Elementary and Intermediate Algebra

FIFTH EDITION

Ron Larson
The Pennsylvania State University
The Behrend College

With the assistance of
Kimberly Nolting
Hillsborough Community College

Australia • Brazil • Japan • Korea • Mexico • Singapore • Spain • United Kingdom • United States

Elementary and Intermediate Algebra, Fifth Edition
Ron Larson

Publisher: Charlie Van Wagner

Associate Development Editor: Laura Localio

Assistant Editor: Shaun Williams

Editorial Assistant: Rebecca Dashiell

Senior Media Editor: Maureen Ross

Executive Marketing Manager: Joe Rogove

Marketing Coordinator: Angela Kim

Marketing Communications Manager: Katherine Malatesta

Project Manager, Editorial Production: Carol Merrigan

Art & Design Manager: Jill Haber

Senior Manufacturing Coordinator: Diane Gibbons

Text Designer: Jerilyn Bockorick

Photo Researcher: Sue McDermott Barlow

Copy Editor: Craig Kirkpatrick

Cover Designer: Irene Morris

Cover Image: Phillip Jarrell/Stone+/Getty Images

Compositor: Larson Texts, Inc.

Chapter Opener and Contents Photo Credits:
p. 1: © Janine Wiedel Photolibrary/Alamy; p. 67: Rubberball/Punchstock; p. 125: Image Source/ Punchstock; p. 215: John Kelly/Getty Images; p. 293: JupiterImages/Banana Stock/Alamy; p. 344: Hemera Technologies/AbleStock/ JupiterImages; p. 345: Photodisc/Punchstock; p. 400: Antonio Scorza/AFP/Getty Images; p. 401: © Digital Vision/Alamy; p. 471: © Richard G. Bingham II/Alamy; p. 554: Purestock/Getty Images; p. 555: © James Marshall/The Image Works; p. 617: © Stock Connection Blue/Alamy; p. 682: Scott T. Baxter/Photodisc/Punchstock; p. 683: Ariel Skelley/Blend Images/Getty Images; p. 761: Purestock/Punchstock; p. 812: © SuperStock/Alamy; p. 813: UpperCut Images/ Alamy.

Library of Congress Control Number: 2008931147

Student Edition

ISBN-13: 978-0-547-10216-0

ISBN-10: 0-547-10216-X

Annotated Instructor's Edition

ISBN-13: 978-0-547-10225-2

ISBN-10: 0-547-10225-9

Brooks/Cole
10 Davis Drive
Belmont, CA 94002-3098
USA

Printed in the United States of America
1 2 3 4 5 6 7 13 12 11 10 09

Common Formulas

Distance

$d = rt$

d = distance traveled
t = time
r = rate

Temperature

$F = \frac{9}{5}C + 32$

F = degrees Fahrenheit
C = degrees Celsius

Simple Interest

$I = Prt$

I = interest
P = principal
r = annual interest rate
t = time in years

Compound Interest

$A = P\left(1 + \frac{r}{n}\right)^{nt}$

A = balance
P = principal
r = annual interest rate
n = compoundings per year
t = time in years

Coordinate Plane: Midpoint Formula

Midpoint of line segment joining (x_1, y_1) and (x_2, y_2)

$$\left(\frac{x_1 + x_2}{2}, \frac{y_1 + y_2}{2}\right)$$

Coordinate Plane: Distance Formula

d = distance between points (x_1, y_1) and (x_2, y_2)

$$d = \sqrt{(x_2 - x_1)^2 + (y_2 - y_1)^2}$$

Quadratic Formula

Solutions of $ax^2 + bx + c = 0$

$$x = \frac{-b \pm \sqrt{b^2 - 4ac}}{2a}$$

Rules of Exponents

(Assume $a \neq 0$ and $b \neq 0$.)

$a^0 = 1$ $\qquad$ $a^m \cdot a^n = a^{m+n}$

$(ab)^m = a^m \cdot b^m$ $\qquad$ $(a^m)^n = a^{mn}$

$\frac{a^m}{a^n} = a^{m-n}$ $\qquad$ $\left(\frac{a}{b}\right)^m = \frac{a^m}{b^m}$

$a^{-n} = \frac{1}{a^n}$ $\qquad$ $\left(\frac{a}{b}\right)^{-n} = \frac{b^n}{a^n}$

Basic Rules of Algebra

Commutative Property of Addition

$a + b = b + a$

Commutative Property of Multiplication

$ab = ba$

Associative Property of Addition

$(a + b) + c = a + (b + c)$

Associative Property of Multiplication

$(ab)c = a(bc)$

Left Distributive Property

$a(b + c) = ab + ac$

Right Distributive Property

$(a + b)c = ac + bc$

Additive Identity Property

$a + 0 = 0 + a = a$

Multiplicative Identity Property

$a \cdot 1 = 1 \cdot a = a$

Additive Inverse Property

$a + (-a) = 0$

Multiplicative Inverse Property

$a \cdot \frac{1}{a} = 1, \quad a \neq 0$

Properties of Equality

Addition Property of Equality

If $a = b$, then $a + c = b + c$.

Multiplication Property of Equality

If $a = b$, then $ac = bc$.

Cancellation Property of Addition

If $a + c = b + c$, then $a = b$.

Cancellation Property of Multiplication

If $ac = bc$, and $c \neq 0$, then $a = b$.

Zero Factor Property

If $ab = 0$, then $a = 0$ or $b = 0$.

Contents

**Appendices C, D, E, F, and G are available on the textbook website. Go to* www.cengage.com/math/larson/algebra *and link to* Elementary and Intermediate Algebra, *Fifth Edition.*

A Word from the Author

Welcome to *Elementary and Intermediate Algebra*, Fifth Edition. In this revision I've focused on laying the groundwork for student success. Each chapter begins with study strategies to help the student do well in the course. Each chapter ends with an interactive summary of what they've learned to prepare them for the chapter test. Throughout the chapter, I've reinforced the skills needed to be successful and check to make sure the student understands the concepts being taught.

In order to address the diverse needs and abilities of students, I offer a straightforward approach to the presentation of difficult concepts. In the Fifth Edition, the emphasis is on helping students learn a variety of techniques—symbolic, numeric, and visual—for solving problems. I am committed to providing students with a successful and meaningful course of study.

Each chapter opens with a *Smart Study Strategy* that will help organize and improve the quality of studying. Mathematics requires students to remember every detail. These study strategies will help students organize, learn, and remember all the details. Each strategy has been student tested.

To improve the usefulness of the text as a study tool, I have a pair of features at the beginning of each section: *What You Should Learn* lists the main objectives that students will encounter throughout the section, and *Why You Should Learn It* provides a motivational explanation for learning the given objectives. To help keep students focused as they read the section, each objective presented in *What You Should Learn* is restated in the margin at the point where the concept is introduced.

In this edition, *Study Tip* features provide hints, cautionary notes, and words of advice for students as they learn the material. *Technology: Tip* features provide point-of-use instruction for using a graphing calculator, whereas *Technology: Discovery* features encourage students to explore mathematical concepts using their graphing or scientific calculators. All technology features are highlighted and can easily be omitted without loss of continuity in coverage of material.

The chapter summary feature *What Did You Learn?* highlights important mathematical vocabulary (*Key Terms*) and primary concepts (*Key Concepts*) from the chapter. For easy reference, the *Key Terms* are correlated to the chapter by page number and the *Key Concepts* by section number.

As students proceed through each chapter, they have many opportunities to assess their understanding and practice skills. A set of *Exercises*, located at the end of each section, correlates to the *Examples* found within the section. *Mid-Chapter Quizzes* and *Chapter Tests* offer students self-assessment tools halfway through and at the conclusion of each chapter. *Review Exercises*, organized by section, restate the *What You Should Learn* objectives so that students may refer back to the appropriate topic discussion when working through the exercises. In addition, the *Concept Check* exercises that precede each exercise set, and the *Cumulative Tests* that follow Chapters 3, 6, 9, and 12, give students more opportunities to revisit and review previously learned concepts.

To show students the practical uses of algebra, I highlight the connections between the mathematical concepts and the real world in the multitude of applications found throughout the text. I believe that students can overcome their difficulties in mathematics if they are encouraged and supported throughout the learning process. Too often, students become frustrated and lose interest in the material when they cannot follow the text. With this in mind, every effort has been made to write a readable text that can be understood by every student. I hope that your students find this approach engaging and effective.

Ron Larson

Ron Larson

Features

Study Skills in Action

Absorbing Details Sequentially

Math is a sequential subject (Nolting, 2008). Learning new math concepts successfully depends on how well you understand all the previous concepts. So, it is important to learn and remember concepts as they are encountered. One way to work through a section sequentially is by following these steps.

1. Work through an example. If you have trouble, consult your notes or seek help from a classmate or instructor.
2. Complete the checkpoint exercise following the example.
3. If you get the checkpoint exercise correct, move on to the next example. If not, make sure you understand your mistake(s) before you move on.
4. When you have finished working through all the examples in the section, take a short break of 5 to 10 minutes. This will give your brain time to process everything.
5. Start the homework exercises.

Kimberly Nolting
VP, Academic Success Press
expert in developmental education

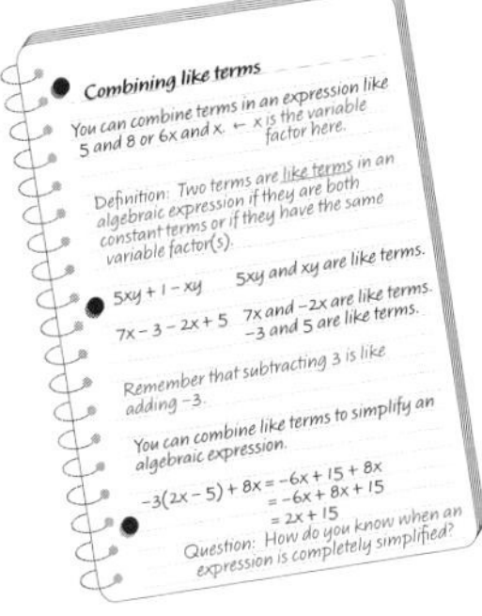

Combining like terms
You can combine terms in an expression like 5 and 8 or 6x and x. ← x is the variable factor here.
Definition: Two terms are like terms in an algebraic expression if they are both constant terms or if they have the same variable factor(s).
5xy + 1 − xy   5xy and xy are like terms.
7x − 3 − 2x + 5   7x and −2x are like terms. −3 and 5 are like terms.
Remember that subtracting 3 is like adding −3.
You can combine like terms to simplify an algebraic expression.
−3(2x − 5) + 8x = −6x + 15 + 8x
= −6x + 8x + 15
= 2x + 15
Question: How do you know when an expression is completely simplified?

Smart Study Strategy

Rework Your Notes

It is almost impossible to write down in your notes all the detailed information you are taught in class. A good way to reinforce the concepts and put them into your long-term memory is to rework your notes. When you take notes, leave extra space on the pages. You can go back after class and fill in:

- important definitions and rules
- additional examples
- questions you have about the material

66

Chapter 2

Fundamentals of Algebra

2.1 Writing and Evaluating Algebraic Expressions
2.2 Simplifying Algebraic Expressions
2.3 Algebra and Problem Solving
2.4 Introduction to Equations

IT WORKED FOR ME!

"When I am in math class, I struggle with keeping all my notes neat. I get everything I need written down in them but they're just sloppy. I rewrite my notes so that I can review what we have covered and so I have a neat set of notes to keep using for reference. Another friend of mine in class is doing it with me now."

Joel
Psychology/film

67

Chapter Opener

Each chapter opener presents a study skill essential to success in mathematics. Following is a *Smart Study Strategy,* which gives concrete ways that students can help themselves with the study skill. In each chapter, there is a *Smart Study Strategy* note in the side column pointing out an appropriate time to use this strategy. Quotes from real students who have successfully used the strategy are given in *It Worked for Me!*

Section 1.2 Adding and Subtracting Integers 11

1.2 Adding and Subtracting Integers

What You Should Learn

1. Add integers using a number line.
2. Add integers with like signs and with unlike signs.
3. Subtract integers with like signs and with unlike signs.

Nancy R. Cohen/Getty Images

Why You Should Learn It

Real numbers are used to represent many real-life quantities. For instance, in Exercise 107 on page 18, you will use real numbers to find the change in digital camera sales.

1 ► Add integers using a number line.

Adding Integers Using a Number Line

In this and the next section, you will study the four operations of arithmetic (addition, subtraction, multiplication, and division) on the set of integers. There are many examples of these operations in real life. For example, your business had a gain of \$550 during one week and a loss of \$600 the next week. Over the two-week period, your business had a combined profit of

$$550 + (-600) = -50$$

which represents an overall loss of \$50.

The number line is a good visual model for demonstrating addition of integers. To add two integers, $a + b$, using a number line, start at 0. Then move right or left a units depending on whether a is positive or negative. From that position, move right or left b units depending on whether b is positive or negative. The final position is called the **sum.**

EXAMPLE 1 Adding Integers with Like Signs Using a Number Line

Find each sum.

a. $5 + 2$ **b.** $-3 + (-5)$

Solution

a. Start at zero and move five units to the right. Then move two more units to the right, as shown in Figure 1.20. So, $5 + 2 = 7$.

Figure 1.20

b. Start at zero and move three units to the left. Then move five more units to the left, as shown in Figure 1.21. So, $-3 + (-5) = -8$.

Figure 1.21

✓ CHECKPOINT Now try Exercise 3.

Section Opener

Every section begins with a list of learning objectives called *What You Should Learn*. Each objective is restated in the margin at the point where it is covered. *Why You Should Learn It* provides a motivational explanation for learning the given objectives.

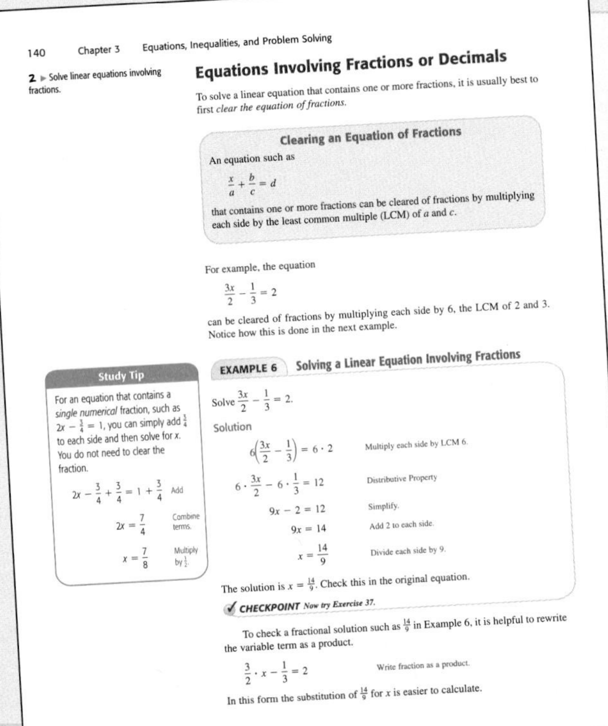

140 Chapter 3 Equations, Inequalities, and Problem Solving

2 ▶ Solve linear equations involving fractions.

Equations Involving Fractions or Decimals

To solve a linear equation that contains one or more fractions, it is usually best to first *clear the equation of fractions*.

Clearing an Equation of Fractions

An equation such as

$$\frac{x}{a} + \frac{b}{c} = d$$

that contains one or more fractions can be cleared of fractions by multiplying each side by the least common multiple (LCM) of a and c.

For example, the equation

$$\frac{3x}{2} - \frac{1}{3} = 2$$

can be cleared of fractions by multiplying each side by 6, the LCM of 2 and 3. Notice how this is done in the next example.

Study Tip

For an equation that contains a *single numerical* fraction, such as $2x - \frac{3}{4} = 1$, you can simply add $\frac{3}{4}$ to each side and then solve for x. You do not need to clear the fraction.

$2x - \frac{3}{4} + \frac{3}{4} = 1 + \frac{3}{4}$ Add $\frac{3}{4}$.

$2x = \frac{7}{4}$ Combine terms.

$x = \frac{7}{8}$ Multiply by $\frac{1}{2}$.

EXAMPLE 6 Solving a Linear Equation Involving Fractions

Solve $\frac{3x}{2} - \frac{1}{3} = 2$.

Solution

$6\left(\frac{3x}{2} - \frac{1}{3}\right) = 6 \cdot 2$ Multiply each side by LCM 6.

$6 \cdot \frac{3x}{2} - 6 \cdot \frac{1}{3} = 12$ Distributive Property

$9x - 2 = 12$ Simplify.

$9x = 14$ Add 2 to each side.

$x = \frac{14}{9}$ Divide each side by 9.

The solution is $x = \frac{14}{9}$. Check this in the original equation.

✓ CHECKPOINT *Now try Exercise 37.*

To check a fractional solution such as $\frac{14}{9}$ in Example 6, it is helpful to rewrite the variable term as a product.

$\frac{3}{2} \cdot x - \frac{1}{3} = 2$ Write fraction as a product.

In this form the substitution of $\frac{14}{9}$ for x is easier to calculate.

Examples

Each example has been carefully chosen to illustrate a particular mathematical concept or problem-solving technique. The examples cover a wide variety of problems and are titled for easy reference. Many examples include detailed, step-by-step solutions with side comments, which explain the key steps of the solution process.

Checkpoints

Each example is followed by a checkpoint exercise. After working through an example, students can try the checkpoint exercise in the exercise set to check their understanding of the concepts presented in the example. Checkpoint exercises are marked with a ✓ in the exercise set for easy reference.

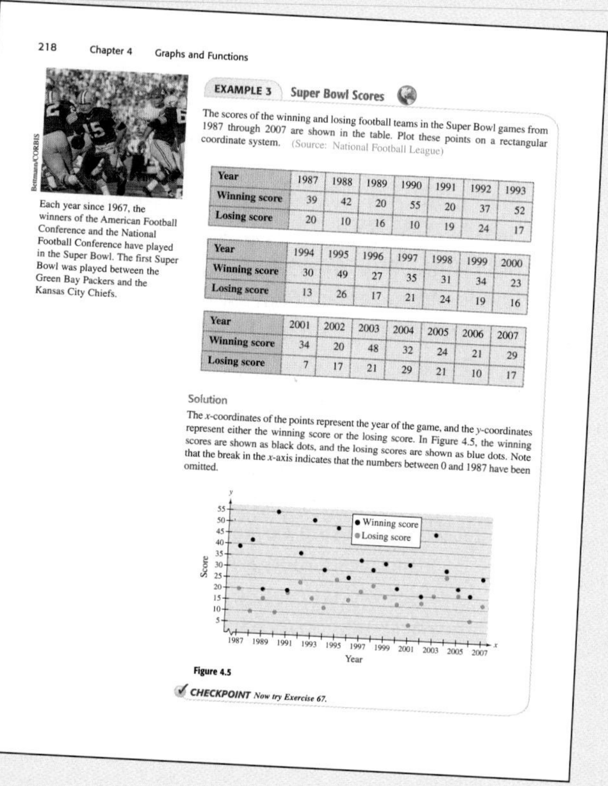

218 Chapter 4 Graphs and Functions

Bettmann/CORBIS

Each year since 1967, the winners of the American Football Conference and the National Football Conference have played in the Super Bowl. The first Super Bowl was played between the Green Bay Packers and the Kansas City Chiefs.

EXAMPLE 3 Super Bowl Scores

The scores of the winning and losing football teams in the Super Bowl games from 1987 through 2007 are shown in the table. Plot these points on a rectangular coordinate system. (Source: National Football League)

Year	1987	1988	1989	1990	1991	1992	1993
Winning score	39	42	20	55	20	37	52
Losing score	20	10	16	10	19	24	17

Year	1994	1995	1996	1997	1998	1999	2000
Winning score	30	49	27	35	31	34	23
Losing score	13	26	17	21	24	19	16

Year	2001	2002	2003	2004	2005	2006	2007
Winning score	34	20	48	32	24	21	29
Losing score	7	17	21	29	21	10	17

Solution

The x-coordinates of the points represent the year of the game, and the y-coordinates represent either the winning score or the losing score. In Figure 4.5, the winning scores are shown as black dots, and the losing scores are shown as blue dots. Note that the break in the x-axis indicates that the numbers between 0 and 1987 have been omitted.

Figure 4.5

✓ CHECKPOINT *Now try Exercise 67.*

Applications

A wide variety of real-life applications are integrated throughout the text in examples and exercises. These applications demonstrate the relevance of algebra in the real world. Many of the applications use current, real data. The icon 🌎 indicates an example involving a real-life application.

Problem Solving

This text provides many opportunities for students to sharpen their problem-solving skills. In both the examples and the exercises, students are asked to apply verbal, numerical, analytical, and graphical approaches to problem solving. In the spirit of the AMATYC and NCTM standards, students are taught a five-step strategy for solving applied problems, which begins with constructing a verbal model and ends with checking the answer.

460 Chapter 7 Rational Expressions, Equations, and Functions

✓ **61.** *Stopping Distance* The stopping distance d of an automobile is directly proportional to the square of its speed s. On one road, a car requires 75 feet to stop from a speed of 30 miles per hour. How many feet does the car require to stop from a speed of 48 miles per hour on the same road?

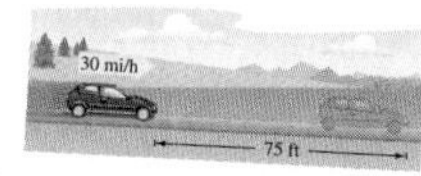

62. *Frictional Force* The frictional force F (between the tires of a car and the road) that is required to keep a car on a curved section of a highway is directly proportional to the square of the speed s of the car. By what factor does the force F change when the speed of the car is doubled on the same curve?

63. *Power Generation* The power P generated by a wind turbine varies directly as the cube of the wind speed w. The turbine generates 400 watts of power in a 20-mile-per-hour wind. Find the power it generates in a 30-mile-per-hour wind.

64. *Weight of an Astronaut* A person's weight on the moon varies directly as his or her weight on Earth. An astronaut weighs 360 pounds on Earth, including heavy equipment. On the moon the astronaut weighs only 60 pounds with the equipment. If the first woman in space, Valentina Tereshkova, had landed on the moon and weighed 54 pounds with equipment, how much would she have weighed on Earth with her equipment?

✓ **65.** *Demand* A company has found that the daily demand x for its boxes of chocolates is inversely proportional to the price p. When the price is \$5, the demand is 800 boxes. Approximate the demand when the price is increased to \$6.

66. *Pressure* When a person walks, the pressure P on each sole varies inversely as the area A of the sole. A person is trudging through deep snow, wearing boots that have a sole area of 29 square inches each. The sole pressure is 4 pounds per square inch. If the person was wearing snowshoes, each with an area 11 times that of their boot soles, what would be the pressure on each snowshoe? The constant of variation in this problem is the weight of the person. How much does the person weigh?

67. *Environment* The graph shows the percent p of oil that remained in Chedabucto Bay, Nova Scotia, after an oil spill. The cleaning of the spill was left primarily to natural actions. After about a year, the percent that remained varied inversely as time. Find a model that relates p and t, where t is the number of years since the spill. Then use it to find the percent of oil that remained $6\frac{1}{2}$ years after the spill, and compare the result with the graph.

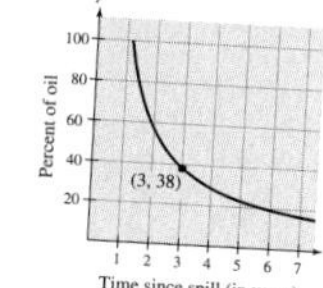

68. *Meteorology* The graph shows the water temperature in relation to depth in the north central Pacific Ocean. At depths greater than 900 meters, the water temperature varies inversely with the water depth. Find a model that relates the temperature T to the depth d. Then use it to find the water temperature at a depth of 4385 meters, and compare the result with the graph.

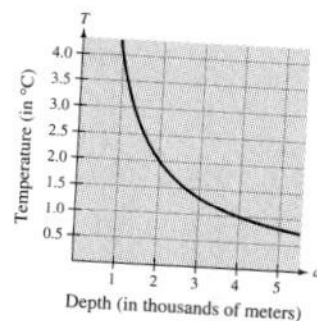

✓ **69.** *Revenue* The weekly demand for a company's frozen pizzas varies directly as the amount spent on advertising and inversely as the price per pizza. At \$5 per pizza, when \$500 is spent each week on ads, the demand is 2000 pizzas. If advertising is increased to \$600, what price will yield a demand of 2000 pizzas? Is this increase worthwhile in terms of revenue?

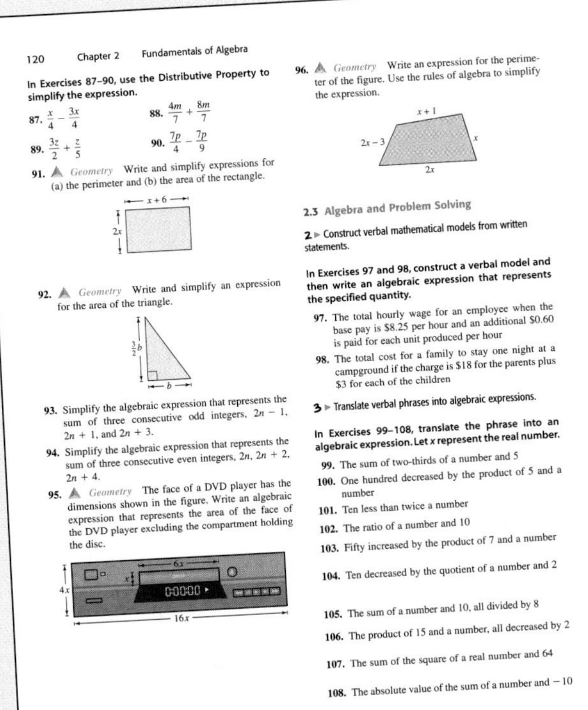

120 Chapter 2 Fundamentals of Algebra

In Exercises 87–90, use the Distributive Property to simplify the expression.

87. $\frac{x}{4} - \frac{3x}{4}$ **88.** $\frac{4m}{7} + \frac{8m}{7}$

89. $\frac{3z}{2} + \frac{z}{5}$ **90.** $\frac{7p}{4} - \frac{7p}{9}$

91. ▲ *Geometry* Write and simplify expressions for (a) the perimeter and (b) the area of the rectangle.

[figure: rectangle, $x + 6$ by $2x$]

92. ▲ *Geometry* Write and simplify an expression for the area of the triangle.

[figure: right triangle, height $\frac{3}{2}b$, base b]

93. Simplify the algebraic expression that represents the sum of three consecutive odd integers, $2n - 1$, $2n + 1$, and $2n + 3$.

94. Simplify the algebraic expression that represents the sum of three consecutive even integers, $2n$, $2n + 2$, $2n + 4$.

95. ▲ *Geometry* The face of a DVD player has the dimensions shown in the figure. Write an algebraic expression that represents the area of the face of the DVD player excluding the compartment holding the disc.

[figure: DVD player, $16x$ by $4x$; compartment $6x$ by x; display 0:00:00]

96. ▲ *Geometry* Write an expression for the perimeter of the figure. Use the rules of algebra to simplify the expression.

[figure: quadrilateral with sides $x + 1$, x, $2x$, $2x - 3$]

2.3 Algebra and Problem Solving

2 ▶ Construct verbal mathematical models from written statements.

In Exercises 97 and 98, construct a verbal model and then write an algebraic expression that represents the specified quantity.

97. The total hourly wage for an employee when the base pay is \$8.25 per hour and an additional \$0.60 is paid for each unit produced per hour

98. The total cost for a family to stay one night at a campground if the charge is \$18 for the parents plus \$3 for each of the children

3 ▶ Translate verbal phrases into algebraic expressions.

In Exercises 99–108, translate the phrase into an algebraic expression. Let x represent the real number.

99. The sum of two-thirds of a number and 5

100. One hundred decreased by the product of 5 and a number

101. Ten less than twice a number

102. The ratio of a number and 10

103. Fifty increased by the product of 7 and a number

104. Ten decreased by the quotient of a number and 2

105. The sum of a number and 10, all divided by 8

106. The product of 15 and a number, all decreased by 2

107. The sum of the square of a real number and 64

108. The absolute value of the sum of a number and -10

Geometry

The Fifth Edition continues to provide coverage and integration of geometry in examples and exercises. The icon ▲ indicates an exercise involving geometry.

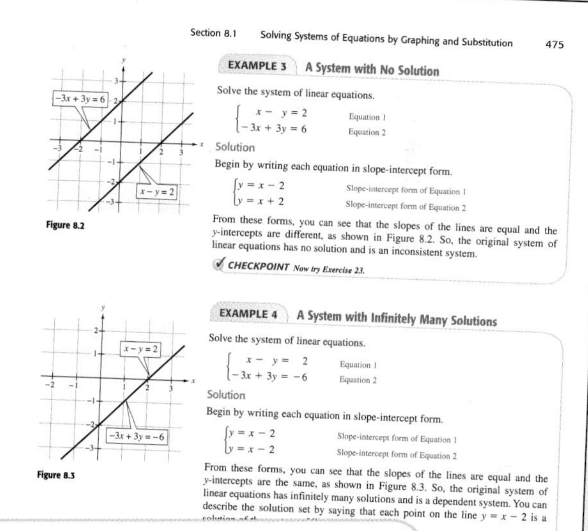

Section 8.1 Solving Systems of Equations by Graphing and Substitution 475

EXAMPLE 3 A System with No Solution

Solve the system of linear equations.

$$\begin{cases} x - y = 2 & \text{Equation 1} \\ -3x + 3y = 6 & \text{Equation 2} \end{cases}$$

Solution

Begin by writing each equation in slope-intercept form.

$$\begin{cases} y = x - 2 & \text{Slope-intercept form of Equation 1} \\ y = x + 2 & \text{Slope-intercept form of Equation 2} \end{cases}$$

From these forms, you can see that the slopes of the lines are equal and the y-intercepts are different, as shown in Figure 8.2. So, the original system of linear equations has no solution and is an inconsistent system.

✓ *CHECKPOINT Now try Exercise 23.*

[graph labels: $-3x + 3y = 6$; $x - y = 2$]

Figure 8.2

EXAMPLE 4 A System with Infinitely Many Solutions

Solve the system of linear equations.

$$\begin{cases} x - y = 2 & \text{Equation 1} \\ -3x + 3y = -6 & \text{Equation 2} \end{cases}$$

Solution

Begin by writing each equation in slope-intercept form.

$$\begin{cases} y = x - 2 & \text{Slope-intercept form of Equation 1} \\ y = x - 2 & \text{Slope-intercept form of Equation 2} \end{cases}$$

From these forms, you can see that the slopes of the lines are equal and the y-intercepts are the same, as shown in Figure 8.3. So, the original system of linear equations has infinitely many solutions and is a dependent system. You can describe the solution set by saying that each point on the line $y = x - 2$ is a ...

[graph labels: $x - y = 2$; $-3x + 3y = -6$]

Figure 8.3

...s representing a system of ...st have either no solution or ...o lines have different slopes, ...onding system has a single

...through Examples 5 and 6. ...f solving a system of linear ...ond, once you have made a ...itical that you check in the ...sen is the correct solution.

Graphics

Visualization is a critical problem-solving skill. To encourage the development of this skill, students are shown how to use graphs to reinforce algebraic and numeric solutions and to interpret data. The numerous figures in examples and exercises throughout the text were computer-generated for accuracy.

Section 7.1 Rational Expressions and Functions

2 ▸ Simplify rational expressions.

Simplifying Rational Expressions

As with numerical fractions, a rational expression is said to be i… **reduced) form** if its numerator and denominator have no comm… than ±1). To simplify rational expressions, you can apply the …

Simplifying Rational Expressions

Let u, v, and w represent real numbers, variables, or algebraic expressions such that $v \neq 0$ and $w \neq 0$. Then the following is valid.

$$\frac{uw}{vw} = \frac{u\cancel{w}}{v\cancel{w}} = \frac{u}{v}$$

Be sure you divide out only *factors*, not *terms*. For instance, consider the expressions below.

$\frac{2 \cdot 2}{2(x + 5)}$ You *can* divide out the common factor 2.

$\frac{3 + x}{3 + 2x}$ You *cannot* divide out the common term 3.

Simplifying a rational expression requires two steps: (1) completely factor the numerator and denominator and (2) divide out any *factors* that are common to both the numerator and denominator. So, your success in simplifying rational expressions actually lies in your ability to *factor completely* the polynomials in both the numerator and denominator.

EXAMPLE 4 Simplifying a Rational Expression

Simplify the rational expression $\frac{2x^3 - 6x}{6x^2}$.

Solution

First note that the domain of the rational expression is all real values of x such that $x \neq 0$. Then, completely factor both the numerator and denominator.

$$\frac{2x^3 - 6x}{6x^2} = \frac{2x(x^2 - 3)}{2x(3x)} \quad \text{Factor numerator and denominator.}$$

$$= \frac{\cancel{2x}(x^2 - 3)}{\cancel{2x}(3x)} \quad \text{Divide out common factor } 2x.$$

$$= \frac{x^2 - 3}{3x} \quad \text{Simplified form}$$

In simplified form, the domain of the rational expression is the same as that of the original expression—all real values of x such that $x \neq 0$.

✓ CHECKPOINT *Now try Exercise 43.*

Simplifying Rational Expressions

Let u, v, and w represent real numbers, variables, or algebraic expressions such that $v \neq 0$ and $w \neq 0$. Then the following is valid.

$$\frac{uw}{vw} = \frac{u\cancel{w}}{v\cancel{w}} = \frac{u}{v}$$

Definitions and Rules

All important definitions, rules, formulas, properties, and summaries of solution methods are highlighted for emphasis. Each of these features is also titled for easy reference.

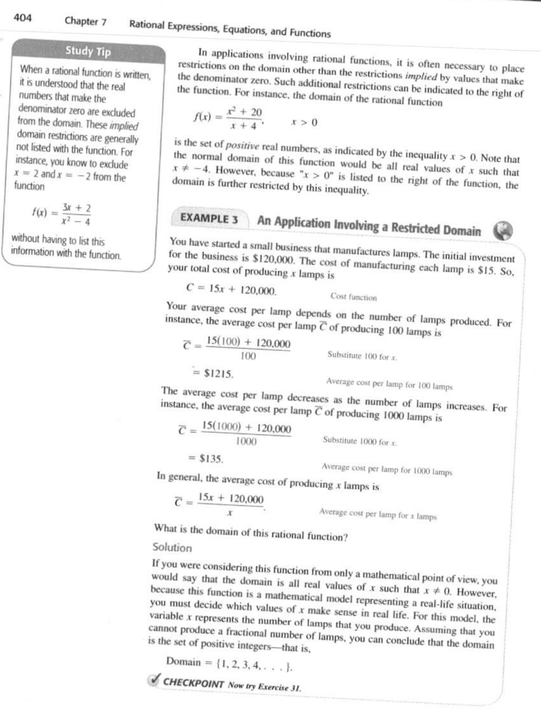

404 Chapter 7 Rational Expressions, Equations, and Functions

Study Tip

When a rational function is written, it is understood that the real numbers that make the denominator zero are excluded from the domain. These *implied* domain restrictions are generally not listed with the function. For instance, you know to exclude $x = 2$ and $x = -2$ from the function

$$f(x) = \frac{3x + 2}{x^2 - 4}$$

without having to list this information with the function.

In applications involving rational functions, it is often necessary to place restrictions on the domain other than the restrictions *implied* by values that make the denominator zero. Such additional restrictions can be indicated to the right of the function. For instance, the domain of the rational function

$$f(x) = \frac{x^2 + 20}{x + 4}, \quad x > 0$$

is the set of *positive* real numbers, as indicated by the inequality $x > 0$. Note that the normal domain of this function would be all real values of x such that $x \neq -4$. However, because "$x > 0$" is listed to the right of the function, the domain is further restricted by this inequality.

EXAMPLE 3 An Application Involving a Restricted Domain

You have started a small business that manufactures lamps. The initial investment for the business is \$120,000. The cost of manufacturing each lamp is \$15. So, your total cost of producing x lamps is

$C = 15x + 120{,}000.$ Cost function

Your average cost per lamp depends on the number of lamps produced. For instance, the average cost per lamp $\overline{C}$ of producing 100 lamps is

$$\overline{C} = \frac{15(100) + 120{,}000}{100} \quad \text{Substitute 100 for } x.$$

$$= \$1215. \quad \text{Average cost per lamp for 100 lamps}$$

The average cost per lamp decreases as the number of lamps increases. For instance, the average cost per lamp $\overline{C}$ of producing 1000 lamps is

$$\overline{C} = \frac{15(1000) + 120{,}000}{1000} \quad \text{Substitute 1000 for } x.$$

$$= \$135. \quad \text{Average cost per lamp for 1000 lamps}$$

In general, the average cost of producing x lamps is

$$\overline{C} = \frac{15x + 120{,}000}{x}. \quad \text{Average cost per lamp for } x \text{ lamps}$$

What is the domain of this rational function?

Solution

If you were considering this function from only a mathematical point of view, you would say that the domain is all real values of x such that $x \neq 0$. However, because this function is a mathematical model representing a real-life situation, you must decide which values of x make sense in real life. For this model, the variable x represents the number of lamps that you produce. Assuming that you cannot produce a fractional number of lamps, you can conclude that the domain is the set of positive integers—that is,

Domain $= \{1, 2, 3, 4, \ldots\}$.

✓ CHECKPOINT *Now try Exercise 31.*

Study Tip

When a rational function is written, it is understood that the real numbers that make the denominator zero are excluded from the domain. These *implied* domain restrictions are generally not listed with the function. For instance, you know to exclude $x = 2$ and $x = -2$ from the function

$$f(x) = \frac{3x + 2}{x^2 - 4}$$

without having to list this information with the function.

Study Tips

Study Tips offer students specific point-of-use suggestions for studying algebra, as well as pointing out common errors and discussing alternative solution methods. They appear in the margins.

Technology Tips

Point-of-use instructions for using graphing calculators appear in the margins. These features encourage the use of graphing technology as a tool for visualization of mathematical concepts, for verification of other solution methods, and for facilitation of computations. The *Technology: Tips* can easily be omitted without loss of continuity in coverage. Answers to questions posed within these features are located in the back of the Annotated Instructor's Edition.

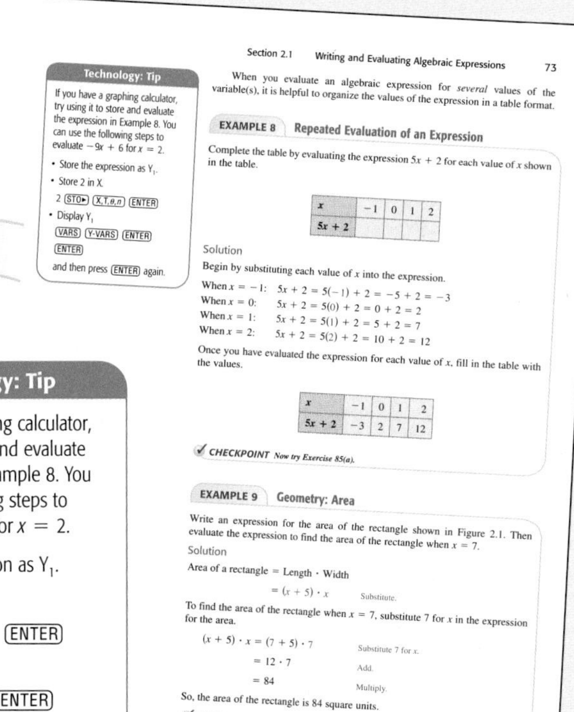
Section 2.1 Writing and Evaluating Algebraic Expressions 73

Technology: Tip

If you have a graphing calculator, try using it to store and evaluate the expression in Example 8. You can use the following steps to evaluate $-9x + 6$ for $x = 2$.

- Store the expression as Y_1.
- Store 2 in X.

  2 (STO▸) (X,T,θ,n) (ENTER)
- Display Y_1

  (VARS) (Y-VARS) (ENTER) (ENTER)

and then press (ENTER) again.

When you evaluate an algebraic expression for *several* values of the variable(s), it is helpful to organize the values of the expression in a table format.

EXAMPLE 8 **Repeated Evaluation of an Expression**

Complete the table by evaluating the expression $5x + 2$ for each value of x shown in the table.

x	-1	0	1	2
$5x + 2$				

Solution

Begin by substituting each value of x into the expression.

When $x = -1$: $5x + 2 = 5(-1) + 2 = -5 + 2 = -3$

When $x = 0$: $5x + 2 = 5(0) + 2 = 0 + 2 = 2$

When $x = 1$: $5x + 2 = 5(1) + 2 = 5 + 2 = 7$

When $x = 2$: $5x + 2 = 5(2) + 2 = 10 + 2 = 12$

Once you have evaluated the expression for each value of x, fill in the table with the values.

x	-1	0	1	2
$5x + 2$	-3	2	7	12

✓ CHECKPOINT *Now try Exercise 85(a).*

EXAMPLE 9 **Geometry: Area**

Write an expression for the area of the rectangle shown in Figure 2.1. Then evaluate the expression to find the area of the rectangle when $x = 7$.

Solution

Area of a rectangle = Length · Width

$= (x + 5) \cdot x$ Substitute.

To find the area of the rectangle when $x = 7$, substitute 7 for x in the expression for the area.

$(x + 5) \cdot x = (7 + 5) \cdot 7$ Substitute 7 for x.

$= 12 \cdot 7$ Add.

$= 84$ Multiply.

So, the area of the rectangle is 84 square units.

✓ CHECKPOINT *Now try Exercise 91.*

Technology: Tip

If you have a graphing calculator, try using it to store and evaluate the expression in Example 8. You can use the following steps to evaluate $-9x + 6$ for $x = 2$.

- Store the expression as Y_1.
- Store 2 in X.

  2 (STO▸) (X,T,θ,n) (ENTER)
- Display Y_1

  (VARS) (Y-VARS) (ENTER) (ENTER)

  and then press (ENTER) again.

Technology: Discovery

Use a graphing calculator to graph the following second-degree equations, and note the numbers of x-intercepts.

$y = x^2 - 10x + 25$

$y = 5x^2 + 60x + 175$

$y = -2x^2 - 4x - 5$

Use a graphing calculator to graph the following third-degree equations, and note the numbers of x-intercepts.

$y = x^3 - 12x^2 + 48x - 60$

$y = x^3 - 4x$

$y = x^3 + 13x^2 + 55x + 75$

Use your results to write a conjecture about how the degree of a polynomial equation is related to the possible number of solutions.

6 Chapter 6 Factoring and Solving Equations

▸ Solve higher-degree polynomial [e]quations by factoring.

Solving Higher-Degree Equations by Factoring

EXAMPLE 5 **Solving a Polynomial Equation with Three Factors**

Solve $3x^3 = 15x^2 + 18x$.

Solution

$3x^3 = 15x^2 + 18x$ Write original equation.

$3x^3 - 15x^2 - 18x = 0$ Write in general form.

$3x(x^2 - 5x - 6) = 0$ Factor out common factor.

$3x(x - 6)(x + 1) = 0$ Factor.

$3x = 0 \Rightarrow x = 0$ Set 1st factor equal to 0.

$x - 6 = 0 \Rightarrow x = 6$ Set 2nd factor equal to 0.

$x + 1 = 0 \Rightarrow x = -1$ Set 3rd factor equal to 0.

The solutions are $x = 0$, $x = 6$, and $x = -1$. Check these three solutions.

✓ CHECKPOINT *Now try Exercise 65.*

Technology: Discovery

Use a graphing calculator to graph the following second-degree equations, and note the numbers of x-intercepts.

$y = x^2 - 10x + 25$

$y = 5x^2 + 60x + 175$

$y = -2x^2 - 4x - 5$

Use a graphing calculator to graph the following third-degree equations, and note the numbers of x-intercepts.

$y = x^3 - 12x^2 + 48x - 60$

$y = x^3 - 4x$

$y = x^3 + 13x^2 + 55x + 75$

Use your results to write a conjecture about how the degree of a polynomial equation is related to the possible number of solutions.

Notice that the equation in Example 5 is a third-degree equation and has three solutions. This is not a coincidence. In general, a polynomial equation can have *at most* as many solutions as its degree. For instance, a second-degree equation can have zero, one, or two solutions. Notice that the equation in Example 6 is a fourth-degree equation and has four solutions.

EXAMPLE 6 **Solving a Polynomial Equation with Four Factors**

Solve $x^4 + x^3 - 4x^2 - 4x = 0$.

Solution

$x^4 + x^3 - 4x^2 - 4x = 0$ Write original equation.

$x(x^3 + x^2 - 4x - 4) = 0$ Factor out common factor.

$x[(x^3 + x^2) + (-4x - 4)] = 0$ Group terms.

$x[x^2(x + 1) - 4(x + 1)] = 0$ Factor grouped terms.

$x[(x + 1)(x^2 - 4)] = 0$ Distributive Property

$x(x + 1)(x + 2)(x - 2) = 0$ Difference of two squares

$x = 0 \Rightarrow x = 0$

$x + 1 = 0 \Rightarrow x = -1$

$x + 2 = 0 \Rightarrow x = -2$

$x - 2 = 0 \Rightarrow x = 2$

The solutions are $x = 0$, $x = -1$, $x = -2$, and $x = 2$. Check these four solutions.

✓ CHECKPOINT *Now try Exercise 75.*

Technology: Discovery

Technology: Discovery features invite students to engage in active exploration of mathematical concepts and discovery of mathematical relationships through the use of scientific or graphing calculators. These activities encourage students to utilize their critical thinking skills and help them develop an intuitive understanding of theoretical concepts. *Technology: Discovery* features can easily be omitted without loss of continuity in coverage. Answers to questions posed within these features are located in the back of the Annotated Instructor's Edition.

Concept Check

Each exercise set is preceded by four exercises that check students' understanding of the main concepts of the section. These exercises could be completed in class to make sure that students are ready to start the exercise set.

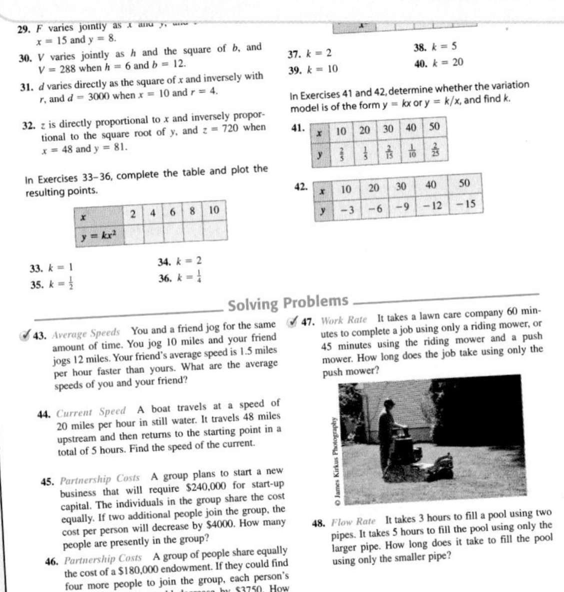

29. F varies jointly as x and y, and ... $x = 15$ and $y = 8$.

30. V varies jointly as h and the square of b, and $V = 288$ when $h = 6$ and $b = 12$.

31. d varies directly as the square of x and inversely with r, and $d = 3000$ when $x = 10$ and $r = 4$.

32. z is directly proportional to x and inversely proportional to the square root of y, and $z = 720$ when $x = 48$ and $y = 81$.

In Exercises 33–36, complete the table and plot the resulting points.

x	2	4	6	8	10
$y = kx^2$					

33. $k = 1$ 34. $k = 2$

35. $k = \frac{1}{2}$ 36. $k = \frac{1}{4}$

37. $k = 2$ 38. $k = 5$

39. $k = 10$ 40. $k = 20$

In Exercises 41 and 42, determine whether the variation model is of the form $y = kx$ or $y = k/x$, and find k.

41.

x	10	20	30	40	50
y	$\frac{2}{5}$	$\frac{1}{5}$	$\frac{2}{15}$	$\frac{1}{10}$	$\frac{2}{25}$

42.

x	10	20	30	40	50
y	−3	−6	−9	−12	−15

Solving Problems

✓ 43. *Average Speeds* You and a friend jog for the same amount of time. You jog 10 miles and your friend jogs 12 miles. Your friend's average speed is 1.5 miles per hour faster than yours. What are the average speeds of you and your friend?

44. *Current Speed* A boat travels at a speed of 20 miles per hour in still water. It travels 48 miles upstream and then returns to the starting point in a total of 5 hours. Find the speed of the current.

45. *Partnership Costs* A group plans to start a new business that will require $240,000 for start-up capital. The individuals in the group share the cost equally. If two additional people join the group, the cost per person will decrease by $4000. How many people are presently in the group?

46. *Partnership Costs* A group of people share equally the cost of a $180,000 endowment. If they could find four more people to join the group, each person's share of the cost would decrease by $3750. How ...

✓ 47. *Work Rate* It takes a lawn care company 60 minutes to complete a job using only a riding mower, or 45 minutes using the riding mower and a push mower. How long does the job take using only the push mower?

© James Kirkus Photography

48. *Flow Rate* It takes 3 hours to fill a pool using two pipes. It takes 5 hours to fill the pool using only the larger pipe. How long does it take to fill the pool using only the smaller pipe?

Section 7.6 Applications and Variation 457

Concept Check

1. In a problem, y varies directly as x and the constant of proportionality is positive. If one of the variables increases, how does the other change? Explain.

2. In a problem, y varies inversely as x and the constant of proportionality is positive. If one of the variables increases, how does the other change? Explain.

3. Are the following statements equivalent? Explain.
(a) y varies directly as x.
(b) y is directly proportional to the square of x.

4. Describe the difference between *combined variation* and *joint variation*.

7.6 EXERCISES Go to pages 462–463 to record your assignments.

Developing Skills

In Exercises 1–14, write a model for the statement.

1. I varies directly as V.
2. C varies directly as r.
3. V is directly proportional to t.
4. A is directly proportional to w.
5. u is directly proportional to the square of v.
6. s varies directly as the cube of t.
7. p varies inversely as d.
8. S varies inversely as the square of v.
9. A is inversely proportional to the fourth power of t.
10. P is inversely proportional to the square root of $1 + r$.
11. A varies jointly as l and w.
12. V varies jointly as h and the square of r.
13. *Boyle's Law* If the temperature of a gas is not allowed to change, its absolute pressure P is inversely proportional to its volume V.
14. *Newton's Law of Universal Gravitation* The gravitational attraction F between two particles of masses m_1 and m_2 is directly proportional to the product of the masses and inversely proportional to the square of the distance r between the particles.

In Exercises 15–20, write a verbal sentence using variation terminology to describe the formula.

15. *Area of a Triangle:* $A = \frac{1}{2}bh$
16. *Area of a Rectangle:* $A = lw$
17. *Volume of a Right Circular Cylinder:* $V = \pi r^2 h$
18. *Volume of a Sphere:* $V = \frac{4}{3}\pi r^3$
19. *Average Speed:* $r = \frac{d}{t}$
20. *Height of a Cylinder:* $h = \frac{V}{\pi r^2}$

In Exercises 21–32, find the constant of proportionality and write an equation that relates the variables.

21. s varies directly as t, and $s = 20$ when $t = 4$.
22. h is directly proportional to r, and $h = 28$ when $r = 12$.
23. F is directly proportional to the square of x, and $F = 500$ when $x = 40$.
24. M varies directly as the cube of n, and $M = 0.012$ when $n = 0.2$.
25. n varies inversely as m, and $n = 32$ when $m = 1.5$.
26. q is inversely proportional to p, and $q = \frac{3}{2}$ when $p = 50$.

Exercises

The exercise sets are grouped into three categories: *Developing Skills, Solving Problems,* and *Explaining Concepts.* The exercise sets offer a diverse variety of computational, conceptual, and applied problems to accommodate many learning styles. Designed to build competence, skill, and understanding, each exercise set is graded in difficulty to allow students to gain confidence as they progress. Detailed solutions to all odd-numbered exercises are given in the *Student Solutions Guide*, and answers to all odd-numbered exercises are given in the back of the student text. Answers are located in place in the Annotated Instructor's Edition.

Section 7.6 Applications and Variation 461

70. *Revenue* The monthly demand for a company's sports caps varies directly as the amount spent on advertising and inversely as the square of the price per cap. At $15 per cap, when $2500 is spent each week on ads, the demand is 300 caps. If advertising is increased to $3000, what price will yield a demand of 300 caps? Is this increase worthwhile in terms of revenue?

✓ 71. *Simple Interest* The simple interest earned by an account varies jointly as the time and the principal. A principal of $600 earns $10 interest in 4 months. How much would $900 earn in 6 months?

72. *Simple Interest* The simple interest earned by an account varies jointly as the time and the principal. In 2 years, a principal of $5000 earns $650 interest. How much would $1000 earn in 1 year?

73. *Engineering* The load P that can be safely supported by a horizontal beam varies jointly as the product of the width W of the beam and the square of the depth D, and inversely as the length L (see figure).
(a) Write a model for the statement.
(b) How does P change when the width and length of the beam are both doubled?
(c) How does P change when the width and depth of the beam are doubled?
(d) How does P change when all three of the dimensions are doubled?
(e) How does P change when the depth of the beam is cut in half?
(f) A beam with width 3 inches, depth 8 inches, and length 120 inches can safely support 2000 pounds. Determine the safe load of a beam made from the same material if its depth is increased to 10 inches.

Explaining Concepts

True or False? In Exercises 74 and 75, determine whether the statement is true or false. Explain your reasoning.

74. In a situation involving combined variation, y can vary directly as x and inversely as x at the same time.

75. In a joint variation problem where z varies jointly as x and y, if x increases, then z and y must both increase.

76. If y varies directly as the square of x and x is doubled, how does y change? Use the rules of exponents to explain your answer.

77. If y varies inversely as the square of x and x is doubled, how does y change? Use the rules of exponents to explain your answer.

78. Describe a real-life problem for each type of variation (direct, inverse, and joint).

Cumulative Review

In Exercises 79–82, write the expression using exponential notation.

79. $(6)(6)(6)(6)$
80. $(-4)(-4)(-4)$
81. $(\frac{1}{2})(\frac{1}{2})(\frac{1}{2})(\frac{1}{2})$
82. $-(-\frac{3}{4})(-\frac{3}{4})(-\frac{3}{4})$

In Exercises 83–86, use synthetic division to divide.

83. $(x^2 - 5x - 14) \div (x + 2)$
84. $(3x^2 - 5x + 2) \div (x + 1)$
85. $\frac{4x^5 - 14x^4 + 6x^3}{x - 3}$
86. $\frac{x^5 - 3x^2 - 5x + 1}{x - 2}$

Cumulative Review

Each exercise set (except those in Chapter 1) is followed by exercises that cover concepts from previous sections. This serves as a review for students and also helps students connect old concepts with new concepts.

206 Chapter 3 Equations, Inequalities, and Problem Solving

What Did You Learn?

Use these two pages to help prepare for a test on this chapter. Check off the key terms and key concepts you know. You can also use this section to record your assignments.

Plan for Test Success

Date of test: / /

Study dates and times: / / at : A.M./P.M.
/ / at : A.M./P.M.

Things to review:

- ☐ Key Terms, *p. 206*
- ☐ Key Concepts, *pp. 206–207*
- ☐ Your class notes
- ☐ Your assignments
- ☐ Study Tips, *pp. 129, 130, 132, 137, 140, 141, 143, 147, 148, 150, 151, 163, 164, 171, 178, 185, 187, 188, 197, 199, 201*
- ☐ Technology Tips, *pp. 128, 139, 173, 188, 202*
- ☐ Mid-Chapter Quiz, *p. 170*
- ☐ Review Exercises, *pp. 208–211*
- ☐ Chapter Test, *p. 212*
- ☐ Video Explanations Online
- ☐ Tutorial Online

Key Terms

- ☐ linear equation, *p. 126*
- ☐ first-degree equation, *p. 126*
- ☐ identity, *p. 131*
- ☐ consecutive integers, *p. 132*
- ☐ equivalent fractions, *p. 142*
- ☐ cross-multiplication, *p. 142*
- ☐ markup, *p. 152*
- ☐ discount, *p. 153*
- ☐ ratio, *p. 159*
- ☐ unit price, *p. 161*
- ☐ proportion, *p. 162*
- ☐ algebraic inequalities, *p. 184*
- ☐ solve an inequality, *p. 184*
- ☐ graph an inequality, *p. 184*
- ☐ bounded intervals, *p. 184*
- ☐ endpoints of an interval, *p. 184*
- ☐ unbounded (infinite) intervals, *p. 185*
- ☐ positive infinity, *p. 185*
- ☐ negative infinity, *p. 185*
- ☐ equivalent inequalities, *p. 186*
- ☐ linear inequality, *p. 187*
- ☐ compound inequality, *p. 189*
- ☐ intersection, *p. 190*
- ☐ union, *p. 190*
- ☐ absolute value equation, *p. 197*
- ☐ standard form of an absolute value equation, *p. 198*

Key Concepts

3.1 Solving Linear Equations

Assignment: ______ Due date: ______

☐ **Solve a linear equation.**
Solve a linear equation by using inverse operations to isolate the variable.

☐ **Write expressions for special types of integers.**
Let n be an integer.
1. $2n$ denotes an *even* integer.
2. $2n - 1$ and $2n + 1$ denote *odd* integers.
3. The set $\{n, n + 1, n + 2\}$ denotes three *consecutive* integers.

What Did You Learn? 207

3.2 Equations That Reduce to Linear Form

Assignment: ______ Due date: ______

☐ **Solve equations containing symbols of grouping.**
Remove symbols of grouping using the Distributive Property, combine like terms, isolate the variable using properties of equality, and check your solution in the original equation.

☐ **Solve equations involving fractions.**
To clear an equation of fractions, multiply each side by the least common multiple (LCM) of the denominators.
Use cross-multiplication to solve a linear equation that equates two fractions.

3.3 Problem Solving with Percents

Assignment: ______ Due date: ______

☐ **Use the percent equation $a = p \cdot b$.**
b = base number
p = percent (in decimal form)
a = number being compared to b

☐ **Use guidelines for solving word problems.**
See page 154.

3.4 Ratios and Proportions

Assignment: ______ Due date: ______

☐ **Define ratio.**
The ratio of the real number a to the real number b is given by a/b, or $a : b$.

☐ **Solve a proportion.**
A proportion equates two ratios.
If $\frac{a}{b} = \frac{c}{d}$, then $ad = bc$.

3.5 Geometric and Scientific Applications

Assignment: ______ Due date: ______

☐ **Use common formulas.**
See pages 171 and 173.

☐ **Solve mixture and work-rate problems.**
Mixture and work-rate problems are composed of the sum of two or more "hidden products" that involve rate factors.

3.6 Linear Inequalities

Assignment: ______ Due date: ______

☐ **Graph solutions on a number line.**
A parenthesis excludes an endpoint from the solution interval. A square bracket includes an endpoint in the solution interval.

☐ **Use properties of inequalities.**
See page 186.

3.7 Absolute Value Equations and Inequalities

Assignment: ______ Due date: ______

☐ **Solve absolute value equations.**
Let x be a variable or an algebraic expression and let a be a real number such that $a \ge 0$. The solutions of the equation $|x| = a$ are given by $x = a$ and $x = -a$.

☐ **Solve an absolute value inequality.**
See page 200.

What Did You Learn? (Chapter Summary)

The *What Did You Learn?* at the end of each chapter has been reorganized and expanded in the Fifth Edition. The *Plan for Test Success* provides a place for students to plan their studying for a test and includes a checklist of things to review. Students are also able to check off the *Key Terms* and *Key Concepts* of the chapter as these are reviewed. A space to record assignments for each section of the chapter is also provided.

Review Exercises

The *Review Exercises* at the end of each chapter contain skill-building and application exercises that are first ordered by section, and then grouped according to the objectives stated within *What You Should Learn*. This organization allows students to easily identify the appropriate sections and concepts for study and review.

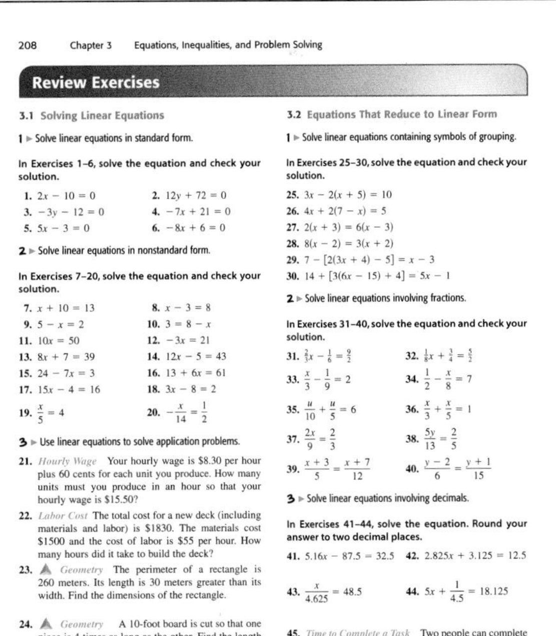

208 Chapter 3 Equations, Inequalities, and Problem Solving

Review Exercises

3.1 Solving Linear Equations

1 ▸ Solve linear equations in standard form.

In Exercises 1–6, solve the equation and check your solution.

1. $2x - 10 = 0$
2. $12y + 72 = 0$
3. $-3y - 12 = 0$
4. $-7x + 21 = 0$
5. $5x - 3 = 0$
6. $-8x + 6 = 0$

2 ▸ Solve linear equations in nonstandard form.

In Exercises 7–20, solve the equation and check your solution.

7. $x + 10 = 13$
8. $x - 3 = 8$
9. $5 - x = 2$
10. $3 = 8 - x$
11. $10x = 50$
12. $-3x = 21$
13. $8x + 7 = 39$
14. $12x - 5 = 43$
15. $24 - 7x = 3$
16. $13 + 6x = 61$
17. $15x - 4 = 16$
18. $3x - 8 = 2$
19. $\frac{x}{5} = 4$
20. $-\frac{x}{14} = \frac{1}{2}$

3 ▸ Use linear equations to solve application problems.

21. *Hourly Wage* Your hourly wage is \$8.30 per hour plus 60 cents for each unit you produce. How many units must you produce in an hour so that your hourly wage is \$15.50?
22. *Labor Cost* The total cost for a new deck (including materials and labor) is \$1830. The materials cost \$1500 and the cost of labor is \$55 per hour. How many hours did it take to build the deck?
23. ▲ *Geometry* The perimeter of a rectangle is 260 meters. Its length is 30 meters greater than its width. Find the dimensions of the rectangle.
24. ▲ *Geometry* A 10-foot board is cut so that one piece is 4 times as long as the other. Find the length of each piece.

3.2 Equations That Reduce to Linear Form

1 ▸ Solve linear equations containing symbols of grouping.

In Exercises 25–30, solve the equation and check your solution.

25. $3x - 2(x + 5) = 10$
26. $4x + 2(7 - x) = 5$
27. $2(x + 3) = 6(x - 3)$
28. $8(x - 2) = 3(x + 2)$
29. $7 - [2(3x + 4) - 5] = x - 3$
30. $14 + [3(6x - 15) + 4] = 5x - 1$

2 ▸ Solve linear equations involving fractions.

In Exercises 31–40, solve the equation and check your solution.

31. $\frac{2}{3}x - \frac{1}{6} = \frac{9}{2}$
32. $\frac{1}{8}x + \frac{3}{4} = \frac{5}{2}$
33. $\frac{x}{3} - \frac{1}{9} = 2$
34. $\frac{1}{2} - \frac{x}{8} = 7$
35. $\frac{u}{10} + \frac{u}{5} = 6$
36. $\frac{x}{3} + \frac{x}{5} = 1$
37. $\frac{2x}{9} = \frac{2}{3}$
38. $\frac{5y}{13} = \frac{2}{5}$
39. $\frac{x + 3}{5} = \frac{x + 7}{12}$
40. $\frac{y - 2}{6} = \frac{y + 1}{15}$

3 ▸ Solve linear equations involving decimals.

In Exercises 41–44, solve the equation. Round your answer to two decimal places.

41. $5.16x - 87.5 = 32.5$
42. $2.825x + 3.125 = 12.5$
43. $\frac{x}{4.625} = 48.5$
44. $5x + \frac{1}{4.5} = 18.125$

45. *Time to Complete a Task* Two people can complete 50% of a task in t hours, where t must satisfy the equation $\frac{t}{10} + \frac{t}{15} = 0.5$. How long will it take for the two people to complete 50% of the task?

170 Chapter 3 Equations, Inequalities, and Problem Solving

Mid-Chapter Quiz

Take this quiz as you would take a quiz in class. After you are done, check your work against the answers in the back of the book.

In Exercises 1–10, solve the equation.

1. $74 - 12x = 2$
2. $10(y - 8) = 0$
3. $3x + 1 = x + 20$
4. $6x + 8 = 8 - 2x$
5. $-10x + \frac{2}{3} = \frac{7}{3} - 5x$
6. $\frac{x}{5} + \frac{x}{7} = 1$
7. $\frac{9 + x}{3} = 15$
8. $3 - 5(4 - x) = -6$
9. $\frac{x + 3}{6} = \frac{4}{3}$
10. $\frac{x + 7}{5} = \frac{x + 9}{7}$

In Exercises 11 and 12, solve the equation. Round your answer to two decimal places. In your own words, explain how to check the solution.

11. $32.86 - 10.5x = 11.25$
12. $\frac{x}{5.45} + 3.2 = 12.6$
13. What number is 62% of 25?
14. What number is $\frac{1}{2}$% of 8400?
15. 300 is what percent of 150?
16. 145.6 is 32% of what number?
17. You work 40 hours a week at a candy store and earn \$7.50 per hour. You also earn \$7.00 per hour baby-sitting and can work as many hours as you want. You want to earn \$370 a week. How many hours must you baby-sit?
18. A region has an area of 42 square meters. It must be divided into three subregions so that the second has twice the area of the first, and the third has twice the area of the second. Find the area of each subregion.
19. To get an A in a psychology course, you must have an average of at least 90 points for 3 tests of 100 points each. For the first 2 tests, your scores are 84 and 93. What must you score on the third test to earn a 90% average for the course?
20. The circle graph at the left shows the numbers of endangered wildlife and plant species as of October 2007. What percent of the total number of endangered wildlife and plant species were birds? (Source: U.S. Fish and Wildlife Service)
21. Two people can paint a room in t hours, where t must satisfy the equation $t/4 + t/12 = 1$. How long will it take for the two people to paint the room?
22. A large round pizza has a radius of $r = 15$ inches, and a small round pizza has a radius of $r = 8$ inches. Find the ratio of the area of the large pizza to the area of the small pizza. (*Hint:* The area of a circle is $A = \pi r^2$.)
23. A car uses 30 gallons of gasoline for a trip of 800 miles. How many gallons would be used on a trip of 700 miles?

Endangered Wildlife and Plant Species

Plants 599
Mammals 325
Birds 251
Other 232
Fishes 85
Reptiles 78

Figure for 20

Mid-Chapter Quiz

Each chapter contains a *Mid-Chapter Quiz.* Answers to all questions in the *Mid-Chapter Quiz* are given in the back of the student text and are located in place in the Annotated Instructor's Edition.

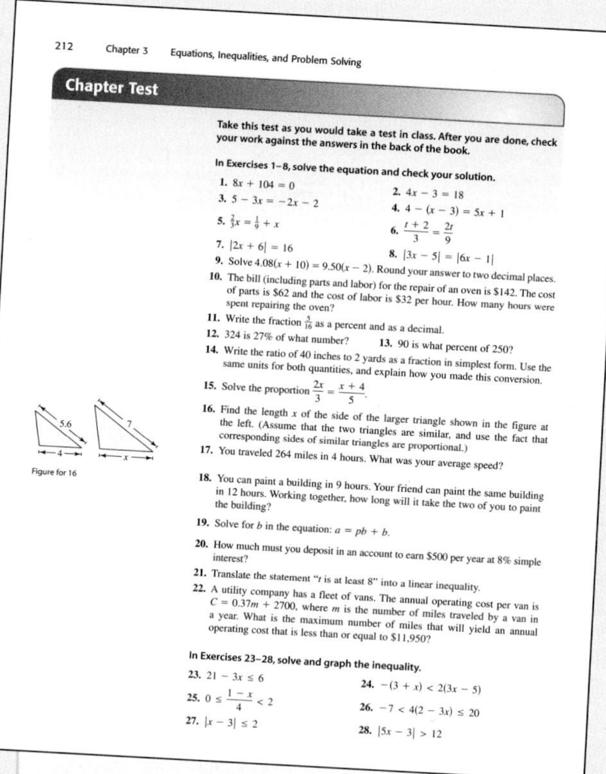

212 Chapter 3 Equations, Inequalities, and Problem Solving

Chapter Test

Take this test as you would take a test in class. After you are done, check your work against the answers in the back of the book.

In Exercises 1–8, solve the equation and check your solution.

1. $8x + 104 = 0$
2. $4x - 3 = 18$
3. $5 - 3x = -2x - 2$
4. $4 - (x - 3) = 5x + 1$
5. $\frac{2}{3}x = \frac{1}{9} + x$
6. $\frac{t + 2}{3} = \frac{2t}{9}$
7. $|2x + 6| = 16$
8. $|3x - 5| = |6x - 1|$
9. Solve $4.08(x + 10) = 9.50(x - 2)$. Round your answer to two decimal places.
10. The bill (including parts and labor) for the repair of an oven is \$142. The cost of parts is \$62 and the cost of labor is \$32 per hour. How many hours were spent repairing the oven?
11. Write the fraction $\frac{5}{16}$ as a percent and as a decimal.
12. 324 is 27% of what number?
13. 90 is what percent of 250?
14. Write the ratio of 40 inches to 2 yards as a fraction in simplest form. Use the same units for both quantities, and explain how you made this conversion.
15. Solve the proportion $\frac{2x}{3} = \frac{x + 4}{5}$.
16. Find the length x of the side of the larger triangle shown in the figure at the left. (Assume that the two triangles are similar, and use the fact that corresponding sides of similar triangles are proportional.)
17. You traveled 264 miles in 4 hours. What was your average speed?
18. You can paint a building in 9 hours. Your friend can paint the same building in 12 hours. Working together, how long will it take the two of you to paint the building?
19. Solve for b in the equation: $a = pb + b$.
20. How much must you deposit in an account to earn \$500 per year at 8% simple interest?
21. Translate the statement "t is at least 8" into a linear inequality.
22. A utility company has a fleet of vans. The annual operating cost per van is $C = 0.37m + 2700$, where m is the number of miles traveled by a van in a year. What is the maximum number of miles that will yield an annual operating cost that is less than or equal to \$11,950?

In Exercises 23–28, solve and graph the inequality.

23. $21 - 3x \le 6$
24. $-(3 + x) < 2(3x - 5)$
25. $0 \le \frac{1 - x}{4} < 2$
26. $-7 < 4(2 - 3x) \le 20$
27. $|x - 3| \le 2$
28. $|5x - 3| > 12$

Figure for 16

Chapter Test

Each chapter ends with a *Chapter Test.* Answers to all questions in the *Chapter Test* are given in the back of the student text and are located in place in the Annotated Instructor's Edition.

Cumulative Test 213

Cumulative Test: Chapters 1–3

Take this test as you would take a test in class. After you are done, check your work against the answers in the back of the book.

1. Place the correct symbol (< or >) between the numbers: $-\frac{3}{4}$ ▢ $\left|-\frac{7}{8}\right|$.

In Exercises 2–7, evaluate the expression.

2. $(-200)(2)(-3)$
3. $\frac{3}{8} - \frac{5}{6}$
4. $-\frac{2}{9} \div \frac{8}{75}$
5. $-(-2)^3$
6. $3 + 2(6) - 1$
7. $24 + 12 \div 3$

In Exercises 8 and 9, evaluate the expression when $x = -2$ and $y = 3$.

8. $-3x - (2y)^2$
9. $\frac{5}{6}y + x^3$
10. Use exponential form to write the product $3 \cdot (x + y) \cdot (x + y) \cdot 3 \cdot 3$.
11. Use the Distributive Property to expand $-2x(x - 3)$.
12. Identify the property of real numbers illustrated by

$$2 + (3 + x) = (2 + 3) + x.$$

In Exercises 13–15, simplify the expression.

13. $(3x^3)(5x^4)$
14. $2x^2 - 3x + 5x^2 - (2 + 3x)$
15. $4(x^2 + x) + 7(2x - x^2)$

In Exercises 16–18, solve the equation and check your solution.

16. $12x - 3 = 7x + 27$
17. $2x - \frac{5x}{4} = 13$
18. $5(x + 8) = -2x - 9$
19. Solve and graph the inequality.

$$-8(x + 5) \le 16$$

20. The sticker on a new car gives the fuel efficiency as 28.3 miles per gallon. In your own words, explain how to estimate the annual fuel cost for the buyer if the car will be driven approximately 15,000 miles per year and the fuel cost is \$2.759 per gallon.
21. Write the ratio "24 ounces to 2 pounds" as a fraction in simplest form.
22. The suggested retail price of a digital camcorder is \$1150. The camcorder is on sale for "20% off" the list price. Find the sale price.
23. The figure at the left shows two pieces of property. The assessed values of the properties are proportional to their areas. The value of the larger piece is \$95,000. What is the value of the smaller piece?

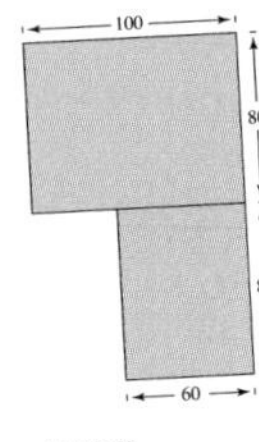

Figure for 23

Cumulative Test

The *Cumulative Tests* that follow Chapters 3, 6, 9, and 12 provide a comprehensive self-assessment tool that helps students check their mastery of previously covered material. Answers to all questions in the *Cumulative Tests* are given in the back of the student text and are located in place in the Annotated Instructor's Edition.

Supplements

Elementary and Intermediate Algebra, Fifth Edition, by Ron Larson is accompanied by a comprehensive supplements package, which includes resources for both students and instructors. All items are keyed to the text.

Printed Resources

For Students

Student Solutions Manual by Carolyn Neptune, Johnson County Community College, and Gerry Fitch, Louisiana State University
(0547140347)

- Detailed, step-by-step solutions to all odd-numbered exercises in the section exercise sets and in the review exercises
- Detailed, step-by-step solutions to all Mid-Chapter Quiz, Chapter Test, and Cumulative Test questions

For Instructors

Annotated Instructor's Edition
(0547102259)

- Includes answers in place for Exercise sets, Review Exercises, Mid-Chapter Quizzes, Chapter Tests, and Cumulative Tests
- Additional Answers section in the back of the text lists those answers that contain large graphics or lengthy exposition
- Answers to the Technology: Tip and Technology: Discovery questions are provided in the back of the book
- Annotations at point of use that offer strategies and suggestions for teaching the course and point out common student errors

Complete Solutions Manual by Carolyn Neptune, Johnson County Community College, and Gerry Fitch, Louisiana State University
(0547140290)

- Chapter and Final Exam test forms with answer key
- Individual test items and answers for Chapters 1–13
- Notes to the instructor including tips and strategies on student assessment, cooperative learning, classroom management, study skills, and problem solving

Technology Resources

For Students

Website *(www.cengage.com/math/larson/algebra)*

Instructional DVDs by Dana Mosely to accompany Larson, Developmental Math Series, 5e (05471402074)

Personal Tutor An easy-to-use and effective live, online tutoring service. *Whiteboard Simulations* and Practice Area promote real-time visual interaction.

For Instructors

Power Lecture CD-ROM with Diploma® (0547140207) This CD-ROM provides the instructor with dynamic media tools for teaching. Create, deliver, and customize tests (both print and online) in minutes with Diploma® computerized testing featuring algorithmic equations. Easily build solution sets for homework or exams using *Solution Builder*'s online solutions manual. Microsoft® PowerPoint® lecture slides, figures from the book, and Test Bank, in electronic format, are also included on this CD-ROM.

WebAssign Instant feedback and ease of use are just two reasons why WebAssign is the most widely used homework system in higher education. WebAssign's homework delivery system allows you to assign, collect, grade, and record homework assignments via the web. And now, this proven system has been enhanced to include links to textbook sections, video examples, and problem-specific tutorials.

Website *(www.cengage.com/math/larson/algebra)*

Solution Builder This online tool lets instructors build customized solution sets in three simple steps and then print and hand out in class or post to a password-protected class website.

Acknowledgments

I would like to thank the many people who have helped me revise the various editions of this text. Their encouragement, criticisms, and suggestions have been invaluable.

Reviewers

Tom Anthony, Central Piedmont Community College; Tina Cannon, Chattanooga State Technical Community College; LeAnne Conaway, Harrisburg Area Community College and Penn State University; Mary Deas, Johnson County Community College; Jeremiah Gilbert, San Bernadino Valley College; Jason Pallett, Metropolitan Community College-Longview; Laurence Small, L.A. Pierce College; Dr. Azar Raiszadeh, Chattanooga State Technical Community College; Patrick Ward, Illinois Central College.

My thanks to Kimberly Nolting, Hillsborough Community College, for her contributions to this project. My thanks also to Robert Hostetler, The Behrend College, The Pennsylvania State University, and Patrick M. Kelly, Mercyhurst College, for their significant contributions to previous editions of this text.

I would also like to thank the staff of Larson Texts, Inc., who assisted in preparing the manuscript, rendering the art package, and typesetting and proofreading the pages and the supplements.

On a personal level, I am grateful to my spouse, Deanna Gilbert Larson, for her love, patience, and support. Also, a special thanks goes to R. Scott O'Neil.

If you have suggestions for improving this text, please feel free to write to me. Over the past two decades I have received many useful comments from both instructors and students, and I value these comments very much.

Ron Larson

Study Skills in Action

Keeping a Positive Attitude

A student's experiences during the first three weeks in a math course often determine whether the student sticks with it or not. You can get yourself off to a good start by immediately acquiring a positive attitude and the study behaviors to support it.

Using Study Strategies

In each *Study Skills in Action* feature, you will learn a new study strategy that will help you progress through the course. Each strategy will help you:

- set up good study habits;
- organize information into smaller pieces;
- create review tools;
- memorize important definitions and rules;
- learn the math at hand.

Kimberly Nolting
VP, Academic Success Press
expert in developmental education

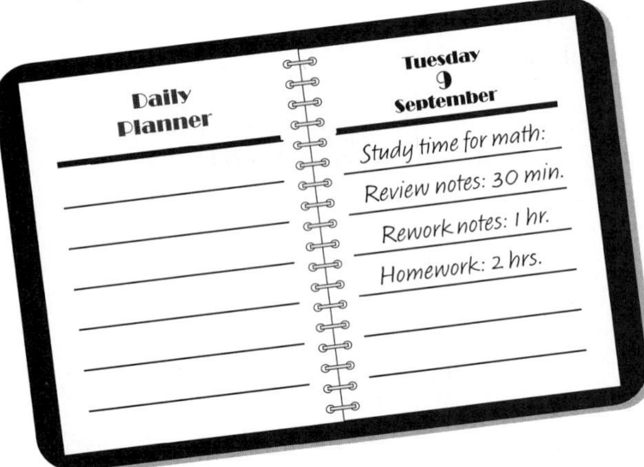

Daily
Planner
Wednesday
10
September
Important information:
Tutoring Center hours:
7:00 a.m. to 11:00 p.m.
Instructor's office hours:
M 4:30 to 6:00 p.m.
Th 9:00 to 10:30 a.m.

Smart Study Strategy

Create a Positive Study Environment

1 ► After the first math class, set aside time for reviewing your notes and the textbook, reworking your notes, and completing homework.

2 ► Find a productive study environment on campus. Most colleges have a tutoring center where students can study and receive assistance as needed.

3 ► Set up a place for studying at home that is comfortable, but not too comfortable. It needs to be away from all potential distractions.

4 ► Make at least two other *collegial friends* in class. Collegial friends are students who study well together, help each other out when someone gets sick, and keep each other's attitudes positive.

5 ► Meet with your instructor at least once during the first two weeks. Ask the instructor what he or she advises for study strategies in the class. This will help you and let the instructor know that you really want to do well.

Chapter 1

The Real Number System

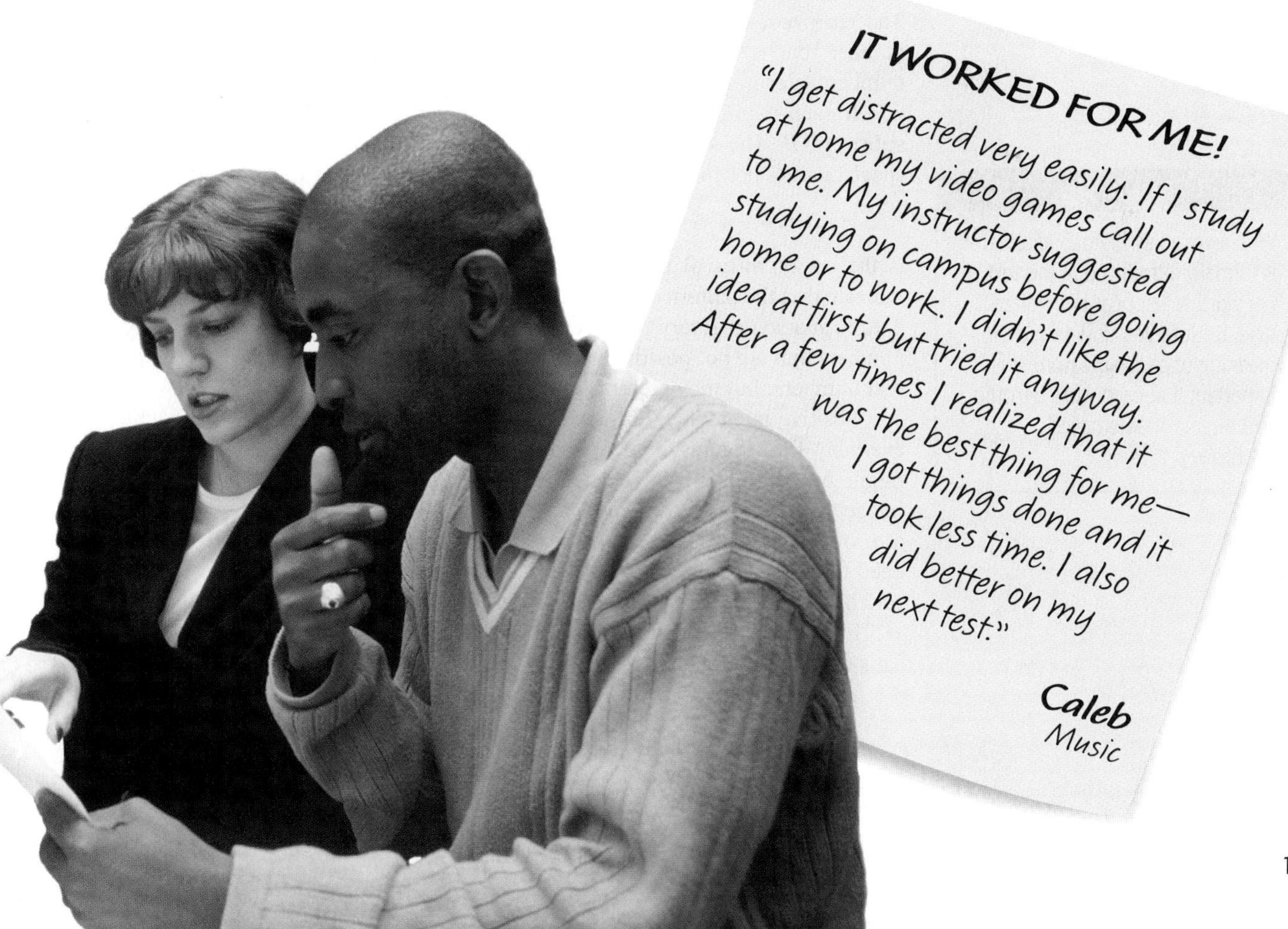

IT WORKED FOR ME!
"I get distracted very easily. If I study at home my video games call out to me. My instructor suggested studying on campus before going home or to work. I didn't like the idea at first, but tried it anyway. After a few times I realized that it was the best thing for me—I got things done and it took less time. I also did better on my next test."
Caleb
Music

1.1 Real Numbers: Order and Absolute Value

Chris Collins/Veer

What You Should Learn

1 ▶ Define sets and use them to classify numbers as natural, integer, rational, or irrational.
2 ▶ Plot numbers on the real number line.
3 ▶ Use the real number line and inequality symbols to order real numbers.
4 ▶ Find the absolute value of a number.

Why You Should Learn It

Understanding sets and subsets of real numbers will help you to analyze real-life situations accurately.

1 ▶ Define sets and use them to classify numbers as natural, integer, rational, or irrational.

Sets and Real Numbers

The ability to communicate precisely is an essential part of a modern society, and it is the primary goal of this text. Specifically, this section introduces the language used to communicate numerical concepts.

The formal term that is used in mathematics to refer to a collection of objects is the word **set.** For instance, the set $\{1, 2, 3\}$ contains the three numbers 1, 2, and 3. Note that a pair of braces $\{\ \ \}$ is used to list the members of the set. Parentheses ( ) and brackets [ ] are used to represent other ideas.

The set of numbers that is used in arithmetic is called the set of **real numbers.** The term *real* distinguishes real numbers from *imaginary* numbers—a type of number that is used in some mathematics courses. You will study imaginary numbers later in the text.

If each member of a set A is also a member of a set B, then A is called a **subset** of B. The set of real numbers has many important subsets, each with a special name. For instance, the set

$\{1, 2, 3, 4, . . .\}$ A subset of the set of real numbers

is the set of **natural numbers** or **positive integers.** Note that the three dots indicate that the pattern continues. For instance, the set also contains the numbers 5, 6, 7, and so on. Every positive integer is a real number, but there are many real numbers that are not positive integers. For example, the numbers -2, 0, and $\frac{1}{2}$ are real numbers, but they are not positive integers.

Positive integers can be used to describe many things that you encounter in everyday life. For instance, you might be taking four classes this term, or you might be paying \$480 a month for rent. But even in everyday life, positive integers cannot describe some concepts accurately. For instance, you could have a zero balance in your checking account. To describe a quantity such as this, you need to expand the set of positive integers to include zero. The expanded set is called the set of **whole numbers.** To describe a quantity such as a temperature of $-5°$F, you need to expand the set of whole numbers to include **negative integers.** This expanded set is called the set of **integers.**

$$\{\underbrace{. . . , -3, -2, -1}_{\text{Negative integers}}, \overset{\text{Zero}}{0}, \underbrace{1, 2, 3, . . .}_{\text{Positive integers}}\}$$ Set of integers

The set of integers is also a subset of the set of real numbers.

Study Tip

Whenever a key mathematical term is formally introduced in this text, the word will appear in boldface type. Be sure you understand the meaning of each new word; it is important that each word become part of your mathematical vocabulary.

Even with the set of integers, there are still many quantities in everyday life that you cannot describe accurately. The costs of many items are not in whole-dollar amounts, but in parts of dollars, such as \$1.19 or \$39.98. You might work $8\frac{1}{2}$ hours, or you might miss the first half of a movie. To describe such quantities, you can expand the set of integers to include **fractions.** The expanded set is called the set of **rational numbers.** In the formal language of mathematics, a real number is **rational** if it can be written as a ratio of two integers. So, $\frac{3}{4}$ is a rational number; so is 0.5 $\left(\text{it can be written as } \frac{1}{2}\right)$; and so is every integer. A real number that is not rational is called **irrational** and cannot be written as the ratio of two integers. One example of an irrational number is $\sqrt{2}$, which is read as the positive square root of 2. Another example is π (the Greek letter pi), which represents the ratio of the circumference of a circle to its diameter. Each of the sets of numbers mentioned—natural numbers, whole numbers, integers, rational numbers, and irrational numbers—is a subset of the set of real numbers, as shown in Figure 1.1.

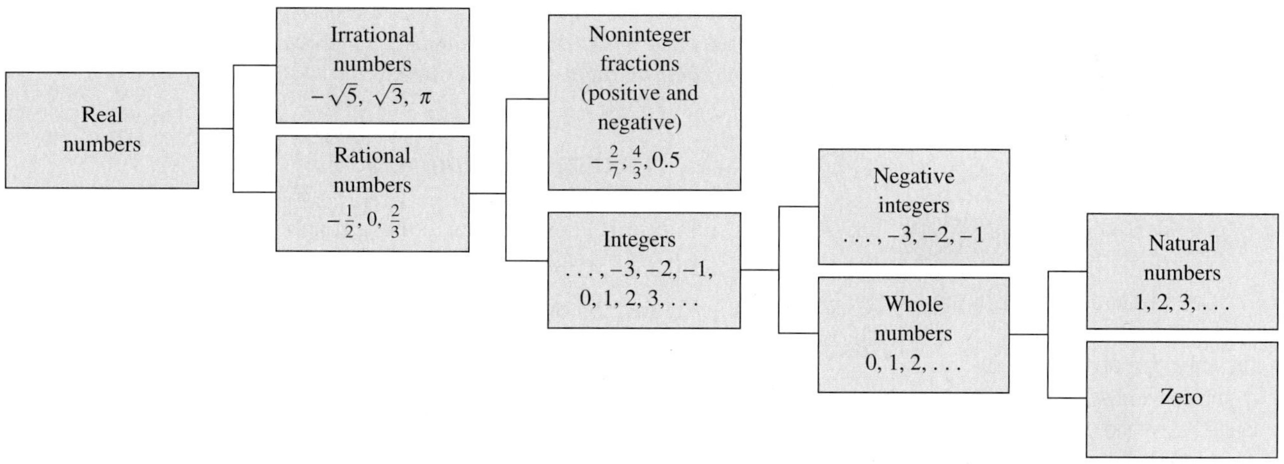

Figure 1.1 Subsets of Real Numbers

Study Tip

In *decimal form,* you can recognize rational numbers as decimals that terminate

$\frac{1}{2} = 0.5$ or $\frac{3}{8} = 0.375$

or repeat

$\frac{4}{3} = 1.\overline{3}$ or $\frac{2}{11} = 0.\overline{18}.$

Irrational numbers are represented by decimals that neither terminate nor repeat, as in

$\sqrt{2} = 1.414213562\ldots$

or

$\pi = 3.141592654\ldots.$

EXAMPLE 1 Classifying Real Numbers

Which of the numbers in the following set are (a) natural numbers, (b) integers, (c) rational numbers, and (d) irrational numbers?

$$\left\{\frac{1}{2}, -1, 0, 4, -\frac{5}{8}, \frac{4}{2}, -\frac{3}{1}, 0.86, \sqrt{2}, \sqrt{9}\right\}$$

Solution

a. Natural numbers: $\left\{4, \frac{4}{2} = 2, \sqrt{9} = 3\right\}$

b. Integers: $\left\{-1, 0, 4, \frac{4}{2} = 2, -\frac{3}{1} = -3, \sqrt{9} = 3\right\}$

c. Rational numbers: $\left\{\frac{1}{2}, -1, 0, 4, -\frac{5}{8}, \frac{4}{2}, -\frac{3}{1}, 0.86, \sqrt{9} = 3\right\}$

d. Irrational number: $\left\{\sqrt{2}\right\}$

✓ ***CHECKPOINT*** *Now try Exercise 3.*

2 ▶ Plot numbers on the real number line.

The Real Number Line

The diagram used to represent the real numbers is called the **real number line.** It consists of a horizontal line with a point (the **origin**) labeled 0. Numbers to the left of 0 are **negative** and numbers to the right of 0 are **positive,** as shown in Figure 1.2. The real number zero is neither positive nor negative. So, the term **nonnegative** implies that a number may be positive or zero.

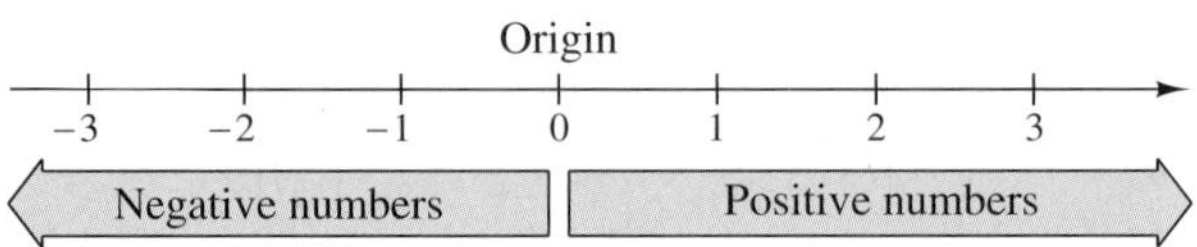

Figure 1.2 The Real Number Line

Drawing the point on the real number line that corresponds to a real number is called **plotting** the real number.

Example 2 illustrates the following principle. *Each point on the real number line corresponds to exactly one real number, and each real number corresponds to exactly one point on the real number line.*

Technology: Tip

The Greek letter pi, denoted by the symbol π, is the ratio of the circumference of a circle to its diameter. Because π cannot be written as a ratio of two integers, it is an irrational number. You can obtain an approximation of π on a scientific or graphing calculator by using the following keystroke.

Keystroke	*Display*
π	3.141592654

Between which two integers would you plot π on the real number line?

EXAMPLE 2 Plotting Real Numbers

a. The point in Figure 1.3 corresponds to the real number $-\frac{1}{2}$.

b. The point in Figure 1.4 corresponds to the real number 2.

c. The point in Figure 1.5 corresponds to the real number $-\frac{3}{2}$.

d. The point in Figure 1.6 corresponds to the real number 1.

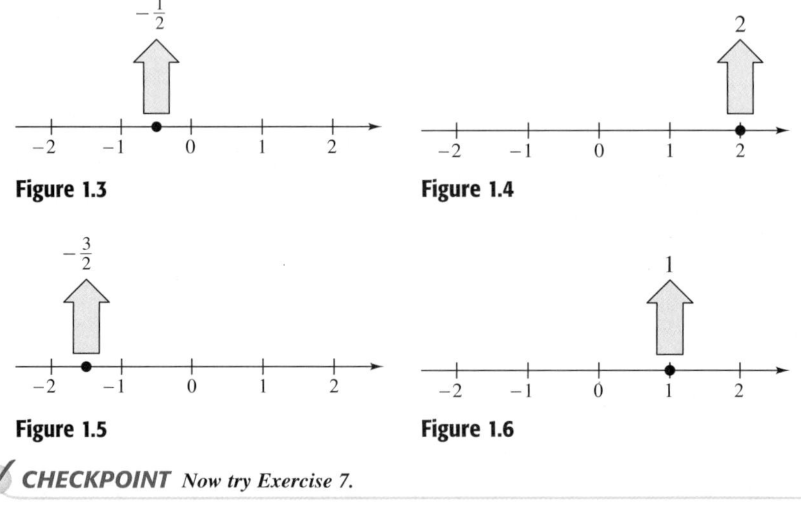

Figure 1.3

Figure 1.4

Figure 1.5

Figure 1.6

✓ **CHECKPOINT** *Now try Exercise 7.*

3 ▶ Use the real number line and inequality symbols to order real numbers.

Ordering Real Numbers

The real number line provides you with a way of comparing any two real numbers. For instance, if you choose any two (different) numbers on the real number line, one of the numbers must be to the left of the other number. The number to the left is **less than** the number to the right. Similarly, the number to the right is **greater than** the number to the left. For example, in Figure 1.7 you can see that -3 is less than 2 because -3 lies to the left of 2 on the number line. A "less than" comparison is denoted by the **inequality symbol** $<$. For instance, "-3 is less than 2" is denoted by $-3 < 2$.

Similarly, the inequality symbol $>$ is used to denote a "greater than" comparison. For instance, "2 is greater than -3" is denoted by $2 > -3$. The inequality symbol $\leq$ means **less than or equal to,** and the inequality symbol $\geq$ means **greater than or equal to.**

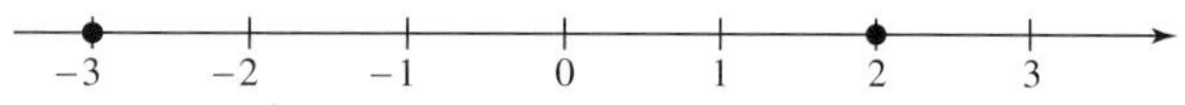

Figure 1.7 -3 lies to the left of 2.

When you are asked to **order** two numbers, you are simply being asked to say which of the two numbers is greater.

EXAMPLE 3 Ordering Integers

Place the correct inequality symbol ($<$ or $>$) between each pair of numbers.

a. 3 ▭ 5 **b.** -3 ▭ -5 **c.** 4 ▭ 0

d. -2 ▭ 2 **e.** 1 ▭ -4

Solution

a. $3 < 5$, because 3 lies to the *left* of 5. See Figure 1.8.

b. $-3 > -5$, because -3 lies to the *right* of -5. See Figure 1.9.

c. $4 > 0$, because 4 lies to the *right* of 0. See Figure 1.10.

d. $-2 < 2$, because -2 lies to the *left* of 2. See Figure 1.11.

e. $1 > -4$, because 1 lies to the *right* of -4. See Figure 1.12.

Figure 1.8

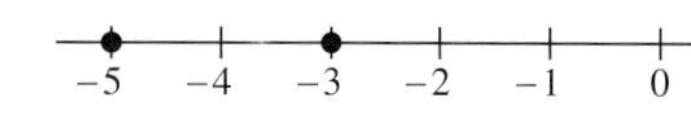

Figure 1.9

Figure 1.10

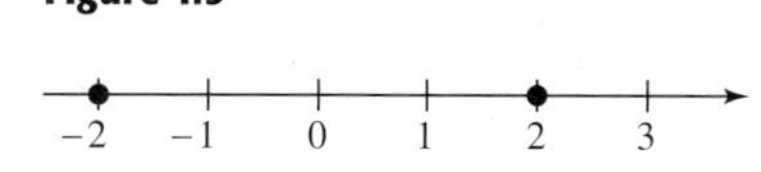

Figure 1.11

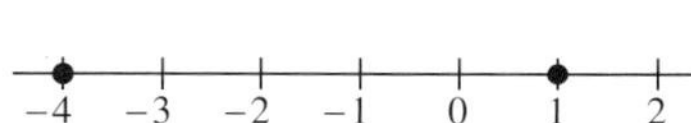

Figure 1.12

✓ **CHECKPOINT** *Now try Exercise 9.*

To order two fractions, you can write both fractions with the same denominator, or you can rewrite both fractions in decimal form. Here are two examples.

$$\frac{1}{3} = \frac{4}{12} \text{ and } \frac{1}{4} = \frac{3}{12} \quad \Rightarrow \quad \frac{1}{3} > \frac{1}{4}$$

$$\frac{11}{131} \approx 0.084 \text{ and } \frac{19}{209} \approx 0.091 \quad \Rightarrow \quad \frac{11}{131} < \frac{19}{209}$$

The symbol $\approx$ means "is approximately equal to."

EXAMPLE 4 Ordering Fractions

Place the correct inequality symbol (< or >) between each pair of numbers.

a. $\frac{1}{3}$ ☐ $\frac{1}{5}$ **b.** $-\frac{3}{2}$ ☐ $-\frac{1}{7}$

Solution

a. Write both fractions with the same denominator.

$$\frac{1}{3} = \frac{1 \cdot 5}{3 \cdot 5} = \frac{5}{15} \qquad \frac{1}{5} = \frac{1 \cdot 3}{5 \cdot 3} = \frac{3}{15}$$

Because $\frac{5}{15} > \frac{3}{15}$, you can conclude that $\frac{1}{3} > \frac{1}{5}$. (See Figure 1.13.)

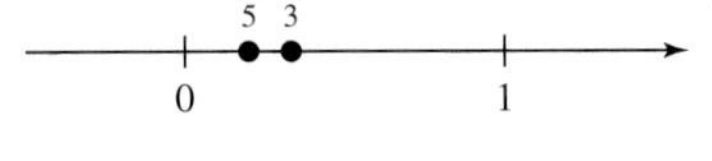

Figure 1.13

b. Write both fractions with the same denominator.

$$-\frac{3}{2} = -\frac{3 \cdot 7}{2 \cdot 7} = -\frac{21}{14} \qquad -\frac{1}{7} = -\frac{1 \cdot 2}{7 \cdot 2} = -\frac{2}{14}$$

Because $-\frac{21}{14} < -\frac{2}{14}$, you can conclude that $-\frac{3}{2} < -\frac{1}{7}$. (See Figure 1.14.)

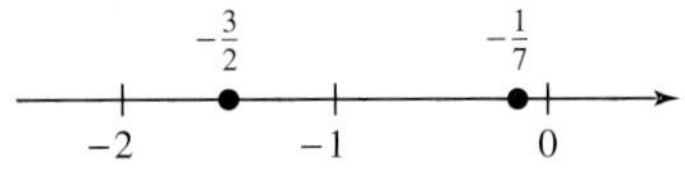

Figure 1.14

✓ ***CHECKPOINT*** *Now try Exercise 15.*

EXAMPLE 5 Ordering Decimals

Place the correct inequality symbol (< or >) between each pair of numbers.

a. -3.1 ☐ 2.8 **b.** -1.09 ☐ -1.90

Solution

a. $-3.1 < 2.8$, because -3.1 lies to the *left* of 2.8. (See Figure 1.15.)

b. $-1.09 > -1.90$, because -1.09 lies to the *right* of -1.90. (See Figure 1.16.)

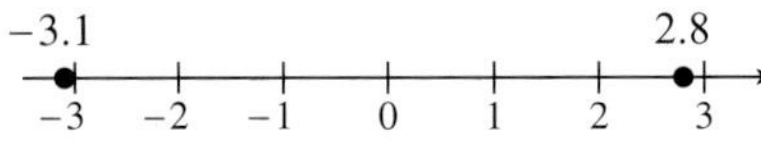

Figure 1.15

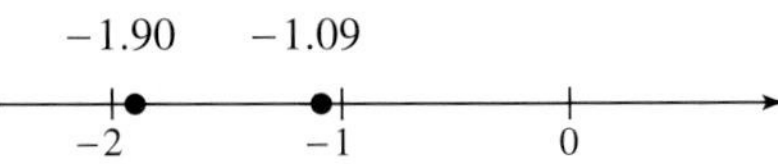

Figure 1.16

✓ ***CHECKPOINT*** *Now try Exercise 19.*

4 ▶ Find the absolute value of a number.

Smart Study Strategy

Go to page xxii for ways to *Create a Positive Study Environment.*

Absolute Value

Two real numbers are **opposites** of each other if they lie the same distance from, but on opposite sides of, zero. For example, -2 is the opposite of 2, and 4 is the opposite of -4, as shown in Figure 1.17.

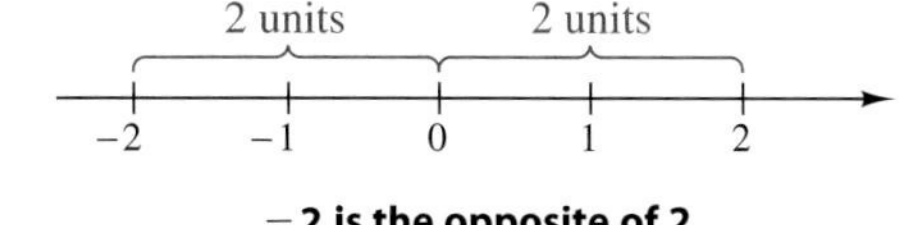

−2 is the opposite of 2.

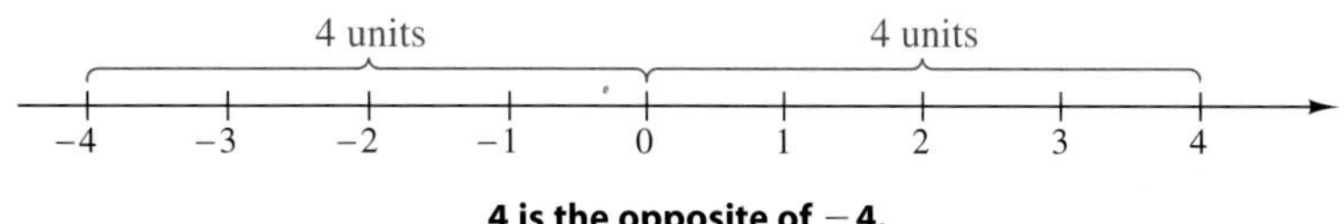

4 is the opposite of −4.

Figure 1.17

Parentheses are useful for denoting the opposite of a negative number. For example, $-(-3)$ means the opposite of -3, which is 3. That is,

$-(-3) = 3.$ The opposite of -3 is 3.

For any real number, its distance from zero on the real number line is its **absolute value.** A pair of vertical bars, $|\ \ |$, is used to denote absolute value. Here are two examples.

$|5| =$ "distance between 5 and 0" $= 5$

$|-8| =$ "distance between -8 and 0" $= 8$ See Figure 1.18.

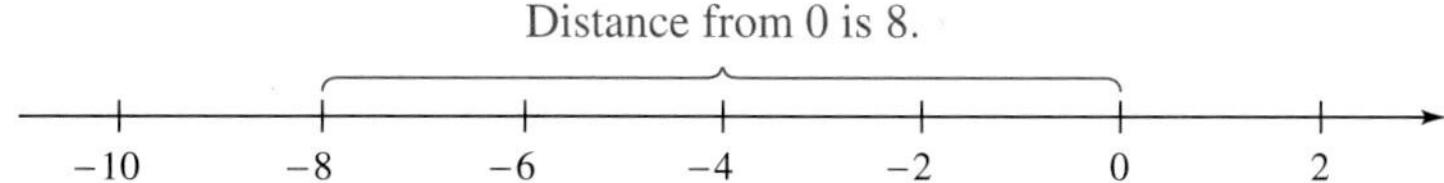

Figure 1.18

Because opposite numbers lie the same distance from zero on the real number line, they have the same absolute value. So, $|5| = 5$ and $|-5| = 5$ (see Figure 1.19).

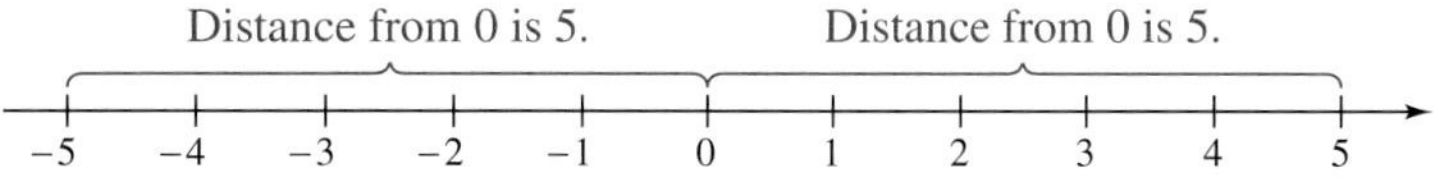

Figure 1.19

You can write this more simply as $|5| = |-5| = 5$.

Definition of Absolute Value

If a is a real number, then the **absolute value** of a is

$$|a| = \begin{cases} a, & \text{if } a \geq 0 \\ -a, & \text{if } a < 0 \end{cases}.$$

The absolute value of a real number is either positive or zero (never negative). For instance, by definition, $|-3| = -(-3) = 3$. Moreover, zero is the only real number whose absolute value is 0. That is, $|0| = 0$.

The word **expression** means a collection of numbers and symbols such as $3 + 5$ or $|-4|$. When asked to **evaluate** an expression, you are to find the *number* that is equal to the expression.

EXAMPLE 6 Evaluating Absolute Values

Evaluate each expression.

a. $|-10|$ **b.** $\left|\frac{3}{4}\right|$

c. $|-3.2|$ **d.** $-|-6|$

Solution

a. $|-10| = 10$, because the distance between -10 and 0 is 10.

b. $\left|\frac{3}{4}\right| = \frac{3}{4}$, because the distance between $\frac{3}{4}$ and 0 is $\frac{3}{4}$.

c. $|-3.2| = 3.2$, because the distance between -3.2 and 0 is 3.2.

d. $-|-6| = -(6) = -6$

✓ ***CHECKPOINT*** *Now try Exercise 37.*

Note in Example 6(d) that $-|-6| = -6$ does not contradict the fact that the absolute value of a real number cannot be negative. The expression $-|-6|$ calls for the *opposite* of an absolute value and so it must be negative.

EXAMPLE 7 Comparing Absolute Values

Place the correct symbol ($<$, $>$, or $=$) between each pair of numbers.

a. $|-9| \quad |9|$

b. $|-3| \quad 5$

c. $0 \quad |-7|$

d. $-4 \quad -|-4|$

e. $|12| \quad |-15|$

f. $2 \quad -|-2|$

Solution

a. $|-9| = |9|$, because $|-9| = 9$ and $|9| = 9$.

b. $|-3| < 5$, because $|-3| = 3$ and 3 is less than 5.

c. $0 < |-7|$, because $|-7| = 7$ and 0 is less than 7.

d. $-4 = -|-4|$, because $-|-4| = -4$ and -4 is equal to -4.

e. $|12| < |-15|$, because $|12| = 12$, $|-15| = 15$, and 12 is less than 15.

f. $2 > -|-2|$, because $-|-2| = -2$ and 2 is greater than -2.

✓ ***CHECKPOINT*** *Now try Exercise 49.*

Concept Check

1. In your own words, define rational and irrational numbers. Give an example of each.

2. Explain the difference between plotting the numbers 4 and -4 on the real number line.

3. Explain how to determine the smaller of two different real numbers.

4. How many numbers are three units from 0 on the real number line? Explain your answer.

1.1 EXERCISES

Go to pages 58–59 to record your assignments.

Developing Skills

In Exercises 1–4, determine which of the numbers in the set are (a) natural numbers, (b) integers, (c) rational numbers, and (d) irrational numbers. *See Example 1.*

1. $\left\{-3, 20, \pi, -\frac{3}{2}, \frac{9}{3}, 4.5\right\}$

2. $\left\{\sqrt{16}, 10, -82, -\frac{24}{3}, -8.2, \sqrt{5}, \frac{1}{5}\right\}$

3. $\left\{\sqrt{13}, -\frac{5}{2}, 6.5, -4.5, \frac{8}{4}, \frac{3}{4}\right\}$

4. $\left\{8, \sqrt{25}, -1, \frac{4}{3}, -3.25, \sqrt{49}, -\frac{10}{2}\right\}$

In Exercises 5–8, plot the numbers on the real number line. *See Example 2.*

5. $-7, 1.5$
6. $4, -3.2$
7. $\frac{1}{4}, 0, -2$
8. $-\frac{3}{2}, 5, 1$

In Exercises 9–20, plot each real number as a point on the real number line and place the correct inequality symbol (< or >) between the pair of real numbers. *See Examples 3, 4, and 5.*

9. $3 \quad \square \quad -4$
10. $6 \quad \square \quad -2$
11. $4 \quad \square \quad -\frac{7}{2}$
12. $2 \quad \square \quad \frac{3}{2}$
13. $0 \quad \square \quad -\frac{7}{16}$
14. $-\frac{7}{3} \quad \square \quad -\frac{5}{2}$
15. $\frac{9}{16} \quad \square \quad \frac{5}{8}$
16. $-\frac{3}{8} \quad \square \quad -\frac{5}{4}$
17. $-4.6 \quad \square \quad 1.5$
18. $28.60 \quad \square \quad -3.75$
19. $-6.58 \quad \square \quad -7.66$
20. $20.156 \quad \square \quad 54.235$

In Exercises 21–24, find the distance between a and zero on the real number line.

21. $a = 2$
22. $a = 5$
23. $a = -8$
24. $a = -17$

In Exercises 25–30, find the opposite of the number. Plot the number and its opposite on the real number line. What is the distance of each from 0?

25. 3
26. -6
27. -3.8
28. 7.5
29. $\frac{5}{2}$
30. $-\frac{3}{4}$

In Exercises 31–34, find the absolute value of the real number and its distance from 0.

31.

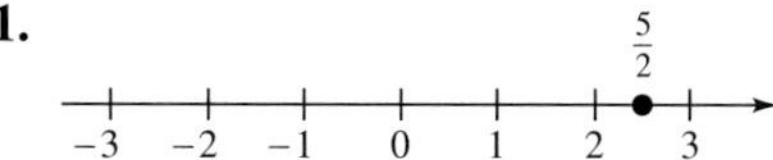

32.

2.4
−3 −2 −1 0 1 2 3

33.

−3 −2 −1 0 1 2 3

34.

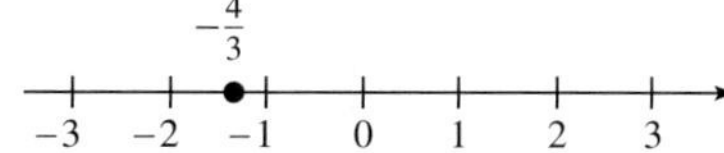

In Exercises 35–48, evaluate the expression. *See Example 6.*

35. $|10|$
36. $|1|$
37. $|-3|$
38. $|-19|$
39. $|-3.4|$
40. $|-16.2|$
41. $\left|-\frac{7}{2}\right|$
42. $\left|-\frac{9}{16}\right|$
43. $-|4.09|$
44. $-|91.3|$
45. $-|-23.6|$
46. $-|-0.08|$
47. $|0|$
48. $|\pi|$

In Exercises 49–58, place the correct symbol (<, >, or =) between the pair of real numbers. *See Example 7.*

49. $|-16| \quad |16|$

50. $|525| \quad |-525|$

51. $|-4| \quad |3|$

52. $|16| \quad |-25|$

53. $\left|\frac{3}{16}\right| \quad \left|\frac{3}{2}\right|$

54. $\left|\frac{7}{8}\right| \quad \left|\frac{4}{3}\right|$

55. $-|-48.5| \quad |-48.5|$

56. $-|-64| \quad |-64|$

57. $|-\pi| \quad -|-2\pi|$

58. $-|-4.9| \quad |-10.2|$

In Exercises 59–62, plot the numbers on the real number line.

59. $\frac{5}{2}, \pi, -1, -|-3|$

60. $3.7, \frac{16}{3}, -|-1.9|, -\frac{1}{2}$

61. $-5, \frac{7}{3}, |-3|, 0, -|4.5|$

62. $|-2.3|, 3.2, -2.3, -|3.2|$

In Exercises 63–68, find all real numbers whose distance from a is given by d.

63. $a = 8, d = 12$

64. $a = 6, d = 7$

65. $a = 21.3, d = 6$

66. $a = 42.5, d = 7$

67. $a = -2, d = 3.5$

68. $a = -7, d = 7.2$

Solving Problems

In Exercises 69–77, give three examples of numbers that satisfy the given conditions.

69. A real number that is a negative integer

70. A real number that is a whole number

71. A real number that is not a rational number

72. A real number that is not an irrational number

73. An integer that is a rational number

74. A rational number that is not an integer

75. A rational number that is not a negative number

76. A real number that is not a positive rational number

77. An integer that is not a whole number

Explaining Concepts

78. Explain why $\frac{8}{4}$ is a natural number, but $\frac{7}{4}$ is not.

79. Which real number lies farther from 0 on the real number line, -15 or 10? Explain.

80. Which real number lies farther from -4 on the real number line, 3 or -10? Explain.

81. Which real number is smaller, $\frac{3}{8}$ or 0.37? Explain.

True or False? In Exercises 82–87, decide whether the statement is true or false. Justify your answer.

82. The absolute value of any real number is always positive.

83. The absolute value of a number is equal to the absolute value of its opposite.

84. The absolute value of a rational number is a rational number.

85. A given real number corresponds to exactly one point on the real number line.

86. The opposite of a positive number is a negative number.

87. Every rational number is an integer.

1.2 Adding and Subtracting Integers

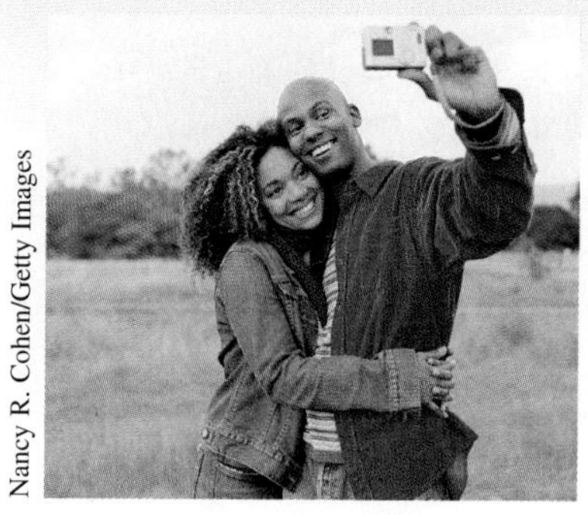
Nancy R. Cohen/Getty Images

What You Should Learn

1 ▶ Add integers using a number line.
2 ▶ Add integers with like signs and with unlike signs.
3 ▶ Subtract integers with like signs and with unlike signs.

Why You Should Learn It

Real numbers are used to represent many real-life quantities. For instance, in Exercise 107 on page 18, you will use real numbers to find the change in digital camera sales.

1 ▶ Add integers using a number line.

Adding Integers Using a Number Line

In this and the next section, you will study the four operations of arithmetic (addition, subtraction, multiplication, and division) on the set of integers. There are many examples of these operations in real life. For example, your business had a gain of $550 during one week and a loss of $600 the next week. Over the two-week period, your business had a combined profit of

$$550 + (-600) = -50$$

which represents an overall loss of $50.

The number line is a good visual model for demonstrating addition of integers. To add two integers, $a + b$, using a number line, start at 0. Then move right or left a units depending on whether a is positive or negative. From that position, move right or left b units depending on whether b is positive or negative. The final position is called the **sum.**

EXAMPLE 1 **Adding Integers with Like Signs Using a Number Line**

Find each sum.

a. $5 + 2$ **b.** $-3 + (-5)$

Solution

a. Start at zero and move five units to the right. Then move two more units to the right, as shown in Figure 1.20. So, $5 + 2 = 7$.

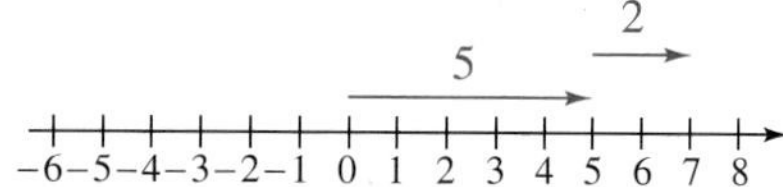

Figure 1.20

b. Start at zero and move three units to the left. Then move five more units to the left, as shown in Figure 1.21. So, $-3 + (-5) = -8$.

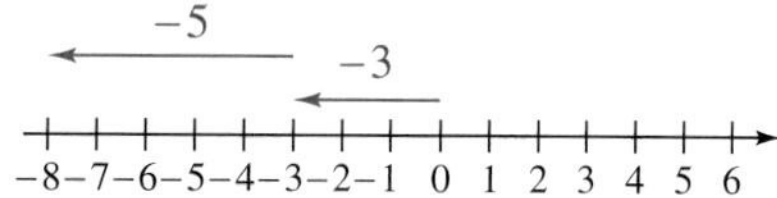

Figure 1.21

✓ ***CHECKPOINT*** *Now try Exercise 3.*

EXAMPLE 2 Adding Integers with Unlike Signs Using a Number Line

Find each sum.

a. $-5 + 2$ **b.** $7 + (-3)$ **c.** $-4 + 4$

Solution

a. Start at zero and move five units to the left. Then move two units to the right, as shown in Figure 1.22. So, $-5 + 2 = -3$.

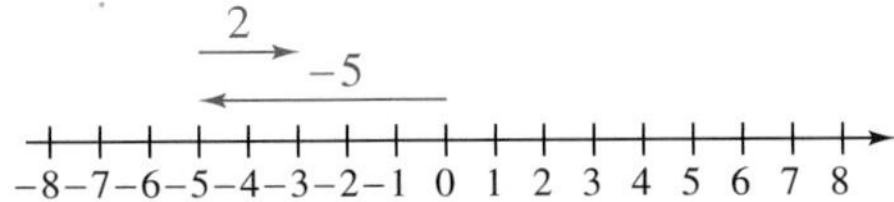

Figure 1.22

b. Start at zero and move seven units to the right. Then move three units to the left, as shown in Figure 1.23. So, $7 + (-3) = 4$.

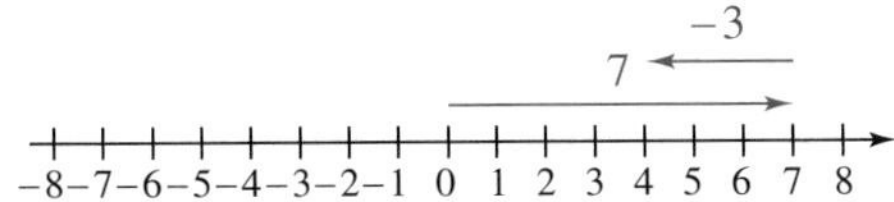

Figure 1.23

c. Start at zero and move four units to the left. Then move four units to the right, as shown in Figure 1.24. So, $-4 + 4 = 0$.

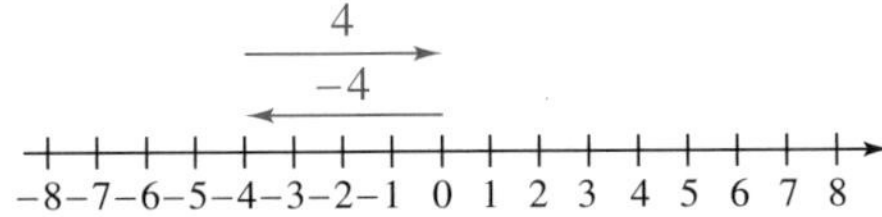

Figure 1.24

✓ ***CHECKPOINT*** *Now try Exercise 7.*

In Example 2(c), notice that the sum of -4 and 4 is 0. Two numbers whose sum is zero are called **opposites** (or **additive inverses**) of each other, So, -4 is the opposite of 4 and 4 is the opposite of -4.

2 ▶ Add integers with like signs and with unlike signs.

Adding Integers Algebraically

Examples 1 and 2 illustrated a *graphical* approach to adding integers. It is more common to use an *algebraic* approach to adding integers, as summarized below.

Addition of Integers

1. To **add** two integers with *like* signs, add their absolute values and attach the common sign to the result.
2. To **add** two integers with *unlike* signs, subtract the smaller absolute value from the larger absolute value and attach the sign of the integer with the larger absolute value.

EXAMPLE 3 Adding Integers

a. Like signs: $-18 + (-62) = -(|-18| + |-62|) = -(18 + 62) = -80$

b. Unlike signs: $22 + (-17) = |22| - |-17| = 22 - 17 = 5$

c. Unlike signs: $-84 + 14 = -(|-84| - |14|) = -(84 - 14) = -70$

✓ ***CHECKPOINT*** *Now try Exercise 17.*

$$\begin{array}{r} {\scriptstyle 1\,1} \\ 148 \\ 62 \\ +\,536 \\ \hline 746 \end{array}$$

Figure 1.25 Carrying Algorithm

There are different ways to add three or more integers. You can use the **carrying algorithm** with a vertical format with nonnegative integers, as shown in Figure 1.25, or you can add them two at a time, as illustrated in Example 4.

EXAMPLE 4 Account Balance 

At the beginning of a month, your account balance was \$28. During the month, you deposited \$60 and withdrew \$40. What was your balance at the end of the month?

Solution

$$\$28 + \$60 + (-\$40) = (\$28 + \$60) + (-\$40)$$
$$= \$88 + (-\$40) = \$48 \quad \text{Balance}$$

 ***CHECKPOINT*** *Now try Exercise 103.*

3 ▶ Subtract integers with like signs and with unlike signs.

Subtracting Integers Algebraically

Subtraction can be thought of as "taking away." For instance, $8 - 5$ can be thought of as "8 take away 5," which leaves 3. Moreover, note that $8 + (-5) = 3$, which means that

$$8 - 5 = 8 + (-5).$$

In other words, $8 - 5$ can also be accomplished by "adding the opposite of 5 to 8."

Subtraction of Integers

To **subtract** one integer from another, add the opposite of the integer being subtracted to the other integer. The result is called the **difference** of the two integers.

EXAMPLE 5 Subtracting Integers

a. $3 - 8 = 3 + (-8) = -5$ — Add opposite of 8.

b. $10 - (-13) = 10 + 13 = 23$ — Add opposite of -13.

c. $-5 - 12 = -5 + (-12) = -17$ — Add opposite of 12.

✓ ***CHECKPOINT*** *Now try Exercise 47.*

```
 3 10 15
 4  1  5
-2  7  6
--------
 1  3  9
```

Figure 1.26 Borrowing Algorithm

Be sure you understand that the terminology of subtraction is not the same as that used for negative numbers. For instance, -5 is read as "negative 5," but $8 - 5$ is read as "8 subtract 5." It is important to distinguish between the operation and the signs of the numbers involved. For instance, in $-3 - 5$ the operation is subtraction and the numbers are -3 and 5.

For subtraction problems involving two nonnegative integers, you can use the **borrowing algorithm** shown in Figure 1.26.

EXAMPLE 6 Subtracting Integers

a. Subtract 10 from -4 means: $-4 - 10 = -4 + (-10) = -14$.

b. -3 subtract -8 means: $-3 - (-8) = -3 + 8 = 5$.

CHECKPOINT *Now try Exercise 77.*

To evaluate an expression that contains a series of additions and subtractions, write the subtractions as equivalent additions and simplify from left to right, as shown in Example 7.

EXAMPLE 7 Evaluating Expressions

Evaluate each expression.

a. $-13 - 7 + 11 - (-4)$ **b.** $5 - (-9) - 12 + 2$

c. $-1 - 3 - 4 + 6$ **d.** $5 - 1 - 8 + 7 - (-10)$

Solution

a. $-13 - 7 + 11 - (-4) = -13 + (-7) + 11 + 4$ Add opposites.

$= -20 + 11 + 4$ Add -13 and -7.

$= -9 + 4$ Add -20 and 11.

$= -5$ Add.

b. $5 - (-9) - 12 + 2 = 5 + 9 + (-12) + 2$ Add opposites.

$= 14 + (-12) + 2$ Add 5 and 9.

$= 2 + 2$ Add 14 and -12.

$= 4$ Add.

c. $-1 - 3 - 4 + 6 = -1 + (-3) + (-4) + 6$ Add opposites.

$= -4 + (-4) + 6$ Add -1 and -3.

$= -8 + 6$ Add -4 and -4.

$= -2$ Add.

d. $5 - 1 - 8 + 7 - (-10) = 5 + (-1) + (-8) + 7 + 10$ Add opposites.

$= 4 + (-8) + 7 + 10$ Add 5 and -1.

$= -4 + 7 + 10$ Add 4 and -8.

$= 3 + 10$ Add -4 and 7.

$= 13$ Add.

CHECKPOINT *Now try Exercise 87.*

EXAMPLE 8 Temperature Change 

The temperature in Minneapolis, Minnesota at 4 P.M. was 15°F. By midnight, the temperature had decreased by 18°. What was the temperature in Minneapolis at midnight?

Solution

To find the temperature at midnight, subtract 18 from 15.

$$15 - 18 = 15 + (-18)$$
$$= -3$$

The temperature in Minneapolis at midnight was -3°F.

 **CHECKPOINT** ***Now try Exercise 97.***

This text includes several examples and exercises that use a calculator. As each new calculator application is encountered, you will be given general instructions for using a calculator. These instructions, however, may not agree precisely with the steps required by *your* calculator, so be sure you are familiar with the use of the keys on your own calculator.

For each of the calculator examples in the text, two possible keystroke sequences are given: one for a standard *scientific* calculator and one for a *graphing* calculator.

EXAMPLE 9 Evaluating Expressions with a Calculator

Evaluate each expression with a calculator.

a. $-4 - 5$ **b.** $2 - (-3) + 9$

	Keystrokes	*Display*	
a.	4 [+/−] [−] 5 [=]	−9	Scientific
	[(−)] 4 [−] 5 [ENTER]	−9	Graphing
	Keystrokes	*Display*	
b.	2 [−] [(] 3 [+/−] [)] [+] 9 [=]	14	Scientific
	2 [−] [(] [(−)] 3 [)] [+] 9 [ENTER]	14	Graphing

CHECKPOINT ***Now try Exercise 93.***

Technology: Tip

The keys [+/−] and [(−)] change a number to its opposite, and [−] is the subtraction key. For instance, the keystrokes [−] 4 [−] 5 [ENTER] will not produce the result shown in Example 9(a).

Concept Check

1. Explain how to use a number line to add three negative integers.
2. In your own words, write the rule for adding two integers with opposite signs. How do you determine the sign of the sum?
3. Explain how to find the difference of two integers.
4. When is the difference of two integers equal to zero?

1.2 EXERCISES

Go to pages 58–59 to record your assignments.

Developing Skills

In Exercises 1–8, find the sum and demonstrate the addition on the real number line. *See Examples 1 and 2.*

1. $2 + 7$
2. $3 + 9$
✓ **3.** $-8 + (-3)$
4. $-4 + (-7)$
5. $10 + (-3)$
6. $14 + (-8)$
✓ **7.** $-6 + 4$
8. $-12 + 5$

In Exercises 9–42, find the sum. *See Example 3.*

9. $6 + 10$
10. $8 + 3$
11. $14 + (-14)$
12. $10 + (-10)$
13. $-45 + 45$
14. $-23 + 23$
15. $14 + 13$
16. $20 + 19$
✓ **17.** $-23 + (-4)$
18. $-32 + (-16)$
19. $18 + (-12)$
20. $34 + (-16)$
21. $75 + 100$
22. $54 + 68$
23. $9 + (-14)$
24. $18 + (-26)$
25. $10 + (-6) + 34$
26. $7 + (-4) + 1$
27. $-15 + (-3) + 8$
28. $-82 + (-36) + 82$
29. $9 + (-18) + 4$
30. $2 + (-51) + 13$
31. $16 + 2 + (-7)$
32. $24 + 1 + (-19)$
33. $-13 + 12 + 4$
34. $-31 + 20 + 15$
35. $75 + (-75) + (-15)$
36. $32 + (-32) + (-16)$
37. $803 + (-104) + (-613) + 214$
38. $4365 + (-2145) + (-1873) + 40{,}084$
39. $312 + (-564) + (-100)$
40. $1200 + (-1300) + (-275)$
41. $-890 + (-90) + 62$
42. $-770 + (-383) + 492$

In Exercises 43–76, find the difference. *See Example 5.*

43. $21 - 18$
44. $47 - 12$
45. $51 - 25$
46. $37 - 37$
✓ **47.** $1 - (-4)$
48. $7 - (-8)$
49. $15 - (-10)$
50. $8 - (-31)$
51. $18 - (-18)$
52. $62 - (-28)$
53. $19 - (-31)$
54. $12 - (-5)$
55. $27 - 57$
56. $18 - 32$
57. $61 - 85$
58. $53 - 74$
59. $22 - 131$
60. $48 - 222$
61. $2 - 11$
62. $3 - 15$
63. $13 - 24$
64. $26 - 34$
65. $-135 - (-114)$
66. $-63 - (-8)$
67. $-4 - (-4)$
68. $-942 - (-942)$
69. $-10 - (-4)$
70. $-12 - (-7)$
71. $-71 - 32$
72. $-84 - 106$
73. $-210 - 400$
74. $-120 - 142$
75. $-110 - (-30)$
76. $-2500 - (-600)$

In Exercises 77–82, find the difference. ***See Example 6.***

77. Subtract 15 from -6.

78. Subtract 24 from -17.

79. Subtract -120 from 380.

80. Subtract -80 from 140.

81. -43 subtract -22

82. -77 subtract -110

83. *Think About It* What number must be added to 10 to obtain -5?

84. *Think About It* What number must be added to 36 to obtain -12?

85. *Think About It* What number must be subtracted from -12 to obtain 24?

86. *Think About It* What number must be subtracted from -20 to obtain 15?

In Exercises 87–92, evaluate the expression. ***See Example 7.***

87. $-1 + 3 - (-4) + 10$

88. $12 - 6 + 3 - (-8)$

89. $6 + 7 - 12 - 5$

90. $-3 + 2 - 20 + 9$

91. $-(-5) + 7 - 18 + 4$

92. $-15 - (-2) + 4 - 6$

In Exercises 93–96, write the keystrokes used to evaluate the expression with a calculator (either scientific or graphing). Then evaluate the expression. ***See Example 9.***

93. $-3 - 7$

94. $9 - (-2)$

95. $6 + 5 - (-7)$

96. $4 - 3 - (-9)$

Solving Problems

97. *Temperature Change* The temperature at 6 A.M. was $-10°$F. By noon, the temperature had increased by 22°F. What was the temperature at noon?

98. *Account Balance* A credit card owner charged \$142 worth of goods on her account that had an initial balance of \$0. Find the balance after a payment of \$87 was made.

99. *Outdoor Recreation* A hiker descended 847 meters into the Grand Canyon. He climbed back up 385 meters and then rested. Find the distance between where the hiker rested and where he started his descent.

John Burcham/Getty Images

The Grand Canyon, located in Arizona, is 277 river miles long.

100. *Outdoor Recreation* A fisherman dropped his line 27 meters below the surface of the water. Because the fish were not biting, he raised his line by 8 meters. How far below the surface of the water was his line?

101. *Profit* A telephone company lost \$650,000 during the first half of the year. By the end of the year, the company had an overall profit of \$362,000. What was the company's profit during the second half of the year?

102. *Altitude* An airplane flying at a cruising altitude of 31,000 feet is instructed to descend as shown in the diagram below. How many feet must the airplane descend?

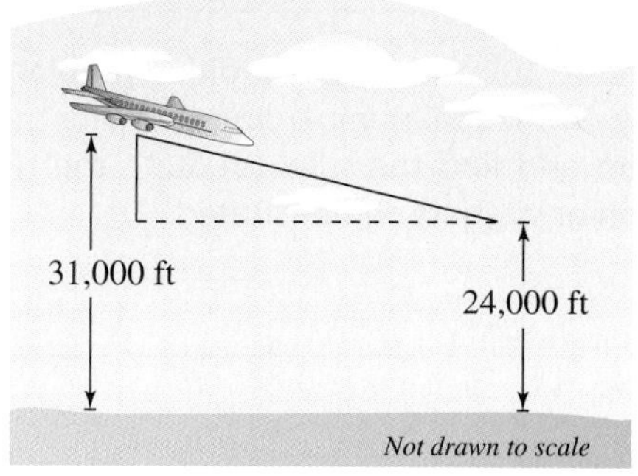

Not drawn to scale

The symbol indicates an exercise in which you are instructed to use a graphing calculator.

✓ **103.** *Account Balance* At the beginning of a month, your account balance was \$2750. During the month, you withdrew \$350 and \$500, deposited \$450, and earned \$6.42 in interest. What was your balance at the end of the month?

104. *Account Balance* At the beginning of a month, your account balance was \$1204. During the month, you withdrew \$425 and \$621, deposited \$150 and \$80, and earned \$2.02 in interest. What was your balance at the end of the month?

105. *Temperature Change* When you left for class in the morning, the temperature was 25°C. By the time class ended, the temperature had increased by 4°. While you studied, the temperature increased by 3°. During your soccer practice, the temperature decreased by 9°. What was the temperature after your soccer practice?

106. *Temperature Change* When you left for class in the morning, the temperature was 40°F. By the time class ended, the temperature had increased by 13°. While you studied, the temperature decreased by 5°. During your club meeting, the temperature decreased by 6°. What was the temperature after your club meeting?

107. *Digital Cameras* The bar graph shows the factory sales (in millions of dollars) of digital cameras in the United States for the years 2000 to 2005. (Source: Consumer Electronics Association)

(a) Find the change in factory sales of digital cameras from 2000 to 2001.

(b) Find the change in factory sales of digital cameras from 2004 to 2005.

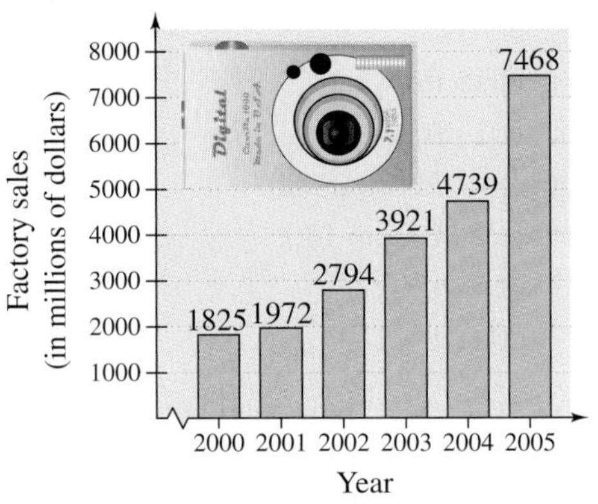

Figure for 107

108. *Population* The bar graph shows the estimated populations (in thousands) of Cleveland, OH for the years 2000 to 2006. (Source: U.S. Census Bureau)

(a) Find the change in population of Cleveland from 2000 to 2003.

(b) Find the change in population of Cleveland from 2005 to 2006.

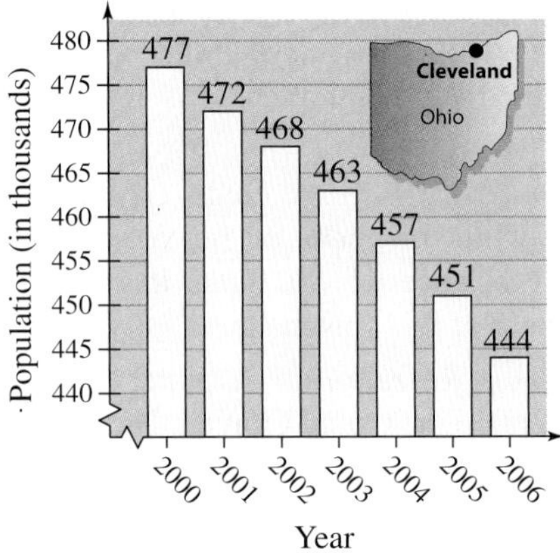

Explaining Concepts

In Exercises 109 and 110, an addition problem is shown visually on the real number line. (a) Write the addition problem and find the sum. (b) State the rule for the addition of integers demonstrated.

109.

110.

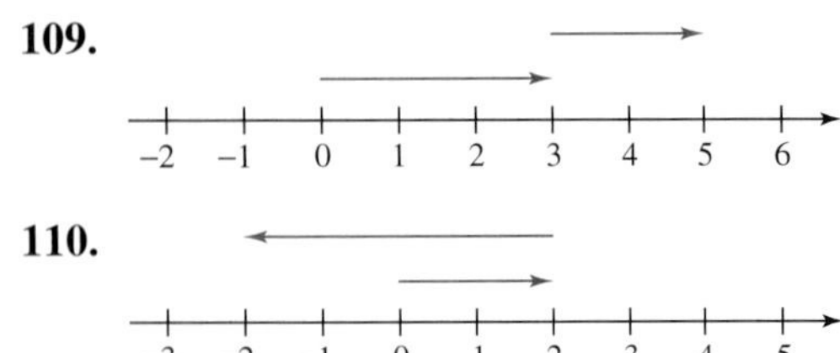

111. ✎ Explain why the sum of two negative integers is a negative integer.

112. ✎ When is the sum of a positive integer and a negative integer a positive integer?

113. ✎ Is it possible that the sum of two positive integers is a negative integer? Explain.

114. ✎ Is it possible that the difference of two negative integers is a positive integer? Explain.

1.3 Multiplying and Dividing Integers

NASA

What You Should Learn

1 ▶ Multiply integers with like signs and with unlike signs.
2 ▶ Divide integers with like signs and with unlike signs.
3 ▶ Find factors and prime factors of an integer.
4 ▶ Represent the definitions and rules of arithmetic symbolically.

Why You Should Learn It

You can multiply and divide integers to solve real-life problems. For instance, in Exercise 109 on page 30, you will divide integers to find the average speed of a space shuttle.

1 ▶ Multiply integers with like signs and with unlike signs.

Multiplying Integers

Multiplication of two integers can be described as repeated addition or subtraction. The result of multiplying one number by another is called a **product.** Here are three examples.

Multiplication	*Repeated Addition or Subtraction*
$3 \times 5 = 15$	$\underbrace{5 + 5 + 5}_{\text{Add 5 three times.}} = 15$
$4 \times (-2) = -8$	$\underbrace{(-2) + (-2) + (-2) + (-2)}_{\text{Add } -2 \text{ four times.}} = -8$
$(-3) \times (-4) = 12$	$\underbrace{-(-4) - (-4) - (-4)}_{\text{Subtract } -4 \text{ three times.}} = 12$

Multiplication is denoted in a variety of ways. For instance,

$$7 \times 3, \quad 7 \cdot 3, \quad 7(3), \quad (7)3, \quad \text{and} \quad (7)(3)$$

all denote the product of "7 times 3," which is 21.

Rules for Multiplying Integers

1. The product of an integer and zero is 0.
2. The product of two integers with *like* signs is *positive.*
3. The product of two integers with *unlike* signs is *negative.*

To find the product of more than two numbers, first find the product of their absolute values. If the number of negative factors is even, then the product is positive. If the number of negative factors is odd, then the product is negative. For instance, the expression

$5(-3)(-4)(7)$ Expression has 2 negative factors.

is positive because it has an even number of negative factors.

EXAMPLE 1 Multiplying Integers

a. $4(10) = 40$	(Positive) · (positive) = positive
b. $-6 \cdot 9 = -54$	(Negative) · (positive) = negative
c. $-5(-7) = 35$	(Negative) · (negative) = positive
d. $3(-12) = -36$	(Positive) · (negative) = negative
e. $-12 \cdot 0 = 0$	(Negative) · (zero) = zero
f. $-2(8)(-3)(-1) = -(2 \cdot 8 \cdot 3 \cdot 1)$	Odd number of negative factors
$= -48$	Answer is negative.

CHECKPOINT *Now try Exercise 11.*

Be careful to distinguish properly between expressions such as $3(-5)$ and $3 - 5$ or $-3(-5)$ and $-3 - 5$. The first of each pair is a *multiplication* problem, whereas the second is a *subtraction* problem.

Multiplication	*Subtraction*
$3(-5) = -15$	$3 - 5 = -2$
$-3(-5) = 15$	$-3 - 5 = -8$

$$\begin{array}{rl} 47 & \\ \times\ 23 & \\ \hline 141 & \Leftarrow \text{Multiply 3 times 47.} \\ 94\ \ & \Leftarrow \text{Multiply 2 times 47.} \\ \hline 1081 & \Leftarrow \text{Add columns.} \end{array}$$

Figure 1.27 Vertical Multiplication Algorithm

To multiply two integers having two or more digits, we suggest the **vertical multiplication algorithm** demonstrated in Figure 1.27. The sign of the product is determined by the usual multiplication rule.

EXAMPLE 2 Geometry: Volume of a Box

Find the volume of the rectangular box shown in Figure 1.28.

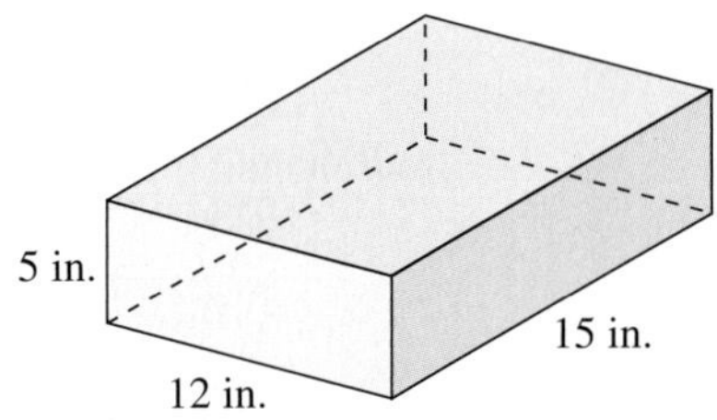

Figure 1.28

Solution

To find the volume, multiply the length, width, and height of the box.

$$\begin{aligned} \text{Volume} &= (\text{Length}) \cdot (\text{Width}) \cdot (\text{Height}) \\ &= (15 \text{ inches}) \cdot (12 \text{ inches}) \cdot (5 \text{ inches}) \\ &= 900 \text{ cubic inches} \end{aligned}$$

So, the box has a volume of 900 cubic inches.

CHECKPOINT *Now try Exercise 113.*

2 ▶ Divide integers with like signs and with unlike signs.

Dividing Integers

Just as subtraction can be expressed in terms of addition, you can express division in terms of multiplication. Here are some examples.

Division		*Related Multiplication*
$15 \div 3 = 5$	because	$15 = 5 \cdot 3$
$-15 \div 3 = -5$	because	$-15 = -5 \cdot 3$
$15 \div (-3) = -5$	because	$15 = (-5) \cdot (-3)$
$-15 \div (-3) = 5$	because	$-15 = 5 \cdot (-3)$

The result of dividing one integer by another is called the **quotient** of the integers. **Division** is denoted by the symbol ÷, or by /, or by a horizontal line. For instance,

$$30 \div 6, \quad 30/6, \quad \text{and} \quad \frac{30}{6}$$

all denote the quotient of 30 and 6, which is 5. Using the form $30 \div 6$, 30 is called the **dividend** and 6 is the **divisor.** In the forms $30/6$ and $\frac{30}{6}$, 30 is the **numerator** and 6 is the **denominator.**

It is important to know how to use 0 in a division problem. Zero divided by a nonzero integer is always 0. For instance,

$$\frac{0}{13} = 0 \quad \text{because} \quad 0 = 0 \cdot 13.$$

On the other hand, division by zero, such as $13 \div 0$, is *undefined.*

Because division can be described in terms of multiplication, the rules for dividing two integers with like or unlike signs are the same as those for multiplying such integers.

Technology: Discovery

Does $\frac{1}{0} = 0$? Does $\frac{2}{0} = 0$? Write each division above in terms of multiplication. What does this tell you about division by zero? What does your calculator display when you perform the division?

Rules for Dividing Integers

1. Zero divided by a nonzero integer is 0, whereas a nonzero integer divided by zero is *undefined.*
2. The quotient of two nonzero integers with *like* signs is *positive.*
3. The quotient of two nonzero integers with *unlike* signs is *negative.*

EXAMPLE 3 Dividing Integers

a. $\dfrac{-42}{-6} = 7$ because $-42 = 7(-6)$.

b. $36 \div (-9) = -4$ because $(-4)(-9) = 36$.

c. $0 \div (-13) = 0$ because $(0)(-13) = 0$.

d. $-105 \div 7 = -15$ because $(-15)(7) = -105$.

e. $-97 \div 0$ is undefined.

✓ ***CHECKPOINT*** ***Now try Exercise 43.***

$$\begin{array}{r} 27 \\ 13\overline{)351} \\ \underline{26} \\ 91 \\ \underline{91} \end{array}$$

Figure 1.29 Long Division Algorithm

When dividing large numbers, the **long division algorithm** can be used. For instance, the long division algorithm in Figure 1.29 shows that

$$351 \div 13 = 27.$$

Remember that division can be checked by multiplying the answer by the divisor. So, it is true that

$$351 \div 13 = 27 \quad \text{because} \quad 27(13) = 351.$$

All four operations on integers (addition, subtraction, multiplication, and division) are used in the following real-life example.

EXAMPLE 4 Stock Purchase 

On Monday you bought \$500 worth of stock in a company. During the rest of the week, you recorded the gains and losses in your stock's value, as shown in the table.

Tuesday	Wednesday	Thursday	Friday
Gained \$15	Lost \$18	Lost \$23	Gained \$10

a. What was the value of the stock at the close of Wednesday?

b. What was the value of the stock at the end of the week?

c. What would the total loss have been if Thursday's loss had occurred on each of the four days?

d. What was the average daily gain (or loss) for the four days recorded?

Solution

a. The value at the close of Wednesday was

$$500 + 15 - 18 = \$497.$$

b. The value of the stock at the end of the week was

$$500 + 15 - 18 - 23 + 10 = \$484.$$

c. The loss on Thursday was \$23. If this loss had occurred each day, the total loss would have been

$$4(23) = \$92.$$

d. To find the average daily gain (or loss), add the gains and losses of the four days and divide by 4.

$$\text{Average} = \frac{15 + (-18) + (-23) + 10}{4} = \frac{-16}{4} = -4$$

This means that during the four days, the stock had an average loss of \$4 per day.

CHECKPOINT *Now try Exercise 103.*

Study Tip

To find the **average** of n numbers, add the numbers and divide the result by n.

3 ▶ Find factors and prime factors of an integer.

Factors and Prime Numbers

The set of positive integers

$$\{1, 2, 3, \ldots\}$$

is one subset of the real numbers that has intrigued mathematicians for many centuries.

Historically, an important number concept has been *factors* of positive integers. In a multiplication problem such as $3 \cdot 7 = 21$, the numbers 3 and 7 are called *factors* of 21.

$$\underbrace{3 \cdot 7}_{\text{Factors}} = \underbrace{21}_{\text{Product}}$$

It is also correct to call the numbers 3 and 7 *divisors* of 21, because 3 and 7 each divide evenly into 21.

Definition of Factor (or Divisor)

If a and b are positive integers, then a is a **factor** (or **divisor**) of b if and only if there is a positive integer c such that $a \cdot c = b$.

The concept of factors allows you to classify positive integers into three groups: *prime* numbers, *composite* numbers, and the number 1.

Definitions of Prime and Composite Numbers

1. A positive integer greater than 1 with no factors other than itself and 1 is called a **prime number,** or simply a **prime.**
2. A positive integer greater than 1 with more than two factors is called a **composite number,** or simply a **composite.**

The numbers 2, 3, 5, 7, and 11 are primes because they have only themselves and 1 as factors. The numbers 4, 6, 8, 9, and 10 are composites because each has more than two factors. The number 1 is neither prime nor composite because 1 is its only factor.

Every composite number can be expressed as a *unique* product of prime factors. Here are some examples.

$$6 = 2 \cdot 3, \;\; 15 = 3 \cdot 5, \;\; 18 = 2 \cdot 3 \cdot 3, \;\; 42 = 2 \cdot 3 \cdot 7, \;\; 124 = 2 \cdot 2 \cdot 31$$

According to the definition of a prime number, is it possible for any negative number to be prime? Consider the number -2. Is it prime? Are its only factors 1 and itself? No, because

$$-2 = 1(-2),$$

$$-2 = (-1)(2),$$

or $-2 = (-1)(1)(2)$.

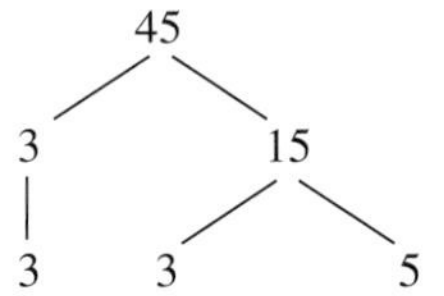

Figure 1.30 Tree Diagram

One strategy for factoring a composite number into prime factors is to begin by finding the smallest prime number that is a factor of the composite number. Dividing this factor into the number yields a *companion* factor. For instance, 3 is the smallest prime number that is a factor of 45, and its companion factor is 15 because $15 = 45 \div 3$. Continue identifying factors and companion factors until each factor is prime. As shown in Figure 1.30, a *tree diagram* is a nice way to record your work. From the tree diagram, you can see that the prime factorization of 45 is $45 = 3 \cdot 3 \cdot 5$.

EXAMPLE 5 Prime Factorization

Write the prime factorization of each number.

a. 84 **b.** 78 **c.** 133 **d.** 43

Solution

a. 2 is a recognized divisor of 84. So,

$$84 = 2 \cdot 42 = 2 \cdot 2 \cdot 21 = 2 \cdot 2 \cdot 3 \cdot 7.$$

b. 2 is a recognized divisor of 78. So,

$$78 = 2 \cdot 39 = 2 \cdot 3 \cdot 13.$$

c. If you do not recognize a divisor of 133, you can start by dividing any of the prime numbers 2, 3, 5, 7, 11, 13, etc., into 133. You will find 7 to be the first prime to divide 133. So,

$$133 = 7 \cdot 19.$$

d. In this case, none of the primes less than 43 divides 43. So, 43 is prime.

CHECKPOINT *Now try Exercise 85.*

To say that a number is **divisible** by n means that n divides into the number without leaving a remainder. Other aids to finding prime factors of a number include the following divisibility tests.

Divisibility Tests

Test	*Example*
1. A number is divisible by 2 if it is *even*.	364 is divisible by 2 because it is even.
2. A number is divisible by 3 if the sum of its digits is divisible by 3.	261 is divisible by 3 because $2 + 6 + 1 = 9$.
3. A number is divisible by 9 if the sum of its digits is divisible by 9.	738 is divisible by 9 because $7 + 3 + 8 = 18$.
4. A number is divisible by 5 if its units digit is 0 or 5.	325 is divisible by 5 because its units digit is 5.
5. A number is divisible by 10 if its units digit is 0.	120 is divisible by 10 because its units digit is 0.

4 ▶ Represent the definitions and rules of arithmetic symbolically.

Summary of Definitions and Rules

So far in this chapter, rules and procedures have been described more with words than with symbols. For instance, subtraction is verbally defined as "adding the opposite of the number being subtracted." As you move to higher and higher levels of mathematics, it becomes more and more convenient to use symbols to describe rules and procedures. For instance, subtraction is symbolically defined as

$$a - b = a + (-b).$$

At its simplest level, algebra is a symbolic form of arithmetic. This arithmetic–algebra connection can be illustrated in the following way.

Arithmetic		*Algebra*
Verbal rules and definitions	→	Symbolic rules and definitions
Specific examples of rules and definitions		

An illustration of this connection is shown in Example 6.

EXAMPLE 6 Writing a Rule of Arithmetic in Symbolic Form

Write an example and an algebraic description of the arithmetic rule: *The product of two integers with unlike signs is negative.*

Solution

Example

For the integers -3 and 7,

$$-3 \cdot 7 = -21.$$

Also, for the integers 3 and -7,

$$3 \cdot (-7) = -21.$$

Algebraic Description

If a and b are positive integers, then

$$\underbrace{(-a) \cdot b}_{\text{Unlike signs}} = \underbrace{-(a \cdot b)}_{\text{Negative product}}$$

and

$$\underbrace{a \cdot (-b)}_{\text{Unlike signs}} = \underbrace{-(a \cdot b)}_{\text{Negative product}}.$$

CHECKPOINT Now try Exercise 95.

The list on the following page summarizes the algebraic versions of important definitions and rules of arithmetic. In each case, a specific example is included for clarification.

Arithmetic Summary

Definitions: Let a, b, and c be integers.

	Definition	*Example*
1. Subtraction:	$a - b = a + (-b)$	$5 - 7 = 5 + (-7)$
2. Multiplication: (a is a positive integer)	$a \cdot b = \underbrace{b + b + \cdots + b}_{a \text{ terms}}$	$3 \cdot 5 = 5 + 5 + 5$
3. Division: ($b \neq 0$)	$a \div b = c$ if and only if $a = c \cdot b$.	$12 \div 4 = 3$ because $12 = 3 \cdot 4$.
4. Less than:	$a < b$ if there is a positive real number c such that $a + c = b$.	$-2 < 1$ because $-2 + 3 = 1$.
5. Absolute value:	$\lvert a \rvert = \begin{cases} a, & \text{if } a \geq 0 \\ -a, & \text{if } a < 0 \end{cases}$	$\lvert -3 \rvert = -(-3) = 3$
6. Divisor:	a is a divisor of b if and only if there is an integer c such that $a \cdot c = b$.	7 is a divisor of 21 because $7 \cdot 3 = 21$.

Rules: Let a and b be integers.

	Rule	*Example*
1. Addition:		
	(a) To add two integers with *like* signs, add their absolute values and attach the common sign to the result.	$3 + 7 = \lvert 3 \rvert + \lvert 7 \rvert = 10$
	(b) To add two integers with *unlike* signs, subtract the smaller absolute value from the larger absolute value and attach the sign of the integer with the larger absolute value.	$-5 + 8 = \lvert 8 \rvert - \lvert -5 \rvert = 8 - 5 = 3$
2. Multiplication:		
	(a) $a \cdot 0 = 0 = 0 \cdot a$	$3 \cdot 0 = 0 = 0 \cdot 3$
	(b) Like signs: $a \cdot b > 0$	$(-2)(-5) = 10$
	(c) Unlike signs: $a \cdot b < 0$	$(2)(-5) = -10$
3. Division:		
	(a) $\dfrac{0}{a} = 0$	$\dfrac{0}{4} = 0$
	(b) $\dfrac{a}{0}$ is undefined.	$\dfrac{6}{0}$ is undefined.
	(c) Like signs: $\dfrac{a}{b} > 0$	$\dfrac{-2}{-3} = \dfrac{2}{3}$
	(d) Unlike signs: $\dfrac{a}{b} < 0$	$\dfrac{-5}{7} = -\dfrac{5}{7}$

EXAMPLE 7 Using Definitions and Rules

a. Use the definition of subtraction to complete the statement.

$4 - 9 = \square$

b. Use the definition of multiplication to complete the statement.

$6 + 6 + 6 + 6 = \square$

c. Use the definition of absolute value to complete the statement.

$|-9| = \square$

d. Use the rule for adding integers with unlike signs to complete the statement.

$-7 + 3 = \square$

e. Use the rule for multiplying integers with unlike signs to complete the statement.

$-9 \times 2 = \square$

Solution

a. $4 - 9 = 4 + (-9) = -5$

b. $6 + 6 + 6 + 6 = 4 \cdot 6 = 24$

c. $|-9| = -(-9) = 9$

d. $-7 + 3 = -(|-7| - |3|) = -4$

e. $-9 \times 2 = -18$

✓ **CHECKPOINT** *Now try Exercise 97.*

EXAMPLE 8 Finding a Pattern

Complete each pattern. Decide which rules the patterns demonstrate.

a.
$3 \cdot (3) = 9$
$3 \cdot (2) = 6$
$3 \cdot (1) = 3$
$3 \cdot (0) = 0$
$3 \cdot (-1) = \square$
$3 \cdot (-2) = \square$
$3 \cdot (-3) = \square$

b.
$-3 \cdot (3) = -9$
$-3 \cdot (2) = -6$
$-3 \cdot (1) = -3$
$-3 \cdot (0) = 0$
$-3 \cdot (-1) = \square$
$-3 \cdot (-2) = \square$
$-3 \cdot (-3) = \square$

Solution

a.
$3 \cdot (-1) = -3$
$3 \cdot (-2) = -6$
$3 \cdot (-3) = -9$

b.
$-3 \cdot (-1) = 3$
$-3 \cdot (-2) = 6$
$-3 \cdot (-3) = 9$

The product of integers with unlike signs is negative, and the product of integers with like signs is positive.

✓ **CHECKPOINT** *Now try Exercise 101.*

Concept Check

1. You multiply a nonzero number by itself 6 times. Explain why the product is positive.

2. In your own words, write the rule for determining the sign of the quotient of integers.

3. Do prime numbers have composite factors? Do composite numbers have prime factors? Explain.

4. Explain the meaning of the algebraic description below.

   Let a, b, and c be positive integers. If a is a factor of b, then a is a factor of cb.

1.3 EXERCISES

Go to pages 58–59 to record your assignments.

Developing Skills

In Exercises 1–4, write each multiplication as repeated addition or subtraction and find the product.

1. $3 \cdot 2$
2. 4×5
3. $5 \times (-3)$
4. $(-6)(-2)$

In Exercises 5–30, find the product. ***See Example 1.***

5. 5×7
6. 8×3
7. $0 \cdot 4$
8. $-12 \cdot 0$
9. $2(-16)$
10. $8(-7)$
11. ✓ $-9(4)$
12. $-6(5)$
13. $230(-3)$
14. $175(-2)$
15. $-7(-13)$
16. $-40(-4)$
17. $-200(-8)$
18. $-150(-4)$
19. $3(-5)(6)$
20. $4(-2)(-6)$
21. $-7(3)(-1)$
22. $-2(5)(-3)$
23. $-2(-3)(-5)$
24. $-10(-4)(-2)$
25. $|(-3)4|$
26. $|8(-9)|$
27. $|3(-5)(6)|$
28. $|8(-3)(5)|$
29. $|6(20)(4)|$
30. $|9(12)(2)|$

In Exercises 31–40, use the vertical multiplication algorithm to find the product.

31. 26×13
32. 14×9
33. -14×24
34. -8×30
35. $75(-63)$
36. $-72(866)$
37. $-13(-20)$
38. $-11(-24)$
39. $-21(-429)$
40. $-14(-585)$

In Exercises 41–60, perform the division, if possible. If not possible, state the reason. ***See Example 3.***

41. $27 \div 9$
42. $35 \div 7$
43. ✓ $72 \div (-12)$
44. $54 \div (-9)$
45. $-28 \div 4$
46. $-108 \div 9$
47. $-56 \div (-8)$
48. $-68 \div (-4)$
49. $\frac{8}{0}$
50. $\frac{17}{0}$
51. $\frac{0}{8}$
52. $\frac{0}{17}$
53. $\dfrac{-81}{-3}$
54. $\dfrac{-125}{-25}$
55. $\dfrac{6}{-1}$
56. $\dfrac{-33}{1}$
57. $\dfrac{-28}{4}$
58. $\dfrac{72}{-12}$
59. $-27 \div (-27)$
60. $-83 \div (-83)$

In Exercises 61–70, use the long division algorithm to find the quotient.

61. $1440 \div 45$
62. $936 \div 52$
63. $1440 \div (-9)$
64. $936 \div (-8)$
65. $-1312 \div 16$
66. $-5152 \div 23$
67. $2750 \div 25$
68. $22{,}010 \div 71$
69. $-9268 \div (-28)$
70. $-6804 \div (-36)$

In Exercises 71–74, use a calculator to find the quotient.

71. $\dfrac{44,290}{515}$

72. $\dfrac{33,511}{47}$

73. $\dfrac{169,290}{162}$

74. $\dfrac{1,027,500}{250}$

In Exercises 75–84, decide whether the number is prime or composite.

75. 240

76. 533

77. 643

78. 257

79. 3911

80. 1321

81. 1281

82. 1323

83. 3555

84. 8324

In Exercises 85–94, write the prime factorization of the number. ***See Example 5.***

✓ **85.** 12

86. 52

87. 561

88. 245

89. 210

90. 525

91. 2535

92. 1521

93. 192

94. 264

In Exercises 95 and 96, write an example and an algebraic description of the arithmetic rule. ***See Example 6.***

✓ **95.** The product of 1 and any real number is the real number itself.

96. Any nonzero real number divided by itself is 1.

In Exercises 97–100, complete the statement using the indicated definition or rule. ***See Example 7.***

✓ **97.** Definition of division: $12 \div 4 = \square$

98. Definition of absolute value: $|-8| = \square$

99. Rule for multiplying integers by 0:

$6 \cdot 0 = \square = 0 \cdot 6$

100. Rule for dividing integers with unlike signs:

$\dfrac{30}{-10} = \square$

In Exercises 101 and 102, complete the pattern. Decide which rule the pattern demonstrates. ***See Example 8.***

✓ **101.** $2(0) = 0$
$1(0) = 0$
$-1(0) = \square$
$-2(0) = \square$

102. $0 \div 2 = 0$
$0 \div 1 = 0$
$0 \div (-1) = \square$
$0 \div (-2) = \square$

Solving Problems

✓ **103.** *Stock Price* The price per share of a technology stock drops \$0.29 on each of four consecutive days. What is the total price change per share during the four days?

104. *Loss Leaders* To attract customers, a grocery store runs a sale on bananas. The bananas are *loss leaders,* which means the store loses money on the bananas but hopes to make it up on other items. The store sells 800 pounds at a loss of \$0.26 per pound. What is the total loss?

105. *Savings Plan* A homeowner saves \$250 per month for home improvements. After 2 years, how much money has the homeowner saved?

106. *Temperature Change* The temperature measured by a weather balloon is decreasing approximately 3° for each 1000-foot increase in altitude. The balloon rises 8000 feet. What is the total temperature change?

107. *Geometry* Find the area of the football field.

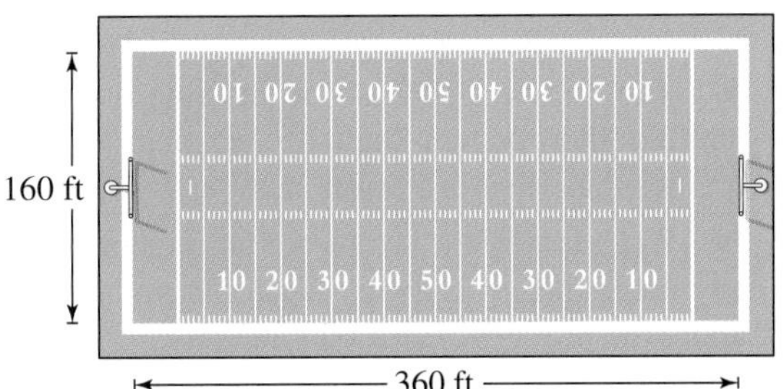

108. *Geometry* Find the area of the city park.

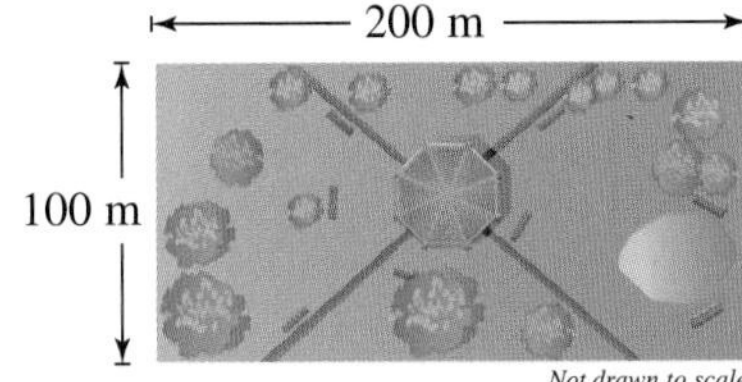

109. *Average Speed* A space shuttle orbiting Earth travels a distance of about 45 miles in 9 seconds. What is the average speed of the space shuttle in miles per second?

110. *Average Speed* A hiker jogs a mountain trail that is 6 miles long in 54 minutes. How many minutes does the hiker average per mile?

111. *Exam Scores* A student has a total of 328 points after four 100-point exams.

(a) What is the average number of points scored per exam?

(b) The scores on the four exams are 87, 73, 77, and 91. Plot each of the scores and the average score on the real number line.

(c) Find the difference between each score and the average score. Find the sum of these differences and give a possible explanation of the result.

112. *Sports* A football team gains a total of 20 yards after four downs.

(a) What is the average number of yards gained per down?

(b) The gains on the four downs are 8 yards, 4 yards, 2 yards, and 6 yards. Plot each of the gains and the average gain on the real number line.

(c) Find the difference between each gain and the average gain. Find the sum of these differences and give a possible explanation of the result.

Geometry In Exercises 113 and 114, find the volume of the shipping box. The volume is found by multiplying the length, width, and height of the box. ***See Example 2.***

✓ **113.**

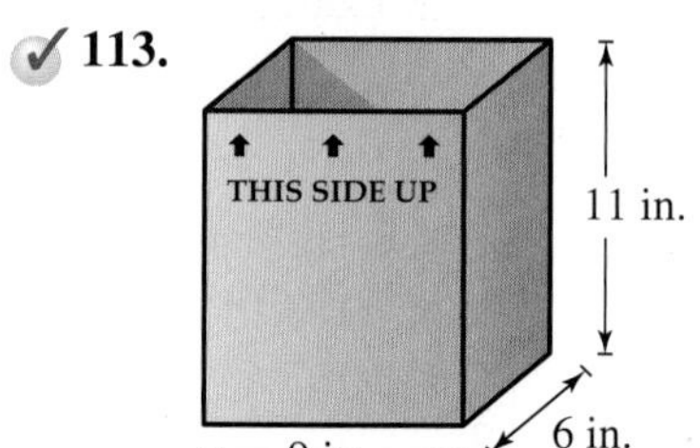

114.

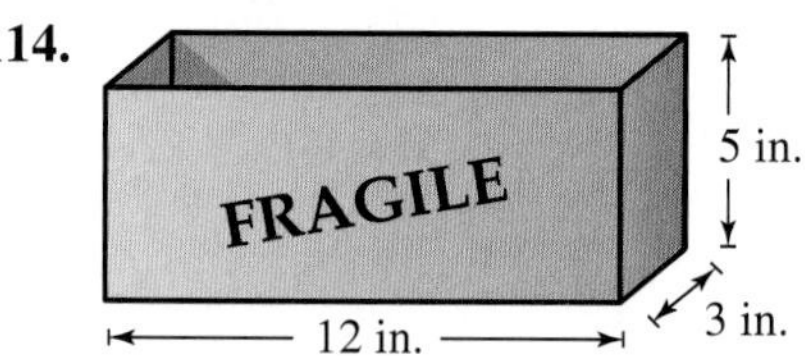

Explaining Concepts

115. What is the only even prime number? Explain why there are no other even prime numbers.

116. The number 1997 is not divisible by a prime number that is less than 45. Explain why this implies that 1997 is a prime number.

117. Explain why the product of an even integer and any other integer is even. What can you conclude about the product of two odd integers?

118. Explain how to check the result of a division problem.

119. *Investigation* Twin primes are prime numbers that differ by 2. For instance, 3 and 5 are twin primes. What are the other twin primes less than 100?

120. *Investigation* The **proper factors** of a number are all its factors less than the number itself. A number is **perfect** if the sum of its proper factors is equal to the number. A number is **abundant** if the sum of its proper factors is greater than the number. Which numbers less than 25 are perfect? Which are abundant? Try to find the first perfect number greater than 25.

121. *The Sieve of Eratosthenes* Write the integers from 1 through 100 in 10 lines of 10 numbers each.

(a) Cross out the number 1. Cross out all multiples of 2 other than 2 itself. Do the same for 3, 5, and 7.

(b) Of what type are the remaining numbers? Explain why this is the only type of number left.

122. *Think About It* An integer n is divided by 2 and the quotient is an even integer. What does this tell you about n? Give an example.

Mid-Chapter Quiz

Take this quiz as you would take a quiz in class. After you are done, check your work against the answers in the back of the book.

In Exercises 1–4, plot each real number as a point on the real number line and place the correct inequality symbol (< or >) between the real numbers.

1. $\frac{3}{16} \quad \square \quad \frac{3}{8}$ 2. $-2.5 \quad \square \quad -4$ 3. $-7 \quad \square \quad 3$ 4. $2\pi \quad \square \quad 6$

In Exercises 5 and 6, evaluate the expression.

5. $-|-0.75|$
6. $\left|-\frac{17}{19}\right|$

In Exercises 7 and 8, place the correct symbol (<, >, or =) between the real numbers.

7. $\left|\frac{7}{2}\right| \quad \square \quad |-3.5|$
8. $\left|\frac{3}{4}\right| \quad \square \quad -|0.75|$
9. Subtract -13 from -22.
10. Find the absolute value of the sum of -54 and 26.

In Exercises 11–22, evaluate the expression.

11. $52 + 47$
12. $-18 + (-35)$
13. $-15 - 12$
14. $-35 - (-10)$
15. $25 + (-75)$
16. $72 - 134$
17. $12 + (-6) - 8 + 10$
18. $-9 - 17 + 36 + (-15)$
19. $-6(10)$
20. $-7(-13)$
21. $\dfrac{-45}{-3}$
22. $\dfrac{-24}{6}$

In Exercises 23–26, decide whether the number is prime or composite. If it is composite, write its prime factorization.

23. 23
24. 91
25. 111
26. 144

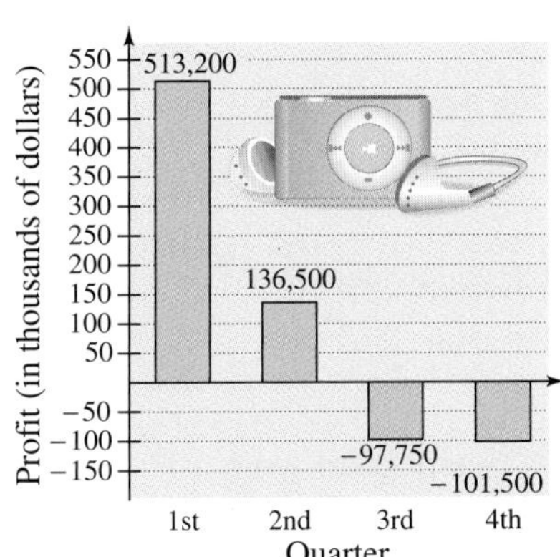

Figure for 27

27. An electronics manufacturer's quarterly profits are shown in the bar graph at the left. What is the manufacturer's total profit for the year?
28. A cord of wood is a pile 8 feet long, 4 feet wide, and 4 feet high. The volume of a rectangular solid is its length times its width times its height. Find the number of cubic feet in a cord of wood.
29. It is necessary to cut a 90-foot rope into six pieces of equal length. What is the length of each piece?
30. At the beginning of a month, your account balance was \$738. During the month, you withdrew \$550, deposited \$189, and paid a fee of \$10. What was your balance at the end of the month?

1.4 Operations with Rational Numbers

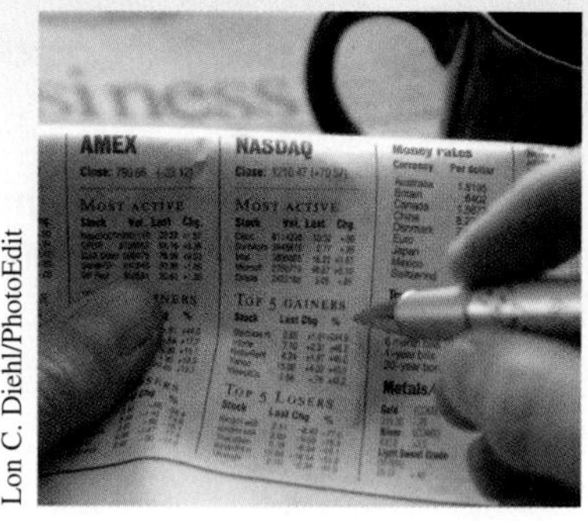
Lon C. Diehl/PhotoEdit

What You Should Learn

1 ▶ Rewrite fractions as equivalent fractions.
2 ▶ Add and subtract fractions.
3 ▶ Multiply and divide fractions.
4 ▶ Add, subtract, multiply, and divide decimals.

Why You Should Learn It

Rational numbers are used to represent many real-life quantities. For instance, in Exercise 151 on page 44, you will use rational numbers to find the increase in the Dow Jones Industrial Average.

1 ▶ Rewrite fractions as equivalent fractions.

Rewriting Fractions

A **fraction** is a number that is written as a quotient, with a *numerator* and a *denominator*. The terms *fraction* and *rational number* are related, but are not exactly the same. The term *fraction* refers to a number's form, whereas the term *rational number* refers to its classification. For instance, the number 2 is a fraction when it is written as $\frac{2}{1}$, but it is a rational number regardless of how it is written.

Rules of Signs for Fractions

1. If the numerator and denominator of a fraction have like signs, the value of the fraction is positive.
2. If the numerator and denominator of a fraction have unlike signs, the value of the fraction is negative.

All of the following fractions are positive and are equivalent to $\frac{2}{3}$.

$$\frac{2}{3}, \frac{-2}{-3}, -\frac{-2}{3}, -\frac{2}{-3}$$

All of the following fractions are negative and are equivalent to $-\frac{2}{3}$.

$$-\frac{2}{3}, \frac{-2}{3}, \frac{2}{-3}, -\frac{-2}{-3}$$

In both arithmetic and algebra, it is often beneficial to write a fraction in **simplest form** or reduced form, which means that the numerator and denominator have no common factors (other than 1). By finding the prime factors of the numerator and the denominator, you can determine what common factor(s) to divide out. The product of the common factors is the **greatest common factor** (or **GCF**).

Study Tip

To find the GCF of two natural numbers, write the prime factorization of each number. For instance, from the prime factorizations

$$18 = 2 \cdot 3 \cdot 3$$

and

$$42 = 2 \cdot 3 \cdot 7$$

you can see that the common factors of 18 and 42 are 2 and 3. So, it follows that the greatest common factor is $2 \cdot 3$, or 6.

Writing a Fraction in Simplest Form

To write a fraction in simplest form, divide both the numerator and denominator by their greatest common factor (GCF).

EXAMPLE 1 Writing Fractions in Simplest Form

Write each fraction in simplest form.

a. $\dfrac{18}{24} = \dfrac{\overset{1}{\cancel{2}} \cdot \overset{1}{\cancel{3}} \cdot 3}{2 \cdot 2 \cdot \underset{1}{\cancel{2}} \cdot \underset{1}{\cancel{3}}} = \dfrac{3}{4}$ Divide out GCF of 6.

b. $\dfrac{35}{21} = \dfrac{5 \cdot \overset{1}{\cancel{7}}}{3 \cdot \underset{1}{\cancel{7}}} = \dfrac{5}{3}$ Divide out GCF of 7.

c. $\dfrac{24}{72} = \dfrac{\overset{1}{\cancel{2}} \cdot \overset{1}{\cancel{2}} \cdot \overset{1}{\cancel{2}} \cdot \overset{1}{\cancel{3}}}{\underset{1}{\cancel{2}} \cdot \underset{1}{\cancel{2}} \cdot \underset{1}{\cancel{2}} \cdot \underset{1}{\cancel{3}} \cdot 3} = \dfrac{1}{3}$ Divide out GCF of 24.

✓ **CHECKPOINT** *Now try Exercise 15.*

You can obtain an **equivalent fraction** by multiplying the numerator and denominator by the same nonzero number or by dividing the numerator and denominator by the same nonzero number. Here are some examples.

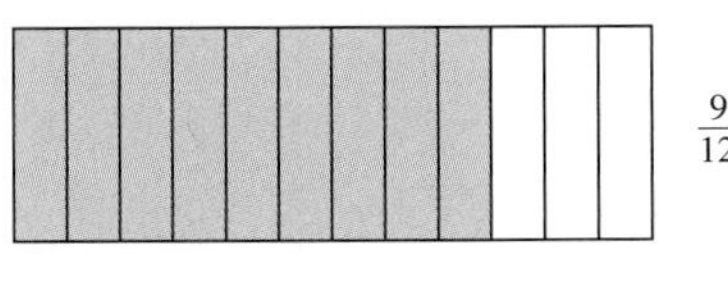

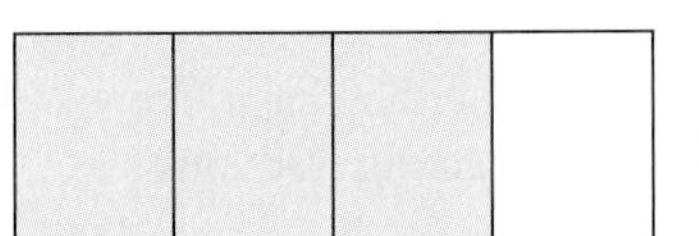

Figure 1.31 Equivalent Fractions

Fraction	*Equivalent Fraction*	*Operation*
$\dfrac{9}{12} = \dfrac{\overset{1}{\cancel{3}} \cdot 3}{\underset{1}{\cancel{3}} \cdot 4}$	$\dfrac{3}{4}$	Divide numerator and denominator by 3. (See Figure 1.31.)
$\dfrac{6}{5} = \dfrac{6 \cdot 2}{5 \cdot 2}$	$\dfrac{12}{10}$	Multiply numerator and denominator by 2.
$\dfrac{-8}{12} = -\dfrac{\overset{1}{\cancel{2}} \cdot \overset{1}{\cancel{2}} \cdot 2}{\underset{1}{\cancel{2}} \cdot \underset{1}{\cancel{2}} \cdot 3}$	$-\dfrac{2}{3}$	Divide numerator and denominator by GCF of 4.

EXAMPLE 2 Writing Equivalent Fractions

Write an equivalent fraction with the indicated denominator.

a. $\dfrac{2}{3} = \dfrac{\quad}{15}$ **b.** $\dfrac{4}{7} = \dfrac{\quad}{42}$ **c.** $\dfrac{9}{15} = \dfrac{\quad}{35}$

Solution

a. $\dfrac{2}{3} = \dfrac{2 \cdot 5}{3 \cdot 5} = \dfrac{10}{15}$ Multiply numerator and denominator by 5.

b. $\dfrac{4}{7} = \dfrac{4 \cdot 6}{7 \cdot 6} = \dfrac{24}{42}$ Multiply numerator and denominator by 6.

c. $\dfrac{9}{15} = \dfrac{\cancel{3} \cdot 3}{\cancel{3} \cdot 5} = \dfrac{3 \cdot 7}{5 \cdot 7} = \dfrac{21}{35}$ Reduce first, then multiply numerator and denominator by 7.

✓ **CHECKPOINT** *Now try Exercise 25.*

2 ▶ Add and subtract fractions.

Adding and Subtracting Fractions

You can use models to add and subtract fractions, as shown in Figure 1.32.

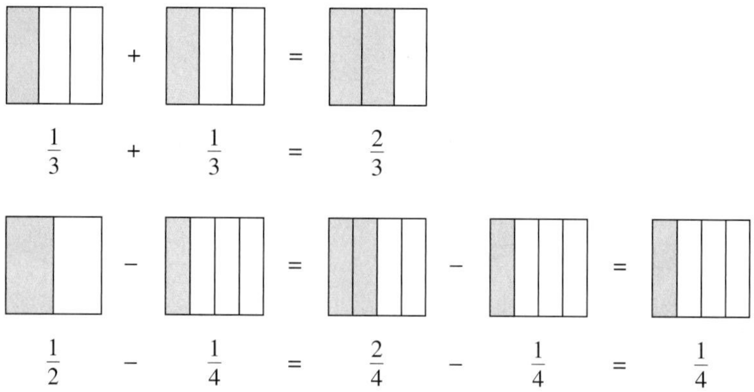

$$\frac{1}{3} + \frac{1}{3} = \frac{2}{3}$$

$$\frac{1}{2} - \frac{1}{4} = \frac{2}{4} - \frac{1}{4} = \frac{1}{4}$$

Figure 1.32

The models suggest the following rules about adding and subtracting fractions.

Study Tip

Adding fractions with unlike denominators is an example of a basic problem-solving strategy that is used in mathematics—rewriting a given problem in a simpler or more familiar form.

Addition and Subtraction of Fractions

Let a, b, and c be integers with $c \neq 0$.

1. *With like denominators:*

$$\frac{a}{c} + \frac{b}{c} = \frac{a+b}{c} \quad \text{or} \quad \frac{a}{c} - \frac{b}{c} = \frac{a-b}{c}$$

2. *With unlike denominators:* Rewrite the fractions so that they have like denominators. Then use the rule for adding and subtracting fractions with like denominators.

EXAMPLE 3 Adding and Subtracting Fractions with Like Denominators

a. $\frac{3}{12} + \frac{4}{12} = \frac{3+4}{12} = \frac{7}{12}$ Add numerators.

b. $\frac{7}{9} - \frac{2}{9} = \frac{7-2}{9} = \frac{5}{9}$ Subtract numerators.

✓ ***CHECKPOINT*** ***Now try Exercise 29.***

To find a like denominator for two or more fractions, find the **least common multiple** (or **LCM**) of their denominators. For instance, the LCM of the denominators of $\frac{3}{8}$ and $-\frac{5}{12}$ is 24. To see this, consider all multiples of 8 (8, 16, 24, 32, 40, 48, . . .) and all multiples of 12 (12, 24, 36, 48, . . .). The numbers 24 and 48 are common multiples, and the number 24 is the smallest of the common multiples.

EXAMPLE 4 Adding Fractions with Unlike Denominators

$$\frac{4}{5} + \frac{11}{15} = \frac{4(3)}{5(3)} + \frac{11}{15}$$ LCM of 5 and 15 is 15.

$$= \frac{12}{15} + \frac{11}{15}$$ Rewrite with like denominators.

$$= \frac{23}{15}$$ Add numerators.

✓ **CHECKPOINT** *Now try Exercise 45.*

Study Tip

In Example 5, a common shortcut for writing $1\frac{7}{9}$ as $\frac{16}{9}$ is to multiply 1 by 9, add the result to 7, and then divide by 9, as follows.

$$1\frac{7}{9} = \frac{1(9) + 7}{9} = \frac{16}{9}$$

EXAMPLE 5 Subtracting Fractions with Unlike Denominators

Evaluate $1\frac{7}{9} - \frac{11}{12}$.

Solution

To begin, rewrite the **mixed number** $1\frac{7}{9}$ as a fraction.

$$1\frac{7}{9} = 1 + \frac{7}{9} = \frac{9}{9} + \frac{7}{9} = \frac{16}{9}$$

Then subtract the two fractions as follows.

$$1\frac{7}{9} - \frac{11}{12} = \frac{16}{9} - \frac{11}{12}$$ Rewrite $1\frac{7}{9}$ as $\frac{16}{9}$.

$$= \frac{16(4)}{9(4)} + \frac{-11(3)}{12(3)}$$ LCM of 9 and 12 is 36.

$$= \frac{64}{36} + \frac{-33}{36}$$ Rewrite with like denominators.

$$= \frac{31}{36}$$ Add numerators.

✓ **CHECKPOINT** *Now try Exercise 61.*

You can add or subtract two fractions, without first finding a common denominator, by using the following rule.

Alternative Rule for Adding and Subtracting Two Fractions

If a, b, c, and d are integers with $b \neq 0$ and $d \neq 0$, then

$$\frac{a}{b} + \frac{c}{d} = \frac{ad + bc}{bd} \quad \text{or} \quad \frac{a}{b} - \frac{c}{d} = \frac{ad - bc}{bd}.$$

Technology: Tip

When you use a scientific or graphing calculator to add or subtract fractions, your answer may appear in decimal form. An answer such as 0.583333333 is not as exact as $\frac{7}{12}$ and may introduce roundoff error. Refer to the user's manual for your calculator for instructions on adding and subtracting fractions and displaying answers in fraction form.

In Example 5, the difference between $1\frac{7}{9}$ and $\frac{11}{12}$ was found using the least common multiple of 9 and 12. Compare those solution steps with the following steps, which use the alternative rule for adding or subtracting two fractions.

$$1\frac{7}{9} - \frac{11}{12} = \frac{16(12) - 9(11)}{9(12)} \qquad \text{Apply alternative rule.}$$

$$= \frac{192 - 99}{108} \qquad \text{Multiply.}$$

$$= \frac{93}{108} = \frac{31}{36} \qquad \text{Simplify.}$$

EXAMPLE 6 Subtracting Fractions

$$\frac{5}{16} - \left(-\frac{7}{30}\right) = \frac{5}{16} + \frac{7}{30} \qquad \text{Add the opposite.}$$

$$= \frac{5(30) + 16(7)}{16(30)} \qquad \text{Apply alternative rule.}$$

$$= \frac{150 + 112}{480} \qquad \text{Multiply.}$$

$$= \frac{262}{480} = \frac{131}{240} \qquad \text{Simplify.}$$

✓ **CHECKPOINT** *Now try Exercise 57.*

EXAMPLE 7 Combining Three or More Fractions

Evaluate $\frac{5}{6} - \frac{7}{15} + \frac{3}{10} - 1$.

Solution

The least common denominator of 6, 15, and 10 is 30. So, you can rewrite the original expression as follows.

$$\frac{5}{6} - \frac{7}{15} + \frac{3}{10} - 1 = \frac{5(5)}{6(5)} + \frac{(-7)(2)}{15(2)} + \frac{3(3)}{10(3)} + \frac{(-1)(30)}{30}$$

$$= \frac{25}{30} + \frac{-14}{30} + \frac{9}{30} + \frac{-30}{30} \qquad \text{Rewrite with like denominators.}$$

$$= \frac{25 - 14 + 9 - 30}{30} \qquad \text{Add numerators.}$$

$$= \frac{-10}{30} = -\frac{1}{3} \qquad \text{Simplify.}$$

✓ **CHECKPOINT** *Now try Exercise 69.*

3 ▶ Multiply and divide fractions.

Multiplying and Dividing Fractions

The procedure for multiplying fractions is simpler than those for adding and subtracting fractions. Regardless of whether the fractions have like or unlike denominators, you can find the product of two fractions by multiplying the numerators and multiplying the denominators.

Multiplication of Fractions

Let a, b, c, and d be integers with $b \neq 0$ and $d \neq 0$. Then the product of $\frac{a}{b}$ and $\frac{c}{d}$ is

$$\frac{a}{b} \cdot \frac{c}{d} = \frac{a \cdot c}{b \cdot d}.$$ Multiply numerators and denominators.

EXAMPLE 8 Multiplying Fractions

a. $\frac{5}{8} \cdot \frac{3}{2} = \frac{5(3)}{8(2)}$ Multiply numerators and denominators.

$= \frac{15}{16}$ Simplify.

b. $\left(-\frac{7}{9}\right)\left(-\frac{5}{21}\right) = \frac{7}{9} \cdot \frac{5}{21}$ Product of two negatives is positive.

$= \frac{7(5)}{9(21)}$ Multiply numerators and denominators.

$= \frac{\cancel{7}(5)}{9(3)(\cancel{7})}$ Divide out common factor.

$= \frac{5}{27}$ Write in simplest form.

CHECKPOINT *Now try Exercise 89.*

EXAMPLE 9 Multiplying Three Fractions

$\left(3\frac{1}{5}\right)\left(-\frac{7}{6}\right)\left(\frac{5}{3}\right) = \left(\frac{16}{5}\right)\left(-\frac{7}{6}\right)\left(\frac{5}{3}\right)$ Rewrite mixed number as a fraction.

$= \frac{16(-7)(5)}{5(6)(3)}$ Multiply numerators and denominators.

$= -\frac{(8)(\cancel{2})(7)(\cancel{5})}{(\cancel{5})(3)(\cancel{2})(3)}$ Divide out common factors.

$= -\frac{56}{9}$ Write in simplest form.

CHECKPOINT *Now try Exercise 97.*

The **reciprocal** or **multiplicative inverse** of a number is the number by which it must be multiplied to obtain 1. For instance, the reciprocal of 3 is $\frac{1}{3}$ because $3\left(\frac{1}{3}\right) = 1$. Similarly, the reciprocal of $-\frac{2}{3}$ is $-\frac{3}{2}$ because

$$\left(-\frac{2}{3}\right)\left(-\frac{3}{2}\right) = 1.$$

To divide two fractions, multiply the first fraction by the *reciprocal* of the second fraction. Another way of saying this is "invert the divisor and multiply."

Division of Fractions

Let a, b, c, and d be integers with $b \neq 0$, $c \neq 0$, and $d \neq 0$. Then the quotient of $\frac{a}{b}$ and $\frac{c}{d}$ is

$$\frac{a}{b} \div \frac{c}{d} = \frac{a}{b} \cdot \frac{d}{c}.$$ Invert divisor and multiply.

EXAMPLE 10 Dividing Fractions

a. $\frac{5}{8} \div \frac{20}{12} = \frac{5}{8} \cdot \frac{12}{20}$ Invert divisor and multiply.

$= \frac{(5)(12)}{(8)(20)}$ Multiply numerators and denominators.

$= \frac{\cancel{(5)}(3)\cancel{(4)}}{(8)\cancel{(4)}\cancel{(5)}}$ Divide out common factors.

$= \frac{3}{8}$ Write in simplest form.

b. $\frac{6}{13} \div \left(-\frac{9}{26}\right) = \frac{6}{13} \cdot \left(-\frac{26}{9}\right)$ Invert divisor and multiply.

$= -\frac{(6)(26)}{(13)(9)}$ Multiply numerators and denominators.

$= -\frac{(2)\cancel{(3)}(2)\cancel{(13)}}{\cancel{(13)}\cancel{(3)}(3)}$ Divide out common factors.

$= -\frac{4}{3}$ Write in simplest form.

c. $-\frac{1}{4} \div (-3) = -\frac{1}{4} \cdot \left(-\frac{1}{3}\right)$ Invert divisor and multiply.

$= \frac{(-1)(-1)}{(4)(3)}$ Multiply numerators and denominators.

$= \frac{1}{12}$ Write in simplest form.

CHECKPOINT *Now try Exercise 107.*

4 ▶ Add, subtract, multiply, and divide decimals.

Operations with Decimals

Rational numbers can be represented as **terminating** or **repeating decimals.** Here are some examples.

Terminating Decimals	*Repeating Decimals*
$\frac{1}{4} = 0.25$	$\frac{1}{6} = 0.1666 \ldots$ or $0.1\overline{6}$
$\frac{3}{8} = 0.375$	$\frac{1}{3} = 0.3333 \ldots$ or $0.\overline{3}$
$\frac{2}{10} = 0.2$	$\frac{1}{12} = 0.0833 \ldots$ or $0.08\overline{3}$
$\frac{5}{16} = 0.3125$	$\frac{8}{33} = 0.2424 \ldots$ or $0.\overline{24}$

Note that bar notation is used to indicate the *repeated* digit (or digits) in the decimal notation. You can obtain the decimal representation of any fraction by long division. For instance, the decimal representation of $\frac{5}{12}$ is $0.41\overline{6}$, which can be obtained using the following long division algorithm.

$$\begin{array}{r} 0.4166 \\ 12\overline{)5.0000} \\ \underline{4\,8} \\ 20 \\ \underline{12} \\ 80 \\ \underline{72} \\ 80 \end{array} \quad \ldots = 0.41\overline{6}$$

For calculations involving decimals such as 0.4166 . . . , you must **round the decimal.** For instance, rounded to two decimal places, the number 0.4166 . . . is 0.42. Similarly, rounded to three decimal places, the number 0.4166 . . . is 0.417.

Rounding a Decimal

1. Determine the number of digits of accuracy you wish to keep. The digit in the last position you keep is called the **rounding digit,** and the digit in the first position you discard is called the **decision digit.**
2. If the decision digit is 5 or greater, round up by adding 1 to the rounding digit.
3. If the decision digit is 4 or less, round down by leaving the rounding digit unchanged.

Given Decimal	*Rounded to Three Places*
0.9768	0.977
0.9765	0.977
0.9763	0.976

EXAMPLE 11 Adding and Multiplying Decimals

a. Add 0.583, 1.06, and 2.9104. **b.** Multiply -3.57 and 0.032.

Solution

a. To add decimals, align the decimal points and proceed as in integer addition.

$$\begin{array}{r@{}l} & \text{1 1} \\ & 0.583 \\ & 1.06 \\ + & 2.9104 \\ \hline & 4.5534 \end{array}$$

b. To multiply decimals, use integer multiplication and then place the decimal point (in the product) so that the number of decimal places equals the sum of the decimal places in the two factors.

$$\begin{array}{rrl} & -3.57 & \text{Two decimal places} \\ \times & 0.032 & \text{Three decimal places} \\ \hline & 714 & \\ & 1071 & \\ \hline & -0.11424 & \text{Five decimal places} \end{array}$$

 **CHECKPOINT** *Now try Exercise 145.*

EXAMPLE 12 Dividing Decimals

Divide 1.483 by 0.56. Round the answer to two decimal places.

Solution

To divide 1.483 by 0.56, convert the divisor to an integer by moving its decimal point to the right. Move the decimal point in the dividend an equal number of places to the right. Place the decimal point in the quotient directly above the new decimal point in the dividend and then divide as with integers.

$$\begin{array}{r} 2.648 \\ 56\,\overline{)\,148.300} \\ \underline{112} \\ 36\,3 \\ \underline{33\,6} \\ 2\,70 \\ \underline{2\,24} \\ 460 \\ \underline{448} \end{array}$$

Rounded to two decimal places, the answer is 2.65. This answer can be written as

$$\frac{1.483}{0.56} \approx 2.65$$

where the symbol $\approx$ means **is approximately equal to.**

CHECKPOINT *Now try Exercise 147.*

EXAMPLE 13 Physical Fitness

To satisfy your health and fitness requirement, you decide to take a tennis class. You learn that you burn about 400 calories per hour playing tennis. In one week, you played tennis for $\frac{3}{4}$ hour on Tuesday, 2 hours on Wednesday, and $1\frac{1}{2}$ hours on Thursday. How many total calories did you burn playing tennis during that week? What was the average number of calories you burned playing tennis for the three days?

Solution

The total number of calories you burned playing tennis during the week was

$$400\left(\frac{3}{4}\right) + 400(2) + 400\left(1\frac{1}{2}\right) = 300 + 800 + 600 = 1700 \text{ calories.}$$

The average number of calories you burned playing tennis for the three days was

$$\frac{1700}{3} \approx 566.67 \text{ calories.}$$

CHECKPOINT *Now try Exercise 161.*

Summary of Rules for Fractions

Let a, b, c, and d be real numbers.

Rule	*Example*
1. Rules of signs for fractions:	
$\frac{-a}{-b} = \frac{a}{b}$	$\frac{-12}{-4} = \frac{12}{4}$
$\frac{-a}{b} = \frac{a}{-b} = -\frac{a}{b}$	$\frac{-12}{4} = \frac{12}{-4} = -\frac{12}{4}$
2. Equivalent fractions:	
$\frac{a}{b} = \frac{a \cdot c}{b \cdot c}, b \neq 0, c \neq 0$	$\frac{1}{4} = \frac{3}{12}$ because $\frac{1}{4} = \frac{1 \cdot 3}{4 \cdot 3} = \frac{3}{12}$
3. Addition of fractions:	
$\frac{a}{b} + \frac{c}{d} = \frac{ad + bc}{bd}, b \neq 0, d \neq 0$	$\frac{1}{3} + \frac{2}{7} = \frac{1 \cdot 7 + 3 \cdot 2}{3 \cdot 7} = \frac{13}{21}$
4. Subtraction of fractions:	
$\frac{a}{b} - \frac{c}{d} = \frac{ad - bc}{bd}, b \neq 0, d \neq 0$	$\frac{1}{3} - \frac{2}{7} = \frac{1 \cdot 7 - 3 \cdot 2}{3 \cdot 7} = \frac{1}{21}$
5. Multiplication of fractions:	
$\frac{a}{b} \cdot \frac{c}{d} = \frac{a \cdot c}{b \cdot d}, b \neq 0, d \neq 0$	$\frac{1}{3} \cdot \frac{2}{7} = \frac{1(2)}{3(7)} = \frac{2}{21}$
6. Division of fractions:	
$\frac{a}{b} \div \frac{c}{d} = \frac{a}{b} \cdot \frac{d}{c}, b \neq 0, c \neq 0, d \neq 0$	$\frac{1}{3} \div \frac{2}{7} = \frac{1}{3} \cdot \frac{7}{2} = \frac{7}{6}$

Concept Check

1. Explain how to write an equivalent fraction.
2. In your own words, explain the procedure for adding and subtracting fractions with unlike denominators.
3. What is the reciprocal of the number n?
4. Explain the procedure for rounding a decimal.

1.4 EXERCISES

Go to pages 58–59 to record your assignments.

Developing Skills

In Exercises 1–12, find the greatest common factor.

1. 5, 10 **2.** 3, 9

3. 20, 45 **4.** 48, 64

5. 45, 90 **6.** 27, 54

7. 18, 84, 90 **8.** 84, 98, 192

9. 240, 300, 360 **10.** 117, 195, 507

11. 134, 225, 315, 945 **12.** 80, 144, 214, 504

In Exercises 13–20, write the fraction in simplest form. ***See Example 1.***

13. $\frac{2}{4}$ **14.** $\frac{4}{16}$

15. $\frac{12}{15}$ **16.** $\frac{14}{35}$

17. $\frac{60}{192}$ **18.** $\frac{90}{225}$

19. $\frac{28}{350}$ **20.** $\frac{88}{154}$

In Exercises 21–24, each figure is divided into regions of equal area. Write a fraction that represents the shaded portion of the figure. Then write the fraction in simplest form.

21.

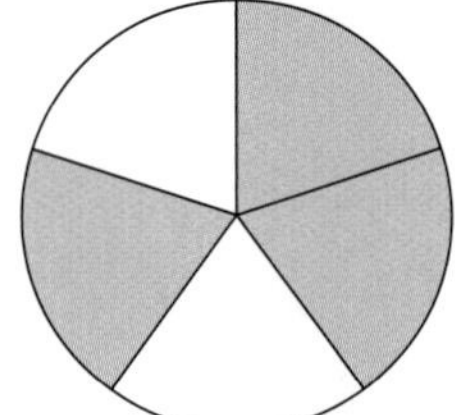

22.

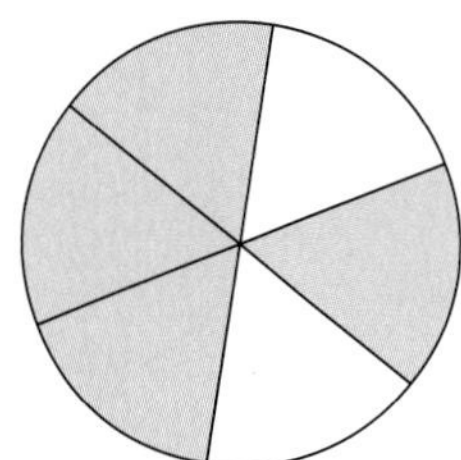

23.

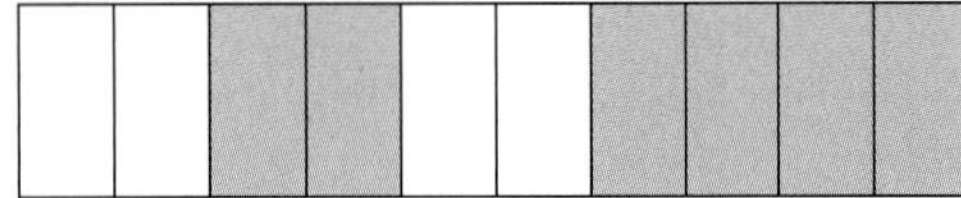

24.

In Exercises 25–28, write an equivalent fraction with the indicated denominator. ***See Example 2.***

25. $\frac{3}{8} = \frac{\square}{16}$

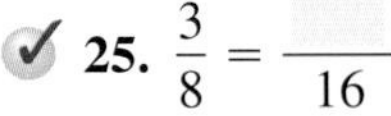

26. $\frac{4}{5} = \frac{\square}{15}$

27. $\frac{6}{15} = \frac{\square}{25}$ **28.** $\frac{21}{49} = \frac{\square}{28}$

In Exercises 29–42, find the sum or difference. Write the result in simplest form. ***See Example 3.***

29. $\frac{7}{15} + \frac{1}{15}$ **30.** $\frac{13}{35} + \frac{5}{35}$

31. $\frac{3}{2} + \frac{5}{2}$ **32.** $\frac{5}{6} + \frac{13}{6}$

33. $\frac{9}{16} - \frac{3}{16}$ **34.** $\frac{15}{32} - \frac{7}{32}$

35. $-\frac{23}{11} + \frac{12}{11}$ **36.** $-\frac{39}{23} + \frac{11}{23}$

37. $\frac{3}{4} - \frac{5}{4}$ **38.** $\frac{7}{8} - \frac{9}{8}$

39. $\frac{7}{10} + \left(-\frac{3}{10}\right)$ **40.** $\frac{11}{15} + \left(-\frac{2}{15}\right)$

41. $\frac{2}{5} + \frac{4}{5} + \frac{1}{5}$ **42.** $\frac{2}{9} + \frac{4}{9} + \frac{1}{9}$

In Exercises 43–66, evaluate the expression. Write the result in simplest form. ***See Examples 4, 5, and 6.***

43. $\frac{1}{2} + \frac{1}{3}$ **44.** $\frac{3}{5} + \frac{1}{2}$

45. $\frac{3}{16} + \frac{3}{8}$ **46.** $\frac{2}{3} + \frac{4}{9}$

47. $\frac{1}{4} - \frac{1}{3}$ **48.** $\frac{2}{3} - \frac{1}{6}$

49. $-\frac{1}{8} - \frac{1}{6}$ **50.** $-\frac{13}{8} - \frac{3}{4}$

51. $4 + \frac{8}{3}$ **52.** $2 + \frac{17}{25}$

53. $-\frac{3}{8} - \frac{1}{12}$ **54.** $-\frac{5}{12} - \frac{1}{9}$

55. $\frac{3}{4} - \frac{2}{5}$ **56.** $\frac{5}{6} - \frac{2}{7}$

57. $-\frac{5}{6} - \left(-\frac{3}{4}\right)$ **58.** $-\frac{1}{9} - \left(-\frac{3}{5}\right)$

59. $3\frac{1}{2} + 5\frac{2}{3}$ **60.** $5\frac{3}{4} + 8\frac{1}{10}$

61. $1\frac{3}{16} - 2\frac{1}{4}$ **62.** $5\frac{7}{8} - 2\frac{1}{2}$

63. $15 - 20\frac{1}{4}$ **64.** $6 - 3\frac{5}{8}$

65. $-5\frac{1}{3} - 4\frac{5}{12}$ **66.** $-2\frac{3}{4} - 3\frac{1}{5}$

In Exercises 67–72, evaluate the expression. Write the result in simplest form. ***See Example 7.***

67. $\frac{5}{12} - \frac{3}{8} + \frac{5}{4}$ **68.** $-\frac{3}{7} + \frac{5}{14} + \frac{3}{4}$

69. $3 + \frac{12}{3} + \frac{1}{9}$ **70.** $1 + \frac{2}{3} - \frac{5}{6}$

71. $2 - \frac{25}{6} - \frac{3}{4}$ **72.** $2 - \frac{15}{16} - \frac{7}{6}$

In Exercises 73–76, determine the unknown fractional part of the circle graph.

73.

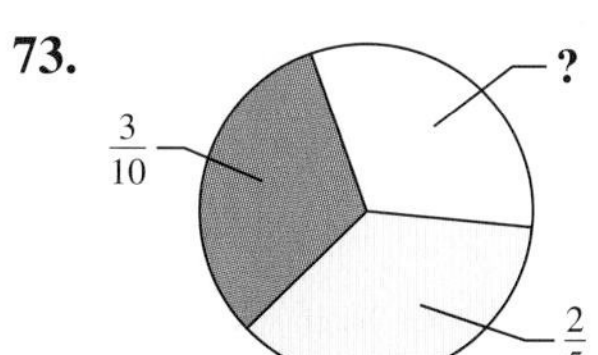

74.

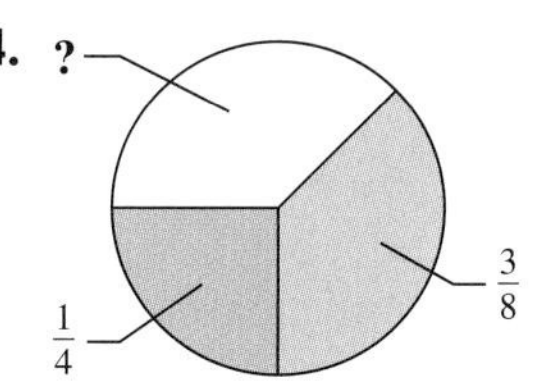

75.

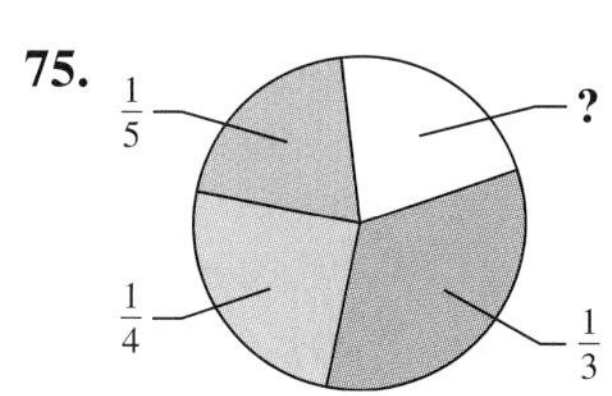

76.

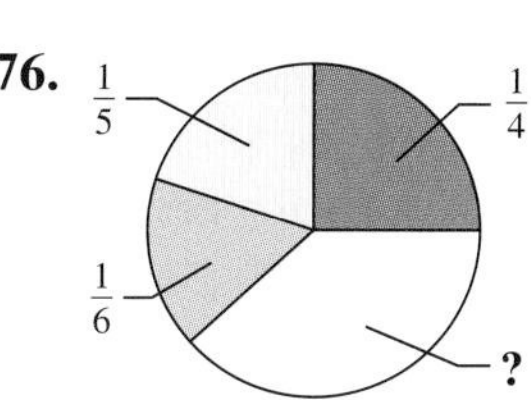

In Exercises 77–100, evaluate the expression. Write the result in simplest form. ***See Examples 8 and 9.***

77. $\frac{1}{2} \cdot \frac{3}{4}$ **78.** $\frac{3}{5} \cdot \frac{1}{2}$

79. $-\frac{2}{3} \cdot \frac{5}{7}$ **80.** $-\frac{5}{6} \cdot \frac{1}{2}$

81. $\frac{2}{3}\left(-\frac{9}{16}\right)$ **82.** $\frac{5}{3}\left(-\frac{3}{5}\right)$

83. $-\frac{3}{4}\left(-\frac{4}{9}\right)$ **84.** $-\frac{15}{16}\left(-\frac{12}{5}\right)$

85. $\frac{5}{18}\left(\frac{3}{4}\right)$ **86.** $\frac{3}{28}\left(\frac{7}{8}\right)$

87. $\frac{11}{12}\left(-\frac{9}{44}\right)$ **88.** $\frac{5}{12}\left(-\frac{6}{25}\right)$

89. $-\frac{3}{11}\left(-\frac{11}{3}\right)$ **90.** $-\frac{7}{15}\left(-\frac{15}{7}\right)$

91. $9\left(\frac{4}{15}\right)$ **92.** $24\left(\frac{7}{18}\right)$

93. $2\frac{3}{4} \cdot 3\frac{2}{3}$ **94.** $-8\frac{1}{2} \cdot 3\frac{2}{5}$

95. $-5\frac{2}{3} \cdot 4\frac{1}{2}$ **96.** $2\frac{4}{5} \cdot 6\frac{2}{3}$

97. $-\frac{3}{2}\left(-\frac{15}{16}\right)\left(\frac{12}{25}\right)$ **98.** $\frac{1}{2}\left(-\frac{4}{15}\right)\left(-\frac{5}{24}\right)$

99. $6\left(\frac{3}{4}\right)\left(\frac{2}{9}\right)$ **100.** $8\left(\frac{5}{12}\right)\left(\frac{3}{10}\right)$

In Exercises 101–104, find the reciprocal of the number. Show that the product of the number and its reciprocal is 1.

101. 7 **102.** 14

103. $\frac{4}{7}$ **104.** $-\frac{5}{9}$

In Exercises 105–124, evaluate the expression and write the result in simplest form. If it is not possible, explain why. ***See Example 10.***

105. $\frac{3}{8} \div \frac{3}{4}$ **106.** $\frac{5}{16} \div \frac{25}{8}$

107. $-\frac{5}{12} \div \frac{45}{32}$ **108.** $-\frac{16}{21} \div \frac{12}{27}$

109. $\frac{8}{3} \div \frac{8}{3}$ **110.** $\frac{5}{7} \div \frac{5}{7}$

111. $\frac{3}{5} \div \frac{7}{5}$ **112.** $\frac{7}{8} \div \frac{3}{8}$

113. $-\frac{5}{6} \div \left(-\frac{8}{10}\right)$ **114.** $-\frac{14}{15} \div \left(-\frac{24}{25}\right)$

115. $-10 \div \frac{1}{9}$ **116.** $-6 \div \frac{1}{3}$

117. $0 \div (-21)$ **118.** $0 \div (-33)$

119. $\frac{3}{5} \div 0$ **120.** $\frac{11}{13} \div 0$

121. $3\frac{3}{4} \div 1\frac{1}{2}$ **122.** $2\frac{4}{9} \div 5\frac{1}{3}$

123. $3\frac{1}{4} \div 2\frac{5}{8}$ **124.** $1\frac{5}{6} \div 2\frac{1}{3}$

In Exercises 125–134, write the fraction in decimal form. (Use bar notation for repeating digits.)

125. $\frac{1}{4}$ **126.** $\frac{7}{8}$

127. $\frac{5}{16}$ **128.** $\frac{9}{20}$

129. $\frac{2}{9}$ **130.** $\frac{1}{6}$

131. $\frac{5}{12}$ **132.** $\frac{5}{18}$

133. $\frac{4}{11}$ **134.** $\frac{4}{15}$

In Exercises 135–150, evaluate the expression. Round your answer to two decimal places. ***See Examples 11 and 12.***

135. $12.33 + 14.76$ **136.** $6.983 + 241.5$

137. $132.1 + (-25.45)$

138. $408.9 + (-13.12)$

139. $1.21 + 4.06 - 3.00$

140. $3.4 + 1.062 - 5.13$

141. $-0.0005 - 2.01 + 0.111$

142. $-1.0012 - 3.25 + 0.2$

143. $-6.3(9.05)$

144. $3.7(-14.8)$

✓ **145.** $-0.05(-85.95)$

146. $-0.09(-0.45)$

✓ **147.** $4.69 \div 0.12$

148. $7.14 \div 0.94$

149. $1.062 \div (-2.1)$

150. $2.011 \div (-3.3)$

Solving Problems

151. *Stock Price* On September 26, 2007, the Dow Jones Industrial Average closed at 13,878.15 points. On September 27, 2007, it closed at 13,912.94 points. Determine the increase in the Dow Jones Industrial Average.

152. *Clothing Design* A designer uses $3\frac{1}{6}$ yards of material to make a skirt and $2\frac{3}{4}$ yards to make a shirt. Find the total amount of material required.

153. *Agriculture* During the months of January, February, and March, a farmer bought $8\frac{3}{4}$ tons, $7\frac{1}{5}$ tons, and $9\frac{3}{8}$ tons of feed, respectively. Find the total amount of feed purchased during the first quarter of the year.

154. *Cooking* You are making a batch of cookies. You have placed 2 cups of flour, $\frac{1}{3}$ cup butter, $\frac{1}{2}$ cup brown sugar, and $\frac{1}{3}$ cup granulated sugar in a mixing bowl. How many cups of ingredients are in the mixing bowl?

155. *Construction Project* The highway workers have a sign beside a construction project indicating what fraction of the work has been completed. At the beginnings of May and June, the fractions of work completed were $\frac{5}{16}$ and $\frac{2}{3}$, respectively. What fraction of the work was completed during the month of May?

156. *Fund Drive* During a fund drive, a charity has a display showing how close it is to reaching its goal. At the end of the first week, the display shows $\frac{1}{8}$ of the goal. At the end of the second week, the display shows $\frac{3}{5}$ of the goal. What fraction of the goal was gained during the second week?

157. *Cooking* You make 60 ounces of dough for breadsticks. Each breadstick requires $\frac{5}{4}$ ounces of dough. How many breadsticks can you make?

158. *Unit Price* A $2\frac{1}{2}$-pound can of food costs $4.95. What is the cost per pound?

159. *Consumer Awareness* The sticker on a new car gives the fuel efficiency as 22.3 miles per gallon. The average cost of fuel is $2.859 per gallon. Estimate the annual fuel cost for a car that will be driven approximately 12,000 miles per year.

160. *Walking Time* Your apartment is $\frac{3}{4}$ mile from the subway. You walk at the rate of $3\frac{1}{4}$ miles per hour. How long does it take you to walk to the subway?

✓ **161.** *Consumer Awareness* At a convenience store, you buy two gallons of milk at $3.75 per gallon and three loaves of bread at $1.68 per loaf. You give the clerk a 20-dollar bill. How much change will you receive? (Assume there is no sales tax.)

162. *Consumer Awareness* A cellular phone company charges $5.35 for the first 250 text messages and $0.10 for each additional text message. Find the cost of 263 text messages.

163. *Stock Purchase* You buy 200 shares of stock at $23.63 per share and 300 shares at $86.25 per share.

(a) Estimate the total cost of the stock.

(b) Use a calculator to find the total cost of the stock.

164. *Music* Each day for a week, you practiced the saxophone for $\frac{2}{3}$ hour.

(a) Explain how to use mental math to estimate the number of hours of practice in a week.

(b) Determine the actual number of hours you practiced during the week. Write the result in decimal form, rounding to one decimal place.

165. *Consumer Awareness* The prices per gallon of regular unleaded gasoline at three service stations are \$2.859, \$2.969, and \$3.079, respectively. Find the average price per gallon.

166. *Consumer Awareness* The prices of a 16-ounce bottle of soda at three different convenience stores are \$1.09, \$1.25, and \$1.10, respectively. Find the average price for the bottle of soda.

Explaining Concepts

167. Is it true that the sum of two fractions of like signs is positive? If not, give an example that shows the statement is false.

168. Does $\frac{2}{3} + \frac{3}{2} = (2 + 3)/(3 + 2) = 1$? Explain your answer.

169. In your own words, describe the rule for determining the sign of the product of two fractions.

170. Is it true that $\frac{2}{3} = 0.67$? Explain your answer.

171. Use the figure to determine how many one-fourths are in 3. Explain how to obtain the same result by division.

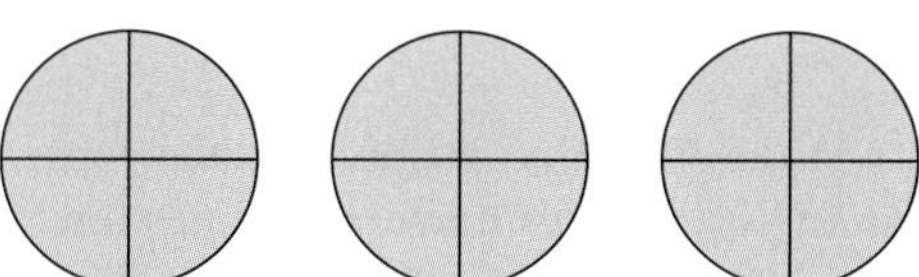

172. Use the figure to determine how many one-sixths are in $\frac{2}{3}$. Explain how to obtain the same result by division.

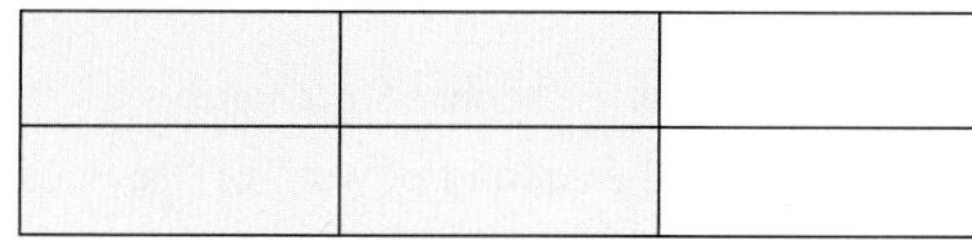

173. *Investigation* When using a calculator to perform operations with decimals, you should try to get in the habit of rounding your answers *only* after all the calculations are done. If you round the answer at a preliminary stage, you can introduce unnecessary roundoff error. The dimensions of a box are $l = 5.24$, $w = 3.03$, and $h = 2.749$. Find the volume, $l \cdot w \cdot h$, by multiplying the numbers and then rounding the answer to one decimal place. Now use a second method, first rounding each dimension to one decimal place and then multiplying the numbers. Compare your answers, and explain which of these techniques produces the more accurate answer.

True or False? In Exercises 174–179, decide whether the statement is true or false. Justify your answer.

174. The reciprocal of every nonzero integer is an integer.

175. The reciprocal of every nonzero rational number is a rational number.

176. The product of two nonzero rational numbers is a rational number.

177. The product of two positive rational numbers is greater than either factor.

178. If $u > v$, then $u - v > 0$.

179. If $u > 0$ and $v > 0$, then $u - v > 0$.

180. *Estimation* Use mental math to determine whether $\left(5\frac{3}{4}\right) \times \left(4\frac{1}{8}\right)$ is less than 20. Explain your reasoning.

181. Determine the placement of the digits 3, 4, 5, and 6 in the following addition problem so that you obtain the specified sum. Use each number only once.

$$\frac{\square}{\square} + \frac{\square}{\square} = \frac{13}{10}$$

182. If the fractions represented by the points P and R are multiplied, what point on the number line best represents their product: M, S, N, P, or T? (Source: National Council of Teachers of Mathematics)

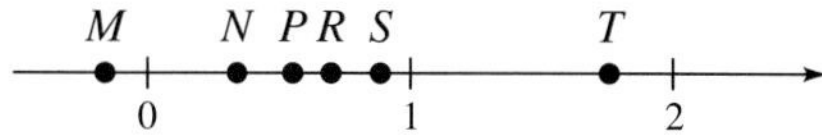

1.5 Exponents, Order of Operations, and Properties of Real Numbers

Getty Images

Why You Should Learn It

Properties of real numbers can be used to solve real-life problems. For instance, in Exercise 130 on page 56, you will use the Distributive Property to find the amount paid for a new truck.

What You Should Learn

1 ▶ Rewrite repeated multiplication in exponential form and evaluate exponential expressions.

2 ▶ Evaluate expressions using order of operations.

3 ▶ Identify and use the properties of real numbers.

Exponents

1 ▶ Rewrite repeated multiplication in exponential form and evaluate exponential expressions.

In Section 1.3, you learned that multiplication by a positive integer can be described as repeated addition.

Multiplication	*Repeated Addition*
4×7	$\underbrace{7 + 7 + 7 + 7}_{\text{4 terms of 7}}$

In a similar way, repeated multiplication can be described in **exponential form.**

Repeated Multiplication	*Exponential Form*
$\underbrace{7 \cdot 7 \cdot 7 \cdot 7}_{\text{4 factors of 7}}$	7^4

In the exponential form 7^4, 7 is the **base** and it specifies the repeated factor. The number 4 is the **exponent** and it indicates how many times the base occurs as a factor.

When you write the exponential form 7^4, you can say that you are raising 7 to the fourth **power.** When a number is raised to the *first* power, you usually do not write the exponent 1. For instance, you would usually write 5 rather than 5^1. Here are some examples of how exponential expressions are read.

Exponential Expression	*Verbal Statement*
7^2	"seven to the second power" or "seven squared"
4^3	"four to the third power" or "four cubed"
$(-2)^4$	"negative two to the fourth power"
-2^4	"the opposite of two to the fourth power"

It is important to recognize how exponential forms such as $(-2)^4$ and -2^4 differ.

$(-2)^4 = (-2)(-2)(-2)(-2)$ The negative sign is part of the base.

$= 16$ The value of the expression is positive.

$-2^4 = -(2 \cdot 2 \cdot 2 \cdot 2)$ The negative sign is not part of the base.

$= -16$ The value of the expression is negative.

Technology: Discovery

When a negative number is raised to a power, the use of parentheses is very important. To discover why, use a calculator to evaluate $(-5)^4$ and -5^4. Write a statement explaining the results. Then use a calculator to evaluate $(-5)^3$ and -5^3. If necessary, write a new statement explaining your discoveries.

Keep in mind that an exponent applies only to the factor (number) directly preceding it. Parentheses are needed to include a negative sign or other factors as part of the base.

EXAMPLE 1 Evaluating Exponential Expressions

a. $2^5 = 2 \cdot 2 \cdot 2 \cdot 2 \cdot 2$ Rewrite expression as a product.

$= 32$ Simplify.

b. $\left(\frac{2}{3}\right)^4 = \frac{2}{3} \cdot \frac{2}{3} \cdot \frac{2}{3} \cdot \frac{2}{3}$ Rewrite expression as a product.

$= \frac{2 \cdot 2 \cdot 2 \cdot 2}{3 \cdot 3 \cdot 3 \cdot 3}$ Multiply fractions.

$= \frac{16}{81}$ Simplify.

✓ **CHECKPOINT** *Now try Exercise 17.*

EXAMPLE 2 Evaluating Exponential Expressions

a. $(-4)^3 = (-4)(-4)(-4)$ Rewrite expression as a product.

$= -64$ Simplify.

b. $(-3)^4 = (-3)(-3)(-3)(-3)$ Rewrite expression as a product.

$= 81$ Simplify.

c. $-3^4 = -(3 \cdot 3 \cdot 3 \cdot 3)$ Rewrite expression as a product.

$= -81$ Simplify.

✓ **CHECKPOINT** *Now try Exercise 23.*

In parts (a) and (b) of Example 2, note that when a negative number is raised to an odd power, the result is *negative,* and when a negative number is raised to an even power, the result is *positive.*

EXAMPLE 3 Transporting Capacity

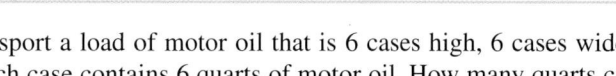

A truck can transport a load of motor oil that is 6 cases high, 6 cases wide, and 6 cases long. Each case contains 6 quarts of motor oil. How many quarts can the truck transport?

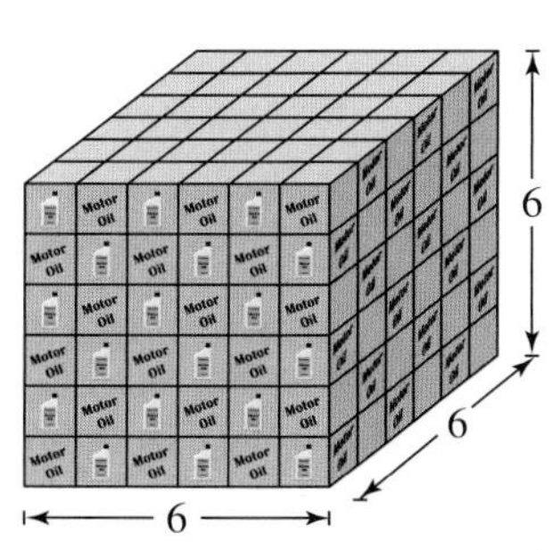

Figure 1.33

Solution

A sketch can help you solve this problem. From Figure 1.33, there are $6 \cdot 6 \cdot 6$ cases of motor oil, and each case contains 6 quarts. You can see that 6 occurs as a factor four times, which implies that the total number of quarts is

$$(6 \cdot 6 \cdot 6) \cdot 6 = 6^4 = 1296.$$

So, the truck can transport 1296 quarts of oil.

✓ **CHECKPOINT** *Now try Exercise 125.*

2 ▶ Evaluate expressions using order of operations.

Order of Operations

Up to this point in the text, you have studied five operations of arithmetic—addition, subtraction, multiplication, division, and exponentiation (repeated multiplication). When you use more than one operation in a given problem, you face the question of which operation to perform first. For example, without further guidelines, you could evaluate $4 + 3 \cdot 5$ in two ways.

Add First

$$4 + 3 \cdot 5 \stackrel{?}{=} 7 \cdot 5$$
$$= 35$$

Multiply First

$$4 + 3 \cdot 5 \stackrel{?}{=} 4 + 15$$
$$= 19$$

Technology: Discovery

To discover if your calculator performs the established order of operations, evaluate $7 + 5 \cdot 3 - 2^4 \div 4$ exactly as it appears. Does your calculator display 5 or 18? If your calculator performs the established order of operations, it will display 18.

According to the established **order of operations,** the second evaluation is correct. The reason for this is that multiplication has a higher priority than addition. The accepted priorities for order of operations are summarized below.

Order of Operations

1. Perform operations inside *symbols of grouping*—( ) or [ ]— or *absolute value symbols,* starting with the innermost symbols.
2. Evaluate all *exponential* expressions.
3. Perform all *multiplications* and *divisions* from left to right.
4. Perform all *additions* and *subtractions* from left to right.

In the priorities for order of operations, note that the highest priority is given to symbols of grouping such as parentheses or brackets. This means that when you want to be sure that you are communicating an expression correctly, you can insert symbols of grouping to specify which operations you intend to be performed first. For instance, if you want to make sure that $4 + 3 \cdot 5$ will be evaluated correctly, you can write it as $4 + (3 \cdot 5)$.

Study Tip

When you use symbols of grouping in an expression, you should alternate between parentheses and brackets. For instance, the expression

$$10 - (3 - [4 - (5 + 7)])$$

is easier to understand than

$$10 - (3 - (4 - (5 + 7))).$$

EXAMPLE 4 Order of Operations

a. $7 - [(5 \cdot 3) + 2^3] = 7 - [15 + 2^3]$ Multiply inside the parentheses.

$= 7 - [15 + 8]$ Evaluate exponential expression.

$= 7 - 23$ Add inside the brackets.

$= -16$ Subtract.

b. $36 \div (3^2 \cdot 2) - 6 = 36 \div (9 \cdot 2) - 6$ Evaluate exponential expression.

$= 36 \div 18 - 6$ Multiply inside the parentheses.

$= 2 - 6$ Divide.

$= -4$ Subtract.

✓ ***CHECKPOINT*** *Now try Exercise 45.*

EXAMPLE 5 **Order of Operations**

a. $\frac{3}{7} \div \frac{8}{7} + \left(-\frac{3}{5}\right)\left(\frac{1}{3}\right) = \frac{3}{7} \cdot \frac{7}{8} + \left(-\frac{3}{5}\right)\left(\frac{1}{3}\right)$ Invert divisor and multiply.

$= \frac{3}{8} + \left(-\frac{1}{5}\right)$ Multiply fractions.

$= \frac{15}{40} + \frac{-8}{40}$ Find common denominator.

$= \frac{7}{40}$ Add fractions.

b. $\frac{8}{3}\left(\frac{1}{6} + \frac{1}{4}\right) = \frac{8}{3}\left(\frac{2}{12} + \frac{3}{12}\right)$ Find common denominator.

$= \frac{8}{3}\left(\frac{5}{12}\right)$ Add inside the parentheses.

$= \frac{40}{36}$ Multiply fractions.

$= \frac{10}{9}$ Simplify.

✓ **CHECKPOINT** ***Now try Exercise 55.***

EXAMPLE 6 **Order of Operations**

Evaluate the expression $6 + \frac{8 + 7}{3^2 - 4} - (-5)$.

Solution

$6 + \frac{8 + 7}{3^2 - 4} - (-5) = 6 + \frac{8 + 7}{9 - 4} - (-5)$ Evaluate exponential expression.

$= 6 + \frac{15}{9 - 4} - (-5)$ Add in numerator.

$= 6 + \frac{15}{5} - (-5)$ Subtract in denominator.

$= 6 + 3 - (-5)$ Divide.

$= 9 + 5$ Add.

$= 14$ Add.

✓ **CHECKPOINT** ***Now try Exercise 71.***

In Example 6, note that a fraction bar acts as a symbol of grouping. For instance,

$$\frac{8 + 7}{3^2 - 4} \quad \text{means} \quad (8 + 7) \div (3^2 - 4), \quad \text{not} \quad 8 + 7 \div 3^2 - 4.$$

3 ▶ Identify and use the properties of real numbers.

Properties of Real Numbers

You are now ready for the symbolic versions of the properties that are true about operations with real numbers. These properties are referred to as **properties of real numbers.** The table shows a verbal description and an illustrative example for each property. Keep in mind that the letters a, b, and c represent real numbers, even though only rational numbers have been used to this point.

Properties of Real Numbers: Let *a*, *b*, and *c* be real numbers.

Property	*Example*
1. *Commutative Property of Addition:* Two real numbers can be added in either order. $a + b = b + a$	$3 + 5 = 5 + 3$
2. *Commutative Property of Multiplication:* Two real numbers can be multiplied in either order. $ab = ba$	$4 \cdot (-7) = -7 \cdot 4$
3. *Associative Property of Addition:* When three real numbers are added, it makes no difference which two are added first. $(a + b) + c = a + (b + c)$	$(2 + 6) + 5 = 2 + (6 + 5)$
4. *Associative Property of Multiplication:* When three real numbers are multiplied, it makes no difference which two are multiplied first. $(ab)c = a(bc)$	$(3 \cdot 5) \cdot 2 = 3 \cdot (5 \cdot 2)$
5. *Distributive Property:* Multiplication distributes over addition. $a(b + c) = ab + ac$ $(a + b)c = ac + bc$	$3(8 + 5) = 3 \cdot 8 + 3 \cdot 5$ $(3 + 8)5 = 3 \cdot 5 + 8 \cdot 5$
6. *Additive Identity Property:* The sum of zero and a real number equals the number itself. $a + 0 = 0 + a = a$	$3 + 0 = 0 + 3 = 3$
7. *Multiplicative Identity Property:* The product of 1 and a real number equals the number itself. $a \cdot 1 = 1 \cdot a = a$	$4 \cdot 1 = 1 \cdot 4 = 4$
8. *Additive Inverse Property:* The sum of a real number and its opposite is zero. $a + (-a) = 0$	$3 + (-3) = 0$
9. *Multiplicative Inverse Property:* The product of a nonzero real number and its reciprocal is 1. $a \cdot \frac{1}{a} = 1,\ a \neq 0$	$8 \cdot \frac{1}{8} = 1$

EXAMPLE 7 Identifying Properties of Real Numbers

Identify the property of real numbers illustrated by each statement.

a. $3(6 + 2) = 3 \cdot 6 + 3 \cdot 2$

b. $5 \cdot \frac{1}{5} = 1$

c. $7 + (5 + 4) = (7 + 5) + 4$

d. $(12 + 3) + 0 = 12 + 3$

e. $4(11) = 11(4)$

Solution

a. This statement illustrates the Distributive Property.

b. This statement illustrates the Multiplicative Inverse Property.

c. This statement illustrates the Associative Property of Addition.

d. This statement illustrates the Additive Identity Property.

e. This statement illustrates the Commutative Property of Multiplication.

✓ **CHECKPOINT** *Now try Exercise 79.*

EXAMPLE 8 Using the Properties of Real Numbers

Complete each statement using the specified property of real numbers.

a. Commutative Property of Addition:

$$5 + 9 = \quad$$

b. Associative Property of Multiplication:

$$6(5 \cdot 13) = \quad$$

c. Distributive Property:

$$4 \cdot 3 + 4 \cdot 7 = \quad$$

Solution

a. By the Commutative Property of Addition, you can write

$$5 + 9 = 9 + 5.$$

b. By the Associative Property of Multiplication, you can write

$$6(5 \cdot 13) = (6 \cdot 5)13.$$

c. By the Distributive Property, you can write

$$4 \cdot 3 + 4 \cdot 7 = 4(3 + 7).$$

✓ **CHECKPOINT** *Now try Exercise 101.*

One of the distinctive things about algebra is that its rules make sense. You don't have to accept them on "blind faith"—instead, you can learn the reasons that the rules work. For instance, there are some basic differences among the operations of addition, multiplication, subtraction, and division.

In the summary of properties of real numbers on page 50, all the properties are listed in terms of addition and multiplication. The reason for this is that subtraction and division lack many of the properties listed in the summary. For instance, subtraction and division are not commutative. To see this, consider the following.

$$7 - 5 \neq 5 - 7 \quad \text{and} \quad 12 \div 4 \neq 4 \div 12$$

Similarly, subtraction and division are not associative.

$$9 - (5 - 3) \neq (9 - 5) - 3 \quad \text{and} \quad 12 \div (4 \div 2) \neq (12 \div 4) \div 2$$

EXAMPLE 9 Geometry: Area 

You measure the width of a billboard and find that it is 60 feet. You are told that its height is 22 feet less than its width.

a. Write an expression for the area of the billboard.

b. Use the Distributive Property to rewrite the expression.

c. Find the area of the billboard.

Solution

Figure 1.34

a. Begin by drawing and labeling a diagram, as shown in Figure 1.34. To find an expression for the area of the billboard, multiply the width by the height.

$$\begin{aligned} \text{Area} &= \text{Width} \times \text{Height} \\ &= 60(60 - 22) \end{aligned}$$

b. To rewrite the expression $60(60 - 22)$ using the Distributive Property, distribute 60 over the subtraction.

$$60(60 - 22) = 60(60) - 60(22)$$

c. To find the area of the billboard, evaluate the expression in part (b) as follows.

$$\begin{aligned} 60(60) - 60(22) &= 3600 - 1320 && \text{Multiply.} \\ &= 2280 && \text{Subtract.} \end{aligned}$$

So, the area of the billboard is 2280 square feet.

✓ ***CHECKPOINT*** *Now try Exercise 131.*

From Example 9(b) you can see that the Distributive Property is also true for subtraction. For instance, the "subtraction form" of $a(b + c) = ab + ac$ is

$$\begin{aligned} a(b - c) &= a[b + (-c)] \\ &= ab + a(-c) \\ &= ab - ac. \end{aligned}$$

Concept Check

1. Consider the expression 3^5.
   (a) What part of the exponential expression is the number 3?
   (b) What part of the exponential expression is the number 5?
2. In your own words, describe the priorities for the established order of operations.
3. In your own words, state the Associative Property of Addition and the Associative Property of Multiplication. Give an example of each.
4. In your own words, state the Commutative Property of Addition and the Commutative Property of Multiplication. Give an example of each.

1.5 EXERCISES

Go to pages 58–59 to record your assignments.

Developing Skills

In Exercises 1–8, rewrite in exponential form.

1. $2 \cdot 2 \cdot 2 \cdot 2 \cdot 2$

2. $4 \cdot 4 \cdot 4 \cdot 4 \cdot 4 \cdot 4$

3. $(-5) \cdot (-5) \cdot (-5) \cdot (-5)$

4. $(-3) \cdot (-3) \cdot (-3)$

5. $\left(-\frac{1}{4}\right) \cdot \left(-\frac{1}{4}\right)$

6. $\left(-\frac{3}{5}\right) \cdot \left(-\frac{3}{5}\right) \cdot \left(-\frac{3}{5}\right) \cdot \left(-\frac{3}{5}\right)$

7. $-[(1.6) \cdot (1.6) \cdot (1.6) \cdot (1.6) \cdot (1.6)]$

8. $-[(8.7) \cdot (8.7) \cdot (8.7)]$

In Exercises 9–16, rewrite as a product.

9. $(-3)^6$

10. $(-8)^2$

11. $\left(\frac{3}{8}\right)^5$

12. $\left(\frac{3}{11}\right)^4$

13. $\left(-\frac{1}{2}\right)^7$

14. $\left(-\frac{4}{5}\right)^6$

15. $-(9.8)^3$

16. $-(0.01)^8$

In Exercises 17–28, evaluate the expression. ***See Examples 1 and 2.***

✓ **17.** 3^2

18. 4^3

19. 2^6

20. 5^3

21. $\left(\frac{1}{4}\right)^3$

22. $\left(\frac{4}{5}\right)^4$

✓ **23.** $(-5)^3$

24. $(-4)^2$

25. -4^2

26. $-(-6)^3$

27. $(-1.2)^3$

28. $(-1.5)^4$

In Exercises 29–72, evaluate the expression. If it is not possible, state the reason. Write fractional answers in simplest form. ***See Examples 4, 5, and 6.***

29. $4 - 6 + 10$

30. $8 + 9 - 12$

31. $5 - (8 - 15)$

32. $13 - (12 - 3)$

33. $17 - |2 - (6 + 5)|$

34. $125 - |10 - (25 - 3)|$

35. $15 + 3 \cdot 4$

36. $9 - 5 \cdot 2$

37. $25 - 32 \div 4$

38. $16 + 24 \div 8$

39. $(16 - 5) \div (3 - 5)$

40. $(19 - 4) \div (7 - 2)$

41. $(10 - 16) \cdot (20 - 26)$

42. $(14 - 17) \cdot (13 - 19)$

43. $(45 \div 10) \cdot 2$

44. $(38 \div 5) \cdot 4$

✓ **45.** $[360 - (8 + 12)] \div 5$

46. $[127 - (13 + 4)] \div 10$

47. $5 + (2^2 \cdot 3)$

48. $181 - (13 \cdot 3^2)$

49. $(-6)^2 - (48 \div 4^2)$

50. $(-3)^3 + (12 \div 2^2)$

51. $\left(3 \cdot \frac{5}{9}\right) + 1 - \frac{1}{3}$

52. $\frac{2}{3}\left(\frac{3}{4}\right) + 2 - \frac{3}{2}$

53. $18\left(\frac{1}{2} + \frac{2}{3}\right)$

54. $4\left(-\frac{2}{3} + \frac{4}{3}\right)$

55. $\frac{7}{25}\left(\frac{7}{16} - \frac{1}{8}\right)$

56. $\frac{3}{2}\left(\frac{2}{3} + \frac{1}{6}\right)$

57. $\frac{7}{3}\left(\frac{2}{3}\right) \div \frac{28}{15}$

58. $\frac{3}{8}\left(\frac{1}{5}\right) \div \frac{25}{32}$

59. $\dfrac{3 + [15 \div (-3)]}{16}$

60. $\dfrac{5 + [(-12) \div 4]}{24}$

61. $\dfrac{1 - 3^2}{-2}$

62. $\dfrac{2^2 + 4^2}{5}$

63. $\dfrac{7^2 - 4^2}{0}$

64. $\dfrac{0}{3^2 - 1^2}$

65. $\dfrac{0}{6^2 + 1}$

66. $\dfrac{3^3 + 1}{0}$

67. $\dfrac{5^2 + 12^2}{13}$

68. $\dfrac{8^2 - 2^3}{4}$

69. $\dfrac{3 \cdot 6 - 4 \cdot 6}{5 + 1}$

70. $\dfrac{5 \cdot 3 + 5 \cdot 6}{7 - 2}$

71. $7 - \dfrac{4 + 6}{2^2 + 1} + 5$

72. $11 - \dfrac{3^3 - 30}{8 + 1} + 1$

In Exercises 73–76, use a calculator to evaluate the expression. Round your answer to two decimal places.

73. $300\left(1 + \dfrac{0.1}{12}\right)^{24}$

74. $1000 \div \left(1 + \dfrac{0.09}{4}\right)^8$

75. $\dfrac{1.32 + 4(3.68)}{1.5}$

76. $\dfrac{4.19 - 7(2.27)}{14.8}$

In Exercises 77–96, identify the property of real numbers illustrated by the statement. *See Example 7.*

77. $6(-3) = -3(6)$

78. $16 + 10 = 10 + 16$

79. $5 + 10 = 10 + 5$

80. $-2(8) = 8(-2)$

81. $6(3 + 13) = 6 \cdot 3 + 6 \cdot 13$

82. $1 \cdot 4 = 4$

83. $-16 + 16 = 0$

84. $(14 + 2)3 = 14 \cdot 3 + 2 \cdot 3$

85. $(10 + 3) + 2 = 10 + (3 + 2)$

86. $25 + (-25) = 0$

87. $12 \cdot 9 - 12 \cdot 3 = 12(9 - 3)$

88. $(32 + 8) + 5 = 32 + (8 + 5)$

89. $7\left(\frac{1}{7}\right) = 1$

90. $-14 + 0 = -14$

91. $0 + 15 = 15$

92. $(2 \cdot 3)4 = 2(3 \cdot 4)$

93. $\frac{1}{4}(3 + 8) = \frac{1}{4}(3) + \frac{1}{4}(8)$

94. $7 \cdot 12 + 7 \cdot 8 = 7(12 + 8)$

95. $4(3 \cdot 10) = (4 \cdot 3)10$

96. $[(7 + 8)6]5 = (7 + 8)(6 \cdot 5)$

In Exercises 97–108, complete the statement using the specified property of real numbers. *See Example 8.*

97. Commutative Property of Addition:
$18 + 5 =$

98. Commutative Property of Addition:
$3 + 12 =$

99. Commutative Property of Multiplication:
$10(-3) =$

100. Commutative Property of Multiplication:
$5(8 + 3) =$

101. Distributive Property:
$6(19 + 2) =$

102. Distributive Property:
$5(7 - 16) =$

103. Distributive Property:
$3 \cdot 4 + 5 \cdot 4 =$

104. Distributive Property:
$(4 - 9)12 =$

105. Associative Property of Addition:
$18 + (12 + 9) =$

106. Associative Property of Addition:
$10 + (8 + 7) =$

107. Associative Property of Multiplication:
$12(3 \cdot 4) =$

108. Associative Property of Multiplication:
$(4 \cdot 11)10 =$

In Exercises 109–116, find (a) the additive inverse and (b) the multiplicative inverse of the quantity.

109. 50

110. 12

111. -1

112. -8

113. $-\frac{1}{2}$

114. $\frac{3}{4}$

115. 0.2

116. 0.45

In Exercises 117–120, simplify the expression using (a) the Distributive Property and (b) order of operations.

117. $3(6 + 10)$

118. $4(8 - 3)$

119. $\frac{2}{3}(9 + 24)$

120. $\frac{1}{2}(4 - 2)$

In Exercises 121–124, justify each step.

121. $7 \cdot 4 + 9 + 2 \cdot 4$

$= 7 \cdot 4 + 2 \cdot 4 + 9$

$= (7 \cdot 4 + 2 \cdot 4) + 9$

$= (7 + 2)4 + 9$

$= 9 \cdot 4 + 9$

$= 9(4 + 1)$

$= 9(5)$

$= 45$

122. $24 + 39 + (-24)$

$= 24 + (-24) + 39$

$= 0 + 39$

$= 39$

123. $\left(\frac{7}{9} + 6\right) + \frac{2}{9}$

$= \frac{7}{9} + \left(6 + \frac{2}{9}\right)$

$= \frac{7}{9} + \left(\frac{2}{9} + 6\right)$

$= \left(\frac{7}{9} + \frac{2}{9}\right) + 6$

$= 1 + 6$

$= 7$

124. $\left(\frac{2}{3} \cdot 7\right) \cdot 21$

$= \frac{2}{3} \cdot (7 \cdot 21)$

$= \frac{2}{3} \cdot (21 \cdot 7)$

$= \left(\frac{2}{3} \cdot 21\right) \cdot 7$

$= 14 \cdot 7$

$= 98$

Solving Problems

125. *Capacity* A truck can transport a load of propane tanks that is 4 cases high, 4 cases wide, and 4 cases long. Each case contains 4 propane tanks. How many tanks can the truck transport?

126. *Capacity* A grocery store has a cereal display that is 8 boxes high, 8 boxes wide, and 8 boxes long. How many cereal boxes are in the display?

Geometry In Exercises 127 and 128, find the area of the region.

127.

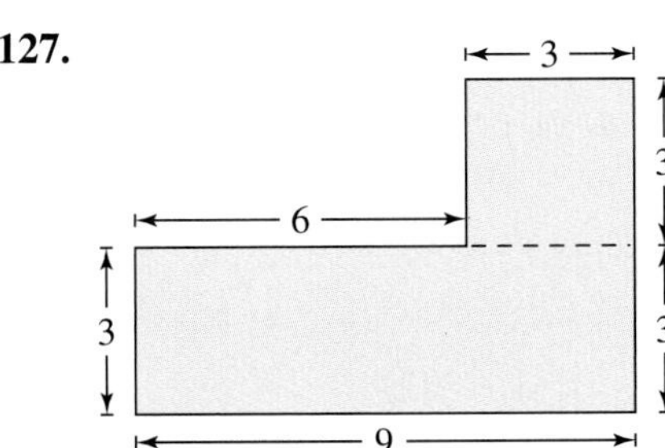

128.

8
12
8
4
8
8

129. *Sales Tax* You purchase a sweater for \$35.95. There is a 6% sales tax, which means that the total amount you must pay is $35.95 + 0.06(35.95)$.

(a) Use the Distributive Property to rewrite the expression.

(b) How much must you pay for the sweater including sales tax?

130. *Cost of a Truck* A new truck can be paid for by 48 monthly payments of \$665 each plus a down payment of 2.5 times the amount of the monthly payment. This means that the total amount paid for the truck is $2.5(665) + 48(665)$.

(a) Use the Distributive Property to rewrite the expression.

(b) What is the total amount paid for the truck?

131. *Geometry* The width of a movie screen is 30 feet and its height is 8 feet less than its width.

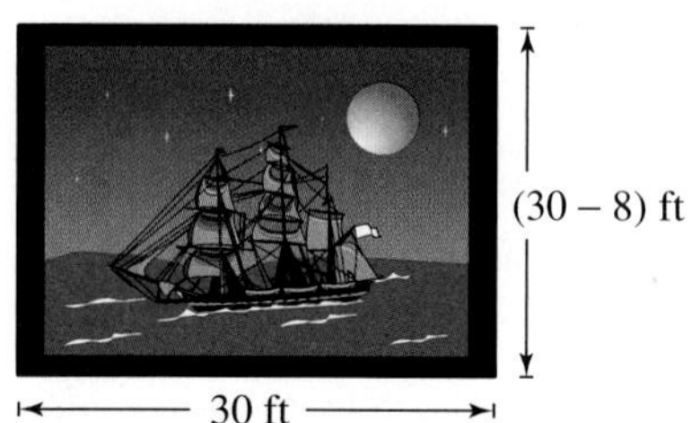

(a) Write an expression for the area of the movie screen.

(b) Use the Distributive Property to rewrite the expression.

(c) Find the area of the movie screen.

132. *Geometry* A picture frame is 36 inches wide and its height is 9 inches less than its width.

(a) Write an expression for the area of the picture frame.

(b) Use the Distributive Property to rewrite the expression.

(c) Find the area of the picture frame.

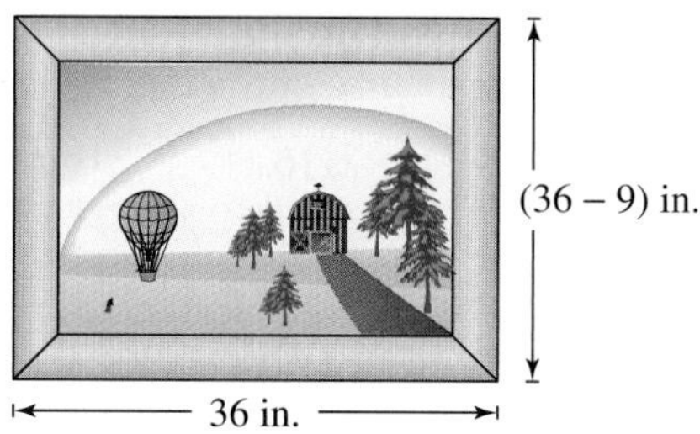

Figure for 132

Geometry In Exercises 133 and 134, write an expression for the perimeter of the triangle shown in the figure. Use the properties of real numbers to simplify the expression.

133.

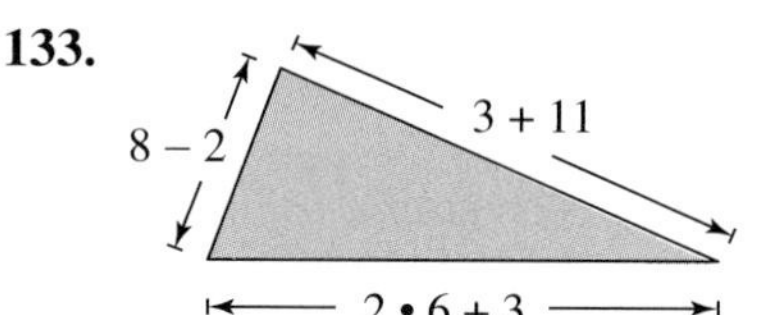

134.

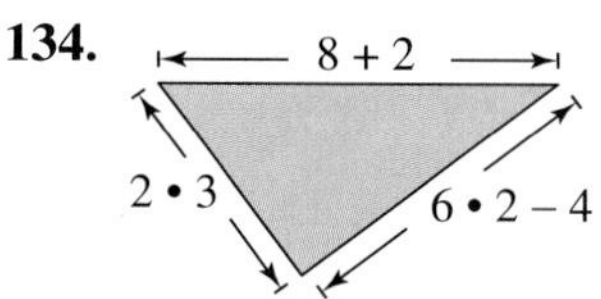

Think About It In Exercises 135 and 136, determine whether the order in which the two activities are performed is "commutative." That is, do you obtain the same result regardless of which activity is performed first?

135. (a) "Put on your socks."

(b) "Put on your shoes."

136. (a) "Weed the flower beds."

(b) "Mow the lawn."

Explaining Concepts

137. Are -6^2 and $(-6)^2$ equal? Explain.

138. Are $2 \cdot 5^2$ and 10^2 equal? Explain.

In Exercises 139–148, explain why the statement is true. (The symbol $\neq$ means "is not equal to.")

139. $4 \cdot 6^2 \neq 24^2$

140. $-3^2 \neq (-3)(-3)$

141. $4 - (6 - 2) \neq 4 - 6 - 2$

142. $\dfrac{8 - 6}{2} \neq 4 - 6$

143. $100 \div 2 \times 50 \neq 1$

144. $\dfrac{16}{2} \cdot 2 \neq 4$

145. $5(7 + 3) \neq 5(7) + 3$

146. $-7(5 - 2) \neq -7(5) - 7(2)$

147. $\frac{8}{0} \neq 0$

148. $5\left(\frac{1}{5}\right) \neq 0$

149. *Error Analysis* Describe and correct the error.

$$-9 + \frac{9+20}{3(5)} - (-3) = -9 + \frac{9}{3} + \frac{20}{5} - (-3)$$
$$= -9 + 3 + 4 - (-3)$$
$$= 1$$

150. *Error Analysis* Describe and correct the error.

$$7 - 3(8+1) - 15 = 4(8+1) - 15$$
$$= 4(9) - 15$$
$$= 36 - 15$$
$$= 21$$

151. Match each expression in the first column with its value in the second column.

Expression	*Value*
$(6+2) \cdot (5+3)$	19
$(6+2) \cdot 5 + 3$	22
$6 + 2 \cdot 5 + 3$	64
$6 + 2 \cdot (5+3)$	43

152. Using the established order of operations, which of the following expressions has a value of 72? For those that don't, decide whether you can insert parentheses into the expression so that its value is 72.

(a) $4 + 2^3 - 7$ (b) $4 + 8 \cdot 6$
(c) $93 - 25 - 4$ (d) $70 + 10 \div 5$
(e) $60 + 20 \div 2 + 32$
(f) $35 \cdot 2 + 2$

153. Consider the rectangle shown in the figure.

(a) Find the area of the rectangle by adding the areas of regions I and II.

(b) Find the area of the rectangle by multiplying its length by its width.

(c) Explain how the results of parts (a) and (b) relate to the Distributive Property.

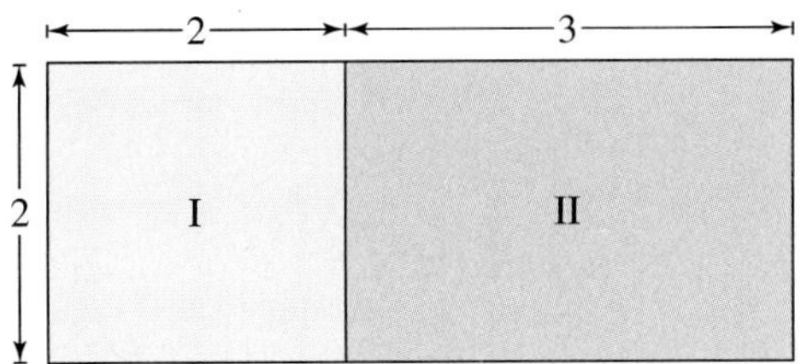

Geometry In Exercises 154 and 155, find the area of the shaded rectangle in two ways. Explain how the results are related to the Distributive Property.

154.

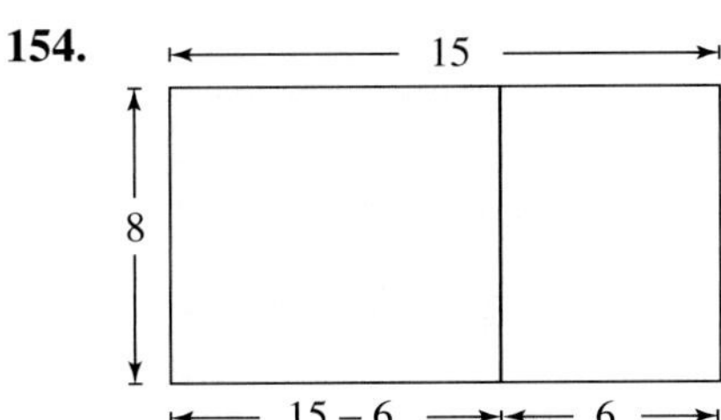

155.

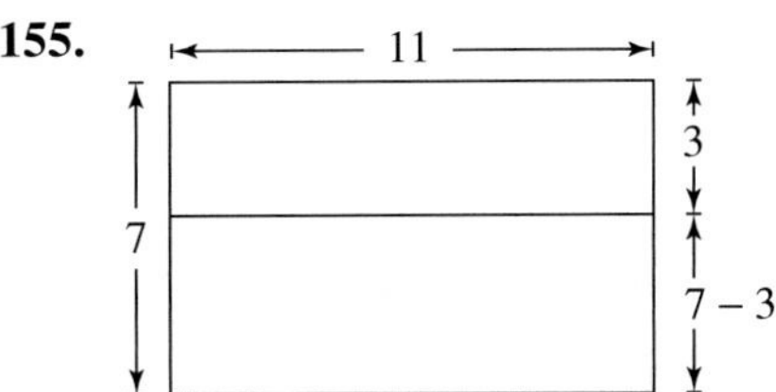

What Did You Learn?

Use these two pages to help prepare for a test on this chapter. Check off the key terms and key concepts you know. You can also use this section to record your assignments.

Plan for Test Success

Date of test: / /

Study dates and times: / / at : A.M./P.M.

/ / at : A.M./P.M.

Things to review:

- ☐ Key Terms, *p. 58*
- ☐ Key Concepts, *pp. 58–59*
- ☐ Your class notes
- ☐ Your assignments
- ☐ Study Tips, *pp. 2, 3, 22, 32, 34, 35, 48*
- ☐ Technology Tips, *pp. 4, 15, 36*
- ☐ Mid-Chapter Quiz, *p. 31*
- ☐ Review Exercises, *pp. 60–64*
- ☐ Chapter Test, *p. 65*
- ☐ Video Explanations Online
- ☐ Tutorial Online

Key Terms

- ☐ set, *p. 2*
- ☐ real numbers, *p. 2*
- ☐ subset, *p. 2*
- ☐ natural numbers, *p. 2*
- ☐ whole numbers, *p. 2*
- ☐ integers, *p. 2*
- ☐ rational numbers, *p. 3*
- ☐ rational, *p. 3*
- ☐ irrational, *p. 3*
- ☐ real number line, *p. 4*
- ☐ origin, *p. 4*
- ☐ negative, *p. 4*
- ☐ positive, *p. 4*
- ☐ nonnegative, *p. 4*
- ☐ inequality symbol, *p. 5*
- ☐ opposites, *p. 7*
- ☐ absolute value, *p. 7*
- ☐ expression, *p. 8*
- ☐ evaluate, *p. 8*
- ☐ sum, *p. 11*
- ☐ additive inverse, *p. 12*
- ☐ difference, *p. 13*
- ☐ product, *p. 19*
- ☐ quotient, *p. 21*
- ☐ dividend, *p. 21*
- ☐ divisor, *p. 21*
- ☐ numerator, *p. 21*
- ☐ denominator, *p. 21*
- ☐ factor, *p. 23*
- ☐ prime, *p. 23*
- ☐ composite, *p. 23*
- ☐ fraction, *p. 32*
- ☐ simplest form, *p. 32*
- ☐ greatest common factor, *p. 32*
- ☐ equivalent fraction, *p. 33*
- ☐ least common multiple, *p. 34*
- ☐ mixed number, *p. 35*
- ☐ reciprocal, *p. 38*
- ☐ terminating decimal, *p. 39*
- ☐ repeating decimal, *p. 39*
- ☐ exponential form, *p. 46*
- ☐ base, *p. 46*
- ☐ exponent, *p. 46*
- ☐ order of operations, *p. 48*

Key Concepts

1.1 Real Numbers: Order and Absolute Value

Assignment: ______________________________ Due date: ________

☐ **Order real numbers.**

Use the real number line and an inequality symbol ($<$, $>$, $\leq$, or $\geq$) to order real numbers.

☐ **Define absolute value.**

The absolute value of a number is its distance from zero on the real number line. The absolute value is either positive or zero.

−4 −3 −2 −1 0 1 2 3 4

Distance from 0 is 3.

$-3 < 2$ because -3 lies to the left of 2 on the real number line. $|-3| = 3$ because the distance between -3 and 0 is 3.

1.2 Adding and Subtracting Integers

Assignment:________________________________ Due date:________

☐ **Add integers.**

To add two integers with like signs, add their absolute values and attach the common sign to the result.

To add two integers with unlike signs, subtract the smaller absolute value from the larger absolute value and attach the sign of the integer with the larger absolute value.

☐ **Subtract integers.**

To subtract one integer from another, add the opposite of the integer being subtracted to the other integer.

1.3 Multiplying and Dividing Integers

Assignment:________________________________ Due date:________

☐ **Use rules for multiplying and dividing integers.**

The product of an integer and zero is 0.

Zero divided by a nonzero integer is 0, whereas a nonzero integer divided by zero is undefined.

The product or quotient of two nonzero integers with like signs is positive.

The product or quotient of two nonzero integers with unlike signs is negative.

☐ **Define factor.**

If a and b are positive integers, then a is a factor (or divisor) of b if and only if there is a positive integer c such that $a \cdot c = b$.

☐ **Define prime and composite numbers.**

A positive integer greater than 1 with no factors other than itself and 1 is called a prime number.

A positive integer greater than 1 with more than two factors is called a composite number.

1.4 Operations with Rational Numbers

Assignment:________________________________ Due date:________

☐ **Add and subtract fractions.**

To combine fractions with like denominators, add (or subtract) the numerators and write the result over the like denominator.

To combine fractions with unlike denominators, rewrite the fractions so they have like denominators. Then use the rule for combining fractions with like denominators.

☐ **Multiply fractions.**

To multiply fractions, multiply the numerators and multiply the denominators.

☐ **Divide fractions.**

To divide fractions, invert the divisor and multiply.

☐ **Add, subtract, multiply, and divide decimals.**

See pages 39 and 40.

1.5 Exponents, Order of Operations, and Properties of Real Numbers

Assignment:________________________________ Due date:________

☐ **Use exponential form.**

Repeated multiplication can be described in exponential form.

$3 \cdot 3 \cdot 3 = 3^3 = 27$

☐ **Use order of operations.**

1. Perform operations inside symbols of grouping.
2. Evaluate all exponential expressions.
3. Multiply and divide from left to right.
4. Add and subtract from left to right.

☐ **Use properties of real numbers.**

Let a, b, and c be real numbers.

Commutative Property of Addition $a + b = b + a$

Commutative Property of Multiplication $ab = ba$

Associative Property of Addition $(a + b) + c = a + (b + c)$

Associative Property of Multiplication $(ab)c = a(bc)$

Distributive Property

$a(b + c) = ab + ac$ $(a + b)c = ac + bc$

Additive Identity Property $a + 0 = 0 + a = a$

Multiplicative Identity Property $a \cdot 1 = 1 \cdot a = a$

Additive Inverse Property $a + (-a) = 0$

Multiplicative Inverse Property $a \cdot \frac{1}{a} = 1, a \neq 0$

Review Exercises

1.1 Real Numbers: Order and Absolute Value

1 ▶ Define sets and use them to classify numbers as natural, integer, rational, or irrational.

In Exercises 1–4, determine which of the numbers in the set are (a) natural numbers, (b) integers, (c) rational numbers, and (d) irrational numbers.

1. $\left\{-1, 4.5, \frac{2}{5}, -\frac{1}{7}, \sqrt{4}, \sqrt{5}\right\}$

2. $\left\{10, -3, \frac{4}{5}, \pi, -3.\overline{16}, -\frac{19}{11}\right\}$

3. $\left\{\frac{30}{2}, 2, -\sqrt{3}, 1.5, -\pi, -\frac{10}{7}\right\}$

4. $\left\{3.75, 33, \dfrac{2}{3}, -2.\overline{6}, -92, -\dfrac{\pi}{4}\right\}$

2 ▶ Plot numbers on the real number line.

In Exercises 5–10, plot the numbers on the real number line.

5. $-3, 5$ **6.** $-8, 11$

7. $-6, \frac{5}{4}$ **8.** $-\frac{7}{2}, 9$

9. $-1, 0, \frac{1}{2}$ **10.** $-2, -\frac{1}{3}, 5$

3 ▶ Use the real number line and inequality symbols to order real numbers.

In Exercises 11–16, plot each real number as a point on the real number line and place the correct inequality symbol ($<$ or $>$) between the pair of real numbers.

11. $-\frac{1}{10}$ ▒ 4 **12.** $\frac{25}{3}$ ▒ $\frac{5}{3}$

13. -3 ▒ -7 **14.** 10.6 ▒ -3.5

15. 5 ▒ $\frac{7}{2}$ **16.** $\frac{3}{8}$ ▒ $\frac{4}{9}$

17. Which is smaller: $\frac{2}{3}$ or 0.6?

18. Which is smaller: $-\frac{1}{3}$ or -0.3?

4 ▶ Find the absolute value of a number.

In Exercises 19–22, find the opposite of the number, and determine the distance of the number and its opposite from 0.

19. 152 **20.** -10.4

21. $-\frac{7}{3}$ **22.** $\frac{2}{3}$

In Exercises 23–30, evaluate the expression.

23. $|-8.5|$ **24.** $|-9.6|$

25. $|3.4|$ **26.** $|4|$

27. $-|-6.2|$ **28.** $|5.98|$

29. $-\left|\frac{8}{5}\right|$ **30.** $-\left|-\frac{7}{9}\right|$

In Exercises 31–36, place the correct symbol ($<, >,$ or $=$) between the pair of real numbers.

31. $|-84|$ ▒ $|84|$

32. $|-10|$ ▒ $|4|$

33. $\left|\frac{5}{2}\right|$ ▒ $\left|\frac{8}{9}\right|$

34. $-|-1.8|$ ▒ $|5.7|$

35. $\left|\frac{3}{10}\right|$ ▒ $-\left|\frac{4}{5}\right|$

36. $|2.3|$ ▒ $-|2.3|$

In Exercises 37–40, find all real numbers whose distance from *a* is given by *d*.

37. $a = 5, d = 7$ **38.** $a = -1, d = 4$

39. $a = 2.6, d = 5$ **40.** $a = -3, d = 6.5$

1.2 Adding and Subtracting Integers

1 ▶ Add integers using a number line.

In Exercises 41–44, find the sum and demonstrate the addition on the real number line.

41. $4 + 3$ **42.** $15 + (-6)$

43. $-1 + (-4)$ **44.** $-6 + (-2)$

2 ▶ Add integers with like signs and with unlike signs.

In Exercises 45–54, find the sum.

45. $16 + (-5)$

46. $25 + (-10)$

47. $-125 + 30$

48. $-54 + 12$

49. $-13 + (-76)$

50. $-24 + (-25)$

51. $-10 + 21 + (-6)$

52. $-23 + 4 + (-11)$

53. $-17 + (-3) + (-9)$

54. $-16 + (-2) + (-8)$

55. *Profit* A small software company had a profit of \$95,000 in January, a loss of \$64,400 in February, and a profit of \$51,800 in March. What was the company's overall profit (or loss) for the three months?

56. *Account Balance* At the beginning of a month, your account balance was \$3090. During the month, you withdrew \$870 and \$465, deposited \$109, and earned \$10.05 in interest. What was your balance at the end of the month?

57. Is the sum of two integers, one negative and one positive, negative? Explain.

58. Is the sum of two negative integers negative? Explain.

3 ▶ Subtract integers with like signs and with unlike signs.

In Exercises 59–68, find the difference.

59. $28 - 7$

60. $43 - 12$

61. $8 - 15$

62. $17 - 26$

63. $14 - (-19)$

64. $28 - (-4)$

65. $-18 - 4$

66. $-37 - 14$

67. $-12 - (-7) - 4$

68. $-26 - (-8) - (-10)$

69. Subtract -549 from 613.

70. What number must be subtracted from -6 to obtain 13?

71. *Account Balance* At the beginning of a month, your account balance was \$1560. During the month you withdrew \$50, \$255, and \$490. What was your balance at the end of the month?

72. *Gasoline Prices* At the beginning of a month, gas cost \$2.89 per gallon. During the month, the price increased by \$0.05 and \$0.02, decreased by \$0.10, and then increased again by \$0.07. How much did gas cost at the end of the month?

1.3 Multiplying and Dividing Integers

1 ▶ Multiply integers with like signs and with unlike signs.

In Exercises 73–84, find the product.

73. $15 \cdot 3$

74. $21 \cdot 4$

75. $-3 \cdot 24$

76. $-2 \cdot 44$

77. $6(-8)$

78. $12(-5)$

79. $-5(-9)$

80. $-10(-81)$

81. $3(-6)(3)$

82. $15(-2)(7)$

83. $-4(-5)(-2)$

84. $-12(-2)(-6)$

85. *Savings Plan* You save \$150 per month for 2 years. What is the total amount you have saved?

86. *Average Speed* A truck drives 65 miles per hour for 3 hours. How far has the truck traveled?

2 ▶ Divide integers with like signs and with unlike signs.

In Exercises 87–98, perform the division, if possible. If not possible, state the reason.

87. $72 \div 8$

88. $63 \div 9$

89. $\dfrac{-72}{6}$

90. $\dfrac{-162}{9}$

91. $75 \div (-5)$

92. $48 \div (-4)$

93. $\dfrac{-52}{-4}$

94. $\dfrac{-64}{-4}$

95. $0 \div 815$

96. $0 \div 25$

97. $135 \div 0$

98. $26 \div 0$

99. *Average Speed* A commuter train travels a distance of 195 miles between two cities in 3 hours. What is the average speed of the train in miles per hour?

100. *Unit Price* At a garage sale, you buy a box of six glass canisters for a total of \$78. All the canisters are of equal value. How much is each one worth?

3 ▶ Find factors and prime factors of an integer.

In Exercises 101–106, decide whether the number is prime or composite.

101. 137

102. 296

103. 839

104. 909

105. 1764

106. 1847

In Exercises 107–112, write the prime factorization of the number.

107. 264

108. 195

109. 378

110. 858

111. 1612

112. 1787

4 ▶ Represent the definitions and rules of arithmetic symbolically.

In Exercises 113–116, complete the statement using the indicated definition or rule.

113. Rule for multiplying integers with unlike signs: $12 \times (-3) = \square$

114. Definition of multiplication: $(-4) + (-4) + (-4) = \square$

115. Definition of absolute value: $|-7| = \square$

116. Rule for adding integers with unlike signs: $-9 + 5 = \square$

1.4 Operations with Rational Numbers

1 ▶ Rewrite fractions as equivalent fractions.

In Exercises 117–122, find the greatest common factor.

117. 54, 90 **118.** 154, 220

119. 2, 6, 9 **120.** 8, 12, 24

121. 63, 84, 441 **122.** 99, 132, 253

In Exercises 123–126, write the fraction in simplest form.

123. $\frac{3}{12}$ **124.** $\frac{15}{25}$

125. $\frac{30}{48}$ **126.** $\frac{126}{162}$

In Exercises 127–130, write an equivalent fraction with the indicated denominator.

127. $\frac{2}{3} = \frac{\square}{15}$ **128.** $\frac{3}{7} = \frac{\square}{28}$

129. $\frac{6}{10} = \frac{\square}{25}$ **130.** $\frac{9}{12} = \frac{\square}{16}$

2 ▶ Add and subtract fractions.

In Exercises 131–144, evaluate the expression. Write the result in simplest form.

131. $\frac{3}{25} + \frac{7}{25}$ **132.** $\frac{9}{64} + \frac{7}{64}$

133. $\frac{27}{16} - \frac{15}{16}$ **134.** $-\frac{5}{12} + \frac{1}{12}$

135. $\frac{3}{8} + \frac{1}{2}$ **136.** $\frac{7}{12} + \frac{5}{18}$

137. $-\frac{5}{9} + \frac{2}{3}$ **138.** $\frac{7}{15} - \frac{2}{25}$

139. $-\frac{25}{32} + \left(-\frac{7}{24}\right)$ **140.** $-\frac{7}{8} - \frac{11}{12}$

141. $5 - \frac{15}{4}$ **142.** $\frac{12}{5} - 3$

143. $5\frac{3}{4} - 3\frac{5}{8}$ **144.** $-3\frac{7}{10} + 1\frac{1}{20}$

145. *Meteorology* The table shows the daily amounts of rainfall (in inches) during a five-day period. What was the total amount of rainfall for the five days?

Day	Mon	Tue	Wed	Thu	Fri
Rainfall (in inches)	$\frac{3}{8}$	$\frac{1}{2}$	$\frac{1}{8}$	$1\frac{1}{4}$	$\frac{1}{2}$

146. *Fuel Consumption* The morning and evening readings of the fuel gauge on a car were $\frac{7}{8}$ and $\frac{1}{3}$, respectively. What fraction of the tank of fuel was used that day?

3 ▸ Multiply and divide fractions.

In Exercises 147–160, evaluate the expression and write the result in simplest form. If it is not possible, explain why.

147. $\frac{5}{8} \cdot \frac{-2}{15}$

148. $\frac{3}{32} \cdot \frac{32}{3}$

149. $35\left(\frac{1}{35}\right)$

150. $-6\left(\frac{5}{36}\right)$

151. $\frac{3}{8}\left(-\frac{2}{27}\right)$

152. $-\frac{5}{12}\left(-\frac{4}{25}\right)$

153. $\frac{5}{14} \div \frac{15}{28}$

154. $-\frac{7}{10} \div \frac{4}{15}$

155. $-\frac{3}{4} \div \left(-\frac{7}{8}\right)$

156. $\frac{15}{32} \div \left(-\frac{5}{4}\right)$

157. $-\frac{5}{9} \div 0$

158. $0 \div \frac{1}{12}$

159. $-5 \cdot 0$

160. $0 \cdot \frac{1}{2}$

161. *Meteorology* During an eight-hour period, $6\frac{3}{4}$ inches of snow fell. What was the average rate of snowfall per hour?

162. *Sports* In three strokes on the golf course, you hit your ball a total distance of $64\frac{7}{8}$ meters. What is your average distance per stroke?

In Exercises 163–166, write the fraction in decimal form. (Use bar notation for repeating digits.)

163. $\frac{5}{8}$

164. $\frac{9}{16}$

165. $\frac{8}{15}$

166. $\frac{5}{11}$

4 ▸ Add, subtract, multiply, and divide decimals.

In Exercises 167–174, evaluate the expression. Round your answer to two decimal places.

167. $4.89 + 0.76$

168. $1.29 + 0.44$

169. $3.815 - 5.19$

170. $7.234 - 8.16$

171. $1.49(-0.5)$

172. $2.34(-1.2)$

173. $5.25 \div 0.25$

174. $10.18 \div 1.6$

175. *Consumer Awareness* A DJ charges \$600 for 4 hours and \$200 for each additional hour. Find the cost to hire the DJ for an 8-hour event.

176. *Consumer Awareness* A plasma television costs \$599.99 plus \$32.96 per month for 18 months. Find the total cost of the television.

1.5 Exponents, Order of Operations, and Properties of Real Numbers

1 ▸ Rewrite repeated multiplication in exponential form and evaluate exponential expressions.

In Exercises 177–180, rewrite in exponential form.

177. $6 \cdot 6 \cdot 6 \cdot 6 \cdot 6$

178. $(-3) \cdot (-3) \cdot (-3)$

179. $\left(\frac{6}{7}\right) \cdot \left(\frac{6}{7}\right) \cdot \left(\frac{6}{7}\right) \cdot \left(\frac{6}{7}\right)$

180. $-[(3.3) \cdot (3.3)]$

In Exercises 181–184, rewrite as a product.

181. $(-7)^4$

182. $\left(\frac{1}{2}\right)^5$

183. $(1.25)^3$

184. $\left(-\frac{4}{9}\right)^6$

In Exercises 185–190, evaluate the expression.

185. 2^4

186. $(-6)^2$

187. $\left(-\frac{3}{4}\right)^3$

188. $\left(\frac{2}{3}\right)^2$

189. -7^2

190. $-(-3)^3$

2 ▸ Evaluate expressions using order of operations.

In Exercises 191–210, evaluate the expression. Write fractional answers in simplest form.

191. $12 - 2 \cdot 3$

192. $1 + 7 \cdot 3 - 10$

193. $18 \div 6 \cdot 7$

194. $3^2 \cdot 4 \div 2$

195. $20 + (8^2 \div 2)$

196. $(8 - 3) \div 15$

197. $240 - (4^2 \cdot 5)$

198. $5^2 - (625 \cdot 5^2)$

199. $3^2(5 - 2)^2$

200. $-5(10 - 7)^3$

201. $\frac{3}{4}\left(\frac{5}{6}\right) + 4$

202. $75 - 24 \div 2^3$

203. $122 - [45 - (32 + 8) - 23]$

204. $-58 - (48 - 12) - (-30 - 4)$

205. $\dfrac{6 \cdot 4 - 36}{4}$

206. $\dfrac{144}{2 \cdot 3 \cdot 3}$

207. $\dfrac{54 - 4 \cdot 3}{6}$

208. $\dfrac{3 \cdot 5 + 125}{10}$

209. $\dfrac{78 - |-78|}{5}$

210. $\dfrac{300}{15 - |-15|}$

In Exercises 211–214, use a calculator to evaluate the expression. Round your answer to two decimal places.

211. $(5.8)^4 - (3.2)^5$

212. $\dfrac{(15.8)^3}{(2.3)^8}$

213. $\dfrac{3000}{(1.05)^{10}}$

214. $500\left(1 + \dfrac{0.07}{4}\right)^{40}$

215. *Depreciation* After 3 years, the value of a $25,000 car is given by $25,000\left(\frac{3}{4}\right)^3$.

(a) What is the value of the car after 3 years?

(b) How much has the car depreciated during the 3 years?

216. *Geometry* The volume of water in a hot tub is given by $V = 6^2 \cdot 3$ (see figure). How many cubic feet of water will the hot tub hold? Find the total weight of the water in the tub. (Use the fact that 1 cubic foot of water weighs 62.4 pounds.)

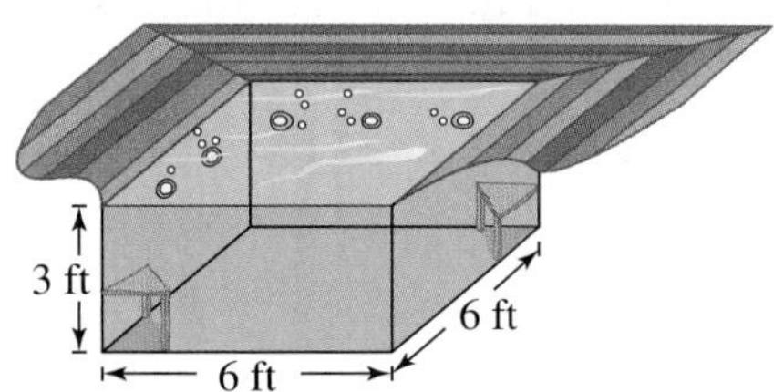

3 ▸ Identify and use the properties of real numbers.

In Exercises 217–224, identify the property of real numbers illustrated by the statement.

217. $123 - 123 = 0$

218. $9 \cdot \frac{1}{9} = 1$

219. $14(3) = 3(14)$

220. $5(3 \cdot 8) = (5 \cdot 3)8$

221. $17 \cdot 1 = 17$

222. $10 + 6 = 6 + 10$

223. $-2(7 + 12) = (-2)7 + (-2)12$

224. $2 + (3 + 19) = (2 + 3) + 19$

In Exercises 225–228, complete the statement using the specified property of real numbers.

225. Additive Identity Property:
$-16 + 0 =$

226. Distributive Property:
$8(7 + 2) =$

227. Commutative Property of Addition:
$24 + 1 =$

228. Associative Property of Multiplication:
$8(5 \cdot 7) =$

229. *Geometry* Find the area of the shaded rectangle in two ways. Explain how the results are related to the Distributive Property.

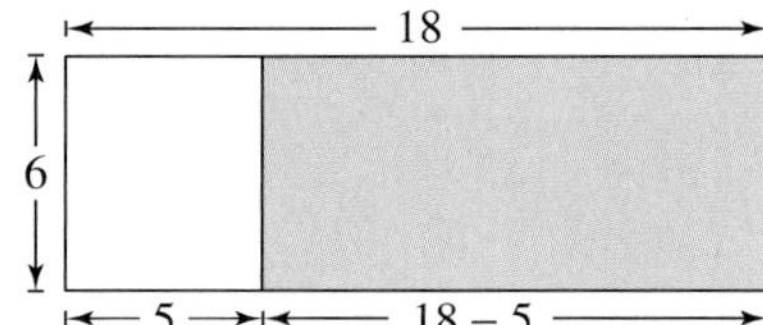

Chapter Test

Take this test as you would take a test in class. After you are done, check your work against the answers in the back of the book.

1. Which of the following are (a) natural numbers, (b) integers, (c) rational numbers, and (d) irrational numbers?
$\{4, -6, \frac{1}{2}, 0, \pi, \frac{7}{9}\}$
2. Place the correct inequality symbol (< or >) between the real numbers.
$-\frac{3}{5} \quad \square \quad -|-2|$

In Exercises 3–20, evaluate the expression. Write fractional answers in simplest form.

3. $|-13|$
4. $-|-6.8|$
5. $16 + (-20)$
6. $-50 - (-60)$
7. $7 + |-3|$
8. $64 - (25 - 8)$
9. $-5(32)$
10. $\frac{-72}{-9}$
11. $\frac{15(-6)}{3}$
12. $-\frac{(-2)(5)}{10}$
13. $\frac{5}{6} - \frac{1}{8}$
14. $-\frac{9}{50}\left(-\frac{20}{27}\right)$
15. $\frac{7}{16} \div \frac{21}{28}$
16. $\frac{-8.1}{0.3}$
17. $-(0.8)^2$
18. $35 - (50 \div 5^2)$
19. $5(3 + 4)^2 - 10$
20. $18 - 7 \cdot 4 + 2^3$

In Exercises 21–24, identify the property of real numbers illustrated by the statement.

21. $3(4 + 6) = 3 \cdot 4 + 3 \cdot 6$
22. $5 \cdot \frac{1}{5} = 1$
23. $3 + (4 + 8) = (3 + 4) + 8$
24. $3(7 + 2) = (7 + 2)3$
25. Write the fraction $\frac{36}{162}$ in simplest form.
26. Write the prime factorization of 216.
27. An electric railway travels a distance of 1218 feet in 21 seconds. What is the average speed of the railway in feet per second?
28. At the grocery store, you buy five cartons of eggs at \$1.49 a carton and two gallons of orange juice at \$3.06 a gallon. You give the clerk a 20-dollar bill. How much change will you receive? (Assume there is no sales tax.)

Study Skills in Action

Absorbing Details Sequentially

Math is a sequential subject (Nolting, 2008). Learning new math concepts successfully depends on how well you understand all the previous concepts. So, it is important to learn and remember concepts as they are encountered. One way to work through a section sequentially is by following these steps.

1 ▶ Work through an example. If you have trouble, consult your notes or seek help from a classmate or instructor.

2 ▶ Complete the checkpoint exercise following the example.

3 ▶ If you get the checkpoint exercise correct, move on to the next example. If not, make sure you understand your mistake(s) before you move on.

4 ▶ When you have finished working through all the examples in the section, take a short break of 5 to 10 minutes. This will give your brain time to process everything.

5 ▶ Start the homework exercises.

Kimberly Nolting
VP, Academic Success Press
expert in developmental education

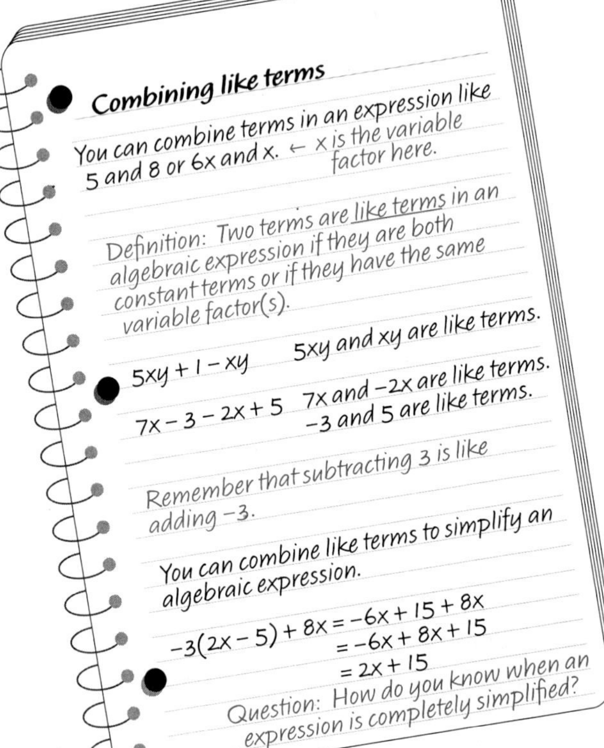

Combining like terms
You can combine terms in an expression like 5 and 8 or 6x and x. ← x is the variable factor here.
Definition: Two terms are like terms in an algebraic expression if they are both constant terms or if they have the same variable factor(s).
5xy + 1 − xy
5xy and xy are like terms.
7x − 3 − 2x + 5
7x and −2x are like terms.
−3 and 5 are like terms.
Remember that subtracting 3 is like adding −3.
You can combine like terms to simplify an algebraic expression.
−3(2x − 5) + 8x = −6x + 15 + 8x
= −6x + 8x + 15
= 2x + 15
Question: How do you know when an expression is completely simplified?

Smart Study Strategy

Rework Your Notes

It is almost impossible to write down in your notes all the detailed information you are taught in class. A good way to reinforce the concepts and put them into your long-term memory is to rework your notes. When you take notes, leave extra space on the pages. You can go back after class and fill in:

- important definitions and rules
- additional examples
- questions you have about the material

Chapter 2

Fundamentals of Algebra

2.1 Writing and Evaluating Algebraic Expressions

Comstock/Photolibrary

Why You Should Learn It

Algebraic expressions can be used to represent real-life quantities. For instance, in Exercise 88 on page 76, you will write and evaluate an expression that can be used to determine how far you travel on the highway.

What You Should Learn

1 ▶ Define and identify terms, variables, and coefficients of algebraic expressions.

2 ▶ Define exponential form and interpret exponential expressions.

3 ▶ Evaluate algebraic expressions using real numbers.

1 ▶ Define and identify terms, variables, and coefficients of algebraic expressions.

Variables and Algebraic Expressions

One of the distinguishing characteristics of algebra is its use of symbols to represent quantities whose numerical values are unknown. Here is a simple example.

EXAMPLE 1 Writing an Algebraic Expression

You accept a part-time job for \$9 per hour. The job offer states that you will be expected to work between 15 and 30 hours a week. Because you don't know how many hours you will work during a week, your total income for a week is unknown. Moreover, your income will probably *vary* from week to week. By representing the variable quantity (the number of hours worked) by the letter x, you can represent the weekly income by the following algebraic expression.

\$9 per hour ↘ ↙ Number of hours worked

$$9x$$

In the product $9x$, the number 9 is a *constant* and the letter x is a *variable.*

CHECKPOINT *Now try Exercise 3.*

Definition of Algebraic Expression

A collection of letters (**variables**) and real numbers (**constants**) combined by using addition, subtraction, multiplication, or division is an **algebraic expression.**

Some examples of algebraic expressions are

$$3x + y, \quad -5a^3, \quad 2W - 7, \quad \frac{x}{y + 3}, \quad \text{and} \quad x^2 - 4x + 5.$$

The **terms** of an algebraic expression are those parts that are separated by *addition.* For example, the expression $x^2 - 4x + 5$ has three terms: x^2, $-4x$, and 5. Note that $-4x$, rather than $4x$, is a term of $x^2 - 4x + 5$ because

$$x^2 - 4x + 5 = x^2 + (-4x) + 5. \qquad \text{To subtract, add the opposite.}$$

For terms such as x^2, $-4x$, and 5, the numerical factor is called the **coefficient** of the term. Here, the coefficients are 1, -4, and 5.

EXAMPLE 2 Identifying the Terms of an Algebraic Expression

Identify the terms of each algebraic expression.

a. $x + 2$

b. $3x + \frac{1}{2}$

c. $2y - 5x - 7$

d. $5(x - 3) + 3x - 4$

e. $4 - 6x + \frac{x + 9}{3}$

Solution

	Algebraic Expression	*Terms*
a.	$x + 2$	$x, 2$
b.	$3x + \frac{1}{2}$	$3x, \frac{1}{2}$
c.	$2y - 5x - 7$	$2y, -5x, -7$
d.	$5(x - 3) + 3x - 4$	$5(x - 3), 3x, -4$
e.	$4 - 6x + \frac{x + 9}{3}$	$4, -6x, \frac{x + 9}{3}$

 **CHECKPOINT** *Now try Exercise 13.*

The terms of an algebraic expression depend on the way the expression is written. Rewriting the expression can (and, in fact, usually does) change its terms. For instance, the expression $2 + 4 - x$ has three terms, but the equivalent expression $6 - x$ has only two terms.

EXAMPLE 3 Identifying Coefficients

Identify the coefficient of each term.

a. $-5x^2$

b. x^3

c. $\frac{2x}{3}$

d. $-\frac{x}{4}$

e. $-x^3$

f. $3\pi x$

Solution

	Term	*Coefficient*	*Comment*
a.	$-5x^2$	-5	Note that $-5x^2 = (-5)x^2$.
b.	x^3	1	Note that $x^3 = 1 \cdot x^3$.
c.	$\frac{2x}{3}$	$\frac{2}{3}$	Note that $\frac{2x}{3} = \frac{2}{3}(x)$.
d.	$-\frac{x}{4}$	$-\frac{1}{4}$	Note that $-\frac{x}{4} = -\frac{1}{4}(x)$.
e.	$-x^3$	-1	Note that $-x^3 = (-1)x^3$.
f.	$3\pi x$	3π	Note that $3\pi x = 3\pi \cdot x$.

CHECKPOINT *Now try Exercise 27.*

2 ▶ Define exponential form and interpret exponential expressions.

Exponential Form

You know from Section 1.5 that a number raised to a power can be evaluated by repeated multiplication. For example, 7^4 represents the product obtained by multiplying 7 by itself four times.

$$\underbrace{7}_{\text{Base}}^{\overset{\text{Exponent}}{4}} = \underbrace{7 \cdot 7 \cdot 7 \cdot 7}_{4 \text{ factors}}$$

In general, for any positive integer n and any real number a, you have

$$a^n = \underbrace{a \cdot a \cdot a \cdot \cdot \cdot a}_{n \text{ factors}}.$$

This rule applies to factors that are variables as well as factors that are *algebraic expressions.*

Study Tip

Be sure you understand the difference between repeated addition

$$\underbrace{x + x + x + x}_{4 \text{ terms}} = 4x$$

and repeated multiplication

$$\underbrace{x \cdot x \cdot x \cdot x}_{4 \text{ factors}} = x^4.$$

Definition of Exponential Form

Let n be a positive integer and let a be a real number, a variable, or an algebraic expression.

$$a^n = \underbrace{a \cdot a \cdot a \cdot \cdot \cdot a}_{n \text{ factors}}$$

In this definition, remember that the letter a can be a number, a variable, or an algebraic expression. It may be helpful to think of a as a box into which you can place any algebraic expression.

$$\square^n = \square \cdot \square \cdot \cdot \cdot \square$$ The box may contain a number, a variable, or an algebraic expression.

EXAMPLE 4 Interpreting Exponential Expressions

a. $3^4 = 3 \cdot 3 \cdot 3 \cdot 3$ **b.** $3x^4 = 3 \cdot x \cdot x \cdot x \cdot x$

c. $(-3x)^4 = (-3x)(-3x)(-3x)(-3x) = (-3)(-3)(-3)(-3) \cdot x \cdot x \cdot x \cdot x$

d. $(y + 2)^3 = (y + 2)(y + 2)(y + 2)$

e. $(5x)^2y^3 = (5x)(5x) \cdot y \cdot y \cdot y = 5 \cdot 5 \cdot x \cdot x \cdot y \cdot y \cdot y$

✓ **CHECKPOINT** *Now try Exercise 37.*

Be sure you understand the priorities for order of operations involving exponents. Here are some examples that tend to cause problems.

Expression	*Correct Evaluation*	*Incorrect Evaluation*
-3^2	$-(3 \cdot 3) = -9$	~~$(-3)(-3) = 9$~~
$(-3)^2$	$(-3)(-3) = 9$	~~$-(3 \cdot 3) = -9$~~
$3x^2$	$3 \cdot x \cdot x$	~~$(3x)(3x)$~~
$-3x^2$	$-3 \cdot x \cdot x$	~~$-(3x)(3x)$~~
$(-3x)^2$	$(-3x)(-3x)$	~~$-(3x)(3x)$~~

3 ▶ Evaluate algebraic expressions using real numbers.

Evaluating Algebraic Expressions

In applications of algebra, you are often required to **evaluate** an algebraic expression. This means you are to find the value of an expression when its variables are replaced by real numbers. For instance, when $x = 2$, the value of the expression $2x + 3$ is as follows.

Expression	*Substitute 2 for x.*	*Value of Expression*
$2x + 3$	$2(2) + 3$	7

When finding the value of an algebraic expression, be sure to replace every occurrence of the specified variable with the appropriate real number. For instance, when $x = -2$, the value of $x^2 - x + 3$ is

$$(-2)^2 - (-2) + 3 = 4 + 2 + 3 = 9.$$

EXAMPLE 5 Evaluating Algebraic Expressions

Evaluate each expression for $x = -3$ and $y = 5$.

a. $-x$

b. $x - y$

c. $3x + 2y$

d. $y - 2(x + y)$

e. $y^2 - 3y$

Solution

a. When $x = -3$, the value of $-x$ is

$-x = -(-3)$ Substitute -3 for x.

$= 3.$ Simplify.

b. When $x = -3$ and $y = 5$, the value of $x - y$ is

$x - y = (-3) - 5$ Substitute -3 for x and 5 for y.

$= -8.$ Subtract.

c. When $x = -3$ and $y = 5$, the value of $3x + 2y$ is

$3x + 2y = 3(-3) + 2(5)$ Substitute -3 for x and 5 for y.

$= -9 + 10$ Multiply.

$= 1.$ Add.

d. When $x = -3$ and $y = 5$, the value of $y - 2(x + y)$ is

$y - 2(x + y) = 5 - 2[(-3) + 5]$ Substitute -3 for x and 5 for y.

$= 5 - 2(2)$ Add.

$= 1.$ Simplify.

e. When $y = 5$, the value of $y^2 - 3y$ is

$y^2 - 3y = (5)^2 - 3(5)$ Substitute 5 for y.

$= 25 - 15$ Simplify.

$= 10.$ Subtract.

✓ ***CHECKPOINT*** *Now try Exercise 67.*

Study Tip

As shown in parts (a), (b), (c), and (d) of Example 5, it is a good idea to use parentheses when substituting a negative number for a variable.

Technology: Tip

Absolute value expressions can be evaluated on a graphing calculator. When evaluating an expression such as $|3 - 6|$, parentheses should surround the entire expression, as in abs(3 − 6).

EXAMPLE 6 Evaluating Algebraic Expressions

Evaluate each expression for $x = 4$ and $y = -6$.

a. y^2 **b.** $-y^2$ **c.** $y - x$ **d.** $|y - x|$ **e.** $|x - y|$

Solution

a. When $y = -6$, the value of y^2 is

$$y^2 = (-6)^2 = 36.$$

b. When $y = -6$, the value of $-y^2$ is

$$-y^2 = -(y^2) = -(-6)^2 = -36.$$

c. When $x = 4$ and $y = -6$, the value of $y - x$ is

$$y - x = (-6) - 4 = -10.$$

d. When $x = 4$ and $y = -6$, the value of $|y - x|$ is

$$|y - x| = |(-6) - 4| = |-10| = 10.$$

e. When $x = 4$ and $y = -6$, the value of $|x - y|$ is

$$|x - y| = |4 - (-6)| = |4 + 6| = |10| = 10.$$

✓ **CHECKPOINT** *Now try Exercise 69.*

EXAMPLE 7 Evaluating Algebraic Expressions

Evaluate each expression for $x = -5$, $y = -2$, and $z = 3$.

a. $\dfrac{y + 2z}{5y - xz}$ **b.** $(y + 2z)(z - 3y)$

Solution

a. When $x = -5$, $y = -2$, and $z = 3$, the value of the expression is

$$\frac{y + 2z}{5y - xz} = \frac{-2 + 2(3)}{5(-2) - (-5)(3)} \qquad \text{Substitute for } x, y, \text{ and } z.$$

$$= \frac{-2 + 6}{-10 + 15} \qquad \text{Simplify.}$$

$$= \frac{4}{5}. \qquad \text{Simplify.}$$

b. When $y = -2$ and $z = 3$, the value of the expression is

$$(y + 2z)(z - 3y) = [(-2) + 2(3)][3 - 3(-2)] \qquad \text{Substitute for } y \text{ and } z.$$

$$= (-2 + 6)(3 + 6) \qquad \text{Simplify.}$$

$$= (4)(9) \qquad \text{Add.}$$

$$= 36. \qquad \text{Multiply.}$$

✓ **CHECKPOINT** *Now try Exercise 79.*

When you evaluate an algebraic expression for *several* values variable(s), it is helpful to organize the values of the expression in a table fo

> **Technology: Tip**
>
> If you have a graphing calculator, try using it to store and evaluate the expression in Example 8. You can use the following steps to evaluate $-9x + 6$ for $x = 2$.
>
> - Store the expression as Y_1.
> - Store 2 in X.
>
>   2 (STO▶) (X,T,θ,n) (ENTER)
> - Display Y_1
>
>   (VARS) (Y-VARS) (ENTER) (ENTER)
>
>   and then press (ENTER) again.

EXAMPLE 8 Repeated Evaluation of an Expression

Complete the table by evaluating the expression $5x + 2$ for each value of x shown in the table.

x	-1	0	1	2
$5x + 2$				

Solution

Begin by substituting each value of x into the expression.

When $x = -1$: $5x + 2 = 5(-1) + 2 = -5 + 2 = -3$

When $x = 0$: $5x + 2 = 5(0) + 2 = 0 + 2 = 2$

When $x = 1$: $5x + 2 = 5(1) + 2 = 5 + 2 = 7$

When $x = 2$: $5x + 2 = 5(2) + 2 = 10 + 2 = 12$

Once you have evaluated the expression for each value of x, fill in the table with the values.

x	-1	0	1	2
$5x + 2$	-3	2	7	12

CHECKPOINT *Now try Exercise 85(a).*

EXAMPLE 9 Geometry: Area

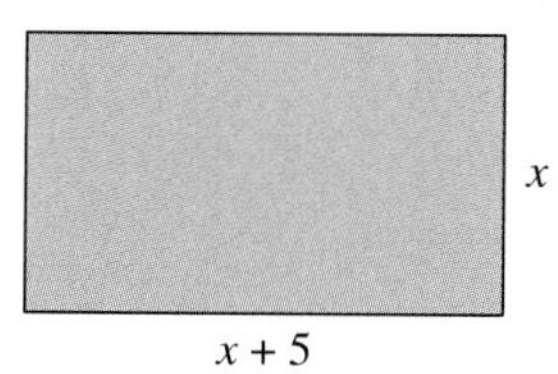

Figure 2.1

Write an expression for the area of the rectangle shown in Figure 2.1. Then evaluate the expression to find the area of the rectangle when $x = 7$.

Solution

Area of a rectangle = Length · Width

$= (x + 5) \cdot x$ Substitute.

To find the area of the rectangle when $x = 7$, substitute 7 for x in the expression for the area.

$(x + 5) \cdot x = (7 + 5) \cdot 7$ Substitute 7 for x.

$= 12 \cdot 7$ Add.

$= 84$ Multiply.

So, the area of the rectangle is 84 square units.

✓ **CHECKPOINT** *Now try Exercise 91.*

Concept Check

... e between terms and factors.

... the base and the exponent of the expression $(x + 1)^3$.

3. Explain why $4x^4$ is not equal to $(4x)(4x)(4x)(4x)$.

4. What value of y would cause $3y + 2$ to equal 8? Explain.

2.1 EXERCISES

Go to pages 116–117 to record your assignments.

Developing Skills

In Exercises 1–4, write an algebraic expression for the statement. ***See Example 1.***

1. The income earned at \$7.55 per hour for w hours
2. The cost for a family of n people to see a movie if the cost per person is \$8.25
3. ✓ The cost of m pounds of meat if the cost per pound is \$3.79
4. The total weight of x bags of fertilizer if each bag weighs 50 pounds

In Exercises 5–10, identify the variable(s) in the expression.

5. $x + 3$ **6.** $y - 1$

7. $m + n$ **8.** $a + b$

9. $2^3 - k$ **10.** $3^2 + z$

In Exercises 11–24, identify the terms of the expression. ***See Example 2.***

11. $4x + 3$ **12.** $3x^2 + 5$

✓ **13.** $6x - 1$ **14.** $5 - 3t^2$

15. $\frac{5}{3} - 3y^3$ **16.** $6x + \frac{2}{3}$

17. $a^2 + 4ab + b^2$ **18.** $x^2 + 18xy + y^2$

19. $3(x + 5) + 10$ **20.** $16 - (x + 1)$

21. $15 + \dfrac{5}{x}$ **22.** $\dfrac{6}{t} - 22$

23. $\dfrac{3}{x + 2} - 3x + 4$ **24.** $\dfrac{5}{x - 5} - 7x^2 + 18$

In Exercises 25–34, identify the coefficient of the term. ***See Example 3.***

25. $14x$ **26.** $25y$

✓ **27.** $-\frac{1}{3}y$ **28.** $\frac{2}{3}n$

29. $\dfrac{2x}{5}$ **30.** $-\dfrac{3x}{4}$

31. $2\pi x^2$ **32.** πt^4

33. $3.06u$ **34.** $-5.32b$

In Exercises 35–52, expand the expression as a product of factors. ***See Example 4.***

35. y^5

36. $(-x)^6$

✓ **37.** 2^2x^4

38. $(-5)^3x^2$

39. $4y^2z^3$

40. $3uv^4$

41. $(a^2)^3$

42. $(z^3)^3$

43. $-4x^3 \cdot x^4$

44. $a^2y^2 \cdot y^3$

45. $-9(ab)^3$

46. $2(xz)^4$

47. $(x + y)^2$

48. $(s - t)^5$

49. $\left(\dfrac{a}{3s}\right)^4$

50. $\left(-\dfrac{2}{5x}\right)^3$

51. $[2(a-b)^3][2(a-b)^2]$

52. $[3(r+s)^2][3(r+s)]^2$

In Exercises 53–62, rewrite the product in exponential form.

53. $-2 \cdot u \cdot u \cdot u \cdot u$

54. $\frac{1}{3} \cdot x \cdot x \cdot x \cdot x \cdot x$

55. $(2u) \cdot (2u) \cdot (2u) \cdot (2u)$

56. $\frac{1}{3}x \cdot \frac{1}{3}x \cdot \frac{1}{3}x \cdot \frac{1}{3}x \cdot \frac{1}{3}x$

57. $-a \cdot (-a) \cdot (-a) \cdot b \cdot b$

58. $y \cdot y \cdot z \cdot z \cdot z \cdot z$

59. $-3 \cdot (x-y) \cdot (x-y) \cdot (-3) \cdot (-3)$

60. $(u-v) \cdot (u-v) \cdot 8 \cdot 8 \cdot 8 \cdot (u-v)$

61. $\frac{x+y}{4} \cdot \frac{x+y}{4} \cdot \frac{x+y}{4}$

62. $\frac{r-s}{5} \cdot \frac{r-s}{5} \cdot \frac{r-s}{5} \cdot \frac{r-s}{5}$

In Exercises 63–84, evaluate the algebraic expression for the given values of the variable(s). If it is not possible, state the reason. ***See Examples 5, 6, and 7.***

	Expression	*Values*
63.	$2x - 1$	(a) $x = \frac{1}{2}$ (b) $x = -4$
64.	$3x - 2$	(a) $x = \frac{4}{3}$ (b) $x = -1$
65.	$2x^2 - 5$	(a) $x = -2$ (b) $x = 3$
66.	$64 - 16t^2$	(a) $t = 2$ (b) $t = -3$
✓ **67.**	$3x - 2y$	(a) $x = 4, y = 3$ (b) $x = \frac{2}{3}, y = -1$
68.	$10u - 3v$	(a) $u = 3, v = 10$ (b) $u = -2, v = \frac{4}{7}$
✓ **69.**	$\lvert 2x - 3y \rvert$	(a) $x = 2, y = 3$ (b) $x = -1, y = 4$
70.	$y - \lvert -3x + y \rvert$	(a) $x = -2, y = -1$ (b) $x = 7, y = 3$

	Expression	*Values*
71.	$x - 3(x - y)$	(a) $x = 3, y = 3$ (b) $x = 4, y = -4$
72.	$-3x + 2(x + y)$	(a) $x = -2, y = 2$ (b) $x = 0, y = 5$
73.	$b^2 - 4ab$	(a) $a = 2, b = -3$ (b) $a = 6, b = -4$
74.	$a^2 + 2ab$	(a) $a = -2, b = 3$ (b) $a = 4, b = -2$
75.	$\frac{x - 2y}{x + 2y}$	(a) $x = 4, y = 2$ (b) $x = 4, y = -2$
76.	$\frac{5x}{y - 3}$	(a) $x = 2, y = 4$ (b) $x = 2, y = 3$
77.	$\frac{-y}{x^2 + y^2}$	(a) $x = 0, y = 5$ (b) $x = 1, y = -3$
78.	$\frac{2x - y}{y^2 + 1}$	(a) $x = 1, y = 2$ (b) $x = 1, y = 3$
✓ **79.**	$(x + 2y)(-3x - z)$	(a) $x = 2, y = -1, z = -1$ (b) $x = -3, y = 2, z = -2$
80.	$\frac{yz - 3}{x + 2z}$	(a) $x = 0, y = -7, z = 3$ (b) $x = -2, y = -3, z = 3$

81. *Area of a Triangle*

$\frac{1}{2}bh$ (a) $b = 3, h = 5$ (b) $b = 2, h = 10$

82. *Distance Traveled*

rt (a) $r = 50, t = 3.5$ (b) $r = 35, t = 4$

83. *Volume of a Rectangular Prism*

lwh (a) $l = 4, w = 2, h = 9$ (b) $l = 100, w = 0.8, h = 4$

84. *Simple Interest*

Prt (a) $P = 1000, r = 0.08, t = 3$ (b) $P = 500, r = 0.07, t = 5$

85. *Finding a Pattern*

(a) Complete the table by evaluating the expression $3x - 2$ for each value of x.

x	-1	0	1	2	3	4
$3x - 2$						

(b) Use the table to find the change in the value of the expression for each one-unit increase in x.

(c) From the pattern of parts (a) and (b), predict the change in the algebraic expression $\frac{2}{3}x + 4$ for each one-unit increase in x. Then verify your prediction by completing a table.

86. *Finding a Pattern*

(a) Complete the table by evaluating the expression $3 - 2x$ for each value of x.

x	-1	0	1	2	3	4
$3 - 2x$						

(b) Use the table to find the change in the value of the expression for each one-unit increase in x.

(c) From the pattern of parts (a) and (b), predict the change in the algebraic expression $4 - \frac{3}{2}x$ for each one-unit increase in x. Then verify your prediction by completing a table.

Solving Problems

87. *Advertising* An advertisement for a new pair of basketball shoes claims that the shoes will help you jump six inches higher than without shoes.

(a) Let x represent the height (in inches) jumped without shoes. Write an expression that represents the height of a jump while wearing the new shoes.

(b) You can jump 23 inches without shoes. How high can you jump while wearing the new shoes?

88. *Distance* You are driving 60 miles per hour on the highway.

(a) Write an expression that represents the distance you travel in t hours.

(b) How far will you travel in 2.75 hours?

Geometry In Exercises 89–92, write an expression for the area of the figure. Then evaluate the expression for the given value(s) of the variable(s). ***See Example 9.***

89. $n = 8$ **90.** $x = 10, y = 3$

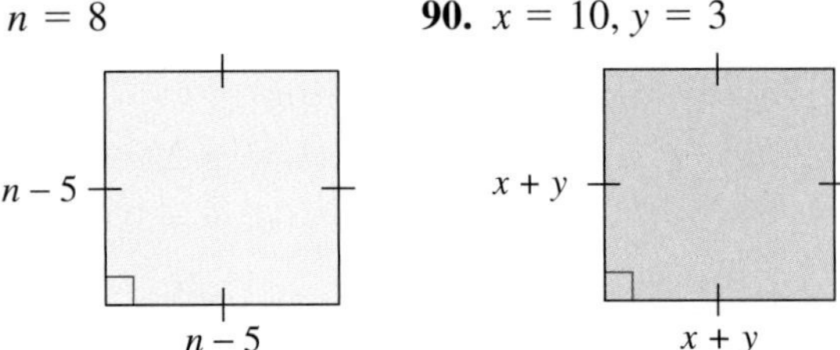

91. $a = 5, b = 4$ **92.** $x = 9$

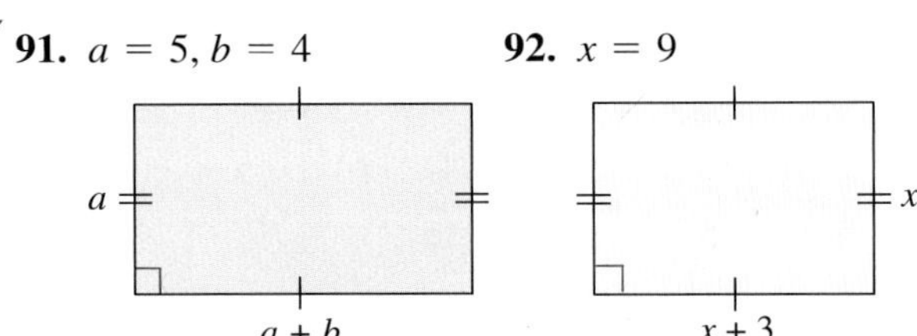

93. *Exploration* For any natural number n, the sum of the numbers $1, 2, 3, \ldots, n$ is equal to

$$\frac{n(n+1)}{2}, \quad n \geq 1.$$

Verify the formula for (a) $n = 3$, (b) $n = 6$, and (c) $n = 10$.

94. *Exploration* A convex polygon with n sides has

$$\frac{n(n-3)}{2}, \quad n \geq 4$$

diagonals. Verify the formula for (a) a square (two diagonals), (b) a pentagon (five diagonals), and (c) a hexagon (nine diagonals).

95. *Iteration and Exploration* Once an expression has been evaluated for a specified value, the expression can be repeatedly evaluated by using the result of the preceding evaluation as the input for the next evaluation.

(a) The procedure for repeated evaluation of the algebraic expression $\frac{1}{2}x + 3$ can be accomplished on a graphing calculator, as follows.

- Clear the display.
- Enter 2 in the display and press ENTER.
- Enter $\frac{1}{2} * \text{ANS} + 3$ and press ENTER.
- Each time ENTER is pressed, the calculator will evaluate the expression at the value of x obtained in the preceding computation. Continue the process six more times. What value does the expression appear to be approaching? If necessary, round your answers to three decimal places.

(b) Repeat part (a) starting with $x = 12$.

96. *Exploration* Repeat Exercise 95 using the expression $\frac{3}{4}x + 2$. If necessary, round your answers to three decimal places.

Explaining Concepts

97. Is $3x$ a term of $4 - 3x$? Explain.

98. Is it possible to evaluate the expression

$$\frac{x+2}{y-3}$$

when $x = 5$ and $y = 3$? Explain.

99. Explain why the formulas in Exercises 93 and 94 will always yield natural numbers.

100. *Error Analysis* Describe and correct the error in evaluating $y - 2(x - y)$ for $x = 2$ and $y = -4$.

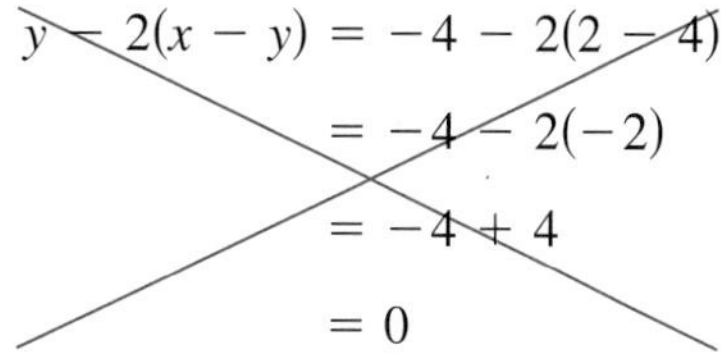

Cumulative Review

In Exercises 101–108, evaluate the expression.

101. $10 - (-7)$

102. $6 - 10 - (-12) + 3$

103. $-5 + 10 - (-9) - 4$

104. $-(-8) + 6 - 4 - 2$

105. $(-6)(-4)$

106. $\frac{-56}{7}$

107. $\frac{-144}{-12}$

108. $5(-7)$

In Exercises 109–112, identify the property of real numbers illustrated by the statement.

109. $3(4) = 4(3)$

110. $10 - 10 = 0$

111. $3(6 + 2) = 3 \cdot 6 + 3 \cdot 2$

112. $7 + (8 + 5) = (7 + 8) + 5$

2.2 Simplifying Algebraic Expressions

Bill Pogue/Getty Images

What You Should Learn

1 ▶ Use the properties of algebra.
2 ▶ Combine like terms of an algebraic expression.
3 ▶ Simplify an algebraic expression by rewriting the terms.
4 ▶ Use the Distributive Property to remove symbols of grouping.

Why You Should Learn It

You can use an algebraic expression to find the area of a house lot, as shown in Exercise 155 on page 89.

Properties of Algebra

1 ▶ Use the properties of algebra.

You are now ready to combine algebraic expressions using the properties below.

Properties of Algebra

Let a, b, and c represent real numbers, variables, or algebraic expressions.

Property	*Example*
Commutative Property of Addition:	
$a + b = b + a$	$3x + x^2 = x^2 + 3x$
Commutative Property of Multiplication:	
$ab = ba$	$(5 + x)x = x(5 + x)$
Associative Property of Addition:	
$(a + b) + c = a + (b + c)$	$(2x + 7) + x^2 = 2x + (7 + x^2)$
Associative Property of Multiplication:	
$(ab)c = a(bc)$	$(2x \cdot 5y) \cdot 7 = 2x \cdot (5y \cdot 7)$
Distributive Property:	
$a(b + c) = ab + ac$	$4x(7 + 3x) = 4x \cdot 7 + 4x \cdot 3x$
$(a + b)c = ac + bc$	$(2y + 5)y = 2y \cdot y + 5 \cdot y$
Additive Identity Property:	
$a + 0 = 0 + a = a$	$3y^2 + 0 = 0 + 3y^2 = 3y^2$
Multiplicative Identity Property:	
$a \cdot 1 = 1 \cdot a = a$	$(-2x^3) \cdot 1 = 1 \cdot (-2x^3) = -2x^3$
Additive Inverse Property:	
$a + (-a) = 0$	$3y^2 + (-3y^2) = 0$
Multiplicative Inverse Property:	
$a \cdot \frac{1}{a} = 1, \quad a \neq 0$	$(x^2 + 2) \cdot \frac{1}{x^2 + 2} = 1$

Study Tip

You'll discover as you review the table of properties at the right that they are the same as the properties of real numbers on page 50. The only difference is that the input for algebra rules can be real numbers, variables, or algebraic expressions.

EXAMPLE 1 Applying the Basic Rules of Algebra

Use the indicated rule to complete each statement.

a. Additive Identity Property:	$(x - 2) + \quad = x - 2$
b. Commutative Property of Multiplication:	$5(y + 6) =$
c. Commutative Property of Addition:	$5(y + 6) =$
d. Distributive Property:	$5(y + 6) =$
e. Associative Property of Addition:	$(x^2 + 3) + 7 =$
f. Additive Inverse Property:	$\quad + 4x = 0$

Solution

a. $(x - 2) + 0 = x - 2$

b. $5(y + 6) = (y + 6)5$

c. $5(y + 6) = 5(6 + y)$

d. $5(y + 6) = 5y + 5(6)$

e. $(x^2 + 3) + 7 = x^2 + (3 + 7)$

f. $-4x + 4x = 0$

✓ ***CHECKPOINT*** *Now try Exercise 23.*

Example 2 illustrates some common uses of the Distributive Property. Study this example carefully. Such uses of the Distributive Property are very important in algebra. Applying the Distributive Property as illustrated in Example 2 is called **expanding** an algebraic expression.

EXAMPLE 2 Using the Distributive Property

Use the Distributive Property to expand each expression.

a. $2(7 - x)$ **b.** $(10 - 2y)3$ **c.** $2x(x + 4y)$ **d.** $-(1 - 2y + x)$

Solution

a. $2(7 - x) = 2 \cdot 7 - 2 \cdot x$
$= 14 - 2x$

b. $(10 - 2y)3 = 10(3) - 2y(3)$
$= 30 - 6y$

c. $2x(x + 4y) = 2x(x) + 2x(4y)$
$= 2x^2 + 8xy$

d. $-(1 - 2y + x) = (-1)(1 - 2y + x)$
$= (-1)(1) - (-1)(2y) + (-1)(x)$
$= -1 + 2y - x$

✓ ***CHECKPOINT*** *Now try Exercise 35.*

Study Tip

In Example 2(d), the negative sign is distributed over each term in the parentheses by multiplying each term by -1.

In the next example, note how area can be used to demonstrate the Distributive Property.

EXAMPLE 3 **The Distributive Property and Area**

Write the area of each component of the figure. Then demonstrate the Distributive Property by writing the total area of each figure in two ways.

a. 2 4; 3; $2 + 4$

b. a b; a; $a + b$

Solution

a. 2 4; 3 | 6 | 12

The total area is $3(2 + 4) = 3 \cdot 2 + 3 \cdot 4 = 6 + 12 = 18$.

b. a b; a | a^2 | ab

The total area is $a(a + b) = a \cdot a + a \cdot b = a^2 + ab$.

CHECKPOINT *Now try Exercise 63.*

2 ▶ Combine like terms of an algebraic expression.

Combining Like Terms

Two or more terms of an algebraic expression can be combined only if they are **like terms.**

Definition of Like Terms

In an algebraic expression, two terms are said to be **like terms** if they are both constant terms or if they have the same *variable factor(s)*. Factors such as x in $5x$ and ab in $6ab$ are called **variable factors.**

The terms $5x$ and $-3x$ are like terms because they have the same variable factor, x. Similarly, $3x^2y$, $-x^2y$, and $\frac{1}{3}(x^2y)$ are like terms because they have the same variable factors, x^2 and y.

Study Tip

Notice in Example 4(b) that $-x^2$ and $3x$ are *not* like terms because the variable x is not raised to the same power in both terms.

EXAMPLE 4 **Identifying Like Terms in Expressions**

	Expression	*Like Terms*
a.	$5xy + 1 - xy$	$5xy$ and $-xy$
b.	$12 - x^2 + 3x - 5$	12 and -5
c.	$7x - 3 - 2x + 5$	$7x$ and $-2x$, -3 and 5

CHECKPOINT *Now try Exercise 67.*

To combine like terms in an algebraic expression, you can simply add their respective coefficients and attach the common variable factor(s). This is actually an application of the Distributive Property, as shown in Example 5.

EXAMPLE 5 Combining Like Terms

Simplify each expression by combining like terms.

a. $5x + 2x - 4$ **b.** $-5 + 8 + 7y - 5y$ **c.** $2y - 3x - 4x$

Solution

a. $5x + 2x - 4 = (5 + 2)x - 4$ Distributive Property

$= 7x - 4$ Add.

b. $-5 + 8 + 7y - 5y = (-5 + 8) + (7 - 5)y$ Distributive Property

$= 3 + 2y$ Simplify.

c. $2y - 3x - 4x = 2y - x(3 + 4)$ Distributive Property

$= 2y - x(7)$ Add.

$= 2y - 7x$ Simplify.

✓ **CHECKPOINT** *Now try Exercise 77.*

Often, you need to use other rules of algebra before you can apply the Distributive Property to combine like terms. This is illustrated in the next example.

EXAMPLE 6 Using Rules of Algebra to Combine Like Terms

Simplify each expression by combining like terms.

a. $7x + 3y - 4x$ **b.** $12a - 5 - 3a + 7$ **c.** $y - 4x - 7y + 9y$

Solution

a. $7x + 3y - 4x = 3y + 7x - 4x$ Commutative Property

$= 3y + (7x - 4x)$ Associative Property

$= 3y + (7 - 4)x$ Distributive Property

$= 3y + 3x$ Subtract.

b. $12a - 5 - 3a + 7 = 12a - 3a - 5 + 7$ Commutative Property

$= (12a - 3a) + (-5 + 7)$ Associative Property

$= (12 - 3)a + (-5 + 7)$ Distributive Property

$= 9a + 2$ Simplify.

c. $y - 4x - 7y + 9y = -4x + (y - 7y + 9y)$ Group like terms.

$= -4x + (1 - 7 + 9)y$ Distributive Property

$= -4x + 3y$ Simplify.

✓ **CHECKPOINT** *Now try Exercise 79.*

Study Tip

As you gain experience with the rules of algebra, you may want to combine some of the steps in your work. For instance, you might feel comfortable listing only the following steps to solve part (b) of Example 6.

$12a - 5 - 3a + 7$

$= (12a - 3a) + (-5 + 7)$

$= 9a + 2$

3 ▶ Simplify an algebraic expression by rewriting the terms.

Smart Study Strategy

Go to page 66 for ways to *Rework Your Notes.*

Simplifying Algebraic Expressions

Simplifying an algebraic expression by rewriting it in a more usable form is one of the most frequently used skills in algebra. You will study two others—solving an equation and sketching the graph of an equation—later in this text. To **simplify an algebraic expression** generally means to remove symbols of grouping and combine like terms. For instance, the expression $x + (3 + x)$ can be simplified as $2x + 3$.

EXAMPLE 7 Simplifying Algebraic Expressions

Simplify each expression.

a. $-3(-5x)$ **b.** $7(-x)$

Solution

a. $-3(-5x) = (-3)(-5)x$ Associative Property

$= 15x$ Multiply.

b. $7(-x) = 7(-1)(x)$ Coefficient of $-x$ is -1.

$= -7x$ Multiply.

✓ **CHECKPOINT** *Now try Exercise 107.*

EXAMPLE 8 Simplifying Algebraic Expressions

Simplify each expression.

a. $\frac{5x}{3} \cdot \frac{3}{5}$ **b.** $x^2(-2x^3)$ **c.** $(-2x)(4x)$ **d.** $(2rs)(r^2s)$

Solution

a. $\frac{5x}{3} \cdot \frac{3}{5} = \left(\frac{5}{3} \cdot x\right) \cdot \frac{3}{5}$ Coefficient of $\frac{5x}{3}$ is $\frac{5}{3}$.

$= \left(\frac{5}{3} \cdot \frac{3}{5}\right) \cdot x$ Commutative and Associative Properties

$= 1 \cdot x$ Multiplicative Inverse

$= x$ Multiplicative Identity

b. $x^2(-2x^3) = (-2)(x^2 \cdot x^3)$ Commutative and Associative Properties

$= -2 \cdot x \cdot x \cdot x \cdot x \cdot x$ Repeated multiplication

$= -2x^5$ Exponential form

c. $(-2x)(4x) = (-2 \cdot 4)(x \cdot x)$ Commutative and Associative Properties

$= -8x^2$ Exponential form

d. $(2rs)(r^2s) = 2(r \cdot r^2)(s \cdot s)$ Commutative and Associative Properties

$= 2 \cdot r \cdot r \cdot r \cdot s \cdot s$ Repeated multiplication

$= 2r^3s^2$ Exponential form

✓ **CHECKPOINT** *Now try Exercise 113.*

4 ▶ Use the Distributive Property to remove symbols of grouping.

Symbols of Grouping

The main tool for removing symbols of grouping is the Distributive Property, as illustrated in Example 9. You may want to review order of operations in Section 1.5.

Study Tip

When a parenthetical expression is preceded by a *plus* sign, you can remove the parentheses without changing the signs of the terms inside.

$3y + (-2y + 7)$

$= 3y - 2y + 7$

When a parenthetical expression is preceded by a *minus* sign, however, you must change the sign of each term to remove the parentheses.

$3y - (2y - 7)$

$= 3y - 2y + 7$

Remember that $-(2y - 7)$ is equal to $(-1)(2y - 7)$, and the Distributive Property can be used to "distribute the minus sign" to obtain $-2y + 7$.

EXAMPLE 9 Removing Symbols of Grouping

Simplify each expression.

a. $-(3y + 5)$ **b.** $5x + (x - 7)2$

c. $-2(4x - 1) + 3x$ **d.** $3(y - 5) - (2y - 7)$

Solution

a. $-(3y + 5) = -3y - 5$ Distributive Property

b. $5x + (x - 7)2 = 5x + 2x - 14$ Distributive Property

$= 7x - 14$ Combine like terms.

c. $-2(4x - 1) + 3x = -8x + 2 + 3x$ Distributive Property

$= -8x + 3x + 2$ Commutative Property

$= -5x + 2$ Combine like terms.

d. $3(y - 5) - (2y - 7) = 3y - 15 - 2y + 7$ Distributive Property

$= (3y - 2y) + (-15 + 7)$ Group like terms.

$= y - 8$ Combine like terms.

✓ **CHECKPOINT** *Now try Exercise 123.*

EXAMPLE 10 Removing Nested Symbols of Grouping

Simplify each expression.

a. $5x - 2[4x + 3(x - 1)]$ **b.** $-7y + 3[2y - (3 - 2y)] - 5y + 4$

Solution

a. $5x - 2[4x + 3(x - 1)]$

$= 5x - 2[4x + 3x - 3]$ Distributive Property

$= 5x - 2[7x - 3]$ Combine like terms.

$= 5x - 14x + 6$ Distributive Property

$= -9x + 6$ Combine like terms.

b. $-7y + 3[2y - (3 - 2y)] - 5y + 4$

$= -7y + 3[2y - 3 + 2y] - 5y + 4$ Distributive Property

$= -7y + 3[4y - 3] - 5y + 4$ Combine like terms.

$= -7y + 12y - 9 - 5y + 4$ Distributive Property

$= (-7y + 12y - 5y) + (-9 + 4)$ Group like terms.

$= -5$ Combine like terms.

✓ **CHECKPOINT** *Now try Exercise 131.*

EXAMPLE 11 Simplifying an Algebraic Expression

Simplify $2x(x + 3y) + 4(5 - xy)$.

Solution

$$\begin{aligned} 2x(x + 3y) + 4(5 - xy) &= 2x^2 + 6xy + 20 - 4xy && \text{Distributive Property} \\ &= 2x^2 + 6xy - 4xy + 20 && \text{Commutative Property} \\ &= 2x^2 + 2xy + 20 && \text{Combine like terms.} \end{aligned}$$

✓ **CHECKPOINT** *Now try Exercise 137.*

EXAMPLE 12 Simplifying a Fractional Expression

Simplify $\frac{x}{4} + \frac{2x}{7}$.

Solution

$$\begin{aligned} \frac{x}{4} + \frac{2x}{7} &= \frac{1}{4}x + \frac{2}{7}x && \text{Write with fractional coefficients.} \\ &= \left(\frac{1}{4} + \frac{2}{7}\right)x && \text{Distributive Property} \\ &= \left[\frac{1(7)}{4(7)} + \frac{2(4)}{7(4)}\right]x && \text{Common denominator} \\ &= \frac{15x}{28} && \text{Simplify.} \end{aligned}$$

✓ **CHECKPOINT** *Now try Exercise 145.*

EXAMPLE 13 Geometry: Perimeter and Area

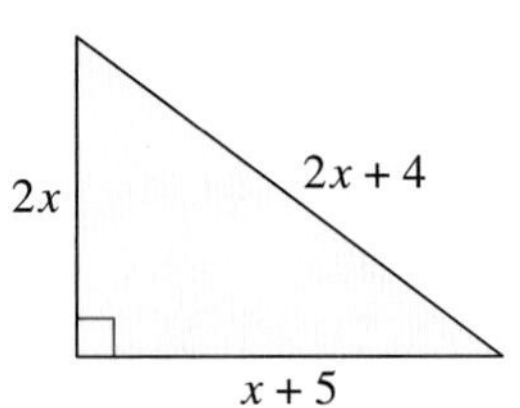

Figure 2.2

Using Figure 2.2, write and simplify an expression for (a) the perimeter and (b) the area of the triangle.

Solution

a. Perimeter of a Triangle = Sum of the Three Sides

$$\begin{aligned} &= 2x + (2x + 4) + (x + 5) && \text{Substitute.} \\ &= (2x + 2x + x) + (4 + 5) && \text{Group like terms.} \\ &= 5x + 9 && \text{Combine like terms.} \end{aligned}$$

b. Area of a Triangle $= \frac{1}{2} \cdot \text{Base} \cdot \text{Height}$

$$\begin{aligned} &= \tfrac{1}{2}(x + 5)(2x) && \text{Substitute.} \\ &= \tfrac{1}{2}(2x)(x + 5) && \text{Commutative Property} \\ &= x(x + 5) && \text{Multiply.} \\ &= x^2 + 5x && \text{Distributive Property} \end{aligned}$$

✓ **CHECKPOINT** *Now try Exercise 151.*

Concept Check

1. Explain the Additive and Multiplicative Inverse Properties. Give an example of each.

2. In your own words, state the definition of like terms. Give an example of like terms and an example of unlike terms.

3. Describe how to combine like terms. Give an example of an expression that can be simplified by combining like terms.

4. In your own words, describe the procedure for removing nested symbols of grouping.

2.2 EXERCISES

Go to pages 116–117 to record your assignments.

Developing Skills

In Exercises 1–22, identify the property (or properties) of algebra illustrated by the statement.

1. $3a + 5b = 5b + 3a$
2. $x + 2y = 2y + x$
3. $-10(xy^2) = (-10x)y^2$
4. $(9x)y = 9(xy)$
5. $rt + 0 = rt$
6. $-8x + 0 = -8x$
7. $(x^2 + y^2) \cdot 1 = x^2 + y^2$
8. $1 \cdot (5z + 12) = 5z + 12$
9. $(3x + 2y) + z = 3x + (2y + z)$
10. $-4a + (b^2 + 2c) = (-4a + b^2) + 2c$
11. $2zy = 2yz$
12. $-7a^2c = -7ca^2$
13. $-5x(y + z) = -5xy - 5xz$
14. $x(y + z) = xy + xz$
15. $(5m + 3) - (5m + 3) = 0$
16. $(2x - 10) - (2x - 10) = 0$
17. $16xy \cdot \dfrac{1}{16xy} = 1, \quad xy \neq 0$
18. $(x + y) \cdot \dfrac{1}{(x + y)} = 1, \quad x + y \neq 0$
19. $(x + 2)(x + y) = x(x + y) + 2(x + y)$
20. $(a + 6)(b + 2c) = (a + 6)b + (a + 6)2c$
21. $x^2 + (y^2 - y^2) = x^2$
22. $3y + (z^3 - z^3) = 3y$

In Exercises 23–34, complete the statement. Then state the property of algebra that you used. ***See Example 1.***

23. $(-5r)s = -5(\quad)$
24. $7(xy^2) = (\quad)y^2$
25. $v \cdot 2 =$
26. $(2x - y)(-3) = -3$
27. $5(t - 2) = 5(\quad) + 5(\quad)$
28. $x(y + 4) = x(\quad) + x(\quad)$
29. $(2z - 3) + \quad = 0$

30. $-(x + 10) + \underline{\qquad\qquad} = 0$

31. $-5x(\underline{\qquad\qquad}) = 1, \quad x \neq 0$

32. $\frac{1}{4z^2}(\underline{\qquad\qquad}) = 1, z \neq 0$

33. $12 + (8 - x) = \underline{\qquad\qquad} - x$

34. $-11 + (5 + 2y) = \underline{\qquad\qquad} + 2y$

In Exercises 35–62, use the Distributive Property to expand the expression. ***See Example 2.***

35. $2(16 + 8z)$

36. $5(7 + 3x)$

37. $8(-3 + 5m)$

38. $12(-2 + y)$

39. $10(9 - 6x)$

40. $3(7 - 4a)$

41. $-8(2 + 5t)$

42. $-9(4 + 2b)$

43. $-5(2x - y)$

44. $-3(11y - 6)$

45. $(x + 1)8$

46. $(r + 10)2$

47. $(4 - t)(-6)$

48. $(3 - x)(-5)$

49. $4(x + xy + y^2)$

50. $6(r - t + s)$

51. $3(x^2 + x)$

52. $9(a^2 + a)$

53. $4(2y^2 - y)$

54. $7(3x^2 - x)$

55. $-z(5 - 2z)$

56. $-t(12 - 4t)$

57. $-4y(3y - 4)$

58. $-6s(6s - 1)$

59. $-(u - v)$

60. $-(x + y)$

61. $x(3x - 4y)$

62. $r(2r^2 - t)$

In Exercises 63–66, write the area of each component of the figure. Then demonstrate the Distributive Property by writing the total area of each figure in two ways. ***See Example 3.***

63.

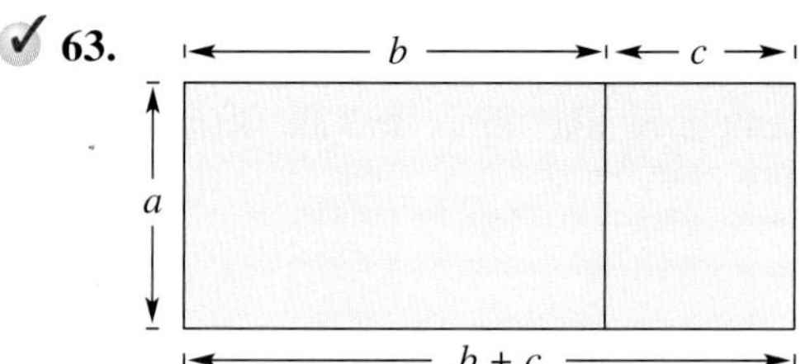

64.

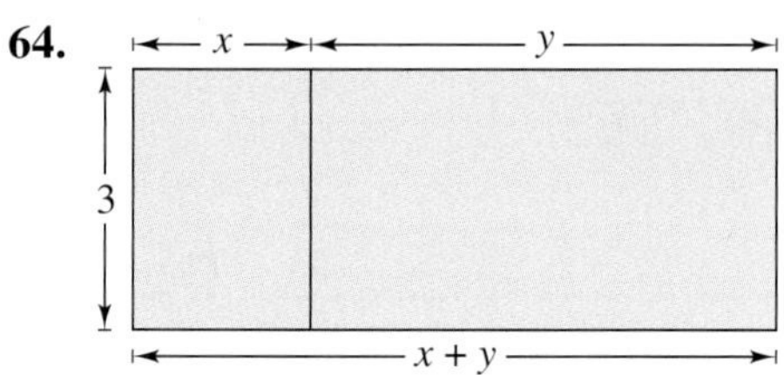

65.

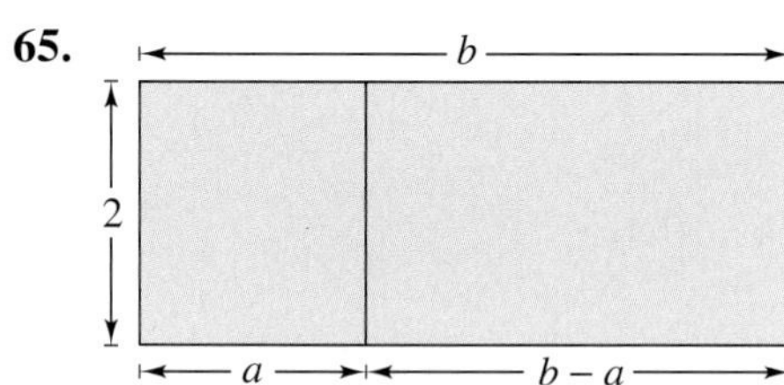

66.

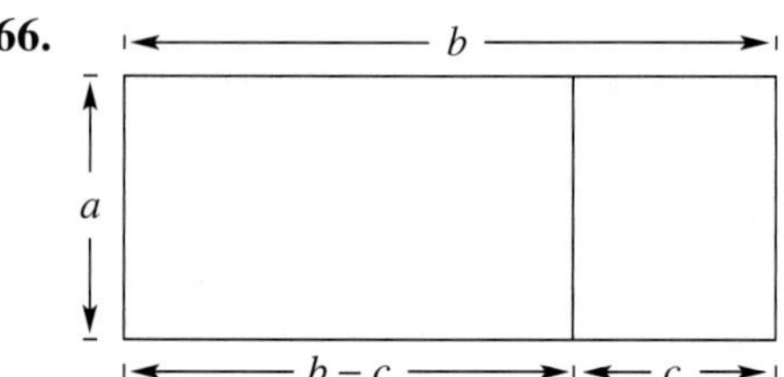

In Exercises 67–72, identify the like terms. ***See Example 4.***

67. $16t^3 + 4t - 5t + 3t^3$

68. $-\frac{1}{4}x^2 - 3x + \frac{3}{4}x^2 + x$

69. $4rs^2 - 5 - 2r^2s + 12rs^2 + 1$

70. $3 + 6x^2y + 2xy - 2 - 4x^2y$

71. $x^3 + 4x^2y - 2y^2 + 5xy^2 + 10x^2y + 3x^3$

72. $a^2 + 5ab^2 - 3b^2 + 7a^2b - ab^2 + a^2$

In Exercises 73–92, simplify the expression by combining like terms. ***See Examples 5 and 6.***

73. $3y - 5y$

74. $-16x + 25x$

75. $x + 5 - 3x$

76. $7s + 3 - 3s$

✓ **77.** $2x + 9x + 4$

78. $10x - 6 - 5x$

✓ **79.** $5r + 6 - 2r + 1$

80. $2t - 4 + 8t + 9$

81. $x^2 - 2xy + 4 + xy$

82. $r^2 + 3rs - 6 - rs$

83. $5z - 5 + 10z + 2z + 16$

84. $7x - 4x + 8 + 3x - 6$

85. $z^3 + 2z^2 + z + z^2 + 2z + 1$

86. $3x^2 - x^2 + 4x + 2x^2 - x + x^2$

87. $2x^2y + 5xy^2 - 3x^2y + 4xy + 7xy^2$

88. $6rt - 3r^2t + 2rt^2 - 4rt - 2r^2t$

89. $3\left(\frac{1}{x}\right) - \frac{1}{x} + 8$

90. $1.2\left(\frac{1}{x}\right) + 3.8\left(\frac{1}{x}\right) - 4x$

91. $5\left(\frac{1}{t}\right) - 3t + 6\left(\frac{1}{t}\right) - 2t$

92. $16\left(\frac{a}{b}\right) - 6\left(\frac{a}{b}\right) + \frac{3}{2} - \frac{1}{2}$

True or False? In Exercises 93–98, determine whether the statement is true or false. Justify your answer.

93. $3(x - 4) \stackrel{?}{=} 3x - 4$

94. $-3(x - 4) \stackrel{?}{=} -3x - 12$

95. $6x - 4x \stackrel{?}{=} 2x$

96. $12y^2 + 3y^2 \stackrel{?}{=} 36y^2$

97. $2 - (x + 4) \stackrel{?}{=} -2x - 8$

98. $-(3 + 2y) - y \stackrel{?}{=} -3 - 3y$

Mental Math In Exercises 99–106, use the Distributive Property to perform the required arithmetic *mentally*. For example, suppose you work as a mechanic where the wage is \$14 per hour and time-and-one-half for overtime. So, your hourly wage for overtime is

$14(1.5) = 14\left(1 + \frac{1}{2}\right) = 14 + 7 = \$21.$

99. $8(52) = 8(50 + 2)$

100. $7(33) = 7(30 + 3)$

101. $9(48) = 9(50 - 2)$

102. $6(29) = 6(30 - 1)$

103. $-4(56) = -4(60 - 4)$

104. $-6(27) = -6(30 - 3)$

105. $5(7.02) = 5(7 + 0.02)$

106. $12(11.95) = 12(12 - 0.05)$

In Exercises 107–120, simplify the expression. ***See Examples 7 and 8.***

✓ **107.** $2(6x)$

108. $-7(5a)$

109. $-(4x)$

110. $-(5t)$

111. $(-2x)(-3x)$

112. $(-3y)(-4y)$

✓ **113.** $(-5z)(2z^2)$

114. $(10t)(-4t^2)$

115. $\frac{18a}{5} \cdot \frac{15}{6}$

116. $\frac{5x}{8} \cdot \frac{16}{5}$

117. $\left(-\frac{3x^2}{2}\right)\left(\frac{4x}{18}\right)$

118. $\left(\frac{4x}{3}\right)\left(\frac{3x}{16}\right)$

119. $(12xy^2)(-2x^3y^2)$

120. $(7r^2s^3)(3rs)$

In Exercises 121–140, simplify the expression by removing symbols of grouping and combining like terms. ***See Examples 9, 10, and 11.***

121. $2(x - 2) + 4$

122. $3(x - 5) - 2$

✓ **123.** $6(2s - 1) + s + 4$

124. $(2x - 1)2 + x + 9$

125. $m - 3(m - 7)$

126. $8l - (3l - 7)$

127. $-6(2 - 3x) + 10(5 - x)$

128. $3(r - 2s) - 5(3r - 5s)$

129. $\frac{2}{3}(12x + 15) + 16$

130. $\frac{3}{8}(4 - y) - \frac{5}{2} + 10$

131. $3 - 2[6 + (4 - x)]$

132. $10x + 5[6 - (2x + 3)]$

133. $7x(2 - x) - 4x$

134. $-6x(x - 1) + x^2$

135. $4x^2 + x(5 - x) - 3$

136. $-z(z - 2) + 3z^2 + 5$

137. $-3t(4 - t) + t(t + 1)$

138. $-2x(x - 1) + x(3x - 2)$

139. $3t[4 - (t - 3)] + t(t + 5)$

140. $4y[5 - (y + 1)] + 3y(y + 1)$

In Exercises 141–148, use the Distributive Property to simplify the expression. ***See Example 12.***

141. $\dfrac{2x}{3} - \dfrac{x}{3}$

142. $\dfrac{4y}{5} - \dfrac{2y}{5}$

143. $\dfrac{3z}{8} + \dfrac{7z}{8}$

144. $\dfrac{5t}{12} + \dfrac{7t}{12}$

145. $\dfrac{x}{3} - \dfrac{5x}{4}$

146. $\dfrac{5x}{7} + \dfrac{2x}{3}$

147. $\dfrac{3x}{10} - \dfrac{x}{15} + \dfrac{4x}{5}$

148. $\dfrac{3z}{4} - \dfrac{z}{2} - \dfrac{z}{3}$

Solving Problems

Geometry In Exercises 149 and 150, write an expression for the perimeter of the triangle shown in the figure. Use the properties of algebra to simplify the expression.

149.

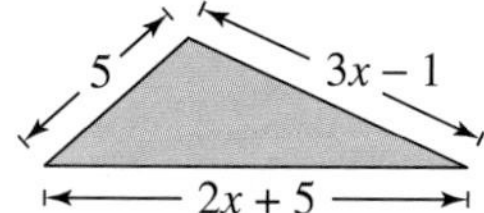

150.

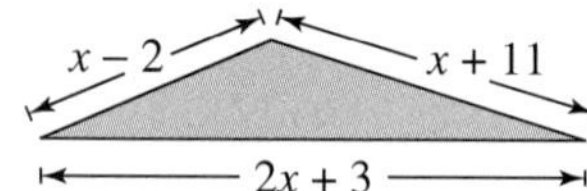

Geometry In Exercises 151 and 152, write and simplify expressions for (a) the perimeter and (b) the area of the rectangle. ***See Example 13.***

151.

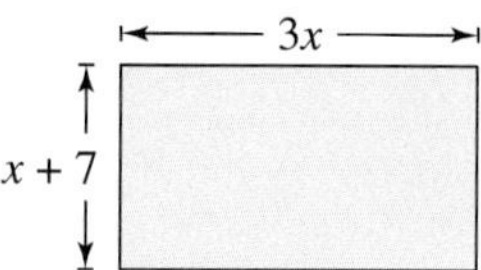

152.

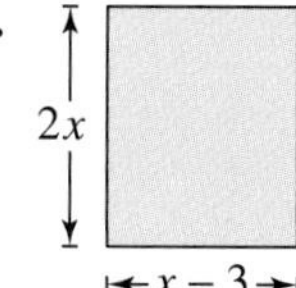

153. *Geometry* The area of a trapezoid with parallel bases of lengths b_1 and b_2 and height h (see figure) is $\frac{1}{2}h(b_1 + b_2)$.

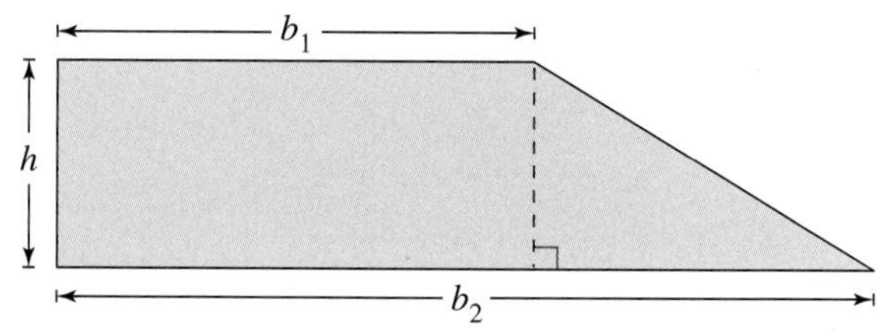

Figure for 153

(a) Show that the area can also be expressed as $b_1h + \frac{1}{2}(b_2 - b_1)h$, and give a geometric explanation for the area represented by each term in this expression.

(b) Find the area of a trapezoid with $b_1 = 7$, $b_2 = 12$, and $h = 3$.

154. *Geometry* The remaining area of a square with side length x after a smaller square with side length y has been removed (see figure) is $(x + y)(x - y)$.

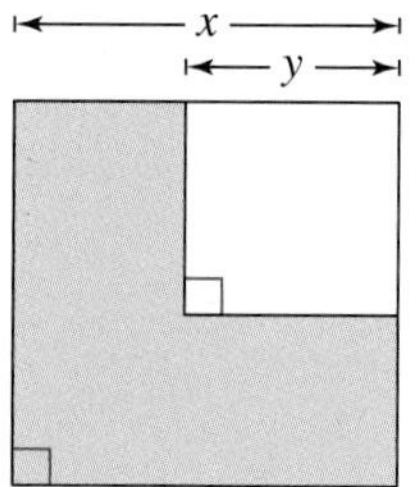

(a) Show that the remaining area can also be expressed as $x(x - y) + y(x - y)$, and give a geometric explanation for the area represented by each term in this expression.

(b) Find the remaining area of a square with side length 9 after a square with side length 5 has been removed.

Geometry In Exercises 155 and 156, use the formula for the area of a trapezoid, $\frac{1}{2}h(b_1 + b_2)$, to write an expression for the area of the trapezoidal house lot and park.

155.

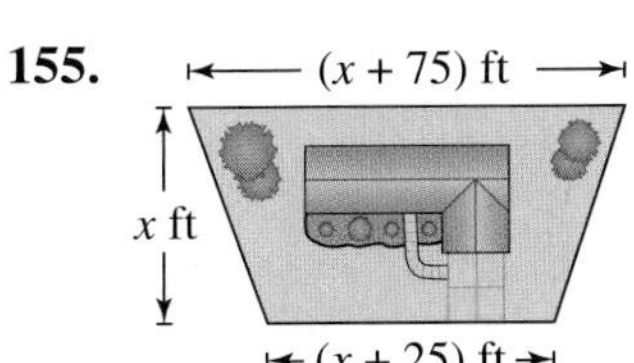

156.

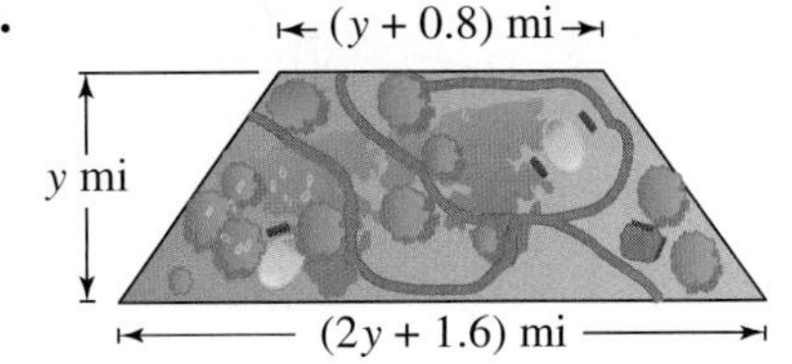

Explaining Concepts

157. Discuss the difference between $(6x)^4$ and $6x^4$.

In Exercises 158 and 159, explain why the two expressions are not like terms.

158. $\frac{1}{2}x^2y, \frac{5}{2}xy^2$

159. $-16x^2y^3, 7x^2y$

160. *Error Analysis* Describe and correct the error.

$$\frac{x}{3} + \frac{4x}{3} = \frac{5x}{6}$$

161. *Error Analysis* Describe and correct the error.

$$\begin{aligned} 4x - 3(x - 1) &= 4x - 3(x) - 3(1) \\ &= 4x - 3x - 3 \\ &= x - 3 \end{aligned}$$

162. Does the expression $[x - (3 \cdot 4)] \div 5$ change if the parentheses are removed? Does it change if the brackets are removed? Explain.

Cumulative Review

In Exercises 163–168, evaluate the expression.

163. $0 - (-12)$

164. $5 - 4 \div 2 + 6$

165. $-12 - 2 + |-3|$

166. $6 + 3(4 + 2)$

167. $\frac{5}{16} - \frac{3}{10}$

168. $\frac{9}{16} + 2\frac{3}{12}$

In Exercises 169–172, evaluate the algebraic expression for the given values of the variable(s).

169. $3x - 2$

(a) $x = 2$

(b) $x = -1$

170. $2x^2 + 3$

(a) $x = 3$

(b) $x = -4$

171. $2y - x$

(a) $x = 1, y = 5$

(b) $x = -6, y = 3$

172. $x + 3(y - 2x)$

(a) $x = 2, y = -3$

(b) $x = -6, y = 4$

Mid-Chapter Quiz

Take this quiz as you would take a quiz in class. After you are done, check your work against the answers in the back of the book.

In Exercises 1 and 2, evaluate the algebraic expression for the given values of the variable(s). If it is not possible, state the reason.

1. $x^2 - 3x$ (a) $x = 3$ (b) $x = -2$

2. $\dfrac{x}{y-3}$ (a) $x = 5, y = 3$ (b) $x = 0, y = -1$

In Exercises 3 and 4, identify the terms of the expression and their coefficients.

3. $4x^2 - 2x$

4. $5x + 3y - z$

In Exercises 5 and 6, rewrite the expression in exponential form.

5. $(-3y)(-3y)(-3y)(-3y)$

6. $2 \cdot (x - 3) \cdot (x - 3) \cdot 2 \cdot 2$

In Exercises 7–10, identify the property of algebra illustrated by the statement.

7. $-3(2y) = (-3 \cdot 2)y$

8. $(x + 2)y = xy + 2y$

9. $3y \cdot \dfrac{1}{3y} = 1, \quad y \neq 0$

10. $x - x^2 + 2 = -x^2 + x + 2$

In Exercises 11 and 12, use the Distributive Property to expand the expression.

11. $2x(3x - 1)$

12. $-6(2y + 3y^2 - 6)$

In Exercises 13–20, simplify the expression.

13. $-4(-5y^2)$

14. $\dfrac{x}{3}\left(-\dfrac{3x}{5}\right)$

15. $(-3y)^2y^3$

16. $\dfrac{2z^2}{3y} \cdot \dfrac{5z}{7}$

17. $y^2 - 3xy + y + 7xy$

18. $10\left(\dfrac{1}{u}\right) - 7\left(\dfrac{1}{u}\right) + 3u$

19. $5(a - 2b) + 3(a + b)$

20. $4x + 3[2 - 4(x + 6)]$

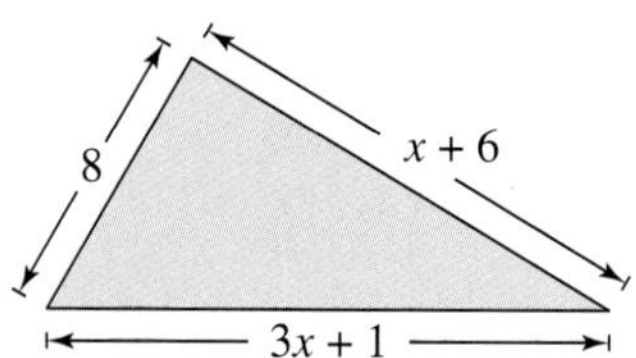

Figure for 21

21. Write and simplify an expression for the perimeter of the triangle (see figure).

22. Your teacher divides your class of x students into 6 teams.

(a) Write an expression representing the number of students on each team.

(b) There are 30 students in your class. How many students are on each team?

2.3 Algebra and Problem Solving

© dmac/Alamy

What You Should Learn

1 ▶ Define algebra as a problem-solving language.

2 ▶ Construct verbal mathematical models from written statements.

3 ▶ Translate verbal phrases into algebraic expressions.

4 ▶ Identify hidden operations when constructing algebraic expressions.

5 ▶ Use problem-solving strategies to solve application problems.

Why You Should Learn It

Translating verbal phrases into algebraic expressions enables you to solve real-life problems. For instance, in Exercise 59 on page 102, you will find an expression for the total camping fee at a campground.

1 ▶ Define algebra as a problem-solving language.

What Is Algebra?

Algebra is a problem-solving language that is used to solve real-life problems. It has four basic components, which tend to nest within each other, as indicated in Figure 2.3.

1. Symbolic representations and applications of the rules of arithmetic
2. Rewriting (reducing, simplifying, factoring) algebraic expressions into equivalent forms
3. Creating and solving equations
4. Studying relationships among variables by the use of functions and graphs

1. Rules of arithmetic

2. Algebraic expressions: rewriting into equivalent forms

3. Algebraic equations: creating and solving

4. Functions and graphs: relationships among variables

Figure 2.3

Study Tip

As you study this text, it is helpful to view algebra from the "big picture," as shown in Figure 2.3. The ability to write algebraic expressions and equations is needed in the major components of algebra—*simplifying* expressions, *solving* equations, and *graphing* functions.

Notice that one of the components deals with expressions and another deals with equations. As you study algebra, it is important to understand the difference between simplifying or rewriting an algebraic *expression,* and solving an algebraic *equation.* In general, remember that a mathematical expression *has no equal sign,* whereas a mathematical equation *must have an equal sign.*

When you use an equal sign to *rewrite* an expression, you are merely indicating the *equivalence* of the new expression and the previous one.

Original Expression	*equals*	*Equivalent Expression*
$(a + b)c$	$=$	$ac + bc$

2 ▶ Construct verbal mathematical models from written statements.

Constructing Verbal Models

In the first two sections of this chapter, you studied techniques for rewriting and simplifying algebraic expressions. In this section you will study ways to construct algebraic expressions from written statements by first constructing a **verbal mathematical model.**

Take another look at Example 1 in Section 2.1 (page 68). In that example, you are paid \$9 per hour and your weekly pay can be represented by the verbal model

Pay per hour · Number of hours = 9 dollars · x hours = $9x$.

Note the hidden operation of multiplication in this expression. Nowhere in the verbal problem does it say you are to multiply 9 times x. It is *implied* in the problem. This is often the case when algebra is used to solve real-life problems.

Telegraph Colour Library/FPG International

In 2005, about 690,000 tons of aluminum containers and packaging were recycled. This accounted for about 36.3% of all aluminum containers and packaging produced. (Source: U.S. Environmental Protection Agency)

EXAMPLE 1 Constructing an Algebraic Expression

You are paid 5 cents for each aluminum soda can and 3 cents for each plastic soda bottle you collect. Write an algebraic expression that represents the total weekly income for this recycling activity.

Solution

Before writing an algebraic expression for the weekly income, it is helpful to construct an informal verbal model. For instance, the following verbal model could be used.

Pay per can · Number of cans + Pay per bottle · Number of bottles

Note that the word *and* in the problem indicates addition. Because both the number of cans and the number of bottles can vary from week to week, you can use the two variables c and b, respectively, to write the following algebraic expression.

5 cents · c cans + 3 cents · b bottles = $5c + 3b$

✓ **CHECKPOINT** *Now try Exercise 53.*

In Example 1, notice that c is used to represent the number of *cans* and b is used to represent the number of *bottles.* When writing algebraic expressions, choose variables that can be identified with the unknown quantities.

The number of one kind of item can sometimes be expressed in terms of the number of another kind of item. Suppose the number of cans in Example 1 was said to be "three times the number of bottles." In this case, only one variable would be needed and the model could be written as

5 cents · $3 \cdot b$ cans + 3 cents · b bottles $= 5(3b) + 3b$

$$= 15b + 3b$$

$$= 18b.$$

3 ▶ Translate verbal phrases into algebraic expressions.

Translating Phrases

When translating verbal phrases into algebraic expressions, it is helpful to watch for key words and phrases that indicate the four different operations of arithmetic. The following list shows several examples.

Translating Phrases into Algebraic Expressions

Key Words and Phrases	*Verbal Description*	*Expression*
Addition:		
Sum, plus, greater than, increased by, more than,	The sum of 6 and x	$6 + x$
exceeds, total of	Eight more than y	$y + 8$
Subtraction:		
Difference, minus, less than, decreased by,	Five decreased by a	$5 - a$
subtracted from, reduced by	Four less than z	$z - 4$
Multiplication:		
Product, multiplied by, twice, times, percent of	Seven times x	$7x$
Division:		
Quotient, divided by, ratio, per	The ratio of x and 3	$\frac{x}{3}$

Study Tip

Order is important when writing subtraction and division expressions. For instance, *three less than m* means $m - 3$, not $3 - m$, and *the quotient of n* and 7 means $\frac{n}{7}$, not $\frac{7}{n}$.

EXAMPLE 2 Translating Phrases Having Specified Variables

Translate each phrase into an algebraic expression.

a. Three less than m **b.** y decreased by 10

c. The product of 5 and x **d.** The quotient of n and 7

Solution

a. Three less than m

$m - 3$ "Less than" indicates subtraction.

b. y decreased by 10

$y - 10$ "Decreased by" indicates subtraction.

c. The product of 5 and x

$5x$ "Product" indicates multiplication.

d. The quotient of n and 7

$\frac{n}{7}$ "Quotient" indicates division.

✓ **CHECKPOINT** *Now try Exercise 1.*

EXAMPLE 3 Translating Phrases Having Specified Variables

Translate each phrase into an algebraic expression.

a. Six times the sum of x and 7

b. The product of 4 and x, all divided by 3

c. k decreased by the product of 8 and m

Solution

a. Six times the sum of x and 7

$6(x + 7)$ Think: 6 multiplied by what?

b. The product of 4 and x, all divided by 3

$\frac{4x}{3}$ Think: What is divided by 3?

c. k decreased by the product of 8 and m

$k - 8m$ Think: What is subtracted from k?

CHECKPOINT *Now try Exercise 3.*

In most applications of algebra, the variables are not specified and it is your task to assign variables to the *appropriate* quantities. Although similar to the translations in Examples 2 and 3, the translations in the next example may seem more difficult because variables have not been assigned to the unknown quantities.

EXAMPLE 4 Translating Phrases Having No Specified Variables

Study Tip

Any variable, such as b, k, n, r, or x, can be chosen to represent an unspecified number. The choice is a matter of preference. In Example 4, x was chosen as the variable.

Translate each phrase into an algebraic expression.

a. The sum of 3 and a number

b. Five decreased by the product of 3 and a number

c. The difference of a number and 3, all divided by 12

Solution

In each case, let x be the unspecified number.

a. The sum of 3 and a number

$3 + x$ Think: 3 added to what?

b. Five decreased by the product of 3 and a number

$5 - 3x$ Think: What is subtracted from 5?

c. The difference of a number and 3, all divided by 12

$\frac{x - 3}{12}$ Think: What is divided by 12?

CHECKPOINT *Now try Exercise 23.*

A good way to learn algebra is to do it *forward* and *backward.* In the next example, algebraic expressions are translated into verbal phrases. Keep in mind that other key words could be used to describe the operation(s) in each expression. Your goal is to use key words or phrases that keep the verbal descriptions clear and concise.

EXAMPLE 5 Translating Algebraic Expressions into Verbal Phrases

Without using a variable, write a verbal description for each expression.

a. $x - 12$ **b.** $7(x + 12)$ **c.** $5 + \frac{x}{2}$ **d.** $\frac{5 + x}{2}$ **e.** $(3x)^2$

Solution

a. *Algebraic expression:* $x - 12$

Operation: Subtraction

Key Phrase: Less than

Verbal description: Twelve less than a number

b. *Algebraic expression:* $7(x + 12)$

Operations: Multiplication, addition

Key Words: Times, sum

Verbal description: Seven times the sum of a number and 12

c. *Algebraic expression:* $5 + \frac{x}{2}$

Operations: Addition, division

Key Words: Plus, quotient

Verbal description: Five plus the quotient of a number and 2

d. *Algebraic expression:* $\frac{5 + x}{2}$

Operations: Addition, division

Key Words: Sum, divided by

Verbal description: The sum of 5 and a number, all divided by 2

e. *Algebraic expression:* $(3x)^2$

Operations: Raise to a power, multiplication

Key Words: Square, product

Verbal description: The square of the product of 3 and x

✓ ***CHECKPOINT*** ***Now try Exercise 33.***

Translating algebraic expressions into verbal phrases is more difficult than it may appear. It is easy to write a phrase that is ambiguous. For instance, what does the phrase "the sum of 5 and a number times 2" mean? Without further information, this phrase could mean

$$5 + 2x \quad \text{or} \quad 2(5 + x).$$

4 ▶ Identify hidden operations when constructing algebraic expressions.

Verbal Models with Hidden Operations

Most real-life problems do not contain verbal expressions that clearly identify all the arithmetic operations involved. You need to rely on past experience and the physical nature of the problem in order to identify the operations hidden in the problem statement. Multiplication is the operation most commonly hidden in real-life applications. Watch for *hidden operations* in the next two examples.

EXAMPLE 6 Discovering Hidden Operations

a. A cash register contains n nickels and d dimes. Write an expression for this amount of money in cents.

b. A person riding a bicycle travels at a constant rate of 12 miles per hour. Write an expression showing how far the person can ride in t hours.

c. A person paid x dollars plus 6% sales tax for an automobile. Write an expression for the total cost of the automobile.

Solution

a. The amount of money is a sum of products.

Verbal Model: [Value of nickel] · [Number of nickels] + [Value of dime] · [Number of dimes]

Labels: Value of nickel = 5 (cents)
Number of nickels = n (nickels)
Value of dime = 10 (cents)
Number of dimes = d (dimes)

Expression: $5n + 10d$ (cents)

b. The distance traveled is a product.

Verbal Model: [Rate of travel] · [Time traveled]

Labels: Rate of travel = 12 (miles per hour)
Time traveled = t (hours)

Expression: $12t$ (miles)

c. The total cost is a sum.

Verbal Model: [Cost of automobile] + [Percent of sales tax] · [Cost of automobile]

Labels: Percent of sales tax = 0.06 (decimal form)
Cost of automobile = x (dollars)

Expression: $x + 0.06x = (1 + 0.06)x$
$= 1.06x$

✓ ***CHECKPOINT*** ***Now try Exercise 55.***

Study Tip

In Example 6(b), the final answer is listed in terms of miles. This unit is found as shown below.

$$12\frac{\text{miles}}{\cancel{\text{hours}}} \cdot t\ \cancel{\text{hours}}$$

Note that the hours "divide out," leaving miles as the unit of measure. This technique is called *unit analysis* and can be very helpful in determining the final unit of measure.

Notice in part (c) of Example 6 that the equal signs are used to denote the equivalence of the three expressions. It is not an equation to be solved.

5 ▶ Use problem-solving strategies to solve application problems.

Additional Problem-Solving Strategies

In addition to constructing verbal models, there are other problem-solving strategies that can help you succeed in this course.

Summary of Additional Problem-Solving Strategies

1. **Guess, Check, and Revise** Guess a reasonable solution based on the given data. Check the guess, and revise it, if necessary. Continue guessing, checking, and revising until a correct solution is found.
2. **Make a Table/Look for a Pattern** Make a table using the data in the problem. Look for a number pattern. Then use the pattern to complete the table or find a solution.
3. **Draw a Diagram** Draw a diagram that shows the facts of the problem. Use the diagram to visualize the action of the problem. Use algebra to find a solution. Then check the solution against the facts.
4. **Solve a Simpler Problem** Construct a simpler problem that is similar to the original problem. Solve the simpler problem. Then use the same procedure to solve the original problem.

Study Tip

The most common errors made by students when solving algebraic problems are arithmetic errors. Be sure to check your arithmetic when solving algebraic problems.

EXAMPLE 7 Guess, Check, and Revise

You deposit \$500 in an account that earns 6% interest compounded annually. The balance A in the account after t years is $A = 500(1 + 0.06)^t$. How long will it take for your investment to double?

Solution

You can solve this problem using a guess, check, and revise strategy. For instance, you might guess that it will take 10 years for your investment to double. The balance after 10 years is

$$A = 500(1 + 0.06)^{10} \approx \$895.42.$$

Because the amount has not yet doubled, you increase your guess to 15 years.

$$A = 500(1 + 0.06)^{15} \approx \$1198.28$$

Because this amount is greater than double the investment, your next guess should be a number between 10 and 15. After trying several more numbers, you can determine that your balance will double in about 11.9 years.

✓ **CHECKPOINT** *Now try Exercise 61.*

Another strategy that works well for a problem such as Example 7 is to make a table of data values. You can use a calculator to create the following table.

t	2	4	6	8	10	12
A	561.80	631.24	709.26	796.92	895.42	1006.10

EXAMPLE 8 Make a Table/Look for a Pattern

Find each product. Then describe the pattern and use your description to find the product of 14 and 16.

$$1 \cdot 3,\ 2 \cdot 4,\ 3 \cdot 5,\ 4 \cdot 6,\ 5 \cdot 7,\ 6 \cdot 8,\ 7 \cdot 9$$

Solution

One way to help find a pattern is to organize the results in a table.

Numbers	$1 \cdot 3$	$2 \cdot 4$	$3 \cdot 5$	$4 \cdot 6$	$5 \cdot 7$	$6 \cdot 8$	$7 \cdot 9$
Product	3	8	15	24	35	48	63

From the table, you can see that each of the products is 1 less than the square of the mean of the numbers. For instance, 3 is 1 less than 2^2 or 4, 8 is 1 less than 3^2 or 9, 15 is 1 less than 4^2 or 16, and so on.

If this pattern continues for other numbers, you can hypothesize that the product of 14 and 16 is 1 less than 15^2 or 225. That is,

$$14 \cdot 16 = 15^2 - 1$$
$$= 224.$$

You can confirm this result by actually multiplying 14 and 16.

CHECKPOINT *Now try Exercise 65.*

EXAMPLE 9 Draw a Diagram 

The outer dimensions of a rectangular apartment are 25 feet by 40 feet. The combination living room, dining room, and kitchen areas occupy two-fifths of the apartment's area. Find the total area of the remaining rooms.

Solution

For this problem, it helps to draw a diagram, as shown in Figure 2.4. From the figure, you can see that the total area of the apartment is

$$\text{Area} = (\text{Length})(\text{Width})$$
$$= (40)(25)$$
$$= 1000 \text{ square feet.}$$

The area occupied by the living room, dining room, and kitchen is

$$\frac{2}{5}(1000) = 400 \text{ square feet.}$$

This implies that the remaining rooms must have a total area of

$$1000 - 400 = 600 \text{ square feet.}$$

CHECKPOINT *Now try Exercise 69.*

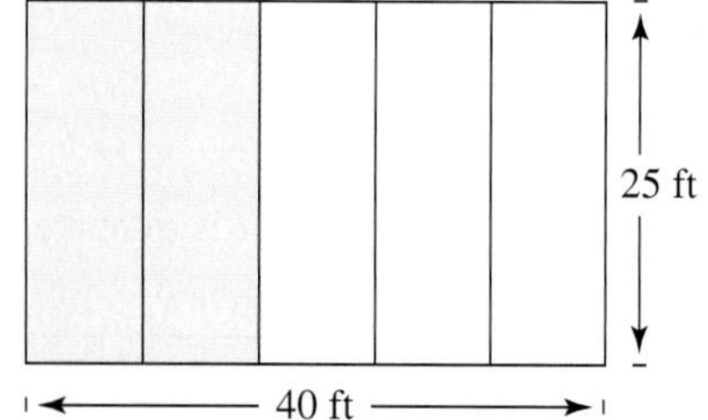

Figure 2.4

EXAMPLE 10 Solve a Simpler Problem 

You are driving on an interstate highway at an average speed of 60 miles per hour. How far will you travel in $12\frac{1}{2}$ hours?

Solution

One way to solve this problem is to use the formula that relates distance, rate, and time. Suppose, however, that you have forgotten the formula. To help you remember, you could solve some simpler problems.

- If you travel 60 miles per hour for 1 hour, you will travel 60 miles.
- If you travel 60 miles per hour for 2 hours, you will travel 120 miles.
- If you travel 60 miles per hour for 3 hours, you will travel 180 miles.

From these examples, it appears that you can find the total miles traveled by multiplying the rate by the time. So, if you travel 60 miles per hour for $12\frac{1}{2}$ hours, you will travel a distance of

$$(60)(12.5) = 750 \text{ miles.}$$

 **CHECKPOINT** *Now try Exercise 71.*

Hidden operations are often involved when variable names (labels) are assigned to unknown quantities. A good strategy is to use a *specific* case to help you write a model for the *general* case. For instance, a specific case of finding three consecutive integers

$$3,\ 3 + 1, \text{ and } 3 + 2$$

may help you write a general case for finding three consecutive integers n, $n + 1$, and $n + 2$. This strategy is illustrated in Examples 11 and 12.

EXAMPLE 11 Using a Specific Case to Find a General Case 

In each of the following, use the variable to label the unknown quantity.

a. A person's weekly salary is d dollars. What is the annual salary?

b. A person's annual salary is y dollars. What is the monthly salary?

Solution

a. There are 52 weeks in a year.

Specific case: If the weekly salary is \$500, then the annual salary (in dollars) is $52 \cdot 500$.

General case: If the weekly salary is d dollars, then the annual salary (in dollars) is $52 \cdot d$ or $52d$.

b. There are 12 months in a year.

Specific case: If the annual salary is \$34,000, then the monthly salary (in dollars) is $34{,}000 \div 12$.

General case: If the annual salary is y dollars, then the monthly salary (in dollars) is $y \div 12$ or $y/12$.

CHECKPOINT *Now try Exercise 75.*

EXAMPLE 12 Using a Specific Case to Find a General Case

In each of the following, use the variable to label the unknown quantity.

a. You are k inches shorter than a friend. You are 60 inches tall. How tall is your friend?

b. A consumer buys g gallons of gasoline for a total of d dollars. What is the price per gallon?

c. A person has driven on the highway at an average speed of 60 miles per hour for t hours. How far has the person traveled?

Solution

a. You are k inches shorter than a friend.

Specific case: If you are 10 inches shorter than your friend, then your friend is $60 + 10$ inches tall.

General case: If you are k inches shorter than your friend, then your friend is $60 + k$ inches tall.

b. To obtain the price per gallon, divide the price by the number of gallons.

Specific case: If the total price is \$26.50 and the total number of gallons is 10, then the price per gallon is $26.50 \div 10$ dollars per gallon.

General case: If the total price is d dollars and the total number of gallons is g, then the price per gallon is $d \div g$ or d/g dollars per gallon.

c. To obtain the distance driven, multiply the speed by the number of hours.

Specific case: If the person has driven for 2 hours at a speed of 60 miles per hour, then the person has traveled $60 \cdot 2$ miles.

General case: If the person has driven for t hours at a speed of 60 miles per hour, then the person has traveled $60t$ miles.

✓ **CHECKPOINT** ***Now try Exercise 77.***

Most of the verbal problems you encounter in a mathematics text have precisely the right amount of information necessary to solve the problem. In real life, however, you may need to collect additional information, as shown in Example 13.

EXAMPLE 13 Enough Information?

Decide what additional information is needed to solve the following problem.

During a given week, a person worked 48 hours for the same employer. The hourly rate for overtime is \$14. Write an expression for the person's gross pay for the week, including any pay received for overtime.

Solution

To solve this problem, you would need to know how much the person is normally paid per hour. You would also need to be sure that the person normally works 40 hours per week and that overtime is paid on time worked beyond 40 hours.

✓ **CHECKPOINT** ***Now try Exercise 79.***

Concept Check

1. The word *difference* indicates what operation?
2. Two unknown quantities in a verbal model are "Number of cherries" and "Number of strawberries." What variables would you use to represent these quantities? Explain.
3. What is a *hidden operation* in a verbal phrase? Explain how to identify hidden operations.
4. Explain how to use a guess, check, and revise problem-solving strategy.

2.3 EXERCISES

Go to pages 116–117 to record your assignments.

Developing Skills

In Exercises 1–6, match the verbal phrase with the correct algebraic expression. ***See Examples 2 and 3.***

(a) $x - 12$ (b) $3x$
(c) $12 - x$ (d) $12 + x$
(e) $3(x - 12)$ (f) $3x - 12$

✓ **1.** Twelve increased by x
2. Twelve reduced by x
✓ **3.** Three times the difference of x and 12
4. Twelve less than the product of 3 and x
5. The product of 3 and x
6. The difference of x and 12

In Exercises 7–30, translate the phrase into an algebraic expression. Let x represent the real number. ***See Example 4.***

7. A number increased by 5
8. 17 more than a number
9. A number decreased by 25
10. A number decreased by 7
11. Six less than a number
12. Ten more than a number
13. Twice a number
14. The product of 30 and a number
15. A number divided by 3
16. A number divided by 100
17. The ratio of a number and 50
18. One-half of a number
19. Three-tenths of a number
20. Twenty-five hundredths of a number
21. Five less than triple a number
22. Eight more than 5 times a number
✓ **23.** Three times the difference of a number and 5
24. Ten times the sum of a number and 4
25. Fifteen more than the quotient of a number and 5
26. Seventeen less than 4 times a number
27. The absolute value of the sum of a number and 4
28. The absolute value of 4 less than twice a number
29. The square of a number, increased by 1
30. Twice the square of a number, increased by 4

In Exercises 31–44, write a verbal description of the algebraic expression, without using a variable. (There is more than one correct answer.) ***See Example 5.***

31. $x - 10$
32. $x + 9$
✓ **33.** $3x + 2$
34. $4 - 7x$
35. $\frac{1}{2}x - 6$

36. $9 - \frac{1}{4}x$

37. $3(2 - x)$

38. $-10(t - 6)$

39. $\frac{t + 1}{2}$

40. $\frac{y - 3}{4}$

41. $\frac{1}{2} - \frac{t}{5}$

42. $\frac{1}{4} + \frac{x}{8}$

43. $x^2 + 5$

44. $x^3 - 1$

In Exercises 45–52, translate the phrase into an algebraic expression. Simplify the expression.

45. x times the sum of x and 3

46. n times the difference of 6 and n

47. x minus the sum of 25 and x

48. The sum of 4 and x added to the sum of x and -8

49. The square of x decreased by the product of x and $2x$

50. The square of x added to the product of x and $x + 1$

51. Eight times the sum of x and 24, all divided by 2

52. Four times the difference of x and 15, all divided by 2

Solving Problems

✓ **53.** *Money* A cash register contains d dimes. Write an algebraic expression that represents the total amount of money (in dollars).

54. *Money* A cash register contains d dimes and q quarters. Write an algebraic expression that represents the total amount of money (in dollars).

✓ **55.** *Sales Tax* The sales tax on a purchase of L dollars is 6%. Write an algebraic expression that represents the total amount of sales tax. (*Hint:* Use the decimal form of 6%.)

56. *Income Tax* The state income tax on a gross income of I dollars in Pennsylvania is 3.07%. Write an algebraic expression that represents the total amount of income tax. (*Hint:* Use the decimal form of 3.07%.)

57. *Travel Time* A truck travels 100 miles at an average speed of r miles per hour. Write an algebraic expression that represents the total travel time.

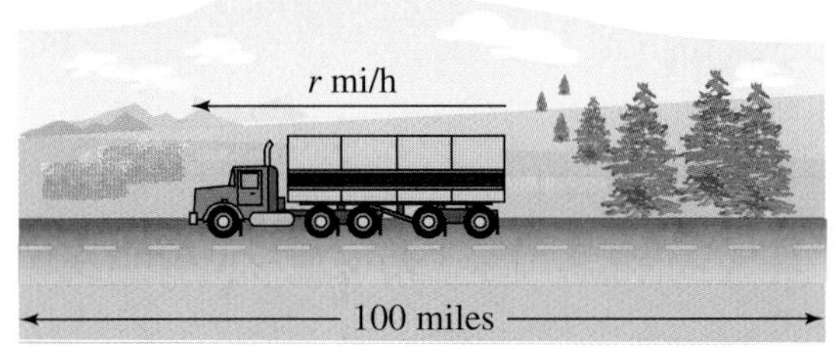

58. *Distance* An airplane travels at a rate of r miles per hour for 3 hours. Write an algebraic expression that represents the total distance traveled by the airplane.

59. *Consumerism* A campground charges \$15 for adults and \$2 for children. Write an algebraic expression that represents the total camping fee for m adults and n children.

60. *Hourly Wage* The hourly wage for an employee is \$12.50 per hour plus 75 cents for each of the q units produced during the hour. Write an algebraic expression that represents the total hourly earnings for the employee.

Guess, Check, and Revise In Exercises 61–64, an expression for the balance in an account is given. Use a guess, check, and revise strategy to determine the time (in years) necessary for the investment of \$1000 to double. ***See Example 7.***

✓ **61.** Interest rate: 7%
$1000(1 + 0.07)^t$

62. Interest rate: 5%
$1000(1 + 0.05)^t$

63. Interest rate: 6%
$1000(1 + 0.06)^t$

64. Interest rate: 8%
$1000(1 + 0.08)^t$

Finding a Pattern In Exercises 65 and 66, describe the pattern and use your description to find the value of the expression when $n = 20$. **See Example 8.**

65.

n	0	1	2	3	4	5
Value of expression	−1	1	3	5	7	9

66.

n	0	1	2	3	4	5
Value of expression	5	12	19	26	33	40

Exploration In Exercises 67 and 68, find values for a and b such that the expression $an + b$ yields the values in the table.

67.

n	0	1	2	3	4	5
$an + b$	4	9	14	19	24	29

68.

n	0	1	2	3	4	5
$an + b$	1	5	9	13	17	21

Drawing a Diagram In Exercises 69 and 70, draw figures satisfying the specified conditions. **See Example 9.**

69. The sides of a square have length a centimeters. Draw the square. Draw the rectangle obtained by extending two parallel sides of the square 6 centimeters. Find expressions for the perimeter and area of each figure.

70. The dimensions of a rectangular lawn are 150 feet by 250 feet. The property owner buys a rectangular strip x feet wide along one 250-foot side of the lawn. Draw diagrams representing the lawn before and after the purchase. Write an expression for the area of each.

Solving a Simpler Problem In Exercises 71 and 72, solve the problem. **See Example 10.**

71. A bubble rises through water at a rate of about 1.15 feet per second. How far will the bubble rise in 5 seconds?

72. A train travels at an average speed of 50 miles per hour. How long will it take the train to travel 350 miles?

Geometry In Exercises 73 and 74, use simpler shapes to write an algebraic expression that represents the area of the trapezoid.

73.

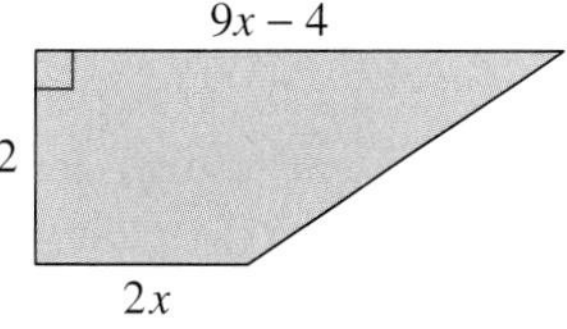

74.

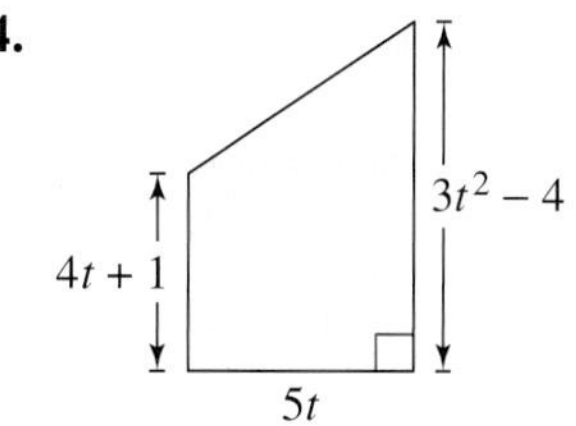

Finding a General Case In Exercises 75–78, use the variable to label the unknown quantity. **See Examples 11 and 12.**

75. A person's monthly cost for satellite television service is d dollars. What is the annual cost?

76. You buy t tickets to a baseball game for 9 dollars per ticket. What is the total cost?

77. The height of a rectangular picture frame is 1.5 times the width w. What is the perimeter of the picture frame?

78. A square dance floor has a side length of s. What is the area of the square dance floor?

In Exercises 79–82, decide what additional information is needed to solve the problem. (Do not solve the problem.) ***See Example 13.***

79. *Distance* A family taking a Sunday drive through the country travels at an average speed of 45 miles per hour. How far have they traveled by 3:00 P.M.?

80. *Painting* You paint a rectangular room that is twice as long as it is wide. One gallon of paint covers 100 square feet. How much money do you spend on paint?

81. *Consumerism* You want to buy a high-definition plasma television that costs \$1250. So, you put half of your weekly earnings into your savings account to save up for it. How many hours will you have to work in order to be able to buy the television?

82. *Consumer Awareness* You purchase an MP3 player during a sale at an electronics store. The MP3 player is discounted by 15%. What is the sale price of the player?

© Ruaridh Stewart/ZUMA/CORBIS

The first MP3 player was sold in Korea in 1998. They now come in different shapes, styles, and storage capacities.

Explaining Concepts

83. Determine which phrase(s) is (are) equivalent to the expression $n + 4$.

(a) 4 more than n
(b) the sum of n and 4
(c) n less than 4
(d) the ratio of n to 4
(e) the total of 4 and n

84. Determine whether order is important when translating each phrase into an algebraic expression. Explain.

(a) x increased by 10
(b) 10 decreased by x
(c) The product of x and 10
(d) The quotient of x and 10

85. Give two interpretations of "the quotient of 5 and a number times 3." Explain why $\frac{3n}{5}$ is not a possible interpretation.

86. Give two interpretations of "the difference of 6 and a number divided by 3." Explain why $\frac{n-6}{3}$ is not a possible interpretation.

Cumulative Review

In Exercises 87–92, evaluate the expression.

87. $(-6)(-13)$

88. $|4(-6)(5)|$

89. $\left(-\frac{4}{3}\right)\left(-\frac{9}{16}\right)$

90. $\frac{7}{8} \div \frac{3}{16}$

91. $\left|-\frac{5}{9}\right| + 2$

92. $-7\frac{3}{5} - 3\frac{1}{2}$

In Exercises 93–96, identify the property of algebra illustrated by the statement.

93. $2a + b = b + 2a$

94. $-4x(1) = -4x$

95. $2(c - d) = 2c - 2d$

96. $-3y^3 + 3y^3 = 0$

2.4 Introduction to Equations

Byron Aughenbaugh/Getty Images

What You Should Learn

1 ▶ Distinguish between an algebraic expression and an algebraic equation.
2 ▶ Check whether a given value is a solution of an equation.
3 ▶ Use properties of equality to solve equations.
4 ▶ Use a verbal model to construct an algebraic equation.

Why You Should Learn It

You can use verbal models to write algebraic equations that model real-life situations. For instance, in Exercise 75 on page 114, you will write an equation for the distance from a lightning strike to an observer based on how long after the strike the observer hears the thunder.

1 ▶ Distinguish between an algebraic expression and an algebraic equation.

Equations

An **equation** is a statement that two algebraic expressions are equal. For example,

$$x = 3, \quad 5x - 2 = 8, \quad \frac{x}{4} = 7, \quad \text{and} \quad x^2 - 9 = 0$$

are equations. To **solve** an equation involving the variable x means to find all values of x for which the equation is true. Such values are called **solutions.** For instance, $x = 2$ is a solution of the equation

$$5x - 2 = 8$$

because

$$5(2) - 2 = 8$$

is a true statement. The solutions of an equation are said to **satisfy** the equation.

Be sure that you understand the distinction between an algebraic expression and an algebraic equation. The differences are summarized in the following table.

Algebraic Expression	Algebraic Equation
• Example: $4(x - 1)$ • Contains *no* equal sign • Can be evaluated for any real number for which the expression is defined • Can sometimes be simplified to an equivalent form: $4(x - 1)$ simplifies to $4x - 4$	• Example: $4(x - 1) = 12$ • Contains an equal sign and is true for only certain values of the variable • Solution is found by forming equivalent equations using the properties of equality: $4(x - 1) = 12$ $4x - 4 = 12$ $4x = 16$ $x = 4$

2 ▶ Check whether a given value is a solution of an equation.

Checking Solutions of Equations

To **check** whether a given value is a solution of an equation, substitute the value into the original equation. If the substitution results in a true statement, then the value is a solution of the equation. If the substitution results in a false statement, then the value is not a solution of the equation. This process is illustrated in Examples 1 and 2.

EXAMPLE 1 Checking a Solution of an Equation

Determine whether $x = -2$ is a solution of $x^2 - 5 = 4x + 7$.

Solution

$$\begin{aligned} x^2 - 5 &= 4x + 7 && \text{Write original equation.} \\ (-2)^2 - 5 &\stackrel{?}{=} 4(-2) + 7 && \text{Substitute } -2 \text{ for } x. \\ 4 - 5 &\stackrel{?}{=} -8 + 7 && \text{Simplify.} \\ -1 &= -1 && \text{Solution checks. } \checkmark \end{aligned}$$

Because the substitution results in a true statement, you can conclude that $x = -2$ is a solution of the original equation.

✓ **CHECKPOINT** *Now try Exercise 1.*

Study Tip

When checking a solution, you should write a question mark over the equal sign to indicate that you are not sure of the validity of the equation.

The fact that you have found one solution of an equation does not mean that you have found all of the solutions. For instance, you can check that $x = 6$ is also a solution of the equation in Example 1, as follows.

$$\begin{aligned} x^2 - 5 &= 4x + 7 && \text{Write original equation.} \\ (6)^2 - 5 &\stackrel{?}{=} 4(6) + 7 && \text{Substitute 6 for } x. \\ 36 - 5 &\stackrel{?}{=} 24 + 7 && \text{Simplify.} \\ 31 &= 31 && \text{Solution checks. } \checkmark \end{aligned}$$

EXAMPLE 2 A Trial Solution That Does Not Check

Determine whether $x = 2$ is a solution of $x^2 - 5 = 4x + 7$.

Solution

$$\begin{aligned} x^2 - 5 &= 4x + 7 && \text{Write original equation.} \\ (2)^2 - 5 &\stackrel{?}{=} 4(2) + 7 && \text{Substitute 2 for } x. \\ 4 - 5 &\stackrel{?}{=} 8 + 7 && \text{Simplify.} \\ -1 &\neq 15 && \text{Solution does not check. } \times \end{aligned}$$

Because the substitution results in a false statement, you can conclude that $x = 2$ is not a solution of the original equation.

✓ **CHECKPOINT** *Now try Exercise 5.*

3 ▶ Use properties of equality to solve equations.

Forming Equivalent Equations

It is helpful to think of an equation as having two sides that are in balance. Consequently, when you try to solve an equation, you must be careful to maintain that balance by performing the same operation(s) on each side.

Two equations that have the same set of solutions are called **equivalent.** For instance, the equations

$$x = 3 \quad \text{and} \quad x - 3 = 0$$

are equivalent because both have only one solution—the number 3. When any one of the operations in the following list is applied to an equation, the resulting equation is equivalent to the original equation.

Forming Equivalent Equations: Properties of Equality

An equation can be transformed into an *equivalent equation* using one or more of the following procedures.

	Original Equation	*Equivalent Equation(s)*
1. *Simplify either side:* Remove symbols of grouping, combine like terms, or simplify fractions on one or both sides of the equation.	$3x - x = 8$	$2x = 8$
2. *Apply the Addition Property of Equality:* Add (or subtract) the same quantity to (from) *each* side of the equation.	$x - 2 = 5$	$x - 2 + 2 = 5 + 2$ $x = 7$
3. *Apply the Multiplication Property of Equality:* Multiply (or divide) each side of the equation by the same *nonzero* quantity.	$3x = 9$	$\frac{3x}{3} = \frac{9}{3}$ $x = 3$
4. *Interchange the two sides of the equation.*	$7 = x$	$x = 7$

The second and third operations in this list can be used to eliminate terms or factors in an equation. For example, to solve the equation $x - 5 = 1$, you need to eliminate the term -5 on the left side. This is accomplished by adding its opposite, 5, to each side.

$$x - 5 = 1 \qquad \text{Write original equation.}$$

$$x - 5 + 5 = 1 + 5 \qquad \text{Add 5 to each side.}$$

$$x + 0 = 6 \qquad \text{Combine like terms.}$$

$$x = 6 \qquad \text{Solution}$$

These four equations are equivalent, and they are called the **steps** of the solution.

The next example shows how the properties of equality can be used to solve equations. You will get many more opportunities to practice these skills in the next chapter. For now, your goal should be to understand why each step in the solution is valid. For instance, the second step in part (a) of Example 3 is valid because the Addition Property of Equality states that you can add the same quantity to each side of an equation.

EXAMPLE 3 Operations Used to Solve Equations

Identify the property of equality used to solve each equation.

a. $x - 5 = 0$ Original equation

$x - 5 + 5 = 0 + 5$ Add 5 to each side.

$x = 5$ Solution

b. $\frac{x}{5} = -2$ Original equation

$\frac{x}{5}(5) = -2(5)$ Multiply each side by 5.

$x = -10$ Solution

c. $4x = 9$ Original equation

$\frac{4x}{4} = \frac{9}{4}$ Divide each side by 4.

$x = \frac{9}{4}$ Solution

d. $\frac{5}{3}x = 7$ Original equation

$\frac{3}{5} \cdot \frac{5}{3}x = \frac{3}{5} \cdot 7$ Multiply each side by $\frac{3}{5}$.

$x = \frac{21}{5}$ Solution

Study Tip

In Example 3(c), each side of the equation is divided by 4 to eliminate the coefficient 4 on the left side. You could just as easily *multiply* each side by $\frac{1}{4}$. Both techniques are legitimate—which one you decide to use is a matter of personal preference.

Solution

a. The Addition Property of Equality is used to add 5 to each side of the equation in the second step. Adding 5 eliminates the term -5 from the left side of the equation.

b. The Multiplication Property of Equality is used to multiply each side of the equation by 5 in the second step. Multiplying by 5 eliminates the denominator from the left side of the equation.

c. The Multiplication Property of Equality is used to divide each side of the equation by 4 $\left(\text{or multiply each side by } \frac{1}{4}\right)$ in the second step. Dividing by 4 eliminates the coefficient from the left side of the equation.

d. The Multiplication Property of Equality is used to multiply each side of the equation by $\frac{3}{5}$ in the second step. Multiplying by the *reciprocal* of the fraction $\frac{5}{3}$ eliminates the fraction from the left side of the equation.

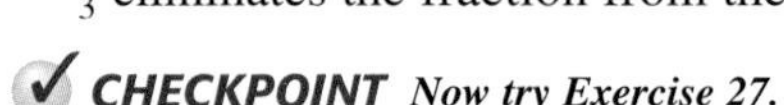
CHECKPOINT Now try Exercise 27.

4 ▶ Use a verbal model to construct an algebraic equation.

Constructing Equations

It is helpful to use two phases in constructing equations that model real-life situations, as shown below.

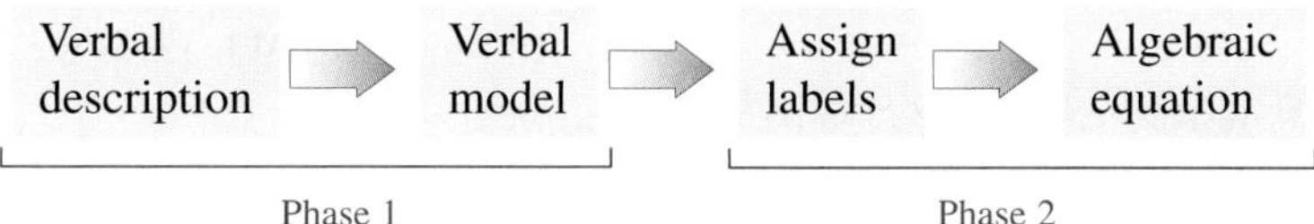

In the first phase, you translate the verbal description into a *verbal model.* In the second phase, you assign labels and translate the verbal model into a *mathematical model* or an *algebraic equation.* Here are two examples of verbal models.

1. The sale price of a basketball is \$28. The sale price is \$7 less than the original price. What is the original price?

   Verbal Model: Sale price = Original price − Discount

   \$28 = Original price − \$7

2. The original price of a basketball is \$35. The original price is discounted by \$7. What is the sale price?

   Verbal Model: Sale price = Original price − Discount

   Sale price = \$35 − \$7

EXAMPLE 4 Using a Verbal Model to Construct an Equation

Write an algebraic equation for the following problem.

The total income that an employee received in a year was \$40,950. How much was the employee paid each week? Assume that each weekly paycheck contained the same amount, and that the year consisted of 52 weeks.

Solution

Verbal Model: Income for year = Number of weeks in a year · Weekly pay

Labels:
Income for year = 40,950 (dollars)
Weekly pay = x (dollars per week)
Number of weeks = 52 (weeks)

Algebraic Model: $40{,}950 = 52x$

✓ **CHECKPOINT** *Now try Exercise 67.*

Study Tip

When you construct an equation, be sure to check that both sides of the equation represent the *same* unit of measure. For instance, in Example 4, both sides of the equation $40{,}950 = 52x$ represent dollar amounts.

EXAMPLE 5 Using a Verbal Model to Construct an Equation

Write an algebraic equation for the following problem.

Returning to college after spring break, you travel 3 hours and stop for lunch. You know that it takes 45 minutes to complete the last 36 miles of the 180-mile trip. What was the average speed during the first 3 hours of the trip?

Study Tip

In Example 5, the information that it takes 45 minutes to complete the last part of the trip is unnecessary information. This type of unnecessary information in an applied problem is sometimes called a *red herring*.

Solution

Verbal Model: Distance = Rate · Time

Labels:
Distance = 180 − 36 = 144 (miles)
Rate = r (miles per hour)
Time = 3 (hours)

Algebraic Model: $144 = 3r$

CHECKPOINT *Now try Exercise 79.*

EXAMPLE 6 Using a Verbal Model to Construct an Equation

Live Earth was a series of concerts held on July 7, 2007 to raise awareness of the climate crisis. More than 150 musical acts performed during the 24 hours of music across 7 continents.
(Source: Live Earth, LLC)

Write an algebraic equation for the following problem.

Tickets for a concert cost \$45 for each floor seat and \$30 for each stadium seat. There were 800 seats on the main floor, and these were sold out. The total revenue from ticket sales was \$54,000. How many stadium seats were sold?

Solution

Verbal Model: Total revenue = Revenue from floor seats + Revenue from stadium seats

Labels:
Total revenue = 54,000 (dollars)
Price per floor seat = 45 (dollars per seat)
Number of floor seats = 800 (seats)
Price per stadium seat = 30 (dollars per seat)
Number of stadium seats = x (seats)

Algebraic Model: $54{,}000 = 45(800) + 30x$

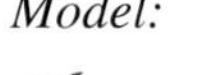

CHECKPOINT *Now try Exercise 81.*

In Example 6, you can use the following *unit analysis* to check that both sides of the equation are measured in dollars.

$$54{,}000 \text{ dollars} = \left(\frac{45 \text{ dollars}}{\cancel{\text{seat}}}\right)(800 \cancel{\text{ seats}}) + \left(\frac{30 \text{ dollars}}{\cancel{\text{seat}}}\right)(x \cancel{\text{ seats}})$$

In Section 3.1, you will study techniques for solving the equations constructed in Examples 4, 5, and 6.

Concept Check

1. In your own words, explain what is meant by the term *equivalent equations.*

2. Describe the steps that can be used to transform an equation into an equivalent equation.

3. Is there more than one way to write a verbal model? Explain.

4. Explain how to decide whether a real number is a solution of an equation. Give an example of an equation with a solution that checks and one that does not check.

2.4 EXERCISES

Go to pages 116–117 to record your assignments.

Developing Skills

In Exercises 1–16, determine whether each value of x is a solution of the equation. ***See Examples 1 and 2.***

	Equation	*Values*	
✓ 1.	$2x - 18 = 0$	(a) $x = 0$	(b) $x = 9$
2.	$3x - 3 = 0$	(a) $x = 4$	(b) $x = 1$
3.	$6x + 1 = -11$	(a) $x = 2$	(b) $x = -2$
4.	$2x + 5 = -15$	(a) $x = -10$	(b) $x = 5$
✓ 5.	$x + 5 = 2x$	(a) $x = -1$	(b) $x = 5$
6.	$15 - 2x = 3x$	(a) $x = 3$	(b) $x = 5$
7.	$7x + 1 = 4(x - 2)$	(a) $x = 1$	(b) $x = 12$
8.	$5x - 1 = 3(x + 5)$	(a) $x = 8$	(b) $x = -2$
9.	$2x + 10 = 7(x + 1)$	(a) $x = \frac{3}{5}$	(b) $x = -\frac{2}{3}$
10.	$3(3x + 2) = 9 - x$	(a) $x = -\frac{3}{4}$	(b) $x = \frac{3}{10}$
11.	$x^2 - 4 = x + 2$	(a) $x = 3$	(b) $x = -2$
12.	$x^2 = 8 - 2x$	(a) $x = 2$	(b) $x = -4$
13.	$\frac{2}{x} - \frac{1}{x} = 1$	(a) $x = 0$	(b) $x = \frac{1}{3}$
14.	$\frac{4}{x} + \frac{2}{x} = 1$	(a) $x = 0$	(b) $x = 6$
15.	$\frac{5}{x - 1} + \frac{1}{x} = 5$	(a) $x = 3$	(b) $x = \frac{1}{6}$
16.	$\frac{3}{x - 2} = x$	(a) $x = -1$	(b) $x = 3$

In Exercises 17–26, use a calculator to determine whether the value of x is a solution of the equation.

	Equation	*Values*
17.	$x + 1.7 = 6.5$	(a) $x = -3.1$ (b) $x = 4.8$
18.	$7.9 - x = 14.6$	(a) $x = -6.7$ (b) $x = 5.4$
19.	$40x - 490 = 0$	(a) $x = 12.25$ (b) $x = -12.25$
20.	$20x - 550 = 0$	(a) $x = 27.5$ (b) $x = -27.5$
21.	$2x^2 - x - 10 = 0$	(a) $x = \frac{5}{2}$ (b) $x = -1.09$
22.	$22x - 5x^2 = 17$	(a) $x = 1$ (b) $x = 3.4$
23.	$\frac{1}{x} - \frac{9}{x - 4} = 1$	(a) $x = 0$ (b) $x = -2$

24. $x = \dfrac{3}{4x + 1}$ (a) $x = -0.25$ (b) $x = 0.75$

25. $x^3 - 1.728 = 0$ (a) $x = \frac{6}{5}$ (b) $x = -\frac{6}{5}$

26. $4x^2 - 10.24 = 0$ (a) $x = \frac{8}{5}$ (b) $x = -\frac{8}{5}$

In Exercises 27–36, justify each step of the solution. ***See Example 3.***

27.
$$\begin{aligned} x - 8 &= 3 \\ x - 8 + 8 &= 3 + 8 \\ x &= 11 \end{aligned}$$

28.
$$\begin{aligned} 6x &= 17 \\ \frac{6x}{6} &= \frac{17}{6} \\ x &= \frac{17}{6} \end{aligned}$$

29.
$$\begin{aligned} \tfrac{2}{3}x &= 12 \\ \tfrac{3}{2}\left(\tfrac{2}{3}x\right) &= \tfrac{3}{2}(12) \\ x &= 18 \end{aligned}$$

30.
$$\begin{aligned} \tfrac{4}{5}x &= -28 \\ \tfrac{5}{4}\left(\tfrac{4}{5}x\right) &= \tfrac{5}{4}(-28) \\ x &= -35 \end{aligned}$$

31.
$$\begin{aligned} 5x + 12 &= 22 \\ 5x + 12 - 12 &= 22 - 12 \\ 5x &= 10 \\ \frac{5x}{5} &= \frac{10}{5} \\ x &= 2 \end{aligned}$$

32.
$$\begin{aligned} 14 - 3x &= 5 \\ 14 - 3x - 14 &= 5 - 14 \\ 14 - 14 - 3x &= 5 - 14 \\ -3x &= -9 \\ \frac{-3x}{-3} &= \frac{-9}{-3} \\ x &= 3 \end{aligned}$$

33.
$$\begin{aligned} 2(x - 1) &= x + 3 \\ 2x - 2 &= x + 3 \\ 2x - 2 - x &= x + 3 - x \\ 2x - x - 2 &= x - x + 3 \\ x - 2 &= 3 \\ x - 2 + 2 &= 3 + 2 \\ x &= 5 \end{aligned}$$

34.
$$\begin{aligned} x + 6 &= -6(4 - x) \\ x + 6 &= -24 + 6x \\ x + 6 - x &= -24 + 6x - x \\ x - x + 6 &= 6x - x - 24 \\ 6 &= 5x - 24 \\ 6 + 24 &= 5x - 24 + 24 \\ 30 &= 5x \\ \frac{30}{5} &= \frac{5x}{5} \\ 6 &= x \end{aligned}$$

35.
$$\begin{aligned} \frac{x}{3} &= x + 1 \\ 3\left(\frac{x}{3}\right) &= 3(x + 1) \\ x &= 3x + 3 \\ x - 3x &= 3x + 3 - 3x \\ x - 3x &= 3x - 3x + 3 \\ -2x &= 3 \\ \frac{-2x}{-2} &= \frac{3}{-2} \\ x &= -\frac{3}{2} \end{aligned}$$

36.
$$\begin{aligned} \frac{4}{5}x &= 4x - 16 \\ \frac{5}{4}\left(\frac{4}{5}x\right) &= \frac{5}{4}(4x - 16) \\ x &= 5x - 20 \\ x - 5x &= 5x - 20 - 5x \\ x - 5x &= 5x - 5x - 20 \\ -4x &= -20 \\ \frac{-4x}{-4} &= \frac{-20}{-4} \\ x &= 5 \end{aligned}$$

In Exercises 37–44, write a verbal description of the algebraic equation without using a variable. (There is more than one correct answer.)

37. $x - 6 = 32$

38. $\frac{x}{2} = 4$

39. $2x + 5 = 21$

40. $3x - 2 = 7$

41. $10(x - 3) = 8x$

42. $2(x - 5) = 12$

43. $\frac{x + 1}{3} = 8$

44. $\frac{x - 2}{10} = 6$

In Exercises 45–54, write an algebraic equation. Do *not* solve the equation.

45. The sum of a number and 12 is 45.

46. The difference of 24 and a number is 3.

47. The quotient of a number and 8 is 6.

48. The product of 7 and a number is 49.

49. The sum of 3 times a number and 4 is 16.

50. Four times the sum of a number and 6 is 100.

51. Six times a number subtracted from 120 is 96.

52. Two times a number decreased by 14 equals the number divided by 3.

53. Four divided by the sum of a number and 5 is 2.

54. The sum of a number and 8, all divided by 4, is 32.

Unit Analysis In Exercises 55–62, simplify the expression. State the units of the simplified value.

55. $\frac{3 \text{ dollars}}{\text{unit}} \cdot (5 \text{ units})$

56. $\frac{25 \text{ miles}}{\text{gallon}} \cdot (15 \text{ gallons})$

57. $\frac{50 \text{ pounds}}{\text{foot}} \cdot (3 \text{ feet})$

58. $\frac{3 \text{ dollars}}{\text{pound}} \cdot (5 \text{ pounds})$

59. $\frac{5 \text{ feet}}{\text{second}} \cdot \frac{60 \text{ seconds}}{\text{minute}} \cdot (20 \text{ minutes})$

60. $\frac{12 \text{ dollars}}{\text{hour}} \cdot \frac{1 \text{ hour}}{60 \text{ minutes}} \cdot (45 \text{ minutes})$

61. $\frac{100 \text{ centimeters}}{\text{meter}} \cdot (2.4 \text{ meters})$

62. $\frac{1000 \text{ milliliters}}{\text{liter}} \cdot (5.6 \text{ liters})$

In Exercises 63–66, use a property of equality to solve the equation. Check your solution.

63. $x - 8 = 5$

64. $x + 3 = 19$

65. $3x = 30$

66. $\frac{x}{4} = 12$

Solving Problems

In Exercises 67–84, write an algebraic equation. Do *not* solve the equation. ***See Examples 4, 5, and 6.***

✓ **67.** *Test Score* After your instructor added 6 points to each student's test score, your score is 94. What was your original score?

68. *Meteorology* With the 1.2-inch rainfall today, the total for the month is 4.5 inches. How much had been recorded for the month before today's rainfall?

69. *Consumerism* You have \$1044 saved for the purchase of a new computer that will cost \$1926. How much more must you save?

70. *Original Price* The sale price of a coat is \$225.98. The discount is \$64. What is the original price?

71. *Consumer Awareness* The price of a gold ring has increased by $45 over the past year. It is now selling for $375. What was the price one year ago?

72. *Travel Costs* A company pays its sales representatives 35 cents per mile if they use their personal cars. A sales representative submitted a bill to be reimbursed for $148.05 for driving. How many miles did the sales representative drive?

73. *Football* During a football game, the running back carried the ball 18 times and his average number of yards per carry was 4.5. How many yards did the running back gain for the game?

74. *Aquarium* The total cost of admission for 6 adults at an aquarium is $132. What is the cost per adult?

Stephen Marks/Getty Images

The Georgia Aquarium houses approximately 80,000 animals from 500 species in more than 8 million gallons of water.

75. *Meteorology* You hear thunder 3 seconds after seeing a lightning strike. The speed of sound is 1100 feet per second. How far away is the lightning?

76. *Fund Raising* A student group is selling boxes of greeting cards at a profit of $1.75 each. The group needs $2000 more to have enough money for a trip to Washington, D.C. How many boxes does the group need to sell to earn $2000?

77. *Geometry* The base of a rectangular trunk used for transporting concert equipment is 4 feet by 6 feet, and its volume is 72 cubic feet (see figure). What is the height of the trunk?

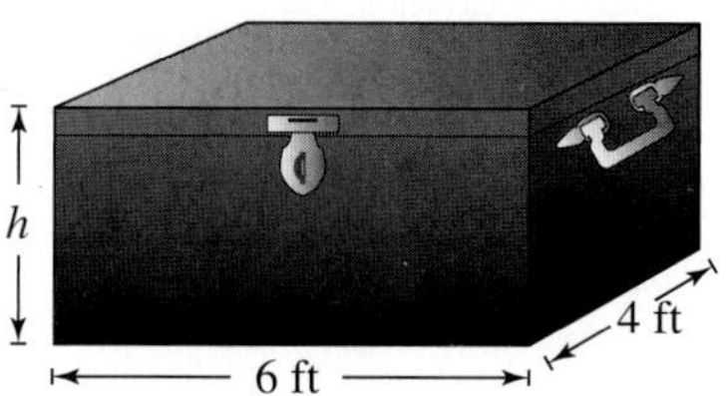

Figure for 77

78. *Geometry* The width of a rectangular mirror is one-third its length (see figure). The perimeter of the mirror is 96 inches. What are the dimensions of the mirror?

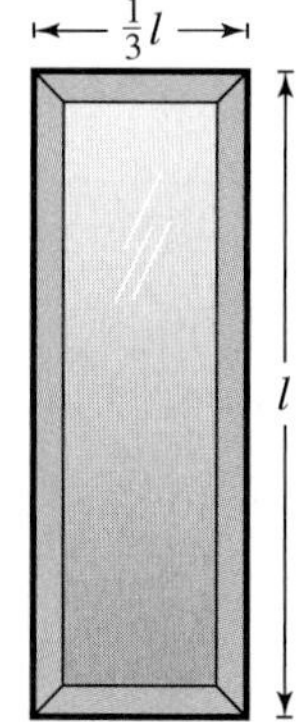

79. *Average Speed* After traveling for 3 hours, your family is still 25 miles from completing a 160-mile trip (see figure). What was the average speed during the first 3 hours of the trip?

80. *Average Speed* A group of students plans to take two cars to a soccer tournament. The first car leaves on time, travels at an average speed of 45 miles per hour, and arrives at the destination in 3 hours. The second car leaves one-half hour after the first car and arrives at the tournament at the same time as the first car. What is the average speed of the second car?

✓ **81.** *Volunteering* You want to volunteer at a soup kitchen for 150 hours over a 15-week period. After 8 weeks, you have volunteered for 72 hours. How many hours will you have to work per week over the remaining 7 weeks to reach your goal?

82. *Money* A student has n quarters and seven \$1 bills totaling \$8.75. How many quarters does the student have?

83. *Depreciation* A textile corporation buys equipment with an initial purchase price of \$750,000. It is estimated that its useful life will be 3 years and at that time its value will be \$75,000. The total depreciation is divided equally among the three years. (Depreciation is the difference between the initial price of an item and its current value.) What is the total amount of depreciation declared each year?

84. *Car Wash* A school science club conducts a car wash to raise money. The club spends \$12 on supplies and charges \$5 per car. After the car wash, the club has a profit of \$113. How many cars did the members of the science club wash?

In Exercises 85 and 86, write an algebraic equation. Simplify the equation, but do not solve the equation.

85. *Attendance* A high school made \$986 in revenue for a play. Tickets for the play cost \$10 for adults and \$6 for students. The number of students attending the play was $\frac{3}{4}$ the number of adults attending the play. How many adults and students attended the play?

86. *Ice Show* An ice show earns a revenue of \$11,041 one night. Tickets for the ice show cost \$18 for adults and \$13 for children. The number of adults attending the ice show was 33 more than the number of children attending the show. How many adults and children attended the show?

Explaining Concepts

87. Are there any equations of the form $ax = b$ $(a \neq 0)$ that are true for more than one value of x? Explain.

88. Determine which equations are equivalent to $14 = x + 8$.

(a) $x + 8 = 14$ (b) $8x = 14$

(c) $x - 8 = 14$ (d) $8 + x = 14$

(e) $2(x + 4) - x = 14$

(f) $3(x + 6) - 2x + 5 = 14$

89. Describe a real-life problem that uses the following verbal model.

Revenue of \$840	=	\$35 per case	·	Number of cases

90. Explain the difference between simplifying an expression and solving an equation. Give an example of each.

Cumulative Review

In Exercises 91–96, simplify the expression.

91. $t^2 \cdot t^5$ **92.** $(-3y^3)y^2$

93. $6x + 9x$ **94.** $4 - 3t + t$

95. $-(-8b)$ **96.** $7(-10x)$

In Exercises 97–100, translate the phrase into an algebraic expression. Let x represent the real number.

97. 23 more than a number

98. A number divided by 6

99. Seven more than 4 times a number

100. Nine times the difference of a number and 3

What Did You Learn?

Use these two pages to help prepare for a test on this chapter. Check off the key terms and key concepts you know. You can also use this section to record your assignments.

Plan for Test Success

Date of test: / /

Study dates and times: / / at : A.M./P.M.

/ / at : A.M./P.M.

Things to review:

- ☐ Key Terms, *p. 116*
- ☐ Key Concepts, *pp. 116–117*
- ☐ Your class notes
- ☐ Your assignments
- ☐ Study Tips, *pp. 70, 71, 78, 79, 80, 81, 83, 91, 93, 94, 96, 97, 106, 108, 109, 110*
- ☐ Technology Tips, *pp. 72, 73*
- ☐ Mid-Chapter Quiz, *p. 90*
- ☐ Review Exercises, *pp. 118–122*
- ☐ Chapter Test, *p. 123*
- ☐ Video Explanations Online
- ☐ Tutorial Online

Key Terms

- ☐ variables, *p. 68*
- ☐ constants, *p. 68*
- ☐ algebraic expression, *p. 68*
- ☐ terms, *p. 68*
- ☐ coefficient, *p. 68*
- ☐ evaluate an algebraic expression, *p. 71*
- ☐ expanding an algebraic expression, *p. 79*
- ☐ like terms, *p. 80*
- ☐ variable factors, *p. 80*
- ☐ simplify an algebraic expression, *p. 82*
- ☐ verbal mathematical model, *p. 92*
- ☐ equation, *p. 105*
- ☐ solve, *p. 105*
- ☐ solutions, *p. 105*
- ☐ satisfy, *p. 105*
- ☐ check, *p. 106*
- ☐ equivalent equations, *p. 107*
- ☐ steps of the solution, *p. 107*

Key Concepts

2.1 Writing and Evaluating Algebraic Expressions

Assignment: ______________________________ Due date: ________

☐ **Identify terms, variables, and coefficients.**

In the expression $4x + 3y$, $4x$ and $3y$ are the terms of the expression, x and y are the variables, and 4 and 3 are the coefficients.

☐ **Use exponential form.**

Repeated multiplication can be expressed in exponential form using a base a and an exponent n, where a is a real number, variable, or algebraic expression and n is a positive integer.

$$a^n = \underbrace{a \cdot a \cdot a \cdot a \cdot \cdot \cdot a}_{n \text{ factors}}$$

☐ **Evaluate algebraic expressions.**

To evaluate an algebraic expression, replace every occurrence of the specified variable in the expression with the appropriate real number, and perform the operation(s).

2.2 Simplifying Algebraic Expressions

Assignment: ______________________ Due date: ________

☐ **Use the Properties of Algebra.**

Let a, b, and c represent real numbers, variables, or algebraic expressions.

Commutative Property of Addition $a + b = b + a$

Commutative Property of Multiplication $ab = ba$

Associative Property of Addition $(a + b) + c = a + (b + c)$

Associative Property of Multiplication $(ab)c = a(bc)$

Distributive Property

$a(b + c) = ab + ac$ $a(b - c) = ab - ac$

$(a + b)c = ac + bc$ $(a - b)c = ac - bc$

Additive Identity Property $a + 0 = 0 + a = a$

Multiplicative Identity Property $a \cdot 1 = 1 \cdot a = a$

Additive Inverse Property $a + (-a) = 0$

Multiplicative Inverse Property $a \cdot \frac{1}{a} = 1, a \neq 0$

☐ **Define like terms.**

Two or more terms of an algebraic expression can be combined only if they are like terms. Two terms are said to be like terms if they are both constant terms or if they have the same variable factor(s).

☐ **Combine like terms.**

To combine like terms in an algebraic expression, add their respective coefficients and attach the common variable factor(s).

☐ **Simplify an algebraic expression.**

To simplify an algebraic expression, remove symbols of grouping and combine like terms.

2.3 Algebra and Problem Solving

Assignment: ______________________ Due date: ________

☐ **Translate verbal phrases into algebraic expressions.**

When translating verbal phrases into algebraic expressions, look for key words and phrases that indicate the four different operations of arithmetic.

☐ **Use problem-solving strategies.**

1. Guess, check, and revise.
2. Make a table/look for a pattern.
3. Draw a diagram.
4. Solve a simpler problem.

2.4 Introduction to Equations

Assignment: ______________________ Due date: ________

☐ **Check a solution of an equation.**

To check whether a given value is a solution of an equation, substitute the value into the original equation. If the substitution results in a true statement, then the value is a solution of the equation. If the substitution results in a false statement, then the value is not a solution of the equation.

☐ **Construct an equation.**

First, write a verbal model. Then, assign labels to the known and unknown quantities and translate the verbal model into an algebraic equation.

☐ **Use Properties of Equality.**

An equation can be transformed into an equivalent equation using one or more of the following procedures.

1. Remove symbols of grouping, combine like terms, or simplify fractions on one or both sides of the equation.
2. Add (or subtract) the same quantity to (from) each side of the equation.
3. Multiply (or divide) each side of the equation by the same nonzero quantity.
4. Interchange the two sides of the equation.

Review Exercises

2.1 Writing and Evaluating Algebraic Expressions

1 ▶ Define and identify terms, variables, and coefficients of algebraic expressions.

In Exercises 1 and 2, write an algebraic expression for the statement.

1. The distance traveled in t hours if the average speed is 60 miles per hour
2. The cost of x pounds of coffee if the cost per pound is \$1.99

In Exercises 3-6, identify the variable(s) in the expression.

3. $15 - x$
4. $t - 5^2$
5. $a - 3b$
6. $y + z$

In Exercises 7–12, identify the terms of the expression and their coefficients.

7. $12y + y^2$
8. $4x - \frac{1}{2}x^3$
9. $5x^2 - 3xy + 10y^2$
10. $y^2 - 10yz + \frac{2}{3}z^2$
11. $\frac{2y}{3} - \frac{4x}{y}$
12. $-\frac{4b}{9} + \frac{11a}{b}$

2 ▶ Define exponential form and interpret exponential expressions.

In Exercises 13–18, rewrite the product in exponential form.

13. $5z \cdot 5z \cdot 5z$
14. $\frac{3}{8}y \cdot \frac{3}{8}y \cdot \frac{3}{8}y \cdot \frac{3}{8}y$
15. $(-3x) \cdot (-3x) \cdot (-3x) \cdot (-3x) \cdot (-3x)$
16. $\left(-\frac{2}{7}\right) \cdot \left(-\frac{2}{7}\right) \cdot \left(-\frac{2}{7}\right)$
17. $(b - c) \cdot (b - c) \cdot 6 \cdot 6$
18. $2 \cdot (a + b) \cdot 2 \cdot (a + b) \cdot 2$

3 ▶ Evaluate algebraic expressions using real numbers.

In Exercises 19–26, evaluate the algebraic expression at the given values of the variable(s).

	Expression	*Values*	
19.	$x^2 - 2x + 5$	(a) $x = 0$	(b) $x = 2$
20.	$x^3 - 8$	(a) $x = 2$	(b) $x = 4$
21.	$x^2 - x(y + 1)$	(a) $x = 2, y = -1$	(b) $x = 1, y = 2$
22.	$2r + r(t^2 - 3)$	(a) $r = 3, t = -2$	(b) $r = -2, t = 3$
23.	$\frac{x + 5}{y}$	(a) $x = -5, y = 3$	(b) $x = 2, y = -1$
24.	$\frac{a - 9}{2b}$	(a) $a = 7, b = -3$	(b) $a = -4, b = 5$
25.	$x^2 - 2y + z$	(a) $x = 1, y = 2, z = 0$	(b) $x = 2, y = -3, z = -4$
26.	$\frac{m + 2n}{-p}$	(a) $m = -1, n = 2, p = 3$	(b) $m = 4, n = -2, p = -7$

2.2 Simplifying Algebraic Expressions

1 ▶ Use the properties of algebra.

In Exercises 27–32, identify the property of algebra illustrated by the statement.

27. $xy \cdot \frac{1}{xy} = 1$
28. $u(vw) = (uv)w$
29. $(x - y)(2) = 2(x - y)$
30. $(a + b) + 0 = a + b$
31. $2x + (3y - z) = (2x + 3y) - z$
32. $x(y + z) = xy + xz$

In Exercises 33–36, complete the statement. Then state the property of algebra that you used.

33. $3(m^2n) = 3m^2(\quad)$

34. $b(c + 1) = b(\quad) + b(\quad)$

35. $(3x + 8) + \quad = 0$

36. $10p(\quad) = 1$

In Exercises 37–46, use the Distributive Property to expand the expression.

37. $4(x + 3y)$

38. $3(8s - 12t)$

39. $-5(2u - 3v)$

40. $-3(-2x - 8y)$

41. $x(8x + 5y)$

42. $-u(3u - 10v)$

43. $-(-a + 3b)$

44. $(7 - 2j)(-6)$

45. $2(x + 3 - 2y)$

46. $-7(4 - 2m - 5n^2)$

2 ▶ Combine like terms of an algebraic expression.

In Exercises 47–50, identify the like terms.

47. $3x - 4 + 2x$

48. $-4y + y^2 - 9y$

49. $10 - z + z^2 - 2$

50. $\frac{2}{5}x - 5 + x^3 - \frac{5}{7}x$

In Exercises 51–62, simplify the expression by combining like terms.

51. $3a - 5a$

52. $6c - 2c$

53. $3p - 4q + q + 8p$

54. $10x - 4y - 25x + 6y$

55. $\frac{1}{4}s - 6t + \frac{7}{2}s + t$

56. $\frac{2}{3}a + \frac{3}{5}a - \frac{1}{2}b + \frac{2}{3}b$

57. $x^2 + 3xy - xy + 4$

58. $uv^2 + 10 - 2uv^2 + 2$

59. $5x - 5y + 3xy - 2x + 2y$

60. $y^3 + 2y^2 + 2y^3 - 3y^2 + 1$

61. $5\left(1 + \frac{r}{n}\right)^2 - 2\left(1 + \frac{r}{n}\right)^2$

62. $-7\left(\frac{1}{u}\right) + 4\left(\frac{1}{u^2}\right) + 3\left(\frac{1}{u}\right)$

3 ▶ Simplify an algebraic expression by rewriting the terms.

In Exercises 63–70, simplify the expression.

63. $12(4t)$

64. $8(7x)$

65. $-5(-9x^2)$

66. $-10(-3b^3)$

67. $(-6x)(2x^2)$

68. $(-3y^2)(15y)$

69. $\frac{12x}{5} \cdot \frac{10}{3}$

70. $\frac{4z}{15} \cdot \frac{9}{2}$

Mental Math **In Exercises 71–74, use the Distributive Property to perform the required arithmetic mentally.**

71. $3(61) = 3(60 + 1)$

72. $-5(29) = -5(30 - 1)$

73. $7(98) = 7(100 - 2)$

74. $-4(41) = -4(40 + 1)$

4 ▶ Use the Distributive Property to remove symbols of grouping.

In Exercises 75–86, simplify the expression by removing symbols of grouping and combining like terms.

75. $5(u - 4) + 10$

76. $16 - 3(v + 2)$

77. $3s - (r - 2s)$

78. $50x - (30x + 100)$

79. $-3(1 - 10z) + 2(1 - 10z)$

80. $8(15 - 3y) - 5(15 - 3y)$

81. $\frac{1}{3}(42 - 18z) - 2(8 - 4z)$

82. $\frac{1}{4}(100 + 36s) - (15 - 4s)$

83. $10 - [8(5 - x) + 2]$

84. $3[2(4x - 5) + 4] - 3$

85. $2[x + 2(y - x)]$

86. $2t[4 - (3 - t)] + 5t$

In Exercises 87–90, use the Distributive Property to simplify the expression.

87. $\frac{x}{4} - \frac{3x}{4}$

88. $\frac{4m}{7} + \frac{8m}{7}$

89. $\frac{3z}{2} + \frac{z}{5}$

90. $\frac{7p}{4} - \frac{7p}{9}$

91. *Geometry* Write and simplify expressions for (a) the perimeter and (b) the area of the rectangle.

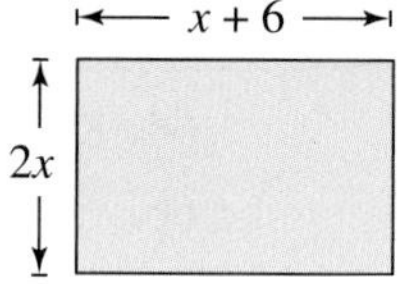

92. *Geometry* Write and simplify an expression for the area of the triangle.

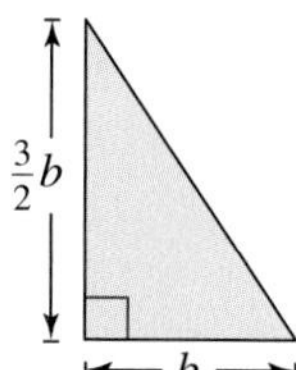

93. Simplify the algebraic expression that represents the sum of three consecutive odd integers, $2n - 1$, $2n + 1$, and $2n + 3$.

94. Simplify the algebraic expression that represents the sum of three consecutive even integers, $2n$, $2n + 2$, $2n + 4$.

95. *Geometry* The face of a DVD player has the dimensions shown in the figure. Write an algebraic expression that represents the area of the face of the DVD player excluding the compartment holding the disc.

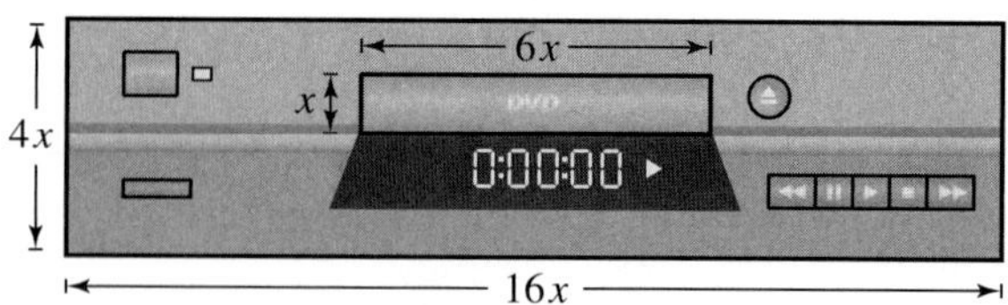

96. *Geometry* Write an expression for the perimeter of the figure. Use the rules of algebra to simplify the expression.

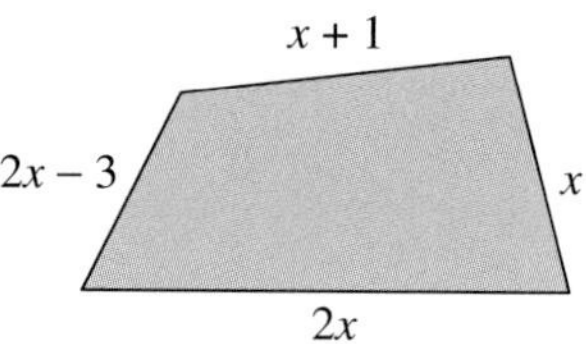

2.3 Algebra and Problem Solving

2 ▶ Construct verbal mathematical models from written statements.

In Exercises 97 and 98, construct a verbal model and then write an algebraic expression that represents the specified quantity.

97. The total hourly wage for an employee when the base pay is $8.25 per hour and an additional $0.60 is paid for each unit produced per hour

98. The total cost for a family to stay one night at a campground if the charge is $18 for the parents plus $3 for each of the children

3 ▶ Translate verbal phrases into algebraic expressions.

In Exercises 99–108, translate the phrase into an algebraic expression. Let *x* represent the real number.

99. The sum of two-thirds of a number and 5

100. One hundred decreased by the product of 5 and a number

101. Ten less than twice a number

102. The ratio of a number and 10

103. Fifty increased by the product of 7 and a number

104. Ten decreased by the quotient of a number and 2

105. The sum of a number and 10, all divided by 8

106. The product of 15 and a number, all decreased by 2

107. The sum of the square of a real number and 64

108. The absolute value of the sum of a number and -10

In Exercises 109–112, write a verbal description of the expression without using a variable. (There is more than one correct answer.)

109. $x + 3$

110. $3x - 2$

111. $\frac{y - 2}{3}$

112. $4(x + 5)$

4 ▶ Identify hidden operations when constructing algebraic expressions.

113. *Commission* A salesperson earns 5% commission on his total weekly sales, x. Write an algebraic expression that represents the amount in commissions that the salesperson earns in a week.

114. *Sale Price* A cordless phone is advertised for 20% off the list price of L dollars. Write an algebraic expression that represents the sale price of the phone.

115. *Rent* The monthly rent for your apartment is $625 for n months. Write an algebraic expression that represents the total rent.

116. *Distance* A car travels for 10 hours at an average speed of s miles per hour. Write an algebraic expression that represents the total distance traveled by the car.

5 ▶ Use problem-solving strategies to solve application problems.

117. *Finding a Pattern* Describe the pattern, and use your description to find the value of the expression when $n = 20$.

n	0	1	2	3	4	5
Value of expression	4	7	10	13	16	19

118. *Finding a Pattern* Find values of a and b such that the expression $an + b$ yields the values in the table.

n	0	1	2	3	4	5
$an + b$	4	9	14	19	24	29

2.4 Introduction to Equations

2 ▶ Check whether a given value is a solution of an equation.

In Exercises 119–128, determine whether each value of x is a solution of the equation.

	Equation	*Values*	
119.	$5x + 6 = 36$	(a) $x = 3$	(b) $x = 6$
120.	$17 - 3x = 8$	(a) $x = 3$	(b) $x = -3$
121.	$3x - 12 = x$	(a) $x = -1$	(b) $x = 6$
122.	$8x + 24 = 2x$	(a) $x = 0$	(b) $x = -4$
123.	$4(2 - x) = 3(2 + x)$	(a) $x = \frac{2}{7}$	(b) $x = -\frac{2}{3}$
124.	$5x + 2 = 3(x + 10)$	(a) $x = 14$	(b) $x = -10$
125.	$\frac{4}{x} - \frac{2}{x} = 5$	(a) $x = -1$	(b) $x = \frac{2}{5}$
126.	$\frac{x}{3} + \frac{x}{6} = 1$	(a) $x = \frac{2}{9}$	(b) $x = -\frac{2}{9}$
127.	$x(x - 7) = -12$	(a) $x = 3$	(b) $x = 4$
128.	$x(x + 1) = 2$	(a) $x = 1$	(b) $x = -2$

3 ▶ Use properties of equality to solve equations.

In Exercises 129–132, justify each step of the solution.

129.
$$\begin{aligned} -7x + 20 &= -1 \\ -7x + 20 - 20 &= -1 - 20 \\ -7x &= -21 \\ \frac{-7x}{-7} &= \frac{-21}{-7} \\ x &= 3 \end{aligned}$$

130.
$$\begin{aligned}
3(x-2) &= x+2\\
3x-6 &= x+2\\
3x-6-x &= x+2-x\\
3x-x-6 &= x-x+2\\
2x-6 &= 2\\
2x-6+6 &= 2+6\\
2x &= 8\\
\frac{2x}{2} &= \frac{8}{2}\\
x &= 4
\end{aligned}$$

131.
$$\begin{aligned}
x &= -(x-14)\\
x &= -x+14\\
x+x &= -x+14+x\\
x+x &= -x+x+14\\
2x &= 14\\
\frac{2x}{2} &= \frac{14}{2}\\
x &= 7
\end{aligned}$$

132.
$$\begin{aligned}
\frac{x}{4} &= x-2\\
4\left(\frac{x}{4}\right) &= 4(x-2)\\
x &= 4x-8\\
x-4x &= 4x-8-4x\\
x-4x &= 4x-4x-8\\
-3x &= -8\\
\frac{-3x}{-3} &= \frac{-8}{-3}\\
x &= \frac{8}{3}
\end{aligned}$$

4 ▶ Use a verbal model to construct an algebraic equation.

In Exercises 133–136, write an algebraic equation. Do *not* solve the equation.

133. The sum of a number and its reciprocal is $\frac{37}{6}$. What is the number?

134. *Distance* A car travels 135 miles in t hours with an average speed of 45 miles per hour (see figure). How many hours did the car travel?

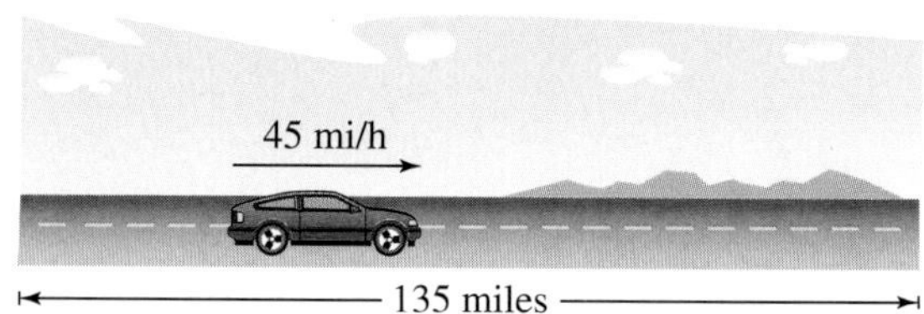

135. *Geometry* The area of the shaded region in the figure is 24 square inches. What is the length of the rectangle?

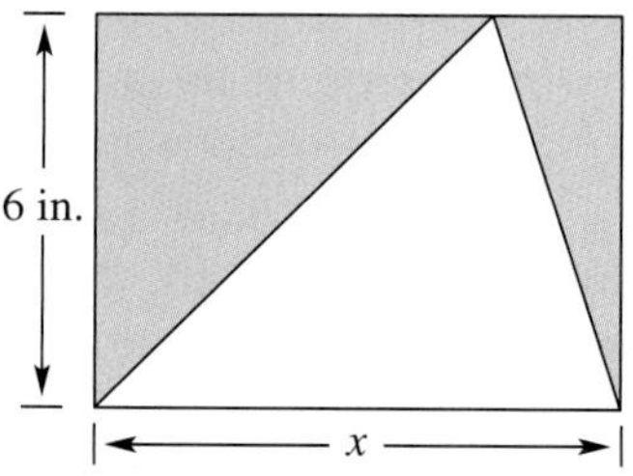

136. *Geometry* The perimeter of the face of a rectangular traffic light is 72 inches (see figure). What are the dimensions of the traffic light?

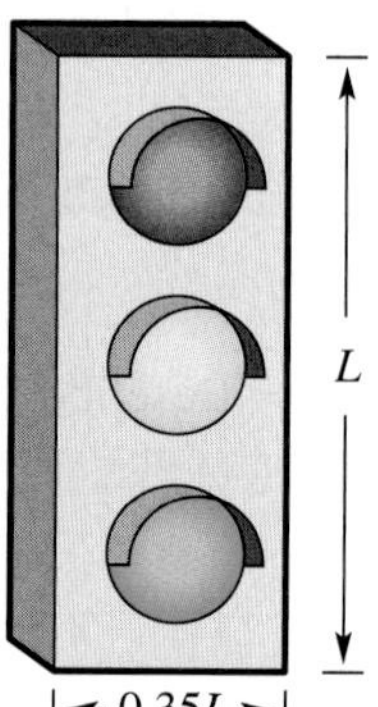

Chapter Test

Take this test as you would take a test in class. After you are done, check your work against the answers in the back of the book.

1. Identify the terms of the expression and their coefficients.

$2x^2 - 7xy + 3y^3$

2. Rewrite the product in exponential form.
$x \cdot (x + y) \cdot x \cdot (x + y) \cdot x$

In Exercises 3–6, identify the property of algebra illustrated by the statement.

3. $(5x)y = 5(xy)$

4. $2 + (x - y) = (x - y) + 2$

5. $7xy + 0 = 7xy$

6. $(x + 5) \cdot \dfrac{1}{(x + 5)} = 1$

In Exercises 7–10, use the Distributive Property to expand the expression.

7. $3(x + 8)$

8. $5(4r - s)$

9. $-y(3 - 2y)$

10. $-9(4 - 2x + x^2)$

In Exercises 11–14, simplify the expression.

11. $3b - 2a + a - 10b$

12. $15(u - v) - 7(u - v)$

13. $3z - (4 - z)$

14. $2[10 - (t + 1)]$

In Exercises 15 and 16, evaluate the expression for $x = 2$ and $y = -10$.

15. $x^3 - 2$

16. $x^2 + 4(y + 2)$

17. Explain why it is not possible to evaluate $\dfrac{a + 2b}{3a - b}$ when $a = 2$ and $b = 6$.

18. Translate the phrase "four less than one-third of a number" into an algebraic expression. Let n represent the number.

19. (a) Write expressions for the perimeter and area of the rectangle at the left. Simplify each expression.

(b) Evaluate each expression for $w = 7$.

w

$2w - 4$

Figure for 19

20. The prices of concert tickets for adults and children are \$25 and \$20, respectively.

(a) Write an algebraic expression that represents the total income from the concert for m adults and n children.

(b) How much will it cost two adults and three children to attend the concert?

21. Determine whether the values of x are solutions of $6(3 - x) - 5(2x - 1) = 7$.

(a) $x = -2$ (b) $x = 1$

Study Skills in Action

Knowing Your Preferred Learning Modality

Math is a specific system of rules, properties, and calculations used to solve problems. However, you can take different approaches to learning this specific system based on learning modalities. A learning modality is a preferred way of taking in information that is then transferred into the brain for processing. The three modalities are *visual, auditory,* and *kinesthetic*. The following are brief descriptions of these modalities.

- **Visual** You take in information more productively if you can see the information.
- **Auditory** You take in information more productively when you listen to an explanation and talk about it.
- **Kinesthetic** You take in information more productively if you can experience it or use physical activity in studying.

You may find that one approach, or even a combination of approaches, works best for you.

Kimberly Nolting
VP, Academic Success Press
expert in developmental education

Smart Study Strategy

Use Your Preferred Learning Modality

Visual *Draw a picture of a word problem.*

- Draw a picture of a word problem before writing a verbal model. You do not have to be an artist.
- When making a review card for a word problem, include a picture. This will help you recall the information while taking a test.
- Make sure your notes are visually neat for easy recall.

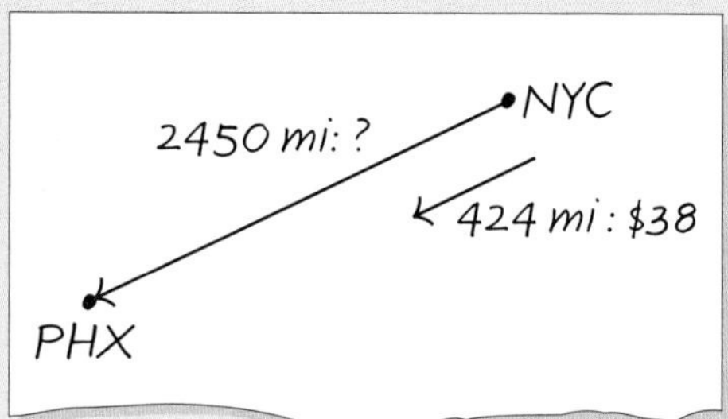

Auditory *Talk about a word problem.*

- Explain how to do a word problem to another student. This is a form of thinking out loud. Write the instructions down on a review card.
- Find several students as serious as you are about math and form a study group.
- Teach the material to an imaginary person when studying alone.

"It takes $38 worth of gas to travel 424 miles. To find the cost of traveling 2450 miles, I can set up and solve a proportion."

Kinesthetic *Incorporate physical activity.*

- Act out a word problem as much as possible. Use props when you can.
- Solve a word problem on a large whiteboard—the physical action of writing is more kinesthetic when the writing is larger and you can move around while doing it.
- Make a review card.

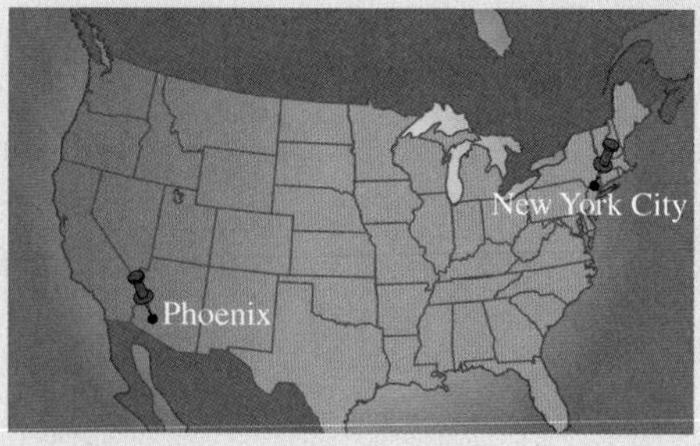

Chapter 3

Equations, Inequalities, and Problem Solving

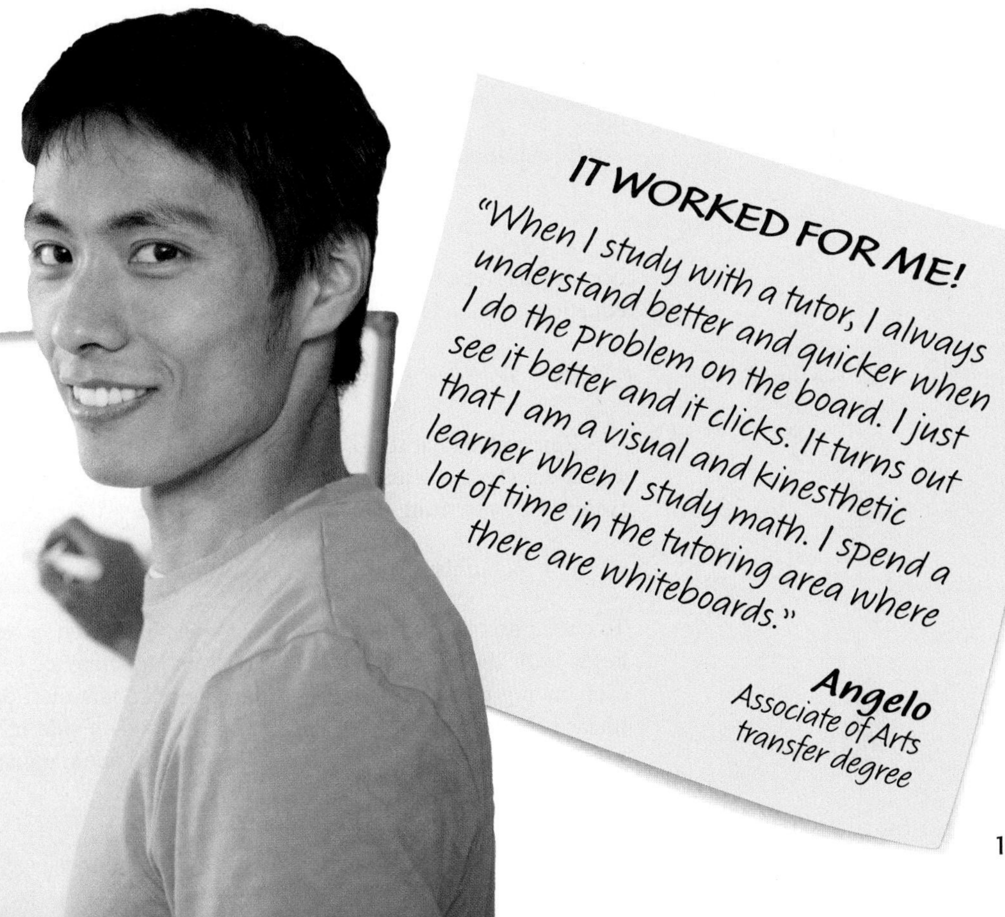

IT WORKED FOR ME!
"When I study with a tutor, I always understand better and quicker when I do the problem on the board. I just see it better and it clicks. It turns out that I am a visual and kinesthetic learner when I study math. I spend a lot of time in the tutoring area where there are whiteboards."
Angelo
Associate of Arts transfer degree

3.1 Solving Linear Equations

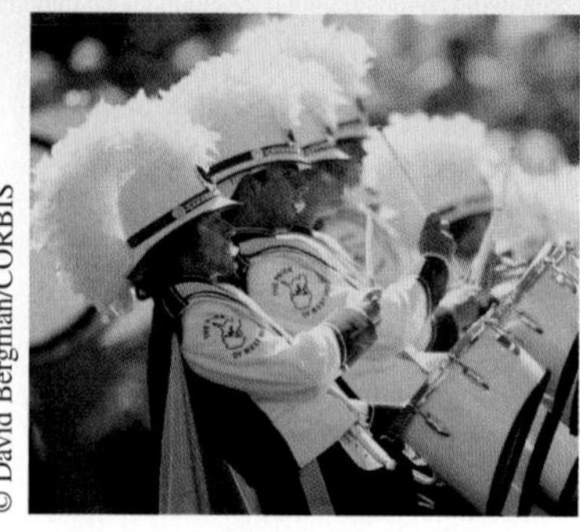
© David Bergman/CORBIS

What You Should Learn

1 ▶ Solve linear equations in standard form.

2 ▶ Solve linear equations in nonstandard form.

3 ▶ Use linear equations to solve application problems.

Why You Should Learn It

Linear equations are used in many real-life applications. For instance, in Exercise 78 on page 135, you will use a linear equation to determine the number of advance tickets sold for a drumline competition.

1 ▶ Solve linear equations in standard form.

Linear Equations in the Standard Form $ax + b = 0$

This is an important step in your study of algebra. In the first two chapters, you were introduced to the rules of algebra, and you learned to use these rules to rewrite and simplify algebraic expressions. In Sections 2.3 and 2.4, you gained experience in translating verbal phrases and problems into algebraic forms. You are now ready to use these skills and experiences to *solve equations.*

In this section, you will learn how the rules of algebra and the properties of equality can be used to solve the most common type of equation—a linear equation in one variable.

Definition of Linear Equation

A **linear equation** in one variable x is an equation that can be written in the standard form

$$ax + b = 0$$

where a and b are real numbers with $a \neq 0$.

A linear equation in one variable is also called a **first-degree equation** because its variable has an (implied) exponent of 1. Some examples of linear equations in standard form are

$$2x = 0, \quad x - 7 = 0, \quad 4x + 6 = 0, \quad \text{and} \quad \frac{x}{2} - 1 = 0.$$

Remember that to *solve* an equation involving x means to find all values of x that satisfy the equation. For the linear equation $ax + b = 0$, the goal is to *isolate* x by rewriting the equation in the form

$$x = \boxed{\text{a number}}. \qquad \text{Isolate the variable } x.$$

To obtain this form, you use the techniques discussed in Section 2.4. That is, you begin with the original equation and write a sequence of equivalent equations, each having the same solution as the original equation. For instance, to solve the linear equation $x - 2 = 0$, you can add 2 to each side of the equation to obtain $x = 2$. As mentioned in Section 2.4, each equivalent equation is called a **step** of the solution.

EXAMPLE 1 Solving a Linear Equation in Standard Form

Solve $3x - 15 = 0$. Then check the solution.

Solution

$3x - 15 = 0$	Write original equation.
$3x - 15 + 15 = 0 + 15$	Add 15 to each side.
$3x = 15$	Combine like terms.
$\frac{3x}{3} = \frac{15}{3}$	Divide each side by 3.
$x = 5$	Simplify.

It appears that the solution is $x = 5$. You can check this as follows.

Check

$3x - 15 = 0$	Write original equation.
$3(5) - 15 \stackrel{?}{=} 0$	Substitute 5 for x.
$15 - 15 \stackrel{?}{=} 0$	Multiply.
$0 = 0$	Solution checks. ✓

So, the solution is $x = 5$.

✓ **CHECKPOINT** *Now try Exercise 17.*

In Example 1, be sure you see that solving an equation has two basic stages. The first stage is to *find* the solution (or solutions). The second stage is to *check* that each solution you find actually satisfies the original equation. You can improve your accuracy in algebra by developing the habit of checking each solution.

A common question in algebra is

"How do I know which step to do *first* to isolate x?"

The answer is that you need practice. By solving many linear equations, you will find that your skill will improve. The key thing to remember is that you can "get rid of" terms and factors by using *inverse* operations. Here are some guidelines and examples.

Guideline	*Equation*	*Inverse Operation*
1. Subtract to remove a sum.	$x + 3 = 0$	Subtract 3 from each side.
2. Add to remove a difference.	$x - 5 = 0$	Add 5 to each side.
3. Divide to remove a product.	$4x = 20$	Divide each side by 4.
4. Multiply to remove a quotient.	$\frac{x}{8} = 2$	Multiply each side by 8.

For additional examples, review Example 3 on page 108. In each case of that example, note how inverse operations are used to isolate the variable.

EXAMPLE 2 Solving a Linear Equation in Standard Form

Solve $2x + 18 = 0$. Then check the solution.

Solution

$2x + 18 = 0$	Write original equation.
$2x + 18 - 18 = 0 - 18$	Subtract 18 from each side.
$2x = -18$	Combine like terms.
$\frac{2x}{2} = -\frac{18}{2}$	Divide each side by 2.
$x = -9$	Simplify.

Check

$2x + 18 = 0$	Write original equation.
$2(-9) + 18 \stackrel{?}{=} 0$	Substitute -9 for x.
$-18 + 18 \stackrel{?}{=} 0$	Multiply.
$0 = 0$	Solution checks. ✓

The solution is $x = -9$.

 **CHECKPOINT** ***Now try Exercise 19.***

EXAMPLE 3 Solving a Linear Equation in Standard Form

Solve $5x - 12 = 0$. Then check the solution.

Solution

$5x - 12 = 0$	Write original equation.
$5x - 12 + 12 = 0 + 12$	Add 12 to each side.
$5x = 12$	Combine like terms.
$\frac{5x}{5} = \frac{12}{5}$	Divide each side by 5.
$x = \frac{12}{5}$	Simplify.

Check

$5x - 12 = 0$	Write original equation.
$5\left(\frac{12}{5}\right) - 12 \stackrel{?}{=} 0$	Substitute $\frac{12}{5}$ for x.
$12 - 12 \stackrel{?}{=} 0$	Multiply.
$0 = 0$	Solution checks. ✓

The solution is $x = \frac{12}{5}$.

CHECKPOINT ***Now try Exercise 31.***

Technology: Tip

Remember to check your solution in the original equation. This can be done efficiently with a graphing calculator.

You know that $x = \frac{12}{5}$ is a solution of the equation in Example 3, but at this point you might be asking, "How can I be sure that the equation does not have other solutions?" The answer is that a linear equation in one variable always has *exactly one* solution. You can show this with the following steps.

$$ax + b = 0 \quad \text{Original equation, with } a \neq 0$$

$$ax + b - b = 0 - b \quad \text{Subtract } b \text{ from each side.}$$

$$ax = -b \quad \text{Combine like terms.}$$

$$\frac{ax}{a} = \frac{-b}{a} \quad \text{Divide each side by } a.$$

$$x = -\frac{b}{a} \quad \text{Simplify.}$$

It is clear that the last equation has only one solution, $x = -b/a$. Because the last equation is equivalent to the original equation, you can conclude that every linear equation in one variable written in standard form has exactly one solution.

Study Tip

To eliminate a fractional coefficient, it may be easier to multiply each side by the *reciprocal* of the fraction than to divide by the fraction itself. Here is an example.

$$-\frac{2}{3}x = 4$$

$$\left(-\frac{3}{2}\right)\left(-\frac{2}{3}\right)x = \left(-\frac{3}{2}\right)4$$

$$x = -\frac{12}{2}$$

$$x = -6$$

EXAMPLE 4 Solving a Linear Equation in Standard Form

Solve $\frac{x}{3} + 3 = 0$.

Solution

$$\frac{x}{3} + 3 = 0 \quad \text{Write original equation.}$$

$$\frac{x}{3} + 3 - 3 = 0 - 3 \quad \text{Subtract 3 from each side.}$$

$$\frac{x}{3} = -3 \quad \text{Combine like terms.}$$

$$3\left(\frac{x}{3}\right) = 3(-3) \quad \text{Multiply each side by 3.}$$

$$x = -9 \quad \text{Simplify.}$$

The solution is $x = -9$. Check this in the original equation.

✓ **CHECKPOINT** *Now try Exercise 35.*

As you gain experience in solving linear equations, you will probably find that you can perform some of the solution steps in your head. For instance, you might solve the equation in Example 4 by writing only the following steps.

$$\frac{x}{3} + 3 = 0 \quad \text{Write original equation.}$$

$$\frac{x}{3} = -3 \quad \text{Subtract 3 from each side.}$$

$$x = -9 \quad \text{Multiply each side by 3.}$$

2 ▶ Solve linear equations in nonstandard form.

Solving a Linear Equation in Nonstandard Form

The definition of linear equation contains the phrase "that can be written in the standard form $ax + b = 0$." This suggests that some linear equations may come in nonstandard or disguised form.

A common type of linear equation is one in which the variable terms are not combined into one term. In such cases, you can begin the solution by *combining like terms.* Note how this is done in the next two examples.

Study Tip

In Example 5, note that the variable in the equation doesn't always have to be x. Any letter can be used.

EXAMPLE 5 Solving a Linear Equation in Nonstandard Form

Solve $3y + 8 - 5y = 4$.

Solution

$3y + 8 - 5y = 4$	Write original equation.
$3y - 5y + 8 = 4$	Group like terms.
$-2y + 8 = 4$	Combine like terms.
$-2y + 8 - 8 = 4 - 8$	Subtract 8 from each side.
$-2y = -4$	Combine like terms.
$\dfrac{-2y}{-2} = \dfrac{-4}{-2}$	Divide each side by -2.
$y = 2$	Simplify.

The solution is $y = 2$. Check this in the original equation.

✓ ***CHECKPOINT*** *Now try Exercise 63.*

Study Tip

You can isolate the variable term on either side of the equal sign. For instance, Example 6 could have been solved in the following way.

$$x + 6 = 2(x - 3)$$
$$x + 6 = 2x - 6$$
$$x - x + 6 = 2x - x - 6$$
$$6 = x - 6$$
$$6 + 6 = x - 6 + 6$$
$$12 = x$$

EXAMPLE 6 Using the Distributive Property

Solve $x + 6 = 2(x - 3)$.

Solution

$x + 6 = 2(x - 3)$	Write original equation.
$x + 6 = 2x - 6$	Distributive Property
$x - 2x + 6 = 2x - 2x - 6$	Subtract $2x$ from each side.
$-x + 6 = -6$	Combine like terms.
$-x + 6 - 6 = -6 - 6$	Subtract 6 from each side.
$-x = -12$	Combine like terms.
$(-1)(-x) = (-1)(-12)$	Multiply each side by -1.
$x = 12$	Simplify.

The solution is $x = 12$. Check this in the original equation.

✓ ***CHECKPOINT*** *Now try Exercise 67.*

EXAMPLE 7 Solving Linear Equations: Special Cases

a.

$2x + 3 = 2(x + 4)$ Original equation

$2x + 3 = 2x + 8$ Distributive Property

$2x - 2x + 3 = 2x - 2x + 8$ Subtract $2x$ from each side.

$3 \neq 8$ Simplify.

Because 3 does not equal 8, you can conclude that the original equation has no solution.

b.

$4(x + 3) = 4x + 12$ Original equation

$4x + 12 = 4x + 12$ Distributive Property

$4x - 4x + 12 = 4x - 4x + 12$ Subtract $4x$ from each side.

$12 = 12$ Simplify.

Because the last equation is true for any value of x, you can conclude that the original equation has infinitely many solutions. This type of equation is called an **identity.**

CHECKPOINT *Now try Exercise 55.*

3 ▶ Use linear equations to solve application problems.

Applications

EXAMPLE 8 Geometry: Dimensions of a Dog Pen

You have 96 feet of fencing to enclose a rectangular pen for your dog. To provide sufficient running space for the dog to exercise, the pen is to be three times as long as it is wide. Find the dimensions of the pen.

Solution

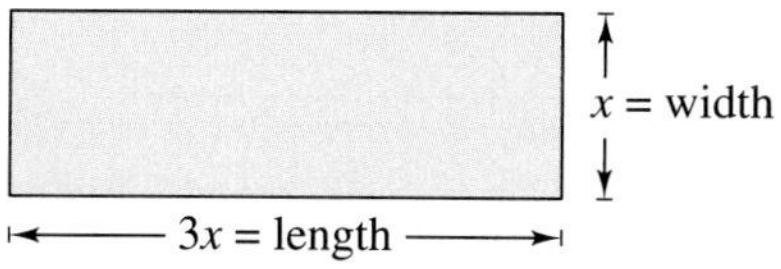

x = width

$3x$ = length

Figure 3.1

Begin by drawing and labeling a diagram, as shown in Figure 3.1. The perimeter of a rectangle is the sum of twice its length and twice its width.

Verbal Model: Perimeter $= 2 \cdot$ Length $+ 2 \cdot$ Width

Algebraic Model: $96 = 2(3x) + 2x$

You can solve this equation as follows.

$96 = 6x + 2x$ Multiply.

$96 = 8x$ Combine like terms.

$\frac{96}{8} = \frac{8x}{8}$ Divide each side by 8.

$12 = x$ Simplify.

So, the width of the pen is 12 feet, and its length is $3(12) = 36$ feet.

CHECKPOINT *Now try Exercise 73.*

Study Tip

When solving a word problem, be sure to ask yourself whether your solution makes sense. For instance, suppose the answer you obtain in Example 9 is -3000 stadium seats sold. This answer does not make sense because the number of stadium seats sold cannot be negative.

EXAMPLE 9 Ticket Sales 

Tickets for a concert cost \$40 for each floor seat and \$20 for each stadium seat. There are 800 seats on the main floor, and these are sold out. The total revenue from ticket sales is \$92,000. How many stadium seats were sold?

Solution

Verbal Model: Total revenue = Revenue from floor seats + Revenue from stadium seats

Labels:		
	Total revenue $= 92{,}000$	(dollars)
	Price per floor seat $= 40$	(dollars per seat)
	Number of floor seats $= 800$	(seats)
	Price per stadium seat $= 20$	(dollars per seat)
	Number of stadium seats $= x$	(seats)

Algebraic Model: $92{,}000 = 40(800) + 20x$

Now that you have written an algebraic equation to represent the problem, you can solve the equation as follows.

$$92{,}000 = 40(800) + 20x \quad \text{Write equation.}$$

$$92{,}000 = 32{,}000 + 20x \quad \text{Simplify.}$$

$$92{,}000 - 32{,}000 = 32{,}000 - 32{,}000 + 20x \quad \text{Subtract 32,000 from each side.}$$

$$60{,}000 = 20x \quad \text{Combine like terms.}$$

$$\frac{60{,}000}{20} = \frac{20x}{20} \quad \text{Divide each side by 20.}$$

$$3000 = x \quad \text{Simplify.}$$

There were 3000 stadium seats sold. To check this solution, you should go back to the original statement of the problem and substitute 3000 stadium seats into the equation. You will find that the total revenue is \$92,000.

✓ ***CHECKPOINT*** *Now try Exercise 77.*

Two integers are called **consecutive integers** if they differ by 1. So, for any integer n, the next two larger consecutive integers are $n + 1$ and $(n + 1) + 1$, or $n + 2$. You can denote three consecutive integers by n, $n + 1$, and $n + 2$.

Expressions for Special Types of Integers

Let n be an integer. Then the following expressions can be used to denote even integers, odd integers, and consecutive integers, respectively.

1. $2n$ denotes an *even* integer.
2. $2n - 1$ and $2n + 1$ denote *odd* integers.
3. The set $\{n, n + 1, n + 2\}$ denotes three *consecutive* integers.

EXAMPLE 10 Consecutive Integers

Find three consecutive integers whose sum is 48.

Solution

Verbal Model: First integer + Second integer + Third integer = 48

Labels: First integer $= n$
Second integer $= n + 1$
Third integer $= n + 2$

Equation:

$$n + (n + 1) + (n + 2) = 48 \qquad \text{Write equation.}$$

$$3n + 3 = 48 \qquad \text{Combine like terms.}$$

$$3n + 3 - 3 = 48 - 3 \qquad \text{Subtract 3 from each side.}$$

$$3n = 45 \qquad \text{Combine like terms.}$$

$$\frac{3n}{3} = \frac{45}{3} \qquad \text{Divide each side by 3.}$$

$$n = 15 \qquad \text{Simplify.}$$

So, the first integer is 15, the second integer is $15 + 1 = 16$, and the third integer is $15 + 2 = 17$. Check this in the original statement of the problem.

✓ **CHECKPOINT** *Now try Exercise 83.*

EXAMPLE 11 Consecutive Even Integers

Find two consecutive even integers such that the sum of the first even integer and three times the second is 78.

Solution

Verbal Model: First even integer + 3 · Second even integer = 78

Labels: First even integer $= 2n$
Second even integer $= 2n + 2$

Equation:

$$2n + 3(2n + 2) = 78 \qquad \text{Write equation.}$$

$$2n + 6n + 6 = 78 \qquad \text{Distributive Property}$$

$$8n + 6 = 78 \qquad \text{Combine like terms.}$$

$$8n + 6 - 6 = 78 - 6 \qquad \text{Subtract 6 from each side.}$$

$$8n = 72 \qquad \text{Combine like terms.}$$

$$n = 9 \qquad \text{Divide each side by 8.}$$

So, the first even integer is $2 \cdot 9 = 18$, and the second even integer is $2 \cdot 9 + 2 = 20$. Check this in the original statement of the problem.

✓ **CHECKPOINT** *Now try Exercise 85.*

Concept Check

1. Let n be an integer. Explain why the expression $2n + 1$ represents an odd integer.

2. Are $x + 3 = 5$ and $x = 2$ equivalent equations? Explain.

3. Without solving the equation, explain why $x + 7 = x + 5$ has no solution.

4. Without solving the equation, explain why $2x + 1 = 2x + 1$ has infinitely many solutions.

3.1 EXERCISES

Go to pages 206–207 to record your assignments.

Developing Skills

Mental Math In Exercises 1–8, solve the equation mentally.

1. $x + 6 = 0$

2. $a + 5 = 0$

3. $x - 9 = 4$

4. $u - 3 = 8$

5. $7y = 28$

6. $4s = 12$

7. $4z = -36$

8. $9z = -63$

In Exercises 9–12, justify each step of the solution. ***See Examples 1–4.***

9.
$$5x + 15 = 0$$
$$5x + 15 - 15 = 0 - 15$$
$$5x = -15$$
$$\frac{5x}{5} = \frac{-15}{5}$$
$$x = -3$$

10.
$$7x - 14 = 0$$
$$7x - 14 + 14 = 0 + 14$$
$$7x = 14$$
$$\frac{7x}{7} = \frac{14}{7}$$
$$x = 2$$

11.
$$-2x - 8 = 0$$
$$-2x - 8 + 8 = 0 + 8$$
$$-2x = 8$$
$$\frac{-2x}{-2} = \frac{8}{-2}$$
$$x = -4$$

12.
$$-3x + 12 = 0$$
$$-3x + 12 - 12 = 0 - 12$$
$$-3x = -12$$
$$\frac{-3x}{-3} = \frac{-12}{-3}$$
$$x = 4$$

In Exercises 13–16, describe the operation that would be used to solve the equation.

13. $x - 8 = 12$

14. $x + 1 = -3$

15. $-\frac{x}{3} = 1$

16. $2x = -11$

In Exercises 17–70, solve the equation and check your solution. (Some equations have no solution.) ***See Examples 1–7.***

✓ **17.** $8x - 16 = 0$

18. $4x - 24 = 0$

✓ **19.** $3x + 21 = 0$

20. $2x + 52 = 0$

21. $5x = 30$

22. $12x = 18$

23. $9x = -21$

24. $-14x = 42$

25. $-8x + 4 = -20$

26. $-7x + 24 = 3$

27. $25x - 4 = 46$

28. $15x - 18 = 12$

29. $10 - 4x = -6$

30. $15 - 3x = -15$

✓ **31.** $6x - 4 = 0$

32. $8z - 2 = 0$

33. $\frac{x}{3} = 10$

34. $-\frac{x}{2} = 3$

✓ 35. $\frac{x}{2} + 1 = 0$

36. $\frac{x}{5} - 2 = 4$

37. $x - \frac{1}{3} = \frac{4}{3}$

38. $x + \frac{5}{2} = \frac{9}{2}$

39. $t - \frac{1}{3} = \frac{1}{2}$

40. $z + \frac{2}{5} = -\frac{3}{10}$

41. $3y - 2 = 2y$

42. $2s - 13 = 28s$

43. $4 - 7x = 5x$

44. $24 - 5x = x$

45. $4 - 5t = 16 + t$

46. $3x + 4 = x + 10$

47. $-3t + 5 = -3t$

48. $4z + 2 = 4z$

49. $15x - 3 = 15 - 3x$

50. $2x - 5 = 7x + 10$

51. $7a - 18 = 3a - 2$

52. $4x - 2 = 3x + 1$

53. $7x + 9 = 3x + 1$

54. $6t - 3 = 8t + 1$

✓ 55. $4x - 6 = 4x - 6$

56. $5 - 3x = 5 - 3x$

57. $2x + 4 = -3(x - 2)$

58. $4(y + 1) = -y + 5$

59. $5(3 - x) = x - 12$

60. $12 - w = -2(3w - 1)$

61. $2x = -3x$

62. $6t = 9t$

✓ 63. $2x - 5 + 10x = 3$

64. $-4x + 10 + 10x = 4$

65. $5t - 4 + 3t = 8t - 4$

66. $7z - 5z - 8 = 2z - 8$

✓ 67. $2(y - 9) = -5y - 4$

68. $-9z + 3 = -(2z - 3)$

69. $3(2 - 7x) = 3(4 - 7x)$

70. $2(5 + 6x) = 4(3x - 1)$

Solving Problems

71. *Geometry* The sides of a yield sign all have the same length (see figure). The perimeter of a roadway yield sign is 225 centimeters. Find the length of each side.

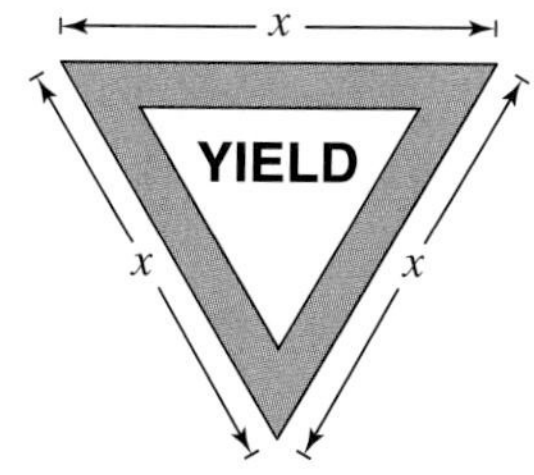

Figure for 71

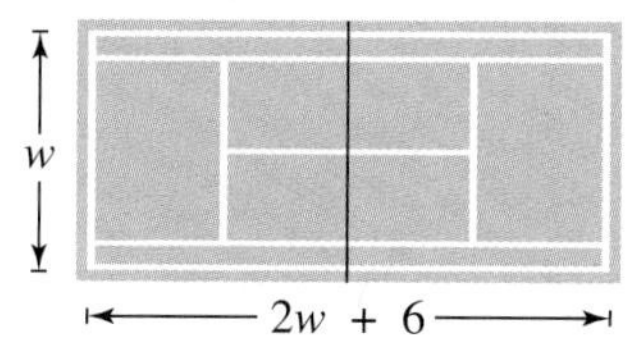

Figure for 72

72. *Geometry* The length of a tennis court is 6 feet more than twice the width (see figure). Find the width of the court if the length is 78 feet.

✓ 73. *Geometry* The perimeter of a Jamaican flag is 120 inches. Its length is twice its width. Find the dimensions of the flag.

74. *Geometry* You are asked to cut a 12-foot board into three pieces. Two pieces are to have the same length and the third is to be twice as long as the others. How long are the pieces?

75. *Car Repair* The bill (including parts and labor) for the repair of your car is shown. Some of the bill is unreadable. From what is given, can you determine how many hours were spent on labor? Explain.

Parts	$285.00
Labor ($44 per hour)	$
Total	**$384.00**

Bill for 75

76. *Car Repair* The bill for the repair of your car was $553. The cost for parts was $265. The cost for labor was $48 per hour. How many hours did the repair work take?

✓ 77. *Ticket Sales* Tickets for a community theater cost $10 for main floor seats and $8 for balcony seats. There are 400 seats on the main floor, and these were sold out for the evening performance. The total revenue from ticket sales was $5200. How many balcony seats were sold?

78. *Ticket Sales* Tickets for a drumline competition cost $5 at the gate and $3 in advance. Eight hundred tickets were sold at the gate. The total revenue from ticket sales was $5500. How many advance tickets were sold?

79. *Summer Jobs* During the summer, you work 40 hours a week and earn $9.25 an hour at a coffee shop. You also tutor for $10.00 an hour and can work as many hours as you want. You want to earn a combined total of $425 a week. How many hours must you tutor?

80. *Summer Jobs* During the summer, you work 30 hours a week and earn \$8.75 an hour at a gas station. You also work as a landscaper for \$11.00 an hour and can work as many hours as you want. You want to earn a combined total of \$400 a week. How many hours must you work at the landscaping job?

81. *Number Problem* Five times the sum of a number and 16 is 100. Find the number.

82. *Number Problem* The sum of twice a number and 27 is 73. Find the number.

✓ **83.** *Number Problem* The sum of two consecutive odd integers is 72. Find the two integers.

84. *Number Problem* The sum of two consecutive even integers is 154. Find the two integers.

✓ **85.** *Number Problem* The sum of three consecutive odd integers is 159. Find the three integers.

86. *Number Problem* The sum of three consecutive even integers is 192. Find the three integers.

Explaining Concepts

87. The scale below is balanced. Each blue box weighs 1 ounce. How much does the red box weigh? If you removed three blue boxes from each side, would the scale still balance? What property of equality does this illustrate?

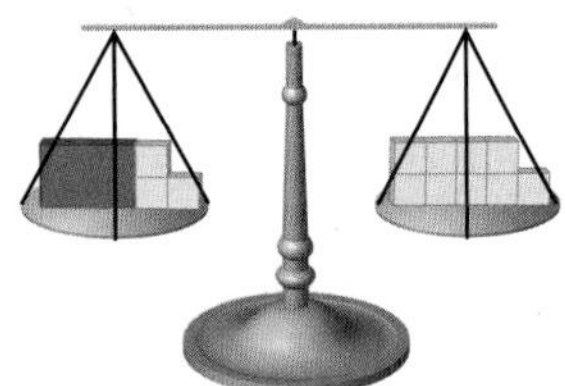

In Exercises 88–91, determine whether the statement is true or false. Justify your answer.

88. Subtracting 0 from each side of an equation yields an equivalent equation.

89. Multiplying each side of an equation by 0 yields an equivalent equation.

90. The sum of two odd integers is even.

91. The sum of an odd integer and an even integer is even.

92. *Finding a Pattern* The length of a rectangle is t times its width (see figure). The rectangle has a perimeter of 1200 meters, which implies that $2w + 2(tw) = 1200$, where w is the width of the rectangle.

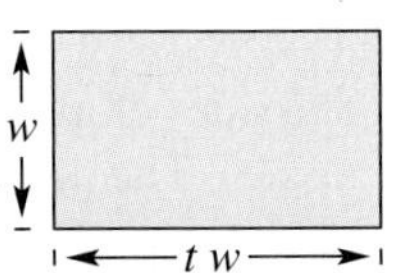

(a) Complete the table.

t	1	1.5	2
Width			
Length			
Area			

t	3	4	5
Width			
Length			
Area			

(b) Use the completed table to draw a conclusion concerning the area of a rectangle of given perimeter as its length increases relative to its width.

Cumulative Review

In Exercises 93–96, plot the numbers on the real number line.

93. $2, -3$

94. $-2.5, 0$

95. $\frac{3}{2}, 1, -1$

96. $4, -\frac{1}{2}, 2.6$

In Exercises 97–102, determine whether (a) $x = -1$ or (b) $x = 2$ is a solution of the equation.

97. $x - 8 = -9$

98. $x + 1.5 = 3.5$

99. $2x - 1 = 3$

100. $3x + 4 = 1$

101. $x + 4 = 2x$

102. $-2(x - 1) = 2 - 2x$

3.2 Equations That Reduce to Linear Form

Robert Brenner/PhotoEdit, Inc.

What You Should Learn

1 ▶ Solve linear equations containing symbols of grouping.
2 ▶ Solve linear equations involving fractions.
3 ▶ Solve linear equations involving decimals.

Why You Should Learn It

Many real-life applications can be modeled with linear equations involving decimals. For instance, Exercise 87 on page 145 shows how a linear equation can model the projected population of the United States.

1 ▶ Solve linear equations containing symbols of grouping.

Equations Containing Symbols of Grouping

In this section you will continue your study of linear equations by looking at more complicated types of linear equations. To solve a linear equation that contains symbols of grouping, use the following guidelines.

1. Remove symbols of grouping from each side by using the Distributive Property.
2. Combine like terms.
3. Isolate the variable using properties of equality.
4. Check your solution in the original equation.

EXAMPLE 1 Solving a Linear Equation Involving Parentheses

Solve $4(x - 3) = 8$. Then check your solution.

Solution

$4(x-3) = 8$	Write original equation.
$4 \cdot x - 4 \cdot 3 = 8$	Distributive Property
$4x - 12 = 8$	Simplify.
$4x - 12 + 12 = 8 + 12$	Add 12 to each side.
$4x = 20$	Combine like terms.
$\frac{4x}{4} = \frac{20}{4}$	Divide each side by 4.
$x = 5$	Simplify.

Check

$4(5-3) \stackrel{?}{=} 8$	Substitute 5 for x in original equation.
$4(2) \stackrel{?}{=} 8$	Simplify.
$8 = 8$	Solution checks. ✓

The solution is $x = 5$.

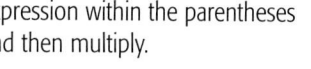

✓ **CHECKPOINT** *Now try Exercise 7.*

Study Tip

Notice in the check of Example 1 that you do not need to use the Distributive Property to remove the parentheses. Simply evaluate the expression within the parentheses and then multiply.

EXAMPLE 2 Solving a Linear Equation Involving Parentheses

Solve $3(2x - 1) + x = 11$. Then check your solution.

Solution

$3(2x - 1) + x = 11$	Write original equation.
$6x - 3 + x = 11$	Distributive Property
$6x + x - 3 = 11$	Group like terms.
$7x - 3 = 11$	Combine like terms.
$7x - 3 + 3 = 11 + 3$	Add 3 to each side.
$7x = 14$	Combine like terms.
$x = 2$	Divide each side by 7.

Check

$3(2x - 1) + x = 11$	Write original equation.
$3[2(2) - 1] + 2 \stackrel{?}{=} 11$	Substitute 2 for x.
$3(4 - 1) + 2 \stackrel{?}{=} 11$	Multiply.
$3(3) + 2 \stackrel{?}{=} 11$	Subtract.
$9 + 2 \stackrel{?}{=} 11$	Multiply.
$11 = 11$	Solution checks. ✓

The solution is $x = 2$.

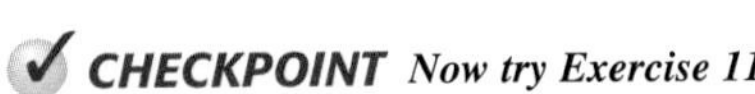
✓ **CHECKPOINT** *Now try Exercise 11.*

EXAMPLE 3 Solving a Linear Equation Involving Parentheses

Solve $5(x + 2) = 2(x - 1)$.

Solution

$5(x + 2) = 2(x - 1)$	Write original equation.
$5x + 10 = 2x - 2$	Distributive Property
$5x - 2x + 10 = 2x - 2x - 2$	Subtract $2x$ from each side.
$3x + 10 = -2$	Combine like terms.
$3x + 10 - 10 = -2 - 10$	Subtract 10 from each side.
$3x = -12$	Combine like terms.
$x = -4$	Divide each side by 3.

The solution is $x = -4$. Check this in the original equation.

✓ **CHECKPOINT** *Now try Exercise 15.*

EXAMPLE 4 Solving a Linear Equation Involving Parentheses

Solve $2(x - 7) - 3(x + 4) = 4 - (5x - 2)$.

Solution

$$\begin{aligned}
2(x - 7) - 3(x + 4) &= 4 - (5x - 2) && \text{Write original equation.}\\
2x - 14 - 3x - 12 &= 4 - 5x + 2 && \text{Distributive Property}\\
-x - 26 &= -5x + 6 && \text{Combine like terms.}\\
-x + 5x - 26 &= -5x + 5x + 6 && \text{Add } 5x \text{ to each side.}\\
4x - 26 &= 6 && \text{Combine like terms.}\\
4x - 26 + 26 &= 6 + 26 && \text{Add 26 to each side.}\\
4x &= 32 && \text{Combine like terms.}\\
x &= 8 && \text{Divide each side by 4.}
\end{aligned}$$

The solution is $x = 8$. Check this in the original equation.

CHECKPOINT *Now try Exercise 25.*

The linear equation in the next example involves both brackets and parentheses. Watch out for nested symbols of grouping such as these. The *innermost symbols of grouping* should be removed first.

Technology: Tip

Try using your graphing calculator to check the solution found in Example 5. You will need to nest some parentheses inside other parentheses. This will give you practice working with nested parentheses on a graphing calculator.

Left side of equation

$$5\left(-\frac{1}{3}\right) - 2\left(4\left(-\frac{1}{3}\right) + 3\left(\left(-\frac{1}{3}\right) - 1\right)\right)$$

Right side of equation

$$8 - 3\left(-\frac{1}{3}\right)$$

EXAMPLE 5 An Equation Involving Nested Symbols of Grouping

Solve $5x - 2[4x + 3(x - 1)] = 8 - 3x$.

Solution

$$\begin{aligned}
5x - 2[4x + 3(x - 1)] &= 8 - 3x && \text{Write original equation.}\\
5x - 2[4x + 3x - 3] &= 8 - 3x && \text{Distributive Property}\\
5x - 2[7x - 3] &= 8 - 3x && \text{Combine like terms inside brackets.}\\
5x - 14x + 6 &= 8 - 3x && \text{Distributive Property}\\
-9x + 6 &= 8 - 3x && \text{Combine like terms.}\\
-9x + 3x + 6 &= 8 - 3x + 3x && \text{Add } 3x \text{ to each side.}\\
-6x + 6 &= 8 && \text{Combine like terms.}\\
-6x + 6 - 6 &= 8 - 6 && \text{Subtract 6 from each side.}\\
-6x &= 2 && \text{Combine like terms.}\\
x &= -\tfrac{2}{6} = -\tfrac{1}{3} && \text{Divide each side by } -6.
\end{aligned}$$

The solution is $x = -\frac{1}{3}$. Check this in the original equation.

CHECKPOINT *Now try Exercise 31.*

2 ▶ Solve linear equations involving fractions.

Equations Involving Fractions or Decimals

To solve a linear equation that contains one or more fractions, it is usually best to first *clear the equation of fractions.*

Clearing an Equation of Fractions

An equation such as

$$\frac{x}{a} + \frac{b}{c} = d$$

that contains one or more fractions can be cleared of fractions by multiplying each side by the least common multiple (LCM) of a and c.

For example, the equation

$$\frac{3x}{2} - \frac{1}{3} = 2$$

can be cleared of fractions by multiplying each side by 6, the LCM of 2 and 3. Notice how this is done in the next example.

Study Tip

For an equation that contains a *single numerical* fraction, such as $2x - \frac{3}{4} = 1$, you can simply add $\frac{3}{4}$ to each side and then solve for x. You do not need to clear the fraction.

$$2x - \frac{3}{4} + \frac{3}{4} = 1 + \frac{3}{4} \qquad \text{Add } \tfrac{3}{4}.$$

$$2x = \frac{7}{4} \qquad \text{Combine terms.}$$

$$x = \frac{7}{8} \qquad \text{Multiply by } \tfrac{1}{2}.$$

EXAMPLE 6 Solving a Linear Equation Involving Fractions

Solve $\frac{3x}{2} - \frac{1}{3} = 2$.

Solution

$$6\left(\frac{3x}{2} - \frac{1}{3}\right) = 6 \cdot 2 \qquad \text{Multiply each side by LCM 6.}$$

$$6 \cdot \frac{3x}{2} - 6 \cdot \frac{1}{3} = 12 \qquad \text{Distributive Property}$$

$$9x - 2 = 12 \qquad \text{Simplify.}$$

$$9x = 14 \qquad \text{Add 2 to each side.}$$

$$x = \frac{14}{9} \qquad \text{Divide each side by 9.}$$

The solution is $x = \frac{14}{9}$. Check this in the original equation.

✓ **CHECKPOINT** *Now try Exercise 37.*

To check a fractional solution such as $\frac{14}{9}$ in Example 6, it is helpful to rewrite the variable term as a product.

$$\frac{3}{2} \cdot x - \frac{1}{3} = 2 \qquad \text{Write fraction as a product.}$$

In this form the substitution of $\frac{14}{9}$ for x is easier to calculate.

EXAMPLE 7 Solving a Linear Equation Involving Fractions

Solve $\frac{x}{5} + \frac{3x}{4} = 19$. Then check your solution.

Solution

$$\frac{x}{5} + \frac{3x}{4} = 19$$ Write original equation.

$$20\left(\frac{x}{5}\right) + 20\left(\frac{3x}{4}\right) = 20(19)$$ Multiply each side by LCM 20.

$$4x + 15x = 380$$ Simplify.

$$19x = 380$$ Combine like terms.

$$x = 20$$ Divide each side by 19.

Check

$$\frac{20}{5} + \frac{3(20)}{4} \stackrel{?}{=} 19$$ Substitute 20 for x in original equation.

$$4 + 15 \stackrel{?}{=} 19$$ Simplify.

$$19 = 19$$ Solution checks. 

The solution is $x = 20$.

✓ CHECKPOINT *Now try Exercise 41.*

Study Tip

Notice in Example 8 that to clear all fractions in the equation, you multiply by 12, which is the LCM of 3, 4, and 2.

EXAMPLE 8 Solving a Linear Equation Involving Fractions

Solve $\frac{2}{3}\left(x + \frac{1}{4}\right) = \frac{1}{2}$.

Solution

$$\frac{2}{3}\left(x + \frac{1}{4}\right) = \frac{1}{2}$$ Write original equation.

$$\frac{2}{3}x + \frac{2}{12} = \frac{1}{2}$$ Distributive Property

$$12 \cdot \frac{2}{3}x + 12 \cdot \frac{2}{12} = 12 \cdot \frac{1}{2}$$ Multiply each side by LCM 12.

$$8x + 2 = 6$$ Simplify.

$$8x = 4$$ Subtract 2 from each side.

$$x = \frac{4}{8} = \frac{1}{2}$$ Divide each side by 8.

The solution is $x = \frac{1}{2}$. Check this in the original equation.

 **CHECKPOINT** *Now try Exercise 49.*

A common type of linear equation is one that equates two fractions. To solve such an equation, consider the fractions to be **equivalent** and use **cross-multiplication.** That is, if

$$\frac{a}{b} = \frac{c}{d}, \quad \text{then} \quad a \cdot d = b \cdot c.$$

Note how cross-multiplication is used in the next example.

EXAMPLE 9 Using Cross-Multiplication

Use cross-multiplication to solve $\frac{x+2}{3} = \frac{8}{5}$. Then check your solution.

Solution

$$\frac{x+2}{3} = \frac{8}{5} \qquad \text{Write original equation.}$$

$$5(x+2) = 3(8) \qquad \text{Cross-multiply.}$$

$$5x + 10 = 24 \qquad \text{Distributive Property}$$

$$5x = 14 \qquad \text{Subtract 10 from each side.}$$

$$x = \frac{14}{5} \qquad \text{Divide each side by 5.}$$

Check

$$\frac{x+2}{3} = \frac{8}{5} \qquad \text{Write original equation.}$$

$$\frac{\left(\frac{14}{5} + 2\right)}{3} \stackrel{?}{=} \frac{8}{5} \qquad \text{Substitute } \tfrac{14}{5} \text{ for } x.$$

$$\frac{\left(\frac{14}{5} + \frac{10}{5}\right)}{3} \stackrel{?}{=} \frac{8}{5} \qquad \text{Write 2 as } \tfrac{10}{5}.$$

$$\frac{\frac{24}{5}}{3} \stackrel{?}{=} \frac{8}{5} \qquad \text{Simplify.}$$

$$\frac{24}{5}\left(\frac{1}{3}\right) \stackrel{?}{=} \frac{8}{5} \qquad \text{Invert and multiply.}$$

$$\frac{8}{5} = \frac{8}{5} \qquad \text{Solution checks. ✓}$$

The solution is $x = \frac{14}{5}$.

CHECKPOINT *Now try Exercise 57.*

Bear in mind that cross-multiplication can be used only with equations written in a form that equates two fractions. Try rewriting the equation in Example 6 in this form, and then use cross-multiplication to solve for x.

More extensive applications of cross-multiplication will be discussed when you study ratios and proportions later in this chapter.

3 ▶ Solve linear equations involving decimals.

Many real-life applications of linear equations involve decimal coefficients. To solve such an equation, you can clear it of decimals in much the same way you clear an equation of fractions. Multiply each side by a power of 10 that converts all decimal coefficients to integers, as shown in the next example.

Study Tip

There are other ways to solve the decimal equation in Example 10. You could first clear the equation of decimals by multiplying each side by 100. Or, you could keep the decimals and use a graphing calculator to do the arithmetic operations. The method you choose is a matter of personal preference.

EXAMPLE 10 Solving a Linear Equation Involving Decimals

Solve $0.3x + 0.2(10 - x) = 0.15(30)$. Then check your solution.

Solution

$$\begin{aligned} 0.3x + 0.2(10 - x) &= 0.15(30) && \text{Write original equation.} \\ 0.3x + 2 - 0.2x &= 4.5 && \text{Distributive Property} \\ 0.1x + 2 &= 4.5 && \text{Combine like terms.} \\ 10(0.1x + 2) &= 10(4.5) && \text{Multiply each side by 10.} \\ x + 20 &= 45 && \text{Simplify.} \\ x &= 25 && \text{Subtract 20 from each side.} \end{aligned}$$

Check

$$\begin{aligned} 0.3(25) + 0.2(10 - 25) &\stackrel{?}{=} 0.15(30) && \text{Substitute 25 for } x \text{ in original equation.} \\ 0.3(25) + 0.2(-15) &\stackrel{?}{=} 0.15(30) && \text{Perform subtraction within parentheses.} \\ 7.5 - 3.0 &\stackrel{?}{=} 4.5 && \text{Multiply.} \\ 4.5 &= 4.5 && \text{Solution checks. ✓} \end{aligned}$$

The solution is $x = 25$.

✓ ***CHECKPOINT*** *Now try Exercise 75.*

EXAMPLE 11 Financial Assistance 

The amount y (in billions) of financial assistance awarded to students from 2001 to 2005 can be approximated by the linear model $y = 6.06t + 43.3$, where t represents the year, with $t = 1$ corresponding to 2001. Use the model to predict the year in which the financial assistance awarded to students will be \$100 billion. (Source: U.S. Department of Education)

Solution

To find the year in which there will be \$100 billion awarded to students, substitute 100 for y in the original equation and solve the equation for t.

$$\begin{aligned} 100 &= 6.06t + 43.3 && \text{Substitute 100 for } y \text{ in original equation.} \\ 56.7 &= 6.06t && \text{Subtract 43.3 from each side.} \\ 9.4 &\approx t && \text{Divide each side by 6.06.} \end{aligned}$$

Because $t = 1$ corresponds to 2001, the financial assistance awarded to students will be \$100 billion during 2009. Check this in the original statement of the problem.

✓ ***CHECKPOINT*** *Now try Exercise 87.*

Concept Check

1. In your own words, describe the procedure for removing symbols of grouping. Give some examples.
2. What is meant by the least common multiple of the denominators of two or more fractions? Discuss one method for finding the least common multiple of the denominators of fractions.
3. When solving an equation that contains fractions, explain what is accomplished by multiplying each side of the equation by the least common multiple of the denominators.
4. Explain the procedure for eliminating decimals in an equation.

3.2 EXERCISES

Go to pages 206–207 to record your assignments.

Developing Skills

In Exercises 1–54, solve the equation and check your solution. (Some of the equations have no solution.) ***See Examples 1–8.***

1. $2(y - 4) = 0$
2. $9(y - 7) = 0$
3. $-5(t + 3) = 10$
4. $-3(x + 1) = 18$
5. $25(z - 2) = 60$
6. $2(x - 3) = 4$
7. ✓ $7(x + 5) = 49$
8. $4(x + 1) = 24$
9. $4 - (z + 6) = 8$
10. $25 - (y + 3) = 10$
11. ✓ $3 - (2x - 18) = 3$
12. $16 - (3x - 10) = 5$
13. $12(x - 3) = 0$
14. $4(z - 8) = 0$
15. ✓ $5(x - 4) = 2(2x + 5)$
16. $-(4x + 10) = 6(x + 2)$
17. $3(2x - 1) = 3(2x + 5)$
18. $4(z - 2) = 2(2z - 4)$
19. $-3(x + 4) = 4(x + 4)$
20. $-8(x - 6) = 3(x - 6)$
21. $7 = 3(x + 2) - 3(x - 5)$
22. $24 = 12(z + 1) - 3(4z - 2)$
23. $7x - 2(x - 2) = 12$
24. $15(x + 1) - 8x = 29$
25. ✓ $4 - (y - 3) = 3(y + 1) - 4(1 - y)$
26. $12 - 2(y + 3) = 4(y - 6) - (y - 1)$
27. $-6(3 + x) + 2(3x + 5) = 0$
28. $-3(5x + 2) + 5(1 + 3x) = 0$
29. $2[(3x + 5) - 7] = 3(4x - 3)$
30. $3[(5x + 1) - 4] = 4(2x - 3)$
31. ✓ $4x + 3[x - 2(2x - 1)] = 4 - 3x$
32. $16 + 4[5x - 4(x + 2)] = 7 - 2x$
33. $\frac{y}{5} = \frac{3}{5}$
34. $\frac{z}{3} = \frac{10}{3}$
35. $\frac{y}{5} = -\frac{3}{10}$
36. $\frac{v}{4} = -\frac{7}{8}$
37. ✓ $\frac{6x}{25} = \frac{3}{5}$
38. $\frac{8x}{9} = \frac{2}{3}$
39. $\frac{5x}{4} + \frac{1}{2} = 0$
40. $\frac{3z}{7} + \frac{5}{14} = 0$
41. ✓ $\frac{x}{5} - \frac{1}{2} = 3$
42. $\frac{y}{6} - \frac{5}{8} = 2$
43. $\frac{x}{5} - \frac{x}{2} = 1$
44. $\frac{x}{3} + \frac{x}{4} = 1$
45. $2s + \frac{3}{2} = 2s + 2$
46. $\frac{3}{4} + 5s = -2 + 5s$
47. $3x + \frac{1}{4} = \frac{3}{4}$
48. $2x - \frac{3}{8} = \frac{5}{8}$
49. ✓ $\frac{1}{5}x + 1 = \frac{3}{10}x - 4$
50. $\frac{1}{8}x + 3 = \frac{1}{4}x + 5$
51. $\frac{2}{3}(z + 5) - \frac{1}{4}(z + 24) = 0$
52. $\frac{3x}{2} + \frac{1}{4}(x - 2) = 10$
53. $\frac{100 - 4u}{3} = \frac{5u + 6}{4} + 6$
54. $\frac{8 - 3x}{2} - 4 = \frac{x}{6}$

In Exercises 55 and 56, the perimeter of the figure is 15. Find the value of x.

55. 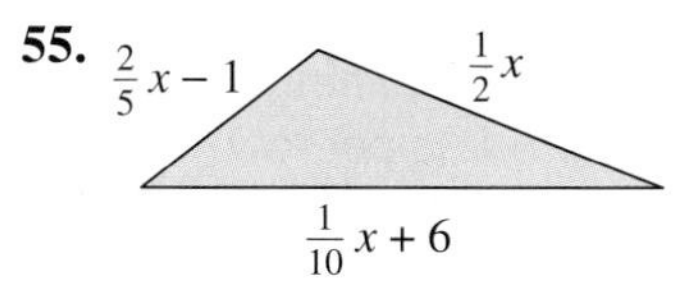

56. 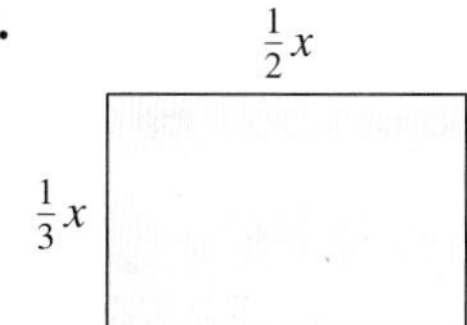

In Exercises 57–66, use cross-multiplication to solve the equation. ***See Example 9.***

57. $\frac{t+4}{6} = \frac{2}{3}$ **58.** $\frac{x-6}{10} = \frac{3}{5}$

59. $\frac{x-2}{5} = \frac{2}{3}$ **60.** $\frac{2x+1}{3} = \frac{5}{2}$

61. $\frac{5x-4}{4} = \frac{2}{3}$ **62.** $\frac{10x+3}{6} = \frac{3}{2}$

63. $\frac{x}{4} = \frac{1-2x}{3}$ **64.** $\frac{x+1}{6} = \frac{3x}{10}$

65. $\frac{10-x}{2} = \frac{x+4}{5}$ **66.** $\frac{2x+3}{5} = \frac{3-4x}{8}$

In Exercises 67–70, state the power of 10 that is needed to clear the decimals in the equation.

67. $3 + 0.03x = 5$ **68.** $0.4x - 0.1 = 2$

69. $1.205x - 0.003 = 0.5$

70. $5.225 + 3.001x = 10.275$

In Exercises 71–80, solve the equation. Round your answer to two decimal places. ***See Example 10.***

71. $0.2x + 5 = 6$ **72.** $4 - 0.3x = 1$

73. $0.234x + 1 = 2.805$ **74.** $2.75x - 3.13 = 5.12$

75. $0.42x - 0.4(x + 2.4) = 0.3(5)$

76. $1.6x + 0.25(12 - x) = 0.43(-12)$

77. $\frac{x}{3.25} + 1 = 2.08$ **78.** $\frac{x}{4.08} + 7.2 = 5.14$

79. $\frac{x}{3.155} = 2.850$ **80.** $\frac{3x}{4.5} = \frac{1}{8}$

Solving Problems

81. *Time to Complete a Task* Two people can complete 80% of a task in t hours, where t must satisfy the equation $t/10 + t/15 = 0.8$. How long will it take for the two people to complete 80% of the task?

82. *Time to Complete a Task* Two machines can complete a task in t hours, where t must satisfy the equation $t/10 + t/15 = 1$. How long will it take for the two machines to complete the task?

83. *Course Grade* To get an A in a course, you must have an average of at least 90 points for four tests of 100 points each. For the first three tests, your scores are 87, 92, and 84. What must you score on the fourth exam to earn a 90% average for the course?

84. *Course Grade* Repeat Exercise 83 under the condition that the fourth test is weighted so that it counts for twice as much as each of the first three tests.

In Exercises 85 and 86, use the equation and solve for x.

$p_1x + p_2(a - x) = p_3a$

85. *Mixture Problem* Determine the number of quarts of a 10% solution that must be mixed with a 30% solution to obtain 100 quarts of a 25% solution. ($p_1 = 0.1$, $p_2 = 0.3$, $p_3 = 0.25$, and $a = 100$.)

86. *Mixture Problem* Determine the number of gallons of a 25% solution that must be mixed with a 50% solution to obtain 5 gallons of a 30% solution. ($p_1 = 0.25$, $p_2 = 0.5$, $p_3 = 0.3$, and $a = 5$.)

87. *Data Analysis* The table shows the projected numbers N (in millions) of persons living in the United States. (Source: U.S. Census Bureau)

Year	2010	2015	2020	2025
N	308.9	322.4	335.8	349.4

A model for the data is $N = 2.70t + 281.9$, where t represents time in years, with $t = 10$ corresponding to the year 2010. According to the model, in what year will the population exceed 360 million?

88. *Data Analysis* The table shows the net sales N (in millions) of Coach for the years 2003 through 2007. (Source: Coach, Inc.)

Year	2003	2004	2005	2006	2007
N	0.93	1.27	1.65	2.04	2.61

A model for the data is

$$N = 0.413t - 0.37$$

where t represents time in years, with $t = 3$ corresponding to the year 2003. According to the model, in what year will the net sales exceed 4 million?

Fireplace Construction In Exercises 89 and 90, use the following information. A fireplace is 93 inches wide. Each brick in the fireplace has a length of 8 inches, and there is $\frac{1}{2}$ inch of mortar between adjoining bricks (see figure). Let n be the number of bricks per row.

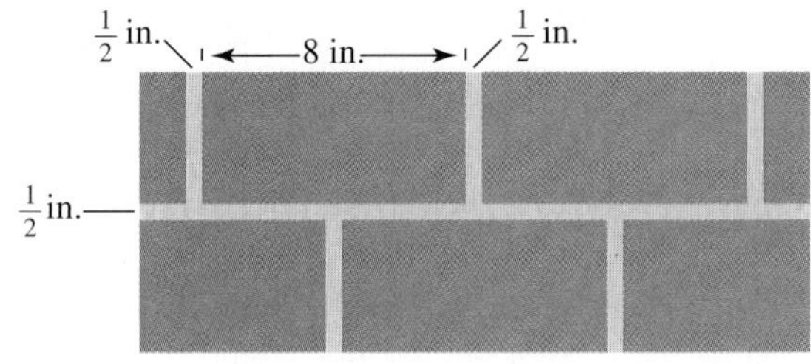

89. Explain why the number of bricks per row is the solution of the equation $8n + \frac{1}{2}(n - 1) = 93$.

90. Find the number of bricks per row in the fireplace.

Explaining Concepts

91. You could solve $3(x - 7) = 15$ by applying the Distributive Property as the first step. However, there is another way to begin. What is it?

92. *Error Analysis* Describe and correct the error.

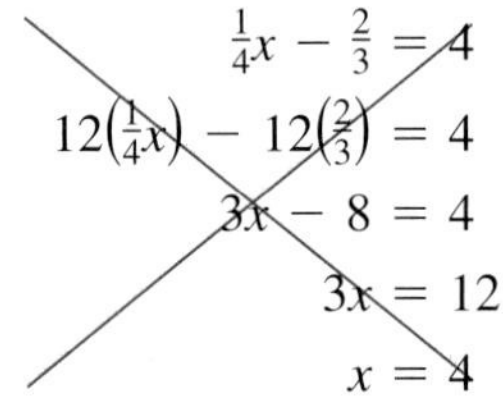

$$\frac{1}{4}x - \frac{2}{3} = 4$$
$$12\left(\frac{1}{4}x\right) - 12\left(\frac{2}{3}\right) = 4$$
$$3x - 8 = 4$$
$$3x = 12$$
$$x = 4$$

93. Explain what happens when you divide each side of a linear equation by a variable factor.

94. When simplifying an algebraic *expression* involving fractions, why can't you simplify the expression by multiplying by the least common multiple of the denominators?

Cumulative Review

In Exercises 95–102, simplify the expression.

95. $(-2x)^2x^4$

96. $-y^2(-2y)^3$

97. $5z^3(z^2)$

98. $a^2 + 3a + 4 - 2a - 6$

99. $\dfrac{5x}{3} - \dfrac{2x}{3} - 4$

100. $2x^2 - 4 + 5 - 3x^2$

101. $-y^2(y^2 + 4) + 6y^2$

102. $5t(2 - t) + t^2$

In Exercises 103–106, solve the equation and check your solution.

103. $3x - 5 = 12$

104. $-5x + 9 = 9$

105. $4 - 2x = 22$

106. $x - 3 + 7x = 29$

3.3 Problem Solving with Percents

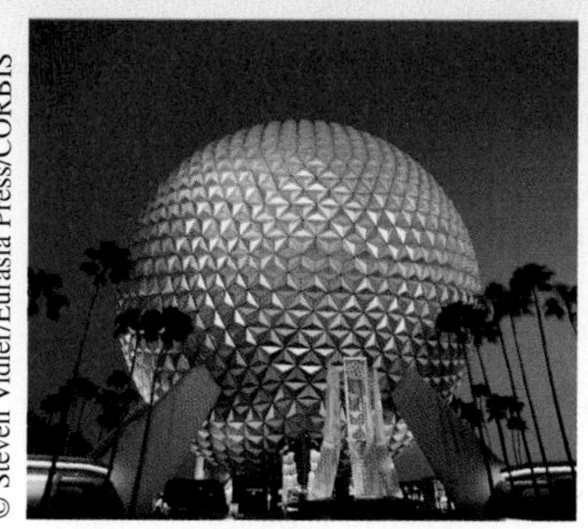
© Steven Vidler/Eurasia Press/CORBIS

Why You Should Learn It

Real-life data can be organized using circle graphs and percents. For instance, in Exercise 95 on page 157, a circle graph is used to classify the revenue of the Walt Disney Company.

What You Should Learn

1 ▶ Convert percents to decimals and fractions, and convert decimals and fractions to percents.
2 ▶ Solve linear equations involving percents.
3 ▶ Solve application problems involving markups and discounts.

1 ▶ Convert percents to decimals and fractions, and convert decimals and fractions to percents.

Percents

In applications involving percents, you usually must convert the percents to decimal (or fractional) form before performing any arithmetic operations. Consequently, you need to be able to convert from percents to decimals (or fractions), and vice versa. The following verbal model can be used to perform the conversions.

Decimal or fraction · 100% = Percent

For example, the decimal 0.38 corresponds to 38 percent. That is,

$$0.38(100\%) = 38\%.$$

Study Tip

Note in Example 1(a) that it is possible to have percents that are larger than 100%. It is also possible to have percents that are less than 1%, such as $\frac{1}{2}\%$ or 0.78%.

EXAMPLE 1 Converting Decimals and Fractions to Percents

Convert each number to a percent.

a. 1.20 **b.** $\frac{3}{5}$

Solution

a. *Verbal Model:* Decimal · 100% = Percent

Equation: $(1.20)(100\%) = 120\%$

So, the decimal 1.20 corresponds to 120%.

b. *Verbal Model:* Fraction · 100% = Percent

Equation: $\frac{3}{5}(100\%) = \frac{300}{5}\%$

$= 60\%$

So, the fraction $\frac{3}{5}$ corresponds to 60%.

✓ **CHECKPOINT** *Now try Exercise 13.*

Study Tip

In Examples 1 and 2, there is a quick way to convert between percent form and decimal form.

- To convert from percent form to decimal form, move the decimal point two places to the *left*. For instance,

$$3.5\% = 0.035.$$

- To convert from decimal form to percent form, move the decimal point two places to the *right*. For instance,

$$1.20 = 120\%.$$

- Decimal-to-fraction or fraction-to-decimal conversions can be done on a calculator. Consult your user's guide.

EXAMPLE 2 Converting Percents to Decimals and Fractions

a. Convert 3.5% to a decimal.

b. Convert 55% to a fraction.

Solution

a. *Verbal Model:* Decimal $\cdot$ 100% $=$ Percent

Label: $x =$ decimal

Equation:

$x(100\%) = 3.5\%$ Write equation.

$x = \dfrac{3.5\%}{100\%}$ Divide each side by 100%.

$x = 0.035$ Simplify.

So, 3.5% corresponds to the decimal 0.035.

b. *Verbal Model:* Fraction $\cdot$ 100% $=$ Percent

Label: $x =$ fraction

Equation:

$x(100\%) = 55\%$ Write equation.

$x = \dfrac{55\%}{100\%}$ Divide each side by 100%.

$x = \dfrac{11}{20}$ Simplify.

So, 55% corresponds to the fraction $\frac{11}{20}$.

CHECKPOINT *Now try Exercise 21.*

Some percents occur so commonly that it is helpful to memorize their conversions. For instance, 100% corresponds to 1 and 200% corresponds to 2. The table below shows the decimal and fraction conversions for several percents.

Percent	10%	$12\frac{1}{2}\%$	20%	25%	$33\frac{1}{3}\%$	50%	$66\frac{2}{3}\%$	75%
Decimal	0.1	0.125	0.2	0.25	$0.\overline{3}$	0.5	$0.\overline{6}$	0.75
Fraction	$\frac{1}{10}$	$\frac{1}{8}$	$\frac{1}{5}$	$\frac{1}{4}$	$\frac{1}{3}$	$\frac{1}{2}$	$\frac{2}{3}$	$\frac{3}{4}$

Percent means *per hundred* or *parts of 100.* (The Latin word for 100 is *centum.*) For example, 20% means 20 parts of 100, which is equivalent to the fraction $20/100$ or $\frac{1}{5}$. In applications involving percent, many people like to state percent in terms of a portion. For instance, the statement "20% of the population lives in apartments" is often stated as "1 out of every 5 people lives in an apartment."

2 ▶ Solve linear equations involving percents.

The Percent Equation

The primary use of percents is to compare two numbers. For example, 2 is 50% of 4, and 5 is 25% of 20. The following model is helpful.

Verbal Model: $a \quad = \quad p$ percent of b

Labels: b = base number
p = percent (in decimal form)
a = number being compared to b

Equation: $a = p \cdot b$

EXAMPLE 3 Solving Percent Equations

a. What number is 30% of 70?

b. Fourteen is 25% of what number?

c. One hundred thirty-five is what percent of 27?

Solution

a. *Verbal Model:* What number = 30% of 70

Label: a = unknown number

Equation: $a = (0.3)(70) = 21$

So, 21 is 30% of 70.

b. *Verbal Model:* 14 = 25% of what number

Label: b = unknown number

Equation:
$$14 = 0.25b$$
$$\frac{14}{0.25} = b$$
$$56 = b$$

So, 14 is 25% of 56.

c. *Verbal Model:* 135 = What percent of 27

Label: p = unknown percent (in decimal form)

Equation:
$$135 = p(27)$$
$$\frac{135}{27} = p$$
$$5 = p$$

So, 135 is 500% of 27.

CHECKPOINT *Now try Exercise 51.*

Study Tip

In most real-life applications, a and b are more disguised than they are in Example 3. It may help to think of a as a "new" amount and b as the "original" amount.

From Example 3, you can see that there are three basic types of percent problems. Each can be solved by substituting the two given quantities into the percent equation and solving for the third quantity.

Question	*Given*	*Percent Equation*
What number is p percent of b?	p and b	Solve for a.
a is p percent of what number?	a and p	Solve for b.
a is what percent of b?	a and b	Solve for p.

EXAMPLE 4 Real Estate Commission 

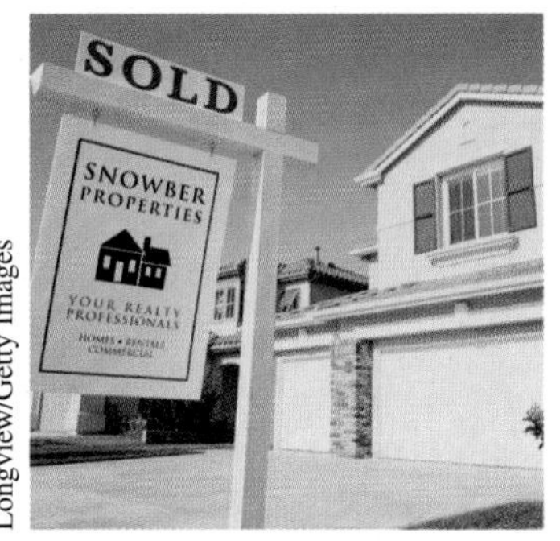

Longview/Getty Images

In 2007, approximately 774,000 new houses were sold, with a median price of $246,900. (Source: U.S. Census Bureau)

A real estate agency receives a commission of $8092.50 for the sale of a $124,500 house. What percent commission is this?

Solution

Verbal Model: Commission = Percent (in decimal form) · Sale price

Labels:
Commission = 8092.50 (dollars)
Percent = p (decimal form)
Sale price = 124,500 (dollars)

Equation:

$$8092.50 = p(124{,}500) \qquad \text{Write equation.}$$

$$\frac{8092.50}{124{,}500} = p \qquad \text{Divide each side by 124,500.}$$

$$0.065 = p \qquad \text{Simplify.}$$

So, the real estate agency receives a commission of 6.5%.

 **CHECKPOINT** *Now try Exercise 83.*

EXAMPLE 5 Cost-of-Living Raise 

A union negotiates for a cost-of-living raise of 7%. What is the raise for a union member whose salary is $33,240? What is this person's new salary?

Solution

Verbal Model: Raise = Percent (in decimal form) · Salary

Labels:
Raise = a (dollars)
Percent = 7% = 0.07 (decimal form)
Salary = 33,240 (dollars)

Equation: $a = 0.07(33{,}240) = 2326.80$

So, the raise is $2326.80 and the new salary is

$$33{,}240.00 + 2326.80 = \$35{,}566.80.$$

CHECKPOINT *Now try Exercise 88.*

EXAMPLE 6 Course Grade 

You missed an A in your chemistry course by only three points. Your point total for the course is 402. How many points were possible in the course? (Assume that you needed 90% of the course total for an A.)

Solution

Verbal Model:

$$\boxed{\text{Your points}} + \boxed{\text{3 points}} = \boxed{\text{Percent (in decimal form)}} \cdot \boxed{\text{Total points}}$$

Labels:
Your points $= 402$ (points)
Percent $= 90\% = 0.9$ (decimal form)
Total points for course $= b$ (points)

Equation:

$402 + 3 = 0.9b$ Write equation.

$405 = 0.9b$ Add.

$450 = b$ Divide each side by 0.9.

So, there were 450 possible points in the course. You can check your solution as follows.

$402 + 3 = 0.9b$ Write original equation.

$402 + 3 \stackrel{?}{=} 0.9(450)$ Substitute 450 for b.

$405 = 405$ Solution checks. ✓

✓ **CHECKPOINT** *Now try Exercise 89.*

EXAMPLE 7 Percent Increase 

The average monthly price for expanded basic cable programming packages was \$10.67 in 1986 and \$41.17 in 2006. Find the percent increase in the price from 1986 to 2006. (Source: National Cable & Telecommunications Association)

Solution

Verbal Model:

$$\boxed{\text{2006 price}} = \boxed{\text{1986 price}} \cdot \boxed{\text{Percent increase (in decimal form)}} + \boxed{\text{1986 price}}$$

Labels:
2006 price $= 41.17$ (dollars)
Percent increase $= p$ (decimal form)
1986 price $= 10.67$ (dollars)

Equation:

$41.17 = 10.67p + 10.67$ Write equation.

$30.5 = 10.67p$ Subtract 10.67 from each side.

$2.86 \approx p$ Divide each side by 10.67.

So, the percent increase in the average monthly price for expanded basic cable programming packages from 1986 to 2006 is approximately 286%. Check this in the original statement of the problem.

 **CHECKPOINT** *Now try Exercise 86.*

Study Tip

Recall from Section 1.4 that the symbol $\approx$ means "is approximately equal to."

3 ▶ Solve application problems involving markups and discounts.

Markups and Discounts

You may have had the experience of buying an item at one store and later finding that you could have paid less for the same item at another store. The basic reason for this price difference is **markup,** which is the difference between the **cost** (the amount a retailer pays for the item) and the **price** (the amount at which the retailer sells the item to the consumer). A verbal model for this problem is as follows.

Selling price = Cost + Markup

In such a problem, the markup may be known or it may be expressed as a percent of the cost. This percent is called the **markup rate.**

Markup = Markup rate · Cost

Markup is one of those "hidden operations" referred to in Section 2.3.

In business and economics, the terms *cost* and *price* do not mean the same thing. The cost of an item is the amount a business pays for the item. The price of an item is the amount for which the business sells the item.

EXAMPLE 8 Finding the Selling Price

A sporting goods store uses a markup rate of 55% on all items. The cost of a golf bag is \$45. What is the selling price of the bag?

Solution

Verbal Model: Selling price = Cost + Markup

Labels:
Selling price $= x$ (dollars)
Cost $= 45$ (dollars)
Markup rate $= 0.55$ (decimal form)
Markup $= (0.55)(45)$ (dollars)

Equation:
$x = 45 + (0.55)(45)$ Write equation.
$= 45 + 24.75$ Multiply.
$= \$69.75$ Add.

The selling price is \$69.75. You can check your solution as follows.

$x = 45 + (0.55)(45)$ Write original equation.
$69.75 \stackrel{?}{=} 45 + (0.55)(45)$ Substitute 69.75 for x.
$69.75 = 69.75$ Solution checks. ✓

Now try Exercise 67.

In Example 8, you are given the cost and are asked to find the selling price. Example 9 illustrates the reverse problem. That is, in Example 9 you are given the selling price and are asked to find the cost.

EXAMPLE 9 Finding the Cost of an Item

The selling price of a pair of ski boots is \$98. The markup rate is 60%. What is the cost of the boots?

Solution

Verbal Model: Selling price = Cost + Markup

Labels:
Selling price = 98 (dollars)
Cost = x (dollars)
Markup rate = 0.60 (decimal form)
Markup = $0.60x$ (dollars)

Equation:
$98 = x + 0.60x$ Write equation.
$98 = 1.60x$ Combine like terms.
$61.25 = x$ Divide each side by 1.60.

The cost is \$61.25. Check this in the original statement of the problem.

 **CHECKPOINT** *Now try Exercise 61.*

EXAMPLE 10 Finding the Markup Rate

A pair of walking shoes sells for \$60. The cost of the walking shoes is \$24. What is the markup rate?

Solution

Verbal Model: Selling price = Cost + Markup

Labels:
Selling price = 60 (dollars)
Cost = 24 (dollars)
Markup rate = p (decimal form)
Markup = $p(24)$ (dollars)

Equation:
$60 = 24 + p(24)$ Write equation.
$36 = 24p$ Subtract 24 from each side.
$1.5 = p$ Divide each side by 24.

Because $p = 1.5$, it follows that the markup rate is 150%.

 **CHECKPOINT** *Now try Exercise 59.*

The mathematics of discounts is similar to that of markups. The model for this situation is

Sale price = List price − Discount

where the **discount** is given in dollars, and the **discount rate** is given as a percent of the list price. Notice the "hidden operation" in the discount.

Discount = Discount rate · List price

EXAMPLE 11 Finding the Discount Rate 

During a midsummer sale, a lawn mower listed at \$199.95 is on sale for \$139.95. What is the discount rate?

Solution

Verbal Model: Discount $=$ Discount rate $\cdot$ List price

Labels:
Discount $= 199.95 - 139.95 = 60$ (dollars)
List price $= 199.95$ (dollars)
Discount rate $= p$ (decimal form)

Equation:
$60 = p(199.95)$ Write equation.
$0.30 \approx p$ Divide each side by 199.95.

Because $p \approx 0.30$, it follows that the discount rate is approximately 30%.

 **CHECKPOINT** *Now try Exercise 69.*

EXAMPLE 12 Finding the Sale Price 

A drug store advertises 40% off the prices of all summer tanning products. A bottle of suntan oil lists for \$3.49. What is the sale price?

Solution

Verbal Model: Sale price $=$ List price $-$ Discount

Labels:
List price $= 3.49$ (dollars)
Discount rate $= 0.4$ (decimal form)
Discount $= 0.4(3.49)$ (dollars)
Sale price $= x$ (dollars)

Equation: $x = 3.49 - (0.4)(3.49) \approx 2.09$

The sale price is \$2.09. Check this in the original statement of the problem.

 **CHECKPOINT** *Now try Exercise 75.*

The following guidelines summarize the problem-solving strategy that you should use when solving word problems.

Guidelines for Solving Word Problems

1. Write a *verbal model* that describes the problem.
2. Assign *labels* to fixed quantities and variable quantities.
3. Rewrite the verbal model as an *algebraic equation* using the assigned labels.
4. *Solve* the resulting algebraic equation.
5. *Check* to see that your solution satisfies the original problem as stated.

Concept Check

1. Explain what is meant by the word *percent*.
2. Can any positive terminating decimal be written as a percent? Explain.
3. Write an equation that can be used to find the number x that is 25% of a number y.
4. Explain the difference between a markup and a discount. Give an example of each.

3.3 EXERCISES

Go to pages 206–207 to record your assignments.

Developing Skills

In Exercises 1–8, convert the decimal to a percent. ***See Example 1.***

1. 0.62
2. 0.57
3. 0.20
4. 0.38
5. 0.075
6. 0.005
7. 2.38
8. 1.75

In Exercises 9–16, convert the fraction to a percent. ***See Example 1.***

9. $\frac{4}{5}$
10. $\frac{1}{4}$
11. $\frac{5}{4}$
12. $\frac{6}{5}$
13. ✓ $\frac{5}{6}$
14. $\frac{2}{3}$
15. $\frac{21}{20}$
16. $\frac{5}{2}$

In Exercises 17–24, convert the percent to a decimal. ***See Example 2.***

17. 12.5%
18. 95%
19. 125%
20. 250%
21. ✓ 8.5%
22. 0.3%
23. $\frac{3}{4}\%$
24. $4\frac{4}{5}\%$

In Exercises 25–32, convert the percent to a fraction. ***See Example 2.***

25. 37.5%
26. 85%
27. 130%
28. 350%
29. 1.4%
30. 0.7%
31. $\frac{1}{2}\%$
32. $2\frac{3}{10}\%$

In Exercises 33–42, complete the table showing the equivalent forms of percents. ***See Examples 1 and 2.***

	Percent	*Parts out of 100*	*Decimal*	*Fraction*
33.	40%			
34.	16%			
35.	7.5%			
36.	75%			
37.		63		
38.		10.5		
39.			0.155	
40.			0.80	
41.				$\frac{3}{5}$
42.				$\frac{3}{20}$

In Exercises 43–46, what percent of the figure is shaded? (There are a total of 360° in a circle.)

43.

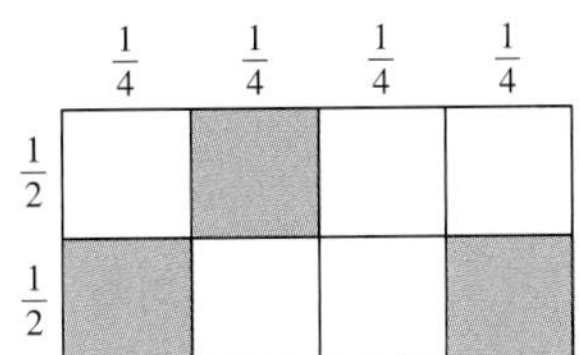

44.

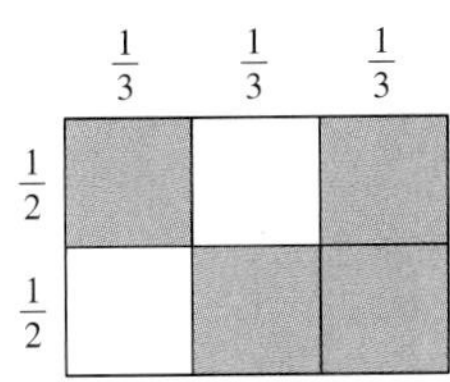

45.

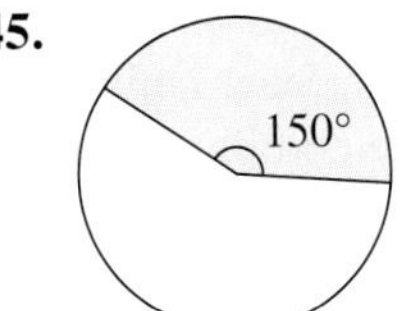

46.

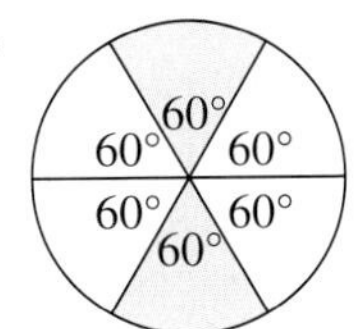

In Exercises 47–58, solve the percent equation. ***See Example 3.***

47. What number is 30% of 150?

48. What number is $66\frac{2}{3}\%$ of 816?

49. What number is 0.75% of 56?

50. What number is 325% of 450?

✓ 51. 903 is 43% of what number?

52. 275 is $12\frac{1}{2}\%$ of what number?

53. 594 is 450% of what number?

54. 51.2 is 0.08% of what number?

55. 576 is what percent of 800?

56. 38 is what percent of 5700?

57. 22 is what percent of 800?

58. 148.8 is what percent of 960?

In Exercises 59–68, find the missing quantities. ***See Examples 8, 9, and 10.***

	Cost	Selling Price	Markup	Markup Rate
✓ 59.	$26.97	$49.95		
60.	$71.97	$119.95		
✓ 61.		$74.38		81.5%
62.		$69.99		55.5%
63.		$125.98	$56.69	
64.		$350.00	$80.77	
65.		$15,900.00	$2650.00	
66.		$224.87	$75.08	
✓ 67.	$107.97			85.2%
68.	$680.00			$33\frac{1}{3}\%$

In Exercises 69–78, find the missing quantities. ***See Examples 11 and 12.***

	List Price	Sale Price	Discount	Discount Rate
✓ 69.	$39.95	$29.95		
70.	$50.99	$45.99		
71.		$18.95		20%
72.		$189.00		40%
73.	$189.99		$30.00	
74.	$18.95		$8.00	
✓ 75.	$119.96			50%
76.	$84.95			65%
77.		$695.00	$300.00	
78.		$259.97	$135.00	

Solving Problems

79. *Rent* You spend 17% of your monthly income of $3200 for rent. What is your monthly payment?

80. *Cost of Housing* You budget 30% of your annual after-tax income for housing. Your after-tax income is $38,500. What amount can you spend on housing?

81. *Retirement Plan* You budget $7\frac{1}{2}\%$ of your gross income for an individual retirement plan. Your annual gross income is $45,800. How much will you put in your retirement plan each year?

82. *Enrollment* In the fall of 2005, about 45% of the undergraduates enrolled at Alabama State University were freshmen. The undergraduate enrollment of the college was 4485. Find the approximate number of freshmen enrolled in the fall of 2005. (Source: Alabama State University)

✓ 83. *Commission* A real estate agency receives a commission of $14,506.50 for the sale of a $152,700 house. What percent commission is this?

84. *Commission* A car salesman receives a commission of $1145 for the sale of a $45,800 car. What percent commission is this?

85. *Lawn Tractor* You purchase a lawn tractor for $3750 and 1 year later you note that the price has increased to $3900. Find the percent increase in the price of the lawn tractor.

✓ 86. *Monthly Bill* The average monthly cell phone bill was $80.90 in 1990 and $49.98 in 2005. Find the percent decrease in price from 1990 to 2005. (Source: Cellular Telecommunications & Internet Association)

87. *Salary Raise* You accept a job with a salary of $35,600. After 6 months, you receive a 5% raise. What is your new salary?

✓ **88.** *Salary Raise* A union negotiates for a cost-of-living raise of 4.5%. What is the raise for a union member whose salary is $37,380? What is this person's new salary?

✓ **89.** *Course Grade* You were six points shy of a B in your mathematics course. Your point total for the course was 394. How many points were possible in the course? (Assume that you needed 80% of the course total for a B.)

90. *Eligible Voters* In 2004, 125,736,000 votes were cast in the presidential election. This represented 63.8% of the eligible voters. How many eligible voters were there in 2004? (Source: U.S. Census Bureau)

The 55th Inaugural Ceremony took place on January 20, 2005 at the United States Capitol.

91. *Original Price* A coat sells for $250 during a 20% off storewide clearance sale. What was the original price of the coat?

92. *Membership Drive* Because of a membership drive for a public television station, the current membership is 125% of what it was a year ago. The current number of members is 7815. How many members did the station have last year?

93. *Geometry* A rectangular plot of land measures 650 feet by 825 feet. A square garage with sides of length 24 feet is built on the plot of land. What percentage of the plot of land is occupied by the garage?

94. *Geometry* A circular target is attached to a rectangular board, as shown in the figure. The radius of the circle is $4\frac{1}{2}$ inches, and the measurements of the board are 12 inches by 15 inches. What percentage of the board is covered by the target? (The area of a circle is $A = \pi r^2$, where r is the radius of the circle.)

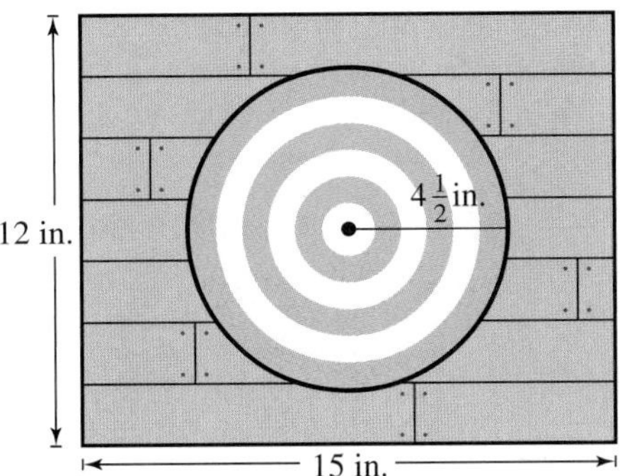

Figure for 94

95. *Revenue* In 2007, the Walt Disney Company had a revenue of $35.5 billion. The circle graph classifies the revenue of the company. Approximate the revenue in each of the classifications. (Source: The Walt Disney Company)

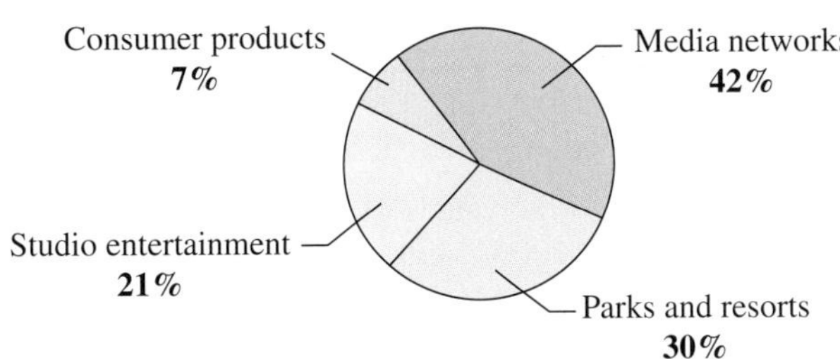

96. *Graphical Estimation* The bar graph shows the numbers (in thousands) of criminal cases for major offenses commenced in the United States District Courts from 2002 through 2006. (Source: Administrative Office of the U.S. Courts)

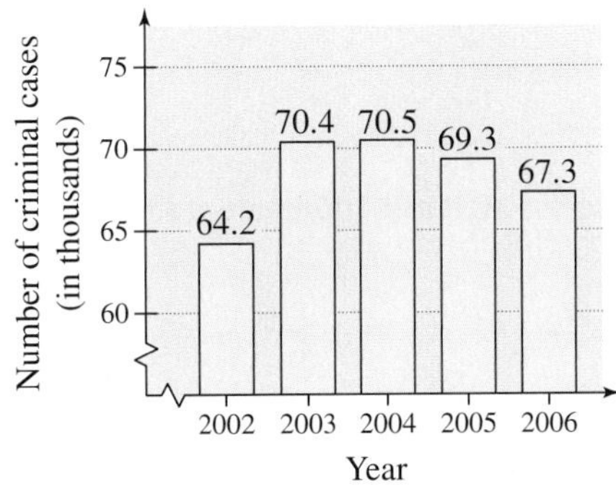

(a) Determine the percent increase in cases from 2002 to 2003.

(b) Determine the percent decrease in cases from 2003 to 2006.

97. *Interpreting a Table* The table shows the numbers and percents of women scientists in the United States in three fields for the years 1983 and 2005. (Source: U.S. Bureau of Labor Statistics)

	1983		2005	
	Number	**%**	**Number**	**%**
Math/ Computer	137,048	29.6	876,420	27.0
Chemistry	22,834	23.3	40,948	35.3
Biology	22,440	40.8	57,953	48.7

(a) Find the total number of mathematicians and computer scientists (men and women) in 2005.

(b) Find the total number of chemists (men and women) in 1983.

(c) Find the total number of biologists (men and women) in 2005.

98. *Data Analysis* The table shows the approximate population (in millions) of Bangladesh for each decade from 1960 through 2000. Approximate the percent growth rate for each decade. If the growth rate of the 2000s continued until the year 2020, approximate the population in 2020. (Source: U.S. Bureau of the Census, International Data Base)

Year	1960	1970	1980	1990	2000
Population	54.6	67.4	88.1	109.9	129.2

Explaining Concepts

99. The fraction $\frac{a}{b}$ $(a > 0, b > 0)$ is converted to a percent. For what values of a and b is the percent greater than 100%? Less than 100%? Equal to 100%? Explain.

100. Would you rather receive a 3% raise followed by a 9% raise or a 9% raise followed by a 3% raise? Explain.

True or False? In Exercises 101–104, decide whether the statement is true or false. Justify your answer.

101. $1 = 1\%$

102. Every percent can be written as a fraction.

103. The question "What is 68% of 50?" can be answered by solving the equation $a = 68(50)$.

104. $\frac{1}{2}\% = 50\%$

Cumulative Review

In Exercises 105 and 106, evaluate the expression.

105. $8 - |-7 + 11| + (-4)$

106. $34 - [54 - (-16 + 4) + 6]$

In Exercises 107 and 108, evaluate the algebraic expression for the specified values of the variables.

107. $x^2 - y^2$

(a) $x = 4, y = 3$ (b) $x = -5, y = 3$

108. $\dfrac{z^2 + 2}{x^2 - 1}$

(a) $x = 1, z = 2$ (b) $x = 2, z = 1$

In Exercises 109 and 110, use the Distributive Property to expand the expression.

109. $4(2x - 5)$ **110.** $-z(xz - 2y^2)$

In Exercises 111–114, solve the equation and check your solution.

111. $4(x + 3) = 0$ **112.** $-3(y - 2) = 21$

113. $22 - (z + 1) = 33$ **114.** $\dfrac{w}{3} = \dfrac{8}{12}$

3.4 Ratios and Proportions

Eunice Harris/Researchers, Inc.

What You Should Learn

1 ▶ Compare relative sizes using ratios.
2 ▶ Find the unit price of a consumer item.
3 ▶ Solve proportions that equate two ratios.
4 ▶ Solve application problems using the Consumer Price Index.

Why You Should Learn It

Ratios can be used to represent many real-life quantities. For instance, in Exercise 64 on page 168, you will find the gear ratios for a five-speed bicycle.

1 ▶ Compare relative sizes using ratios.

Setting Up Ratios

A **ratio** is a comparison of one number with another by division. For example, in a class of 29 students made up of 16 women and 13 men, the ratio of women to men is 16 to 13 or $\frac{16}{13}$. Some other ratios for this class are as follows.

Men to women: $\dfrac{13}{16}$ Men to students: $\dfrac{13}{29}$ Students to women: $\dfrac{29}{16}$

Note the order implied by a ratio. The ratio of a to b means a/b, whereas the ratio of b to a means b/a.

Definition of Ratio

The **ratio** of the real number a to the real number b is given by

$$\frac{a}{b}.$$

The ratio of a to b is sometimes written as $a : b$.

EXAMPLE 1 Writing Ratios in Fractional Form

a. The ratio of 7 to 5 is given by $\frac{7}{5}$.

b. The ratio of 12 to 8 is given by $\frac{12}{8} = \frac{3}{2}$.

Note that the fraction $\frac{12}{8}$ can be written in simplest form as $\frac{3}{2}$.

c. The ratio of $3\frac{1}{2}$ to $5\frac{1}{4}$ is given by

$$\frac{3\frac{1}{2}}{5\frac{1}{4}} = \frac{\frac{7}{2}}{\frac{21}{4}} \qquad \text{Rewrite mixed numbers as fractions.}$$

$$= \frac{7}{2} \cdot \frac{4}{21} \qquad \text{Invert divisor and multiply.}$$

$$= \frac{2}{3}. \qquad \text{Simplify.}$$

✓ **CHECKPOINT** *Now try Exercise 5.*

There are many real-life applications of ratios. For instance, ratios are used to describe opinion surveys (for/against), populations (male/female, unemployed/employed), and mixtures (oil/gasoline, water/alcohol).

When comparing two *measurements* by a ratio, you should use the same unit of measurement in both the numerator and the denominator. For example, to find the ratio of 4 feet to 8 inches, you could convert 4 feet to 48 inches (by multiplying by 12) to obtain

$$\frac{4 \text{ feet}}{8 \text{ inches}} = \frac{48 \text{ \cancel{inches}}}{8 \text{ \cancel{inches}}} = \frac{48}{8} = \frac{6}{1}$$

or you could convert 8 inches to $\frac{8}{12}$ foot (by dividing by 12) to obtain

$$\frac{4 \text{ feet}}{8 \text{ inches}} = \frac{4 \text{ \cancel{feet}}}{\frac{8}{12} \text{ \cancel{foot}}} = 4 \cdot \frac{12}{8} = \frac{6}{1}.$$

If you use different units of measurement in the numerator and denominator, then you *must* include the units. If you use the same units of measurement in the numerator and denominator, then it is not necessary to write the units.

EXAMPLE 2 Comparing Measurements

Find ratios to compare the relative sizes of the following.

a. 5 gallons to 7 gallons **b.** 3 meters to 40 centimeters

c. 200 cents to 3 dollars **d.** 30 months to $1\frac{1}{2}$ years

Solution

a. Because the units of measurement are the same, the ratio is $\frac{5}{7}$.

b. Because the units of measurement are different, begin by converting meters to centimeters *or* centimeters to meters. Here, it is easier to convert meters to centimeters by multiplying by 100.

$$\frac{3 \text{ meters}}{40 \text{ centimeters}} = \frac{3(100) \text{ \cancel{centimeters}}}{40 \text{ \cancel{centimeters}}} \quad \text{Convert meters to centimeters.}$$

$$= \frac{300}{40} \quad \text{Multiply in numerator.}$$

$$= \frac{15}{2} \quad \text{Simplify.}$$

c. Because 200 cents is the same as 2 dollars, the ratio is

$$\frac{200 \text{ cents}}{3 \text{ dollars}} = \frac{2 \text{ \cancel{dollars}}}{3 \text{ \cancel{dollars}}} = \frac{2}{3}.$$

d. Because $1\frac{1}{2}$ years $= 18$ months, the ratio is

$$\frac{30 \text{ months}}{1\frac{1}{2} \text{ years}} = \frac{30 \text{ \cancel{months}}}{18 \text{ \cancel{months}}} = \frac{30}{18} = \frac{5}{3}.$$

Now try Exercise 21.

2 ▶ Find the unit price of a consumer item.

Unit Prices

As a consumer, you must be able to determine the unit prices of items you buy in order to make the best use of your money. The **unit price** of an item is given by the ratio of the total price to the total number of units.

$$\text{Unit price} = \frac{\text{Total price}}{\text{Total units}}$$

The word *per* is used to state unit prices. For instance, the unit price for a particular brand of coffee might be \$4.69 *per* pound.

EXAMPLE 3 Finding a Unit Price 

Find the unit price (in dollars per ounce) for a five-pound, four-ounce box of detergent that sells for \$7.14.

Solution

Begin by writing the weight in ounces. That is,

$$\begin{aligned} 5 \text{ pounds} + 4 \text{ ounces} &= 5 \cancel{\text{pounds}}\left(\frac{16 \text{ ounces}}{1 \cancel{\text{pound}}}\right) + 4 \text{ ounces} \\ &= 80 \text{ ounces} + 4 \text{ ounces} \\ &= 84 \text{ ounces}. \end{aligned}$$

Next, determine the unit price as follows.

$$\text{Unit price} = \frac{\text{Total price}}{\text{Total units}} = \frac{\$7.14}{84 \text{ ounces}} = \$0.085 \text{ per ounce}$$

CHECKPOINT *Now try Exercise 33.*

EXAMPLE 4 Comparing Unit Prices

Which has the lower unit price: a 12-ounce box of breakfast cereal for \$2.69 or a 16-ounce box of the same cereal for \$3.49?

Solution

The unit price for the smaller box is

$$\text{Unit price} = \frac{\text{Total price}}{\text{Total units}} = \frac{\$2.69}{12 \text{ ounces}} \approx \$0.224 \text{ per ounce.}$$

The unit price for the larger box is

$$\text{Unit price} = \frac{\text{Total price}}{\text{Total units}} = \frac{\$3.49}{16 \text{ ounces}} \approx \$0.218 \text{ per ounce.}$$

So, the larger box has a slightly lower unit price.

✓ **CHECKPOINT** *Now try Exercise 37.*

3 ▶ Solve proportions that equate two ratios.

Solving Proportions

A **proportion** is a statement that equates two ratios. For example, if the ratio of a to b is the same as the ratio of c to d, you can write the proportion as

$$\frac{a}{b} = \frac{c}{d}.$$

In typical applications, you know three of the values and are required to find the fourth. To solve such a fractional equation, you can use the *cross-multiplication* procedure introduced in Section 3.2.

Solving a Proportion

If

$$\frac{a}{b} = \frac{c}{d}$$

then $ad = bc$. The quantities a and d are called the **extremes** of the proportion, whereas b and c are called the **means** of the proportion.

EXAMPLE 5 Solving Proportions

Solve each proportion.

a. $\frac{50}{x} = \frac{2}{28}$ **b.** $\frac{x-2}{5} = \frac{4}{3}$

Solution

a.

$\frac{50}{x} = \frac{2}{28}$	Write original proportion.
$50(28) = 2x$	Cross-multiply.
$\frac{1400}{2} = x$	Divide each side by 2.
$700 = x$	Simplify.

So, the ratio of 50 to 700 is the same as the ratio of 2 to 28.

b.

$\frac{x-2}{5} = \frac{4}{3}$	Write original proportion.
$3(x-2) = 20$	Cross-multiply.
$3x - 6 = 20$	Distributive Property
$3x = 26$	Add 6 to each side.
$x = \frac{26}{3}$	Divide each side by 3.

✓ **CHECKPOINT** *Now try Exercise 41.*

100 ft

210 ft

Triangular lot

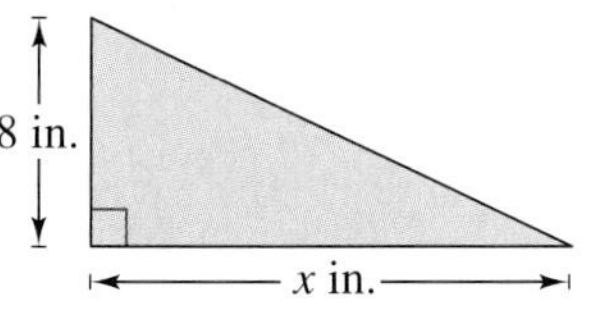

Sketch

Figure 3.2

EXAMPLE 6 Geometry: Similar Triangles 

A triangular lot has perpendicular sides of lengths 100 feet and 210 feet. You are to make a proportional sketch of this lot using 8 inches as the length of the shorter side. How long should you make the longer side?

Solution

This is a case of similar triangles in which the ratios of the corresponding sides are equal. The triangles are shown in Figure 3.2.

$$\frac{\text{Shorter side of lot}}{\text{Longer side of lot}} = \frac{\text{Shorter side of sketch}}{\text{Longer side of sketch}} \qquad \text{Proportion for similar triangles}$$

$$\frac{100}{210} = \frac{8}{x} \qquad \text{Substitute.}$$

$$x \cdot 100 = 210 \cdot 8 \qquad \text{Cross-multiply.}$$

$$100x = 1680 \qquad \text{Simplify.}$$

$$x = \frac{1680}{100} \qquad \text{Divide each side by 100.}$$

$$= 16.8 \qquad \text{Simplify.}$$

So, the length of the longer side of the sketch should be 16.8 inches.

 ***CHECKPOINT*** *Now try Exercise 81.*

EXAMPLE 7 Resizing a Graph 

You have a 7-by-8-inch graph that must be reduced to a size of 5.25 inches by 6 inches for a research paper. What percent does the graph need to be reduced to in order to fit in the allotted space?

Solution

Because the longer side must be reduced from 8 inches to 6 inches, consider the following proportion.

$$\frac{\text{New length}}{\text{Old length}} = \frac{\text{New percent}}{\text{Old percent}} \qquad \text{Write proportion.}$$

$$\frac{6}{8} = \frac{x}{100} \qquad \text{Substitute.}$$

$$\frac{6}{8} \cdot 100 = x \qquad \text{Multiply each side by 100.}$$

$$75 = x \qquad \text{Simplify.}$$

So, the graph needs to be reduced to 75% of its original size.

✓ ***CHECKPOINT*** *Now try Exercise 85.*

Study Tip

To solve an equation, you want to isolate the variable. In Example 7, this was done by multiplying each side by 100 instead of cross-multiplying. In this case, multiplying each side by 100 was the only step needed to isolate the x-variable. However, either method is valid for solving the equation.

Smart Study Strategy

Go to page 124 for ways to *Use Your Preferred Learning Modality.*

EXAMPLE 8 **Gasoline Cost** 

You are driving from New York to Phoenix, a trip of 2450 miles. You begin the trip with a full tank of gas, and after traveling 424 miles, you refill the tank for $38.00. How much should you plan to spend on gasoline for the entire trip?

Solution

Verbal Model: $\dfrac{\text{Cost for entire trip}}{\text{Cost for one tank}} = \dfrac{\text{Miles for entire trip}}{\text{Miles for one tank}}$

Labels:
Cost of gas for entire trip $= x$ (dollars)
Cost of gas for one tank $= 38$ (dollars)
Miles for entire trip $= 2450$ (miles)
Miles for one tank $= 424$ (miles)

Proportion: $\dfrac{x}{38} = \dfrac{2450}{424}$ Write proportion.

$x = 38\left(\dfrac{2450}{424}\right)$ Multiply each side by 38.

$x \approx 219.58$ Simplify.

You should plan to spend approximately $219.58 for gasoline on the trip. Check this in the original statement of the problem.

 **CHECKPOINT** *Now try Exercise 67.*

4 ▶ Solve application problems using the Consumer Price Index.

The Consumer Price Index

The rate of inflation is important to all of us. Simply stated, *inflation* is an economic condition in which the price of a fixed amount of goods or services increases. So, a fixed amount of money buys less in a given year than in previous years.

The most widely used measurement of inflation in the United States is the *Consumer Price Index* (CPI). The table below shows the "All Items" or general index for the years 1975 to 2006. (Source: Bureau of Labor Statistics)

Study Tip

To determine (from the CPI) the change in the buying power of a dollar from one year to another, use the following proportion.

$$\frac{\text{Price in year } n}{\text{Price in year } m} = \frac{\text{Index in year } n}{\text{Index in year } m}$$

Year	CPI	Year	CPI	Year	CPI	Year	CPI
1975	53.8	1983	99.6	1991	136.2	1999	166.6
1976	56.9	1984	103.9	1992	140.3	2000	172.2
1977	60.6	1985	107.6	1993	144.5	2001	177.1
1978	65.2	1986	109.6	1994	148.2	2002	179.9
1979	72.6	1987	113.6	1995	152.4	2003	184.0
1980	82.4	1988	118.3	1996	156.9	2004	188.9
1981	90.9	1989	124.0	1997	160.5	2005	195.3
1982	96.5	1990	130.7	1998	163.0	2006	201.6

EXAMPLE 9 Using the Consumer Price Index 

You paid \$43,000 for a house in 1976. What is the amount you would pay for the same house in 2006?

Solution

Verbal Model: $\dfrac{\text{Price in 2006}}{\text{Price in 1976}} = \dfrac{\text{Index in 2006}}{\text{Index in 1976}}$

Labels: Price in 2006 $= x$ (dollars)
Price in 1976 $= 43{,}000$ (dollars)
Index in 2006 $= 201.6$
Index in 1976 $= 56.9$

Proportion: $\dfrac{x}{43{,}000} = \dfrac{201.6}{56.9}$ Write proportion.

$x = \dfrac{201.6}{56.9} \cdot 43{,}000$ Multiply each side by 43,000.

$x \approx 152{,}351$ Simplify.

So, you would pay approximately \$152,351 for the house in 2006. Check this in the original statement of the problem.

 **CHECKPOINT** *Now try Exercise 87.*

EXAMPLE 10 Using the Consumer Price Index 

You inherited a diamond pendant from your grandmother in 2005. The pendant was appraised at \$1600. What was the value of the pendant when your grandmother bought it in 1978?

Solution

Verbal Model: $\dfrac{\text{Price in 2005}}{\text{Price in 1978}} = \dfrac{\text{Index in 2005}}{\text{Index in 1978}}$

Labels: Price in 2005 $= 1600$ (dollars)
Price in 1978 $= x$ (dollars)
Index in 2005 $= 195.3$
Index in 1978 $= 65.2$

Proportion: $\dfrac{1600}{x} = \dfrac{195.3}{65.2}$ Write proportion.

$104{,}320 = 195.3x$ Cross-multiply.

$534 \approx x$ Divide each side by 195.3.

So, the value of the pendant in 1978 was approximately \$534. Check this in the original statement of the problem.

CHECKPOINT *Now try Exercise 89.*

The four C's—cut, color, carat weight, and clarity—are used to classify the rarity of diamonds.

Concept Check

1. Does the order in which a ratio is written have any significance? Explain.
2. Explain how to find the unit price of an item.
3. Explain how to solve a proportion.
4. State the proportion you would use to find the current price of an item using the Consumer Price Index.

3.4 EXERCISES

Go to pages 206–207 to record your assignments.

Developing Skills

In Exercises 1–12, write the ratio as a fraction in simplest form. ***See Example 1.***

1. 36 to 9
2. 45 to 15
3. 27 to 54
4. 27 to 63
✓ **5.** $5\frac{2}{3}$ to $1\frac{1}{3}$
6. $2\frac{1}{4}$ to $3\frac{3}{8}$
7. 14 : 21
8. 12 : 30
9. 144 : 16
10. 60 : 45
11. $3\frac{1}{5} : 5\frac{3}{10}$
12. $1\frac{2}{7} : \frac{1}{2}$

In Exercises 13–30, find a ratio that compares the relative sizes of the quantities. (Use the same units of measurement for both quantities.) ***See Example 2.***

13. Forty-two inches to 21 inches
14. Eighty-one feet to 27 feet
15. Forty dollars to $60
16. Twenty-four pounds to 30 pounds
17. One quart to 1 gallon
18. Three inches to 2 feet
19. Seven nickels to 3 quarters
20. Twenty-four ounces to 3 pounds
✓ **21.** Three hours to 90 minutes
22. Twenty-one feet to 35 yards
23. Seventy-five centimeters to 2 meters
24. Three miles to 2000 feet
25. Sixty milliliters to 1 liter
26. Two weeks to 7 days
27. Two kilometers to 2500 meters
28. Five and one-half pints to 2 quarts
29. Three thousand pounds to 5 tons
30. Four days to 30 hours

In Exercises 31–34, find the unit price (in dollars per ounce). ***See Example 3.***

31. A 20-ounce can of pineapple for $0.98

32. An 18-ounce box of cereal for $4.29

✓ **33.** A one-pound, four-ounce loaf of bread for $1.46

34. A one-pound package of cheese for $3.08

In Exercises 35–40, determine which product has the lower unit price. ***See Example 4.***

35. (a) A gallon of orange juice for $3.49
(b) A quart of orange juice for $1.39

36. (a) A 16-ounce bag of chocolates for $1.99
(b) An 18-ounce bag of chocolates for $2.29

✓ **37.** (a) A 4-pound bag of sugar for $1.89
(b) A 10-pound bag of sugar for $4.49

38. (a) An 18-ounce jar of peanut butter for $1.92
(b) A 28-ounce jar of peanut butter for $3.18

39. (a) A two-liter bottle (67.6 ounces) of soft drink for $1.09
(b) Six 12-ounce cans of soft drink for $1.69

40. (a) A one-quart container of oil for $2.12
(b) A 2.5-gallon container of oil for $19.99

In Exercises 41–56, solve the proportion. ***See Example 5.***

41. $\frac{5}{3} = \frac{20}{y}$

42. $\frac{9}{x} = \frac{18}{5}$

43. $\frac{5}{x} = \frac{3}{2}$

44. $\frac{4}{t} = \frac{2}{25}$

45. $\frac{z}{35} = \frac{5}{8}$

46. $\frac{y}{25} = \frac{12}{10}$

47. $\frac{8}{3} = \frac{t}{6}$

48. $\frac{x}{6} = \frac{7}{12}$

49. $\frac{0.5}{0.8} = \frac{n}{0.3}$

50. $\frac{2}{4.5} = \frac{t}{0.5}$

51. $\frac{x+1}{5} = \frac{3}{10}$

52. $\frac{z-3}{8} = \frac{3}{10}$

53. $\frac{x+6}{3} = \frac{x-5}{2}$

54. $\frac{x-2}{4} = \frac{x+10}{10}$

55. $\frac{x+2}{8} = \frac{x-1}{3}$

56. $\frac{x-4}{5} = \frac{x+2}{7}$

Solving Problems

In Exercises 57–66, express the statement as a ratio in simplest form. (Use the same units of measurement for both quantities.)

57. *Study Hours* You study 4 hours per day and are in class 6 hours per day. Find the ratio of the number of study hours to class hours.

58. *Income Tax* You have \$22 of state tax withheld from your paycheck per week when your gross pay is \$750. Find the ratio of tax to gross pay.

59. *Sports* Last football season, a player had 806 rushing yards and 217 receiving yards. Find the ratio of rushing yards to receiving yards.

60. *Education* There are 2921 students and 127 faculty members at your school. Find the ratio of the number of students to the number of faculty members.

61. *Compression Ratio* The *compression ratio* of an engine is the ratio of the expanded volume of gas in one of its cylinders to the compressed volume of gas in the cylinder (see figure). A cylinder in a diesel engine has an expanded volume of 345 cubic centimeters and a compressed volume of 17.25 cubic centimeters. What is the compression ratio of this engine?

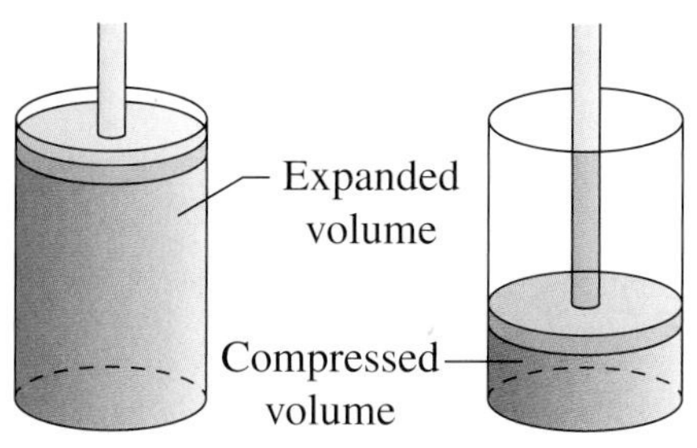

62. *Turn Ratio* The *turn ratio* of a transformer is the ratio of the number of turns on the secondary winding to the number of turns on the primary winding (see figure). A transformer has a primary winding with 250 turns and a secondary winding with 750 turns. What is its turn ratio?

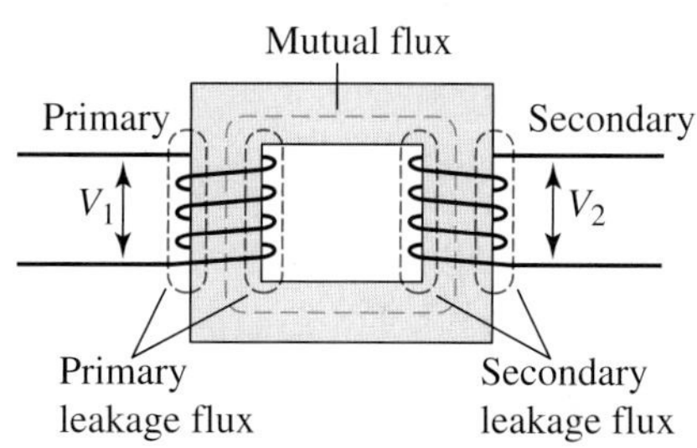

63. *Gear Ratio* The gear ratio of two gears is the ratio of the number of teeth on one gear to the number of teeth on the other gear. Find the gear ratio of the larger gear to the smaller gear for the gears shown in the figure.

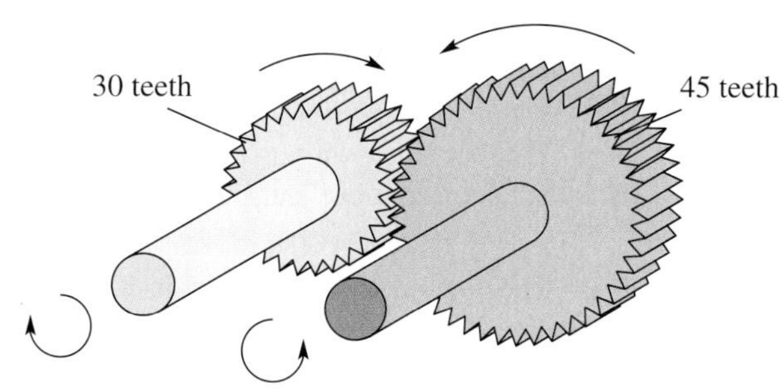

64. *Gear Ratio* On a five-speed bicycle, the ratio of the pedal gear to the axle gear depends on which axle gear is engaged. Use the table to find the gear ratios for the five different gears. For which gear is it easiest to pedal? Why?

Gear	1st	2nd	3rd	4th	5th
Teeth on pedal gear	52	52	52	52	52
Teeth on axle gear	28	24	20	17	14

65. *Geometry* A large pizza has a radius of 10 inches and a small pizza has a radius of 7 inches. Find the ratio of the area of the large pizza to the area of the small pizza. (*Note:* The area of a circle is $A = \pi r^2$.)

66. *Specific Gravity* The *specific gravity* of a substance is the ratio of its weight to the weight of an equal volume of water. Kerosene weighs 0.82 gram per cubic centimeter and water weighs 1 gram per cubic centimeter. What is the specific gravity of kerosene?

✓ **67.** *Amount of Fuel* A car uses 20 gallons of gasoline for a trip of 500 miles. How many gallons would be used on a trip of 400 miles?

68. *Amount of Fuel* A tractor requires 4 gallons of diesel fuel to plow for 90 minutes. How many gallons of fuel would be required to plow for 8 hours?

69. *Building Material* One hundred cement blocks are required to build a 16-foot wall. How many blocks are needed to build a 40-foot wall?

70. *Force on a Spring* A force of 50 pounds stretches a spring 4 inches. How much force is required to stretch the spring 6 inches?

71. *Sales Tax* The tax on a shirt priced at \$19.99 is \$1.20. Find the tax on a pair of jeans priced at \$34.99.

72. *Sales Tax* The tax on a CD priced at \$15.99 is \$0.64. Find the tax on an MP3 player priced at \$149.99.

73. *Polling Results* In a poll, 624 people from a sample of 1100 indicated they would vote for the Republican candidate. How many votes can the candidate expect to receive from 40,000 votes cast?

74. *Quality Control* A quality control engineer found two defective units in a sample of 50. At this rate, what is the expected number of defective units in a shipment of 10,000 units?

75. *Pumping Time* A pump can fill a 750-gallon tank in 35 minutes. How long will it take to fill a 1000-gallon tank with this pump?

76. *Recipe* Two cups of flour are required to make one batch of cookies. How many cups are required for $2\frac{1}{2}$ batches?

77. *Map Scale* On a map, $1\frac{1}{4}$ inches represents 80 miles. Estimate the distance between two cities that are 6 inches apart on the map.

78. *Map Scale* On a map, $1\frac{1}{2}$ inches represents 40 miles. Estimate the distance between two cities that are 4 inches apart on the map.

79. *Salt Water* The fresh water to salt ratio for a mixture is 25 to 1. How much fresh water is required to produce a mixture that contains one-half pound of salt?

80. *Building Material* The ratio of cement to sand in an 80-pound bag of dry mix is 1 to 4. Find the number of pounds of sand in the bag. (*Note:* Dry mix is composed of only cement and sand.)

Geometry In Exercises 81 and 82, find the length x of the side of the larger triangle. (Assume that the two triangles are similar, and use the fact that corresponding sides of similar triangles are proportional.) ***See Example 6.***

✓ **81.**

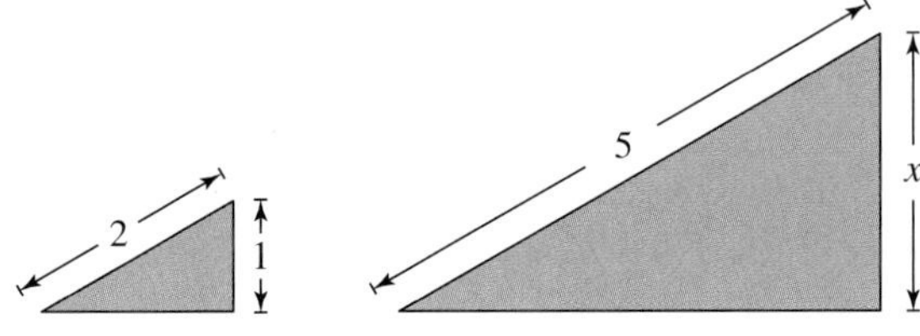

82.

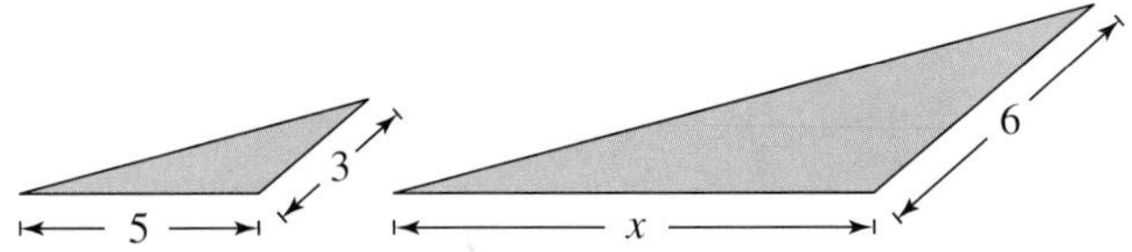

83. *Geometry* Find the length of the shadow of the man shown in the figure. (*Hint:* Use similar triangles to create a proportion.)

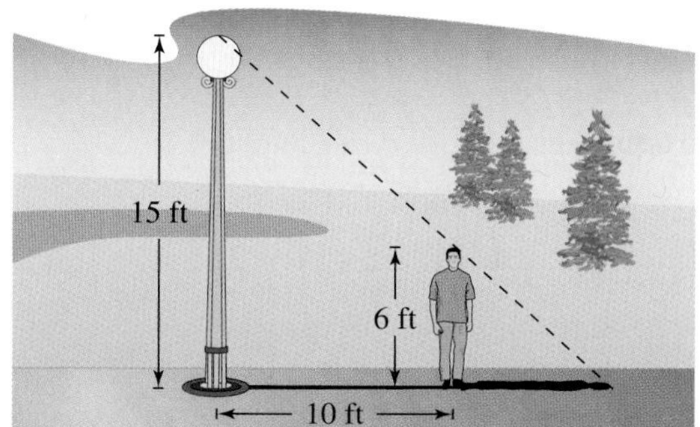

84. *Geometry* Find the height of the tree shown in the figure. (*Hint:* Use similar triangles to create a proportion.)

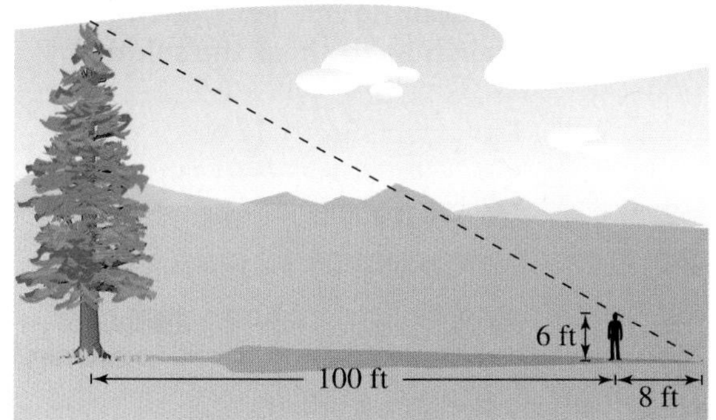

85. *Resizing a Picture* You have an 8-by-10-inch photo of a soccer player that must be reduced to a size of 1.6 inches by 2 inches for the school newsletter. What percent does the photo need to be reduced to in order for it to fit in the allotted space?

86. *Resizing a Picture* You have a 7-by-5-inch photo of the math club that must be reduced to a size of 5.6 inches by 4 inches for the school yearbook. What percent does the photo need to be reduced to in order for it to fit in the allotted space?

In Exercises 87–90, use the Consumer Price Index table on page 164 to estimate the price of the item in the indicated year. ***See Examples 9 and 10.***

87. The 2006 price of a lawn tractor that cost \$2875 in 1979

88. The 2000 price of a watch that cost \$158 in 1988

89. The 1975 price of a gallon of milk that cost \$2.75 in 1996

90. The 1980 price of a coat that cost \$225 in 2001

Explaining Concepts

91. You are told that the ratio of men to women in a class is 2 to 1. Does this information tell you the total number of people in the class? Explain.

92. Explain the following statement. "When setting up a ratio, be sure you are comparing apples to apples and not apples to oranges."

93. Create a proportion problem. Exchange problems with another student and solve the problem you receive.

Cumulative Review

In Exercises 94–99, evaluate the expression.

94. $3^2 - (-4)$

95. $(-5)^3 + 3$

96. 9.3×10^6

97. $\dfrac{-|7 + 3^2|}{4}$

98. $(-4)^2 - (30 \div 50)$

99. $(8 \cdot 9) + (-4)^3$

In Exercises 100–103, solve the percent equation.

100. What number is 25% of 250?

101. What number is 45% of 90?

102. 150 is 250% of what number?

103. 465 is what percent of 500?

Mid-Chapter Quiz

Take this quiz as you would take a quiz in class. After you are done, check your work against the answers in the back of the book.

In Exercises 1–10, solve the equation.

1. $74 - 12x = 2$

2. $10(y - 8) = 0$

3. $3x + 1 = x + 20$

4. $6x + 8 = 8 - 2x$

5. $-10x + \frac{2}{3} = \frac{7}{3} - 5x$

6. $\frac{x}{5} + \frac{x}{7} = 1$

7. $\frac{9 + x}{3} = 15$

8. $3 - 5(4 - x) = -6$

9. $\frac{x + 3}{6} = \frac{4}{3}$

10. $\frac{x + 7}{5} = \frac{x + 9}{7}$

In Exercises 11 and 12, solve the equation. Round your answer to two decimal places. In your own words, explain how to check the solution.

11. $32.86 - 10.5x = 11.25$

12. $\frac{x}{5.45} + 3.2 = 12.6$

13. What number is 62% of 25?

14. What number is $\frac{1}{2}\%$ of 8400?

15. 300 is what percent of 150?

16. 145.6 is 32% of what number?

17. You work 40 hours a week at a candy store and earn \$7.50 per hour. You also earn \$7.00 per hour baby-sitting and can work as many hours as you want. You want to earn \$370 a week. How many hours must you baby-sit?

18. A region has an area of 42 square meters. It must be divided into three subregions so that the second has twice the area of the first, and the third has twice the area of the second. Find the area of each subregion.

19. To get an A in a psychology course, you must have an average of at least 90 points for 3 tests of 100 points each. For the first 2 tests, your scores are 84 and 93. What must you score on the third test to earn a 90% average for the course?

Endangered Wildlife and Plant Species

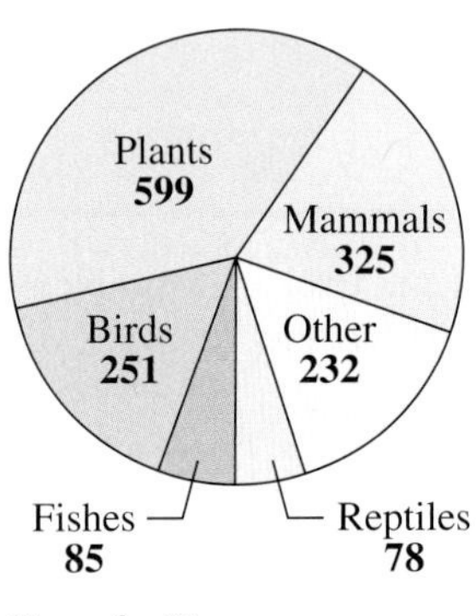

Figure for 20

20. The circle graph at the left shows the numbers of endangered wildlife and plant species as of October 2007. What percent of the total number of endangered wildlife and plant species were birds? (Source: U.S. Fish and Wildlife Service)

21. Two people can paint a room in t hours, where t must satisfy the equation $t/4 + t/12 = 1$. How long will it take for the two people to paint the room?

22. A large round pizza has a radius of $r = 15$ inches, and a small round pizza has a radius of $r = 8$ inches. Find the ratio of the area of the large pizza to the area of the small pizza. (*Hint:* The area of a circle is $A = \pi r^2$.)

23. A car uses 30 gallons of gasoline for a trip of 800 miles. How many gallons would be used on a trip of 700 miles?

3.5 Geometric and Scientific Applications

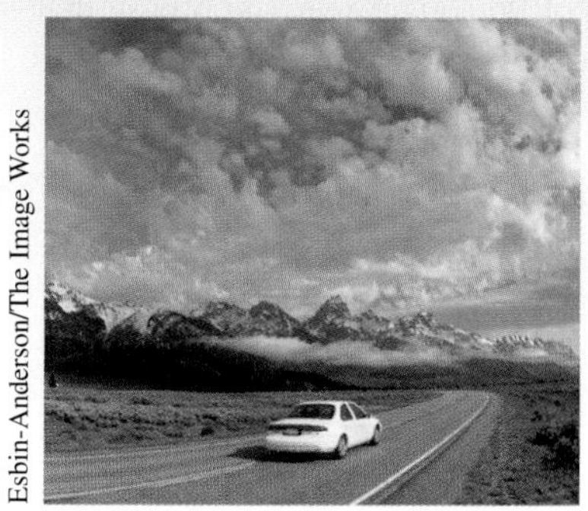
Esbin-Anderson/The Image Works

What You Should Learn

1 ▶ Use common formulas to solve application problems.

2 ▶ Solve mixture problems involving hidden products.

3 ▶ Solve work-rate problems.

Why You Should Learn It

The formula for distance can be used whenever you decide to take a road trip. For instance, in Exercise 69 on page 183, you will use the formula for distance to find the travel time for an automobile trip.

Using Formulas

1 ▶ Use common formulas to solve application problems.

Some formulas occur so frequently in problem solving that it is to your benefit to memorize them. For instance, the following formulas for area, perimeter, and volume are often used to create verbal models for word problems. In the geometric formulas below, A represents area, P represents perimeter, C represents circumference, and V represents volume.

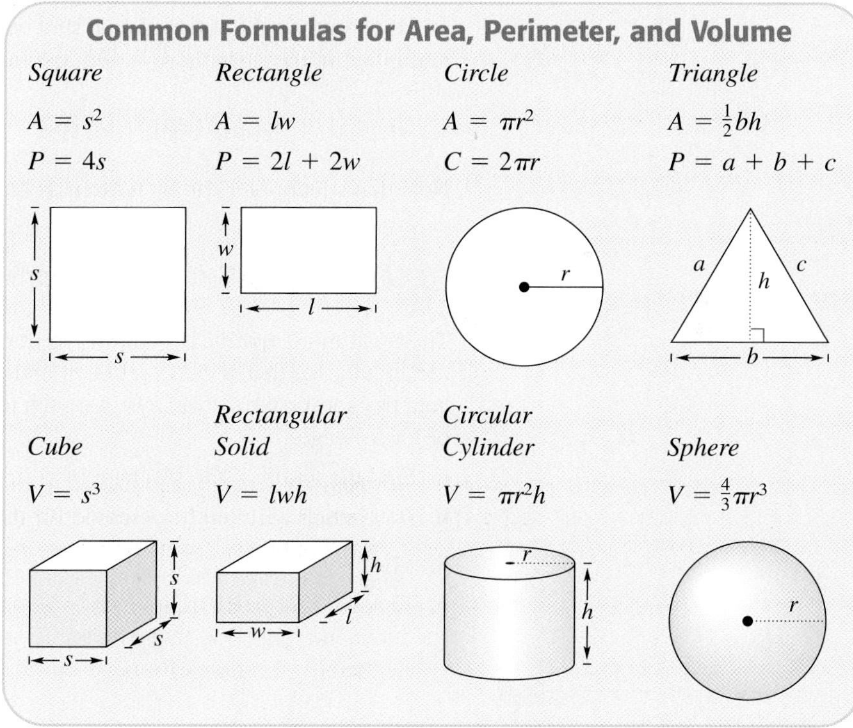

Common Formulas for Area, Perimeter, and Volume
Square
$A = s^2$
$P = 4s$
s
s
Rectangle
$A = lw$
$P = 2l + 2w$
w
l
Circle
$A = \pi r^2$
$C = 2\pi r$
r
Triangle
$A = \frac{1}{2}bh$
$P = a + b + c$
a
c
h
b
Cube
$V = s^3$
s
s
s
Rectangular Solid
$V = lwh$
h
l
w
Circular Cylinder
$V = \pi r^2 h$
r
h
Sphere
$V = \frac{4}{3}\pi r^3$
r

- *Perimeter* is always measured in linear units, such as inches, feet, miles, centimeters, meters, and kilometers.
- *Area* is always measured in square units, such as square inches, square feet, square centimeters, and square meters.
- *Volume* is always measured in cubic units, such as cubic inches, cubic feet, cubic centimeters, and cubic meters.

Study Tip

When solving problems involving perimeter, area, or volume, be sure you list the units of measurement in your answers.

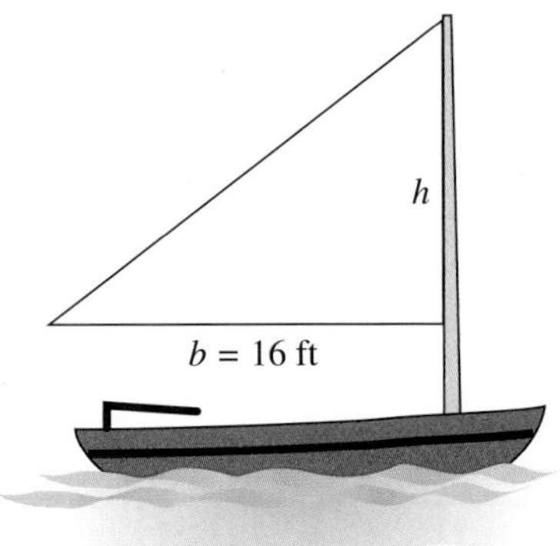

Figure 3.3

EXAMPLE 1 Using a Geometric Formula 

A sailboat has a triangular sail with an area of 96 square feet and a base that is 16 feet long, as shown in Figure 3.3. What is the height of the sail?

Solution

Because the sail is triangular and you are given its area, you should begin with the formula for the area of a triangle.

$A = \frac{1}{2}bh$	Area of a triangle
$96 = \frac{1}{2}(16)h$	Substitute 96 for A and 16 for b.
$96 = 8h$	Simplify.
$12 = h$	Divide each side by 8.

The height of the sail is 12 feet.

CHECKPOINT *Now try Exercise 29.*

In Example 1, notice that b and h are measured in feet. When they are multiplied in the formula $A = \frac{1}{2}bh$, the resulting area is measured in *square* feet.

$$A = \frac{1}{2}(16 \text{ feet})(12 \text{ feet}) = 96 \text{ feet}^2$$

Note that square feet can be written as feet^2.

EXAMPLE 2 Using a Geometric Formula 

The local municipality is planning to develop the street along which you own a rectangular lot that is 500 feet deep and has an area of 100,000 square feet. To help pay for the new sewer system, each lot owner will be assessed \$5.50 per foot of lot frontage.

a. Find the width of the frontage of your lot.

b. How much will you be assessed for the new sewer system?

Solution

a. To solve this problem, it helps to begin by drawing a diagram such as the one shown in Figure 3.4. In the diagram, label the depth of the property as $l = 500$ feet and the unknown frontage as w.

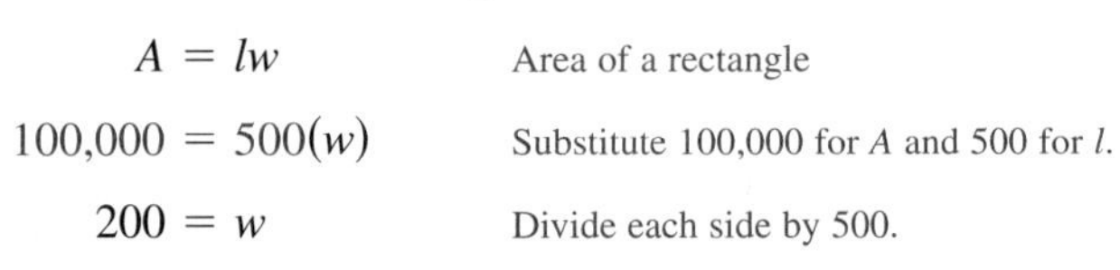

$A = lw$	Area of a rectangle
$100{,}000 = 500(w)$	Substitute 100,000 for A and 500 for l.
$200 = w$	Divide each side by 500.

The frontage of the rectangular lot is 200 feet.

b. If each foot of frontage costs \$5.50, then your total assessment will be $200(5.50) = \$1100$.

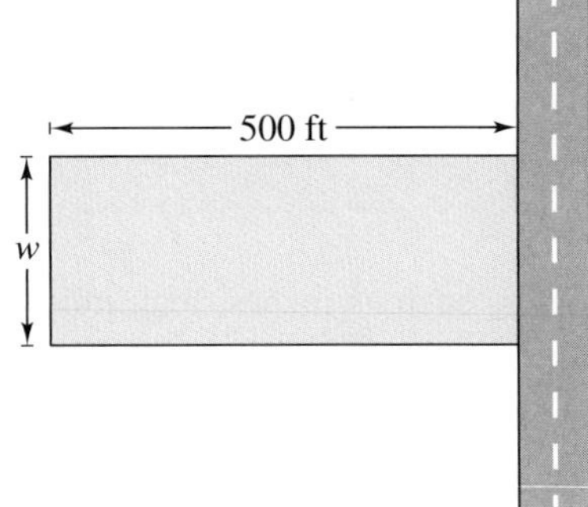

Figure 3.4

CHECKPOINT *Now try Exercise 33.*

Miscellaneous Common Formulas

Temperature: F = degrees Fahrenheit, C = degrees Celsius

$$F = \frac{9}{5}C + 32$$

Simple Interest: I = interest, P = principal, r = interest rate (decimal form), t = time (years)

$$I = Prt$$

Distance: d = distance traveled, r = rate, t = time

$$d = rt$$

In some applications, it helps to rewrite a common formula by solving for a different variable. For instance, using the common formula for temperature, you can obtain a formula for C (degrees Celsius) in terms of F (degrees Fahrenheit) as follows.

$$F = \frac{9}{5}C + 32 \qquad \text{Temperature formula}$$

$$F - 32 = \frac{9}{5}C \qquad \text{Subtract 32 from each side.}$$

$$\frac{5}{9}(F - 32) = C \qquad \text{Multiply each side by } \tfrac{5}{9}.$$

$$C = \frac{5}{9}(F - 32) \qquad \text{Formula}$$

EXAMPLE 3 Simple Interest 

An amount of \$5000 is deposited in an account paying simple interest. After 6 months, the account has earned \$162.50 in interest. What is the annual interest rate for this account?

Solution

$$I = Prt \qquad \text{Simple interest formula}$$

$$162.50 = 5000(r)\left(\frac{1}{2}\right) \qquad \text{Substitute for } I, P, \text{ and } t.$$

$$162.50 = 2500r \qquad \text{Simplify.}$$

$$\frac{162.50}{2500} = r \qquad \text{Divide each side by 2500.}$$

$$0.065 = r \qquad \text{Simplify.}$$

The annual interest rate is $r = 0.065$ (or 6.5%). Check this solution in the original statement of the problem.

✓ **CHECKPOINT** *Now try Exercise 21.*

Technology: Tip

You can use a graphing calculator to solve simple interest problems by using the program found at our website, *www.cengage.com/math/larson/algebra.* Use the program to check the results of Example 3. Then use the program and the guess, check, and revise method to find P when $I = \$5269$, $r = 11\%$, and $t = 5$ years.

One of the most familiar rate problems and most often used formulas in real life is the one that relates distance, rate (or speed), and time: $d = rt$. For instance, if you travel at a constant (or average) rate of 50 miles per hour for 45 minutes, the total distance traveled is given by

$$\left(50\,\frac{\text{miles}}{\text{hour}}\right)\left(\frac{45}{60}\text{ hour}\right) = 37.5 \text{ miles.}$$

As with all problems involving applications, be sure to check that the units in the model make sense. For instance, in this problem the rate is given in *miles per hour.* So, in order for the solution to be given in *miles,* you must convert the time (from minutes) to *hours.* In the model, you can think of dividing out the 2 "hours," as follows.

$$\left(50\,\frac{\text{miles}}{\cancel{\text{hour}}}\right)\left(\frac{45}{60}\,\cancel{\text{hour}}\right) = 37.5 \text{ miles}$$

EXAMPLE 4 A Distance-Rate-Time Problem 

You jog at an average rate of 8 kilometers per hour. How long will it take you to jog 14 kilometers?

Solution

Verbal Model: Distance = Rate · Time

Labels:
Distance = 14 (kilometers)
Rate = 8 (kilometers per hour)
Time = t (hours)

Equation:

$14 = 8(t)$ Write equation.

$\frac{14}{8} = t$ Divide each side by 8.

$1.75 = t$ Simplify.

It will take you 1.75 hours (or 1 hour and 45 minutes). Check this in the original statement of the problem.

 **CHECKPOINT** *Now try Exercise 25.*

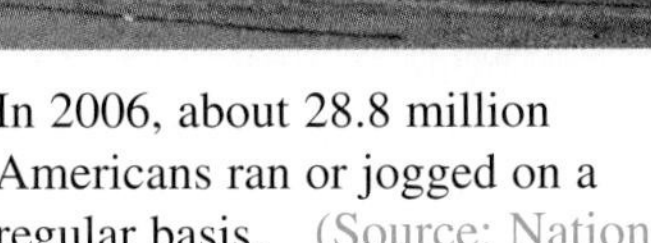

Karl Weatherly/Getty Images

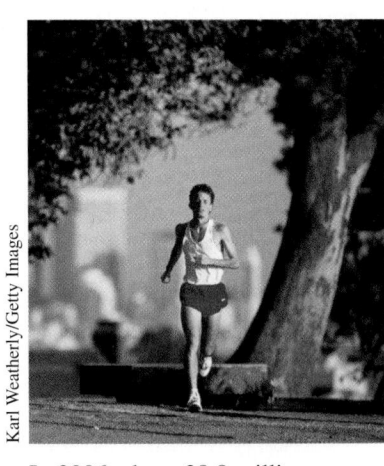

In 2006, about 28.8 million Americans ran or jogged on a regular basis. (Source: National Sporting Goods Association)

If you are having trouble solving a distance-rate-time problem, consider making a table like the one shown below for Example 4.

Distance = Rate · Time

Rate (km/hr)	8	8	8	8	8	8	8	8
Time (hours)	0.25	0.50	0.75	1.00	1.25	1.50	1.75	2.00
Distance (kilometers)	2	4	6	8	10	12	14	16

2 ▶ Solve mixture problems involving hidden products.

Solving Mixture Problems

Many real-world problems involve combinations of two or more quantities that make up a new or different quantity. Such problems are called **mixture problems.** They are usually composed of the sum of two or more "hidden products" that involve *rate factors*. Here is the generic form of the verbal model for mixture problems.

$$\underbrace{\boxed{\text{First rate}} \cdot \boxed{\text{Amount}}}_{\text{First component}} + \underbrace{\boxed{\text{Second rate}} \cdot \boxed{\text{Amount}}}_{\text{Second component}} = \underbrace{\boxed{\text{Final rate}} \cdot \boxed{\text{Final amount}}}_{\text{Final mixture}}$$

The rate factors are usually expressed as *percents* or *percents of measure* such as dollars per pound, jobs per hour, or gallons per minute.

EXAMPLE 5 A Nut Mixture Problem 

A grocer wants to mix cashew nuts worth \$7 per pound with 15 pounds of peanuts worth \$2.50 per pound. To obtain a nut mixture worth \$4 per pound, how many pounds of cashews are needed? How many pounds of mixed nuts will be produced for the grocer to sell?

Solution

In this problem, the rates are the *unit prices* of the nuts.

Verbal Model: $\boxed{\text{Total cost of cashews}} + \boxed{\text{Total cost of peanuts}} = \boxed{\text{Total cost of mixed nuts}}$

Labels:
Unit price of cashews $= 7$ (dollars per pound)
Unit price of peanuts $= 2.5$ (dollars per pound)
Unit price of mixed nuts $= 4$ (dollars per pound)
Amount of cashews $= x$ (pounds)
Amount of peanuts $= 15$ (pounds)
Amount of mixed nuts $= x + 15$ (pounds)

Equation:

$7(x) + 2.5(15) = 4(x + 15)$ Write equation.

$7x + 37.5 = 4x + 60$ Simplify.

$3x = 22.5$ Simplify.

$x = \dfrac{22.5}{3} = 7.5$ Divide each side by 3.

The grocer needs 7.5 pounds of cashews. This will result in $x + 15 = 7.5 + 15 = 22.5$ pounds of mixed nuts. You can check these results as follows.

$$\underbrace{(\$7.00/\text{lb})(7.5\text{ lb})}_{\text{Cashews}} + \underbrace{(\$2.50/\text{lb})(15\text{ lb})}_{\text{Peanuts}} \stackrel{?}{=} \underbrace{(\$4.00/\text{lb})(22.5\text{ lb})}_{\text{Mixed Nuts}}$$

$$\$52.50 + \$37.50 \stackrel{?}{=} \$90.00$$

$\$90.00 = \$90.00$ Solution checks. ✓

✓ **CHECKPOINT** *Now try Exercise 43.*

EXAMPLE 6 A Solution Mixture Problem 

A pharmacist needs to strengthen a 15% alcohol solution with a pure alcohol solution to obtain a 32% solution. How much pure alcohol should be added to 100 milliliters of the 15% solution? (See Figure 3.5.)

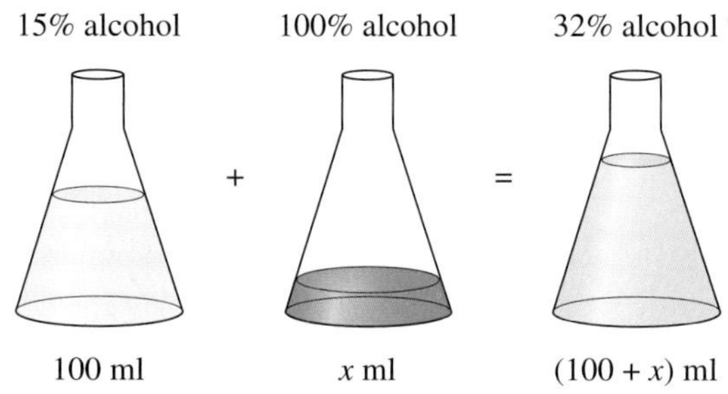

Figure 3.5

Solution

In this problem, the rates are the alcohol *percents* of the solutions.

Verbal Model: $\boxed{\text{Amount of alcohol in 15\% alcohol solution}} + \boxed{\text{Amount of alcohol in 100\% alcohol solution}} = \boxed{\text{Amount of alcohol in final alcohol solution}}$

Labels:
15% solution: Percent alcohol $= 0.15$ (decimal form)
Amount of alcohol solution $= 100$ (milliliters)
100% solution: Percent alcohol $= 1.00$ (decimal form)
Amount of alcohol solution $= x$ (milliliters)
32% solution: Percent alcohol $= 0.32$ (decimal form)
Amount of alcohol solution $= 100 + x$ (milliliters)

Equation:
$0.15(100) + 1.00(x) = 0.32(100 + x)$ Write equation.
$15 + x = 32 + 0.32x$ Simplify.
$0.68x = 17$ Simplify.
$x = \dfrac{17}{0.68}$ Divide each side by 0.68.
$= 25$ ml Simplify.

So, the pharmacist should add 25 milliliters of pure alcohol to the 15% solution. This will result in $100 + x = 100 + 25 = 125$ milliliters of the 32% solution. You can check these results as follows.

$$\overbrace{0.15(100)}^{\text{15\% solution}} + \overbrace{1.00(25)}^{\text{100\% solution}} \stackrel{?}{=} \overbrace{0.32(125)}^{\text{Final solution}}$$

$$15 + 25 \stackrel{?}{=} 40$$

$$40 = 40$$ Solution checks. ✓

✓ ***CHECKPOINT*** *Now try Exercise 45.*

Remember that mixture problems are sums of two or more hidden products that involve different rates. Watch for such problems in the exercises.

Mixture problems can also involve a "mix" of investments, as shown in the next example.

EXAMPLE 7 **Investment Mixture** 

You invested a total of \$10,000 in two funds earning $4\frac{1}{2}\%$ and $5\frac{1}{2}\%$ simple interest. There is more risk in the $5\frac{1}{2}\%$ fund. During 1 year, the two funds earned a total of \$508.75 in interest. How much did you invest in each fund?

Solution

Verbal Model: [Interest earned at $4\frac{1}{2}\%$] + [Interest earned at $5\frac{1}{2}\%$] = [Total interest earned]

Labels:
Amount invested at $4\frac{1}{2}\% = x$ (dollars)
Amount invested at $5\frac{1}{2}\% = 10{,}000 - x$ (dollars)
Interest earned at $4\frac{1}{2}\% = (x)(0.045)(1)$ (dollars)
Interest earned at $5\frac{1}{2}\% = (10{,}000 - x)(0.055)(1)$ (dollars)
Total interest earned $= 508.75$ (dollars)

Equation:

$0.045x + 0.055(10{,}000 - x) = 508.75$ Write equation.

$0.045x + 550 - 0.055x = 508.75$ Distributive Property

$550 - 0.01x = 508.75$ Simplify.

$-0.01x = -41.25$ Subtract 550 from each side.

$x = 4125$ Divide each side by -0.01.

So, you invested \$4125 at $4\frac{1}{2}\%$ and $10{,}000 - x = 10{,}000 - 4125 = \$5875$ at $5\frac{1}{2}\%$. Check this in the original statement of the problem.

 **CHECKPOINT** *Now try Exercise 51.*

3 ▶ Solve work-rate problems.

Solving Work-Rate Problems

Although not generally referred to as such, most **work-rate problems** are actually *mixture* problems because they involve two or more rates. Here is the generic form of the verbal model for work-rate problems.

[First rate] · [Time] + [Second rate] · [Time] = [1 (one whole job completed)]

In work-rate problems, the work rate is the reciprocal of the time needed to do the entire job. For instance, if it takes 7 hours to complete a job, the per-hour work rate is $\frac{1}{7}$ job per hour. Similarly, if it takes $4\frac{1}{2}$ minutes to complete a job, the per-minute rate is

$$\frac{1}{4\frac{1}{2}} = \frac{1}{\frac{9}{2}} = \frac{2}{9} \text{ job per minute.}$$

EXAMPLE 8 A Work-Rate Problem 

Consider two machines in a paper manufacturing plant. Machine 1 can complete one job (2000 pounds of paper) in 3 hours. Machine 2 is newer and can complete one job in $2\frac{1}{2}$ hours. How long will it take the two machines working together to complete one job?

Solution

Verbal Model: [Portion done by machine 1] + [Portion done by machine 2] = [1 (one whole job completed)]

Labels:

One whole job completed = 1 (job)

Rate (machine 1) $= \frac{1}{3}$ (job per hour)

Time (machine 1) $= t$ (hours)

Rate (machine 2) $= \frac{2}{5}$ (job per hour)

Time (machine 2) $= t$ (hours)

Equation:

$\left(\frac{1}{3}\right)(t) + \left(\frac{2}{5}\right)(t) = 1$ Write equation.

$\left(\frac{11}{15}\right)(t) = 1$ Combine like terms.

$t = \frac{15}{11}$ Multiply each side by $\frac{15}{11}$.

It will take $\frac{15}{11}$ hours (or about 1.36 hours) for the machines to complete the job working together. Check this solution in the original statement of the problem.

 **CHECKPOINT** *Now try Exercise 53.*

Study Tip

Note in Example 8 that "2000 pounds of paper" is unnecessary information. The 2000 pounds is represented as "one complete job."

EXAMPLE 9 A Fluid-Rate Problem 

An above-ground swimming pool has a capacity of 15,600 gallons, as shown in Figure 3.6. A drain pipe can empty the pool in $6\frac{1}{2}$ hours. At what rate (in gallons per minute) does the water flow through the drain pipe?

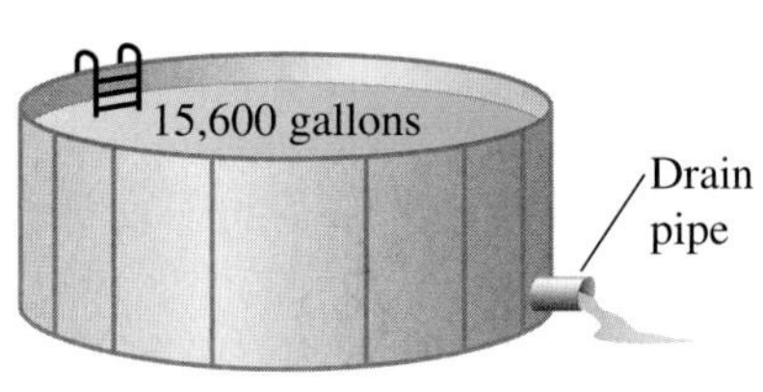

Figure 3.6

Solution

To begin, change the time from hours to minutes by multiplying by 60. That is, $6\frac{1}{2}$ hours is equal to (6.5)(60) or 390 minutes.

Verbal Model: [Volume of pool] = [Rate] · [Time]

Labels:

Volume = 15,600 (gallons)

Rate $= r$ (gallons per minute)

Time = 390 (minutes)

Equation:

$15{,}600 = r(390)$ Write equation.

$40 = r$ Divide each side by 390.

The water is flowing through the drain pipe at a rate of 40 gallons per minute. Check this solution in the original statement of the problem.

Concept Check

1. In your own words, describe the units of measure used for perimeter, area, and volume. Give examples of each.
2. Rewrite the formula for simple interest by solving for P.
3. A chemist wants to determine how much of a 20% acid solution should be added to a 60% acid solution to obtain a 35% acid solution. What additional information do you need to know to solve the problem?
4. It takes you 5 hours to complete a job. What portion of the job do you complete each hour? Explain.

3.5 EXERCISES

Go to pages 206–207 to record your assignments.

Developing Skills

In Exercises 1–14, solve for the specified variable.

1. Solve for h: $A = \frac{1}{2}bh$
2. Solve for R: $E = IR$
3. Solve for r: $A = P + Prt$
4. Solve for l: $P = 2l + 2w$
5. Solve for l: $V = lwh$
6. Solve for r: $C = 2\pi r$
7. Solve for C: $S = C + RC$
8. Solve for L: $S = L - RL$
9. Solve for m_2: $F = \alpha\frac{m_1m_2}{r^2}$
10. Solve for b: $V = \frac{4}{3}\pi a^2b$
11. Solve for b: $A = \frac{1}{2}(a + b)h$
12. Solve for r: $V = \frac{1}{3}\pi h^2(3r - h)$
13. Solve for a: $h = v_0t + \frac{1}{2}at^2$
14. Solve for a: $S = \frac{n}{2}[2a + (n - 1)d]$

In Exercises 15–18, evaluate the formula for the specified values of the variables. (List the *units* of the answer.)

15. *Volume of a Right Circular Cylinder:* $V = \pi r^2h$
    $r = 5$ meters, $h = 4$ meters
16. *Volume of a Rectangular Solid:* $V = lwh$
    $l = 7$ inches, $w = 4$ inches, $h = 2$ inches
17. *Body Mass Index:* $B = \frac{703w}{h^2}$
    $w = 127$ pounds, $h = 61$ inches
18. *Electric Power:* $I = \frac{P}{V}$
    $P = 1500$ watts, $V = 110$ volts

In Exercises 19–22, find the missing interest, principal, interest rate, or time. ***See Example 3.***

19. $I =$
    $P = \$870$
    $r = 3.8\%$
    $t = 18$ months
20. $I = \$180$
    $P =$
    $r = 4.5\%$
    $t = 3$ years
21. ✓ $I = \$54$
    $P = \$450$
    $r =$
    $t = 2$ years
22. $I = \$97.50$
    $P = \$1200$
    $r = 4.5\%$
    $t =$

In Exercises 23–28, find the missing distance, rate, or time. ***See Example 4.***

	Distance, d	*Rate, r*	*Time, t*
23.		4 m/min	12 min
24.		62 mi/hr	$2\frac{1}{2}$ hr
25.	128 km	8 km/hr	
26.	210 mi	50 mi/hr	
27.	2054 m		18 sec
28.	482 ft		40 min

Solving Problems

In Exercises 29–34, use a common geometric formula to solve the problem. ***See Examples 1 and 2.***

29. *Geometry* A triangular piece of stained glass has an area of 6 square inches and a height of 3 inches. What is the length of the base?

30. *Geometry* A dime has a circumference of about 56.27 millimeters. What is the radius of a dime? Round your answer to two decimal places.

31. *Geometry* An Olympic-size swimming pool in the shape of a rectangular solid has a volume of 3125 cubic meters, a length of 50 meters, and a width of 25 meters. What is the depth of the pool?

32. *Geometry* A cylindrical bass drum has a volume of about 3054 cubic inches and a radius of 9 inches. What is the height of the drum? Round your answer to one decimal place.

33. *Flooring Costs* You want to install a hardwood floor in a rectangular room. The floor has a perimeter of 66 feet and a length of 18 feet.

(a) What is the width of the floor?

(b) Hardwood flooring costs $12 per square foot. What is the cost of the floor?

34. *Fencing Costs* A fence borders a rectangular piece of land. The land has an area of 11,250 square feet and a width of 90 feet.

(a) What is the length of the piece of land?

(b) Fencing costs $8 per foot. What is the cost of the fence?

Simple Interest In Exercises 35–38, use the formula for simple interest. ***See Example 3.***

35. Find the annual interest rate on a certificate of deposit that earned $128.98 interest in 1 year on a principal of $1500.

36. How long must $700 be invested at an annual interest rate of 6.25% to earn $460 interest?

37. You borrow $15,000 for $\frac{1}{2}$ year. You promise to pay back the principal and the interest in one lump sum. The annual interest rate is 13%. What is your payment?

38. You have a balance of $650 on your credit card that you cannot pay this month. The annual interest rate on an unpaid balance is 19%. Find the lump sum of principal and interest due in 1 month.

In Exercises 39–42, use the formula for distance to solve the problem. ***See Example 4.***

39. *Average Speed* Determine the average speed of an experimental plane that can travel 3000 miles in 2.6 hours.

40. *Average Speed* Determine the average speed of an Olympic runner who completes the 10,000-meter race in 27 minutes and 45 seconds.

41. *Space Shuttle* The speed of the space shuttle (see figure) is 17,500 miles per hour. How long will it take the shuttle to travel a distance of 3000 miles? Round your answer to two decimal places.

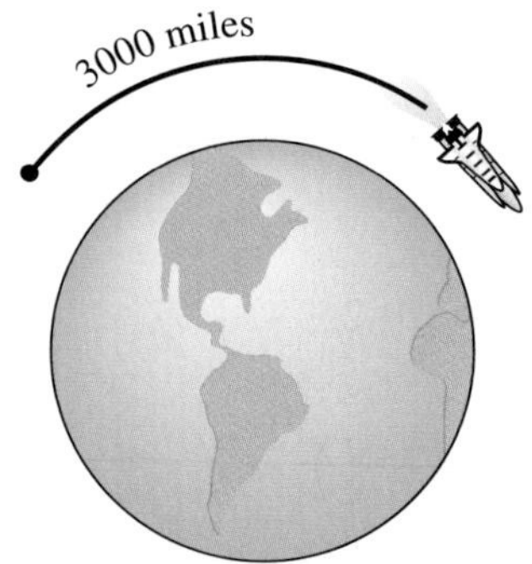

42. *Speed of Light* The speed of light is about 670,616,629.4 miles per hour, and the distance between Earth and the Sun is about 93,000,000 miles. How long does it take light from the Sun to reach Earth? Round your answer to two decimal places.

✓ **43.** *Numbers of Stamps* You have 100 stamps that have a total value of $35.10. Some of the stamps are worth $0.27 each and the others are worth $0.42 each. How many stamps of each type do you have?

44. *Numbers of Coins* A person has 50 coins in dimes and quarters with a combined value of $7.70. Determine the number of coins of each type.

Mixture Problem In Exercises 45–48, determine the numbers of units of solution 2 required to obtain the desired percent alcohol concentration of the final solution. Then find the amount of the final solution. ***See Example 6.***

	Concentration Solution 1	*Amount of Solution 1*	*Concentration Solution 2*	*Concentration Final Solution*
✓ **45.**	10%	25 gal	30%	25%
46.	25%	4 L	50%	30%
47.	15%	5 qt	45%	30%
48.	70%	18.75 gal	90%	75%

49. *Antifreeze* The cooling system in a truck contains 4 gallons of coolant that is 30% antifreeze. How much must be withdrawn and replaced with 100% antifreeze to bring the coolant in the system to 50% antifreeze?

50. *Chemistry* You need 175 milliliters of a 6% hydrochloric acid solution for an experiment. Your chemistry lab has a bottle of 3% hydrochloric acid solution and a bottle of 10% hydrochloric acid solution. How many milliliters of each solution should you mix together?

✓ **51.** *Investment Mixture* You invested a total of $6000 in two funds earning 7% and 9% simple interest. During 1 year, the two funds earned a total of $500 in interest. How much did you invest in each fund?

52. *Investment Mixture* You invested a total of $30,000 in two funds earning 8.5% and 10% simple interest. During 1 year, the two funds earned a total of $2700 in interest. How much did you invest in each fund?

✓ **53.** *Work Rate* You can mow a lawn in 2 hours using a riding mower, and your friend can mow the same lawn in 3 hours using a push mower. Using both machines together, how long will it take you and your friend to mow the lawn?

54. *Work Rate* One person can complete a typing project in 6 hours, and another can complete the same project in 8 hours. If they both work on the project, in how many hours can it be completed?

55. *Work Rate* One worker can complete a task in m minutes while a second can complete the task in $9m$ minutes. Show that by working together they can complete the task in $t = \frac{9}{10}m$ minutes.

56. *Work Rate* One worker can complete a task in h hours while a second can complete the task in $3h$ hours. Show that by working together they can complete the task in $t = \frac{3}{4}h$ hours.

✓ **57.** *Intravenous Bag* A 1000-milliliter intravenous bag is attached to a patient with a tube and is empty after 8 hours. At what rate does the solution flow through the tube?

58. *Swimming Pool* A swimming pool has a capacity of 10,800 gallons. A drain pipe empties the pool at a rate of 12 gallons per minute. How long, in hours, will it take for the pool to empty?

59. *Flower Order* A floral shop receives a $384 order for roses and carnations. The prices per dozen for the roses and carnations are $18 and $12, respectively. The order contains twice as many roses as carnations. How many of each type of flower are in the order?

60. *Ticket Sales* Ticket sales for a play total $1700. The number of tickets sold to adults is three times the number sold to children. The prices of the tickets for adults and children are $5 and $2, respectively. How many of each type were sold?

61. *Interpreting a Table* An agricultural corporation must purchase 100 tons of cattle feed. The feed is to be a mixture of soybeans, which cost \$200 per ton, and corn, which costs \$125 per ton.

(a) Complete the table, where x is the number of tons of corn in the mixture.

Corn, x	Soybeans, $100 - x$	Price per ton of the mixture
0		
20		
40		
60		
80		
100		

(b) How does an increase in the number of tons of corn affect the number of tons of soybeans in the mixture?

(c) How does an increase in the number of tons of corn affect the price per ton of the mixture?

(d) If there were equal weights of corn and soybeans in the mixture, how would the price of the mixture relate to the price of each component?

62. *Interpreting a Table* A metallurgist is making 5 ounces of an alloy of metal A, which costs \$52 per ounce, and metal B, which costs \$16 per ounce.

(a) Complete the table, where x is the number of ounces of metal A in the alloy.

Metal A, x	Metal B, $5 - x$	Price per ounce of the alloy
0		
1		
2		
3		
4		
5		

(b) How does an increase in the number of ounces of metal A in the alloy affect the number of ounces of metal B in the alloy?

(c) How does an increase in the number of ounces of metal A in the alloy affect the price of the alloy?

(d) If there were equal amounts of metal A and metal B in the alloy, how would the price of the alloy relate to the price of each of the components?

63. *Geometry* The length of a rectangle is 8 inches more than the width. The perimeter is 36 inches. What are the length and width of the rectangle?

64. *Geometry* The length of a rectangle is 11 feet more than twice the width. The perimeter is 118 feet. What are the length and width of the rectangle?

65. *Geometry* Two sides of a triangle have the same length. The third side is 7 meters less than 4 times that length. The perimeter is 83 meters. What are the lengths of the three sides of the triangle?

66. *Geometry* The longest side of a triangle is 3 times the length of the shortest side. The third side of the triangle is 4 inches longer than the shortest side. The perimeter is 49 inches. What are the lengths of the three sides of the triangle?

67. *Distance* Two cars start at a given point and travel in the same direction at average speeds of 45 miles per hour and 52 miles per hour (see figure). How far apart will they be in 4 hours?

68. *Distance* Two planes leave Orlando International Airport at approximately the same time and fly in opposite directions (see figure). Their speeds are 510 miles per hour and 600 miles per hour. How far apart will the planes be after $1\frac{1}{2}$ hours?

69. *Travel Time* On the first part of a 225-mile automobile trip you averaged 55 miles per hour. On the last part of the trip you averaged 48 miles per hour because of increased traffic congestion. The total trip took 4 hours and 15 minutes. Find the travel time for each part of the trip.

70. *Time* A jogger leaves a point on a fitness trail running at a rate of 4 miles per hour. Ten minutes later, a second jogger leaves from the same location running at 5 miles per hour. How long will it take the second jogger to overtake the first? How far will each have run at that point?

Explaining Concepts

71. It takes you 4 hours to drive 180 miles. Explain how to use mental math to find your average speed. Then explain how your method is related to the formula $d = rt$.

72. *Error Analysis* A student solves the equation $S = 2lw + 2lh + 2wh$ for w and his answer is

$$w = \frac{S - 2lw - 2lh}{2h}.$$

Describe and correct the student's error.

73. Write three equations that are equivalent to $A = \frac{1}{2}(x + y)h$ by solving for each variable, where A is the area, h is the height, and x and y are the bases of a trapezoid. Explain when you would use each equation.

74. If the height of a triangle is doubled, does the area of the triangle double? Explain.

75. If the radius of a circle is doubled, does its circumference double? Does its area double? Explain.

Cumulative Review

In Exercises 76–79, determine which of the numbers in the set are (a) natural numbers, (b) integers, (c) rational numbers, and (d) irrational numbers.

76. $\left\{-6, \frac{7}{4}, 2.1, \sqrt{49}, -8, \frac{4}{3}\right\}$

77. $\left\{1.8, \frac{1}{10}, 7, -2.75, 1, -3\right\}$

78. $\left\{0, -1, \frac{7}{12}, \frac{18}{5}, \sqrt{8}, 1.6\right\}$

79. $\left\{-2.2, 9, \frac{1}{3}, \frac{3}{5}, -6, \sqrt{13}\right\}$

In Exercises 80–85, solve the proportion.

80. $\frac{x}{3} = \frac{28}{12}$

81. $\frac{1}{4} = \frac{y}{36}$

82. $\frac{z}{18} = \frac{8}{12}$

83. $\frac{3}{2} = \frac{9}{x}$

84. $\frac{5}{t} = \frac{75}{165}$

85. $\frac{34}{x} = \frac{102}{48}$

3.6 Linear Inequalities

© Jose Fuste Raga/CORBIS

Why You Should Learn It

You can use linear inequalities to model and solve real-life problems. For instance, you will use inequalities to analyze the consumption of wind energy in Exercises 119 and 120 on page 196.

What You Should Learn

1 ▶ Sketch the graphs of inequalities.
2 ▶ Identify the properties of inequalities that can be used to create equivalent inequalities.
3 ▶ Solve linear inequalities.
4 ▶ Solve compound inequalities.
5 ▶ Solve application problems involving inequalities.

1 ▶ Sketch the graphs of inequalities.

Intervals on the Real Number Line

In this section you will study **algebraic inequalities,** which are inequalities that contain one or more variable terms. Some examples are

$$x \le 4, \quad x \ge -3, \quad x + 2 < 7, \quad \text{and} \quad 4x - 6 < 3x + 8.$$

As with an equation, you **solve** an inequality in the variable x by finding all values of x for which the inequality is true. Such values are called **solutions** and are said to *satisfy* the inequality. The set of all solutions of an inequality is the **solution set** of the inequality. The **graph** of an inequality is obtained by plotting its solution set on the real number line. Often, these graphs are intervals—either bounded or unbounded.

Bounded Intervals on the Real Number Line

Let a and b be real numbers such that $a < b$. The following intervals on the real number line are called **bounded intervals.** The numbers a and b are the **endpoints** of each interval. A bracket indicates that the endpoint is included in the interval, and a parenthesis indicates that the endpoint is excluded.

Notation	*Interval Type*	*Inequality*	*Graph*
$[a, b]$	Closed	$a \le x \le b$	
(a, b)	Open	$a < x < b$	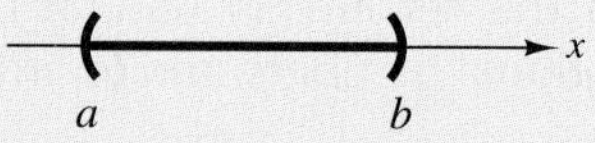
$[a, b)$		$a \le x < b$	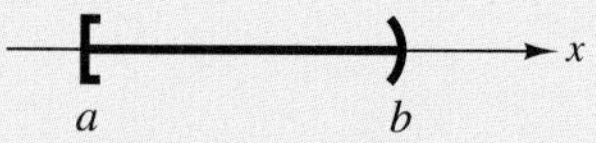
$(a, b]$		$a < x \le b$	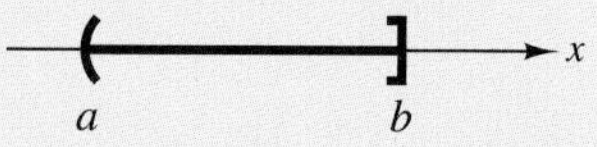

The **length** of the interval $[a, b]$ is the distance between its endpoints: $b - a$. The lengths of $[a, b]$, (a, b), $[a, b)$, and $(a, b]$ are the same. The reason that these four types of intervals are called "bounded" is that each has a finite length. An interval that *does not* have a finite length is **unbounded** (or **infinite**).

Unbounded Intervals on the Real Number Line

Let a and b be real numbers. The following intervals on the real number line are called **unbounded intervals**.

Notation	*Interval Type*	*Inequality*	*Graph*
$[a, \infty)$		$x \geq a$	
(a, ∞)	Open	$x > a$	
$(-\infty, b]$		$x \leq b$	
$(-\infty, b)$	Open	$x < b$	
$(-\infty, \infty)$	Entire real line		

The symbols ∞ (**positive infinity**) and $-\infty$ (**negative infinity**) do not represent real numbers. They are simply convenient symbols used to describe the unboundedness of an interval such as $(-5, \infty)$. This is read as the interval from -5 to infinity.

Study Tip

In Example 1(c), the inequality $-3 < x$ can also be written as $x > -3$. In other words, saying "-3 is less than x" is the same as saying "x is greater than -3."

EXAMPLE 1 Graphing Inequalities

Sketch the graph of each inequality.

a. $-3 < x \leq 1$ **b.** $0 < x < 2$

c. $-3 < x$ **d.** $x \leq 2$

Solution

a. The graph of $-3 < x \leq 1$ is a bounded interval.

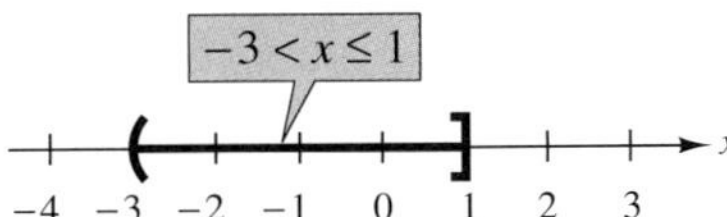

b. The graph of $0 < x < 2$ is a bounded interval.

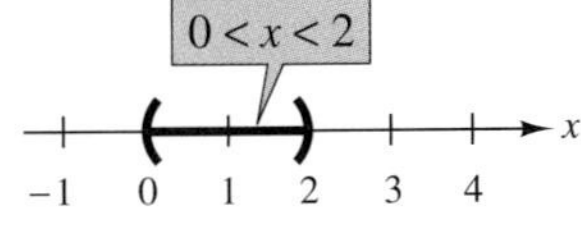

c. The graph of $-3 < x$ is an unbounded interval.

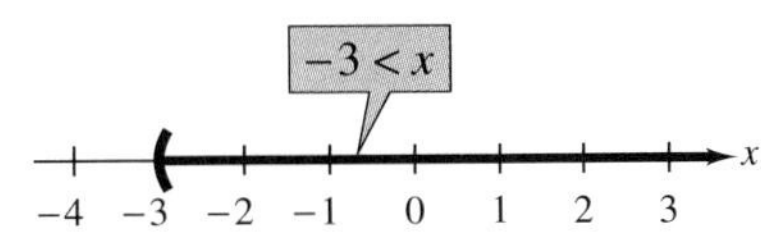

d. The graph of $x \leq 2$ is an unbounded interval.

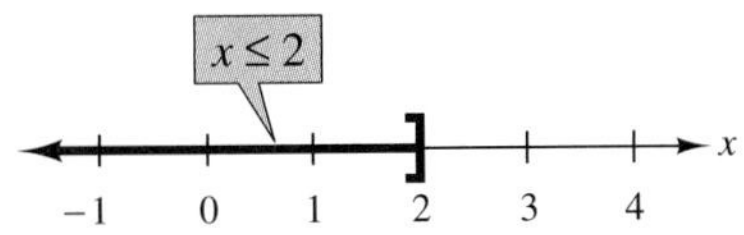

✓ ***CHECKPOINT*** *Now try Exercise 11.*

2 ▶ Identify the properties of inequalities that can be used to create equivalent inequalities.

Properties of Inequalities

Solving a linear inequality is much like solving a linear equation. You isolate the variable by using the **properties of inequalities.** These properties are similar to the properties of equality, but there are two important exceptions. *When each side of an inequality is multiplied or divided by a negative number, the direction of the inequality symbol must be reversed.* Here is an example.

$-2 < 5$	Original inequality
$(-3)(-2) > (-3)(5)$	Multiply each side by -3 and reverse the inequality.
$6 > -15$	Simplify.

Two inequalities that have the same solution set are **equivalent inequalities.** The following list of operations can be used to create equivalent inequalities.

Properties of Inequalities

1. *Addition and Subtraction Properties*

   Adding the same quantity to, or subtracting the same quantity from, each side of an inequality produces an equivalent inequality.

   If $a < b$, then $a + c < b + c$.

   If $a < b$, then $a - c < b - c$.

2. *Multiplication and Division Properties: Positive Quantities*

   Multiplying or dividing each side of an inequality by a positive quantity produces an equivalent inequality.

   If $a < b$ and c is positive, then $ac < bc$.

   If $a < b$ and c is positive, then $\frac{a}{c} < \frac{b}{c}$.

3. *Multiplication and Division Properties: Negative Quantities*

   Multiplying or dividing each side of an inequality by a negative quantity produces an equivalent inequality in which the inequality symbol is reversed.

   If $a < b$ and c is negative, then $ac > bc$. Reverse inequality.

   If $a < b$ and c is negative, then $\frac{a}{c} > \frac{b}{c}$. Reverse inequality.

4. *Transitive Property*

   Consider three quantities for which the first quantity is less than the second, and the second is less than the third. It follows that the first quantity must be less than the third quantity.

   If $a < b$ and $b < c$, then $a < c$.

These properties remain true if the symbols $<$ and $>$ are replaced by $\leq$ and $\geq$. Moreover, a, b, and c can represent real numbers, variables, or expressions. Note that you cannot multiply or divide each side of an inequality by zero.

3 ▶ Solve linear inequalities.

Solving a Linear Inequality

An inequality in one variable is a **linear inequality** if it can be written in one of the following forms.

$$ax + b \le 0, \quad ax + b < 0, \quad ax + b \ge 0, \quad ax + b > 0$$

The solution set of a linear inequality can be written in set notation. For the solution $x > 1$, the set notation is $\{x|x > 1\}$ and is read "the set of all x such that x is greater than 1."

As you study the following examples, *remember that when you multiply or divide an inequality by a negative number, you must reverse the inequality symbol.*

Study Tip

Checking the solution set of an inequality is not as simple as checking the solution set of an equation. (There are usually too many x-values to substitute back into the original inequality.) You can, however, get an indication of the validity of a solution set by substituting a few convenient values of x. For instance, in Example 2, try checking that $x = 0$ satisfies the original inequality, whereas $x = 4$ does not.

EXAMPLE 2 Solving a Linear Inequality

$x + 6 < 9$	Original inequality
$x + 6 - 6 < 9 - 6$	Subtract 6 from each side.
$x < 3$	Combine like terms.

The solution set consists of all real numbers that are less than 3. The solution set in interval notation is $(-\infty, 3)$ and in set notation is $\{x|x < 3\}$. The graph is shown in Figure 3.7.

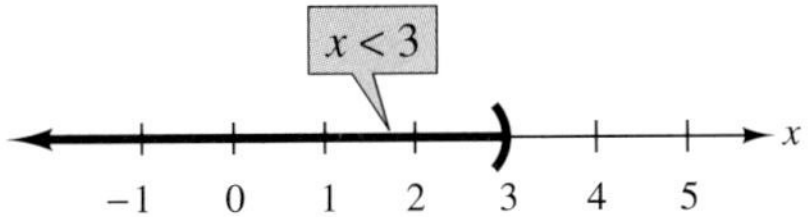

Figure 3.7

✓ ***CHECKPOINT*** *Now try Exercise 37.*

EXAMPLE 3 Solving a Linear Inequality

$8 - 3x \le 20$	Original inequality
$8 - 8 - 3x \le 20 - 8$	Subtract 8 from each side.
$-3x \le 12$	Combine like terms.
$\dfrac{-3x}{-3} \ge \dfrac{12}{-3}$	Divide each side by -3 and reverse the inequality symbol.
$x \ge -4$	Simplify.

The solution set in interval notation is $[-4, \infty)$ and in set notation is $\{x|x \ge -4\}$. The graph is shown in Figure 3.8.

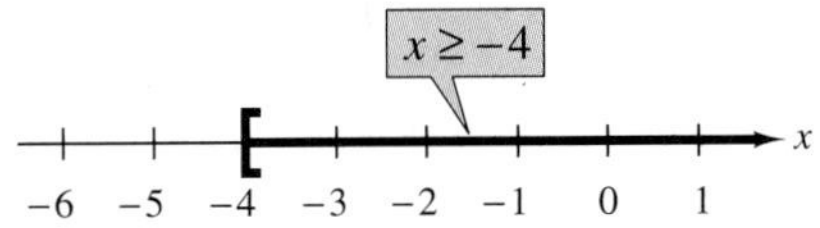

Figure 3.8

✓ ***CHECKPOINT*** *Now try Exercise 41.*

Technology: Tip

Most graphing calculators can graph a linear inequality. Consult your user's guide for specific instructions. The graph below shows the solution of the inequality in Example 4. Notice that the graph representing the solution interval lies above the x-axis.

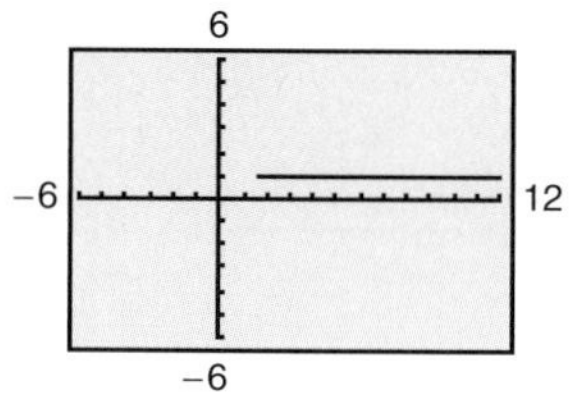

EXAMPLE 4 Solving a Linear Inequality

$7x - 3 > 3(x + 1)$ Original inequality

$7x - 3 > 3x + 3$ Distributive Property

$7x - 3x - 3 > 3x - 3x + 3$ Subtract $3x$ from each side.

$4x - 3 > 3$ Combine like terms.

$4x - 3 + 3 > 3 + 3$ Add 3 to each side.

$4x > 6$ Combine like terms.

$\frac{4x}{4} > \frac{6}{4}$ Divide each side by 4.

$x > \frac{3}{2}$ Simplify.

The solution set consists of all real numbers that are greater than $\frac{3}{2}$. The solution set in interval notation is $\left(\frac{3}{2}, \infty\right)$ and in set notation is $\left\{x \middle| x > \frac{3}{2}\right\}$. The graph is shown in Figure 3.9.

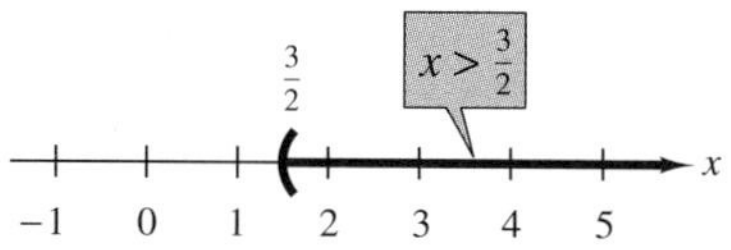

Figure 3.9

✓ ***CHECKPOINT*** *Now try Exercise 79.*

Study Tip

An inequality can be cleared of fractions in the same way an equation can be cleared of fractions—by multiplying each side by the least common denominator. This is shown in Example 5.

EXAMPLE 5 Solving a Linear Inequality

$\frac{2x}{3} + 12 < \frac{x}{6} + 18$ Original inequality

$6 \cdot \left(\frac{2x}{3} + 12\right) < 6 \cdot \left(\frac{x}{6} + 18\right)$ Multiply each side by LCD of 6.

$4x + 72 < x + 108$ Distributive Property

$4x - x < 108 - 72$ Subtract x and 72 from each side.

$3x < 36$ Combine like terms.

$x < 12$ Divide each side by 3.

The solution set consists of all real numbers that are less than 12. The solution set in interval notation is $(-\infty, 12)$ and in set notation is $\{x | x < 12\}$. The graph is shown in Figure 3.10.

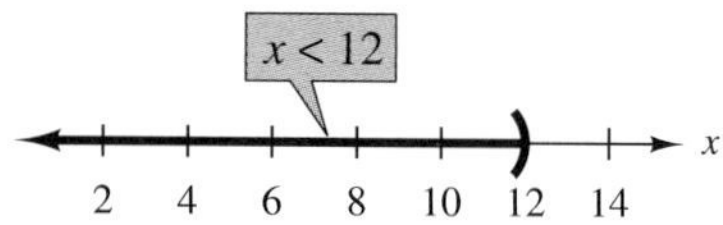

Figure 3.10

✓ ***CHECKPOINT*** *Now try Exercise 55.*

4 ▶ Solve compound inequalities.

Solving a Compound Inequality

Two inequalities joined by the word *and* or the word *or* constitute a **compound inequality.** When two inequalities are joined by the word *and*, the solution set consists of all real numbers that satisfy *both* inequalities. The solution set for the compound inequality $-4 \le 5x - 2$ *and* $5x - 2 < 7$ can be written more simply as the **double inequality**

$$-4 \le 5x - 2 < 7.$$

A compound inequality formed by the word *and* is called **conjunctive** and is the only kind that has the potential to form a double inequality. A compound inequality joined by the word *or* is called **disjunctive** and cannot be re-formed into a double inequality.

EXAMPLE 6 Solving a Double Inequality

Solve the double inequality $-7 \le 5x - 2 < 8$.

Solution

$-7 \le 5x - 2 < 8$	Write original inequality.
$-7 + 2 \le 5x - 2 + 2 < 8 + 2$	Add 2 to all three parts.
$-5 \le 5x < 10$	Combine like terms.
$\frac{-5}{5} \le \frac{5x}{5} < \frac{10}{5}$	Divide each part by 5.
$-1 \le x < 2$	Simplify.

The solution set consists of all real numbers that are greater than or equal to -1 and less than 2. The solution set in interval notation is $[-1, 2)$ and in set notation is $\{x|-1 \le x < 2\}$. The graph is shown in Figure 3.11.

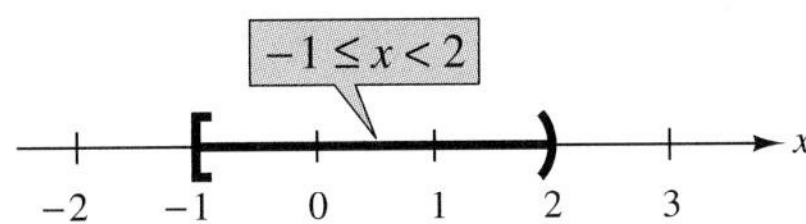

Figure 3.11

✓ **CHECKPOINT** *Now try Exercise 61.*

The double inequality in Example 6 could have been solved in two parts, as follows.

$-7 \le 5x - 2$	and	$5x - 2 < 8$
$-5 \le 5x$		$5x < 10$
$-1 \le x$		$x < 2$

The solution set consists of all real numbers that satisfy both inequalities. In other words, the solution set is the set of all values of x for which

$$-1 \le x < 2.$$

EXAMPLE 7 Solving a Conjunctive Inequality

Solve the compound inequality $-1 \le 2x - 3$ and $2x - 3 < 5$.

Solution

Begin by writing the conjunctive inequality as a double inequality.

$$-1 \le 2x - 3 < 5$$ Write as double inequality.

$$-1 + 3 \le 2x - 3 + 3 < 5 + 3$$ Add 3 to all three parts.

$$2 \le 2x < 8$$ Combine like terms.

$$\frac{2}{2} \le \frac{2x}{2} < \frac{8}{2}$$ Divide each part by 2.

$$1 \le x < 4$$ Solution set (See Figure 3.12.)

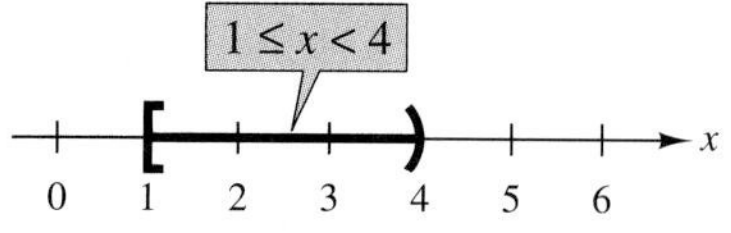

Figure 3.12

The solution set in interval notation is $[1, 4)$ and in set notation is $\{x|1 \le x < 4\}$.

✓ **CHECKPOINT** *Now try Exercise 71.*

EXAMPLE 8 Solving a Disjunctive Inequality

Solve the compound inequality $-3x + 6 \le 2$ or $-3x + 6 \ge 7$.

Solution

$$-3x + 6 \le 2 \quad \text{or} \quad -3x + 6 \ge 7$$ Write original inequality.

$$-3x + 6 - 6 \le 2 - 6 \qquad -3x + 6 - 6 \ge 7 - 6$$ Subtract 6 from all parts.

$$-3x \le -4 \qquad -3x \ge 1$$ Combine like terms.

$$\frac{-3x}{-3} \ge \frac{-4}{-3} \qquad \frac{-3x}{-3} \le \frac{1}{-3}$$ Divide all parts by -3 and reverse both inequality symbols.

$$x \ge \frac{4}{3} \qquad x \le -\frac{1}{3}$$ Solution set (See Figure 3.13.)

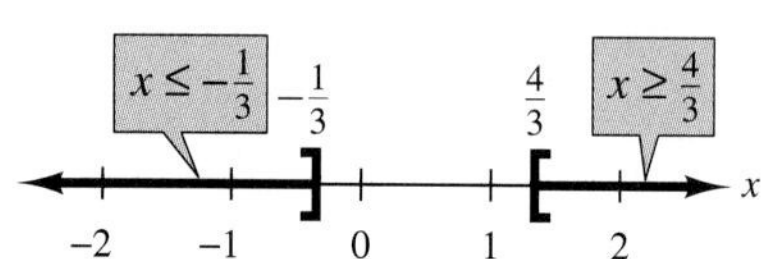

Figure 3.13

The solution set in set notation is $\{x|x \le -\frac{1}{3} \text{ or } x \ge \frac{4}{3}\}$.

✓ **CHECKPOINT** *Now try Exercise 75.*

Compound inequalities can be written using *symbols*. For compound inequalities, the word *and* is represented by the symbol $\cap$, which is read as **intersection.** The word *or* is represented by the symbol $\cup$, which is read as **union.** Graphical representations are shown in Figure 3.14. If A and B are sets, then x is in $A \cap B$ if it is in both A and B. Similarly, x is in $A \cup B$ if it is in A, B, or both A and B.

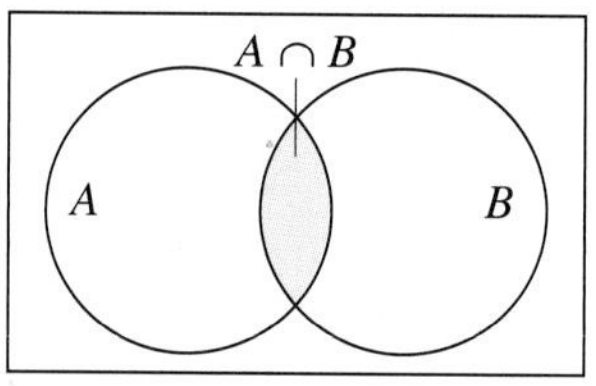

Intersection of two sets

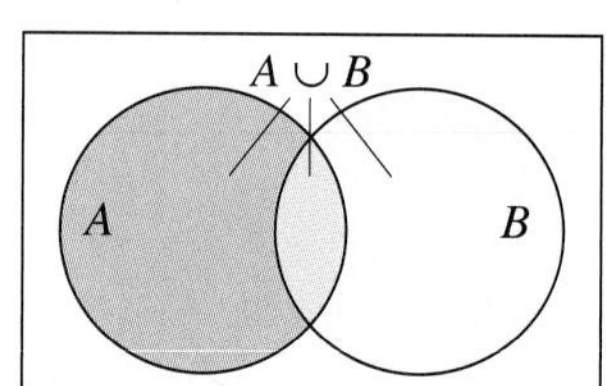

Union of two sets

Figure 3.14

EXAMPLE 9 Writing a Solution Set Using Union

A solution set is shown on the number line in Figure 3.15.

a. Write the solution set as a compound inequality.

b. Write the solution set using the union symbol.

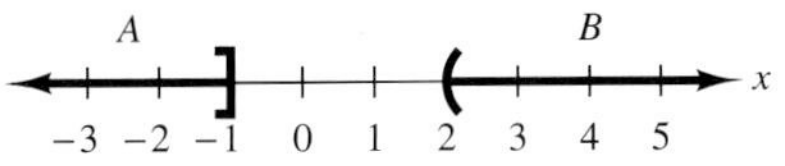

Figure 3.15

Solution

a. As a compound inequality, you can write the solution set as $x \le -1$ or $x > 2$.

b. Using set notation, you can write the left interval as $A = \{x|x \le -1\}$ and the right interval as $B = \{x|x > 2\}$. So, using the union symbol, the entire solution set can be written as $A \cup B$.

✓ ***CHECKPOINT*** *Now try Exercise 83.*

EXAMPLE 10 Writing a Solution Set Using Intersection

Write the compound inequality using the intersection symbol.

$$-3 \le x \le 4$$

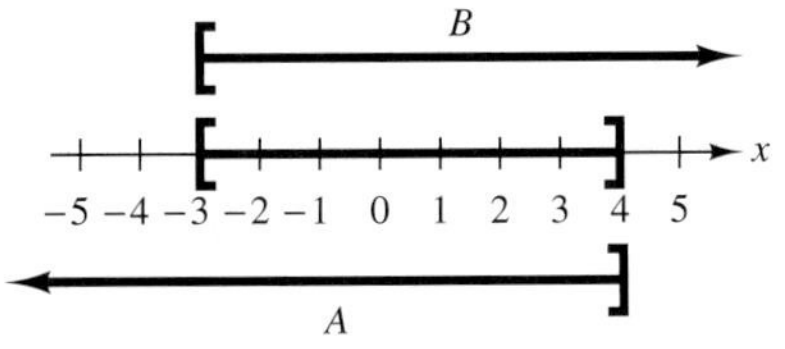

Figure 3.16

Solution

Consider the two sets $A = \{x|x \le 4\}$ and $B = \{x|x \ge -3\}$. These two sets overlap, as shown on the number line in Figure 3.16. The compound inequality $-3 \le x \le 4$ consists of all numbers that are in both $x \le 4$ and $x \ge -3$, which means that it can be written as $A \cap B$.

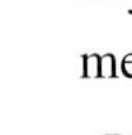

✓ ***CHECKPOINT*** *Now try Exercise 89.*

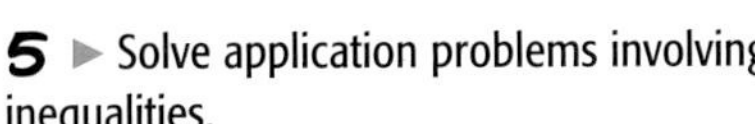

5 ▶ Solve application problems involving inequalities.

Applications

Linear inequalities in real-life problems arise from statements that involve phrases such as "at least," "no more than," "minimum value," and so on. Study the meanings of the key phrases in the next example.

EXAMPLE 11 Translating Verbal Statements

Verbal Statement	*Inequality*	
a. x is at most 3.	$x \le 3$	"at most" means "less than or equal to."
b. x is no more than 3.	$x \le 3$	
c. x is at least 3.	$x \ge 3$	"at least" means "greater than or equal to."
d. x is no less than 3.	$x \ge 3$	
e. x is more than 3.	$x > 3$	
f. x is less than 3.	$x < 3$	
g. x is a minimum of 3.	$x \ge 3$	
h. x is at least 2, but less than 7.	$2 \le x < 7$	
i. x is greater than 2, but no more than 7.	$2 < x \le 7$	

✓ ***CHECKPOINT*** *Now try Exercise 97.*

To solve real-life problems involving inequalities, you can use the same "verbal-model approach" you use with equations.

EXAMPLE 12 Finding the Maximum Width of a Package 

Figure 3.17

An overnight delivery service will not accept any package with a combined length and girth (perimeter of a cross section perpendicular to the length) exceeding 132 inches. Consider a rectangular box that is 68 inches long and has square cross sections. What is the maximum acceptable width of such a box?

Solution

First make a sketch (see Figure 3.17). The length of the box is 68 inches, and because a cross section is square, the width and height are each x inches long.

Verbal Model: Length + Girth $\leq$ 132 inches

Labels: Width of a side $= x$ (inches)
Length $= 68$ (inches)
Girth $= 4x$ (inches)

Inequality:
$$68 + 4x \leq 132$$
$$4x \leq 64$$
$$x \leq 16$$

The width of the box can be at most 16 inches.

 **CHECKPOINT** *Now try Exercise 113.*

EXAMPLE 13 Comparing Costs 

Company A rents a subcompact car for \$240 per week with no extra charge for mileage. Company B rents a similar car for \$100 per week plus an additional 25 cents for each mile driven. How many miles must you drive in a week so that the rental fee of Company B is more than that of Company A?

Solution

Verbal Model: Weekly cost for Company B > Weekly cost for Company A

Labels: Number of miles driven in one week $= m$ (miles)
Weekly cost for Company A $= 240$ (dollars)
Weekly cost for Company B $= 100 + 0.25m$ (dollars)

Inequality:
$$100 + 0.25m > 240$$
$$0.25m > 140$$
$$m > 560$$

So, the car from Company B is more expensive if you drive more than 560 miles in a week. The table shown at the left helps confirm this conclusion.

Miles driven	Company A	Company B
520	\$240.00	\$230.00
530	\$240.00	\$232.50
540	\$240.00	\$235.00
550	\$240.00	\$237.50
560	\$240.00	\$240.00
570	\$240.00	\$242.50

CHECKPOINT *Now try Exercise 117.*

Concept Check

1. Is dividing each side of an inequality by 5 the same as multiplying each side by $\frac{1}{5}$? Explain.

2. State whether each inequality is equivalent to $x > 3$. Explain your reasoning in each case.

   (a) $x < 3$ (b) $3 < x$

   (c) $-x < -3$ (d) $-3 < x$

3. Describe two types of situations involving application of properties of inequalities for which you must reverse the inequality symbol.

4. Explain the distinction between using the word *or* and using the word *and* to form a compound inequality.

3.6 EXERCISES

Go to pages 206–207 to record your assignments.

Developing Skills

In Exercises 1–4, determine whether each value of x satisfies the inequality.

	Inequality	*Values*	
1.	$7x - 10 > 0$	(a) $x = 3$	(b) $x = -2$
		(c) $x = \frac{5}{2}$	(d) $x = \frac{1}{2}$
2.	$3x + 2 < \frac{7x}{5}$	(a) $x = 0$	(b) $x = 4$
		(c) $x = -4$	(d) $x = -1$
3.	$0 < \frac{x + 4}{5} < 2$	(a) $x = 10$	(b) $x = -4$
		(c) $x = 0$	(d) $x = 6$
4.	$-3 < \frac{2 - x}{2} \leq 3$	(a) $x = 0$	(b) $x = 7$
		(c) $x = 9$	(d) $x = -1$

In Exercises 5–10, match the inequality with its graph. [The graphs are labeled (a), (b), (c), (d), (e), and (f).]

(a)
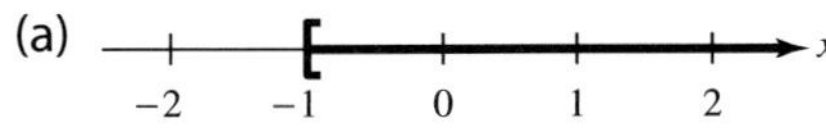

(b)
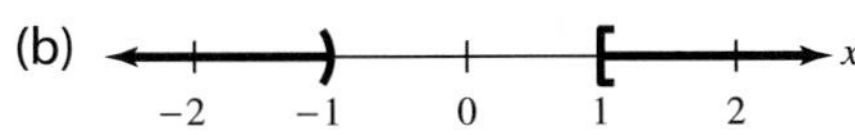

(c)

(d)

(e)

(f)

5. $x \geq -1$
6. $-1 < x \leq 1$
7. $x \leq -1$ or $x \geq 2$
8. $x < -1$ or $x \geq 1$
9. $-2 < x < 1$
10. $x < 2$

In Exercises 11–24, sketch the graph of the inequality. *See Example 1.*

11. $x \leq 4$
12. $x > -6$
13. $x > 3.5$
14. $x \leq -2.5$
15. $-5 < x \leq 3$
16. $-1 < x \leq 5$
17. $4 > x \geq 1$
18. $9 \geq x \geq 3$
19. $\frac{3}{2} \geq x > 0$
20. $-\frac{15}{4} < x < -\frac{5}{2}$
21. $x < -5$ or $x \geq -1$
22. $x \leq -4$ or $x > 0$
23. $x \leq 3$ or $x > 7$
24. $x \leq -1$ or $x \geq 1$
25. Write an inequality equivalent to $5 - \frac{1}{3}x > 8$ by multiplying each side by -3.
26. Write an inequality equivalent to $5 - \frac{1}{3}x > 8$ by adding $\frac{1}{3}x$ to each side.

In Exercises 27–34, determine whether the inequalities are equivalent.

27. $3x - 2 < 12, \quad 3x < 10$

28. $6x + 7 \geq 11, \quad 6x \geq 18$

29. $-5(x + 12) > 25, \quad x + 12 > -5$

30. $-4(5 - x) < 32, \quad 5 - x < -8$

31. $7x - 6 \leq 3x + 12, \quad 4x \leq 18$

32. $11 - 3x \geq 7x + 1, \quad 10 \geq 10x$

33. $3x > 5x, \quad 3 > 5$

34. $4x > -8x, \quad -4 < 8$

In Exercises 35–82, solve the inequality and sketch the solution on the real number line. ***See Examples 2–8.***

35. $x - 4 \geq 0$

36. $x + 1 < 0$

37. $x + 7 \leq 9$

38. $z - 4 > 0$

39. $2x < 8$

40. $3x \geq 12$

41. $-9x \geq 36$

42. $-6x \leq 24$

43. $-\frac{3}{4}x < -6$

44. $-\frac{1}{5}x > -2$

45. $5 - x \leq -2$

46. $1 - y \geq -5$

47. $2x - 5.3 > 9.8$

48. $1.6x + 4 \leq 12.4$

49. $5 - 3x < 7$

50. $12 - 5x > 5$

51. $3x - 11 > -x + 7$

52. $21x - 11 \leq 6x + 19$

53. $-3x + 7 < 8x - 13$

54. $6x - 1 > 3x - 11$

55. $\dfrac{x}{4} > 2 - \dfrac{x}{2}$

56. $\dfrac{x}{6} - 1 \leq \dfrac{x}{4}$

57. $\dfrac{x - 4}{3} + 3 \leq \dfrac{x}{8}$

58. $\dfrac{x + 3}{6} + \dfrac{x}{8} \geq 1$

59. $\dfrac{3x}{5} - 4 < \dfrac{2x}{3} - 3$

60. $\dfrac{4x}{7} + 1 > \dfrac{x}{2} + \dfrac{5}{7}$

61. $0 < 2x - 5 < 9$

62. $-6 \leq 3x - 9 < 0$

63. $8 < 6 - 2x \leq 12$

64. $-10 \leq 4 - 7x < 10$

65. $-1 < -0.2x < 1$

66. $-2 < -0.5s \leq 0$

67. $-3 < \dfrac{2x - 3}{2} < 3$

68. $0 \leq \dfrac{x - 5}{2} < 4$

69. $1 > \dfrac{x - 4}{-3} > -2$

70. $-\dfrac{2}{3} < \dfrac{x - 4}{-6} \leq \dfrac{1}{3}$

71. $2x - 4 \leq 4$ and $2x + 8 > 6$

72. $7 + 4x < -5 + x$ and $2x + 10 \leq -2$

73. $8 - 3x > 5$ and $x - 5 \geq 10$

74. $9 - x \leq 3 + 2x$ and $3x - 7 \leq -22$

75. $7.2 - 1.1x > 1$ or $1.2x - 4 > 2.7$

76. $0.4x - 3 \leq 8.1$ or $4.2 - 1.6x \leq 3$

77. $7x + 11 < 3 + 4x$ or $\dfrac{5}{2}x - 1 \geq 9 - \dfrac{3}{2}x$

78. $3x + 10 \leq -x - 6$ or $\dfrac{1}{2}x + 5 < \dfrac{5}{2}x - 4$

79. $-3(y + 10) \geq 4(y + 10)$

80. $2(4 - z) \geq 8(1 + z)$

81. $-4 \leq 2 - 3(x + 2) < 11$

82. $16 < 4(y + 2) - 5(2 - y) \leq 24$

In Exercises 83–88, write the solution set as a compound inequality. Then write the solution using set notation and the union or intersection symbol. ***See Example 9.***

83.
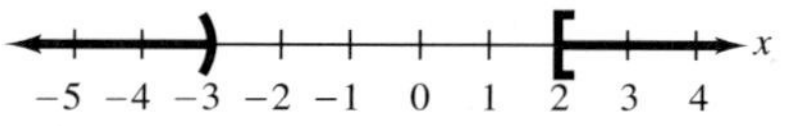

84.
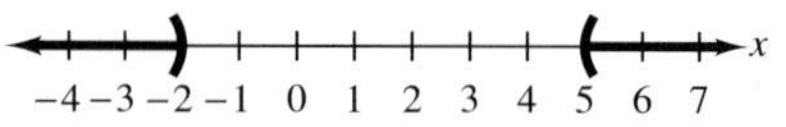

85.
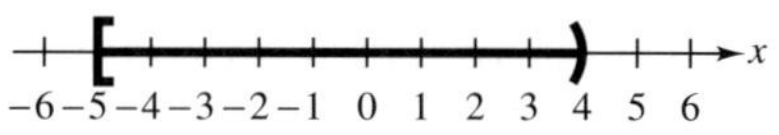

86.
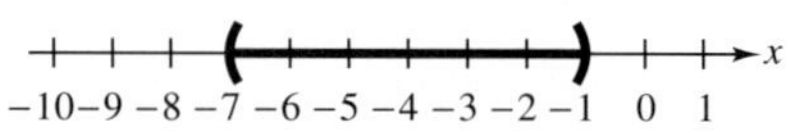

87. 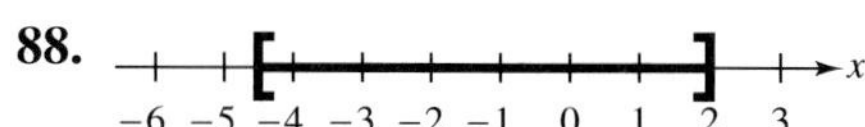

88.

−6 −5 −4 −3 −2 −1 0 1 2 3 x

In Exercises 89–94, write the compound inequality using set notation and the union or intersection symbol. ***See Example 10.***

89. $-7 \le x < 0$

90. $2 < x < 8$

91. $-\frac{9}{2} < x \le -\frac{3}{2}$

92. $-\frac{4}{5} \le x < \frac{1}{5}$

93. $x < 0$ or $x \ge \frac{2}{3}$

94. $-3 > x$ or $x > 8$

In Exercises 95–100, rewrite the statement using inequality notation. ***See Example 11.***

95. x is nonnegative.

96. y is more than -2.

97. z is at least 8.

98. m is at least 4.

99. n is at least 10, but no more than 16.

100. x is at least 450, but no more than 500.

In Exercises 101–104, write a verbal description of the inequality.

101. $x \ge \frac{5}{2}$

102. $t < 4$

103. $3 \le y < 5$

104. $0 < z \le \pi$

Solving Problems

105. *Budget* A student group has \$4500 budgeted for a field trip. The cost of transportation for the trip is \$1900. To stay within the budget, all other costs C must be no more than what amount?

106. *Budget* You have budgeted \$1800 per month for your total expenses. The cost of rent per month is \$600 and the cost of food is \$350. To stay within your budget, all other costs C must be no more than what amount?

107. *Meteorology* Miami's average temperature is greater than the average temperature in Washington, DC, and the average temperature in Washington, DC is greater than the average temperature in New York City. How does the average temperature in Miami compare with the average temperature in New York City?

108. *Elevation* The elevation (above sea level) of San Francisco is less than the elevation of Dallas, and the elevation of Dallas is less than the elevation of Denver. How does the elevation of San Francisco compare with the elevation of Denver?

109. *Operating Costs* A utility company has a fleet of vans. The annual operating cost per van is $C = 0.35m + 2900$, where m is the number of miles traveled by a van in a year. What is the maximum number of miles that will yield an annual operating cost that is no more than \$12,000?

110. *Operating Costs* A fuel company has a fleet of trucks. The annual operating cost per truck is $C = 0.58m + 7800$, where m is the number of miles traveled by a truck in a year. What is the maximum number of miles that will yield an annual operating cost that is less than \$25,000?

Cost, Revenue, and Profit In Exercises 111 and 112, the revenue R from selling x units and the cost C of producing x units of a product are given. In order to obtain a profit, the revenue must be greater than the cost. For what values of x will this product produce a profit?

111. $R = 89.95x$
$C = 61x + 875$

112. $R = 105.45x$
$C = 78x + 25{,}850$

113. *Geometry* The width of a rectangle is 22 meters. The perimeter of the rectangle must be at least 90 meters and not more than 120 meters. Find the interval for the length x (see figure on page 196).

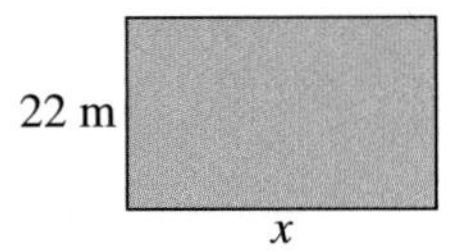

Figure for 113

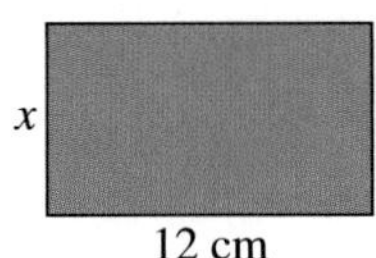

Figure for 114

114. *Geometry* The length of a rectangle is 12 centimeters. The perimeter of the rectangle must be at least 30 centimeters and not more than 42 centimeters. Find the interval for the width x.

115. *Number Problem* Four times a number n must be at least 12 and no more than 30. What interval represents the possible values of n?

116. *Number Problem* What interval represents the values of n for which $\frac{1}{3}n$ is no more than 9?

✓ **117.** *Hourly Wage* Your company requires you to select one of two payment plans. One plan pays a straight \$12.50 per hour. The second plan pays \$8.00 per hour plus \$0.75 per unit produced per hour. Write an inequality for the number of units that must be produced per hour so that the second option yields the greater hourly wage. Solve the inequality.

118. *Monthly Wage* Your company requires you to select one of two payment plans. One plan pays a straight \$3000 per month. The second plan pays \$1000 per month plus a commission of 4% of your gross sales. Write an inequality for the gross sales per month for which the second option yields the greater monthly wage. Solve the inequality.

Energy In Exercises 119 and 120, use the equation $y = 21.8t - 160$, for $9 \le t \le 15$, which models the annual consumption of energy produced by wind (in trillions of British thermal units) in the United States from 1999 to 2005. In this model, t represents the year, with $t = 9$ corresponding to 1999. (Source: U.S. Energy Information Administration)

119. During which years was the consumption of energy produced by wind less than 100 trillion Btu?

120. During which years was the consumption of energy produced by wind greater than 130 trillion Btu?

Explaining Concepts

121. Describe any differences between properties of equalities and properties of inequalities.

122. If $-3 \le x \le 10$, then $-x$ must be in what interval? Explain.

123. Discuss whether the solution set of a linear inequality is a *bounded* interval or an *unbounded* interval.

124. Two linear inequalities are joined by the word *or* to form a compound inequality. Discuss whether the solution set is a bounded interval.

In Exercises 125–128, let a and b be real numbers such that $a < b$. Use a and b to write a compound algebraic inequality in x with the given type of solution. Explain your reasoning.

125. A bounded interval

126. Two unbounded intervals

127. The set of all real numbers

128. No solution

Cumulative Review

In Exercises 129–132, place the correct symbol ($<$, $>$, or $=$) between the pair of real numbers.

129. $|4| \quad \square \quad |-5|$

130. $|-4| \quad \square \quad |-6|$

131. $|-7| \quad \square \quad |7|$

132. $-|5| \quad \square \quad -(5)$

In Exercises 133–136, determine whether each value of the variable is a solution of the equation.

133. $3x = 27$; $x = 6, x = 9$

134. $x - 14 = 8$; $x = 6, x = 22$

135. $7x - 5 = 7 + x$; $x = 2, x = 6$

136. $2 + 5x = 8x - 13$; $x = 3, x = 5$

In Exercises 137–140, solve the equation.

137. $2x - 17 = 0$

138. $x - 17 = 4$

139. $32x = -8$

140. $14x + 5 = 2 - x$

3.7 Absolute Value Equations and Inequalities

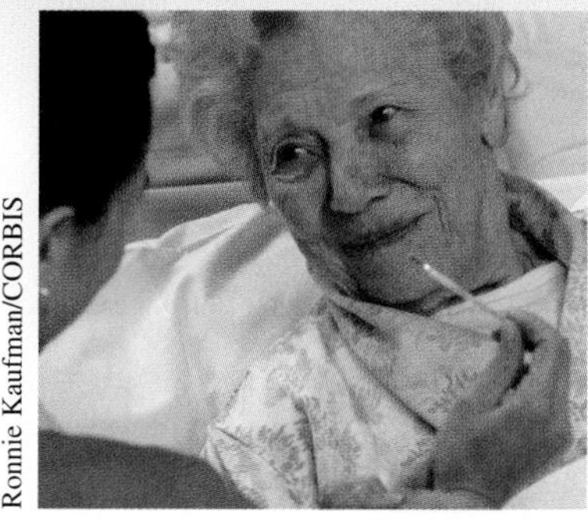

Ronnie Kaufman/CORBIS

What You Should Learn

1 ▶ Solve absolute value equations.
2 ▶ Solve inequalities involving absolute value.

Why You Should Learn It

Absolute value equations and inequalities can be used to model and solve real-life problems. For instance, in Exercise 92 on page 205, you will use an absolute value inequality to describe the normal body temperature range.

1 ▶ Solve absolute value equations.

Solving Equations Involving Absolute Value

Consider the **absolute value equation**

$$|x| = 3.$$

The only solutions of this equation are $x = -3$ and $x = 3$, because these are the only two real numbers whose distance from zero is 3. (See Figure 3.18.) In other words, the absolute value equation $|x| = 3$ has exactly two solutions:

$$x = -3 \quad \text{and} \quad x = 3.$$

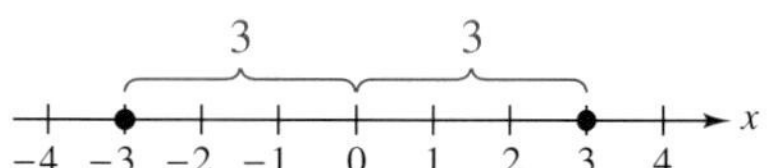

Figure 3.18

Solving an Absolute Value Equation

Let x be a variable or an algebraic expression and let a be a real number such that $a \geq 0$. The solutions of the equation $|x| = a$ are given by $x = -a$ and $x = a$. That is,

$$|x| = a \quad \Longrightarrow \quad x = -a \quad \text{or} \quad x = a.$$

EXAMPLE 1 Solving Absolute Value Equations

Solve each absolute value equation.

a. $|x| = 10$ **b.** $|x| = 0$ **c.** $|y| = -1$

Solution

a. This equation is equivalent to the two linear equations

$x = -10$ and $x = 10$. Equivalent linear equations

So, the absolute value equation has two solutions: $x = -10$ and $x = 10$.

b. This equation is equivalent to the two linear equations

$x = -(0) = 0$ and $x = 0$. Equivalent linear equations

Because both equations are the same, you can conclude that the absolute value equation has only one solution: $x = 0$.

c. This absolute value equation has *no solution* because it is not possible for the absolute value of a real number to be negative.

✓ **CHECKPOINT** *Now try Exercise 13.*

Study Tip

The strategy for solving an absolute value equation is to *rewrite* the equation in *equivalent forms* that can be solved by previously learned methods. This is a common strategy in mathematics. That is, when you encounter a new type of problem, you try to rewrite the problem so that it can be solved by techniques you already know.

EXAMPLE 2 Solving an Absolute Value Equation

Solve $|3x + 4| = 10$.

Solution

$\|3x + 4\| = 10$			Write original equation.
$3x + 4 = -10$	or	$3x + 4 = 10$	Equivalent equations
$3x + 4 - 4 = -10 - 4$		$3x + 4 - 4 = 10 - 4$	Subtract 4 from each side.
$3x = -14$		$3x = 6$	Combine like terms.
$x = -\frac{14}{3}$		$x = 2$	Divide each side by 3.

Check

$$|3x + 4| = 10 \qquad |3x + 4| = 10$$
$$\left|3\left(-\tfrac{14}{3}\right) + 4\right| \stackrel{?}{=} 10 \qquad |3(2) + 4| \stackrel{?}{=} 10$$
$$|-14 + 4| \stackrel{?}{=} 10 \qquad |6 + 4| \stackrel{?}{=} 10$$
$$|-10| = 10 \checkmark \qquad |10| = 10 \checkmark$$

✓ ***CHECKPOINT*** *Now try Exercise 21.*

When solving absolute value equations, remember that it is possible that they have no solution. For instance, the equation $|3x + 4| = -10$ has no solution because the absolute value of a real number cannot be negative. Do not make the mistake of trying to solve such an equation by writing the "equivalent" linear equations as $3x + 4 = -10$ and $3x + 4 = 10$. These equations have solutions, but they are both extraneous.

The equation in the next example is not given in the **standard form**

$$|ax + b| = c, \quad c \geq 0.$$

Notice that the first step in solving such an equation is to write it in standard form.

EXAMPLE 3 An Absolute Value Equation in Nonstandard Form

Solve $|2x - 1| + 3 = 8$.

Solution

$\|2x - 1\| + 3 = 8$			Write original equation.
$\|2x - 1\| = 5$			Write in standard form.
$2x - 1 = -5$	or	$2x - 1 = 5$	Equivalent equations
$2x = -4$		$2x = 6$	Add 1 to each side.
$x = -2$		$x = 3$	Divide each side by 2.

The solutions are $x = -2$ and $x = 3$. Check these in the original equation.

✓ ***CHECKPOINT*** *Now try Exercise 27.*

If two algebraic expressions are equal in absolute value, they must either be equal to each other or be the *opposites* of each other. So, you can solve equations of the form $|ax + b| = |cx + d|$ by forming the two linear equations

$$\overbrace{ax + b = cx + d}^{\text{Expressions equal}} \quad \text{and} \quad \overbrace{ax + b = -(cx + d)}^{\text{Expressions opposite}}.$$

EXAMPLE 4 Solving an Equation Involving Two Absolute Values

Solve $|3x - 4| = |7x - 16|$.

Solution

$|3x - 4| = |7x - 16|$ Write original equation.

$3x - 4 = 7x - 16$ or $3x - 4 = -(7x - 16)$ Equivalent equations

$-4x - 4 = -16$ $\quad$ $3x - 4 = -7x + 16$

$-4x = -12$ $\quad$ $10x = 20$

$x = 3$ $\quad$ $x = 2$ Solutions

The solutions are $x = 3$ and $x = 2$. Check these in the original equation.

✓ **CHECKPOINT** *Now try Exercise 35.*

Study Tip

When solving an equation of the form

$$|ax + b| = |cx + d|$$

it is possible that one of the resulting equations will not have a solution. Note this occurrence in Example 5.

EXAMPLE 5 Solving an Equation Involving Two Absolute Values

Solve $|x + 5| = |x + 11|$.

Solution

By equating the expression $(x + 5)$ to the opposite of $(x + 11)$, you obtain

$x + 5 = -(x + 11)$ Equivalent equation

$x + 5 = -x - 11$ Distributive Property

$2x + 5 = -11$ Add x to each side.

$2x = -16$ Subtract 5 from each side.

$x = -8.$ Divide each side by 2.

However, by setting the two expressions equal to each other, you obtain

$x + 5 = x + 11$ Equivalent equation

$x = x + 6$ Subtract 5 from each side.

$0 = 6$ Subtract x from each side.

which is a false statement. So, the original equation has only one solution: $x = -8$. Check this solution in the original equation.

✓ **CHECKPOINT** *Now try Exercise 37.*

2 ▶ Solve inequalities involving absolute value.

Solving Inequalities Involving Absolute Value

To see how to solve inequalities involving absolute value, consider the following comparisons.

$|x| = 2$

$x = -2$ and $x = 2$

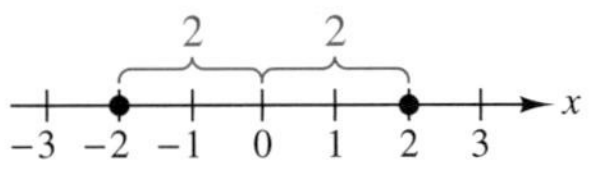

$|x| < 2$

$-2 < x < 2$

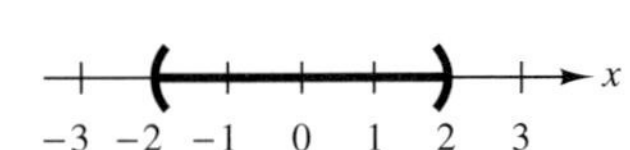

$|x| > 2$

$x < -2$ or $x > 2$

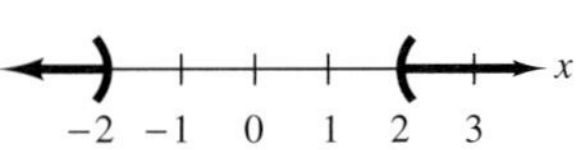

These comparisons suggest the following rules for solving inequalities involving absolute value.

Solving an Absolute Value Inequality

Let x be a variable or an algebraic expression and let a be a real number such that $a > 0$.

1. The solutions of $|x| < a$ are all values of x that lie between $-a$ and a. That is,

   $|x| < a$ if and only if $-a < x < a$.

2. The solutions of $|x| > a$ are all values of x that are less than $-a$ or greater than a. That is,

   $|x| > a$ if and only if $x < -a$ or $x > a$.

These rules are also valid if $<$ is replaced by $\leq$ and $>$ is replaced by $\geq$.

EXAMPLE 6 Solving an Absolute Value Inequality

Solve $|x - 5| < 2$.

Solution

$|x - 5| < 2$ Write original inequality.

$-2 < x - 5 < 2$ Equivalent double inequality

$-2 + 5 < x - 5 + 5 < 2 + 5$ Add 5 to all three parts.

$3 < x < 7$ Combine like terms.

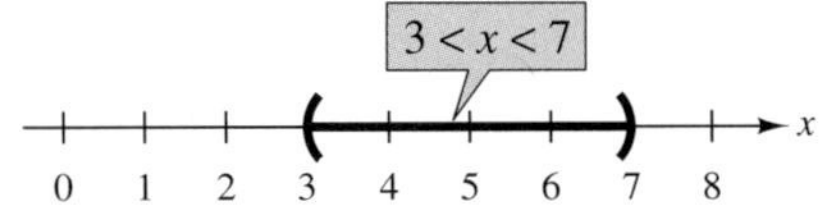

Figure 3.19

The solution set consists of all real numbers that are greater than 3 and less than 7. The solution set in interval notation is $(3, 7)$ and in set notation is $\{x|3 < x < 7\}$. The graph of this solution set is shown in Figure 3.19.

✓ ***CHECKPOINT*** *Now try Exercise 53.*

To verify the solution of an absolute value inequality, you must check values in the solution set and outside of the solution set. In Example 6 you can check that $x = 4$ is in the solution set and that $x = 2$ and $x = 8$ are not in the solution set.

Study Tip

In Example 6, note that an absolute value inequality of the form $|x| < a$ (or $|x| \le a$) can be solved with a double inequality. An inequality of the form $|x| > a$ (or $|x| \ge a$) cannot be solved with a double inequality. Instead, you must solve two separate inequalities, as demonstrated in Example 7.

EXAMPLE 7 Solving an Absolute Value Inequality

Solve $|3x - 4| \ge 5$.

Solution

$|3x - 4| \ge 5$			Write original inequality.
$3x - 4 \le -5$	or	$3x - 4 \ge 5$	Equivalent inequalities
$3x - 4 + 4 \le -5 + 4$		$3x - 4 + 4 \ge 5 + 4$	Add 4 to all parts.
$3x \le -1$		$3x \ge 9$	Combine like terms.
$\frac{3x}{3} \le \frac{-1}{3}$		$\frac{3x}{3} \ge \frac{9}{3}$	Divide each side by 3.
$x \le -\frac{1}{3}$		$x \ge 3$	Simplify.

The solution set consists of all real numbers that are less than or equal to $-\frac{1}{3}$ or greater than or equal to 3. The solution set in interval notation is $\left(-\infty, -\frac{1}{3}\right] \cup [3, \infty)$ and in set notation is $\left\{x \middle| x \le -\frac{1}{3} \text{ or } x \ge 3\right\}$. The graph is shown in Figure 3.20.

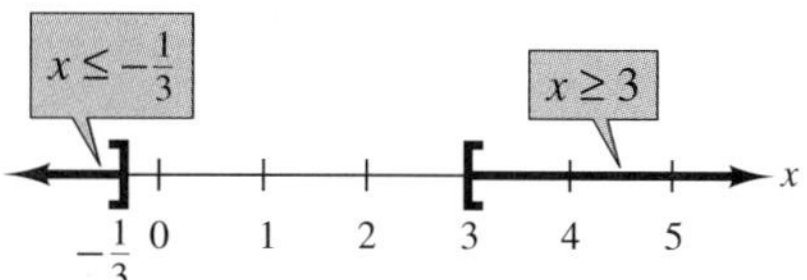

Figure 3.20

✓ ***CHECKPOINT*** *Now try Exercise 59.*

EXAMPLE 8 Solving an Absolute Value Inequality

$\left|2 - \frac{x}{3}\right| \le 0.01$	Original inequality
$-0.01 \le 2 - \frac{x}{3} \le 0.01$	Equivalent double inequality
$-2.01 \le -\frac{x}{3} \le -1.99$	Subtract 2 from all three parts.
$-2.01(-3) \ge -\frac{x}{3}(-3) \ge -1.99(-3)$	Multiply all three parts by -3 and reverse both inequality symbols.
$6.03 \ge x \ge 5.97$	Simplify.
$5.97 \le x \le 6.03$	Solution set in standard form

The solution set consists of all real numbers that are greater than or equal to 5.97 *and* less than or equal to 6.03. The solution set in interval notation is $[5.97, 6.03]$ and in set notation is $\{x|5.97 \le x \le 6.03\}$. The graph is shown in Figure 3.21.

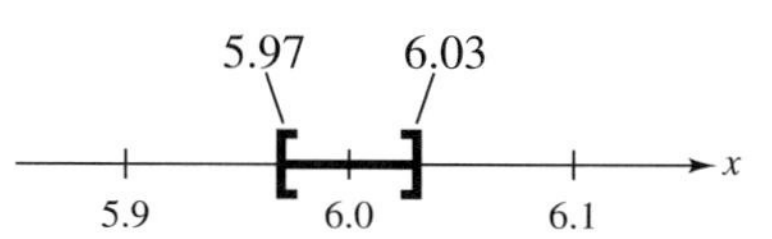

Figure 3.21

✓ ***CHECKPOINT*** *Now try Exercise 65.*

Technology: Tip

Most graphing calculators can graph absolute value inequalities. Consult your user's guide for specific instructions. The graph below shows the solution of the inequality in Example 6 on page 200, which is $|x - 5| < 2$. Notice that the graph representing the solution interval lies above the x-axis.

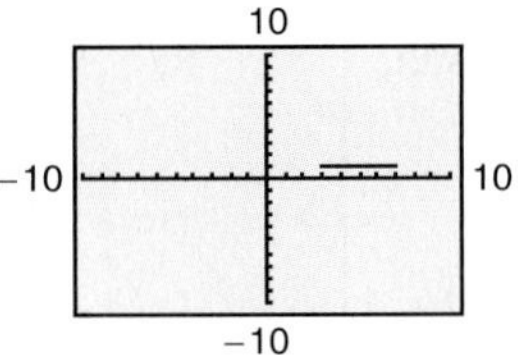

EXAMPLE 9 Production 

The estimated daily production at an oil refinery is given by the absolute value inequality $|x - 200{,}000| \le 25{,}000$, where x is measured in barrels of oil. Solve the inequality to determine the maximum and minimum production levels.

Solution

$|x - 200{,}000| \le 25{,}000$ Write original inequality.

$-25{,}000 \le x - 200{,}000 \le 25{,}000$ Equivalent double inequality

$175{,}000 \le x \le 225{,}000$ Add 200,000 to all three parts.

So, the oil refinery produces a maximum of 225,000 barrels of oil and a minimum of 175,000 barrels of oil per day.

 **CHECKPOINT** *Now try Exercise 89.*

EXAMPLE 10 Creating a Model 

To test the accuracy of a rattlesnake's "pit-organ sensory system," a biologist blindfolds a rattlesnake and presents the snake with a warm "target." Of 36 strikes, the snake is on target 17 times. Let A represent the number of degrees by which the snake is off target. Then $A = 0$ represents a strike that is aimed directly at the target. Positive values of A represent strikes to the right of the target, and negative values of A represent strikes to the left of the target. Use the diagram shown in Figure 3.22 to write an absolute value inequality that describes the interval in which the 36 strikes occurred.

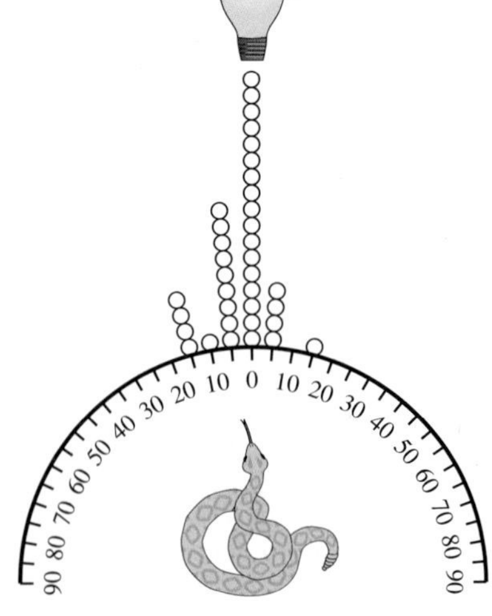

Figure 3.22

Solution

From the diagram, you can see that in the 36 strikes, the snake is never off by more than 15 degrees in either direction. As a compound inequality, this can be represented by $-15 \le A \le 15$. As an absolute value inequality, this interval can be represented by $|A| \le 15$.

CHECKPOINT *Now try Exercise 91.*

Concept Check

1. In your own words, explain how to solve an absolute value equation. Illustrate your explanation with an example.
2. In the equation $|x| = b$, b is a positive real number. How many solutions does this equation have? Explain.
3. In the inequality $|x| < b$, b is a positive real number. Describe the solution set of this inequality.
4. Can you use a double inequality to solve the inequality $|5x - 4| \geq 3$? Explain your reasoning.

3.7 EXERCISES

Go to pages 206–207 to record your assignments.

Developing Skills

In Exercises 1–4, determine whether the value is a solution of the equation.

	Equation	*Value*
1.	$\lvert 4x + 5\rvert = 10$	$x = -3$
2.	$\lvert 2x - 16\rvert = 10$	$x = 3$
3.	$\lvert 6 - 2w\rvert = 2$	$w = 4$
4.	$\left\lvert \frac{1}{2}t + 4\right\rvert = 8$	$t = 6$

In Exercises 5–8, transform the absolute value equation into two linear equations.

5. $|x - 10| = 17$

6. $|7 - 2t| = 5$

7. $|4x + 1| = \frac{1}{2}$

8. $|22k + 6| = 9$

In Exercises 9–12, write the absolute value equation in standard form.

9. $|3x| + 7 = 8$ **10.** $|5x| - 6 = -3$

11. $3|2x| - 1 = 5$ **12.** $\frac{1}{4}|3x + 1| = 4$

In Exercises 13–40, solve the equation. (Some equations have no solution.) ***See Examples 1–5.***

✓ **13.** $|x| = 4$ **14.** $|x| = 3$

15. $|t| = -45$ **16.** $|s| = 16$

17. $|h| = 0$ **18.** $|x| = -82$

19. $|5x| = 15$ **20.** $\left|\frac{1}{3}x\right| = 2$

✓ **21.** $|x + 1| = 5$ **22.** $|x + 5| = 7$

23. $\left|\frac{2s + 3}{5}\right| = 5$ **24.** $\left|\frac{7a + 6}{4}\right| = 2$

25. $|4 - 3x| = 0$ **26.** $|3x - 2| = -5$

✓ **27.** $|5x - 3| + 8 = 22$ **28.** $|5 - 2x| + 10 = 6$

29. $\left|\frac{x - 2}{3}\right| + 6 = 6$ **30.** $\left|\frac{x - 2}{5}\right| + 4 = 4$

31. $-2|7 - 4x| = -16$ **32.** $4|5x + 1| = 24$

33. $3|2x - 5| + 4 = 7$

34. $2|4 - 3x| - 6 = -2$

✓ **35.** $|x + 8| = |2x + 1|$

36. $|10 - 3x| = |x + 7|$

✓ **37.** $|3x + 1| = |3x - 3|$

38. $|2x + 7| = |2x + 9|$

39. $|4x - 10| = 2|2x + 3|$

40. $3|2 - 3x| = |9x + 21|$

Think About It In Exercises 41 and 42, write an absolute value equation that represents the verbal statement.

41. The distance between x and 4 is 9.

42. The distance between -3 and t is 5.

In Exercises 43–46, determine whether the x-value is a solution of the inequality.

	Inequality	*Value*
43.	$\lvert x\rvert < 3$	$x = 2$
44.	$\lvert x\rvert \leq 5$	$x = -7$
45.	$\lvert x - 7\rvert \geq 3$	$x = 9$
46.	$\lvert x - 3\rvert > 5$	$x = 16$

In Exercises 47–50, transform the absolute value inequality into a double inequality or two separate inequalities.

47. $|y + 5| < 3$

48. $|6x + 7| \leq 5$

49. $|7 - 2h| \geq 9$

50. $|8 - x| > 25$

In Exercises 51–70, solve the inequality. ***See Examples 6–8.***

51. $|y| < 4$ **52.** $|x| < 6$

53. $|x| \geq 6$ **54.** $|y| \geq 4$

55. $|2x| < 14$ **56.** $|4z| \leq 9$

57. $\left|\frac{y}{3}\right| \leq \frac{1}{3}$ **58.** $\left|\frac{t}{5}\right| < \frac{3}{5}$

59. $|x + 6| > 10$ **60.** $|y - 2| \leq 4$

61. $|2x - 1| \leq 7$ **62.** $|6t + 15| \geq 30$

63. $|3x + 10| < -1$ **64.** $|4x - 5| > -3$

65. $\frac{|y - 16|}{4} < 30$ **66.** $\frac{|a + 6|}{2} \geq 16$

67. $|0.2x - 3| < 4$ **68.** $|1.5t - 8| \leq 16$

69. $\left|\frac{3x - 2}{4}\right| + 5 \geq 5$ **70.** $\left|\frac{2x - 4}{5}\right| - 9 \leq 3$

In Exercises 71–76, use a graphing calculator to solve the inequality.

71. $|3x + 2| < 4$ **72.** $|2x - 1| \leq 3$

73. $|2x + 3| > 9$ **74.** $|7r - 3| > 11$

75. $|x - 5| + 3 \leq 5$ **76.** $|a + 1| - 4 < 0$

In Exercises 77–80, match the inequality with its graph. [The graphs are labeled (a), (b), (c), and (d).]

(a)

(b)

(c)

(d)

77. $|x - 4| \leq 4$ **78.** $|x - 4| < 1$

79. $\frac{1}{2}|x - 4| > 4$ **80.** $|2(x - 4)| \geq 4$

In Exercises 81–84, write an absolute value inequality that represents the interval.

81.

82.

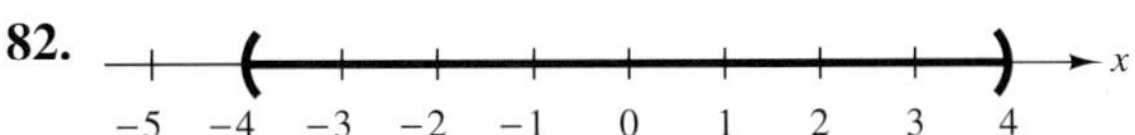

83.

84.

In Exercises 85–88, write an absolute value inequality that represents the verbal statement.

85. The set of all real numbers x whose distance from 0 is less than 3.

86. The set of all real numbers x whose distance from 0 is more than 2.

87. The set of all real numbers x for which the distance from 0 to 3 less than twice x is more than 5.

88. The set of all real numbers x for which the distance from 0 to 5 more than half of x is less than 13.

Solving Problems

89. *Speed Skating* In the 2006 Winter Olympics, each skater in the 500-meter short track speed skating final had a time that satisfied the inequality $|t - 42.238| \le 0.412$, where t is the time in seconds. Sketch the graph of the solution of the inequality. What are the fastest and slowest times?

90. *Time Study* A time study was conducted to determine the length of time required to perform a task in a manufacturing process. The times required by approximately two-thirds of the workers in the study satisfied the inequality

$$\left|\frac{t - 15.6}{1.9}\right| \le 1$$

where t is time in minutes. Sketch the graph of the solution of the inequality. What are the maximum and minimum times?

✓ **91.** *Accuracy of Measurements* In woodshop class, you must cut several pieces of wood to within $\frac{3}{16}$ inch of the teacher's specifications. Let $(s - x)$ represent the difference between the specification s and the measured length x of a cut piece.

(a) Write an absolute value inequality that describes the values of x that are within specifications.

(b) The length of one piece of wood is specified to be $s = 5\frac{1}{8}$ inches. Describe the acceptable lengths for this piece.

92. *Body Temperature* Physicians generally consider an adult's body temperature x to be normal if it is within 1°F of the temperature 98.6°F.

(a) Write an absolute value inequality that describes the values of x that are considered normal.

(b) Describe the range of body temperatures that are considered normal.

Explaining Concepts

93. The graph of the inequality $|x - 3| < 2$ can be described as *all real numbers that are within two units of 3*. Give a similar description of $|x - 4| < 1$.

94. Write an absolute value inequality to represent all the real numbers that are more than $|a|$ units from b. Then write an example showing the solution of the inequality for sample values of a and b.

95. Complete $|2x - 6| \le$ ▢ so that the solution is $0 \le x \le 6$.

96. Describe and correct the error. Explain how you can recognize that the solution is wrong without solving the inequality.

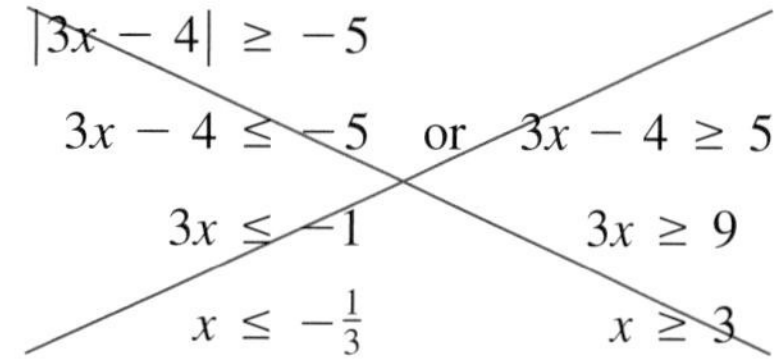

$|3x - 4| \ge -5$

$3x - 4 \le -5$ or $3x - 4 \ge 5$

$3x \le -1$ $\quad$ $3x \ge 9$

$x \le -\frac{1}{3}$ $\quad$ $x \ge 3$

Cumulative Review

In Exercises 97 and 98, translate the verbal phrase into an algebraic expression.

97. Four times the sum of a number n and 3

98. Eight less than two times a number n

In Exercises 99 and 100, find the missing quantities.

	Cost	*Selling Price*	*Markup*	*Markup Rate*
99.	\$80.00			40%
100.	\$74.00			62%

In Exercises 101–104, solve the inequality.

101. $x - 7 > 13$

102. $x + 7 \le 13$

103. $4x + 11 \ge 27$

104. $-4 < x + 2 < 12$

What Did You Learn?

Use these two pages to help prepare for a test on this chapter. Check off the key terms and key concepts you know. You can also use this section to record your assignments.

Plan for Test Success

Date of test: / /

Study dates and times: / / at : A.M./P.M.
/ / at : A.M./P.M.

Things to review:

- ☐ Key Terms, *p. 206*
- ☐ Key Concepts, *pp. 206–207*
- ☐ Your class notes
- ☐ Your assignments
- ☐ Study Tips, *pp. 129, 130, 132, 137, 140, 141, 143, 147, 148, 150, 151, 163, 164, 171, 178, 185, 187, 188, 197, 199, 201*
- ☐ Technology Tips, *pp. 128, 139, 173, 188, 202*
- ☐ Mid-Chapter Quiz, *p. 170*
- ☐ Review Exercises, *pp. 208–211*
- ☐ Chapter Test, *p. 212*
- ☐ Video Explanations Online
- ☐ Tutorial Online

Key Terms

- ☐ linear equation, *p. 126*
- ☐ first-degree equation, *p. 126*
- ☐ identity, *p. 131*
- ☐ consecutive integers, *p. 132*
- ☐ equivalent fractions, *p. 142*
- ☐ cross-multiplication, *p. 142*
- ☐ markup, *p. 152*
- ☐ discount, *p. 153*
- ☐ ratio, *p. 159*
- ☐ unit price, *p. 161*
- ☐ proportion, *p. 162*
- ☐ algebraic inequalities, *p. 184*
- ☐ solve an inequality, *p. 184*
- ☐ graph an inequality, *p. 184*
- ☐ bounded intervals, *p. 184*
- ☐ endpoints of an interval, *p. 184*
- ☐ unbounded (infinite) intervals, *p. 185*
- ☐ positive infinity, *p. 185*
- ☐ negative infinity, *p. 185*
- ☐ equivalent inequalities, *p. 186*
- ☐ linear inequality, *p. 187*
- ☐ compound inequality, *p. 189*
- ☐ intersection, *p. 190*
- ☐ union, *p. 190*
- ☐ absolute value equation, *p. 197*
- ☐ standard form of an absolute value equation, *p. 198*

Key Concepts

3.1 Solving Linear Equations

Assignment: ______________________________ Due date: ________

☐ **Solve a linear equation.**

Solve a linear equation by using inverse operations to isolate the variable.

☐ **Write expressions for special types of integers.**

Let n be an integer.

1. $2n$ denotes an *even* integer.
2. $2n - 1$ and $2n + 1$ denote *odd* integers.
3. The set $\{n, n + 1, n + 2\}$ denotes three *consecutive* integers.

3.2 Equations That Reduce to Linear Form

Assignment:______________________________ Due date:________

☐ **Solve equations containing symbols of grouping.**

Remove symbols of grouping using the Distributive Property, combine like terms, isolate the variable using properties of equality, and check your solution in the original equation.

☐ **Solve equations involving fractions.**

To clear an equation of fractions, multiply each side by the least common multiple (LCM) of the denominators.

Use cross-multiplication to solve a linear equation that equates two fractions.

3.3 Problem Solving with Percents

Assignment:______________________________ Due date:________

☐ **Use the percent equation $a = p \cdot b$.**

$b =$ base number

$p =$ percent (in decimal form)

$a =$ number being compared to b

☐ **Use guidelines for solving word problems.**

See page 154.

3.4 Ratios and Proportions

Assignment:______________________________ Due date:________

☐ **Define ratio.**

The ratio of the real number a to the real number b is given by a/b, or $a : b$.

☐ **Solve a proportion.**

A proportion equates two ratios.

If $\frac{a}{b} = \frac{c}{d}$, then $ad = bc$.

3.5 Geometric and Scientific Applications

Assignment:______________________________ Due date:________

☐ **Use common formulas.**

See pages 171 and 173.

☐ **Solve mixture and work-rate problems.**

Mixture and work-rate problems are composed of the sum of two or more "hidden products" that involve rate factors.

3.6 Linear Inequalities

Assignment:______________________________ Due date:________

☐ **Graph solutions on a number line.**

A parenthesis excludes an endpoint from the solution interval. A square bracket includes an endpoint in the solution interval.

☐ **Use properties of inequalities.**

See page 186.

3.7 Absolute Value Equations and Inequalities

Assignment:______________________________ Due date:________

☐ **Solve absolute value equations.**

Let x be a variable or an algebraic expression and let a be a real number such that $a \geq 0$. The solutions of the equation $|x| = a$ are given by $x = a$ and $x = -a$.

☐ **Solve an absolute value inequality.**

See page 200.

Review Exercises

3.1 Solving Linear Equations

1 ▶ Solve linear equations in standard form.

In Exercises 1–6, solve the equation and check your solution.

1. $2x - 10 = 0$
2. $12y + 72 = 0$
3. $-3y - 12 = 0$
4. $-7x + 21 = 0$
5. $5x - 3 = 0$
6. $-8x + 6 = 0$

2 ▶ Solve linear equations in nonstandard form.

In Exercises 7–20, solve the equation and check your solution.

7. $x + 10 = 13$
8. $x - 3 = 8$
9. $5 - x = 2$
10. $3 = 8 - x$
11. $10x = 50$
12. $-3x = 21$
13. $8x + 7 = 39$
14. $12x - 5 = 43$
15. $24 - 7x = 3$
16. $13 + 6x = 61$
17. $15x - 4 = 16$
18. $3x - 8 = 2$
19. $\frac{x}{5} = 4$
20. $-\frac{x}{14} = \frac{1}{2}$

3 ▶ Use linear equations to solve application problems.

21. *Hourly Wage* Your hourly wage is \$8.30 per hour plus 60 cents for each unit you produce. How many units must you produce in an hour so that your hourly wage is \$15.50?

22. *Labor Cost* The total cost for a new deck (including materials and labor) is \$1830. The materials cost \$1500 and the cost of labor is \$55 per hour. How many hours did it take to build the deck?

23. *Geometry* The perimeter of a rectangle is 260 meters. Its length is 30 meters greater than its width. Find the dimensions of the rectangle.

24. *Geometry* A 10-foot board is cut so that one piece is 4 times as long as the other. Find the length of each piece.

3.2 Equations That Reduce to Linear Form

1 ▶ Solve linear equations containing symbols of grouping.

In Exercises 25–30, solve the equation and check your solution.

25. $3x - 2(x + 5) = 10$
26. $4x + 2(7 - x) = 5$
27. $2(x + 3) = 6(x - 3)$
28. $8(x - 2) = 3(x + 2)$
29. $7 - [2(3x + 4) - 5] = x - 3$
30. $14 + [3(6x - 15) + 4] = 5x - 1$

2 ▶ Solve linear equations involving fractions.

In Exercises 31–40, solve the equation and check your solution.

31. $\frac{2}{3}x - \frac{1}{6} = \frac{9}{2}$
32. $\frac{1}{8}x + \frac{3}{4} = \frac{5}{2}$
33. $\frac{x}{3} - \frac{1}{9} = 2$
34. $\frac{1}{2} - \frac{x}{8} = 7$
35. $\frac{u}{10} + \frac{u}{5} = 6$
36. $\frac{x}{3} + \frac{x}{5} = 1$
37. $\frac{2x}{9} = \frac{2}{3}$
38. $\frac{5y}{13} = \frac{2}{5}$
39. $\frac{x + 3}{5} = \frac{x + 7}{12}$
40. $\frac{y - 2}{6} = \frac{y + 1}{15}$

3 ▶ Solve linear equations involving decimals.

In Exercises 41–44, solve the equation. Round your answer to two decimal places.

41. $5.16x - 87.5 = 32.5$
42. $2.825x + 3.125 = 12.5$
43. $\frac{x}{4.625} = 48.5$
44. $5x + \frac{1}{4.5} = 18.125$

45. *Time to Complete a Task* Two people can complete 50% of a task in t hours, where t must satisfy the equation $\frac{t}{10} + \frac{t}{15} = 0.5$. How long will it take for the two people to complete 50% of the task?

46. *Course Grade* To get an A in a course, you must have an average of at least 90 points for four tests of 100 points each. For the first three tests, your scores are 85, 96, and 89. What must you score on the fourth exam to earn a 90% average for the course?

3.3 Problem Solving with Percents

1 ▶ Convert percents to decimals and fractions, and convert decimals and fractions to percents.

In Exercises 47-54, complete the table showing the equivalent forms of a percent.

	Percent	*Parts out of 100*	*Decimal*	*Fraction*
47.	60%			
48.	35%			
49.				$\frac{4}{5}$
50.				$\frac{5}{8}$
51.			0.20	
52.			1.35	
53.		55		
54.		12.5		

2 ▶ Solve linear equations involving percents.

In Exercises 55–60, solve the percent equation.

55. What number is 125% of 16?

56. What number is 0.8% of 3250?

57. 150 is $37\frac{1}{2}\%$ of what number?

58. 323 is 95% of what number?

59. 150 is what percent of 250?

60. 130.6 is what percent of 3265?

3 ▶ Solve application problems involving markups and discounts.

61. *Selling Price* An electronics store uses a markup rate of 78% on all items. The cost of a CD player is $48. What is the selling price of the CD player?

62. *Sale Price* A sporting goods store advertises 30% off the list price of all golf equipment. A set of golf clubs has a list price of $229.99. What is the sale price?

63. *Sales* The sales (in millions) for the Yankee Candle Company for the years 2001 to 2005 are shown in the bar graph below. (Source: The Yankee Candle Company)

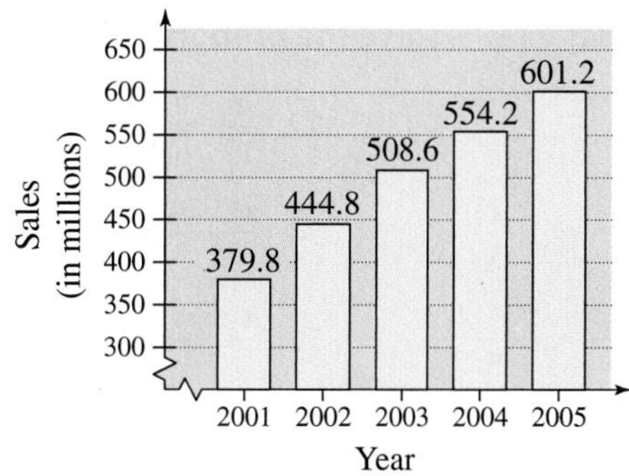

(a) Determine the percent increase in sales from 2004 to 2005.

(b) Determine the percent increase in sales from 2001 to 2005.

64. *Price Increase* The manufacturer's suggested retail price for a car is $18,459. Estimate the price of a comparably equipped car for the next model year if the price will increase by $4\frac{1}{2}\%$.

3.4 Ratios and Proportions

1 ▶ Compare relative sizes using ratios.

In Exercises 65 and 66, find a ratio that compares the relative sizes of the quantities. (Use the same units of measurement for both quantities.)

65. Eighteen inches to 4 yards

66. Four meters to 150 centimeters

67. *Hours* You are in school for 7.5 hours per day and you sleep for 6 hours per day. Find the ratio of the number of hours you sleep to the number of hours you are in school.

68. *Grandchildren* A grandmother has two grandsons and eight granddaughters. Find the ratio of the number of grandsons to the number of granddaughters.

2 ▶ Find the unit price of a consumer item.

In Exercises 69 and 70, which product has the lower unit price?

69. (a) An 18-ounce container of cooking oil for $0.89

(b) A 24-ounce container of cooking oil for $1.12

70. (a) A 17.4-ounce box of pasta noodles for $1.32
(b) A 32-ounce box of pasta noodles for $2.62

3 ▶ Solve proportions that equate two ratios.

In Exercises 71–76, solve the proportion.

71. $\frac{7}{16} = \frac{z}{8}$

72. $\frac{x}{12} = \frac{5}{4}$

73. $\frac{x+2}{4} = -\frac{1}{3}$

74. $\frac{x-4}{1} = \frac{9}{4}$

75. $\frac{x-3}{2} = \frac{x+6}{5}$

76. $\frac{x+1}{3} = \frac{x+2}{4}$

77. *Entertainment* A band charges $200 to play for 3 hours. How much would it charge to play for 2 hours?

78. *Resizing a Picture* You have a 4-by-6 inch photo of the Student Council that must be reduced to a size of 3.2 inches by 4.8 inches for the school yearbook. What percent does the photo need to be reduced by in order for it to fit in the allotted space?

4 ▶ Solve application problems using the Consumer Price Index.

In Exercises 79 and 80, use the Consumer Price Index table on page 164 to estimate the price of the item in the indicated year.

79. The 2006 price of a chair that cost $78 in 1984

80. The 1986 price of an oven that cost $120 in 1999

3.5 Geometric and Scientific Applications

1 ▶ Use common formulas to solve application problems.

In Exercises 81 and 82, solve for the specified variable.

81. Solve for x: $z = \frac{x-m}{s}$

82. Solve for h: $V = \pi r^2 h$

In Exercises 83–88, find the missing distance, rate, or time.

	Distance, d	*Rate, r*	*Time, t*
83.		65 mi/hr	8 hr
84.		45 mi/hr	2 hr
85.	855 m	5 m/min	
86.	205 mi	60 mi/hr	
87.	3000 mi		50 hr
88.	1000 km		25 hr

89. *Distance* An airplane has an average speed of 475 miles per hour. How far will it travel in $2\frac{1}{3}$ hours?

90. *Average Speed* You can walk 20 kilometers in 3 hours and 47 minutes. What is your average speed?

91. *Geometry* The width of a rectangular swimming pool is 4 feet less than its length. The perimeter of the pool is 112 feet. Find the dimensions of the pool.

92. *Geometry* The perimeter of an isosceles triangle is 65 centimeters. Find the length of the two equal sides if each is 10 centimeters longer than the third side. (An isosceles triangle has two sides of equal length.)

Simple Interest **In Exercises 93 and 94, use the simple interest formula.**

93. Find the total interest you will earn on a $1000 corporate bond that matures in 5 years and has an annual interest rate of 9.5%.

94. Find the annual interest rate on a certificate of deposit that pays $60 per year in interest on a principal of $750.

2 ▶ Solve mixture problems involving hidden products.

95. *Number of Coins* You have 30 coins in dimes and quarters with a combined value of $5.55. Determine the number of coins of each type.

96. *Birdseed Mixture* A pet store owner mixes two types of birdseed that cost $1.25 per pound and $2.20 per pound to make 20 pounds of a mixture that costs $1.82 per pound. How many pounds of each kind of birdseed are in the mixture?

3 ▶ Solve work-rate problems.

97. *Work-Rate* One person can complete a task in 5 hours, and another can complete the same task in 6 hours. How long will it take both people working together to complete the task?

98. *Work-Rate* The person in Exercise 97 who can complete the task in 5 hours has already worked 1 hour when the second person starts. How long will they work together to complete the task?

3.6 Linear Inequalities

1 ▶ Sketch the graphs of inequalities.

In Exercises 99–102, sketch the graph of the inequality.

99. $-3 \le x < 1$

100. $-2.5 \le x < 4$

101. $-7 < x$

102. $x \ge -2$

3 ▶ Solve linear inequalities.

In Exercises 103–108, solve the inequality and sketch the solution on the real number line.

103. $-6x < -24$

104. $-16x \ge -48$

105. $8x + 1 \ge 10x - 11$

106. $12 - 3x < 4x - 2$

107. $\frac{1}{3} - \frac{1}{2}y < 12$

108. $\dfrac{x}{4} - 2 < \dfrac{3x}{8} + 5$

4 ▶ Solve compound inequalities.

In Exercises 109–112, solve the compound inequality and sketch the solution on the real number line.

109. $-6 \le 2x + 8 < 4$

110. $-13 \le 3 - 4x < 13$

111. $5x - 4 < 6$ and $3x + 1 > -8$

112. $6 - 2x \le 1$ or $10 - 4x > -6$

5 ▶ Solve application problems involving inequalities.

113. *Earnings* A waiter earns \$6 per hour plus tips of at least 15% of each customer's restaurant tab. What amount of restaurant tabs assures the waiter of making at least \$150 in a five-hour shift?

114. *Long-Distance Charges* The cost of an international long-distance telephone call is \$0.99 for the first minute and \$0.49 for each additional minute. Your prepaid calling card has \$22.50 left to pay for a call. How many minutes can you talk?

3.7 Absolute Value Equations and Inequalities

1 ▶ Solve absolute value equations.

In Exercises 115–120, solve the equation.

115. $|4 - 3x| = 8$

116. $|2x + 3| = 7$

117. $|5x + 4| - 10 = -6$

118. $|x - 2| - 2 = 4$

119. $|3x - 4| = |x + 2|$

120. $|5x + 6| = |2x - 1|$

2 ▶ Solve inequalities involving absolute value.

In Exercises 121–126, solve the inequality.

121. $|x - 4| > 3$

122. $|t + 3| > 2$

123. $|3x| < 12$

124. $|4x - 1| > 7$

125. $|b + 2| - 6 > 1$

126. $|2y - 1| + 4 < -1$

In Exercises 127 and 128, use a graphing calculator to solve the inequality.

127. $|4(x - 3)| \ge 8$

128. $|5(1 - x)| \le 25$

In Exercises 129 and 130, write an absolute value inequality that represents the interval.

129.

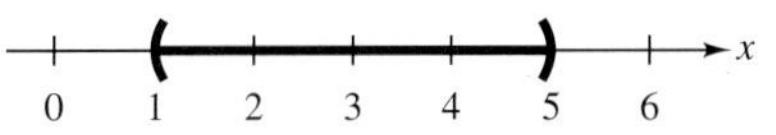

130.

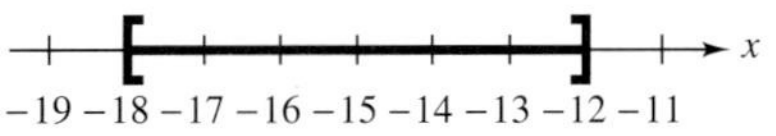

131. *Temperature* The storage temperature (in degrees Fahrenheit) of a computer must satisfy the inequality $|t - 78.3| \le 38.3$, where t is the temperature. Sketch the graph of the solution of the inequality. What are the maximum and minimum temperatures?

132. *Temperature* The operating temperature (in degrees Fahrenheit) of a computer must satisfy the inequality $|t - 77| \le 27$, where t is the temperature. Sketch the graph of the solution of the inequality. What are the maximum and minimum temperatures?

Chapter Test

Take this test as you would take a test in class. After you are done, check your work against the answers in the back of the book.

In Exercises 1–8, solve the equation and check your solution.

1. $8x + 104 = 0$

2. $4x - 3 = 18$

3. $5 - 3x = -2x - 2$

4. $4 - (x - 3) = 5x + 1$

5. $\frac{2}{3}x = \frac{1}{9} + x$

6. $\dfrac{t + 2}{3} = \dfrac{2t}{9}$

7. $|2x + 6| = 16$

8. $|3x - 5| = |6x - 1|$

9. Solve $4.08(x + 10) = 9.50(x - 2)$. Round your answer to two decimal places.

10. The bill (including parts and labor) for the repair of an oven is \$142. The cost of parts is \$62 and the cost of labor is \$32 per hour. How many hours were spent repairing the oven?

11. Write the fraction $\frac{5}{16}$ as a percent and as a decimal.

12. 324 is 27% of what number?

13. 90 is what percent of 250?

14. Write the ratio of 40 inches to 2 yards as a fraction in simplest form. Use the same units for both quantities, and explain how you made this conversion.

15. Solve the proportion $\dfrac{2x}{3} = \dfrac{x + 4}{5}$.

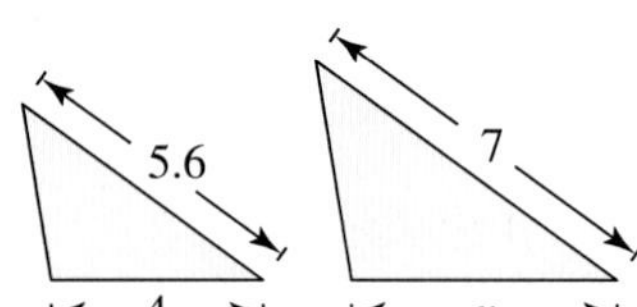

Figure for 16

16. Find the length x of the side of the larger triangle shown in the figure at the left. (Assume that the two triangles are similar, and use the fact that corresponding sides of similar triangles are proportional.)

17. You traveled 264 miles in 4 hours. What was your average speed?

18. You can paint a building in 9 hours. Your friend can paint the same building in 12 hours. Working together, how long will it take the two of you to paint the building?

19. Solve for b in the equation: $a = pb + b$.

20. How much must you deposit in an account to earn \$500 per year at 8% simple interest?

21. Translate the statement "t is at least 8" into a linear inequality.

22. A utility company has a fleet of vans. The annual operating cost per van is $C = 0.37m + 2700$, where m is the number of miles traveled by a van in a year. What is the maximum number of miles that will yield an annual operating cost that is less than or equal to \$11,950?

In Exercises 23–28, solve and graph the inequality.

23. $21 - 3x \le 6$

24. $-(3 + x) < 2(3x - 5)$

25. $0 \le \dfrac{1 - x}{4} < 2$

26. $-7 < 4(2 - 3x) \le 20$

27. $|x - 3| \le 2$

28. $|5x - 3| > 12$

Cumulative Test: Chapters 1–3

Take this test as you would take a test in class. After you are done, check your work against the answers in the back of the book.

1. Place the correct symbol (< or >) between the numbers: $-\frac{3}{4}$ $\square$ $\left|-\frac{7}{8}\right|$.

In Exercises 2–7, evaluate the expression.

2. $(-200)(2)(-3)$ **3.** $\frac{3}{8} - \frac{5}{6}$ **4.** $-\frac{2}{9} \div \frac{8}{75}$

5. $-(-2)^3$ **6.** $3 + 2(6) - 1$ **7.** $24 + 12 \div 3$

In Exercises 8 and 9, evaluate the expression when $x = -2$ and $y = 3$.

8. $-3x - (2y)^2$ **9.** $\frac{5}{6}y + x^3$

10. Use exponential form to write the product $3 \cdot (x + y) \cdot (x + y) \cdot 3 \cdot 3$.

11. Use the Distributive Property to expand $-2x(x - 3)$.

12. Identify the property of real numbers illustrated by

$$2 + (3 + x) = (2 + 3) + x.$$

In Exercises 13–15, simplify the expression.

13. $(3x^3)(5x^4)$

14. $2x^2 - 3x + 5x^2 - (2 + 3x)$

15. $4(x^2 + x) + 7(2x - x^2)$

In Exercises 16–18, solve the equation and check your solution.

16. $12x - 3 = 7x + 27$ **17.** $2x - \dfrac{5x}{4} = 13$

18. $5(x + 8) = -2x - 9$

19. Solve and graph the inequality.

$$-8(x + 5) \leq 16$$

20. The sticker on a new car gives the fuel efficiency as 28.3 miles per gallon. In your own words, explain how to estimate the annual fuel cost for the buyer if the car will be driven approximately 15,000 miles per year and the fuel cost is \$2.759 per gallon.

21. Write the ratio "24 ounces to 2 pounds" as a fraction in simplest form.

22. The suggested retail price of a digital camcorder is \$1150. The camcorder is on sale for "20% off" the list price. Find the sale price.

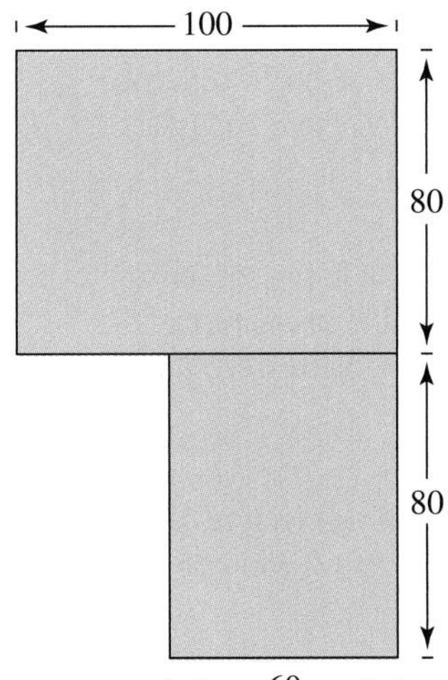

Figure for 23

23. The figure at the left shows two pieces of property. The assessed values of the properties are proportional to their areas. The value of the larger piece is \$95,000. What is the value of the smaller piece?

Study Skills in Action

Reading Your Textbook Like a Manual

Many students avoid opening their textbooks for the same reason many people avoid opening their checkbooks—anxiety and frustration. The truth? Not opening your math textbook will cause more anxiety and frustration! Your textbook is a manual designed to help you master skills and understand and remember concepts. It contains many features and resources that can help you be successful in your course.

For more information about reading a textbook, refer to the *Math Study Skills Workbook* (Nolting, 2008).

Kimberly Nolting
VP, Academic Success Press
expert in developmental education

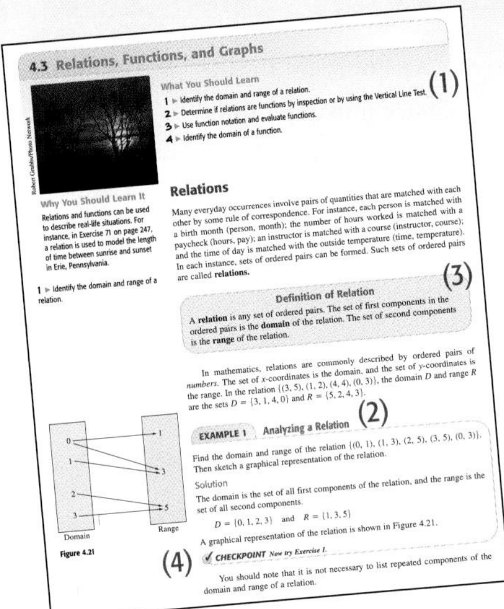

4.3 Relations, Functions, and Graphs

What You Should Learn (1)

1 ▶ Identify the domain and range of a relation.
2 ▶ Determine if relations are functions by inspection or by using the Vertical Line Test.
3 ▶ Use function notation and evaluate functions.
4 ▶ Identify the domain of a function.

Robert Grubbs/Photo Network

Why You Should Learn It

Relations and functions can be used to describe real-life situations. For instance, in Exercise 71 on page 247, a relation is used to model the length of time between sunrise and sunset in Erie, Pennsylvania.

1 ▶ Identify the domain and range of a relation.

Relations

Many everyday occurrences involve pairs of quantities that are matched with each other by some rule of correspondence. For instance, each person is matched with a birth month (person, month); the number of hours worked is matched with a paycheck (hours, pay); an instructor is matched with a course (instructor, course); and the time of day is matched with the outside temperature (time, temperature). In each instance, sets of ordered pairs can be formed. Such sets of ordered pairs are called **relations.** (3)

Definition of Relation

A **relation** is any set of ordered pairs. The set of first components in the ordered pairs is the **domain** of the relation. The set of second components is the **range** of the relation.

In mathematics, relations are commonly described by ordered pairs of *numbers*. The set of x-coordinates is the domain, and the set of y-coordinates is the range. In the relation $\{(3, 5), (1, 2), (4, 4), (0, 3)\}$, the domain D and range R are the sets $D = \{3, 1, 4, 0\}$ and $R = \{5, 2, 4, 3\}$.

EXAMPLE 1 Analyzing a Relation (2)

Find the domain and range of the relation $\{(0, 1), (1, 3), (2, 5), (3, 5), (0, 3)\}$. Then sketch a graphical representation of the relation.

Solution

The domain is the set of all first components of the relation, and the range is the set of all second components.

$D = \{0, 1, 2, 3\}$ and $R = \{1, 3, 5\}$

A graphical representation of the relation is shown in Figure 4.21.

(4) ✓ CHECKPOINT Now try Exercise 1.

You should note that it is not necessary to list repeated components of the domain and range of a relation.

Domain Range

Figure 4.21

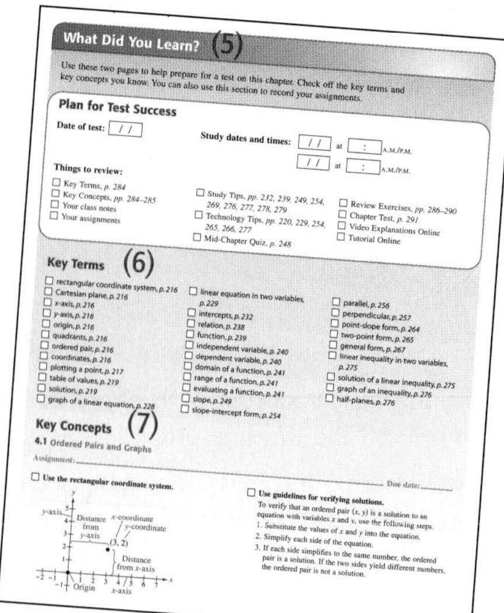

What Did You Learn? (5)

Use these two pages to help prepare for a test on this chapter. Check off the key terms and key concepts you know. You can also use this section to record your assignments.

Plan for Test Success

Date of test: / /

Study dates and times: / / at : A.M./P.M.
/ / at : A.M./P.M.

Things to review:

- ☐ Key Terms, *p. 284*
- ☐ Key Concepts, *pp. 284–285*
- ☐ Your class notes
- ☐ Your assignments
- ☐ Study Tips, *pp. 232, 239, 249, 254, 269, 276, 277, 278, 279*
- ☐ Technology Tips, *pp. 220, 229, 254, 265, 266, 277*
- ☐ Mid-Chapter Quiz, *p. 248*
- ☐ Review Exercises, *pp. 286–290*
- ☐ Chapter Test, *p. 291*
- ☐ Video Explanations Online
- ☐ Tutorial Online

Key Terms (6)

- ☐ rectangular coordinate system, *p. 216*
- ☐ Cartesian plane, *p. 216*
- ☐ *x*-axis, *p. 216*
- ☐ *y*-axis, *p. 216*
- ☐ origin, *p. 216*
- ☐ quadrants, *p. 216*
- ☐ ordered pair, *p. 216*
- ☐ coordinates, *p. 216*
- ☐ plotting a point, *p. 217*
- ☐ table of values, *p. 219*
- ☐ solution, *p. 219*
- ☐ graph of a linear equation, *p. 228*
- ☐ linear equation in two variables, *p. 229*
- ☐ intercepts, *p. 232*
- ☐ relation, *p. 238*
- ☐ function, *p. 239*
- ☐ independent variable, *p. 240*
- ☐ dependent variable, *p. 240*
- ☐ domain of a function, *p. 241*
- ☐ range of a function, *p. 241*
- ☐ evaluating a function, *p. 241*
- ☐ slope, *p. 249*
- ☐ slope-intercept form, *p. 254*
- ☐ parallel, *p. 256*
- ☐ perpendicular, *p. 257*
- ☐ point-slope form, *p. 264*
- ☐ two-point form, *p. 265*
- ☐ general form, *p. 267*
- ☐ linear inequality in two variables, *p. 275*
- ☐ solution of a linear inequality, *p. 275*
- ☐ graph of an inequality, *p. 276*
- ☐ half-planes, *p. 276*

Key Concepts (7)

4.1 Ordered Pairs and Graphs

Assignment: Due date:

☐ Use the rectangular coordinate system.

☐ Use guidelines for verifying solutions.

To verify that an ordered pair (x, y) is a solution to an equation with variables x and y, use the following steps.

1. Substitute the values of x and y into the equation.
2. Simplify each side of the equation.
3. If each side simplifies to the same number, the ordered pair is a solution. If the two sides yield different numbers, the ordered pair is not a solution.

Smart Study Strategy

Use the Features of Your Textbook

To review what you learned in a previous class:

- Read the list of What You Should Learn (1) at the beginning of the section. If you cannot remember how to perform a skill, review the appropriate example (2) in the section.
- Read and understand the contents of all tinted concept boxes (3)—these contain important definitions and rules.

To prepare for homework:

- Complete the checkpoint exercises (4) in the section. If you have difficulty with a checkpoint exercise, reread the example or seek help from a peer or instructor.

To review for quizzes and tests:

- Make use of the What Did You Learn? (5) feature. Check off the key terms (6) and key concepts (7) you know, and review those you do not know.
- Complete the Review Exercises. Then take the Mid-Chapter Quiz, Chapter Test, or Cumulative Test, as appropriate.

Chapter 4

Graphs and Functions

IT WORKED FOR ME!

"I used to be really afraid of math, so reading the textbook was torture. I have learned that it just takes different strategies to read the textbook. It's my resource book when I do homework. I lug it to class because it helps me follow along. I'm not afraid of math anymore because I know how to study it—finally."

Jodie
Elementary education

4.1 Ordered Pairs and Graphs

What You Should Learn

1 ▶ Plot and find the coordinates of a point on a rectangular coordinate system.

2 ▶ Construct tables of values for equations and determine whether ordered pairs are solutions of equations.

3 ▶ Use the verbal problem-solving method to plot points on a rectangular coordinate system.

Why You Should Learn It

The Cartesian plane can be used to represent relationships between two variables. For instance, Exercises 69–72 on page 226 show how to represent graphically the average price of regular unleaded gasoline in the United States.

1 ▶ Plot and find the coordinates of a point on a rectangular coordinate system.

The Rectangular Coordinate System

Just as you can represent real numbers by points on the real number line, you can represent **ordered pairs** of real numbers by points in a plane. This plane is called a **rectangular coordinate system** or the **Cartesian plane,** after the French mathematician René Descartes (1596–1650).

A rectangular coordinate system is formed by two real lines intersecting at right angles, as shown in Figure 4.1. The horizontal number line is usually called the ***x*-axis** and the vertical number line is usually called the ***y*-axis.** (The plural of axis is *axes.*) The point of intersection of the two axes is called the **origin,** and the axes separate the plane into four regions called **quadrants.**

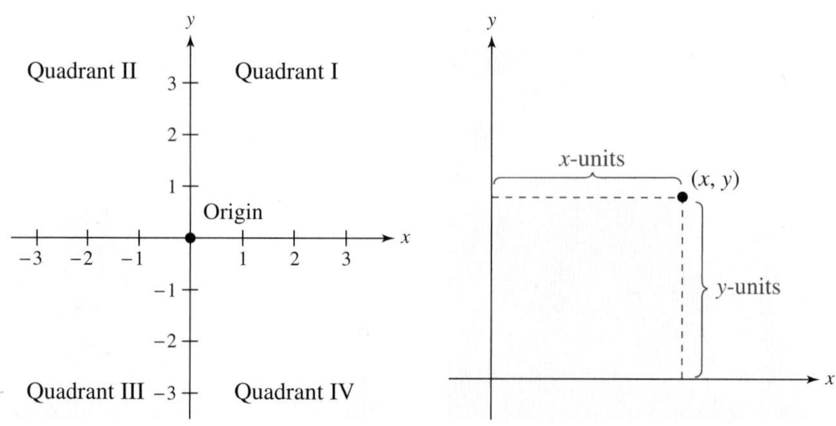

Figure 4.1 **Figure 4.2**

Each point in the plane corresponds to an **ordered pair** (x, y) of real numbers x and y, called the **coordinates** of the point. The first number (or ***x*-coordinate**) tells how far to the left or right the point is from the vertical axis, and the second number (or ***y*-coordinate**) tells how far above or below the point is from the horizontal axis, as shown in Figure 4.2.

A positive x-coordinate implies that the point lies to the *right* of the vertical axis; a negative x-coordinate implies that the point lies to the *left* of the vertical axis; and an x-coordinate of zero implies that the point lies *on* the vertical axis. Similarly, a positive y-coordinate implies that the point lies *above* the horizontal axis, and a negative y-coordinate implies that the point lies *below* the horizontal axis.

Locating a point in a plane is called **plotting** the point. This procedure is demonstrated in Example 1.

EXAMPLE 1 Plotting Points on a Rectangular Coordinate System

Plot the points $(-1, 2)$, $(3, 0)$, $(2, -1)$, $(3, 4)$, $(0, 0)$, and $(-2, -3)$ on a rectangular coordinate system.

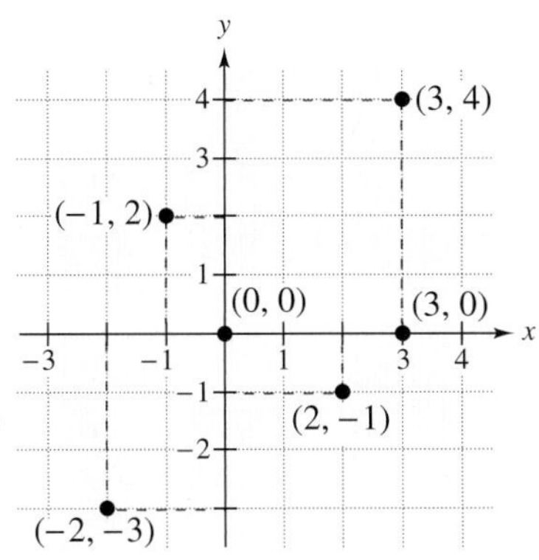

Figure 4.3

Solution

The point $(-1, 2)$ is one unit to the *left* of the vertical axis and two units *above* the horizontal axis.

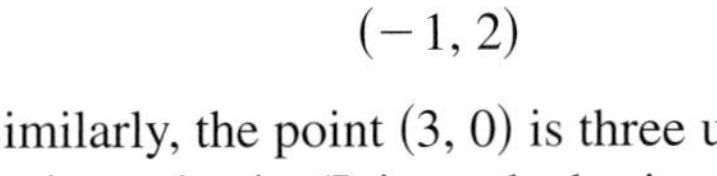

Similarly, the point $(3, 0)$ is three units to the *right* of the vertical axis and *on* the horizontal axis. (It is on the horizontal axis because the y-coordinate is zero.) The other four points can be plotted in a similar way, as shown in Figure 4.3.

✓ **CHECKPOINT** *Now try Exercise 5.*

In Example 1 you were given the coordinates of several points and were asked to plot the points on a rectangular coordinate system. Example 2 looks at the reverse problem—that is, you are given points on a rectangular coordinate system and are asked to determine their coordinates.

EXAMPLE 2 Finding Coordinates of Points

Determine the coordinates of each of the points shown in Figure 4.4.

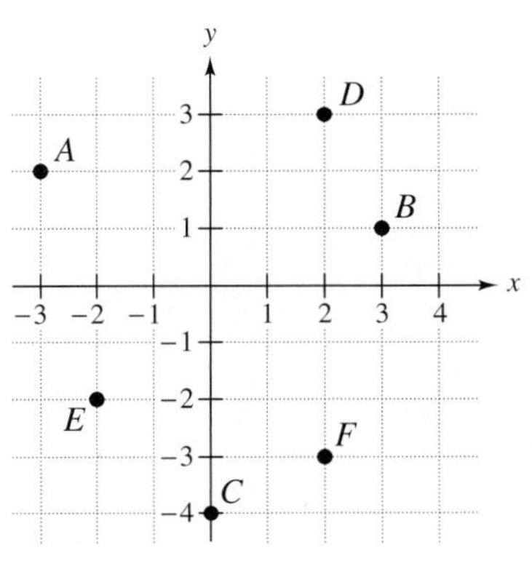

Figure 4.4

Solution

Point A lies three units to the *left* of the vertical axis and two units *above* the horizontal axis. So, point A must be given by the ordered pair $(-3, 2)$. The coordinates of the other four points can be determined in a similar way, and the results are summarized as follows.

Point	*Position*	*Coordinates*
A	Three units *left,* two units *up*	$(-3, 2)$
B	Three units *right,* one unit *up*	$(3, 1)$
C	Zero units *left* (or right), four units *down*	$(0, -4)$
D	Two units *right,* three units *up*	$(2, 3)$
E	Two units *left,* two units *down*	$(-2, -2)$
F	Two units *right,* three units *down*	$(2, -3)$

 **CHECKPOINT** *Now try Exercise 11.*

In Example 2, note that point $A(-3, 2)$ and point $F(2, -3)$ are different points. The order in which the numbers appear in an ordered pair is important. Notice that because point C lies on the y-axis, it has an x-coordinate of 0.

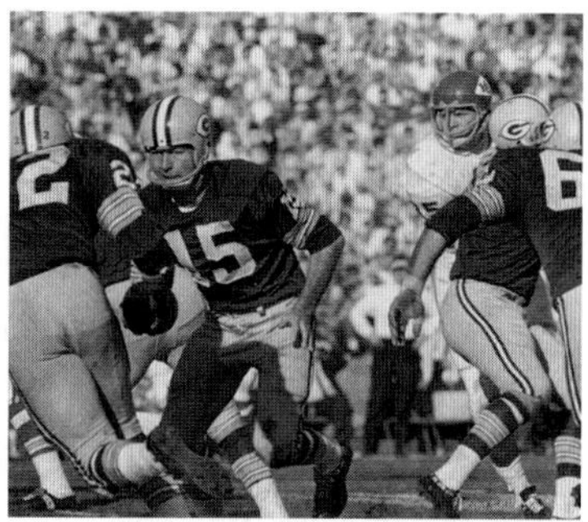

Bettmann/CORBIS

Each year since 1967, the winners of the American Football Conference and the National Football Conference have played in the Super Bowl. The first Super Bowl was played between the Green Bay Packers and the Kansas City Chiefs.

EXAMPLE 3 Super Bowl Scores

The scores of the winning and losing football teams in the Super Bowl games from 1987 through 2007 are shown in the table. Plot these points on a rectangular coordinate system. (Source: National Football League)

Year	1987	1988	1989	1990	1991	1992	1993
Winning score	39	42	20	55	20	37	52
Losing score	20	10	16	10	19	24	17

Year	1994	1995	1996	1997	1998	1999	2000
Winning score	30	49	27	35	31	34	23
Losing score	13	26	17	21	24	19	16

Year	2001	2002	2003	2004	2005	2006	2007
Winning score	34	20	48	32	24	21	29
Losing score	7	17	21	29	21	10	17

Solution

The x-coordinates of the points represent the year of the game, and the y-coordinates represent either the winning score or the losing score. In Figure 4.5, the winning scores are shown as black dots, and the losing scores are shown as blue dots. Note that the break in the x-axis indicates that the numbers between 0 and 1987 have been omitted.

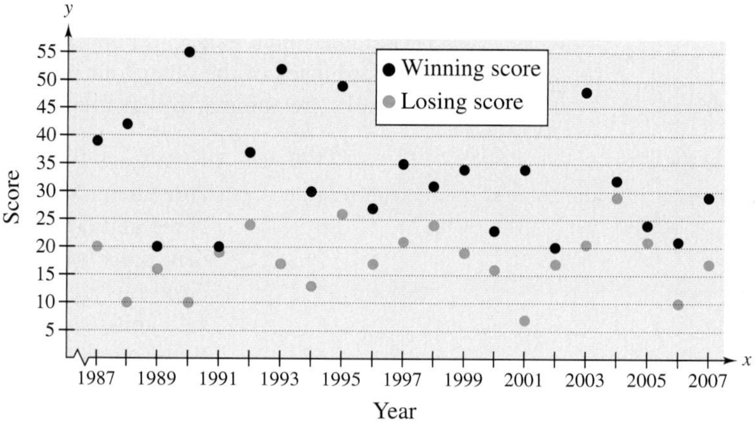

Figure 4.5

✓ **CHECKPOINT** *Now try Exercise 67.*

2 ▶ Construct tables of values for equations and determine whether ordered pairs are solutions of equations.

Ordered Pairs as Solutions of Equations

In Example 3, the relationship between the year and the Super Bowl scores was illustrated by a **table of values.** In mathematics, the relationship between the variables x and y is often given by an equation. From the equation, you can construct your own table of values. For instance, consider the equation

$$y = 2x + 1.$$

To construct a table of values for this equation, choose several x-values and then calculate the corresponding y-values. For example, if you choose $x = 1$, the corresponding y-value is

$y = 2(1) + 1$ Substitute 1 for x.

$y = 3.$ Simplify.

The corresponding ordered pair $(x, y) = (1, 3)$ is a **solution point** (or **solution**) of the equation. The table below is a table of values (and the corresponding solution points) using x-values of -3, -2, -1, 0, 1, 2, and 3. These x-values are arbitrary. You should try to use x-values that are convenient and simple to use.

Choose x	Calculate y using $y = 2x + 1$	Solution point
$x = -3$	$y = 2(-3) + 1 = -5$	$(-3, -5)$
$x = -2$	$y = 2(-2) + 1 = -3$	$(-2, -3)$
$x = -1$	$y = 2(-1) + 1 = -1$	$(-1, -1)$
$x = 0$	$y = 2(0) + 1 = 1$	$(0, 1)$
$x = 1$	$y = 2(1) + 1 = 3$	$(1, 3)$
$x = 2$	$y = 2(2) + 1 = 5$	$(2, 5)$
$x = 3$	$y = 2(3) + 1 = 7$	$(3, 7)$

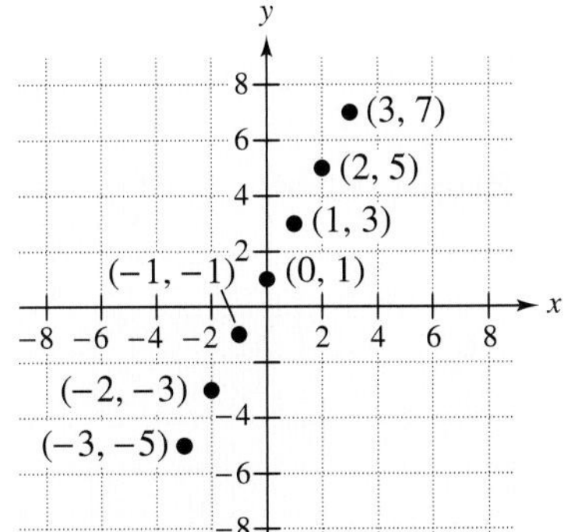

Figure 4.6

Once you have constructed a table of values, you can get a visual idea of the relationship between the variables x and y by plotting the solution points on a rectangular coordinate system. For instance, the solution points shown in the table are plotted in Figure 4.6.

In many places throughout this course, you will see that approaching a problem in different ways can help you understand the problem better. For instance, the discussion above looks at solutions of an equation in three ways.

Three Approaches to Problem Solving

1. **Algebraic Approach** Use algebra to find several solutions.
2. **Numerical Approach** Construct a table that shows several solutions.
3. **Graphical Approach** Draw a graph that shows several solutions.

Technology: Tip

Consult the user's guide for your graphing calculator to see if your graphing calculator has a *table* feature. By using the *table* feature in the *ask* mode, you can create a table of values for an equation.

When constructing a table of values for an equation, it is helpful first to solve the equation for y. For instance, the equation $4x + 2y = -8$ can be solved for y as follows.

$$4x + 2y = -8 \quad \text{Write original equation.}$$

$$4x - 4x + 2y = -8 - 4x \quad \text{Subtract } 4x \text{ from each side.}$$

$$2y = -8 - 4x \quad \text{Combine like terms.}$$

$$\frac{2y}{2} = \frac{-8 - 4x}{2} \quad \text{Divide each side by 2.}$$

$$y = -4 - 2x \quad \text{Simplify.}$$

This procedure is further demonstrated in Example 4.

EXAMPLE 4 Constructing a Table of Values

Construct a table of values showing five solution points for the equation

$6x - 2y = 4.$

Then plot the solution points on a rectangular coordinate system. Choose x-values of $-2, -1, 0, 1,$ and $2.$

Solution

$$6x - 2y = 4 \quad \text{Write original equation.}$$

$$6x - 6x - 2y = 4 - 6x \quad \text{Subtract } 6x \text{ from each side.}$$

$$-2y = -6x + 4 \quad \text{Combine like terms.}$$

$$\frac{-2y}{-2} = \frac{-6x + 4}{-2} \quad \text{Divide each side by } -2.$$

$$y = 3x - 2 \quad \text{Simplify.}$$

Now, using the equation $y = 3x - 2$, you can construct a table of values, as shown below.

x	-2	-1	0	1	2
$y = 3x - 2$	-8	-5	-2	1	4
Solution point	$(-2, -8)$	$(-1, -5)$	$(0, -2)$	$(1, 1)$	$(2, 4)$

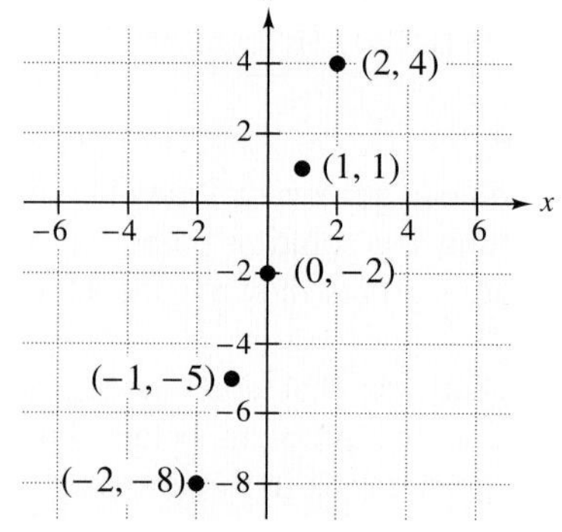

Figure 4.7

Finally, from the table you can plot the five solution points on a rectangular coordinate system, as shown in Figure 4.7.

✓ **CHECKPOINT** *Now try Exercise 37.*

In the next example, you are given several ordered pairs and are asked to determine whether they are solutions of the original equation. To do this, you need to substitute the values of x and y into the equation. If the substitution produces a true statement, the ordered pair (x, y) is a solution and is said to **satisfy** the equation.

Smart Study Strategy

Go to page 214 for ways to *Use the Features of Your Textbook.*

Guidelines for Verifying Solutions

To verify that an ordered pair (x, y) is a solution of an equation with variables x and y, use the following steps.

1. Substitute the values of x and y into the equation.
2. Simplify each side of the equation.
3. If each side simplifies to the same number, the ordered pair is a solution. If the two sides yield different numbers, the ordered pair is not a solution.

EXAMPLE 5 Verifying Solutions of an Equation

Determine whether each of the ordered pairs is a solution of $x + 3y = 6$.

a. $(1, 2)$ **b.** $\left(-2, \frac{8}{3}\right)$ **c.** $(0, 2)$

Solution

a. For the ordered pair $(1, 2)$, substitute $x = 1$ and $y = 2$ into the original equation.

$$x + 3y = 6 \qquad \text{Write original equation.}$$

$$1 + 3(2) \stackrel{?}{=} 6 \qquad \text{Substitute 1 for } x \text{ and 2 for } y.$$

$$7 \neq 6 \qquad \text{Not a solution ✗}$$

Because the substitution does not satisfy the original equation, you can conclude that the ordered pair $(1, 2)$ *is not* a solution of the original equation.

b. For the ordered pair $\left(-2, \frac{8}{3}\right)$, substitute $x = -2$ and $y = \frac{8}{3}$ into the original equation.

$$x + 3y = 6 \qquad \text{Write original equation.}$$

$$(-2) + 3\left(\frac{8}{3}\right) \stackrel{?}{=} 6 \qquad \text{Substitute } -2 \text{ for } x \text{ and } \frac{8}{3} \text{ for } y.$$

$$-2 + 8 \stackrel{?}{=} 6 \qquad \text{Simplify.}$$

$$6 = 6 \qquad \text{Solution ✓}$$

Because the substitution satisfies the original equation, you can conclude that the ordered pair $\left(-2, \frac{8}{3}\right)$ *is* a solution of the original equation.

c. For the ordered pair $(0, 2)$, substitute $x = 0$ and $y = 2$ into the original equation.

$$x + 3y = 6 \qquad \text{Write original equation.}$$

$$0 + 3(2) \stackrel{?}{=} 6 \qquad \text{Substitute 0 for } x \text{ and 2 for } y.$$

$$6 = 6 \qquad \text{Solution ✓}$$

Because the substitution satisfies the original equation, you can conclude that the ordered pair $(0, 2)$ *is* a solution of the original equation.

✓ **CHECKPOINT** *Now try Exercise 53.*

3 ▶ Use the verbal problem-solving method to plot points on a rectangular coordinate system.

Application

EXAMPLE 6 Total Cost

You set up a small business to assemble computer keyboards. Your initial cost is \$120,000, and your unit cost of assembling each keyboard is \$40. Write an equation that relates your total cost to the number of keyboards produced. Then plot the total costs of producing 1000, 2000, 3000, 4000, and 5000 keyboards.

Solution

The total cost equation must represent both the unit cost and the initial cost. A verbal model for this problem is as follows.

Verbal Model: Total cost = Unit cost · Number of keyboards + Initial cost

Labels: Total cost $= C$ (dollars)
Unit cost $= 40$ (dollars per keyboard)
Number of keyboards $= x$ (keyboards)
Initial cost $= 120{,}000$ (dollars)

Algebraic Model: $C = 40x + 120{,}000$

Using this equation, you can construct the following table of values.

x	1000	2000	3000	4000	5000
$C = 40x + 120{,}000$	160,000	200,000	240,000	280,000	320,000

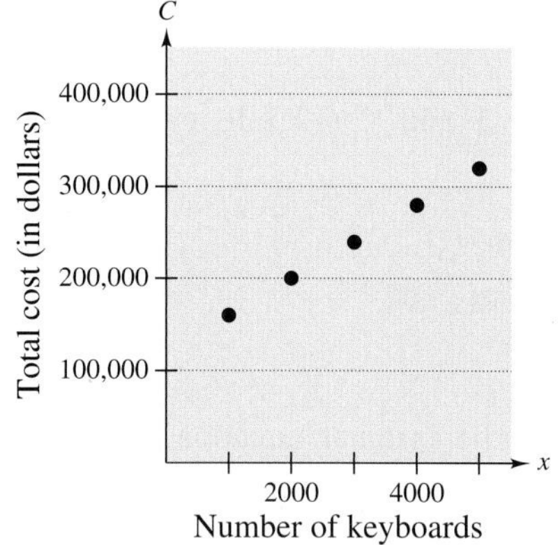

Figure 4.8

From the table, you can plot the ordered pairs, as shown in Figure 4.8.

✓ ***CHECKPOINT*** *Now try Exercise 63.*

Although graphs can help you visualize relationships between two variables, they can also be misleading. The graphs shown in Figure 4.9 and Figure 4.10 represent the yearly profits for a truck rental company. The graph in Figure 4.9 is misleading. The scale on the vertical axis makes it appear that the change in profits from 2003 to 2007 is dramatic, but the total change is only \$3000, which is small in comparison with \$3,000,000.

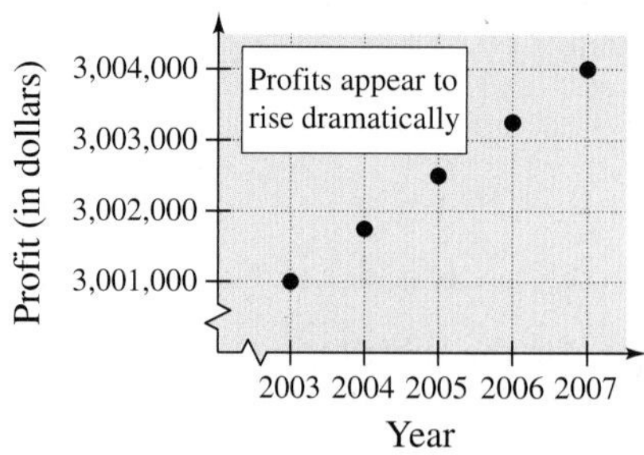

Figure 4.9

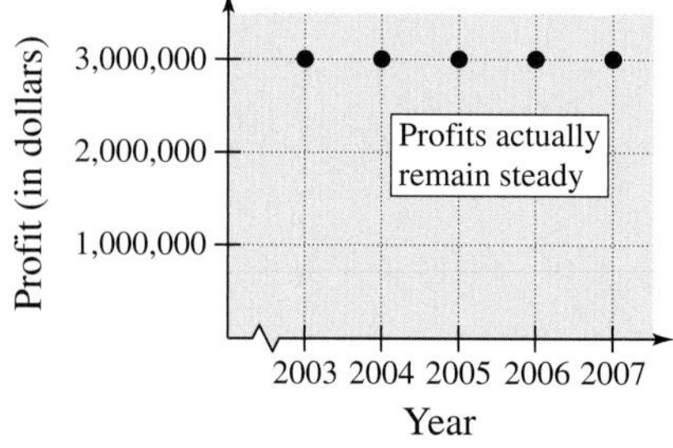

Figure 4.10

Concept Check

1. Describe the signs of the x- and y-coordinates of points that lie in each of the four quadrants.

2. When the point (x, y) is plotted, what does the x-coordinate represent? What does the y-coordinate represent?

3. What is the x-coordinate of any point on the y-axis? What is the y-coordinate of any point on the x-axis?

4. How do you verify that an ordered pair is a solution of an equation?

4.1 EXERCISES

Go to pages 284–285 to record your assignments.

Developing Skills

In Exercises 1–10, plot the points on a rectangular coordinate system. *See Example 1.*

1. $(3, 2), (-4, 2), (2, -4)$
2. $(-1, 6), (-1, -6), (4, 6)$
3. $(-10, -4), (4, -4), (0, 0)$
4. $(-6, 4), (0, 0), (3, -2)$

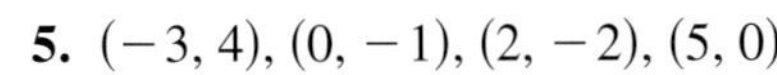

5. $(-3, 4), (0, -1), (2, -2), (5, 0)$
6. $(-1, 3), (0, 2), (-4, -4), (-1, 0)$
7. $\left(\frac{3}{2}, -1\right), \left(-3, \frac{3}{4}\right), \left(\frac{1}{2}, -\frac{1}{2}\right)$
8. $\left(-\frac{2}{3}, 4\right), \left(\frac{1}{2}, -\frac{5}{2}\right), \left(-4, -\frac{5}{4}\right)$
9. $(3, -4), \left(\frac{5}{2}, 0\right), (0, 3)$
10. $\left(\frac{5}{2}, 2\right), \left(-3, \frac{4}{3}\right), \left(\frac{3}{4}, \frac{9}{4}\right)$

In Exercises 11–14, determine the coordinates of the points. *See Example 2.*

11.

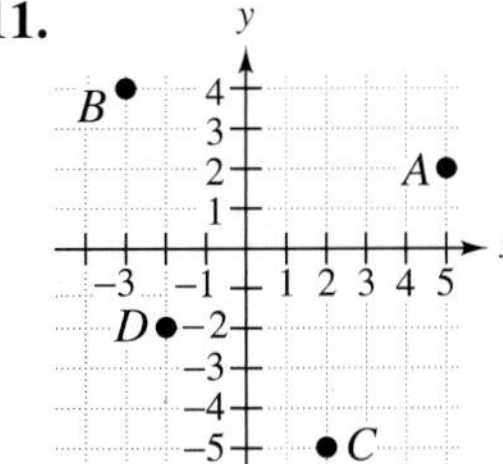

12.

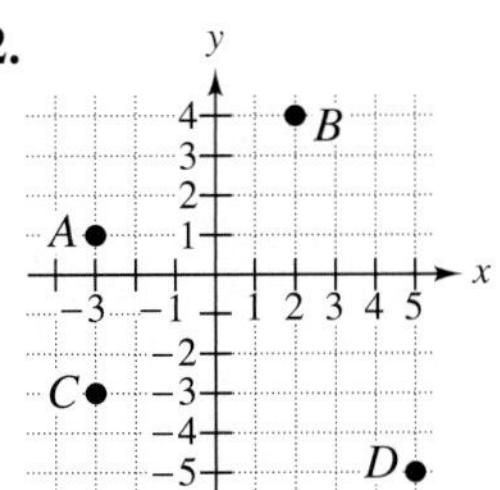

13.

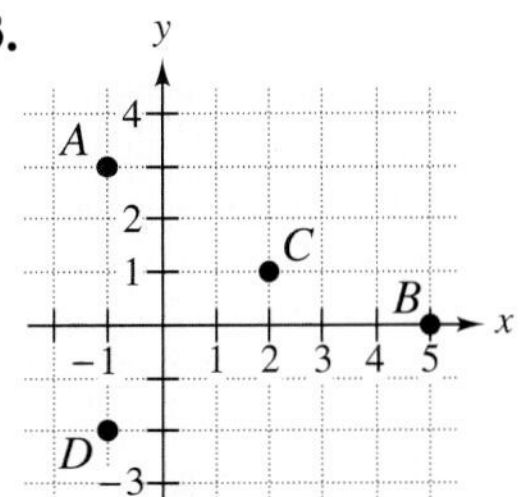

14.

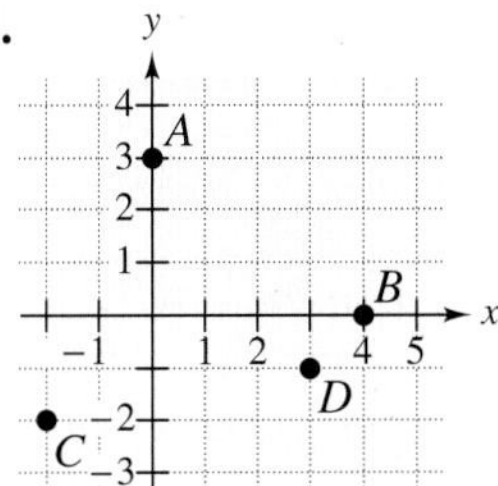

In Exercises 15–20, determine the quadrant in which the point is located without plotting it.

15. $(-3, 1)$
16. $(4, -3)$
17. $\left(-\frac{1}{8}, -\frac{2}{7}\right)$
18. $\left(\frac{3}{11}, \frac{7}{8}\right)$
19. $(-100, -365.6)$
20. $(-157.4, 305.6)$

In Exercises 21–26, determine the quadrant(s) in which the point is located without plotting it. Assume $x \neq 0$ and $y \neq 0$.

21. $(-5, y)$, y is a real number.
22. $(6, y)$, y is a real number.
23. $(x, -2)$, x is a real number.
24. $(x, 3)$, x is a real number.
25. (x, y), $xy < 0$
26. (x, y), $xy > 0$

In Exercises 27–34, plot the points and connect them with line segments to form the figure.

27. Triangle: $(-1, 1), (2, -1), (3, 4)$
28. Triangle: $(0, 3), (-1, -2), (4, 8)$
29. Square: $(2, 4), (5, 1), (2, -2), (-1, 1)$
30. Rectangle: $(2, 1), (4, 2), (1, 8), (-1, 7)$
31. Parallelogram: $(5, 2), (7, 0), (1, -2), (-1, 0)$
32. Parallelogram: $(-1, 1), (0, 4), (5, 1), (4, -2)$
33. Rhombus: $(0, 0), (3, 2), (5, 5), (2, 3)$
34. Rhombus: $(0, 0), (1, 2), (3, 3), (2, 1)$

In Exercises 35–40, complete the table of values. Then plot the solution points on a rectangular coordinate system. ***See Example 4.***

35.

x	-2	0	2	4	6
$y = 3x - 4$					

36.

x	-2	0	2	4	6
$y = 2x + 1$					

✓ **37.**

x	-2	0	4	6	8
$y = -\frac{3}{2}x + 5$					

38.

x	-4	-2	0	2	4
$y = -\frac{1}{2}x + 3$					

39.

x	-2	-1	0	1	2
$y = -4x - 5$					

40.

x	-2	0	$\frac{1}{2}$	2	4
$y = -\frac{7}{2}x + 3$					

In Exercises 41–50, solve the equation for *y*.

41. $7x + y = 8$ **42.** $2x + y = 1$

43. $10x - y = 2$ **44.** $12x - y = 7$

45. $6x - 3y = 3$ **46.** $15x - 5y = 25$

47. $x + 4y = 8$ **48.** $x - 2y = -6$

49. $4x - 5y = 3$ **50.** $9x - 2y = 5$

In Exercises 51–58, determine whether each ordered pair is a solution of the equation. ***See Example 5.***

51. $y = 2x + 4$ (a) $(3, 10)$ (b) $(-1, 3)$ (c) $(0, 0)$ (d) $(-2, 0)$

52. $y = 5x - 2$ (a) $(2, 0)$ (b) $(-2, -12)$ (c) $(6, 28)$ (d) $(1, 1)$

✓ **53.** $2y - 3x + 1 = 0$ (a) $(1, 1)$ (b) $(5, 7)$ (c) $(-3, -1)$ (d) $(-3, -5)$

54. $x - 8y + 10 = 0$ (a) $(-2, 1)$ (b) $(6, 2)$ (c) $(0, -1)$ (d) $(2, -4)$

55. $y = \frac{2}{3}x$ (a) $(6, 6)$ (b) $(-9, -6)$ (c) $(0, 0)$ (d) $\left(-1, \frac{2}{3}\right)$

56. $y = -\frac{7}{8}x$ (a) $(-5, -2)$ (b) $(0, 0)$ (c) $(8, 8)$ (d) $\left(\frac{3}{5}, 1\right)$

57. $y = 3 - 4x$ (a) $\left(-\frac{1}{2}, 5\right)$ (b) $(1, 7)$ (c) $(0, 0)$ (d) $\left(-\frac{3}{4}, 0\right)$

58. $y = \frac{3}{2}x + 1$ (a) $\left(0, \frac{3}{2}\right)$ (b) $(4, 7)$ (c) $\left(\frac{2}{3}, 2\right)$ (d) $(-2, -2)$

In Exercises 59 and 60, complete each ordered pair so that it satisfies the equation.

59. $y = 3x + 4$ (a) $(\quad, 0)$ (b) $(4, \quad)$ (c) $(\quad, -2)$

60. $y = -2x - 7$ (a) $(0, \quad)$ (b) $(-1, \quad)$ (c) $(\quad, 1)$

Solving Problems

61. *Organizing Data* The distance y (in centimeters) a spring is compressed by a force x (in kilograms) is given by $y = 0.066x$. Complete a table of values for $x = 20, 40, 60, 80,$ and 100 to determine the distance the spring is compressed for each of the specified forces. Plot the results on a rectangular coordinate system.

62. *Organizing Data* A company buys a new copier for \$9500. Its value y after x years is given by $y = -800x + 9500$. Complete a table of values for $x = 0, 2, 4, 6,$ and 8 to determine the value of the copier at each specified time. Plot the results on a rectangular coordinate system.

✓ 63. *Organizing Data* With an initial cost of \$5000, a company will produce x units of a video game at \$25 per unit. Write an equation that relates the total cost of producing x units to the number of units produced. Plot the cost of producing 100, 150, 200, 250, and 300 units.

64. *Organizing Data* An employee earns \$10 plus \$0.50 for every x units produced per hour. Write an equation that relates the employee's total hourly wage to the number of units produced. Plot the hourly wages for producing 2, 5, 8, 10, and 20 units per hour.

65. *Organizing Data* The table shows the normal average temperatures y (in degrees Fahrenheit) in Anchorage, Alaska for each month x of the year, with $x = 1$ corresponding to January. (Source: National Climatic Data Center)

x	1	2	3	4	5	6
y	16	19	26	36	47	55

x	7	8	9	10	11	12
y	58	56	48	34	22	18

(a) Plot the data in the table. Did you use the same scale on both axes? Explain.

(b) Using the graph, find the month for which the normal average temperature changed the least from the previous month.

66. *Organizing Data* The table shows the speeds of a car x (in miles per hour) and the approximate stopping distances y (in feet).

x	20	30	40	50	60
y	63	109	164	229	303

(a) Plot the data in the table.

(b) The x-coordinates increase in equal increments of 10 miles per hour. Describe the pattern of the y-coordinates. What are the implications for the driver?

✓ 67. *Graphical Interpretation* The table shows the numbers of hours x that a student studied for five different algebra exams, and the resulting scores y.

x	3.5	1	8	4.5	0.5
y	72	67	95	81	53

(a) Plot the data in the table.

(b) Use the graph to describe the relationship between the number of hours studied and the resulting exam score.

68. *Graphical Interpretation* The table shows the lowest prices per share of common stock y (in dollars) for the Dow Chemical Company for the years 1997 through 2006, where x represents the year. (Source: Dow Chemical Company 2006 Annual Report)

x	1997	1998	1999	2000	2001
y	25.25	24.90	28.50	23.00	25.06

x	2002	2003	2004	2005	2006
y	23.66	24.83	36.35	40.18	33.00

(a) Plot the data in the table.

(b) Use the graph to determine the two consecutive years between which the greatest increase occurred and the two consecutive years between which the greatest decrease occurred in the lowest price per share of common stock.

Graphical Estimation In Exercises 69–72, use the scatter plot, which shows the average prices (in dollars) of a gallon of regular unleaded gasoline from 1998 through 2005. (Source: U.S. Energy Information Administration)

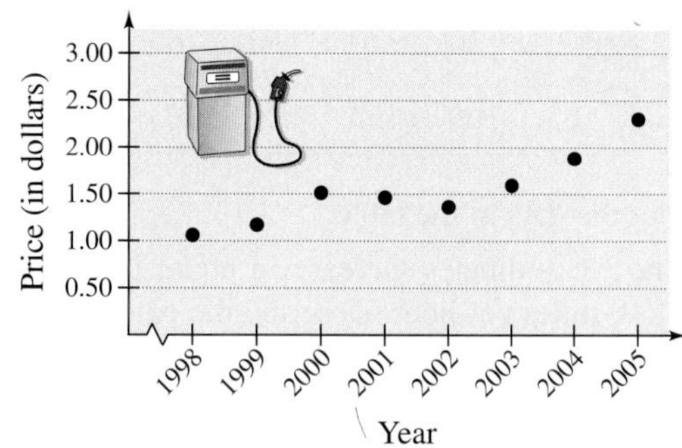

69. Estimate the price of gasoline in 1999.

70. Estimate the price of gasoline in 2002.

71. Estimate the increase in the price of gasoline from 2003 to 2004.

72. Estimate the decrease in the price of gasoline from 2000 to 2001.

Graphical Estimation In Exercises 73–76, use the scatter plot, which shows the per capita personal income in the United States from 1999 through 2006. (Source: U.S. Bureau of Economic Analysis)

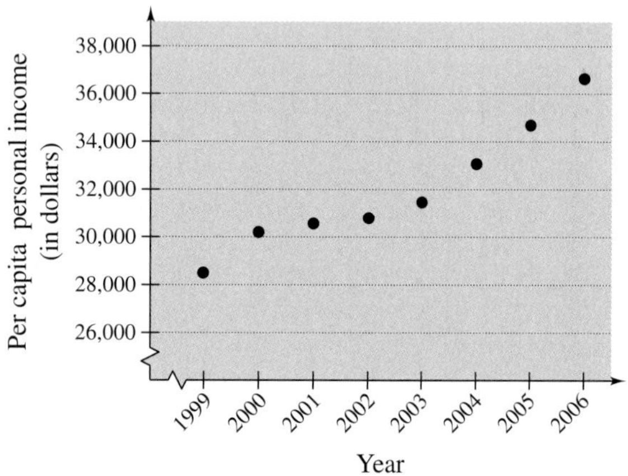

73. Estimate the per capita personal income in 2002.

74. Estimate the per capita personal income in 2003.

75. Estimate the percent increase in per capita personal income from 2005 to 2006.

76. The per capita personal income in 1980 was $10,205. Estimate the percent increase in per capita personal income from 1980 to 1999.

Graphical Estimation In Exercises 77 and 78, use the bar graph, which shows the percents of gross domestic product spent on health care in several countries in 2005. (Source: Organization for Economic Cooperation and Development)

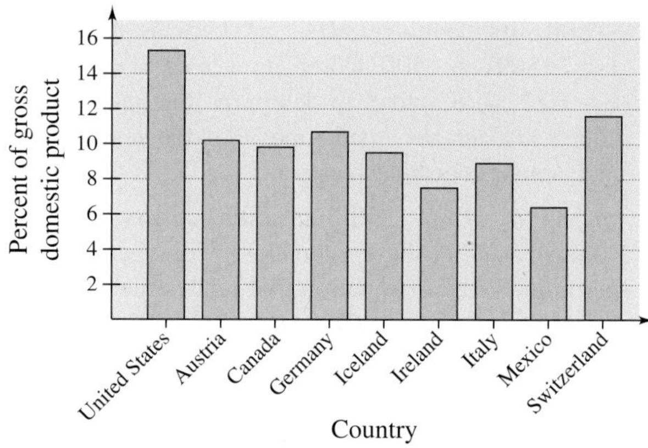

77. Estimate the percent of gross domestic product spent on health care in Mexico.

78. Estimate the percent of gross domestic product spent on health care in the United States.

Graphical Estimation In Exercises 79 and 80, use the bar graph, which shows the highest recorded temperatures during January in select cities. (Source: U.S. Census Bureau)

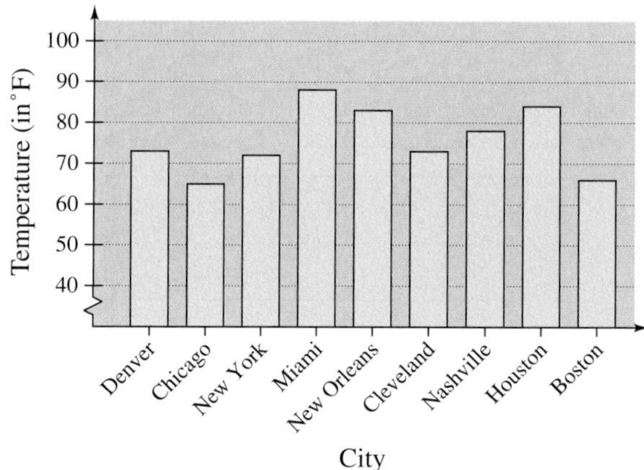

79. Estimate the highest recorded temperature in Chicago.

80. Estimate the highest recorded temperature in Houston.

Explaining Concepts

81. (a) Plot the points $(3, 2)$, $(-5, 4)$, and $(6, -4)$ on a rectangular coordinate system.

(b) Change the sign of the y-coordinate of each point plotted in part (a). Plot the three new points on the same rectangular coordinate system used in part (a).

(c) What can you infer about the location of a point when the sign of its y-coordinate is changed?

82. (a) Plot the points $(3, 2)$, $(-5, 4)$, and $(6, -4)$ on a rectangular coordinate system.

(b) Change the sign of the x-coordinate of each point plotted in part (a). Plot the three new points on the same rectangular coordinate system used in part (a).

(c) What can you infer about the location of a point when the sign of its x-coordinate is changed?

83. The points $(6, -1)$, $(-2, -1)$, and $(-2, 4)$ are three vertices of a rectangle. Find the coordinates of the fourth vertex.

84. Discuss the significance of the word "ordered" when referring to an ordered pair (x, y).

85. In a rectangular coordinate system, must the scales on the x-axis and y-axis be the same? If not, give an example in which the scales differ.

86. Review the tables in Exercises 35–40 and observe that in some cases the y-coordinates of the solution points increase and in others the y-coordinates decrease. What factor in the equation causes this? Explain.

Cumulative Review

In Exercises 87–94, solve the equation.

87. $-y = 10$

88. $10 - t = 6$

89. $3x - 42 = 0$

90. $64 - 16x = 0$

91. $125(r - 1) = 625$

92. $2(3 - y) = 7y + 5$

93. $20 - \frac{1}{9}x = 4$

94. $0.35x = 70$

In Exercises 95–98, solve the inequality.

95. $x + 3 > 2$

96. $y - 4 < -8$

97. $3x < 12$

98. $2(z - 4) > 10z$

4.2 Graphs of Equations in Two Variables

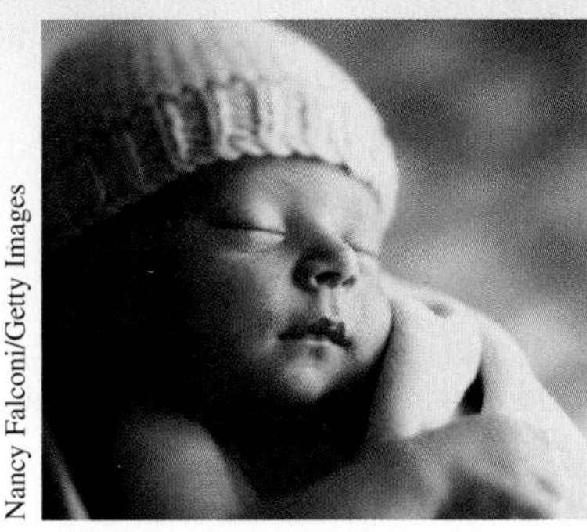

Nancy Falconi/Getty Images

Why You Should Learn It

The graph of an equation can help you see relationships between real-life quantities. For instance, in Exercise 69 on page 236, a graph can be used to illustrate the change over time in the life expectancy for a male child at birth.

What You Should Learn

1 ▶ Sketch graphs of equations using the point-plotting method.

2 ▶ Find and use x- and y-intercepts as aids to sketching graphs.

3 ▶ Use the verbal problem-solving method to write an equation and sketch its graph.

1 ▶ Sketch graphs of equations using the point-plotting method.

The Graph of an Equation in Two Variables

You have already seen that the solutions of an equation involving two variables can be represented by points on a rectangular coordinate system. The set of *all* such points is called the **graph** of the equation.

To see how to sketch a graph, consider the equation

$$y = 2x - 1.$$

To begin sketching the graph of this equation, construct a table of values, as shown at the left. Next, plot the solution points on a rectangular coordinate system, as shown in Figure 4.11. Finally, find a pattern for the plotted points and use the pattern to connect the points with a smooth curve or line, as shown in Figure 4.12.

x	$y = 2x - 1$	Solution point
-3	-7	$(-3, -7)$
-2	-5	$(-2, -5)$
-1	-3	$(-1, -3)$
0	-1	$(0, -1)$
1	1	$(1, 1)$
2	3	$(2, 3)$
3	5	$(3, 5)$

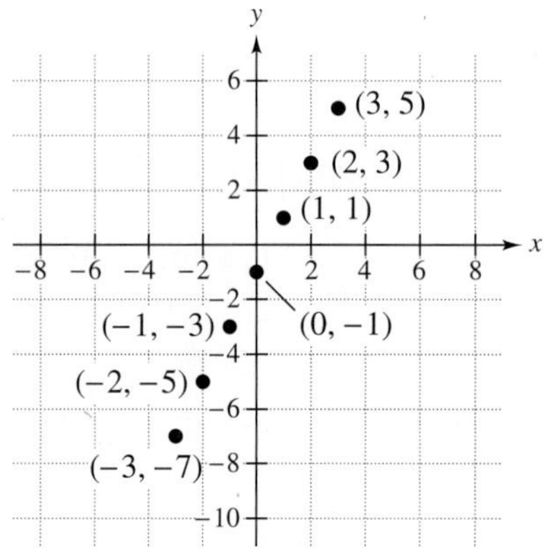

Figure 4.11

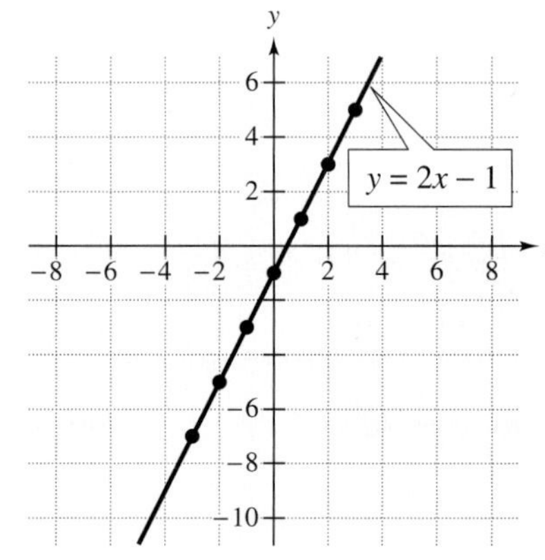

Figure 4.12

The Point-Plotting Method of Sketching a Graph

1. If possible, rewrite the equation by isolating one of the variables.
2. Make a table of values showing several solution points.
3. Plot these points on a rectangular coordinate system.
4. Connect the points with a smooth curve or line.

> **Technology: Tip**
>
> To graph an equation using a graphing calculator, use the following steps.
>
> (1) Select a viewing window.
>
> (2) Solve the original equation for y in terms of x.
>
> (3) Enter the equation in the equation editor.
>
> (4) Display the graph.
>
> Consult the user's guide for your graphing calculator for specific instructions.

EXAMPLE 1 Sketching the Graph of an Equation

Sketch the graph of $3x + y = 5$.

Solution

Begin by solving the equation for y, so that y is isolated on the left.

$$3x + y = 5 \qquad \text{Write original equation.}$$

$$3x - 3x + y = -3x + 5 \qquad \text{Subtract } 3x \text{ from each side.}$$

$$y = -3x + 5 \qquad \text{Simplify.}$$

Next, create a table of values, as shown below.

x	-2	-1	0	1	2	3
$y = -3x + 5$	11	8	5	2	-1	-4
Solution point	$(-2, 11)$	$(-1, 8)$	$(0, 5)$	$(1, 2)$	$(2, -1)$	$(3, -4)$

Now, plot the solution points, as shown in Figure 4.13. It appears that all six points lie on a line, so complete the sketch by drawing a line through the points, as shown in Figure 4.14.

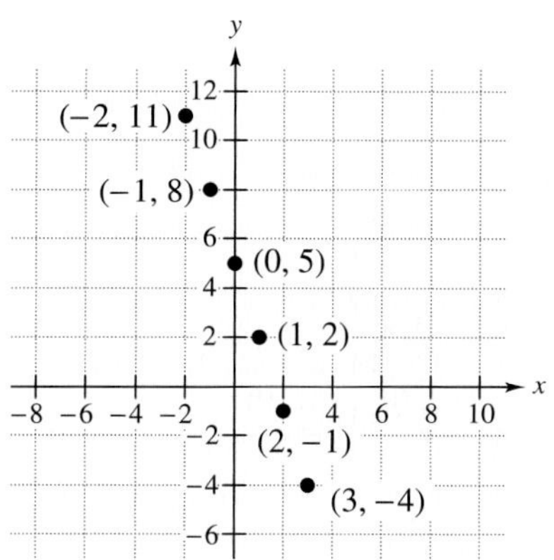

Figure 4.13

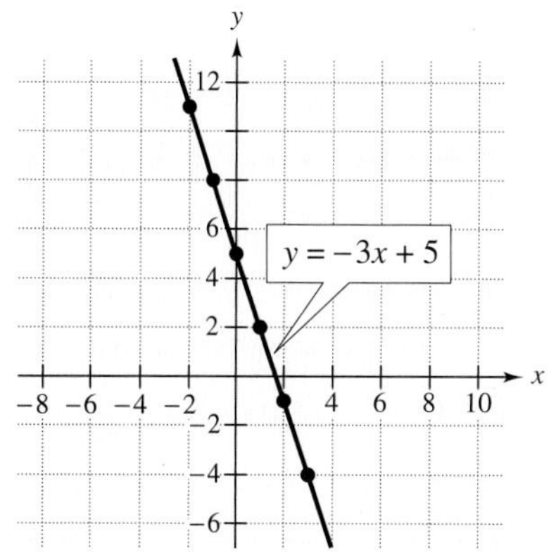

Figure 4.14

✓ ***CHECKPOINT*** *Now try Exercise 9.*

When creating a table of values, you are generally free to choose any x-values. When doing this, however, remember that the more x-values you choose, the easier it will be to recognize a pattern.

The equation in Example 1 is an example of a **linear equation in two variables**—the variables are raised to the first power, and the graph of the equation is a line. As shown in the next two examples, graphs of nonlinear equations are not lines.

Technology: Discovery

Most graphing calculators have the following standard viewing window.

Xmin = -10
Xmax = 10
Xscl = 1
Ymin = -10
Ymax = 10
Yscl = 1

What happens when the equation $x + y = 12$ is graphed using a standard viewing window?

To see where the graph crosses the x- and y-axes, you need to change the viewing window. What changes would you make in the viewing window to see where the line intersects the axes?

Graph each equation using a graphing calculator and describe the viewing window used.

a. $y = \frac{1}{2}x + 6$

b. $y = 2x^2 + 5x + 10$

c. $y = 10 - x$

d. $y = -3x^3 + 5x + 8$

EXAMPLE 2 Sketching the Graph of a Nonlinear Equation

Sketch the graph of $x^2 + y = 4$.

Solution

Begin by solving the equation for y, so that y is isolated on the left.

$$x^2 + y = 4 \qquad \text{Write original equation.}$$

$$x^2 - x^2 + y = -x^2 + 4 \qquad \text{Subtract } x^2 \text{ from each side.}$$

$$y = -x^2 + 4 \qquad \text{Simplify.}$$

Next, create a table of values, as shown below. Be careful with the signs of the numbers when creating the table. For instance, when $x = -3$, the value of y is

$$y = -(-3)^2 + 4$$
$$= -9 + 4$$
$$= -5.$$

x	-3	-2	-1	0	1	2	3
$y = -x^2 + 4$	-5	0	3	4	3	0	-5
Solution point	$(-3, -5)$	$(-2, 0)$	$(-1, 3)$	$(0, 4)$	$(1, 3)$	$(2, 0)$	$(3, -5)$

Now, plot the solution points, as shown in Figure 4.15. Finally, connect the points with a smooth curve, as shown in Figure 4.16.

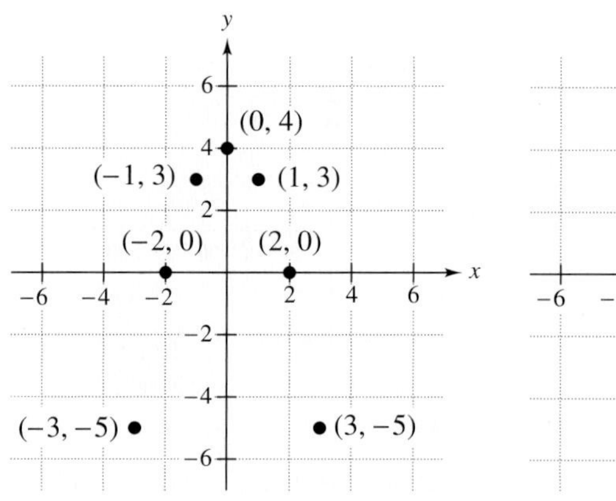

Figure 4.15

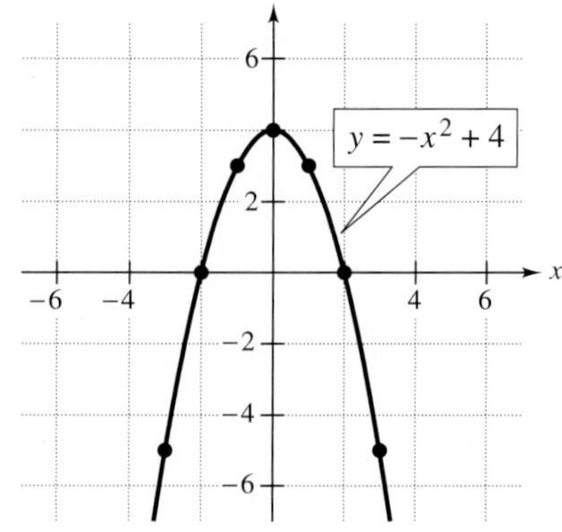

Figure 4.16

CHECKPOINT *Now try Exercise 13.*

The graph of the equation in Example 2 is called a *parabola*. You will study this type of graph in a later chapter.

Example 3 examines the graph of an equation that involves an absolute value. Remember that the absolute value of a number is its distance from zero on the real number line. For instance, $|-5| = 5$, $|2| = 2$, and $|0| = 0$.

EXAMPLE 3 The Graph of an Absolute Value Equation

Sketch the graph of $y = |x - 1|$.

Solution

This equation is already written in a form with y isolated on the left. You can begin by creating a table of values, as shown below. Notice that because of the absolute value, all of the y-values are nonnegative. For instance, when $x = -2$, the value of y is

$$\begin{aligned} y &= |-2 - 1| \\ &= |-3| \\ &= 3 \end{aligned}$$

and when $x = 2$, the value of y is $|2 - 1| = 1$.

x	-2	-1	0	1	2	3	4
$y = \lvert x - 1 \rvert$	3	2	1	0	1	2	3
Solution point	$(-2, 3)$	$(-1, 2)$	$(0, 1)$	$(1, 0)$	$(2, 1)$	$(3, 2)$	$(4, 3)$

Plot the solution points, as shown in Figure 4.17. It appears that the points lie in a "V-shaped" pattern, with the point $(1, 0)$ lying at the bottom of the "V." Following this pattern, connect the points to form the graph shown in Figure 4.18.

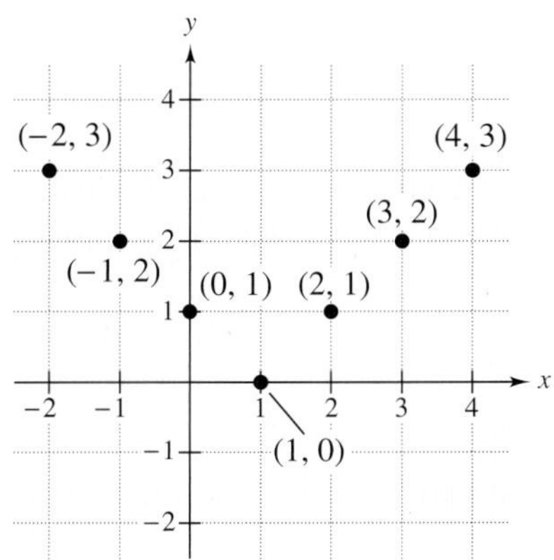

Figure 4.17

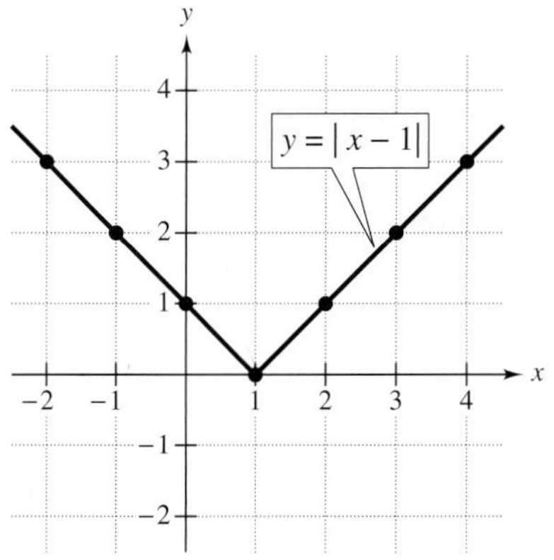

Figure 4.18

✓ **CHECKPOINT** *Now try Exercise 15.*

2 ▶ Find and use x- and y-intercepts as aids to sketching graphs.

Intercepts: Aids to Sketching Graphs

Two types of solution points that are especially useful are those having zero as either the x- or the y-coordinate. Such points are called **intercepts** because they are the points at which the graph intersects the x- or y-axis.

Study Tip

Some texts denote the x-intercept as the x-coordinate at the point $(a, 0)$, rather than the point itself. Unless it is necessary to make a distinction, we will use the term *intercept* to mean either the point or the coordinate.

Definitions of Intercepts

The point $(a, 0)$ is called an ***x*-intercept** of the graph of an equation if it is a solution point of the equation. To find the x-intercept(s), let $y = 0$ and solve the equation for x.

The point $(0, b)$ is called a ***y*-intercept** of the graph of an equation if it is a solution point of the equation. To find the y-intercept(s), let $x = 0$ and solve the equation for y.

EXAMPLE 4 Finding the Intercepts of a Graph

Find the intercepts and sketch the graph of $y = 2x - 5$.

Solution

To find any x-intercepts, let $y = 0$ and solve the resulting equation for x.

$y = 2x - 5$ Write original equation.

$0 = 2x - 5$ Let $y = 0$.

$\frac{5}{2} = x$ Solve equation for x.

To find any y-intercepts, let $x = 0$ and solve the resulting equation for y.

$y = 2x - 5$ Write original equation.

$y = 2(0) - 5$ Let $x = 0$.

$y = -5$ Solve equation for y.

So, the graph has one x-intercept, which occurs at the point $\left(\frac{5}{2}, 0\right)$, and one y-intercept, which occurs at the point $(0, -5)$. To sketch the graph of the equation, create a table of values. (Include the intercepts in the table.) Then plot the points and connect the points with a line, as shown in Figure 4.19.

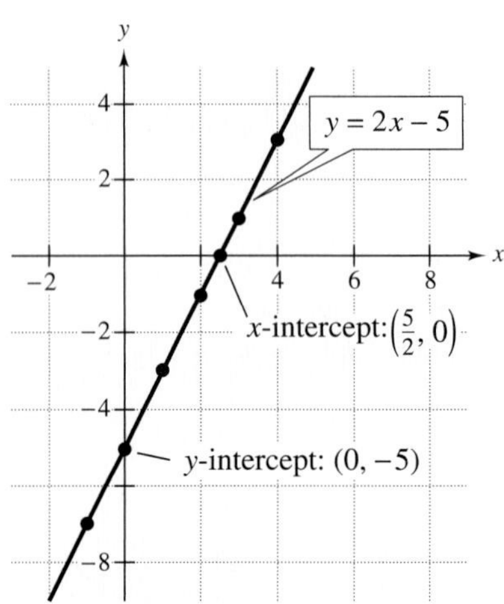

Figure 4.19

x	-1	0	1	2	$\frac{5}{2}$	3	4
$y = 2x - 5$	-7	-5	-3	-1	0	1	3
Solution point	$(-1, -7)$	$(0, -5)$	$(1, -3)$	$(2, -1)$	$\left(\frac{5}{2}, 0\right)$	$(3, 1)$	$(4, 3)$

✓ ***CHECKPOINT*** *Now try Exercise 27.*

When you create a table of values, include any intercepts you have found. You should also include points to the left and to the right of the intercepts. This helps to give a more complete view of the graph.

3 ▶ Use the verbal problem-solving method to write an equation and sketch its graph.

Application

EXAMPLE 5 Depreciation 

The value of a \$35,500 sport utility vehicle (SUV) depreciates over 10 years (the depreciation is the same each year). At the end of the 10 years, the salvage value is expected to be \$5500.

a. Write an equation that relates the value of the SUV to its age in years.

b. Sketch the graph of the equation.

c. What is the y-intercept of the graph, and what does it represent in the context of the problem?

Solution

a. The total depreciation over the 10 years is $35{,}500 - 5500 = \$30{,}000$. Because the same amount is depreciated each year, it follows that the annual depreciation is $30{,}000/10 = \$3000$.

Verbal Model: Value after t years = Original value − Annual depreciation · Number of years

Labels:

Value after t years $= y$	(dollars)
Original value $= 35{,}500$	(dollars)
Annual depreciation $= 3000$	(dollars per year)
Number of years $= t$	(years)

Algebraic Model: $y = 35{,}500 - 3000t$

b. A sketch of the graph of the depreciation equation is shown in Figure 4.20.

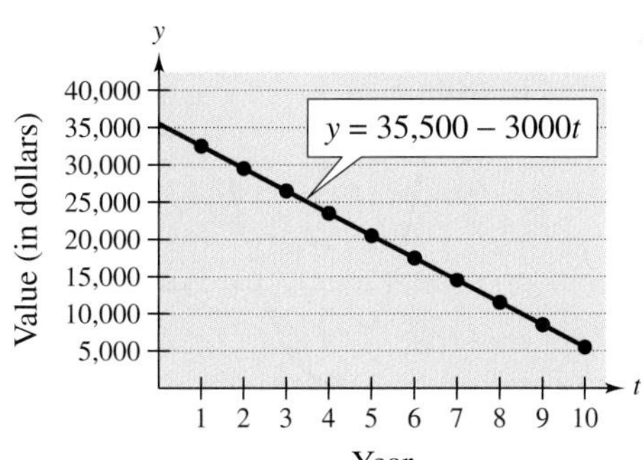

Figure 4.20

c. To find the y-intercept of the graph, let $t = 0$ and solve the equation for y.

$y = 35{,}500 - 3000t$ Write original equation.

$y = 35{,}500 - 3000(0)$ Substitute 0 for t.

$y = 35{,}500$ Simplify.

So, the y-intercept is $(0, 35{,}500)$, and it corresponds to the original value of the SUV.

CHECKPOINT *Now try Exercise 67.*

Concept Check

1. In your own words, explain what is meant by the *graph* of an equation.

2. In your own words, describe the point-plotting method of sketching the graph of an equation.

3. Describe how you can check whether an ordered pair (x, y) is a solution of an equation.

4. Explain how to find the x- and y-intercepts of a graph.

4.2 EXERCISES

Go to pages 284–285 to record your assignments.

Developing Skills

In Exercises 1–8, match the equation with its graph. [The graphs are labeled (a), (b), (c), (d), (e), (f), (g), and (h).]

(a)

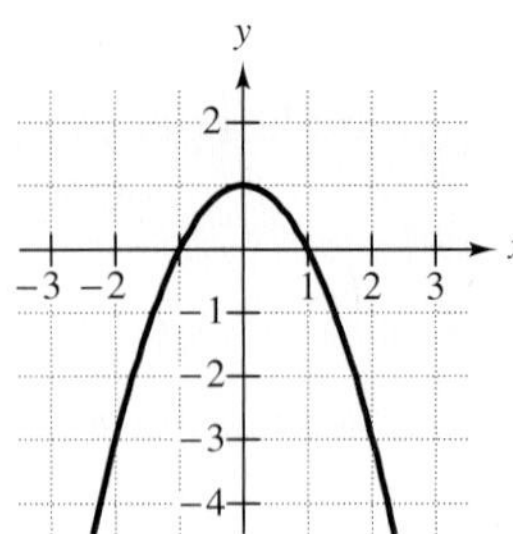

(b)

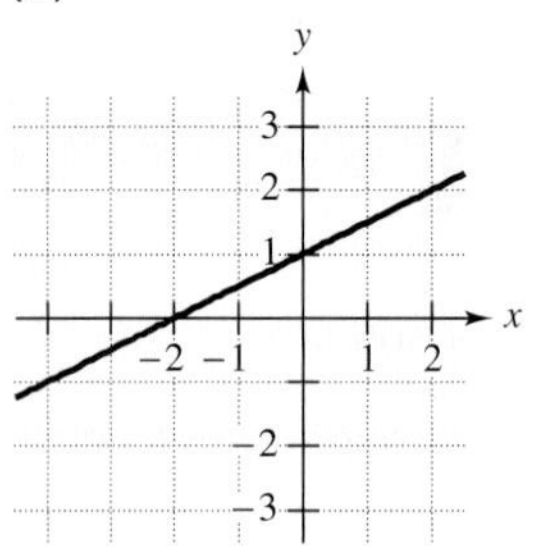

(c)

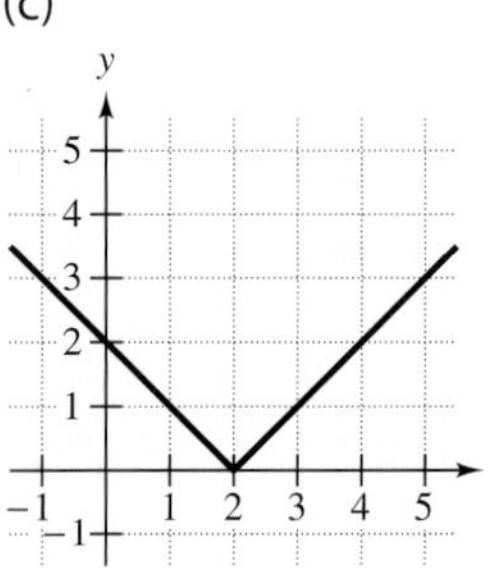

(d)

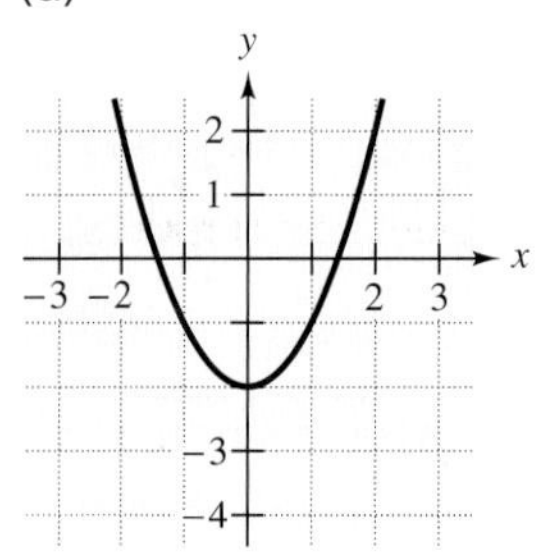

(e)

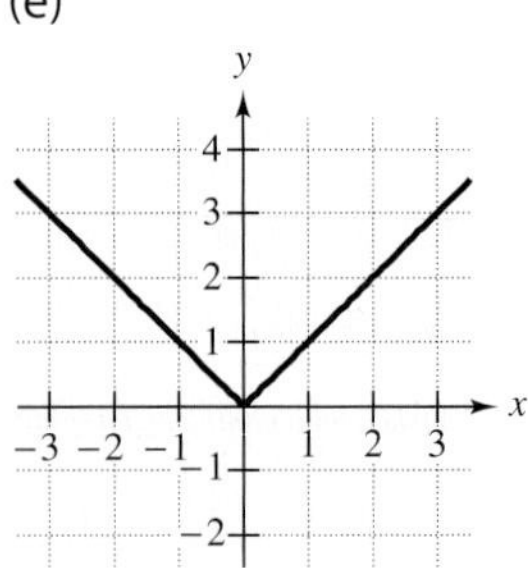

(f)

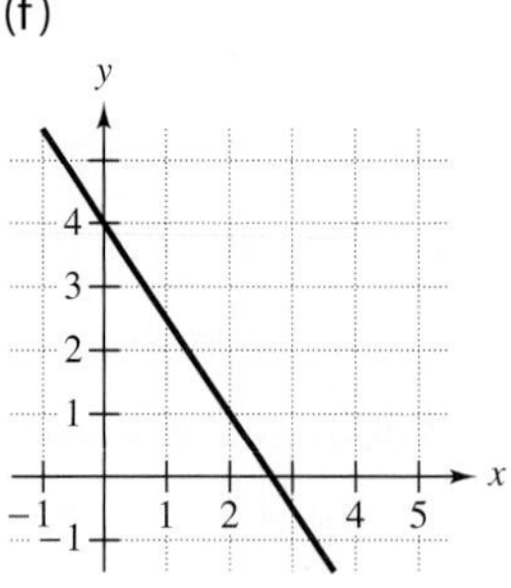

(g)

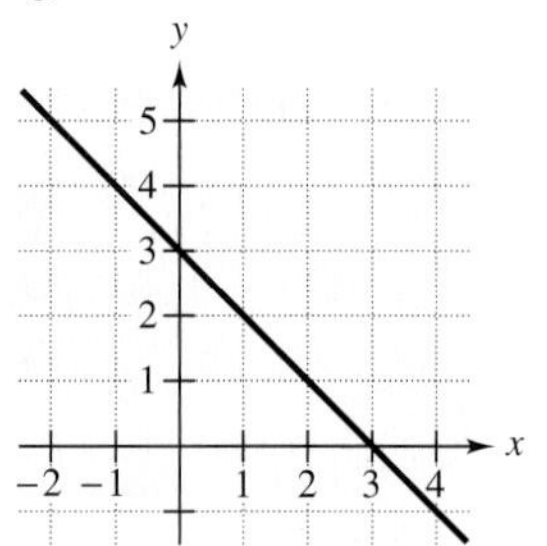

(h)

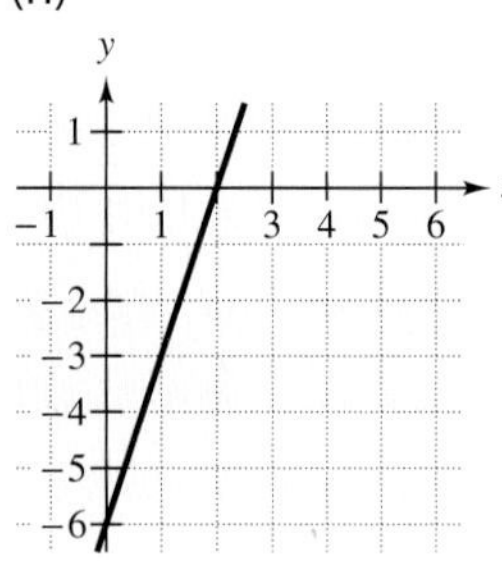

1. $y = 3 - x$
2. $y = \frac{1}{2}x + 1$
3. $y = -x^2 + 1$
4. $y = |x|$
5. $y = 3x - 6$
6. $y = |x - 2|$
7. $y = x^2 - 2$
8. $y = 4 - \frac{3}{2}x$

In Exercises 9–16, complete the table and use the results to sketch the graph of the equation. ***See Examples 1–3.***

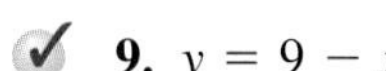

9. $y = 9 - x$

x	-2	-1	0	1	2
y					

10. $y = x - 1$

x	-2	-1	0	1	2
y					

11. $x + 2y = 4$

x	-2	0	2	4	6
y					

12. $3x - 2y = 6$

x	-2	0	2	4	6
y					

✓ **13.** $y = x^2 + 3$

x	-2	-1	0	1	2
y					

14. $x^2 + y = -1$

x	-2	-1	0	1	2
y					

✓ **15.** $y = |x + 1|$

x	-3	-2	-1	0	1
y					

16. $y = |x| - 2$

x	-2	-1	0	1	2
y					

In Exercises 17–24, graphically estimate the x- and y-intercepts of the graph. Then check your results algebraically.

17. $4x - 2y = -8$

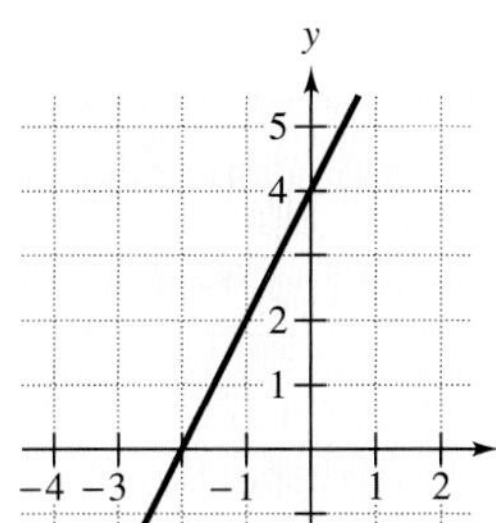

18. $5y - 2x = 10$

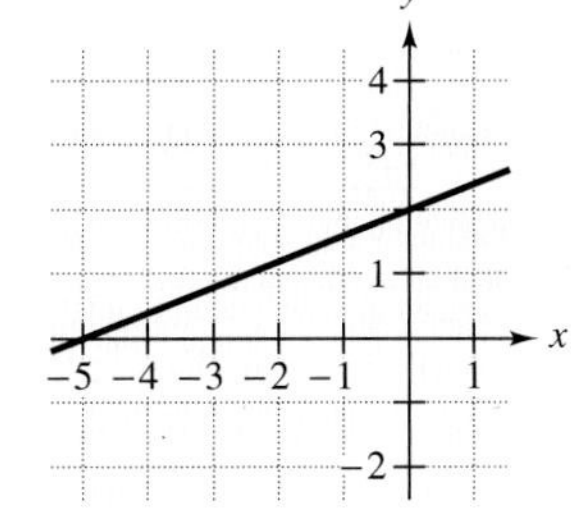

19. $x + 3y = 6$

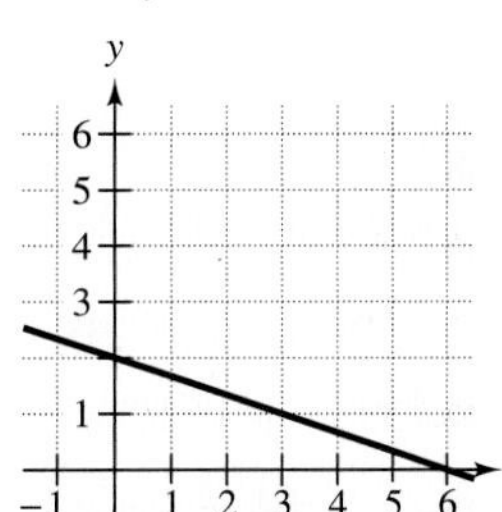

20. $4x + 3y = 12$

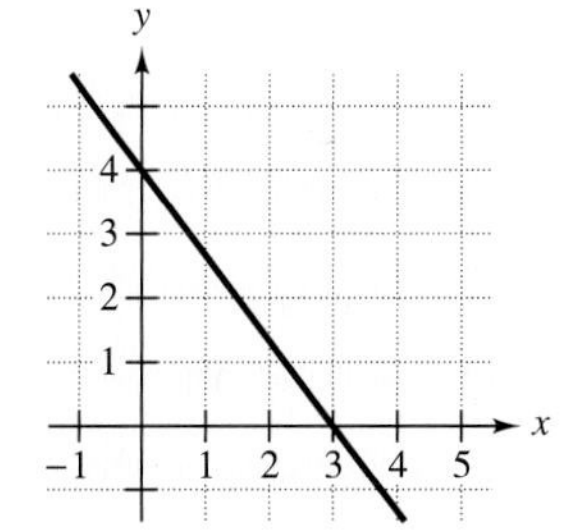

21. $y = |x| - 3$

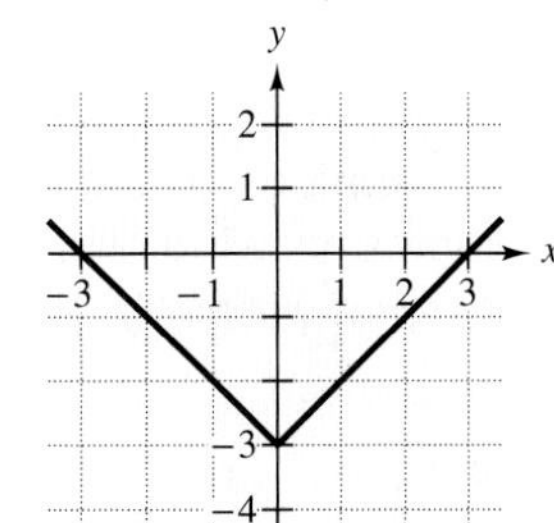

22. $y = 4 - |x|$

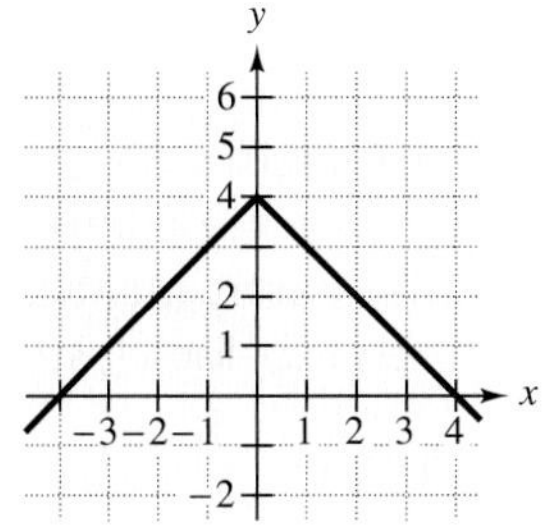

23. $y = 16 - x^2$

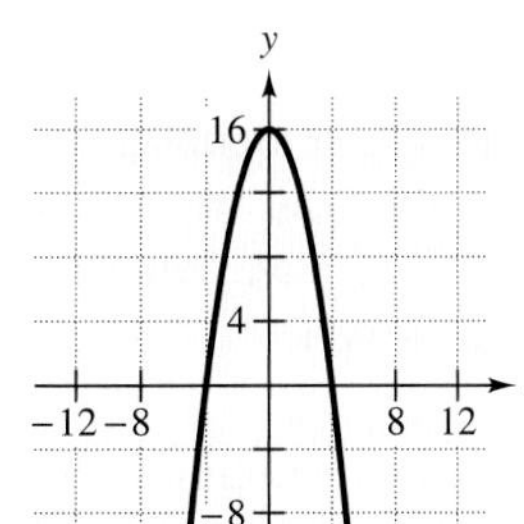

24. $y = x^2 - 4$

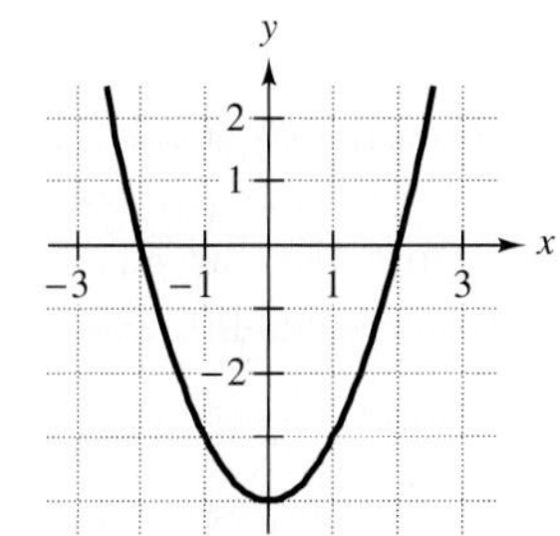

In Exercises 25–36, find the x- and y-intercepts (if any) of the graph of the equation. ***See Example 4.***

25. $y = -2x + 7$

26. $y = 5x - 3$

✓ **27.** $y = \frac{1}{2}x - 1$

28. $y = -\frac{1}{2}x + 3$

29. $x - y = 1$

30. $x + y = 10$

31. $2x + y = -2$

32. $3x - 2y = 1$

33. $2x + 6y - 9 = 0$

34. $2x - 5y + 50 = 0$

35. $\frac{3}{4}x - \frac{1}{2}y = 3$

36. $\frac{1}{2}x + \frac{2}{3}y = 1$

In Exercises 37–52, sketch the graph of the equation and label the coordinates of at least three solution points.

37. $y = 2 - x$

38. $y = x + 8$

39. $y = 3x$

40. $y = -2x$

41. $4x + y = 6$

42. $10x + 5y = 20$

43. $7x + 7y = 14$

44. $2x - y = 5$

45. $y = \frac{3}{8}x + 15$

46. $y = 14 - \frac{2}{3}x$

47. $y = -x^2 + 9$

48. $y = x^2 - 1$

49. $y = |x - 5|$

50. $y = |x + 3|$

51. $y = 5 - |x|$

52. $y = |x| + 3$

In Exercises 53–56, use a graphing calculator to graph both equations in the same viewing window. Are the graphs identical? If so, what property of real numbers is being illustrated?

53. $y_1 = \frac{1}{3}x - 1$
$y_2 = 1 - \frac{1}{3}x$

54. $y_1 = 3\left(\frac{1}{4}x\right)$
$y_2 = \left(3 \cdot \frac{1}{4}\right)x$

55. $y_1 = 2(x - 2)$
$y_2 = 2x - 4$

56. $y_1 = 2 + (x + 4)$
$y_2 = (2 + x) + 4$

In Exercises 57–64, use a graphing calculator to graph the equation. Is it appropriate to use a standard viewing window? If not, describe an appropriate viewing window.

57. $y = 25 - 5x$

58. $y = \frac{2}{3}x - 6$

59. $y = 1.25x + 1.8$

60. $y = 1.7 - 0.1x$

61. $y = \frac{1}{4}x^2 - 4x + 12$

62. $y = 16 - 4x - x^2$

63. $y = |x + 3| - 4$

64. $y = 4 - |x - 7|$

Solving Problems

65. *Creating a Model* Let y represent the distance traveled by a car that is moving at a constant speed of 35 miles per hour. Let t represent the number of hours the car has traveled. Write an equation that relates y and t, and sketch its graph.

66. *Creating a Model* The cost of printing a book is \$500, plus \$5 per book. Let C represent the total cost and let x represent the number of books. Write an equation that relates C and x, and sketch its graph.

✓ **67.** *Hot-Air Balloon* A hot-air balloon at 1120 feet descends at a rate of 80 feet per minute. Let y represent the height of the balloon and let x represent the number of minutes the balloon descends.

(a) Write an equation that relates the height of the hot-air balloon and the number of minutes it descends.

(b) Sketch the graph of the equation.

(c) What is the y-intercept of the graph, and what does it represent in the context of the problem?

68. *Fitness* You run and walk on a trail that is 6 miles long. You run 4 miles per hour and walk 3 miles per hour. Let y be the number of hours you walk and let x be the number of hours you run.

(a) Write an equation that relates the number of hours you run and the number of hours you walk to the total length of the trail.

(b) Sketch the graph of the equation.

(c) What is the y-intercept of the graph, and what does it represent in the context of the problem?

69. *Life Expectancy* The table shows the life expectancies y (in years) in the United States for a male child at birth for the years 1999 through 2004.

Year	1999	2000	2001	2002	2003	2004
y	73.9	74.3	74.4	74.5	74.7	75.2

A model for this data is $y = 0.22t + 71.9$, where t is the time in years, with $t = 9$ corresponding to 1999. (Source: U.S. National Center for Health Statistics)

(a) Plot the data and graph the model on the same set of coordinate axes.

(b) Use the model to predict the life expectancy for a male child born in 2015.

70. *Life Expectancy* The table shows the life expectancies y (in years) in the United States for a female child at birth for the years 1999 through 2004.

Year	1999	2000	2001	2002	2003	2004
y	79.4	79.7	79.8	79.9	80.0	80.4

A model for this data is $y = 0.17t + 77.9$, where t is the time in years, with $t = 9$ corresponding to 1999. (Source: U.S. National Center for Health Statistics)

(a) Plot the data and graph the model on the same set of coordinate axes.

(b) Use the model to predict the life expectancy for a female child born in 2015.

Explaining Concepts

71. Do all graphs of linear equations in two variables have a y-intercept? Explain.

72. If the graph of a linear equation in two variables has a negative x-intercept and a positive y-intercept, does the line rise or fall from left to right? Through which quadrant(s) does the line pass? Use a graph to illustrate your answer.

73. You walk toward a tree at a constant speed. Let x represent the time (in seconds) and let y represent the distance (in feet) between you and the tree. Sketch a possible graph of this situation. Explain how x and y are related. What does the x-intercept mean?

74. How many solution points does a linear equation in two variables have? Explain.

Cumulative Review

In Exercises 75–78, evaluate the expression.

75. $-4 + (-7) - 3 + 1$

76. $-6 + 3 - (-1) + 11$

77. $-(-3) + 5 - 4 + 9$

78. $-18 - (-6) + 2 - 8$

In Exercises 79–84, evaluate the expression and write the result in simplest form.

79. $\frac{3}{4}\left(-\frac{2}{9}\right)$

80. $\left(-\frac{1}{6}\right)\left(-\frac{8}{15}\right)$

81. $\left(-\frac{7}{12}\right)\left(-\frac{18}{35}\right)$

82. $-\frac{6}{7} \div \frac{5}{21}$

83. $\frac{12}{5} \div \left(-\frac{1}{3}\right)$

84. $\left(-\frac{16}{25}\right) \div \left(-\frac{4}{5}\right)$

In Exercises 85–88, determine whether the ordered pairs are solutions of the equation.

85. $y = 3x - 5$ (a) $(0, 5)$ (b) $(-1, -2)$ (c) $(3, 4)$ (d) $(-2, -11)$

86. $y = 2x + 1$ (a) $(-3, -5)$ (b) $(-1, 3)$ (c) $(5, 8)$ (d) $(2, 5)$

87. $3y - 4x = 7$ (a) $(1, 1)$ (b) $(-5, 9)$ (c) $(4, -3)$ (d) $(7, 7)$

88. $x - 2y = -2$ (a) $(-6, 2)$ (b) $(-2, 2)$ (c) $(4, 3)$ (d) $(2, 0)$

4.3 Relations, Functions, and Graphs

Robert Grubbs/Photo Network

What You Should Learn

1 ▶ Identify the domain and range of a relation.
2 ▶ Determine if relations are functions by inspection or by using the Vertical Line Test.
3 ▶ Use function notation and evaluate functions.
4 ▶ Identify the domain of a function.

Why You Should Learn It

Relations and functions can be used to describe real-life situations. For instance, in Exercise 71 on page 247, a relation is used to model the length of time between sunrise and sunset in Erie, Pennsylvania.

1 ▶ Identify the domain and range of a relation.

Relations

Many everyday occurrences involve pairs of quantities that are matched with each other by some rule of correspondence. For instance, each person is matched with a birth month (person, month); the number of hours worked is matched with a paycheck (hours, pay); an instructor is matched with a course (instructor, course); and the time of day is matched with the outside temperature (time, temperature). In each instance, sets of ordered pairs can be formed. Such sets of ordered pairs are called **relations.**

Definition of Relation

A **relation** is any set of ordered pairs. The set of first components in the ordered pairs is the **domain** of the relation. The set of second components is the **range** of the relation.

In mathematics, relations are commonly described by ordered pairs of *numbers*. The set of x-coordinates is the domain, and the set of y-coordinates is the range. In the relation $\{(3, 5), (1, 2), (4, 4), (0, 3)\}$, the domain D and range R are the sets $D = \{3, 1, 4, 0\}$ and $R = \{5, 2, 4, 3\}$.

EXAMPLE 1 Analyzing a Relation

Find the domain and range of the relation $\{(0, 1), (1, 3), (2, 5), (3, 5), (0, 3)\}$. Then sketch a graphical representation of the relation.

Solution

The domain is the set of all first components of the relation, and the range is the set of all second components.

$$D = \{0, 1, 2, 3\} \quad \text{and} \quad R = \{1, 3, 5\}$$

A graphical representation of the relation is shown in Figure 4.21.

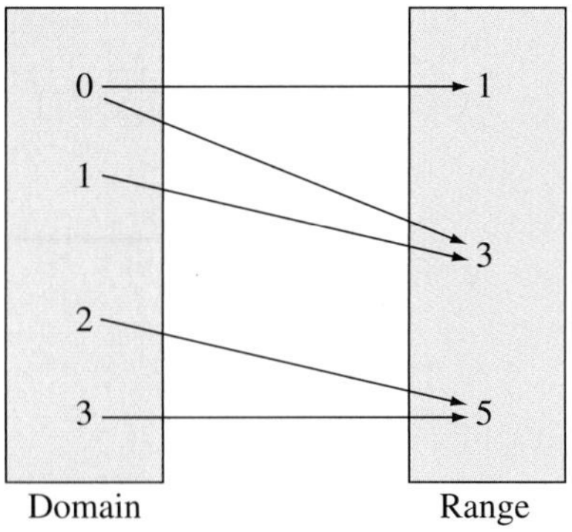

Figure 4.21

✓ ***CHECKPOINT*** *Now try Exercise 1.*

You should note that it is not necessary to list repeated components of the domain and range of a relation.

2 ▶ Determine if relations are functions by inspection or by using the Vertical Line Test.

Functions

In the study of mathematics and its applications, the focus is mainly on a special type of relation called a **function.**

Definition of Function

A **function** is a relation in which no two ordered pairs have the same first component and different second components.

This definition means that a given first component cannot be paired with two different second components. For instance, the pairs (1, 3) and (1, −1) could not be ordered pairs of a function.

Consider the relations described at the beginning of this section.

Relation	*Ordered Pairs*	*Sample Relation*
1	(person, month)	{(A, May), (B, Dec), (C, Oct), . . .}
2	(hours, pay)	{(12, 84), (4, 28), (6, 42), (15, 105), . . .}
3	(instructor, course)	{(A, MATH001), (A, MATH002), . . .}
4	(time, temperature)	{(8, 70°), (10, 78°), (12, 78°), . . .}

The first relation *is* a function because each person has only one birth month. The second relation *is* a function because the number of hours worked at a particular job can yield only *one* paycheck amount. The third relation *is not* a function because an instructor can teach more than one course. The fourth relation *is* a function. Note that the ordered pairs (10, 78°) and (12, 78°) do not violate the definition of a function.

Study Tip

The ordered pairs of a relation can be thought of in the form (input, output). For a *function*, a given input cannot yield two different outputs. For instance, if the input is a person's name and the output is that person's month of birth, then your name as the input can yield only your month of birth as the output.

EXAMPLE 2 Testing Whether a Relation Is a Function

Decide whether each relation represents a function.

a. Input: a, b, c

Output: 2, 3, 4

$\{(a, 2), (b, 3), (c, 4)\}$

b.

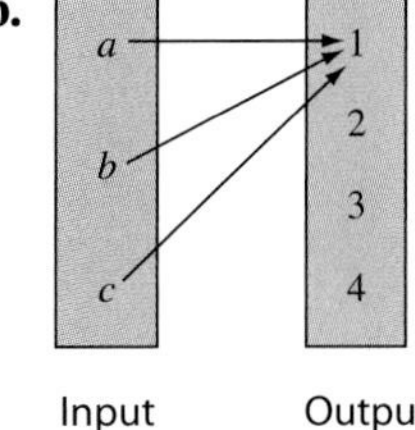

c.

Input x	Output y	(x, y)
3	1	(3, 1)
4	3	(4, 3)
5	4	(5, 4)
3	2	(3, 2)

Solution

a. This set of ordered pairs *does* represent a function. No first component has two different second components.

b. This diagram *does* represent a function. No first component has two different second components.

c. This table *does not* represent a function. The first component 3 is paired with two different second components, 1 and 2.

Now try Exercise 7.

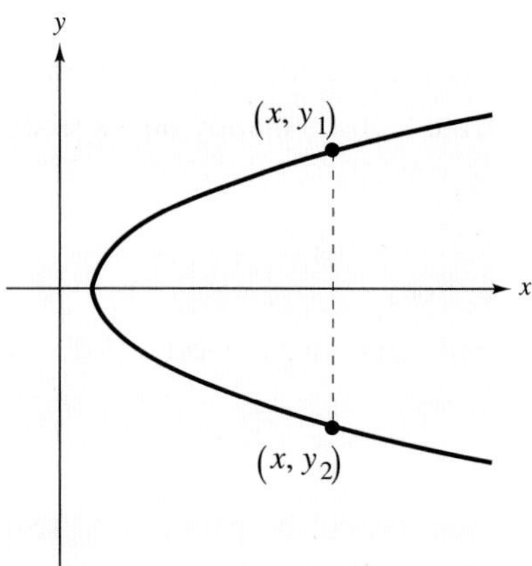

Figure 4.22

In algebra, it is common to represent functions by equations in two variables rather than by ordered pairs. For instance, the equation $y = x^2$ represents the variable y as a function of x. The variable x is the **independent variable** (the input) and y is the **dependent variable** (the output). In this context, the domain of the function is the set of all *allowable* values of x, and the range is the *resulting* set of all values taken on by the dependent variable y.

From the graph of an equation, it is easy to determine whether the equation represents y as a function of x. The graph in Figure 4.22 *does not* represent a function of x because the indicated value of x is paired with two y-values. Graphically, this means that a vertical line intersects the graph more than once.

Vertical Line Test

A set of points on a rectangular coordinate system is the graph of y as a function of x if and only if no vertical line intersects the graph at more than one point.

EXAMPLE 3 Using the Vertical Line Test for Functions

Use the Vertical Line Test to determine whether y is a function of x.

a.

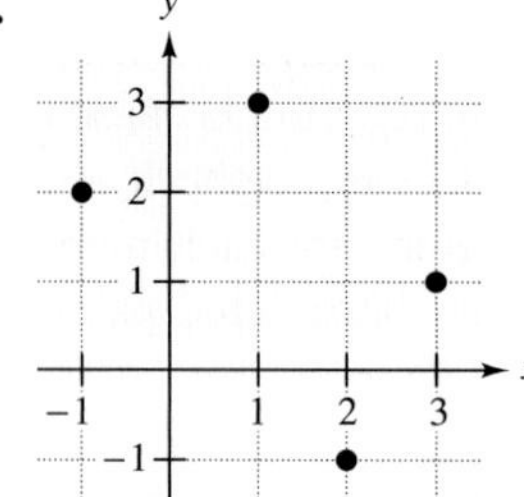

b.

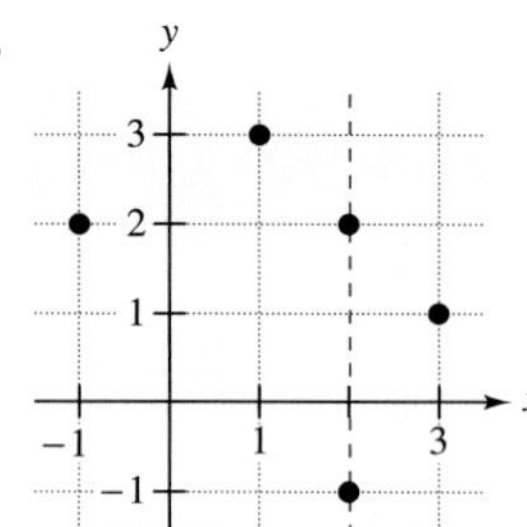

c.

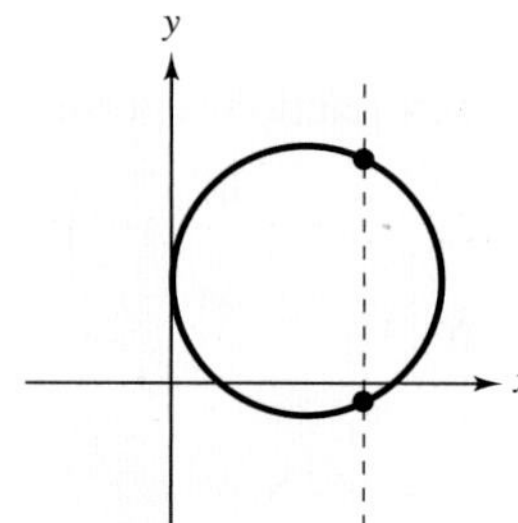

d.

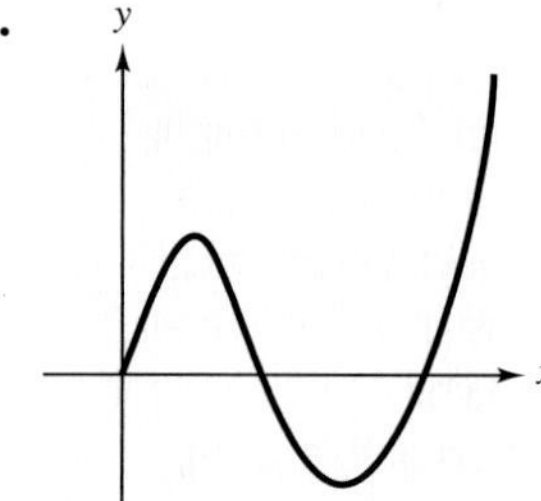

Solution

a. From the graph, you can see that no vertical line intersects more than one point on the graph. So, the relation *does* represent y as a function of x.

b. From the graph, you can see that a vertical line intersects more than one point on the graph. So, the relation *does not* represent y as a function of x.

c. From the graph, you can see that a vertical line intersects more than one point on the graph. So, the relation *does not* represent y as a function of x.

d. From the graph, you can see that no vertical line intersects more than one point on the graph. So, the relation *does* represent y as a function of x.

3 ▶ Use function notation and evaluate functions.

Function Notation

To discuss functions represented by equations, it is common practice to give them names using **function notation.** For instance, the function

$$y = 2x - 6$$

can be given the name "f" and written in function notation as

$$f(x) = 2x - 6.$$

Function Notation

In the notation $f(x)$:

f is the **name** of the function.
x is a **domain** (or input) value.
$f(x)$ is a **range** (or output) value y for a given x.

The symbol $f(x)$ is read as *the value of f at x* or simply *f of x.*

The process of finding the value of $f(x)$ for a given value of x is called **evaluating a function.** This is accomplished by substituting a given x-value (input) into the equation to obtain the value of $f(x)$ (output). Here is an example.

Function	*x-Values (input)*	*Function Values (output)*
$f(x) = 4 - 3x$	$x = -2$	$f(-2) = 4 - 3(-2) = 4 + 6 = 10$
	$x = -1$	$f(-1) = 4 - 3(-1) = 4 + 3 = 7$
	$x = 0$	$f(0) = 4 - 3(0) = 4 - 0 = 4$
	$x = 2$	$f(2) = 4 - 3(2) = 4 - 6 = -2$
	$x = 3$	$f(3) = 4 - 3(3) = 4 - 9 = -5$

Although f and x are often used as a convenient function name and independent (input) variable, you can use other letters. For instance, the equations

$$f(x) = x^2 - 3x + 5, \quad f(t) = t^2 - 3t + 5, \quad \text{and} \quad g(s) = s^2 - 3s + 5$$

all define the same function. In fact, the letters used are just "placeholders" and this same function is well described by the form

$$f(\quad) = (\quad)^2 - 3(\quad) + 5$$

where the parentheses are used in place of a letter. To evaluate $f(-2)$, simply place -2 in each set of parentheses, as follows.

$$\begin{aligned} f(-2) &= (-2)^2 - 3(-2) + 5 \\ &= 4 + 6 + 5 \\ &= 15 \end{aligned}$$

It is important to put parentheses around the x-value (input) and then simplify the result.

EXAMPLE 4 Evaluating a Function

Let $f(x) = x^2 + 1$. Find each value of the function.

a. $f(-2)$ **b.** $f(0)$

Solution

a. $f(x) = x^2 + 1$ Write original function.

$f(-2) = (-2)^2 + 1$ Substitute -2 for x.

$= 4 + 1 = 5$ Simplify.

b. $f(x) = x^2 + 1$ Write original function.

$f(0) = (0)^2 + 1$ Substitute 0 for x.

$= 0 + 1 = 1$ Simplify.

✓ ***CHECKPOINT*** *Now try Exercise 45.*

EXAMPLE 5 Evaluating a Function

Let $g(x) = 3x - x^2$. Find each value of the function.

a. $g(2)$ **b.** $g(0)$

Solution

a. Substituting 2 for x produces $g(2) = 3(2) - (2)^2 = 6 - 4 = 2$.

b. Substituting 0 for x produces $g(0) = 3(0) - (0)^2 = 0 - 0 = 0$.

✓ ***CHECKPOINT*** *Now try Exercise 47.*

4 ▶ Identify the domain of a function.

Finding the Domain of a Function

The domain of a function may be explicitly described along with the function, or it may be *implied* by the context in which the function is used. For instance, if weekly pay is a function of hours worked (for a 40-hour work week), the implied domain is $0 \le x \le 40$. Certainly x cannot be negative in this context.

EXAMPLE 6 Finding the Domain of a Function

Find the domain of each function.

a. f:$\{(-3, 0), (-1, 2), (0, 4), (2, 4), (4, -1)\}$

b. Area of a square: $A = s^2$

Solution

a. The domain of f consists of all first components in the set of ordered pairs. So, the domain is $\{-3, -1, 0, 2, 4\}$.

b. For the area of a square, you must choose positive values for the side s. So, the domain is the set of all real numbers s such that $s > 0$.

✓ ***CHECKPOINT*** *Now try Exercise 53.*

Concept Check

1. Explain the difference between a relation and a function.

2. Explain the meanings of the terms *domain* and *range* in the context of a function.

3. In your own words, explain how to use the Vertical Line Test.

4. What is the meaning of the notation $f(3)$?

4.3 EXERCISES

Go to pages 284–285 to record your assignments.

Developing Skills

In Exercises 1–6, find the domain and range of the relation. ***See Example 1.***

✓ **1.** $\{(-4, 3), (2, 5), (1, 2), (4, -3)\}$

2. $\{(-1, 5), (8, 3), (4, 6), (-5, -2)\}$

3. $\{(2, 16), (-9, -10), (\frac{1}{2}, 0)\}$

4. $\{(\frac{2}{3}, -4), (-6, \frac{1}{4}), (0, 0)\}$

5. $\{(-1, 3), (5, -7), (-1, 4), (8, -2), (1, -7)\}$

6. $\{(1, 1), (2, 4), (3, 9), (-2, 4), (-1, 1)\}$

In Exercises 7–26, determine whether the relation represents a function. ***See Example 2.***

✓ **7.** **8.**

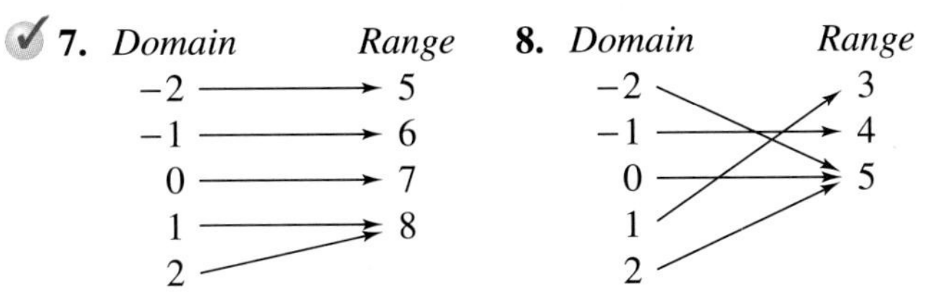

9. **10.**

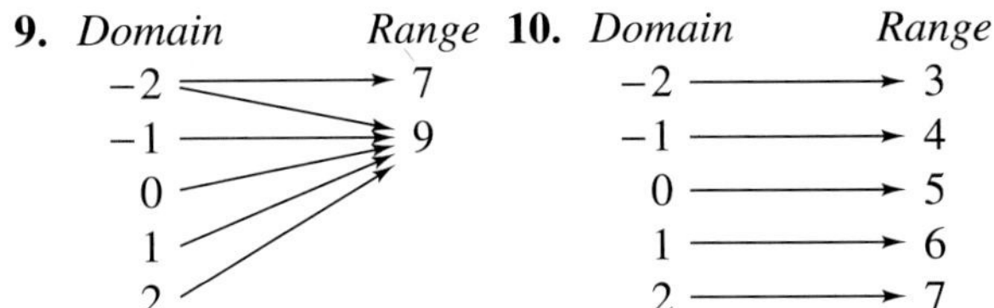

11. **12.**

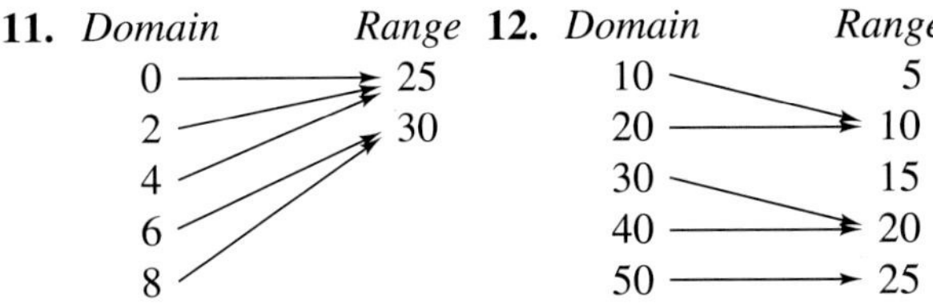

13. **14.**

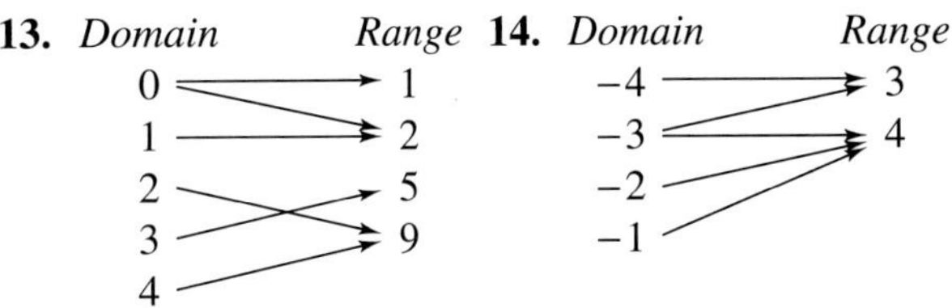

15.

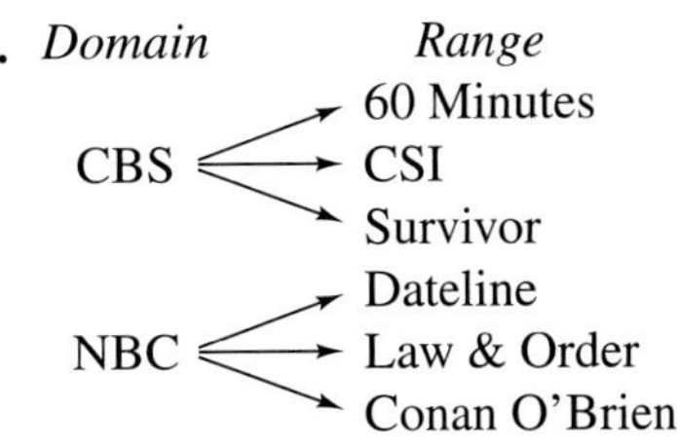

16.

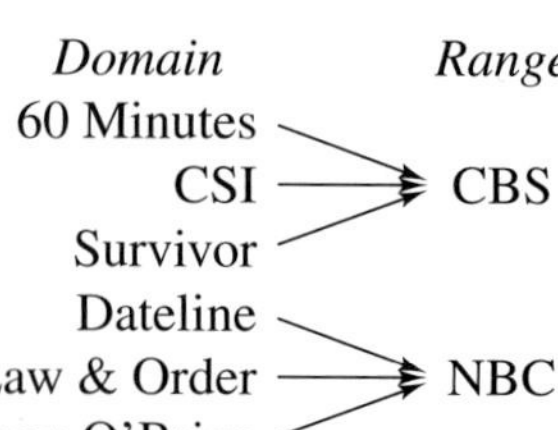

17. *Domain* *Range*

Year	Single women in the labor force (in percent)
2002	67.4
2003	66.2
2004	65.9
2005	66.0

(Source: U.S. Bureau of Labor Statistics)

18. *Domain* — Percent daily value of vitamin C per serving; *Range* — Cereal

10% → Corn Flakes, Wheaties, Cheerios
100% → Cheerios, Total

19.

Input x	Output y	(x, y)
0	2	(0, 2)
1	4	(1, 4)
2	6	(2, 6)
3	8	(3, 8)
4	10	(4, 10)

20.

Input x	Output y	(x, y)
0	2	(0, 2)
1	4	(1, 4)
2	6	(2, 6)
1	8	(1, 8)
0	10	(0, 10)

21.

Input x	Output y	(x, y)
1	1	(1, 1)
3	2	(3, 2)
5	3	(5, 3)
3	4	(3, 4)
1	5	(1, 5)

22.

Input x	Output y	(x, y)
2	1	(2, 1)
4	1	(4, 1)
6	1	(6, 1)
8	1	(8, 1)
10	1	(10, 1)

23. $\{(0, 25), (2, 25), (4, 30), (6, 30), (8, 30)\}$

24. $\{(10, 5), (20, 10), (30, 15), (40, 20), (50, 25)\}$

25. Input: a, b, c
Output: 0, 4, 9
$\{(a, 0), (b, 4), (c, 9)\}$

26. Input: 3, 5, 7
Output: d, e, f
$\{(3, d)\ (5, e), (7, f), (7, d)\}$

In Exercises 27–36, use the Vertical Line Test to determine whether y is a function of x. ***See Example 3.***

✓ **27.**

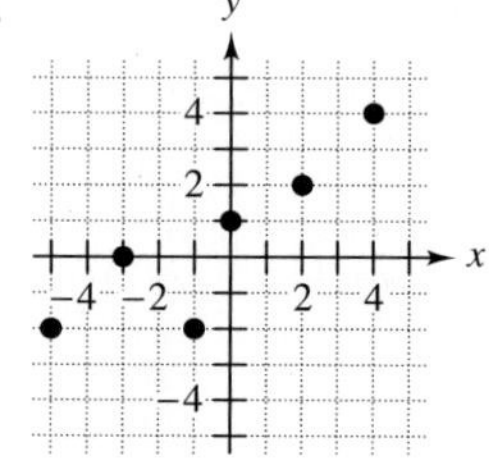

28.

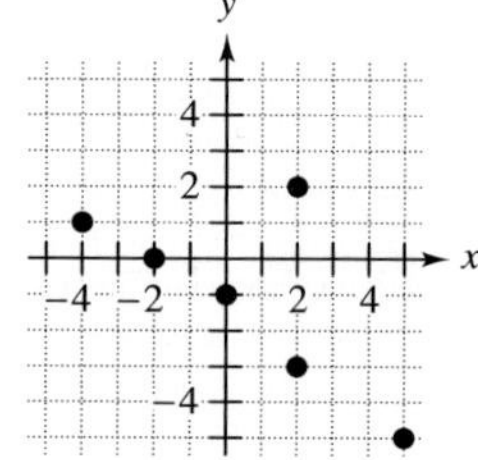

29.

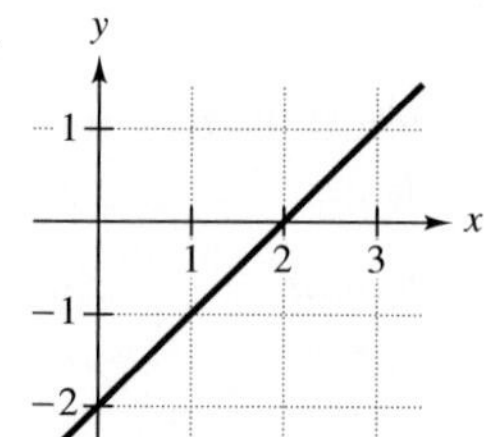

30.

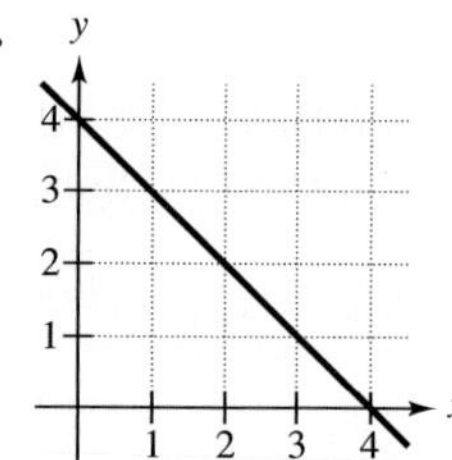

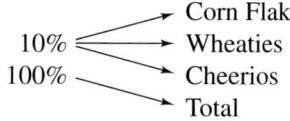

31.

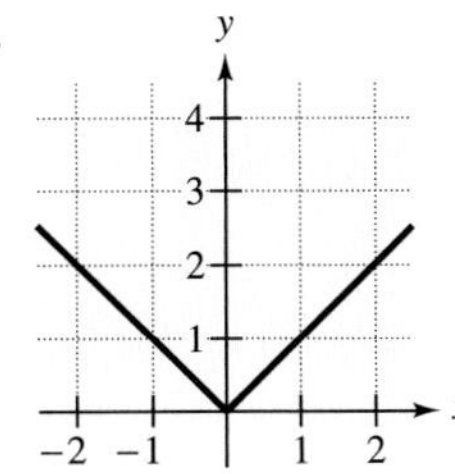

32.

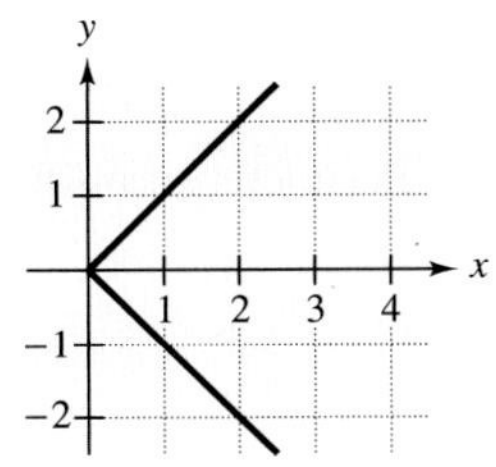

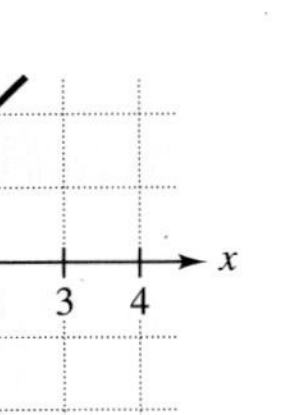

33.

34.

35.

36.

In Exercises 37–52, evaluate the function as indicated, and simplify. ***See Examples 4 and 5.***

37. $f(x) = \frac{1}{2}x$ (a) $f(2)$ (b) $f(5)$ (c) $f(-4)$ (d) $f\left(-\frac{2}{3}\right)$

38. $g(x) = -\frac{4}{5}x$ (a) $g(5)$ (b) $g(0)$ (c) $g(-3)$ (d) $g\left(-\frac{5}{4}\right)$

39. $f(x) = 2x - 1$ (a) $f(0)$ (b) $f(3)$ (c) $f(-3)$ (d) $f\left(-\frac{1}{2}\right)$

40. $f(t) = 3 - 4t$ (a) $f(0)$ (b) $f(1)$ (c) $f(-2)$ (d) $f\left(\frac{3}{4}\right)$

41. $h(t) = \frac{1}{4}t - 1$ (a) $h(200)$ (b) $h(-12)$ (c) $h(8)$ (d) $h\left(-\frac{5}{2}\right)$

42. $f(s) = 4 - \frac{2}{3}s$ (a) $f(60)$ (b) $f(-15)$ (c) $f(-18)$ (d) $f\left(\frac{1}{2}\right)$

43. $f(v) = \frac{1}{2}v^2$ (a) $f(-4)$ (b) $f(4)$ (c) $f(0)$ (d) $f(2)$

44. $g(u) = -2u^2$ (a) $g(0)$ (b) $g(2)$ (c) $g(3)$ (d) $g(-4)$

✓ **45.** $f(x) = 4x^2 + 2$ (a) $f(1)$ (b) $f(-1)$ (c) $f(-4)$ (d) $f\left(-\frac{3}{2}\right)$

46. $g(t) = 5 - 2t^2$ (a) $g\left(\frac{5}{2}\right)$ (b) $g(-10)$ (c) $g(0)$ (d) $g\left(\frac{3}{4}\right)$

✓ **47.** $g(x) = 2x^2 - 3x + 1$ (a) $g(0)$ (b) $g(-2)$ (c) $g(1)$ (d) $g\left(\frac{1}{2}\right)$

48. $h(x) = 1 - 4x - x^2$ (a) $h(0)$ (b) $h(-4)$ (c) $h(10)$ (d) $h\left(\frac{3}{2}\right)$

49. $g(u) = |u + 2|$ (a) $g(2)$ (b) $g(-2)$ (c) $g(10)$ (d) $g\left(-\frac{5}{2}\right)$

50. $h(s) = |s| + 2$ (a) $h(4)$ (b) $h(-10)$ (c) $h(-2)$ (d) $h\left(\frac{3}{2}\right)$

51. $h(x) = x^3 - 1$ (a) $h(0)$ (b) $h(1)$ (c) $h(3)$ (d) $h\left(\frac{1}{2}\right)$

52. $f(x) = 16 - x^4$ (a) $f(-2)$ (b) $f(2)$ (c) $f(1)$ (d) $f(3)$

In Exercises 53–60, find the domain of the function. ***See Example 6.***

✓ **53.** $f:\{(0, 4), (1, 3), (2, 2), (3, 1), (4, 0)\}$

54. $f:\{(-2, -1), (-1, 0), (0, 1), (1, 2), (2, 3)\}$

55. $g:\{(-8, -1), (-6, 0), (2, 7), (5, 0), (12, 10)\}$

56. $g:\{(-4, 4), (3, 8), (4, 5), (9, -2), (10, -7)\}$

57. $h:\{(-5, 2), (-4, 2), (-3, 2), (-2, 2), (-1, 2)\}$

58. $h:\{(10, 100), (20, 200), (30, 300), (40, 400)\}$

59. Area of a circle: $A = \pi r^2$

60. Perimeter of a square: $P = 4s$

Solving Problems

61. *Demand* The demand for a product is a function of its price. Consider the demand function

$f(p) = 20 - 0.5p$

where p is the price in dollars.

(a) Find $f(10)$ and $f(15)$.

(b) Describe the effect a price increase has on demand.

62. *Maximum Load* The maximum safe load L (in pounds) for a wooden beam 2 inches wide and d inches high is $L(d) = 100d^2$.

(a) Complete the table.

d	2	4	6	8
$L(d)$				

(b) Describe the effect of an increase in height on the maximum safe load.

63. *Distance* The function $d(t) = 50t$ gives the distance (in miles) that a car will travel in t hours at an average speed of 50 miles per hour. Find the distance traveled for (a) $t = 2$, (b) $t = 4$, and (c) $t = 10$.

64. *Speed of Sound* The function $S(h) = 1116 - 4.04h$ approximates the speed of sound (in feet per second) at altitude h (in thousands of feet). Use the function to approximate the speed of sound for (a) $h = 0$, (b) $h = 10$, and (c) $h = 30$.

Interpreting a Graph In Exercises 65–68, use the information in the graph. (Source: U.S. National Center for Education Statistics)

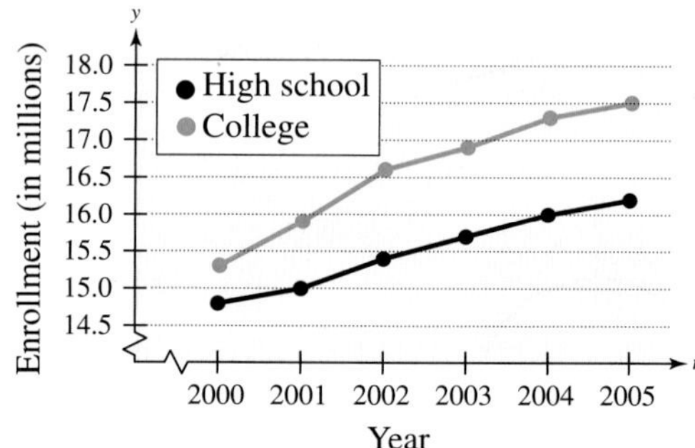

65. Is the high school enrollment a function of the year?

66. Is the college enrollment a function of the year?

67. Let $f(t)$ represent the number of high school students in year t. Find $f(2001)$.

68. Let $g(t)$ represent the number of college students in year t. Find $g(2005)$.

69. *Geometry* Write the formula for the perimeter P of a square with sides of length s. Is P a function of s? Explain.

70. *Geometry* Write the formula for the volume V of a cube with sides of length t. Is V a function of t? Explain.

71. *Sunrise and Sunset* The graph approximates the length of time L (in hours) between sunrise and sunset in Erie, Pennsylvania for the year 2007. The variable t represents the day of the year.

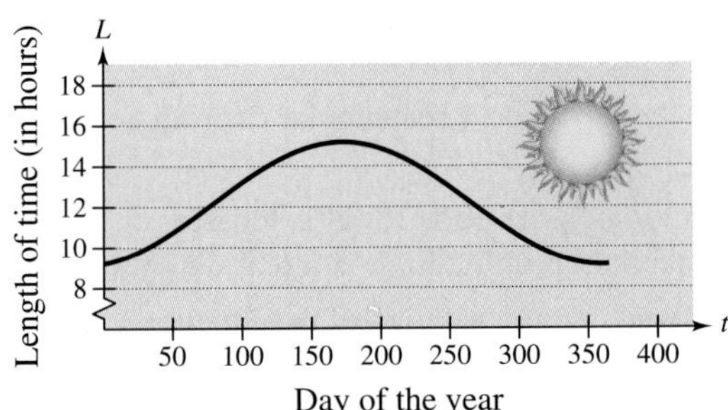

(a) Is the length of time L a function of the day of the year t?

(b) Estimate the range of this relation.

72. *SAT Scores and Grade-Point Average* The graph shows the SAT scores x and the grade-point averages (GPA) y for 12 students.

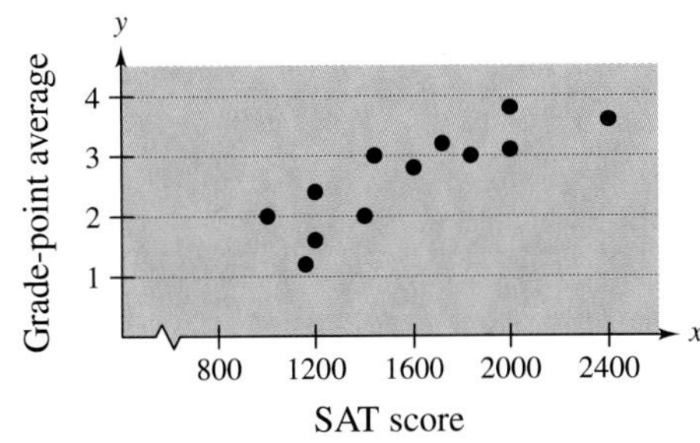

(a) Is the GPA y a function of the SAT score x?

(b) Estimate the range of this relation.

Explaining Concepts

73. Is it possible to find a relation that is not a function? If it is, find one.

74. Is it possible to find a function that is not a relation? If it is, find one.

75. Is it possible for the number of elements in the domain of a relation to be greater than the number of elements in the range of the relation? Explain.

76. Determine whether the statement uses the word *function* in a way that is mathematically correct. Explain your reasoning.

(a) The amount of money in your savings account is a function of your salary.

(b) The speed at which a free-falling baseball strikes the ground is a function of the height from which it is dropped.

Cumulative Review

In Exercises 77–80, rewrite the statement using inequality notation.

77. x is negative.

78. m is at least -3.

79. z is at least 85, but no more than 100.

80. n is less than 20, but no less than 16.

In Exercises 81–88, solve the equation.

81. $|x| = 8$

82. $|g| = -4$

83. $|4h| = 24$

84. $\left|\frac{1}{5}m\right| = 2$

85. $|x + 4| = 5$

86. $|2t - 3| = 11$

87. $|6b + 8| = 2b$

88. $|n - 2| = |2n + 9|$

Mid-Chapter Quiz

Take this quiz as you would take a quiz in class. After you are done, check your work against the answers in the back of the book.

1. Plot the points $(4, -2)$ and $\left(-1, -\frac{5}{2}\right)$ on a rectangular coordinate system.

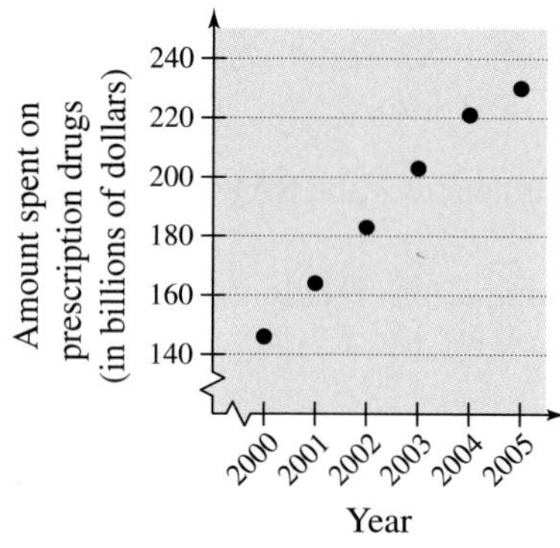

Figure for 4

2. Determine the quadrant(s) in which the point $(3, y)$ is located, or the axis on which the point is located, without plotting it. (y is a real number.)
3. Determine whether each ordered pair is a solution of the equation $y = 9 - |x|$.
   (a) $(2, 7)$ (b) $(-3, 12)$ (c) $(-9, 0)$ (d) $(0, -9)$
4. The scatter plot at the left shows the amounts (in billions of dollars) spent on prescription drugs in the United States for the years 2000 through 2005. Estimate the amount spent on prescription drugs for each year from 2000 to 2005. (Source: National Association of Chain Drug Stores)

In Exercises 5 and 6, find the *x*- and *y*-intercepts of the graph of the equation.

5. $x - 3y = 12$
6. $y = -7x + 2$

In Exercises 7–9, sketch the graph of the equation.

7. $y = 5 - 2x$
8. $y = (x + 2)^2$
9. $y = |x + 3|$

In Exercises 10 and 11, find the domain and range of the relation.

10. $\{(1, 4), (2, 6), (3, 10), (2, 14), (1, 0)\}$
11. $\{(-3, 6), (-2, 6), (-1, 6), (0, 6)\}$
12. Determine whether the relation in the figure is a function of x using the Vertical Line Test.

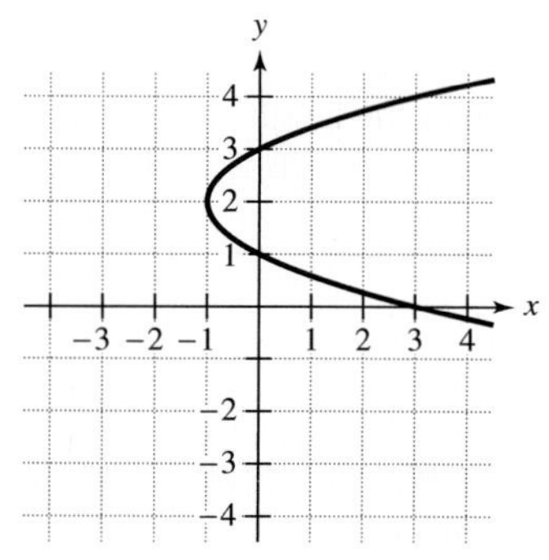

Figure for 12

In Exercises 13 and 14, evaluate the function as indicated, and simplify.

13. $f(x) = 3(x + 2) - 4$
    (a) $f(0)$ (b) $f(-3)$
14. $g(x) = 4 - x^2$
    (a) $g(-1)$ (b) $g(8)$
15. Find the domain of the function f: $\{(10, 1), (15, 3), (20, 9), (25, 27)\}$.
16. Use a graphing calculator to graph $y = 3.6x - 2.4$. Graphically estimate the intercepts of the graph. Explain how to verify your estimates algebraically.
17. A new computer system sells for approximately \$2000 and depreciates at the rate of \$500 per year.
    (a) Find an equation that relates the value of the computer system to the number of years.
    (b) Sketch the graph of the equation.
    (c) What is the *y*-intercept of the graph, and what does it represent in the context of the problem?

4.4 Slope and Graphs of Linear Equations

Stockbyte/Getty Images

What You Should Learn

1 ▶ Determine the slope of a line through two points.

2 ▶ Write linear equations in slope-intercept form and graph the equations.

3 ▶ Use slopes to determine whether lines are parallel, perpendicular, or neither.

Why You Should Learn It

Slopes of lines can be used in many business applications. For instance, in Exercise 94 on page 261, you will interpret the meaning of the slope of a line segment that represents the average price of a troy ounce of gold.

1 ▶ Determine the slope of a line through two points.

The Slope of a Line

The **slope** of a nonvertical line is the number of units the line rises or falls vertically for each unit of horizontal change from left to right. For example, the line in Figure 4.23 rises two units for each unit of horizontal change from left to right, and so this line has a slope of $m = 2$.

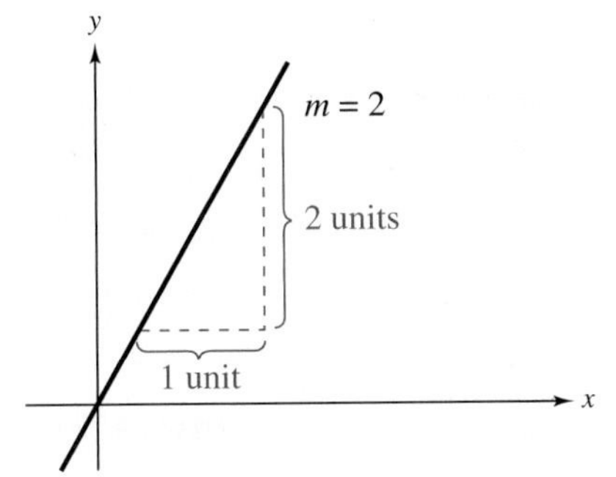

Figure 4.23

Figure 4.24

Study Tip

In the definition at the right, the *rise* is the vertical change between the points and the *run* is the horizontal change between the points.

Definition of the Slope of a Line

The **slope** m of a nonvertical line passing through the points (x_1, y_1) and (x_2, y_2) is

$$m = \frac{y_2 - y_1}{x_2 - x_1} = \frac{\text{Change in } y}{\text{Change in } x} = \frac{\text{Rise}}{\text{Run}}$$

where $x_1 \neq x_2$. (See Figure 4.24.)

When the formula for slope is used, the *order of subtraction* is important. Given two points on a line, you are free to label either of them (x_1, y_1) and the other (x_2, y_2). However, once this has been done, you must form the numerator and denominator using the same order of subtraction.

$m = \frac{y_2 - y_1}{x_2 - x_1}$ Correct

$m = \frac{y_1 - y_2}{x_1 - x_2}$ Correct

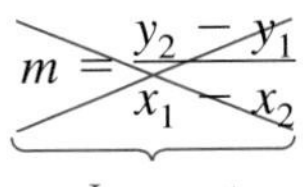

Incorrect

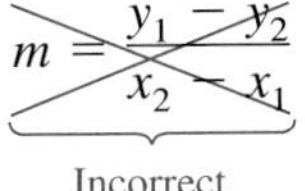

Incorrect

EXAMPLE 1 Finding the Slope of a Line Through Two Points

Find the slope of the line passing through each pair of points.

a. $(-2, 0)$ and $(3, 1)$ **b.** $(0, 0)$ and $(1, -1)$

Solution

a. Let $(x_1, y_1) = (-2, 0)$ and $(x_2, y_2) = (3, 1)$. The slope of the line through these points is

$$m = \frac{y_2 - y_1}{x_2 - x_1}$$

$$= \frac{1 - 0}{3 - (-2)}$$ Difference in y-values / Difference in x-values

$$= \frac{1}{5}.$$ Simplify.

The graph of the line is shown in Figure 4.25.

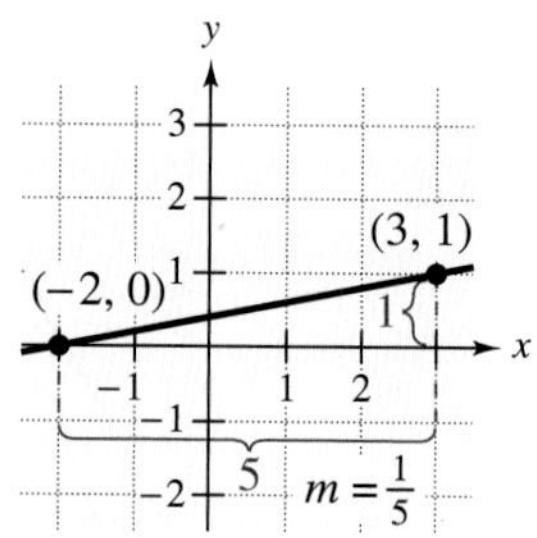

Figure 4.25

b. The slope of the line through $(0, 0)$ and $(1, -1)$ is

$$m = \frac{-1 - 0}{1 - 0}$$ Difference in y-values / Difference in x-values

$$= \frac{-1}{1}$$ Simplify.

$$= -1.$$ Simplify.

The graph of the line is shown in Figure 4.26.

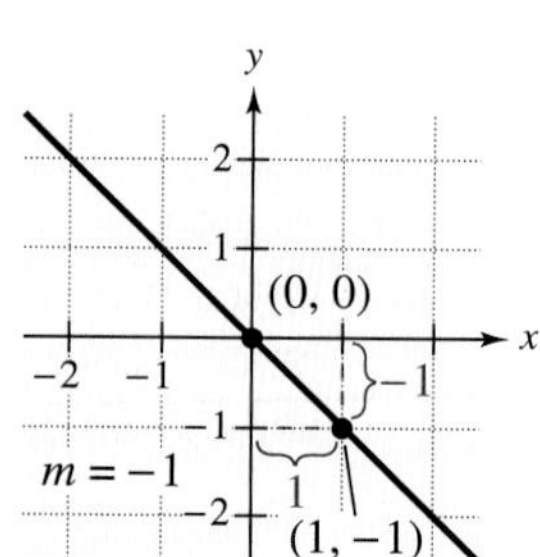

Figure 4.26

✓ ***CHECKPOINT*** *Now try Exercise 11.*

EXAMPLE 2 Horizontal and Vertical Lines and Slope

Find the slope of the line passing through each pair of points.

a. $(-1, 2)$ and $(2, 2)$ **b.** $(2, 4)$ and $(2, 1)$

Solution

a. The line through $(-1, 2)$ and $(2, 2)$ is horizontal because its y-coordinates are the same. The slope of this horizontal line is

$$m = \frac{2 - 2}{2 - (-1)} \qquad \text{Difference in } y\text{-values} \over \text{Difference in } x\text{-values}$$

$$= \frac{0}{3} \qquad \text{Simplify.}$$

$$= 0. \qquad \text{Simplify.}$$

The graph of the line is shown in Figure 4.27.

b. The line through $(2, 4)$ and $(2, 1)$ is vertical because its x-coordinates are the same. Applying the formula for slope, you have

$$\frac{4 - 1}{2 - 2} = \frac{3}{0}. \qquad \text{Division by 0 is undefined.}$$

Because division by zero is not defined, the slope of a vertical line is not defined. The graph of the line is shown in Figure 4.28.

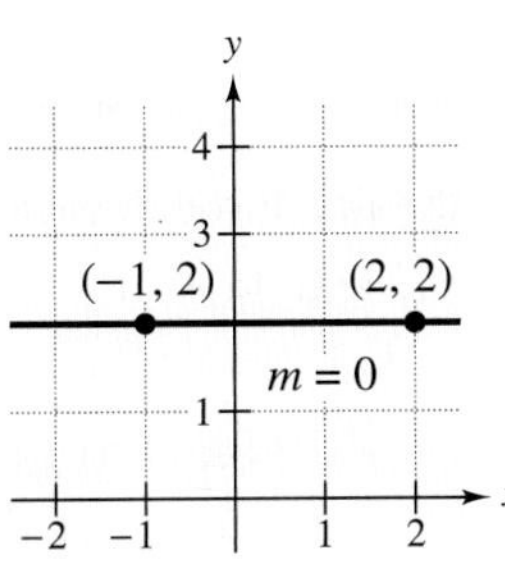

Figure 4.27

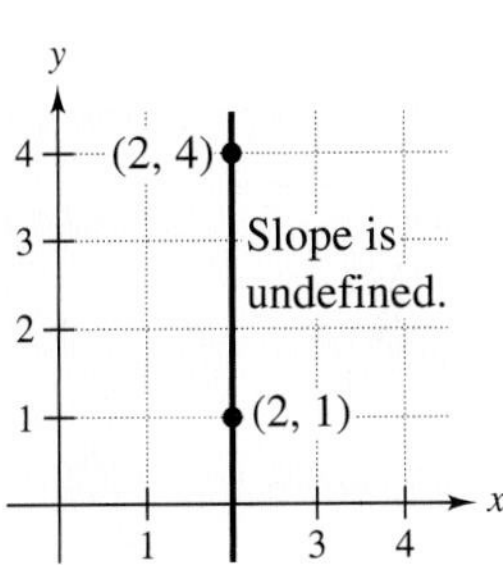

Figure 4.28

CHECKPOINT *Now try Exercise 17.*

From the slopes of the lines shown in Figures 4.25–4.28, you can make several generalizations about the slope of a line.

Slope of a Line

1. A line with positive slope $(m > 0)$ *rises* from left to right.
2. A line with negative slope $(m < 0)$ *falls* from left to right.
3. A line with zero slope $(m = 0)$ is *horizontal.*
4. A line with undefined slope is *vertical.*

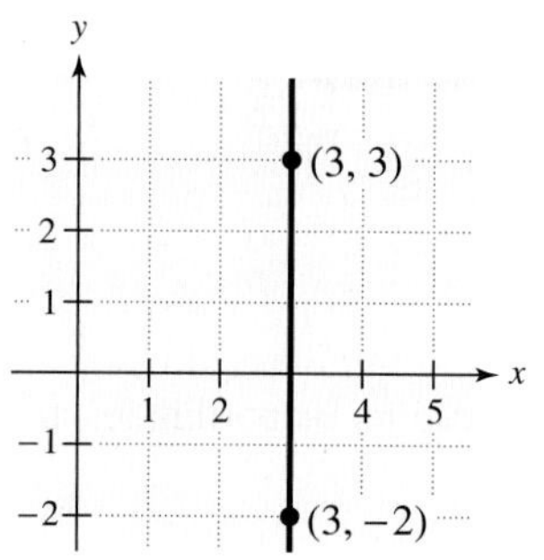

Vertical line: undefined slope
Figure 4.29

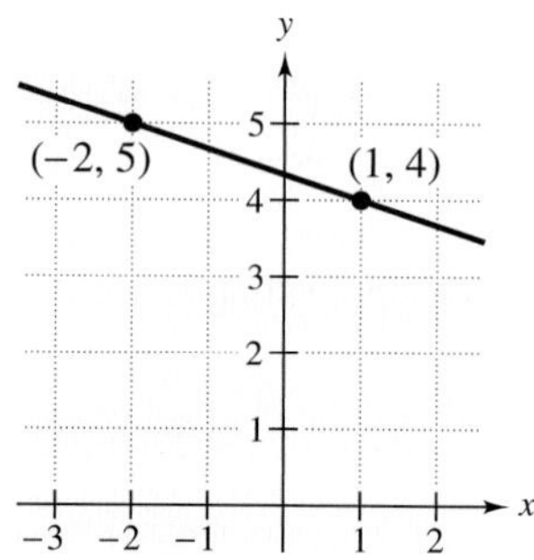

Line falls: negative slope
Figure 4.30

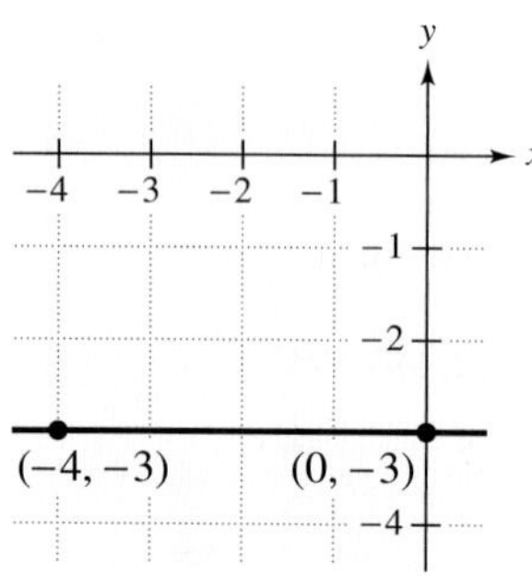

Horizontal line: zero slope
Figure 4.31

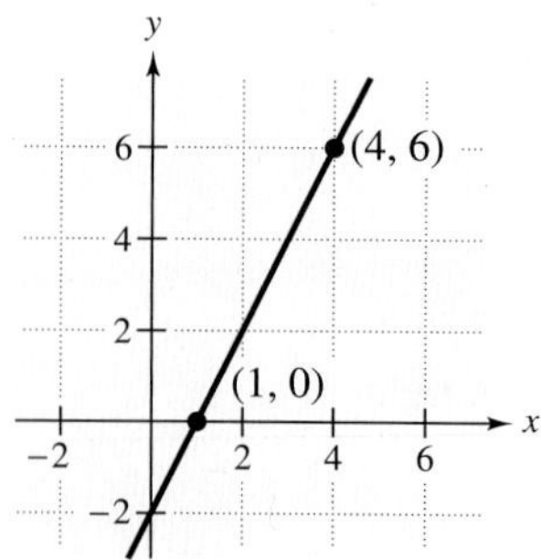

Line rises: positive slope
Figure 4.32

EXAMPLE 3 Using Slope to Describe Lines

Describe the line through each pair of points.

a. $(3, -2)$ and $(3, 3)$ **b.** $(-2, 5)$ and $(1, 4)$

Solution

a. Let $(x_1, y_1) = (3, -2)$ and $(x_2, y_2) = (3, 3)$.

$$m = \frac{3 - (-2)}{3 - 3}$$ Definition of slope

$$= \frac{5}{0}$$ Undefined slope (See Figure 4.29.)

Because the slope is undefined, the line is vertical.

b. Let $(x_1, y_1) = (-2, 5)$ and $(x_2, y_2) = (1, 4)$.

$$m = \frac{4 - 5}{1 - (-2)}$$ Definition of slope

$$= -\frac{1}{3}$$ Negative slope (See Figure 4.30.)

Because the slope is negative, the line falls from left to right.

✓ ***CHECKPOINT*** *Now try Exercise 13.*

EXAMPLE 4 Using Slope to Describe Lines

Describe the line through each pair of points.

a. $(-4, -3)$ and $(0, -3)$ **b.** $(1, 0)$ and $(4, 6)$

Solution

a. Let $(x_1, y_1) = (-4, -3)$ and $(x_2, y_2) = (0, -3)$.

$$m = \frac{-3 - (-3)}{0 - (-4)}$$ Definition of slope

$$= \frac{0}{4} = 0$$ Zero slope (See Figure 4.31.)

Because the slope is zero, the line is horizontal.

b. Let $(x_1, y_1) = (1, 0)$ and $(x_2, y_2) = (4, 6)$.

$$m = \frac{6 - 0}{4 - 1}$$ Definition of slope

$$= \frac{6}{3} = 2$$ Positive slope (See Figure 4.32.)

Because the slope is positive, the line rises from left to right.

✓ ***CHECKPOINT*** *Now try Exercise 19.*

Any two points on a nonvertical line can be used to calculate its slope. This is demonstrated in the next two examples.

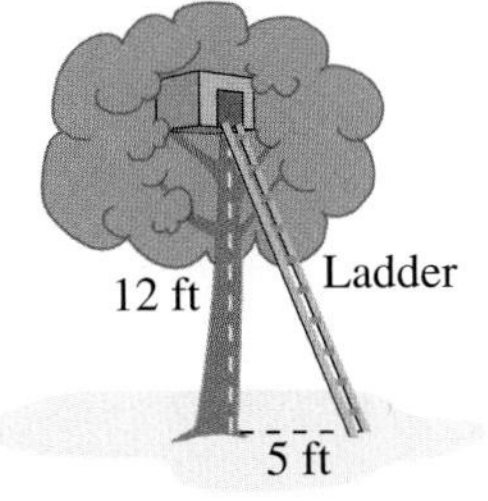

Figure 4.33

EXAMPLE 5 Finding the Slope of a Ladder 

Find the slope of the ladder leading up to the tree house in Figure 4.33.

Solution

Consider the tree trunk as the y-axis and the level ground as the x-axis. The endpoints of the ladder are (0, 12) and (5, 0). So, the slope of the ladder is

$$m = \frac{y_2 - y_1}{x_2 - x_1} = \frac{0 - 12}{5 - 0} = -\frac{12}{5}.$$

✓ **CHECKPOINT** *Now try Exercise 87.*

EXAMPLE 6 Finding the Slope of a Line

Sketch the graph of the line $3x - 2y = 4$. Then find the slope of the line. (Choose two different pairs of points on the line and show that the same slope is obtained from either pair.)

Solution

Begin by solving the equation for y.

$$y = \frac{3}{2}x - 2$$

Then, construct a table of values, as shown below.

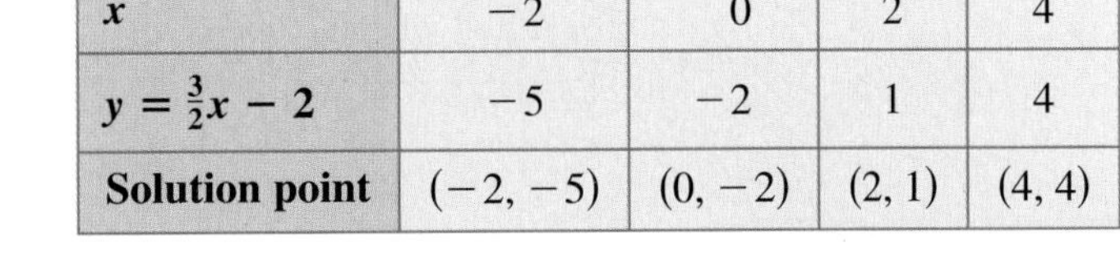

x	-2	0	2	4
$y = \frac{3}{2}x - 2$	-5	-2	1	4
Solution point	$(-2, -5)$	$(0, -2)$	$(2, 1)$	$(4, 4)$

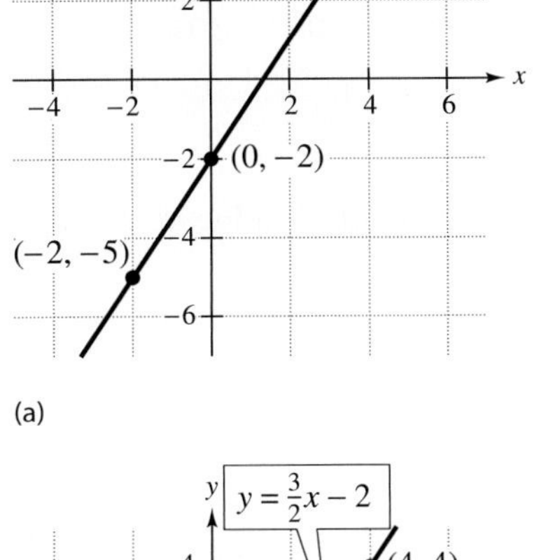

(a)

(b)

Figure 4.34

From the solution points shown in the table, sketch the line, as shown in Figure 4.34. To calculate the slope of the line using two different sets of points, first use the points $(-2, -5)$ and $(0, -2)$, as shown in Figure 4.34(a), and obtain a slope of

$$m = \frac{-2 - (-5)}{0 - (-2)} = \frac{3}{2}.$$

Next, use the points (2, 1) and (4, 4), as shown in Figure 4.34(b), and obtain a slope of

$$m = \frac{4 - 1}{4 - 2} = \frac{3}{2}.$$

Try some other pairs of points on the line to see that you obtain a slope of $m = \frac{3}{2}$ regardless of which two points you use.

✓ **CHECKPOINT** *Now try Exercise 29.*

2 ▶ Write linear equations in slope-intercept form and graph the equations.

Slope as a Graphing Aid

You saw in Section 4.1 that before creating a table of values for an equation, it is helpful first to solve the equation for y. When you do this for a linear equation, you obtain some very useful information. Consider the results of Example 6.

$$3x - 2y = 4 \quad \text{Write original equation.}$$

$$3x - 3x - 2y = -3x + 4 \quad \text{Subtract } 3x \text{ from each side.}$$

$$-2y = -3x + 4 \quad \text{Simplify.}$$

$$\frac{-2y}{-2} = \frac{-3x + 4}{-2} \quad \text{Divide each side by } -2.$$

$$y = \frac{3}{2}x - 2 \quad \text{Simplify.}$$

Observe that the coefficient of x is the slope of the graph of this equation. (See Example 6.) Moreover, the constant term, -2, gives the y-intercept of the graph.

$$y = \frac{3}{2}x + -2$$

slope: $\frac{3}{2}$ y-intercept $(0, -2)$

This form is called the **slope-intercept form** of the equation of the line.

Technology: Tip

Setting the viewing window on a graphing calculator affects the appearance of a line's slope. When you are using a graphing calculator, remember that you cannot judge whether a slope is steep or shallow unless you use a *square* setting—a setting that shows equal spacing of the units on both axes. For many graphing calculators, a square setting is obtained by using the ratio of 10 vertical units to 15 horizontal units.

Slope-Intercept Form of the Equation of a Line

The graph of the equation

$$y = mx + b$$

is a line whose slope is m and whose y-intercept is $(0, b)$. (See Figure 4.35.)

Study Tip

Remember that slope is a *rate of change.* In the slope-intercept equation

$$y = mx + b$$

the slope m is the rate of change of y with respect to x.

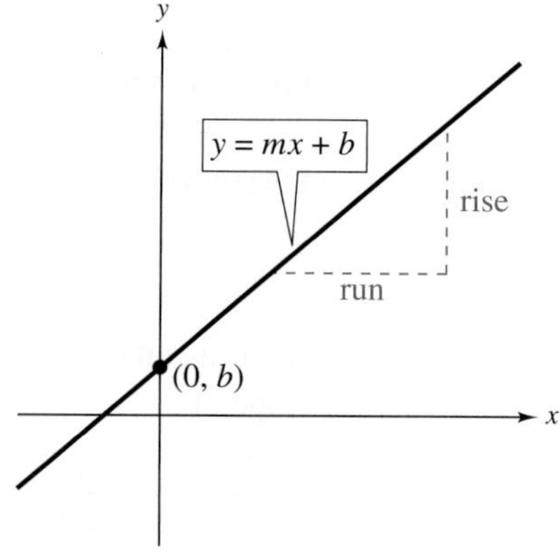

Figure 4.35

So far, you have been plotting several points to sketch the equation of a line. However, now that you can recognize equations of lines, you don't have to plot as many points—two points are enough. (You might remember from geometry that *two points are all that are necessary to determine a line.*) The next example shows how to use the slope to help sketch a line.

EXAMPLE 7 Using the Slope and y-Intercept to Sketch a Line

Use the slope and y-intercept to sketch the graph of

$$x - 3y = -6.$$

Solution

First, write the equation in slope-intercept form.

$x - 3y = -6$ Write original equation.

$-3y = -x - 6$ Subtract x from each side.

$y = \dfrac{-x - 6}{-3}$ Divide each side by -3.

$y = \dfrac{1}{3}x + 2$ Simplify to slope-intercept form.

So, the slope of the line is $m = \frac{1}{3}$ and the y-intercept is $(0, b) = (0, 2)$. Now you can sketch the graph of the equation. First, plot the y-intercept, as shown in Figure 4.36. Then, using a slope of $\frac{1}{3}$,

$$m = \frac{1}{3} = \frac{\text{Change in } y}{\text{Change in } x}$$

locate a second point on the line by moving three units to the right and one unit up (or one unit up and three units to the right), also shown in Figure 4.36. Finally, obtain the graph by drawing a line through the two points. (See Figure 4.37.)

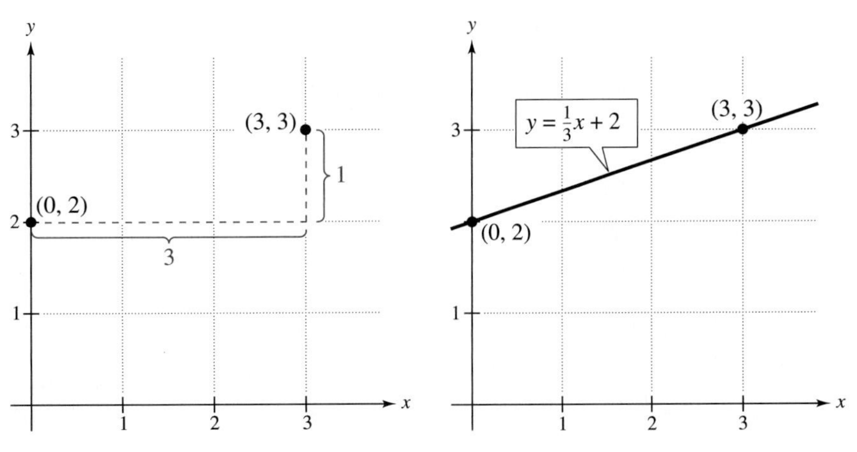

Figure 4.36 **Figure 4.37**

✓ **CHECKPOINT** *Now try Exercise 59.*

3 ▶ Use slopes to determine whether lines are parallel, perpendicular, or neither.

Parallel and Perpendicular Lines

You know from geometry that two lines in a plane are **parallel** if they do not intersect. What this means in terms of their slopes is shown in Example 8.

EXAMPLE 8 Lines That Have the Same Slope

On the same set of coordinate axes, sketch the graphs of the lines $y = 3x$ and $y = 3x - 4$.

Solution

For the line

$$y = 3x$$

the slope is $m = 3$ and the y-intercept is $(0, 0)$. For the line

$$y = 3x - 4$$

the slope is also $m = 3$ and the y-intercept is $(0, -4)$. See Figure 4.38.

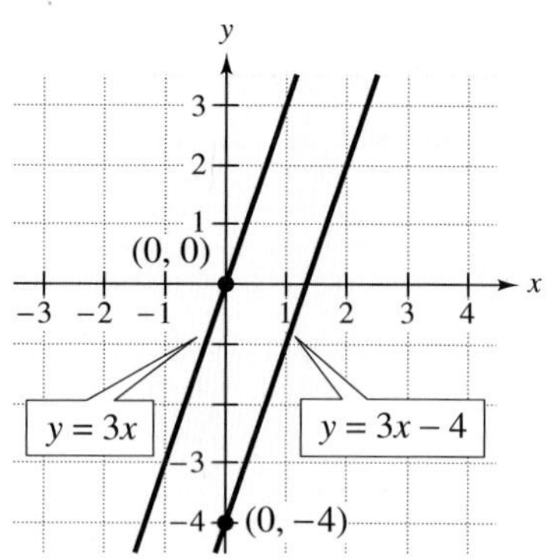

Figure 4.38

✓ ***CHECKPOINT*** *Now try Exercise 73.*

In Example 8, notice that the two lines have the same slope and that the two lines appear to be parallel. The following rule states that this is always the case.

Parallel Lines

Two distinct nonvertical lines are parallel if and only if they have the same slope.

The phrase "if and only if" in this rule is used in mathematics as a way to write two statements in one. The first statement says that *if two distinct nonvertical lines have the same slope, they must be parallel.* The second (or reverse) statement says that *if two distinct nonvertical lines are parallel, they must have the same slope.*

EXAMPLE 9 Lines That Have Negative Reciprocal Slopes

On the same set of coordinate axes, sketch the graphs of the lines $y = 5x + 2$ and $y = -\frac{1}{5}x - 4$.

Solution

For the line

$$y = 5x + 2$$

the slope is $m = 5$ and the y-intercept is $(0, 2)$. For the line

$$y = -\tfrac{1}{5}x - 4$$

the slope is $m = -\frac{1}{5}$ and the y-intercept is $(0, -4)$. See Figure 4.39.

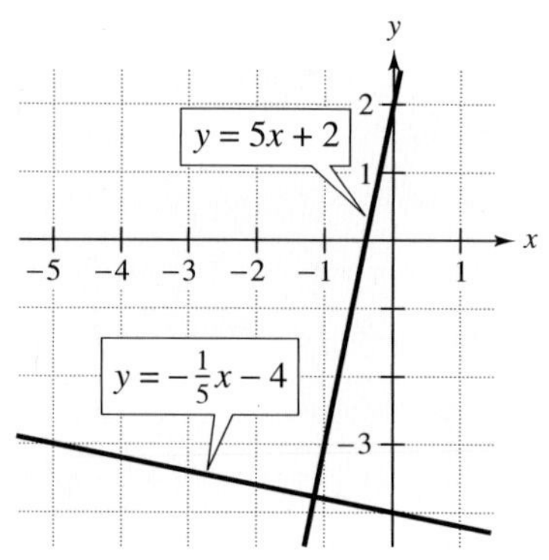

Figure 4.39

✓ ***CHECKPOINT*** *Now try Exercise 75.*

In Example 9, notice that the two lines have slopes that are negative reciprocals of each other and that the two lines appear to be perpendicular. Another rule from geometry is that two lines in a plane are **perpendicular** if they intersect at right angles. In terms of their slopes, this means that two nonvertical lines are perpendicular if their slopes are negative reciprocals of each other.

Perpendicular Lines

Consider two nonvertical lines whose slopes are m_1 and m_2. The two lines are perpendicular if and only if their slopes are *negative reciprocals* of each other. That is,

$$m_1 = -\frac{1}{m_2}$$

or, equivalently,

$$m_1 \cdot m_2 = -1.$$

EXAMPLE 10 Parallel or Perpendicular?

Determine whether the pairs of lines are parallel, perpendicular, or neither.

a. $y = -3x - 2,\ y = \frac{1}{3}x + 1$

b. $y = \frac{1}{2}x + 1,\ y = \frac{1}{2}x - 1$

Solution

a. The first line has a slope of $m_1 = -3$ and the second line has a slope of $m_2 = \frac{1}{3}$. Because these slopes are negative reciprocals of each other, the two lines must be perpendicular, as shown in Figure 4.40.

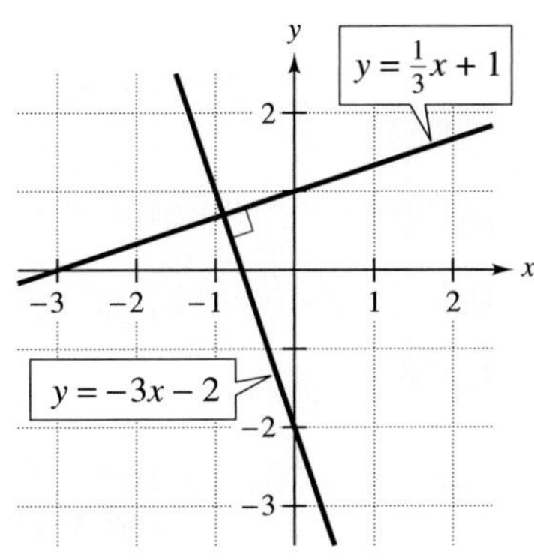

Figure 4.40

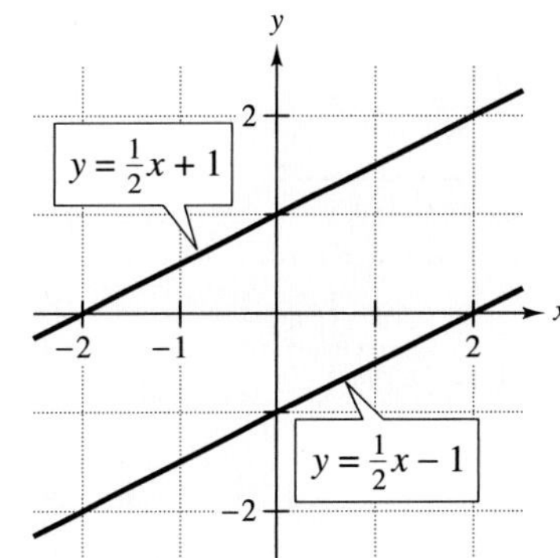

Figure 4.41

b. Both lines have a slope of $m = \frac{1}{2}$. So, the two lines must be parallel, as shown in Figure 4.41.

✓ **CHECKPOINT** *Now try Exercise 79.*

Concept Check

1. Is the slope of a line a ratio? Explain.

2. Explain how you can visually determine the sign of the slope of a line by observing the graph of the line.

3. Which slope is steeper: -5 or 2? Explain.

4. Determine whether the following statement is true or false. Justify your answer.

   If both the x- and y-intercepts of a line are positive, then the slope of the line is positive.

4.4 EXERCISES

Go to pages 284–285 to record your assignments.

Developing Skills

In Exercises 1–6, estimate the slope (if it exists) of the line from its graph.

1.

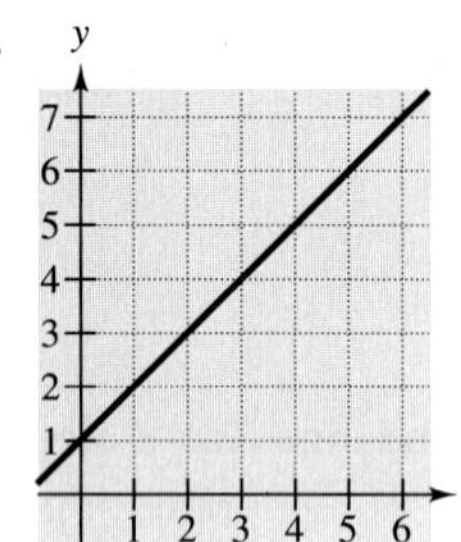

2.

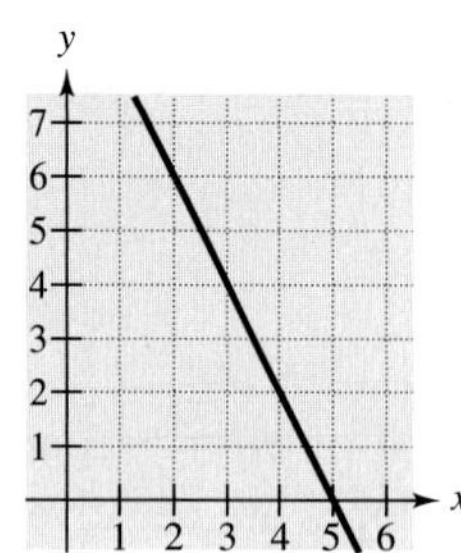

3.

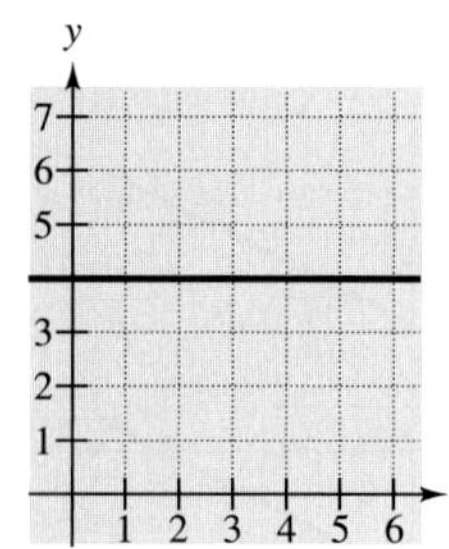

4.

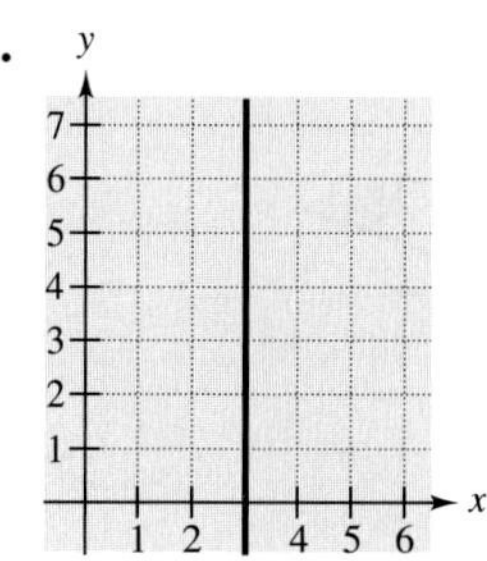

5.

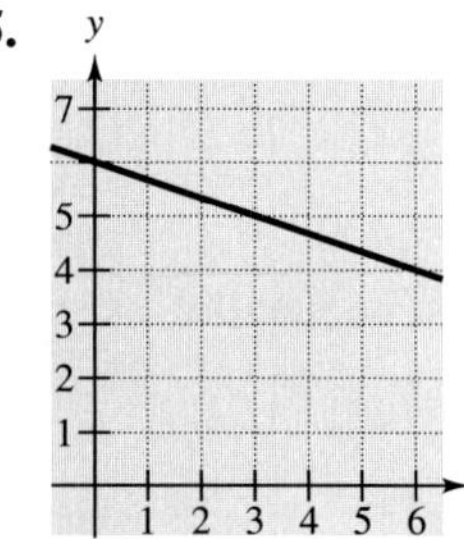

6.

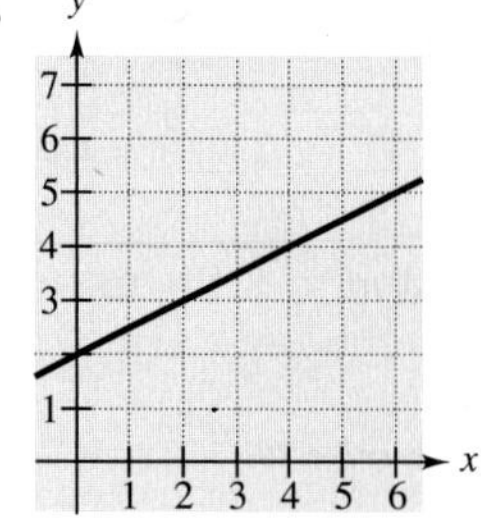

In Exercises 7 and 8, identify the line in the figure that has each slope.

7. (a) $m = \frac{3}{2}$

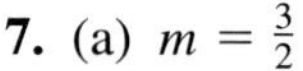

(b) $m = 0$

(c) $m = -\frac{2}{3}$

(d) $m = -2$

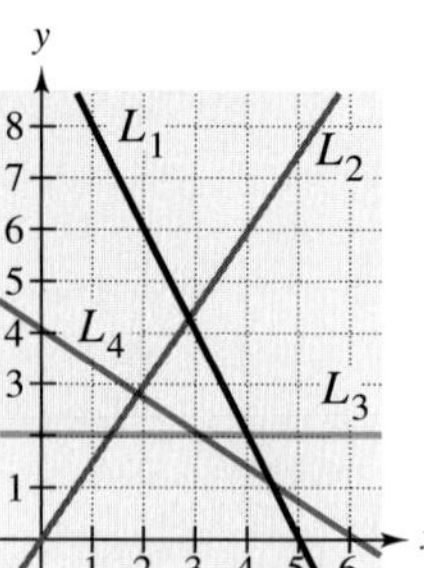

8. (a) $m = -\frac{3}{4}$

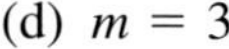

(b) $m = \frac{1}{2}$

(c) m is undefined.

(d) $m = 3$

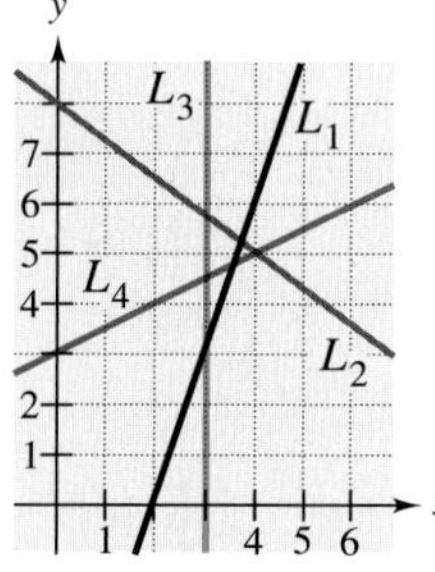

In Exercises 9–28, plot the points and find the slope of the line passing through the points. State whether the line rises, falls, is horizontal, or is vertical. ***See Examples 1–4.***

9. (0, 0), (4, 5)

10. (0, 0), (3, 6)

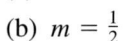

✓ **11.** (0, 0), (8, -4)

12. (0, 0), (-1, 3)

✓ **13.** $(6, 0), (0, 4)$ **14.** $(0, -3), (5, 0)$

15. $(-4, -1), (2, 6)$ **16.** $(5, 3), (-3, 1)$

✓ **17.** $(-6, -1), (-6, 4)$

18. $(-4, -10), (-4, 0)$

✓ **19.** $(3, -4), (8, -4)$

20. $(1, -2), (-2, -2)$

21. $\left(\frac{1}{4}, \frac{3}{2}\right), \left(\frac{9}{2}, -3\right)$ **22.** $\left(\frac{5}{4}, \frac{1}{4}\right), \left(\frac{7}{8}, 2\right)$

23. $(3.2, -1), (-3.2, 4)$ **24.** $(1.4, 3), (-1.4, 5)$

25. $(3.5, -1), (5.75, 4.25)$ **26.** $(6, 6.4), (-3.1, 5.2)$

27. $(a, 3), (4, 3),\ a \neq 4$

28. $(4, a), (4, 2),\ a \neq 2$

In Exercises 29 and 30, complete the table. Use two different pairs of solution points to show that the same slope is obtained using either pair. ***See Example 6.***

x	-2	0	2	4
y				
Solution point				

✓ **29.** $y = -2x - 2$ **30.** $y = 3x + 4$

In Exercises 31–34, use the formula for slope to find the value of y such that the line through the points has the given slope.

31. Points: $(4, -1), (0, y)$
Slope: $m = -4$

32. Points: $(-2, y), (1, 6)$
Slope: $m = 6$

33. Points: $(-4, y), (7, 6)$
Slope: $m = \frac{5}{2}$

34. Points: $(0, 10), (6, y)$
Slope: $m = -\frac{1}{3}$

In Exercises 35–46, a point on a line and the slope of the line are given. Plot the point and use the slope to find two additional points on the line. (There are many correct answers.)

35. $(2, 1),\ m = 0$ **36.** $(5, 10),\ m = 0$

37. $(1, -6),\ m = 2$ **38.** $(-2, 4),\ m = 1$

39. $(0, 1),\ m = -2$ **40.** $(5, 6),\ m = -3$

41. $(-4, 0),\ m = \frac{2}{3}$ **42.** $(-1, -1),\ m = \frac{1}{4}$

43. $(3, 5),\ m = -\frac{1}{2}$ **44.** $(1, 3),\ m = -\frac{4}{3}$

45. $(-8, 1)$
m is undefined.

46. $(6, -4)$
m is undefined.

In Exercises 47–52, sketch the graph of the line through the point $(3, 0)$ having the given slope.

47. $m = 0$ **48.** m is undefined.

49. $m = 2$ **50.** $m = -1$

51. $m = -\frac{2}{3}$ **52.** $m = \frac{3}{5}$

In Exercises 53–58, plot the x- and y-intercepts and sketch the graph of the line.

53. $2x + 3y + 6 = 0$ **54.** $3x + 4y + 12 = 0$

55. $5x - 2y - 10 = 0$ **56.** $3x - 7y - 21 = 0$

57. $6x - 4y + 12 = 0$ **58.** $2x - 5y - 20 = 0$

In Exercises 59–72, write the equation in slope-intercept form. Use the slope and y-intercept to sketch the graph of the line. ***See Example 7.***

✓ **59.** $x - y = 0$ **60.** $x + y = 0$

61. $\frac{1}{2}x + y = 0$ **62.** $\frac{3}{4}x - y = 0$

63. $2x - y - 3 = 0$ **64.** $x - y + 2 = 0$

65. $x - 2y + 2 = 0$ **66.** $x + 3y - 12 = 0$

67. $2x - 6y - 15 = 0$ **68.** $10x + 6y - 3 = 0$

69. $3x - 4y + 2 = 0$ **70.** $2x - 3y + 1 = 0$

71. $y + 5 = 0$ **72.** $y - 3 = 0$

In Exercises 73–78, sketch the graphs of the two lines on the same rectangular coordinate system. Identify the slope and y-intercept of each line. ***See Examples 8 and 9.***

73. $y_1 = 4x$
$y_2 = 4x - 3$

74. $y_1 = -\frac{2}{5}x + 1$
$y_2 = -\frac{2}{5}x - 3$

75. $y_1 = \frac{1}{3}x - 5$
$y_2 = -3x + 2$

76. $y_1 = -2x + 6$
$y_2 = -\frac{1}{2}x + 2$

77. $y_1 = \frac{4}{5}x - 2$
$y_2 = -\frac{4}{5}x - 5$

78. $y_1 = \frac{3}{2}x - 5$
$y_2 = -\frac{2}{3}x - 3$

In Exercises 79–82, sketch the graphs of the two lines on the same rectangular coordinate system. Determine whether the lines are parallel, perpendicular, or neither. Use a graphing calculator to verify your result. (Use a square setting.) ***See Example 10.***

79. $y_1 = 2x - 3$
$y_2 = 2x + 1$

80. $y_1 = -\frac{1}{3}x - 3$
$y_2 = \frac{1}{3}x + 1$

81. $y_1 = 2x - 3$
$y_2 = -\frac{1}{2}x + 1$

82. $y_1 = -\frac{1}{3}x - 3$
$y_2 = 3x + 1$

In Exercises 83–86, determine whether the lines L_1 and L_2 passing through the pairs of points are parallel, perpendicular, or neither.

83. L_1: $(0, -1), (5, 9)$
L_2: $(0, 3), (4, 1)$

84. L_1: $(-2, -1), (1, 5)$
L_2: $(1, 3), (5, 5)$

85. L_1: $(3, 6), (-6, 0)$
L_2: $(0, -1), \left(5, \frac{7}{3}\right)$

86. L_1: $(4, 8), (-4, 2)$
L_2: $(3, -5), \left(-1, \frac{1}{3}\right)$

Solving Problems

87. *Slide* The ladder of a straight slide in a playground is 8 feet high. The distance along the ground from the ladder to the foot of the slide is 12 feet. Approximate the slope of the slide.

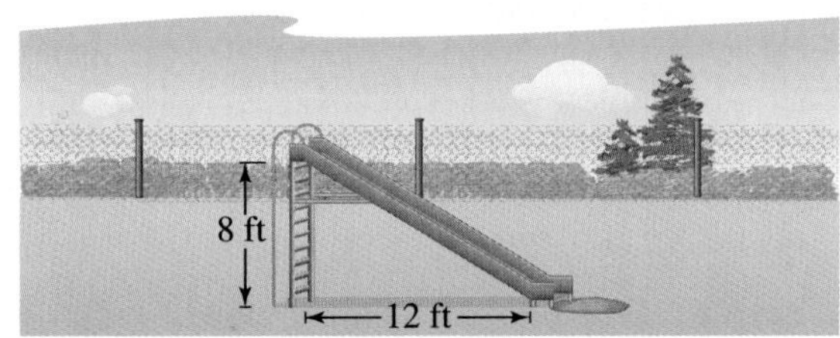

88. *Ladder* Find the slope of the ladder shown in the figure.

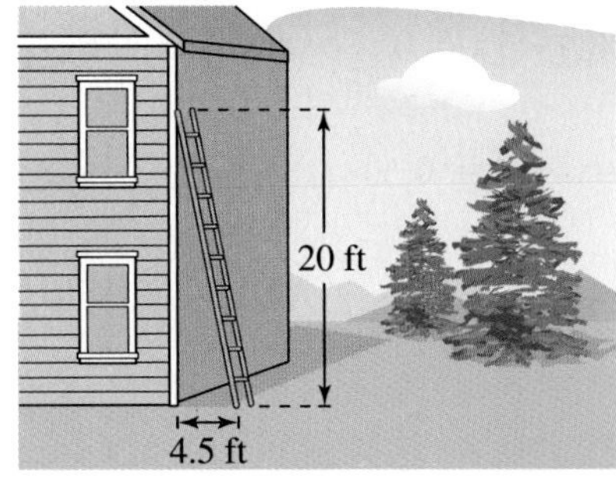

89. *Flight Path* An airplane leaves an airport. As it flies over a town, its altitude is 4 miles. The town is about 20 miles from the airport. Approximate the slope of the linear path followed by the airplane during takeoff.

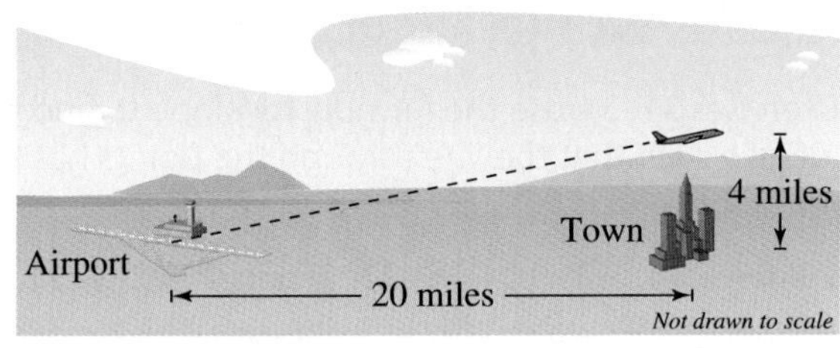

90. *Roof Pitch* Determine the slope, or pitch, of the roof of the house shown in the figure.

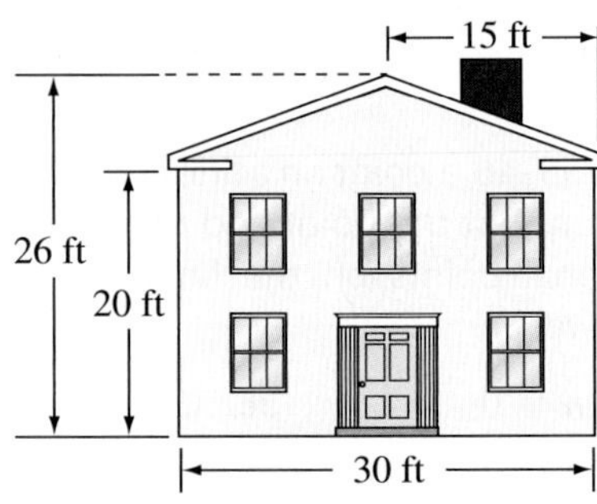

91. *Subway Track* A subway track rises 3 feet over a 200-foot horizontal distance.

(a) Draw a diagram of the track and label the rise and run.

(b) Find the slope of the track.

(c) Would the slope be steeper if the track rose 3 feet over a distance of 100 feet? Explain.

Piotr Redlinski/CORBIS

Approximately 660 miles of track are used for passenger service in the New York City subway system. (Source: Metropolitan Transit Authority)

92. *Skateboarding Ramp* A wedge-shaped skateboarding ramp rises to a height of 12 inches over a 50-inch horizontal distance.

(a) Draw a diagram of the ramp and label the rise and run.

(b) Find the slope of the ramp.

(c) Would the slope be steeper if the ramp rose 12 inches over a distance of 60 inches? Explain.

93. *Theatrical Films* The graph shows the numbers of theatrical films released in the years 2002 through 2006. (Source: Motion Picture Association of America)

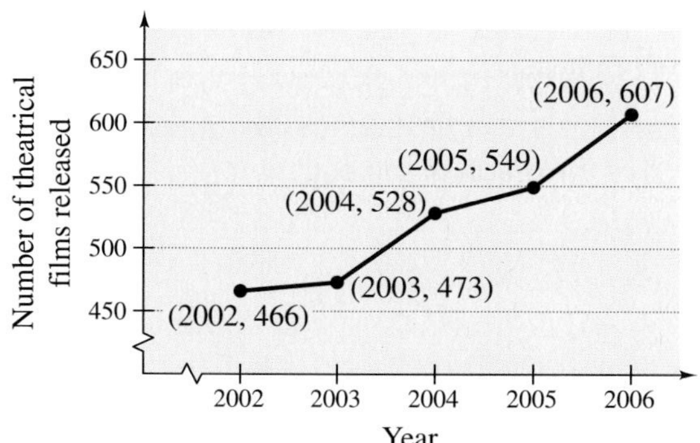

(a) Find the slopes of the four line segments.

(b) Find the slope of the line segment connecting the years 2002 and 2006. Interpret the meaning of this slope in the context of the problem.

94. *Gold Prices* The graph shows the average prices (in dollars) of a troy ounce of gold for the years 2000 through 2006.

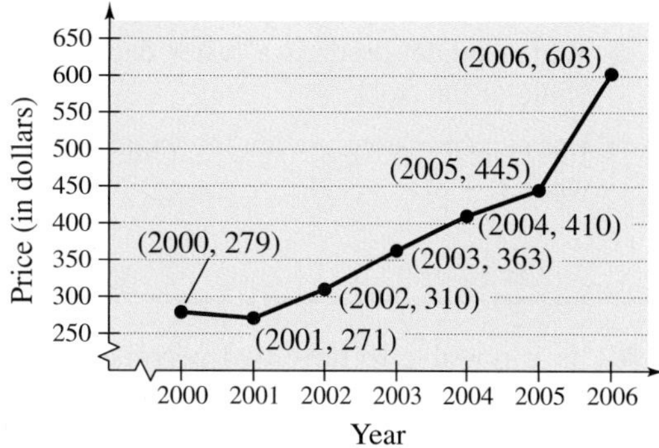

(a) Find the slopes of the six line segments.

(b) Find the slope of the line segment connecting the years 2000 and 2006. Interpret the meaning of this slope in the context of the problem.

95. *Profit* Based on different assumptions, the marketing department of an outerwear manufacturer develops two linear models to predict the annual profits of the company over the next 10 years. The models are $P_1 = 0.2t + 2.4$ and $P_2 = 0.3t + 2.4$, where P_1 and P_2 represent profit in millions of dollars and t is time in years $(0 \le t \le 10)$.

(a) Interpret the slopes of the two linear models in the context of the problem.

(b) Which model predicts a faster increase in profits?

(c) Use each model to predict profits when $t = 10$.

(d) Use a graphing calculator to graph the models in the same viewing window.

96. *Stock Market* Based on different assumptions, a research firm develops two linear models to predict the value of a stock over the next 12 months. The models are $S_1 = -2.6t + 40.2$ and $S_2 = -1.8t + 40.2$, where S_1 and S_2 are the values of the stock in dollars and t is the time in months $(1 \le t \le 12)$.

(a) Interpret the slopes of the two linear models in the context of the problem.

(b) Which model predicts a faster decrease in the value of the stock?

(c) Use each model to predict the value of the stock when $t = 8$.

(d) Use a graphing calculator to graph the models in the same viewing window.

Rate of Change In Exercises 97–102, the slopes of lines representing annual sales y in terms of time t in years are given. Use the slopes to determine any change in annual sales for a 1-year increase in time t.

97. $m = 76$

98. $m = 32$

99. $m = 0$

100. $m = 0.5$

101. $m = -14$

102. $m = -4$

Explaining Concepts

103. Is it possible to have two perpendicular lines with positive slopes? Explain.

104. The slope of a line is $\frac{3}{2}$. How much will y change if x is increased by eight units? Explain.

105. When a quantity y is increasing or decreasing at a constant rate over time t, the graph of y versus t is a line. What is another name for the rate of change?

106. Explain how to use slopes to determine if the points $(-2, -3)$, $(1, 1)$, and $(3, 4)$ lie on the same line.

107. When determining the slope of a line through two points, does the order of subtracting the coordinates of the points matter? Explain.

108. *Misleading Graphs*

(a) Use a graphing calculator to graph the line $y = 0.75x - 2$ for each viewing window.

Window 1	Window 2
Xmin = -10	Xmin = 0
Xmax = 10	Xmax = 1
Xscl = 2	Xscl = 0.5
Ymin = -100	Ymin = -2
Ymax = 100	Ymax = -1.5
Yscl = 10	Yscl = 0.1

(b) Do the lines appear to have the same slope?

(c) Does either of the lines appear to have a slope of 0.75? If not, find a viewing window that will make the line appear to have a slope of 0.75.

Cumulative Review

In Exercises 109–116, simplify the expression.

109. $x^2 \cdot x^3$

110. $z^2 \cdot z^2$

111. $(-y^2)y$

112. $5x^2(x^5)$

113. $(25x^3)(2x^2)$

114. $(3yz)(6yz^3)$

115. $x^2 - 2x - x^2 + 3x + 2$

116. $x^2 - 5x - 2 + x$

In Exercises 117–122, find the x- and y-intercepts (if any) of the graph of the equation.

117. $y = 6x - 3$

118. $y = -\frac{4}{3}x - 8$

119. $2x + y = -3$

120. $3x - 5y = 15$

121. $x + 6y - 9 = 0$

122. $4x - 9y + 18 = 0$

4.5 Equations of Lines

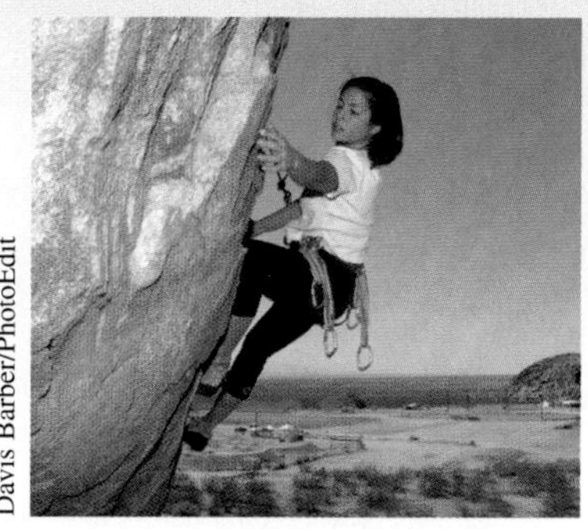

Davis Barber/PhotoEdit

Why You Should Learn It

Real-life problems can be modeled and solved using linear equations. For instance, in Example 8 on page 269, a linear equation is used to model the relationship between the time and the height of a rock climber.

What You Should Learn

1 ▶ Write equations of lines using the point-slope form.
2 ▶ Write the equations of horizontal and vertical lines.
3 ▶ Use linear models to solve application problems.

1 ▶ Write equations of lines using the point-slope form.

The Point-Slope Form of the Equation of a Line

In Sections 4.1 through 4.4, you studied analytic (or coordinate) geometry. Analytic geometry uses a coordinate plane to give visual representations of algebraic concepts, such as equations and functions.

There are two basic types of problems in analytic geometry.

1. Given an equation, sketch its graph.

   Algebra ⇨ Geometry

2. Given a graph, write its equation.

   Geometry 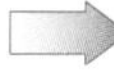 Algebra

In Section 4.4, you worked primarily with the first type of problem. In this section, you will study the second type. Specifically, you will learn how to write the equation of a line when you are given its slope and a point on the line. Before a general formula for doing this is given, consider the following example.

EXAMPLE 1 Writing an Equation of a Line

A line has a slope of $\frac{5}{3}$ and passes through the point $(2, 1)$. Find its equation.

Solution

Begin by sketching the line, as shown in Figure 4.42. The slope of a line is the same through any two points on the line. So, using *any* representative point (x, y) and the given point $(2, 1)$, it follows that the slope of the line is

$$m = \frac{y - 1}{x - 2}.$$ Difference in y-coordinates / Difference in x-coordinates

By substituting $\frac{5}{3}$ for m, you obtain the equation of the line.

$$\frac{5}{3} = \frac{y - 1}{x - 2}$$ Slope formula

$$5(x - 2) = 3(y - 1)$$ Cross-multiply.

$$5x - 10 = 3y - 3$$ Distributive Property

$$5x - 3y = 7$$ Equation of line

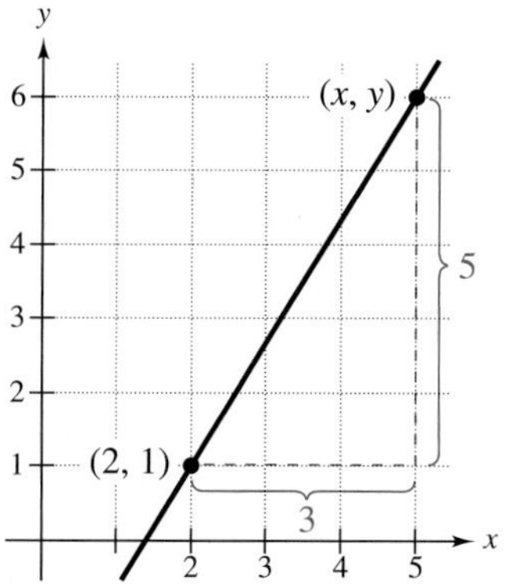

✓ **CHECKPOINT** *Now try Exercise 5.*

Figure 4.42

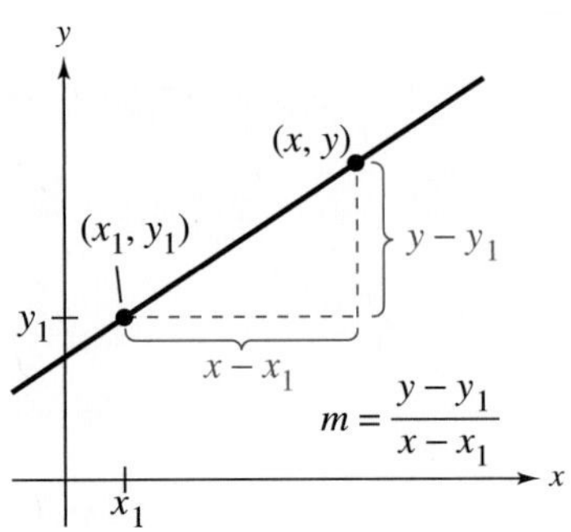

Figure 4.43

The procedure in Example 1 can be used to derive a *formula* for the equation of a line given its slope and a point on the line. In Figure 4.43, let (x_1, y_1) be a given point on a line whose slope is m. If (x, y) is any *other* point on the line, it follows that

$$\frac{y - y_1}{x - x_1} = m.$$

This equation in variables x and y can be rewritten in the form

$$y - y_1 = m(x - x_1)$$

which is called the **point-slope form** of the equation of a line.

Point-Slope Form of the Equation of a Line

The **point-slope form** of the equation of a line with slope m and passing through the point (x_1, y_1) is

$$y - y_1 = m(x - x_1).$$

EXAMPLE 2 The Point-Slope Form of the Equation of a Line

Write an equation of the line that passes through the point $(1, -2)$ and has slope $m = 3$.

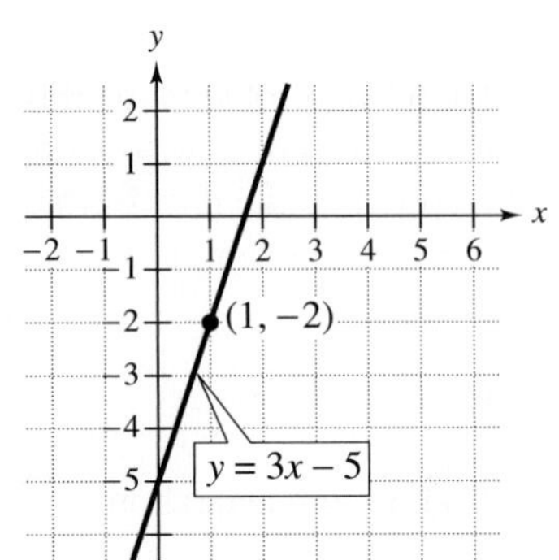

Figure 4.44

Solution

Use the point-slope form with $(x_1, y_1) = (1, -2)$ and $m = 3$.

$$y - y_1 = m(x - x_1) \qquad \text{Point-slope form}$$

$$y - (-2) = 3(x - 1) \qquad \text{Substitute } -2 \text{ for } y_1\text{, } 1 \text{ for } x_1\text{, and } 3 \text{ for } m.$$

$$y + 2 = 3x - 3 \qquad \text{Simplify.}$$

$$y = 3x - 5 \qquad \text{Equation of line}$$

So, an equation of the line is $y = 3x - 5$. Note that this is the slope-intercept form of the equation. The graph of this line is shown in Figure 4.44.

✓ **CHECKPOINT** *Now try Exercise 15.*

In Example 2, note that it was concluded that $y = 3x - 5$ is "an" equation of the line rather than "the" equation of the line. The reason for this is that every equation can be written in many equivalent forms. For instance,

$$y = 3x - 5, \quad 3x - y = 5, \quad \text{and} \quad 3x - y - 5 = 0$$

are all equations of the line in Example 2. The first of these equations $(y = 3x - 5)$ is in the slope-intercept form

$$y = mx + b \qquad \text{Slope-intercept form}$$

and provides the most information about the line. The last of these equations $(3x - y - 5 = 0)$ is in the general form of the equation of a line.

$$ax + by + c = 0 \qquad \text{General form}$$

Technology: Tip

A program for several models of graphing calculators that uses the two-point form to find the equation of a line is available at our website *www.cengage.com/math/larson/algebra.*

The program prompts for the coordinates of the two points and then outputs the slope and the y-intercept of the line that passes through the two points. Verify Example 3 using this program.

The point-slope form can be used to find an equation of a line passing through any two points (x_1, y_1) and (x_2, y_2). First, use the formula for the slope of a line passing through these two points.

$$m = \frac{y_2 - y_1}{x_2 - x_1}$$

Then, knowing the slope, use the point-slope form to obtain the equation

$$y - y_1 = \frac{y_2 - y_1}{x_2 - x_1}(x - x_1). \qquad \text{Two-point form}$$

This is sometimes called the **two-point form** of the equation of a line.

EXAMPLE 3 An Equation of a Line Passing Through Two Points

Write an equation of the line that passes through the points $(3, 1)$ and $(-3, 4)$.

Solution

Let $(x_1, y_1) = (3, 1)$ and $(x_2, y_2) = (-3, 4)$. The slope of a line passing through these points is

$$m = \frac{y_2 - y_1}{x_2 - x_1} \qquad \text{Formula for slope}$$

$$= \frac{4 - 1}{-3 - 3} \qquad \text{Substitute for } x_1, y_1, x_2, \text{ and } y_2.$$

$$= \frac{3}{-6} \qquad \text{Subtract.}$$

$$= -\frac{1}{2}. \qquad \text{Simplify.}$$

Now, use the point-slope form to find an equation of the line.

$$y - y_1 = m(x - x_1) \qquad \text{Point-slope form}$$

$$y - 1 = -\frac{1}{2}(x - 3) \qquad \text{Substitute 1 for } y_1, 3 \text{ for } x_1, \text{ and } -\tfrac{1}{2} \text{ for } m.$$

$$y - 1 = -\frac{1}{2}x + \frac{3}{2} \qquad \text{Distributive Property}$$

$$y = -\frac{1}{2}x + \frac{5}{2} \qquad \text{Equation of line}$$

The graph of this line is shown in Figure 4.45.

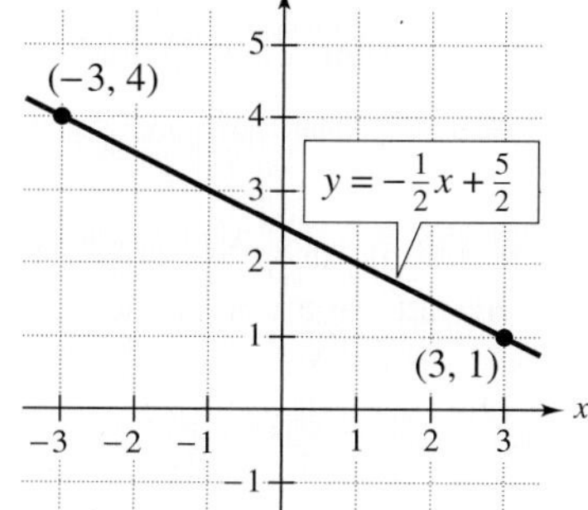

Figure 4.45

✓ **CHECKPOINT** *Now try Exercise 51.*

In Example 3, it does not matter which of the two points is labeled (x_1, y_1) and which is labeled (x_2, y_2). Try switching these labels to $(x_1, y_1) = (-3, 4)$ and $(x_2, y_2) = (3, 1)$ and reworking the problem to see that you obtain the same equation.

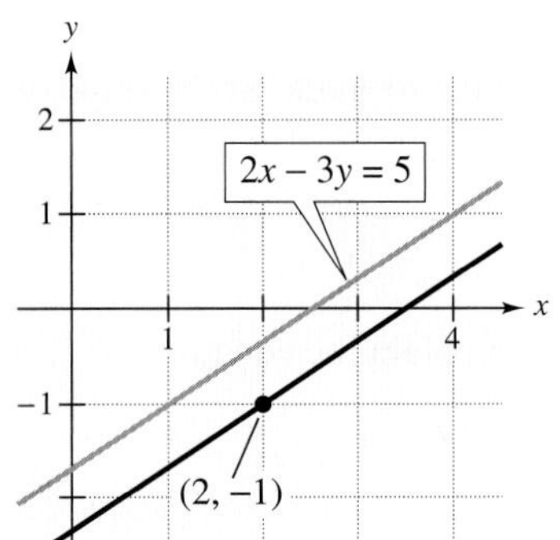

Figure 4.46

Technology: Tip

With a graphing calculator, parallel lines appear to be parallel in both *square* and *nonsquare* window settings. Verify this by graphing $y = 2x - 3$ and $y = 2x + 1$ in both a square and a nonsquare window.

Such is not the case with perpendicular lines, as you can see by graphing $y = 2x - 3$ and $y = -\frac{1}{2}x + 1$ in a square and a nonsquare window.

EXAMPLE 4 Equations of Parallel Lines

Write an equation of the line that passes through the point $(2, -1)$ and is parallel to the line $2x - 3y = 5$, as shown in Figure 4.46.

Solution

To begin, write the original equation in slope-intercept form.

$$2x - 3y = 5 \qquad \text{Write original equation.}$$

$$-3y = -2x + 5 \qquad \text{Subtract } 2x \text{ from each side.}$$

$$y = \frac{2}{3}x - \frac{5}{3} \qquad \text{Divide each side by } -3.$$

Because the line has a slope of $m = \frac{2}{3}$, it follows that any parallel line must have the same slope. So, an equation of the line through $(2, -1)$, parallel to the original line, is

$$y - y_1 = m(x - x_1) \qquad \text{Point-slope form}$$

$$y - (-1) = \frac{2}{3}(x - 2) \qquad \text{Substitute } -1 \text{ for } y_1,\ 2 \text{ for } x_1, \text{ and } \tfrac{2}{3} \text{ for } m.$$

$$y + 1 = \frac{2}{3}x - \frac{4}{3} \qquad \text{Simplify.}$$

$$y = \frac{2}{3}x - \frac{7}{3}. \qquad \text{Equation of parallel line}$$

✓ ***CHECKPOINT*** *Now try Exercise 73(a).*

EXAMPLE 5 Equations of Perpendicular Lines

Write an equation of the line that passes through the point $(2, -1)$ and is perpendicular to the line $2x - 3y = 5$, as shown in Figure 4.47.

Solution

From Example 4, you know that the original line has a slope of $\frac{2}{3}$. So, any line perpendicular to this line must have a slope of $-\frac{3}{2}$. So, an equation of the line through $(2, -1)$, perpendicular to the original line, is

$$y - y_1 = m(x - x_1) \qquad \text{Point-slope form}$$

$$y - (-1) = -\frac{3}{2}(x - 2) \qquad \text{Substitute } -1 \text{ for } y_1,\ 2 \text{ for } x_1, \text{ and } -\tfrac{3}{2} \text{ for } m.$$

$$y + 1 = -\frac{3}{2}x + 3 \qquad \text{Simplify.}$$

$$y = -\frac{3}{2}x + 2. \qquad \text{Equation of perpendicular line}$$

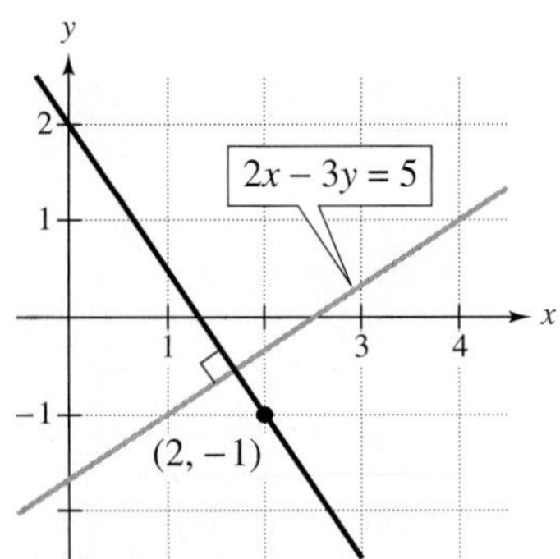

Figure 4.47

✓ ***CHECKPOINT*** *Now try Exercise 73(b).*

2 ▶ Write the equations of horizontal and vertical lines.

Equations of Horizontal and Vertical Lines

Recall from Section 4.4 that a horizontal line has a slope of zero. From the slope-intercept form of the equation of a line, you can see that a horizontal line has an equation of the form

$$y = (0)x + b \quad \text{or} \quad y = b. \qquad \text{Horizontal line}$$

This is consistent with the fact that each point on a horizontal line through $(0, b)$ has a y-coordinate of b. Similarly, each point on a vertical line through $(a, 0)$ has an x-coordinate of a. Because you know that a vertical line has an undefined slope, you know that it has an equation of the form

$$x = a. \qquad \text{Vertical line}$$

Every line has an equation that can be written in the **general form**

$$ax + by + c = 0 \qquad \text{General form}$$

where a and b are not *both* zero.

EXAMPLE 6 Writing Equations of Horizontal and Vertical Lines

Write an equation for each line.

a. Vertical line through $(-3, 2)$

b. Line passing through $(-1, 2)$ and $(4, 2)$

c. Line passing through $(0, 2)$ and $(0, -2)$

d. Horizontal line through $(0, -4)$

Solution

a. Because the line is vertical and passes through the point $(-3, 2)$, every point on the line has an x-coordinate of -3. So, the equation of the line is

$$x = -3. \qquad \text{Vertical line}$$

b. Because both points have the same y-coordinate, the line through $(-1, 2)$ and $(4, 2)$ is horizontal. So, its equation is

$$y = 2. \qquad \text{Horizontal line}$$

c. Because both points have the same x-coordinate, the line through $(0, 2)$ and $(0, -2)$ is vertical. So, its equation is

$$x = 0. \qquad \text{Vertical line (y-axis)}$$

d. Because the line is horizontal and passes through the point $(0, -4)$, every point on the line has a y-coordinate of -4. So, the equation of the line is

$$y = -4. \qquad \text{Horizontal line}$$

The graphs of the lines are shown in Figure 4.48.

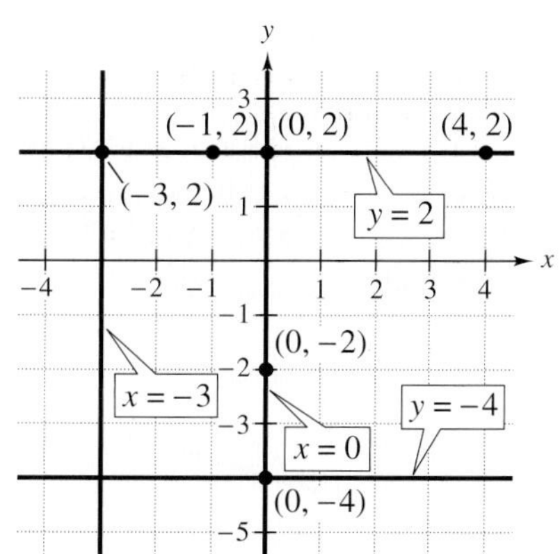

Figure 4.48

✓ **CHECKPOINT** *Now try Exercise 83.*

In Example 6(c), note that the equation $x = 0$ represents the y-axis. In a similar way, you can show that the equation $y = 0$ represents the x-axis.

3 ▶ Use linear models to solve application problems.

Applications

EXAMPLE 7 Predicting Net Revenue

The net revenue of Harley-Davidson, Inc. was \$5.02 billion in 2004 and \$5.34 billion in 2005. Using only this information, write a linear equation that models the net revenue in terms of the year. Then predict the net revenue for 2006. (Source: Harley-Davidson, Inc.)

Solution

Let $t = 4$ represent 2004. Then the two given values are represented by the data points (4, 5.02) and (5, 5.34). The slope of the line through these points is

$$m = \frac{5.34 - 5.02}{5 - 4}$$

$$= 0.32.$$

Using the point-slope form, you can find the equation that relates the net revenue y and the year t to be

$$y - y_1 = m(t - t_1) \quad \text{Point-slope form}$$

$$y - 5.34 = 0.32(t - 5) \quad \text{Substitute 5.34 for } y_1, \text{ 5 for } t_1, \text{ and 0.32 for } m.$$

$$y - 5.34 = 0.32t - 1.6 \quad \text{Distributive Property}$$

$$y = 0.32t + 3.74. \quad \text{Write in slope-intercept form.}$$

Using this equation, a prediction of the net revenue in 2006 ($t = 6$) is

$$y = 0.32(6) + 3.74 = \$5.66 \text{ billion.}$$

In this case, the prediction is fairly good—the actual net revenue in 2006 was \$5.80 billion. The graph of this equation is shown in Figure 4.49.

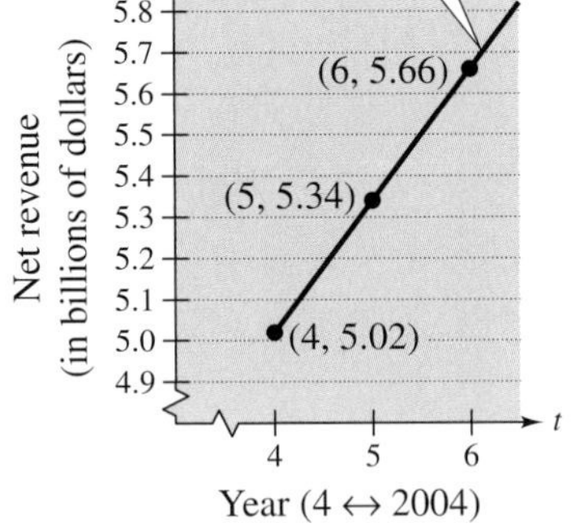

Figure 4.49

✓ **CHECKPOINT** *Now try Exercise 107.*

The prediction method illustrated in Example 7 is called **linear extrapolation.** Note in Figure 4.50 that for linear extrapolation, the estimated point lies *to the right* of the given points. When the estimated point lies *between* two given points, the method is called **linear interpolation,** as shown in Figure 4.51.

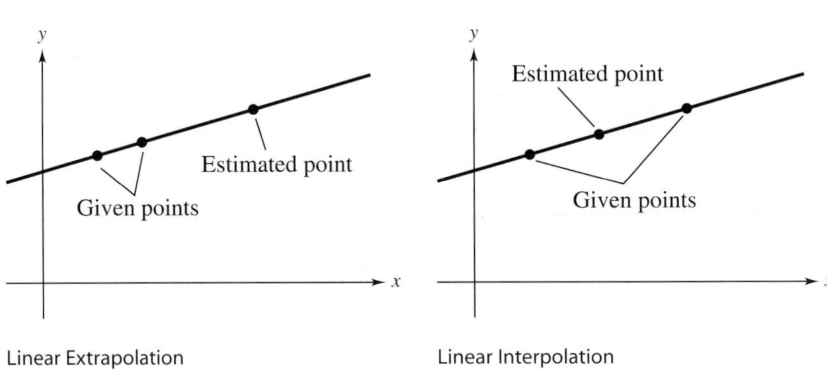

Linear Extrapolation
Figure 4.50

Linear Interpolation
Figure 4.51

In the linear equation $y = mx + b$, you know that m represents the slope of the line. In applications, the slope of a line can often be interpreted as the *rate of change of y with respect to x.* Rates of change should always be described with appropriate units of measure.

EXAMPLE 8 Using Slope as a Rate of Change 

A rock climber is climbing up a 500-foot cliff. By 1 P.M., the rock climber has climbed 115 feet up the cliff. By 4 P.M., the climber has reached a height of 280 feet, as shown in Figure 4.52. Find the average rate of change of the climber and use this rate of change to find a linear model that relates the height of the climber to the time.

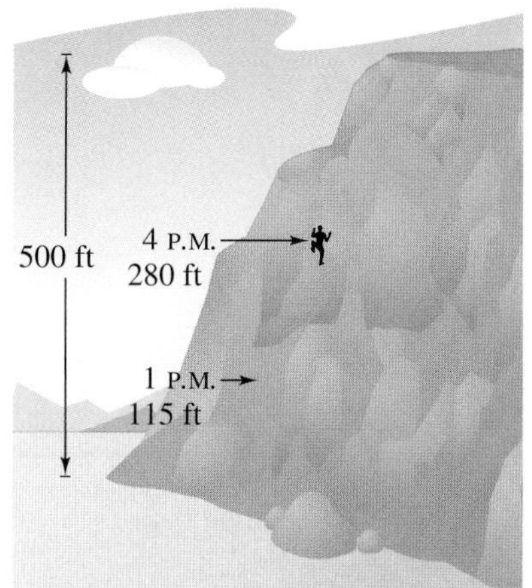

Figure 4.52

Solution

Let y represent the height of the climber and let t represent the time. Then the two points that represent the climber's two positions are $(t_1, y_1) = (1, 115)$ and $(t_2, y_2) = (4, 280)$. So, the average rate of change of the climber is

$$\text{Average rate of change} = \frac{y_2 - y_1}{t_2 - t_1} = \frac{280 - 115}{4 - 1} = 55 \text{ feet per hour.}$$

So, an equation that relates the height of the climber to the time is

$$y - y_1 = m(t - t_1) \qquad \text{Point-slope form}$$

$$y - 115 = 55(t - 1) \qquad \text{Substitute 115 for } y_1, \text{ 1 for } t_1, \text{ and 55 for } m.$$

$$y = 55t + 60. \qquad \text{Linear model}$$

✓ **CHECKPOINT** *Now try Exercise 103.*

You have now studied several formulas that relate to equations of lines. In the summary below, remember that the formulas that deal with slope cannot be applied to vertical lines. For instance, the lines $x = 2$ and $y = 3$ are perpendicular, but they do not follow the "negative reciprocal property" of perpendicular lines because the line $x = 2$ is vertical (and has no slope).

Study Tip

The slope-intercept form of the equation of a line is better suited for *sketching a line.* On the other hand, the point-slope form of the equation of a line is better suited for *creating the equation of a line,* given its slope and a point on the line.

Summary of Equations of Lines

1. Slope of the line through (x_1, y_1) and (x_2, y_2): $m = \dfrac{y_2 - y_1}{x_2 - x_1}$
2. General form of the equation of a line: $ax + by + c = 0$
3. Equation of a vertical line: $x = a$
4. Equation of a horizontal line: $y = b$
5. Slope-intercept form of the equation of a line: $y = mx + b$
6. Point-slope form of the equation of a line: $y - y_1 = m(x - x_1)$
7. Parallel lines have *equal* slopes: $m_1 = m_2$
8. Perpendicular lines have *negative reciprocal* slopes: $m_1 = -\dfrac{1}{m_2}$

Concept Check

1. In the equation $y = mx + b$, what do m and b represent?

2. When is it most convenient to use the slope-intercept form of the equation of a line? When is it most convenient to use the point-slope form of the equation of a line?

3. Explain how to find the equation of a line perpendicular to a given line through the point (a, b).

4. What is the slope of a vertical line? What is the slope of a horizontal line?

4.5 EXERCISES

Go to pages 284–285 to record your assignments.

Developing Skills

In Exercises 1–14, write an equation of the line that passes through the point and has the specified slope. Sketch the line. ***See Example 1.***

1. $(0, 0), m = -2$
2. $(0, -2), m = 3$
3. $(6, 0), m = \frac{1}{2}$
4. $(0, 10), m = -\frac{1}{4}$
5. ✓ $(-2, 1), m = 2$
6. $(3, -5), m = -1$
7. $(-8, -1), m = -\frac{1}{5}$
8. $(12, 4), m = \frac{2}{3}$
9. $\left(\frac{1}{2}, -3\right), m = 0$
10. $\left(-\frac{5}{4}, 6\right), m = 0$
11. $\left(0, \frac{3}{2}\right), m = \frac{2}{3}$
12. $\left(-\frac{5}{2}, 0\right), m = \frac{3}{4}$
13. $(2, 4), m = -0.8$
14. $(6, -3), m = 0.67$

In Exercises 15–26, use the point-slope form to write an equation of the line that passes through the point and has the specified slope. Write the equation in slope-intercept form. ***See Example 2.***

15. ✓ $(0, -4), m = 4$
16. $(-7, 0), m = 1$
17. $(-3, 6), m = -3$
18. $(-4, 1), m = -4$
19. $(9, 0), m = -\frac{1}{3}$
20. $(0, 2), m = \frac{3}{5}$
21. $(-10, 4), m = 0$
22. $(-2, -5), m = 0$
23. $(8, -1), m = -\frac{3}{4}$
24. $(1, 10), m = -\frac{1}{6}$
25. $(-2, 1), m = \frac{3}{8}$
26. $(-1, -3), m = \frac{1}{2}$

In Exercises 27–38, determine the slope of the line.

27. $y = \frac{3}{8}x - 4$
28. $y = -\frac{3}{5}x - 2$
29. $y - 2 = 5(x + 3)$
30. $y + 3 = -2(x - 6)$
31. $y + \frac{5}{6} = \frac{2}{3}(x + 4)$
32. $y - \frac{1}{4} = \frac{5}{8}\left(x - \frac{13}{5}\right)$
33. $y + 9 = 0$
34. $y - 6 = 0$
35. $x - 12 = 0$
36. $x + 5 = 0$
37. $3x - 2y + 10 = 0$
38. $5x + 4y - 8 = 0$

In Exercises 39–42, write the slope-intercept form of the line that has the specified y-intercept and slope.

39.

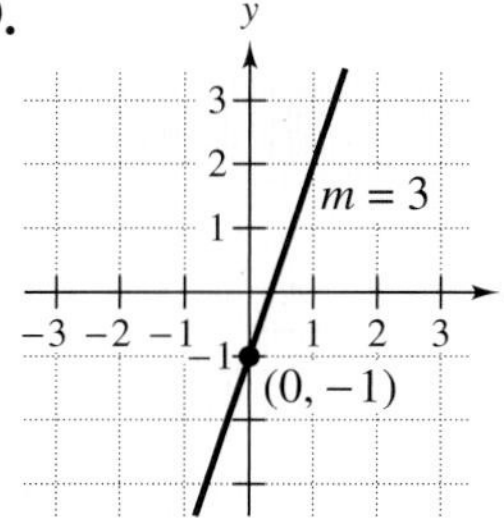

40.

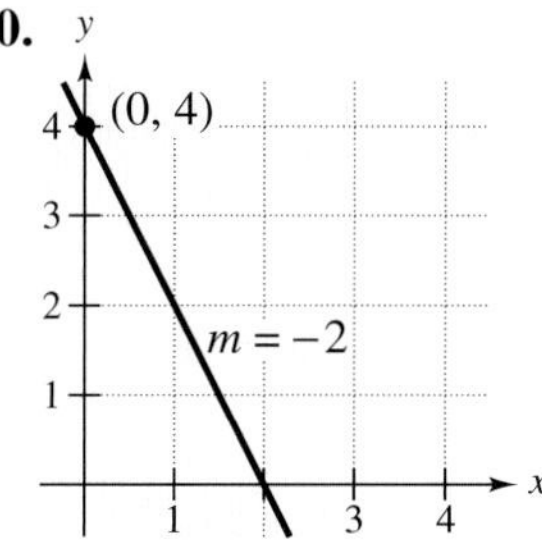

41.

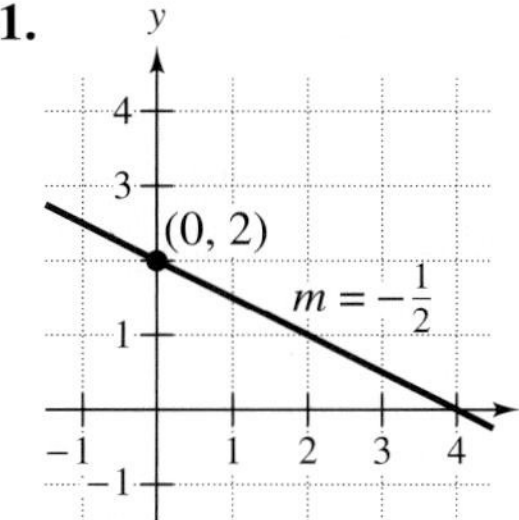

42.

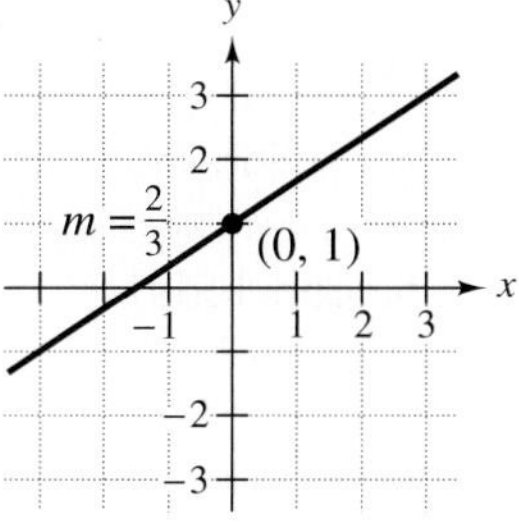

In Exercises 43–46, write the point-slope form of the equation of the line.

43.

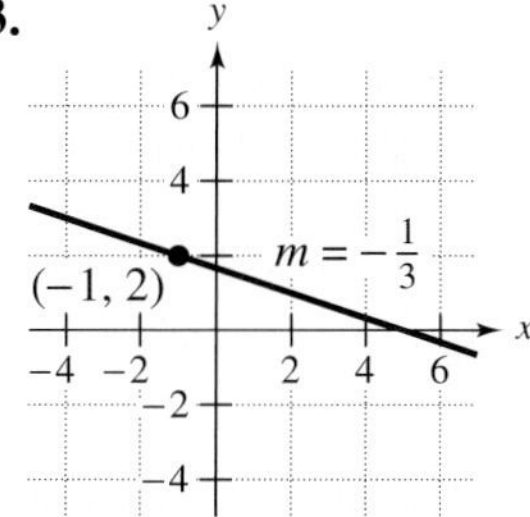

44.

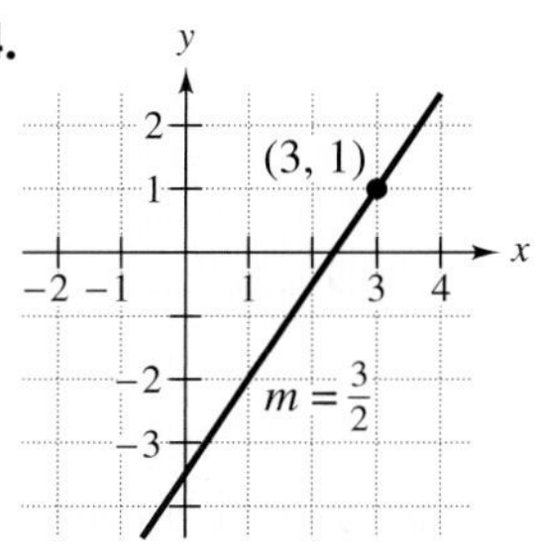

45.

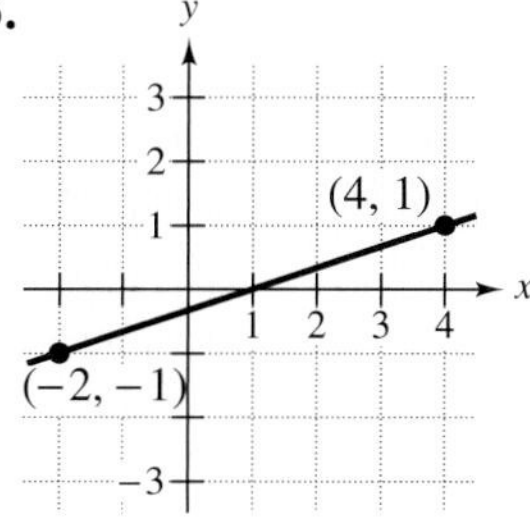

46.

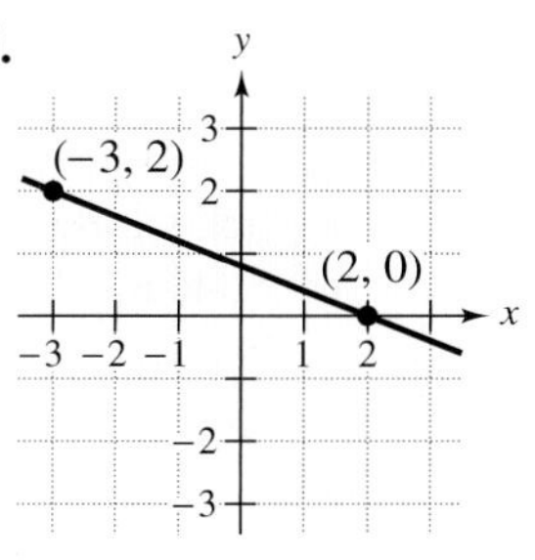

In Exercises 47–58, write an equation of the line that passes through the points. When possible, write the equation in slope-intercept form. Sketch the line. ***See Example 3.***

47. $(0, 0), (4, -4)$

48. $(0, 0), (-2, 4)$

49. $(4, 0), (2, -4)$

50. $(-1, 7), (0, 3)$

✓ **51.** $(-2, 3), (6, -5)$

52. $(-4, 6), (-2, 3)$

53. $(-6, 2), (3, 5)$

54. $(-9, 7), (-4, 4)$

55. $(5, -1), (5, 2)$

56. $(0, 3), (5, 3)$

57. $\left(\frac{5}{2}, -1\right), \left(\frac{9}{2}, 7\right)$

58. $\left(4, \frac{5}{3}\right), \left(-1, \frac{2}{3}\right)$

In Exercises 59–72, write an equation of the line passing through the points. Write the equation in general form.

59. $(0, 3), (3, 0)$

60. $(0, -2), (-2, 0)$

61. $(5, -1), (-5, 5)$

62. $(2, 10), (-5, -4)$

63. $(2, 13), (-6, 1)$

64. $(-5, 7), (-2, 1)$

65. $(5, -1), (-7, -4)$

66. $(3, 5), (1, 6)$

67. $(-3, 8), (2, 5)$

68. $(-9, -9), (-7, -5)$

69. $\left(2, \frac{1}{2}\right), \left(\frac{1}{2}, \frac{5}{2}\right)$

70. $\left(\frac{1}{4}, 1\right), \left(-\frac{3}{4}, -\frac{2}{3}\right)$

71. $(1, 0.6), (2, -0.6)$

72. $(-8, 0.6), (2, -2.4)$

In Exercises 73–82, write an equation of the line that passes through the point and is (a) parallel and (b) perpendicular to the given line. ***See Examples 4 and 5.***

✓ **73.** $(2, 1)$
$x - y = 3$

74. $(-3, 2)$
$x + y = 7$

75. $(-12, 4)$
$3x + 4y = 7$

76. $(15, -2)$
$5x - 3y = 0$

77. $(1, 3)$
$2x + y = 0$

78. $(-5, -2)$
$x + 5y = 3$

79. $(-1, 0)$
$y + 3 = 0$

80. $(2, 5)$
$x - 4 = 0$

81. $(4, -1)$
$3y - 2x = 7$

82. $(-6, 5)$
$4x - 5y = 2$

In Exercises 83–90, write an equation of the line. ***See Example 6.***

✓ **83.** Vertical line through $(-2, 4)$

84. Horizontal line through $(7, 3)$

85. Horizontal line through $\left(\frac{1}{2}, \frac{2}{3}\right)$

86. Vertical line through $\left(\frac{1}{4}, 0\right)$

87. Line passing through $(4, 1)$ and $(4, 8)$

88. Line passing through $(-1, 5)$ and $(6, 5)$

89. Line passing through $(1, -8)$ and $(7, -8)$

90. Line passing through $(3, 0)$ and $(3, 5)$

Graphical Exploration In Exercises 91–94, use a graphing calculator to graph the lines in the same viewing window. Use the square setting. Are the lines parallel, perpendicular, or neither?

91. $y_1 = -0.4x + 3$
$y_2 = \frac{5}{2}x - 1$

92. $y_1 = \frac{2x - 3}{3}$
$y_2 = \frac{4x + 3}{6}$

93. $y_1 = 0.4x + 1$
$y_2 = x + 2.5$

94. $y_1 = \frac{3}{4}x - 5$
$y_2 = -\frac{3}{4}x + 2$

Graphical Exploration In Exercises 95 and 96, use a graphing calculator to graph the equations in the same viewing window. Use the square setting. What can you conclude?

95. $y_1 = \frac{1}{3}x + 2$
$y_2 = -3x + 2$

96. $y_1 = 4x + 2$
$y_2 = -\frac{1}{4}x + 2$

Solving Problems

97. *Wages* A sales representative receives a monthly salary of \$2000 plus a commission of 2% of the total monthly sales. Write a linear model that relates total monthly wages W to sales S.

98. *Reimbursed Expenses* A sales representative is reimbursed \$250 per day for lodging and meals plus \$0.30 per mile driven. Write a linear model that relates the daily cost C to the number of miles driven x.

99. *Discount* A department store is offering a 20% discount on all items in its inventory.

(a) Write a linear model that relates the sale price S to the list price L.

(b) Use a graphing calculator to graph the model.

(c) Use the graph to estimate the sale price of a coffee maker whose list price is \$49.98. Verify your estimate algebraically.

100. *Discount* A car dealership is offering a 15% discount on all trucks in its inventory.

(a) Write a linear model that relates the sale price S to the list price L.

(b) Use a graphing calculator to graph the model.

(c) Use the graph to estimate the sale price of a truck whose list price is \$29,500. Verify your estimate algebraically.

101. *Depreciation* A school district purchases a high-volume printer, copier, and scanner for \$25,000. After 1 year, its depreciated value is \$22,700. The depreciation is linear.

(a) Write a linear model that relates the value V of the equipment to the time t in years.

(b) Use the model to estimate the value of the equipment after 3 years.

102. *Depreciation* A sub shop purchases a used pizza oven for \$875. After 1 year, its depreciated value is \$790. The depreciation is linear.

(a) Write a linear model that relates the value V of the oven to the time t in years.

(b) Use the model to estimate the value of the oven after 5 years.

✓ **103.** *Swimming Pool* You use a garden hose to fill a swimming pool at a constant rate. The pool contains 45 gallons of water after 5 minutes. After 30 minutes, the pool contains 120 gallons of water. Find the average rate of change of the number of gallons of water in the pool and use this rate of change to find a linear model that relates the number of gallons of water in the pool to the time.

104. *Bike Path* A city is paving a bike path. The same length of path is paved each day. After 4 days, there is 14 miles of path remaining to be paved. After 6 more days, there is 11 miles of path remaining to be paved. Find the average rate of change of the distance (in miles) remaining to be paved and use this rate of change to find a linear model that relates the distance (in miles) remaining to be paved to the number of days.

105. *Rental Demand* A real estate office handles an apartment complex with 50 units. When the rent per unit is \$580 per month, all 50 units are occupied. However, when the rent is \$625 per month, the average number of occupied units drops to 47. Assume that the relationship between the monthly rent p and the demand x is linear.

(a) Represent the given information as two ordered pairs of the form (x, p). Plot these ordered pairs.

(b) Write a linear model that relates the monthly rent p to the demand x. Graph the model and describe the relationship between the rent and the demand.

(c) *Linear Extrapolation* Use the model in part (b) to predict the number of units occupied if the rent is raised to \$655.

(d) *Linear Interpolation* Use the model in part (b) to estimate the number of units occupied if the rent is \$595.

106. *Soft Drink Demand* When soft drinks sold for \$0.80 per can at football games, approximately 6000 cans were sold. When the price was raised to \$1.00 per can, the demand dropped to 4000. Assume that the relationship between the price p and the demand x is linear.

(a) Represent the given information as two ordered pairs of the form (x, p). Plot these ordered pairs.

(b) Write a linear model that relates the price p to the demand x. Graph the model and describe the relationship between the price and the demand.

(c) *Linear Extrapolation* Use the model in part (b) to predict the number of soft drinks sold if the price is raised to \$1.10.

(d) *Linear Interpolation* Use the model in part (b) to estimate the number of soft drinks sold if the price is \$0.90.

✓ **107.** *Net Revenue* The net revenue of the Adidas Group was \$6.6 billion in 2005 and \$10.1 billion in 2006. Using only this information, write a linear equation that models the net revenue s in terms of the year t. Then predict the net revenue for 2007. (Let $t = 5$ represent 2005.) (Source: The Adidas Group)

108. *Net Revenue* The net revenue of Cabela's Inc. was \$1.8 billion in 2005 and \$2.1 billion in 2006. Using only this information, write a linear equation that models the net revenue R in terms of the year t. Then predict the net revenue for 2007. (Let $t = 5$ represent 2005.) (Source: Cabela's Inc.)

109. *Rate of Change* You are given the dollar value of a product in 2007 and the rate at which the value is expected to change during the next 5 years. Use this information to write a linear equation that gives the dollar value V of the product in terms of the year t. (Let $t = 7$ represent 2007.)

	2007 Value	*Rate*
(a)	\$2540	\$125 increase per year
(b)	\$156	\$4.50 increase per year
(c)	\$20,400	\$2000 decrease per year
(d)	\$45,000	\$2300 decrease per year

110. *Graphical Interpretation* Match each of the situations labeled (a), (b), (c), and (d) with one of the graphs labeled (e), (f), (g), and (h). Then determine the slope of each line and interpret the slope in the context of the real-life situation.

(a) A friend is paying you \$10 per week to repay a \$100 loan.

(b) An employee is paid \$12.50 per hour plus \$1.50 for each unit produced per hour.

(c) A sales representative receives \$40 per day for food plus \$0.32 for each mile traveled.

(d) A television purchased for \$600 depreciates \$100 per year.

(e)

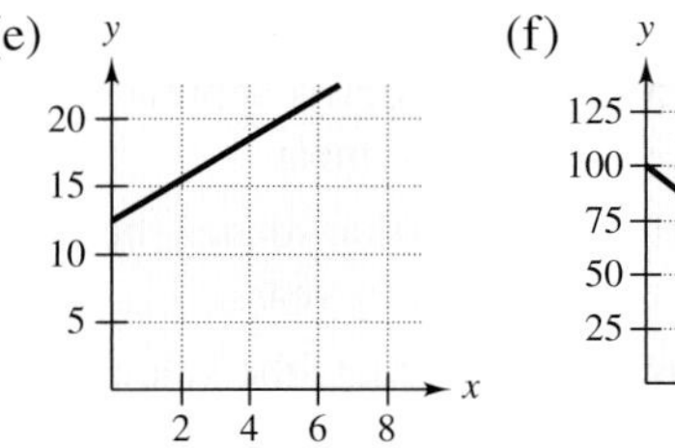

(f)

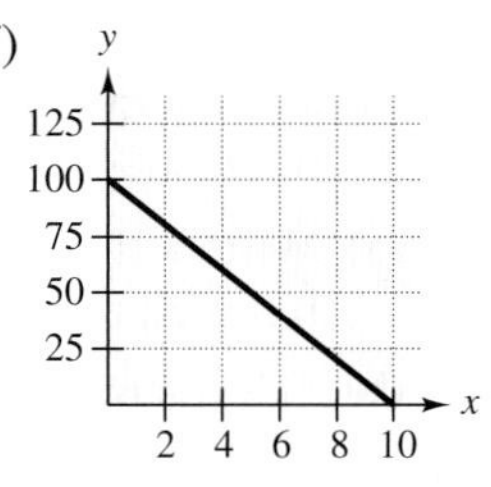

(g)

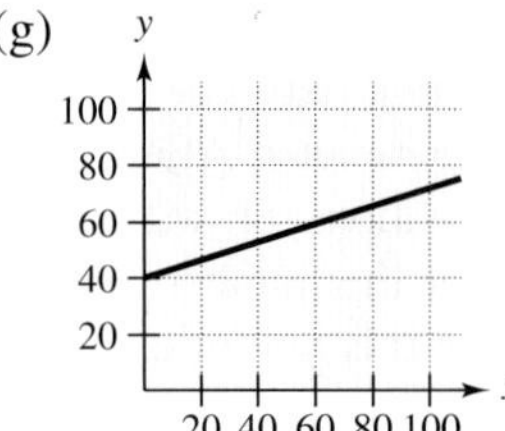

(h)

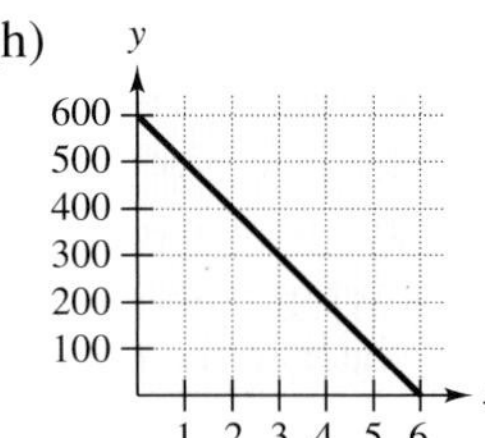

Explaining Concepts

111. Can any pair of points on a line be used to calculate the slope of the line? Explain.

112. Can the equation of a vertical line be written in slope-intercept form? Explain.

113. Explain how to find analytically the x-intercept of the line given by $y = mx + b$.

114. *Think About It* Find the slope of the line given by $5x + 7y - 21 = 0$. Use the same process to find a formula for the slope of the line $ax + by + c = 0$ where $b \neq 0$.

115. What is implied about the graphs of the lines $a_1x + b_1y + c_1 = 0$ and $a_2x + b_2y + c_2 = 0$ if $\frac{a_1}{b_1} = \frac{a_2}{b_2}$?

116. *Research Project* Use a newspaper or weekly news magazine to find an example of data that is *increasing* linearly with time. Find a linear model that relates the data to time. Repeat the project for data that is *decreasing*.

Cumulative Review

In Exercises 117–120, simplify the expression.

117. $4(3 - 2x)$

118. $x^2(xy^3)$

119. $3x - 2(x - 5)$

120. $u - [3 + (u - 4)]$

In Exercises 121–124, solve for y in terms of x.

121. $3x + y = 4$

122. $4 - y + x = 0$

123. $4x - 5y = -2$

124. $3x + 4y - 5 = 0$

In Exercises 125–128, determine the slope of the line passing through the given points.

125. $(3, 0)$ and $(4, 2)$

126. $(-1, 3)$ and $(2, -5)$

127. $(0, -2)$ and $(-7, -1)$

128. $(-2, 4)$ and $(-5, 10)$

4.6 Graphs of Linear Inequalities

Eric Mencher/*Philadelphia Inquirer*/MCT/Newscom

What You Should Learn

1 ▶ Determine whether an ordered pair is a solution of a linear inequality in two variables.
2 ▶ Sketch graphs of linear inequalities in two variables.
3 ▶ Use linear inequalities to model and solve real-life problems.

Why You Should Learn It

Linear inequalities can be used to model and solve real-life problems. For instance, in Exercise 61 on page 283, you will use a linear inequality to analyze the number of pieces assembled by a furniture company in one day.

1 ▶ Determine whether an ordered pair is a solution of a linear inequality in two variables.

Linear Inequalities in Two Variables

A **linear inequality in two variables,** x and y, is an inequality that can be written in one of the forms below (where a and b are not both zero).

$$ax + by < c, \quad ax + by > c, \quad ax + by \le c, \quad ax + by \ge c$$

Some examples include: $x - y > 2$, $3x - 2y \le 6$, $x \ge 5$, and $y < -1$. An ordered pair (x_1, y_1) is a **solution** of a linear inequality in x and y if the inequality is true when x_1 and y_1 are substituted for x and y, respectively. For instance, the ordered pair $(3, 2)$ is a solution of the inequality $x - y > 0$ because $3 - 2 > 0$ is a true statement.

EXAMPLE 1 Verifying Solutions of Linear Inequalities

Determine whether each point is a solution of $3x - y \ge -1$.

a. $(0, 0)$ **b.** $(1, 4)$ **c.** $(-1, 2)$

Solution

a. $3x - y \ge -1$ Write original inequality.

$3(0) - 0 \overset{?}{\ge} -1$ Substitute 0 for x and 0 for y.

$0 \ge -1$ Inequality is satisfied. ✓

Because the inequality is satisfied, the point $(0, 0)$ *is* a solution.

b. $3x - y \ge -1$ Write original inequality.

$3(1) - 4 \overset{?}{\ge} -1$ Substitute 1 for x and 4 for y.

$-1 \ge -1$ Inequality is satisfied. ✓

Because the inequality is satisfied, the point $(1, 4)$ *is* a solution.

c. $3x - y \ge -1$ Write original inequality.

$3(-1) - 2 \overset{?}{\ge} -1$ Substitute -1 for x and 2 for y.

$-5 \not\ge -1$ Inequality is not satisfied. ✗

Because the inequality is not satisfied, the point $(-1, 2)$ *is not* a solution.

✓ **CHECKPOINT** *Now try Exercise 1.*

2 ▶ Sketch graphs of linear inequalities in two variables.

The Graph of a Linear Inequality in Two Variables

Study Tip

When the inequality is less than ($<$) or greater than ($>$), the graph of the corresponding equation is a dashed line because the points on the line are *not* solutions of the inequality. When the inequality is less than or equal to ($\le$) or greater than or equal to ($\ge$), the graph of the corresponding equation is a solid line because the points on the line *are* solutions of the inequality.

The **graph** of an inequality is the collection of all solution points of the inequality. To sketch the graph of a linear inequality such as

$3x - 2y < 6$ Original linear inequality

begin by sketching the graph of the *corresponding linear equation*

$3x - 2y = 6.$ Corresponding linear equation

Use *dashed* lines for the inequalities $<$ and $>$ and *solid* lines for the inequalities $\le$ and $\ge$. The graph of the equation separates the plane into two regions, called **half-planes.** In each half-plane, one of the following *must* be true.

1. All points in the half-plane are solutions of the inequality.

2. No point in the half-plane is a solution of the inequality.

So, you can determine whether the points in an entire half-plane satisfy the inequality by simply testing *one* point in the region. This graphing procedure is summarized as follows.

Sketching the Graph of a Linear Inequality in Two Variables

1. Replace the inequality sign by an equal sign and sketch the graph of the resulting equation. (Use a dashed line for $<$ or $>$ and a solid line for $\le$ or $\ge$.)

2. Test one point in each of the half-planes formed by the graph in Step 1. If the point satisfies the inequality, then shade the entire half-plane to denote that every point in the region satisfies the inequality.

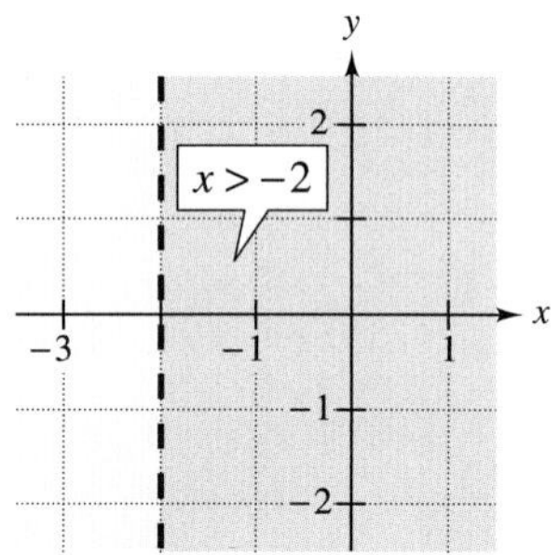

Figure 4.53

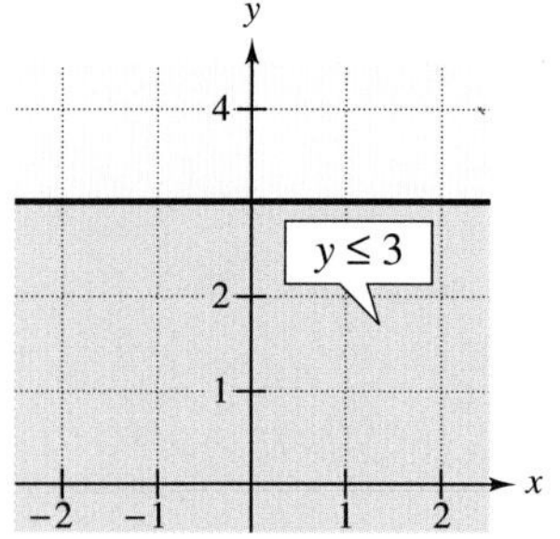

Figure 4.54

EXAMPLE 2 Sketching the Graph of a Linear Inequality

Sketch the graph of each linear inequality.

a. $x > -2$

b. $y \le 3$

Solution

a. The graph of the corresponding equation $x = -2$ is a vertical line. The points (x, y) that satisfy the inequality $x > -2$ are those lying to the right of this line, as shown in Figure 4.53.

b. The graph of the corresponding equation $y = 3$ is a horizontal line. The points (x, y) that satisfy the inequality $y \le 3$ are those lying on or below this line, as shown in Figure 4.54.

✓ ***CHECKPOINT*** *Now try Exercise 21.*

Notice that in Example 2 a dashed line is used for the graph of $x > -2$ and a solid line is used for the graph of $y \le 3$.

Study Tip

You can use any point that is not on the line as a test point. However, the origin is often the most convenient test point because it is easy to evaluate expressions in which 0 is substituted for each variable.

EXAMPLE 3 Sketching the Graph of a Linear Inequality

Sketch the graph of the linear inequality

$x - y < 2.$

Solution

The graph of the corresponding equation

$x - y = 2$ Write corresponding equation.

is the line shown in Figure 4.55. Because the origin (0, 0) does not lie on the line, use it as the test point.

$x - y < 2$ Write original inequality.

$0 - 0 \stackrel{?}{<} 2$ Substitute 0 for x and 0 for y.

$0 < 2$ Inequality is satisfied. ✓

Because (0, 0) satisfies the inequality, the graph consists of the half-plane lying above the line. Try checking a point below the line. Regardless of which point you choose, you will see that it does not satisfy the inequality.

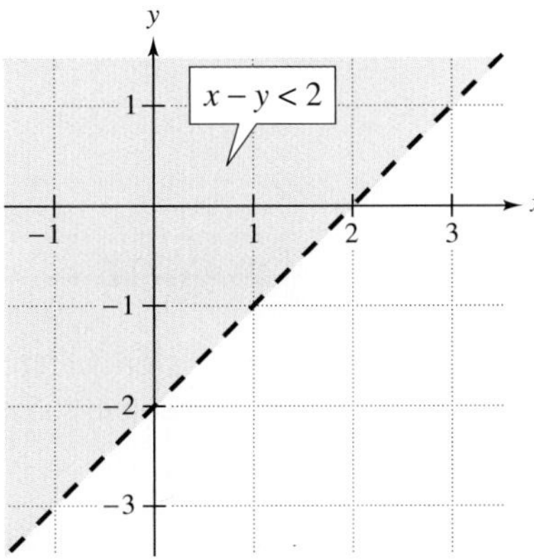

Figure 4.55

✓ **CHECKPOINT** *Now try Exercise 25.*

Technology: Tip

Many graphing calculators are capable of graphing linear inequalities. Consult the user's guide for your graphing calculator for specific instructions.

The graph of $y \leq -x + 2$ shown at the right.

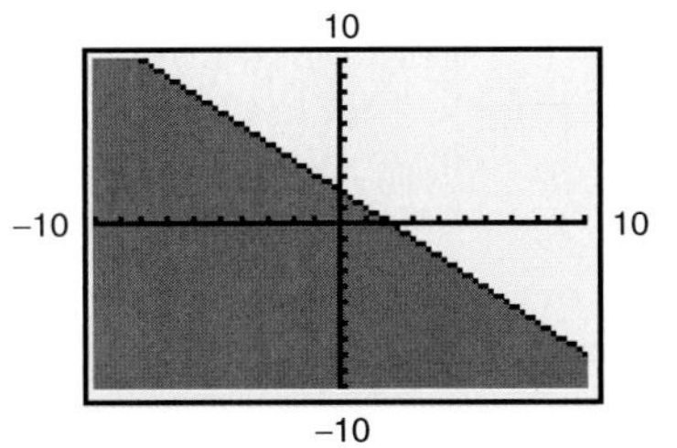

> **Study Tip**
>
> The solution of the inequality
>
> $$y < \tfrac{3}{2}x - \tfrac{5}{2}$$
>
> is a half-plane with the line
>
> $$y = \tfrac{3}{2}x - \tfrac{5}{2}$$
>
> as its boundary. The y-values that are less than those yielded by this equation make the inequality true. So, you want to shade the half-plane with the smaller y-values, as shown in Figure 4.56.

For a linear inequality in two variables, you can sometimes simplify the graphing procedure by writing the inequality in *slope-intercept* form. For instance, by writing $x - y < 2$ in the form $y > x - 2$, you can see that the solution points lie *above* the line $y = x - 2$, as shown in Figure 4.55. Similarly, by writing the inequality $3x - 2y > 5$ in the form

$$y < \frac{3}{2}x - \frac{5}{2}$$

you can see that the solutions lie *below* the line $y = \frac{3}{2}x - \frac{5}{2}$, as shown in Figure 4.56.

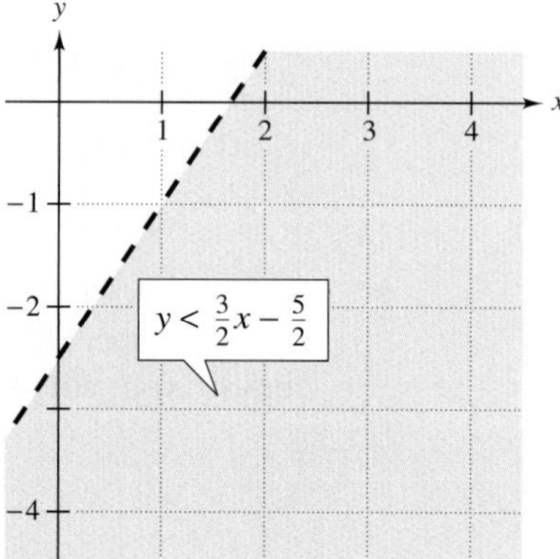

Figure 4.56

EXAMPLE 4 Sketching the Graph of a Linear Inequality

Use the slope-intercept form of a linear equation as an aid in sketching the graph of the inequality $5x + 4y \le 12$.

Solution

To begin, rewrite the inequality in slope-intercept form.

$$5x + 4y \le 12 \qquad \text{Write original inequality.}$$

$$4y \le -5x + 12 \qquad \text{Subtract } 5x \text{ from each side.}$$

$$y \le -\frac{5}{4}x + 3 \qquad \text{Divide each side by 4.}$$

From this form, you can conclude that the solution is the half-plane lying *on* or *below* the line $y = -\frac{5}{4}x + 3$. The graph is shown in Figure 4.57. You can verify this by testing the solution point $(0, 0)$.

$$5x + 4y \le 12 \qquad \text{Write original inequality.}$$

$$5(0) + 4(0) \overset{?}{\le} 12 \qquad \text{Substitute 0 for } x \text{ and 0 for } y.$$

$$0 \le 12 \qquad \text{Inequality is satisfied. ✓}$$

✓ **CHECKPOINT** *Now try Exercise 31.*

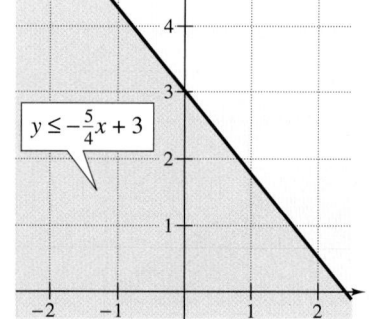

Figure 4.57

3 ▶ Use linear inequalities to model and solve real-life problems.

Application

EXAMPLE 5 Working to Meet a Budget

Your budget requires you to earn *at least* \$250 per week. You work two part-time jobs. One is at a fast-food restaurant, which pays \$8 per hour, and the other is tutoring for \$10 per hour. Let x represent the number of hours you work at the restaurant and let y represent the number of hours you work as a tutor.

a. Write an inequality that represents the numbers of hours you can work at each job in order to meet your budget requirements.

b. Graph the inequality and find at least two ordered pairs (x, y) that identify the numbers of hours you can work at each job in order to meet your budget requirements.

Solution

a. To write the inequality, use the problem-solving method.

Verbal Model: Hourly pay at fast-food restaurant · Number of hours at fast-food restaurant + Hourly pay tutoring · Number of hours tutoring ≥ Minimum weekly earnings

Labels:		
	Hourly pay at fast-food restaurant $= 8$	(dollars per hour)
	Number of hours at fast-food restaurant $= x$	(hours)
	Hourly pay for tutoring $= 10$	(dollars per hour)
	Number of hours tutoring $= y$	(hours)
	Minimum weekly earnings $= 250$	(dollars)

Algebraic Inequality: $8x + 10y \geq 250$

b. To sketch the graph, rewrite the inequality in slope-intercept form.

$$8x + 10y \geq 250 \quad \text{Write original inequality.}$$

$$10y \geq -8x + 250 \quad \text{Subtract } 8x \text{ from each side.}$$

$$y \geq -\frac{4}{5}x + 25 \quad \text{Divide each side by 10.}$$

Graph the corresponding equation

$$y = -\frac{4}{5}x + 25$$

and shade the half-plane lying above the line, as shown in Figure 4.58. From the graph, you can see that two solutions that will yield the desired weekly earnings of at least \$250 are (10, 17) and (20, 15). In other words, you can work 10 hours at the restaurant and 17 hours as a tutor, or 20 hours at the restaurant and 15 hours as a tutor, to meet your budget requirements. There are many other solutions.

✓ **CHECKPOINT** *Now try Exercise 59.*

Study Tip

The variables in Example 5 cannot represent negative numbers. So, the graph of the inequality does not include points in Quadrants II, III, or IV.

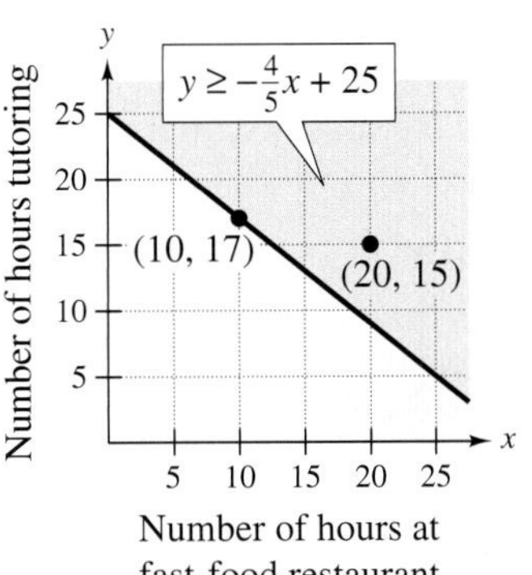

Figure 4.58

Concept Check

1. List the four forms of a linear inequality in variables x and y.

2. What is meant by saying that (x_1, y_1) is a solution of a linear inequality in x and y?

3. When sketching the graph of a linear inequality, should you test a point that is on the graph of the corresponding linear equation? Explain.

4. What point is often the most convenient test point to use when sketching the graph of a linear inequality?

4.6 EXERCISES

Go to pages 284–285 to record your assignments.

Developing Skills

In Exercises 1–8, determine whether the points are solutions of the inequality. ***See Example 1.***

	Inequality	*Points*
✓	1. $x + 4y > 10$	(a) $(0, 0)$
		(b) $(3, 2)$
		(c) $(1, 2)$
		(d) $(-2, 4)$
	2. $2x + 3y > 9$	(a) $(0, 0)$
		(b) $(1, 1)$
		(c) $(2, 2)$
		(d) $(-2, 5)$
	3. $-3x + 5y \le 12$	(a) $(1, 2)$
		(b) $(2, -3)$
		(c) $(1, 3)$
		(d) $(2, 8)$
	4. $5x + 3y < 100$	(a) $(25, 10)$
		(b) $(6, 10)$
		(c) $(0, -12)$
		(d) $(4, 5)$
	5. $3x - 2y < 2$	(a) $(1, 3)$
		(b) $(2, 0)$
		(c) $(0, 0)$
		(d) $(3, -5)$
	6. $y - 2x > 5$	(a) $(4, 13)$
		(b) $(8, 1)$
		(c) $(0, 7)$
		(d) $(1, -3)$

Inequality	*Points*
7. $6x + 4y \ge -4$	(a) $(2, -4)$
	(b) $(6, -9)$
	(c) $(-3, 4)$
	(d) $(3, -1)$
8. $5y + 8x \le 14$	(a) $(-3, 8)$
	(b) $(7, -6)$
	(c) $(1, 1)$
	(d) $(3, 0)$

In Exercises 9–12, state whether the boundary of the graph of the inequality should be dashed or solid.

9. $2x + 3y < 6$
10. $2x + 3y \le 6$
11. $2x + 3y \ge 6$
12. $2x + 3y > 6$

In Exercises 13–16, match the inequality with its graph. [The graphs are labeled (a), (b), (c), and (d).]

(a)

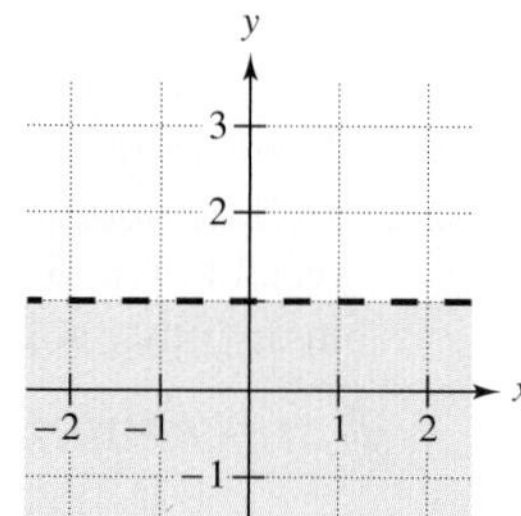

(b)

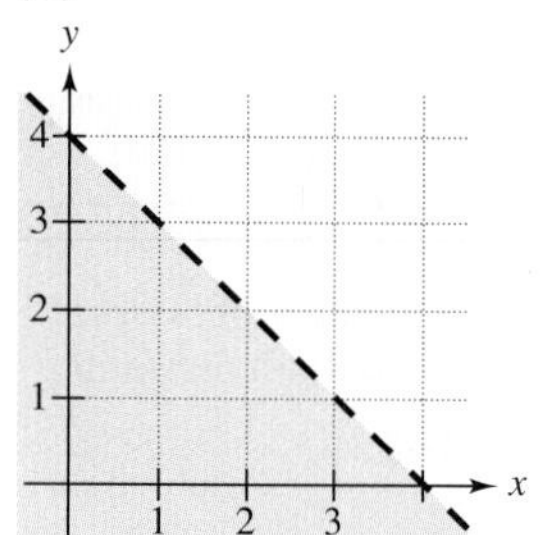

(c) (d)

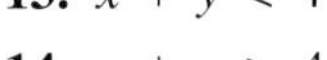

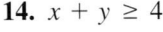

13. $x + y < 4$
14. $x + y \geq 4$
15. $x > 1$
16. $y < 1$

In Exercises 17–20, match the inequality with its graph. [The graphs are labeled (a), (b), (c), and (d).]

(a)

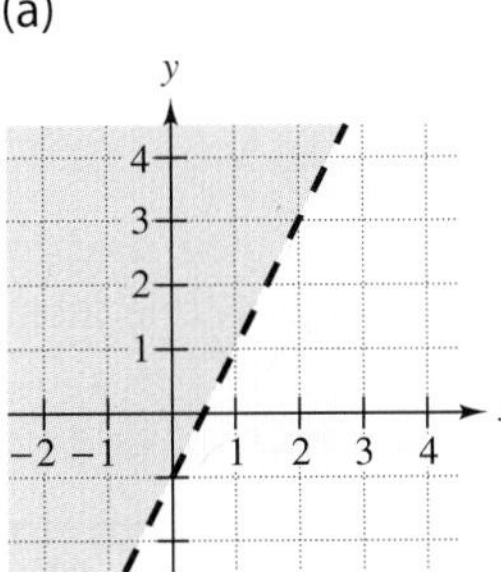

(b)

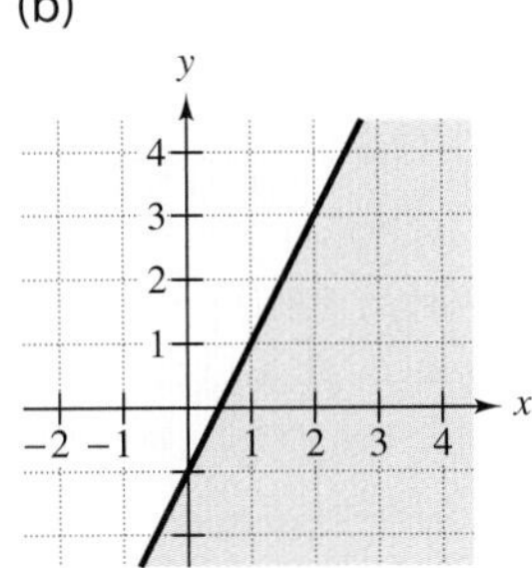

(c)

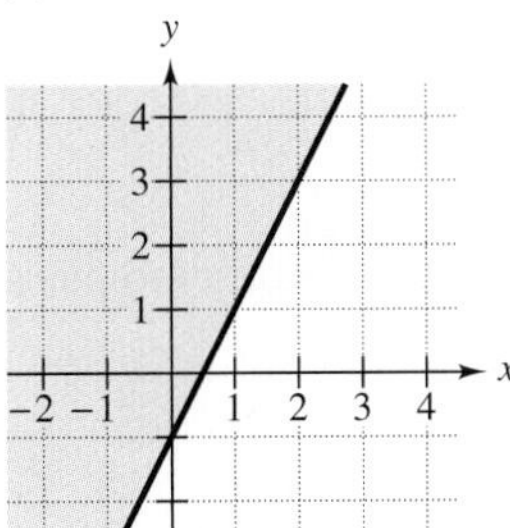

(d)

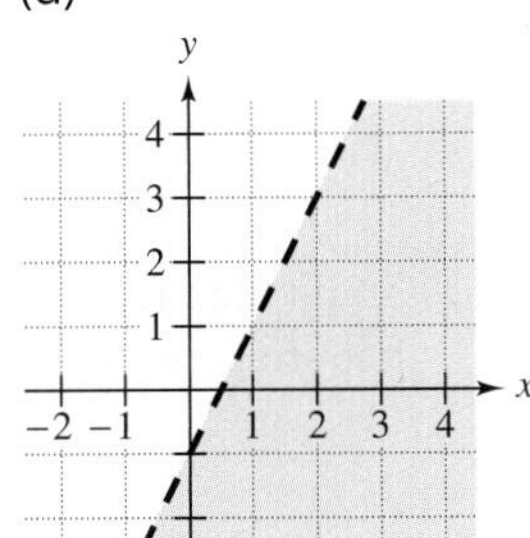

17. $2x - y \leq 1$
18. $2x - y < 1$
19. $2x - y \geq 1$
20. $2x - y > 1$

In Exercises 21–36, sketch the graph of the linear inequality. ***See Examples 2–4.***

21. $y \geq 3$
22. $x > -4$
23. $y \leq 3x$
24. $y \geq 5x$
25. $y \leq 2x - 1$
26. $y \geq 2x - 1$
27. $x - y < 0$
28. $x + y > 0$
29. $y > -2x + 10$
30. $y < 3x + 1$
31. $-3x + 2y < 6$
32. $x - 2y \leq -6$
33. $x \geq 3y - 5$
34. $x \leq -2y + 10$
35. $y - 3 > \frac{1}{2}(x - 4)$
36. $y + 1 < -2(x - 3)$

In Exercises 37–44, use a graphing calculator to graph the linear inequality.

37. $y \geq 2x - 1$
38. $y \leq 4 - 0.5x$
39. $y \leq -2x + 4$
40. $y \geq x - 3$
41. $y \geq \frac{1}{2}x + 2$
42. $y \leq -\frac{2}{3}x + 6$
43. $6x + 10y - 15 \leq 0$
44. $3x - 2y + 4 \geq 0$

In Exercises 45–50, write the statement as a linear inequality. Then sketch the graph of the inequality.

45. y is more than six times x.
46. x is at most three times y.
47. The sum of x and y is at least 9.
48. The difference of x and y is less than 20.
49. y is no more than the sum of x and 3.
50. The sum of x and 7 is more than three times y.

In Exercises 51–56, write an inequality for the shaded region shown in the figure.

51.

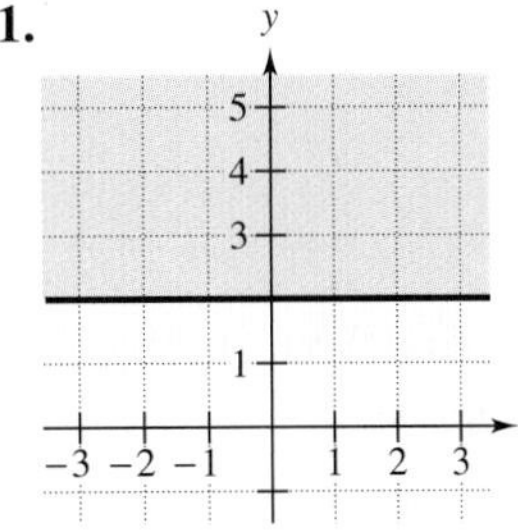

52.

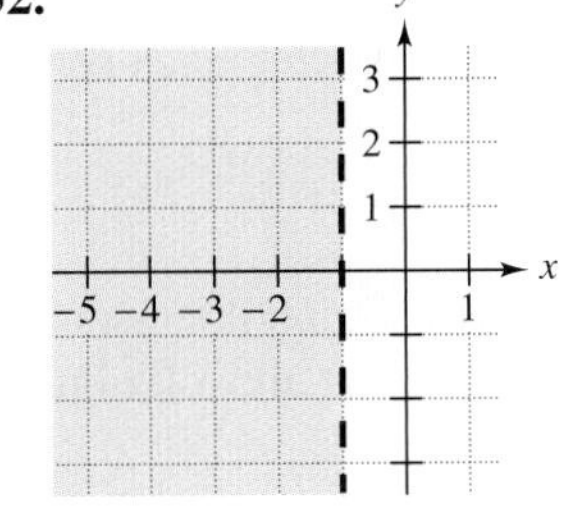

53.

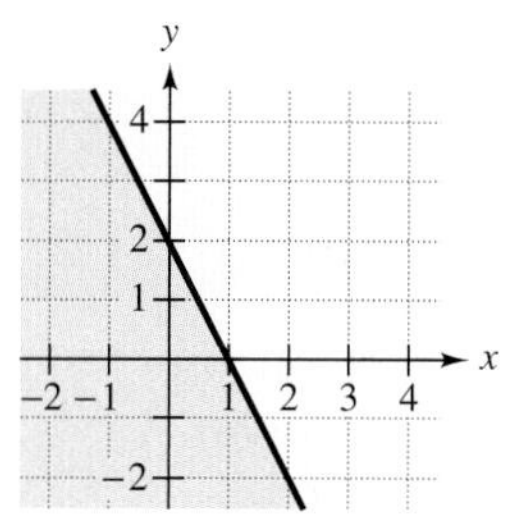

54.

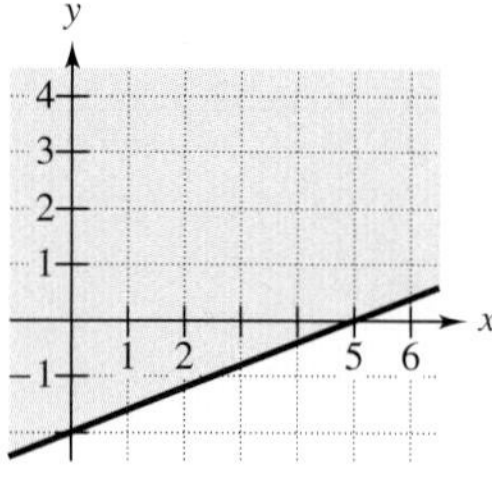

55.

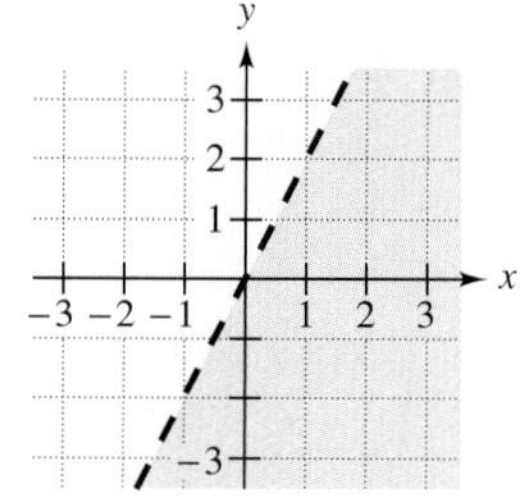

56.

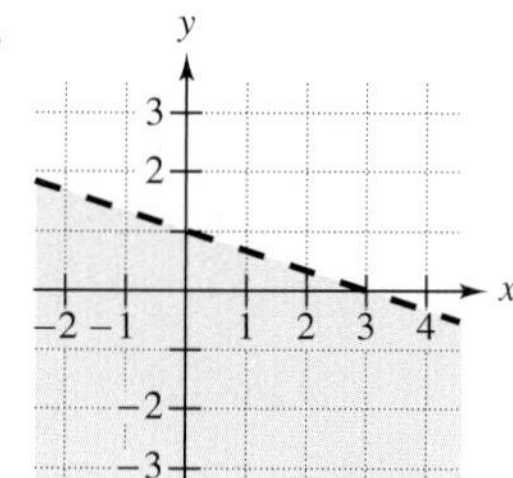

Solving Problems

57. *Nutrition* A nutritionist recommends that the fat calories consumed per day should be at most 35% of the total calories consumed per day.

(a) Write a linear inequality that represents the different numbers of total calories and fat calories that are recommended for one day.

(b) Graph the inequality and find three ordered pairs that are solutions of the inequality.

58. *Music* A jazz band holds tryouts for new members. For the tryout, you need to perform two songs that are no longer than 10 minutes combined.

(a) Write a linear inequality that represents the different numbers of minutes that can be spent playing each song.

(b) Graph the inequality and find three ordered pairs that are solutions of the inequality.

✓ 59. *Part-Time Jobs* Your budget requires you to earn at least $210 per week. You work two part-time jobs. One is at a grocery store, which pays $9 per hour, and the other is mowing lawns, which pays $12 per hour.

(a) Write a linear inequality that represents the different numbers of hours you can work at each job in order to meet your budget requirements.

(b) Graph the inequality and find three ordered pairs that are solutions of the inequality.

60. *Money* A cash register must have at least $25 in change consisting of d dimes and q quarters.

(a) Write a linear inequality that represents the different numbers of dimes and quarters that can satisfy the requirement.

(b) Graph the inequality and find three ordered pairs that are solutions of the inequality.

61. *Manufacturing* Each table produced by a furniture company requires 1 hour in the assembly center. The matching chair requires $1\frac{1}{2}$ hours in the assembly center. A total of 12 hours per day is available in the assembly center.

(a) Write a linear inequality that represents the different numbers of hours that can be spent assembling tables and chairs.

(b) Graph the inequality and find three ordered pairs that are solutions of the inequality.

62. *Inventory* A store sells two models of computers. The costs to the store of the two models are \$2000 and \$3000, and the owner of the store does not want more than \$30,000 invested in the inventory for these two models.

(a) Write a linear inequality that represents the different numbers of each model that can be held in inventory.

(b) Graph the inequality and find three ordered pairs that are solutions of the inequality.

Explaining Concepts

63. Write the inequality whose graph consists of all points above the x-axis.

64. Write an inequality whose graph has no points in the first quadrant.

65. Does $2x < 2y$ have the same graph as $y > x$? Explain.

66. Are there any points in the coordinate plane that are not solutions of either $y > x - 2$ or $y < x - 2$? Explain.

67. Explain the difference between graphing the solution of the inequality $x \geq 1$ (a) on the real number line and (b) on a rectangular coordinate system.

Cumulative Review

In Exercises 68–73, find a ratio that compares the relative sizes of the quantities. (Use the same units of measurement for both quantities.)

68. Thirty-six feet to 8 feet
69. Eight dimes to 36 nickels
70. Four hours to 45 minutes
71. Fifty centimeters to 3 meters
72. Nine pounds to 8 ounces
73. Forty-six inches to 4 feet

In Exercises 74–79, write an equation of the line that passes through the points. When possible, write the equation in slope-intercept form.

74. $(0, 0), (-3, 6)$
75. $(5, 8), (1, -4)$
76. $(5, 0), (0, -2)$
77. $(4, 3), (-4, 5)$
78. $(6, -1), (-3, -3)$
79. $\left(\frac{5}{6}, -2\right), \left(\frac{1}{6}, 4\right)$

What Did You Learn?

Use these two pages to help prepare for a test on this chapter. Check off the key terms and key concepts you know. You can also use this section to record your assignments.

Plan for Test Success

Date of test: / /

Study dates and times: / / at : A.M./P.M.
/ / at : A.M./P.M.

Things to review:

- ☐ Key Terms, *p. 284*
- ☐ Key Concepts, *pp. 284–285*
- ☐ Your class notes
- ☐ Your assignments
- ☐ Study Tips, *pp. 232, 239, 249, 254, 269, 276, 277, 278, 279*
- ☐ Technology Tips, *pp. 220, 229, 254, 265, 266, 277*
- ☐ Mid-Chapter Quiz, *p. 248*
- ☐ Review Exercises, *pp. 286–290*
- ☐ Chapter Test, *p. 291*
- ☐ Video Explanations Online
- ☐ Tutorial Online

Key Terms

- ☐ rectangular coordinate system, *p. 216*
- ☐ Cartesian plane, *p. 216*
- ☐ *x*-axis, *p. 216*
- ☐ *y*-axis, *p. 216*
- ☐ origin, *p. 216*
- ☐ quadrants, *p. 216*
- ☐ ordered pair, *p. 216*
- ☐ coordinates, *p. 216*
- ☐ plotting a point, *p. 217*
- ☐ table of values, *p. 219*
- ☐ solution, *p. 219*
- ☐ graph of a linear equation, *p. 228*
- ☐ linear equation in two variables, *p. 229*
- ☐ intercepts, *p. 232*
- ☐ relation, *p. 238*
- ☐ function, *p. 239*
- ☐ independent variable, *p. 240*
- ☐ dependent variable, *p. 240*
- ☐ domain of a function, *p. 241*
- ☐ range of a function, *p. 241*
- ☐ evaluating a function, *p. 241*
- ☐ slope, *p. 249*
- ☐ slope-intercept form, *p. 254*
- ☐ parallel, *p. 256*
- ☐ perpendicular, *p. 257*
- ☐ point-slope form, *p. 264*
- ☐ two-point form, *p. 265*
- ☐ general form, *p. 267*
- ☐ linear inequality in two variables, *p. 275*
- ☐ solution of a linear inequality, *p. 275*
- ☐ graph of an inequality, *p. 276*
- ☐ half-planes, *p. 276*

Key Concepts

4.1 Ordered Pairs and Graphs

Assignment: ______________________ Due date: ________

☐ **Use the rectangular coordinate system.**

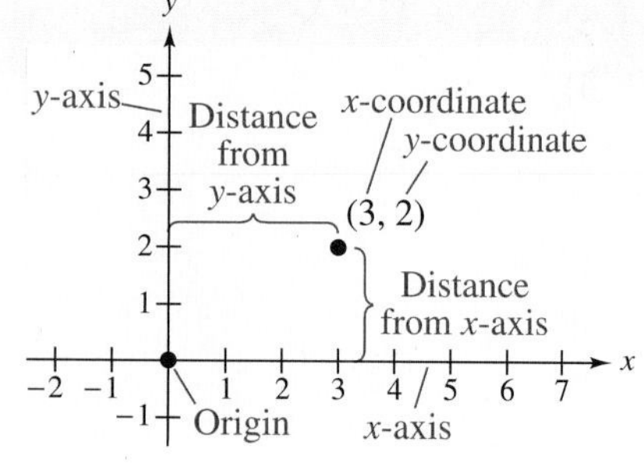

☐ **Use guidelines for verifying solutions.**

To verify that an ordered pair (x, y) is a solution to an equation with variables x and y, use the following steps.

1. Substitute the values of x and y into the equation.
2. Simplify each side of the equation.
3. If each side simplifies to the same number, the ordered pair is a solution. If the two sides yield different numbers, the ordered pair is not a solution.

4.2 Graphs of Equations in Two Variables

Assignment: ____________________ Due date: ________

☐ **Use the point-plotting method of sketching a graph.**

See page 228.

☐ **Find the x- and y-intercepts of a graph.**

To find the x-intercept(s), let $y = 0$ and solve the equation for x.

To find the y-intercept(s), let $x = 0$ and solve the equation for y.

4.3 Relations, Functions, and Graphs

Assignment: ____________________ Due date: ________

☐ **Define function.**

A function is a set of ordered pairs in which no two ordered pairs have the same first component and different second components.

☐ **Use the Vertical Line Test.**

A set of points on a rectangular coordinate system is the graph of y as a function of x if and only if no vertical line intersects the graph at more than one point.

4.4 Slope and Graphs of Linear Equations

Assignment: ____________________ Due date: ________

☐ **Determine the slope of a line.**

The slope m of a nonvertical line passing through points (x_1, y_1) and (x_2, y_2) is

$$m = \frac{y_2 - y_1}{x_2 - x_1} = \frac{\text{Change in } y}{\text{Change in } x} = \frac{\text{Rise}}{\text{Run}}$$

where $x_1 \neq x_2$.

☐ **Write linear equations in slope-intercept form.**

The graph of the equation $y = mx + b$ is a line whose slope is m and whose y-intercept is $(0, b)$.

☐ **Define parallel and perpendicular lines.**

Consider two nonvertical lines whose slopes are m_1 and m_2.
Parallel: $m_1 = m_2$

Perpendicular: $m_1 = -\dfrac{1}{m_2}$

4.5 Equations of Lines

Assignment: ____________________ Due date: ________

☐ **Write equations of lines.**

The point-slope form of the equation of line with slope m and passing through the point (x_1, y_1) is $y - y_1 = m(x - x_1)$.

The general form of the equation of a line is $ax + by + c = 0$.

☐ **Write equations of horizontal and vertical lines.**

A horizontal line has a slope of zero and an equation of the form $y = b$. A vertical line has an undefined slope and an equation of the form $x = a$.

4.6 Graphs of Linear Inequalities

Assignment: ____________________ Due date: ________

☐ **Check a solution of a linear inequality.**

An ordered pair (x_1, y_1) is a solution of a linear inequality in x and y if the inequality is true when x_1 and y_1 are substituted for x and y, respectively.

☐ **Sketch the graph of a linear inequality in two variables.**

See page 276.

Review Exercises

4.1 Ordered Pairs and Graphs

1 ▶ Plot and find the coordinates of a point on a rectangular coordinate system.

In Exercises 1 and 2, plot the points on a rectangular coordinate system.

1. $(-1, 6), (4, -3), (-2, 2), (3, 5)$
2. $(0, -1), (-4, 2), (5, 1), (3, -4)$

In Exercises 3 and 4, determine the coordinates of the points.

3.

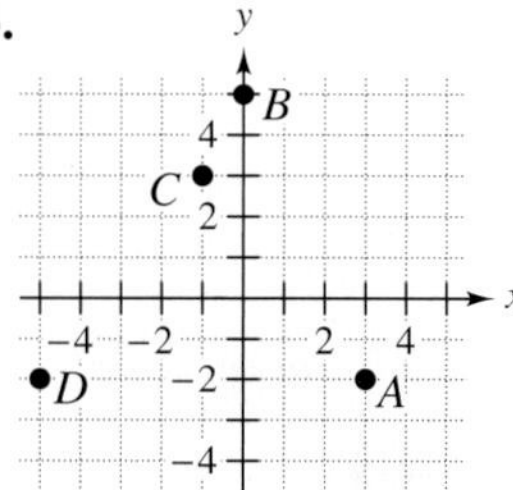

4.

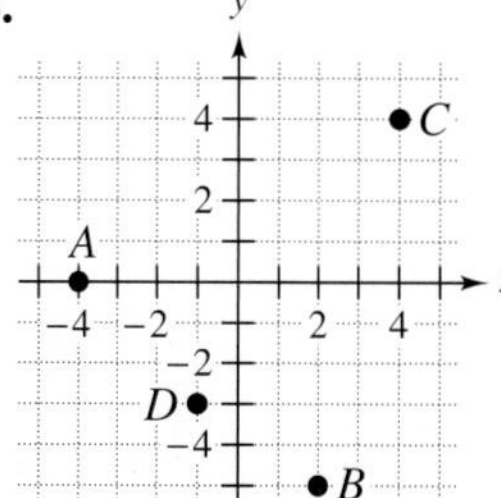

In Exercises 5–8, determine the quadrant(s) in which the point is located, or the axis on which the point is located, without plotting it.

5. $(x, 5),\ x < 0$
6. $(-3, y),\ y > 0$
7. $(-6, y)$, y is a real number.
8. $(x, -1)$, x is a real number.

2 ▶ Construct a table of values for equations and determine whether ordered pairs are solutions of equations.

In Exercises 9 and 10, complete the table of values. Then plot the solution points on a rectangular coordinate system.

9.

x	-1	0	1	2
$y = -\frac{1}{2}x - 1$				

10.

x	-1	0	1	2
$y = \frac{3}{2}x + 5$				

In Exercises 11–16, solve the equation for *y*.

11. $3x + 4y = 12$
12. $2x + 3y = 6$
13. $9x - 3y = 12$
14. $-7x + y = -22$
15. $x - 2y = 8$
16. $-x - 3y = 9$

In Exercises 17–20, determine whether the ordered pairs are solutions of the equation.

17. $x - 3y = 4$
    (a) $(1, -1)$
    (b) $(0, 0)$
    (c) $(2, 1)$
    (d) $(5, -2)$
18. $y - 2x = -1$
    (a) $(3, 7)$
    (b) $(0, -1)$
    (c) $(-2, -5)$
    (d) $(-1, 0)$
19. $y = \frac{2}{3}x + 3$
    (a) $(3, 5)$
    (b) $(-3, 1)$
    (c) $(-6, 0)$
    (d) $(0, 3)$
20. $y = \frac{1}{4}x + 2$
    (a) $(-4, 1)$
    (b) $(-8, 0)$
    (c) $(12, 5)$
    (d) $(0, 2)$

3 ▶ Use the verbal problem-solving method to plot points on a rectangular coordinate system.

21. *Organizing Data* The data from a study measuring the relationship between the wattage x of a standard 120-volt light bulb and the energy rate y (in lumens) is shown in the table.

x	25	40	60	100	150	200
y	235	495	840	1675	2650	3675

(a) Plot the data in the table.

(b) Use the graph to describe the relationship between the wattage and energy rate.

22. *Organizing Data* The table shows the numbers of employees (in millions) in the health services industry in the United States for the years 2000 through 2005, where x represents the year. (Source: U.S. Bureau of Labor Statistics)

x	2000	2001	2002	2003	2004	2005
y	12.7	13.1	13.6	13.9	14.2	14.5

(a) Plot the data in the table.

(b) Use the graph to describe the relationship between the year and the number of employees.

(c) Find the percent increase in the number of employees in the health services industry from 2000 to 2005.

4.2 Graphs of Equations in Two Variables

1 ▶ Sketch graphs of equations using the point-plotting method.

In Exercises 23 and 24, complete the table and use the results to sketch the graph of the equation.

23. $y = x^2 - 1$

x	-2	-1	0	1	2
y					

24. $y = |x - 2|$

x	0	1	2	3	4
y					

In Exercises 25–36, sketch the graph of the equation using the point-plotting method.

25. $y = 7$

26. $x = -2$

27. $y = 3x$

28. $y = -2x$

29. $y = 4 - \frac{1}{2}x$

30. $y = \frac{3}{2}x - 3$

31. $y - 2x - 4 = 0$

32. $3x + 2y + 6 = 0$

33. $y = 2x - 1$

34. $y = 5 - 4x$

35. $y = \frac{1}{4}x + 2$

36. $y = -\frac{2}{3}x - 2$

2 ▶ Find and use x- and y-intercepts as aids to sketching graphs.

In Exercises 37–42, find the x- and y-intercepts (if any) of the graph of the equation. Then sketch the graph of the equation and label the x- and y-intercepts.

37. $y = \frac{2}{5}x - 2$

38. $y = \frac{1}{3}x + 1$

39. $2x - y = 4$

40. $3x - y = 10$

41. $4x + 2y = 8$

42. $9x + 3y = 6$

3 ▶ Use the verbal problem-solving method to write an equation and sketch its graph.

43. *Creating a Model* The cost of producing a DVD is \$125, plus \$3 per DVD. Let C represent the total cost and let x represent the number of DVDs. Write an equation that relates C and x and sketch its graph.

44. *Creating a Model* Let y represent the distance traveled by a train that is moving at a constant speed of 80 miles per hour. Let t represent the number of hours the train has traveled. Write an equation that relates y and t, and sketch its graph.

4.3 Relations, Functions, and Graphs

1 ▶ Identify the domain and range of a relation.

In Exercises 45–48, find the domain and range of the relation.

45. $\{(8, 3), (-2, 7), (5, 1), (3, 8)\}$

46. $\{(0, 1), (-1, 3), (4, 6), (-7, 5)\}$

47. $\{(2, -3), (-2, 3), (2, 4), (4, 0)\}$

48. $\{(1, 7), (-3, 4), (6, 4), (-2, 4)\}$

2 ▶ Determine if relations are functions by inspection or by using the Vertical Line Test.

In Exercises 49 and 50, determine whether the relation represents a function.

49. $\{(-1, 3), (3, 3), (0, 3), (7, 9), (10, 9)\}$

50. $\{(a, 4), (b, 4), (b, 8), (c, 9)\}$

In Exercises 51–54, use the Vertical Line Test to determine whether *y* is a function of *x*.

51.

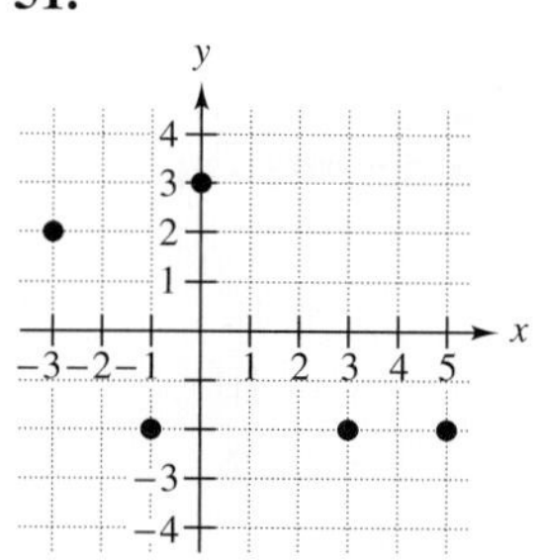

52.

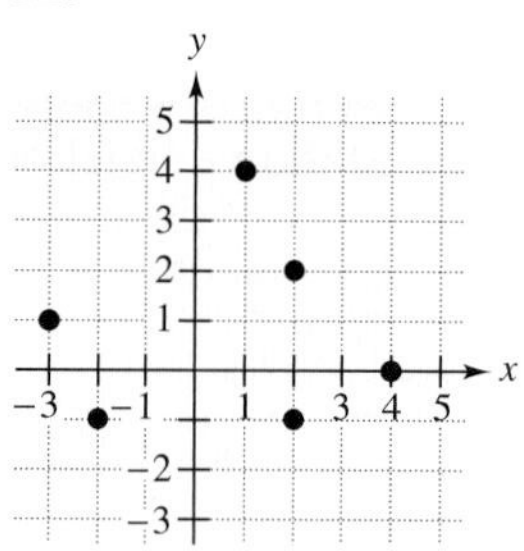

53.

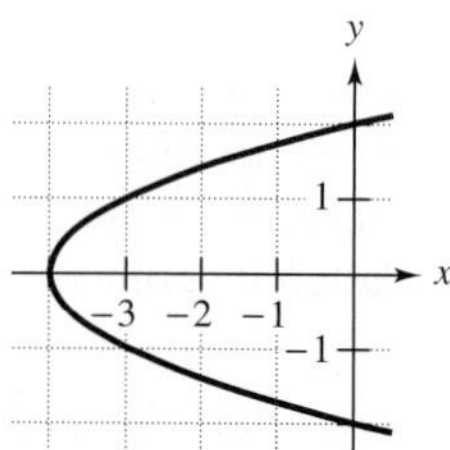

54.

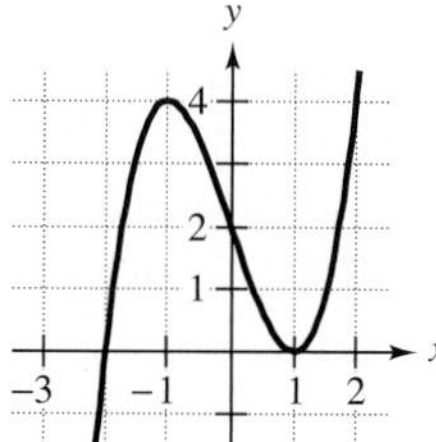

3 ▶ Use function notation and evaluate functions.

In Exercises 55–60, evaluate the function as indicated, and simplify.

55. $f(x) = \frac{3}{4}x$ (a) $f(-1)$ (b) $f(4)$ (c) $f(10)$ (d) $f\left(-\frac{4}{3}\right)$

56. $f(x) = 2x - 7$ (a) $f(-1)$ (b) $f(3)$ (c) $f\left(\frac{1}{2}\right)$ (d) $f(-4)$

57. $g(t) = -16t^2 + 64$ (a) $g(0)$ (b) $g\left(\frac{1}{4}\right)$ (c) $g(1)$ (d) $g(2)$

58. $h(u) = u^3 + 2u^2 - 4$ (a) $h(0)$ (b) $h(3)$ (c) $h(-1)$ (d) $h(-2)$

59. $f(x) = |2x + 3|$ (a) $f(0)$ (b) $f(5)$ (c) $f(-4)$ (d) $f\left(-\frac{3}{2}\right)$

60. $f(x) = |x| - 4$ (a) $f(-1)$ (b) $f(1)$ (c) $f(-4)$ (d) $f(2)$

61. *Demand* The demand for a product is a function of its price. Consider the demand function $f(p) = 40 - 0.2p$, where p is the price in dollars. Find the demand for (a) $p = 10$, (b) $p = 50$, and (c) $p = 100$.

62. *Profit* The profit for a product is a function of the amount spent on advertising for the product. In the profit function

$f(x) = 8000 + 2000x - 50x^2$

x is the amount (in hundreds of dollars) spent on advertising. Find the profit for (a) $x = 5$, (b) $x = 10$, and (c) $x = 20$.

4 ▶ Identify the domain of a function.

In Exercises 63–66, find the domain of the function.

63. $f{:}\{(1, 5), (2, 10), (3, 15), (4, -10), (5, -15)\}$

64. $g{:}\{(-3, 6), (-2, 4), (-1, 2), (0, 0), (1, -2)\}$

65. $f{:}\ \{(3, -1), (4, 6), (-2, -1), (0, -2), (7, 0)\}$

66. $g{:}\{(-8, 0), (3, -2), (10, 3), (-5, 1), (0, 0)\}$

4.4 Slope and Graphs of Linear Equations

1 ▶ Determine the slope of a line through two points.

In Exercises 67–70, estimate the slope of the line from its graph.

67.

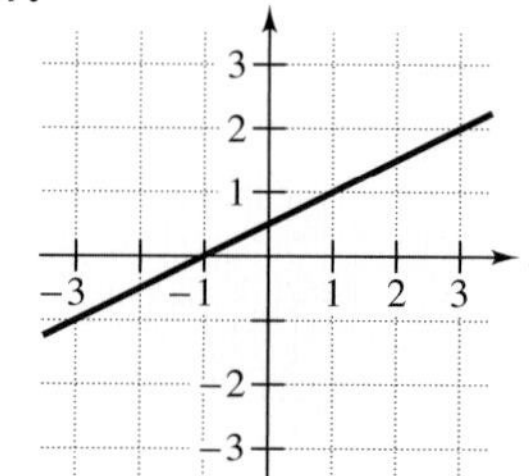

68.

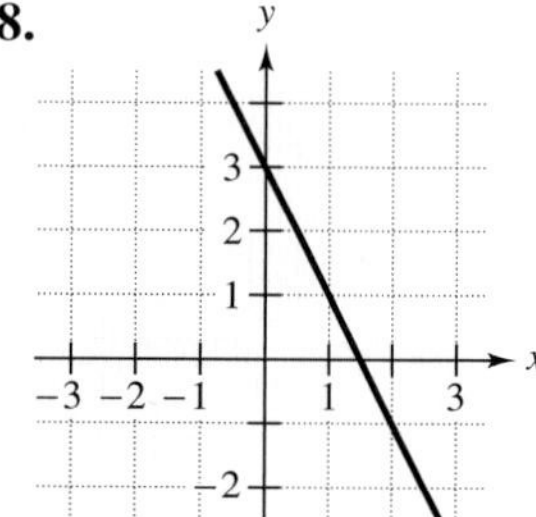

69.

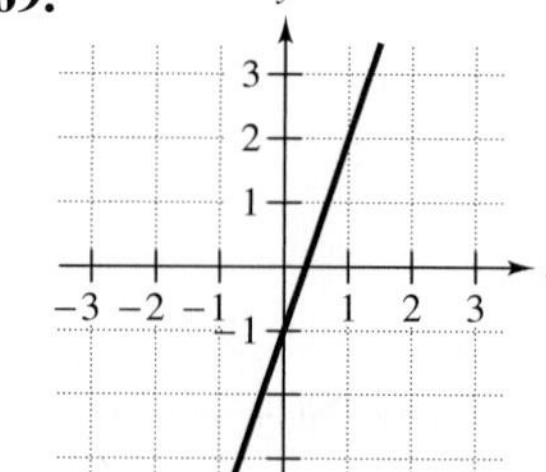

70.

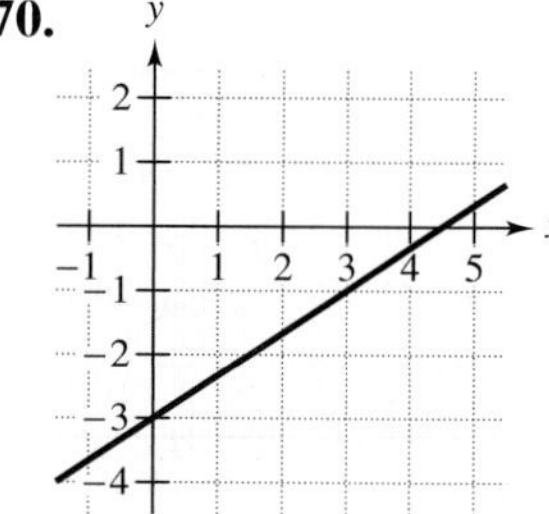

In Exercises 71–76, plot the points and find the slope of the line passing through the points. State whether the line rises, falls, is horizontal, or is vertical.

71. $(2, 1), (14, 6)$
72. $(-2, 2), (3, -10)$
73. $(4, 0), (4, 6)$
74. $(1, 3), (4, 3)$
75. $(-1, -4), (-5, -10)$
76. $(-3, -3), (-8, -6)$
77. *Truck* The floor of a truck is 4 feet above ground level. The end of the ramp used in loading the truck rests on the ground 6 feet behind the truck. Determine the slope of the ramp.
78. *Flight Path* An aircraft is on its approach to an airport. As it flies over a town, its altitude is 15,000 feet. The town is about 10 miles from the airport. Approximate the slope of the linear path followed by the aircraft during landing.

2 ▶ Write linear equations in slope-intercept form and graph the equations.

In Exercises 79–86, write the equation in slope-intercept form. Use the slope and y-intercept to sketch the line.

79. $x + y = 6$
80. $x - y = -3$
81. $2x - y = -1$
82. $-4x + y = -2$
83. $3x + 6y = 12$
84. $7x + 21y = -14$
85. $5y - 2x = 5$
86. $3y - x = 6$

3 ▶ Use slopes to determine whether lines are parallel, perpendicular, or neither.

In Exercises 87–90, determine whether the lines L_1 and L_2 passing through the pairs of points are parallel, perpendicular, or neither.

87. L_1: $(0, 3), (-2, 1)$
L_2: $(-8, -3), (4, 9)$

88. L_1: $(-3, -1), (2, 5)$
L_2: $(2, 11), (8, 6)$

89. L_1: $(3, 6), (-1, -5)$
L_2: $(-2, 3), (4, 7)$

90. L_1: $(-1, 2), (-1, 4)$
L_2: $(7, 3), (4, 7)$

4.5 Equations of Lines

1 ▶ Write equations of lines using the point-slope form.

In Exercises 91–98, use the point-slope form to write an equation of the line that passes through the point and has the specified slope. Write the equation in slope-intercept form.

91. $(4, -1),\ m = 2$
92. $(-5, 2),\ m = 3$
93. $(1, 2),\ m = -4$
94. $(7, -3),\ m = -1$
95. $(-5, -2),\ m = \frac{4}{5}$
96. $(12, -4),\ m = -\frac{1}{6}$
97. $(3, 8)$, m is undefined.
98. $(-4, 6),\ m = 0$

In Exercises 99–106, write an equation of the line passing through the points. Write the equation in general form.

99. $(-4, 0), (0, -2)$
100. $(-4, -2), (4, 6)$
101. $(0, 8), (6, 8)$
102. $(2, -6), (2, 5)$
103. $(-1, -2), (-4, -7)$
104. $\left(0, \frac{4}{3}\right), (3, 0)$
105. $(2.4, 3.3), (6, 7.8)$
106. $(-1.4, 0), (3.2, 9.2)$

In Exercises 107–110, write an equation of the line that passes through the point and is (a) parallel and (b) perpendicular to the given line.

107. $(-6, 3)$
$x - y = -2$

108. $\left(\frac{1}{5}, -\frac{4}{5}\right)$
$5x + y = 2$

109. $\left(\frac{3}{8}, 4\right)$
$y - 9 = 0$

110. $(-2, 1)$
$5x = 2$

2 ▶ Write the equations of horizontal and vertical lines.

In Exercises 111–114, write an equation of the line.

111. Horizontal line through $(-4, 5)$
112. Horizontal line through $(3, -7)$

113. Vertical line through $(5, -1)$

114. Vertical line through $(-10, 4)$

3 ▶ Use linear models to solve application problems.

115. *Wages* A pharmaceutical salesperson receives a monthly salary of \$5500 plus a commission of 7% of the total monthly sales. Write a linear model that relates total monthly wages W to sales S.

116. *Rental Demand* An apartment complex has 50 units. When the rent per unit is \$425 per month, all 50 units are occupied. When the rent is \$480 per month, the average number of occupied units drops to 47. Assume that the relationship between the monthly rent p and the demand x is linear.

(a) Represent the given information as two ordered pairs of the form (x, p). Plot these ordered pairs.

(b) Write a linear model that relates the monthly rent p to the demand x. Graph the model and describe the relationship between the rent and the demand.

(c) *Linear Extrapolation* Use the model in part (b) to predict the number of units occupied if the rent is raised to \$530.

(d) *Linear Interpolation* Use the model in part (b) to estimate the number of units occupied if the rent is \$475.

4.6 Graphs of Linear Inequalities

1 ▶ Determine whether an ordered pair is a solution of a linear inequality in two variables.

In Exercises 117–120, determine whether the points are solutions of the inequality.

117. $x - y > 4$

(a) $(-1, -5)$ (b) $(0, 0)$

(c) $(3, -2)$ (d) $(8, 1)$

118. $y - 2x \leq -1$

(a) $(0, 0)$ (b) $(-2, 1)$

(c) $(-3, 4)$ (d) $(-1, -6)$

119. $3x - 2y < -1$

(a) $(3, 4)$ (b) $(-1, 2)$

(c) $(1, 8)$ (d) $(0, 0)$

120. $-4y + 5x > 3$

(a) $(1, 2)$ (b) $(-3, 6)$

(c) $(-1, -3)$ (d) $(4, 4)$

2 ▶ Sketch graphs of linear inequalities in two variables.

In Exercises 121–126, sketch the graph of the linear inequality.

121. $x - 2 \geq 0$ **122.** $y + 3 < 0$

123. $2x + y < 1$ **124.** $3x - 4y > 2$

125. $x \leq 4y - 2$ **126.** $x \geq 3 - 2y$

In Exercises 127 and 128, write an inequality for the shaded region shown in the figure.

127.

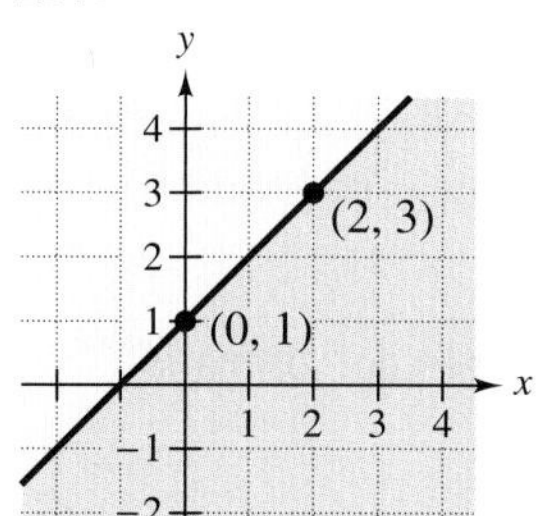

128.

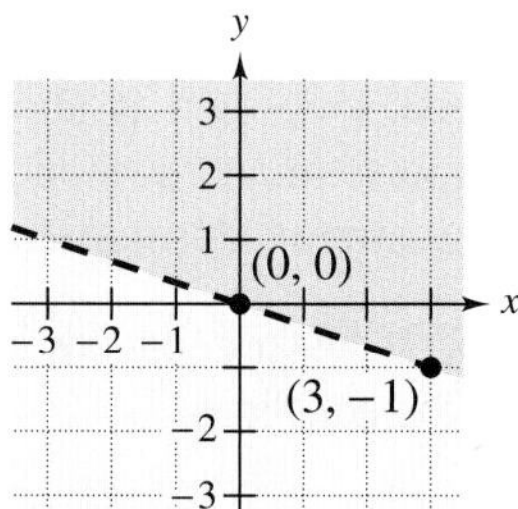

3 ▶ Use linear inequalities to model and solve real-life problems.

129. *Manufacturing* Each DVD player produced by an electronics manufacturer requires 2 hours in the assembly center. Each camcorder produced by the same manufacturer requires 3 hours in the assembly center. A total of 120 hours per week is available in the assembly center. Write a linear inequality that represents the different numbers of hours that can be spent assembling DVD players and camcorders. Graph the inequality and find three ordered pairs that are solutions of the inequality.

130. *Manufacturing* A company produces two types of lawn mowers, Economy and Deluxe. The Deluxe model requires 3 hours in the assembly center and the Economy model requires $1\frac{1}{2}$ hours in the assembly center. A total of 24 hours per day is available in the assembly center. Write a linear inequality that represents the different numbers of hours that can be spent assembling the two models. Graph the inequality and find three ordered pairs that are solutions of the inequality.

Chapter Test

Take this test as you would take a test in class. After you are done, check your work against the answers in the back of the book.

1. Plot the points $(-1, 2)$, $(1, 4)$, and $(2, -1)$ on a rectangular coordinate system. Connect the points with line segments to form a right triangle.

2. Determine whether the ordered pairs are solutions of $y = |x| + |x - 2|$.
   (a) $(0, -2)$ (b) $(0, 2)$ (c) $(-4, 10)$ (d) $(-2, -2)$

3. What is the y-coordinate of any point on the x-axis?

4. Find the x- and y-intercepts of the graph of $8x - 2y = -16$.

5. Complete the table and use the results to sketch the graph of the equation.
   $3x + y = -4$

x	-2	-1	0	1	2
y					

Input, x	Output, y
0	4
1	5
2	8
1	-3
0	-1

Table for 9

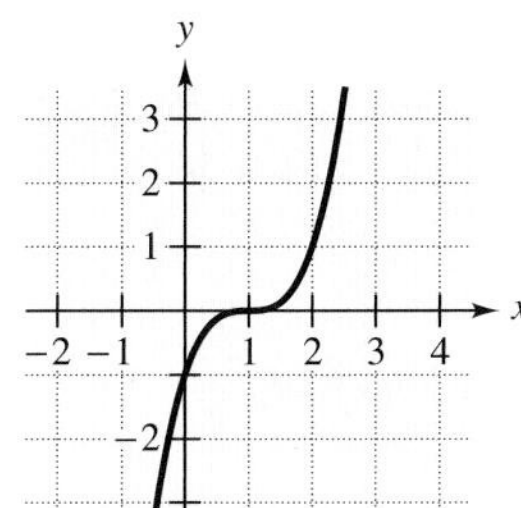

Figure for 10

In Exercises 6–8, sketch the graph of the equation.

6. $x + 2y = 6$
7. $y = |x + 2|$
8. $y = (x - 3)^2$
9. Does the table at the left represent y as a function of x? Explain.
10. Does the graph at the left represent y as a function of x?
11. Evaluate $f(x) = x^3 - 2x^2$ as indicated, and simplify.
    (a) $f(0)$ (b) $f(2)$ (c) $f(-2)$ (d) $f\left(\frac{1}{2}\right)$
12. Find the slope of the line passing through the points $(-5, 0)$ and $\left(2, \frac{3}{2}\right)$. Then write an equation of the line in slope-intercept form.
13. A line with slope $m = -2$ passes through the point $(-3, 4)$. Plot the point and use the slope to find two additional points on the line. (There are many correct answers.)
14. Find the slope of a line *perpendicular* to the line $7x - 8y + 5 = 0$.
15. Find an equation of the line that passes through the point $(0, 6)$ with slope $m = -\frac{3}{8}$.
16. Write an equation of the vertical line that passes through the point $(3, -7)$.

17. Determine whether the points are solutions of $3x + 5y \le 16$.
    (a) $(2, 2)$ (b) $(6, -1)$ (c) $(-2, 4)$ (d) $(7, -1)$

In Exercises 18–20, sketch the graph of the linear inequality.

18. $y \ge -2$
19. $y < 5 - 2x$
20. $-y + 4x > 3$
21. The sales y of a product are modeled by $y = 230x + 5000$, where x is time in years. Interpret the meaning of the slope in this model.

Study Skills in Action

Managing Test Anxiety

Test anxiety is different from the typical nervousness that usually occurs during tests. It interferes with the thinking process. After leaving the classroom, have you suddenly been able to recall what you could not remember during the test? It is likely that this was a result of test anxiety. Test anxiety is a learned reaction or response–no one is born with it. The good news is that most students can learn to manage test anxiety.

It is important to get as much information as you can into your long-term memory and to practice retrieving the information before you take a test. The more you practice retrieving information, the easier it will be during the test.

Kimberly Nolting
VP, Academic Success Press
expert in developmental education

Smart Study Strategy

Make Mental Cheat Sheets

No, we are not asking you to cheat! Just prepare as if you were going to and then memorize the information you've gathered.

1 ► Write down important information on note cards. This can include:

- formulas
- examples of problems you find difficult
- concepts that always trip you up

2 ► Memorize the information on the note cards. Flash through the cards, placing the ones containing information you know in one stack and the ones containing information you do not know in another stack. Keep working on the information you do not know.

3 ► As soon as you receive your test, turn it over and write down all the information you remember, starting with things you have the greatest difficulty remembering. Having this information available should boost your confidence and free up mental energy for focusing on the test.

The FOIL Method

*To multiply two binomials, you can combine the products of the **F**irst, **O**utside, **I**nside and **L**ast terms.*

$(2x+1)(x-5)$

F O I L

$= 2x(x) + 2x(-5) + 1(x) + 1(-5)$

$= 2x^2 - 10x + x - 5$

$= 2x^2 - 9x - 5$

Special Products:

$(a+b)(a-b) = a^2 - b^2$

$(a+b)^2 = a^2 + 2ab + b^2$

$(a-b)^2 = a^2 - 2ab + b^2$

Do not wait until the night before the test to make note cards. Make them after you study each section. Then review them two or three times a week.

Chapter 5

Exponents and Polynomials

IT WORKED FOR ME!
"I used to waste time reviewing for a test by just reading through my notes over and over again. Then I learned how to make note cards and keep working through them, saving the important ones for my mental cheat sheets. Now a couple of my friends and I go through our notes and make mental cheat sheets together. We are all doing well in our math class."
Stephanie
Associate of Arts transfer degree

5.1 Integer Exponents and Scientific Notation

© Blend Images/Alamy

Why You Should Learn It

Scientific notation can be used to represent very large real-life quantities. For instance, in Exercise 152 on page 303, you will use scientific notation to determine the total amount of rice consumed in the world in 1 year.

What You Should Learn

1 ▶ Use the rules of exponents to simplify expressions.
2 ▶ Rewrite exponential expressions involving negative and zero exponents.
3 ▶ Write very large and very small numbers in scientific notation.

1 ▶ Use the rules of exponents to simplify expressions.

Rules of Exponents

Recall from Section 1.5 that *repeated multiplication* can be written in what is called **exponential form.** Let n be a positive integer and let a be a real number. Then the product of n factors of a is given by

$$a^n = \underbrace{a \cdot a \cdot a \cdot \cdot \cdot a}_{n \text{ factors}}. \qquad a \text{ is the base and } n \text{ is the exponent.}$$

When multiplying two exponential expressions that have the *same base*, you add exponents. To see why this is true, consider the product $a^3 \cdot a^2$. Because the first expression represents three factors of a and the second represents two factors of a, the product of the two expressions represents five factors of a, as follows.

$$a^3 \cdot a^2 = \underbrace{(a \cdot a \cdot a)}_{3 \text{ factors}} \cdot \underbrace{(a \cdot a)}_{2 \text{ factors}} = \underbrace{(a \cdot a \cdot a \cdot a \cdot a)}_{5 \text{ factors}} = a^{3+2} = a^5$$

Rules of Exponents

Let m and n be positive integers, and let a and b represent real numbers, variables, or algebraic expressions.

Rule	*Example*
1. Product: $a^m \cdot a^n = a^{m+n}$	$x^5(x^4) = x^{5+4} = x^9$
2. Product-to-Power: $(ab)^m = a^m \cdot b^m$	$(2x)^3 = 2^3(x^3) = 8x^3$
3. Power-to-Power: $(a^m)^n = a^{mn}$	$(x^2)^3 = x^{2 \cdot 3} = x^6$
4. Quotient: $\dfrac{a^m}{a^n} = a^{m-n}, m > n, a \neq 0$	$\dfrac{x^5}{x^3} = x^{5-3} = x^2, x \neq 0$
5. Quotient-to-Power: $\left(\dfrac{a}{b}\right)^m = \dfrac{a^m}{b^m}, b \neq 0$	$\left(\dfrac{x}{4}\right)^2 = \dfrac{x^2}{4^2} = \dfrac{x^2}{16}$

The product rule and the product-to-power rule can be extended to three or more factors. For example,

$$a^m \cdot a^n \cdot a^k = a^{m+n+k} \quad \text{and} \quad (abc)^m = a^m b^m c^m.$$

Study Tip

In the expression $x + 5$, the coefficient of x is understood to be 1. Similarly, the power (or exponent) of x is also understood to be 1. So

$$x^4 \cdot x \cdot x^2 = x^{4+1+2} = x^7.$$

Note such occurrences in Examples 1(a) and 2(b).

EXAMPLE 1 Using Rules of Exponents

Simplify: **a.** $(x^2y^4)(3x)$ **b.** $-2(y^2)^3$ **c.** $(-2y^2)^3$ **d.** $(3x^2)(-5x)^3$

Solution

a. $(x^2y^4)(3x) = 3(x^2 \cdot x)(y^4) = 3(x^{2+1})(y^4) = 3x^3y^4$

b. $-2(y^2)^3 = (-2)(y^{2\cdot 3}) = -2y^6$

c. $(-2y^2)^3 = (-2)^3(y^2)^3 = -8(y^{2\cdot 3}) = -8y^6$

d. $(3x^2)(-5x)^3 = 3(-5)^3(x^2 \cdot x^3) = 3(-125)(x^{2+3}) = -375x^5$

✓ **CHECKPOINT** *Now try Exercise 1.*

EXAMPLE 2 Using Rules of Exponents

Simplify: **a.** $\dfrac{14a^5b^3}{7a^2b^2}$ **b.** $\left(\dfrac{x^2}{2y}\right)^3$ **c.** $\dfrac{x^ny^{3n}}{x^2y^4}$ **d.** $\dfrac{(2a^2b^3)^2}{a^3b^2}$

Solution

a. $\dfrac{14a^5b^3}{7a^2b^2} = 2(a^{5-2})(b^{3-2}) = 2a^3b$

b. $\left(\dfrac{x^2}{2y}\right)^3 = \dfrac{(x^2)^3}{(2y)^3} = \dfrac{x^{2\cdot 3}}{2^3y^3} = \dfrac{x^6}{8y^3}$

c. $\dfrac{x^ny^{3n}}{x^2y^4} = x^{n-2}y^{3n-4}$

d. $\dfrac{(2a^2b^3)^2}{a^3b^2} = \dfrac{2^2(a^{2\cdot 2})(b^{3\cdot 2})}{a^3b^2} = \dfrac{4a^4b^6}{a^3b^2} = 4(a^{4-3})(b^{6-2}) = 4ab^4$

✓ **CHECKPOINT** *Now try Exercise 11.*

2 ▶ Rewrite exponential expressions involving negative and zero exponents.

Integer Exponents

The definition of an exponent can be extended to include zero and negative integers. If a is a real number such that $a \neq 0$, then a^0 is defined as 1. Moreover, if m is an integer, then a^{-m} is defined as the reciprocal of a^m.

Definitions of Zero Exponents and Negative Exponents

Let a and b be real numbers such that $a \neq 0$ and $b \neq 0$, and let m be an integer.

1. $a^0 = 1$ **2.** $a^{-m} = \dfrac{1}{a^m}$ **3.** $\left(\dfrac{a}{b}\right)^{-m} = \left(\dfrac{b}{a}\right)^m$

These definitions are consistent with the rules of exponents given on page 294. For instance, consider the following.

$$x^0 \cdot x^m = x^{0+m} = x^m = 1 \cdot x^m$$

(x^0 is the same as 1)

EXAMPLE 3 Zero Exponents and Negative Exponents

Evaluate each expression.

a. 3^0 **b.** 3^{-2} **c.** $\left(\frac{3}{4}\right)^{-1}$

Solution

a. $3^0 = 1$ Definition of zero exponents

b. $3^{-2} = \dfrac{1}{3^2} = \dfrac{1}{9}$ Definition of negative exponents

c. $\left(\dfrac{3}{4}\right)^{-1} = \left(\dfrac{4}{3}\right)^1 = \dfrac{4}{3}$ Definition of negative exponents

 ***CHECKPOINT*** *Now try Exercise 21.*

Study Tip

Notice that by definition, $a^0 = 1$ for all real *nonzero* values of a. Zero cannot have a zero exponent, because the expression 0^0 is undefined.

Summary of Rules of Exponents

Let m and n be integers, and let a and b represent real numbers, variables, or algebraic expressions. (All denominators and bases are nonzero.)

Product and Quotient Rules	*Example*
1. $a^m \cdot a^n = a^{m+n}$	$x^4(x^3) = x^{4+3} = x^7$
2. $\dfrac{a^m}{a^n} = a^{m-n}$	$\dfrac{x^3}{x} = x^{3-1} = x^2$
Power Rules	
3. $(ab)^m = a^m \cdot b^m$	$(3x)^2 = 3^2(x^2) = 9x^2$
4. $(a^m)^n = a^{mn}$	$(x^3)^3 = x^{3 \cdot 3} = x^9$
5. $\left(\dfrac{a}{b}\right)^m = \dfrac{a^m}{b^m}$	$\left(\dfrac{x}{3}\right)^2 = \dfrac{x^2}{3^2} = \dfrac{x^2}{9}$
Zero and Negative Exponent Rules	
6. $a^0 = 1$	$(x^2 + 1)^0 = 1$
7. $a^{-m} = \dfrac{1}{a^m}$	$x^{-2} = \dfrac{1}{x^2}$
8. $\left(\dfrac{a}{b}\right)^{-m} = \left(\dfrac{b}{a}\right)^m$	$\left(\dfrac{x}{3}\right)^{-2} = \left(\dfrac{3}{x}\right)^2 = \dfrac{3^2}{x^2} = \dfrac{9}{x^2}$

Study Tip

Notice that the first five rules of exponents were first listed on page 294 for *positive* values of m and n, and the quotient rule included the restriction $m > n$.

Because the rules shown here allow the use of zero exponents and negative exponents, the restriction $m > n$ is no longer necessary for the quotient rule.

EXAMPLE 4 Using Rules of Exponents

a. $2x^{-1} = 2(x^{-1}) = 2\left(\dfrac{1}{x}\right) = \dfrac{2}{x}$ Use negative exponent rule and simplify.

b. $(2x)^{-1} = \dfrac{1}{(2x)^1} = \dfrac{1}{2x}$ Use negative exponent rule and simplify.

 ***CHECKPOINT*** *Now try Exercise 55.*

Study Tip

As you become accustomed to working with negative exponents, you will probably not write as many steps as shown in Example 5. For instance, to rewrite a fraction involving exponents, you might use the following simplified rule. *To move a factor from the numerator to the denominator or vice versa, change the sign of its exponent.* You can apply this rule to the expression in Example 5(a) by "moving" the factor x^{-2} to the numerator and changing the exponent to 2. That is,

$$\frac{3}{x^{-2}} = 3x^2.$$

Remember, you can move only *factors* in this manner, not terms.

EXAMPLE 5 Using Rules of Exponents

Rewrite each expression using only positive exponents. (Assume that $x \neq 0$.)

a. $\dfrac{3}{x^{-2}} = \dfrac{3}{\left(\dfrac{1}{x^2}\right)}$ Negative exponent rule

$= 3\left(\dfrac{x^2}{1}\right) = 3x^2$ Invert divisor and multiply.

b. $\dfrac{1}{(3x)^{-2}} = \dfrac{1}{\left[\dfrac{1}{(3x)^2}\right]}$ Use negative exponent rule.

$= \dfrac{1}{\left(\dfrac{1}{9x^2}\right)}$ Use product-to-power rule and simplify.

$= (1)\left(\dfrac{9x^2}{1}\right) = 9x^2$ Invert divisor and multiply.

CHECKPOINT *Now try Exercise 63.*

EXAMPLE 6 Using Rules of Exponents

Rewrite each expression using only positive exponents. (Assume that $x \neq 0$ and $y \neq 0$.)

a. $(-5x^{-3})^2 = (-5)^2(x^{-3})^2$ Product-to-power rule

$= 25x^{-6}$ Power-to-product rule

$= \dfrac{25}{x^6}$ Negative exponent rule

b. $-\left(\dfrac{7x}{y^2}\right)^{-2} = -\left(\dfrac{y^2}{7x}\right)^2$ Negative exponent rule

$= -\dfrac{(y^2)^2}{(7x)^2}$ Quotient-to-power rule

$= -\dfrac{y^4}{49x^2}$ Power-to-power and product-to-power rules

c. $\dfrac{12x^2y^{-4}}{6x^{-1}y^2} = 2(x^{2-(-1)})(y^{-4-2})$ Quotient rule

$= 2x^3y^{-6}$ Simplify.

$= \dfrac{2x^3}{y^6}$ Negative exponent rule

CHECKPOINT *Now try Exercise 73.*

EXAMPLE 7 Using Rules of Exponents

Rewrite each expression using only positive exponents. (Assume that $x \neq 0$ and $y \neq 0$.)

a. $\left(\dfrac{8x^{-1}y^4}{4x^3y^2}\right)^{-3} = \left(\dfrac{2y^2}{x^4}\right)^{-3}$ Simplify.

$= \left(\dfrac{x^4}{2y^2}\right)^3$ Negative exponent rule

$= \dfrac{x^{12}}{2^3y^6} = \dfrac{x^{12}}{8y^6}$ Quotient-to-power rule

b. $\dfrac{3xy^0}{x^2(5y)^0} = \dfrac{3x(1)}{x^2(1)} = \dfrac{3}{x}$ Zero exponent rule

✓ ***CHECKPOINT*** *Now try Exercise 75.*

3 ▶ Write very large and very small numbers in scientific notation.

Scientific Notation

Exponents provide an efficient way of writing and computing with very large and very small numbers. For instance, a drop of water contains more than 33 billion billion molecules—that is, 33 followed by 18 zeros. It is convenient to write such numbers in **scientific notation.** This notation has the form $c \times 10^n$, where $1 \leq c < 10$ and n is an integer. So, the number of molecules in a drop of water can be written in scientific notation as follows.

$$33{,}000{,}000{,}000{,}000{,}000{,}000 = 3.3 \times 10^{19}$$

19 places

The *positive* exponent 19 indicates that the number being written in scientific notation is *large* (10 or more) and that the decimal point has been moved 19 places. A *negative* exponent in scientific notation indicates that the number is *small* (less than 1).

EXAMPLE 8 Writing in Scientific Notation

Write each number in scientific notation.

a. 0.0000684 **b.** 937,200,000

Solution

a. $0.0000684 = 6.84 \times 10^{-5}$ Small number ⇨ negative exponent

Five places

b. $937{,}200{,}000.0 = 9.372 \times 10^8$ Large number ⇨ positive exponent

Eight places

✓ ***CHECKPOINT*** *Now try Exercise 101.*

EXAMPLE 9 Writing in Decimal Notation

a. $2.486 \times 10^2 = 248.6$ — Two places. Positive exponent  large number

b. $1.81 \times 10^{-6} = 0.00000181$ — Six places. Negative exponent  small number

 ***CHECKPOINT*** *Now try Exercise 115.*

EXAMPLE 10 Using Scientific Notation

Rewrite the factors in scientific notation and then evaluate

$$\frac{(2{,}400{,}000{,}000)(0.0000045)}{(0.00003)(1500)}.$$

Solution

$$\frac{(2{,}400{,}000{,}000)(0.0000045)}{(0.00003)(1500)} = \frac{(2.4 \times 10^9)(4.5 \times 10^{-6})}{(3.0 \times 10^{-5})(1.5 \times 10^3)}$$

$$= \frac{(2.4)(4.5)(10^3)}{(4.5)(10^{-2})}$$

$$= (2.4)(10^5) = 2.4 \times 10^5$$

 ***CHECKPOINT*** *Now try Exercise 133.*

EXAMPLE 11 Using Scientific Notation with a Calculator

Use a calculator to evaluate each expression.

a. $65{,}000 \times 3{,}400{,}000{,}000$ **b.** $0.000000348 \div 870$

Solution

a. 6.5 [EXP] 4 [×] 3.4 [EXP] 9 [=] Scientific

6.5 [EE] 4 [×] 3.4 [EE] 9 [ENTER] Graphing

The calculator display should read [2.21E 14], which implies that

$$(6.5 \times 10^4)(3.4 \times 10^9) = 2.21 \times 10^{14} = 221{,}000{,}000{,}000{,}000.$$

b. 3.48 [EXP] 7 [±] [÷] 8.7 [EXP] 2 [=] Scientific

3.48 [EE] [(−)] 7 [÷] 8.7 [EE] 2 [ENTER] Graphing

The calculator display should read [4E −10], which implies that

$$\frac{3.48 \times 10^{-7}}{8.7 \times 10^2} = 4.0 \times 10^{-10} = 0.0000000004.$$

 ***CHECKPOINT*** *Now try Exercise 135.*

Technology: Tip

Most scientific and graphing calculators automatically switch to scientific notation to show large or small numbers that exceed the display range.

To *enter* numbers in scientific notation, your calculator should have a key labeled [EE] or [EXP]. Consult the user's guide for your calculator for specific instructions.

Concept Check

1. In your own words, describe how to simplify the expression.
   (a) $x^a \cdot x^b$ (b) $(xy)^m$
   (c) $(x^a)^b$ (d) $\frac{c^m}{c^n}$

2. Let x represent a real number such that $x > 1$. What can you say about the value of each of the expressions x^2, x^1, x^0, x^{-1}, and x^{-2}?

3. To write a decimal in scientific notation, you write the number as $c \times 10^n$, where $1 \leq c < 10$. Explain in your own words how to write the number c for a given decimal.

4. Explain why you would not use scientific notation to list the ingredient amounts in a cooking recipe.

5.1 EXERCISES

Go to pages 338–339 to record your assignments.

Developing Skills

In Exercises 1–20, use the rules of exponents to simplify the expression (if possible). ***See Examples 1 and 2.***

✓ 1. (a) $-3x^3 \cdot x^5$ (b) $(-3x)^2 \cdot x^5$
2. (a) $5^2y^4 \cdot y^2$ (b) $(5y)^2 \cdot y^4$
3. (a) $(-5z^2)^3$ (b) $(-5z^4)^2$
4. (a) $(-5z^3)^2$ (b) $(-5z)^4$
5. (a) $(u^3v)(2v^2)$ (b) $(-4u^4)(u^5v)$
6. (a) $(6xy^7)(-x)$ (b) $(x^5y^3)(2y^3)$
7. (a) $5u^2 \cdot (-3u^6)$ (b) $(2u)^4(4u)$
8. (a) $(3y)^3(2y^2)$ (b) $3y^3 \cdot 2y^2$
9. (a) $-(m^5n)^3(-m^2n^2)^2$ (b) $(-m^5n)(m^2n^2)$
10. (a) $-(m^3n^2)(mn^3)$ (b) $-(m^3n^2)^2(-mn^3)$
✓ 11. (a) $\frac{27m^5n^6}{9mn^3}$ (b) $\frac{-18m^3n^6}{-6mn^3}$
12. (a) $\frac{28x^2y^3}{2xy^2}$ (b) $\frac{24xy^2}{8y}$
13. (a) $\left(\frac{3x}{4y}\right)^2$ (b) $\left(\frac{5u}{3v}\right)^3$
14. (a) $\left(\frac{2a}{3y}\right)^5$ (b) $-\left(\frac{2a}{3y}\right)^2$
15. (a) $-\frac{(-2x^2y)^3}{9x^2y^2}$ (b) $-\frac{(-2xy^3)^2}{6y^2}$
16. (a) $\frac{(-4xy)^3}{8xy^2}$ (b) $\frac{(-xy)^4}{-3(xy)^2}$
17. (a) $\left[\frac{(-5u^3v)^2}{10u^2v}\right]^2$ (b) $\left[\frac{-5(u^3v)^2}{10u^2v}\right]^2$
18. (a) $\left[\frac{(3x^2)(2x)^2}{(-2x)(6x)}\right]^2$ (b) $\left[\frac{(3x^2)(2x)^4}{(-2x)^2(6x)}\right]^2$
19. (a) $\frac{x^{2n+4}\,y^{4n}}{x^5\,y^{2n+1}}$ (b) $\frac{x^{6n}\,y^{n-7}}{x^{4n+2}\,y^5}$
20. (a) $\frac{x^{3n}\,y^{2n-1}}{x^n\,y^{n+3}}$ (b) $\frac{x^{4n-6}\,y^{n+10}}{x^{2n-5}\,y^{n-2}}$

In Exercises 21–50, evaluate the expression. ***See Example 3.***

✓ 21. 5^{-2}
22. 2^{-4}
23. -10^{-3}
24. -20^{-2}
25. $(-3)^0$
26. 25^0
27. $\frac{1}{4^{-3}}$
28. $\frac{1}{-8^{-2}}$
29. $\frac{1}{(-2)^{-5}}$
30. $-\frac{1}{6^{-2}}$
31. $\left(\frac{2}{3}\right)^{-1}$
32. $\left(\frac{4}{5}\right)^{-3}$
33. $\left(\frac{3}{16}\right)^0$
34. $\left(-\frac{5}{8}\right)^{-2}$
35. $27 \cdot 3^{-3}$
36. $16 \cdot 4^{-4}$
37. $\frac{3^4}{3^{-2}}$
38. $\frac{5^{-1}}{5^2}$
39. $\frac{10^3}{10^{-2}}$
40. $\frac{10^{-5}}{10^{-6}}$
41. $(4^2 \cdot 4^{-1})^{-2}$
42. $(5^3 \cdot 5^{-4})^{-3}$

43. $(2^{-3})^2$
44. $(-4^{-1})^{-2}$
45. $2^{-3} + 2^{-4}$
46. $4 - 3^{-2}$
47. $\left(\frac{3}{4} + \frac{5}{8}\right)^{-2}$
48. $\left(\frac{1}{2} - \frac{2}{3}\right)^{-1}$
49. $(5^0 - 4^{-2})^{-1}$
50. $(32 + 4^{-3})^0$

In Exercises 51–90, rewrite the expression using only positive exponents, and simplify. (Assume that any variables in the expression are nonzero.) ***See Examples 4–7.***

51. $y^4 \cdot y^{-2}$
52. $x^{-2} \cdot x^{-5}$
53. $z^5 \cdot z^{-3}$
54. $t^{-1} \cdot t^{-6}$
55. $7x^{-4}$
56. $3y^{-3}$
57. $(4x)^{-3}$
58. $(5u)^{-2}$
59. $\dfrac{1}{x^{-6}}$
60. $\dfrac{4}{y^{-1}}$
61. $\dfrac{8a^{-6}}{6a^{-7}}$
62. $\dfrac{6u^{-2}}{15u^{-1}}$
63. $\dfrac{(4t)^0}{t^{-2}}$
64. $\dfrac{(5u)^{-4}}{(5u)^0}$
65. $(2x^2)^{-2}$
66. $(4a^{-2})^{-3}$
67. $(-3x^{-3}y^2)(4x^2y^{-5})$
68. $(5s^5t^{-5})(-6s^{-2}t^4)$
69. $(3x^2y^{-2})^{-2}$
70. $(-4y^{-3}z)^{-3}$
71. $\left(\dfrac{x}{10}\right)^{-1}$
72. $\left(\dfrac{4}{z}\right)^{-2}$
73. $\dfrac{6x^3y^{-3}}{12x^{-2}y}$
74. $\dfrac{2y^{-1}z^{-3}}{4yz^{-3}}$
75. $\left(\dfrac{3u^2v^{-1}}{3^3u^{-1}v^3}\right)^{-2}$
76. $\left(\dfrac{5^2x^3y^{-3}}{125xy}\right)^{-1}$
77. $\left(\dfrac{a^{-2}}{b^{-2}}\right)\left(\dfrac{b}{a}\right)^3$
78. $\left(\dfrac{a^{-3}}{b^{-3}}\right)\left(\dfrac{b}{a}\right)^3$
79. $(2x^3y^{-1})^{-3}(4xy^{-6})$
80. $(ab)^{-2}(a^2b^2)^{-1}$
81. $u^4(6u^{-3}v^0)(7v)^0$
82. $x^5(3x^0y^4)(7y)^0$
83. $[(x^{-4}y^{-6})^{-1}]^2$
84. $[(2x^{-3}y^{-2})^2]^{-2}$
85. $\dfrac{(2a^{-2}b^4)^3b}{(10a^3b)^2}$
86. $\dfrac{(5x^2y^{-5})^{-1}}{2x^{-5}y^4}$
87. $(u + v^{-2})^{-1}$
88. $x^{-2}(x^2 + y^2)$
89. $\dfrac{a + b}{b^{-1}a + 1}$
90. $\dfrac{u^{-1} - v^{-1}}{u^{-1} + v^{-1}}$

In Exercises 91–100, evaluate the expression when $x = -3$ and $y = 4$.

91. $x^2 \cdot x^{-3} \cdot x \cdot y$
92. $x^4 \cdot x^{-1} \cdot x^{-1} \cdot y$
93. $\dfrac{x^2}{y^{-2}}$
94. $\dfrac{y^2}{x^{-2}}$
95. $(x + y)^{-4}$
96. $(-x - y)^{-2}$
97. $\left(\dfrac{5x}{3y}\right)^{-1}$
98. $\left(\dfrac{3y}{12x}\right)^{-2}$
99. $(xy)^{-2}$
100. $(x^2y)^{-1}$

In Exercises 101–114, write the number in scientific notation. ***See Example 8.***

101. 3,600,000
102. 98,100,000
103. 47,620,000
104. 841,000,000,000
105. 0.00031
106. 0.00625
107. 0.0000000381
108. 0.0000000000692
109. *Land Area of Earth*: 57,300,000 square miles
110. *Water Area of Earth*: 139,500,000 square miles
111. *Light Year*: 9,460,800,000,000 kilometers
112. *Thickness of a Soap Bubble*: 0.0000001 meter
113. *Relative Density of Hydrogen*: 0.0899 gram per milliliter
114. *One Micron (Millionth of a Meter)*: 0.00003937 inch

In Exercises 115–124, write the number in decimal notation. ***See Example 9.***

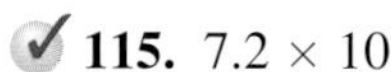

115. 7.2×10^8

116. 7.413×10^{11}

117. 1.359×10^{-7}

118. 8.6×10^{-9}

119. *2006 Merrill Lynch Revenues:* $\$3.4659 \times 10^{10}$ (Source: 2006 Merrill Lynch Annual Report)

120. *Number of Air Sacs in the Lungs*: 3.5×10^8

121. *Interior Temperature of the Sun*: 1.5×10^7 degrees Celsius

122. *Width of an Air Molecule*: 9.0×10^{-9} meter

123. *Charge of an Electron*: 4.8×10^{-10} electrostatic unit

124. *Width of a Human Hair*: 9.0×10^{-4} meter

In Exercises 125–134, evaluate the expression without a calculator. ***See Example 10.***

125. $(2 \times 10^9)(3.4 \times 10^{-4})$

126. $(6.5 \times 10^6)(2 \times 10^4)$

127. $(5 \times 10^4)^2$

128. $(4 \times 10^6)^3$

129. $\dfrac{3.6 \times 10^{12}}{6 \times 10^5}$

130. $\dfrac{2.5 \times 10^{-3}}{5 \times 10^2}$

131. $(4{,}500{,}000)(2{,}000{,}000{,}000)$

132. $(62{,}000{,}000)(0.0002)$

133. $\dfrac{64{,}000{,}000}{0.00004}$

134. $\dfrac{72{,}000{,}000{,}000}{0.00012}$

In Exercises 135–142, evaluate with a calculator. Write the answer in scientific notation, $c \times 10^n$, with c rounded to two decimal places. ***See Example 11.***

135. $\dfrac{(0.0000565)(2{,}850{,}000{,}000{,}000)}{0.00465}$

136. $\dfrac{(3{,}450{,}000{,}000)(0.000125)}{(52{,}000{,}000)(0.000003)}$

137. $\dfrac{1.357 \times 10^{12}}{(4.2 \times 10^2)(6.87 \times 10^{-3})}$

138. $\dfrac{(3.82 \times 10^5)^2}{(8.5 \times 10^4)(5.2 \times 10^{-3})}$

139. $(2.58 \times 10^6)^4$

140. $(8.67 \times 10^4)^7$

141. $\dfrac{(5{,}000{,}000)^3(0.000037)^2}{(0.005)^4}$

142. $\dfrac{(6{,}200{,}000)(0.005)^3}{(0.00035)^5}$

Solving Problems

143. *Distance* The distance from Earth to the Sun is approximately 93 million miles. Write this distance in scientific notation.

144. *Stars* A study by Australian astronomers estimated the number of stars within range of modern telescopes to be 70,000,000,000,000,000,000,000. Write this number in scientific notation. (Source: The Australian National University)

Larry Landolfi/ Photo Researchers, Inc.

145. *Electrons* A cube of copper with an edge of 1 centimeter has approximately 8.483×10^{22} free electrons. Write this real number in decimal notation.

146. *Lumber Consumption* The total volume of the lumber consumed in the United States in 2005 was about 1.0862×10^{10} cubic feet. Write this volume in decimal notation. (Source: U.S. Forest Service)

147. *Light Year* One light year (the distance light can travel in 1 year) is approximately 9.46×10^{15} meters. Approximate the time to the nearest minute for light to travel from the Sun to Earth if that distance is approximately 1.50×10^{11} meters.

148. *Masses of Earth and the Sun* The masses of Earth and the Sun are approximately 5.98×10^{24} kilograms and 1.99×10^{30} kilograms, respectively. The mass of the Sun is approximately how many times that of Earth?

149. *Distance* The star Alpha Andromeda is approximately 95 light years from Earth. Determine this distance in meters. (See Exercise 147 for the definition of a light year.)

150. *Metal Expansion* When the temperature of an iron steam pipe 200 feet long is increased by 75°C, the length of the pipe will increase by an amount $75(200)(1.1 \times 10^{-5})$. Find this amount and write the answer in decimal notation.

151. *Federal Debt* In 2005, the resident population of the United States was about 296 million people, and it would have cost each resident about \$26,600 to pay off the federal debt. Use these two numbers to approximate the federal debt in 2005. (Source: U.S. Census Bureau and U.S. Office of Management and Budget)

152. *Rice Consumption* In 2005, the population of the world was about 6.451 billion people, and the average person consumed about 141.8 pounds of milled rice. Use these two numbers to approximate the total amount (in pounds) of milled rice consumed in the world in 2005. (Source: U.S. Census Bureau and U.S. Department of Agriculture)

Explaining Concepts

153. *Think About It* Discuss whether you feel that using scientific notation to multiply or divide very large or very small numbers makes the process *easier* or *more difficult*. Support your position with an example.

154. You multiply an expression by a^5. The product is a^{12}. What was the original expression? Explain how you found your answer.

True or False? In Exercises 155 and 156, determine whether the statement is true or false. Justify your reasoning.

155. The value of $\frac{1}{3^{-3}}$ is less than 1.

156. The expression 0.142×10^{10} is in scientific notation.

In Exercises 157–160, use the rules of exponents to explain why the statement is *false*.

157. $a^m \cdot b^n = ab^{m+n}$ ✗

158. $(ab)^m = a^m + b^m$ ✗

159. $(a^m)^n = a^{m+n}$ ✗

160. $\frac{a^m}{a^n} = a^m - a^n$ ✗

Cumulative Review

In Exercises 161–164, simplify the expression by combining like terms.

161. $3x + 4x - x$

162. $y - 3x + 4y - 2$

163. $a^2 + 2ab - b^2 + ab + 4b^2$

164. $x^2 + 5x^2y - 3x^2y + 4x^2$

In Exercises 165–170, sketch the graph of the linear inequality.

165. $y < 7$

166. $x > -2$

167. $y \geq 8x$

168. $y > 3 + x$

169. $y \leq -2x + 4$

170. $x + 3y < -1$

5.2 Adding and Subtracting Polynomials

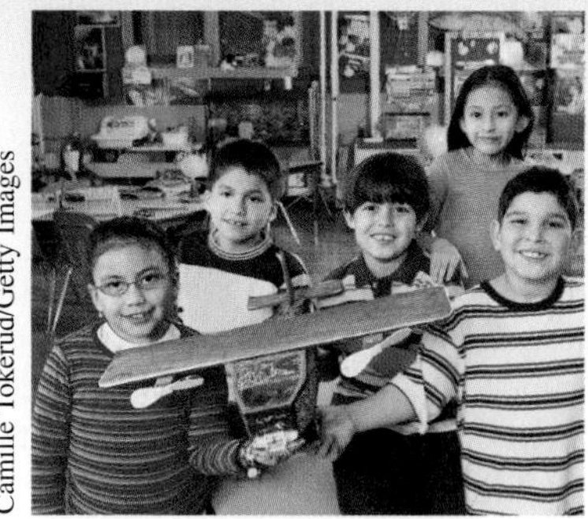
Camille Tokerud/Getty Images

Why You Should Learn It

Polynomials can be used to model and solve real-life problems. For instance, in Exercise 103 on page 312, polynomials are used to model the projected enrollments of public and private schools in the United States.

What You Should Learn

1 ▶ Identify the degrees and leading coefficients of polynomials.
2 ▶ Add polynomials using a horizontal or vertical format.
3 ▶ Subtract polynomials using a horizontal or vertical format.

Basic Definitions

1 ▶ Identify the degrees and leading coefficients of polynomials.

Recall from Section 2.1 that the *terms* of an algebraic expression are those parts separated by addition. An algebraic expression whose terms are all of the form ax^k, where a is any real number and k is a nonnegative integer, is called a **polynomial in one variable,** or simply a **polynomial.** Here are some examples of polynomials in one variable.

$$2x + 5, \quad x^2 - 3x + 7, \quad 9x^5, \quad \text{and} \quad x^3 + 8$$

In the term ax^k, a is the **coefficient** of the term and k is the **degree** of the term. Note that the degree of the term ax is 1, and the degree of a constant term is 0. Because a polynomial is an algebraic *sum*, the coefficients take on the signs between the terms. For instance,

$$x^4 + 2x^3 - 5x^2 + 7 = (1)x^4 + 2x^3 + (-5)x^2 + (0)x + 7$$

has coefficients 1, 2, -5, 0, and 7. For this polynomial, the last term, 7, is the **constant term.** Polynomials are usually written in order of descending powers of the variable. This is called **standard form.** Here are three examples.

Nonstandard Form	*Standard Form*
$4 + x$	$x + 4$
$3x^2 - 5 - x^3 + 2x$	$-x^3 + 3x^2 + 2x - 5$
$18 - x^2 + 3$	$-x^2 + 21$

The **degree of a polynomial** is the degree of the term with the highest power, and the coefficient of this term is the **leading coefficient** of the polynomial. For instance, the polynomial

Degree

$$-3x^4 + 4x^2 + x + 7$$

Leading coefficient

is of fourth degree, and its leading coefficient is -3. The reasons why the degree of a polynomial is important will become clear as you study factoring and problem solving in Chapter 6.

Definition of a Polynomial in x

Let $a_n, a_{n-1}, \ldots, a_2, a_1, a_0$ be real numbers and let n be a nonnegative integer. A **polynomial in x** is an expression of the form

$$a_nx^n + a_{n-1}x^{n-1} + \cdots + a_2x^2 + a_1x + a_0$$

where $a_n \neq 0$. The polynomial is of **degree n,** and the number a_n is called the **leading coefficient.** The number a_0 is called the **constant term.**

EXAMPLE 1 Identifying Polynomials

Determine whether each expression is a polynomial. If it is not, explain why.

a. $3x^4 - 8x + x^{-1}$ **b.** $x^2 - 3x + 1$

c. $x^3 + 3x^{1/2}$ **d.** $-\frac{1}{3}x + \frac{x^3}{4}$

Solution

a. $3x^4 - 8x + x^{-1}$ *is not* a polynomial because the third term, x^{-1}, has a negative exponent.

b. $x^2 - 3x + 1$ *is* a polynomial of degree 2 with integer coefficients.

c. $x^3 + 3x^{1/2}$ *is not* a polynomial because the exponent in the second term, $3x^{1/2}$, is not an integer.

d. $-\frac{1}{3}x + \frac{x^3}{4}$ *is* a polynomial of degree 3 with rational coefficients.

CHECKPOINT *Now try Exercise 3.*

EXAMPLE 2 Determining Degrees and Leading Coefficients

Write each polynomial in standard form and identify the degree and leading coefficient.

Polynomial	*Standard Form*	*Degree*	*Leading Coefficient*
a. $4x^2 - 5x^7 - 2 + 3x$	$-5x^7 + 4x^2 + 3x - 2$	7	-5
b. $4 - 9x^2$	$-9x^2 + 4$	2	-9
c. 8	8	0	8
d. $2 + x^3 - 5x^2$	$x^3 - 5x^2 + 2$	3	1

In part (c), note that a polynomial with only a constant term has a degree of zero.

CHECKPOINT *Now try Exercise 13.*

A polynomial with only one term is called a **monomial.** A polynomial with two *unlike* terms is called a **binomial,** and a polynomial with three *unlike* terms is called a **trinomial.** For example, $3x^2$ is a *monomial,* $-3x + 1$ is a *binomial,* and $4x^3 - 5x + 6$ is a *trinomial.*

2 ▶ Add polynomials using a horizontal or vertical format.

Adding Polynomials

As with algebraic expressions, the key to adding two polynomials is to recognize *like* terms—those having the *same degree.* By the Distributive Property, you can then combine the like terms using either a horizontal or a vertical format of terms. For instance, the polynomials $2x^2 + 3x + 1$ and $x^2 - 2x + 2$ can be added horizontally to obtain

$$(2x^2 + 3x + 1) + (x^2 - 2x + 2) = (2x^2 + x^2) + (3x - 2x) + (1 + 2)$$
$$= 3x^2 + x + 3$$

or they can be added vertically to obtain the same result.

$$\begin{array}{r} 2x^2 + 3x + 1 \\ \underline{x^2 - 2x + 2} \\ 3x^2 + x + 3 \end{array}$$ Vertical format

Technology: Tip

You can use a graphing calculator to check the results of adding or subtracting polynomials. For instance, try graphing

$y_1 = (2x + 1) + (-3x - 4)$

and

$y_2 = -x - 3$

in the same viewing window, as shown below. Because both graphs are the same, you can conclude that

$(2x + 1) + (-3x - 4) = -x - 3.$

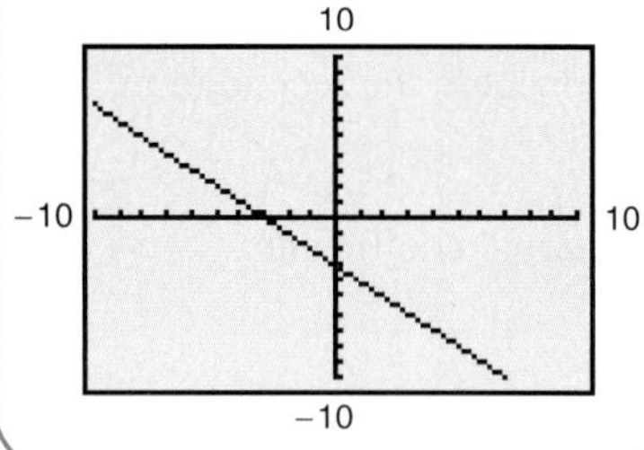

EXAMPLE 3 Adding Polynomials Horizontally

Use a horizontal format to find each sum.

a. $(2x^2 + 4x - 1) + (x^2 - 3)$ Original polynomials

$= (2x^2 + x^2) + (4x) + (-1 - 3)$ Group like terms.

$= 3x^2 + 4x - 4$ Combine like terms.

b. $(x^3 + 2x^2 + 4) + (3x^2 - x + 5)$ Original polynomials

$= (x^3) + (2x^2 + 3x^2) + (-x) + (4 + 5)$ Group like terms.

$= x^3 + 5x^2 - x + 9$ Combine like terms.

c. $(2x^2 - x + 3) + (4x^2 - 7x + 2) + (-x^2 + x - 2)$ Original polynomials

$= (2x^2 + 4x^2 - x^2) + (-x - 7x + x) + (3 + 2 - 2)$ Group like terms.

$= 5x^2 - 7x + 3$ Combine like terms.

✓ ***CHECKPOINT*** *Now try Exercise 33.*

EXAMPLE 4 Adding Polynomials Vertically

Use a vertical format to find each sum.

a.
$$\begin{array}{r} -4x^3 - 2x^2 + x - 5 \\ \underline{2x^3 + 3x + 4} \\ -2x^3 - 2x^2 + 4x - 1 \end{array}$$

b.
$$\begin{array}{r} 5x^3 + 2x^2 - x + 7 \\ 3x^2 - 4x + 7 \\ \underline{-x^3 + 4x^2 - 2x - 8} \\ 4x^3 + 9x^2 - 7x + 6 \end{array}$$

✓ ***CHECKPOINT*** *Now try Exercise 47.*

Study Tip

When you use a vertical format to add polynomials, be sure that you line up the *like terms.*

3 ▶ Subtract polynomials using a horizontal or vertical format.

Subtracting Polynomials

To subtract one polynomial from another, you *add the opposite* by changing the sign of each term of the polynomial that is being subtracted and then adding the resulting like terms.

Be especially careful when subtracting polynomials. One of the most common mistakes in algebra is not changing signs correctly when subtracting one expression from another. Here is an example.

Wrong sign ↓ (on $-x^2$); Wrong sign ↑ (on -2)

$$(x^2 + 3) - (x^2 + 2x - 2) \neq x^2 + 3 - x^2 + 2x - 2$$ Common error

Note that the error is forgetting to change *all* of the signs in the polynomial that is being subtracted. Here is the correct way to perform the subtraction.

Correct sign ↓ (on $-2x$); Correct sign ↑ (on $+2$)

$$(x^2 + 3) - (x^2 + 2x - 2) = x^2 + 3 - x^2 - 2x + 2$$ Correct

$$= (x^2 - x^2) + (-2x) + (3 + 2)$$ Group like terms.

$$= -2x + 5$$ Combine like terms.

Recall that by the Distributive Property

$$-(x^2 + 2x - 2) = (-1)(x^2 + 2x - 2) = -x^2 - 2x + 2.$$

EXAMPLE 5 Subtracting Polynomials Horizontally

Use a horizontal format to find each difference.

a. $(2x^2 + 3) - (3x^2 - 4)$

b. $(3x^3 - 4x^2 + 3) - (x^3 + 3x^2 - x - 4)$

Solution

a. $(2x^2 + 3) - (3x^2 - 4) = 2x^2 + 3 - 3x^2 + 4$ Distributive Property

$= (2x^2 - 3x^2) + (3 + 4)$ Group like terms.

$= -x^2 + 7$ Combine like terms.

b. $(3x^3 - 4x^2 + 3) - (x^3 + 3x^2 - x - 4)$ Original polynomials

$= 3x^3 - 4x^2 + 3 - x^3 - 3x^2 + x + 4$ Distributive Property

$= (3x^3 - x^3) + (-4x^2 - 3x^2) + (x) + (3 + 4)$ Group like terms.

$= 2x^3 - 7x^2 + x + 7$ Combine like terms.

✓ **CHECKPOINT** *Now try Exercise 59.*

Just as you did for addition, you can use a vertical format to subtract one polynomial from another. (The vertical format does not work well with subtractions involving three or more polynomials.) When using a vertical format, write the polynomial being subtracted underneath the one from which it is being subtracted. Be sure to line up like terms in vertical columns.

EXAMPLE 6 Subtracting Polynomials Vertically

Use a vertical format to find each difference.

a. $(3x^2 + 7x - 6) - (3x^2 + 7x)$

b. $(5x^3 - 2x^2 + x) - (4x^2 - 3x + 2)$

c. $(4x^4 - 2x^3 + 5x^2 - x + 8) - (3x^4 - 2x^3 + 3x - 4)$

Solution

a.
$$\begin{array}{r} (3x^2 + 7x - 6) \\ -(3x^2 + 7x \quad\;\;) \\ \hline \end{array} \Rightarrow \begin{array}{r} 3x^2 + 7x - 6 \\ -3x^2 - 7x \quad\;\; \\ \hline -6 \end{array}$$
Change signs and add.

b.
$$\begin{array}{r} (5x^3 - 2x^2 + x \quad\;\;) \\ -(\quad 4x^2 - 3x + 2) \\ \hline \end{array} \Rightarrow \begin{array}{r} 5x^3 - 2x^2 + x \quad\;\; \\ - 4x^2 + 3x - 2 \\ \hline 5x^3 - 6x^2 + 4x - 2 \end{array}$$
Change signs and add.

c.
$$\begin{array}{r} (4x^4 - 2x^3 + 5x^2 - x + 8) \\ -(3x^4 - 2x^3 \qquad + 3x - 4) \\ \hline \end{array} \Rightarrow \begin{array}{r} 4x^4 - 2x^3 + 5x^2 - x + 8 \\ -3x^4 + 2x^3 \qquad - 3x + 4 \\ \hline x^4 \qquad + 5x^2 - 4x + 12 \end{array}$$

✓ **CHECKPOINT** *Now try Exercise 67.*

EXAMPLE 7 Combining Polynomials Horizontally

Use a horizontal format to perform the indicated operations.

$$(x^2 - 2x + 1) - [(x^2 + x - 3) + (-2x^2 - 4x)]$$

Solution

$(x^2 - 2x + 1) - [(x^2 + x - 3) + (-2x^2 - 4x)]$ Original polynomials

$= (x^2 - 2x + 1) - [(x^2 - 2x^2) + (x - 4x) + (-3)]$ Group like terms.

$= (x^2 - 2x + 1) - [-x^2 - 3x - 3]$ Combine like terms.

$= x^2 - 2x + 1 + x^2 + 3x + 3$ Distributive Property

$= (x^2 + x^2) + (-2x + 3x) + (1 + 3)$ Group like terms.

$= 2x^2 + x + 4$ Combine like terms.

✓ **CHECKPOINT** *Now try Exercise 89.*

EXAMPLE 8 Combining Polynomials

Perform the indicated operations and simplify.

$$(3x^2 - 7x + 2) - (4x^2 + 6x - 1) + (-x^2 + 4x + 5)$$

Solution

$$\begin{aligned}(3x^2 - 7x + 2) &- (4x^2 + 6x - 1) + (-x^2 + 4x + 5)\\ &= 3x^2 - 7x + 2 - 4x^2 - 6x + 1 - x^2 + 4x + 5\\ &= (3x^2 - 4x^2 - x^2) + (-7x - 6x + 4x) + (2 + 1 + 5)\\ &= -2x^2 - 9x + 8\end{aligned}$$

 **CHECKPOINT** *Now try Exercise 93.*

EXAMPLE 9 Combining Polynomials

Perform the indicated operations and simplify.

$$(-2x^2 + 4x - 3) - [(4x^2 - 5x + 8) - 2(-x^2 + x + 3)]$$

Solution

$$\begin{aligned}(-2x^2 + 4x - 3) &- [(4x^2 - 5x + 8) - 2(-x^2 + x + 3)]\\ &= (-2x^2 + 4x - 3) - [4x^2 - 5x + 8 + 2x^2 - 2x - 6]\\ &= (-2x^2 + 4x - 3) - [(4x^2 + 2x^2) + (-5x - 2x) + (8 - 6)]\\ &= (-2x^2 + 4x - 3) - [6x^2 - 7x + 2]\\ &= -2x^2 + 4x - 3 - 6x^2 + 7x - 2\\ &= (-2x^2 - 6x^2) + (4x + 7x) + (-3 - 2)\\ &= -8x^2 + 11x - 5\end{aligned}$$

CHECKPOINT *Now try Exercise 95.*

EXAMPLE 10 Geometry: Area of a Region

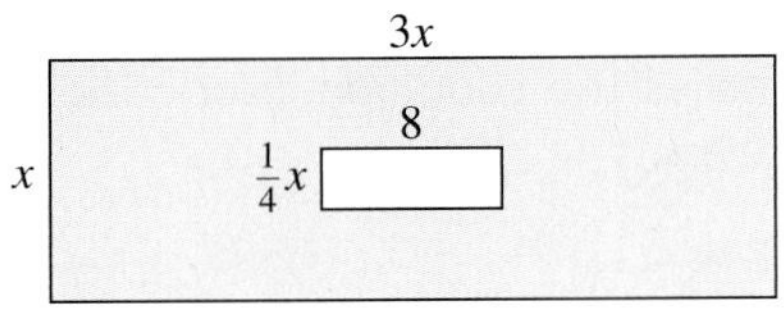

Find an expression for the area of the shaded region shown in Figure 5.1.

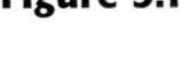

Figure 5.1

Solution

To find a polynomial that represents the area of the shaded region, subtract the area of the inner rectangle from the area of the outer rectangle, as follows.

Area of shaded region	=	Area of outer rectangle	−	Area of inner rectangle

$$\begin{aligned}&= 3x(x) - 8\left(\frac{1}{4}x\right)\\ &= 3x^2 - 2x\end{aligned}$$

CHECKPOINT *Now try Exercise 99.*

Concept Check

1. Explain the difference between the degree of a term of a polynomial and the degree of a polynomial.

2. Determine which of the two statements is always true. Is the other statement always false? Explain.
   a. A polynomial is a trinomial.
   b. A trinomial is a polynomial.

3. Is the sum of two binomials always a binomial? Explain.

4. In your own words, explain how to subtract polynomials using a horizontal or vertical format.

5.2 EXERCISES

Go to pages 338–339 to record your assignments.

Developing Skills

In Exercises 1–8, determine whether the expression is a polynomial. If it is not, explain why. ***See Example 1.***

1. $9 - z$
2. $t^2 - 4$
3. $p^{3/4} - 16$
4. $9 - z^{1/2}$
5. $6x^{-1}$
6. $4 - 9x^4$
7. $z^2 - 3z + \frac{1}{4}$
8. $t^3 - 3t + 4$

In Exercises 9–18, write the polynomial in standard form. Then identify its degree and leading coefficient. ***See Example 2.***

9. $12x + 9$
10. $15 - 6c$
11. $7x - 5x^2 + 10$
12. $1 - 4z + 12z^3$
13. $6m - 3m^5 - m^2 + 12$
14. $5x^3 - 3x^2 + 10$
15. 10
16. -32
17. $v_0t - 16t^2$ (v_0 is a constant.)
18. $64 - \frac{1}{2}at^2$ (a is a constant.)

In Exercises 19–24, determine whether the polynomial is a monomial, a binomial, or a trinomial.

19. $14y - 2$
20. $z^2 + 5z - 14$
21. -32
22. $8 - t^3$
23. $4x + 18x^2 - 5$
24. $45v^4$

In Exercises 25–30, give an example of a polynomial in one variable that satisfies the condition. (*Note:* There are many correct answers.)

25. A binomial of degree 3
26. A trinomial of degree 4
27. A monomial of degree 2
28. A binomial of degree 5
29. A trinomial of degree 6
30. A monomial of degree 0

In Exercises 31–42, use a horizontal format to find the sum. ***See Example 3.***

31. $(4w + 5) + (16w - 9)$

32. $(-2x + 4) + (x - 6)$

✓ **33.** $(3z^2 - z + 2) + (z^2 - 4)$

34. $(6x^4 + 8x) + (4x - 6)$

35. $\left(\frac{2}{3}y^2 - \frac{3}{4}\right) + \left(\frac{5}{6}y^2 + 2\right)$

36. $\left(\frac{3}{4}x^3 - \frac{1}{2}\right) + \left(\frac{1}{8}x^3 + 3\right)$

37. $(0.1t^3 - 3.4t^2) + (1.5t^3 - 7.3)$

38. $(0.7x^2 - 0.2x + 2.5) + (7.4x - 3.9)$

39. $b^2 + (b^3 - 2b^2 + 3) + (b^3 - 3)$

40. $(3x^2 - x) + 5x^3 + (-4x^3 + x^2 - 8)$

41. $(2ab - 3) + (a^2 - 2ab) + (4b^2 - a^2)$

42. $(uv - 3) + (4uv - v^2) + (u^2 - 8uv)$

In Exercises 43–56, use a vertical format to find the sum. ***See Example 4.***

43. $\begin{array}{r} 2x + 5 \\ \underline{3x + 8} \end{array}$

44. $\begin{array}{r} 11x + 5 \\ \underline{7x - 6} \end{array}$

45. $\begin{array}{r} -2x + 10 \\ \underline{x - 38} \end{array}$

46. $\begin{array}{r} 4x^2 + 13 \\ \underline{3x^2 - 11} \end{array}$

✓ **47.** $(x^2 - 4) + (2x^2 + 6)$

48. $(x^3 + 2x - 3) + (4x + 5)$

49. $(-x^3 + 3) + (3x^3 + 2x^2 + 5)$

50. $(2z^3 + 3z - 2) + (z^2 - 2z)$

51. $(3x^4 - 2x^3 - 4x^2 + 2x - 5) + (x^2 - 7x + 5)$

52. $(x^5 - 4x^3 + x + 9) + (2x^4 + 3x^3 - 3)$

53. $(x^2 - 2x + 2) + (x^2 + 4x) + 2x^2$

54. $(5y + 10) + (y^2 - 3y - 2) + (2y^2 + 4y - 3)$

55. Add $8y^3 + 7$ to $5 - 3y^3$.

56. Add $2z - 8z^2 - 3$ to $z^2 + 5z$.

In Exercises 57–66, use a horizontal format to find the difference. ***See Example 5.***

57. $(11x - 8) - (2x + 3)$

58. $(5x + 1) - (18x - 7)$

✓ **59.** $(x^2 - x) - (x - 2)$

60. $(x^2 - 4) - (x^2 - 4)$

61. $(4 - 2x - x^3) - (3 - 2x + 2x^3)$

62. $(t^4 - 2t^2) - (3t^2 - t^4 - 5)$

63. $8 - (w^3 - w)$

64. $(z^3 + z^2 + 1) - z^2$

65. $(x^5 - 3x^4 + x^3 - 5x + 1) - (4x^5 - x^3 + x - 5)$

66. $(t^4 + 5t^3 - t^2 + 8t - 10) - (t^4 + t^3 + 2t^2 + 4t - 7)$

In Exercises 67–80, use a vertical format to find the difference. ***See Example 6.***

✓ **67.** $\begin{array}{r} 2x - 2 \\ \underline{-\,(x - 1)} \end{array}$

68. $\begin{array}{r} 9x + 7 \\ \underline{-\,(3x + 9)} \end{array}$

69. $\begin{array}{r} 2x^2 - x + 2 \\ \underline{-\,(3x^2 + x - 1)} \end{array}$

70. $\begin{array}{r} y^4 - 2 \\ \underline{-\,(y^4 + 2)} \end{array}$

71. $(-3x^3 - 4x^2 + 2x - 5) - (2x^4 + 2x^3 - 4x + 5)$

72. $(12x^3 + 25x^2 - 15) - (-2x^3 + 18x^2 - 3x)$

73. $(2 - x^3) - (2 + x^3)$

74. $(4z^3 - 6) - (-z^3 + z - 2)$

75. $(4t^3 - 3t + 5) - (3t^2 - 3t - 10)$

76. $(-s^2 - 3) - (2s^2 + 10s)$

77. $(7x^2 - x) - (x^3 - 2x^2 + 10)$

78. $(y^2 - 3y + 8) - (y^3 + y^2 - 3)$

79. Subtract $7x^3 - 4x + 5$ from $10x^3 + 15$.

80. Subtract $y^5 - y^4$ from $y^2 + 3y^4$.

In Exercises 81–96, perform the indicated operations and simplify. ***See Examples 7, 8, and 9.***

81. $(6x - 5) - (8x + 15)$

82. $(2x^2 + 1) + (x^2 - 2x + 1)$

83. $-(x^3 - 2) + (4x^3 - 2x)$

84. $-(5x^2 - 1) - (-3x^2 + 5)$

85. $2(x^4 + 2x) + (5x + 2)$

86. $(z^4 - 2z^2) + 3(z^4 + 4)$

87. $(15x^2 - 6) - (-8x^3 - 14x^2 - 17)$

88. $(15x^4 - 18x - 19) - (-13x^4 - 5x + 15)$

89. $5z - [3z - (10z + 8)]$

90. $9w^2 - [2w - (w^2 + 3w)]$

91. $(y^3 + 1) - [(y^2 + 1) + (3y - 7)]$

92. $(a^2 - a) - [(2a^2 + 3a) - (5a^2 - 12)]$

93. $2(t^2 + 5) - 3(t^2 + 5) + 5(t^2 + 5)$

94. $-10(u + 1) + 8(u - 1) - 3(u + 6)$

95. $8v - 6(3v - v^2) + 10(10v + 3)$

96. $3(x^2 - 2x + 3) - 4(4x + 1) - (3x^2 - 2x)$

Solving Problems

Geometry In Exercises 97 and 98, find an expression for the perimeter of the figure.

97.

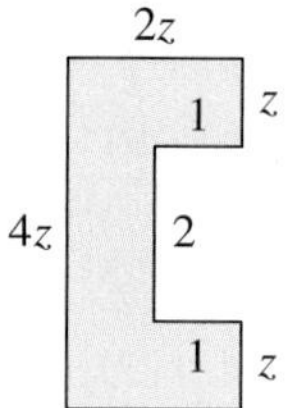

98.

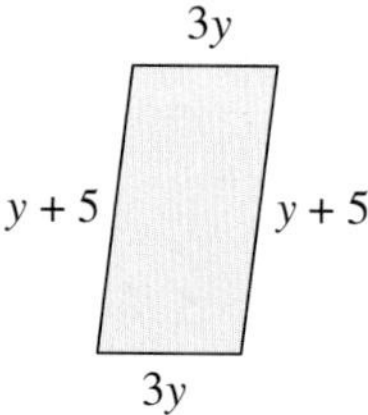

Geometry In Exercises 99–102, find an expression for the area of the shaded region of the figure. ***See Example 10.***

99.

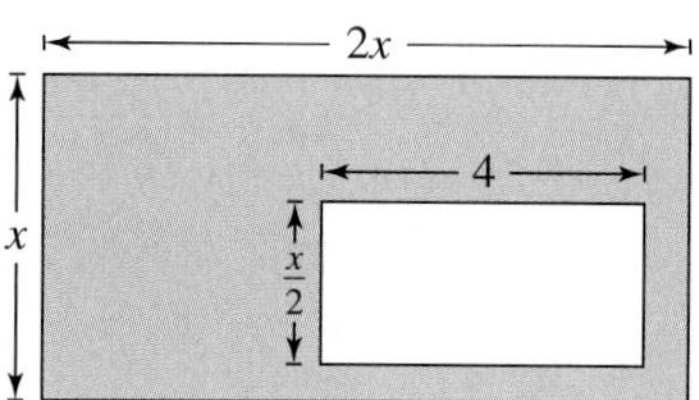

100.

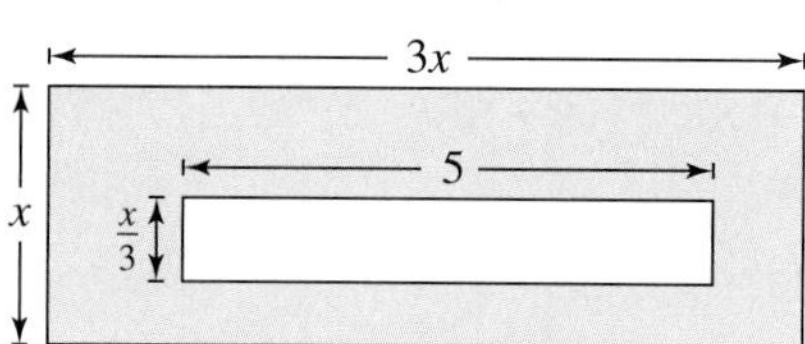

101.

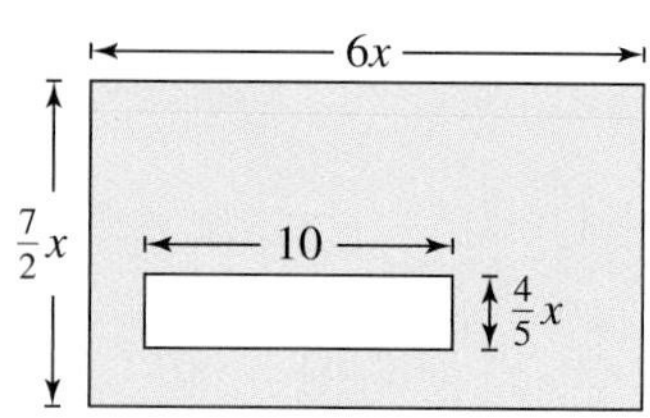

102.

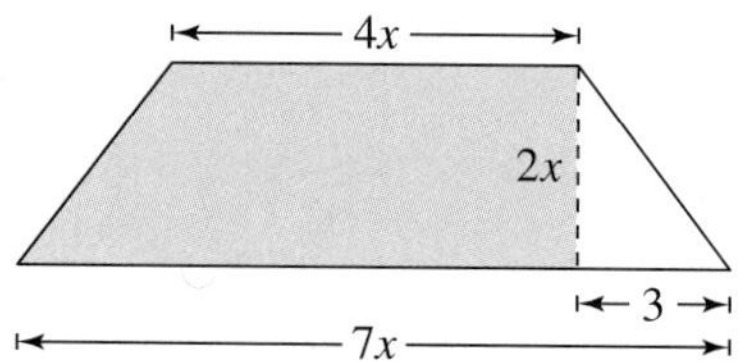

103. *Enrollment* The projected enrollments (in millions) of students in public schools P and private schools R for the years 2010 through 2015 can be modeled by

$P = 0.53t + 58.1, \quad 10 \le t \le 15$

and

$R = 0.007t^2 - 0.06t + 11.0, \quad 10 \le t \le 15$

where t represents the year, with $t = 10$ corresponding to 2010. (Source: U.S. Center for Education Statistics)

(a) Add the polynomials to find a model for the projected total enrollment T of students in public and private schools.

(b) Use a graphing calculator to graph all three models.

(c) Use the graphs from part (b) to determine whether the numbers of public, private, and total school enrollments are increasing or decreasing.

104. *Cost, Revenue, and Profit* The cost C of producing x units of a product is $C = 200 + 45x$. The revenue R for selling x units is $R = 120x - x^2$, where $0 \le x \le 60$. The profit P is the difference between revenue and cost.

(a) Perform the subtraction required to find the polynomial representing profit P.

(b) Use a graphing calculator to graph the polynomial representing profit.

(c) Determine the profit when 40 units are produced and sold. Use the graph in part (b) to predict the change in profit when x is some value other than 40.

Explaining Concepts

105. In your own words, define "like terms." What is the only factor of like terms that can differ?

106. Describe how to combine like terms. What operations are used?

107. Is a polynomial an algebraic expression? Explain.

108. Write a paragraph that explains how the adage "You can't add apples and oranges" might relate to adding two polynomials. Include several examples to illustrate the applicability of this statement.

Cumulative Review

In Exercises 109–112, solve the equation and check your solution.

109. $\frac{4x}{27} = \frac{8}{9}$

110. $\frac{x}{6} - \frac{x}{18} = 3$

111. $\frac{x+3}{6} = \frac{2}{5}$

112. $\frac{x-5}{2} = \frac{4x}{3}$

In Exercises 113 and 114, graph the equation. Use a graphing calculator to verify your graph.

113. $y = 2 - \frac{3}{2}x$

114. $y = |x - 1|$

In Exercises 115 and 116, determine the exponent that makes the statement true.

115. $2^{\quad} = \frac{1}{32}$

116. $(x^{\quad}y^2)^{-3} = \frac{1}{x^{12}y^6}$

In Exercises 117 and 118, use your calculator to evaluate the expression. Write your answer in scientific notation. Round your answer to four decimal places.

117. $(4.15 \times 10^3)^{-4}$

118. $\frac{1.5 \times 10^8}{2.3 \times 10^5}$

Mid-Chapter Quiz

Take this quiz as you would take a quiz in class. After you are done, check your work against the answers in the back of the book.

In Exercises 1–4, simplify the expression. (Assume that no denominator is zero.)

1. $(4m^3n^2)^4$
2. $(-3xy)^2(2x^2y)^3$
3. $\dfrac{-12x^3y}{9x^5y^2}$
4. $\dfrac{3t^3}{(-6t)^2}$

In Exercises 5 and 6, rewrite the expression using only positive exponents.

5. $5x^{-2}y^{-3}$
6. $\dfrac{3x^{-2}y}{5z^{-1}}$

In Exercises 7 and 8, use the rules of exponents to simplify the expression using only positive exponents. (Assume that no variable is zero.)

7. $(3a^{-3}b^2)^{-2}$
8. $(4t^{-3})^0$

9. Write the number 8,168,000,000,000 in scientific notation.
10. Write the number 5.021×10^{-8} in decimal notation.
11. Explain why $x^2 + 2x - 3x^{-1}$ is not a polynomial.
12. Determine the degree and the leading coefficient of the polynomial $10 + x^2 - 4x^3$.
13. Give an example of a trinomial in one variable of degree 5.

In Exercises 14–17, perform the indicated operations and simplify.

14. $(y^2 + 3y - 1) + (4 + 3y)$
15. $(3v^2 - 5) - (v^3 + 2v^2 - 6v)$
16. $9s - [6 - (s - 5) + 7s]$
17. $-3(4 - x) + 4(x^2 + 2) - (x^2 - 2x)$

In Exercises 18 and 19, use a vertical format to find the sum.

18. $$\begin{array}{r} 5x^4 \qquad\qquad + 2x^2 + x - 3 \\ \underline{3x^3 - 2x^2 - 3x + 5} \end{array}$$

19. $$\begin{array}{r} x^3 - 3x^2 \qquad\qquad - 15 \\ \underline{2x^2 + 5x - 4} \end{array}$$

In Exercises 20 and 21, use a vertical format to find the difference.

20. $$\begin{array}{r} x^2 - x + 2 \\ \underline{-\quad (x - 4)} \end{array}$$

21. $$\begin{array}{r} 6x^4 + 3x^3 \qquad\qquad + 8 \\ \underline{-(x^4 \qquad\qquad + 4x^2 + 2)} \end{array}$$

Figure for 22

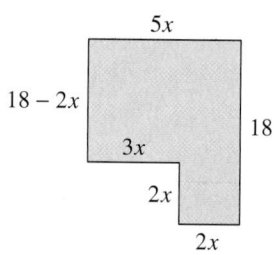

22. Find an expression for the perimeter of the figure at the left.

5.3 Multiplying Polynomials: Special Products

Holly Harris/Getty Images

Why You Should Learn It

Multiplying polynomials enables you to model and solve real-life problems. For instance, in Exercise 131 on page 326, you will multiply polynomials to find the total consumption of milk in the United States.

What You Should Learn

1 ▶ Find products with monomial multipliers.
2 ▶ Multiply binomials using the Distributive Property and the FOIL Method.
3 ▶ Multiply polynomials using a horizontal or vertical format.
4 ▶ Identify and use special binomial products.

1 ▶ Find products with monomial multipliers.

Monomial Multipliers

To multiply polynomials, you use many of the rules for simplifying algebraic expressions. You may want to review these rules in Section 2.2 and Section 5.1.

1. The Distributive Property
2. Combining like terms
3. Removing symbols of grouping
4. Rules of exponents

The simplest type of polynomial multiplication involves a monomial multiplier. The product is obtained by direct application of the Distributive Property. For instance, to multiply the monomial x by the polynomial $(2x + 5)$, multiply *each* term of the polynomial by x.

$$(x)(2x + 5) = (x)(2x) + (x)(5) = 2x^2 + 5x$$

EXAMPLE 1 Finding Products with Monomial Multipliers

Find each product.

a. $(3x - 7)(-2x)$ **b.** $3x^2(5x - x^3 + 2)$ **c.** $(-x)(2x^2 - 3x)$

Solution

a. $(3x - 7)(-2x) = 3x(-2x) - 7(-2x)$ — Distributive Property

$= -6x^2 + 14x$ — Write in standard form.

b. $3x^2(5x - x^3 + 2)$

$= (3x^2)(5x) - (3x^2)(x^3) + (3x^2)(2)$ — Distributive Property

$= 15x^3 - 3x^5 + 6x^2$ — Rules of exponents

$= -3x^5 + 15x^3 + 6x^2$ — Write in standard form.

c. $(-x)(2x^2 - 3x) = (-x)(2x^2) - (-x)(3x)$ — Distributive Property

$= -2x^3 + 3x^2$ — Write in standard form.

✓ **CHECKPOINT** *Now try Exercise 13.*

2 ▶ Multiply binomials using the Distributive Property and the FOIL Method.

Multiplying Binomials

To multiply two binomials, you can use both (left and right) forms of the Distributive Property. For example, if you treat the binomial $(5x + 7)$ as a single quantity, you can multiply $(3x - 2)$ by $(5x + 7)$ as follows.

$$(3x - 2)(5x + 7) = 3x(5x + 7) - 2(5x + 7)$$
$$= (3x)(5x) + (3x)(7) - (2)(5x) - 2(7)$$
$$= 15x^2 + 21x - 10x - 14$$

Product of First terms | Product of Outer terms | Product of Inner terms | Product of Last terms

$$= 15x^2 + 11x - 14$$

With practice, you should be able to multiply two binomials without writing out all of the steps above. In fact, the four products in the boxes above suggest that you can write the product of two binomials in just one step. This is called the **FOIL Method.** Note that the words *first, outer, inner,* and *last* refer to the positions of the terms in the original product.

First
Outer

$$(3x - 2)(5x + 7)$$

Inner
Last

Technology: Tip

Remember that you can use a graphing calculator to check whether you have performed a polynomial operation correctly. For instance, to check if

$$(x - 1)(x + 5) = x^2 + 4x - 5$$

you can graph the left side of the equation and graph the right side of the equation in the same viewing window, as shown below. Because both graphs are the same, you can conclude that the multiplication was performed correctly.

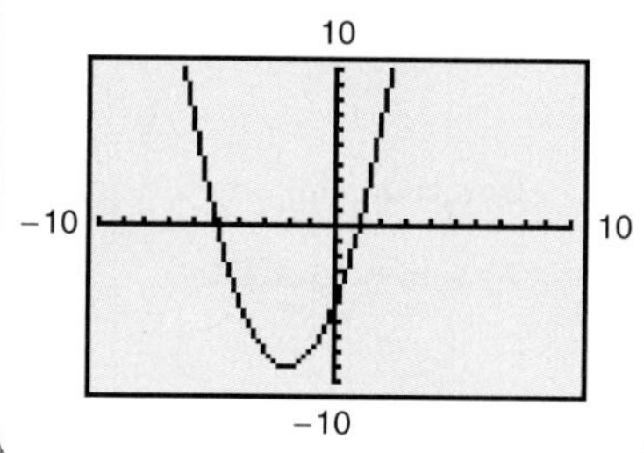

EXAMPLE 2 Multiplying with the Distributive Property

Use the Distributive Property to find each product.

a. $(x - 1)(x + 5)$

b. $(2x + 3)(x - 2)$

Solution

a. $(x - 1)(x + 5) = x(x + 5) - 1(x + 5)$ Right Distributive Property

$= x^2 + 5x - x - 5$ Left Distributive Property

$= x^2 + (5x - x) - 5$ Group like terms.

$= x^2 + 4x - 5$ Combine like terms.

b. $(2x + 3)(x - 2) = 2x(x - 2) + 3(x - 2)$ Right Distributive Property

$= 2x^2 - 4x + 3x - 6$ Left Distributive Property

$= 2x^2 + (-4x + 3x) - 6$ Group like terms.

$= 2x^2 - x - 6$ Combine like terms.

✓ ***CHECKPOINT*** *Now try Exercise 31.*

EXAMPLE 3 Multiplying Binomials Using the FOIL Method

Use the FOIL Method to find each product.

a. $(x + 4)(x - 4)$

b. $(3x + 5)(2x + 1)$

Solution

$$\begin{aligned} &\quad\ \ \text{F}\quad\ \text{O}\quad\ \ \text{I}\quad\ \ \text{L} \\ \textbf{a. } (x + 4)(x - 4) &= x^2 - 4x + 4x - 16 \\ &= x^2 - 16 \end{aligned}$$

Combine like terms.

$$\begin{aligned} &\quad\ \ \text{F}\quad\ \ \text{O}\quad\ \ \text{I}\quad\ \ \text{L} \\ \textbf{b. } (3x + 5)(2x + 1) &= 6x^2 + 3x + 10x + 5 \\ &= 6x^2 + 13x + 5 \end{aligned}$$

Combine like terms.

CHECKPOINT *Now try Exercise 39.*

In Example 3(a), note that the outer and inner products add up to zero.

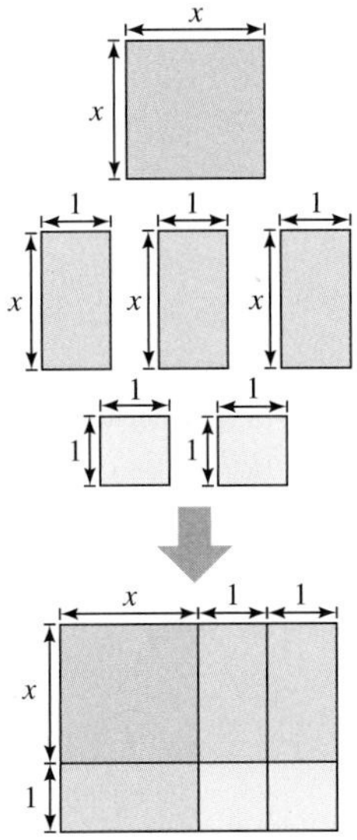

EXAMPLE 4 A Geometric Model of a Polynomial Product

Use the geometric model shown in Figure 5.2 to show that

$$x^2 + 3x + 2 = (x + 1)(x + 2).$$

Solution

The upper part of the figure shows that the sum of the areas of the six rectangles is

$$x^2 + (x + x + x) + (1 + 1) = x^2 + 3x + 2.$$

The lower part of the figure shows that the area of the rectangle is

$$\begin{aligned} (x + 1)(x + 2) &= x^2 + 2x + x + 2 \\ &= x^2 + 3x + 2. \end{aligned}$$

So, $x^2 + 3x + 2 = (x + 1)(x + 2)$.

CHECKPOINT *Now try Exercise 125.*

Figure 5.2

EXAMPLE 5 Simplifying a Polynomial Expression

Simplify the expression and write the result in standard form.

$$(4x + 5)^2$$

Solution

$$\begin{aligned} (4x + 5)^2 &= (4x + 5)(4x + 5) && \text{Repeated multiplication} \\ &= 16x^2 + 20x + 20x + 25 && \text{Use FOIL Method.} \\ &= 16x^2 + 40x + 25 && \text{Combine like terms.} \end{aligned}$$

CHECKPOINT *Now try Exercise 45.*

EXAMPLE 6 **Simplifying a Polynomial Expression**

Simplify the expression and write the result in standard form.

$(3x^2 - 2)(4x + 7) - (4x)^2$

Solution

$(3x^2 - 2)(4x + 7) - (4x)^2$

$= 12x^3 + 21x^2 - 8x - 14 - (4x)^2$ Use FOIL Method.

$= 12x^3 + 21x^2 - 8x - 14 - 16x^2$ Square monomial.

$= 12x^3 + 5x^2 - 8x - 14$ Combine like terms.

✓ **CHECKPOINT** *Now try Exercise 49.*

3 ▶ Multiply polynomials using a horizontal or vertical format.

Multiplying Polynomials

The FOIL Method for multiplying two binomials is simply a device for guaranteeing that *each term of one binomial is multiplied by each term of the other binomial.*

$$(ax + b)(cx + d) = \overset{F}{ax(cx)} + \overset{O}{ax(d)} + \overset{I}{b(cx)} + \overset{L}{b(d)}$$

This same rule applies to the product of any two polynomials: *each term of one polynomial must be multiplied by each term of the other polynomial.* This can be accomplished using either a horizontal or a vertical format.

EXAMPLE 7 **Multiplying Polynomials (Horizontal Format)**

Use a horizontal format to find each product.

a. $(x - 4)(x^2 - 4x + 2)$ **b.** $(2x^2 - 7x + 1)(4x + 3)$

Solution

a. $(x - 4)(x^2 - 4x + 2)$

$= x(x^2 - 4x + 2) - 4(x^2 - 4x + 2)$ Distributive Property

$= x^3 - 4x^2 + 2x - 4x^2 + 16x - 8$ Distributive Property

$= x^3 - 8x^2 + 18x - 8$ Combine like terms.

b. $(2x^2 - 7x + 1)(4x + 3)$

$= (2x^2 - 7x + 1)(4x) + (2x^2 - 7x + 1)(3)$ Distributive Property

$= 8x^3 - 28x^2 + 4x + 6x^2 - 21x + 3$ Distributive Property

$= 8x^3 - 22x^2 - 17x + 3$ Combine like terms.

✓ **CHECKPOINT** *Now try Exercise 59.*

EXAMPLE 8 Multiplying Polynomials (Vertical Format)

Use a vertical format to find the product of $(3x^2 + x - 5)$ and $(2x - 1)$.

Solution

With a vertical format, line up like terms in the same vertical columns, just as you align digits in whole number multiplication.

$$
\begin{array}{rrrrl}
 & 3x^2 & + \; x & - \; 5 & \text{Place polynomial with most terms on top.} \\
\times & & 2x & - \; 1 & \text{Line up like terms.} \\
\hline
 & -3x^2 & - \; x & + \; 5 & \Leftarrow \; -1(3x^2 + x - 5) \\
6x^3 & + \; 2x^2 & - \; 10x & & \Leftarrow \; 2x(3x^2 + x - 5) \\
\hline
6x^3 & - \; x^2 & - \; 11x & + \; 5 & \text{Combine like terms in columns.}
\end{array}
$$

 **CHECKPOINT** *Now try Exercise 69.*

EXAMPLE 9 Multiplying Polynomials (Vertical Format)

Use a vertical format to find the product of $(4x^3 + 8x - 1)$ and $(2x^2 + 3)$.

Solution

$$
\begin{array}{rrrrrl}
 & 4x^3 & & + \; 8x & - \; 1 & \text{Place polynomial with most terms on top.} \\
\times & & 2x^2 & & + \; 3 & \text{Line up like terms.} \\
\hline
 & 12x^3 & & + \; 24x & - \; 3 & 3(4x^3 + 8x - 1) \\
8x^5 & + \; 16x^3 & - \; 2x^2 & & & 2x^2(4x^3 + 8x - 1) \\
\hline
8x^5 & + \; 28x^3 & - \; 2x^2 & + \; 24x & - \; 3 & \text{Combine like terms in columns.}
\end{array}
$$

 **CHECKPOINT** *Now try Exercise 71.*

Study Tip

When multiplying two polynomials, it is best to write each in standard form before using either the horizontal or the vertical format.

EXAMPLE 10 Multiplying Polynomials (Vertical Format)

Write the polynomials in standard form and use a vertical format to find the product of $(x + 3x^2 - 4)$ and $(5 + 3x - x^2)$.

Solution

$$
\begin{array}{rrrrrl}
 & & 3x^2 & + \; x & - \; 4 & \text{Write in standard form.} \\
\times & & -x^2 & + \; 3x & + \; 5 & \text{Write in standard form.} \\
\hline
 & & 15x^2 & + \; 5x & - \; 20 & \Leftarrow \; 5(3x^2 + x - 4) \\
 & 9x^3 & + \; 3x^2 & - \; 12x & & \Leftarrow \; 3x(3x^2 + x - 4) \\
-3x^4 & - \; x^3 & + \; 4x^2 & & & \Leftarrow \; -x^2(3x^2 + x - 4) \\
\hline
-3x^4 & + \; 8x^3 & + \; 22x^2 & - \; 7x & - \; 20 & \text{Combine like terms.}
\end{array}
$$

CHECKPOINT *Now try Exercise 75.*

EXAMPLE 11 Multiplying Polynomials

Multiply $(x - 3)^3$.

Solution

To raise $(x - 3)$ to the third power, you can use two steps. First, because $(x - 3)^3 = (x - 3)^2(x - 3)$, find the product $(x - 3)^2$.

$$(x - 3)^2 = (x - 3)(x - 3) \qquad \text{Repeated multiplication}$$

$$= x^2 - 3x - 3x + 9 \qquad \text{Use FOIL Method.}$$

$$= x^2 - 6x + 9 \qquad \text{Combine like terms.}$$

Now multiply $x^2 - 6x + 9$ by $x - 3$, as follows.

$$(x^2 - 6x + 9)(x - 3) = (x^2 - 6x + 9)(x) - (x^2 - 6x + 9)(3)$$

$$= x^3 - 6x^2 + 9x - 3x^2 + 18x - 27$$

$$= x^3 - 9x^2 + 27x - 27$$

So, $(x - 3)^3 = x^3 - 9x^2 + 27x - 27$.

✓ **CHECKPOINT** *Now try Exercise 77.*

4 ▶ Identify and use special binomial products.

Special Products

Some binomial products, such as those in Examples 3(a) and 5, have special forms that occur frequently in algebra. The product

$$(x + 4)(x - 4)$$

is called a **product of the sum and difference of two terms.** With such products, the two middle terms cancel, as follows.

$$(x + 4)(x - 4) = x^2 - 4x + 4x - 16 \qquad \text{Sum and difference of two terms}$$

$$= x^2 - 16 \qquad \text{Product has no middle term.}$$

Study Tip

You should learn to recognize the patterns of the two special products at the right. The FOIL Method can be used to verify each rule.

Another common type of product is the **square of a binomial.**

$$(4x + 5)^2 = (4x + 5)(4x + 5) \qquad \text{Square of a binomial}$$

$$= 16x^2 + 20x + 20x + 25 \qquad \text{Use FOIL Method.}$$

$$= 16x^2 + 40x + 25 \qquad \text{Middle term is twice the product of the terms of the binomial.}$$

In general, when a binomial is squared, the resulting middle term is always twice the product of the two terms.

$$(a + b)^2 = a^2 + 2(ab) + b^2$$

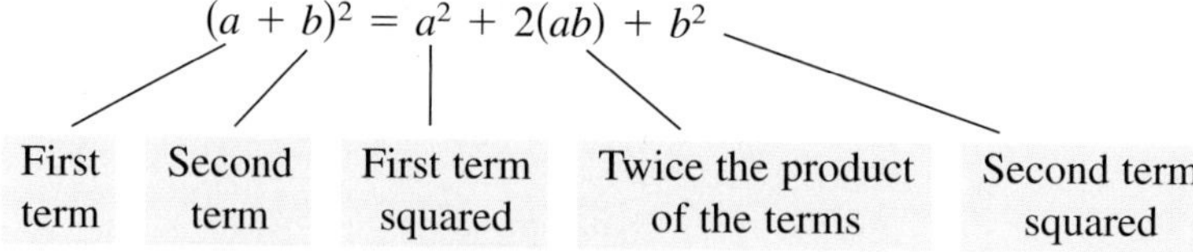

Be sure to include the middle term. For instance, $(a + b)^2$ is not equal to $a^2 + b^2$.

Smart Study Strategy

Go to page 292 for ways to *Make Mental Cheat Sheets.*

Special Products

Let a and b be real numbers, variables, or algebraic expressions.

Special Product	*Example*
Sum and Difference of Two Terms:	
$(a + b)(a - b) = a^2 - b^2$	$(2x - 5)(2x + 5) = 4x^2 - 25$
Square of a Binomial:	
$(a + b)^2 = a^2 + 2ab + b^2$	$(3x + 4)^2 = 9x^2 + 2(3x)(4) + 16$ $= 9x^2 + 24x + 16$
$(a - b)^2 = a^2 - 2ab + b^2$	$(x - 7)^2 = x^2 - 2(x)(7) + 49$ $= x^2 - 14x + 49$

EXAMPLE 12 Finding the Product of the Sum and Difference of Two Terms

Multiply $(x + 2)(x - 2)$.

Solution

Sum Difference (1st term)^2 (2nd term)^2

$$(x + 2)(x - 2) = (x)^2 - (2)^2 = x^2 - 4$$

✓ **CHECKPOINT** ***Now try Exercise 85.***

EXAMPLE 13 Finding the Product of the Sum and Difference of Two Terms

Multiply $(5x - 6)(5x + 6)$.

Solution

Difference Sum (1st term)^2 (2nd term)^2

$$(5x - 6)(5x + 6) = (5x)^2 - (6)^2 = 25x^2 - 36$$

✓ **CHECKPOINT** ***Now try Exercise 91.***

EXAMPLE 14 Squaring a Binomial

Multiply $(4x - 9)^2$.

Solution

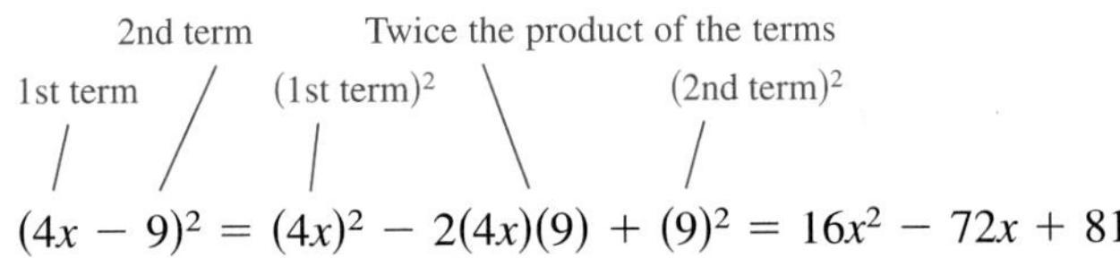

✓ **CHECKPOINT** ***Now try Exercise 101.***

EXAMPLE 15 Squaring a Binomial

Multiply $(3x + 7)^2$.

Solution

1st term; 2nd term; (1st term)^2; Twice the product of the terms; (2nd term)^2

$$(3x + 7)^2 = (3x)^2 + 2(3x)(7) + (7)^2 = 9x^2 + 42x + 49$$

CHECKPOINT Now try Exercise 103.

EXAMPLE 16 Squaring a Binomial

Multiply $(6 - 5x^2)^2$.

Solution

1st term; 2nd term; (1st term)^2; Twice the product of the terms; (2nd term)^2

$$(6 - 5x^2)^2 = (6)^2 - 2(6)(5x^2) + (5x^2)^2$$

$$= 36 - 60x^2 + (5)^2(x^2)^2 = 36 - 60x^2 + 25x^4$$

CHECKPOINT Now try Exercise 107.

EXAMPLE 17 Finding the Dimensions of a Golf Tee

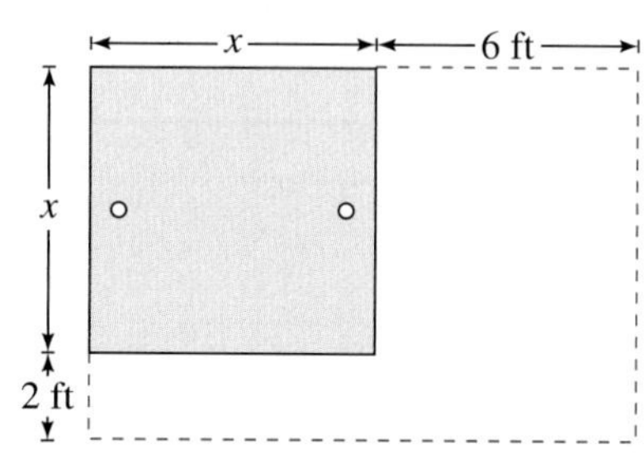
x
6 ft
x
2 ft

Figure 5.3

A landscaper wants to reshape a square tee area for the ninth hole of a golf course. The new tee area is to have one side 2 feet longer and the adjacent side 6 feet longer than the original tee. (See Figure 5.3.) The new tee has 204 square feet more area than the original tee. What are the dimensions of the original tee?

Solution

Verbal Model: New area = Old area + 204

Labels:
Original length = original width = x (feet)
Original area = x^2 (square feet)
New length = $x + 6$ (feet)
New width = $x + 2$ (feet)

Equation:

$(x + 6)(x + 2) = x^2 + 204$ Write equation.

$x^2 + 8x + 12 = x^2 + 204$ Multiply factors.

$8x + 12 = 204$ Subtract x^2 from each side.

$8x = 192$ Subtract 12 from each side.

$x = 24$ Divide each side by 8.

The original tee measured 24 feet by 24 feet.

CHECKPOINT Now try Exercise 123.

Concept Check

1. Describe the rules of exponents that are used to multiply polynomials. Give examples.

2. Explain the meaning of each letter of "FOIL" as it relates to multiplying two binomials.

3. Describe how to multiply polynomials using a vertical format.

4. State the two special products and explain how to evaluate them.

5.3 EXERCISES

Go to pages 338–339 to record your assignments.

Developing Skills

In Exercises 1–52, perform the multiplication and simplify. ***See Examples 1–3, 5, and 6.***

1. $x(-2x)$

2. $y(-3y)$

3. $t^2(4t)$

4. $3u(u^4)$

5. $\left(\frac{x}{4}\right)(10x)$

6. $9x\left(\frac{x}{12}\right)$

7. $(-2b^2)(-3b)$

8. $(-4x^3)(-5x^4)$

9. $y(3 - y)$

10. $z(z - 1)$

11. $-x(x^2 - 4)$

12. $-t(10 - 9t^2)$

✓ **13.** $-3x(2x^2 + 5)$

14. $-5u(u^2 + 4)$

15. $-4x(3 + 3x^2 - 6x^3)$

16. $-5v(5 - 4v + 5v^2)$

17. $3x(x^2 - 2x + 1)$

18. $4y(4y^2 + 2y - 3)$

19. $2x(x^2 - 2x + 8)$

20. $7y(y^2 - y + 5)$

21. $4t^3(t - 3)$

22. $-2t^4(t + 6)$

23. $x^2(4x^2 - 3x + 1)$

24. $y^2(2y^2 + y - 5)$

25. $-3x^3(4x^2 - 6x + 2)$

26. $5u^4(2u^3 - 3u + 3)$

27. $-2x(-3x)(5x + 2)$

28. $4x(-2x)(x^2 - 1)$

29. $-2x(-6x^4) - 3x^2(2x^2)$

30. $-8y(-5y^4) - 2y^2(5y^3)$

✓ **31.** $(x + 3)(x + 4)$

32. $(x - 5)(x + 10)$

33. $(3x - 5)(x + 1)$

34. $(7x - 2)(x - 3)$

35. $(2x - y)(x - 2y)$

36. $(x + y)(x + 2y)$

37. $(5x + 6)(3x + 1)$

38. $(4x + 3)(2x - 1)$

✓ **39.** $(6 - 2x)(4x + 3)$

40. $(8x - 6)(5 - 4x)$

41. $(3x - 2y)(x - y)$

42. $(7x + 5y)(x + y)$

43. $(3x^2 - 4)(x + 2)$

44. $(5x^2 - 2)(x - 1)$

✓ **45.** $(2x + 4)^2$

46. $(7x - 3)^2$

47. $(8x + 2)^2$

48. $(5x - 1)^2$

✓ **49.** $(3s + 1)(3s + 4) - (3s)^2$

50. $(2t + 5)(4t - 2) - (2t)^2$

51. $(4x^2 - 1)(2x + 8) + (-x^2)^3$

52. $(3 - 3x^2)(4 - 5x^2) - (-x^4)^2$

In Exercises 53–66, use a horizontal format to find the product. ***See Example 7.***

53. $(x + 10)(x + 2)$ **54.** $(x - 1)(x + 3)$

55. $(2x - 5)(7x + 2)$ **56.** $(3x - 2)(2x - 3)$

57. $(x + 1)(x^2 + 2x - 1)$ **58.** $(x - 3)(x^2 - 3x + 4)$

✓ **59.** $(x^3 - 2x + 1)(x - 5)$

60. $(x^2 - x + 1)(x + 4)$

61. $(x - 2)(5x^2 + 2x + 4)$ **62.** $(x + 9)(2x^2 - x - 4)$

63. $(x^2 + 3)(x^2 - 6x + 2)$

64. $(x^2 - 5)(x^2 - 2x + 3)$

65. $(3x^2 - 4x - 2)(3x^2 + 1)$

66. $(8x^2 + 2x + 5)(4x^3 - 2)$

In Exercises 67–76, use a vertical format to find the product. ***See Examples 8–10.***

67. $\begin{array}{r} x + 3 \\ \times\ x - 2 \\ \hline \end{array}$ **68.** $\begin{array}{r} 2x - 1 \\ \times\ 5x + 1 \\ \hline \end{array}$

✓ **69.** $\begin{array}{r} 4x^4 - 6x^2 \quad + 9 \\ \times \qquad\quad 2x + 3 \\ \hline \end{array}$

70. $\begin{array}{r} x^2 - 3x + 9 \\ \times \qquad x + 3 \\ \hline \end{array}$

✓ **71.** $(3x^3 + x + 7)(x^2 + 1)$

72. $(5x^4 - 3x + 2)(2x^2 - 4)$

73. $(x^2 - x + 2)(x^2 + x - 2)$

74. $(x^2 + 2x + 5)(2x^2 - x - 1)$

✓ **75.** $(x + 3 - 2x^2)(5x + x^2 - 4)$

76. $(1 - x - x^2)(x + 1 - x^3)$

In Exercises 77–84, perform the multiplication and simplify. ***See Example 11.***

✓ **77.** $(x - 2)^3$

78. $(x + 3)^3$

79. $(x - 1)^2(x - 1)^2$

80. $(x + 4)^2(x + 4)^2$

81. $(x + 2)^2(x - 4)$ **82.** $(x - 4)^2(x - 1)$

83. $(u - 1)(2u + 3)(2u + 1)$

84. $(2x + 5)(x - 2)(5x - 3)$

In Exercises 85–114, use a special product pattern to find the product. ***See Examples 12–16.***

✓ **85.** $(x + 3)(x - 3)$ **86.** $(x - 5)(x + 5)$

87. $(x + 20)(x - 20)$ **88.** $(y + 9)(y - 9)$

89. $(2u + 3)(2u - 3)$ **90.** $(3z + 4)(3z - 4)$

✓ **91.** $(4t - 6)(4t + 6)$ **92.** $(3u + 7)(3u - 7)$

93. $(2x^2 + 5)(2x^2 - 5)$

94. $(6t + 1)(6t - 1)$

95. $(4x + y)(4x - y)$

96. $(5u + 12v)(5u - 12v)$

97. $(9u + 7v)(9u - 7v)$

98. $(8a - 5b)(8a + 5b)$

99. $(x + 6)^2$

100. $(a - 2)^2$

✓ **101.** $(t - 3)^2$

102. $(x + 10)^2$

✓ **103.** $(3x + 2)^2$

104. $(2x - 8)^2$

105. $(8 - 3z)^2$

106. $(1 - 5t)^2$

107. $(4 + 7s^2)^2$

108. $(3 + 8v^2)^2$

109. $(2x - 5y)^2$

110. $(4s + 3t)^2$

111. $[(x + 1) + y]^2$

112. $[(x - 3) - y]^2$

113. $[u - (3 + v)]^2$

114. $[2u + (v - 2)]^2$

In Exercises 115 and 116, perform the multiplication and simplify.

115. $(x + 2)^2 - (x - 2)^2$ **116.** $(u + 5)^2 + (u - 5)^2$

Think About It In Exercises 117 and 118, is the equation an identity? Explain.

117. $(x + y)^3 = x^3 + 3x^2y + 3xy^2 + y^3$

118. $(x - y)^3 = x^3 - 3x^2y + 3xy^2 - y^3$

In Exercises 119 and 120, use the result of Exercise 117 to find the product.

119. $(x + 2)^3$ **120.** $(x + 1)^3$

Solving Problems

121. *Geometry* The base of a triangular sail is $2x$ feet and its height is $(x + 10)$ feet (see figure). Find an expression for the area of the sail.

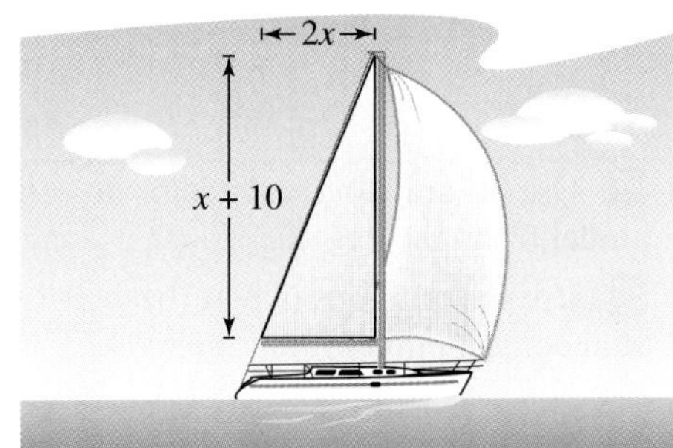

122. *Geometry* The height of a rectangular sign is twice its width w (see figure). Find expressions for (a) the perimeter and (b) the area of the sign.

123. *Geometry* A park recreation manager wants to reshape a square sandbox. The new sandbox will have one side 2 feet longer and the adjacent side 3 feet longer than the original sandbox. The new sandbox will have 26 square feet more area than the original sandbox. What are the dimensions of the original sandbox?

124. *Geometry* A carpenter wants to expand a square room. The new room will have one side 4 feet longer and the adjacent side 6 feet longer than the original room. The new room will have 144 square feet more area than the original room. What are the dimensions of the original room?

Geometry In Exercises 125 and 126, what polynomial product is represented by the geometric model? Explain.

125.

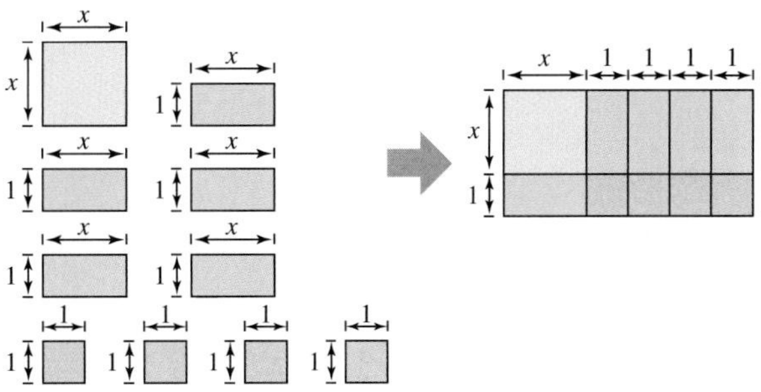

126.

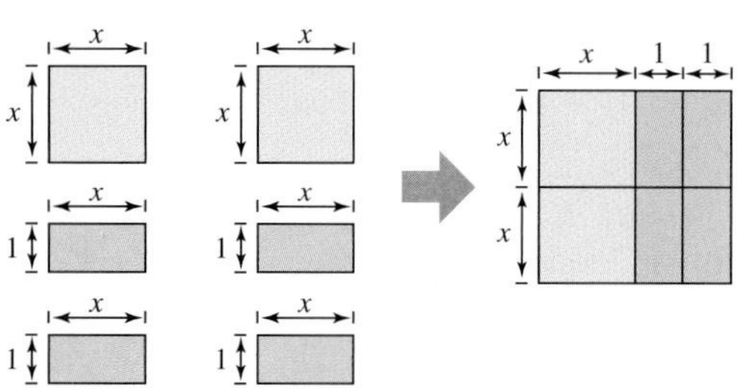

127. *Geometry* Add the areas of the four rectangular regions shown in the figure. What special product does the geometric model represent?

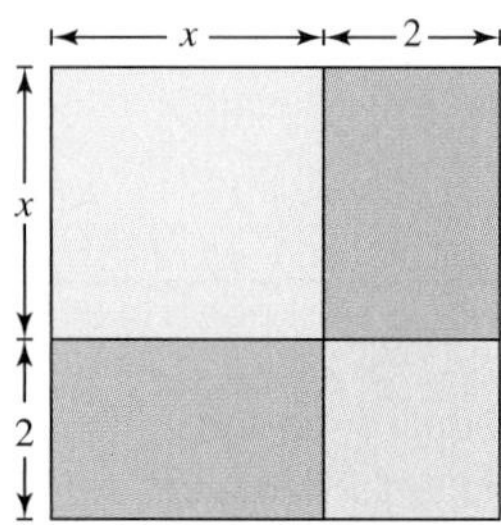

128. *Geometry* Add the areas of the four rectangular regions shown in the figure. What special method does the geometric model represent?

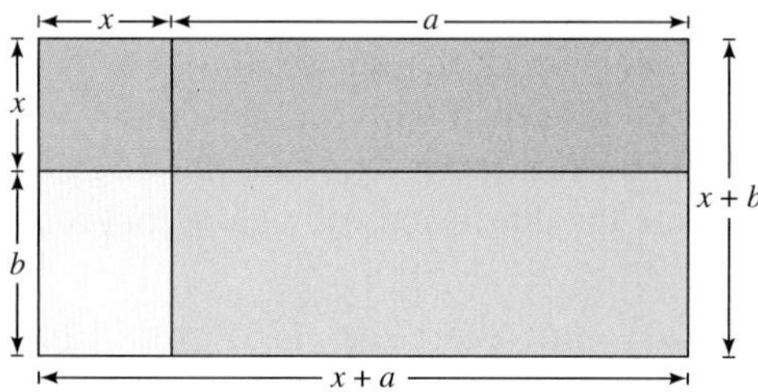

Geometry In Exercises 129 and 130, find a polynomial product that represents the area of the region. Then simplify the product.

129.

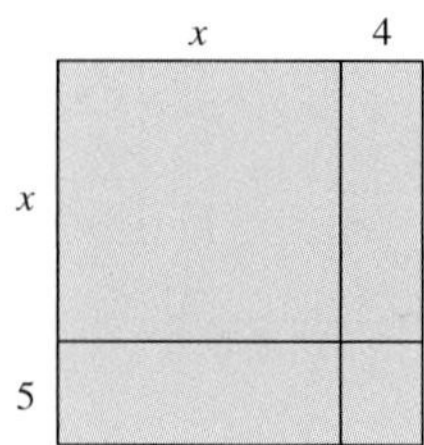

130.

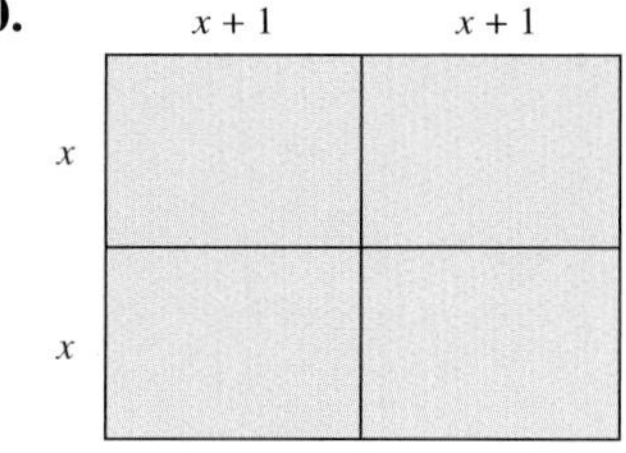

131. *Milk Consumption* The per capita consumption (average consumption per person) of milk M (in gallons) in the United States in the years 1996 through 2005 is given by

$M = -0.31t + 25.6, \quad 6 \le t \le 15.$

The population P (in millions) of the United States during the same time period is given by

$P = -0.034t^2 + 3.72t + 248.6, \quad 6 \le t \le 15.$

In both models, t represents the year, with $t = 6$ corresponding to 1996. (Source: USDA/Economic Research Service and U.S. Census Bureau)

(a) Multiply the polynomials to find a model for the total consumption T of milk in the United States.

(b) Use a graphing calculator to graph the model from part (a).

(c) Use the graph from part (b) to estimate the total consumption of milk in 2004.

132. *Interpreting Graphs* When x units of a home theater system are sold, the revenue R is given by

$R = x(900 - 0.5x).$

(a) Use a graphing calculator to graph the equation.

(b) Multiply the factors in the expression for revenue and use a graphing calculator to graph the product in the same viewing window you used in part (a). Verify that the graph is the same as in part (a).

(c) Find the revenue when 500 units are sold. Use the graph to determine if revenue would increase or decrease if more units were sold.

133. *Compound Interest* After 2 years, an investment of \$500 compounded annually at interest rate r will yield an amount $500(1 + r)^2$. Find this product.

134. *Compound Interest* After 3 years, an investment of \$1200 compounded annually at interest rate r will yield an amount $1200(1 + r)^3$. Find this product.

Explaining Concepts

135. Explain why an understanding of the Distributive Property is essential in multiplying polynomials. Illustrate your explanation with an example.

136. Discuss any differences between the expressions $(3x)^2$ and $3x^2$.

137. What is the degree of the product of two polynomials of degrees m and n? Explain.

138. A polynomial with m terms is multiplied by a polynomial with n terms. How many monomial-by-monomial products must be found? Explain.

139. *True or False?* Because the product of two monomials is a monomial, it follows that the product of two binomials is a binomial. Justify your answer.

140. *Finding a Pattern* Perform each multiplication.

(a) $(x - 1)(x + 1)$

(b) $(x - 1)(x^2 + x + 1)$

(c) $(x - 1)(x^3 + x^2 + x + 1)$

(d) From the pattern formed in the first three products, can you predict the product of $(x - 1)(x^4 + x^3 + x^2 + x + 1)$?

Verify your prediction by multiplying.

Cumulative Review

In Exercises 141–144, perform the indicated operations and simplify.

141. $(12x - 3) + (3x - 4)$

142. $(9x - 5) - (x + 7)$

143. $(-8x + 11) - (-4x - 6)$

144. $-(5x - 10) + (-13x + 40)$

In Exercises 145–148, solve the percent equation.

145. What number is 25% of 45?

146. 78 is 10% of what number?

147. 20 is what percent of 60?

148. What number is 55% of 62?

In Exercises 149–152, solve the proportion.

149. $\frac{2}{5} = \frac{x}{10}$

150. $\frac{3}{2} = \frac{15}{y}$

151. $\frac{z}{6} = \frac{5}{8}$

152. $\frac{9}{w} = \frac{6}{7}$

5.4 Dividing Polynomials and Synthetic Division

What You Should Learn

1 ▶ Divide polynomials by monomials and write in simplest form.
2 ▶ Use long division to divide polynomials by polynomials.
3 ▶ Use synthetic division to divide polynomials by polynomials of the form $x - k$.
4 ▶ Use synthetic division to factor polynomials.

Why You Should Learn It

Division of polynomials is useful in higher-level mathematics when factoring and finding zeros of polynomials.

1 ▶ Divide polynomials by monomials and write in simplest form.

Dividing a Polynomial by a Monomial

To divide a polynomial by a monomial, *reverse* the procedure used to add or subtract two rational expressions. Here is an example.

$$2 + \frac{1}{x} = \frac{2x}{x} + \frac{1}{x} = \frac{2x + 1}{x}$$ Add fractions.

$$\frac{2x + 1}{x} = \frac{2x}{x} + \frac{1}{x} = 2 + \frac{1}{x}$$ Divide by monomial.

Dividing a Polynomial by a Monomial

Let u, v, and w represent real numbers, variables, or algebraic expressions such that $w \neq 0$.

1. $\dfrac{u + v}{w} = \dfrac{u}{w} + \dfrac{v}{w}$ **2.** $\dfrac{u - v}{w} = \dfrac{u}{w} - \dfrac{v}{w}$

When dividing a polynomial by a monomial, remember to write the resulting expressions in simplest form, as illustrated in Example 1.

EXAMPLE 1 Dividing a Polynomial by a Monomial

Perform the division and simplify.

$$\frac{12x^2 - 20x + 8}{4x}$$

Solution

$$\frac{12x^2 - 20x + 8}{4x} = \frac{12x^2}{4x} - \frac{20x}{4x} + \frac{8}{4x}$$ Divide each term in the numerator by $4x$.

$$= \frac{3(\cancel{4x})(x)}{\cancel{4x}} - \frac{5(\cancel{4x})}{\cancel{4x}} + \frac{2(\cancel{4})}{\cancel{4}x}$$ Divide out common factors.

$$= 3x - 5 + \frac{2}{x}$$ Simplified form

✓ **CHECKPOINT** *Now try Exercise 5.*

2 ▶ Use long division to divide polynomials by polynomials.

Long Division

In Example 1, you learned how to divide a polynomial by a monomial by factoring and dividing out common factors. You can also use this procedure when dividing one polynomial by another. For instance, you can divide $x^2 - 2x - 3$ by $x - 3$ as follows.

$$(x^2 - 2x - 3) \div (x - 3) = \frac{x^2 - 2x - 3}{x - 3} \qquad \text{Write as a fraction.}$$

$$= \frac{(x + 1)(x - 3)}{x - 3} \qquad \text{Factor numerator.}$$

$$= \frac{(x + 1)\cancel{(x - 3)}}{\cancel{x - 3}} \qquad \text{Divide out common factor.}$$

$$= x + 1, \quad x \neq 3 \qquad \text{Simplified form}$$

This procedure works well for polynomials that factor easily. For those that do not, you can use a more general procedure that follows a "long division algorithm" similar to the algorithm used for dividing positive integers, which is reviewed in Example 2.

EXAMPLE 2 Long Division Algorithm for Positive Integers

Use the long division algorithm to divide 6584 by 28.

Solution

Think $\frac{65}{28} \approx 2$.

Think $\frac{98}{28} \approx 3$.

Think $\frac{144}{28} \approx 5$.

$$\begin{array}{r} 235 \\ 28 \overline{)6584} \end{array}$$

56	Multiply 2 by 28.
98	Subtract and bring down 8.
84	Multiply 3 by 28.
144	Subtract and bring down 4.
140	Multiply 5 by 28.
4	Remainder

So, you have

$$6584 \div 28 = 235 + \frac{4}{28}$$

$$= 235 + \frac{1}{7}.$$

✔ ***CHECKPOINT*** *Now try Exercise 15.*

In Example 2, 6584 is the **dividend,** 28 is the **divisor,** 235 is the **quotient,** and 4 is the **remainder.**

In the next several examples, you will see how the long division algorithm can be extended to cover the division of one polynomial by another.

When you use long division to divide polynomials, follow the steps below.

Long Division of Polynomials

1. Write the dividend and divisor in descending powers of the variable.
2. Insert placeholders with zero coefficients for missing powers of the variable. (See Example 5.)
3. Perform the long division of the polynomials as you would with integers.
4. Continue the process until the degree of the remainder is less than that of the divisor.

EXAMPLE 3 Long Division Algorithm for Polynomials

Think $x^2/x = x$.

Think $3x/x = 3$.

$$\begin{array}{r l}
x + 3 & \\
x - 1 \overline{) x^2 + 2x + 4} & \\
\underline{x^2 - x} & \text{Multiply } x \text{ by } (x - 1). \\
3x + 4 & \text{Subtract and bring down 4.} \\
\underline{3x - 3} & \text{Multiply 3 by } (x - 1). \\
7 & \text{Subtract.}
\end{array}$$

The remainder is a fractional part of the divisor, so you can write

$$\overbrace{\frac{x^2 + 2x + 4}{\underbrace{x - 1}_{\text{Divisor}}}}^{\text{Dividend}} = \overbrace{x + 3}^{\text{Quotient}} + \frac{\overset{\text{Remainder}}{7}}{\underbrace{x - 1}_{\text{Divisor}}}.$$

✓ **CHECKPOINT** *Now try Exercise 19.*

Study Tip

Note that in Example 3, the division process requires $3x - 3$ to be subtracted from $3x + 4$. The difference

$$\begin{array}{r} 3x + 4 \\ \underline{-(3x - 3)} \end{array}$$

is implied and written simply as

$$\begin{array}{r} 3x + 4 \\ \underline{3x - 3} \\ 7. \end{array}$$

You can check a long division problem by multiplying by the divisor. For instance, you can check the result of Example 3 as follows.

$$\frac{x^2 + 2x + 4}{x - 1} \stackrel{?}{=} x + 3 + \frac{7}{x - 1}$$

$$(x - 1)\left(\frac{x^2 + 2x + 4}{x - 1}\right) \stackrel{?}{=} (x - 1)\left(x + 3 + \frac{7}{x - 1}\right)$$

$$x^2 + 2x + 4 \stackrel{?}{=} (x + 3)(x - 1) + 7$$

$$x^2 + 2x + 4 \stackrel{?}{=} (x^2 + 2x - 3) + 7$$

$$x^2 + 2x + 4 = x^2 + 2x + 4 \quad ✓$$

Technology: Tip

You can check the result of a division problem *graphically* with a graphing calculator by comparing the graphs of the original quotient and the simplified form. The graphical check for Example 4 is shown below. Because the graphs coincide, you can conclude that the solution checks.

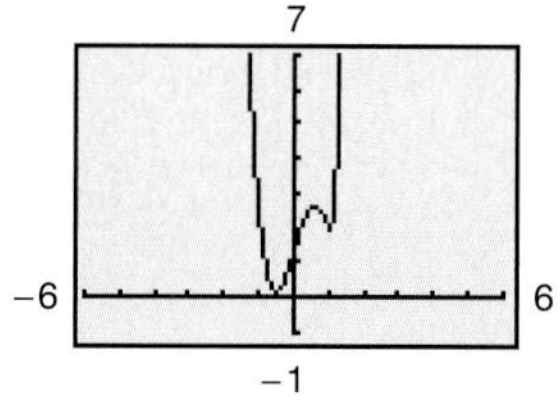

EXAMPLE 4 Writing in Standard Form Before Dividing

Divide $-13x^3 + 10x^4 + 8x - 7x^2 + 4$ by $3 - 2x$.

Solution

First write the divisor and dividend in standard polynomial form.

$$\begin{array}{rll}
 & -5x^3 - x^2 + 2x - 1 & \\
-2x + 3\,) & 10x^4 - 13x^3 - 7x^2 + 8x + 4 & \\
 & \underline{10x^4 - 15x^3} & \text{Multiply } -5x^3 \text{ by } (-2x + 3). \\
 & 2x^3 - 7x^2 & \text{Subtract and bring down } -7x^2. \\
 & \underline{2x^3 - 3x^2} & \text{Multiply } -x^2 \text{ by } (-2x + 3). \\
 & -4x^2 + 8x & \text{Subtract and bring down } 8x. \\
 & \underline{-4x^2 + 6x} & \text{Multiply } 2x \text{ by } (-2x + 3). \\
 & 2x + 4 & \text{Subtract and bring down } 4. \\
 & \underline{2x - 3} & \text{Multiply } -1 \text{ by } (-2x + 3). \\
 & 7 & \text{Subtract.}
\end{array}$$

This shows that

$$\underbrace{\frac{\overbrace{10x^4 - 13x^3 - 7x^2 + 8x + 4}^{\text{Dividend}}}{\underbrace{-2x + 3}_{\text{Divisor}}}} = \overbrace{-5x^3 - x^2 + 2x - 1}^{\text{Quotient}} + \frac{\overset{\text{Remainder}}{7}}{\underbrace{-2x + 3}_{\text{Divisor}}}.$$

✓ **CHECKPOINT** *Now try Exercise 25.*

When the dividend is missing one or more powers of x, the long division algorithm requires that you account for the missing powers, as shown in Example 5.

EXAMPLE 5 Accounting for Missing Powers of *x*

Divide $x^3 - 2$ by $x - 1$.

Solution

To account for the missing x^2- and x-terms, insert $0x^2$ and $0x$.

$$\begin{array}{rll}
 & x^2 + x + 1 & \\
x - 1\,) & x^3 + 0x^2 + 0x - 2 & \text{Insert } 0x^2 \text{ and } 0x. \\
 & \underline{x^3 - x^2} & \text{Multiply } x^2 \text{ by } (x - 1). \\
 & x^2 + 0x & \text{Subtract and bring down } 0x. \\
 & \underline{x^2 - x} & \text{Multiply } x \text{ by } (x - 1). \\
 & x - 2 & \text{Subtract and bring down } -2. \\
 & \underline{x - 1} & \text{Multiply } 1 \text{ by } (x - 1). \\
 & -1 & \text{Subtract.}
\end{array}$$

So, you have

$$\frac{x^3 - 2}{x - 1} = x^2 + x + 1 - \frac{1}{x - 1}.$$

✓ **CHECKPOINT** *Now try Exercise 41.*

In each of the long division examples presented so far, the divisor has been a first-degree polynomial. The long division algorithm works just as well with polynomial divisors of degree two or more, as shown in Example 6.

EXAMPLE 6 A Second-Degree Divisor

Divide $x^4 + 6x^3 + 6x^2 - 10x - 3$ by $x^2 + 2x - 3$.

Solution

$$\begin{array}{rll}
 & x^2 + 4x + 1 & \\
x^2 + 2x - 3\,\overline{)\,x^4 + 6x^3 + 6x^2 - 10x - 3} & & \\
\underline{x^4 + 2x^3 - 3x^2} & & \text{Multiply } x^2 \text{ by } (x^2 + 2x - 3). \\
4x^3 + 9x^2 - 10x & & \text{Subtract and bring down } -10x. \\
\underline{4x^3 + 8x^2 - 12x} & & \text{Multiply } 4x \text{ by } (x^2 + 2x - 3). \\
x^2 + 2x - 3 & & \text{Subtract and bring down } -3. \\
\underline{x^2 + 2x - 3} & & \text{Multiply 1 by } (x^2 + 2x - 3). \\
0 & & \text{Subtract.}
\end{array}$$

So, $x^2 + 2x - 3$ divides evenly into $x^4 + 6x^3 + 6x^2 - 10x - 3$. That is,

$$\frac{x^4 + 6x^3 + 6x^2 - 10x - 3}{x^2 + 2x - 3} = x^2 + 4x + 1, x \neq -3, x \neq 1.$$

CHECKPOINT ***Now try Exercise 49.***

Study Tip

If the remainder of a division problem is zero, the divisor is said to **divide evenly** into the dividend.

3 ▶ Use synthetic division to divide polynomials by polynomials of the form $x - k$.

Synthetic Division

There is a nice shortcut for division by polynomials of the form $x - k$. It is called **synthetic division** and is outlined for a third-degree polynomial as follows.

Synthetic Division of a Third-Degree Polynomial

Use synthetic division to divide $ax^3 + bx^2 + cx + d$ by $x - k$, as follows.

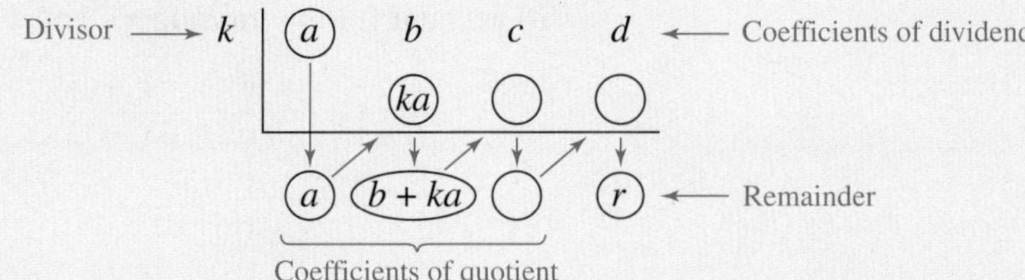

Vertical Pattern: Add terms.

Diagonal Pattern: Multiply by k.

Keep in mind that this algorithm for synthetic division works *only* for divisors of the form $x - k$. Remember that $x + k = x - (-k)$. Moreover, the degree of the quotient is always one less than the degree of the dividend.

EXAMPLE 7 Using Synthetic Division

Use synthetic division to divide $x^3 + 3x^2 - 4x - 10$ by $x - 2$.

Solution

The coefficients of the dividend form the top row of the synthetic division array. Because you are dividing by $x - 2$, write 2 at the top left of the array. To begin the algorithm, bring down the first coefficient. Then multiply this coefficient by 2, write the result in the second row, and add the two numbers in the second column. By continuing this pattern, you obtain the following.

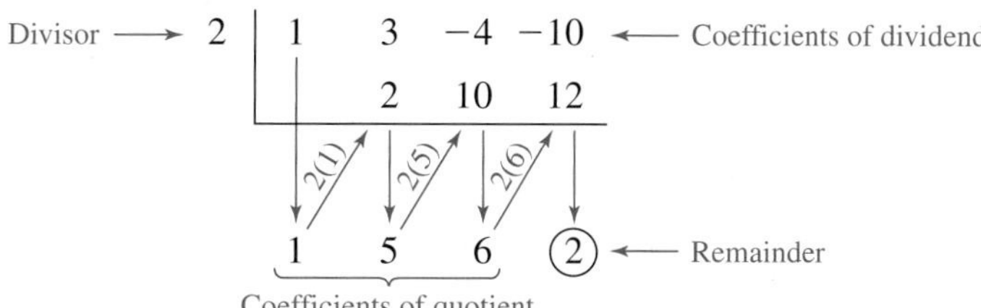

The bottom row shows the coefficients of the quotient. So, the quotient is

$$1x^2 + 5x + 6$$

and the remainder is 2. So, the result of the division problem is

$$\frac{x^3 + 3x^2 - 4x - 10}{x - 2} = x^2 + 5x + 6 + \frac{2}{x - 2}.$$

 ***CHECKPOINT*** *Now try Exercise 61.*

4 ▶ Use synthetic division to factor polynomials.

Factoring and Division

Synthetic division (or long division) can be used to factor polynomials. If the remainder in a synthetic division problem is zero, you know that the divisor divides *evenly* into the dividend.

EXAMPLE 8 Factoring a Polynomial

Study Tip

In Example 8, synthetic division is used to divide the polynomial by the factor $x - 1$. Long division could be used also.

The polynomial $x^3 - 7x + 6$ can be factored completely using synthetic division. Because $x - 1$ is a factor of the polynomial, you can divide as follows.

1	1	0	−7	6	
		1	1	−6	
	1	1	−6	(0)	← Remainder

Because the remainder is zero, the divisor divides evenly into the dividend:

$$\frac{x^3 - 7x + 6}{x - 1} = x^2 + x - 6.$$

From this result, you can factor the original polynomial as follows.

$$x^3 - 7x + 6 = (x - 1)(x^2 + x - 6) = (x - 1)(x + 3)(x - 2)$$

CHECKPOINT *Now try Exercise 73.*

Concept Check

1. Consider the equation $1253 \div 12 = 104 + \frac{5}{12}$. Identify the dividend, divisor, quotient, and remainder.

2. Explain what it means for a divisor to divide *evenly* into a dividend.

3. Explain how you can check polynomial division.

4. For synthetic division, what form must the divisor have?

5.4 EXERCISES

Go to pages 338–339 to record your assignments.

Developing Skills

In Exercises 1–14, perform the division. ***See Example 1.***

1. $(7x^3 - 2x^2) \div x$
2. $(3w^2 - 6w) \div w$
3. $(4x^2 - 2x) \div (-x)$
4. $(5y^3 + 6y^2 - 3y) \div (-y)$
5. $(m^4 + 2m^2 - 7) \div m$
6. $(x^3 + x - 2) \div x$
7. $\dfrac{50z^3 + 30z}{-5z}$
8. $\dfrac{18c^4 - 24c^2}{-6c}$
9. $\dfrac{4v^4 + 10v^3 - 8v^2}{4v^2}$
10. $\dfrac{6x^4 + 8x^3 - 18x^2}{3x^2}$
11. $\dfrac{4x^5 - 6x^4 + 12x^3 - 8x^2}{4x^2}$
12. $\dfrac{15x^{12} - 5x^9 + 30x^6}{5x^6}$
13. $(5x^2y - 8xy + 7xy^2) \div 2xy$
14. $(-14s^4t^2 + 7s^2t^2 - 18t) \div 2s^2t$

In Exercises 15–18, use the long division algorithm to perform the division. ***See Example 2.***

15. Divide 1013 by 9.
16. Divide 3713 by 22.
17. $3235 \div 15$
18. $6344 \div 28$

In Exercises 19–56, perform the division. ***See Examples 3–6.***

19. $\dfrac{x^2 - 8x + 15}{x - 3}$
20. $\dfrac{t^2 - 18t + 72}{t - 6}$
21. $(x^2 + 15x + 50) \div (x + 5)$
22. $(y^2 - 6y - 16) \div (y + 2)$
23. Divide $x^2 - 5x + 8$ by $x - 2$.
24. Divide $x^2 + 10x - 9$ by $x - 3$.
25. Divide $21 - 4x - x^2$ by $3 - x$.
26. Divide $5 + 4x - x^2$ by $1 + x$.
27. $\dfrac{5x^2 + 2x + 3}{x + 2}$
28. $\dfrac{2x^2 + 13x + 15}{x + 5}$
29. $\dfrac{12x^2 + 17x - 5}{3x + 2}$
30. $\dfrac{8x^2 + 2x + 3}{4x - 1}$
31. $(12 - 17t + 6t^2) \div (2t - 3)$
32. $(15 - 14u - 8u^2) \div (5 + 2u)$
33. Divide $2y^2 + 7y + 3$ by $2y + 1$.
34. Divide $10t^2 - 7t - 12$ by $2t - 3$.

35. $\dfrac{x^3 - 2x^2 + 4x - 8}{x - 2}$

36. $\dfrac{x^3 + 4x^2 + 7x + 28}{x + 4}$

37. $\dfrac{9x^3 - 3x^2 - 3x + 4}{3x + 2}$

38. $\dfrac{4y^3 + 12y^2 + 7y - 3}{2y + 3}$

39. $(2x + 9) \div (x + 2)$

40. $(12x - 5) \div (2x + 3)$

✓ 41. $\dfrac{x^2 + 16}{x + 4}$

42. $\dfrac{y^2 + 8}{y + 2}$

43. $\dfrac{6z^2 + 7z}{5z - 1}$

44. $\dfrac{8y^2 - 2y}{3y + 5}$

45. $\dfrac{16x^2 - 1}{4x + 1}$

46. $\dfrac{25y^2 - 9}{5y - 3}$

47. $\dfrac{x^3 + 125}{x + 5}$

48. $\dfrac{x^3 - 27}{x - 3}$

✓ 49. $(x^3 + 4x^2 + 7x + 7) \div (x^2 + 2x + 3)$

50. $(2x^3 + 2x^2 - 2x - 15) \div (2x^2 + 4x + 5)$

51. $(4x^4 - 3x^2 + x - 5) \div (x^2 - 3x + 2)$

52. $(8x^5 + 6x^4 - x^3 + 1) \div (2x^3 - x^2 - 3)$

53. Divide $x^4 - 1$ by $x - 1$.

54. Divide $x^6 - 1$ by $x - 1$.

55. $x^5 \div (x^2 + 1)$

56. $x^6 \div (x^3 - 1)$

In Exercises 57–60, simplify the expression.

57. $\dfrac{8u^2v}{2u} + \dfrac{3(uv)^2}{uv}$

58. $\dfrac{15x^3y}{10x^2} + \dfrac{3xy^2}{2y}$

59. $\dfrac{x^2 + 3x + 2}{x + 2} + (2x + 3)$

60. $\dfrac{x^2 + 2x - 3}{x - 1} - (3x - 4)$

In Exercises 61–72, use synthetic division to divide. ***See Example 7.***

✓ 61. $(x^2 + x - 6) \div (x - 2)$

62. $(x^2 + 5x - 6) \div (x + 6)$

63. $\dfrac{x^3 + 3x^2 - 1}{x + 4}$

64. $\dfrac{x^3 - 4x + 7}{x - 1}$

65. $\dfrac{x^4 - 4x^3 + x + 10}{x - 2}$

66. $\dfrac{2x^5 - 3x^3 + x}{x - 3}$

67. $\dfrac{5x^3 - 6x^2 + 8}{x - 4}$

68. $\dfrac{5x^3 + 6x + 8}{x + 2}$

69. $\dfrac{10x^4 - 50x^3 - 800}{x - 6}$

70. $\dfrac{x^5 - 13x^4 - 120x + 80}{x + 3}$

71. $\dfrac{0.1x^2 + 0.8x + 1}{x - 0.2}$

72. $\dfrac{x^3 - 0.8x + 2.4}{x + 0.1}$

In Exercises 73–80, completely factor the polynomial given one of its factors. **See Example 8.**

	Polynomial	*Factor*
✓ **73.**	$x^3 - x^2 - 14x + 24$	$x - 3$
74.	$x^3 + x^2 - 32x - 60$	$x + 5$
75.	$4x^3 - 3x - 1$	$x - 1$
76.	$9x^3 + 51x^2 + 88x + 48$	$x + 3$
77.	$x^4 + 7x^3 + 3x^2 - 63x - 108$	$x + 4$
78.	$x^4 - 6x^3 - 8x^2 + 96x - 128$	$x - 4$
79.	$15x^2 - 2x - 8$	$x - \frac{4}{5}$
80.	$18x^2 - 9x - 20$	$x + \frac{5}{6}$

In Exercises 81 and 82, find the constant c such that the denominator divides evenly into the numerator.

81. $\dfrac{x^3 + 2x^2 - 4x + c}{x - 2}$ **82.** $\dfrac{x^4 - 3x^2 + c}{x + 6}$

In Exercises 83 and 84, use a graphing calculator to graph the two equations in the same viewing window. Use the graphs to verify that the expressions are equivalent. Verify the results algebraically.

83. $y_1 = \dfrac{x + 4}{2x}$

$y_2 = \dfrac{1}{2} + \dfrac{2}{x}$

84. $y_1 = \dfrac{x^2 + 2}{x + 1}$

$y_2 = x - 1 + \dfrac{3}{x + 1}$

In Exercises 85 and 86, perform the division assuming that n is a positive integer.

85. $\dfrac{x^{3n} + 3x^{2n} + 6x^n + 8}{x^n + 2}$ **86.** $\dfrac{x^{3n} - x^{2n} + 5x^n - 5}{x^n - 1}$

Think About It In Exercises 87 and 88, the divisor, quotient, and remainder are given. Find the dividend.

	Divisor	*Quotient*	*Remainder*
87.	$x - 6$	$x^2 + x + 1$	-4
88.	$x + 3$	$x^2 - 2x - 5$	8

Finding a Pattern In Exercises 89 and 90, complete the table for the function. The first row is completed for Exercise 89. What conclusion can you draw as you compare the values of $f(k)$ with the remainders? (Use synthetic division to find the remainders.)

89. $f(x) = x^3 - x^2 - 2x$

90. $f(x) = 2x^3 - x^2 - 2x + 1$

k	$f(k)$	Divisor $(x - k)$	Remainder
-2	-8	$x + 2$	-8
-1			
0			
$\frac{1}{2}$			
1			
2			

Solving Problems

91. *Geometry* The volume of a cube is $x^3 + 3x^2 + 3x + 1$. The height of the cube is $x + 1$. Use division to find the area of the base.

92. *Geometry* A rectangular house has a volume of $x^3 + 55x^2 + 650x + 2000$ cubic feet (the space in the attic is not included). The height of the house is $x + 5$ feet (see figure). Find the number of square feet of floor space *on the first floor* of the house.

Figure for 92

Geometry In Exercises 93 and 94, you are given the expression for the volume of the solid shown. Find the expression for the missing dimension.

93. $V = x^3 + 18x^2 + 80x + 96$

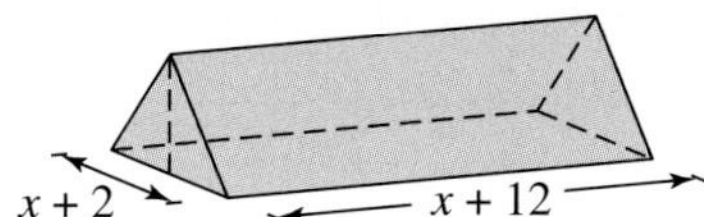

94. $V = 2h^3 + 3h^2 + h$

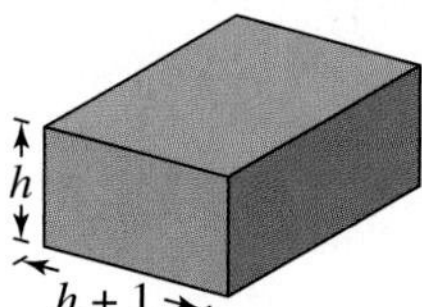

Explaining Concepts

95. *Error Analysis* Describe and correct the error.

$\frac{6x + 5y}{x} = \frac{6\cancel{x} + 5y}{\cancel{x}} = 6 + 5y$ (crossed out)

96. *Error Analysis* Describe and correct the error.

$\frac{x^2}{x+1} = \frac{x^{\cancel{2}}}{\cancel{x}} + \frac{x^2}{1} = x + x^2$ (crossed out)

97. Create a polynomial division problem and identify the dividend, divisor, quotient, and remainder.

98. *True or False?* If the divisor divides evenly into the dividend, then the divisor and quotient are factors of the dividend. Justify your answer.

99. Use a graphing calculator to graph each polynomial in the same viewing window using the standard setting. Use the *zero* or *root* feature to find the x-intercepts. What can you conclude about the polynomials? Verify your conclusion algebraically.

(a) $y = (x - 4)(x - 2)(x + 1)$

(b) $y = (x^2 - 6x + 8)(x + 1)$

(c) $y = x^3 - 5x^2 + 2x + 8$

100. Use a graphing calculator to graph the function

$$f(x) = \frac{x^3 - 5x^2 + 2x + 8}{x - 2}.$$

Use the *zero* or *root* feature to find the x-intercepts. Why does this function have only two x-intercepts? To what other function does the graph of $f(x)$ appear to be equivalent? What is the difference between the two graphs?

Cumulative Review

In Exercises 101–106, solve the inequality.

101. $7 - 3x > 4 - x$

102. $2(x + 6) - 20 < 2$

103. $|x - 3| < 2$

104. $|x - 5| > 3$

105. $\left|\frac{1}{4}x - 1\right| \geq 3$

106. $\left|2 - \frac{1}{3}x\right| \leq 10$

In Exercises 107 and 108, determine the quadrants in which the point must be located.

107. $(-3, y)$, y is a real number.

108. $(x, 7)$, x is a real number.

109. Describe the location of the set of points whose x-coordinates are 0.

110. Find the coordinates of the point five units to the right of the y-axis and seven units below the x-axis.

What Did You Learn?

Use these two pages to help prepare for a test on this chapter. Check off the key terms and key concepts you know. You can also use this section to record your assignments.

Plan for Test Success

Date of test: / /

Study dates and times: / / at : A.M./P.M.

/ / at : A.M./P.M.

Things to review:

- ☐ Key Terms, *p. 338*
- ☐ Key Concepts, *pp. 338–339*
- ☐ Your class notes
- ☐ Your assignments
- ☐ Study Tips, *pp. 295, 296, 297, 306, 319, 320, 330, 332, 333*
- ☐ Technology Tips, *pp. 299, 306, 316, 331*
- ☐ Mid-Chapter Quiz, *p. 314*
- ☐ Review Exercises, *pp. 340–342*
- ☐ Chapter Test, *p. 343*
- ☐ Video Explanations Online
- ☐ Tutorial Online

Key Terms

- ☐ exponential form, *p. 294*
- ☐ scientific notation, *p. 298*
- ☐ polynomial in one variable, *p. 304*
- ☐ polynomial, *p. 304*
- ☐ coefficient, *p. 304*
- ☐ degree, *p. 304*
- ☐ constant term, *p. 304*
- ☐ standard form, *p. 304*
- ☐ degree of a polynomial, *p. 304*
- ☐ leading coefficient, *p. 304*
- ☐ polynomial in *x*, *p. 305*
- ☐ monomial, *p. 305*
- ☐ binomial, *p. 305*
- ☐ trinomial, *p. 305*
- ☐ FOIL Method, *p. 316*
- ☐ product of the sum and difference of two terms, *p. 320*
- ☐ square of a binomial, *p. 320*
- ☐ dividend, *p. 329*
- ☐ divisor, *p. 329*
- ☐ quotient, *p. 329*
- ☐ remainder, *p. 329*
- ☐ synthetic division, *p. 332*

Key Concepts

5.1 Integer Exponents and Scientific Notation

Assignment: ______________________________ **Due date:** ________

☐ **Use the rules of exponents.**

Let m and n be integers, and let a and b be real numbers, variables, or algebraic expressions, such that $a \neq 0$ and $b \neq 0$.

1. $a^m \cdot a^n = a^{m+n}$
2. $\dfrac{a^m}{a^n} = a^{m-n}$
3. $(ab)^m = a^m b^m$
4. $(a^m)^n = a^{mn}$
5. $\left(\dfrac{a}{b}\right)^m = \dfrac{a^m}{b^m}$
6. $a^0 = 1$
7. $a^{-n} = \dfrac{1}{a^n}$
8. $\left(\dfrac{a}{b}\right)^{-m} = \left(\dfrac{b}{a}\right)^m$

☐ **Write numbers in scientific notation.**

Scientific notation has the form $c \times 10^n$, where $1 \leq c < 10$ and n is an integer. If $n > 0$, the number is large (10 or more). If $n < 0$, the number is small (less than 1).

5.2 Adding and Subtracting Polynomials

Assignment: ______________________ Due date: ________

☐ **Define a polynomial in *x*.**

Let $a_n, a_{n-1}, \ldots, a_2, a_1, a_0$ be real numbers and let n be a nonnegative integer. A polynomial in x is an expression of the form

$a_nx^n + a_{n-1}x^{n-1} + \cdots + a_2x^2 + a_1x + a_0$

where $a_n \neq 0$. The polynomial is of degree n, and the number a_n is called the leading coefficient. The number a_0 is called the constant term.

☐ **Add polynomials.**

To add polynomials, combine like terms (those having the same degree) by using the Distributive Property.

☐ **Subtract polynomials.**

To subtract one polynomial from another, you add the opposite by changing the sign of each term of the polynomial that is being subtracted and then adding the resulting like terms.

5.3 Multiplying Polynomials: Special Products

Assignment: ______________________ Due date: ________

☐ **Multiply polynomials.**

1. To multiply a polynomial by a monomial, apply the Distributive Property.
2. To multiply two binomials, use the FOIL Method. Combine the product of the **F**irst terms, the product of the **O**uter terms, the product of the **I**nner terms, and the product of the **L**ast terms.
3. To multiply two polynomials, use the Distributive Property to multiply each term of one polynomial by each term of the other polynomial.

☐ **Use special products.**

Let a and b be real numbers, variables, or algebraic expressions.

Sum and Difference of Two Terms:

$(a + b)(a - b) = a^2 - b^2$

Square of a Binomial:

$(a + b)^2 = a^2 + 2ab + b^2$

$(a - b)^2 = a^2 - 2ab + b^2$

5.4 Dividing Polynomials and Synthetic Division

Assignment: ______________________ Due date: ________

☐ **Divide a polynomial by a monomial.**

Let a, b, and c be real numbers, variables, or algebraic expressions, such that $c \neq 0$.

1. $\dfrac{a + b}{c} = \dfrac{a}{c} + \dfrac{b}{c}$
2. $\dfrac{a - b}{c} = \dfrac{a}{c} - \dfrac{b}{c}$

☐ **Divide polynomials.**

Along with the long division algorithm, follow these steps when performing long division of polynomials.

1. Write the dividend and divisor in descending powers of the variable.
2. Insert placeholders with zero coefficients for missing powers of the variable.
3. Perform the long division of the polynomials as you would with integers. Continue the process until the degree of the remainder is less than the degree of the divisor.
4. Use synthetic division to divide a polynomial by a binomial of the form $x - k$. [Remember that $x + k = x - (-k)$.]

Review Exercises

5.1 Integer Exponents and Scientific Notation

1 ▶ Use the rules of exponents to simplify expressions.

In Exercises 1–14, use the rules of exponents to simplify the expression (if possible).

1. $x^4 \cdot x^5$ **2.** $-3y^2 \cdot y^4$

3. $(u^2)^3$ **4.** $(v^4)^2$

5. $(-2z)^3$ **6.** $(-3y)^2(2)$

7. $-(u^2v)^2(-4u^3v)$ **8.** $(12x^2y)(3x^2y^4)^2$

9. $\dfrac{12z^5}{6z^2}$ **10.** $\dfrac{15m^3}{25m}$

11. $\dfrac{25g^4d^2}{80g^2d^2}$ **12.** $-\dfrac{-48u^8v^6}{(-2u^2v)^3}$

13. $\left(\dfrac{72x^4}{6x^2}\right)^2$ **14.** $\left(-\dfrac{y^2}{2}\right)^3$

2 ▶ Rewrite exponential expressions involving negative and zero exponents.

In Exercises 15–18, evaluate the expression.

15. $(2^3 \cdot 3^2)^{-1}$ **16.** $(2^{-2} \cdot 5^2)^{-2}$

17. $\left(\dfrac{3}{4}\right)^{-3}$ **18.** $\left(\dfrac{1}{3^{-2}}\right)^2$

In Exercises 19–30, rewrite the expression using only positive exponents, and simplify. (Assume that any variables in the expression are nonzero.)

19. $(6y^4)(2y^{-3})$ **20.** $4(-3x)^{-3}$

21. $\dfrac{4x^{-2}}{2x}$ **22.** $\dfrac{15t^5}{24t^{-3}}$

23. $(x^3y^{-4})^0$ **24.** $(5x^{-2}y^4)^{-2}$

25. $\dfrac{7a^6b^{-2}}{14a^{-1}b^4}$ **26.** $\dfrac{2u^0v^{-2}}{10u^{-1}v^{-3}}$

27. $\left(\dfrac{3x^{-1}y^2}{12x^5y^{-3}}\right)^{-1}$ **28.** $\left(\dfrac{4x^{-3}z^{-1}}{8x^4z}\right)^{-2}$

29. $u^3(5u^0v^{-1})(9u)^2$ **30.** $a^4(16a^{-2}b^4)(2b)^{-3}$

3 ▶ Write very large and very small numbers in scientific notation.

In Exercises 31 and 32, write the number in scientific notation.

31. 0.0000538 **32.** 30,296,000,000

In Exercises 33 and 34, write the number in decimal form.

33. 4.833×10^8 **34.** 2.74×10^{-4}

In Exercises 35–38, evaluate the expression without a calculator.

35. $(6 \times 10^3)^2$ **36.** $(3 \times 10^{-3})(8 \times 10^7)$

37. $\dfrac{3.5 \times 10^7}{7 \times 10^4}$ **38.** $\dfrac{1}{(6 \times 10^{-3})^2}$

5.2 Adding and Subtracting Polynomials

1 ▶ Identify the degrees and leading coefficients of polynomials.

In Exercises 39–44, write the polynomial in standard form. Then identify its degree and leading coefficient.

39. $10x - 4 - 5x^3$

40. $2x^2 + 9$

41. $4x^3 - 2x + 5x^4 - 7x^2$

42. $6 - 3x + 6x^2 - x^3$

43. $7x^4 - 1 + 11x^2$

44. $12x^2 + 2x - 8x^5 + 1$

2 ▶ Add polynomials using a horizontal or vertical format.

In Exercises 45–52, find the sum.

45. $(8x + 4) + (x - 4)$ **46.** $\left(\frac{1}{2}x + \frac{2}{3}\right) + \left(4x + \frac{1}{3}\right)$

47. $(3y^3 + 5y^2 - 9y) + (2y^3 - 3y + 10)$

48. $(6 - x + x^2) + (3x^2 + x)$

49. $(3u + 4u^2) + 5(u + 1) + 3u^2$

50. $6(u^2 + 2) + 12u + (u^2 - 5u + 2)$

51. $\begin{array}{r} -x^4 - 2x^2 + 3 \\ \underline{3x^4 - 5x^2 } \end{array}$

52. $\begin{array}{r} 5z^3 - 4z - 7 \\ \underline{-z^2 - 4z } \end{array}$

53. *Geometry* The length of a rectangular wall is x units, and its height is $(x - 3)$ units (see figure). Find an expression for the perimeter of the wall.

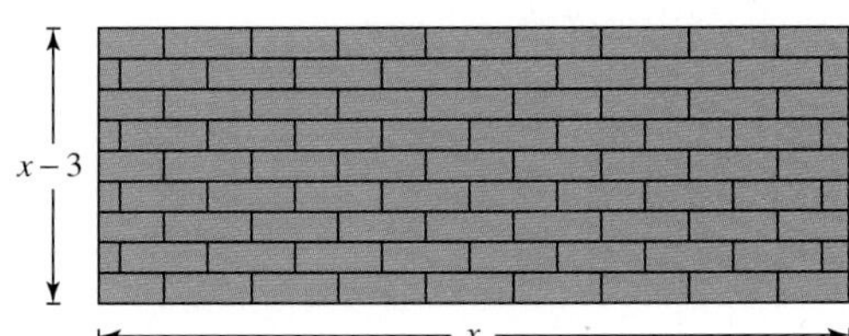

54. *Geometry* A rectangular garden has length $(t + 5)$ feet and width $2t$ feet (see figure). Find an expression for the perimeter of the garden.

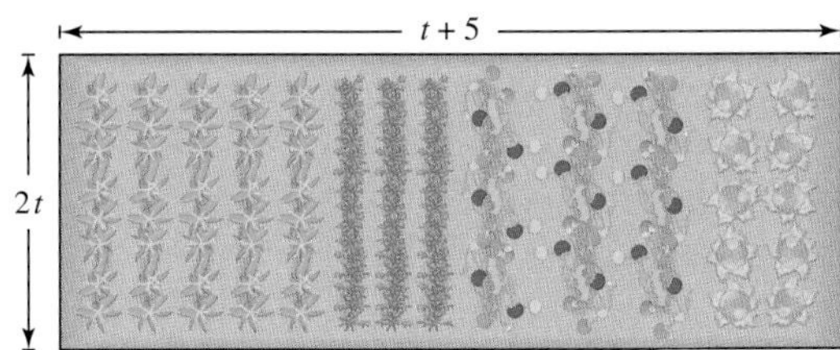

3 ▶ Subtract polynomials using a horizontal or vertical format.

In Exercises 55–62, find the difference.

55. $(3t - 5) - (3t - 9)$

56. $\left(2x - \frac{1}{5}\right) - \left(\frac{1}{4}x + \frac{1}{4}\right)$

57. $(6x^2 - 9x - 5) - 3(4x^2 - 6x + 1)$

58. $(5t^2 + 2) - 2(4t^2 + 1)$

59. $4y^2 - [y - 3(y^2 + 2)]$

60. $(6a^3 + 3a) - 2[a - (a^3 + 2)]$

61. $\begin{array}{r} 5x^2 + 2x - 27 \\ \underline{-(2x^2 - 2x - 13)} \end{array}$

62. $\begin{array}{r} 12y^4 - 15y^2 + 7 \\ \underline{-(18y^4 - 9)} \end{array}$

63. *Cost, Revenue, and Profit* The cost C of producing x units of a product is $C = 15 + 26x$. The revenue R for selling x units is $R = 40x - \frac{1}{2}x^2$, where $0 \le x \le 20$. The profit P is the difference between revenue and cost.

(a) Perform the subtraction required to find the polynomial representing profit P.

(b) Use a graphing calculator to graph the polynomial representing profit.

(c) Determine the profit when 14 units are produced and sold. Use the graph in part (b) to describe the profit when x is less than or greater than 14.

5.3 Multiplying Polynomials: Special Products

1 ▶ Find products with monomial multipliers.

In Exercises 64–67, perform the multiplication and simplify.

64. $2x(x + 4)$

65. $3y(y - 1)$

66. $(4x + 2)(-3x^2)$

67. $(5 - 7y)(-6y^2)$

2 ▶ Multiply binomials using the Distributive Property and the FOIL Method.

In Exercises 68–71, perform the multiplication and simplify.

68. $(x + 4)(x + 6)$

69. $(u + 5)(u - 2)$

70. $(4x - 3)(3x + 4)$

71. $(6 - 2x)(7x - 10)$

3 ▶ Multiply polynomials using a horizontal or vertical format.

In Exercises 72–81, perform the multiplication and simplify.

72. $(x^2 + 5x + 2)(x - 6)$

73. $(s^2 + 4s - 3)(s - 3)$

74. $(2t - 1)(t^2 - 3t + 3)$

75. $(4x + 2)(x^2 + 6x - 5)$

76. $\begin{array}{r} 3x^2 + x - 2 \\ \underline{\times 4x - 5} \end{array}$

77. $\begin{array}{r} 5y^2 - 2y + 9 \\ \underline{\times 3y + 4} \end{array}$

78. $\begin{array}{r} y^2 - 4y + 5 \\ \underline{\times \; y^2 + 2y - 3} \end{array}$

79. $\begin{array}{r} x^2 + 8x - 12 \\ \underline{\times \; x^2 - 9x + 2} \end{array}$

80. $(2x + 1)^3$

81. $(3y - 2)^3$

82. *Geometry* The width of a rectangular window is $(2x + 6)$ inches and its height is $(3x + 10)$ inches (see figure). Find an expression for the area of the window.

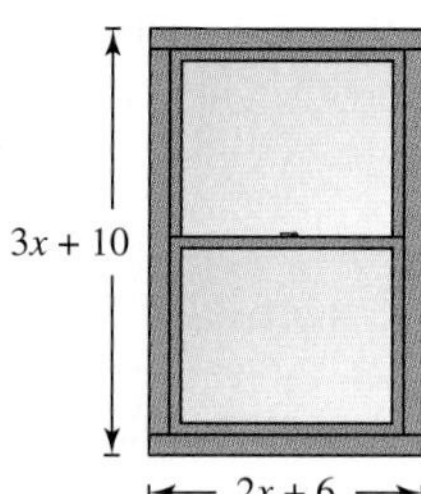

83. *Geometry* The width of a rectangular parking lot is $(x + 25)$ meters and its length is $(x + 30)$ meters (see figure). Find an expression for the area of the parking lot.

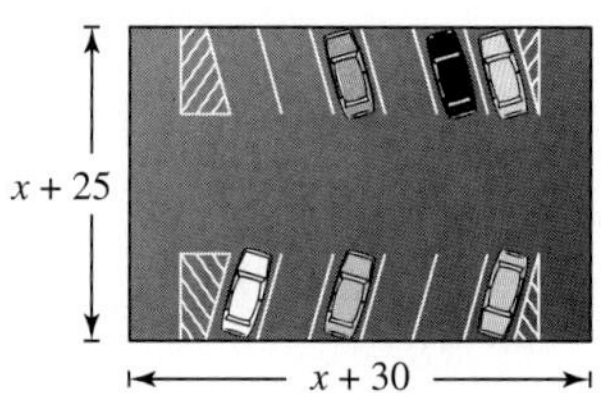

4 ▸ Identify and use special binomial products.

In Exercises 84–91, use a special product pattern to find the product.

84. $(x + 3)^2$

85. $(x - 5)^2$

86. $\left(\frac{1}{2}x - 4\right)^2$

87. $(4 + 3b)^2$

88. $(u - 6)(u + 6)$

89. $(r + 7)(r - 7)$

90. $(2x - 4y)(2x + 4y)$

91. $(4u + 5v)(4u - 5v)$

92. *Garden* A square garden with side length s is to be reconfigured so that its length is 2 feet longer and its width is 2 feet shorter. Find an expression for the new area of the garden.

5.4 Dividing Polynomials and Synthetic Division

1 ▸ Divide polynomials by monomials and write in simplest form.

In Exercises 93–96, perform the division.

93. $(4x^3 - x) \div (2x)$

94. $(10x + 15) \div (5x)$

95. $\dfrac{3x^3y^2 - x^2y^2 + x^2y}{x^2y}$

96. $\dfrac{6a^3b^3 + 2a^2b - 4ab^2}{2ab}$

2 ▸ Use long division to divide polynomials by polynomials.

In Exercises 97–102, perform the division.

97. $\dfrac{6x^3 + 2x^2 - 4x + 2}{3x - 1}$

98. $\dfrac{4x^4 - x^3 - 7x^2 + 18x}{x - 2}$

99. $\dfrac{x^4 - 3x^2 + 2}{x^2 - 1}$

100. $\dfrac{x^4 - 4x^3 + 3x}{x^2 - 1}$

101. $\dfrac{x^5 - 3x^4 + x^2 + 6}{x^3 - 2x^2 + x - 1}$

102. $\dfrac{x^6 + 4x^5 - 3x^2 + 5x}{x^3 + x^2 - 4x + 3}$

3 ▸ Use synthetic division to divide polynomials by polynomials of the form $x - k$.

In Exercises 103–106, use synthetic division to divide.

103. $\dfrac{x^3 + 7x^2 + 3x - 14}{x + 2}$

104. $\dfrac{x^4 - 2x^3 - 15x^2 - 2x + 10}{x - 5}$

105. $(x^4 - 3x^2 - 25) \div (x - 3)$

106. $(2x^3 + 5x - 2) \div \left(x + \frac{1}{2}\right)$

4 ▸ Use synthetic division to factor polynomials.

In Exercises 107 and 108, completely factor the polynomial given one of its factors.

	Polynomial	*Factor*
107.	$x^3 + 2x^2 - 5x - 6$	$x - 2$
108.	$2x^3 + x^2 - 2x - 1$	$x + 1$

Chapter Test

Take this test as you would take a test in class. After you are done, check your work against the answers in the back of the book.

In Exercises 1–6, rewrite the expression using only positive exponents, and simplify. (Assume that no variable is zero.)

1. $(3x^{-2}y^3)^{-2}$

2. $(4u^{-3}v^2)^{-2}(8u^{-1}v^{-2})^0$

3. $\dfrac{12x^{-3}y^5}{4x^{-2}y^{-1}}$

4. $\left(\dfrac{-2x^2y}{z^{-3}}\right)^{-2}$

5. $\left(-\dfrac{2u^2}{v^{-1}}\right)^3\left(\dfrac{3v^2}{u^{-3}}\right)$

6. $\dfrac{(-3x^2y^{-1})^4}{6x^2y^0}$

7. (a) Write 0.00015 in scientific notation.

(b) Write 8×10^7 in decimal notation.

8. Explain why the expression is not a polynomial: $\dfrac{4}{x^2 + 2}$.

9. Determine the degree and the leading coefficient of $-3x^4 - 5x^2 + 2x - 10$.

10. Give an example of a trinomial in one variable of degree 5.

In Exercises 11–22, simplify the expression. (Assume that no variable or denominator is zero.)

11. $(3z^2 - 3z + 7) + (8 - z^2)$

12. $(8u^3 + 3u^2 - 2u - 1) - (u^3 + 3u^2 - 2u)$

13. $6y - [2y - (3 + 4y - y^2)]$

14. $-5(x^2 - 1) + 3(4x + 7) - (x^2 + 26)$

15. $(x - 7)^2$

16. $(2x - 3)(2x + 3)$

17. $(z + 2)(2z^2 - 3z + 5)$

18. $(y + 3)(y^4 + y^2 - 4)$

19. $\dfrac{4z^3 + z}{2z}$

20. $\dfrac{16x^2 - 12}{-8}$

21. $\dfrac{x^3 - x - 6}{x - 2}$

22. $\dfrac{4x^3 + 10x^2 - 2x - 5}{2x - 4}$

3x + 2
2x + 5
2x
x

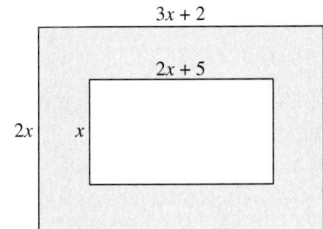

Figure for 23

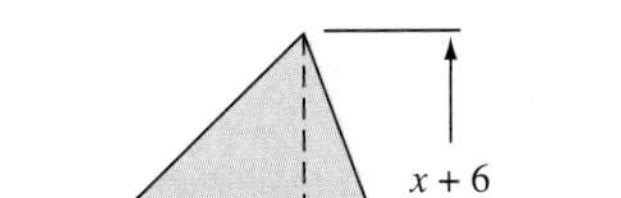
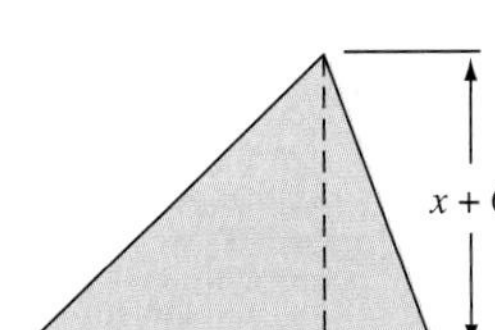

Figure for 24

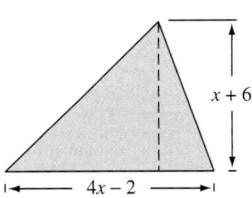

23. Find an expression for the area of the shaded region shown in the figure.

24. Find an expression for the area of the triangle shown in the figure.

25. The revenue R from the sale of x guitars is given by $R = x^2 - 35x$. The cost C of producing x guitars is given by $C = 150 + 12x$. Perform the subtraction required to find the polynomial representing the profit P.

26. The area of a rectangle is $x^2 - 2x - 3$ and its length is $x + 1$. Find the width of the rectangle.

Study Skills in Action

Being Confident

How does someone "get" confidence? Confidence is linked to another attribute called self-efficacy. Self-efficacy is the belief that one has the ability to accomplish a specific task. It is possible for a student to have high self-efficacy when it comes to writing a personal essay, but to have low self-efficacy when it comes to learning math.

A good way to foster self-efficacy is by building a support system. A support system should include faculty and staff who can encourage and guide you, and other students who can help you study and stay focused.

Kimberly Nolting
VP, Academic Success Press
expert in developmental education

Collegial friends, who share the same desire to do well, can be the best type of support.

Smart Study Strategy

Build a Support System

1 ► **Surround yourself with positive collegial friends.** Find another student in class with whom to study. Make sure this person is not anxious about math because you do not want another student's anxiety to increase your own. Arrange to meet on campus and compare notes, homework, and so on at least two times per week. Collegial friends can encourage each other.

2 ► **Find a place on campus to study where other students are also studying.** Libraries, learning centers, and tutoring centers are great places to study. While studying in such places, you will be able to ask for assistance when you have questions. You do not want to study alone if you typically get down on yourself with lots of negative self-talk.

3 ► **Establish a relationship with a learning assistant.** Get to know someone who can help you find assistance for any type of academic issue. Learning assistants, tutors, and instructors are excellent resources.

4 ► **Seek out assistance before you are overwhelmed.** Visit your instructor when you need help. Instructors are more than willing to help their students, particularly during office hours. Go with a friend if you are nervous about visiting your instructor.

5 ► **Be your own support.** Listen to what you tell yourself when frustrated with studying math. Replace any negative self-talk dialog with more positive statements. Here are some examples of positive statements:

"I may not have done well in the past, but I'm learning how to study math, and will get better."

"It does not matter what others believe—I know that I can get through this course."

"Wow, I messed up on a quiz. I need to talk to someone and figure out what I need to do differently."

Chapter 6

Factoring and Solving Equations

IT WORKED FOR ME!
"I always get to know the instructor because then he or she is more willing to help you. I usually talk to the instructor before class starts, just to visit. I used to avoid going to instructors' offices, but now I realize that they are used to helping students, and it really is no big deal. I'm able to ask more questions about what is confusing me. I feel more connected in class too."
Laura
Business

6.1 Factoring Polynomials with Common Factors

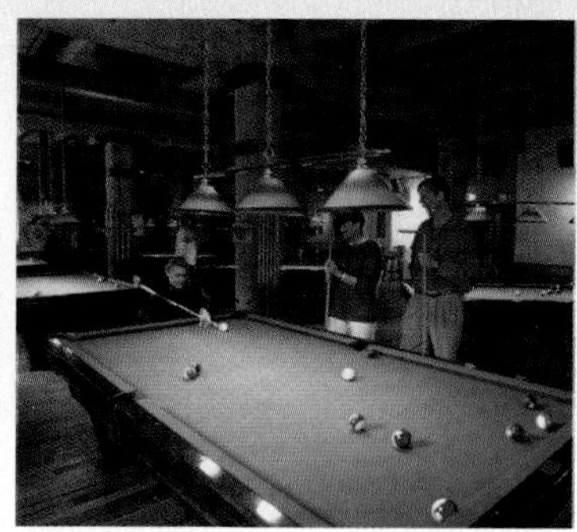

Dave G. Houser/CORBIS

Why You Should Learn It

In some cases, factoring a polynomial enables you to determine unknown quantities. For example, in Exercise 108 on page 353, you will factor the expression for the revenue from selling pool tables to determine an expression for the price of the pool tables.

What You Should Learn

1 ▶ Find the greatest common factor of two or more expressions.

2 ▶ Factor out the greatest common monomial factor from polynomials.

3 ▶ Factor polynomials by grouping.

Greatest Common Factor

1 ▶ Find the greatest common factor of two or more expressions.

In Chapter 5, you used the Distributive Property to multiply polynomials. In this chapter, you will study the *reverse* process, which is **factoring.**

Multiplying Polynomials

$\underbrace{2x}_{\text{Factor}}\underbrace{(7-3x)}_{\text{Factor}} \Rightarrow \underbrace{14x - 6x^2}_{\text{Product}}$

Factoring Polynomials

$\underbrace{14x - 6x^2}_{\text{Product}} \Rightarrow \underbrace{2x}_{\text{Factor}}\underbrace{(7-3x)}_{\text{Factor}}$

To factor an expression efficiently, you need to understand the concept of the *greatest common factor* of two (or more) integers or terms. In Section 1.4, you learned that the **greatest common factor** of two or more integers is the greatest integer that is a factor of each integer. For example, the greatest common factor of $12 = 2 \cdot 2 \cdot 3$ and $30 = 2 \cdot 3 \cdot 5$ is $2 \cdot 3 = 6$.

EXAMPLE 1 Finding the Greatest Common Factor

To find the greatest common factor of $5x^2y^2$ and $30x^3y$, first factor each term.

$$5x^2y^2 = 5 \cdot x \cdot x \cdot y \cdot y = (5x^2y)(y)$$

$$30x^3y = 2 \cdot 3 \cdot 5 \cdot x \cdot x \cdot x \cdot y = (5x^2y)(6x)$$

So, you can conclude that the greatest common factor is $5x^2y$.

CHECKPOINT *Now try Exercise 7.*

EXAMPLE 2 Finding the Greatest Common Factor

To find the greatest common factor of $8x^5$, $20x^3$, and $16x^4$, first factor each term.

$$8x^5 = 2 \cdot 2 \cdot 2 \cdot x \cdot x \cdot x \cdot x \cdot x = (4x^3)(2x^2)$$

$$20x^3 = 2 \cdot 2 \cdot 5 \cdot x \cdot x \cdot x = (4x^3)(5)$$

$$16x^4 = 2 \cdot 2 \cdot 2 \cdot 2 \cdot x \cdot x \cdot x \cdot x = (4x^3)(4x)$$

So, you can conclude that the greatest common factor is $4x^3$.

CHECKPOINT *Now try Exercise 11.*

2 ▶ Factor out the greatest common monomial factor from polynomials.

Common Monomial Factors

Consider the three terms listed in Example 2 as terms of the polynomial

$$8x^5 + 16x^4 + 20x^3.$$

The greatest common factor, $4x^3$, of these terms is the **greatest common monomial factor** of the polynomial. When you use the Distributive Property to remove this factor from each term of the polynomial, you are **factoring out** the greatest common monomial factor.

$$8x^5 + 16x^4 + 20x^3 = 4x^3(2x^2) + 4x^3(4x) + 4x^3(5)$$ Factor each term.

$$= 4x^3(2x^2 + 4x + 5)$$ Factor out common monomial factor.

Study Tip

To find the greatest common monomial factor of a polynomial, answer these two questions.

1. What is the greatest integer factor common to each coefficient of the polynomial?
2. What is the highest-power variable factor common to each term of the polynomial?

EXAMPLE 3 Greatest Common Monomial Factor

Factor out the greatest common monomial factor from $6x - 18$.

Solution

The greatest common integer factor of $6x$ and 18 is 6. There is no common variable factor.

$$6x - 18 = 6(x) - 6(3)$$ Greatest common monomial factor is 6.

$$= 6(x - 3)$$ Factor 6 out of each term.

✓ ***CHECKPOINT*** *Now try Exercise 21.*

EXAMPLE 4 Greatest Common Monomial Factor

Factor out the greatest common monomial factor from $10y^3 - 25y^2$.

Solution

For the terms $10y^3$ and $25y^2$, 5 is the greatest common integer factor and y^2 is the highest-power common variable factor.

$$10y^3 - 25y^2 = (5y^2)(2y) - (5y^2)(5)$$ Greatest common factor is $5y^2$.

$$= 5y^2(2y - 5)$$ Factor $5y^2$ out of each term.

✓ ***CHECKPOINT*** *Now try Exercise 31.*

EXAMPLE 5 Greatest Common Monomial Factor

Factor out the greatest common monomial factor from $45x^3 - 15x^2 - 15$.

Solution

The greatest common integer factor of $45x^3$, $15x^2$, and 15 is 15. There is no common variable factor.

$$45x^3 - 15x^2 - 15 = 15(3x^3) - 15(x^2) - 15(1)$$

$$= 15(3x^3 - x^2 - 1)$$

✓ ***CHECKPOINT*** *Now try Exercise 43.*

EXAMPLE 6 Greatest Common Monomial Factor

Factor out the greatest common monomial factor from $35y^3 - 7y^2 - 14y$.

Solution

$$35y^3 - 7y^2 - 14y = 7y(5y^2) - 7y(y) - 7y(2)$$ Greatest common factor is $7y$.

$$= 7y(5y^2 - y - 2)$$ Factor $7y$ out of each term.

 **CHECKPOINT** *Now try Exercise 47.*

EXAMPLE 7 Greatest Common Monomial Factor

Factor out the greatest common monomial factor from $3xy^2 - 15x^2y + 12xy$.

Solution

$$3xy^2 - 15x^2y + 12xy = 3xy(y) - 3xy(5x) + 3xy(4)$$ Greatest common factor is $3xy$.

$$= 3xy(y - 5x + 4)$$ Factor $3xy$ out of each term.

 **CHECKPOINT** *Now try Exercise 51.*

The greatest common monomial factor of the terms of a polynomial is usually considered to have a positive coefficient. However, sometimes it is convenient to factor a negative number out of a polynomial.

EXAMPLE 8 A Negative Common Monomial Factor

Factor the polynomial $-2x^2 + 8x - 12$ in two ways.

a. Factor out a common monomial factor of 2.

b. Factor out a common monomial factor of -2.

Solution

a. To factor out the common monomial factor of 2, write the following.

$$-2x^2 + 8x - 12 = 2(-x^2) + 2(4x) + 2(-6)$$ Factor each term.

$$= 2(-x^2 + 4x - 6)$$ Factored form

b. To factor -2 out of the polynomial, write the following.

$$-2x^2 + 8x - 12 = -2(x^2) + (-2)(-4x) + (-2)(6)$$ Factor each term.

$$= -2(x^2 - 4x + 6)$$ Factored form

Check this result by multiplying $(x^2 - 4x + 6)$ by -2. When you do, you will obtain the original polynomial.

 **CHECKPOINT** *Now try Exercise 59.*

Smart Study Strategy

Go to page 344 for ways to *Build a Support System.*

With experience, you should be able to omit writing the first step shown in Examples 6, 7, and 8. For instance, to factor -2 out of $-2x^2 + 8x - 12$, you could simply write

$$-2x^2 + 8x - 12 = -2(x^2 - 4x + 6).$$

3 ▶ Factor polynomials by grouping.

Factoring by Grouping

There are occasions when the common factor of an expression is not simply a monomial. For instance, the expression

$$x^2(x - 2) + 3(x - 2)$$

has the common *binomial* factor $(x - 2)$. Factoring out this common factor produces

$$x^2(x - 2) + 3(x - 2) = (x - 2)(x^2 + 3).$$

This type of factoring is part of a more general procedure called **factoring by grouping.**

EXAMPLE 9 Common Binomial Factors

Factor each expression.

a. $5x^2(7x - 1) - 3(7x - 1)$ **b.** $2x(3x - 4) + (3x - 4)$

c. $3y^2(y - 3) + 4(3 - y)$

Solution

a. Each of the terms of this expression has a binomial factor of $(7x - 1)$.

$$5x^2(7x - 1) - 3(7x - 1) = (7x - 1)(5x^2 - 3)$$

b. Each of the terms of this expression has a binomial factor of $(3x - 4)$.

$$2x(3x - 4) + (3x - 4) = (3x - 4)(2x + 1)$$

Be sure you see that when $(3x - 4)$ is factored out of itself, you are left with the factor 1. This follows from the fact that $(3x - 4)(1) = (3x - 4)$.

c. $3y^2(y - 3) + 4(3 - y) = 3y^2(y - 3) - 4(y - 3)$ Write $4(3 - y)$ as $-4(y - 3)$.

$= (y - 3)(3y^2 - 4)$ Common factor is $(y - 3)$.

✓ **CHECKPOINT** ***Now try Exercise 61.***

In Example 9, the polynomials were already grouped so that it was easy to determine the common binomial factors. In practice, you will have to do the grouping as well as the factoring. To see how this works, consider the expression

$$x^3 + 2x^2 + 3x + 6$$

and try to factor it. Note first that there is no common monomial factor to take out of all four terms. But suppose you *group* the first two terms together and the last two terms together.

$x^3 + 2x^2 + 3x + 6 = (x^3 + 2x^2) + (3x + 6)$ Group terms.

$= x^2(x + 2) + 3(x + 2)$ Factor out common monomial factor in each group.

$= (x + 2)(x^2 + 3)$ Factored form

When factoring by grouping, be sure to group terms that have a common monomial factor. For example, in the polynomial above, you should not group the first term x^3 with the fourth term 6.

EXAMPLE 10 Factoring by Grouping

Factor $x^3 + 2x^2 + x + 2$.

Solution

$$x^3 + 2x^2 + x + 2 = (x^3 + 2x^2) + (x + 2) \quad \text{Group terms.}$$

$$= x^2(x + 2) + (x + 2) \quad \text{Factor out common monomial factor in each group.}$$

$$= (x + 2)(x^2 + 1) \quad \text{Factored form}$$

✓ **CHECKPOINT** *Now try Exercise 69.*

Note that in Example 10 the polynomial is factored by grouping the first and second terms and the third and fourth terms. You could just as easily have grouped the first and third terms and the second and fourth terms, as follows.

$$x^3 + 2x^2 + x + 2 = (x^3 + x) + (2x^2 + 2)$$

$$= x(x^2 + 1) + 2(x^2 + 1) = (x^2 + 1)(x + 2)$$

Study Tip

Notice in Example 12 that the polynomial is not written in standard form. You could have rewritten the polynomial before factoring and still obtained the same result.

$$2x^3 + 4x - x^2 - 2$$
$$= 2x^3 - x^2 + 4x - 2$$
$$= (2x^3 - x^2) + (4x - 2)$$
$$= x^2(2x - 1) + 2(2x - 1)$$
$$= (2x - 1)(x^2 + 2)$$

EXAMPLE 11 Factoring by Grouping

Factor $3x^2 - 12x - 5x + 20$.

Solution

$$3x^2 - 12x - 5x + 20 = (3x^2 - 12x) + (-5x + 20) \quad \text{Group terms.}$$

$$= 3x(x - 4) - 5(x - 4) \quad \text{Factor out common monomial factor in each group.}$$

$$= (x - 4)(3x - 5) \quad \text{Factored form}$$

Note how a -5 is factored out so that the common binomial factor $x - 4$ appears.

✓ **CHECKPOINT** *Now try Exercise 75.*

You can always check to see that you have factored an expression correctly by multiplying the factors and comparing the result with the original expression. Try using multiplication to check the results of Examples 10 and 11.

EXAMPLE 12 Geometry: Area of a Rectangle

The area of a rectangle of width $2x - 1$ is given by the polynomial $2x^3 + 4x - x^2 - 2$, as shown in Figure 6.1. Factor this expression to determine the length of the rectangle.

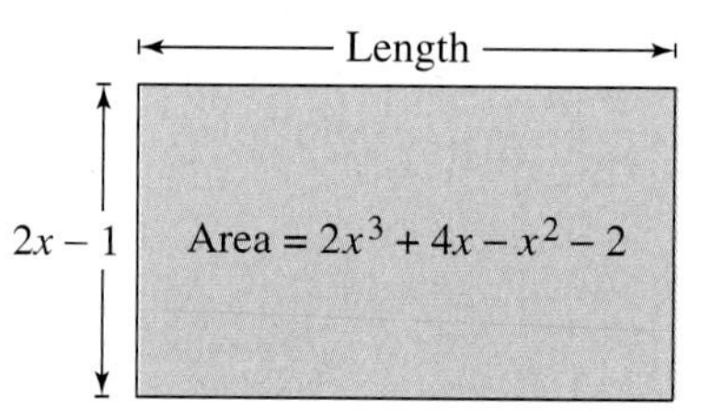

Figure 6.1

Solution

$$2x^3 + 4x - x^2 - 2 = (2x^3 + 4x) + (-x^2 - 2) \quad \text{Group terms.}$$

$$= 2x(x^2 + 2) - (x^2 + 2) \quad \text{Factor out common monomial factor in each group.}$$

$$= (x^2 + 2)(2x - 1) \quad \text{Factored form}$$

You can see that the length of the rectangle is $x^2 + 2$.

✓ **CHECKPOINT** *Now try Exercise 101.*

Concept Check

1. How do you check your result when factoring a polynomial?
2. In your own words, describe a method for finding the greatest common factor of a polynomial.
3. How do you determine which terms to group together when factoring by grouping?
4. Explain how the word *factor* can be used as a noun and a verb.

6.1 EXERCISES

Go to pages 392–393 to record your assignments.

Developing Skills

In Exercises 1–16, find the greatest common factor of the expressions. ***See Examples 1 and 2.***

1. $z^2, -z^6$
2. t^4, t^7
3. $2x^2, 12x$
4. $36x^4, 18x^3$
5. u^2v, u^3v^2
6. $r^6s^4, -rs$
✓ **7.** $9y^8z^4, -12y^5z^4$
8. $-15x^6y^3, 45xy^3$
9. $14x^2, 1, 7x^4$
10. $5y^4, 10x^2y^2, 1$
✓ **11.** $28a^4b^2, 14a^3, 42a^2b^5$
12. $16x^2y, 12xy^2, 36x^2$
13. $2(x + 3), 3(x + 3)$
14. $4(x - 5), 3x(x - 5)$
15. $x(7x + 5), 7x + 5$
16. $x - 4, y(x - 4)$

In Exercises 17–52, factor the polynomial. (*Note:* Some of the polynomials have no common monomial factor.) ***See Examples 3–7.***

17. $3x + 3$
18. $5y - 5$
19. $6z + 36$
20. $4x - 28$
✓ **21.** $8t - 16$
22. $4u + 12$
23. $-25x - 10$
24. $-14y - 7$
25. $24y^2 - 18$
26. $8z^3 + 12$
27. $x^2 + x$
28. $-s^3 - s$
29. $25u^2 - 14u$
30. $36t^4 + 24t^2$
✓ **31.** $2x^4 + 6x^3$
32. $9z^6 + 27z^4$
33. $7s^2 + 9t^2$
34. $12x^2 - 5y^3$
35. $12x^2 - 2x$
36. $12u^7 + 9u^5$
37. $-10r^3 - 35r$
38. $-144a^8 + 24a^6$
39. $12x^2 + 16x - 8$
40. $9 - 3y - 15y^2$
41. $100 + 75z - 50z^2$
42. $42t^3 - 21t^2 + 7$
✓ **43.** $9x^4 + 6x^3 + 18x^2$
44. $32a^5 - 2a^3 + 6a$
45. $5u^2 + 5u^2 + 5u$
46. $11y^3 - 22y^2 + 11y^2$
✓ **47.** $16a^3b^3 + 24a^4b^3$
48. $9x^4y + 24x^2y$
49. $10ab + 10a^2b$
50. $21x^2z^5 + 35x^6z$
✓ **51.** $4xy + 8x^2y - 24x^4y^5$
52. $15m^4n^3 - 25m^7n + 30m^4n^8$

In Exercises 53–60, factor a negative real number from the polynomial and then write the polynomial factor in standard form. ***See Example 8.***

53. $5 - 10x$
54. $3 - 6x$
55. $-10x - 3000$
56. $-3x^2 + 4$
57. $-x^2 + 5x + 10$
58. $-4x^2 - 8x + 20$
✓ **59.** $4 + 12x - 2x^2$
60. $8 - 4x - 12x^2$

In Exercises 61–94, factor the polynomial by grouping. ***See Examples 9–11.***

61. $x(x - 3) + 5(x - 3)$

62. $x(x + 6) + 3(x + 6)$

63. $y(q - 5) - 10(q - 5)$

64. $a^2(b + 2) - b(b + 2)$

65. $x^3(y + 4) + y(y + 4)$

66. $x^3(x - 2) + 6(x - 2)$

67. $(a + b)(a - b) + a(a + b)$

68. $(x + y)(x - y) - x(x - y)$

69. $x^2 + 10x + x + 10$

70. $x^2 - 5x + x - 5$

71. $x^2 + 3x + 4x + 12$

72. $x^2 - 6x + 5x - 30$

73. $x^2 + 3x - 5x - 15$

74. $x^2 + 4x + 2x + 8$

75. $4x^2 - 14x + 14x - 49$

76. $4x^2 - 6x + 6x - 9$

77. $6x^2 + 3x - 2x - 1$

78. $4x^2 + 20x - x - 5$

79. $8x^2 + 32x + x + 4$

80. $8x^2 - 4x - 2x + 1$

81. $9x^2 - 21x + 6x - 14$

82. $12x^2 + 42x - 10x - 35$

83. $2x^2 - 4x - 3x + 6$

84. $35x^2 - 40x + 21x - 24$

85. $t^3 - 3t^2 + 2t - 6$

86. $s^3 + 6s^2 + 2s + 12$

87. $16z^3 + 8z^2 + 2z + 1$

88. $4u^3 - 2u^2 - 6u + 3$

89. $x^3 - 3x - x^2 + 3$

90. $x^3 + 7x - 3x^2 - 21$

91. $3x^2 + x^3 - 18 - 6x$

92. $5x^2 + 10x^3 + 4 + 8x$

93. $ky^2 - 4ky + 2y - 8$

94. $ay^2 + 3ay + 3y + 9$

In Exercises 95–100, fill in the missing factor.

95. $\frac{1}{4}x + \frac{3}{4} = \frac{1}{4}(\quad)$

96. $\frac{5}{6}x - \frac{1}{6} = \frac{1}{6}(\quad)$

97. $2y - \frac{1}{5} = \frac{1}{5}(\quad)$

98. $3z + \frac{5}{4} = \frac{1}{4}(\quad)$

99. $\frac{7}{8}x + \frac{5}{16}y = \frac{1}{16}(\quad)$

100. $\frac{5}{12}u - \frac{5}{8}v = \frac{1}{24}(\quad)$

Solving Problems

Geometry In Exercises 101 and 102, factor the polynomial to find an expression for the length of the rectangle. ***See Example 12.***

101. Area $= 2x^2 + 2x$

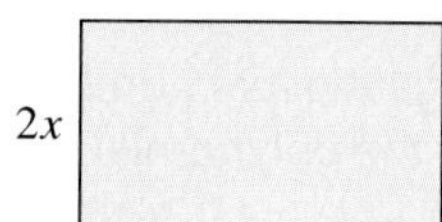

102. Area $= x^2 + 2x + 10x + 20$

$x + 2$

Geometry In Exercises 103 and 104, write an expression for the area of the shaded region and factor the expression if possible.

103.

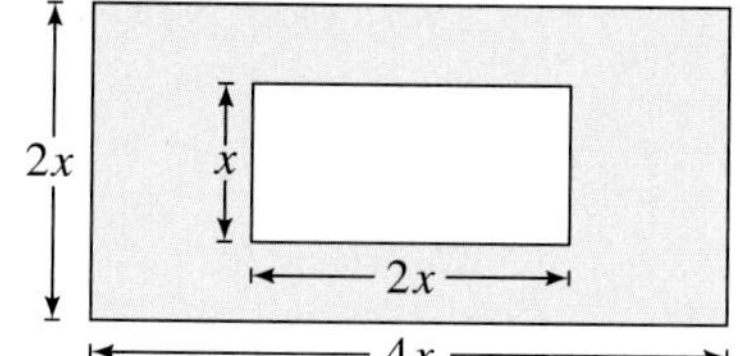

104.

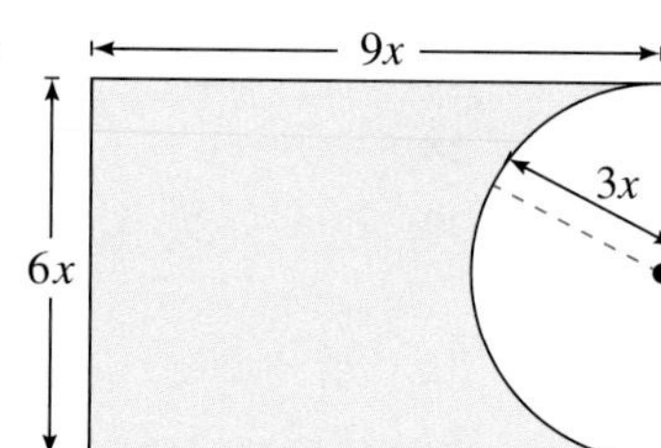

105. *Geometry* The surface area of a right circular cylinder is given by

$2\pi r^2 + 2\pi rh$

where r is the radius of the base of the cylinder and h is the height of the cylinder. Factor this expression.

106. *Simple Interest* The amount after t years when a principal of P dollars is invested at $r\%$ simple interest is given by

$P + Prt.$

Factor this expression.

107. *Chemical Reaction* The rate of change in a chemical reaction is

$kQx - kx^2$

where Q is the original amount, x is the new amount, and k is a constant of proportionality. Factor this expression.

108. *Unit Price* The revenue R from selling x units of a product at a price of p dollars per unit is given by $R = xp$. For a pool table the revenue is

$R = 900x - 0.1x^2.$

Factor the revenue model and determine an expression that represents the price p in terms of x.

Explaining Concepts

109. Explain why $x^2(2x + 1)$ is in factored form.

110. Explain why $3x$ is the greatest common monomial factor of $3x^3 + 3x^2 + 3x$.

111. Give several examples of the use of the Distributive Property in factoring.

112. Give an example of a polynomial with four terms that can be factored by grouping.

Cumulative Review

In Exercises 113–116, determine whether each value is a solution of the equation.

113. $x + 2 = 3$ (a) $x = 5$ (b) $x = 1$

114. $x - 5 = 10$ (a) $x = 0$ (b) $x = 15$

115. $2x - 4 = 0$ (a) $x = 2$ (b) $x = -2$

116. $-7x - 8 = 6$ (a) $x = 1$ (b) $x = -2$

In Exercises 117–120, simplify the expression.

117. $\left(\frac{3y}{2x^3}\right)^2$ **118.** $\left(\frac{b^{-4}}{4a^{-5}}\right)^2$

119. $z^2 \cdot z^{-6}$ **120.** $(x^2y)^5(x^7y^6)^2$

In Exercises 121–124, perform the division and simplify.

121. $\frac{3m^2n}{m}$ **122.** $\frac{12x^3y^5}{3x^2y^5}$

123. $\frac{x^2 + 9x + 8}{x + 8}$ **124.** $\frac{2x^2 + 5x - 12}{2x - 3}$

6.2 Factoring Trinomials

What You Should Learn

1 ▶ Factor trinomials of the form $x^2 + bx + c$.
2 ▶ Factor trinomials in two variables.
3 ▶ Factor trinomials completely.

Why You Should Learn It

The techniques for factoring trinomials will help you in solving polynomial equations in Section 6.5.

1 ▶ Factor trinomials of the form $x^2 + bx + c$.

Factoring Trinomials of the Form $x^2 + bx + c$

From Section 5.3, you know that the product of two binomials is often a trinomial. Here are some examples.

Factored Form		*F*	*O*	*I*	*L*		*Trinomial Form*
$(x-1)(x+5)$	$=$	x^2	$+\ 5x$	$-\ x$	$-\ 5$	$=$	$x^2 + 4x - 5$
$(x-3)(x-3)$	$=$	x^2	$-\ 3x$	$-\ 3x$	$+\ 9$	$=$	$x^2 - 6x + 9$
$(x+5)(x+1)$	$=$	x^2	$+\ x$	$+\ 5x$	$+\ 5$	$=$	$x^2 + 6x + 5$
$(x-2)(x-4)$	$=$	x^2	$-\ 4x$	$-\ 2x$	$+\ 8$	$=$	$x^2 - 6x + 8$

Try covering the factored forms in the left-hand column above. Can you determine the factored forms from the trinomial forms? In this section, you will learn how to factor trinomials of the form $x^2 + bx + c$. To begin, consider the following factorization.

$$\begin{aligned}(x+m)(x+n) &= x^2 + nx + mx + mn \\ &= x^2 + \underbrace{(n+m)}_{\text{Sum of terms}}x + \underbrace{mn}_{\text{Product of terms}} \\ &= x^2 + \quad b \quad x + \quad c\end{aligned}$$

So, to *factor* a trinomial $x^2 + bx + c$ into a product of two binomials, you must find two numbers m and n whose product is c and whose sum is b.

There are many different techniques that can be used to factor trinomials. The most common technique is to use *guess, check, and revise* with mental math. For example, try factoring the trinomial

$$x^2 + 5x + 6.$$

You need to find two numbers whose product is 6 and whose sum is 5. Using mental math, you can determine that the numbers are 2 and 3.

$$x^2 + 5x + 6 = (x+2)(x+3)$$

The product of 2 and 3 is 6.

The sum of 2 and 3 is 5.

EXAMPLE 1 Factoring a Trinomial

Factor the trinomial $x^2 + 5x - 6$.

Solution

You need to find two numbers whose product is -6 and whose sum is 5.

The product of -1 and 6 is -6.

$$x^2 + 5x - 6 = (x - 1)(x + 6)$$

The sum of -1 and 6 is 5.

✓ **CHECKPOINT** *Now try Exercise 15.*

EXAMPLE 2 Factoring a Trinomial

Factor the trinomial $x^2 - x - 6$.

Solution

The product of -3 and 2 is -6.

$$x^2 - x - 6 = (x - 3)(x + 2)$$

The sum of -3 and 2 is -1.

✓ **CHECKPOINT** *Now try Exercise 17.*

Study Tip

Use a list to help you find the two numbers with the required product and sum. For Example 2:

Factors of -6	*Sum*
$1, -6$	-5
$-1, 6$	5
$2, -3$	-1
$-2, 3$	1

Because -1 is the required sum, the correct factorization is

$x^2 - x - 6 = (x - 3)(x + 2)$.

EXAMPLE 3 Factoring a Trinomial

Factor the trinomial $x^2 - 5x + 6$.

Solution

The product of -2 and -3 is 6.

$$x^2 - 5x + 6 = (x - 2)(x - 3)$$

The sum of -2 and -3 is -5.

✓ **CHECKPOINT** *Now try Exercise 19.*

EXAMPLE 4 Factoring a Trinomial

Factor the trinomial $14 + 5x - x^2$.

Solution

First, factor out -1 and write the polynomial factor in standard form. Then find two numbers whose product is -14 and whose sum is -5. So,

The product of -7 and 2 is -14.

$$14 + 5x - x^2 = -(x^2 - 5x - 14) = -(x - 7)(x + 2).$$

The sum of -7 and 2 is -5.

✓ **CHECKPOINT** *Now try Exercise 21.*

If you have trouble factoring a trinomial, it helps to make a list of all the distinct pairs of factors and then check each sum. For instance, consider the trinomial

$$x^2 - 5x - 24 = (x + \quad)(x - \quad).$$ Opposite signs

In this trinomial the constant term is negative, so you need to find two numbers with opposite signs whose product is -24 and whose sum is -5.

Factors of -24	*Sum*	
$1, -24$	-23	
$-1, 24$	23	
$2, -12$	-10	
$-2, 12$	10	
$3, -8$	-5	Correct choice
$-3, 8$	5	
$4, -6$	-2	
$-4, 6$	2	

So, $x^2 - 5x - 24 = (x + 3)(x - 8)$.

With experience, you will be able to narrow the list of possible factors *mentally* to only two or three possibilities whose sums can then be tested to determine the correct factorization. Here are some suggestions for narrowing the list.

Guidelines for Factoring $x^2 + bx + c$

To factor $x^2 + bx + c$, you need to find two numbers m and n whose product is c and whose sum is b.

$$x^2 + bx + c = (x + m)(x + n)$$

1. If c is positive, then m and n have like signs that match the sign of b.
2. If c is negative, then m and n have unlike signs.
3. If $|b|$ is small relative to $|c|$, first try those factors of c that are closest to each other in absolute value.

Study Tip

Notice that factors may be written in any order. For example,

$(x - 2)(x + 12) =$
$(x + 12)(x - 2)$

and

$(x + 4)(x - 6) =$
$(x - 6)(x + 4)$

because of the Commutative Property of Multiplication.

EXAMPLE 5 Factoring a Trinomial

Factor the trinomial $x^2 - 8x - 48$.

Solution

You need to find two numbers whose product is -48 and whose sum is -8.

The product of -12 and 4 is -48.

$$x^2 - 8x - 48 = (x - 12)(x + 4)$$

The sum of -12 and 4 is -8.

✓ **CHECKPOINT** *Now try Exercise 23.*

Study Tip

Not all trinomials are factorable using integer factors. For instance, $x^2 - 2x - 6$ is not factorable using integer factors because there is no pair of factors of -6 whose sum is -2. Such nonfactorable trinomials are called **prime polynomials.**

EXAMPLE 6 Factoring a Trinomial

Factor the trinomial $x^2 + 7x - 30$.

Solution

You need to find two numbers whose product is -30 and whose sum is 7.

The product of -3 and 10 is -30.

$$x^2 + 7x - 30 = (x - 3)(x + 10)$$

The sum of -3 and 10 is 7.

✓ **CHECKPOINT** *Now try Exercise 25.*

2 ▶ Factor trinomials in two variables.

Factoring Trinomials in Two Variables

The next three examples show how to factor trinomials of the form

$$x^2 + bxy + cy^2. \qquad \text{Trinomial in two variables}$$

Note that this trinomial has two variables, x and y. However, from the factorization

$$x^2 + bxy + cy^2 = x^2 + (m + n)xy + mny^2 = (x + my)(x + ny)$$

you can see that you still need to find two factors of c whose sum is b.

Study Tip

With *any* factoring problem, remember that you can check your result by multiplying. For instance, in Example 7, you can check the result by multiplying $(x - 4y)$ by $(x + 3y)$ to obtain $x^2 - xy - 12y^2$.

EXAMPLE 7 Factoring a Trinomial in Two Variables

Factor the trinomial $x^2 - xy - 12y^2$.

Solution

You need to find two numbers whose product is -12 and whose sum is -1.

The product of -4 and 3 is -12.

$$x^2 - xy - 12y^2 = (x - 4y)(x + 3y)$$

The sum of -4 and 3 is -1.

✓ **CHECKPOINT** *Now try Exercise 33.*

EXAMPLE 8 Factoring a Trinomial in Two Variables

Factor the trinomial $y^2 - 6xy + 8x^2$.

Solution

You need to find two numbers whose product is 8 and whose sum is -6.

The product of -2 and -4 is 8.

$$y^2 - 6xy + 8x^2 = (y - 2x)(y - 4x)$$

The sum of -2 and -4 is -6.

✓ **CHECKPOINT** *Now try Exercise 35.*

EXAMPLE 9 Factoring a Trinomial in Two Variables

Factor the trinomial $x^2 + 11xy + 10y^2$.

Solution

You need to find two numbers whose product is 10 and whose sum is 11.

The product of 1 and 10 is 10.

$$x^2 + 11xy + 10y^2 = (x + y)(x + 10y)$$

The sum of 1 and 10 is 11.

✓ **CHECKPOINT** ***Now try Exercise 37.***

3 ▶ Factor trinomials completely.

Factoring Completely

Some trinomials have a common monomial factor. In such cases you should first factor out the common monomial factor. Then you can try to factor the resulting trinomial by the methods of this section. This "multiple-stage factoring process" is called **factoring completely.** The trinomial below is completely factored.

$$2x^2 - 4x - 6 = 2(x^2 - 2x - 3)$$ Factor out common monomial factor 2.

$$= 2(x - 3)(x + 1)$$ Factor trinomial.

EXAMPLE 10 Factoring Completely

Factor the trinomial $2x^2 - 12x + 10$ completely.

Solution

$$2x^2 - 12x + 10 = 2(x^2 - 6x + 5)$$ Factor out common monomial factor 2.

$$= 2(x - 5)(x - 1)$$ Factor trinomial.

✓ **CHECKPOINT** ***Now try Exercise 41.***

EXAMPLE 11 Factoring Completely

Factor the trinomial $3x^3 - 27x^2 + 54x$ completely.

Solution

$$3x^3 - 27x^2 + 54x = 3x(x^2 - 9x + 18)$$ Factor out common monomial factor $3x$.

$$= 3x(x - 3)(x - 6)$$ Factor trinomial.

✓ **CHECKPOINT** ***Now try Exercise 47.***

EXAMPLE 12 Factoring Completely

$$4y^4 + 32y^3 + 28y^2 = 4y^2(y^2 + 8y + 7)$$ Factor out common monomial factor $4y^2$.

$$= 4y^2(y + 1)(y + 7)$$ Factor trinomial.

✓ **CHECKPOINT** ***Now try Exercise 51.***

Concept Check

1. In your own words, explain how to factor a trinomial of the form $x^2 + bx + c$. Give examples with your explanation.
2. What does it mean for a trinomial to be *prime*?
3. Can you completely factor a trinomial into two different sets of prime factors? Explain.
4. After factoring a trinomial, how can you check your result?

6.2 EXERCISES

Go to pages 392–393 to record your assignments.

Developing Skills

In Exercises 1–6, fill in the missing factor. Then check your answer by multiplying the factors.

1. $x^2 + 8x + 7 = (x + 7)(\quad)$
2. $x^2 + 2x - 3 = (x + 3)(\quad)$
3. $y^2 + 11y - 12 = (y - 1)(\quad)$
4. $y^2 - 2y - 8 = (y + 2)(\quad)$
5. $z^2 - 7z + 12 = (z - 4)(\quad)$
6. $z^2 - 4z + 3 = (z - 3)(\quad)$

In Exercises 7–12, find all possible products of the form $(x + m)(x + n)$, where $m \cdot n$ is the specified product. (Assume that m and n are integers.)

7. $m \cdot n = 11$
8. $m \cdot n = 5$
9. $m \cdot n = 14$
10. $m \cdot n = -10$
11. $m \cdot n = -12$
12. $m \cdot n = 18$

In Exercises 13–40, factor the trinomial. (*Note:* Some of the trinomials may be prime.) ***See Examples 1–9.***

13. $x^2 + 6x + 8$
14. $x^2 + 12x + 35$
15. ✓ $x^2 + 2x - 15$
16. $x^2 + 4x - 21$
17. ✓ $x^2 - 9x - 22$
18. $x^2 - 9x - 10$
19. ✓ $x^2 - 9x + 14$
20. $y^2 - 6y + 10$
21. ✓ $2x + 15 - x^2$
22. $3x + 18 - x^2$
23. ✓ $u^2 - 22u - 48$
24. $x^2 - 8x - 240$
25. ✓ $x^2 + 3x - 70$
26. $x^2 - 13x + 40$
27. $x^2 + 19x + 60$
28. $x^2 + 10x + 24$
29. $x^2 - 17x + 72$
30. $r^2 - 30r + 216$
31. $y^2 + 5y + 11$
32. $x^2 - x - 36$
33. ✓ $x^2 - 7xz - 18z^2$
34. $u^2 - 4uv - 5v^2$
35. ✓ $x^2 - 5xy + 6y^2$
36. $x^2 + xy - 2y^2$
37. ✓ $x^2 + 8xy + 15y^2$
38. $x^2 + 15xy + 50y^2$
39. $a^2 + 2ab - 15b^2$
40. $y^2 + 4yz - 60z^2$

In Exercises 41–60, factor the trinomial completely. (*Note*: Some of the trinomials may be prime.) ***See Examples 10–12.***

41. $4x^2 - 32x + 60$

42. $4y^2 - 8y - 12$

43. $3z^2 + 5z + 6$

44. $7x^2 + 5x + 10$

45. $9x^2 + 18x - 18$

46. $6x^2 - 24x - 6$

47. $x^3 - 13x^2 + 30x$

48. $x^3 + x^2 - 2x$

49. $3x^3 + 18x^2 + 24x$

50. $4x^3 + 8x^2 - 12x$

51. $x^4 - 5x^3 + 6x^2$

52. $x^4 + 3x^3 - 10x^2$

53. $2x^4 - 20x^3 + 42x^2$

54. $5x^4 - 10x^3 - 240x^2$

55. $x^3 + 5x^2y + 6xy^2$

56. $x^2y - 6xy^2 + y^3$

57. $-3y^2x - 9yx + 54x$

58. $-5x^2z + 15xz + 50z$

59. $2x^3y + 4x^2y^2 - 6xy^3$

60. $2x^3y - 10x^2y^2 + 6xy^3$

In Exercises 61–66, find all integers b such that the trinomial can be factored.

61. $x^2 + bx + 18$

62. $x^2 + bx + 10$

63. $x^2 + bx - 35$

64. $x^2 + bx - 11$

65. $x^2 + bx + 36$

66. $x^2 + bx - 48$

In Exercises 67–72, find two integers c such that the trinomial can be factored. (There are many correct answers.)

67. $x^2 + 3x + c$

68. $x^2 + 5x + c$

69. $x^2 - 4x + c$

70. $x^2 - 15x + c$

71. $x^2 - 9x + c$

72. $x^2 + 12x + c$

Solving Problems

73. *Exploration* An open box is to be made from a four-foot-by-six-foot sheet of metal by cutting equal squares from the corners and turning up the sides (see figure). The volume of the box can be modeled by $V = 4x^3 - 20x^2 + 24x, \quad 0 < x < 2$.

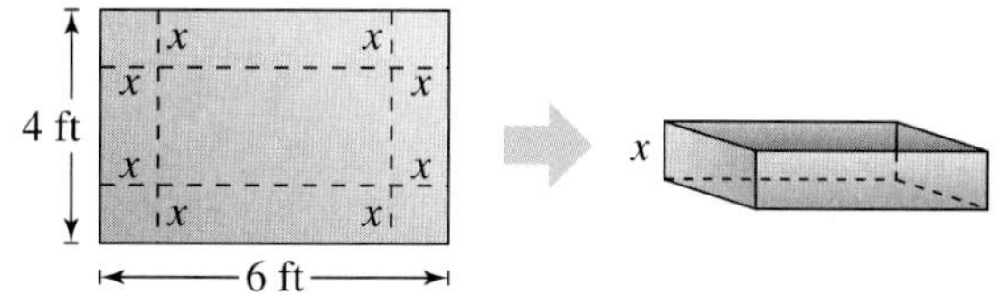

(a) Factor the trinomial that models the volume of the box. Use the factored form to explain how the model was found.

(b) Use a graphing calculator to graph the trinomial over the specified interval. Use the graph to approximate the size of the squares to be cut from the corners so that the box has the maximum volume.

74. *Exploration* If the box in Exercise 73 is to be made from a six-foot-by-eight-foot sheet of metal, the volume of the box would be modeled by

$V = 4x^3 - 28x^2 + 48x, \quad 0 < x < 3.$

(a) Factor the trinomial that models the volume of the box. Use the factored form to explain how the model was found.

(b) Use a graphing calculator to graph the trinomial over the specified interval. Use the graph to approximate the size of the squares to be cut from the corners so that the box has the maximum volume.

75. *Geometry* The area of the rectangle shown in the figure is $x^2 + 30x + 200$. What is the area of the shaded region?

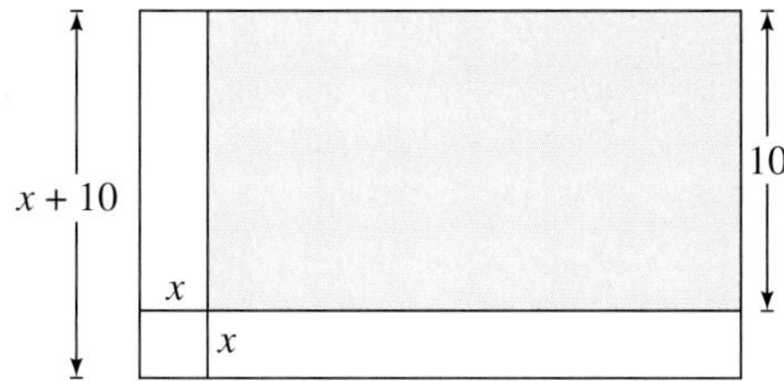

76. *Geometry* The area of the rectangle shown in the figure is $x^2 + 17x + 70$. What is the area of the shaded region?

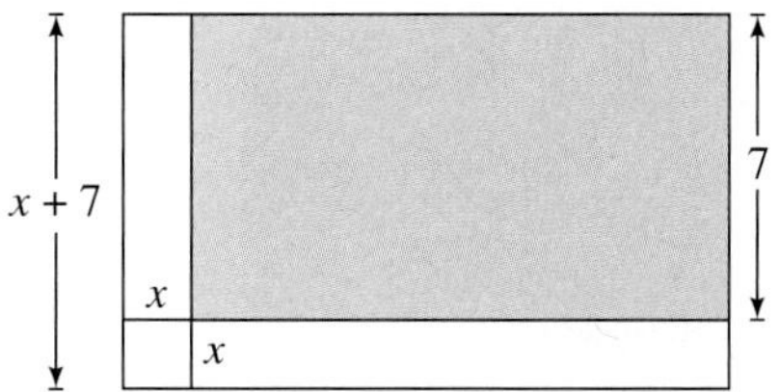

Explaining Concepts

77. State which of the following are factorizations of $2x^2 + 6x - 20$. For each correct factorization, state whether or not it is complete.

(a) $(2x - 4)(x + 5)$

(b) $(2x - 4)(2x + 10)$

(c) $(x - 2)(x + 5)$

(d) $2(x - 2)(x + 5)$

78. In factoring $x^2 - 4x + 3$, why is it unnecessary to test $(x - 1)(x + 3)$ and $(x + 1)(x - 3)$?

79. In factoring the trinomial $x^2 + bx + c$, is the process easier if c is a prime number such as 5 or if c is a composite number such as 120? Explain.

Cumulative Review

In Exercises 80–87, solve the equation and check your solution.

80. $5x - 9 = 26$

81. $3x - 5 = 16$

82. $5 - 3x = 6$

83. $7 - 2x = 9$

84. $7x - 12 = 3x$

85. $10x + 24 = 2x$

86. $5x - 16 = 7x - 9$

87. $3x - 8 = 9x + 4$

In Exercises 88–95, find the greatest common factor of the expressions.

88. $6x^3, 52x^6$

89. $35t^2, 7t^8$

90. a^5b^4, a^3b^7

91. xy^3, x^2y^2

92. $18r^3s^3, -54r^5s^3$

93. $12xy^3, 28x^3y^4$

94. $16u^2v^5, 8u^4v^3, 2u^3v^7$

95. $21xy^4, 42x^2y^2, 9x^4y$

6.3 More About Factoring Trinomials

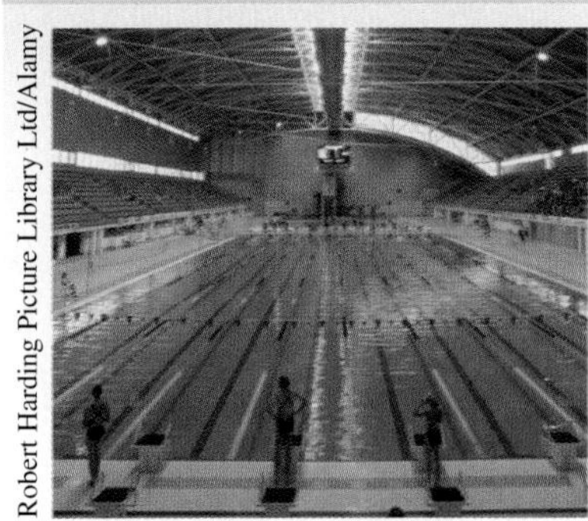
Robert Harding Picture Library Ltd/Alamy

Why You Should Learn It

Trinomials can be used in many geometric applications. For example, in Exercise 112 on page 369, you will factor the expression for the volume of a swimming pool to determine an expression for the width of the swimming pool.

What You Should Learn

1 ▶ Factor trinomials of the form $ax^2 + bx + c$.
2 ▶ Factor trinomials completely.
3 ▶ Factor trinomials by grouping.

1 ▶ Factor trinomials of the form $ax^2 + bx + c$.

Factoring Trinomials of the Form $ax^2 + bx + c$

In this section, you will learn how to factor trinomials whose leading coefficients are *not* 1. To see how this works, consider the following.

Factors of a

$$ax^2 + bx + c = (\;\square\; x + \;\square\;)(\;\square\; x + \;\square\;)$$

Factors of c

The goal is to find a combination of factors of a and c such that the outer and inner products add up to the middle term bx.

EXAMPLE 1 Factoring a Trinomial of the Form $ax^2 + bx + c$

Factor the trinomial $4x^2 - 4x - 3$.

Solution

First, observe that $4x^2 - 4x - 3$ has no common monomial factor. For this trinomial, $a = 4$ and $c = -3$. You need to find a combination of the factors of 4 and -3 such that the outer and inner products add up to $-4x$. The possible combinations are as follows.

Factors	$O + I$	
$(x + 1)(4x - 3)$ (Inner product $= 4x$; Outer product $= -3x$)	$-3x + 4x = x$	x does not equal $-4x$.
$(x - 1)(4x + 3)$	$3x - 4x = -x$	$-x$ does not equal $-4x$.
$(x + 3)(4x - 1)$	$-x + 12x = 11x$	$11x$ does not equal $-4x$.
$(x - 3)(4x + 1)$	$x - 12x = -11x$	$-11x$ does not equal $-4x$.
$(2x + 1)(2x - 3)$	$-6x + 2x = -4x$	$-4x$ equals $-4x$. ✓
$(2x - 1)(2x + 3)$	$6x - 2x = 4x$	$4x$ does not equal $-4x$.

So, the correct factorization is $4x^2 - 4x - 3 = (2x + 1)(2x - 3)$.

✓ **CHECKPOINT** *Now try Exercise 23.*

EXAMPLE 2 Factoring a Trinomial of the Form $ax^2 + bx + c$

Factor the trinomial $6x^2 + 5x - 4$.

Solution

First, observe that $6x^2 + 5x - 4$ has no common monomial factor. For this trinomial, $a = 6$ and $c = -4$. You need to find a combination of the factors of 6 and -4 such that the outer and inner products add up to $5x$.

Factors	*O + I*	
$(x + 1)(6x - 4)$	$-4x + 6x = 2x$	$2x$ does not equal $5x$.
$(x - 1)(6x + 4)$	$4x - 6x = -2x$	$-2x$ does not equal $5x$.
$(x + 4)(6x - 1)$	$-x + 24x = 23x$	$23x$ does not equal $5x$.
$(x - 4)(6x + 1)$	$x - 24x = -23x$	$-23x$ does not equal $5x$.
$(x + 2)(6x - 2)$	$-2x + 12x = 10x$	$10x$ does not equal $5x$.
$(x - 2)(6x + 2)$	$2x - 12x = -10x$	$-10x$ does not equal $5x$.
$(2x + 1)(3x - 4)$	$-8x + 3x = -5x$	$-5x$ does not equal $5x$.
$(2x - 1)(3x + 4)$	$8x - 3x = 5x$	$5x$ equals $5x$. ✓
$(2x + 4)(3x - 1)$	$-2x + 12x = 10x$	$10x$ does not equal $5x$.
$(2x - 4)(3x + 1)$	$2x - 12x = -10x$	$-10x$ does not equal $5x$.
$(2x + 2)(3x - 2)$	$-4x + 6x = 2x$	$2x$ does not equal $5x$.
$(2x - 2)(3x + 2)$	$4x - 6x = -2x$	$-2x$ does not equal $5x$.

So, the correct factorization is $6x^2 + 5x - 4 = (2x - 1)(3x + 4)$.

✓ **CHECKPOINT** *Now try Exercise 37.*

Study Tip

If the original trinomial has no common monomial factors, then its binomial factors cannot have common monomial factors. So, in Example 2, you don't have to test factors, such as $(6x - 4)$, that have a common monomial factor of 2.

The following guidelines can help shorten the list of possible factorizations.

Guidelines for Factoring $ax^2 + bx + c$ $(a > 0)$

1. If the trinomial has a common monomial factor, you should factor out the common factor before trying to find the binomial factors.
2. Because the resulting trinomial has no common monomial factors, you do not have to test any binomial factors that have a common monomial factor.
3. Switch the signs of the factors of c when the middle term (O + I) is correct except in sign.

Using these guidelines, you can shorten the list in Example 2 to the following.

$(x + 4)(6x - 1) = 6x^2 + 23x - 4$	$23x$ does not equal $5x$.
$(2x + 1)(3x - 4) = 6x^2 - 5x - 4$	Opposite sign
$(2x - 1)(3x + 4) = 6x^2 + 5x - 4$ ⇦	Correct factorization

Technology: Tip

As with other types of factoring, you can use a graphing calculator to check your results. For instance, graph

$y_1 = 2x^2 + x - 15$ and

$y_2 = (2x - 5)(x + 3)$

in the same viewing window, as shown below. Because both graphs are the same, you can conclude that

$2x^2 + x - 15$

$= (2x - 5)(x + 3).$

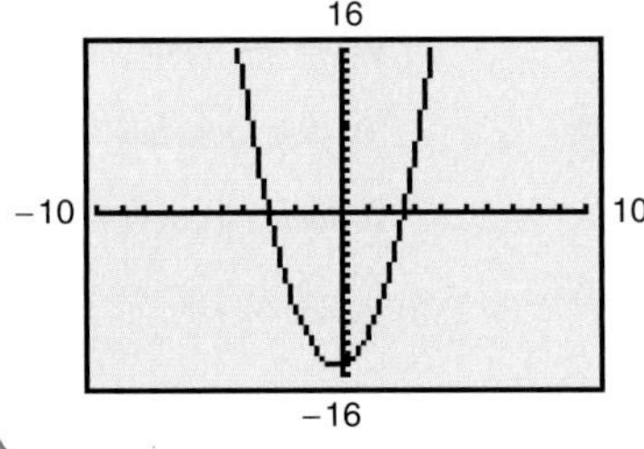

EXAMPLE 3 Factoring a Trinomial of the Form $ax^2 + bx + c$

Factor the trinomial $2x^2 + x - 15$.

Solution

First, observe that $2x^2 + x - 15$ has no common monomial factor. For this trinomial, $a = 2$, which factors as $(1)(2)$, and $c = -15$, which factors as $(1)(-15)$, $(-1)(15)$, $(3)(-5)$, and $(-3)(5)$.

$$(2x + 1)(x - 15) = 2x^2 - 29x - 15$$

$$(2x + 15)(x - 1) = 2x^2 + 13x - 15$$

$$(2x + 3)(x - 5) = 2x^2 - 7x - 15$$

$$(2x + 5)(x - 3) = 2x^2 - x - 15$$ Middle term has opposite sign.

$$(2x - 5)(x + 3) = 2x^2 + x - 15$$ 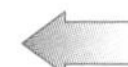 Correct factorization

So, the correct factorization is

$$2x^2 + x - 15 = (2x - 5)(x + 3).$$

✓ **CHECKPOINT** *Now try Exercise 43.*

Notice in Example 3 that when the middle term has the incorrect sign, you need to change only the signs of the second terms of the two factors.

2 ▶ Factor trinomials completely.

Factoring Completely

Remember that if a trinomial has a common monomial factor, the common monomial factor should be factored out first. The complete factorization will then show all monomial and binomial factors.

EXAMPLE 4 Factoring Completely

Factor $4x^3 - 30x^2 + 14x$ completely.

Solution

Begin by factoring out the common monomial factor.

$$4x^3 - 30x^2 + 14x = 2x(2x^2 - 15x + 7)$$

Now, for the new trinomial $2x^2 - 15x + 7$, $a = 2$ and $c = 7$. The possible factorizations of this trinomial are as follows.

$$(2x - 7)(x - 1) = 2x^2 - 9x + 7$$

$$(2x - 1)(x - 7) = 2x^2 - 15x + 7$$ Correct factorization

So, the complete factorization of the original trinomial is

$$4x^3 - 30x^2 + 14x = 2x(2x^2 - 15x + 7)$$

$$= 2x(2x - 1)(x - 7).$$

✓ **CHECKPOINT** *Now try Exercise 61.*

In factoring a trinomial with a negative leading coefficient, first factor -1 out of the trinomial, as demonstrated in Example 5.

EXAMPLE 5 A Negative Leading Coefficient

Factor the trinomial $-5x^2 + 7x + 6$.

Solution

This trinomial has a negative leading coefficient, so you should begin by factoring -1 out of the trinomial.

$$-5x^2 + 7x + 6 = (-1)(5x^2 - 7x - 6)$$

Now, for the new trinomial $5x^2 - 7x - 6$, you have $a = 5$ and $c = -6$. After testing the possible factorizations, you can conclude that

$$(x - 2)(5x + 3) = 5x^2 - 7x - 6.$$ Correct factorization

So, a correct factorization is

$$-5x^2 + 7x + 6 = (-1)(x - 2)(5x + 3)$$

$$= (-x + 2)(5x + 3).$$ Distributive Property

Another correct factorization is $(x - 2)(-5x - 3)$.

✓ **CHECKPOINT** *Now try Exercise 73.*

3 ▶ Factor trinomials by grouping.

Factoring by Grouping

The examples in this section and the preceding section have shown how to use the *guess, check, and revise* strategy to factor trinomials. An alternative technique to use is factoring by grouping. Recall from Section 6.1 that the polynomial

$$x^3 + 2x^2 + 3x + 6$$

was factored by first grouping terms and then applying the Distributive Property.

$$x^3 + 2x^2 + 3x + 6 = (x^3 + 2x^2) + (3x + 6)$$ Group terms.

$$= x^2(x + 2) + 3(x + 2)$$ Factor out common monomial factor in each group.

$$= (x + 2)(x^2 + 3)$$ Distributive Property

By rewriting the middle term of the trinomial in Example 3 as

$$2x^2 + x - 15 = 2x^2 + 6x - 5x - 15$$

you can group the first two terms and the last two terms and factor the trinomial as follows.

$$2x^2 + x - 15 = 2x^2 + 6x - 5x - 15$$ Rewrite middle term.

$$= (2x^2 + 6x) + (-5x - 15)$$ Group terms.

$$= 2x(x + 3) - 5(x + 3)$$ Factor out common monomial factor in each group.

$$= (x + 3)(2x - 5)$$ Distributive Property

Guidelines for Factoring $ax^2 + bx + c$ by Grouping

1. If necessary, write the trinomial in standard form.
2. Choose factors of the product ac that add up to b.
3. Use these factors to rewrite the middle term as a sum or difference.
4. Group and remove any common monomial factors from the first two terms and the last two terms.
5. If possible, factor out the common binomial factor.

EXAMPLE 6 Factoring a Trinomial by Grouping

Use factoring by grouping to factor the trinomial $2x^2 + 5x - 3$.

Solution

In the trinomial $2x^2 + 5x - 3$, $a = 2$ and $c = -3$, which implies that the product ac is -6. Now, because -6 factors as $(6)(-1)$, and $6 - 1 = 5 = b$, you can rewrite the middle term as $5x = 6x - x$. This produces the following.

$$2x^2 + 5x - 3 = 2x^2 + 6x - x - 3 \quad \text{Rewrite middle term.}$$

$$= (2x^2 + 6x) + (-x - 3) \quad \text{Group terms.}$$

$$= 2x(x + 3) - (x + 3) \quad \text{Factor out common monomial factor in each group.}$$

$$= (x + 3)(2x - 1) \quad \text{Factor out common binomial factor.}$$

So, the trinomial factors as

$$2x^2 + 5x - 3 = (x + 3)(2x - 1).$$

CHECKPOINT *Now try Exercise 95.*

EXAMPLE 7 Factoring a Trinomial by Grouping

Use factoring by grouping to factor the trinomial $6x^2 - 11x - 10$.

Solution

In the trinomial $6x^2 - 11x - 10$, $a = 6$ and $c = -10$, which implies that the product ac is -60. Now, because -60 factors as $(-15)(4)$ and $-15 + 4 = -11 = b$, you can rewrite the middle term as $-11x = -15x + 4x$. This produces the following.

$$6x^2 - 11x - 10 = 6x^2 - 15x + 4x - 10 \quad \text{Rewrite middle term.}$$

$$= (6x^2 - 15x) + (4x - 10) \quad \text{Group terms.}$$

$$= 3x(2x - 5) + 2(2x - 5) \quad \text{Factor out common monomial factor in each group.}$$

$$= (2x - 5)(3x + 2) \quad \text{Factor out common binomial factor.}$$

So, the trinomial factors as

$$6x^2 - 11x - 10 = (2x - 5)(3x + 2).$$

CHECKPOINT *Now try Exercise 99.*

Concept Check

1. When creating a list of possible factorizations, the middle term of a trinomial is correct except in sign. What can you do to find the correct factorization?

2. How is the Distributive Property used in factoring by grouping?

3. How do you approach factoring a polynomial with a negative leading coefficient?

4. What is the first step in factoring $ax^2 + bx + c$ by grouping?

6.3 EXERCISES

Go to pages 392–393 to record your assignments.

Developing Skills

In Exercises 1–18, fill in the missing factor.

1. $2x^2 + 7x - 4 = (2x - 1)(\quad)$
2. $3x^2 + x - 4 = (3x + 4)(\quad)$
3. $3t^2 + 4t - 15 = (3t - 5)(\quad)$
4. $5t^2 + t - 18 = (5t - 9)(\quad)$
5. $7x^2 + 15x + 2 = (7x + 1)(\quad)$
6. $3x^2 + 4x + 1 = (3x + 1)(\quad)$
7. $5x^2 + 18x + 9 = (5x + 3)(\quad)$
8. $5x^2 + 19x + 12 = (5x + 4)(\quad)$
9. $5a^2 + 12a - 9 = (a + 3)(\quad)$
10. $7c^2 + 32c - 15 = (c + 5)(\quad)$
11. $4z^2 - 13z + 3 = (z - 3)(\quad)$
12. $8z^2 - 19z + 6 = (z - 2)(\quad)$
13. $6x^2 - 23x + 7 = (3x - 1)(\quad)$
14. $6x^2 - 13x + 6 = (2x - 3)(\quad)$
15. $9a^2 - 6a - 8 = (3a + 2)(\quad)$
16. $4a^2 - 4a - 15 = (2a + 3)(\quad)$
17. $18t^2 + 3t - 10 = (6t + 5)(\quad)$
18. $12x^2 - 31x + 20 = (3x - 4)(\quad)$

In Exercises 19–22, find all possible products of the form $(5x + m)(x + n)$, where $m \cdot n$ is the specified product. (Assume that m and n are integers.)

19. $m \cdot n = 3$

20. $m \cdot n = 15$

21. $m \cdot n = 28$

22. $m \cdot n = 36$

In Exercises 23–50, factor the trinomial. (*Note:* Some of the trinomials may be prime.) ***See Examples 1–3.***

23. ✓ $2x^2 + 5x + 3$
24. $3x^2 + 7x + 2$
25. $4y^2 + 5y + 1$
26. $3x^2 + 5x - 2$
27. $6y^2 - 7y + 1$
28. $7a^2 - 9a + 2$
29. $12x^2 + 7x - 5$
30. $5z^2 - 24z - 5$
31. $5x^2 - 2x + 1$
32. $4z^2 - 8z + 1$
33. $2x^2 + x + 3$
34. $6x^2 - 10x + 5$
35. $8s^2 - 14s + 3$
36. $6v^2 + v - 2$

37. $4x^2 + 13x - 12$
38. $16y^2 + 24y - 27$
39. $9x^2 - 18x + 8$
40. $25a^2 - 40a + 12$
41. $18u^2 - 3u - 28$
42. $24s^2 + 37s - 5$
43. $15a^2 + 14a - 8$
44. $12x^2 - 8x - 15$
45. $10t^2 - 3t - 18$
46. $14t^2 + 61t - 9$
47. $15m^2 + 13m - 20$
48. $21b^2 - 40b - 21$
49. $16z^2 - 34z + 15$
50. $12x^2 - 41x + 24$

In Exercises 51–72, factor the polynomial completely. (*Note:* Some of the polynomials may be prime.) ***See Example 4.***

51. $6x^2 - 3x$
52. $7a^4 - 49a^3$
53. $15y^2 - 40y$
54. $24y^3 - 16y$
55. $u(u - 3) + 9(u - 3)$
56. $x(x - 8) - 2(x - 8)$
57. $2v^2 + 8v - 42$
58. $4z^2 - 12z - 40$
59. $-3x^2 - 3x - 60$
60. $5y^2 + 25y + 35$
61. $9z^2 - 24z + 15$
62. $6x^2 + 8x - 8$
63. $4x^2 - 4x - 8$
64. $6x^2 - 6x - 36$
65. $-15x^4 - 2x^3 + 8x^2$
66. $15y^2 - 7y^3 - 2y^4$
67. $3x^3 + 4x^2 + 2x$
68. $5x^3 - 3x^2 - 4x$
69. $6x^3 + 24x^2 - 192x$
70. $35x + 28x^2 - 7x^3$
71. $18u^4 + 18u^3 - 27u^2$
72. $12x^5 - 16x^4 + 8x^3$

In Exercises 73–82, factor the trinomial. (*Note:* The leading coefficient is negative.) ***See Example 5.***

73. $-2x^2 + 7x + 9$
74. $-5x^2 + x + 4$
75. $4 - 4x - 3x^2$
76. $-4x^2 + 17x + 15$
77. $-6x^2 + 7x + 10$
78. $3 + 2x - 8x^2$
79. $1 - 4x - 60x^2$
80. $2 + 5x - 12x^2$
81. $16 - 8x - 15x^2$
82. $20 + 17x - 10x^2$

In Exercises 83–88, find all integers b such that the trinomial can be factored.

83. $3x^2 + bx + 10$
84. $4x^2 + bx + 3$
85. $2x^2 + bx - 6$
86. $5x^2 + bx - 6$
87. $6x^2 + bx + 20$
88. $8x^2 + bx - 18$

In Exercises 89–94, find two integers c such that the trinomial can be factored. (There are many correct answers.)

89. $4x^2 + 3x + c$
90. $2x^2 + 5x + c$
91. $3x^2 - 10x + c$
92. $8x^2 - 3x + c$
93. $6x^2 - 5x + c$
94. $4x^2 - 9x + c$

In Exercises 95–110, factor the trinomial by grouping. *See Examples 6 and 7.*

✓ **95.** $3x^2 + 4x + 1$

96. $2x^2 + 5x + 2$

97. $7x^2 + 20x - 3$

98. $5x^2 - 14x - 3$

✓ **99.** $6x^2 + 5x - 4$

100. $12y^2 + 11y + 2$

101. $15x^2 - 11x + 2$

102. $12x^2 - 13x + 1$

103. $3a^2 + 11a + 10$

104. $7z^2 - 18z - 9$

105. $16x^2 + 2x - 3$

106. $20c^2 + 19c - 1$

107. $12x^2 - 17x + 6$

108. $10y^2 - 13y - 30$

109. $6u^2 - 5u - 14$

110. $12x^2 + 28x + 15$

Solving Problems

111. *Geometry* The sandbox shown in the figure has a height of x and a width of $x + 2$. The volume of the sandbox is $2x^3 + 7x^2 + 6x$. Find the length l of the sandbox.

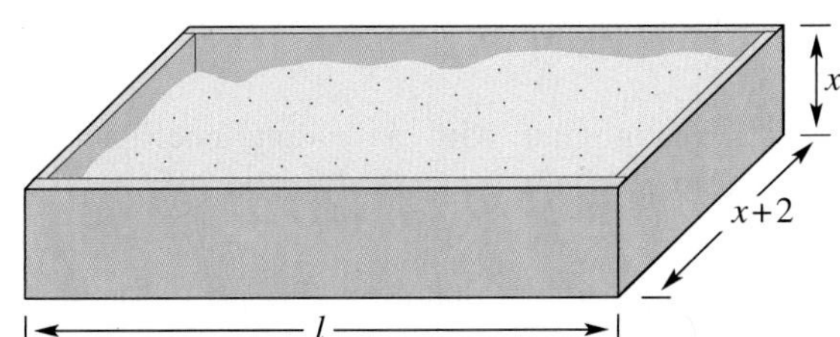

112. *Geometry* The swimming pool shown in the figure has a depth of d and a length of $5d + 2$. The volume of the swimming pool is $15d^3 - 14d^2 - 8d$. Find the width w of the swimming pool.

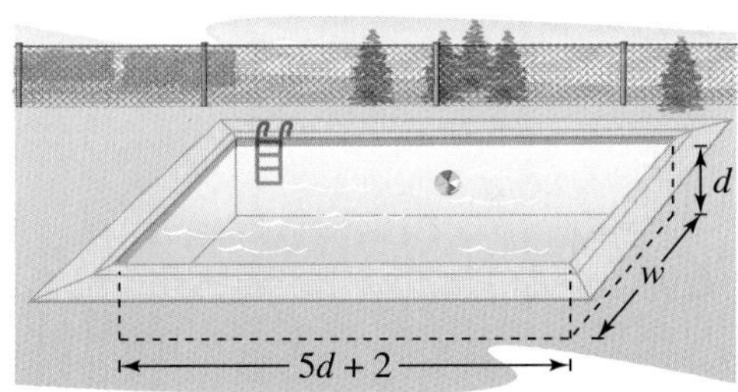

113. *Geometry* The area of the rectangle shown in the figure is $2x^2 + 9x + 10$. What is the area of the shaded region?

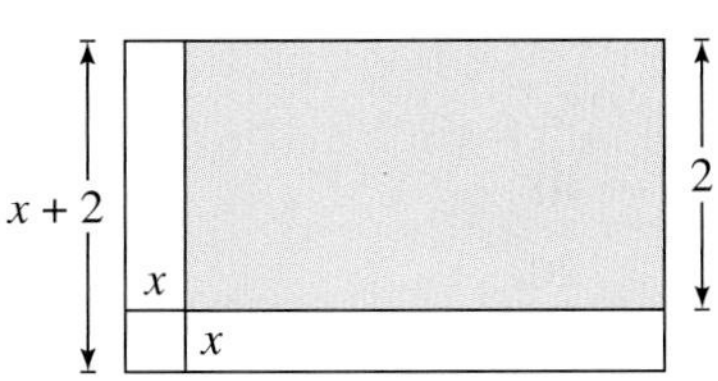

114. *Geometry* The area of the rectangle shown in the figure is $3x^2 + 10x + 3$. What is the area of the shaded region?

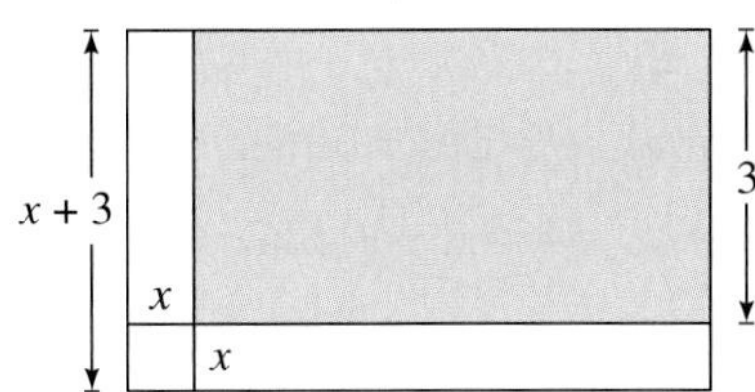

115. *Graphical Exploration* Consider the equations

$$y_1 = 2x^3 + 3x^2 - 5x$$

and

$$y_2 = x(2x + 5)(x - 1).$$

(a) Factor the trinomial represented by y_1. What is the relationship between y_1 and y_2?

(b) Demonstrate your answer to part (a) graphically by using a graphing calculator to graph y_1 and y_2 in the same viewing window.

(c) Identify the x- and y-intercepts of the graphs of y_1 and y_2.

116. *Beam Deflection* A cantilever beam of length l is fixed at the origin. A load weighing W pounds is attached to the end of the beam (see figure). The deflection y of the beam x units from the origin is given by

$$y = -\frac{1}{10}x^2 - \frac{1}{120}x^3, \quad 0 \le x \le 3.$$

Factor the expression for the deflection. (Write the binomial factor with positive integer coefficients.)

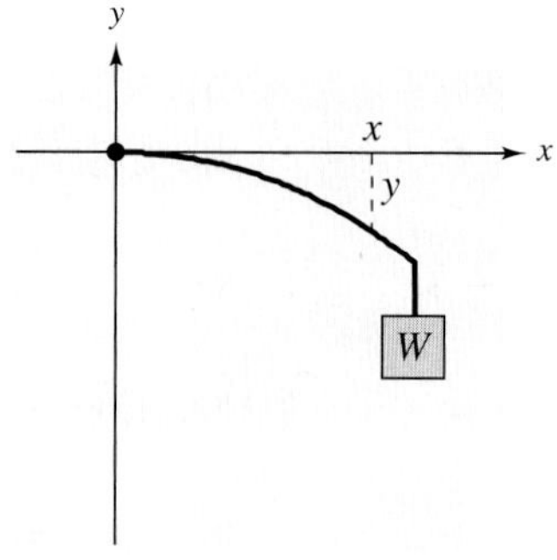

Figure for 116

Explaining Concepts

117. Without multiplying the factors, explain why $(2x + 3)(x + 5)$ is not a factorization of $2x^2 + 7x - 15$.

118. *Error Analysis* Describe and correct the error.

$$9x^2 - 9x - 54 = (3x + 6)(3x - 9)$$
$$= 3(x + 2)(x - 3)$$

119. In factoring $ax^2 + bx + c$, how many possible factorizations must be tested if a and c are prime? Explain your reasoning.

120. Give an example of a prime trinomial that is of the form $ax^2 + bx + c$.

121. Give an example of a trinomial of the form $ax^3 + bx^2 + cx$ that has a common monomial factor of $2x$.

122. Can a trinomial with a leading coefficient not equal to 1 have two identical factors? If so, give an example.

123. Many people think the technique of factoring a trinomial by grouping is more efficient than the *guess, check, and revise* strategy, especially when the coefficients a and c have many factors. Try factoring $6x^2 - 13x + 6$, $2x^2 + 5x - 12$, and $3x^2 + 11x - 4$ using both methods. Which method do you prefer? Explain the advantages and disadvantages of each method.

Cumulative Review

In Exercises 124–127, write the prime factorization of the number.

124. 500

125. 315

126. 792

127. 2275

In Exercises 128–131, perform the multiplication and simplify.

128. $(2x - 5)(x + 7)$

129. $(3x - 2)^2$

130. $(7y + 2)(7y - 2)$

131. $(3y + 8)(9y + 3)$

In Exercises 132–135, factor the trinomial.

132. $x^2 - 4x - 45$

133. $y^2 + 2y - 15$

134. $z^2 + 22z + 40$

135. $x^2 - 12x + 20$

Mid-Chapter Quiz

Take this quiz as you would take a quiz in class. After you are done, check your work against the answers in the back of the book.

In Exercises 1–4, fill in the missing factor.

1. $\frac{2}{3}x - 1 = \frac{1}{3}(\quad)$

2. $x^2y - xy^2 = xy(\quad)$

3. $y^2 + y - 42 = (y + 7)(\quad)$

4. $3y^2 - y - 30 = (3y - 10)(\quad)$

In Exercises 5–16, factor the polynomial completely.

5. $9x^2 + 21$

6. $5a^3b - 25a^2b^2$

7. $x(x + 7) - 6(x + 7)$

8. $t^3 - 3t^2 + t - 3$

9. $y^2 + 11y + 30$

10. $u^2 + u - 56$

11. $x^3 - x^2 - 30x$

12. $2x^2y + 8xy - 64y$

13. $2y^2 - 3y - 27$

14. $6 - 13z - 5z^2$

15. $12x^2 - 5x - 2$

16. $10s^4 - 14s^3 + 2s^2$

17. Find all integers b such that the trinomial

$x^2 + bx + 12$

can be factored. Describe the method you used.

18. Find two integers c such that the trinomial

$x^2 - 10x + c$

can be factored. Describe the method you used. (There are many correct answers.)

19. Find all possible products of the form

$(3x + m)(x + n)$

such that $m \cdot n = 6$. Describe the method you used.

20. The area of the rectangle shown in the figure is $3x^2 + 38x + 80$. What is the area of the shaded region?

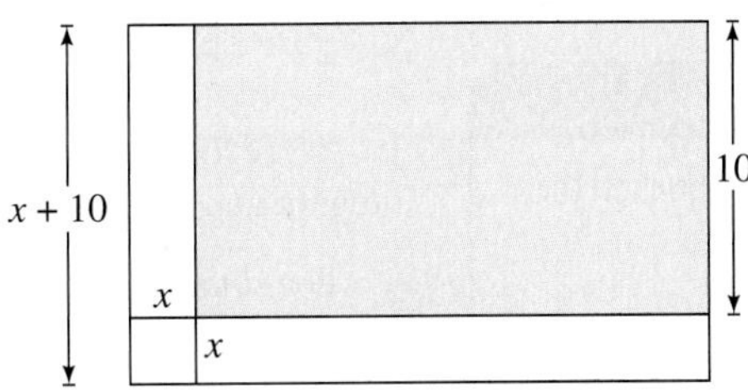

21. Use a graphing calculator to graph $y_1 = -2x^2 + 11x - 12$ and $y_2 = (3 - 2x)(x - 4)$ in the same viewing window. What can you conclude?

6.4 Factoring Polynomials with Special Forms

© Paul Collis/Alamy

What You Should Learn

1 ▶ Factor the difference of two squares.
2 ▶ Recognize repeated factorization.
3 ▶ Identify and factor perfect square trinomials.
4 ▶ Factor the sum or difference of two cubes.

Why You Should Learn It

You can factor polynomials with special forms that model real-life situations. For instance, in Example 12 on page 377, you will factor an expression that models the safe working load for a piano lifted by a rope.

1 ▶ Factor the difference of two squares.

Difference of Two Squares

One of the easiest special polynomial forms to recognize and to factor is the form $a^2 - b^2$. It is called a **difference of two squares** and it factors according to the following pattern.

Difference of Two Squares

Let a and b be real numbers, variables, or algebraic expressions.

$$a^2 - b^2 = (a + b)(a - b)$$

Difference ($a^2 - b^2$); Opposite signs ($(a+b)$ and $(a-b)$)

Technology: Discovery

Use your calculator to verify the special polynomial form called the "difference of two squares." To do so, evaluate the equation when $a = 16$ and $b = 9$. Try more values, including negative values. What can you conclude?

This pattern can be illustrated geometrically, as shown in Figure 6.2. The area of the shaded region on the left is represented by $a^2 - b^2$ (the area of the larger square minus the area of the smaller square). On the right, the *same* area is represented by a rectangle whose width is $a + b$ and whose length is $a - b$.

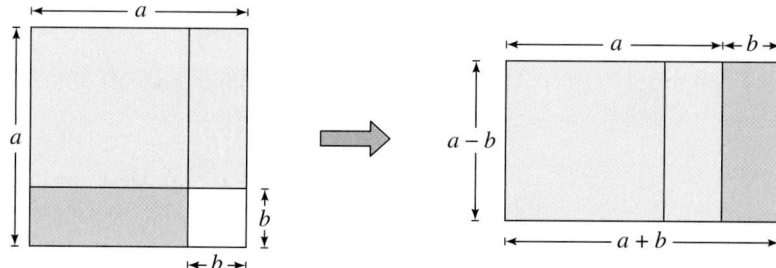

Figure 6.2

Study Tip

Note in the following equation that x-terms of higher powers can be perfect squares.

$$25 - 64x^4 = (5)^2 - (8x^2)^2$$
$$= (5 + 8x^2)(5 - 8x^2)$$

To recognize perfect square terms, look for coefficients that are squares of integers and for variables raised to *even* powers. Here are some examples.

Original Polynomial		*Difference of Squares*		*Factored Form*
$x^2 - 1$	⇨	$(x)^2 - (1)^2$	⇨	$(x + 1)(x - 1)$
$4x^2 - 9$	⇨	$(2x)^2 - (3)^2$	⇨	$(2x + 3)(2x - 3)$

Study Tip

When factoring a polynomial, remember that you can check your result by multiplying the factors. For instance, you can check the factorization in Example 1(a) as follows.

$(x + 6)(x - 6)$

$= x^2 - 6x + 6x - 36$

$= x^2 - 36$

EXAMPLE 1 Factoring the Difference of Two Squares

Factor each polynomial.

a. $x^2 - 36$ **b.** $x^2 - \frac{4}{25}$ **c.** $81x^2 - 49$

Solution

a. $x^2 - 36 = x^2 - 6^2$ Write as difference of two squares.

$= (x + 6)(x - 6)$ Factored form

b. $x^2 - \frac{4}{25} = x^2 - \left(\frac{2}{5}\right)^2$ Write as difference of two squares.

$= \left(x + \frac{2}{5}\right)\left(x - \frac{2}{5}\right)$ Factored form

c. $81x^2 - 49 = (9x)^2 - 7^2$ Write as difference of two squares.

$= (9x + 7)(9x - 7)$ Factored form

Check your results by using the FOIL Method.

CHECKPOINT *Now try Exercise 3.*

The rule $a^2 - b^2 = (a + b)(a - b)$ applies to polynomials or expressions in which a and b are themselves expressions.

EXAMPLE 2 Factoring the Difference of Two Squares

Factor the polynomial $(x + 1)^2 - 4$.

Solution

$(x + 1)^2 - 4 = (x + 1)^2 - 2^2$ Write as difference of two squares.

$= [(x + 1) + 2][(x + 1) - 2]$ Factored form

$= (x + 3)(x - 1)$ Simplify.

Check your result by using the FOIL Method.

CHECKPOINT *Now try Exercise 15.*

Sometimes the difference of two squares can be hidden by the presence of a common monomial factor. Remember that with all factoring techniques, you should first remove any common monomial factors.

EXAMPLE 3 Removing a Common Monomial Factor First

Factor the polynomial $20x^3 - 5x$.

Solution

$20x^3 - 5x = 5x(4x^2 - 1)$ Factor out common monomial factor $5x$.

$= 5x[(2x)^2 - 1^2]$ Write as difference of two squares.

$= 5x(2x + 1)(2x - 1)$ Factored form

CHECKPOINT *Now try Exercise 27.*

2 ▶ Recognize repeated factorization.

Repeated Factorization

To factor a polynomial completely, you should always check to see whether the factors obtained might themselves be factorable. That is, can any of the factors be factored? For instance, after factoring the polynomial $(x^4 - 1)$ once as the difference of two squares

$$x^4 - 1 = (x^2)^2 - 1^2 \quad \text{Write as difference of two squares.}$$

$$= (x^2 + 1)(x^2 - 1) \quad \text{Factored form}$$

you can see that the second factor is itself the difference of two squares. So, to factor the polynomial *completely*, you must continue the factoring process.

$$x^4 - 1 = (x^2 + 1)(x^2 - 1) \quad \text{Factor as difference of two squares.}$$

$$= (x^2 + 1)(x + 1)(x - 1) \quad \text{Factor completely.}$$

Study Tip

Note in Examples 4 and 5 that no attempt is made to factor the *sum of two squares.* A second-degree polynomial that is the sum of two squares cannot be factored as the product of binomials (using integer coefficients). In general, *the sum of two squares is not factorable.*

EXAMPLE 4 Factoring Completely

Factor the polynomial $x^4 - 16$ completely.

Solution

Recognizing $x^4 - 16$ as a difference of two squares, you can write

$$x^4 - 16 = (x^2)^2 - 4^2 \quad \text{Write as difference of two squares.}$$

$$= (x^2 + 4)(x^2 - 4). \quad \text{Factored form}$$

Note that the second factor $(x^2 - 4)$ is itself a difference of two squares, and so

$$x^4 - 16 = (x^2 + 4)(x^2 - 4) \quad \text{Factor as difference of two squares.}$$

$$= (x^2 + 4)(x + 2)(x - 2). \quad \text{Factor completely.}$$

 **CHECKPOINT** *Now try Exercise 29.*

EXAMPLE 5 Factoring Completely

Factor the polynomial $48x^4 - 3$ completely.

Solution

Start by removing the common monomial factor.

$$48x^4 - 3 = 3(16x^4 - 1) \quad \text{Remove common monomial factor 3.}$$

Recognizing $16x^4 - 1$ as the difference of two squares, you can write

$$48x^4 - 3 = 3(16x^4 - 1) \quad \text{Factor out common monomial.}$$

$$= 3[(4x^2)^2 - 1^2] \quad \text{Write as difference of two squares.}$$

$$= 3(4x^2 + 1)(4x^2 - 1) \quad \text{Recognize } 4x^2 - 1 \text{ as a difference of two squares.}$$

$$= 3(4x^2 + 1)[(2x)^2 - 1^2] \quad \text{Write as difference of two squares.}$$

$$= 3(4x^2 + 1)(2x + 1)(2x - 1). \quad \text{Factor completely.}$$

✓ **CHECKPOINT** *Now try Exercise 33.*

3 ▶ Identify and factor perfect square trinomials.

Perfect Square Trinomials

A **perfect square trinomial** is the square of a binomial. For instance,

$$x^2 + 4x + 4 = (x + 2)(x + 2)$$
$$= (x + 2)^2$$

is the square of the binomial $(x + 2)$. Perfect square trinomials come in two forms: one in which the middle term is positive and the other in which the middle term is negative. In both cases, the first and last terms are positive perfect squares.

Perfect Square Trinomials

Let a and b be real numbers, variables, or algebraic expressions.

1. $a^2 + 2ab + b^2 = (a + b)^2$ (Same sign) **2.** $a^2 - 2ab + b^2 = (a - b)^2$ (Same sign)

Study Tip

To recognize a perfect square trinomial, remember that the first and last terms must be perfect squares and positive, and the middle term must be twice the product of a and b. (The middle term can be positive or negative.) Watch for squares of fractions.

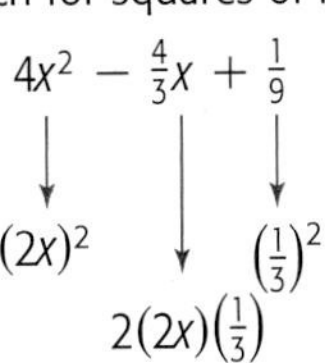

EXAMPLE 6 Identifying Perfect Square Trinomials

Which of the following are perfect square trinomials?

a. $m^2 - 4m + 4$

b. $4x^2 - 2x + 1$

c. $y^2 + 6y - 9$

d. $x^2 + x + \frac{1}{4}$

Solution

a. This polynomial *is* a perfect square trinomial. It factors as $(m - 2)^2$.

b. This polynomial *is not* a perfect square trinomial because the middle term is not twice the product of $2x$ and 1.

c. This polynomial *is not* a perfect square trinomial because the last term, -9, is not positive.

d. This polynomial *is* a perfect square trinomial. It factors as $\left(x + \frac{1}{2}\right)^2$.

✓ ***CHECKPOINT*** *Now try Exercise 39.*

EXAMPLE 7 Factoring a Perfect Square Trinomial

Factor the trinomial $y^2 - 6y + 9$.

Solution

$$y^2 - 6y + 9 = y^2 - 2(3y) + 3^2 \qquad \text{Recognize the pattern.}$$
$$= (y - 3)^2 \qquad \text{Write in factored form.}$$

✓ ***CHECKPOINT*** *Now try Exercise 45.*

EXAMPLE 8 Factoring a Perfect Square Trinomial

Factor the trinomial $16x^2 + 40x + 25$.

Solution

$$16x^2 + 40x + 25 = (4x)^2 + 2(4x)(5) + 5^2 \quad \text{Recognize the pattern.}$$

$$= (4x + 5)^2 \quad \text{Write in factored form.}$$

✓ **CHECKPOINT** *Now try Exercise 51.*

EXAMPLE 9 Factoring a Perfect Square Trinomial

$$9x^2 - 24xy + 16y^2 = (3x)^2 - 2(3x)(4y) + (4y)^2 \quad \text{Recognize the pattern.}$$

$$= (3x - 4y)^2 \quad \text{Write in factored form.}$$

✓ **CHECKPOINT** *Now try Exercise 57.*

4 ▶ Factor the sum or difference of two cubes.

Sum or Difference of Two Cubes

The last type of special factoring presented in this section is the sum or difference of two *cubes*. The patterns for these two special forms are summarized below.

Study Tip

When using either of the factoring patterns at the right, pay special attention to the signs. Remembering the "like" and "unlike" patterns for the signs is helpful.

Sum or Difference of Two Cubes

Let a and b be real numbers, variables, or algebraic expressions.

Like signs

1. $a^3 + b^3 = (a + b)(a^2 - ab + b^2)$

Unlike signs

Like signs

2. $a^3 - b^3 = (a - b)(a^2 + ab + b^2)$

Unlike signs

EXAMPLE 10 Factoring a Sum of Two Cubes

Factor the polynomial $y^3 + 27$.

Solution

$$y^3 + 27 = y^3 + 3^3 \quad \text{Write as sum of two cubes.}$$

$$= (y + 3)[y^2 - (y)(3) + 3^2] \quad \text{Factored form}$$

$$= (y + 3)(y^2 - 3y + 9) \quad \text{Simplify.}$$

✓ **CHECKPOINT** *Now try Exercise 73.*

Study Tip

It is easy to make arithmetic errors when applying the patterns for factoring the sum or difference of two cubes. When you use these patterns, be sure to check your work by multiplying the factors.

EXAMPLE 11 Factoring Differences of Two Cubes

Factor each polynomial.

a. $64 - x^3$ **b.** $2x^3 - 16$

Solution

a. $64 - x^3 = 4^3 - x^3$ Write as difference of two cubes.

$= (4 - x)[4^2 + (4)(x) + x^2]$ Factored form

$= (4 - x)(16 + 4x + x^2)$ Simplify.

b. $2x^3 - 16 = 2(x^3 - 8)$ Factor out common monomial factor 2.

$= 2(x^3 - 2^3)$ Write as difference of two cubes.

$= 2(x - 2)[x^2 + (x)(2) + 2^2]$ Factored form

$= 2(x - 2)(x^2 + 2x + 4)$ Simplify.

✓ ***CHECKPOINT*** *Now try Exercise 77.*

EXAMPLE 12 Safe Working Load

An object lifted with a rope should not weigh more than the safe working load for the rope. To lift a 600-pound piano, the safe working load for a natural fiber rope is given by $150c^2 - 600$, where c is the circumference of the rope (in inches). Factor this expression.

Solution

$150c^2 - 600 = 150(c^2 - 4)$ Factor out common monomial factor.

$= 150(c^2 - 2^2)$ Write as difference of two squares.

$= 150(c + 2)(c - 2)$ Factored form

✓ ***CHECKPOINT*** *Now try Exercise 129.*

The following guidelines are steps for applying the various procedures involved in factoring polynomials.

Guidelines for Factoring Polynomials

1. Factor out any common factors.
2. Factor according to one of the special polynomial forms: difference of two squares, sum or difference of two cubes, or perfect square trinomials.
3. Factor trinomials, $ax^2 + bx + c$, with $a = 1$ or $a \neq 1$.
4. Factor by grouping—for polynomials with four terms.
5. Check to see whether the factors themselves can be factored.
6. Check the results by multiplying the factors.

Concept Check

1. Explain how to identify and factor the difference of two squares.

2. Explain how to identify and factor a perfect square trinomial.

3. What is the factorization of the sum of two cubes, $x^3 + y^3$? What is the factorization of the difference of two cubes, $x^3 - y^3$?

4. In your own words, state the guidelines for factoring polynomials.

6.4 EXERCISES

Go to pages 392–393 to record your assignments.

Developing Skills

In Exercises 1–22, factor the difference of two squares. *See Examples 1 and 2.*

1. $x^2 - 9$
2. $y^2 - 49$
3. $u^2 - 64$
4. $x^2 - 4$
5. $144 - x^2$
6. $81 - x^2$
7. $u^2 - \frac{1}{4}$
8. $t^2 - \frac{1}{16}$
9. $v^2 - \frac{4}{9}$
10. $u^2 - \frac{25}{81}$
11. $16y^2 - 9$
12. $36z^2 - 121$
13. $100 - 49x^2$
14. $16 - 81x^2$
15. $(x - 1)^2 - 4$
16. $(t + 2)^2 - 9$
17. $25 - (z + 5)^2$
18. $(x + 6)^2 - 36$
19. $16 - (a + 2)^2$
20. $(x + 7)^2 - 64$
21. $9y^2 - 25z^2$
22. $100x^2 - 81y^2$

In Exercises 23–36, factor the polynomial completely. *See Examples 3–5.*

23. $2x^2 - 72$
24. $3x^2 - 27$
25. $4x - 25x^3$
26. $a^3 - 16a$
27. $8y^3 - 50y$
28. $20x^3 - 180x$
29. $y^4 - 81$
30. $z^4 - 625$
31. $1 - x^4$
32. $256 - u^4$
33. $2x^4 - 162$
34. $5x^4 - 80$
35. $81x^4 - 16y^4$
36. $81x^4 - z^4$

In Exercises 37–42, determine whether the polynomial is a perfect square trinomial. ***See Example 6.***

37. $9b^2 + 24b + 16$

38. $y^2 - 2y + 6$

✓ **39.** $m^2 - 2m - 1$

40. $16n^2 + 2n + 1$

41. $4k^2 - 20k + 25$

42. $x^2 + 20x + 100$

In Exercises 43–60, factor the perfect square trinomial. ***See Examples 7–9.***

43. $x^2 - 8x + 16$

44. $x^2 + 10x + 25$

✓ **45.** $x^2 + 14x + 49$

46. $a^2 - 12a + 36$

47. $b^2 + b + \frac{1}{4}$

48. $x^2 + \frac{2}{5}x + \frac{1}{25}$

49. $4t^2 + 4t + 1$

50. $9x^2 - 12x + 4$

✓ **51.** $25y^2 - 10y + 1$

52. $16z^2 + 24z + 9$

53. $4x^2 - x + \frac{1}{16}$

54. $4t^2 - \frac{4}{3}t + \frac{1}{9}$

55. $x^2 - 6xy + 9y^2$

56. $16x^2 - 8xy + y^2$

✓ **57.** $4y^2 + 20yz + 25z^2$

58. $36u^2 + 84uv + 49v^2$

59. $9a^2 - 12ab + 4b^2$

60. $49m^2 - 28mn + 4n^2$

Think About It In Exercises 61–66, find two real numbers b such that the expression is a perfect square trinomial.

61. $x^2 + bx + 1$

62. $x^2 + bx + 100$

63. $x^2 + bx + \frac{16}{25}$

64. $y^2 + by + \frac{1}{9}$

65. $4x^2 + bx + 81$

66. $4x^2 + bx + 9$

In Exercises 67–70, find a real number c such that the expression is a perfect square trinomial.

67. $x^2 + 6x + c$

68. $x^2 + 10x + c$

69. $y^2 - 4y + c$

70. $z^2 - 14z + c$

In Exercises 71–82, factor the sum or difference of two cubes. ***See Examples 10 and 11.***

71. $x^3 - 8$

72. $x^3 - 27$

✓ **73.** $y^3 + 64$

74. $z^3 + 125$

75. $1 + 8t^3$

76. $1 + 27s^3$

✓ **77.** $27u^3 - 8$

78. $64v^3 - 125$

79. $4x^3 - 32s^3$

80. $2a^3 - 250b^3$

81. $27x^3 + 64y^3$

82. $27y^3 + 125z^3$

In Exercises 83–124, factor the polynomial completely. (*Note:* Some of the polynomials may be prime.)

83. $4x - 28$

84. $9t + 56$

85. $u^2 + 3u$

86. $x^3 - 4x^2$

87. $5y^2 - 25y$

88. $12a^2 - 24a$

89. $5y^2 - 125$

90. $6x^2 - 54y^2$

91. $y^4 - 25y^2$

92. $x^6 - 49x^4$

93. $x^2 - 4xy + 4y^2$

94. $9y^2 - 6yz + z^2$

95. $x^2 - 2x + 1$

96. $81 + 18x + x^2$

97. $9x^2 + 10x + 1$

98. $8x^2 + 10x + 3$

99. $2x^3 - 2x^2y - 4xy^2$

100. $2y^3 - 7y^2z - 15yz^2$

101. $9t^2 - 16$

102. $25t^2 - 144$

103. $36 - (z + 6)^2$

104. $(t - 10)^2 - 100$

105. $(t - 1)^2 - 121$

106. $(x - 3)^2 - 100$

107. $u^3 + 2u^2 + 3u$

108. $u^3 + 2u^2 - 3u$

109. $x^2 + 81$

110. $x^2 + 16$

111. $2t^3 - 16$

112. $24x^3 - 3$

113. $3a^3 + 24b^3$

114. $54x^3 - 2y^3$

115. $x^4 - 81$

116. $2x^4 - 32$

117. $x^4 - y^4$

118. $81y^4 - z^4$

119. $x^3 - 4x^2 - x + 4$

120. $y^3 + 3y^2 - 4y - 12$

121. $x^4 + 3x^3 - 16x^2 - 48x$

122. $36x + 18x^2 - 4x^3 - 2x^4$

123. $64 - y^6$

124. $1 - y^8$

Mental Math In Exercises 125–128, evaluate the quantity mentally using the two samples as models.

$$29^2 = (30 - 1)^2$$
$$= 30^2 - 2 \cdot 30 \cdot 1 + 1^2$$
$$= 900 - 60 + 1 = 841$$
$$48 \cdot 52 = (50 - 2)(50 + 2)$$
$$= 50^2 - 2^2 = 2496$$

125. 21^2

126. 49^2

127. $59 \cdot 61$

128. $28 \cdot 32$

Solving Problems

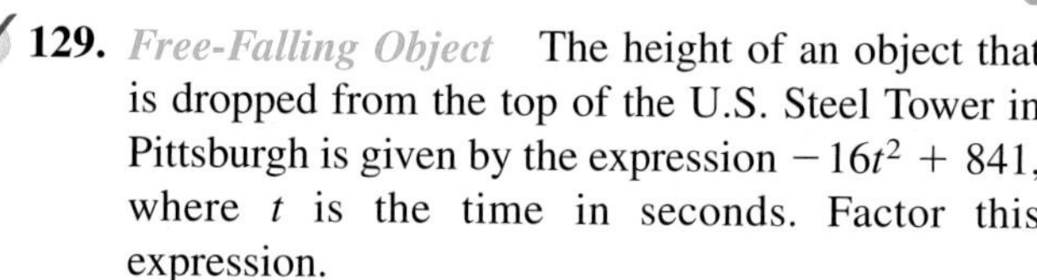

129. *Free-Falling Object* The height of an object that is dropped from the top of the U.S. Steel Tower in Pittsburgh is given by the expression $-16t^2 + 841$, where t is the time in seconds. Factor this expression.

130. *Geometry* An annulus is the region between two concentric circles. The area of the annulus shown in the figure is $\pi R^2 - \pi r^2$. Give the complete factorization of this expression.

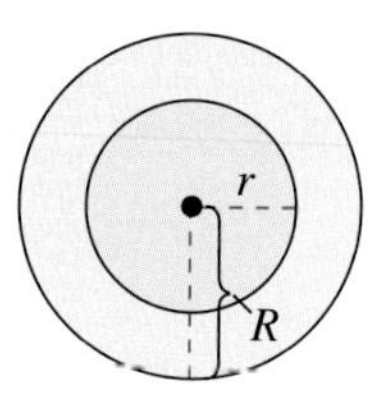

In Exercises 131 and 132, write the polynomial as the difference of two squares. Use the result to factor the polynomial completely.

131. $x^2 + 6x + 8 = (x^2 + 6x + 9) - 1$

$= (\quad)^2 - (\quad)^2$

132. $x^2 + 8x + 12 = (x^2 + 8x + 16) - 4$

$= (\quad)^2 - (\quad)^2$

133. The figure below shows two cubes: a large cube whose volume is a^3 and a smaller cube whose volume is b^3. If the smaller cube is removed from the larger, the remaining solid has a volume of $a^3 - b^3$ and is composed of three rectangular boxes, labeled Box 1, Box 2, and Box 3. Find the volume of each box and describe how these results are related to the following special product pattern.

$$a^3 - b^3 = (a - b)(a^2 + ab + b^2)$$
$$= (a - b)a^2 + (a - b)ab + (a - b)b^2$$

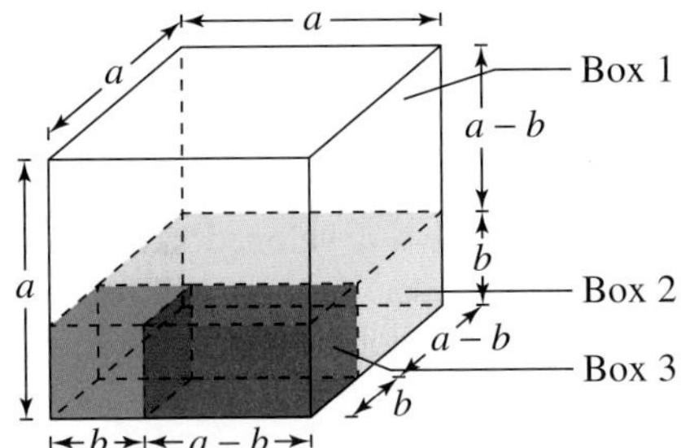

134. *Geometry* From the eight vertices of a cube of dimension x, cubes of dimension y are removed (see figure).

(a) Write an expression for the volume of the solid that remains after the eight cubes at the vertices are removed.

(b) Factor the expression for the volume in part (a).

(c) In the context of this problem, y must be less than what multiple of x? Explain your answer geometrically and from the result of part (b).

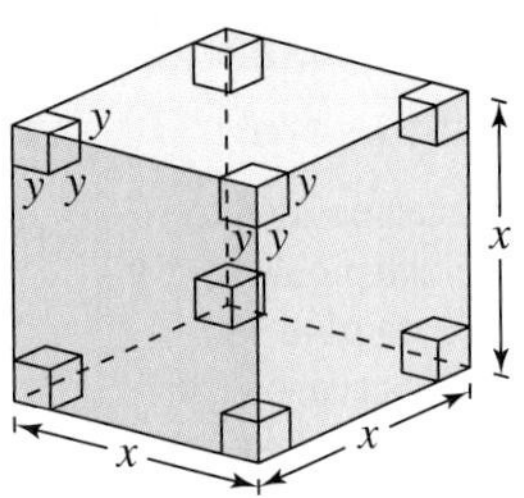

Explaining Concepts

135. Is the expression $x(x + 2) - 2(x + 2)$ completely factored? If not, rewrite it in factored form.

136. Is $x^2 + 4$ equal to $(x + 2)^2$? Explain.

137. Is $x^3 - 27$ equal to $(x - 3)^3$ Explain.

138. *True or False?* Because the sum of two squares cannot be factored, it follows that the sum of two cubes cannot be factored. Justify your answer.

Cumulative Review

In Exercises 139–144, solve the equation and check your solution.

139. $7 + 5x = 7x - 1$

140. $2 - 5(x - 1) = 2[x + 10(x - 1)]$

141. $2(x + 1) = 0$

142. $\frac{3}{4}(12x - 8) = 10$

143. $\frac{x}{5} + \frac{1}{5} = \frac{7}{10}$

144. $\frac{3x}{4} + \frac{1}{2} = 8$

In Exercises 145–148, factor the trinomial.

145. $2x^2 + 7x + 3$

146. $3y^2 - 5y - 12$

147. $6m^2 + 7m - 20$

148. $15x^2 - 28x + 12$

6.5 Solving Polynomial Equations by Factoring

Carol Havens/CORBIS

What You Should Learn

1 ▶ Use the Zero-Factor Property to solve equations.
2 ▶ Solve quadratic equations by factoring.
3 ▶ Solve higher-degree polynomial equations by factoring.
4 ▶ Solve application problems by factoring.

Why You Should Learn It

Quadratic equations can be used to model and solve real-life problems. For instance, Exercise 105 on page 390 shows how a quadratic equation can be used to model the time it takes an object thrown from the Royal Gorge Bridge to reach the ground.

1 ▶ Use the Zero-Factor Property to solve equations.

The Zero-Factor Property

In the first four sections of this chapter, you have developed skills for *rewriting* (simplifying and factoring) polynomials. In this section you will use these skills, together with the **Zero-Factor Property,** to solve polynomial equations.

Zero-Factor Property

Let a and b be real numbers, variables, or algebraic expressions. If a and b are factors such that

$$ab = 0$$

then $a = 0$ or $b = 0$. This property also applies to three or more factors.

Study Tip

The Zero-Factor Property is basically a formal way of saying that the only way the product of two or more factors can be zero is if one (or more) of the factors is zero.

The Zero-Factor Property is the primary property for solving equations in algebra. For instance, to solve the equation

$$(x - 1)(x + 2) = 0 \qquad \text{Original equation}$$

you can use the Zero-Factor Property to conclude that either $(x - 1)$ or $(x + 2)$ must be zero. Setting the first factor equal to zero implies that $x = 1$ is a solution.

$$x - 1 = 0 \implies x = 1 \qquad \text{First solution}$$

Similarly, setting the second factor equal to zero implies that $x = -2$ is a solution.

$$x + 2 = 0 \implies x = -2 \qquad \text{Second solution}$$

So, the equation $(x - 1)(x + 2) = 0$ has exactly two solutions: $x = 1$ and $x = -2$. Check these solutions by substituting them into the original equation.

$$(x - 1)(x + 2) = 0 \qquad \text{Write original equation.}$$

$$(1 - 1)(1 + 2) \stackrel{?}{=} 0 \qquad \text{Substitute 1 for } x.$$

$$(0)(3) = 0 \qquad \text{First solution checks. } \checkmark$$

$$(-2 - 1)(-2 + 2) \stackrel{?}{=} 0 \qquad \text{Substitute } -2 \text{ for } x.$$

$$(-3)(0) = 0 \qquad \text{Second solution checks. } \checkmark$$

2 ▶ Solve quadratic equations by factoring.

Solving Quadratic Equations by Factoring

Definition of Quadratic Equation

A **quadratic equation** is an equation that can be written in the general form

$ax^2 + bx + c = 0$ Quadratic equation

where a, b, and c are real numbers with $a \neq 0$.

Here are some examples of quadratic equations.

$$x^2 - 2x - 3 = 0, \quad 2x^2 + x - 1 = 0, \quad x^2 - 5x = 0$$

In the next four examples, note how you can combine your factoring skills with the Zero-Factor Property to solve quadratic equations.

EXAMPLE 1 Solving a Quadratic Equation by Factoring

Solve $x^2 - x - 6 = 0$.

Solution

First, make sure that the right side of the equation is zero. Next, factor the left side of the equation. Finally, apply the Zero-Factor Property to find the solutions.

$x^2 - x - 6 = 0$ Write original equation.

$(x + 2)(x - 3) = 0$ Factor left side of equation.

$x + 2 = 0 \Rightarrow x = -2$ Set 1st factor equal to 0 and solve for x.

$x - 3 = 0 \Rightarrow x = 3$ Set 2nd factor equal to 0 and solve for x.

The equation has two solutions: $x = -2$ and $x = 3$.

Check

$(-2)^2 - (-2) - 6 \stackrel{?}{=} 0$ Substitute -2 for x in original equation.

$4 + 2 - 6 \stackrel{?}{=} 0$ Simplify.

$0 = 0$ Solution checks. ✓

$(3)^2 - 3 - 6 \stackrel{?}{=} 0$ Substitute 3 for x in original equation.

$9 - 3 - 6 \stackrel{?}{=} 0$ Simplify.

$0 = 0$ Solution checks. ✓

✓ ***CHECKPOINT*** *Now try Exercise 25.*

Study Tip

In Section 3.1, you learned that the basic idea in solving a linear equation is to *isolate the variable.* Notice in Example 1 that the basic idea in solving a quadratic equation is to factor the left side so that the equation can be converted into two linear equations.

Factoring and the Zero-Factor Property allow you to solve a quadratic equation by converting it into two *linear* equations, which you already know how to solve. This is a common strategy of algebra—to break down a given problem into simpler parts, each of which can be solved by previously learned methods.

In order for the Zero-Factor Property to be used, a polynomial equation *must* be written in **general form.** That is, the polynomial must be on one side of the equation and zero must be the only term on the other side of the equation. To write $x^2 - 3x = 10$ in general form, subtract 10 from each side of the equation.

$$x^2 - 3x = 10 \qquad \text{Write original equation.}$$

$$x^2 - 3x - 10 = 10 - 10 \qquad \text{Subtract 10 from each side.}$$

$$x^2 - 3x - 10 = 0 \qquad \text{General form}$$

To solve this equation, factor the left side as $(x - 5)(x + 2)$, then form the linear equations $x - 5 = 0$ and $x + 2 = 0$ to obtain $x = 5$ and $x = -2$, respectively.

Guidelines for Solving Quadratic Equations

1. Write the quadratic equation in general form.
2. Factor the left side of the equation.
3. Set each factor with a variable equal to zero.
4. Solve each linear equation.
5. Check each solution in the original equation.

EXAMPLE 2 Solving a Quadratic Equation by Factoring

Solve $2x^2 + 5x = 12$.

Solution

$$2x^2 + 5x = 12 \qquad \text{Write original equation.}$$

$$2x^2 + 5x - 12 = 0 \qquad \text{Write in general form.}$$

$$(2x - 3)(x + 4) = 0 \qquad \text{Factor left side of equation.}$$

$$2x - 3 = 0 \Rightarrow x = \tfrac{3}{2} \qquad \text{Set 1st factor equal to 0 and solve for } x.$$

$$x + 4 = 0 \Rightarrow x = -4 \qquad \text{Set 2nd factor equal to 0 and solve for } x.$$

The solutions are $x = \frac{3}{2}$ and $x = -4$.

Check

$$2\left(\tfrac{3}{2}\right)^2 + 5\left(\tfrac{3}{2}\right) \stackrel{?}{=} 12 \qquad \text{Substitute } \tfrac{3}{2} \text{ for } x \text{ in original equation.}$$

$$\tfrac{9}{2} + \tfrac{15}{2} \stackrel{?}{=} 12 \qquad \text{Simplify.}$$

$$12 = 12 \qquad \text{Solution checks. ✓}$$

$$2(-4)^2 + 5(-4) \stackrel{?}{=} 12 \qquad \text{Substitute } -4 \text{ for } x \text{ in original equation.}$$

$$32 - 20 \stackrel{?}{=} 12 \qquad \text{Simplify.}$$

$$12 = 12 \qquad \text{Solution checks. ✓}$$

✓ **CHECKPOINT** ***Now try Exercise 29.***

Study Tip

Be sure you see that one side of an equation must be zero to apply the Zero-Factor Property. For instance, in Example 2, you cannot simply factor the left side to obtain $x(2x + 5) = 12$ and assume that $x = 12$ and $2x + 5 = 12$ yield correct solutions. In fact, neither of the resulting solutions satisfies the original equation.

In Examples 1 and 2, the original equations each involved a second-degree (quadratic) polynomial, and each had *two different* solutions. You will sometimes encounter second-degree polynomial equations that have only one (repeated) solution. This occurs when the left side of the general form of the equation is a perfect square trinomial, as shown in Example 3.

Technology: Discovery

Use the general forms of the equations in Examples 3 and 4 to write the following equations in two variables.

$$y = x^2 - 8x + 16$$

$$y = x^2 + 9x + 14$$

Use your graphing calculator to graph each equation and determine the x-intercepts.

How are the x-intercepts of the graphs related to the solutions of the equations in Examples 3 and 4? Explain why this relationship makes sense by comparing the general equations in Examples 3 and 4 with the equations you graphed.

EXAMPLE 3 A Quadratic Equation with a Repeated Solution

Solve $x^2 - 2x + 16 = 6x$.

Solution

$x^2 - 2x + 16 = 6x$	Write original equation.
$x^2 - 8x + 16 = 0$	Write in general form.
$(x - 4)^2 = 0$	Factor.
$x - 4 = 0$ or $x - 4 = 0$	Set factors equal to 0.
$x = 4$	Solve for x.

Note that even though the left side of this equation has two factors, the factors are the same. So, the only solution of the equation is $x = 4$. This solution is called a **repeated solution.**

Check

$x^2 - 2x + 16 = 6x$	Write original equation.
$(4)^2 - 2(4) + 16 \stackrel{?}{=} 6(4)$	Substitute 4 for x.
$16 - 8 + 16 \stackrel{?}{=} 24$	Simplify.
$24 = 24$	Solution checks. ✓

✓ **CHECKPOINT** ***Now try Exercise 37.***

EXAMPLE 4 Solving a Quadratic Equation by Factoring

Solve $(x + 3)(x + 6) = 4$.

Solution

Begin by multiplying the factors on the left side.

$(x + 3)(x + 6) = 4$		Write original equation.
$x^2 + 9x + 18 = 4$		Multiply factors.
$x^2 + 9x + 14 = 0$		Write in general form.
$(x + 2)(x + 7) = 0$		Factor.
$x + 2 = 0$	$\Rightarrow x = -2$	Set 1st factor equal to 0 and solve for x.
$x + 7 = 0$	$\Rightarrow x = -7$	Set 2nd factor equal to 0 and solve for x.

The equation has two solutions: $x = -2$ and $x = -7$. Check these in the original equation.

✓ **CHECKPOINT** ***Now try Exercise 47.***

3 ▶ Solve higher-degree polynomial equations by factoring.

Solving Higher-Degree Equations by Factoring

EXAMPLE 5 Solving a Polynomial Equation with Three Factors

Solve $3x^3 = 15x^2 + 18x$.

Solution

$3x^3 = 15x^2 + 18x$	Write original equation.
$3x^3 - 15x^2 - 18x = 0$	Write in general form.
$3x(x^2 - 5x - 6) = 0$	Factor out common factor.
$3x(x - 6)(x + 1) = 0$	Factor.
$3x = 0 \Rightarrow x = 0$	Set 1st factor equal to 0.
$x - 6 = 0 \Rightarrow x = 6$	Set 2nd factor equal to 0.
$x + 1 = 0 \Rightarrow x = -1$	Set 3rd factor equal to 0.

The solutions are $x = 0$, $x = 6$, and $x = -1$. Check these three solutions.

✓ **CHECKPOINT** *Now try Exercise 65.*

Technology: Discovery

Use a graphing calculator to graph the following second-degree equations, and note the numbers of x-intercepts.

$y = x^2 - 10x + 25$

$y = 5x^2 + 60x + 175$

$y = -2x^2 - 4x - 5$

Use a graphing calculator to graph the following third-degree equations, and note the numbers of x-intercepts.

$y = x^3 - 12x^2 + 48x - 60$

$y = x^3 - 4x$

$y = x^3 + 13x^2 + 55x + 75$

Use your results to write a conjecture about how the degree of a polynomial equation is related to the possible number of solutions.

Notice that the equation in Example 5 is a third-degree equation and has three solutions. This is not a coincidence. In general, a polynomial equation can have *at most* as many solutions as its degree. For instance, a second-degree equation can have zero, one, or two solutions. Notice that the equation in Example 6 is a fourth-degree equation and has four solutions.

EXAMPLE 6 Solving a Polynomial Equation with Four Factors

Solve $x^4 + x^3 - 4x^2 - 4x = 0$.

Solution

$x^4 + x^3 - 4x^2 - 4x = 0$	Write original equation.
$x(x^3 + x^2 - 4x - 4) = 0$	Factor out common factor.
$x[(x^3 + x^2) + (-4x - 4)] = 0$	Group terms.
$x[x^2(x + 1) - 4(x + 1)] = 0$	Factor grouped terms.
$x[(x + 1)(x^2 - 4)] = 0$	Distributive Property
$x(x + 1)(x + 2)(x - 2) = 0$	Difference of two squares
$x = 0 \Rightarrow x = 0$	
$x + 1 = 0 \Rightarrow x = -1$	
$x + 2 = 0 \Rightarrow x = -2$	
$x - 2 = 0 \Rightarrow x = 2$	

The solutions are $x = 0$, $x = -1$, $x = -2$, and $x = 2$. Check these four solutions.

✓ **CHECKPOINT** *Now try Exercise 75.*

4 ▶ Solve application problems by factoring.

Applications

EXAMPLE 7 Geometry: Dimensions of a Room 

A rectangular room has an area of 192 square feet. The length of the room is 4 feet more than its width, as shown in Figure 6.3. Find the dimensions of the room.

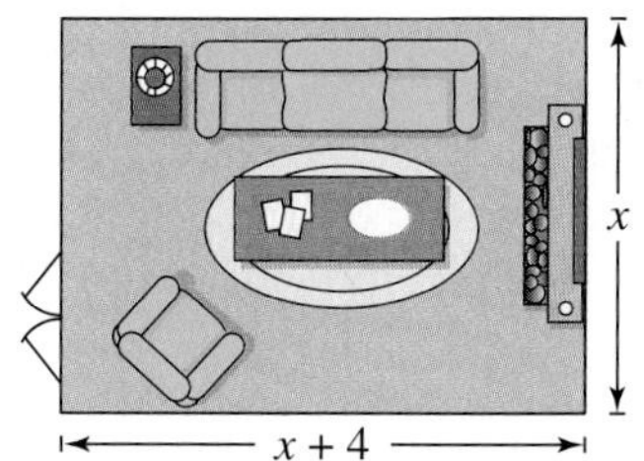

Figure 6.3

Solution

Verbal Model: Length · Width = Area

Labels:
Length $= x + 4$ (feet)
Width $= x$ (feet)
Area $= 192$ (square feet)

Equation:

$$(x + 4)x = 192$$

$$x^2 + 4x - 192 = 0$$

$$(x + 16)(x - 12) = 0$$

$$x = -16 \quad \text{or} \quad x = 12$$

Because the negative solution does not make sense, choose the positive solution $x = 12$. When the width of the room is 12 feet, the length of the room is

$$\text{Length} = x + 4 = 12 + 4 = 16 \text{ feet.}$$

So, the dimensions of the room are 12 feet by 16 feet.

 **CHECKPOINT** *Now try Exercise 97.*

EXAMPLE 8 Free-Falling Object 

A rock is dropped into a well from a height of 64 feet above the water. (See Figure 6.4.) The rock's height (in feet) relative to the water surface is given by the position function $h(t) = -16t^2 + 64$, where t is the time (in seconds) since the rock was dropped. How long does it take for the rock to hit the water?

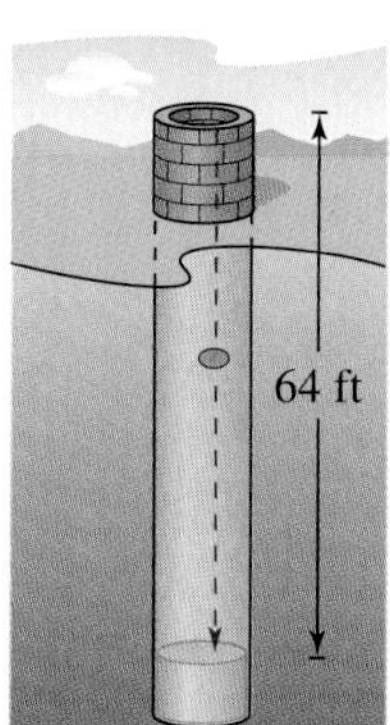

Figure 6.4

Solution

The water surface corresponds to a height of 0 feet. So, substitute a height of 0 for $h(t)$ in the equation, and solve for t.

$0 = -16t^2 + 64$ Substitute 0 for $h(t)$.

$16t^2 - 64 = 0$ Write in general form.

$16(t^2 - 4) = 0$ Factor out common factor.

$16(t + 2)(t - 2) = 0$ Difference of two squares

$t = -2$ or $t = 2$ Solutions using Zero-Factor Property

Because a time of -2 seconds does not make sense, choose the positive solution $t = 2$, and conclude that the rock hits the water 2 seconds after it is dropped.

 **CHECKPOINT** *Now try Exercise 101.*

Concept Check

1. **Fill in the blanks to complete the statement:** In order to apply the Zero-Factor Property to an equation, one side of the equation must consist of a ______ of two or more ______, and the other side must consist of the number ______.

2. *True or False?* If $(2x - 5)(x + 4) = 1$, then $2x - 5 = 1$ or $x + 4 = 1$. Justify your answer.

3. Is it possible for a quadratic equation to have just one solution? Explain.

4. You want to solve an equation of the form $ax^2 + bx + c = d$, where a, b, c, and d are nonzero integers. What step(s) must you perform before you can apply the Zero-Factor Property?

6.5 EXERCISES

Go to pages 392–393 to record your assignments.

Developing Skills

In Exercises 1–12, use the Zero-Factor Property to solve the equation.

1. $x(x - 4) = 0$
2. $z(z + 6) = 0$
3. $(y - 3)(y + 10) = 0$
4. $(s - 7)(s + 4) = 0$
5. $25(a + 4)(a - 2) = 0$
6. $17(t - 3)(t + 8) = 0$
7. $(2t + 5)(3t + 1) = 0$
8. $(5x - 3)(2x - 8) = 0$
9. $4x(2x - 3)(2x + 25) = 0$
10. $\frac{1}{5}x(x - 2)(3x + 4) = 0$
11. $(x - 3)(2x + 1)(x + 4) = 0$
12. $(y - 39)(2y + 7)(y + 12) = 0$

In Exercises 13–78, solve the equation by factoring. *See Examples 1–6.*

13. $5y - y^2 = 0$
14. $3x^2 + 9x = 0$
15. $9x^2 + 15x = 0$
16. $4x^2 - 6x = 0$
17. $2x^2 = 32x$
18. $8x^2 = 5x$
19. $5y^2 = 15y$
20. $5x^2 = 7x$
21. $x^2 - 25 = 0$
22. $x^2 - 121 = 0$
23. $3y^2 - 48 = 0$
24. $5z^2 - 45 = 0$
25. ✓ $x^2 - 3x - 10 = 0$
26. $x^2 - x - 12 = 0$
27. $x^2 - 10x + 24 = 0$
28. $x^2 - 13x + 42 = 0$
29. ✓ $4x^2 + 15x = 25$
30. $14x^2 + 9x = -1$
31. $7 + 13x - 2x^2 = 0$
32. $11 + 32y - 3y^2 = 0$
33. $3y^2 - 2 = -y$
34. $-2x - 15 = -x^2$
35. $-13x + 36 = -x^2$
36. $x^2 - 15 = -2x$
37. ✓ $m^2 - 8m + 18 = 2$
38. $a^2 + 4a + 10 = 6$
39. $x^2 + 16x + 57 = -7$
40. $x^2 - 12x + 21 = -15$
41. $4z^2 - 12z + 15 = 6$
42. $16t^2 + 48t + 40 = 4$
43. $x(x + 2) - 10(x + 2) = 0$
44. $x(x - 15) + 3(x - 15) = 0$
45. $u(u - 3) + 3(u - 3) = 0$
46. $x(x + 10) - 2(x + 10) = 0$
47. ✓ $x(x - 5) = 36$
48. $s(s + 4) = 96$
49. $y(y + 6) = 72$
50. $x(x - 4) = 12$
51. $3t(2t - 3) = 15$
52. $3u(3u + 1) = 20$
53. $(a + 2)(a + 5) = 10$
54. $(x - 8)(x - 7) = 20$
55. $(x - 4)(x + 5) = 10$
56. $(u - 6)(u + 4) = -21$
57. $(t - 2)^2 = 16$
58. $(s + 4)^2 = 49$
59. $9 = (x + 2)^2$
60. $1 = (y + 3)^2$

61. $(x - 3)^2 - 25 = 0$

62. $1 - (x + 1)^2 = 0$

63. $81 - (x + 4)^2 = 0$

64. $(s + 5)^2 - 49 = 0$

✓ 65. $x^3 - 19x^2 + 84x = 0$

66. $x^3 + 18x^2 + 45x = 0$

67. $6t^3 = t^2 + t$

68. $3u^3 = 5u^2 + 2u$

69. $z^2(z + 2) - 4(z + 2) = 0$

70. $16(3 - u) - u^2(3 - u) = 0$

71. $a^3 + 2a^2 - 9a - 18 = 0$

72. $x^3 - 2x^2 - 4x + 8 = 0$

73. $c^3 - 3c^2 - 9c + 27 = 0$

74. $v^3 + 4v^2 - 4v - 16 = 0$

✓ 75. $x^4 - 3x^3 - x^2 + 3x = 0$

76. $x^4 + 2x^3 - 9x^2 - 18x = 0$

77. $8x^4 + 12x^3 - 32x^2 - 48x = 0$

78. $9x^4 - 15x^3 - 9x^2 + 15x = 0$

Graphical Reasoning In Exercises 79–82, determine the x-intercepts of the graph and explain how the x-intercepts correspond to the solutions of the polynomial equation when $y = 0$.

79. $y = x^2 - 9$

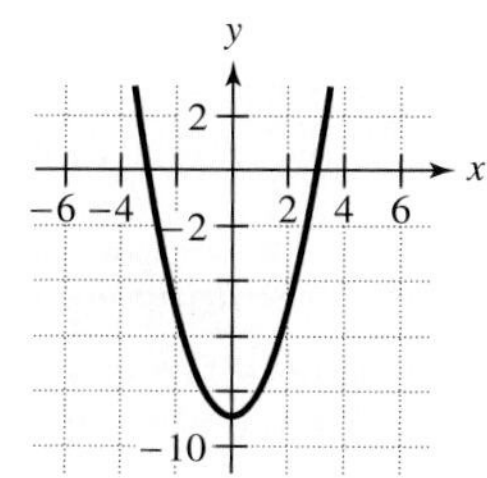

80. $y = x^2 - 4x + 4$

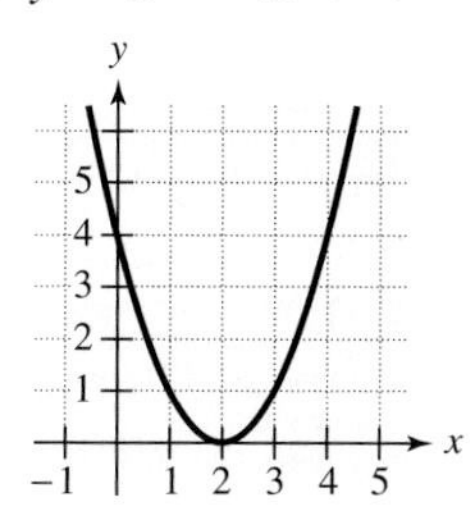

81. $y = x^3 - 6x^2 + 9x$

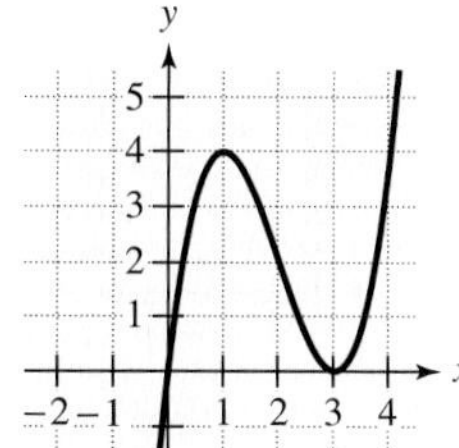

82. $y = x^3 - 3x^2 - x + 3$

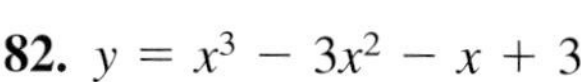

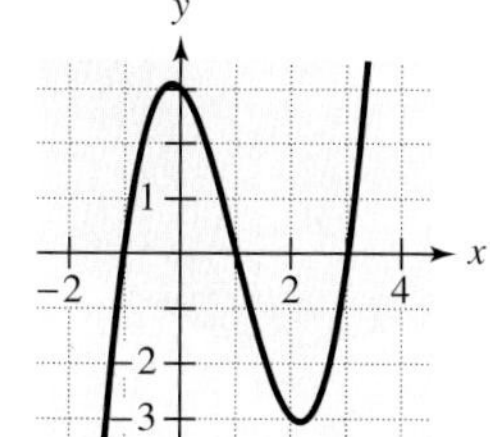

In Exercises 83–90, use a graphing calculator to graph the equation and find any x-intercepts of the graph. Verify algebraically that any x-intercepts are solutions of the polynomial equation when $y = 0$.

83. $y = x^2 + 5x$

84. $y = x^2 - 11x + 28$

85. $y = x^2 - 8x + 12$

86. $y = (x - 2)^2 - 9$

87. $y = 2x^2 + 5x - 12$

88. $y = x^3 - 9x$

89. $y = 2x^3 - 5x^2 - 12x$

90. $y = 2 + x - 2x^2 - x^3$

91. Let a and b be real numbers such that $a \neq 0$. Find the solutions of $ax^2 + bx = 0$.

92. Let a be a nonzero real number. Find the solutions of $ax^2 - ax = 0$.

Solving Problems

Think About It In Exercises 93 and 94, find a quadratic equation with the given solutions.

93. $x = -2, \quad x = 6$

94. $x = -2, \quad x = 4$

95. *Number Problem* The sum of a positive number and its square is 240. Find the number.

96. *Number Problem* Find two consecutive positive integers whose product is 132.

✓ 97. *Geometry* The rectangular floor of a storage shed has an area of 540 square feet. The length of the floor is 7 feet more than its width (see figure on the next page). Find the dimensions of the floor.

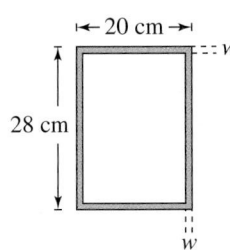

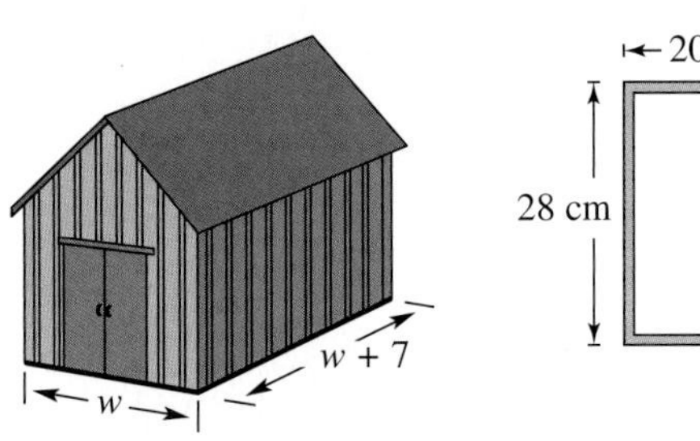

Figure for 97

20 cm
28 cm
w
w

Figure for 98

98. *Geometry* The outside dimensions of a picture frame are 28 centimeters and 20 centimeters (see figure). The area of the exposed part of the picture is 468 square centimeters. Find the width w of the frame.

99. *Geometry* A triangle has an area of 27 square inches. The height of the triangle is $1\frac{1}{2}$ times its base. Find the base and height of the triangle.

100. *Geometry* The height of a triangle is 2 inches less than its base. The area of the triangle is 60 square inches. Find the base and height of the triangle.

✓ **101.** *Free-Falling Object* A hammer is dropped from a construction project 400 feet above the ground. The height h (in feet) of the hammer is modeled by the position equation $h = -16t^2 + 400$, where t is the time in seconds. How long does it take for the hammer to reach the ground?

102. *Free-Falling Object* A penny is dropped from the roof of a building 256 feet above the ground. The height h (in feet) of the penny after t seconds is modeled by the equation $h = -16t^2 + 256$. How long does it take for the penny to reach the ground?

103. *Free-Falling Object* An object falls from the roof of a building 80 feet above the ground toward a balcony 16 feet above the ground. The object's height h (in feet, relative to the ground) after t seconds is modeled by the equation $h = -16t^2 + 80$. How long does it take for the object to reach the balcony?

104. *Free-Falling Object* You throw a baseball upward with an initial velocity of 30 feet per second. The baseball's height h (in feet) relative to your glove after t seconds is modeled by the equation $h = -16t^2 + 30t$. How long does it take for the ball to reach your glove?

105. *Free-Falling Object* An object is thrown upward from the Royal Gorge Bridge in Colorado, 1053 feet above the Arkansas River, with an initial velocity of 48 feet per second. The height h (in feet) of the object is modeled by the position equation $h = -16t^2 + 48t + 1053$, where t is the time measured in seconds. How long does it take for the object to reach the river?

106. *Free-Falling Object* Your friend stands 96 feet above you on a cliff. You throw an object upward with an initial velocity of 80 feet per second. The height h (in feet) of the object after t seconds is modeled by the equation $h = -16t^2 + 80t$. How long does it take for the object to reach your friend on the way up? On the way down?

107. *Break-Even Analysis* The revenue R from the sale of x home theater systems is given by $R = 140x - x^2$. The cost of producing x systems is given by $C = 2000 + 50x$. How many home theater systems must be produced and sold in order to break even?

108. *Break-Even Analysis* The revenue R from the sale of x digital cameras is given by $R = 120x - x^2$. The cost of producing x digital cameras is given by $C = 1200 + 40x$. How many cameras must be produced and sold in order to break even?

109. *Investigation* Solve the equation

$2(x + 3)^2 + (x + 3) - 15 = 0$ in the following two ways.

(a) Let $u = x + 3$, and solve the resulting equation for u. Then find the corresponding values of x that are solutions of the original equation.

(b) Expand and collect like terms in the original equation, and solve the resulting equation for x.

(c) Which method is easier? Explain.

110. *Investigation* Solve each equation using both methods described in Exercise 109.

(a) $3(x + 6)^2 - 10(x + 6) - 8 = 0$

(b) $8(x + 2)^2 - 18(x + 2) + 9 = 0$

111. *Geometry* An open box is to be made from a rectangular piece of material that is 5 meters long and 4 meters wide. The box is made by cutting squares of dimension x from the corners and turning up the sides, as shown in the figure. The volume V of a rectangular solid is the product of its length, width, and height.

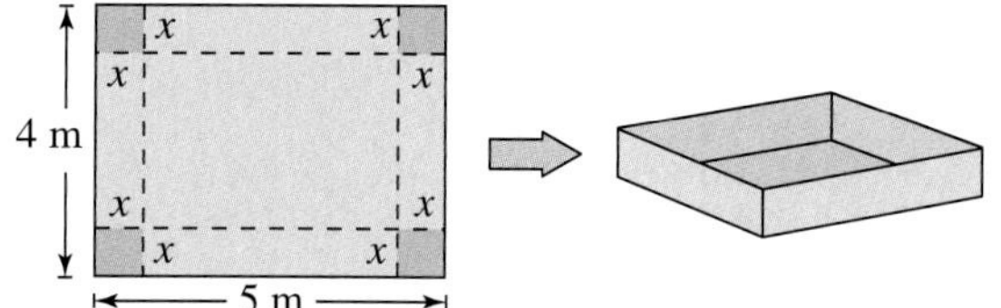

(a) Show algebraically that the volume of the box is given by $V = (5 - 2x)(4 - 2x)x$.

(b) Determine the values of x for which $V = 0$. Determine an appropriate domain for the function V in the context of this problem.

(c) Complete the table.

x	0.25	0.50	0.75	1.00	1.25	1.50	1.75
V							

(d) Use the table to determine x when $V = 3$. Verify the result algebraically.

(e) Use a graphing calculator to graph the volume function. Use the graph to approximate the value of x that yields the box of greatest volume.

112. *Geometry* An open box with a square base stands 5 inches tall. The total surface area of the outside of the box is 525 square inches. What are the dimensions of the base?

Explaining Concepts

113. What is the maximum number of solutions of an nth-degree polynomial equation? Give an example of a third-degree equation that has only one real number solution.

114. What is the maximum number of first-degree factors that an nth-degree polynomial equation can have? Explain.

115. *Think About It* A quadratic equation has a repeated solution. Describe the x-intercept(s) of the graph of the equation formed by replacing 0 with y in the general form of the equation.

116. A third-degree polynomial equation has two solutions. What must be special about one of the solutions? Explain.

117. There are some polynomial equations that have real number solutions but cannot be solved by factoring. Explain how this can be.

118. The polynomial equation $x^3 - x - 3 = 0$ *cannot* be solved algebraically using any of the techniques described in this book. It does, however, have one solution that is a real number.

(a) *Graphical Solution:* Use a graphing calculator to graph the equation and estimate the solution.

(b) *Numerical Solution:* Use the *table* feature of a graphing calculator to create a table and estimate the solution.

Cumulative Review

In Exercises 119–122, find the unit price (in dollars per ounce) of the product.

119. A 12-ounce soda for \$0.75

120. A 12-ounce package of brown-and-serve rolls for \$1.89

121. A 30-ounce can of pumpkin pie filling for \$2.13

122. Turkey meat priced at \$0.94 per pound

In Exercises 123–126, find the domain of the function.

123. $f(x) = \dfrac{x + 3}{x + 1}$

124. $f(x) = \dfrac{12}{x - 2}$

125. $g(x) = \sqrt{3 - x}$

126. $h(x) = \sqrt{x^2 - 4}$

What Did You Learn?

Use these two pages to help prepare for a test on this chapter. Check off the key terms and key concepts you know. You can also use this section to record your assignments.

Plan for Test Success

Date of test: / /

Study dates and times: / / at : A.M./P.M.

/ / at : A.M./P.M.

Things to review:

- ☐ Key Terms, *p. 392*
- ☐ Key Concepts, *pp. 392–393*
- ☐ Your class notes
- ☐ Your assignments
- ☐ Study Tips, *pp. 347, 350, 355, 356, 357, 363, 372, 373, 374, 375, 376, 377, 382, 383, 384*
- ☐ Technology Tips, *p. 364*
- ☐ Mid-Chapter Quiz, *p. 371*
- ☐ Review Exercises, *pp. 394–397*
- ☐ Chapter Test, *p. 398*
- ☐ Video Explanations Online
- ☐ Tutorial Online

Key Terms

- ☐ factoring, *p. 346*
- ☐ greatest common factor, *p. 346*
- ☐ greatest common monomial factor, *p. 347*
- ☐ factoring out, *p. 347*
- ☐ factoring by grouping, *p. 349*
- ☐ prime polynomials, *p. 357*
- ☐ factoring completely, *p. 358*
- ☐ difference of two squares, *p. 372*
- ☐ perfect square trinomial, *p. 375*
- ☐ Zero-Factor Property, *p. 382*
- ☐ quadratic equation, *p. 383*
- ☐ general form of a quadratic equation, *p. 384*
- ☐ repeated solution, *p. 385*

Key Concepts

6.1 Factoring Polynomials with Common Factors

Assignment: ______________________________ Due date: ________

☐ **Factor out common monomial factors.**

Use the Distributive Property to remove the greatest common monomial factor from each term of a polynomial.

☐ **Factor polynomials by grouping.**

For polynomials with four terms, group terms that have a common monomial factor. Factor the two groupings and then look for a common binomial factor.

6.2 Factoring Trinomials

Assignment: ______________________________ Due date: ________

☐ **Use guidelines for factoring $x^2 + bx + c$.**

To factor $x^2 + bx + c$, you need to find two numbers m and n whose product is c and whose sum is b.

$x^2 + bx + c = (x + m)(x + n)$

1. If c is positive, then m and n have like signs that match the sign of b.
2. If c is negative, then m and n have unlike signs.
3. If $|b|$ is small relative to $|c|$, first try those factors of c that are closest to each other in absolute value.

6.3 More About Factoring Trinomials

Assignment:______________________________ Due date:__________

☐ **Use guidelines for factoring $ax^2 + bx + c$ $(a > 0)$.**

1. If the trinomial has a common monomial factor, you should factor out the common factor before trying to find the binomial factors.
2. You do not have to test any binomial factors that have a common monomial factor.
3. Switch the signs of the factors of c when the middle term (O + I) is correct except in sign.

☐ **Use guidelines for factoring $ax^2 + bx + c$ by grouping.**

1. If necessary, write the trinomial in standard form.
2. Choose factors of the product ac that add up to b.
3. Use these factors to rewrite the middle term as a sum or difference.
4. Group and remove any common monomial factors from the first two terms and the last two terms.
5. If possible, factor out the common binomial factor.

6.4 Factoring Polynomials with Special Forms

Assignment:______________________________ Due date:__________

☐ **Factor special products.**

Let a and b be real numbers, variables, or algebraic expressions.

Difference of Two Squares:

$a^2 - b^2 = (a + b)(a - b)$

Perfect Square Trinomials:

$a^2 + 2ab + b^2 = (a + b)^2$

$a^2 - 2ab + b^2 = (a - b)^2$

Sum or Difference of Two Cubes:

$a^3 + b^3 = (a + b)(a^2 - ab + b^2)$

$a^3 - b^3 = (a - b)(a^2 + ab + b^2)$

☐ **Use guidelines for factoring polynomials.**

1. Factor out any common factors.
2. Factor according to one of the special polynomial forms: difference of two squares, sum or difference of two cubes, or perfect square trinomials.
3. Factor trinomials, $ax^2 + bx + c$, with $a = 1$ or $a \neq 1$.
4. Factor by grouping—for polynomials with four terms.
5. Check to see whether the factors themselves can be factored.
6. Check the results by multiplying the factors.

6.5 Solving Polynomial Equations by Factoring

Assignment:______________________________ Due date:__________

☐ **Use the Zero-Factor Property.**

Let a and b be real numbers, variables, or algebraic expressions. If a and b are factors such that $ab = 0$, then $a = 0$ or $b = 0$. This property also applies to three or more factors.

☐ **Use guidelines for solving quadratic equations.**

1. Write the quadratic equation in general form.
2. Factor the left side of the equation.
3. Set each factor with a variable equal to zero.
4. Solve each linear equation.
5. Check each solution in the original equation.

Review Exercises

6.1 Factoring Polynomials with Common Factors

1 ▶ Find the greatest common factor of two or more expressions.

In Exercises 1–8, find the greatest common factor of the expressions.

1. t^2, t^5

2. $-y^3, y^8$

3. $3x^4, 21x^2$

4. $14z^2, 21z$

5. $14x^2y^3, -21x^3y^5$

6. $-15y^2z^2, 5y^2z$

7. $8x^2y, 24xy^2, 4xy$

8. $27ab^5, 9ab^6, 18a^2b^3$

2 ▶ Factor out the greatest common monomial factor from polynomials.

In Exercises 9–22, factor the polynomial.

9. $3x - 6$

10. $7 + 21x$

11. $3t - t^2$

12. $u^2 - 6u$

13. $5x^2 + 10x^3$

14. $7y - 21y^4$

15. $8a^2 - 12a^3$

16. $14x - 26x^4$

17. $5x^3 + 5x^2 - 5x$

18. $6u - 9u^2 + 15u^3$

19. $8y^2 + 4y + 12$

20. $3z^4 - 21z^3 + 10z$

21. $p(p - 4) - 2(p - 4)$

22. $3x(x + 2) + 5(x + 2)$

 *Geometry* **In Exercises 23 and 24, write an expression for the area of the shaded region and factor the expression.**

23.

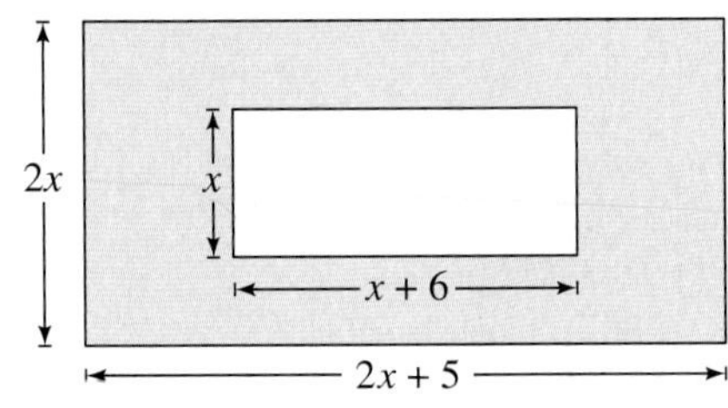

24.

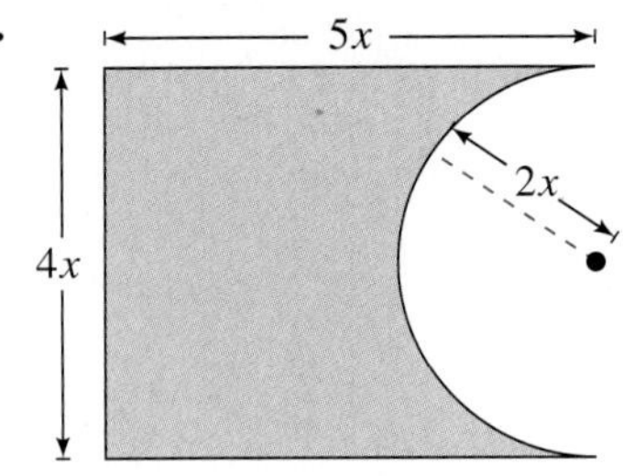

3 ▶ Factor polynomials by grouping.

In Exercises 25–34, factor the polynomial by grouping.

25. $x(x + 1) - 3(x + 1)$

26. $5(y - 3) - y(y - 3)$

27. $2u(u - 2) + 5(u - 2)$

28. $7(x + 8) + 3x(x + 8)$

29. $y^3 + 3y^2 + 2y + 6$

30. $z^3 - 5z^2 + z - 5$

31. $x^3 + 2x^2 + x + 2$

32. $x^3 - 5x^2 + 5x - 25$

33. $x^2 - 4x + 3x - 12$

34. $2x^2 + 6x - 5x - 15$

6.2 Factoring Trinomials

1 ▶ Factor trinomials of the form $x^2 + bx + c$.

In Exercises 35–46, factor the trinomial.

35. $x^2 - 3x - 28$

36. $x^2 - 3x - 40$

37. $u^2 + 5u - 36$

38. $y^2 + 15y + 56$

39. $x^2 - 2x - 24$

40. $x^2 + 8x + 15$

41. $y^2 + 10y + 21$

42. $a^2 - 7a + 12$

43. $b^2 + 13b - 30$

44. $z^2 - 9z + 18$

45. $w^2 + 3w - 40$

46. $x^2 - 7x - 8$

In Exercises 47–50, find all integers *b* such that the trinomial can be factored.

47. $x^2 + bx + 9$

48. $y^2 + by + 25$

49. $z^2 + bz + 11$

50. $x^2 + bx + 14$

2 ▶ Factor trinomials in two variables.

In Exercises 51–56, factor the trinomial.

51. $x^2 + 9xy - 10y^2$

52. $u^2 + uv - 5v^2$

53. $y^2 - 6xy - 27x^2$

54. $v^2 + 18uv + 32u^2$

55. $x^2 - 2xy - 8y^2$

56. $a^2 - ab - 30b^2$

3 ▶ Factor trinomials completely.

In Exercises 57–64, factor the trinomial completely.

57. $4x^2 - 24x + 32$

58. $3u^2 - 6u - 72$

59. $x^3 + 9x^2 + 18x$

60. $y^3 - 8y^2 + 15y$

61. $4x^3 + 36x^2 + 56x$

62. $2y^3 - 4y^2 - 30y$

63. $3x^2 + 18x - 81$

64. $8x^2 - 48x + 64$

6.3 More About Factoring Trinomials

1 ▶ Factor trinomials of the form $ax^2 + bx + c$.

In Exercises 65–78, factor the trinomial.

65. $5 - 2x - 3x^2$

66. $8x^2 - 18x + 9$

67. $50 - 5x - x^2$

68. $7 + 5x - 2x^2$

69. $6x^2 + 7x + 2$

70. $16x^2 + 13x - 3$

71. $4y^2 - 3y - 1$

72. $5x^2 - 12x + 7$

73. $3x^2 + 7x - 6$

74. $45y^2 - 8y - 4$

75. $3x^2 + 5x - 2$

76. $7x^2 - 4x - 3$

77. $2x^2 - 3x + 1$

78. $3x^2 + 8x + 4$

In Exercises 79 and 80, find all integers *b* such that the trinomial can be factored.

79. $x^2 + bx - 24$

80. $2x^2 + bx - 16$

In Exercises 81 and 82, find two integers *c* such that the trinomial can be factored. (There are many correct answers.)

81. $2x^2 - 4x + c$

82. $5x^2 + 6x + c$

2 ▶ Factor trinomials completely.

In Exercises 83–92, factor the trinomial completely.

83. $3x^2 + 33x + 90$

84. $4x^2 + 12x - 16$

85. $6y^2 + 39y - 21$

86. $10b^2 - 38b + 24$

87. $6u^3 + 3u^2 - 30u$

88. $8x^3 - 8x^2 - 30x$

89. $8y^3 - 20y^2 + 12y$

90. $14x^3 + 26x^2 - 4x$

91. $6x^3 + 14x^2 - 12x$

92. $12y^3 + 36y^2 + 15y$

93. *Geometry* The cake box shown in the figure has a height of x and a width of $x + 1$. The volume of the box is $3x^3 + 4x^2 + x$. Find the length l of the box.

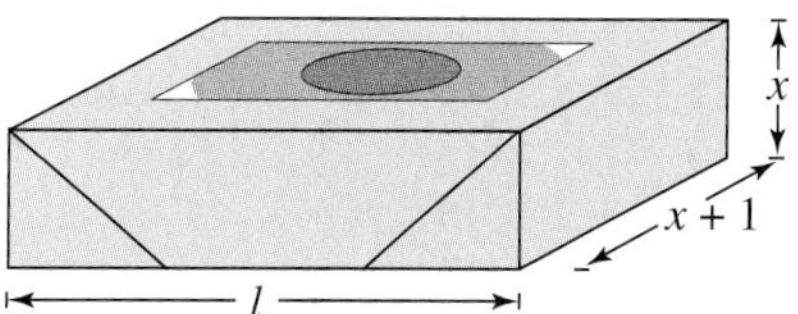

94. *Geometry* The area of the rectangle shown in the figure is $2x^2 + 5x + 3$. What is the area of the shaded region?

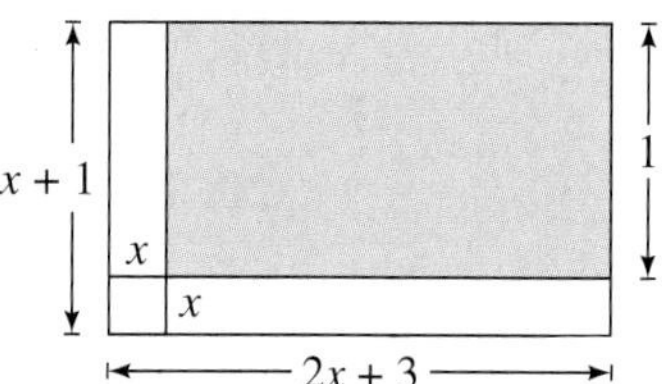

3 ▶ Factor trinomials by grouping.

In Exercises 95–102, factor the trinomial by grouping.

95. $2x^2 - 13x + 21$ **96.** $3a^2 - 13a - 10$

97. $4y^2 + y - 3$ **98.** $6z^2 - 43z + 7$

99. $6x^2 + 11x - 10$ **100.** $21x^2 - 25x - 4$

101. $14x^2 + 17x + 5$ **102.** $5t^2 + 27t - 18$

6.4 Factoring Polynomials with Special Forms

1 ▶ Factor the difference of two squares.

In Exercises 103–112, factor the difference of two squares.

103. $a^2 - 100$ **104.** $36 - b^2$

105. $25 - 4y^2$ **106.** $16b^2 - 1$

107. $12x^2 - 27$ **108.** $100x^2 - 64$

109. $(u + 1)^2 - 4$ **110.** $(y - 2)^2 - 9$

111. $16 - (z - 5)^2$ **112.** $81 - (x + 9)^2$

2 ▶ Recognize repeated factorization.

In Exercises 113 and 114, fill in the missing factors.

113. $x^3 - x = x(\quad)(\quad)$

114. $u^4 - v^4 = (u^2 + v^2)(\quad)(\quad)$

In Exercises 115–122, factor the polynomial completely.

115. $3y^3 - 75y$ **116.** $16b^3 - 36b$

117. $s^3t - st^3$ **118.** $5x^3 - 20xy^2$

119. $x^4 - 81$ **120.** $2a^4 - 32$

121. $x^3 - 2x^2 + 4x - 8$ **122.** $b^3 - 3b^2 + 9b - 27$

3 ▶ Identify and factor perfect square trinomials.

In Exercises 123–130, factor the perfect square trinomial.

123. $x^2 - 8x + 16$ **124.** $y^2 + 24y + 144$

125. $9s^2 + 12s + 4$ **126.** $16x^2 - 40x + 25$

127. $y^2 + 4yz + 4z^2$ **128.** $u^2 - 2uv + v^2$

129. $x^2 + \frac{2}{3}x + \frac{1}{9}$ **130.** $y^2 - \frac{4}{3}y + \frac{4}{9}$

4 ▶ Factor the sum or difference of two cubes.

In Exercises 131–136, factor the sum or difference of two cubes.

131. $a^3 + 1$

132. $z^3 + 8$

133. $27 - 8t^3$

134. $z^3 - 125$

135. $8x^3 + y^3$

136. $125a^3 - 27b^3$

6.5 Solving Polynomial Equations by Factoring

1 ▶ Use the Zero-Factor Property to solve equations.

In Exercises 137–142, use the Zero-Factor Property to solve the equation.

137. $4x(x - 2) = 0$

138. $-3x(2x + 6) = 0$

139. $(2x + 1)(x - 3) = 0$

140. $(x - 7)(3x - 8) = 0$

141. $(x + 10)(4x - 1)(5x + 9) = 0$

142. $3x(x + 8)(2x - 7) = 0$

2 ▶ Solve quadratic equations by factoring.

In Exercises 143–150, solve the quadratic equation by factoring.

143. $3s^2 - 2s - 8 = 0$

144. $5v^2 - 12v - 9 = 0$

145. $m(2m - 1) + 3(2m - 1) = 0$

146. $4w(2w + 8) - 7(2w + 8) = 0$

147. $z(5 - z) + 36 = 0$

148. $(x + 3)^2 - 25 = 0$

149. $v^2 - 100 = 0$

150. $x^2 - 121 = 0$

3 ▶ Solve higher-degree polynomial equations by factoring.

In Exercises 151–158, solve the polynomial equation by factoring.

151. $2y^4 + 2y^3 - 24y^2 = 0$

152. $9x^4 - 15x^3 - 6x^2 = 0$

153. $x^3 - 11x^2 + 18x = 0$

154. $x^3 + 20x^2 + 36x = 0$

155. $b^3 - 6b^2 - b + 6 = 0$

156. $q^3 + 3q^2 - 4q - 12 = 0$

157. $x^4 - 5x^3 - 9x^2 + 45x = 0$

158. $2x^4 + 6x^3 - 50x^2 - 150x = 0$

4 ▶ Solve application problems by factoring.

159. *Number Problem* Find two consecutive positive odd integers whose product is 99.

160. *Number Problem* Find two consecutive positive even integers whose product is 168.

161. *Geometry* A rectangle has an area of 900 square inches. The length of the rectangle is $2\frac{1}{4}$ times its width. Find the dimensions of the rectangle.

162. *Geometry* A rectangle has an area of 432 square inches. The width of the rectangle is $\frac{3}{4}$ times its length. Find the dimensions of the rectangle.

163. *Geometry* A closed box with a square base stands 12 inches tall. The total surface area of the outside of the box is 512 square inches. What are the dimensions of the base? (*Hint:* The surface area is given by $S = 2x^2 + 4xh$.)

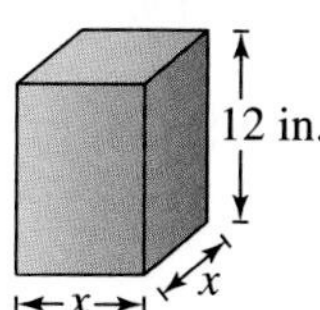

164. *Geometry* An open box with a square base stands 10 inches tall. The total surface area of the outside of the box is 225 square inches. What are the dimensions of the base? (*Hint:* The surface area is given by $S = x^2 + 4xh$.)

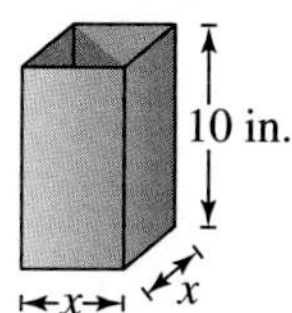

165. *Free-Falling Object* An object is dropped from a weather balloon 3600 feet above the ground. The height h (in feet) of the object is modeled by the position equation $h = -16t^2 + 3600$, where t is the time (in seconds). How long will it take for the object to reach the ground?

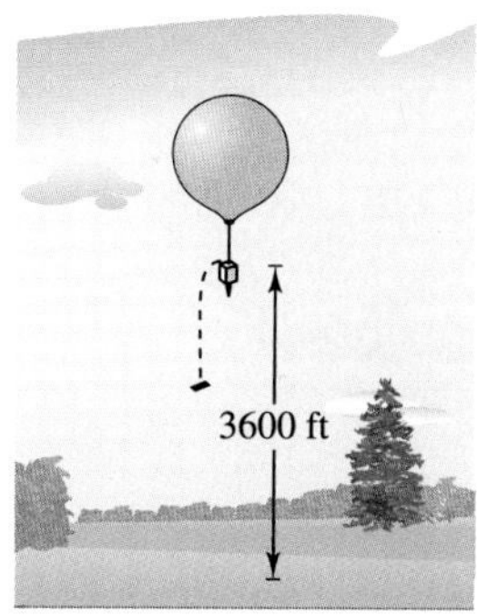

166. *Free-Falling Object* An object is thrown upward from the Trump Tower in New York City, which is 664 feet tall, with an initial velocity of 45 feet per second. The height h (in feet) of the object is modeled by the position equation $h = -16t^2 + 45t + 664$, where t is the time (in seconds). How long will it take for the object to reach the ground?

Chapter Test

Take this test as you would take a test in class. After you are done, check your work against the answers in the back of the book.

In Exercises 1–10, factor the polynomial completely.

1. $9x^2 - 63x^5$

2. $z(z + 17) - 10(z + 17)$

3. $t^2 - 2t - 80$

4. $6x^2 - 11x + 4$

5. $3y^3 + 72y^2 - 75y$

6. $4 - 25v^2$

7. $x^3 + 8$

8. $100 - (z + 11)^2$

9. $x^3 + 2x^2 - 9x - 18$

10. $16 - z^4$

11. Fill in the missing factor: $\frac{2}{5}x - \frac{3}{5} = \frac{1}{5}(\quad)$.

12. Find all integers b such that $x^2 + bx + 5$ can be factored.

13. Find a real number c such that $x^2 + 12x + c$ is a perfect square trinomial.

14. Explain why $(x + 1)(3x - 6)$ is not a complete factorization of $3x^2 - 3x - 6$.

In Exercises 15–20, solve the equation.

15. $(x + 4)(2x - 3) = 0$

16. $3x^2 + 7x - 6 = 0$

17. $y(2y - 1) = 6$

18. $2x^2 - 3x = 8 + 3x$

19. $2x^3 - 8x^2 - 24x = 0$

20. $y^4 + 7y^3 - 3y^2 - 21y = 0$

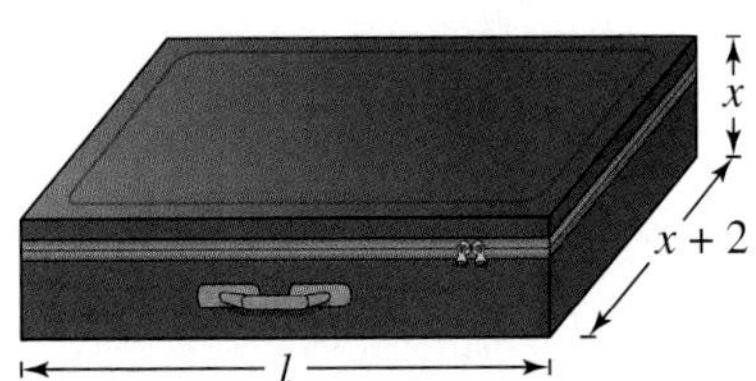

Figure for 21

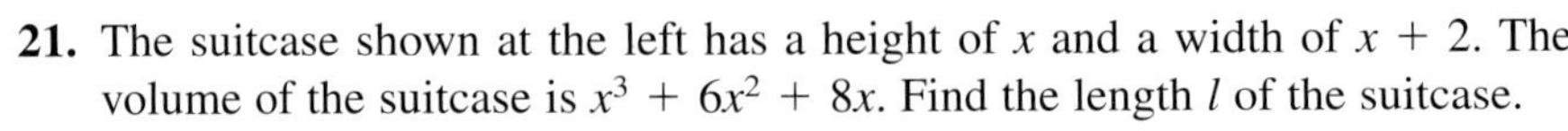

21. The suitcase shown at the left has a height of x and a width of $x + 2$. The volume of the suitcase is $x^3 + 6x^2 + 8x$. Find the length l of the suitcase.

22. The width of a rectangle is 5 inches less than its length. The area of the rectangle is 84 square inches. Find the dimensions of the rectangle.

23. An object is thrown upward from the top of the Aon Center in Chicago at a height of 1136 feet, with an initial velocity of 14 feet per second. The height h (in feet) of the object is modeled by the position equation

$$h = -16t^2 + 14t + 1136$$

where t is the time measured in seconds. How long will it take for the object to reach the ground? How long will it take the object to fall to a height of 806 feet?

24. Find two consecutive positive even integers whose product is 624.

25. The perimeter of a rectangular storage lot at a car dealership is 800 feet. The lot is surrounded by fencing that costs \$15 per foot for the front side and \$10 per foot for the remaining three sides. The total cost of the fencing is \$9500. Find the dimensions of the storage lot.

Cumulative Test: Chapters 4–6

Take this test as you would take a test in class. After you are done, check your work against the answers in the back of the book.

1. Describe how to identify the quadrants in which the points $(-2, y)$ must be located. (y is a real number.)

2. Determine whether the ordered pairs are solution points of the equation $9x - 4y + 36 = 0$.

(a) $(-1, -1)$ (b) $(8, 27)$ (c) $(-4, 0)$ (d) $(3, -2)$

In Exercises 3 and 4, sketch the graph of the equation and determine any intercepts of the graph.

3. $y = 3 + |x|$ **4.** $x + 2y = 6$

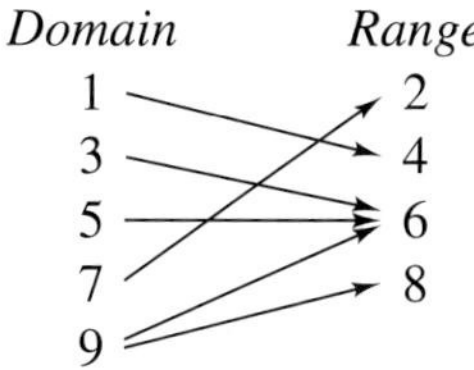

Figure for 5

5. Determine whether the relation at the left represents a function.

6. The slope of a line is $-\frac{1}{4}$ and a point on the line is $(2, 1)$. Find the coordinates of a second point on the line. Explain why there are many correct answers.

7. Write the slope-intercept form of the equation of the line that passes through the point $\left(0, -\frac{3}{2}\right)$ and has slope $m = \frac{2}{5}$.

In Exercises 8 and 9, sketch the lines and determine whether they are parallel, perpendicular, or neither.

8. $y_1 = \frac{2}{3}x - 3, y_2 = -\frac{3}{2}x + 1$ **9.** $y_1 = 2 - 0.4x, y_2 = -\frac{2}{5}x$

10. Subtract: $(x^3 - 3x^2) - (x^3 + 2x^2 - 5)$.

11. Multiply: $(6z)(-7z)(z^2)$.

12. Multiply: $(3x + 5)(x - 4)$. **13.** Multiply: $(5x - 3)(5x + 3)$.

14. Expand: $(5x + 6)^2$. **15.** Divide: $(6x^2 + 72x) \div 6x$.

16. Divide: $\dfrac{x^2 - 3x - 2}{x - 4}$. **17.** Simplify: $\dfrac{(3xy^2)^{-2}}{6x^{-3}}$.

18. Factor: $2u^2 - 6u$. **19.** Factor and simplify: $(x - 4)^2 - 36$.

20. Factor completely: $x^3 + 8x^2 + 16x$.

21. Factor completely: $x^3 + 2x^2 - 4x - 8$.

22. Solve: $u(u - 12) = 0$. **23.** Solve: $5x^2 - 12x - 9 = 0$.

24. Rewrite the expression $\left(\dfrac{x}{2}\right)^{-2}$ using only positive exponents and simplify.

25. A sales representative is reimbursed \$150 per day for lodging and meals, plus \$0.45 per mile driven. Write a linear equation giving the daily cost C to the company in terms of x, the number of miles driven. Explain the reasoning you used in writing the model. Find the cost for a day when the representative drives 70 miles.

26. The cost of operating a pizza delivery car is \$0.70 per mile after an initial investment of \$9000. What mileage on the car will keep the cost at or below \$36,400?

Study Skills in Action

Using a Test-Taking Strategy

What do runners do before a race? They design a strategy for running their best. They make sure they get enough rest, eat sensibly, and get to the track early to warm up. In the same way, it is important for students to get a good night's sleep, eat a healthy meal, and get to class early to allow time to focus before a test.

The biggest difference between a runner's race and a math test is that a math student does not have to reach the finish line first! In fact, many students would increase their scores if they used all the test time instead of worrying about being the last student left in the class. This is why it is important to have a strategy for taking the test.

Kimberly Nolting
VP, Academic Success Press
expert in developmental education

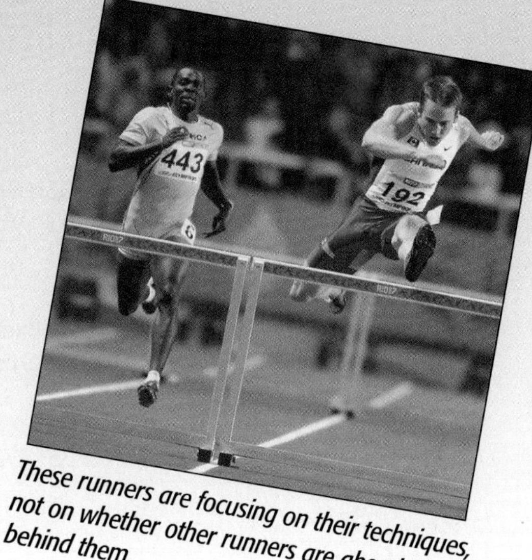

These runners are focusing on their techniques, not on whether other runners are ahead of or behind them.

Smart Study Strategy

Use Ten Steps for Test-Taking

1 ► **Do a memory data dump.** As soon as you get the test, turn it over and write down anything that you still have trouble remembering sometimes (formulas, calculations, rules).

2 ► **Preview the test.** Look over the test and mark the questions you know how to do easily. These are the problems you should do first.

3 ► **Do a second memory data dump.** As you previewed the test, you may have remembered other information. Write this information on the back of the test.

4 ► **Develop a test progress schedule.** Based on how many points each question is worth, decide on a progress schedule. You should always have more than half the test done before half the time has elapsed.

5 ► **Answer the easiest problems first.** Solve the problems you marked while previewing the test.

6 ► **Skip difficult problems.** Skip the problems that you suspect will give you trouble.

7 ► **Review the skipped problems.** After solving all the problems that you know how to do easily, go back and reread the problems you skipped.

8 ► **Try your best at the remaining problems that confuse you.** Even if you cannot completely solve a problem, you may be able to get partial credit for a few correct steps.

9 ► **Review the test.** Look for any careless errors you may have made.

10 ► **Use all the allowed test time.** The test is not a race against the other students.

Chapter 7

Rational Expressions, Equations, and Functions

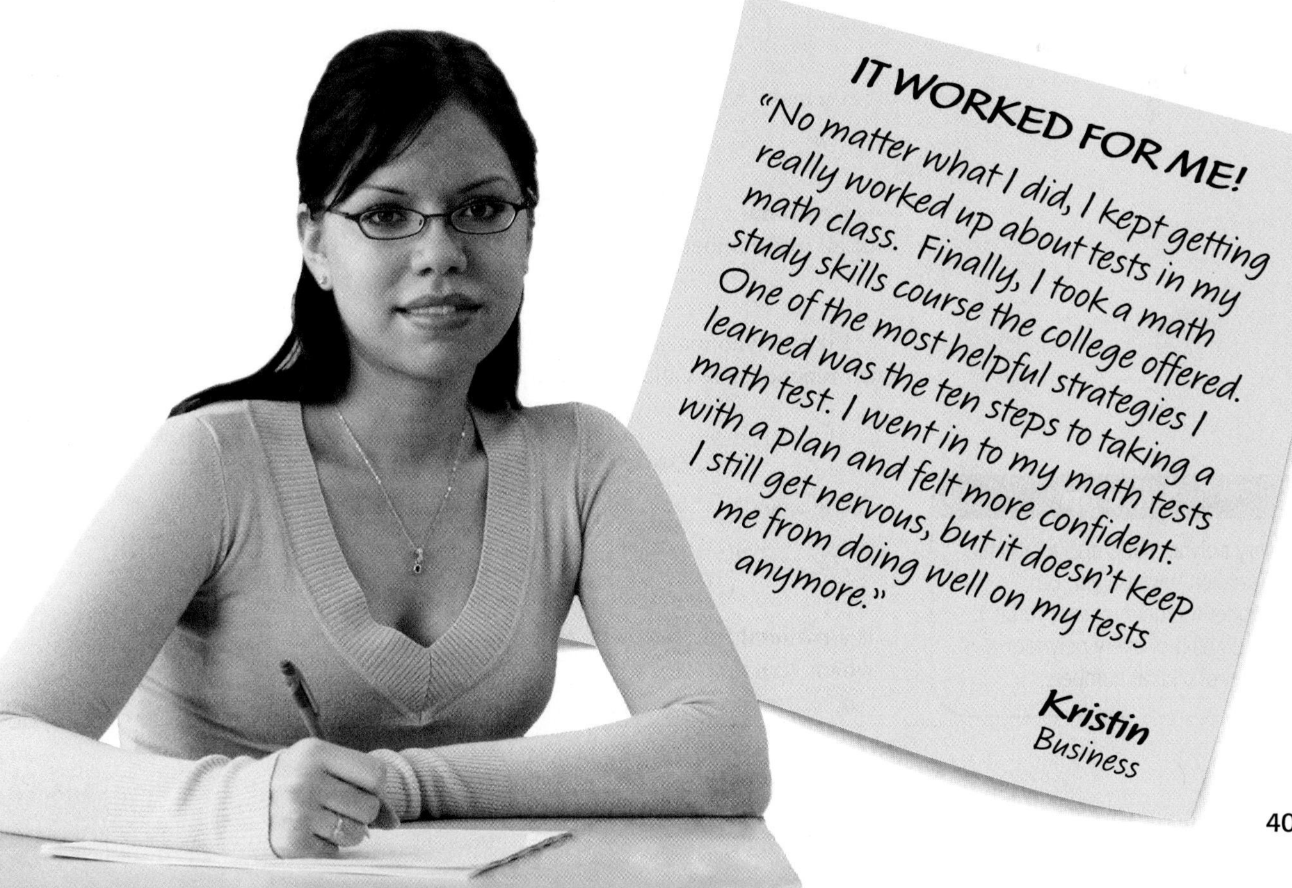

IT WORKED FOR ME!
"No matter what I did, I kept getting really worked up about tests in my math class. Finally, I took a math study skills course the college offered. One of the most helpful strategies I learned was the ten steps to taking a math test. I went in to my math tests with a plan and felt more confident. I still get nervous, but it doesn't keep me from doing well on my tests anymore."
Kristin
Business

7.1 Rational Expressions and Functions

Paul Barton/CORBIS

Why You Should Learn It

Rational expressions can be used to solve real-life problems. For instance, in Exercise 93 on page 413, you will find a rational expression that models the average cable television revenue per subscriber.

What You Should Learn

1 ▶ Find the domain of a rational function.

2 ▶ Simplify rational expressions.

1 ▶ Find the domain of a rational function.

The Domain of a Rational Function

A fraction whose numerator and denominator are polynomials is called a **rational expression.** Some examples are

$$\frac{3}{x+4}, \quad \frac{2x}{x^2-4x+4}, \quad \text{and} \quad \frac{x^2-5x}{x^2+2x-3}.$$

In Section 1.3, you learned that because division by zero is undefined, the denominator of a rational expression cannot be zero. So, in your work with rational expressions, you must assume that all real number values of the variable that make the denominator zero are excluded. For the three fractions above, $x = -4$ is excluded from the first fraction, $x = 2$ from the second, and both $x = 1$ and $x = -3$ from the third. The set of *usable* values of the variable is called the **domain** of the rational expression.

Definition of a Rational Expression

Let u and v be polynomials. The algebraic expression

$$\frac{u}{v}$$

is a **rational expression.** The **domain** of this rational expression is the set of all real numbers for which $v \neq 0$.

Like polynomials, rational expressions can be used to describe functions. Such functions are called **rational functions.**

Study Tip

Every polynomial is also a rational expression because you can consider the denominator to be 1. The domain of every polynomial is the set of all real numbers.

Definition of a Rational Function

Let $u(x)$ and $v(x)$ be polynomial functions. The function

$$f(x) = \frac{u(x)}{v(x)}$$

is a **rational function.** The **domain** of f is the set of all real numbers for which $v(x) \neq 0$.

EXAMPLE 1 Finding the Domains of Rational Functions

Find the domain of each rational function.

a. $f(x) = \dfrac{4}{x - 2}$ **b.** $g(x) = \dfrac{2x + 5}{8}$

Solution

a. The denominator is zero when $x - 2 = 0$ or $x = 2$. So, the domain is all real values of x such that $x \neq 2$. In interval notation, you can write the domain as

$$\text{Domain} = (-\infty, 2) \cup (2, \infty).$$

b. The denominator, 8, is never zero, so the domain is the set of *all* real numbers. In interval notation, you can write the domain as

$$\text{Domain} = (-\infty, \infty).$$

 ***CHECKPOINT*** *Now try Exercise 3.*

Technology: Discovery

Use a graphing calculator to graph the equation that corresponds to part (a) of Example 1. Then use the *trace* or *table* feature of the calculator to determine the behavior of the graph near $x = 2$. Graph the equation that corresponds to part (b) of Example 1. How does this graph differ from the graph in part (a)?

EXAMPLE 2 Finding the Domains of Rational Functions

Find the domain of each rational function.

a. $f(x) = \dfrac{5x}{x^2 - 16}$ **b.** $h(x) = \dfrac{3x - 1}{x^2 - 2x - 3}$

Study Tip

Remember that when interval notation is used, the symbol $\cup$ means *union* and the symbol $\cap$ means *intersection*.

Solution

a. The denominator is zero when $x^2 - 16 = 0$. Solving this equation by factoring, you find that the denominator is zero when $x = -4$ or $x = 4$. So, the domain is all real values of x such that $x \neq -4$ and $x \neq 4$. In interval notation, you can write the domain as

$$\text{Domain} = (-\infty, -4) \cup (-4, 4) \cup (4, \infty).$$

b. The denominator is zero when $x^2 - 2x - 3 = 0$. Solving this equation by factoring, you find that the denominator is zero when $x = 3$ or when $x = -1$. So, the domain is all real values of x such that $x \neq 3$ and $x \neq -1$. In interval notation, you can write the domain as

$$\text{Domain} = (-\infty, -1) \cup (-1, 3) \cup (3, \infty).$$

CHECKPOINT *Now try Exercise 15.*

Study Tip

When a rational function is written, it is understood that the real numbers that make the denominator zero are excluded from the domain. These *implied* domain restrictions are generally not listed with the function. For instance, you know to exclude $x = 2$ and $x = -2$ from the function

$$f(x) = \frac{3x + 2}{x^2 - 4}$$

without having to list this information with the function.

In applications involving rational functions, it is often necessary to place restrictions on the domain other than the restrictions *implied* by values that make the denominator zero. Such additional restrictions can be indicated to the right of the function. For instance, the domain of the rational function

$$f(x) = \frac{x^2 + 20}{x + 4}, \qquad x > 0$$

is the set of *positive* real numbers, as indicated by the inequality $x > 0$. Note that the normal domain of this function would be all real values of x such that $x \neq -4$. However, because "$x > 0$" is listed to the right of the function, the domain is further restricted by this inequality.

EXAMPLE 3 An Application Involving a Restricted Domain 

You have started a small business that manufactures lamps. The initial investment for the business is \$120,000. The cost of manufacturing each lamp is \$15. So, your total cost of producing x lamps is

$$C = 15x + 120{,}000. \qquad \text{Cost function}$$

Your average cost per lamp depends on the number of lamps produced. For instance, the average cost per lamp $\overline{C}$ of producing 100 lamps is

$$\overline{C} = \frac{15(100) + 120{,}000}{100} \qquad \text{Substitute 100 for } x.$$

$$= \$1215. \qquad \text{Average cost per lamp for 100 lamps}$$

The average cost per lamp decreases as the number of lamps increases. For instance, the average cost per lamp $\overline{C}$ of producing 1000 lamps is

$$\overline{C} = \frac{15(1000) + 120{,}000}{1000} \qquad \text{Substitute 1000 for } x.$$

$$= \$135. \qquad \text{Average cost per lamp for 1000 lamps}$$

In general, the average cost of producing x lamps is

$$\overline{C} = \frac{15x + 120{,}000}{x}. \qquad \text{Average cost per lamp for } x \text{ lamps}$$

What is the domain of this rational function?

Solution

If you were considering this function from only a mathematical point of view, you would say that the domain is all real values of x such that $x \neq 0$. However, because this function is a mathematical model representing a real-life situation, you must decide which values of x make sense in real life. For this model, the variable x represents the number of lamps that you produce. Assuming that you cannot produce a fractional number of lamps, you can conclude that the domain is the set of positive integers—that is,

$$\text{Domain} = \{1, 2, 3, 4, \ldots\}.$$

CHECKPOINT *Now try Exercise 31.*

2 ▶ Simplify rational expressions.

Simplifying Rational Expressions

As with numerical fractions, a rational expression is said to be in **simplified** (or **reduced**) **form** if its numerator and denominator have no common factors (other than ± 1). To simplify rational expressions, you can apply the rule below.

Simplifying Rational Expressions

Let u, v, and w represent real numbers, variables, or algebraic expressions such that $v \neq 0$ and $w \neq 0$. Then the following is valid.

$$\frac{uw}{vw} = \frac{u\cancel{w}}{v\cancel{w}} = \frac{u}{v}$$

Be sure you divide out only *factors*, not *terms*. For instance, consider the expressions below.

$$\frac{\cancel{2} \cdot 2}{\cancel{2}(x + 5)}$$ You *can* divide out the common factor 2.

$$\frac{3 + x}{3 + 2x}$$ You *cannot* divide out the common term 3.

Simplifying a rational expression requires two steps: (1) completely factor the numerator and denominator and (2) divide out any *factors* that are common to both the numerator and denominator. So, your success in simplifying rational expressions actually lies in your ability to *factor completely* the polynomials in both the numerator and denominator.

EXAMPLE 4 Simplifying a Rational Expression

Simplify the rational expression $\frac{2x^3 - 6x}{6x^2}$.

Solution

First note that the domain of the rational expression is all real values of x such that $x \neq 0$. Then, completely factor both the numerator and denominator.

$$\frac{2x^3 - 6x}{6x^2} = \frac{2x(x^2 - 3)}{2x(3x)}$$ Factor numerator and denominator.

$$= \frac{\cancel{2x}(x^2 - 3)}{\cancel{2x}(3x)}$$ Divide out common factor $2x$.

$$= \frac{x^2 - 3}{3x}$$ Simplified form

In simplified form, the domain of the rational expression is the same as that of the original expression—all real values of x such that $x \neq 0$.

CHECKPOINT *Now try Exercise 43.*

Technology: Tip

Use the *table* feature of a graphing calculator to compare the original and simplified forms of the expression in Example 5.

$$y_1 = \frac{x^2 + 2x - 15}{3x - 9}$$

$$y_2 = \frac{x + 5}{3}$$

Set the increment value of the table to 1 and compare the values at $x = 0, 1, 2, 3, 4$, and 5. Next set the increment value to 0.1 and compare the values at $x = 2.8, 2.9, 3.0, 3.1$, and 3.2. From the table you can see that the functions differ only at $x = 3$. This shows why $x \neq 3$ must be written as part of the simplified form of the original expression.

EXAMPLE 5 Simplifying a Rational Expression

Simplify the rational expression $\dfrac{x^2 + 2x - 15}{3x - 9}$.

Solution

The domain of the rational expression is all real values of x such that $x \neq 3$.

$$\frac{x^2 + 2x - 15}{3x - 9} = \frac{(x + 5)(x - 3)}{3(x - 3)} \qquad \text{Factor numerator and denominator.}$$

$$= \frac{(x + 5)\cancel{(x - 3)}}{3\cancel{(x - 3)}} \qquad \text{Divide out common factor } (x - 3).$$

$$= \frac{x + 5}{3}, \ x \neq 3 \qquad \text{Simplified form}$$

✓ **CHECKPOINT** *Now try Exercise 51.*

Dividing out common factors from the numerator and denominator of a rational expression can change the implied domain. For instance, in Example 5 the domain restriction $x \neq 3$ must be listed because it is no longer implied in the simplified expression. With this restriction, the new expression is equal to the original expression.

EXAMPLE 6 Simplifying a Rational Expression

Simplify the rational expression $\dfrac{x^3 - 16x}{x^2 - 2x - 8}$.

Solution

The domain of the rational expression is all real values of x such that $x \neq -2$ and $x \neq 4$.

$$\frac{x^3 - 16x}{x^2 - 2x - 8} = \frac{x(x^2 - 16)}{(x + 2)(x - 4)} \qquad \text{Partially factor.}$$

$$= \frac{x(x + 4)(x - 4)}{(x + 2)(x - 4)} \qquad \text{Factor completely.}$$

$$= \frac{x(x + 4)\cancel{(x - 4)}}{(x + 2)\cancel{(x - 4)}} \qquad \text{Divide out common factor } (x - 4).$$

$$= \frac{x(x + 4)}{x + 2}, \ x \neq 4 \qquad \text{Simplified form}$$

✓ **CHECKPOINT** *Now try Exercise 61.*

When you simplify a rational expression, keep in mind that you must list any domain restrictions that are no longer implied in the simplified expression. For instance, in Example 6 the restriction $x \neq 4$ is listed so that the domains agree for the original and simplified expressions. The example does not list $x \neq -2$ because this restriction is apparent by looking at either expression.

Study Tip

Be sure to *factor completely* the numerator and denominator of a rational expression in order to find any common factors. You may need to use a change in signs. Remember that the Distributive Property allows you to write $(b - a)$ as $-(a - b)$. Watch for this in Example 7.

EXAMPLE 7 Simplification Involving a Change in Sign

Simplify the rational expression $\dfrac{2x^2 - 9x + 4}{12 + x - x^2}$.

Solution

The domain of the rational expression is all real values of x such that $x \neq -3$ and $x \neq 4$.

$$\frac{2x^2 - 9x + 4}{12 + x - x^2} = \frac{(2x - 1)(x - 4)}{(4 - x)(3 + x)}$$ Factor numerator and denominator.

$$= \frac{(2x - 1)(x - 4)}{-(x - 4)(3 + x)}$$ $(4 - x) = -(x - 4)$

$$= \frac{(2x - 1)\cancel{(x - 4)}}{-\cancel{(x - 4)}(3 + x)}$$ Divide out common factor $(x - 4)$.

$$= -\frac{2x - 1}{3 + x},\quad x \neq 4$$ Simplified form

The simplified form is equivalent to the original expression for all values of x such that $x \neq 4$. Note that by implied restriction, $x = -3$ is excluded from the domains of both the original and simplified expressions.

✓ **CHECKPOINT** *Now try Exercise 65.*

In Example 7, be sure you see that when dividing the numerator and denominator by the common factor of $(x - 4)$, you keep the minus sign. In the simplified form of the fraction, this text uses the convention of moving the minus sign out in front of the fraction. However, this is a personal preference. All of the following forms are equivalent.

$$-\frac{2x - 1}{3 + x} = \frac{-(2x - 1)}{3 + x} = \frac{-2x + 1}{3 + x} = \frac{2x - 1}{-3 - x} = \frac{2x - 1}{-(3 + x)}$$

EXAMPLE 8 A Rational Expression Involving Two Variables

Simplify the rational expression $\dfrac{3xy + y^2}{2y}$.

Solution

The domain of the rational expression is all real values of y such that $y \neq 0$.

$$\frac{3xy + y^2}{2y} = \frac{y(3x + y)}{2y}$$ Factor numerator and denominator.

$$= \frac{\cancel{y}(3x + y)}{2\cancel{y}}$$ Divide out common factor y.

$$= \frac{3x + y}{2},\quad y \neq 0$$ Simplified form

✓ **CHECKPOINT** *Now try Exercise 71.*

EXAMPLE 9 **A Rational Expression Involving Two Variables**

$$\frac{2x^2 + 2xy - 4y^2}{5x^3 - 5xy^2} = \frac{2(x - y)(x + 2y)}{5x(x - y)(x + y)} \qquad \text{Factor numerator and denominator.}$$

$$= \frac{2\cancel{(x - y)}(x + 2y)}{5x\cancel{(x - y)}(x + y)} \qquad \text{Divide out common factor } (x - y).$$

$$= \frac{2(x + 2y)}{5x(x + y)}, \quad x \neq y \qquad \text{Simplified form}$$

The domain of the original rational expression is all real values of x and y such that $x \neq 0$ and $x \neq \pm y$.

 **CHECKPOINT** *Now try Exercise 73.*

Study Tip

As you study the examples and work the exercises in this section and the next three sections, keep in mind that you are *rewriting expressions in simpler forms*. You are not solving equations. Equal signs are used in the steps of the simplification process only to indicate that the new form of the expression is *equivalent* to the original form.

EXAMPLE 10 **A Rational Expression Involving Two Variables**

$$\frac{4x^2y - y^3}{2x^2y - xy^2} = \frac{(2x - y)(2x + y)y}{(2x - y)xy} \qquad \text{Factor numerator and denominator.}$$

$$= \frac{\cancel{(2x - y)}(2x + y)\cancel{y}}{\cancel{(2x - y)}x\cancel{y}} \qquad \text{Divide out common factors } (2x - y) \text{ and } y.$$

$$= \frac{2x + y}{x}, \quad y \neq 0, \; y \neq 2x \qquad \text{Simplified form}$$

The domain of the original rational expression is all real values of x and y such that $x \neq 0$, $y \neq 0$, and $y \neq 2x$.

 **CHECKPOINT** *Now try Exercise 75.*

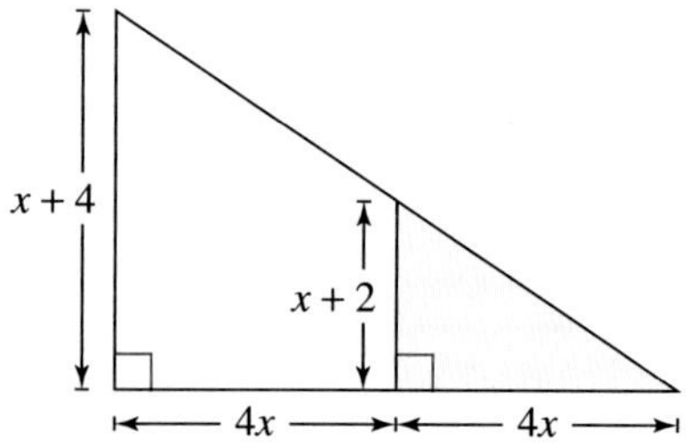

Figure 7.1

EXAMPLE 11 **Geometry: Area**

Find the ratio of the area of the shaded portion of the triangle to the total area of the triangle. (See Figure 7.1.)

Solution

The area of the shaded portion of the triangle is given by

$$\text{Area} = \tfrac{1}{2}(4x)(x + 2) = \tfrac{1}{2}(4x^2 + 8x) = 2x^2 + 4x.$$

The total area of the triangle is given by

$$\text{Area} = \tfrac{1}{2}(4x + 4x)(x + 4) = \tfrac{1}{2}(8x)(x + 4) = \tfrac{1}{2}(8x^2 + 32x) = 4x^2 + 16x.$$

So, the ratio of the area of the shaded portion of the triangle to the total area of the triangle is

$$\frac{2x^2 + 4x}{4x^2 + 16x} = \frac{2x(x + 2)}{4x(x + 4)} = \frac{x + 2}{2(x + 4)}, \quad x > 0.$$

 **CHECKPOINT** *Now try Exercise 85.*

Concept Check

1. Describe the process for finding the implied domain restrictions of a rational function.

2. Describe a situation in which you would need to indicate a domain restriction to the right of a rational function.

3. What expression(s) must you factor completely in order to simplify a rational function of the form $f(x) = \frac{u(x)}{v(x)}$, and why?

4. After factoring completely, what is one additional step that is sometimes needed to find common factors in the numerator and denominator of a rational expression?

7.1 EXERCISES

Go to pages 462–463 to record your assignments.

Developing Skills

In Exercises 1–22, find the domain of the rational function. ***See Examples 1 and 2.***

1. $f(x) = \frac{x^2 + 9}{4}$

2. $f(y) = \frac{y^2 - 3}{7}$

✓ 3. $f(x) = \frac{4}{x - 3}$

4. $g(x) = \frac{-2}{x - 7}$

5. $f(x) = \frac{12x}{9 - x}$

6. $h(y) = \frac{2y}{1 - y}$

7. $g(x) = \frac{2x}{x + 10}$

8. $f(x) = \frac{4x}{x + 1}$

9. $h(x) = \frac{x}{x^2 + 4}$

10. $h(x) = \frac{4x}{x^2 + 16}$

11. $f(y) = \frac{y - 4}{y(y + 3)}$

12. $f(z) = \frac{z + 2}{z(z - 4)}$

13. $f(x) = \frac{x^2}{x(x - 1)}$

14. $g(x) = \frac{x^3}{x(x + 2)}$

✓ 15. $f(t) = \frac{5t}{t^2 - 16}$

16. $f(x) = \frac{x}{x^2 - 4}$

17. $g(y) = \frac{y + 5}{y^2 - 3y}$

18. $g(t) = \frac{t - 6}{t^2 + 5t}$

19. $g(x) = \frac{x + 1}{x^2 - 5x + 6}$

20. $h(t) = \frac{3t^2}{t^2 - 2t - 3}$

21. $f(u) = \frac{u^2}{3u^2 - 2u - 5}$

22. $g(y) = \frac{y + 5}{4y^2 - 5y - 6}$

In Exercises 23–28, evaluate the rational function as indicated, and simplify. If not possible, state the reason.

23. $f(x) = \frac{4x}{x + 3}$

(a) $f(1)$ (b) $f(-2)$

(c) $f(-3)$ (d) $f(0)$

24. $f(x) = \frac{x - 5}{4x}$

(a) $f(10)$ (b) $f(0)$

(c) $f(-3)$ (d) $f(5)$

25. $g(x) = \dfrac{x^2 - 4x}{x^2 - 9}$

(a) $g(0)$ (b) $g(4)$

(c) $g(3)$ (d) $g(-3)$

26. $g(t) = \dfrac{t^2 + 4t}{t^2 - 4}$

(a) $g(2)$ (b) $g(1)$

(c) $g(-2)$ (d) $g(-4)$

27. $h(s) = \dfrac{s^2}{s^2 - s - 2}$

(a) $h(10)$ (b) $h(0)$

(c) $h(-1)$ (d) $h(2)$

28. $f(x) = \dfrac{x^3 + 1}{x^2 - 6x + 9}$

(a) $f(-1)$ (b) $f(3)$

(c) $f(-2)$ (d) $f(2)$

In Exercises 29–34, describe the domain. ***See Example 3.***

29. *Geometry* A rectangle of length x inches has an area of 500 square inches. The perimeter P of the rectangle is given by

$$P = 2\left(x + \frac{500}{x}\right).$$

30. *Cost* The cost C in millions of dollars for the government to seize $p\%$ of an illegal drug as it enters the country is given by

$$C = \frac{528p}{100 - p}.$$

✓ **31.** *Inventory Cost* The inventory cost I when x units of a product are ordered from a supplier is given by

$$I = \frac{0.25x + 2000}{x}.$$

32. *Average Cost* The average cost $\overline{C}$ for a manufacturer to produce x units of a product is given by

$$\overline{C} = \frac{1.35x + 4570}{x}.$$

33. *Pollution Removal* The cost C in dollars of removing $p\%$ of the air pollutants in the stack emission of a utility company is given by the rational function

$$C = \frac{60{,}000p}{100 - p}.$$

34. *Consumer Awareness* The average cost of a movie rental $\overline{M}$ when you consider the cost of purchasing a DVD player and renting x DVDs at \$3.49 per movie is

$$\overline{M} = \frac{75 + 3.49x}{x}.$$

In Exercises 35–42, fill in the missing factor.

35. $\dfrac{5(\quad)}{6(x + 3)} = \dfrac{5}{6}, \quad x \neq -3$

36. $\dfrac{7(\quad)}{15(x - 10)} = \dfrac{7}{15}, \quad x \neq 10$

37. $\dfrac{3x(x + 16)^2}{2(\quad)} = \dfrac{x}{2}, \quad x \neq -16$

38. $\dfrac{25x^2(x - 10)}{12(\quad)} = \dfrac{5x}{12}, \quad x \neq 10, \quad x \neq 0$

39. $\dfrac{(x + 5)(\quad)}{3x^2(x - 2)} = \dfrac{x + 5}{3x}, \quad x \neq 2$

40. $\dfrac{(3y - 7)(\quad)}{y^2 - 4} = \dfrac{3y - 7}{y + 2}, \quad y \neq 2$

41. $\dfrac{8x(\quad)}{x^2 - 2x - 15} = \dfrac{8x}{x - 5}, \quad x \neq -3$

42. $\dfrac{(3 - z)(\quad)}{z^3 + 2z^2} = \dfrac{3 - z}{z^2}, \quad z \neq -2$

In Exercises 43–80, simplify the rational expression. ***See Examples 4–10.***

✓ **43.** $\dfrac{5x}{25}$

44. $\dfrac{32y}{24}$

45. $\dfrac{12x^2}{12x}$

46. $\dfrac{15z^3}{15z^3}$

47. $\dfrac{18x^2y}{15xy^4}$

48. $\dfrac{24xz^4}{16x^3z}$

49. $\dfrac{3x^2 - 9x}{12x^2}$

50. $\dfrac{8x^3 + 4x^2}{20x}$

✓ **51.** $\dfrac{x^2(x - 8)}{x(x - 8)}$

52. $\dfrac{a^2b(b - 3)}{b^3(b - 3)^2}$

53. $\dfrac{2x - 3}{4x - 6}$

54. $\dfrac{5 - x}{3x - 15}$

55. $\dfrac{y^2 - 49}{2y - 14}$

56. $\dfrac{x^2 - 36}{6 - x}$

57. $\dfrac{a + 3}{a^2 + 6a + 9}$

58. $\dfrac{u^2 - 12u + 36}{u - 6}$

59. $\dfrac{x^2 - 7x}{x^2 - 14x + 49}$

60. $\dfrac{z^2 + 22z + 121}{3z + 33}$

✓ **61.** $\dfrac{y^3 - 4y}{y^2 + 4y - 12}$

62. $\dfrac{x^3 - 4x}{x^2 - 5x + 6}$

63. $\dfrac{y^4 - 16y^2}{y^2 + y - 12}$

64. $\dfrac{x^4 - 25x^2}{x^2 + 2x - 15}$

✓ **65.** $\dfrac{3x^2 - 7x - 20}{12 + x - x^2}$

66. $\dfrac{2x^2 + 3x - 5}{7 - 6x - x^2}$

67. $\dfrac{2x^2 + 19x + 24}{2x^2 - 3x - 9}$

68. $\dfrac{2y^2 + 13y + 20}{2y^2 + 17y + 30}$

69. $\dfrac{15x^2 + 7x - 4}{25x^2 - 16}$

70. $\dfrac{56z^2 - 3z - 20}{49z^2 - 16}$

✓ **71.** $\dfrac{3xy^2}{xy^2 + x}$

72. $\dfrac{x + 3x^2y}{3xy + 1}$

✓ **73.** $\dfrac{y^2 - 64x^2}{5(3y + 24x)}$

74. $\dfrac{x^2 - 25z^2}{2(3x + 15z)}$

✓ **75.** $\dfrac{5xy + 3x^2y^2}{xy^3}$

76. $\dfrac{4u^2v - 12uv^2}{18uv}$

77. $\dfrac{u^2 - 4v^2}{u^2 + uv - 2v^2}$

78. $\dfrac{x^2 + 4xy}{x^2 - 16y^2}$

79. $\dfrac{3m^2 - 12n^2}{m^2 + 4mn + 4n^2}$

80. $\dfrac{x^2 + xy - 2y^2}{x^2 + 3xy + 2y^2}$

In Exercises 81 and 82, complete the table. What can you conclude?

81.

x	-2	-1	0	1	2	3	4
$\dfrac{x^2 - x - 2}{x - 2}$							
$x + 1$							

82.

x	-2	-1	0	1	2	3	4
$\dfrac{x^2 + 5x}{x}$							
$x + 5$							

Solving Problems

▲ *Geometry* In Exercises 83–86, find the ratio of the area of the shaded portion to the total area of the figure. ***See Example 11.***

83.

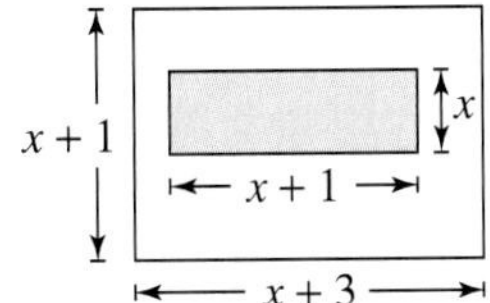

84.

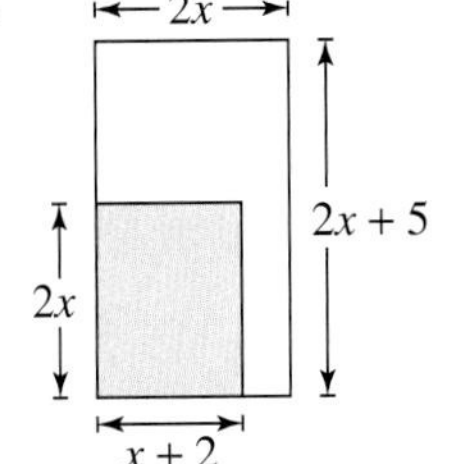

85. 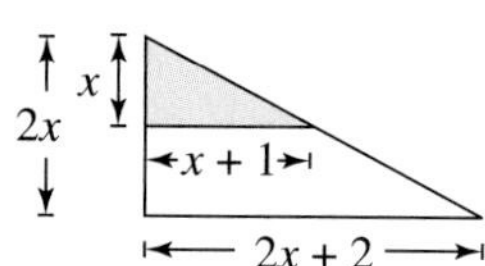

86. 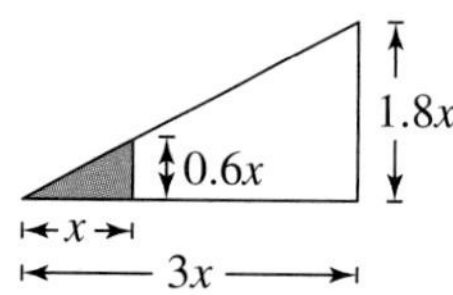

87. *Average Cost* A machine shop has a setup cost of \$2500 for the production of a new product. The cost of labor and material for producing each unit is \$9.25.

(a) Write the total cost C as a function of x, the number of units produced.

(b) Write the average cost per unit $\overline{C} = C/x$ as a function of x, the number of units produced.

(c) Determine the domain of the function in part (b).

(d) Find the value of $\overline{C}(100)$.

88. *Average Cost* A greeting card company has an initial investment of \$60,000. The cost of producing one dozen cards is \$6.50.

(a) Write the total cost C as a function of x, the number of dozens of cards produced.

(b) Write the average cost per dozen $\overline{C} = C/x$ as a function of x, the number of dozens of cards produced.

(c) Determine the domain of the function in part (b).

(d) Find the value of $\overline{C}(11{,}000)$.

89. *Distance Traveled* A van starts on a trip and travels at an average speed of 45 miles per hour. Three hours later, a car starts on the same trip and travels at an average speed of 60 miles per hour.

(a) Find the distance each vehicle has traveled when the car has been on the road for t hours.

(b) Use the result of part (a) to write the distance between the van and the car as a function of t.

(c) Write the ratio of the distance the car has traveled to the distance the van has traveled as a function of t.

90. *Distance Traveled* A car starts on a trip and travels at an average speed of 55 miles per hour. Two hours later, a second car starts on the same trip and travels at an average speed of 65 miles per hour.

(a) Find the distance each vehicle has traveled when the second car has been on the road for t hours.

(b) Use the result of part (a) to write the distance between the first car and the second car as a function of t.

(c) Write the ratio of the distance the second car has traveled to the distance the first car has traveled as a function of t.

91. *Geometry* One swimming pool is circular and another is rectangular. The rectangular pool's width is three times its depth. Its length is 6 feet more than its width. The circular pool has a diameter that is twice the width of the rectangular pool, and it is 2 feet deeper. Find the ratio of the circular pool's volume to the rectangular pool's volume.

92. *Geometry* A circular pool has a radius five times its depth. A rectangular pool has the same depth as the circular pool. Its width is 4 feet more than three times its depth and its length is 2 feet less than six times its depth. Find the ratio of the rectangular pool's volume to the circular pool's volume.

Cable TV Revenue In Exercises 93 and 94, use the following polynomial models, which give the total basic cable television revenue R (in millions of dollars) and the number of basic cable subscribers S (in millions) for the years 2001 through 2005 (see figures).

$R = 1189.2t + 25{,}266, \quad 1 \le t \le 5$

$S = -0.35t + 67.1, \quad 1 \le t \le 5$

In these models, t represents the year, with $t = 1$ corresponding to 2001. (Source: Kagan Research, LLC)

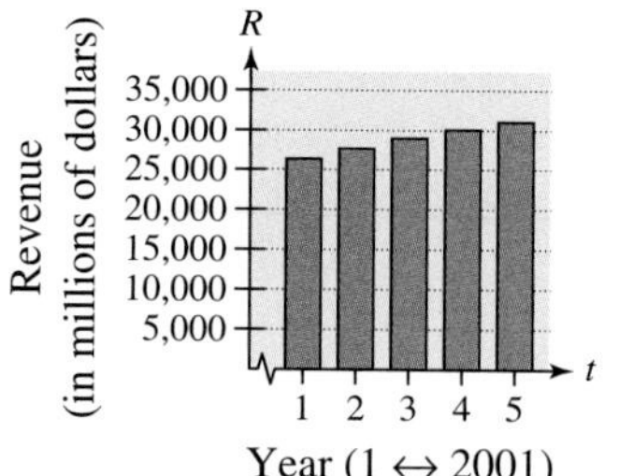

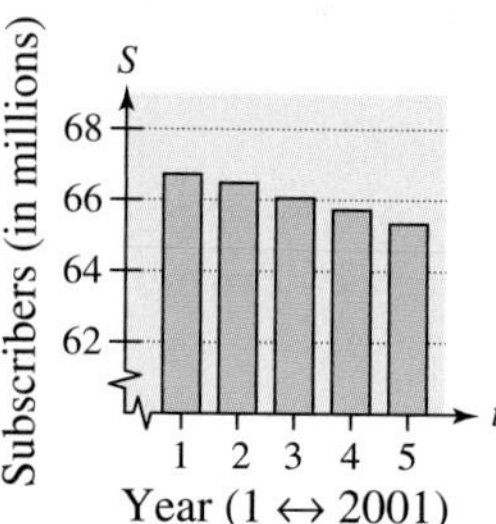

Figures for 93 and 94

93. Find a rational model that represents the average basic cable television revenue per subscriber during the years 2001 to 2005.

94. Use the model found in Exercise 93 to complete the table, which shows the average basic cable television revenue per subscriber.

Year	2001	2002	2003	2004	2005
Average revenue					

Explaining Concepts

95. How do you determine whether a rational expression is in simplified form?

96. Can you divide out common terms from the numerator and denominator of a rational expression? Explain.

97. Give an example of a rational function whose domain is the set of all real numbers and whose denominator is a second-degree polynomial function.

98. *Error Analysis* Describe the error.

$$\frac{2x^2}{x^2+4} = \frac{2x^2}{x^2+4} = \frac{2}{1+4} = \frac{2}{5}$$

99. A student writes the following incorrect solution for simplifying a rational expression. Discuss the student's errors and misconceptions, and construct a correct solution.

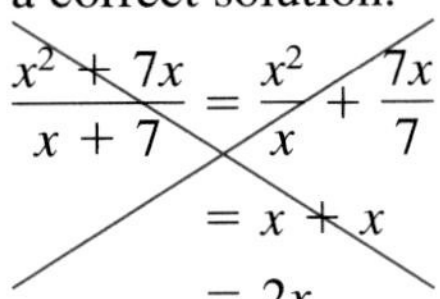

$\frac{x^2+7x}{x+7} = \frac{x^2}{x} + \frac{7x}{7}$
$= x + x$
$= 2x$

100. Is the following statement true? Explain.

$$\frac{6x-5}{5-6x} = -1$$

101. Explain how you can use a given polynomial function $f(x)$ to write a rational function $g(x)$ that is equivalent to $f(x)$, $x \neq 2$.

102. Is it possible for a rational function $f(x)$ (without added domain restrictions) to be undefined on an interval $[a, b]$, where a and b are real numbers such that $a < b$? Explain.

Cumulative Review

In Exercises 103–106, find the product.

103. $\frac{1}{4}\left(\frac{3}{4}\right)$

104. $\frac{2}{3}\left(-\frac{5}{6}\right)$

105. $\frac{1}{3}\left(\frac{3}{5}\right)(5)$

106. $\left(-\frac{3}{7}\right)\left(\frac{2}{5}\right)\left(-\frac{1}{6}\right)$

In Exercises 107–110, perform the indicated multiplication.

107. $(-2a^3)(-2a)$

108. $6x^2(-3x)$

109. $(-3b)(b^2 - 3b + 5)$

110. $ab^2(3a - 4ab + 6a^2b^2)$

7.2 Multiplying and Dividing Rational Expressions

Steven E. Frischling/Bloomberg News/Landov

What You Should Learn

1 ▶ Multiply rational expressions and simplify.

2 ▶ Divide rational expressions and simplify.

Why You Should Learn It

Multiplication and division of rational expressions can be used to solve real-life applications. For instance, Example 9 on page 418 shows how to divide rational expressions to find a model for the annual amount the average American spent on meals away from home from 2000 to 2006.

1 ▶ Multiply rational expressions and simplify.

Multiplying Rational Expressions

The rule for multiplying rational expressions is the same as the rule for multiplying numerical fractions. That is, you *multiply numerators, multiply denominators, and write the new fraction in simplified form.*

$$\frac{3}{4}\cdot\frac{7}{6}=\frac{21}{24}=\frac{\cancel{3}\cdot 7}{\cancel{3}\cdot 8}=\frac{7}{8}$$

Multiplying Rational Expressions

Let u, v, w, and z represent real numbers, variables, or algebraic expressions such that $v \neq 0$ and $z \neq 0$. Then the product of u/v and w/z is

$$\frac{u}{v}\cdot\frac{w}{z}=\frac{uw}{vz}.$$

In order to recognize common factors when simplifying the product, use factoring in the numerator and denominator, as demonstrated in Example 1.

EXAMPLE 1 Multiplying Rational Expressions

Multiply the rational expressions $\dfrac{4x^3y}{3xy^4}\cdot\dfrac{-6x^2y^2}{10x^4}$.

Solution

$$\frac{4x^3y}{3xy^4}\cdot\frac{-6x^2y^2}{10x^4}=\frac{(4x^3y)\cdot(-6x^2y^2)}{(3xy^4)\cdot(10x^4)}$$ Multiply numerators and denominators.

$$=\frac{-24x^5y^3}{30x^5y^4}$$ Simplify.

$$=\frac{-4\cancel{(6)}\cancel{(x^5)}\cancel{(y^3)}}{5\cancel{(6)}\cancel{(x^5)}\cancel{(y^3)}(y)}$$ Factor and divide out common factors.

$$=-\frac{4}{5y},\ x\neq 0$$ Simplified form

✓ **CHECKPOINT** *Now try Exercise 11.*

EXAMPLE 2 Multiplying Rational Expressions

Multiply the rational expressions.

$$\frac{x}{5x^2 - 20x} \cdot \frac{x - 4}{2x^2 + x - 3}$$

Solution

$$\frac{x}{5x^2 - 20x} \cdot \frac{x - 4}{2x^2 + x - 3}$$

$$= \frac{x \cdot (x - 4)}{(5x^2 - 20x) \cdot (2x^2 + x - 3)}$$ Multiply numerators and denominators.

$$= \frac{x(x - 4)}{5x(x - 4)(x - 1)(2x + 3)}$$ Factor.

$$= \frac{\cancel{x}\cancel{(x - 4)}}{5\cancel{x}\cancel{(x - 4)}(x - 1)(2x + 3)}$$ Divide out common factors.

$$= \frac{1}{5(x - 1)(2x + 3)}, \; x \neq 0, \; x \neq 4$$ Simplified form

✓ **CHECKPOINT** *Now try Exercise 23.*

EXAMPLE 3 Multiplying Rational Expressions

Multiply the rational expressions.

$$\frac{4x^2 - 4x}{x^2 + 2x - 3} \cdot \frac{x^2 + x - 6}{4x}$$

Solution

$$\frac{4x^2 - 4x}{x^2 + 2x - 3} \cdot \frac{x^2 + x - 6}{4x}$$

$$= \frac{4x(x - 1)(x + 3)(x - 2)}{(x - 1)(x + 3)(4x)}$$ Multiply and factor.

$$= \frac{\cancel{4x}\cancel{(x - 1)}\cancel{(x + 3)}(x - 2)}{\cancel{(x - 1)}\cancel{(x + 3)}\cancel{(4x)}}$$ Divide out common factors.

$$= x - 2, \; x \neq 0, \; x \neq 1, \; x \neq -3$$ Simplified form

✓ **CHECKPOINT** *Now try Exercise 25.*

The rule for multiplying rational expressions can be extended to cover products involving expressions that are not in fractional form. To do this, rewrite each expression that is not in fractional form as a fraction whose denominator is 1. Here is a simple example.

$$\frac{x + 3}{x - 2} \cdot (5x) = \frac{x + 3}{x - 2} \cdot \frac{5x}{1} = \frac{(x + 3)(5x)}{x - 2} = \frac{5x(x + 3)}{x - 2}$$

Technology: Tip

You can use a graphing calculator to check your results when multiplying rational expressions. For instance, in Example 3, try graphing the equations

$$y_1 = \frac{4x^2 - 4x}{x^2 + 2x - 3} \cdot \frac{x^2 + x - 6}{4x}$$

and

$$y_2 = x - 2$$

in the same viewing window and use the *table* feature to create a table of values for the two equations. If the two graphs coincide, and the values of y_1 and y_2 are the same in the table except where a common factor has been divided out, as shown below, you can conclude that the solution checks.

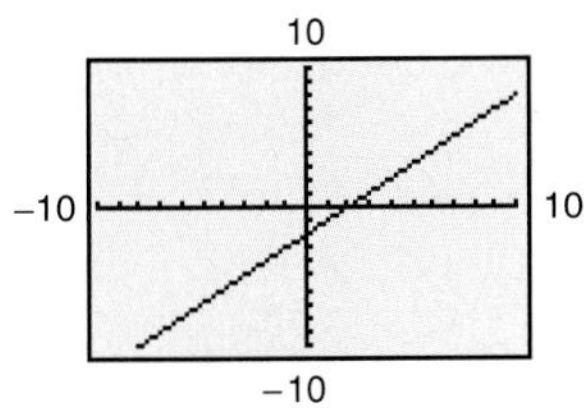

X	Y1	Y2
-4	-6	-6
-3	ERROR	-5
-2	-4	-4
-1	-3	-3
0	ERROR	-2
1	ERROR	-1
2	0	0

X= -4

In the next example, note how to divide out a factor that differs only in sign. The Distributive Property is used in the step in which $(y - x)$ is rewritten as $(-1)(x - y)$.

EXAMPLE 4 Multiplying Rational Expressions

Multiply the rational expressions.

$$\frac{x - y}{y^2 - x^2} \cdot \frac{x^2 - xy - 2y^2}{3x - 6y}$$

Solution

$$\frac{x - y}{y^2 - x^2} \cdot \frac{x^2 - xy - 2y^2}{3x - 6y}$$

$$= \frac{(x - y)(x - 2y)(x + y)}{(y + x)(y - x)(3)(x - 2y)} \qquad \text{Multiply and factor.}$$

$$= \frac{(x - y)(x - 2y)(x + y)}{(y + x)(-1)(x - y)(3)(x - 2y)} \qquad (y - x) = -1(x - y)$$

$$= \frac{\cancel{(x - y)}\cancel{(x - 2y)}\cancel{(x + y)}}{\cancel{(x + y)}(-1)\cancel{(x - y)}(3)\cancel{(x - 2y)}} \qquad \text{Divide out common factors.}$$

$$= -\frac{1}{3},\ x \neq y,\ x \neq -y,\ x \neq 2y \qquad \text{Simplified form}$$

✓ **CHECKPOINT** *Now try Exercise 27.*

The rule for multiplying rational expressions can be extended to cover products of three or more expressions, as shown in Example 5.

EXAMPLE 5 Multiplying Three Rational Expressions

Multiply the rational expressions.

$$\frac{x^2 - 3x + 2}{x + 2} \cdot \frac{3x}{x - 2} \cdot \frac{2x + 4}{x^2 - 5x}$$

Solution

$$\frac{x^2 - 3x + 2}{x + 2} \cdot \frac{3x}{x - 2} \cdot \frac{2x + 4}{x^2 - 5x}$$

$$= \frac{(x - 1)(x - 2)(3)(x)(2)(x + 2)}{(x + 2)(x - 2)(x)(x - 5)} \qquad \text{Multiply and factor.}$$

$$= \frac{(x - 1)\cancel{(x - 2)}(3)\cancel{(x)}(2)\cancel{(x + 2)}}{\cancel{(x + 2)}\cancel{(x - 2)}\cancel{(x)}(x - 5)} \qquad \text{Divide out common factors.}$$

$$= \frac{6(x - 1)}{x - 5},\ x \neq 0,\ x \neq 2,\ x \neq -2 \qquad \text{Simplified form}$$

✓ **CHECKPOINT** *Now try Exercise 31.*

2 ▶ Divide rational expressions and simplify.

Dividing Rational Expressions

To divide two rational expressions, multiply the first expression by the *reciprocal* of the second. That is, *invert the divisor and multiply.*

Dividing Rational Expressions

Let u, v, w, and z represent real numbers, variables, or algebraic expressions such that $v \neq 0$, $w \neq 0$, and $z \neq 0$. Then the quotient of u/v and w/z is

$$\frac{u}{v} \div \frac{w}{z} = \frac{u}{v} \cdot \frac{z}{w} = \frac{uz}{vw}.$$

Study Tip

Don't forget to add domain restrictions as needed in division problems. In Example 6, an implied domain restriction in the original expression is $x \neq 1$. Because this restriction is not implied by the final expression, it must be added as a written restriction.

EXAMPLE 6 Dividing Rational Expressions

Divide the rational expressions.

$$\frac{x}{x+3} \div \frac{4}{x-1}$$

Solution

$$\frac{x}{x+3} \div \frac{4}{x-1} = \frac{x}{x+3} \cdot \frac{x-1}{4}$$ Invert divisor and multiply.

$$= \frac{x(x-1)}{(x+3)(4)}$$ Multiply numerators and denominators.

$$= \frac{x(x-1)}{4(x+3)}, \quad x \neq 1$$ Simplify.

✓ **CHECKPOINT** *Now try Exercise 37.*

EXAMPLE 7 Dividing Rational Expressions

$$\frac{2x}{3x-12} \div \frac{x^2-2x}{x^2-6x+8}$$ Original expressions

$$= \frac{2x}{3x-12} \cdot \frac{x^2-6x+8}{x^2-2x}$$ Invert divisor and multiply.

$$= \frac{(2)(x)(x-2)(x-4)}{(3)(x-4)(x)(x-2)}$$ Factor.

$$= \frac{(2)\cancel{(x)}\cancel{(x-2)}\cancel{(x-4)}}{(3)\cancel{(x-4)}\cancel{(x)}\cancel{(x-2)}}$$ Divide out common factors.

$$= \frac{2}{3}, \quad x \neq 0, \; x \neq 2, \; x \neq 4$$ Simplified form

Remember that the original expression is equivalent to $\frac{2}{3}$ except for $x = 0$, $x = 2$, and $x = 4$.

✓ **CHECKPOINT** *Now try Exercise 47.*

EXAMPLE 8 Dividing Rational Expressions

Divide the rational expressions.

$$\frac{x^2 - y^2}{2x + 2y} \div \frac{2x^2 - 3xy + y^2}{6x + 2y}$$

Solution

$$\frac{x^2 - y^2}{2x + 2y} \div \frac{2x^2 - 3xy + y^2}{6x + 2y}$$

$$= \frac{x^2 - y^2}{2x + 2y} \cdot \frac{6x + 2y}{2x^2 - 3xy + y^2}$$ Invert divisor and multiply.

$$= \frac{(x + y)(x - y)(2)(3x + y)}{(2)(x + y)(2x - y)(x - y)}$$ Factor.

$$= \frac{\cancel{(x + y)}\cancel{(x - y)}\cancel{(2)}(3x + y)}{\cancel{(2)}\cancel{(x + y)}(2x - y)\cancel{(x - y)}}$$ Divide out common factors.

$$= \frac{3x + y}{2x - y}, \; x \neq y, x \neq -y, y \neq -3x$$ Simplified form

✓ **CHECKPOINT** *Now try Exercise 49.*

EXAMPLE 9 Amount Spent on Meals and Beverages 

The annual amount A (in millions of dollars) Americans spent on meals and beverages purchased away from home, and the population P (in millions) of the United States, for the years 2000 through 2006 can be modeled by

$$A = \frac{-8242.58t + 348{,}299.6}{-0.06t + 1}, \quad 0 \leq t \leq 6$$

and

$$P = 2.71t + 282.7, \quad 0 \leq t \leq 6$$

where t represents the year, with $t = 0$ corresponding to 2000. Find a model T for the amount Americans spent *per person* on meals and beverages. (Source: U.S. Bureau of Economic Analysis and U.S. Census Bureau)

Solution

To find a model T for the amount Americans spent per person on meals and beverages, divide the total amount by the population.

$$T = \frac{-8242.58t + 348{,}299.6}{-0.06t + 1} \div (2.71t + 282.7)$$ Divide amount spent by population.

$$= \frac{-8242.58t + 348{,}299.6}{-0.06t + 1} \cdot \frac{1}{2.71t + 282.7}$$ Invert divisor and multiply.

$$= \frac{-8242.58t + 348{,}299.6}{(-0.06t + 1)(2.71t + 282.7)}, \quad 0 \leq t \leq 6$$ Model

✓ **CHECKPOINT** *Now try Exercise 69.*

Concept Check

1. Why is factoring used in multiplying rational expressions?
2. In your own words, explain how to divide rational expressions.
3. Explain how to divide a rational expression by a polynomial.
4. In dividing rational expressions, explain how you can lose implied domain restrictions when you invert the divisor.

7.2 EXERCISES

Go to pages 462–463 to record your assignments.

Developing Skills

In Exercises 1–8, fill in the missing factor.

1. $\dfrac{7x^2}{3y(\quad)} = \dfrac{7}{3y}, \quad x \neq 0$
2. $\dfrac{14x(x-3)^2}{(x-3)(\quad)} = \dfrac{2x}{x-3}$
3. $\dfrac{3x(x+2)^2}{(x-4)(\quad)} = \dfrac{3x}{x-4}, \quad x \neq -2$
4. $\dfrac{(x+1)^3}{x(\quad)} = \dfrac{x+1}{x}, \quad x \neq -1$
5. $\dfrac{3u(\quad)}{7v(u+1)} = \dfrac{3u}{7v}, \quad u \neq -1$
6. $\dfrac{(3t+5)(\quad)}{5t^2(3t-5)} = \dfrac{3t+5}{t}, \quad t \neq \dfrac{5}{3}$
7. $\dfrac{13x(\quad)}{4-x^2} = \dfrac{13x}{x-2}, \quad x \neq -2$
8. $\dfrac{x^2(\quad)}{x^2-10x} = \dfrac{x^2}{10-x}, \quad x \neq 0$

In Exercises 9–36, multiply and simplify. ***See Examples 1–5.***

9. $4x \cdot \dfrac{7}{12x}$
10. $\dfrac{8}{7y} \cdot (42y)$
11. ✓ $\dfrac{8s^3}{9s} \cdot \dfrac{6s^2}{32s}$
12. $\dfrac{3x^4}{7x} \cdot \dfrac{8x^2}{9}$
13. $16u^4 \cdot \dfrac{12}{8u^2}$
14. $18x^4 \cdot \dfrac{4}{15x}$
15. $\dfrac{8}{3+4x} \cdot (9+12x)$
16. $(6-4x) \cdot \dfrac{10}{3-2x}$
17. $\dfrac{8u^2v}{3u+v} \cdot \dfrac{u+v}{12u}$
18. $\dfrac{1-3xy}{4x^2y} \cdot \dfrac{46x^4y^2}{15-45xy}$
19. $\dfrac{12-r}{3} \cdot \dfrac{3}{r-12}$
20. $\dfrac{8-z}{8+z} \cdot \dfrac{z+8}{z-8}$
21. $\dfrac{(2x-3)(x+8)}{x^3} \cdot \dfrac{x}{3-2x}$
22. $\dfrac{x+14}{x^3(10-x)} \cdot \dfrac{x(x-10)}{5}$
23. ✓ $\dfrac{4r-12}{r-2} \cdot \dfrac{r^2-4}{r-3}$
24. $\dfrac{5y-20}{5y+15} \cdot \dfrac{2y+6}{y-4}$
25. ✓ $\dfrac{2t^2-t-15}{t+2} \cdot \dfrac{t^2-t-6}{t^2-6t+9}$
26. $\dfrac{y^2-16}{y^2+8y+16} \cdot \dfrac{3y^2-5y-2}{y^2-6y+8}$
27. ✓ $(4y^2-x^2) \cdot \dfrac{xy}{(x-2y)^2}$
28. $(u-2v)^2 \cdot \dfrac{u+2v}{2v-u}$
29. $\dfrac{x^2+2xy-3y^2}{(x+y)^2} \cdot \dfrac{x^2-y^2}{x+3y}$
30. $\dfrac{(x-2y)^2}{x+2y} \cdot \dfrac{x^2+7xy+10y^2}{x^2-4y^2}$

31. $\frac{x+5}{x-5} \cdot \frac{2x^2-9x-5}{3x^2+x-2} \cdot \frac{x^2-1}{x^2+7x+10}$

32. $\frac{t^2+4t+3}{2t^2-t-10} \cdot \frac{t}{t^2+3t+2} \cdot \frac{2t^2+4t^3}{t^2+3t}$

33. $\frac{9-x^2}{2x+3} \cdot \frac{4x^2+8x-5}{4x^2-8x+3} \cdot \frac{6x^4-2x^3}{8x^2+4x}$

34. $\frac{16x^2-1}{4x^2+9x+5} \cdot \frac{5x^2-9x-18}{x^2-12x+36} \cdot \frac{12+4x-x^2}{4x^2-13x+3}$

35. $\frac{x^3+3x^2-4x-12}{x^3-3x^2-4x+12} \cdot \frac{x^2-9}{x}$

36. $\frac{xu-yu+xv-yv}{xu+yu-xv-yv} \cdot \frac{xu+yu+xv+yv}{xu-yu-xv+yv}$

In Exercises 37–52, divide and simplify. ***See Examples 6–8.***

37. $\frac{x}{x+2} \div \frac{3}{x+1}$

38. $\frac{x+3}{4} \div \frac{x-2}{x}$

39. $x^2 \div \frac{3x}{4}$

40. $\frac{u}{10} \div u^2$

41. $\frac{2x}{5} \div \frac{x^2}{15}$

42. $\frac{3y^2}{20} \div \frac{y}{15}$

43. $\frac{7xy^2}{10u^2v} \div \frac{21x^3}{45uv}$

44. $\frac{25x^2y}{60x^3y^2} \div \frac{5x^4y^3}{16x^2y}$

45. $\frac{3(a+b)}{4} \div \frac{(a+b)^2}{2}$

46. $\frac{x^2+9}{5(x+2)} \div \frac{x+3}{5(x^2-4)}$

47. $\frac{4x}{3x-3} \div \frac{x^2+2x}{x^2+x-2}$

48. $\frac{5x+5}{2x} \div \frac{x^2-3x}{x^2-2x-3}$

49. $\frac{(x^3y)^2}{(x+2y)^2} \div \frac{x^2y}{(x+2y)^3}$

50. $\frac{x^2-y^2}{2x^2-8x} \div \frac{(x-y)^2}{2xy}$

51. $\frac{x^2+2x-15}{x^2+11x+30} \div \frac{x^2-8x+15}{x^2+2x-24}$

52. $\frac{y^2+5y-14}{y^2+10y+21} \div \frac{y^2+5y+6}{y^2+7y+12}$

In Exercises 53–60, perform the operations and simplify. (In Exercises 59 and 60, n is a positive integer.)

53. $\left[\frac{x^2}{9} \cdot \frac{3(x+4)}{x^2+2x}\right] \div \frac{x}{x+2}$

54. $\left(\frac{x^2+6x+9}{x^2} \cdot \frac{2x+1}{x^2-9}\right) \div \frac{4x^2+4x+1}{x^2-3x}$

55. $\left[\frac{xy+y}{4x} \div (3x+3)\right] \div \frac{y}{3x}$

56. $\frac{3u^2-u-4}{u^2} \div \frac{3u^2+12u+4}{u^4-3u^3}$

57. $\frac{2x^2+5x-25}{3x^2+5x+2} \cdot \frac{3x^2+2x}{x+5} \div \left(\frac{x}{x+1}\right)^2$

58. $\frac{t^2-100}{4t^2} \cdot \frac{t^3-5t^2-50t}{t^4+10t^3} \div \frac{(t-10)^2}{5t}$

59. $x^3 \cdot \frac{x^{2n}-9}{x^{2n}+4x^n+3} \div \frac{x^{2n}-2x^n-3}{x}$

60. $\dfrac{x^{n+1} - 8x}{x^{2n} + 2x^n + 1} \cdot \dfrac{x^{2n} - 4x^n - 5}{x} \div x^n$

In Exercises 61 and 62, use a graphing calculator to graph the two equations in the same viewing window. Use the graphs and a table of values to verify that the expressions are equivalent. Verify the results algebraically.

61. $y_1 = \dfrac{x^2 - 10x + 25}{x^2 - 25} \cdot \dfrac{x + 5}{2}$

$y_2 = \dfrac{x - 5}{2}, \quad x \neq \pm 5$

62. $y_1 = \dfrac{3x + 15}{x^4} \div \dfrac{x + 5}{x^2}$

$y_2 = \dfrac{3}{x^2}, \quad x \neq -5$

Solving Problems

Geometry In Exercises 63 and 64, write and simplify an expression for the area of the shaded region.

63.

$\frac{2w + 3}{3}$ $\frac{2w + 3}{3}$ $\frac{2w + 3}{3}$

$\frac{w}{2}$ $\frac{w}{2}$

64.

$\frac{2w - 1}{2}$ $\frac{2w - 1}{2}$

$\frac{w}{3}$ $\frac{w}{3}$ $\frac{w}{3}$

Probability In Exercises 65–68, consider an experiment in which a marble is tossed into a rectangular box with dimensions $2x$ centimeters by $4x + 2$ centimeters. The probability that the marble will come to rest in the unshaded portion of the box is equal to the ratio of the unshaded area to the total area of the figure. Find the probability in simplified form.

65.

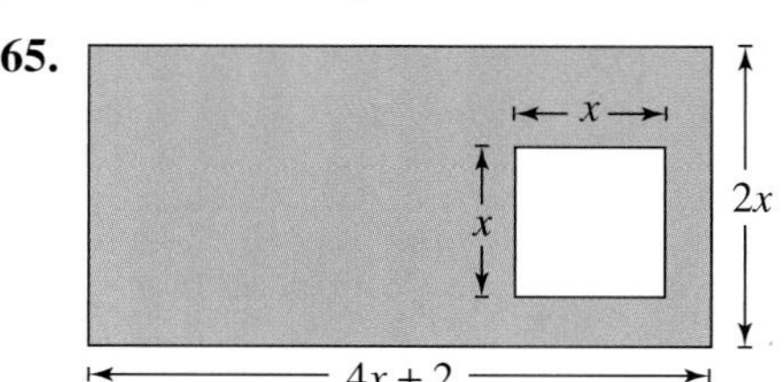

66.

x $2x$ $2x$ $4x + 2$

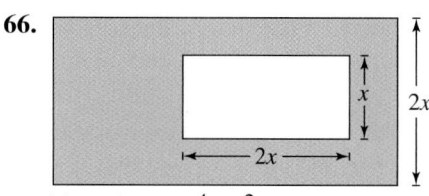

67.

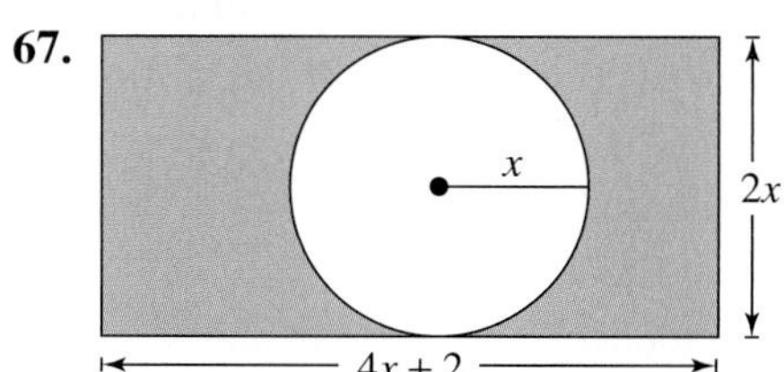

68.

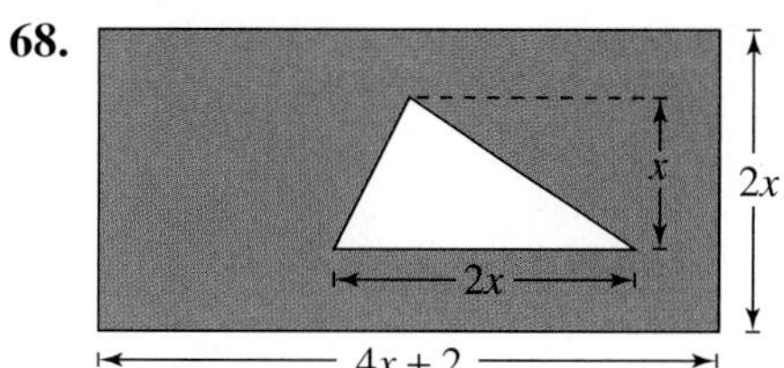

69. *Employment* The number of jobs J (in millions) in Florida, and the population P (in millions) of Florida, for the years 2001 through 2006 can be modeled by

$J = \dfrac{-0.696t + 8.94}{-0.092t + 1}, \quad 1 \leq t \leq 6$ and

$P = 0.352t + 15.97, \quad 1 \leq t \leq 6$

where t represents the year, with $t = 1$ corresponding to 2001. Find a model Y for the number of jobs per person during these years. (Source: U.S. Bureau of Economic Analysis)

70. *Per Capita Income* The total annual amount I (in millions of dollars) of personal income earned in Alabama, and its population P (in millions), for the years 2001 through 2006 can be modeled by

$$I = \frac{-4.665t + 106.48}{-0.075t + 1}, \quad 1 \le t \le 6 \quad \text{and}$$

$$P = \frac{-0.467t + 4.46}{-0.107t + 1}, \quad 1 \le t \le 6$$

where t represents the year, with $t = 1$ corresponding to 2001. Find a model Y for the annual per capita income for these years. (Source: U.S. Bureau of Economic Analysis)

Explaining Concepts

71. Describe how the operation of division is used in the process of simplifying a product of rational expressions.

72. In a quotient of two rational expressions, the denominator of the divisor is x. Describe a set of circumstances in which you will *not* need to list $x \neq 0$ as a domain restriction after dividing.

73. Explain what is missing in the following statement.

$$\frac{x-a}{x-b} \div \frac{x-a}{x-b} = 1$$

74. When two rational expressions are multiplied, the resulting expression is a polynomial. Explain how the total number of factors in the numerators of the expressions you multiplied compares to the total number of factors in the denominators.

75. *Error Analysis* Describe and correct the errors.

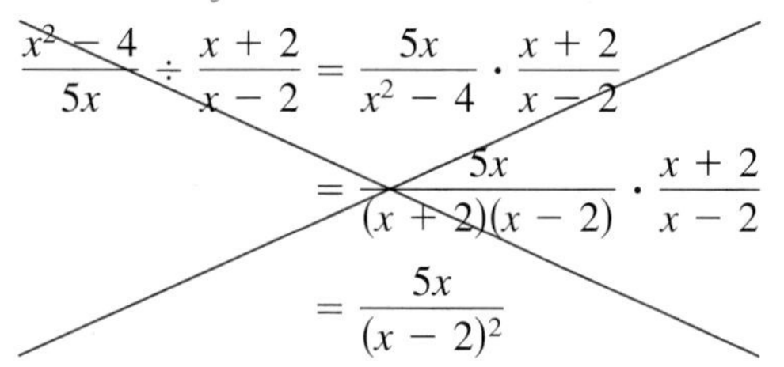

76. Complete the table for the given values of x. Round your answers to five decimal places.

x	60	100	1000
$\frac{x-10}{x+10}$			
$\frac{x+50}{x-50}$			
$\frac{x-10}{x+10} \cdot \frac{x+50}{x-50}$			

x	10,000	100,000	1,000,000
$\frac{x-10}{x+10}$			
$\frac{x+50}{x-50}$			
$\frac{x-10}{x+10} \cdot \frac{x+50}{x-50}$			

What kind of pattern do you see? Try to explain what is going on. Can you see why?

Cumulative Review

In Exercises 77–80, evaluate the expression.

77. $\frac{1}{8} + \frac{3}{8} + \frac{5}{8}$

78. $\frac{3}{7} - \frac{2}{7}$

79. $\frac{3}{5} + \frac{4}{15}$

80. $\frac{7}{6} - \frac{9}{7}$

In Exercises 81–84, solve the equation by factoring.

81. $x^2 + 3x = 0$

82. $x^2 + 3x - 10 = 0$

83. $4x^2 - 25 = 0$

84. $x(x - 4) + 2(x - 12) = 0$

7.3 Adding and Subtracting Rational Expressions

What You Should Learn

1 ▶ Add or subtract rational expressions with like denominators, and simplify.

2 ▶ Add or subtract rational expressions with unlike denominators, and simplify.

Why You Should Learn It

Addition and subtraction of rational expressions can be used to solve real-life applications. For instance, in Exercise 89 on page 431, you will find a rational expression that models the total number of people enrolled as undergraduate students.

Adding or Subtracting with Like Denominators

1 ▶ Add or subtract rational expressions with like denominators, and simplify.

As with numerical fractions, the procedure used to add or subtract two rational expressions depends on whether the expressions have *like* or *unlike* denominators. To add or subtract two rational expressions with *like* denominators, simply combine their numerators and place the result over the common denominator.

Adding or Subtracting with Like Denominators

If u, v, and w are real numbers, variables, or algebraic expressions, and $w \neq 0$, the following rules are valid.

1. $\dfrac{u}{w} + \dfrac{v}{w} = \dfrac{u+v}{w}$ — Add fractions with like denominators.

2. $\dfrac{u}{w} - \dfrac{v}{w} = \dfrac{u-v}{w}$ — Subtract fractions with like denominators.

EXAMPLE 1 Adding and Subtracting with Like Denominators

a. $\dfrac{x}{4} + \dfrac{5-x}{4} = \dfrac{x+(5-x)}{4} = \dfrac{5}{4}$ — Add numerators.

b. $\dfrac{7}{2x-3} - \dfrac{3x}{2x-3} = \dfrac{7-3x}{2x-3}$ — Subtract numerators.

CHECKPOINT *Now try Exercise 1.*

Study Tip

After adding or subtracting two (or more) rational expressions, check the resulting fraction to see if it can be simplified, as illustrated in Example 2.

EXAMPLE 2 Subtracting Rational Expressions and Simplifying

$$\frac{x}{x^2-2x-3} - \frac{3}{x^2-2x-3} = \frac{x-3}{x^2-2x-3} \qquad \text{Subtract numerators.}$$

$$= \frac{(1)(x-3)}{(x-3)(x+1)} \qquad \text{Factor.}$$

$$= \frac{1}{x+1}, \quad x \neq 3 \qquad \text{Simplified form}$$

CHECKPOINT *Now try Exercise 17.*

The rules for adding and subtracting rational expressions with like denominators can be extended to sums and differences involving three or more rational expressions, as illustrated in Example 3.

EXAMPLE 3 **Combining Three Rational Expressions**

$$\frac{x^2 - 26}{x - 5} - \frac{2x + 4}{x - 5} + \frac{10 + x}{x - 5}$$ Original expressions

$$= \frac{(x^2 - 26) - (2x + 4) + (10 + x)}{x - 5}$$ Write numerator over common denominator.

$$= \frac{x^2 - x - 20}{x - 5}$$ Use the Distributive Property and combine like terms.

$$= \frac{\cancel{(x - 5)}(x + 4)}{\cancel{x - 5}}$$ Factor and divide out common factor.

$$= x + 4, \quad x \neq 5$$ Simplified form

✓ **CHECKPOINT** *Now try Exercise 21.*

2 ▶ Add or subtract rational expressions with unlike denominators, and simplify.

Adding or Subtracting with Unlike Denominators

The **least common multiple (LCM)** of two (or more) polynomials can be helpful when adding or subtracting rational expressions with *unlike* denominators. The least common multiple of two (or more) polynomials is the simplest polynomial that is a multiple of each of the original polynomials. This means that the LCM must contain all the *different* factors in each polynomial, with each factor raised to the greatest power of its occurrence in any one of the polynomials.

EXAMPLE 4 **Finding Least Common Multiples**

a. The least common multiple of

$$6x = 2 \cdot 3 \cdot x, \quad 2x^2 = 2 \cdot x^2, \quad \text{and} \quad 9x^3 = 3^2 \cdot x^3$$

is $2 \cdot 3^2 \cdot x^3 = 18x^3$.

b. The least common multiple of

$$x^2 - x = x(x - 1) \quad \text{and} \quad 2x - 2 = 2(x - 1)$$

is $2x(x - 1)$.

c. The least common multiple of

$$3x^2 + 6x = 3x(x + 2) \quad \text{and} \quad x^2 + 4x + 4 = (x + 2)^2$$

is $3x(x + 2)^2$.

✓ **CHECKPOINT** *Now try Exercise 23.*

To add or subtract rational expressions with *unlike* denominators, you must first rewrite the rational expressions so that they have a common denominator. You can always find a common denominator of two (or more) rational expressions by multiplying their denominators. However, if you use the **least common denominator (LCD),** which is the least common multiple of the denominators, you may have less simplifying to do. After the rational expressions have been written with a common denominator, you can simply add or subtract using the rules given at the beginning of this section.

Technology: Tip

You can use a graphing calculator to check your results when adding or subtracting rational expressions. For instance, in Example 5, try graphing the equations

$$y_1 = \frac{7}{6x} + \frac{5}{8x}$$

and

$$y_2 = \frac{43}{24x}$$

in the same viewing window. If the two graphs coincide, as shown below, you can conclude that the solution checks.

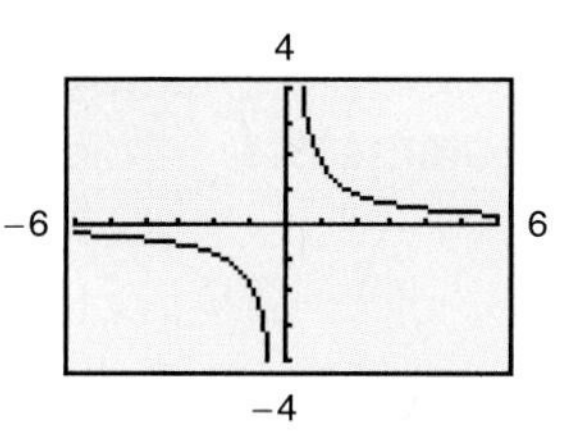

EXAMPLE 5 Adding with Unlike Denominators

Add the rational expressions: $\frac{7}{6x} + \frac{5}{8x}$.

Solution

By factoring the denominators, $6x = 2 \cdot 3 \cdot x$ and $8x = 2^3 \cdot x$, you can conclude that the least common denominator is $2^3 \cdot 3 \cdot x = 24x$.

$$\frac{7}{6x} + \frac{5}{8x} = \frac{7(4)}{6x(4)} + \frac{5(3)}{8x(3)}$$ Rewrite expressions using LCD of $24x$.

$$= \frac{28}{24x} + \frac{15}{24x}$$ Like denominators

$$= \frac{28 + 15}{24x} = \frac{43}{24x}$$ Add fractions and simplify.

✓ **CHECKPOINT** *Now try Exercise 51.*

EXAMPLE 6 Subtracting with Unlike Denominators

Subtract the rational expressions: $\frac{3}{x-3} - \frac{5}{x+2}$.

Solution

The only factors of the denominators are $x - 3$ and $x + 2$. So, the least common denominator is $(x - 3)(x + 2)$.

$$\frac{3}{x-3} - \frac{5}{x+2}$$ Write original expressions.

$$= \frac{3(x+2)}{(x-3)(x+2)} - \frac{5(x-3)}{(x-3)(x+2)}$$ Rewrite expressions using LCD of $(x - 3)(x + 2)$.

$$= \frac{3x+6}{(x-3)(x+2)} - \frac{5x-15}{(x-3)(x+2)}$$ Distributive Property

$$= \frac{3x + 6 - 5x + 15}{(x-3)(x+2)}$$ Subtract fractions and use the Distributive Property.

$$= \frac{-2x + 21}{(x-3)(x+2)}$$ Simplified form

✓ **CHECKPOINT** *Now try Exercise 63.*

Study Tip

In Example 7, notice that the denominator $2 - x$ is rewritten as $(-1)(x - 2)$ and then the problem is changed from addition to subtraction.

EXAMPLE 7 Adding with Unlike Denominators

$\frac{6x}{x^2 - 4} + \frac{3}{2 - x}$ Original expressions

$= \frac{6x}{(x + 2)(x - 2)} + \frac{3}{(-1)(x - 2)}$ Factor denominators.

$= \frac{6x}{(x + 2)(x - 2)} - \frac{3(x + 2)}{(x + 2)(x - 2)}$ Rewrite expressions using LCD of $(x + 2)(x - 2)$.

$= \frac{6x}{(x + 2)(x - 2)} - \frac{3x + 6}{(x + 2)(x - 2)}$ Distributive Property

$= \frac{6x - (3x + 6)}{(x + 2)(x - 2)}$ Subtract.

$= \frac{6x - 3x - 6}{(x + 2)(x - 2)}$ Distributive Property

$= \frac{3x - 6}{(x + 2)(x - 2)}$ Simplify.

$= \frac{3\cancel{(x - 2)}}{(x + 2)\cancel{(x - 2)}}$ Factor and divide out common factor.

$= \frac{3}{x + 2}, \quad x \neq 2$ Simplified form

✓ **CHECKPOINT** *Now try Exercise 55.*

EXAMPLE 8 Subtracting with Unlike Denominators

$\frac{x}{x^2 - 5x + 6} - \frac{1}{x^2 - x - 2}$ Original expressions

$= \frac{x}{(x - 3)(x - 2)} - \frac{1}{(x - 2)(x + 1)}$ Factor denominators.

$= \frac{x(x + 1)}{(x - 3)(x - 2)(x + 1)} - \frac{1(x - 3)}{(x - 3)(x - 2)(x + 1)}$ Rewrite expressions using LCD of $(x - 3)(x - 2)(x + 1)$.

$= \frac{x^2 + x}{(x - 3)(x - 2)(x + 1)} - \frac{x - 3}{(x - 3)(x - 2)(x + 1)}$ Distributive Property

$= \frac{(x^2 + x) - (x - 3)}{(x - 3)(x - 2)(x + 1)}$ Subtract fractions.

$= \frac{x^2 + x - x + 3}{(x - 3)(x - 2)(x + 1)}$ Distributive Property

$= \frac{x^2 + 3}{(x - 3)(x - 2)(x + 1)}$ Simplified form

✓ **CHECKPOINT** *Now try Exercise 71.*

EXAMPLE 9 **Combining Three Rational Expressions**

$$\frac{4x}{x^2 - 16} + \frac{x}{x + 4} - \frac{2}{x} = \frac{4x}{(x + 4)(x - 4)} + \frac{x}{x + 4} - \frac{2}{x}$$

$$= \frac{4x(x)}{x(x + 4)(x - 4)} + \frac{x(x)(x - 4)}{x(x + 4)(x - 4)} - \frac{2(x + 4)(x - 4)}{x(x + 4)(x - 4)}$$

$$= \frac{4x^2 + x^2(x - 4) - 2(x^2 - 16)}{x(x + 4)(x - 4)}$$

$$= \frac{4x^2 + x^3 - 4x^2 - 2x^2 + 32}{x(x + 4)(x - 4)}$$

$$= \frac{x^3 - 2x^2 + 32}{x(x + 4)(x - 4)}$$

✓ **CHECKPOINT** *Now try Exercise 77.*

To add or subtract two rational expressions, you can use the LCD method or the basic definition

$$\frac{a}{b} \pm \frac{c}{d} = \frac{ad \pm bc}{bd}, \quad b \neq 0, d \neq 0.$$ Basic definition

This definition provides an efficient way of adding or subtracting two rational expressions that have no common factors in their denominators.

EXAMPLE 10 **Motor Vehicle and Parts Sales** 

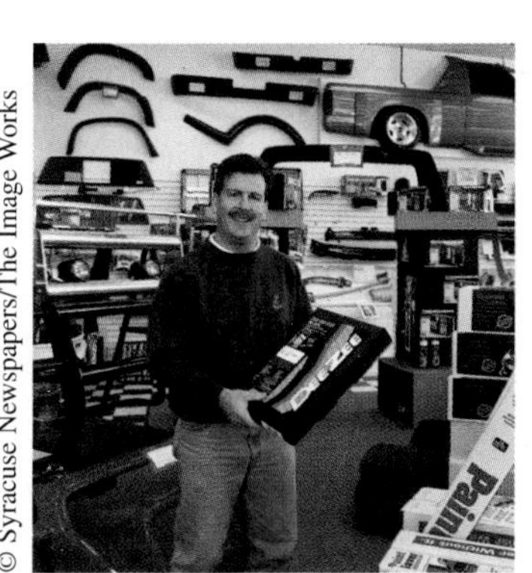

© Syracuse Newspapers/The Image Works

In 2005, E-commerce accounted for only about 2 percent of the sales of vehicles and vehicle parts in the United States.

For the years 2000 through 2005, the total annual retail sales T (in billions of dollars) and the E-commerce annual retail sales E (in billions of dollars) of motor vehicles and vehicle parts in the United States can be modeled by

$$T = \frac{-57.6t + 800}{-0.085t + 1} \quad \text{and} \quad E = \frac{t^2 + 4.3}{0.03t^2 + 1}, \quad 0 \leq t \leq 5$$

where t represents the year, with $t = 0$ corresponding to 2000. Find a rational model N for the annual retail sales *not* from E-commerce during this time period. (Source: U.S. Department of Labor)

Solution

To find a model for N, find the difference of T and E.

$$N = \frac{-57.6t + 800}{-0.085t + 1} - \frac{t^2 + 4.3}{0.03t^2 + 1}$$ Subtract E from T.

$$= \frac{(-57.6t + 800)(0.03t^2 + 1) - (-0.085t + 1)(t^2 + 4.3)}{(-0.085t + 1)(0.03t^2 + 1)}$$ Basic definition

$$= \frac{-1.643t^3 + 23t^2 - 57.2345t + 795.7}{(-0.085t + 1)(0.03t^2 + 1)}$$ Use FOIL Method and combine like terms.

✓ **CHECKPOINT** *Now try Exercise 89.*

Concept Check

1. *True or False?* Two rational expressions with *like* denominators have a common denominator.
2. When adding or subtracting rational expressions, how do you rewrite each rational expression as an equivalent expression whose denominator is the LCD?
3. In your own words, describe how to add or subtract rational expressions with *like* denominators.
4. In your own words, describe how to add or subtract rational expressions with *unlike* denominators.

7.3 EXERCISES

Go to pages 462–463 to record your assignments.

Developing Skills

In Exercises 1–22, combine and simplify. ***See Examples 1–3.***

✓ 1. $\frac{5x}{6} + \frac{4x}{6}$

2. $\frac{7y}{12} + \frac{9y}{12}$

3. $\frac{2}{3a} - \frac{11}{3a}$

4. $\frac{6}{19x} - \frac{7}{19x}$

5. $\frac{x}{9} - \frac{x+2}{9}$

6. $\frac{4-y}{4} + \frac{3y}{4}$

7. $\frac{z^2}{3} + \frac{z^2-2}{3}$

8. $\frac{10x^2+1}{3} - \frac{10x^2}{3}$

9. $\frac{2x+5}{3x} + \frac{1-x}{3x}$

10. $\frac{16+z}{5z} - \frac{11-z}{5z}$

11. $\frac{3y-22}{y-6} - \frac{2y-16}{y-6}$

12. $\frac{5x-1}{x+4} + \frac{5-4x}{x+4}$

13. $\frac{2x-1}{x(x-3)} + \frac{1-x}{x(x-3)}$

14. $\frac{3-2n}{n(n+2)} - \frac{1-3n}{n(n+2)}$

15. $\frac{w}{w^2-4} + \frac{2}{w^2-4}$

16. $\frac{d}{d^2-36} - \frac{6}{d^2-36}$

✓ 17. $\frac{c}{c^2+3c-4} - \frac{1}{c^2+3c-4}$

18. $\frac{2v}{2v^2-5v-12} + \frac{3}{2v^2-5v-12}$

19. $\frac{3y}{3} - \frac{3y-3}{3} - \frac{7}{3}$

20. $\frac{-16u}{9} - \frac{27-16u}{9} + \frac{2}{9}$

✓ 21. $\frac{x^2-4x}{x-3} + \frac{10-4x}{x-3} - \frac{x-8}{x-3}$

22. $\frac{6-7z}{z+4} + \frac{z^2-14}{z+4} + \frac{4z-20}{z+4}$

In Exercises 23–34, find the least common multiple of the expressions. ***See Example 4.***

✓ 23. $5x^2, 20x^3$

24. $14t^2, 42t^5$

25. $9y^3, 12y$

26. $18m^2, 45m$

27. $15x^2, 3(x+5)$

28. $6x^2, 15x(x-1)$

29. $63z^2(z+1), 14(z+1)^4$

30. $18y^3, 27y(y-3)^2$

31. $8t(t+2), 14(t^2-4)$

32. $6(x^2-4), 2x(x+2)$

33. $2y^2+y-1, 4y^2-2y$

34. $t^3+3t^2+9t, 2t^2(t^2-9)$

In Exercises 35–40, fill in the missing factor.

35. $\frac{7x^2}{4a(\quad)} = \frac{7}{4a}, \quad x \neq 0$

36. $\frac{8y^2}{(b+2)(\quad)} = \frac{2y}{b+2}, \quad y \neq 0$

37. $\dfrac{5r(\quad)}{3v(u+1)} = \dfrac{5r}{3v}, \quad u \neq -1$

38. $\dfrac{(3t+5)(\quad)}{10t^2(3t-5)} = \dfrac{3t+5}{2t}, \quad t \neq \dfrac{5}{3}$

39. $\dfrac{7y(\quad)}{4-x^2} = \dfrac{7y}{x-2}, \quad x \neq -2$

40. $\dfrac{4x^2(\quad)}{x^2-10x} = \dfrac{4x^2}{10-x}, \quad x \neq 0$

In Exercises 41–48, find the least common denominator of the two fractions and rewrite each fraction using the least common denominator.

41. $\dfrac{n+8}{3n-12}, \dfrac{10}{6n^2}$

42. $\dfrac{y-4}{2y+14}, \dfrac{3y}{10y^3}$

43. $\dfrac{2}{x^2(x-3)}, \dfrac{5}{x(x+3)}$

44. $\dfrac{5t}{2t(t-3)^2}, \dfrac{4}{t(t-3)}$

45. $\dfrac{v}{2v^2+2v}, \dfrac{4}{3v^2}$

46. $\dfrac{4x}{(x+5)^2}, \dfrac{x-2}{x^2-25}$

47. $\dfrac{x-8}{x^2-25}, \dfrac{9x}{x^2-10x+25}$

48. $\dfrac{3y}{y^2-y-12}, \dfrac{y-4}{y^2+3y}$

In Exercises 49–82, combine and simplify. ***See Examples 5–9.***

49. $\dfrac{5}{4x} - \dfrac{3}{5}$

50. $\dfrac{10}{b} + \dfrac{1}{10b}$

✓ 51. $\dfrac{7}{a} + \dfrac{14}{a^2}$

52. $\dfrac{1}{6u^2} - \dfrac{2}{9u}$

53. $25 + \dfrac{10}{x+4}$

54. $\dfrac{30}{x-6} - 4$

✓ 55. $\dfrac{20}{x-4} + \dfrac{20}{4-x}$

56. $\dfrac{15}{2-t} - \dfrac{7}{t-2}$

57. $\dfrac{3x}{x-8} - \dfrac{6}{8-x}$

58. $\dfrac{1}{y-6} + \dfrac{y}{6-y}$

59. $\dfrac{3x}{3x-2} + \dfrac{2}{2-3x}$

60. $\dfrac{y}{5y-3} - \dfrac{3}{3-5y}$

61. $\dfrac{9}{5v} + \dfrac{3}{v-1}$

62. $\dfrac{3}{y-1} + \dfrac{5}{4y}$

✓ 63. $\dfrac{x}{x+3} - \dfrac{5}{x-2}$

64. $\dfrac{1}{x+4} - \dfrac{1}{x+2}$

65. $\dfrac{12}{x^2-9} - \dfrac{2}{x-3}$

66. $\dfrac{12}{x^2-4} - \dfrac{3}{x+2}$

67. $\dfrac{3}{x-5} + \dfrac{2}{x+5}$

68. $\dfrac{7}{2x-3} + \dfrac{3}{2x+3}$

69. $\dfrac{4}{x^2} - \dfrac{4}{x^2+1}$

70. $\dfrac{3}{y^2-3} + \dfrac{2}{3y^2}$

✓ 71. $\dfrac{x}{x^2-x-30} - \dfrac{1}{x+5}$

72. $\dfrac{x}{x^2-9} + \dfrac{3}{x^2-5x+6}$

73. $\dfrac{4}{x-4} + \dfrac{16}{(x-4)^2}$

74. $\dfrac{3}{x-2} - \dfrac{1}{(x-2)^2}$

75. $\frac{y}{x^2 + xy} - \frac{x}{xy + y^2}$

76. $\frac{5}{x + y} + \frac{5}{x^2 - y^2}$

77. $\frac{4}{x} - \frac{2}{x^2} + \frac{4}{x + 3}$

78. $\frac{5}{2} - \frac{1}{2x} - \frac{3}{x + 1}$

79. $\frac{3u}{u^2 - 2uv + v^2} + \frac{2}{u - v} - \frac{u}{u - v}$

80. $\frac{1}{x - y} - \frac{3}{x + y} + \frac{3x - y}{x^2 - y^2}$

81. $\frac{x + 2}{x - 1} - \frac{2}{x + 6} - \frac{14}{x^2 + 5x - 6}$

82. $\frac{-2x - 10}{x^2 + 8x + 15} + \frac{2}{x + 3} + \frac{x}{x + 5}$

In Exercises 83 and 84, use a graphing calculator to graph the two equations in the same viewing window. Use the graphs to verify that the expressions are equivalent. Verify the results algebraically.

83. $y_1 = \frac{2}{x} + \frac{4}{x - 2}, \quad y_2 = \frac{6x - 4}{x(x - 2)}$

84. $y_1 = \frac{x}{3} - \frac{2}{x + 3}, \quad y_2 = \frac{x^2 + 3x - 6}{3x + 9}$

Solving Problems

85. *Work Rate* After working together for t hours on a common task, two workers have completed fractional parts of the job equal to $t/4$ and $t/6$. What fractional part of the task has been completed?

86. *Work Rate* After working together for t hours on a common task, two workers have completed fractional parts of the job equal to $t/3$ and $t/5$. What fractional part of the task has been completed?

87. *Rewriting a Fraction* The fraction $4/(x^3 - x)$ can be rewritten as a sum of three fractions, as follows.

$$\frac{4}{x^3 - x} = \frac{A}{x} + \frac{B}{x + 1} + \frac{C}{x - 1}$$

The numbers A, B, and C are the solutions of the system

$$\begin{cases} A + B + C = 0 \\ \quad - B + C = 0 \\ -A \qquad\quad = 4. \end{cases}$$

Solve the system and verify that the sum of the three resulting fractions is the original fraction.

88. *Rewriting a Fraction* The fraction

$$\frac{x + 1}{x^3 - x^2}$$

can be rewritten as a sum of three fractions, as follows.

$$\frac{x + 1}{x^3 - x^2} = \frac{A}{x} + \frac{B}{x^2} + \frac{C}{x - 1}$$

The numbers A, B, and C are the solutions of the system

$$\begin{cases} A \qquad + C = 0 \\ -A + B \qquad = 1 \\ \quad - B \qquad = 1. \end{cases}$$

Solve the system and verify that the sum of the three resulting fractions is the original fraction.

Undergraduate Students In Exercises 89 and 90, use the following models, which give the numbers (in thousands) of males M and females F enrolled as undergraduate students from 2000 through 2005.

$$M = \frac{1434.4t + 5797.28}{0.205t + 1}, \quad 0 \le t \le 5$$

and $F = \frac{1809.8t + 7362.51}{0.183t + 1}, \quad 0 \le t \le 5$

In these models, t represents the year, with $t = 0$ corresponding to 2000. (Source: U.S. National Center for Education Statistics)

89. Find a rational model T for the total number of undergraduate students (in thousands) from 2000 through 2005.

Year	2000	2001	2002
Undergraduates (in thousands)			

90. Use the model you found in Exercise 89 to complete the table showing the total number of undergraduate students (rounded to the nearest thousand) each year from 2000 through 2005.

Year	2003	2004	2005
Undergraduates (in thousands)			

Explaining Concepts

91. *Error Analysis* Describe the error.

$$\frac{x-1}{x+4} - \frac{4x-11}{x+4} = \frac{x-1-4x-11}{x+4}$$
$$= \frac{-3x-12}{x+4}$$
$$= \frac{-3(x+4)}{x+4}$$
$$= -3$$

92. *Error Analysis* Describe the error.

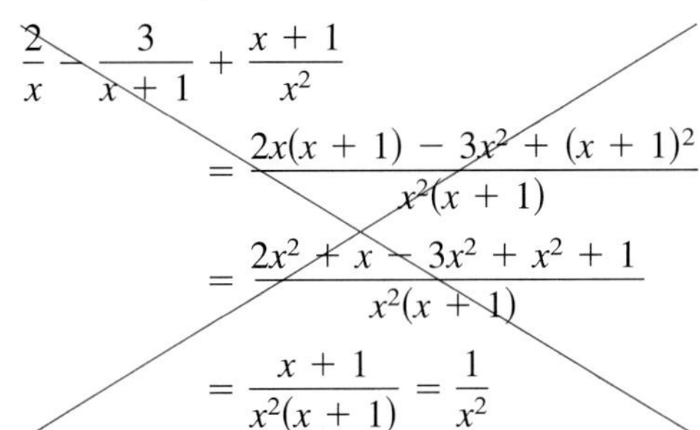

$$\frac{2}{x} - \frac{3}{x+1} + \frac{x+1}{x^2}$$
$$= \frac{2x(x+1) - 3x^2 + (x+1)^2}{x^2(x+1)}$$
$$= \frac{2x^2 + x - 3x^2 + x^2 + 1}{x^2(x+1)}$$
$$= \frac{x+1}{x^2(x+1)} = \frac{1}{x^2}$$

93. Is it possible for the least common denominator of two fractions to be the same as one of the fraction's denominators? If so, give an example.

94. Evaluate each expression at the given value of the variable in two different ways: (1) combine and simplify the rational expressions first and then evaluate the simplified expression at the given value of the variable, and (2) substitute the given value of the variable first and then simplify the resulting expression. Do you get the same result with each method? Discuss which method you prefer and why. List the advantages and/or disadvantages of each method.

(a) $\dfrac{1}{m-4} - \dfrac{1}{m+4} + \dfrac{3m}{m^2-16}, \quad m = 2$

(b) $\dfrac{x-2}{x^2-9} + \dfrac{3x+2}{x^2-5x+6}, \quad x = 4$

(c) $\dfrac{3y^2+16y-8}{y^2+2y-8} - \dfrac{y-1}{y-2} + \dfrac{y}{y+4}, \quad y = 3$

Cumulative Review

In Exercises 95–98, find the sum or difference.

95. $5v + (4 - 3v)$

96. $(2v + 7) + (9v + 8)$

97. $(x^2 - 4x + 3) - (6 - 2x)$

98. $(5y + 2) - (2y^2 + 8y - 5)$

In Exercises 99–102, factor the trinomial, if possible.

99. $x^2 - 7x + 12$

100. $c^2 + 6c + 10$

101. $2a^2 - 9a - 18$

102. $6w^2 + 14w - 12$

Mid-Chapter Quiz

Take this quiz as you would take a quiz in class. After you are done, check your work against the answers in the back of the book.

1. Determine the domain of $f(x) = \dfrac{x}{x^2 + x}$.

2. Evaluate $f(x) = \dfrac{2x - 1}{x^2 + 1}$ for the indicated values of x, and simplify. If it is not possible, state the reason.

(a) $f(3)$ (b) $f(1)$ (c) $f(-1)$ (d) $f\left(\frac{1}{2}\right)$

3. Evaluate $h(x) = (x^2 - 9)/(x^2 - x - 2)$ for the indicated values of x, and simplify. If it is not possible, state the reason.

(a) $h(-3)$ (b) $h(0)$ (c) $h(-1)$ (d) $h(5)$

In Exercises 4–9, simplify the rational expression.

4. $\dfrac{9y^2}{6y}$

5. $\dfrac{6u^4v^3}{15uv^3}$

6. $\dfrac{4x^2 - 1}{x - 2x^2}$

7. $\dfrac{(z + 3)^2}{2z^2 + 5z - 3}$

8. $\dfrac{5a^2b + 3ab^3}{a^2b^2}$

9. $\dfrac{2mn^2 - n^3}{2m^2 + mn - n^2}$

In Exercises 10–20, perform the indicated operations and simplify.

10. $\dfrac{11t^2}{6} \cdot \dfrac{9}{33t}$

11. $(x^2 + 2x) \cdot \dfrac{5}{x^2 - 4}$

12. $\dfrac{4}{3(x - 1)} \cdot \dfrac{12x}{6(x^2 + 2x - 3)}$

13. $\dfrac{32z^4}{5x^5y^5} \div \dfrac{80z^5}{25x^8y^6}$

14. $\dfrac{a - b}{9a + 9b} \div \dfrac{a^2 - b^2}{a^2 + 2a + 1}$

15. $\dfrac{10}{x^2 + 2x} \div \dfrac{15}{x^2 + 3x + 2}$

16. $\dfrac{3x}{x + 5} \cdot \dfrac{x + 4x - 5}{x^2} \div \dfrac{x - 1}{2x}$

17. $\dfrac{5u}{3(u + v)} \cdot \dfrac{2(u^2 - v^2)}{3v} \div \dfrac{25u^2}{18(u - v)}$

18. $\dfrac{5x - 6}{x - 2} + \dfrac{2x - 5}{x - 2}$

19. $\dfrac{x}{x^2 - 9} - \dfrac{4(x - 3)}{x + 3}$

20. $\dfrac{x^2 + 2}{x^2 - x - 2} + \dfrac{1}{x + 1} - \dfrac{x}{x - 2}$

21. You open a floral shop with a setup cost of \$25,000. The cost of creating one dozen floral arrangements is \$144.

(a) Write the total cost C as a function of x, the number of floral arrangements (in dozens) created.

(b) Write the average cost per dozen $\overline{C} = C/x$ as a function of x, the number of floral arrangements (in dozens) created.

(c) Find the value of $\overline{C}(500)$.

7.4 Complex Fractions

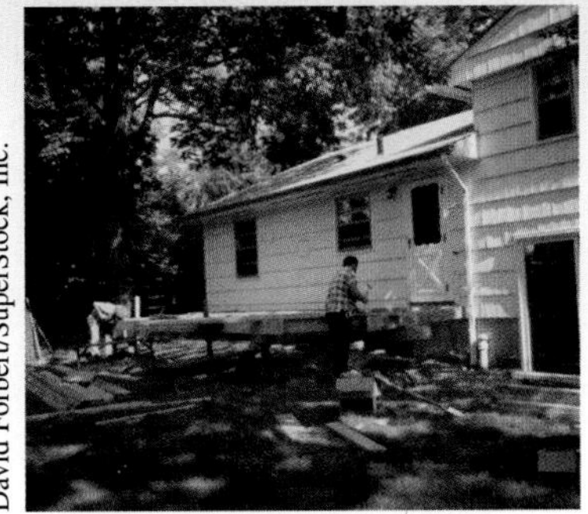
David Forbert/SuperStock, Inc.

Why You Should Learn It

Complex fractions can be used to model real-life situations. For instance, in Exercise 64 on page 439, a complex fraction is used to model the annual percent rate for a home-improvement loan.

What You Should Learn

1 ▶ Simplify complex fractions using rules for dividing rational expressions.

2 ▶ Simplify complex fractions having a sum or difference in the numerator and/or denominator.

1 ▶ Simplify complex fractions using rules for dividing rational expressions.

Complex Fractions

Problems involving the division of two rational expressions are sometimes written as **complex fractions.** A complex fraction is a fraction that has a fraction in its numerator or denominator, or both. The rules for dividing rational expressions still apply. For instance, consider the following complex fraction.

$$\frac{\left(\frac{x+2}{3}\right)}{\left(\frac{x-2}{x}\right)}$$

Numerator fraction: $\left(\frac{x+2}{3}\right)$

Main fraction line

Denominator fraction: $\left(\frac{x-2}{x}\right)$

To perform the division implied by this complex fraction, invert the denominator fraction (the divisor) and multiply, as follows.

$$\frac{\left(\frac{x+2}{3}\right)}{\left(\frac{x-2}{x}\right)} = \frac{x+2}{3} \cdot \frac{x}{x-2}$$

$$= \frac{x(x+2)}{3(x-2)}, \quad x \neq 0$$

Note that for complex fractions, you make the main fraction line slightly longer than the fraction lines in the numerator and denominator.

EXAMPLE 1 Simplifying a Complex Fraction

$$\frac{\left(\frac{5}{14}\right)}{\left(\frac{25}{8}\right)} = \frac{5}{14} \cdot \frac{8}{25} \qquad \text{Invert divisor and multiply.}$$

$$= \frac{\cancel{5} \cdot \cancel{2} \cdot 2 \cdot 2}{\cancel{2} \cdot 7 \cdot \cancel{5} \cdot 5} \qquad \text{Multiply, factor, and divide out common factors.}$$

$$= \frac{4}{35} \qquad \text{Simplified form}$$

✓ **CHECKPOINT** *Now try Exercise 1.*

EXAMPLE 2 Simplifying a Complex Fraction

Simplify the complex fraction.

$$\frac{\left(\frac{4y^3}{(5x)^2}\right)}{\left(\frac{(2y)^2}{10x^3}\right)}$$

Solution

$$\frac{\left(\frac{4y^3}{(5x)^2}\right)}{\left(\frac{(2y)^2}{10x^3}\right)} = \frac{4y^3}{25x^2} \cdot \frac{10x^3}{4y^2}$$ Invert divisor and multiply.

$$= \frac{4y^2 \cdot y \cdot 2 \cdot 5x^2 \cdot x}{5 \cdot 5x^2 \cdot 4y^2}$$ Multiply and factor.

$$= \frac{\cancel{4y^2} \cdot y \cdot 2 \cdot \cancel{5x^2} \cdot x}{5 \cdot \cancel{5x^2} \cdot \cancel{4y^2}}$$ Divide out common factors.

$$= \frac{2xy}{5}, \quad x \neq 0, \; y \neq 0$$ Simplified form

✓ CHECKPOINT *Now try Exercise 5.*

Study Tip

Domain restrictions result from the values that make any denominator zero in a complex fraction. In Example 2, note that the original expression has three denominators: $(5x)^2$, $10x^3$, and $(2y)^2/10x^3$. The domain restrictions that result from these denominators are $x \neq 0$ and $y \neq 0$.

EXAMPLE 3 Simplifying a Complex Fraction

Simplify the complex fraction.

$$\frac{\left(\frac{x+1}{x+2}\right)}{\left(\frac{x+1}{x+5}\right)}$$

Solution

$$\frac{\left(\frac{x+1}{x+2}\right)}{\left(\frac{x+1}{x+5}\right)} = \frac{x+1}{x+2} \cdot \frac{x+5}{x+1}$$ Invert divisor and multiply.

$$= \frac{(x+1)(x+5)}{(x+2)(x+1)}$$ Multiply numerators and denominators.

$$= \frac{\cancel{(x+1)}(x+5)}{(x+2)\cancel{(x+1)}}$$ Divide out common factors.

$$= \frac{x+5}{x+2}, \quad x \neq -1, \; x \neq -5$$ Simplified form

✓ CHECKPOINT *Now try Exercise 7.*

EXAMPLE 4 **Simplifying a Complex Fraction**

$$\frac{\left(\frac{x^2+4x+3}{x-2}\right)}{2x+6} = \frac{\left(\frac{x^2+4x+3}{x-2}\right)}{\left(\frac{2x+6}{1}\right)}$$ Rewrite denominator.

$$= \frac{x^2+4x+3}{x-2} \cdot \frac{1}{2x+6}$$ Invert divisor and multiply.

$$= \frac{(x+1)(x+3)}{(x-2)(2)(x+3)}$$ Multiply and factor.

$$= \frac{(x+1)\cancel{(x+3)}}{(x-2)(2)\cancel{(x+3)}}$$ Divide out common factor.

$$= \frac{x+1}{2(x-2)}, \quad x \neq -3$$ Simplified form

✓ **CHECKPOINT** *Now try Exercise 11.*

2 ▶ Simplify complex fractions having a sum or difference in the numerator and/or denominator.

Complex Fractions with Sums or Differences

Complex fractions can have numerators and/or denominators that are sums or differences of fractions. One way to simplify such a complex fraction is to combine the terms so that the numerator and denominator each consist of a single fraction. Then divide by inverting the denominator and multiplying.

Study Tip

Another way of simplifying the complex fraction in Example 5 is to multiply the numerator and denominator by $3x$, the least common denominator of all the fractions in the numerator and denominator. This produces the same result, as shown below.

$$\frac{\left(\frac{x}{3}+\frac{2}{3}\right)}{\left(1-\frac{2}{x}\right)} = \frac{\left(\frac{x}{3}+\frac{2}{3}\right)}{\left(1-\frac{2}{x}\right)} \cdot \frac{3x}{3x}$$

$$= \frac{\frac{x}{3}(3x)+\frac{2}{3}(3x)}{(1)(3x)-\frac{2}{x}(3x)}$$

$$= \frac{x^2+2x}{3x-6}$$

$$= \frac{x(x+2)}{3(x-2)}, \quad x \neq 0$$

EXAMPLE 5 **Simplifying a Complex Fraction**

$$\frac{\left(\frac{x}{3}+\frac{2}{3}\right)}{\left(1-\frac{2}{x}\right)} = \frac{\left(\frac{x}{3}+\frac{2}{3}\right)}{\left(\frac{x}{x}-\frac{2}{x}\right)}$$

$$= \frac{\left(\frac{x+2}{3}\right)}{\left(\frac{x-2}{x}\right)}$$ Add fractions. Rewrite with least common denominators.

$$= \frac{x+2}{3} \cdot \frac{x}{x-2}$$ Invert divisor and multiply.

$$= \frac{x(x+2)}{3(x-2)}, \quad x \neq 0$$ Simplified form

✓ **CHECKPOINT** *Now try Exercise 25.*

Study Tip

In Example 6, you might wonder about the domain restrictions that result from the main denominator

$$\left(\frac{3}{x+2}+\frac{2}{x}\right)$$

of the original expression. By setting this expression equal to zero and solving for x, you can see that it leads to the domain restriction $x \neq -\frac{4}{5}$. Notice that this restriction is implied by the denominator of the simplified expression.

EXAMPLE 6 Simplifying a Complex Fraction

$$\frac{\left(\frac{2}{x+2}\right)}{\left(\frac{3}{x+2}+\frac{2}{x}\right)} = \frac{\left(\frac{2}{x+2}\right)(x)(x+2)}{\left(\frac{3}{x+2}\right)(x)(x+2)+\left(\frac{2}{x}\right)(x)(x+2)}$$ $x(x + 2)$ is the least common denominator.

$$= \frac{2x}{3x+2(x+2)}$$ Multiply and simplify.

$$= \frac{2x}{3x+2x+4}$$ Distributive Property

$$= \frac{2x}{5x+4},\quad x \neq -2,\quad x \neq 0$$ Simplify.

Notice that the numerator and denominator of the complex fraction were multiplied by $(x)(x + 2)$, which is the least common denominator of the fractions in the original complex fraction.

✓ **CHECKPOINT** *Now try Exercise 41.*

When simplifying a rational expression containing negative exponents, first rewrite the expression with positive exponents and then proceed with simplifying the expression. This is demonstrated in Example 7.

EXAMPLE 7 Simplifying a Complex Fraction

$$\frac{5+x^{-2}}{8x^{-1}+x} = \frac{\left(5+\frac{1}{x^2}\right)}{\left(\frac{8}{x}+x\right)}$$ Rewrite with positive exponents.

$$= \frac{\left(\frac{5x^2}{x^2}+\frac{1}{x^2}\right)}{\left(\frac{8}{x}+\frac{x^2}{x}\right)}$$ Rewrite with least common denominators.

$$= \frac{\left(\frac{5x^2+1}{x^2}\right)}{\left(\frac{x^2+8}{x}\right)}$$ Add fractions.

$$= \frac{5x^2+1}{x^2}\cdot\frac{x}{x^2+8}$$ Invert divisor and multiply.

$$= \frac{\cancel{x}(5x^2+1)}{\cancel{x}(x)(x^2+8)}$$ Divide out common factor.

$$= \frac{5x^2+1}{x(x^2+8)}$$ Simplified form

✓ **CHECKPOINT** *Now try Exercise 45.*

Concept Check

1. What kind of division problem can be represented by a complex fraction?

2. Describe a method for simplifying complex fractions that uses the process for the "Division of Fractions" on page 38.

3. Describe the method for simplifying complex fractions that involves the use of a least common denominator.

4. Explain how you can find the implied domain restrictions for a complex fraction.

7.4 EXERCISES

Go to pages 462–463 to record your assignments.

Developing Skills

In Exercises 1–22, simplify the complex fraction. ***See Examples 1–4.***

✓ 1. $\dfrac{\left(\dfrac{3}{16}\right)}{\left(\dfrac{9}{12}\right)}$

2. $\dfrac{\left(\dfrac{20}{21}\right)}{\left(\dfrac{8}{7}\right)}$

3. $\dfrac{\left(\dfrac{8x^2y}{3z^2}\right)}{\left(\dfrac{4xy}{9z^5}\right)}$

4. $\dfrac{\left(\dfrac{36x^4}{5y^4z^5}\right)}{\left(\dfrac{9xy^2}{20z^5}\right)}$

✓ 5. $\dfrac{\left(\dfrac{6x^3}{(5y)^2}\right)}{\left(\dfrac{(3x)^2}{15y^4}\right)}$

6. $\dfrac{\left(\dfrac{(3r)^3}{10t^4}\right)}{\left(\dfrac{9r}{(2t)^2}\right)}$

✓ 7. $\dfrac{\left(\dfrac{y}{3-y}\right)}{\left(\dfrac{y^2}{y-3}\right)}$

8. $\dfrac{\left(\dfrac{x}{x-4}\right)}{\left(\dfrac{x}{4-x}\right)}$

9. $\dfrac{\left(\dfrac{25x^2}{x-5}\right)}{\left(\dfrac{10x}{5+4x-x^2}\right)}$

10. $\dfrac{\left(\dfrac{5x}{x+7}\right)}{\left(\dfrac{10}{x^2+8x+7}\right)}$

✓ 11. $\dfrac{\left(\dfrac{x^2+3x-10}{x+4}\right)}{3x-6}$

12. $\dfrac{\left(\dfrac{x^2-2x-8}{x-1}\right)}{5x-20}$

13. $\dfrac{2x-14}{\left(\dfrac{x^2-9x+14}{x+3}\right)}$

14. $\dfrac{4x+16}{\left(\dfrac{x^2+9x+20}{x-1}\right)}$

15. $\dfrac{\left(\dfrac{6x^2-17x+5}{3x^2+3x}\right)}{\left(\dfrac{3x-1}{3x+1}\right)}$

16. $\dfrac{\left(\dfrac{6x^2-13x-5}{5x^2+5x}\right)}{\left(\dfrac{2x-5}{5x+1}\right)}$

17. $\dfrac{\left(\dfrac{16x^2+8x+1}{3x^2+8x-3}\right)}{\left(\dfrac{4x^2-3x-1}{x^2+6x+9}\right)}$

18. $\dfrac{\left(\dfrac{9x^2 - 24x + 16}{x^2 + 10x + 25}\right)}{\left(\dfrac{6x^2 - 5x - 4}{2x^2 + 3x - 35}\right)}$

19. $\dfrac{x^2 + x - 6}{x^2 - 4} \div \dfrac{x + 3}{x^2 + 4x + 4}$

20. $\dfrac{t^3 + t^2 - 9t - 9}{t^2 - 5t + 6} \div \dfrac{t^2 + 6t + 9}{t - 2}$

21. $\dfrac{\left(\dfrac{x^2 - 3x - 10}{x^2 - 4x + 4}\right)}{\left(\dfrac{21 + 4x - x^2}{x^2 - 5x - 14}\right)}$

22. $\dfrac{\left(\dfrac{x^2 + 5x + 6}{4x^2 - 20x + 25}\right)}{\left(\dfrac{x^2 - 5x - 24}{4x^2 - 25}\right)}$

31. $\dfrac{\left(\dfrac{3}{x^2} + \dfrac{1}{x}\right)}{\left(2 - \dfrac{4}{5x}\right)}$

32. $\dfrac{\left(16 - \dfrac{1}{x^2}\right)}{\left(\dfrac{1}{4x^2} - 4\right)}$

33. $\dfrac{\left(\dfrac{y}{x} - \dfrac{x}{y}\right)}{\left(\dfrac{x + y}{xy}\right)}$

34. $\dfrac{\left(\dfrac{x}{y} - \dfrac{y}{x}\right)}{\left(\dfrac{x - y}{xy}\right)}$

35. $\dfrac{\left(x - \dfrac{2y^2}{x - y}\right)}{x - 2y}$

36. $\dfrac{\left(x - \dfrac{6y^2}{x - y}\right)}{x - 3y}$

In Exercises 23–44, simplify the complex fraction. ***See Examples 5 and 6.***

23. $\dfrac{\left(1 + \dfrac{4}{y}\right)}{y}$

24. $\dfrac{x}{\left(\dfrac{3}{x} + 2\right)}$

✓ 25. $\dfrac{\left(\dfrac{4}{x} + 3\right)}{\left(\dfrac{4}{x} - 3\right)}$

26. $\dfrac{\left(\dfrac{1}{t} - 1\right)}{\left(\dfrac{1}{t} + 1\right)}$

27. $\dfrac{\left(\dfrac{x}{2}\right)}{\left(2 + \dfrac{3}{x}\right)}$

28. $\dfrac{\left(1 - \dfrac{2}{x}\right)}{\left(\dfrac{x}{2}\right)}$

29. $\dfrac{\left(3 + \dfrac{9}{x - 3}\right)}{\left(4 + \dfrac{12}{x - 3}\right)}$

30. $\dfrac{\left(4 + \dfrac{16}{x - 4}\right)}{\left(5 + \dfrac{20}{x - 4}\right)}$

37. $\dfrac{\left(1 - \dfrac{1}{y}\right)}{\left(\dfrac{1 - 4y}{y - 3}\right)}$

38. $\dfrac{\left(\dfrac{x + 1}{x + 2} - \dfrac{1}{x}\right)}{\left(\dfrac{2}{x + 2}\right)}$

39. $\dfrac{\left(\dfrac{10}{x + 1}\right)}{\left(\dfrac{1}{2x + 2} + \dfrac{3}{x + 1}\right)}$

40. $\dfrac{\left(\dfrac{2}{x + 5}\right)}{\left(\dfrac{2}{x + 5} + \dfrac{1}{4x + 20}\right)}$

✓ 41. $\dfrac{\left(\dfrac{1}{x} - \dfrac{1}{x + 1}\right)}{\left(\dfrac{1}{x + 1}\right)}$

42. $\dfrac{\left(\dfrac{5}{y} - \dfrac{6}{2y + 1}\right)}{\left(\dfrac{5}{2y + 1}\right)}$

43. $\dfrac{\left(\dfrac{x}{x - 3} - \dfrac{2}{3}\right)}{\left(\dfrac{10}{3x} + \dfrac{x^2}{x - 3}\right)}$

44. $\dfrac{\left(\dfrac{1}{2x} - \dfrac{6}{x + 5}\right)}{\left(\dfrac{x}{x - 5} + \dfrac{1}{x}\right)}$

In Exercises 45–52, simplify the expression. *See Example 7.*

✓ **45.** $\dfrac{2y - y^{-1}}{10 - y^{-2}}$

46. $\dfrac{9x - x^{-1}}{3 + x^{-1}}$

47. $\dfrac{7x^2 + 2x^{-1}}{5x^{-3} + x}$

48. $\dfrac{3x^{-2} - x}{4x^{-1} + 6x}$

49. $\dfrac{x^{-1} + y^{-1}}{x^{-1} - y^{-1}}$

50. $\dfrac{x^{-1} - y^{-1}}{x^{-2} - y^{-2}}$

51. $\dfrac{x^{-2} - y^{-2}}{(x + y)^2}$

52. $\dfrac{x - y}{x^{-2} - y^{-2}}$

In Exercises 53 and 54, use the function to find and simplify the expression for

$$\frac{f(2 + h) - f(2)}{h}.$$

53. $f(x) = \dfrac{1}{x}$

54. $f(x) = \dfrac{x}{x - 1}$

Solving Problems

55. *Average of Two Numbers* Determine the average of two real numbers $x/5$ and $x/6$.

56. *Average of Two Numbers* Determine the average of two real numbers $2x/3$ and $3x/5$.

57. *Average of Two Numbers* Determine the average of two real numbers $2x/3$ and $x/4$.

58. *Average of Two Numbers* Determine the average of two real numbers $4/a^2$ and $2/a$.

59. *Average of Two Numbers* Determine the average of two real numbers $(b + 5)/4$ and $2/b$.

60. *Average of Two Numbers* Determine the average of two real numbers $5/2s$ and $(s + 1)/5$.

61. *Number Problem* Find three real numbers that divide the real number line between $x/9$ and $x/6$ into four equal parts (see figure).

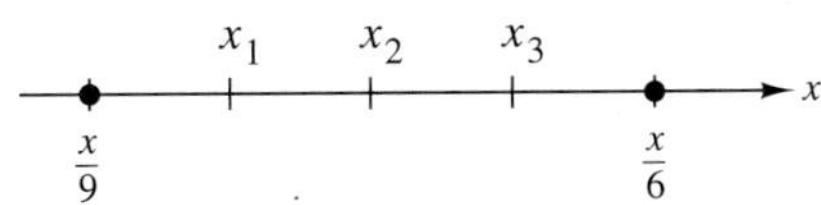

62. *Number Problem* Find two real numbers that divide the real number line between $x/3$ and $5x/4$ into three equal parts (see figure).

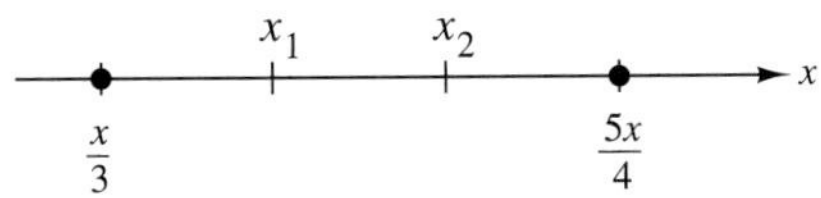

63. *Electrical Resistance* When two resistors of resistances R_1 and R_2 are connected in parallel, the total resistance is modeled by

$$\frac{1}{\left(\dfrac{1}{R_1} + \dfrac{1}{R_2}\right)}.$$

Simplify this complex fraction.

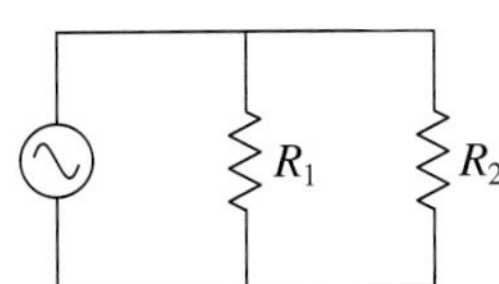

64. *Monthly Payment* The approximate annual percent interest rate r of a monthly installment loan is

$$r = \frac{\left[\dfrac{24(MN - P)}{N}\right]}{\left(P + \dfrac{MN}{12}\right)}$$

where N is the total number of payments, M is the monthly payment, and P is the amount financed.

(a) Simplify the expression.

(b) Approximate the annual percent interest rate for a four-year home-improvement loan of \$15,000 with monthly payments of \$350.

In Exercises 65 and 66, use the following models, which give the number N (in thousands) of cellular telephone subscribers and the annual revenue R (in millions of dollars) from cell phone subscriptions in the United States from 2000 through 2005.

$$N = \frac{6433.62t + 111{,}039.2}{-0.06t + 1}, \quad 0 \le t \le 5$$

and $R = \dfrac{8123.73t + 60{,}227.5}{-0.04t + 1}, \quad 0 \le t \le 5$

In these models, t represents the year, with $t = 0$ corresponding to 2000. (Source: Cellular Telecommunications and Internet Association)

© Strauss/Curtis/CORBIS

65. (a) Use a graphing calculator to graph the two models in the same viewing window.

(b) Find a model for the average monthly bill per subscriber. (*Note:* Modify the revenue model from years to months.)

66. (a) Use the model in Exercise 65 (b) to complete the table.

Year, t	0	1	2	3	4	5
Monthly bill						

(b) Use the model in Exercise 65(b) to predict the average monthly bill per subscriber in 2006, 2007, and 2008. Notice that, according to the model, N and R increase in these years, but the average bill decreases. Explain how this is possible.

Explaining Concepts

67. Is the simplified form of a complex fraction a complex fraction? Explain.

68. Describe the effect of multiplying two rational expressions by their least common denominator.

Error Analysis In Exercises 69 and 70, describe and correct the error.

69. $\dfrac{\left(\dfrac{a}{b}\right)}{b} = \dfrac{a}{\left(\dfrac{b}{b}\right)} = \dfrac{a}{1} = a, \quad b \neq 0$

70. $\dfrac{\left(\dfrac{a}{b}\right)}{\left(\dfrac{b}{c}\right)} = \dfrac{a}{b} \cdot \dfrac{b}{c} = \dfrac{a}{c}, \quad b \neq 0$

Cumulative Review

In Exercises 71 and 72, use the rules of exponents to simplify the expression.

71. $(2y)^3(3y)^2$

72. $\dfrac{27x^4y^2}{9x^3y}$

In Exercises 73 and 74, factor the trinomial.

73. $3x^2 + 5x - 2$

74. $x^2 + xy - 2y^2$

In Exercises 75–78, divide and simplify.

75. $\dfrac{x^2}{2} \div 4x$

76. $\dfrac{4x^3}{3} \div \dfrac{2x^2}{9}$

77. $\dfrac{(x + 1)^2}{x + 2} \div \dfrac{x + 1}{(x + 2)^3}$

78. $\dfrac{x^2 - 4x + 4}{x - 3} \div \dfrac{x^2 - 3x + 2}{x^2 - 6x + 9}$

7.5 Solving Rational Equations

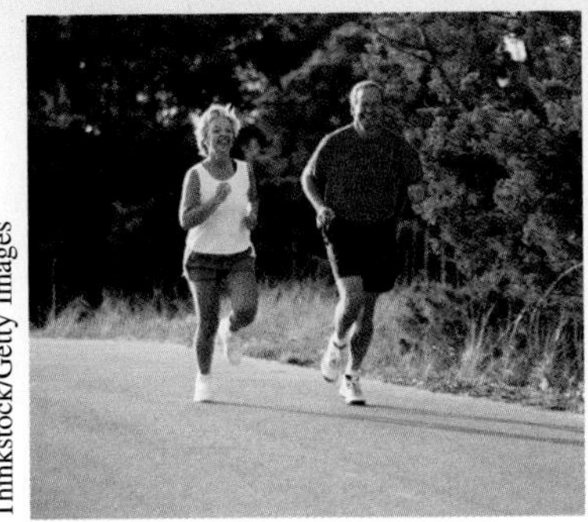
Thinkstock/Getty Images

What You Should Learn

1 ▶ Solve rational equations containing constant denominators.

2 ▶ Solve rational equations containing variable denominators.

Why You Should Learn It

Rational equations can be used to model and solve real-life applications. For instance, in Exercise 86 on page 448, you will use a rational equation to determine the speeds of two runners.

Equations Containing Constant Denominators

1 ▶ Solve rational equations containing constant denominators.

In Section 3.2, you studied a strategy for solving equations that contain fractions with *constant* denominators. That procedure is reviewed here because it is the basis for solving more general equations involving fractions. Recall from Section 3.2 that you can "clear an equation of fractions" by multiplying each side of the equation by the least common denominator (LCD) of the fractions in the equation. Note how this is done in the next three examples.

Study Tip

A *rational equation* is an equation containing one or more rational expressions.

EXAMPLE 1 An Equation Containing Constant Denominators

Solve $\frac{3}{5} = \frac{x}{2} + 1$.

Solution

The least common denominator of the fractions is 10, so begin by multiplying each side of the equation by 10.

$$\frac{3}{5} = \frac{x}{2} + 1$$ Write original equation.

$$10\left(\frac{3}{5}\right) = 10\left(\frac{x}{2} + 1\right)$$ Multiply each side by LCD of 10.

$$6 = 5x + 10$$ Distribute and simplify.

$$-4 = 5x \quad \Rightarrow \quad -\frac{4}{5} = x$$ Subtract 10 from each side, then divide each side by 5.

The solution is $x = -\frac{4}{5}$. You can check this in the original equation as follows.

Check

$$\frac{3}{5} \stackrel{?}{=} \frac{-4/5}{2} + 1$$ Substitute $-\frac{4}{5}$ for x in the original equation.

$$\frac{3}{5} \stackrel{?}{=} -\frac{4}{5} \cdot \frac{1}{2} + 1$$ Invert divisor and multiply.

$$\frac{3}{5} = -\frac{2}{5} + 1$$ Solution checks. ✓

✓ **CHECKPOINT** *Now try Exercise 5.*

EXAMPLE 2 An Equation Containing Constant Denominators

Solve $\frac{x-3}{6} = 7 - \frac{x}{12}$.

Solution

The least common denominator of the fractions is 12, so begin by multiplying each side of the equation by 12.

$$\frac{x-3}{6} = 7 - \frac{x}{12}$$ Write original equation.

$$12\left(\frac{x-3}{6}\right) = 12\left(7 - \frac{x}{12}\right)$$ Multiply each side by LCD of 12.

$$2x - 6 = 84 - x$$ Distribute and simplify.

$$3x - 6 = 84$$ Add x to each side.

$$3x = 90 \quad \Rightarrow \quad x = 30$$ Add 6 to each side, then divide each side by 3.

The solution is $x = 30$. Check this in the original equation.

✓ **CHECKPOINT** *Now try Exercise 11.*

EXAMPLE 3 An Equation That Has Two Solutions

Solve $\frac{x^2}{3} + \frac{x}{2} = \frac{5}{6}$.

Solution

The least common denominator of the fractions is 6, so begin by multiplying each side of the equation by 6.

$$\frac{x^2}{3} + \frac{x}{2} = \frac{5}{6}$$ Write original equation.

$$6\left(\frac{x^2}{3} + \frac{x}{2}\right) = 6\left(\frac{5}{6}\right)$$ Multiply each side by LCD of 6.

$$\frac{6x^2}{3} + \frac{6x}{2} = \frac{30}{6}$$ Distributive Property

$$2x^2 + 3x = 5$$ Simplify.

$$2x^2 + 3x - 5 = 0$$ Subtract 5 from each side.

$$(2x + 5)(x - 1) = 0$$ Factor.

$$2x + 5 = 0 \quad \Rightarrow \quad x = -\tfrac{5}{2}$$ Set 1st factor equal to 0.

$$x - 1 = 0 \quad \Rightarrow \quad x = 1$$ Set 2nd factor equal to 0.

The solutions are $x = -\frac{5}{2}$ and $x = 1$. Check these in the original equation.

✓ **CHECKPOINT** *Now try Exercise 15.*

2 ▶ Solve rational equations containing variable denominators.

Equations Containing Variable Denominators

In a rational expression, remember that the variable(s) cannot take on values that make the denominator zero. This is especially critical in solving rational equations that contain variable denominators.

Technology: Tip

You can use a graphing calculator to approximate the solution of the equation in Example 4. To do this, graph the left side of the equation and the right side of the equation in the same viewing window.

$$y_1 = \frac{7}{x} - \frac{1}{3x} \text{ and } y_2 = \frac{8}{3}$$

The solution of the equation is the x-coordinate of the point at which the two graphs intersect, as shown below. You can use the *intersect* feature of the graphing calculator to approximate the point of intersection to be $\left(\frac{5}{2}, \frac{8}{3}\right)$. So, the solution is $x = \frac{5}{2}$, which is the same solution obtained in Example 4.

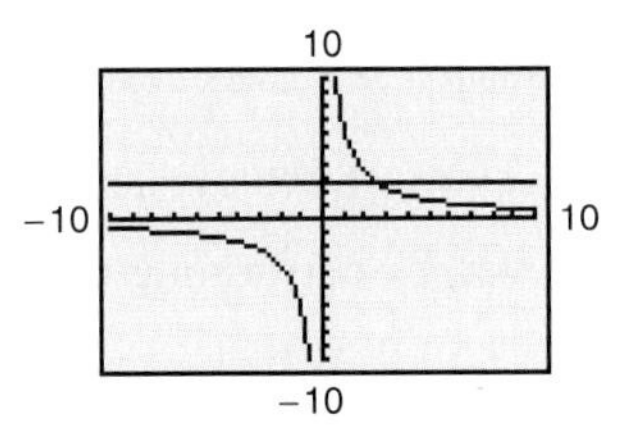

EXAMPLE 4 An Equation Containing Variable Denominators

Solve the equation.

$$\frac{7}{x} - \frac{1}{3x} = \frac{8}{3}$$

Solution

The least common denominator of the fractions is $3x$, so begin by multiplying each side of the equation by $3x$.

$$\frac{7}{x} - \frac{1}{3x} = \frac{8}{3}$$ Write original equation.

$$3x\left(\frac{7}{x} - \frac{1}{3x}\right) = 3x\left(\frac{8}{3}\right)$$ Multiply each side by LCD of $3x$.

$$\frac{21x}{x} - \frac{3x}{3x} = \frac{24x}{3}$$ Distributive Property

$$21 - 1 = 8x$$ Simplify.

$$\frac{20}{8} = x$$ Combine like terms and divide each side by 8.

$$\frac{5}{2} = x$$ Simplify.

The solution is $x = \frac{5}{2}$. You can check this in the original equation as follows.

Check

$$\frac{7}{x} - \frac{1}{3x} = \frac{8}{3}$$ Write original equation.

$$\frac{7}{5/2} - \frac{1}{3(5/2)} \stackrel{?}{=} \frac{8}{3}$$ Substitute $\frac{5}{2}$ for x.

$$7\left(\frac{2}{5}\right) - \left(\frac{1}{3}\right)\left(\frac{2}{5}\right) \stackrel{?}{=} \frac{8}{3}$$ Invert divisors and multiply.

$$\frac{14}{5} - \frac{2}{15} \stackrel{?}{=} \frac{8}{3}$$ Simplify.

$$\frac{40}{15} \stackrel{?}{=} \frac{8}{3}$$ Combine like terms.

$$\frac{8}{3} = \frac{8}{3}$$ Solution checks. ✓

✓ ***CHECKPOINT*** *Now try Exercise 25.*

Throughout the text, the importance of checking solutions is emphasized. Up to this point, the main reason for checking has been to make sure that you did not make arithmetic errors in the solution process. In the next example, you will see that there is another reason for checking solutions in the *original* equation. That is, even with no mistakes in the solution process, it can happen that a "trial solution" does not satisfy the original equation. This type of solution is called an **extraneous solution.** An extraneous solution of an equation *must not* be listed as an actual solution.

EXAMPLE 5 An Equation with No Solution

Solve $\frac{5x}{x-2} = 7 + \frac{10}{x-2}$.

Solution

The least common denominator of the fractions is $x - 2$, so begin by multiplying each side of the equation by $x - 2$.

$$\frac{5x}{x-2} = 7 + \frac{10}{x-2}$$ Write original equation.

$$(x-2)\left(\frac{5x}{x-2}\right) = (x-2)\left(7 + \frac{10}{x-2}\right)$$ Multiply each side by $x - 2$.

$$5x = 7(x-2) + 10$$ Distribute and simplify.

$$5x = 7x - 14 + 10$$ Distributive Property

$$5x = 7x - 4$$ Combine like terms.

$$-2x = -4$$ Subtract $7x$ from each side.

$$x = 2$$ Divide each side by -2.

At this point, the solution appears to be $x = 2$. However, by performing a check, you can see that this "trial solution" is extraneous.

Check

$$\frac{5x}{x-2} = 7 + \frac{10}{x-2}$$ Write original equation.

$$\frac{5(2)}{2-2} \stackrel{?}{=} 7 + \frac{10}{2-2}$$ Substitute 2 for x.

$$\frac{10}{0} \stackrel{?}{=} 7 + \frac{10}{0}$$ Solution does not check. ✗

Because the check results in *division by zero*, you can conclude that 2 is extraneous. So, the original equation has no solution.

✓ **CHECKPOINT** ***Now try Exercise 45.***

Notice that $x = 2$ is excluded from the domains of the two fractions in the original equation in Example 5. You may find it helpful when solving these types of equations to list the domain restrictions *before* beginning the solution process.

Study Tip

Although cross-multiplication can be a little quicker than multiplying by the least common denominator, remember that *it can be used only with equations that have a single fraction on each side* of the equation.

EXAMPLE 6 Cross-Multiplying

Solve $\dfrac{2x}{x+4} = \dfrac{3}{x-1}$.

Solution

The domain is all real values of x such that $x \neq -4$ and $x \neq 1$. You can use cross-multiplication to solve this equation.

$$\begin{aligned}
\frac{2x}{x+4} &= \frac{3}{x-1} && \text{Write original equation.}\\
2x(x-1) &= 3(x+4) && \text{Cross-multiply.}\\
2x^2 - 2x &= 3x + 12 && \text{Distributive Property}\\
2x^2 - 5x - 12 &= 0 && \text{Subtract } 3x \text{ and } 12 \text{ from each side.}\\
(2x+3)(x-4) &= 0 && \text{Factor.}\\
2x + 3 = 0 \quad &\Longrightarrow \quad x = -\tfrac{3}{2} && \text{Set 1st factor equal to 0.}\\
x - 4 = 0 \quad &\Longrightarrow \quad x = 4 && \text{Set 2nd factor equal to 0.}
\end{aligned}$$

The solutions are $x = -\frac{3}{2}$ and $x = 4$. Check these in the original equation.

✓ ***CHECKPOINT*** *Now try Exercise 47.*

Technology: Discovery

Use a graphing calculator to graph the equation

$$y = \frac{3x}{x+1} - \frac{12}{x^2-1} - 2.$$

Then use the *zero* or *root* feature of the calculator to determine the x-intercepts. How do the x-intercepts compare with the solutions to Example 7? What can you conclude?

EXAMPLE 7 An Equation That Has Two Solutions

Solve $\dfrac{3x}{x+1} = \dfrac{12}{x^2-1} + 2$.

Solution

The domain is all real values of x such that $x \neq 1$ and $x \neq -1$. The least common denominator is $(x+1)(x-1) = x^2 - 1$.

$$\begin{aligned}
(x^2-1)\left(\frac{3x}{x+1}\right) &= (x^2-1)\left(\frac{12}{x^2-1} + 2\right) && \text{Multiply each side of original equation by LCD of } x^2-1.\\
(x-1)(3x) &= 12 + 2(x^2-1) && \text{Simplify.}\\
3x^2 - 3x &= 12 + 2x^2 - 2 && \text{Distributive Property}\\
x^2 - 3x - 10 &= 0 && \text{Subtract } 2x^2 \text{ and } 10 \text{ from each side.}\\
(x+2)(x-5) &= 0 && \text{Factor.}\\
x + 2 = 0 \quad &\Longrightarrow \quad x = -2 && \text{Set 1st factor equal to 0.}\\
x - 5 = 0 \quad &\Longrightarrow \quad x = 5 && \text{Set 2nd factor equal to 0.}
\end{aligned}$$

The solutions are $x = -2$ and $x = 5$. Check these in the original equation.

✓ ***CHECKPOINT*** *Now try Exercise 61.*

Concept Check

1. What is a rational equation?

2. Describe how to solve a rational equation.

3. Explain the domain restrictions that may exist for a rational equation.

4. When can you use cross-multiplication to solve a rational equation? Explain.

7.5 EXERCISES

Go to pages 462–463 to record your assignments.

Developing Skills

In Exercises 1–4, determine whether each value of x is a solution of the equation.

Equation	*Values*
1. $\frac{x}{3} - \frac{x}{5} = \frac{4}{3}$	(a) $x = 0$ (b) $x = -2$ (c) $x = \frac{1}{8}$ (d) $x = 10$
2. $\frac{x}{4} + \frac{3}{4x} = 1$	(a) $x = -1$ (b) $x = 1$ (c) $x = 3$ (d) $x = \frac{1}{2}$
3. $x = 4 + \frac{21}{x}$	(a) $x = 0$ (b) $x = -3$ (c) $x = 7$ (d) $x = -1$
4. $5 - \frac{1}{x - 3} = 2$	(a) $x = \frac{10}{3}$ (b) $x = -\frac{1}{3}$ (c) $x = 0$ (d) $x = 1$

In Exercises 5–22, solve the equation. ***See Examples 1–3.***

✓ 5. $\frac{x}{6} - 1 = \frac{2}{3}$

6. $\frac{y}{8} + 7 = -\frac{1}{2}$

7. $\frac{1}{4} = \frac{z + 1}{8}$

8. $\frac{a}{2} = \frac{a + 2}{3}$

9. $\frac{x}{4} + \frac{x}{2} = \frac{2x}{3}$

10. $\frac{x}{4} - \frac{x}{6} = \frac{1}{4}$

✓ 11. $\frac{z + 2}{3} = 4 - \frac{z}{12}$

12. $\frac{2y - 9}{6} = 3y - \frac{3}{4}$

13. $\frac{x - 5}{5} + 3 = -\frac{x}{4}$

14. $\frac{4x - 2}{7} - \frac{5}{14} = 2x$

✓ 15. $\frac{t}{2} = 12 - \frac{3t^2}{2}$

16. $\frac{x^2}{2} - \frac{3x}{5} = -\frac{1}{10}$

17. $\frac{5y - 1}{12} + \frac{y}{3} = -\frac{1}{4}$

18. $\frac{z - 4}{9} - \frac{3z + 1}{18} = \frac{3}{2}$

19. $\frac{h + 2}{5} - \frac{h - 1}{9} = \frac{2}{3}$

20. $\frac{u - 2}{6} + \frac{2u + 5}{15} = 3$

21. $\frac{x + 5}{4} - \frac{3x - 8}{3} = \frac{4 - x}{12}$

22. $\frac{2x - 7}{10} - \frac{3x + 1}{5} = \frac{6 - x}{5}$

In Exercises 23–66, solve the equation. (Check for extraneous solutions.) ***See Examples 4–7.***

23. $\frac{9}{25 - y} = -\frac{1}{4}$

24. $-\frac{6}{u + 3} = \frac{2}{3}$

✓ 25. $5 - \frac{12}{a} = \frac{5}{3}$

26. $\frac{5}{b} - 18 = 21$

27. $\frac{4}{x} - \frac{7}{5x} = -\frac{1}{2}$

28. $\frac{5}{3} = \frac{6}{7x} + \frac{2}{x}$

29. $\frac{12}{y + 5} + \frac{1}{2} = 2$

30. $\frac{7}{8} - \frac{16}{t - 2} = \frac{3}{4}$

31. $\frac{5}{x} = \frac{25}{3(x + 2)}$

32. $\frac{10}{x + 4} = \frac{15}{4(x + 1)}$

33. $\frac{8}{3x + 5} = \frac{1}{x + 2}$

34. $\frac{500}{3x + 5} = \frac{50}{x - 3}$

35. $\dfrac{3}{x+2} - \dfrac{1}{x} = \dfrac{1}{5x}$

36. $\dfrac{12}{x+5} + \dfrac{5}{x} = \dfrac{20}{x}$

37. $\dfrac{1}{2} = \dfrac{18}{x^2}$

38. $\dfrac{1}{4} = \dfrac{16}{z^2}$

39. $\dfrac{t}{4} = \dfrac{4}{t}$

40. $\dfrac{20}{u} = \dfrac{u}{5}$

41. $x + 1 = \dfrac{72}{x}$

42. $\dfrac{48}{x} = x - 2$

43. $y + \dfrac{18}{y} = 9$

44. $x - \dfrac{24}{x} = 5$

✓ **45.** $\dfrac{4}{x(x-1)} + \dfrac{3}{x} = \dfrac{4}{x-1}$

46. $\dfrac{x-2}{2} - \dfrac{15}{2x} = 0$

✓ **47.** $\dfrac{2x}{5} = \dfrac{x^2-5x}{5x}$

48. $\dfrac{3x}{4} = \dfrac{x^2+3x}{8x}$

49. $\dfrac{y+1}{y+10} = \dfrac{y-2}{y+4}$

50. $\dfrac{x-3}{x+1} = \dfrac{x-6}{x+5}$

51. $\dfrac{15}{x} + \dfrac{9x-7}{x+2} = 9$

52. $\dfrac{3z-2}{z+1} = 4 - \dfrac{z+2}{z-1}$

53. $\dfrac{2}{6q+5} - \dfrac{3}{4(6q+5)} = \dfrac{1}{28}$

54. $\dfrac{10}{x(x-2)} + \dfrac{4}{x} = \dfrac{5}{x-2}$

55. $\dfrac{4}{2x+3} + \dfrac{17}{5x-3} = 3$

56. $\dfrac{5}{3x+1} + \dfrac{3}{2x+2} = 2$

57. $\dfrac{2}{x-10} - \dfrac{3}{x-2} = \dfrac{6}{x^2-12x+20}$

58. $\dfrac{5}{x+2} + \dfrac{2}{x^2-6x-16} = -\dfrac{4}{x-8}$

59. $\dfrac{x+3}{x^2-9} + \dfrac{4}{3-x} - 2 = 0$

60. $1 - \dfrac{6}{4-x} = \dfrac{x+2}{x^2-16}$

✓ **61.** $\dfrac{x}{x-2} + \dfrac{3x}{x-4} = -\dfrac{2(x-6)}{x^2-6x+8}$

62. $\dfrac{2(x+1)}{x^2-4x+3} + \dfrac{6x}{x-3} = \dfrac{3x}{x-1}$

63. $\dfrac{5}{x^2+4x+3} + \dfrac{2}{x^2+x-6} = \dfrac{3}{x^2-x-2}$

64. $\dfrac{2}{x^2+2x-8} - \dfrac{1}{x^2+9x+20} = \dfrac{4}{x^2+3x-10}$

65. $\dfrac{x}{3} = \dfrac{1+\frac{4}{x}}{1+\frac{2}{x}}$

66. $\dfrac{2x}{3} = \dfrac{1+\frac{2}{x}}{1+\frac{1}{x}}$

In Exercises 67–70, (a) use the graph to determine any x-intercepts of the graph and (b) set $y = 0$ and solve the resulting rational equation to confirm the result of part (a).

67. $y = \dfrac{x+2}{x-2}$

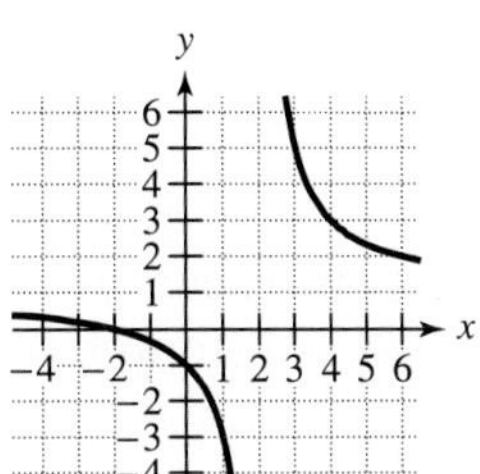

68. $y = \dfrac{2x}{x+4}$

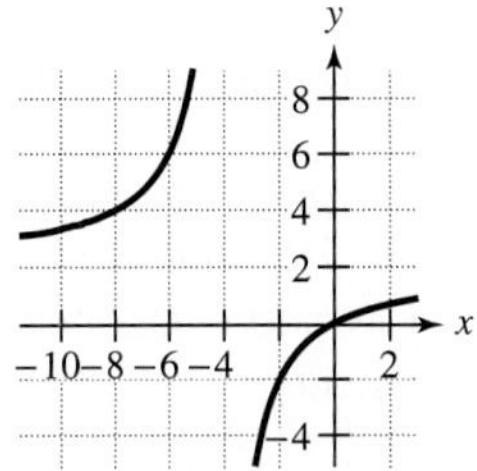

69. $y = x - \dfrac{1}{x}$

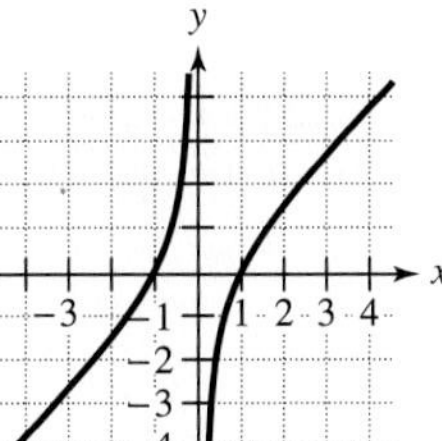

70. $y = x - \dfrac{2}{x} - 1$

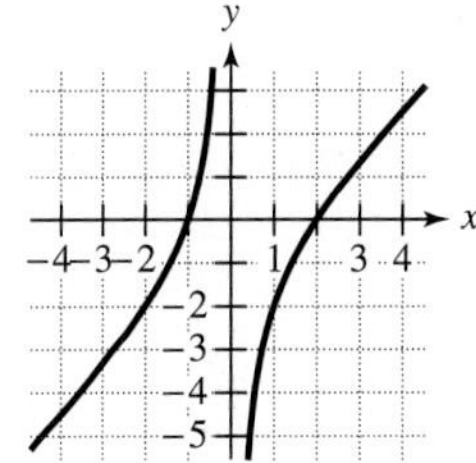

In Exercises 71–76, (a) use a graphing calculator to graph the equation and determine any x-intercepts of the graph and (b) set $y = 0$ and solve the resulting rational equation to confirm the result of part (a).

71. $y = \dfrac{x-4}{x+5}$

72. $y = \dfrac{1}{x} - \dfrac{3}{x+4}$

73. $y = x + 3 + \dfrac{7}{x}$

74. $y = 20\left(\dfrac{2}{x} - \dfrac{3}{x-1}\right)$

75. $y = (x+1) - \dfrac{6}{x}$

76. $y = \dfrac{x^2+9}{x}$

Think About It In Exercises 77–80, if the exercise is an equation, solve it; if it is an expression, simplify it.

77. $\dfrac{16}{x^2 - 16} + \dfrac{x}{2x - 8} = \dfrac{1}{2}$

78. $\dfrac{5}{x + 3} + \dfrac{5}{3} + 3$

79. $\dfrac{16}{x^2 - 16} + \dfrac{x}{2x - 8} + \dfrac{1}{2}$

80. $\dfrac{5}{x + 3} + \dfrac{5}{3} = 3$

Solving Problems

81. *Number Problem* Find a number such that the sum of the number and its reciprocal is $\frac{37}{6}$.

82. *Number Problem* Find a number such that the sum of two times the number and three times its reciprocal is $\frac{203}{10}$.

83. *Painting* A painter can paint a fence in 4 hours, while his partner can paint the fence in 6 hours. How long would it take to paint the fence if both worked together?

84. *Roofing* A roofer requires 15 hours to shingle a roof, while an apprentice requires 21 hours. How long would it take to shingle the roof if both worked together?

85. *Wind Speed* A plane has a speed of 300 miles per hour in still air. The plane travels a distance of 680 miles with a tail wind in the same time it takes to travel 520 miles into a head wind. Find the speed of the wind.

86. *Speed* One person runs 2 miles per hour faster than a second person. The first person runs 5 miles in the same time the second person runs 4 miles. Find the speed of each person.

87. *Saves* A hockey goalie has faced 799 shots and saved 707 of them. How many additional consecutive saves does the goalie need to obtain a save percent (in decimal form) of .900?

Nick Didlick/Getty Images

88. *Batting Average* A softball player has been up to bat 47 times and has hit the ball safely 8 times. How many additional consecutive times must the player hit the ball safely to obtain a batting average of .250?

Explaining Concepts

89. Define the term *extraneous solution*. How do you identify an extraneous solution?

90. Explain how you can use a graphing calculator to estimate the solution of a rational equation.

91. Explain why the equation $\dfrac{n}{x} + n = \dfrac{n}{x}$ has no solution if n is any real nonzero number.

92. Does multiplying a rational equation by its LCD produce an equivalent equation? Explain.

Cumulative Review

In Exercises 93–96, factor the expression.

93. $x^2 - 81$

94. $x^2 - 121$

95. $4x^2 - \frac{1}{4}$

96. $49 - (x - 2)^2$

In Exercises 97 and 98, find the domain of the rational function.

97. $f(x) = \dfrac{2x^2}{5}$

98. $f(x) = \dfrac{4}{x - 6}$

7.6 Applications and Variation

NASA

What You Should Learn

1 ▶ Solve application problems involving rational equations.
2 ▶ Solve application problems involving direct variation.
3 ▶ Solve application problems involving inverse variation.
4 ▶ Solve application problems involving joint variation.

Why You Should Learn It

You can use mathematical models in a wide variety of applications involving variation. For instance, in Exercise 64 on page 460, you will use direct variation to model the weight of a person on the moon.

1 ▶ Solve application problems involving rational equations.

Rational Equation Applications

The first three examples in this section are types of application problems that you have seen earlier in the text. The difference now is that the variable appears in the denominator of a rational expression.

EXAMPLE 1 Average Speeds 

You and your friend travel to separate colleges in the same amount of time. You drive 380 miles and your friend drives 400 miles. Your friend's average speed is 3 miles per hour faster than your average speed. What is your average speed and what is your friend's average speed?

Solution

Begin by setting your time equal to your friend's time. Then use an alternative version of the formula for distance that gives the time in terms of the distance and the rate.

Verbal Model: Your time = Your friend's time

$$\frac{\text{Your distance}}{\text{Your rate}} = \frac{\text{Friend's distance}}{\text{Friend's rate}}$$

Labels:
Your distance = 380 (miles)
Your rate = r (miles per hour)
Friend's distance = 400 (miles)
Friend's rate = $r + 3$ (miles per hour)

Equation:

$$\frac{380}{r} = \frac{400}{r+3}$$ Original equation.

$$380(r+3) = 400(r), \quad r \neq 0, r \neq -3$$ Cross-multiply.

$$380r + 1140 = 400r$$ Distributive Property

$$1140 = 20r$$  $$57 = r$$ Simplify.

Your average speed is 57 miles per hour and your friend's average speed is $57 + 3 = 60$ miles per hour. Check this in the original statement of the problem.

✓ **CHECKPOINT** *Now try Exercise 43.*

Study Tip

When determining the domain of a real-life problem, you must also consider the context of the problem. For instance, in Example 2, the time it takes to fill the tub with water could not be a negative number. The problem implies that the domain must be all real numbers greater than zero.

EXAMPLE 2 A Work-Rate Problem 

With the cold water valve open, it takes 8 minutes to fill a washing machine tub. With both the hot and cold water valves open, it takes 5 minutes to fill the tub. How long will it take to fill the tub with only the hot water valve open?

Solution

Verbal Model: Rate for cold water + Rate for hot water = Rate for warm water

Labels: Rate for cold water $= \frac{1}{8}$ (tub per minute)

Rate for hot water $= \frac{1}{t}$ (tub per minute)

Rate for warm water $= \frac{1}{5}$ (tub per minute)

Equation:

$\frac{1}{8} + \frac{1}{t} = \frac{1}{5}$ Original equation

$5t + 40 = 8t$ Multiply each side by LCD of $40t$ and simplify.

$40 = 3t$  $\frac{40}{3} = t$ Simplify.

So, it takes $13\frac{1}{3}$ minutes to fill the tub with hot water. Check this solution.

CHECKPOINT *Now try Exercise 47.*

EXAMPLE 3 Cost-Benefit Model 

A utility company burns coal to generate electricity. The cost C (in dollars) of removing $p\%$ of the pollutants from smokestack emissions is modeled by

$$C = \frac{80{,}000p}{100 - p}, \quad 0 \le p < 100.$$

What percent of air pollutants in the stack emissions can be removed for \$420,000?

Solution

To determine the percent of air pollutants in the stack emissions that can be removed for \$420,000, substitute 420,000 for C in the model.

$420{,}000 = \frac{80{,}000p}{100 - p}$ Substitute 420,000 for C.

$420{,}000(100 - p) = 80{,}000p$ Cross-multiply.

$42{,}000{,}000 - 420{,}000p = 80{,}000p$ Distributive Property

$42{,}000{,}000 = 500{,}000p$ Add $420{,}000p$ to each side.

$84 = p$ Divide each side by 500,000.

So, 84% of air pollutants in the stack emissions can be removed for \$420,000.

CHECKPOINT *Now try Exercise 49.*

2 ▶ Solve application problems involving direct variation.

Direct Variation

In the mathematical model for **direct variation,** y is a *linear* function of x. Specifically,

$$y = kx.$$

To use this mathematical model in applications involving direct variation, you need to use given values of x and y to find the value of the constant k.

> **Direct Variation**
>
> The following statements are equivalent.
>
> **1.** y **varies directly** as x.
>
> **2.** y is **directly proportional** to x.
>
> **3.** $y = kx$ for some constant k.
>
> The number k is called the **constant of proportionality.**

EXAMPLE 4 Direct Variation

The total revenue R (in dollars) obtained from selling x ice show tickets is directly proportional to the number of tickets sold x. When 10,000 tickets are sold, the total revenue is \$142,500.

a. Find a mathematical model that relates the total revenue R to the number of tickets sold x.

b. Find the total revenue obtained from selling 12,000 tickets.

Solution

a. Because the total revenue is directly proportional to the number of tickets sold, the linear model is $R = kx$. To find the value of the constant k, use the fact that $R = 142{,}500$ when $x = 10{,}000$. Substituting these values into the model produces

$$142{,}500 = k(10{,}000) \qquad \text{Substitute for } R \text{ and } x.$$

which implies that

$$k = \frac{142{,}500}{10{,}000} = 14.25.$$

So, the equation relating the total revenue to the total number of tickets sold is

$$R = 14.25x. \qquad \text{Direct variation model}$$

The graph of this equation is shown in Figure 7.2.

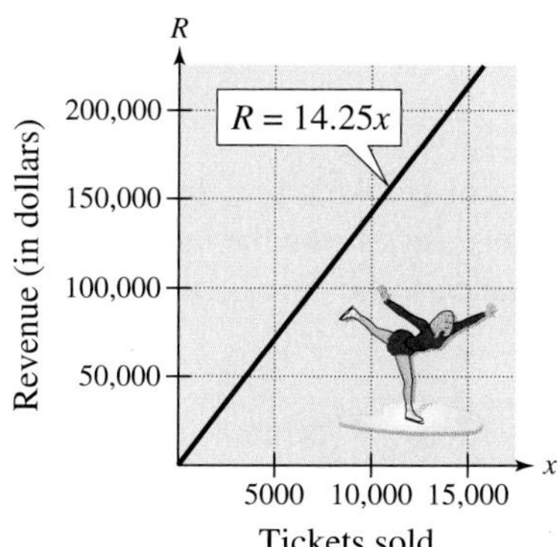

Figure 7.2

b. When $x = 12{,}000$, the total revenue is

$$R = 14.25(12{,}000) = \$171{,}000.$$

CHECKPOINT *Now try Exercise 53.*

EXAMPLE 5 Direct Variation 

Hooke's Law for springs states that the distance a spring is stretched (or compressed) is directly proportional to the force on the spring. A force of 20 pounds stretches a spring 5 inches.

a. Find a mathematical model that relates the distance the spring is stretched to the force applied to the spring.

b. How far will a force of 30 pounds stretch the spring?

Solution

a. For this problem, let d represent the distance (in inches) that the spring is stretched and let F represent the force (in pounds) that is applied to the spring. Because the distance d is directly proportional to the force F, the model is

$$d = kF.$$

To find the value of the constant k, use the fact that $d = 5$ when $F = 20$. Substituting these values into the model produces

$$5 = k(20) \qquad \text{Substitute 5 for } d \text{ and 20 for } F.$$

$$\frac{5}{20} = k \qquad \text{Divide each side by 20.}$$

$$\frac{1}{4} = k. \qquad \text{Simplify.}$$

So, the equation relating distance and force is

$$d = \frac{1}{4}F. \qquad \text{Direct variation model}$$

b. When $F = 30$, the distance is

$$d = \frac{1}{4}(30) = 7.5 \text{ inches.} \qquad \text{See Figure 7.3.}$$

Figure 7.3

CHECKPOINT *Now try Exercise 55.*

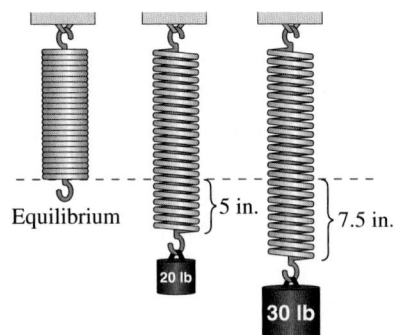

In Example 5, you can get a clearer understanding of Hooke's Law by using the model $d = \frac{1}{4}F$ to create a table or a graph (see Figure 7.4). From the table or from the graph, you can see what it means for the distance to be "proportional to the force."

Force, F	10 lb	20 lb	30 lb	40 lb	50 lb	60 lb
Distance, d	2.5 in.	5.0 in.	7.5 in.	10.0 in.	12.5 in.	15.0 in.

Figure 7.4

In Examples 4 and 5, the direct variations are such that an *increase* in one variable corresponds to an *increase* in the other variable. There are, however, other applications of direct variation in which an increase in one variable corresponds to a *decrease* in the other variable. For instance, in the model $y = -2x$, an increase in x will yield a decrease in y.

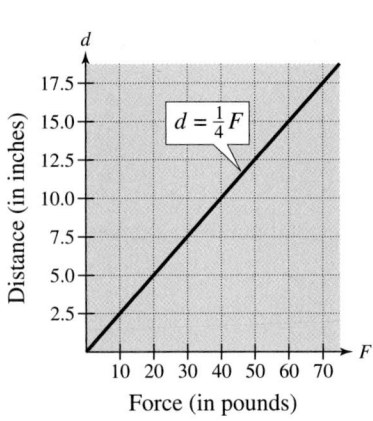

Another type of direct variation relates one variable to a power of another.

Direct Variation as nth Power

The following statements are equivalent.

1. y **varies directly as the nth power** of x.
2. y is **directly proportional to the nth power** of x.
3. $y = kx^n$ for some constant k.

EXAMPLE 6 **Direct Variation as a Power** 

The distance a ball rolls down an inclined plane is directly proportional to the square of the time it rolls. During the first second, a ball rolls down a plane a distance of 6 feet.

a. Find a mathematical model that relates the distance traveled to the time.

b. How far will the ball roll during the first 2 seconds?

Solution

a. Letting d be the distance (in feet) that the ball rolls and letting t be the time (in seconds), you obtain the model

$$d = kt^2.$$

Because $d = 6$ when $t = 1$, you obtain

$d = kt^2$ Write original equation.

$6 = k(1)^2 \Rightarrow 6 = k.$ Substitute 6 for d and 1 for t.

So, the equation relating distance to time is

$d = 6t^2.$ Direct variation as 2nd power model

The graph of this equation is shown in Figure 7.5.

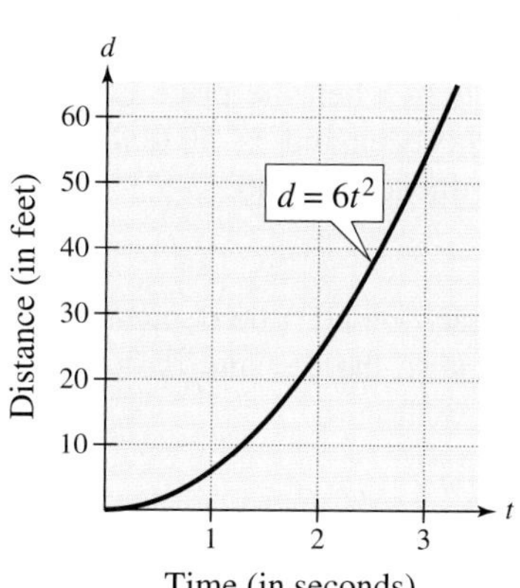

Figure 7.5

b. When $t = 2$, the distance traveled is

$d = 6(2)^2 = 6(4) = 24$ feet. See Figure 7.6.

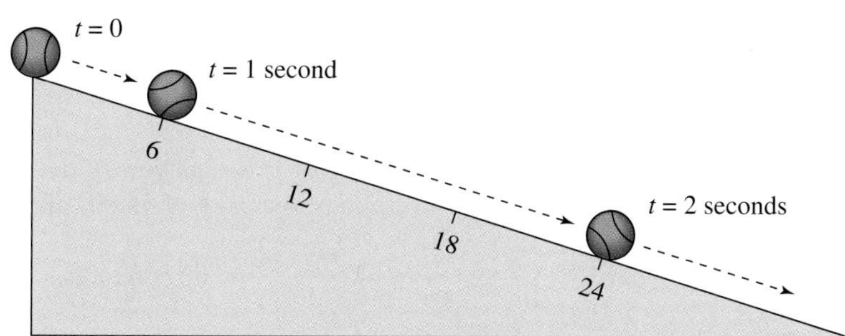

Figure 7.6

✓ ***CHECKPOINT*** *Now try Exercise 61.*

3 ▶ Solve application problems involving inverse variation.

Inverse Variation

A second type of variation is called **inverse variation.** With this type of variation, one of the variables is said to be inversely proportional to the other variable.

> **Inverse Variation**
>
> **1.** The following three statements are equivalent.
>
> **a.** y **varies inversely** as x.
>
> **b.** y is **inversely proportional** to x.
>
> **c.** $y = \dfrac{k}{x}$ for some constant k.
>
> **2.** If $y = \dfrac{k}{x^n}$, then y is inversely proportional to the nth power of x.

EXAMPLE 7 Inverse Variation 

The marketing department of a large company has found that the demand for one of its hand tools varies inversely as the price of the product. (When the price is low, more people are willing to buy the product than when the price is high.) When the price of the tool is \$7.50, the monthly demand is 50,000 tools. Approximate the monthly demand if the price is reduced to \$6.00.

Solution

Let x represent the number of tools that are sold each month (the demand), and let p represent the price per tool (in dollars). Because the demand is inversely proportional to the price, the model is

$$x = \frac{k}{p}.$$

By substituting $x = 50,000$ when $p = 7.50$, you obtain

$$50,000 = \frac{k}{7.50} \qquad \text{Substitute 50,000 for } x \text{ and 7.50 for } p.$$

$$375,000 = k. \qquad \text{Multiply each side by 7.50.}$$

So, the inverse variation model is $x = \dfrac{375,000}{p}$.

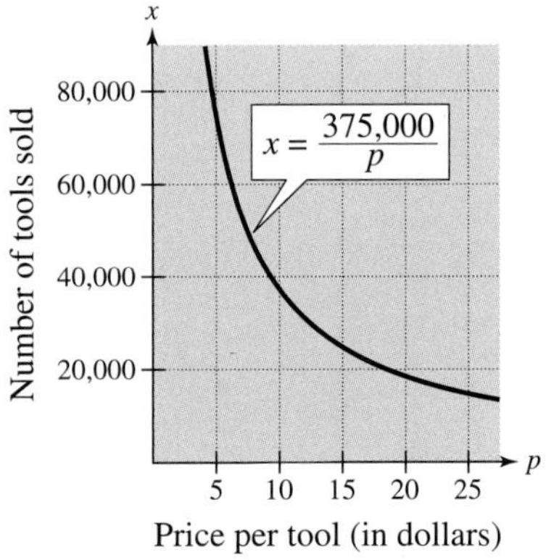

The graph of this equation is shown in Figure 7.7. To find the demand that corresponds to a price of \$6.00, substitute 6 for p in the equation and obtain

$$x = \frac{375,000}{6} = 62,500 \text{ tools.}$$

So, if the price is lowered from \$7.50 per tool to \$6.00 per tool, you can expect the monthly demand to increase from 50,000 tools to 62,500 tools.

 ***CHECKPOINT*** *Now try Exercise 65.*

Figure 7.7

Some applications of variation involve problems with *both* direct and inverse variation in the same model. These types of models are said to have **combined variation.**

EXAMPLE 8 Direct and Inverse Variation 

A computer hardware manufacturer determines that the demand for its USB flash drive is directly proportional to the amount spent on advertising and inversely proportional to the price of the flash drive. When \$40,000 is spent on advertising and the price per unit is \$20, the monthly demand is 10,000 flash drives.

a. If the amount of advertising were increased to \$50,000, how much could the price be increased to maintain a monthly demand of 10,000 flash drives?

b. If you were in charge of the advertising department, would you recommend this increased expense in advertising?

Solution

a. Let x represent the number of flash drives that are sold each month (the demand), let a represent the amount spent on advertising (in dollars), and let p represent the price per unit (in dollars). Because the demand is directly proportional to the advertising expense and inversely proportional to the price, the model is

$$x = \frac{ka}{p}.$$

By substituting 10,000 for x when $a = 40{,}000$ and $p = 20$, you obtain

$$10{,}000 = \frac{k(40{,}000)}{20}$$ Substitute 10,000 for x, 40,000 for a, and 20 for p.

$$200{,}000 = 40{,}000k$$ Multiply each side by 20.

$$5 = k.$$ Divide each side by 40,000.

So, the model is

$$x = \frac{5a}{p}.$$ Direct and inverse variation model

To find the price that corresponds to a demand of 10,000 and an advertising expense of \$50,000, substitute 10,000 for x and 50,000 for a into the model and solve for p.

$$10{,}000 = \frac{5(50{,}000)}{p} \qquad p = \frac{5(50{,}000)}{10{,}000} = \$25$$

So, the price increase would be $\$25 - \$20 = \$5$.

b. The total revenue for selling 10,000 units at \$20 each is \$200,000, and the revenue for selling 10,000 units at \$25 each is \$250,000. So, increasing the advertising expense from \$40,000 to \$50,000 would increase the revenue by \$50,000. This implies that you should recommend the increased expense in advertising.

✓ ***CHECKPOINT*** ***Now try Exercise 69.***

4 ▶ Solve application problems involving joint variation.

Joint Variation

The model used in Example 8 involved both direct and inverse variation, and the word "and" was used to couple the two types of variation together. To describe two different *direct* variations in the same statement, the word "jointly" is used. For instance, the model $z = kxy$ can be described by saying that z is *jointly* proportional to x and y. So, in **joint variation,** one variable varies directly with the product of two variables.

Joint Variation

1. The following three statements are equivalent.
   a. z **varies jointly** as x and y.
   b. z is **jointly proportional** to x and y.
   c. $z = kxy$ for some constant k.
2. If $z = kx^n y^m$, then z is jointly proportional to the nth power of x and the mth power of y.

EXAMPLE 9 Joint Variation 

The *simple interest* earned by a savings account is jointly proportional to the time and the principal. After one quarter (3 months), the interest for a principal of \$6000 is \$120. How much interest would a principal of \$7500 earn in 5 months?

Solution

To begin, let I represent the interest earned (in dollars), let P represent the principal (in dollars), and let t represent the time (in years). Because the interest is jointly proportional to the time and the principal, the model is

$$I = ktP.$$

Because $I = 120$ when $P = 6000$ and $t = \frac{1}{4}$, you have

$$120 = k\left(\frac{1}{4}\right)(6000) \qquad \text{Substitute 120 for } I, \tfrac{1}{4} \text{ for } t, \text{ and 6000 for } P.$$

$$120 = 1500\,k \qquad \text{Simplify.}$$

$$0.08 = k. \qquad \text{Divide each side by 1500.}$$

So, the model that relates interest to time and principal is

$$I = 0.08tP. \qquad \text{Joint variation model}$$

To find the interest earned on a principal of \$7500 over a five-month period of time, substitute $P = 7500$ and $t = \frac{5}{12}$ into the model to obtain an interest of

$$I = 0.08\left(\frac{5}{12}\right)(7500) = \$250.$$

CHECKPOINT Now try Exercise 71.

Concept Check

1. In a problem, y varies directly as x and the constant of proportionality is positive. If one of the variables increases, how does the other change? Explain.

2. In a problem, y varies inversely as x and the constant of proportionality is positive. If one of the variables increases, how does the other change? Explain.

3. Are the following statements equivalent? Explain.
   (a) y varies directly as x.
   (b) y is directly proportional to the square of x.

4. Describe the difference between *combined variation* and *joint variation.*

7.6 EXERCISES

Go to pages 462–463 to record your assignments.

Developing Skills

In Exercises 1–14, write a model for the statement.

1. I varies directly as V.
2. C varies directly as r.
3. V is directly proportional to t.
4. A is directly proportional to w.
5. u is directly proportional to the square of v.
6. s varies directly as the cube of t.
7. p varies inversely as d.
8. S varies inversely as the square of v.
9. A is inversely proportional to the fourth power of t.
10. P is inversely proportional to the square root of $1 + r$.
11. A varies jointly as l and w.
12. V varies jointly as h and the square of r.
13. *Boyle's Law* If the temperature of a gas is not allowed to change, its absolute pressure P is inversely proportional to its volume V.
14. *Newton's Law of Universal Gravitation* The gravitational attraction F between two particles of masses m_1 and m_2 is directly proportional to the product of the masses and inversely proportional to the square of the distance r between the particles.

In Exercises 15–20, write a verbal sentence using variation terminology to describe the formula.

15. *Area of a Triangle:* $A = \frac{1}{2}bh$
16. *Area of a Rectangle:* $A = lw$
17. *Volume of a Right Circular Cylinder:* $V = \pi r^2 h$
18. *Volume of a Sphere:* $V = \frac{4}{3}\pi r^3$
19. *Average Speed:* $r = \dfrac{d}{t}$
20. *Height of a Cylinder:* $h = \dfrac{V}{\pi r^2}$

In Exercises 21–32, find the constant of proportionality and write an equation that relates the variables.

21. s varies directly as t, and $s = 20$ when $t = 4$.
22. h is directly proportional to r, and $h = 28$ when $r = 12$.
23. F is directly proportional to the square of x, and $F = 500$ when $x = 40$.
24. M varies directly as the cube of n, and $M = 0.012$ when $n = 0.2$.
25. n varies inversely as m, and $n = 32$ when $m = 1.5$.
26. q is inversely proportional to p, and $q = \frac{3}{2}$ when $p = 50$.

27. g varies inversely as the square root of z, and $g = \frac{4}{5}$ when $z = 25$.

28. u varies inversely as the square of v, and $u = 40$ when $v = \frac{1}{2}$.

29. F varies jointly as x and y, and $F = 500$ when $x = 15$ and $y = 8$.

30. V varies jointly as h and the square of b, and $V = 288$ when $h = 6$ and $b = 12$.

31. d varies directly as the square of x and inversely with r, and $d = 3000$ when $x = 10$ and $r = 4$.

32. z is directly proportional to x and inversely proportional to the square root of y, and $z = 720$ when $x = 48$ and $y = 81$.

In Exercises 33–36, complete the table and plot the resulting points.

x	2	4	6	8	10
$y = kx^2$					

33. $k = 1$

34. $k = 2$

35. $k = \frac{1}{2}$

36. $k = \frac{1}{4}$

In Exercises 37–40, complete the table and plot the resulting points.

x	2	4	6	8	10
$y = \dfrac{k}{x^2}$					

37. $k = 2$

38. $k = 5$

39. $k = 10$

40. $k = 20$

In Exercises 41 and 42, determine whether the variation model is of the form $y = kx$ or $y = k/x$, and find k.

41.

x	10	20	30	40	50
y	$\frac{2}{5}$	$\frac{1}{5}$	$\frac{2}{15}$	$\frac{1}{10}$	$\frac{2}{25}$

42.

x	10	20	30	40	50
y	-3	-6	-9	-12	-15

Solving Problems

✓ **43.** *Average Speeds* You and a friend jog for the same amount of time. You jog 10 miles and your friend jogs 12 miles. Your friend's average speed is 1.5 miles per hour faster than yours. What are the average speeds of you and your friend?

44. *Current Speed* A boat travels at a speed of 20 miles per hour in still water. It travels 48 miles upstream and then returns to the starting point in a total of 5 hours. Find the speed of the current.

45. *Partnership Costs* A group plans to start a new business that will require \$240,000 for start-up capital. The individuals in the group share the cost equally. If two additional people join the group, the cost per person will decrease by \$4000. How many people are presently in the group?

46. *Partnership Costs* A group of people share equally the cost of a \$180,000 endowment. If they could find four more people to join the group, each person's share of the cost would decrease by \$3750. How many people are presently in the group?

✓ **47.** *Work Rate* It takes a lawn care company 60 minutes to complete a job using only a riding mower, or 45 minutes using the riding mower and a push mower. How long does the job take using only the push mower?

48. *Flow Rate* It takes 3 hours to fill a pool using two pipes. It takes 5 hours to fill the pool using only the larger pipe. How long does it take to fill the pool using only the smaller pipe?

49. *Pollution Removal* The cost C in dollars of removing $p\%$ of the air pollutants in the stack emissions of a utility company is modeled by the equation below. Determine the percent of air pollutants in the stack emissions that can be removed for \$680,000.

$$C = \frac{120,000p}{100 - p}$$

50. *Population Growth* A biologist starts a culture with 100 bacteria. The population P of the culture is approximated by the model below, where t is the time in hours. Find the time required for the population to increase to 800 bacteria.

$$P = \frac{500(1 + 3t)}{5 + t}$$

51. *Nail Sizes* The unit for determining the size of a nail is the *penny*. For example, 8d represents an 8-penny nail. The number N of finishing nails per pound can be modeled by

$$N = -139.1 + \frac{2921}{x}$$

where x is the size of the nail.

(a) What is the domain of the function?

(b) Use a graphing calculator to graph the function.

(c) Use the graph to determine the size of the finishing nail if there are 153 nails per pound.

(d) Verify the result of part (c) algebraically.

52. *Learning Curve* A psychologist observes that the number of lines N of a poem that a four-year-old child can memorize depends on the number x of short sessions spent on the task, according to the model

$$N = \frac{20x}{x + 1}.$$

(a) What is the domain of the function?

(b) Use a graphing calculator to graph the function.

(c) Use the graph to determine the number of sessions needed for a child to memorize 15 lines of the poem.

(d) Verify the result of part (c) algebraically.

53. *Revenue* The total revenue R is directly proportional to the number of units sold x. When 500 units are sold, the revenue is \$4825. Find the revenue when 620 units are sold. Then interpret the constant of proportionality.

54. *Revenue* The total revenue R is directly proportional to the number of units sold x. When 25 units are sold, the revenue is \$300. Find the revenue when 42 units are sold. Then interpret the constant of proportionality.

55. *Hooke's Law* A force of 50 pounds stretches a spring 5 inches.

(a) How far will a force of 20 pounds stretch the spring?

(b) What force is required to stretch the spring 1.5 inches?

56. *Hooke's Law* A force of 50 pounds stretches a spring 3 inches.

(a) How far will a force of 20 pounds stretch the spring?

(b) What force is required to stretch the spring 1.5 inches?

57. *Hooke's Law* A baby weighing $10\frac{1}{2}$ pounds compresses the spring of a baby scale 7 millimeters. Determine the weight of a baby that compresses the spring 12 millimeters.

58. *Hooke's Law* An apple weighing 14 ounces compresses the spring of a produce scale 3 millimeters. Determine the weight of a grapefruit that compresses the spring 5 millimeters.

59. *Free-Falling Object* The velocity v of a free-falling object is directly proportional to the time t (in seconds) that the object has fallen. The velocity of a falling object is -64 feet per second after the object has fallen for 2 seconds. Find the velocity of the object after it has fallen for a total of 4 seconds.

60. *Free-Falling Object* Neglecting air resistance, the distance d that an object falls varies directly as the square of the time t it has fallen. An object falls 64 feet in 2 seconds. Determine the distance it will fall in 6 seconds.

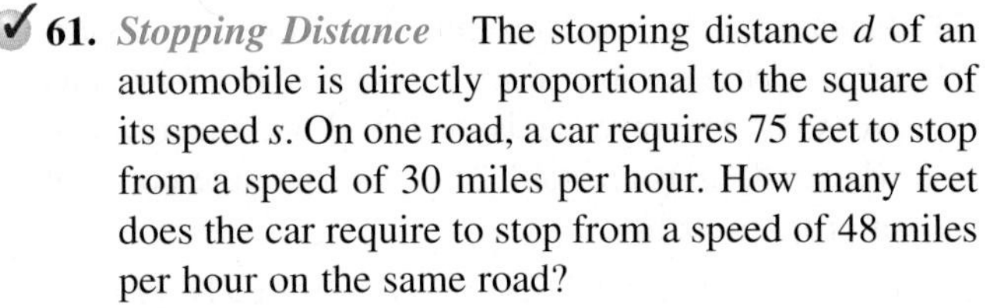

✓ **61.** *Stopping Distance* The stopping distance d of an automobile is directly proportional to the square of its speed s. On one road, a car requires 75 feet to stop from a speed of 30 miles per hour. How many feet does the car require to stop from a speed of 48 miles per hour on the same road?

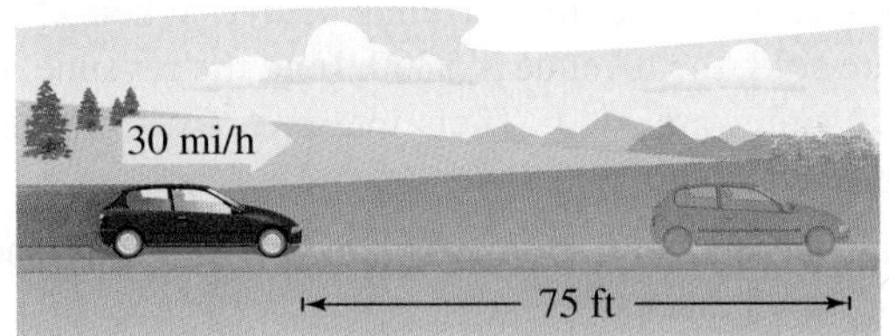

62. *Frictional Force* The frictional force F (between the tires of a car and the road) that is required to keep a car on a curved section of a highway is directly proportional to the square of the speed s of the car. By what factor does the force F change when the speed of the car is doubled on the same curve?

63. *Power Generation* The power P generated by a wind turbine varies directly as the cube of the wind speed w. The turbine generates 400 watts of power in a 20-mile-per-hour wind. Find the power it generates in a 30-mile-per-hour wind.

64. *Weight of an Astronaut* A person's weight on the moon varies directly as his or her weight on Earth. An astronaut weighs 360 pounds on Earth, including heavy equipment. On the moon the astronaut weighs only 60 pounds with the equipment. If the first woman in space, Valentina Tereshkova, had landed on the moon and weighed 54 pounds with equipment, how much would she have weighed on Earth with her equipment?

✓ **65.** *Demand* A company has found that the daily demand x for its boxes of chocolates is inversely proportional to the price p. When the price is \$5, the demand is 800 boxes. Approximate the demand when the price is increased to \$6.

66. *Pressure* When a person walks, the pressure P on each sole varies inversely as the area A of the sole. A person is trudging through deep snow, wearing boots that have a sole area of 29 square inches each. The sole pressure is 4 pounds per square inch. If the person was wearing snowshoes, each with an area 11 times that of their boot soles, what would be the pressure on each snowshoe? The constant of variation in this problem is the weight of the person. How much does the person weigh?

67. *Environment* The graph shows the percent p of oil that remained in Chedabucto Bay, Nova Scotia, after an oil spill. The cleaning of the spill was left primarily to natural actions. After about a year, the percent that remained varied inversely as time. Find a model that relates p and t, where t is the number of years since the spill. Then use it to find the percent of oil that remained $6\frac{1}{2}$ years after the spill, and compare the result with the graph.

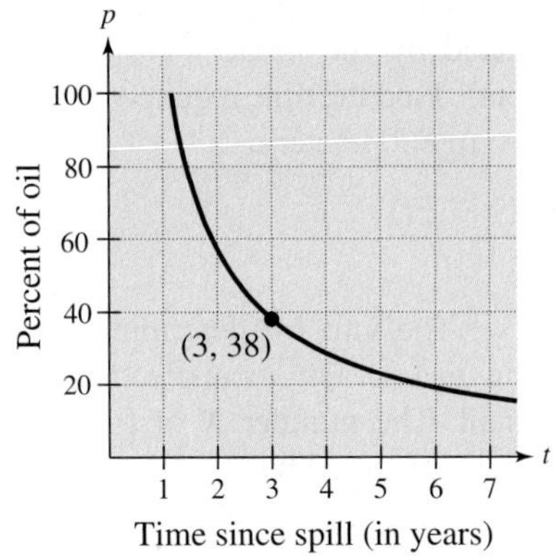

68. *Meteorology* The graph shows the water temperature in relation to depth in the north central Pacific Ocean. At depths greater than 900 meters, the water temperature varies inversely with the water depth. Find a model that relates the temperature T to the depth d. Then use it to find the water temperature at a depth of 4385 meters, and compare the result with the graph.

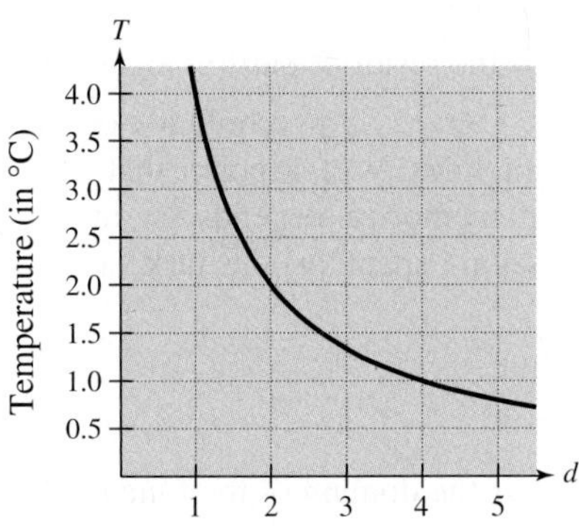

✓ **69.** *Revenue* The weekly demand for a company's frozen pizzas varies directly as the amount spent on advertising and inversely as the price per pizza. At \$5 per pizza, when \$500 is spent each week on ads, the demand is 2000 pizzas. If advertising is increased to \$600, what price will yield a demand of 2000 pizzas? Is this increase worthwhile in terms of revenue?

70. *Revenue* The monthly demand for a company's sports caps varies directly as the amount spent on advertising and inversely as the square of the price per cap. At \$15 per cap, when \$2500 is spent each week on ads, the demand is 300 caps. If advertising is increased to \$3000, what price will yield a demand of 300 caps? Is this increase worthwhile in terms of revenue?

✓ **71.** *Simple Interest* The simple interest earned by an account varies jointly as the time and the principal. A principal of \$600 earns \$10 interest in 4 months. How much would \$900 earn in 6 months?

72. *Simple Interest* The simple interest earned by an account varies jointly as the time and the principal. In 2 years, a principal of \$5000 earns \$650 interest. How much would \$1000 earn in 1 year?

73. *Engineering* The load P that can be safely supported by a horizontal beam varies jointly as the product of the width W of the beam and the square of the depth D, and inversely as the length L (see figure).

(a) Write a model for the statement.

(b) How does P change when the width and length of the beam are both doubled?

(c) How does P change when the width and depth of the beam are doubled?

(d) How does P change when all three of the dimensions are doubled?

(e) How does P change when the depth of the beam is cut in half?

(f) A beam with width 3 inches, depth 8 inches, and length 120 inches can safely support 2000 pounds. Determine the safe load of a beam made from the same material if its depth is increased to 10 inches.

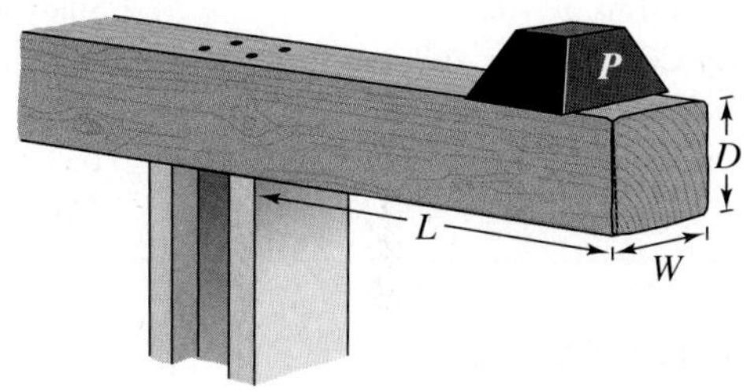

Explaining Concepts

True or False? In Exercises 74 and 75, determine whether the statement is true or false. Explain your reasoning.

74. In a situation involving combined variation, y can vary directly as x and inversely as x at the same time.

75. In a joint variation problem where z varies jointly as x and y, if x increases, then z and y must both increase.

76. ✎ If y varies directly as the square of x and x is doubled, how does y change? Use the rules of exponents to explain your answer.

77. ✎ If y varies inversely as the square of x and x is doubled, how does y change? Use the rules of exponents to explain your answer.

78. ✎ Describe a real-life problem for each type of variation (direct, inverse, and joint).

Cumulative Review

In Exercises 79–82, write the expression using exponential notation.

79. $(6)(6)(6)(6)$

80. $(-4)(-4)(-4)$

81. $\left(\frac{1}{5}\right)\left(\frac{1}{5}\right)\left(\frac{1}{5}\right)\left(\frac{1}{5}\right)\left(\frac{1}{5}\right)$

82. $-\left(-\frac{3}{4}\right)\left(-\frac{3}{4}\right)\left(-\frac{3}{4}\right)$

In Exercises 83–86, use synthetic division to divide.

83. $(x^2 - 5x - 14) \div (x + 2)$

84. $(3x^2 - 5x + 2) \div (x + 1)$

85. $\dfrac{4x^5 - 14x^4 + 6x^3}{x - 3}$

86. $\dfrac{x^5 - 3x^2 - 5x + 1}{x - 2}$

What Did You Learn?

Use these two pages to help prepare for a test on this chapter. Check off the key terms and key concepts you know. You can also use this section to record your assignments.

Plan for Test Success

Date of test: / /

Study dates and times: / / at : A.M./P.M.

/ / at : A.M./P.M.

Things to review:

- ☐ Key Terms, *p. 462*
- ☐ Key Concepts, *pp. 462–463*
- ☐ Your class notes
- ☐ Your assignments
- ☐ Study Tips, *pp. 402, 403, 404, 407, 408, 417, 423, 426, 434, 435, 436, 441, 445, 450*
- ☐ Technology Tips, *pp. 406, 415, 425, 443*
- ☐ Mid-Chapter Quiz, *p. 432*
- ☐ Review Exercises, *pp. 464–468*
- ☐ Chapter Test, *p. 469*
- ☐ Video Explanations Online
- ☐ Tutorial Online

Key Terms

- ☐ rational expression, *p. 402*
- ☐ rational function, *p. 402*
- ☐ domain (of a rational function), *p. 402*
- ☐ simplified form, *p. 405*
- ☐ least common multiple, *p. 424*
- ☐ least common denominator, *p. 425*
- ☐ complex fraction, *p. 433*
- ☐ extraneous solution, *p. 444*
- ☐ direct variation, *p. 451*
- ☐ constant of proportionality, *p. 451*
- ☐ inverse variation, *p. 454*
- ☐ combined variation, *p. 455*
- ☐ joint variation, *p. 456*

Key Concepts

7.1 Rational Expressions and Functions

Assignment: ______________________ **Due date:** ________

☐ **Find the domain of a rational function.**

The set of *usable* values of the variable (values that do not make the denominator zero) is called the **domain** of a rational expression.

☐ **Simplify a rational expression.**

Divide out common factors: $\frac{uw}{vw} = \frac{u\not{w}}{v\not{w}} = \frac{u}{v},\ w \neq 0$

Domain restrictions of the original expression that are not implied by the simplified form must be listed.

7.2 Multiplying and Dividing Rational Expressions

Assignment: ______________________ **Due date:** ________

☐ **Multiply rational expressions.**

1. Multiply the numerators and the denominators.
   $\frac{u}{v} \cdot \frac{w}{z} = \frac{uw}{vz}$
2. Factor the numerator and the denominator.
3. Simplify by dividing out the common factors.

☐ **Divide rational expressions.**

Invert the divisor and multiply using the steps for multiplying rational expressions.

$\frac{u}{v} \div \frac{w}{z} = \frac{u}{v} \cdot \frac{z}{w} = \frac{uz}{vw}$

7.3 Adding and Subtracting Rational Expressions

Assignment: ______________________________ Due date: __________

☐ **Add or subtract rational expressions with like denominators.**

1. Combine the numerators: $\frac{u}{w} \pm \frac{v}{w} = \frac{u \pm v}{w}$
2. Simplify the resulting rational expression.

☐ **Add or subtract rational expressions with unlike denominators.**

Multiply the rational expressions by their LCD, then use the process for adding or subtracting rational expressions with like denominators.

7.4 Complex Fractions

Assignment: ______________________________ Due date: __________

☐ **Simplify a complex fraction.**

When the numerator and denominator of the complex fraction each consist of a single fraction, use the rules for dividing rational expressions.

When a sum or difference is present in the numerator or denominator of the complex fraction, first combine the terms so that the numerator and the denominator each consist of a single fraction.

7.5 Solving Rational Equations

Assignment: ______________________________ Due date: __________

☐ **Solve a rational equation containing constant denominators.**

1. Multiply each side of the equation by the LCD of all the fractions in the equation.
2. Solve the resulting equation.

☐ **Solve a rational equation containing variable denominators.**

1. Determine the domain restrictions of the equation.
2. Multiply each side by the LCD of all the fractions in the equation.
3. Solve the resulting equation.

☐ **Solve a rational equation by cross-multiplying.**

This method can be used when each side of the equation consists of a single fraction.

7.6 Applications and Variation

Assignment: ______________________________ Due date: __________

☐ **Solve application problems involving rational equations.**

☐ **Solve application problems involving direct variation.**

Direct variation: $y = kx$

Direct variation as nth power: $y = kx^n$

☐ **Solve application problems involving inverse variation.**

Inverse variation: $y = \frac{k}{x}$

Inverse variation as nth power: $y = \frac{k}{x^n}$

☐ **Solve application problems involving joint variation.**

Joint variation: $z = kxy$

Joint variation as nth and mth powers: $z = kx^n y^m$

Review Exercises

7.1 Rational Expressions and Functions

1 ▶ Find the domain of a rational function.

In Exercises 1–6, find the domain of the rational function.

1. $f(y) = \dfrac{3y}{y - 8}$

2. $g(t) = \dfrac{t + 4}{t + 12}$

3. $f(x) = \dfrac{2x}{x^2 + 1}$

4. $g(t) = \dfrac{t + 2}{t^2 + 4}$

5. $g(u) = \dfrac{u}{u^2 - 7u + 6}$

6. $f(x) = \dfrac{x - 12}{x(x^2 - 16)}$

7. *Geometry* A rectangle with a width of w inches has an area of 36 square inches. The perimeter P of the rectangle is given by

$$P = 2\left(w + \frac{36}{w}\right).$$

Describe the domain of the function.

8. *Average Cost* The average cost $\overline{C}$ for a manufacturer to produce x units of a product is given by

$$\overline{C} = \frac{15{,}000 + 0.75x}{x}.$$

Describe the domain of the function.

2 ▶ Simplify rational expressions.

In Exercises 9–16, simplify the rational expression.

9. $\dfrac{6x^4y^2}{15xy^2}$

10. $\dfrac{2(y^3z)^2}{28(yz^2)^2}$

11. $\dfrac{5b - 15}{30b - 120}$

12. $\dfrac{4a}{10a^2 + 26a}$

13. $\dfrac{9x - 9y}{y - x}$

14. $\dfrac{x + 3}{x^2 - x - 12}$

15. $\dfrac{x^2 - 5x}{2x^2 - 50}$

16. $\dfrac{x^2 + 3x + 9}{x^3 - 27}$

In Exercises 17 and 18, find the ratio of the area of the shaded region to the area of the whole figure.

17.

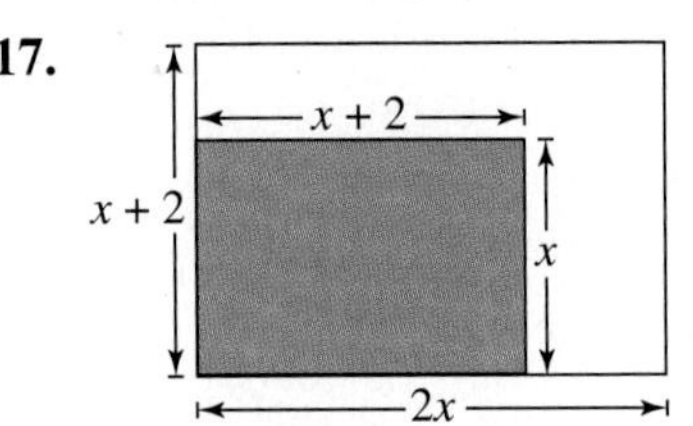

18.

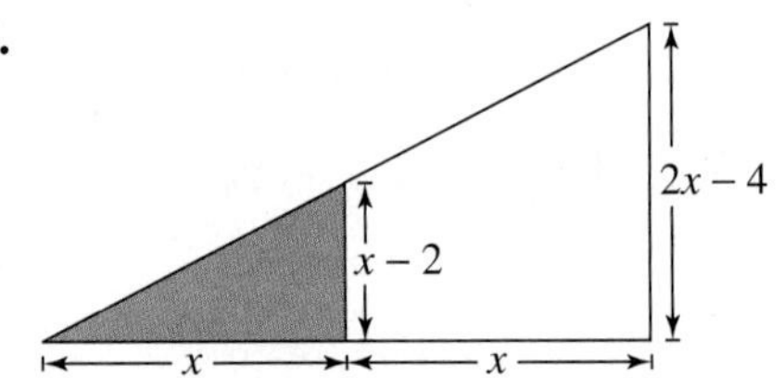

7.2 Multiplying and Dividing Rational Expressions

1 ▶ Multiply rational expressions and simplify.

In Exercises 19–26, multiply and simplify.

19. $\dfrac{4}{x} \cdot \dfrac{x^2}{12}$

20. $\dfrac{3}{y^3} \cdot 5y^3$

21. $\dfrac{7}{8} \cdot \dfrac{2x}{y} \cdot \dfrac{y^2}{14x^2}$

22. $\dfrac{15(x^2y)^3}{3y^3} \cdot \dfrac{12y}{x}$

23. $\dfrac{60z}{z + 6} \cdot \dfrac{z^2 - 36}{5}$

24. $\dfrac{x^2 - 16}{6} \cdot \dfrac{3}{x^2 - 8x + 16}$

25. $\dfrac{u}{u - 3} \cdot \dfrac{3u - u^2}{4u^2}$

26. $x^2 \cdot \dfrac{x + 1}{x^2 - x} \cdot \dfrac{(5x - 5)^2}{x^2 + 6x + 5}$

2 ▶ Divide rational expressions and simplify.

In Exercises 27–34, divide and simplify.

27. $24x^4 \div \dfrac{6x}{5}$

28. $\dfrac{8u^2}{3} \div \dfrac{u}{9}$

29. $25y^2 \div \dfrac{xy}{5}$

30. $\dfrac{6}{z^2} \div 4z^2$

31. $\dfrac{x^2 + 3x + 2}{3x^2 + x - 2} \div (x + 2)$

32. $\dfrac{x^2 - 14x + 48}{x^2 - 6x} \div (3x - 24)$

33. $\dfrac{x^2 - 7x}{x + 1} \div \dfrac{x^2 - 14x + 49}{x^2 - 1}$

34. $\dfrac{x^2 - x}{x + 1} \div \dfrac{5x - 5}{x^2 + 6x + 5}$

7.3 Adding and Subtracting Rational Expressions

1 ▶ Add or subtract rational expressions with like denominators, and simplify.

In Exercises 35–44, combine and simplify.

35. $\dfrac{4x}{5} + \dfrac{11x}{5}$

36. $\dfrac{7y}{12} - \dfrac{4y}{12}$

37. $\dfrac{15}{3x} - \dfrac{3}{3x}$

38. $\dfrac{4}{5x} + \dfrac{1}{5x}$

39. $\dfrac{8 - x}{4x} + \dfrac{5}{4x}$

40. $\dfrac{3}{5x} - \dfrac{x - 1}{5x}$

41. $\dfrac{2(3y + 4)}{2y + 1} + \dfrac{3 - y}{2y + 1}$

42. $\dfrac{4x - 2}{3x + 1} - \dfrac{x + 1}{3x + 1}$

43. $\dfrac{4x}{x + 2} + \dfrac{3x - 7}{x + 2} - \dfrac{9}{x + 2}$

44. $\dfrac{3}{2y - 3} - \dfrac{y - 10}{2y - 3} + \dfrac{5y}{2y - 3}$

2 ▶ Add or subtract rational expressions with unlike denominators, and simplify.

In Exercises 45–54, combine and simplify.

45. $\dfrac{3}{5x^2} + \dfrac{4}{10x}$

46. $\dfrac{3}{z} - \dfrac{5}{2z^2}$

47. $\dfrac{1}{x + 5} + \dfrac{3}{x - 12}$

48. $\dfrac{2}{x - 10} + \dfrac{3}{4 - x}$

49. $5x + \dfrac{2}{x - 3} - \dfrac{3}{x + 2}$

50. $4 - \dfrac{4x}{x + 6} + \dfrac{7}{x - 5}$

51. $\dfrac{6}{x - 5} - \dfrac{4x + 7}{x^2 - x - 20}$

52. $\dfrac{5}{x + 2} + \dfrac{25 - x}{x^2 - 3x - 10}$

53. $\dfrac{5}{x + 3} - \dfrac{4x}{(x + 3)^2} - \dfrac{1}{x - 3}$

54. $\dfrac{8}{y} - \dfrac{3}{y + 5} + \dfrac{4}{y - 2}$

In Exercises 55 and 56, use a graphing calculator to graph the two equations in the same viewing window. Use the graphs to verify that the expressions are equivalent. Verify the results algebraically.

55. $y_1 = \dfrac{1}{x} - \dfrac{3}{x + 3}$

$y_2 = \dfrac{3 - 2x}{x(x + 3)}$

56. $y_1 = \dfrac{5x}{x-5} + \dfrac{7}{x+1}$

$y_2 = \dfrac{5x^2 + 12x - 35}{x^2 - 4x - 5}$

7.4 Complex Fractions

1 ▸ Simplify complex fractions using rules for dividing rational expressions.

In Exercises 57–62, simplify the complex fraction.

57. $\dfrac{\left(\dfrac{6}{x}\right)}{\left(\dfrac{2}{x^3}\right)}$

58. $\dfrac{xy}{\left(\dfrac{5x^2}{2y}\right)}$

59. $\dfrac{\left(\dfrac{x}{x-2}\right)}{\left(\dfrac{2x}{2-x}\right)}$

60. $\dfrac{\left(\dfrac{y^2}{5-y}\right)}{\left(\dfrac{y}{y-5}\right)}$

61. $\dfrac{\left(\dfrac{6x^2}{x^2+2x-35}\right)}{\left(\dfrac{x^3}{x^2-25}\right)}$

62. $\dfrac{\left[\dfrac{24-18x}{(2-x)^2}\right]}{\left(\dfrac{60-45x}{x^2-4x+4}\right)}$

2 ▸ Simplify complex fractions having a sum or difference in the numerator and/or denominator.

In Exercises 63–68, simplify the complex fraction.

63. $\dfrac{3t}{\left(5 - \dfrac{2}{t}\right)}$

64. $\dfrac{\left(\dfrac{1}{x} - \dfrac{1}{2}\right)}{2x}$

65. $\dfrac{\left(x - 3 + \dfrac{2}{x}\right)}{\left(1 - \dfrac{2}{x}\right)}$

66. $\dfrac{3x-1}{\left(\dfrac{2}{x^2} + \dfrac{5}{x}\right)}$

67. $\dfrac{\left(\dfrac{1}{a^2-16} - \dfrac{1}{a}\right)}{\left(\dfrac{1}{a^2+4a} + 4\right)}$

68. $\dfrac{\left(\dfrac{1}{x^2} - \dfrac{1}{y^2}\right)}{\left(\dfrac{1}{x} + \dfrac{1}{y}\right)}$

7.5 Solving Rational Equations

1 ▸ Solve rational equations containing constant denominators.

In Exercises 69–74, solve the equation.

69. $\dfrac{x}{15} + \dfrac{3}{5} = 1$

70. $\dfrac{x}{6} + \dfrac{5}{3} = 3$

71. $\dfrac{3x}{8} = -15 + \dfrac{x}{4}$

72. $\dfrac{t+1}{6} = \dfrac{1}{2} - 2t$

73. $\dfrac{x^2}{6} - \dfrac{x}{12} = \dfrac{1}{2}$

74. $\dfrac{x^2}{4} = -\dfrac{x}{12} + \dfrac{1}{6}$

2 ▸ Solve rational equations containing variable denominators.

In Exercises 75–90, solve the equation.

75. $8 - \dfrac{12}{t} = \dfrac{1}{3}$

76. $5 + \dfrac{2}{x} = \dfrac{1}{4}$

77. $\dfrac{2}{y} - \dfrac{1}{3y} = \dfrac{1}{3}$

78. $\dfrac{7}{4x} - \dfrac{6}{8x} = 1$

79. $r = 2 + \dfrac{24}{r}$

80. $\dfrac{2}{x} - \dfrac{x}{6} = \dfrac{2}{3}$

81. $8\left(\dfrac{6}{x} - \dfrac{1}{x+5}\right) = 15$

82. $16\left(\dfrac{5}{x} - \dfrac{3}{x+2}\right) = 4$

83. $\dfrac{3}{y+1} - \dfrac{8}{y} = 1$

84. $\dfrac{4x}{x-5} + \dfrac{2}{x} = -\dfrac{4}{x-5}$

85. $\dfrac{2x}{x-3} - \dfrac{3}{x} = 0$

86. $\dfrac{6x}{x-3} = 9 + \dfrac{18}{x-3}$

87. $\dfrac{12}{x^2+x-12} - \dfrac{1}{x-3} = -1$

88. $\dfrac{3}{x-1} + \dfrac{6}{x^2-3x+2} = 2$

89. $\dfrac{5}{x^2-4} - \dfrac{6}{x-2} = -5$

90. $\dfrac{3}{x^2-9} + \dfrac{4}{x+3} = 1$

7.6 Applications and Variation

1 ▶ Solve application problems involving rational equations.

91. *Average Speeds* You and a friend ride bikes for the same amount of time. You ride 24 miles and your friend rides 15 miles. Your friend's average speed is 6 miles per hour slower than yours. What are the average speeds of you and your friend?

92. *Average Speed* You drive 220 miles to see a friend. The return trip takes 20 minutes less than the original trip, and your average speed is 5 miles per hour faster. What is your average speed on the return trip?

93. *Partnership Costs* A group of people starting a business agree to share equally in the cost of a $60,000 piece of machinery. If they could find two more people to join the group, each person's share of the cost would decrease by $5000. How many people are presently in the group?

94. *Work Rate* One painter works $1\frac{1}{2}$ times as fast as another painter. It takes them 4 hours working together to paint a room. Find the time it takes each painter to paint the room working alone.

95. *Population Growth* The Parks and Wildlife Commission introduces 80,000 fish into a large lake. The population P (in thousands) of the fish is approximated by the model

$$P = \frac{20(4 + 3t)}{1 + 0.05t}$$

where t is the time in years. Find the time required for the population to increase to 400,000 fish.

96. *Average Cost* The average cost $\overline{C}$ of producing x units of a product is given by

$$\overline{C} = 1.5 + \frac{4200}{x}.$$

Determine the number of units that must be produced to obtain an average cost of $2.90 per unit.

2 ▶ Solve application problems involving direct variation.

97. *Hooke's Law* A force of 100 pounds stretches a spring 4 inches. Find the force required to stretch the spring 6 inches.

98. *Stopping Distance* The stopping distance d of an automobile is directly proportional to the square of its speed s. How will the stopping distance be changed by doubling the speed of the car?

3 ▸ Solve application problems involving inverse variation.

99. *Travel Time* The travel time between two cities is inversely proportional to the average speed. A train travels between the cities in 3 hours at an average speed of 65 miles per hour. How long would it take to travel between the cities at an average speed of 80 miles per hour?

100. *Demand* A company has found that the daily demand x for its cordless telephones is inversely proportional to the price p. When the price is \$25, the demand is 1000 telephones. Approximate the demand when the price is increased to \$28.

101. *Revenue* The monthly demand for Brand X athletic shoes varies directly as the amount spent on advertising and inversely as the square of the price per pair. When \$20,000 is spent on monthly advertising and the price per pair of shoes is \$55, the demand is 900 pairs. If advertising is increased to \$25,000, what price will yield a demand of 900 pairs? Is this increase worthwhile in terms of revenue?

102. *Revenue* The seasonal demand for Ace brand sunglasses varies directly as the amount spent on advertising and inversely as the square of the price per pair. When \$125,000 is spent on advertising and the price per pair is \$35, the demand is 5000 pairs. If advertising is increased to \$135,000, what price will yield a demand of 5000 pairs? Is this increase worthwhile in terms of revenue?

4 ▸ Solve application problems involving joint variation.

103. *Simple Interest* The simple interest earned on a savings account is jointly proportional to the time and the principal. After three quarters (9 months), the interest for a principal of \$12,000 is \$675. How much interest would a principal of \$8200 earn in 18 months?

104. *Cost* The cost of constructing a wooden box with a square base varies jointly as the height of the box and the square of the width of the box. A box of height 16 inches and of width 6 inches costs \$28.80. How much would a box of height 14 inches and of width 8 inches cost?

Chapter Test

Take this test as you would take a test in class. After you are done, check your work against the answers in the back of the book.

1. Find the domain of $f(x) = \dfrac{x+1}{x^2 - 6x + 5}$.

In Exercises 2 and 3, simplify the rational expression.

2. $\dfrac{4 - 2x}{x - 2}$

3. $\dfrac{2a^2 - 5a - 12}{5a - 20}$

4. Find the least common multiple of x^2, $3x^3$, and $(x + 4)^2$.

Smart Study Strategy

Go to page 400 for ways to *Use Ten Steps for Test-Taking.*

In Exercises 5–16, perform the operation and simplify.

5. $\dfrac{4z^3}{5} \cdot \dfrac{25}{12z^2}$

6. $\dfrac{y^2 + 8y + 16}{2(y - 2)} \cdot \dfrac{8y - 16}{(y + 4)^3}$

7. $\dfrac{(2xy^2)^3}{15} \div \dfrac{12x^3}{21}$

8. $(4x^2 - 9) \div \dfrac{2x + 3}{2x^2 - x - 3}$

9. $\dfrac{3}{x - 3} + \dfrac{x - 2}{x - 3}$

10. $2x + \dfrac{1 - 4x^2}{x + 1}$

11. $\dfrac{5x}{x + 2} - \dfrac{2}{x^2 - x - 6}$

12. $\dfrac{3}{x} - \dfrac{5}{x^2} + \dfrac{2x}{x^2 + 2x + 1}$

13. $\dfrac{\left(\dfrac{3x}{x + 2}\right)}{\left(\dfrac{12}{x^3 + 2x^2}\right)}$

14. $\dfrac{\left(9x - \dfrac{1}{x}\right)}{\left(\dfrac{1}{x} - 3\right)}$

15. $\dfrac{3x^{-2} + y^{-1}}{(x + y)^{-1}}$

16. $\dfrac{6x^2 - 4x + 8}{2x}$

In Exercises 17–19, solve the equation.

17. $\dfrac{3}{h + 2} = \dfrac{1}{6}$

18. $\dfrac{2}{x + 5} - \dfrac{3}{x + 3} = \dfrac{1}{x}$

19. $\dfrac{1}{x + 1} + \dfrac{1}{x - 1} = \dfrac{2}{x^2 - 1}$

20. Find a mathematical model that relates u and v if v varies directly as the square root of u, and $v = \frac{3}{2}$ when $u = 36$.

21. If the temperature of a gas is not allowed to change, the absolute pressure P of the gas is inversely proportional to its volume V, according to Boyle's Law. A large balloon is filled with 180 cubic meters of helium at atmospheric pressure (1 atm) at sea level. What is the volume of the helium if the balloon rises to an altitude at which the atmospheric pressure is 0.75 atm? (Assume that the temperature does not change.)

Study Skills in Action

Viewing Math as a Foreign Language

Learning math requires more than just completing homework problems. For instance, learning the material in a chapter may require using approaches similar to those used for learning a foreign language (Nolting, 2008) in that you must:

- understand and memorize vocabulary words;
- understand and memorize mathematical rules (as you would memorize grammatical rules); and
- apply rules to mathematical expressions or equations (like creating sentences using correct grammar rules).

You should understand the vocabulary words and rules in a chapter as well as memorize and say them out loud. Strive to speak the mathematical language with fluency, just as a student learning a foreign language must.

Kimberly Nolting
VP, Academic Success Press
expert in developmental education

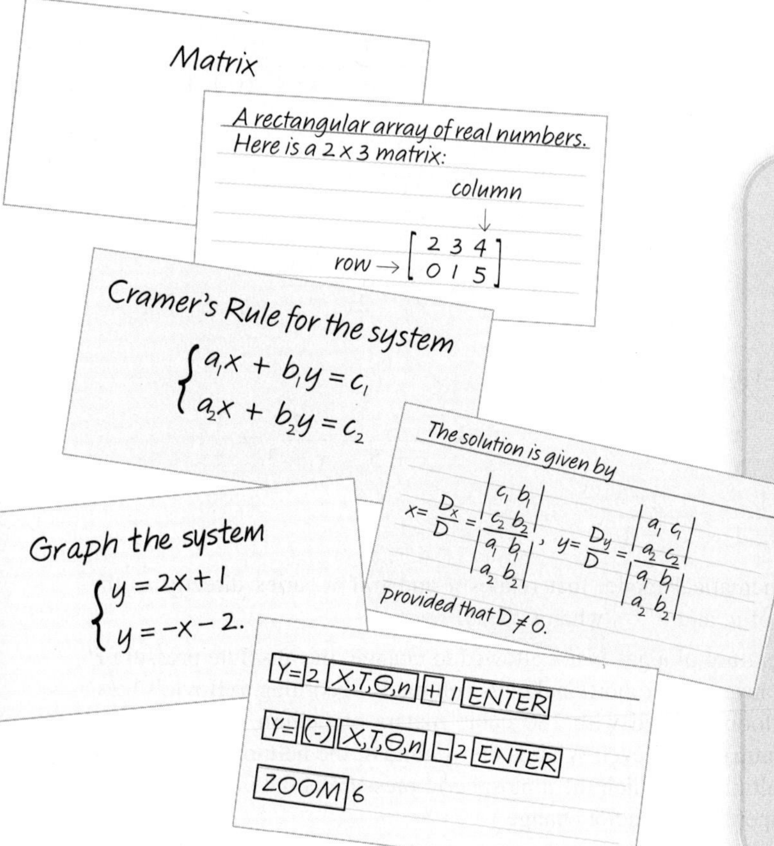

Smart Study Strategy

Make Note Cards

Invest in three different colors of 4 × 6 note cards. Use one color for each of the following: vocabulary words; rules; and graphing keystrokes.

1 ► Write vocabulary words on note cards, one word per card. Write the definition and an example on the other side. If possible, put definitions in your own words.

2 ► Write rules on note cards, one per card. Include an example and an explanation on the other side.

3 ► Write each kind of calculation on a separate note card. Include the keystrokes required to perform the calculation on the other side.

Use the note cards as references while doing your homework. Quiz yourself once a day.

Chapter 8

Systems of Equations and Inequalities

IT WORKED FOR ME!
"When I trained to be a tutor we learned how note cards can be used to help review and memorize math concepts. I started using them in my own math class and it really helps me, especially when I am caught with short amounts of time on campus. I don't have to pull out all of my books to study. I just keep my note cards in my purse."
Di Iris
Computer science

8.1 Solving Systems of Equations by Graphing and Substitution

Ronald Martinez/Getty Images

Why You Should Learn It

Many businesses use systems of equations to help determine their sales goals. For instance, Exercise 106 on page 486 uses the graph of a system of equations to determine the break-even point of producing and selling hockey sticks.

What You Should Learn

1 ▶ Determine if an ordered pair is a solution of a system of equations.

2 ▶ Use a coordinate system to solve systems of linear equations graphically.

3 ▶ Use the method of substitution to solve systems of equations algebraically.

4 ▶ Solve application problems using systems of equations.

1 ▶ Determine if an ordered pair is a solution of a system of equations.

Systems of Equations

Many problems in business and science involve **systems of equations.** These systems consist of two or more equations, each containing two or more variables.

$$\begin{cases} ax + by = c & \text{Equation 1} \\ dx + ey = f & \text{Equation 2} \end{cases}$$

A **solution** of such a system is an ordered pair (x, y) of real numbers that satisfies *each* equation in the system. When you find the set of all solutions of the system of equations, you are finding the **solution of the system of equations.**

EXAMPLE 1 Checking Solutions of a System of Equations

Which of the ordered pairs is a solution of the system: (a) (3, 3) or (b) (4, 2)?

$$\begin{cases} x + y = 6 & \text{Equation 1} \\ 2x - 5y = -2 & \text{Equation 2} \end{cases}$$

Solution

a. To determine whether the ordered pair (3, 3) is a solution of the system of equations, you should substitute 3 for x and 3 for y in *each* of the equations. Substituting into Equation 1 produces

$3 + 3 = 6.$ ✓ Substitute 3 for x and 3 for y.

Similarly, substituting into Equation 2 produces

$2(3) - 5(3) \neq -2.$ ✗ Substitute 3 for x and 3 for y.

Because the ordered pair (3, 3) fails to check in *both* equations, you can conclude that it *is not* a solution of the system of equations.

b. By substituting 4 for x and 2 for y in the original equations, you can determine that the ordered pair (4, 2) *is* a solution of *both* equations.

$4 + 2 = 6$ ✓ Substitute 4 for x and 2 for y in Equation 1.

$2(4) - 5(2) = -2$ ✓ Substitute 4 for x and 2 for y in Equation 2.

So, (4, 2) *is* a solution of the original system of equations.

✓ **CHECKPOINT** ***Now try Exercise 3.***

2 ▶ Use a coordinate system to solve systems of linear equations graphically.

Solving a System of Linear Equations by Graphing

In this chapter you will study three methods of solving a system of linear equations. The first method is *solution by graphing*. With this method, you first sketch the lines representing the equations. Then you try to determine whether the lines intersect at a point, as illustrated in Example 2.

EXAMPLE 2 Solving a System of Linear Equations

Use the graphical method to solve the system of linear equations.

$$\begin{cases} 2x + 3y = 7 & \text{Equation 1} \\ 2x - 5y = -1 & \text{Equation 2} \end{cases}$$

Solution

Because both equations in the system are linear, you know that they have graphs that are lines. To sketch these lines, first write each equation in slope-intercept form, as follows.

$$\begin{cases} y = -\frac{2}{3}x + \frac{7}{3} & \text{Slope-intercept form of Equation 1} \\ y = \frac{2}{5}x + \frac{1}{5} & \text{Slope-intercept form of Equation 2} \end{cases}$$

The lines corresponding to these two equations are shown in Figure 8.1.

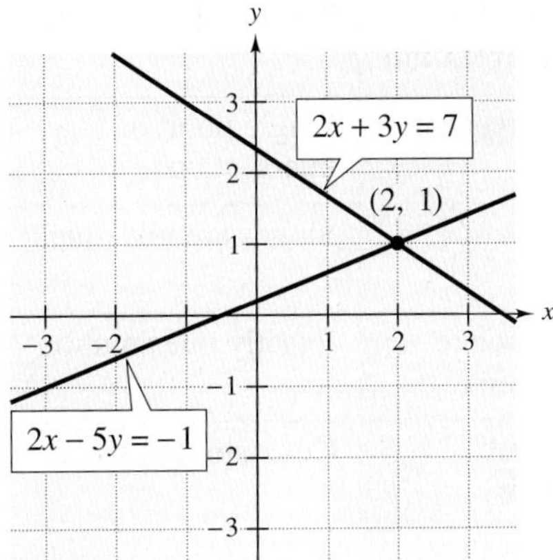

Figure 8.1

It appears that the two lines intersect at a single point, (2, 1). To verify this, substitute the coordinates of the point into each of the two original equations.

Substitute in 1st Equation

$2x + 3y = 7$ Equation 1

$2(2) + 3(1) \stackrel{?}{=} 7$ Substitute 2 for x and 1 for y.

$7 = 7$ Solution checks. ✓

Substitute in 2nd Equation

$2x - 5y = -1$ Equation 2

$2(2) - 5(1) \stackrel{?}{=} -1$ Substitute 2 for x and 1 for y.

$-1 = -1$ Solution checks. ✓

Because *both* equations are satisfied, the point (2, 1) is the solution of the system.

✓ **CHECKPOINT** *Now try Exercise 21.*

Technology: Discovery

Rewrite each system of equations in slope-intercept form and graph the equations using a graphing calculator. What is the relationship between the slopes of the two lines and the number of points of intersection?

a. $\begin{cases} 3x + 4y = 12 \\ 2x - 3y = -9 \end{cases}$ b. $\begin{cases} -x + 2y = 8 \\ 2x - 4y = 5 \end{cases}$ c. $\begin{cases} x + y = 6 \\ 3x + 3y = 18 \end{cases}$

A system of linear equations can have exactly one solution, infinitely many solutions, or no solution. To see why this is true, consider the graphical interpretations of three systems of two linear equations shown below.

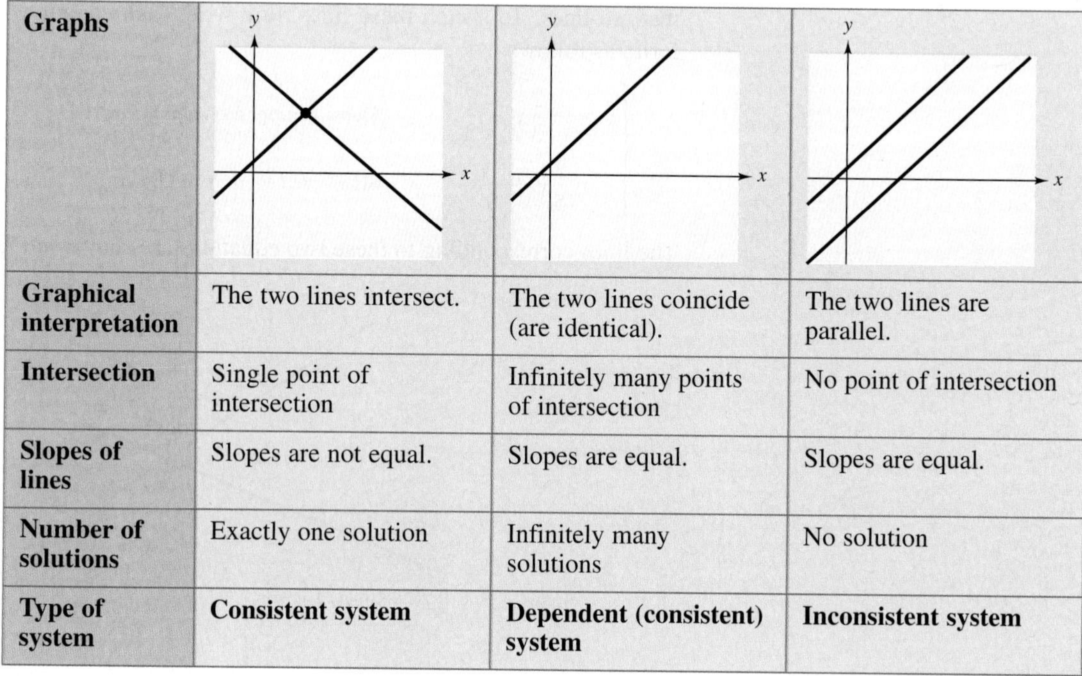

Graphs			
Graphical interpretation	The two lines intersect.	The two lines coincide (are identical).	The two lines are parallel.
Intersection	Single point of intersection	Infinitely many points of intersection	No point of intersection
Slopes of lines	Slopes are not equal.	Slopes are equal.	Slopes are equal.
Number of solutions	Exactly one solution	Infinitely many solutions	No solution
Type of system	**Consistent system**	**Dependent (consistent) system**	**Inconsistent system**

Note that for dependent systems, the slopes of the lines are equal and the y-intercepts are the same. For inconsistent systems, the slopes of the lines are equal, but the y-intercepts are different. Also, note that the word *consistent* is used to mean that the system of linear equations has at least one solution, whereas the word *inconsistent* is used to mean that the system of linear equations has no solution.

You can see from the graphs above that a comparison of the slopes of two lines gives useful information about the number of solutions of the corresponding system of equations. So, to solve a system of equations graphically, it helps to begin by writing the equations in slope-intercept form,

$$y = mx + b. \qquad \text{Slope-intercept form}$$

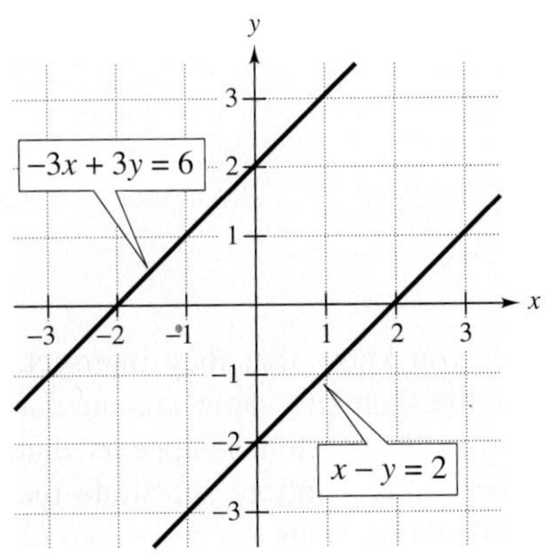

Figure 8.2

EXAMPLE 3 A System with No Solution

Solve the system of linear equations.

$$\begin{cases} x - y = 2 & \text{Equation 1} \\ -3x + 3y = 6 & \text{Equation 2} \end{cases}$$

Solution

Begin by writing each equation in slope-intercept form.

$$\begin{cases} y = x - 2 & \text{Slope-intercept form of Equation 1} \\ y = x + 2 & \text{Slope-intercept form of Equation 2} \end{cases}$$

From these forms, you can see that the slopes of the lines are equal and the y-intercepts are different, as shown in Figure 8.2. So, the original system of linear equations has no solution and is an inconsistent system.

 ***CHECKPOINT*** *Now try Exercise 23.*

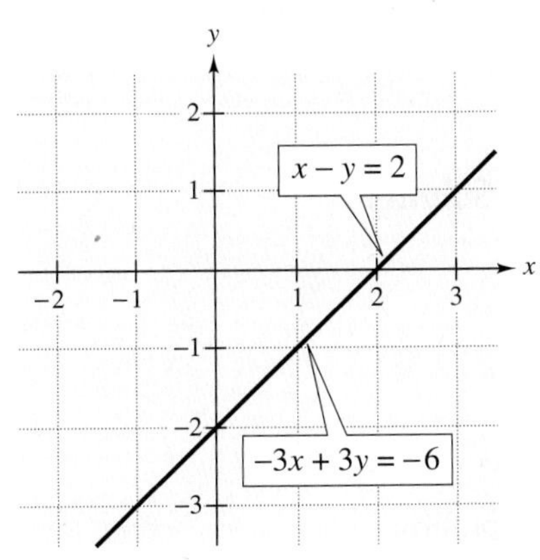

Figure 8.3

EXAMPLE 4 A System with Infinitely Many Solutions

Solve the system of linear equations.

$$\begin{cases} x - y = 2 & \text{Equation 1} \\ -3x + 3y = -6 & \text{Equation 2} \end{cases}$$

Solution

Begin by writing each equation in slope-intercept form.

$$\begin{cases} y = x - 2 & \text{Slope-intercept form of Equation 1} \\ y = x - 2 & \text{Slope-intercept form of Equation 2} \end{cases}$$

From these forms, you can see that the slopes of the lines are equal and the y-intercepts are the same, as shown in Figure 8.3. So, the original system of linear equations has infinitely many solutions and is a dependent system. You can describe the solution set by saying that each point on the line $y = x - 2$ is a solution of the system of linear equations.

 ***CHECKPOINT*** *Now try Exercise 35.*

Note in Examples 3 and 4 that if the two lines representing a system of linear equations have the same slope, the system must have either no solution or infinitely many solutions. On the other hand, if the two lines have different slopes, they must intersect at a single point and the corresponding system has a single solution.

There are two things you should note as you read through Examples 5 and 6. First, your success in applying the graphical method of solving a system of linear equations depends on sketching accurate graphs. Second, once you have made a graph and estimated the point of intersection, it is critical that you check in the original system to see whether the point you have chosen is the correct solution.

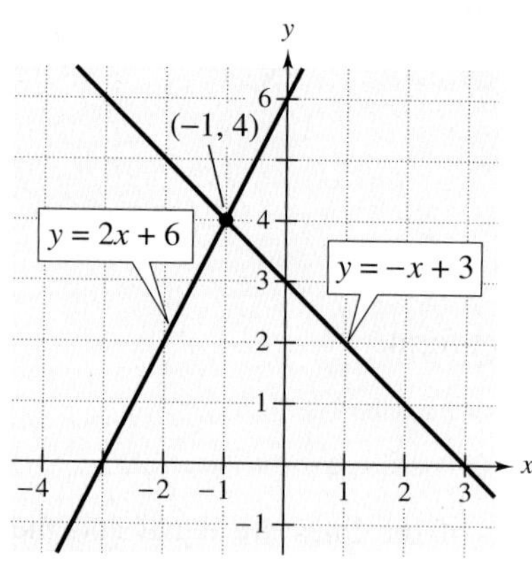

Figure 8.4

EXAMPLE 5 A System with a Single Solution

Solve the system of linear equations.

$$\begin{cases} y = -x + 3 & \text{Equation 1} \\ y = 2x + 6 & \text{Equation 2} \end{cases}$$

Solution

Because the lines do not have the same slope, you know that they intersect. To find the point of intersection, sketch both lines on the same rectangular coordinate system, as shown in Figure 8.4. From this sketch, it appears that the solution occurs at the point $(-1, 4)$. To check this solution, substitute the coordinates of the point into each of the two original equations.

Substitute in 1st Equation

$$y = -x + 3$$
$$4 \stackrel{?}{=} -(-1) + 3$$
$$4 = 4 \checkmark$$

Substitute in 2nd Equation

$$y = 2x + 6$$
$$4 \stackrel{?}{=} 2(-1) + 6$$
$$4 = 4 \checkmark$$

Because *both* equations are satisfied, the point $(-1, 4)$ is the solution of the system.

✓ **CHECKPOINT** *Now try Exercise 25.*

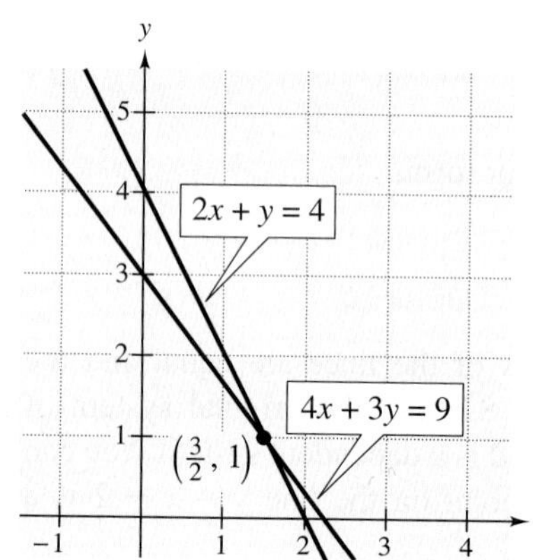

Figure 8.5

EXAMPLE 6 A System with a Single Solution

Solve the system of linear equations.

$$\begin{cases} 2x + y = 4 & \text{Equation 1} \\ 4x + 3y = 9 & \text{Equation 2} \end{cases}$$

Solution

Begin by writing each equation in slope-intercept form.

$$\begin{cases} y = -2x + 4 & \text{Slope-intercept form of Equation 1} \\ y = -\frac{4}{3}x + 3 & \text{Slope-intercept form of Equation 2} \end{cases}$$

Because the lines do not have the same slope, you know that they intersect. To find the point of intersection, sketch both lines on the same rectangular coordinate system, as shown in Figure 8.5. From this sketch, it appears that the solution occurs at the point $\left(\frac{3}{2}, 1\right)$. To check this solution, substitute the coordinates of the point into each of the two original equations.

Substitute in 1st Equation

$$2x + y = 4$$
$$2\left(\tfrac{3}{2}\right) + 1 \stackrel{?}{=} 4$$
$$3 + 1 = 4 \checkmark$$

Substitute in 2nd Equation

$$4x + 3y = 9$$
$$4\left(\tfrac{3}{2}\right) + 3(1) \stackrel{?}{=} 9$$
$$6 + 3 = 9 \checkmark$$

Because *both* equations are satisfied, the point $\left(\frac{3}{2}, 1\right)$ is the solution of the system.

✓ **CHECKPOINT** *Now try Exercise 27.*

3 ▶ Use the method of substitution to solve systems of equations algebraically.

The Method of Substitution

Solving systems of equations by graphing is useful but less accurate than using algebraic methods. In this section, you will study an algebraic method called the **method of substitution.** The goal of the method of substitution is to *reduce a system of two linear equations in two variables to a single equation in one variable.* Examples 7 and 8 illustrate the basic steps of the method.

EXAMPLE 7 The Method of Substitution

Solve the system of linear equations.

$$\begin{cases} -x + y = 1 & \text{Equation 1} \\ 2x + y = -2 & \text{Equation 2} \end{cases}$$

Solution

Begin by solving for y in Equation 1.

$-x + y = 1$ Original Equation 1

$y = x + 1$ Revised Equation 1

Next, substitute this expression for y in Equation 2.

$2x + y = -2$ Equation 2

$2x + (x + 1) = -2$ Substitute $x + 1$ for y.

$3x + 1 = -2$ Combine like terms.

$3x = -3$ Subtract 1 from each side.

$x = -1$ Divide each side by 3.

At this point, you know that the x-coordinate of the solution is -1. To find the y-coordinate, *back-substitute* the x-value in the revised Equation 1.

$y = x + 1$ Revised Equation 1

$y = -1 + 1$ Substitute -1 for x.

$y = 0$ Simplify.

So, the solution is $(-1, 0)$. Check this solution by substituting $x = -1$ and $y = 0$ in both of the original equations.

✓ **CHECKPOINT** *Now try Exercise 65.*

Study Tip

The term *back-substitute* implies that you work backwards. After solving for one of the variables, substitute that value back into one of the equations in the original (or revised) system to find the value of the other variable.

When you use substitution, it does not matter which variable you choose to solve for first. You should choose the variable and equation that are easier to work with. For instance, in the system below on the left, it is best to solve for x in Equation 2, whereas for the system on the right, it is best to solve for y in Equation 1.

$$\begin{cases} 3x - 2y = 1 & \text{Equation 1} \\ x + 4y = 3 & \text{Equation 2} \end{cases} \qquad \begin{cases} 2x + y = 5 & \text{Equation 1} \\ 3x - 2y = 11 & \text{Equation 2} \end{cases}$$

EXAMPLE 8 The Method of Substitution

Solve the system of linear equations.

$$\begin{cases} 5x + 7y = 1 & \text{Equation 1} \\ x + 4y = -5 & \text{Equation 2} \end{cases}$$

Solution

For this system, it is convenient to begin by solving for x in the second equation.

$x + 4y = -5$ Original Equation 2

$x = -4y - 5$ Revised Equation 2

Substituting this expression for x in the first equation produces the following.

$5(-4y - 5) + 7y = 1$ Substitute $-4y - 5$ for x in Equation 1.

$-20y - 25 + 7y = 1$ Distributive Property

$-13y - 25 = 1$ Combine like terms.

$-13y = 26$ Add 25 to each side.

$y = -2$ Divide each side by -13.

Finally, back-substitute this y-value into the revised second equation.

$x = -4(-2) - 5 = 3$ Substitute -2 for y in revised Equation 2.

The solution is $(3, -2)$. Check this by substituting $x = 3$ and $y = -2$ into both of the original equations, as follows.

Substitute into Equation 1

$5x + 7y = 1$

$5(3) + 7(-2) \stackrel{?}{=} 1$

$15 - 14 = 1$ ✓

Substitute into Equation 2

$x + 4y = -5$

$(3) + 4(-2) \stackrel{?}{=} -5$

$3 - 8 = -5$ ✓

✓ **CHECKPOINT** *Now try Exercise 71.*

The steps for using the method of substitution to solve a system of two equations involving two variables are summarized as follows.

The Method of Substitution

1. Solve one of the equations for one variable in terms of the other.
2. Substitute the expression obtained in Step 1 in the other equation to obtain an equation in one variable.
3. Solve the equation obtained in Step 2.
4. Back-substitute the solution from Step 3 in the expression obtained in Step 1 to find the value of the other variable.
5. Check the solution to see that it satisfies *both* of the original equations.

EXAMPLE 9 The Method of Substitution

Solve the system of linear equations.

$$\begin{cases} 5x + 3y = 18 & \text{Equation 1} \\ 2x - 7y = -1 & \text{Equation 2} \end{cases}$$

Solution

Step 1 Because neither variable has a coefficient of 1, you can choose to solve for either variable. For instance, you can begin by solving for x in Equation 1 to obtain $x = -\frac{3}{5}y + \frac{18}{5}$.

Step 2 Substitute for x in Equation 2 and solve for y.

$$2x - 7y = -1 \qquad \text{Equation 2}$$

$$2\left(-\tfrac{3}{5}y + \tfrac{18}{5}\right) - 7y = -1 \qquad \text{Substitute } -\tfrac{3}{5}y + \tfrac{18}{5} \text{ for } x.$$

Step 3

$$-\tfrac{6}{5}y + \tfrac{36}{5} - 7y = -1 \qquad \text{Distributive Property}$$

$$-6y + 36 - 35y = -5 \qquad \text{Multiply each side by 5.}$$

$$y = 1 \qquad \text{Solve for } y.$$

Step 4 Back-substitute this y-value into the revised first equation.

$$x = -\tfrac{3}{5}y + \tfrac{18}{5} \qquad \text{Revised Equation 1}$$

$$x = -\tfrac{3}{5}(1) + \tfrac{18}{5} = 3 \qquad \text{Substitute 1 for } y.$$

Step 5 The solution is $(3, 1)$. Check this in the original system.

CHECKPOINT *Now try Exercise 73.*

EXAMPLE 10 The Method of Substitution: No-Solution Case

Solve the system of linear equations.

$$\begin{cases} x - 3y = 2 & \text{Equation 1} \\ -2x + 6y = 2 & \text{Equation 2} \end{cases}$$

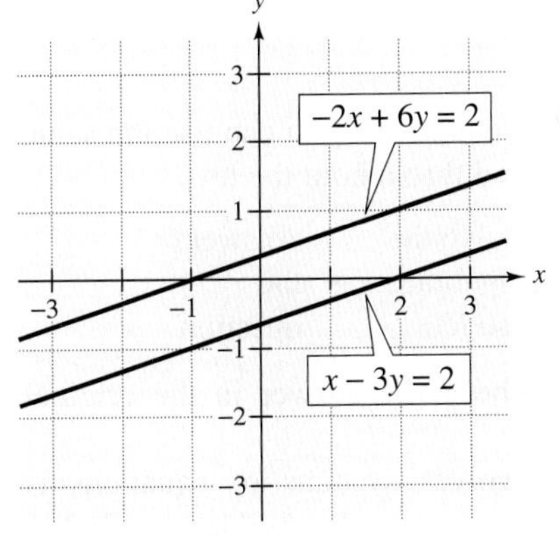

Figure 8.6

Solution

Begin by solving for x in Equation 1 to obtain $x = 3y + 2$. Then substitute this expression for x in Equation 2.

$$-2x + 6y = 2 \qquad \text{Equation 2}$$

$$-2(3y + 2) + 6y = 2 \qquad \text{Substitute } 3y + 2 \text{ for } x.$$

$$-6y - 4 + 6y = 2 \qquad \text{Distributive Property}$$

$$-4 \neq 2 \qquad \text{Simplify.}$$

Because -4 does not equal 2, you can conclude that the original system is inconsistent and has no solution. The graphs in Figure 8.6 confirm this result.

CHECKPOINT *Now try Exercise 85.*

EXAMPLE 11 The Method of Substitution: Many-Solution Case

Solve the system of linear equations.

$$\begin{cases} 9x + 3y = 15 & \text{Equation 1} \\ 3x + y = 5 & \text{Equation 2} \end{cases}$$

Solution

Begin by solving for y in Equation 2 to obtain $y = -3x + 5$. Then, substitute this expression for y in Equation 1.

$$9x + 3y = 15 \qquad \text{Equation 1}$$

$$9x + 3(-3x + 5) = 15 \qquad \text{Substitute } -3x + 5 \text{ for } y.$$

$$9x - 9x + 15 = 15 \qquad \text{Distributive Property}$$

$$15 = 15 \qquad \text{Simplify.}$$

The equation $15 = 15$ is true for any value of x. This implies that any solution of Equation 2 is also a solution of Equation 1. In other words, the original system of linear equations is *dependent* and has infinitely many solutions. The solutions consist of all ordered pairs (x, y) lying on the line $3x + y = 5$. Some sample solutions are $(-1, 8)$, $(0, 5)$, and $(1, 2)$. Check these as follows:

Solution Point	*Substitute into* $3x + y = 5$
$(-1, 8)$	$3(-1) + 8 = -3 + 8 = 5$ ✓
$(0, 5)$	$3(0) + 5 = 0 + 5 = 5$ ✓
$(1, 2)$	$3(1) + 2 = 3 + 2 = 5$ ✓

CHECKPOINT *Now try Exercise 81.*

By writing both equations in Example 11 in slope-intercept form, you will get identical equations. This means that the lines coincide and the system has infinitely many solutions.

4 ▶ Solve application problems using systems of equations.

Applications

To model a real-life situation with a system of equations, you can use the same basic problem-solving strategy that has been used throughout the text.

Write a verbal model. → Assign labels. → Write an algebraic model. → Solve the algebraic model. → Answer the question.

After answering the question, remember to check the answer in the original statement of the problem.

A common business application that involves systems of equations is break-even analysis. The total cost C of producing x units of a product usually has two components—the *initial cost* and the *cost per unit*. When enough units have been sold so that the total revenue R equals the total cost C, the sales have reached the **break-even point.** You can find this break-even point by finding the point of intersection of the cost and revenue graphs.

EXAMPLE 12 Break-Even Analysis 

A small business invests \$14,000 to produce a new energy bar. Each bar costs \$0.80 to produce and is sold for \$1.50. How many energy bars must be sold before the business breaks even?

Solution

Verbal Model: Total cost = Cost per bar · Number of bars + Initial cost

Total revenue = Price per bar · Number of bars

Labels:
Total cost $= C$ (dollars)
Cost per bar $= 0.80$ (dollars per bar)
Number of bars $= x$ (bars)
Initial cost $= 14{,}000$ (dollars)
Total revenue $= R$ (dollars)
Price per bar $= 1.50$ (dollars per bar)

System:
$$\begin{cases} C = 0.80x + 14{,}000 & \text{Equation 1} \\ R = 1.50x & \text{Equation 2} \end{cases}$$

The two equations are in slope-intercept form and because the lines do not have the same slope, you know that they intersect. So, to find the break-even point, graph both equations and determine the point of intersection of the two graphs, as shown in Figure 8.7. From this graph, it appears that the break-even point occurs at the point (20,000, 30,000). To check this solution, substitute the coordinates of the point in each of the two original equations.

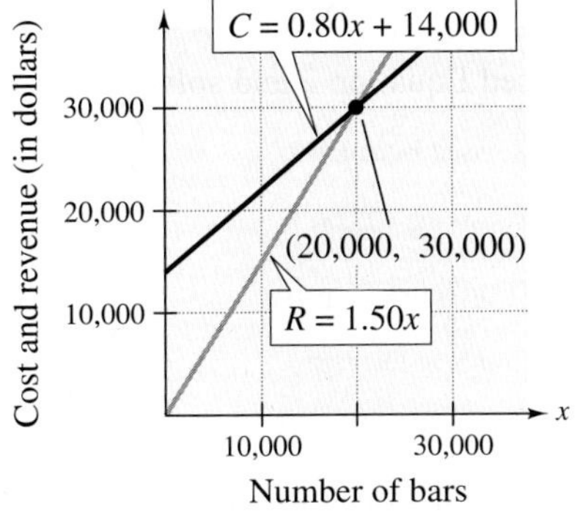

Figure 8.7

Substitute in Equation 1

$C = 0.80x + 14{,}000$ — Equation 1

$30{,}000 \stackrel{?}{=} 0.80(20{,}000) + 14{,}000$ — Substitute 20,000 for x and 30,000 for C.

$30{,}000 \stackrel{?}{=} 16{,}000 + 14{,}000$ — Multiply.

$30{,}000 = 30{,}000$ ✓ — Simplify.

Substitute in Equation 2

$R = 1.50x$ — Equation 2

$30{,}000 \stackrel{?}{=} 1.50(20{,}000)$ — Substitute 20,000 for x and 30,000 for R.

$30{,}000 = 30{,}000$ ✓ — Simplify.

Because both equations are satisfied, the business must sell 20,000 energy bars before it breaks even.

CHECKPOINT *Now try Exercise 105.*

Profit P (or loss) for the business can be determined by the equation $P = R - C$. Note in Figure 8.7 that sales less than the break-even point correspond to a loss for the business, whereas sales greater than the break-even point correspond to a profit for the business.

EXAMPLE 13 Simple Interest 

A total of \$12,000 is invested in two funds paying 6% and 8% simple interest. The combined annual interest for the two funds is \$880. How much of the \$12,000 is invested at each rate?

Solution

Verbal Model: [Amount in 6% fund] + [Amount in 8% fund] = [12,000]

[6%] · [Amount in 6% fund] + [8%] · [Amount in 8% fund] = [880]

Labels: Amount in 6% fund $= x$ (dollars)
Amount in 8% fund $= y$ (dollars)

System:
$$\begin{cases} x + y = 12{,}000 & \text{Equation 1} \\ 0.06x + 0.08y = 880 & \text{Equation 2} \end{cases}$$

To begin, it is convenient to multiply each side of the second equation by 100. This eliminates the need to work with decimals.

$0.06x + 0.08y = 880$ Equation 2

$6x + 8y = 88{,}000$ Multiply each side by 100.

Then solve for x in Equation 1.

$x = 12{,}000 - y$ Revised Equation 1

Next, substitute this expression for x in the revised Equation 2 and solve for y.

$6x + 8y = 88{,}000$ Revised Equation 2

$6(12{,}000 - y) + 8y = 88{,}000$ Substitute $12{,}000 - y$ for x.

$72{,}000 - 6y + 8y = 88{,}000$ Distributive Property

$72{,}000 + 2y = 88{,}000$ Combine like terms.

$y = 8000$ Solve for y.

Back-substitute the value $y = 8000$ in the revised Equation 1 and solve for x.

$x = 12{,}000 - y$ Revised Equation 1

$x = 12{,}000 - 8000$ Substitute 8000 for y.

$x = 4000$ Simplify.

So, \$4000 was invested at 6% simple interest and \$8000 was invested at 8% simple interest. Check this in the original statement of the problem as follows.

Substitute in Equation 1

$x + y = 12{,}000$

$4000 + 8000 = 12{,}000$ ✓

Substitute in Equation 2

$0.06x + 0.08y = 880$

$0.06(4000) + 0.08(8000) = 880$

$240 + 640 = 800$ ✓

CHECKPOINT *Now try Exercise 109.*

Concept Check

1. Give graphical descriptions of the three cases for a system of linear equations in two variables.

2. Explain how you can check the solution of a system of linear equations algebraically and graphically.

3. In your own words, explain the basic steps in solving a system of linear equations by the method of substitution.

4. Is it possible for a consistent system of linear equations to have exactly two solutions? Explain.

8.1 EXERCISES

Go to pages 546–547 to record your assignments.

Developing Skills

In Exercises 1–6, determine whether each ordered pair is a solution of the system of equations. ***See Example 1.***

	System	*Ordered Pairs*	
1.	$\begin{cases} x + 3y = 11 \\ -x + 3y = 7 \end{cases}$	(a) $(2, 3)$	(b) $(5, 4)$
2.	$\begin{cases} 3x - y = -2 \\ x - 3y = 2 \end{cases}$	(a) $(0, 2)$	(b) $(-1, -1)$
✓ **3.**	$\begin{cases} 2x - 3y = -8 \\ x + y = 1 \end{cases}$	(a) $(5, -3)$	(b) $(-1, 2)$
4.	$\begin{cases} 5x - 3y = -12 \\ x - 4y = 1 \end{cases}$	(a) $(-3, -1)$	(b) $(3, 1)$
5.	$\begin{cases} 5x - 3y = 3 \\ -10x + 6y = -6 \end{cases}$	(a) $(0, -1)$	(b) $(3, 4)$
6.	$\begin{cases} 8x + 2y = 0 \\ 24x + 6y = 15 \end{cases}$	(a) $(1, -4)$	(b) $(3, -9)$

In Exercises 7–14, use the graphs of the equations to determine the solution (if any) of the system of linear equations. Check your solution.

7. $\begin{cases} 2x + y = 4 \\ x - y = 2 \end{cases}$

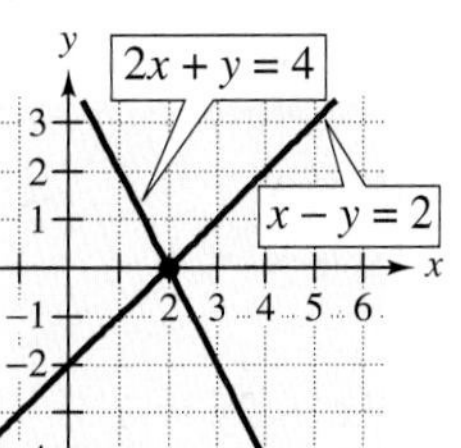

8. $\begin{cases} x + 3y = 2 \\ -x + 2y = 3 \end{cases}$

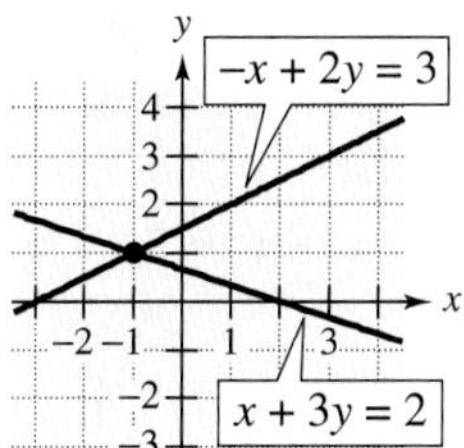

9. $\begin{cases} x - y = 0 \\ 3x - 2y = -1 \end{cases}$

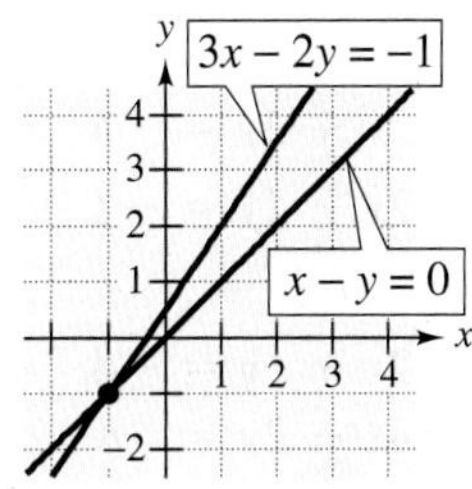

10. $\begin{cases} 2x - y = 2 \\ 4x + 3y = 24 \end{cases}$

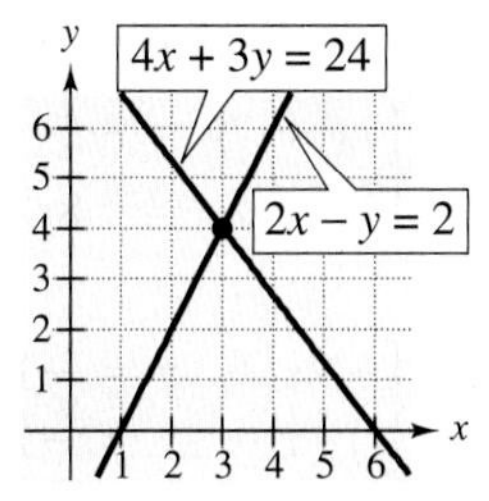

11. $\begin{cases} x - 2y = -4 \\ -0.5x + y = 2 \end{cases}$

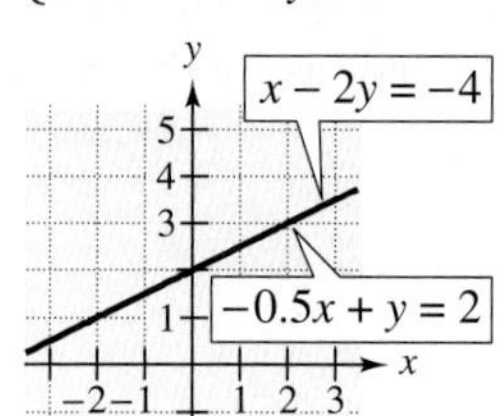

12. $\begin{cases} 2x - 5y = 10 \\ 6x - 15y = 75 \end{cases}$

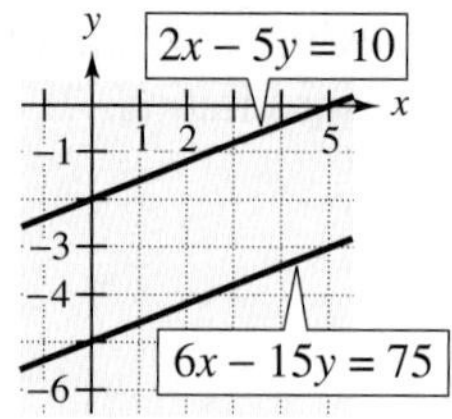

13. $\begin{cases} x + 4y = 8 \\ 3x + 12y = 12 \end{cases}$

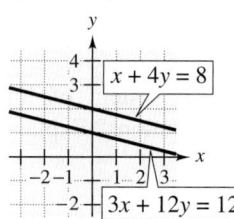

14. $\begin{cases} 2x - y = -3 \\ -4x + 2y = 6 \end{cases}$

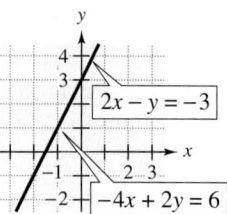

In Exercises 15–40, sketch the graphs of the equations and approximate any solutions of the system of linear equations. *See Examples 2–6.*

15. $\begin{cases} y = -x + 3 \\ y = x + 1 \end{cases}$

16. $\begin{cases} y = 2x - 1 \\ y = x + 1 \end{cases}$

17. $\begin{cases} y = 2x - 4 \\ y = -\frac{1}{2}x + 1 \end{cases}$

18. $\begin{cases} y = \frac{1}{2}x + 2 \\ y = -x + 8 \end{cases}$

19. $\begin{cases} x - y = 3 \\ x + y = 3 \end{cases}$

20. $\begin{cases} x - y = 0 \\ x + y = 4 \end{cases}$

✓ **21.** $\begin{cases} x + y = 4 \\ 4x + 4y = 2 \end{cases}$

22. $\begin{cases} 3x + 3y = 1 \\ x + y = 3 \end{cases}$

✓ **23.** $\begin{cases} x + 2y = 3 \\ x - 3y = 13 \end{cases}$

24. $\begin{cases} -x + 10y = 30 \\ x + 10y = 10 \end{cases}$

✓ **25.** $\begin{cases} x + 7y = -5 \\ 3x - 2y = 8 \end{cases}$

26. $\begin{cases} x + 2y = 4 \\ 2x - 2y = -1 \end{cases}$

✓ **27.** $\begin{cases} -2x + y = -1 \\ x - 2y = -1 \end{cases}$

28. $\begin{cases} 2x + y = -4 \\ 4x - 2y = 8 \end{cases}$

29. $\begin{cases} x - 2y = 4 \\ 2x - 4y = 8 \end{cases}$

30. $\begin{cases} -4x + 2y = 12 \\ 2x + y = 6 \end{cases}$

31. $\begin{cases} 3x + 2y = -6 \\ 3x - 2y = 6 \end{cases}$

32. $\begin{cases} 2x + 3y = 6 \\ 4x + 6y = 12 \end{cases}$

33. $\begin{cases} 4x - 5y = 0 \\ 6x - 5y = 10 \end{cases}$

34. $\begin{cases} 4x - 3y = 3 \\ 4x - 3y = 0 \end{cases}$

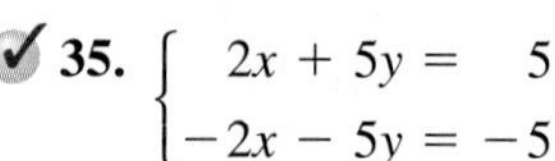

✓ **35.** $\begin{cases} 2x + 5y = 5 \\ -2x - 5y = -5 \end{cases}$

36. $\begin{cases} \frac{1}{2}x + 2y = -4 \\ -3x + y = 11 \end{cases}$

37. $\begin{cases} x + \frac{5}{4}y = 5 \\ 4x + 5y = 20 \end{cases}$

38. $\begin{cases} 7x + 3y = 21 \\ \frac{7}{3}x + y = 7 \end{cases}$

39. $\begin{cases} 8x - 6y = -12 \\ x - \frac{3}{4}y = -2 \end{cases}$

40. $\begin{cases} -x + \frac{2}{3}y = 5 \\ 9x - 6y = 6 \end{cases}$

In Exercises 41–44, use a graphing calculator to graph the equations and approximate any solutions of the system of linear equations. Check your solution.

41. $\begin{cases} y = 2x - 1 \\ y = -3x + 9 \end{cases}$

42. $\begin{cases} y = \frac{3}{4}x + 2 \\ y = x + 1 \end{cases}$

43. $\begin{cases} y = x - 1 \\ y = -2x + 8 \end{cases}$

44. $\begin{cases} y = 2x + 3 \\ y = -x - 3 \end{cases}$

In Exercises 45–52, write the equations of the lines in slope-intercept form. What can you conclude about the number of solutions of the system?

45. $\begin{cases} 2x - 3y = -12 \\ -8x + 12y = -12 \end{cases}$

46. $\begin{cases} -5x + 8y = 8 \\ -5x + 8y = -28 \end{cases}$

47. $\begin{cases} -x + 4y = 7 \\ 3x - 12y = -21 \end{cases}$

48. $\begin{cases} 3x + 8y = 28 \\ -4x + 9y = 1 \end{cases}$

49. $\begin{cases} -2x + 3y = 4 \\ 2x + 3y = 8 \end{cases}$

50. $\begin{cases} 2x + 5y = 15 \\ 2x - 5y = 5 \end{cases}$

51. $\begin{cases} -6x + 8y = 9 \\ 3x - 4y = -6 \end{cases}$

52. $\begin{cases} 7x + 6y = -4 \\ 3.5x + 3y = -2 \end{cases}$

In Exercises 53–58, solve the system by the method of substitution. Use the graph to check the solution.

53. $\begin{cases} x + y = 1 \\ 2x - y = 2 \end{cases}$

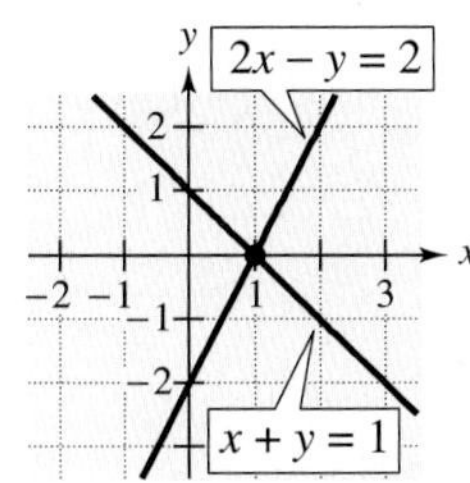

54. $\begin{cases} -2x + y = 4 \\ -x + y = 3 \end{cases}$

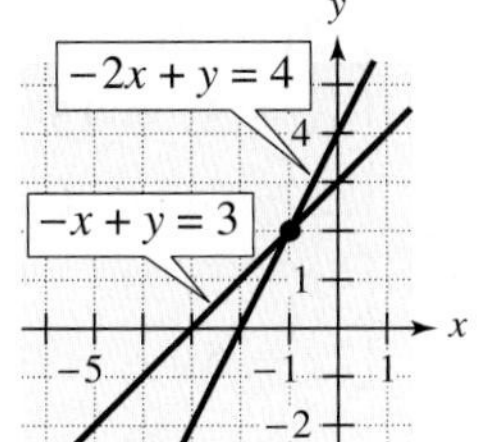

55. $\begin{cases} -x + y = 1 \\ x - y = 1 \end{cases}$

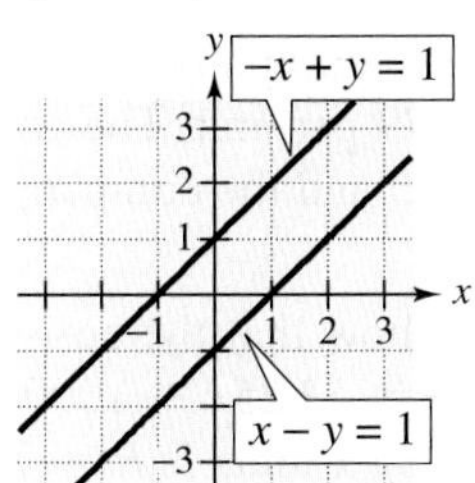

56. $\begin{cases} x + 2y = 6 \\ x + 2y = 2 \end{cases}$

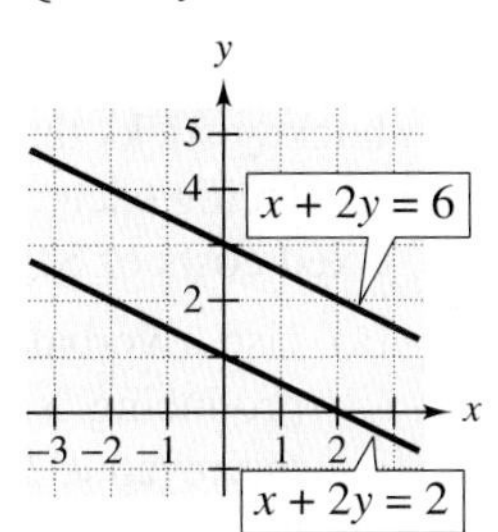

57. $\begin{cases} 2x + y = 3 \\ 4x + 2y = 6 \end{cases}$

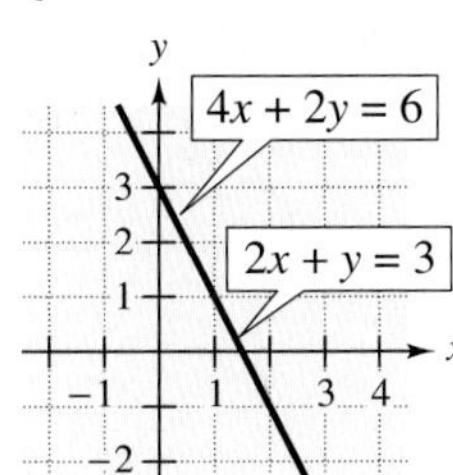

58. $\begin{cases} -4x + 3y = 6 \\ 8x - 6y = -12 \end{cases}$

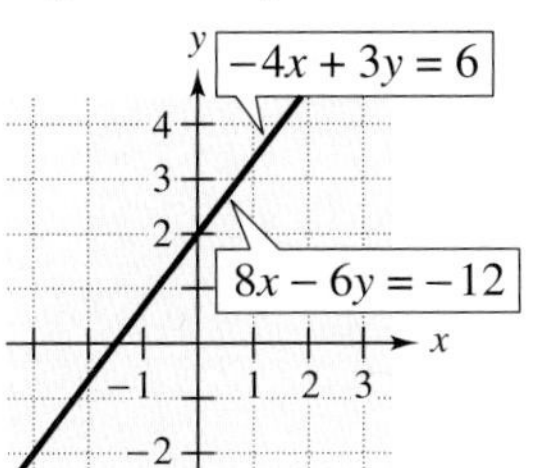

In Exercises 59–96, solve the system by the method of substitution. ***See Examples 7–11.***

59. $\begin{cases} y = 2x - 1 \\ y = -x + 5 \end{cases}$

60. $\begin{cases} y = -2x + 9 \\ y = 3x - 1 \end{cases}$

61. $\begin{cases} x = 4y - 5 \\ x = 3y \end{cases}$

62. $\begin{cases} x = -5y - 2 \\ x = 2y - 23 \end{cases}$

63. $\begin{cases} 2x = 8 \\ x - 2y = 12 \end{cases}$

64. $\begin{cases} 2x - y = 0 \\ 3y = 6 \end{cases}$

65. $\begin{cases} x - y = 0 \\ 2x + y = 0 \end{cases}$

66. $\begin{cases} x - y = 0 \\ 5x - 3y = 10 \end{cases}$

67. $\begin{cases} x - 2y = -10 \\ 3x - y = 0 \end{cases}$

68. $\begin{cases} x - 2y = 5 \\ 3x - y = 0 \end{cases}$

69. $\begin{cases} 2x - y = -2 \\ 4x + y = 5 \end{cases}$

70. $\begin{cases} x + 6y = 7 \\ -x + 4y = -2 \end{cases}$

71. $\begin{cases} x + 2y = 1 \\ 5x - 4y = -23 \end{cases}$

72. $\begin{cases} -2x + 4y = -18 \\ 3x + 3y = 9 \end{cases}$

73. $\begin{cases} 8x + 4y = 7 \\ -12x + 3y = -11 \end{cases}$

74. $\begin{cases} -3x + 6y = 4 \\ 2x + y = 4 \end{cases}$

75. $\begin{cases} 5x + 3y = 11 \\ x - 5y = 5 \end{cases}$

76. $\begin{cases} -3x + y = 4 \\ -9x + 5y = 10 \end{cases}$

77. $\begin{cases} 5x + 2y = 0 \\ x - 3y = 0 \end{cases}$

78. $\begin{cases} -2x + 3y = 17 \\ 2x - y = -7 \end{cases}$

79. $\begin{cases} 2x + 5y = -4 \\ 3x - y = 11 \end{cases}$

80. $\begin{cases} 2x + 5y = 1 \\ -x + 6y = 8 \end{cases}$

81. $\begin{cases} 4x - y = 2 \\ 2x - \frac{1}{2}y = 1 \end{cases}$

82. $\begin{cases} 3x - y = 6 \\ 4x - \frac{2}{3}y = -4 \end{cases}$

83. $\begin{cases} \frac{1}{5}x + \frac{1}{2}y = 8 \\ 2x + y = 20 \end{cases}$

84. $\begin{cases} \frac{1}{2}x + \frac{3}{4}y = 10 \\ 4x - y = 4 \end{cases}$

85. $\begin{cases} -5x + 4y = 14 \\ 5x - 4y = 4 \end{cases}$

86. $\begin{cases} 3x - 2y = 3 \\ -6x + 4y = -6 \end{cases}$

87. $\begin{cases} x - 4y = 2 \\ 5x + 2.5y = 10 \end{cases}$

88. $\begin{cases} 0.5x + 0.3y = 2 \\ 4x - 3y = -2 \end{cases}$

89. $\begin{cases} -6x + 1.5y = 6 \\ 8x - 2y = -8 \end{cases}$

90. $\begin{cases} 0.3x - 0.3y = 0 \\ x - y = 4 \end{cases}$

91. $\begin{cases} \dfrac{x}{3} - \dfrac{y}{4} = 2 \\ \dfrac{x}{2} + \dfrac{y}{6} = 3 \end{cases}$

92. $\begin{cases} -\dfrac{x}{5} + \dfrac{y}{2} = -3 \\ \dfrac{x}{4} - \dfrac{y}{4} = 0 \end{cases}$

93. $\begin{cases} \dfrac{x}{4} + \dfrac{y}{2} = 1 \\ \dfrac{x}{2} - \dfrac{y}{3} = 1 \end{cases}$

94. $\begin{cases} -\dfrac{x}{6} + \dfrac{y}{12} = 1 \\ \dfrac{x}{2} + \dfrac{y}{8} = 1 \end{cases}$

95. $\begin{cases} 2(x - 5) = y + 2 \\ 3x = 4(y + 2) \end{cases}$

96. $\begin{cases} 3(x - 2) + 5 = 4(y + 3) - 2 \\ 2x + 7 = 2y + 8 \end{cases}$

In Exercises 97–102, solve the system by the method of substitution. Use a graphing calculator to verify the solution graphically.

97. $\begin{cases} y = \frac{1}{4}x + \frac{19}{4} \\ y = \frac{8}{5}x - 1 \end{cases}$

98. $\begin{cases} y = \frac{5}{4}x + 3 \\ y = \frac{1}{2}x + 6 \end{cases}$

99. $\begin{cases} 3x + 2y = 12 \\ x - y = 3 \end{cases}$

100. $\begin{cases} 16x - 8y = 5 \\ 32x + 8y = 19 \end{cases}$

101. $\begin{cases} 5x + 3y = 15 \\ 2x - 3y = 6 \end{cases}$

102. $\begin{cases} 4x - 5y = 0 \\ 2x - 5y = -10 \end{cases}$

Solving Problems

103. *Number Problem* The sum of two numbers x and y is 20 and the difference of the two numbers is 2. Find the two numbers.

104. *Number Problem* The sum of two numbers x and y is 35 and the difference of the two numbers is 11. Find the two numbers.

105. *Break-Even Analysis* A small company produces bird feeders that sell for \$23 per unit. The cost of producing each unit is \$16.75, and the company has fixed costs of \$400.

(a) Use a verbal model to show that the cost C of producing x units is $C = 16.75x + 400$ and the revenue R from selling x units is $R = 23x$.

(b) Use a graphing calculator to graph the cost and revenue functions in the same viewing window. Approximate the point of intersection of the graphs and interpret the result.

106. *Break-Even Analysis* A company produces hockey sticks that sell for \$79 per unit. The cost of producing each unit is \$53.25 and the company has fixed costs of \$1000.

(a) Use a verbal model to show that the cost C of producing x units is $C = 53.25x + 1000$ and the revenue R from selling x units is $R = 79x$.

(b) Use a graphing calculator to graph the cost and revenue functions in the same viewing window. Approximate the point of intersection of the graphs and interpret the result.

Think About It In Exercises 107 and 108, the graphs of the two equations appear parallel. Are the two lines actually parallel? Does the system have a solution? If so, find the solution.

107. $\begin{cases} x - 200y = -200 \\ x - 199y = 198 \end{cases}$

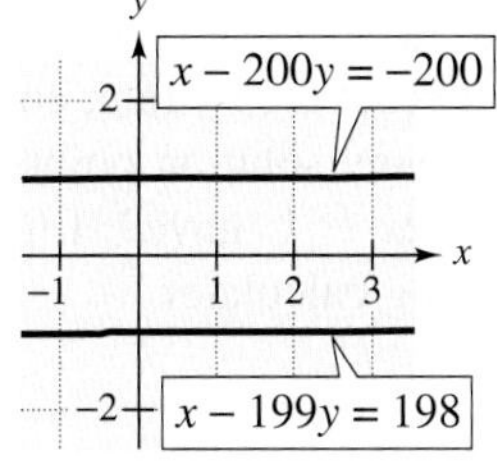

108. $\begin{cases} 25x - 24y = 0 \\ 13x - 12y = 24 \end{cases}$

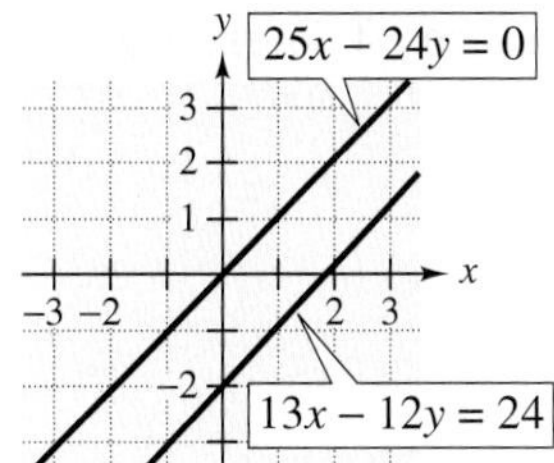

109. *Investment* A total of \$15,000 is invested in two funds paying 5% and 8% simple interest. (There is more risk in the 8% fund.) The combined annual interest for the two funds is \$900. Determine how much of the \$15,000 is invested at each rate.

110. *Investment* A total of \$10,000 is invested in two funds paying 7% and 10% simple interest. (There is more risk in the 10% fund.) The combined annual interest for the two funds is \$775. Determine how much of the \$10,000 is invested at each rate.

111. *Dinner Price* Six people ate dinner for \$63.90. The price for an adult was \$16.95 and the price for a child was \$7.50. Determine how many adults attended the dinner.

112. *Ticket Sales* You are selling football tickets. Student tickets cost \$3 and general admission tickets cost \$5. You sell 1957 tickets and collect \$8113. Determine how many of each type of ticket were sold.

113. *Comparing Costs* Car model ES costs \$18,000 and costs an average of \$0.26 per mile to maintain. Car model LS costs \$22,000 and costs an average of \$0.22 per mile to maintain. Determine after how many miles the total costs of the two models will be the same (the two models are driven the same number of miles).

114. *Comparing Costs* A solar heating system for a three-bedroom home costs \$28,500 for installation and \$125 per year to operate. An electric heating system for the same home costs \$5750 for installation and \$1000 per year to operate. Determine after how many years the total costs of solar heating and electric heating will be the same. What will be the cost at that time?

115. *Geometry* Find an equation of the line with slope $m = 2$ passing through the intersection of the lines $x - 2y = 3$ and $3x + y = 16$.

116. *Geometry* Find an equation of the line with slope $m = -3$ passing through the intersection of the lines $4x + 6y = 26$ and $5x - 2y = -15$.

Explaining Concepts

117. In your own words, explain what is meant by a dependent system of linear equations.

118. In your own words, explain what is meant by an inconsistent system of linear equations.

119. When solving a system of linear equations by the method of substitution, how do you recognize that it has no solution?

120. When solving a system of linear equations by the method of substitution, how do you recognize that it has infinitely many solutions?

121. Describe any advantages of the method of substitution over the graphical method of solving a system of linear equations.

122. *Creating a System* Write a system of linear equations with integer coefficients that has the unique solution $(3, -4)$. (There are many correct answers.)

123. *Creating an Example* Write an example of a system of linear equations that has no solution. (There are many correct answers.)

124. *Creating an Example* Write an example of a system of linear equations that has infinitely many solutions. (There are many correct answers.)

125. When solving a linear equation in two variables by graphing, why should the solution be checked in both equations?

126. Your instructor says, "An equation (not in standard form) such as $2x - 3 = 5x - 9$ can be considered a system of equations." Create the system, and find the solution point. How many solution points does the "system" $x^2 - 1 = 2x - 1$ have? Illustrate your results with a graphing calculator.

Think About It In Exercises 127–130, find the value of *a* or *b* such that the system of linear equations is inconsistent.

127. $\begin{cases} x + by = 1 \\ x + 2y = 2 \end{cases}$

128. $\begin{cases} ax + 3y = 6 \\ 5x - 5y = 2 \end{cases}$

129. $\begin{cases} -6x + y = 4 \\ 2x + by = 3 \end{cases}$

130. $\begin{cases} 6x - 3y = 4 \\ ax - y = -2 \end{cases}$

Cumulative Review

In Exercises 131–134, evaluate the expression.

131. $\frac{2}{3} + \frac{1}{3}$

132. $\frac{5}{8} + \frac{3}{2}$

133. $\frac{2}{7} - \frac{1}{3}$

134. $\frac{3}{5} - \frac{7}{6}$

In Exercises 135–138, solve the equation and check your solution.

135. $x - 6 = 5x$

136. $2 - 3x = 14 + x$

137. $y - 3(4y - 2) = 1$

138. $y + 6(3 - 2y) = 4$

In Exercises 139–142, solve the rational equation.

139. $\frac{x}{5} + \frac{2x}{5} = 3$

140. $\frac{3x}{5} + \frac{4x}{8} = \frac{11}{10}$

141. $\frac{x - 3}{x + 1} = \frac{4}{3}$

142. $\frac{3}{x} = \frac{9}{2(x + 2)}$

8.2 Solving Systems of Equations by Elimination

Staffan Widstrand/CORBIS

Why You Should Learn It

The method of elimination is one method of solving a system of linear equations. For instance, in Exercise 70 on page 497, this method is convenient for solving a system of linear equations used to find the focal length of a camera.

What You Should Learn

1 ▶ Solve systems of linear equations algebraically using the method of elimination.

2 ▶ Choose a method for solving systems of equations.

The Method of Elimination

1 ▶ Solve systems of linear equations algebraically using the method of elimination.

In this section, you will study another way to solve a system of linear equations algebraically—the **method of elimination.** The key step in this method is to obtain opposite coefficients for one of the variables so that *adding* the two equations eliminates this variable. For instance, by adding the equations

$$\begin{cases} 3x + 5y = 7 & \text{Equation 1} \\ -3x - 2y = -1 & \text{Equation 2} \end{cases}$$

$$3y = 6 \qquad \text{Add equations.}$$

you eliminate the variable x and obtain a single equation in one variable, y.

EXAMPLE 1 The Method of Elimination

Solve the system of linear equations.

$$\begin{cases} 4x + 3y = 1 & \text{Equation 1} \\ 2x - 3y = 5 & \text{Equation 2} \end{cases}$$

Solution

Begin by noting that the coefficients of y are opposites. So, by adding the two equations, you can eliminate y.

$$\begin{cases} 4x + 3y = 1 & \text{Equation 1} \\ 2x - 3y = 5 & \text{Equation 2} \end{cases}$$

$$6x \qquad = 6 \qquad \text{Add equations.}$$

So, $x = 1$. By back-substituting this value into the first equation, you can solve for y, as follows.

$$4(1) + 3y = 1 \qquad \text{Substitute 1 for } x \text{ in Equation 1.}$$

$$3y = -3 \qquad \text{Subtract 4 from each side.}$$

$$y = -1 \qquad \text{Divide each side by 3.}$$

The solution is $(1, -1)$. Check this in both of the original equations.

✓ **CHECKPOINT** *Now try Exercise 9.*

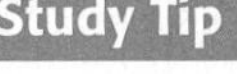

Study Tip

Try solving the system in Example 1 by substitution. Notice that the method of elimination is more efficient for this system.

To obtain opposite coefficients for one of the variables, you often need to multiply one or both of the equations by a suitable constant. This is demonstrated in the following example.

EXAMPLE 2 The Method of Elimination

Solve the system of linear equations.

$$\begin{cases} 2x - 3y = -7 & \text{Equation 1} \\ 3x + y = -5 & \text{Equation 2} \end{cases}$$

Solution

For this system, you can obtain opposite coefficients of y by multiplying the second equation by 3.

$$\begin{cases} 2x - 3y = -7 \\ 3x + y = -5 \end{cases} \Rightarrow \begin{aligned} 2x - 3y &= -7 && \text{Equation 1} \\ 9x + 3y &= -15 && \text{Multiply Equation 2 by 3.} \\ \hline 11x \quad &= -22 && \text{Add equations.} \end{aligned}$$

So, $x = -2$. By back-substituting this value of x into the second equation, you can solve for y.

$$\begin{aligned} 3x + y &= -5 && \text{Equation 2} \\ 3(-2) + y &= -5 && \text{Substitute } -2 \text{ for } x. \\ -6 + y &= -5 && \text{Simplify.} \\ y &= 1 && \text{Add 6 to each side.} \end{aligned}$$

The solution is $(-2, 1)$. Check this in the original equations, as follows.

Substitute into Equation 1

$$\begin{aligned} 2x - 3y &= -7 \\ 2(-2) - 3(1) &\stackrel{?}{=} -7 \\ -4 - 3 &= -7 \checkmark \end{aligned}$$

Substitute into Equation 2

$$\begin{aligned} 3x + y &= -5 \\ 3(-2) + 1 &\stackrel{?}{=} -5 \\ -6 + 1 &= -5 \checkmark \end{aligned}$$

✓ **CHECKPOINT** *Now try Exercise 13.*

This method is called "elimination" because the first step in the process is to "eliminate" one of the variables. This method is summarized as follows.

The Method of Elimination

1. Obtain opposite coefficients of x (or y) by multiplying all terms of one or both equations by suitable constants.
2. Add the equations to eliminate one variable, and solve the resulting equation.
3. Back-substitute the value obtained in Step 2 in either of the original equations and solve for the other variable.
4. Check your solution in *both* of the original equations.

EXAMPLE 3 The Method of Elimination

Solve the system of linear equations.

$$\begin{cases} 5x + 3y = 6 & \text{Equation 1} \\ 2x - 4y = 5 & \text{Equation 2} \end{cases}$$

Solution

You can obtain opposite coefficients of y by multiplying the first equation by 4 and the second equation by 3.

$$\begin{cases} 5x + 3y = 6 \\ 2x - 4y = 5 \end{cases} \Rightarrow \begin{array}{rl} 20x + 12y = 24 & \text{Multiply Equation 1 by 4.} \\ 6x - 12y = 15 & \text{Multiply Equation 2 by 3.} \\ \hline 26x \qquad = 39 & \text{Add equations.} \end{array}$$

From this equation, you can see that $x = \frac{3}{2}$. By back-substituting this value of x into the second equation, you can solve for y, as follows.

$$2x - 4y = 5 \qquad \text{Equation 2}$$

$$2\left(\frac{3}{2}\right) - 4y = 5 \qquad \text{Substitute } \tfrac{3}{2} \text{ for } x.$$

$$3 - 4y = 5 \qquad \text{Simplify.}$$

$$-4y = 2 \qquad \text{Subtract 3 from each side.}$$

$$y = -\frac{1}{2} \qquad \text{Divide each side by } -4.$$

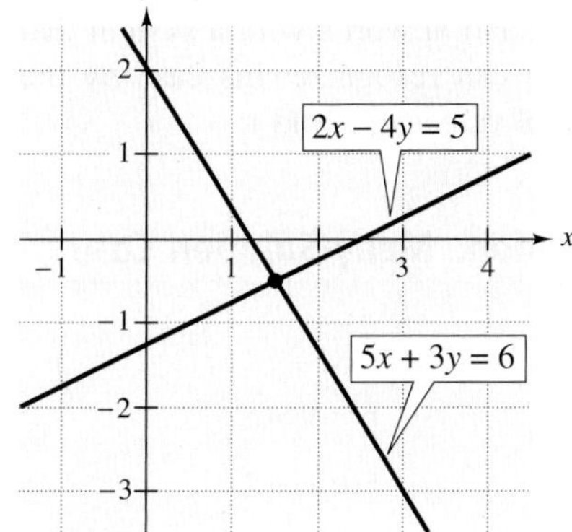

Figure 8.8

The solution is $\left(\frac{3}{2}, -\frac{1}{2}\right)$. You can check this as follows.

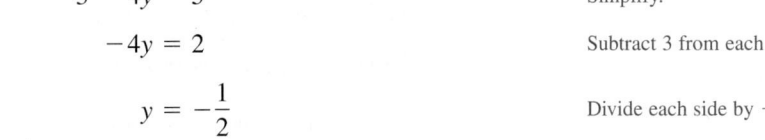

Substitute into Equation 1

$$5x + 3y = 6$$

$$5\left(\frac{3}{2}\right) + 3\left(-\frac{1}{2}\right) \stackrel{?}{=} 6$$

$$\frac{15}{2} - \frac{3}{2} = 6 \checkmark$$

Substitute into Equation 2

$$2x - 4y = 5$$

$$2\left(\frac{3}{2}\right) - 4\left(-\frac{1}{2}\right) \stackrel{?}{=} 5$$

$$3 + 2 = 5 \checkmark$$

The graph of this system is shown in Figure 8.8. From the graph it appears that the solution $\left(\frac{3}{2}, -\frac{1}{2}\right)$ is reasonable.

✓ **CHECKPOINT** *Now try Exercise 19.*

In Example 3, the y-variable was eliminated first. You could have just as easily solved the system by eliminating the x-variable first, as follows.

$$\begin{cases} 5x + 3y = 6 \\ 2x - 4y = 5 \end{cases} \Rightarrow \begin{array}{rl} 10x + 6y = 12 & \text{Multiply Equation 1 by 2.} \\ -10x + 20y = -25 & \text{Multiply Equation 2 by } -5. \\ \hline 26y = -13 & \text{Add equations.} \end{array}$$

From this equation, $y = -\frac{1}{2}$. By back-substituting this value of y into the second equation, you can solve for x to obtain $x = \frac{3}{2}$.

In the next example, note how the method of elimination can be used to determine that a system of linear equations has no solution. As with substitution, notice that the key is recognizing the occurrence of a *false statement.*

EXAMPLE 4 The Method of Elimination: No-Solution Case

Solve the system of linear equations.

$$\begin{cases} 2x - 6y = 5 & \text{Equation 1} \\ 3x - 9y = 2 & \text{Equation 2} \end{cases}$$

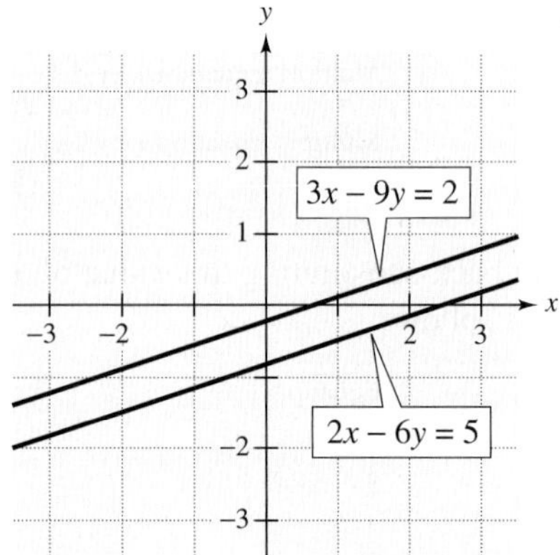

Figure 8.9

Solution

To obtain coefficients that differ only in sign, multiply the first equation by 3 and multiply the second equation by -2.

$$\begin{cases} 2x - 6y = 5 \\ 3x - 9y = 2 \end{cases} \Rightarrow \begin{array}{rl} 6x - 18y = 15 & \text{Multiply Equation 1 by 3.} \\ -6x + 18y = -4 & \text{Multiply Equation 2 by } -2. \\ \hline 0 = 11 & \text{Add equations.} \end{array}$$

Because $0 = 11$ is a false statement, you can conclude that the system is inconsistent and has no solution. The lines corresponding to the two equations of this system are shown in Figure 8.9. Note that the two lines are parallel and so have no point of intersection.

✓ **CHECKPOINT** *Now try Exercise 25.*

Example 5 shows how the method of elimination works with a system that has infinitely many solutions. Notice that you can recognize this case by the occurrence of an equation that is true for all real values of x and y.

EXAMPLE 5 The Method of Elimination: Many-Solution Case

Solve the system of linear equations.

$$\begin{cases} 2x - 6y = -5 & \text{Equation 1} \\ -4x + 12y = 10 & \text{Equation 2} \end{cases}$$

Solution

To obtain coefficients of x that differ only in sign, multiply the first equation by 2.

$$\begin{cases} 2x - 6y = -5 \\ -4x + 12y = 10 \end{cases} \Rightarrow \begin{array}{rl} 4x - 12y = -10 & \text{Multiply Equation 1 by 2.} \\ -4x + 12y = 10 & \text{Equation 2} \\ \hline 0 = 0 & \text{Add equations.} \end{array}$$

Because $0 = 0$ is a true statement, you can conclude that the system is dependent and has infinitely many solutions. The solution set consists of all ordered pairs (x, y) lying on the line $2x - 6y = -5$.

✓ **CHECKPOINT** *Now try Exercise 27.*

Study Tip

By writing both equations in Example 5 in slope-intercept form, you will obtain identical equations. This shows that the system has infinitely many solutions.

The next example shows how the method of elimination works with a system of linear equations with decimal coefficients.

EXAMPLE 6 Solving a System with Decimal Coefficients

Solve the system of linear equations.

$$\begin{cases} 0.02x - 0.05y = -0.38 & \text{Equation 1} \\ 0.03x + 0.04y = 1.04 & \text{Equation 2} \end{cases}$$

Solution

Because the coefficients in this system have two decimal places, begin by multiplying each equation by 100. This produces a system in which the coefficients are all integers.

$$\begin{cases} 2x - 5y = -38 & \text{Revised Equation 1} \\ 3x + 4y = 104 & \text{Revised Equation 2} \end{cases}$$

Study Tip

When multiplying an equation by a negative number, be sure to distribute the negative sign to each term of the equation.

Now, to obtain coefficients of x that differ only in sign, multiply the first equation by 3 and multiply the second equation by -2.

$$\begin{cases} 2x - 5y = -38 \\ 3x + 4y = 104 \end{cases} \quad \Rightarrow \quad \begin{aligned} 6x - 15y &= -114 && \text{Multiply Equation 1 by 3.} \\ -6x - 8y &= -208 && \text{Multiply Equation 2 by } -2. \\ -23y &= -322 && \text{Add equations.} \end{aligned}$$

So, the y-coordinate of the solution is

$$y = \frac{-322}{-23} = 14.$$

Back-substituting this value into revised Equation 2 produces the following.

$$\begin{aligned} 3x + 4(14) &= 104 && \text{Substitute 14 for } y \text{ in revised Equation 2.} \\ 3x + 56 &= 104 && \text{Simplify.} \\ 3x &= 48 && \text{Subtract 56 from each side.} \\ x &= 16 && \text{Divide each side by 3.} \end{aligned}$$

So, the solution is (16, 14). You can check this solution as follows.

Substitute into Equation 1

$$\begin{aligned} 0.02x - 0.05y &= -0.38 && \text{Equation 1} \\ 0.02(16) - 0.05(14) &\stackrel{?}{=} -0.38 && \text{Substitute 16 for } x \text{ and 14 for } y. \\ 0.32 - 0.70 &= -0.38 && \text{Solution checks. ✓} \end{aligned}$$

Substitute into Equation 2

$$\begin{aligned} 0.03x + 0.04y &= 1.04 && \text{Equation 2} \\ 0.03(16) + 0.04(14) &\stackrel{?}{=} 1.04 && \text{Substitute 16 for } x \text{ and 14 for } y. \\ 0.48 + 0.56 &= 1.04 && \text{Solution checks. ✓} \end{aligned}$$

✓ CHECKPOINT Now try Exercise 35.

EXAMPLE 7 An Application of a System of Linear Equations 

A fundraising dinner was held on two consecutive nights. On the first night, 100 adult tickets and 175 children's tickets were sold, for a total of \$1225. On the second night, 200 adult tickets and 316 children's tickets were sold, for a total of \$2348. The system of linear equations that represents this problem is

$$\begin{cases} 100x + 175y = 1225 & \text{Equation 1} \\ 200x + 316y = 2348 & \text{Equation 2} \end{cases}$$

where x represents the price of an adult ticket and y represents the price of a child's ticket. Solve this system to find the price of each type of ticket.

Solution

To obtain coefficients of x that differ only in sign, multiply Equation 1 by -2.

$$\begin{cases} 100x + 175y = 1225 \\ 200x + 316y = 2348 \end{cases}$$

$$\begin{aligned} -200x - 350y &= -2450 && \text{Multiply Equation 1 by } -2. \\ 200x + 316y &= 2348 && \text{Equation 2} \\ \hline -34y &= -102 && \text{Add equations.} \end{aligned}$$

So, the y-coordinate of the solution is $y = -102/-34 = 3$. Back-substituting this value into Equation 2 produces the following.

$$\begin{aligned} 200x + 316(3) &= 2348 && \text{Substitute 3 for } y \text{ in Equation 2.} \\ 200x &= 1400 && \text{Simplify.} \\ x &= 7 && \text{Divide each side by 200.} \end{aligned}$$

The solution is $(7, 3)$. So the price of an adult ticket was \$7 and the price of a child's ticket was \$3. Check this solution in both of the original equations.

✓ **CHECKPOINT** *Now try Exercise 61.*

2 ▶ Choose a method for solving systems of equations.

Choosing Methods

To decide which of the three methods (graphing, substitution, or elimination) to use to solve a system of two linear equations, use the following guidelines.

Guidelines for Solving a System of Linear Equations

To decide whether to use graphing, substitution, or elimination, consider the following.

1. The graphing method is useful for approximating the solution and for giving an overall picture of how one variable changes with respect to the other.
2. To find exact solutions, use either substitution or elimination.
3. For systems of equations in which one variable has a coefficient of 1, substitution may be more efficient than elimination.
4. Elimination is usually more efficient. This is especially true when the coefficients of one of the variables are opposites.

Concept Check

1. Explain how to solve a system of linear equations by elimination.
2. When solving a system by the method of elimination, how do you recognize that it has no solution?
3. When solving a system by the method of elimination, how do you recognize that it has infinitely many solutions?
4. Both $(-2, 3)$ and $(8, 1)$ are solutions of a system of linear equations. How many solutions does the system have? Explain.

8.2 EXERCISES

Go to pages 546–547 to record your assignments.

Developing Skills

In Exercises 1–4, solve the system by the method of elimination. Use the graph to check your solution.

1. $\begin{cases} 2x + y = 4 \\ x - y = 2 \end{cases}$

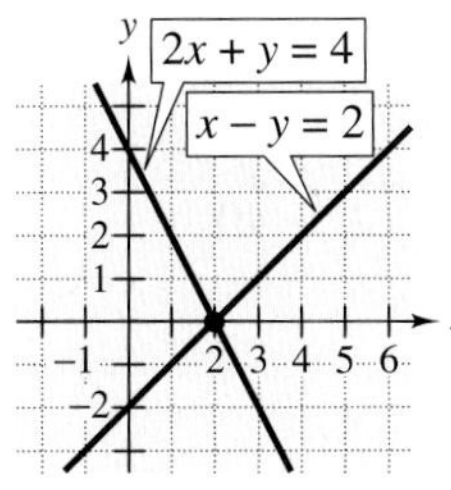

2. $\begin{cases} x + 3y = 2 \\ -x + 2y = 3 \end{cases}$

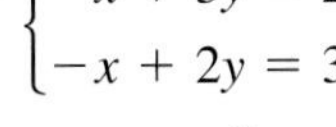

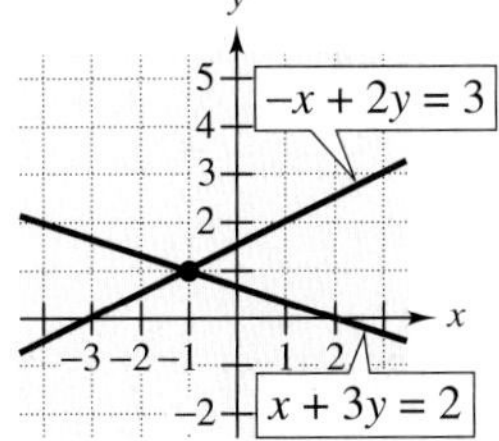

3. $\begin{cases} x - y = 0 \\ 3x - 2y = -1 \end{cases}$

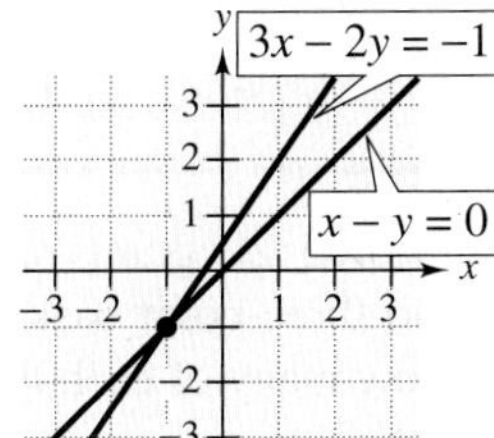

4. $\begin{cases} 2x - y = 2 \\ 4x + 3y = 24 \end{cases}$

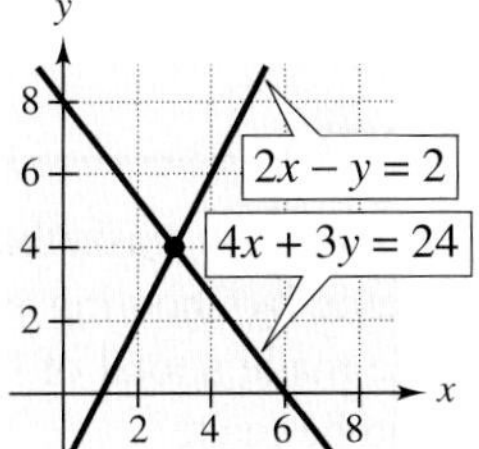

In Exercises 5–36, solve the system by the method of elimination. ***See Examples 1–6.***

5. $\begin{cases} x - y = 4 \\ x + y = 12 \end{cases}$

6. $\begin{cases} x + y = 7 \\ x - y = 3 \end{cases}$

7. $\begin{cases} -x + 2y = 12 \\ x + 6y = 20 \end{cases}$

8. $\begin{cases} x + 2y = 14 \\ x - 2y = 10 \end{cases}$

✓ 9. $\begin{cases} 3x - 5y = 1 \\ 2x + 5y = 9 \end{cases}$

10. $\begin{cases} -2x + 3y = -4 \\ 2x - 4y = 6 \end{cases}$

11. $\begin{cases} 2a + 5b = 3 \\ 2a + b = 9 \end{cases}$

12. $\begin{cases} 4a + 5b = 9 \\ 2a + 5b = 7 \end{cases}$

✓ 13. $\begin{cases} -x + 2y = 6 \\ 2x + 5y = 6 \end{cases}$

14. $\begin{cases} -4x + 8y = 0 \\ 3x - 2y = 2 \end{cases}$

15. $\begin{cases} 3x - 4y = 11 \\ 2x + 3y = -4 \end{cases}$

16. $\begin{cases} 2x + 3y = 16 \\ 5x - 10y = 30 \end{cases}$

17. $\begin{cases} 3x + 2y = -1 \\ -2x + 7y = 9 \end{cases}$

18. $\begin{cases} 5x + 3y = 27 \\ 7x - 2y = 13 \end{cases}$

✓ 19. $\begin{cases} 3x - 4y = 1 \\ 4x + 3y = 1 \end{cases}$

20. $\begin{cases} 2x - 5y = -1 \\ 2x - y = 1 \end{cases}$

21. $\begin{cases} 3x + 2y = 10 \\ 2x + 5y = 3 \end{cases}$

22. $\begin{cases} 4x + 5y = 7 \\ 6x - 2y = -18 \end{cases}$

23. $\begin{cases} 5u + 6v = 14 \\ 3u + 5v = 7 \end{cases}$

24. $\begin{cases} 5x + 3y = 18 \\ 2x - 7y = -1 \end{cases}$

✓ 25. $\begin{cases} 2x - 4y = 1 \\ 5x - 10y = -3 \end{cases}$

26. $\begin{cases} 4x + 8y = 0 \\ 3x + 6y = 2 \end{cases}$

27. $\begin{cases} -3x - 12y = 3 \\ 5x + 20y = -5 \end{cases}$

28. $\begin{cases} 7x + 10y = 0 \\ 21x + 30y = 0 \end{cases}$

29. $\begin{cases} 6r + 5s = 3 \\ \frac{3}{2}r - \frac{5}{4}s = \frac{3}{4} \end{cases}$

30. $\begin{cases} \frac{1}{4}x - y = \frac{1}{2} \\ 4x + 4y = 3 \end{cases}$

31. $\begin{cases} \frac{1}{2}s - t = \frac{3}{2} \\ 4s + 2t = 27 \end{cases}$

32. $\begin{cases} 3u + 4v = 14 \\ \frac{1}{6}u - v = -2 \end{cases}$

33. $\begin{cases} 0.4a + 0.7b = 3 \\ 0.8a + 1.4b = 6 \end{cases}$

34. $\begin{cases} 0.2u - 0.1v = 1 \\ -0.8u + 0.4v = 3 \end{cases}$

35. $\begin{cases} 0.02x - 0.05y = -0.19 \\ 0.03x + 0.04y = 0.52 \end{cases}$

36. $\begin{cases} 0.05x - 0.03y = 0.21 \\ 0.01x + 0.01y = 0.09 \end{cases}$

In Exercises 37–42, solve the system by the method of elimination. Use a graphing calculator to verify your solution.

37. $\begin{cases} 3x + 2y = 10 \\ x - 2y = 14 \end{cases}$

38. $\begin{cases} -2x + 2y = 7 \\ 2x + y = 8 \end{cases}$

39. $\begin{cases} 7x + 8y = 6 \\ 3x - 4y = 10 \end{cases}$

40. $\begin{cases} 10x - 11y = 7 \\ 2x - y = 5 \end{cases}$

41. $\begin{cases} 5x + 2y = 7 \\ 3x - 6y = -3 \end{cases}$

42. $\begin{cases} -4x + 5y = 8 \\ 2x + 3y = 18 \end{cases}$

In Exercises 43–54, use the most convenient method (graphing, substitution, or elimination) to solve the system of linear equations. State which method you used.

43. $\begin{cases} x - y = 2 \\ y = 3 \end{cases}$

44. $\begin{cases} y = 2x - 1 \\ y = x + 1 \end{cases}$

45. $\begin{cases} 6x + 21y = 132 \\ 6x - 4y = 32 \end{cases}$

46. $\begin{cases} -2x + y = 12 \\ 2x + 3y = 20 \end{cases}$

47. $\begin{cases} -4x + 3y = 11 \\ 3x - 10y = 15 \end{cases}$

48. $\begin{cases} -3x + 5y = -11 \\ 5x - 9y = 19 \end{cases}$

49. $\begin{cases} 0.1x - 0.1y = 0 \\ 0.8x + 0.3y = 1.5 \end{cases}$

50. $\begin{cases} x - 2y = 0 \\ 0.2x + 0.8y = 2.4 \end{cases}$

51. $\begin{cases} -\frac{x}{4} + y = 1 \\ \frac{x}{4} + \frac{y}{2} = 1 \end{cases}$

52. $\begin{cases} \frac{x}{3} - \frac{y}{5} = 1 \\ \frac{x}{12} + \frac{y}{40} = 1 \end{cases}$

53. $\begin{cases} 3(x + 5) - 7 = 2(3 - 2y) \\ 2x + 1 = 4(y + 2) \end{cases}$

54. $\begin{cases} \frac{1}{2}(x - 4) + 9 = y - 10 \\ -5(x + 3) = 8 - 2(y - 3) \end{cases}$

Solving Problems

55. *Number Problem* The sum of two numbers x and y is 40 and the difference of the two numbers is 10. Find the two numbers.

56. *Number Problem* The sum of two numbers x and y is 50 and the difference of the two numbers is 20. Find the two numbers.

57. *Number Problem* The sum of two numbers x and y is 82 and the difference of the numbers is 14. Find the two numbers.

58. *Number Problem* The sum of two numbers x and y is 154 and the difference of the numbers is 38. Find the two numbers.

59. *Sports* A basketball player scored 20 points in a game by shooting two-point and three-point baskets. He made a total of 9 baskets. How many of each type did he make?

60. *Sports* A basketball team scored 84 points in a game by shooting two-point and three-point baskets. The team made a total of 36 baskets. How many of each type did the team make?

✓ **61.** *Ticket Sales* Ticket sales for a play were \$3799 on the first night and \$4905 on the second night. On the first night, 213 student tickets were sold and 632 general admission tickets were sold. On the second night, 275 student tickets were sold and 816 general admission tickets were sold. Determine the price of each type of ticket.

62. *Ticket Sales* Ticket sales for an annual variety show were \$540 on the first night and \$850 on the second night. On the first night, 150 student tickets were sold and 80 general admission tickets were sold. On the second night, 200 student tickets were sold and 150 general admission tickets were sold. Determine the price of each type of ticket.

63. *Investment* You invest a total of \$10,000 in two investments earning 7.5% and 10% simple interest. (There is more risk in the 10% fund.) Your goal is to have a total annual interest income of \$850. Determine the smallest amount that you can invest at 10% in order to meet your objective.

64. *Investment* You invest a total of \$12,000 in two investments earning 8% and 11.5% simple interest. (There is more risk in the 11.5% fund.) Your goal is to have a total annual interest income of \$1065. Determine the smallest amount that you can invest at 11.5% in order to meet your objective.

65. *Music* A music instructor charges \$25 for a private flute lesson and \$18 per student for a group flute lesson. In one day, the instructor earns \$265 from 12 students. How many students of each type did the instructor teach?

66. *Dance* A tap dance instructor charges \$20 for a private lesson and \$12 per student for a group lesson. In one day, the instructor earns \$216 from 14 students. How many students of each type did the instructor teach?

67. *Jewelry* A bracelet that is supposed to be 18-karat gold weighs 277.92 grams. The volume of the bracelet is 18.52 cubic centimeters. The bracelet is made of gold and copper. Gold weighs 19.3 grams per cubic centimeter and copper weighs 9 grams per cubic centimeter. Determine whether or not the bracelet is 18-karat gold. (An 18-karat gold bracelet is $3/4$ gold by weight.)

Photographer's Choice/Punchstock

In 2006, the United States produced about 260 metric tons of gold, which was about 10% of the world production. (Source: U.S. Geological Survey)

68. *Geometry* Find an equation of the line of slope $m = -2$ passing through the intersection of the lines

$2x + 5y = 11$ and $4x - y = 11$.

69. *Geometry* Find an equation of the line of slope $m = 1/3$ passing through the intersection of the lines $3x + 4y = 7$ and $5x - 4y = 1$.

70. *Focal Length* When parallel rays of light pass through a convex lens, they are bent inward and meet at a *focus* (see figure). The distance from the center of the lens to the focus is called the *focal length.* The equations of the lines that represent the two bent rays in the camera are

$$\begin{cases} x + 3y = 1 \\ -x + 3y = -1 \end{cases}$$

where x and y are measured in inches. Which equation is the upper ray? What is the focal length?

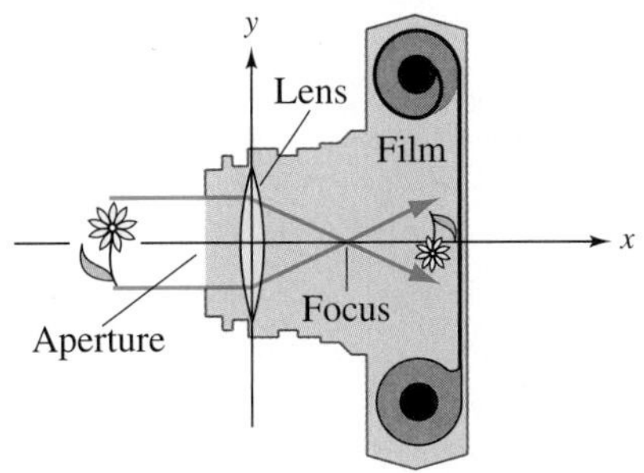

Explaining Concepts

71. *Creating an Example* Explain how to "clear" a system of decimals. (There are many correct answers.)

72. *Creating a System* Write a system of linear equations that is better solved by the method of elimination than by the method of substitution. (There are many correct answers.)

73. *Creating a System* Write a system of linear equations that is better solved by the method of substitution than by the method of elimination. (There are many correct answers.)

74. Consider the system of linear equations.

$$\begin{cases} x + y = 8 \\ 2x + 2y = k \end{cases}$$

(a) Find the value(s) of k for which the system has an infinite number of solutions.

(b) Find one value of k for which the system has no solution. (There are many correct answers.)

(c) Can the system have a single solution for some value of k? Why or why not?

Cumulative Review

In Exercises 75–80, plot the points and find the slope (if possible) of the line passing through the points. If not possible, state why.

75. $(-6, 4), (-3, -4)$

76. $(4, 6), (8, -2)$

77. $\left(\frac{7}{2}, \frac{9}{2}\right), \left(\frac{4}{3}, -3\right)$

78. $\left(-\frac{3}{4}, -\frac{7}{4}\right), \left(-1, \frac{5}{2}\right)$

79. $(-3, 6), (-3, 2)$

80. $(6, 2), (10, 2)$

In Exercises 81–84, solve and graph the inequality.

81. $x \le 3$

82. $x > -4$

83. $x + 5 < 6$

84. $3x - 7 \ge 2x + 9$

In Exercises 85–88, solve the system by the method of substitution.

85. $\begin{cases} y = x \\ x + 3y = 20 \end{cases}$

86. $\begin{cases} x + y = 9 \\ 2x + 2y = 18 \end{cases}$

87. $\begin{cases} 2x + y = 5 \\ 5x + 3y = 12 \end{cases}$

88. $\begin{cases} 5x + 6y = 21 \\ 25x + 30y = 10 \end{cases}$

8.3 Linear Systems in Three Variables

Frank Whitney/Getty Images

What You Should Learn

1 ▶ Solve systems of linear equations in row-echelon form using back-substitution.
2 ▶ Solve systems of linear equations using the method of Gaussian elimination.
3 ▶ Solve application problems using the method of Gaussian elimination.

Why You Should Learn It

Systems of linear equations in three variables can be used to model and solve real-life problems. For instance, in Exercise 47 on page 509, a system of linear equations can be used to determine a chemical mixture for a pesticide.

1 ▶ Solve systems of linear equations in row-echelon form using back-substitution.

Row-Echelon Form

The method of elimination can be applied to a system of linear equations in more than two variables. In fact, this method easily adapts to computer use for solving systems of linear equations with dozens of variables.

When the method of elimination is used to solve a system of linear equations, the goal is to rewrite the system in a form to which back-substitution can be applied. For instance, consider the following two systems of linear equations.

$$\begin{cases} x - 2y + 2z = 9 \\ -x + 3y = -4 \\ 2x - 5y + z = 10 \end{cases} \qquad \begin{cases} x - 2y + 2z = 9 \\ y + 2z = 5 \\ z = 3 \end{cases}$$

Which of these two systems do you think is easier to solve? After comparing the two systems, it should be clear that it is easier to solve the system on the right because the value of z is already shown and back-substitution will readily yield the values of x and y. The system on the right is said to be in **row-echelon form,** which means that it has a "stair-step" pattern with leading coefficients of 1.

EXAMPLE 1 Using Back-Substitution

In the following system of linear equations, you know the value of z from Equation 3.

$$\begin{cases} x - 2y + 2z = 9 & \text{Equation 1} \\ y + 2z = 5 & \text{Equation 2} \\ z = 3 & \text{Equation 3} \end{cases}$$

To solve for y, substitute $z = 3$ in Equation 2 to obtain

$$y + 2(3) = 5 \quad \Rightarrow \quad y = -1. \qquad \text{Substitute 3 for } z.$$

Finally, substitute $y = -1$ and $z = 3$ in Equation 1 to obtain

$$x - 2(-1) + 2(3) = 9 \quad \Rightarrow \quad x = 1. \qquad \text{Substitute } -1 \text{ for } y \text{ and 3 for } z.$$

The solution is $x = 1$, $y = -1$, and $z = 3$, which can also be written as the **ordered triple** $(1, -1, 3)$. Check this in the original system of equations.

✓ **CHECKPOINT** *Now try Exercise 3.*

Study Tip

When checking a solution, remember that the solution must satisfy each equation in the original system.

2 ▶ Solve systems of linear equations using the method of Gaussian elimination.

The Method of Gaussian Elimination

Two systems of equations are **equivalent systems** if they have the same solution set. To solve a system that is not in row-echelon form, first convert it to an *equivalent* system that is in row-echelon form. To see how this is done, take another look at the method of elimination, as applied to a system of two linear equations.

EXAMPLE 2 The Method of Elimination

Solve the system of linear equations.

$$\begin{cases} 3x - 2y = -1 & \text{Equation 1} \\ x - y = 0 & \text{Equation 2} \end{cases}$$

Solution

$$\begin{cases} x - y = 0 \\ 3x - 2y = -1 \end{cases}$$ Interchange the two equations in the system.

$$\begin{aligned} -3x + 3y &= 0 \\ 3x - 2y &= -1 \\ \hline y &= -1 \end{aligned}$$ Multiply new Equation 1 by -3 and add it to new Equation 2.

$$\begin{cases} x - y = 0 \\ y = -1 \end{cases}$$ New system in row-echelon form

Using back-substitution, you can determine that the solution is $(-1, -1)$. Check the solution in each equation in the original system, as follows.

Equation 1

$$3x - 2y \stackrel{?}{=} -1$$

$$3(-1) - 2(-1) = -1 \ \checkmark$$

Equation 2

$$x - y \stackrel{?}{=} 0$$

$$(-1) - (-1) = 0 \ \checkmark$$

 ***CHECKPOINT*** *Now try Exercise 7.*

Rewriting a system of linear equations in row-echelon form usually involves a chain of equivalent systems, each of which is obtained by using one of the three basic row operations. This process is called **Gaussian elimination.**

Operations That Produce Equivalent Systems

Each of the following **row operations** on a system of linear equations produces an *equivalent* system of linear equations.

1. Interchange two equations.
2. Multiply one of the equations by a nonzero constant.
3. Add a multiple of one of the equations to another equation to replace the latter equation.

EXAMPLE 3 Using Gaussian Elimination to Solve a System

Solve the system of linear equations.

$$\begin{cases} x - 2y + 2z = 9 & \text{Equation 1} \\ -x + 3y = -4 & \text{Equation 2} \\ 2x - 5y + z = 10 & \text{Equation 3} \end{cases}$$

Solution

Because the leading coefficient of the first equation is 1, you can begin by keeping the x in the upper left position and eliminating the other x terms from the first column, as follows.

$$\begin{cases} x - 2y + 2z = 9 \\ y + 2z = 5 \\ 2x - 5y + z = 10 \end{cases}$$

Adding the first equation to the second equation produces a new second equation.

$$\begin{cases} x - 2y + 2z = 9 \\ y + 2z = 5 \\ -y - 3z = -8 \end{cases}$$

Adding -2 times the first equation to the third equation produces a new third equation.

Now that the x terms are eliminated from all but the first row, work on the second column. (You need to eliminate y from the third equation.)

$$\begin{cases} x - 2y + 2z = 9 \\ y + 2z = 5 \\ -z = -3 \end{cases}$$

Adding the second equation to the third equation produces a new third equation.

Finally, you need a coefficient of 1 for z in the third equation.

$$\begin{cases} x - 2y + 2z = 9 \\ y + 2z = 5 \\ z = 3 \end{cases}$$

Multiplying the third equation by -1 produces a new third equation.

This is the same system that was solved in Example 1, and, as in that example, you can conclude by back-substitution that the solution is

$x = 1, \quad y = -1, \quad \text{and} \quad z = 3.$

The solution can be written as the ordered triple

$(1, -1, 3).$

You can check the solution by substituting 1 for x, -1 for y, and 3 for z in each equation of the original system, as follows.

Check

Equation 1: $x - 2y + 2z \stackrel{?}{=} 9$

$1 - 2(-1) + 2(3) = 9$ ✓

Equation 2: $-x + 3y \stackrel{?}{=} -4$

$-(1) + 3(-1) = -4$ ✓

Equation 3: $2x - 5y + z \stackrel{?}{=} 10$

$2(1) - 5(-1) + 3 = 10$ ✓

✓ **CHECKPOINT** *Now try Exercise 11.*

EXAMPLE 4 Using Gaussian Elimination to Solve a System

Solve the system of linear equations.

$$\begin{cases} 4x + y - 3z = 11 & \text{Equation 1} \\ 2x - 3y + 2z = 9 & \text{Equation 2} \\ x + y + z = -3 & \text{Equation 3} \end{cases}$$

Solution

$$\begin{cases} x + y + z = -3 \\ 2x - 3y + 2z = 9 \\ 4x + y - 3z = 11 \end{cases}$$

Interchange the first and third equations.

$$\begin{cases} x + y + z = -3 \\ -5y = 15 \\ 4x + y - 3z = 11 \end{cases}$$

Adding -2 times the first equation to the second equation produces a new second equation.

$$\begin{cases} x + y + z = -3 \\ -5y = 15 \\ -3y - 7z = 23 \end{cases}$$

Adding -4 times the first equation to the third equation produces a new third equation.

$$\begin{cases} x + y + z = -3 \\ y = -3 \\ -3y - 7z = 23 \end{cases}$$

Multiplying the second equation by $-\frac{1}{5}$ produces a new second equation.

$$\begin{cases} x + y + z = -3 \\ y = -3 \\ -7z = 14 \end{cases}$$

Adding 3 times the second equation to the third equation produces a new third equation.

$$\begin{cases} x + y + z = -3 \\ y = -3 \\ z = -2 \end{cases}$$

Multiplying the third equation by $-\frac{1}{7}$ produces a new third equation.

Now you can see that $z = -2$ and $y = -3$. Moreover, by back-substituting these values in Equation 1, you can determine that $x = 2$. So, the solution is

$$x = 2, \quad y = -3, \quad \text{and} \quad z = -2$$

which can be written as the ordered triple $(2, -3, -2)$. You can check this solution as follows.

Check

Equation 1:

$$4x + y - 3z \stackrel{?}{=} 11$$
$$4(2) + (-3) - 3(-2) \stackrel{?}{=} 11$$
$$11 = 11 \checkmark$$

Equation 2:

$$2x - 3y + 2z \stackrel{?}{=} 9$$
$$2(2) - 3(-3) + 2(-2) \stackrel{?}{=} 9$$
$$9 = 9 \checkmark$$

Equation 3:

$$x + y + z \stackrel{?}{=} -3$$
$$(2) + (-3) + (-2) \stackrel{?}{=} -3$$
$$-3 = -3 \checkmark$$

✓ ***CHECKPOINT*** *Now try Exercise 15.*

The next example involves an inconsistent system—one that has no solution. The key to recognizing an inconsistent system is that at some stage in the elimination process, you obtain a false statement such as $0 = 6$. Watch for such statements as you do the exercises for this section.

EXAMPLE 5 An Inconsistent System

Solve the system of linear equations.

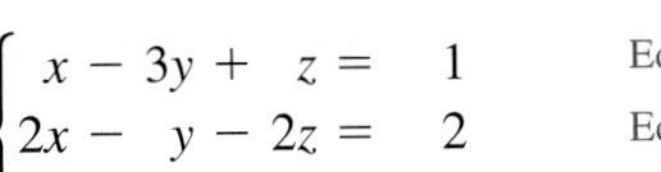

$$\begin{cases} x - 3y + z = 1 & \text{Equation 1} \\ 2x - y - 2z = 2 & \text{Equation 2} \\ x + 2y - 3z = -1 & \text{Equation 3} \end{cases}$$

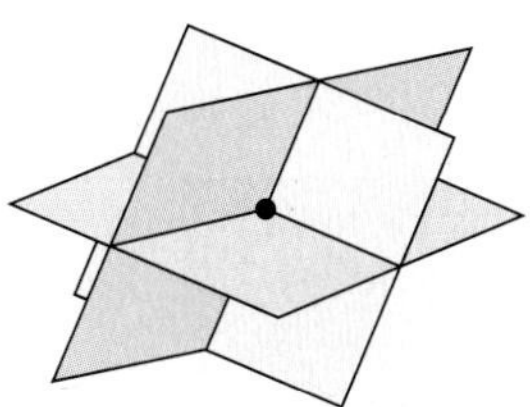

Solution: one point
Figure 8.10

Solution

$$\begin{cases} x - 3y + z = 1 \\ 5y - 4z = 0 \\ x + 2y - 3z = -1 \end{cases}$$

Adding -2 times the first equation to the second equation produces a new second equation.

$$\begin{cases} x - 3y + z = 1 \\ 5y - 4z = 0 \\ 5y - 4z = -2 \end{cases}$$

Adding -1 times the first equation to the third equation produces a new third equation.

$$\begin{cases} x - 3y + z = 1 \\ 5y - 4z = 0 \\ 0 = -2 \end{cases}$$

Adding -1 times the second equation to the third equation produces a new third equation.

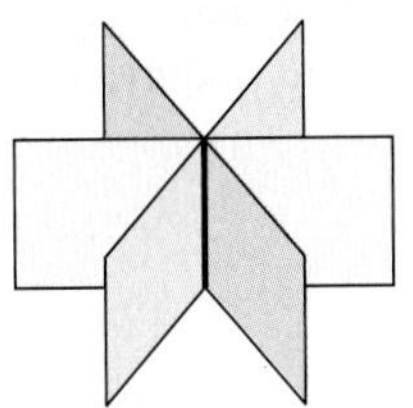

Solution: one line

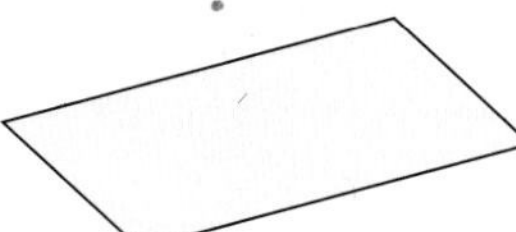

Solution: one plane
Figure 8.11

Because the third "equation" is a false statement, you can conclude that this system is inconsistent and therefore has no solution. Moreover, because this system is equivalent to the original system, you can conclude that the original system also has no solution.

✓ ***CHECKPOINT*** *Now try Exercise 17.*

As with a system of linear equations in two variables, the number of solutions of a system of linear equations in more than two variables must fall into one of three categories.

The Number of Solutions of a Linear System

For a system of linear equations, exactly one of the following is true.

1. There is exactly one solution.
2. There are infinitely many solutions.
3. There is no solution.

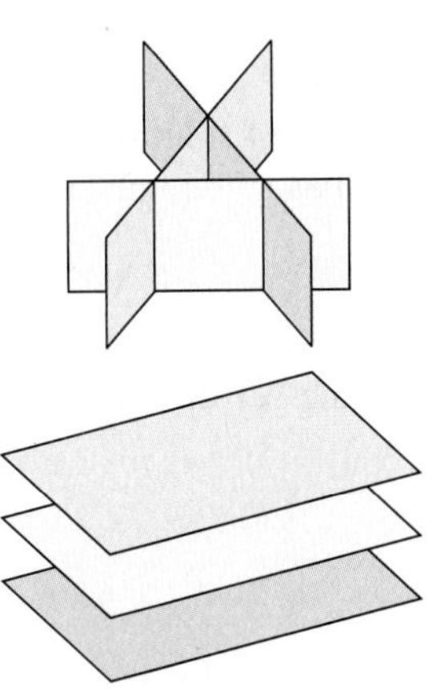

Solution: none
Figure 8.12

The graph of a system of three linear equations in three variables consists of *three planes*. When these planes intersect in a single point, the system has exactly one solution. (See Figure 8.10.) When the three planes intersect in a line or a plane, the system has infinitely many solutions. (See Figure 8.11.) When the three planes have no point in common, the system has no solution. (See Figure 8.12.)

EXAMPLE 6 A System with Infinitely Many Solutions

Solve the system of linear equations.

$$\begin{cases} x + y - 3z = -1 & \text{Equation 1} \\ y - z = 0 & \text{Equation 2} \\ -x + 2y = 1 & \text{Equation 3} \end{cases}$$

Solution

Begin by rewriting the system in row-echelon form.

$$\begin{cases} x + y - 3z = -1 \\ y - z = 0 \\ 3y - 3z = 0 \end{cases}$$

Adding the first equation to the third equation produces a new third equation.

$$\begin{cases} x + y - 3z = -1 \\ y - z = 0 \\ 0 = 0 \end{cases}$$

Adding -3 times the second equation to the third equation produces a new third equation.

This means that Equation 3 depends on Equations 1 and 2 in the sense that it gives no additional information about the variables. So, the original system is equivalent to the system

$$\begin{cases} x + y - 3z = -1 \\ y - z = 0. \end{cases}$$

In the last equation, solve for y in terms of z to obtain $y = z$. Back-substituting for y in the previous equation produces $x = 2z - 1$. Finally, letting $z = a$, where a is any real number, you can see that there are an infinite number of solutions to the original system, all of the form

$$x = 2a - 1,\ y = a, \text{ and } z = a.$$

So, every ordered triple of the form

$$(2a - 1, a, a), \qquad a \text{ is a real number}$$

is a solution of the system.

CHECKPOINT *Now try Exercise 27.*

In Example 6, there are other ways to write the same infinite set of solutions. For instance, letting $x = b$, the solutions could have been written as

$$\left(b, \frac{1}{2}(b + 1), \frac{1}{2}(b + 1)\right), \qquad b \text{ is a real number.}$$

To convince yourself that this description produces the same set of solutions, consider the comparison shown below.

Substitution	*Solution*	
$a = 0$	$(2(0) - 1, 0, 0) = (-1, 0, 0)$	
$b = -1$	$\left(-1, \frac{1}{2}(-1 + 1), \frac{1}{2}(-1 + 1)\right) = (-1, 0, 0)$	Same solution
$a = 1$	$(2(1) - 1, 1, 1) = (1, 1, 1)$	
$b = 1$	$\left(1, \frac{1}{2}(1 + 1), \frac{1}{2}(1 + 1)\right) = (1, 1, 1)$	Same solution

Study Tip

When comparing descriptions of an infinite solution set, keep in mind that there is more than one way to describe the set.

3 ▶ Solve application problems using the method of Gaussian elimination.

Applications

EXAMPLE 7 Vertical Motion 

The height at time t of an object that is moving in a (vertical) line with constant acceleration a is given by the **position equation**

$$s = \frac{1}{2}at^2 + v_0t + s_0.$$

The height s is measured in feet, the acceleration a is measured in feet per second squared, the time t is measured in seconds, v_0 is the initial velocity (at time $t = 0$), and s_0 is the initial height. Find the values of a, v_0, and s_0, if $s = 164$ feet at 1 second, $s = 180$ feet at 2 seconds, and $s = 164$ feet at 3 seconds.

© Peter Brogden/Alamy

The "Big Shot" zero-gravity ride looks like part of a needle atop the Stratosphere Tower in Las Vegas. The "Big Shot" lets riders experience zero gravity by allowing them to free-fall after first catapulting them upward to a height of nearly 1100 feet above the ground.

Solution

By substituting the three values of t and s into the position equation, you obtain three linear equations in a, v_0, and s_0.

When $t = 1$, $s = 164$: $\frac{1}{2}a(1)^2 + v_0(1) + s_0 = 164$

When $t = 2$, $s = 180$: $\frac{1}{2}a(2)^2 + v_0(2) + s_0 = 180$

When $t = 3$, $s = 164$: $\frac{1}{2}a(3)^2 + v_0(3) + s_0 = 164$

By multiplying the first and third equations by 2, this system can be rewritten as

$$\begin{cases} a + 2v_0 + 2s_0 = 328 & \text{Equation 1} \\ 2a + 2v_0 + s_0 = 180 & \text{Equation 2} \\ 9a + 6v_0 + 2s_0 = 328 & \text{Equation 3} \end{cases}$$

and you can apply Gaussian elimination to obtain

$$\begin{cases} a + 2v_0 + 2s_0 = 328 & \text{Equation 1} \\ \quad -2v_0 - 3s_0 = -476 & \text{Equation 2} \\ \qquad\quad 2s_0 = 232. & \text{Equation 3} \end{cases}$$

From the third equation, $s_0 = 116$, so back-substitution in Equation 2 yields

$$\begin{aligned} -2v_0 - 3(116) &= -476 \\ -2v_0 &= -128 \\ v_0 &= 64. \end{aligned}$$

Finally, back-substituting $s_0 = 116$ and $v_0 = 64$ in Equation 1 yields

$$\begin{aligned} a + 2(64) + 2(116) &= 328 \\ a &= -32. \end{aligned}$$

So, the position equation for this object is $s = -16t^2 + 64t + 116$.

✓ **CHECKPOINT** *Now try Exercise 37.*

EXAMPLE 8 A Geometry Application

The sum of the measures of two angles of a triangle is twice the measure of the third angle. The measure of the first angle is 18° more than the measure of the third angle. Find the measures of the three angles.

Solution

Let x, y, and z represent the measures of the first, second, and third angles, respectively. The sum of the measures of the three angles of a triangle is 180°. From the given information, you can write a system of equations as follows.

$$\begin{cases} x + y + z = 180 & \text{Equation 1} \\ x + y = 2z & \text{Equation 2} \\ x = z + 18 & \text{Equation 3} \end{cases}$$

By rewriting this system with the variable terms on the left side, you obtain

$$\begin{cases} x + y + z = 180 & \text{Equation 1} \\ x + y - 2z = 0 & \text{Equation 2} \\ x - z = 18. & \text{Equation 3} \end{cases}$$

Using Gaussian elimination to solve this system yields $x = 78$, $y = 42$, and $z = 60$. So, the measures of the three angles are 78°, 42°, and 60°, respectively.

Check

Equation 1: $78 + 42 + 60 = 180$ ✓

Equation 2: $78 + 42 - 2(60) = 0$ ✓

Equation 3: $78 - 60 = 18$ ✓

✓ **CHECKPOINT** *Now try Exercise 41.*

EXAMPLE 9 Grades of Paper

A paper manufacturer sells a 50-pound package that consists of three grades of computer paper. Grade A costs \$6.00 per pound, grade B costs \$4.50 per pound, and grade C costs \$3.50 per pound. Half of the 50-pound package consists of the two cheaper grades. The cost of the 50-pound package is \$252.50. How many pounds of each grade of paper are there in the 50-pound package?

Solution

Let A, B, and C represent the numbers of pounds of grade A, grade B, and grade C paper, respectively. From the given information, you can write a system of equations as follows.

$$\begin{cases} A + B + C = 50 & \text{Equation 1} \\ 6A + 4.50B + 3.50C = 252.50 & \text{Equation 2} \\ B + C = 25 & \text{Equation 3} \end{cases}$$

Using Gaussian elimination to solve this system yields $A = 25$, $B = 15$, and $C = 10$. So, there are 25 pounds of grade A paper, 15 pounds of grade B paper, and 10 pounds of grade C paper in the 50-pound package. Check this solution.

✓ **CHECKPOINT** *Now try Exercise 49.*

Concept Check

1. How can the process of Gaussian elimination help you to solve a system of equations? In general, after applying Gaussian elimination in a system of equations, what are the next steps you take to find the solution of the system?
2. Describe the three row operations that you can use to produce an equivalent system of equations while applying Gaussian elimination.
3. Give an example of a system of three linear equations in three variables that is in row-echelon form.
4. Show how to use back-substitution to solve the system of equations you wrote in Concept Check 3.

8.3 EXERCISES

Go to pages 546–547 to record your assignments.

Developing Skills

In Exercises 1 and 2, determine whether each ordered triple is a solution of the system of linear equations.

1. $$\begin{cases} x + 3y + 2z = 1 \\ 5x - y + 3z = 16 \\ -3x + 7y + z = -14 \end{cases}$$

(a) $(0, 3, -2)$ (b) $(12, 5, -13)$
(c) $(1, -2, 3)$ (d) $(-2, 5, -3)$

2. $$\begin{cases} 3x - y + 4z = -10 \\ -x + y + 2z = 6 \\ 2x - y + z = -8 \end{cases}$$

(a) $(-2, 4, 0)$ (b) $(0, -3, 10)$
(c) $(1, -1, 5)$ (d) $(7, 19, -3)$

In Exercises 3–6, use back-substitution to solve the system of linear equations. *See Example 1.*

✓ 3. $$\begin{cases} x - 2y + 4z = 4 \\ 3y - z = 2 \\ z = -5 \end{cases}$$

4. $$\begin{cases} 5x + 4y - z = 0 \\ 10y - 3z = 11 \\ z = 3 \end{cases}$$

5. $$\begin{cases} x - 2y + 4z = 4 \\ y = 3 \\ y + z = 2 \end{cases}$$

6. $$\begin{cases} x = 10 \\ 3x + 2y = 2 \\ x + y + 2z = 0 \end{cases}$$

In Exercises 7 and 8, perform the row operation and write the equivalent system of linear equations. ***See Example 2.***

✓ 7. Add Equation 1 to Equation 2.

$$\begin{cases} x - 2y = 8 & \text{Equation 1} \\ -x + 3y = 6 & \text{Equation 2} \end{cases}$$

What did this operation accomplish?

8. Add -2 times Equation 1 to Equation 3.

$$\begin{cases} x - 2y + 3z = 5 & \text{Equation 1} \\ -x + y + 5z = 4 & \text{Equation 2} \\ 2x - 3z = 0 & \text{Equation 3} \end{cases}$$

What did this operation accomplish?

In Exercises 9 and 10, determine whether the two systems of linear equations are equivalent. Give reasons for your answer.

9. $$\begin{cases} x + 3y - z = 6 \\ 2x - y + 2z = 1 \\ 3x + 2y - z = 2 \end{cases} \qquad \begin{cases} x + 3y - z = 6 \\ -7y + 4z = -11 \\ -7y + 2z = -16 \end{cases}$$

10. $$\begin{cases} x - 2y + 3z = 9 \\ -x + 3y = -4 \\ 2x - 5y + 5z = 17 \end{cases} \qquad \begin{cases} x - 2y + 3z = 9 \\ y + 3z = 5 \\ -y - z = -1 \end{cases}$$

In Exercises 11–34, solve the system of linear equations. ***See Examples 3–6.***

✓ **11.** $\begin{cases} x + z = 4 \\ y = 2 \\ 4x + z = 7 \end{cases}$

12. $\begin{cases} x + y = 6 \\ 3x - y = 2 \\ z = 3 \end{cases}$

13. $\begin{cases} x + y + z = 6 \\ 2x - y + z = 3 \\ 3x - z = 0 \end{cases}$

14. $\begin{cases} x + y + z = 2 \\ -x + 3y + 2z = 8 \\ 4x + y = 4 \end{cases}$

✓ **15.** $\begin{cases} x + y + z = -3 \\ 4x + y - 3z = 11 \\ 2x - 3y + 2z = 9 \end{cases}$

16. $\begin{cases} x - y + 2z = -4 \\ 3x + y - 4z = -6 \\ 2x + 3y - 4z = 4 \end{cases}$

✓ **17.** $\begin{cases} x + 2y + 6z = 5 \\ -x + y - 2z = 3 \\ x - 4y - 2z = 1 \end{cases}$

18. $\begin{cases} x + 6y + 2z = 9 \\ 3x - 2y + 3z = -1 \\ 5x - 5y + 2z = 7 \end{cases}$

19. $\begin{cases} 2x + 2z = 2 \\ 5x + 3y = 4 \\ 3y - 4z = 4 \end{cases}$

20. $\begin{cases} x + y + 8z = 3 \\ 2x + y + 11z = 4 \\ x + 3z = 0 \end{cases}$

21. $\begin{cases} 6y + 4z = -12 \\ 3x + 3y = 9 \\ 2x - 3z = 10 \end{cases}$

22. $\begin{cases} 2x - 4y + z = 0 \\ 3x + 2z = -1 \\ -6x + 3y + 2z = -10 \end{cases}$

23. $\begin{cases} 2x + y + 3z = 1 \\ 2x + 6y + 8z = 3 \\ 6x + 8y + 18z = 5 \end{cases}$

24. $\begin{cases} 3x - y - 2z = 5 \\ 2x + y + 3z = 6 \\ 6x - y - 4z = 9 \end{cases}$

25. $\begin{cases} y + z = 5 \\ 2x + 4z = 4 \\ 2x - 3y = -14 \end{cases}$

26. $\begin{cases} 5x + 2y = -8 \\ z = 5 \\ 3x - y + z = 9 \end{cases}$

✓ **27.** $\begin{cases} 2x + z = 1 \\ 5y - 3z = 2 \\ 6x + 20y - 9z = 11 \end{cases}$

28. $\begin{cases} 2x + y - z = 4 \\ y + 3z = 2 \\ 3x + 2y = 4 \end{cases}$

29. $\begin{cases} 2x + 3z = 4 \\ 5x + y + z = 2 \\ 11x + 3y - 3z = 0 \end{cases}$

30. $\begin{cases} 3x + y + z = 2 \\ 4x + 2z = 1 \\ 5x - y + 3z = 0 \end{cases}$

31. $\begin{cases} 0.2x + 1.3y + 0.6z = 0.1 \\ 0.1x + 0.3z = 0.7 \\ 2x + 10y + 8z = 8 \end{cases}$

32. $\begin{cases} 0.3x - 0.1y + 0.2z = 0.35 \\ 2x + y - 2z = -1 \\ 2x + 4y + 3z = 10.5 \end{cases}$

33. $\begin{cases} x + 4y - 2z = 2 \\ -3x + y + z = -2 \\ 5x + 7y - 5z = 6 \end{cases}$

34. $\begin{cases} x - 2y - z = 3 \\ 2x + y - 3z = 1 \\ x + 8y - 3z = -7 \end{cases}$

In Exercises 35 and 36, find a system of linear equations in three variables with integer coefficients that has the given point as a solution. (There are many correct answers.)

35. $(4, -3, 2)$

36. $(5, 7, -10)$

Solving Problems

Vertical Motion In Exercises 37–40, find the position equation $s = \frac{1}{2}at^2 + v_0t + s_0$ for an object that has the indicated heights at the specified times. ***See Example 7.***

✓ **37.** $s = 128$ feet at $t = 1$ second
$s = 80$ feet at $t = 2$ seconds
$s = 0$ feet at $t = 3$ seconds

38. $s = 48$ feet at $t = 1$ second
$s = 64$ feet at $t = 2$ seconds
$s = 48$ feet at $t = 3$ seconds

39. $s = 32$ feet at $t = 1$ second
$s = 32$ feet at $t = 2$ seconds
$s = 0$ feet at $t = 3$ seconds

40. $s = 10$ feet at $t = 0$ seconds
$s = 54$ feet at $t = 1$ second
$s = 46$ feet at $t = 3$ seconds

✓ **41.** *Geometry* The sum of the measures of two angles of a triangle is twice the measure of the third angle. The measure of the second angle is 28° less than the measure of the third angle. Find the measures of the three angles.

42. *Geometry* The measure of one angle of a triangle is two-thirds the measure of a second angle, and the measure of the second angle is 12° greater than the measure of the third angle. Find the measures of the three angles.

43. *Investment* An inheritance of $80,000 is divided among three investments yielding a total of $8850 in simple interest in 1 year. The interest rates for the three investments are 6%, 10%, and 15%. The amount invested at 10% is $750 more than the amount invested at 15%. Find the amount invested at each rate.

44. *Investment* An inheritance of $16,000 is divided among three investments yielding a total of $940 in simple interest in 1 year. The interest rates for the three investments are 5%, 6%, and 7%. The amount invested at 6% is $3000 less than the amount invested at 5%. Find the amount invested at each rate.

45. *Investment* You receive a total of $1150 in simple interest in 1 year from three investments. The interest rates for the three investments are 6%, 8%, and 9%. The 8% investment is twice the 6% investment, and the 9% investment is $1000 less than the 6% investment. What is the amount of each investment?

46. *Investment* You receive a total of $620 in simple interest in 1 year from three investments. The interest rates for the three investments are 5%, 7%, and 8%. The 5% investment is twice the 7% investment, and the 7% investment is $1500 less than the 8% investment. What is the amount of each investment?

47. *Chemical Mixture* A mixture of 12 gallons of chemical A, 16 gallons of chemical B, and 26 gallons of chemical C is required to kill a destructive crop insect. Commercial spray X contains one, two, and two parts of these chemicals. Spray Y contains only chemical C. Spray Z contains only chemicals A and B in equal amounts. How much of each type of commercial spray is needed to obtain the desired mixture?

48. *Fertilizer Mixture* A mixture of 5 pounds of fertilizer A, 13 pounds of fertilizer B, and 4 pounds of fertilizer C provides the optimal nutrients for a plant. Commercial brand X contains equal parts of fertilizer B and fertilizer C. Brand Y contains one part of fertilizer A and two parts of fertilizer B. Brand Z contains two parts of fertilizer A, five parts of fertilizer B, and two parts of fertilizer C. How much of each fertilizer brand is needed to obtain the desired mixture?

✓ **49.** *Hot Dogs* A vendor sells three sizes of hot dogs at prices of $1.50, $2.50, and $3.25. On a day when the vendor had a total revenue of $289.25 from sales of 143 hot dogs, four times as many $1.50 hot dogs were sold as $3.25 hot dogs. How many hot dogs were sold at each price?

50. *Coffee* A coffee manufacturer sells a 10-pound package that consists of three flavors of coffee. Vanilla flavored coffee costs $6 per pound, Hazelnut flavored coffee costs $6.50 per pound, and French Roast flavored coffee costs $7 per pound. The package contains the same amount of Hazelnut coffee as French Roast coffee. The cost of the 10-pound package is $66. How many pounds of each type of coffee are in the package?

51. *Mixture Problem* A chemist needs 12 gallons of a 20% acid solution. It is mixed from three solutions whose concentrations are 10%, 15%, and 25%. How many gallons of each solution will satisfy each condition?

(a) Use 4 gallons of the 25% solution.

(b) Use as little as possible of the 25% solution.

(c) Use as much as possible of the 25% solution.

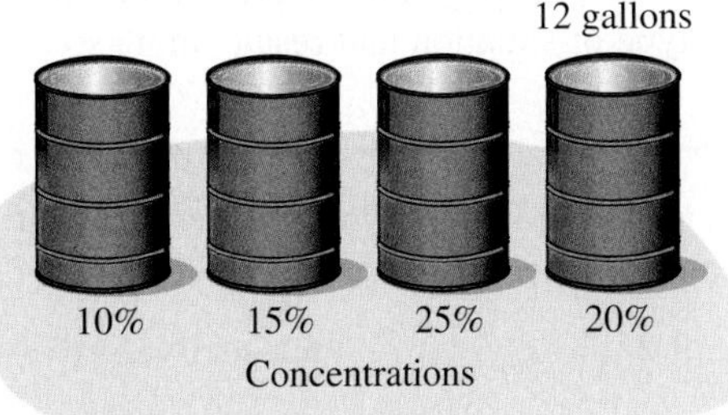

52. *Mixture Problem* A chemist needs 10 liters of a 25% acid solution. It is mixed from three solutions whose concentrations are 10%, 20%, and 50%. How many liters of each solution will satisfy each condition?

(a) Use 2 liters of the 50% solution.

(b) Use as little as possible of the 50% solution.

(c) Use as much as possible of the 50% solution.

53. *School Orchestra* The table shows the percents of each section of the North High School orchestra that were chosen to participate in the city orchestra, the county orchestra, and the state orchestra. Thirty members of the city orchestra, 17 members of the county orchestra, and 10 members of the state orchestra are from North High. How many members are in each section of North High's orchestra?

Orchestra	String	Wind	Percussion
City orchestra	40%	30%	50%
County orchestra	20%	25%	25%
State orchestra	10%	15%	25%

54. *Sports* The table shows the percents of each unit of the North High School football team that were chosen for academic honors, as city all-stars, and as county all-stars. Of all the players on the football team, 5 were awarded with academic honors, 13 were named city all-stars, and 4 were named county all-stars. How many members of each unit are there on the football team?

	Defense	Offense	Special teams
Academic honors	0%	10%	20%
City all-stars	10%	20%	50%
County all-stars	10%	0%	20%

Explaining Concepts

55. You apply Gaussian elimination to a system of three equations in the variables x, y, and z. From the row-echelon form, the solution $(1, -3, 4)$ is apparent *without* applying back-substitution or any other calculations. Explain why.

56. A system of three linear equations in three variables has an infinite number of solutions. Is it possible that the graphs of two of the three equations are parallel planes? Explain.

57. Two ways that a system of three linear equations in three variables can have no solution are shown in Figure 8.12 on page 503. Describe the graph for a third type of situation that results in no solution.

58. Describe the graphs and numbers of solutions possible for a system of three linear equations in three variables in which at least two of the equations are dependent.

59. Describe the graphs and numbers of solutions possible for a system of three linear equations in three variables if each pair of equations is consistent and *not* dependent.

60. Write a system of four linear equations in four unknowns, and use Gaussian elimination with back-substitution to solve it.

Cumulative Review

In Exercises 61–64, identify the terms and coefficients of the algebraic expression.

61. $3x + 2$

62. $4x^2 + 5x - 4$

63. $14t^5 - t + 25$

64. $5s^2 + 3st + 2t^2$

In Exercises 65–68, solve the system of linear equations by the method of elimination.

65. $\begin{cases} 2x + 3y = 17 \\ 4y = 12 \end{cases}$

66. $\begin{cases} x - 2y = 11 \\ 3x + 3y = 6 \end{cases}$

67. $\begin{cases} 3x - 4y = -30 \\ 5x + 4y = 14 \end{cases}$

68. $\begin{cases} 3x + 5y = 1 \\ 4x + 15y = 5 \end{cases}$

Mid-Chapter Quiz

Take this quiz as you would take a quiz in class. After you are done, check your work against the answers in the back of the book.

1. Is (4, 2) a solution of $3x + 4y = 4$ *and* $5x - 3y = 14$? Explain.

2. Is (2, −1) a solution of $2x - 3y = 7$ *and* $3x + 5y = 1$? Explain.

In Exercises 3–5, use the given graphs to solve the system of linear equations.

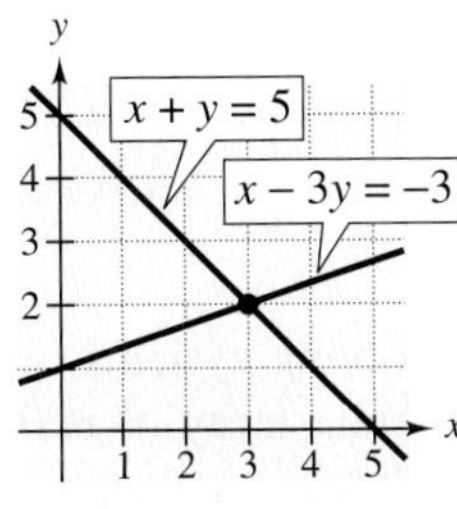

Figure for 3

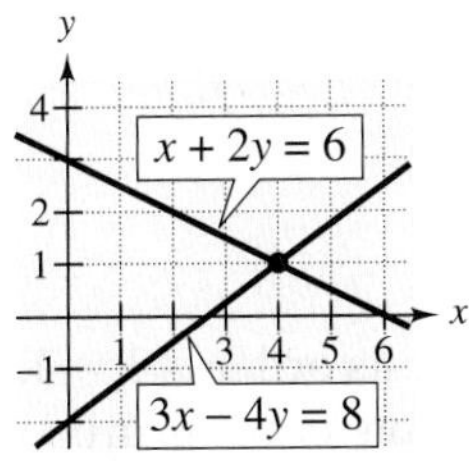

Figure for 4

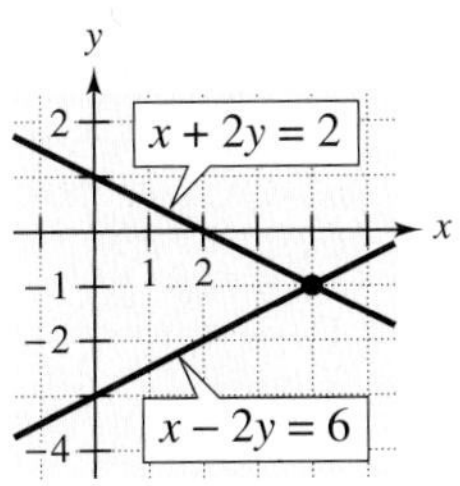

Figure for 5

3. $\begin{cases} x + y = 5 \\ x - 3y = -3 \end{cases}$ **4.** $\begin{cases} x + 2y = 6 \\ 3x - 4y = 8 \end{cases}$ **5.** $\begin{cases} x + 2y = 2 \\ x - 2y = 6 \end{cases}$

In Exercises 6–8, sketch the graphs of the equations and approximate any solutions of the system of linear equations. Check your solution.

6. $\begin{cases} x = -3 \\ x + y = 8 \end{cases}$ **7.** $\begin{cases} y = \frac{3}{2}x - 1 \\ y = -x + 4 \end{cases}$ **8.** $\begin{cases} 4x + y = 0 \\ -x + y = 5 \end{cases}$

In Exercises 9–11, solve the system by the method of substitution.

9. $\begin{cases} x - y = 4 \\ y = 2 \end{cases}$ **10.** $\begin{cases} y = -\frac{2}{3}x + 5 \\ y = 2x - 3 \end{cases}$ **11.** $\begin{cases} 2x - y = -7 \\ 4x + 3y = 16 \end{cases}$

In Exercises 12–15, use elimination or Gaussian elimination to solve the system.

12. $\begin{cases} 2x + y = 1 \\ 6x + 5y = 13 \end{cases}$ **13.** $\begin{cases} -x + 3y = 10 \\ 9x - 4y = 5 \end{cases}$

14. $\begin{cases} a + b + c = 1 \\ 4a + 2b + c = 2 \\ 9a + 3b + c = 4 \end{cases}$ **15.** $\begin{cases} x + 4z = 17 \\ -3x + 2y - z = -20 \\ x - 5y + 3z = 19 \end{cases}$

In Exercises 16 and 17, find a system of linear equations that has the ordered pair as its only solution. (There are many correct answers.)

16. (−1, 9) **17.** (3, 0)

18. A small company produces one-time-use cameras that sell for $5.95 per unit. The cost of producing each camera is $3.45, and the company has fixed costs of $16,000. Use a graphing calculator to graph the cost and revenue functions in the same viewing window. Approximate the point of intersection of the graphs and interpret the results.

19. The measure of the second angle of a triangle is 10° less than twice the measure of the first angle. The measure of the third angle is 10° greater than the measure of the first angle. Find the measures of the three angles.

8.4 Matrices and Linear Systems

© South West Images Scotland/Alamy

What You Should Learn

1 ► Determine the orders of matrices.

2 ► Form coefficient and augmented matrices, and form linear systems from augmented matrices.

3 ► Perform elementary row operations to solve systems of linear equations in matrix form.

4 ► Use matrices and Gaussian elimination with back-substitution to solve systems of linear equations.

Why You Should Learn It

Systems of linear equations that model real-life situations can be solved using matrices. For instance, in Exercise 85 on page 523, the numbers of computer parts a company produces can be found using a matrix.

1 ► Determine the orders of matrices.

Matrices

In this section, you will study a streamlined technique for solving systems of linear equations. This technique involves the use of a rectangular array of real numbers called a **matrix.** (The plural of matrix is *matrices.*) Here is an example of a matrix.

$$\begin{array}{r} \\ \text{Row 1} \\ \text{Row 2} \\ \text{Row 3} \end{array} \begin{array}{c} \begin{array}{cccc} \text{Column 1} & \text{Column 2} & \text{Column 3} & \text{Column 4} \end{array} \\ \begin{bmatrix} 3 & -2 & 4 & 1 \\ 0 & 1 & -1 & 2 \\ 2 & 0 & -3 & 0 \end{bmatrix} \end{array}$$

This matrix has three rows and four columns, which means that its **order** is 3×4, which is read as "3 by 4." Each number in the matrix is an **entry** of the matrix.

EXAMPLE 1 Orders of Matrices

Determine the order of each matrix.

a. $\begin{bmatrix} 1 & -2 & 4 \\ 0 & 1 & -2 \end{bmatrix}$ **b.** $\begin{bmatrix} 0 & 0 \\ 0 & 0 \end{bmatrix}$ **c.** $\begin{bmatrix} 1 & -3 \\ -2 & 0 \\ 4 & -2 \end{bmatrix}$

Solution

a. This matrix has two rows and three columns, so the order is 2×3.

b. This matrix has two rows and two columns, so the order is 2×2.

c. This matrix has three rows and two columns, so the order is 3×2.

✓ **CHECKPOINT** *Now try Exercise 1.*

Smart Study Strategy

Go to page 470 for ways to *Make Note Cards.*

Study Tip

The order of a matrix is always given as *row by column.* A matrix with the same number of rows as columns is called a **square matrix.** For instance, the 2×2 matrix in Example 1(b) is square.

2 ▶ Form coefficient and augmented matrices, and form linear systems from augmented matrices.

Augmented and Coefficient Matrices

A matrix derived from a system of linear equations (each written in standard form with the constant term on the right) is the **augmented matrix** of the system. Moreover, the matrix derived from the coefficients of the system (but not including the constant terms) is the **coefficient matrix** of the system. Here is an example.

Study Tip

Note the use of 0 for the missing y-variable in the third equation, and also note the fourth column of constant terms in the augmented matrix.

System / *Coefficient Matrix* / *Augmented Matrix*

$$\begin{cases} x - 4y + 3z = 5 \\ -x + 3y - z = -3 \\ 2x - 4z = 6 \end{cases} \quad \begin{bmatrix} 1 & -4 & 3 \\ -1 & 3 & -1 \\ 2 & 0 & -4 \end{bmatrix} \quad \left[\begin{array}{ccc|c} 1 & -4 & 3 & 5 \\ -1 & 3 & -1 & -3 \\ 2 & 0 & -4 & 6 \end{array}\right]$$

When forming either the coefficient matrix or the augmented matrix of a system, you should begin by vertically aligning the variables in the equations.

Given System / *Align Variables* / *Form Augmented Matrix*

$$\begin{cases} x + 3y = 9 \\ -y + 4z = -2 \\ x - 5z = 0 \end{cases} \quad \begin{cases} x + 3y = 9 \\ -y + 4z = -2 \\ x - 5z = 0 \end{cases} \quad \left[\begin{array}{ccc|c} 1 & 3 & 0 & 9 \\ 0 & -1 & 4 & -2 \\ 1 & 0 & -5 & 0 \end{array}\right]$$

EXAMPLE 2 Forming Coefficient and Augmented Matrices

Form the coefficient matrix and the augmented matrix for each system.

a. $\begin{cases} -x + 5y = 2 \\ 7x - 2y = -6 \end{cases}$ **b.** $\begin{cases} 3x + 2y - z = 1 \\ x + 2z = -3 \\ -2x - y = 4 \end{cases}$

Solution

System / *Coefficient Matrix* / *Augmented Matrix*

a. $\begin{cases} -x + 5y = 2 \\ 7x - 2y = -6 \end{cases} \quad \begin{bmatrix} -1 & 5 \\ 7 & -2 \end{bmatrix} \quad \left[\begin{array}{cc|c} -1 & 5 & 2 \\ 7 & -2 & -6 \end{array}\right]$

b. $\begin{cases} 3x + 2y - z = 1 \\ x + 2z = -3 \\ -2x - y = 4 \end{cases} \quad \begin{bmatrix} 3 & 2 & -1 \\ 1 & 0 & 2 \\ -2 & -1 & 0 \end{bmatrix} \quad \left[\begin{array}{ccc|c} 3 & 2 & -1 & 1 \\ 1 & 0 & 2 & -3 \\ -2 & -1 & 0 & 4 \end{array}\right]$

✓ **CHECKPOINT** *Now try Exercise 11.*

EXAMPLE 3 Forming Linear Systems from Their Matrices

Write the system of linear equations represented by each matrix.

a. $\left[\begin{array}{cc|c} 3 & -5 & 4 \\ -1 & 2 & 0 \end{array}\right]$ **b.** $\left[\begin{array}{cc|c} 1 & 3 & 2 \\ 0 & 1 & -3 \end{array}\right]$ **c.** $\left[\begin{array}{ccc|c} 2 & 0 & -8 & 1 \\ -1 & 1 & 1 & 2 \\ 5 & -1 & 7 & 3 \end{array}\right]$

Solution

a. $\begin{cases} 3x - 5y = 4 \\ -x + 2y = 0 \end{cases}$ **b.** $\begin{cases} x + 3y = 2 \\ y = -3 \end{cases}$ **c.** $\begin{cases} 2x - 8z = 1 \\ -x + y + z = 2 \\ 5x - y + 7z = 3 \end{cases}$

✓ **CHECKPOINT** *Now try Exercise 17.*

3 ▶ Perform elementary row operations to solve systems of linear equations in matrix form.

Elementary Row Operations

In Section 8.3, you studied three operations that can be used on a system of linear equations to produce an equivalent system: (1) interchange two equations, (2) multiply an equation by a nonzero constant, and (3) add a multiple of an equation to another equation. In matrix terminology, these three operations correspond to **elementary row operations.**

Study Tip

Although elementary row operations are simple to perform, they involve a lot of arithmetic. So that you can check your work, you should get in the habit of noting the elementary row operations performed in each step. People use different schemes to do this. The scheme that is used in this text is to write an abbreviated version of the row operation at the left of the row that has been changed, as shown in Example 4. .

Elementary Row Operations

Any of the following **elementary row operations** performed on an augmented matrix will produce a matrix that is row-equivalent to the original matrix. Two matrices are **row-equivalent** if one can be obtained from the other by a sequence of elementary row operations.

1. Interchange two rows.
2. Multiply a row by a nonzero constant.
3. Add a multiple of a row to another row.

EXAMPLE 4 Performing Elementary Row Operations

a. Interchange the first and second rows.

Original Matrix / *New Row-Equivalent Matrix*

$$\begin{bmatrix} 0 & 1 & 3 & 4 \\ -1 & 2 & 0 & 3 \\ 2 & -3 & 4 & 1 \end{bmatrix} \qquad \begin{matrix} R_2 \\ R_1 \\ \end{matrix} \begin{bmatrix} -1 & 2 & 0 & 3 \\ 0 & 1 & 3 & 4 \\ 2 & -3 & 4 & 1 \end{bmatrix}$$

b. Multiply the first row by $\frac{1}{2}$.

Original Matrix / *New Row-Equivalent Matrix*

$$\begin{bmatrix} 2 & -4 & 6 & -2 \\ 1 & 3 & -3 & 0 \\ 5 & -2 & 1 & 2 \end{bmatrix} \qquad \begin{matrix} \frac{1}{2}R_1 \rightarrow \\ \\ \end{matrix} \begin{bmatrix} 1 & -2 & 3 & -1 \\ 1 & 3 & -3 & 0 \\ 5 & -2 & 1 & 2 \end{bmatrix}$$

c. Add -2 times the first row to the third row.

Original Matrix / *New Row-Equivalent Matrix*

$$\begin{bmatrix} 1 & 2 & -4 & 3 \\ 0 & 3 & -2 & -1 \\ 2 & 1 & 5 & -2 \end{bmatrix} \qquad \begin{matrix} \\ \\ -2R_1 + R_3 \rightarrow \end{matrix} \begin{bmatrix} 1 & 2 & -4 & 3 \\ 0 & 3 & -2 & -1 \\ 0 & -3 & 13 & -8 \end{bmatrix}$$

d. Add 6 times the first row to the second row.

Original Matrix / *New Row-Equivalent Matrix*

$$\begin{bmatrix} 1 & 2 & 2 & -4 \\ -6 & -11 & 3 & 18 \\ 0 & 0 & 4 & 7 \end{bmatrix} \qquad \begin{matrix} \\ 6R_1 + R_2 \rightarrow \\ \end{matrix} \begin{bmatrix} 1 & 2 & 2 & -4 \\ 0 & 1 & 15 & -6 \\ 0 & 0 & 4 & 7 \end{bmatrix}$$

✓ **CHECKPOINT** *Now try Exercise 31.*

In Section 8.3, Gaussian elimination was used with back-substitution to solve systems of linear equations. Example 5 demonstrates the matrix version of Gaussian elimination. The two methods are essentially the same. The basic difference is that with matrices you do not need to keep writing the variables.

Technology: Tip

Most graphing calculators can be used to perform row operations on matrices. Some have a function that gives the reduced row-echelon form of a matrix. When you perform a sequence of row operations on a graphing calculator, it is a good idea to store the resulting matrix from each step in a matrix variable. Consult your user's guide to learn how to perform elementary row operations on your graphing calculator.

Enter the matrix from Example 5 in your graphing calculator and perform the indicated row operations.

EXAMPLE 5 Solving a System of Linear Equations

Linear System / *Associated Augmented Matrix*

$$\begin{cases} x - 2y + 2z = 9 \\ -x + 3y = -4 \\ 2x - 5y + z = 10 \end{cases} \qquad \left[\begin{array}{rrr:r} 1 & -2 & 2 & 9 \\ -1 & 3 & 0 & -4 \\ 2 & -5 & 1 & 10 \end{array}\right]$$

Add the first equation to the second equation. / Add the first row to the second row.

$$\begin{cases} x - 2y + 2z = 9 \\ y + 2z = 5 \\ 2x - 5y + z = 10 \end{cases} \qquad R_1 + R_2 \rightarrow \left[\begin{array}{rrr:r} 1 & -2 & 2 & 9 \\ 0 & 1 & 2 & 5 \\ 2 & -5 & 1 & 10 \end{array}\right]$$

Add -2 times the first equation to the third equation. / Add -2 times the first row to the third row.

$$\begin{cases} x - 2y + 2z = 9 \\ y + 2z = 5 \\ -y - 3z = -8 \end{cases} \qquad -2R_1 + R_3 \rightarrow \left[\begin{array}{rrr:r} 1 & -2 & 2 & 9 \\ 0 & 1 & 2 & 5 \\ 0 & -1 & -3 & -8 \end{array}\right]$$

Add the second equation to the third equation. / Add the second row to the third row.

$$\begin{cases} x - 2y + 2z = 9 \\ y + 2z = 5 \\ -z = -3 \end{cases} \qquad R_2 + R_3 \rightarrow \left[\begin{array}{rrr:r} 1 & -2 & 2 & 9 \\ 0 & 1 & 2 & 5 \\ 0 & 0 & -1 & -3 \end{array}\right]$$

Multiply the third equation by -1. / Multiply the third row by -1.

$$\begin{cases} x - 2y + 2z = 9 \\ y + 2z = 5 \\ z = 3 \end{cases} \qquad -R_3 \rightarrow \left[\begin{array}{rrr:r} 1 & -2 & 2 & 9 \\ 0 & 1 & 2 & 5 \\ 0 & 0 & 1 & 3 \end{array}\right]$$

At this point, you can use back-substitution to find that the solution is $x = 1$, $y = -1$, and $z = 3$. The solution can be written as the ordered triple $(1, -1, 3)$.

✓ **CHECKPOINT** *Now try Exercise 53.*

Study Tip

The last matrix in Example 5 is in **row-echelon form.** The term *echelon* refers to the stair-step pattern formed by the nonzero elements of the matrix.

Definition of Row-Echelon Form of a Matrix

A matrix in **row-echelon form** has the following properties.

1. All rows consisting entirely of zeros occur at the bottom of the matrix.
2. For each row that does not consist entirely of zeros, the first nonzero entry is 1 (called a **leading 1**).
3. For two successive (nonzero) rows, the leading 1 in the higher row is farther to the left than the leading 1 in the lower row.

4 ▶ Use matrices and Gaussian elimination with back-substitution to solve systems of linear equations.

Solving a System of Linear Equations

Gaussian Elimination with Back-Substitution

To use matrices and Gaussian elimination to solve a system of linear equations, use the following steps.

1. Write the augmented matrix of the system of linear equations.
2. Use elementary row operations to rewrite the augmented matrix in row-echelon form.
3. Write the system of linear equations corresponding to the matrix in row-echelon form, and use back-substitution to find the solution.

When you perform Gaussian elimination with back-substitution, you should operate from *left to right by columns*, using elementary row operations to obtain zeros in all entries directly below the leading 1's.

EXAMPLE 6 Gaussian Elimination with Back-Substitution

Solve the system of linear equations.

$$\begin{cases} 2x - 3y = -2 \\ x + 2y = 13 \end{cases}$$

Solution

$$\begin{bmatrix} 2 & -3 & \vdots & -2 \\ 1 & 2 & \vdots & 13 \end{bmatrix}$$ Augmented matrix for system of linear equations

$$\begin{matrix} R_2 \\ R_1 \end{matrix} \begin{bmatrix} 1 & 2 & \vdots & 13 \\ 2 & -3 & \vdots & -2 \end{bmatrix}$$ First column has leading 1 in upper left corner.

$$-2R_1 + R_2 \rightarrow \begin{bmatrix} 1 & 2 & \vdots & 13 \\ 0 & -7 & \vdots & -28 \end{bmatrix}$$ First column has a zero under its leading 1.

$$-\tfrac{1}{7}R_2 \rightarrow \begin{bmatrix} 1 & 2 & \vdots & 13 \\ 0 & 1 & \vdots & 4 \end{bmatrix}$$ Second column has leading 1 in second row.

The system of linear equations that corresponds to the (row-echelon) matrix is

$$\begin{cases} x + 2y = 13 \\ y = 4. \end{cases}$$

Using back-substitution, you can find that the solution of the system is $x = 5$ and $y = 4$, which can be written as the ordered pair $(5, 4)$. Check this solution in the original system, as follows.

Check

Equation 1: $2(5) - 3(4) = -2$ ✓

Equation 2: $5 + 2(4) = 13$ ✓

✓ ***CHECKPOINT*** *Now try Exercise 55.*

EXAMPLE 7 Gaussian Elimination with Back-Substitution

Solve the system of linear equations.

$$\begin{cases} 3x + 3y \qquad = 9 \\ 2x \qquad - 3z = 10 \\ \qquad 6y + 4z = -12 \end{cases}$$

Solution

$$\begin{bmatrix} 3 & 3 & 0 & \vdots & 9 \\ 2 & 0 & -3 & \vdots & 10 \\ 0 & 6 & 4 & \vdots & -12 \end{bmatrix}$$ Augmented matrix for system of linear equations

$$\tfrac{1}{3}R_1 \rightarrow \begin{bmatrix} 1 & 1 & 0 & \vdots & 3 \\ 2 & 0 & -3 & \vdots & 10 \\ 0 & 6 & 4 & \vdots & -12 \end{bmatrix}$$ First column has leading 1 in upper left corner.

$$-2R_1 + R_2 \rightarrow \begin{bmatrix} 1 & 1 & 0 & \vdots & 3 \\ 0 & -2 & -3 & \vdots & 4 \\ 0 & 6 & 4 & \vdots & -12 \end{bmatrix}$$ First column has zeros under its leading 1.

$$-\tfrac{1}{2}R_2 \rightarrow \begin{bmatrix} 1 & 1 & 0 & \vdots & 3 \\ 0 & 1 & \frac{3}{2} & \vdots & -2 \\ 0 & 6 & 4 & \vdots & -12 \end{bmatrix}$$ Second column has leading 1 in second row.

$$-6R_2 + R_3 \rightarrow \begin{bmatrix} 1 & 1 & 0 & \vdots & 3 \\ 0 & 1 & \frac{3}{2} & \vdots & -2 \\ 0 & 0 & -5 & \vdots & 0 \end{bmatrix}$$ Second column has zero under its leading 1.

$$-\tfrac{1}{5}R_3 \rightarrow \begin{bmatrix} 1 & 1 & 0 & \vdots & 3 \\ 0 & 1 & \frac{3}{2} & \vdots & -2 \\ 0 & 0 & 1 & \vdots & 0 \end{bmatrix}$$ Third column has leading 1 in third row.

The system of linear equations that corresponds to this (row-echelon) matrix is

$$\begin{cases} x + y \qquad = 3 \\ \quad y + \dfrac{3}{2}z = -2 \\ \qquad\quad z = 0. \end{cases}$$

Using back-substitution, you can find that the solution is

$$x = 5, \quad y = -2, \quad \text{and} \quad z = 0$$

which can be written as the ordered triple $(5, -2, 0)$. Check this in the original system, as follows.

Check

Equation 1: $3(5) + 3(-2) = 9$ ✓

Equation 2: $2(5) - 3(0) = 10$ ✓

Equation 3: $6(-2) + 4(0) = -12$ ✓

✓ **CHECKPOINT** *Now try Exercise 65.*

EXAMPLE 8 A System with No Solution

Solve the system of linear equations.

$$\begin{cases} 6x - 10y = -4 \\ 9x - 15y = 5 \end{cases}$$

Solution

$$\begin{bmatrix} 6 & -10 & \vdots & -4 \\ 9 & -15 & \vdots & 5 \end{bmatrix}$$ Augmented matrix for system of linear equations

$$\tfrac{1}{6}R_1 \rightarrow \begin{bmatrix} 1 & -\frac{5}{3} & \vdots & -\frac{2}{3} \\ 9 & -15 & \vdots & 5 \end{bmatrix}$$ First column has leading 1 in upper left corner.

$$\begin{matrix} \\ -9R_1 + R_2 \rightarrow \end{matrix} \begin{bmatrix} 1 & -\frac{5}{3} & \vdots & -\frac{2}{3} \\ 0 & 0 & \vdots & 11 \end{bmatrix}$$ First column has a zero under its leading 1.

The "equation" that corresponds to the second row of this matrix is $0 = 11$. Because this is a false statement, the system of equations has no solution.

CHECKPOINT *Now try Exercise 59.*

EXAMPLE 9 A System with Infinitely Many Solutions

Solve the system of linear equations.

$$\begin{cases} 12x - 6y = -3 \\ -8x + 4y = 2 \end{cases}$$

Solution

$$\begin{bmatrix} 12 & -6 & \vdots & -3 \\ -8 & 4 & \vdots & 2 \end{bmatrix}$$ Augmented matrix for system of linear equations

$$\tfrac{1}{12}R_1 \rightarrow \begin{bmatrix} 1 & -\frac{1}{2} & \vdots & -\frac{1}{4} \\ -8 & 4 & \vdots & 2 \end{bmatrix}$$ First column has leading 1 in upper left corner.

$$\begin{matrix} \\ 8R_1 + R_2 \rightarrow \end{matrix} \begin{bmatrix} 1 & -\frac{1}{2} & \vdots & -\frac{1}{4} \\ 0 & 0 & \vdots & 0 \end{bmatrix}$$ First column has a zero under its leading 1.

Because the second row of the matrix is all zeros, the system of equations has an infinite number of solutions, represented by all points (x, y) on the line

$$x - \frac{1}{2}y = -\frac{1}{4}.$$

Because this line can be written as

$$x = \frac{1}{2}y - \frac{1}{4}$$

you can write the solution set as

$$\left(\frac{1}{2}a - \frac{1}{4}, a\right), \text{ where } a \text{ is any real number.}$$

CHECKPOINT *Now try Exercise 63.*

EXAMPLE 10 Investment Portfolio 

You have a portfolio totaling \$219,000 and want to invest in municipal bonds, blue-chip stocks, and growth or speculative stocks. The municipal bonds pay 6% annually. Over a five-year period, you expect blue-chip stocks to return 10% annually and growth stocks to return 15% annually. You want a combined annual return of 8%, and you also want to have only one-fourth of the portfolio invested in stocks. How much should be allocated to each type of investment?

Solution

Let M, B, and G represent the amounts invested in municipal bonds, blue-chip stocks, and growth stocks, respectively. This situation is represented by the following system.

$$\begin{cases} M + B + G = 219{,}000 \\ 0.06M + 0.10B + 0.15G = 17{,}520 \\ B + G = 54{,}750 \end{cases}$$

Equation 1: Total investment is \$219,000.
Equation 2: Combined annual return is 8%.
Equation 3: $\frac{1}{4}$ of investment is allocated to stocks.

Form the augmented matrix for this system of equations, and then use elementary row operations to obtain the row-echelon form of the matrix.

$$\begin{bmatrix} 1 & 1 & 1 & \vdots & 219{,}000 \\ 0.06 & 0.10 & 0.15 & \vdots & 17{,}520 \\ 0 & 1 & 1 & \vdots & 54{,}750 \end{bmatrix}$$

Augmented matrix for system of linear equations

$$-0.06R_1 + R_2 \rightarrow \begin{bmatrix} 1 & 1 & 1 & \vdots & 219{,}000 \\ 0 & 0.04 & 0.09 & \vdots & 4{,}380 \\ 0 & 1 & 1 & \vdots & 54{,}750 \end{bmatrix}$$

First column has zeros under its leading 1.

$$25R_2 \rightarrow \begin{bmatrix} 1 & 1 & 1 & \vdots & 219{,}000 \\ 0 & 1 & 2.25 & \vdots & 109{,}500 \\ 0 & 1 & 1 & \vdots & 54{,}750 \end{bmatrix}$$

Second column has leading 1 in second row.

$$\begin{matrix} \\ \\ -R_2 + R_3 \rightarrow \end{matrix} \begin{bmatrix} 1 & 1 & 1 & \vdots & 219{,}000 \\ 0 & 1 & 2.25 & \vdots & 109{,}500 \\ 0 & 0 & -1.25 & \vdots & -54{,}750 \end{bmatrix}$$

Second column has zero under its leading 1.

$$\begin{matrix} \\ \\ -0.8R_3 \rightarrow \end{matrix} \begin{bmatrix} 1 & 1 & 1 & \vdots & 219{,}000 \\ 0 & 1 & 2.25 & \vdots & 109{,}500 \\ 0 & 0 & 1 & \vdots & 43{,}800 \end{bmatrix}$$

Third column has leading 1 in third row and matrix is in row-echelon form.

From the row-echelon form, you can see that $G = 43{,}800$. By back-substituting G into the revised second equation, you can determine the value of B.

$$B + 2.25(43{,}800) = 109{,}500 \quad \Rightarrow \quad B = 10{,}950$$

By back-substituting B and G into Equation 1, you can solve for M.

$$M + 10{,}950 + 43{,}800 = 219{,}000 \quad \Rightarrow \quad M = 164{,}250$$

So, you should invest \$164,250 in municipal bonds, \$10,950 in blue-chip stocks, and \$43,800 in growth or speculative stocks. Check this solution by substituting these values into the original system of equations.

CHECKPOINT *Now try Exercise 81.*

Concept Check

1. A matrix contains exactly four entries. What are the possible orders of the matrix? State the numbers of rows and columns in each possible order.

2. For a given system of equations, which has more entries, the coefficient matrix or the augmented matrix? Explain.

3. What is the primary difference between performing row operations on a system of equations and performing elementary row operations?

4. After using matrices to perform Gaussian elimination, what steps are generally needed to find the solution of the original system of equations?

8.4 EXERCISES

Go to pages 546–547 to record your assignments.

Developing Skills

In Exercises 1–10, determine the order of the matrix. ***See Example 1.***

✓ **1.** $\begin{bmatrix} 3 & -2 \\ -4 & 0 \\ 2 & -7 \\ -1 & -3 \end{bmatrix}$

2. $\begin{bmatrix} 3 & 4 \\ 2 & -1 \\ 8 & 10 \\ -6 & -6 \\ 12 & 50 \end{bmatrix}$

3. $\begin{bmatrix} -2 & 5 \\ 0 & -1 \end{bmatrix}$

4. $\begin{bmatrix} 5 & -8 & 32 \\ 7 & 15 & 28 \end{bmatrix}$

5. $\begin{bmatrix} 4 \\ -2 \\ 0 \\ 1 \end{bmatrix}$

6. $\begin{bmatrix} 4 & 0 & -5 \\ -1 & 8 & 9 \\ 0 & -3 & 4 \end{bmatrix}$

7. $[5]$

8. $[1 \quad -1 \quad 2 \quad 3]$

9. $[13 \quad 12 \quad -9 \quad 0]$

10. $\begin{bmatrix} 6 \\ -13 \\ 22 \end{bmatrix}$

In Exercises 11–16, form (a) the coefficient matrix and (b) the augmented matrix for the system of linear equations. ***See Example 2.***

✓ **11.** $\begin{cases} 4x - 5y = -2 \\ -x + 8y = 10 \end{cases}$

12. $\begin{cases} 8x + 3y = 25 \\ 3x - 9y = 12 \end{cases}$

13. $\begin{cases} x + y = 0 \\ 5x - 2y - 2z = 12 \\ 2x + 4y + z = 5 \end{cases}$

14. $\begin{cases} 9x - 3y + z = 13 \\ 12x - 8z = 5 \\ 3x + 4y - z = 6 \end{cases}$

15. $\begin{cases} 5x + y - 3z = 7 \\ 2y + 4z = 12 \end{cases}$

16. $\begin{cases} 10x + 6y - 8z = -4 \\ -4x - 7y = 9 \end{cases}$

In Exercises 17–24, write the system of linear equations represented by the augmented matrix. (Use variables x, y, z, and w.) ***See Example 3.***

✓ **17.** $\left[\begin{array}{cc:c} 4 & 3 & 8 \\ 1 & -2 & 3 \end{array}\right]$

18. $\left[\begin{array}{cc:c} 9 & -4 & 0 \\ 6 & 1 & -4 \end{array}\right]$

19. $\begin{bmatrix} 1 & 0 & 2 & \vdots & -10 \\ 0 & 3 & -1 & \vdots & 5 \\ 4 & 2 & 0 & \vdots & 3 \end{bmatrix}$

20. $\begin{bmatrix} 4 & -1 & 3 & \vdots & 5 \\ 2 & 0 & -2 & \vdots & -1 \\ -1 & 6 & 0 & \vdots & 3 \end{bmatrix}$

21. $\begin{bmatrix} 5 & 8 & 2 & 0 & \vdots & -1 \\ -2 & 15 & 5 & 1 & \vdots & 9 \\ 1 & 6 & -7 & 0 & \vdots & -3 \end{bmatrix}$

22. $\begin{bmatrix} 0 & 1 & -5 & 8 & \vdots & 10 \\ 2 & 4 & -1 & 0 & \vdots & 15 \\ 1 & 1 & 7 & 9 & \vdots & -8 \end{bmatrix}$

23. $\begin{bmatrix} 13 & 1 & 4 & -2 & \vdots & -4 \\ 5 & 4 & 0 & -1 & \vdots & 0 \\ 1 & 2 & 6 & 8 & \vdots & 5 \\ -10 & 12 & 3 & 1 & \vdots & -2 \end{bmatrix}$

24. $\begin{bmatrix} 7 & 3 & -2 & 4 & \vdots & 2 \\ -1 & 0 & 4 & -1 & \vdots & 6 \\ 8 & 3 & 0 & 0 & \vdots & -4 \\ 0 & 2 & -4 & 3 & \vdots & 12 \end{bmatrix}$

In Exercises 25–30, describe the elementary row operation used to transform the first matrix into the second matrix. ***See Examples 4 and 5.***

25. $\begin{bmatrix} 0 & 3 & -2 \\ 2 & 5 & -7 \end{bmatrix} \Rightarrow \begin{bmatrix} 2 & 5 & -7 \\ 0 & 3 & -2 \end{bmatrix}$

26. $\begin{bmatrix} 1 & 6 & 7 \\ 0 & 1 & 2 \\ 0 & -3 & -4 \end{bmatrix} \Rightarrow \begin{bmatrix} 1 & 6 & 7 \\ 0 & 1 & 2 \\ 0 & 0 & 2 \end{bmatrix}$

27. $\begin{bmatrix} -3 & 6 & 9 \\ 5 & 6 & 7 \end{bmatrix} \Rightarrow \begin{bmatrix} 1 & -2 & -3 \\ 5 & 6 & 7 \end{bmatrix}$

28. $\begin{bmatrix} \frac{1}{3} & 1 & 4 \\ -7 & 2 & 5 \end{bmatrix} \Rightarrow \begin{bmatrix} 1 & 3 & 12 \\ -7 & 2 & 5 \end{bmatrix}$

29. $\begin{bmatrix} 1 & 3 & 2 \\ -3 & 4 & 2 \\ -5 & 6 & -7 \end{bmatrix} \Rightarrow \begin{bmatrix} 1 & 3 & 2 \\ -3 & 4 & 2 \\ 0 & 21 & 3 \end{bmatrix}$

30. $\begin{bmatrix} 0 & -2 & 6 \\ 2 & 5 & 3 \end{bmatrix} \Rightarrow \begin{bmatrix} 2 & 5 & 3 \\ 0 & -2 & 6 \end{bmatrix}$

In Exercises 31–36, fill in the entries of the row-equivalent matrix formed by performing the indicated elementary row operation. ***See Example 4.***

31. $\begin{bmatrix} 1 & 1 & -4 & 2 \\ 0 & 0 & 8 & 3 \\ 0 & 4 & 5 & 5 \end{bmatrix} \begin{matrix} \\ R_3 \\ R_2 \end{matrix} \begin{bmatrix} \square & \square & \square & \square \\ \square & \square & \square & \square \\ \square & \square & \square & \square \end{bmatrix}$

32. $\begin{bmatrix} 0 & 0 & -5 & 2 \\ 0 & -7 & -3 & 3 \\ 1 & 4 & 5 & 4 \end{bmatrix} \begin{matrix} R_3 \\ \\ R_1 \end{matrix} \begin{bmatrix} \square & \square & \square & \square \\ \square & \square & \square & \square \\ \square & \square & \square & \square \end{bmatrix}$

33. $\begin{bmatrix} 9 & -18 & 27 \\ 3 & 4 & 5 \end{bmatrix} \quad \frac{1}{9}R_1 \rightarrow \begin{bmatrix} \square & \square & \square \\ \square & \square & \square \end{bmatrix}$

34. $\begin{bmatrix} 1 & 21 & 7 \\ 0 & -7 & 14 \end{bmatrix} \quad -\frac{1}{7}R_2 \rightarrow \begin{bmatrix} \square & \square & \square \\ \square & \square & \square \end{bmatrix}$

35. $\begin{bmatrix} 1 & 4 & 3 \\ 2 & 8 & 6 \end{bmatrix} \quad -2R_1 + R_2 \rightarrow \begin{bmatrix} \square & \square & \square \\ \square & \square & \square \end{bmatrix}$

36. $\begin{bmatrix} 1 & 4 & 5 \\ 4 & -7 & 3 \end{bmatrix} \quad -4R_1 + R_2 \rightarrow \begin{bmatrix} \square & \square & \square \\ \square & \square & \square \end{bmatrix}$

In Exercises 37–42, convert the matrix to row-echelon form. (There are many correct answers.)

37. $\begin{bmatrix} 1 & 2 & 3 \\ 2 & -1 & -4 \end{bmatrix}$

38. $\begin{bmatrix} 1 & 3 & 6 \\ -4 & -9 & 3 \end{bmatrix}$

39. $\begin{bmatrix} 4 & 6 & 1 \\ -2 & 2 & 5 \end{bmatrix}$

40. $\begin{bmatrix} 3 & 2 & 6 \\ 2 & 3 & -3 \end{bmatrix}$

41. $\begin{bmatrix} 1 & 1 & 0 & 5 \\ -2 & -1 & 2 & -10 \\ 3 & 6 & 7 & 14 \end{bmatrix}$

42. $\begin{bmatrix} 1 & 2 & -1 & 3 \\ 3 & 7 & -5 & 14 \\ -2 & -1 & -3 & 8 \end{bmatrix}$

In Exercises 43–46, use the matrix capabilities of a graphing calculator to write the matrix in row-echelon form. (There are many correct answers.)

43. $\begin{bmatrix} 1 & -1 & -1 & 1 \\ 4 & -4 & 1 & 8 \\ -6 & 8 & 18 & 0 \end{bmatrix}$

44. $\begin{bmatrix} 1 & -3 & 0 & -7 \\ -3 & 10 & 1 & 23 \\ 4 & -10 & 2 & -24 \end{bmatrix}$

45. $\begin{bmatrix} 1 & 1 & -1 & 3 \\ 2 & 1 & 2 & 5 \\ 3 & 2 & 1 & 8 \end{bmatrix}$

46. $\begin{bmatrix} 1 & -3 & -2 & -8 \\ 1 & 3 & -2 & 17 \\ 1 & 2 & -2 & -5 \end{bmatrix}$

In Exercises 47–52, write the system of linear equations represented by the augmented matrix. Then use back-substitution to find the solution. (Use variables x, y, and z.)

47. $\left[\begin{array}{cc|c} 1 & -2 & 4 \\ 0 & 1 & -3 \end{array}\right]$

48. $\left[\begin{array}{cc|c} 1 & 5 & 0 \\ 0 & 1 & -1 \end{array}\right]$

49. $\left[\begin{array}{cc|c} 1 & 5 & 3 \\ 0 & 1 & -2 \end{array}\right]$

50. $\left[\begin{array}{ccc|c} 1 & 5 & -3 & 0 \\ 0 & 1 & 0 & 6 \\ 0 & 0 & 1 & -5 \end{array}\right]$

51. $\left[\begin{array}{ccc|c} 1 & -1 & 2 & 4 \\ 0 & 1 & -1 & 2 \\ 0 & 0 & 1 & -2 \end{array}\right]$

52. $\left[\begin{array}{ccc|c} 1 & 2 & -2 & -1 \\ 0 & 1 & 1 & 9 \\ 0 & 0 & 1 & -3 \end{array}\right]$

In Exercises 53–78, use matrices to solve the system of linear equations. ***See Examples 5–9.***

✓ **53.** $\begin{cases} x + 2y = 7 \\ 3x - 7y = 8 \end{cases}$

54. $\begin{cases} 2x + 6y = 16 \\ 2x + 3y = 7 \end{cases}$

✓ **55.** $\begin{cases} 6x - 4y = 2 \\ 5x + 2y = 7 \end{cases}$

56. $\begin{cases} x - 3y = 5 \\ -2x + 6y = -10 \end{cases}$

57. $\begin{cases} 12x + 10y = -14 \\ 4x - 3y = -11 \end{cases}$

58. $\begin{cases} -x - 5y = -10 \\ 2x - 3y = 7 \end{cases}$

✓ **59.** $\begin{cases} -x + 2y = 1.5 \\ 2x - 4y = 3 \end{cases}$

60. $\begin{cases} 2x - y = -0.1 \\ 3x + 2y = 1.6 \end{cases}$

61. $\begin{cases} x - 2y - z = 6 \\ y + 4z = 5 \\ 4x + 2y + 3z = 8 \end{cases}$

62. $\begin{cases} x - 3z = -2 \\ 3x + y - 2z = 5 \\ 2x + 2y + z = 4 \end{cases}$

✓ **63.** $\begin{cases} x + y - 5z = 3 \\ x - 2z = 1 \\ 2x - y - z = 0 \end{cases}$

64. $\begin{cases} 2y + z = 3 \\ -4y - 2z = 0 \\ x + y + z = 2 \end{cases}$

✓ **65.** $\begin{cases} 2x + 4y = 10 \\ 2x + 2y + 3z = 3 \\ -3x + y + 2z = -3 \end{cases}$

66. $\begin{cases} 2x - y + 3z = 24 \\ 2y - z = 14 \\ 7x - 5y = 6 \end{cases}$

67. $\begin{cases} x - 3y + 2z = 8 \\ 2y - z = -4 \\ x + z = 3 \end{cases}$

68. $\begin{cases} 2x + 3z = 3 \\ 4x - 3y + 7z = 5 \\ 8x - 9y + 15z = 9 \end{cases}$

69. $\begin{cases} -2x - 2y - 15z = 0 \\ x + 2y + 2z = 18 \\ 3x + 3y + 22z = 2 \end{cases}$

70. $\begin{cases} 2x + 4y + 5z = 5 \\ x + 3y + 3z = 2 \\ 2x + 4y + 4z = 2 \end{cases}$

71. $\begin{cases} 2x + 4z = 1 \\ x + y + 3z = 0 \\ x + 3y + 5z = 0 \end{cases}$

72. $\begin{cases} 3x + y - 2z = 2 \\ 6x + 2y - 4z = 1 \\ -3x - y + 2z = 1 \end{cases}$

73. $\begin{cases} x + 3y = 2 \\ 2x + 6y = 4 \\ 2x + 5y + 4z = 3 \end{cases}$

74. $\begin{cases} 4x + 3y = 10 \\ 2x - y = 10 \\ -2x + z = -9 \end{cases}$

75. $\begin{cases} 4x - y + z = 4 \\ -6x + 3y - 2z = -5 \\ 2x + 5y - z = 7 \end{cases}$

76. $\begin{cases} 2x + 2y + z = 8 \\ 2x + 3y + z = 7 \\ 6x + 8y + 3z = 22 \end{cases}$

77. $\begin{cases} 2x + y - 2z = 4 \\ 3x - 2y + 4z = 6 \\ -4x + y + 6z = 12 \end{cases}$

78. $\begin{cases} 3x + 3y + z = 4 \\ 2x + 6y + z = 5 \\ -x - 3y + 2z = -5 \end{cases}$

Solving Problems

79. *Investment* A corporation borrows \$1,500,000 to expand its line of clothing. Some of the money is borrowed at 8%, some at 9%, and the remainder at 12%. The annual interest payment to the lenders is \$133,000. The amount borrowed at 8% is four times the amount borrowed at 12%. How much is borrowed at each rate?

80. *Investment* An inheritance of \$25,000 is divided among three investments yielding a total of \$1890 in simple interest per year. The interest rates for the three investments are 5%, 7%, and 10%. The 5% and 7% investments are \$2000 and \$3000 less than the 10% investment, respectively. Find the amount placed in each investment.

✓ **81.** *Ticket Sales* A theater owner wants to sell 1500 total tickets at his three theaters for a total revenue of \$10,050. Tickets cost \$1.50 at Theater A, \$7.50 at Theater B, and \$8.50 at Theater C. Theaters B and C each have twice as many seats as Theater A. How many tickets must be sold at each theater to reach the owner's goal?

82. *Nut Mixture* A grocer wishes to mix three kinds of nuts to obtain 50 pounds of a mixture priced at \$4.10 per pound. Peanuts cost \$3.00 per pound, pecans cost \$4.00 per pound, and cashews cost \$6.00 per pound. Three-quarters of the mixture is composed of peanuts and pecans. How many pounds of each variety should the grocer use?

83. *Number Problem* The sum of three positive numbers is 33. The second number is 3 greater than the first, and the third is four times the first. Find the three numbers.

84. *Number Problem* The sum of three positive numbers is 24. The second number is 4 greater than the first, and the third is three times the first. Find the three numbers.

85. *Production* A company produces computer chips, resistors, and transistors. Each computer chip requires 2 units of copper, 2 units of zinc, and 1 unit of glass. Each resistor requires 1 unit of copper, 3 units of zinc, and 2 units of glass. Each transistor requires 3 units of copper, 2 units of zinc, and 2 units of glass. There are 70 units of copper, 80 units of zinc, and 55 units of glass available for use. Find the numbers of computer chips, resistors, and transistors the company can produce.

86. *Production* A gourmet baked goods company specializes in chocolate muffins, chocolate cookies, and chocolate brownies. Each muffin requires 2 units of chocolate, 3 units of flour, and 2 units of sugar. Each cookie requires 1 unit of chocolate, 1 unit of flour, and 1 unit of sugar. Each brownie requires 2 units of chocolate, 1 unit of flour, and 1.5 units of sugar. There are 550 units of chocolate, 525 units of flour, and 500 units of sugar available for use. Find the numbers of chocolate muffins, chocolate cookies, and chocolate brownies the company can produce.

Investment Portfolio In Exercises 87 and 88, consider an investor with a portfolio totaling \$500,000 that is to be allocated among the following types of investments: certificates of deposit, municipal bonds, blue-chip stocks, and growth or speculative stocks. Use the given conditions to find expressions for the amounts that can be invested in each type of stock. Then find the other amounts when the amount invested in growth stocks is \$100,000.

87. The certificates of deposit pay 10% annually, and the municipal bonds pay 8% annually. Over a five-year period, the investor expects the blue-chip stocks to return 12% annually and the growth stocks to return 13% annually. The investor wants a combined annual return of 10% and also wants to have only one-fourth of the portfolio invested in stocks.

88. The certificates of deposit pay 9% annually, and the municipal bonds pay 5% annually. Over a five-year period, the investor expects the blue-chip stocks to return 12% annually and the growth stocks to return 14% annually. The investor wants a combined annual return of 10% and also wants to have only one-fourth of the portfolio invested in stocks.

Explaining Concepts

89. The entries in a matrix consist of the whole numbers from 1 to 15. The matrix has more than one row and there are more columns than rows. What is the order of the matrix? Explain.

90. Give an example of a matrix in *row-echelon form*. There are many correct answers.

91. Describe the row-echelon form of an augmented matrix that corresponds to a system of linear equations that is inconsistent.

92. Describe the row-echelon form of an augmented matrix that corresponds to a system of linear equations that has an infinite number of solutions.

93. An augmented matrix in row-echelon form represents a system of three variables in three equations that has exactly one solution. The matrix has six nonzero entries, and three of them are in the last column. Discuss the possible entries in the first three columns of this matrix.

94. An augmented matrix in row-echelon form represents a system of three variables in three equations with exactly one solution. What is the smallest number of nonzero entries that this matrix can have? Explain.

Cumulative Review

In Exercises 95–98, evaluate the expression.

95. $6(-7)$

96. $45 \div (-5)$

97. $5(4) - 3(-2)$

98. $\dfrac{(-45) - (-20)}{-5}$

In Exercises 99 and 100, solve the system of linear equations.

99. $\begin{cases} x = 4 \\ 3y + 2z = -4 \\ x + y + z = 3 \end{cases}$

100. $\begin{cases} x - 2y - 3z = 4 \\ 2x + 2y + z = -4 \\ -2x + z = 0 \end{cases}$

8.5 Determinants and Linear Systems

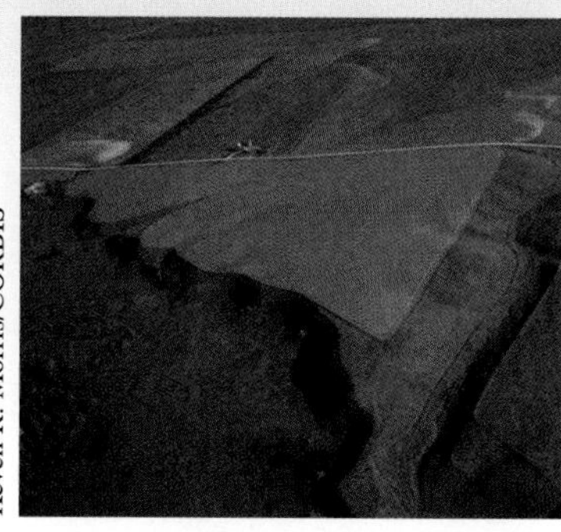
Keven R. Morris/CORBIS

What You Should Learn

1 ▶ Find determinants of 2×2 matrices and 3×3 matrices.

2 ▶ Use determinants and Cramer's Rule to solve systems of linear equations.

3 ▶ Use determinants to find areas of triangles, to test for collinear points, and to find equations of lines.

Why You Should Learn It

You can use determinants and matrices to model and solve real-life problems. For instance, in Exercise 71 on page 535, you can use a matrix to estimate the area of a region of land.

1 ▶ Find determinants of 2×2 matrices and 3×3 matrices.

The Determinant of a Matrix

Associated with each square matrix is a real number called its **determinant.** The use of determinants arose from special number patterns that occur during the solution of systems of linear equations. For instance, the system

$$\begin{cases} a_1x + b_1y = c_1 \\ a_2x + b_2y = c_2 \end{cases}$$

has a solution given by

$$x = \frac{b_2c_1 - b_1c_2}{a_1b_2 - a_2b_1} \quad \text{and} \quad y = \frac{a_1c_2 - a_2c_1}{a_1b_2 - a_2b_1}$$

provided that $a_1b_2 - a_2b_1 \neq 0$. Note that the denominator of each fraction is the same. This denominator is called the **determinant** of the coefficient matrix of the system.

Coefficient Matrix $\quad$ *Determinant*

$$A = \begin{bmatrix} a_1 & b_1 \\ a_2 & b_2 \end{bmatrix} \qquad \det(A) = a_1b_2 - a_2b_1$$

The determinant of the matrix A can also be denoted by vertical bars on both sides of the matrix, as indicated in the following definition.

Study Tip

Note that det(A) and $|A|$ are used interchangeably to represent the determinant of A. Although vertical bars are also used to denote the absolute value of a real number, the context will show which use is intended.

Definition of the Determinant of a 2 × 2 Matrix

$$\det(A) = |A| = \begin{vmatrix} a_1 & b_1 \\ a_2 & b_2 \end{vmatrix} = a_1b_2 - a_2b_1$$

A convenient method for remembering the formula for the determinant of a 2×2 matrix is shown in the diagram below.

$$\det(A) = \begin{vmatrix} a_1 & b_1 \\ a_2 & b_2 \end{vmatrix} = a_1b_2 - a_2b_1$$

Note that the determinant is given by the difference of the products of the two diagonals of the matrix.

EXAMPLE 1 The Determinant of a 2 × 2 Matrix

Find the determinant of each matrix.

a. $A = \begin{bmatrix} 2 & -3 \\ 1 & 4 \end{bmatrix}$ **b.** $B = \begin{bmatrix} -1 & 2 \\ 2 & -4 \end{bmatrix}$ **c.** $C = \begin{bmatrix} 1 & 3 \\ 2 & 5 \end{bmatrix}$

Solution

a. $\det(A) = \begin{vmatrix} 2 & -3 \\ 1 & 4 \end{vmatrix} = 2(4) - 1(-3) = 8 + 3 = 11$

b. $\det(B) = \begin{vmatrix} -1 & 2 \\ 2 & -4 \end{vmatrix} = (-1)(-4) - 2(2) = 4 - 4 = 0$

c. $\det(C) = \begin{vmatrix} 1 & 3 \\ 2 & 5 \end{vmatrix} = 1(5) - 2(3) = 5 - 6 = -1$

✓ ***CHECKPOINT*** *Now try Exercise 1.*

Technology: Tip

A graphing calculator with matrix capabilities can be used to evaluate the determinant of a square matrix. Consult the user's guide of your graphing calculator to learn how to evaluate a determinant. Use the graphing calculator to check the result in Example 1(a). Then try to evaluate the determinant of the 3×3 matrix at the right using a graphing calculator. Finish the evaluation of the determinant by expanding by minors to check the result.

Notice in Example 1 that the determinant of a matrix can be positive, zero, or negative.

One way to evaluate the determinant of a 3×3 matrix, called **expanding by minors,** allows you to write the determinant of a 3×3 matrix in terms of three 2×2 determinants. The **minor** of an entry in a 3×3 matrix is the determinant of the 2×2 matrix that remains after deletion of the row and column in which the entry occurs. Here are three examples.

Determinant	*Entry*	*Minor of Entry*	*Value of Minor*
$\begin{vmatrix} 1 & -1 & 3 \\ 0 & 2 & 5 \\ -2 & 4 & -7 \end{vmatrix}$	1	$\begin{vmatrix} 2 & 5 \\ 4 & -7 \end{vmatrix}$	$2(-7) - 4(5) = -34$
$\begin{vmatrix} 1 & -1 & 3 \\ 0 & 2 & 5 \\ -2 & 4 & -7 \end{vmatrix}$	-1	$\begin{vmatrix} 0 & 5 \\ -2 & -7 \end{vmatrix}$	$0(-7) - (-2)(5) = 10$
$\begin{vmatrix} 1 & -1 & 3 \\ 0 & 2 & 5 \\ -2 & 4 & -7 \end{vmatrix}$	3	$\begin{vmatrix} 0 & 2 \\ -2 & 4 \end{vmatrix}$	$0(4) - (-2)(2) = 4$

Expanding by Minors

$$\det(A) = \begin{vmatrix} a_1 & b_1 & c_1 \\ a_2 & b_2 & c_2 \\ a_3 & b_3 & c_3 \end{vmatrix}$$

$$= a_1(\text{minor of } a_1) - b_1(\text{minor of } b_1) + c_1(\text{minor of } c_1)$$

$$= a_1\begin{vmatrix} b_2 & c_2 \\ b_3 & c_3 \end{vmatrix} - b_1\begin{vmatrix} a_2 & c_2 \\ a_3 & c_3 \end{vmatrix} + c_1\begin{vmatrix} a_2 & b_2 \\ a_3 & b_3 \end{vmatrix}$$

This pattern is called **expanding by minors** along the first row. A similar pattern can be used to expand by minors along any row or column.

$$\begin{bmatrix} + & - & + \\ - & + & - \\ + & - & + \end{bmatrix}$$

Figure 8.13 Sign Pattern for a 3×3 Matrix

The *signs* of the terms used in expanding by minors follow the alternating pattern shown in Figure 8.13. For instance, the signs used to expand by minors along the second row are $-$, $+$, $-$, as shown below.

$$\det(A) = \begin{vmatrix} a_1 & b_1 & c_1 \\ a_2 & b_2 & c_2 \\ a_3 & b_3 & c_3 \end{vmatrix}$$

$$= -a_2(\text{minor of } a_2) + b_2(\text{minor of } b_2) - c_2(\text{minor of } c_2)$$

EXAMPLE 2 Finding the Determinant of a 3×3 Matrix

Find the determinant of $A = \begin{bmatrix} -1 & 1 & 2 \\ 0 & 2 & 3 \\ 3 & 4 & 2 \end{bmatrix}$.

Solution

By expanding by minors along the *first column*, you obtain

$$\det(A) = \begin{vmatrix} -1 & 1 & 2 \\ 0 & 2 & 3 \\ 3 & 4 & 2 \end{vmatrix}$$

$$= (-1)\begin{vmatrix} 2 & 3 \\ 4 & 2 \end{vmatrix} - (0)\begin{vmatrix} 1 & 2 \\ 4 & 2 \end{vmatrix} + (3)\begin{vmatrix} 1 & 2 \\ 2 & 3 \end{vmatrix}$$

$$= (-1)(4 - 12) - (0)(2 - 8) + (3)(3 - 4)$$

$$= 8 - 0 - 3 = 5.$$

✓ ***CHECKPOINT*** *Now try Exercise 13.*

EXAMPLE 3 Finding the Determinant of a 3×3 Matrix

Find the determinant of $A = \begin{bmatrix} 1 & 2 & 1 \\ 3 & 0 & 2 \\ 4 & 0 & -1 \end{bmatrix}$.

Solution

By expanding by minors along the *second column*, you obtain

$$\det(A) = \begin{vmatrix} 1 & 2 & 1 \\ 3 & 0 & 2 \\ 4 & 0 & -1 \end{vmatrix}$$

$$= -(2)\begin{vmatrix} 3 & 2 \\ 4 & -1 \end{vmatrix} + (0)\begin{vmatrix} 1 & 1 \\ 4 & -1 \end{vmatrix} - (0)\begin{vmatrix} 1 & 1 \\ 3 & 2 \end{vmatrix}$$

$$= -(2)(-3 - 8) + 0 - 0 = 22.$$

✓ ***CHECKPOINT*** *Now try Exercise 17.*

A zero entry in a matrix will always yield a zero term when expanding by minors. So, when you are evaluating the determinant of a matrix, choose to expand along the row or column that has the most zero entries.

2 ▶ Use determinants and Cramer's Rule to solve systems of linear equations.

Cramer's Rule

So far in this chapter, you have studied four methods for solving a system of linear equations: graphing, substitution, elimination with equations, and elimination with matrices. You will now learn one more method, called **Cramer's Rule,** which is named after Gabriel Cramer (1704–1752). This rule uses determinants to write the solution of a system of linear equations.

In Cramer's Rule, the value of a variable is expressed as the quotient of two determinants of the coefficient matrix of the system. The numerator is the determinant of the matrix formed by using the column of constants as replacements for the coefficients of the variable. In the definition below, note the notation for the different determinants.

Study Tip

Cramer's Rule is not as general as the elimination method because Cramer's Rule requires that the coefficient matrix of the system be square *and* that the system have exactly one solution.

Cramer's Rule

1. For the system of linear equations

$$\begin{cases} a_1x + b_1y = c_1 \\ a_2x + b_2y = c_2 \end{cases}$$

the solution is given by

$$x = \frac{D_x}{D} = \frac{\begin{vmatrix} c_1 & b_1 \\ c_2 & b_2 \end{vmatrix}}{\begin{vmatrix} a_1 & b_1 \\ a_2 & b_2 \end{vmatrix}}, \qquad y = \frac{D_y}{D} = \frac{\begin{vmatrix} a_1 & c_1 \\ a_2 & c_2 \end{vmatrix}}{\begin{vmatrix} a_1 & b_1 \\ a_2 & b_2 \end{vmatrix}}$$

provided that $D \neq 0$.

2. For the system of linear equations

$$\begin{cases} a_1x + b_1y + c_1z = d_1 \\ a_2x + b_2y + c_2z = d_2 \\ a_3x + b_3y + c_3z = d_3 \end{cases}$$

the solution is given by

$$x = \frac{D_x}{D} = \frac{\begin{vmatrix} d_1 & b_1 & c_1 \\ d_2 & b_2 & c_2 \\ d_3 & b_3 & c_3 \end{vmatrix}}{\begin{vmatrix} a_1 & b_1 & c_1 \\ a_2 & b_2 & c_2 \\ a_3 & b_3 & c_3 \end{vmatrix}}, \qquad y = \frac{D_y}{D} = \frac{\begin{vmatrix} a_1 & d_1 & c_1 \\ a_2 & d_2 & c_2 \\ a_3 & d_3 & c_3 \end{vmatrix}}{\begin{vmatrix} a_1 & b_1 & c_1 \\ a_2 & b_2 & c_2 \\ a_3 & b_3 & c_3 \end{vmatrix}},$$

$$z = \frac{D_z}{D} = \frac{\begin{vmatrix} a_1 & b_1 & d_1 \\ a_2 & b_2 & d_2 \\ a_3 & b_3 & d_3 \end{vmatrix}}{\begin{vmatrix} a_1 & b_1 & c_1 \\ a_2 & b_2 & c_2 \\ a_3 & b_3 & c_3 \end{vmatrix}}, \ D \neq 0.$$

EXAMPLE 4 Using Cramer's Rule for a 2 × 2 System

Use Cramer's Rule to solve the system of linear equations.

$$\begin{cases} 4x - 2y = 10 \\ 3x - 5y = 11 \end{cases}$$

Solution

The determinant of the coefficient matrix is

$$D = \begin{vmatrix} 4 & -2 \\ 3 & -5 \end{vmatrix} = -20 - (-6) = -14.$$

$$x = \frac{D_x}{D} = \frac{\begin{vmatrix} 10 & -2 \\ 11 & -5 \end{vmatrix}}{-14} = \frac{(-50) - (-22)}{-14} = \frac{-28}{-14} = 2$$

$$y = \frac{D_y}{D} = \frac{\begin{vmatrix} 4 & 10 \\ 3 & 11 \end{vmatrix}}{-14} = \frac{44 - 30}{-14} = \frac{14}{-14} = -1$$

The solution is $(2, -1)$. Check this in the original system of equations.

 **CHECKPOINT** *Now try Exercise 37.*

EXAMPLE 5 Using Cramer's Rule for a 3 × 3 System

Use Cramer's Rule to solve the system of linear equations.

$$\begin{cases} -x + 2y - 3z = 1 \\ 2x \qquad\;\; + z = 0 \\ 3x - 4y + 4z = 2 \end{cases}$$

Solution

The determinant of the coefficient matrix is $D = 10$.

$$x = \frac{D_x}{D} = \frac{\begin{vmatrix} 1 & 2 & -3 \\ 0 & 0 & 1 \\ 2 & -4 & 4 \end{vmatrix}}{10} = \frac{8}{10} = \frac{4}{5}$$

$$y = \frac{D_y}{D} = \frac{\begin{vmatrix} -1 & 1 & -3 \\ 2 & 0 & 1 \\ 3 & 2 & 4 \end{vmatrix}}{10} = \frac{-15}{10} = -\frac{3}{2}$$

$$z = \frac{D_z}{D} = \frac{\begin{vmatrix} -1 & 2 & 1 \\ 2 & 0 & 0 \\ 3 & -4 & 2 \end{vmatrix}}{10} = \frac{-16}{10} = -\frac{8}{5}$$

The solution is $\left(\frac{4}{5}, -\frac{3}{2}, -\frac{8}{5}\right)$. Check this in the original system of equations.

CHECKPOINT *Now try Exercise 47.*

Study Tip

When using Cramer's Rule, remember that the method *does not* apply if the determinant of the coefficient matrix is zero.

3 ▶ Use determinants to find areas of triangles, to test for collinear points, and to find equations of lines.

Applications of Determinants

In addition to Cramer's Rule, determinants have many other practical applications. For instance, you can use a determinant to find the area of a triangle whose vertices are given by three points on a rectangular coordinate system.

Area of a Triangle

The area of a triangle with vertices (x_1, y_1), (x_2, y_2), and (x_3, y_3) is

$$\text{Area} = \pm\frac{1}{2}\begin{vmatrix} x_1 & y_1 & 1 \\ x_2 & y_2 & 1 \\ x_3 & y_3 & 1 \end{vmatrix}$$

where the symbol $(\pm)$ indicates that the appropriate sign should be chosen to yield a positive area.

EXAMPLE 6 Finding the Area of a Triangle

Find the area of the triangle whose vertices are $(2, 0)$, $(1, 3)$, and $(3, 2)$, as shown in Figure 8.14.

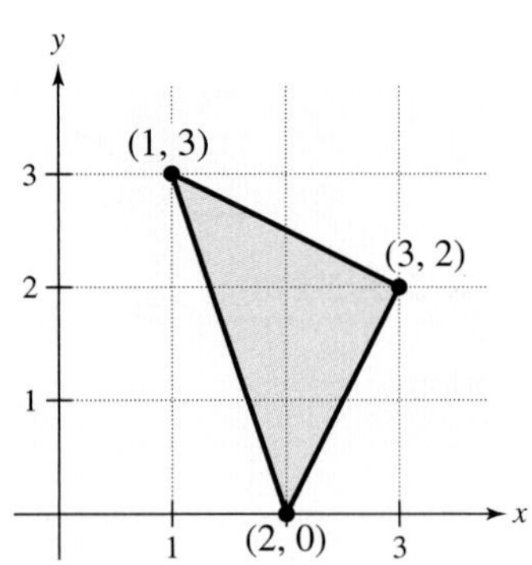

Figure 8.14

Solution

Choose $(x_1, y_1) = (2, 0)$, $(x_2, y_2) = (1, 3)$, and $(x_3, y_3) = (3, 2)$. To find the area of the triangle, evaluate the determinant by expanding by minors along the first row.

$$\begin{vmatrix} x_1 & y_1 & 1 \\ x_2 & y_2 & 1 \\ x_3 & y_3 & 1 \end{vmatrix} = \begin{vmatrix} 2 & 0 & 1 \\ 1 & 3 & 1 \\ 3 & 2 & 1 \end{vmatrix}$$

$$= 2\begin{vmatrix} 3 & 1 \\ 2 & 1 \end{vmatrix} - 0\begin{vmatrix} 1 & 1 \\ 3 & 1 \end{vmatrix} + 1\begin{vmatrix} 1 & 3 \\ 3 & 2 \end{vmatrix}$$

$$= 2(1) - 0 + 1(-7)$$

$$= -5$$

Using this value, you can conclude that the area of the triangle is

$$\text{Area} = -\frac{1}{2}\begin{vmatrix} 2 & 0 & 1 \\ 1 & 3 & 1 \\ 3 & 2 & 1 \end{vmatrix}$$

$$= -\frac{1}{2}(-5) = \frac{5}{2}.$$

✔ ***CHECKPOINT*** *Now try Exercise 59.*

To see the benefit of the "determinant formula," try finding the area of the triangle in Example 6 by using the standard formula:

$$\text{Area} = \frac{1}{2}(\text{Base})(\text{Height}).$$

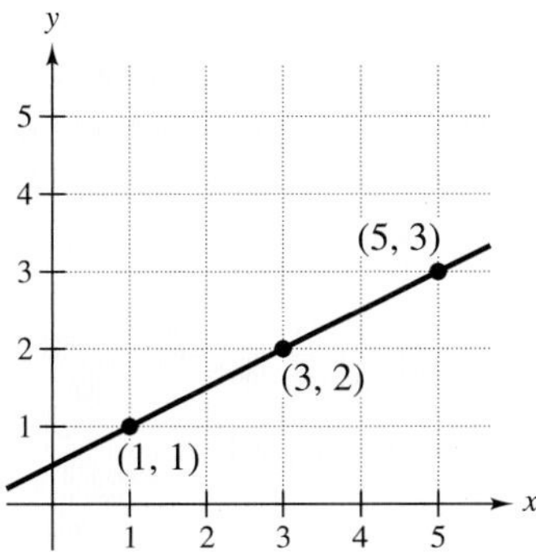

Figure 8.15

Suppose the three points in Example 6 had been on the same line. What would have happened in applying the area formula to three such points? The answer is that the determinant would have been zero. Consider, for instance, the three collinear points (1, 1), (3, 2), and (5, 3), as shown in Figure 8.15. The area of the "triangle" that has these three points as vertices is

$$\frac{1}{2}\begin{vmatrix} 1 & 1 & 1 \\ 3 & 2 & 1 \\ 5 & 3 & 1 \end{vmatrix} = \frac{1}{2}\left(1\begin{vmatrix} 2 & 1 \\ 3 & 1 \end{vmatrix} - 1\begin{vmatrix} 3 & 1 \\ 5 & 1 \end{vmatrix} + 1\begin{vmatrix} 3 & 2 \\ 5 & 3 \end{vmatrix}\right)$$

$$= \frac{1}{2}[-1 - (-2) + (-1)]$$

$$= 0.$$

This result is generalized as follows.

Test for Collinear Points

Three points (x_1, y_1), (x_2, y_2), and (x_3, y_3) are collinear (lie on the same line) if and only if

$$\begin{vmatrix} x_1 & y_1 & 1 \\ x_2 & y_2 & 1 \\ x_3 & y_3 & 1 \end{vmatrix} = 0.$$

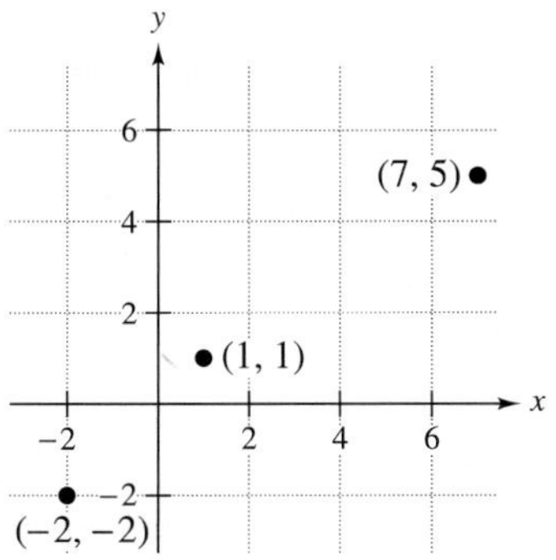

EXAMPLE 7 Testing for Collinear Points

Determine whether the points $(-2, -2)$, $(1, 1)$, and $(7, 5)$ are collinear. (See Figure 8.16.)

Figure 8.16

Solution

Letting $(x_1, y_1) = (-2, -2)$, $(x_2, y_2) = (1, 1)$, and $(x_3, y_3) = (7, 5)$, you have

$$\begin{vmatrix} x_1 & y_1 & 1 \\ x_2 & y_2 & 1 \\ x_3 & y_3 & 1 \end{vmatrix} = \begin{vmatrix} -2 & -2 & 1 \\ 1 & 1 & 1 \\ 7 & 5 & 1 \end{vmatrix}$$

$$= -2\begin{vmatrix} 1 & 1 \\ 5 & 1 \end{vmatrix} - (-2)\begin{vmatrix} 1 & 1 \\ 7 & 1 \end{vmatrix} + 1\begin{vmatrix} 1 & 1 \\ 7 & 5 \end{vmatrix}$$

$$= -2(-4) - (-2)(-6) + 1(-2)$$

$$= -6.$$

Because the value of this determinant *is not* zero, you can conclude that the three points *do not* lie on the same line and so are not collinear.

CHECKPOINT *Now try Exercise 73.*

As a good review, look at how the slope can be used to verify the result in Example 7. Label the points $A(-2, -2)$, $B(1, 1)$, and $C(7, 5)$. Because the slopes from A to B and from A to C are different, the points are not collinear.

You can also use determinants to find the equation of a line through two points. In this case, the first row consists of the variables x and y and the number 1. By expanding by minors along the first row, the resulting 2×2 determinants are the coefficients of the variables x and y and the constant of the linear equation, as shown in Example 8.

Two-Point Form of the Equation of a Line

An equation of the line passing through the distinct points (x_1, y_1) and (x_2, y_2) is given by

$$\begin{vmatrix} x & y & 1 \\ x_1 & y_1 & 1 \\ x_2 & y_2 & 1 \end{vmatrix} = 0.$$

EXAMPLE 8 Finding an Equation of a Line

Find an equation of the line passing through $(-2, 1)$ and $(3, -2)$.

Solution

Applying the determinant formula for the equation of a line produces

$$\begin{vmatrix} x & y & 1 \\ -2 & 1 & 1 \\ 3 & -2 & 1 \end{vmatrix} = 0.$$

To evaluate this determinant, you can expand by minors along the first row to obtain the following.

$$x\begin{vmatrix} 1 & 1 \\ -2 & 1 \end{vmatrix} - y\begin{vmatrix} -2 & 1 \\ 3 & 1 \end{vmatrix} + 1\begin{vmatrix} -2 & 1 \\ 3 & -2 \end{vmatrix} = 0$$

$$3x + 5y + 1 = 0$$

So, an equation of the line is $3x + 5y + 1 = 0$.

✓ **CHECKPOINT** *Now try Exercise 79.*

Note that this method of finding the equation of a line works for all lines, including horizontal and vertical lines, as shown below.

Vertical Line Through $(2, 0)$ *and* $(2, 2)$:

$$\begin{vmatrix} x & y & 1 \\ 2 & 0 & 1 \\ 2 & 2 & 1 \end{vmatrix} = 0$$

$$-2x - 0y + 4 = 0$$

$$-2x = -4$$

$$x = 2$$

Horizontal Line Through $(-3, 4)$ *and* $(2, 4)$:

$$\begin{vmatrix} x & y & 1 \\ -3 & 4 & 1 \\ 2 & 4 & 1 \end{vmatrix} = 0$$

$$0x + 5y - 20 = 0$$

$$5y = 20$$

$$y = 4$$

Concept Check

1. The determinant of a matrix can be represented by vertical bars, similar to the vertical bars used for absolute value. Does this mean that every determinant is nonnegative?

2. Is it possible to find the determinant of a 2×3 matrix? Explain.

3. If one column of a 3×3 matrix is all zeros, what is the determinant of the matrix? Explain.

4. Can Cramer's Rule be used to solve any system of linear equations? Explain.

8.5 EXERCISES

Go to pages 546–547 to record your assignments.

Developing Skills

In Exercises 1–12, find the determinant of the matrix. *See Example 1.*

✓ **1.** $\begin{bmatrix} 2 & 1 \\ 3 & 4 \end{bmatrix}$

2. $\begin{bmatrix} -3 & 1 \\ 5 & 2 \end{bmatrix}$

3. $\begin{bmatrix} 5 & 2 \\ -6 & 3 \end{bmatrix}$

4. $\begin{bmatrix} 2 & -2 \\ 4 & 3 \end{bmatrix}$

5. $\begin{bmatrix} -4 & 0 \\ 9 & 0 \end{bmatrix}$

6. $\begin{bmatrix} 4 & -3 \\ 0 & 0 \end{bmatrix}$

7. $\begin{bmatrix} 3 & -3 \\ -6 & 6 \end{bmatrix}$

8. $\begin{bmatrix} -2 & 3 \\ 6 & -9 \end{bmatrix}$

9. $\begin{bmatrix} -7 & 6 \\ \frac{1}{2} & 3 \end{bmatrix}$

10. $\begin{bmatrix} \frac{2}{3} & \frac{5}{6} \\ 14 & -2 \end{bmatrix}$

11. $\begin{bmatrix} 0.4 & 0.7 \\ 0.7 & 0.4 \end{bmatrix}$

12. $\begin{bmatrix} -1.2 & 4.5 \\ 0.4 & -0.9 \end{bmatrix}$

In Exercises 13–30, evaluate the determinant of the matrix. Expand by minors along the row or column that appears to make the computation easiest. *See Examples 2 and 3.*

✓ **13.** $\begin{bmatrix} 2 & 3 & -1 \\ 6 & 0 & 0 \\ 4 & 1 & 1 \end{bmatrix}$

14. $\begin{bmatrix} 10 & 2 & -4 \\ 8 & 0 & -2 \\ 4 & 0 & 2 \end{bmatrix}$

15. $\begin{bmatrix} 1 & 1 & 2 \\ 3 & 1 & 0 \\ -2 & 0 & 3 \end{bmatrix}$

16. $\begin{bmatrix} 2 & 1 & 3 \\ 1 & 4 & 4 \\ 1 & 0 & 2 \end{bmatrix}$

✓ **17.** $\begin{bmatrix} 2 & 4 & 6 \\ 0 & 3 & 1 \\ 0 & 0 & -5 \end{bmatrix}$

18. $\begin{bmatrix} 2 & 3 & 1 \\ 0 & 5 & -2 \\ 0 & 0 & -2 \end{bmatrix}$

19. $\begin{bmatrix} -2 & 2 & 3 \\ 1 & -1 & 0 \\ 0 & 1 & 4 \end{bmatrix}$

20. $\begin{bmatrix} -2 & 3 & 0 \\ 3 & 1 & -4 \\ 0 & 4 & 2 \end{bmatrix}$

21. $\begin{bmatrix} 1 & 4 & -2 \\ 3 & 6 & -6 \\ -2 & 1 & 4 \end{bmatrix}$

22. $\begin{bmatrix} 2 & -1 & 0 \\ 4 & 2 & 1 \\ 4 & 2 & 1 \end{bmatrix}$

23. $\begin{bmatrix} 2 & -2 & 7 \\ 1 & -3 & -2 \\ -2 & 6 & 4 \end{bmatrix}$

24. $\begin{bmatrix} 6 & 8 & -7 \\ 0 & 0 & 0 \\ 4 & -6 & 22 \end{bmatrix}$

25. $\begin{bmatrix} 2 & -5 & 0 \\ 4 & 7 & 0 \\ -7 & 25 & 3 \end{bmatrix}$

26. $\begin{bmatrix} 8 & 7 & 6 \\ -4 & 0 & 0 \\ 5 & 1 & 4 \end{bmatrix}$

27. $\begin{bmatrix} 0.1 & 0.2 & 0.3 \\ -0.3 & 0.2 & 0.2 \\ 5 & 4 & 4 \end{bmatrix}$

28. $\begin{bmatrix} -0.4 & 0.4 & 0.3 \\ 0.2 & 0.2 & 0.2 \\ 0.3 & 0.2 & 0.2 \end{bmatrix}$

29. $\begin{bmatrix} x & y & 1 \\ 3 & 1 & 1 \\ -2 & 0 & 1 \end{bmatrix}$

30. $\begin{bmatrix} x & y & 1 \\ -2 & -2 & 1 \\ 1 & 5 & 1 \end{bmatrix}$

In Exercises 31–36, use a graphing calculator to evaluate the determinant of the matrix.

31. $\begin{bmatrix} 5 & -3 & 2 \\ 7 & 5 & -7 \\ 0 & 6 & -1 \end{bmatrix}$

32. $\begin{bmatrix} 3 & -1 & 2 \\ 1 & -1 & 2 \\ -2 & 3 & 10 \end{bmatrix}$

33. $\begin{bmatrix} -\frac{1}{2} & -1 & 6 \\ 8 & -\frac{1}{4} & -4 \\ 1 & 2 & 1 \end{bmatrix}$

34. $\begin{bmatrix} \frac{1}{2} & \frac{3}{2} & \frac{1}{2} \\ 4 & 8 & 10 \\ -2 & -6 & 12 \end{bmatrix}$

35. $\begin{bmatrix} 0.6 & 0.4 & -0.6 \\ 0.1 & 0.5 & -0.3 \\ 8 & -2 & 12 \end{bmatrix}$

36. $\begin{bmatrix} 0.4 & 0.3 & 0.3 \\ -0.2 & 0.6 & 0.6 \\ 3 & 1 & 1 \end{bmatrix}$

In Exercises 37–52, use Cramer's Rule to solve the system of linear equations. (If not possible, state the reason.) ***See Examples 4 and 5.***

37. $\begin{cases} x + 2y = 5 \\ -x + y = 1 \end{cases}$

38. $\begin{cases} 2x - y = -10 \\ 3x + 2y = -1 \end{cases}$

39. $\begin{cases} 3x + 4y = -2 \\ 5x + 3y = 4 \end{cases}$

40. $\begin{cases} 3x + 2y = -3 \\ 4x + 5y = -11 \end{cases}$

41. $\begin{cases} 20x + 8y = 11 \\ 12x - 24y = 21 \end{cases}$

42. $\begin{cases} 13x - 6y = 17 \\ 26x - 12y = 8 \end{cases}$

43. $\begin{cases} -0.4x + 0.8y = 1.6 \\ 2x - 4y = 5 \end{cases}$

44. $\begin{cases} -0.4x + 0.8y = 1.6 \\ 0.2x + 0.3y = 2.2 \end{cases}$

45. $\begin{cases} 3u + 6v = 5 \\ 6u + 14v = 11 \end{cases}$

46. $\begin{cases} 3x_1 + 2x_2 = 1 \\ 2x_1 + 10x_2 = 6 \end{cases}$

47. $\begin{cases} 4x - y + z = -5 \\ 2x + 2y + 3z = 10 \\ 5x - 2y + 6z = 1 \end{cases}$

48. $\begin{cases} 4x - 2y + 3z = -2 \\ 2x + 2y + 5z = 16 \\ 8x - 5y - 2z = 4 \end{cases}$

49. $\begin{cases} 4a + 3b + 4c = 1 \\ 4a - 6b + 8c = 8 \\ -a + 9b - 2c = -7 \end{cases}$

50. $\begin{cases} 2x + 3y + 5z = 4 \\ 3x + 5y + 9z = 7 \\ 5x + 9y + 17z = 13 \end{cases}$

51. $\begin{cases} 5x - 3y + 2z = 2 \\ 2x + 2y - 3z = 3 \\ x - 7y + 8z = -4 \end{cases}$

52. $\begin{cases} 5x + 4y - 6z = -10 \\ -4x + 2y + 3z = -1 \\ 8x + 4y + 12z = 2 \end{cases}$

In Exercises 53–56, solve the system of linear equations using a graphing calculator and Cramer's Rule. ***See Examples 4 and 5.***

53. $\begin{cases} -3x + 10y = 22 \\ 9x - 3y = 0 \end{cases}$

54. $\begin{cases} 3x + 7y = 3 \\ 7x + 25y = 11 \end{cases}$

55. $\begin{cases} 3x - 2y + 3z = 8 \\ x + 3y + 6z = -3 \\ x + 2y + 9z = -5 \end{cases}$

56. $\begin{cases} 6x + 4y - 8z = -22 \\ -2x + 2y + 3z = 13 \\ -2x + 2y - z = 5 \end{cases}$

In Exercises 57 and 58, solve the equation.

57. $\begin{vmatrix} -3x & x \\ 4 & 3 \end{vmatrix} = 26$

58. $\begin{vmatrix} -8 & x \\ 6 & -x \end{vmatrix} = 6$

Solving Problems

Area of a Triangle In Exercises 59–66, use a determinant to find the area of the triangle with the given vertices. ***See Example 6.***

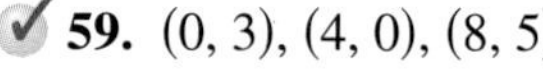

59. $(0, 3), (4, 0), (8, 5)$

60. $(2, 0), (0, 5), (6, 3)$

61. $(-3, 4), (1, -2), (6, 1)$

62. $(-2, -3), (2, -3), (0, 4)$

63. $(-2, 1), (3, -1), (1, 6)$

64. $(-1, 4), (-4, 0), (1, 3)$

65. $\left(0, \frac{1}{2}\right), \left(\frac{5}{2}, 0\right), (4, 3)$

66. $\left(\frac{1}{4}, 0\right), \left(0, \frac{3}{4}\right), (8, -2)$

Area of a Region In Exercises 67–70, find the area of the shaded region of the figure.

67. **68.**

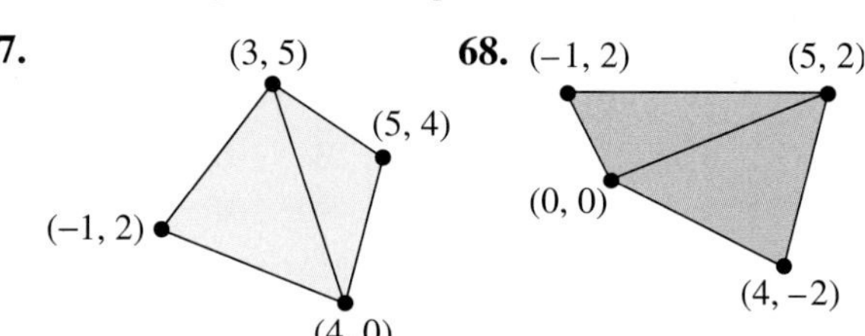

69. **70.**

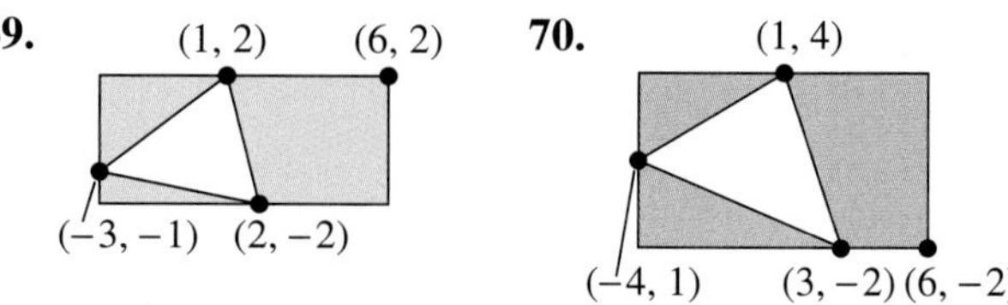

71. *Area of a Region* A large region of forest has been infested with gypsy moths. The region is roughly triangular, as shown in the figure. Find the area of this region. (*Note:* The measurements in the figure are in miles.)

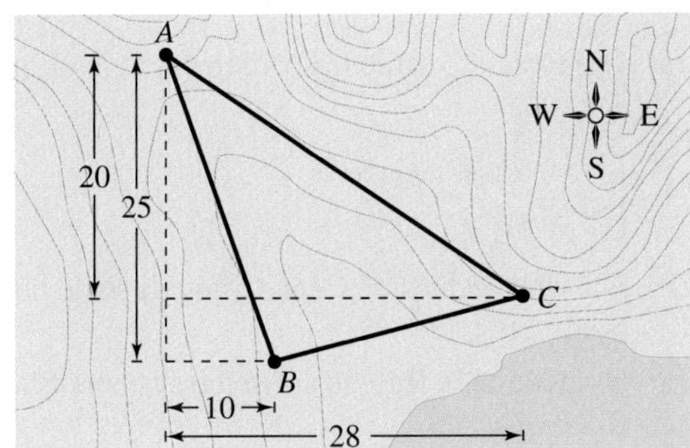

72. *Area of a Region* You have purchased a triangular tract of land, as shown in the figure. What is the area of this tract of land? (*Note:* The measurements in the figure are in feet.)

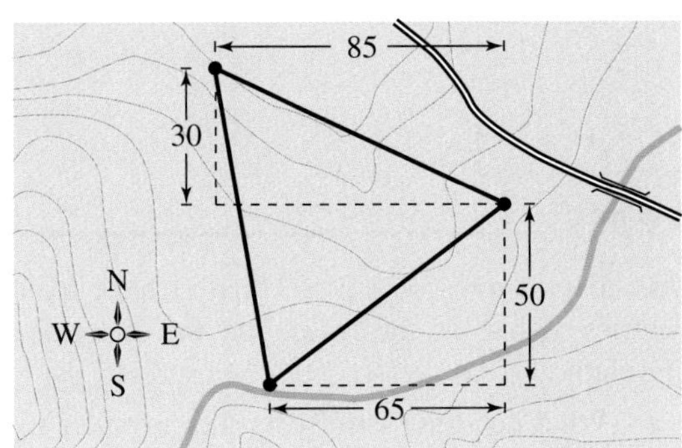

Collinear Points In Exercises 73–78, determine whether the points are collinear. ***See Example 7.***

73. $(-1, 11), (0, 8), (2, 2)$

74. $(-1, -1), (1, 9), (2, 13)$

75. $(2, -4), (5, 2), (10, 10)$

76. $(1, 8), (3, 2), (6, -7)$

77. $\left(-2, \frac{1}{3}\right), (2, 1), \left(3, \frac{1}{5}\right)$

78. $\left(0, \frac{1}{2}\right), \left(1, \frac{7}{6}\right), \left(9, \frac{13}{2}\right)$

Equation of a Line In Exercises 79–86, use a determinant to find the equation of the line through the points. ***See Example 8.***

79. $(-2, -1), (4, 2)$

80. $(-1, 3), (2, -6)$

81. $(10, 7), (-2, -7)$

82. $(-8, 3), (4, 6)$

83. $\left(-2, \frac{3}{2}\right), (3, -3)$

84. $\left(-\frac{1}{2}, 3\right), \left(\frac{5}{2}, 1\right)$

85. $(2, 3.6), (8, 10)$

86. $(3, 1.6), (5, -2.2)$

87. *Electrical Networks* Laws that deal with electrical currents are known as *Kirchhoff's Laws.* When Kirchhoff's Laws are applied to the electrical network shown in the figure, the currents I_1, I_2, and I_3 are the solution of the system

$$\begin{cases} I_1 - I_2 + I_3 = 0 \\ 3I_1 + 2I_2 = 7 \\ 2I_2 + 4I_3 = 8. \end{cases}$$

Find the currents.

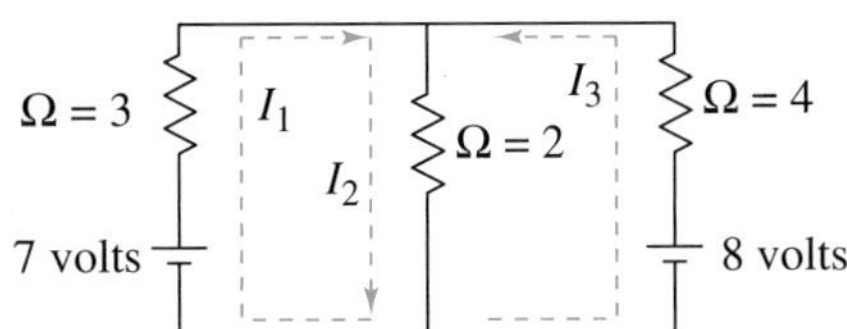

© Jose Luis Pelaez, Inc./CORBIS

88. *Force* When three forces are applied to a beam, Newton's Laws suggest that the forces F_1, F_2, and F_3 are the solution of the system

$$\begin{cases} 3F_1 + F_2 - F_3 = 2 \\ F_1 - 2F_2 + F_3 = 0 \\ 4F_1 - F_2 + F_3 = 0. \end{cases}$$

Find the forces.

89. *Electrical Networks* When Kirchhoff's Laws are applied to the electrical network shown in the figure, the currents I_1, I_2, and I_3 are the solution of the system

$$\begin{cases} I_1 + I_2 - I_3 = 0 \\ I_1 + 2I_3 = 12 \\ I_1 - 2I_2 = -4. \end{cases}$$

Find the currents.

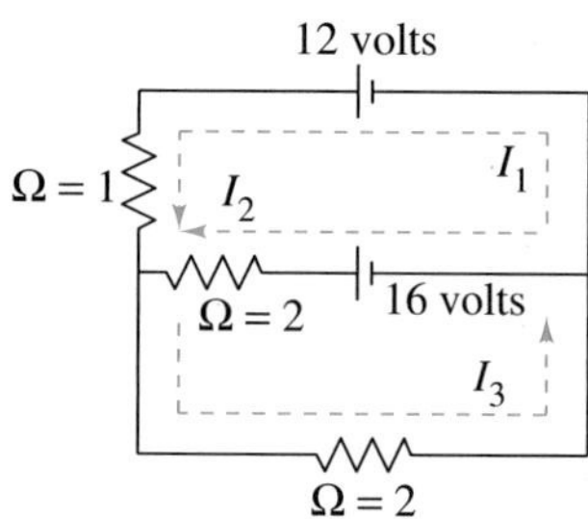

90. *Electrical Networks* When Kirchhoff's Laws are applied to the electrical network shown in the figure, the currents I_1, I_2, and I_3 are the solution of the system

$$\begin{cases} I_1 - I_2 + I_3 = 0 \\ I_2 + 4I_3 = 8 \\ 4I_1 + I_2 = 16. \end{cases}$$

Find the currents.

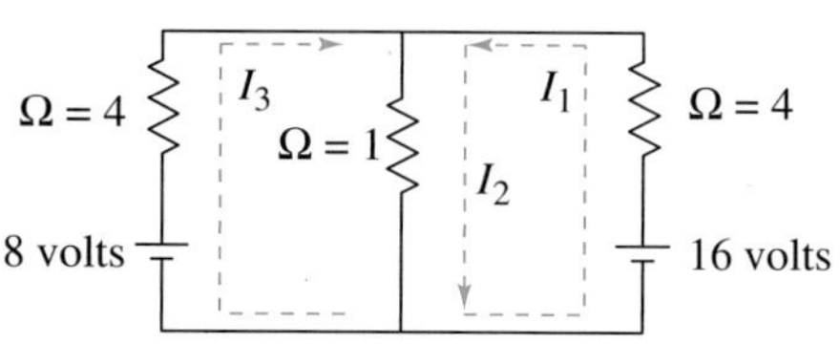

Figure for 90

91. (a) Use Cramer's Rule to solve the system of linear equations.

$$\begin{cases} kx + 3ky = 2 \\ (2 + k)x + ky = 5 \end{cases}$$

(b) For what values of k can Cramer's Rule not be used?

92. (a) Use Cramer's Rule to solve the system of linear equations.

$$\begin{cases} kx + (1 - k)y = 1 \\ (1 - k)x + ky = 3 \end{cases}$$

(b) For what value(s) of k will the system be inconsistent?

Explaining Concepts

93. Explain the difference between a square matrix and its determinant.

94. What is meant by the minor of an entry of a square matrix?

95. If two rows of a 3×3 matrix have identical entries, what is the value of the determinant? Explain.

96. What conditions must be met in order to use Cramer's Rule to solve a system of linear equations?

Cumulative Review

In Exercises 97–100, sketch the graph of the solution of the linear inequality.

97. $4x - 2y < 0$

98. $2x + 8y \geq 0$

99. $-x + 3y > 12$

100. $-3x - y \leq 2$

In Exercises 101–104, write the general form of the equation of the line that passes through the two points.

101. $(0, 0), (4, 2)$

102. $(1, 2), (6, 3)$

103. $(-1, 2), (5, 2)$

104. $(-3, 3), (8, -6)$

In Exercises 105–108, determine if the set of ordered pairs is a function.

105. $\{(0, 0), (2, 1), (4, 2), (6, 3)\}$

106. $\{(0, 2), (1, 4), (4, 1), (0, 4)\}$

107. $\{(-4, 5), (-1, 0), (3, -2), (3, -4)\}$

108. $\{(-3, 1), (-1, 3), (1, 3), (3, 1)\}$

8.6 Systems of Linear Inequalities

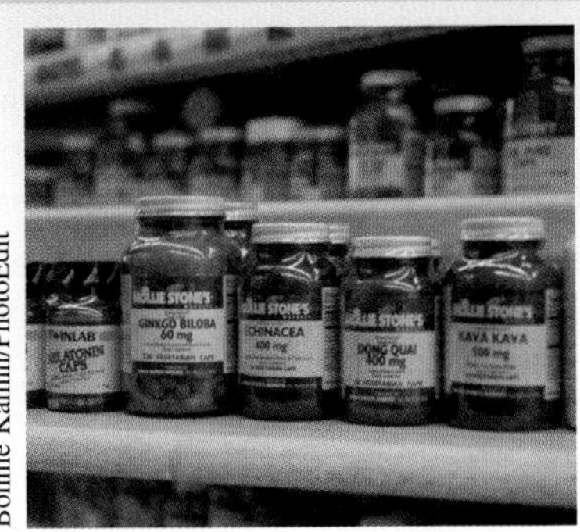

Bonnie Kamin/PhotoEdit

What You Should Learn

1 ▶ Solve systems of linear inequalities in two variables.

2 ▶ Use systems of linear inequalities to model and solve real-life problems.

Why You Should Learn It

Systems of linear inequalities can be used to model and solve real-life problems. For instance, in Exercise 61 on page 544, a system of linear inequalities can be used to analyze the compositions of dietary supplements.

1 ▶ Solve systems of linear inequalities in two variables.

Systems of Linear Inequalities in Two Variables

You have already graphed linear inequalities in two variables. However, many practical problems in business, science, and engineering involve **systems of linear inequalities.** This type of system arises in problems that have *constraint* statements that contain phrases such as "more than," "less than," "at least," "no more than," "a minimum of," and "a maximum of." A **solution** of a system of linear inequalities in x and y is a point (x, y) that satisfies each inequality in the system.

To sketch the graph of a system of inequalities in two variables, first sketch (on the same coordinate system) the graph of each individual inequality. The **solution set** is the region that is *common* to every graph in the system.

EXAMPLE 1 Graphing a System of Linear Inequalities

Sketch the graph of the system of linear inequalities.

$$\begin{cases} 2x - y \le 5 \\ x + 2y \ge 2 \end{cases}$$

Solution

Begin by rewriting each inequality in slope-intercept form. Then sketch the line for the corresponding equation of each inequality. See Figures 8.17–8.19.

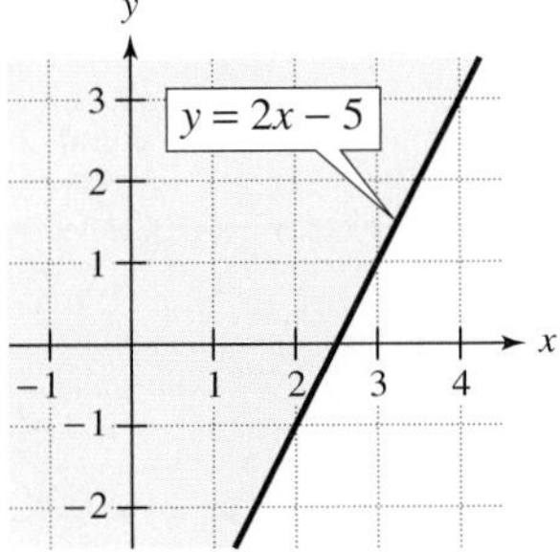

Graph of $2x - y \le 5$ is all points on and above $y = 2x - 5$.

Figure 8.17

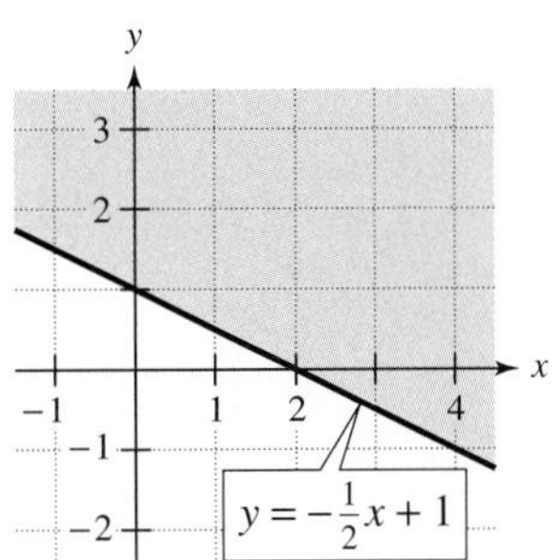

Graph of $x + 2y \ge 2$ is all points on and above $y = -\frac{1}{2}x + 1$.

Figure 8.18

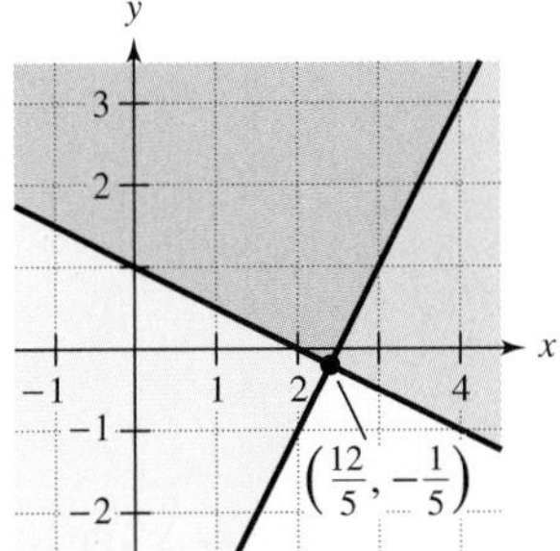

Graph of system is the purple wedge-shaped region.

Figure 8.19

✓ **CHECKPOINT** *Now try Exercise 13.*

In Figure 8.19, note that the two borderlines of the region

$$y = 2x - 5 \quad \text{and} \quad y = -\frac{1}{2}x + 1$$

intersect at the point $\left(\frac{12}{5}, -\frac{1}{5}\right)$. Such a point is called a **vertex** of the region. The region shown in the figure has only one vertex. Some regions, however, have several vertices. When you are sketching the graph of a system of linear inequalities, it is helpful to find and label any vertices of the region.

Graphing a System of Linear Inequalities

1. Sketch the line that corresponds to each inequality. (Use dashed lines for inequalities with $<$ or $>$ and solid lines for inequalities with $\le$ or $\ge$.)
2. Lightly shade the half-plane that is the graph of each linear inequality. (Colored pencils may help distinguish different half-planes.)
3. The graph of the system is the intersection of the half-planes. (If you use colored pencils, it is the region that is selected with *every* color.)

EXAMPLE 2 Graphing a System of Linear Inequalities

Sketch the graph of the system of linear inequalities: $\begin{cases} y < 4 \\ y > 1 \end{cases}$.

Solution

The graph of the first inequality is the half-plane below the horizontal line

$y = 4.$ Upper boundary

The graph of the second inequality is the half-plane above the horizontal line

$y = 1.$ Lower boundary

The graph of the system is the horizontal band that lies *between* the two horizontal lines (where $y < 4$ *and* $y > 1$), as shown in Figure 8.20.

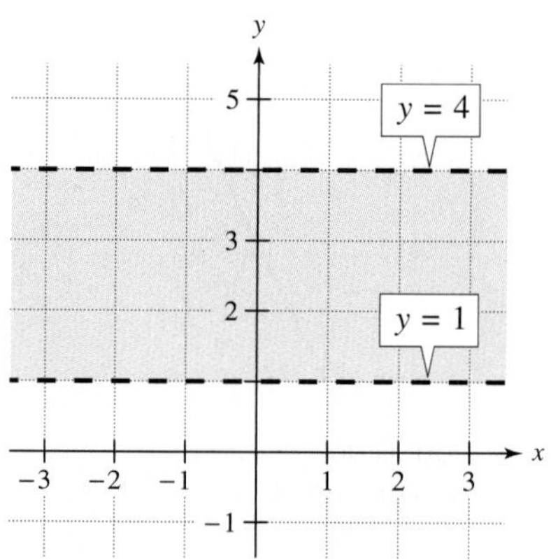

Figure 8.20

✓ ***CHECKPOINT*** *Now try Exercise 11.*

EXAMPLE 3 Graphing a System of Linear Inequalities

Sketch the graph of the system of linear inequalities, and label the vertices.

$$\begin{cases} x - y < 2 \\ x > -2 \\ y \leq 3 \end{cases}$$

Solution

Begin by sketching the half-planes represented by the three linear inequalities. The graph of

$$x - y < 2$$

is the half-plane lying above the line $y = x - 2$, the graph of

$$x > -2$$

is the half-plane lying to the right of the line $x = -2$, and the graph of

$$y \leq 3$$

is the half-plane lying on and below the line $y = 3$. As shown in Figure 8.21, the region that is common to all three of these half-planes is a triangle. The vertices of the triangle are found as follows.

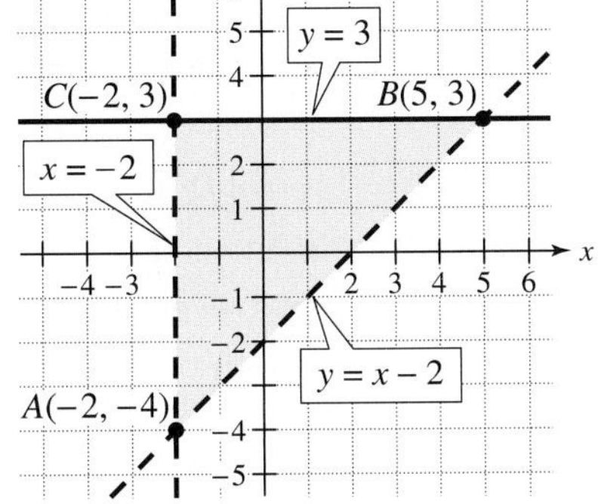

Figure 8.21

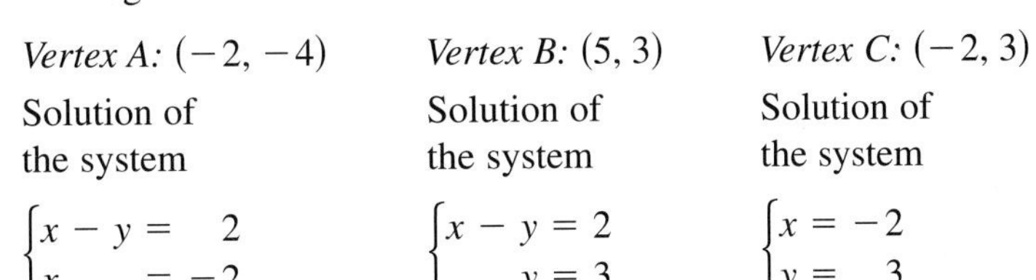

Vertex A: $(-2, -4)$	*Vertex B:* $(5, 3)$	*Vertex C:* $(-2, 3)$
Solution of the system	Solution of the system	Solution of the system
$\begin{cases} x - y = 2 \\ x = -2 \end{cases}$	$\begin{cases} x - y = 2 \\ y = 3 \end{cases}$	$\begin{cases} x = -2 \\ y = 3 \end{cases}$

✓ **CHECKPOINT** *Now try Exercise 31.*

For the triangular region shown in Figure 8.21, each point of intersection of a pair of boundary lines corresponds to a vertex. With more complicated regions, two border lines can sometimes intersect at a point that is not a vertex of the region, as shown in Figure 8.22. To keep track of which points of intersection are actually vertices of the region, you should sketch the region and refer to your sketch as you find each point of intersection.

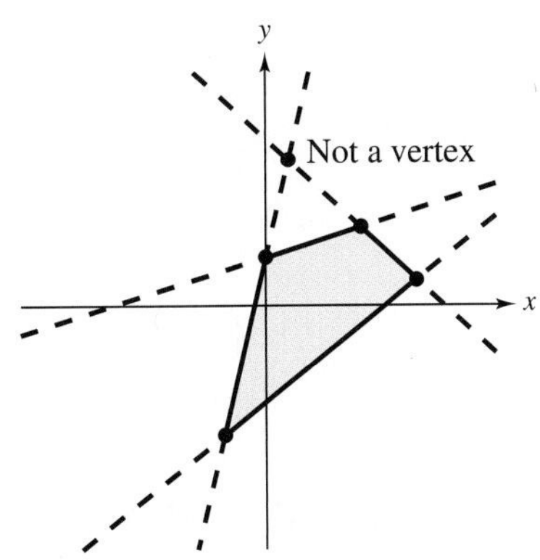

Figure 8.22

EXAMPLE 4 Graphing a System of Linear Inequalities

Sketch the graph of the system of linear inequalities, and label the vertices.

$$\begin{cases} x + y \le 5 \\ 3x + 2y \le 12 \\ x \ge 0 \\ y \ge 0 \end{cases}$$

Solution

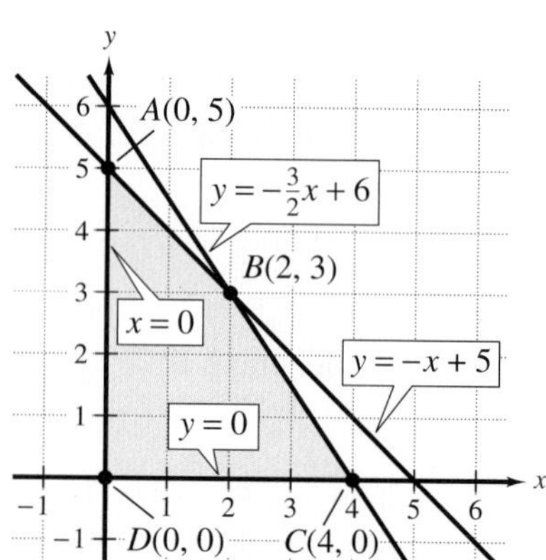

Figure 8.23

Begin by sketching the half-planes represented by the four linear inequalities. The graph of $x + y \le 5$ is the half-plane lying on and below the line $y = -x + 5$. The graph of $3x + 2y \le 12$ is the half-plane lying on and below the line $y = -\frac{3}{2}x + 6$. The graph of $x \ge 0$ is the half-plane lying on and to the right of the y-axis, and the graph of $y \ge 0$ is the half-plane lying on and above the x-axis. As shown in Figure 8.23, the region that is common to all four of these half-planes is a four-sided polygon. The vertices of the region are found as follows.

Vertex A: (0, 5)	*Vertex B:* (2, 3)	*Vertex C:* (4, 0)	*Vertex D:* (0, 0)
Solution of the system	Solution of the system	Solution of the system	Solution of the system
$\begin{cases} x + y = 5 \\ x = 0 \end{cases}$	$\begin{cases} x + y = 5 \\ 3x + 2y = 12 \end{cases}$	$\begin{cases} 3x + 2y = 12 \\ y = 0 \end{cases}$	$\begin{cases} x = 0 \\ y = 0 \end{cases}$

EXAMPLE 5 Finding the Boundaries of a Region

Find a system of inequalities that defines the region shown in Figure 8.24.

Solution

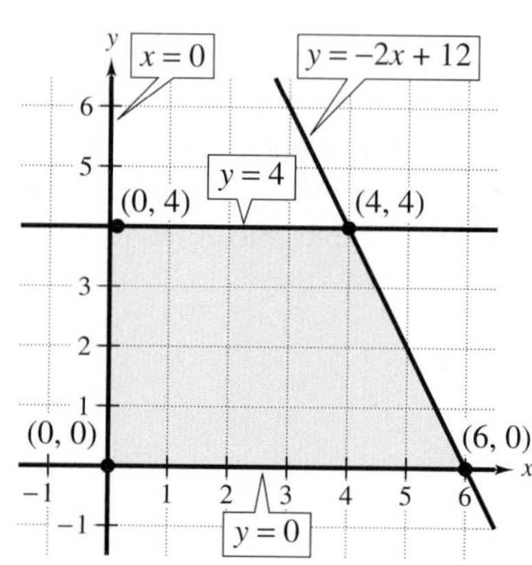

Figure 8.24

Three of the boundaries of the region are horizontal or vertical—they are easy to find. To find the diagonal boundary line, you can use the techniques of Section 4.5 to find the equation of the line passing through the points (4, 4) and (6, 0). Use the formula for slope to find $m = -2$, and then use the point-slope form with point (6, 0) and $m = -2$ to obtain

$$y - 0 = -2(x - 6).$$

So, the equation is $y = -2x + 12$. The system of linear inequalities that describes the region is as follows.

$$\begin{cases} y \le 4 & \text{Region lies on and below line } y = 4. \\ y \ge 0 & \text{Region lies on and above } x\text{-axis.} \\ x \ge 0 & \text{Region lies on and to the right of } y\text{-axis.} \\ y \le -2x + 12 & \text{Region lies on and below line } y = -2x + 12. \end{cases}$$

Technology: Tip

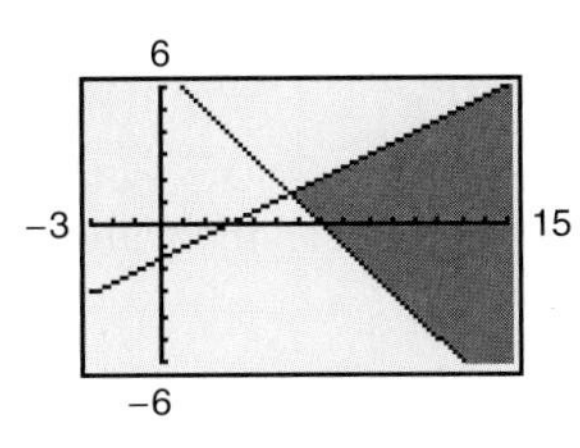

A graphing calculator can be used to graph a system of linear inequalities. The graph of

$$\begin{cases} 4y < 2x - 6 \\ x + y \geq 7 \end{cases}$$

is shown at the left. The grey shaded region, in which all points satisfy both inequalities, is the solution of the system. Try using a graphing calculator to graph

$$\begin{cases} 3x + y < 1 \\ -2x - 2y < 8. \end{cases}$$

2 ▶ Use systems of linear inequalities to model and solve real-life problems.

Application

EXAMPLE 6 Nutrition 

The minimum daily requirements for the liquid portion of a diet are 300 calories, 36 units of vitamin A, and 90 units of vitamin C. A cup of dietary drink X provides 60 calories, 12 units of vitamin A, and 10 units of vitamin C. A cup of dietary drink Y provides 60 calories, 6 units of vitamin A, and 30 units of vitamin C. Write a system of linear inequalities that describes how many cups of each drink should be consumed each day to meet the minimum daily requirements for calories and vitamins.

Solution

Begin by letting x and y represent the following.

$x =$ number of cups of dietary drink X

$y =$ number of cups of dietary drink Y

To meet the minimum daily requirements, the following inequalities must be satisfied.

$$\begin{cases} 60x + 60y \geq 300 & \text{Calories} \\ 12x + 6y \geq 36 & \text{Vitamin A} \\ 10x + 30y \geq 90 & \text{Vitamin C} \\ x \geq 0 \\ y \geq 0 \end{cases}$$

The last two inequalities are included because x and y cannot be negative. The graph of this system of inequalities is shown in Figure 8.25.

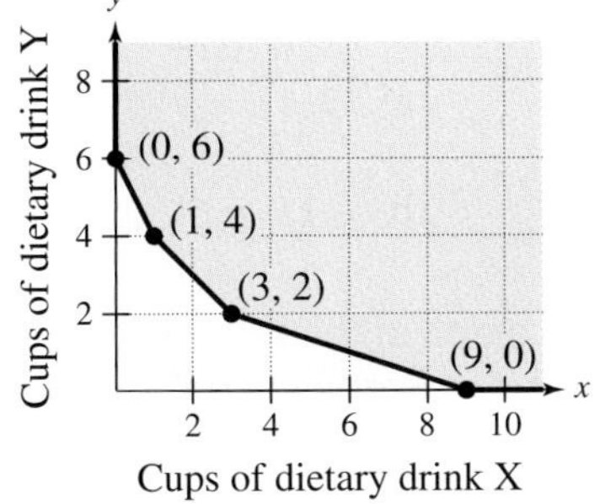

Figure 8.25

✓ **CHECKPOINT** *Now try Exercise 57.*

Concept Check

1. What is a system of linear inequalities in two variables?

2. Explain when you should use dashed lines and when you should use solid lines in sketching a system of linear inequalities.

3. Does the point of intersection of each pair of boundary lines correspond to a vertex? Explain.

4. Is it possible for a system of linear inequalities to have no solution? Explain.

8.6 EXERCISES

Go to pages 546–547 to record your assignments.

Developing Skills

In Exercises 1–6, match the system of linear inequalities with its graph. [The graphs are labeled (a), (b), (c), (d), (e), and (f).]

(a)

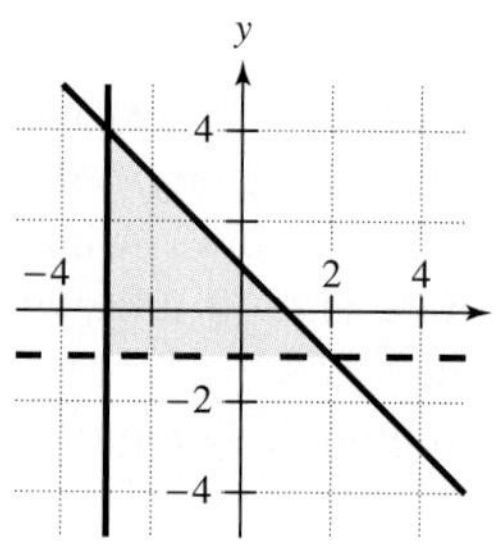

(b)

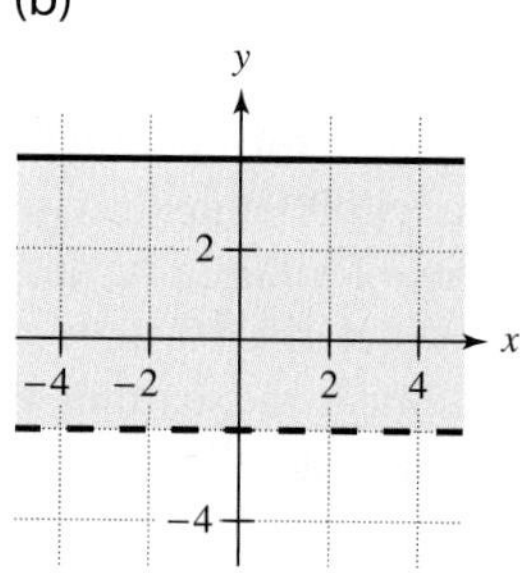

(c)

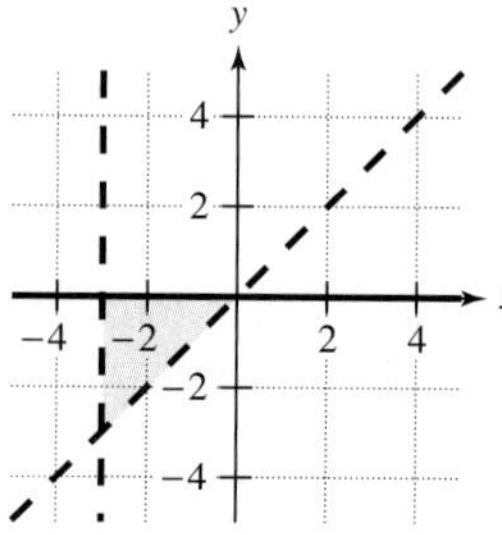

(d)

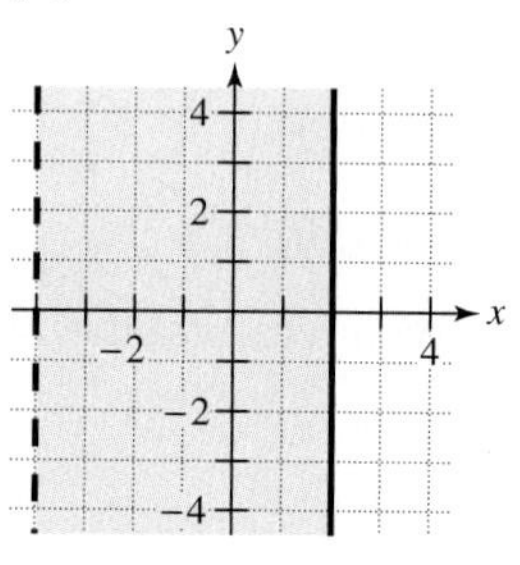

(e)

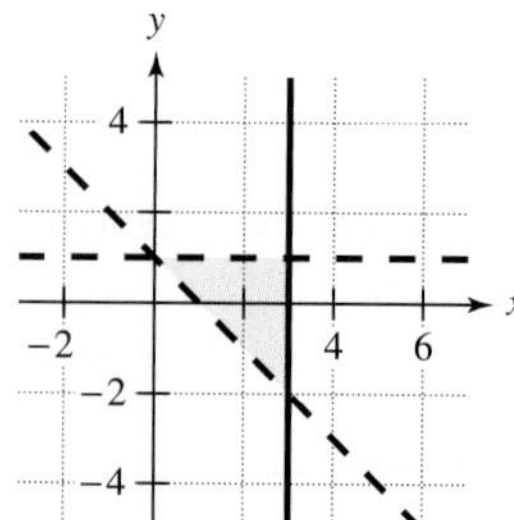

(f)

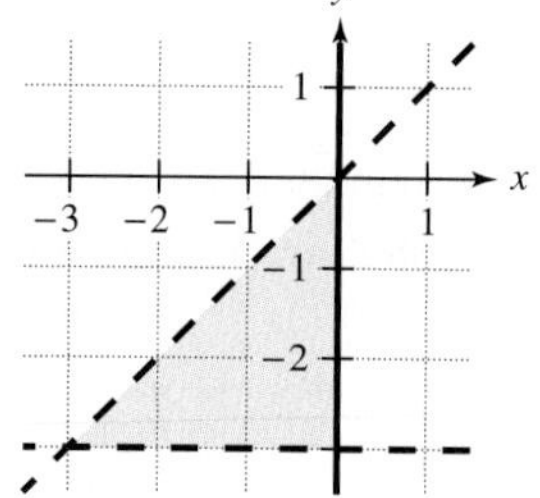

1. $\begin{cases} x > -4 \\ x \le 2 \end{cases}$

2. $\begin{cases} y \le 4 \\ y > -2 \end{cases}$

3. $\begin{cases} y < x \\ y > -3 \\ x \le 0 \end{cases}$

4. $\begin{cases} y > x \\ x > -3 \\ y \le 0 \end{cases}$

5. $\begin{cases} y > -1 \\ x \ge -3 \\ y \le -x + 1 \end{cases}$

6. $\begin{cases} x \le 3 \\ y < 1 \\ y > -x + 1 \end{cases}$

In Exercises 7–10, determine if each ordered pair is a solution of the system of linear inequalities.

7. $\begin{cases} 2x - y > 4 \\ x + 3y \le 6 \end{cases}$

(a) $(2, 0)$

(b) $(4, -2)$

8. $\begin{cases} x + 4y > 4 \\ 3x + 2y \ge -6 \end{cases}$

(a) $(0, 2)$

(b) $(-1, 1)$

9. $\begin{cases} -x + y < -2 \\ 4x + y < -3 \end{cases}$

(a) $(-3, 4)$

(b) $(-1, -3)$

10. $\begin{cases} 5x - 3y \le 12 \\ -3x + 5y \ge -15 \end{cases}$

(a) $(3, 1)$

(b) $(-5, -6)$

In Exercises 11–44, sketch a graph of the solution of the system of linear inequalities. ***See Examples 1–4.***

✓ 11. $\begin{cases} x < 3 \\ x > -2 \end{cases}$

12. $\begin{cases} y > -1 \\ y \le 2 \end{cases}$

✓ 13. $\begin{cases} x + y \le 3 \\ x - 1 \le 1 \end{cases}$

14. $\begin{cases} x + y \ge 2 \\ x - y \le 2 \end{cases}$

15. $\begin{cases} 2x - 4y \le 6 \\ x + y \ge 2 \end{cases}$

16. $\begin{cases} 4x + 10y \le 5 \\ x - y \le 4 \end{cases}$

17. $\begin{cases} x + 2y \le 6 \\ x - 2y \le 0 \end{cases}$

18. $\begin{cases} 2x + y \le 0 \\ x - y \le 8 \end{cases}$

19. $\begin{cases} x - 2y > 4 \\ 2x + y > 6 \end{cases}$

20. $\begin{cases} 3x + y < 6 \\ x + 2y > 2 \end{cases}$

21. $\begin{cases} x + y > -1 \\ x + y < 3 \end{cases}$

22. $\begin{cases} x - y > 2 \\ x - y < -4 \end{cases}$

23. $\begin{cases} y \ge \frac{4}{3}x + 1 \\ y \le 5x - 2 \end{cases}$

24. $\begin{cases} y \ge \frac{1}{2}x + \frac{1}{2} \\ y \le 4x - \frac{1}{2} \end{cases}$

25. $\begin{cases} y > x - 2 \\ y > -\frac{1}{3}x + 5 \end{cases}$

26. $\begin{cases} y > x - 4 \\ y > \frac{2}{3}x + \frac{1}{3} \end{cases}$

27. $\begin{cases} y \ge 3x - 3 \\ y \le -x + 1 \end{cases}$

28. $\begin{cases} y \ge 2x - 3 \\ y \le 3x + 1 \end{cases}$

29. $\begin{cases} x + 2y \le -4 \\ y \ge x + 5 \end{cases}$

30. $\begin{cases} x + y \le -3 \\ y \ge 3x - 4 \end{cases}$

✓ **31.** $\begin{cases} x + y \le 4 \\ x \ge 0 \\ y \ge 0 \end{cases}$

32. $\begin{cases} 2x + y \le 6 \\ x \ge 0 \\ y \ge 0 \end{cases}$

33. $\begin{cases} 4x - 2y > 8 \\ x \ge 0 \\ y \le 0 \end{cases}$

34. $\begin{cases} 2x - 6y > 6 \\ x \le 0 \\ y \le 0 \end{cases}$

35. $\begin{cases} y > -5 \\ x \le 2 \\ y \le x + 2 \end{cases}$

36. $\begin{cases} y \ge -1 \\ x < 3 \\ y \ge x - 1 \end{cases}$

37. $\begin{cases} x + y \le 1 \\ -x + y \le 1 \\ y \ge 0 \end{cases}$

38. $\begin{cases} 3x + 2y < 6 \\ x - 3y \ge 1 \\ y \ge 0 \end{cases}$

39. $\begin{cases} x + y \le 5 \\ x - 2y \ge 2 \\ y \ge 3 \end{cases}$

40. $\begin{cases} 2x + y \ge 2 \\ x - 3y \le 2 \\ y \le 1 \end{cases}$

41. $\begin{cases} -3x + 2y < 6 \\ x - 4y > -2 \\ 2x + y < 3 \end{cases}$

42. $\begin{cases} x + 2y > 14 \\ -2x + 3y > 15 \\ x + 3y < 3 \end{cases}$

✓ **43.** $\begin{cases} x \ge 1 \\ x - 2y \le 3 \\ 3x + 2y \ge 9 \\ x + y \le 6 \end{cases}$

44. $\begin{cases} x + y \le 4 \\ x + y \ge -1 \\ x - y \ge -2 \\ x - y \le 2 \end{cases}$

In Exercises 45–50, use a graphing calculator to graph the solution of the system of linear inequalities.

45. $\begin{cases} 2x - 3y \le 6 \\ y \le 4 \end{cases}$

46. $\begin{cases} 6x + 3y \ge 12 \\ y \le 4 \end{cases}$

47. $\begin{cases} 2x - 2y \le 5 \\ y \le 6 \end{cases}$

48. $\begin{cases} 2x + 3y \ge 12 \\ y \ge 2 \end{cases}$

49. $\begin{cases} 2x + y \le 2 \\ y \ge -4 \end{cases}$

50. $\begin{cases} 4x - 3y \ge -3 \\ y \ge -1 \end{cases}$

In Exercises 51–56, write a system of linear inequalities that describes the shaded region. ***See Example 5.***

✓ **51.**

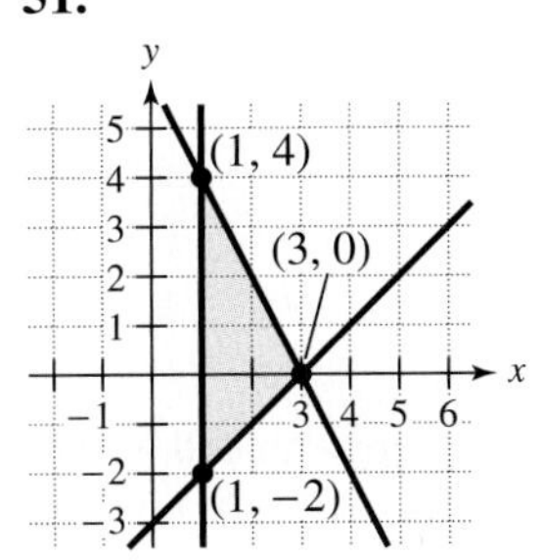

52.

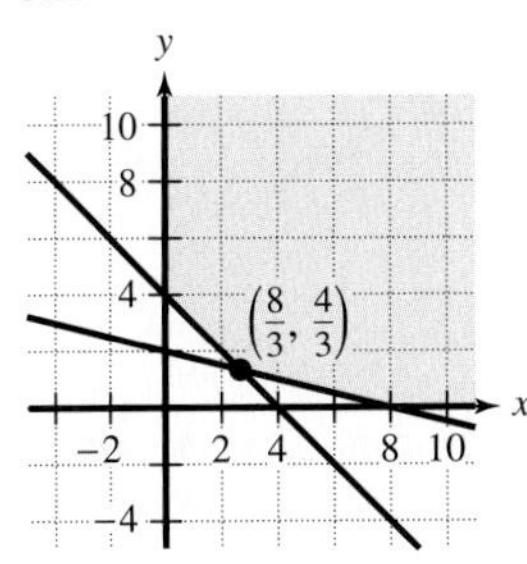

53.

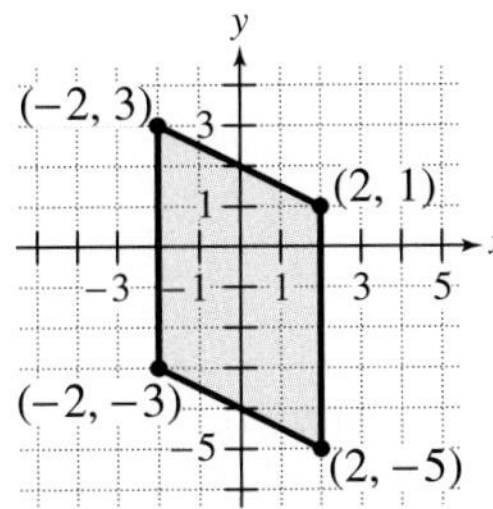

54.

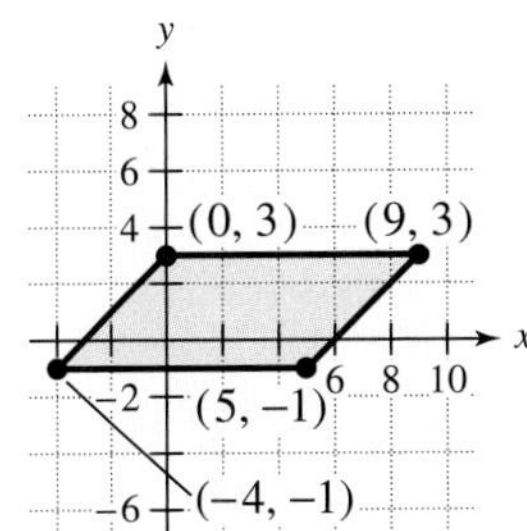

55.

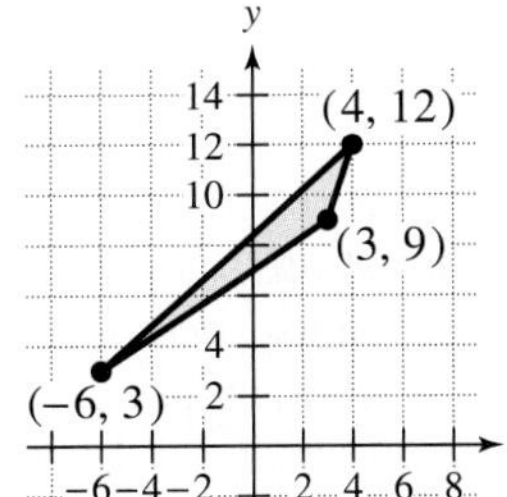

56.

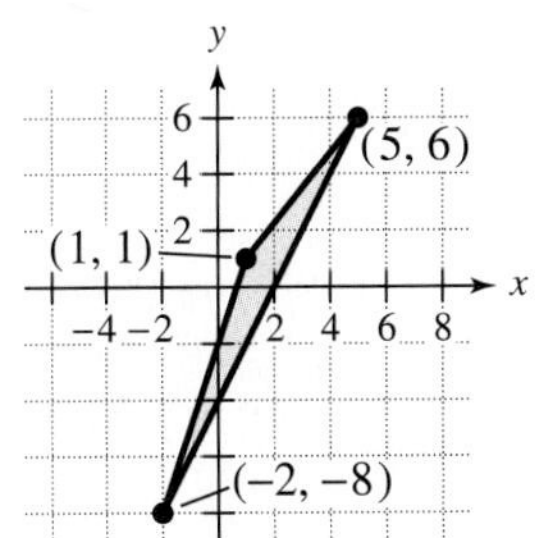

Solving Problems

✓ **57.** *Production* A furniture company can sell all the tables and chairs it produces. Each table requires 1 hour in the assembly center and $1\frac{1}{3}$ hours in the finishing center. Each chair requires $1\frac{1}{2}$ hours in the assembly center and $\frac{3}{4}$ hour in the finishing center. The company's assembly center is available 12 hours per day, and its finishing center is available 16 hours per day. Write a system of linear inequalities describing the different production levels. Graph the system.

58. *Production* An electronics company can sell all the HD TVs and DVD players it produces. Each HD TV requires 3 hours on the assembly line and $1\frac{1}{4}$ hours on the testing line. Each DVD player requires $2\frac{1}{2}$ hours on the assembly line and 1 hour on the testing line. The company's assembly line is available 20 hours per day, and its testing line is available 16 hours per day. Write a system of linear inequalities describing the different production levels. Graph the system.

59. *Investment* A person plans to invest up to $40,000 in two different interest-bearing accounts, account X and account Y. Account X is to contain at least $10,000. Moreover, account Y should have at least twice the amount in account X. Write a system of linear inequalities describing the various amounts that can be deposited in each account. Graph the system.

60. *Investment* A person plans to invest up to $25,000 in two different interest-bearing accounts, account X and account Y. Account Y is to contain at least $4000. Moreover, account X should have at least three times the amount in account Y. Write a system of linear inequalities describing the various amounts that can be deposited in each account. Graph the system.

61. *Nutrition* A dietitian is asked to design a special dietary supplement using two different foods. Each ounce of food X contains 20 units of calcium, 15 units of iron, and 10 units of vitamin B. Each ounce of food Y contains 10 units of calcium, 10 units of iron, and 20 units of vitamin B. The minimum daily requirements in the diet are 280 units of calcium, 160 units of iron, and 180 units of vitamin B. Write a system of linear inequalities describing the different amounts of food X and food Y that can be used in the diet. Use a graphing calculator to graph the system.

62. *Nutrition* A veterinarian is asked to design a special canine dietary supplement using two different dog foods. Each ounce of food X contains 12 units of calcium, 8 units of iron, and 6 units of protein. Each ounce of food Y contains 10 units of calcium, 10 units of iron, and 8 units of protein. The minimum daily requirements of the diet are 200 units of calcium, 100 units of iron, and 120 units of protein. Write a system of linear inequalities describing the different amounts of dog food X and dog food Y that can be used. Use a graphing calculator to graph the system.

63. *Ticket Sales* Two types of tickets are to be sold for a concert. General admission tickets cost $30 per ticket and stadium seat tickets cost $45 per ticket. The promoter of the concert must sell at least 15,000 tickets, including at least 8000 general admission tickets and at least 4000 stadium seat tickets. Moreover, the gross receipts must total at least $525,000 in order for the concert to be held. Write a system of linear inequalities describing the different numbers of tickets that can be sold. Use a graphing calculator to graph the system.

64. *Ticket Sales* For a concert event, there are \$30 reserved seat tickets and \$20 general admission tickets. There are 2000 reserved seats available, and fire regulations limit the number of paid ticket holders to 3000. The promoter must take in at least \$75,000 in ticket sales. Write a system of linear inequalities describing the different numbers of tickets that can be sold. Use a graphing calculator to graph the system.

65. *Geometry* The figure shows a cross section of a roped-off swimming area at a beach. Write a system of linear inequalities describing the cross section. (Each unit in the coordinate system represents 1 foot.)

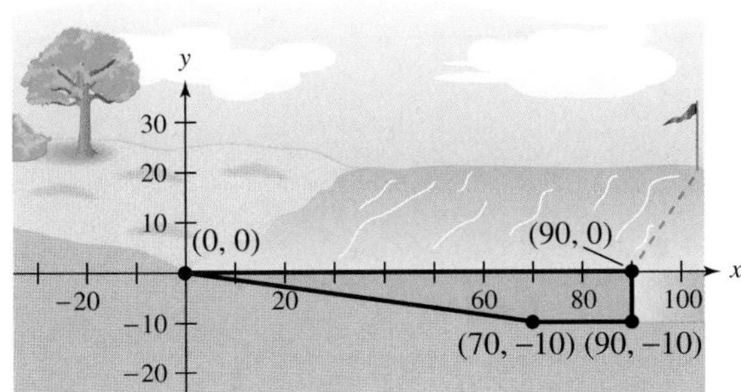

66. *Geometry* The figure shows the chorus platform on a stage. Write a system of linear inequalities describing the part of the audience that can see the full chorus. (Each unit in the coordinate system represents 1 meter.)

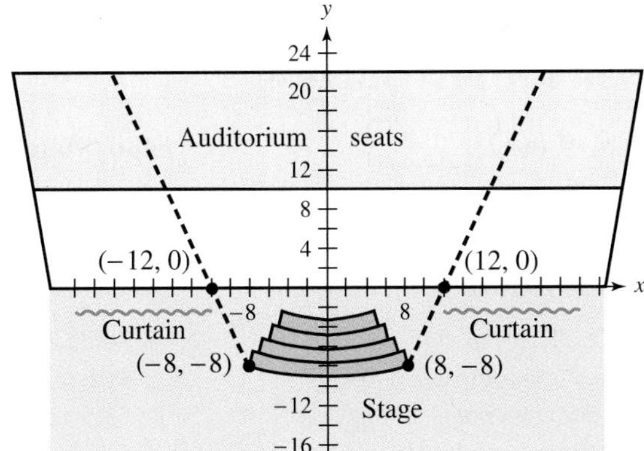

Explaining Concepts

67. Explain the meaning of the term *half-plane*. Give an example of an inequality whose graph is a half-plane.

68. Explain how you can check any single point (x_1, y_1) to determine whether the point is a solution of a system of linear inequalities.

69. Explain how to determine the vertices of the solution region for a system of linear inequalities.

70. Describe the difference between the solution set of a system of linear equations and the solution set of a system of linear inequalities.

Cumulative Review

In Exercises 71–76, find the *x*- and *y*-intercepts (if any) of the graph of the equation.

71. $y = 4x + 2$

72. $y = 8 - 3x$

73. $-x + 3y = -3$

74. $3x - 6y = 12$

75. $y = |x + 2|$

76. $y = |x - 1| - 2$

In Exercises 77–80, evaluate the function as indicated, and simplify.

77. $f(x) = 3x - 7$
(a) $f(-1)$
(b) $f\left(\frac{2}{3}\right)$

78. $f(x) = x^2 + x$
(a) $f(3)$
(b) $f(-2)$

79. $f(x) = 3x - x^2$
(a) $f(0)$ (b) $f(2m)$

80. $f(x) = \dfrac{x + 3}{x - 1}$
(a) $f(8)$ (b) $f(k - 2)$

What Did You Learn?

Use these two pages to help prepare for a test on this chapter. Check off the key terms and key concepts you know. You can also use this section to record your assignments.

Plan for Test Success

Date of test: / /

Study dates and times: / / at : A.M./P.M.
/ / at : A.M./P.M.

Things to review:

- ☐ Key Terms, *p. 546*
- ☐ Key Concepts, *pp. 546–547*
- ☐ Your class notes
- ☐ Your assignments
- ☐ Study Tips, *pp. 477, 489, 492, 493, 499, 504, 512, 513, 514, 515, 525, 528, 529*
- ☐ Technology Tips, *pp. 515, 526, 541*
- ☐ Mid-Chapter Quiz, *p. 511*
- ☐ Review Exercises, *pp. 548–552*
- ☐ Chapter Test, *p. 553*
- ☐ Video Explanations Online
- ☐ Tutorial Online

Key Terms

- ☐ systems of equations, *p. 472*
- ☐ solution of a system of equations, *p. 472*
- ☐ consistent system, *p. 474*
- ☐ dependent system, *p. 474*
- ☐ inconsistent system, *p. 474*
- ☐ method of substitution, *p. 477*
- ☐ break-even point, *p. 480*
- ☐ method of elimination, *p. 489*
- ☐ row-echelon form, *pp. 499, 515*
- ☐ ordered triple, *p. 499*
- ☐ equivalent systems, *p. 500*
- ☐ Gaussian elimination, *p. 500*
- ☐ row operations, *p. 500*
- ☐ position equation, *p. 505*
- ☐ matrix, *p. 512*
- ☐ order (of a matrix), *p. 512*
- ☐ entry (of a matrix), *p. 512*
- ☐ square matrix, *p. 512*
- ☐ augmented matrix, *p. 513*
- ☐ coefficient matrix, *p. 513*
- ☐ elementary row operations, *p. 514*
- ☐ row-equivalent matrices, *p. 514*
- ☐ determinant, *p. 525*
- ☐ expanding by minors, *p. 526*
- ☐ minor (of an entry), *p. 526*
- ☐ Cramer's Rule, *p. 528*
- ☐ system of linear inequalities, *p. 537*
- ☐ solution of a system of linear inequalities, *p. 537*
- ☐ solution set, *p. 537*
- ☐ vertex, *p. 538*

Key Concepts

8.1 Solving Systems of Equations by Graphing and Substitution

Assignment: ________ **Due date:** ________

☐ **Solve systems of equations graphically.**

A system of equations can have one solution, infinitely many solutions, or no solution.

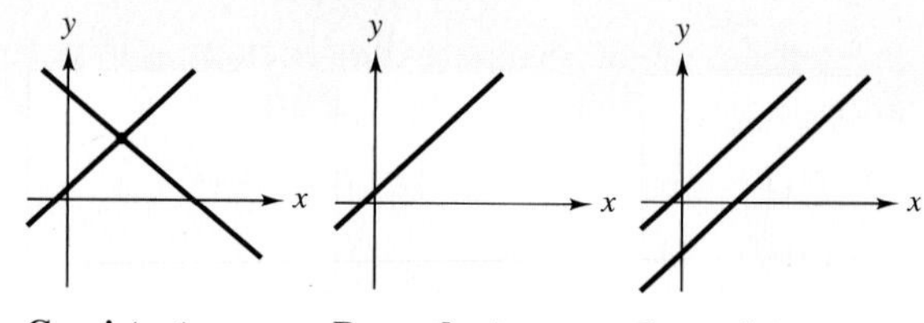

Consistent system (one solution) **Dependent (consistent) system** (infinitely many solutions) **Inconsistent system** (no solution)

☐ **Solve systems of equations algebraically using the method of substitution.**

1. Solve one equation for one variable in terms of the other variable.
2. Substitute the expression obtained in Step 1 in the other equation to obtain an equation in one variable.
3. Solve the equation obtained in Step 2.
4. Back-substitute the Step 3 solution in the expression found in Step 1 to find the value of the other variable.
5. Check the solution to see that it satisfies both of the original equations.

8.2 Solving Systems of Equations by Elimination

Assignment: ______________________________ **Due date:** __________

☐ **Use the method of elimination.**

1. Obtain opposite coefficients of x (or y) by multiplying all terms of one or both equations by suitable constants.
2. Add the equations to eliminate one variable, and solve the resulting equation.
3. Back-substitute the value obtained in Step 2 into either of the original equations and solve for the other variable.
4. Check your solution in both of the original equations.

8.3 Linear Systems in Three Variables

Assignment: ______________________________ **Due date:** __________

☐ **Solve systems of linear equations in row-echelon form using back-substitution.**

A system of equations in *row-echelon form* has a stair-step pattern with leading coefficients of 1. You can use back-substitution to solve a system in row-echelon form.

☐ **Use Gaussian elimination to write a system of linear equations in row-echelon form.**

Two systems of equations are *equivalent systems* if they have the same solution set.

Each of the following *row operations* produces an equivalent system of linear equations.

1. Interchange two equations.
2. Multiply one of the equations by a nonzero constant.
3. Add a multiple of one of the equations to another equation to replace the latter equation.

Gaussian elimination is the process of forming a chain of equivalent systems by performing one row operation at a time to obtain an equivalent system in row-echelon form.

8.4 Matrices and Linear Systems

Assignment: ______________________________ **Due date:** __________

☐ **Perform elementary row operations on a matrix.**

1. Interchange two rows.
2. Multiply a row by a nonzero constant.
3. Add a multiple of a row to another row.

☐ **Use Gaussian elimination with back-substitution.**

1. Write the augmented matrix of the system of equations.
2. Use elementary row operations to rewrite the augmented matrix in row-echelon form.
3. Write the system of equations corresponding to the matrix in row-echelon form. Then use back-substitution to find the solution.

8.5 Determinants and Linear Systems

Assignment: ______________________________ **Due date:** __________

☐ **Find the determinant of a 2 × 2 matrix.**

$$\begin{vmatrix} a_1 & b_1 \\ a_2 & b_2 \end{vmatrix} = a_1b_2 - a_2b_1$$

☐ **Use expanding by minors to find the determinant of a 3 × 3 matrix.**

$$\begin{vmatrix} a_1 & b_1 & c_1 \\ a_2 & b_2 & c_2 \\ a_3 & b_3 & c_3 \end{vmatrix} = a_1\begin{vmatrix} b_2 & c_2 \\ b_3 & c_3 \end{vmatrix} - b_1\begin{vmatrix} a_2 & c_2 \\ a_3 & c_3 \end{vmatrix} + c_1\begin{vmatrix} a_2 & b_2 \\ a_3 & b_3 \end{vmatrix}$$

☐ **Use Cramer's Rule to solve a system.**

See page 528.

8.6 Systems of Linear Inequalities

Assignment: ______________________________ **Due date:** __________

☐ **Graph a system of linear inequalities.**

1. Sketch a dashed or solid line corresponding to each inequality.
2. Shade the half-plane for each inequality.
3. The intersection of all half-planes represents the system.

Review Exercises

8.1 Solving Systems of Equations by Graphing and Substitution

1 ▶ Determine if an ordered pair is a solution of a system of equations.

In Exercises 1–4, determine whether each ordered pair is a solution of the system of equations.

	System	*Ordered Pairs*	
1.	$\begin{cases} 3x - 5y = 11 \\ -x + 2y = -4 \end{cases}$	(a) $(2, -1)$	(b) $(3, -2)$
2.	$\begin{cases} 10x + 8y = -2 \\ 2x - 5y = 26 \end{cases}$	(a) $(4, -4)$	(b) $(3, -4)$
3.	$\begin{cases} 0.2x + 0.4y = 5 \\ x + 3y = 30 \end{cases}$	(a) $(0.5, -0.7)$	(b) $(15, 5)$
4.	$\begin{cases} -\frac{1}{2}x - \frac{2}{3}y = \frac{1}{2} \\ x + y = 1 \end{cases}$	(a) $(-5, 6)$	(b) $(7, -3)$

2 ▶ Use a coordinate system to solve systems of linear equations graphically.

In Exercises 5–8, use the graphs of the equations to determine the solution (if any) of the system of linear equations. Check your solution.

5. $\begin{cases} 2x + y = 4 \\ 2x - y = 0 \end{cases}$

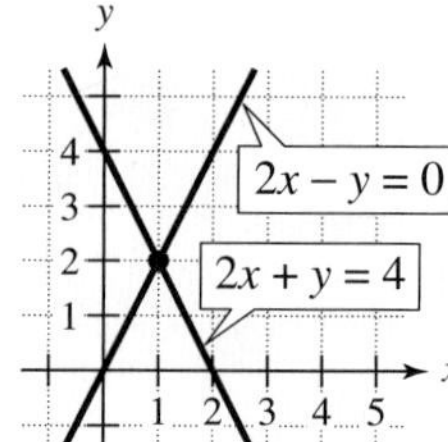

6. $\begin{cases} -x + y = 1 \\ x + y = 5 \end{cases}$

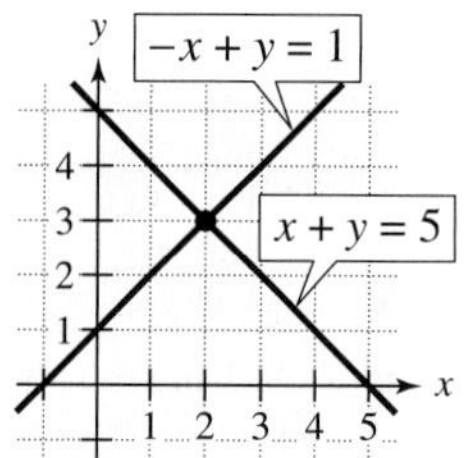

7. $\begin{cases} 3x - 2y = 6 \\ 2x - y = 3 \end{cases}$

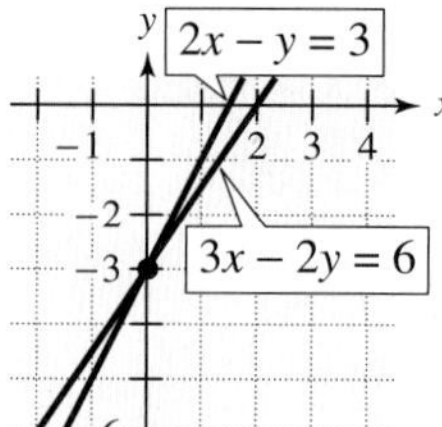

8. $\begin{cases} x + 2y = 6 \\ x + 2y = 2 \end{cases}$

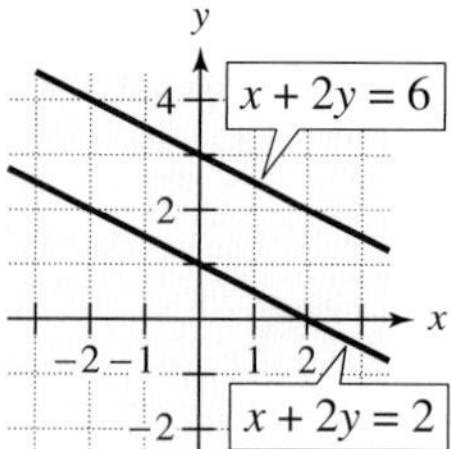

In Exercises 9–18, sketch the graphs of the equations and approximate any solutions of the system of linear equations.

9. $\begin{cases} y = x - 4 \\ y = 2x - 9 \end{cases}$

10. $\begin{cases} y = -\frac{5}{3}x + 6 \\ y = x - 10 \end{cases}$

11. $\begin{cases} x + y = 2 \\ x - y = 0 \end{cases}$

12. $\begin{cases} x - y = 9 \\ -x + y = 1 \end{cases}$

13. $\begin{cases} 2x + 3 = 3y \\ y = \frac{2}{3}x \end{cases}$

14. $\begin{cases} x + y = -1 \\ 3x + 2y = 0 \end{cases}$

15. $\begin{cases} -3x - 3y = -6 \\ x + y = 2 \end{cases}$

16. $\begin{cases} 2x - 3y = 9 \\ x + y = 3 \end{cases}$

17. $\begin{cases} x + 7y = 6 \\ -3x + y = 4 \end{cases}$

18. $\begin{cases} 5x - 4y = 12 \\ -x + 3y = 2 \end{cases}$

3 ▶ Use the method of substitution to solve systems of equations algebraically.

In Exercises 19–32, solve the system by the method of substitution.

19. $\begin{cases} y = 2x \\ y = x + 4 \end{cases}$

20. $\begin{cases} x = -2y + 13 \\ x = \dfrac{y}{2} + 3 \end{cases}$

21. $\begin{cases} x = 3y - 2 \\ x = 6 - y \end{cases}$

22. $\begin{cases} y = -4x + 1 \\ y = x - 4 \end{cases}$

23. $\begin{cases} x = y + 3 \\ x = y + 1 \end{cases}$

24. $\begin{cases} y = 3x + 4 \\ 9x = 3y - 12 \end{cases}$

25. $\begin{cases} x - 2y = 6 \\ 3x + 2y = 10 \end{cases}$ **26.** $\begin{cases} 5x + y = 20 \\ 7x - 5y = -4 \end{cases}$

27. $\begin{cases} 2x - y = 2 \\ 6x + 8y = 39 \end{cases}$ **28.** $\begin{cases} 3x + 4y = 1 \\ x - 7y = -3 \end{cases}$

29. $\begin{cases} -6x + y = -3 \\ 12x - 2y = 6 \end{cases}$ **30.** $\begin{cases} 3x + 4y = 7 \\ 6x + 8y = 10 \end{cases}$

31. $\begin{cases} \frac{3}{5}x - y = 8 \\ 2x - 3y = 25 \end{cases}$ **32.** $\begin{cases} -x + 8y = -115 \\ 2x + \frac{2}{7}y = 2 \end{cases}$

4 ▶ Solve application problems using systems of equations.

33. *Number Problem* The sum of two numbers x and y is 52, and the difference of the two numbers is 20. Find the two numbers.

34. *Break-Even Analysis* A small company produces sunglasses that sell for \$18 per unit. The cost of producing each unit is \$10.25, and the company has fixed costs of \$350.

(a) Use a verbal model to show that the cost C of producing x units is $C = 10.25x + 350$ and the revenue R from selling x units is $R = 18x$.

(b) Use a graphing calculator to graph the cost and revenue functions in the same viewing window. Approximate the point of intersection of the graphs and interpret the result.

35. *Investment* A total of \$12,000 is invested in two funds paying 5% and 10% simple interest. (There is more risk in the 10% fund.) The combined annual interest for the two funds is \$800. Determine how much of the \$12,000 is invested at each rate.

36. *Comparing Costs* An MP3 player costs \$200 plus \$10 for every album purchased. A CD player costs \$50 plus \$15 for every album purchased. Determine after how many albums the total costs for the two music formats will be the same. What will be the cost?

37. *Ticket Sales* You are selling tickets to your school musical. Adult tickets cost \$5 and children's tickets cost \$3. You sell 1510 tickets and collect \$6138. Determine how many of each type of ticket were sold.

8.2 Solving Systems of Equations by Elimination

1 ▶ Solve systems of linear equations algebraically using the method of elimination.

In Exercises 38–47, solve the system by the method of elimination.

38. $\begin{cases} 2x + 4y = 2 \\ -2x - 7y = 4 \end{cases}$ **39.** $\begin{cases} 3x - y = 5 \\ 2x + y = 5 \end{cases}$

40. $\begin{cases} 3x - 2y = 9 \\ x + y = 3 \end{cases}$ **41.** $\begin{cases} 5x + 4y = 2 \\ -x + y = -22 \end{cases}$

42. $\begin{cases} 2x - 5y = 2 \\ 6x - 15y = 4 \end{cases}$ **43.** $\begin{cases} 8x - 6y = 4 \\ -4x + 3y = -2 \end{cases}$

44. $\begin{cases} \frac{1}{2}x - \frac{3}{5}y = \frac{1}{6} \\ -3x + 6y = 1 \end{cases}$ **45.** $\begin{cases} \frac{2}{3}x + \frac{1}{12}y = \frac{3}{4} \\ 3x - 4y = 2 \end{cases}$

46. $\begin{cases} 0.2x - 0.1y = 0.07 \\ 0.4x - 0.5y = -0.01 \end{cases}$

47. $\begin{cases} 0.2x + 0.1y = 0.03 \\ 0.3x - 0.1y = -0.13 \end{cases}$

2 ▶ Choose a method for solving systems of equations.

In Exercises 48–65, use the most convenient method (graphing, substitution, or elimination) to solve the system of linear equations. State which method you used.

48. $\begin{cases} -x + 2y = 2 \\ x = 4 \end{cases}$ **49.** $\begin{cases} 6x - 5y = 0 \\ y = 6 \end{cases}$

50. $\begin{cases} -x + y = 4 \\ x + y = 4 \end{cases}$ **51.** $\begin{cases} -x + 4y = 4 \\ x + y = 6 \end{cases}$

52. $\begin{cases} x + y = 0 \\ 2x + y = 0 \end{cases}$ **53.** $\begin{cases} x - y = 0 \\ x - 6y = 5 \end{cases}$

54. $\begin{cases} -7x + 9y = 9 \\ 2x + 9y = -18 \end{cases}$

55. $\begin{cases} 5x + 8y = 8 \\ x - 8y = 16 \end{cases}$

56. $\begin{cases} -5x + 2y = -4 \\ x - 6y = 4 \end{cases}$

57. $\begin{cases} 6x - 3y = 27 \\ -2x + y = -9 \end{cases}$

58. $\begin{cases} x + 2y = 2 \\ x - 4y = 20 \end{cases}$

59. $\begin{cases} 2x + 6y = 16 \\ 2x + 3y = 7 \end{cases}$

60. $\begin{cases} -\frac{1}{4}x + \frac{2}{3}y = 1 \\ 3x - 8y = 1 \end{cases}$

61. $\begin{cases} \frac{1}{5}x + \frac{3}{2}y = 2 \\ 2x + 13y = 20 \end{cases}$

62. $\begin{cases} \frac{1}{2}x - \frac{1}{3}y = 0 \\ 3x + 2y = 0 \end{cases}$

63. $\begin{cases} \frac{1}{3}x + \frac{4}{7}y = 3 \\ 2x + 3y = 15 \end{cases}$

64. $\begin{cases} 0.2u + 0.3v = 0.14 \\ 0.4u + 0.5v = 0.20 \end{cases}$

65. $\begin{cases} 1.2s + 4.2t = -1.7 \\ 3.0s - 1.8t = 1.9 \end{cases}$

66. *College Credits* Each course at a college is worth either 3 or 4 credits. The members of the student council are taking a total of 47 courses that are worth a total of 156 credits. Determine the number of each type of course being taken.

8.3 Linear Systems in Three Variables

1 ▶ Solve systems of linear equations in row-echelon form using back-substitution.

In Exercises 67–70, use back-substitution to solve the system of linear equations.

67. $\begin{cases} x = 3 \\ x + 2y = 7 \\ -3x - y + 4z = 9 \end{cases}$

68. $\begin{cases} 2x + 3y = 9 \\ 4x - 6z = 12 \\ y = 5 \end{cases}$

69. $\begin{cases} x + 2y = 6 \\ 3y = 9 \\ x + 2z = 12 \end{cases}$

70. $\begin{cases} 3x - 2y + 5z = -10 \\ 3y = 18 \\ 6x - 4y = -6 \end{cases}$

2 ▶ Solve systems of linear equations using the method of Gaussian elimination.

In Exercises 71–74, solve the system of linear equations.

71. $\begin{cases} -x + y + 2z = 1 \\ 2x + 3y + z = -2 \\ 5x + 4y + 2z = 4 \end{cases}$

72. $\begin{cases} 2x + 3y + z = 10 \\ 2x - 3y - 3z = 22 \\ 4x - 2y + 3z = -2 \end{cases}$

73. $\begin{cases} x - y - z = 1 \\ -2x + y + 3z = -5 \\ 3x + 4y - z = 6 \end{cases}$

74. $\begin{cases} -3x + y + 2z = -13 \\ -x - y + z = 0 \\ 2x + 2y - 3z = -1 \end{cases}$

3 ▶ Solve application problems using the method of Gaussian elimination.

75. *Investment* An inheritance of \$20,000 is divided among three investments yielding a total of \$1780 in interest per year. The interest rates for the three investments are 7%, 9%, and 11%. The amounts invested at 9% and 11% are \$3000 and \$1000 less than the amount invested at 7%, respectively. Find the amount invested at each rate.

76. *Vertical Motion* Find the position equation

$$s = \frac{1}{2}at^2 + v_0t + s_0$$

for an object that has the indicated heights at the specified times.

$s = 192$ feet at $t = 1$ second

$s = 152$ feet at $t = 2$ seconds

$s = 80$ feet at $t = 3$ seconds

8.4 Matrices and Linear Systems

1 ▶ Determine the orders of matrices.

In Exercises 77–80, determine the order of the matrix.

77. $\begin{bmatrix} 3 & 6 & -7 & -8 \end{bmatrix}$

78. $\begin{bmatrix} 1 & 5 \\ 3 & -4 \end{bmatrix}$

79. $\begin{bmatrix} 5 & 7 & 9 \\ 11 & -12 & 0 \end{bmatrix}$

80. $\begin{bmatrix} 15 \\ 13 \\ -9 \end{bmatrix}$

2 ▶ Form coefficient and augmented matrices, and form linear systems from augmented matrices.

In Exercises 81 and 82, form (a) the coefficient matrix and (b) the augmented matrix for the system of linear equations.

81. $\begin{cases} 7x - 5y = 11 \\ x - y = -5 \end{cases}$

82. $\begin{cases} x + 2y + z = 4 \\ 3x - z = 2 \\ -x + 5y - 2z = -6 \end{cases}$

In Exercises 83 and 84, write the system of linear equations represented by the matrix. (Use variables *x*, *y*, and *z*.)

83. $\begin{bmatrix} 4 & -1 & 0 & \vdots & 2 \\ 6 & 3 & 2 & \vdots & 1 \\ 0 & 1 & 4 & \vdots & 0 \end{bmatrix}$

84. $\begin{bmatrix} 7 & 8 & \vdots & -26 \\ 4 & -9 & \vdots & -12 \end{bmatrix}$

3 ▶ Perform elementary row operations to solve systems of linear equations in matrix form.

In Exercises 85–88, use matrices and elementary row operations to solve the system.

85. $\begin{cases} 5x + 4y = 2 \\ -x + y = -22 \end{cases}$

86. $\begin{cases} 2x - 5y = 2 \\ 3x - 7y = 1 \end{cases}$

87. $\begin{cases} 0.2x - 0.1y = 0.07 \\ 0.4x - 0.5y = -0.01 \end{cases}$

88. $\begin{cases} 2x + y = 0.3 \\ 3x - y = -1.3 \end{cases}$

4 ▶ Use matrices and Gaussian elimination with back-substitution to solve systems of linear equations.

In Exercises 89–94, use matrices to solve the system of linear equations.

89. $\begin{cases} x + 4y + 4z = 7 \\ -3x + 2y + 3z = 0 \\ 4x - 2z = -2 \end{cases}$

90. $\begin{cases} -x + 3y - z = -4 \\ 2x + 6z = 14 \\ -3x - y + z = 10 \end{cases}$

91. $\begin{cases} 2x_1 + 3x_2 + 3x_3 = 3 \\ 6x_1 + 6x_2 + 12x_3 = 13 \\ 12x_1 + 9x_2 - x_3 = 2 \end{cases}$

92. $\begin{cases} -x_1 + 2x_2 + 3x_3 = 4 \\ 2x_1 - 4x_2 - x_3 = -13 \\ 3x_1 + 2x_2 - 4x_3 = -1 \end{cases}$

93. $\begin{cases} x - 4z = 17 \\ -2x + 4y + 3z = -14 \\ 5x - y + 2z = -3 \end{cases}$

94. $\begin{cases} 2x + 3y - 5z = 3 \\ -x + 2y = 3 \\ 3x + 5y + 2z = 15 \end{cases}$

8.5 Determinants and Linear Systems

1 ▶ Find determinants of 2×2 matrices and 3×3 matrices.

In Exercises 95–100, find the determinant of the matrix using any appropriate method.

95. $\begin{bmatrix} 9 & 8 \\ 10 & 10 \end{bmatrix}$

96. $\begin{bmatrix} -3.4 & 1.2 \\ -5 & 2.5 \end{bmatrix}$

97. $\begin{bmatrix} 8 & 6 & 3 \\ 6 & 3 & 0 \\ 3 & 0 & 2 \end{bmatrix}$

98. $\begin{bmatrix} 7 & -1 & 10 \\ -3 & 0 & -2 \\ 12 & 1 & 1 \end{bmatrix}$

99. $\begin{bmatrix} 8 & 3 & 2 \\ 1 & -2 & 4 \\ 6 & 0 & 5 \end{bmatrix}$

100. $\begin{bmatrix} 4 & 0 & 10 \\ 0 & 10 & 0 \\ 10 & 0 & 34 \end{bmatrix}$

2 ▶ Use determinants and Cramer's Rule to solve systems of linear equations.

In Exercises 101–104, use Cramer's Rule to solve the system of linear equations. (If not possible, state the reason.)

101. $\begin{cases} 7x + 12y = 63 \\ 2x + 3y = 15 \end{cases}$

102. $\begin{cases} 12x + 42y = -17 \\ 30x - 18y = 19 \end{cases}$

103. $\begin{cases} -x + y + 2z = 1 \\ 2x + 3y + z = -2 \\ 5x + 4y + 2z = 4 \end{cases}$

104. $\begin{cases} 2x_1 + x_2 + 2x_3 = 4 \\ 2x_1 + 2x_2 = 5 \\ 2x_1 - x_2 + 6x_3 = 2 \end{cases}$

3 ▶ Use determinants to find areas of triangles, to test for collinear points, and to find equations of lines.

Area of a Triangle **In Exercises 105–108, use a determinant to find the area of the triangle with the given vertices.**

105. $(1, 0), (5, 0), (5, 8)$

106. $(-6, 0), (6, 0), (0, 5)$

107. $(1, 2), (4, -5), (3, 2)$

108. $\left(\frac{3}{2}, 1\right), \left(4, -\frac{1}{2}\right), (4, 2)$

Collinear Points **In Exercises 109 and 110, determine whether the points are collinear.**

109. $(1, 2), (5, 0), (10, -2)$

110. $(-4, 3), (1, 1), (6, -1)$

Equation of a Line **In Exercises 111 and 112, use a determinant to find the equation of the line through the points.**

111. $(-4, 0), (4, 4)$

112. $\left(-\frac{5}{2}, 3\right), \left(\frac{7}{2}, 1\right)$

8.6 Systems of Linear Inequalities

1 ▶ Solve systems of linear inequalities in two variables.

In Exercises 113–116, sketch a graph of the solution of the system of linear inequalities.

113. $\begin{cases} x + y < 5 \\ x > 2 \\ y \geq 0 \end{cases}$

114. $\begin{cases} \frac{1}{2}x + y > 4 \\ x < 6 \\ y < 3 \end{cases}$

115. $\begin{cases} x + 2y \leq 160 \\ 3x + y \leq 180 \\ x \geq 0 \\ y \geq 0 \end{cases}$

116. $\begin{cases} 2x + 3y \leq 24 \\ 2x + y \leq 16 \\ x \geq 0 \\ y \geq 0 \end{cases}$

2 ▶ Use systems of linear inequalities to model and solve real-life problems.

117. *Soup Distribution* A charitable organization can purchase up to 500 cartons of soup to be divided between a soup kitchen and a homeless shelter in the Chicago area. These two organizations need at least 150 cartons and 220 cartons, respectively. Write a system of linear inequalities describing the various numbers of cartons that can go to each organization. Graph the system.

118. *Inventory Costs* A warehouse operator has up to 24,000 square feet of floor space in which to store two products. Each unit of product x requires 20 square feet of floor space and costs \$12 per day to store. Each unit of product y requires 30 square feet of floor space and costs \$8 per day to store. The total storage cost per day cannot exceed \$12,400. Write a system of linear inequalities describing the various ways the two products can be stored. Graph the system.

Chapter Test

Take this test as you would take a test in class. After you are done, check your work against the answers in the back of the book.

$$\begin{cases} 2x - 2y = 2 \\ -x + 2y = 0 \end{cases}$$

System for 1

1. Determine whether each ordered pair is a solution of the system at the left.
   (a) $(2, 1)$ (b) $(4, 3)$

In Exercises 2–10, use the indicated method to solve the system.

2. *Graphical:* $\begin{cases} x - 2y = -1 \\ 2x + 3y = 12 \end{cases}$

3. *Substitution:* $\begin{cases} 4x - y = 1 \\ 4x - 3y = -5 \end{cases}$

4. *Substitution:* $\begin{cases} 2x - 2y = -2 \\ 3x + y = 9 \end{cases}$

5. *Elimination:* $\begin{cases} 3x - 4y = -14 \\ -3x + y = 8 \end{cases}$

6. *Elimination:* $\begin{cases} x + 2y - 4z = 0 \\ 3x + y - 2z = 5 \\ 3x - y + 2z = 7 \end{cases}$

7. *Matrices:* $\begin{cases} x \quad\quad - 3z = -10 \\ \quad -2y + 2z = 0 \\ x - 2y \quad\quad = -7 \end{cases}$

8. *Matrices:* $\begin{cases} x - 3y + z = -3 \\ 3x + 2y - 5z = 18 \\ y + z = -1 \end{cases}$

9. *Cramer's Rule:* $\begin{cases} 2x - 7y = 7 \\ 3x + 7y = 13 \end{cases}$

10. *Any Method:* $\begin{cases} 3x - 2y + z = 12 \\ x - 3y \quad\quad = 2 \\ -3x \quad\quad - 9z = -6 \end{cases}$

$$\begin{bmatrix} 2 & -2 & 0 \\ -1 & 3 & 1 \\ 2 & 8 & 1 \end{bmatrix}$$

Matrix for 11

11. Evaluate the determinant of the matrix shown at the left.

12. Use a determinant to find the area of the triangle with vertices $(0, 0)$, $(5, 4)$, and $(6, 0)$.

13. Graph the solution of the system of linear inequalities.
$$\begin{cases} x - 2y > -3 \\ 2x + 3y \le 22 \\ y \ge 0 \end{cases}$$

14. A midsize car costs $24,000 and costs an average of $0.28 per mile to maintain. A minivan costs $26,000 and costs an average of $0.24 per mile to maintain. Determine after how many miles the total costs of the two vehicles will be the same. (Each model is driven the same number of miles.)

15. An inheritance of $25,000 is divided among three investments yielding a total of $1275 in interest per year. The interest rates for the three investments are 4.5%, 5%, and 8%. The amounts invested at 5% and 8% are $4000 and $10,000 less than the amount invested at 4.5%, respectively. Find the amount invested at each rate.

16. Two types of tickets are sold for a concert. Reserved seat tickets cost $30 per ticket and floor seat tickets cost $40 per ticket. The promoter of the concert can sell at most 9000 reserved seat tickets and 4000 floor seat tickets. Gross receipts must total at least $300,000 in order for the concert to be held. Write a system of linear inequalities describing the different numbers of tickets that can be sold. Graph the system.

Study Skills in Action

Studying in a Group

Many students endure unnecessary frustration because they study by themselves. Studying in a group or with a partner has many benefits. First, the combined memory and comprehension of the members minimizes the likelihood of any member getting "stuck" on a particular problem. Second, discussing math often helps clarify unclear areas. Third, regular study groups keep many students from procrastinating. Finally, study groups often build a camaraderie that helps students stick with the course when it gets tough.

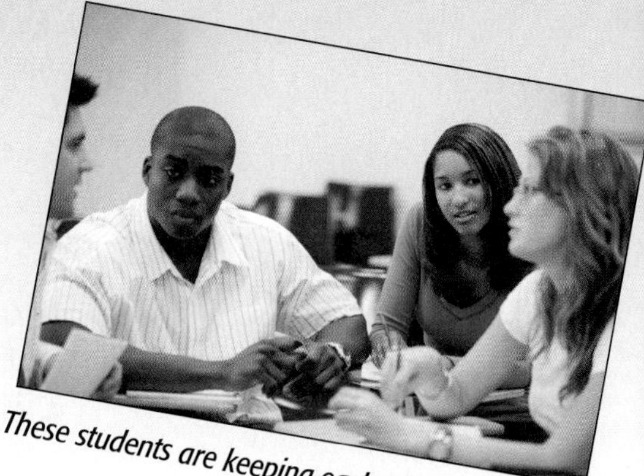

These students are keeping each other motivated.

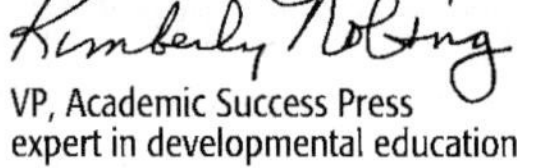

VP, Academic Success Press
expert in developmental education

Smart Study Strategy

Form a Weekly Study Group

1 ▶ Set up the group.

- Select students who are just as dedicated to doing well in the math class as you are.
- Find a regular meeting place on campus that has minimal distractions. Try to find a place that has a white board.
- Compare schedules, and select at least one time a week to meet, allowing at least 1.5 hours for study time.

2 ▶ Organize the study time. If you are unsure about how to structure your time during the first few study sessions, try using the guidelines at the right.

- Review and compare notes - 20 minutes
- Identify and review the key rules, definitions, etc. - 20 minutes
- Demonstrate at least one homework problem for each key concept - 40 minutes
- Make small talk (saving this until the end improves your chances of getting through all the math) - 10 minutes

3 ▶ Set up rules for the group. Consider using the following rules:

- Members must attend regularly, be on time, and participate.
- The sessions will focus on the key math concepts, not on the needs of one student.
- Students who skip classes will not be allowed to participate in the study group.
- Students who keep the group from being productive will be asked to leave the group.

4 ▶ Inform the instructor. Let the instructor know about your study group. Ask for advice about maintaining a productive group.

Chapter 9
Radicals and Complex Numbers

IT WORKED FOR ME!
"I learned about study groups by accident last semester. I started studying in the learning center. When I saw someone else from my class come into the center, I mentioned something to the tutor and she suggested asking her over to study with us. We did, and it actually turned into a group of three or four students every session. I learned a lot more and enjoyed it more too."
Jeremy
Graphic design

9.1 Radicals and Rational Exponents

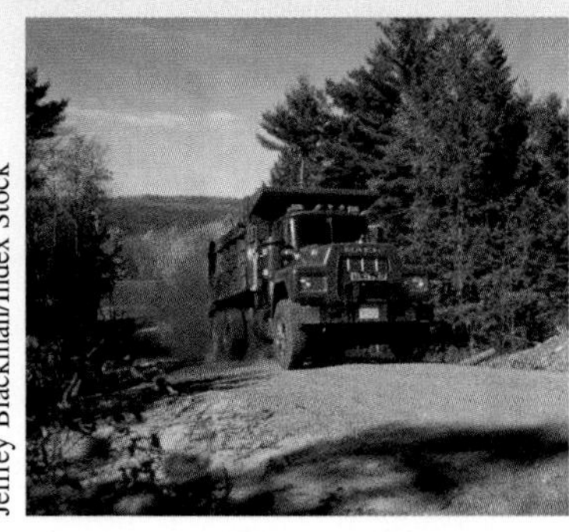
Jeffrey Blackman/Index Stock

What You Should Learn

1 ▶ Determine the *n*th roots of numbers and evaluate radical expressions.

2 ▶ Use the rules of exponents to evaluate or simplify expressions with rational exponents.

3 ▶ Use a calculator to evaluate radical expressions.

4 ▶ Evaluate radical functions and find the domains of radical functions.

Why You Should Learn It

Algebraic equations often involve rational exponents. For instance, in Exercise 167 on page 565, you will use an equation involving a rational exponent to find the depreciation rate for a truck.

1 ▶ Determine the *n*th roots of numbers and evaluate radical expressions.

Roots and Radicals

A **square root** of a number is defined as one of its two equal factors. For example, 5 is a square root of 25 because 5 is one of the two equal factors of 25. In a similar way, a **cube root** of a number is one of its three equal factors.

Number	*Equal Factors*	*Root*	*Type*
$9 = 3^2$	$3 \cdot 3$	3	Square root
$25 = (-5)^2$	$(-5)(-5)$	-5	Square root
$-27 = (-3)^3$	$(-3)(-3)(-3)$	-3	Cube root
$64 = 4^3$	$4 \cdot 4 \cdot 4$	4	Cube root
$16 = 2^4$	$2 \cdot 2 \cdot 2 \cdot 2$	2	Fourth root

Definition of *n*th Root of a Number

Let a and b be real numbers and let n be an integer such that $n \geq 2$. If

$$a = b^n$$

then b is an ***n*th root of *a*.** If $n = 2$, the root is a **square root.** If $n = 3$, the root is a **cube root.**

Study Tip

In the definition at the right, "the *n*th root that has the same sign as *a*" means that the principal *n*th root of *a* is positive if *a* is positive and negative if *a* is negative. For example, $\sqrt{4} = 2$ and $\sqrt[3]{-8} = -2$. Furthermore, to denote the negative square root of a number, you must use a negative sign in front of the radical. For example, $-\sqrt{4} = -2$.

Some numbers have more than one *n*th root. For example, both 5 and -5 are square roots of 25. To avoid ambiguity about which root you are referring to, the **principal *n*th root** of a number is defined in terms of a **radical symbol** $\sqrt[n]{\ }$. So the *principal square root* of 25, written as $\sqrt{25}$, is the positive root, 5.

Principal *n*th Root of a Number

Let a be a real number that has at least one (real number) nth root. The **principal *n*th root of *a*** is the nth root that has the same sign as a, and it is denoted by the **radical symbol**

$$\sqrt[n]{a}. \qquad \text{Principal } n\text{th root}$$

The positive integer n is the **index** of the radical, and the number a is the **radicand.** If $n = 2$, omit the index and write $\sqrt{a}$ rather than $\sqrt[2]{a}$.

EXAMPLE 1 Finding Roots of Numbers

Find each root.

a. $\sqrt{36}$ **b.** $-\sqrt{36}$ **c.** $\sqrt{-4}$ **d.** $\sqrt[3]{8}$ **e.** $\sqrt[3]{-8}$

Solution

a. $\sqrt{36} = 6$ because $6 \cdot 6 = 6^2 = 36$.

b. $-\sqrt{36} = -6$ because $6 \cdot 6 = 6^2 = 36$. So, $(-1)(\sqrt{36}) = (-1)(6) = -6$.

c. $\sqrt{-4}$ is not real because there is no real number that when multiplied by itself yields -4.

d. $\sqrt[3]{8} = 2$ because $2 \cdot 2 \cdot 2 = 2^3 = 8$.

e. $\sqrt[3]{-8} = -2$ because $(-2)(-2)(-2) = (-2)^3 = -8$.

CHECKPOINT *Now try Exercise 1.*

Properties of nth Roots

Property	*Example*
1. If a is a positive real number and n is even, then a has exactly two (real) nth roots, which are denoted by $\sqrt[n]{a}$ and $-\sqrt[n]{a}$.	The two real square roots of 81 are $\sqrt{81} = 9$ and $-\sqrt{81} = -9$.
2. If a is any real number and n is odd, then a has only one (real) nth root, which is denoted by $\sqrt[n]{a}$.	$\sqrt[3]{27} = 3$ $\sqrt[3]{-64} = -4$
3. If a is a negative real number and n is even, then a has no (real) nth root.	$\sqrt{-64}$ is not a real number.

Numbers such as 1, 4, 9, 16, 49, and 81 are called **perfect squares** because they have rational square roots. Similarly, numbers such as 1, 8, 27, 64, and 125 are called **perfect cubes** because they have rational cube roots.

Study Tip

The square roots of perfect squares are rational numbers, so $\sqrt{25}$, $\sqrt{49}$, and $\sqrt{\frac{4}{9}}$ are rational numbers. However, square roots such as $\sqrt{5}$, $\sqrt{6}$, and $\sqrt{\frac{2}{5}}$ are irrational numbers. Similarly, $\sqrt[3]{27}$ and $\sqrt[4]{16}$ are rational numbers, whereas $\sqrt[3]{6}$ and $\sqrt[4]{21}$ are irrational numbers.

EXAMPLE 2 Classifying Perfect nth Powers

State whether each number is a perfect square, a perfect cube, both, or neither.

a. 81 **b.** -125 **c.** 64 **d.** 32

Solution

a. 81 is a perfect square because $9^2 = 81$. It is not a perfect cube.

b. -125 is a perfect cube because $(-5)^3 = -125$. It is not a perfect square.

c. 64 is a perfect square because $8^2 = 64$, and it is also a perfect cube because $4^3 = 64$.

d. 32 is not a perfect square or a perfect cube. (It is, however, a perfect fifth power, because $2^5 = 32$.)

CHECKPOINT *Now try Exercise 9.*

Raising a number to the nth power and taking the principal nth root of a number can be thought of as *inverse* operations. Here are some examples.

$$\left(\sqrt{4}\right)^2 = (2)^2 = 4 \quad \text{and} \quad \sqrt{4} = \sqrt{2^2} = 2$$
$$\left(\sqrt[3]{27}\right)^3 = (3)^3 = 27 \quad \text{and} \quad \sqrt[3]{27} = \sqrt[3]{3^3} = 3$$
$$\left(\sqrt[4]{16}\right)^4 = (2)^4 = 16 \quad \text{and} \quad \sqrt[4]{16} = \sqrt[4]{2^4} = 2$$
$$\left(\sqrt[5]{-243}\right)^5 = (-3)^5 = -243 \quad \text{and} \quad \sqrt[5]{-243} = \sqrt[5]{(-3)^5} = -3$$

Inverse Properties of nth Powers and nth Roots

Let a be a real number, and let n be an integer such that $n \geq 2$.

Property	*Example*
1. If a has a principal nth root, then $\left(\sqrt[n]{a}\right)^n = a.$	$\left(\sqrt{5}\right)^2 = 5$
2. If n is odd, then $\sqrt[n]{a^n} = a.$	$\sqrt[3]{5^3} = 5$
If n is even, then $\sqrt[n]{a^n} = \lvert a \rvert.$	$\sqrt{(-5)^2} = \lvert -5 \rvert = 5$

EXAMPLE 3 Evaluating Radical Expressions

Evaluate each radical expression.

a. $\sqrt[3]{4^3}$ **b.** $\sqrt[3]{(-2)^3}$ **c.** $\left(\sqrt{7}\right)^2$

d. $\sqrt{(-3)^2}$ **e.** $\sqrt{-3^2}$

Solution

a. Because the index of the radical is odd, you can write

$$\sqrt[3]{4^3} = 4.$$

b. Because the index of the radical is odd, you can write

$$\sqrt[3]{(-2)^3} = -2.$$

c. Because the radicand is positive, $\sqrt{7}$ is real and you can write

$$\left(\sqrt{7}\right)^2 = 7.$$

d. Because the index of the radical is even, you must include absolute value signs, and write

$$\sqrt{(-3)^2} = \lvert -3 \rvert = 3.$$

e. Because $\sqrt{-3^2} = \sqrt{-9}$ is an even root of a negative number, its value is not a real number.

✓ ***CHECKPOINT*** ***Now try Exercise 39.***

Study Tip

In parts (d) and (e) of Example 3, notice that the two expressions inside the radical are different. In part (d), the negative sign is part of the base of the exponential expression. In part (e), the negative sign is not part of the base.

2 ▶ Use the rules of exponents to evaluate or simplify expressions with rational exponents.

Rational Exponents

So far in the text you have worked with algebraic expressions involving only integer exponents. Next you will see that algebraic expressions may also contain **rational exponents.**

Definition of Rational Exponents

Let a be a real number, and let n be an integer such that $n \geq 2$. If the principal nth root of a exists, then $a^{1/n}$ is defined as

$$a^{1/n} = \sqrt[n]{a}.$$

If m is a positive integer that has no common factor with n, then

$$a^{m/n} = (a^{1/n})^m = \left(\sqrt[n]{a}\right)^m \quad \text{and} \quad a^{m/n} = (a^m)^{1/n} = \sqrt[n]{a^m}.$$

Study Tip

The numerator of a rational exponent denotes the *power* to which the base is raised, and the denominator denotes the *root* to be taken.

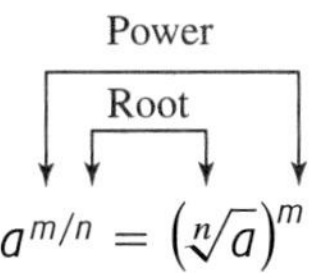

$$a^{m/n} = \left(\sqrt[n]{a}\right)^m$$

It does not matter in which order the two operations are performed, provided the nth root exists. Here is an example.

$8^{2/3} = \left(\sqrt[3]{8}\right)^2 = 2^2 = 4$ Cube root, then second power

$8^{2/3} = \sqrt[3]{8^2} = \sqrt[3]{64} = 4$ Second power, then cube root

The rules of exponents that were listed in Section 5.1 also apply to rational exponents (provided the roots indicated by the denominators exist). These rules are listed below, with different examples.

Technology: Discovery

Use a calculator to evaluate each pair of expressions below.

a. $5.6^{3.5} \cdot 5.6^{0.4}$ and $5.6^{3.9}$

b. $\dfrac{3.4^{4.6}}{3.4^{3.1}}$ and $3.4^{1.5}$

c. $6.2^{-0.75}$ and $\dfrac{1}{6.2^{0.75}}$

What rule is illustrated by each pair of expressions? Use your calculator to illustrate some of the other rules of exponents.

Summary of Rules of Exponents

Let r and s be rational numbers, and let a and b be real numbers, variables, or algebraic expressions. (All denominators and bases are nonzero.)

Product and Quotient Rules	*Example*
1. $a^r \cdot a^s = a^{r+s}$	$4^{1/2}(4^{1/3}) = 4^{5/6}$
2. $\dfrac{a^r}{a^s} = a^{r-s}$	$\dfrac{x^2}{x^{1/2}} = x^{2-(1/2)} = x^{3/2}$
Power Rules	
3. $(ab)^r = a^r \cdot b^r$	$(2x)^{1/2} = 2^{1/2}(x^{1/2})$
4. $(a^r)^s = a^{rs}$	$(x^3)^{1/2} = x^{3/2}$
5. $\left(\dfrac{a}{b}\right)^r = \dfrac{a^r}{b^r}$	$\left(\dfrac{x}{3}\right)^{2/3} = \dfrac{x^{2/3}}{3^{2/3}}$
Zero and Negative Exponent Rules	
6. $a^0 = 1$	$(3x)^0 = 1$
7. $a^{-r} = \dfrac{1}{a^r}$	$4^{-3/2} = \dfrac{1}{4^{3/2}} = \dfrac{1}{(2)^3} = \dfrac{1}{8}$
8. $\left(\dfrac{a}{b}\right)^{-r} = \left(\dfrac{b}{a}\right)^r$	$\left(\dfrac{x}{4}\right)^{-1/2} = \left(\dfrac{4}{x}\right)^{1/2} = \dfrac{2}{x^{1/2}}$

EXAMPLE 4 Evaluating Expressions with Rational Exponents

Evaluate each expression.

a. $8^{4/3}$ **b.** $(4^2)^{3/2}$ **c.** $25^{-3/2}$

d. $\left(\dfrac{64}{125}\right)^{2/3}$ **e.** $-16^{1/2}$ **f.** $(-16)^{1/2}$

Solution

a. $8^{4/3} = (8^{1/3})^4 = (\sqrt[3]{8})^4 = 2^4 = 16$ Root is 3. Power is 4.

b. $(4^2)^{3/2} = 4^{2 \cdot (3/2)} = 4^{6/2} = 4^3 = 64$ Root is 2. Power is 3.

c. $25^{-3/2} = \dfrac{1}{25^{3/2}} = \dfrac{1}{(\sqrt{25})^3} = \dfrac{1}{5^3} = \dfrac{1}{125}$ Root is 2. Power is 3.

d. $\left(\dfrac{64}{125}\right)^{2/3} = \dfrac{64^{2/3}}{125^{2/3}} = \dfrac{(\sqrt[3]{64})^2}{(\sqrt[3]{125})^2} = \dfrac{4^2}{5^2} = \dfrac{16}{25}$ Root is 3. Power is 2.

e. $-16^{1/2} = -\sqrt{16} = -(4) = -4$ Root is 2. Power is 1.

f. $(-16)^{1/2} = \sqrt{-16}$ is not a real number. Root is 2. Power is 1.

CHECKPOINT *Now try Exercise 75.*

Study Tip

In parts (e) and (f) of Example 4, be sure that you see the distinction between the expressions $-16^{1/2}$ and $(-16)^{1/2}$.

EXAMPLE 5 Using Rules of Exponents

Rewrite each expression using rational exponents.

a. $x\sqrt[4]{x^3}$ **b.** $\dfrac{\sqrt[3]{x^2}}{\sqrt{x^3}}$ **c.** $\sqrt[3]{x^2y}$

Solution

a. $x\sqrt[4]{x^3} = x(x^{3/4}) = x^{1+(3/4)} = x^{7/4}$

b. $\dfrac{\sqrt[3]{x^2}}{\sqrt{x^3}} = \dfrac{x^{2/3}}{x^{3/2}} = x^{(2/3)-(3/2)} = x^{-5/6} = \dfrac{1}{x^{5/6}}$

c. $\sqrt[3]{x^2y} = (x^2y)^{1/3} = (x^2)^{1/3}y^{1/3} = x^{2/3}y^{1/3}$

CHECKPOINT *Now try Exercise 91.*

EXAMPLE 6 Using Rules of Exponents

Use rules of exponents to simplify each expression.

a. $\sqrt{\sqrt[3]{x}}$ **b.** $\dfrac{(2x-1)^{4/3}}{\sqrt[3]{2x-1}}$

Solution

a. $\sqrt{\sqrt[3]{x}} = \sqrt{x^{1/3}} = (x^{1/3})^{1/2} = x^{(1/3)(1/2)} = x^{1/6}$

b. $\dfrac{(2x-1)^{4/3}}{\sqrt[3]{2x-1}} = \dfrac{(2x-1)^{4/3}}{(2x-1)^{1/3}} = (2x-1)^{(4/3)-(1/3)} = (2x-1)^{3/3} = 2x-1$

CHECKPOINT *Now try Exercise 107.*

3 ▶ Use a calculator to evaluate radical expressions.

Radicals and Calculators

There are two methods of evaluating radicals on most calculators. For square roots, you can use the *square root key* [√] or [√x]. For other roots, you can first convert the radical to exponential form and then use the *exponential key* [yˣ] or [^].

Technology: Tip

Some calculators have cube root functions $\sqrt[3]{\ }$ or $\sqrt[3]{x}$ and xth root functions $\sqrt[x]{\ }$ or $\sqrt[x]{y}$ that can be used to evaluate roots other than square roots. Consult the user's guide of your calculator for specific keystrokes.

EXAMPLE 7 Evaluating Roots with a Calculator

Evaluate each expression. Round the result to three decimal places.

a. $\sqrt{5}$ **b.** $\sqrt[5]{25}$ **c.** $\sqrt[3]{-4}$ **d.** $(-1.4)^{3/2}$

Solution

a. 5 [√x] Scientific

[√] 5 [ENTER] Graphing

The display is 2.236067977. Rounded to three decimal places, $\sqrt{5} \approx 2.236$.

b. First rewrite the expression as $\sqrt[5]{25} = 25^{1/5}$. Then use one of the following keystroke sequences.

25 [yˣ] [(] 1 [÷] 5 [)] [=] Scientific

25 [^] [(] 1 [÷] 5 [)] [ENTER] Graphing

The display is 1.903653939. Rounded to three decimal places, $\sqrt[5]{25} \approx 1.904$.

c. If your calculator does not have a cube root key, use the fact that

$$\sqrt[3]{-4} = \sqrt[3]{(-1)(4)} = \sqrt[3]{-1}\sqrt[3]{4} = -\sqrt[3]{4} = -4^{1/3}$$

and attach the negative sign as the last keystroke.

4 [yˣ] [(] 1 [÷] 3 [)] [=] [+/−] Scientific

[∛] [(−)] 4 [)] [ENTER] Graphing

The display is −1.587401052. Rounded to three decimal places, $\sqrt[3]{-4} \approx -1.587$.

d. 1.4 [+/−] [yˣ] [(] 3 [÷] 2 [)] [=] Scientific

[(] [(−)] 1.4 [)] [^] [(] 3 [÷] 2 [)] [ENTER] Graphing

The display should indicate an error because an even root of a negative number is not real.

✓ ***CHECKPOINT*** *Now try Exercise 129.*

4 ▶ Evaluate radical functions and find the domains of radical functions.

Radical Functions

A **radical function** is a function that contains a radical such as

$$f(x) = \sqrt{x} \quad \text{or} \quad g(x) = \sqrt[3]{x}.$$

When evaluating a radical function, note that the radical symbol is a symbol of grouping.

Technology: Discovery

Consider the function $f(x) = x^{2/3}$.

a. What is the domain of the function?

b. Use your graphing calculator to graph each equation, in order.

$y_1 = x^{(2 \div 3)}$

$y_2 = (x^2)^{1/3}$ Power, then root

$y_3 = (x^{1/3})^2$ Root, then power

c. Are the graphs all the same? Are their domains all the same?

d. On your graphing calculator, which of the forms properly represent the function $f(x) = x^{m/n}$?

$y_1 = x^{(m \div n)}$

$y_2 = (x^m)^{1/n}$

$y_3 = (x^{1/n})^m$

e. Explain how the domains of $f(x) = x^{2/3}$ and $g(x) = x^{-2/3}$ differ.

EXAMPLE 8 Evaluating Radical Functions

Evaluate each radical function when $x = 4$.

a. $f(x) = \sqrt[3]{x - 31}$ **b.** $g(x) = \sqrt{16 - 3x}$

Solution

a. $f(4) = \sqrt[3]{4 - 31} = \sqrt[3]{-27} = -3$

b. $g(4) = \sqrt{16 - 3(4)} = \sqrt{16 - 12} = \sqrt{4} = 2$

CHECKPOINT *Now try Exercise 143.*

The **domain** of the radical function $f(x) = \sqrt[n]{x}$ is the set of all real numbers such that x has a principal nth root.

Domain of a Radical Function

Let n be an integer that is greater than or equal to 2.

1. If n is odd, the domain of $f(x) = \sqrt[n]{x}$ is the set of all real numbers.
2. If n is even, the domain of $f(x) = \sqrt[n]{x}$ is the set of all nonnegative real numbers.

EXAMPLE 9 Finding the Domains of Radical Functions

Describe the domains of (a) $f(x) = \sqrt[3]{x}$ and (b) $f(x) = \sqrt{x^3}$.

Solution

a. The domain of $f(x) = \sqrt[3]{x}$ is the set of all real numbers because for any real number x, the expression $\sqrt[3]{x}$ is a real number.

b. The domain of $f(x) = \sqrt{x^3}$ is the set of all nonnegative real numbers. For instance, 1 is in the domain but -1 is not because $\sqrt{(-1)^3} = \sqrt{-1}$ is not a real number.

CHECKPOINT *Now try Exercise 149.*

EXAMPLE 10 Finding the Domain of a Radical Function

Find the domain of $f(x) = \sqrt{2x - 1}$.

Solution

The domain of f consists of all x such that $2x - 1 \geq 0$. Using the methods described in Section 3.6, you can solve this inequality as follows.

$2x - 1 \geq 0$ Write original inequality.

$2x \geq 1$ Add 1 to each side.

$x \geq \frac{1}{2}$ Divide each side by 2.

So, the domain is the set of all real numbers x such that $x \geq \frac{1}{2}$.

CHECKPOINT *Now try Exercise 155.*

Study Tip

In general, when the index n of a radical function is even, the domain of the function includes all real values for which the expression under the radical is greater than or equal to zero.

Concept Check

1. Describe all values of n and x for which x has each number of real nth roots.

   (a) 2 (b) 1 (c) 0

2. Describe two ways of performing two operations to evaluate the expression $32^{2/5}$.

3. You rewrite a radical expression using a rational exponent. What part of the rational exponent corresponds to the index of the radical?

4. If n is even, what must be true about the radicand for the nth root to be a real number?

9.1 EXERCISES

Go to pages 608–609 to record your assignments.

Developing Skills

In Exercises 1–8, find the root if it exists. ***See Example 1.***

 **1.** $\sqrt{64}$ **2.** $-\sqrt{100}$

3. $-\sqrt{49}$ **4.** $\sqrt{-25}$

5. $\sqrt[3]{-27}$ **6.** $\sqrt[3]{-64}$

7. $\sqrt{-1}$ **8.** $-\sqrt{-1}$

In Exercises 9–14, state whether the number is a perfect square, a perfect cube, or neither. ***See Example 2.***

9. 49 **10.** -27

11. 1728 **12.** 964

13. 96 **14.** 225

In Exercises 15–22, find all of the square roots of the perfect square.

15. 25 **16.** 121

17. $\frac{9}{16}$ **18.** $\frac{25}{36}$

19. $\frac{1}{49}$ **20.** $\frac{81}{100}$

21. 0.16 **22.** 0.25

In Exercises 23–30, find all of the cube roots of the perfect cube.

23. 8 **24.** -8

25. $-\frac{1}{27}$ **26.** $\frac{27}{64}$

27. $\frac{1}{1000}$ **28.** $-\frac{8}{125}$

29. 0.001 **30.** -0.008

In Exercises 31–38, determine whether the square root is a rational or an irrational number.

31. $\sqrt{3}$ **32.** $\sqrt{6}$

33. $\sqrt{16}$ **34.** $\sqrt{25}$

35. $\sqrt{\frac{4}{25}}$ **36.** $\sqrt{\frac{2}{7}}$

37. $\sqrt{\frac{3}{16}}$ **38.** $\sqrt{\frac{1}{49}}$

In Exercises 39–68, evaluate the radical expression without using a calculator. If not possible, state the reason. ***See Example 3.***

39. $\sqrt{8^2}$ **40.** $-\sqrt{10^2}$

41. $\sqrt{(-10)^2}$ **42.** $\sqrt{(-12)^2}$

43. $\sqrt{-9^2}$ **44.** $\sqrt{-12^2}$

45. $-\sqrt{\left(\frac{2}{3}\right)^2}$ **46.** $\sqrt{\left(\frac{3}{4}\right)^2}$

47. $\sqrt{-\left(\frac{3}{10}\right)^2}$ **48.** $\sqrt{\left(-\frac{3}{5}\right)^2}$

49. $\left(\sqrt{5}\right)^2$ **50.** $-\left(\sqrt{10}\right)^2$

51. $-\left(\sqrt{23}\right)^2$ **52.** $\left(-\sqrt{18}\right)^2$

53. $\sqrt[3]{5^3}$ **54.** $\sqrt[3]{(-7)^3}$

55. $\sqrt[3]{10^3}$ **56.** $\sqrt[3]{4^3}$

57. $-\sqrt[3]{(-6)^3}$ **58.** $-\sqrt[3]{9^3}$

59. $\sqrt[3]{\left(-\frac{1}{4}\right)^3}$ **60.** $-\sqrt[3]{\left(\frac{1}{5}\right)^3}$

61. $\left(\sqrt[3]{11}\right)^3$ **62.** $\left(\sqrt[3]{-6}\right)^3$

63. $\left(-\sqrt[3]{24}\right)^3$ **64.** $\left(\sqrt[3]{21}\right)^3$

65. $\sqrt[4]{3^4}$ **66.** $\sqrt[5]{(-2)^5}$

67. $-\sqrt[4]{2^4}$ **68.** $-\sqrt[4]{-5^4}$

In Exercises 69–72, fill in the missing description.

	Radical Form	*Rational Exponent Form*
69.	$\sqrt{36} = 6$	
70.	$\sqrt[3]{27^2} = 9$	
71.		$256^{3/4} = 64$
72.		$125^{1/3} = 5$

In Exercises 73–88, evaluate without using a calculator. ***See Example 4.***

73. $25^{1/2}$ **74.** $49^{1/2}$

75. $-36^{1/2}$ **76.** $-121^{1/2}$

77. $32^{-2/5}$ **78.** $81^{-3/4}$

79. $(-27)^{-2/3}$ **80.** $(-243)^{-3/5}$

81. $\left(\frac{8}{27}\right)^{2/3}$ **82.** $\left(\frac{256}{625}\right)^{1/4}$

83. $\left(\frac{121}{9}\right)^{-1/2}$ **84.** $\left(\frac{27}{1000}\right)^{-4/3}$

85. $(-3^3)^{2/3}$ **86.** $(8^2)^{3/2}$

87. $-(4^4)^{3/4}$ **88.** $(-2^3)^{5/3}$

In Exercises 89–106, rewrite the expression using rational exponents. ***See Example 5.***

89. $\sqrt{t}$ **90.** $\sqrt[3]{x}$

91. $x\sqrt[3]{x^6}$ **92.** $t\sqrt[5]{t^2}$

93. $u^2\sqrt[3]{u}$ **94.** $y\sqrt[4]{y^2}$

95. $\dfrac{\sqrt{x}}{\sqrt{x^3}}$ **96.** $\dfrac{\sqrt[3]{x^2}}{\sqrt[3]{x^4}}$

97. $\dfrac{\sqrt[4]{t}}{\sqrt{t^5}}$ **98.** $\dfrac{\sqrt[3]{x^4}}{\sqrt{x^3}}$

99. $\sqrt[3]{x^2} \cdot \sqrt[3]{x^7}$ **100.** $\sqrt[5]{z^3} \cdot \sqrt[5]{z^2}$

101. $\sqrt[4]{y^3} \cdot \sqrt[3]{y}$ **102.** $\sqrt[6]{x^5} \cdot \sqrt[3]{x^4}$

103. $\sqrt[4]{x^3y}$ **104.** $\sqrt[3]{u^4v^2}$

105. $z^2\sqrt{y^5z^4}$ **106.** $x^2\sqrt[3]{xy^4}$

In Exercises 107–128, simplify the expression. ***See Example 6.***

107. $3^{1/4} \cdot 3^{3/4}$ **108.** $2^{2/5} \cdot 2^{3/5}$

109. $(2^{1/2})^{2/3}$ **110.** $(4^{1/3})^{9/4}$

111. $\dfrac{2^{1/5}}{2^{6/5}}$ **112.** $\dfrac{5^{-3/4}}{5}$

113. $(c^{3/2})^{1/3}$ **114.** $(k^{-1/3})^{3/2}$

115. $\dfrac{18y^{4/3}z^{-1/3}}{24y^{-2/3}z}$ **116.** $\dfrac{a^{3/4} \cdot a^{1/2}}{a^{5/2}}$

117. $(3x^{-1/3}y^{3/4})^2$ **118.** $(-2u^{3/5}v^{-1/5})^3$

119. $\left(\dfrac{x^{1/4}}{x^{1/6}}\right)^3$ **120.** $\left(\dfrac{3m^{1/6}n^{1/3}}{4n^{-2/3}}\right)^2$

121. $\sqrt{\sqrt[4]{y}}$ **122.** $\sqrt[3]{\sqrt{2x}}$

123. $\sqrt[4]{\sqrt{x^3}}$ **124.** $\sqrt[5]{\sqrt[3]{y^4}}$

125. $\dfrac{(x+y)^{3/4}}{\sqrt[4]{x+y}}$ **126.** $\dfrac{(a-b)^{1/3}}{\sqrt[3]{a-b}}$

127. $\dfrac{(3u-2v)^{2/3}}{\sqrt{(3u-2v)^3}}$ **128.** $\dfrac{\sqrt[4]{2x+y}}{(2x+y)^{3/2}}$

In Exercises 129–142, use a calculator to evaluate the expression. Round your answer to four decimal places. If not possible, state the reason. ***See Example 7.***

129. $\sqrt{35}$ **130.** $\sqrt{-23}$

131. $315^{2/5}$ **132.** $962^{2/3}$

133. $82^{-3/4}$ **134.** $382.5^{-3/2}$

135. $\sqrt[4]{212}$ **136.** $\sqrt[3]{-411}$

137. $\sqrt[3]{545^2}$ **138.** $\sqrt[5]{-35^3}$

139. $\dfrac{8-\sqrt{35}}{2}$ **140.** $\dfrac{-5+\sqrt{3215}}{10}$

141. $\dfrac{3+\sqrt{17}}{9}$ **142.** $\dfrac{7-\sqrt{241}}{12}$

In Exercises 143–148, evaluate the function for each indicated x-value, if possible, and simplify. ***See Example 8.***

143. $f(x) = \sqrt{2x+9}$
(a) $f(0)$ (b) $f(8)$ (c) $f(-6)$ (d) $f(36)$

144. $g(x) = \sqrt{5x-6}$
(a) $g(0)$ (b) $g(2)$ (c) $g(30)$ (d) $g\left(\frac{7}{5}\right)$

145. $g(x) = \sqrt[3]{x+1}$
(a) $g(7)$ (b) $g(26)$ (c) $g(-9)$ (d) $g(-65)$

146. $f(x) = \sqrt[3]{2x-1}$
(a) $f(0)$ (b) $f(-62)$ (c) $f(-13)$ (d) $f(63)$

147. $f(x) = \sqrt[4]{x-3}$
(a) $f(19)$ (b) $f(1)$ (c) $f(84)$ (d) $f(4)$

148. $g(x) = \sqrt[4]{x+1}$
(a) $g(0)$ (b) $g(15)$ (c) $g(-82)$ (d) $g(80)$

In Exercises 149–158, describe the domain of the function. ***See Examples 9 and 10.***

✓ **149.** $f(x) = 3\sqrt{x}$

150. $h(x) = \sqrt[4]{x}$

151. $g(x) = \sqrt{4-9x}$

152. $g(x) = \sqrt{10-2x}$

153. $f(x) = \sqrt[3]{x^4}$

154. $f(x) = \sqrt{-x}$

✓ **155.** $h(x) = \sqrt{2x+9}$

156. $f(x) = \sqrt{3x-5}$

157. $g(x) = \dfrac{2}{\sqrt[4]{x}}$

158. $g(x) = \dfrac{10}{\sqrt[3]{x}}$

In Exercises 159–162, describe the domain of the function. Then check your answer by using a graphing calculator to graph the function.

159. $y = \dfrac{5}{\sqrt[4]{x^3}}$

160. $y = 4\sqrt[3]{x}$

161. $g(x) = 2x^{3/5}$

162. $h(x) = 5x^{2/3}$

In Exercises 163–166, perform the multiplication. Use a graphing calculator to confirm your result.

163. $x^{1/2}(2x-3)$

164. $x^{4/3}(3x^2-4x+5)$

165. $y^{-1/3}(y^{1/3}+5y^{4/3})$

166. $(x^{1/2}-3)(x^{1/2}+3)$

Solving Problems

Mathematical Modeling In Exercises 167 and 168, use the formula for the *declining balances method*

$$r = 1 - \left(\frac{S}{C}\right)^{1/n}$$

to find the depreciation rate r. In the formula, n is the useful life of the item (in years), S is the salvage value (in dollars), and C is the original cost (in dollars).

167. A \$75,000 truck depreciates over an eight-year period, as shown in the graph. Find r. (Round your answer to three decimal places.)

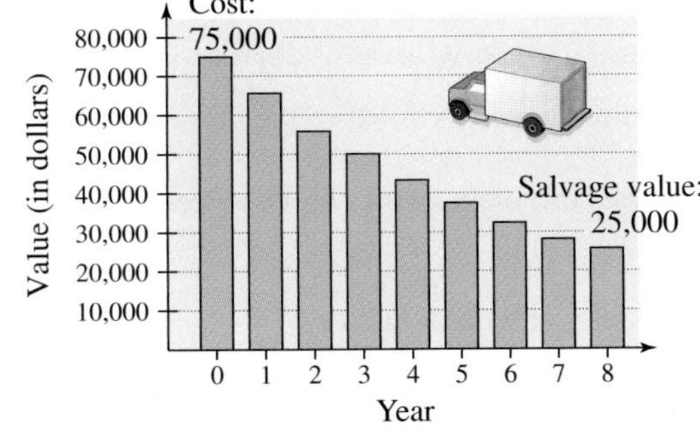

168. A \$125,000 stretch limousine depreciates over a 10-year period, as shown in the graph. Find r. (Round your answer to three decimal places.)

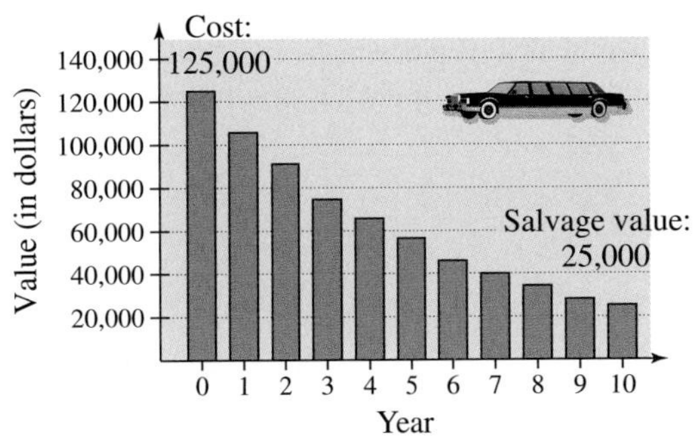

169. *Geometry* Find the dimensions of a piece of carpet for a classroom with 529 square feet of floor space, assuming the floor is square.

170. *Geometry* Find the dimensions of a square mirror with an area of 1024 square inches.

171. *Geometry* The length D of a diagonal of a rectangular solid of length l, width w, and height h is represented by $D = \sqrt{l^2 + w^2 + h^2}$. Approximate to two decimal places the length of the diagonal D of the bed of the pickup truck shown in the figure.

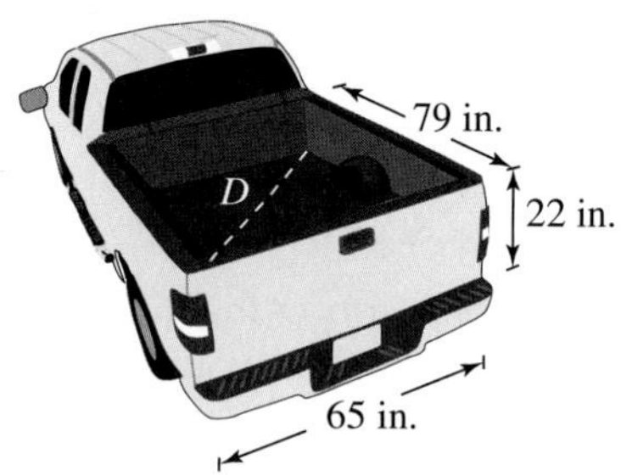

172. *Velocity* A stream of water moving at a rate of v feet per second can carry a particle of a certain type if its diameter is at most $0.03\sqrt{v}$ inches.

(a) Find the largest particle size that can be carried by a stream flowing at the rate of $\frac{3}{4}$ foot per second. Round your answer to three decimal places.

(b) Find the largest particle size that can be carried by a stream flowing at the rate of $\frac{3}{16}$ foot per second. Round your answer to three decimal places.

Explaining Concepts

173. In your own words, explain what the nth root of a number is.

174. Explain how you can determine the domain of a function that has the form of a fraction with radical expressions in both the numerator and the denominator.

175. Is it true that $\sqrt{2} = 1.414$? Explain.

176. Given that x represents a real number, state the conditions on n for each of the following.

(a) $\sqrt[n]{x^n} = x$

(b) $\sqrt[n]{x^n} = |x|$

177. *Investigation* Find all possible "last digits" of perfect squares. (For instance, the last digit of 81 is 1 and the last digit of 64 is 4.) Is it possible that 4,322,788,986 is a perfect square?

178. Use what you know about the domains of radical functions to write a set of rules for the domain of a rational exponent function of the form $f(x) = x^{1/n}$. Can the same rules be used for a function of the form $f(x) = x^{m/n}$? Explain.

Cumulative Review

In Exercises 179–182, solve the equation.

179. $\frac{a}{5} = \frac{a - 3}{2}$

180. $\frac{x}{3} - \frac{3x}{4} = \frac{5x}{12}$

181. $\frac{2}{u + 4} = \frac{5}{8}$

182. $\frac{6}{b} + 22 = 24$

In Exercises 183–186, write a model for the statement.

183. s is directly proportional to the square of t.

184. r varies inversely as the fourth power of x.

185. a varies jointly as b and c.

186. x is directly proportional to y and inversely proportional to z.

9.2 Simplifying Radical Expressions

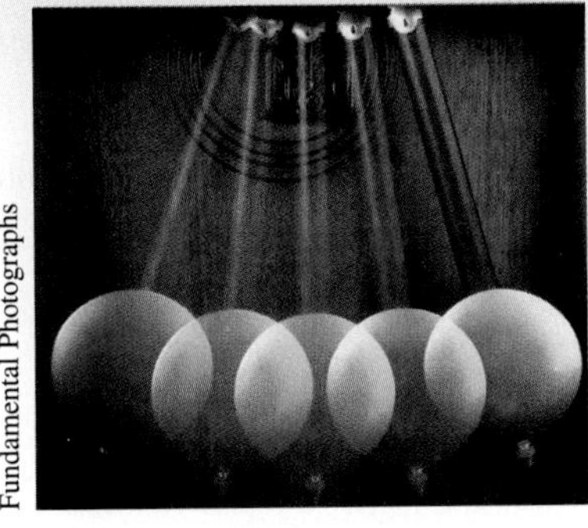

Fundamental Photographs

What You Should Learn

1 ▶ Use the Product and Quotient Rules for Radicals to simplify radical expressions.

2 ▶ Use rationalization techniques to simplify radical expressions.

3 ▶ Use the Pythagorean Theorem in application problems.

Why You Should Learn It

Algebraic equations often involve radicals. For instance, in Exercise 76 on page 573, you will use a radical equation to find the period of a pendulum.

Simplifying Radicals

In this section, you will study ways to simplify radicals. For instance, the expression $\sqrt{12}$ can be simplified as

$$\sqrt{12} = \sqrt{4 \cdot 3} = \sqrt{4}\sqrt{3} = 2\sqrt{3}.$$

This rewritten form is based on the following rules for multiplying and dividing radicals.

1 ▶ Use the Product and Quotient Rules for Radicals to simplify radical expressions.

Product and Quotient Rules for Radicals

Let u and v be real numbers, variables, or algebraic expressions. If the nth roots of u and v are real, the following rules are true.

1. $\sqrt[n]{uv} = \sqrt[n]{u}\sqrt[n]{v}$ — Product Rule for Radicals
2. $\sqrt[n]{\dfrac{u}{v}} = \dfrac{\sqrt[n]{u}}{\sqrt[n]{v}}, \quad v \neq 0$ — Quotient Rule for Radicals

Study Tip

The Product and Quotient Rules for Radicals can be shown to be true by converting the radicals to exponential form and using the rules of exponents on page 559.

Using Rule 3

$$\sqrt[n]{uv} = (uv)^{1/n} = u^{1/n}v^{1/n} = \sqrt[n]{u}\sqrt[n]{v}$$

Using Rule 5

$$\sqrt[n]{\frac{u}{v}} = \left(\frac{u}{v}\right)^{1/n} = \frac{u^{1/n}}{v^{1/n}} = \frac{\sqrt[n]{u}}{\sqrt[n]{v}}$$

You can use the Product Rule for Radicals to *simplify* square root expressions by finding the largest perfect square factor and removing it from the radical, as follows.

$$\sqrt{48} = \sqrt{16 \cdot 3} = \sqrt{16}\sqrt{3} = 4\sqrt{3}$$

This process is called **removing perfect square factors from the radical.**

EXAMPLE 1 Removing Constant Factors from Radicals

Simplify each radical by removing as many factors as possible.

a. $\sqrt{75}$ **b.** $\sqrt{72}$ **c.** $\sqrt{162}$

Solution

a. $\sqrt{75} = \sqrt{25 \cdot 3} = \sqrt{25}\sqrt{3} = 5\sqrt{3}$ — 25 is a perfect square factor of 75.

b. $\sqrt{72} = \sqrt{36 \cdot 2} = \sqrt{36}\sqrt{2} = 6\sqrt{2}$ — 36 is a perfect square factor of 72.

c. $\sqrt{162} = \sqrt{81 \cdot 2} = \sqrt{81}\sqrt{2} = 9\sqrt{2}$ — 81 is a perfect square factor of 162.

✓ **CHECKPOINT** *Now try Exercise 1.*

When removing *variable* factors from a square root radical, remember that it is not valid to write $\sqrt{x^2} = x$ *unless* you happen to know that x is nonnegative. Without knowing anything about x, the only way you can simplify $\sqrt{x^2}$ is to include absolute value signs when you remove x from the radical.

$\sqrt{x^2} = |x|$ — Restricted by absolute value signs

When simplifying the expression $\sqrt{x^3}$, it is not necessary to include absolute value signs because the domain does not include negative numbers.

$\sqrt{x^3} = \sqrt{x^2(x)} = x\sqrt{x}$ — Restricted by domain of radical

EXAMPLE 2 Removing Variable Factors from Radicals

Simplify each radical expression.

a. $\sqrt{25x^2}$ **b.** $\sqrt{12x^3}$ **c.** $\sqrt{144x^4}$ **d.** $\sqrt{72x^3y^2}$

Solution

a. $\sqrt{25x^2} = \sqrt{5^2x^2} = \sqrt{5^2}\sqrt{x^2}$ — Product Rule for Radicals

$= 5|x|$ — $\sqrt{x^2} = |x|$

b. $\sqrt{12x^3} = \sqrt{2^2x^2(3x)} = \sqrt{2^2}\sqrt{x^2}\sqrt{3x}$ — Product Rule for Radicals

$= 2x\sqrt{3x}$ — $\sqrt{2^2}\sqrt{x^2} = 2x, \quad x \geq 0$

c. $\sqrt{144x^4} = \sqrt{12^2(x^2)^2} = \sqrt{12^2}\sqrt{(x^2)^2}$ — Product Rule for Radicals

$= 12x^2$ — $\sqrt{12^2}\sqrt{(x^2)^2} = 12|x^2| = 12x^2$

d. $\sqrt{72x^3y^2} = \sqrt{6^2x^2y^2} \cdot \sqrt{2x}$ — Product Rule for Radicals

$= \sqrt{6^2}\sqrt{x^2}\sqrt{y^2} \cdot \sqrt{2x}$ — Product Rule for Radicals

$= 6x|y|\sqrt{2x}$ — $\sqrt{6^2}\sqrt{x^2}\sqrt{y^2} = 6x|y|, x \geq 0$

✓ **CHECKPOINT** *Now try Exercise 19.*

You can use the inverse properties of nth powers and nth roots described in Section 9.1 to remove perfect nth powers from nth root radicals.

EXAMPLE 3 Removing Factors from Radicals

Simplify each radical expression.

a. $\sqrt[3]{40}$ **b.** $\sqrt[4]{x^5}$

Solution

a. $\sqrt[3]{40} = \sqrt[3]{8(5)} = \sqrt[3]{2^3} \cdot \sqrt[3]{5}$ — Product Rule for Radicals

$= 2\sqrt[3]{5}$ — $\sqrt[3]{2^3} = 2$

b. $\sqrt[4]{x^5} = \sqrt[4]{x^4(x)} = \sqrt[4]{x^4}\sqrt[4]{x}$ — Product Rule for Radicals

$= x\sqrt[4]{x}$ — $\sqrt[4]{x^4} = x, \quad x \geq 0$

✓ **CHECKPOINT** *Now try Exercise 29.*

Study Tip

When you write a number as a product of prime numbers, you are writing its *prime factorization.* To find the perfect nth root factor of 486 in Example 4(a), you can write the prime factorization of 486.

$$486 = 2 \cdot 3 \cdot 3 \cdot 3 \cdot 3 \cdot 3$$
$$= 2 \cdot 3^5$$

From its prime factorization, you can see that 3^5 is a fifth root factor of 486.

$$\sqrt[5]{486} = \sqrt[5]{2 \cdot 3^5}$$
$$= \sqrt[5]{3^5}\sqrt[5]{2}$$
$$= 3\sqrt[5]{2}$$

EXAMPLE 4 Removing Factors from Radicals

Simplify each radical expression.

a. $\sqrt[5]{486x^7}$ **b.** $\sqrt[3]{128x^3y^5}$

Solution

a. $\sqrt[5]{486x^7} = \sqrt[5]{243x^5(2x^2)} = \sqrt[5]{3^5x^5} \cdot \sqrt[5]{2x^2}$ Product Rule for Radicals

$= 3x\sqrt[5]{2x^2}$ $\sqrt[5]{3^5}\sqrt[5]{x^5} = 3x$

b. $\sqrt[3]{128x^3y^5} = \sqrt[3]{64x^3y^3(2y^2)} = \sqrt[3]{4^3x^3y^3} \cdot \sqrt[3]{2y^2}$ Product Rule for Radicals

$= 4xy\sqrt[3]{2y^2}$ $\sqrt[3]{4^3}\sqrt[3]{x^3}\sqrt[3]{y^3} = 4xy$

✓ **CHECKPOINT** *Now try Exercise 35.*

EXAMPLE 5 Removing Factors from Radicals

Simplify each radical expression.

a. $\sqrt{\dfrac{81}{25}}$ **b.** $\dfrac{\sqrt{56x^2}}{\sqrt{8}}$

Solution

a. $\sqrt{\dfrac{81}{25}} = \dfrac{\sqrt{81}}{\sqrt{25}} = \dfrac{9}{5}$ Quotient Rule for Radicals

b. $\dfrac{\sqrt{56x^2}}{\sqrt{8}} = \sqrt{\dfrac{56x^2}{8}}$ Quotient Rule for Radicals

$= \sqrt{7x^2}$ Simplify.

$= \sqrt{7} \cdot \sqrt{x^2}$ Product Rule for Radicals

$= \sqrt{7}|x|$ $\sqrt{x^2} = |x|$

✓ **CHECKPOINT** *Now try Exercise 43.*

EXAMPLE 6 Removing Factors from Radicals

Simplify $-\sqrt[3]{\dfrac{y^5}{27x^3}}$.

Solution

$-\sqrt[3]{\dfrac{y^5}{27x^3}} = -\dfrac{\sqrt[3]{y^3y^2}}{\sqrt[3]{27x^3}}$ Quotient Rule for Radicals

$= -\dfrac{\sqrt[3]{y^3} \cdot \sqrt[3]{y^2}}{\sqrt[3]{27} \cdot \sqrt[3]{x^3}}$ Product Rule for Radicals

$= -\dfrac{y\sqrt[3]{y^2}}{3x}$ Simplify.

CHECKPOINT Now try Exercise 49.

2 ▶ Use rationalization techniques to simplify radical expressions.

Rationalization Techniques

Removing factors from radicals is only one of two techniques used to simplify radicals. Three conditions must be met in order for a radical expression to be in simplest form. These three conditions are summarized as follows.

Simplifying Radical Expressions

A radical expression is said to be in *simplest form* if all three of the statements below are true.

1. All possible *n*th powered factors have been removed from each radical.
2. No radical contains a fraction.
3. No denominator of a fraction contains a radical.

To meet the last two conditions, you can use a second technique for simplifying radical expressions called **rationalizing the denominator.** This involves multiplying both the numerator and the denominator by a *rationalizing factor* that creates a perfect *n*th power in the denominator.

Study Tip

When rationalizing a denominator, remember that for square roots you want a perfect square in the denominator, for cube roots you want a perfect cube, and so on. For instance, to find the rationalizing factor needed to create a perfect square in the denominator of Example 7(c), you can write the prime factorization of 18.

$$18 = 2 \cdot 3 \cdot 3$$
$$= 2 \cdot 3^2$$

From its prime factorization, you can see that 3^2 is a square root factor of 18. You need one more factor of 2 to create a perfect square in the denominator:

$$2 \cdot (2 \cdot 3^2) = 2 \cdot 2 \cdot 3^2$$
$$= 2^2 \cdot 3^2$$
$$= 4 \cdot 9 = 36.$$

EXAMPLE 7 Rationalizing the Denominator

Rationalize the denominator in each expression.

a. $\sqrt{\dfrac{3}{5}}$ **b.** $\dfrac{4}{\sqrt[3]{9}}$ **c.** $\dfrac{8}{3\sqrt{18}}$

Solution

a. $\sqrt{\dfrac{3}{5}} = \dfrac{\sqrt{3}}{\sqrt{5}} = \dfrac{\sqrt{3}}{\sqrt{5}} \cdot \dfrac{\sqrt{5}}{\sqrt{5}} = \dfrac{\sqrt{15}}{\sqrt{5^2}} = \dfrac{\sqrt{15}}{5}$ Multiply by $\sqrt{5}/\sqrt{5}$ to create a perfect square in the denominator.

b. $\dfrac{4}{\sqrt[3]{9}} = \dfrac{4}{\sqrt[3]{9}} \cdot \dfrac{\sqrt[3]{3}}{\sqrt[3]{3}} = \dfrac{4\sqrt[3]{3}}{\sqrt[3]{27}} = \dfrac{4\sqrt[3]{3}}{3}$ Multiply by $\sqrt[3]{3}/\sqrt[3]{3}$ to create a perfect cube in the denominator.

c. $\dfrac{8}{3\sqrt{18}} = \dfrac{8}{3\sqrt{18}} \cdot \dfrac{\sqrt{2}}{\sqrt{2}} = \dfrac{8\sqrt{2}}{3\sqrt{36}} = \dfrac{8\sqrt{2}}{3\sqrt{6^2}}$ Multiply by $\sqrt{2}/\sqrt{2}$ to create a perfect square in the denominator.

$$= \frac{8\sqrt{2}}{3(6)} = \frac{4\sqrt{2}}{9}$$

✓ **CHECKPOINT** *Now try Exercise 55.*

EXAMPLE 8 Rationalizing the Denominator

a. $\sqrt{\dfrac{8x}{12y^5}} = \sqrt{\dfrac{(4)(2)x}{(4)(3)y^5}} = \sqrt{\dfrac{2x}{3y^5}} = \dfrac{\sqrt{2x}}{\sqrt{3y^5}} \cdot \dfrac{\sqrt{3y}}{\sqrt{3y}} = \dfrac{\sqrt{6xy}}{\sqrt{3^2y^6}} = \dfrac{\sqrt{6xy}}{3|y^3|}$

b. $\sqrt[3]{\dfrac{54x^6y^3}{5z^2}} = \dfrac{\sqrt[3]{(3^3)(2)(x^6)(y^3)}}{\sqrt[3]{5z^2}} \cdot \dfrac{\sqrt[3]{25z}}{\sqrt[3]{25z}} = \dfrac{3x^2y\sqrt[3]{50z}}{\sqrt[3]{5^3z^3}} = \dfrac{3x^2y\sqrt[3]{50z}}{5z}$

✓ **CHECKPOINT** *Now try Exercise 69.*

3 ▶ Use the Pythagorean Theorem in application problems.

Applications of Radicals

Radicals commonly occur in applications involving right triangles. Recall that a right triangle is one that contains a right (or 90°) angle, as shown in Figure 9.1. The relationship among the three sides of a right triangle is described by the **Pythagorean Theorem,** which states that if a and b are the lengths of the legs and c is the length of the hypotenuse, then

$$c = \sqrt{a^2 + b^2} \quad \text{and} \quad a = \sqrt{c^2 - b^2}. \qquad \text{Pythagorean Theorem: } a^2 + b^2 = c^2$$

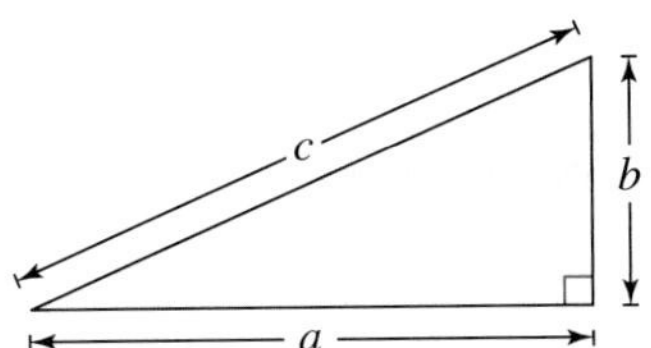

Figure 9.1

EXAMPLE 9 The Pythagorean Theorem

Find the length of the hypotenuse of the right triangle shown in Figure 9.2.

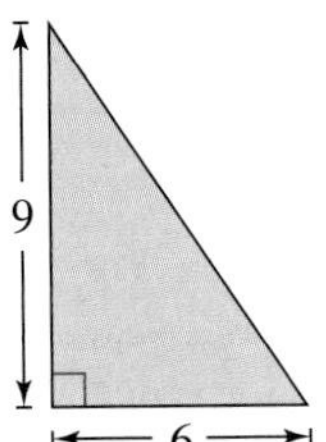

Figure 9.2

Solution

Because you know that $a = 6$ and $b = 9$, you can use the Pythagorean Theorem to find c, as follows.

$c = \sqrt{a^2 + b^2}$	Pythagorean Theorem
$= \sqrt{6^2 + 9^2}$	Substitute 6 for a and 9 for b.
$= \sqrt{117}$	Simplify.
$= \sqrt{9}\sqrt{13}$	Product Rule for Radicals
$= 3\sqrt{13}$	Simplify.

✓ ***CHECKPOINT*** *Now try Exercise 73.*

EXAMPLE 10 An Application of the Pythagorean Theorem

A softball diamond has the shape of a square with 60-foot sides, as shown in Figure 9.3. The catcher is 5 feet behind home plate. How far does the catcher have to throw to reach second base?

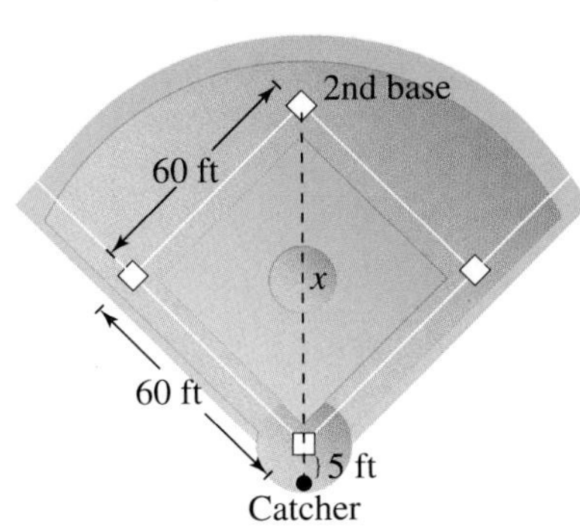

Figure 9.3

Solution

In Figure 9.3, let x be the hypotenuse of a right triangle with 60-foot sides. So, by the Pythagorean Theorem, you have the following.

$x = \sqrt{60^2 + 60^2}$	Pythagorean Theorem
$= \sqrt{7200}$	Simplify.
$= \sqrt{3600}\sqrt{2}$	Product Rule for Radicals
$= 60\sqrt{2}$	Simplify.
≈ 84.9 feet	Use a calculator.

So, the distance from home plate to second base is approximately 84.9 feet. Because the catcher is 5 feet behind home plate, the catcher must make a throw of

$$x + 5 \approx 84.9 + 5 = 89.9 \text{ feet.}$$

✓ ***CHECKPOINT*** *Now try Exercise 77.*

Concept Check

1. When is $\sqrt{x^2} \neq x$? Explain.
2. Explain why $\sqrt{8}$ is not in simplest form.
3. Describe the three conditions that characterize a simplified radical expression.
4. Describe how you would simplify $\frac{1}{\sqrt{3}}$.

9.2 EXERCISES

Go to pages 608–609 to record your assignments.

Developing Skills

In Exercises 1–18, simplify the radical. (Do not use a calculator.) ***See Example 1.***

✓ 1. $\sqrt{18}$
2. $\sqrt{27}$
3. $\sqrt{45}$
4. $\sqrt{125}$
5. $\sqrt{96}$
6. $\sqrt{84}$
7. $\sqrt{153}$
8. $\sqrt{147}$
9. $\sqrt{1183}$
10. $\sqrt{1176}$
11. $\sqrt{0.04}$
12. $\sqrt{0.25}$
13. $\sqrt{0.0072}$
14. $\sqrt{0.0027}$
15. $\sqrt{\frac{60}{3}}$
16. $\sqrt{\frac{208}{4}}$
17. $\sqrt{\frac{13}{25}}$
18. $\sqrt{\frac{15}{36}}$

In Exercises 19–54, simplify the radical expression. ***See Examples 2–6.***

✓ 19. $\sqrt{9x^5}$
20. $\sqrt{64x^3}$
21. $\sqrt{48y^4}$
22. $\sqrt{32x}$
23. $\sqrt{117y^6}$
24. $\sqrt{160x^8}$
25. $\sqrt{120x^2y^3}$
26. $\sqrt{125u^4v^6}$
27. $\sqrt{192a^5b^7}$
28. $\sqrt{363x^{10}y^9}$

✓ 29. $\sqrt[3]{48}$
30. $\sqrt[3]{54}$
31. $\sqrt[3]{112}$
32. $\sqrt[4]{112}$
33. $\sqrt[3]{40x^5}$
34. $\sqrt[3]{81a^7}$
✓ 35. $\sqrt[4]{324y^6}$
36. $\sqrt[5]{160x^8}$
37. $\sqrt[3]{x^4y^3}$
38. $\sqrt[3]{a^5b^6}$
39. $\sqrt[4]{4x^4y^6}$
40. $\sqrt[4]{128u^4v^7}$
41. $\sqrt[5]{32x^5y^6}$
42. $\sqrt[3]{16x^4y^5}$
✓ 43. $\sqrt[3]{\frac{35}{64}}$
44. $\sqrt[4]{\frac{5}{16}}$
45. $\frac{\sqrt{39y^2}}{\sqrt{3}}$
46. $\frac{\sqrt{56w^3}}{\sqrt{2}}$
47. $\sqrt{\frac{32a^4}{b^2}}$
48. $\sqrt{\frac{18x^2}{z^6}}$
✓ 49. $\sqrt[5]{\frac{32x^2}{y^5}}$
50. $\sqrt[3]{\frac{16z^3}{y^6}}$
51. $\sqrt[3]{\frac{54a^4}{b^9}}$
52. $\sqrt[4]{\frac{3u^2}{16v^8}}$
53. $-\sqrt[3]{\frac{3w^4}{8z^3}}$
54. $-\sqrt[4]{\frac{42y^7}{81x^4}}$

In Exercises 55–72, rationalize the denominator and simplify further, if possible. ***See Examples 7 and 8.***

✓ 55. $\sqrt{\frac{1}{3}}$
56. $\sqrt{\frac{1}{5}}$
57. $\frac{1}{\sqrt{7}}$
58. $\frac{12}{\sqrt{3}}$
59. $\sqrt[4]{\frac{5}{4}}$
60. $\sqrt[3]{\frac{9}{25}}$
61. $\frac{6}{\sqrt[3]{32}}$
62. $\frac{10}{\sqrt[5]{16}}$
63. $\frac{1}{\sqrt{y}}$
64. $\frac{2}{\sqrt{3c}}$
65. $\sqrt{\frac{4}{x}}$
66. $\sqrt{\frac{4}{x^3}}$
67. $\frac{1}{x\sqrt{2}}$
68. $\frac{1}{3x\sqrt{x}}$
✓ 69. $\frac{6}{\sqrt{3b^3}}$
70. $\frac{1}{\sqrt{xy}}$
71. $\sqrt[3]{\frac{2x}{3y}}$
72. $\sqrt[3]{\frac{20x^2}{9y^2}}$

Solving Problems

▲ *Geometry* In Exercises 73 and 74, find the length of the hypotenuse of the right triangle. ***See Example 9.***

73.

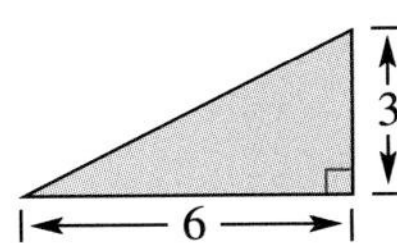

74.

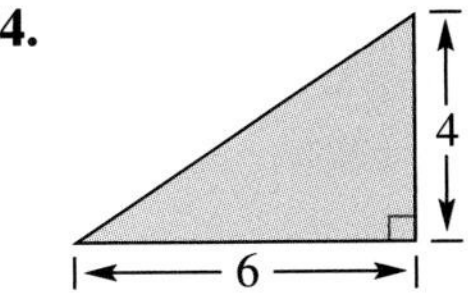

75. *Frequency* The frequency f in cycles per second of a vibrating string is given by

$$f = \frac{1}{100}\sqrt{\frac{400 \times 10^6}{5}}.$$

Use a calculator to approximate this number. (Round the result to two decimal places.)

76. *Period of a Pendulum* The time t (in seconds) for a pendulum of length L (in feet) to go through one complete cycle (its period) is given by

$$t = 2\pi\sqrt{\frac{L}{32}}.$$

Find the period of a pendulum whose length is 4 feet. (Round your answer to two decimal places.)

✓ **77.** ▲ *Geometry* A ladder is to reach a window that is 26 feet high. The ladder is placed 10 feet from the base of the wall (see figure). How long must the ladder be?

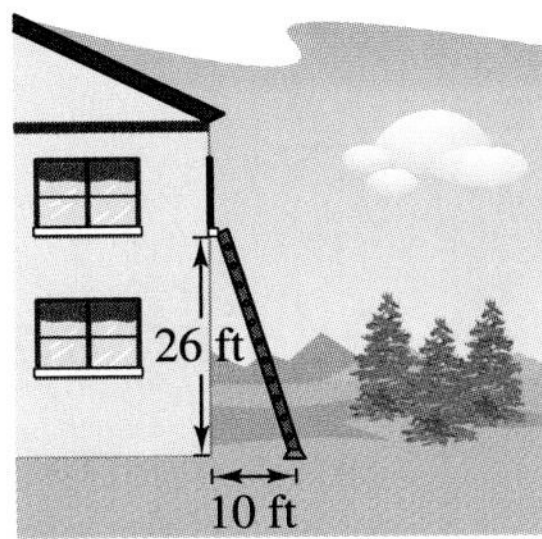

Figure for 77

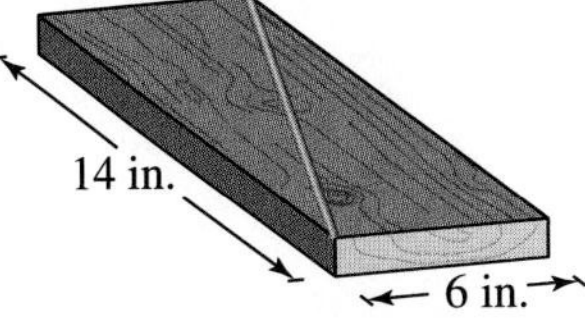

Figure for 78

78. ▲ *Geometry* A string is attached to opposite corners of a piece of wood that is 6 inches wide and 14 inches long (see figure). How long must the string be?

Explaining Concepts

79. Give an example of multiplying two radicals.

80. Enter any positive real number into your calculator and find its square root. Then repeatedly take the square root of the result.

$$\sqrt{x}, \sqrt{\sqrt{x}}, \sqrt{\sqrt{\sqrt{x}}}, . . .$$

What real number does the display appear to be approaching?

81. Square the real number $5/\sqrt{3}$ and note that the radical is eliminated from the denominator. Is this equivalent to rationalizing the denominator? Why or why not?

82. Let u be a positive real number. Explain why $\sqrt[3]{u} \cdot \sqrt[4]{u} \neq \sqrt[12]{u}$.

83. Explain how to find a perfect nth root factor in the radicand of an nth root radical.

Cumulative Review

In Exercises 84 and 85, sketch the graphs of the equations and approximate any solutions of the system of linear equations.

84. $\begin{cases} 3x + 2y = -4 \\ y = 3x + 7 \end{cases}$

85. $\begin{cases} 2x + 3y = 12 \\ 4x - y = 10 \end{cases}$

In Exercises 86 and 87, solve the system by the method of substitution.

86. $\begin{cases} x - 3y = -2 \\ 7y - 4x = 6 \end{cases}$

87. $\begin{cases} y = x + 2 \\ y - x = 8 \end{cases}$

In Exercises 88 and 89, solve the system by the method of elimination.

88. $\begin{cases} 1.5x - 3 = -2y \\ 3x + 4y = 6 \end{cases}$

89. $\begin{cases} x + 4y + 3z = 2 \\ 2x + y + z = 10 \\ -x + y + 2z = 8 \end{cases}$

9.3 Adding and Subtracting Radical Expressions

© James Marshall/The Image Works

What You Should Learn

1 ▶ Use the Distributive Property to add and subtract like radicals.

2 ▶ Use radical expressions in application problems.

Why You Should Learn It

Radical expressions can be used to model and solve real-life problems. For instance, Example 6 on page 576 shows how to find a radical expression that models the out-of-pocket expense of attending college.

1 ▶ Use the Distributive Property to add and subtract like radicals.

Adding and Subtracting Radical Expressions

Two or more radical expressions are called **like radicals** if they have the same index and the same radicand. For instance, the expressions $\sqrt{2}$ and $3\sqrt{2}$ are like radicals, whereas the expressions $\sqrt{3}$ and $\sqrt[3]{3}$ are not. Two radical expressions that are like radicals can be added or subtracted by adding or subtracting their coefficients.

EXAMPLE 1 Combining Radical Expressions

Simplify each expression by combining like radicals.

a. $\sqrt{7} + 5\sqrt{7} - 2\sqrt{7}$

b. $6\sqrt{x} - \sqrt[3]{4} - 5\sqrt{x} + 2\sqrt[3]{4}$

c. $3\sqrt[3]{x} + 2\sqrt[3]{x} + \sqrt{x} - 8\sqrt{x}$

Solution

a. $\sqrt{7} + 5\sqrt{7} - 2\sqrt{7} = (1 + 5 - 2)\sqrt{7}$ Distributive Property

$= 4\sqrt{7}$ Simplify.

b. $6\sqrt{x} - \sqrt[3]{4} - 5\sqrt{x} + 2\sqrt[3]{4}$

$= \left(6\sqrt{x} - 5\sqrt{x}\right) + \left(-\sqrt[3]{4} + 2\sqrt[3]{4}\right)$ Group like radicals.

$= (6 - 5)\sqrt{x} + (-1 + 2)\sqrt[3]{4}$ Distributive Property

$= \sqrt{x} + \sqrt[3]{4}$ Simplify.

c. $3\sqrt[3]{x} + 2\sqrt[3]{x} + \sqrt{x} - 8\sqrt{x}$

$= (3 + 2)\sqrt[3]{x} + (1 - 8)\sqrt{x}$ Distributive Property

$= 5\sqrt[3]{x} - 7\sqrt{x}$ Simplify.

✓ ***CHECKPOINT*** *Now try Exercise 13.*

Before concluding that two radicals cannot be combined, you should first rewrite them in simplest form. This is illustrated in Examples 2 and 3.

Study Tip

It is important to realize that the expression $\sqrt{a} + \sqrt{b}$ is not equal to $\sqrt{a + b}$. For instance, you may be tempted to add $\sqrt{6} + \sqrt{3}$ and get $\sqrt{9} = 3$. But remember, you cannot add unlike radicals. So, $\sqrt{6} + \sqrt{3}$ cannot be simplified further.

EXAMPLE 2 **Simplifying Before Combining Radical Expressions**

Simplify each expression by combining like radicals.

a. $\sqrt{45x} + 3\sqrt{20x}$

b. $5\sqrt{x^3} - x\sqrt{4x}$

Solution

a. $\sqrt{45x} + 3\sqrt{20x} = 3\sqrt{5x} + 6\sqrt{5x}$ — Simplify radicals.

$= 9\sqrt{5x}$ — Combine like radicals.

b. $5\sqrt{x^3} - x\sqrt{4x} = 5x\sqrt{x} - 2x\sqrt{x}$ — Simplify radicals.

$= 3x\sqrt{x}$ — Combine like radicals.

CHECKPOINT *Now try Exercise 25.*

EXAMPLE 3 **Simplifying Before Combining Radical Expressions**

Simplify each expression by combining like radicals.

a. $\sqrt[3]{54y^5} + 4\sqrt[3]{2y^2}$

b. $\sqrt[3]{6x^4} + \sqrt[3]{48x} - \sqrt[3]{162x^4}$

Solution

a. $\sqrt[3]{54y^5} + 4\sqrt[3]{2y^2} = 3y\sqrt[3]{2y^2} + 4\sqrt[3]{2y^2}$ — Simplify radicals.

$= (3y + 4)\sqrt[3]{2y^2}$ — Distributive Property

b. $\sqrt[3]{6x^4} + \sqrt[3]{48x} - \sqrt[3]{162x^4}$ — Write original expression.

$= x\sqrt[3]{6x} + 2\sqrt[3]{6x} - 3x\sqrt[3]{6x}$ — Simplify radicals.

$= (x + 2 - 3x)\sqrt[3]{6x}$ — Distributive Property

$= (2 - 2x)\sqrt[3]{6x}$ — Combine like terms.

CHECKPOINT *Now try Exercise 35.*

It may be necessary to rationalize denominators before combining radicals.

EXAMPLE 4 **Rationalizing Denominators Before Simplifying**

$\sqrt{7} - \dfrac{5}{\sqrt{7}} = \sqrt{7} - \left(\dfrac{5}{\sqrt{7}} \cdot \dfrac{\sqrt{7}}{\sqrt{7}}\right)$ — Multiply by $\sqrt{7}/\sqrt{7}$ to remove the radical from the denominator.

$= \sqrt{7} - \dfrac{5\sqrt{7}}{7}$ — Simplify.

$= \left(1 - \dfrac{5}{7}\right)\sqrt{7}$ — Distributive Property

$= \dfrac{2}{7}\sqrt{7}$ — Simplify.

CHECKPOINT *Now try Exercise 47.*

2 ▶ Use radical expressions in application problems.

Applications

EXAMPLE 5 Geometry: Perimeter of a Triangle

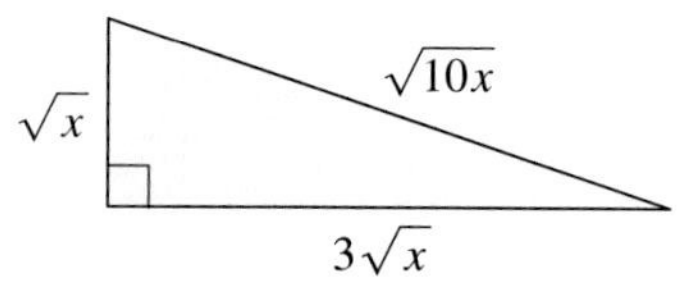

Figure 9.4

Write and simplify an expression for the perimeter of the triangle shown in Figure 9.4.

Solution

$P = a + b + c$ — Formula for perimeter of a triangle

$= \sqrt{x} + 3\sqrt{x} + \sqrt{10x}$ — Substitute.

$= (1 + 3)\sqrt{x} + \sqrt{10x}$ — Distributive Property

$= 4\sqrt{x} + \sqrt{10x}$ — Simplify.

✓ **CHECKPOINT** *Now try Exercise 61.*

EXAMPLE 6 Out-of-Pocket Expense 

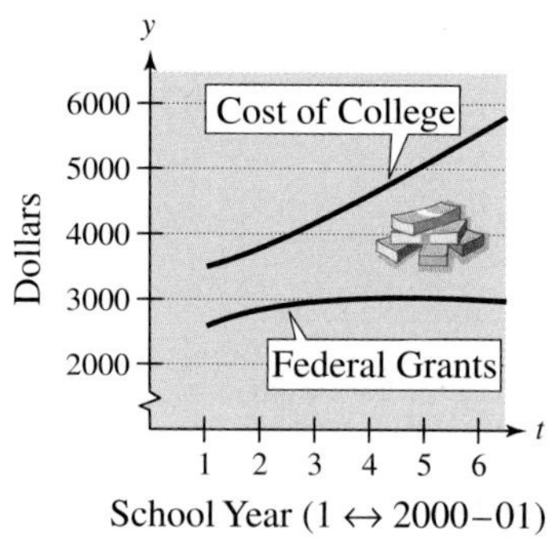

From the graph, what can you deduce about the out-of-pocket expense of a college student?

Figure 9.5

The cost C of attending a four-year public college and the aid A per recipient (for eligible students) from federal grants from 2000 to 2006 can be modeled by the equations

$C = 3783 + 697.7t - 989.2\sqrt{t}, \quad 1 \le t \le 6$ — College cost

$A = 1473 - 329.1t + 1431.1\sqrt{t}, \quad 1 \le t \le 6$ — Grant amount

where t represents the school year, with $t = 1$ corresponding to 2000–01. (See Figure 9.5.) Find a radical expression that models the out-of-pocket expense E incurred by a college student from 2000 to 2006. Estimate the out-of-pocket expense E of a college student during the 2004–05 school year. (Source: *Annual Survey of Colleges*, The College Board)

Solution

The difference of the cost and the aid gives the out-of-pocket expense.

$$\begin{aligned} C - A &= (3783 + 697.7t - 989.2\sqrt{t}) - (1473 - 329.1t + 1431.1\sqrt{t}) \\ &= 3783 + 697.7t - 989.2\sqrt{t} - 1473 + 329.1t - 1431.1\sqrt{t} \\ &= (3783 - 1473) + (697.7t + 329.1t) - (989.2\sqrt{t} + 1431.1\sqrt{t}) \\ &= 2310 + 1026.8t - 2420.3\sqrt{t} \end{aligned}$$

So, the radical expression that models the out-of-pocket expense incurred by a college student is

$$\begin{aligned} E &= C - A \\ &= 2310 + 1026.8t - 2420.3\sqrt{t}. \end{aligned}$$

Using this model, substitute $t = 5$ to estimate the out-of-pocket expense of a college student during the 2004–05 school year.

$$E = 2310 + 1026.8(5) - 2420.3\sqrt{5} \approx 2032$$

✓ **CHECKPOINT** *Now try Exercise 67.*

Concept Check

1. Explain what it means for two radical expressions to be like radicals.
2. Explain how to add or subtract like radicals.
3. Is $\sqrt{2} + \sqrt{18}$ in simplest form? Explain.
4. Is $\sqrt{2} - \dfrac{1}{\sqrt{2}}$ in simplest form? Explain.

9.3 EXERCISES

Go to pages 608–609 to record your assignments.

Developing Skills

In Exercises 1–46, combine the radical expressions, if possible. ***See Examples 1–3.***

1. $3\sqrt{2} - \sqrt{2}$
2. $6\sqrt{5} - 2\sqrt{5}$
3. $2\sqrt{6} + 5\sqrt{6}$
4. $3\sqrt{7} + 2\sqrt{7}$
5. $8\sqrt{5} + 9\sqrt[3]{5}$
6. $3\sqrt[3]{3} + 6\sqrt[3]{3}$
7. $9\sqrt[3]{5} - 6\sqrt[3]{5}$
8. $8\sqrt[4]{5} - 2\sqrt[3]{5}$
9. $4\sqrt[3]{y} + 9\sqrt[3]{y}$
10. $13\sqrt{x} + \sqrt{x}$
11. $15\sqrt[4]{s} - \sqrt[4]{s}$
12. $9\sqrt[4]{t} - 3\sqrt[4]{t}$
13. ✓ $8\sqrt{2} + 6\sqrt{2} - 5\sqrt{2}$
14. $2\sqrt{6} + 8\sqrt{6} - 3\sqrt{6}$
15. $\sqrt[4]{5} - 6\sqrt[4]{13} + 3\sqrt[4]{5} - \sqrt[4]{13}$
16. $9\sqrt[3]{17} + 7\sqrt[3]{2} - 4\sqrt[3]{17} + \sqrt[3]{2}$
17. $9\sqrt[3]{7} - \sqrt{3} + 4\sqrt[3]{7} + 2\sqrt{3}$
18. $5\sqrt{7} - 8\sqrt[4]{11} + \sqrt{7} + 9\sqrt[4]{11}$
19. $8\sqrt{27} - 3\sqrt{3}$
20. $9\sqrt{50} - 4\sqrt{2}$
21. $3\sqrt{45} + 7\sqrt{20}$
22. $5\sqrt{12} + 16\sqrt{27}$
23. $2\sqrt[3]{54} + 12\sqrt[3]{16}$
24. $4\sqrt[4]{48} - \sqrt[4]{243}$
25. ✓ $5\sqrt{9x} - 3\sqrt{x}$
26. $4\sqrt{y} + 2\sqrt{16y}$
27. $3\sqrt{x+1} + 10\sqrt{x+1}$
28. $7\sqrt{2a-3} - 4\sqrt{2a-3}$
29. $\sqrt{25y} + \sqrt{64y}$
30. $\sqrt[3]{16t^4} - \sqrt[3]{54t^4}$
31. $10\sqrt[3]{z} - \sqrt[3]{z^4}$
32. $5\sqrt[3]{24u^2} + 2\sqrt[3]{81u^5}$
33. $\sqrt{5a} + 2\sqrt{45a^3}$
34. $4\sqrt{3x^3} - \sqrt{12x}$
35. ✓ $\sqrt[3]{6x^4} + \sqrt[3]{48x}$
36. $\sqrt[3]{54x} - \sqrt[3]{2x^4}$
37. $\sqrt{9x-9} + \sqrt{x-1}$
38. $\sqrt{4y+12} + \sqrt{y+3}$
39. $\sqrt{x^3-x^2} + \sqrt{4x-4}$
40. $\sqrt{9x-9} - \sqrt{x^3-x^2}$
41. $2\sqrt[3]{a^4b^2} + 3a\sqrt[3]{ab^2}$
42. $3y\sqrt[4]{2x^5y^3} - x\sqrt[4]{162xy^7}$
43. $\sqrt{4r^7s^5} + 3r^2\sqrt{r^3s^5} - 2rs\sqrt{r^5s^3}$
44. $x\sqrt[3]{27x^5y^2} - x^2\sqrt[3]{x^2y^2} + z\sqrt[3]{x^8y^2}$
45. $\sqrt[3]{128x^9y^{10}} - 2x^2y\sqrt[3]{16x^3y^7}$
46. $5\sqrt[3]{320x^5y^8} + 2x\sqrt[3]{135x^2y^8}$

In Exercises 47–56, perform the addition or subtraction and simplify your answer. ***See Example 4.***

47. ✓ $\sqrt{5} - \dfrac{3}{\sqrt{5}}$
48. $\sqrt{10} + \dfrac{5}{\sqrt{10}}$

49. $\sqrt{32} + \sqrt{\dfrac{1}{2}}$

50. $\sqrt{\dfrac{1}{5}} - \sqrt{45}$

51. $\sqrt{12y} - \dfrac{y}{\sqrt{3y}}$

52. $\dfrac{x}{\sqrt{3x}} + \sqrt{27x}$

53. $\dfrac{2}{\sqrt{3x}} + \sqrt{3x}$

54. $2\sqrt{7x} - \dfrac{4}{\sqrt{7x}}$

55. $\sqrt{7y^3} - \sqrt{\dfrac{9}{7y^3}}$

56. $\sqrt{\dfrac{4}{3x^3}} + \sqrt{3x^3}$

In Exercises 57–60, place the correct symbol (<, >, or =) between the numbers.

57. $\sqrt{7} + \sqrt{18}$ ☐ $\sqrt{7 + 18}$

58. $\sqrt{10} - \sqrt{6}$ ☐ $\sqrt{10 - 6}$

59. 5 ☐ $\sqrt{9^2 - 4^2}$

60. 5 ☐ $\sqrt{3^2 + 4^2}$

Solving Problems

Geometry In Exercises 61–64, write a simplified expression for the perimeter of the figure. ***See Example 5.***

61.

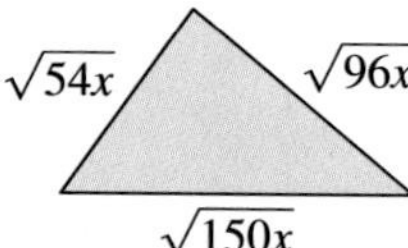

62.

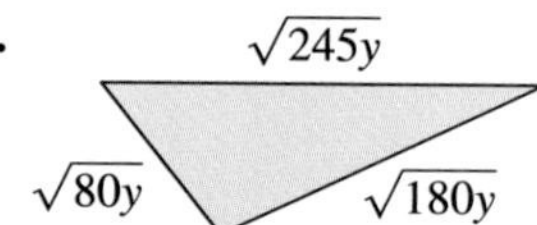

63.

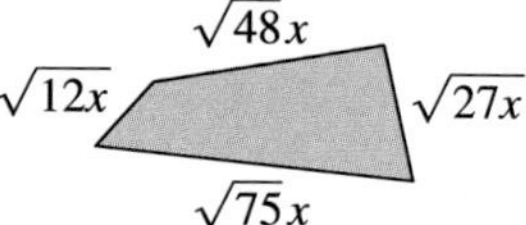

64.

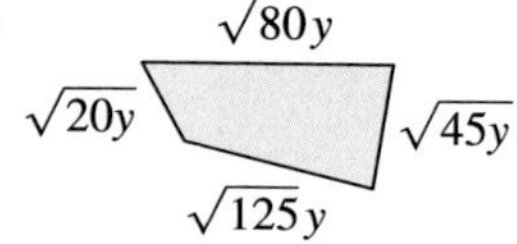

65. *Geometry* The foundation of a house is 40 feet long and 30 feet wide. The height of the attic is 5 feet (see figure).

(a) Use the Pythagorean Theorem to find the length of the hypotenuse of each of the two right triangles formed by the roof line. (Assume there is no overhang.)

(b) Use the result of part (a) to determine the total area of the roof.

66. *Geometry* The four corners are cut from a four-foot-by-eight-foot sheet of plywood, as shown in the figure. Find the perimeter of the remaining piece of plywood.

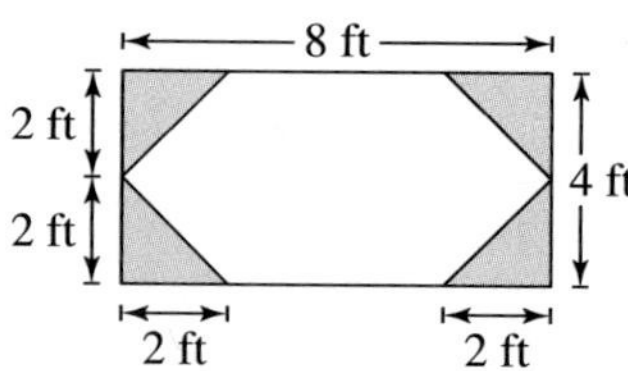

✓ **67.** *Immigration* Legal permanent residents are immigrants who have received a "green card," which grants them the right to live in the United States. The number of immigrants from Colombia C (in thousands) and South America S (in thousands) who received green cards from 2001 to 2006 can be modeled by the equations

$$C = -118 + 286.7\sqrt{t} - 195.1t + 42.8\sqrt{t^3},\quad 1 \le t \le 6$$

$$S = -182 + 565.5\sqrt{t} - 409.2t + 94.6\sqrt{t^3},\quad 1 \le t \le 6$$

where t represents the year, with $t = 1$ corresponding to 2001. Find a radical expression that models the total number T of legal permanent residents from South America, excluding Colombia, from 2001 to 2006. Estimate T in 2003. (Source: U.S. Department of Homeland Security)

68. *Immigration* The number of immigrants from Africa A (in thousands) who received green cards from 2001 to 2006 can be modeled by the equation

$$A = -812 + 2138.9\sqrt{t} - 1881t + 694.2\sqrt{t^3} - 89.8t^2,\quad 1 \le t \le 6$$

where t represents the year, with $t = 1$ corresponding to 2001. Use the South American model from Exercise 67 to find a radical expression that models the total number T of legal permanent residents from Africa and South America from 2001 to 2006. Estimate T in 2005. (Source: U.S. Department of Homeland Security)

Explaining Concepts

69. Will the sum of two radicals always be a radical? Give an example to support your answer.

70. Will the difference of two radicals always be a radical? Give an example to support your answer.

71. Is $\sqrt{2x} + \sqrt{2x}$ equal to $\sqrt{8x}$? Explain.

72. Explain how adding two monomials compares to adding two radicals.

73. You are an algebra instructor, and one of your students hands in the following work. Find and correct the errors, and discuss how you can help your student avoid such errors in the future.

(a) ~~$7\sqrt{3} + 4\sqrt{2} = 11\sqrt{5}$~~

(b) ~~$3\sqrt[3]{k} - 6\sqrt{k} = -3\sqrt{k}$~~

Cumulative Review

In Exercises 74–79, combine the rational expressions and simplify.

74. $\dfrac{2x+1}{3x} + \dfrac{3-4x}{3x}$

75. $\dfrac{7z-2}{2z} - \dfrac{4z+1}{2z}$

76. $\dfrac{4m+6}{m+2} - \dfrac{3m+4}{m+2}$

77. $\dfrac{2x+3}{x-3} + \dfrac{6-5x}{x-3}$

78. $\dfrac{4}{x-4} + \dfrac{2x}{x+1}$

79. $\dfrac{2v}{v-5} - \dfrac{3}{5-v}$

In Exercises 80–83, simplify the complex fraction.

80. $\dfrac{\left(\dfrac{2}{3}\right)}{\left(\dfrac{4}{15}\right)}$

81. $\dfrac{\left(\dfrac{27a^3}{4b^2c}\right)}{\left(\dfrac{9ac^2}{10b^2}\right)}$

82. $\dfrac{\left(\dfrac{x^2+2x-8}{x-8}\right)}{2x+8}$

83. $\dfrac{3w-9}{\left(\dfrac{w^2-10w+21}{w+1}\right)}$

Mid-Chapter Quiz

Take this quiz as you would take a quiz in class. After you are done, check your work against the answers in the back of the book.

In Exercises 1–4, evaluate the expression.

1. $\sqrt{225}$
2. $\sqrt[4]{\frac{81}{16}}$
3. $49^{1/2}$
4. $(-27)^{2/3}$

In Exercises 5 and 6, evaluate the function as indicated, if possible, and simplify.

5. $f(x) = \sqrt{3x - 5}$
   (a) $f(0)$ (b) $f(2)$ (c) $f(10)$
6. $g(x) = \sqrt{9 - x}$
   (a) $g(-7)$ (b) $g(5)$ (c) $g(9)$

In Exercises 7 and 8, describe the domain of the function.

7. $g(x) = \dfrac{12}{\sqrt[3]{x}}$
8. $h(x) = \sqrt{3x + 10}$

In Exercises 9–14, simplify the expression.

9. $\sqrt{27x^2}$
10. $\sqrt[4]{32x^8}$
11. $\sqrt{\dfrac{4u^3}{9}}$
12. $\sqrt[3]{\dfrac{16}{u^6}}$
13. $\sqrt{125x^3y^2z^4}$
14. $2a\sqrt[3]{16a^3b^5}$

In Exercises 15 and 16, rationalize the denominator and simplify further, if possible.

15. $\dfrac{24}{\sqrt{12}}$
16. $\dfrac{21x^2}{\sqrt{7x}}$

In Exercises 17–22, combine the radical expressions, if possible.

17. $2\sqrt{3} - 4\sqrt{7} + \sqrt{3}$
18. $\sqrt{200y} - 3\sqrt{8y}$
19. $5\sqrt{12} + 2\sqrt{3} - \sqrt{75}$
20. $\sqrt{25x + 50} - \sqrt{x + 2}$
21. $6x\sqrt[3]{5x^2} + 2\sqrt[3]{40x^4}$
22. $3\sqrt{x^3y^4z^5} + 2xy^2\sqrt{xz^5} - xz^2\sqrt{xy^4z}$
23. The four corners are cut from an $8\frac{1}{2}$-inch-by-11-inch sheet of paper, as shown in the figure at the left. Find the perimeter of the remaining piece of paper.

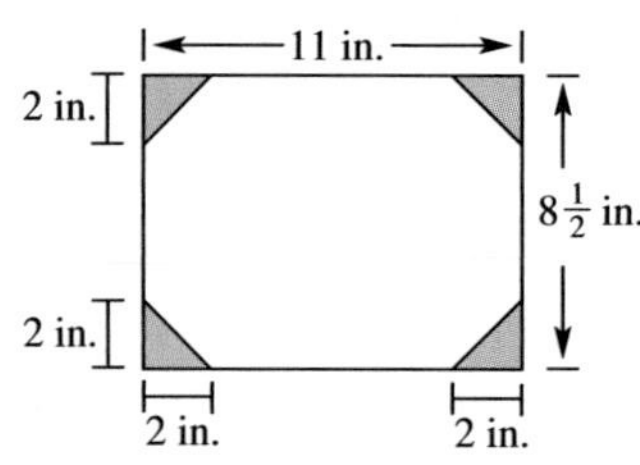

Figure for 23

9.4 Multiplying and Dividing Radical Expressions

Paul A. Souders/CORBIS

What You Should Learn

1 ▶ Use the Distributive Property or the FOIL Method to multiply radical expressions.

2 ▶ Determine the products of conjugates.

3 ▶ Simplify quotients involving radicals by rationalizing the denominators.

Why You Should Learn It

Multiplication of radicals is often used in real-life applications. For instance, in Exercise 111 on page 587, you will multiply two radical expressions to find the area of the cross section of a wooden beam.

1 ▶ Use the Distributive Property or the FOIL Method to multiply radical expressions.

Multiplying Radical Expressions

To multiply radical expressions, you can use the Product Rule for Radicals from Section 9.2, given by $\sqrt[n]{uv} = \sqrt[n]{u}\sqrt[n]{v}$, where u and v are real numbers whose nth roots are also real numbers. When the expressions you are multiplying involve sums or differences, you can also use the Distributive Property or the FOIL Method.

EXAMPLE 1 Multiplying Radical Expressions

Find each product and simplify.

a. $\sqrt{6} \cdot \sqrt{3}$ **b.** $\sqrt[3]{5} \cdot \sqrt[3]{16}$

Solution

a. $\sqrt{6} \cdot \sqrt{3} = \sqrt{6 \cdot 3} = \sqrt{18} = \sqrt{9 \cdot 2} = 3\sqrt{2}$

b. $\sqrt[3]{5} \cdot \sqrt[3]{16} = \sqrt[3]{5 \cdot 16} = \sqrt[3]{80} = \sqrt[3]{8 \cdot 10} = 2\sqrt[3]{10}$

✓ **CHECKPOINT** *Now try Exercise 1.*

EXAMPLE 2 Multiplying Radical Expressions

Find each product and simplify.

a. $\sqrt{3}(2 + \sqrt{5})$ **b.** $\sqrt{2}(4 - \sqrt{8})$ **c.** $\sqrt{6}(\sqrt{12} - \sqrt{3})$

Solution

a. $\sqrt{3}(2 + \sqrt{5}) = 2\sqrt{3} + \sqrt{3}\sqrt{5}$ Distributive Property

$= 2\sqrt{3} + \sqrt{15}$ Product Rule for Radicals

b. $\sqrt{2}(4 - \sqrt{8}) = 4\sqrt{2} - \sqrt{2}\sqrt{8}$ Distributive Property

$= 4\sqrt{2} - \sqrt{16} = 4\sqrt{2} - 4$ Product Rule for Radicals

c. $\sqrt{6}(\sqrt{12} - \sqrt{3}) = \sqrt{6}\sqrt{12} - \sqrt{6}\sqrt{3}$ Distributive Property

$= \sqrt{72} - \sqrt{18}$ Product Rule for Radicals

$= 6\sqrt{2} - 3\sqrt{2} = 3\sqrt{2}$ Find perfect square factors.

✓ **CHECKPOINT** *Now try Exercise 9.*

Smart Study Strategy

Go to page 554 for ways to *Form a Weekly Study Group.*

In Example 2, the Distributive Property was used to multiply radical expressions. In Example 3, note how the FOIL Method is used.

EXAMPLE 3 **Using the FOIL Method**

a. $(2\sqrt{7} - 4)(\sqrt{7} + 1) = \overbrace{2(\sqrt{7})^2}^{F} + \overbrace{2\sqrt{7}}^{O} - \overbrace{4\sqrt{7}}^{I} - \overbrace{4}^{L}$ FOIL Method

$= 2(7) + (2 - 4)\sqrt{7} - 4$ Combine like radicals.

$= 10 - 2\sqrt{7}$ Simplify.

b. $(3 - \sqrt{x})(1 + \sqrt{x}) = 3 + 3\sqrt{x} - \sqrt{x} - (\sqrt{x})^2$ FOIL Method

$= 3 + 2\sqrt{x} - x, \quad x \geq 0$ Combine like radicals and simplify.

✓ **CHECKPOINT** *Now try Exercise 23.*

2 ▶ Determine the products of conjugates.

Conjugates

The expressions $3 + \sqrt{6}$ and $3 - \sqrt{6}$ are called **conjugates** of each other. Notice that they differ only in the sign between the terms. The product of two conjugates is the difference of two squares, which is given by the special product formula $(a + b)(a - b) = a^2 - b^2$. Here are some other examples.

Expression	*Conjugate*	*Product*
$1 - \sqrt{3}$	$1 + \sqrt{3}$	$(1)^2 - (\sqrt{3})^2 = 1 - 3 = -2$
$\sqrt{5} + \sqrt{2}$	$\sqrt{5} - \sqrt{2}$	$(\sqrt{5})^2 - (\sqrt{2})^2 = 5 - 2 = 3$
$\sqrt{10} - 3$	$\sqrt{10} + 3$	$(\sqrt{10})^2 - (3)^2 = 10 - 9 = 1$
$\sqrt{x} + 2$	$\sqrt{x} - 2$	$(\sqrt{x})^2 - (2)^2 = x - 4, x \geq 0$

EXAMPLE 4 **Multiplying Conjugates**

Find the conjugate of the expression and multiply the expression by its conjugate.

a. $2 - \sqrt{5}$ **b.** $\sqrt{3} + \sqrt{x}$

Solution

a. The conjugate of $2 - \sqrt{5}$ is $2 + \sqrt{5}$.

$(2 - \sqrt{5})(2 + \sqrt{5}) = 2^2 - (\sqrt{5})^2$ Special product formula

$= 4 - 5 = -1$ Simplify.

b. The conjugate of $\sqrt{3} + \sqrt{x}$ is $\sqrt{3} - \sqrt{x}$.

$(\sqrt{3} + \sqrt{x})(\sqrt{3} - \sqrt{x}) = (\sqrt{3})^2 - (\sqrt{x})^2$ Special product formula

$= 3 - x, \quad x \geq 0$ Simplify.

✓ **CHECKPOINT** *Now try Exercise 57.*

3 ▶ Simplify quotients involving radicals by rationalizing the denominators.

Dividing Radical Expressions

To simplify a *quotient* involving radicals, you rationalize the denominator. For single-term denominators, you can use the rationalization process described in Section 9.2. To rationalize a denominator involving two terms, multiply both the numerator and denominator by the *conjugate of the denominator.*

EXAMPLE 5 Simplifying Quotients Involving Radicals

Simplify (a) $\dfrac{\sqrt{3}}{1-\sqrt{5}}$ and (b) $\dfrac{4}{2-\sqrt{3}}$.

Solution

a. $\dfrac{\sqrt{3}}{1-\sqrt{5}} = \dfrac{\sqrt{3}}{1-\sqrt{5}} \cdot \dfrac{1+\sqrt{5}}{1+\sqrt{5}}$ Multiply numerator and denominator by conjugate of denominator.

$= \dfrac{\sqrt{3}(1+\sqrt{5})}{1^2-(\sqrt{5})^2}$ Special product formula

$= \dfrac{\sqrt{3}+\sqrt{15}}{1-5}$ Simplify.

$= -\dfrac{\sqrt{3}+\sqrt{15}}{4}$ Simplify.

b. $\dfrac{4}{2-\sqrt{3}} = \dfrac{4}{2-\sqrt{3}} \cdot \dfrac{2+\sqrt{3}}{2+\sqrt{3}}$ Multiply numerator and denominator by conjugate of denominator.

$= \dfrac{4(2+\sqrt{3})}{2^2-(\sqrt{3})^2}$ Special product formula

$= \dfrac{8+4\sqrt{3}}{4-3}$ Simplify.

$= 8+4\sqrt{3}$ Simplify.

✓ **CHECKPOINT** *Now try Exercise 75.*

EXAMPLE 6 Simplifying a Quotient Involving Radicals

$\dfrac{5\sqrt{2}}{\sqrt{7}+\sqrt{2}} = \dfrac{5\sqrt{2}}{\sqrt{7}+\sqrt{2}} \cdot \dfrac{\sqrt{7}-\sqrt{2}}{\sqrt{7}-\sqrt{2}}$ Multiply numerator and denominator by conjugate of denominator.

$= \dfrac{5(\sqrt{14}-\sqrt{4})}{(\sqrt{7})^2-(\sqrt{2})^2}$ Special product formula

$= \dfrac{5(\sqrt{14}-2)}{7-2}$ Simplify.

$= \dfrac{\cancel{5}(\sqrt{14}-2)}{\cancel{5}}$ Divide out common factor.

$= \sqrt{14}-2$ Simplest form

✓ **CHECKPOINT** *Now try Exercise 81.*

EXAMPLE 7 Dividing Radical Expressions

Perform each division and simplify.

a. $6 \div \left(\sqrt{x} - 2\right)$

b. $\left(2 - \sqrt{3}\right) \div \left(\sqrt{6} + \sqrt{2}\right)$

Solution

a. $$\frac{6}{\sqrt{x} - 2} = \frac{6}{\sqrt{x} - 2} \cdot \frac{\sqrt{x} + 2}{\sqrt{x} + 2}$$ Multiply numerator and denominator by conjugate of denominator.

$$= \frac{6\left(\sqrt{x} + 2\right)}{\left(\sqrt{x}\right)^2 - 2^2}$$ Special product formula

$$= \frac{6\sqrt{x} + 12}{x - 4}$$ Simplify.

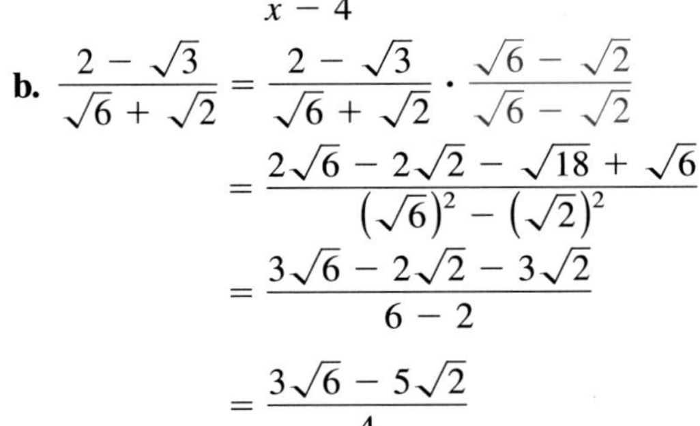

b. $$\frac{2 - \sqrt{3}}{\sqrt{6} + \sqrt{2}} = \frac{2 - \sqrt{3}}{\sqrt{6} + \sqrt{2}} \cdot \frac{\sqrt{6} - \sqrt{2}}{\sqrt{6} - \sqrt{2}}$$ Multiply numerator and denominator by conjugate of denominator.

$$= \frac{2\sqrt{6} - 2\sqrt{2} - \sqrt{18} + \sqrt{6}}{\left(\sqrt{6}\right)^2 - \left(\sqrt{2}\right)^2}$$ FOIL Method and special product formula

$$= \frac{3\sqrt{6} - 2\sqrt{2} - 3\sqrt{2}}{6 - 2}$$ Simplify.

$$= \frac{3\sqrt{6} - 5\sqrt{2}}{4}$$ Simplify.

✓ ***CHECKPOINT*** *Now try Exercise 85.*

EXAMPLE 8 Dividing Radical Expressions

Perform the division and simplify.

$1 \div \left(\sqrt{x} - \sqrt{x + 1}\right)$

Solution

$$\frac{1}{\sqrt{x} - \sqrt{x + 1}} = \frac{1}{\sqrt{x} - \sqrt{x + 1}} \cdot \frac{\sqrt{x} + \sqrt{x + 1}}{\sqrt{x} + \sqrt{x + 1}}$$ Multiply numerator and denominator by conjugate of denominator.

$$= \frac{\sqrt{x} + \sqrt{x + 1}}{\left(\sqrt{x}\right)^2 - \left(\sqrt{x + 1}\right)^2}$$ Special product formula

$$= \frac{\sqrt{x} + \sqrt{x + 1}}{x - (x + 1)}$$ Simplify.

$$= \frac{\sqrt{x} + \sqrt{x + 1}}{-1}$$ Combine like terms.

$$= -\sqrt{x} - \sqrt{x + 1}$$ Simplify.

✓ ***CHECKPOINT*** *Now try Exercise 97.*

Concept Check

1. Give an example of a product of expressions involving radicals in which the Distributive Property can be used to perform the multiplication.
2. Give an example of a product of expressions involving radicals in which the FOIL Method can be used to perform the multiplication.
3. Write a rule that can be used to find the conjugate of the expression $a + b$, where at least one of the expressions a and b is a radical expression.
4. Is the number $\frac{3}{1+\sqrt{5}}$ in simplest form? If not, explain the steps for writing it in simplest form.

9.4 EXERCISES

Go to pages 608–609 to record your assignments.

Developing Skills

In Exercises 1–50, multiply and simplify. ***See Examples 1–3.***

✓ 1. $\sqrt{2}\cdot\sqrt{8}$
2. $\sqrt{6}\cdot\sqrt{18}$
3. $\sqrt{3}\cdot\sqrt{15}$
4. $\sqrt{5}\cdot\sqrt{10}$
5. $\sqrt[3]{12}\cdot\sqrt[3]{6}$
6. $\sqrt[3]{9}\cdot\sqrt[3]{3}$
7. $\sqrt[4]{8}\cdot\sqrt[4]{2}$
8. $\sqrt[4]{54}\cdot\sqrt[4]{3}$
✓ 9. $\sqrt{7}(3-\sqrt{7})$
10. $\sqrt{3}(4+\sqrt{3})$
11. $\sqrt{2}(\sqrt{20}+8)$
12. $\sqrt{7}(\sqrt{14}+3)$
13. $\sqrt{6}(\sqrt{12}-\sqrt{3})$
14. $\sqrt{10}(\sqrt{5}+\sqrt{6})$
15. $4\sqrt{3}(\sqrt{3}-\sqrt{5})$
16. $3\sqrt{5}(\sqrt{5}-\sqrt{2})$
17. $\sqrt{y}(\sqrt{y}+4)$
18. $\sqrt{x}(5-\sqrt{x})$
19. $\sqrt{a}(4-\sqrt{a})$
20. $\sqrt{z}(\sqrt{z}+5)$
21. $\sqrt[3]{4}(\sqrt[3]{2}-7)$
22. $\sqrt[3]{9}(\sqrt[3]{3}+2)$
✓ 23. $(\sqrt{5}+3)(\sqrt{3}-5)$
24. $(\sqrt{7}+6)(\sqrt{2}+6)$
25. $(\sqrt{20}+2)^2$
26. $(4-\sqrt{20})^2$
27. $(\sqrt[3]{6}-3)(\sqrt[3]{4}+3)$
28. $(\sqrt[3]{9}+5)(\sqrt[3]{12}-5)$
29. $(\sqrt{3}+2)(\sqrt{3}-2)$
30. $(3-\sqrt{5})(3+\sqrt{5})$
31. $(6-\sqrt{7})(6+\sqrt{7})$
32. $(\sqrt{8}-5)(\sqrt{8}+5)$
33. $(\sqrt{5}-\sqrt{3})(\sqrt{5}-\sqrt{3})$
34. $(\sqrt{2}+\sqrt{7})(\sqrt{2}+\sqrt{7})$
35. $(10+\sqrt{2x})^2$
36. $(5-\sqrt{3v})^2$
37. $(9\sqrt{x}+2)(5\sqrt{x}-3)$
38. $(16\sqrt{u}-3)(\sqrt{u}-1)$
39. $(2\sqrt{2x}-\sqrt{5})(2\sqrt{2x}+\sqrt{5})$
40. $(\sqrt{7}-3\sqrt{3t})(\sqrt{7}+3\sqrt{3t})$
41. $(\sqrt[3]{2x}+5)^2$
42. $(\sqrt[3]{3x}-4)^2$
43. $(\sqrt[3]{y}+2)(\sqrt[3]{y^2}-5)$
44. $(\sqrt[3]{2y}+10)(\sqrt[3]{4y^2}-10)$
45. $(\sqrt[3]{t}+1)(\sqrt[3]{t^2}+4\sqrt[3]{t}-3)$
46. $(\sqrt[3]{x}-2)(\sqrt[3]{x^2}-2\sqrt[3]{x}+1)$
47. $\sqrt{x^3y^4}(2\sqrt{xy^2}-\sqrt{x^3y})$
48. $3\sqrt{xy^3}(\sqrt{x^3y}+2\sqrt{xy^2})$
49. $2\sqrt[3]{x^4y^5}(\sqrt[3]{8x^{12}y^4}+\sqrt[3]{16xy^9})$
50. $\sqrt[4]{8x^3y^5}(\sqrt[4]{4x^5y^7}-\sqrt[4]{3x^7y^6})$

In Exercises 51–56, complete the statement.

51. $5x\sqrt{3}+15\sqrt{3}=5\sqrt{3}(\quad)$
52. $x\sqrt{7}-x^2\sqrt{7}=x\sqrt{7}(\quad)$
53. $4\sqrt{12}-2x\sqrt{27}=2\sqrt{3}(\quad)$
54. $5\sqrt{50}+10y\sqrt{8}=5\sqrt{2}(\quad)$
55. $6u^2+\sqrt{18u^3}=3u(\quad)$
56. $12s^3-\sqrt{32s^4}=4s^2(\quad)$

In Exercises 57–70, find the conjugate of the expression. Then multiply the expression by its conjugate and simplify. *See Example 4.*

✓ **57.** $2 + \sqrt{5}$

58. $\sqrt{2} - 9$

59. $\sqrt{11} - \sqrt{3}$

60. $\sqrt{10} + \sqrt{7}$

61. $\sqrt{15} + 3$

62. $\sqrt{14} - 3$

63. $\sqrt{x} - 3$

64. $\sqrt{t} + 7$

65. $\sqrt{2u} - \sqrt{3}$

66. $\sqrt{5a} + \sqrt{2}$

67. $2\sqrt{2} + \sqrt{4}$

68. $4\sqrt{3} + \sqrt{2}$

69. $\sqrt{x} + \sqrt{y}$

70. $3\sqrt{u} + \sqrt{3v}$

In Exercises 71–74, evaluate the function as indicated and simplify.

71. $f(x) = x^2 - 6x + 1$

(a) $f(2 - \sqrt{3})$ (b) $f(3 - 2\sqrt{2})$

72. $g(x) = x^2 + 8x + 11$

(a) $g(-4 + \sqrt{5})$ (b) $g(-4\sqrt{2})$

73. $f(x) = x^2 - 2x - 2$

(a) $f(1 + \sqrt{3})$ (b) $f(3 - \sqrt{3})$

74. $g(x) = x^2 - 4x + 1$

(a) $g(1 + \sqrt{5})$ (b) $g(2 - \sqrt{3})$

In Exercises 75–98, simplify the expression. *See Examples 5–8.*

✓ **75.** $\dfrac{6}{\sqrt{11} - 2}$

76. $\dfrac{8}{\sqrt{7} + 3}$

77. $\dfrac{7}{\sqrt{3} + 5}$

78. $\dfrac{5}{9 - \sqrt{6}}$

79. $\dfrac{3}{2\sqrt{10} - 5}$

80. $\dfrac{4}{3\sqrt{5} - 1}$

✓ **81.** $\dfrac{2}{\sqrt{6} + \sqrt{2}}$

82. $\dfrac{10}{\sqrt{9} + \sqrt{5}}$

83. $\dfrac{10}{2\sqrt{3} - \sqrt{7}}$

84. $\dfrac{12}{2\sqrt{5} + \sqrt{8}}$

✓ **85.** $(\sqrt{7} + 2) \div (\sqrt{7} - 2)$

86. $(5 - \sqrt{3}) \div (3 + \sqrt{3})$

87. $(\sqrt{x} - 5) \div (2\sqrt{x} - 1)$

88. $(2\sqrt{t} + 1) \div (2\sqrt{t} - 1)$

89. $\dfrac{3x}{\sqrt{15} - \sqrt{3}}$

90. $\dfrac{5y}{\sqrt{12} + \sqrt{10}}$

91. $\dfrac{\sqrt{5t}}{\sqrt{5} - \sqrt{t}}$

92. $\dfrac{\sqrt{2x}}{\sqrt{x} - \sqrt{2}}$

93. $\dfrac{8a}{\sqrt{3a} + \sqrt{a}}$

94. $\dfrac{7z}{\sqrt{5z} - \sqrt{z}}$

95. $\dfrac{3(x - 4)}{x^2 - \sqrt{x}}$

96. $\dfrac{6(y + 1)}{y^2 + \sqrt{y}}$

✓ **97.** $\dfrac{\sqrt{u + v}}{\sqrt{u - v} - \sqrt{u}}$

98. $\dfrac{z}{\sqrt{u + z} - \sqrt{u}}$

In Exercises 99–102, use a graphing calculator to graph the functions in the same viewing window. Use the graphs to verify that the expressions are equivalent. Verify your results algebraically.

99. $y_1 = \dfrac{10}{\sqrt{x} + 1}$

$y_2 = \dfrac{10(\sqrt{x} - 1)}{x - 1}, \quad x \neq 1$

100. $y_1 = \dfrac{4x}{\sqrt{x} + 4}$

$y_2 = \dfrac{4x(\sqrt{x} - 4)}{x - 16}, \quad x \neq 16$

101. $y_1 = \dfrac{2\sqrt{3x}}{2 - \sqrt{3x}}$

$y_2 = \dfrac{2(2\sqrt{3x} + 3x)}{4 - 3x}$

102. $y_1 = \dfrac{\sqrt{2x} + 6}{\sqrt{2x} - 2}$

$y_2 = \dfrac{x + 6 + 4\sqrt{2x}}{x - 2}$

Rationalizing Numerators In the study of calculus, students sometimes rewrite an expression by rationalizing the numerator. In Exercises 103–110, rationalize the numerator. (*Note:* The results will not be in simplest radical form.)

103. $\dfrac{\sqrt{2}}{7}$

104. $\dfrac{\sqrt{3}}{3}$

105. $\dfrac{\sqrt{10}}{\sqrt{3x}}$

106. $\dfrac{\sqrt{5}}{\sqrt{7x}}$

107. $\dfrac{\sqrt{7} + \sqrt{3}}{5}$

108. $\dfrac{\sqrt{2} - \sqrt{5}}{4}$

109. $\dfrac{\sqrt{y} - 5}{\sqrt{3}}$

110. $\dfrac{\sqrt{x} + 6}{\sqrt{2}}$

Solving Problems

111. *Geometry* The width w and height h of the strongest rectangular beam that can be cut from a log of diameter 24 inches (see figure) are given by

$w = 8\sqrt{3}$ and $h = \sqrt{24^2 - (8\sqrt{3})^2}$.

Find the area of a rectangular cross section of the beam, and write the area in simplest form.

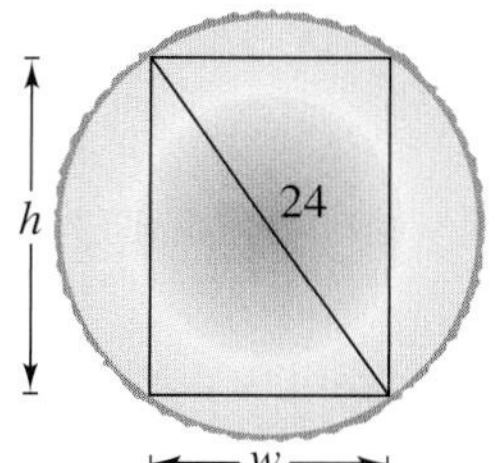

112. *Basketball* The area of the circular cross-section of a basketball is 70 square inches. The area enclosed by a basketball hoop is about 254 square inches. Find the ratio of the diameter of the basketball to the diameter of the hoop.

Hutch Axilrod/Getty Images

113. *Force* The force required to slide a steel block weighing 500 pounds across a milling machine is

$$\frac{500k}{\dfrac{1}{\sqrt{k^2+1}} + \dfrac{k^2}{\sqrt{k^2+1}}}$$

where k is the friction constant (see figure). Simplify this expression.

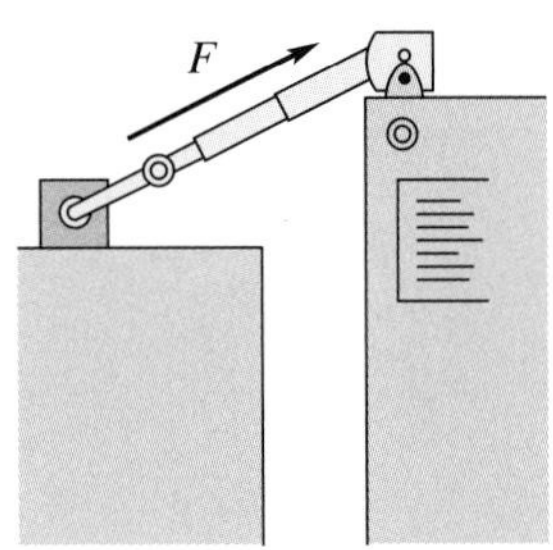

114. The ratio of the width of the Temple of Hephaestus to its height (see figure) is approximately

$$\frac{w}{h} \approx \frac{2}{\sqrt{5}-1}.$$

This number is called the **golden section.** Early Greeks believed that the most aesthetically pleasing rectangles were those whose sides had this ratio.

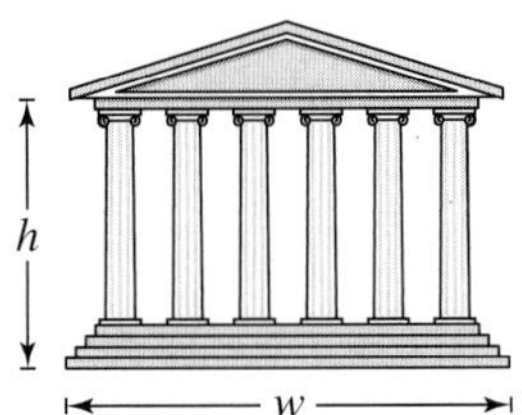

(a) Rationalize the denominator for this expression. Approximate your answer, rounded to two decimal places.

(b) Use the Pythagorean Theorem, a straightedge, and a compass to construct a rectangle whose sides have the golden section as their ratio.

Explaining Concepts

115. Let a and b be integers, but not perfect squares. Describe the circumstances (if any) for which each expression represents a rational number. Explain.

(a) $a\sqrt{b}$

(b) $\sqrt{a}\sqrt{b}$

116. Given that a and b are positive integers, what type of number is the product of the expression $\sqrt{a} + \sqrt{b}$ and its conjugate? Explain.

117. Find the conjugate of $\sqrt{a} + \sqrt{b}$. Multiply the conjugates. Next, find the conjugate of $\sqrt{b} + \sqrt{a}$. Multiply the conjugates. Explain how changing the order of the terms affects the conjugate and the product of the conjugates.

118. Rationalize the denominators of $\dfrac{1}{\sqrt{a}+\sqrt{b}}$ and $\dfrac{1}{\sqrt{b}+\sqrt{a}}$. Explain how changing the order of the terms in the denominator affects the rationalized form of the quotient.

Cumulative Review

In Exercises 119–122, solve the equation. If there is exactly one solution, check your answer. If not, describe the solution.

119. $3x - 18 = 0$

120. $7t - 4 = 4t + 8$

121. $3x - 4 = 3x$

122. $3(2x + 5) = 6x + 15$

In Exercises 123–126, solve the equation by factoring.

123. $x^2 - 144 = 0$

124. $4x^2 - 25 = 0$

125. $x^2 + 2x - 15 = 0$

126. $6x^2 - x - 12 = 0$

In Exercises 127–130, simplify the radical expression.

127. $\sqrt{32x^2y^5}$

128. $\sqrt[3]{32x^2y^5}$

129. $\sqrt[4]{32x^2y^5}$

130. $\sqrt[5]{32x^2y^5}$

9.5 Radical Equations and Applications

Jeff Greenberg/The Image Works

What You Should Learn

1 ▶ Solve a radical equation by raising each side to the nth power.

2 ▶ Solve application problems involving radical equations.

Why You Should Learn It

Radical equations can be used to model and solve real-life applications. For instance, in Exercise 106 on page 598, a radical equation is used to model the total monthly cost of daily flights between Chicago and Denver.

1 ▶ Solve a radical equation by raising each side to the nth power.

Solving Radical Equations

Solving equations involving radicals is somewhat like solving equations that contain fractions—first try to eliminate the radicals and obtain a polynomial equation. Then, solve the polynomial equation using the standard procedures. The following property plays a key role.

Raising Each Side of an Equation to the nth Power

Let u and v be real numbers, variables, or algebraic expressions, and let n be a positive integer. If $u = v$, then it follows that

$$u^n = v^n.$$

This is called **raising each side of an equation to the nth power.**

To use this property to solve a radical equation, first try to isolate one of the radicals on one side of the equation. When using this property to solve radical equations, it is critical that you check your solutions in the original equation.

Technology: Tip

To use a graphing calculator to check the solution in Example 1, graph

$$y = \sqrt{x} - 8$$

as shown below. Notice that the graph crosses the x-axis at $x = 64$, which confirms the solution that was obtained algebraically.

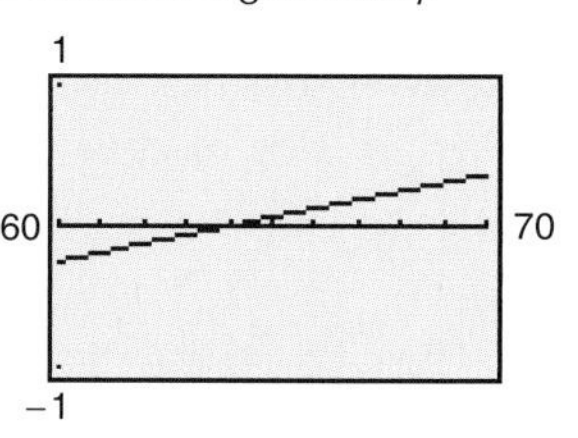

EXAMPLE 1 Solving an Equation Having One Radical

Solve $\sqrt{x} - 8 = 0$.

Solution

$\sqrt{x} - 8 = 0$ Write original equation.

$\sqrt{x} = 8$ Isolate radical.

$(\sqrt{x})^2 = 8^2$ Square each side.

$x = 64$ Simplify.

Check

$\sqrt{64} - 8 \stackrel{?}{=} 0$ Substitute 64 for x in original equation.

$8 - 8 = 0$ Solution checks. ✓

So, the equation has one solution: $x = 64$.

 ***CHECKPOINT*** *Now try Exercise 11.*

Checking solutions of a radical equation is especially important because raising each side of an equation to the nth power to remove the radical(s) often introduces *extraneous* solutions.

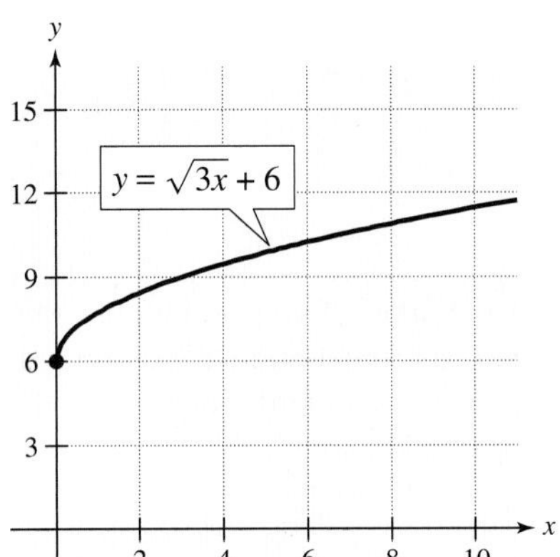

Figure 9.6

EXAMPLE 2 Solving an Equation Having One Radical

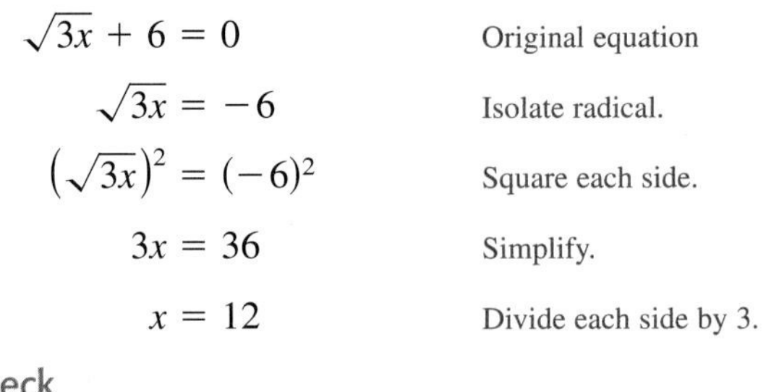

$\sqrt{3x} + 6 = 0$ Original equation

$\sqrt{3x} = -6$ Isolate radical.

$(\sqrt{3x})^2 = (-6)^2$ Square each side.

$3x = 36$ Simplify.

$x = 12$ Divide each side by 3.

Check

$\sqrt{3(12)} + 6 \stackrel{?}{=} 0$ Substitute 12 for x in original equation.

$6 + 6 \neq 0$ Solution does not check. ✗

The solution $x = 12$ is an extraneous solution. So, the original equation has no solution. You can also check this graphically, as shown in Figure 9.6. Notice that the graph does not cross the x-axis and so has no x-intercept.

CHECKPOINT *Now try Exercise 21.*

EXAMPLE 3 Solving an Equation Having One Radical

$\sqrt[3]{2x + 1} - 2 = 3$ Original equation

$\sqrt[3]{2x + 1} = 5$ Isolate radical.

$(\sqrt[3]{2x + 1})^3 = 5^3$ Cube each side.

$2x + 1 = 125$ Simplify.

$2x = 124$ Subtract 1 from each side.

$x = 62$ Divide each side by 2.

Check

$\sqrt[3]{2(62) + 1} - 2 \stackrel{?}{=} 3$ Substitute 62 for x in original equation.

$\sqrt[3]{125} - 2 \stackrel{?}{=} 3$ Simplify.

$5 - 2 = 3$ Solution checks. ✓

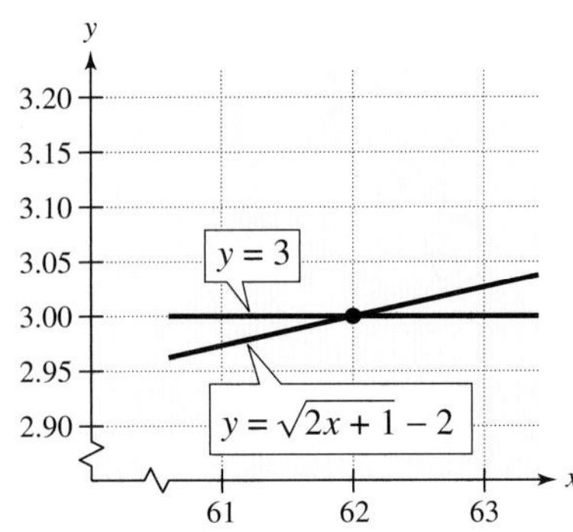

Figure 9.7

So, the equation has one solution: $x = 62$. You can also check the solution graphically by determining the point of intersection of the graphs of $y = \sqrt[3]{2x + 1} - 2$ (left side of equation) and $y = 3$ (right side of equation), as shown in Figure 9.7.

CHECKPOINT *Now try Exercise 27.*

Technology: Tip

In Example 4, you can graphically check the solution of the equation by graphing the left side and right side in the same viewing window. That is, graph the equations

$$y_1 = \sqrt{5x + 3}$$

and

$$y_2 = \sqrt{x + 11}$$

in the same viewing window, as shown below. Using the *intersect* feature of the graphing calculator will enable you to approximate the point(s) at which the graphs intersect. From the figure, you can see that the two graphs intersect at $x = 2$, which is the solution obtained in Example 4.

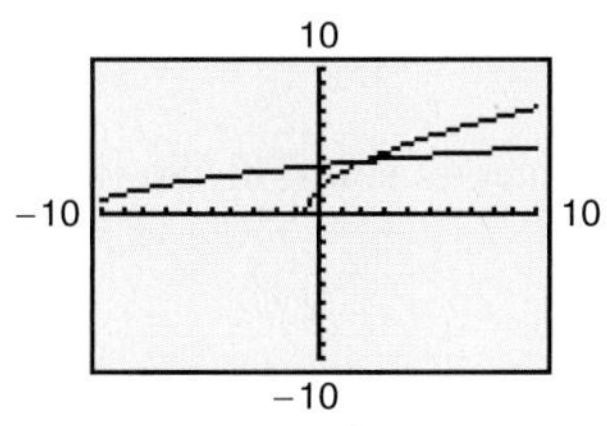

EXAMPLE 4 Solving an Equation Having Two Radicals

Solve $\sqrt{5x + 3} = \sqrt{x + 11}$.

Solution

$\sqrt{5x + 3} = \sqrt{x + 11}$	Write original equation.
$(\sqrt{5x + 3})^2 = (\sqrt{x + 11})^2$	Square each side.
$5x + 3 = x + 11$	Simplify.
$4x + 3 = 11$	Subtract x from each side.
$4x = 8$	Subtract 3 from each side.
$x = 2$	Divide each side by 4.

Check

$\sqrt{5x + 3} = \sqrt{x + 11}$	Write original equation.
$\sqrt{5(2) + 3} \stackrel{?}{=} \sqrt{2 + 11}$	Substitute 2 for x.
$\sqrt{13} = \sqrt{13}$	Solution checks. ✓

So, the equation has one solution: $x = 2$.

 **CHECKPOINT** *Now try Exercise 31.*

EXAMPLE 5 Solving an Equation Having Two Radicals

Solve $\sqrt[4]{3x} + \sqrt[4]{2x - 5} = 0$.

Solution

$\sqrt[4]{3x} + \sqrt[4]{2x - 5} = 0$	Write original equation.
$\sqrt[4]{3x} = -\sqrt[4]{2x - 5}$	Isolate radicals.
$(\sqrt[4]{3x})^4 = (-\sqrt[4]{2x - 5})^4$	Raise each side to fourth power.
$3x = 2x - 5$	Simplify.
$x = -5$	Subtract $2x$ from each side.

Check

$\sqrt[4]{3x} + \sqrt[4]{2x - 5} = 0$	Write original equation.
$\sqrt[4]{3(-5)} + \sqrt[4]{2(-5) - 5} \stackrel{?}{=} 0$	Substitute -5 for x.
$\sqrt[4]{-15} + \sqrt[4]{-15} \neq 0$	Solution does not check. ✗

The solution does not check because it yields fourth roots of negative radicands. So, this equation has no solution. Try checking this graphically. If you graph both sides of the equation, you will discover that the graphs do not intersect.

CHECKPOINT *Now try Exercise 33.*

In the next example you will see that squaring each side of the equation results in a quadratic equation. Remember that you must check the solutions in the *original* radical equation.

EXAMPLE 6 An Equation That Converts to a Quadratic Equation

Solve $\sqrt{x} + 2 = x$.

Solution

$$\begin{aligned}
\sqrt{x} + 2 &= x && \text{Write original equation.}\\
\sqrt{x} &= x - 2 && \text{Isolate radical.}\\
\left(\sqrt{x}\right)^2 &= (x - 2)^2 && \text{Square each side.}\\
x &= x^2 - 4x + 4 && \text{Simplify.}\\
-x^2 + 5x - 4 &= 0 && \text{Write in general form.}\\
(-1)(x - 4)(x - 1) &= 0 && \text{Factor.}\\
x - 4 &= 0 \quad \Rightarrow \quad x = 4 && \text{Set 1st factor equal to 0.}\\
x - 1 &= 0 \quad \Rightarrow \quad x = 1 && \text{Set 2nd factor equal to 0.}
\end{aligned}$$

Check

First Solution	*Second Solution*
$\sqrt{4} + 2 \stackrel{?}{=} 4$	$\sqrt{1} + 2 \stackrel{?}{=} 1$
$2 + 2 = 4$	$1 + 2 \neq 1$

From the check you can see that $x = 1$ is an extraneous solution. So, the only solution is $x = 4$.

CHECKPOINT *Now try Exercise 39.*

When an equation contains two radicals, it may not be possible to isolate both. In such cases, you may have to raise each side of the equation to a power at *two* different stages in the solution.

EXAMPLE 7 Repeatedly Squaring Each Side of an Equation

$$\begin{aligned}
\sqrt{3t + 1} &= 2 - \sqrt{3t} && \text{Original equation}\\
\left(\sqrt{3t + 1}\right)^2 &= \left(2 - \sqrt{3t}\right)^2 && \text{Square each side (1st time).}\\
3t + 1 &= 4 - 4\sqrt{3t} + 3t && \text{Simplify.}\\
-3 &= -4\sqrt{3t} && \text{Isolate radical.}\\
(-3)^2 &= \left(-4\sqrt{3t}\right)^2 && \text{Square each side (2nd time).}\\
9 &= 16(3t) && \text{Simplify.}\\
\frac{3}{16} &= t && \text{Divide each side by 48 and simplify.}
\end{aligned}$$

The solution is $t = \frac{3}{16}$. Check this in the original equation.

CHECKPOINT *Now try Exercise 47.*

2 ▶ Solve application problems involving radical equations.

Applications

EXAMPLE 8 Electricity 

The amount of power consumed by an electrical appliance is given by $I = \sqrt{P/R}$, where I is the current measured in amps, R is the resistance measured in ohms, and P is the power measured in watts. Find the power used by an electric heater for which $I = 10$ amps and $R = 16$ ohms.

Solution

$$10 = \sqrt{\frac{P}{16}}$$ Substitute 10 for I and 16 for R in original equation.

$$10^2 = \left(\sqrt{\frac{P}{16}}\right)^2$$ Square each side.

$$100 = \frac{P}{16} \quad \Rightarrow \quad 1600 = P$$ Simplify and multiply each side by 16.

So, the solution is $P = 1600$ watts. Check this in the original equation.

✓ ***CHECKPOINT*** *Now try Exercise 97.*

Study Tip

An alternative way to solve the problem in Example 8 would be first to solve the equation for P.

$$I = \sqrt{\frac{P}{R}}$$

$$I^2 = \left(\sqrt{\frac{P}{R}}\right)^2$$

$$I^2 = \frac{P}{R}$$

$$I^2R = P$$

At this stage, you can substitute the known values of I and R to obtain

$$P = (10)^2 16 = 1600.$$

EXAMPLE 9 An Application of the Pythagorean Theorem 

The distance between a house on shore and a playground on shore is 40 meters. The distance between the playground and a house on an island is 50 meters. (See Figure 9.8.) What is the distance between the two houses?

Solution

From Figure 9.8, you can see that the distances form a right triangle. So, you can use the Pythagorean Theorem to find the distance between the two houses.

$$c = \sqrt{a^2 + b^2}$$ Pythagorean Theorem

$$50 = \sqrt{40^2 + b^2}$$ Substitute 40 for a and 50 for c.

$$50 = \sqrt{1600 + b^2}$$ Simplify.

$$50^2 = (\sqrt{1600 + b^2})^2$$ Square each side.

$$2500 = 1600 + b^2$$ Simplify.

$$0 = b^2 - 900$$ Write in general form.

$$0 = (b + 30)(b - 30)$$ Factor.

$$b + 30 = 0 \quad \Rightarrow \quad b = -30$$ Set 1st factor equal to 0.

$$b - 30 = 0 \quad \Rightarrow \quad b = 30$$ Set 2nd factor equal to 0.

Choose the positive solution to obtain a distance of 30 meters. Check this solution in the original equation.

✓ ***CHECKPOINT*** *Now try Exercise 85.*

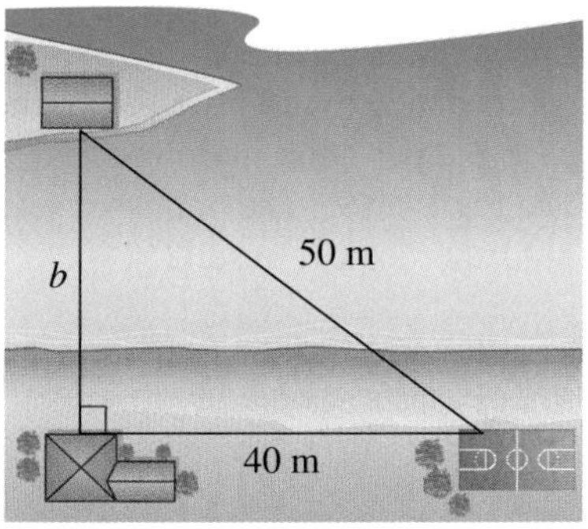

b
50 m
40 m

Figure 9.8

EXAMPLE 10 Velocity of a Falling Object 

The velocity of a free-falling object can be determined from the equation $v = \sqrt{2gh}$, where v is the velocity measured in feet per second, $g = 32$ feet per second per second, and h is the distance (in feet) the object has fallen. Find the height from which a rock has been dropped when it strikes the ground with a velocity of 50 feet per second.

Solution

$$v = \sqrt{2gh} \qquad \text{Write original equation.}$$

$$50 = \sqrt{2(32)h} \qquad \text{Substitute 50 for } v \text{ and 32 for } g.$$

$$50^2 = \left(\sqrt{64h}\right)^2 \qquad \text{Square each side.}$$

$$2500 = 64h \qquad \text{Simplify.}$$

$$39 \approx h \qquad \text{Divide each side by 64.}$$

Check

$$v = \sqrt{2gh} \qquad \text{Write original equation.}$$

$$50 \overset{?}{\approx} \sqrt{2(32)(39)} \qquad \text{Substitute 50 for } v\text{, 32 for } g\text{, and 39 for } h.$$

$$50 \overset{?}{\approx} \sqrt{2496} \qquad \text{Simplify.}$$

$$50 \approx 49.96 \qquad \text{Solution checks. ✓}$$

So, the height from which the rock has been dropped is approximately 39 feet.

✓ CHECKPOINT *Now try Exercise 99.*

Study Tip

When checking a solution with a value that has been rounded, the check will not result in an equality. If the solution is valid, the expressions on each side of the equal sign will be *approximately* equal to each other, as shown in Example 10.

EXAMPLE 11 Market Research 

The marketing department at a publisher determines that the demand for a book depends on the price of the book in accordance with the formula $p = 40 - \sqrt{0.0001x + 1}$, $x \geq 0$, where p is the price per book in dollars and x is the number of books sold at the given price. (See Figure 9.9.) The publisher sets the price at \$12.95. How many copies can the publisher expect to sell?

Figure 9.9

Solution

$$p = 40 - \sqrt{0.0001x + 1} \qquad \text{Write original equation.}$$

$$12.95 = 40 - \sqrt{0.0001x + 1} \qquad \text{Substitute 12.95 for } p.$$

$$\sqrt{0.0001x + 1} = 27.05 \qquad \text{Isolate radical.}$$

$$0.0001x + 1 = 731.7025 \qquad \text{Square each side.}$$

$$0.0001x = 730.7025 \qquad \text{Subtract 1 from each side.}$$

$$x = 7{,}307{,}025 \qquad \text{Divide each side by 0.0001.}$$

So, the publisher can expect to sell about 7.3 million copies.

✓ CHECKPOINT *Now try Exercise 105.*

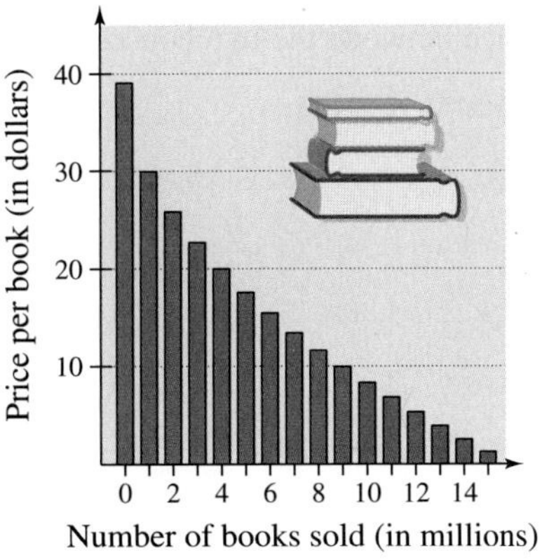

Concept Check

1. Is $1 - x\sqrt{5} = x^2$ a radical equation? Explain your reasoning.
2. One reason for checking a solution in the original equation is to discover errors made when solving the equation. Describe another reason.
3. In your own words, describe the steps used to solve $\sqrt{x} + 2 = x$.
4. The graphs of $f(x) = x - 1$ and $g(x) = \sqrt{x + 5}$ are shown at the right. Explain how you can use the graphs to solve $x - 1 = \sqrt{x + 5}$.

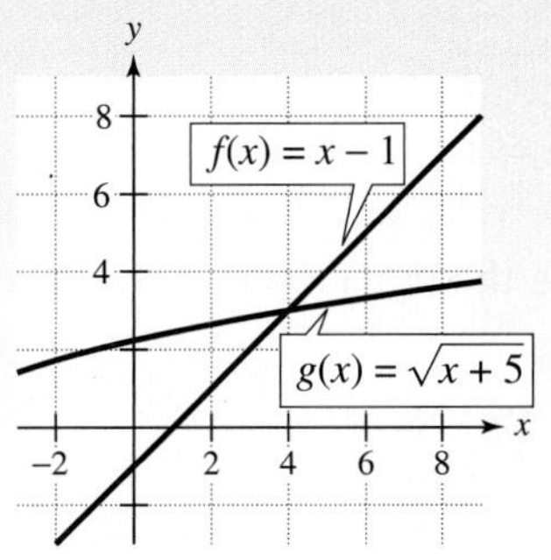

Figure for 4

9.5 EXERCISES

Go to pages 608–609 to record your assignments.

Developing Skills

In Exercises 1–4, determine whether each value of *x* is a solution of the equation.

Equation	*Values of x*	
1. $\sqrt{x} - 10 = 0$	(a) $x = -4$	(b) $x = -100$
	(c) $x = \sqrt{10}$	(d) $x = 100$
2. $\sqrt{3x} - 6 = 0$	(a) $x = \frac{2}{3}$	(b) $x = 2$
	(c) $x = 12$	(d) $x = -\frac{1}{3}\sqrt{6}$
3. $\sqrt[3]{x - 4} = 4$	(a) $x = -60$	(b) $x = 68$
	(c) $x = 20$	(d) $x = 0$
4. $\sqrt[4]{2x} + 2 = 6$	(a) $x = 128$	(b) $x = 2$
	(c) $x = -2$	(d) $x = 0$

In Exercises 5–54, solve the equation and check your solution(s). (Some of the equations have no solution.) ***See Examples 1–7.***

5. $\sqrt{x} = 12$
6. $\sqrt{x} = 5$
7. $\sqrt{y} = 7$
8. $\sqrt{t} = 4$
9. $\sqrt[3]{z} = 3$
10. $\sqrt[4]{x} = 3$
11. $\sqrt{y} - 7 = 0$
12. $\sqrt{t} - 13 = 0$
13. $\sqrt{u} + 13 = 0$
14. $\sqrt{y} + 15 = 0$
15. $\sqrt{x} - 8 = 0$
16. $\sqrt{x} - 10 = 0$
17. $\sqrt{10x} = 30$
18. $\sqrt{8x} = 6$
19. $\sqrt{-3x} = 9$
20. $\sqrt{-4y} = 4$
21. $\sqrt{5t} - 2 = 0$
22. $10 - \sqrt{6x} = 0$
23. $\sqrt{3y + 1} = 4$
24. $\sqrt{3 - 2x} = 2$
25. $\sqrt{9 - 2x} = -9$
26. $\sqrt{2t - 7} = -5$
27. $\sqrt[3]{y - 3} + 4 = 6$
28. $\sqrt[4]{6a - 11} + 8 = -5$
29. $6\sqrt[4]{x + 3} = 15$
30. $4\sqrt[3]{x + 4} = 7$
31. $\sqrt{x + 3} = \sqrt{2x - 1}$
32. $\sqrt{3t + 1} = \sqrt{t + 15}$
33. $\sqrt{3y - 5} - 3\sqrt{y} = 0$
34. $\sqrt{2u + 10} - 2\sqrt{u} = 0$
35. $\sqrt[3]{3x - 4} = \sqrt[3]{x + 10}$
36. $2\sqrt[3]{10 - 3x} = \sqrt[3]{2 - x}$
37. $\sqrt[3]{2x + 15} - \sqrt[3]{x} = 0$
38. $\sqrt[4]{2x} + \sqrt[4]{x + 3} = 0$
39. $\sqrt{x^2 - 2} = x + 4$
40. $\sqrt{x^2 - 4} = x - 2$
41. $\sqrt{2x} = x - 4$
42. $\sqrt{x} = 6 - x$
43. $\sqrt{8x + 1} = x + 2$
44. $\sqrt{3x + 7} = x + 3$
45. $\sqrt{3x + 4} = \sqrt{4x + 3}$
46. $\sqrt{2x - 7} = \sqrt{3x - 12}$
47. $\sqrt{z + 2} = 1 + \sqrt{z}$
48. $\sqrt{2x + 5} = 7 - \sqrt{2x}$
49. $\sqrt{2t + 3} = 3 - \sqrt{2t}$
50. $\sqrt{x} + \sqrt{x + 2} = 2$

51. $\sqrt{x+5} - \sqrt{x} = 1$

52. $\sqrt{x+1} = 2 - \sqrt{x}$

53. $\sqrt{x-6} + 3 = \sqrt{x+9}$

54. $\sqrt{x+3} - \sqrt{x-1} = 1$

In Exercises 55–62, solve the equation and check your solution(s).

55. $t^{3/2} = 8$ **56.** $v^{2/3} = 25$

57. $3y^{1/3} = 18$ **58.** $2x^{3/4} = 54$

59. $(x+4)^{2/3} = 4$ **60.** $(u-2)^{4/3} = 81$

61. $(2x+5)^{1/3} + 3 = 0$ **62.** $(x-6)^{3/2} - 27 = 0$

In Exercises 63–72, use a graphing calculator to graph each side of the equation in the same viewing window. Use the graphs to approximate the solution(s). Verify your answer algebraically.

63. $\sqrt{x} = 2(2-x)$ **64.** $\sqrt{2x+3} = 4x - 3$

65. $\sqrt{x^2+1} = 5 - 2x$ **66.** $\sqrt{8-3x} = x$

67. $\sqrt{x+3} = 5 - \sqrt{x}$ **68.** $\sqrt[3]{5x-8} = 4 - \sqrt[3]{x}$

69. $3\sqrt[4]{x} = 9 - x$ **70.** $\sqrt[3]{x+4} = \sqrt{6-x}$

71. $\sqrt{15-4x} = 2x$ **72.** $\dfrac{4}{\sqrt{x}} = 3\sqrt{x} - 4$

In Exercises 73–76, use the given function to find the indicated value of x.

73. For $f(x) = \sqrt{x} - \sqrt{x-9}$, find x such that $f(x) = 1$.

74. For $g(x) = \sqrt{x} + \sqrt{x-5}$, find x such that $g(x) = 5$.

75. For $h(x) = \sqrt{x-2} - \sqrt{4x+1}$, find x such that $h(x) = -3$.

76. For $f(x) = \sqrt{2x+7} - \sqrt{x+15}$, find x such that $f(x) = -1$.

In Exercises 77–80, find the x-intercept(s) of the graph of the function without graphing the function.

77. $f(x) = \sqrt{x+5} - 3 + \sqrt{x}$

78. $f(x) = \sqrt{6x+7} - 2 - \sqrt{2x+3}$

79. $f(x) = \sqrt{3x-2} - 1 - \sqrt{2x-3}$

80. $f(x) = \sqrt{5x+6} - 1 - \sqrt{3x+3}$

Solving Problems

Geometry In Exercises 81–84, find the length x of the unknown side of the right triangle. (Round your answer to two decimal places.)

81.

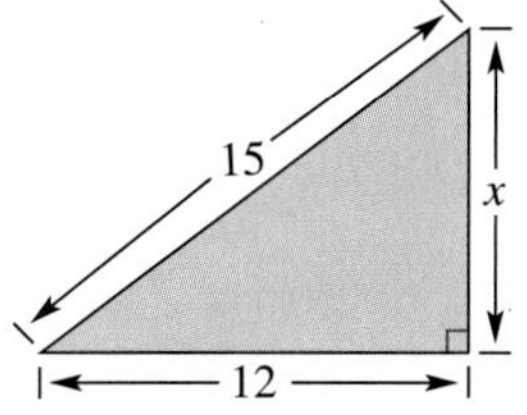

82.

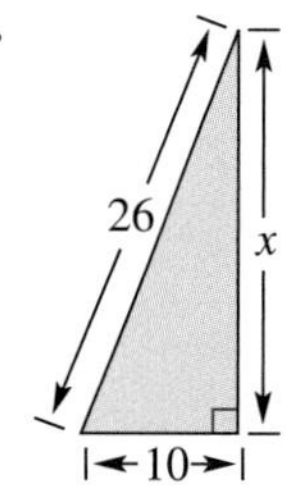

83.

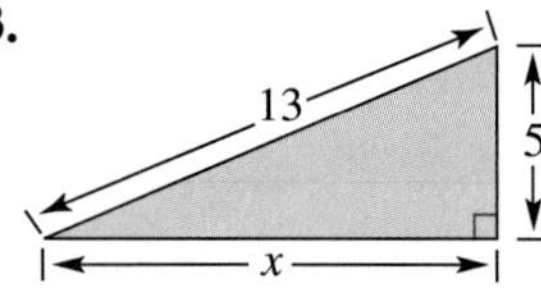

84.

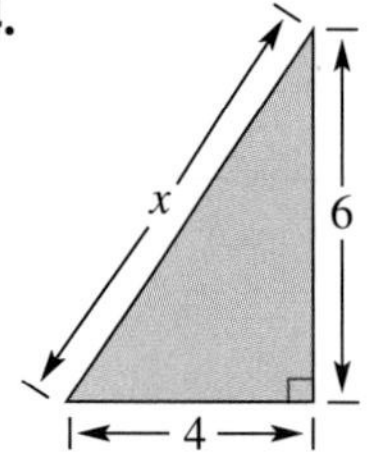

✓ **85.** *Plasma TV* The screen of a plasma television has a diagonal of 50 inches and a width of 43.75 inches. Draw a diagram of the plasma television and find the length of the screen.

86. *Basketball* A basketball court is 50 feet wide and 94 feet long. Draw a diagram of the basketball court and find the length of a diagonal of the court.

87. *Ladders* An extension ladder is placed against the side of a house such that the base of the ladder is 2 meters from the base of the house and the ladder reaches 6 meters up the side of the house. How far is the ladder extended?

88. *Guy Wires* A guy wire on a 100-foot radio tower is attached to the top of the tower and to an anchor 50 feet from the base of the tower. Find the length of the guy wire.

89. *Ladders* A ladder is 17 feet long, and the bottom of the ladder is 8 feet from the side of a house. How far does the ladder reach up the side of the house?

90. *Construction* A 10-foot plank is used to brace a basement wall during construction of a home. The plank is nailed to the wall 6 feet above the floor. Find the slope of the plank.

91. *Geometry* Determine the length and width of a rectangle with a perimeter of 92 inches and a diagonal of 34 inches.

92. *Geometry* Determine the length and width of a rectangle with a perimeter of 68 inches and a diagonal of 26 inches.

93. *Geometry* The lateral surface area of a cone (see figure) is given by $S = \pi r\sqrt{r^2 + h^2}$. Solve the equation for h. Then find the height of a cone with a lateral surface area of $364\pi\sqrt{2}$ square centimeters and a radius of 14 centimeters.

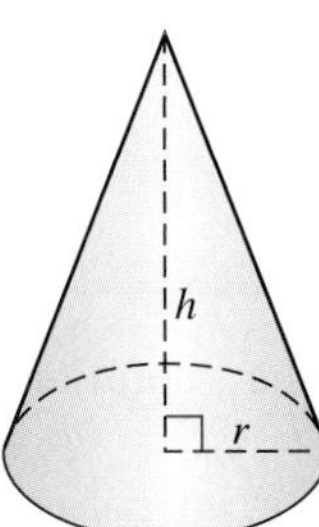

94. *Geometry* The slant height l of a truncated pyramid (see figure) is given by

$l = \sqrt{h^2 + \frac{1}{4}(b_2 - b_1)^2}$.

Solve the equation for h. Then find the height of a truncated pyramid when $l = 2\sqrt{26}$ inches, $b_2 = 8$ inches, and $b_1 = 4$ inches.

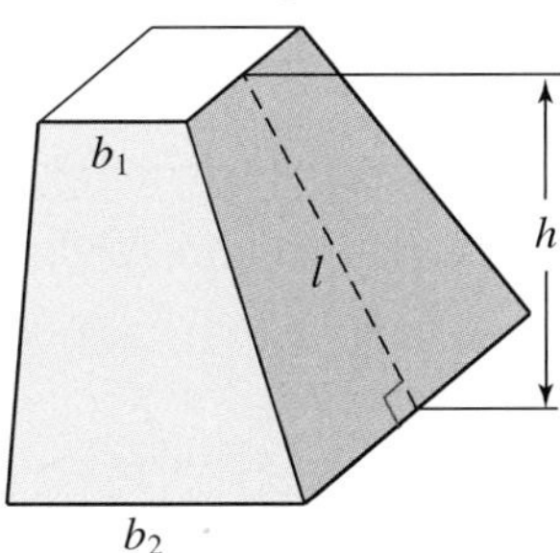

95. *Geometry* Write a function that gives the radius r of a circle in terms of the circle's area A. Use a graphing calculator to graph this function.

96. *Geometry* Write a function that gives the radius r of a sphere in terms of the sphere's volume V. Use a graphing calculator to graph this function.

Height In Exercises 97 and 98, use the formula $t = \sqrt{d/16}$, which gives the time t in seconds for a free-falling object to fall d feet.

✓ **97.** A construction worker drops a nail from a building and observes it strike a water puddle after approximately 2 seconds. Estimate the height from which the nail was dropped.

98. A farmer drops a stone down a well and hears it strike the water after approximately 4.5 seconds. Estimate the depth of the well.

Free-Falling Object In Exercises 99–102, use the equation for the velocity of a free-falling object, $v = \sqrt{2gh}$, as described in Example 10.

✓ **99.** A cliff diver dives from a height of 80 feet. Estimate the velocity of the diver when the diver strikes the water.

100. A coin is dropped from a hot air balloon that is 250 feet above the ground. Estimate the velocity of the coin when the coin strikes the ground.

101. An egg strikes the ground with a velocity of 50 feet per second. Estimate to two decimal places the height from which the egg was dropped.

102. A stone strikes the water with a velocity of 130 feet per second. Estimate to two decimal places the height from which the stone was dropped.

Period of a Pendulum In Exercises 103 and 104, the time t (in seconds) for a pendulum of length L (in feet) to go through one complete cycle (its period) is given by $t = 2\pi\sqrt{L/32}$.

103. How long is the pendulum of a grandfather clock with a period of 1.5 seconds?

104. How long is the pendulum of a mantel clock with a period of 0.75 second?

✓ **105.** *Demand* The demand equation for a sweater is given by

$$p = 50 - \sqrt{0.8(x - 1)}$$

where x is the number of units demanded per day and p is the price per sweater. Find the demand when the price is set at \$30.02.

106. *Airline Passengers* An airline offers daily flights between Chicago and Denver. The total monthly cost C (in millions of dollars) of these flights is

$$C = \sqrt{0.2x + 1}, \quad x \geq 0$$

where x is measured in thousands of passengers (see figure). The total cost of the flights for June is 2.5 million dollars. Approximately how many passengers flew in June?

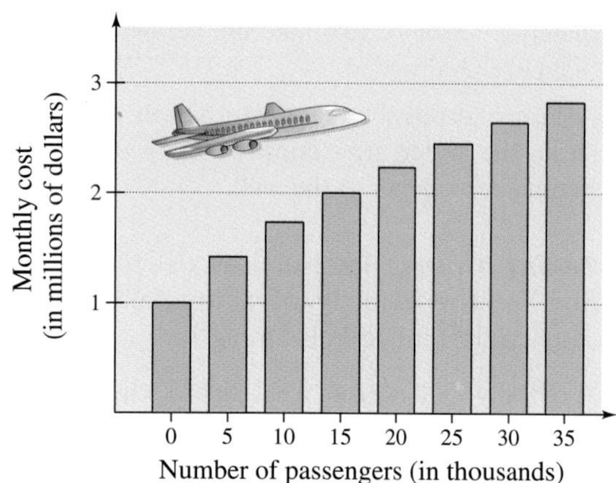

107. *Killer Whales* The weight w (in pounds) of a killer whale can be modeled by

$$w = 280 + 325\sqrt{t}, \quad 0 \leq t \leq 144$$

where t represents the age (in months) of the killer whale.

(a) Use a graphing calculator to graph the model.

(b) At what age did the killer whale weigh about 3400 pounds?

108. *Consumer Spending* The average movie ticket price p (in dollars) consumers paid in theaters in the United States for the years 1997 through 2006 can be modeled by

$$p = 0.518 + 1.52\sqrt{t}, \quad 7 \leq t \leq 16$$

where t represents the year, with $t = 7$ corresponding to 1997. (Source: Veronis, Suhler & Associates Inc.)

(a) Use a graphing calculator to graph the model.

(b) In what year did the average movie ticket price in theaters reach $5.80?

Explaining Concepts

109. *Error Analysis* Describe the error.

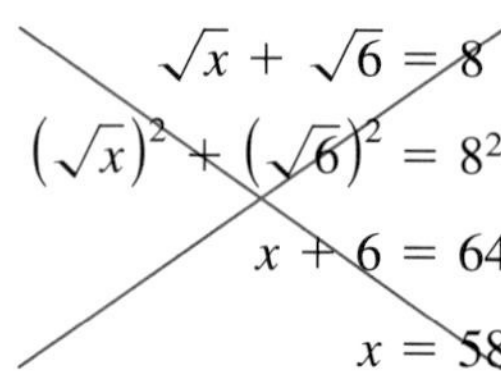

110. Does raising each side of an equation to the nth power always yield an equivalent equation? Explain.

111. *Exploration* The solution of the equation $x + \sqrt{x - a} = b$ is $x = 20$. Discuss how to find a and b. (There are many correct values for a and b.)

112. Explain how you can tell that $\sqrt{x - 9} = -4$ has no solution without solving the equation.

Cumulative Review

In Exercises 113–116, determine whether the two lines are parallel, perpendicular, or neither.

113. $L_1: y = 4x + 2$
$L_2: y = 4x - 1$

114. $L_1: y = 3x - 8$
$L_2: y = -3x - 8$

115. $L_1: y = -x + 5$
$L_2: y = x - 3$

116. $L_1: y = 2x$
$L_2: y = \frac{1}{2}x + 4$

In Exercises 117 and 118, use matrices to solve the system of linear equations.

117. $\begin{cases} 4x - y = 10 \\ -7x - 2y = -25 \end{cases}$

118. $\begin{cases} 3x - 2y = 5 \\ 6x - 5y = 14 \end{cases}$

In Exercises 119–122, simplify the expression.

119. $a^{3/5} \cdot a^{1/5}$

120. $\dfrac{m^2}{m^{2/3}}$

121. $\left(\dfrac{x^{1/2}}{x^{1/8}}\right)^4$

122. $\dfrac{(a + b)^{3/4}}{\sqrt[4]{a + b}}$

9.6 Complex Numbers

What You Should Learn

1 ▶ Write square roots of negative numbers in *i*-form and perform operations on numbers in *i*-form.

2 ▶ Determine the equality of two complex numbers.

3 ▶ Add, subtract, and multiply complex numbers.

4 ▶ Use complex conjugates to write the quotient of two complex numbers in standard form.

Why You Should Learn It

Understanding complex numbers can help you in Section 10.3 to identify quadratic equations that have no real solutions.

1 ▶ Write square roots of negative numbers in *i*-form and perform operations on numbers in *i*-form.

The Imaginary Unit *i*

In Section 9.1, you learned that a negative number has no *real* square root. For instance, $\sqrt{-1}$ is not real because there is no real number x such that $x^2 = -1$. So, as long as you are dealing only with real numbers, the equation $x^2 = -1$ has no solution. To overcome this deficiency, mathematicians have expanded the set of numbers by including the **imaginary unit *i*,** defined as

$$i = \sqrt{-1}. \qquad \text{Imaginary unit}$$

This number has the property that $i^2 = -1$. So, the imaginary unit i is a solution of the equation $x^2 = -1$.

The Square Root of a Negative Number

Let c be a positive real number. Then the square root of $-c$ is given by

$$\sqrt{-c} = \sqrt{c(-1)} = \sqrt{c}\sqrt{-1} = \sqrt{c}\,i.$$

When writing $\sqrt{-c}$ in the ***i*-form,** $\sqrt{c}\,i$, note that i is outside the radical.

Technology: Discovery

Use a calculator to evaluate each radical. Does one result in an error message? Explain why.

a. $\sqrt{121}$

b. $\sqrt{-121}$

c. $-\sqrt{121}$

EXAMPLE 1 Writing Numbers in *i*-Form

Write each number in *i*-form.

a. $\sqrt{-36}$ **b.** $\sqrt{-\dfrac{16}{25}}$ **c.** $\sqrt{-54}$ **d.** $\dfrac{\sqrt{-48}}{\sqrt{-3}}$

Solution

a. $\sqrt{-36} = \sqrt{36(-1)} = \sqrt{36}\sqrt{-1} = 6i$

b. $\sqrt{-\dfrac{16}{25}} = \sqrt{\dfrac{16}{25}(-1)} = \sqrt{\dfrac{16}{25}}\sqrt{-1} = \dfrac{4}{5}i$

c. $\sqrt{-54} = \sqrt{54(-1)} = \sqrt{54}\sqrt{-1} = 3\sqrt{6}i$

d. $\dfrac{\sqrt{-48}}{\sqrt{-3}} = \dfrac{\sqrt{48}\sqrt{-1}}{\sqrt{3}\sqrt{-1}} = \dfrac{\sqrt{48}\,i}{\sqrt{3}\,i} = \sqrt{\dfrac{48}{3}} = \sqrt{16} = 4$

✓ **CHECKPOINT** *Now try Exercise 1.*

To perform operations with square roots of negative numbers, you must *first* write the numbers in i-form. You can then add, subtract, and multiply as follows.

$$ai + bi = (a + b)i \qquad \text{Addition}$$

$$ai - bi = (a - b)i \qquad \text{Subtraction}$$

$$(ai)(bi) = ab(i^2) = ab(-1) = -ab \qquad \text{Multiplication}$$

Study Tip

When performing operations with numbers in i-form, you sometimes need to be able to evaluate powers of the imaginary unit i. The first several powers of i are as follows.

$i^1 = i$

$i^2 = -1$

$i^3 = i(i^2) = i(-1) = -i$

$i^4 = (i^2)(i^2) = (-1)(-1) = 1$

$i^5 = i(i^4) = i(1) = i$

$i^6 = (i^2)(i^4) = (-1)(1) = -1$

$i^7 = (i^3)(i^4) = (-i)(1) = -i$

$i^8 = (i^4)(i^4) = (1)(1) = 1$

Note how the pattern of values $i, -1, -i$, and 1 repeats itself for powers greater than 4.

EXAMPLE 2 Operations with Square Roots of Negative Numbers

Perform each operation.

a. $\sqrt{-9} + \sqrt{-49}$ **b.** $\sqrt{-32} - 2\sqrt{-2}$

Solution

a. $\sqrt{-9} + \sqrt{-49} = \sqrt{9}\sqrt{-1} + \sqrt{49}\sqrt{-1}$ Product Rule for Radicals

$= 3i + 7i$ Write in i-form.

$= 10i$ Simplify.

b. $\sqrt{-32} - 2\sqrt{-2} = \sqrt{32}\sqrt{-1} - 2\sqrt{2}\sqrt{-1}$ Product Rule for Radicals

$= 4\sqrt{2}i - 2\sqrt{2}i$ Write in i-form.

$= 2\sqrt{2}i$ Simplify.

✓ **CHECKPOINT** *Now try Exercise 19.*

EXAMPLE 3 Multiplying Square Roots of Negative Numbers

Find each product.

a. $\sqrt{-15}\sqrt{-15}$ **b.** $\sqrt{-5}(\sqrt{-45} - \sqrt{-4})$

Solution

a. $\sqrt{-15}\sqrt{-15} = (\sqrt{15}i)(\sqrt{15}i)$ Write in i-form.

$= (\sqrt{15})^2 i^2$ Multiply.

$= 15(-1)$ $i^2 = -1$

$= -15$ Simplify.

b. $\sqrt{-5}(\sqrt{-45} - \sqrt{-4}) = \sqrt{5}i(3\sqrt{5}i - 2i)$ Write in i-form.

$= (\sqrt{5}i)(3\sqrt{5}i) - (\sqrt{5}i)(2i)$ Distributive Property

$= 3(5)(-1) - 2\sqrt{5}(-1)$ Multiply.

$= -15 + 2\sqrt{5}$ Simplify.

✓ **CHECKPOINT** *Now try Exercise 27.*

When multiplying square roots of negative numbers, always write them in i-form *before multiplying*. If you do not do this, you can obtain incorrect answers. For instance, in Example 3(a) be sure you see that

$$\sqrt{-15}\sqrt{-15} \neq \sqrt{(-15)(-15)} = \sqrt{225} = 15.$$

2 ▶ Determine the equality of two complex numbers.

Complex Numbers

A number of the form $a + bi$, where a and b are real numbers, is called a **complex number.** The real number a is called the **real part** of the complex number $a + bi$, and the number bi is called the **imaginary part.**

Definition of Complex Number

If a and b are real numbers, the number $a + bi$ is a **complex number,** and it is said to be written in **standard form.** If $b = 0$, the number $a + bi = a$ is a real number. If $b \neq 0$, the number $a + bi$ is called an **imaginary number.** A number of the form bi, where $b \neq 0$, is called a **pure imaginary number.**

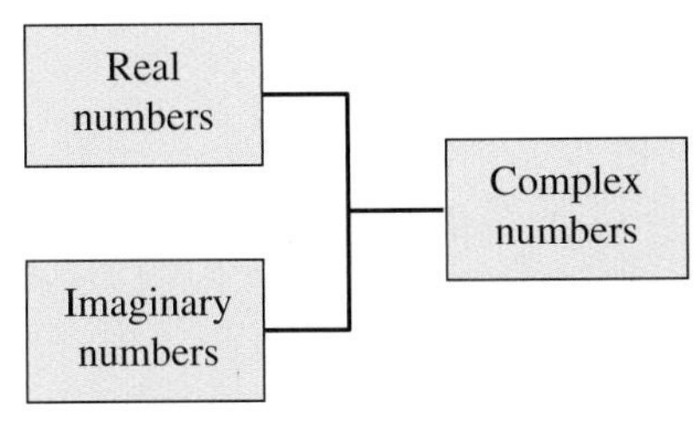

Figure 9.10

A number cannot be both real and imaginary. For instance, the numbers -2, 0, 1, $\frac{1}{2}$, and $\sqrt{2}$ are real numbers, and the numbers $-3i$, $2 + 4i$, and $-1 + i$ are imaginary numbers. The diagram shown in Figure 9.10 shows that the real numbers and the imaginary numbers make up the complex numbers.

Two complex numbers $a + bi$ and $c + di$, in standard form, are equal if and only if $a = c$ and $b = d$.

EXAMPLE 4 Equality of Two Complex Numbers

To determine whether the complex numbers $\sqrt{9} + \sqrt{-48}$ and $3 - 4\sqrt{3}i$ are equal, begin by writing the first number in standard form.

$$\sqrt{9} + \sqrt{-48} = \sqrt{3^2} + \sqrt{4^2(3)(-1)} = 3 + 4\sqrt{3}i$$

The two numbers are not equal because their imaginary parts differ in sign.

✓ **CHECKPOINT** *Now try Exercise 43.*

EXAMPLE 5 Equality of Two Complex Numbers

To find values of x and y that satisfy the equation $3x - \sqrt{-25} = -6 + 3yi$, begin by writing the left side of the equation in standard form.

$$3x - 5i = -6 + 3yi \qquad \text{Each side is in standard form.}$$

For these two numbers to be equal, their real parts must be equal to each other and their imaginary parts must be equal to each other.

Real Parts	*Imaginary Parts*
$3x = -6$	$3yi = -5i$
$x = -2$	$3y = -5$
	$y = -\frac{5}{3}$

So, $x = -2$ and $y = -\frac{5}{3}$.

✓ **CHECKPOINT** *Now try Exercise 51.*

3 ▶ Add, subtract, and multiply complex numbers.

Operations with Complex Numbers

To add or subtract two complex numbers, you add (or subtract) the real and imaginary parts separately. This is similar to combining like terms of a polynomial.

$(a + bi) + (c + di) = (a + c) + (b + d)i$ Addition of complex numbers

$(a + bi) - (c + di) = (a - c) + (b - d)i$ Subtraction of complex numbers

Study Tip

Note in part (b) of Example 6 that the sum of two complex numbers can be a real number.

EXAMPLE 6 Adding and Subtracting Complex Numbers

a. $(3 - i) + (-2 + 4i) = (3 - 2) + (-1 + 4)i = 1 + 3i$

b. $3i + (5 - 3i) = 5 + (3 - 3)i = 5$

c. $4 - (-1 + 5i) + (7 + 2i) = [4 - (-1) + 7] + (-5 + 2)i = 12 - 3i$

d. $(6 + 3i) + \left(2 - \sqrt{-8}\right) - \sqrt{-4} = (6 + 3i) + \left(2 - 2\sqrt{2}i\right) - 2i$

$= (6 + 2) + \left(3 - 2\sqrt{2} - 2\right)i$

$= 8 + \left(1 - 2\sqrt{2}\right)i$

✓ **CHECKPOINT** *Now try Exercise 55.*

The Commutative, Associative, and Distributive Properties of real numbers are also valid for complex numbers, as is the FOIL Method.

EXAMPLE 7 Multiplying Complex Numbers

Perform each operation and write the result in standard form.

a. $(7i)(-3i)$ **b.** $(1 - i)\left(\sqrt{-9}\right)$

c. $(2 - i)(4 + 3i)$ **d.** $(3 + 2i)(3 - 2i)$

Solution

a. $(7i)(-3i) = -21i^2$ Multiply.

$= -21(-1) = 21$ $i^2 = -1$

b. $(1 - i)\left(\sqrt{-9}\right) = (1 - i)(3i)$ Write in i-form.

$= 3i - 3(i^2)$ Distributive Property

$= 3i - 3(-1) = 3 + 3i$ $i^2 = -1$

c. $(2 - i)(4 + 3i) = 8 + 6i - 4i - 3i^2$ FOIL Method

$= 8 + 6i - 4i - 3(-1)$ $i^2 = -1$

$= 11 + 2i$ Combine like terms.

d. $(3 + 2i)(3 - 2i) = 3^2 - (2i)^2$ Special product formula

$= 9 - 4i^2$ Simplify.

$= 9 - 4(-1) = 13$ $i^2 = -1$

✓ **CHECKPOINT** *Now try Exercise 71.*

4 ▶ Use complex conjugates to write the quotient of two complex numbers in standard form.

Complex Conjugates

In Example 7(d), note that the product of two complex numbers can be a real number. This occurs with pairs of complex numbers of the form $a + bi$ and $a - bi$, called **complex conjugates.** In general, the product of complex conjugates has the following form.

$$(a + bi)(a - bi) = a^2 - (bi)^2 = a^2 - b^2i^2 = a^2 - b^2(-1) = a^2 + b^2$$

Here are some examples.

Complex Number	*Complex Conjugate*	*Product*
$4 - 5i$	$4 + 5i$	$4^2 + 5^2 = 41$
$3 + 2i$	$3 - 2i$	$3^2 + 2^2 = 13$
$-2 = -2 + 0i$	$-2 = -2 - 0i$	$(-2)^2 + 0^2 = 4$
$i = 0 + i$	$-i = 0 - i$	$0^2 + 1^2 = 1$

To write the quotient of $a + bi$ and $c + di$ (where $d \neq 0$) in standard form, multiply the numerator and denominator by the *complex conjugate of the denominator*, as shown in Example 8.

EXAMPLE 8 Writing Quotients of Complex Numbers in Standard Form

a. $\dfrac{2 - i}{4i} = \dfrac{2 - i}{4i} \cdot \dfrac{(-4i)}{(-4i)}$ Multiply numerator and denominator by complex conjugate of denominator.

$= \dfrac{-8i + 4i^2}{-16i^2}$ Multiply fractions.

$= \dfrac{-8i + 4(-1)}{-16(-1)}$ $i^2 = -1$

$= \dfrac{-8i - 4}{16}$ Simplify.

$= -\dfrac{1}{4} - \dfrac{1}{2}i$ Write in standard form.

b. $\dfrac{5}{3 - 2i} = \dfrac{5}{3 - 2i} \cdot \dfrac{3 + 2i}{3 + 2i}$ Multiply numerator and denominator by complex conjugate of denominator.

$= \dfrac{5(3 + 2i)}{(3 - 2i)(3 + 2i)}$ Multiply fractions.

$= \dfrac{5(3 + 2i)}{3^2 + 2^2}$ Product of complex conjugates

$= \dfrac{15 + 10i}{13}$ Simplify.

$= \dfrac{15}{13} + \dfrac{10}{13}i$ Write in standard form.

✓ **CHECKPOINT** *Now try Exercise 123.*

EXAMPLE 9 Writing a Quotient of Complex Numbers in Standard Form

$\dfrac{8-i}{8+i} = \dfrac{8-i}{8+i} \cdot \dfrac{8-i}{8-i}$	Multiply numerator and denominator by complex conjugate of denominator.
$= \dfrac{64 - 16i + i^2}{8^2 + 1^2}$	Multiply fractions.
$= \dfrac{64 - 16i + (-1)}{8^2 + 1^2}$	$i^2 = -1$
$= \dfrac{63 - 16i}{65}$	Simplify.
$= \dfrac{63}{65} - \dfrac{16}{65}i$	Write in standard form.

 **CHECKPOINT** *Now try Exercise 135.*

EXAMPLE 10 Writing a Quotient of Complex Numbers in Standard Form

$\dfrac{2+3i}{4-2i} = \dfrac{2+3i}{4-2i} \cdot \dfrac{4+2i}{4+2i}$	Multiply numerator and denominator by complex conjugate of denominator.
$= \dfrac{8 + 16i + 6i^2}{4^2 + 2^2}$	Multiply fractions.
$= \dfrac{8 + 16i + 6(-1)}{4^2 + 2^2}$	$i^2 = -1$
$= \dfrac{2 + 16i}{20} = \dfrac{1}{10} + \dfrac{4}{5}i$	Write in standard form.

CHECKPOINT *Now try Exercise 137.*

EXAMPLE 11 Verifying a Complex Solution of an Equation

Show that $x = 2 + i$ is a solution of the equation $x^2 - 4x + 5 = 0$.

Solution

$x^2 - 4x + 5 = 0$	Write original equation.
$(2 + i)^2 - 4(2 + i) + 5 \stackrel{?}{=} 0$	Substitute $2 + i$ for x.
$4 + 4i + i^2 - 8 - 4i + 5 \stackrel{?}{=} 0$	Expand.
$i^2 + 1 \stackrel{?}{=} 0$	Combine like terms.
$(-1) + 1 \stackrel{?}{=} 0$	$i^2 = -1$
$0 = 0$	Solution checks. ✓

So, $x = 2 + i$ is a solution of the original equation.

CHECKPOINT *Now try Exercise 145.*

Concept Check

1. Write (in words) the steps you can use to write the square root of a negative number in i-form.

2. Describe the values of a and b for which the complex number $a + bi$ is (a) a real number, (b) an imaginary number, and (c) a pure imaginary number. Then (d) explain what you must do to show that two complex numbers are equal.

3. Explain how adding two complex numbers is similar to adding two binomials. Then explain how multiplying two complex numbers is similar to multiplying two binomials.

4. Explain how you can use a complex conjugate to write the quotient of two complex numbers in standard form.

9.6 EXERCISES

Go to pages 608–609 to record your assignments.

Developing Skills

In Exercises 1–18, write the number in i-form. ***See Example 1.***

✓ 1. $\sqrt{-4}$
2. $\sqrt{-9}$
3. $-\sqrt{-144}$
4. $\sqrt{-49}$
5. $\sqrt{-\frac{4}{25}}$
6. $\sqrt{-\frac{9}{64}}$
7. $-\sqrt{-\frac{36}{121}}$
8. $-\sqrt{-\frac{9}{25}}$
9. $\sqrt{-8}$
10. $\sqrt{-75}$
11. $\sqrt{-7}$
12. $\sqrt{-15}$
13. $\dfrac{\sqrt{-12}}{\sqrt{-3}}$
14. $\dfrac{\sqrt{-45}}{\sqrt{-5}}$
15. $\sqrt{-\frac{18}{25}}$
16. $\sqrt{-\frac{20}{49}}$
17. $\sqrt{-0.09}$
18. $\sqrt{-0.0004}$

In Exercises 19–42, perform the operation(s) and write the result in standard form. ***See Examples 2 and 3.***

✓ 19. $\sqrt{-16} + \sqrt{-36}$
20. $\sqrt{-25} - \sqrt{-9}$
21. $\sqrt{-9} - \sqrt{-1}$
22. $\sqrt{-81} + \sqrt{-64}$
23. $\sqrt{-50} - \sqrt{-8}$
24. $\sqrt{-500} + \sqrt{-45}$

25. $\sqrt{-48} + \sqrt{-12} - \sqrt{-27}$
26. $\sqrt{-32} - \sqrt{-18} + \sqrt{-50}$
✓ 27. $\sqrt{-12}\sqrt{-2}$
28. $\sqrt{-25}\sqrt{-6}$
29. $\sqrt{-18}\sqrt{-3}$
30. $\sqrt{-7}\sqrt{-7}$
31. $\sqrt{-0.16}\sqrt{-1.21}$
32. $\sqrt{-0.49}\sqrt{-1.44}$

33. $\sqrt{-3}(\sqrt{-3} + \sqrt{-4})$
34. $\sqrt{-12}(\sqrt{-3} - \sqrt{-12})$
35. $\sqrt{-5}(\sqrt{-16} - \sqrt{-10})$
36. $\sqrt{-3}(\sqrt{-24} + \sqrt{-27})$
37. $\sqrt{-2}(3 - \sqrt{-8})$
38. $\sqrt{-9}(1 + \sqrt{-16})$

39. $(\sqrt{-16})^2$
40. $(\sqrt{-2})^2$
41. $(\sqrt{-4})^3$
42. $(\sqrt{-5})^3$

In Exercises 43–46, determine whether the complex numbers are equal. ***See Example 4.***

✓ 43. $\sqrt{1} + \sqrt{-25}$ and $1 + 5i$
44. $\sqrt{16} + \sqrt{-9}$ and $4 - 3i$
45. $\sqrt{27} - \sqrt{-8}$ and $3\sqrt{3} + 2\sqrt{2}\,i$
46. $\sqrt{18} - \sqrt{-12}$ and $3\sqrt{2} - 2\sqrt{3}\,i$

In Exercises 47–54, determine the values of a and b that satisfy the equation. ***See Examples 4 and 5.***

47. $3 - 4i = a + bi$
48. $-8 + 6i = a + bi$
49. $5 - 4i = (a + 3) + (b - 1)i$
50. $-10 + 12i = 2a + (5b - 3)i$
✓ 51. $-4 - \sqrt{-8} = a + bi$
52. $\sqrt{-36} - 3 = a + bi$
53. $\sqrt{a} + \sqrt{-49} = 8 + bi$
54. $\sqrt{100} + \sqrt{b} = a + 2\sqrt{3}\,i$

In Exercises 55–70, perform the operation(s) and write the result in standard form. ***See Example 6.***

✓ 55. $(4 - 3i) + (6 + 7i)$

56. $(-10 + 2i) + (4 - 7i)$

57. $(-4 - 7i) + (-10 - 33i)$

58. $(15 + 10i) - (2 + 10i)$

59. $13i - (14 - 7i)$ **60.** $17i + (9 - 14i)$

61. $(30 - i) - (18 + 6i) + 3i^2$

62. $(4 + 6i) + (15 + 24i) - 10i^2$

63. $6 - (3 - 4i) + 2i$

64. $22 + (-5 + 8i) + 10i$

65. $\left(\frac{4}{3} + \frac{1}{3}i\right) + \left(\frac{5}{6} + \frac{7}{6}i\right)$ **66.** $\left(\frac{4}{5} + \frac{2}{5}i\right) + \left(\frac{3}{10} - \frac{3}{10}i\right)$

67. $(0.05 + 2.50i) - (6.2 + 11.8i)$

68. $(1.8 + 4.3i) - (0.8 - 0.7i)$

69. $15i - (3 - 25i) + \sqrt{-81}$

70. $(-1 + i) - \sqrt{2} - \sqrt{-2}$

In Exercises 71–98, perform the operation and write the result in standard form. ***See Example 7.***

✓ **71.** $(3i)(12i)$ **72.** $(-5i)(4i)$

73. $(3i)(-8i)$ **74.** $(-2i)(-10i)$

75. $(-5i)(-i)(\sqrt{-49})$ **76.** $(10i)(\sqrt{-36})(-5i)$

77. $(-3i)^3$ **78.** $(8i)^2$

79. $(-3i)^2$ **80.** $(2i)^4$

81. $-5(13 + 2i)$ **82.** $10(8 - 6i)$

83. $4i(-3 - 5i)$ **84.** $-3i(10 - 15i)$

85. $(9 - 2i)(\sqrt{-4})$ **86.** $(11 + 3i)(\sqrt{-25})$

87. $(4 + 3i)(-7 + 4i)$ **88.** $(3 + 5i)(2 + 15i)$

89. $(-7 + 7i)(4 - 2i)$ **90.** $(3 + 5i)(2 - 15i)$

91. $(-2 + \sqrt{-5})(-2 - \sqrt{-5})$

92. $(-3 - \sqrt{-12})(4 - \sqrt{-12})$

93. $(3 - 4i)^2$ **94.** $(7 + i)^2$

95. $(2 + 5i)^2$ **96.** $(8 - 3i)^2$

97. $(3 + i)^3$ **98.** $(2 - 2i)^3$

In Exercises 99–108, simplify the expression.

99. i^7 **100.** i^{11}

101. i^{24} **102.** i^{35}

103. i^{42} **104.** i^{64}

105. i^9 **106.** i^{71}

107. $(-i)^6$ **108.** $(-i)^4$

In Exercises 109–122, multiply the number by its complex conjugate and simplify.

109. $2 + i$ **110.** $3 + 2i$

111. $-2 - 8i$ **112.** $10 - 3i$

113. $5 - \sqrt{6}i$ **114.** $-4 + \sqrt{2}i$

115. $10i$ **116.** 20

117. -12 **118.** $-12i$

119. $1 + \sqrt{-3}$ **120.** $-3 - \sqrt{-5}$

121. $1.5 + \sqrt{-0.25}$ **122.** $3.2 - \sqrt{-0.04}$

In Exercises 123–138, write the quotient in standard form. ***See Examples 8–10.***

✓ **123.** $\dfrac{20}{2i}$ **124.** $\dfrac{-5}{-3i}$

125. $\dfrac{2 + i}{-5i}$ **126.** $\dfrac{1 + i}{3i}$

127. $\dfrac{4}{1 - i}$ **128.** $\dfrac{20}{3 + i}$

129. $\dfrac{7i + 14}{7i}$ **130.** $\dfrac{6i + 3}{3i}$

131. $\dfrac{-12}{2 + 7i}$ **132.** $\dfrac{15}{2(1 - i)}$

133. $\dfrac{3i}{5 + 2i}$ **134.** $\dfrac{4i}{5 - 3i}$

✓ **135.** $\dfrac{5 - i}{5 + i}$ **136.** $\dfrac{9 + i}{9 - i}$

✓ **137.** $\dfrac{4 + 5i}{3 - 7i}$ **138.** $\dfrac{5 + 3i}{7 - 4i}$

In Exercises 139–144, perform the operation by first writing each quotient in standard form.

139. $\dfrac{5}{3 + i} + \dfrac{1}{3 - i}$ **140.** $\dfrac{1}{1 - 2i} + \dfrac{4}{1 + 2i}$

141. $\dfrac{3i}{1 + i} + \dfrac{2}{2 + 3i}$ **142.** $\dfrac{i}{4 - 3i} - \dfrac{5}{2 + i}$

143. $\dfrac{1+i}{i} - \dfrac{3}{5-2i}$ **144.** $\dfrac{3-2i}{i} - \dfrac{1}{7+i}$

In Exercises 145–148, determine whether each number is a solution of the equation. *See Example 11.*

✓ **145.** $x^2 + 2x + 5 = 0$

(a) $x = -1 + 2i$ (b) $x = -1 - 2i$

146. $x^2 - 4x + 13 = 0$

(a) $x = 2 - 3i$ (b) $x = 2 + 3i$

147. $x^3 + 4x^2 + 9x + 36 = 0$

(a) $x = -4$ (b) $x = -3i$

148. $x^3 - 8x^2 + 25x - 26 = 0$

(a) $x = 2$ (b) $x = 3 - 2i$

149. *Cube Roots* The principal cube root of 125, $\sqrt[3]{125}$, is 5. Evaluate the expression x^3 for each value of x.

(a) $x = \dfrac{-5 + 5\sqrt{3}i}{2}$

(b) $x = \dfrac{-5 - 5\sqrt{3}i}{2}$

150. *Cube Roots* The principal cube root of 27, $\sqrt[3]{27}$, is 3. Evaluate the expression x^3 for each value of x.

(a) $x = \dfrac{-3 + 3\sqrt{3}i}{2}$

(b) $x = \dfrac{-3 - 3\sqrt{3}i}{2}$

151. *Pattern Recognition* Compare the results of Exercises 149 and 150. Use the results to list possible cube roots of (a) 1, (b) 8, and (c) 64. Verify your results algebraically.

152. *Algebraic Properties* Consider the complex number $1 + 5i$.

(a) Find the additive inverse of the number.

(b) Find the multiplicative inverse of the number.

In Exercises 153–156, perform the operations.

153. $(a + bi) + (a - bi)$

154. $(a + bi)(a - bi)$

155. $(a + bi) - (a - bi)$

156. $(a + bi)^2 + (a - bi)^2$

Explaining Concepts

157. Look back at Exercises 153–156. Based on your results, write a general rule for each exercise about operations on complex conjugates of the form $a + bi$ and $a - bi$.

158. *True or False?* Some numbers are both real and imaginary. Justify your answer.

159. *Error Analysis* Describe and correct the error.

$\sqrt{-3}\sqrt{-3} = \sqrt{(-3)(-3)} = \sqrt{9} = 3$

160. Explain why the Product Rule for Radicals cannot be used to produce the second expression in Exercise 159.

161. The denominator of a quotient is a pure imaginary number of the form bi. How can you use the complex conjugate of bi to write the quotient in standard form? Can you use the number i instead of the conjugate of bi? Explain.

162. The polynomial $x^2 + 1$ is prime *with respect to the integers*. It is not, however, prime *with respect to the complex numbers*. Show how $x^2 + 1$ can be factored using complex numbers.

Cumulative Review

In Exercises 163–166, use the Zero-Factor Property to solve the equation.

163. $(x - 5)(x + 7) = 0$ **164.** $z(z - 2) = 0$

165. $3y(y - 3)(y + 4) = 0$

166. $(3x - 2)(4x + 1)(x + 9) = 0$

In Exercises 167–170, solve the equation and check your solution.

167. $\sqrt{x} = 9$ **168.** $\sqrt[3]{t} = 8$

169. $\sqrt{x} - 5 = 0$ **170.** $\sqrt{2x + 3} - 7 = 0$

What Did You Learn?

Use these two pages to help prepare for a test on this chapter. Check off the key terms and key concepts you know. You can also use this section to record your assignments.

Plan for Test Success

Date of test: / /

Study dates and times: / / at : A.M./P.M.

/ / at : A.M./P.M.

Things to review:

- ☐ Key Terms, *p. 608*
- ☐ Key Concepts, *pp. 608–609*
- ☐ Your class notes
- ☐ Your assignments
- ☐ Study Tips, *pp. 556, 557, 558, 559, 560, 562, 567, 569, 570, 574, 593, 594, 600, 602*
- ☐ Technology Tips, *pp. 561, 589, 591*
- ☐ Mid-Chapter Quiz, *p. 580*
- ☐ Review Exercises, *pp. 610–612*
- ☐ Chapter Test, *p. 613*
- ☐ Video Explanations Online
- ☐ Tutorial Online

Key Terms

- ☐ square root, *p. 556*
- ☐ cube root, *p. 556*
- ☐ *n*th root of *a*, *p. 556*
- ☐ principal *n*th root of *a*, *p. 556*
- ☐ radical symbol, *p. 556*
- ☐ index, *p. 556*
- ☐ radicand, *p. 556*
- ☐ perfect square, *p. 557*
- ☐ perfect cube, *p. 557*
- ☐ rational exponent, *p. 559*
- ☐ radical function, *p. 561*
- ☐ rationalizing the denominator, *p. 570*
- ☐ Pythagorean Theorem, *p. 571*
- ☐ like radicals, *p. 574*
- ☐ conjugates, *p. 582*
- ☐ imaginary unit *i*, *p. 599*
- ☐ *i*-form, *p. 599*
- ☐ complex number, *p. 601*
- ☐ real part, *p. 601*
- ☐ imaginary part, *p. 601*
- ☐ imaginary number, *p. 601*
- ☐ complex conjugates, *p. 603*

Key Concepts

9.1 Radicals and Rational Exponents

Assignment:______________________________ Due date:________

☐ **Use the properties of *n*th roots.**

1. If a is a positive real number and n is even, then a has exactly two (real) nth roots, which are denoted by $\sqrt[n]{a}$ and $-\sqrt[n]{a}$.
2. If a is any real number and n is odd, then a has only one (real) nth root, denoted by $\sqrt[n]{a}$.
3. If a is a negative real number and n is even, then a has no (real) nth root.

☐ **Use the inverse properties of *n*th powers and *n*th roots.**

Let a be a real number, and let n be an integer such that $n \geq 2$.

1. If a has a principal nth root, then $(\sqrt[n]{a})^n = a$.
2. If n is odd, then $\sqrt[n]{a^n} = a$.
   If n is even, then $\sqrt[n]{a^n} = |a|$.

☐ **Understand rational exponents.**

1. $a^{1/n} = \sqrt[n]{a}$
2. $a^{m/n} = (a^{1/n})^m = (\sqrt[n]{a})^m$
3. $a^{m/n} = (a^m)^{1/n} = \sqrt[n]{a^m}$

☐ **Use the rules of exponents.**

See page 559 for the rules of exponents as they apply to rational exponents.

☐ **Find the domain of a radical function.**

Let n be an integer that is greater than or equal to 2.

1. If n is odd, the domain of $f(x) = \sqrt[n]{x}$ is the set of all real numbers.
2. If n is even, the domain of $f(x) = \sqrt[n]{x}$ is the set of all nonnegative real numbers.

9.2 Simplifying Radical Expressions

Assignment:______________________________ Due date:________

☐ **Use the Product and Quotient Rules for Radicals.**

Let u and v be real numbers, variables, or algebraic expressions. If the nth roots of u and v are real, the following rules are true.

1. $\sqrt[n]{uv} = \sqrt[n]{u}\sqrt[n]{v}$
2. $\sqrt[n]{\dfrac{u}{v}} = \dfrac{\sqrt[n]{u}}{\sqrt[n]{v}}, \; v \neq 0$

☐ **Simplify radical expressions.**

A radical expression is said to be in *simplest form* if all three of the statements below are true.

1. All possible nth powered factors have been removed from each radical.
2. No radical contains a fraction.
3. No denominator of a fraction contains a radical.

9.3 Adding and Subtracting Radical Expressions

Assignment:______________________________ Due date:________

☐ **Add and subtract radical expressions.**

1. *Like radicals* have the same index and the same radicand. Combine like radicals by combining their coefficients.
2. Before concluding that two radicals cannot be combined, simplify each radical expression to see if they become like radicals.

Example:

$\sqrt{4x} + \sqrt{9x}$	Sum of *unlike* radicals
$= 2\sqrt{x} + 3\sqrt{x}$	Simplify radicals.
$= 5\sqrt{x}$	Combine like radicals.

9.4 Multiplying and Dividing Radical Expressions

Assignment:______________________________ Due date:________

☐ **Multiply radical expressions.**

Multiply radical expressions by using the Distributive Property or the FOIL Method.

☐ **Divide radical expressions.**

To simplify a quotient involving radicals, rationalize the denominator by multiplying both the numerator and denominator by the conjugate of the denominator.

9.5 Radical Equations and Applications

Assignment:______________________________ Due date:________

☐ **Raise each side of an equation to the nth power.**

Let u and v be real numbers, variables, or algebraic expressions, and let n be a positive integer. If $u = v$, then it follows that $u^n = v^n$.

Example:

$\sqrt{x} = 9$	Let $u = \sqrt{x}$, $v = 9$.
$(\sqrt{x})^2 = 9^2$	$u^2 = v^2$
$x = 81$	Simplify.

9.6 Complex Numbers

Assignment:______________________________ Due date:________

☐ **Find the square root of a negative number.**

Let c be a positive real number. Then the square root of $-c$ is given by $\sqrt{-c} = \sqrt{c(-1)} = \sqrt{c}\sqrt{-1} = \sqrt{c}\,i$.

When writing $\sqrt{-c}$ in the i-form, $\sqrt{c}\,i$, note that i is outside the radical.

☐ **Perform operations with complex numbers.**

If a and b are real numbers, $a + bi$ is a complex number. To add complex numbers, add the real and imaginary parts separately. Use the FOIL Method or the Distributive Property to multiply complex numbers. To write the quotient of two complex numbers in standard form, multiply the numerator and denominator by the complex conjugate of the denominator, and simplify.

Review Exercises

9.1 Radicals and Rational Exponents

1 ▶ Determine the *n*th roots of numbers and evaluate radical expressions.

In Exercises 1–10, evaluate the radical expression without using a calculator. If not possible, state the reason.

1. $-\sqrt{81}$
2. $\sqrt{-16}$
3. $-\sqrt[3]{64}$
4. $\sqrt[3]{-8}$
5. $-\sqrt{\left(\frac{3}{4}\right)^2}$
6. $\sqrt{\left(-\frac{9}{13}\right)^2}$
7. $\sqrt[3]{-\left(\frac{1}{5}\right)^3}$
8. $-\sqrt[3]{\left(-\frac{27}{64}\right)^3}$
9. $\sqrt{-2^2}$
10. $-\sqrt{-3^2}$

2 ▶ Use the rules of exponents to evaluate or simplify expressions with rational exponents.

In Exercises 11–14, fill in the missing description.

	Radical Form	*Rational Exponent Form*
11.	$\sqrt[3]{27} = 3$	
12.	$\sqrt[3]{0.125} = 0.5$	
13.		$216^{1/3} = 6$
14.		$16^{1/4} = 2$

In Exercises 15–20, evaluate without using a calculator.

15. $27^{4/3}$
16. $16^{3/4}$
17. $(-25)^{3/2}$
18. $-(4^3)^{2/3}$
19. $8^{-4/3}$
20. $243^{-2/5}$

In Exercises 21–32, rewrite the expression using rational exponents.

21. $x^{3/4} \cdot x^{-1/6}$
22. $a^{2/3} \cdot a^{3/5}$
23. $z\sqrt[3]{z^2}$
24. $x^2\sqrt[4]{x^3}$
25. $\dfrac{\sqrt[4]{x^3}}{\sqrt{x^4}}$
26. $\dfrac{\sqrt{x^3}}{\sqrt[3]{x^2}}$
27. $\sqrt[3]{a^3b^2}$
28. $\sqrt[4]{m^3n^8}$
29. $\sqrt[4]{\sqrt{x}}$
30. $\sqrt{\sqrt[3]{x^4}}$
31. $\dfrac{(3x+2)^{2/3}}{\sqrt[3]{3x+2}}$
32. $\dfrac{\sqrt[5]{3x+6}}{(3x+6)^{4/5}}$

3 ▶ Use a calculator to evaluate radical expressions.

In Exercises 33–36, use a calculator to evaluate the expression. Round the answer to four decimal places.

33. $75^{-3/4}$
34. $158^{7/3}$
35. $\sqrt{13^2 - 4(2)(7)}$
36. $\dfrac{-3.7 + \sqrt{15.8}}{2(2.3)}$

4 ▶ Evaluate radical functions and find the domains of radical functions.

In Exercises 37–40, evaluate the function as indicated, if possible, and simplify.

37. $f(x) = \sqrt{x-2}$
    (a) $f(-7)$ (b) $f(51)$
38. $f(x) = \sqrt{6x-5}$
    (a) $f(5)$ (b) $f(-1)$
39. $g(x) = \sqrt[3]{2x-1}$
    (a) $g(0)$ (b) $g(14)$
40. $g(x) = \sqrt[4]{x+5}$
    (a) $g(-4)$ (b) $g(76)$

In Exercises 41 and 42, describe the domain of the function.

41. $f(x) = \sqrt{9-2x}$
42. $g(x) = \sqrt[3]{x+2}$

9.2 Simplifying Radical Expressions

1 ▶ Use the Product and Quotient Rules for Radicals to simplify radical expressions.

In Exercises 43–48, simplify the radical expression.

43. $\sqrt{36u^5v^2}$
44. $\sqrt{24x^3y^4}$
45. $\sqrt{0.25x^4y}$
46. $\sqrt{0.16s^6t^3}$
47. $\sqrt[3]{48a^3b^4}$
48. $\sqrt[4]{48u^4v^6}$

2 ▶ Use rationalization techniques to simplify radical expressions.

In Exercises 49–52, rationalize the denominator and simplify further, if possible.

49. $\sqrt{\dfrac{5}{6}}$
50. $\dfrac{4y}{\sqrt{10z}}$
51. $\dfrac{2}{\sqrt[3]{2x}}$
52. $\sqrt[3]{\dfrac{16t}{s^2}}$

3 ▶ Use the Pythagorean Theorem in application problems.

Geometry **In Exercises 53 and 54, find the length of the hypotenuse of the right triangle.**

53. 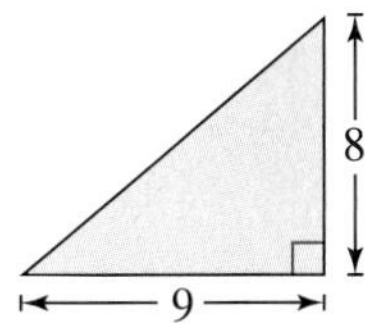

54. 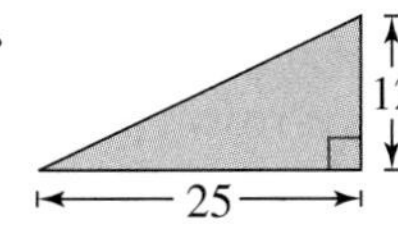

9.3 Adding and Subtracting Radical Expressions

1 ▶ Use the Distributive Property to add and subtract like radicals.

In Exercises 55–62, combine the radical expressions, if possible.

55. $2\sqrt{24} + 7\sqrt{6} - \sqrt{54}$

56. $9\sqrt{50} - 5\sqrt{8} + \sqrt{48}$

57. $5\sqrt{x} - \sqrt[3]{x} + 9\sqrt{x} - 8\sqrt[3]{x}$

58. $\sqrt{3x} - \sqrt[4]{6x^2} + 2\sqrt[4]{6x^2} - 4\sqrt{3x}$

59. $10\sqrt[4]{y+3} - 3\sqrt[4]{y+3}$

60. $5\sqrt[3]{x-3} + 4\sqrt[3]{x-3}$

61. $2x\sqrt[3]{24x^2y} - \sqrt[3]{3x^5y}$

62. $4xy^2\sqrt[4]{243x} + 2y^2\sqrt[4]{48x^5}$

2 ▶ Use radical expressions in application problems.

Dining Hall **In Exercises 63 and 64, a campus dining hall is undergoing renovations. The four corners of the hall are to be walled off and used as storage units (see figure).**

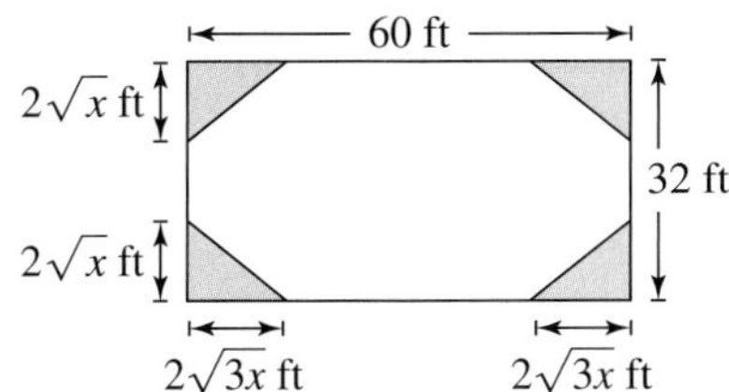

63. Find the perimeter of one of the storage units.

64. Find the perimeter of the newly designed dining hall.

9.4 Multiplying and Dividing Radical Expressions

1 ▶ Use the Distributive Property or the FOIL Method to multiply radical expressions.

In Exercises 65–70, multiply and simplify.

65. $\sqrt{15} \cdot \sqrt{20}$

66. $\sqrt{36} \cdot \sqrt{60}$

67. $\sqrt{10}(\sqrt{2} + \sqrt{5})$

68. $\sqrt{12}(\sqrt{6} - \sqrt{8})$

69. $(\sqrt{3} - \sqrt{x})(\sqrt{3} + \sqrt{x})$

70. $(4 - 3\sqrt{2})^2$

2 ▶ Determine the products of conjugates.

In Exercises 71–74, find the conjugate of the expression. Then multiply the expression by its conjugate and simplify.

71. $3 - \sqrt{7}$

72. $\sqrt{6} + 9$

73. $\sqrt{x} + 20$

74. $9 - \sqrt{2y}$

3 ▶ Simplify quotients involving radicals by rationalizing the denominators.

In Exercises 75–78, rationalize the denominator of the expression and simplify.

75. $\dfrac{\sqrt{2} - 1}{\sqrt{3} - 4}$

76. $\dfrac{2 + \sqrt{20}}{3 + \sqrt{5}}$

77. $(\sqrt{x} + 10) \div (\sqrt{x} - 10)$

78. $(3\sqrt{s} + 4) \div (\sqrt{s} + 2)$

9.5 Radical Equations and Applications

1 ▶ Solve a radical equation by raising each side to the *n*th power.

In Exercises 79–88, solve the equation and check your solution(s).

79. $\sqrt{2x} - 8 = 0$

80. $\sqrt{4x} + 6 = 9$

81. $\sqrt[4]{3x - 1} + 6 = 3$

82. $\sqrt[3]{5x - 7} - 3 = -1$

83. $\sqrt[3]{5x + 2} - \sqrt[3]{7x - 8} = 0$

84. $\sqrt[4]{9x - 2} - \sqrt[4]{8x} = 0$

85. $\sqrt{2(x + 5)} = x + 5$

86. $y - 2 = \sqrt{y + 4}$

87. $\sqrt{1 + 6x} = 2 - \sqrt{6x}$

88. $\sqrt{2 + 9b} + 1 = 3\sqrt{b}$

2 ▶ Solve application problems involving radical equations.

89. *Geometry* Determine the length and width of a rectangle with a perimeter of 46 inches and a diagonal of 17 inches.

90. *Geometry* Determine the length and width of a rectangle with a perimeter of 82 inches and a diagonal of 29 inches.

91. *Period of a Pendulum* The time t (in seconds) for a pendulum of length L (in feet) to go through one complete cycle (its period) is given by

$$t = 2\pi\sqrt{\frac{L}{32}}.$$

How long is the pendulum of a grandfather clock with a period of 1.9 seconds?

92. *Height* The time t (in seconds) for a free-falling object to fall d feet is given by $t = \sqrt{d/16}$. A child drops a pebble from a bridge and observes it strike the water after approximately 4 seconds. Estimate the height from which the pebble was dropped.

Free-Falling Object **In Exercises 93 and 94, the velocity of a free-falling object can be determined from the equation $v = \sqrt{2gh}$, where v is the velocity (in feet per second), $g = 32$ feet per second per second, and h is the distance (in feet) the object has fallen.**

93. Find the height from which a brick has been dropped when it strikes the ground with a velocity of 64 feet per second.

94. Find the height from which a wrench has been dropped when it strikes the ground with a velocity of 112 feet per second.

9.6 Complex Numbers

1 ▶ Write square roots of negative numbers in *i*-form and perform operations on numbers in *i*-form.

In Exercises 95–100, write the number in *i*-form.

95. $\sqrt{-48}$

96. $\sqrt{-0.16}$

97. $10 - 3\sqrt{-27}$

98. $3 + 2\sqrt{-500}$

99. $\frac{3}{4} - 5\sqrt{-\frac{3}{25}}$

100. $-0.5 + 3\sqrt{-1.21}$

In Exercises 101–104, perform the operation(s) and write the result in standard form.

101. $\sqrt{-81} + \sqrt{-36}$

102. $\sqrt{-121} - \sqrt{-84}$

103. $\sqrt{-10}(\sqrt{-4} - \sqrt{-7})$

104. $\sqrt{-5}(\sqrt{-10} + \sqrt{-15})$

2 ▶ Determine the equality of two complex numbers.

In Exercises 105–108, determine the values of *a* and *b* that satisfy the equation.

105. $12 - 5i = (a + 2) + (b - 1)i$

106. $-48 + 9i = (a - 5) + (b + 10)i$

107. $\sqrt{-49} + 4 = a + bi$

108. $-3 - \sqrt{-4} = a + bi$

3 ▶ Add, subtract, and multiply complex numbers.

In Exercises 109–114, perform the operation and write the result in standard form.

109. $(-4 + 5i) - (-12 + 8i)$

110. $(-6 + 3i) + (-1 + i)$

111. $(4 - 3i)(4 + 3i)$

112. $(12 - 5i)(2 + 7i)$

113. $(6 - 5i)^2$

114. $(2 - 9i)^2$

4 ▶ Use complex conjugates to write the quotient of two complex numbers in standard form.

In Exercises 115–120, write the quotient in standard form.

115. $\dfrac{7}{3i}$

116. $\dfrac{4}{5i}$

117. $\dfrac{-3i}{4 - 6i}$

118. $\dfrac{5i}{2 + 9i}$

119. $\dfrac{3 - 5i}{6 + i}$

120. $\dfrac{2 + i}{1 - 9i}$

Chapter Test

Take this test as you would take a test in class. After you are done, check your work against the answers in the back of the book.

In Exercises 1 and 2, evaluate each expression without using a calculator.

1. (a) $16^{3/2}$
(b) $\sqrt{5}\sqrt{20}$

2. (a) $125^{-2/3}$
(b) $\sqrt{3}\sqrt{12}$

3. For $f(x) = \sqrt{9 - 5x}$, find $f(-8)$ and $f(0)$.

4. Find the domain of $g(x) = \sqrt{7x - 3}$.

In Exercises 5–7, simplify each expression.

5. (a) $\left(\dfrac{x^{1/2}}{x^{1/3}}\right)^2$
(b) $5^{1/4} \cdot 5^{7/4}$

6. (a) $\sqrt{\dfrac{32}{9}}$
(b) $\sqrt[3]{24}$

7. (a) $\sqrt{24x^3}$
(b) $\sqrt[4]{16x^5y^8}$

In Exercises 8 and 9, rationalize the denominator of the expression and simplify.

8. $\dfrac{2}{\sqrt[3]{9y}}$

9. $\dfrac{10}{\sqrt{6} - \sqrt{2}}$

10. Subtract: $6\sqrt{18x} - 3\sqrt{32x}$

11. Multiply and simplify: $\sqrt{5}\left(\sqrt{15x} + 3\right)$

12. Expand: $\left(4 - \sqrt{2x}\right)^2$

13. Factor: $7\sqrt{27} + 14y\sqrt{12} = 7\sqrt{3}(\quad)$

In Exercises 14–16, solve the equation.

14. $\sqrt{6z} + 5 = 17$

15. $\sqrt{x^2 - 1} = x - 2$

16. $\sqrt{x} - x + 6 = 0$

In Exercises 17–20, perform the operation(s) and simplify.

17. $(2 + 3i) - \sqrt{-25}$

18. $(3 - 5i)^2$

19. $\sqrt{-16}\left(1 + \sqrt{-4}\right)$

20. $(3 - 2i)(1 + 5i)$

21. Write $\dfrac{5 - 2i}{3 + i}$ in standard form.

22. The velocity v (in feet per second) of an object is given by $v = \sqrt{2gh}$, where $g = 32$ feet per second per second and h is the distance (in feet) the object has fallen. Find the height from which a rock has been dropped when it strikes the ground with a velocity of 96 feet per second.

Cumulative Test: Chapters 7–9

Take this test as you would take a test in class. After you are done, check your work against the answers in the back of the book.

1. Find the domain of $f(x) = \dfrac{3(x-1)}{8x-3}$.

In Exercises 2–5, perform the indicated operation(s) and simplify.

2. $\dfrac{x^2+8x+16}{18x^2} \cdot \dfrac{2x^4+4x^3}{x^2-16}$

3. $\dfrac{x^3-4x}{x^2+10x+24} \div \dfrac{x^2+2x}{x^2+5x+4}$

4. $\dfrac{2}{x} - \dfrac{x}{x^3+3x^2} + \dfrac{1}{x+3}$

5. $\dfrac{\left(\dfrac{x}{y}-\dfrac{y}{x}\right)}{\left(\dfrac{x-y}{xy}\right)}$

6. Determine whether each ordered pair is a solution of the system of linear equations.

$\begin{cases} 2x - y = 2 \\ -x + 3y = 4 \end{cases}$ (a) $(2, 2)$ (b) $(0, 4)$

In Exercises 7–10, match the system of equations with its graph. [The graphs are labeled (a), (b), (c), and (d).]

(a)

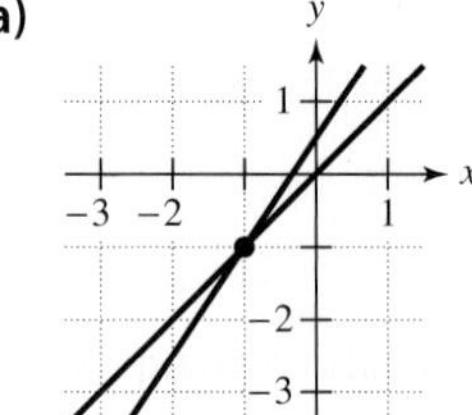

(b)

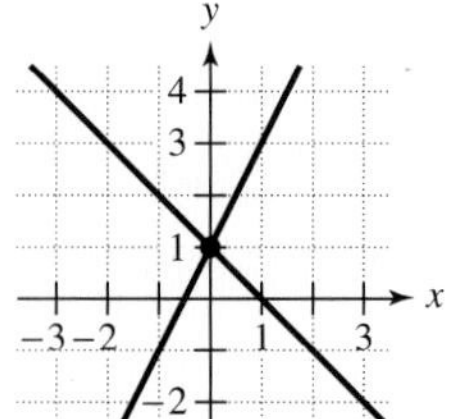

(c)

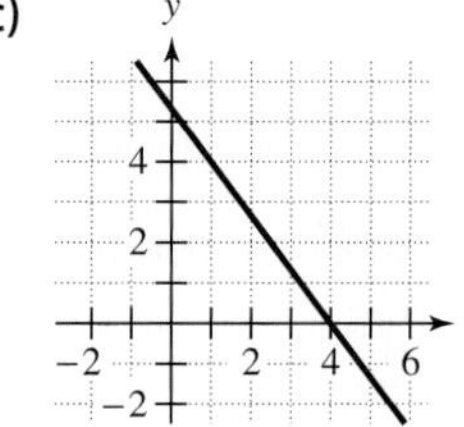

(d)

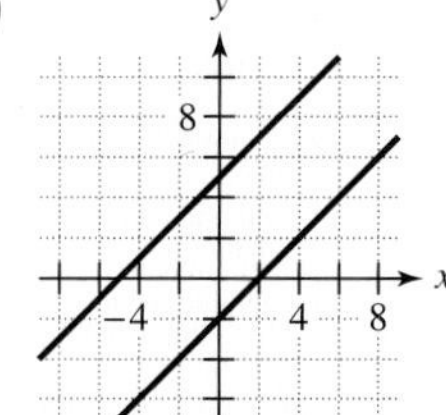

7. $\begin{cases} x + y = 1 \\ 2x - y = -1 \end{cases}$

8. $\begin{cases} 4x + 3y = 16 \\ -8x - 6y = -32 \end{cases}$

9. $\begin{cases} 5x - 5y = 10 \\ -x + y = 5 \end{cases}$

10. $\begin{cases} -x + y = 0 \\ 3x - 2y = -1 \end{cases}$

In Exercises 11–16, use the indicated method to solve the system.

11. *Graphical*: $\begin{cases} x - y = 1 \\ 2x + y = 5 \end{cases}$

12. *Substitution*: $\begin{cases} 4x + 2y = 8 \\ x - 5y = 13 \end{cases}$

13. *Elimination*: $\begin{cases} 4x - 3y = 8 \\ -2x + y = -6 \end{cases}$

14. *Elimination*: $\begin{cases} x + y + z = -1 \\ x = 0 \\ 2x + y = 1 \end{cases}$

15. *Matrices*: $\begin{cases} x + y + z = 1 \\ 5x + 4y + 3z = 0 \\ 6x + 3y + 2z = 1 \end{cases}$

16. *Cramer's Rule*: $\begin{cases} 2x - y = 4 \\ 3x + y = -5 \end{cases}$

17. Graph the solution of the system of inequalities.

$$\begin{cases} x - 2y < 0 \\ -2x + y > 2 \\ y > 0 \end{cases}$$

In Exercises 18–23, perform the indicated operation and simplify.

18. $\sqrt{-2}\left(\sqrt{-8} + 3\right)$

19. $(5 + 2i)^2$

20. $\left(\dfrac{t^{1/2}}{t^{1/4}}\right)^2$

21. $10\sqrt{20x} + 3\sqrt{125x}$

22. $\left(\sqrt{6}x + 7\right)^2$

23. $\dfrac{3}{\sqrt{5} - 2}$

24. Write the quotient in standard form: $\dfrac{2 + 3i}{6 - 2i}$.

In Exercises 25–28, solve the equation.

25. $\dfrac{1}{x} + \dfrac{4}{10 - x} = 1$

26. $\dfrac{x - 3}{x} + 1 = \dfrac{x - 4}{x - 6}$

27. $\sqrt{x} - x + 12 = 0$

28. $\sqrt{5 - x} + 10 = 11$

29. The stopping distance d of a car is directly proportional to the square of its speed s. On a certain type of pavement, a car requires 50 feet to stop when its speed is 25 miles per hour. Estimate the stopping distance when the speed of the car is 40 miles per hour. Explain your reasoning.

30. The number N of prey t months after a predator is introduced into an area is inversely proportional to $t + 1$. If $N = 300$ when $t = 0$, find N when $t = 5$.

31. At a local high school city championship basketball game, 1435 tickets were sold. A student admission ticket cost \$2.50 and an adult admission ticket cost \$5.00. The total ticket sales for the basketball game were \$4587.50. How many of each type of ticket were sold?

32. A total of \$50,000 is invested in two funds paying 8% and 8.5% simple interest. The combined yearly interest is \$4150. How much is invested at each rate?

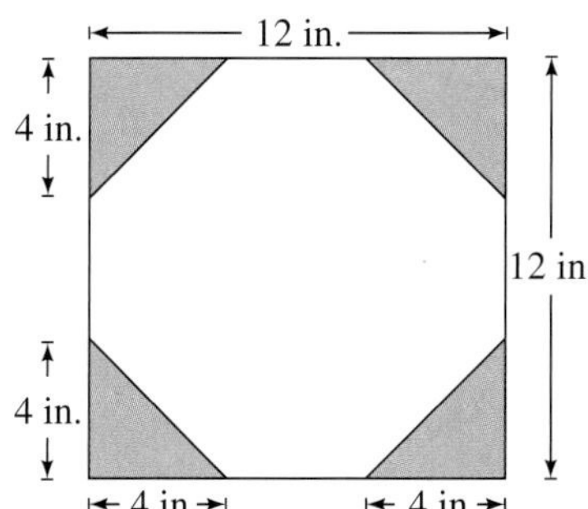

Figure for 33

33. The four corners are cut from a 12-inch-by-12-inch piece of glass, as shown in the figure. Find the perimeter of the remaining piece of glass.

34. The velocity v (in feet per second) of an object is given by $v = \sqrt{2gh}$, where $g = 32$ feet per second per second and h is the distance (in feet) the object has fallen. Find the height from which a rock has dropped if it strikes the ground with a velocity of 65 feet per second.

Study Skills in Action

Improving Your Memory

Have you ever driven on a highway for ten minutes when all of a sudden you kind of woke up and wondered where the last ten miles had gone? The car was on autopilot. The same thing happens to many college students as they sit through back-to-back classes. The longer students sit through classes on autopilot, the more likely they will "crash" when it comes to studying outside of class on their own.

While on autopilot, you do not process and retain new information effectively. Your memory can be improved by learning how to focus during class and while studying on your own.

Kimberly Nolting
VP, Academic Success Press
expert in developmental education

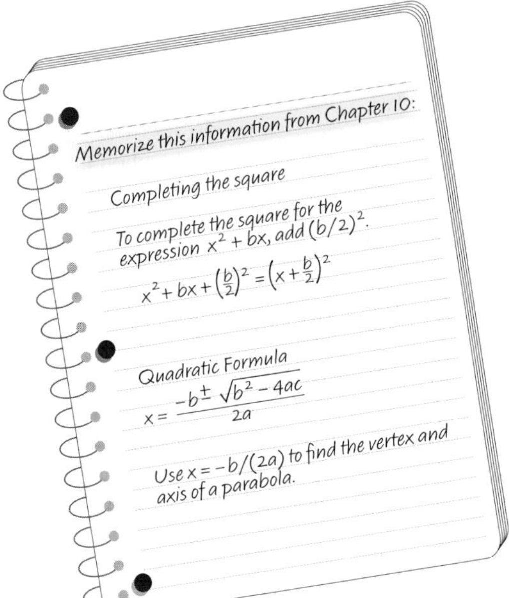

Memorize this information from Chapter 10:
Completing the square
To complete the square for the expression $x^2 + bx$, add $(b/2)^2$.
$x^2 + bx + \left(\frac{b}{2}\right)^2 = \left(x + \frac{b}{2}\right)^2$
Quadratic Formula
$x = \frac{-b \pm \sqrt{b^2 - 4ac}}{2a}$
Use $x = -b/(2a)$ to find the vertex and axis of a parabola.

Smart Study Strategy

Keep Your Mind Focused

During class

- When you sit down at your desk, get all other issues out of your mind by reviewing your notes from the last class and focusing just on math.
- Repeat in your mind what you are writing in your notes.
- When the math is particularly difficult, ask your instructor for another example.

While completing homework

- Before doing homework, review the concept boxes and examples. Talk through the examples out loud.
- Complete homework as though you were also preparing for a quiz. Memorize the different types of problems, formulas, rules, and so on.

Between classes

- Review the concept boxes and check your memory using the checkpoint exercises, Concept Check exercises, and the What Did You Learn? section.

Preparing for a test

- Review all your notes that pertain to the upcoming test. Review examples of each type of problem that could appear on the test.

Chapter 10

Quadratic Equations, Functions, and Inequalities

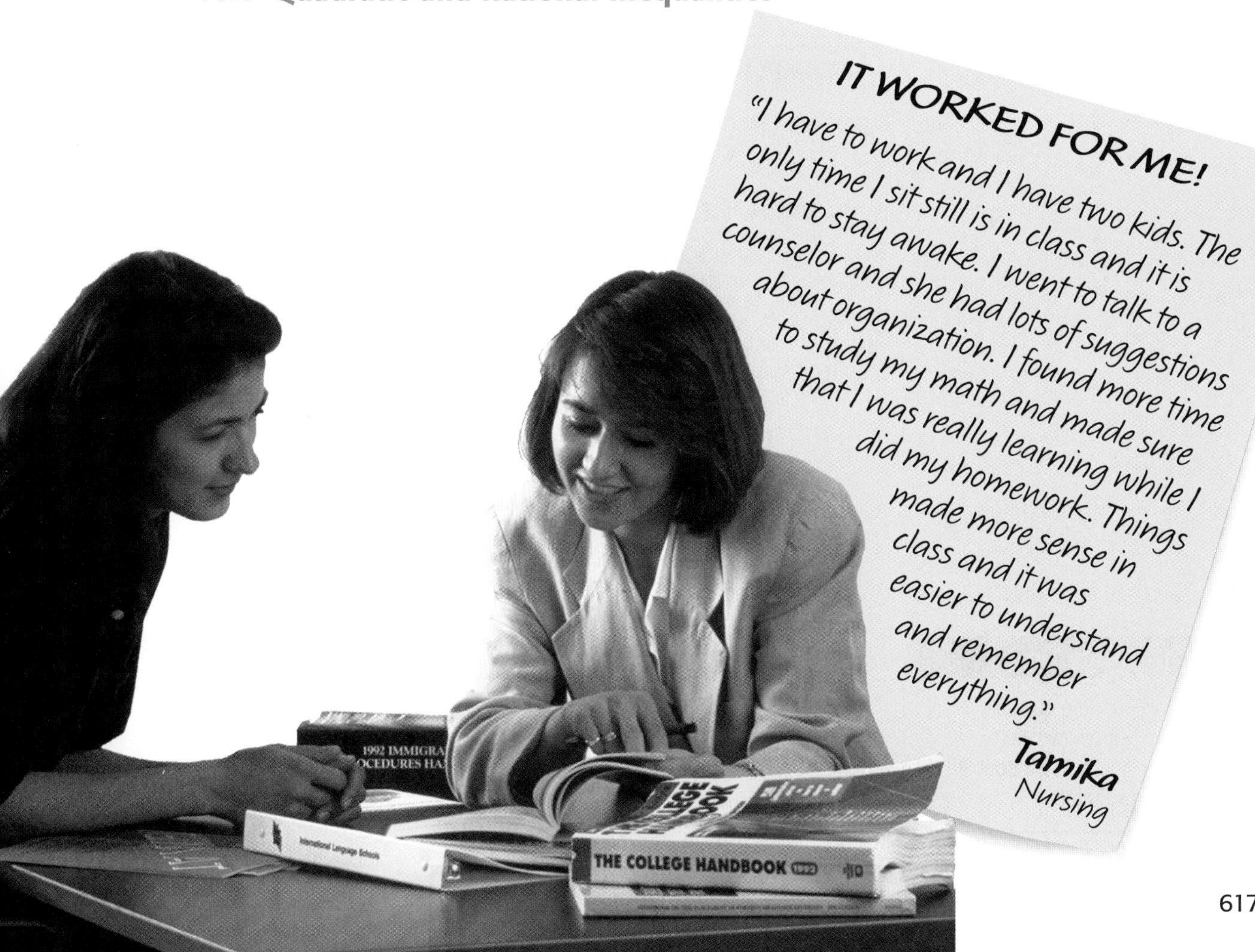

IT WORKED FOR ME!

"I have to work and I have two kids. The only time I sit still is in class and it is hard to stay awake. I went to talk to a counselor and she had lots of suggestions about organization. I found more time to study my math and made sure that I was really learning while I did my homework. Things made more sense in class and it was easier to understand and remember everything."

Tamika
Nursing

10.1 Solving Quadratic Equations: Factoring and Special Forms

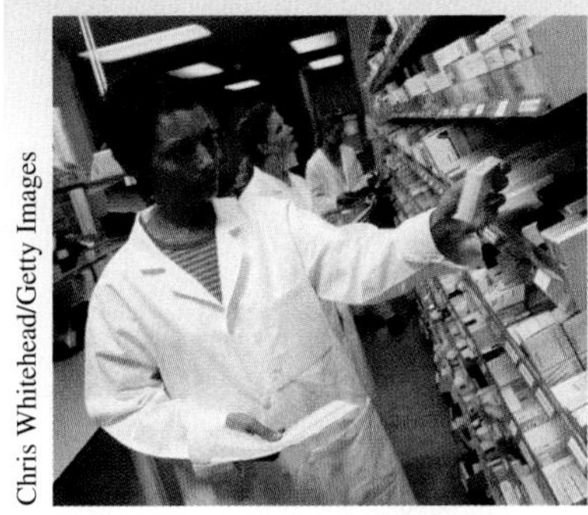
Chris Whitehead/Getty Images

Why You Should Learn It

Quadratic equations can be used to model and solve real-life problems. For instance, in Exercises 141 and 142 on page 626, you will use a quadratic equation to determine national health care expenditures in the United States.

What You Should Learn

1 ▶ Solve quadratic equations by factoring.

2 ▶ Solve quadratic equations by the Square Root Property.

3 ▶ Solve quadratic equations with complex solutions by the Square Root Property.

4 ▶ Use substitution to solve equations of quadratic form.

1 ▶ Solve quadratic equations by factoring.

Solving Quadratic Equations by Factoring

In this chapter, you will study methods for solving quadratic equations and equations of quadratic form. To begin, let's review the method of factoring that you studied in Section 6.5.

Remember that the first step in solving a quadratic equation by factoring is to write the equation in general form. Next, factor the left side. Finally, set each factor equal to zero and solve for x. Be sure to check each solution in the original equation.

EXAMPLE 1 Solving Quadratic Equations by Factoring

a.

$x^2 + 5x = 24$ Original equation

$x^2 + 5x - 24 = 0$ Write in general form.

$(x + 8)(x - 3) = 0$ Factor.

$x + 8 = 0 \Rightarrow x = -8$ Set 1st factor equal to 0.

$x - 3 = 0 \Rightarrow x = 3$ Set 2nd factor equal to 0.

b.

$3x^2 = 4 - 11x$ Original equation

$3x^2 + 11x - 4 = 0$ Write in general form.

$(3x - 1)(x + 4) = 0$ Factor.

$3x - 1 = 0 \Rightarrow x = \frac{1}{3}$ Set 1st factor equal to 0.

$x + 4 = 0 \Rightarrow x = -4$ Set 2nd factor equal to 0.

c.

$9x^2 + 12 = 3 + 12x + 5x^2$ Original equation

$4x^2 - 12x + 9 = 0$ Write in general form.

$(2x - 3)(2x - 3) = 0$ Factor.

$2x - 3 = 0 \Rightarrow x = \frac{3}{2}$ Set factor equal to 0.

Check each solution in its original equation.

✓ **CHECKPOINT** *Now try Exercise 1.*

Study Tip

In Example 1(c), the quadratic equation produces two identical solutions. This is called a **double** or **repeated solution.**

2 ▶ Solve quadratic equations by the Square Root Property.

The Square Root Property

Consider the following equation, where $d > 0$ and u is an algebraic expression.

$$u^2 = d \qquad \text{Original equation}$$

$$u^2 - d = 0 \qquad \text{Write in general form.}$$

$$(u + \sqrt{d})(u - \sqrt{d}) = 0 \qquad \text{Factor.}$$

$$u + \sqrt{d} = 0 \Rightarrow u = -\sqrt{d} \qquad \text{Set 1st factor equal to 0.}$$

$$u - \sqrt{d} = 0 \Rightarrow u = \sqrt{d} \qquad \text{Set 2nd factor equal to 0.}$$

Because the solutions differ only in sign, they can be written together using a "plus or minus sign": $u = \pm\sqrt{d}$. This form of the solution is read as "u is equal to plus or minus the square root of d." Now you can use the **Square Root Property** to solve an equation of the form $u^2 = d$ *without* going through the steps of factoring.

Technology: Tip

To check graphically the solutions of an equation written in general form, graph the left side of the equation and locate its x-intercepts. For instance, in Example 2(b), write the equation as

$$(x - 2)^2 - 10 = 0$$

and then use a graphing calculator to graph

$$y = (x - 2)^2 - 10$$

as shown below. You can use the *zoom* and *trace* features or the *zero* or *root* feature to approximate the x-intercepts of the graph to be $x \approx 5.16$ and $x \approx -1.16$.

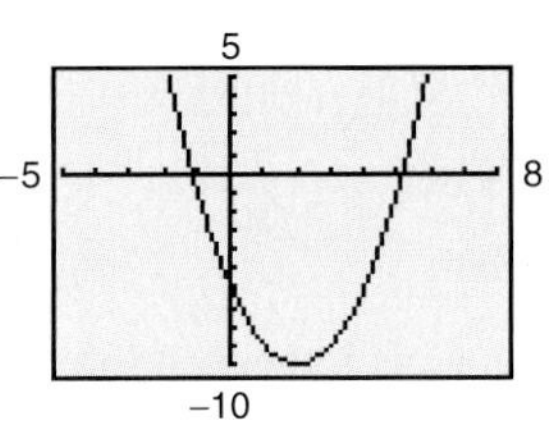

Square Root Property

The equation $u^2 = d$, where $d > 0$, has exactly two solutions:

$$u = \sqrt{d} \quad \text{and} \quad u = -\sqrt{d}.$$

These solutions can also be written as $u = \pm\sqrt{d}$. This solution process is also called **extracting square roots.**

EXAMPLE 2 Square Root Property

a. $3x^2 = 15$ — Original equation

$x^2 = 5$ — Divide each side by 3.

$x = \pm\sqrt{5}$ — Square Root Property

The solutions are $x = \sqrt{5}$ and $x = -\sqrt{5}$. Check these in the original equation.

b. $(x - 2)^2 = 10$ — Original equation

$x - 2 = \pm\sqrt{10}$ — Square Root Property

$x = 2 \pm \sqrt{10}$ — Add 2 to each side.

The solutions are $x = 2 + \sqrt{10} \approx 5.16$ and $x = 2 - \sqrt{10} \approx -1.16$.

c. $(3x - 6)^2 - 8 = 0$ — Original equation

$(3x - 6)^2 = 8$ — Add 8 to each side.

$3x - 6 = \pm 2\sqrt{2}$ — Square Root Property and rewrite $\sqrt{8}$ as $2\sqrt{2}$.

$3x = 6 \pm 2\sqrt{2}$ — Add 6 to each side.

$x = 2 \pm \dfrac{2\sqrt{2}}{3}$ — Divide each side by 3.

The solutions are $x = 2 + 2\sqrt{2}/3 \approx 2.94$ and $x = 2 - 2\sqrt{2}/3 \approx 1.06$.

✓ ***CHECKPOINT*** *Now try Exercise 23.*

3 ▶ Solve quadratic equations with complex solutions by the Square Root Property.

Quadratic Equations with Complex Solutions

Prior to Section 9.6, the only solutions you could find were real numbers. But now that you have studied complex numbers, it makes sense to look for other types of solutions. For instance, although the quadratic equation $x^2 + 1 = 0$ has no solutions that are real numbers, it does have two solutions that are complex numbers: i and $-i$. To check this, substitute i and $-i$ for x.

$(i)^2 + 1 = -1 + 1 = 0$ Solution checks. ✓

$(-i)^2 + 1 = -1 + 1 = 0$ Solution checks. ✓

One way to find complex solutions of a quadratic equation is to extend the Square Root Property to cover the case in which d is a negative number.

Technology: Discovery

Solve each quadratic equation below algebraically. Then use a graphing calculator to check the solutions. Which equations have real solutions and which have complex solutions? Which graphs have x-intercepts and which have no x-intercepts? Compare the type(s) of solution(s) of each quadratic equation with the x-intercept(s) of the graph of the equation.

a. $y = 2x^2 + 3x - 5$

b. $y = 2x^2 + 4x + 2$

c. $y = x^2 + 4$

d. $y = (x + 7)^2 + 2$

Square Root Property (Complex Square Root)

The equation $u^2 = d$, where $d < 0$, has exactly two solutions:

$u = \sqrt{|d|}i$ and $u = -\sqrt{|d|}i$.

These solutions can also be written as $u = \pm\sqrt{|d|}i$.

EXAMPLE 3 Square Root Property

a. $x^2 + 8 = 0$ Original equation

$x^2 = -8$ Subtract 8 from each side.

$x = \pm\sqrt{8}i = \pm 2\sqrt{2}i$ Square Root Property

The solutions are $x = 2\sqrt{2}i$ and $x = -2\sqrt{2}i$. Check these in the original equation.

b. $(x - 4)^2 = -3$ Original equation

$x - 4 = \pm\sqrt{3}i$ Square Root Property

$x = 4 \pm \sqrt{3}i$ Add 4 to each side.

The solutions are $x = 4 + \sqrt{3}i$ and $x = 4 - \sqrt{3}i$. Check these in the original equation.

c. $2(3x - 5)^2 + 32 = 0$ Original equation

$2(3x - 5)^2 = -32$ Subtract 32 from each side.

$(3x - 5)^2 = -16$ Divide each side by 2.

$3x - 5 = \pm 4i$ Square Root Property

$3x = 5 \pm 4i$ Add 5 to each side.

$x = \frac{5}{3} \pm \frac{4}{3}i$ Divide each side by 3.

The solutions are $x = 5/3 + 4/3i$ and $x = 5/3 - 4/3i$. Check these in the original equation.

✓ **CHECKPOINT** *Now try Exercise 45.*

4 ▶ Use substitution to solve equations of quadratic form.

Equations of Quadratic Form

Both the factoring method and the Square Root Property can be applied to nonquadratic equations that are of **quadratic form.** An equation is said to be of quadratic form if it has the form

$$au^2 + bu + c = 0$$

where u is an algebraic expression. Here are some examples.

Equation	*Written in Quadratic Form*
$x^4 + 5x^2 + 4 = 0$	$(x^2)^2 + 5(x^2) + 4 = 0$
$x - 5\sqrt{x} + 6 = 0$	$(\sqrt{x})^2 - 5(\sqrt{x}) + 6 = 0$
$2x^{2/3} + 5x^{1/3} - 3 = 0$	$2(x^{1/3})^2 + 5(x^{1/3}) - 3 = 0$
$18 + 2x^2 + (x^2 + 9)^2 = 8$	$(x^2 + 9)^2 + 2(x^2 + 9) - 8 = 0$

To solve an equation of quadratic form, it helps to make a substitution and rewrite the equation in terms of u, as demonstrated in Examples 4 and 5.

Technology: Tip

You may find it helpful to graph the equation with a graphing calculator before you begin. The graph will indicate the number of real solutions an equation has. For instance, shown below is the graph of the equation in Example 4. You can see from the graph that there are four x-intercepts and so there are four real solutions.

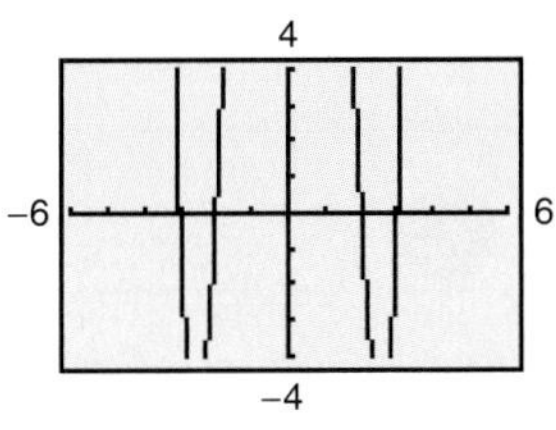

EXAMPLE 4 Solving an Equation of Quadratic Form

Solve $x^4 - 13x^2 + 36 = 0$.

Solution

Begin by writing the original equation in quadratic form, as follows.

$x^4 - 13x^2 + 36 = 0$ Write original equation.

$(x^2)^2 - 13(x^2) + 36 = 0$ Write in quadratic form.

Next, let $u = x^2$ and substitute u into the equation written in quadratic form. Then, factor and solve the equation.

$u^2 - 13u + 36 = 0$ Substitute u for x^2.

$(u - 4)(u - 9) = 0$ Factor.

$u - 4 = 0 \Rightarrow u = 4$ Set 1st factor equal to 0.

$u - 9 = 0 \Rightarrow u = 9$ Set 2nd factor equal to 0.

At this point you have found the "u-solutions." To find the "x-solutions," replace u with x^2 and solve for x.

$u = 4 \Rightarrow x^2 = 4 \Rightarrow x = \pm 2$

$u = 9 \Rightarrow x^2 = 9 \Rightarrow x = \pm 3$

The solutions are $x = 2$, $x = -2$, $x = 3$, and $x = -3$. Check these in the original equation.

✓ **CHECKPOINT** *Now try Exercise 101.*

Be sure you see in Example 4 that the u-solutions of 4 and 9 represent only a temporary step. They are not solutions of the original equation and cannot be substituted into the original equation.

Study Tip

When solving equations involving square roots, be sure to check for extraneous solutions.

EXAMPLE 5 Solving Equations of Quadratic Form

a. $x - 5\sqrt{x} + 6 = 0$ Original equation

This equation is of quadratic form with $u = \sqrt{x}$.

$(\sqrt{x})^2 - 5(\sqrt{x}) + 6 = 0$ Write in quadratic form.

$u^2 - 5u + 6 = 0$ Substitute u for $\sqrt{x}$.

$(u - 2)(u - 3) = 0$ Factor.

$u - 2 = 0 \Rightarrow u = 2$ Set 1st factor equal to 0.

$u - 3 = 0 \Rightarrow u = 3$ Set 2nd factor equal to 0.

Now, using the u-solutions of 2 and 3, you obtain the x-solutions as follows.

$u = 2 \Rightarrow \sqrt{x} = 2 \Rightarrow x = 4$

$u = 3 \Rightarrow \sqrt{x} = 3 \Rightarrow x = 9$

b. $x^{2/3} - x^{1/3} - 6 = 0$ Original equation

This equation is of quadratic form with $u = x^{1/3}$.

$(x^{1/3})^2 - (x^{1/3}) - 6 = 0$ Write in quadratic form.

$u^2 - u - 6 = 0$ Substitute u for $x^{1/3}$.

$(u + 2)(u - 3) = 0$ Factor.

$u + 2 = 0 \Rightarrow u = -2$ Set 1st factor equal to 0.

$u - 3 = 0 \Rightarrow u = 3$ Set 2nd factor equal to 0.

Now, using the u-solutions of -2 and 3, you obtain the x-solutions as follows.

$u = -2 \Rightarrow x^{1/3} = -2 \Rightarrow x = -8$

$u = 3 \Rightarrow x^{1/3} = 3 \Rightarrow x = 27$

CHECKPOINT *Now try Exercise 107.*

EXAMPLE 6 Surface Area of a Softball

The surface area of a sphere of radius r is given by $S = 4\pi r^2$. The surface area of a softball is $144/\pi$ square inches. Find the diameter d of the softball.

Solution

$\frac{144}{\pi} = 4\pi r^2$ Substitute $144/\pi$ for S.

$\frac{36}{\pi^2} = r^2 \Rightarrow \pm\sqrt{\frac{36}{\pi^2}} = r$ Divide each side by 4π and use Square Root Property.

Choose the positive root to obtain $r = 6/\pi$. The diameter of the softball is

$$d = 2r = 2\left(\frac{6}{\pi}\right) = \frac{12}{\pi} \approx 3.82 \text{ inches.}$$

CHECKPOINT *Now try Exercise 131.*

Concept Check

1. Explain the Zero-Factor Property and how it can be used to solve a quadratic equation.

2. Determine whether the following statement is true or false. Justify your answer.

   The only solution of the equation $x^2 = 25$ is $x = 5$.

3. Does the equation $4x^2 + 9 = 0$ have two real solutions or two complex solutions? Explain your reasoning.

4. Is the equation $x^6 - 6x^3 + 9 = 0$ of quadratic form? Explain your reasoning.

10.1 EXERCISES

Go to pages 676–677 to record your assignments.

Developing Skills

In Exercises 1–20, solve the equation by factoring. ***See Example 1.***

✓ **1.** $x^2 - 15x + 54 = 0$ **2.** $x^2 + 15x + 44 = 0$

3. $x^2 - x - 30 = 0$ **4.** $x^2 - 2x - 48 = 0$

5. $x^2 + 4x = 45$ **6.** $x^2 - 7x = 18$

7. $x^2 - 16x + 64 = 0$

8. $x^2 + 60x + 900 = 0$

9. $9x^2 - 10x - 16 = 0$

10. $8x^2 - 10x + 3 = 0$

11. $4x^2 - 12x = 0$ **12.** $25y^2 - 75y = 0$

13. $u(u - 9) - 12(u - 9) = 0$

14. $16x(x - 8) - 12(x - 8) = 0$

15. $2x(x - 5) + 9(x - 5) = 0$

16. $3(4 - x) - 2x(4 - x) = 0$

17. $(y - 4)(y - 3) = 6$

18. $(5 + u)(2 + u) = 4$

19. $2x(3x + 2) = 5 - 6x^2$

20. $(2z + 1)(2z - 1) = -4z^2 - 5z + 2$

In Exercises 21–42, solve the equation by using the Square Root Property. ***See Example 2.***

21. $x^2 = 49$ **22.** $p^2 = 169$

✓ **23.** $6x^2 = 54$ **24.** $5t^2 = 5$

25. $25x^2 = 16$ **26.** $9z^2 = 121$

27. $\frac{w^2}{4} = 49$ **28.** $\frac{x^2}{6} = 24$

29. $4x^2 - 25 = 0$ **30.** $16y^2 - 121 = 0$

31. $4u^2 - 225 = 0$ **32.** $16x^2 - 1 = 0$

33. $(x + 4)^2 = 64$

34. $(m - 12)^2 = 400$

35. $(x - 3)^2 = 0.25$

36. $(x + 2)^2 = 0.81$

37. $(x - 2)^2 = 7$ **38.** $(y + 4)^2 = 27$

39. $(2x + 1)^2 = 50$ **40.** $(3x - 5)^2 = 48$

41. $(9m - 2)^2 - 108 = 0$ **42.** $(5x + 11)^2 - 300 = 0$

In Exercises 43–64, solve the equation by using the Square Root Property. ***See Example 3.***

43. $z^2 = -36$ **44.** $x^2 = -16$

✓ **45.** $x^2 + 4 = 0$ **46.** $p^2 + 9 = 0$

47. $9u^2 + 17 = 0$ **48.** $25x^2 + 4 = 0$

49. $(t - 3)^2 = -25$

50. $(x + 5)^2 = -81$

51. $(3z + 4)^2 + 144 = 0$

52. $(2y - 3)^2 + 25 = 0$

53. $(4m + 1)^2 = -80$ **54.** $(6y - 5)^2 = -8$

55. $36(t + 3)^2 = -100$ **56.** $4(x - 4)^2 = -169$

57. $(x - 1)^2 = -27$ **58.** $(2x + 3)^2 = -54$

59. $(x + 1)^2 + 0.04 = 0$ **60.** $(y - 5)^2 + 6.25 = 0$

61. $\left(c - \frac{2}{3}\right)^2 + \frac{1}{9} = 0$ **62.** $\left(u + \frac{5}{8}\right)^2 + \frac{49}{16} = 0$

63. $\left(x + \frac{7}{3}\right)^2 = -\frac{38}{9}$ **64.** $\left(y - \frac{5}{8}\right)^2 = -\frac{5}{4}$

In Exercises 65–80, find all real and complex solutions of the quadratic equation.

65. $2x^2 - 5x = 0$ **66.** $4t^2 + 20t = 0$

67. $2x^2 + 5x - 12 = 0$ **68.** $3x^2 + 8x - 16 = 0$

69. $x^2 - 900 = 0$ **70.** $z^2 - 256 = 0$

71. $x^2 + 900 = 0$ **72.** $z^2 + 256 = 0$

73. $\frac{2}{3}x^2 = 6$ **74.** $\frac{1}{3}x^2 = 4$

75. $(p - 2)^2 - 108 = 0$ **76.** $(y + 12)^2 - 400 = 0$

77. $(p - 2)^2 + 108 = 0$ **78.** $(y + 12)^2 + 400 = 0$

79. $(x + 2)^2 + 18 = 0$ **80.** $(x + 2)^2 - 18 = 0$

In Exercises 81–90, use a graphing calculator to graph the function. Use the graph to approximate any x-intercepts. Set $y = 0$ and solve the resulting equation. Compare the result with the x-intercepts of the graph.

81. $y = x^2 - 9$

82. $y = 5x - x^2$

83. $y = x^2 - 2x - 15$

84. $y = x^2 + 3x - 40$

85. $y = 4 - (x - 3)^2$

86. $y = 4(x + 1)^2 - 9$

87. $y = 2x^2 - x - 6$

88. $y = 4x^2 - x - 14$

89. $y = 3x^2 - 13x - 10$

90. $y = 5x^2 + 9x - 18$

In Exercises 91–96, use a graphing calculator to graph the function and observe that the graph has no x-intercepts. Set $y = 0$ and solve the resulting equation. Of what type are the solutions of the equation?

91. $y = x^2 + 7$

92. $y = x^2 + 5$

93. $y = (x - 4)^2 + 2$

94. $y = (x + 2)^2 + 3$

95. $y = (x + 3)^2 + 5$

96. $y = (x - 2)^2 + 3$

In Exercises 97–100, solve for y in terms of x. Let f and g be functions representing, respectively, the positive square root and the negative square root. Use a graphing calculator to graph f and g in the same viewing window.

97. $x^2 + y^2 = 4$ **98.** $x^2 - y^2 = 4$

99. $x^2 + 4y^2 = 4$ **100.** $x - y^2 = 0$

In Exercises 101–130, solve the equation of quadratic form. (Find all real and complex solutions.) *See Examples 4 and 5.*

✓ **101.** $x^4 - 5x^2 + 4 = 0$

102. $x^4 - 10x^2 + 25 = 0$

103. $x^4 - 5x^2 + 6 = 0$

104. $x^4 - 10x^2 + 21 = 0$

105. $(x^2 - 4)^2 + 2(x^2 - 4) - 3 = 0$

106. $(x^2 - 1)^2 + (x^2 - 1) - 6 = 0$

✓ **107.** $x - 3\sqrt{x} - 4 = 0$

108. $x - \sqrt{x} - 6 = 0$

109. $x - 7\sqrt{x} + 10 = 0$

110. $x - 11\sqrt{x} + 24 = 0$

111. $x^{2/3} - x^{1/3} - 6 = 0$

112. $x^{2/3} + 3x^{1/3} - 10 = 0$

113. $2x^{2/3} - 7x^{1/3} + 5 = 0$

114. $5x^{2/3} - 13x^{1/3} + 6 = 0$

115. $x^{2/5} - 3x^{1/5} + 2 = 0$

116. $x^{2/5} + 5x^{1/5} + 6 = 0$

117. $2x^{2/5} - 7x^{1/5} + 3 = 0$

118. $2x^{2/5} + 3x^{1/5} + 1 = 0$

119. $x^{1/3} - x^{1/6} - 6 = 0$

120. $x^{1/3} + 2x^{1/6} - 3 = 0$

121. $x^{1/2} - 3x^{1/4} + 2 = 0$

122. $x^{1/2} - 5x^{1/4} + 6 = 0$

123. $\frac{1}{x^2} - \frac{3}{x} + 2 = 0$

124. $\frac{1}{x^2} - \frac{1}{x} - 6 = 0$

125. $4x^{-2} - x^{-1} - 5 = 0$

126. $2x^{-2} - x^{-1} - 1 = 0$

127. $(x^2 - 3x)^2 - 2(x^2 - 3x) - 8 = 0$

128. $(x^2 - 6x)^2 - 2(x^2 - 6x) - 35 = 0$

129. $16\left(\frac{x - 1}{x - 8}\right)^2 + 8\left(\frac{x - 1}{x - 8}\right) + 1 = 0$

130. $9\left(\frac{x + 2}{x + 3}\right)^2 - 6\left(\frac{x + 2}{x + 3}\right) + 1 = 0$

Solving Problems

✓ **131.** *Unisphere* The Unisphere is the world's largest man-made globe. It was built as the symbol of the 1964–1965 New York World's Fair. A sphere with the same diameter as the Unisphere globe would have a surface area of 45,239 square feet. What is the diameter of the Unisphere? (Source: The World's Fair and Exposition Information and Reference Guide)

© Rudy Sulgan/CORBIS

Designing the Unisphere was an engineering challenge that at one point involved simultaneously solving 670 equations.

132. *Geometry* The surface area S of a basketball is $900/\pi$ square inches. Find the radius r of the basketball.

Free-Falling Object In Exercises 133–136, find the time required for an object to reach the ground when it is dropped from a height of s_0 feet. The height h (in feet) is given by $h = -16t^2 + s_0$, where t measures the time (in seconds) after the object is released.

133. $s_0 = 256$

134. $s_0 = 48$

135. $s_0 = 128$

136. $s_0 = 500$

137. *Free-Falling Object* The height h (in feet) of an object thrown vertically upward from the top of a tower 144 feet tall is given by $h = 144 + 128t - 16t^2$, where t measures the time in seconds from the time when the object is released. How long does it take for the object to reach the ground?

138. *Profit* The monthly profit P (in dollars) a company makes depends on the amount x (in dollars) the company spends on advertising according to the model

$$P = 800 + 120x - \frac{1}{2}x^2.$$

Find the amount spent on advertising that will yield a monthly profit of \$8000.

Compound Interest The amount A in an account after 2 years when a principal of P dollars is invested at annual interest rate r compounded annually is given by $A = P(1 + r)^2$. In Exercises 139 and 140, find r.

139. $P = \$1500,\ A = \$1685.40$

140. $P = \$5000,\ A = \$5724.50$

National Health Expenditures In Exercises 141 and 142, the national expenditures for health care in the United States from 1997 through 2006 are given by

$y = 4.95t^2 + 876, \quad 7 \le t \le 16.$

In this model, y represents the expenditures (in billions of dollars) and t represents the year, with $t = 7$ corresponding to 1997 (see figure). (Source: U.S. Centers for Medicare & Medicaid Services)

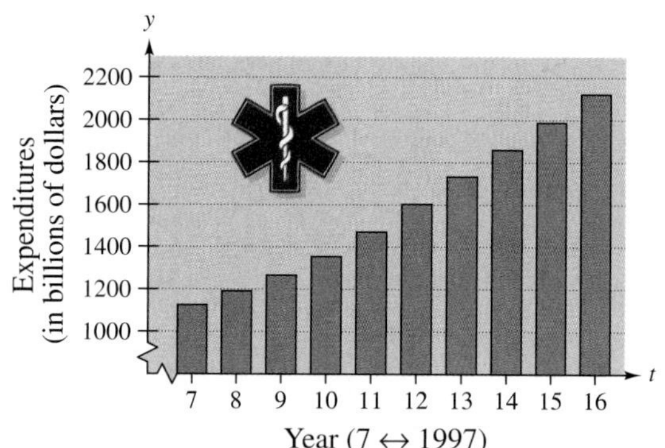

Figure for 141 and 142

141. Algebraically determine the year when expenditures were approximately \$1500 billion. Graphically confirm the result.

142. Algebraically determine the year when expenditures were approximately \$1850 billion. Graphically confirm the result.

Explaining Concepts

143. For a quadratic equation $ax^2 + bx + c = 0$, where a, b, and c are real numbers with $a \neq 0$, explain why b and c can equal 0, but a cannot.

144. Is it possible for a quadratic equation of the form $x^2 = m$ to have one real solution and one complex solution? Explain your reasoning.

145. Describe the steps you would use to solve a quadratic equation when using the Square Root Property.

146. Describe a procedure for solving an equation of quadratic form. Give an example.

Cumulative Review

In Exercises 147–150, solve the inequality and sketch the solution on the real number line.

147. $3x - 8 > 4$

148. $4 - 5x \ge 12$

149. $2x - 6 \le 9 - x$

150. $x - 4 < 6$ or $x + 3 > 8$

In Exercises 151 and 152, solve the system of linear equations.

151. $x + y - z = 4$
$2x + y + 2z = 10$
$x - 3y - 4z = -7$

152. $2x - y + z = -6$
$x + 5y - z = 7$
$-x - 2y - 3z = 8$

In Exercises 153–158, combine the radical expressions, if possible, and simplify.

153. $5\sqrt{3} - 2\sqrt{3}$

154. $8\sqrt{27} + 4\sqrt{27}$

155. $16\sqrt[3]{y} - 9\sqrt[3]{x}$

156. $12\sqrt{x - 1} + 6\sqrt{x - 1}$

157. $\sqrt{16m^4n^3} + m\sqrt{m^2n}$

158. $x^2y\sqrt[4]{32x^2} + x\sqrt[4]{2x^6y^4} - y\sqrt[4]{162x^{10}}$

10.2 Completing the Square

© Sean Cayton/The Image Works

Why You Should Learn It

You can use techniques such as completing the square to solve quadratic equations that model real-life situations. For instance, in Exercise 90 on page 633, you will find the dimensions of an outdoor enclosure of a kennel by completing the square.

What You Should Learn

1 ▶ Rewrite quadratic expressions in completed square form.

2 ▶ Solve quadratic equations by completing the square.

1 ▶ Rewrite quadratic expressions in completed square form.

Constructing Perfect Square Trinomials

Consider the quadratic equation

$$(x-2)^2 = 10. \qquad \text{Completed square form}$$

You know from Example 2(b) in the preceding section that this equation has two solutions: $x = 2 + \sqrt{10}$ and $x = 2 - \sqrt{10}$. Suppose you were given the equation in its general form

$$x^2 - 4x - 6 = 0. \qquad \text{General form}$$

How could you solve this form of the quadratic equation? You could try factoring, but after attempting to do so you would find that the left side of the equation is not factorable using integer coefficients.

In this section, you will study a technique for rewriting an equation in a completed square form. This technique is called **completing the square.** Note that prior to completing the square, the coefficient of the second-degree term must be 1.

Completing the Square

To **complete the square** for the expression $x^2 + bx$, add $(b/2)^2$, which is the square of half the coefficient of x. Consequently,

$$x^2 + bx + \left(\frac{b}{2}\right)^2 = \left(x + \frac{b}{2}\right)^2.$$

(half)^2

EXAMPLE 1 Constructing a Perfect Square Trinomial

What term should be added to $x^2 - 8x$ so that it becomes a perfect square trinomial? To find this term, notice that the coefficient of the x-term is -8. Take half of this coefficient and square the result to get $(-4)^2 = 16$. Add this term to the expression to make it a perfect square trinomial.

$$x^2 - 8x + (-4)^2 = x^2 - 8x + 16 \qquad \text{Add } (-4)^2 = 16 \text{ to the expression.}$$

You can then rewrite the expression as the square of a binomial, $(x-4)^2$.

✓ **CHECKPOINT** *Now try Exercise 3.*

2 ▶ Solve quadratic equations by completing the square.

Solving Equations by Completing the Square

Completing the square can be used to solve quadratic equations. When using this procedure, remember to *preserve the equality* by adding the same constant to each side of the equation.

Study Tip

In Example 2, completing the square is used for the sake of illustration. This particular equation would be easier to solve by factoring. Try reworking the problem by factoring to see that you obtain the same two solutions.

EXAMPLE 2 Completing the Square: Leading Coefficient Is 1

Solve $x^2 + 12x = 0$ by completing the square.

Solution

$x^2 + 12x = 0$ Write original equation.

$x^2 + 12x + 6^2 = 36$ Add $6^2 = 36$ to each side.

$\left(\frac{12}{2}\right)^2$

$(x + 6)^2 = 36$ Completed square form

$x + 6 = \pm\sqrt{36}$ Square Root Property

$x = -6 \pm 6$ Subtract 6 from each side.

$x = -6 + 6$ or $x = -6 - 6$ Separate solutions.

$x = 0$ $\quad x = -12$ Simplify.

The solutions are $x = 0$ and $x = -12$. Check these in the original equation.

✓ **CHECKPOINT** *Now try Exercise 17.*

Technology: Tip

You can use a graphing calculator to check the solution to Example 3. Graph

$y = x^2 - 6x + 7$

as shown below. Then use the *zero* or *root* feature of the graphing calculator to approximate the x-intercepts to be $x \approx 4.41$ and $x \approx 1.59$, which are the same solutions obtained in Example 3.

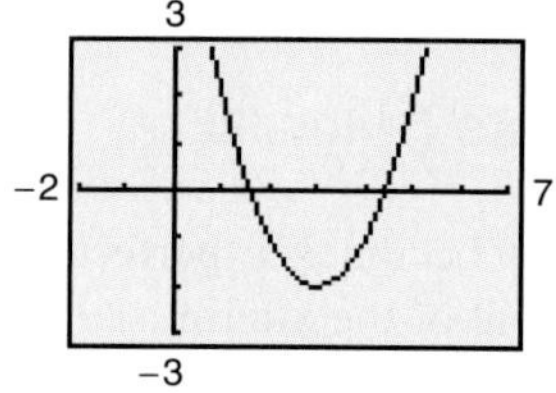

EXAMPLE 3 Completing the Square: Leading Coefficient Is 1

Solve $x^2 - 6x + 7 = 0$ by completing the square.

Solution

$x^2 - 6x + 7 = 0$ Write original equation.

$x^2 - 6x = -7$ Subtract 7 from each side.

$x^2 - 6x + (-3)^2 = -7 + 9$ Add $(-3)^2 = 9$ to each side.

$\left(-\frac{6}{2}\right)^2$

$(x - 3)^2 = 2$ Completed square form

$x - 3 = \pm\sqrt{2}$ Square Root Property

$x = 3 \pm \sqrt{2}$ Add 3 to each side.

$x = 3 + \sqrt{2}$ or $x = 3 - \sqrt{2}$ Separate solutions.

The solutions are $x = 3 + \sqrt{2} \approx 4.41$ and $x = 3 - \sqrt{2} \approx 1.59$. Check these in the original equation.

✓ **CHECKPOINT** *Now try Exercise 33.*

If the leading coefficient of a quadratic equation is not 1, you must divide each side of the equation by this coefficient *before* completing the square.

EXAMPLE 4 Completing the Square: Leading Coefficient Is Not 1

$2x^2 - x - 2 = 0$	Original equation
$2x^2 - x = 2$	Add 2 to each side.
$x^2 - \frac{1}{2}x = 1$	Divide each side by 2.
$x^2 - \frac{1}{2}x + \left(-\frac{1}{4}\right)^2 = 1 + \frac{1}{16}$	Add $\left(-\frac{1}{4}\right)^2 = \frac{1}{16}$ to each side.
$\left(x - \frac{1}{4}\right)^2 = \frac{17}{16}$	Completed square form
$x - \frac{1}{4} = \pm\frac{\sqrt{17}}{4}$	Square Root Property
$x = \frac{1}{4} \pm \frac{\sqrt{17}}{4}$	Add $\frac{1}{4}$ to each side.

The solutions are $x = \frac{1}{4} + \frac{\sqrt{17}}{4} \approx 1.28$ and $x = \frac{1}{4} - \frac{\sqrt{17}}{4} \approx -0.78$.

CHECKPOINT *Now try Exercise 61.*

EXAMPLE 5 Completing the Square: Leading Coefficient Is Not 1

$3x^2 - 6x + 1 = 0$	Original equation
$3x^2 - 6x = -1$	Subtract 1 from each side.
$x^2 - 2x = -\frac{1}{3}$	Divide each side by 3.
$x^2 - 2x + (-1)^2 = -\frac{1}{3} + 1$	Add $(-1)^2 = 1$ to each side.
$(x - 1)^2 = \frac{2}{3}$	Completed square form
$x - 1 = \pm\sqrt{\frac{2}{3}}$	Square Root Property
$x - 1 = \pm\frac{\sqrt{6}}{3}$	Rationalize the denominator.
$x = 1 \pm \frac{\sqrt{6}}{3}$	Add 1 to each side.

The solutions are $x = 1 + \sqrt{6}/3 \approx 1.82$ and $x = 1 - \sqrt{6}/3 \approx 0.18$.

CHECKPOINT *Now try Exercise 59.*

EXAMPLE 6 A Quadratic Equation with Complex Solutions

Solve $x^2 - 4x + 8 = 0$ by completing the square.

Solution

$x^2 - 4x + 8 = 0$	Write original equation.
$x^2 - 4x = -8$	Subtract 8 from each side.
$x^2 - 4x + (-2)^2 = -8 + 4$	Add $(-2)^2 = 4$ to each side.
$(x - 2)^2 = -4$	Completed square form
$x - 2 = \pm 2i$	Square Root Property
$x = 2 \pm 2i$	Add 2 to each side.

The solutions are $x = 2 + 2i$ and $x = 2 - 2i$. Check these in the original equation.

 **CHECKPOINT** ***Now try Exercise 49.***

Tony Avelar/*Bloomberg News*/Landov

Figure 10.1

EXAMPLE 7 Dimensions of an iPhone 

The first generation of the iPhone™ has an approximate volume of 4.968 cubic inches. Its width is 0.46 inch and its face has the dimensions x inches by $x + 2.1$ inches. (See Figure 10.1.) Find the dimensions of the face in inches. (Source: Apple, Inc.)

Solution

$lwh = V$	Formula for volume of a rectangular solid
$(x)(0.46)(x + 2.1) = 4.968$	Substitute 4.968 for V, x for l, 0.46 for w, and $x + 2.1$ for h.
$0.46x^2 + 0.966x = 4.968$	Multiply factors.
$x^2 + 2.1x = 10.8$	Divide each side by 0.46.
$x^2 + 2.1x + \left(\frac{2.1}{2}\right)^2 = 10.8 + 1.1025$	Add $\left(\frac{2.1}{2}\right)^2 = 1.1025$ to each side.
$(x + 1.05)^2 = 11.9025$	Completed square form
$x + 1.05 = \pm\sqrt{11.9025}$	Square Root Property
$x = -1.05 \pm \sqrt{11.9025}$	Subtract 1.05 from each side.

Choosing the positive root, you obtain

$$x = -1.05 + 3.45 = 2.4 \text{ inches} \qquad \text{Length of face}$$

and

$$x + 2.1 = 2.4 + 2.1 = 4.5 \text{ inches.} \qquad \text{Height of face}$$

 **CHECKPOINT** ***Now try Exercise 91.***

Concept Check

1. What is a perfect square trinomial?

2. What term must be added to $x^2 + 5x$ to complete the square? Explain how you found the term.

3. When using the method of completing the square to solve $2x^2 - 7x = 6$, what is the first step? Is the resulting equation equivalent to the original equation? Explain.

4. Is it possible for a quadratic equation to have no real number solution? If so, give an example.

10.2 EXERCISES

Go to pages 676–677 to record your assignments.

Developing Skills

In Exercises 1–16, add a term to the expression so that it becomes a perfect square trinomial. ***See Example 1.***

1. $x^2 + 8x +$
2. $x^2 + 12x +$

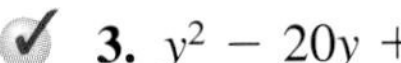

3. $y^2 - 20y +$
4. $y^2 - 2y +$
5. $x^2 + 14x +$
6. $x^2 - 24x +$
7. $t^2 + 5t +$
8. $u^2 + 7u +$
9. $x^2 - 9x +$
10. $y^2 - 11y +$
11. $a^2 - \frac{1}{3}a +$
12. $y^2 + \frac{4}{3}y +$
13. $y^2 + \frac{8}{5}y +$
14. $x^2 - \frac{9}{5}x +$
15. $r^2 - 0.4r +$
16. $s^2 + 4.6s +$

In Exercises 17–32, solve the equation first by completing the square and then by factoring. ***See Examples 2–5.***

17. $x^2 - 20x = 0$
18. $x^2 + 32x = 0$
19. $x^2 + 6x = 0$
20. $t^2 - 10t = 0$
21. $y^2 - 5y = 0$
22. $t^2 - 9t = 0$
23. $t^2 - 8t + 7 = 0$
24. $y^2 - 4y + 4 = 0$
25. $x^2 + 7x + 12 = 0$
26. $z^2 + 3z - 10 = 0$
27. $x^2 - 3x - 18 = 0$
28. $a^2 + 12a + 32 = 0$
29. $2u^2 - 12u + 18 = 0$
30. $3x^2 - 3x - 6 = 0$
31. $4x^2 + 4x - 15 = 0$
32. $6a^2 - 23a + 15 = 0$

In Exercises 33–72, solve the equation by completing the square. Give the solutions in exact form and in decimal form rounded to two decimal places. (The solutions may be complex numbers.) ***See Examples 2–6.***

33. $x^2 - 4x - 3 = 0$
34. $x^2 - 6x + 7 = 0$
35. $x^2 + 4x - 3 = 0$
36. $x^2 + 6x + 7 = 0$
37. $x^2 + 6x = 7$
38. $x^2 - 4x = -3$
39. $x^2 - 12x = -10$
40. $x^2 - 4x = -9$
41. $x^2 + 8x + 7 = 0$
42. $x^2 + 10x + 9 = 0$
43. $x^2 - 10x + 21 = 0$
44. $x^2 - 10x + 24 = 0$
45. $y^2 + 5y + 3 = 0$
46. $y^2 + 8y + 9 = 0$
47. $x^2 + 10 = 6x$
48. $x^2 + 23 = 10x$
49. $z^2 + 4z + 13 = 0$
50. $z^2 - 6z + 18 = 0$
51. $-x^2 + x - 1 = 0$
52. $1 - x - x^2 = 0$

53. $a^2 + 7a + 11 = 0$

54. $y^2 + 5y + 9 = 0$

55. $x^2 - \frac{2}{3}x - 3 = 0$

56. $x^2 + \frac{4}{5}x - 1 = 0$

57. $v^2 + \frac{3}{4}v - 2 = 0$

58. $u^2 - \frac{2}{3}u + 5 = 0$

✓ **59.** $2x^2 + 8x + 3 = 0$

60. $3x^2 - 24x - 5 = 0$

✓ **61.** $3x^2 + 9x + 5 = 0$

62. $5x^2 - 15x + 7 = 0$

63. $4y^2 + 4y - 9 = 0$

64. $4z^2 - 3z + 2 = 0$

65. $5x^2 - 3x + 10 = 0$

66. $7u^2 - 8u - 3 = 0$

67. $x\left(x - \frac{2}{3}\right) = 14$

68. $2x\left(x + \frac{4}{3}\right) = 5$

69. $0.1x^2 + 0.5x = -0.2$

70. $0.2x^2 + 0.1x = -0.5$

71. $0.75x^2 + 1.25x + 1.5 = 0$

72. $0.625x^2 - 0.875x + 0.25 = 0$

In Exercises 73–78, find the real solutions.

73. $\frac{x}{2} - \frac{1}{x} = 1$

74. $\frac{x}{2} + \frac{5}{x} = 4$

75. $\frac{x^2}{8} = \frac{x + 3}{2}$

76. $\frac{x^2 + 2}{24} = \frac{x - 1}{3}$

77. $\sqrt{2x + 1} = x - 3$

78. $\sqrt{3x - 2} = x - 2$

In Exercises 79–86, use a graphing calculator to graph the function. Use the graph to approximate any x-intercepts of the graph. Set $y = 0$ and solve the resulting equation. Compare the result with the x-intercepts of the graph.

79. $y = x^2 + 4x - 1$

80. $y = x^2 + 6x - 4$

81. $y = x^2 - 2x - 5$

82. $y = 2x^2 - 6x - 5$

83. $y = \frac{1}{3}x^2 + 2x - 6$

84. $y = \frac{1}{2}x^2 - 3x + 1$

85. $y = x - 2\sqrt{x} + 1$

86. $y = \sqrt{x} - x + 2$

Solving Problems

87. *Geometric Modeling*

(a) Find the area of the two adjoining rectangles and large square in the figure.

(b) Find the area of the small square in the lower right-hand corner of the figure and add it to the area found in part (a).

(c) Find the dimensions and the area of the entire figure after adjoining the small square in the lower right-hand corner of the figure. Note that you have shown geometrically the technique of completing the square.

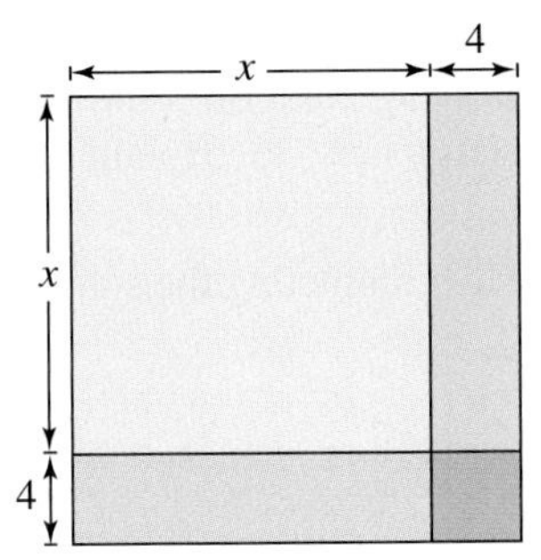

Figure for 87

Figure for 88

88. *Geometric Modeling* Repeat Exercise 87 for the model shown in the figure.

89. *Geometry* You have 200 meters of fencing to enclose two adjacent rectangular corrals (see figure). The total area of the enclosed region is 1400 square meters. What are the dimensions of each corral? (The corrals are the same size.)

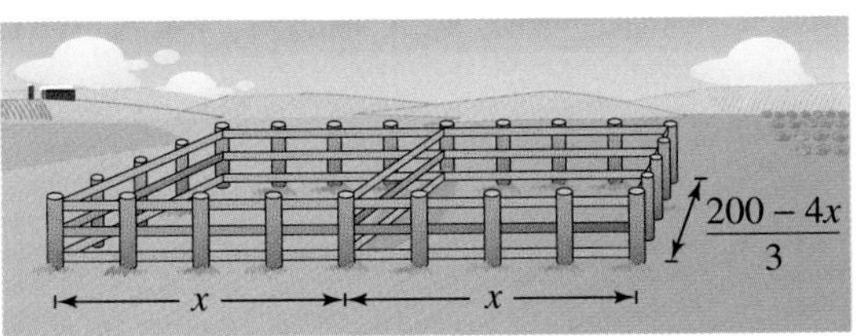

90. *Geometry* A kennel is adding a rectangular outdoor enclosure along one side of the kennel wall (see figure). The other three sides of the enclosure will be formed by a fence. The kennel has 111 feet of fencing and plans to use 1215 square feet of land for the enclosure. What are the dimensions of the enclosure?

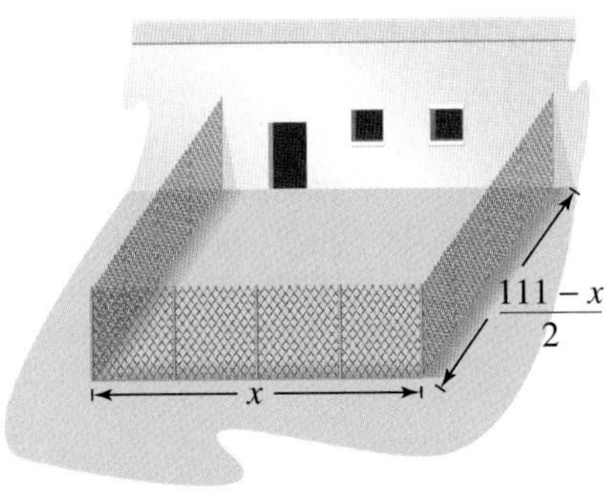

✓ **91.** *Geometry* An open box with a rectangular base of x inches by $x + 4$ inches has a height of 6 inches (see figure). The volume of the box is 840 cubic inches. Find the dimensions of the box.

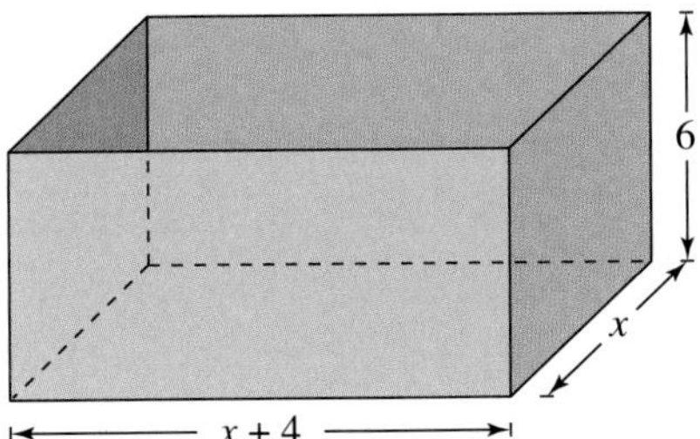

92. *Geometry* An open box with a rectangular base of $2x$ inches by $6x - 2$ inches has a height of 9 inches (see figure). The volume of the box is 1584 cubic inches. Find the dimensions of the box.

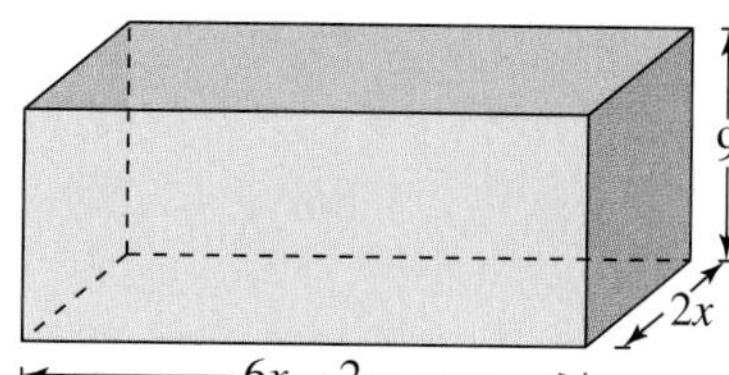

93. *Revenue* The revenue R (in dollars) from selling x pairs of running shoes is given by

$$R = x\left(80 - \frac{1}{2}x\right).$$

Find the number of pairs of running shoes that must be sold to produce a revenue of \$2750.

94. *Revenue* The revenue R (in dollars) from selling x golf clubs is given by

$$R = x\left(150 - \frac{1}{10}x\right).$$

Find the number of golf clubs that must be sold to produce a revenue of \$15,033.60.

Explaining Concepts

95. Explain the use of the Square Root Property when solving a quadratic equation by the method of completing the square.

96. *True or False?* If you solve a quadratic equation by completing the square and obtain solutions that are rational numbers, then you could have solved the equation by factoring. Justify your answer.

97. Consider the quadratic equation $(x - 1)^2 = d$.

(a) What value(s) of d will produce a quadratic equation that has exactly one (repeated) solution?

(b) Describe the value(s) of d that will produce two different solutions, both of which are *rational* numbers.

(c) Describe the value(s) of d that will produce two different solutions, both of which are *irrational* numbers.

(d) Describe the value(s) of d that will produce two different solutions, both of which are *complex* numbers.

98. You teach an algebra class and one of your students hands in the following solution. Find and correct the error(s). Discuss how to explain the error(s) to your student.

Solve $x^2 + 6x - 13 = 0$ by completing the square.

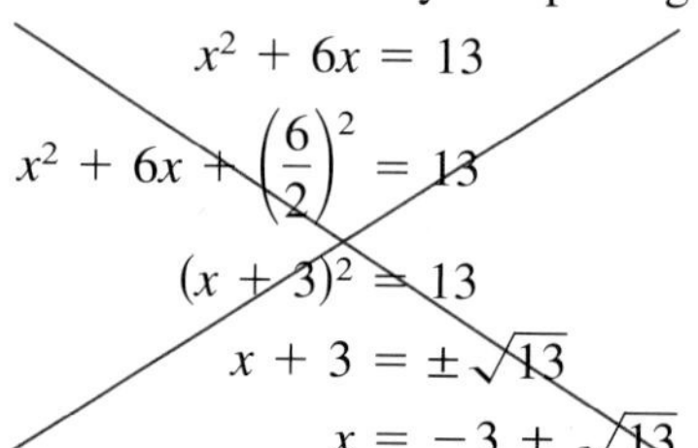

Cumulative Review

In Exercises 99–106, perform the operation and simplify the expression.

99. $3\sqrt{5}\sqrt{500}$

100. $2\sqrt{2x^2}\sqrt{27x}$

101. $(3 + \sqrt{2})(3 - \sqrt{2})$

102. $(\sqrt[3]{6} - 2)(\sqrt[3]{4} + 1)$

103. $(3 + \sqrt{2})^2$

104. $(2\sqrt{x} - 5)^2$

105. $\dfrac{8}{\sqrt{10}}$

106. $\dfrac{5}{\sqrt{12} - 2}$

In Exercises 107 and 108, rewrite the expression using the specified rule, where a and b are nonnegative real numbers.

107. Product Rule: $\sqrt{ab} = \_\_\_\_$.

108. Quotient Rule: $\sqrt{\dfrac{a}{b}} = \_\_\_\_$.

10.3 The Quadratic Formula

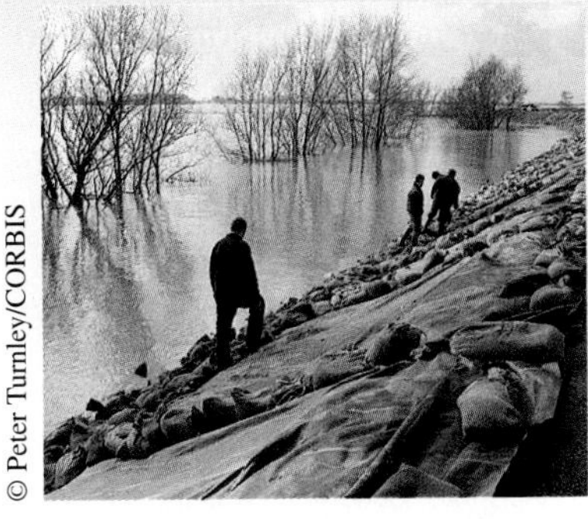
© Peter Turnley/CORBIS

What You Should Learn

1 ▶ Derive the Quadratic Formula by completing the square for a general quadratic equation.
2 ▶ Use the Quadratic Formula to solve quadratic equations.
3 ▶ Determine the types of solutions of quadratic equations using the discriminant.
4 ▶ Write quadratic equations from solutions of the equations.

Why You Should Learn It

Knowing the Quadratic Formula can be helpful in solving quadratic equations that model real-life situations. For instance, in Exercise 99 on page 642, you will use a quadratic equation that models the depth of a river after a heavy rain begins.

1 ▶ Derive the Quadratic Formula by completing the square for a general quadratic equation.

The Quadratic Formula

A fourth technique for solving a quadratic equation involves the **Quadratic Formula.** This formula is derived by completing the square for a general quadratic equation.

$$ax^2 + bx + c = 0$$ General form, $a \neq 0$

$$ax^2 + bx = -c$$ Subtract c from each side.

$$x^2 + \frac{b}{a}x = -\frac{c}{a}$$ Divide each side by a.

$$x^2 + \frac{b}{a}x + \left(\frac{b}{2a}\right)^2 = -\frac{c}{a} + \left(\frac{b}{2a}\right)^2$$ Add $\left(\frac{b}{2a}\right)^2$ to each side.

$$\left(x + \frac{b}{2a}\right)^2 = \frac{b^2 - 4ac}{4a^2}$$ Simplify.

$$x + \frac{b}{2a} = \pm\sqrt{\frac{b^2 - 4ac}{4a^2}}$$ Square Root Property

$$x = -\frac{b}{2a} \pm \frac{\sqrt{b^2 - 4ac}}{2|a|}$$ Subtract $\frac{b}{2a}$ from each side.

$$x = \frac{-b \pm \sqrt{b^2 - 4ac}}{2a}$$ Simplify.

Notice in the derivation of the Quadratic Formula that, because $\pm 2|a|$ represents the same numbers as $\pm 2a$, you can omit the absolute value bars.

Study Tip

The Quadratic Formula is one of the most important formulas in algebra, and you should memorize it. It helps to try to memorize a verbal statement of the rule. For instance, you might try to remember the following verbal statement of the Quadratic Formula: "The opposite of *b*, plus or minus the square root of *b* squared minus 4*ac*, all divided by 2*a*."

The Quadratic Formula

The solutions of $ax^2 + bx + c = 0$, $a \neq 0$, are given by the **Quadratic Formula**

$$x = \frac{-b \pm \sqrt{b^2 - 4ac}}{2a}.$$

2 ▶ Use the Quadratic Formula to solve quadratic equations.

Solving Equations by the Quadratic Formula

When using the Quadratic Formula, remember that *before* the formula can be applied, you must first write the quadratic equation in general form in order to determine the values of a, b, and c.

Study Tip

In Example 1, the solutions are rational numbers, which means that the equation could have been solved by factoring. Try solving the equation by factoring.

EXAMPLE 1 The Quadratic Formula: Two Distinct Solutions

$$x^2 + 6x = 16$$ Original equation

$$x^2 + 6x - 16 = 0$$ Write in general form.

$$x = \frac{-b \pm \sqrt{b^2 - 4ac}}{2a}$$ Quadratic Formula

$$x = \frac{-6 \pm \sqrt{6^2 - 4(1)(-16)}}{2(1)}$$ Substitute 1 for a, 6 for b, and -16 for c.

$$x = \frac{-6 \pm \sqrt{100}}{2}$$ Simplify.

$$x = \frac{-6 \pm 10}{2}$$ Simplify.

$$x = 2 \quad \text{or} \quad x = -8$$ Solutions

The solutions are $x = 2$ and $x = -8$. Check these in the original equation.

✓ ***CHECKPOINT*** *Now try Exercise 5.*

Study Tip

If the leading coefficient of a quadratic equation is negative, you should begin by multiplying each side of the equation by -1, as shown in Example 2. This will produce a positive leading coefficient, which is easier to work with.

EXAMPLE 2 The Quadratic Formula: Two Distinct Solutions

$$-x^2 - 4x + 8 = 0$$ Leading coefficient is negative.

$$x^2 + 4x - 8 = 0$$ Multiply each side by -1.

$$x = \frac{-b \pm \sqrt{b^2 - 4ac}}{2a}$$ Quadratic Formula

$$x = \frac{-4 \pm \sqrt{4^2 - 4(1)(-8)}}{2(1)}$$ Substitute 1 for a, 4 for b, and -8 for c.

$$x = \frac{-4 \pm \sqrt{48}}{2}$$ Simplify.

$$x = \frac{-4 \pm 4\sqrt{3}}{2}$$ Simplify.

$$x = -2 \pm 2\sqrt{3}$$ Solutions

The solutions are $x = -2 + 2\sqrt{3}$ and $x = -2 - 2\sqrt{3}$. Check these in the original equation.

✓ ***CHECKPOINT*** *Now try Exercise 15.*

Study Tip

Example 3 could have been solved as follows, without dividing each side by 2 in the first step.

$$x = \frac{-(-24) \pm \sqrt{(-24)^2 - 4(18)(8)}}{2(18)}$$

$$x = \frac{24 \pm \sqrt{576 - 576}}{36}$$

$$x = \frac{24 \pm 0}{36}$$

$$x = \frac{2}{3}$$

While the result is the same, dividing each side by 2 simplifies the equation before the Quadratic Formula is applied and so allows you to work with smaller numbers.

EXAMPLE 3 The Quadratic Formula: One Repeated Solution

$18x^2 - 24x + 8 = 0$ Original equation

$9x^2 - 12x + 4 = 0$ Divide each side by 2.

$x = \dfrac{-b \pm \sqrt{b^2 - 4ac}}{2a}$ Quadratic Formula

$x = \dfrac{-(-12) \pm \sqrt{(-12)^2 - 4(9)(4)}}{2(9)}$ Substitute 9 for a, -12 for b, and 4 for c.

$x = \dfrac{12 \pm \sqrt{144 - 144}}{18}$ Simplify.

$x = \dfrac{12 \pm \sqrt{0}}{18}$ Simplify.

$x = \dfrac{2}{3}$ Solution

The only solution is $x = \frac{2}{3}$. Check this in the original equation.

✓ **CHECKPOINT** *Now try Exercise 9.*

EXAMPLE 4 The Quadratic Formula: Complex Solutions

$2x^2 - 4x + 5 = 0$ Original equation

$x = \dfrac{-b \pm \sqrt{b^2 - 4ac}}{2a}$ Quadratic Formula

$x = \dfrac{-(-4) \pm \sqrt{(-4)^2 - 4(2)(5)}}{2(2)}$ Substitute 2 for a, -4 for b, and 5 for c.

$x = \dfrac{4 \pm \sqrt{-24}}{4}$ Simplify.

$x = \dfrac{4 \pm 2\sqrt{6}i}{4}$ Write in i-form.

$x = \dfrac{2(2 \pm \sqrt{6}i)}{2 \cdot 2}$ Factor numerator and denominator.

$x = \dfrac{\cancel{2}(2 \pm \sqrt{6}i)}{\cancel{2} \cdot 2}$ Divide out common factor.

$x = 1 \pm \dfrac{\sqrt{6}}{2}i$ Solutions

The solutions are $x = 1 + \frac{\sqrt{6}}{2}i$ and $x = 1 - \frac{\sqrt{6}}{2}i$. Check these in the original equation.

✓ **CHECKPOINT** *Now try Exercise 21.*

3 ▶ Determine the types of solutions of quadratic equations using the discriminant.

The Discriminant

The radicand in the Quadratic Formula, $b^2 - 4ac$, is called the **discriminant** because it allows you to "discriminate" among different types of solutions.

Study Tip

By reexamining Examples 1 through 4, you can see that the equations with rational or repeated solutions could have been solved by *factoring*. In general, quadratic equations (with integer coefficients) for which the discriminant is either zero or a perfect square are factorable using integer coefficients. Consequently, a quick test of the discriminant will help you decide which solution method to use to solve a quadratic equation.

Using the Discriminant

Let a, b, and c be rational numbers such that $a \neq 0$. The discriminant of the quadratic equation $ax^2 + bx + c = 0$ is given by $b^2 - 4ac$, and can be used to classify the solutions of the equation as follows.

Discriminant	*Solution Type*
1. Perfect square	Two distinct rational solutions (Example 1)
2. Positive nonperfect square	Two distinct irrational solutions (Example 2)
3. Zero	One repeated rational solution (Example 3)
4. Negative number	Two distinct complex solutions (Example 4)

Technology: Discovery

Use a graphing calculator to graph each equation.

a. $y = x^2 - x + 2$

b. $y = 2x^2 - 3x - 2$

c. $y = x^2 - 2x + 1$

d. $y = x^2 - 2x - 10$

Describe the solution type of each equation and check your results with those shown in Example 5. Why do you think the discriminant is used to determine solution types?

EXAMPLE 5 Using the Discriminant

Determine the type of solution(s) for each quadratic equation.

a. $x^2 - x + 2 = 0$

b. $2x^2 - 3x - 2 = 0$

c. $x^2 - 2x + 1 = 0$

d. $x^2 - 2x - 1 = 9$

Solution

Equation	*Discriminant*	*Solution Type*
a. $x^2 - x + 2 = 0$	$b^2 - 4ac = (-1)^2 - 4(1)(2)$ $= 1 - 8 = -7$	Two distinct complex solutions
b. $2x^2 - 3x - 2 = 0$	$b^2 - 4ac = (-3)^2 - 4(2)(-2)$ $= 9 + 16 = 25$	Two distinct rational solutions
c. $x^2 - 2x + 1 = 0$	$b^2 - 4ac = (-2)^2 - 4(1)(1)$ $= 4 - 4 = 0$	One repeated rational solution
d. $x^2 - 2x - 1 = 9$	$b^2 - 4ac = (-2)^2 - 4(1)(-10)$ $= 4 + 40 = 44$	Two distinct irrational solutions

✓ **CHECKPOINT** *Now try Exercise 41.*

Summary of Methods for Solving Quadratic Equations

Method	*Example*
1. Factoring	$3x^2 + x = 0$
	$x(3x + 1) = 0 \Rightarrow x = 0$ and $x = -\frac{1}{3}$
2. Square Root Property	$(x + 2)^2 = 7$
	$x + 2 = \pm\sqrt{7} \Rightarrow x = -2 + \sqrt{7}$ and $x = -2 - \sqrt{7}$
3. Completing the square	$x^2 + 6x = 2$
	$x^2 + 6x + 3^2 = 2 + 9$
	$(x + 3)^2 = 11 \Rightarrow x = -3 + \sqrt{11}$ and $x = -3 - \sqrt{11}$
4. Quadratic Formula	$3x^2 - 2x + 2 = 0 \Rightarrow x = \frac{-(-2) \pm \sqrt{(-2)^2 - 4(3)(2)}}{2(3)} = \frac{1}{3} \pm \frac{\sqrt{5}}{3}i$

4 ▶ Write quadratic equations from solutions of the equations.

Writing Quadratic Equations from Solutions

Using the Zero-Factor Property, you know that the equation $(x + 5)(x - 2) = 0$ has two solutions, $x = -5$ and $x = 2$. You can use the Zero-Factor Property in reverse to find a quadratic equation given its solutions. This process is demonstrated in Example 6.

Reverse of Zero-Factor Property

Let a and b be real numbers, variables, or algebraic expressions. If $a = 0$ or $b = 0$, then a and b are factors such that $ab = 0$.

Technology: Tip

A program for several models of graphing calculators that uses the Quadratic Formula to solve quadratic equations can be found at our website, *www.cengage.com/math/larson/algebra.* This program will display real solutions to quadratic equations.

EXAMPLE 6 **Writing a Quadratic Equation from Its Solutions**

Write a quadratic equation that has the solutions $x = 4$ and $x = -7$. Using the solutions $x = 4$ and $x = -7$, you can write the following.

$x = 4$	and	$x = -7$	Solutions
$x - 4 = 0$		$x + 7 = 0$	Obtain zero on one side of each equation.
		$(x - 4)(x + 7) = 0$	Reverse of Zero-Factor Property
		$x^2 + 3x - 28 = 0$	FOIL Method

So, a quadratic equation that has the solutions $x = 4$ and $x = -7$ is

$$x^2 + 3x - 28 = 0.$$

This is not the only quadratic equation with the solutions $x = 4$ and $x = -7$. You can obtain other quadratic equations with these solutions by multiplying $x^2 + 3x - 28 = 0$ by any nonzero real number.

✓ ***CHECKPOINT*** ***Now try Exercise 65.***

Concept Check

1. What method is used to derive the Quadratic Formula from a general quadratic equation?

2. To solve the quadratic equation $3x^2 = 3 - x$ using the Quadratic Formula, what are the values of a, b, and c?

3. The discriminant of a quadratic equation is -25. What type of solution(s) does the equation have?

4. Describe the steps you would use to write a quadratic equation that has the solutions $x = 3$ and $x = -2$.

10.3 EXERCISES

Go to pages 676–677 to record your assignments.

Developing Skills

In Exercises 1–4, write the quadratic equation in general form.

1. $2x^2 = 7 - 2x$
2. $7x^2 + 15x = 5$
3. $x(10 - x) = 5$
4. $x(2x + 9) = 12$

In Exercises 5–14, solve the equation first by using the Quadratic Formula and then by factoring. ***See Examples 1–4.***

✓ 5. $x^2 - 11x + 28 = 0$
6. $x^2 - 12x + 27 = 0$
7. $x^2 + 6x + 8 = 0$
8. $x^2 + 9x + 14 = 0$
✓ 9. $16x^2 + 8x + 1 = 0$
10. $9x^2 + 12x + 4 = 0$
11. $4x^2 + 12x + 9 = 0$
12. $10x^2 - 11x + 3 = 0$
13. $x^2 - 5x - 300 = 0$
14. $x^2 + 20x - 300 = 0$

In Exercises 15–40, solve the equation by using the Quadratic Formula. (Find all real *and* complex solutions.) ***See Examples 1–4.***

✓ 15. $x^2 - 2x - 4 = 0$
16. $x^2 - 2x - 6 = 0$
17. $t^2 + 4t + 1 = 0$
18. $y^2 + 6y - 8 = 0$
19. $x^2 - 10x + 23 = 0$
20. $u^2 - 12u + 29 = 0$
✓ 21. $2x^2 + 3x + 3 = 0$
22. $2x^2 - 2x + 3 = 0$
23. $3v^2 - 2v - 1 = 0$
24. $4x^2 + 6x + 1 = 0$
25. $2x^2 + 4x - 3 = 0$
26. $x^2 - 8x + 19 = 0$
27. $-4x^2 - 6x + 3 = 0$
28. $-5x^2 - 15x + 10 = 0$
29. $8x^2 - 6x + 2 = 0$
30. $6x^2 + 3x - 9 = 0$
31. $-4x^2 + 10x + 12 = 0$
32. $-15x^2 - 10x + 25 = 0$
33. $9x^2 = 1 + 9x$
34. $7x^2 = 3 - 5x$
35. $2x - 3x^2 = 3 - 7x^2$
36. $x - x^2 = 1 - 6x^2$
37. $x^2 - 0.4x - 0.16 = 0$
38. $x^2 + 0.6x - 0.41 = 0$
39. $2.5x^2 + x - 0.9 = 0$
40. $0.09x^2 - 0.12x - 0.26 = 0$

In Exercises 41–48, use the discriminant to determine the type of solution(s) of the quadratic equation. ***See Example 5.***

✓ **41.** $x^2 + x + 1 = 0$

42. $x^2 + x - 1 = 0$

43. $3x^2 - 2x - 5 = 0$

44. $5x^2 + 7x + 3 = 0$

45. $9x^2 - 24x + 16 = 0$

46. $2x^2 + 10x + 6 = 0$

47. $3x^2 - x + 2 = 0$

48. $4x^2 - 16x + 16 = 0$

In Exercises 49–64, solve the quadratic equation by using the most convenient method. (Find all real *and* complex solutions.)

49. $z^2 - 169 = 0$ **50.** $t^2 = 144$

51. $5y^2 + 15y = 0$ **52.** $12u^2 + 30u = 0$

53. $25(x - 3)^2 - 36 = 0$

54. $9(x + 4)^2 + 16 = 0$

55. $2y(y - 18) + 3(y - 18) = 0$

56. $4y(y + 7) - 5(y + 7) = 0$

57. $x^2 + 8x + 25 = 0$

58. $y^2 + 21y + 108 = 0$

59. $3x^2 - 13x + 169 = 0$ **60.** $2x^2 - 15x + 225 = 0$

61. $25x^2 + 80x + 61 = 0$ **62.** $14x^2 + 11x - 40 = 0$

63. $7x(x + 2) + 5 = 3x(x + 1)$

64. $5x(x - 1) - 7 = 4x(x - 2)$

In Exercises 65–74, write a quadratic equation having the given solutions. ***See Example 6.***

✓ **65.** $5, -2$ **66.** $-2, 3$

67. $1, 7$ **68.** $2, 8$

69. $1 + \sqrt{2}, 1 - \sqrt{2}$

70. $-3 + \sqrt{5}, -3 - \sqrt{5}$

71. $5i, -5i$ **72.** $2i, -2i$

73. 12 **74.** -4

In Exercises 75–80, use a graphing calculator to graph the function. Use the graph to approximate any x-intercepts of the graph. Set $y = 0$ and solve the resulting equation. Compare the result with the x-intercepts of the graph.

75. $y = 3x^2 - 6x + 1$ **76.** $y = x^2 + x + 1$

77. $y = x^2 - 4x + 3$ **78.** $y = 5x^2 - 18x + 6$

79. $y = -0.03x^2 + 2x - 0.4$

80. $y = 3.7x^2 - 10.2x + 3.2$

In Exercises 81–84, use a graphing calculator to determine the number of real solutions of the quadratic equation. Verify your answer algebraically.

81. $2x^2 - 5x + 5 = 0$ **82.** $3x^2 - 7x - 6 = 0$

83. $\frac{1}{5}x^2 + \frac{6}{5}x - 8 = 0$ **84.** $\frac{1}{3}x^2 - 5x + 25 = 0$

In Exercises 85–88, determine all real values of x for which the function has the indicated value.

85. $f(x) = 2x^2 - 7x + 1, f(x) = -3$

86. $f(x) = 3x^2 - 7x + 4, f(x) = 0$

87. $g(x) = 2x^2 - 3x + 16, g(x) = 14$

88. $h(x) = 6x^2 + x + 10, h(x) = -2$

In Exercises 89–92, solve the equation.

89. $\frac{x^2}{4} - \frac{2x}{3} = 1$ **90.** $\frac{x^2 - 9x}{6} = \frac{x - 1}{2}$

91. $\sqrt{x + 3} = x - 1$ **92.** $\sqrt{2x - 3} = x - 2$

Think About It In Exercises 93 and 94, describe the values of c such that the equation has (a) two real number solutions, (b) one real number solution, and (c) two complex number solutions.

93. $x^2 - 6x + c = 0$

94. $x^2 + 2x + c = 0$

Solving Problems

95. *Geometry* A rectangle has a width of x inches, a length of $x + 6.3$ inches, and an area of 58.14 square inches. Find its dimensions.

96. *Geometry* A rectangle has a length of $x + 1.5$ inches, a width of x inches, and an area of 18.36 square inches. Find its dimensions.

97. *Free-Falling Object* A stone is thrown vertically upward at a velocity of 40 feet per second from a bridge that is 50 feet above the level of the water (see figure). The height h (in feet) of the stone at time t (in seconds) after it is thrown is

$$h = -16t^2 + 40t + 50.$$

(a) Find the time when the stone is again 50 feet above the water.

(b) Find the time when the stone strikes the water.

(c) Does the stone reach a height of 80 feet? Use the determinant to justify your answer.

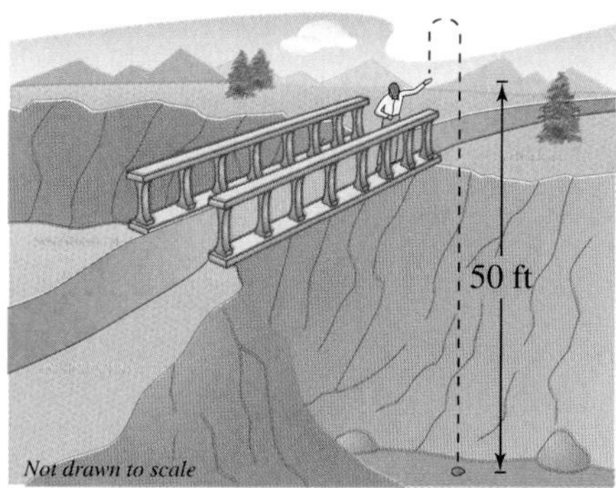

98. *Free-Falling Object* A stone is thrown vertically upward at a velocity of 20 feet per second from a bridge that is 40 feet above the level of the water. The height h (in feet) of the stone at time t (in seconds) after it is thrown is

$$h = -16t^2 + 20t + 40.$$

(a) Find the time when the stone is again 40 feet above the water.

(b) Find the time when the stone strikes the water.

(c) Does the stone reach a height of 50 feet? Use the determinant to justify your answer.

99. *Depth of a River* The depth d (in feet) of a river is given by

$$d = -0.25t^2 + 1.7t + 3.5, \quad 0 \le t \le 7$$

where t is the time (in hours) after a heavy rain begins. When is the river 6 feet deep?

100. *Baseball* The path of a baseball after it has been hit is given by

$$h = -0.003x^2 + 1.19x + 5.2$$

where h is the height (in feet) of the baseball and x is the horizontal distance (in feet) of the ball from home plate. The ball hits the top of the outfield fence that is 10 feet high. How far is the outfield fence from home plate?

Paul Spinelli/MLB Photos via Getty Images

101. *Fuel Economy* The fuel economy y (in miles per gallon) of a car is given by

$$y = -0.013x^2 + 1.25x + 5.6,\ 5 \le x \le 75$$

where x is the speed (in miles per hour) of the car.

(a) Use a graphing calculator to graph the model.

(b) Use the graph in part (a) to find the speeds at which you can travel and have a fuel economy of 32 miles per gallon. Verify your results algebraically.

102. *Cellular Phone Subscribers* The number s (in millions) of cellular phone subscribers in the United States for the years 1997 through 2006 can be modeled by

$$s = 0.510t^2 + 7.74t - 23.5,\ 7 \le t \le 16$$

where $t = 7$ corresponds to 1997.

(Source: Cellular Telecommunications & Internet Association)

(a) Use a graphing calculator to graph the model.

(b) Use the graph in part (a) to determine the year in which there were 145 million cellular phone subscribers. Verify your answer algebraically.

103. *Exploration* Determine the two solutions, x_1 and x_2, of each quadratic equation. Use the values of x_1 and x_2 to fill in the boxes.

Equation	x_1, x_2	$x_1 + x_2$	x_1x_2
(a) $x^2 - x - 6 = 0$			
(b) $2x^2 + 5x - 3 = 0$			
(c) $4x^2 - 9 = 0$			
(d) $x^2 - 10x + 34 = 0$			

104. *Think About It* Consider a general quadratic equation $ax^2 + bx + c = 0$ whose solutions are x_1 and x_2. Use the results of Exercise 103 to determine a relationship among the coefficients a, b, and c, and the sum $(x_1 + x_2)$ and product (x_1x_2) of the solutions.

Explaining Concepts

In Exercises 105–108, which method of solving the quadratic equation would be most convenient? Explain your reasoning.

105. $(x - 3)^2 = 25$

106. $x^2 + 8x - 12 = 0$

107. $2x^2 - 9x + 12 = 0$

108. $8x^2 - 40x = 0$

109. Explain how the discriminant of $ax^2 + bx + c = 0$ is related to the number of x-intercepts of the graph of $y = ax^2 + bx + c$.

110. *Error Analysis* Describe and correct the student's error in writing a quadratic equation that has solutions $x = 2$ and $x = 4$.

$(x + 2)(x + 4) = 0$

$x^2 + 6x + 8 = 0$

Cumulative Review

In Exercises 111–114, use the Distance Formula to determine whether the three points are collinear.

111. $(-1, 11), (2, 2), (1, 5)$

112. $(-2, 4), (3, -3), (5, -1)$

113. $(-6, -2), (-3, -4), (3, -4)$

114. $(-4, 7), (0, 4), (8, -2)$

In Exercises 115–118, sketch the graph of the function.

115. $f(x) = (x - 1)^2$

116. $f(x) = \frac{1}{2}x^2$

117. $f(x) = (x - 2)^2 + 4$

118. $f(x) = (x + 3)^2 - 1$

Mid-Chapter Quiz

Take this quiz as you would take a quiz in class. After you are done, check your work against the answers in the back of the book.

In Exercises 1–8, solve the quadratic equation by the specified method.

1. Factoring:
$2x^2 - 72 = 0$

2. Factoring:
$2x^2 + 3x - 20 = 0$

3. Square Root Property:
$3x^2 = 36$

4. Square Root Property:
$(u - 3)^2 - 16 = 0$

5. Completing the square:
$m^2 + 7m + 2 = 0$

6. Completing the square:
$2y^2 + 6y - 5 = 0$

7. Quadratic Formula:
$x^2 + 4x - 6 = 0$

8. Quadratic Formula:
$6v^2 - 3v - 4 = 0$

In Exercises 9–16, solve the equation by using the most convenient method. (Find all real *and* complex solutions.)

9. $x^2 + 5x + 7 = 0$

10. $36 - (t - 4)^2 = 0$

11. $x(x - 10) + 3(x - 10) = 0$

12. $x(x - 3) = 10$

13. $4b^2 - 12b + 9 = 0$

14. $3m^2 + 10m + 5 = 0$

15. $x - 4\sqrt{x} - 21 = 0$

16. $x^4 + 7x^2 + 12 = 0$

In Exercises 17 and 18, solve the equation of quadratic form. (Find all real *and* complex solutions.)

17. $x - 4\sqrt{x} + 3 = 0$

18. $x^4 - 14x^2 + 24 = 0$

In Exercises 19 and 20, use a graphing calculator to graph the function. Use the graph to approximate any *x*-intercepts of the graph. Set $y = 0$ and solve the resulting equation. Compare the results with the *x*-intercepts of the graph.

19. $y = \frac{1}{2}x^2 - 3x - 1$

20. $y = x^2 + 0.45x - 4$

21. The revenue R from selling x handheld video games is given by

$R = x(180 - 1.5x).$

Find the number of handheld video games that must be sold to produce a revenue of \$5400.

22. A rectangle has a length of x meters, a width of $100 - x$ meters, and an area of 2275 square meters. Find its dimensions.

10.4 Graphs of Quadratic Functions

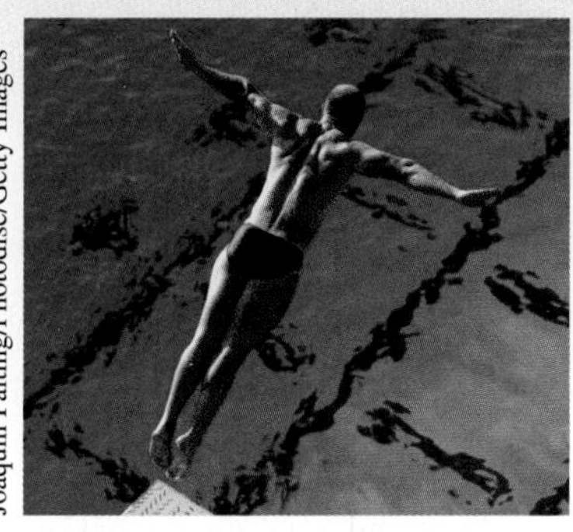
Joaquin Palting/Photodisc/Getty Images

Why You Should Learn It

Real-life situations can be modeled by graphs of quadratic functions. For instance, in Exercise 101 on page 653, a quadratic equation is used to model the maximum height of a diver.

What You Should Learn

1 ▶ Determine the vertices of parabolas by completing the square.

2 ▶ Sketch parabolas.

3 ▶ Write the equation of a parabola given the vertex and a point on the graph.

4 ▶ Use parabolas to solve application problems.

1 ▶ Determine the vertices of parabolas by completing the square.

Graphs of Quadratic Functions

In this section, you will study graphs of quadratic functions of the form

$$f(x) = ax^2 + bx + c. \qquad \text{Quadratic function}$$

Figure 10.2 shows the graph of a simple quadratic function, $f(x) = x^2$.

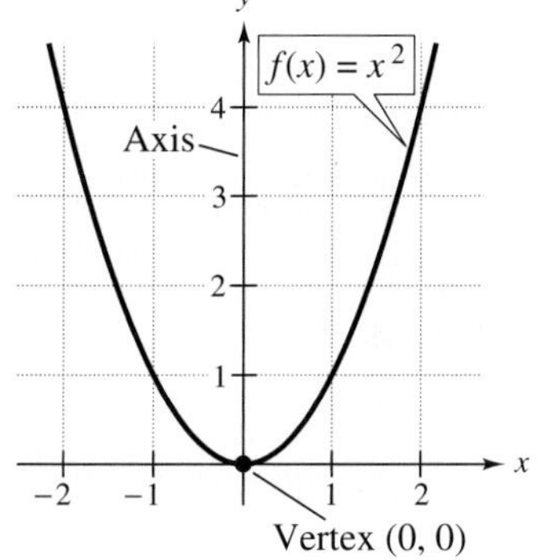

Figure 10.2

Graphs of Quadratic Functions

The graph of $f(x) = ax^2 + bx + c$, $a \neq 0$, is a **parabola.** The completed square form

$$f(x) = a(x - h)^2 + k \qquad \text{Standard form}$$

is the **standard form** of the function. The **vertex** of the parabola occurs at the point (h, k), and the vertical line passing through the vertex is the **axis** of the parabola.

Every parabola is *symmetric* about its axis, which means that if it were folded along its axis, the two parts would match.

If a is positive, the graph of $f(x) = ax^2 + bx + c$ opens upward, and if a is negative, the graph opens downward, as shown in Figure 10.3. Observe in Figure 10.3 that the y-coordinate of the vertex identifies the minimum function value if $a > 0$ and the maximum function value if $a < 0$.

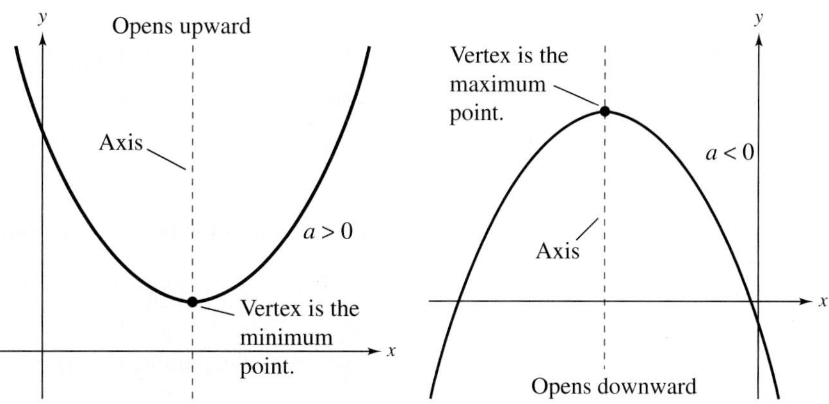

Figure 10.3

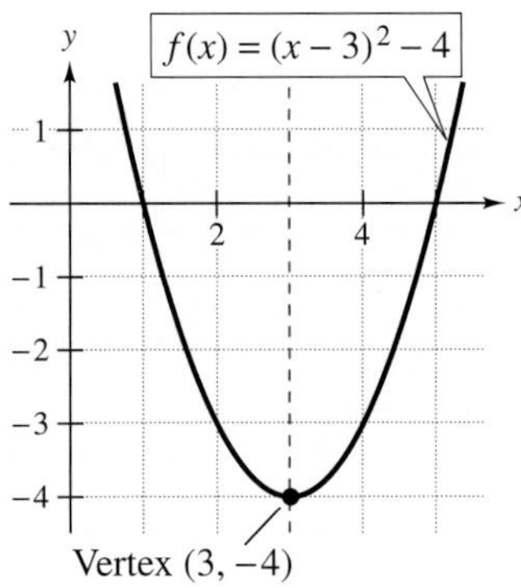

Figure 10.4

EXAMPLE 1 Finding the Vertex by Completing the Square

Find the vertex of the parabola given by $f(x) = x^2 - 6x + 5$.

Solution

Begin by writing the function in standard form.

$$f(x) = x^2 - 6x + 5 \qquad \text{Original function}$$

$$f(x) = x^2 - 6x + (-3)^2 - (-3)^2 + 5 \qquad \text{Complete the square.}$$

$$f(x) = (x^2 - 6x + 9) - 9 + 5 \qquad \text{Regroup terms.}$$

$$f(x) = (x - 3)^2 - 4 \qquad \text{Standard form}$$

From the standard form, you can see that the vertex of the parabola occurs at the point $(3, -4)$, as shown in Figure 10.4. The minimum value of the function is $f(3) = -4$.

✓ **CHECKPOINT** *Now try Exercise 9.*

Study Tip

When a number is added to a function and then that same number is subtracted from the function, the value of the function remains unchanged. Notice in Example 1 that $(-3)^2$ is added to the function to complete the square and then $(-3)^2$ is subtracted from the function so that the value of the function remains the same.

In Example 1, the vertex of the graph was found by *completing the square*. Another approach to finding the vertex is to complete the square once for a general function and then use the resulting formula to find the vertex.

$$f(x) = ax^2 + bx + c \qquad \text{Quadratic function}$$

$$= a\left(x^2 + \frac{b}{a}x\right) + c \qquad \text{Factor } a \text{ out of first two terms.}$$

$$= a\left[x^2 + \frac{b}{a}x + \left(\frac{b}{2a}\right)^2\right] + c - \left(\frac{b}{4a}\right)^2 \qquad \text{Complete the square.}$$

$$= a\left(x + \frac{b}{2a}\right)^2 + c - \frac{b^2}{4a} \qquad \text{Standard form}$$

From this form you can see that the vertex occurs when $x = -b/(2a)$.

EXAMPLE 2 Finding the Vertex Using a Formula

Find the vertex of the parabola given by $f(x) = 3x^2 - 9x$.

Solution

From the original function, it follows that $a = 3$ and $b = -9$. So, the x-coordinate of the vertex is

$$x = \frac{-b}{2a} = \frac{-(-9)}{2(3)} = \frac{3}{2}.$$

Substitute $\frac{3}{2}$ for x in the original equation to find the y-coordinate.

$$f\left(-\frac{b}{2a}\right) = f\left(\frac{3}{2}\right) = 3\left(\frac{3}{2}\right)^2 - 9\left(\frac{3}{2}\right) = -\frac{27}{4}$$

So, the vertex of the parabola is $\left(\frac{3}{2}, -\frac{27}{4}\right)$, the minimum value of the function is $f\left(\frac{3}{2}\right) = -\frac{27}{4}$, and the parabola opens upward, as shown in Figure 10.5.

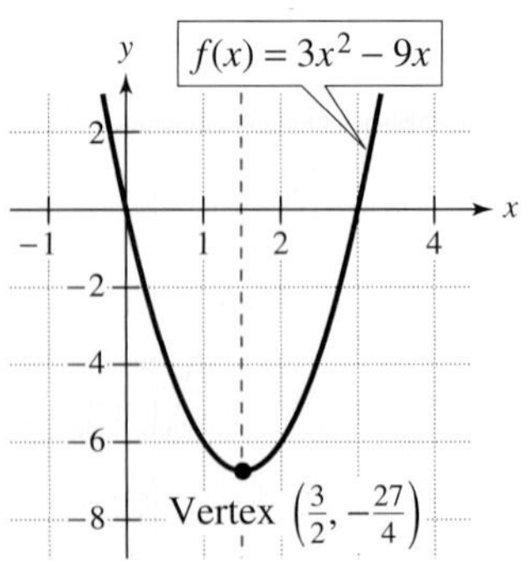

Figure 10.5

✓ **CHECKPOINT** *Now try Exercise 19.*

2 ▶ Sketch parabolas.

Sketching a Parabola

To obtain an accurate sketch of a parabola, the following guidelines are useful.

Sketching a Parabola

1. Determine the vertex and axis of the parabola by completing the square or by using the formula $x = -b/(2a)$.
2. Plot the vertex, axis, x- and y-intercepts, and a few additional points on the parabola. (Using the symmetry about the axis can reduce the number of points you need to plot.)
3. Use the fact that the parabola opens *upward* if $a > 0$ and opens *downward* if $a < 0$ to complete the sketch.

Study Tip

The x- and y-intercepts are useful points to plot. Another convenient fact is that the x-coordinate of the vertex lies halfway between the x-intercepts. Keep this in mind as you study the examples and do the exercises in this section.

EXAMPLE 3 Sketching a Parabola

To sketch the parabola given by $y = x^2 + 6x + 8$, begin by writing the equation in standard form.

$$y = x^2 + 6x + 8 \qquad \text{Write original equation.}$$

$$y = (x^2 + 6x + 3^2 - 3^2) + 8 \qquad \text{Complete the square.}$$

$$\left(\tfrac{6}{2}\right)^2$$

$$y = (x^2 + 6x + 9) - 9 + 8 \qquad \text{Regroup terms.}$$

$$y = (x + 3)^2 - 1 \qquad \text{Standard form}$$

The vertex occurs at the point $(-3, -1)$ and the axis is the line $x = -3$. After plotting this information, calculate a few additional points on the parabola, as shown in the table. Note that the y-intercept is $(0, 8)$ and the x-intercepts are solutions to the equation

$$x^2 + 6x + 8 = (x + 4)(x + 2) = 0.$$

x	-5	-4	-3	-2	-1
$y = (x + 3)^2 - 1$	3	0	-1	0	3
Solution point	$(-5, 3)$	$(-4, 0)$	$(-3, -1)$	$(-2, 0)$	$(-1, 3)$

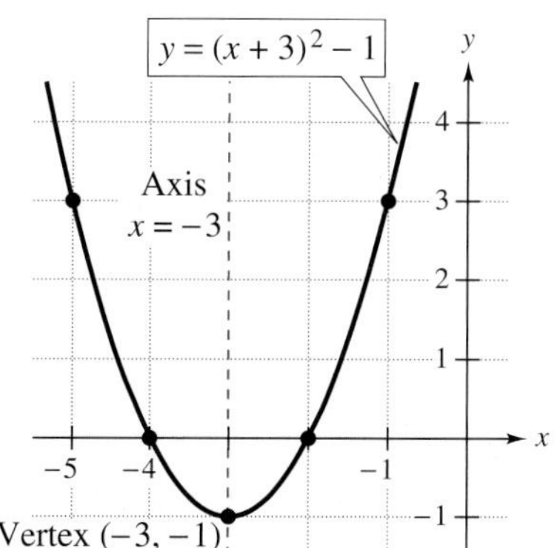

Figure 10.6

The graph of the parabola is shown in Figure 10.6. Note that the parabola opens upward because the leading coefficient (in general form) is positive.

✓ **CHECKPOINT** *Now try Exercise 47.*

3 ▶ Write the equation of a parabola given the vertex and a point on the graph.

Writing the Equation of a Parabola

To write the equation of a parabola with a vertical axis, use the fact that its standard equation has the form $y = a(x - h)^2 + k$, where (h, k) is the vertex.

EXAMPLE 4 Writing the Equation of a Parabola

Write the equation of the parabola with vertex $(-2, 1)$ and y-intercept $(0, -3)$, as shown in Figure 10.7.

Figure 10.7

Solution

Because the vertex occurs at $(h, k) = (-2, 1)$, the equation has the form

$y = a(x - h)^2 + k$	Standard form
$y = a[x - (-2)]^2 + 1$	Substitute -2 for h and 1 for k.
$y = a(x + 2)^2 + 1.$	Simplify.

To find the value of a, use the fact that the y-intercept is $(0, -3)$.

$y = a(x + 2)^2 + 1$	Write standard form.
$-3 = a(0 + 2)^2 + 1$	Substitute 0 for x and -3 for y.
$-1 = a$	Simplify.

So, the standard form of the equation of the parabola is $y = -(x + 2)^2 + 1$.

✓ **CHECKPOINT** *Now try Exercise 83.*

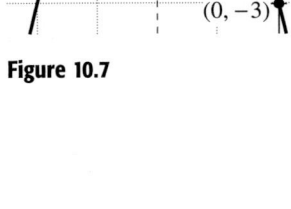

EXAMPLE 5 Writing the Equation of a Parabola

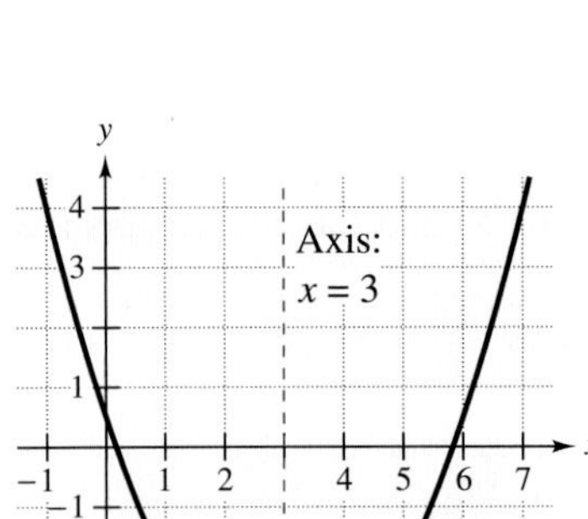

Figure 10.8

Write the equation of the parabola with vertex $(3, -4)$ and that passes through the point $(5, -2)$, as shown in Figure 10.8.

Solution

Because the vertex occurs at $(h, k) = (3, -4)$, the equation has the form

$y = a(x - h)^2 + k$	Standard form
$y = a(x - 3)^2 + (-4)$	Substitute 3 for h and -4 for k.
$y = a(x - 3)^2 - 4.$	Simplify.

To find the value of a, use the fact that the parabola passes through the point $(5, -2)$.

$y = a(x - 3)^2 - 4$	Write standard form.
$-2 = a(5 - 3)^2 - 4$	Substitute 5 for x and -2 for y.
$\frac{1}{2} = a$	Simplify.

So, the standard form of the equation of the parabola is $y = \frac{1}{2}(x - 3)^2 - 4$.

✓ **CHECKPOINT** *Now try Exercise 91.*

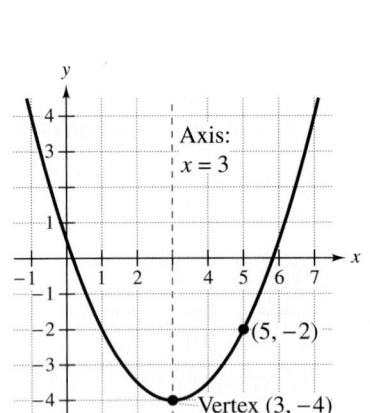

4 ▶ Use parabolas to solve application problems.

Application

EXAMPLE 6 Golden Gate Bridge

Each cable of the Golden Gate Bridge is suspended (in the shape of a parabola) between two towers that are 1280 meters apart. The top of each tower is 152 meters above the roadway. The cables touch the roadway at the midpoint between the towers. (See Figure 10.9.)

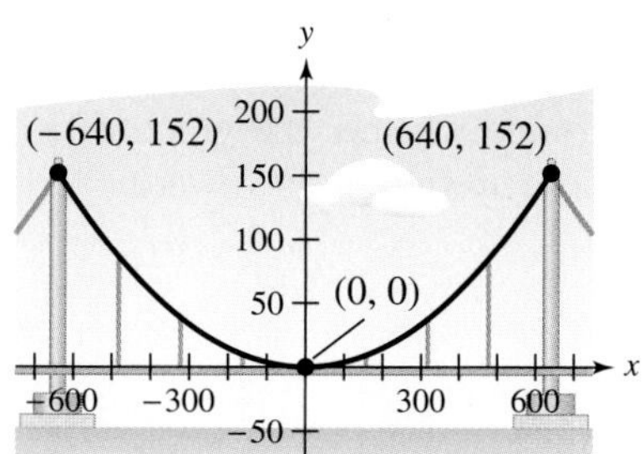

Figure 10.9

a. Write an equation that models the cables of the bridge.

b. Find the height of the suspension cables over the roadway at a distance of 320 meters from the center of the bridge.

Solution

a. From Figure 10.9, you can see that the vertex of the parabola occurs at (0, 0). So, the equation has the form

$y = a(x - h)^2 + k$ Standard form

$y = a(x - 0)^2 + 0$ Substitute 0 for h and 0 for k.

$y = ax^2.$ Simplify.

To find the value of a, use the fact that the parabola passes through the point (640, 152).

$y = ax^2$ Write standard form.

$152 = a(640)^2$ Substitute 640 for x and 152 for y.

$\frac{19}{51{,}200} = a$ Simplify.

So, an equation that models the cables of the bridge is

$$y = \frac{19}{51{,}200}x^2.$$

b. To find the height of the suspension cables over the roadway at a distance of 320 meters from the center of the bridge, evaluate the equation from part (a) for $x = 320$.

$y = \frac{19}{51{,}200}x^2$ Write original equation.

$y = \frac{19}{51{,}200}(320)^2$ Substitute 320 for x.

$y = 38$ Simplify.

So, the height of the suspension cables over the roadway is 38 meters.

CHECKPOINT *Now try Exercise 107.*

Concept Check

1. In your own words, describe the graph of the quadratic function $f(x) = ax^2 + bx + c$.

2. Explain how to find the vertex of the graph of a quadratic function.

3. Explain how to find any x- or y-intercepts of the graph of a quadratic function.

4. Explain how to determine whether the graph of a quadratic function opens upward or downward.

10.4 EXERCISES

Go to pages 676–677 to record your assignments.

Developing Skills

In Exercises 1–6, match the equation with its graph. [The graphs are labeled (a), (b), (c), (d), (e), and (f).]

(a)

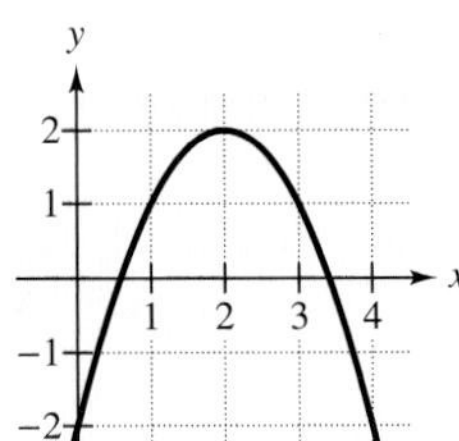

(b)

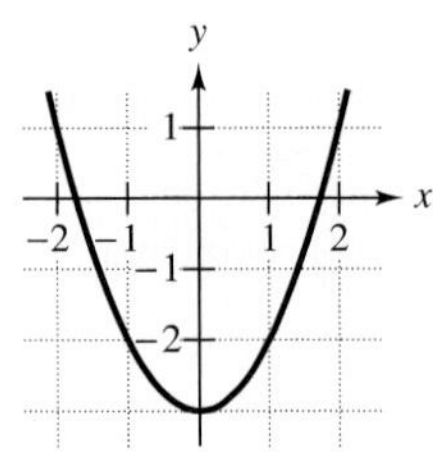

(c)

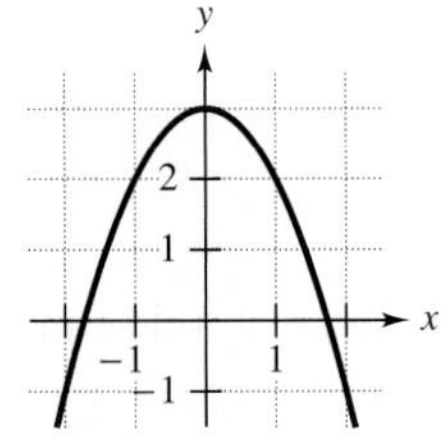

(d)

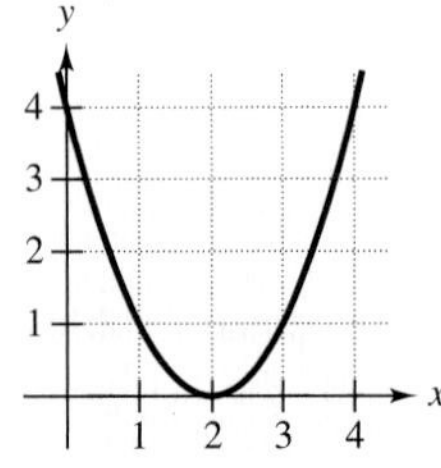

(e)

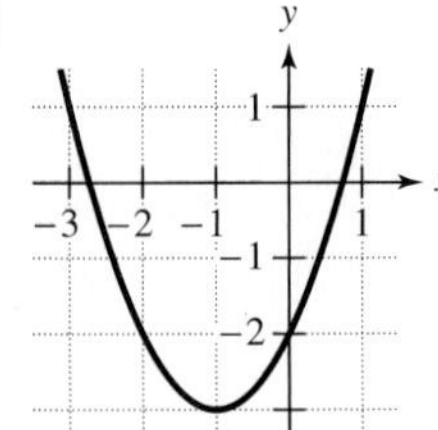
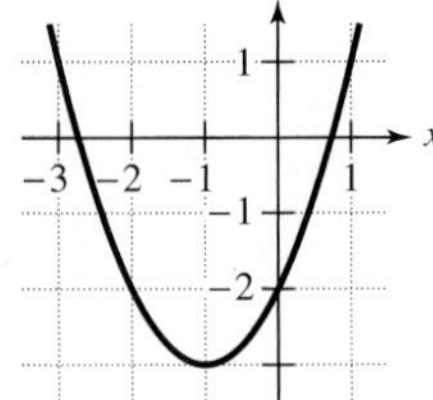

(f)

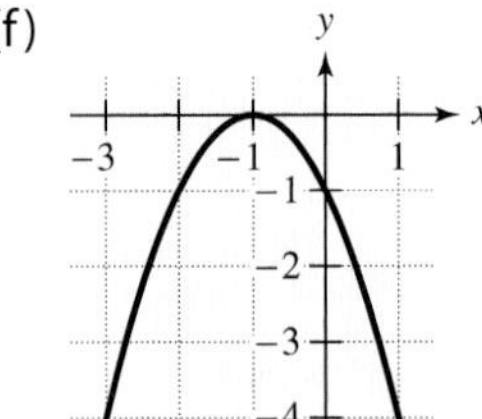

1. $y = (x + 1)^2 - 3$

2. $y = -(x + 1)^2$

3. $y = x^2 - 3$

4. $y = -x^2 + 3$

5. $y = (x - 2)^2$

6. $y = 2 - (x - 2)^2$

In Exercises 7–18, write the equation of the parabola in standard form and find the vertex of its graph. ***See Example 1.***

7. $y = x^2 - 2x$

8. $y = x^2 + 2x$

✓ **9.** $y = x^2 - 4x + 7$

10. $y = x^2 + 6x - 5$

11. $y = x^2 + 6x + 5$

12. $y = x^2 - 4x + 5$

13. $y = -x^2 + 6x - 10$

14. $y = -x^2 + 4x - 8$

15. $y = -x^2 - 8x + 5$

16. $y = -x^2 - 10x + 10$

17. $y = 2x^2 + 6x + 2$

18. $y = 3x^2 - 3x - 9$

In Exercises 19–24, find the vertex of the graph of the function by using the formula $x = -b/(2a)$. ***See Example 2.***

✓ **19.** $f(x) = x^2 - 8x + 15$

20. $f(x) = x^2 + 4x + 1$

21. $g(x) = -x^2 - 2x + 1$

22. $h(x) = -x^2 + 14x - 14$

23. $y = 4x^2 + 4x + 4$

24. $y = 9x^2 - 12x$

In Exercises 25–34, state whether the graph opens upward or downward, and find the vertex.

25. $y = 2(x - 0)^2 + 2$

26. $y = -3(x + 5)^2 - 3$

27. $y = 4 - (x - 10)^2$

28. $y = 2(x - 12)^2 + 3$

29. $y = x^2 - 6$

30. $y = -(x + 1)^2$

31. $y = -(x - 3)^2$

32. $y = x^2 - 6x$

33. $y = -x^2 + 6x$

34. $y = -x^2 - 5$

In Exercises 35–46, find the x- and y-intercepts of the graph.

35. $y = 25 - x^2$

36. $y = x^2 - 49$

37. $y = x^2 - 9x$

38. $y = x^2 + 4x$

39. $y = -x^2 - 6x + 7$

40. $y = -x^2 + 4x - 5$

41. $y = 4x^2 - 12x + 9$

42. $y = 10 - x - 2x^2$

43. $y = x^2 - 3x + 3$

44. $y = x^2 - 3x - 10$

45. $y = -2x^2 - 6x + 5$

46. $y = -4x^2 + 20x + 3$

In Exercises 47–70, sketch the parabola. Identify the vertex and any x-intercepts. Use a graphing calculator to verify your results. ***See Example 3.***

✓ **47.** $g(x) = x^2 - 4$

48. $h(x) = x^2 - 9$

49. $f(x) = -x^2 + 4$

50. $f(x) = -x^2 + 9$

51. $f(x) = x^2 - 3x$

52. $g(x) = x^2 - 4x$

53. $y = -x^2 + 3x$

54. $y = -x^2 + 4x$

55. $y = (x - 4)^2$

56. $y = -(x + 4)^2$

57. $y = x^2 - 9x - 18$

58. $y = x^2 + 4x + 2$

59. $y = -(x^2 + 6x + 5)$

60. $y = -x^2 + 2x + 8$

61. $q(x) = -x^2 + 6x - 7$

62. $g(x) = x^2 + 4x + 7$

63. $y = -2x^2 - 12x - 21$

64. $y = -3x^2 + 6x - 5$

65. $y = \frac{1}{2}(x^2 - 2x - 3)$

66. $y = -\frac{1}{2}(x^2 - 6x + 7)$

67. $y = \frac{1}{5}(3x^2 - 24x + 38)$

68. $y = \frac{1}{5}(2x^2 - 4x + 7)$

69. $f(x) = 5 - \frac{1}{3}x^2$

70. $f(x) = \frac{1}{3}x^2 - 2$

In Exercises 71–78, identify the transformation of the graph of $f(x) = x^2$, and sketch a graph of h.

71. $h(x) = x^2 - 1$

72. $h(x) = x^2 + 3$

73. $h(x) = (x + 2)^2$

74. $h(x) = (x - 4)^2$

75. $h(x) = -(x + 5)^2$

76. $h(x) = -x^2 - 6$

77. $h(x) = -(x - 2)^2 - 3$

78. $h(x) = -(x + 1)^2 + 5$

In Exercises 79–82, use a graphing calculator to approximate the vertex of the graph. Verify the result algebraically.

79. $y = \frac{1}{6}(2x^2 - 8x + 11)$

80. $y = -\frac{1}{4}(4x^2 - 20x + 13)$

81. $y = -0.7x^2 - 2.7x + 2.3$

82. $y = 0.6x^2 + 4.8x + 10.4$

In Exercises 83–86, write an equation of the parabola. *See Example 4.*

83.

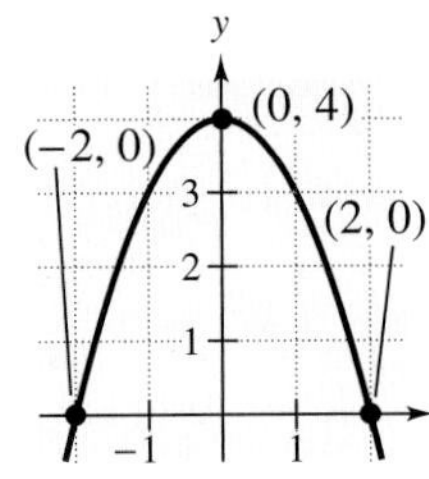

84.

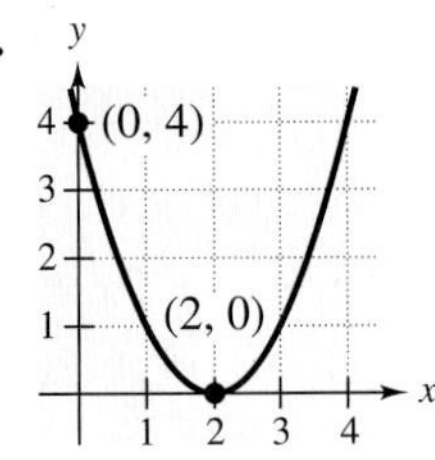

85.

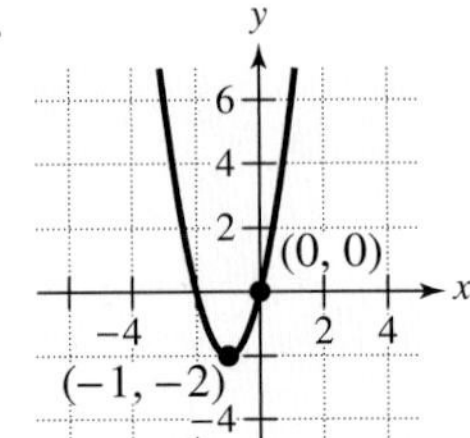

86.

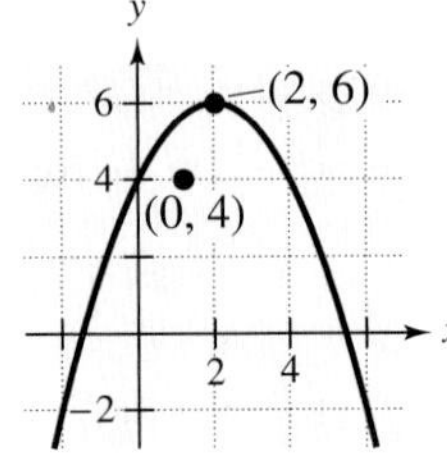

In Exercises 87–94, write an equation of the parabola $y = a(x - h)^2 + k$ that satisfies the given conditions. *See Example 5.*

87. Vertex: $(2, 1)$; $a = 1$

88. Vertex: $(-3, -3)$; $a = 1$

89. Vertex: $(2, -4)$; Point on the graph: $(0, 0)$

90. Vertex: $(-2, -4)$; Point on the graph: $(0, 0)$

91. Vertex: $(-2, -1)$; Point on the graph: $(1, 8)$

92. Vertex: $(4, 2)$; Point on the graph: $(2, -4)$

93. Vertex: $(-1, 1)$; Point on the graph: $(-4, 7)$

94. Vertex: $(5, 2)$; Point on the graph: $(10, 3)$

Solving Problems

95. *Path of a Ball* The height y (in feet) of a ball thrown by a child is given by

$$y = -\tfrac{1}{12}x^2 + 2x + 4$$

where x is the horizontal distance (in feet) from where the ball is thrown.

(a) How high is the ball when it leaves the child's hand?

(b) How high is the ball when it reaches its maximum height?

(c) How far from the child does the ball strike the ground?

96. *Path of a Ball* Repeat Exercise 95 if the path of the ball is modeled by

$$y = -\tfrac{1}{20}x^2 + 2x + 5.$$

97. *Path of an Object* A child launches a toy rocket from a table. The height y (in feet) of the rocket is given by

$$y = -\tfrac{1}{5}x^2 + 6x + 3$$

where x is the horizontal distance (in feet) from where the rocket is launched.

(a) Determine the height from which the rocket is launched.

(b) How high is the rocket at its maximum height?

(c) How far from where it is launched does the rocket land?

98. *Path of an Object* You use a fishing rod to cast a lure into the water. The height y (in feet) of the lure is given by

$$y = -\tfrac{1}{90}x^2 + \tfrac{1}{5}x + 9$$

where x is the horizontal distance (in feet) from the point where the lure is released.

(a) Determine the height from which the lure is released.

(b) How high is the lure at its maximum height?

(c) How far from its release point does the lure land?

99. *Path of a Golf Ball* The height y (in yards) of a golf ball hit by a professional golfer is given by

$$y = -\frac{1}{480}x^2 + \frac{1}{2}x$$

where x is the horizontal distance (in yards) from where the ball is hit.

(a) How high is the ball when it is hit?

(b) How high is the ball at its maximum height?

(c) How far from where the ball is hit does it strike the ground?

100. *Path of a Softball* The height y (in feet) of a softball that you hit is given by

$$y = -\frac{1}{70}x^2 + 2x + 2$$

where x is the horizontal distance (in feet) from where you hit the ball.

(a) How high is the ball when you hit it?

(b) How high is the ball at its maximum height?

(c) How far from where you hit the ball does it strike the ground?

101. *Path of a Diver* The path of a diver is given by

$$y = -\frac{4}{9}x^2 + \frac{24}{9}x + 10$$

where y is the height in feet and x is the horizontal distance from the end of the diving board in feet. What is the maximum height of the diver?

102. *Path of a Diver* Repeat Exercise 101 if the path of the diver is modeled by

$$y = -\frac{4}{3}x^2 + \frac{10}{3}x + 10.$$

103. *Cost* The cost C of producing x units of a product is given by

$$C = 800 - 10x + \frac{1}{4}x^2, \quad 0 < x < 40.$$

Use a graphing calculator to graph this function and approximate the value of x at which C is minimum.

104. *Geometry* The area A of a rectangle is given by the function

$$A = \frac{2}{\pi}(80x - 2x^2), \quad 0 < x < 40$$

where x is the length of the base of the rectangle in feet. Use a graphing calculator to graph the function and to approximate the value of x when A is maximum.

105. *Graphical Estimation* A bridge is to be constructed over a gorge with the main supporting arch being a parabola. The equation of the parabola is

$$y = 6[80 - (x^2/2400)]$$

where x and y are measured in feet. Use a graphing calculator to graph the equation and approximate the maximum height of the arch (relative to its base). Verify the maximum height algebraically.

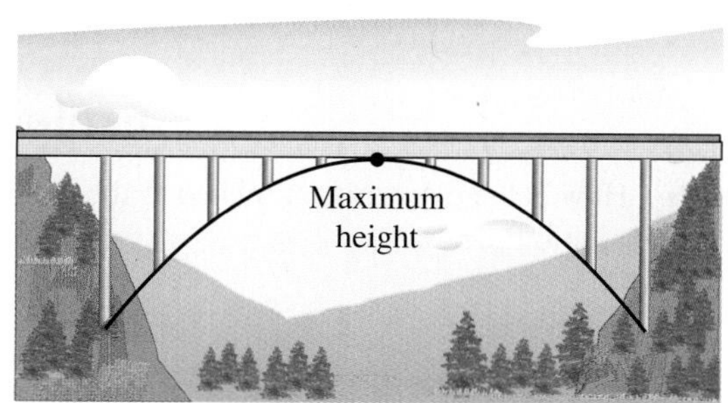

106. *Graphical Estimation* The profit P (in thousands of dollars) for a landscaping company is given by

$$P = 230 + 20s - \frac{1}{2}s^2$$

where s is the amount (in hundreds of dollars) spent on advertising. Use a graphing calculator to graph the profit function and approximate the amount of advertising that yields a maximum profit. Verify the maximum profit algebraically.

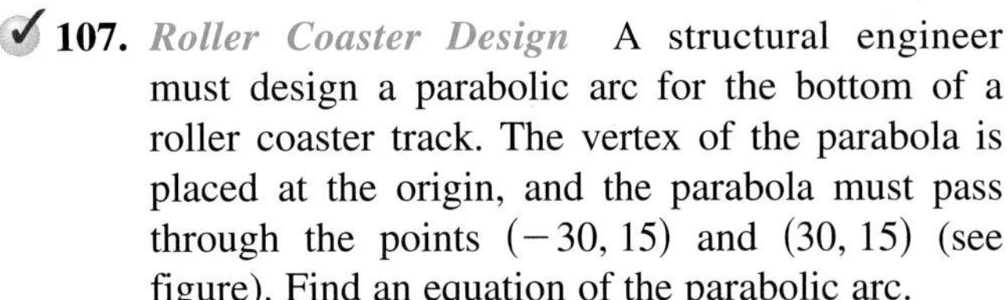

107. *Roller Coaster Design* A structural engineer must design a parabolic arc for the bottom of a roller coaster track. The vertex of the parabola is placed at the origin, and the parabola must pass through the points $(-30, 15)$ and $(30, 15)$ (see figure). Find an equation of the parabolic arc.

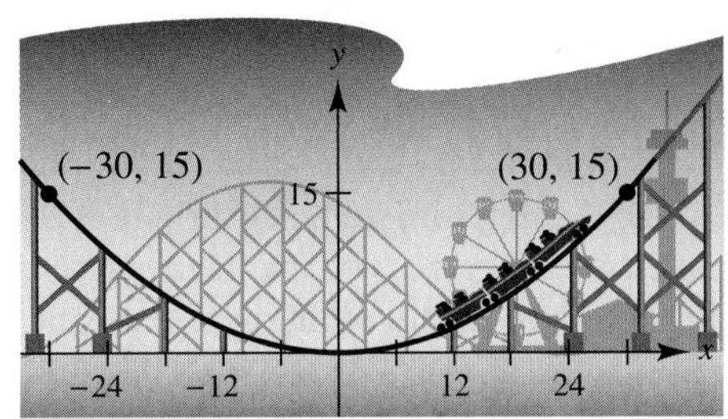

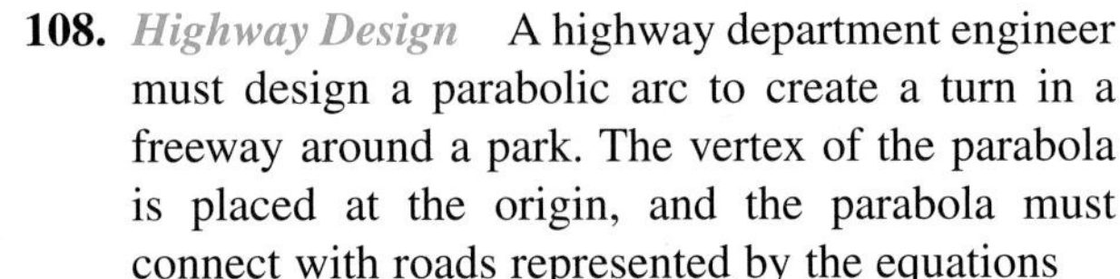

108. *Highway Design* A highway department engineer must design a parabolic arc to create a turn in a freeway around a park. The vertex of the parabola is placed at the origin, and the parabola must connect with roads represented by the equations

$$y = -0.4x - 100, \quad x < -500$$

and

$$y = 0.4x - 100, \quad x > 500$$

(see figure). Find an equation of the parabolic arc.

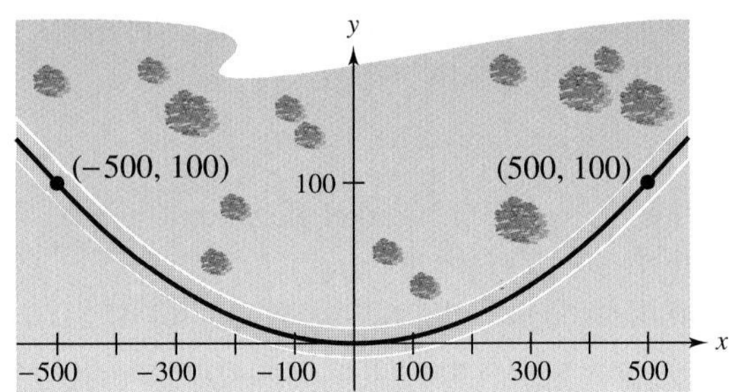

Explaining Concepts

109. How is the discriminant related to the graph of a quadratic function?

110. Is it possible for the graph of a quadratic function to have two y-intercepts? Explain.

111. Explain how to determine the maximum (or minimum) value of a quadratic function.

112. The domain of a quadratic function is the set of real numbers. Explain how to find the range.

Cumulative Review

In Exercises 113–120, find the slope-intercept form of the equation of the line through the two points.

113. $(0, 0), (4, -2)$

114. $(0, 0), (100, 75)$

115. $(-1, -2), (3, 6)$

116. $(1, 5), (6, 0)$

117. $\left(\frac{3}{2}, 8\right), \left(\frac{11}{2}, \frac{5}{2}\right)$

118. $(0, 2), (7.3, 15.4)$

119. $(0, 8), (5, 8)$

120. $(-3, 2), (-3, 5)$

In Exercises 121–124, write the number in i-form.

121. $\sqrt{-64}$

122. $\sqrt{-32}$

123. $\sqrt{-0.0081}$

124. $\sqrt{-\frac{20}{16}}$

10.5 Applications of Quadratic Equations

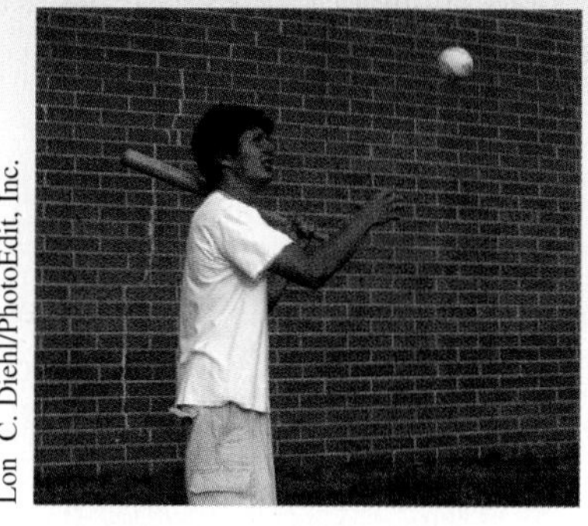

Lon C. Diehl/PhotoEdit, Inc.

Why You Should Learn It

Quadratic equations are used in a wide variety of real-life problems. For instance, in Exercise 41 on page 664, a quadratic equation is used to model the height of a baseball after being hit.

What You Should Learn

1 ▶ Use quadratic equations to solve application problems.

Applications of Quadratic Equations

1 ▶ Use quadratic equations to solve application problems.

EXAMPLE 1 An Investment Problem

A car dealer buys a fleet of cars from a car rental agency for a total of \$120,000. The dealer regains this \$120,000 investment by selling all but four of the cars at an average profit of \$2500 each. How many cars has the dealer sold, and what is the average price per car?

Solution

Although this problem is stated in terms of average price and average profit per car, you can use a model that assumes that each car has sold for the same price.

Verbal Model: Selling price per car = Cost per car + Profit per car

Labels:
Number of cars sold $= x$ (cars)
Number of cars bought $= x + 4$ (cars)
Selling price per car $= 120{,}000/x$ (dollars per car)
Cost per car $= 120{,}000/(x + 4)$ (dollars per car)
Profit per car $= 2500$ (dollars per car)

Equation:

$$\frac{120{,}000}{x} = \frac{120{,}000}{x + 4} + 2500$$

$$120{,}000(x + 4) = 120{,}000x + 2500x(x + 4),\ x \neq 0,\ x \neq -4$$

$$120{,}000x + 480{,}000 = 120{,}000x + 2500x^2 + 10{,}000x$$

$$0 = 2500x^2 + 10{,}000x - 480{,}000$$

$$0 = x^2 + 4x - 192$$

$$0 = (x - 12)(x + 16)$$

$$x - 12 = 0 \quad \Rightarrow \quad x = 12$$

$$x + 16 = 0 \quad \Rightarrow \quad x = -16$$

Choosing the positive value, it follows that the dealer sold 12 cars at an average price of $120{,}000/12 = \$10{,}000$ per car. Check this in the original statement.

✓ **CHECKPOINT** *Now try Exercise 1.*

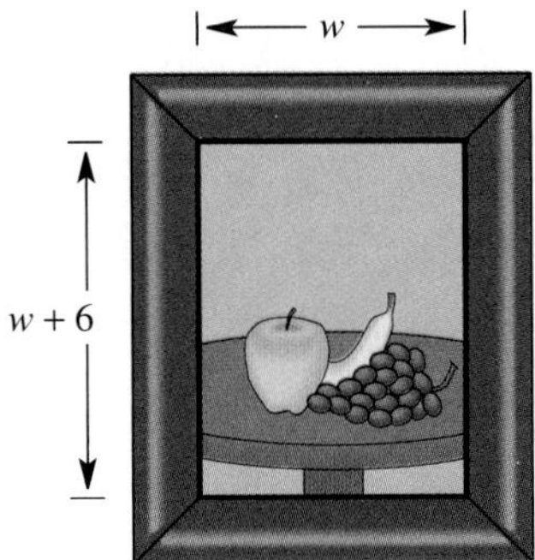

Figure 10.10

EXAMPLE 2 Geometry: Dimensions of a Picture

A picture is 6 inches taller than it is wide and has an area of 216 square inches, as shown in Figure 10.10. What are the dimensions of the picture?

Solution

Verbal Model: Area of picture = Width · Height

Labels:
Picture width $= w$ (inches)
Picture height $= w + 6$ (inches)
Area $= 216$ (square inches)

Equation:

$$216 = w(w + 6)$$

$$0 = w^2 + 6w - 216$$

$$0 = (w + 18)(w - 12)$$

$w + 18 = 0$  $w = -18$

$w - 12 = 0$ ➡ $w = 12$

Choosing the positive value of w, you can conclude that the picture is $w = 12$ inches wide and $w + 6 = 12 + 6 = 18$ inches tall. Check these solutions.

✓ **CHECKPOINT** *Now try Exercise 15.*

EXAMPLE 3 An Interest Problem

The formula $A = P(1 + r)^2$ represents the amount of money A in an account in which P dollars is deposited for 2 years at an annual interest rate of r (in decimal form). Find the interest rate if a deposit of \$6000 increases to \$6933.75 over a two-year period.

Solution

$A = P(1 + r)^2$ Write given formula.

$6933.75 = 6000(1 + r)^2$ Substitute 6933.75 for A and 6000 for P.

$1.155625 = (1 + r)^2$ Divide each side by 6000.

$\pm 1.075 = 1 + r$ Square Root Property

$0.075 = r$ Choose positive solution.

The annual interest rate is $r = 0.075 = 7.5\%$.

Check

$A = P(1 + r)^2$ Write given formula.

$6933.75 \stackrel{?}{=} 6000(1 + 0.075)^2$ Substitute 6933.75 for A, 6000 for P, and 0.075 for r.

$6933.75 \stackrel{?}{=} 6000(1.155625)$ Simplify.

$6933.75 = 6933.75$ Solution checks. ✓

✓ **CHECKPOINT** *Now try Exercise 23.*

EXAMPLE 4 Reduced Rates 

A ski club charters a bus for a ski trip at a cost of \$720. When four nonmembers accept invitations from the club to go on the trip, the bus fare per skier decreases by \$6. How many club members are going on the trip?

Solution

Verbal Model: Fare per skier · Number of skiers = \$720

Labels:
Number of ski club members $= x$ (people)
Number of skiers $= x + 4$ (people)
Original fare per skier $= \dfrac{720}{x}$ (dollars per person)
New fare per skier $= \dfrac{720}{x} - 6$ (dollars per person)

Equation:

$$\left(\frac{720}{x} - 6\right)(x + 4) = 720$$ Original equation

$$\left(\frac{720 - 6x}{x}\right)(x + 4) = 720$$ Rewrite 1st factor.

$$(720 - 6x)(x + 4) = 720x, \ x \neq 0$$ Multiply each side by x.

$$720x + 2880 - 6x^2 - 24x = 720x$$ Multiply factors.

$$-6x^2 - 24x + 2880 = 0$$ Subtract $720x$ from each side.

$$x^2 + 4x - 480 = 0$$ Divide each side by -6.

$$(x + 24)(x - 20) = 0$$ Factor left side of equation.

$$x + 24 = 0 \Rightarrow x = -24$$ Set 1st factor equal to 0.

$$x - 20 = 0 \Rightarrow x = 20$$ Set 2nd factor equal to 0.

Choosing the positive value of x, you can conclude that 20 ski club members are going on the trip. Check this solution in the original statement of the problem, as follows.

Check

Original fare per skier for 20 ski club members:

$$\frac{720}{x} = \frac{720}{20} = \$36$$ Substitute 20 for x.

New fare per skier with 4 nonmembers:

$$\frac{720}{x + 4} = \frac{720}{24} = \$30$$

Decrease in fare per skier with 4 nonmembers:

$$36 - 30 = \$6$$ Solution checks. ✓

CHECKPOINT *Now try Exercise 29.*

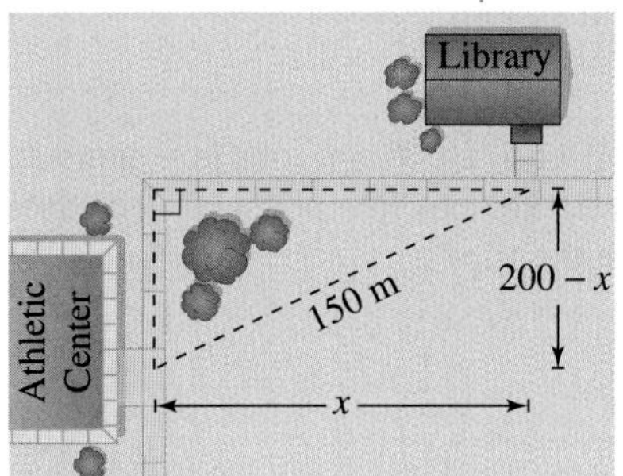

Figure 10.11

EXAMPLE 5 An Application Involving the Pythagorean Theorem

An L-shaped sidewalk from the athletic center to the library on a college campus is 200 meters long, as shown in Figure 10.11. By cutting diagonally across the grass, students shorten the walking distance to 150 meters. What are the lengths of the two legs of the sidewalk?

Solution

Common Formula: $a^2 + b^2 = c^2$ Pythagorean Theorem

Labels: Length of one leg $= x$ (meters)
Length of other leg $= 200 - x$ (meters)
Length of diagonal $= 150$ (meters)

Equation:

$$x^2 + (200 - x)^2 = 150^2$$

$$x^2 + 40{,}000 - 400x + x^2 = 22{,}500$$

$$2x^2 - 400x + 40{,}000 = 22{,}500$$

$$2x^2 - 400x + 17{,}500 = 0$$

$$x^2 - 200x + 8750 = 0$$

Using the Quadratic Formula, you can find the solutions as follows.

$$x = \frac{-(-200) \pm \sqrt{(-200)^2 - 4(1)(8750)}}{2(1)}$$ Substitute 1 for a, -200 for b, and 8750 for c.

$$= \frac{200 \pm \sqrt{5000}}{2}$$

$$= \frac{200 \pm 50\sqrt{2}}{2}$$

$$= \frac{\cancel{2}(100 \pm 25\sqrt{2})}{\cancel{2}}$$

$$= 100 \pm 25\sqrt{2}$$

Both solutions are positive, so it does not matter which you choose. If you let

$$x = 100 + 25\sqrt{2} \approx 135.4 \text{ meters}$$

the length of the other leg is

$$200 - x \approx 200 - 135.4 \approx 64.6 \text{ meters.}$$

✓ **CHECKPOINT** *Now try Exercise 31.*

In Example 5, notice that you obtain the same dimensions if you choose the other value of x. That is, if you let

$$x = 100 - 25\sqrt{2} \approx 64.6 \text{ meters}$$

the length of the other leg is

$$200 - x \approx 200 - 64.6 \approx 135.4 \text{ meters.}$$

EXAMPLE 6 Work-Rate Problem 

An office contains two copy machines. Machine B is known to take 12 minutes longer than machine A to copy the company's monthly report. Using both machines together, it takes 8 minutes to reproduce the report. How long would it take each machine alone to reproduce the report?

Solution

Verbal Model:

[Work done by machine A] + [Work done by machine B] = [1 complete job]

[Rate for A] · [Time for both] + [Rate for B] · [Time for both] = [1]

Labels:

Time for machine A $= t$ (minutes)

Rate for machine A $= \dfrac{1}{t}$ (job per minute)

Time for machine B $= t + 12$ (minutes)

Rate for machine B $= \dfrac{1}{t+12}$ (job per minute)

Time for both machines $= 8$ (minutes)

Rate for both machines $= \dfrac{1}{8}$ (job per minute)

Equation:

$$\frac{1}{t}(8) + \frac{1}{t+12}(8) = 1$$ Original equation

$$8\left(\frac{1}{t} + \frac{1}{t+12}\right) = 1$$ Distributive Property

$$8\left[\frac{t + 12 + t}{t(t+12)}\right] = 1$$ Rewrite with common denominator.

$$8t(t+12)\left[\frac{2t+12}{t(t+12)}\right] = t(t+12)$$ Multiply each side by $t(t + 12)$.

$$8(2t + 12) = t^2 + 12t$$ Simplify.

$$16t + 96 = t^2 + 12t$$ Distributive Property

$$0 = t^2 - 4t - 96$$ Subtract $16t + 96$ from each side.

$$0 = (t - 12)(t + 8)$$ Factor right side of equation.

$t - 12 = 0 \Rightarrow t = 12$ Set 1st factor equal to 0.

$t + 8 = 0 \Rightarrow t = -8$ Set 2nd factor equal to 0.

By choosing the positive value for t, you can conclude that the times for the two machines are

Time for machine A $= t = 12$ minutes

Time for machine B $= t + 12 = 12 + 12 = 24$ minutes.

Check these solutions in the original statement of the problem.

CHECKPOINT *Now try Exercise 35.*

EXAMPLE 7 The Height of a Model Rocket 

A model rocket is projected straight upward from ground level according to the height equation

$$h = -16t^2 + 192t,\ t \geq 0$$

where h is the height in feet and t is the time in seconds.

a. After how many seconds is the height 432 feet?

b. After how many seconds does the rocket hit the ground?

c. What is the maximum height of the rocket?

Solution

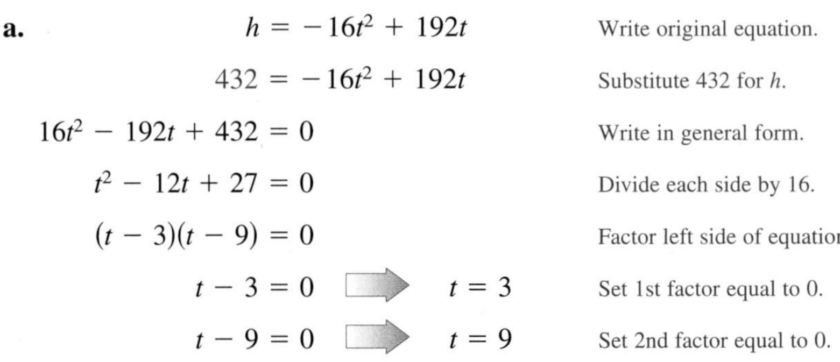

a.

$h = -16t^2 + 192t$ Write original equation.

$432 = -16t^2 + 192t$ Substitute 432 for h.

$16t^2 - 192t + 432 = 0$ Write in general form.

$t^2 - 12t + 27 = 0$ Divide each side by 16.

$(t - 3)(t - 9) = 0$ Factor left side of equation.

$t - 3 = 0 \Rightarrow t = 3$ Set 1st factor equal to 0.

$t - 9 = 0 \Rightarrow t = 9$ Set 2nd factor equal to 0.

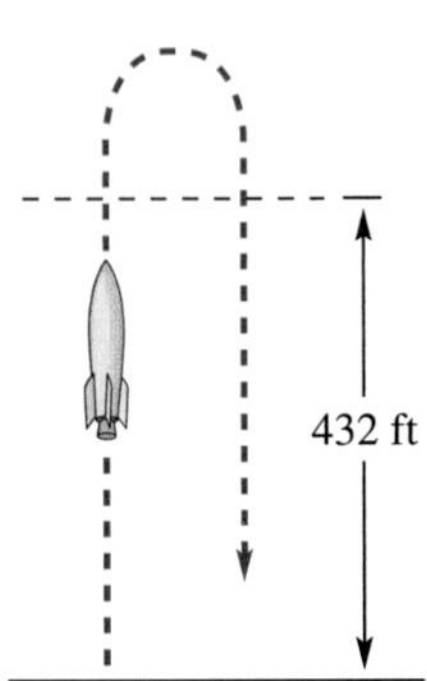

Figure 10.12

The rocket attains a height of 432 feet at two different times—once (going up) after 3 seconds, and again (coming down) after 9 seconds. (See Figure 10.12.)

b. To find the time it takes for the rocket to hit the ground, let the height be 0.

$0 = -16t^2 + 192t$ Substitute 0 for h in original equation.

$0 = t^2 - 12t$ Divide each side by -16.

$0 = t(t - 12)$ Factor right side of equation.

$t = 0$ or $t = 12$ Solutions

The rocket hits the ground after 12 seconds. (Note that the time of $t = 0$ seconds corresponds to the time of lift-off.)

c. The maximum value of h in the equation $h = -16t^2 + 192t$ occurs when $t = -\dfrac{b}{2a}$. So, the t-coordinate is

$$t = \frac{-b}{2a} = \frac{-192}{2(-16)} = 6$$

and the h-coordinate is

$$h = -16(6)^2 + 192(6) = 576.$$

So, the maximum height of the rocket is 576 feet.

 ***CHECKPOINT*** *Now try Exercise 41.*

Concept Check

In Questions 1–4, a problem situation is given. Describe two quantities that can be set equal to each other so as to write an equation that can be used to solve the problem.

1. You know the length of the hypotenuse, and the sum of the lengths of the legs, of a right triangle. You want to find the lengths of the legs.

2. You know the area of a rectangle and you know how many units longer the length is than the width. You want to find the length and width.

3. You know the amount invested in an unknown number of product units. You know the number of units remaining when the investment is regained, and the profit per unit sold. You want to find the number of units sold and the price per unit.

4. You know the time in minutes for two machines to complete a task together and you know how many more minutes it takes one machine than the other to complete the task alone. You want to find the time to complete the task alone for each machine.

10.5 EXERCISES

Go to pages 676–677 to record your assignments.

Solving Problems

1. *Selling Price* A store owner bought a case of eggs for $21.60. By the time all but 6 dozen of the eggs had been sold at a profit of $0.30 per dozen, the original investment of $21.60 had been regained. How many dozen eggs did the owner sell, and what was the selling price per dozen? ***See Example 1.***

2. *Selling Price* A computer store manager buys several computers of the same model for $12,600. The store can regain this $12,600 investment by selling all but four of the computers at a profit of $360 per computer. To do this, how many computers must be sold, and at what price?

3. *Selling Price* A flea market vendor buys a box of DVD movies for $50. After selling several of the DVDs at a profit of $3 each, the vendor still has 15 of the DVDs left by the time she regains her $50 investment. How many DVDs has the vendor sold, and at what price?

4. *Selling Price* A sorority buys a case of sweatshirts for $750 to sell at a mixer. The sorority needs to sell all but 20 of the sweatshirts at a profit of $10 per sweatshirt to regain the $750 investment. How many sweatshirts must be sold, and at what price, to do this?

Geometry In Exercises 5–14, complete the table of widths, lengths, perimeters, and areas of rectangles.

	Width	*Length*	*Perimeter*	*Area*
5.	$1.4l$	l	54 in.	
6.	w	$3.5w$	60 m	
7.	w	$2.5w$		250 ft^2
8.	w	$1.5w$		216 cm^2
9.	$\frac{1}{3}l$	l		192 in.^2
10.	$\frac{3}{4}l$	l		2700 in.^2
11.	w	$w + 3$	54 km	
12.	$l - 6$	l	108 ft	
13.	$l - 20$	l		12,000 m^2
14.	w	$w + 5$		500 ft^2

15. *Geometry* A picture frame is 4 inches taller than it is wide and has an area of 192 square inches. What are the dimensions of the picture frame? ***See Example 2.***

16. *Geometry* The height of a triangle is 8 inches less than its base. The area of the triangle is 192 square inches. Find the dimensions of the triangle.

17. *Storage Area* A retail lumberyard plans to store lumber in a rectangular region adjoining the sales office (see figure). The region will be fenced on three sides, and the fourth side will be bounded by the wall of the office building. There is 350 feet of fencing available, and the area of the region is 12,500 square feet. Find the dimensions of the region.

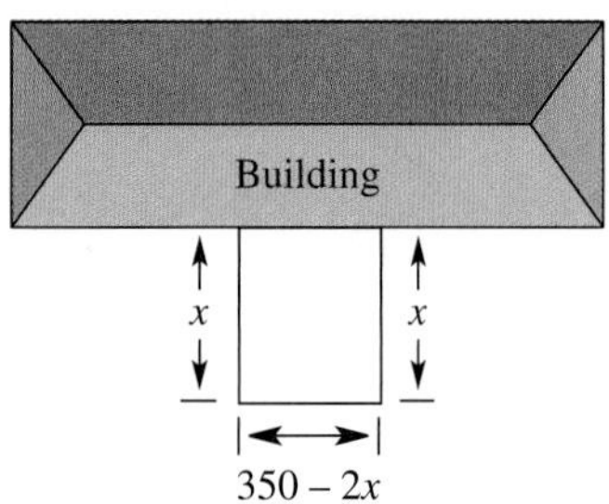

18. *Geometry* Your home is on a square lot. To add more space to your yard, you purchase an additional 20 feet along one side of the property (see figure). The area of the lot is now 25,500 square feet. What are the dimensions of the new lot?

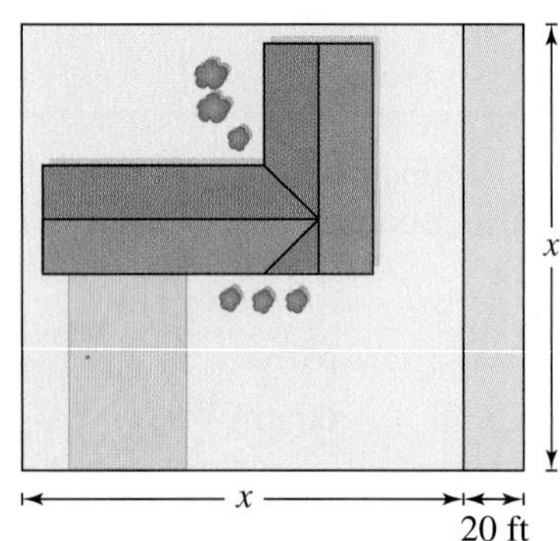

19. *Fenced Area* A family built a fence around three sides of their property (see figure). In total, they used 550 feet of fencing. By their calculations, the lot is 1 acre (43,560 square feet). Is this correct? Explain your reasoning.

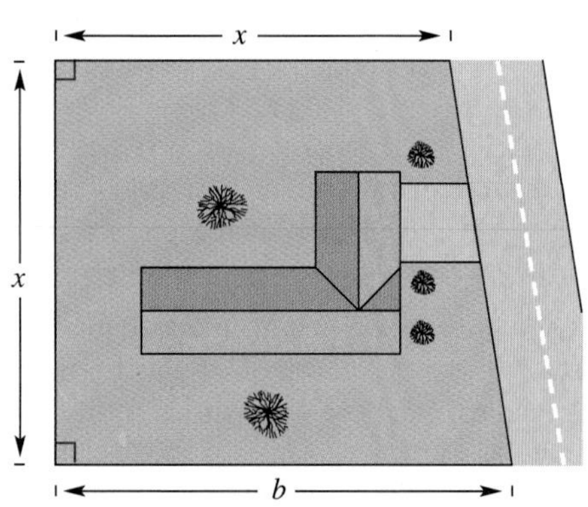

20. *Fenced Area* You have 100 feet of fencing. Do you have enough to enclose a rectangular region whose area is 630 square feet? Is there enough to enclose a circular area of 630 square feet? Explain.

21. *Open Conduit* An open-topped rectangular conduit for carrying water in a manufacturing process is made by folding up the edges of a sheet of aluminum 48 inches wide (see figure). A cross section of the conduit must have an area of 288 square inches. Find the width and height of the conduit.

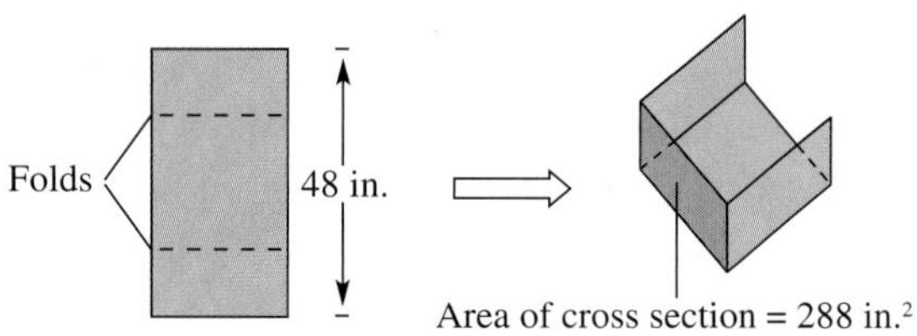

22. *Photography* A photographer has a photograph that is 6 inches by 8 inches. The photographer wants to crop the photo down to half of its original area by trimming equal lengths from each side. How many inches should be trimmed from each side?

Compound Interest In Exercises 23–28, find the interest rate r. Use the formula $A = P(1 + r)^2$, where A is the amount after 2 years in an account earning r percent (in decimal form) compounded annually, and P is the original investment. ***See Example 3.***

23. $P = \$10{,}000$, $A = \$11{,}990.25$

24. $P = \$3000$, $A = \$3499.20$

25. $P = \$500$, $A = \$572.45$

26. $P = \$250$, $A = \$280.90$

27. $P = \$6500$, $A = \$7372.46$

28. $P = \$8000$, $A = \$8421.41$

29. *Reduced Rates* A service organization pays \$210 for a block of tickets to a baseball game. The block contains three more tickets than the organization needs for its members. By inviting three more people to attend (and share in the cost), the organization lowers the price per person by \$3.50. How many people are going to the game? ***See Example 4.***

30. *Reduced Fares* A science club charters a bus to attend a science fair at a cost of \$480. To lower the bus fare per person, the club invites nonmembers to go along. When two nonmembers join the trip, the fare per person is decreased by \$1. How many people are going on the excursion?

31. *Delivery Route* You deliver pizzas to an insurance office and an apartment complex (see figure). Your total mileage in driving to the insurance office and then to the apartment complex is 12 miles. By using a direct route, you are able to drive just 9 miles to return to the pizza shop. Estimate the distance from the pizza shop to the insurance office. ***See Example 5.***

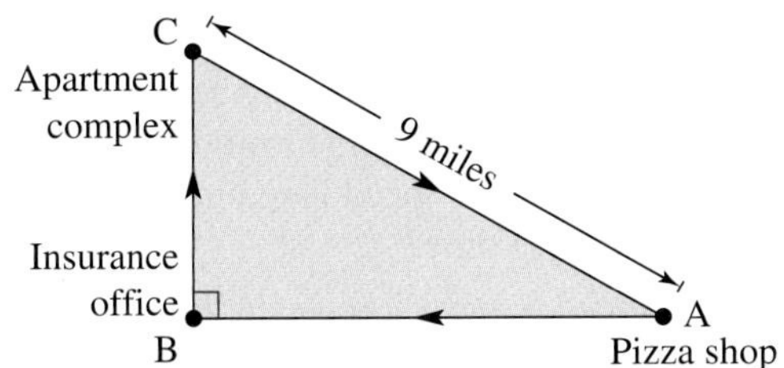

32. *Geometry* An L-shaped sidewalk from the library (point A) to the gym (point B) on a high school campus is 100 yards long, as shown in the figure. By cutting diagonally across the grass, students shorten the walking distance to 80 yards. What are the lengths of the two legs of the sidewalk?

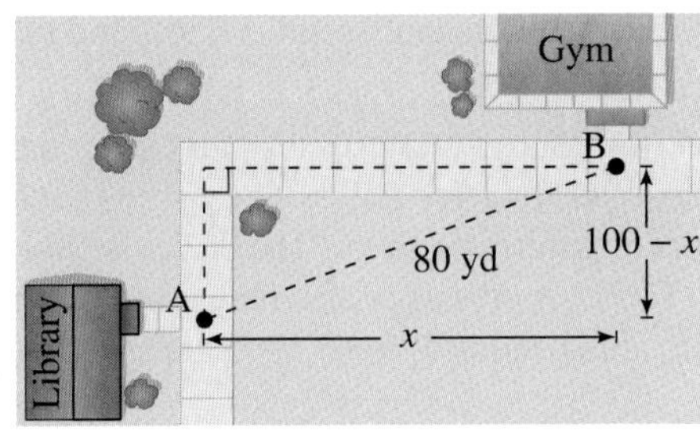

33. *Solving Graphically and Numerically* A meteorologist is positioned 100 feet from the point where a weather balloon is launched (see figure).

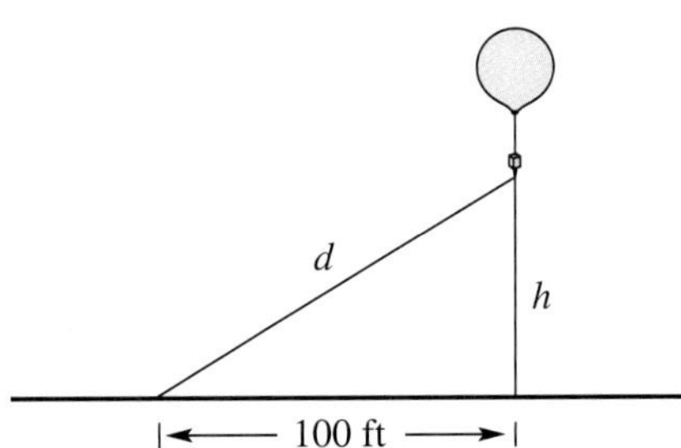

(a) Write an equation relating the distance d between the balloon and the meteorologist to the height h of the balloon.

(b) Use a graphing calculator to graph the equation.

(c) Use the graph to approximate the value of h when $d = 200$ feet.

(d) Complete the table.

h	0	100	200	300
d				

34. *Geometry* An adjustable rectangular form has minimum dimensions of 3 meters by 4 meters. The length and width can be expanded by equal amounts x (see figure).

(a) Write an equation relating the length d of the diagonal to x.

(b) Use a graphing calculator to graph the equation.

(c) Use the graph to approximate the value of x when $d = 10$ meters.

(d) Find x algebraically when $d = 10$.

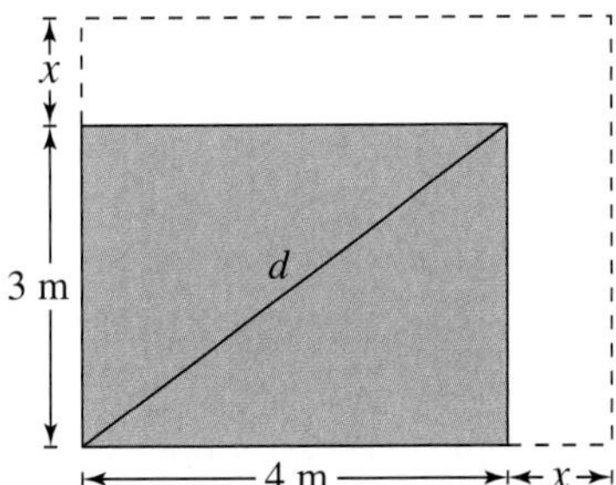

35. *Work Rate* An office contains two printers. Machine B is known to take 3 minutes longer than machine A to produce the company's monthly financial report. Using both machines together, it takes 6 minutes to produce the report. How long would it take each machine to produce the report? ***See Example 6.***

36. *Work Rate* A builder works with two plumbing companies. Company A is known to take 3 days longer than Company B to install the plumbing in a particular style of house. Using both companies, it takes 4 days. How long would it take to install the plumbing using each company individually?

Free-Falling Object In Exercises 37–40, find the time necessary for an object to fall to ground level from an initial height of h_0 feet if its height h at any time t (in seconds) is given by $h = h_0 - 16t^2$.

37. $h_0 = 169$ **38.** $h_0 = 729$

39. $h_0 = 1454$ (height of the Sears Tower)

40. $h_0 = 984$ (height of the Eiffel Tower)

41. *Height* The height h in feet of a baseball t seconds after being hit at a point 3 feet above the ground is given by $h = 3 + 75t - 16t^2$. Find the time when the ball hits the ground. ***See Example 7.***

42. *Height* You are hitting baseballs. When you toss the ball into the air, your hand is 5 feet above the ground (see figure). You hit the ball when it falls back to a height of 3.5 feet. You toss the ball with an initial velocity of 18 feet per second. The height h of the ball t seconds after leaving your hand is given by $h = 5 + 18t - 16t^2$. About how much time passes before you hit the ball?

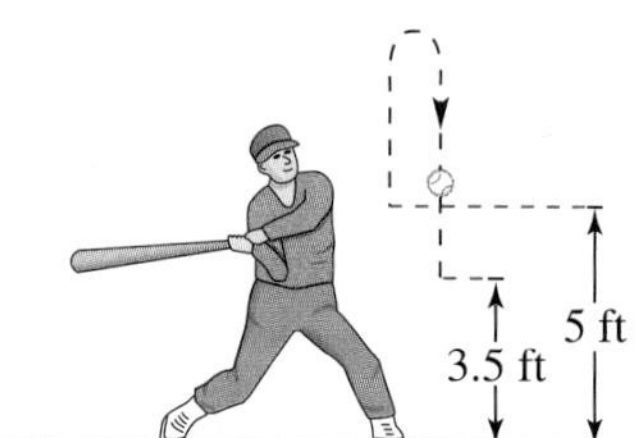

43. *Height* A model rocket is projected straight upward from ground level according to the height equation $h = -16t^2 + 160t$, where h is the height of the rocket in feet and t is the time in seconds.
    (a) After how many seconds is the height 336 feet?
    (b) After how many seconds does the rocket hit the ground?
    (c) What is the maximum height of the rocket?

44. *Height* A tennis ball is tossed vertically upward from a height of 5 feet according to the height equation $h = -16t^2 + 21t + 5$, where h is the height of the tennis ball in feet and t is the time in seconds.
    (a) After how many seconds is the height 11 feet?
    (b) After how many seconds does the tennis ball hit the ground?
    (c) What is the maximum height of the ball?

Number Problems In Exercises 45–50, find two positive integers that satisfy the requirement.

45. The product of two consecutive integers is 182.

46. The product of two consecutive integers is 1806.

47. The product of two consecutive even integers is 168.

48. The product of two consecutive even integers is 2808.

49. The product of two consecutive odd integers is 323.

50. The product of two consecutive odd integers is 1443.

51. *Air Speed* An airline runs a commuter flight between two cities that are 720 miles apart. If the average speed of the planes could be increased by 40 miles per hour, the travel time would be decreased by 12 minutes. What air speed is required to obtain this decrease in travel time?

52. *Average Speed* A truck traveled the first 100 miles of a trip at one speed and the last 135 miles at an average speed of 5 miles per hour less. The entire trip took 5 hours. What was the average speed for the first part of the trip?

53. *Speed* A company uses a pickup truck for deliveries. The cost per hour for fuel is $C = v^2/300$, where v is the speed in miles per hour. The driver is paid \$15 per hour. The cost of wages and fuel for an 80-mile trip at constant speed is \$36. Find the speed.

54. *Speed* A hobby shop uses a small car for deliveries. The cost per hour for fuel is $C = v^2/600$, where v is the speed in miles per hour. The driver is paid \$10 per hour. The cost of wages and fuel for a 110-mile trip at constant speed is \$29.32. Find the speed.

55. *Distance* Find any points on the line $y = 9$ that are 10 units from the point $(2, 3)$.

56. *Distance* Find any points on the line $y = 14$ that are 13 units from the point $(1, 2)$.

57. *Geometry* The area of an ellipse is given by $A = \pi ab$ (see figure on the next page). For a certain ellipse, it is required that $a + b = 20$.
    (a) Show that $A = \pi a(20 - a)$.
    (b) Complete the table.

a	4	7	10	13	16
A					

    (c) Find two values of a such that $A = 300$.

(d) Use a graphing calculator to graph the area equation. Then use the graph to verify the results in part (c).

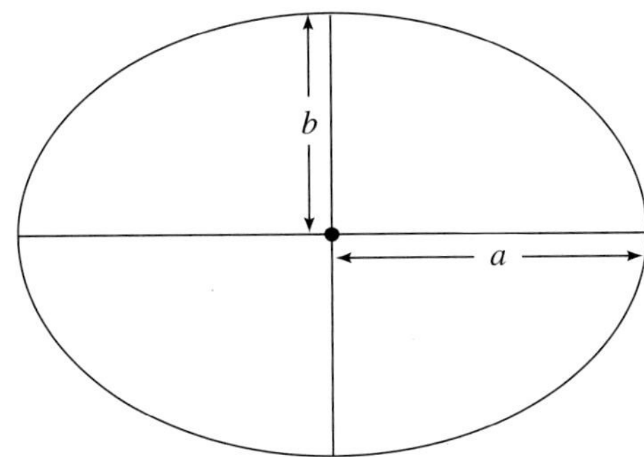

Figure for 57

58. *Data Analysis* For the years 2000 through 2006, the reported numbers of boating accidents A in the United States can be approximated by

$$A = 118.52t^2 - 1140.4t + 7615, \quad 0 \le t \le 6$$

where t is the year, with $t = 0$ corresponding to 2000. (Source: U.S. Coast Guard)

(a) Approximate the numbers of boating accidents in the years 2000 and 2006.

(b) During which year from 2000 to 2006 were there approximately 5800 boating accidents?

(c) Use the Internet to find the Coast Guard's data for boating accidents in the years since 2006. Then use a graphing calculator to graph the model for A given in the problem. Discuss the behavior of the graph for $t > 6$. Use your data and the graph to discuss the appropriateness of the model for making predictions after 2006.

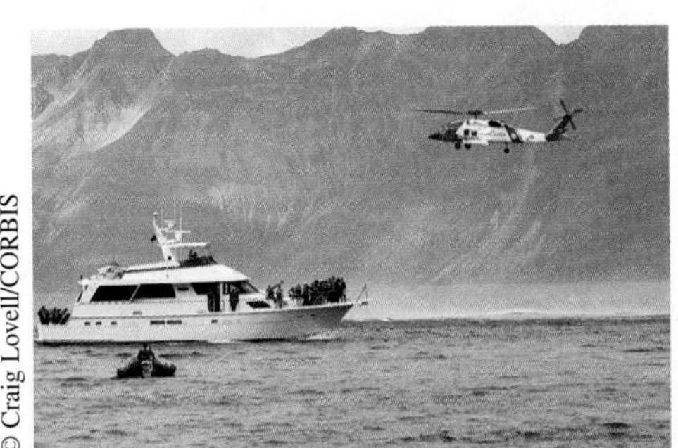

© Craig Lovell/CORBIS

Explaining Concepts

59. To solve some of the problems in this section, you wrote rational equations. Explain why these types of problems are included as applications of quadratic equations.

60. In each of Exercises 5–14, finding the area or perimeter of a rectangle is involved. The solution requires writing an equation that can be solved for the length or width of the rectangle. Explain how you can tell when this equation will be a *quadratic equation* or a *linear equation*.

61. In a *reduced rates* problem such as Example 4, does the cost per person decrease by the same amount for each additional person? Explain.

62. In a *height of an object* problem such as Example 7, suppose you try solving the height equation using a height greater than the maximum height reached by the object. What type of result will you get for t? Explain.

Cumulative Review

In Exercises 63 and 64, solve the inequality and sketch the solution on the real number line.

63. $5 - 3x > 17$

64. $-3 < 2x + 3 < 5$

In Exercises 65 and 66, solve the equation by completing the square.

65. $x^2 - 8x = 0$

66. $x^2 - 2x - 2 = 0$

10.6 Quadratic and Rational Inequalities

Will Hart/PhotoEdit, Inc.

What You Should Learn

1 ▶ Determine test intervals for polynomials.

2 ▶ Use test intervals to solve quadratic inequalities.

3 ▶ Use test intervals to solve rational inequalities.

4 ▶ Use inequalities to solve application problems.

Why You Should Learn It

Rational inequalities can be used to model and solve real-life problems. For instance, in Exercise 120 on page 675, a rational inequality is used to model the temperature of a metal in a laboratory experiment.

1 ▶ Determine test intervals for polynomials.

Finding Test Intervals

When working with polynomial inequalities, it is important to realize that the value of a polynomial can change signs only at its **zeros.** That is, a polynomial can change signs only at the x-values for which the value of the polynomial is zero. For instance, the first-degree polynomial $x + 2$ has a zero at $x = -2$, and it changes signs at that zero. You can picture this result on the real number line, as shown in Figure 10.13.

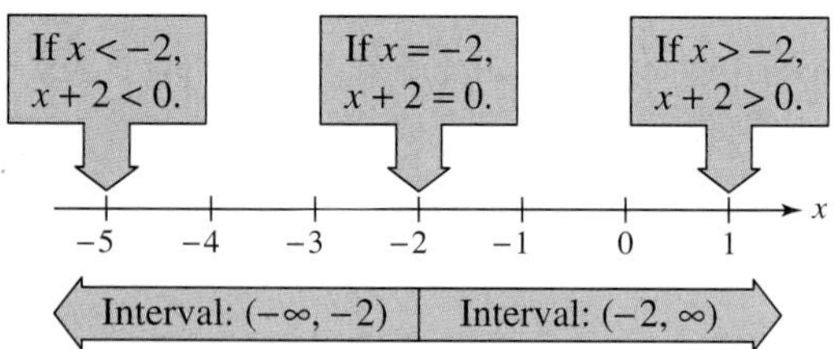

Figure 10.13

Note in Figure 10.13 that the zero of the polynomial partitions the real number line into two **test intervals.** The value of the polynomial is negative for every x-value in the first test interval $(-\infty, -2)$, and positive for every x-value in the second test interval $(-2, \infty)$. You can use the same basic approach to determine the test intervals for any polynomial.

Finding Test Intervals for a Polynomial

1. Find all real zeros of the polynomial, and arrange the zeros in increasing order. The zeros of a polynomial are called its **critical numbers.**
2. Use the critical numbers of the polynomial to determine its test intervals.
3. Choose a representative x-value in each test interval and evaluate the polynomial at that value. If the value of the polynomial is negative, the polynomial will have negative values for *every* x-value in the interval. If the value of the polynomial is positive, the polynomial will have positive values for *every* x-value in the interval.

2 ▶ Use test intervals to solve quadratic inequalities.

Quadratic Inequalities

The concepts of critical numbers and test intervals can be used to solve nonlinear inequalities, as demonstrated in Examples 1, 2, and 4.

Technology: Tip

Most graphing calculators can graph the solution set of a quadratic inequality. Consult the user's guide of your graphing calculator for specific instructions. Notice that the solution set for the quadratic inequality

$$x^2 - 5x < 0$$

shown below appears as a horizontal line above the x-axis.

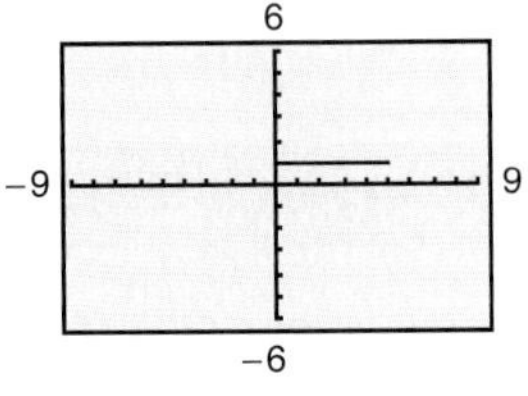

EXAMPLE 1 Solving a Quadratic Inequality

Solve the inequality $x^2 - 5x < 0$.

Solution

First find the *critical numbers* of $x^2 - 5x < 0$ by finding the solutions of the equation $x^2 - 5x = 0$.

$x^2 - 5x = 0$ Write corresponding equation.

$x(x - 5) = 0$ Factor.

$x = 0, x = 5$ Critical numbers

This implies that the test intervals are $(-\infty, 0)$, $(0, 5)$, and $(5, \infty)$. To test an interval, choose a convenient value in the interval and determine if the value satisfies the inequality.

Test interval	Representative x-value	Is inequality satisfied?
$(-\infty, 0)$	$x = -1$	$(-1)^2 - 5(-1) \stackrel{?}{<} 0$ $6 \not< 0$
$(0, 5)$	$x = 1$	$1^2 - 5(1) \stackrel{?}{<} 0$ $-4 < 0$
$(5, \infty)$	$x = 6$	$6^2 - 5(6) \stackrel{?}{<} 0$ $6 \not< 0$

Because the inequality $x^2 - 5x < 0$ is satisfied only by the value $x = 1$ (the value in the middle test interval), you can conclude that the solution set of the inequality is the interval $(0, 5)$, as shown in Figure 10.14.

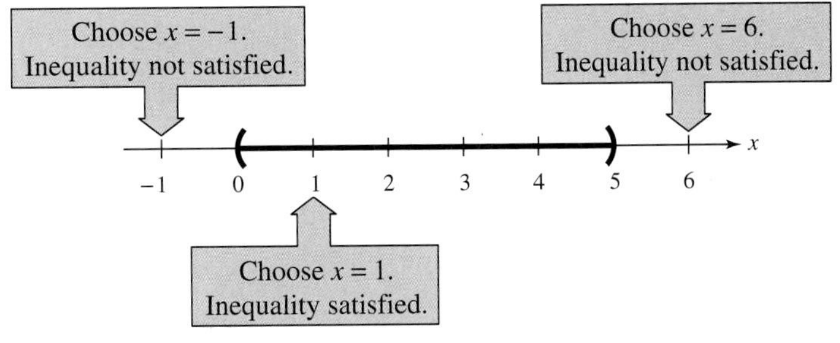

Figure 10.14

✓ **CHECKPOINT** *Now try Exercise 21.*

Study Tip

In Example 1, note that you would have used the same basic procedure if the inequality symbol had been $\leq$, $>$, or $\geq$. For instance, in Figure 10.14, you can see that the solution set of the inequality $x^2 - 5x \geq 0$ consists of the union of the half-open intervals $(-\infty, 0]$ and $[5, \infty)$, which is written as $(-\infty, 0] \cup [5, \infty)$.

Just as in solving quadratic *equations*, the first step in solving a quadratic *inequality* is to write the inequality in **general form,** with the polynomial on the left and zero on the right, as demonstrated in Example 2.

EXAMPLE 2 Solving a Quadratic Inequality

Solve the inequality $2x^2 + 5x \geq 12$.

Solution

Begin by writing the inequality in the general form $2x^2 + 5x - 12 \geq 0$. Next, find the critical numbers by finding the solutions of the equation $2x^2 + 5x - 12 = 0$.

$$2x^2 + 5x - 12 = 0$$ Write corresponding equation.

$$(x + 4)(2x - 3) = 0$$ Factor.

$$x = -4,\ x = \frac{3}{2}$$ Critical numbers

This implies that the test intervals are $(-\infty, -4)$, $\left(-4, \frac{3}{2}\right)$, and $\left(\frac{3}{2}, \infty\right)$. To test an interval, choose a convenient value in the interval and determine if the value satisfies the inequality.

Test interval	Representative x-value	Is inequality satisfied?
$(-\infty, -4)$	$x = -5$	$2(-5)^2 + 5(-5) \overset{?}{\geq} 12$ $25 \geq 12$
$\left(-4, \frac{3}{2}\right)$	$x = 0$	$2(0)^2 + 5(0) \overset{?}{\geq} 12$ $0 \not\geq 12$
$\left(\frac{3}{2}, \infty\right)$	$x = 2$	$2(2)^2 + 5(2) \overset{?}{\geq} 12$ $18 \geq 12$

From this table you can see that the inequality $2x^2 + 5x \geq 12$ is satisfied by x-values in the intervals $(-\infty, -4)$ and $\left(\frac{3}{2}, \infty\right)$. The critical numbers -4 and $\frac{3}{2}$ are also solutions to the inequality. So, the solution set of the inequality is $(-\infty, -4] \cup \left[\frac{3}{2}, \infty\right)$, as shown in Figure 10.15.

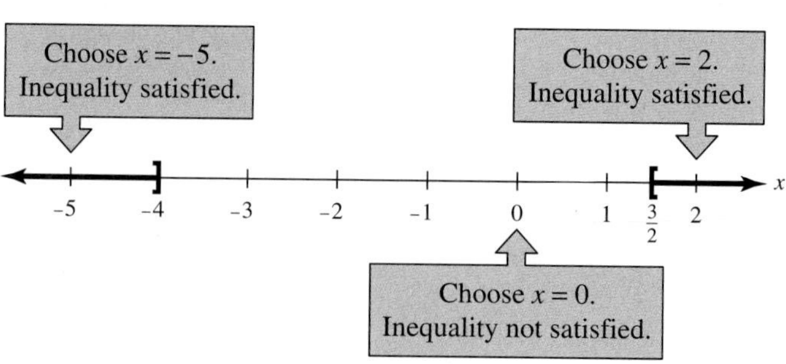

Figure 10.15

✓ **CHECKPOINT** *Now try Exercise 31.*

Study Tip

In Examples 1 and 2, the critical numbers are found by factoring. With quadratic polynomials that do not factor, you can use the Quadratic Formula to find the critical numbers. For instance, to solve the inequality

$$x^2 - 2x - 1 \leq 0$$

you can use the Quadratic Formula to determine that the critical numbers are

$$1 - \sqrt{2} \approx -0.414$$

and

$$1 + \sqrt{2} \approx 2.414.$$

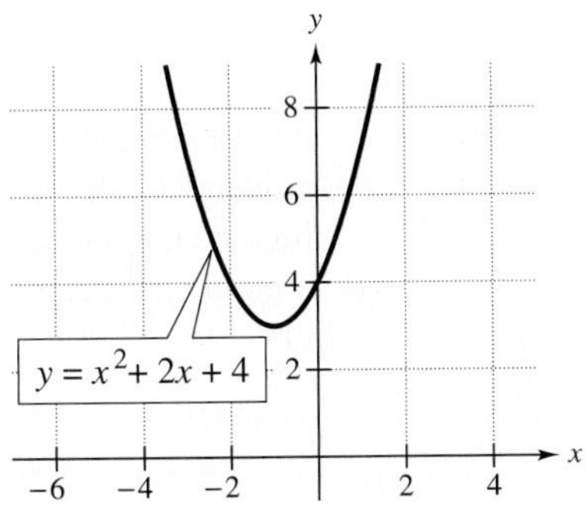

Figure 10.16

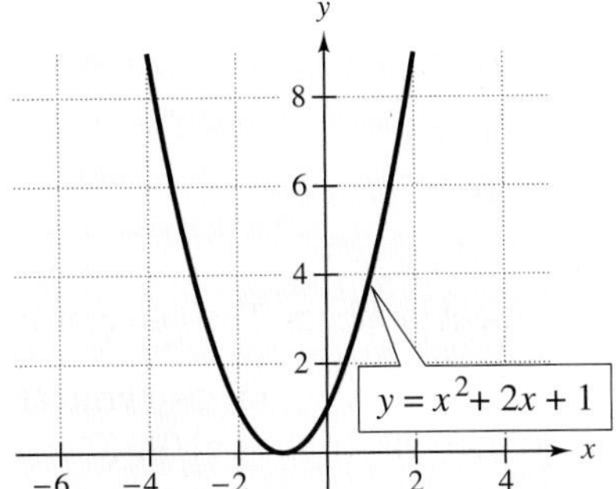

Figure 10.17

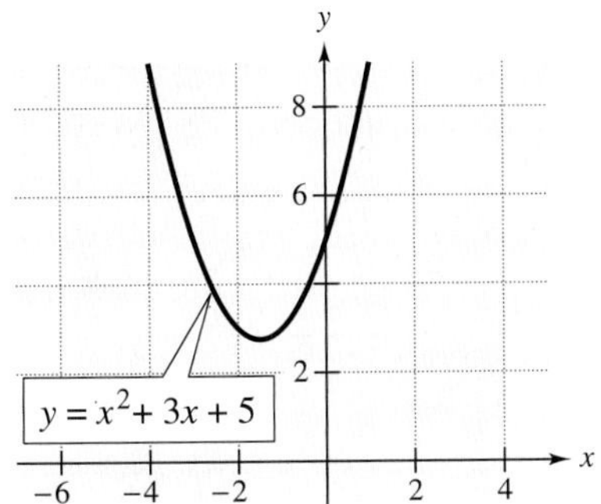

Figure 10.18

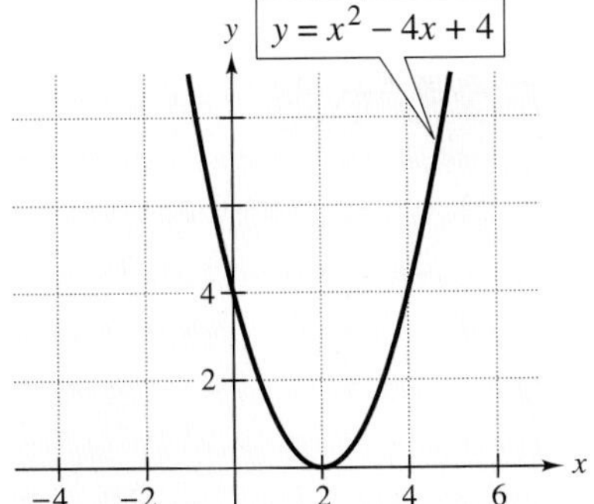

Figure 10.19

The solutions of the quadratic inequalities in Examples 1 and 2 consist, respectively, of a single interval and the union of two intervals. When solving the exercises for this section, you should watch for some unusual solution sets, as illustrated in Example 3.

EXAMPLE 3 Unusual Solution Sets

Solve each inequality.

a. The solution set of the quadratic inequality

$$x^2 + 2x + 4 > 0$$

consists of the entire set of real numbers, $(-\infty, \infty)$. This is true because the value of the quadratic $x^2 + 2x + 4$ is positive for every real value of x. You can see in Figure 10.16 that the entire parabola lies above the x-axis.

b. The solution set of the quadratic inequality

$$x^2 + 2x + 1 \le 0$$

consists of the single number $\{-1\}$. This is true because $x^2 + 2x + 1 = (x + 1)^2$ has just one critical number, $x = -1$, and it is the only value that satisfies the inequality. You can see in Figure 10.17 that the parabola meets the x-axis at $x = -1$.

c. The solution set of the quadratic inequality

$$x^2 + 3x + 5 < 0$$

is empty. This is true because the value of the quadratic $x^2 + 3x + 5$ is not less than zero for any value of x. No point on the parabola lies below the x-axis, as shown in Figure 10.18.

d. The solution set of the quadratic inequality

$$x^2 - 4x + 4 > 0$$

consists of all real numbers *except* the number 2. In interval notation, this solution set can be written as $(-\infty, 2) \cup (2, \infty)$. You can see in Figure 10.19 that the parabola lies above the x-axis *except* at $x = 2$, where it meets the x-axis.

✔ **CHECKPOINT** *Now try Exercise 35.*

Remember that checking the solution set of an inequality is not as straightforward as checking the solutions of an equation, because inequalities tend to have infinitely many solutions. Even so, you should check several x-values in your solution set to confirm that they satisfy the inequality. Also try checking x-values that are not in the solution set to verify that they do not satisfy the inequality.

For instance, the solution set of $x^2 - 5x < 0$ is the interval $(0, 5)$. Try checking some numbers in this interval to verify that they satisfy the inequality. Then check some numbers outside the interval to verify that they do not satisfy the inequality.

3 ▶ Use test intervals to solve rational inequalities.

Rational Inequalities

The concepts of critical numbers and test intervals can be extended to inequalities involving rational expressions. To do this, use the fact that the value of a rational expression can change sign only at its *zeros* (the x-values for which its numerator is zero) and its *undefined values* (the x-values for which its denominator is zero). These two types of numbers make up the **critical numbers** of a rational inequality. For instance, the critical numbers of the inequality

$$\frac{x-2}{(x-1)(x+3)} < 0$$

are $x = 2$ (the numerator is zero), and $x = 1$ and $x = -3$ (the denominator is zero). From these three critical numbers, you can see that the inequality has *four* test intervals: $(-\infty, -3)$, $(-3, 1)$, $(1, 2)$, and $(2, \infty)$.

Study Tip

When solving a rational inequality, you should begin by writing the inequality in general form, with the rational expression (as a single fraction) on the left and zero on the right. For instance, the first step in solving

$$\frac{2x}{x+3} < 4$$

is to write it as

$$\frac{2x}{x+3} - 4 < 0$$

$$\frac{2x - 4(x+3)}{x+3} < 0$$

$$\frac{-2x - 12}{x+3} < 0.$$

Try solving this inequality. You should find that the solution set is $(-\infty, -6) \cup (-3, \infty)$.

EXAMPLE 4 Solving a Rational Inequality

To solve the inequality $\frac{x}{x-2} > 0$, first find the critical numbers. The numerator is zero when $x = 0$, and the denominator is zero when $x = 2$. So, the two critical numbers are 0 and 2, which implies that the test intervals are $(-\infty, 0)$, $(0, 2)$, and $(2, \infty)$. To test an interval, choose a convenient value in the interval and determine if the value satisfies the inequality, as shown in the table.

Test interval	Representative x-value	Is inequality satisfied?
$(-\infty, 0)$	$x = -1$	$\frac{-1}{-1-2} \overset{?}{>} 0 \Rightarrow \frac{1}{3} > 0$
$(0, 2)$	$x = 1$	$\frac{1}{1-2} \overset{?}{>} 0 \Rightarrow -1 \not> 0$
$(2, \infty)$	$x = 3$	$\frac{3}{3-2} \overset{?}{>} 0 \Rightarrow 3 > 0$

You can see that the inequality is satisfied for the intervals $(-\infty, 0)$ and $(2, \infty)$. So, the solution set of the inequality is $(-\infty, 0) \cup (2, \infty)$. (See Figure 10.20.)

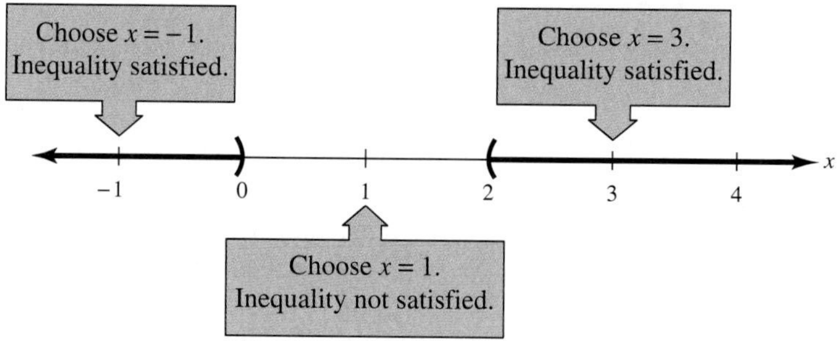

Figure 10.20

✓ ***CHECKPOINT*** *Now try Exercise 77.*

4 ▶ Use inequalities to solve application problems.

Application

EXAMPLE 5 The Height of a Projectile

A projectile is fired straight upward from ground level with an initial velocity of 256 feet per second, as shown in Figure 10.21, so that its height h at any time t is given by

$$h = -16t^2 + 256t$$

where h is measured in feet and t is measured in seconds. During what interval of time will the height of the projectile exceed 960 feet?

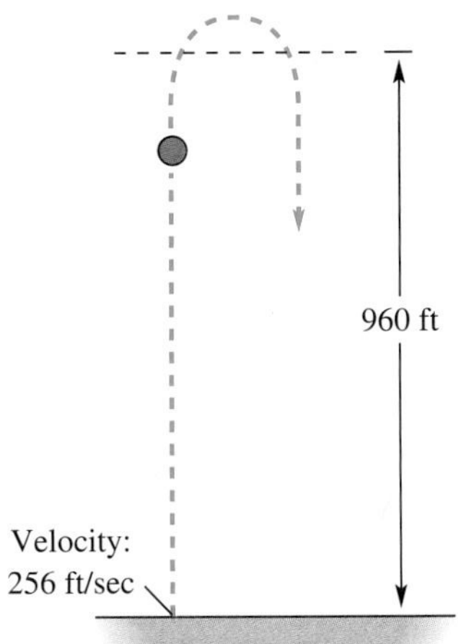

Figure 10.21

Solution

To solve this problem, begin by writing the inequality in general form.

$$-16t^2 + 256t > 960 \qquad \text{Write original inequality.}$$

$$-16t^2 + 256t - 960 > 0 \qquad \text{Write in general form.}$$

Next, find the critical numbers for $-16t^2 + 256t - 960 > 0$ by finding the solution to the equation $-16t^2 + 256t - 960 = 0$.

$$-16t^2 + 256t - 960 = 0 \qquad \text{Write corresponding equation.}$$

$$t^2 - 16t + 60 = 0 \qquad \text{Divide each side by } -16.$$

$$(t - 6)(t - 10) = 0 \qquad \text{Factor.}$$

$$t = 6, t = 10 \qquad \text{Critical numbers}$$

This implies that the test intervals are

$$(-\infty, 6), (6, 10), \text{ and } (10, \infty). \qquad \text{Test intervals}$$

To test an interval, choose a convenient value in the interval and determine if the value satisfies the inequality.

Test interval	Representative t-value	Is inequality satisfied?
$(-\infty, 6)$	$t = 0$	$-16(0)^2 + 256(0) \overset{?}{>} 960$ $0 \not> 960$
$(6, 10)$	$t = 7$	$-16(7)^2 + 256(7) \overset{?}{>} 960$ $1008 > 960$
$(10, \infty)$	$t = 11$	$-16(11)^2 + 256(11) \overset{?}{>} 960$ $880 \not> 960$

So, the height of the projectile will exceed 960 feet for values of t such that $6 < t < 10$.

✓ **CHECKPOINT** *Now try Exercise 111.*

Smart Study Strategy

Go to page 616 for ways to *Keep Your Mind Focused.*

Concept Check

1. The test intervals of a polynomial are $(-\infty, -1)$, $(-1, 3)$, $(3, \infty)$. What are the critical numbers of the polynomial?
2. In your own words, describe a procedure for solving quadratic inequalities.
3. Give a verbal description of the solution set $(-\infty, -5] \cup [10, \infty)$ of the inequality $x^2 - 5x - 50 \ge 0$.
4. How is the procedure for finding the critical numbers of a quadratic inequality different from the procedure for finding the critical numbers of a rational inequality?

10.6 EXERCISES

Go to pages 676–677 to record your assignments.

Developing Skills

In Exercises 1–10, find the critical numbers.

1. $x(2x - 5)$
2. $5x(x - 3)$
3. $4x^2 - 81$
4. $9y^2 - 16$
5. $x(x + 3) - 5(x + 3)$
6. $y(y - 4) - 3(y - 4)$
7. $x^2 - 4x + 3$
8. $3x^2 - 2x - 8$
9. $6x^2 + 13x - 15$
10. $4x^2 - 4x - 3$

In Exercises 11–20, determine the intervals for which the polynomial is entirely negative and entirely positive.

11. $x - 4$
12. $3 - x$
13. $3 - \frac{1}{2}x$
14. $\frac{2}{3}x - 8$
15. $4x(x - 5)$
16. $7x(3 - x)$
17. $4 - x^2$
18. $x^2 - 36$
19. $x^2 - 4x - 5$
20. $2x^2 - 4x - 3$

In Exercises 21–60, solve the inequality and graph the solution on the real number line. (Some of the inequalities have no solutions.) *See Examples 1–3.*

✓ 21. $3x(x - 2) < 0$
22. $5x(x - 8) > 0$
23. $3x(2 - x) \ge 0$
24. $5x(8 - x) > 0$
25. $x^2 > 4$
26. $z^2 \le 9$
27. $x^2 - 3x - 10 \ge 0$
28. $x^2 + 8x + 7 < 0$
29. $x^2 + 4x > 0$
30. $x^2 - 5x \ge 0$
✓ 31. $x^2 + 5x \le 36$
32. $t^2 - 4t > 12$
33. $u^2 + 2u - 2 > 1$
34. $t^2 - 15t + 50 < 0$
✓ 35. $x^2 + 4x + 5 < 0$
36. $x^2 + 6x + 10 > 0$
37. $x^2 + 2x + 1 \ge 0$
38. $y^2 - 5y + 6 > 0$
39. $x^2 - 4x + 2 > 0$
40. $-x^2 + 8x - 11 \le 0$
41. $x^2 - 6x + 9 \ge 0$
42. $x^2 + 14x + 49 < 0$
43. $u^2 - 10u + 25 < 0$
44. $y^2 + 16y + 64 \le 0$
45. $3x^2 + 2x - 8 \le 0$
46. $2t^2 - 3t - 20 \ge 0$
47. $-6u^2 + 19u - 10 > 0$
48. $4x^2 - 4x - 63 < 0$
49. $-2u^2 + 7u + 4 < 0$
50. $-3x^2 - 4x + 4 \le 0$
51. $4x^2 + 28x + 49 \le 0$

52. $9x^2 - 24x + 16 \geq 0$

53. $(x - 2)^2 < 0$

54. $(y + 3)^2 \geq 0$

55. $6 - (x - 2)^2 < 0$

56. $(y + 3)^2 - 6 \geq 0$

57. $16 \leq (u + 5)^2$

58. $25 \geq (x - 3)^2$

59. $x(x - 2)(x + 2) > 0$

60. $x(x - 1)(x + 4) \leq 0$

In Exercises 61–68, use a graphing calculator to solve the inequality. Verify your result algebraically.

61. $x^2 - 6x < 0$

62. $2x^2 + 5x > 0$

63. $0.5x^2 + 1.25x - 3 > 0$

64. $\frac{1}{3}x^2 - 3x < 0$

65. $x^2 + 6x + 5 \geq 8$

66. $x^2 - 6x + 9 < 16$

67. $9 - 0.2(x - 2)^2 < 4$

68. $8x - x^2 > 12$

Graphical Analysis In Exercises 69–72, use a graphing calculator to graph the function. Use the graph to approximate the values of x that satisfy the specified inequalities.

Function	*Inequalities*
69. $f(x) = x^2 - 2x + 3$	(a) $f(x) \geq 0$
	(b) $f(x) \leq 6$
70. $f(x) = -3x^2 + 6x + 2$	(a) $f(x) \leq 0$
	(b) $f(x) \geq 5$
71. $f(x) = -2x^2 + 6x - 9$	(a) $f(x) > -11$
	(b) $f(x) < 10$
72. $f(x) = 4x^2 - 10x - 7$	(a) $f(x) > -1$
	(b) $f(x) < 8$

In Exercises 73–76, find the critical numbers.

73. $\dfrac{5}{x - 3}$

74. $\dfrac{-6}{x + 2}$

75. $\dfrac{2x}{x + 5}$

76. $\dfrac{x - 2}{x - 10}$

In Exercises 77–98, solve the inequality and graph the solution on the real number line. ***See Example 4.***

✓ **77.** $\dfrac{5}{x - 3} > 0$

78. $\dfrac{3}{4 - x} > 0$

79. $\dfrac{-5}{x - 3} > 0$

80. $\dfrac{-3}{4 - x} > 0$

81. $\dfrac{3}{y - 1} \leq -1$

82. $\dfrac{2}{x - 3} \geq -1$

83. $\dfrac{x + 4}{x - 2} > 0$

84. $\dfrac{x - 5}{x + 2} < 0$

85. $\dfrac{y - 4}{y - 1} \leq 0$

86. $\dfrac{y + 7}{y + 3} \geq 0$

87. $\dfrac{4x - 2}{2x - 4} > 0$

88. $\dfrac{3x + 4}{2x - 1} < 0$

89. $\dfrac{x + 2}{4x + 6} \leq 0$

90. $\dfrac{u - 6}{3u - 5} \leq 0$

91. $\dfrac{3(u - 3)}{u + 1} < 0$

92. $\dfrac{2(4 - t)}{4 + t} > 0$

93. $\dfrac{2}{x - 5} \geq 3$

94. $\dfrac{1}{x + 2} > -3$

95. $\dfrac{4x}{x + 2} < -1$

96. $\dfrac{6x}{x - 4} < 5$

97. $\dfrac{x - 3}{x - 6} \leq 4$

98. $\dfrac{x + 4}{x - 5} \geq 10$

In Exercises 99–106, use a graphing calculator to solve the rational inequality. Verify your result algebraically.

99. $\dfrac{1}{x} - x > 0$

100. $\dfrac{1}{x} - 3 < 0$

101. $\frac{x+6}{x+1} - 2 < 0$

102. $\frac{x+12}{x+2} - 3 \geq 0$

103. $\frac{6x-3}{x+5} < 2$

104. $\frac{3x-4}{x-4} < -5$

105. $x + \frac{1}{x} > 3$

106. $4 - \frac{1}{x^2} > 1$

Graphical Analysis In Exercises 107–110, use a graphing calculator to graph the function. Use the graph to approximate the values of x that satisfy the specified inequalities.

	Function	*Inequalities*	
107.	$f(x) = \frac{3x}{x-2}$	(a) $f(x) \leq 0$	(b) $f(x) \geq 6$
108.	$f(x) = \frac{2(x-2)}{x+1}$	(a) $f(x) \leq 0$	(b) $f(x) \geq 8$
109.	$f(x) = \frac{2x^2}{x^2+4}$	(a) $f(x) \geq 1$	(b) $f(x) \leq 2$
110.	$f(x) = \frac{5x}{x^2+4}$	(a) $f(x) \geq 1$	(b) $f(x) \geq 0$

Solving Problems

✓ **111.** *Height* A projectile is fired straight upward from ground level with an initial velocity of 128 feet per second, so that its height h at any time t is given by $h = -16t^2 + 128t$, where h is measured in feet and t is measured in seconds. During what interval of time will the height of the projectile exceed 240 feet?

112. *Height* A projectile is fired straight upward from ground level with an initial velocity of 88 feet per second, so that its height h at any time t is given by $h = -16t^2 + 88t$, where h is measured in feet and t is measured in seconds. During what interval of time will the height of the projectile exceed 50 feet?

113. *Compound Interest* You are investing \$1000 in a certificate of deposit for 2 years, and you want the interest for that time period to exceed \$150. The interest is compounded annually. What interest rate should you have? [*Hint:* Solve the inequality $1000(1 + r)^2 > 1150$.]

114. *Compound Interest* You are investing \$500 in a certificate of deposit for 2 years, and you want the interest for that time to exceed \$50. The interest is compounded annually. What interest rate should you have? [*Hint:* Solve the inequality $500(1 + r)^2 > 550$.]

115. *Geometry* You have 64 feet of fencing to enclose a rectangular region. Determine the interval for the length such that the area will exceed 240 square feet.

116. *Geometry* A rectangular playing field with a perimeter of 100 meters is to have an area of at least 500 square meters. Within what bounds must the length of the field lie?

117. *Cost, Revenue, and Profit* The revenue and cost equations for a computer desk are given by

$$R = x(50 - 0.0002x) \text{ and } C = 12x + 150{,}000$$

where R and C are measured in dollars and x represents the number of desks sold. How many desks must be sold to obtain a profit of at least \$1,650,000?

118. *Cost, Revenue, and Profit* The revenue and cost equations for a digital camera are given by

$$R = x(125 - 0.0005x) \text{ and } C = 3.5x + 185{,}000$$

where R and C are measured in dollars and x represents the number of cameras sold. How many cameras must be sold to obtain a profit of at least \$6,000,000?

119. *Antibiotics* The concentration C (in milligrams per liter) of an antibiotic 30 minutes after it is administered is given by

$$C(t) = \frac{21.9 - 0.043t}{1 + 0.005t}, \quad 30 \le t \le 500$$

where t is the time (in minutes).

(a) Use a graphing calculator to graph the concentration function.

(b) How long does it take for the concentration of the antibiotic to fall below 5 milligrams per liter?

120. *Data Analysis* The temperature T (in degrees Fahrenheit) of a metal in a laboratory experiment was recorded every 2 minutes for a period of 16 minutes. The table shows the experimental data, where t is the time in minutes.

t	0	2	4	6	8
T	250	290	338	410	498

t	10	12	14	16
T	560	530	370	160

A model for this data is

$$T = \frac{248.5 - 13.72t}{1.0 - 0.13t + 0.005t^2}.$$

(a) Use a graphing calculator to plot the data and graph the model in the same viewing window. Does the model fit the data well?

(b) Use the graph to approximate the times when the temperature was at least 400°F.

Explaining Concepts

121. Explain why the critical numbers of a polynomial are not included in its test intervals.

122. Explain the difference in the solution sets of $x^2 - 4 < 0$ and $x^2 - 4 \le 0$.

123. The graph of a quadratic function g lies completely above the x-axis. What is the solution set of the inequality $g(x) < 0$? Explain your reasoning.

124. Explain how you can use the graph of $f(x) = x^2 - x - 6$ to check the solution of $x^2 - x - 6 > 0$.

Cumulative Review

In Exercises 125–130, perform the operation and simplify.

125. $\dfrac{4xy^3}{x^2y} \cdot \dfrac{y}{8x}$

126. $\dfrac{2x^2 - 2}{x^2 - 6x - 7} \cdot (x^2 - 10x + 21)$

127. $\dfrac{x^2 - x - 6}{4x^3} \cdot \dfrac{x + 1}{x^2 + 5x + 6}$

128. $\dfrac{32x^3y}{y^9} \div \dfrac{8x^4}{y^6}$

129. $\dfrac{x^2 + 8x + 16}{x^2 - 6x} \div (3x - 24)$

130. $\dfrac{x^2 + 6x - 16}{3x^2} \div \dfrac{x + 8}{6x}$

In Exercises 131–134, evaluate the expression for the specified value. Round your result to the nearest hundredth, if necessary.

131. x^2; $x = -\dfrac{1}{3}$

132. $1000 - 20x^3$; $x = 4.02$

133. $\dfrac{100}{x^4}$; $x = 1.06$

134. $\dfrac{50}{1 - \sqrt{x}}$; $x = 0.1024$

What Did You Learn?

Use these two pages to help prepare for a test on this chapter. Check off the key terms and key concepts you know. You can also use this section to record your assignments.

Plan for Test Success

Date of test: / /

Study dates and times: / / at : A.M./P.M.
/ / at : A.M./P.M.

Things to review:

- ☐ Key Terms, *p. 676*
- ☐ Key Concepts, *pp. 676–677*
- ☐ Your class notes
- ☐ Your assignments
- ☐ Study Tips, *pp. 618, 622, 628, 635, 636, 637, 638, 646, 647, 667, 668, 670*
- ☐ Technology Tips, *pp. 619, 621, 628, 639, 667*
- ☐ Mid-Chapter Quiz, *p. 644*
- ☐ Review Exercises, *pp. 678–680*
- ☐ Chapter Test, *p. 681*
- ☐ Video Explanations Online
- ☐ Tutorial Online

Key Terms

- ☐ double or repeated solution, *p. 618*
- ☐ Square Root Property, *p. 619*
- ☐ extracting square roots, *p. 619*
- ☐ quadratic form, *p. 621*
- ☐ completing the square, *p. 627*
- ☐ Quadratic Formula, *p. 635*
- ☐ discriminant, *p. 638*
- ☐ parabola, *p. 645*
- ☐ standard form of a quadratic function, *p. 645*
- ☐ vertex of a parabola, *p. 645*
- ☐ axis of a parabola, *p. 645*
- ☐ zeros of a polynomial, *p. 666*
- ☐ test intervals, *p. 666*
- ☐ critical numbers of a polynomial, *p. 666*
- ☐ general form of an inequality, *p. 668*
- ☐ critical numbers of a rational inequality, *p. 670*

Key Concepts

10.1 Solving Quadratic Equations: Factoring and Special Forms

Assignment: ______________________________ **Due date:** ________

☐ **Solve quadratic equations by factoring.**

1. Write the equation in general form.
2. Factor the left side.
3. Set each factor equal to zero and solve for x.

☐ **Solve nonquadratic equations that are of quadratic form.**

An equation is said to be of quadratic form if it has the form $au^2 + bu + c = 0$, where u is an algebraic expression.

☐ **Use the Square Root Property to solve quadratic equations.**

1. The equation $u^2 = d$, where $d > 0$, has exactly two solutions: $u = \pm\sqrt{d}$. This solution process is called *extracting square roots.*
2. The equation $u^2 = d$, where $d < 0$, has exactly two solutions: $u = \pm\sqrt{|d|}i$.

10.2 Completing the Square

Assignment: ______________________________ **Due date:** ________

☐ **Write an expression in completed square form.**

To complete the square for the expression $x^2 + bx$, add $(b/2)^2$, which is the square of half the coefficient of x. Consequently,

$$x^2 + bx + \left(\frac{b}{2}\right)^2 = \left(x + \frac{b}{2}\right)^2.$$

☐ **Solve quadratic equations by completing the square.**

1. Prior to completing the square, the coefficient of the second-degree term must be 1.
2. Preserve the equality by adding the same constant to each side of the equation.
3. Use the Square Root Property to solve the quadratic equation.

10.3 The Quadratic Formula

Assignment: ______ Due date: ______

☐ **Use the Quadratic Formula.**

The solutions of $ax^2 + bx + c = 0$, $a \neq 0$, are given by the *Quadratic Formula*

$$x = \frac{-b \pm \sqrt{b^2 - 4ac}}{2a}.$$

The expression inside the radical, $b^2 - 4ac$, is called the *discriminant.*

☐ **Write quadratic equations from their solutions.**

You can use the Zero-Factor Property in reverse to find a quadratic equation given its solutions.

☐ **Use the discriminant.**

Let a, b, and c be rational numbers such that $a \neq 0$. The discriminant of the quadratic equation $ax^2 + bx + c = 0$ is given by $b^2 - 4ac$, and can be used to classify the solutions of the equation as follows.

Discriminant	*Solution Type*
1. Perfect square	Two distinct rational solutions
2. Positive nonperfect square	Two distinct irrational solutions
3. Zero	One repeated rational solution
4. Negative number	Two distinct complex solutions

10.4 Graphs of Quadratic Functions

Assignment: ______ Due date: ______

☐ **Recognize graphs of quadratic functions.**

The graph of $f(x) = ax^2 + bx + c$, $a \neq 0$, is a *parabola.* The completed square form $f(x) = a(x - h)^2 + k$ is the *standard form* of the function. The *vertex* of the parabola occurs at the point (h, k), and the vertical line passing through the vertex is the *axis* of the parabola.

☐ **Sketch a parabola.**

1. Determine the vertex and axis of the parabola by completing the square or by using the formula $x = -b/(2a)$.
2. Plot the vertex, axis, x- and y-intercepts, and a few additional points on the parabola. (Using the symmetry about the axis can reduce the number of points you need to plot.)
3. Use the fact that the parabola opens upward if $a > 0$ and opens downward if $a < 0$ to complete the sketch.

10.5 Applications of Quadratic Equations

Assignment: ______ Due date: ______

☐ **Use quadratic equations to solve a wide variety of real-life problems.**

The following are samples of applications of quadratic equations.

1. Investment	2. Interest
3. Height of a projectile	4. Geometric dimensions
5. Work rate	6. Structural design

10.6 Quadratic and Rational Inequalities

Assignment: ______ Due date: ______

☐ **Find test intervals for inequalities.**

1. For a polynomial expression, find all the real zeros. For a rational expression, find all the real zeros and those x-values for which the function is undefined.
2. Arrange the numbers found in Step 1 in increasing order. These numbers are called *critical numbers.*
3. Use the critical numbers to determine the test intervals.
4. Choose a representative x-value in each test interval and evaluate the expression at that value. If the value of the expression is negative, the expression will have negative values for every x-value in the interval. If the value of the expression is positive, the expression will have positive values for every x-value in the interval.

Review Exercises

10.1 Solving Quadratic Equations: Factoring and Special Forms

1 ▶ Solve quadratic equations by factoring.

In Exercises 1–10, solve the equation by factoring.

1. $x^2 + 12x = 0$
2. $u^2 - 18u = 0$
3. $3y^2 - 27 = 0$
4. $2z^2 - 72 = 0$
5. $4y^2 + 20y + 25 = 0$
6. $x^2 + \frac{8}{3}x + \frac{16}{9} = 0$
7. $2x^2 - 2x - 180 = 0$
8. $9x^2 + 18x - 135 = 0$
9. $6x^2 - 12x = 4x^2 - 3x + 18$
10. $10x - 8 = 3x^2 - 9x + 12$

2 ▶ Solve quadratic equations by the Square Root Property.

In Exercises 11–16, solve the equation by using the Square Root Property.

11. $z^2 = 144$
12. $2x^2 = 98$
13. $y^2 - 12 = 0$
14. $y^2 - 45 = 0$
15. $(x - 16)^2 = 400$
16. $(x + 3)^2 = 900$

3 ▶ Solve quadratic equations with complex solutions by the Square Root Property.

In Exercises 17–22, solve the equation by using the Square Root Property.

17. $z^2 = -121$
18. $u^2 = -225$
19. $y^2 + 50 = 0$
20. $x^2 + 48 = 0$
21. $(y + 4)^2 + 18 = 0$
22. $(x - 2)^2 + 24 = 0$

4 ▶ Use substitution to solve equations of quadratic form.

In Exercises 23–30, solve the equation of quadratic form. (Find all real *and* complex solutions.)

23. $x^4 - 4x^2 - 5 = 0$
24. $x^4 - 10x^2 + 9 = 0$
25. $x - 4\sqrt{x} + 3 = 0$
26. $x - 4\sqrt{x} + 13 = 0$
27. $(x^2 - 2x)^2 - 4(x^2 - 2x) - 5 = 0$
28. $(\sqrt{x} - 2)^2 + 2(\sqrt{x} - 2) - 3 = 0$
29. $x^{2/3} + 3x^{1/3} - 28 = 0$
30. $x^{2/5} + 4x^{1/5} + 3 = 0$

10.2 Completing the Square

1 ▶ Rewrite quadratic expressions in completed square form.

In Exercises 31–36, add a term to the expression so that it becomes a perfect square trinomial.

31. $z^2 + 18z +$ ▭
32. $y^2 - 80y +$ ▭
33. $x^2 - 15x +$ ▭
34. $x^2 + 21x +$ ▭
35. $y^2 + \frac{2}{5}y +$ ▭
36. $x^2 - \frac{3}{4}x +$ ▭

2 ▶ Solve quadratic equations by completing the square.

In Exercises 37–42, solve the equation by completing the square. Give the solutions in exact form and in decimal form rounded to two decimal places. (The solutions may be complex numbers.)

37. $x^2 - 6x - 3 = 0$
38. $x^2 + 12x + 6 = 0$
39. $v^2 + 5v + 4 = 0$
40. $u^2 - 5u + 6 = 0$
41. $y^2 - \frac{2}{3}y + 2 = 0$
42. $t^2 + \frac{1}{2}t - 1 = 0$

10.3 The Quadratic Formula

2 ▶ Use the Quadratic Formula to solve quadratic equations.

In Exercises 43–48, solve the equation by using the Quadratic Formula. (Find all real *and* complex solutions.)

43. $v^2 + v - 42 = 0$
44. $x^2 - x - 72 = 0$
45. $2y^2 + y - 21 = 0$

46. $2x^2 - 3x - 20 = 0$

47. $5x^2 - 16x + 2 = 0$

48. $3x^2 + 12x + 4 = 0$

3 ▶ Determine the types of solutions of quadratic equations using the discriminant.

In Exercises 49–56, use the discriminant to determine the type of solutions of the quadratic equation.

49. $x^2 + 4x + 4 = 0$

50. $y^2 - 26y + 169 = 0$

51. $s^2 - s - 20 = 0$

52. $r^2 - 5r - 45 = 0$

53. $4t^2 + 16t + 10 = 0$

54. $8x^2 + 85x - 33 = 0$

55. $v^2 - 6v + 21 = 0$

56. $9y^2 + 1 = 0$

4 ▶ Write quadratic equations from solutions of the equations.

In Exercises 57–62, write a quadratic equation having the given solutions.

57. $3, -7$

58. $-2, 8$

59. $5 + \sqrt{7}, 5 - \sqrt{7}$

60. $2 + \sqrt{2}, 2 - \sqrt{2}$

61. $6 + 2i, 6 - 2i$

62. $3 + 4i, 3 - 4i$

10.4 Graphs of Quadratic Functions

1 ▶ Determine the vertices of parabolas by completing the square.

In Exercises 63–66, write the equation of the parabola in standard form, and find the vertex of its graph.

63. $y = x^2 - 8x + 3$

64. $y = 8 - 8x - x^2$

65. $y = 2x^2 - x + 3$

66. $y = 3x^2 + 2x - 6$

2 ▶ Sketch parabolas.

In Exercises 67–70, sketch the parabola. Identify the vertex and any *x*-intercepts. Use a graphing calculator to verify your results.

67. $y = x^2 + 8x$

68. $y = -x^2 + 3x$

69. $f(x) = -x^2 - 2x + 4$

70. $f(x) = x^2 + 3x - 10$

3 ▶ Write the equation of a parabola given the vertex and a point on the graph.

In Exercises 71–74, write an equation of the parabola $y = a(x - h)^2 + k$ that satisfies the conditions.

71. Vertex: $(2, -5)$; Point on the graph: $(0, 3)$

72. Vertex: $(-4, 0)$; Point on the graph: $(0, -6)$

73. Vertex: $(5, 0)$; Point on the graph: $(1, 1)$

74. Vertex: $(-2, 5)$; Point on the graph: $(-4, 11)$

4 ▶ Use parabolas to solve application problems.

75. *Path of a Ball* The height y (in feet) of a ball thrown by a child is given by $y = -\frac{1}{10}x^2 + 3x + 6$, where x is the horizontal distance (in feet) from where the ball is thrown.

(a) Use a graphing calculator to graph the path of the ball.

(b) How high is the ball when it leaves the child's hand?

(c) How high is the ball when it reaches its maximum height?

(d) How far from the child does the ball strike the ground?

76. *Graphical Estimation* The numbers N (in thousands) of bankruptcies filed by businesses in the United States in the years 2001 through 2005 are approximated by $N = -0.736t^2 + 3.12t + 35.0$, $1 \le t \le 5$, where t is the time in years, with $t = 1$ corresponding to 2001. (Source: Administrative Office of the U.S. Courts)

(a) Use a graphing calculator to graph the model.

(b) Use the graph from part (a) to approximate the maximum number of bankruptcies filed by businesses from 2001 through 2005. During what year did this maximum occur?

10.5 Applications of Quadratic Equations

1 ▶ Use quadratic equations to solve application problems.

77. *Selling Price* A car dealer bought a fleet of used cars for a total of \$80,000. By the time all but four of the cars had been sold, at an average profit of \$1000 each, the original investment of \$80,000 had been regained. How many cars were sold, and what was the average price per car?

78. *Selling Price* A manager of a computer store bought several computers of the same model for \$27,000. When all but five of the computers had been sold at a profit of \$900 per computer, the original investment of \$27,000 had been regained. How many computers were sold, and what was the selling price of each computer?

79. *Geometry* The length of a rectangle is 12 inches greater than its width. The area of the rectangle is 85 square inches. Find the dimensions of the rectangle.

80. *Compound Interest* You want to invest \$35,000 for 2 years at an annual interest rate of r (in decimal form). Interest on the account is compounded annually. Find the interest rate if a deposit of \$35,000 increases to \$40,221.44 over a two-year period.

81. *Reduced Rates* A Little League baseball team obtains a block of tickets to a ball game for \$96. After three more people decide to go to the game, the price per ticket is decreased by \$1.60. How many people are going to the game?

82. *Geometry* A corner lot has an L-shaped sidewalk along its sides. The total length of the sidewalk is 69 feet. By cutting diagonally across the lot, the walking distance is shortened to 51 feet. What are the lengths of the two legs of the sidewalk?

83. *Work-Rate Problem* Working together, two people can complete a task in 10 hours. Working alone, one person takes 2 hours longer than the other. How long would it take each person to do the task alone?

84. *Height* An object is projected vertically upward at an initial velocity of 64 feet per second from a height of 192 feet, so that the height h at any time t is given by $h = -16t^2 + 64t + 192$, where t is the time in seconds.

(a) After how many seconds is the height 256 feet?

(b) After how many seconds does the object hit the ground?

10.6 Quadratic and Rational Inequalities

1 ▶ Determine test intervals for polynomials.

In Exercises 85–88, find the critical numbers.

85. $2x(x + 7)$

86. $x(x - 2) + 4(x - 2)$

87. $x^2 - 6x - 27$

88. $2x^2 + 11x + 5$

2 ▶ Use test intervals to solve quadratic inequalities.

In Exercises 89–94, solve the inequality and graph the solution on the real number line.

89. $5x(7 - x) > 0$

90. $-2x(x - 10) \leq 0$

91. $16 - (x - 2)^2 \leq 0$

92. $(x - 5)^2 - 36 > 0$

93. $2x^2 + 3x - 20 < 0$

94. $-3x^2 + 10x + 8 \geq 0$

3 ▶ Use test intervals to solve rational inequalities.

In Exercises 95–98, solve the inequality and graph the solution on the real number line.

95. $\frac{x + 3}{2x - 7} \geq 0$

96. $\frac{3x + 2}{x - 3} > 0$

97. $\frac{x + 4}{x - 1} < 0$

98. $\frac{2x - 9}{x - 1} \leq 0$

4 ▶ Use inequalities to solve application problems.

99. *Height* A projectile is fired straight upward from ground level with an initial velocity of 312 feet per second, so that its height h at any time t is given by $h = -16t^2 + 312t$, where h is measured in feet and t is measured in seconds. During what interval of time will the height of the projectile exceed 1200 feet?

100. *Average Cost* The cost C of producing x notebooks is $C = 100{,}000 + 0.9x$, $x > 0$. Write the average cost $\overline{C} = C/x$ as a function of x. Then determine how many notebooks must be produced if the average cost per unit is to be less than \$2.

Chapter Test

Take this test as you would take a test in class. After you are done, check your work against the answers in the back of the book.

In Exercises 1–6, solve the equation by the specified method.

1. Factoring:
$x(x - 3) - 10(x - 3) = 0$

2. Factoring:
$6x^2 - 34x - 12 = 0$

3. Square Root Property:
$(x - 2)^2 = 0.09$

4. Square Root Property:
$(x + 4)^2 + 100 = 0$

5. Completing the square:
$2x^2 - 6x + 3 = 0$

6. Quadratic Formula:
$2y(y - 2) = 7$

In Exercises 7 and 8, solve the equation of quadratic form.

7. $\frac{1}{x^2} - \frac{6}{x} + 4 = 0$

8. $x^{2/3} - 9x^{1/3} + 8 = 0$

9. Find the discriminant and explain what it means in terms of the type of solutions of the quadratic equation $5x^2 - 12x + 10 = 0$.

10. Find a quadratic equation having the solutions -7 and -3.

In Exercises 11 and 12, sketch the parabola. Identify the vertex and any *x*-intercepts. Use a graphing calculator to verify your results.

11. $y = -x^2 + 2x - 4$

12. $y = x^2 - 2x - 15$

In Exercises 13–15, solve the inequality and sketch its solution.

13. $16 \le (x - 2)^2$

14. $2x(x - 3) < 0$

15. $\frac{x + 1}{x - 5} \le 0$

16. The width of a rectangle is 22 feet less than its length. The area of the rectangle is 240 square feet. Find the dimensions of the rectangle.

17. An English club chartered a bus trip to a Shakespearean festival. The cost of the bus was \$1250. To lower the bus fare per person, the club invited nonmembers to go along. When 10 nonmembers joined the trip, the fare per person decreased by \$6.25. How many club members are going on the trip?

18. An object is dropped from a height of 75 feet. Its height h (in feet) at any time t is given by $h = -16t^2 + 75$, where t is measured in seconds. Find the time required for the object to fall to a height of 35 feet.

19. Two buildings are connected by an L-shaped protected walkway. The total length of the walkway is 155 feet. By cutting diagonally across the grass, the walking distance is shortened to 125 feet. What are the lengths of the two legs of the walkway?

Study Skills in Action

Making the Most of Class Time

Have you ever slumped at your desk while in class and thought "I'll just get the notes down and study later—I'm too tired"? Learning math in college is a team effort, between instructor and student. The more you understand in class, the more you will be able to learn while studying outside of class.

Approach math class with the intensity of a navy pilot during a mission briefing. The pilot has strategic plans to learn during the briefing. He or she listens intensely, takes notes, and memorizes important information. The goal is for the pilot to leave the briefing with a clear picture of the mission. It is the same with a student in a math class.

Kimberly Nolting
VP, Academic Success Press
expert in developmental education

These students are sitting in the front row, where they are more likely to pay attention.

Smart Study Strategy

Take Control of Your Class Time

1 ▶ **Sit where you can easily see and hear the instructor, and the instructor can see you.** The instructor may be able to tell when you are confused just by the look on your face, and may adjust the lesson accordingly. In addition, sitting in this strategic place will keep your mind from wandering.

2 ▶ **Pay attention to what the instructor says about the math, not just what is written on the board.** Write problems on the left side of your notes and what the instructor says about the problems on the right side.

3 ▶ **If the instructor is moving through the material too fast, ask a question.** Questions help to slow the pace for a few minutes and also to clarify what is confusing to you.

4 ▶ **Try to memorize new information while learning it.** Repeat in your head what you are writing in your notes. That way you are reviewing the information twice.

5 ▶ **Ask for clarification.** If you don't understand something at all and do not even know how to phrase a question, just ask for clarification. You might say something like, "Could you please explain the steps in this problem one more time?"

6 ▶ **Think as intensely as if you were going to take a quiz on the material at the end of class.** This kind of mindset will help you to process new information.

7 ▶ **If the instructor asks for someone to go up to the board, volunteer.** The student at the board often receives additional attention and instruction to complete the problem.

8 ▶ **At the end of class, identify concepts or problems on which you still need clarification.** Make sure you see the instructor or a tutor as soon as possible.

Chapter 11

Exponential and Logarithmic Functions

IT WORKED FOR ME!

"I failed my first college math class because, for some reason, I thought just showing up and listening would be enough. I was wrong. Now, I get to class a little early to review my notes, sit where I can see the instructor, and ask questions. I try to learn and remember as much as possible in class because I am so busy juggling work and college."

Aaron
Business

11.1 Exponential Functions

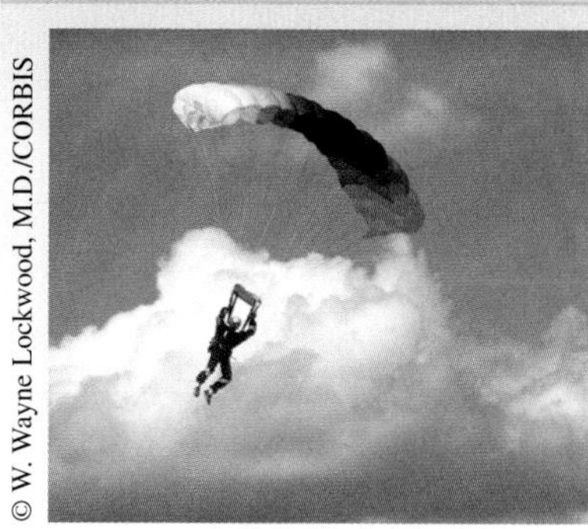

What You Should Learn

1 ▶ Evaluate exponential functions.
2 ▶ Graph exponential functions.
3 ▶ Evaluate the natural base e and graph natural exponential functions.
4 ▶ Use exponential functions to solve application problems.

Why You Should Learn It

Exponential functions can be used to model and solve real-life problems. For instance, in Exercise 89 on page 695, you will use an exponential function to model the descent of a parachutist.

1 ▶ Evaluate exponential functions.

Exponential Functions

In this section, you will study a new type of function called an **exponential function.** Whereas polynomial and rational functions have terms with variable bases and constant exponents, exponential functions have terms with *constant bases* and *variable exponents*. Here are some examples.

Polynomial or Rational Function	*Exponential Function*
Constant Exponents	Variable Exponents
$f(x) = x^2$, $f(x) = x^{-3}$	$f(x) = 2^x$, $f(x) = 3^{-x}$
Variable Bases	Constant Bases

Definition of Exponential Function

The **exponential function *f* with base *a*** is denoted by

$$f(x) = a^x$$

where $a > 0$, $a \neq 1$, and x is any real number.

The base $a = 1$ is excluded because $f(x) = 1^x = 1$ is a constant function, *not* an exponential function.

In Chapters 5 and 9, you learned to evaluate a^x for integer and rational values of x. For example, you know that

$$a^3 = a \cdot a \cdot a, \quad a^{-4} = \frac{1}{a^4}, \quad \text{and} \quad a^{5/3} = \left(\sqrt[3]{a}\right)^5.$$

However, to evaluate a^x for any real number x, you need to interpret forms with *irrational* exponents, such as $a^{\sqrt{2}}$ or a^{π}. For the purposes of this text, it is sufficient to think of a number such as $a^{\sqrt{2}}$, where $\sqrt{2} \approx 1.414214$, as the number that has the successively closer approximations

$$a^{1.4}, a^{1.41}, a^{1.414}, a^{1.4142}, a^{1.41421}, a^{1.414214}, \ldots.$$

The rules of exponents that were discussed in Section 5.1 can be extended to cover exponential functions, as described on the following page.

Study Tip

Rule 4 of the rules of exponential functions indicates that 2^{-x} can be written as

$$2^{-x} = \frac{1}{2^x}.$$

Similarly, $\frac{1}{3^{-x}}$ can be written as

$$\frac{1}{3^{-x}} = 3^x.$$

In other words, you can move a *factor* from the numerator to the denominator (or from the denominator to the numerator) by changing the sign of its exponent.

Rules of Exponential Functions

Let a be a positive real number, and let x and y be real numbers, variables, or algebraic expressions.

1. $a^x \cdot a^y = a^{x+y}$ Product rule
2. $\frac{a^x}{a^y} = a^{x-y}$ Quotient rule
3. $(a^x)^y = a^{xy}$ Power-to-power rule
4. $a^{-x} = \frac{1}{a^x} = \left(\frac{1}{a}\right)^x$ Negative exponent rule

To evaluate exponential functions with a calculator, you can use the exponential key y^x or ^. For example, to evaluate $3^{-1.3}$, you can use the following keystrokes.

Keystrokes	*Display*	
3 y^x 1.3 +/− =	0.239741	Scientific
3 ^ ((−) 1.3) ENTER	0.239741	Graphing

EXAMPLE 1 Evaluating Exponential Functions

Evaluate each function. Use a calculator only if it is necessary or more efficient.

Function	*Values*
a. $f(x) = 2^x$	$x = 3, x = -4, x = \pi$
b. $g(x) = 12^x$	$x = 3, x = -0.1, x = \frac{5}{7}$
c. $h(x) = (1.04)^{2x}$	$x = 0, x = -2, x = \sqrt{2}$

Solution

Evaluation	*Comment*
a. $f(3) = 2^3 = 8$	Calculator is not necessary.
$f(-4) = 2^{-4} = \frac{1}{2^4} = \frac{1}{16}$	Calculator is not necessary.
$f(\pi) = 2^{\pi} \approx 8.825$	Calculator is necessary.
b. $g(3) = 12^3 = 1728$	Calculator is more efficient.
$g(-0.1) = 12^{-0.1} \approx 0.780$	Calculator is necessary.
$g\left(\frac{5}{7}\right) = 12^{5/7} \approx 5.900$	Calculator is necessary.
c. $h(0) = (1.04)^{2 \cdot 0} = (1.04)^0 = 1$	Calculator is not necessary.
$h(-2) = (1.04)^{2(-2)} \approx 0.855$	Calculator is more efficient.
$h(\sqrt{2}) = (1.04)^{2\sqrt{2}} \approx 1.117$	Calculator is necessary.

✓ **CHECKPOINT** *Now try Exercise 17.*

2 ▶ Graph exponential functions.

Graphs of Exponential Functions

The basic nature of the graph of an exponential function can be determined by the point-plotting method or by using a graphing calculator.

Figure 11.1

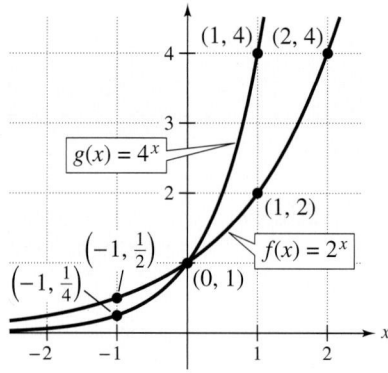

EXAMPLE 2 The Graphs of Exponential Functions

In the same coordinate plane, sketch the graph of each function. Determine the domain and range.

a. $f(x) = 2^x$

b. $g(x) = 4^x$

Solution

The table lists some values of each function, and Figure 11.1 shows the graph of each function. From the graphs, you can see that the domain of each function is the set of all real numbers and that the range of each function is the set of all positive real numbers.

x	-2	-1	0	1	2	3
2^x	$\frac{1}{4}$	$\frac{1}{2}$	1	2	4	8
4^x	$\frac{1}{16}$	$\frac{1}{4}$	1	4	16	64

✓ **CHECKPOINT** *Now try Exercise 31.*

Note in Example 3 that a graph of the form $f(x) = a^x$ (as shown in Example 2) is a reflection in the y-axis of a graph of the form $g(x) = a^{-x}$.

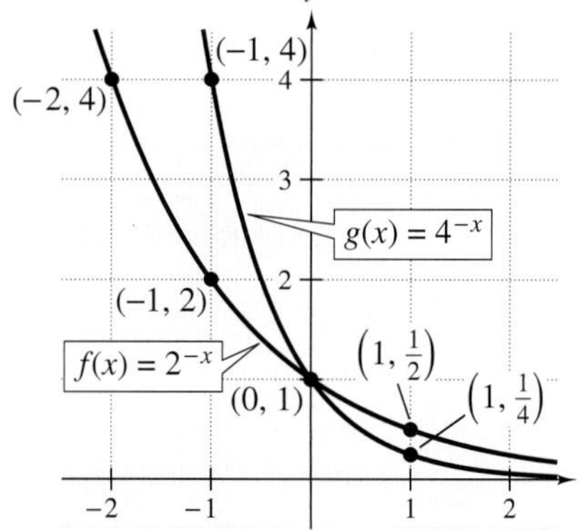

Figure 11.2

EXAMPLE 3 The Graphs of Exponential Functions

In the same coordinate plane, sketch the graph of each function.

a. $f(x) = 2^{-x}$

b. $g(x) = 4^{-x}$

Solution

The table lists some values of each function, and Figure 11.2 shows the graph of each function.

x	-3	-2	-1	0	1	2
2^{-x}	8	4	2	1	$\frac{1}{2}$	$\frac{1}{4}$
4^{-x}	64	16	4	1	$\frac{1}{4}$	$\frac{1}{16}$

✓ **CHECKPOINT** *Now try Exercise 33.*

Study Tip

An **asymptote** of a graph is a line to which the graph becomes arbitrarily close as $|x|$ or $|y|$ increases without bound. In other words, if a graph has an asymptote, it is possible to move far enough out on the graph so that there is almost no difference between the graph and the asymptote.

Examples 2 and 3 suggest that for $a > 1$, the values of the function $y = a^x$ increase as x increases and the values of the function $y = a^{-x} = (1/a)^x$ decrease as x increases. The graphs shown in Figure 11.3 are typical of the graphs of exponential functions. Note that each graph has a y-intercept at $(0, 1)$ and a **horizontal asymptote** of $y = 0$ (the x-axis).

Graph of $y = a^x$

- Domain: $(-\infty, \infty)$
- Range: $(0, \infty)$
- Intercept: $(0, 1)$
- Increasing (moves up to the right)
- Asymptote: x-axis

Graph of $y = a^{-x} = \left(\frac{1}{a}\right)^x$

- Domain: $(-\infty, \infty)$
- Range: $(0, \infty)$
- Intercept: $(0, 1)$
- Decreasing (moves down to the right)
- Asymptote: x-axis

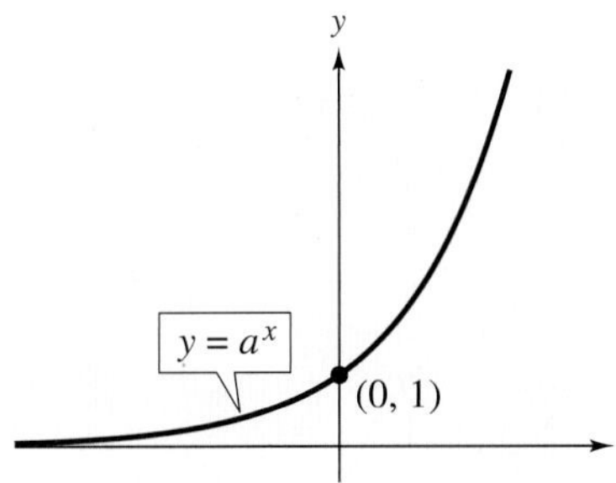

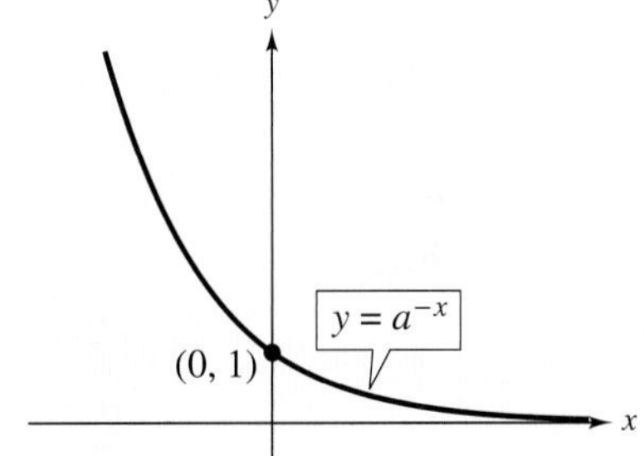

Figure 11.3 Characteristics of the exponential functions $y = a^x$ and $y = a^{-x}$ $(a > 1)$

In the next two examples, notice how the graph of $y = a^x$ can be used to sketch the graphs of functions of the form $f(x) = b \pm a^{x+c}$. Also note that the transformation in Example 4(a) keeps the x-axis as a horizontal asymptote, but the transformation in Example 4(b) yields a new horizontal asymptote of $y = -2$. Also, be sure to note how the y-intercept is affected by each transformation.

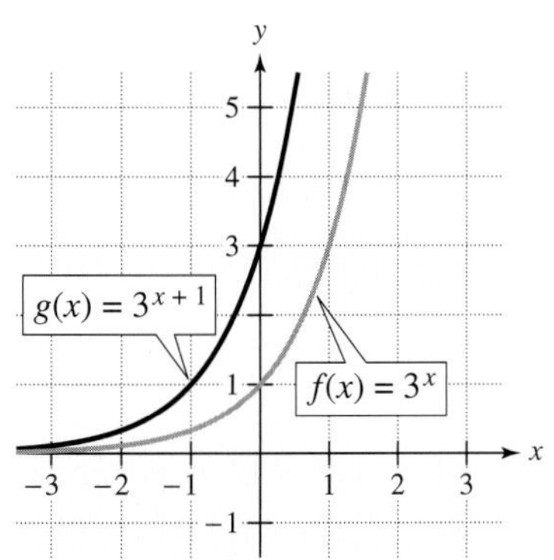

Figure 11.4

Figure 11.5

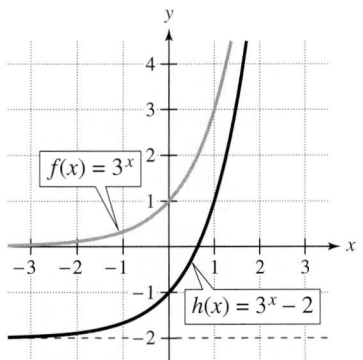

EXAMPLE 4 Transformations of Graphs of Exponential Functions

Use transformations to analyze and sketch the graph of each function.

a. $g(x) = 3^{x+1}$ **b.** $h(x) = 3^x - 2$

Solution

Consider the function $f(x) = 3^x$.

a. The function g is related to f by $g(x) = f(x + 1)$. To sketch the graph of g, shift the graph of f one unit to the left, as shown in Figure 11.4. Note that the y-intercept of g is $(0, 3)$.

b. The function h is related to f by $h(x) = f(x) - 2$. To sketch the graph of g, shift the graph of f two units downward, as shown in Figure 11.5. Note that the y-intercept of h is $(0, -1)$ and the horizontal asymptote is $y = -2$.

✓ **CHECKPOINT** *Now try Exercise 63.*

EXAMPLE 5 Reflections of Graphs of Exponential Functions

Use transformations to analyze and sketch the graph of each function.

a. $g(x) = -3^x$ **b.** $h(x) = 3^{-x}$

Solution

Consider the function $f(x) = 3^x$.

a. The function g is related to f by $g(x) = -f(x)$. To sketch the graph of g, reflect the graph of f about the x-axis, as shown in Figure 11.6. Note that the y-intercept of g is $(0, -1)$.

b. The function h is related to f by $h(x) = f(-x)$. To sketch the graph of h, reflect the graph of f about the y-axis, as shown in Figure 11.7.

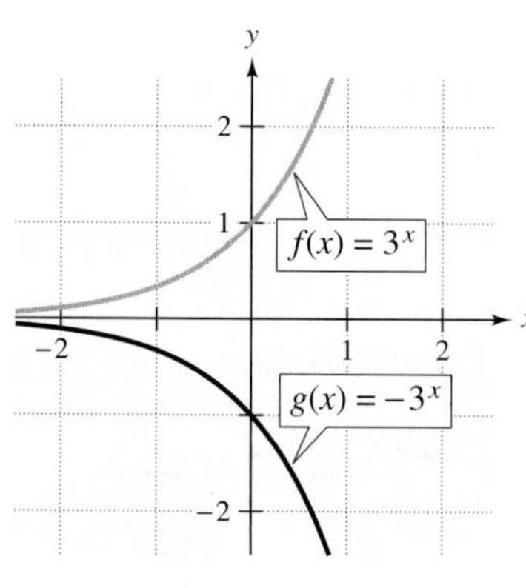

Figure 11.6

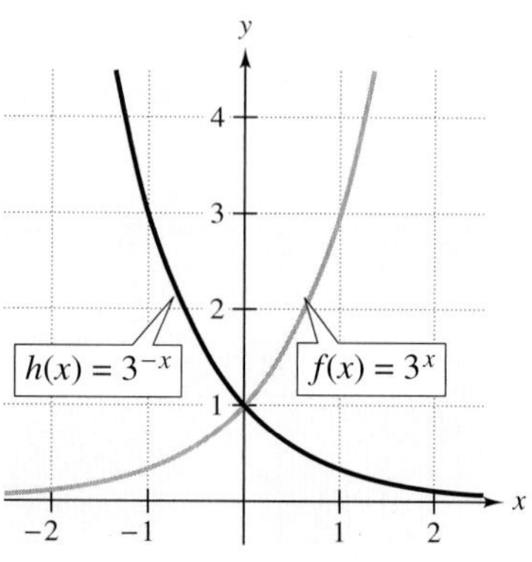

Figure 11.7

 ***CHECKPOINT*** *Now try Exercise 67.*

3 ▶ Evaluate the natural base e and graph natural exponential functions.

The Natural Exponential Function

So far, integers or rational numbers have been used as bases of exponential functions. In many applications of exponential functions, the convenient choice for a base is the following irrational number, denoted by the letter "e."

$e \approx 2.71828\ldots$ Natural base

This number is called the **natural base.** The function

$f(x) = e^x$ Natural exponential function

is called the **natural exponential function.** To evaluate the natural exponential function, you need a calculator, preferably one having a natural exponential key [e^x]. Here are some examples of how to use such a calculator to evaluate the natural exponential function.

Technology: Tip

Some calculators do not have a key labeled [e^x]. If your calculator does not have this key, but does have a key labeled [LN], you may be able to use the two-keystroke sequence [INV] [LN] in place of [e^x].

Value	*Keystrokes*	*Display*	
e^2	2 [e^x]	7.3890561	Scientific
e^2	[e^x] 2 [)] [ENTER]	7.3890561	Graphing
e^{-3}	3 [+/−] [e^x]	0.0497871	Scientific
e^{-3}	[e^x] [(−)] 3 [)] [ENTER]	0.0497871	Graphing

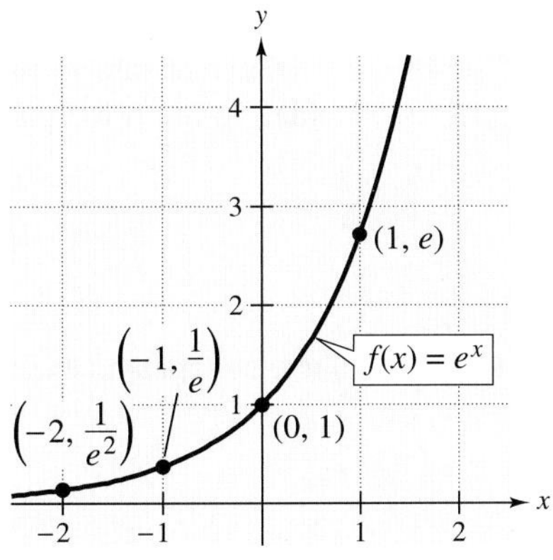

Figure 11.8

When evaluating the natural exponential function, remember that e is the constant number 2.71828 . . . , and x is a variable. After evaluating this function at several values, as shown in the table, you can sketch its graph, as shown in Figure 11.8.

x	−2	−1.5	−1.0	−0.5	0.0	0.5	1.0	1.5
$f(x) = e^x$	0.135	0.223	0.368	0.607	1.000	1.649	2.718	4.482

From the graph, notice the following characteristics of the natural exponential function.

- Domain: $(-\infty, \infty)$
- Range: $(0, \infty)$
- Intercept: $(0, 1)$
- Increasing (moves up to the right)
- Asymptote: x-axis

Notice that these characteristics are consistent with those listed for the exponential function $y = a^x$ on page 687.

4 ▶ Use exponential functions to solve application problems.

Applications

A common scientific application of exponential functions is **radioactive decay.**

EXAMPLE 6 Radioactive Decay 

A particular radioactive element has a half-life of 25 years. For an initial mass of 10 grams, the mass y (in grams) that remains after t years is given by

$$y = 10\left(\frac{1}{2}\right)^{t/25}, \quad t \geq 0.$$

How much of the initial mass remains after 120 years?

Solution

When $t = 120$, the mass is given by

$$y = 10\left(\frac{1}{2}\right)^{120/25} \qquad \text{Substitute 120 for } t.$$

$$= 10\left(\frac{1}{2}\right)^{4.8} \qquad \text{Simplify.}$$

$$\approx 0.359. \qquad \text{Use a calculator.}$$

So, after 120 years, the mass has decayed from an initial amount of 10 grams to only 0.359 gram. Note in Figure 11.9 that the graph of the function shows the 25-year half-life. That is, after 25 years the mass is 5 grams (half of the original), after another 25 years the mass is 2.5 grams, and so on.

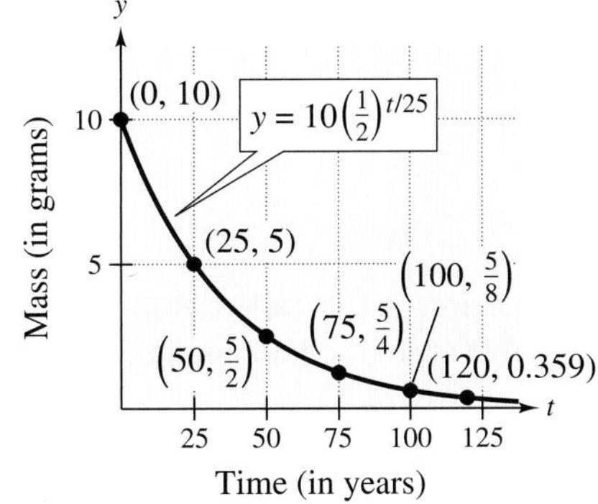

Figure 11.9

✓ ***CHECKPOINT*** *Now try Exercise 69.*

One of the most familiar uses of exponential functions involves **compound interest.** For instance, a principal P is invested at an annual interest rate r (in decimal form), compounded once a year. If the interest is added to the principal at the end of the year, the balance is

$$\begin{aligned} A &= P + Pr \\ &= P(1 + r). \end{aligned}$$

This pattern of multiplying the previous principal by $(1 + r)$ is then repeated each successive year, as shown below.

Time in Years	*Balance at Given Time*
0	$A = P$
1	$A = P(1 + r)$
2	$A = P(1 + r)(1 + r) = P(1 + r)^2$
3	$A = P(1 + r)^2(1 + r) = P(1 + r)^3$
$\vdots$	$\vdots$
t	$A = P(1 + r)^t$

To account for more frequent compounding of interest (such as quarterly or monthly compounding), let n be the number of compoundings per year and let t be the number of years. Then the rate per compounding is r/n and the account balance after t years is

$$A = P\left(1 + \frac{r}{n}\right)^{nt}.$$

EXAMPLE 7 **Finding the Balance for Compound Interest** 

A sum of \$10,000 is invested at an annual interest rate of 7.5%, compounded monthly. Find the balance in the account after 10 years.

Solution

Using the formula for compound interest, with $P = 10{,}000$, $r = 0.075$, $n = 12$ (for monthly compounding), and $t = 10$, you obtain the following balance.

$$\begin{aligned} A &= 10{,}000\left(1 + \frac{0.075}{12}\right)^{12(10)} \\ &\approx \$21{,}120.65 \end{aligned}$$

 **CHECKPOINT** *Now try Exercise 71.*

A second method that banks use to compute interest is called **continuous compounding.** The formula for the balance for this type of compounding is

$$A = Pe^{rt}.$$

The formulas for both types of compounding are summarized on the next page.

Formulas for Compound Interest

After t years, the balance A in an account with principal P and annual interest rate r (in decimal form) is given by one of the following formulas.

1. For n compoundings per year: $A = P\left(1 + \frac{r}{n}\right)^{nt}$

2. For continuous compounding: $A = Pe^{rt}$

Technology: Discovery

Use a graphing calculator to evaluate

$$A = 15{,}000\left(1 + \frac{0.08}{n}\right)^{n(6)}$$

for $n = 1000$, 10,000, and 100,000. Compare these values with those found in parts (a) and (b) of Example 8.

As n gets larger and larger, do you think that the value of A will ever exceed the value found in Example 8(c)? Explain.

EXAMPLE 8 Comparing Three Types of Compounding 

A total of \$15,000 is invested at an annual interest rate of 8%. Find the balance after 6 years for each type of compounding.

a. Quarterly

b. Monthly

c. Continuous

Solution

a. Letting $P = 15{,}000$, $r = 0.08$, $n = 4$, and $t = 6$, the balance after 6 years at quarterly compounding is

$$A = 15{,}000\left(1 + \frac{0.08}{4}\right)^{4(6)}$$

$$\approx \$24{,}126.56.$$

b. Letting $P = 15{,}000$, $r = 0.08$, $n = 12$, and $t = 6$, the balance after 6 years at monthly compounding is

$$A = 15{,}000\left(1 + \frac{0.08}{12}\right)^{12(6)}$$

$$\approx \$24{,}202.53.$$

c. Letting $P = 15{,}000$, $r = 0.08$, and $t = 6$, the balance after 6 years at continuous compounding is

$$A = 15{,}000e^{0.08(6)}$$

$$\approx \$24{,}241.12.$$

Note that the balance is greater with continuous compounding than with quarterly or monthly compounding.

 **CHECKPOINT** *Now try Exercise 73.*

Example 8 illustrates the following general rule. For a given principal, interest rate, and time, the more often the interest is compounded per year, the greater the balance will be. Moreover, the balance obtained by continuous compounding is larger than the balance obtained by compounding n times per year.

Concept Check

1. For the exponential function f with base a, describe the operation you must perform to evaluate f at $x = 2$.

2. Consider the graphs of the functions $f(x) = 3^x$, $g(x) = 4^x$, and $h(x) = 4^{-x}$.
   (a) What point do the graphs of f, g, and h have in common?
   (b) Describe the asymptote for each graph.
   (c) State whether each graph increases or decreases as x increases.
   (d) Compare the graphs of f and g.

3. What special function is approximately equivalent to the function $f(x) = 2.72^x$?

4. The natural base e is used in the formula for which type of interest compounding?

11.1 EXERCISES

Go to pages 752–753 to record your assignments.

Developing Skills

In Exercises 1–8, simplify the expression.

1. $3^x \cdot 3^{x+2}$
2. $e^{3x} \cdot e^{-x}$
3. $\dfrac{e^{x+2}}{e^x}$
4. $\dfrac{3^{2x+3}}{3^{x+1}}$
5. $3(e^x)^{-2}$
6. $4(e^{2x})^{-1}$
7. $\sqrt[3]{-8e^{3x}}$
8. $\sqrt{4e^{6x}}$

In Exercises 9–16, evaluate the expression. (Round your answer to three decimal places.)

9. $5^{\sqrt{2}}$
10. $4^{-\pi}$
11. $e^{1/3}$
12. $e^{-1/3}$
13. $3(2e^{1/2})^3$
14. $(9e^2)^{3/2}$
15. $\dfrac{4e^3}{12e^2}$
16. $\dfrac{6e^5}{10e^7}$

In Exercises 17–30, evaluate the function as indicated. Use a calculator only if it is necessary or more efficient. (Round your answers to three decimal places.) *See Example 1.*

✓ 17. $f(x) = 3^x$
   (a) $x = -2$
   (b) $x = 0$
   (c) $x = 1$

18. $F(x) = 3^{-x}$
   (a) $x = -2$
   (b) $x = 0$
   (c) $x = 1$

19. $g(x) = 2.2^{-x}$
   (a) $x = 1$
   (b) $x = 3$
   (c) $x = \sqrt{6}$

20. $G(x) = 4.2^x$
   (a) $x = -1$
   (b) $x = -2$
   (c) $x = \sqrt{2}$

21. $f(t) = 500\left(\frac{1}{2}\right)^t$
   (a) $t = 0$
   (b) $t = 1$
   (c) $t = \pi$

22. $g(s) = 1200\left(\frac{2}{3}\right)^s$
   (a) $s = 0$
   (b) $s = 2$
   (c) $s = \sqrt{2}$

23. $f(x) = 1000(1.05)^{2x}$
   (a) $x = 0$
   (b) $x = 5$
   (c) $x = 10$

24. $g(t) = 10{,}000(1.03)^{4t}$
   (a) $t = 1$
   (b) $t = 3$
   (c) $t = 5.5$

25. $h(x) = \dfrac{5000}{(1.06)^{8x}}$
   (a) $x = 5$
   (b) $x = 10$
   (c) $x = 20$

26. $P(t) = \dfrac{10{,}000}{(1.01)^{12t}}$
   (a) $t = 2$
   (b) $t = 10$
   (c) $t = 20$

27. $g(x) = 10e^{-0.5x}$
   (a) $x = -4$
   (b) $x = 4$
   (c) $x = 8$

28. $A(t) = 200e^{0.1t}$
   (a) $t = 10$
   (b) $t = 20$
   (c) $t = 40$

29. $g(x) = \dfrac{1000}{2 + e^{-0.12x}}$
   (a) $x = 0$
   (b) $x = 10$
   (c) $x = 50$

30. $f(z) = \dfrac{100}{1 + e^{-0.05z}}$
   (a) $z = 0$
   (b) $z = 10$
   (c) $z = 20$

In Exercises 31–46, sketch the graph of the function. Identify the horizontal asymptote. *See Examples 2 and 3.*

✓ 31. $f(x) = 3^x$

32. $h(x) = \frac{1}{2}(3^x)$

✓ 33. $f(x) = 3^{-x} = \left(\frac{1}{3}\right)^x$

34. $h(x) = \frac{1}{2}(3^{-x})$

35. $g(x) = 3^x - 2$

36. $g(x) = 3^x + 1$

37. $g(x) = 5^{x-1}$

38. $g(x) = 5^{x+3}$

39. $f(t) = 2^{-t^2}$

40. $f(t) = 2^{t^2}$

41. $f(x) = -2^{0.5x}$

42. $h(t) = -2^{-0.5t}$

43. $f(x) = -\left(\frac{1}{3}\right)^x$

44. $f(x) = \left(\frac{3}{4}\right)^x + 1$

45. $g(t) = 200\left(\frac{1}{2}\right)^t$

46. $h(x) = 27\left(\frac{2}{3}\right)^x$

In Exercises 47–58, use a graphing calculator to graph the function.

47. $y = 7^{x/2}$

48. $y = 7^{-x/2}$

49. $y = 7^{-x/2} + 5$

50. $y = 7^{(x-3)/2}$

51. $y = 500(1.06)^t$

52. $y = 100(1.06)^{-t}$

53. $y = 3e^{0.2x}$

54. $y = 50e^{-0.05x}$

55. $P(t) = 100e^{-0.1t}$

56. $N(t) = 10{,}000e^{0.05t}$

57. $y = 6e^{-x^2/3}$

58. $g(x) = 7e^{(x+1)/2}$

In Exercises 59–62, match the function with its graph. [The graphs are labeled (a), (b), (c), and (d).]

(a)

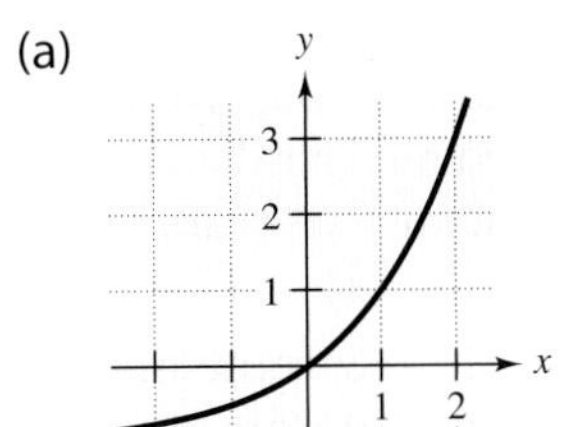

(b)

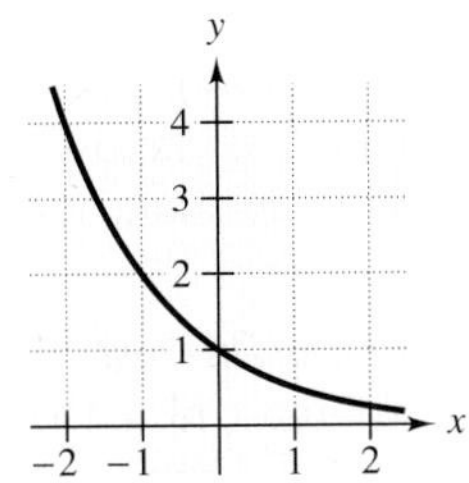

(c)

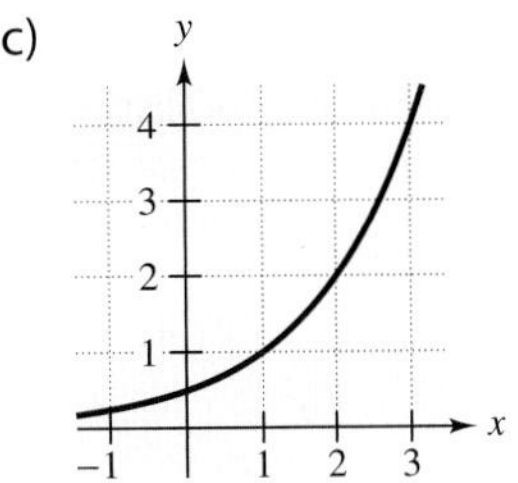

(d)

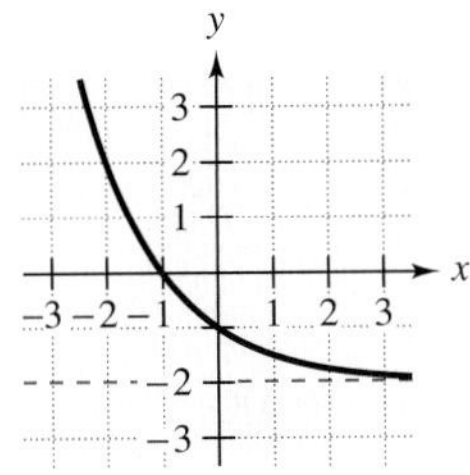

59. $f(x) = 2^{-x}$

60. $f(x) = 2^x - 1$

61. $f(x) = 2^{x-1}$

62. $f(x) = \left(\frac{1}{2}\right)^x - 2$

In Exercises 63–68, identify the transformation of the graph of $f(x) = 4^x$, and sketch the graph of h. *See Examples 4 and 5.*

✓ 63. $h(x) = 4^x - 1$

64. $h(x) = 4^x + 2$

65. $h(x) = 4^{x+2}$

66. $h(x) = 4^{x-4}$

✓ 67. $h(x) = -4^x$

68. $h(x) = 4^{-x}$

Solving Problems

✓ 69. *Radioactive Decay* After t years, 16 grams of a radioactive element with a half-life of 30 years decays to a mass y (in grams) given by

$$y = 16\left(\frac{1}{2}\right)^{t/30}, \quad t \geq 0.$$

How much of the initial mass remains after 80 years?

70. ***Radioactive Substance*** In July of 1999, an individual bought several leaded containers from a metals recycler and found two of them labeled "radioactive." An investigation showed that the containers, originally obtained from Ohio State University, apparently had been used to store iodine-131 starting in January of 1999. Because iodine-131 has a half-life of only 8 days, no elevated radiation levels were detected. (Source: United States Nuclear Regulatory Commission)

Suppose 6 grams of iodine-131 is stored in January. The mass y (in grams) that remains after t days is given by $y = 6\left(\frac{1}{2}\right)^{t/8}, \quad t \geq 0$. How much of the substance is left in July, after 180 days have passed?

71. *Compound Interest* A sum of \$5000 is invested at an annual interest rate of 6%, compounded monthly. Find the balance in the account after 5 years.

72. *Compound Interest* A sum of \$2000 is invested at an annual interest rate of 8%, compounded quarterly. Find the balance in the account after 10 years.

Compound Interest In Exercises 73–76, complete the table to determine the balance A for P dollars invested at rate r for t years, compounded n times per year.

n	1	4	12	365	Continuous compounding
A					

	Principal	*Rate*	*Time*
73.	$P = \$100$	$r = 7\%$	$t = 15$ years
74.	$P = \$600$	$r = 4\%$	$t = 5$ years
75.	$P = \$2000$	$r = 9.5\%$	$t = 10$ years
76.	$P = \$1500$	$r = 6.5\%$	$t = 20$ years

Compound Interest In Exercises 77–80, complete the table to determine the principal P that will yield a balance of A dollars when invested at rate r for t years, compounded n times per year.

n	1	4	12	365	Continuous compounding
P					

	Balance	*Rate*	*Time*
77.	$A = \$5000$	$r = 7\%$	$t = 10$ years
78.	$A = \$100{,}000$	$r = 9\%$	$t = 20$ years
79.	$A = \$1{,}000{,}000$	$r = 10.5\%$	$t = 40$ years
80.	$A = \$10{,}000$	$r = 12.4\%$	$t = 3$ years

81. *Demand* The daily demand x and the price p for a collectible are related by $p = 25 - 0.4e^{0.02x}$. Find the prices for demands of (a) $x = 100$ units and (b) $x = 125$ units.

82. *Population Growth* The populations P (in millions) of the United States from 1980 to 2006 can be approximated by the exponential function $P(t) = 226(1.0110)^t$, where t is the time in years, with $t = 0$ corresponding to 1980. Use the model to estimate the populations in the years (a) 2010 and (b) 2020. (Source: U.S. Census Bureau)

83. *Property Value* The value of a piece of property doubles every 15 years. You buy the property for \$64,000. Its value t years after the date of purchase should be $V(t) = 64{,}000(2)^{t/15}$. Use the model to approximate the values of the property (a) 5 years and (b) 20 years after it is purchased.

84. *Inflation Rate* Suppose the annual rate of inflation is expected to be 4% for the next 5 years. Then the cost C of goods or services t years from now can be approximated by $C(t) = P(1.04)^t$, $0 \le t \le 5$, where P is the present cost. Estimate the yearly cost of tuition four years from now for a college where the tuition is currently \$32,000 per year.

85. *Depreciation* After t years, the value of a car that originally cost \$16,000 depreciates so that each year it is worth $\frac{3}{4}$ of its value from the previous year. Find a model for $V(t)$, the value of the car after t years. Sketch a graph of the model, and determine the value of the car 2 years and 4 years after its purchase.

86. *Depreciation* Suppose the value of the car in Exercise 85 changes by straight-line depreciation of \$3000 per year. Find a model $v(t)$ for the value of the car. Sketch a graph of $v(t)$ and the graph of $V(t)$ from Exercise 85, on the same coordinate axes used in Exercise 85. Which model do you prefer if you sell the car after 2 years? after 4 years?

87. *Match Play* A fraternity sponsors a match play golf tournament to benefit charity. The tournament is bracketed for 1024 golfers, and half of the golfers are eliminated in each round of one-on-one matches. So, 512 golfers remain after the first round, 256 after the second round, and so on, until there is a single champion. Write an exponential function that models this problem. How many golfers remain after the eighth round?

88. *Savings Plan* You decide to start saving pennies according to the following pattern. You save 1 penny the first day, 2 pennies the second day, 4 the third day, 8 the fourth day, and so on. Each day you save twice the number of pennies you saved on the previous day. Write an exponential function that models this problem. How many pennies do you save on the thirtieth day?

89. *Parachute Drop* A parachutist jumps from a plane and releases her parachute 2000 feet above the ground (see figure). From there, her height h (in feet) is given by $h = 1950 + 50e^{-0.4433t} - 22t$, where t is the number of seconds after the parachute is released.

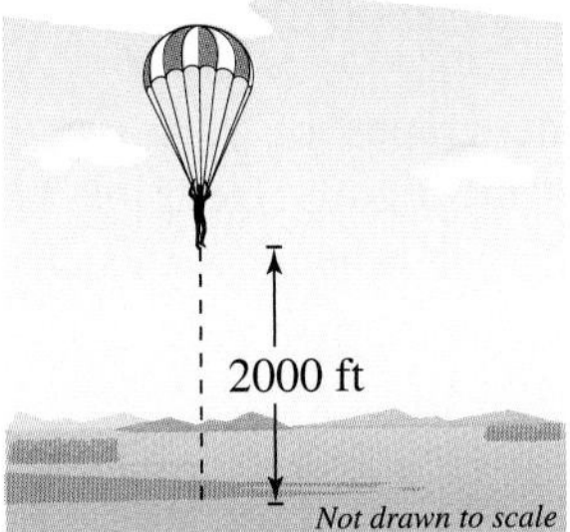

(a) Use a graphing calculator to graph the function.

(b) Use a graphing calculator to find the parachutist's height every 10 seconds from $t = 0$ until the time she reaches the ground. Record your results in a table.

(c) During which 10-second period does the parachutist's height change the most? Use the context of the problem to explain why.

90. *Parachute Drop* A parachutist jumps from a plane and releases his parachute 3000 feet above the ground. From there, his height h (in feet) is given by $h = 2940 + 60e^{-0.4021t} - 24t$, where t is the number of seconds after the parachute is released.

(a) Use a graphing calculator to graph the function.

(b) Use a graphing calculator to find the parachutist's height every 10 seconds from $t = 0$ until the time he reaches the ground. Record your results in a table.

(c) During which 10-second period does the parachutist's height change the most? Use the context of the problem to explain why.

91. *Data Analysis* A meteorologist measures the atmospheric pressure P (in kilograms per square meter) at various altitudes h (in kilometers). The data are shown in the table.

h	0	5	10	15	20
P	10,332	5583	2376	1240	517

(a) Use a graphing calculator to plot the data points.

(b) A model for the data is given by $P = 10{,}958e^{-0.15h}$. Use a graphing calculator to graph the model with the data points from part (a), in the same viewing window. How well does the model fit the data?

(c) Use a graphing calculator to create a table comparing the model with the data points.

(d) Estimate the atmospheric pressure at an altitude of 8 kilometers.

(e) Use the graph to estimate the altitude at which the atmospheric pressure is 2000 kilograms per square meter.

92. *Data Analysis* The median prices of existing one-family homes sold in the United States in the years 1999 through 2006 are shown in the table. (Source: National Association of Realtors)

Year	1999	2000	2001	2002
Price	\$141,200	\$147,300	\$156,600	\$167,600

Year	2003	2004	2005	2006
Price	\$180,200	\$195,200	\$219,000	\$221,900

A model for this data is given by $y = 73{,}482e^{0.0700t}$, where t is time in years, with $t = 9$ representing 1999.

(a) Use the model to complete the table below and compare the results with the actual data.

Year	1999	2000	2001	2002
Price				

Year	2003	2004	2005	2006
Price				

(b) Use a graphing calculator to graph the model with the data points from part (a), in the same viewing window.

(c) Beyond $t = 16$, do the y-values of the model increase at a *higher rate*, a *constant rate*, or a *lower rate* for increasing values of t? Do you think that the model is reliable for making predictions past 2006? Explain.

93. *Calculator Experiment*

(a) Use a calculator to complete the table.

x	1	10	100	1000	10,000
$\left(1 + \frac{1}{x}\right)^x$					

(b) Use the table to sketch the graph of the function

$$f(x) = \left(1 + \frac{1}{x}\right)^x.$$

Does this graph appear to be approaching a horizontal asymptote?

(c) From parts (a) and (b), what conclusions can you make about the value of

$$\left(1 + \frac{1}{x}\right)^x$$

as x gets larger and larger?

94. Identify the graphs of $y_1 = e^{0.2x}$, $y_2 = e^{0.5x}$, and $y_3 = e^x$ in the figure. Describe the effect on the graph of $y = e^{kx}$ when $k > 0$ is changed.

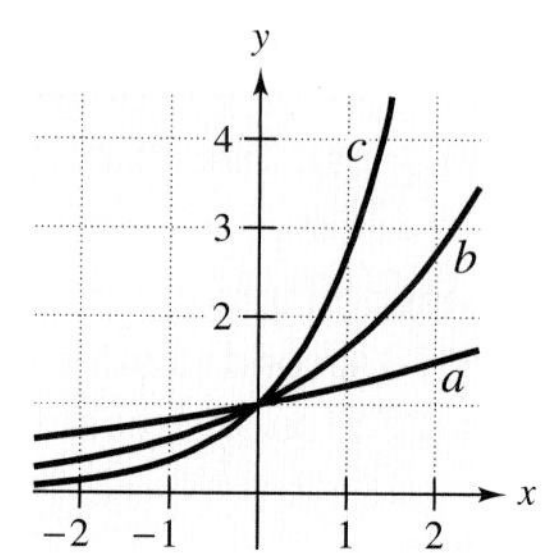

Explaining Concepts

95. Explain why 1^x is not an exponential function.

96. Compare the graphs of $f(x) = 3^x$ and $g(x) = \left(\frac{1}{3}\right)^x$.

97. Does e equal $\dfrac{271{,}801}{99{,}990}$? Explain.

98. Use the characteristics of the exponential function with base 2 to explain why $2^{\sqrt{2}}$ is greater than 2 but less than 4.

99. Consider functions of the form $f(x) = k^x$. Describe the real values of k for which the values of f will *increase*, *decrease*, and *remain constant* as x increases.

100. Look back at your answers to the compound interest problems in Exercises 73–76. In terms of the interest earned, would you say there is much difference between daily compounding and continuous compounding? Explain your reasoning.

Cumulative Review

In Exercises 101 and 102, find the domain of the function.

101. $g(s) = \sqrt{s - 4}$

102. $h(t) = \dfrac{\sqrt{t^2 - 1}}{t - 2}$

In Exercises 103 and 104, sketch a graph of the equation. Use the Vertical Line Test to determine whether y is a function of x.

103. $y^2 = x - 1$

104. $x = y^4 + 1$

11.2 Composite and Inverse Functions

Superstock, Inc.

What You Should Learn

1 ▶ Form compositions of two functions and find the domains of composite functions.
2 ▶ Use the Horizontal Line Test to determine whether functions have inverse functions.
3 ▶ Find inverse functions algebraically.
4 ▶ Graphically verify that two functions are inverse functions of each other.

Why You Should Learn It

Inverse functions can be used to model and solve real-life problems. For instance, in Exercise 109 on page 709, you will use an inverse function to determine the number of units produced for a certain hourly wage.

Composite Functions

1 ▶ Form compositions of two functions and find the domains of composite functions.

Two functions can be combined to form another function called the **composition** of the two functions. For instance, if $f(x) = 2x^2$ and $g(x) = x - 1$, the composition of f with g is denoted by $f \circ g$ and is given by

$$f(g(x)) = f(x - 1) = 2(x - 1)^2.$$

Definition of Composition of Two Functions

The **composition** of the functions f and g is given by

$$(f \circ g)(x) = f(g(x)).$$

The domain of the **composite function** $(f \circ g)$ is the set of all x in the domain of g such that $g(x)$ is in the domain of f. (See Figure 11.10.)

$f \circ g$

x $g(x)$ $f(g(x))$

Domain of g Domain of f

Figure 11.10

Study Tip

A composite function can be viewed as a function within a function, where the composition

$$(f \circ g)(x) = f(g(x))$$

has f as the "outer" function and g as the "inner" function. This is reversed in the composition

$$(g \circ f)(x) = g(f(x)).$$

EXAMPLE 1 Forming the Composition of Two Functions

Given $f(x) = 2x + 4$ and $g(x) = 3x - 1$, find the composition of f with g. Then evaluate the composite function when $x = 1$ and when $x = -3$.

Solution

$(f \circ g)(x) = f(g(x))$	Definition of $f \circ g$
$= f(3x - 1)$	$g(x) = 3x - 1$ is the inner function.
$= 2(3x - 1) + 4$	Input $3x - 1$ into the outer function f.
$= 6x - 2 + 4$	Distributive Property
$= 6x + 2$	Simplify.

When $x = 1$, the value of this composite function is

$$(f \circ g)(1) = 6(1) + 2 = 8.$$

When $x = -3$, the value of this composite function is

$$(f \circ g)(-3) = 6(-3) + 2 = -16.$$

✓ **CHECKPOINT** *Now try Exercise 1.*

The composition of f with g is generally *not* the same as the composition of g with f. This is illustrated in Example 2.

EXAMPLE 2 Comparing the Compositions of Functions

Given $f(x) = 2x - 3$ and $g(x) = x^2 + 1$, find each composition.

a. $(f \circ g)(x)$ **b.** $(g \circ f)(x)$

Solution

a. $(f \circ g)(x) = f(g(x))$ — Definition of $f \circ g$

$= f(x^2 + 1)$ — $g(x) = x^2 + 1$ is the inner function.

$= 2(x^2 + 1) - 3$ — Input $x^2 + 1$ into the outer function f.

$= 2x^2 + 2 - 3$ — Distributive Property

$= 2x^2 - 1$ — Simplify.

b. $(g \circ f)(x) = g(f(x))$ — Definition of $g \circ f$

$= g(2x - 3)$ — $f(x) = 2x - 3$ is the inner function.

$= (2x - 3)^2 + 1$ — Input $2x - 3$ into the outer function g.

$= 4x^2 - 12x + 9 + 1$ — Expand.

$= 4x^2 - 12x + 10$ — Simplify.

Note that $(f \circ g)(x) \neq (g \circ f)(x)$.

✓ **CHECKPOINT** *Now try Exercise 3.*

To determine the domain of a composite function, first write the composite function in simplest form. Then use the fact that its domain *either is equal to or is a restriction of the domain of the "inner" function*. This is demonstrated in Example 3.

EXAMPLE 3 Finding the Domain of a Composite Function

Find the domain of the composition of f with g when $f(x) = x^2$ and $g(x) = \sqrt{x}$.

Solution

$(f \circ g)(x) = f(g(x))$ — Definition of $f \circ g$

$= f(\sqrt{x})$ — $g(x) = \sqrt{x}$ is the inner function.

$= (\sqrt{x})^2$ — Input $\sqrt{x}$ into the outer function f.

$= x, \ x \geq 0$ — Domain of $f \circ g$ is all $x \geq 0$.

The domain of the inner function $g(x) = \sqrt{x}$ is the set of all nonnegative real numbers. The simplified form of $f \circ g$ has no restriction on this set of numbers. So, the restriction $x \geq 0$ must be added to the composition of this function. The domain of $f \circ g$ is the set of all nonnegative real numbers.

✓ **CHECKPOINT** *Now try Exercise 21.*

2 ▶ Use the Horizontal Line Test to determine whether functions have inverse functions.

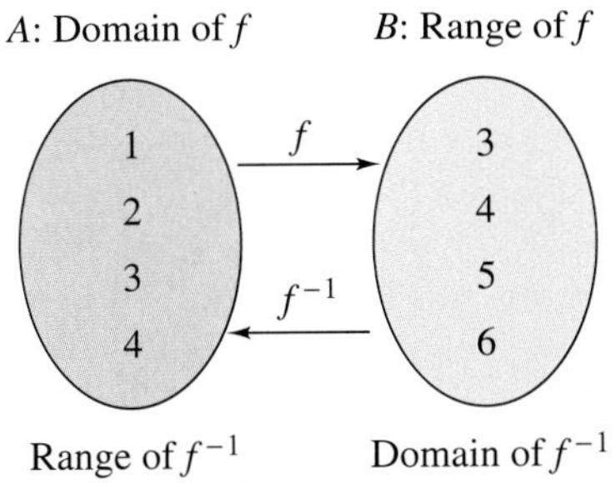

Figure 11.11 The function f is one-to-one and has inverse function f^{-1}.

One-to-One and Inverse Functions

In Section 4.3, you learned that a function can be represented by a set of ordered pairs. For instance, the function $f(x) = x + 2$ from the set $A = \{1, 2, 3, 4\}$ to the set $B = \{3, 4, 5, 6\}$ can be written as follows.

$$f(x) = x + 2: \quad \{(1, 3), (2, 4), (3, 5), (4, 6)\}$$

By interchanging the first and second coordinates of each of these ordered pairs, you can form another function that is called the **inverse function** of f, denoted by f^{-1}. It is a function from the set B to the set A, and can be written as follows.

$$f^{-1}(x) = x - 2: \quad \{(3, 1), (4, 2), (5, 3), (6, 4)\}$$

Interchanging the ordered pairs for a function f will only produce another function when f is one-to-one. A function f is **one-to-one** if each value of the dependent variable corresponds to exactly one value of the independent variable. Figure 11.11 shows that the domain of f is the range of f^{-1} and the range of f is the domain of f^{-1}.

Horizontal Line Test for Inverse Functions

A function f has an inverse function f^{-1} if and only if f is one-to-one. Graphically, a function f has an inverse function f^{-1} if and only if no *horizontal* line intersects the graph of f at more than one point.

EXAMPLE 4 Applying the Horizontal Line Test

Use the Horizontal Line Test to determine if the function is one-to-one and so has an inverse function.

a. The graph of the function $f(x) = x^3 - 1$ is shown in Figure 11.12. Because no horizontal line intersects the graph of f at more than one point, you can conclude that f *is* a one-to-one function and *does* have an inverse function.

b. The graph of the function $f(x) = x^2 - 1$ is shown in Figure 11.13. Because it is possible to find a horizontal line that intersects the graph of f at more than one point, you can conclude that f *is not* a one-to-one function and *does not* have an inverse function.

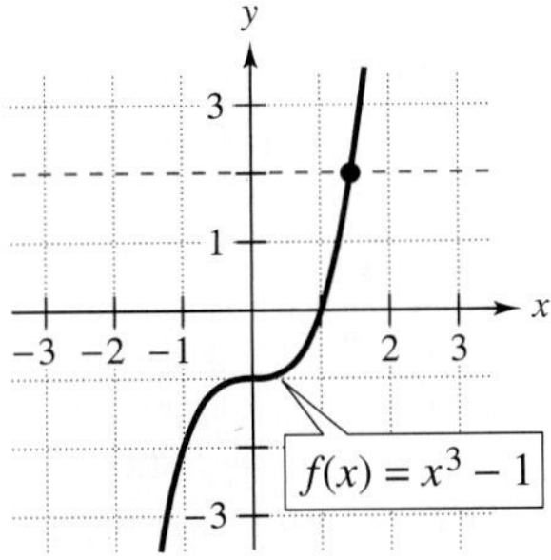

Figure 11.12

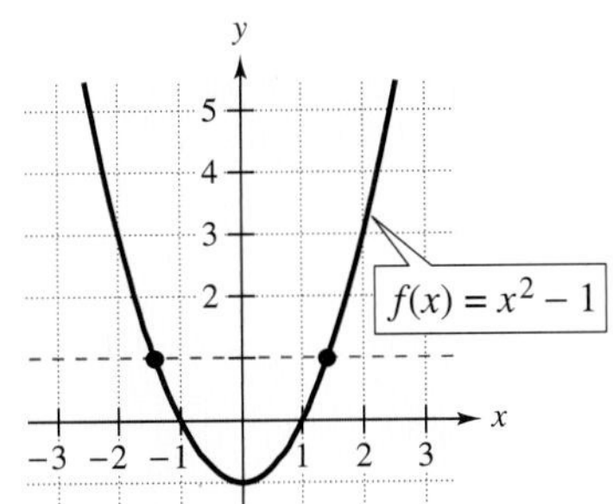

Figure 11.13

✓ ***CHECKPOINT*** *Now try Exercise 35.*

The formal definition of an inverse function is given as follows.

Definition of Inverse Function

Let f and g be two functions such that

$f(g(x)) = x$ for every x in the domain of g

and

$g(f(x)) = x$ for every x in the domain of f.

The function g is called the **inverse function** of the function f, and is denoted by f^{-1} (read "f-inverse"). So, $f(f^{-1}(x)) = x$ and $f^{-1}(f(x)) = x$. The domain of f must be equal to the range of f^{-1}, and vice versa.

Do not be confused by the use of -1 to denote the inverse function f^{-1}. Whenever f^{-1} is written, it *always* refers to the inverse function of f and *not* to the reciprocal of $f(x)$.

If the function g is the inverse function of the function f, it must also be true that the function f is the inverse function of the function g. For this reason, you can refer to the functions f and g as being *inverse functions of each other.*

EXAMPLE 5 Verifying Inverse Functions

Show that $f(x) = 2x - 4$ and $g(x) = \dfrac{x + 4}{2}$ are inverse functions of each other.

Solution

Begin by noting that the domain and range of both functions are the entire set of real numbers. To show that f and g are inverse functions of each other, you need to show that $f(g(x)) = x$ and $g(f(x)) = x$, as follows.

$$f(g(x)) = f\left(\frac{x + 4}{2}\right) \qquad g(x) = (x + 4)/2 \text{ is the inner function.}$$

$$= 2\left(\frac{x + 4}{2}\right) - 4 \qquad \text{Input } (x + 4)/2 \text{ into the outer function } f.$$

$$= x + 4 - 4 = x \qquad \text{Simplify.}$$

$$g(f(x)) = g(2x - 4) \qquad f(x) = 2x - 4 \text{ is the inner function.}$$

$$= \frac{(2x - 4) + 4}{2} \qquad \text{Input } 2x - 4 \text{ into the outer function } g.$$

$$= \frac{2x}{2} = x \qquad \text{Simplify.}$$

Note that the two functions f and g "undo" each other in the following verbal sense. The function f first *multiplies* the input x by 2 and then *subtracts* 4, whereas the function g first *adds* 4 and then *divides* the result by 2.

✓ ***CHECKPOINT*** *Now try Exercise 43.*

EXAMPLE 6 **Verifying Inverse Functions**

Show that the functions

$$f(x) = x^3 + 1 \text{ and } g(x) = \sqrt[3]{x - 1}$$

are inverse functions of each other.

Solution

Begin by noting that the domain and range of both functions are the entire set of real numbers. To show that f and g are inverse functions of each other, you need to show that $f(g(x)) = x$ and $g(f(x)) = x$, as follows.

$f(g(x)) = f\left(\sqrt[3]{x - 1}\right)$	$g(x) = \sqrt[3]{x - 1}$ is the inner function.
$= \left(\sqrt[3]{x - 1}\right)^3 + 1$	Input $\sqrt[3]{x - 1}$ into the outer function f.
$= (x - 1) + 1 = x$	Simplify.
$g(f(x)) = g(x^3 + 1)$	$f(x) = x^3 + 1$ is the inner function.
$= \sqrt[3]{(x^3 + 1) - 1}$	Input $x^3 + 1$ into the outer function g.
$= \sqrt[3]{x^3} = x$	Simplify.

Note that the two functions f and g "undo" each other in the following verbal sense. The function f first *cubes* the input x and then *adds* 1, whereas the function g first *subtracts* 1 and then *takes the cube root* of the result.

CHECKPOINT *Now try Exercise 45.*

3 ▶ Find inverse functions algebraically.

Finding an Inverse Function Algebraically

You can find the inverse function of a simple function by inspection. For instance, the inverse function of $f(x) = 10x$ is $f^{-1}(x) = x/10$. For more complicated functions, however, it is best to use the following steps for finding an inverse function. The key step in these guidelines is switching the roles of x and y. This step corresponds to the fact that inverse functions have ordered pairs with the coordinates reversed.

Study Tip

You can graph a function and use the Horizontal Line Test to see if the function is one-to-one before trying to find its inverse function.

Finding an Inverse Function Algebraically

1. In the equation for $f(x)$, replace $f(x)$ with y.
2. Interchange x and y.
3. If the new equation does not represent y as a function of x, the function f does not have an inverse function. If the new equation does represent y as a function of x, solve the new equation for y.
4. Replace y with $f^{-1}(x)$.
5. Verify that f and f^{-1} are inverse functions of each other by showing that $f(f^{-1}(x)) = x = f^{-1}(f(x))$.

EXAMPLE 7 Finding an Inverse Function

Determine whether each function has an inverse function. If it does, find its inverse function.

a. $f(x) = 2x + 3$ **b.** $f(x) = x^3 + 3$

Solution

a.

$f(x) = 2x + 3$ Write original function.

$y = 2x + 3$ Replace $f(x)$ with y.

$x = 2y + 3$ Interchange x and y.

$y = \dfrac{x - 3}{2}$ Solve for y.

$f^{-1}(x) = \dfrac{x - 3}{2}$ Replace y with $f^{-1}(x)$.

You can verify that $f(f^{-1}(x)) = x = f^{-1}(f(x))$, as follows.

$$f(f^{-1}(x)) = f\left(\frac{x-3}{2}\right) = 2\left(\frac{x-3}{2}\right) + 3 = (x - 3) + 3 = x$$

$$f^{-1}(f(x)) = f^{-1}(2x + 3) = \frac{(2x + 3) - 3}{2} = \frac{2x}{2} = x$$

b.

$f(x) = x^3 + 3$ Write original function.

$y = x^3 + 3$ Replace $f(x)$ with y.

$x = y^3 + 3$ Interchange x and y.

$\sqrt[3]{x - 3} = y$ Solve for y.

$f^{-1}(x) = \sqrt[3]{x - 3}$ Replace y with $f^{-1}(x)$.

You can verify that $f(f^{-1}(x)) = x = f^{-1}(f(x))$, as follows.

$$f(f^{-1}(x)) = f(\sqrt[3]{x - 3}) = (\sqrt[3]{x - 3})^3 + 3 = (x - 3) + 3 = x$$

$$f^{-1}(f(x)) = f^{-1}(x^3 + 3) = \sqrt[3]{(x^3 + 3) - 3} = \sqrt[3]{x^3} = x$$

✓ ***CHECKPOINT*** *Now try Exercise 55.*

Technology: Discovery

Use a graphing calculator to graph $f(x) = x^3 + 1$, $f^{-1}(x) = \sqrt[3]{x - 1}$, and $y = x$ in the same viewing window.

a. Relative to the line $y = x$, how do the graphs of f and f^{-1} compare?

b. For the graph of f, complete the table.

x	-1	0	1
f			

For the graph of f^{-1}, complete the table.

x	0	1	2
f^{-1}			

What can you conclude about the coordinates of the points on the graph of f compared with those on the graph of f^{-1}?

EXAMPLE 8 A Function That Has No Inverse Function

$f(x) = x^2$ Original equation

$y = x^2$ Replace $f(x)$ with y.

$x = y^2$ Interchange x and y.

Recall from Section 4.3 that the equation $x = y^2$ does not represent y as a function of x because you can find two different y-values that correspond to the same x-value. Because the equation does not represent y as a function of x, you can conclude that the original function f does not have an inverse function.

✓ ***CHECKPOINT*** *Now try Exercise 67.*

4 ▶ Graphically verify that two functions are inverse functions of each other.

Graphs of Inverse Functions

The graphs of f and f^{-1} are related to each other in the following way. If the point (a, b) lies on the graph of f, the point (b, a) must lie on the graph of f^{-1}, and vice versa. This means that the graph of f^{-1} is a reflection of the graph of f in the line $y = x$, as shown in Figure 11.14. This "reflective property" of the graphs of f and f^{-1} is illustrated in Examples 9 and 10.

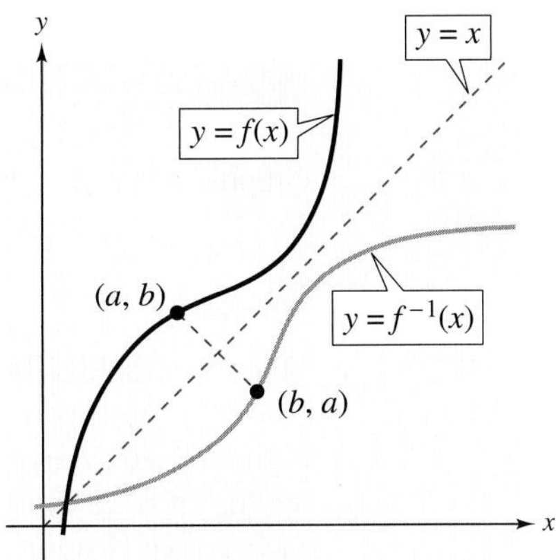

Figure 11.14 The graph of f^{-1} is a reflection of the graph of f in the line $y = x$.

EXAMPLE 9 The Graphs of f and f^{-1}

Sketch the graphs of the inverse functions $f(x) = 2x - 3$ and $f^{-1}(x) = \frac{1}{2}(x + 3)$ on the same rectangular coordinate system, and show that the graphs are reflections of each other in the line $y = x$.

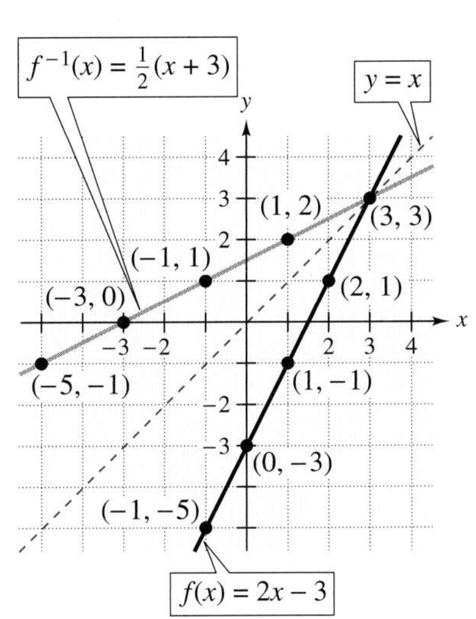

Figure 11.15

Solution

The graphs of f and f^{-1} are shown in Figure 11.15. Visually, it appears that the graphs are reflections of each other. You can verify this reflective property by testing a few points on each graph. Note in the following list that if the point (a, b) is on the graph of f, the point (b, a) is on the graph of f^{-1}.

$f(x) = 2x - 3$	$f^{-1}(x) = \frac{1}{2}(x + 3)$
$(-1, -5)$	$(-5, -1)$
$(0, -3)$	$(-3, 0)$
$(1, -1)$	$(-1, 1)$
$(2, 1)$	$(1, 2)$
$(3, 3)$	$(3, 3)$

✓ ***CHECKPOINT*** *Now try Exercise 79.*

You can sketch the graph of an inverse function without knowing the equation of the inverse function. Simply find the coordinates of points that lie on the original function. By interchanging the x- and y-coordinates, you have points that lie on the graph of the inverse function. Plot these points and sketch the graph of the inverse function.

In Example 8, you saw that the function

$$f(x) = x^2$$

has no inverse function. A more complete way of saying this is "*assuming that the domain of f is the entire real line,* the function $f(x) = x^2$ has no inverse function." If, however, you *restrict* the domain of f to the nonnegative real numbers, then f does have an inverse function, as demonstrated in Example 10.

EXAMPLE 10 Verifying Inverse Functions Graphically

Graphically verify that f and g are inverse functions of each other.

$$f(x) = x^2, \quad x \geq 0 \quad \text{and} \quad g(x) = \sqrt{x}$$

Solution

You can graphically verify that f and g are inverse functions of each other by graphing the functions on the same rectangular coordinate system, as shown in Figure 11.16. Visually, it appears that the graphs are reflections of each other in the line $y = x$. You can verify this reflective property by testing a few points on each graph. Note in the following list that if the point (a, b) is on the graph of f, the point (b, a) is on the graph of g.

$f(x) = x^2, \quad x \geq 0$	$g(x) = f^{-1}(x) = \sqrt{x}$
$(0, 0)$	$(0, 0)$
$(1, 1)$	$(1, 1)$
$(2, 4)$	$(4, 2)$
$(3, 9)$	$(9, 3)$

Technology: Tip

A graphing calculator program for several models of graphing calculators that graphs the function f and its reflection in the line $y = x$ can be found at our website *www.cengage.com/math/larson/algebra.*

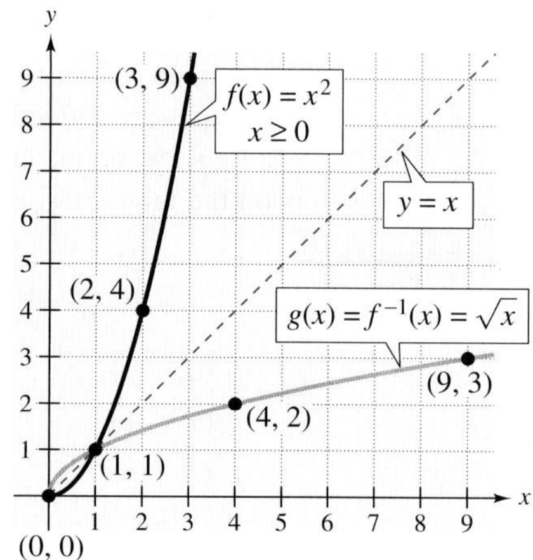

Figure 11.16

So, f and g are inverse functions of each other.

✓ ***CHECKPOINT*** *Now try Exercise 93.*

Concept Check

1. In general, is the composition of f with g equal to the composition of g with f? Give an example.
2. Explain the Horizontal Line Test. What is the relationship between this test and a function being one-to-one?
3. Describe how to find the inverse of a function given by an equation in x and y. Give an example.
4. Describe the relationship between the graphs of a function and its inverse function.

11.2 EXERCISES

Go to pages 752–753 to record your assignments.

Developing Skills

In Exercises 1–10, find the compositions. ***See Examples 1 and 2.***

✓ **1.** $f(x) = 2x + 3, \quad g(x) = x - 6$
(a) $(f \circ g)(x)$ (b) $(g \circ f)(x)$
(c) $(f \circ g)(4)$ (d) $(g \circ f)(7)$

2. $f(x) = x - 5, \quad g(x) = 3x + 2$
(a) $(f \circ g)(x)$ (b) $(g \circ f)(x)$
(c) $(f \circ g)(3)$ (d) $(g \circ f)(3)$

✓ **3.** $f(x) = x^2 + 3, \quad g(x) = x + 2$
(a) $(f \circ g)(x)$ (b) $(g \circ f)(x)$
(c) $(f \circ g)(2)$ (d) $(g \circ f)(-3)$

4. $f(x) = 2x + 1, \quad g(x) = x^2 - 5$
(a) $(f \circ g)(x)$ (b) $(g \circ f)(x)$
(c) $(f \circ g)(-1)$ (d) $(g \circ f)(3)$

5. $f(x) = |x - 3|, \quad g(x) = 3x$
(a) $(f \circ g)(x)$ (b) $(g \circ f)(x)$
(c) $(f \circ g)(1)$ (d) $(g \circ f)(2)$

6. $f(x) = |3x|, \quad g(x) = x - 3$
(a) $(f \circ g)(x)$ (b) $(g \circ f)(x)$
(c) $(f \circ g)(-2)$ (d) $(g \circ f)(-4)$

7. $f(x) = \sqrt{x - 4}, \quad g(x) = x + 5$
(a) $(f \circ g)(x)$ (b) $(g \circ f)(x)$
(c) $(f \circ g)(3)$ (d) $(g \circ f)(8)$

8. $f(x) = \sqrt{x + 6}, \quad g(x) = 2x - 3$
(a) $(f \circ g)(x)$ (b) $(g \circ f)(x)$
(c) $(f \circ g)(3)$ (d) $(g \circ f)(-2)$

9. $f(x) = \dfrac{1}{x - 3}, \quad g(x) = \dfrac{2}{x^2}$
(a) $(f \circ g)(x)$ (b) $(g \circ f)(x)$
(c) $(f \circ g)(-1)$ (d) $(g \circ f)(2)$

10. $f(x) = \dfrac{4}{x^2 - 4}, \quad g(x) = \dfrac{1}{x}$
(a) $(f \circ g)(x)$ (b) $(g \circ f)(x)$
(c) $(f \circ g)(-2)$ (d) $(g \circ f)(1)$

In Exercises 11–14, use the functions f and g to find the indicated values.

$f = \{(-2, 3), (-1, 1), (0, 0), (1, -1), (2, -3)\}$,
$g = \{(-3, 1), (-1, -2), (0, 2), (2, 2), (3, 1)\}$

11. (a) $f(1)$
(b) $g(-1)$
(c) $(g \circ f)(1)$

12. (a) $g(0)$
(b) $f(-1)$
(c) $(f \circ g)(0)$

13. (a) $(f \circ g)(-3)$
(b) $(g \circ f)(-2)$

14. (a) $(f \circ g)(2)$
(b) $(g \circ f)(2)$

In Exercises 15–18, use the functions f and g to find the indicated values.

$f = \{(0, 1), (1, 2), (2, 5), (3, 10), (4, 17)\}$,
$g = \{(5, 4), (10, 1), (2, 3), (17, 0), (1, 2)\}$

15. (a) $f(2)$
(b) $g(10)$
(c) $(g \circ f)(1)$

16. (a) $g(2)$
(b) $f(0)$
(c) $(f \circ g)(10)$

17. (a) $(g \circ f)(4)$
(b) $(f \circ g)(2)$

18. (a) $(f \circ g)(1)$
(b) $(g \circ f)(0)$

In Exercises 19–26, find the compositions (a) $f \circ g$ and (b) $g \circ f$. Then find the domain of each composition. ***See Example 3.***

19. $f(x) = 3x + 4$
$g(x) = x - 7$

20. $f(x) = x + 5$
$g(x) = 4x - 1$

21. $f(x) = \sqrt{x + 2}$
$g(x) = x - 4$

22. $f(x) = \sqrt{x - 5}$
$g(x) = x + 3$

23. $f(x) = x^2 + 3$
$g(x) = \sqrt{x - 1}$

24. $f(x) = \sqrt{2x - 2}$
$g(x) = x^2 - 8$

25. $f(x) = \dfrac{x}{x + 5}$
$g(x) = \sqrt{x - 1}$

26. $f(x) = \dfrac{x}{x - 4}$
$g(x) = \sqrt{x}$

In Exercises 27–34, use a graphing calculator to graph the function and determine whether the function is one-to-one.

27. $f(x) = x^3 - 1$

28. $f(x) = (2 - x)^3$

29. $f(t) = \sqrt[3]{5 - t}$

30. $h(t) = 4 - \sqrt[3]{t}$

31. $g(x) = (x - 3)^4$

32. $f(x) = x^5 + 2$

33. $h(t) = \dfrac{5}{t}$

34. $g(t) = \dfrac{5}{t^2}$

In Exercises 35–40, use the Horizontal Line Test to determine if the function is one-to-one and so has an inverse function. ***See Example 4.***

35. $f(x) = x^2 - 2$

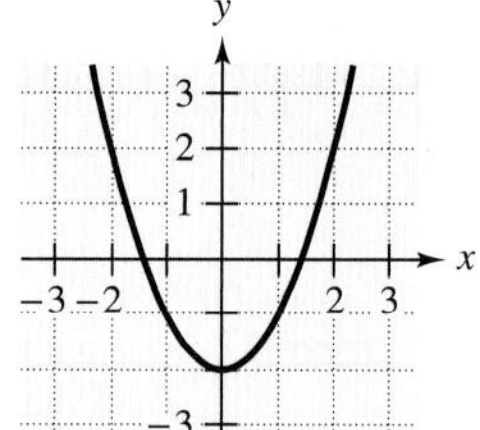

36. 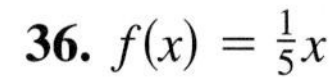

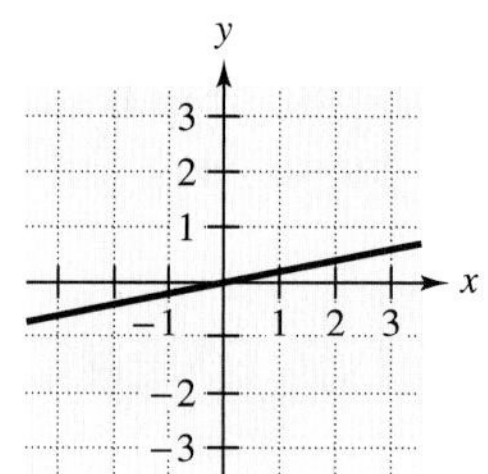

37. $f(x) = x^2, \quad x \ge 0$

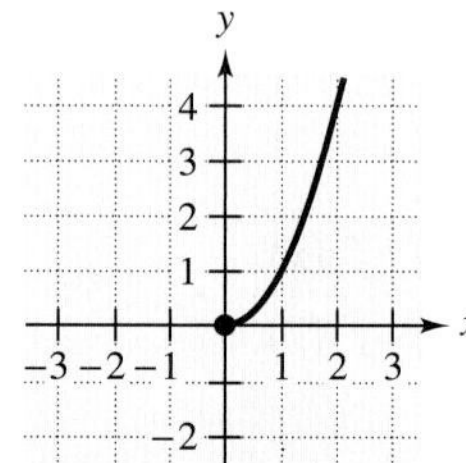

38. 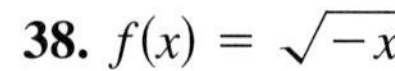

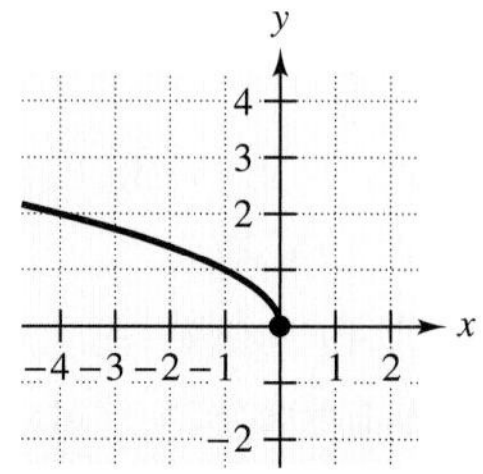

39. $g(x) = \sqrt{25 - x^2}$

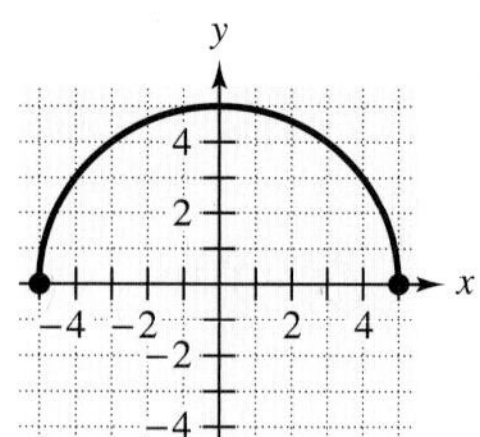

40. 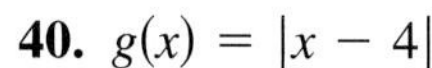

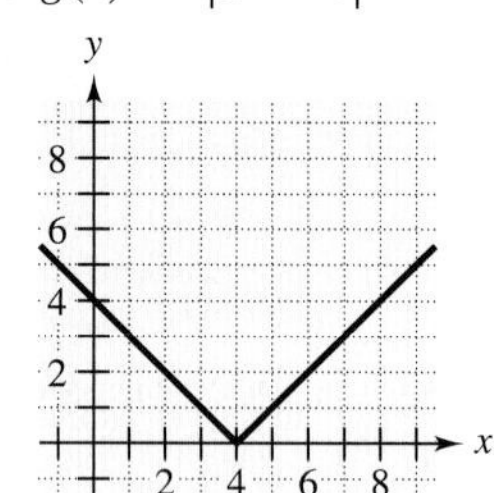

In Exercises 41–48, verify algebraically that the functions f and g are inverse functions of each other. ***See Examples 5 and 6.***

41. $f(x) = -6x, \; g(x) = -\frac{1}{6}x$

42. $f(x) = \frac{2}{3}x, \; g(x) = \frac{3}{2}x$

43. $f(x) = 1 - 2x, \; g(x) = \frac{1}{2}(1 - x)$

44. $f(x) = 2x - 1, \; g(x) = \frac{1}{2}(x + 1)$

45. $f(x) = \sqrt[3]{x + 1}, \; g(x) = x^3 - 1$

46. $f(x) = x^7, \; g(x) = \sqrt[7]{x}$

47. $f(x) = \frac{1}{x}, \; g(x) = \frac{1}{x}$

48. $f(x) = \frac{1}{x + 1}, \; g(x) = \frac{1 - x}{x}$

In Exercises 49–60, find the inverse function of f. Verify that $f(f^{-1}(x))$ and $f^{-1}(f(x))$ are equal to the identity function. ***See Example 7.***

49. $f(x) = 5x$

50. $f(x) = -8x$

51. $f(x) = -\frac{2}{5}x$

52. $f(x) = \frac{1}{3}x$

53. $f(x) = x + 10$

54. $f(x) = x - 5$

55. $f(x) = 5 - x$

56. $f(x) = 8 - x$

57. $f(x) = x^9$

58. $f(x) = x^5$

59. $f(x) = \sqrt[3]{x}$

60. $f(x) = x^{1/7}$

In Exercises 61–74, find the inverse function (if it exists). ***See Examples 7 and 8.***

61. $g(x) = x + 25$

62. $f(x) = 7 - x$

63. $g(x) = 3 - 4x$

64. $g(t) = 6t + 1$

65. $g(t) = \frac{1}{4}t + 2$

66. $h(s) = 5 - \frac{3}{2}s$

67. $g(x) = x^2 + 4$

68. $h(x) = (4 - x)^2$

69. $h(x) = \sqrt{x}$

70. $h(x) = \sqrt{x + 5}$

71. $f(t) = t^3 - 1$

72. $h(t) = t^5 + 8$

73. $f(x) = \sqrt{x + 3}$

74. $f(x) = \sqrt{x^2 - 4}, \quad x \geq 2$

In Exercises 75–78, match the graph with the graph of its inverse function. [The graphs of the inverse functions are labeled (a), (b), (c), and (d).]

(a)

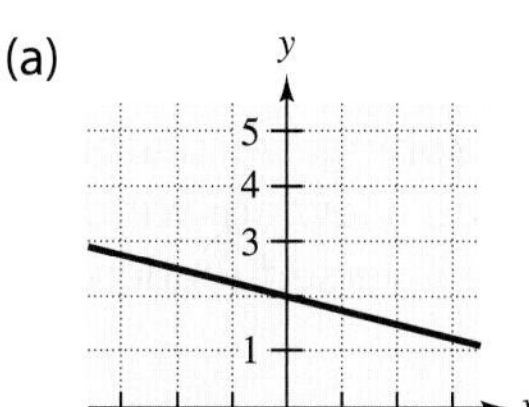

(b)

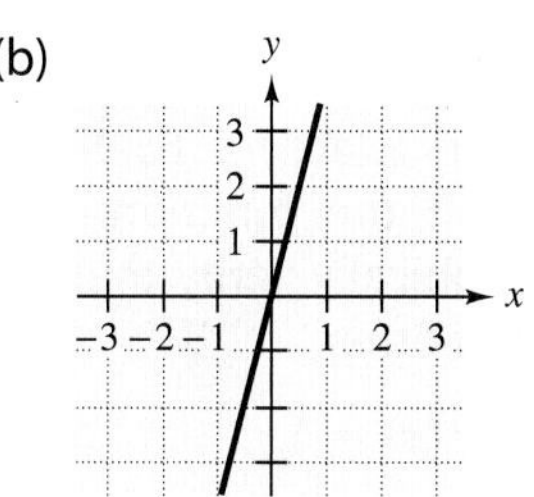

(c)

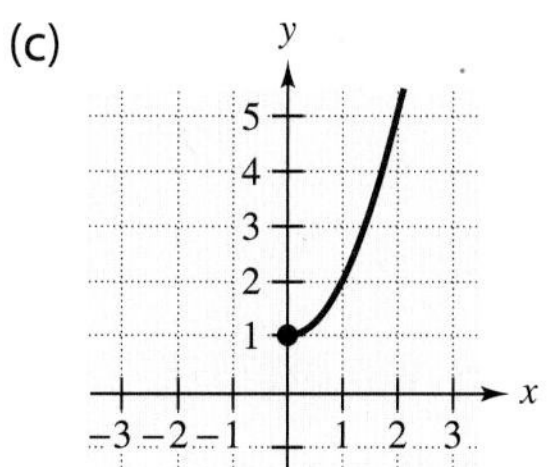

(d)

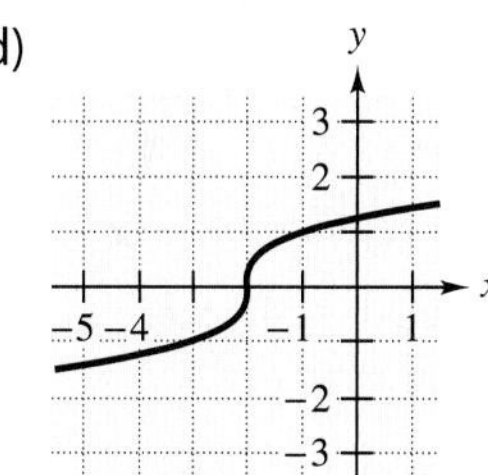

75.

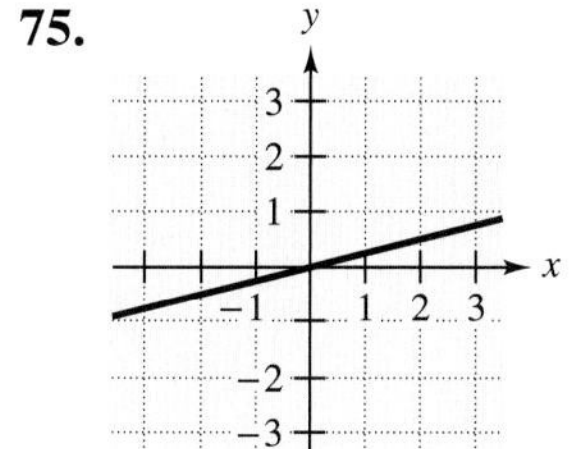

76.

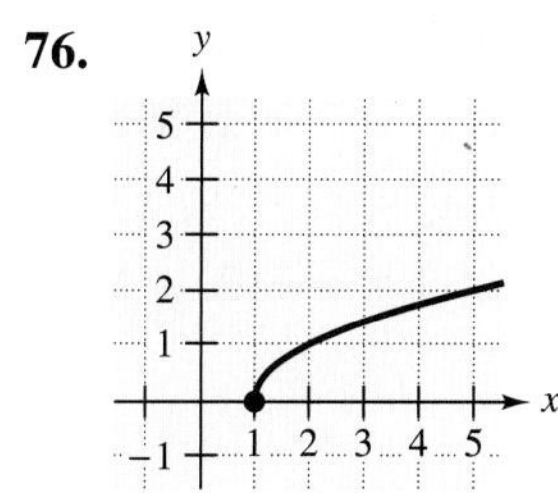

77.

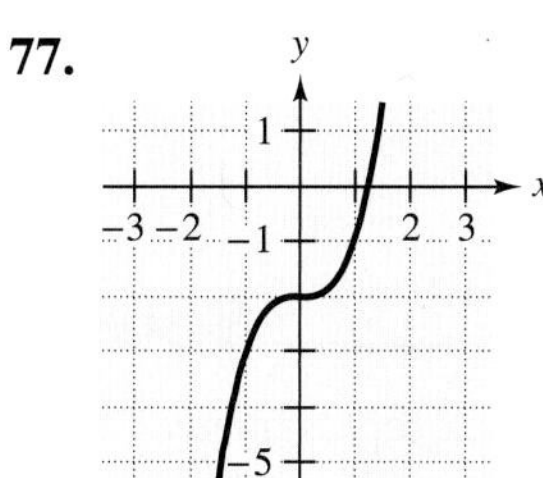

78.

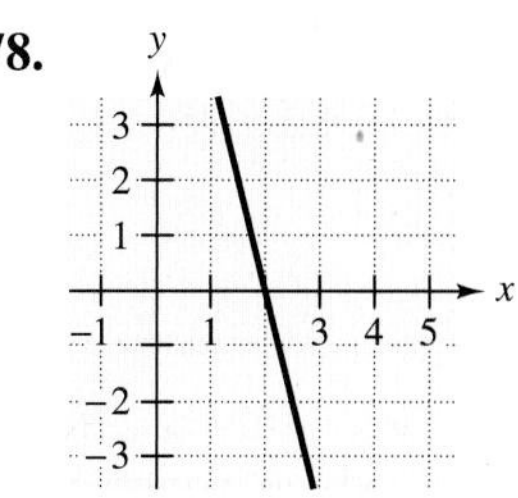

In Exercises 79–84, sketch the graphs of f and f^{-1} on the same rectangular coordinate system. Show that the graphs are reflections of each other in the line $y = x$. ***See Example 9.***

✓ **79.** $f(x) = x + 4,\ f^{-1}(x) = x - 4$

80. $f(x) = x - 7,\ f^{-1}(x) = x + 7$

81. $f(x) = 3x - 1,\ f^{-1}(x) = \frac{1}{3}(x + 1)$

82. $f(x) = 5 - 4x,\ f^{-1}(x) = -\frac{1}{4}(x - 5)$

83. $f(x) = x^2 - 1,\ x \ge 0$

$f^{-1}(x) = \sqrt{x + 1}$

84. $f(x) = (x + 2)^2,\ x \ge -2$

$f^{-1}(x) = \sqrt{x} - 2$

In Exercises 85–92, use a graphing calculator to graph the functions in the same viewing window. Graphically verify that f and g are inverse functions of each other.

85. $f(x) = \frac{1}{3}x$
$g(x) = 3x$

86. $f(x) = \frac{1}{5}x - 1$
$g(x) = 5x + 5$

87. $f(x) = \sqrt{x - 4}$
$g(x) = x^2 + 4,\ x \ge 0$

88. $f(x) = \sqrt{4 - x}$
$g(x) = 4 - x^2,\ x \ge 0$

89. $f(x) = \frac{1}{8}x^3$
$g(x) = 2\sqrt[3]{x}$

90. $f(x) = \sqrt[3]{x + 2}$
$g(x) = x^3 - 2$

91. $f(x) = |3 - x|,\ x \ge 3$
$g(x) = 3 + x,\ x \ge 0$

92. $f(x) = |x - 2|,\ x \ge 2$
$g(x) = x + 2,\ x \ge 0$

In Exercises 93–96, delete part of the graph of the function so that the remaining part is one-to-one. Find the inverse function of the remaining part and find the domain of the inverse function. (*Note:* There is more than one correct answer.) ***See Example 10.***

✓ **93.** $f(x) = (x - 2)^2$

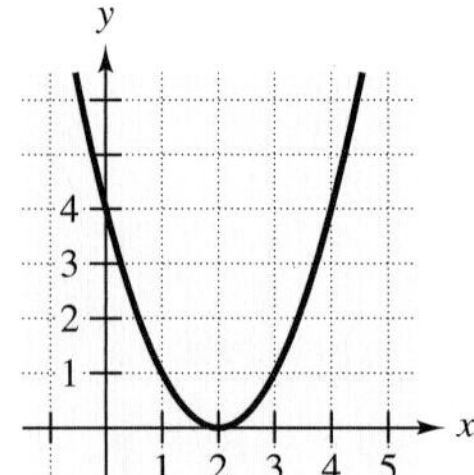

94. $f(x) = 9 - x^2$

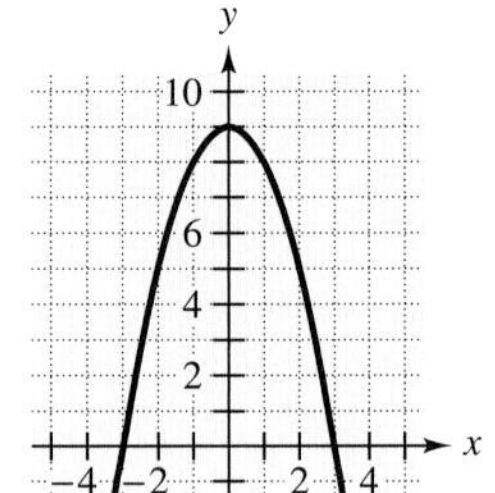

95. $f(x) = |x| + 1$

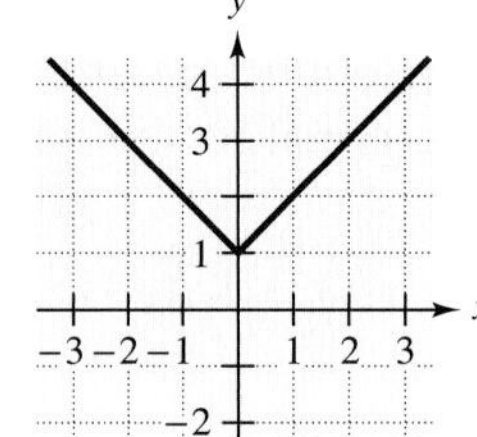

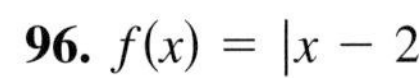

96. $f(x) = |x - 2|$

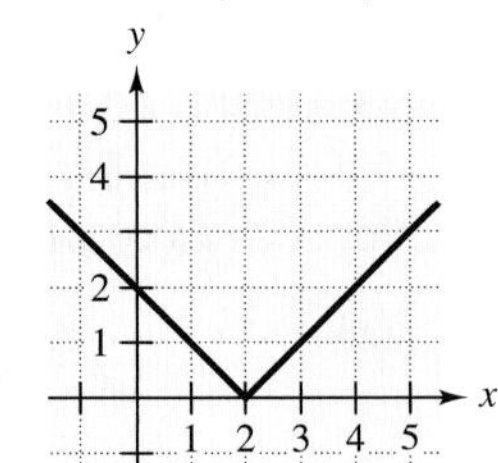

In Exercises 97 and 98, consider the function $f(x) = 3 - 2x$.

97. Find $f^{-1}(x)$.

98. Find $(f^{-1})^{-1}(x)$.

In Exercises 99–102, f is a one-to-one function such that $f(a) = b$, $f(b) = c$, and $f(c) = a$.

99. Find $f^{-1}(a)$.

100. Find $f^{-1}(b)$.

101. Find $f^{-1}(f^{-1}(c))$.

102. Find $f^{-1}(f^{-1}(a))$.

Solving Problems

103. *Sales Bonus* You are a sales representative for a clothing manufacturer. You are paid an annual salary plus a bonus of 3% of your sales over \$300,000. Consider the two functions $s(x) = x - 300{,}000$ and $p(s) = 0.03s$. If x is greater than \$300,000, find and interpret $p(s(x))$.

104. *Daily Production Cost* The daily cost of producing x units in a manufacturing process is $C(x) = 8.5x + 300$. The number of units produced in t hours during a day is given by $x(t) = 12t$, $0 \le t \le 8$. Find, simplify, and interpret $(C \circ x)(t)$.

105. *Ripples* You are standing on a bridge over a calm pond and drop a pebble, causing ripples of concentric circles in the water. The radius (in feet) of the outermost ripple is given by $r(t) = 0.6t$, where t is time in seconds after the pebble hits the water. The area of the circle is given by the function $A(r) = \pi r^2$. Find an equation for the composition $A(r(t))$. What are the input and output of this composite function? What is the area of the circle after 3 seconds?

106. *Oil Spill* An oil tanker hits a reef and begins to spill crude oil into the water. The oil forms a circular region around the ship with the radius (in feet) given by $r(t) = 15t$, where t is the time in hours after the hull is breached. The area of the circle is given by the function $A(r) = \pi r^2$. Find an equation for the composition $A(r(t))$. What are the input and output of this composite function? What is the area of the circle after 4 hours?

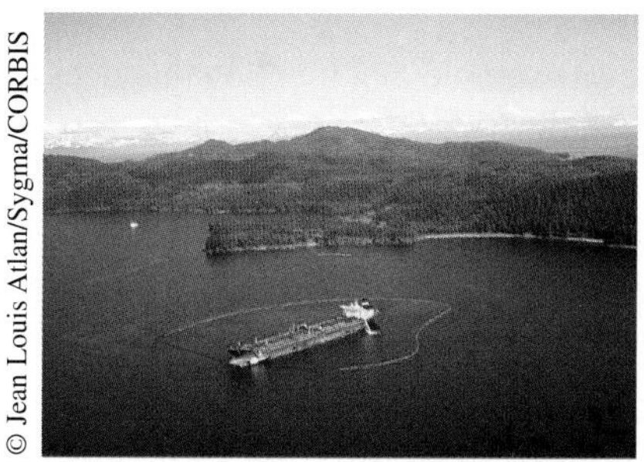

107. *Rebate and Discount* The suggested retail price of a new car is p dollars. The dealership advertised a factory rebate of \$2000 and a 5% discount.

(a) Write a function R in terms of p, giving the cost of the car after receiving the factory rebate.

(b) Write a function S in terms of p, giving the cost of the car after receiving the dealership discount.

(c) Form the composite functions $(R \circ S)(p)$ and $(S \circ R)(p)$ and interpret each.

(d) Find $(R \circ S)(26{,}000)$ and $(S \circ R)(26{,}000)$. Which yields the smaller cost for the car? Explain.

108. *Rebate and Discount* The suggested retail price of a plasma television is p dollars. The electronics store is offering a manufacturer's rebate of \$200 and a 10% discount.

(a) Write a function R in terms of p, giving the cost of the television after receiving the manufacturer's rebate.

(b) Write a function S in terms of p, giving the cost of the television after receiving the 10% discount.

(c) Form the composite functions $(R \circ S)(p)$ and $(S \circ R)(p)$ and interpret each.

(d) Find $(R \circ S)(1600)$ and $(S \circ R)(1600)$. Which yields the smaller cost for the plasma television? Explain.

109. *Hourly Wage* Your wage is \$9.00 per hour plus \$0.65 for each unit produced per hour. So, your hourly wage y in terms of the number of units produced x is $y = 9 + 0.65x$.

(a) Find the inverse function.

(b) What does each variable represent in the inverse function?

(c) Use the context of the problem to determine the domain of the inverse function.

(d) Determine the number of units produced when your hourly wage averages \$14.20.

110. *Federal Income Tax* In 2007, the function $T = 0.15(x - 7825) + 782.5$ represented the federal income tax owed by a single person whose adjusted gross income x was between \$7825 and \$31,850. (Source: Internal Revenue Service)

(a) Find the inverse function.

(b) What does each variable represent in the inverse function?

(c) Use the context of the problem to determine the domain of the inverse function.

(d) Determine a single person's adjusted gross income if they owed \$3808.75 in federal income taxes in 2007.

111. *Exploration* Consider the functions $f(x) = 4x$ and $g(x) = x + 6$.

(a) Find $(f \circ g)(x)$.

(b) Find $(f \circ g)^{-1}(x)$.

(c) Find $f^{-1}(x)$ and $g^{-1}(x)$.

(d) Find $(g^{-1} \circ f^{-1})(x)$ and compare the result with that of part (b).

(e) Make a conjecture about $(f \circ g)^{-1}(x)$ and $(g^{-1} \circ f^{-1})(x)$.

112. *Exploration* Repeat Exercise 111 for $f(x) = x^2 + 1, x \geq 0$ and $g(x) = 3x$.

Explaining Concepts

True or False? In Exercises 113–116, decide whether the statement is true or false. If true, explain your reasoning. If false, give an example.

113. If the inverse function of f exists, the y-intercept of f is an x-intercept of f^{-1}. Explain.

114. There exists no function f such that $f = f^{-1}$.

115. If the inverse function of f exists, the domains of f and f^{-1} are the same.

116. If the inverse function of f exists and its graph passes through the point (2, 2), the graph of f^{-1} also passes through the point (2, 2).

117. Describe how to find the inverse of a function given by a set of ordered pairs. Give an example.

118. Give an example of a function that does not have an inverse function.

119. Why must a function be one-to-one in order for its inverse to be a function?

Cumulative Review

In Exercises 120–123, identify the transformation of the graph of $f(x) = x^2$.

120. $g(x) = (x - 4)^2$

121. $h(x) = -x^2$

122. $v(x) = x^2 + 1$

123. $k(x) = (x + 3)^2 - 5$

In Exercises 124–127, factor the expression completely.

124. $2x^3 - 6x$

125. $16 - (y + 2)^2$

126. $t^2 + 10t + 25$

127. $5 - u + 5u^2 - u^3$

In Exercises 128–131, graph the equation.

128. $y = 3 - \frac{1}{2}x$

129. $3x - 4y = 6$

130. $y = x^2 - 6x + 5$

131. $y = -(x - 2)^2 + 1$

11.3 Logarithmic Functions

A.T. Willett/Alamy

What You Should Learn

1 ▶ Evaluate logarithmic functions.

2 ▶ Graph logarithmic functions.

3 ▶ Graph and evaluate natural logarithmic functions.

4 ▶ Use the change-of-base formula to evaluate logarithms.

Why You Should Learn It

Logarithmic functions can be used to model and solve real-life problems. For instance, in Exercise 128 on page 721, you will use a logarithmic function to determine the speed of the wind near the center of a tornado.

1 ▶ Evaluate logarithmic functions.

Logarithmic Functions

In Section 11.2, you learned the concept of an inverse function. Moreover, you saw that if a function has the property that no horizontal line intersects the graph of the function more than once, the function must have an inverse function. By looking back at the graphs of the exponential functions introduced in Section 11.1, you will see that every function of the form

$$f(x) = a^x$$

passes the Horizontal Line Test and so must have an inverse function. To describe the inverse function of $f(x) = a^x$, follow the steps used in Section 11.2.

$y = a^x$ Replace $f(x)$ by y.

$x = a^y$ Interchange x and y.

At this point, there is no way to solve for y. A verbal description of y in the equation $x = a^y$ is "y equals the power to which a must be raised to obtain x." This inverse of $f(x) = a^x$ is denoted by the **logarithmic function with base *a***

$$f^{-1}(x) = \log_a x.$$

Definition of Logarithmic Function

Let a and x be positive real numbers such that $a \neq 1$. The **logarithm of *x* with base *a*** is denoted by $\log_a x$ and is defined as follows.

$$y = \log_a x \quad \text{if and only if} \quad x = a^y$$

The function $f(x) = \log_a x$ is the **logarithmic function with base *a*.**

From the definition it is clear that

Logarithmic Equation		*Exponential Equation*
$y = \log_a x$	is equivalent to	$x = a^y$.

So, to find the value of $\log_a x$, *think*

"$\log_a x$ = the power to which a must be raised to obtain x."

Smart Study Strategy

Go to page 682 for ways to *Take Control of Your Class Time.*

For instance,

$y = \log_2 8$ Think: "The power to which 2 must be raised to obtain 8."

$y = 3.$

That is,

$3 = \log_2 8.$ This is equivalent to $2^3 = 8$.

By now it should be clear that *a logarithm is an exponent.*

EXAMPLE 1 Evaluating Logarithms

Evaluate each logarithm.

a. $\log_2 16$ **b.** $\log_3 9$ **c.** $\log_4 2$

Solution

In each case you should answer the question, "To what power must the base be raised to obtain the given number?"

a. The power to which 2 must be raised to obtain 16 is 4. That is,

$$2^4 = 16 \Rightarrow \log_2 16 = 4.$$

b. The power to which 3 must be raised to obtain 9 is 2. That is,

$$3^2 = 9 \Rightarrow \log_3 9 = 2.$$

c. The power to which 4 must be raised to obtain 2 is $\frac{1}{2}$. That is,

$$4^{1/2} = 2 \Rightarrow \log_4 2 = \frac{1}{2}.$$

✔ **CHECKPOINT** *Now try Exercise 25.*

Study Tip

Study the results of Example 2 carefully. Each of the logarithms illustrates an important special property of logarithms that you should know.

EXAMPLE 2 Evaluating Logarithms

Evaluate each logarithm.

a. $\log_5 1$ **b.** $\log_{10} \frac{1}{10}$ **c.** $\log_3(-1)$ **d.** $\log_4 0$

Solution

a. The power to which 5 must be raised to obtain 1 is 0. That is,

$$5^0 = 1 \Rightarrow \log_5 1 = 0.$$

b. The power to which 10 must be raised to obtain $\frac{1}{10}$ is -1. That is,

$$10^{-1} = \frac{1}{10} \Rightarrow \log_{10} \frac{1}{10} = -1.$$

c. There is no power to which 3 can be raised to obtain -1. The reason for this is that for any value of x, 3^x is a positive number. So, $\log_3(-1)$ is undefined.

d. There is no power to which 4 can be raised to obtain 0. So, $\log_4 0$ is undefined.

✔ **CHECKPOINT** *Now try Exercise 35.*

The following properties of logarithms follow directly from the definition of the logarithmic function with base a.

Properties of Logarithms

Let a and x be positive real numbers such that $a \neq 1$. Then the following properties are true.

1. $\log_a 1 = 0$ because $a^0 = 1$.

2. $\log_a a = 1$ because $a^1 = a$.

3. $\log_a a^x = x$ because $a^x = a^x$.

The logarithmic function with base 10 is called the **common logarithmic function.** On most calculators, this function can be evaluated with the common logarithmic key LOG, as illustrated in the next example.

EXAMPLE 3 Evaluating Common Logarithms

Evaluate each logarithm. Use a calculator only if necessary.

a. $\log_{10} 100$ **b.** $\log_{10} 0.01$

c. $\log_{10} 5$ **d.** $\log_{10} 2.5$

Solution

a. The power to which 10 must be raised to obtain 100 is 2. That is,

$$10^2 = 100 \quad \Rightarrow \quad \log_{10} 100 = 2.$$

b. The power to which 10 must be raised to obtain 0.01 or $\frac{1}{100}$ is -2. That is,

$$10^{-2} = \frac{1}{100} \quad \Rightarrow \quad \log_{10} 0.01 = -2.$$

c. There is no simple power to which 10 can be raised to obtain 5, so you should use a calculator to evaluate $\log_{10} 5$.

Keystrokes	*Display*	
5 LOG	0.69897	Scientific
LOG 5) ENTER	0.69897	Graphing

So, rounded to three decimal places, $\log_{10} 5 \approx 0.699$.

d. There is no simple power to which 10 can be raised to obtain 2.5, so you should use a calculator to evaluate $\log_{10} 2.5$.

Keystrokes	*Display*	
2.5 LOG	0.39794	Scientific
LOG 2.5) ENTER	0.39794	Graphing

So, rounded to three decimal places, $\log_{10} 2.5 \approx 0.398$.

CHECKPOINT Now try Exercise 47.

Study Tip

Be sure you see that the value of a logarithm can be zero or negative, as in Example 3(b), but you *cannot* take the logarithm of zero or a negative number. This means that the logarithms $\log_{10}(-10)$ and $\log_5 0$ are undefined.

2 ▶ Graph logarithmic functions.

Graphs of Logarithmic Functions

To sketch the graph of

$$y = \log_a x$$

you can use the fact that the graphs of inverse functions are reflections of each other in the line $y = x$.

EXAMPLE 4 Graphs of Exponential and Logarithmic Functions

On the same rectangular coordinate system, sketch the graph of each function.

a. $f(x) = 2^x$ **b.** $g(x) = \log_2 x$

Solution

a. Begin by making a table of values for $f(x) = 2^x$.

x	-2	-1	0	1	2	3
$f(x) = 2^x$	$\frac{1}{4}$	$\frac{1}{2}$	1	2	4	8

By plotting these points and connecting them with a smooth curve, you obtain the graph shown in Figure 11.17.

b. Because $g(x) = \log_2 x$ is the inverse function of $f(x) = 2^x$, the graph of g is obtained by reflecting the graph of f in the line $y = x$, as shown in Figure 11.17.

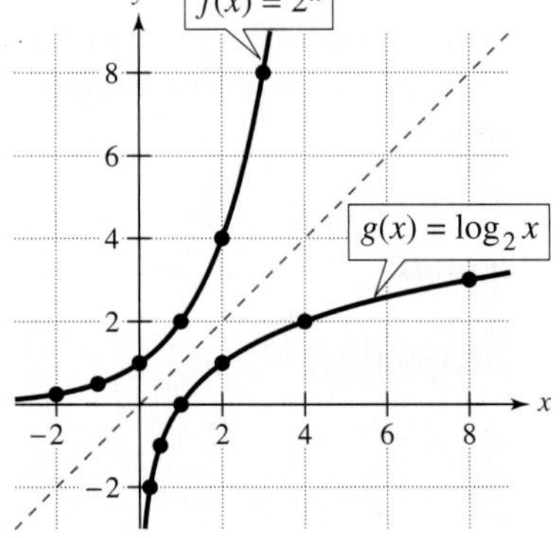

Figure 11.17 Inverse Functions

✓ ***CHECKPOINT*** *Now try Exercise 57.*

Study Tip

In Example 4, the inverse property of logarithmic functions is used to sketch the graph of $g(x) = \log_2 x$. You could also use a standard point-plotting approach or a graphing calculator.

Notice from the graph of $g(x) = \log_2 x$, shown in Figure 11.17, that the domain of the function is the set of positive numbers and the range is the set of all real numbers. The basic characteristics of the graph of a logarithmic function are summarized in Figure 11.18. In this figure, note that the graph has one x-intercept at $(1, 0)$. Also note that $x = 0$ (y-axis) is a vertical asymptote of the graph.

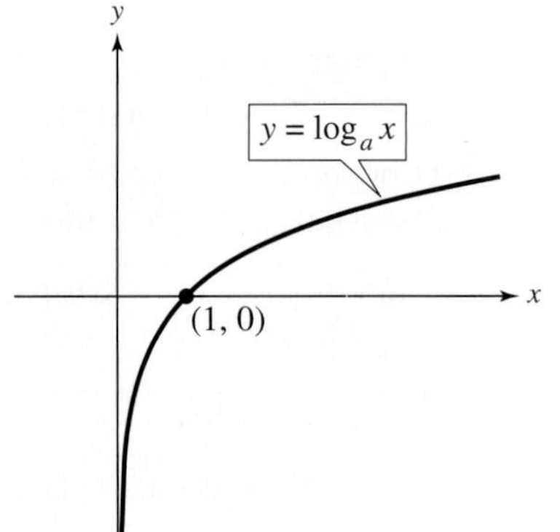

Graph of $y = \log_a x$, $a > 1$

- Domain: $(0, \infty)$
- Range: $(-\infty, \infty)$
- Intercept: $(1, 0)$
- Increasing (moves up to the right)
- Asymptote: y-axis

Figure 11.18 Characteristics of logarithmic function $y = \log_a x$ $(a > 1)$

In the following example, the graph of $\log_a x$ is used to sketch the graphs of functions of the form $y = b \pm \log_a(x + c)$. Notice how each transformation affects the vertical asymptote.

EXAMPLE 5 Sketching the Graphs of Logarithmic Functions

The graph of each function is similar to the graph of $f(x) = \log_{10} x$, as shown in Figure 11.19. From the graph you can determine the domain of the function.

a. Because $g(x) = \log_{10}(x - 1) = f(x - 1)$, the graph of g can be obtained by shifting the graph of f one unit to the right. The vertical asymptote of the graph of g is $x = 1$. The domain of g is $(1, \infty)$.

b. Because $h(x) = 2 + \log_{10} x = 2 + f(x)$, the graph of h can be obtained by shifting the graph of f two units upward. The vertical asymptote of the graph of h is $x = 0$. The domain of h is $(0, \infty)$.

c. Because $k(x) = -\log_{10} x = -f(x)$, the graph of k can be obtained by reflecting the graph of f in the x-axis. The vertical asymptote of the graph of k is $x = 0$. The domain of k is $(0, \infty)$.

d. Because $j(x) = \log_{10}(-x) = f(-x)$, the graph of j can be obtained by reflecting the graph of f in the y-axis. The vertical asymptote of the graph of j is $x = 0$. The domain of j is $(-\infty, 0)$.

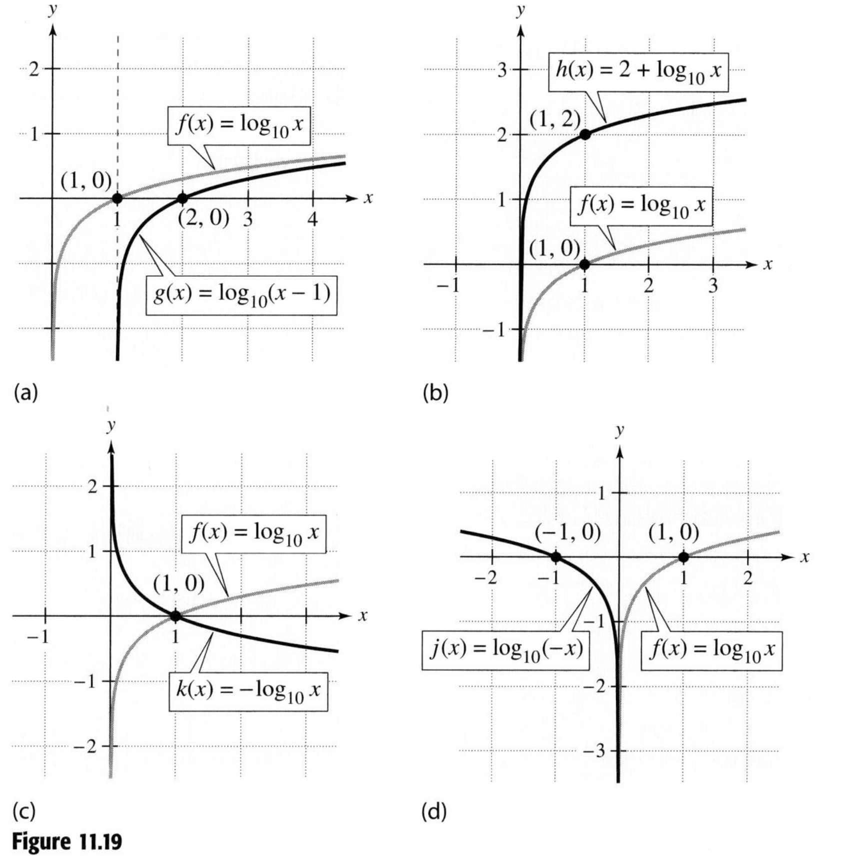

Figure 11.19

✓ **CHECKPOINT** *Now try Exercise 61.*

3 ▶ Graph and evaluate natural logarithmic functions.

The Natural Logarithmic Function

As with exponential functions, the most widely used base for logarithmic functions is the number e. The logarithmic function with base e is the **natural logarithmic function** and is denoted by the special symbol $\ln x$, which is read as "el en of x."

The Natural Logarithmic Function

The function defined by

$$f(x) = \log_e x = \ln x$$

where $x > 0$, is called the **natural logarithmic function.**

Graph of $g(x) = \ln x$
- Domain: $(0, \infty)$
- Range: $(-\infty, \infty)$
- Intercept: $(1, 0)$
- Increasing (moves up to the right)
- Asymptote: y-axis

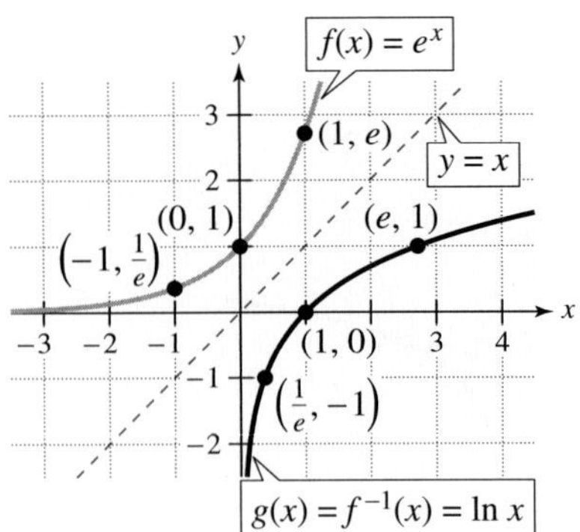

Figure 11.20 Characteristics of the natural logarithmic function $g(x) = \ln x$

The definition above implies that the natural logarithmic function and the natural exponential function are inverse functions of each other. So, every logarithmic equation can be written in an equivalent exponential form, and every exponential equation can be written in logarithmic form.

Because the functions $f(x) = e^x$ and $g(x) = \ln x$ are inverse functions of each other, their graphs are reflections of each other in the line $y = x$. This reflective property is illustrated in Figure 11.20. The figure also contains a summary of several characteristics of the graph of the natural logarithmic function.

Notice that the domain of the natural logarithmic function, as with every other logarithmic function, is the set of *positive real numbers*—be sure you see that $\ln x$ is not defined for zero or for negative numbers.

The three properties of logarithms listed earlier in this section are also valid for natural logarithms.

Properties of Natural Logarithms

Let x be a positive real number. Then the following properties are true.

1. $\ln 1 = 0$ because $e^0 = 1$.

2. $\ln e = 1$ because $e^1 = e$.

3. $\ln e^x = x$ because $e^x = e^x$.

Technology: Tip

On most calculators, the natural logarithm key is denoted by LN.

For instance, on a scientific calculator, you can evaluate ln 2 as 2 LN and on a graphing calculator, you can evaluate it as LN 2) ENTER. In either case, you should obtain a display of 0.6931472.

EXAMPLE 6 Evaluating Natural Logarithmic Functions

Evaluate each expression.

a. $\ln e^2$ **b.** $\ln \dfrac{1}{e}$

Solution

Using the property that $\ln e^x = x$, you obtain the following.

a. $\ln e^2 = 2$ **b.** $\ln \dfrac{1}{e} = \ln e^{-1} = -1$

✓ **CHECKPOINT** *Now try Exercise 89.*

4 ▶ Use the change-of-base formula to evaluate logarithms.

Change of Base

Although 10 and e are the most frequently used bases, you occasionally need to evaluate logarithms with other bases. In such cases, the following **change-of-base formula** is useful.

Technology: Tip

You can use a graphing calculator to graph logarithmic functions that do not have a base of 10 by using the change-of-base formula. Use the change-of-base formula to rewrite $g(x) = \log_2 x$ in Example 4 on page 714 (with $b = 10$) and graph the function. You should obtain a graph similar to the one below.

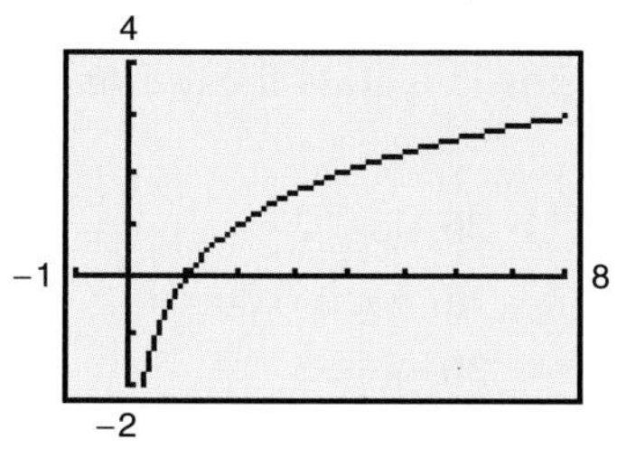

Change-of-Base Formula

Let a, b, and x be positive real numbers such that $a \neq 1$ and $b \neq 1$. Then $\log_a x$ is given as follows.

$$\log_a x = \frac{\log_b x}{\log_b a} \quad \text{or} \quad \log_a x = \frac{\ln x}{\ln a}$$

The usefulness of this change-of-base formula is that you can use a calculator that has only the common logarithm key (LOG) and the natural logarithm key (LN) to evaluate logarithms to any base.

EXAMPLE 7 Changing Bases to Evaluate Logarithms

a. Use *common* logarithms to evaluate $\log_3 5$.

b. Use *natural* logarithms to evaluate $\log_6 2$.

Solution

Using the change-of-base formula, you can convert to common and natural logarithms by writing

$$\log_3 5 = \frac{\log_{10} 5}{\log_{10} 3} \quad \text{and} \quad \log_6 2 = \frac{\ln 2}{\ln 6}.$$

Now, use the following keystrokes.

a. *Keystrokes*	*Display*	
5 (LOG) (÷) 3 (LOG) (=)	1.4649735	Scientific
(LOG) 5 ()) (÷) (LOG) 3 ()) (ENTER)	1.4649735	Graphing

So, $\log_3 5 \approx 1.465$.

b. *Keystrokes*	*Display*	
2 (LN) (÷) 6 (LN) (=)	0.3868528	Scientific
(LN) 2 ()) (÷) (LN) 6 ()) (ENTER)	0.3868528	Graphing

So, $\log_6 2 \approx 0.387$.

✓ **CHECKPOINT** ***Now try Exercise 111.***

Study Tip

In Example 7(a), $\log_3 5$ could have been evaluated using natural logarithms in the change-of-base formula.

$$\log_3 5 = \frac{\ln 5}{\ln 3} \approx 1.465$$

Notice that you get the same answer whether you use natural logarithms or common logarithms in the change-of-base formula.

At this point, you have been introduced to all the basic types of functions that are covered in this course: polynomial functions, radical functions, rational functions, exponential functions, and logarithmic functions. The only other common types of functions are *trigonometric functions*, which you will study if you go on to take a course in trigonometry or precalculus.

Concept Check

In Concept Check Exercises 1 and 2, determine whether the statement is true or false. Justify your answer.

1. The statement $8 = 2^3$ is equivalent to $2 = \log_8 3$.

2. The graph of $f(x) = \ln x$ is the reflection of the graph of $f(x) = e^x$ in the x-axis.

3. Explain how to use the graph of $y = \log_{10} x$ to graph $y = 3 + \log_{10}(x - 1)$.

4. Describe and correct the error in evaluating $\log_4 10$.

$$\log_4 10 = \frac{\ln 4}{\ln 10} \approx 0.6021$$

11.3 EXERCISES

Go to pages 752–753 to record your assignments.

Developing Skills

In Exercises 1–12, write the logarithmic equation in exponential form.

1. $\log_7 49 = 2$
2. $\log_{11} 121 = 2$
3. $\log_2 \frac{1}{32} = -5$
4. $\log_3 \frac{1}{27} = -3$
5. $\log_3 \frac{1}{243} = -5$
6. $\log_{10} 10{,}000 = 4$
7. $\log_{36} 6 = \frac{1}{2}$
8. $\log_{64} 4 = \frac{1}{3}$
9. $\log_8 4 = \frac{2}{3}$
10. $\log_{16} 8 = \frac{3}{4}$
11. $\log_2 5.278 \approx 2.4$
12. $\log_3 1.179 \approx 0.15$

In Exercises 13–24, write the exponential equation in logarithmic form.

13. $6^2 = 36$
14. $3^5 = 243$
15. $5^{-3} = \frac{1}{125}$
16. $6^{-4} = \frac{1}{1296}$
17. $8^{2/3} = 4$
18. $81^{3/4} = 27$
19. $25^{-1/2} = \frac{1}{5}$
20. $6^{-3} = \frac{1}{216}$
21. $4^0 = 1$
22. $6^1 = 6$
23. $5^{1.4} \approx 9.518$
24. $10^{0.36} \approx 2.291$

In Exercises 25–46, evaluate the logarithm without using a calculator. (If not possible, state the reason.) *See Examples 1 and 2.*

25. $\log_2 8$
26. $\log_3 27$
27. $\log_{10} 1000$
28. $\log_{10} 0.00001$
29. $\log_2 \frac{1}{16}$
30. $\log_3 \frac{1}{9}$
31. $\log_4 \frac{1}{64}$
32. $\log_6 \frac{1}{216}$
33. $\log_{10} \frac{1}{10{,}000}$
34. $\log_{10} \frac{1}{100}$
35. $\log_2(-3)$
36. $\log_4(-4)$
37. $\log_4 1$
38. $\log_3 1$
39. $\log_5(-6)$
40. $\log_2 0$
41. $\log_9 3$
42. $\log_{125} 5$
43. $\log_{16} 8$
44. $\log_{81} 9$
45. $\log_7 7^4$
46. $\log_5 5^3$

In Exercises 47–52, use a calculator to evaluate the common logarithm. (Round your answer to four decimal places.) *See Example 3.*

47. $\log_{10} 42$
48. $\log_{10} 7561$
49. $\log_{10} 0.023$
50. $\log_{10} 0.149$
51. $\log_{10}\left(\sqrt{5} + 3\right)$
52. $\log_{10} \frac{\sqrt{3}}{2}$

In Exercises 53–56, match the function with its graph. [The graphs are labeled (a), (b), (c), and (d).]

(a)

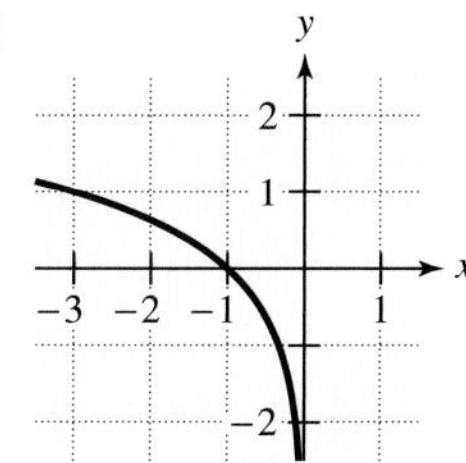

(b)

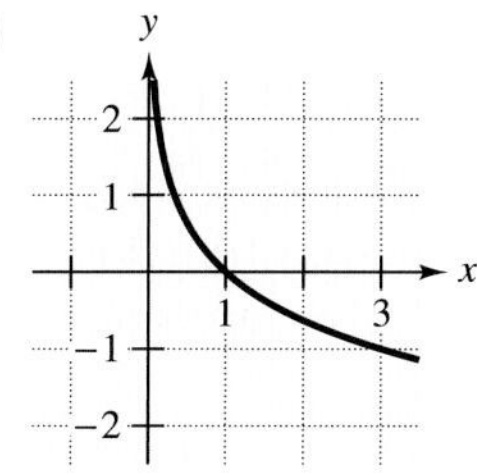

(c)

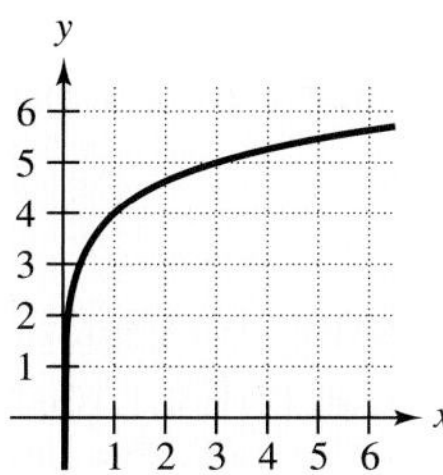

(d)

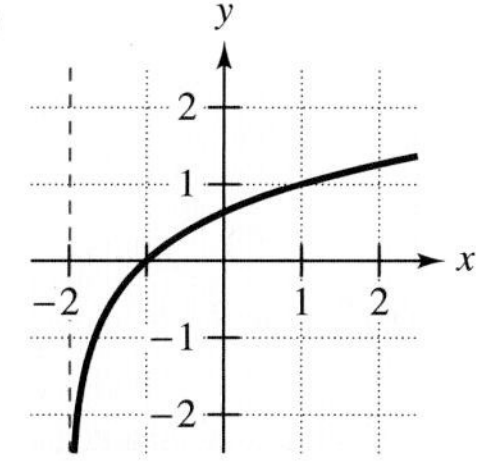

53. $f(x) = 4 + \log_3 x$ **54.** $f(x) = -\log_3 x$

55. $f(x) = \log_3(-x)$ **56.** $f(x) = \log_3(x + 2)$

In Exercises 57–60, sketch the graph of *f*. Then use the graph of *f* to sketch the graph of *g*. ***See Example 4.***

✓ **57.** $f(x) = 3^x$, $g(x) = \log_3 x$ **58.** $f(x) = 4^x$, $g(x) = \log_4 x$

59. $f(x) = 6^x$, $g(x) = \log_6 x$ **60.** $f(x) = \left(\frac{1}{2}\right)^x$, $g(x) = \log_{1/2} x$

In Exercises 61–66, identify the transformation of the graph of $f(x) = \log_2 x$. Then sketch the graph of *h*. ***See Example 5.***

✓ **61.** $h(x) = 3 + \log_2 x$ **62.** $h(x) = -5 + \log_2 x$

63. $h(x) = \log_2(x - 2)$ **64.** $h(x) = \log_2(x + 5)$

65. $h(x) = \log_2(-x)$ **66.** $h(x) = -\log_2 x$

In Exercises 67–76, sketch the graph of the function. Identify the vertical asymptote.

67. $f(x) = \log_5 x$ **68.** $g(x) = \log_8 x$

69. $g(t) = -\log_9 t$ **70.** $h(s) = -2 \log_3 s$

71. $f(x) = 2 + \log_4 x$ **72.** $f(x) = -2 + \log_3 x$

73. $g(x) = \log_2(x - 3)$ **74.** $h(x) = \log_3(x + 1)$

75. $f(x) = \log_{10}(10x)$ **76.** $g(x) = \log_4(4x)$

In Exercises 77–82, find the domain and vertical asymptote of the function. Then sketch its graph.

77. $f(x) = \log_4 x$ **78.** $g(x) = \log_6 x$

79. $h(x) = \log_5(x - 4)$ **80.** $f(x) = -\log_6(x + 2)$

81. $y = -\log_3 x + 2$ **82.** $y = \log_4(x - 2) + 3$

In Exercises 83–88, use a graphing calculator to graph the function. Determine the domain and the vertical asymptote.

83. $y = 5 \log_{10} x$ **84.** $y = 5 \log_{10}(x - 3)$

85. $y = -3 + 5 \log_{10} x$ **86.** $y = 5 \log_{10}(3x)$

87. $y = \log_{10}\left(\frac{x}{5}\right)$ **88.** $y = \log_{10}(-x)$

In Exercises 89–94, use a calculator to evaluate the natural logarithm. (Round your answer to four decimal places.) ***See Example 6.***

✓ **89.** $\ln 38$ **90.** $\ln 18.6$

91. $\ln 0.15$ **92.** $\ln 0.002$

93. $\ln\left(\frac{3 - \sqrt{2}}{5}\right)$ **94.** $\ln\left(1 + \frac{0.10}{12}\right)$

In Exercises 95–98, match the function with its graph. [The graphs are labeled (a), (b), (c), and (d).]

(a)

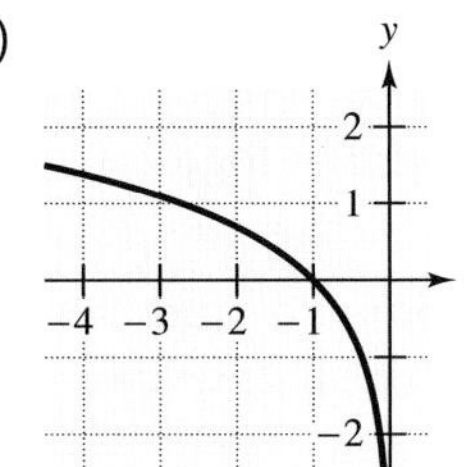

(b)

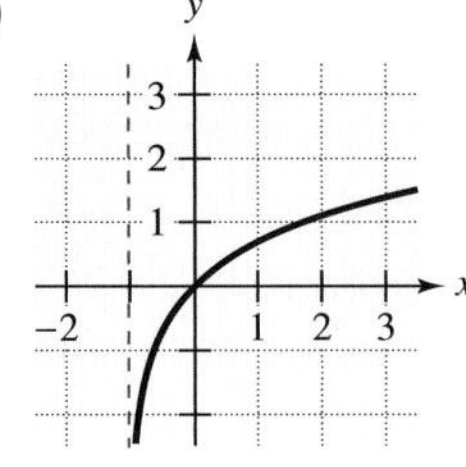

(c)

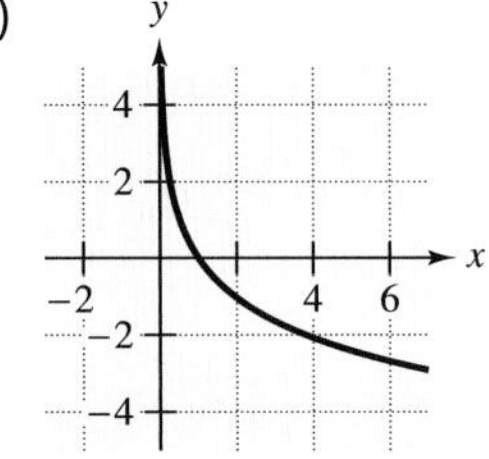

(d)

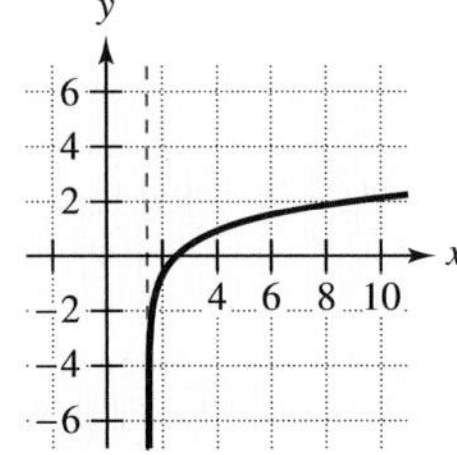

95. $f(x) = \ln(x + 1)$ **96.** $f(x) = \ln(-x)$

97. $f(x) = \ln\left(x - \frac{3}{2}\right)$ **98.** $f(x) = -\frac{3}{2} \ln x$

In Exercises 99–106, sketch the graph of the function. Identify the vertical asymptote.

99. $f(x) = -\ln x$ **100.** $f(x) = -5 \ln x$

101. $f(x) = 3 \ln x$ **102.** $h(t) = 4 \ln t$

103. $f(x) = 3 + \ln x$ **104.** $h(x) = 2 + \ln x$

105. $g(t) = 2 \ln(t - 4)$ **106.** $g(x) = -3 \ln(x + 3)$

In Exercises 107–110, use a graphing calculator to graph the function. Determine the domain and the vertical asymptote.

107. $g(x) = -\ln(x + 1)$ **108.** $h(x) = \ln(x + 4)$

109. $f(t) = 7 + 3 \ln t$ **110.** $g(t) = \ln(5 - t)$

In Exercises 111–124, use a calculator to evaluate the logarithm by means of the change-of-base formula. Use (a) the common logarithm key and (b) the natural logarithm key. (Round your answer to four decimal places.) ***See Example 7.***

✓ **111.** $\log_9 36$ **112.** $\log_7 411$

113. $\log_5 14$ **114.** $\log_6 9$

115. $\log_2 0.72$ **116.** $\log_{12} 0.6$

117. $\log_{15} 1250$ **118.** $\log_{20} 125$

119. $\log_{1/4} 16$ **120.** $\log_{1/3} 18$

121. $\log_4 \sqrt{42}$ **122.** $\log_5 \sqrt{21}$

123. $\log_2(1 + e)$ **124.** $\log_4(2 + e^3)$

Solving Problems

125. *American Elk* The antler spread a (in inches) and shoulder height h (in inches) of an adult male American elk are related by the model

$$h = 116 \log_{10}(a + 40) - 176.$$

Approximate to one decimal place the shoulder height of a male American elk with an antler spread of 55 inches.

126. *Sound Intensity* The relationship between the number of decibels B and the intensity of a sound I in watts per centimeter squared is given by

$$B = 10 \log_{10}\left(\frac{I}{10^{-16}}\right).$$

Determine the number of decibels of an alarm clock with an intensity of 10^{-8} watt per centimeter squared.

127. *Compound Interest* The time t in years for an investment to double in value when compounded continuously at interest rate r is given by

$$t = \frac{\ln 2}{r}.$$

Complete the table, which shows the "doubling times" for several interest rates.

r	0.07	0.08	0.09	0.10	0.11	0.12
t						

128. *Meteorology* Most tornadoes last less than 1 hour and travel about 20 miles. The speed of the wind S (in miles per hour) near the center of the tornado and the distance d (in miles) the tornado travels are related by the model $S = 93 \log_{10} d + 65$. On March 18, 1925, a large tornado struck portions of Missouri, Illinois, and Indiana, covering a distance of 220 miles. Approximate to one decimal place the speed of the wind near the center of this tornado.

129. *Tractrix* A person walking along a dock (the y-axis) drags a boat by a 10-foot rope (see figure). The boat travels along a path known as a *tractrix*. The equation of the path is

$$y = 10 \ln\left(\frac{10 + \sqrt{100 - x^2}}{x}\right) - \sqrt{100 - x^2}.$$

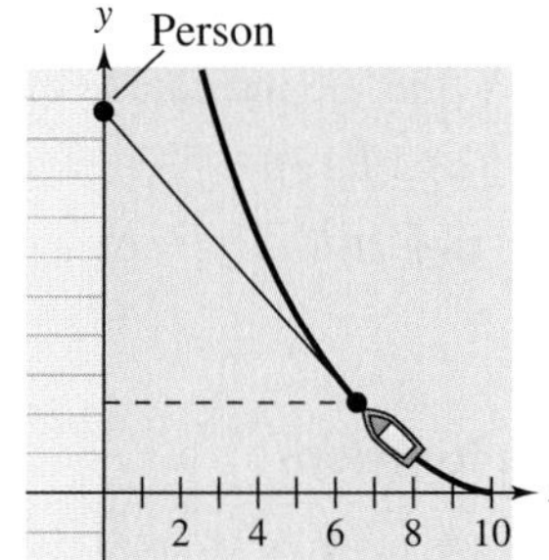

(a) Use a graphing calculator to graph the function. What is the domain of the function?

(b) Identify any asymptotes.

(c) Determine the position of the person when the x-coordinate of the position of the boat is $x = 2$.

130. *Home Mortgage* The model

$$t = 10.042 \ln\left(\frac{x}{x - 1250}\right), \quad x > 1250$$

approximates the length t (in years) of a home mortgage of \$150,000 at 10% interest in terms of the monthly payment x.

(a) Use a graphing calculator to graph the model. Describe the change in the length of the mortgage as the monthly payment increases.

(b) Use the graph in part (a) to approximate the length of the mortgage when the monthly payment is \$1316.35.

(c) Use the result of part (b) to find the total amount paid over the term of the mortgage. What amount of the total is interest costs?

Think About It In Exercises 131–136, answer the question for the function $f(x) = \log_{10} x$. (Do not use a calculator.)

131. What is the domain of f?

132. Find the inverse function of f.

133. Describe the values of $f(x)$ for $1000 \le x \le 10{,}000$.

134. Describe the values of x, given that $f(x)$ is negative.

135. By what amount will x increase, given that $f(x)$ is increased by 1 unit?

136. Find the ratio of a to b when $f(a) = 3 + f(b)$.

Explaining Concepts

137. Explain the difference between common logarithms and natural logarithms.

138. Explain the relationship between the domain of the graph of $f(x) = \log_5 x$ and the range of the graph of $g(x) = 5^x$.

139. Discuss how shifting or reflecting the graph of a logarithmic function affects the domain and the range.

140. Explain why $\log_a x$ is defined only when $0 < a < 1$ and $a > 1$.

Cumulative Review

In Exercises 141–144, use the rules of exponents to simplify the expression.

141. $(-m^6n)(m^4n^3)$

142. $(m^2n^4)^3(mn^2)$

143. $\dfrac{36x^4y}{8xy^3}$

144. $-\left(\dfrac{3x}{5y}\right)^5$

In Exercises 145–148, perform the indicated operation(s) and simplify. (Assume all variables are positive.)

145. $25\sqrt{3x} - 3\sqrt{12x}$

146. $(\sqrt{x} + 3)(\sqrt{x} - 3)$

147. $\sqrt{u}(\sqrt{20} - \sqrt{5})$

148. $(2\sqrt{t} + 3)^2$

Mid-Chapter Quiz

Take this quiz as you would take a quiz in class. After you are done, check your work against the answers in the back of the book.

1. Given $f(x) = \left(\frac{4}{3}\right)^x$, find (a) $f(2)$, (b) $f(0)$, (c) $f(-1)$, and (d) $f(1.5)$.

2. Identify the horizontal asymptote of the graph of $g(x) = -3^{-0.5x}$.

In Exercises 3–6, sketch the graph of the function. Identify the horizontal asymptote. Use a graphing calculator for Exercises 5 and 6.

3. $y = \frac{1}{2}(4^x)$

4. $y = 5(2^{-x})$

5. $f(t) = 12e^{-0.4t}$

6. $g(x) = 100(1.08)^x$

7. Given $f(x) = 2x - 3$ and $g(x) = x^3$, find the indicated composition.

(a) $(f \circ g)(x)$ (b) $(g \circ f)(x)$ (c) $(f \circ g)(-2)$ (d) $(g \circ f)(4)$

8. Verify algebraically and graphically that $f(x) = 5 - 2x$ and $g(x) = \frac{1}{2}(5 - x)$ are inverse functions of each other.

In Exercises 9 and 10, find the inverse function.

9. $h(x) = 10x + 3$

10. $g(t) = \frac{1}{2}t^3 + 2$

11. Write the logarithmic equation $\log_9 \frac{1}{81} = -2$ in exponential form.

12. Write the exponential equation $2^6 = 64$ in logarithmic form.

13. Evaluate $\log_5 125$ without a calculator.

In Exercises 14 and 15, use a graphing calculator to graph the function. Identify the vertical asymptote.

14. $f(t) = -2\ln(t + 3)$

15. $h(x) = 5 + \frac{1}{2}\ln x$

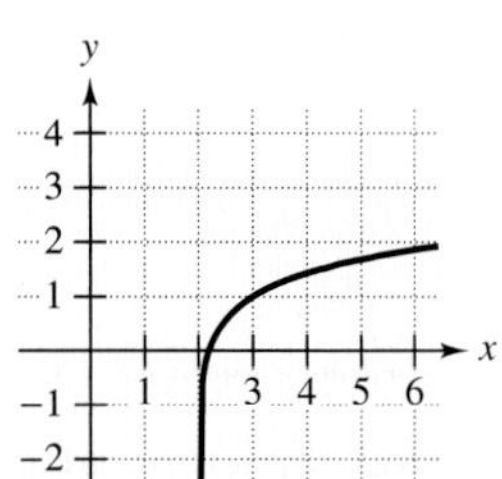

Figure for 16

16. Use the graph of f shown at the left to determine h and k if $f(x) = \log_5(x - h) + k$.

17. Use a calculator and the change-of-base formula to evaluate $\log_3 782$.

18. You deposit \$1200 in an account at an annual interest rate of $6\frac{1}{4}\%$. Complete the table showing the balances A in the account after 15 years for several types of compounding.

n	1	4	12	365	Continuous compounding
A					

19. After t years, the remaining mass y (in grams) of 14 grams of a radioactive element whose half-life is 40 years is given by $y = 14\left(\frac{1}{2}\right)^{t/40}$, $t \geq 0$. How much of the initial mass remains after 125 years?

11.4 Properties of Logarithms

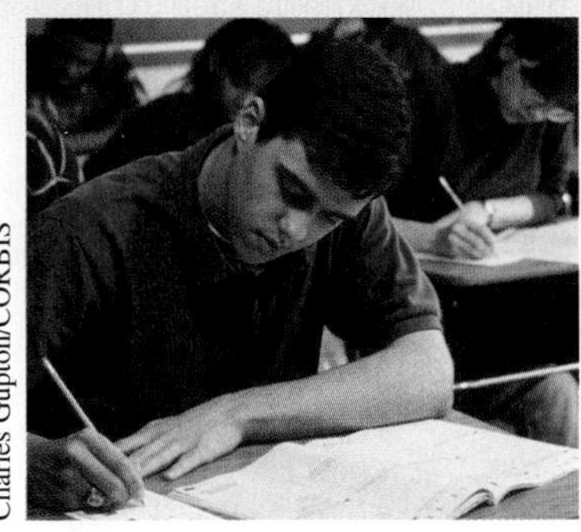
Charles Gupton/CORBIS

What You Should Learn

1 ▶ Use the properties of logarithms to evaluate logarithms.

2 ▶ Use the properties of logarithms to rewrite, expand, or condense logarithmic expressions.

3 ▶ Use the properties of logarithms to solve application problems.

Why You Should Learn It

Logarithmic equations are often used to model scientific observations. For instance, in Example 5 on page 726, a logarithmic equation is used to model human memory.

1 ▶ Use the properties of logarithms to evaluate logarithms.

Properties of Logarithms

You know from the preceding section that the logarithmic function with base a is the *inverse function* of the exponential function with base a. So, it makes sense that each property of exponents should have a corresponding property of logarithms. For instance, the exponential property

$a^0 = 1$ Exponential property

has the corresponding logarithmic property

$\log_a 1 = 0.$ Corresponding logarithmic property

In this section you will study the logarithmic properties that correspond to the following three exponential properties.

	Base a	*Natural Base*
1.	$a^m a^n = a^{m+n}$	$e^m e^n = e^{m+n}$
2.	$\dfrac{a^m}{a^n} = a^{m-n}$	$\dfrac{e^m}{e^n} = e^{m-n}$
3.	$(a^m)^n = a^{mn}$	$(e^m)^n = e^{mn}$

Properties of Logarithms

Let a be a positive real number such that $a \neq 1$, and let n be a real number. If u and v are real numbers, variables, or algebraic expressions such that $u > 0$ and $v > 0$, the following properties are true.

	Logarithm with Base a	*Natural Logarithm*
1. Product Property:	$\log_a(uv) = \log_a u + \log_a v$	$\ln(uv) = \ln u + \ln v$
2. Quotient Property:	$\log_a \dfrac{u}{v} = \log_a u - \log_a v$	$\ln \dfrac{u}{v} = \ln u - \ln v$
3. Power Property:	$\log_a u^n = n \log_a u$	$\ln u^n = n \ln u$

There is no general property of logarithms that can be used to simplify $\log_a(u + v)$. Specifically,

$\log_a(u + v)$ *does not equal* $\log_a u + \log_a v$.

EXAMPLE 1 Using Properties of Logarithms

Use $\ln 2 \approx 0.693$, $\ln 3 \approx 1.099$, and $\ln 5 \approx 1.609$ to approximate each expression.

a. $\ln \frac{2}{3}$ **b.** $\ln 10$ **c.** $\ln 30$

Solution

a. $\ln \frac{2}{3} = \ln 2 - \ln 3$ Quotient Property

$\approx 0.693 - 1.099 = -0.406$ Substitute for ln 2 and ln 3.

b. $\ln 10 = \ln(2 \cdot 5)$ Factor.

$= \ln 2 + \ln 5$ Product Property

$\approx 0.693 + 1.609$ Substitute for ln 2 and ln 5.

$= 2.302$ Simplify.

c. $\ln 30 = \ln(2 \cdot 3 \cdot 5)$ Factor.

$= \ln 2 + \ln 3 + \ln 5$ Product Property

$\approx 0.693 + 1.099 + 1.609$ Substitute for ln 2, ln 3, and ln 5.

$= 3.401$ Simplify.

***CHECKPOINT** Now try Exercise 25.*

When using the properties of logarithms, it helps to state the properties *verbally*. For instance, the verbal form of the Product Property

$$\ln(uv) = \ln u + \ln v$$

is: *The log of a product is the sum of the logs of the factors.* Similarly, the verbal form of the Quotient Property

$$\ln \frac{u}{v} = \ln u - \ln v$$

is: *The log of a quotient is the difference of the logs of the numerator and denominator.*

Study Tip

Remember that you can verify results such as those given in Examples 1 and 2 with a calculator.

EXAMPLE 2 Using Properties of Logarithms

Use the properties of logarithms to verify that $-\ln 2 = \ln \frac{1}{2}$.

Solution

Using the Power Property, you can write the following.

$-\ln 2 = (-1) \ln 2$ Rewrite coefficient as -1.

$= \ln 2^{-1}$ Power Property

$= \ln \frac{1}{2}$ Rewrite 2^{-1} as $\frac{1}{2}$.

***CHECKPOINT** Now try Exercise 41.*

2 ▶ Use the properties of logarithms to rewrite, expand, or condense logarithmic expressions.

Rewriting Logarithmic Expressions

In Examples 1 and 2, the properties of logarithms were used to rewrite logarithmic expressions involving the log of a *constant*. A more common use of these properties is to rewrite the log of a *variable expression*.

EXAMPLE 3 Expanding Logarithmic Expressions

Use the properties of logarithms to expand each expression.

a. $\log_{10} 7x^3 = \log_{10} 7 + \log_{10} x^3$ Product Property

$= \log_{10} 7 + 3 \log_{10} x$ Power Property

b. $\log_6 \dfrac{8x^3}{y} = \log_6 8x^3 - \log_6 y$ Quotient Property

$= \log_6 8 + \log_6 x^3 - \log_6 y$ Product Property

$= \log_6 8 + 3 \log_6 x - \log_6 y$ Power Property

c. $\ln \dfrac{\sqrt{3x - 5}}{7} = \ln\left[\dfrac{(3x - 5)^{1/2}}{7}\right]$ Rewrite using rational exponent.

$= \ln(3x - 5)^{1/2} - \ln 7$ Quotient Property

$= \dfrac{1}{2} \ln(3x - 5) - \ln 7$ Power Property

✓ ***CHECKPOINT*** *Now try Exercise 47.*

When you rewrite a logarithmic expression as in Example 3, you are **expanding** the expression. The reverse procedure is demonstrated in Example 4, and is called **condensing** a logarithmic expression.

EXAMPLE 4 Condensing Logarithmic Expressions

Use the properties of logarithms to condense each expression.

a. $\ln x - \ln 3 = \ln \dfrac{x}{3}$ Quotient Property

b. $\dfrac{1}{2} \log_3 x + \log_3 5 = \log_3 x^{1/2} + \log_3 5$ Power Property

$= \log_3 5\sqrt{x}$ Product Property

c. $3(\ln 4 + \ln x) = 3(\ln 4x)$ Product Property

$= \ln (4x)^3$ Power Property

$= \ln 64x^3$ Simplify.

✓ ***CHECKPOINT*** *Now try Exercise 75.*

When you expand or condense a logarithmic expression, it is possible to change the domain of the expression. For instance, the domain of the function

$$f(x) = 2 \ln x \qquad \text{Domain is the set of positive real numbers.}$$

is the set of positive real numbers, whereas the domain of

$$g(x) = \ln x^2 \qquad \text{Domain is the set of nonzero real numbers.}$$

is the set of nonzero real numbers. So, when you expand or condense a logarithmic expression, you should check to see whether the rewriting has changed the domain of the expression. In such cases, you should restrict the domain appropriately. For instance, you can write

$$f(x) = 2 \ln x$$
$$= \ln x^2, x > 0.$$

3 ▶ Use the properties of logarithms to solve application problems.

Application

EXAMPLE 5 Human Memory Model 

In an experiment, students attended several lectures on a subject. Every month for a year after that, the students were tested to see how much of the material they remembered. The average scores for the group are given by the **human memory model**

$$f(t) = 80 - \ln(t + 1)^9, \quad 0 \le t \le 12$$

where t is the time in months. Find the average scores for the group after 8 months.

Solution

To make the calculations easier, rewrite the model using the Power Property, as follows.

$$f(t) = 80 - 9 \ln(t + 1), \quad 0 \le t \le 12$$

After 8 months, the average score was

$$f(8) = 80 - 9 \ln(8 + 1) \qquad \text{Substitute 8 for } t.$$
$$\approx 80 - 19.8 \qquad \text{Simplify.}$$
$$= 60.2. \qquad \text{Average score after 8 months}$$

The graph of the function is shown in Figure 11.21.

✓ **CHECKPOINT** *Now try Exercise 113.*

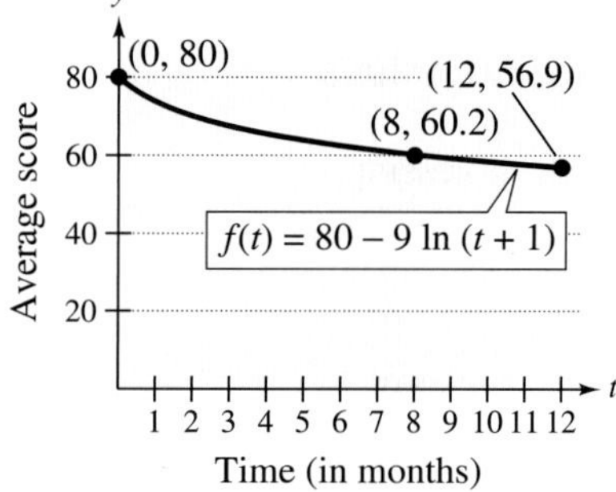

Human Memory Model
Figure 11.21

Concept Check

1. In your own words, give a verbal description of the Power Property $\ln u^n = n \ln u$.

2. Explain the steps you would take to evaluate the expression $\log_6 9 + \log_6 4$.

3. Explain the steps you would take to expand the expression

$$\log_{10}\frac{x^3}{y^5}.$$

4. For $x > 0$, is $2 \ln x$ *greater than, less than,* or *equal to* $\ln x^3 + \ln x^4 - \ln x^5$? Justify your answer.

11.4 EXERCISES

Go to pages 752–753 to record your assignments.

Developing Skills

In Exercises 1–24, use properties of logarithms to evaluate the expression without a calculator. (If not possible, state the reason.)

1. $\log_{12} 12^3$
2. $\log_3 81$
3. $\log_4\left(\frac{1}{16}\right)^2$
4. $\log_7\left(\frac{1}{49}\right)^3$
5. $\log_5 \sqrt[3]{5}$
6. $\ln \sqrt{e}$
7. $\ln 14^0$
8. $\ln \left(\dfrac{7.14}{7.14}\right)$
9. $\ln e^{-9}$
10. $\ln e^7$
11. $\log_4 8 + \log_4 2$
12. $\log_6 2 + \log_6 3$
13. $\log_8 4 + \log_8 16$
14. $\log_{10} 5 + \log_{10} 20$
15. $\log_3 54 - \log_3 2$
16. $\log_5 50 - \log_5 2$
17. $\log_6 72 - \log_6 2$
18. $\log_3 324 - \log_3 4$
19. $\log_2 5 - \log_2 40$
20. $\log_4\left(\frac{3}{16}\right) + \log_4\left(\frac{1}{3}\right)$
21. $\ln e^8 + \ln e^4$
22. $\ln e^5 - \ln e^2$
23. $\ln \dfrac{e^3}{e^2}$
24. $\ln(e^2 \cdot e^4)$

In Exercises 25–32, use $\log_4 2 = 0.5000$, $\log_4 3 \approx 0.7925$, and the properties of logarithms to approximate the expression. Do not use a calculator. ***See Example 1.***

25. $\log_4 8$
26. $\log_4 24$
27. $\log_4 \frac{3}{2}$
28. $\log_4 \frac{9}{2}$
29. $\log_4 \sqrt[3]{9}$
30. $\log_4 \sqrt{3 \cdot 2^5}$
31. $\log_4 3^0$
32. $\log_4 4^3$

In Exercises 33–40, use $\ln 3 \approx 1.0986$, $\ln 5 \approx 1.6094$, and the properties of logarithms to approximate the expression. Use a calculator to verify your result.

33. $\ln 9$
34. $\ln 75$
35. $\ln \frac{5}{3}$
36. $\ln \frac{27}{125}$
37. $\ln \sqrt{45}$
38. $\ln \sqrt[3]{25}$
39. $\ln(3^5 \cdot 5^2)$
40. $\ln \sqrt{3^2 \cdot 5^3}$

In Exercises 41–46, use the properties of logarithms to verify the statement. ***See Example 2.***

41. $-3 \log_4 2 = \log_4 \frac{1}{8}$

42. $-2 \ln \frac{1}{3} = \ln 9$

43. $-3 \log_{10} 3 + \log_{10} \frac{3}{2} = \log_{10} \frac{1}{18}$

44. $-2 \ln 2 + \ln 24 = \ln 6$

45. $-\ln \frac{1}{7} = \ln 56 - \ln 8$

46. $-\log_5 10 = \log_5 10 - \log_5 100$

In Exercises 47–74, use the properties of logarithms to expand the expression. (Assume all variables are positive.) ***See Example 3.***

47. $\log_3 11x$

48. $\log_2 3x$

49. $\ln 3y$

50. $\ln 5x$

51. $\log_7 x^2$

52. $\log_3 x^3$

53. $\log_4 x^{-3}$

54. $\log_2 s^{-4}$

55. $\log_4 \sqrt{3x}$

56. $\log_3 \sqrt[3]{5y}$

57. $\log_2 \dfrac{z}{17}$

58. $\log_{10} \dfrac{7}{y}$

59. $\log_9 \dfrac{\sqrt{x}}{12}$

60. $\ln \dfrac{\sqrt{x}}{x+9}$

61. $\ln x^2(y + 2)$

62. $\ln y(y + 1)^2$

63. $\log_4[x^6(x + 7)^2]$

64. $\log_8[(x - y)^3 z^6]$

65. $\log_3 \sqrt[3]{x + 1}$

66. $\log_5 \sqrt{xy}$

67. $\ln \sqrt{x(x + 2)}$

68. $\ln \sqrt[3]{x(x + 5)}$

69. $\ln\left(\dfrac{x+1}{x+4}\right)^2$

70. $\log_2\left(\dfrac{x^2}{x-3}\right)^3$

71. $\ln \sqrt[3]{\dfrac{x^2}{x+1}}$

72. $\ln \sqrt{\dfrac{3x}{x-5}}$

73. $\ln \dfrac{xy^2}{z^3}$

74. $\log_5 \dfrac{x^2y^5}{z^7}$

In Exercises 75–102, use the properties of logarithms to condense the expression. ***See Example 4.***

75. $\log_{12} x - \log_{12} 3$

76. $\log_6 12 - \log_6 y$

77. $\log_3 5 + \log_3 x$

78. $\log_5 2x + \log_5 3y$

79. $\log_{10} 4 - \log_{10} x$

80. $\ln 10x - \ln z$

81. $4 \ln b$

82. $12 \log_4 z$

83. $-2 \log_5 2x$

84. $-5 \ln(x + 3)$

85. $7 \log_2 x + 3 \log_2 z$

86. $2 \log_{10} x + \frac{1}{2} \log_{10} y$

87. $\log_3 2 + \frac{1}{2} \log_3 y$

88. $\ln 6 - 3 \ln z$

89. $3 \ln x + \ln y - 2 \ln z$

90. $4 \ln 2 + 2 \ln x - \frac{1}{2} \ln y$

91. $4(\ln x + \ln y)$

92. $\frac{1}{3}(\ln 10 + \ln 4x)$

93. $2[\ln x - \ln(x + 1)]$

94. $5\left[\ln x - \frac{1}{2} \ln(x + 4)\right]$

95. $\log_4(x + 8) - 3 \log_4 x$

96. $5 \log_3 x + \log_3(x - 6)$

97. $\frac{1}{3} \log_5(x + 3) - \log_5(x - 6)$

98. $\frac{1}{4} \log_6(x + 1) - 5 \log_6(x - 4)$

99. $5 \log_6(c + d) - \frac{1}{2} \log_6(m - n)$

100. $2 \log_5(x + y) + 3 \log_5 w$

101. $\frac{1}{5}(3 \log_2 x - 4 \log_2 y)$

102. $\frac{1}{3}[\ln(x - 6) - 4 \ln y - 2 \ln z]$

In Exercises 103–108, simplify the expression.

103. $\ln 3e^2$

104. $\log_3(3^2 \cdot 4)$

105. $\log_5 \sqrt{50}$

106. $\log_2 \sqrt{22}$

107. $\log_8 \dfrac{8}{x^3}$

108. $\ln \dfrac{6}{e^5}$

In Exercises 109–112, use a graphing calculator to graph the two equations in the same viewing window. Use the graphs to verify that the expressions are equivalent. Assume $x > 0$.

109. $y_1 = \ln\left(\dfrac{10}{x^2 + 1}\right)^2$

$y_2 = 2[\ln 10 - \ln(x^2 + 1)]$

110. $y_1 = \ln \sqrt{x(x + 1)}$

$y_2 = \frac{1}{2}[\ln x + \ln(x + 1)]$

111. $y_1 = \ln[x^2(x + 2)]$

$y_2 = 2 \ln x + \ln(x + 2)$

112. $y_1 = \ln \dfrac{\sqrt{x}}{x - 3}$

$y_2 = \frac{1}{2} \ln x - \ln(x - 3)$

Solving Problems

✓ **113.** *Sound Intensity* The relationship between the number of decibels B and the intensity of a sound I in watts per centimeter squared is given by

$$B = 10 \log_{10}\left(\frac{I}{10^{-16}}\right).$$

Use properties of logarithms to write the formula in simpler form, and determine the number of decibels of a thunderclap with an intensity of 10^{-3} watt per centimeter squared.

© Scott Stulberg/CORBIS

114. *Human Memory Model* Students participating in an experiment attended several lectures on a subject. Every month for a year after that, the students were tested to see how much of the material they remembered. The average scores for the group are given by the human memory model

$$f(t) = 80 - \log_{10}(t + 1)^{12}, \quad 0 \le t \le 12$$

where t is the time in months.

(a) Find the average scores for the group after 2 months and 8 months.

(b) Use a graphing calculator to graph the function.

Molecular Transport In Exercises 115 and 116, use the following information. The energy E (in kilocalories per gram molecule) required to transport a substance from the outside to the inside of a living cell is given by

$$E = 1.4(\log_{10} C_2 - \log_{10} C_1)$$

where C_1 and C_2 are the concentrations of the substance outside and inside the cell, respectively.

115. Condense the expression.

116. The concentration of a substance inside a cell is twice the concentration outside the cell. How much energy is required to transport the substance from outside to inside the cell?

Explaining Concepts

True or False? In Exercises 117–122, use properties of logarithms to determine whether the equation is true or false. Justify your answer.

117. $\log_2 8x = 3 + \log_2 x$

118. $\log_3(u + v) = \log_3 u + \log_3 v$

119. $\log_3(u + v) = \log_3 u \cdot \log_3 v$

120. $\dfrac{\log_6 10}{\log_6 3} = \log_6 10 - \log_6 3$

121. If $f(x) = \log_a x$, then $f(ax) = 1 + f(x)$.

122. If $f(x) = \log_a x$, then $f(a^n) = n$.

True or False? In Exercises 123–127, determine whether the statement is true or false given that $f(x) = \ln x$. Justify your answer.

123. $f(0) = 0$

124. $f(2x) = \ln 2 + \ln x$

125. $f(x - 3) = \ln x - \ln 3, \quad x > 3$

126. $\sqrt{f(x)} = \frac{1}{2} \ln x$

127. If $f(x) > 0$, then $x > 1$.

128. *Think About It* Without a calculator, approximate the natural logarithms of as many integers as possible between 1 and 20 using $\ln 2 \approx 0.6931$, $\ln 3 \approx 1.0986$, $\ln 5 \approx 1.6094$, and $\ln 7 \approx 1.9459$. Explain the method you used. Then verify your results with a calculator and explain any differences in the results.

129. *Think About It* Explain how you can show that

$$\frac{\ln x}{\ln y} \neq \ln \frac{x}{y}.$$

Cumulative Review

In Exercises 130–135, solve the equation.

130. $\frac{2}{3}x + \frac{2}{3} = 4x - 6$

131. $x^2 - 10x + 17 = 0$

132. $\dfrac{5}{2x} - \dfrac{4}{x} = 3$

133. $\dfrac{1}{x} + \dfrac{2}{x - 5} = 0$

134. $|x - 4| = 3$

135. $\sqrt{x + 2} = 7$

In Exercises 136–139, sketch the parabola. Identify the vertex and any x-intercepts.

136. $g(x) = -(x + 2)^2$

137. $f(x) = x^2 - 16$

138. $g(x) = -2x^2 + 4x - 7$

139. $h(x) = x^2 + 6x + 14$

In Exercises 140–143, find the compositions (a) $f \circ g$ and (b) $g \circ f$. Then find the domain of each composition.

140. $f(x) = 4x + 9$
$g(x) = x - 5$

141. $f(x) = \sqrt{x}$
$g(x) = x - 3$

142. $f(x) = \dfrac{1}{x}$
$g(x) = x + 2$

143. $f(x) = \dfrac{5}{x^2 - 4}$
$g(x) = x + 1$

11.5 Solving Exponential and Logarithmic Equations

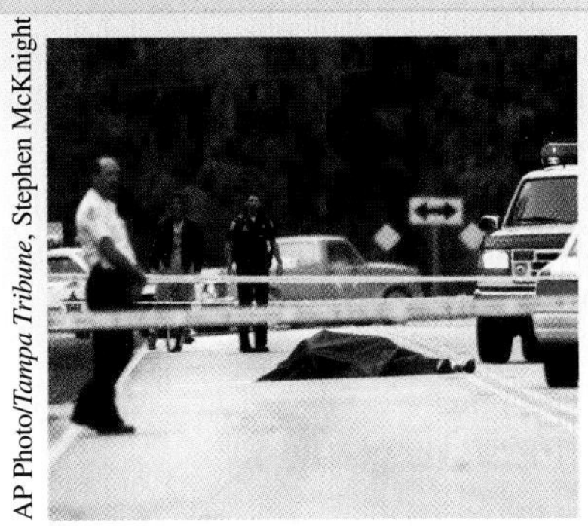
AP Photo/*Tampa Tribune*, Stephen McKnight

Why You Should Learn It

Exponential and logarithmic equations occur in many scientific applications. For instance, in Exercise 131 on page 739, you will use a logarithmic equation to determine a person's time of death.

What You Should Learn

1 ▶ Solve basic exponential and logarithmic equations.

2 ▶ Use inverse properties to solve exponential equations.

3 ▶ Use inverse properties to solve logarithmic equations.

4 ▶ Use exponential or logarithmic equations to solve application problems.

1 ▶ Solve basic exponential and logarithmic equations.

Exponential and Logarithmic Equations

In this section, you will study procedures for *solving equations* that involve exponential or logarithmic expressions. As a simple example, consider the exponential equation $2^x = 16$. By rewriting this equation in the form $2^x = 2^4$, you can see that the solution is $x = 4$. To solve this equation, you can use one of the following properties, which result from the fact that exponential and logarithmic functions are one-to-one functions.

One-to-One Properties of Exponential and Logarithmic Equations

Let a be a positive real number such that $a \neq 1$, and let x and y be real numbers. Then the following properties are true.

1. $a^x = a^y$ if and only if $x = y$.

2. $\log_a x = \log_a y$ if and only if $x = y$ $(x > 0, y > 0)$.

EXAMPLE 1 Solving Exponential and Logarithmic Equations

Solve each equation.

a.

$$4^{x+2} = 64 \qquad \text{Original equation}$$

$$4^{x+2} = 4^3 \qquad \text{Rewrite with like bases.}$$

$$x + 2 = 3 \qquad \text{One-to-one property}$$

$$x = 1 \qquad \text{Subtract 2 from each side.}$$

The solution is $x = 1$. Check this in the original equation.

b.

$$\ln(2x - 3) = \ln 11 \qquad \text{Original equation}$$

$$2x - 3 = 11 \qquad \text{One-to-one property}$$

$$2x = 14 \qquad \text{Add 3 to each side.}$$

$$x = 7 \qquad \text{Divide each side by 2.}$$

The solution is $x = 7$. Check this in the original equation.

✓ **CHECKPOINT** *Now try Exercise 21.*

2 ▶ Use inverse properties to solve exponential equations.

Solving Exponential Equations

In Example 1(a), you were able to use a one-to-one property to solve the original equation because each side of the equation was written in exponential form with the same base. However, if only one side of the equation is written in exponential form or if both sides cannot be written with the same base, it is more difficult to solve the equation. For example, to solve the equation $2^x = 7$, you must find the power to which 2 can be raised to obtain 7. To do this, *rewrite the exponential equation in logarithmic form* by taking the logarithm of each side, and use one of the inverse properties of exponents and logarithms listed below.

Solving Exponential Equations

To solve an exponential equation, first isolate the exponential expression. Then **take the logarithm of each side of the equation** (or write the equation in logarithmic form) and solve for the variable.

Technology: Discovery

Use a graphing calculator to graph each side of each equation. What does this tell you about the inverse properties of exponents and logarithms?

1. (a) $\log_{10}(10^x) = x$
   (b) $10^{(\log_{10} x)} = x$
2. (a) $\ln(e^x) = x$
   (b) $e^{(\ln x)} = x$

Inverse Properties of Exponents and Logarithms

	Base a	*Natural Base e*
1.	$\log_a(a^x) = x$	$\ln(e^x) = x$
2.	$a^{(\log_a x)} = x$	$e^{(\ln x)} = x$

Study Tip

Remember that to evaluate a logarithm such as $\log_2 7$ you need to use the change-of-base formula.

$$\log_2 7 = \frac{\ln 7}{\ln 2} \approx 2.807$$

Similarly,

$$\log_4 9 + 3 = \frac{\ln 9}{\ln 4} + 3$$

$$\approx 1.585 + 3$$

$$= 4.585$$

EXAMPLE 2 Solving Exponential Equations

Solve each exponential equation.

a. $2^x = 7$ **b.** $4^{x-3} = 9$ **c.** $2e^x = 10$

Solution

a. To isolate x, take the $\log_2$ of each side of the equation or write the equation in logarithmic form, as follows.

$2^x = 7$ Write original equation.

$x = \log_2 7$ Inverse property

The solution is $x = \log_2 7 \approx 2.807$. Check this in the original equation.

b. $4^{x-3} = 9$ Write original equation.

$x - 3 = \log_4 9$ Inverse property

$x = \log_4 9 + 3$ Add 3 to each side.

The solution is $x = \log_4 9 + 3 \approx 4.585$. Check this in the original equation.

c. $2e^x = 10$ Write original equation.

$e^x = 5$ Divide each side by 2.

$x = \ln 5$ Inverse property

The solution is $x = \ln 5 \approx 1.609$. Check this in the original equation.

✓ ***CHECKPOINT*** *Now try Exercise 39.*

Technology: Tip

Remember that you can use a graphing calculator to solve equations graphically or to check solutions that are obtained algebraically. For instance, to check the solutions in Examples 2(a) and 2(c), graph each side of each equation, as shown below.

Graph $y_1 = 2^x$ and $y_2 = 7$. Then use the *intersect* feature of the graphing calculator to approximate the intersection of the two graphs to be $x \approx 2.807$.

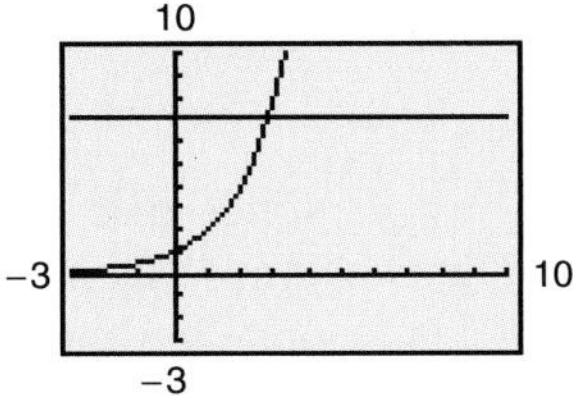

Graph $y_1 = 2e^x$ and $y_2 = 10$. Then use the *intersect* feature of the graphing calculator to approximate the intersection of the two graphs to be $x \approx 1.609$.

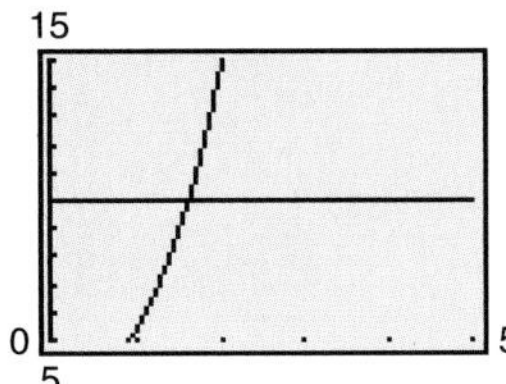

EXAMPLE 3 Solving an Exponential Equation

Solve $5 + e^{x+1} = 20$.

Solution

$5 + e^{x+1} = 20$	Write original equation.
$e^{x+1} = 15$	Subtract 5 from each side.
$\ln e^{x+1} = \ln 15$	Take the logarithm of each side.
$x + 1 = \ln 15$	Inverse property
$x = -1 + \ln 15$	Subtract 1 from each side.

The solution is $x = -1 + \ln 15 \approx 1.708$. You can check this as follows.

Check

$5 + e^{x+1} = 20$	Write original equation.
$5 + e^{-1+\ln 15+1} \stackrel{?}{=} 20$	Substitute $-1 + \ln 15$ for x.
$5 + e^{\ln 15} \stackrel{?}{=} 20$	Simplify.
$5 + 15 = 20$	Solution checks. ✓

CHECKPOINT *Now try Exercise 67.*

3 ▶ Use inverse properties to solve logarithmic equations.

Solving Logarithmic Equations

You know how to solve an exponential equation by *taking the logarithm of each side*. To solve a logarithmic equation, you need to **exponentiate** each side. For instance, to solve a logarithmic equation such as $\ln x = 2$, you can exponentiate each side of the equation as follows.

$\ln x = 2$ Write original equation.

$e^{\ln x} = e^2$ Exponentiate each side.

$x = e^2$ Inverse property

Notice that you obtain the same result by writing the equation in exponential form. This procedure is demonstrated in the next three examples. The following guideline can be used for solving logarithmic equations.

Solving Logarithmic Equations

To solve a logarithmic equation, first isolate the logarithmic expression. Then **exponentiate each side of the equation** (or write the equation in exponential form) and solve for the variable.

EXAMPLE 4 Solving Logarithmic Equations

a. $2 \log_4 x = 5$ Original equation

$\log_4 x = \frac{5}{2}$ Divide each side by 2.

$4^{\log_4 x} = 4^{5/2}$ Exponentiate each side.

$x = 4^{5/2}$ Inverse property

$x = 32$ Simplify.

The solution is $x = 32$. Check this in the original equation, as follows.

Check

$2 \log_4 x = 5$ Original equation

$2 \log_4(32) \stackrel{?}{=} 5$ Substitute 32 for x.

$2(2.5) \stackrel{?}{=} 5$ Use a calculator.

$5 = 5$ Solution checks. ✓

b. $\frac{1}{4} \log_2 x = \frac{1}{2}$ Original equation

$\log_2 x = 2$ Multiply each side by 4.

$2^{\log_2 x} = 2^2$ Exponentiate each side.

$x = 4$ Inverse property

The solution is $x = 4$. Check this in the original equation.

✓ CHECKPOINT *Now try Exercise 83.*

EXAMPLE 5 Solving a Logarithmic Equation

$3 \log_{10} x = 6$	Original equation
$\log_{10} x = 2$	Divide each side by 3.
$x = 10^2$	Exponential form
$x = 100$	Simplify.

The solution is $x = 100$. Check this in the original equation.

 **CHECKPOINT** ***Now try Exercise 87.***

Study Tip

When checking approximate solutions to exponential and logarithmic equations, be aware that the check will not be exact because the solutions are approximate.

EXAMPLE 6 Solving a Logarithmic Equation

$20 \ln 0.2x = 30$	Original equation
$\ln 0.2x = 1.5$	Divide each side by 20.
$e^{\ln 0.2x} = e^{1.5}$	Exponentiate each side.
$0.2x = e^{1.5}$	Inverse property
$x = 5e^{1.5}$	Divide each side by 0.2.

The solution is $x = 5e^{1.5} \approx 22.408$. Check this in the original equation.

 **CHECKPOINT** ***Now try Exercise 93.***

The next two examples use logarithmic properties as part of the solutions.

EXAMPLE 7 Solving a Logarithmic Equation

$\log_3 2x - \log_3(x - 3) = 1$	Original equation
$\log_3 \dfrac{2x}{x - 3} = 1$	Condense the left side.
$\dfrac{2x}{x - 3} = 3^1$	Exponential form
$2x = 3x - 9$	Multiply each side by $x - 3$.
$-x = -9$	Subtract $3x$ from each side.
$x = 9$	Divide each side by -1.

The solution is $x = 9$. Check this in the original equation.

CHECKPOINT ***Now try Exercise 105.***

Study Tip

Recall from Section 11.3 that you cannot take the logarithm of a negative number. So, another way to recognize extraneous solutions of a logarithmic equation is to identify the restrictions on the domain before you solve the equation. Because the domain of $y = \log_b x$ is $x > 0$, the equation in Example 8 must have $x > 0$ and $x - 5 > 0$, which means that $x > 5$. Any solution that is less than or equal to 5 is extraneous.

EXAMPLE 8 Checking For Extraneous Solutions

$\log_6 x + \log_6(x - 5) = 2$	Original equation
$\log_6 [x(x - 5)] = 2$	Condense the left side.
$x(x - 5) = 6^2$	Exponential form
$x^2 - 5x - 36 = 0$	Write in general form.
$(x - 9)(x + 4) = 0$	Factor.
$x - 9 = 0 \Rightarrow x = 9$	Set 1st factor equal to 0.
$x + 4 = 0 \Rightarrow x = -4$	Set 2nd factor equal to 0.

Check the possible solutions $x = 9$ and $x = -4$ in the original equation.

First Solution

$$\log_6 (9) + \log_6 (9 - 5) \stackrel{?}{=} 2$$
$$\log_6 (9 \cdot 4) \stackrel{?}{=} 2$$
$$\log_6 36 = 2 \checkmark$$

Second Solution

$$\log_6 (-4) + \log_6 (-4 - 5) \stackrel{?}{=} 2$$
$$\log_6 (-4) + \log_6 (-9) \neq 2 \text{ ✗}$$

Of the two possible solutions, only $x = 9$ checks. So, $x = -4$ is extraneous.

✓ CHECKPOINT *Now try Exercise 109.*

4 ▶ Use exponential or logarithmic equations to solve application problems.

Application

EXAMPLE 9 Compound Interest

A deposit of \$5000 is placed in a savings account for 2 years. The interest on the account is compounded continuously. At the end of 2 years, the balance in the account is \$5416.44. What is the annual interest rate for this account?

Solution

Formula: $A = Pe^{rt}$

Labels:		
	Principal $= P = 5000$	(dollars)
	Amount $= A = 5416.44$	(dollars)
	Time $= t = 2$	(years)
	Annual interest rate $= r$	(percent in decimal form)

Equation:

$5416.44 = 5000e^{2r}$	Substitute for A, P, and t.
$1.083288 = e^{2r}$	Divide each side by 5000 and simplify.
$\ln 1.083288 - \ln(e^{2r})$	Takc logarithm of cach side.
$0.08 \approx 2r \Rightarrow 0.04 \approx r$	Inverse property

The annual interest rate is approximately 4%. Check this solution.

✓ CHECKPOINT *Now try Exercise 123.*

Concept Check

1. Can the one-to-one property of logarithms be used to solve the equation $\log_2 x = \log_3 9$? Explain.

2. Explain how to solve $5^{x+2} = 5^4$.

3. Which equation requires logarithms for its solution: $2^{x-1} = 32$ or $2^{x-1} = 30$? Explain.

4. If a solution of a logarithmic equation is negative, does this imply that the solution is extraneous? Explain.

11.5 EXERCISES

Go to pages 752–753 to record your assignments.

Developing Skills

In Exercises 1–6, determine whether each value of *x* is a solution of the equation.

1. $3^{2x-5} = 27$
(a) $x = 1$
(b) $x = 4$

2. $2^{x+5} = 16$
(a) $x = -1$
(b) $x = 0$

3. $e^{x+5} = 45$
(a) $x = -5 + \ln 45$
(b) $x \approx -2.1933$

4. $4^{x-2} = 250$
(a) $x = 2 + \log_4 250$
(b) $x \approx 4.9829$

5. $\log_9(6x) = \frac{3}{2}$
(a) $x = 27$
(b) $x = \frac{9}{2}$

6. $\ln(x + 3) = 2.5$
(a) $x = -3 + e^{2.5}$
(b) $x \approx 9.1825$

In Exercises 7–34, solve the equation. (Do not use a calculator.) ***See Example 1.***

7. $7^x = 7^3$
8. $4^x = 4^6$
9. $e^{1-x} = e^4$
10. $e^{x+3} = e^8$
11. $5^{x+6} = 25^5$
12. $2^{x-4} = 8^2$
13. $6^{2x} = 36$
14. $5^{3x} = 25$
15. $3^{2-x} = 81$
16. $4^{2x-1} = 64$
17. $5^x = \frac{1}{125}$
18. $3^x = \frac{1}{243}$
19. $2^{x+2} = \frac{1}{16}$
20. $3^{x+2} = \frac{1}{27}$
✓ **21.** $4^{x+3} = 32^x$
22. $9^{x-2} = 243^{x+1}$
23. $\ln 5x = \ln 22$
24. $\ln 4x = \ln 30$
25. $\log_6 3x = \log_6 18$
26. $\log_5 2x = \log_5 36$
27. $\ln(3 - x) = \ln 10$
28. $\ln(2x - 3) = \ln 17$
29. $\log_2(x + 3) = \log_2 7$
30. $\log_4(x - 8) = \log_4(-4)$
31. $\log_5(2x - 3) = \log_5(4x - 5)$
32. $\log_3(4 - 3x) = \log_3(2x + 9)$
33. $\log_3(2 - x) = 2$
34. $\log_2(3x - 1) = 5$

In Exercises 35–38, simplify the expression.

35. $\ln e^{2x-1}$
36. $\log_3 3^{x^2}$
37. $10^{\log_{10} 2x}$
38. $e^{\ln(x+1)}$

In Exercises 39–82, solve the exponential equation. (Round your answer to two decimal places.) ***See Examples 2 and 3.***

✓ **39.** $3^x = 91$
40. $4^x = 40$
41. $5^x = 8.2$
42. $2^x = 3.6$
43. $6^{2x} = 205$
44. $4^{3x} = 168$
45. $7^{3y} = 126$
46. $5^{5y} = 305$
47. $3^{2-x} = 8$
48. $5^{3-x} = 15$
49. $10^{x+6} = 250$
50. $12^{x-1} = 324$
51. $4e^{-x} = 24$
52. $6e^{-x} = 3$
53. $\frac{1}{4}e^x = 5$
54. $\frac{2}{3}e^x = 1$
55. $\frac{1}{2}e^{-2x} = 9$
56. $4e^{-3x} = 6$
57. $250(1.04)^x = 1000$
58. $32(1.5)^x = 640$
59. $300e^{x/2} = 9000$
60. $7500e^{x/3} = 1500$
61. $1000^{0.12x} = 25{,}000$
62. $1800^{0.2x} = 225$

63. $\frac{1}{5}(4^{x+2}) = 300$

64. $3(2^{t+4}) = 350$

65. $6 + 2^{x-1} = 1$

66. $5^{x+6} - 4 = 12$

✓ **67.** $7 + e^{2-x} = 28$

68. $24 + e^{4-x} = 22$

69. $8 - 12e^{-x} = 7$

70. $6 - 3e^{-x} = -15$

71. $4 + e^{2x} = 10$

72. $10 + e^{4x} = 18$

73. $17 - e^{x/4} = 14$

74. $50 - e^{x/2} = 35$

75. $23 - 5e^{x+1} = 3$

76. $3e^{1-x} - 5 = 72$

77. $4(1 + e^{x/3}) = 84$

78. $50(3 - e^{2x}) = 125$

79. $\dfrac{8000}{(1.03)^t} = 6000$

80. $\dfrac{5000}{(1.05)^x} = 250$

81. $\dfrac{300}{2 - e^{-0.15t}} = 200$

82. $\dfrac{500}{1 + e^{-0.1t}} = 400$

In Exercises 83–118, solve the logarithmic equation. (Round your answer to two decimal places.) ***See Examples 4–8.***

✓ **83.** $\log_{10} x = -1$

84. $\log_{10} x = 3$

85. $\log_3 x = 4.7$

86. $\log_3 x = -1.8$

✓ **87.** $4 \log_3 x = 28$

88. $6 \log_2 x = 18$

89. $16 \ln x = 30$

90. $12 \ln x = 20$

91. $\log_{10} 4x = 2$

92. $\log_3 6x = 4$

✓ **93.** $\ln 2x = \frac{1}{5}$

94. $\ln(0.5t) = \frac{1}{4}$

95. $\ln x^2 = 6$

96. $\ln \sqrt{x} = 1.3$

97. $2 \log_4(x + 5) = 3$

98. $5 \log_{10}(x + 2) = 15$

99. $\frac{3}{4} \ln(x + 4) = -2$

100. $\frac{2}{3} \ln(x + 1) = -1$

101. $7 - 2 \log_2 x = 4$

102. $5 - 4 \log_2 x = 2$

103. $-1 + 3 \log_{10} \dfrac{x}{2} = 8$

104. $-5 + 2 \ln 3x = 5$

✓ **105.** $\log_4 x + \log_4 5 = 2$

106. $\log_5 x - \log_5 4 = 2$

107. $\log_6(x + 8) + \log_6 3 = 2$

108. $\log_7(x - 1) - \log_7 4 = 1$

✓ **109.** $\log_5(x + 3) - \log_5 x = 1$

110. $\log_3(x - 2) + \log_3 5 = 3$

111. $\log_{10} x + \log_{10}(x - 3) = 1$

112. $\log_{10} x + \log_{10}(x + 1) = 0$

113. $\log_2(x - 1) + \log_2(x + 3) = 3$

114. $\log_6(x - 5) + \log_6 x = 2$

115. $\log_{10} 4x - \log_{10}(x - 2) = 1$

116. $\log_2 3x - \log_2(x + 4) = 3$

117. $\log_2 x + \log_2(x + 2) - \log_2 3 = 4$

118. $\log_3 2x + \log_3(x - 1) - \log_3 4 = 1$

In Exercises 119–122, use a graphing calculator to approximate the x-intercept of the graph.

119. $y = 10^{x/2} - 5$

120. $y = 2e^x - 21$

121. $y = 6 \ln(0.4x) - 13$

122. $y = 5 \log_{10}(x + 1) - 3$

Solving Problems

✓ **123.** *Compound Interest* A deposit of $10,000 is placed in a savings account for 2 years. The interest for the account is compounded continuously. At the end of 2 years, the balance in the account is $11,051.71. What is the annual interest rate for this account?

124. *Compound Interest* A deposit of $2500 is placed in a savings account for 2 years. The interest for the account is compounded continuously. At the end of 2 years, the balance in the account is $2847.07. What is the annual interest rate for this account?

125. *Doubling Time* Solve the exponential equation $5000 = 2500e^{0.09t}$ for t to determine the number of years for an investment of \$2500 to double in value when compounded continuously at the rate of 9%.

126. *Doubling Rate* Solve the exponential equation $10{,}000 = 5000e^{10r}$ for r to determine the interest rate required for an investment of \$5000 to double in value when compounded continuously for 10 years.

127. *Sound Intensity* The relationship between the number of decibels B and the intensity of a sound I in watts per centimeter squared is given by

$$B = 10\log_{10}\left(\frac{I}{10^{-16}}\right).$$

Determine the intensity of a sound I if it registers 80 decibels on a decibel meter.

128. *Sound Intensity* The relationship between the number of decibels B and the intensity of a sound I in watts per centimeter squared is given by

$$B = 10\log_{10}\left(\frac{I}{10^{-16}}\right).$$

Determine the intensity of a sound I if it registers 110 decibels on a decibel meter.

129. *Friction* In order to restrain an untrained horse, a trainer partially wraps a rope around a cylindrical post in a corral (see figure). The horse is pulling on the rope with a force of 200 pounds. The force F (in pounds) needed to hold back the horse is $F = 200e^{-0.5\pi\theta/180}$, where θ is the angle of wrap (in degrees). The trainer needs to know the smallest value of θ for which a force of 80 pounds will hold the horse.

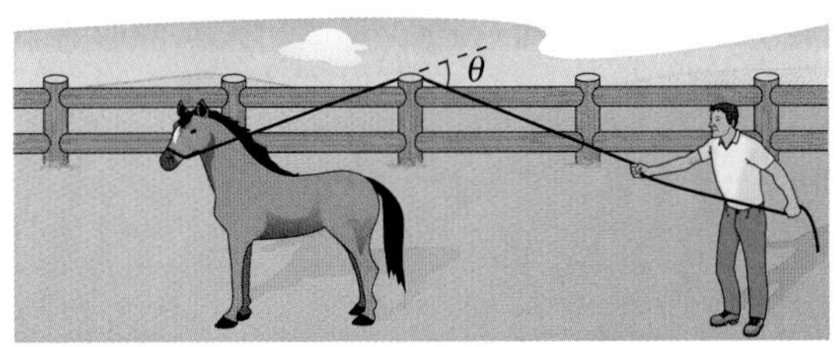

(a) Find the smallest value of θ algebraically by letting $F = 80$ and solving the resulting equation.

(b) Find the smallest value of θ graphically by using a graphing calculator to graph the equations $y_1 = 200e^{-0.5\pi\theta/180}$ and $y_2 = 80$ in order to find the point of intersection.

130. *Online Retail* The projected online retail sales S (in billions of dollars) in the United States for the years 2006 through 2011 are modeled by the equation $S = 59.8e^{0.1388t}$, for $6 \le t \le 11$, where t is the time in years, with $t = 6$ corresponding to 2006. You want to know the year when S is about \$210 billion. (Source: Forrester Research, Inc.)

(a) Find the year algebraically by letting $S = 210$ and solving the resulting equation.

(b) Find the year graphically by using a graphing calculator to graph the equations $y_1 = 59.8e^{0.1388t}$ and $y_2 = 210$ in order to find the point of intersection.

Newton's Law of Cooling In Exercises 131 and 132, use Newton's Law of Cooling

$$kt = \ln\frac{T - S}{T_0 - S}$$

where T is the temperature of a body (in °F), t is the number of hours elapsed, S is the temperature of the environment, and T_0 is the initial temperature of the body.

131. *Time of Death* A corpse was discovered in a motel room at 10 P.M., and its temperature was 85°F. Three hours later, the temperature of the corpse was 78°F. The temperature of the motel room is a constant 65°F.

(a) What is the constant k?

(b) Find the time of death using the fact that the temperature of the corpse at the time of death was 98.6°F.

(c) What is the temperature of the corpse two hours after death?

132. *Time of Death* A corpse was discovered in the bedroom of a home at 7 A.M., and its temperature was 92°F. Two hours later, the temperature of the corpse was 88°F. The temperature of the bedroom is a constant 68°F.

(a) What is the constant k?

(b) Find the time of death using the fact that the temperature of the corpse at the time of death was 98.6°F.

(c) What is the temperature of the corpse three hours after death?

Oceanography In Exercises 133 and 134, use the following information. Oceanographers use the density d (in grams per cubic centimeter) of seawater to obtain information about the circulation of water masses and the rates at which waters of different densities mix. For water with a salinity of 30%, the water temperature T (in °C) is related to the density by

$T = 7.9 \ln(1.0245 - d) + 61.84.$

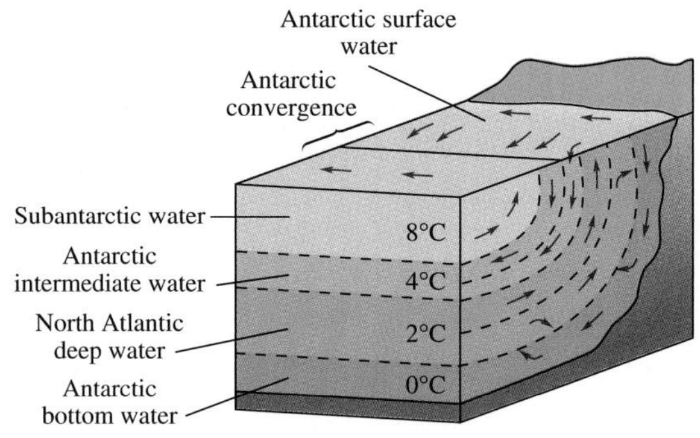

Figure for 133 and 134

This cross section shows complex currents at various depths in the South Atlantic Ocean off Antarctica.

133. Find the densities of the Subantarctic water and the Antarctic bottom water shown in the figure.

134. Find the densities of the Antarctic intermediate water and the North Atlantic deep water shown in the figure.

Explaining Concepts

135. State the three basic properties of logarithms.

136. Explain how to solve $10^{2x-1} = 5316$.

137. In your own words, state the guidelines for solving exponential and logarithmic equations.

138. Why is it possible for a logarithmic equation to have an extraneous solution?

Cumulative Review

In Exercises 139–142, solve the equation by using the Square Root Property.

139. $x^2 = -25$

140. $x^2 - 49 = 0$

141. $9n^2 - 16 = 0$

142. $(2a + 3)^2 = 18$

In Exercises 143 and 144, solve the equation of quadratic form.

143. $t^4 - 13t^2 + 36 = 0$

144. $u + 2\sqrt{u} - 15 = 0$

In Exercises 145–148, complete the table of widths, lengths, perimeters, and areas of rectangles.

	Width	*Length*	*Perimeter*	*Area*
145.	$2.5x$	x	42 in.	
146.	w	$1.6w$	78 ft	
147.	w	$w + 4$		192 km^2
148.	$x - 3$	x		270 cm^2

11.6 Applications

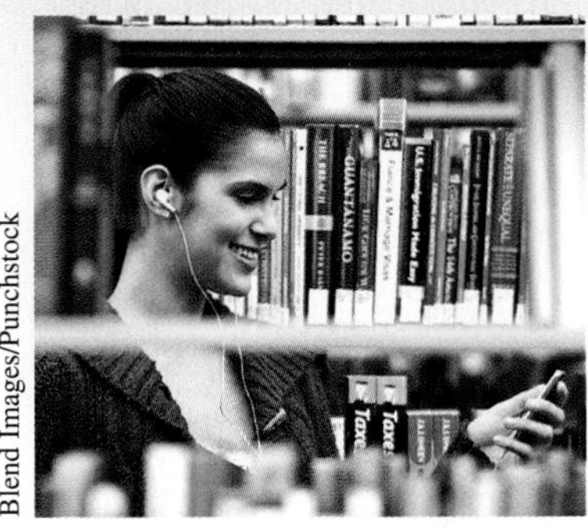

Blend Images/Punchstock

Why You Should Learn It

Exponential growth and decay models can be used in many real-life situations. For instance, in Exercise 62 on page 750, you will use an exponential growth model to represent the number of music albums downloaded in the United States.

What You Should Learn

1 ▶ Use exponential equations to solve compound interest problems.

2 ▶ Use exponential equations to solve growth and decay problems.

3 ▶ Use logarithmic equations to solve intensity problems.

1 ▶ Use exponential equations to solve compound interest problems.

Compound Interest

In Section 11.1, you were introduced to two formulas for compound interest. Recall that in these formulas, A is the balance, P is the principal, r is the annual interest rate (in decimal form), and t is the time in years.

n Compoundings per Year

$$A = P\left(1 + \frac{r}{n}\right)^{nt}$$

Continuous Compounding

$$A = Pe^{rt}$$

EXAMPLE 1 Finding the Annual Interest Rate

An investment of $50,000 is made in an account that compounds interest quarterly. After 4 years, the balance in the account is $71,381.07. What is the annual interest rate for this account?

Solution

Formula: $A = P\left(1 + \frac{r}{n}\right)^{nt}$

Labels:
Principal $= P = 50{,}000$ (dollars)
Amount $= A = 71{,}381.07$ (dollars)
Time $= t = 4$ (years)
Number of compoundings per year $= n = 4$
Annual interest rate $= r$ (percent in decimal form)

Equation:

$71{,}381.07 = 50{,}000\left(1 + \frac{r}{4}\right)^{(4)(4)}$ Substitute for A, P, n, and t.

$1.42762 \approx \left(1 + \frac{r}{4}\right)^{16}$ Divide each side by 50,000.

$(1.42762)^{1/16} \approx 1 + \frac{r}{4}$ Raise each side to $\frac{1}{16}$ power.

$1.0225 \approx 1 + \frac{r}{4}$ Simplify.

$0.09 \approx r$ Subtract 1 from each side and then multiply each side by 4.

The annual interest rate is approximately 9%. Check this in the original problem.

✓ **CHECKPOINT** *Now try Exercise 1.*

Study Tip

To remove an exponent from one side of an equation, you can often raise each side of the equation to the *reciprocal* power. For instance, in Example 1, the exponent 16 is eliminated from the right side by raising each side to the reciprocal power $\frac{1}{16}$.

EXAMPLE 2 Doubling Time for Continuous Compounding

An investment is made in a trust fund at an annual interest rate of 8.75%, compounded continuously. How long will it take for the investment to double?

Study Tip

In "doubling time" problems, you do not need to know the value of the principal P to find the doubling time. As shown in Example 2, the factor P divides out of the equation and so does not affect the doubling time.

Solution

$A = Pe^{rt}$ Formula for continuous compounding

$2P = Pe^{0.0875t}$ Substitute known values.

$2 = e^{0.0875t}$ Divide each side by P.

$\ln 2 = 0.0875t$ Inverse property

$\dfrac{\ln 2}{0.0875} = t$ Divide each side by 0.0875.

$7.92 \approx t$ Use a calculator.

It will take approximately 7.92 years for the investment to double.

Check

$2P \stackrel{?}{=} Pe^{0.0875(7.92)}$ Substitute $2P$ for A, 0.0875 for r, and 7.92 for t.

$2P \stackrel{?}{=} Pe^{0.693}$ Simplify.

$2P \approx 1.9997P$ Solution checks. ✓

✓ **CHECKPOINT** *Now try Exercise 9.*

EXAMPLE 3 Finding the Type of Compounding

You deposit \$1000 in an account. At the end of 1 year, your balance is \$1077.63. The bank tells you that the annual interest rate for the account is 7.5%. How was the interest compounded?

Solution

If the interest had been compounded continuously at 7.5%, the balance would have been $A = 1000e^{(0.075)(1)} = \$1077.88$. Because the actual balance is slightly less than this, you should use the formula for interest that is compounded n times per year.

$$A = 1000\left(1 + \frac{0.075}{n}\right)^n = 1000\left(1 + \frac{0.075}{12}\right)^{12} \approx 1077.63$$

At this point, it is not clear what you should do to solve the equation for n. However, by completing a table like the one shown below, you can see that $n = 12$. So, the interest was compounded monthly.

n	1	4	12	365
$1000\left(1 + \frac{0.075}{n}\right)^n$	1075	1077.14	1077.63	1077.88

✓ **CHECKPOINT** *Now try Exercise 13.*

In Example 3, notice that an investment of \$1000 compounded monthly produced a balance of \$1077.63 at the end of 1 year. Because \$77.63 of this amount is interest, the **effective yield** for the investment is

$$\text{Effective yield} = \frac{\text{Year's interest}}{\text{Amount invested}} = \frac{77.63}{1000} = 0.07763 = 7.763\%.$$

In other words, the effective yield for an investment collecting compound interest is the *simple interest rate* that would yield the same balance at the end of 1 year.

EXAMPLE 4 Finding the Effective Yield 

An investment is made in an account that pays 6.75% interest, compounded continuously. What is the effective yield for this investment?

Solution

Notice that you do not have to know the principal or the time that the money will be left in the account. Instead, you can choose an arbitrary principal, such as \$1000. Then, because effective yield is based on the balance at the end of 1 year, you can use the following formula.

$$\begin{aligned} A &= Pe^{rt} \\ &= 1000e^{0.0675(1)} \\ &\approx 1069.83 \end{aligned}$$

Now, because the account would earn \$69.83 in interest after 1 year for a principal of \$1000, you can conclude that the effective yield is

$$\text{Effective yield} = \frac{69.83}{1000} = 0.06983 = 6.983\%.$$

 ***CHECKPOINT*** *Now try Exercise 19.*

2 ▶ Use exponential equations to solve growth and decay problems.

Growth and Decay

The balance in an account earning *continuously* compounded interest is one example of a quantity that increases over time according to the **exponential growth model** $y = Ce^{kt}$.

Exponential Growth and Decay

The mathematical model for exponential growth or decay is given by

$$y = Ce^{kt}.$$

For this model, t is the time, C is the original amount of the quantity, and y is the amount after time t. The number k is a constant that is determined by the rate of growth (or decay). If $k > 0$, the model represents **exponential growth,** and if $k < 0$, it represents **exponential decay.**

One common application of exponential growth is in modeling the growth of a population. Example 5 illustrates the use of the growth model

$$y = Ce^{kt}, \quad k > 0.$$

EXAMPLE 5 Population Growth

The population of Texas was 17 million in 1990 and 21 million in 2000. What would you estimate the population of Texas to be in 2010? (Source: U.S. Census Bureau)

Solution

If you assumed a *linear growth model*, you would simply estimate the population in the year 2010 to be 25 million because the population would increase by 4 million every 10 years. However, social scientists and demographers have discovered that *exponential growth models* are better than linear growth models for representing population growth. So, you can use the exponential growth model

$$y = Ce^{kt}.$$

In this model, let $t = 0$ represent 1990. The given information about the population can be described by the following table.

t (year)	0	10	20
Ce^{kt} (million)	$Ce^{k(0)} = 17$	$Ce^{k(10)} = 21$	$Ce^{k(20)} = ?$

To find the population when $t = 20$, you must first find the values of C and k. From the table, you can use the fact that $Ce^{k(0)} = Ce^0 = 17$ to conclude that $C = 17$. Then, using this value of C, you can solve for k as follows.

$$Ce^{k(10)} = 21 \qquad \text{From table}$$

$$17e^{10k} = 21 \qquad \text{Substitute value of } C.$$

$$e^{10k} = \frac{21}{17} \qquad \text{Divide each side by 17.}$$

$$10k = \ln\frac{21}{17} \qquad \text{Inverse property}$$

$$k = \frac{1}{10}\ln\frac{21}{17} \qquad \text{Divide each side by 10.}$$

$$k \approx 0.0211 \qquad \text{Simplify.}$$

Finally, you can use this value of k in the model from the table for 2010 (for $t = 20$) to estimate the population in the year 2010 to be

$$17e^{0.0211(20)} \approx 17(1.53) = 26.01 \text{ million}.$$

Figure 11.22 graphically compares the exponential growth model with a linear growth model.

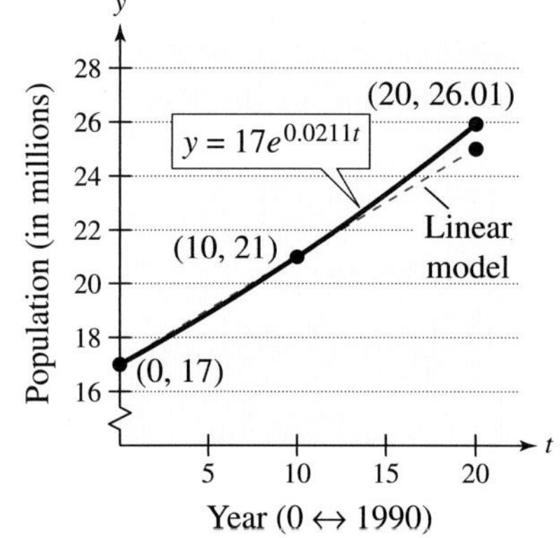

Population Models
Figure 11.22

✓ ***CHECKPOINT*** *Now try Exercise 47.*

EXAMPLE 6 Radioactive Decay 

Radioactive iodine-125 is a by-product of some types of nuclear reactors. Its **half-life** is 60 days. That is, after 60 days, a given amount of radioactive iodine-125 will have decayed to half the original amount. A nuclear accident occurs and releases 20 grams of radioactive iodine-125. How long will it take for the radioactive iodine to decay to a level of 1 gram?

Solution

Use the model for exponential decay, $y = Ce^{kt}$, and the information given in the problem to set up the following table.

t (days)	0	60	?
Ce^{kt} (grams)	$Ce^{k(0)} = 20$	$Ce^{k(60)} = 10$	$Ce^{k(t)} = 1$

Because $Ce^{k(0)} = Ce^0 = 20$, you can conclude that $C = 20$. Then, using this value of C, you can solve for k as follows.

$$Ce^{k(60)} = 10 \qquad \text{From table}$$

$$20e^{60k} = 10 \qquad \text{Substitute value of } C.$$

$$e^{60k} = \frac{1}{2} \qquad \text{Divide each side by 20.}$$

$$60k = \ln\frac{1}{2} \qquad \text{Inverse property}$$

$$k = \frac{1}{60}\ln\frac{1}{2} \qquad \text{Divide each side by 60.}$$

$$k \approx -0.01155 \qquad \text{Simplify.}$$

Finally, you can use this value of k in the model from the table to find the time when the amount is 1 gram, as follows.

$$Ce^{kt} = 1 \qquad \text{From table}$$

$$20e^{-0.01155t} = 1 \qquad \text{Substitute values of } C \text{ and } k.$$

$$e^{-0.01155t} = \frac{1}{20} \qquad \text{Divide each side by 20.}$$

$$-0.01155t = \ln\frac{1}{20} \qquad \text{Inverse property}$$

$$t = \frac{1}{-0.01155}\ln\frac{1}{20} \qquad \text{Divide each side by } -0.01155.$$

$$t \approx 259.4 \text{ days} \qquad \text{Simplify.}$$

So, 20 grams of radioactive iodine-125 will have decayed to 1 gram after about 259.4 days. This solution is shown graphically in Figure 11.23.

✓ **CHECKPOINT** *Now try Exercise 65.*

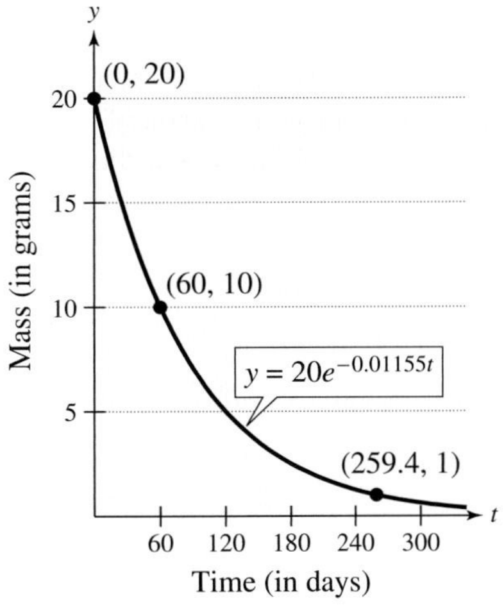

Radioactive Decay
Figure 11.23

EXAMPLE 7 Website Growth 

Your college created an algebra tutoring website in 2000. The number of hits per year at the website has grown exponentially. The website had 4080 hits in 2000 and 22,440 hits in 2008. Predict the number of hits in 2014.

Solution

t (year)	Ce^{kt}
0	$Ce^{k(0)} = 4080$
8	$Ce^{k(8)} = 22{,}440$
14	$Ce^{k(14)} = ?$

In the exponential growth model $y = Ce^{kt}$, let $t = 0$ represent 2000. Next, use the information given in the problem to set up the table shown at the left. Because $Ce^{k(0)} = Ce^0 = 4080$, you can conclude that $C = 4080$. Then, using this value of C, you can solve for k as follows.

$$Ce^{k(8)} = 22{,}440 \quad \text{From table}$$

$$4080e^{8k} = 22{,}440 \quad \text{Substitute value of } C.$$

$$e^{8k} = 5.5 \quad \text{Divide each side by 4080.}$$

$$8k = \ln 5.5 \quad \text{Inverse property}$$

$$k = \tfrac{1}{8}\ln 5.5 \approx 0.2131 \quad \text{Divide each side by 8 and simplify.}$$

Finally, you can use this value of k in the model from the table to predict the number of hits in 2014 to be $4080e^{0.2131(14)} \approx 80{,}600$.

 *CHECKPOINT* *Now try Exercise 61.*

3 ▶ Use logarithmic equations to solve intensity problems.

Intensity Models

On the **Richter scale,** the magnitude R of an earthquake can be measured by the **intensity model** $R = \log_{10} I$, where I is the intensity of the shock wave.

EXAMPLE 8 Earthquake Intensity 

In 2007, an earthquake near the coast of Peru measured 8.0 on the Richter scale. Weeks later, an earthquake in San Francisco that was felt by thousands of people measured 5.4 on the Richter scale. Compare the intensities of the two earthquakes.

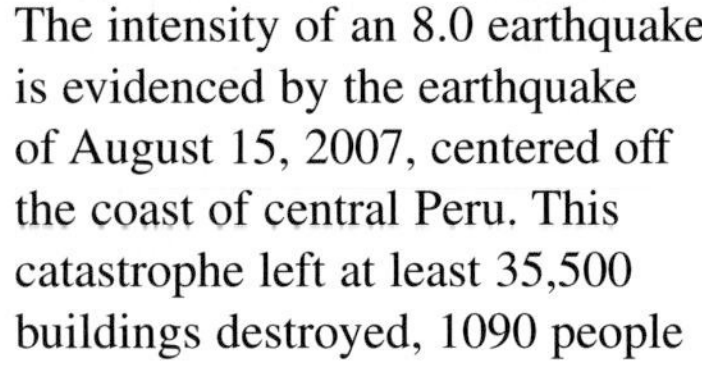

AP Photo/Martin Mejia

The intensity of an 8.0 earthquake is evidenced by the earthquake of August 15, 2007, centered off the coast of central Peru. This catastrophe left at least 35,500 buildings destroyed, 1090 people injured, and 514 people dead.

Solution

The intensity of Peru's earthquake is given as follows.

$$8.0 = \log_{10} I \quad \Rightarrow \quad 10^8 = I \qquad \text{Inverse property}$$

The intensity of San Francisco's earthquake can be found in a similar way.

$$5.4 = \log_{10} I \quad \Rightarrow \quad 10^{5.4} = I \qquad \text{Inverse property}$$

The ratio of these two intensities is

$$\frac{I \text{ for Peru}}{I \text{ for San Francisco}} = \frac{10^{8.0}}{10^{5.4}} = 10^{8.0-5.4} = 10^{2.6} \approx 398.$$

So, Peru's earthquake had an intensity that was about 398 times greater than the intensity of San Francisco's earthquake.

 *CHECKPOINT* *Now try Exercise 75.*

Concept Check

1. *True or False?* If an account earns simple interest for one year, then the effective yield for the account is the simple interest rate. Justify your answer.

2. *True or False?* The exponential model $y = Ce^{0.1t}$ represents exponential decay. Justify your answer.

3. In a radioactive decay problem, you are given the initial amount and the half-life of a radioactive substance. To write the decay model $y = Ce^{kt}$ for the problem, explain how you can find the values of C and k.

4. Explain how you can use the Richter scale measurements R_1 and R_2 of two earthquakes to compare the intensities of the two earthquakes.

11.6 EXERCISES

Go to pages 752–753 to record your assignments.

Solving Problems

Compound Interest In Exercises 1–6, find the annual interest rate. ***See Example 1.***

	Principal	*Balance*	*Time*	*Compounding*
✓ **1.**	\$500	\$1004.83	10 years	Monthly
2.	\$3000	\$21,628.70	20 years	Quarterly
3.	\$1000	\$36,581.00	40 years	Daily
4.	\$200	\$314.85	5 years	Yearly
5.	\$750	\$8267.38	30 years	Continuous
6.	\$2000	\$4234.00	10 years	Continuous

Doubling Time In Exercises 7–12, find the time for the investment to double. Use a graphing calculator to verify the result graphically. ***See Example 2.***

	Principal	*Rate*	*Compounding*
7.	\$2500	7.5%	Monthly
8.	\$900	$5\frac{3}{4}\%$	Quarterly
✓ **9.**	\$18,000	8%	Continuous
10.	\$250	6.5%	Yearly
11.	\$1500	$7\frac{1}{4}\%$	Monthly
12.	\$600	9.75%	Continuous

Compound Interest In Exercises 13–18, determine the type of compounding. Solve the problem by trying the more common types of compounding. ***See Example 3.***

	Principal	*Balance*	*Time*	*Rate*
✓ **13.**	\$5000	\$8954.24	10 years	6%
14.	\$5000	\$9096.98	10 years	6%
15.	\$750	\$1587.75	10 years	7.5%
16.	\$10,000	\$73,890.56	20 years	10%
17.	\$100	\$141.48	5 years	7%
18.	\$4000	\$4788.76	2 years	9%

Effective Yield In Exercises 19–26, find the effective yield. ***See Example 4.***

	Rate	*Compounding*
✓ **19.**	8%	Continuous
20.	9.5%	Daily
21.	7%	Monthly
22.	8%	Yearly
23.	6%	Quarterly
24.	9%	Quarterly
25.	8%	Monthly
26.	$5\frac{1}{4}\%$	Daily

27. *Doubling Time* Is it necessary to know the principal P to find the doubling time in Exercises 7–12? Explain.

28. *Effective Yield*

(a) Is it necessary to know the principal P to find the effective yield in Exercises 19–26? Explain.

(b) When the interest is compounded more frequently, what inference can you make about the difference between the effective yield and the stated annual percentage rate?

Compound Interest In Exercises 29–36, find the principal that must be deposited under the specified conditions to obtain the given balance.

	Balance	*Rate*	*Time*	*Compounding*
29.	\$10,000	9%	20 years	Continuous
30.	\$5000	8%	5 years	Continuous
31.	\$750	6%	3 years	Daily
32.	\$3000	7%	10 years	Monthly
33.	\$25,000	7%	30 years	Monthly
34.	\$8000	6%	2 years	Monthly
35.	\$1000	5%	1 year	Daily
36.	\$100,000	9%	40 years	Daily

Monthly Deposits In Exercises 37–40, you make monthly deposits of P dollars in a savings account at an annual interest rate r, compounded continuously. Find the balance A after t years given that

$$A = \frac{P(e^{rt} - 1)}{e^{r/12} - 1}.$$

	Principal	*Rate*	*Time*
37.	$P = 30$	$r = 8\%$	$t = 10$ years
38.	$P = 100$	$r = 9\%$	$t = 30$ years
39.	$P = 50$	$r = 10\%$	$t = 40$ years
40.	$P = 20$	$r = 7\%$	$t = 20$ years

Monthly Deposits In Exercises 41 and 42, you make monthly deposits of \$30 in a savings account at an annual interest rate of 8%, compounded continuously (see figure).

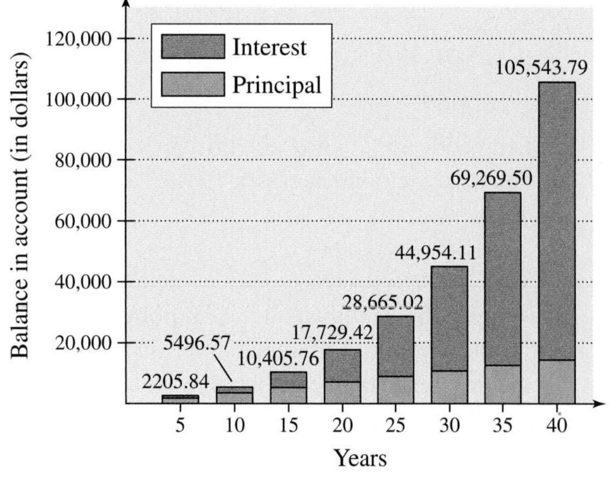

41. Find the total amount that has been deposited in the account after 20 years, and the total interest earned.

42. Find the total amount that has been deposited in the account after 40 years, and the total interest earned.

Exponential Growth and Decay In Exercises 43–46, find the constant k such that the graph of $y = Ce^{kt}$ passes through the points.

43.

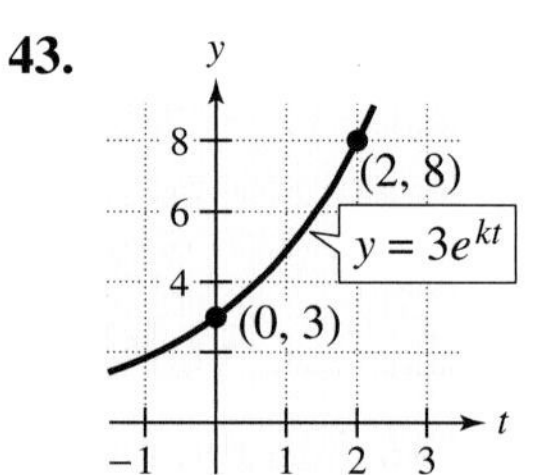

44.

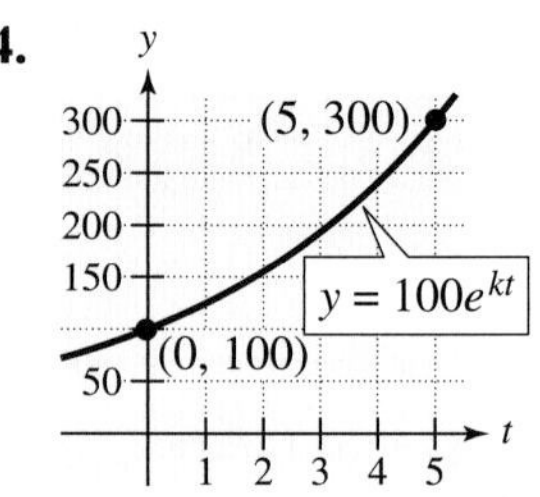

45.

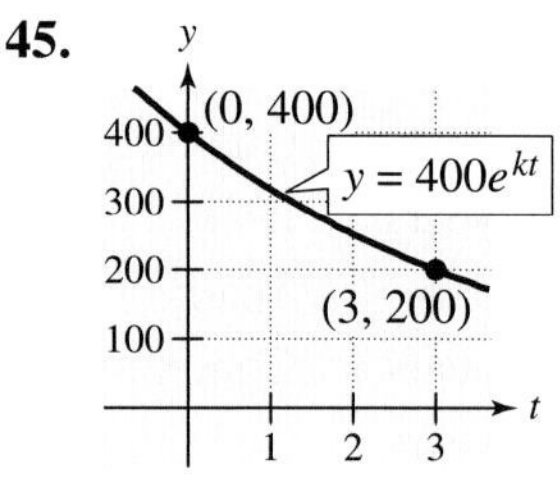

46.

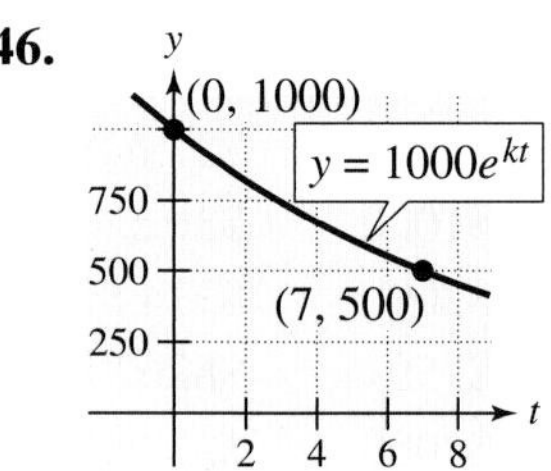

Population of a Country In Exercises 47–54, the population (in thousands) of a Caribbean locale in 2000 and the predicted population (in thousands) for 2020 are given. Find the constants C and k to obtain the exponential growth model $y = Ce^{kt}$ for the population. (Let $t = 0$ correspond to the year 2000.) Use the model to predict the population in the year 2025. ***See Example 5.*** (Source: United Nations)

	Country	*2000*	*2020*
✓ **47.**	Aruba	90	106
48.	Bahamas	303	381
49.	Barbados	286	303
50.	Belize	245	363
51.	Jamaica	2589	2872

	Country	*2000*	*2020*
52.	Haiti	8573	11,584
53.	Puerto Rico	3834	4252
54.	Saint Lucia	153	188

55. *Rate of Growth* Compare the values of k in Exercises 47 and 51. Which is larger? Explain.

56. *Exponential Growth Models* What variable in the continuous compound interest formula is equivalent to k in the model for population growth? Use your answer to give an interpretation of k.

57. *World Population* The figure shows the population P (in billions) of the world as projected by the U.S. Census Bureau. The bureau's projection can be modeled by

$$P = \frac{11.7}{1 + 1.21e^{-0.0269t}}$$

where $t = 0$ represents 1990. Use the model to predict the population in 2025.

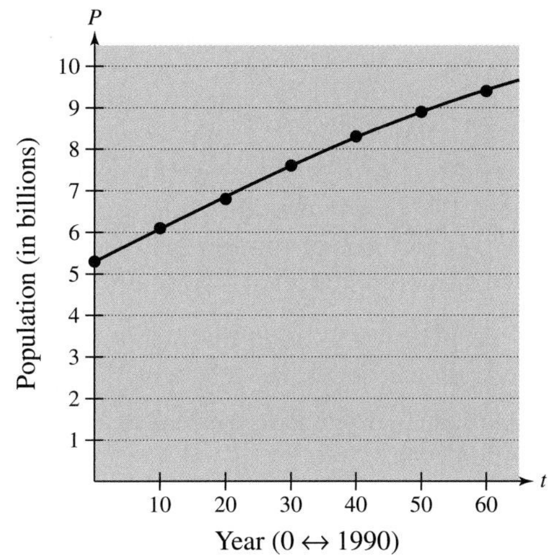

58. *World Population* Use the model P given in Exercise 57 to predict the world population in 2045.

59. *Computer Virus* In 2005, a computer worm called "Samy" interrupted the operations of a social networking website by inserting the payload message "but most of all, Samy is my hero" in the personal profile pages of the website's users. It is said that the "Samy" worm's message spread from 73 users to 1 million users within 20 hours.

(a) Find the constants C and k to obtain an exponential growth model $y = Ce^{kt}$ for the "Samy" worm.

(b) Use your model from part (a) to estimate how long it took the "Samy" worm to drop its payload message in 5300 personal profile pages.

60. *Stamp Collecting* The most expensive stamp in the world is the "Treskilling Yellow," a stamp issued in Sweden in 1855. The Treskilling Yellow sold in 1990 for \$1.3 million and again in 2008 for \$2.3 million.

AP Photo/Donald Stampfli

(a) Find the constants C and k to obtain the exponential growth model $y = Ce^{kt}$ for the value of the Treskilling Yellow.

(b) Use your model from part (a) to estimate the year in which the value of the Treskilling Yellow will reach \$3 million.

✓ **61.** *Cellular Phones* In 2000, there were 109,478,000 cellular telephone users in the United States. By 2006, the number had grown to 233,041,000. Use an exponential growth model to predict the number of cell phone users in 2013. (Source: CTIA–The Wireless Association®)

62. *Album Downloads* In 2004, about 4.6 million albums were purchased through downloading in the United States. In 2006, the number had increased to about 27.6 million. Use an exponential growth model to predict the number of albums that will be purchased through downloading in 2013. (Source: Recording Industry Association of America)

Radioactive Decay In Exercises 63–68, complete the table for the radioactive isotopes. ***See Example 6.***

	Isotope	Half-Life (Years)	Initial Quantity	Amount After 1000 Years
63.	^{226}Ra	1620	6 g	▒ g
64.	^{226}Ra	1620	▒ g	0.25 g
65.	^{14}C	5730	▒ g	4.0 g
66.	^{14}C	5730	10 g	▒ g
67.	^{239}Pu	24,100	4.2 g	▒ g
68.	^{239}Pu	24,100	▒ g	1.5 g

69. *Radioactive Decay* Radioactive radium (^{226}Ra) has a half-life of 1620 years. If you start with 5 grams of the isotope, how much will remain after 1000 years?

70. *Carbon 14 Dating* Carbon 14 dating assumes that all living organisms contain ^{14}C (radioactive carbon) in the same relative proportion. When an organism dies, however, its ^{14}C begins to decay according to its half-life of 5730 years. A piece of charcoal from an ancient tree contains only 15% as much ^{14}C as a piece of modern charcoal. How long ago did the ancient tree die? (Round your answer to the nearest 100 years.)

71. *Radioactive Decay* The isotope ^{230}Pu has a half-life of 24,360 years. If you start with 10 grams of this isotope, how much will remain after 10,000 years?

72. *Radioactive Decay* Carbon 14 (^{14}C) has a half-life of 5730 years. If you start with 5 grams of this isotope, how much will remain after 1000 years?

73. *Depreciation* A sport utility vehicle that cost $34,000 new has a depreciated value of $26,000 after 1 year. Find the value of the sport utility vehicle when it is 3 years old by using the exponential model $y = Ce^{kt}$.

74. *Depreciation* After x years, the value y of a recreational vehicle that cost $8000 new is given by

$y = 8000(0.8)^x$.

(a) Use a graphing calculator to graph the model.

(b) Graphically approximate the value of the recreational vehicle after 1 year.

(c) Graphically approximate the time when the recreational vehicle's value will be $4000.

Earthquake Intensity In Exercises 75–78, compare the intensities of the two earthquakes. ***See Example 8.***

	Location	Date	Magnitude
75.	Chile	5/22/1960	9.5
	Chile	1/22/2008	5.2
76.	Southern Alaska	3/28/1964	9.2
	Southern Alaska	12/31/2007	3.5
77.	Fiji	1/15/2008	6.5
	Philippines	1/15/2008	4.7
78.	India	1/14/2008	5.8
	Taiwan	1/14/2008	3.5

Acidity In Exercises 79–82, use the acidity model $pH = -\log_{10}[H^+]$, where acidity (pH) is a measure of the hydrogen ion concentration $[H^+]$ (measured in moles of hydrogen per liter) of a solution.

79. Find the pH of a solution that has a hydrogen ion concentration of 9.2×10^{-8}.

80. Compute the hydrogen ion concentration if the pH of a solution is 4.7.

81. A blueberry has a pH of 2.5 and an antacid tablet has a pH of 9.5. The hydrogen ion concentration of the fruit is how many times the concentration of the tablet?

82. If the pH of a solution is decreased by 1 unit, the hydrogen ion concentration is increased by what factor?

83. *Population Growth* The population p of a species of wild rabbit t years after it is introduced into a new habitat is given by

$$p(t) = \frac{5000}{1 + 4e^{-t/6}}.$$

(a) Use a graphing calculator to graph the population function.

(b) Determine the size of the population of rabbits that was introduced into the habitat.

(c) Determine the size of the population of rabbits after 9 years.

(d) After how many years will the size of the population of rabbits be 2000?

84. *Sales Growth* Annual sales y of a personal digital assistant x years after it is introduced are approximated by

$$y = \frac{2000}{1 + 4e^{-x/2}}.$$

(a) Use a graphing calculator to graph the model.

(b) Use the graph in part (a) to approximate annual sales of this personal digital assistant model when $x = 4$.

(c) Use the graph in part (a) to approximate the time when annual sales of this personal digital assistant model are $y = 1100$ units.

(d) Use the graph in part (a) to estimate the maximum level that annual sales of this model will approach.

85. *Advertising Effect* The sales S (in thousands of units) of a brand of jeans after the company spent x hundred dollars in advertising are given by

$$S = 10(1 - e^{kx}).$$

(a) Write S as a function of x if 2500 pairs of jeans are sold when \$500 is spent on advertising.

(b) How many pairs of jeans will be sold if advertising expenditures are raised to \$700?

86. *Advertising Effect* The sales S of a video game after the company spent x thousand dollars in advertising are given by

$$S = 4500(1 - e^{kx}).$$

(a) Write S as a function of x if 2030 copies of the video game are sold when \$10,000 is spent on advertising.

(b) How many copies of the video game will be sold if advertising expenditures are raised to \$25,000?

Explaining Concepts

87. Explain how to determine whether an exponential model of the form $y = Ce^{kt}$ models growth or decay.

88. The formulas for periodic and continuous compounding have the four variables A, P, r, and t in common. Explain what each variable measures.

89. For what types of compounding is the effective yield of an investment greater than the annual interest rate? Explain.

90. If the reading on the Richter scale is increased by 1, the intensity of the earthquake is increased by what factor? Explain.

Cumulative Review

In Exercises 91–94, solve the equation by using the Quadratic Formula.

91. $x^2 - 7x - 5 = 0$

92. $x^2 + 5x - 3 = 0$

93. $3x^2 + 9x + 4 = 0$

94. $3x^2 + 4x = -2x + 5$

In Exercises 95–98, solve the inequality and graph the solution on the real number line.

95. $\frac{4}{x - 4} > 0$

96. $\frac{x - 1}{x + 2} < 0$

97. $\frac{2x}{x - 3} > 1$

98. $\frac{x - 5}{x + 2} \le -1$

What Did You Learn?

Use these two pages to help prepare for a test on this chapter. Check off the key terms and key concepts you know. You can also use this section to record your assignments.

Plan for Test Success

Date of test: / /

Study dates and times: / / at : A.M./P.M.

/ / at : A.M./P.M.

Things to review:

- ☐ Key Terms, *p. 752*
- ☐ Key Concepts, *pp. 752–753*
- ☐ Your class notes
- ☐ Your assignments
- ☐ Study Tips, *pp. 685, 687, 697, 701, 712, 713, 714, 717, 724, 732, 735, 736, 741, 742*
- ☐ Technology Tips, *pp. 688, 704, 716, 717, 733*
- ☐ Mid-Chapter Quiz, *p. 722*
- ☐ Review Exercises, *pp. 754–758*
- ☐ Chapter Test, *p. 759*
- ☐ Video Explanations Online
- ☐ Tutorial Online

Key Terms

- ☐ exponential function, *p. 684*
- ☐ asymptote, *p. 687*
- ☐ horizontal asymptote, *p. 687*
- ☐ natural base, *p. 688*
- ☐ natural exponential function, *p. 688*
- ☐ composition, *p. 697*
- ☐ inverse function, *p. 699*
- ☐ one-to-one function, *p. 699*
- ☐ logarithmic function with base *a*, *p. 711*
- ☐ common logarithmic function, *p. 713*
- ☐ natural logarithmic function, *p. 716*
- ☐ change-of-base formula, *p. 717*
- ☐ exponentiate, *p. 734*
- ☐ effective yield, *p. 743*
- ☐ exponential growth, *p. 743*
- ☐ exponential decay, *p. 743*

Key Concepts

11.1 Exponential Functions

Assignment: ______________________ Due date: ________

☐ **Evaluate exponential functions of the form $f(x) = a^x$.**

☐ **Graph exponential functions using the characteristics of the graph of $y = a^x$:**

Domain: $(-\infty, \infty)$ Range: $(0, \infty)$

Intercept: $(0, 1)$ Increases from left to right.

The x-axis is a horizontal asymptote.

☐ **Evaluate and graph natural exponential functions of the form $f(x) = e^x$.**

The natural exponential function is simply an exponential function with a special base, the natural base $e \approx 2.71828$.

11.2 Composite and Inverse Functions

Assignment: ______________________ Due date: ________

☐ **Form the composition of two functions.**

$(f \circ g)(x) = f(g(x))$

The domain of $(f \circ g)$ is the set of all x in the domain of g such that $g(x)$ is in the domain of f.

☐ **Use the Horizontal Line Test to determine whether a function has an inverse.**

☐ **Find an inverse function algebraically.**

1. In the equation for $f(x)$, replace $f(x)$ with y.
2. Interchange x and y.
3. Solve for y. (If y is not a function of x, the original equation does not have an inverse.)
4. Replace y with $f^{-1}(x)$.
5. Verify that $f(f^{-1}(x)) = x = f^{-1}(f(x))$.

11.3 Logarithmic Functions

Assignment: ______________________________ Due date: ________

☐ **Evaluate logarithmic functions.**

A logarithm is an exponent. In other words, $y = \log_a x$ if and only if $x = a^y$.

Properties of logarithms Let a and x be positive real numbers such that $a \neq 1$. Then:

1. $\log_a 1 = 0$ because $a^0 = 1$.
2. $\log_a a = 1$ because $a^1 = a$.
3. $\log_a a^x = x$ because $a^x = a^x$.

☐ **Graph logarithmic functions using the characteristics of the graph of $y = \log_a x$:**

Domain: $(0, \infty)$ Range: $(-\infty, \infty)$

Intercept: $(1, 0)$ Increases from left to right.

The y-axis is a vertical asymptote.

☐ **Use a Change-of-Base Formula.**

$$\log_a x = \frac{\log_b x}{\log_b a} \quad \text{or} \quad \log_a x = \frac{\ln x}{\ln a}$$

11.4 Properties of Logarithms

Assignment: ______________________________ Due date: ________

☐ **Use properties of logarithms.**

Let a be a positive real number such that $a \neq 1$, and let n be a real number. If u and v are real numbers, variables, or algebraic expressions such that $u > 0$ and $v > 0$, then the following properties are true.

Logarithm with base a	*Natural logarithm*
1. $\log_a(uv) = \log_a u + \log_a v$	$\ln(uv) = \ln u + \ln v$
2. $\log_a \frac{u}{v} = \log_a u - \log_a v$	$\ln \frac{u}{v} = \ln u - \ln v$
3. $\log_a u^n = n \log_a u$	$\ln u^n = n \ln u$

11.5 Solving Exponential and Logarithmic Equations

Assignment: ______________________________ Due date: ________

☐ **Solve exponential and logarithmic equations using the one-to-one properties.**

Let a be a positive real number such that $a \neq 1$, and let x and y be real numbers. Then the following properties are true.

1. $a^x = a^y$ if and only if $x = y$.
2. $\log_a x = \log_a y$ if and only if $x = y$ $(x > 0, y > 0)$.

☐ **Solve exponential and logarithmic equations using the inverse properties.**

Base a	*Natural base e*
1. $\log_a(a^x) = x$	$\ln(e^x) = x$
2. $a^{(\log_a x)} = x$	$e^{(\ln x)} = x$

11.6 Applications

Assignment: ______________________________ Due date: ________

☐ **Solve compound interest problems.**

The following compound interest formulas are for the balance A, principal P, annual interest rate r (in decimal form), and time t (in years).

n Compoundings per Year: $A = P\left(1 + \frac{r}{n}\right)^{nt}$

Continuous Compounding: $A = Pe^{rt}$

☐ **Solve growth and decay problems.**

The mathematical model for growth or decay is

$y = Ce^{kt}$

where y is the amount of an initial quantity C that remains after time t. The number k is a constant. The model represents growth if $k > 0$ or decay if $k < 0$.

☐ **Solve earthquake intensity problems.**

On the Richter scale, the magnitude R of an earthquake can be measured by the intensity model $R = \log_{10} I$, where I is the intensity of the shock wave.

Review Exercises

11.1 Exponential Functions

1 ▶ Evaluate exponential functions.

In Exercises 1–4, evaluate the exponential function as indicated. (Round your answer to three decimal places.)

1. $f(x) = 4^x$
(a) $x = -3$
(b) $x = 1$
(c) $x = 2$

2. $g(x) = 4^{-x}$
(a) $x = -2$
(b) $x = 0$
(c) $x = 2$

3. $g(t) = 5^{-t/3}$
(a) $t = -3$
(b) $t = \pi$
(c) $t = 6$

4. $h(s) = 1 - 3^{0.2s}$
(a) $s = 0$
(b) $s = 2$
(c) $s = \sqrt{10}$

2 ▶ Graph exponential functions.

In Exercises 5–14, sketch the graph of the function. Identify the horizontal asymptote.

5. $f(x) = 3^x$

6. $f(x) = 3^{-x}$

7. $f(x) = 3^x - 3$

8. $f(x) = 3^x + 5$

9. $f(x) = 3^{x+1}$

10. $f(x) = 3^{x-1}$

11. $f(x) = 3^{x/2}$

12. $f(x) = 3^{-x/2}$

13. $f(x) = 3^{x/2} - 2$

14. $f(x) = 3^{x/2} + 3$

In Exercises 15–18, use a graphing calculator to graph the function.

15. $f(x) = 2^{-x^2}$

16. $g(x) = 2^{|x|}$

17. $y = 10(1.09)^t$

18. $y = 250(1.08)^t$

3 ▶ Evaluate the natural base e and graph natural exponential functions.

In Exercises 19 and 20, evaluate the exponential function as indicated. (Round your answers to three decimal places.)

19. $f(x) = 3e^{-2x}$
(a) $x = 3$
(b) $x = 0$
(c) $x = -19$

20. $g(x) = e^{x/5} + 11$
(a) $x = 12$
(b) $x = -8$
(c) $x = 18.4$

In Exercises 21–24, use a graphing calculator to graph the function.

21. $y = 4e^{-x/3}$

22. $y = 6 - e^{x/2}$

23. $f(x) = e^{x+2}$

24. $h(t) = \dfrac{8}{1 + e^{-t/5}}$

4 ▶ Use exponential functions to solve application problems.

Compound Interest **In Exercises 25 and 26, complete the table to determine the balance *A* for *P* dollars invested at interest rate *r* for *t* years, compounded *n* times per year.**

n	1	4	12	365	Continuous compounding
A					

	Principal	*Rate*	*Time*
25.	$P = \$5000$	$r = 10\%$	$t = 40$ years
26.	$P = \$10{,}000$	$r = 9.5\%$	$t = 30$ years

27. *Radioactive Decay* After t years, the remaining mass y (in grams) of 21 grams of a radioactive element whose half-life is 25 years is given by $y = 21\left(\frac{1}{2}\right)^{t/25}$, $t \geq 0$. How much of the initial mass remains after 58 years?

28. *Depreciation* After t years, a truck that originally cost \$38,000 depreciates in value so that each year it is worth $\frac{2}{3}$ of its value for the previous year. Find a model for $V(t)$, the value of the truck after t years. Sketch a graph of the model and determine the value of the truck 6 years after it was purchased.

11.2 Composite and Inverse Functions

1 ▶ Form compositions of two functions and find the domains of composite functions.

In Exercises 29–32, find the compositions.

29. $f(x) = x + 2$, $g(x) = x^2$

 (a) $(f \circ g)(2)$ (b) $(g \circ f)(-1)$

30. $f(x) = \sqrt[3]{x}$, $g(x) = x + 2$

 (a) $(f \circ g)(6)$ (b) $(g \circ f)(64)$

31. $f(x) = \sqrt{x + 1}$, $g(x) = x^2 - 1$

 (a) $(f \circ g)(5)$ (b) $(g \circ f)(-1)$

32. $f(x) = \dfrac{1}{x - 4}$, $g(x) = \dfrac{x + 1}{2x}$

 (a) $(f \circ g)(1)$ (b) $(g \circ f)\left(\dfrac{1}{5}\right)$

In Exercises 33 and 34, find the compositions (a) $f \circ g$ and (b) $g \circ f$. Then find the domain of each composition.

33. $f(x) = \sqrt{x + 6}$, $g(x) = 2x$

34. $f(x) = \dfrac{2}{x - 4}$, $g(x) = x^2$

2 ▶ Use the Horizontal Line Test to determine whether functions have inverse functions.

In Exercises 35–38, use the Horizontal Line Test to determine if the function is one-to-one and so has an inverse function.

35. $f(x) = x^2 - 25$

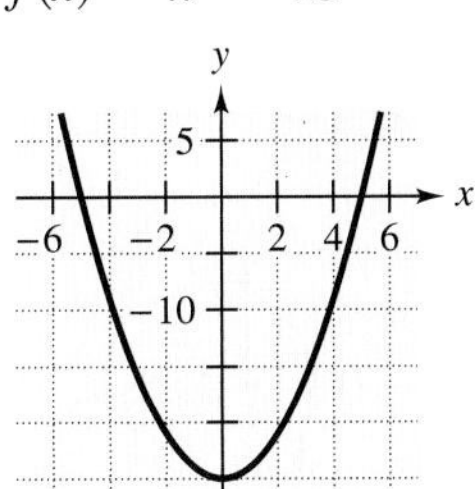

36. $f(x) = \frac{1}{4}x^3$

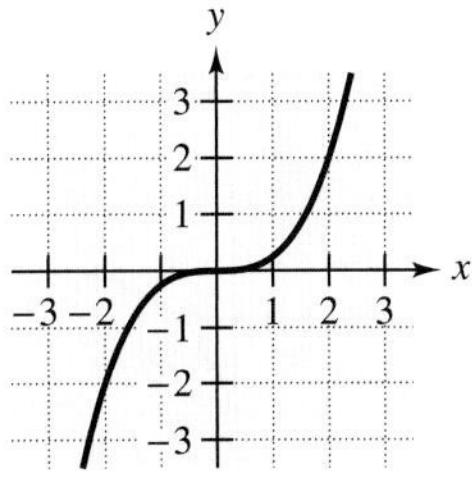

37. $h(x) = 4\sqrt[3]{x}$

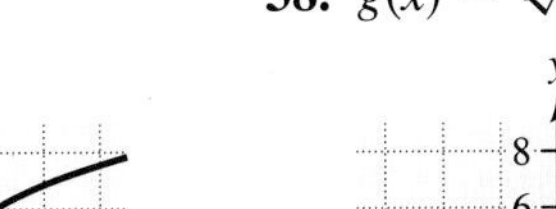

38. $g(x) = \sqrt{16 - x^2}$

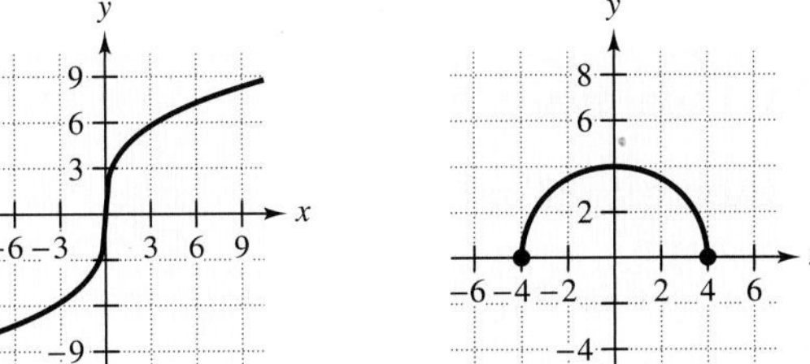

3 ▶ Find inverse functions algebraically.

In Exercises 39–44, find the inverse function.

39. $f(x) = 3x + 4$

40. $f(x) = 2x - 3$

41. $h(x) = \sqrt{5x}$

42. $g(x) = x^2 + 2$, $x \geq 0$

43. $f(t) = t^3 + 4$

44. $h(t) = \sqrt[3]{t - 1}$

4 ▶ Graphically verify that two functions are inverse functions of each other.

In Exercises 45 and 46, use a graphing calculator to graph the functions in the same viewing window. Graphically verify that f and g are inverse functions of each other.

45. $f(x) = 3x + 4$

 $g(x) = \frac{1}{3}(x - 4)$

46. $f(x) = \frac{1}{3}\sqrt[3]{x}$

 $g(x) = 27x^3$

In Exercises 47–50, use the graph of f to sketch the graph of f^{-1}.

47.

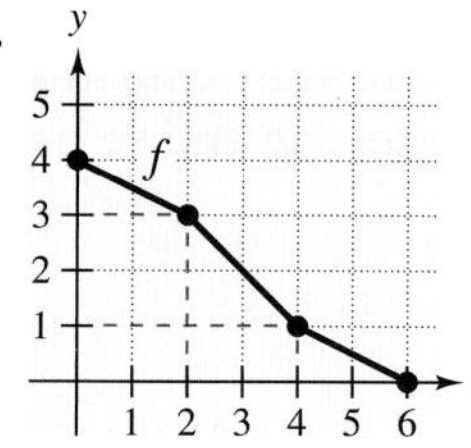

48.

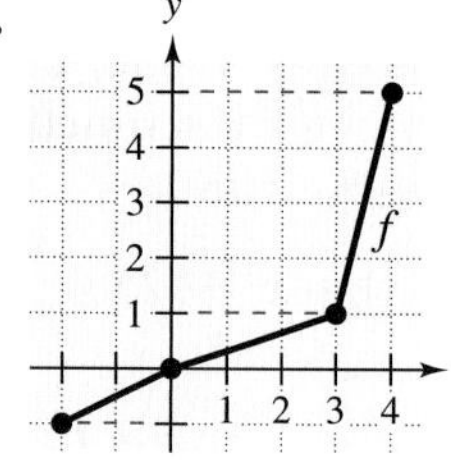

49.

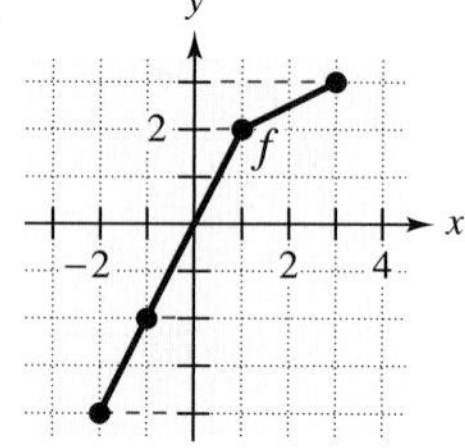

50.

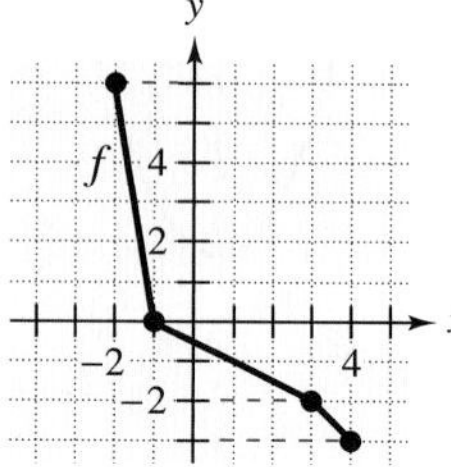

11.3 Logarithmic Functions

1 ▶ Evaluate logarithmic functions.

In Exercises 51–58, evaluate the logarithm.

51. $\log_{10} 1000$

52. $\log_{27} 3$

53. $\log_3 \frac{1}{9}$

54. $\log_4 \frac{1}{16}$

55. $\log_2 64$

56. $\log_{10} 0.01$

57. $\log_3 1$

58. $\log_2 \sqrt{4}$

2 ▶ Graph logarithmic functions.

In Exercises 59–64, sketch the graph of the function. Identify the vertical asymptote.

59. $f(x) = \log_3 x$

60. $f(x) = -\log_3 x$

61. $f(x) = -1 + \log_3 x$

62. $f(x) = 1 + \log_3 x$

63. $y = \log_2(x - 4)$

64. $y = \log_4(x + 1)$

3 ▶ Graph and evaluate natural logarithmic functions.

In Exercises 65 and 66, use your calculator to evaluate the natural logarithm. (Round your answer to four decimal places.)

65. $\ln 50$

66. $\ln\left(\frac{5 - \sqrt{3}}{2}\right)$

In Exercises 67–70, sketch the graph of the function. Identify the vertical asymptote.

67. $y = \ln(x - 3)$

68. $y = -\ln(x + 2)$

69. $y = 5 - \ln x$

70. $y = 3 + \ln x$

4 ▶ Use the change-of-base formula to evaluate logarithms.

In Exercises 71–74, use a calculator to evaluate the logarithm by means of the change-of-base formula. (Round your answer to four decimal places.)

71. $\log_4 9$

72. $\log_{1/2} 5$

73. $\log_8 160$

74. $\log_3 0.28$

11.4 Properties of Logarithms

1 ▶ Use the properties of logarithms to evaluate logarithms.

In Exercises 75–80, use $\log_5 2 \approx 0.4307$ and $\log_5 3 \approx 0.6826$ to approximate the expression. Do not use a calculator.

75. $\log_5 18$

76. $\log_5 \sqrt{6}$

77. $\log_5 \frac{1}{2}$

78. $\log_5 \frac{2}{3}$

79. $\log_5(12)^{2/3}$

80. $\log_5(5^2 \cdot 6)$

2 ▶ Use the properties of logarithms to rewrite, expand, or condense logarithmic expressions.

In Exercises 81–88, use the properties of logarithms to expand the expression. (Assume all variables are positive.)

81. $\log_4 6x^4$

82. $\log_{12} 2x^{-5}$

83. $\log_5 \sqrt{x+2}$

84. $\ln \sqrt[3]{\dfrac{x}{5}}$

85. $\ln \dfrac{x+2}{x+3}$

86. $\ln x(x+4)^2$

87. $\ln\left[\sqrt{2x}(x+3)^5\right]$

88. $\log_3 \dfrac{a^2\sqrt{b}}{cd^5}$

In Exercises 89–98, use the properties of logarithms to condense the expression.

89. $-\frac{2}{3}\ln 3y$

90. $5 \log_2 y$

91. $\log_8 16x + \log_8 2x^2$

92. $\log_4 6x - \log_4 10$

93. $-2(\ln 2x - \ln 3)$

94. $5(1 + \ln x + \ln 2)$

95. $4[\log_2 k - \log_2(k-t)]$

96. $\frac{1}{3}(\log_8 a + 2\log_8 b)$

97. $3\ln x + 4\ln y + \ln z$

98. $\ln(x+4) - 3\ln x - \ln y$

True or False? **In Exercises 99–104, use properties of logarithms to determine whether the equation is true or false. If it is false, state why or give an example to show that it is false.**

99. $\log_2 4x = 2\log_2 x$

100. $\dfrac{\ln 5x}{\ln 10x} = \ln \dfrac{1}{2}$

101. $\log_{10} 10^{2x} = 2x$

102. $e^{\ln t} = t,\ t > 0$

103. $\log_4 \dfrac{16}{x} = 2 - \log_4 x$

104. $6\ln x + 6\ln y = \ln(xy)^6,\ x > 0,\ y > 0$

3 ▶ Use the properties of logarithms to solve application problems.

105. *Light Intensity* The intensity of light y as it passes through a medium is given by

$$y = \ln\left(\frac{I_0}{I}\right)^{0.83}.$$

Use properties of logarithms to write the formula in simpler form, and determine the intensity of light passing through this medium when $I_0 = 4.2$ and $I = 3.3$.

106. *Human Memory Model* A psychologist finds that the percent p of retention in a group of subjects can be modeled by

$$p = \frac{\log_{10}(10^{68})}{\log_{10}(t+1)^{20}}$$

where t is the time in months after the subjects' initial testing. Use properties of logarithms to write the formula in simpler form, and determine the percent of retention after 5 months.

11.5 Solving Exponential and Logarithmic Equations

1 ▶ Solve basic exponential and logarithmic equations.

In Exercises 107–112, solve the equation.

107. $2^x = 64$

108. $6^x = 216$

109. $4^{x-3} = \frac{1}{16}$

110. $3^{x-2} = 81$

111. $\log_7(x+6) = \log_7 12$

112. $\ln(8-x) = \ln 3$

2 ▶ Use inverse properties to solve exponential equations.

In Exercises 113–118, solve the exponential equation. (Round your answer to two decimal places.)

113. $3^x = 500$

114. $8^x = 1000$

115. $2e^{0.5x} = 45$

116. $125e^{-0.4x} = 40$

117. $12(1 - 4^x) = 18$

118. $25(1 - e^t) = 12$

3 ▶ Use inverse properties to solve logarithmic equations.

In Exercises 119–128, solve the logarithmic equation. (Round your answer to two decimal places.)

119. $\ln x = 7.25$

120. $\ln x = -0.5$

121. $\log_{10} 4x = 2.1$

122. $\log_2 2x = -0.65$

123. $\log_3(2x + 1) = 2$

124. $\log_5(x - 10) = 2$

125. $\frac{1}{3}\log_2 x + 5 = 7$

126. $4\log_5(x + 1) = 4.8$

127. $\log_3 x + \log_3 7 = 4$

128. $2\log_4 x - \log_4(x - 1) = 1$

4 ▶ Use exponential or logarithmic equations to solve application problems.

129. *Compound Interest* A deposit of \$5000 is placed in a savings account for 2 years. The interest for the account is compounded continuously. At the end of 2 years, the balance in the account is \$5751.37. What is the annual interest rate for this account?

130. *Sound Intensity* The relationship between the number of decibels B and the intensity of a sound I in watts per centimeter squared is given by

$$B = 10\log_{10}\left(\frac{I}{10^{-16}}\right).$$

Determine the intensity of a firework display I if it registers 130 decibels on a decibel meter.

11.6 Applications

1 ▶ Use exponential equations to solve compound interest problems.

Annual Interest Rate **In Exercises 131–136, find the annual interest rate.**

	Principal	*Balance*	*Time*	*Compounding*
131.	\$250	\$410.90	10 years	Quarterly
132.	\$1000	\$1348.85	5 years	Monthly
133.	\$5000	\$15,399.30	15 years	Daily
134.	\$10,000	\$35,236.45	20 years	Yearly
135.	\$1800	\$46,422.61	50 years	Continuous
136.	\$7500	\$15,877.50	15 years	Continuous

Effective Yield **In Exercises 137–142, find the effective yield.**

	Rate	*Compounding*
137.	5.5%	Daily
138.	6%	Monthly
139.	7.5%	Quarterly
140.	8%	Yearly
141.	7.5%	Continuously
142.	3.75%	Continuously

2 ▶ Use exponential equations to solve growth and decay problems.

Radioactive Decay **In Exercises 143–148, complete the table for the radioactive isotopes.**

	Isotope	*Half-Life (Years)*	*Initial Quantity*	*Amount After 1000 Years*
143.	^{226}Ra	1620	3.5 g	▭ g
144.	^{226}Ra	1620	▭ g	0.5 g
145.	^{14}C	5730	▭ g	2.6 g
146.	^{14}C	5730	10 g	▭ g
147.	^{239}Pu	24,100	5 g	▭ g
148.	^{239}Pu	24,100	▭ g	2.5 g

3 ▶ Use logarithmic equations to solve intensity problems.

In Exercises 149 and 150, compare the intensities of the two earthquakes.

	Location	*Date*	*Magnitude*
149.	San Francisco, California	4/18/1906	8.3
	Napa, California	9/3/2000	4.9
150.	Mexico	1/23/2008	5.8
	Puerto Rico	1/23/2008	3.3

Chapter Test

Take this test as you would take a test in class. After you are done, check your work against the answers in the back of the book.

1. Evaluate $f(t) = 54\left(\frac{2}{3}\right)^t$ when $t = -1, 0, \frac{1}{2}$, and 2.
2. Sketch a graph of the function $f(x) = 2^{x/3}$ and identify the horizontal asymptote.
3. Find the compositions (a) $f \circ g$ and (b) $g \circ f$. Then find the domain of each composition.
   $f(x) = 2x^2 + x \qquad g(x) = 5 - 3x$
4. Find the inverse function of $f(x) = 9x - 4$.
5. Verify algebraically that the functions f and g are inverse functions of each other.
   $f(x) = -\frac{1}{2}x + 3, \qquad g(x) = -2x + 6$
6. Evaluate $\log_4 \frac{1}{256}$ without a calculator.
7. Describe the relationship between the graphs of $f(x) = \log_5 x$ and $g(x) = 5^x$.
8. Use the properties of logarithms to expand $\log_8\left(4\sqrt{x}/y^4\right)$.
9. Use the properties of logarithms to condense $\ln x - 4 \ln y$.

In Exercises 10–17, solve the equation. Round your answer to two decimal places, if necessary.

10. $\log_2 x = 5$
11. $9^{2x} = 182$
12. $400e^{0.08t} = 1200$
13. $3 \ln(2x - 3) = 10$
14. $12(7 - 2^x) = -300$
15. $\log_2 x + \log_2 4 = 5$
16. $\ln x - \ln 2 = 4$
17. $30(e^x + 9) = 300$
18. Determine the balance after 20 years if \$2000 is invested at 7% compounded (a) quarterly and (b) continuously.
19. Determine the principal that will yield \$100,000 when invested at 9% compounded quarterly for 25 years.
20. A principal of \$500 yields a balance of \$1006.88 in 10 years when the interest is compounded continuously. What is the annual interest rate?
21. A car that cost \$20,000 new has a depreciated value of \$15,000 after 1 year. Find the value of the car when it is 5 years old by using the exponential model $y = Ce^{kt}$.

In Exercises 22–24, the population p of a species of fox t years after it is introduced into a new habitat is given by

$$p(t) = \frac{2400}{1 + 3e^{-t/4}}.$$

22. Determine the size of the population that was introduced into the habitat.
23. Determine the population after 4 years.
24. After how many years will the population be 1200?

Study Skills in Action

Avoiding Test-Taking Errors

For some students, the day they get their math tests back is just as nerve-racking as the day they take the test. Do you look at your grade, sigh hopelessly, and stuff the test in your book bag? This kind of response is not going to help you to do better on the next test. When professional football players lose a game, the coach does not let them just forget about it. They review all their mistakes and discuss how to correct them. That is what you need to do with every math test.

There are six types of test errors (Nolting, 2008), as listed below. Look at your test and see what types of errors you make. Then decide what you can do to avoid making them again. Many students need to do this with a tutor or instructor the first time through.

Kimberly Nolting
VP, Academic Success Press
expert in developmental education

Smart Study Strategy

Analyze Your Errors

Type of error	Corrective action
1 ► **Misreading Directions:** You do not correctly read or understand directions.	Read the instructions in the textbook exercises at least twice and make sure you understand what they mean. Make this a habit in time for the next test.
2 ► **Careless Errors:** You understand how to do a problem but make careless errors, such as not carrying a sign, miscopying numbers, and so on.	Pace yourself during a test to avoid hurrying. Also, make sure you write down every step of a solution neatly. Use a finger to move from one step to the next, looking for errors.
3 ► **Concept Errors:** You do not understand how to apply the properties and rules needed to solve a problem.	Find a tutor who will work with you on the next chapter. Visit the instructor to make sure you understand the math.
4 ► **Application Errors:** You can do numerical problems that are similar to your homework problems but struggle with problems that vary, such as application problems.	Do not just mimic the steps of solving an application problem. Explain out loud why you are doing each step. Ask the instructor or tutor for different types of problems.
5 ► **Test-Taking Errors:** You hurry too much, do not use all of the allowed time, spend too much time on one problem, and so on.	Refer to the *Ten Steps for Test-Taking* on page 400.
6 ► **Study Errors:** You do not study the right material or do not learn it well enough to remember it on a test without resources such as notes.	Take a practice test. Work with a study group. Confer with your instructor. Don't try to learn a whole chapter's worth of material in one night—cramming does not work in math!

Chapter 12

Conics

IT WORKED FOR ME!

"My instructor told me that if I just put a little more effort into studying and getting ready for the final, I could possibly get an A. So, I pulled out my old tests. That is when I noticed that most of my mistakes involved word problems and not reading directions carefully. I got help from a tutor on word problems and made sure I correctly read the instructions on the final. It worked."

Maribeth
Education

12.1 Circles and Parabolas

What You Should Learn

1 ▶ Recognize the four basic conics: circles, parabolas, ellipses, and hyperbolas.
2 ▶ Graph and write equations of circles centered at the origin.
3 ▶ Graph and write equations of circles centered at (h, k).
4 ▶ Graph and write equations of parabolas.

Why You Should Learn It

Circles can be used to model and solve scientific problems. For instance, in Exercise 93 on page 772, you will write an equation that represents the circular orbit of a satellite.

1 ▶ Recognize the four basic conics: circles, parabolas, ellipses, and hyperbolas.

The Conics

In Section 10.4, you saw that the graph of a second-degree equation of the form $y = ax^2 + bx + c$ is a parabola. A parabola is one of four types of **conics** or **conic sections.** The other three types are circles, ellipses, and hyperbolas. All four types have equations of second degree. Each figure can be obtained by intersecting a plane with a double-napped cone, as shown in Figure 12.1.

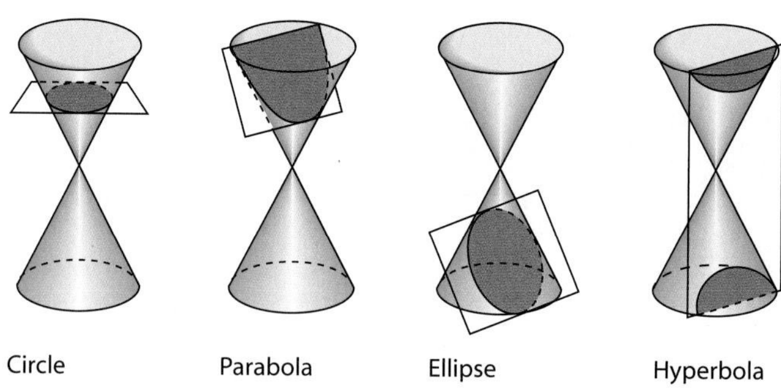

Figure 12.1

2 ▶ Graph and write equations of circles centered at the origin.

Conics occur in many practical applications. Reflective surfaces in satellite dishes, flashlights, and telescopes often have a parabolic shape. The orbits of planets are elliptical, and the orbits of comets are usually elliptical or hyperbolic. Ellipses and parabolas are also used in building archways and bridges.

Circles Centered at the Origin

Definition of a Circle

A **circle** in the rectangular coordinate system consists of all points (x, y) that are a given positive distance r from a fixed point, called the **center** of the circle. The distance r is called the **radius** of the circle.

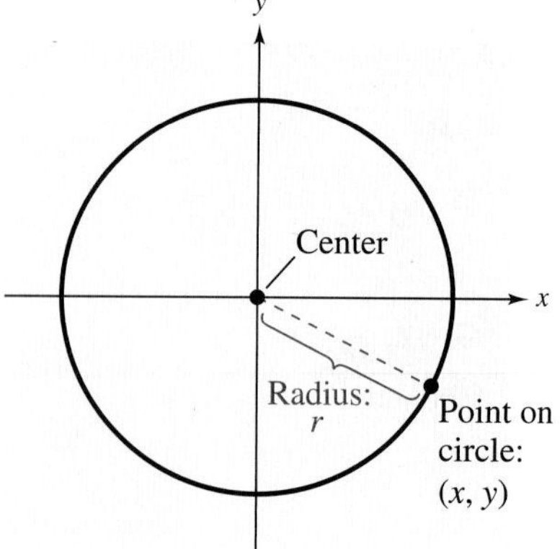

Figure 12.2

If the center of the circle is the origin, as shown in Figure 12.2, the relationship between the coordinates of any point (x, y) on the circle and the radius r is

$$r = \sqrt{(x - 0)^2 + (y - 0)^2} = \sqrt{x^2 + y^2}. \qquad \text{Distance Formula}$$

By squaring each side of this equation, you obtain the equation below, which is called the **standard form of the equation of a circle centered at the origin.**

> **Standard Equation of a Circle (Center at Origin)**
>
> The **standard form of the equation of a circle centered at the origin** is
>
> $x^2 + y^2 = r^2.$ Circle with center at (0, 0)
>
> The positive number r is called the **radius** of the circle.

EXAMPLE 1 Writing an Equation of a Circle

Write an equation of the circle that is centered at the origin and has a radius of 2, as shown in Figure 12.3.

Solution

Using the standard form of the equation of a circle (with center at the origin) and $r = 2$, you obtain

$x^2 + y^2 = r^2$ Standard form with center at (0, 0)

$x^2 + y^2 = 2^2$ Substitute 2 for r.

$x^2 + y^2 = 4.$ Equation of circle

CHECKPOINT *Now try Exercise 7.*

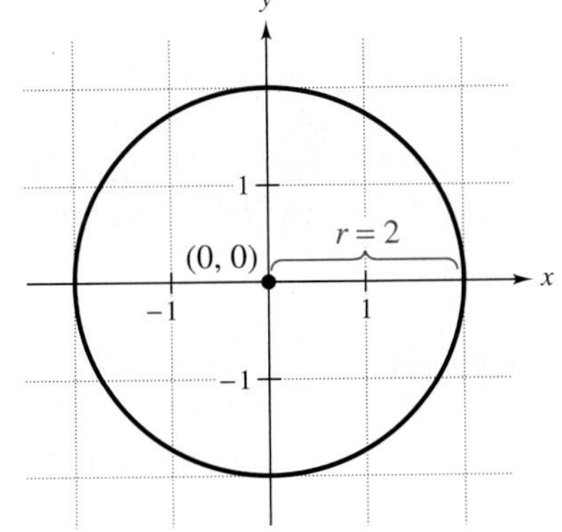

Figure 12.3

To sketch the graph of the equation of a given circle, write the equation in standard form, which will allow you to identify the radius of the circle.

EXAMPLE 2 Sketching a Circle

Identify the radius of the circle given by the equation $4x^2 + 4y^2 - 25 = 0$. Then sketch the circle.

Solution

Begin by writing the equation in standard form.

$4x^2 + 4y^2 - 25 = 0$ Write original equation.

$4x^2 + 4y^2 = 25$ Add 25 to each side.

$x^2 + y^2 = \frac{25}{4}$ Divide each side by 4.

$x^2 + y^2 = \left(\frac{5}{2}\right)^2$ Standard form

From the standard form of the equation of this circle centered at the origin, you can see that the radius is $\frac{5}{2}$. The graph of the circle is shown in Figure 12.4.

CHECKPOINT *Now try Exercise 23.*

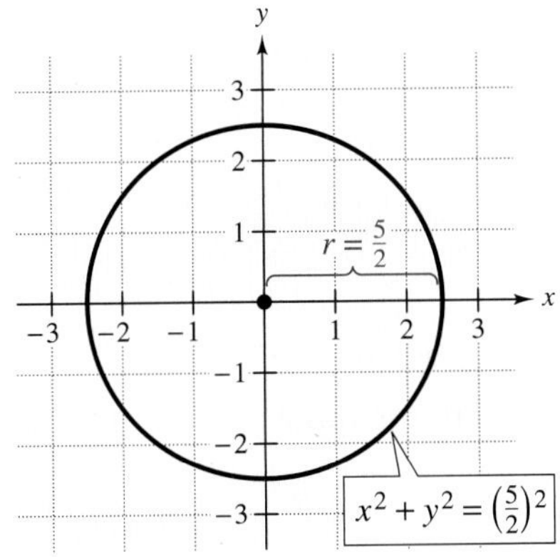

Figure 12.4

3 ▶ Graph and write equations of circles centered at (h, k).

Circles Centered at (h, k)

Consider a circle whose radius is r and whose center is the point (h, k), as shown in Figure 12.5. Let (x, y) be any point on the circle. To find an equation for this circle, you can use a variation of the Distance Formula and write

$$\text{Radius} = r = \sqrt{(x-h)^2 + (y-k)^2}. \qquad \text{Distance Formula}$$

By squaring each side of this equation, you obtain the equation shown below, which is called the **standard form of the equation of a circle centered at (h, k).**

Center: (h, k)

Radius: r

Point on circle: (x, y)

Figure 12.5

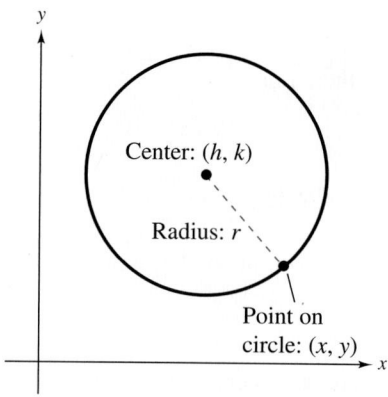

> **Standard Equation of a Circle [Center at (h, k)]**
>
> The **standard form of the equation of a circle centered at (h, k)** is
>
> $$(x-h)^2 + (y-k)^2 = r^2.$$

When $h = 0$ and $k = 0$, the circle is centered at the origin. Otherwise, you can shift the center of the circle h units horizontally and k units vertically from the origin.

EXAMPLE 3 Writing an Equation of a Circle

The point $(2, 5)$ lies on a circle whose center is $(5, 1)$, as shown in Figure 12.6. Write the standard form of the equation of this circle.

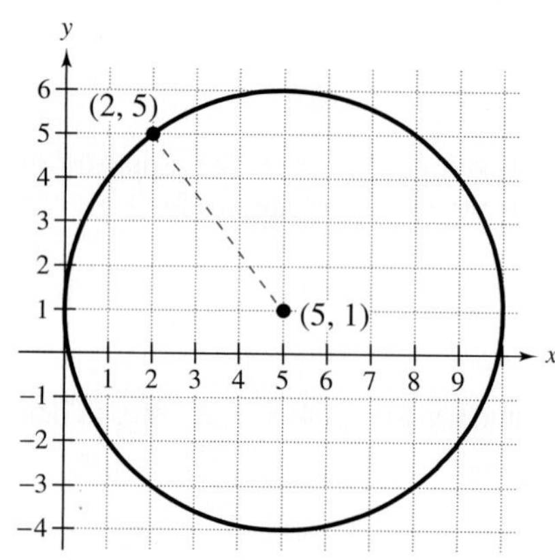

Figure 12.6

Solution

The radius r of the circle is the distance between $(2, 5)$ and $(5, 1)$.

$r = \sqrt{(2-5)^2 + (5-1)^2}$	Distance Formula
$= \sqrt{(-3)^2 + 4^2}$	Simplify.
$= \sqrt{9 + 16}$	Simplify.
$= \sqrt{25}$	Simplify.
$= 5$	Radius

Using $(h, k) = (5, 1)$ and $r = 5$, the equation of the circle is

$(x-h)^2 + (y-k)^2 = r^2$	Standard form
$(x-5)^2 + (y-1)^2 = 5^2$	Substitute for h, k, and r.
$(x-5)^2 + (y-1)^2 = 25.$	Equation of circle

From the graph, you can see that the center of the circle is shifted five units to the right and one unit upward from the origin.

✓ ***CHECKPOINT*** *Now try Exercise 15.*

Writing the equation of a circle in standard form helps you to determine both the radius and center of the circle. To write the standard form of the equation of a circle, you may need to complete the square, as demonstrated in Example 4.

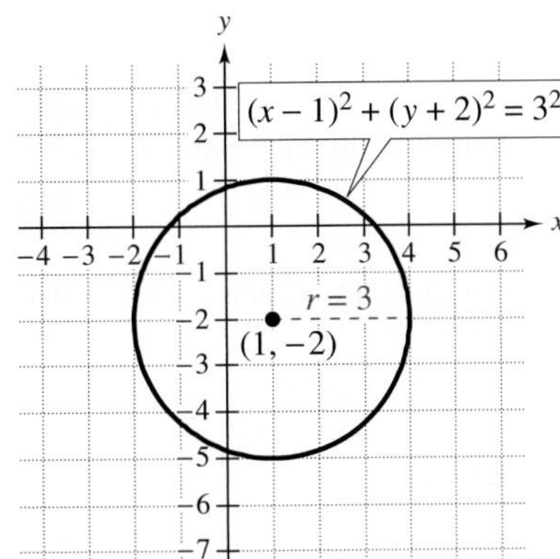

Figure 12.7

EXAMPLE 4 Writing an Equation in Standard Form

Write the equation $x^2 + y^2 - 2x + 4y - 4 = 0$ in standard form. Then sketch the circle represented by the equation.

Solution

$$x^2 + y^2 - 2x + 4y - 4 = 0 \qquad \text{Write original equation.}$$

$$(x^2 - 2x + \quad) + (y^2 + 4y + \quad) = 4 \qquad \text{Group terms.}$$

$$[x^2 - 2x + (-1)^2] + (y^2 + 4y + 2^2) = 4 + 1 + 4 \qquad \text{Complete the squares.}$$

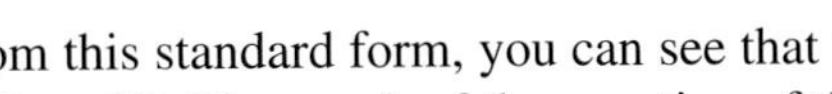

$$(x - 1)^2 + (y + 2)^2 = 3^2 \qquad \text{Standard form}$$

From this standard form, you can see that the circle is centered at $(1, -2)$ with a radius of 3. The graph of the equation of the circle is shown in Figure 12.7.

✓ **CHECKPOINT** *Now try Exercise 31.*

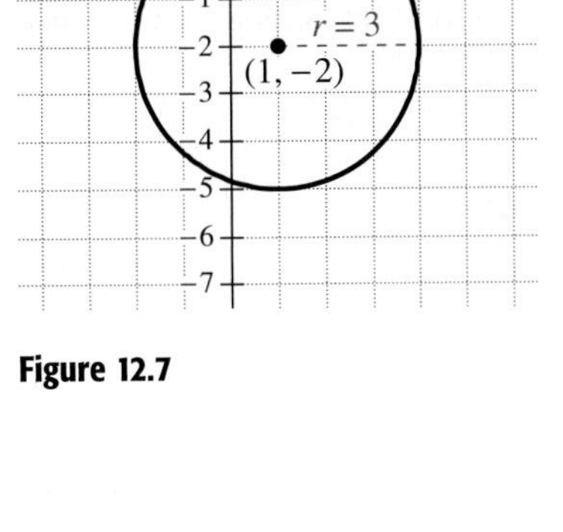

Figure 12.8

EXAMPLE 5 An Application: Mechanical Drawing

In a mechanical drawing class, you have to program a computer to model the metal piece shown in Figure 12.8. Part of the assignment is to find an equation that represents the semicircular portion of the hole in the metal piece. What is the equation?

Solution

From the drawing, you can see that the center of the circle is $(h, k) = (5, 2)$ and that the radius of the circle is $r = 1.5$. This implies that the equation of the entire circle is

$$(x - h)^2 + (y - k)^2 = r^2 \qquad \text{Standard form}$$

$$(x - 5)^2 + (y - 2)^2 = 1.5^2 \qquad \text{Substitute for } h, k, \text{ and } r.$$

$$(x - 5)^2 + (y - 2)^2 = 2.25. \qquad \text{Equation of circle}$$

To find the equation of the upper portion of the circle, solve this standard equation for y.

$$(x - 5)^2 + (y - 2)^2 = 2.25$$

$$(y - 2)^2 = 2.25 - (x - 5)^2$$

$$y - 2 = \pm\sqrt{2.25 - (x - 5)^2}$$

$$y = 2 \pm \sqrt{2.25 - (x - 5)^2}$$

Finally, take the positive square root to obtain the equation of the upper portion of the circle.

$$y = 2 + \sqrt{2.25 - (x - 5)^2}$$

✓ **CHECKPOINT** *Now try Exercise 95.*

Study Tip

In Example 5, if you had wanted the equation of the lower portion of the circle, you would have taken the negative square root

$$y = 2 - \sqrt{2.25 - (x - 5)^2}.$$

4 ▶ Graph and write equations of parabolas.

Equations of Parabolas

The second basic type of conic is a **parabola.** In Section 10.4, you studied some of the properties of parabolas. There you saw that the graph of a quadratic function of the form $y = ax^2 + bx + c$ is a parabola that opens upward if a is positive and downward if a is negative. You also learned that each parabola has a vertex and that the vertex of the graph of $y = ax^2 + bx + c$ occurs when $x = -b/(2a)$.

In this section, you will study a general definition of a parabola in the sense that it is independent of the orientation of the parabola.

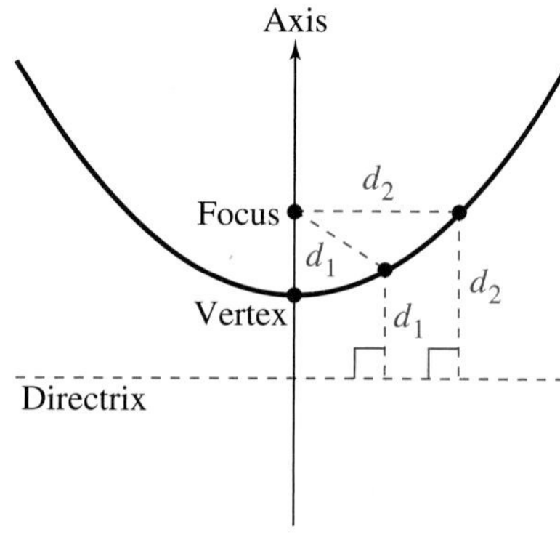

Figure 12.9

Definition of a Parabola

A **parabola** is the set of all points (x, y) that are equidistant from a fixed line (**directrix**) and a fixed point (**focus**) not on the line.

The midpoint between the focus and the directrix is called the **vertex,** and the line passing through the focus and the vertex is called the **axis** of the parabola. Note in Figure 12.9 that a parabola is symmetric with respect to its axis. Using the definition of a parabola, you can derive the **standard form of the equation of a parabola** whose directrix is parallel to the x-axis or to the y-axis.

Study Tip

If the focus of a parabola is above or to the right of the vertex, p is positive. If the focus is below or to the left of the vertex, p is negative.

Standard Equation of a Parabola

The **standard form of the equation of a parabola** with vertex at the origin $(0, 0)$ is

$x^2 = 4py, \quad p \neq 0$ Vertical axis

$y^2 = 4px, \quad p \neq 0.$ Horizontal axis

The focus lies on the axis p units (*directed distance*) from the vertex. If the vertex is at (h, k), then the standard form of the equation is

$(x - h)^2 = 4p(y - k), \quad p \neq 0$ Vertical axis; directrix: $y = k - p$

$(y - k)^2 = 4p(x - h), \quad p \neq 0.$ Horizontal axis; directrix: $x = h - p$

(See Figure 12.10.)

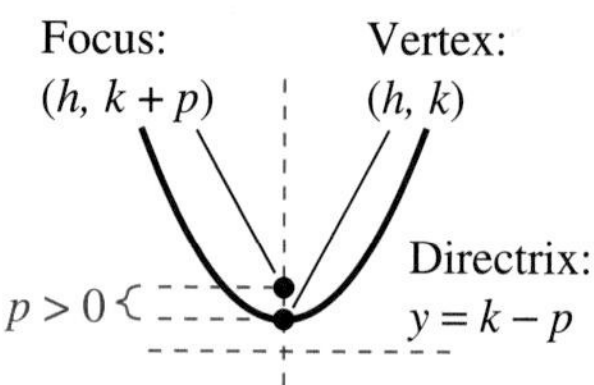

Parabola with vertical axis

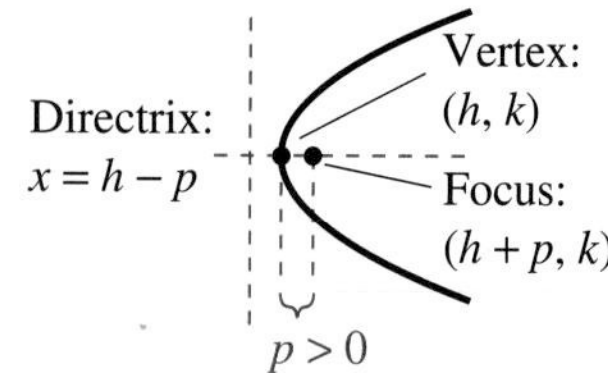

Parabola with horizontal axis

Figure 12.10

EXAMPLE 6 Writing the Standard Equation of a Parabola

Write the standard form of the equation of the parabola with vertex (0, 0) and focus (0, −2), as shown in Figure 12.11.

Figure 12.11

Solution

Because the vertex is at the origin and the axis of the parabola is vertical, use the equation

$$x^2 = 4py$$

where p is the directed distance from the vertex to the focus. Because the focus is two units *below* the vertex, $p = -2$. So, the equation of the parabola is

$x^2 = 4py$	Standard form
$x^2 = 4(-2)y$	Substitute for p.
$x^2 = -8y.$	Equation of parabola

 ***CHECKPOINT*** *Now try Exercise 55.*

EXAMPLE 7 Writing the Standard Equation of a Parabola

Write the standard form of the equation of the parabola with vertex (3, −2) and focus (4, −2), as shown in Figure 12.12.

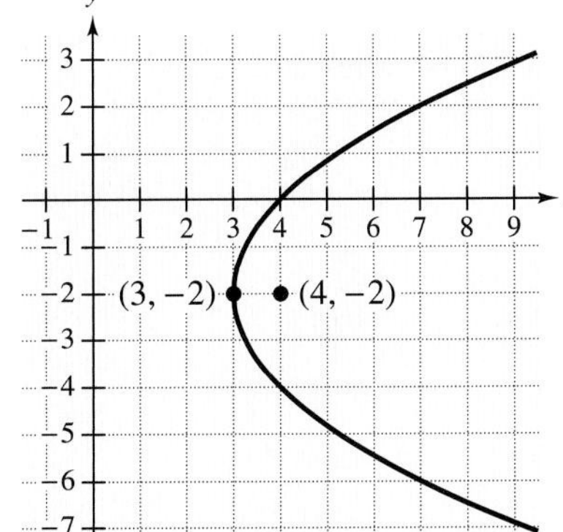

Figure 12.12

Solution

Because the vertex is at $(h, k) = (3, -2)$ and the axis of the parabola is horizontal, use the equation

$$(y - k)^2 = 4p(x - h)$$

where $h = 3$, $k = -2$, and $p = 1$. So, the equation of the parabola is

$(y - k)^2 = 4p(x - h)$	Standard form
$[y - (-2)]^2 = 4(1)(x - 3)$	Substitute for h, k, and p.
$(y + 2)^2 = 4(x - 3).$	Equation of parabola

 ***CHECKPOINT*** *Now try Exercise 69.*

Technology: Tip

You cannot represent a circle or a parabola with a horizontal axis as a single function of x. You can, however, represent it by two functions of x. For instance, try using a graphing calculator to graph the equations below in the same viewing window. Use a viewing window in which $-1 \le x \le 10$ and $-8 \le y \le 4$. Do you obtain a parabola? Does the graphing calculator connect the two portions of the parabola?

$y_1 = -2 + 2\sqrt{x - 3}$	Upper portion of parabola
$y_2 = -2 - 2\sqrt{x - 3}$	Lower portion of parabola

EXAMPLE 8 Analyzing a Parabola

Sketch the graph of the parabola $y = \frac{1}{8}x^2$ and identify its vertex and focus.

Solution

Because the equation can be written in the standard form $x^2 = 4py$, it is a parabola whose vertex is at the origin. You can identify the focus of the parabola by writing its equation in standard form.

$y = \frac{1}{8}x^2$ Write original equation.

$\frac{1}{8}x^2 = y$ Interchange sides of the equation.

$x^2 = 8y$ Multiply each side by 8.

$x^2 = 4(2)y$ Rewrite 8 in the form $4p$.

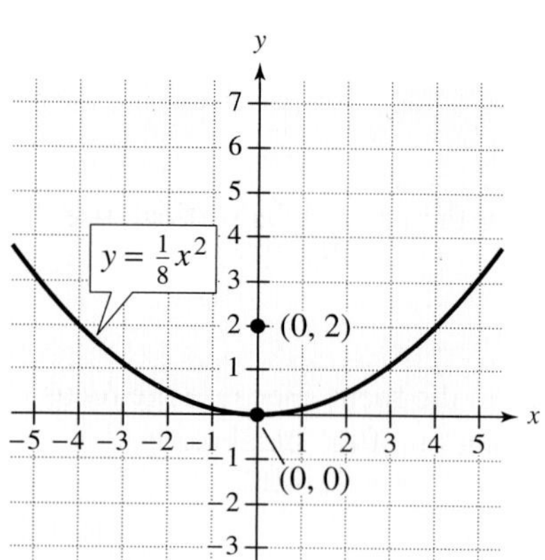

Figure 12.13

From this standard form, you can see that $p = 2$. Because the parabola opens upward, as shown in Figure 12.13, you can conclude that the focus lies $p = 2$ units above the vertex. So, the focus is $(0, 2)$.

✓ **CHECKPOINT** *Now try Exercise 75.*

EXAMPLE 9 Analyzing a Parabola

Sketch the parabola $x = \frac{1}{3}y^2 - 2y + 5$ and identify its vertex and focus.

Solution

This equation can be written in the standard form $(y - k)^2 = 4p(x - h)$. To do this, you can complete the square, as follows.

$x = \frac{1}{3}y^2 - 2y + 5$ Write original equation.

$\frac{1}{3}y^2 - 2y + 5 = x$ Interchange sides of the equation.

$y^2 - 6y + 15 = 3x$ Multiply each side by 3.

$y^2 - 6y = 3x - 15$ Subtract 15 from each side.

$y^2 - 6y + 9 = 3x - 15 + 9$ Complete the square.

$(y - 3)^2 = 3x - 6$ Simplify.

$(y - 3)^2 = 3(x - 2)$ Factor.

$(y - 3)^2 = 4\left(\frac{3}{4}\right)(x - 2)$ Rewrite 3 in the form $4p$.

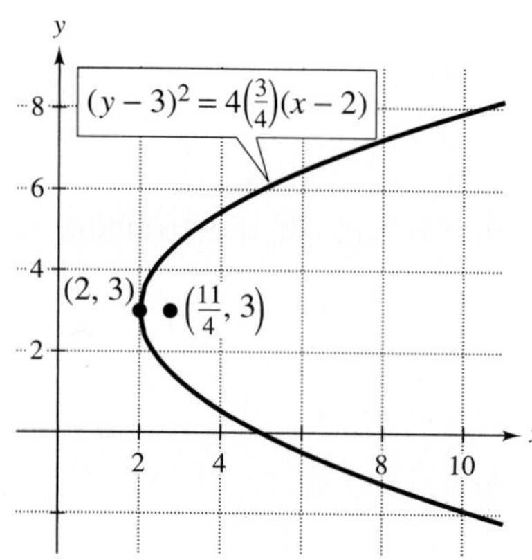

Figure 12.14

From this standard form, you can see that the vertex is $(h, k) = (2, 3)$ and $p = \frac{3}{4}$. Because the parabola opens to the right, as shown in Figure 12.14, the focus lies $p = \frac{3}{4}$ unit to the right of the vertex. So, the focus is $\left(\frac{11}{4}, 3\right)$.

✓ **CHECKPOINT** *Now try Exercise 85.*

Parabolas occur in a wide variety of applications. For instance, a parabolic reflector can be formed by revolving a parabola around its axis. The light rays emanating from the focus of a parabolic reflector used in a flashlight are all parallel to one another, as shown in Figure 12.15.

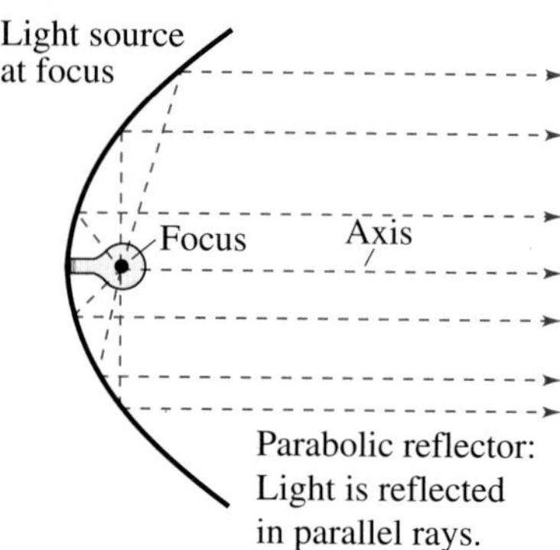

Figure 12.15

Concept Check

1. Explain how you would identify the center and radius of the circle given by the equation $x^2 + y^2 - 36 = 0$.

2. Is the center of the circle given by the equation $(x + 2)^2 + (y + 4)^2 = 20$ shifted two units to the right and four units upward? Explain your reasoning.

3. Which standard form of the equation of a parabola would you use to write an equation for a parabola with vertex $(2, -3)$ and focus $(2, 1)$? Explain your reasoning.

4. Given the equation of a parabola, explain how to determine if the parabola opens upward, downward, to the right, or to the left.

12.1 EXERCISES

Go to pages 804–805 to record your assignments.

Developing Skills

In Exercises 1–6, match the equation with its graph. [The graphs are labeled (a), (b), (c), (d), (e), and (f).]

(a)

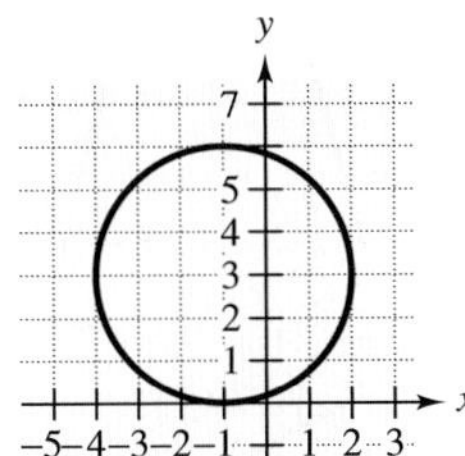

(b)

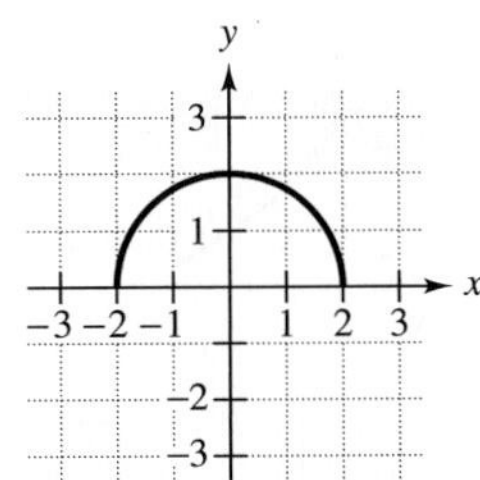

(c)

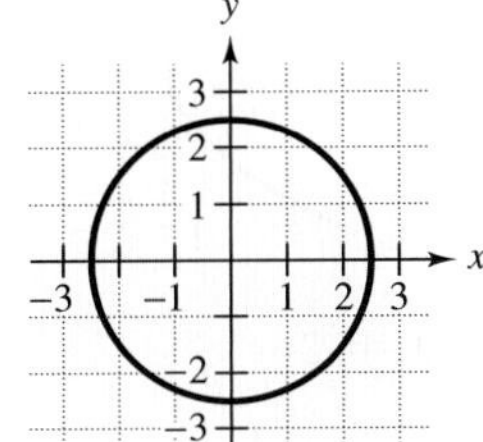

(d)

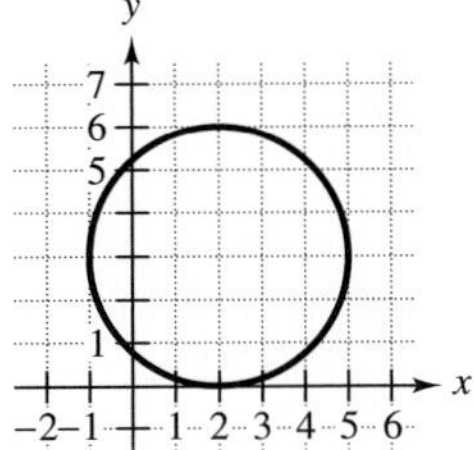

(e)

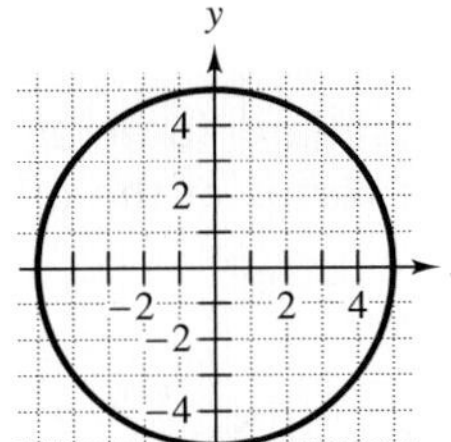

(f)

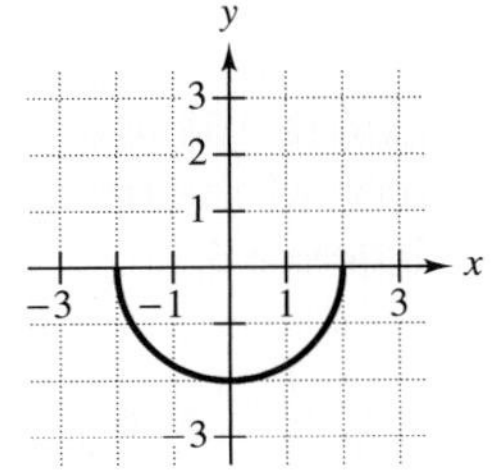

1. $x^2 + y^2 = 25$
2. $4x^2 + 4y^2 = 25$
3. $(x - 2)^2 + (y - 3)^2 = 9$
4. $(x + 1)^2 + (y - 3)^2 = 9$
5. $y = -\sqrt{4 - x^2}$
6. $y = \sqrt{4 - x^2}$

In Exercises 7–14, write the standard form of the equation of the circle with center at $(0, 0)$ that satisfies the criterion. *See Example 1.*

7. Radius: 5
8. Radius: 9
9. Radius: $\frac{2}{3}$
10. Radius: $\frac{5}{2}$
11. Passes through the point $(0, 6)$
12. Passes through the point $(-2, 0)$
13. Passes through the point $(5, 2)$
14. Passes through the point $(-1, -4)$

In Exercises 15–22, write the standard form of the equation of the circle with center at (h, k) that satisfies the criteria. *See Example 3.*

15. Center: $(4, 3)$
    Radius: 10

16. Center: $(-4, 8)$
    Radius: 7

17. Center: $(6, -5)$
    Radius: 3

18. Center: $(-5, -2)$
Radius: $\frac{5}{2}$

19. Center: $(-2, 1)$
Passes through the point $(0, 1)$

20. Center: $(8, 2)$
Passes through the point $(8, 0)$

21. Center: $(3, 2)$
Passes through the point $(4, 6)$

22. Center: $(-3, -5)$
Passes through the point $(0, 0)$

In Exercises 23–42, identify the center and radius of the circle and sketch the circle. ***See Examples 2 and 4.***

23. $x^2 + y^2 = 16$ **24.** $x^2 + y^2 = 1$

25. $x^2 + y^2 = 36$ **26.** $x^2 + y^2 = 15$

27. $4x^2 + 4y^2 = 1$ **28.** $9x^2 + 9y^2 = 64$

29. $25x^2 + 25y^2 - 144 = 0$

30. $\frac{x^2}{4} + \frac{y^2}{4} - 1 = 0$

31. $(x + 1)^2 + (y - 5)^2 = 64$

32. $(x - 9)^2 + (y + 2)^2 = 144$

33. $(x - 2)^2 + (y - 3)^2 = 4$

34. $(x + 4)^2 + (y - 3)^2 = 25$

35. $\left(x + \frac{9}{4}\right)^2 + (y - 4)^2 = 16$

36. $(x - 5)^2 + \left(y + \frac{3}{4}\right)^2 = 1$

37. $x^2 + y^2 - 4x - 2y + 1 = 0$

38. $x^2 + y^2 + 6x - 4y - 3 = 0$

39. $x^2 + y^2 + 2x + 6y + 6 = 0$

40. $x^2 + y^2 - 2x + 6y - 15 = 0$

41. $x^2 + y^2 + 10x - 4y - 7 = 0$

42. $x^2 + y^2 - 14x + 8y + 56 = 0$

In Exercises 43–46, use a graphing calculator to graph the circle. (*Note:* Solve for y. Use a square setting so that the circles display correctly.)

43. $x^2 + y^2 = 30$ **44.** $4x^2 + 4y^2 = 45$

45. $(x - 2)^2 + y^2 = 10$ **46.** $x^2 + (y - 5)^2 = 21$

In Exercises 47–52, match the equation with its graph. [The graphs are labeled (a), (b), (c), (d), (e), and (f).]

(a)
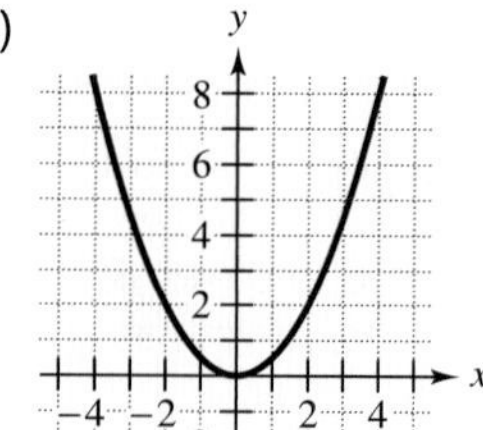
(b)
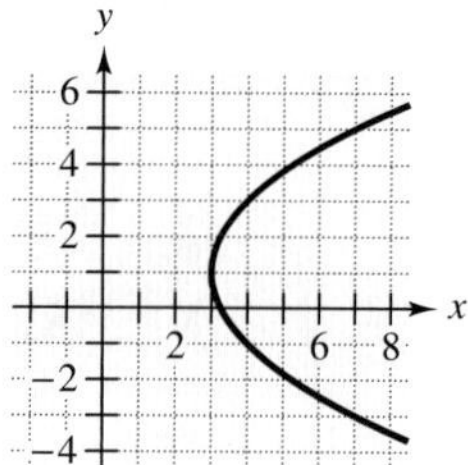
(c)
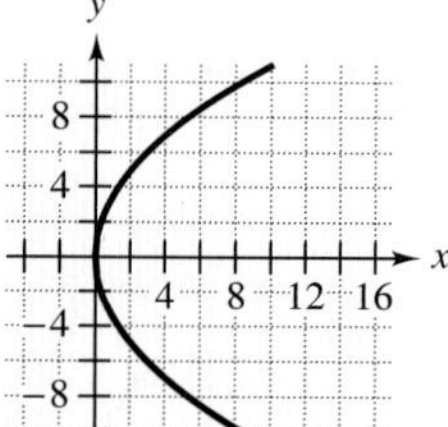
(d)
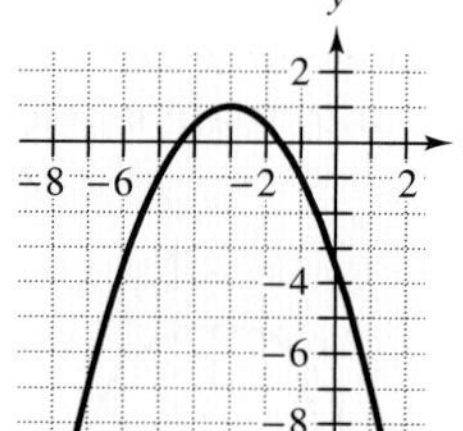
(e)
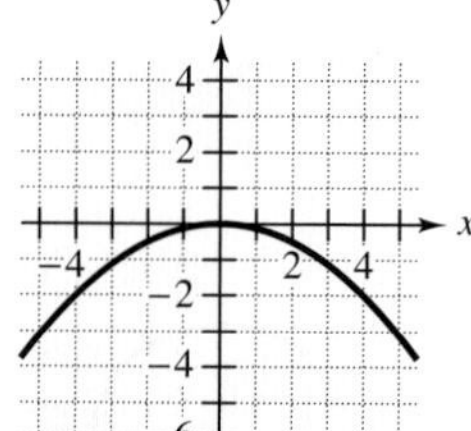
(f)
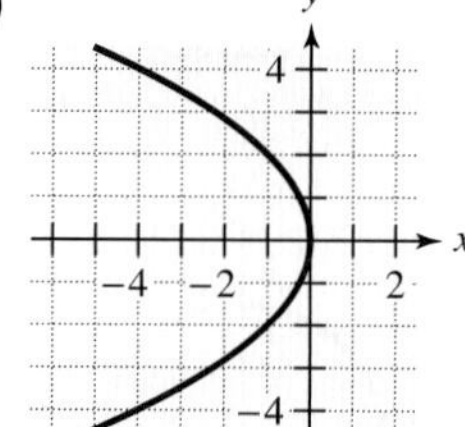

47. $y^2 = -4x$ **48.** $x^2 = 2y$

49. $x^2 = -8y$ **50.** $y^2 = 12x$

51. $(y - 1)^2 = 4(x - 3)$ **52.** $(x + 3)^2 = -2(y - 1)$

In Exercises 53–64, write the standard form of the equation of the parabola with its vertex at the origin. ***See Example 6.***

53.
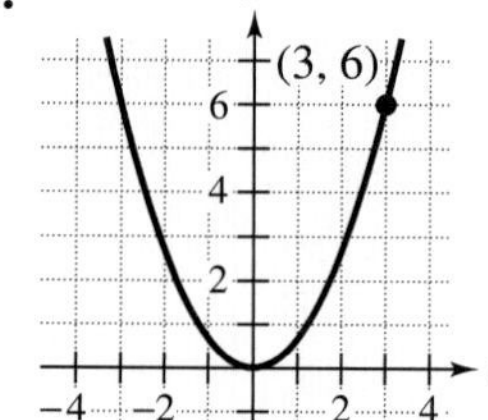

54.
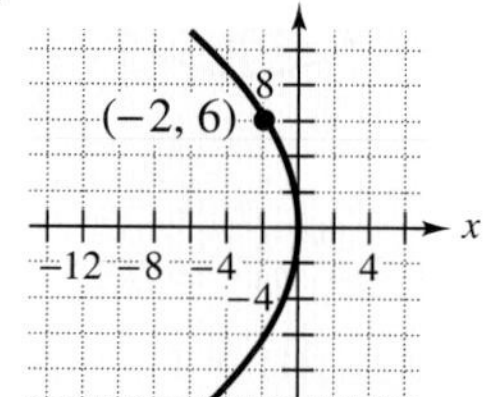

55. Focus: $\left(0, -\frac{3}{2}\right)$

56. Focus: $\left(\frac{5}{4}, 0\right)$

57. Focus: $(-2, 0)$

58. Focus: $(0, -2)$

59. Focus: $(0, 1)$

60. Focus: $(-3, 0)$

61. Focus: $(6, 0)$

62. Focus: $(0, 2)$

63. Passes through $(4, 6)$; Horizontal axis

64. Passes through $(-6, -12)$; Vertical axis

In Exercises 65–74, write the standard form of the equation of the parabola with its vertex at (h, k). ***See Example 7.***

65.

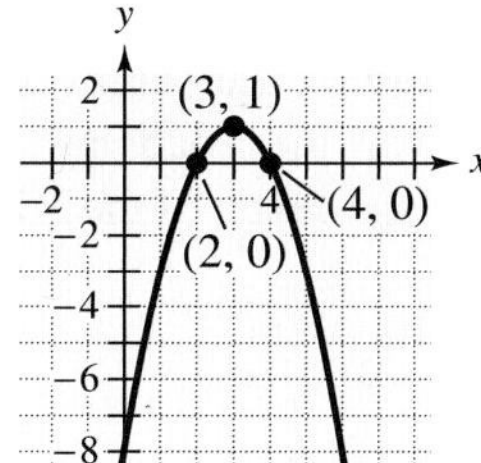

66.

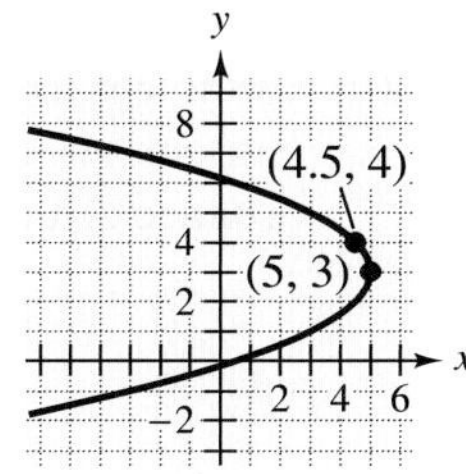

67.

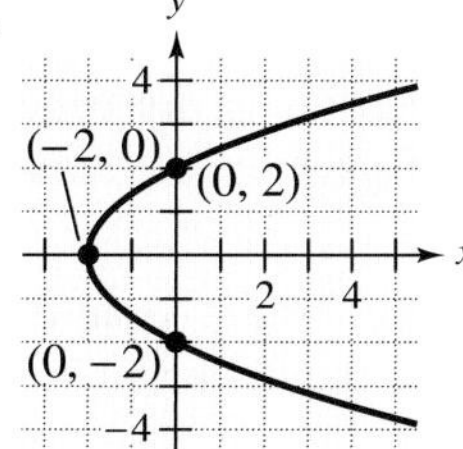

68.

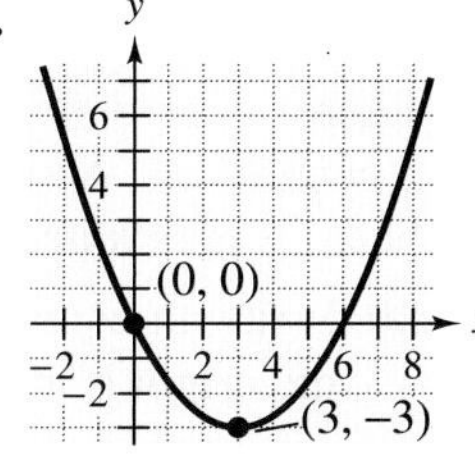

69. Vertex: $(3, 2)$; Focus: $(1, 2)$

70. Vertex: $(-1, 2)$; Focus: $(-1, 0)$

71. Vertex: $(0, -4)$; Focus: $(0, -1)$

72. Vertex: $(-2, 1)$; Focus: $(-5, 1)$

73. Vertex: $(0, 2)$;
Passes through $(1, 3)$; Horizontal axis

74. Vertex: $(0, 2)$;
Passes through $(6, 0)$; Vertical axis

In Exercises 75–88, identify the vertex and focus of the parabola and sketch the parabola. ***See Examples 8 and 9.***

75. $y = \frac{1}{2}x^2$

76. $y = 2x^2$

77. $y^2 = -10x$

78. $y^2 = 3x$

79. $x^2 + 8y = 0$

80. $x + y^2 = 0$

81. $(x - 1)^2 + 8(y + 2) = 0$

82. $(x + 3) + (y - 2)^2 = 0$

83. $\left(y + \frac{1}{2}\right)^2 = 2(x - 5)$

84. $\left(x + \frac{1}{2}\right)^2 = 4(y - 3)$

85. $y = \frac{1}{3}(x^2 - 2x + 10)$

86. $4x - y^2 - 2y - 33 = 0$

87. $y^2 + 6y + 8x + 25 = 0$

88. $y^2 - 4y - 4x = 0$

In Exercises 89–92, use a graphing calculator to graph the parabola. Identify the vertex and focus.

89. $y = -\frac{1}{6}(x^2 + 4x - 2)$

90. $x^2 - 2x + 8y + 9 = 0$

91. $y^2 + x + y = 0$

92. $3y^2 - 10x + 25 = 0$

Solving Problems

93. *Satellite Orbit* Write an equation that represents the circular orbit of a satellite 500 miles above the surface of Earth. Place the origin of the rectangular coordinate system at the center of Earth, and assume the radius of Earth is 4000 miles.

94. *Observation Wheel* Write an equation that represents the circular wheel of the Singapore Flyer in Singapore, which has a diameter of 150 meters. Place the origin of the rectangular coordinate system at the center of the wheel.

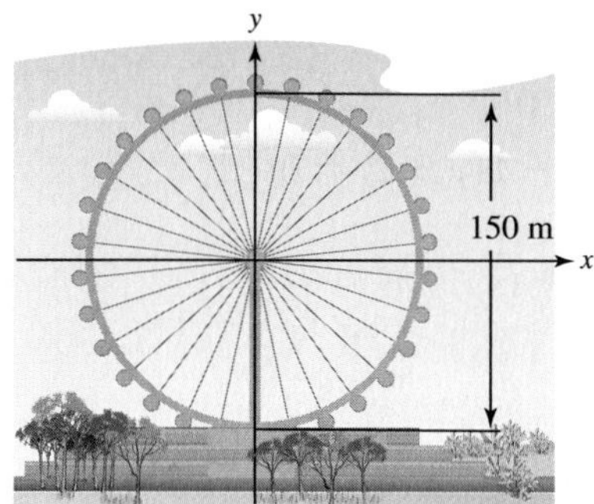

95. *Mirror* Write an equation that represents the circular mirror, with a diameter of 3 feet, shown in the figure. The wall hangers of the mirror are shown as two points on the circle. Use the equation to determine the height of the left wall hanger.

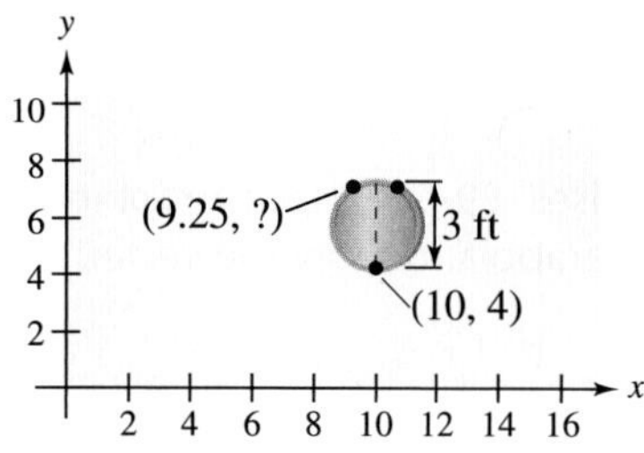

96. *Dog Leash* A leash allows a dog a semicircular boundary that has a diameter of 80 feet. Write an equation that represents the semicircle. The dog is located on the semicircle, 10 feet from the fence at the right. How far is the dog from the house?

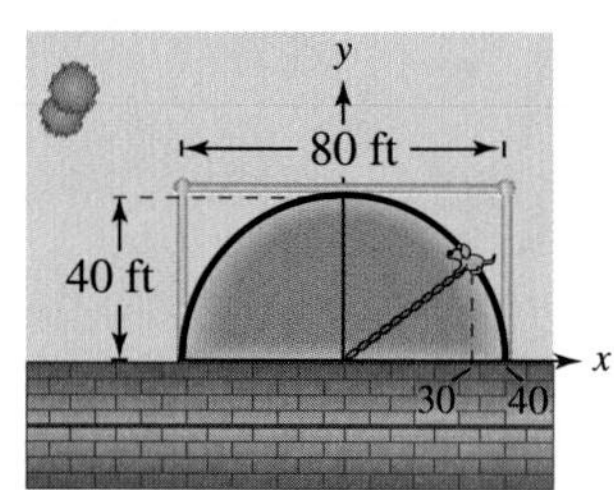

97. *Suspension Bridge* Each cable of a suspension bridge is suspended (in the shape of a parabola) between two towers that are 120 meters apart, and the top of each tower is 20 meters above the roadway. The cables touch the roadway at the midpoint between the two towers (see figure).

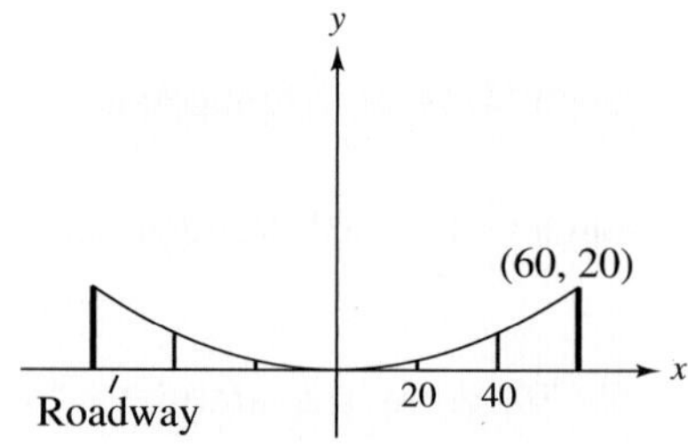

(a) Write an equation that represents the parabolic shape of each cable.

(b) Complete the table by finding the height of the suspension cables y over the roadway at a distance of x meters from the center of the bridge.

x	0	20	40	60
y				

98. *Beam Deflection* A simply supported beam is 16 meters long and has a load at the center (see figure). The deflection of the beam at its center is 3 centimeters. Assume that the shape of the deflected beam is parabolic.

(a) Write an equation of the parabola. (Assume that the origin is at the center of the deflected beam.)

(b) How far from the center of the beam is the deflection equal to 1 centimeter?

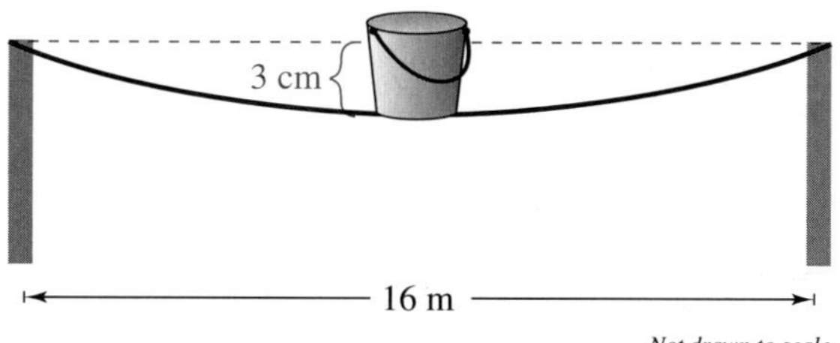

Not drawn to scale

99. *Revenue* The revenue R generated by the sale of x video game systems is given by $R = 575x - \frac{5}{4}x^2$.

(a) Use a graphing calculator to graph the function.

(b) Use the graph to approximate the number of sales that will maximize revenue.

100. *Path of a Softball* The path of a softball is given by $y = -0.08x^2 + x + 4$. The coordinates x and y are measured in feet, with $x = 0$ corresponding to the position from which the ball was thrown.

(a) Use a graphing calculator to graph the path of the softball.

(b) Move the cursor along the path to approximate the highest point and the range of the path.

101. *Graphical Estimation* A rectangle centered at the origin with sides parallel to the coordinate axes is placed in a circle of radius 25 inches centered at the origin (see figure). The length of the rectangle is $2x$ inches.

(a) Show that the width and area of the rectangle are given by $2\sqrt{625 - x^2}$ and $4x\sqrt{625 - x^2}$, respectively.

(b) Use a graphing calculator to graph the area function. Approximate the value of x for which the area is maximum.

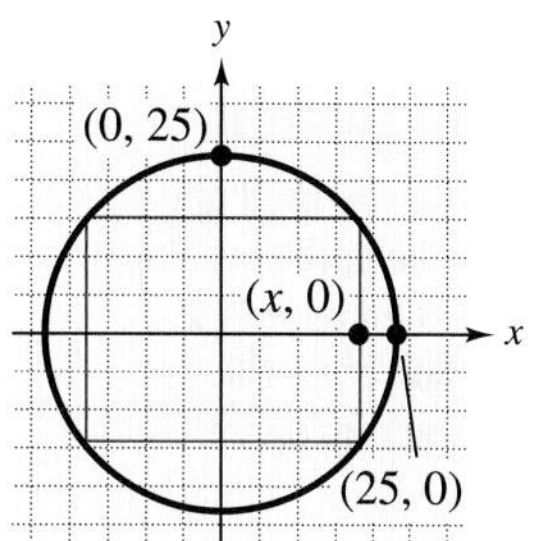

Explaining Concepts

102. The point $(-4, 3)$ lies on a circle with center $(-1, 1)$. Does the point $(3, 2)$ lie on the same circle? Explain your reasoning.

103. A student claims that

$$x^2 + y^2 - 6y = -5$$

does not represent a circle. Is the student correct? Explain your reasoning.

104. Is y a function of x in the equation $y^2 = 6x$? Explain.

105. Is it possible for a parabola to intersect its directrix? Explain.

106. If the vertex and focus of a parabola are on a horizontal line, is the directrix of the parabola vertical? Explain.

Cumulative Review

In Exercises 107–112, solve the equation by completing the square.

107. $x^2 + 4x = 6$

108. $x^2 + 6x = -4$

109. $x^2 - 2x - 3 = 0$

110. $4x^2 - 12x - 10 = 0$

111. $2x^2 + 5x - 8 = 0$ **112.** $9x^2 - 12x = 14$

In Exercises 113–116, use the properties of logarithms to expand the expression. (Assume all variables are positive.)

113. $\log_8 x^{10}$ **114.** $\log_{10} \sqrt{xy^3}$

115. $\ln 5x^2y$ **116.** $\ln \frac{x}{y^4}$

In Exercises 117–120, use the properties of logarithms to condense the expression.

117. $\log_{10} x + \log_{10} 6$

118. $2 \log_3 x - \log_3 y$

119. $3 \ln x + \ln y - \ln 9$

120. $4(\ln x + \ln y) - \ln(x^4 + y^4)$

12.2 Ellipses

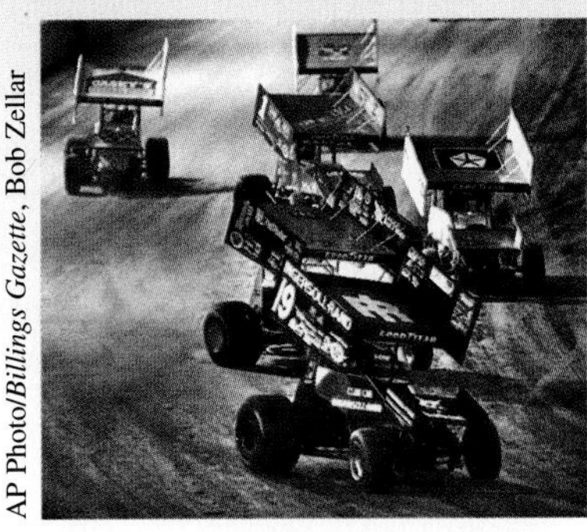
AP Photo/*Billings Gazette*, Bob Zellar

What You Should Learn

1 ▶ Graph and write equations of ellipses centered at the origin.

2 ▶ Graph and write equations of ellipses centered at (h, k).

Why You Should Learn It

Equations of ellipses can be used to model and solve real-life problems. For instance, in Exercise 59 on page 782, you will use an equation of an ellipse to model a dirt race track for sprint cars.

1 ▶ Graph and write equations of ellipses centered at the origin.

Ellipses Centered at the Origin

The third type of conic is called an *ellipse* and is defined as follows.

Definition of an Ellipse

An **ellipse** in the rectangular coordinate system consists of all points (x, y) such that the sum of the distances between (x, y) and two distinct fixed points is a constant, as shown in Figure 12.16. Each of the two fixed points is called a **focus** of the ellipse. (The plural of focus is *foci.*)

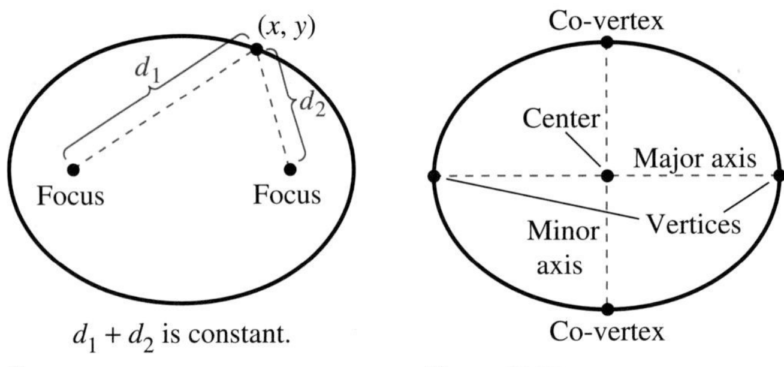

Figure 12.16 **Figure 12.17**

The line through the foci intersects the ellipse at two points, called the **vertices,** as shown in Figure 12.17. The line segment joining the vertices is called the **major axis,** and its midpoint is called the **center** of the ellipse. The line segment perpendicular to the major axis at the center is called the **minor axis** of the ellipse, and the points at which the minor axis intersects the ellipse are called **co-vertices.**

To trace an ellipse, place two thumbtacks at the foci, as shown in Figure 12.18. If the ends of a fixed length of string are fastened to the thumbtacks and the string is drawn taut with a pencil, the path traced by the pencil will be an ellipse.

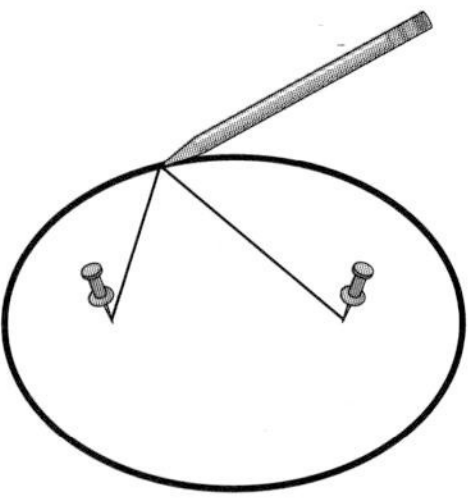

Figure 12.18

The standard form of the equation of an ellipse takes one of two forms, depending on whether the major axis is horizontal or vertical.

Standard Equation of an Ellipse (Center at Origin)

The **standard form of the equation of an ellipse centered at the origin** with major and minor axes of lengths $2a$ and $2b$ is

$$\frac{x^2}{a^2} + \frac{y^2}{b^2} = 1 \quad \text{or} \quad \frac{x^2}{b^2} + \frac{y^2}{a^2} = 1, \qquad 0 < b < a.$$

The vertices lie on the major axis, a units from the center, and the co-vertices lie on the minor axis, b units from the center, as shown in Figure 12.19.

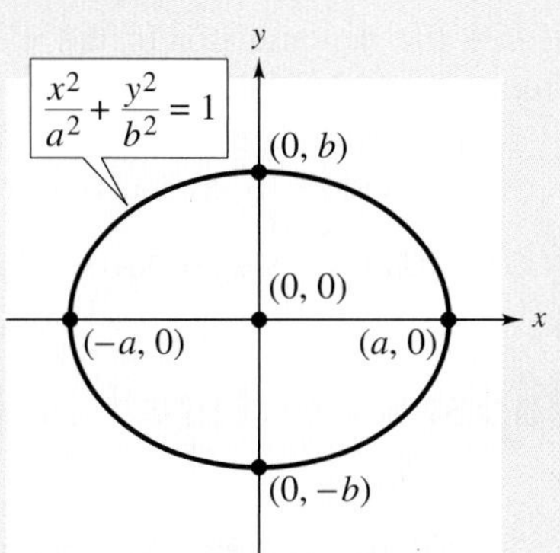

Major axis is horizontal.
Minor axis is vertical.

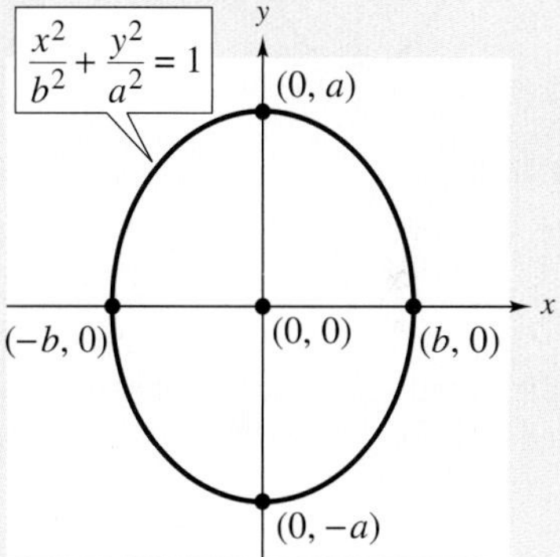

Major axis is vertical.
Minor axis is horizontal.

Figure 12.19

Study Tip

Notice that the standard form of the equation of an ellipse centered at the origin with a horizontal major axis has x-intercepts of $(\pm a, 0)$ and y-intercepts of $(0, \pm b)$.

EXAMPLE 1 Writing the Standard Equation of an Ellipse

Write an equation of the ellipse that is centered at the origin, with vertices $(-3, 0)$ and $(3, 0)$ and co-vertices $(0, -2)$ and $(0, 2)$.

Solution

Begin by plotting the vertices and co-vertices, as shown in Figure 12.20. The center of the ellipse is $(0, 0)$. So, the equation of the ellipse has the form

$$\frac{x^2}{a^2} + \frac{y^2}{b^2} = 1. \qquad \text{Major axis is horizontal.}$$

For this ellipse, the major axis is horizontal. So, a is the distance between the center and either vertex, which implies that $a = 3$. Similarly, b is the distance between the center and either co-vertex, which implies that $b = 2$. So, the standard form of the equation of the ellipse is

$$\frac{x^2}{3^2} + \frac{y^2}{2^2} = 1. \qquad \text{Standard form}$$

✓ **CHECKPOINT** *Now try Exercise 7.*

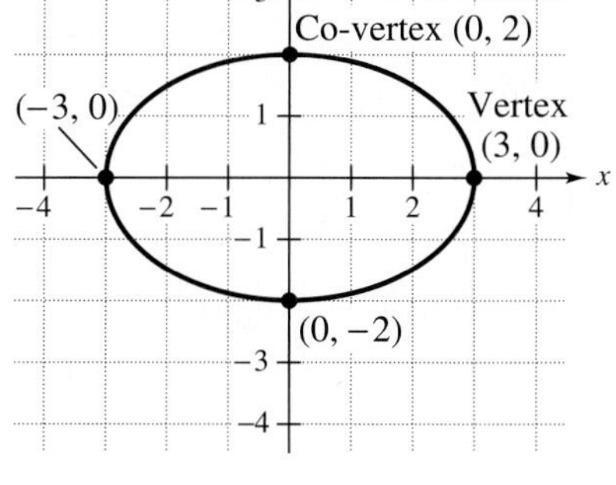

Figure 12.20

EXAMPLE 2 Sketching an Ellipse

Sketch the ellipse given by $4x^2 + y^2 = 36$. Identify the vertices and co-vertices.

Solution

To sketch an ellipse, it helps first to write its equation in standard form.

$$4x^2 + y^2 = 36 \qquad \text{Write original equation.}$$

$$\frac{x^2}{9} + \frac{y^2}{36} = 1 \qquad \text{Divide each side by 36 and simplify.}$$

$$\frac{x^2}{3^2} + \frac{y^2}{6^2} = 1 \qquad \text{Standard form}$$

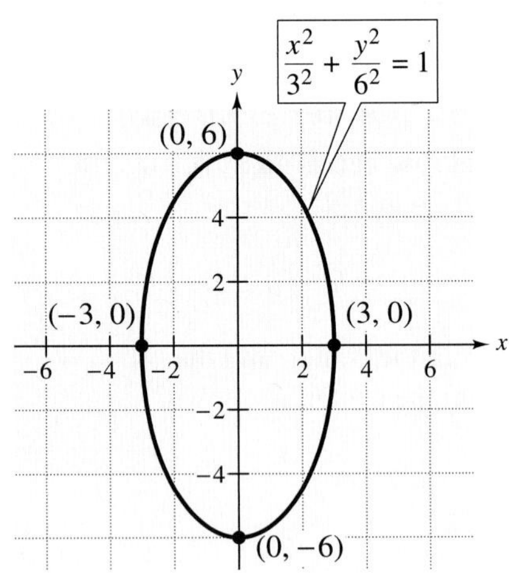

Figure 12.21

Because the denominator of the y^2-term is larger than the denominator of the x^2-term, you can conclude that the major axis is vertical. Moreover, because $a = 6$, the vertices are $(0, -6)$ and $(0, 6)$. Finally, because $b = 3$, the co-vertices are $(-3, 0)$ and $(3, 0)$, as shown in Figure 12.21.

CHECKPOINT *Now try Exercise 19.*

2 ▶ Graph and write equations of ellipses centered at (h, k).

Ellipses Centered at (h, k)

Standard Equation of an Ellipse [Center at (h, k)]

The **standard form of the equation of an ellipse centered at (h, k)** with major and minor axes of lengths $2a$ and $2b$, where $0 < b < a$, is

$$\frac{(x-h)^2}{a^2} + \frac{(y-k)^2}{b^2} = 1 \qquad \text{Major axis is horizontal.}$$

or

$$\frac{(x-h)^2}{b^2} + \frac{(y-k)^2}{a^2} = 1. \qquad \text{Major axis is vertical.}$$

The foci lie on the major axis, c units from the center, with $c^2 = a^2 - b^2$.

Figure 12.22 shows the horizontal and vertical orientations for an ellipse.

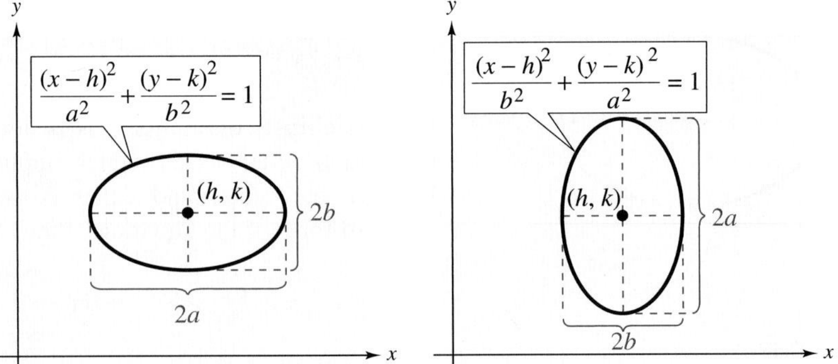

Figure 12.22

When $h = 0$ and $k = 0$, the ellipse is centered at the origin. Otherwise, you can shift the center of the ellipse h units horizontally and k units vertically from the origin.

EXAMPLE 3 Writing the Standard Equation of an Ellipse

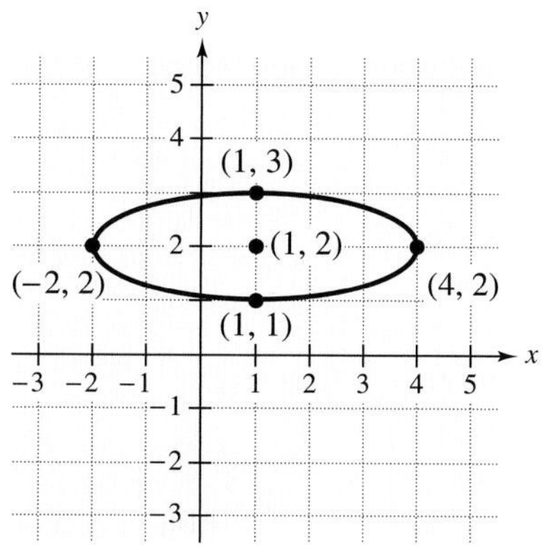

Figure 12.23

Write the standard form of the equation of the ellipse with vertices $(-2, 2)$ and $(4, 2)$ and co-vertices $(1, 3)$ and $(1, 1)$, as shown in Figure 12.23.

Solution

Because the vertices are $(-2, 2)$ and $(4, 2)$, the center of the ellipse is $(h, k) = (1, 2)$. The distance from the center to either vertex is $a = 3$, and the distance to either co-vertex is $b = 1$. Because the major axis is horizontal, the standard form of the equation is

$$\frac{(x-h)^2}{a^2} + \frac{(y-k)^2}{b^2} = 1. \quad \text{Major axis is horizontal.}$$

Substitute the values of h, k, a, and b to obtain

$$\frac{(x-1)^2}{3^2} + \frac{(y-2)^2}{1^2} = 1. \quad \text{Standard form}$$

From the graph, you can see that the center of the ellipse is shifted one unit to the right and two units upward from the origin.

✓ ***CHECKPOINT*** *Now try Exercise 37.*

Technology: Tip

You can use a graphing calculator to graph an ellipse by graphing the upper and lower portions in the same viewing window. For instance, to graph the ellipse $x^2 + 4y^2 = 4$, first solve for y to obtain

$$y_1 = \frac{1}{2}\sqrt{4 - x^2}$$

and

$$y_2 = -\frac{1}{2}\sqrt{4 - x^2}.$$

Use a viewing window in which $-3 \le x \le 3$ and $-2 \le y \le 2$. You should obtain the graph shown below.

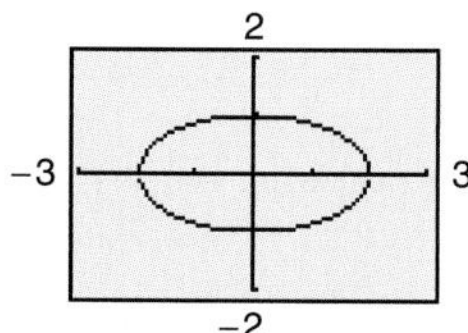

Use this procedure to graph the ellipse in Example 3 on your graphing calculator.

To write an equation of an ellipse in standard form, you must group the x-terms and the y-terms and then complete each square, as shown in Example 4.

EXAMPLE 4 Sketching an Ellipse

Sketch the ellipse given by $4x^2 + y^2 - 8x + 6y + 9 = 0$.

Solution

Begin by writing the equation in standard form. In the fourth step, note that 9 and 4 are added to *each* side of the equation.

$$4x^2 + y^2 - 8x + 6y + 9 = 0$$ Write original equation.

$$(4x^2 - 8x + \quad) + (y^2 + 6y + \quad) = -9$$ Group terms.

$$4(x^2 - 2x + \quad) + (y^2 + 6y + \quad) = -9$$ Factor 4 out of x-terms.

$$4(x^2 - 2x + 1) + (y^2 + 6y + 9) = -9 + 4(1) + 9$$ Complete the squares.

$$4(x - 1)^2 + (y + 3)^2 = 4$$ Simplify.

$$\frac{(x-1)^2}{1} + \frac{(y+3)^2}{4} = 1$$ Divide each side by 4.

$$\frac{(x-1)^2}{1^2} + \frac{(y+3)^2}{2^2} = 1$$ Standard form

You can see that the center of the ellipse is at $(h, k) = (1, -3)$. Because the denominator of the y^2-term is larger than the denominator of the x^2-term, you can conclude that the major axis is vertical. Because the denominator of the x^2-term is $b^2 = 1^2$, you can locate the endpoints of the minor axis one unit to the right of the center and one unit to the left of the center. Because the denominator of the y^2-term is $a^2 = 2^2$, you can locate the endpoints of the major axis two units up from the center and two units down from the center, as shown in Figure 12.24. To complete the graph, sketch an oval shape that is determined by the vertices and co-vertices, as shown in Figure 12.25.

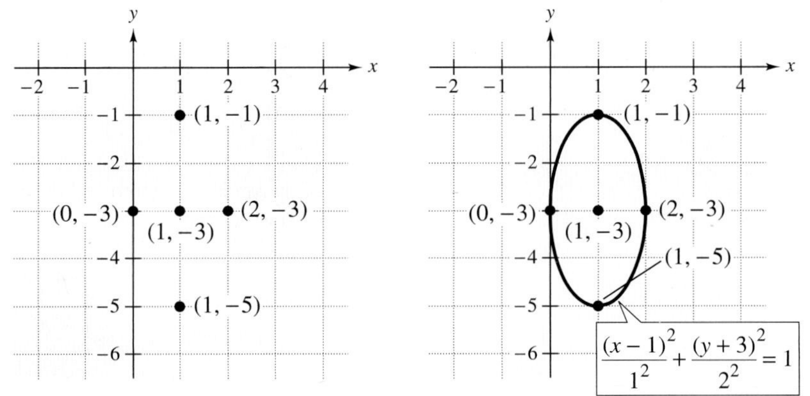

Figure 12.24 **Figure 12.25**

From Figure 12.25, you can see that the center of the ellipse is shifted one unit to the right and three units down from the origin.

✓ **CHECKPOINT** *Now try Exercise 45.*

EXAMPLE 5 An Application: Semielliptical Archway 

10 ft

30 ft

Figure 12.26

You are responsible for designing a semielliptical archway, as shown in Figure 12.26. The height of the archway is 10 feet, and its width is 30 feet. Write an equation of the ellipse and use the equation to sketch an accurate diagram of the archway.

Solution

To make the equation simple, place the origin at the center of the ellipse. This means that the standard form of the equation is

$$\frac{x^2}{a^2} + \frac{y^2}{b^2} = 1. \qquad \text{Major axis is horizontal.}$$

Because the major axis is horizontal, it follows that $a = 15$ and $b = 10$, which implies that the equation is

$$\frac{x^2}{15^2} + \frac{y^2}{10^2} = 1. \qquad \text{Standard form}$$

To make an accurate sketch of the ellipse, solve this equation for y as follows.

$$\frac{x^2}{225} + \frac{y^2}{100} = 1 \qquad \text{Simplify denominators.}$$

$$\frac{y^2}{100} = 1 - \frac{x^2}{225} \qquad \text{Subtract } \frac{x^2}{225} \text{ from each side.}$$

$$y^2 = 100\left(1 - \frac{x^2}{225}\right) \qquad \text{Multiply each side by 100.}$$

$$y = 10\sqrt{1 - \frac{x^2}{225}} \qquad \text{Take the positive square root of each side.}$$

Next, calculate several y-values for the archway, as shown in the table. Then use the values in the table to sketch the archway, as shown in Figure 12.27.

x	± 15	± 12.5	± 10	± 7.5	± 5	± 2.5	0
y	0	5.53	7.45	8.66	9.43	9.86	10

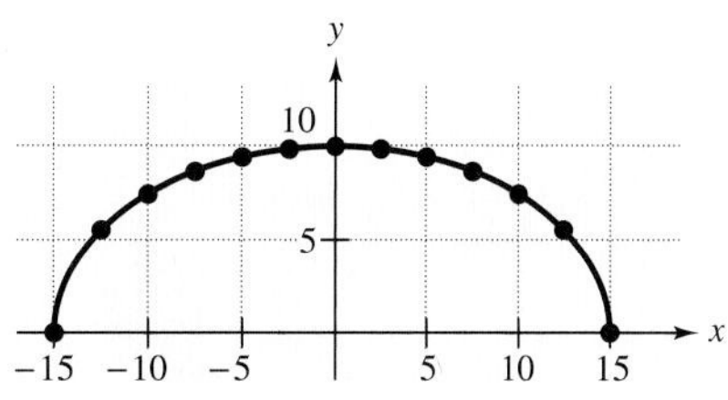

Figure 12.27

✓ **CHECKPOINT** *Now try Exercise 57.*

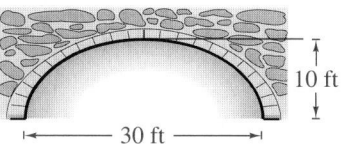

Concept Check

1. Define an ellipse and write the standard form of the equation of an ellipse centered at the origin.

2. What points do you need to know in order to write the equation of an ellipse?

3. From the standard equation, how can you determine the lengths of the major and minor axes of an ellipse?

4. From the standard equation, how can you determine the orientation of the major and minor axes of an ellipse?

12.2 EXERCISES

Go to pages 804–805 to record your assignments.

Developing Skills

In Exercises 1–6, match the equation with its graph. [The graphs are labeled (a), (b), (c), (d), (e), and (f).]

(a)

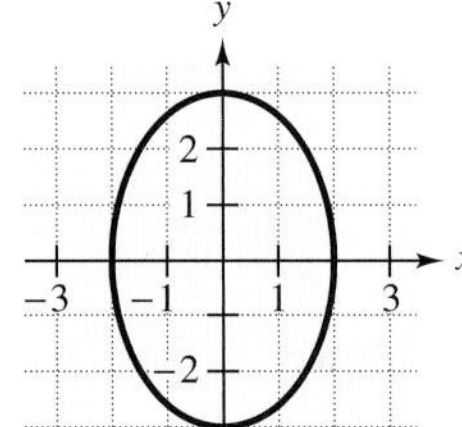

(b)

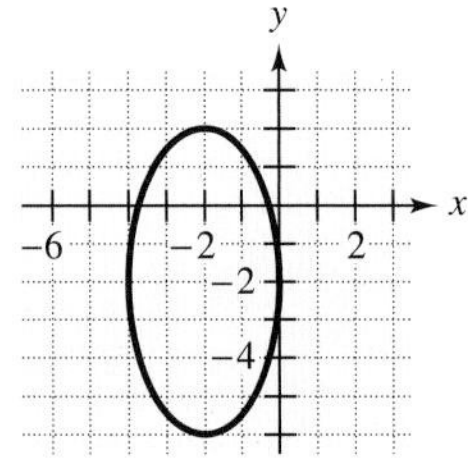

(c)

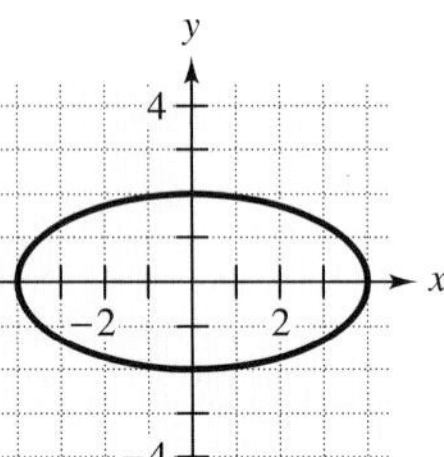

(d)

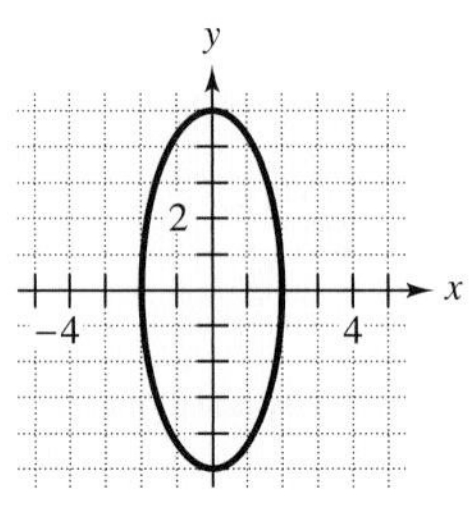

(e)

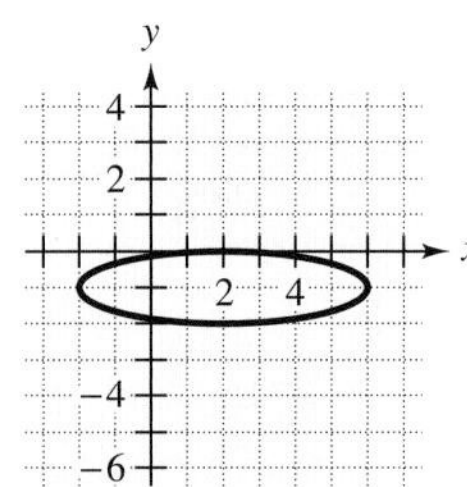

(f)

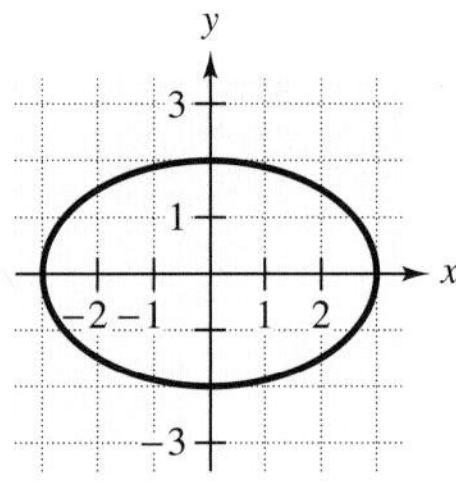

1. $\frac{x^2}{4} + \frac{y^2}{9} = 1$

2. $\frac{x^2}{9} + \frac{y^2}{4} = 1$

3. $\frac{x^2}{4} + \frac{y^2}{25} = 1$

4. $\frac{y^2}{4} + \frac{x^2}{16} = 1$

5. $\frac{(x-2)^2}{16} + \frac{(y+1)^2}{1} = 1$

6. $\frac{(x+2)^2}{4} + \frac{(y+2)^2}{16} = 1$

In Exercises 7–18, write the standard form of the equation of the ellipse centered at the origin. ***See Example 1.***

	Vertices	*Co-vertices*
✓ 7.	$(-4, 0), (4, 0)$	$(0, -3), (0, 3)$
8.	$(-4, 0), (4, 0)$	$(0, -1), (0, 1)$
9.	$(-2, 0), (2, 0)$	$(0, -1), (0, 1)$
10.	$(-7, 0), (7, 0)$	$(0, -4), (0, 4)$
11.	$(0, -6), (0, 6)$	$(-3, 0), (3, 0)$
12.	$(0, -5), (0, 5)$	$(-1, 0), (1, 0)$
13.	$(0, -2), (0, 2)$	$(-1, 0), (1, 0)$
14.	$(0, -8), (0, 8)$	$(-4, 0), (4, 0)$

15. Major axis (vertical) 10 units, minor axis 6 units

16. Major axis (horizontal) 24 units, minor axis 10 units

17. Major axis (horizontal) 20 units, minor axis 12 units

18. Major axis (vertical) 40 units, minor axis 30 units

In Exercises 19–32, sketch the ellipse. Identify the vertices and co-vertices. *See Example 2.*

19. $\dfrac{x^2}{16} + \dfrac{y^2}{4} = 1$

20. $\dfrac{x^2}{25} + \dfrac{y^2}{9} = 1$

21. $\dfrac{x^2}{4} + \dfrac{y^2}{16} = 1$

22. $\dfrac{x^2}{9} + \dfrac{y^2}{25} = 1$

23. $\dfrac{x^2}{25/9} + \dfrac{y^2}{16/9} = 1$

24. $\dfrac{x^2}{1} + \dfrac{y^2}{1/4} = 1$

25. $\dfrac{9x^2}{4} + \dfrac{25y^2}{16} = 1$

26. $\dfrac{36x^2}{49} + \dfrac{16y^2}{9} = 1$

27. $16x^2 + 25y^2 - 9 = 0$

28. $64x^2 + 36y^2 - 49 = 0$

29. $4x^2 + y^2 - 4 = 0$

30. $4x^2 + 9y^2 - 36 = 0$

31. $10x^2 + 16y^2 - 160 = 0$

32. $15x^2 + 3y^2 - 75 = 0$

In Exercises 33–36, use a graphing calculator to graph the ellipse. Identify the vertices. (*Note:* Solve for *y*.)

33. $x^2 + 2y^2 = 4$

34. $9x^2 + y^2 = 64$

35. $3x^2 + y^2 - 12 = 0$

36. $5x^2 + 2y^2 - 10 = 0$

In Exercises 37–40, write the standard form of the equation of the ellipse. *See Example 3.*

37.

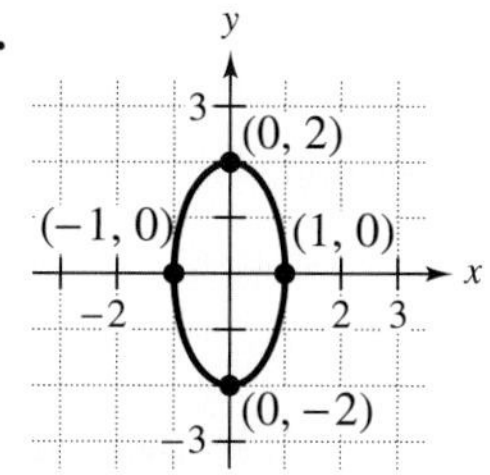

38.

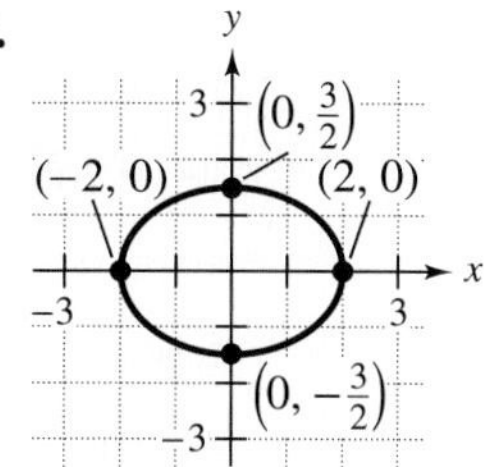

39.

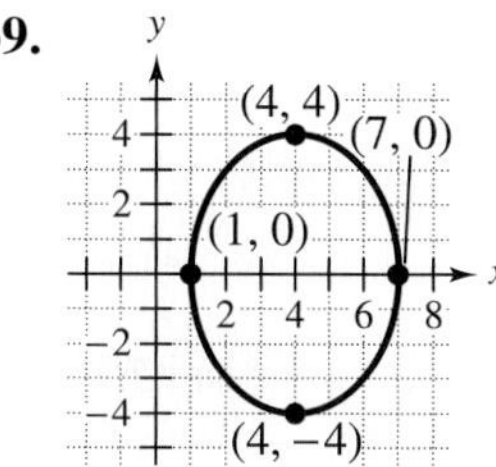

40.

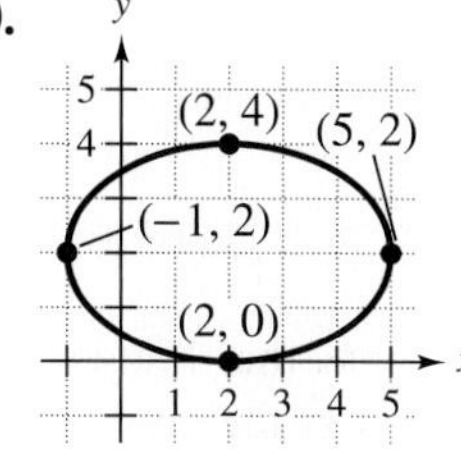

In Exercises 41–54, find the center and vertices of the ellipse and sketch the ellipse. *See Example 4.*

41. $\dfrac{(x + 5)^2}{16} + y^2 = 1$

42. $x^2 + \dfrac{(y - 3)^2}{9} = 1$

43. $\dfrac{(x - 1)^2}{9} + \dfrac{(y - 5)^2}{25} = 1$

44. $\dfrac{(x + 2)^2}{1/4} + \dfrac{(y + 4)^2}{1} = 1$

45. $4(x - 2)^2 + 9(y + 2)^2 = 36$

46. $2(x + 5)^2 + 8(y - 2)^2 = 72$

47. $12(x + 4)^2 + 3(y - 1)^2 = 48$

48. $16(x - 2)^2 + 4(y + 3)^2 = 16$

49. $9x^2 + 4y^2 + 36x - 24y + 36 = 0$

50. $9x^2 + 4y^2 - 36x + 8y + 31 = 0$

51. $25x^2 + 9y^2 - 200x + 54y + 256 = 0$

52. $25x^2 + 16y^2 - 150x - 128y + 81 = 0$

53. $x^2 + 4y^2 - 4x - 8y - 92 = 0$

54. $x^2 + 4y^2 + 6x + 16y - 11 = 0$

Solving Problems

55. *Wading Pool* You are building a wading pool that is in the shape of an ellipse. Your plans give the following equation for the elliptical shape of the pool, measured in feet.

$$\frac{x^2}{324} + \frac{y^2}{196} = 1$$

Find the longest distance and shortest distance across the pool.

56. *Oval Office* In the White House, the Oval Office is in the shape of an ellipse. The perimeter of the floor can be modeled in meters by the equation

$$\frac{x^2}{19.36} + \frac{y^2}{30.25} = 1.$$

Find the longest distance and shortest distance across the office.

✓ **57.** *Architecture* A semielliptical arch for a tunnel under a river has a width of 100 feet and a height of 40 feet (see figure). Determine the height of the arch 5 feet from the edge of the tunnel.

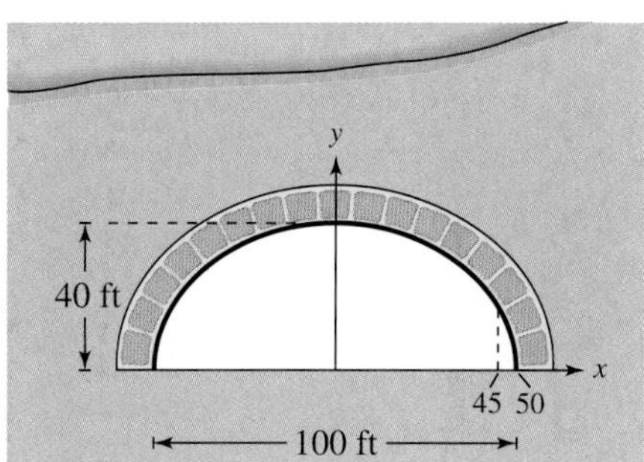

58. *Architecture* A semielliptical arch for a fireplace has a width of 54 inches and a height of 30 inches. Determine the height of the arch 10 inches from the edge of the fireplace.

59. *Motorsports* Most sprint car dirt tracks are elliptical in shape. Write an equation of an elliptical race track with a major axis that is 1230 feet long and a minor axis that is 580 feet long.

60. *Bicycle Chainwheel* The pedals of a bicycle drive a chainwheel, which drives a smaller sprocket wheel on the rear axle (see figure). Many chainwheels are circular. Some, however, are slightly elliptical, which tends to make pedaling easier. Write an equation of an elliptical chainwheel with a major axis that is 8 inches long and a minor axis that is $7\frac{1}{2}$ inches long.

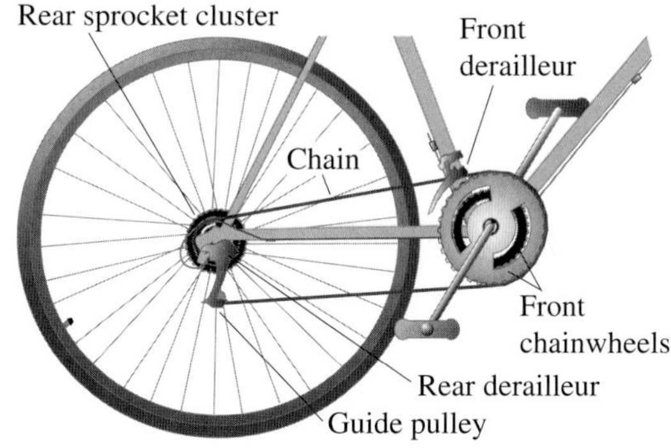

Airplane In Exercises 61 and 62, an airplane with enough fuel to fly 800 miles safely will take off from airport A and land at airport B. Answer the following questions given the situation in each exercise. Round answers to two decimal places, if necessary.

(a) *Explain* why the region in which the airplane can fly is bounded by an ellipse (see figure).

(b) Let $(0, 0)$ represent the center of the ellipse. Find the coordinates of each airport.

(c) Suppose the plane flies from airport A straight past airport B to a vertex of the ellipse, and then straight back to airport B. How far does the plane fly? Use your answer to find the coordinates of the vertices.

(d) Write an equation of the ellipse. (*Hint:* $c^2 = a^2 - b^2$)

(e) The area of an ellipse is given by $A = \pi ab$. Find the area of the ellipse.

61. Airport A is 500 miles from airport B.

62. Airport A is 650 miles from airport B.

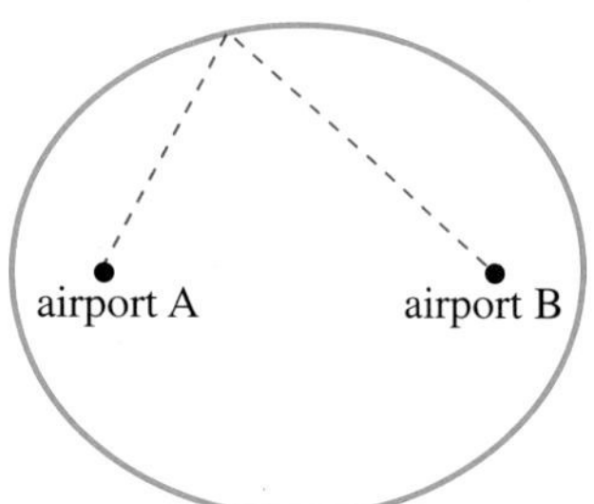

Figure for 61 and 62

Explaining Concepts

63. Describe the relationship between circles and ellipses. How are they similar? How do they differ?

64. The area of an ellipse is given by $A = \pi ab$. Explain how this area is related to the area of a circle.

65. Explain the significance of the foci in an ellipse.

66. Explain how to write an equation of an ellipse if you know the coordinates of the vertices and co-vertices.

67. From the standard form of the equation, explain how you can determine if the graph of an ellipse intersects the x- or y-axis.

Cumulative Review

In Exercises 68–75, evaluate the function as indicated and sketch the graph of the function.

68. $f(x) = 4^x$
(a) $x = 3$
(b) $x = -1$

69. $f(x) = 3^{-x}$
(a) $x = -2$
(b) $x = 2$

70. $g(x) = 5^{x-1}$
(a) $x = 4$
(b) $x = 0$

71. $g(x) = 6e^{0.5x}$
(a) $x = -1$
(b) $x = 2$

72. $h(x) = \log_{10} 2x$
(a) $x = 5$
(b) $x = 500$

73. $h(x) = \log_{16} 4x$
(a) $x = 4$
(b) $x = 64$

74. $f(x) = \ln(-x)$
(a) $x = -6$
(b) $x = 3$

75. $f(x) = \log_4(x - 3)$
(a) $x = 3$
(b) $x = 35$

Mid-Chapter Quiz

Take this quiz as you would take a quiz in class. After you are done, check your work against the answers in the back of the book.

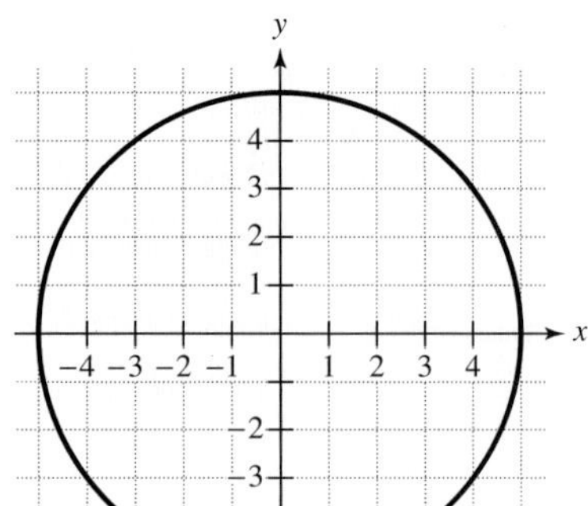

Figure for 1

1. Write the standard form of the equation of the circle shown in the figure.

2. Write the standard form of the equation of the parabola shown in the figure.

3. Write the standard form of the equation of the ellipse shown in the figure.

4. Write the standard form of the equation of the circle with center $(3, -5)$ and passing through the point $(0, -1)$.

5. Write the standard form of the equation of the parabola with vertex $(2, 3)$ and focus $(2, 1)$.

6. Write the standard form of the equation of the ellipse with vertices $(0, -10)$ and $(0, 10)$ and co-vertices $(-6, 0)$ and $(6, 0)$.

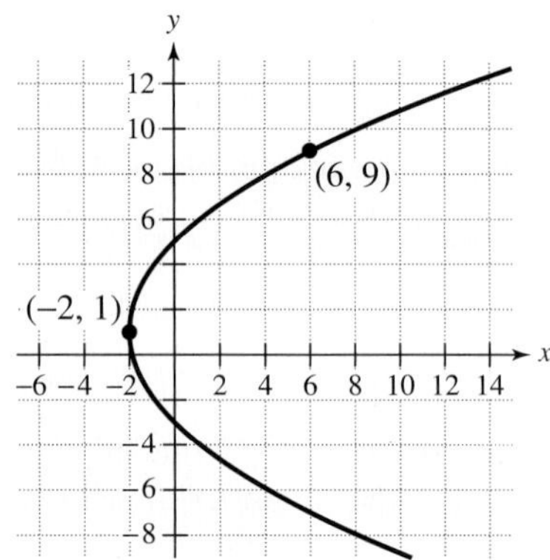

Figure for 2

In Exercises 7 and 8, write the equation of the circle in standard form. Then find the center and the radius of the circle.

7. $x^2 + y^2 + 6y - 7 = 0$

8. $x^2 + y^2 + 2x - 4y + 4 = 0$

In Exercises 9 and 10, write the equation of the parabola in standard form. Then find the vertex and the focus of the parabola.

9. $x = y^2 - 6y - 7$

10. $x^2 - 8x + y + 12 = 0$

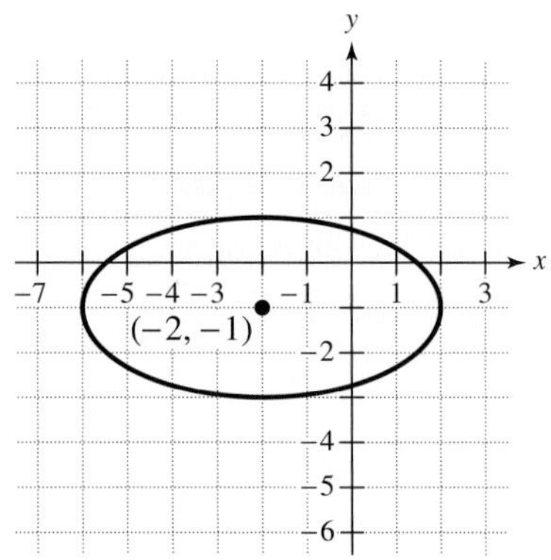

Figure for 3

In Exercises 11 and 12, write the equation of the ellipse in standard form. Then find the center and the vertices of the ellipse.

11. $4x^2 + y^2 - 16x - 20 = 0$

12. $4x^2 + 9y^2 - 48x + 36y + 144 = 0$

In Exercises 13–18, sketch the graph of the equation.

13. $(x + 5)^2 + (y - 1)^2 = 9$

14. $9x^2 + y^2 = 81$

15. $x = -y^2 - 4y$

16. $x^2 + (y + 4)^2 = 1$

17. $y = x^2 - 2x + 1$

18. $4(x + 3)^2 + (y - 2)^2 = 16$

12.3 Hyperbolas

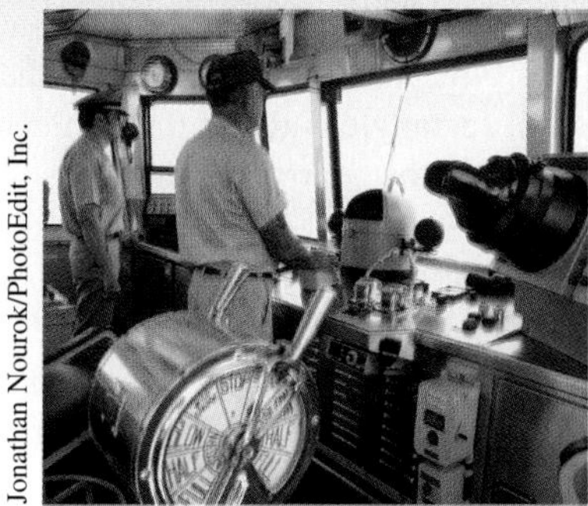

Why You Should Learn It

Equations of hyperbolas are often used in navigation. For instance, in Exercise 49 on page 792, a hyperbola is used to model long-distance radio navigation for a ship.

What You Should Learn

1 ▶ Graph and write equations of hyperbolas centered at the origin.

2 ▶ Graph and write equations of hyperbolas centered at (h, k).

Hyperbolas Centered at the Origin

1 ▶ Graph and write equations of hyperbolas centered at the origin.

The fourth basic type of conic is called a **hyperbola** and is defined as follows.

Definition of a Hyperbola

A **hyperbola** in the rectangular coordinate system consists of all points (x, y) such that the *difference* of the distances between (x, y) and two fixed points is a positive constant, as shown in Figure 12.28. The two fixed points are called the **foci** of the hyperbola. The line on which the foci lie is called the **transverse axis** of the hyperbola.

d_2 (x, y) d_1 Focus Focus

$d_2 - d_1$ is a positive constant.

Figure 12.28

Standard Equation of a Hyperbola (Center at Origin)

The **standard form of the equation of a hyperbola centered at the origin** is

$$\frac{x^2}{a^2} - \frac{y^2}{b^2} = 1 \qquad \text{or} \qquad \frac{y^2}{a^2} - \frac{x^2}{b^2} = 1$$

Transverse axis is horizontal. Transverse axis is vertical.

where a and b are positive real numbers. The **vertices** of the hyperbola lie on the transverse axis, a units from the center, as shown in Figure 12.29.

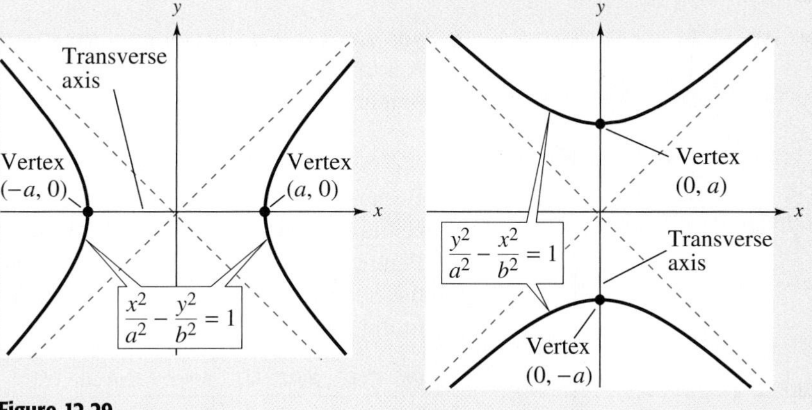

Figure 12.29

A hyperbola has two disconnected parts, each of which is called a **branch** of the hyperbola. The two branches approach a pair of intersecting lines called the **asymptotes** of the hyperbola. The two asymptotes intersect at the center of the hyperbola. To sketch a hyperbola, form a **central rectangle** that is centered at the origin and has side lengths of $2a$ and $2b$. Note in Figure 12.30 that the asymptotes pass through the corners of the central rectangle and that the vertices of the hyperbola lie at the centers of opposite sides of the central rectangle.

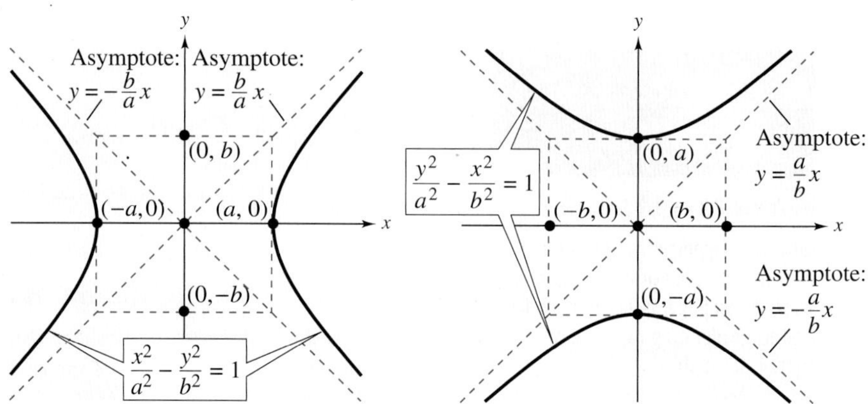

Figure 12.30

EXAMPLE 1 Sketching a Hyperbola

Identify the vertices of the hyperbola given by the equation, and sketch the hyperbola.

$$\frac{x^2}{36} - \frac{y^2}{16} = 1$$

Solution

From the standard form of the equation

$$\frac{x^2}{6^2} - \frac{y^2}{4^2} = 1$$

you can see that the center of the hyperbola is the origin and the transverse axis is horizontal. So, the vertices lie six units to the left and right of the center at the points

$(-6, 0)$ and $(6, 0)$.

Because $a = 6$ and $b = 4$, you can sketch the hyperbola by first drawing a central rectangle with a width of $2a = 12$ and a height of $2b = 8$, as shown in Figure 12.31. Next, draw the asymptotes of the hyperbola through the corners of the central rectangle, and plot the vertices. Finally, draw the hyperbola, as shown in Figure 12.32.

CHECKPOINT *Now try Exercise 11.*

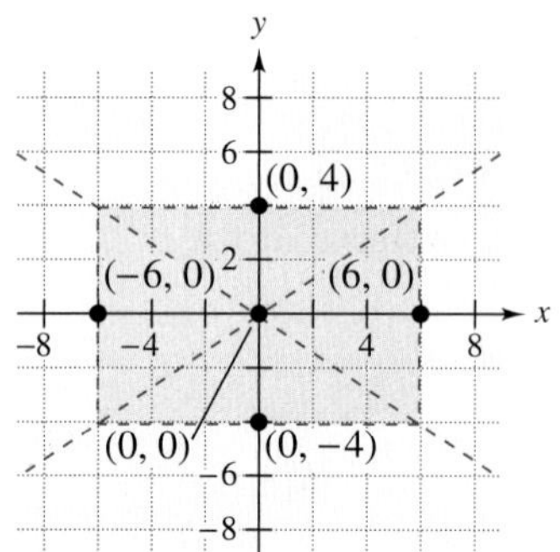

Figure 12.31

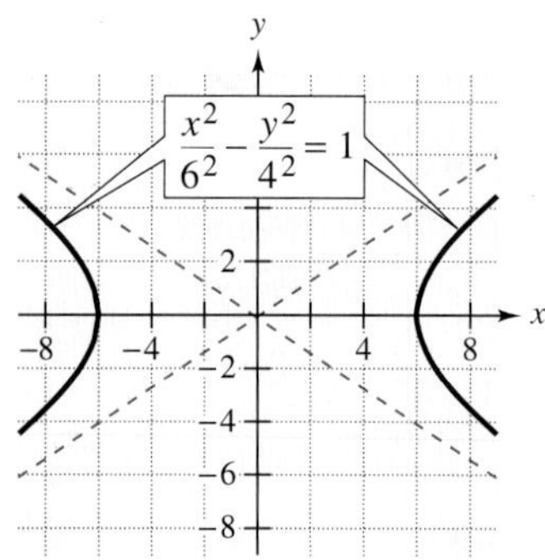

Figure 12.32

Writing the equation of a hyperbola is a little more difficult than writing equations of the other three types of conics. However, if you know the vertices and the asymptotes, you can find the values of a and b, which enable you to write the equation. Notice in Example 2 that the key to this procedure is knowing that the central rectangle has a width of $2b$ and a height of $2a$.

EXAMPLE 2 Writing the Equation of a Hyperbola

Write the standard form of the equation of the hyperbola with a vertical transverse axis and vertices $(0, 3)$ and $(0, -3)$. The equations of the asymptotes of the hyperbola are $y = \frac{3}{5}x$ and $y = -\frac{3}{5}x$.

Solution

To begin, sketch the lines that represent the asymptotes, as shown in Figure 12.33. Note that these two lines intersect at the origin, which implies that the center of the hyperbola is $(0, 0)$. Next, plot the two vertices at the points $(0, 3)$ and $(0, -3)$. You can use the vertices and asymptotes to sketch the central rectangle of the hyperbola, as shown in Figure 12.33. Note that the corners of the central rectangle occur at the points

$$(-5, 3),\ (5, 3),\ (-5, -3), \text{ and } (5, -3).$$

Because the width of the central rectangle is $2b = 10$, it follows that $b = 5$. Similarly, because the height of the central rectangle is $2a = 6$, it follows that $a = 3$. Now that you know the values of a and b, you can use the standard form of the equation of a hyperbola to write an equation.

$$\frac{y^2}{a^2} - \frac{x^2}{b^2} = 1 \qquad \text{Transverse axis is vertical.}$$

$$\frac{y^2}{3^2} - \frac{x^2}{5^2} = 1 \qquad \text{Substitute 3 for } a \text{ and 5 for } b.$$

$$\frac{y^2}{9} - \frac{x^2}{25} = 1 \qquad \text{Simplify.}$$

The graph is shown in Figure 12.34.

Study Tip

For a hyperbola, note that a and b are not determined in the same way as for an ellipse, where a is always greater than b. In the standard form of the equation of a hyperbola, a^2 is always the denominator of the positive term.

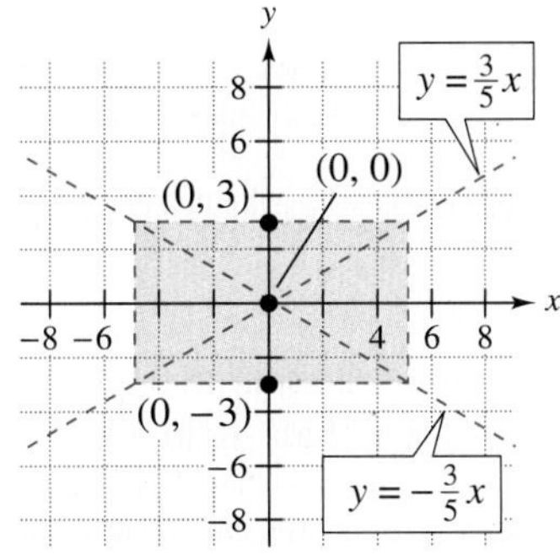

Figure 12.33

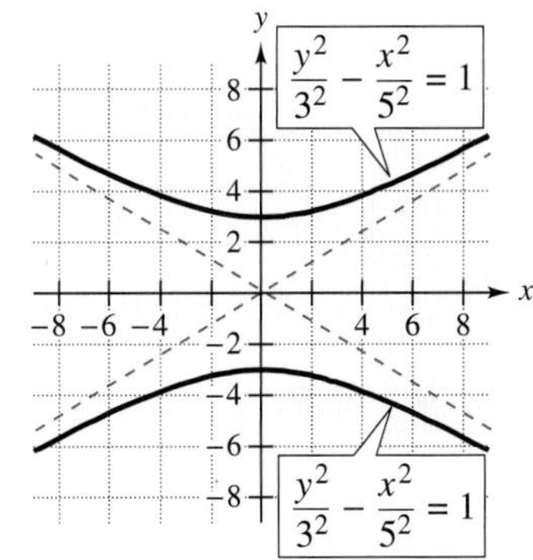

Figure 12.34

✓ **CHECKPOINT** *Now try Exercise 19.*

2 ▶ Graph and write equations of hyperbolas centered at (h, k).

Hyperbolas Centered at (h, k)

Standard Equation of a Hyperbola [Center at (h, k)]

The **standard form of the equation of a hyperbola centered at (h, k)** is

$$\frac{(x-h)^2}{a^2} - \frac{(y-k)^2}{b^2} = 1 \qquad \text{Transverse axis is horizontal.}$$

or

$$\frac{(y-k)^2}{a^2} - \frac{(x-h)^2}{b^2} = 1 \qquad \text{Transverse axis is vertical.}$$

where a and b are positive real numbers. The vertices lie on the transverse axis, a units from the center, as shown in Figure 12.35.

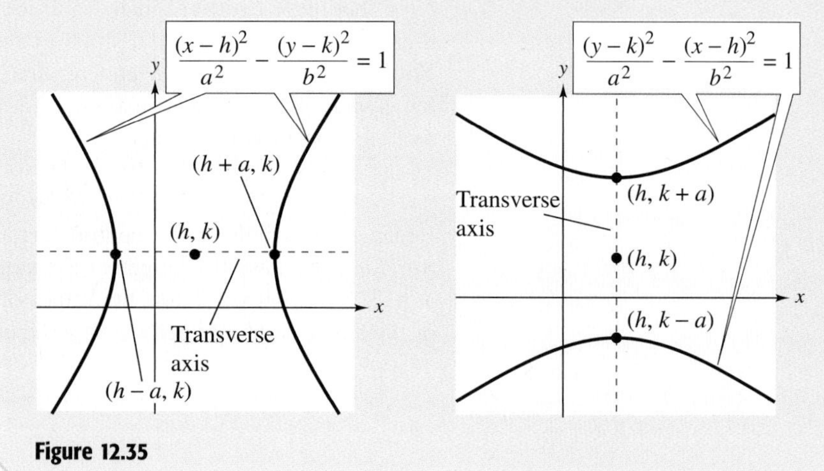

Figure 12.35

When $h = 0$ and $k = 0$, the hyperbola is centered at the origin. Otherwise, you can shift the center of the hyperbola h units horizontally and k units vertically from the origin.

EXAMPLE 3 Sketching a Hyperbola

Sketch the hyperbola given by $\frac{(y-1)^2}{9} - \frac{(x+2)^2}{4} = 1$.

Solution

From the form of the equation, you can see that the transverse axis is vertical. The center of the hyperbola is $(h, k) = (-2, 1)$. Because $a = 3$ and $b = 2$, you can begin by sketching a central rectangle that is six units high and four units wide, centered at $(-2, 1)$. Then, sketch the asymptotes by drawing lines through the corners of the central rectangle. Sketch the hyperbola, as shown in Figure 12.36. From the graph, you can see that the center of the hyperbola is shifted two units to the left and one unit upward from the origin.

✓ **CHECKPOINT** *Now try Exercise 33.*

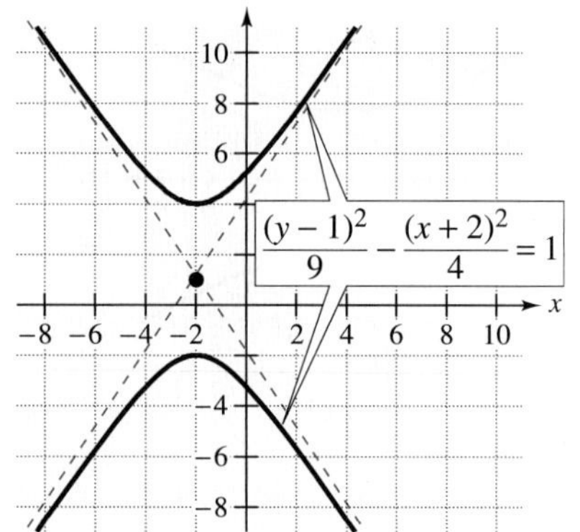

Figure 12.36

EXAMPLE 4 Sketching a Hyperbola

Sketch the hyperbola given by $x^2 - 4y^2 + 8x + 16y - 4 = 0$.

Solution

Complete the square to write the equation in standard form.

$$x^2 - 4y^2 + 8x + 16y - 4 = 0 \qquad \text{Write original equation.}$$

$$(x^2 + 8x + \boxed{\quad}) - (4y^2 - 16y + \boxed{\quad}) = 4 \qquad \text{Group terms.}$$

$$(x^2 + 8x + \boxed{\quad}) - 4(y^2 - 4y + \boxed{\quad}) = 4 \qquad \text{Factor 4 out of } y\text{-terms.}$$

$$(x^2 + 8x + 16) - 4(y^2 - 4y + 4) = 4 + 16 - 4(4) \qquad \text{Complete the squares.}$$

$$(x + 4)^2 - 4(y - 2)^2 = 4 \qquad \text{Simplify.}$$

$$\frac{(x + 4)^2}{4} - \frac{(y - 2)^2}{1} = 1 \qquad \text{Divide each side by 4.}$$

$$\frac{(x + 4)^2}{2^2} - \frac{(y - 2)^2}{1^2} = 1 \qquad \text{Standard form}$$

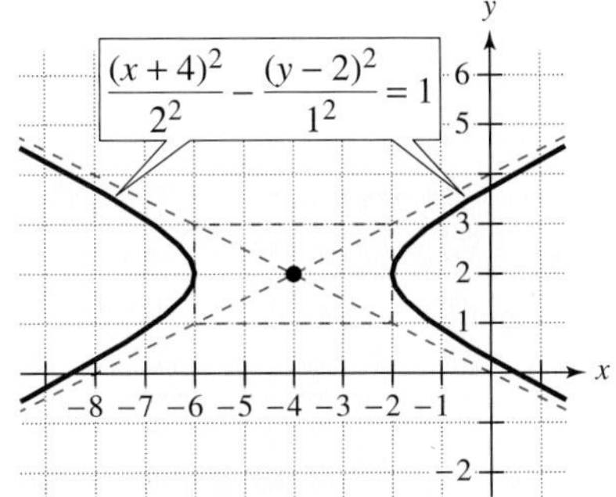

Figure 12.37

From this standard form, you can see that the transverse axis is horizontal and the center of the hyperbola is $(h, k) = (-4, 2)$. Because $a = 2$ and $b = 1$, you can begin by sketching a central rectangle that is four units wide and two units high, centered at $(-4, 2)$. Then, sketch the asymptotes by drawing lines through the corners of the central rectangle. Sketch the hyperbola, as shown in Figure 12.37. From the graph, you can see that the center of the hyperbola is shifted four units to the left and two units upward from the origin.

 ***CHECKPOINT*** *Now try Exercise 35.*

Technology: Tip

You can use a graphing calculator to graph a hyperbola. For instance, to graph the hyperbola $4y^2 - 9x^2 = 36$, first solve for y to obtain

$$y_1 = 3\sqrt{\frac{x^2}{4} + 1}$$

and

$$y_2 = -3\sqrt{\frac{x^2}{4} + 1}.$$

Use a viewing window in which $-6 \le x \le 6$ and $-8 \le y \le 8$. You should obtain the graph shown below.

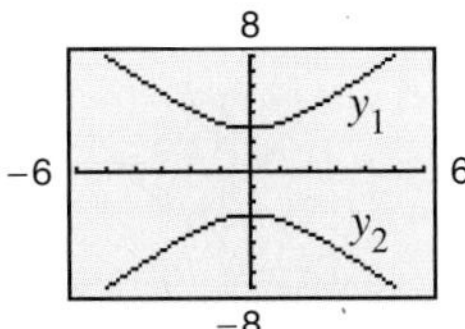

Concept Check

1. You are given the equation of a hyperbola in the standard form

$$\frac{x^2}{a^2} - \frac{y^2}{b^2} = 1.$$

Explain how you can sketch the central rectangle for the hyperbola. Explain how you can use the central rectangle to sketch the asymptotes of the hyperbola.

2. You are given the vertices and the equations of the asymptotes of a hyperbola. Explain how you can determine the values of a and b in the standard form of the equation of the hyperbola.

3. What are the dimensions of the central rectangle and the coordinates of the center of the hyperbola whose equation in standard form is

$$\frac{(y-k)^2}{a^2} - \frac{(x-h)^2}{b^2} = 1?$$

4. Given the equation of a hyperbola in the general polynomial form

$$ax^2 - by^2 + cx + dy + e = 0$$

what process can you use to find the center of the hyperbola?

12.3 EXERCISES

Go to pages 804–805 to record your assignments.

Developing Skills

In Exercises 1–6, match the equation with its graph. [The graphs are labeled (a), (b), (c), (d), (e), and (f).]

(a)

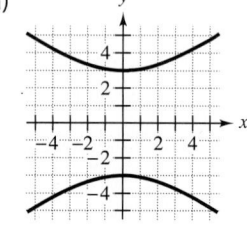

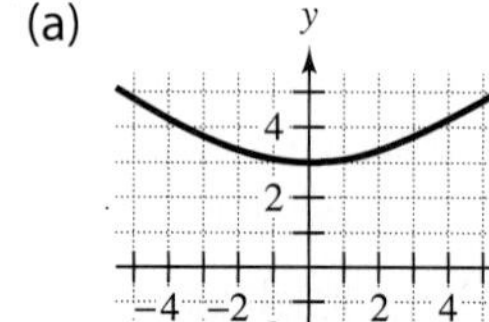

(b)

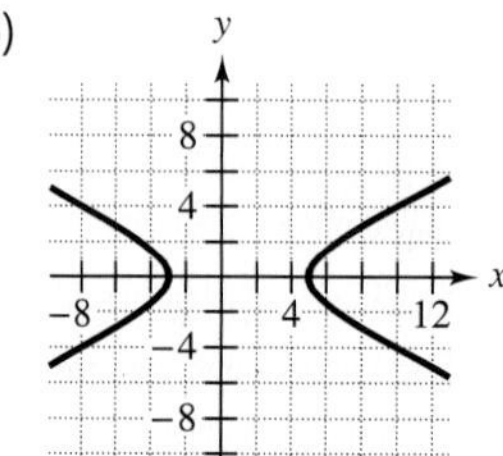

(c)

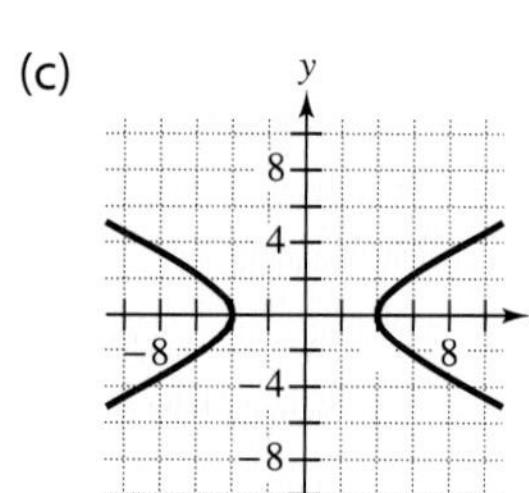

(d)

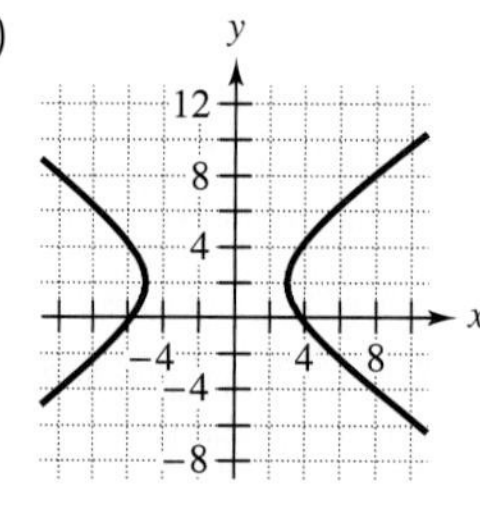

(e)

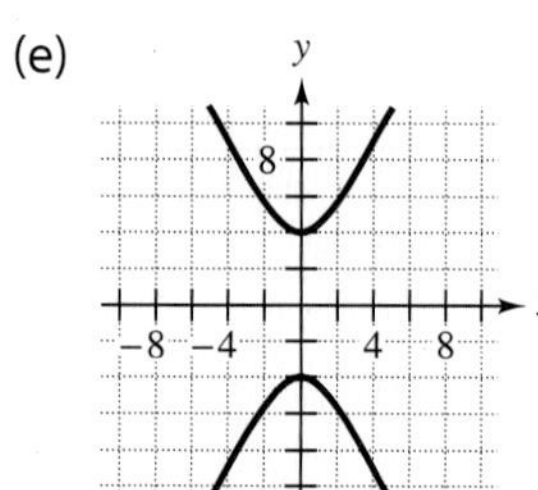

(f)

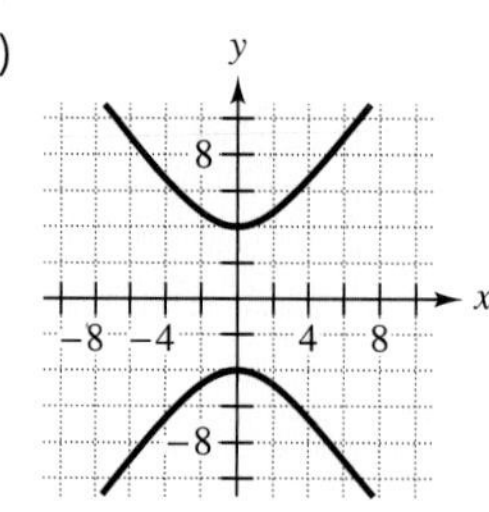

1. $\frac{x^2}{16} - \frac{y^2}{4} = 1$

2. $\frac{y^2}{16} - \frac{x^2}{4} = 1$

3. $\frac{y^2}{9} - \frac{x^2}{16} = 1$

4. $\frac{y^2}{16} - \frac{x^2}{9} = 1$

5. $\frac{(x-1)^2}{16} - \frac{y^2}{4} = 1$

6. $\frac{(x+1)^2}{16} - \frac{(y-2)^2}{9} = 1$

In Exercises 7–18, sketch the hyperbola. Identify the vertices and asymptotes. ***See Example 1.***

7. $x^2 - y^2 = 9$

8. $x^2 - y^2 = 1$

9. $y^2 - x^2 = 9$

10. $y^2 - x^2 = 1$

✓ **11.** $\frac{x^2}{9} - \frac{y^2}{25} = 1$

12. $\frac{x^2}{4} - \frac{y^2}{9} = 1$

13. $\frac{y^2}{9} - \frac{x^2}{25} = 1$

14. $\frac{y^2}{4} - \frac{x^2}{9} = 1$

15. $\frac{x^2}{1} - \frac{y^2}{9/4} = 1$

16. $\frac{y^2}{1/4} - \frac{x^2}{25/4} = 1$

17. $4y^2 - x^2 + 16 = 0$

18. $4y^2 - 9x^2 - 36 = 0$

In Exercises 19–26, write the standard form of the equation of the hyperbola centered at the origin. ***See Example 2.***

	Vertices	*Asymptotes*	
✓ **19.**	$(-4, 0), (4, 0)$	$y = 2x$	$y = -2x$
20.	$(-2, 0), (2, 0)$	$y = \frac{1}{3}x$	$y = -\frac{1}{3}x$
21.	$(0, -4), (0, 4)$	$y = \frac{1}{2}x$	$y = -\frac{1}{2}x$
22.	$(0, -2), (0, 2)$	$y = 3x$	$y = -3x$
23.	$(-9, 0), (9, 0)$	$y = \frac{2}{3}x$	$y = -\frac{2}{3}x$
24.	$(-1, 0), (1, 0)$	$y = \frac{1}{2}x$	$y = -\frac{1}{2}x$
25.	$(0, -1), (0, 1)$	$y = 2x$	$y = -2x$
26.	$(0, -5), (0, 5)$	$y = x$	$y = -x$

In Exercises 27–30, use a graphing calculator to graph the equation. (*Note:* Solve for *y*.)

27. $\frac{x^2}{16} - \frac{y^2}{4} = 1$

28. $\frac{y^2}{16} - \frac{x^2}{4} = 1$

29. $5x^2 - 2y^2 + 10 = 0$

30. $x^2 - 2y^2 - 4 = 0$

In Exercises 31–38, find the center and vertices of the hyperbola and sketch the hyperbola. ***See Examples 3 and 4.***

31. $(y + 4)^2 - (x - 3)^2 = 25$

32. $(y + 6)^2 - (x - 2)^2 = 1$

✓ **33.** $\frac{(x - 1)^2}{4} - \frac{(y + 2)^2}{1} = 1$

34. $\frac{(x - 2)^2}{4} - \frac{(y - 3)^2}{9} = 1$

✓ **35.** $9x^2 - y^2 - 36x - 6y + 18 = 0$

36. $x^2 - 9y^2 + 36y - 72 = 0$

37. $4x^2 - y^2 + 24x + 4y + 28 = 0$

38. $25x^2 - 4y^2 + 100x + 8y + 196 = 0$

In Exercises 39–42, write the standard form of the equation of the hyperbola.

39.

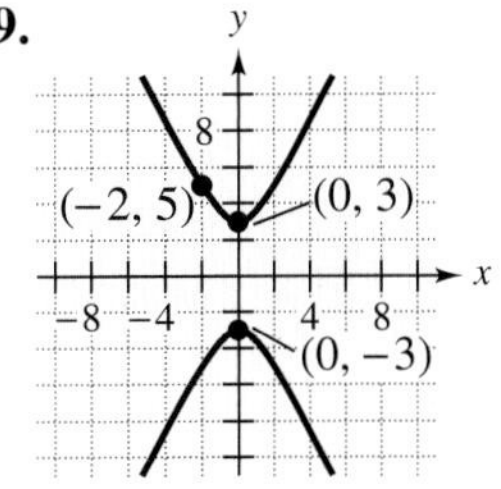

40.

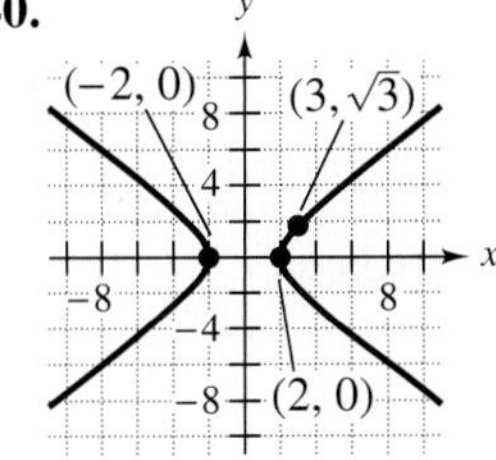

41.

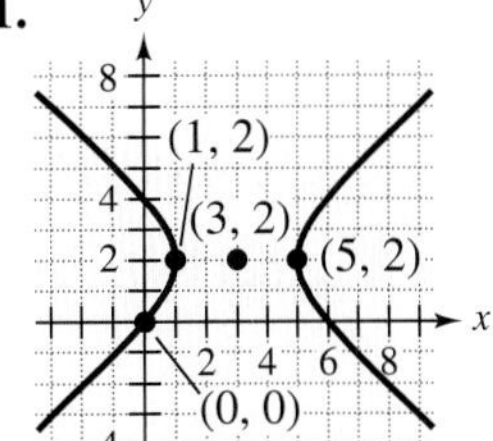

42.

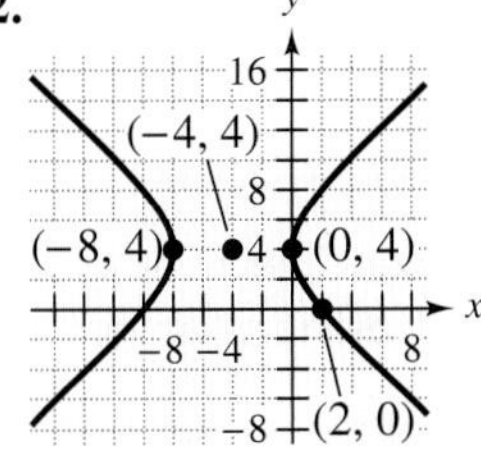

In Exercises 43–48, determine whether the graph represented by the equation is a *circle*, a *parabola*, an *ellipse*, or a *hyperbola*.

43. $\dfrac{(x-3)^2}{4^2} + \dfrac{(y-4)^2}{6^2} = 1$

44. $\dfrac{(x+2)^2}{25} + \dfrac{(y-2)^2}{25} = 1$

45. $x^2 - y^2 = 1$

46. $2x + y^2 = 0$

47. $y^2 - x^2 - 2y + 8x - 19 = 0$

48. $9x^2 + y^2 - 18x - 8y + 16 = 0$

Solving Problems

49. *Navigation* Long-distance radio navigation for aircraft and ships uses synchronized pulses transmitted by widely separated transmitting stations. The locations of two transmitting stations that are 300 miles apart are represented by the points $(-150, 0)$ and $(150, 0)$ (see figure). A ship's location is given by $(x, 75)$. The difference in the arrival times of pulses transmitted simultaneously to the ship from the two stations is constant at any point on the hyperbola given by

$$\frac{x^2}{93^2} - \frac{y^2}{13{,}851} = 1$$

which passes through the ship's location and has the two stations as foci. Use the equation to find the x-coordinate of the ship's location.

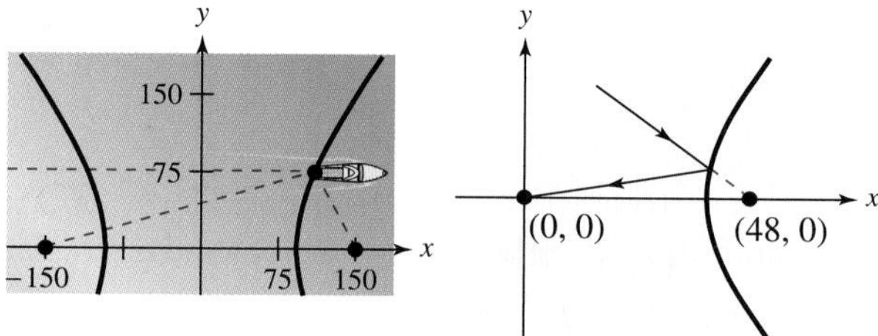

Figure for 49

Figure for 50

50. *Optics* Hyperbolic mirrors are used in some telescopes. The figure shows a cross section of a hyperbolic mirror as the right branch of a hyperbola. A property of the mirror is that a light ray directed at the focus $(48, 0)$ is reflected to the other focus $(0, 0)$. Use the equation of the hyperbola

$$89x^2 - 55y^2 - 4272x + 31{,}684 = 0$$

to find the coordinates of the mirror's vertex.

Explaining Concepts

51. *Think About It* Describe the part of the hyperbola

$$\frac{(x-3)^2}{4} - \frac{(y-1)^2}{9} = 1$$

given by each equation.

(a) $x = 3 - \frac{2}{3}\sqrt{9 + (y-1)^2}$

(b) $y = 1 + \frac{3}{2}\sqrt{(x-3)^2 - 4}$

52. Consider the definition of a hyperbola. How many hyperbolas have a given pair of points as foci? Explain your reasoning.

53. How many hyperbolas pass through a given point and have a given pair of points as foci? Explain your reasoning.

54. Cut cone-shaped pieces of styrofoam to demonstrate how to obtain each type of conic section: circle, parabola, ellipse, and hyperbola. Discuss how you could write directions for someone else to form each conic section. Compile a list of real-life situations and/or everyday objects in which conic sections may be seen.

Cumulative Review

In Exercises 55 and 56, determine whether the system is consistent or inconsistent.

55. $\begin{cases} -x + 3y = 8 \\ 4x - 12y = -32 \end{cases}$

56. $\begin{cases} x - 3y = 5 \\ 2x - 6y = -5 \end{cases}$

In Exercises 57 and 58, solve the system of linear equations by the method of elimination.

57. $\begin{cases} x + y = 3 \\ x - y = 2 \end{cases}$

58. $\begin{cases} 4x + 3y = 3 \\ x - 2y = 9 \end{cases}$

12.4 Solving Nonlinear Systems of Equations

StockShot/Alamy

Why You Should Learn It

Nonlinear systems of equations can be used in real-life applications. For instance, in Example 7 on page 799, nonlinear equations are used to assist rescuers in their search for victims buried by an avalanche.

What You Should Learn

1 ▶ Solve nonlinear systems of equations graphically.

2 ▶ Solve nonlinear systems of equations by substitution.

3 ▶ Solve nonlinear systems of equations by elimination.

4 ▶ Use nonlinear systems of equations to model and solve real-life problems.

1 ▶ Solve nonlinear systems of equations graphically.

Solving Nonlinear Systems of Equations by Graphing

In Chapter 8, you studied several methods for solving systems of linear equations. For instance, the following linear system has one solution, $(2, -1)$, which means that $(2, -1)$ is a point of intersection of the two lines represented by the system.

$$\begin{cases} 2x - 3y = 7 \\ x + 4y = -2 \end{cases}$$

In Chapter 8, you also learned that a linear system can have no solution, exactly one solution, or infinitely many solutions. A **nonlinear system of equations** is a system that contains at least one nonlinear equation. Nonlinear systems of equations can have no solution, one solution, or two or more solutions. For instance, the hyperbola and line in Figure 12.38(a) have no point of intersection, the circle and line in Figure 12.38(b) have one point of intersection, and the parabola and line in Figure 12.38(c) have two points of intersection.

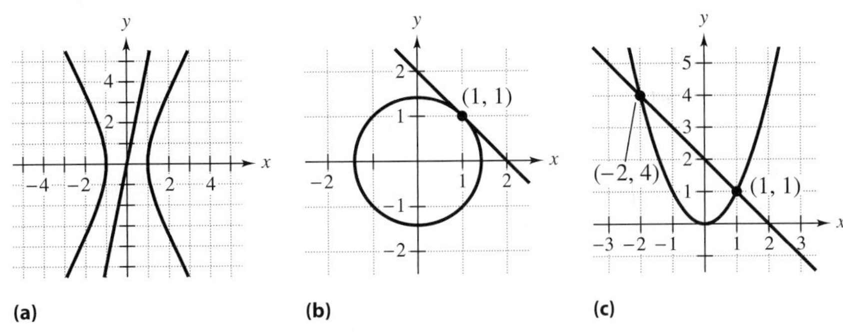

Figure 12.38

You can solve a nonlinear system of equations graphically, as follows.

Solving a Nonlinear System Graphically

1. Sketch the graph of each equation in the system.
2. Locate the point(s) of intersection of the graphs (if any) and graphically approximate the coordinates of the point(s).
3. Check the coordinates by substituting them into each equation in the original system. If the coordinates do not check, you may have to use an algebraic approach, as discussed later in this section.

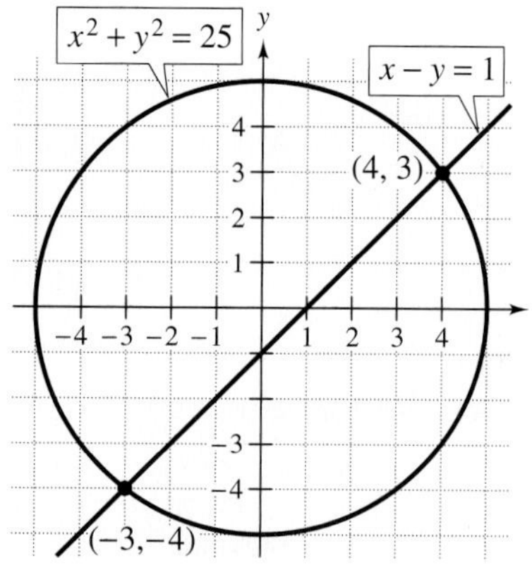

Figure 12.39

Technology: Tip

Try using a graphing calculator to solve the system in Example 1. When you do this, remember that the circle needs to be entered as two separate equations.

$y_1 = \sqrt{25 - x^2}$ Top half of circle

$y_2 = -\sqrt{25 - x^2}$ Bottom half of circle

$y_3 = x - 1$ Line

EXAMPLE 1 Solving a Nonlinear System Graphically

Find all solutions of the nonlinear system of equations.

$$\begin{cases} x^2 + y^2 = 25 & \text{Equation 1} \\ x - y = 1 & \text{Equation 2} \end{cases}$$

Solution

Begin by sketching the graph of each equation. The first equation graphs as a circle centered at the origin and having a radius of 5. The second equation graphs as a line with a slope of 1 and a y-intercept of $(0, -1)$. The system appears to have two solutions: $(-3, -4)$ and $(4, 3)$, as shown in Figure 12.39.

Check

To check $(-3, -4)$, substitute -3 for x and -4 for y in each equation.

$(-3)^2 + (-4)^2 \stackrel{?}{=} 25$ Substitute -3 for x and -4 for y in Equation 1.

$9 + 16 = 25$ Solution checks in Equation 1. ✓

$(-3) - (-4) \stackrel{?}{=} 1$ Substitute -3 for x and -4 for y in Equation 2.

$-3 + 4 = 1$ Solution checks in Equation 2. ✓

To check $(4, 3)$, substitute 4 for x and 3 for y in each equation.

$4^2 + 3^2 \stackrel{?}{=} 25$ Substitute 4 for x and 3 for y in Equation 1.

$16 + 9 = 25$ Solution checks in Equation 1. ✓

$4 - 3 \stackrel{?}{=} 1$ Substitute 4 for x and 3 for y in Equation 2.

$1 = 1$ Solution checks in Equation 2. ✓

✓ **CHECKPOINT** *Now try Exercise 1.*

EXAMPLE 2 Solving a Nonlinear System Graphically

Find all solutions of the nonlinear system of equations.

$$\begin{cases} x = (y - 3)^2 & \text{Equation 1} \\ x + y = 5 & \text{Equation 2} \end{cases}$$

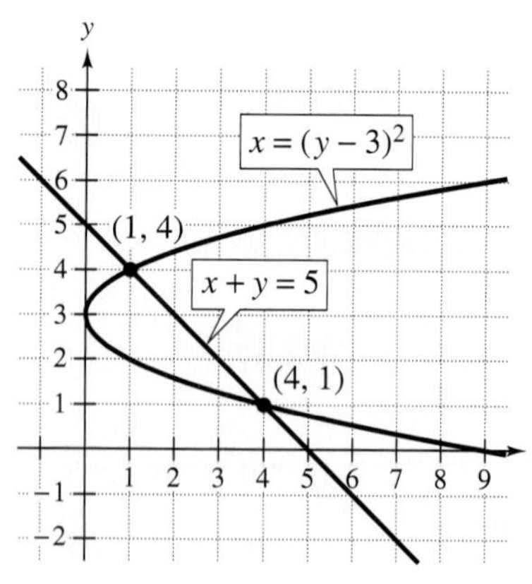

Figure 12.40

Solution

Begin by sketching the graph of each equation. Solve the first equation for y.

$x = (y - 3)^2$ Write original equation.

$\pm\sqrt{x} = y - 3$ Take the square root of each side.

$3 \pm \sqrt{x} = y$ Add 3 to each side.

The graph of $y = 3 \pm \sqrt{x}$ is a parabola with its vertex at $(0, 3)$. The second equation graphs as a line with a slope of -1 and a y-intercept of $(0, 5)$. The system appears to have two solutions: $(4, 1)$ and $(1, 4)$, as shown in Figure 12.40. Check these solutions in the original system.

✓ **CHECKPOINT** *Now try Exercise 5.*

2 ▶ Solve nonlinear systems of equations by substitution.

Solving Nonlinear Systems of Equations by Substitution

The graphical approach to solving any type of system (linear or nonlinear) in two variables is very useful. For systems with solutions having "messy" coordinates, however, a graphical approach is usually not accurate enough to produce exact solutions. In such cases, you should use an algebraic approach. (With an algebraic approach, you should still sketch the graph of each equation in the system.)

As with systems of *linear* equations, there are two basic algebraic approaches: substitution and elimination. Substitution usually works well for systems in which one of the equations is linear, as shown in Example 3.

EXAMPLE 3 Using Substitution to Solve a Nonlinear System

Solve the nonlinear system of equations.

$$\begin{cases} 4x^2 + y^2 = 4 & \text{Equation 1} \\ -2x + y = 2 & \text{Equation 2} \end{cases}$$

Solution

Begin by solving for y in Equation 2 to obtain $y = 2x + 2$. Next, substitute this expression for y into Equation 1.

$$4x^2 + y^2 = 4 \qquad \text{Write Equation 1.}$$

$$4x^2 + (2x + 2)^2 = 4 \qquad \text{Substitute } 2x + 2 \text{ for } y.$$

$$4x^2 + 4x^2 + 8x + 4 = 4 \qquad \text{Expand.}$$

$$8x^2 + 8x = 0 \qquad \text{Simplify.}$$

$$8x(x + 1) = 0 \qquad \text{Factor.}$$

$$8x = 0 \Rightarrow x = 0 \qquad \text{Set 1st factor equal to 0.}$$

$$x + 1 = 0 \Rightarrow x = -1 \qquad \text{Set 2nd factor equal to 0.}$$

Finally, back-substitute these values of x into the revised Equation 2 to solve for y.

For $x = 0$: $\quad y = 2(0) + 2 = 2$

For $x = -1$: $\quad y = 2(-1) + 2 = 0$

So, the system of equations has two solutions: $(0, 2)$ and $(-1, 0)$. Figure 12.41 shows the graph of the system. You can check the solutions as follows.

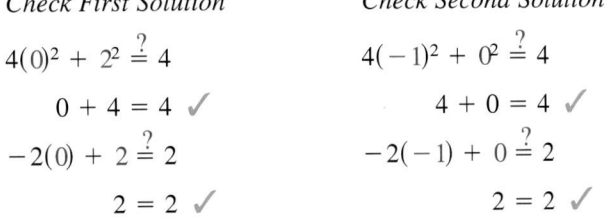

Check First Solution

$$4(0)^2 + 2^2 \stackrel{?}{=} 4$$

$$0 + 4 = 4 \checkmark$$

$$-2(0) + 2 \stackrel{?}{=} 2$$

$$2 = 2 \checkmark$$

Check Second Solution

$$4(-1)^2 + 0^2 \stackrel{?}{=} 4$$

$$4 + 0 = 4 \checkmark$$

$$-2(-1) + 0 \stackrel{?}{=} 2$$

$$2 = 2 \checkmark$$

Figure 12.41

✓ **CHECKPOINT** *Now try Exercise 31.*

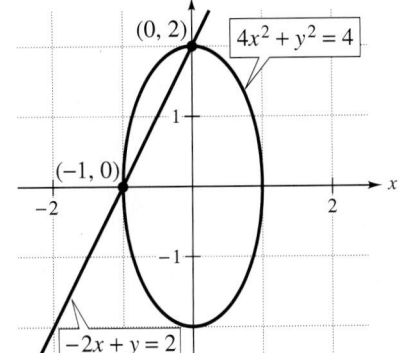

The steps for using the method of substitution to solve a system of two equations involving two variables are summarized as follows.

Method of Substitution

To solve a system of two equations in two variables, use the steps below.

1. Solve one of the equations for one variable in terms of the other variable.
2. Substitute the expression found in Step 1 into the other equation to obtain an equation in one variable.
3. Solve the equation obtained in Step 2.
4. Back-substitute the solution from Step 3 into the expression obtained in Step 1 to find the value of the other variable.
5. Check the solution to see that it satisfies *both* of the original equations.

Example 4 shows how the method of substitution and graphing can be used to determine that a nonlinear system of equations has no solution.

EXAMPLE 4 Solving a Nonlinear System: No-Solution Case

Solve the nonlinear system of equations.

$$\begin{cases} x^2 - y = 0 & \text{Equation 1} \\ x - y = 1 & \text{Equation 2} \end{cases}$$

Solution

Begin by solving for y in Equation 2 to obtain $y = x - 1$. Next, substitute this expression for y into Equation 1.

$$x^2 - y = 0 \qquad \text{Write Equation 1.}$$

$$x^2 - (x - 1) = 0 \qquad \text{Substitute } x - 1 \text{ for } y.$$

$$x^2 - x + 1 = 0 \qquad \text{Distributive Property}$$

Use the Quadratic Formula, because this equation cannot be factored.

$$x = \frac{-(-1) \pm \sqrt{(-1)^2 - 4(1)(1)}}{2(1)} \qquad \text{Use Quadratic Formula.}$$

$$= \frac{1 \pm \sqrt{1 - 4}}{2} = \frac{1 \pm \sqrt{-3}}{2} \qquad \text{Simplify.}$$

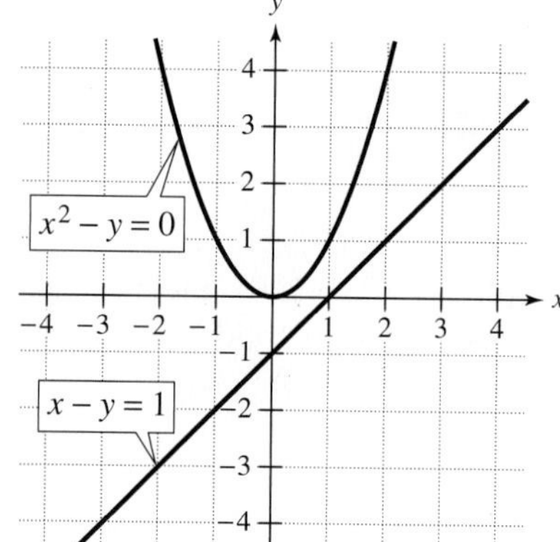

Figure 12.42

Now, because the Quadratic Formula yields a negative number inside the radical, you can conclude that the equation $x^2 - x + 1 = 0$ has no (real) solution. So, the system has no (real) solution. Figure 12.42 shows the graph of this system. From the graph, you can see that the parabola and the line have no point of intersection.

CHECKPOINT *Now try Exercise 51.*

3 ▶ Solve nonlinear systems of equations by elimination.

Solving Nonlinear Systems of Equations by Elimination

In Section 8.2, you learned how to use the method of elimination to solve a system of linear equations. This method can also be used with special types of nonlinear systems, as demonstrated in Example 5.

EXAMPLE 5 Using Elimination to Solve a Nonlinear System

Solve the nonlinear system of equations.

$$\begin{cases} 4x^2 + y^2 = 64 & \text{Equation 1} \\ x^2 + y^2 = 52 & \text{Equation 2} \end{cases}$$

Solution

Because both equations have y^2 as a term (and no other terms containing y), you can eliminate y by subtracting Equation 2 from Equation 1.

$$\begin{array}{rcl} 4x^2 + y^2 &=& 64 \\ -x^2 - y^2 &=& -52 \\ \hline 3x^2 \qquad &=& 12 \end{array} \qquad \text{Subtract Equation 2 from Equation 1.}$$

After eliminating y, solve the remaining equation for x.

$3x^2 = 12$ — Write resulting equation.

$x^2 = 4$ — Divide each side by 3.

$x = \pm 2$ — Take square root of each side.

By substituting $x = 2$ into Equation 2, you obtain

$x^2 + y^2 = 52$ — Write Equation 2.

$(2)^2 + y^2 = 52$ — Substitute 2 for x.

$y^2 = 48$ — Subtract 4 from each side.

$y = \pm 4\sqrt{3}.$ — Take square root of each side and simplify.

By substituting $x = -2$, you obtain the same values of y, as follows.

$x^2 + y^2 = 52$ — Write Equation 2.

$(-2)^2 + y^2 = 52$ — Substitute -2 for x.

$y^2 = 48$ — Subtract 4 from each side.

$y = \pm 4\sqrt{3}$ — Take square root of each side and simplify.

This implies that the system has four solutions:

$$\left(2, 4\sqrt{3}\right), \quad \left(2, -4\sqrt{3}\right), \quad \left(-2, 4\sqrt{3}\right), \quad \left(-2, -4\sqrt{3}\right).$$

Check these solutions in the original system. Figure 12.43 shows the graph of the system. Notice that the graph of Equation 1 is an ellipse and the graph of Equation 2 is a circle.

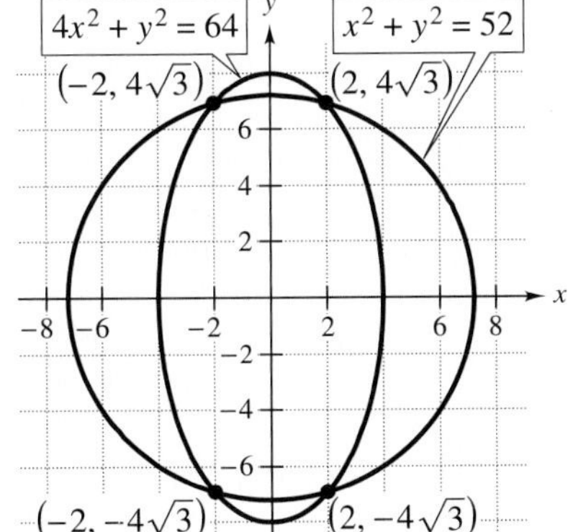

Figure 12.43

✓ **CHECKPOINT** *Now try Exercise 59.*

Smart Study Strategy

Go to page 760 for ways to *Analyze Your Errors.*

EXAMPLE 6 Using Elimination to Solve a Nonlinear System

Solve the nonlinear system of equations.

$$\begin{cases} x^2 - 2y = 4 & \text{Equation 1} \\ x^2 - y^2 = 1 & \text{Equation 2} \end{cases}$$

Solution

Because both equations have x^2 as a term (and no other terms containing x), you can eliminate x by subtracting Equation 2 from Equation 1.

$$\begin{aligned} x^2 - 2y &= 4 \\ -x^2 + y^2 &= -1 \quad \text{Subtract Equation 2 from Equation 1.} \\ \hline y^2 - 2y &= 3 \end{aligned}$$

After eliminating x, solve the remaining equation for y.

$y^2 - 2y = 3$ Write resulting equation.

$y^2 - 2y - 3 = 0$ Write in general form.

$(y + 1)(y - 3) = 0$ Factor.

$y + 1 = 0 \Rightarrow y = -1$ Set 1st factor equal to 0.

$y - 3 = 0 \Rightarrow y = 3$ Set 2nd factor equal to 0.

When $y = -1$, you obtain

$x^2 - y^2 = 1$ Write Equation 2.

$x^2 - (-1)^2 = 1$ Substitute -1 for y.

$x^2 - 1 = 1$ Simplify.

$x^2 = 2$ Add 1 to each side.

$x = \pm\sqrt{2}.$ Take square root of each side.

When $y = 3$, you obtain

$x^2 - y^2 = 1$ Write Equation 2.

$x^2 - (3)^2 = 1$ Substitute 3 for y.

$x^2 - 9 = 1$ Simplify.

$x^2 = 10$ Add 9 to each side.

$x = \pm\sqrt{10}.$ Take square root of each side.

This implies that the system has four solutions:

$$(\sqrt{2}, -1), \quad (-\sqrt{2}, -1), \quad (\sqrt{10}, 3), \quad (-\sqrt{10}, 3).$$

Check these solutions in the original system. Figure 12.44 shows the graph of the system. Notice that the graph of Equation 1 is a parabola and the graph of Equation 2 is a hyperbola.

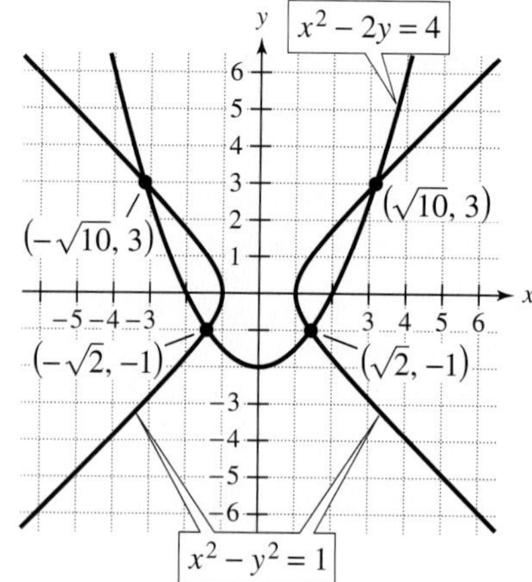

Figure 12.44

CHECKPOINT *Now try Exercise 63.*

In Example 6, the method of elimination yields the four exact solutions $(\sqrt{2}, -1)$, $(-\sqrt{2}, -1)$, $(\sqrt{10}, 3)$, and $(-\sqrt{10}, 3)$. You can use a calculator to approximate these solutions as $(1.41, -1)$, $(-1.41, -1)$, $(3.16, 3)$, and $(-3.16, 3)$. If you use the decimal approximations to check your solutions in the original system, be aware that they may not check.

4 ▶ Use nonlinear systems of equations to model and solve real-life problems.

Application

There are many examples of the use of nonlinear systems of equations in business and science. For instance, in Example 7 a nonlinear system of equations is used to help rescue avalanche victims.

EXAMPLE 7 Avalanche Rescue System

RECCO® is an avalanche rescue system utilized by rescue organizations worldwide. RECCO technology enables quick directional pinpointing of a victim's exact location using harmonic radar. The two-part system consists of a detector used by rescuers, and reflectors that are integrated into apparel, helmets, protection gear, or boots. The range of the detector through snow is 30 meters. Two rescuers are 30 meters apart on the surface. What is the maximum depth of a reflector that is in range of both rescuers?

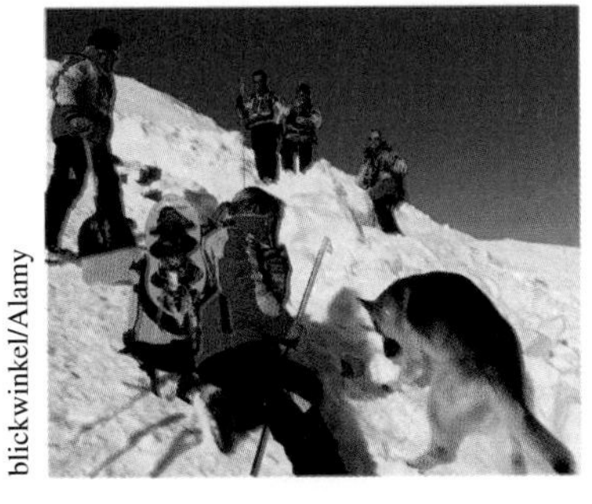

RECCO technology is often used in conjunction with other rescue methods such as avalanche dogs, transceivers, and probe lines.

Solution

Let the first rescuer be located at the origin and let the second rescuer be located 30 meters (units) to the right. The range of each detector is circular and can be modeled by the following equations.

$$\begin{cases} x^2 + y^2 = 30^2 & \text{Range of first rescuer} \\ (x - 30)^2 + y^2 = 30^2 & \text{Range of second rescuer} \end{cases}$$

Using methods demonstrated earlier in this section, you'll find that these two equations intersect when $x = 15$ and $y \approx \pm 25.98$. You are concerned only about the lower portions of the circles. So, the maximum depth in range of both rescuers is point R, as shown in Figure 12.45, which is about 26 meters beneath the surface.

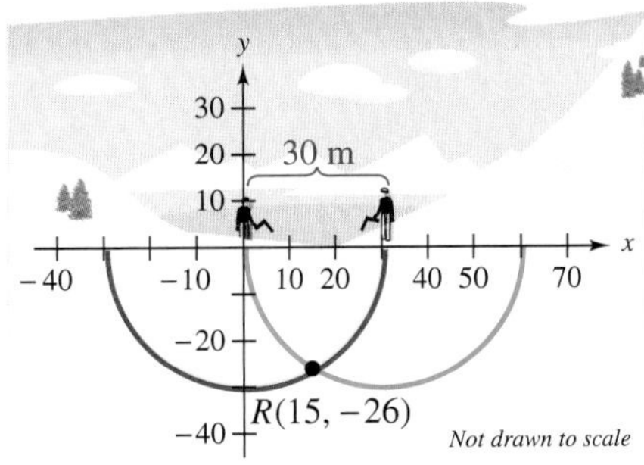

Figure 12.45

✓ **CHECKPOINT** *Now try Exercise 87.*

Concept Check

1. How is a system of nonlinear equations different from a system of linear equations?

2. Identify the different methods you can use to solve a system of nonlinear equations.

3. If one of the equations in a system is linear, which algebraic method usually works best for solving the system?

4. If both of the equations in a system are conics, which algebraic method usually works best for solving the system?

12.4 EXERCISES

Go to pages 804–805 to record your assignments.

Developing Skills

In Exercises 1–16, graph the equations to determine whether the system has any solutions. Find any solutions that exist. ***See Examples 1 and 2.***

✓ **1.** $\begin{cases} y = 1 \\ x^2 + y = 0 \end{cases}$

2. $\begin{cases} x^2 + y^2 = 1 \\ x + y = 4 \end{cases}$

3. $\begin{cases} x = 0 \\ x^2 + y^2 = 9 \end{cases}$

4. $\begin{cases} y = 4 \\ x^2 - y = 0 \end{cases}$

✓ **5.** $\begin{cases} x + y = 2 \\ x^2 - y = 0 \end{cases}$

6. $\begin{cases} 2x + y = 10 \\ x^2 + y^2 = 25 \end{cases}$

7. $\begin{cases} x^2 + y = 9 \\ x - y = -3 \end{cases}$

8. $\begin{cases} x - y^2 = 0 \\ x - y = 2 \end{cases}$

9. $\begin{cases} y = \sqrt{x - 2} \\ x - 2y = 1 \end{cases}$

10. $\begin{cases} x - 2y = 4 \\ x^2 - y = 0 \end{cases}$

11. $\begin{cases} x^2 + y^2 = 100 \\ x + y = 2 \end{cases}$

12. $\begin{cases} x^2 + y^2 = 169 \\ x + y = 7 \end{cases}$

13. $\begin{cases} x^2 + y^2 = 25 \\ 2x - y = -5 \end{cases}$

14. $\begin{cases} x^2 - y^2 = 16 \\ 3x - y = 12 \end{cases}$

15. $\begin{cases} 9x^2 - 4y^2 = 36 \\ 5x - 2y = 0 \end{cases}$

16. $\begin{cases} 9x^2 + 4y^2 = 36 \\ y + 1 = \sqrt{x - 1} \end{cases}$

In Exercises 17–30, use a graphing calculator to graph the equations and find any solutions of the system.

17. $\begin{cases} y = 2x^2 \\ y = -2x + 12 \end{cases}$

18. $\begin{cases} y = 5x^2 \\ y = -15x - 10 \end{cases}$

19. $\begin{cases} y = x \\ y = x^3 \end{cases}$

20. $\begin{cases} y = x^4 \\ y = 5x + 6 \end{cases}$

21. $\begin{cases} y = x^2 \\ y = -x^2 + 4x \end{cases}$

22. $\begin{cases} y = 8 - x^2 \\ y = 6 - x \end{cases}$

23. $\begin{cases} x^2 - y = 2 \\ 3x + y = 2 \end{cases}$

24. $\begin{cases} x^2 + 2y = 6 \\ x - y = -4 \end{cases}$

25. $\begin{cases} x - 3y = 1 \\ \sqrt{x} - 1 = y \end{cases}$

26. $\begin{cases} \sqrt{x} + 1 = y \\ 2x + y = 4 \end{cases}$

27. $\begin{cases} y = x^3 \\ y = x^3 - 3x^2 + 3x \end{cases}$

28. $\begin{cases} y = -2(x^2 - 1) \\ y = 2(x^4 - 2x^2 + 1) \end{cases}$

29. $\begin{cases} y = \ln x - 2 \\ y = x - 2 \end{cases}$

30. $\begin{cases} y = -x^2 + 1 \\ y = \ln(x + 1) + 1 \end{cases}$

In Exercises 31–58, solve the system by the method of substitution. ***See Examples 3 and 4.***

31. $\begin{cases} y = 2x^2 \\ y = 6x - 4 \end{cases}$

32. $\begin{cases} y = 5x^2 \\ y = -5x + 10 \end{cases}$

33. $\begin{cases} x^2 + y = 5 \\ 2x + y = 5 \end{cases}$

34. $\begin{cases} x - y^2 = 0 \\ x - y = 2 \end{cases}$

35. $\begin{cases} x^2 + y^2 = 4 \\ x + y = 2 \end{cases}$

36. $\begin{cases} x^2 + y^2 = 36 \\ x = 6 \end{cases}$

37. $\begin{cases} x^2 + y^2 = 25 \\ y = 5 \end{cases}$

38. $\begin{cases} x^2 + y^2 = 1 \\ x + y = 7 \end{cases}$

39. $\begin{cases} x^2 + y^2 = 64 \\ -3x + y = 8 \end{cases}$

40. $\begin{cases} x^2 + y^2 = 81 \\ x + 3y = 27 \end{cases}$

41. $\begin{cases} 4x + y^2 = 2 \\ 2x - y = -11 \end{cases}$

42. $\begin{cases} x^2 + y^2 = 10 \\ 2x - y = 5 \end{cases}$

43. $\begin{cases} x^2 + y^2 = 9 \\ x + 2y = 3 \end{cases}$

44. $\begin{cases} x^2 + y^2 = 4 \\ x - 2y = 4 \end{cases}$

45. $\begin{cases} 2x^2 - y^2 = -8 \\ x - y = 6 \end{cases}$

46. $\begin{cases} y^2 = -x + 4 \\ x^2 + y^2 = 6 \end{cases}$

47. $\begin{cases} y = x^2 - 5 \\ 3x + 2y = 10 \end{cases}$

48. $\begin{cases} x + y = 4 \\ x^2 - y^2 = 4 \end{cases}$

49. $\begin{cases} y = \sqrt{4 - x} \\ x + 3y = 6 \end{cases}$

50. $\begin{cases} y = \sqrt{25 - x^2} \\ x + y = 7 \end{cases}$

51. $\begin{cases} x^2 - 4y^2 = 16 \\ x^2 + y^2 = 1 \end{cases}$

52. $\begin{cases} 2x^2 - y^2 = 12 \\ 3x^2 - y^2 = -4 \end{cases}$

53. $\begin{cases} y = x^2 - 3 \\ x^2 + y^2 = 9 \end{cases}$

54. $\begin{cases} x^2 + y^2 = 25 \\ x - 3y = -5 \end{cases}$

55. $\begin{cases} 16x^2 + 9y^2 = 144 \\ 4x + 3y = 12 \end{cases}$

56. $\begin{cases} 4x^2 + 16y^2 = 64 \\ x + 2y = 4 \end{cases}$

57. $\begin{cases} x^2 - y^2 = 9 \\ x^2 + y^2 = 1 \end{cases}$

58. $\begin{cases} x^2 - y^2 = 4 \\ x - y = 2 \end{cases}$

In Exercises 59–80, solve the system by the method of elimination. ***See Examples 5 and 6.***

59. $\begin{cases} x^2 + 2y = 1 \\ x^2 + y^2 = 4 \end{cases}$

60. $\begin{cases} x + y^2 = 5 \\ 2x^2 + y^2 = 6 \end{cases}$

61. $\begin{cases} -x + y^2 = 10 \\ x^2 - y^2 = -8 \end{cases}$

62. $\begin{cases} x^2 + y = 9 \\ x^2 - y^2 = 7 \end{cases}$

63. $\begin{cases} x^2 + y^2 = 7 \\ x^2 - y^2 = 1 \end{cases}$

64. $\begin{cases} x^2 + y^2 = 25 \\ y^2 - x^2 = 7 \end{cases}$

65. $\begin{cases} x^2 - y^2 = 4 \\ x^2 + y^2 = 4 \end{cases}$

66. $\begin{cases} x^2 + y^2 = 25 \\ x^2 - y^2 = -36 \end{cases}$

67. $\begin{cases} x^2 + y^2 = 13 \\ 2x^2 + 3y^2 = 30 \end{cases}$

68. $\begin{cases} 3x^2 - y^2 = 4 \\ x^2 + 4y^2 = 10 \end{cases}$

69. $\begin{cases} 4x^2 + 9y^2 = 36 \\ 2x^2 - 9y^2 = 18 \end{cases}$

70. $\begin{cases} 5x^2 - 2y^2 = -13 \\ 3x^2 + 4y^2 = 39 \end{cases}$

71. $\begin{cases} 2x^2 + 3y^2 = 21 \\ x^2 + 2y^2 = 12 \end{cases}$

72. $\begin{cases} 2x^2 + y^2 = 11 \\ x^2 + 3y^2 = 28 \end{cases}$

73. $\begin{cases} -x^2 - 2y^2 = 6 \\ 5x^2 + 15y^2 = 20 \end{cases}$

74. $\begin{cases} x^2 - 2y^2 = 7 \\ x^2 + y^2 = 34 \end{cases}$

75. $\begin{cases} x^2 + y^2 = 9 \\ 16x^2 - 4y^2 = 64 \end{cases}$

76. $\begin{cases} 3x^2 + 4y^2 = 35 \\ 2x^2 + 5y^2 = 42 \end{cases}$

77. $\begin{cases} \dfrac{x^2}{4} + y^2 = 1 \\ x^2 + \dfrac{y^2}{4} = 1 \end{cases}$

78. $\begin{cases} x^2 - y^2 = 1 \\ \dfrac{x^2}{2} + y^2 = 1 \end{cases}$

79. $\begin{cases} y^2 - x^2 = 10 \\ x^2 + y^2 = 16 \end{cases}$

80. $\begin{cases} x^2 + y^2 = 25 \\ x^2 + 2y^2 = 36 \end{cases}$

Solving Problems

81. *Ice Rink* A rectangular ice rink has an area of 3000 square feet. The diagonal across the rink is 85 feet. Find the dimensions of the rink.

82. *Cell Phone* A cell phone has a rectangular external display that contains 19,200 pixels with a diagonal of 200 pixels. Find the resolution (the dimensions in pixels) of the external display.

83. *Dog Park* A rectangular dog park has a diagonal sidewalk that measures 290 feet. The perimeter of each triangle formed by the diagonal is 700 feet. Find the dimensions of the dog park.

84. *Sailboat* A sail for a sailboat is shaped like a right triangle that has a perimeter of 36 meters and a hypotenuse of 15 meters. Find the dimensions of the sail.

85. *Hyperbolic Mirror* In a hyperbolic mirror, light rays directed to one focus are reflected to the other focus. The mirror in the figure has the equation

$$\frac{x^2}{9} - \frac{y^2}{16} = 1.$$

At which point on the mirror will light from the point (0, 10) reflect to the focus?

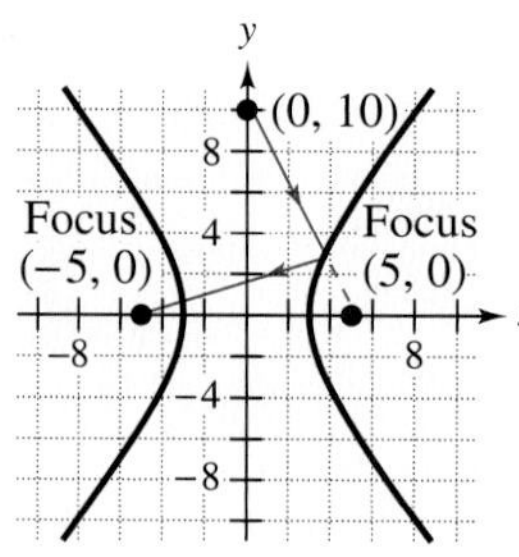

Figure for 85

86. *Miniature Golf* You are playing miniature golf and your golf ball is at $(-15, 25)$ (see figure). A wall at the end of the enclosed area is part of a hyperbola whose equation is

$$\frac{x^2}{19} - \frac{y^2}{81} = 1.$$

Using the reflective property of hyperbolas given in Exercise 85, at which point on the wall must your ball hit for it to go into the hole? (The ball bounces off the wall only once.)

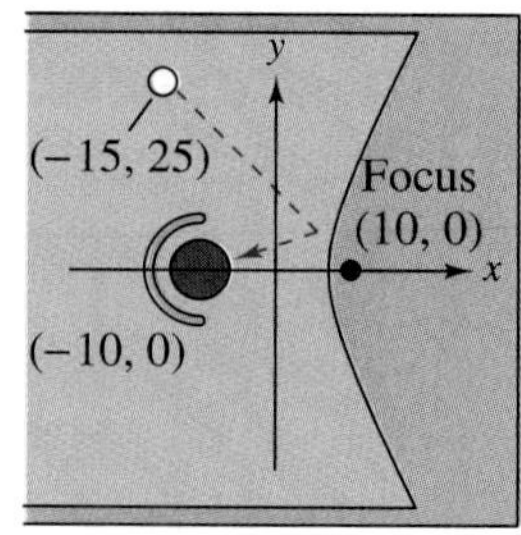

87. *Busing Boundary* To be eligible to ride the school bus to East High School, a student must live at least 1 mile from the school (see figure). Describe the portion of Clarke Street for which the residents are *not* eligible to ride the school bus. Use a coordinate system in which the school is at (0, 0) and each unit represents 1 mile.

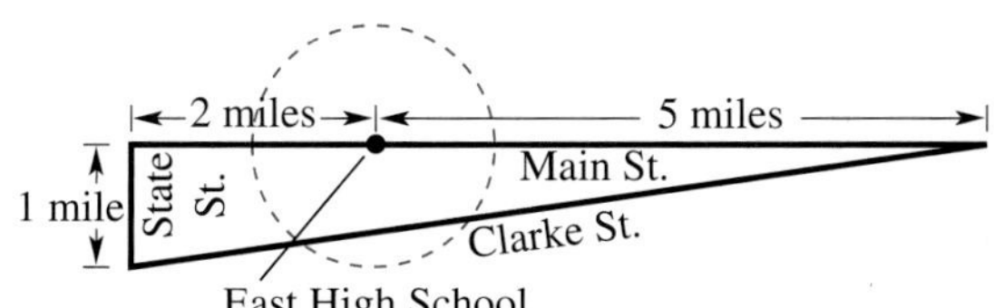

88. *Search Team* A search team of three members splits to search an area in the woods. Each member carries a family service radio with a circular range of 3 miles. The team members agree to communicate from their bases every hour. The second member sets up base 3 miles north of the first member. Where should the third member set up base to be as far east as possible but within direct communication range of each of the other two searchers? Use a coordinate system in which the first member is at (0, 0) and each unit represents 1 mile.

Explaining Concepts

89. Explain how to solve a nonlinear system of equations using the method of substitution.

90. Explain how to solve a nonlinear system of equations using the method of elimination.

91. What is the maximum number of points of intersection of a line and a hyperbola? Explain.

92. A circle and a parabola can have 0, 1, 2, 3, or 4 points of intersection. Sketch the circle given by $x^2 + y^2 = 4$. Discuss how this circle could intersect a parabola with an equation of the form $y = x^2 + C$. Then find the values of C for each of the five cases described below.

(a) No points of intersection

(b) One point of intersection

(c) Two points of intersection

(d) Three points of intersection

(e) Four points of intersection

Use a graphing calculator to confirm your results.

Cumulative Review

In Exercises 93–104, solve the equation and check your solution(s).

93. $\sqrt{6 - 2x} = 4$

94. $\sqrt{x + 3} = -9$

95. $\sqrt{x} = x - 6$

96. $\sqrt{x + 14} = \sqrt{x} + 3$

97. $3^x = 243$

98. $4^x = 256$

99. $5^{x-1} = 310$

100. $e^{0.5x} = 8$

101. $\log_{10} x = 0.01$

102. $\log_4 8x = 3$

103. $2 \ln(x + 1) = -2$

104. $\ln(x + 3) - \ln x = \ln 1$

What Did You Learn?

Use these two pages to help prepare for a test on this chapter. Check off the key terms and key concepts you know. You can also use this section to record your assignments.

Plan for Test Success

Date of test: / /

Study dates and times: / / at : A.M./P.M.

/ / at : A.M./P.M.

Things to review:

- ☐ Key Terms, *p. 804*
- ☐ Key Concepts, *pp. 804–805*
- ☐ Your class notes
- ☐ Your assignments
- ☐ Study Tips, *pp. 765, 766, 775, 787*
- ☐ Technology Tips, *pp. 767, 777, 789, 794*
- ☐ Mid-Chapter Quiz, *p. 784*
- ☐ Review Exercises, *pp. 806–808*
- ☐ Chapter Test, *p. 809*
- ☐ Video Explanations Online
- ☐ Tutorial Online

Key Terms

- ☐ conic (conic section), *p. 762*
- ☐ circle, *p. 762*
- ☐ center (of a circle), *p. 762*
- ☐ radius, *pp. 762, 763*
- ☐ standard form of equation of circle centered at (0, 0), *p. 763*
- ☐ standard form of equation of circle centered at (h, k), *p. 764*
- ☐ parabola, *p. 766*
- ☐ directrix, *p. 766*
- ☐ focus (of a parabola), *p. 766*
- ☐ vertex (of a parabola), *p. 766*
- ☐ axis (of a parabola), *p. 766*
- ☐ ellipse, *p. 774*
- ☐ focus (of an ellipse), *p. 774*
- ☐ vertices (of an ellipse), *p. 774*
- ☐ major axis, *p. 774*
- ☐ center (of an ellipse), *p. 774*
- ☐ minor axis, *p. 774*
- ☐ co-vertices, *p. 774*
- ☐ standard form of equation of ellipse centered at (0, 0), *p. 775*
- ☐ standard form of equation of ellipse centered at (h, k), *p. 776*
- ☐ hyperbola, *p. 785*
- ☐ foci (of a hyperbola), *p. 785*
- ☐ transverse axis (of a hyperbola), *p. 785*
- ☐ standard form of equation of hyperbola centered at (0, 0), *p. 785*
- ☐ vertices (of a hyperbola), *p. 785*
- ☐ branch (of a hyperbola), *p. 786*
- ☐ asymptotes (of a hyperbola), *p. 786*
- ☐ central rectangle, *p. 786*
- ☐ standard form of equation of hyperbola centered at (h, k), *p. 788*
- ☐ nonlinear system of equations, *p. 793*

Key Concepts

12.1 Circles and Parabolas

Assignment: ______ **Due date:** ______

☐ **Graph and write equations of circles centered at the origin.**

Standard equation of a circle with radius r and center (0, 0):

$x^2 + y^2 = r^2$

☐ **Graph and write equations of circles centered at (h, k).**

Standard equation of a circle with radius r and center (h, k):

$(x - h)^2 + (y - k)^2 = r^2$

☐ **Graph and write equations of parabolas.**

Standard equation of a parabola with vertex at (0, 0):

$x^2 = 4py$ Vertical axis

$y^2 = 4px$ Horizontal axis

Standard equation of a parabola with vertex at (h, k):

$(x - h)^2 = 4p(y - k)$ Vertical axis

$(y - k)^2 = 4p(x - h)$ Horizontal axis

The focus of a parabola lies on the axis, a directed distance of p units from the vertex.

12.2 Ellipses

Assignment: ______ **Due date:** ______

☐ **Graph and write equations of ellipses centered at the origin.**

Standard equation of an ellipse centered at the origin:

$\frac{x^2}{a^2} + \frac{y^2}{b^2} = 1$ Major axis is horizontal.

$\frac{x^2}{b^2} + \frac{y^2}{a^2} = 1$ Major axis is vertical.

☐ **Graph and write equations of ellipses centered at (h, k).**

Standard equation of an ellipse centered at (h, k):

$\frac{(x-h)^2}{a^2} + \frac{(y-k)^2}{b^2} = 1$ Major axis is horizontal.

$\frac{(x-h)^2}{b^2} + \frac{(y-k)^2}{a^2} = 1$ Major axis is vertical.

In all of the standard equations for ellipses, $0 < b < a$. The vertices of the ellipse lie on the major axis, a units from the center (the major axis has length $2a$). The co-vertices lie on the minor axis, b units from the center.

12.3 Hyperbolas

Assignment: ______ **Due date:** ______

☐ **Graph and write equations of hyperbolas centered at the origin.**

Standard equation of a hyperbola centered at the origin:

$\frac{x^2}{a^2} - \frac{y^2}{b^2} = 1$ Transverse axis is horizontal.

$\frac{y^2}{a^2} - \frac{x^2}{b^2} = 1$ Transverse axis is vertical.

☐ **Graph and write equations of hyperbolas centered at (h, k).**

Standard equation of a hyperbola centered at (h, k):

$\frac{(x-h)^2}{a^2} - \frac{(y-k)^2}{b^2} = 1$ Transverse axis is horizontal.

$\frac{(y-k)^2}{a^2} - \frac{(x-h)^2}{b^2} = 1$ Transverse axis is vertical.

A hyperbola's vertices lie on the transverse axis, a units from the center. A hyperbola's central rectangle has side lengths of $2a$ and $2b$. A hyperbola's asymptotes pass through opposite corners of its central rectangle.

12.4 Solving Nonlinear Systems of Equations

Assignment: ______ **Due date:** ______

☐ **Solve a nonlinear system graphically.**

1. Sketch the graph of each equation in the system.
2. Graphically approximate the coordinates of any points of intersection of the graphs.
3. Check the coordinates by substituting them into each equation in the original system. If the coordinates do not check, it may be necessary to use an algebraic approach such as substitution or elimination.

☐ **Use substitution to solve a nonlinear system of two equations in two variables.**

1. Solve one equation for one variable in terms of the other.
2. Substitute the expression found in Step 1 into the other equation to obtain an equation in one variable.
3. Solve the equation obtained in Step 2.
4. Back-substitute the solution from Step 3 into the expression obtained in Step 1 to find the value of the other variable.
5. Check that the solution satisfies *both* original equations.

☐ **Use elimination to solve a nonlinear system of equations.**

Review Exercises

12.1 Circles and Parabolas

1 ▶ Recognize the four basic conics: circles, parabolas, ellipses, and hyperbolas.

In Exercises 1–6, identify the conic.

1.

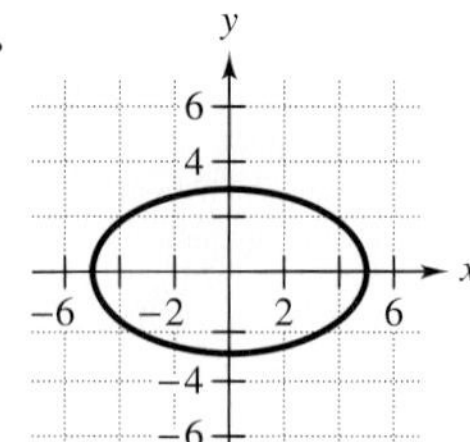

2.

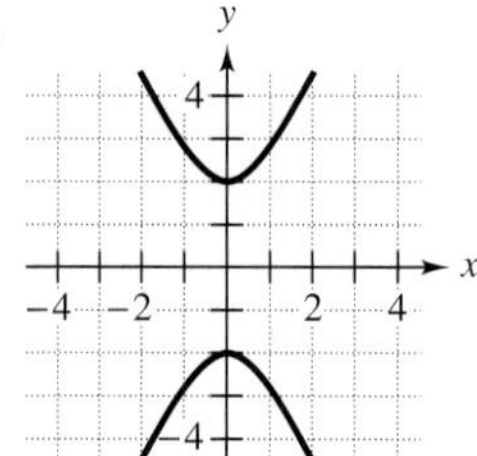

3.

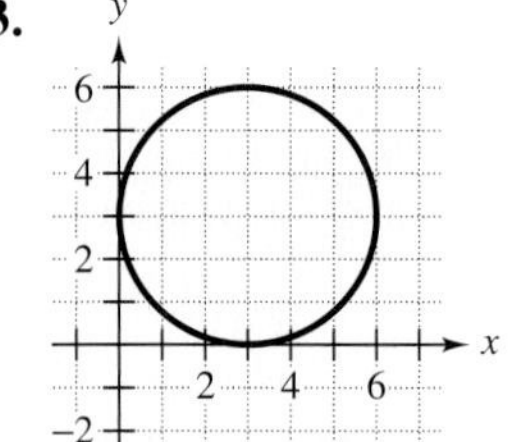

4.

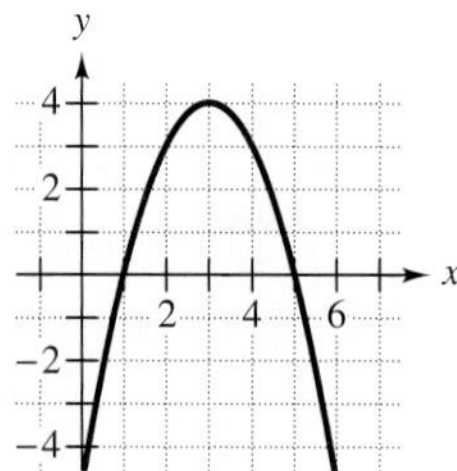

5.

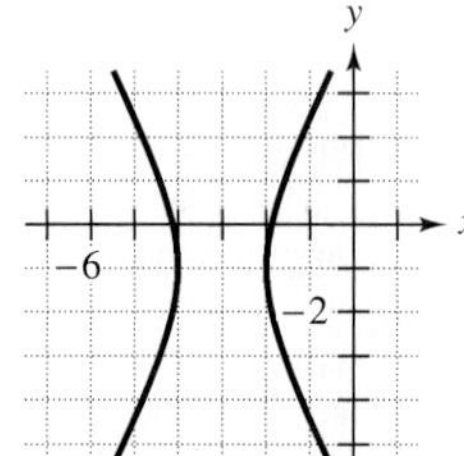

6.

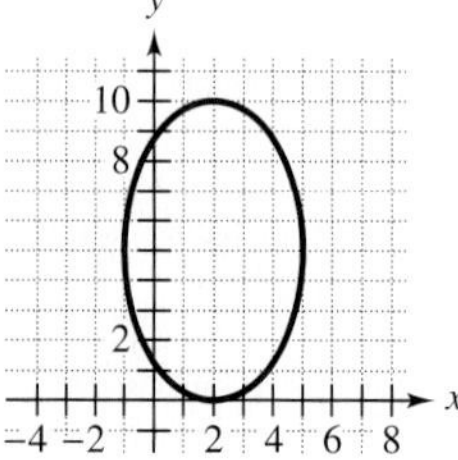

2 ▶ Graph and write equations of circles centered at the origin.

In Exercises 7 and 8, write the standard form of the equation of the circle with center at $(0, 0)$ that satisfies the criterion.

7. Radius: 6

8. Passes through the point $(-1, 3)$

In Exercises 9 and 10, identify the center and radius of the circle and sketch the circle.

9. $x^2 + y^2 = 64$

10. $9x^2 + 9y^2 - 49 = 0$

3 ▶ Graph and write equations of circles centered at (h, k).

In Exercises 11 and 12, write the standard form of the equation of the circle with center at (h, k) that satisfies the criteria.

11. Center: $(2, 6)$; Radius: 3

12. Center: $(-2, 3)$; Passes through the point $(1, 1)$

In Exercises 13 and 14, identify the center and radius of the circle and sketch the circle.

13. $x^2 + y^2 + 6x + 8y + 21 = 0$

14. $x^2 + y^2 - 8x + 16y + 75 = 0$

4 ▶ Graph and write equations of parabolas.

In Exercises 15–18, write the standard form of the equation of the parabola. Then sketch the parabola.

15. Vertex: $(0, 0)$; Focus: $(-2, 0)$

16. Vertex: $(0, 5)$; Focus: $(2, 5)$

17. Vertex: $(-1, 3)$;
Passes through the point $(-2, 5)$; Vertical axis

18. Vertex: $(8, 0)$;
Passes through the point $(2, -2)$; Horizontal axis

In Exercises 19 and 20, identify the vertex and focus of the parabola and sketch the parabola.

19. $y = \frac{1}{2}x^2 - 8x + 7$

20. $x = y^2 + 10y - 4$

12.2 Ellipses

1 ▶ Graph and write equations of ellipses centered at the origin.

In Exercises 21–24, write the standard form of the equation of the ellipse centered at the origin.

21. Vertices: $(0, -5), (0, 5)$;
Co-vertices: $(-2, 0), (2, 0)$

22. Vertices: $(-8, 0), (8, 0)$;
Co-vertices: $(0, -3), (0, 3)$

23. Major axis (vertical) 6 units, minor axis 4 units

24. Major axis (horizontal) 12 units, minor axis 2 units

In Exercises 25–28, sketch the ellipse. Identify the vertices and co-vertices.

25. $\frac{x^2}{64} + \frac{y^2}{16} = 1$

26. $\frac{x^2}{9} + y^2 = 1$

27. $36x^2 + 9y^2 - 36 = 0$

28. $100x^2 + 4y^2 - 4 = 0$

2 ▶ Graph and write equations of ellipses centered at (h, k).

In Exercises 29–32, write the standard form of the equation of the ellipse.

29. Vertices: $(-2, 4), (8, 4)$; Co-vertices: $(3, 0), (3, 8)$

30. Vertices: $(0, 5), (12, 5)$; Co-vertices: $(6, 2), (6, 8)$

31. Vertices: $(0, 0), (0, 8)$; Co-vertices: $(-3, 4), (3, 4)$

32. Vertices: $(5, -3), (5, 13)$;
Co-vertices: $(3, 5), (7, 5)$

In Exercises 33–36, find the center and vertices of the ellipse and sketch the ellipse.

33. $9(x + 1)^2 + 4(y - 2)^2 = 144$

34. $x^2 + 25y^2 - 4x - 21 = 0$

35. $16x^2 + y^2 + 6y - 7 = 0$

36. $x^2 + 4y^2 + 10x - 24y + 57 = 0$

12.3 Hyperbolas

1 ▶ Graph and write equations of hyperbolas centered at the origin.

In Exercises 37–40, sketch the hyperbola. Identify the vertices and asymptotes.

37. $x^2 - y^2 = 25$

38. $y^2 - x^2 = 16$

39. $\frac{y^2}{25} - \frac{x^2}{4} = 1$

40. $\frac{x^2}{16} - \frac{y^2}{25} = 1$

In Exercises 41–44, write the standard form of the equation of the hyperbola centered at the origin.

	Vertices	*Asymptotes*	
41.	$(-2, 0), (2, 0)$	$y = \frac{3}{2}x$	$y = -\frac{3}{2}x$
42.	$(0, -6), (0, 6)$	$y = 3x$	$y = -3x$
43.	$(0, -8), (0, 8)$	$y = \frac{4}{5}x$	$y = -\frac{4}{5}x$
44.	$(-3, 0), (3, 0)$	$y = \frac{4}{3}x$	$y = -\frac{4}{3}x$

2 ▶ Graph and write equations of hyperbolas centered at (h, k).

In Exercises 45–48, find the center and vertices of the hyperbola and sketch the hyperbola.

45. $\dfrac{(x-3)^2}{9} - \dfrac{(y+1)^2}{4} = 1$

46. $\dfrac{(x+4)^2}{25} - \dfrac{(y-7)^2}{64} = 1$

47. $8y^2 - 2x^2 + 48y + 16x + 8 = 0$

48. $25x^2 - 4y^2 - 200x - 40y = 0$

In Exercises 49 and 50, write the standard form of the equation of the hyperbola.

49.

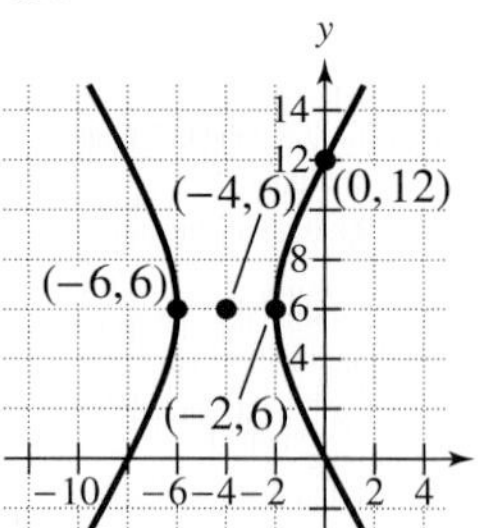

50.

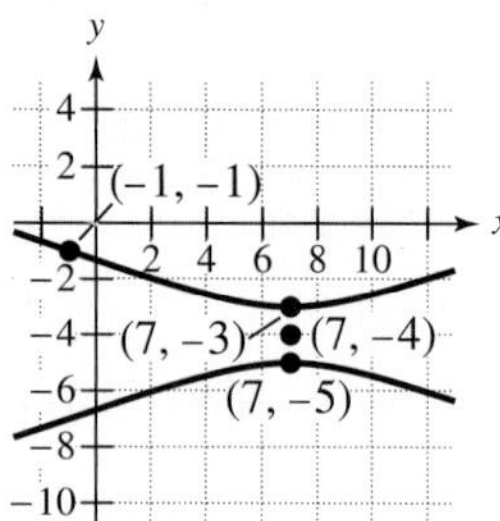

12.4 Solving Nonlinear Systems of Equations

1 ▶ Solve nonlinear systems of equations graphically.

In Exercises 51–54, use a graphing calculator to graph the equations and find any solutions of the system.

51. $\begin{cases} y = x^2 \\ y = 3x \end{cases}$

52. $\begin{cases} y = 2 + x^2 \\ y = 8 - x \end{cases}$

53. $\begin{cases} x^2 + y^2 = 16 \\ -x + y = 4 \end{cases}$

54. $\begin{cases} 2x^2 - y^2 = -8 \\ y = x + 6 \end{cases}$

2 ▶ Solve nonlinear systems of equations by substitution.

In Exercises 55–58, solve the system by the method of substitution.

55. $\begin{cases} y = 5x^2 \\ y = -15x - 10 \end{cases}$

56. $\begin{cases} y^2 = 16x \\ 4x - y = -24 \end{cases}$

57. $\begin{cases} x^2 + y^2 = 1 \\ x + y = -1 \end{cases}$

58. $\begin{cases} y^2 - x^2 = 9 \\ x + y = 1 \end{cases}$

3 ▶ Solve nonlinear systems of equations by elimination.

In Exercises 59 and 60, solve the system by the method of elimination.

59. $\begin{cases} 6x^2 - y^2 = 15 \\ x^2 + y^2 = 13 \end{cases}$

60. $\begin{cases} x^2 + y^2 = 16 \\ -x^2 + \dfrac{y^2}{16} = 1 \end{cases}$

4 ▶ Use nonlinear systems of equations to model and solve real-life problems.

61. *Geometry* A computer manufacturer needs a circuit board with a perimeter of 28 centimeters and a diagonal of length 10 centimeters. What should the dimensions of the circuit board be?

62. *Geometry* A home interior decorator wants to find a ceramic tile with a perimeter of 6 inches and a diagonal of length $\sqrt{5}$ inches. What should the dimensions of the tile be?

63. *Geometry* A piece of wire 100 inches long is to be cut into two pieces. Each of the two pieces is then to be bent into a square. The area of one square is to be 144 square inches greater than the area of the other square. How should the wire be cut?

64. *Geometry* You have 250 feet of fencing to enclose two corrals of equal size (see figure). The combined area of the corrals is 2400 square feet. Find the dimensions of each corral.

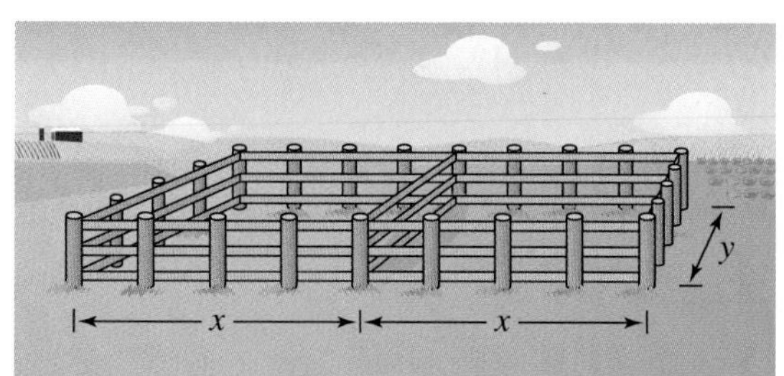

Chapter Test

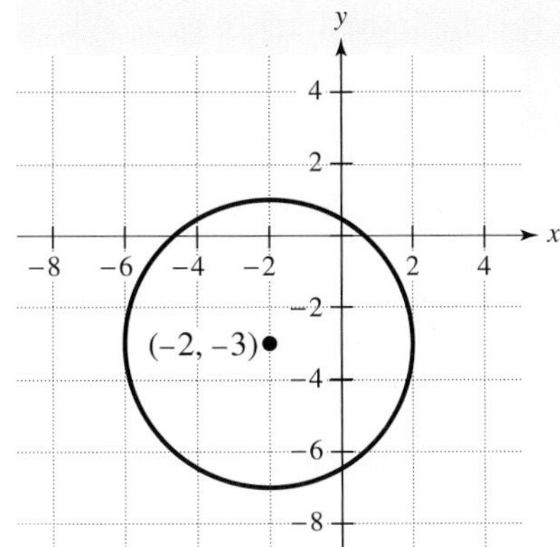

Figure for 1

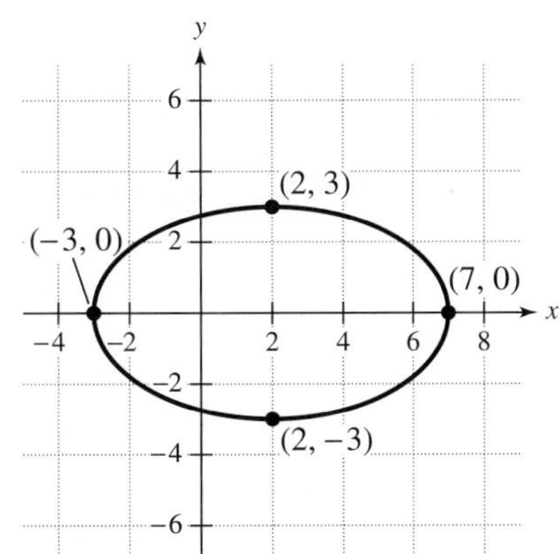

Figure for 6

Take this test as you would take a test in class. After you are done, check your work against the answers in the back of the book.

1. Write the standard form of the equation of the circle shown in the figure.

In Exercises 2 and 3, write the equation of the circle in standard form. Then sketch the circle.

2. $x^2 + y^2 - 2x - 6y + 1 = 0$
3. $x^2 + y^2 + 4x - 6y + 4 = 0$
4. Identify the vertex and the focus of the parabola $x = -3y^2 + 12y - 8$. Then sketch the parabola.
5. Write the standard form of the equation of the parabola with vertex $(7, -2)$ and focus $(7, 0)$.
6. Write the standard form of the equation of the ellipse shown in the figure.

In Exercises 7 and 8, find the center and vertices of the ellipse. Then sketch the ellipse.

7. $16x^2 + 4y^2 = 64$
8. $25x^2 + 4y^2 - 50x - 24y - 39 = 0$

In Exercises 9 and 10, write the standard form of the equation of the hyperbola.

9. Vertices: $(-3, 0), (3, 0)$; Asymptotes: $y = \pm\frac{2}{3}x$
10. Vertices: $(0, -5), (0, 5)$; Asymptotes: $y = \pm\frac{5}{2}x$

In Exercises 11 and 12, find the center and vertices of the hyperbola. Then sketch the hyperbola.

11. $4x^2 - 2y^2 - 24x + 20 = 0$
12. $16y^2 - 25x^2 + 64y + 200x - 736 = 0$

In Exercises 13–15, solve the nonlinear system of equations.

13. $\begin{cases} x^2/16 + y^2/9 = 1 \\ 3x + 4y = 12 \end{cases}$ 14. $\begin{cases} x^2 + y^2 = 16 \\ x^2/16 - y^2/9 = 1 \end{cases}$ 15. $\begin{cases} x^2 + y^2 = 10 \\ x^2 = y^2 + 2 \end{cases}$

16. Write the equation of the circular orbit of a satellite 1000 miles above the surface of Earth. Place the origin of the rectangular coordinate system at the center of Earth and assume the radius of Earth to be 4000 miles.
17. A rectangle has a perimeter of 56 inches and a diagonal of length 20 inches. Find the dimensions of the rectangle.

Cumulative Test: Chapters 10–12

Take this test as you would take a test in class. After you are done, check your work against the answers in the back of the book.

In Exercises 1–4, solve the equation by the specified method.

1. Factoring:
$4x^2 - 9x - 9 = 0$

2. Square Root Property:
$(x - 5)^2 - 64 = 0$

3. Completing the square:
$x^2 - 10x - 25 = 0$

4. Quadratic Formula:
$3x^2 + 6x + 2 = 0$

5. Solve the equation of quadratic form: $x^4 - 8x^2 + 15 = 0$.

In Exercises 6 and 7, solve the inequality and graph the solution on the real number line.

6. $3x^2 + 8x \le 3$

7. $\dfrac{3x + 4}{2x - 1} < 0$

8. Find a quadratic equation having the solutions -2 and 6.

9. Find the compositions (a) $f \circ g$ and (b) $g \circ f$ for $f(x) = 2x^2 - 3$ and $g(x) = 5x - 1$. Then find the domain of each composition.

10. Find the inverse function of $f(x) = \dfrac{5 - 3x}{4}$.

11. Evaluate $f(x) = 7 + 2^{-x}$ when $x = 1, 0.5$, and 3.

12. Sketch the graph of $f(x) = 4^{x-1}$ and identify the horizontal asymptote.

13. Describe the relationship between the graphs of $f(x) = e^x$ and $g(x) = \ln x$.

14. Sketch the graph of $\log_3(x - 1)$ and identify the vertical asymptote.

15. Evaluate $\log_4 \frac{1}{16}$ without using a calculator.

16. Use the properties of logarithms to condense $3(\log_2 x + \log_2 y) - \log_2 z$.

17. Use the properties of logarithms to expand $\log_{10} \dfrac{\sqrt{x + 1}}{x^4}$.

In Exercises 18–21, solve the equation.

18. $\log_x\left(\frac{1}{9}\right) = -2$

19. $4 \ln x = 10$

20. $500(1.08)^t = 2000$

21. $3(1 + e^{2x}) = 20$

22. If the inflation rate averages 2.8% over the next 5 years, the approximate cost C of goods and services t years from now is given by

$C(t) = P(1.028)^t,\ 0 \le t \le 5$

where P is the present cost. The price of an oil change is presently \$29.95. Estimate the price 5 years from now.

23. Determine the effective yield of an 8% interest rate compounded continuously.

24. Determine the length of time for an investment of \$1500 to quadruple in value if the investment earns 7% compounded continuously.

25. Write the equation of the circle in standard form and sketch the circle:

$x^2 + y^2 - 6x + 14y - 6 = 0.$

26. Identify the vertex and focus of the parabola and sketch the parabola:

$y = 2x^2 - 20x + 5.$

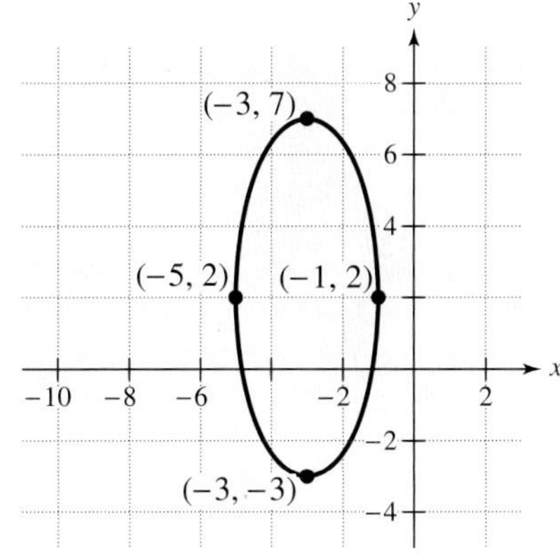

Figure for 27

27. Write the standard form of the equation of the ellipse shown in the figure.

28. Find the center and vertices of the ellipse and sketch the ellipse:

$4x^2 + y^2 = 4.$

29. Write the standard form of the equation of the hyperbola with vertices $(0, -3)$ and $(0, 3)$ and asymptotes $y = \pm 3x$.

30. Find the center and vertices of the hyperbola and sketch the hyperbola:

$x^2 - 9y^2 + 18y = 153.$

In Exercises 31 and 32, solve the nonlinear system of equations.

31. $\begin{cases} y = x^2 - x - 1 \\ 3x - y = 4 \end{cases}$

32. $\begin{cases} x^2 + 5y^2 = 21 \\ -x + y^2 = 5 \end{cases}$

33. A rectangle has an area of 32 square feet and a perimeter of 24 feet. Find the dimensions of the rectangle.

34. The path of a ball is given by $y = -0.1x^2 + 3x + 6$. The coordinates x and y are measured in feet, with $x = 0$ corresponding to the position from which the ball was thrown.

(a) Use a graphing calculator to graph the path of the ball.

(b) Move the cursor along the path to approximate the highest point and the range of the path.

Study Skills in Action

Preparing for the Final Exam

At the end of the semester, most students are inundated with projects, papers, and tests. Instructors may speed up the pace in lectures to get through all the material. If something unexpected is going to happen to a student, it often happens during this time.

Getting through the last couple of weeks of a math course can be challenging. This is why it is important to plan your review time for the final exam at least three weeks before the exam.

Kimberly Nolting
VP, Academic Success Press
expert in developmental education

These students are planning how they will study for the final exam.

Smart Study Strategy

Form a Final Exam Study Group

1 ► **Form a study group of three or four students several weeks before the final exam.** The intent of this group is to review what you have already learned while continuing to learn new material.

2 ► **Find out what material you must know for the final, even if the instructor has not yet covered it.** As a group, meet with the instructor outside of class. A group is likely to receive more attention and can ask more questions.

3 ► **Ask for or create a practice final and have the instructor look at it.** Make sure the problems are on an appropriate level of difficulty. Look for sample problems in old tests and in cumulative tests in the textbook. Review what the textbook and your notes say as you look for problems. This will refresh your memory.

4 ► **Have each group member take the practice final exam.** Then have each member identify what he or she needs to study. Make sure you can complete the problems with the speed and accuracy that are necessary to complete the real final exam.

5 ► **Decide when the group is going to meet during the next couple of weeks and what you will cover during each session.** The tutoring or learning center on campus is an ideal setting in which to meet. Many libraries have small study rooms that study groups can reserve. Set up several study times for each week. If you live at home, make sure your family knows that this is a busy time.

6 ► **During the study group sessions, make sure you stay on track.** Prepare for each study session by knowing what material in the textbook you are going to cover and having the class notes for that material. When you have questions, assign a group member to go to the instructor for answers. Then this member can relay the correct information to the other group members. Save socializing for after the final exam.

Chapter 13

Sequences, Series, and the Binomial Theorem

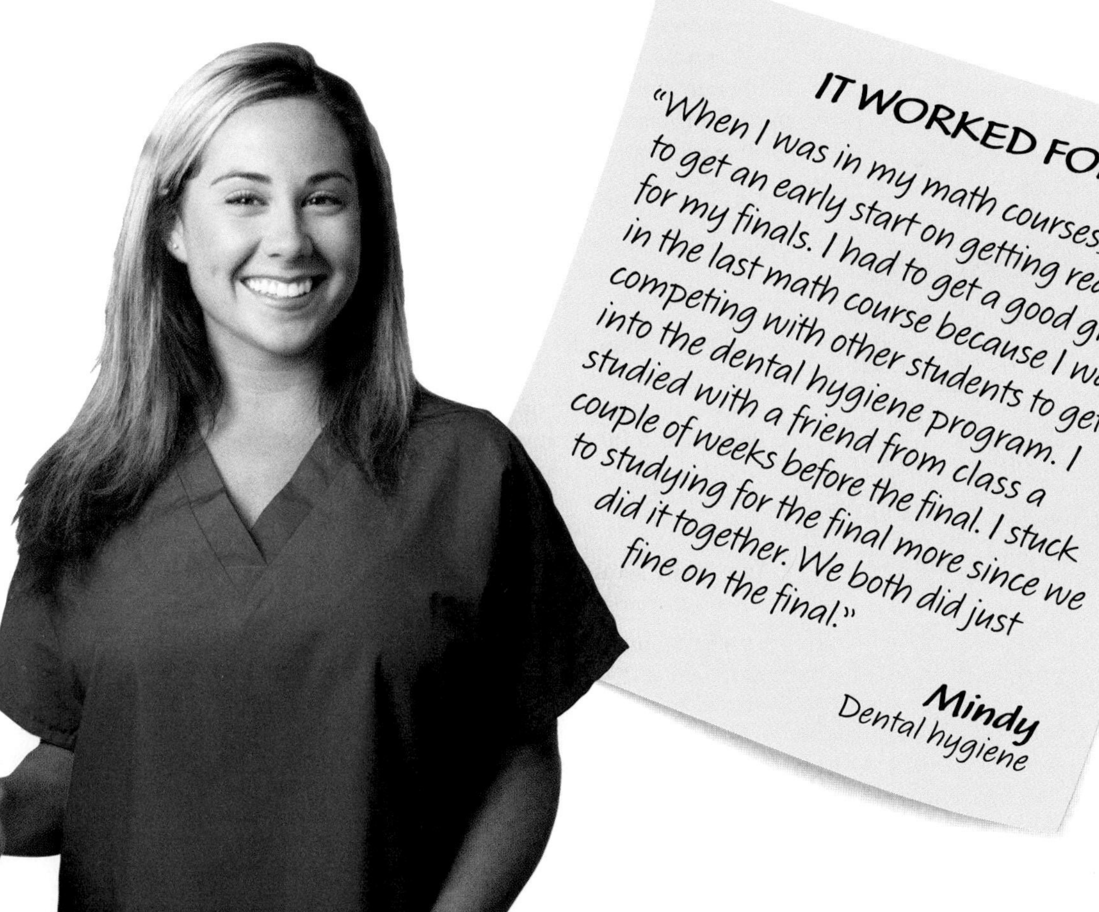

IT WORKED FOR ME!
"When I was in my math courses, I had to get an early start on getting ready for my finals. I had to get a good grade in the last math course because I was competing with other students to get into the dental hygiene program. I studied with a friend from class a couple of weeks before the final. I stuck to studying for the final more since we did it together. We both did just fine on the final."
Mindy
Dental hygiene

13.1 Sequences and Series

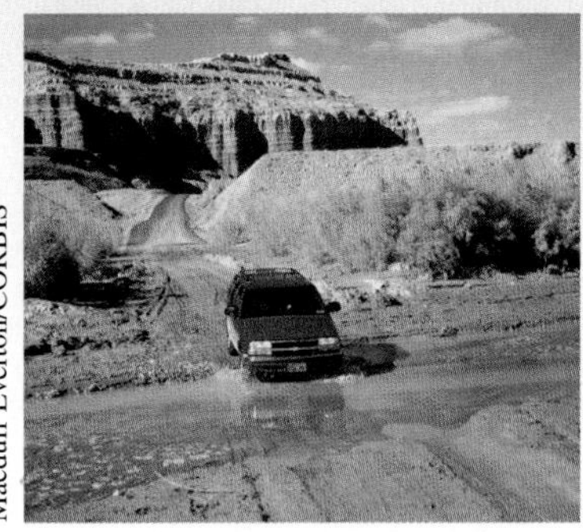
Macduff Everton/CORBIS

Why You Should Learn It

Sequences and series are useful in modeling sets of values in order to identify patterns. For instance, in Exercise 117 on page 824, you will use a sequence to model the depreciation of a sport utility vehicle.

What You Should Learn

1 ▶ Use sequence notation to write the terms of sequences.

2 ▶ Write the terms of sequences involving factorials.

3 ▶ Find the apparent *n*th term of a sequence.

4 ▶ Sum the terms of sequences to obtain series, and use sigma notation to represent partial sums.

Sequences

1 ▶ Use sequence notation to write the terms of sequences.

You are given the following choice of contract offers for the next 5 years of employment.

Contract A $30,000 the first year and a $3300 raise each year

Contract B $30,000 the first year and a 10% raise each year

Which contract offers the largest salary over the five-year period? The salaries for each contract are shown in the table at the left. Notice that after 4 years contract B represents a better contract offer than contract A. The salaries for each contract option represent a sequence.

Year	Contract A	Contract B
1	$30,000	$30,000
2	$33,300	$33,000
3	$36,600	$36,300
4	$39,900	$39,930
5	$43,200	$43,923
Total	$183,000	$183,153

A mathematical **sequence** is simply an ordered list of numbers. Each number in the list is a **term** of the sequence. A sequence can have a finite number of terms or an infinite number of terms. For instance, the sequence of positive odd integers that are less than 15 is a *finite* sequence

$1, 3, 5, 7, 9, 11, 13$ Finite sequence

whereas the sequence of positive odd integers is an *infinite* sequence.

$1, 3, 5, 7, 9, 11, 13, . . .$ Infinite sequence

Note that the three dots indicate that the sequence continues and has an infinite number of terms.

Because each term of a sequence is matched with its location, a sequence can be defined as a function whose domain is a subset of positive integers.

Sequences

An **infinite sequence** $a_1, a_2, a_3, . . . , a_n, . . .$ is a function whose domain is the set of positive integers.

A **finite sequence** $a_1, a_2, a_3, . . . , a_n$ is a function whose domain is the finite set $\{1, 2, 3, . . ., n\}$.

In some cases it is convenient to begin subscripting a sequence with 0 instead of 1. Then the domain of the infinite sequence is the set of nonnegative integers and the domain of the finite sequence is the set $\{0, 1, 2, . . . , n\}$. The terms of the sequence are denoted by $a_0, a_1, a_2, a_3, a_4, . . . , a_n, ...$

$a_{(\ )} = 2(\ ) + 1$

$a_{(1)} = 2(1) + 1 = 3$

$a_{(2)} = 2(2) + 1 = 5$

$\vdots$

$a_{(51)} = 2(51) + 1 = 103$

The subscripts of a sequence are used in place of function notation. For instance, if parentheses replaced the n in $a_n = 2n + 1$, the notation would be similar to function notation, as shown at the left.

Technology: Tip

Most graphing calculators have a "sequence graphing mode" that allows you to plot the terms of a sequence as points on a rectangular coordinate system. For instance, the graph of the first six terms of the sequence given by

$$a_n = n^2 - 1$$

is shown below.

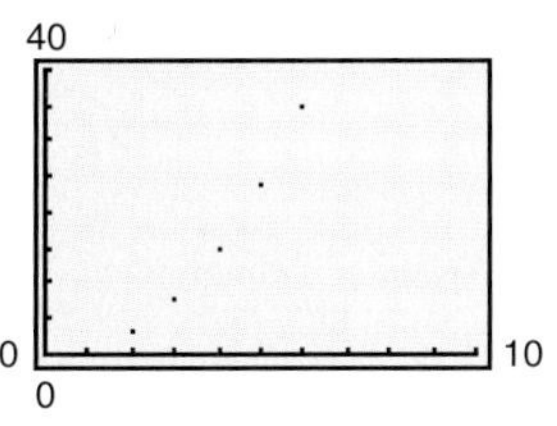

EXAMPLE 1 Writing the Terms of a Sequence

Write the first six terms of the sequence whose nth term is

$a_n = n^2 - 1.$ Begin sequence with $n = 1$.

Solution

$a_1 = (1)^2 - 1 = 0$ $\quad a_2 = (2)^2 - 1 = 3$ $\quad a_3 = (3)^2 - 1 = 8$

$a_4 = (4)^2 - 1 = 15$ $\quad a_5 = (5)^2 - 1 = 24$ $\quad a_6 = (6)^2 - 1 = 35$

The sequence can be written as $0, 3, 8, 15, 24, 35, \ldots, n^2 - 1, \ldots$.

CHECKPOINT *Now try Exercise 1.*

EXAMPLE 2 Writing the Terms of a Sequence

Write the first six terms of the sequence whose nth term is

$a_n = 3(2^n).$ Begin sequence with $n = 0$.

Solution

$a_0 = 3(2^0) = 3 \cdot 1 = 3$ $\quad a_1 = 3(2^1) = 3 \cdot 2 = 6$

$a_2 = 3(2^2) = 3 \cdot 4 = 12$ $\quad a_3 = 3(2^3) = 3 \cdot 8 = 24$

$a_4 = 3(2^4) = 3 \cdot 16 = 48$ $\quad a_5 = 3(2^5) = 3 \cdot 32 = 96$

The sequence can be written as $3, 6, 12, 24, 48, 96, \ldots, 3(2^n), \ldots$.

CHECKPOINT *Now try Exercise 3.*

EXAMPLE 3 A Sequence Whose Terms Alternate in Sign

Write the first six terms of the sequence whose nth term is

$a_n = \dfrac{(-1)^n}{2n - 1}.$ Begin sequence with $n = 1$.

Solution

$a_1 = \dfrac{(-1)^1}{2(1) - 1} = -\dfrac{1}{1}$ $\quad a_2 = \dfrac{(-1)^2}{2(2) - 1} = \dfrac{1}{3}$ $\quad a_3 = \dfrac{(-1)^3}{2(3) - 1} = -\dfrac{1}{5}$

$a_4 = \dfrac{(-1)^4}{2(4) - 1} = \dfrac{1}{7}$ $\quad a_5 = \dfrac{(-1)^5}{2(5) - 1} = -\dfrac{1}{9}$ $\quad a_6 = \dfrac{(-1)^6}{2(6) - 1} = \dfrac{1}{11}$

The sequence can be written as $-1, \dfrac{1}{3}, -\dfrac{1}{5}, \dfrac{1}{7}, -\dfrac{1}{9}, \dfrac{1}{11}, \ldots, \dfrac{(-1)^n}{2n - 1}, \ldots$.

✓ **CHECKPOINT** *Now try Exercise 5.*

Smart Study Strategy

Go to page 812 for ways to *Form a Final Exam Study Group.*

2 ▶ Write the terms of sequences involving factorials.

Factorial Notation

Some very important sequences in mathematics involve terms that are defined with special types of products called **factorials.**

Definition of Factorial

If n is a positive integer, n **factorial** is defined as

$$n! = 1 \cdot 2 \cdot 3 \cdot 4 \cdot \cdot \cdot \cdot \cdot (n - 1) \cdot n.$$

As a special case, zero factorial is defined as $0! = 1$.

The first several factorial values are as follows.

$0! = 1$ $\qquad$ $1! = 1$

$2! = 1 \cdot 2 = 2$ $\qquad$ $3! = 1 \cdot 2 \cdot 3 = 6$

$4! = 1 \cdot 2 \cdot 3 \cdot 4 = 24$ $\qquad$ $5! = 1 \cdot 2 \cdot 3 \cdot 4 \cdot 5 = 120$

Many calculators have a factorial key, denoted by $\boxed{n!}$. If your calculator has such a key, try using it to evaluate $n!$ for several values of n. You will see that as n increases, the value of $n!$ becomes very large. For instance,

$$10! = 3{,}628{,}800.$$

EXAMPLE 4 A Sequence Involving Factorials

Write the first six terms of the sequence with the given nth term.

a. $a_n = \dfrac{1}{n!}$ $\qquad$ Begin sequence with $n = 0$.

b. $a_n = \dfrac{2^n}{n!}$ $\qquad$ Begin sequence with $n = 0$.

Solution

a. $a_0 = \dfrac{1}{0!} = \dfrac{1}{1} = 1$ $\qquad$ $a_1 = \dfrac{1}{1!} = \dfrac{1}{1} = 1$

$a_2 = \dfrac{1}{2!} = \dfrac{1}{1 \cdot 2} = \dfrac{1}{2}$ $\qquad$ $a_3 = \dfrac{1}{3!} = \dfrac{1}{1 \cdot 2 \cdot 3} = \dfrac{1}{6}$

$a_4 = \dfrac{1}{4!} = \dfrac{1}{1 \cdot 2 \cdot 3 \cdot 4} = \dfrac{1}{24}$ $\qquad$ $a_5 = \dfrac{1}{5!} = \dfrac{1}{1 \cdot 2 \cdot 3 \cdot 4 \cdot 5} = \dfrac{1}{120}$

b. $a_0 = \dfrac{2^0}{0!} = \dfrac{1}{1} = 1$ $\qquad$ $a_1 = \dfrac{2^1}{1!} = \dfrac{2}{1} = 2$

$a_2 = \dfrac{2^2}{2!} = \dfrac{2 \cdot 2}{1 \cdot 2} = \dfrac{4}{2} = 2$ $\qquad$ $a_3 = \dfrac{2^3}{3!} = \dfrac{2 \cdot 2 \cdot 2}{1 \cdot 2 \cdot 3} = \dfrac{8}{6} = \dfrac{4}{3}$

$a_4 = \dfrac{2^4}{4!} = \dfrac{2 \cdot 2 \cdot 2 \cdot 2}{1 \cdot 2 \cdot 3 \cdot 4} = \dfrac{2}{3}$ $\qquad$ $a_5 = \dfrac{2^5}{5!} = \dfrac{2 \cdot 2 \cdot 2 \cdot 2 \cdot 2}{1 \cdot 2 \cdot 3 \cdot 4 \cdot 5} = \dfrac{4}{15}$

✓ ***CHECKPOINT*** *Now try Exercise 19.*

3 ▶ Find the apparent nth term of a sequence.

Finding the nth Term of a Sequence

Sometimes you have the first several terms of a sequence and need to find a formula (the nth term) to generate those terms. Pattern recognition is crucial in finding a form for the nth term.

Study Tip

Simply listing the first few terms is not sufficient to define a unique sequence—the nth term must be given. Consider the sequence

$$\frac{1}{2}, \frac{1}{4}, \frac{1}{8}, \frac{1}{15}, ...$$

The first three terms are identical to the first three terms of the sequence in Example 5(a). However, the nth term of this sequence is defined as

$$a_n = \frac{6}{(n+1)(n^2 - n + 6)}.$$

EXAMPLE 5 Finding the nth Term of a Sequence

Write an expression for the nth term of each sequence.

a. $\frac{1}{2}, \frac{1}{4}, \frac{1}{8}, \frac{1}{16}, \frac{1}{32}, . . .$ **b.** $1, -4, 9, -16, 25, . . .$

Solution

a.

n:	1	2	3	4	5	. . .	n
Terms:	$\frac{1}{2}$	$\frac{1}{4}$	$\frac{1}{8}$	$\frac{1}{16}$	$\frac{1}{32}$	. . .	a_n

Pattern: The numerators are 1 and the denominators are increasing powers of 2.

So, an expression for the nth term is $\frac{1}{2^n}$.

b.

n:	1	2	3	4	5	. . .	n
Terms:	1	-4	9	-16	25	. . .	a_n

Pattern: The terms have alternating signs, with those in the even positions being negative. The absolute value of each term is the square of n.

So, an expression for the nth term is $(-1)^{n+1}n^2$.

✓ **CHECKPOINT** *Now try Exercise 55.*

4 ▶ Sum the terms of sequences to obtain series, and use sigma notation to represent partial sums.

Series

In the table of salaries at the beginning of this section, the terms of the finite sequence were *added*. If you add all the terms of an *infinite* sequence, you obtain a **series.**

Definition of Series

For an infinite sequence $a_1, a_2, a_3, . . . , a_n, . . .$

1. the sum of the first n terms

$$S_n = a_1 + a_2 + a_3 + \cdot \cdot \cdot + a_n$$

is called a **partial sum,** and

2. the sum of all the terms

$$a_1 + a_2 + a_3 + \cdot \cdot \cdot + a_n + \cdot \cdot \cdot$$

is called an **infinite series,** or simply a **series.**

Technology: Tip

Most graphing calculators have a built-in program that will calculate the partial sum of a sequence. Consult the user's guide for your graphing calculator for specific instructions.

EXAMPLE 6 Finding Partial Sums

Find the indicated partial sums for each sequence.

a. Find S_1, S_2, and S_5 for $a_n = 3n - 1$.

b. Find S_2, S_3, and S_4 for $a_n = \dfrac{(-1)^n}{n+1}$.

Solution

a. The first five terms of the sequence $a_n = 3n - 1$ are

$$a_1 = 2,\ a_2 = 5,\ a_3 = 8,\ a_4 = 11, \text{ and } a_5 = 14.$$

So, the partial sums are

$$S_1 = 2,\ S_2 = 2 + 5 = 7, \text{ and } S_5 = 2 + 5 + 8 + 11 + 14 = 40.$$

b. The first four terms of the sequence $a_n = \dfrac{(-1)^n}{n+1}$ are

$$a_1 = -\frac{1}{2},\ a_2 = \frac{1}{3},\ a_3 = -\frac{1}{4}, \text{ and } a_4 = \frac{1}{5}.$$

So, the partial sums are

$$S_2 = -\frac{1}{2} + \frac{1}{3} = -\frac{1}{6},\ S_3 = -\frac{1}{2} + \frac{1}{3} - \frac{1}{4} = -\frac{5}{12}, \text{ and}$$

$$S_4 = -\frac{1}{2} + \frac{1}{3} - \frac{1}{4} + \frac{1}{5} = -\frac{13}{60}.$$

✓ ***CHECKPOINT*** *Now try Exercise 67.*

A convenient shorthand notation for denoting a partial sum is called **sigma notation.** This name comes from the use of the uppercase Greek letter sigma, written as Σ.

Definition of Sigma Notation

The sum of the first n terms of the sequence whose nth term is a_n is

$$\sum_{i=1}^{n} a_i = a_1 + a_2 + a_3 + a_4 + \cdot \cdot \cdot + a_n$$

where i is the **index of summation,** n is the **upper limit of summation,** and 1 is the **lower limit of summation.**

Sigma (summation) notation is an instruction to add the terms of a sequence. From the definition above, the upper limit of summation tells you where to end the sum. Sigma notation helps you generate the appropriate terms of the sequence prior to finding the actual sum.

EXAMPLE 7 Finding a Sum in Sigma Notation

$$\sum_{i=1}^{6} 2i = 2(1) + 2(2) + 2(3) + 2(4) + 2(5) + 2(6)$$
$$= 2 + 4 + 6 + 8 + 10 + 12$$
$$= 42$$

CHECKPOINT *Now try Exercise 71.*

Study Tip

In Example 7, the index of summation is i and the summation begins with $i = 1$. Any letter can be used as the index of summation, and the summation can begin with any integer. For instance, in Example 8, the index of summation is k and the summation begins with $k = 0$.

EXAMPLE 8 Finding a Sum in Sigma Notation

$$\sum_{k=0}^{8} \frac{1}{k!} = \frac{1}{0!} + \frac{1}{1!} + \frac{1}{2!} + \frac{1}{3!} + \frac{1}{4!} + \frac{1}{5!} + \frac{1}{6!} + \frac{1}{7!} + \frac{1}{8!}$$
$$= 1 + 1 + \frac{1}{2} + \frac{1}{6} + \frac{1}{24} + \frac{1}{120} + \frac{1}{720} + \frac{1}{5040} + \frac{1}{40{,}320}$$
$$\approx 2.71828$$

Note that this sum is approximately $e = 2.71828\ldots$.

CHECKPOINT *Now try Exercise 77.*

EXAMPLE 9 Writing a Sum in Sigma Notation

Write each sum in sigma notation.

a. $\frac{2}{2} + \frac{2}{3} + \frac{2}{4} + \frac{2}{5} + \frac{2}{6}$ **b.** $1 - \frac{1}{3} + \frac{1}{9} - \frac{1}{27} + \frac{1}{81}$

Solution

a. To write this sum in sigma notation, you must find a pattern for the terms. You can see that the terms have numerators of 2 and denominators that range over the integers from 2 to 6. So, one possible sigma notation is

$$\sum_{i=1}^{5} \frac{2}{i+1} = \frac{2}{2} + \frac{2}{3} + \frac{2}{4} + \frac{2}{5} + \frac{2}{6}.$$

b. To write this sum in sigma notation, you must find a pattern for the terms. You can see that their numerators alternate in sign and their denominators are integer powers of 3, starting with 3^0 and ending with 3^4. So, one possible sigma notation is

$$\sum_{i=0}^{4} \frac{(-1)^i}{3^i} = \frac{1}{3^0} + \frac{-1}{3^1} + \frac{1}{3^2} + \frac{-1}{3^3} + \frac{1}{3^4}.$$

CHECKPOINT *Now try Exercise 95.*

Concept Check

1. Explain how to find the sixth term of the sequence $a_n = \frac{n-3}{4}$.

2. Explain the difference between $a_n = 5n$ and $a_n = 5n!$.

3. Determine whether the statement is true or false. Justify your answer. The only expression for the nth term of the sequence 1, 3, 9, . . . is $2n^2 + 1$.

4. How many terms are in the sum $\sum_{i=1}^{5} i^2$? In the sum $\sum_{i=0}^{5} 10i$?

13.1 EXERCISES

Go to pages 852–853 to record your assignments.

Developing Skills

In Exercises 1–22, write the first five terms of the sequence. (Assume that n begins with 1.) ***See Examples 1–4.***

✓ 1. $a_n = 2n$

2. $a_n = 3n$

✓ 3. $a_n = \left(\frac{1}{4}\right)^n$

4. $a_n = \left(\frac{1}{3}\right)^n$

✓ 5. $a_n = (-1)^n 2n$

6. $a_n = (-1)^{n+1} 3n$

7. $a_n = \left(-\frac{1}{2}\right)^{n+1}$

8. $a_n = \left(\frac{2}{3}\right)^{n-1}$

9. $a_n = 5n - 2$

10. $a_n = 2n + 3$

11. $a_n = \frac{4}{n+3}$

12. $a_n = \frac{9}{5+n}$

13. $a_n = \frac{3n}{5n-1}$

14. $a_n = \frac{2n}{6n-3}$

15. $a_n = \frac{(-1)^n}{n^2}$

16. $a_n = \frac{1}{\sqrt{n}}$

17. $a_n = 2 + \frac{1}{4^n}$

18. $a_n = 10 - \frac{1}{5^n}$

✓ 19. $a_n = \frac{(n+1)!}{n!}$

20. $a_n = \frac{n!}{(n-1)!}$

21. $a_n = \frac{2 + (-2)^n}{n!}$

22. $a_n = \frac{1 + (-1)^n}{n^2}$

In Exercises 23–26, find the indicated term of the sequence.

23. $a_n = (-1)^n(5n - 3)$

$a_{15} =$ ▢

24. $a_n = (-1)^{n+1}(3n + 10)$

$a_{20} =$ ▢

25. $a_n = \frac{n^2 - 2}{(n-1)!}$

$a_8 =$ ▢

26. $a_n = \frac{n^2}{n!}$

$a_{12} =$ ▢

In Exercises 27–38, simplify the expression.

27. $\frac{5!}{4!}$

28. $\frac{6!}{8!}$

29. $\frac{10!}{12!}$

30. $\frac{16!}{13!}$

31. $\dfrac{25!}{20!5!}$

32. $\dfrac{20!}{15!5!}$

33. $\dfrac{n!}{(n+1)!}$

34. $\dfrac{(n+2)!}{n!}$

35. $\dfrac{(n+1)!}{(n-1)!}$

36. $\dfrac{(3n)!}{(3n+2)!}$

37. $\dfrac{(2n)!}{(2n-1)!}$

38. $\dfrac{(5n+2)!}{(5n)!}$

In Exercises 39–42, match the sequence with the graph of its first 10 terms. [The graphs are labeled (a), (b), (c), and (d).]

(a)

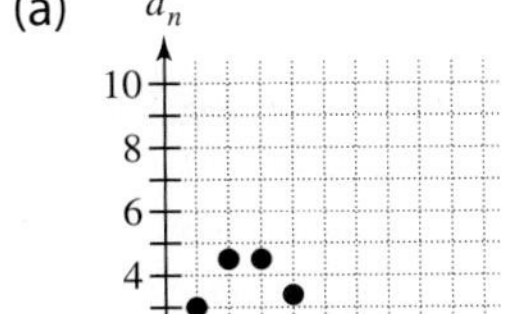

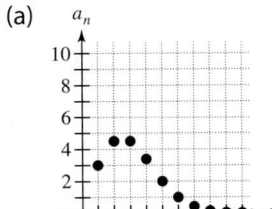

(b)

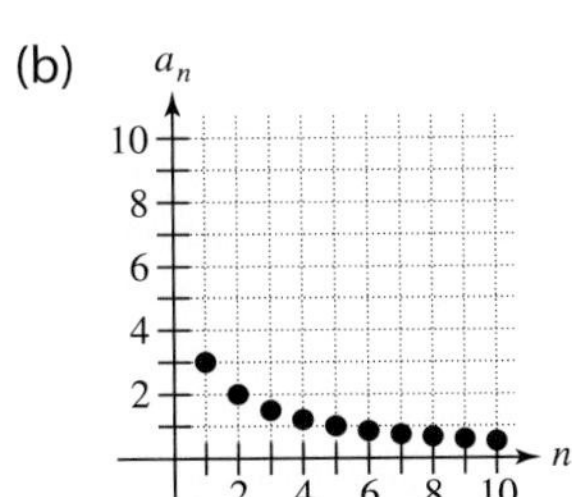

(c)

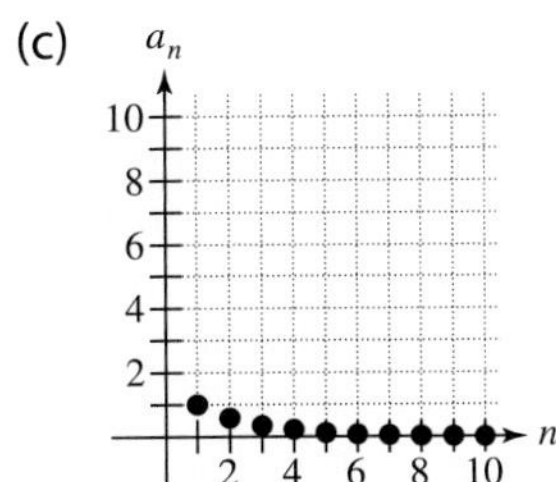

(d)

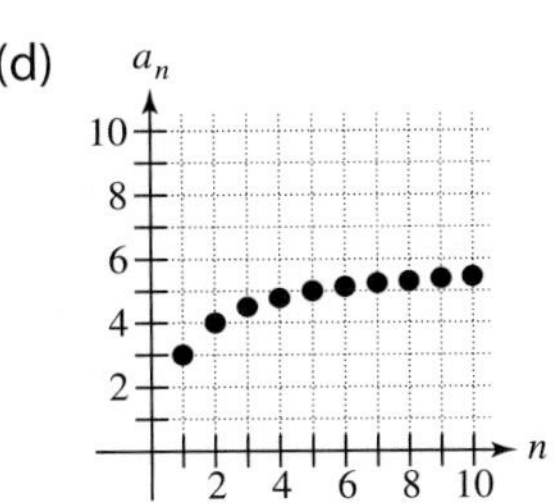

39. $a_n = \dfrac{6}{n+1}$

40. $a_n = \dfrac{6n}{n+1}$

41. $a_n = (0.6)^{n-1}$

42. $a_n = \dfrac{3^n}{n!}$

In Exercises 43–48, use a graphing calculator to graph the first 10 terms of the sequence.

43. $a_n = \dfrac{4n^2}{n^2 - 2}$

44. $a_n = \dfrac{2n^2}{n^2 + 1}$

45. $a_n = 3 - \dfrac{4}{n}$

46. $a_n = 5 + \dfrac{3}{n}$

47. $a_n = 100(-0.4)^{n+1}$

48. $a_n = 10\left(\dfrac{3}{4}\right)^{n-1}$

In Exercises 49–66, write an expression for the nth term of the sequence. (Assume that n begins with 1.) ***See Example 5.***

49. 1, 3, 5, 7, 9, . . .

50. 2, 4, 6, 8, 10, . . .

51. 2, 6, 10, 14, 18, . . .

52. 6, 11, 16, 21, 26, . . .

53. 0, 3, 8, 15, 24, . . .

54. 1, 8, 27, 64, 125, . . .

✓ 55. 2, −4, 6, −8, 10, . . .

56. 1, −1, 1, −1, 1, . . .

57. $\frac{2}{3}, \frac{3}{4}, \frac{4}{5}, \frac{5}{6}, \frac{6}{7}, \ldots$

58. $\frac{2}{1}, \frac{3}{3}, \frac{4}{5}, \frac{5}{7}, \frac{6}{9}, \ldots$

59. $\frac{-1}{5}, \frac{1}{25}, \frac{-1}{125}, \frac{1}{625}, \frac{-1}{3125}, \ldots$

60. $1, \frac{1}{4}, \frac{1}{9}, \frac{1}{16}, \frac{1}{25}, \ldots$

61. $1, \frac{1}{2}, \frac{1}{4}, \frac{1}{8}, \ldots$

62. $\frac{1}{3}, \frac{2}{9}, \frac{4}{27}, \frac{8}{81}, \ldots$

63. $1 + \frac{1}{1}, 1 + \frac{1}{2}, 1 + \frac{1}{3}, 1 + \frac{1}{4}, 1 + \frac{1}{5}, \ldots$

64. $1 + \frac{1}{2}, 1 + \frac{3}{4}, 1 + \frac{7}{8}, 1 + \frac{15}{16}, 1 + \frac{31}{32}, \ldots$

65. $-\frac{1}{2}, \frac{1}{6}, -\frac{1}{24}, \frac{1}{120}, -\frac{1}{720}, \ldots$

66. $1, 2, \dfrac{2^2}{2}, \dfrac{2^3}{6}, \dfrac{2^4}{24}, \dfrac{2^5}{120}, \ldots$

In Exercises 67–70, find the indicated partial sums for the sequence. ***See Example 6.***

✓ 67. Find S_1, S_2, and S_6 for $a_n = 2n + 5$.

68. Find S_3, S_4, and S_{10} for $a_n = n^3 - 1$.

69. Find S_2, S_3, and S_9 for $a_n = \dfrac{1}{n}$.

70. Find S_1, S_3, and S_5 for $a_n = \dfrac{(-1)^{n+1}}{n+1}$.

In Exercises 71–86, find the partial sum. ***See Examples 7 and 8.***

71. $\sum_{k=1}^{5} 6k$

72. $\sum_{k=1}^{4} 5k$

73. $\sum_{i=0}^{6} (2i + 5)$

74. $\sum_{i=0}^{4} (2i + 3)$

75. $\sum_{j=3}^{7} (6j - 10)$

76. $\sum_{i=2}^{7} (4i - 1)$

77. $\sum_{j=1}^{5} \frac{(-1)^{j+1}}{j^2}$

78. $\sum_{j=0}^{3} \frac{1}{j^2 + 1}$

79. $\sum_{m=1}^{8} \frac{m}{m + 1}$

80. $\sum_{k=1}^{6} \frac{k - 3}{k + 2}$

81. $\sum_{k=1}^{6} (-8)$

82. $\sum_{n=3}^{12} 10$

83. $\sum_{i=1}^{8} \left(\frac{1}{i} - \frac{1}{i + 1}\right)$

84. $\sum_{k=1}^{5} \left(\frac{2}{k} - \frac{2}{k + 2}\right)$

85. $\sum_{n=0}^{5} \left(-\frac{1}{3}\right)^n$

86. $\sum_{k=1}^{4} \left(\frac{5}{3}\right)^{k-1}$

In Exercises 87–94, use a graphing calculator to find the partial sum.

87. $\sum_{n=1}^{8} 10n^2$

88. $\sum_{n=0}^{5} 2n^2$

89. $\sum_{j=2}^{6} (j! - j)$

90. $\sum_{i=0}^{4} (i! + 4)$

91. $\sum_{j=0}^{4} \frac{6}{j!}$

92. $\sum_{k=1}^{6} \left(\frac{1}{2k} - \frac{1}{2k - 1}\right)$

93. $\sum_{k=1}^{6} \ln k$

94. $\sum_{k=3}^{5} \frac{\log_{10} k}{k}$

In Exercises 95–112, write the sum using sigma notation. (Begin with $k = 0$ or $k = 1$.) ***See Example 9.***

95. $1 + 2 + 3 + 4 + 5$

96. $8 + 9 + 10 + 11 + 12 + 13 + 14$

97. $5 + 10 + 15 + 20 + 25 + 30$

98. $24 + 30 + 36 + 42$

99. $\frac{3}{1 + 1} + \frac{3}{1 + 2} + \frac{3}{1 + 3} + \cdots + \frac{3}{1 + 50}$

100. $\frac{1}{2(1) + 1} + \frac{1}{2(2) + 1} + \frac{1}{2(3) + 1} + \cdots + \frac{1}{2(30) + 1}$

101. $\frac{1}{2(1)} + \frac{1}{2(2)} + \frac{1}{2(3)} + \frac{1}{2(4)} + \cdots + \frac{1}{2(10)}$

102. $\frac{1}{1^2} + \frac{1}{2^2} + \frac{1}{3^2} + \frac{1}{4^2} + \cdots + \frac{1}{20^2}$

103. $\frac{1}{2^0} + \frac{1}{2^1} + \frac{1}{2^2} + \frac{1}{2^3} + \cdots + \frac{1}{2^{12}}$

104. $\left(-\frac{2}{3}\right)^0 + \left(-\frac{2}{3}\right)^1 + \left(-\frac{2}{3}\right)^2 + \cdots + \left(-\frac{2}{3}\right)^{20}$

105. $\frac{1}{1(1 + 1)} + \frac{1}{2(2 + 1)} + \frac{1}{3(3 + 1)} + \cdots + \frac{1}{20(20 + 1)}$

106. $\frac{1}{2^3} - \frac{1}{4^3} + \frac{1}{6^3} - \frac{1}{8^3} + \cdots + \frac{1}{14^3}$

107. $\frac{1}{2} + \frac{2}{3} + \frac{3}{4} + \frac{4}{5} + \frac{5}{6} + \cdots + \frac{11}{12}$

108. $\frac{2}{4} + \frac{4}{7} + \frac{6}{10} + \frac{8}{13} + \frac{10}{16} + \cdots + \frac{20}{31}$

109. $\frac{2}{4} + \frac{4}{5} + \frac{6}{6} + \frac{8}{7} + \cdots + \frac{40}{23}$

110. $\left(2 + \frac{1}{1}\right) + \left(2 + \frac{1}{2}\right) + \left(2 + \frac{1}{3}\right) + \cdots + \left(2 + \frac{1}{25}\right)$

111. $1 + 1 + 2 + 6 + 24 + 120 + 720$

112. $-\frac{1}{10} + \frac{1}{20} - \frac{1}{60} + \frac{1}{240} - \frac{1}{1200} + \frac{1}{7200} - \frac{1}{50{,}400}$

Solving Problems

113. *Compound Interest* A deposit of $500 is made in an account that earns 7% interest compounded yearly. The balance in the account after N years is given by

$$A_N = 500(1 + 0.07)^N, \quad N = 1, 2, 3, \ldots.$$

(a) Compute the first eight terms of the sequence.

(b) Find the balance in this account after 40 years by computing A_{40}.

(c) Use a graphing calculator to graph the first 40 terms of the sequence.

(d) The terms are increasing. Is the rate of growth of the terms increasing? Explain.

114. *Sports* The number of degrees a_n in each angle of a regular n-sided polygon is

$$a_n = \frac{180(n - 2)}{n}, \quad n \geq 3.$$

The surface of a soccer ball is made of regular hexagons and pentagons. When a soccer ball is taken apart and flattened, as shown in the figure, the sides don't meet each other. Use the terms a_5 and a_6 to explain why there are gaps between adjacent hexagons.

115. *Stars* Stars are formed by placing n equally spaced points on a circle and connecting each point with a second point on the circle (see figure). The measure in degrees d_n of the angle at each tip of the star is given by

$$d_n = \frac{180(n - 4)}{n}, \quad n \geq 5.$$

Write the first six terms of this sequence.

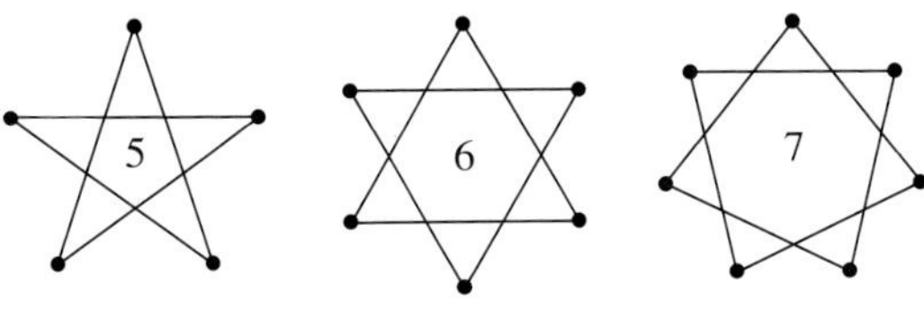

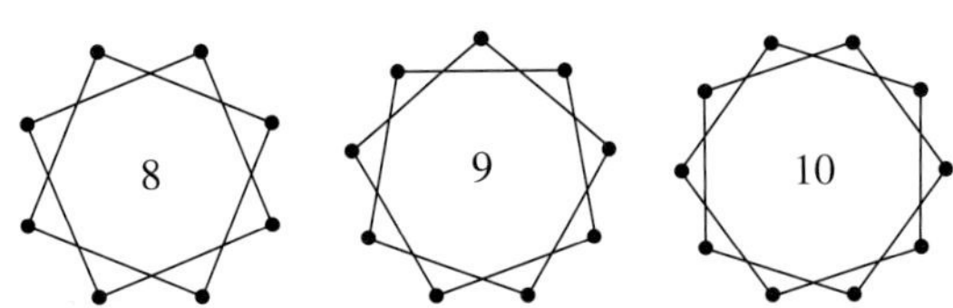

116. *Stars* The stars in Exercise 115 were formed by placing n equally spaced points on a circle and connecting each point with the second point from it on the circle. The stars in the figure for this exercise were formed in a similar way except that each point was connected with the third point from it. For these stars, the measure in degrees d_n of the angle at each point is given by

$$d_n = \frac{180(n - 6)}{n}, \quad n \geq 7.$$

Write the first five terms of this sequence.

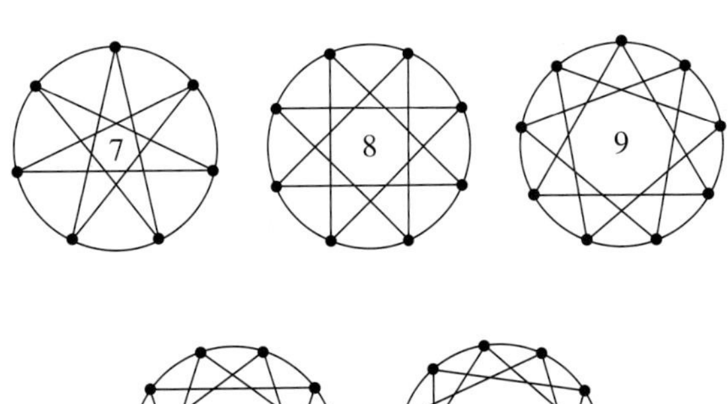

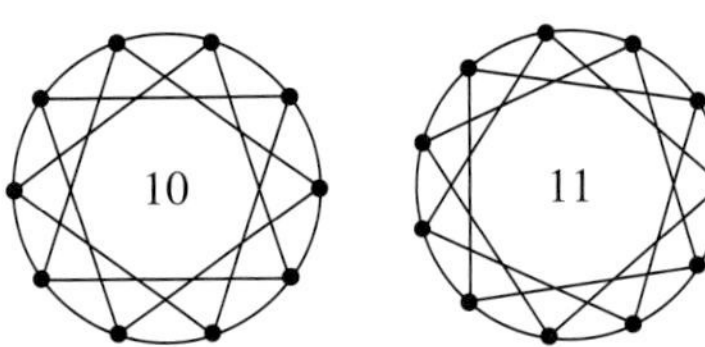

117. *Depreciation* At the end of each year, the value of a sport utility vehicle with an initial cost of \$32,000 is three-fourths what it was at the beginning of the year. After n years, its value is given by

$$a_n = 32{,}000\left(\frac{3}{4}\right)^n, \quad n = 1, 2, 3, \ldots.$$

(a) Find the value of the sport utility vehicle 3 years after it was purchased by computing a_3.

(b) Find the value of the sport utility vehicle 6 years after it was purchased by computing a_6. Is this value half of what it was after 3 years? Explain.

118. *Annual Revenue* The revenue a_n (in millions of dollars) of Netflix, Inc., for the years 2001 through 2007 is modeled by

$$a_n = 18.054n^2 + 51.47n - 11.1, \quad n = 1, 2, \ldots, 7$$

where n is the year, with $n = 1$ corresponding to 2001. (Source: Netflix, Inc.)

(a) Find the terms of this finite sequence.

(b) Use a graphing calculator to construct a bar graph that represents the sequence.

(c) Find $\sum_{i=1}^{7} a_i$. What does this number represent?

Explaining Concepts

119. In your own words, explain why a sequence is a function.

120. The nth term of a sequence is $a_n = (-1)^n n$. Which terms of the sequence are negative? Explain.

121. Explain the difference between $a_n = 4n!$ and $a_n = (4n)!$.

In Exercises 122–124, decide whether the statement is true or false. Justify your answer.

122. $\sum_{i=1}^{4} (i^2 + 2i) = \sum_{i=1}^{4} i^2 + \sum_{i=1}^{4} 2i$

123. $\sum_{k=1}^{4} 3k = 3\sum_{k=1}^{4} k$

124. $\sum_{j=1}^{4} 2^j = \sum_{j=3}^{6} 2^{j-2}$

Cumulative Review

In Exercises 125–128, evaluate the expression for the specified value of the variable.

125. $-2n + 15;\ n = 3$

126. $-20n + 100;\ n = 4$

127. $25 - 3(n + 4);\ n = 8$

128. $-\frac{3}{2}(n - 1) + 6;\ n = 10$

In Exercises 129–132, identify the center and radius of the circle, and sketch the circle.

129. $x^2 + y^2 = 36$

130. $4x^2 + 4y^2 = 9$

131. $x^2 + y^2 + 4x - 12 = 0$

132. $x^2 + y^2 - 10x - 2y - 199 = 0$

In Exercises 133–136, identify the vertex and focus of the parabola, and sketch the parabola.

133. $x^2 = 6y$

134. $y^2 = 9x$

135. $x^2 + 8y + 32 = 0$

136. $y^2 - 10x + 6y + 29 = 0$

13.2 Arithmetic Sequences

Lynn Goldsmith/CORBIS

What You Should Learn

1 ▶ Recognize, write, and find the nth terms of arithmetic sequences.

2 ▶ Find the nth partial sum of an arithmetic sequence.

3 ▶ Use arithmetic sequences to solve application problems.

Why You Should Learn It

An arithmetic sequence can reduce the amount of time it takes to find the sum of a sequence of numbers with a common difference. For instance, in Exercise 125 on page 831, you will use an arithmetic sequence to determine how much to charge for tickets to a concert at an outdoor arena.

1 ▶ Recognize, write, and find the nth terms of arithmetic sequences.

Arithmetic Sequences

A sequence whose consecutive terms have a common difference is called an **arithmetic sequence.**

> **Definition of Arithmetic Sequence**
>
> A sequence is called **arithmetic** if the differences between consecutive terms are the same. So, the sequence
>
> $$a_1, a_2, a_3, a_4, \ldots, a_n, \ldots$$
>
> is arithmetic if there is a number d such that
>
> $$a_2 - a_1 = d, \quad a_3 - a_2 = d, \quad a_4 - a_3 = d$$
>
> and so on. The number d is the **common difference** of the sequence.

EXAMPLE 1 Examples of Arithmetic Sequences

a. The sequence whose nth term is $3n + 2$ is arithmetic. For this sequence, the common difference between consecutive terms is 3.

$$\underbrace{5, 8}_{8-5=3}, 11, 14, \ldots, 3n + 2, \ldots \qquad \text{Begin with } n = 1.$$

b. The sequence whose nth term is $7 - 5n$ is arithmetic. For this sequence, the common difference between consecutive terms is -5.

$$\underbrace{2, -3}_{-3-2=-5}, -8, -13, \ldots, 7 - 5n, \ldots \qquad \text{Begin with } n = 1.$$

c. The sequence whose nth term is $\frac{1}{4}(n + 3)$ is arithmetic. For this sequence, the common difference between consecutive terms is $\frac{1}{4}$.

$$\underbrace{1, \frac{5}{4}}_{\frac{5}{4}-1=\frac{1}{4}}, \frac{3}{2}, \frac{7}{4}, \ldots, \frac{1}{4}(n + 3), \ldots \qquad \text{Begin with } n = 1.$$

✓ ***CHECKPOINT*** *Now try Exercise 1.*

Study Tip

The *n*th term of an arithmetic sequence can be derived from the following pattern.

$a_1 = a_1$ 1st term

$a_2 = a_1 + d$ 2nd term

$a_3 = a_1 + 2d$ 3rd term

$a_4 = a_1 + 3d$ 4th term

$a_5 = a_1 + 4d$ 5th term

1 less

⋮ ⋮

$a_n = a_1 + (n - 1)d$ *n*th term

1 less

The *n*th Term of an Arithmetic Sequence

The *n*th term of an arithmetic sequence has the form

$$a_n = a_1 + (n - 1)d$$

where d is the common difference between the terms of the sequence, and a_1 is the first term.

EXAMPLE 2 Finding the *n*th Term of an Arithmetic Sequence

Find a formula for the *n*th term of the arithmetic sequence whose common difference is 2 and whose first term is 5.

Solution

You know that the formula for the *n*th term is of the form $a_n = a_1 + (n - 1)d$. Moreover, because the common difference is $d = 2$ and the first term is $a_1 = 5$, the formula must have the form

$$a_n = 5 + 2(n - 1).$$

So, the formula for the *n*th term is $a_n = 2n + 3$, and the sequence has the following form.

$$5, 7, 9, 11, 13, \ldots, 2n + 3, \ldots$$

CHECKPOINT *Now try Exercise 53.*

If you know the *n*th term and the common difference of an arithmetic sequence, you can find the $(n + 1)$th term by using the **recursion formula**

$$a_{n+1} = a_n + d.$$

EXAMPLE 3 Using a Recursion Formula

The 12th term of an arithmetic sequence is 52 and the common difference is 3.

a. What is the 13th term of the sequence? **b.** What is the first term?

Solution

a. You know that $a_{12} = 52$ and $d = 3$. So, using the recursion formula $a_{13} = a_{12} + d$, you can determine that the 13th term of the sequence is

$$a_{13} = 52 + 3 = 55.$$

b. Using $n = 12$, $d = 3$, and $a_{12} = 52$ in the formula $a_n = a_1 + (n - 1)d$ yields

$$52 = a_1 + (12 - 1)(3)$$

$$19 = a_1.$$

CHECKPOINT *Now try Exercise 71.*

2 ▶ Find the nth partial sum of an arithmetic sequence.

The Partial Sum of an Arithmetic Sequence

The sum of the first n terms of an arithmetic sequence is called the ***n*th partial sum** of the sequence. For instance, the fifth partial sum of the arithmetic sequence whose nth term is $3n + 4$ is

$$\sum_{i=1}^{5} (3i + 4) = 7 + 10 + 13 + 16 + 19 = 65.$$

To find a formula for the nth partial sum S_n of an arithmetic sequence, write out S_n forwards and backwards and then add the two forms, as follows.

$$S_n = a_1 + (a_1 + d) + (a_1 + 2d) + \cdots + [a_1 + (n - 1)d] \quad \text{Forwards}$$

$$S_n = a_n + (a_n - d) + (a_n - 2d) + \cdots + [a_n - (n - 1)d] \quad \text{Backwards}$$

$$2S_n = (a_1 + a_n) + (a_1 + a_n) + (a_1 + a_n) + \cdots + [a_1 + a_n] \quad \text{Sum of two equations}$$

$$= n(a_1 + a_n) \quad n \text{ groups of } (a_1 + a_n)$$

Dividing each side by 2 yields the following formula.

Study Tip

You can use the formula for the nth partial sum of an arithmetic sequence to find the sum of consecutive numbers. For instance, the sum of the integers from 1 to 100 is

$$\sum_{i=1}^{100} i = \frac{100}{2}(1 + 100)$$

$$= 50(101)$$

$$= 5050.$$

The *n*th Partial Sum of an Arithmetic Sequence

The nth partial sum of the arithmetic sequence whose nth term is a_n is

$$\sum_{i=1}^{n} a_i = a_1 + a_2 + a_3 + a_4 + \cdots + a_n$$

$$= \frac{n}{2}(a_1 + a_n).$$

Or, equivalently, you can find the sum of the first n terms of an arithmetic sequence by multiplying the average of the first and nth terms by n.

EXAMPLE 4 Finding the *n*th Partial Sum

Find the sum of the first 20 terms of the arithmetic sequence whose nth term is $4n + 1$.

Solution

The first term of this sequence is $a_1 = 4(1) + 1 = 5$ and the 20th term is $a_{20} = 4(20) + 1 = 81$. So, the sum of the first 20 terms is given by

$$\sum_{i=1}^{n} a_i = \frac{n}{2}(a_1 + a_n) \quad n\text{th partial sum formula}$$

$$\sum_{i=1}^{20} (4i + 1) = \frac{20}{2}(a_1 + a_{20}) \quad \text{Substitute 20 for } n.$$

$$= 10(5 + 81) \quad \text{Substitute 5 for } a_1 \text{ and 81 for } a_{20}.$$

$$= 10(86) \quad \text{Simplify.}$$

$$= 860. \quad n\text{th partial sum}$$

✓ **CHECKPOINT** *Now try Exercise 79.*

EXAMPLE 5 Finding the *n*th Partial Sum

Find the sum of the even integers from 2 to 100.

Solution

Because the integers

2, 4, 6, 8, . . . , 100

form an arithmetic sequence, you can find the sum as follows.

$$\sum_{i=1}^{n} a_i = \frac{n}{2}(a_1 + a_n)$$ *nth partial sum formula*

$$\sum_{i=1}^{50} 2i = \frac{50}{2}(a_1 + a_{50})$$ Substitute 50 for n.

$$= 25(2 + 100)$$ Substitute 2 for a_1 and 100 for a_{50}.

$$= 25(102)$$ Simplify.

$$= 2550$$ *nth partial sum*

✓ **CHECKPOINT** *Now try Exercise 89.*

3 ▶ Use arithmetic sequences to solve application problems.

Application

EXAMPLE 6 Total Sales 

Your business sells \$100,000 worth of handmade furniture during its first year. You have a goal of increasing annual sales by \$25,000 each year for 9 years. If you meet this goal, how much will you sell during your first 10 years of business?

Solution

The annual sales during the first 10 years form the following arithmetic sequence.

\$100,000, \$125,000, \$150,000, \$175,000, \$200,000,
\$225,000, \$250,000, \$275,000, \$300,000, \$325,000

Using the formula for the nth partial sum of an arithmetic sequence, you can find the total sales during the first 10 years as follows.

$$\text{Total sales} = \frac{n}{2}(a_1 + a_n)$$ *nth partial sum formula*

$$= \frac{10}{2}(100{,}000 + 325{,}000)$$ Substitute for n, a_1, and a_n.

$$= 5(425{,}000)$$ Simplify.

$$= \$2{,}125{,}000$$ Simplify.

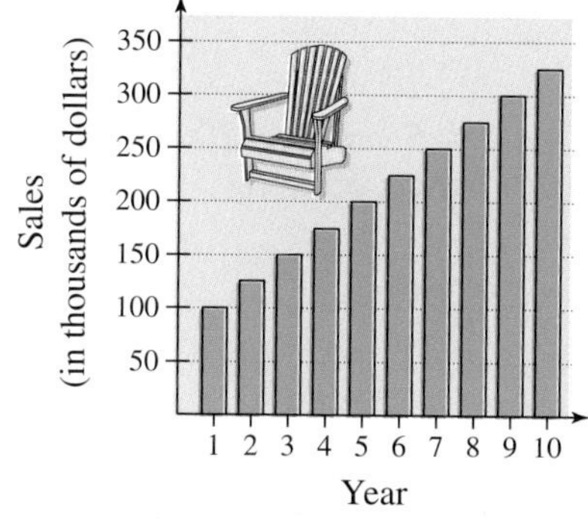

Figure 13.1

From the bar graph shown in Figure 13.1, notice that the annual sales for your company follow a *linear growth* pattern. In other words, saying that a quantity increases arithmetically is the same as saying that it increases linearly.

✓ **CHECKPOINT** *Now try Exercise 123.*

Concept Check

1. In an arithmetic sequence, the common difference between consecutive terms is 3. How can you use the value of one term to find the value of the next term in the sequence?

2. Explain how you can use the first two terms of an arithmetic sequence to write a formula for the nth term of the sequence.

3. Explain how you can use the average of the first term and the nth term of an arithmetic sequence to find the nth partial sum of the sequence.

4. In an arithmetic sequence, you know the common difference d between consecutive terms. What else do you need to know to find a_6?

13.2 EXERCISES

Go to pages 852–853 to record your assignments.

Developing Skills

In Exercises 1–10, find the common difference of the arithmetic sequence. *See Example 1.*

✓ **1.** $2, 5, 8, 11, \ldots$

2. $-8, 0, 8, 16, \ldots$

3. $100, 94, 88, 82, \ldots$

4. $3200, 2800, 2400, 2000, \ldots$

5. $10, -2, -14, -26, -38, \ldots$

6. $4, \frac{9}{2}, 5, \frac{11}{2}, 6, \ldots$

7. $1, \frac{5}{3}, \frac{7}{3}, 3, \ldots$

8. $\frac{1}{2}, \frac{5}{4}, 2, \frac{11}{4}, \ldots$

9. $\frac{7}{2}, \frac{9}{4}, 1, -\frac{1}{4}, -\frac{3}{2}, \ldots$

10. $\frac{5}{2}, \frac{11}{6}, \frac{7}{6}, \frac{1}{2}, -\frac{1}{6}, \ldots$

In Exercises 11–16, find the common difference of the arithmetic sequence with the given nth term. *See Example 1.*

11. $a_n = 4n + 5$

12. $a_n = 7n + 6$

13. $a_n = 8 - 3n$

14. $a_n = 12 - 4n$

15. $a_n = \frac{1}{2}(n + 1)$

16. $a_n = \frac{1}{3}(n + 4)$

In Exercises 17–32, determine whether the sequence is arithmetic. If so, find the common difference.

17. $2, 4, 6, 8, \ldots$

18. $1, 2, 4, 8, 16, \ldots$

19. $10, 8, 6, 4, 2, \ldots$

20. $2, 6, 10, 14, \ldots$

21. $32, 16, 0, -16, \ldots$

22. $32, 16, 8, 4, \ldots$

23. $3.2, 4, 4.8, 5.6, \ldots$

24. $8, 4, 2, 1, 0.5, 0.25, \ldots$

25. $2, \frac{7}{2}, 5, \frac{13}{2}, \ldots$

26. $3, \frac{5}{2}, 2, \frac{3}{2}, 1, \ldots$

27. $\frac{1}{3}, \frac{2}{3}, \frac{4}{3}, \frac{8}{3}, \frac{16}{3}, \ldots$

28. $\frac{9}{4}, 2, \frac{7}{4}, \frac{3}{2}, \frac{5}{4}, \ldots$

29. $1, \sqrt{2}, \sqrt{3}, 2, \sqrt{5}, \ldots$

30. $1, 4, 9, 16, 25, \ldots$

31. $\ln 4, \ln 8, \ln 12, \ln 16, \ldots$

32. $e, e^2, e^3, e^4, \ldots$

In Exercises 33–42, write the first five terms of the arithmetic sequence.

33. $a_1 = 7, d = 5$

34. $a_1 = 8, d = 3$

35. $a_1 = 11, d = 4$

36. $a_1 = 18, d = 10$

37. $a_1 = 20, d = -4$

38. $a_1 = 16, d = -3$

39. $a_1 = 6, a_2 = 11$

40. $a_1 = 9, a_2 = 11$

41. $a_1 = 22, a_2 = 18$

42. $a_1 = 30, a_2 = 20$

In Exercises 43–52, write the first five terms of the arithmetic sequence. (Assume that n begins with 1.)

43. $a_n = 3n + 4$

44. $a_n = 5n - 4$

45. $a_n = -2n + 8$

46. $a_n = -10n + 100$

47. $a_n = \frac{5}{2}n - 1$

48. $a_n = \frac{2}{3}n + 2$

49. $a_n = \frac{3}{5}n + 1$

50. $a_n = \frac{3}{4}n - 2$

51. $a_n = -\frac{1}{4}(n - 1) + 4$

52. $a_n = 4(n + 2) + 24$

In Exercises 53–70, find a formula for the *n*th term of the arithmetic sequence. *See Example 2.*

✓ **53.** $a_1 = 4, \quad d = 3$

54. $a_1 = 7, \quad d = 2$

55. $a_1 = \frac{1}{2}, \quad d = \frac{3}{2}$

56. $a_1 = \frac{5}{3}, \quad d = \frac{1}{3}$

57. $a_1 = 100, \quad d = -5$

58. $a_1 = -6, \quad d = -1$

59. $a_3 = 6, \quad d = \frac{3}{2}$

60. $a_6 = 5, \quad d = \frac{3}{2}$

61. $a_1 = 5, \quad a_5 = 15$

62. $a_2 = 93, \quad a_6 = 65$

63. $a_3 = 16, \quad a_4 = 20$

64. $a_5 = 30, \quad a_4 = 25$

65. $a_1 = 50, \quad a_3 = 30$

66. $a_{10} = 32, \quad a_{12} = 48$

67. $a_2 = 10, \quad a_6 = 8$

68. $a_7 = 8, \quad a_{13} = 6$

69. $a_1 = 0.35, \quad a_2 = 0.30$

70. $a_1 = 0.08, \quad a_2 = 0.082$

In Exercises 71–78, write the first five terms of the arithmetic sequence defined recursively. *See Example 3.*

✓ **71.** $a_1 = 14$, $a_{k+1} = a_k + 6$

72. $a_1 = 3$, $a_{k+1} = a_k - 2$

73. $a_1 = 23$, $a_{k+1} = a_k - 5$

74. $a_1 = 12$, $a_{k+1} = a_k + 6$

75. $a_1 = -16$, $a_{k+1} = a_k + 5$

76. $a_1 = -22$, $a_{k+1} = a_k - 4$

77. $a_1 = 3.4$, $a_{k+1} = a_k - 1.1$

78. $a_1 = 10.9$, $a_{k+1} = a_k + 0.7$

In Exercises 79–88, find the partial sum. *See Example 4.*

✓ **79.** $\sum_{k=1}^{20} k$

80. $\sum_{k=1}^{30} 4k$

81. $\sum_{k=1}^{50} (k + 3)$

82. $\sum_{n=1}^{30} (n + 2)$

83. $\sum_{k=1}^{10} (5k - 2)$

84. $\sum_{k=1}^{100} (4k - 1)$

85. $\sum_{n=1}^{500} \frac{n}{2}$

86. $\sum_{n=1}^{300} \frac{n}{3}$

87. $\sum_{n=1}^{30} \left(\frac{1}{3}n - 4\right)$

88. $\sum_{n=1}^{75} (0.3n + 5)$

In Exercises 89–100, find the *n*th partial sum of the arithmetic sequence. *See Example 5.*

✓ **89.** $5, 12, 19, 26, 33, \ldots, \quad n = 12$

90. $2, 12, 22, 32, 42, \ldots, \quad n = 20$

91. $2, 8, 14, 20, \ldots, \quad n = 25$

92. $500, 480, 460, 440, \ldots, \quad n = 20$

93. $200, 175, 150, 125, 100, \ldots, \quad n = 8$

94. $800, 785, 770, 755, 740, \ldots, \quad n = 25$

95. $-50, -38, -26, -14, -2, \ldots, \quad n = 50$

96. $-16, -8, 0, 8, 16, \ldots, \quad n = 30$

97. $1, 4.5, 8, 11.5, 15, \ldots, \quad n = 12$

98. $2.2, 2.8, 3.4, 4.0, 4.6, \ldots, \quad n = 12$

99. $a_1 = 0.5, \ a_4 = 1.7, \ldots, \quad n = 10$

100. $a_1 = 15, \ a_{100} = 307, \ldots, \quad n = 100$

In Exercises 101–106, match the arithmetic sequence with its graph. [The graphs are labeled (a), (b), (c), (d), (e), and (f).]

(a)

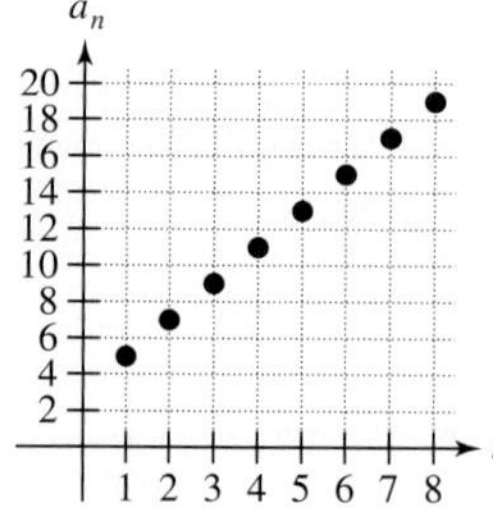

(b)

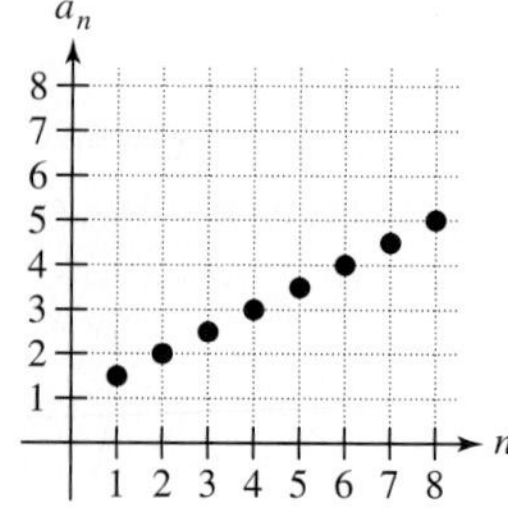

(c)

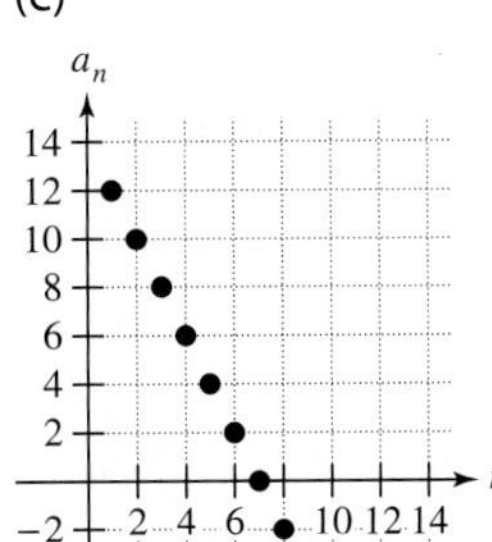

(d)

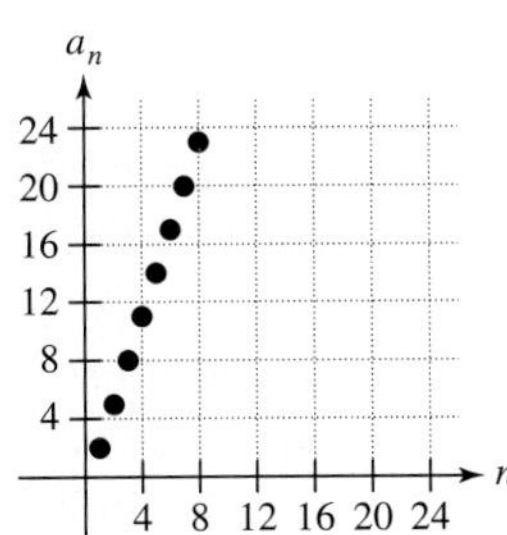

(e)

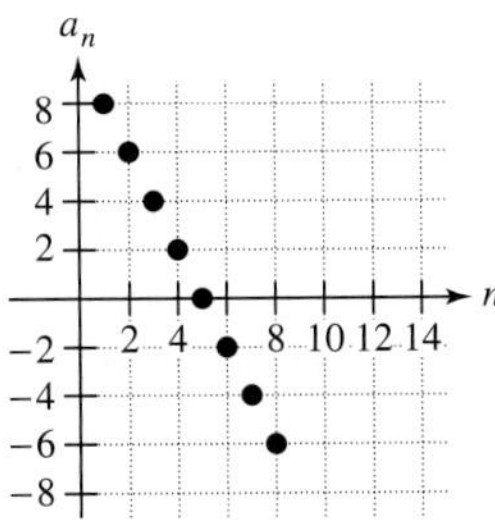

(f)

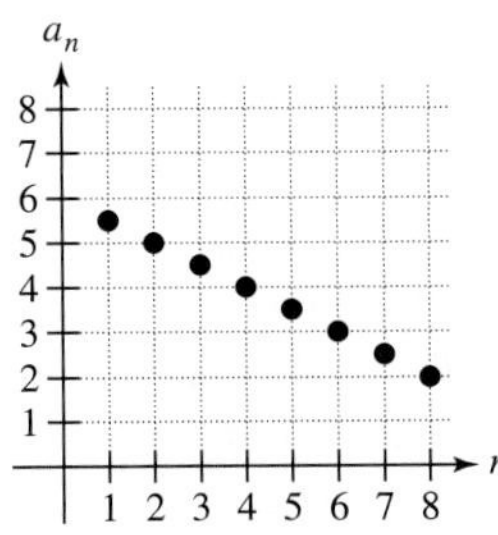

101. $a_n = \frac{1}{2}n + 1$

102. $a_n = -\frac{1}{2}n + 6$

103. $a_n = -2n + 10$

104. $a_n = 2n + 3$

105. $a_1 = 12$

$a_{n+1} = a_n - 2$

106. $a_1 = 2$

$a_{n+1} = a_n + 3$

In Exercises 107–112, use a graphing calculator to graph the first 10 terms of the sequence.

107. $a_n = -2n + 21$

108. $a_n = -25n + 500$

109. $a_n = \frac{3}{5}n + \frac{3}{2}$

110. $a_n = \frac{3}{2}n + 1$

111. $a_n = 2.5n - 8$

112. $a_n = 6.2n + 3$

In Exercises 113–118, use a graphing calculator to find the partial sum.

113. $\sum_{j=1}^{25} (750 - 30j)$

114. $\sum_{n=1}^{40} (1000 - 25n)$

115. $\sum_{i=1}^{60} \left(300 - \frac{8}{3}i\right)$

116. $\sum_{n=1}^{20} \left(500 - \frac{1}{10}n\right)$

117. $\sum_{n=1}^{50} (2.15n + 5.4)$

118. $\sum_{n=1}^{60} (200 - 3.4n)$

Solving Problems

119. *Number Problem* Find the sum of the first 75 positive integers.

120. *Number Problem* Find the sum of the integers from 35 to 100.

121. *Number Problem* Find the sum of the first 50 positive odd integers.

122. *Number Problem* Find the sum of the first 100 positive even integers.

✓ **123.** *Salary* In your new job as an actuary, your starting salary will be \$54,000 with an increase of \$3000 at the end of each of the first 5 years. How much will you be paid through the end of your first 6 years of employment with the company?

124. *Wages* You earn 5 dollars on the first day of the month, 10 dollars on the second day, 15 dollars on the third day, and so on. Determine the total amount that you will earn during a 30-day month.

125. *Ticket Prices* There are 20 rows of seats on the main floor of an outdoor arena: 20 seats in the first row, 21 seats in the second row, 22 seats in the third row, and so on (see figure). How much should you charge per ticket in order to obtain \$15,000 for the sale of all the seats on the main floor?

126. *Pile of Logs* Logs are stacked in a pile as shown in the figure. The top row has 15 logs and the bottom row has 21 logs. How many logs are in the pile?

127. *Baling Hay* In the first two trips baling hay around a large field (see figure), a farmer obtains 93 bales and 89 bales, respectively. The farmer estimates that the same pattern will continue. Estimate the total number of bales obtained if there are another six trips around the field.

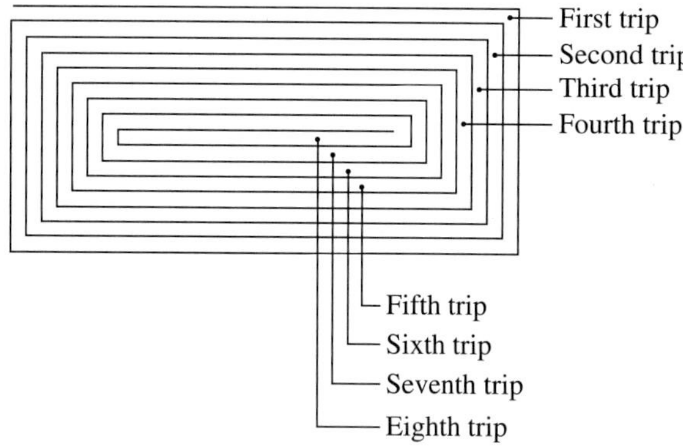

128. *Baling Hay* In the first two trips baling hay around a field (see figure), a farmer obtains 64 bales and 60 bales, respectively. The farmer estimates that the same pattern will continue. Estimate the total number of bales obtained if there are another four trips around the field.

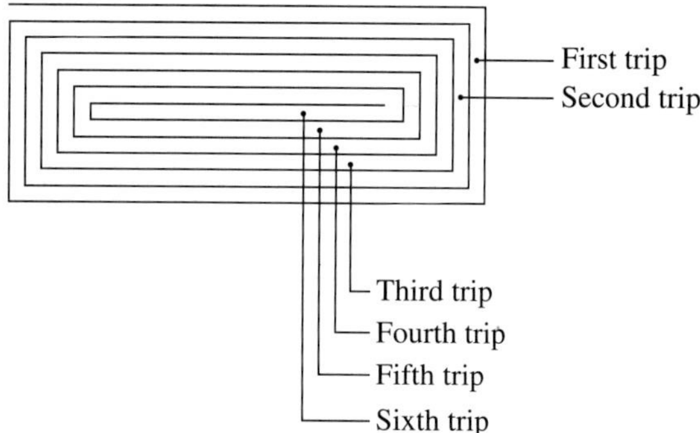

129. *Clock Chimes* A clock chimes once at 1:00, twice at 2:00, three times at 3:00, and so on. The clock also chimes once at 15-minute intervals that are not on the hour. How many times does the clock chime in a 12-hour period?

130. *Clock Chimes* A clock chimes once at 1:00, twice at 2:00, three times at 3:00, and so on. The clock also chimes once on the half-hour. How many times does the clock chime in a 12-hour period?

131. *Free-Falling Object* A free-falling object falls 16 feet during the first second, 48 feet during the second second, 80 feet during the third second, and so on. What total distance does the object fall in 8 seconds?

132. *Free-Falling Object* A free-falling object falls 4.9 meters during the first second, 14.7 meters during the second second, 24.5 meters during the third second, and so on. What total distance does the object fall in 5 seconds?

133. *Craft Beer Sales* Craft beers are produced by small, independent brewers using innovative techniques. The amount of craft beer (in barrels) sold in the United States each year from 2003 through 2007 can be modeled by the arithmetic sequence whose nth term is given by

$$a_n = 4.41 + 0.59n$$

where $n = 1$ corresponds to 2003. What is the total amount of craft beer sold in these five years? (Source: Brewer's Association)

AP Photo/*The Journal Record*, Mark Hancock

Explaining Concepts

134. *Growth Rate* A growth chart shows how a typical child with certain characteristics is expected to grow. A growth chart for Chinese girls shows that a 6-year-old Chinese girl is in the 50th percentile for her weight. From the data in the chart, the girl's expected yearly weight gains (in pounds) over the next five years can be modeled by the arithmetic sequence whose nth term is given by

$a_n = 2.3 + 0.86n$

where $n = 1$ corresponds to the weight gained from age 6 to age 7. How much total weight is the girl expected to gain in the next five years?

135. Explain what a recursion formula does.

136. Explain how to use the nth term a_n and the common difference d of an arithmetic sequence to write a recursion formula for the term a_{n+2} of the sequence.

137. Is it possible to use the nth term a_n and the common difference d of an arithmetic sequence to write a recursion formula for the term a_{2n}? Explain.

138. Explain why you cannot use the formula $a_n = a_1 + (n - 1)d$ to find the nth term of a sequence whose first term is a_0. Discuss the changes that can be made in the formula to create a new formula that can be used.

139. *Pattern Recognition*

(a) Compute the sums of positive odd integers.

$1 + 3 = \square$

$1 + 3 + 5 = \square$

$1 + 3 + 5 + 7 = \square$

$1 + 3 + 5 + 7 + 9 = \square$

$1 + 3 + 5 + 7 + 9 + 11 = \square$

(b) Do the partial sums of the positive odd integers form an arithmetic sequence? Explain.

(c) Use the sums in part (a) to make a conjecture about the sums of positive odd integers. Check your conjecture for the sum

$1 + 3 + 5 + 7 + 9 + 11 + 13 = \square$.

(d) Verify your conjecture in part (c) analytically.

140. Each term of an arithmetic sequence is multiplied by a constant C. Is the resulting sequence arithmetic? If so, how does the common difference compare with the common difference of the original sequence?

Cumulative Review

In Exercises 141–144, find the center and vertices of the ellipse.

141. $\dfrac{(x - 4)^2}{25} + \dfrac{(y + 5)^2}{9} = 1$

142. $\dfrac{(x + 2)}{4} + (y - 8)^2 = 1$

143. $9x^2 + 4y^2 - 18x + 24y + 9 = 0$

144. $x^2 + 4y^2 - 8x + 12 = 0$

In Exercises 145–148, write the sum using sigma notation. (Begin with $k = 1$.)

145. $3 + 4 + 5 + 6 + 7 + 8 + 9$

146. $3 + 6 + 9 + 12 + 15$

147. $12 + 15 + 18 + 21 + 24$

148. $2 + 2^2 + 2^3 + 2^4 + 2^5$

Mid-Chapter Quiz

Take this quiz as you would take a quiz in class. After you are done, check your work against the answers in the back of the book.

In Exercises 1–4, write the first five terms of the sequence. (Assume that *n* begins with 1.)

1. $a_n = 4n$

2. $a_n = 2n + 5$

3. $a_n = 32\left(\frac{1}{4}\right)^{n-1}$

4. $a_n = \frac{(-3)^n n}{n + 4}$

In Exercises 5–10, find the sum.

5. $\sum_{k=1}^{4} 10k$

6. $\sum_{i=1}^{10} 4$

7. $\sum_{j=1}^{5} \frac{60}{j + 1}$

8. $\sum_{n=1}^{4} \frac{12}{n}$

9. $\sum_{n=1}^{5} (3n - 1)$

10. $\sum_{k=1}^{4} (k^2 - 1)$

In Exercises 11–14, write the sum using sigma notation. (Begin with $k = 1$.)

11. $\frac{2}{3(1)} + \frac{2}{3(2)} + \frac{2}{3(3)} + \cdots + \frac{2}{3(20)}$

12. $\frac{1}{1^3} - \frac{1}{2^3} + \frac{1}{3^3} - \cdots + \frac{1}{25^3}$

13. $0 + \frac{1}{2} + \frac{2}{3} + \frac{3}{4} + \cdots + \frac{19}{20}$

14. $\frac{1}{2} + \frac{4}{2} + \frac{9}{2} + \cdots + \frac{100}{2}$

In Exercises 15 and 16, find the common difference of the arithmetic sequence.

15. $1, \frac{3}{2}, 2, \frac{5}{2}, 3, \ldots$

16. $100, 94, 88, 82, 76, \ldots$

In Exercises 17 and 18, find a formula for the *n*th term of the arithmetic sequence.

17. $a_1 = 20, \quad a_4 = 11$

18. $a_1 = 32, \quad d = -4$

19. Find the sum of the first 200 positive even numbers.

20. You save \$.50 on one day, \$1.00 the next day, \$1.50 the next day, and so on. How much will you have accumulated at the end of one year (365 days)?

13.3 Geometric Sequences and Series

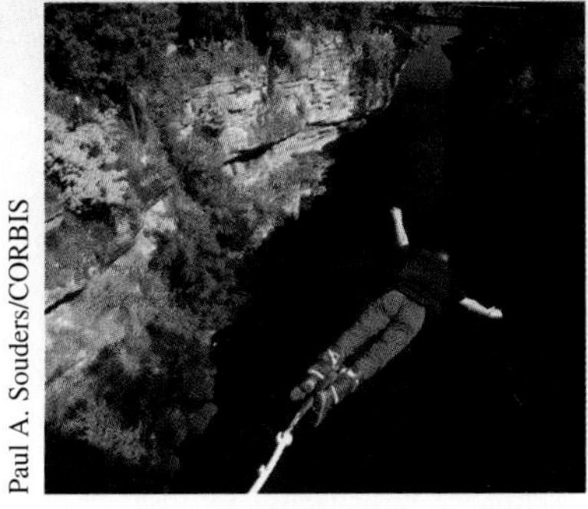
Paul A. Souders/CORBIS

Why You Should Learn It

A geometric sequence can reduce the amount of time it takes to find the sum of a sequence of numbers with a common ratio. For instance, in Exercise 121 on page 843, you will use a geometric sequence to find the total distance traveled by a bungee jumper.

What You Should Learn

1 ▶ Recognize, write, and find the nth terms of geometric sequences.
2 ▶ Find the nth partial sum of a geometric sequence.
3 ▶ Find the sum of an infinite geometric series.
4 ▶ Use geometric sequences to solve application problems.

1 ▶ Recognize, write, and find the nth terms of geometric sequences.

Geometric Sequences

In Section 13.2, you studied sequences whose consecutive terms have a common *difference*. In this section, you will study sequences whose consecutive terms have a common *ratio*.

Definition of Geometric Sequence

A sequence is called **geometric** if the ratios of consecutive terms are the same. So, the sequence $a_1, a_2, a_3, a_4, \ldots, a_n, \ldots$ is geometric if there is a number r, with $r \neq 0$, such that

$$\frac{a_2}{a_1} = r, \quad \frac{a_3}{a_2} = r, \quad \frac{a_4}{a_3} = r$$

and so on. The number r is the **common ratio** of the sequence.

EXAMPLE 1 Examples of Geometric Sequences

a. The sequence whose nth term is 2^n is geometric. For this sequence, the common ratio between consecutive terms is 2.

$$\underbrace{2, 4}_{\frac{4}{2} = 2}, 8, 16, \ldots, 2^n, \ldots \qquad \text{Begin with } n = 1.$$

b. The sequence whose nth term is $4(3^n)$ is geometric. For this sequence, the common ratio between consecutive terms is 3.

$$\underbrace{12, 36}_{\frac{36}{12} = 3}, 108, 324, \ldots, 4(3^n), \ldots \qquad \text{Begin with } n = 1.$$

c. The sequence whose nth term is $\left(-\frac{1}{3}\right)^n$ is geometric. For this sequence, the common ratio between consecutive terms is $-\frac{1}{3}$.

$$\underbrace{-\frac{1}{3}, \frac{1}{9}}_{\frac{1/9}{-1/3} = -\frac{1}{3}}, -\frac{1}{27}, \frac{1}{81}, \ldots, \left(-\frac{1}{3}\right)^n, \ldots \qquad \text{Begin with } n = 1.$$

✓ **CHECKPOINT** *Now try Exercise 1.*

Study Tip

If you know the nth term of a geometric sequence, the $(n + 1)$th term can be found by multiplying by r. That is, $a_{n+1} = ra_n$.

The nth Term of a Geometric Sequence

The nth term of a geometric sequence has the form

$$a_n = a_1 r^{n-1}$$

where r is the common ratio of consecutive terms of the sequence. So, every geometric sequence can be written in the following form.

$$a_1, a_1r, a_1r^2, a_1r^3, a_1r^4, \ldots, a_1r^{n-1}, \ldots$$

EXAMPLE 2 Finding the nth Term of a Geometric Sequence

a. Find a formula for the nth term of the geometric sequence whose common ratio is 3 and whose first term is 1.

b. What is the eighth term of the sequence found in part (a)?

Solution

a. The formula for the nth term is of the form $a_n = a_1r^{n-1}$. Moreover, because the common ratio is $r = 3$ and the first term is $a_1 = 1$, the formula must have the form

$$a_n = a_1r^{n-1}$$ Formula for geometric sequence

$$= (1)(3)^{n-1}$$ Substitute 1 for a_1 and 3 for r.

$$= 3^{n-1}.$$ Simplify.

The sequence has the form $1, 3, 9, 27, 81, \ldots, 3^{n-1}, \ldots$.

b. The eighth term of the sequence is $a_8 = 3^{8-1} = 3^7 = 2187$.

✓ **CHECKPOINT** *Now try Exercise 39.*

EXAMPLE 3 Finding the nth Term of a Geometric Sequence

Find a formula for the nth term of the geometric sequence whose first two terms are 4 and 2.

Solution

Because the common ratio is

$$r = \frac{a_2}{a_1} = \frac{2}{4} = \frac{1}{2}$$

the formula for the nth term must be

$$a_n = a_1r^{n-1}$$ Formula for geometric sequence

$$= 4\left(\frac{1}{2}\right)^{n-1}.$$ Substitute 4 for a_1 and $\frac{1}{2}$ for r.

The sequence has the form $4, 2, 1, \frac{1}{2}, \frac{1}{4}, \ldots, 4\left(\frac{1}{2}\right)^{n-1}, \ldots$.

✓ **CHECKPOINT** *Now try Exercise 47.*

2 ▶ Find the nth partial sum of a geometric sequence.

The Partial Sum of a Geometric Sequence

The nth Partial Sum of a Geometric Sequence

The nth partial sum of the geometric sequence whose nth term is $a_n = a_1 r^{n-1}$ is given by

$$\sum_{i=1}^{n} a_1 r^{i-1} = a_1 + a_1 r + a_1 r^2 + a_1 r^3 + \cdot \cdot \cdot + a_1 r^{n-1} = a_1\left(\frac{r^n - 1}{r - 1}\right).$$

EXAMPLE 4 **Finding the nth Partial Sum**

Find the sum $1 + 2 + 4 + 8 + 16 + 32 + 64 + 128$.

Solution

This is a geometric sequence whose common ratio is $r = 2$. Because the first term of the sequence is $a_1 = 1$, it follows that the sum is

$$\sum_{i=1}^{8} 2^{i-1} = (1)\left(\frac{2^8 - 1}{2 - 1}\right) = \frac{256 - 1}{2 - 1} = 255.$$ Substitute 1 for a_1 and 2 for r.

✓ **CHECKPOINT** *Now try Exercise 71.*

EXAMPLE 5 **Finding the nth Partial Sum**

Find the sum of the first five terms of the geometric sequence whose nth term is $a_n = \left(\frac{2}{3}\right)^n$.

Solution

$$\sum_{i=1}^{5} \left(\frac{2}{3}\right)^i = \frac{2}{3}\left[\frac{(2/3)^5 - 1}{(2/3) - 1}\right]$$ Substitute $\frac{2}{3}$ for a_1 and $\frac{2}{3}$ for r.

$$= \frac{2}{3}\left[\frac{(32/243) - 1}{-1/3}\right]$$ Simplify.

$$= \frac{422}{243} \approx 1.737$$ Use a calculator to simplify.

✓ **CHECKPOINT** *Now try Exercise 77.*

3 ▶ Find the sum of an infinite geometric series.

Geometric Series

Suppose that in Example 5, you were to find the sum of all the terms of the infinite geometric sequence

$$\frac{2}{3}, \frac{4}{9}, \frac{8}{27}, \frac{16}{81}, \ldots, \left(\frac{2}{3}\right)^n, \ldots.$$

The sum of all the terms of an infinite geometric sequence is called an **infinite geometric series,** or simply a **geometric series.**

Technology: Discovery

Evaluate $\left(\frac{1}{2}\right)^n$ for $n = 1$, 10, and 100. What happens to the value of $\left(\frac{1}{2}\right)^n$ as n increases? Make a conjecture about the value of $\left(\frac{1}{2}\right)^n$ as n approaches infinity.

In your mind, would this sum be infinitely large or would it be a finite number? Consider the formula for the nth partial sum of a geometric sequence.

$$S_n = a_1\left(\frac{r^n - 1}{r - 1}\right) = a_1\left(\frac{1 - r^n}{1 - r}\right)$$

Suppose that $|r| < 1$. As you let n become larger, it follows that r^n approaches 0, so that the term r^n drops out of the formula above. The sum becomes

$$S = a_1\left(\frac{1}{1 - r}\right) = \frac{a_1}{1 - r}.$$

Because n is not involved, you can use this formula to evaluate the sum. In the case of Example 5, $r = \left(\frac{2}{3}\right) < 1$, and so the sum of the infinite geometric sequence is

$$S = \sum_{i=1}^{\infty}\left(\frac{2}{3}\right)^i = \frac{a_1}{1 - r} = \frac{2/3}{1 - (2/3)} = \frac{2/3}{1/3} = 2.$$

Sum of an Infinite Geometric Series

If $a_1, a_1r, a_1r^2, \ldots, a_1r^n, \ldots$ is an infinite geometric sequence and $|r| < 1$, the sum of the terms of the corresponding infinite geometric series is

$$S = \sum_{i=0}^{\infty} a_1r^i = \frac{a_1}{1 - r}.$$

EXAMPLE 6 Finding the Sum of an Infinite Geometric Series

Find each sum.

a. $\displaystyle\sum_{i=1}^{\infty} 5\left(\frac{3}{4}\right)^{i-1}$ **b.** $\displaystyle\sum_{n=0}^{\infty} 4\left(\frac{3}{10}\right)^n$ **c.** $\displaystyle\sum_{i=0}^{\infty}\left(-\frac{3}{5}\right)^i$

Solution

a. The series is geometric, with $a_1 = 5\left(\frac{3}{4}\right)^{1-1} = 5$ and $r = \frac{3}{4}$. So,

$$\sum_{i=1}^{\infty} 5\left(\frac{3}{4}\right)^{i-1} = \frac{5}{1 - (3/4)} = \frac{5}{1/4} = 20.$$

b. The series is geometric, with $a_1 = 4\left(\frac{3}{10}\right)^0 = 4$ and $r = \frac{3}{10}$. So,

$$\sum_{n=0}^{\infty} 4\left(\frac{3}{10}\right)^n = \frac{4}{1 - (3/10)} = \frac{4}{7/10} = \frac{40}{7}.$$

c. The series is geometric, with $a_1 = \left(-\frac{3}{5}\right)^0 = 1$ and $r = -\frac{3}{5}$. So,

$$\sum_{i=0}^{\infty}\left(-\frac{3}{5}\right)^i = \frac{1}{1 - (-3/5)} = \frac{1}{1 + (3/5)} = \frac{5}{8}.$$

✓ **CHECKPOINT** *Now try Exercise 93.*

4 ▶ Use geometric sequences to solve application problems.

Applications

EXAMPLE 7 A Lifetime Salary 

You have accepted a job as a meteorologist that pays a salary of \$45,000 the first year. During the next 39 years, suppose you receive a 6% raise each year. What will your total salary be over the 40-year period?

Solution

Using a geometric sequence, your salary during the first year will be $a_1 = 45{,}000$. Then, with a 6% raise each year, your salary for the next 2 years will be as follows.

$$a_2 = 45{,}000 + 45{,}000(0.06) = 45{,}000(1.06)^1$$

$$a_3 = 45{,}000(1.06) + 45{,}000(1.06)(0.06) = 45{,}000(1.06)^2$$

From this pattern, you can see that the common ratio of the geometric sequence is $r = 1.06$. Using the formula for the nth partial sum of a geometric sequence, you will find that the total salary over the 40-year period is given by

$$\text{Total salary} = a_1\left(\frac{r^n - 1}{r - 1}\right)$$

$$= 45{,}000\left[\frac{(1.06)^{40} - 1}{1.06 - 1}\right]$$

$$= 45{,}000\left[\frac{(1.06)^{40} - 1}{0.06}\right] \approx \$6{,}964{,}288.$$

✓ ***CHECKPOINT*** *Now try Exercise 107.*

AP Photo/*The Journal Record*, Mark Hancock

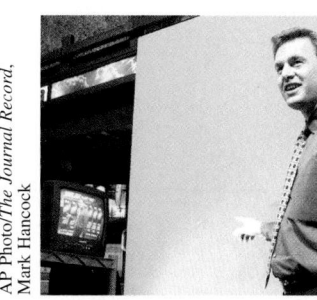

In 2007, meteorologists employed by the federal government earned an average salary of \$84,882.

EXAMPLE 8 Increasing Annuity 

You deposit \$100 in an account each month for 2 years. The account pays an annual interest rate of 9%, compounded monthly. What is your balance at the end of 2 years? (This type of savings plan is called an **increasing annuity.**)

Solution

The first deposit would earn interest for the full 24 months, the second deposit would earn interest for 23 months, the third deposit would earn interest for 22 months, and so on. Using the formula for compound interest, you can see that the total of the 24 deposits would be

$$\text{Total} = a_1 + a_2 + \cdot\cdot\cdot + a_{24}$$

$$= 100\left(1 + \frac{0.09}{12}\right)^1 + 100\left(1 + \frac{0.09}{12}\right)^2 + \cdot\cdot\cdot + 100\left(1 + \frac{0.09}{12}\right)^{24}$$

$$= 100(1.0075)^1 + 100(1.0075)^2 + \cdot\cdot\cdot + 100(1.0075)^{24}$$

$$= 100(1.0075)\left(\frac{1.0075^{24} - 1}{1.0075 - 1}\right) = \$2638.49.$$

✓ ***CHECKPOINT*** *Now try Exercise 109.*

Concept Check

1. Explain the difference between an arithmetic sequence and a geometric sequence.

2. How can you determine whether a sequence is geometric?

3. What is the general formula for the nth term of a geometric sequence?

4. Can you find the sum of an infinite geometric series if the common ratio is $\frac{3}{2}$? Explain.

13.3 EXERCISES

Go to pages 852–853 to record your assignments.

Developing Skills

In Exercises 1–12, find the common ratio of the geometric sequence. ***See Example 1.***

1. $3, 6, 12, 24, \ldots$
2. $2, 6, 18, 54, \ldots$
3. $5, -5, 5, -5, \ldots$
4. $-5, -0.5, -0.05, -0.005, \ldots$
5. $\frac{1}{2}, -\frac{1}{4}, \frac{1}{8}, -\frac{1}{16}, \ldots$
6. $\frac{2}{3}, -\frac{4}{3}, \frac{8}{3}, -\frac{16}{3}, \ldots$
7. $75, 15, 3, \frac{3}{5}, \ldots$
8. $12, -4, \frac{4}{3}, -\frac{4}{9}, \ldots$
9. $1, \pi, \pi^2, \pi^3, \ldots$
10. $e, e^2, e^3, e^4, \ldots$
11. $50(1.04), 50(1.04)^2, 50(1.04)^3, 50(1.04)^4, \ldots$
12. $25(1.07), 25(1.07)^2, 25(1.07)^3, 25(1.07)^4, \ldots$

In Exercises 13–24, determine whether the sequence is geometric. If so, find the common ratio.

13. $64, 32, 16, 8, \ldots$
14. $64, 32, 0, -32, \ldots$
15. $10, 15, 20, 25, \ldots$
16. $10, 20, 40, 80, \ldots$
17. $5, 10, 20, 40, \ldots$
18. $270, 90, 30, 10, \ldots$
19. $1, 8, 27, 64, 125, \ldots$
20. $2, 4, 8, 14, 22, \ldots$
21. $1, -\frac{2}{3}, \frac{4}{9}, -\frac{8}{27}, \ldots$
22. $\frac{1}{3}, -\frac{2}{3}, \frac{4}{3}, -\frac{8}{3}, \ldots$
23. $1, 1.1, 1.21, 1.331, \ldots$
24. $1, 0.2, 0.04, 0.008, \ldots$

In Exercises 25–38, write the first five terms of the geometric sequence. If necessary, round your answers to two decimal places.

25. $a_1 = 4, \quad r = 2$
26. $a_1 = 2, \quad r = 4$
27. $a_1 = 6, \quad r = \frac{1}{2}$
28. $a_1 = 90, \quad r = \frac{1}{3}$
29. $a_1 = 5, \quad r = -2$
30. $a_1 = -12, \quad r = -1$
31. $a_1 = -4, \quad r = -\frac{1}{2}$
32. $a_1 = 3, \quad r = -\frac{3}{2}$
33. $a_1 = 10, \quad r = 1.02$
34. $a_1 = 200, \quad r = 1.07$
35. $a_1 = 10, \quad r = \frac{3}{5}$
36. $a_1 = 36, \quad r = \frac{2}{3}$
37. $a_1 = \frac{3}{2}, \quad r = \frac{2}{3}$
38. $a_1 = \frac{4}{5}, \quad r = \frac{1}{2}$

In Exercises 39–52, find a formula for the nth term of the geometric sequence. (Assume that n begins with 1.) ***See Examples 2 and 3.***

39. $a_1 = 1, \quad r = 2$
40. $a_1 = 5, \quad r = 4$

41. $a_1 = 2, \quad r = 2$

42. $a_1 = 25, \quad r = 5$

43. $a_1 = 10, \quad r = -\frac{1}{5}$

44. $a_1 = 1, \quad r = -\frac{4}{3}$

45. $a_1 = 4, \quad r = -\frac{1}{2}$

46. $a_1 = 9, \quad r = \frac{2}{3}$

47. $a_1 = 8, \quad a_2 = 2$

48. $a_1 = 18, \quad a_2 = 8$

49. $a_1 = 14, \quad a_2 = \frac{21}{2}$

50. $a_1 = 36, \quad a_2 = \frac{27}{2}$

51. $4, -6, 9, -\frac{27}{2}, \ldots$

52. $1, \frac{3}{2}, \frac{9}{4}, \frac{27}{8}, \ldots$

In Exercises 53–66, find the specified term of the geometric sequence. Round to the nearest hundredth, if necessary.

53. $a_1 = 6, \quad r = \frac{1}{2}, \quad a_{10} =$

54. $a_1 = 8, \quad r = \frac{3}{4}, \quad a_8 =$

55. $a_1 = 3, \quad r = \sqrt{2}, \quad a_{10} =$

56. $a_1 = 5, \quad r = \sqrt{3}, \quad a_9 =$

57. $a_1 = 200, \quad r = 1.2, \quad a_{12} =$

58. $a_1 = 500, \quad r = 1.06, \quad a_{40} =$

59. $a_1 = 120, \quad r = -\frac{1}{3}, \quad a_{10} =$

60. $a_1 = 240, \quad r = -\frac{1}{4}, \quad a_{13} =$

61. $a_1 = 4, \quad a_2 = 3, \quad a_5 =$

62. $a_1 = 1, \quad a_2 = 9, \quad a_5 =$

63. $a_3 = 3, \quad a_4 = 6, \quad a_5 =$

64. $a_2 = 5, \quad a_3 = 7, \quad a_4 =$

65. $a_2 = 12, \quad a_3 = 16, \quad a_5 =$

66. $a_4 = 100, \quad a_5 = -25, \quad a_7 =$

In Exercises 67–70, match the geometric sequence with its graph. [The graphs are labeled (a), (b), (c), and (d).]

(a)

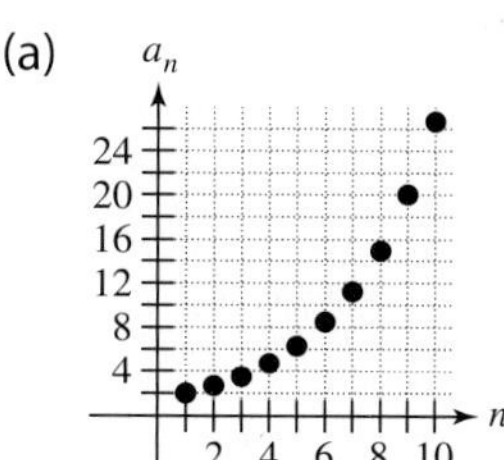

(b)

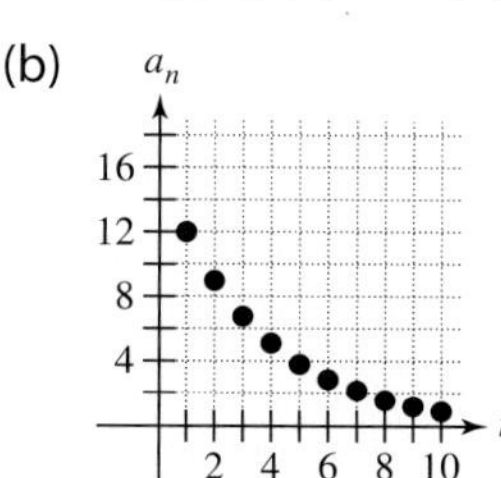

(c)

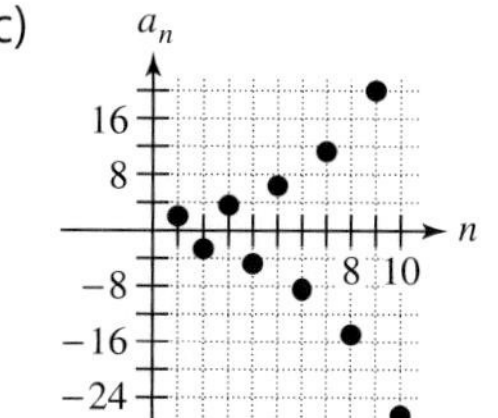

(d)

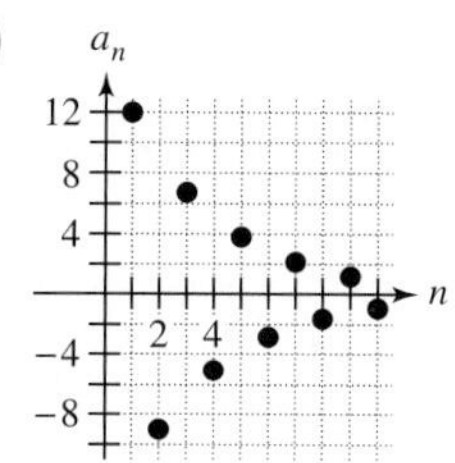

67. $a_n = 12\left(\frac{3}{4}\right)^{n-1}$

68. $a_n = 12\left(-\frac{3}{4}\right)^{n-1}$

69. $a_n = 2\left(-\frac{4}{3}\right)^{n-1}$

70. $a_n = 2\left(\frac{4}{3}\right)^{n-1}$

In Exercises 71–80, find the partial sum. Round to the nearest hundredth, if necessary. ***See Examples 4 and 5.***

71. $\sum_{i=1}^{10} 2^{i-1}$

72. $\sum_{i=1}^{6} 3^{i-1}$

73. $\sum_{i=1}^{12} 3\left(\frac{3}{2}\right)^{i-1}$

74. $\sum_{i=1}^{20} 12\left(\frac{2}{3}\right)^{i-1}$

75. $\sum_{i=1}^{15} 3\left(-\frac{1}{3}\right)^{i-1}$

76. $\sum_{i=1}^{8} 8\left(-\frac{1}{4}\right)^{i-1}$

77. $\sum_{i=1}^{6} \left(\frac{3}{4}\right)^{i}$

78. $\sum_{i=1}^{4} \left(\frac{5}{6}\right)^{i}$

79. $\sum_{i=1}^{8} 6(0.1)^{i-1}$

80. $\sum_{i=1}^{12} 1000(1.06)^{i-1}$

In Exercises 81–92, find the nth partial sum of the geometric sequence. Round to the nearest hundredth, if necessary.

81. $1, -3, 9, -27, 81, \ldots, \quad n = 10$

82. $3, -6, 12, -24, 48, \ldots, \quad n = 12$

83. $8, 4, 2, 1, \frac{1}{2}, \ldots, \quad n = 15$

84. $9, 6, 4, \frac{8}{3}, \frac{16}{9}, \ldots, \quad n = 10$

85. $4, 12, 36, 108, \ldots, \quad n = 8$

86. $\frac{1}{36}, -\frac{1}{12}, \frac{1}{4}, -\frac{3}{4}, \ldots, \quad n = 20$

87. $60, -15, \frac{15}{4}, -\frac{15}{16}, \ldots, \quad n = 12$

88. $40, -10, \frac{5}{2}, -\frac{5}{8}, \frac{5}{32}, \ldots, n = 10$

89. $30, 30(1.06), 30(1.06)^2, 30(1.06)^3, \ldots, n = 20$

90. $100, 100(1.08), 100(1.08)^2, 100(1.08)^3, \ldots,$ $n = 40$

91. $1, \sqrt{3}, 3, 3\sqrt{3}, 9, \ldots, n = 18$

92. $1, \sqrt{2}, 2, 2\sqrt{2}, 4, \ldots, n = 12$

In Exercises 93–100, find the sum. ***See Example 6.***

 **93.** $\sum_{n=1}^{\infty} \left(\frac{1}{2}\right)^{n-1}$

94. $\sum_{n=1}^{\infty} \left(-\frac{1}{2}\right)^{n-1}$

95. $\sum_{n=0}^{\infty} 2\left(-\frac{2}{3}\right)^{n}$

96. $\sum_{n=0}^{\infty} 2\left(\frac{2}{3}\right)^{n}$

97. $\sum_{n=0}^{\infty} \left(\frac{1}{10}\right)^{n}$

98. $\sum_{n=0}^{\infty} 4\left(\frac{1}{4}\right)^{n}$

99. $8 + 6 + \frac{9}{2} + \frac{27}{8} + \cdots$

100. $3 - 1 + \frac{1}{3} - \frac{1}{9} + \cdots$

In Exercises 101–104, use a graphing calculator to graph the first 10 terms of the sequence.

101. $a_n = 20(-0.6)^{n-1}$

102. $a_n = 4(1.4)^{n-1}$

103. $a_n = 15(0.6)^{n-1}$

104. $a_n = 8(-0.6)^{n-1}$

Solving Problems

105. *Depreciation* A company buys a machine for \$250,000. During the next 5 years, the machine depreciates at the rate of 25% per year. (That is, at the end of each year, the depreciated value is 75% of what it was at the beginning of the year.)

(a) Find a formula for the nth term of the geometric sequence that gives the value of the machine n full years after it was purchased.

(b) Find the depreciated value of the machine at the end of 5 full years.

(c) During which year did the machine depreciate the most?

106. *Population Increase* A city of 350,000 people is growing at the rate of 1% per year. (That is, at the end of each year, the population is 1.01 times the population at the beginning of the year.)

(a) Find a formula for the nth term of the geometric sequence that gives the population after n years.

(b) Estimate the population after 10 years.

(c) During which year did the population grow the least?

107. *Salary* You accept a job as an archaeologist that pays a salary of \$30,000 the first year. During the next 39 years, you receive a 5% raise each year. What would your total salary be over the 40-year period?

108. *Salary* You accept a job as a marine biologist that pays a salary of \$45,000 the first year. During the next 39 years, you receive a 5.5% raise each year. What would your total salary be over the 40-year period?

Increasing Annuity In Exercises 109–114, find the balance A in an increasing annuity in which a principal of P dollars is invested each month for t years, compounded monthly at rate r.

109. $P = \$50$ $t = 10$ years $r = 9\%$

110. $P = \$50$ $t = 5$ years $r = 7\%$

111. $P = \$30$ $t = 40$ years $r = 8\%$

112. $P = \$200$ $t = 30$ years $r = 10\%$

113. $P = \$100$ $t = 30$ years $r = 6\%$

114. $P = \$100$ $t = 25$ years $r = 8\%$

115. *Wages* You start work at a company that pays \$0.01 for the first day, \$0.02 for the second day, \$0.04 for the third day, and so on. The daily wage keeps doubling. What would your total income be for working (a) 29 days and (b) 30 days?

116. *Wages* You start work at a company that pays \$0.01 for the first day, \$0.03 for the second day, \$0.09 for the third day, and so on. The daily wage keeps tripling. What would your total income be for working (a) 25 days and (b) 26 days?

117. *Power Supply* The electrical power for an implanted medical device decreases by 0.1% each day.

(a) Find a formula for the nth term of the geometric sequence that gives the percent of the initial power n days after the device is implanted.

(b) What percent of the initial power is still available 1 year after the device is implanted?

(c) The power supply needs to be changed when half the power is depleted. Use a graphing calculator to graph the first 750 terms of the sequence. Estimate when the power source should be changed.

118. *Cooling* The temperature of water in an ice cube tray is 70°F when it is placed in a freezer. Its temperature n hours after being placed in the freezer is 20% less than 1 hour earlier.

(a) Find a formula for the nth term of the geometric sequence that gives the temperature of the water n hours after being placed in the freezer.

(b) Find the temperature of the water 6 hours after it is placed in the freezer.

(c) Use a graphing calculator to estimate the time when the water freezes. Explain how you found your answer.

119. *Geometry* An equilateral triangle has an area of 1 square unit. The triangle is divided into four smaller triangles and the center triangle is shaded (see figure). Each of the three unshaded triangles is then divided into four smaller triangles and each center triangle is shaded. This process is repeated one more time. What is the total area of the shaded region?

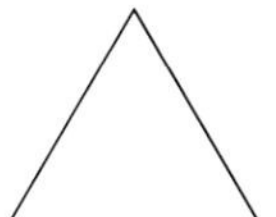
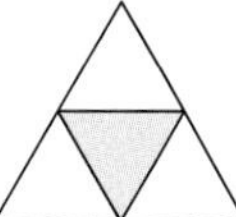
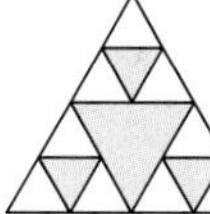

120. *Geometry* A square has an area of 1 square unit. The square is divided into nine smaller squares and the center square is shaded (see figure). Each of the eight unshaded squares is then divided into nine smaller squares and each center square is shaded. This process is repeated one more time. What is the total area of the shaded region?

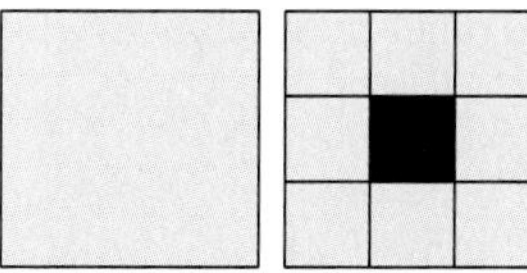
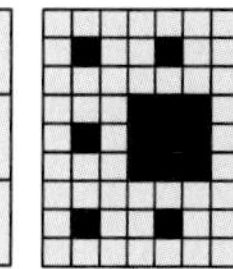
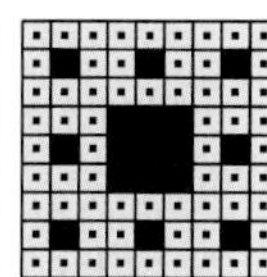

121. *Bungee Jumping* A bungee jumper jumps from a bridge and stretches a cord 100 feet. Each successive bounce stretches the cord 75% of its length for the preceding bounce (see figure). Find the total distance traveled by the bungee jumper during 10 bounces.

$100 + 2(100)(0.75) + \cdot \cdot \cdot + 2(100)(0.75)^{10}$

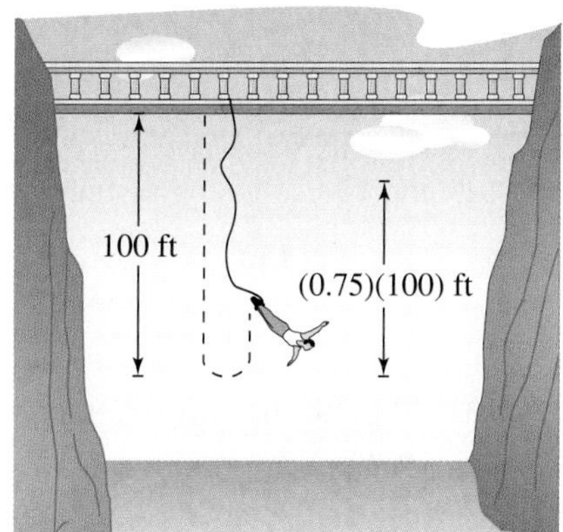

122. *Distance* A ball is dropped from a height of 16 feet. Each time it drops h feet, it rebounds $0.81h$ feet.

(a) Find the total distance traveled by the ball.

(b) The ball takes the following times for each fall.

$$s_1 = -16t^2 + 16, \qquad s_1 = 0 \text{ if } t = 1$$
$$s_2 = -16t^2 + 16(0.81), \qquad s_2 = 0 \text{ if } t = 0.9$$
$$s_3 = -16t^2 + 16(0.81)^2, \qquad s_3 = 0 \text{ if } t = (0.9)^2$$
$$s_4 = -16t^2 + 16(0.81)^3, \qquad s_4 = 0 \text{ if } t = (0.9)^3$$
$$\vdots \qquad \vdots$$
$$s_n = -16t^2 + 16(0.81)^{n-1}, \; s_n = 0 \text{ if } t = (0.9)^{n-1}$$

Beginning with s_2, the ball takes the same amount of time to bounce up as it does to fall, and so the total time elapsed before it comes to rest is

$$t = 1 + 2\sum_{n=1}^{\infty} (0.9)^n.$$

Find this total.

Explaining Concepts

123. The second and third terms of a geometric sequence are 6 and 3, respectively. What is the first term?

124. Give an example of a geometric sequence whose terms alternate in sign.

125. ✎ Explain why the terms of a geometric sequence decrease when $a_1 > 0$ and $0 < r < 1$.

126. ✎ In your own words, describe an increasing annuity.

127. ✎ Explain what is meant by the nth partial sum of a sequence.

128. ✎ A unit square is divided into two equal rectangles. One of the resulting rectangles is then divided into two equal rectangles, as shown in the figure. This process is repeated indefinitely.

(a) Explain why the areas of the rectangles (from largest to smallest) form a geometric sequence.

(b) Find a formula for the nth term of the geometric sequence.

(c) Use the formula for the sum of an infinite geometric series to show that the combined area of the rectangles is 1.

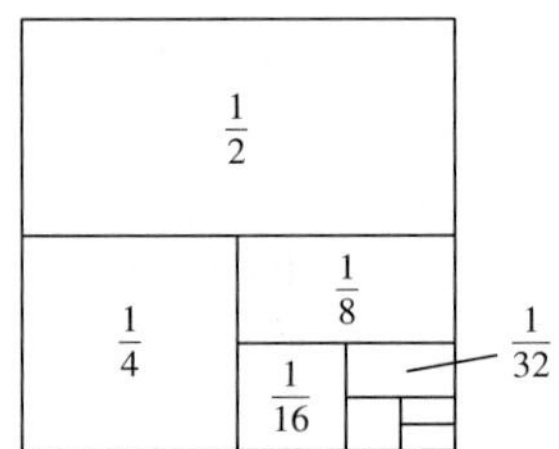

Cumulative Review

In Exercises 129 and 130, solve the system.

129. $\begin{cases} y = 2x^2 \\ y = 2x + 4 \end{cases}$

130. $\begin{cases} x^2 + y^2 = 1 \\ x^2 + y = 1 \end{cases}$

In Exercises 131–134, find the annual interest rate.

	Principal	*Balance*	*Time*	*Compounding*
131.	\$1000	\$2219.64	10 years	Monthly
132.	\$2000	\$3220.65	8 years	Quarterly
133.	\$2500	\$10,619.63	20 years	Yearly
134.	\$3500	\$25,861.70	40 years	Continuous

In Exercises 135 and 136, sketch the hyperbola. Identify the vertices and asymptotes.

135. $\dfrac{x^2}{16} - \dfrac{y^2}{9} = 1$

136. $\dfrac{y^2}{1} - \dfrac{x^2}{4} = 1$

13.4 The Binomial Theorem

What You Should Learn

1 ▶ Use the Binomial Theorem to calculate binomial coefficients.

2 ▶ Use Pascal's Triangle to calculate binomial coefficients.

3 ▶ Expand binomial expressions.

Why You Should Learn It

You can use the Binomial Theorem to expand quantities used in probability. See Exercises 55–58 on page 851.

1 ▶ Use the Binomial Theorem to calculate binomial coefficients.

Binomial Coefficients

Recall that a **binomial** is a polynomial that has two terms. In this section, you will study a formula that provides a quick method of raising a binomial to a power. To begin, let's look at the expansion of $(x + y)^n$ for several values of n.

$$(x + y)^0 = 1$$

$$(x + y)^1 = x + y$$

$$(x + y)^2 = x^2 + 2xy + y^2$$

$$(x + y)^3 = x^3 + 3x^2y + 3xy^2 + y^3$$

$$(x + y)^4 = x^4 + 4x^3y + 6x^2y^2 + 4xy^3 + y^4$$

$$(x + y)^5 = x^5 + 5x^4y + 10x^3y^2 + 10x^2y^3 + 5xy^4 + y^5$$

There are several observations you can make about these expansions.

1. In each expansion, there are $n + 1$ terms.
2. In each expansion, x and y have symmetrical roles. The powers of x decrease by 1 in successive terms, whereas the powers of y increase by 1.
3. The sum of the powers of each term is n. For instance, in the expansion of $(x + y)^5$, the sum of the powers of each term is 5.

$$(x + y)^5 = x^5 + 5\overbrace{x^4y^1}^{4+1=5} + 10\overbrace{x^3y^2}^{3+2=5} + 10x^2y^3 + 5xy^4 + y^5$$

4. The coefficients increase and then decrease in a symmetrical pattern.

The coefficients of a binomial expansion are called **binomial coefficients.** To find them, you can use the **Binomial Theorem.**

Study Tip

Other notations that are commonly used for ${}_nC_r$ are

$\binom{n}{r}$ and $C(n, r)$.

The Binomial Theorem

In the expansion of $(x + y)^n$

$$(x + y)^n = x^n + nx^{n-1}y + \cdot \cdot \cdot + {}_nC_rx^{n-r}y^r + \cdot \cdot \cdot + nxy^{n-1} + y^n$$

the coefficient of $x^{n-r}y^r$ is given by

$${}_nC_r = \frac{n!}{(n - r)!r!}.$$

EXAMPLE 1 Finding Binomial Coefficients

Find each binomial coefficient.

a. ${}_8C_2$ **b.** ${}_{10}C_3$ **c.** ${}_7C_0$ **d.** ${}_8C_8$

Solution

a. $${}_8C_2 = \frac{8!}{6! \cdot 2!} = \frac{(8 \cdot 7) \cdot \cancel{6!}}{\cancel{6!} \cdot 2!} = \frac{8 \cdot 7}{2 \cdot 1} = 28$$

b. $${}_{10}C_3 = \frac{10!}{7! \cdot 3!} = \frac{(10 \cdot 9 \cdot 8) \cdot \cancel{7!}}{\cancel{7!} \cdot 3!} = \frac{10 \cdot 9 \cdot 8}{3 \cdot 2 \cdot 1} = 120$$

c. $${}_7C_0 = \frac{7!}{7! \cdot 0!} = 1$$ **d.** $${}_8C_8 = \frac{8!}{0! \cdot 8!} = 1$$

CHECKPOINT *Now try Exercise 1.*

Technology: Tip

The formula for the binomial coefficient is the same as the formula for combinations in the study of probability. Most graphing calculators have the capability to evaluate a binomial coefficient. Consult the user's guide for your graphing calculator.

When $r \neq 0$ and $r \neq n$, as in parts (a) and (b) of Example 1, there is a simple pattern for evaluating binomial coefficients that results from dividing a common factorial expression out of the numerator and denominator.

$${}_8C_2 = \frac{\overbrace{8 \cdot 7}^{\text{2 factors}}}{\underbrace{2 \cdot 1}_{\text{2 factors}}} \quad \text{and} \quad {}_{10}C_3 = \frac{\overbrace{10 \cdot 9 \cdot 8}^{\text{3 factors}}}{\underbrace{3 \cdot 2 \cdot 1}_{\text{3 factors}}}$$

EXAMPLE 2 Finding Binomial Coefficients

Find each binomial coefficient.

a. ${}_7C_3$ **b.** ${}_7C_4$ **c.** ${}_{12}C_1$ **d.** ${}_{12}C_{11}$

Solution

a. $${}_7C_3 = \frac{7 \cdot 6 \cdot 5}{3 \cdot 2 \cdot 1} = 35$$

b. $${}_7C_4 = \frac{7 \cdot 6 \cdot 5 \cdot 4}{4 \cdot 3 \cdot 2 \cdot 1} = 35$$ ${}_7C_4 = {}_7C_3$

c. $${}_{12}C_1 = \frac{12!}{11! \cdot 1!} = \frac{(12) \cdot \cancel{11!}}{\cancel{11!} \cdot 1!} = \frac{12}{1} = 12$$

d. $${}_{12}C_{11} = \frac{12!}{1! \cdot 11!} = \frac{(12) \cdot \cancel{11!}}{1! \cdot \cancel{11!}} = \frac{12}{1} = 12$$ ${}_{12}C_{11} = {}_{12}C_1$

CHECKPOINT *Now try Exercise 7.*

In Example 2, it is not a coincidence that the answers to parts (a) and (b) are the same and that the answers to parts (c) and (d) are the same. In general,

$${}_nC_r = {}_nC_{n-r}.$$

This shows the symmetric property of binomial coefficients.

2 ▶ Use Pascal's Triangle to calculate binomial coefficients.

Pascal's Triangle

There is a convenient way to remember a pattern for binomial coefficients. By arranging the coefficients in a triangular pattern, you obtain the following array, which is called **Pascal's Triangle.** This triangle is named after the famous French mathematician Blaise Pascal (1623–1662).

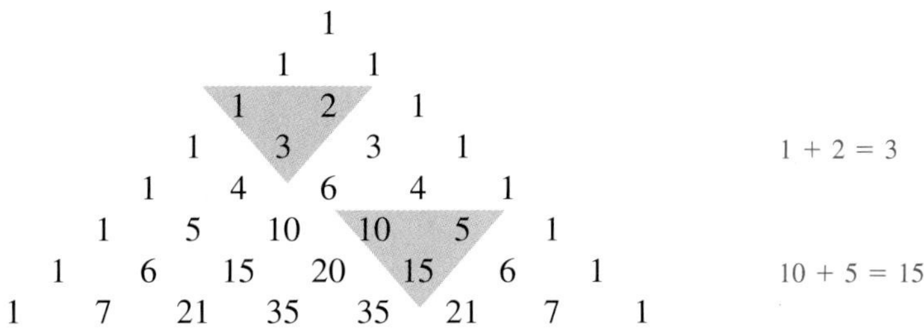

Study Tip

The top row in Pascal's Triangle is called the *zeroth row* because it corresponds to the binomial expansion

$(x + y)^0 = 1.$

Similarly, the next row is called the *first row* because it corresponds to the binomial expansion

$(x + y)^1 = 1(x) + 1(y).$

In general, the *nth row* in Pascal's Triangle gives the coefficients of $(x + y)^n$.

The first and last numbers in each row of Pascal's Triangle are 1. Every other number in each row is formed by adding the two numbers immediately above the number. Pascal noticed that numbers in this triangle are precisely the same numbers that are the coefficients of binomial expansions, as follows.

$$(x + y)^0 = 1 \qquad \text{0th row}$$
$$(x + y)^1 = 1x + 1y \qquad \text{1st row}$$
$$(x + y)^2 = 1x^2 + 2xy + 1y^2 \qquad \text{2nd row}$$
$$(x + y)^3 = 1x^3 + 3x^2y + 3xy^2 + 1y^3 \qquad \text{3rd row}$$
$$(x + y)^4 = 1x^4 + 4x^3y + 6x^2y^2 + 4xy^3 + 1y^4 \qquad \vdots$$
$$(x + y)^5 = 1x^5 + 5x^4y + 10x^3y^2 + 10x^2y^3 + 5xy^4 + 1y^5$$
$$(x + y)^6 = 1x^6 + 6x^5y + 15x^4y^2 + 20x^3y^3 + 15x^2y^4 + 6xy^5 + 1y^6$$
$$(x + y)^7 = 1x^7 + 7x^6y + 21x^5y^2 + 35x^4y^3 + 35x^3y^4 + 21x^2y^5 + 7xy^6 + 1y^7$$

Use the seventh row to find the binomial coefficients of the eighth row.

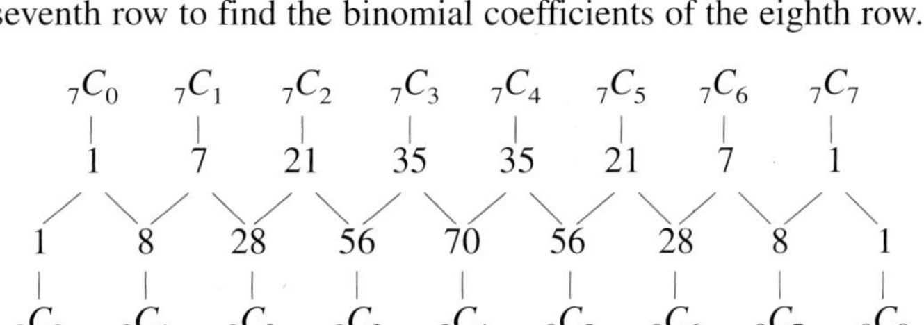

EXAMPLE 3 Using Pascal's Triangle

Use the fifth row of Pascal's Triangle to evaluate $_5C_2$.

Solution

1	5	10	10	5	1
$_5C_0$	$_5C_1$	$_5C_2$	$_5C_3$	$_5C_4$	$_5C_5$

So, $_5C_2 = 10$.

✓ **CHECKPOINT** *Now try Exercise 17.*

3 ▶ Expand binomial expressions.

Binomial Expansions

As mentioned at the beginning of this section, when you write out the coefficients of a binomial raised to a power, you are **expanding a binomial.** The formulas for binomial coefficients give you an easy way to expand binomials.

EXAMPLE 4 Expanding a Binomial

Write the expansion of the expression $(x + 1)^5$.

Solution

The binomial coefficients from the fifth row of Pascal's Triangle are

1, 5, 10, 10, 5, 1.

So, the expansion is as follows.

$$(x + 1)^5 = (1)x^5 + (5)x^4(1) + (10)x^3(1^2) + (10)x^2(1^3) + (5)x(1^4) + (1)(1^5)$$
$$= x^5 + 5x^4 + 10x^3 + 10x^2 + 5x + 1$$

✓ **CHECKPOINT** *Now try Exercise 23.*

To expand binomials representing *differences*, rather than sums, you alternate signs. Here are two examples.

$$(x - 1)^3 = x^3 - 3x^2 + 3x - 1$$
$$(x - 1)^4 = x^4 - 4x^3 + 6x^2 - 4x + 1$$

EXAMPLE 5 Expanding a Binomial

Write the expansion of each expression.

a. $(x - 3)^4$ **b.** $(2x - 1)^3$

Solution

a. The binomial coefficients from the fourth row of Pascal's Triangle are

1, 4, 6, 4, 1.

So, the expansion is as follows.

$$(x - 3)^4 = (1)x^4 - (4)x^3(3) + (6)x^2(3^2) - (4)x(3^3) + (1)(3^4)$$
$$= x^4 - 12x^3 + 54x^2 - 108x + 81$$

b. The binomial coefficients from the third row of Pascal's Triangle are

1, 3, 3, 1.

So, the expansion is as follows.

$$(2x - 1)^3 = (1)(2x)^3 - (3)(2x)^2(1) + (3)(2x)(1^2) - (1)(1^3)$$
$$= 8x^3 - 12x^2 + 6x - 1$$

✓ **CHECKPOINT** *Now try Exercise 25.*

EXAMPLE 6 Expanding a Binomial

Write the expansion of the expression $(x - 2y)^4$.

Solution

Use the fourth row of Pascal's Triangle, as follows.

$$(x - 2y)^4 = (1)x^4 - (4)x^3(2y) + (6)x^2(2y)^2 - (4)x(2y)^3 + (1)(2y)^4$$
$$= x^4 - 8x^3y + 24x^2y^2 - 32xy^3 + 16y^4$$

✓ **CHECKPOINT** *Now try Exercise 27.*

EXAMPLE 7 Expanding a Binomial

Write the expansion of the expression $(x^2 + 4)^3$.

Solution

Use the third row of Pascal's Triangle, as follows.

$$(x^2 + 4)^3 = (1)(x^2)^3 + (3)(x^2)^2(4) + (3)x^2(4^2) + (1)(4^3)$$
$$= x^6 + 12x^4 + 48x^2 + 64$$

✓ **CHECKPOINT** *Now try Exercise 31.*

Sometimes you will need to find a specific term in a binomial expansion. Instead of writing out the entire expansion, you can use the fact that from the Binomial Theorem, the $(r + 1)$th term is ${}_nC_r\, x^{n-r}y^r$.

EXAMPLE 8 Finding a Term in a Binomial Expansion

a. Find the sixth term in the expansion of $(a + 2b)^8$.

b. Find the coefficient of the term a^6b^5 in the expansion of $(3a - 2b)^{11}$.

Solution

a. In this case, $6 = r + 1$ means that $r = 5$. Because $n = 8$, $x = a$, and $y = 2b$, the sixth term in the binomial expansion is

$${}_8C_5 a^{8-5}(2b)^5 = 56 \cdot a^3 \cdot (2b)^5$$
$$= 56(2^5)a^3b^5$$
$$= 1792a^3b^5.$$

b. In this case, $n = 11$, $r = 5$, $x = 3a$, and $y = -2b$. Substitute these values to obtain

$${}_nC_r\, x^{n-r}y^r = {}_{11}C_5(3a)^6(-2b)^5$$
$$= 462(729a^6)(-32b^5)$$
$$= -10{,}777{,}536a^6b^5.$$

So, the coefficient is $-10{,}777{,}536$.

✓ **CHECKPOINT** *Now try Exercise 43.*

Concept Check

1. How many terms are in the expansion of $(x + y)^{10}$?

2. In the expansion of $(x + y)^{10}$, is ${}_6C_4$ the coefficient of the x^4y^6 term? Explain.

3. Which row of Pascal's Triangle would you use to evaluate ${}_{10}C_3$?

4. When finding the seventh term of a binomial expansion by evaluating ${}_nC_r x^{n-r}y^r$, what value should you substitute for r? Explain.

13.4 EXERCISES

Go to pages 852–853 to record your assignments.

Developing Skills

In Exercises 1–10, evaluate the binomial coefficient ${}_nC_r$. ***See Examples 1 and 2.***

✓ 1. ${}_6C_4$
2. ${}_9C_3$
3. ${}_{10}C_5$
4. ${}_{12}C_9$
5. ${}_{12}C_{12}$
6. ${}_8C_1$
✓ 7. ${}_{20}C_6$
8. ${}_{15}C_{10}$
9. ${}_{20}C_{14}$
10. ${}_{15}C_5$

In Exercises 11–16, use a graphing calculator to evaluate ${}_nC_r$.

11. ${}_{30}C_6$
12. ${}_{40}C_8$
13. ${}_{52}C_5$
14. ${}_{100}C_4$
15. ${}_{800}C_{797}$
16. ${}_{1000}C_2$

In Exercises 17–22, use Pascal's Triangle to evaluate ${}_nC_r$. ***See Example 3.***

✓ 17. ${}_6C_2$
18. ${}_9C_3$
19. ${}_7C_3$
20. ${}_9C_5$
21. ${}_8C_4$
22. ${}_{10}C_6$

In Exercises 23–32, use Pascal's Triangle to expand the expression. ***See Examples 4–7.***

✓ 23. $(t + 5)^3$
24. $(y + 2)^4$
✓ 25. $(m - n)^5$
26. $(r - s)^7$
✓ 27. $(3a - 1)^5$
28. $(1 - 4b)^3$
29. $(2y + z)^6$
30. $(3c + d)^6$
✓ 31. $(x^2 + 2)^4$
32. $(5 + y^2)^5$

In Exercises 33–42, use the Binomial Theorem to expand the expression.

33. $(x + 3)^6$
34. $(m - 4)^4$
35. $(u - 2v)^3$
36. $(2x + y)^5$
37. $(3a + 2b)^4$
38. $(4u - 3v)^3$
39. $\left(x + \frac{2}{y}\right)^4$
40. $\left(s + \frac{1}{t}\right)^5$
41. $(2x^2 - y)^5$
42. $(x - 4y^3)^4$

In Exercises 43–46, find the specified term in the expansion of the binomial. ***See Example 8.***

✓ 43. $(x + y)^{10}$, 4th term
44. $(x - y)^6$, 7th term
45. $(a + 6b)^9$, 5th term
46. $(3a - b)^{12}$, 10th term

In Exercises 47–50, find the coefficient of the given term in the expansion of the binomial. ***See Example 8.***

	Expression	*Term*
47.	$(x + 1)^{10}$	x^7
48.	$(x + 3)^{12}$	x^9
49.	$(x^2 - 3)^4$	x^4
50.	$(3 - y^3)^5$	y^9

In Exercises 51–54, use the Binomial Theorem to approximate the quantity rounded to three decimal places. For example:

$$(1.02)^{10} = (1 + 0.02)^{10} \approx 1 + 10(0.02) + 45(0.02)^2.$$

51. $(1.02)^8$ **52.** $(2.005)^{10}$

53. $(2.99)^{12}$ **54.** $(1.98)^9$

Solving Problems

Probability In Exercises 55–58, use the Binomial Theorem to expand the expression. In the study of probability, it is sometimes necessary to use the expansion $(p + q)^n$, where $p + q = 1$.

55. $\left(\frac{1}{2} + \frac{1}{2}\right)^5$

56. $\left(\frac{2}{3} + \frac{1}{3}\right)^4$

57. $\left(\frac{1}{4} + \frac{3}{4}\right)^4$

58. $\left(\frac{2}{5} + \frac{3}{5}\right)^3$

59. *Pascal's Triangle* Rows 0 through 6 of Pascal's Triangle are shown. Find the sum of the numbers in each row. Describe the pattern.

1
1 1
1 2 1
1 3 3 1
1 4 6 4 1
1 5 10 10 5 1
1 6 15 20 15 6 1

60. *Pascal's Triangle* Use each encircled group of numbers to form a 2×2 matrix. Find the determinant of each matrix. Describe the pattern.

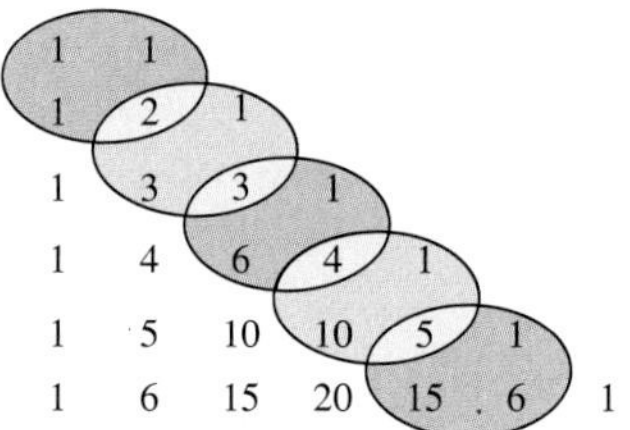

Explaining Concepts

61. How do the expansions of $(x + y)^n$ and $(x - y)^n$ differ?

62. Which of the following is equal to ${}_{11}C_5$? Explain.

(a) $\dfrac{11 \cdot 10 \cdot 9 \cdot 8 \cdot 7}{5 \cdot 4 \cdot 3 \cdot 2 \cdot 1}$ (b) $\dfrac{11 \cdot 10 \cdot 9 \cdot 8 \cdot 7}{6 \cdot 5 \cdot 4 \cdot 3 \cdot 2 \cdot 1}$

63. ✎ In your own words, explain how to form the rows in Pascal's Triangle.

64. In the expansion of $(x + 2)^9$, are the coefficients of the x^3-term and the x^6-term identical? Explain.

Cumulative Review

In Exercises 65 and 66, find the partial sum of the arithmetic sequence.

65. $\displaystyle\sum_{i=1}^{15} (2 + 3i)$ **66.** $\displaystyle\sum_{k=1}^{25} (9k - 5)$

In Exercises 67 and 68, find the partial sum of the geometric sequence. Round to the nearest hundredth.

67. $\displaystyle\sum_{k=1}^{8} 5^{k-1}$ **68.** $\displaystyle\sum_{i=1}^{20} 10\left(\tfrac{3}{4}\right)^{i-1}$

What Did You Learn?

Use these two pages to help prepare for a test on this chapter. Check off the key terms and key concepts you know. You can also use this section to record your assignments.

Plan for Test Success

Date of test: / /

Study dates and times: / / at : A.M./P.M.
/ / at : A.M./P.M.

Things to review:

- ☐ Key Terms, *p. 852*
- ☐ Key Concepts, *pp. 852–853*
- ☐ Your class notes
- ☐ Your assignments
- ☐ Study Tips, *pp. 817, 819, 826, 827, 836, 845, 847*
- ☐ Technology Tips, *pp. 815, 818, 846*
- ☐ Mid-Chapter Quiz, *p. 834*
- ☐ Review Exercises, *pp. 854–856*
- ☐ Chapter Test, *p. 857*
- ☐ Video Explanations Online
- ☐ Tutorial Online

Key Terms

- ☐ sequence, *p. 814*
- ☐ term (of a sequence), *p. 814*
- ☐ infinite sequence, *p. 814*
- ☐ finite sequence, *p. 814*
- ☐ factorials, *p. 816*
- ☐ series, *p. 817*
- ☐ partial sum, *p. 817*
- ☐ infinite series, *p. 817*
- ☐ sigma notation, *p. 818*
- ☐ index of summation, *p. 818*
- ☐ upper limit of summation, *p. 818*
- ☐ lower limit of summation, *p. 818*
- ☐ arithmetic sequence, *p. 825*
- ☐ common difference, *p. 825*
- ☐ recursion formula, *p. 826*
- ☐ *n*th partial sum, *pp. 827, 837*
- ☐ geometric sequence, *p. 835*
- ☐ common ratio, *p. 835*
- ☐ infinite geometric series, *p. 837*
- ☐ increasing annuity, *p. 839*
- ☐ binomial coefficients, *p. 845*
- ☐ Pascal's Triangle, *p. 847*
- ☐ expanding a binomial, *p. 848*

Key Concepts

13.1 Sequences and Series

Assignment: ______________________________ **Due date:** ________

☐ **Write the terms of a sequence.**

An infinite sequence $a_1, a_2, a_3, \ldots, a_n, \ldots$ is a function whose domain is the set of positive integers.

A finite sequence $a_1, a_2, a_3, \ldots, a_n$ is a function whose domain is the finite set $\{1, 2, 3, \ldots, n\}$.

☐ **Write the terms of a sequence involving factorials.**

If n is a positive integer, n factorial is defined as

$n! = 1 \cdot 2 \cdot 3 \cdot 4 \cdot \ldots \cdot (n-1) \cdot n.$

As a special case, zero factorial is defined as $0! = 1$.

☐ **Sum the terms of a sequence to obtain a series.**

For an infinite sequence $a_1, a_2, a_3, \ldots, a_n, \ldots$

1. The sum of the first n terms is called a partial sum.
2. The sum of all the terms is called an infinite series, or simply a series.

☐ **Use sigma notation to represent a partial sum.**

The sum of the first n terms of the sequence whose nth term is a_n is

$$\sum_{i=1}^{n} a_i = a_1 + a_2 + a_3 + a_4 + \cdots + a_n$$

where i is the index of summation, n is the upper limit of summation, and 1 is the lower limit of summation.

13.2 Arithmetic Sequences

Assignment:________________________________ Due date:________

☐ **Recognize, write, and find the *n*th term of an arithmetic sequence.**

The *n*th term of an arithmetic sequence has the form

$a_n = a_1 + (n - 1)d$

where d is the common difference between the terms of the sequence, and a_1 is the first term.

☐ **Find the *n*th partial sum of an arithmetic sequence.**

The *n*th partial sum of the arithmetic sequence whose *n*th term is a_n is

$$\sum_{i=1}^{n} a_i = a_1 + a_2 + a_3 + a_4 + \cdots + a_n$$
$$= \frac{n}{2}(a_1 + a_n).$$

13.3 Geometric Sequences and Series

Assignment:________________________________ Due date:________

☐ **Recognize, write, and find the *n*th term of a geometric sequence.**

The *n*th term of a geometric sequence has the form

$a_n = a_1 r^{n-1}$

where r is the common ratio of consecutive terms of the sequence. So, every geometric sequence can be written in the following form.

$a_1, a_1 r, a_1 r^2, a_1 r^3, a_1 r^4, \ldots, a_1 r^{n-1}, \ldots$

☐ **Find the *n*th partial sum of a geometric sequence.**

The *n*th partial sum of the geometric sequence whose *n*th term is $a_n = a_1 r^{n-1}$ is given by

$$\sum_{i=1}^{n} a_1 r^{i-1} = a_1 + a_1 r + a_1 r^2 + a_1 r^3 + \cdots + a_1 r^{n-1}$$
$$= a_1\left(\frac{r^n - 1}{r - 1}\right).$$

☐ **Find the sum of an infinite geometric series.**

If $a_1, a_1 r, a_1 r^2, \ldots, a_1 r^n, \ldots$ is an infinite geometric sequence and $|r| < 1$, the sum of the terms of the corresponding infinite geometric series is

$$S = \sum_{i=0}^{\infty} a_1 r^i = \frac{a_1}{1 - r}.$$

13.4 The Binomial Theorem

Assignment:________________________________ Due date:________

☐ **Use the Binomial Theorem to calculate binomial coefficients.**

In the expansion of $(x + y)^n$

$$(x + y)^n = x^n + nx^{n-1}y + \cdots + {}_nC_r x^{n-r}y^r + \cdots + nxy^{n-1} + y^n$$

the coefficient of $x^{n-r}y^r$ is given by

$${}_nC_r = \frac{n!}{(n - r)!r!}.$$

☐ **Use Pascal's Triangle to calculate binomial coefficients.**

The first and last numbers in each row of Pascal's Triangle are 1. Every other number in each row is the sum of the two numbers immediately above it.

1
1 1
1 2 1
1 3 3 1
1 4 6 4 1
1 5 10 10 5 1
1 6 15 20 15 6 1

Review Exercises

13.1 Sequences and Series

1 ▸ Use sequence notation to write the terms of sequences.

In Exercises 1–4, write the first five terms of the sequence. (Assume that n begins with 1.)

1. $a_n = 3n + 5$

2. $a_n = \frac{1}{2}n - 4$

3. $a_n = \dfrac{n}{3n - 1}$

4. $a_n = 3^n + n$

2 ▸ Write the terms of sequences involving factorials.

In Exercises 5–8, write the first five terms of the sequence. (Assume that n begins with 1.)

5. $a_n = (n + 1)!$

6. $a_n = (-1)^n n!$

7. $a_n = \dfrac{n!}{2n}$

8. $a_n = \dfrac{(n + 1)!}{(2n)!}$

3 ▸ Find the apparent nth term of a sequence.

In Exercises 9–16, write an expression for the nth term of the sequence. (Assume that n begins with 1.)

9. $4, 7, 10, 13, 16, \ldots$

10. $3, -6, 9, -12, 15, \ldots$

11. $\frac{1}{2}, \frac{1}{5}, \frac{1}{10}, \frac{1}{17}, \frac{1}{26}, \ldots$

12. $\frac{0}{2}, \frac{1}{3}, \frac{2}{4}, \frac{3}{5}, \frac{4}{6}, \ldots$

13. $3, 1, -1, -3, -5, \ldots$

14. $3, 7, 11, 15, 19, \ldots$

15. $\frac{3}{2}, \frac{12}{5}, \frac{27}{10}, \frac{48}{17}, \frac{75}{26}, \ldots$

16. $-1, \frac{1}{2}, -\frac{1}{4}, \frac{1}{8}, -\frac{1}{16}, \ldots$

4 ▸ Sum the terms of sequences to obtain series, and use sigma notation to represent partial sums.

In Exercises 17–20, find the partial sum.

17. $\displaystyle\sum_{k=1}^{4} 7$

18. $\displaystyle\sum_{k=1}^{4} \frac{(-1)^k}{k}$

19. $\displaystyle\sum_{i=1}^{5} \frac{i - 2}{i + 1}$

20. $\displaystyle\sum_{n=1}^{4} \left(\frac{1}{n} - \frac{1}{n + 2}\right)$

In Exercises 21–24, write the sum using sigma notation. (Begin with $k = 0$ or $k = 1$.)

21. $[5(1) - 3] + [5(2) - 3] + [5(3) - 3] + [5(4) - 3]$

22. $(1)[(1) - 5] + (2)[(2) - 5] + (3)[(3) - 5] + (4)[(4) - 5] + (5)[(5) - 5]$

23. $\dfrac{1}{3(1)} + \dfrac{1}{3(2)} + \dfrac{1}{3(3)} + \dfrac{1}{3(4)} + \dfrac{1}{3(5)} + \dfrac{1}{3(6)}$

24. $\left(-\frac{1}{3}\right)^0 + \left(-\frac{1}{3}\right)^1 + \left(-\frac{1}{3}\right)^2 + \left(-\frac{1}{3}\right)^3 + \left(-\frac{1}{3}\right)^4$

13.2 Arithmetic Sequences

1 ▸ Recognize, write, and find the nth terms of arithmetic sequences.

In Exercises 25 and 26, find the common difference of the arithmetic sequence.

25. $50, 44.5, 39, 33.5, 28, \ldots$

26. $9, 12, 15, 18, 21, \ldots$

In Exercises 27–32, write the first five terms of the arithmetic sequence. (Assume that n begins with 1.)

27. $a_n = 132 - 5n$

28. $a_n = 2n + 3$

29. $a_n = \frac{1}{3}n + \frac{5}{3}$

30. $a_n = -\frac{3}{5}n + 1$

31. $a_1 = 80$
$a_{k+1} = a_k - \frac{5}{2}$

32. $a_1 = 30$
$a_{k+1} = a_k - 12$

In Exercises 33–36, find a formula for the nth term of the arithmetic sequence.

33. $a_1 = 10, \quad d = 4$

34. $a_1 = 32, \quad d = -2$

35. $a_1 = 1000, \quad a_2 = 950$

36. $a_2 = 150, \quad a_5 = 201$

2 ▶ Find the *n*th partial sum of an arithmetic sequence.

In Exercises 37–40, find the partial sum.

37. $\sum_{k=1}^{12} (7k - 5)$

38. $\sum_{k=1}^{10} (100 - 10k)$

39. $\sum_{j=1}^{120} \left(\frac{1}{4}j + 1\right)$

40. $\sum_{j=1}^{50} \frac{3j}{2}$

In Exercises 41 and 42, use a graphing calculator to find the partial sum.

41. $\sum_{i=1}^{60} (1.25i + 4)$

42. $\sum_{i=1}^{150} \frac{i + 4}{2}$

3 ▶ Use arithmetic sequences to solve application problems.

43. *Number Problem* Find the sum of the first 50 positive integers that are multiples of 4.

44. *Number Problem* Find the sum of the integers from 225 to 300.

45. *Auditorium Seating* Each row in a small auditorium has three more seats than the preceding row. The front row seats 22 people and there are 12 rows of seats. Find the seating capacity of the auditorium.

46. *Wages* You earn \$25 on the first day of the month and \$100 on the last day of the month. Each day you are paid \$2.50 more than the previous day. How much do you earn in a 31-day month?

13.3 Geometric Sequences and Series

1 ▶ Recognize, write, and find the *n*th terms of geometric sequences.

In Exercises 47 and 48, find the common ratio of the geometric sequence.

47. $8, 20, 50, 125, \frac{625}{2}, \ldots$

48. $27, -18, 12, -8, \frac{16}{3}, \ldots$

In Exercises 49–54, write the first five terms of the geometric sequence.

49. $a_1 = 10, \quad r = 3$

50. $a_1 = 2, \quad r = -5$

51. $a_1 = 100, \quad r = -\frac{1}{2}$

52. $a_1 = 20, \quad r = \frac{1}{5}$

53. $a_1 = 4, \quad r = \frac{3}{2}$

54. $a_1 = 32, \quad r = -\frac{3}{4}$

In Exercises 55–60, find a formula for the nth term of the geometric sequence. (Assume that n begins with 1.)

55. $a_1 = 1, \quad r = -\frac{2}{3}$

56. $a_1 = 100, \quad r = 1.07$

57. $a_1 = 24, \quad a_2 = 72$

58. $a_1 = 16, \quad a_2 = -4$

59. $a_1 = 12, \quad a_4 = -\frac{3}{2}$

60. $a_2 = 1, \quad a_3 = \frac{1}{3}$

2 ▶ Find the *n*th partial sum of a geometric sequence.

In Exercises 61–66, find the partial sum. Round to the nearest thousandth, if necessary.

61. $\sum_{n=1}^{12} 2^n$

62. $\sum_{n=1}^{12} (-2)^n$

63. $\sum_{k=1}^{8} 5\left(-\frac{3}{4}\right)^k$

64. $\sum_{k=1}^{12} (-0.6)^{k-1}$

65. $\sum_{n=1}^{120} 500(1.01)^n$

66. $\sum_{n=1}^{40} 1000(1.1)^n$

In Exercises 67 and 68, use a graphing calculator to find the partial sum. Round to the nearest thousandth, if necessary.

67. $\sum_{k=1}^{75} 200(1.4)^{k-1}$

68. $\sum_{j=1}^{60} 25(0.9)^{j-1}$

3 ▶ Find the sum of an infinite geometric series.

In Exercises 69–72, find the sum.

69. $\sum_{i=1}^{\infty} \left(\frac{7}{8}\right)^{i-1}$

70. $\sum_{i=1}^{\infty} \left(\frac{3}{5}\right)^{i-1}$

71. $\sum_{k=0}^{\infty} 4\left(\frac{2}{3}\right)^k$

72. $\sum_{k=0}^{\infty} 1.3\left(\frac{1}{10}\right)^k$

4 ▶ Use geometric sequences to solve application problems.

73. *Depreciation* A company pays $120,000 for a machine. During the next 5 years, the machine depreciates at the rate of 30% per year. (That is, at the end of each year, the depreciated value is 70% of what it was at the beginning of the year.)

(a) Find a formula for the nth term of the geometric sequence that gives the value of the machine n full years after it was purchased.

(b) Find the depreciated value of the machine at the end of 5 full years.

74. *Population Increase* A city of 85,000 people is growing at the rate of 1.2% per year. (That is, at the end of each year, the population is 1.012 times what it was at the beginning of the year.)

(a) Find a formula for the nth term of the geometric sequence that gives the population after n years.

(b) Estimate the population after 50 years.

75. *Internet* On its first day, a website has 1000 visits. During the next 89 days, the number of visits increases by 12.5% each day. What is the total number of visits during the 90-day period?

76. *Increasing Annuity* You deposit $200 in an account each month for 10 years. The account pays an annual interest rate of 8%, compounded monthly. What is your balance at the end of 10 years?

13.4 The Binomial Theorem

1 ▶ Use the Binomial Theorem to calculate binomial coefficients.

In Exercises 77–80, evaluate $_nC_r$.

77. $_8C_3$

78. $_{12}C_2$

79. $_{15}C_4$

80. $_{100}C_1$

In Exercises 81–84, use a graphing calculator to evaluate $_nC_r$.

81. $_{40}C_4$

82. $_{32}C_8$

83. $_{25}C_6$

84. $_{48}C_5$

2 ▶ Use Pascal's Triangle to calculate binomial coefficients.

In Exercises 85–88, use Pascal's Triangle to evaluate $_nC_r$.

85. $_5C_3$

86. $_9C_9$

87. $_8C_4$

88. $_7C_2$

3 ▶ Expand binomial expressions.

In Exercises 89–92, use Pascal's Triangle to expand the expression.

89. $(x - 5)^4$

90. $(x + y)^7$

91. $(5x + 2)^3$

92. $(x - 3y)^4$

In Exercises 93–98, use the Binomial Theorem to expand the expression.

93. $(x + 1)^{10}$

94. $(y - 2)^6$

95. $(3x - 2y)^4$

96. $(4u + v)^5$

97. $(u^2 + v^3)^5$

98. $(x^4 - y^5)^4$

In Exercises 99 and 100, find the specified term in the expansion of the binomial.

99. $(x + 2)^{10}$, 7th term

100. $(2x - 3y)^5$, 4th term

In Exercises 101 and 102, find the coefficient of the given term in the expansion of the binomial.

	Expression	*Term*
101.	$(x - 3)^{10}$	x^5
102.	$(3x + 4y)^6$	x^2y^4

Chapter Test

Take this test as you would take a test in class. After you are done, check your work against the answers in the back of the book.

1. Write the first five terms of the sequence $a_n = \left(-\frac{3}{5}\right)^{n-1}$. (Assume that n begins with 1.)

2. Write the first five terms of the sequence $a_n = 3n^2 - n$. (Assume that n begins with 1.)

In Exercises 3–5, find the partial sum.

3. $\sum_{n=1}^{12} 5$

4. $\sum_{k=0}^{8} (2k - 3)$

5. $\sum_{n=1}^{5} (3 - 4n)$

6. Use sigma notation to write $\dfrac{2}{3(1) + 1} + \dfrac{2}{3(2) + 1} + \cdot \cdot \cdot + \dfrac{2}{3(12) + 1}$.

7. Use sigma notation to write

$$\left(\frac{1}{2}\right)^0 + \left(\frac{1}{2}\right)^2 + \left(\frac{1}{2}\right)^4 + \left(\frac{1}{2}\right)^6 + \left(\frac{1}{2}\right)^8 + \left(\frac{1}{2}\right)^{10}.$$

8. Write the first five terms of the arithmetic sequence whose first term is $a_1 = 12$ and whose common difference is $d = 4$.

9. Find a formula for the nth term of the arithmetic sequence whose first term is $a_1 = 5000$ and whose common difference is $d = -100$.

10. Find the sum of the first 50 positive integers that are multiples of 3.

11. Find the common ratio of the geometric sequence: $-4, 3, -\frac{9}{4}, \frac{27}{16}, \ldots$

12. Find a formula for the nth term of the geometric sequence whose first term is $a_1 = 4$ and whose common ratio is $r = \frac{1}{2}$.

In Exercises 13 and 14, find the partial sum.

13. $\sum_{n=1}^{8} 2(2^n)$

14. $\sum_{n=1}^{10} 3\left(\frac{1}{2}\right)^n$

In Exercises 15 and 16, find the sum of the infinite geometric series.

15. $\sum_{i=1}^{\infty} \left(\frac{1}{2}\right)^i$

16. $\sum_{i=1}^{\infty} 10(0.4)^{i-1}$

17. Evaluate: ${}_{20}C_3$

18. Use Pascal's Triangle to expand $(x - 2)^5$.

19. Find the coefficient of the term x^3y^5 in the expansion of $(x + y)^8$.

20. A free-falling object will fall 4.9 meters during the first second, 14.7 more meters during the second second, 24.5 more meters during the third second, and so on. What is the total distance the object will fall in 10 seconds if this pattern continues?

21. You deposit \$80 each month in an increasing annuity that pays 4.8% compounded monthly. What is the balance after 45 years?

Appendix A Review of Elementary Algebra Topics

- Sets and Real Numbers
- Operations with Real Numbers
- Properties of Real Numbers

A.1 The Real Number System

Sets and Real Numbers

Real numbers are used in everyday life to describe quantities such as age, miles per gallon, container size, and population. Real numbers are represented by symbols such as

$$-5, 9, 0, \tfrac{4}{3}, 0.666\ldots, 28.21, \sqrt{2}, \pi, \text{ and } \sqrt[3]{-32}.$$

Here are some important subsets of the set of real numbers.

$\{1, 2, 3, 4, \ldots\}$ Set of natural numbers

$\{0, 1, 2, 3, 4, \ldots\}$ Set of whole numbers

$\{\ldots, -3, -2, -1, 0, 1, 2, 3, \ldots\}$ Set of integers

A real number is rational if it can be written as the ratio p/q of two integers, where $q \neq 0$. For instance, the numbers

$$\tfrac{1}{3} = 0.3333\ldots = 0.\overline{3}, \tfrac{1}{8} = 0.125, \text{ and } \tfrac{125}{111} = 1.126126\ldots = 1.\overline{126}$$

are rational. The decimal representation of a rational number either repeats or terminates.

$\frac{173}{55} = 3.1\overline{45}$ A rational number that repeats

$\frac{1}{2} = 0.5$ A rational number that terminates

A real number that cannot be written as the ratio of two integers is called irrational. Irrational numbers have infinite nonrepeating decimal representations. For instance, the numbers

$$\sqrt{2} \approx 1.4142136 \quad \text{and} \quad \pi \approx 3.1415927$$

are irrational. (The symbol $\approx$ means "is approximately equal to.")

Real numbers are represented graphically by a real number line. The point 0 on the real number line is the origin. Numbers to the right of 0 are positive, and numbers to the left are negative, as shown in Figure A.1. The term "nonnegative" describes a number that is either positive or zero.

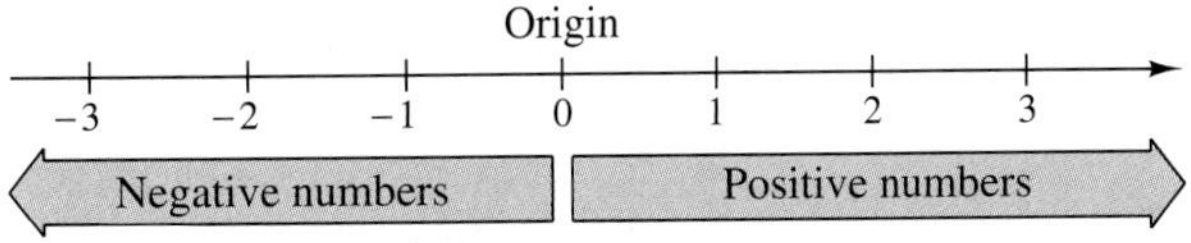

Figure A.1 The Real Number Line

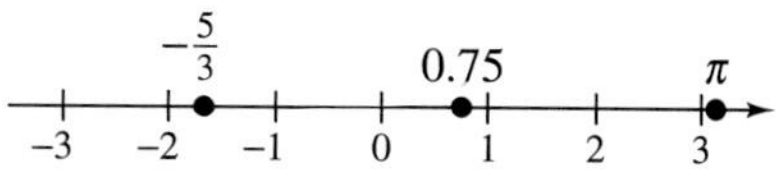

Every real number corresponds to exactly one point on the real number line.

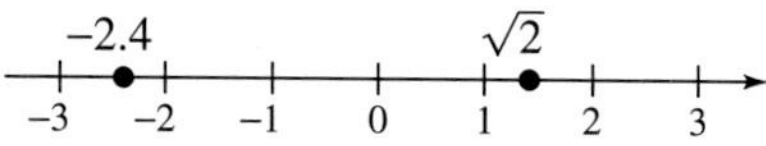

Every point on the real number line corresponds to exactly one real number.

Figure A.2

As illustrated in Figure A.2, there is a *one-to-one correspondence* between real numbers and points on the real number line.

The real number line provides you with a way of comparing any two real numbers. For any two (different) numbers on the real number line, one of the numbers must be to the left of the other number. A "less than" comparison is denoted by the inequality symbol $<$, a "greater than" comparison is denoted by $>$, a "less than or equal to" comparison is denoted by $\leq$, and a "greater than or equal to" comparison is denoted by $\geq$. When you are asked to order two numbers, you are simply being asked to say which of the two numbers is greater.

EXAMPLE 1 Ordering Real Numbers

Place the correct inequality symbol ($<$ or $>$) between each pair of numbers.

a. $2 \quad\square\quad -1$ **b.** $-\frac{1}{2} \quad\square\quad \frac{1}{4}$ **c.** $-1.1 \quad\square\quad -1.2$

Solution

a. $2 > -1$, because 2 lies to the *right* of -1. See Figure A.3.

b. $-\frac{1}{2} < \frac{1}{4}$, because $-\frac{1}{2} = -\frac{2}{4}$ lies to the *left* of $\frac{1}{4}$. See Figure A.4.

c. $-1.1 > -1.2$, because -1.1 lies to the *right* of -1.2. See Figure A.5.

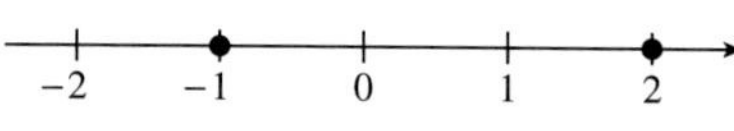

Figure A.3

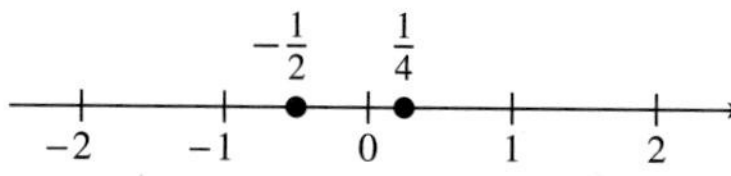

Figure A.4

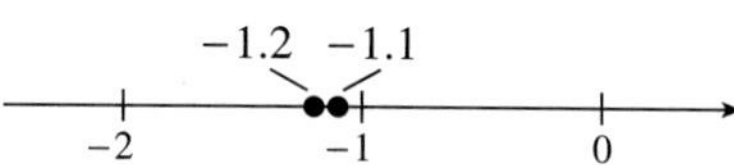

Figure A.5

Two real numbers are opposites of each other if they lie the same distance from, but on opposite sides of, zero. For instance, -1 is the opposite of 1, as shown in Figure A.6.

The absolute value of a real number is its distance from zero on the real number line. A pair of vertical bars, $|\ \ |$, is used to denote absolute value. The absolute value of a real number is either positive or zero (never negative).

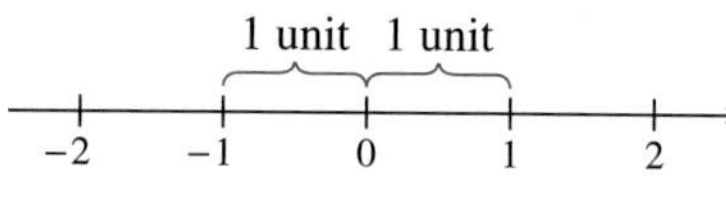

-1 is the opposite of 1.

Figure A.6

EXAMPLE 2 Evaluating Absolute Values

a. $|5| = 5$, because the distance between 5 and 0 is 5.

b. $|0| = 0$, because the distance between 0 and itself is 0.

c. $\left|-\frac{2}{3}\right| = \frac{2}{3}$, because the distance between $-\frac{2}{3}$ and 0 is $\frac{2}{3}$.

Operations with Real Numbers

There are four basic arithmetic operations with real numbers: addition, subtraction, multiplication, and division.

The result of adding two real numbers is called the sum of the two numbers. Subtraction of one real number from another can be described as adding the opposite of the second number to the first number. For instance,

$$7 - 5 = 7 + (-5) = 2 \text{ and } 10 - (-13) = 10 + 13 = 23.$$

The result of subtracting one real number from another is called the difference of the two numbers.

Study Tip

In the fraction

$$\frac{a}{b}$$

a is the numerator and b is the denominator.

EXAMPLE 3 Adding and Subtracting Real Numbers

a. $-25 + 12 = -13$

b. $5 + (-10) = -5$

c. $-13.8 - 7.02 = -13.8 + (-7.02) = -20.82$

d. To add two fractions with unlike denominators, you must first rewrite one (or both) of the fractions so that they have a common denominator. To do this, find the least common multiple (LCM) of the denominators.

$$\frac{1}{3} + \frac{2}{9} = \frac{1(3)}{3(3)} + \frac{2}{9} \qquad \text{LCM of 3 and 9 is 9.}$$

$$= \frac{3}{9} + \frac{2}{9} = \frac{5}{9} \qquad \text{Rewrite with like denominators and add numerators.}$$

The result of multiplying two real numbers is called their product, and each of the numbers is called a factor of the product. The product of zero and any other number is zero. Multiplication is denoted in a variety of ways. For instance,

3×2, $3 \cdot 2$, $3(2)$, and $(3)(2)$

all denote the product of "3 times 2," which you know is 6.

EXAMPLE 4 Multiplying Real Numbers

a. $(6)(-4) = -24$ **b.** $(-1.2)(-0.4) = 0.48$

c. To find the product of more than two numbers, find the product of their absolute values. If there is an *even* number of negative factors, the product is positive. If there is an *odd* number of negative factors, the product is negative. For instance, in the product $6(2)(-5)(-8)$, there are two negative factors, so the product must be positive, and you can write $6(2)(-5)(-8) = 480$.

d. To multiply two fractions, multiply their numerators and their denominators. For instance, the product of $\frac{2}{3}$ and $\frac{4}{5}$ is

$$\left(\frac{2}{3}\right)\left(\frac{4}{5}\right) = \frac{(2)(4)}{(3)(5)} = \frac{8}{15}.$$

The reciprocal of a nonzero real number a is defined as the number by which a must be multiplied to obtain 1. The reciprocal of the fraction a/b is b/a.

To divide one real number by a second (nonzero) real number, multiply the first number by the reciprocal of the second number. The result of dividing two real numbers is called the quotient of the two numbers. Division is denoted in a variety of ways. For instance,

$12 \div 4$, $12/4$, $\frac{12}{4}$, and $4\overline{)12}$

all denote the quotient of "12 divided by 4," which you know is 3.

EXAMPLE 5 **Dividing Real Numbers**

a. $-30 \div 5 = -30\left(\frac{1}{5}\right) = -\frac{30}{5} = -6$ **b.** $-\frac{9}{14} \div -\frac{1}{3} = -\frac{9}{14}\left(-\frac{3}{1}\right) = \frac{27}{14}$

c. $\frac{5}{16} \div 2\frac{3}{4} = \frac{5}{16} \div \frac{11}{4} = \frac{5}{16}\left(\frac{4}{11}\right) = \frac{5(\cancel{4})}{4(\cancel{4})(11)} = \frac{5}{44}$

Let n be a positive integer and let a be a real number. Then the product of n factors of a is given by

$$a^n = \underbrace{a \cdot a \cdot a \cdot \cdots \cdot a}_{n \text{ factors}}.$$

In the exponential form a^n, a is called the base and n is called the exponent.

EXAMPLE 6 **Evaluating Exponential Expressions**

a. $(-2)^5 = (-2)(-2)(-2)(-2)(-2) = -32$

b. $\left(\frac{1}{5}\right)^3 = \left(\frac{1}{5}\right)\left(\frac{1}{5}\right)\left(\frac{1}{5}\right) = \frac{1}{125}$ **c.** $(-7)^2 = (-7)(-7) = 49$

One way to help avoid confusion when communicating algebraic ideas is to establish an order of operations. This order is summarized below.

Order of Operations

1. Perform operations inside *symbols of grouping*—( ) or [ ]—or *absolute value symbols*, starting with the innermost set of symbols.
2. Evaluate all *exponential* expressions.
3. Perform all *multiplications* and *divisions* from left to right.
4. Perform all *additions* and *subtractions* from left to right.

EXAMPLE 7 **Order of Operations**

a. $20 - 2 \cdot 3^2 = 20 - 2 \cdot 9 = 20 - 18 = 2$

b. $-4 + 2(-2 + 5)^2 = -4 + 2(3)^2 = -4 + 2(9) = -4 + 18 = 14$

c. $\frac{2 \cdot 5^2 - 10}{3^2 - 4} = (2 \cdot 5^2 - 10) \div (3^2 - 4)$ Rewrite using parentheses.

$= (50 - 10) \div (9 - 4)$ Evaluate exponential expressions and multiply within symbols of grouping.

$= 40 \div 5 = 8$ Simplify.

Study Tip

For more review of the real number system, refer to Chapter 1.

Properties of Real Numbers

Below is a review of the properties of real numbers. In this list, a verbal description of each property is given, as well as an example.

Properties of Real Numbers: Let *a*, *b*, and *c* be real numbers.

Property	*Example*
1. *Commutative Property of Addition:* Two real numbers can be added in either order. $a + b = b + a$	$3 + 5 = 5 + 3$
2. *Commutative Property of Multiplication:* Two real numbers can be multiplied in either order. $ab = ba$	$4 \cdot (-7) = -7 \cdot 4$
3. *Associative Property of Addition:* When three real numbers are added, it makes no difference which two are added first. $(a + b) + c = a + (b + c)$	$(2 + 6) + 5 = 2 + (6 + 5)$
4. *Associative Property of Multiplication:* When three real numbers are multiplied, it makes no difference which two are multiplied first. $(ab)c = a(bc)$	$(3 \cdot 5) \cdot 2 = 3 \cdot (5 \cdot 2)$
5. *Distributive Property:* Multiplication distributes over addition. $a(b + c) = ab + ac$; $(a + b)c = ac + bc$	$3(8 + 5) = 3 \cdot 8 + 3 \cdot 5$; $(3 + 8)5 = 3 \cdot 5 + 8 \cdot 5$
6. *Additive Identity Property:* The sum of zero and a real number equals the number itself. $a + 0 = 0 + a = a$	$3 + 0 = 0 + 3 = 3$
7. *Multiplicative Identity Property:* The product of 1 and a real number equals the number itself. $a \cdot 1 = 1 \cdot a = a$	$4 \cdot 1 = 1 \cdot 4 = 4$
8. *Additive Inverse Property:* The sum of a real number and its opposite is zero. $a + (-a) = 0$	$3 + (-3) = 0$
9. *Multiplicative Inverse Property:* The product of a nonzero real number and its reciprocal is 1. $a \cdot \frac{1}{a} = 1,\ a \neq 0$	$8 \cdot \frac{1}{8} = 1$

A.2 Fundamentals of Algebra

Algebraic Expressions

One characteristic of algebra is the use of letters to represent numbers. The letters are variables, and combinations of letters and numbers are algebraic expressions. The terms of an algebraic expression are those parts separated by addition. For example, in the expression $-x^2 + 5x + 8$, $-x^2$ and $5x$ are the variable terms and 8 is the constant term. The coefficient of the variable term $-x^2$ is -1 and the coefficient of $5x$ is 5.

To evaluate an algebraic expression, substitute numerical values for each of the variables in the expression.

EXAMPLE 1 Evaluating Algebraic Expressions

a. Evaluate the expression $-3x + 5$ when $x = 3$.

$$-3(3) + 5 = -9 + 5 = -4$$

b. Evaluate the expression $3x^2 + 2xy - y^2$ when $x = 3$ and $y = -1$.

$$3(3)^2 + 2(3)(-1) - (-1)^2 = 3(9) + (-6) - 1 = 20$$

The properties of real numbers listed on page A5 can be used to rewrite and simplify algebraic expressions. To simplify an algebraic expression generally means to remove symbols of grouping such as parentheses or brackets and to combine like terms. In an algebraic expression, two terms are said to be like terms if they are both constant terms or if they have the same variable factor(s). To combine like terms in an algebraic expression, add their respective coefficients and attach the common variable factor.

EXAMPLE 2 Combining Like Terms

a. $2x + 3y - 6x - y = (2x - 6x) + (3y - y)$ Group like terms.

$= (2 - 6)x + (3 - 1)y$ Distributive Property

$= -4x + 2y$ Simplest form

b. $4x^2 + 5x - x^2 - 8x = (4x^2 - x^2) + (5x - 8x)$ Group like terms.

$= (4 - 1)x^2 + (5 - 8)x$ Distributive Property

$= 3x^2 - 3x$ Simplest form

EXAMPLE 3 **Removing Symbols of Grouping**

a. $-2(a + 5) + 4(a - 8) = -2a - 10 + 4a - 32$ Distributive Property

$= (-2a + 4a) + (-10 - 32)$ Group like terms.

$= 2a - 42$ Combine like terms.

b. $3x^2 - [9x + 3x(2x - 1)] = 3x^2 - [9x + 6x^2 - 3x]$ Distributive Property

$= 3x^2 - [6x^2 + 6x]$ Combine like terms.

$= 3x^2 - 6x^2 - 6x$ Distributive Property

$= -3x^2 - 6x$ Combine like terms.

Constructing Verbal Models

When you translate a verbal sentence or phrase into an algebraic expression, watch for key words and phrases that indicate the four different operations of arithmetic.

EXAMPLE 4 **Translating Verbal Phrases**

a. *Verbal Description:* Seven more than 3 times x

Algebraic Expression: $3x + 7$

b. *Verbal Description:* Four times the sum of y and 9

Algebraic Expression: $4(y + 9)$

c. *Verbal Description:* Five decreased by the product of 2 and a number

Label: The number $= x$ *Algebraic Expression:* $5 - 2x$

d. *Verbal Description:* One more than the product of 8 and a number, all divided by 6

Label: The number $= x$ *Algebraic Expression:* $\dfrac{8x + 1}{6}$

Study Tip

When verbal phrases are translated into algebraic expressions, products are often overlooked, as demonstrated in Example 5.

EXAMPLE 5 **Constructing Verbal Models** 

A cash register contains x quarters. Write an expression for this amount of money in dollars.

Solution

Verbal Model: Value of coin · Number of coins

Labels: Value of coin $= 0.25$ (dollars per quarter)
Number of coins $= x$ (quarters)

Expression: $0.25x$ (dollars)

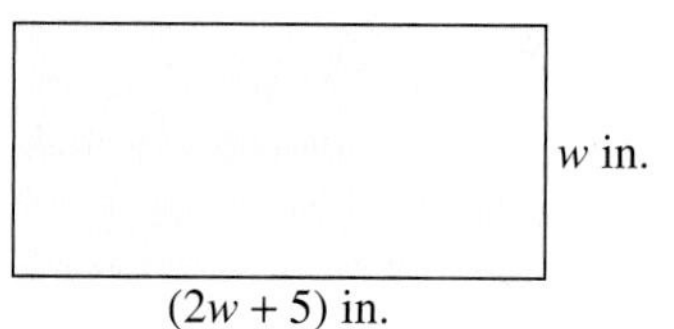

Figure A.7

EXAMPLE 6 Constructing Verbal Models

The width of a rectangle is w inches. The length of the rectangle is 5 inches more than twice its width. Write an expression for the perimeter of the rectangle.

Solution

Draw a rectangle, as shown in Figure A.7. Next, use a verbal model to solve the problem. Use the formula (perimeter) = 2(length) + 2(width).

Verbal Model: 2 · Length + 2 · Width

Labels: Length $= 2w + 5$ (inches)
Width $= w$ (inches)

Expression: $2(2w + 5) + 2w = 4w + 10 + 2w = 6w + 10$ (inches)

Equations

An equation is a statement that equates two algebraic expressions. Solving an equation involving x means finding all values of x for which the equation is true. Such values are solutions and are said to satisfy the equation. Example 7 shows how to check whether a given value is a solution of an equation.

EXAMPLE 7 Checking a Solution of an Equation

Determine whether $x = -3$ is a solution of $-3x - 5 = 4x + 16$.

$-3(-3) - 5 \stackrel{?}{=} 4(-3) + 16$ Substitute -3 for x in original equation.

$9 - 5 \stackrel{?}{=} -12 + 16$ Simplify.

$4 = 4$ Solution checks. ✓

EXAMPLE 8 Using a Verbal Model to Construct an Equation 

You are given a speeding ticket for \$80 for speeding on a road where the speed limit is 45 miles per hour. You are fined \$10 for each mile per hour over the speed limit. How fast were you driving? Write an algebraic equation that models the situation.

Solution

Verbal Model: Fine · Speed over limit = Amount of ticket

Labels: Fine $= 10$ (dollars per mile per hour)
Your speed $= x$ (miles per hour)
Speed over limit $= x - 45$ (miles per hour)
Amount of ticket $= 80$ (dollars)

Algebraic Model: $10(x - 45) = 80$

Study Tip

For more review on the fundamentals of algebra, refer to Chapter 2.

- Equations
- Inequalities
- Problem Solving

A.3 Equations, Inequalities, and Problem Solving

Equations

A linear equation in one variable x is an equation that can be written in the standard form

$$ax + b = 0$$

where a and b are real numbers with $a \neq 0$. To solve a linear equation, you want to isolate x on one side of the equation by a sequence of equivalent equations, each having the same solution(s) as the original equation. The operations that yield equivalent equations are as follows.

Operations That Yield Equivalent Equations

1. Remove symbols of grouping, combine like terms, or simplify fractions on one or both sides of the equation.
2. Add (or subtract) the same quantity to (from) each side of the equation.
3. Multiply (or divide) each side of the equation by the same nonzero quantity.
4. Interchange the two sides of the equation.

EXAMPLE 1 Solving a Linear Equation in Standard Form

Solve $3x - 6 = 0$. Then check the solution.

Solution

$3x - 6 = 0$	Write original equation.
$3x - 6 + 6 = 0 + 6$	Add 6 to each side.
$3x = 6$	Combine like terms.
$\dfrac{3x}{3} = \dfrac{6}{3}$	Divide each side by 3.
$x = 2$	Simplify.

Check

$3x - 6 = 0$	Write original equation.
$3(2) - 6 \stackrel{?}{=} 0$	Substitute 2 for x.
$0 = 0$	Solution checks. ✓

So, the solution is $x = 2$.

EXAMPLE 2 **Solving a Linear Equation in Nonstandard Form**

Solve $5x + 4 = 3x - 8$.

Solution

$$5x + 4 = 3x - 8 \quad \text{Write original equation.}$$

$$5x - 3x + 4 = 3x - 3x - 8 \quad \text{Subtract } 3x \text{ from each side.}$$

$$2x + 4 = -8 \quad \text{Combine like terms.}$$

$$2x + 4 - 4 = -8 - 4 \quad \text{Subtract 4 from each side.}$$

$$2x = -12 \quad \text{Combine like terms.}$$

$$\frac{2x}{2} = \frac{-12}{2} \quad \text{Divide each side by 2.}$$

$$x = -6 \quad \text{Simplify.}$$

The solution is $x = -6$. Check this in the original equation.

Linear equations often contain parentheses or other symbols of grouping. In most cases, it helps to remove symbols of grouping as a first step in solving an equation. This is illustrated in Example 3.

EXAMPLE 3 **Solving a Linear Equation Involving Parentheses**

Solve $2(x + 4) = 5(x - 8)$.

Solution

$$2(x + 4) = 5(x - 8) \quad \text{Write original equation.}$$

$$2x + 8 = 5x - 40 \quad \text{Distributive Property}$$

$$2x - 5x + 8 = 5x - 5x - 40 \quad \text{Subtract } 5x \text{ from each side.}$$

$$-3x + 8 = -40 \quad \text{Combine like terms.}$$

$$-3x + 8 - 8 = -40 - 8 \quad \text{Subtract 8 from each side.}$$

$$-3x = -48 \quad \text{Combine like terms.}$$

$$\frac{-3x}{-3} = \frac{-48}{-3} \quad \text{Divide each side by } -3.$$

$$x = 16 \quad \text{Simplify.}$$

The solution is $x = 16$. Check this in the original equation.

Study Tip

Recall that when finding the least common multiple of a set of numbers, you should first consider all multiples of each number. Then, you should choose the smallest of the common multiples of the numbers.

To solve an equation involving fractional expressions, find the least common multiple (LCM) of the denominators and multiply each side by the LCM.

EXAMPLE 4 **Solving a Linear Equation Involving Fractions**

Solve $\frac{x}{3} + \frac{3x}{4} = 2$.

Solution

$$12\left(\frac{x}{3} + \frac{3x}{4}\right) = 12(2)$$ Multiply each side of original equation by LCM 12.

$$12 \cdot \frac{x}{3} + 12 \cdot \frac{3x}{4} = 24$$ Distributive Property

$$4x + 9x = 24$$ Clear fractions.

$$13x = 24$$ Combine like terms.

$$x = \frac{24}{13}$$ Divide each side by 13.

The solution is $x = \frac{24}{13}$. Check this in the original equation.

To solve an equation involving an absolute value, remember that the expression inside the absolute value signs can be positive or negative. This results in two separate equations, each of which must be solved.

EXAMPLE 5 **Solving an Equation Involving Absolute Value**

Solve $|4x - 3| = 13$.

Solution

$$|4x - 3| = 13$$ Write original equation.

$$4x - 3 = -13 \quad \text{or} \quad 4x - 3 = 13$$ Equivalent equations

$$4x = -10 \qquad\qquad 4x = 16$$ Add 3 to each side.

$$x = -\frac{5}{2} \qquad\qquad x = 4$$ Divide each side by 4.

The solutions are $x = -\frac{5}{2}$ and $x = 4$. Check these in the original equation.

Inequalities

The simplest type of inequality is a linear inequality in one variable. For instance, $2x + 3 > 4$ is a linear inequality in x. The procedures for solving linear inequalities in one variable are much like those for solving linear equations, as described on page A9. The exception is that when each side of an inequality is multiplied or divided by a negative number, the direction of the inequality symbol *must be reversed.*

EXAMPLE 6 **Solving a Linear Inequality**

Solve and graph the inequality $-5x - 7 > 3x + 9$.

Solution

$-5x - 7 > 3x + 9$	Write original inequality.
$-8x - 7 > 9$	Subtract $3x$ from each side.
$-8x > 16$	Add 7 to each side.
$x < -2$	Divide each side by -8 and reverse the direction of the inequality symbol.

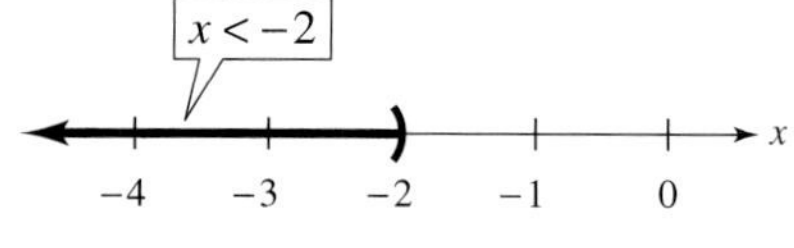

Figure A.8

The solution set in interval notation is $(-\infty, -2)$ and in set notation is $\{x|x < -2\}$. The graph of the solution set is shown in Figure A.8.

Two inequalities joined by the word *and* or the word *or* constitute a compound inequality. Sometimes it is possible to write a compound inequality as a double inequality. For instance, you can write $-3 < 6x - 1$ *and* $6x - 1 < 3$ more simply as $-3 < 6x - 1 < 3$. A compound inequality formed by the word *and* is called conjunctive and may be rewritten as a double inequality. A compound inequality joined by the word *or* is called disjunctive and cannot be rewritten as a double inequality.

EXAMPLE 7 **Solving a Conjunctive Inequality**

Solve and graph the inequality $2x + 3 \geq 4$ and $3x - 8 < -2$.

Solution

$2x + 3 \geq 4$	and	$3x - 8 < -2$
$2x \geq 1$		$3x < 6$
$x \geq \frac{1}{2}$		$x < 2$

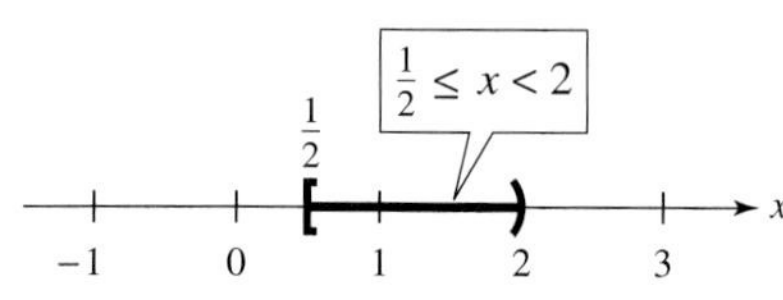

Figure A.9

The solution set in interval notation is $\left[\frac{1}{2}, 2\right)$ and in set notation is $\left\{x\middle|\frac{1}{2} \leq x < 2\right\}$. The graph of the solution set is shown in Figure A.9.

Study Tip

Recall that the word *or* is represented by the symbol $\cup$, which is read as *union*.

EXAMPLE 8 **Solving a Disjunctive Inequality**

Solve and graph the inequality $x - 8 > -3$ or $-6x + 1 \geq -5$.

Solution

$x - 8 > -3$	or	$-6x + 1 \geq -5$
$x > 5$		$-6x \geq -6$
		$x \leq 1$

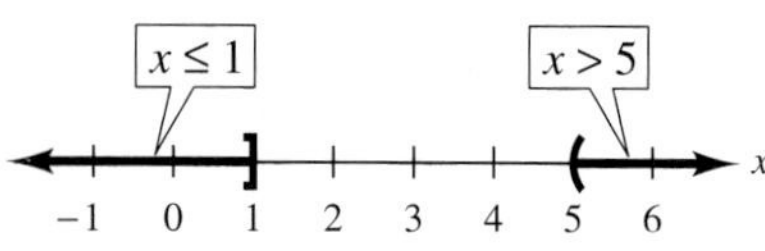

Figure A.10

The solution set in interval notation is $(-\infty, 1] \cup (5, \infty)$ and in set notation is $\{x|x > 5 \text{ or } x \leq 1\}$. The graph of the solution set is shown in Figure A.10.

To solve an absolute value inequality, use the following rules.

Solving an Absolute Value Inequality

Let x be a variable or an algebraic expression and let a be a real number such that $a > 0$.

1. The solutions of $|x| < a$ are all values of x that lie between $-a$ and a.

$$|x| < a \text{ if and only if } -a < x < a$$

2. The solutions of $|x| > a$ are all values of x that are less than $-a$ or greater than a.

$$|x| > a \text{ if and only if } x < -a \text{ or } x > a$$

These rules are also valid if $<$ is replaced by $\leq$ and $>$ is replaced by $\geq$.

EXAMPLE 9 Solving Absolute Value Inequalities

Solve and graph each inequality.

a. $|4x + 3| > 9$ **b.** $|2x - 7| \leq 1$

Solution

a. $|4x + 3| > 9$ Write original inequality.

$4x + 3 < -9$ or $4x + 3 > 9$ Equivalent inequalities

$4x < -12$ $\quad 4x > 6$ Subtract 3 from each side.

$x < -3$ $\quad x > \frac{3}{2}$ Divide each side by 4.

The solution set consists of all real numbers that are less than -3 or greater than $\frac{3}{2}$. The solution set in interval notation is $(-\infty, -3) \cup \left(\frac{3}{2}, \infty\right)$ and in set notation is $\left\{x \mid x < -3 \text{ or } x > \frac{3}{2}\right\}$. The graph is shown in Figure A.11.

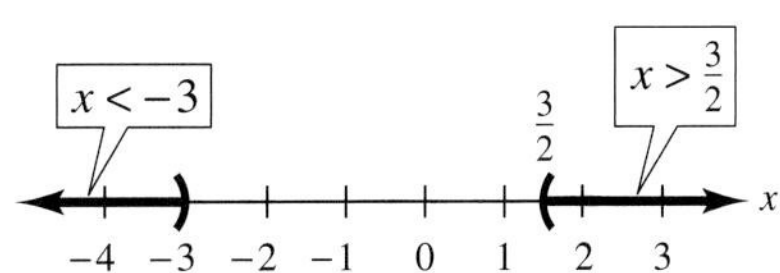

Figure A.11

b. $|2x - 7| \leq 1$ Write original inequality.

$-1 \leq 2x - 7 \leq 1$ Equivalent double inequality

$6 \leq 2x \leq 8$ Add 7 to all three parts.

$3 \leq x \leq 4$ Divide all three parts by 2.

The solution set consists of all real numbers that are greater than or equal to 3 and less than or equal to 4. The solution set in interval notation is $[3, 4]$ and in set notation is $\{x \mid 3 \leq x \leq 4\}$. The graph is shown in Figure A.12.

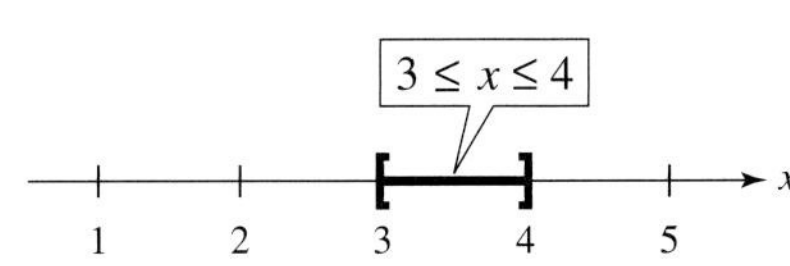

Figure A.12

Problem Solving

Algebra is used to solve word problems that relate to real-life situations. The following guidelines summarize the problem-solving strategy that you should use when solving word problems.

Guidelines for Solving Word Problems

1. Write a *verbal model* that describes the problem.
2. Assign *labels* to fixed quantities and variable quantities.
3. Rewrite the verbal model as an *algebraic equation* using the assigned labels.
4. *Solve* the resulting algebraic equation.
5. *Check* to see that your solution satisfies the original problem as stated.

EXAMPLE 10 Finding the Percent of Monthly Expenses 

Your family has an annual income of \$77,520 and the following monthly expenses: mortgage (\$1500), car payment (\$510), food (\$400), utilities (\$325), and credit cards (\$300). The total expenses for one year represent what percent of your family's annual income?

Solution

The total amount of your family's monthly expenses is

$$1500 + 510 + 400 + 325 + 300 = \$3035.$$

The total monthly expenses for one year are

$$3035 \cdot 12 = \$36{,}420.$$

Verbal Model: Expenses = Percent · Income

Labels:
Expenses = 36,420 (dollars)
Percent = p (in decimal form)
Income = 77,520 (dollars)

Equation:

$$36{,}420 = p \cdot 77{,}520 \qquad \text{Original equation}$$

$$\frac{36{,}420}{77{,}520} = p \qquad \text{Divide each side by 77,520.}$$

$$0.470 \approx p \qquad \text{Use a calculator.}$$

Your family's total expenses for one year are approximately 0.470 or 47.0% of your family's annual income.

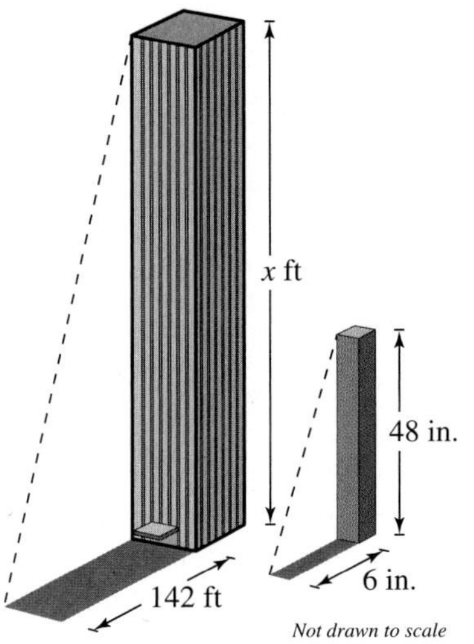

Figure A.13

EXAMPLE 11 Geometry: Similar Triangles 

To determine the height of the Aon Center Building (in Chicago), you measure the shadow cast by the building and find it to be 142 feet long, as shown in Figure A.13. Then you measure the shadow cast by a four-foot post and find it to be 6 inches long. Estimate the building's height.

Solution

To solve this problem, you use a property from geometry that states that the ratios of corresponding sides of similar triangles are equal.

Verbal Model:

$$\frac{\text{Height of building}}{\text{Length of building's shadow}} = \frac{\text{Height of post}}{\text{Length of post's shadow}}$$

Labels:
Height of building $= x$ (feet)
Length of building's shadow $= 142$ (feet)
Height of post $= 4$ feet $= 48$ inches (inches)
Length of post's shadow $= 6$ (inches)

Proportion:

$$\frac{x}{142} = \frac{48}{6}$$ Original proportion

$$x \cdot 6 = 142 \cdot 48$$ Cross-multiply

$$x = 1136$$ Divide each side by 6.

So, you can estimate the Aon Center Building to be 1136 feet high.

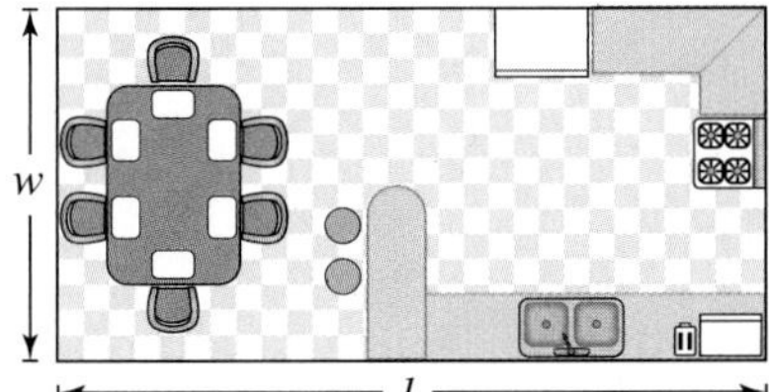

Figure A.14

EXAMPLE 12 Geometry: Dimensions of a Room 

A rectangular kitchen is twice as long as it is wide, and its perimeter is 84 feet. Find the dimensions of the kitchen.

Solution

For this problem, it helps to sketch a diagram, as shown in Figure A.14.

Verbal Model: $2 \cdot$ Length $+ 2 \cdot$ Width $=$ Perimeter

Labels:
Length $= l = 2w$ (feet)
Width $= w$ (feet)
Perimeter $= 84$ (feet)

Equation:

$$2(2w) + 2w = 84$$ Original equation

$$6w = 84$$ Combine like terms.

$$w = 14$$ Divide each side by 6.

Because the length is twice the width, you have $l = 2w = 2(14) = 28$. So, the dimensions of the room are 14 feet by 28 feet.

Study Tip

For more review on equations and inequalities, refer to Chapter 3.

A.4 Graphs and Functions

The Rectangular Coordinate System

You can represent ordered pairs of real numbers by points in a plane. This plane is called a rectangular coordinate system. A rectangular coordinate system is formed by two real lines, the x-axis (horizontal line) and the y-axis (vertical line), intersecting at right angles. The point of intersection of the axes is called the origin, and the axes divide the plane into four regions called quadrants.

Each point in the plane corresponds to an ordered pair (x, y) of real numbers x and y, called the coordinates of the point. The x-coordinate tells how far to the left or right the point is from the vertical axis, and the y-coordinate tells how far up or down the point is from the horizontal axis, as shown in Figure A.15.

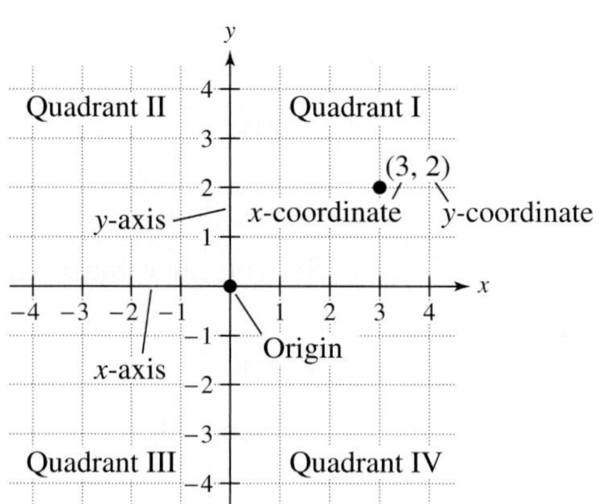

Figure A.15

EXAMPLE 1 Finding Coordinates of Points

Determine the coordinates of each of the points shown in Figure A.16, and then determine the quadrant in which each point is located.

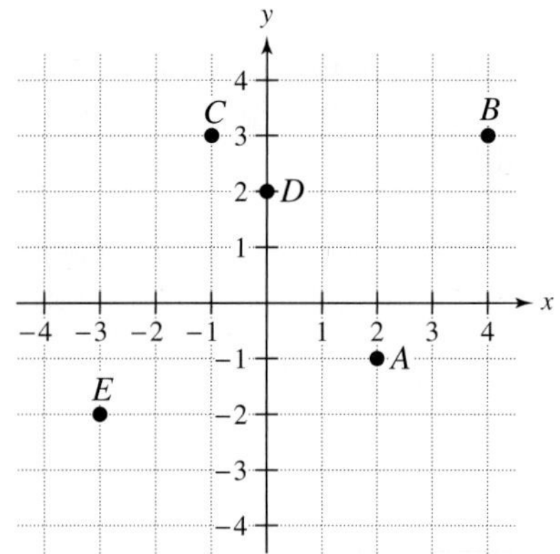

Figure A.16

Solution

Point A lies two units to the *right* of the vertical axis and one unit *below* the horizontal axis. So, point A must be given by $(2, -1)$. The coordinates of the other four points can be determined in a similar way. The results are as follows.

Point	*Coordinates*	*Quadrant*
A	$(2, -1)$	IV
B	$(4, 3)$	I
C	$(-1, 3)$	II
D	$(0, 2)$	None
E	$(-3, -2)$	III

Graphs of Equations

The solutions of an equation involving two variables can be represented by points on a rectangular coordinate system. The graph of an equation is the set of all points that are solutions of the equation.

The simplest way to sketch the graph of an equation is the point-plotting method. With this method, you construct a table of values consisting of several solution points of the equation, plot these points, and then connect the points with a smooth curve or line.

EXAMPLE 2 Sketching the Graph of an Equation

Sketch the graph of $y = x^2 - 2$.

Solution

Begin by choosing several x-values and then calculating the corresponding y-values. For example, if you choose $x = -2$, the corresponding y-value is

$y = x^2 - 2$ Original equation

$y = (-2)^2 - 2$ Substitute -2 for x.

$y = 4 - 2 = 2.$ Simplify.

Then, create a table using these values, as shown below.

x	-2	-1	0	1	2	3
$y = x^2 - 2$	2	-1	-2	-1	2	7
Solution point	$(-2, 2)$	$(-1, -1)$	$(0, -2)$	$(1, -1)$	$(2, 2)$	$(3, 7)$

Next, plot the solution points, as shown in Figure A.17. Finally, connect the points with a smooth curve, as shown in Figure A.18.

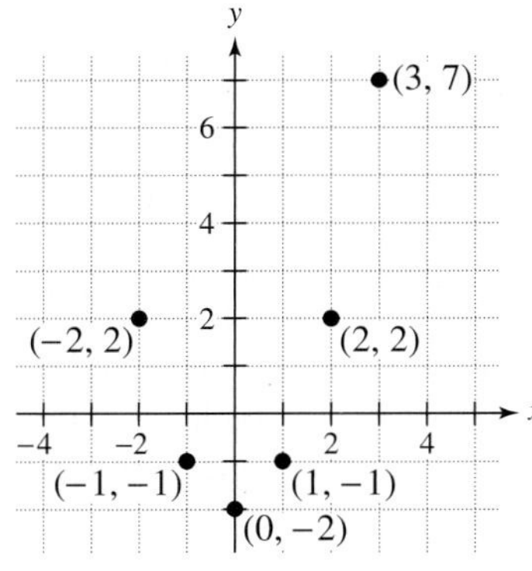

Figure A.17

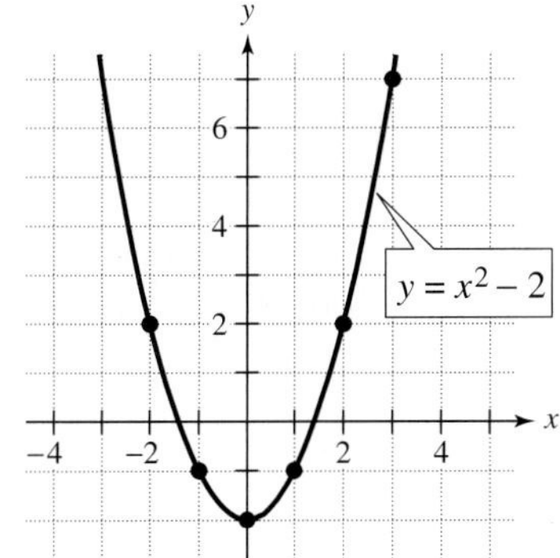

Figure A.18

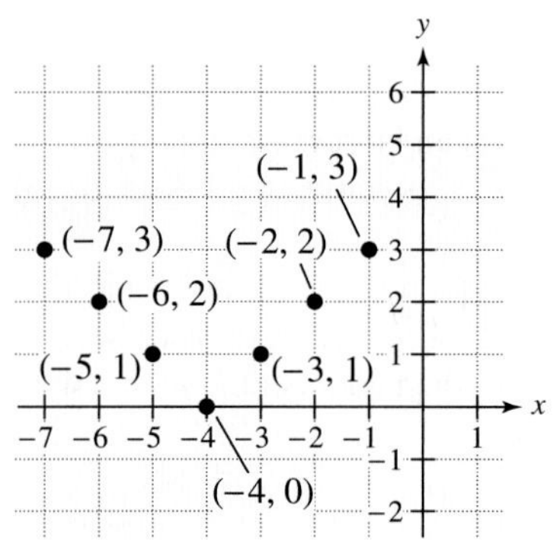

Figure A.19

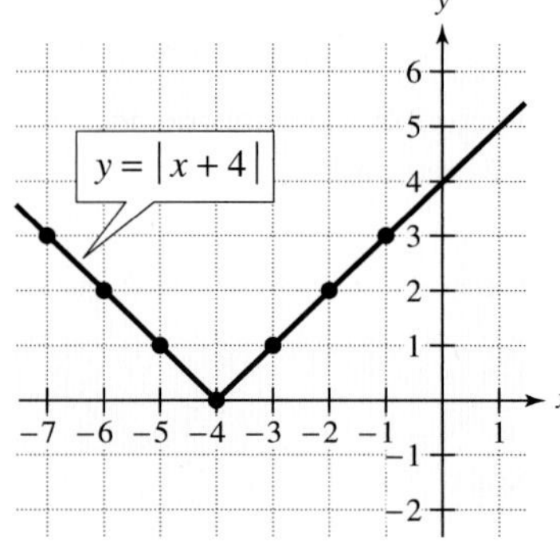

Figure A.20

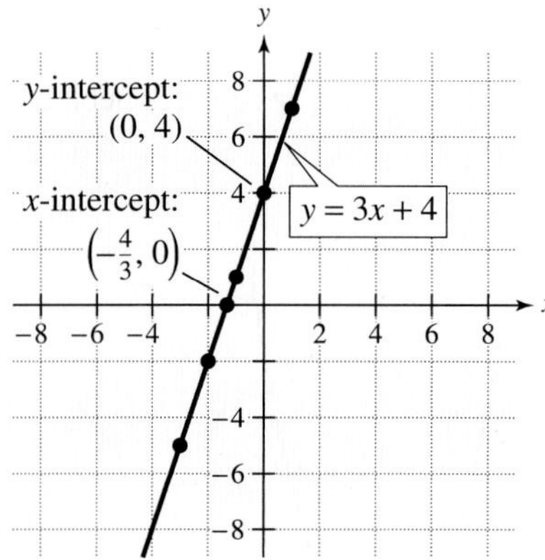

Figure A.21

EXAMPLE 3 Sketching the Graph of an Equation

Sketch the graph of $y = |x + 4|$.

Solution

Begin by creating a table of values, as shown below. Plot the solution points, as shown in Figure A.19. It appears that the points lie in a "V-shaped" pattern, with the point $(-4, 0)$ lying at the bottom of the "V." Following this pattern, connect the points to form the graph shown in Figure A.20.

x	-7	-6	-5	-4	-3	-2	-1
$y = \|x + 4\|$	3	2	1	0	1	2	3
Solution point	$(-7, 3)$	$(-6, 2)$	$(-5, 1)$	$(-4, 0)$	$(-3, 1)$	$(-2, 2)$	$(-1, 3)$

Intercepts of a graph are the points at which the graph intersects the x- or y-axis. To find x-intercepts, let $y = 0$ and solve the equation for x. To find y-intercepts, let $x = 0$ and solve the equation for y.

EXAMPLE 4 Finding the Intercepts of a Graph

Find the intercepts and sketch the graph of $y = 3x + 4$.

Solution

To find any x-intercepts, let $y = 0$ and solve the resulting equation for x.

$y = 3x + 4$ Write original equation.

$0 = 3x + 4$ Let $y = 0$.

$-\dfrac{4}{3} = x$ Solve equation for x.

To find any y-intercepts, let $x = 0$ and solve the resulting equation for y.

$y = 3x + 4$ Write original equation.

$y = 3(0) + 4$ Let $x = 0$.

$y = 4$ Solve equation for y.

So, the x-intercept is $\left(-\frac{4}{3}, 0\right)$ and the y-intercept is $(0, 4)$. To sketch the graph of the equation, create a table of values (including intercepts), as shown below. Then plot the points and connect them with a line, as shown in Figure A.21.

x	-3	-2	$-\frac{4}{3}$	-1	0	1
$y = 3x + 4$	-5	-2	0	1	4	7
Solution point	$(-3, -5)$	$(-2, -2)$	$\left(-\frac{4}{3}, 0\right)$	$(-1, 1)$	$(0, 4)$	$(1, 7)$

Functions

A relation is any set of ordered pairs, which can be thought of as (input, output). A function is a relation in which no two ordered pairs have the same first component and different second components.

EXAMPLE 5 **Testing Whether Relations Are Functions**

Decide whether the relation represents a function.

a.

Input: a, b, c

Output: 1, 2, 3, 4

$a \to 1$, $a \to 2$, $b \to 3$, $c \to 4$

b. Input: 2, 5, 7
Output: 1, 2, 3
$\{(2, 1), (5, 2), (7, 3)\}$

Solution

a. This diagram *does not* represent a function. The first component a is paired with two different second components, 1 and 2.

b. This set of ordered pairs *does* represent a function. No first component has two different second components.

The graph of an equation represents y as a function of x if and only if no vertical line intersects the graph more than once. This is called the Vertical Line Test.

EXAMPLE 6 **Using the Vertical Line Test for Functions**

Use the Vertical Line Test to determine whether y is a function of x.

a.

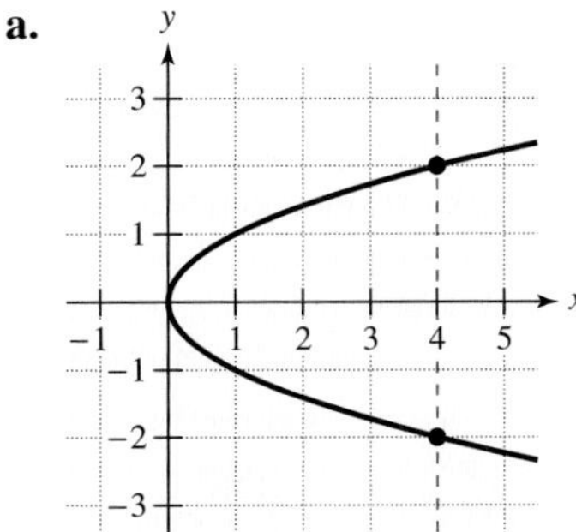

b.

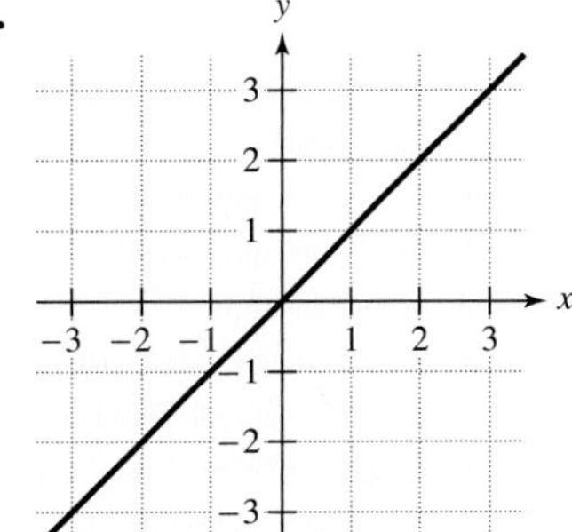

Solution

a. From the graph, you can see that a vertical line intersects more than one point on the graph. So, the relation *does not* represent y as a function of x.

b. From the graph, you can see that no vertical line intersects more than one point on the graph. So, the relation *does* represent y as a function of x.

Slope and Linear Equations

The graph in Figure A.21 on page A18 is an example of a graph of a linear equation. The equation is written in slope-intercept form, $y = mx + b$, where m is the slope and $(0, b)$ is the y-intercept. Linear equations can be written in other forms, as shown below.

Forms of Linear Equations

1. General form: $ax + by + c = 0$
2. Slope-intercept form: $y = mx + b$
3. Point-slope form: $y - y_1 = m(x - x_1)$

The slope of a nonvertical line is the number of units the line rises or falls vertically for each unit of horizontal change from left to right. To find the slope m of the line through (x_1, y_1) and (x_2, y_2), use the following formula.

$$m = \frac{y_2 - y_1}{x_2 - x_1} = \frac{\text{Change in } y}{\text{Change in } x}$$

EXAMPLE 7 Finding the Slope of a Line Through Two Points

Find the slope of the line passing through $(3, 1)$ and $(-6, 0)$.

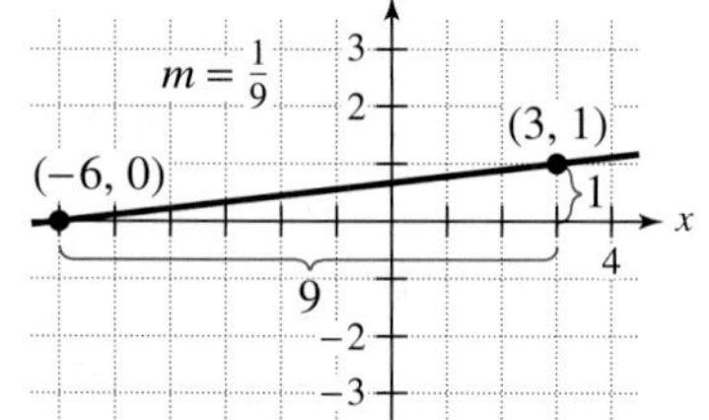

Figure A.22

Solution

Let $(x_1, y_1) = (3, 1)$ and $(x_2, y_2) = (-6, 0)$. The slope of the line through these points is

$$m = \frac{y_2 - y_1}{x_2 - x_1} = \frac{0 - 1}{-6 - 3} = \frac{-1}{-9} = \frac{1}{9}.$$

The graph of the line is shown in Figure A.22.

You can make several generalizations about the slopes of lines.

Slope of a Line

1. A line with positive slope $(m > 0)$ rises from left to right.
2. A line with negative slope $(m < 0)$ falls from left to right.
3. A line with zero slope $(m = 0)$ is horizontal.
4. A line with undefined slope is vertical.
5. Parallel lines have equal slopes: $m_1 = m_2$
6. Perpendicular lines have negative reciprocal slopes: $m_1 = -\frac{1}{m_2}$

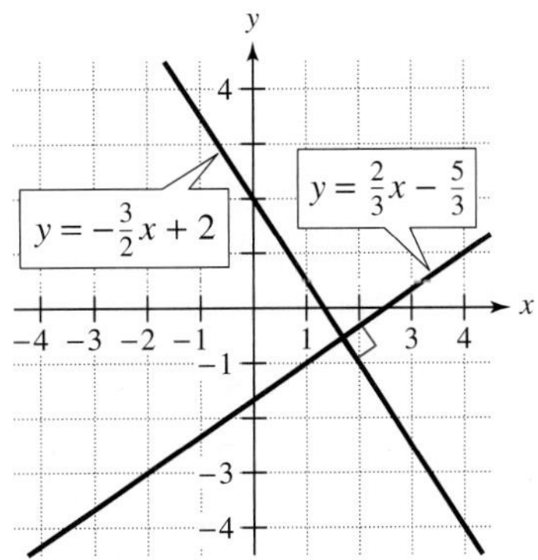

Figure A.23

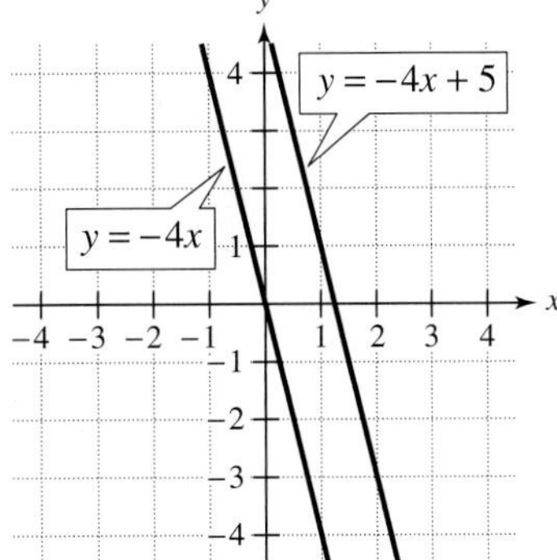

Figure A.24

EXAMPLE 8 Parallel or Perpendicular?

Determine whether the pairs of lines are parallel, perpendicular, or neither.

a. $y = \frac{2}{3}x - \frac{5}{3}$

$y = -\frac{3}{2}x + 2$

b. $4x + y = 5$

$-8x - 2y = 0$

Solution

a. The first line has a slope of $m_1 = \frac{2}{3}$ and the second line has a slope of $m_2 = -\frac{3}{2}$. Because these slopes are negative reciprocals of each other, the two lines must be perpendicular, as shown in Figure A.23.

b. To begin, write each equation in slope-intercept form.

$4x + y = 5$ Write first equation.

$y = -4x + 5$ Slope-intercept form

So, the first line has a slope of $m_1 = -4$.

$-8x - 2y = 0$ Write second equation.

$-2y = 8x$ Add $8x$ to each side.

$y = -4x$ Slope-intercept form

So, the second line has a slope of $m_2 = -4$. Because both lines have the same slope, they must be parallel, as shown in Figure A.24.

You can use the point-slope form of the equation of a line to write the equation of a line when you are given its slope and a point on the line.

EXAMPLE 9 Writing an Equation of a Line

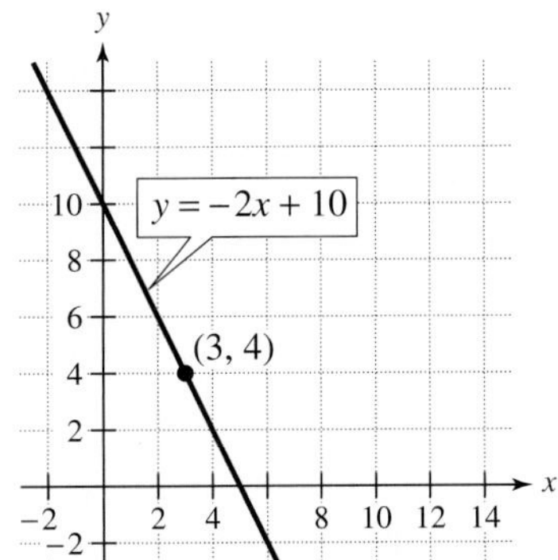

Figure A.25

Write an equation of the line that passes through the point (3, 4) and has slope $m = -2$.

Solution

Use the point-slope form with $(x_1, y_1) = (3, 4)$ and $m = -2$.

$y - y_1 = m(x - x_1)$ Point-slope form

$y - 4 = -2(x - 3)$ Substitute 4 for y_1, 3 for x_1, and -2 for m.

$y - 4 = -2x + 6$ Simplify.

$y = -2x + 10$ Equation of line

So, an equation of the line in slope-intercept form is $y = -2x + 10$. The graph of this line is shown in Figure A.25.

The point-slope form can also be used to write the equation of a line passing through any two points. To use this form, substitute the formula for slope into the point-slope form, as follows.

$y - y_1 = m(x - x_1)$ Point-slope form

$y - y_1 = \dfrac{y_2 - y_1}{x_2 - x_1}(x - x_1)$ Substitute formula for slope.

EXAMPLE 10 An Equation of a Line Passing Through Two Points

Write an equation of the line that passes through the points $(5, -1)$ and $(2, 0)$.

Solution

Let $(x_1, y_1) = (5, -1)$ and $(x_2, y_2) = (2, 0)$. The slope of a line passing through these points is

$$m = \frac{y_2 - y_1}{x_2 - x_1} = \frac{0 - (-1)}{2 - 5} = \frac{1}{-3} = -\frac{1}{3}.$$

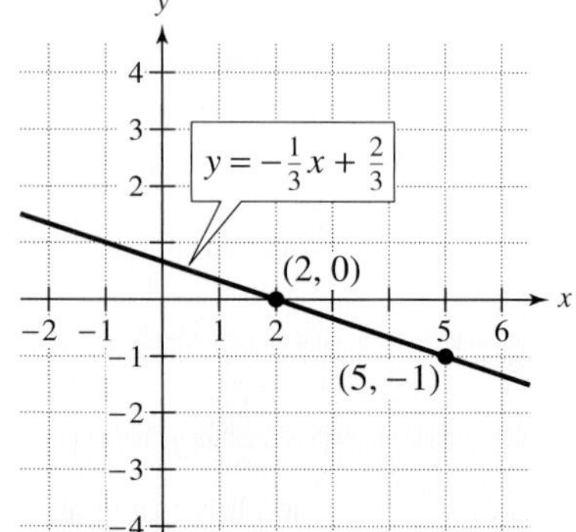

Figure A.26

Now, use the point-slope form to find an equation of the line.

$y - y_1 = m(x - x_1)$ Point-slope form

$y - (-1) = -\frac{1}{3}(x - 5)$ Substitute -1 for y_1, 5 for x_1, and $-\frac{1}{3}$ for m.

$y + 1 = -\frac{1}{3}x + \frac{5}{3}$ Simplify.

$y = -\frac{1}{3}x + \frac{2}{3}$ Equation of line

The graph of this line is shown in Figure A.26.

The slope and y-intercept of a line can be used as an aid when you are sketching a line.

EXAMPLE 11 Using the Slope and y-Intercept to Sketch a Line

Use the slope and y-intercept to sketch the graph of $-x + 2y = -4$.

Solution

First, write the equation in slope-intercept form.

$-x + 2y = -4$ Write original equation.

$2y = x - 4$ Add x to each side.

$y = \frac{1}{2}x - 2$ Slope-intercept form

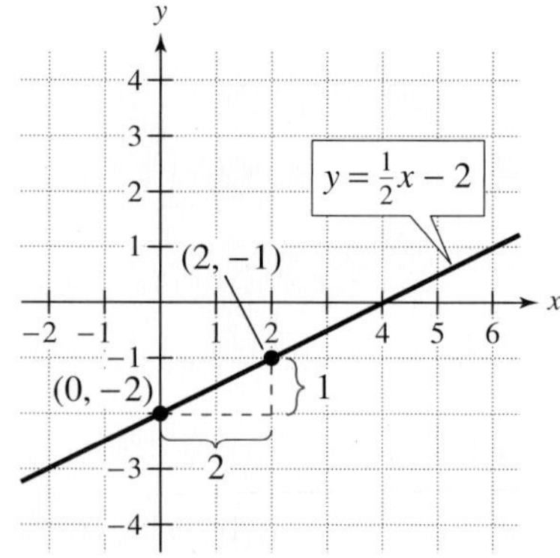

Figure A.27

So, the slope of the line is $m = \frac{1}{2}$ and the y-intercept is $(0, b) = (0, -2)$. Now, plot the y-intercept and locate a second point by using the slope. Because the slope is $m = \frac{1}{2}$, move two units to the right and one unit upward from the y-intercept. Then draw a line through the two points, as shown in Figure A.27.

You know that a horizontal line has a slope of $m = 0$. So, the equation of a horizontal line is $y = b$. A vertical line has an undefined slope, so it has an equation of the form $x = a$.

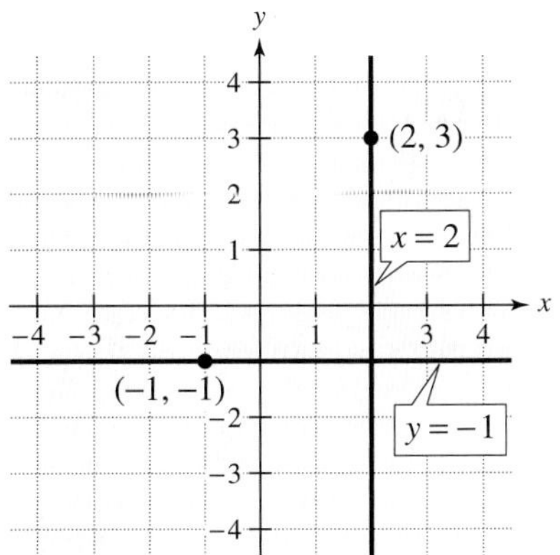

Figure A.28

EXAMPLE 12 **Equations of Horizontal and Vertical Lines**

a. Write an equation of the horizontal line passing through $(-1, -1)$.

b. Write an equation of the vertical line passing through $(2, 3)$.

Solution

a. The line is horizontal and passes through the point $(-1, -1)$, so every point on the line has a y-coordinate of -1. The equation of the line is $y = -1$.

b. The line is vertical and passes through the point $(2, 3)$, so every point on the line has an x-coordinate of 2. The equation of the line is $x = 2$.

The graphs of these two lines are shown in Figure A.28.

Graphs of Linear Inequalities

The statements $3x - 2y < 6$ and $2x + 3y \geq 1$ are linear inequalities in two variables. An ordered pair (x_1, y_1) is a solution of a linear inequality in x and y if the inequality is true when x_1 and y_1 are substituted for x and y, respectively. The graph of a linear inequality is the collection of all solution points of the inequality. To sketch the graph of a linear inequality, begin by sketching the graph of the corresponding linear equation (use a dashed line for $<$ and $>$ and a solid line for $\leq$ and $\geq$). The graph of the equation separates the plane into two regions, called half-planes. In each half-plane, either *all* points in the half-plane are solutions of the inequality or *no* point in the half-plane is a solution of the inequality. To determine whether the points in an entire half-plane satisfy the inequality, simply test one point in the region. If the point satisfies the inequality, then shade the entire half-plane to denote that every point in the region satisfies the inequality.

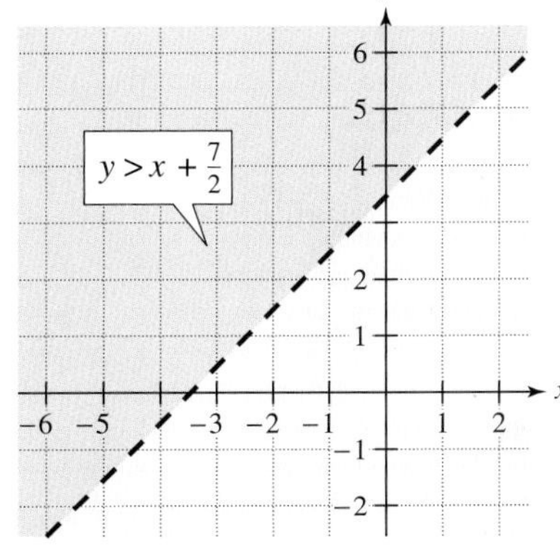

Figure A.29

Study Tip

For more review on graphs and functions, refer to Chapter 4.

EXAMPLE 13 **Sketching the Graph of a Linear Inequality**

Use the slope-intercept form of a linear equation to graph $-2x + 2y > 7$.

Solution

To begin, rewrite the inequality in slope-intercept form.

$$2y > 2x + 7 \qquad \text{Add } 2x \text{ to each side.}$$

$$y > x + \frac{7}{2} \qquad \text{Write in slope-intercept form.}$$

From this form, you can conclude that the solution is the half-plane lying above the line $y = x + \frac{7}{2}$. The graph is shown in Figure A.29.

- Exponents
- Polynomials
- Operations with Polynomials

A.5 Exponents and Polynomials

Exponents

Repeated multiplication can be written in exponential form. In general, if a is a real number and n is a positive integer, then

$$a^n = \underbrace{a \cdot a \cdot a \cdot \cdots \cdot a}_{n \text{ factors}}$$

where n is the exponent and a is the base. The following is a summary of the rules of exponents. In Rule 6 below, be sure you see how to use a negative exponent.

Summary of Rules of Exponents

Let m and n be integers, and let a and b be real numbers, variables, or algebraic expressions, such that $a \neq 0$ and $b \neq 0$.

Rule	*Example*
1. *Product Rule*: $a^m a^n = a^{m+n}$	$y^2 \cdot y^4 = y^{2+4} = y^6$
2. *Quotient Rule*: $\dfrac{a^m}{a^n} = a^{m-n}$	$\dfrac{x^7}{x^4} = x^{7-4} = x^3$
3. *Product-to-Power Rule*: $(ab)^m = a^m b^m$	$(5x)^4 = 5^4x^4$
4. *Quotient-to-Power Rule*: $\left(\dfrac{a}{b}\right)^m = \dfrac{a^m}{b^m}$	$\left(\dfrac{2}{x}\right)^3 = \dfrac{2^3}{x^3}$
5. *Power-to-Power Rule*: $(a^m)^n = a^{mn}$	$(y^3)^{-4} = y^{3(-4)} = y^{-12}$
6. *Negative Exponent Rule*: $a^{-n} = \dfrac{1}{a^n}$	$y^{-4} = \dfrac{1}{y^4}$
7. *Zero Exponent Rule*: $a^0 = 1$	$(x^2 + 1)^0 = 1$

EXAMPLE 1 **Using Rules of Exponents**

Use the rules of exponents to simplify each expression.

a. $(a^2b^4)(3ab^{-2})$ **b.** $(2xy^2)^3$ **c.** $3a(-4a^2)^0$ **d.** $\left(\dfrac{4x}{y^3}\right)^3$

Solution

a.
$$\begin{aligned}(a^2b^4)(3ab^{-2}) &= (3)(a^2)(a)(b^4)(b^{-2}) && \text{Regroup factors.}\\ &= (3)(a^{2+1})(b^{4-2}) && \text{Apply rules of exponents.}\\ &= 3a^3b^2 && \text{Simplify.}\end{aligned}$$

b. $(2xy^2)^3 = (2)^3(x)^3(y^2)^3$ Apply rules of exponents.

$= 8x^3y^{2 \cdot 3}$ Apply rules of exponents.

$= 8x^3y^6$ Simplify.

c. $3a(-4a^2)^0 = (3a)(-4)^0(a^2)^0$ Apply rules of exponents.

$= 3a(1)(a^{2 \cdot 0})$ Apply rules of exponents.

$= 3a, \ a \neq 0$ Simplify.

d. $\left(\frac{4x}{y^3}\right)^3 = \frac{4^3x^3}{(y^3)^3}$ Apply rules of exponents.

$= \frac{64x^3}{y^{3 \cdot 3}}$ Apply rules of exponents.

$= \frac{64x^3}{y^9}$ Simplify.

EXAMPLE 2 Rewriting with Positive Exponents

Use rules of exponents to simplify each expression using only positive exponents. (Assume that no variable is equal to zero.)

a. x^{-1} **b.** $\frac{1}{3x^{-2}}$

c. $\frac{25a^3b^{-4}}{5a^{-2}b}$ **d.** $\left(\frac{2x^{-1}}{xy^0}\right)^{-2}$

Solution

a. $x^{-1} = \frac{1}{x}$ Apply rules of exponents.

b. $\frac{1}{3x^{-2}} = \frac{x^2}{3}$ Apply rules of exponents.

c. $\frac{25a^3b^{-4}}{5a^{-2}b} = 5a^{3-(-2)}b^{-4-1}$ Apply rules of exponents.

$= 5a^5b^{-5}$ Simplify.

$= \frac{5a^5}{b^5}$ Apply rules of exponents.

d. $\left(\frac{2x^{-1}}{xy^0}\right)^{-2} = \frac{(2)^{-2}x^2}{x^{-2}y^0}$ Apply rules of exponents.

$= \frac{x^2 \cdot x^2}{2^2}$ Simplify.

$= \frac{x^4}{4}$ Apply rules of exponents.

It is convenient to write very large or very small numbers in scientific notation. This notation has the form $c \times 10^n$, where $1 \le c < 10$ and n is an integer. A positive exponent indicates that the number is large (10 or more) and a negative exponent indicates that the number is small (less than 1).

EXAMPLE 3 Scientific Notation

Write each number in scientific notation.

a. 0.0000782 **b.** 836,100,000

Solution

a. $0.0000782 = 7.82 \times 10^{-5}$ **b.** $836{,}100{,}000 = 8.361 \times 10^8$

EXAMPLE 4 Decimal Notation

Write each number in decimal notation.

a. 9.36×10^{-6} **b.** 1.345×10^2

Solution

a. $9.36 \times 10^{-6} = 0.00000936$ **b.** $1.345 \times 10^2 = 134.5$

Polynomials

The most common type of algebraic expression is the polynomial. Some examples are

$$-x + 1, \quad 2x^2 - 5x + 4, \quad \text{and} \quad 3x^3.$$

A polynomial in x is an expression of the form

$$a_nx^n + a_{n-1}x^{n-1} + \cdot \cdot \cdot + a_2x^2 + a_1x + a_0$$

where $a_n, a_{n-1}, \ldots, a_2, a_1, a_0$ are real numbers, n is a nonnegative integer, and $a_n \neq 0$. The polynomial is of degree n, a_n is called the leading coefficient, and a_0 is called the constant term.

Polynomials with one, two, and three terms are called monomials, binomials, and trinomials, respectively. In standard form, a polynomial is written with descending powers of x.

EXAMPLE 5 Determining Degrees and Leading Coefficients

	Polynomial	*Standard Form*	*Degree*	*Leading Coefficient*
a.	$3x - x^2 + 4$	$-x^2 + 3x + 4$	2	-1
b.	-5	-5	0	-5
c.	$8 - 4x^3 + 7x + 2x^5$	$2x^5 - 4x^3 + 7x + 8$	5	2

Operations with Polynomials

You can add and subtract polynomials in much the same way that you add and subtract real numbers. Simply add or subtract the like terms (terms having the same variables to the same powers) by adding their coefficients. For instance, $-3xy^2$ and $5xy^2$ are like terms and their sum is

$$-3xy^2 + 5xy^2 = (-3 + 5)xy^2 = 2xy^2.$$

To subtract one polynomial from another, add the opposite by changing the sign of each term of the polynomial that is being subtracted and then adding the resulting like terms. You can add and subtract polynomials using either a horizontal or a vertical format.

EXAMPLE 6 Adding and Subtracting Polynomials

a.
$$\begin{array}{r} 3x^2 - 2x + 4 \\ \underline{-x^2 + 7x - 9} \\ 2x^2 + 5x - 5 \end{array}$$

b. $(5x^3 - 7x^2 - 3) + (x^3 + 2x^2 - x + 8)$ Original polynomials

$= (5x^3 + x^3) + (-7x^2 + 2x^2) - x + (-3 + 8)$ Group like terms.

$= 6x^3 - 5x^2 - x + 5$ Combine like terms.

c.
$$\begin{array}{rcr} (4x^3 + 3x - 6) & \Rightarrow & 4x^3 + 3x - 6 \\ \underline{-(3x^3 + x + 10)} & \Rightarrow & \underline{-3x^3 - x - 10} \\ & & x^3 + 2x - 16 \end{array}$$

d. $(7x^4 - x^2 - x + 2) - (3x^4 - 4x^2 + 3x)$ Original polynomials

$= 7x^4 - x^2 - x + 2 - 3x^4 + 4x^2 - 3x$ Distributive Property

$= (7x^4 - 3x^4) + (-x^2 + 4x^2) + (-x - 3x) + 2$ Group like terms.

$= 4x^4 + 3x^2 - 4x + 2$ Combine like terms.

The simplest type of polynomial multiplication involves a monomial multiplier. The product is obtained by direct application of the Distributive Property.

EXAMPLE 7 Finding a Product with a Monomial Multiplier

Find the product of $4x^2$ and $-2x^3 + 3x + 1$.

Solution

$$\begin{aligned} 4x^2(-2x^3 + 3x + 1) &= (4x^2)(-2x^3) + (4x^2)(3x) + (4x^2)(1) \\ &= -8x^5 + 12x^3 + 4x^2 \end{aligned}$$

To multiply two binomials, use the FOIL Method illustrated below.

First
Outer

$$(ax + b)(cx + d) = \underset{\text{F}}{ax(cx)} + \underset{\text{O}}{ax(d)} + \underset{\text{I}}{b(cx)} + \underset{\text{L}}{b(d)}$$

Inner
Last

EXAMPLE 8 Using the FOIL Method

Use the FOIL Method to find the product of $x - 3$ and $x - 9$.

Solution

$$(x - 3)(x - 9) = \overset{\text{F}}{x^2} - \overset{\text{O}}{9x} - \overset{\text{I}}{3x} + \overset{\text{L}}{27}$$

$$= x^2 - 12x + 27$$ Combine like terms.

EXAMPLE 9 Using the FOIL Method

Use the FOIL Method to find the product of $2x - 4$ and $x + 5$.

Solution

$$(2x - 4)(x + 5) = \overset{\text{F}}{2x^2} + \overset{\text{O}}{10x} - \overset{\text{I}}{4x} - \overset{\text{L}}{20}$$

$$= 2x^2 + 6x - 20$$ Combine like terms.

To multiply two polynomials that have three or more terms, you can use the same basic principle that you use when multiplying monomials and binomials. That is, each term of one polynomial must be multiplied by each term of the other polynomial. This can be done using either a vertical or a horizontal format.

EXAMPLE 10 Multiplying Polynomials (Vertical Format)

Multiply $x^2 - 2x + 2$ by $x^2 + 3x + 4$ using a vertical format.

Solution

$$\begin{array}{rrrrrl}
 & & x^2 & - 2x & + 2 & \\
\times & & x^2 & + 3x & + 4 & \\
\hline
 & & 4x^2 & - 8x & + 8 & \Leftarrow 4(x^2 - 2x + 2) \\
 & 3x^3 & - 6x^2 & + 6x & & \Leftarrow 3x(x^2 - 2x + 2) \\
x^4 & - 2x^3 & + 2x^2 & & & \Leftarrow x^2(x^2 - 2x + 2) \\
\hline
x^4 & + x^3 & + 0x^2 & - 2x & + 8 & \text{Combine like terms.}
\end{array}$$

So, $(x^2 - 2x + 2)(x^2 + 3x + 4) = x^4 + x^3 - 2x + 8$.

EXAMPLE 11 Multiplying Polynomials (Horizontal Format)

$$\begin{aligned}(4x^2 - 3x - 1)(2x - 5)&\\ &= 4x^2(2x - 5) - 3x(2x - 5) - 1(2x - 5) && \text{Distributive Property}\\ &= 8x^3 - 20x^2 - 6x^2 + 15x - 2x + 5 && \text{Distributive Property}\\ &= 8x^3 - 26x^2 + 13x + 5 && \text{Combine like terms.}\end{aligned}$$

Some binomial products have special forms that occur frequently in algebra. These special products are listed below.

1. Sum and Difference of Two Terms: $(a + b)(a - b) = a^2 - b^2$
2. Square of a Binomial: $(a + b)^2 = a^2 + 2ab + b^2$
   $(a - b)^2 = a^2 - 2ab + b^2$

EXAMPLE 12 Finding Special Products

a. $(3x - 2)(3x + 2) = (3x)^2 - 2^2$ Special product

$= 9x^2 - 4$ Simplify.

b. $(2x - 7)^2 = (2x)^2 - 2(2x)(7) + 7^2$ Special product

$= 4x^2 - 28x + 49$ Simplify.

c. $(4a + 5b)^2 = (4a)^2 + 2(4a)(5b) + (5b)^2$ Special product

$= 16a^2 + 40ab + 25b^2$ Simplify.

To divide a polynomial by a monomial, separate the original division problem into multiple division problems, each involving the division of a monomial by a monomial.

EXAMPLE 13 Dividing a Polynomial by a Monomial

Perform the division and simplify.

$$\frac{7x^3 - 12x^2 + 4x + 1}{4x}$$

Solution

$$\frac{7x^3 - 12x^2 + 4x + 1}{4x} = \frac{7x^3}{4x} - \frac{12x^2}{4x} + \frac{4x}{4x} + \frac{1}{4x}$$ Divide each term separately.

$$= \frac{7x^2}{4} - 3x + 1 + \frac{1}{4x}$$ Use rules of exponents.

Polynomial division is similar to long division of integers. To use polynomial long division, write the dividend and divisor in descending powers of the variable, insert placeholders with zero coefficients for missing powers of the variable, and divide as you would with integers. Continue this process until the degree of the remainder is less than that of the divisor.

EXAMPLE 14 Long Division Algorithm for Polynomials

Use the long division algorithm to divide $x^2 + 2x + 4$ by $x - 1$.

Solution

Think $\frac{x^2}{x} = x$.

Think $\frac{3x}{x} = 3$.

$$
\begin{array}{r l}
x + 3 & \\
x - 1\,\big)\overline{x^2 + 2x + 4} & \\
\underline{x^2 - x\phantom{{}+4}} & \text{Multiply } x \text{ by } (x - 1). \\
3x + 4 & \text{Subtract and bring down 4.} \\
\underline{3x - 3} & \text{Multiply 3 by } (x - 1). \\
7 & \text{Remainder}
\end{array}
$$

Considering the remainder as a fractional part of the divisor, the result is

$$\overbrace{\frac{x^2 + 2x + 4}{\underbrace{x - 1}_{\text{Divisor}}}}^{\text{Dividend}} = \overbrace{x + 3}^{\text{Quotient}} + \frac{\overset{\text{Remainder}}{7}}{\underbrace{x - 1}_{\text{Divisor}}}.$$

EXAMPLE 15 Accounting for Missing Powers of *x*

Divide $x^3 - 2$ by $x - 1$.

Solution

Note how the missing x^2- and x-terms are accounted for.

$$
\begin{array}{r l}
x^2 + x + 1 & \\
x - 1\,\big)\overline{x^3 + 0x^2 + 0x - 2} & \text{Insert } 0x^2 \text{ and } 0x. \\
\underline{x^3 - x^2} & \text{Multiply } x^2 \text{ by } (x - 1). \\
x^2 + 0x & \text{Subtract and bring down } 0x. \\
\underline{x^2 - x} & \text{Multiply } x \text{ by } (x - 1). \\
x - 2 & \text{Subtract and bring down } -2. \\
\underline{x - 1} & \text{Multiply 1 by } (x - 1). \\
-1 & \text{Remainder}
\end{array}
$$

So, you have $\frac{x^3 - 2}{x - 1} = x^2 + x + 1 - \frac{1}{x - 1}$.

Synthetic division is a nice shortcut for dividing by polynomials of the form $x - k$.

Synthetic Division of a Third-Degree Polynomial

Use synthetic division to divide $ax^3 + bx^2 + cx + d$ by $x - k$, as follows.

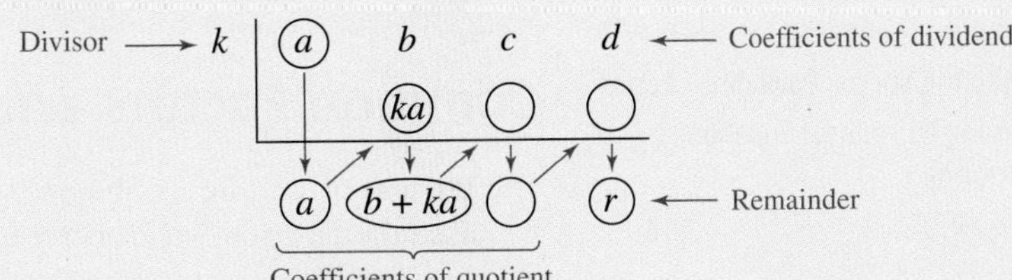

Vertical Pattern: Add terms.
Diagonal Pattern: Multiply by k.

EXAMPLE 16 **Using Synthetic Division**

Use synthetic division to divide $x^3 - 6x^2 + 4$ by $x - 3$.

Solution

You should set up the array as follows. Note that a zero is included for the missing x-term in the dividend.

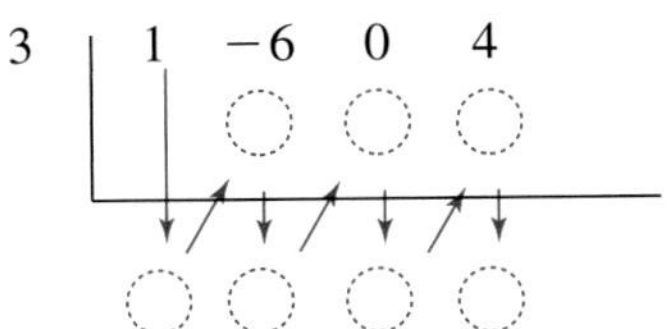

Then, use the synthetic division pattern by adding terms in columns and multiplying the results by 3.

Divisor: $x - 3$ Dividend: $x^3 - 6x^2 + 4$

3	1	−6	0	4
		3	−9	−27
	1	−3	−9	−23

Remainder: −23

Quotient: $x^2 - 3x - 9$

So, you have

$$\frac{x^3 - 6x^2 + 4}{x - 3} = x^2 - 3x - 9 - \frac{23}{x - 3}.$$

For more review on exponents and polynomials, refer to Chapter 5.

A.6 Factoring and Solving Equations

Common Factors and Factoring by Grouping

The process of writing a polynomial as a product is called factoring. Previously, you used the Distributive Property to *multiply* and *remove* parentheses. Now, you will use the Distributive Property in the reverse direction to *factor* and *create* parentheses.

Removing the common monomial factor is the first step in completely factoring a polynomial. When you use the Distributive Property to remove this factor from each term of the polynomial, you are factoring out the greatest common monomial factor.

EXAMPLE 1 Common Monomial Factors

Factor out the greatest common monomial factor from each polynomial.

a. $3x + 9$ **b.** $6x^3 - 4x$ **c.** $-4y^2 + 12y - 16$

Solution

a. $3x + 9 = 3(x) + 3(3) = 3(x + 3)$ Greatest common monomial factor is 3.

b. $6x^3 - 4x = 2x(3x^2) - 2x(2) = 2x(3x^2 - 2)$ Greatest common monomial factor is $2x$.

c. $-4y^2 + 12y - 16 = -4(y^2) + (-4)(-3y) + (-4)(4)$ Greatest common monomial factor is -4.

$= -4(y^2 - 3y + 4)$ Factor -4 out of each term.

Some expressions have common factors that are not simple monomials. For instance, the expression $2x(x - 2) + 3(x - 2)$ has the common binomial factor $(x - 2)$. Factoring out this common factor produces

$$2x(x - 2) + 3(x - 2) = (x - 2)(2x + 3).$$

This type of factoring is called factoring by grouping.

EXAMPLE 2 Common Binomial Factor

Factor $7a(3a + 4b) + 2(3a + 4b)$.

Solution

Each of the terms of this expression has a binomial factor of $(3a + 4b)$.

$$7a(3a + 4b) + 2(3a + 4b) = (3a + 4b)(7a + 2)$$

In Example 2, the expression was already grouped, and so it was easy to determine the common binomial factor. In practice, you will have to do the grouping *and* the factoring.

EXAMPLE 3 **Factoring by Grouping**

Factor $x^3 - 5x^2 + x - 5$.

Solution

$$x^3 - 5x^2 + x - 5 = (x^3 - 5x^2) + (x - 5) \qquad \text{Group terms.}$$

$$= x^2(x - 5) + 1(x - 5) \qquad \text{Factoring out common monomial factor in each group.}$$

$$= (x - 5)(x^2 + 1) \qquad \text{Factored form}$$

Factoring Trinomials

To factor a trinomial $x^2 + bx + c$ into a product of two binomials, you must find two numbers m and n whose product is c and whose sum is b. If c is positive, then m and n have like signs that match the sign of b. If c is negative, then m and n have unlike signs. If $|b|$ is small relative to $|c|$, first try those factors of c that are close to each other in absolute value.

EXAMPLE 4 **Factoring Trinomials**

Factor each trinomial.

a. $x^2 - 7x + 12$ **b.** $x^2 - 2x - 8$

Solution

a. You need to find two numbers whose product is 12 and whose sum is -7.

The product of -3 and -4 is 12.

$$x^2 - 7x + 12 = (x - 3)(x - 4)$$

The sum of -3 and -4 is -7.

b. You need to find two numbers whose product is -8 and whose sum is -2.

The product of -4 and 2 is -8.

$$x^2 - 2x - 8 = (x - 4)(x + 2)$$

The sum of -4 and 2 is -2.

Applications of algebra sometimes involve trinomials that have a common monomial factor. To factor such trinomials completely, first factor out the common monomial factor. Then try to factor the resulting trinomial by the methods given in this section.

EXAMPLE 5 Factoring Completely

Factor the trinomial $5x^3 + 20x^2 + 15x$ completely.

Solution

$$5x^3 + 20x^2 + 15x = 5x(x^2 + 4x + 3)$$ Factor out common monomial factor $5x$.

$$= 5x(x + 1)(x + 3)$$ Factor trinomial.

To factor a trinomial whose leading coefficient is not 1, use the following pattern.

$$ax^2 + bx + c = (\underbrace{\square}_{\text{Factors of } a} x + \square)(\square x + \square)$$

Factors of a (first terms of each binomial); Factors of c (last terms of each binomial)

Use the following guidelines to help shorten the list of possible factorizations of a trinomial.

Guidelines for Factoring $ax^2 + bx + c$ $(a > 0)$

1. If the trinomial has a common monomial factor, you should factor out the common factor before trying to find the binomial factors.
2. Because the resulting trinomial has no common monomial factors, you do not have to test any binomial factors that have a common monomial factor.
3. Switch the signs of the factors of c when the middle term $(O + I)$ is correct except in sign.

EXAMPLE 6 Factor a Trinomial of the Form $ax^2 + bx + c$

Factor the trinomial $6x^2 + 17x + 5$.

Solution

First, observe that $6x^2 + 17x + 5$ has no common monomial factor. For this trinomial, $a = 6$, which factors as (1)(6) or (2)(3), and $c = 5$, which factors as (1)(5).

$$(x + 1)(6x + 5) = 6x^2 + 11x + 5$$

$$(x + 5)(6x + 1) = 6x^2 + 31x + 5$$

$$(2x + 1)(3x + 5) = 6x^2 + 13x + 5$$

$$(2x + 5)(3x + 1) = 6x^2 + 17x + 5$$ ⟸ Correct factorization

So, the correct factorization is $6x^2 + 17x + 5 = (2x + 5)(3x + 1)$.

EXAMPLE 7 Factoring a Trinomial of the Form $ax^2 + bx + c$

Factor the trinomial.

$$3x^2 - 16x - 35$$

Solution

First, observe that $3x^2 - 16x - 35$ has no common monomial factor. For this trinomial,

$$a = 3 \text{ and } c = -35.$$

The possible factorizations of this trinomial are as follows.

$$(3x - 1)(x + 35) = 3x^2 + 104x - 35$$

$$(3x - 35)(x + 1) = 3x^2 - 32x - 35$$

$$(3x - 5)(x + 7) = 3x^2 + 16x - 35$$ Middle term has opposite sign.

$$(3x + 5)(x - 7) = 3x^2 - 16x - 35$$ Correct factorization

So, the correct factorization is

$$3x^2 - 16x - 35 = (3x + 5)(x - 7).$$

EXAMPLE 8 Factoring Completely

Factor the trinomial completely.

$$6x^2y + 16xy + 10y$$

Solution

Begin by factoring out the common monomial factor $2y$.

$$6x^2y + 16xy + 10y = 2y(3x^2 + 8x + 5)$$

Now, for the new trinomial $3x^2 + 8x + 5$,

$$a = 3 \text{ and } c = 5.$$

The possible factorizations of this trinomial are as follows.

$$(3x + 1)(x + 5) = 3x^2 + 16x + 5$$

$$(3x + 5)(x + 1) = 3x^2 + 8x + 5$$ Correct factorization

So, the correct factorization is

$$6x^2y + 16xy + 10y = 2y(3x^2 + 8x + 5) = 2y(3x + 5)(x + 1).$$

Factoring a trinomial can involve quite a bit of trial and error. Some of this trial and error can be lessened by using factoring by grouping. The key to this method of factoring is knowing how to rewrite the middle term. In general, to factor a trinomial $ax^2 + bx + c$ by grouping, choose factors of the product ac that add up to b and use these factors to rewrite the middle term. This technique is illustrated in Example 9.

EXAMPLE 9 **Factoring a Trinomial by Grouping**

Use factoring by grouping to factor the trinomial $6y^2 + 5y - 4$.

Solution

In the trinomial $6y^2 + 5y - 4$, $a = 6$ and $c = -4$, which implies that the product of ac is -24. Now, because -24 factors as $(8)(-3)$, and $8 - 3 = 5 = b$, you can rewrite the middle term as $5y = 8y - 3y$. This produces the following result.

$$6y^2 + 5y - 4 = 6y^2 + 8y - 3y - 4$$ Rewrite middle term.

$$= (6y^2 + 8y) - (3y + 4)$$ Group terms.

$$= 2y(3y + 4) - (3y + 4)$$ Factor out common monomial factor in first group.

$$= (3y + 4)(2y - 1)$$ Distributive Property

So, the trinomial factors as $6y^2 + 5y - 4 = (3y + 4)(2y - 1)$.

Factoring Special Polynomial Forms

Some polynomials have special forms. You should learn to recognize these forms so that you can factor such polynomials easily.

Factoring Special Polynomial Forms

Let a and b be real numbers, variables, or algebraic expressions.

1. Difference of Two Squares: $a^2 - b^2 = (a + b)(a - b)$
2. Perfect Square Trinomial: $a^2 + 2ab + b^2 = (a + b)^2$
   $a^2 - 2ab + b^2 = (a - b)^2$
3. Sum or Difference of Two Cubes: $a^3 + b^3 = (a + b)(a^2 - ab + b^2)$
   $a^3 - b^3 = (a - b)(a^2 + ab + b^2)$

EXAMPLE 10 **Factoring the Difference of Two Squares**

Factor each polynomial.

a. $x^2 - 144$ **b.** $4a^2 - 9b^2$

Solution

a. $x^2 - 144 = x^2 - 12^2$ Write as difference of two squares.

$$= (x + 12)(x - 12)$$ Factored form

b. $4a^2 - 9b^2 = (2a)^2 - (3b)^2$ Write as difference of two squares.

$$= (2a + 3b)(2a - 3b)$$ Factored form

To recognize perfect square terms, look for coefficients that are squares of integers and for variables raised to even powers.

EXAMPLE 11 Factoring Perfect Square Trinomials

Factor each trinomial.

a. $x^2 - 10x + 25$

b. $4y^2 + 4y + 1$

Solution

a. $x^2 - 10x + 25 = x^2 - 2(5x) + 5^2$ Recognize the pattern.

$= (x - 5)^2$ Write in factored form.

b. $4y^2 + 4y + 1 = (2y)^2 + 2(2y)(1) + 1^2$ Recognize the pattern.

$= (2y + 1)^2$ Write in factored form.

EXAMPLE 12 Factoring Sum or Difference of Two Cubes

Factor each polynomial.

a. $x^3 + 1$

b. $27x^3 - 64y^3$

Solution

a. $x^3 + 1 = x^3 + 1^3$ Write as sum of two cubes.

$= (x + 1)[x^2 - (x)(1) + 1^2]$ Factored form.

$= (x + 1)(x^2 - x + 1)$ Simplify.

b. $27x^3 - 64y^3 = (3x)^3 - (4y)^3$ Write as difference of two cubes.

$= (3x - 4y)[(3x)^2 + (3x)(4y) + (4y)^2]$ Factored form

$= (3x - 4y)(9x^2 + 12xy + 16y^2)$ Simplify.

Solving Polynomial Equations by Factoring

A quadratic equation is an equation that can be written in the general form

$$ax^2 + bx + c = 0, \quad a \neq 0.$$

You can combine your factoring skills with the Zero-Factor Property to solve quadratic equations.

Zero-Factor Property

Let a and b be real numbers, variables, or algebraic expressions. If a and b are factors such that

$$ab = 0$$

then $a = 0$ or $b = 0$. This property also applies to three or more factors.

In order for the Zero-Factor Property to be used, a quadratic equation must be written in general form.

EXAMPLE 13 Using Factoring to Solve a Quadratic Equation

Solve the equation.

$$x^2 - x - 12 = 0$$

Solution

First, check to see that the right side of the equation is zero. Next, factor the left side of the equation. Finally, apply the Zero-Factor Property to find the solutions.

$x^2 - x - 12 = 0$ Write original equation.

$(x + 3)(x - 4) = 0$ Factor left side of equation.

$x + 3 = 0 \Rightarrow x = -3$ Set 1st factor equal to 0 and solve for x.

$x - 4 = 0 \Rightarrow x = 4$ Set 2nd factor equal to 0 and solve for x.

So, the equation has two solutions: $x = -3$ and $x = 4$.

Remember to check your solutions in the original equation, as follows.

Check First Solution

$x^2 - x - 12 = 0$ Write original equation.

$(-3)^2 - (-3) - 12 \stackrel{?}{=} 0$ Substitute -3 for x.

$9 + 3 - 12 \stackrel{?}{=} 0$ Simplify.

$0 = 0$ Solution checks. ✓

Check Second Solution

$x^2 - x - 12 = 0$ Write original equation.

$4^2 - 4 - 12 \stackrel{?}{=} 0$ Substitute 4 for x.

$16 - 4 - 12 \stackrel{?}{=} 0$ Simplify.

$0 = 0$ Solution checks. ✓

EXAMPLE 14 Using Factoring to Solve a Quadratic Equation

Solve $2x^2 - 3 = 7x + 1$.

Solution

$2x^2 - 3 = 7x + 1$ Write original equation.

$2x^2 - 7x - 4 = 0$ Write in general form.

$(2x + 1)(x - 4) = 0$ Factor.

$2x + 1 = 0 \Rightarrow x = -\frac{1}{2}$ Set 1st factor equal to 0 and solve for x.

$x - 4 = 0 \Rightarrow x = 4$ Set 2nd factor equal to 0 and solve for x.

So, the equation has two solutions: $x = -\frac{1}{2}$ and $x = 4$. Check these in the original equation, as follows.

Check First Solution

$2\left(-\frac{1}{2}\right)^2 - 3 \stackrel{?}{=} 7\left(-\frac{1}{2}\right) + 1$ Substitute $-\frac{1}{2}$ for x in original equation.

$\frac{1}{2} - 3 \stackrel{?}{=} -\frac{7}{2} + 1$ Simplify.

$-\frac{5}{2} = -\frac{5}{2}$ Solution checks. ✓

Check Second Solution

$2(4)^2 - 3 \stackrel{?}{=} 7(4) + 1$ Substitute 4 for x in original equation.

$32 - 3 \stackrel{?}{=} 28 + 1$ Simplify.

$29 = 29$ Solution checks. ✓

The Zero-Factor Property can be used to solve polynomial equations of degree 3 or higher. To do this, use the same strategy you used with quadratic equations.

Study Tip

The solution $x = -6$ in Example 15 is called a repeated solution.

EXAMPLE 15 Solving a Polynomial Equation with Three Factors

Solve $x^3 + 12x^2 + 36x = 0$.

Solution

$x^3 + 12x^2 + 36x = 0$ Write original equation.

$x(x^2 + 12x + 36) = 0$ Factor out common monomial factor.

$x(x + 6)(x + 6) = 0$ Factor perfect square trinomial.

$x = 0$ Set 1st factor equal to 0.

$x + 6 = 0 \Rightarrow x = -6$ Set 2nd factor equal to 0 and solve for x.

$x + 6 = 0 \Rightarrow x = -6$ Set 3rd factor equal to 0 and solve for x.

Note that even though the left side of the equation has three factors, two of the factors are the same. So, you conclude that the solutions of the equation are $x = 0$ and $x = -6$. Check these in the original equation.

Study Tip

For more review on factoring and solving equations, refer to Chapter 6.

Appendix B Introduction to Graphing Calculators

Introduction

You previously studied the point-plotting method for sketching the graph of an equation. One of the disadvantages of the point-plotting method is that to get a good idea about the shape of a graph, you need to plot *many* points. By plotting only a few points, you can badly misrepresent the graph.

For instance, consider the equation $y = x^3$. To graph this equation, suppose you calculated only the following three points.

x	-1	0	1
$y = x^3$	-1	0	1
Solution point	$(-1, -1)$	$(0, 0)$	$(1, 1)$

By plotting these three points, as shown in Figure B.1, you might assume that the graph of the equation is a line. This, however, is not correct. By plotting several more points, as shown in Figure B.2, you can see that the actual graph is not straight at all.

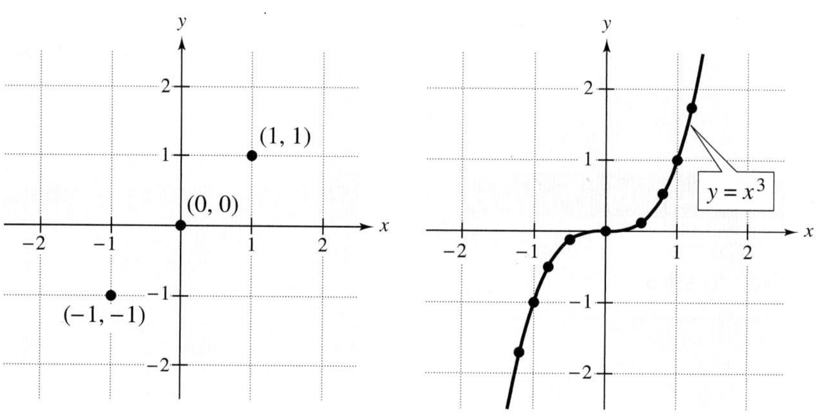

Figure B.1 **Figure B.2**

So, the point-plotting method leaves you with a dilemma. On the one hand, the method can be very inaccurate if only a few points are plotted. But, on the other hand, it is very time-consuming to plot a dozen (or more) points. Technology can help you solve this dilemma. Plotting several points (or even hundreds of points) on a rectangular coordinate system is something that a computer or graphing calculator can do easily.

Using a Graphing Calculator

There are many different graphing utilities: some are graphing programs for computers and some are hand-held graphing calculators. In this appendix, the steps used to graph an equation with a *TI-84* or *TI-84 Plus* graphing calculator are described. Keystroke sequences are often given for illustration; however, these sequences may not agree precisely with the steps required by *your* calculator.*

Graphing an Equation with a *TI-84* or *TI-84 Plus* Graphing Calculator

Before performing the following steps, set your calculator so that all of the standard defaults are active. For instance, all of the options at the left of the (MODE) screen should be highlighted.

1. Set the viewing window for the graph. (See Example 3.) To set the standard viewing window, press (ZOOM) 6.
2. Rewrite the equation so that y is isolated on the left side of the equation.
3. Press the (Y=) key. Then enter the right side of the equation on the first line of the display. (The first line is labeled $Y_1 =$.)
4. Press the (GRAPH) key.

EXAMPLE 1 Graphing a Linear Equation

Use a graphing calculator to graph $2y + x = 4$.

Solution

To begin, solve the equation for y in terms of x.

$$2y + x = 4 \quad \text{Write original equation.}$$

$$2y = -x + 4 \quad \text{Subtract } x \text{ from each side.}$$

$$y = -\frac{1}{2}x + 2 \quad \text{Divide each side by 2.}$$

Press the (Y=) key, and enter the following keystrokes.

((−)) (X,T,θ,n) (÷) 2 (+) 2

The top row of the display should now be as follows.

Y_1 = -X/2 + 2

Press the (GRAPH) key, and the screen should look like the one in Figure B.3.

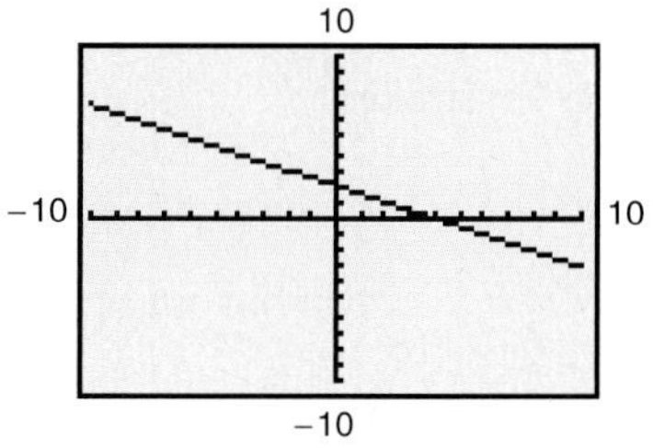

10
-10 10
-10

Figure B.3

*The graphing calculator keystrokes given in this appendix correspond to the *TI-84* and *TI-84 Plus* graphing calculators by Texas Instruments. For other graphing calculators, the keystrokes may differ. Consult your user's guide.

In Figure B.3, notice that the calculator screen does not label the tick marks on the x-axis or the y-axis. To see what the tick marks represent, you can press (WINDOW). If you set your calculator to the standard graphing defaults before working Example 1, the screen should show the following values.

Xmin = -10	The minimum x-value is -10.
Xmax = 10	The maximum x-value is 10.
Xscl = 1	The x-scale is 1 unit per tick mark.
Ymin = -10	The minimum y-value is -10.
Ymax = 10	The maximum y-value is 10.
Yscl = 1	The y-scale is 1 unit per tick mark.
Xres = 1	Sets the pixel resolution

These settings are summarized visually in Figure B.4.

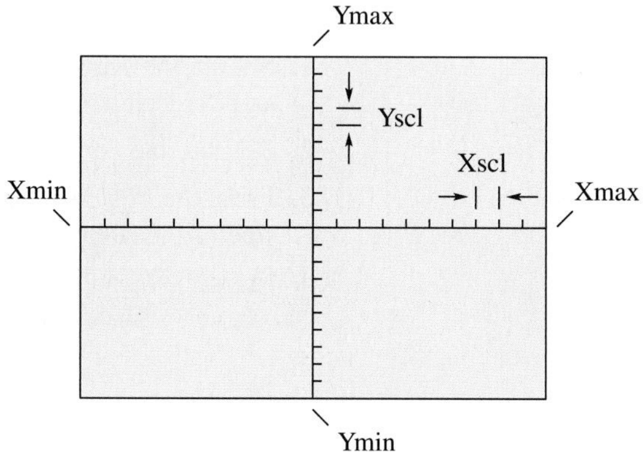

Figure B.4

EXAMPLE 2 Graphing an Equation Involving Absolute Value

Use a graphing calculator to graph

$$y = |x - 3|.$$

Solution

This equation is already written so that y is isolated on the left side of the equation. Press the (Y=) key, and enter the following keystrokes.

(MATH) (▶) 1 (X,T,θ,n) (−) 3 ())

The top row of the display should now be as follows.

$$Y_1 = \text{abs}(X - 3)$$

Press the (GRAPH) key, and the screen should look like the one shown in Figure B.5.

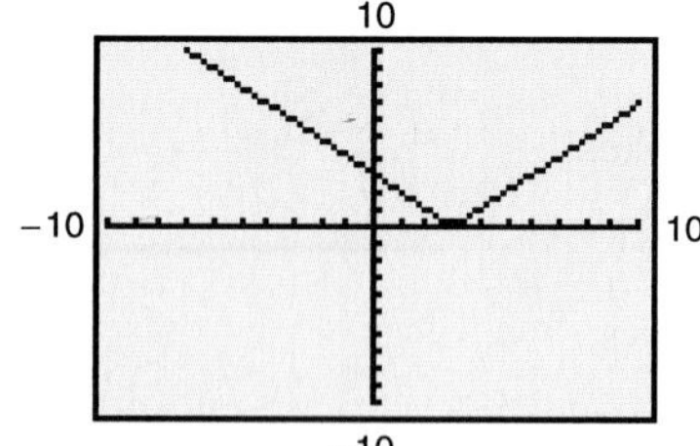

Figure B.5

Using Special Features of a Graphing Calculator

To use your graphing calculator to its best advantage, you must learn to set the viewing window, as illustrated in the next example.

EXAMPLE 3 Setting the Viewing Window

Use a graphing calculator to graph

$y = x^2 + 12.$

Solution

Press (Y=) and enter $x^2 + 12$ on the first line.

(X,T,θ,n) (x²) (+) 12

Press the (GRAPH) key. If your calculator is set to the standard viewing window, nothing will appear on the screen. The reason for this is that the lowest point on the graph of $y = x^2 + 12$ occurs at the point (0, 12). Using the standard viewing window, you obtain a screen whose largest y-value is 10. In other words, none of the graph is visible on a screen whose y-values vary between -10 and 10, as shown in Figure B.6. To change these settings, press (WINDOW) and enter the following values.

Xmin = -10	The minimum x-value is -10.
Xmax = 10	The maximum x-value is 10.
Xscl = 1	The x-scale is 1 unit per tick mark.
Ymin = -10	The minimum y-value is -10.
Ymax = 30	The maximum y-value is 30.
Yscl = 5	The y-scale is 5 units per tick mark.
Xres = 1	Sets the pixel resolution

Press (GRAPH) and you will obtain the graph shown in Figure B.7. On this graph, note that each tick mark on the y-axis represents five units because you changed the y-scale to 5. Also note that the highest point on the y-axis is now 30 because you changed the maximum value of y to 30.

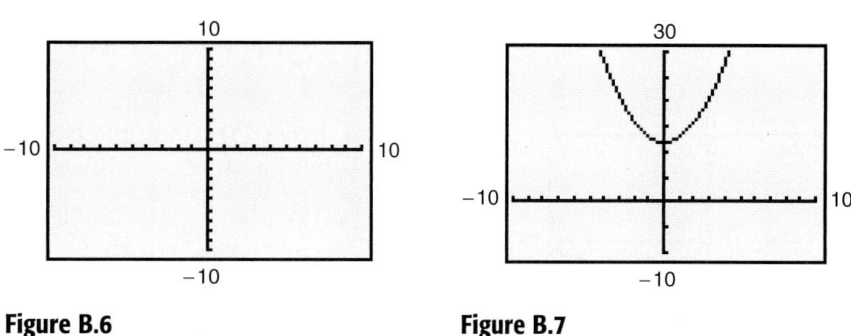

Figure B.6 **Figure B.7**

If you changed the y-maximum and y-scale on your calculator as indicated in Example 3, you should return to the standard setting before working Example 4. To do this, press (ZOOM) 6.

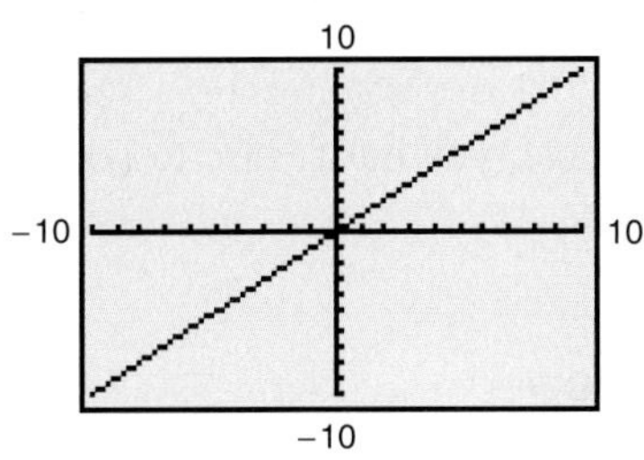

Figure B.8

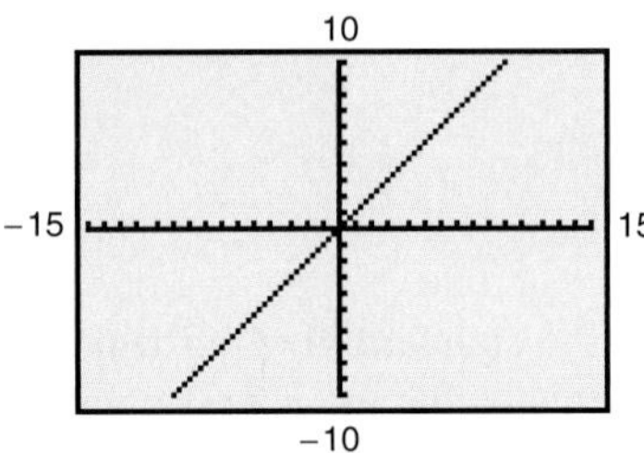

Figure B.9

EXAMPLE 4 Using a Square Setting

Use a graphing calculator to graph $y = x$. The graph of this equation is a line that makes a 45° angle with the x-axis and with the y-axis. From the graph on your calculator, does the angle appear to be 45°?

Solution

Press (Y=) and enter x on the first line.

$$Y_1 = X$$

Press the (GRAPH) key and you will obtain the graph shown in Figure B.8. Notice that the angle the line makes with the x-axis doesn't appear to be 45°. The reason for this is that the screen is wider than it is tall. This makes the tick marks on the x-axis farther apart than the tick marks on the y-axis. To obtain the same distance between tick marks on both axes, you can change the graphing settings from "standard" to "square." To do this, press the following keys.

(ZOOM) 5 Square setting

The screen should look like the one shown in Figure B.9. Note in this figure that the square setting has changed the viewing window so that the x-values vary from -15 to 15.

There are many possible square settings on a graphing calculator. To create a square setting, you need the following ratio to be $\frac{2}{3}$.

$$\frac{\text{Ymax} - \text{Ymin}}{\text{Xmax} - \text{Xmin}}$$

For instance, the setting in Example 4 is square because

$$\frac{\text{Ymax} - \text{Ymin}}{\text{Xmax} - \text{Xmin}} = \frac{10 - (-10)}{15 - (-15)} = \frac{20}{30} = \frac{2}{3}.$$

EXAMPLE 5 Graphing More than One Equation

Use a graphing calculator to graph each equation in the same viewing window.

$$y = -x + 4, \quad y = -x, \quad \text{and} \quad y = -x - 4$$

Solution

To begin, press (Y=) and enter all three equations on the first three lines. The display should now be as follows.

$Y_1 = -X + 4$ (−) (X,T,θ,n) (+) 4

$Y_2 = -X$ (−) (X,T,θ,n)

$Y_3 = -X - 4$ (−) (X,T,θ,n) (−) 4

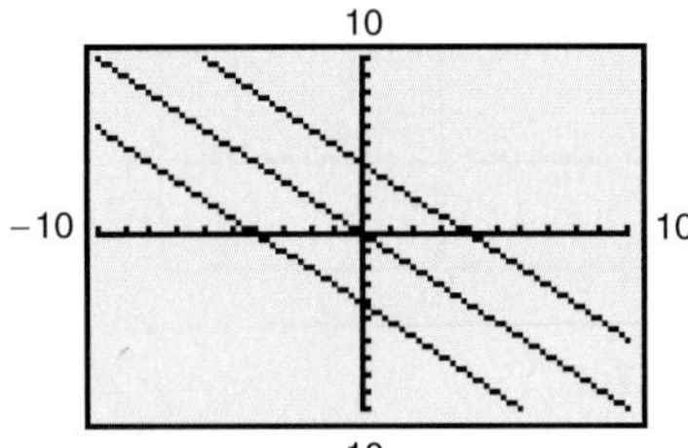

Figure B.10

Press the (GRAPH) key and you will obtain the graph shown in Figure B.10. Note that the graph of each equation is a line and that the lines are parallel to each other.

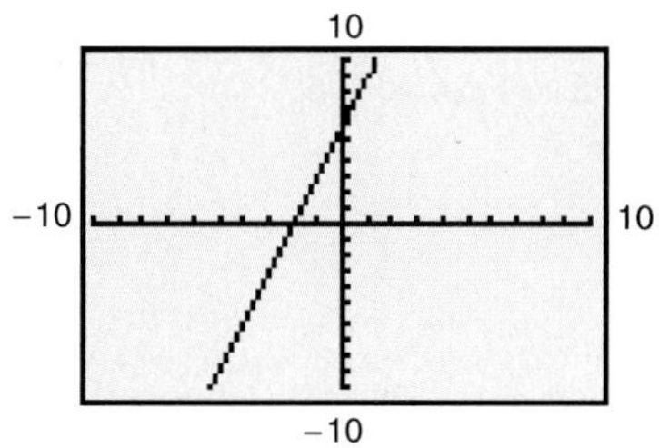

Figure B.11

EXAMPLE 6 Using the Trace Feature

Approximate the x- and y-intercepts of $y = 3x + 6$ by using the *trace* feature of a graphing calculator.

Solution

Press (Y=) and enter $3x + 6$ on the first line.

3 (X,T,θ,n) (+) 6

Press the (GRAPH) key and you will obtain the graph shown in Figure B.11. Then press the (TRACE) key and use the (◄) (►) keys to move along the graph. To get a better approximation of a solution point, you can use the following keystrokes repeatedly.

(ZOOM) 2 (ENTER)

As you can see in Figures B.12 and B.13, the x-intercept is $(-2, 0)$ and the y-intercept is $(0, 6)$.

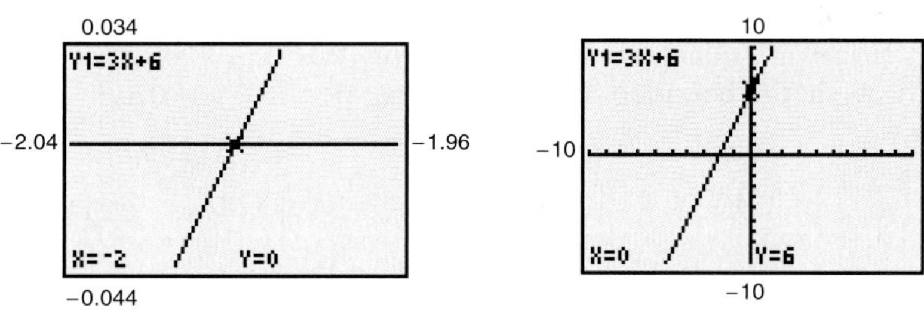

Figure B.12 **Figure B.13**

Appendix B Exercises

In Exercises 1–12, use a graphing calculator to graph the equation. (Use a standard setting.) ***See Examples 1 and 2.***

1. $y = -3x$

2. $y = x - 4$

3. $y = \frac{3}{4}x - 6$

4. $y = -3x + 2$

5. $y = \frac{1}{2}x^2$

6. $y = -\frac{2}{3}x^2$

7. $y = x^2 - 4x + 2$

8. $y = -0.5x^2 - 2x + 2$

9. $y = |x - 5|$

10. $y = |x + 4|$

11. $y = |x^2 - 4|$

12. $y = |x - 2| - 5$

In Exercises 13–16, use a graphing calculator to graph the equation using the given window settings. ***See Example 3.***

13. $y = 27x + 100$

Xmin = 0
Xmax = 5
Xscl = .5
Ymin = 75
Ymax = 250
Yscl = 25
Xres = 1

14. $y = 50{,}000 - 6000x$

Xmin = 0
Xmax = 7
Xscl = .5
Ymin = 0
Ymax = 50000
Yscl = 5000
Xres = 1

15. $y = 0.001x^2 + 0.5x$

Xmin = -500
Xmax = 200
Xscl = 50
Ymin = -100
Ymax = 100
Yscl = 20
Xres = 1

16. $y = 100 - 0.5|x|$

Xmin = -300
Xmax = 300
Xscl = 60
Ymin = -100
Ymax = 100
Yscl = 20
Xres = 1

In Exercises 17–20, find a viewing window that shows the important characteristics of the graph.

17. $y = 15 + |x - 12|$

18. $y = 15 + (x - 12)^2$

19. $y = -15 + |x + 12|$

20. $y = -15 + (x + 12)^2$

Geometry In Exercises 21–24, use a graphing calculator to graph the equations in the same viewing window. Using a "square setting," determine the geometrical shape bounded by the graphs. ***See Example 4.***

21. $y = -4, \quad y = -|x|$

22. $y = |x|, \quad y = 5$

23. $y = |x| - 8, \quad y = -|x| + 8$

24. $y = -\frac{1}{2}x + 7, \quad y = \frac{8}{3}(x + 5), \quad y = \frac{2}{7}(3x - 4)$

In Exercises 25–28, use a graphing calculator to graph both equations in the same viewing window. Are the graphs identical? If so, what basic rule of algebra is being illustrated? ***See Example 5.***

25. $y_1 = 2x + (x + 1)$
$y_2 = (2x + x) + 1$

26. $y_1 = \frac{1}{2}(3 - 2x)$
$y_2 = \frac{3}{2} - x$

27. $y_1 = 2\left(\frac{1}{2}\right)$
$y_2 = 1$

28. $y_1 = x(0.5x)$
$y_2 = (0.5x)x$

In Exercises 29–36, use the *trace* feature of a graphing calculator to approximate the x- and y-intercepts of the graph. ***See Example 6.***

29. $y = 9 - x^2$

30. $y = 3x^2 - 2x - 5$

31. $y = 6 - |x + 2|$

32. $y = (x - 2)^2 - 3$

33. $y = 2x - 5$

34. $y = 4 - |x|$

35. $y = x^2 + 1.5x - 1$

36. $y = x^3 - 4x$

Modeling Data In Exercises 37 and 38, use the following models, which give the number of pieces of first-class mail and the number of pieces of standard mail handled by the U.S. Postal Service.

First Class

$y = 0.5x^2 - 5.06x + 110.3, \quad 2 \le x \le 6$

Standard

$y = -0.221x^2 + 5.88x + 75.8, \quad 2 \le x \le 6$

In these models, y is the number of pieces handled (in billions) and x is the year, with $x = 2$ corresponding to 2002. (Source: U.S. Postal Service)

37. Use the following window settings to graph both models in the same viewing window on a graphing calculator.

Xmin = 2
Xmax = 6
Xscl = 1
Ymin = 50
Ymax = 150
Yscl = 25
Xres = 1

38. (a) Were the numbers of pieces of first-class mail and standard mail increasing or decreasing over time?

(b) In what year were the numbers of pieces of first-class mail and standard mail the same?

(c) After the year in part (b), was more first-class mail or more standard mail handled?

Answers to Odd-Numbered Exercises, Quizzes, and Tests

CHAPTER 1

Section 1.1 *(page 9)*

1. (a) $20, \frac{9}{3}$ (b) $-3, 20, \frac{9}{3}$ (c) $-3, 20, -\frac{3}{2}, \frac{9}{3}, 4.5$ (d) π

3. (a) $\frac{8}{4}$ (b) $\frac{8}{4}$ (c) $-\frac{5}{2}, 6.5, -4.5, \frac{8}{4}, \frac{3}{4}$ (d) $\sqrt{13}$

5. **7.**

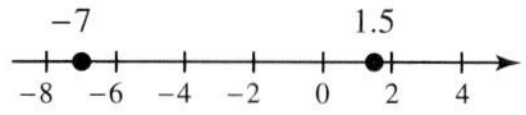

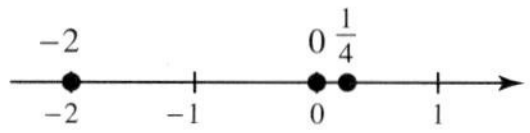

9. 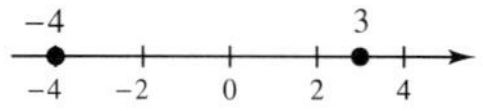 $>$ **11.** $>$

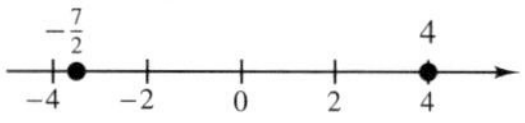

13. 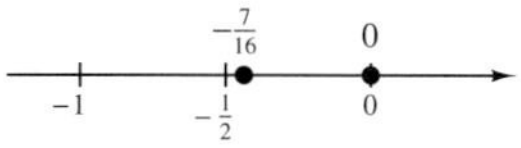 $>$ **15.** $<$

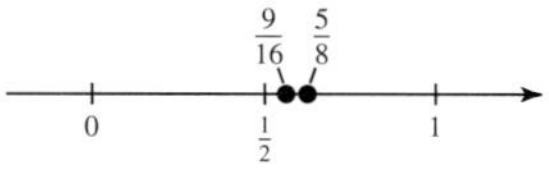

17. 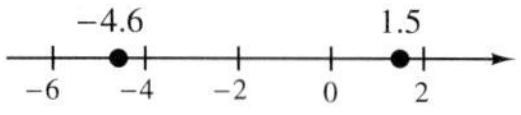 $<$ **19.** $>$

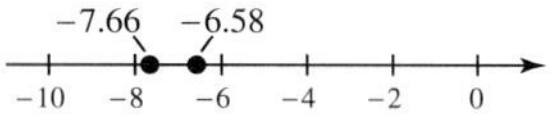

21. 2 **23.** 8

25. -3; Distance: 3 **27.** 3.8; Distance: 3.8

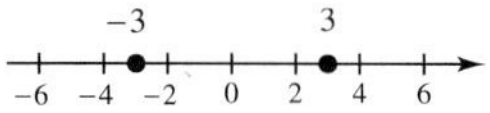

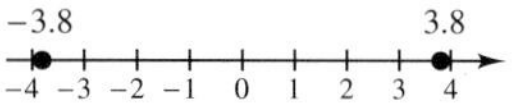

29. $-\frac{5}{2}$; Distance: $\frac{5}{2}$

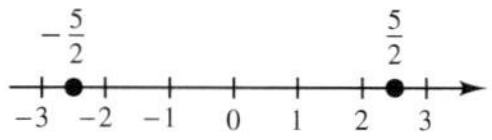

31. $\frac{5}{2}, \frac{5}{2}$ **33.** 3, 3 **35.** 10 **37.** 3 **39.** 3.4

41. $\frac{7}{2}$ **43.** -4.09 **45.** -23.6 **47.** 0 **49.** $=$

51. $>$ **53.** $<$ **55.** $<$ **57.** $>$

59. **61.**

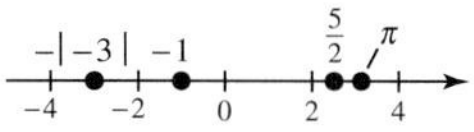

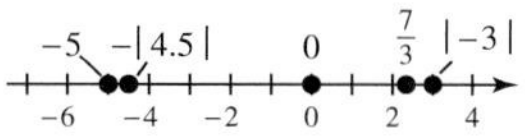

63. $-4, 20$ **65.** 15.3, 27.3 **67.** $-5.5, 1.5$

69. Sample answer: $-3, -100, -\frac{4}{1}$

71. Sample answer: $\sqrt{2}, \pi, -\sqrt{3}$

73. Sample answer: $-7, 1, 341$

75. Sample answer: $\frac{1}{2}, 10, 20\frac{1}{5}$

77. Sample answer: $-7, -\frac{4}{2}, -\frac{5}{1}$

79. -15; $|-15| > |10|$

81. 0.37; 0.37 lies to the left of $\frac{3}{8} = 0.375$.

83. True. For example, $|-4| = |4|$.

85. True. By definition, each point on the real number line corresponds to exactly one real number, and each real number corresponds to exactly one point on the real number line.

87. False. $-\frac{3}{2}$ is a rational number, but not an integer.

Section 1.2 *(page 16)*

1. 9 **3.** -11

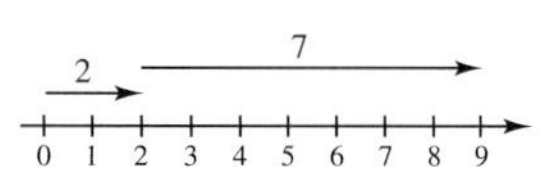

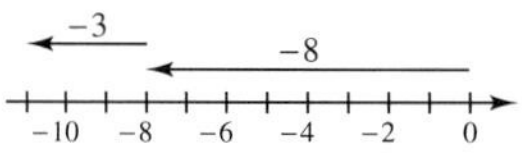

5. 7 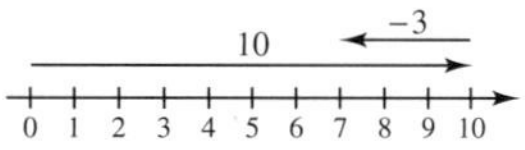 **7.** -2

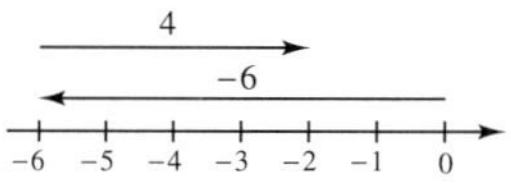

9. 16 **11.** 0 **13.** 0 **15.** 27 **17.** -27

19. 6 **21.** 175 **23.** -5 **25.** 38 **27.** -10

29. -5 **31.** 11 **33.** 3 **35.** -15 **37.** 300

39. -352 **41.** -918 **43.** 3 **45.** 26 **47.** 5

49. 25 **51.** 36 **53.** 50 **55.** -30 **57.** -24

59. -109 **61.** -9 **63.** -11 **65.** -21 **67.** 0

69. -6 **71.** -103 **73.** -610 **75.** -80

77. -21 **79.** 500 **81.** -21 **83.** -15

85. -36 **87.** 16 **89.** -4 **91.** -2

93. -10

3 [+/−] [−] 7 [=]

[(−)] 3 [−] 7 [ENTER]

95. 18

6 [+] 5 [−] [(] 7 [+/−] [)] [=]

6 [+] 5 [−] [(] [(−)] 7 [)] [ENTER]

97. 12°F **99.** 462 meters **101.** \$1,012,000

103. \$2356.42 **105.** 23°C

107. (a) \$147 million (b) \$2729 million

109. (a) $3 + 2 = 5$ (b) Adding two integers with like signs

111. To add two negative integers, add their absolute values and attach the negative sign.

113. No. To add two positive integers, add their absolute values and attach the common sign, which is always the positive sign.

Section 1.3 *(page 28)*

1. $2 + 2 + 2 = 6$

3. $(-3) + (-3) + (-3) + (-3) + (-3) = -15$

5. 35 **7.** 0 **9.** -32 **11.** -36

13. -690 **15.** 91 **17.** 1600 **19.** -90

21. 21 **23.** -30 **25.** 12 **27.** 90

29. 480 **31.** 338 **33.** -336 **35.** -4725

37. 260 **39.** 9009 **41.** 3 **43.** -6

45. -7 **47.** 7 **49.** Division by zero is undefined.

51. 0 **53.** 27 **55.** -6 **57.** -7

59. 1 **61.** 32 **63.** -160 **65.** -82

67. 110 **69.** 331 **71.** 86 **73.** 1045

75. Composite **77.** Prime **79.** Prime

81. Composite **83.** Composite **85.** $2 \cdot 2 \cdot 3$

87. $3 \cdot 11 \cdot 17$ **89.** $2 \cdot 3 \cdot 5 \cdot 7$

91. $3 \cdot 5 \cdot 13 \cdot 13$ **93.** $2 \cdot 2 \cdot 2 \cdot 2 \cdot 2 \cdot 2 \cdot 3$

95. Example: For the number -4, $1 \cdot (-4) = -4$. Also, for the number 4, $1 \cdot 4 = 4$.

Algebraic Description: If a is a real number, then $1 \cdot a = a$.

97. 3 **99.** 0

101. $-1(0) = 0$

$-2(0) = 0$

The product of an integer and zero is 0.

103. $-\$1.16$ **105.** \$6000 **107.** 57,600 square feet

109. 5 miles per second

111. (a) 82

(b) 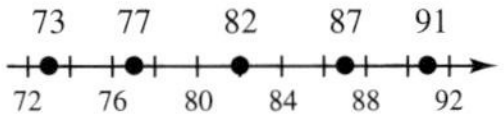

(c) 5, -9, -5, and 9; Sum is 0; Explanations will vary.

113. 594 cubic inches

115. 2; it is divisible only by 1 and itself. Any other even number is divisible by 1, itself, and 2.

117. $(2m)n = 2(mn)$. The product of two odd integers is odd.

119. 5 and 7; 11 and 13; 17 and 19; 29 and 31; 41 and 43; 59 and 61; 71 and 73

121. (a)

~~01~~	02	03	~~04~~	05	~~06~~	07	~~08~~	~~09~~	~~10~~
11	~~12~~	13	~~14~~	~~15~~	~~16~~	17	~~18~~	19	~~20~~
~~21~~	~~22~~	23	~~24~~	~~25~~	~~26~~	~~27~~	~~28~~	29	~~30~~
31	~~32~~	~~33~~	~~34~~	~~35~~	~~36~~	37	~~38~~	~~39~~	~~40~~
41	~~42~~	43	~~44~~	~~45~~	~~46~~	47	~~48~~	~~49~~	~~50~~
~~51~~	~~52~~	53	~~54~~	~~55~~	~~56~~	~~57~~	~~58~~	59	~~60~~
61	~~62~~	~~63~~	~~64~~	~~65~~	~~66~~	67	~~68~~	~~69~~	~~70~~
71	~~72~~	73	~~74~~	~~75~~	~~76~~	~~77~~	~~78~~	79	~~80~~
~~81~~	~~82~~	83	~~84~~	~~85~~	~~86~~	~~87~~	~~88~~	89	~~90~~
~~91~~	~~92~~	~~93~~	~~94~~	~~95~~	~~96~~	97	~~98~~	~~99~~	~~100~~

(b) Prime numbers; The multiples of 2, 3, 5, and 7, other than the numbers themselves, cannot be prime because they have 2, 3, 5, and 7 as factors.

Mid-Chapter Quiz *(page 31)*

1. $<$

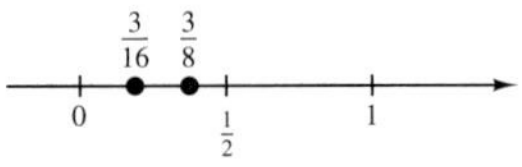

2. $>$

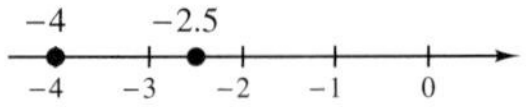

3. $<$

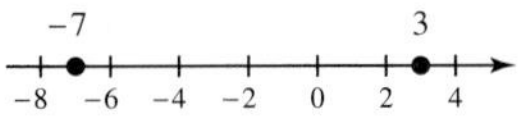

4. $>$

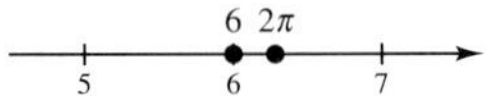

5. -0.75 **6.** $\frac{17}{19}$ **7.** $=$ **8.** $>$ **9.** -9

10. 28 **11.** 99 **12.** -53 **13.** -27 **14.** -25

15. -50 **16.** -62 **17.** 8 **18.** -5 **19.** -60

20. 91 **21.** 15 **22.** -4 **23.** Prime

24. Composite; $7 \cdot 13$ **25.** Composite; $3 \cdot 37$

26. Composite; $2 \cdot 2 \cdot 2 \cdot 2 \cdot 3 \cdot 3$ **27.** \$450,450

28. 128 cubic feet **29.** 15 feet **30.** \$367

Section 1.4 *(page 42)*

1. 5 **3.** 5 **5.** 45 **7.** 6 **9.** 60 **11.** 1

13. $\frac{1}{2}$ **15.** $\frac{4}{5}$ **17.** $\frac{5}{16}$ **19.** $\frac{2}{25}$ **21.** $\frac{3}{5}$

23. $\frac{6}{10} = \frac{3}{5}$ **25.** 6 **27.** 10 **29.** $\frac{8}{15}$ **31.** 4

33. $\frac{3}{8}$ **35.** -1 **37.** $-\frac{1}{2}$ **39.** $\frac{2}{5}$ **41.** $\frac{7}{5}$

43. $\frac{5}{6}$ **45.** $\frac{9}{16}$ **47.** $-\frac{1}{12}$ **49.** $-\frac{7}{24}$ **51.** $\frac{20}{3}$

53. $-\frac{11}{24}$ **55.** $\frac{7}{20}$ **57.** $-\frac{1}{12}$ **59.** $\frac{55}{6}$ **61.** $-\frac{17}{16}$

63. $-\frac{21}{4}$ **65.** $-\frac{39}{4}$ **67.** $\frac{31}{24}$ **69.** $\frac{64}{9}$ **71.** $-\frac{35}{12}$

73. $\frac{3}{10}$ **75.** $\frac{13}{60}$ **77.** $\frac{3}{8}$ **79.** $-\frac{10}{21}$ **81.** $-\frac{3}{8}$

83. $\frac{1}{3}$ **85.** $\frac{5}{24}$ **87.** $-\frac{3}{16}$ **89.** 1 **91.** $\frac{12}{5}$

93. $\frac{121}{12}$ **95.** $-\frac{51}{2}$ **97.** $\frac{27}{40}$ **99.** 1

101. $\frac{1}{7}$; $7 \cdot \frac{1}{7} = 1$ **103.** $\frac{7}{4}$; $\frac{4}{7} \cdot \frac{7}{4} = 1$ **105.** $\frac{1}{2}$

107. $-\frac{8}{27}$ **109.** 1 **111.** $\frac{3}{7}$ **113.** $\frac{25}{24}$ **115.** -90

117. 0 **119.** Division by zero is undefined. **121.** $\frac{5}{2}$

123. $\frac{26}{21}$ **125.** 0.25 **127.** 0.3125 **129.** $0.\overline{2}$

131. $0.41\overline{6}$ **133.** $0.\overline{36}$ **135.** 27.09 **137.** 106.65

139. 2.27 **141.** -1.90 **143.** -57.02

145. 4.30 **147.** 39.08 **149.** -0.51

151. 34.79 points **153.** $25\frac{13}{40}$ tons **155.** $\frac{17}{48}$

157. 48 breadsticks **159.** \$1538.48 **161.** \$7.46

163. (a) Answers will vary. (b) \$30,601 **165.** \$2.969

167. No; $-\frac{1}{2} + \left(-\frac{3}{4}\right) = -\frac{5}{4}$

169. The product of two fractions with like signs is positive. The product of two fractions with different signs is negative.

171. 12; $3 \div \frac{1}{4} = 12$

173. 43.6; 42.12

The first method produces the more accurate answer because you round only the answer, while in the second method you round each dimension before you multiply to get the answer.

175. True. The reciprocal of a rational number can always be written as a ratio of two integers.

177. False. $\frac{1}{2} \cdot \frac{1}{3} = \frac{1}{6}$

179. False. If $u = 1$ and $v = 2$, then $u - v = 1 - 2 = -1 \not> 0$.

181. $\frac{4}{5} + \frac{3}{6}$

Section 1.5 *(page 53)*

1. 2^5 **3.** $(-5)^4$ **5.** $\left(-\frac{1}{4}\right)^2$ **7.** $-(1.6)^5$

9. $(-3)(-3)(-3)(-3)(-3)(-3)$ **11.** $\left(\frac{3}{8}\right)\left(\frac{3}{8}\right)\left(\frac{3}{8}\right)\left(\frac{3}{8}\right)\left(\frac{3}{8}\right)$

13. $\left(-\frac{1}{2}\right)\left(-\frac{1}{2}\right)\left(-\frac{1}{2}\right)\left(-\frac{1}{2}\right)\left(-\frac{1}{2}\right)\left(-\frac{1}{2}\right)\left(-\frac{1}{2}\right)$

15. $-[(9.8)(9.8)(9.8)]$ **17.** 9 **19.** 64 **21.** $\frac{1}{64}$

23. -125 **25.** -16 **27.** -1.728 **29.** 8

31. 12 **33.** 8 **35.** 27 **37.** 17 **39.** $-\frac{11}{2}$

41. 36 **43.** 9 **45.** 68 **47.** 17 **49.** 33

51. $\frac{7}{3}$ **53.** 21 **55.** $\frac{7}{80}$ **57.** $\frac{5}{6}$ **59.** $-\frac{1}{8}$

61. 4 **63.** Division by zero is undefined. **65.** 0

67. 13 **69.** -1 **71.** 10 **73.** 366.12

75. 10.69 **77.** Commutative Property of Multiplication

79. Commutative Property of Addition

81. Distributive Property **83.** Additive Inverse Property

85. Associative Property of Addition

87. Distributive Property

89. Multiplicative Inverse Property

91. Additive Identity Property **93.** Distributive Property

95. Associative Property of Multiplication

97. $5 + 18$ **99.** $-3(10)$

101. $6 \cdot 19 + 6 \cdot 2$ **103.** $(3 + 5)4$

105. $(18 + 12) + 9$ **107.** $(12 \cdot 3)4$

109. (a) -50 (b) $\frac{1}{50}$ **111.** (a) 1 (b) -1

113. (a) $\frac{1}{2}$ (b) -2 **115.** (a) -0.2 (b) 5

117. (a) 48 (b) 48 **119.** (a) 22 (b) 22

121. $7 \cdot 4 + 9 + 2 \cdot 4$

$= 7 \cdot 4 + 2 \cdot 4 + 9$ Commutative Property of Addition

$= (7 \cdot 4 + 2 \cdot 4) + 9$ Associative Property of Addition

$= (7 + 2)4 + 9$ Distributive Property

$= 9 \cdot 4 + 9$ Add.

$= 9(4 + 1)$ Distributive Property

$= 9(5)$ Add.

$= 45$ Multiply.

123. $\left(\frac{7}{9} + 6\right) + \frac{2}{9}$

$= \frac{7}{9} + \left(6 + \frac{2}{9}\right)$ Associative Property of Addition

$= \frac{7}{9} + \left(\frac{2}{9} + 6\right)$ Commutative Property of Addition

$= \left(\frac{7}{9} + \frac{2}{9}\right) + 6$ Associative Property of Addition

$= 1 + 6$ Add.

$= 7$ Add.

125. 256 tanks **127.** 36 square units

129. (a) $35.95(1 + 0.06) = 35.95(1.06)$ (b) \$38.11

131. (a) $30(30 - 8)$ (b) $30(30) - 30(8)$ (c) 660 square feet

133. $8 - 2 + 3 + 11 + 2 \cdot 6 + 3 = 35$ **135.** No

137. No. -6^2 is the opposite of six squared, which is -36. $(-6)^2$ is negative six squared, which is 36.

139. $24^2 = (4 \cdot 6)^2 = 4^2 \cdot 6^2$

141. $4 - (6 - 2) = 4 - 6 + 2$

143. $100 \div 2 \times 50 = 50 \times 50 = 2500$

145. $5(7 + 3) = 5(7) + 5(3)$

147. Division by zero is undefined.

149. Fraction was simplified incorrectly.

$$-9 + \frac{9 + 20}{3(5)} - (-3) = -9 + \frac{29}{15} + 3$$
$$= -6 + \frac{29}{15}$$
$$= \frac{-90 + 29}{15}$$
$$= -\frac{61}{15}$$

151.

Expression	*Value*
$(6+2)\cdot(5+3)$	$=64$
$(6+2)\cdot 5+3$	$=43$
$6+2\cdot 5+3$	$=19$
$6+2\cdot(5+3)$	$=22$

153. (a) $2\cdot 2+2\cdot 3=4+6=10$

(b) $2\cdot 5=10$

(c) Explanations will vary.

155. $11\cdot 7-11\cdot 3=77-33=44$

$11(7-3)=11(4)=44$

Explanations will vary.

Review Exercises *(page 60)*

1. (a) $\sqrt{4}$ (b) $-1, \sqrt{4}$

(c) $-1, 4.5, \frac{2}{5}, -\frac{1}{7}, \sqrt{4}$ (d) $\sqrt{5}$

3. (a) $\frac{30}{2}, 2$ (b) $\frac{30}{2}, 2$ (c) $\frac{30}{2}, 2, 1.5, \frac{10}{7}$ (d) $-\sqrt{3}, -\pi$

5.

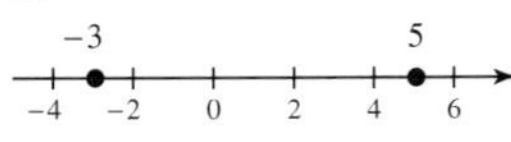

7.

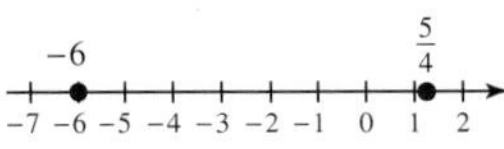

9.

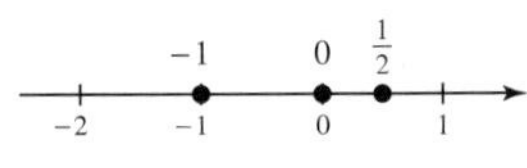

11. <

13. >

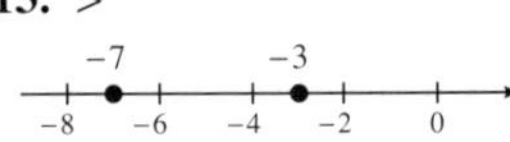

15. >

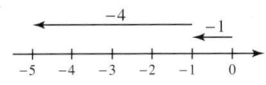

17. 0.6 **19.** $-152, 152$ **21.** $\frac{7}{3}, \frac{7}{3}$ **23.** 8.5

25. 3.4 **27.** -6.2 **29.** $-\frac{8}{5}$ **31.** =

33. > **35.** > **37.** $-2, 12$ **39.** $-2.4, 7.6$

41. 7 **43.** -5

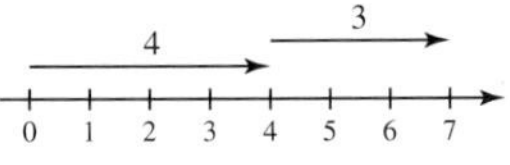

45. 11 **47.** -95 **49.** -89

51. 5 **53.** -29 **55.** $82,400

57. The sum can be positive or negative. The sign is determined by the integer with the greater absolute value.

59. 21 **61.** -7 **63.** 33 **65.** -22 **67.** -9

69. 1162 **71.** $765 **73.** 45 **75.** -72

77. -48 **79.** 45 **81.** -54 **83.** -40

85. $3600 **87.** 9 **89.** -12 **91.** -15

93. 13 **95.** 0 **97.** Division by zero is undefined.

99. 65 miles per hour **101.** Prime **103.** Prime

105. Composite **107.** $2\cdot 2\cdot 2\cdot 3\cdot 11$

109. $2\cdot 3\cdot 3\cdot 3\cdot 7$ **111.** $2\cdot 2\cdot 13\cdot 31$ **113.** -36

115. 7 **117.** 18 **119.** 1 **121.** 21 **123.** $\frac{1}{4}$

125. $\frac{5}{8}$ **127.** 10 **129.** 15 **131.** $\frac{2}{5}$ **133.** $\frac{3}{4}$

135. $\frac{7}{8}$ **137.** $\frac{1}{9}$ **139.** $-\frac{103}{96}$ **141.** $\frac{5}{4}$ **143.** $\frac{17}{8}$

145. $2\frac{3}{4}$ inches **147.** $-\frac{1}{12}$ **149.** 1 **151.** $-\frac{1}{36}$

153. $\frac{2}{3}$ **155.** $\frac{6}{7}$ **157.** Division by zero is undefined.

159. 0 **161.** $\frac{27}{32}$ inch per hour **163.** 0.625

165. $0.5\overline{3}$ **167.** 5.65 **169.** -1.38 **171.** -0.75

173. 21 **175.** $1400 **177.** 6^5 **179.** $\left(\frac{6}{7}\right)^4$

181. $(-7)\cdot(-7)\cdot(-7)\cdot(-7)$

183. $(1.25)\cdot(1.25)\cdot(1.25)$ **185.** 16 **187.** $-\frac{27}{64}$

189. -49 **191.** 6 **193.** 21 **195.** 52 **197.** 160

199. 81 **201.** $\frac{37}{8}$ **203.** 140 **205.** -3 **207.** 7

209. 0 **211.** 796.11 **213.** 1841.74

215. (a) $10,546.88 (b) $14,453.12

217. Additive Inverse Property

219. Commutative Property of Multiplication

221. Multiplicative Identity Property

223. Distributive Property

225. -16 **227.** $1+24$

229. $6\cdot 18-6\cdot 5=108-30=78$

$6(18-5)=6(13)=78$

Explanations will vary.

Chapter Test *(page 65)*

1. (a) 4 (b) $4, -6, 0$ (c) $4, -6, \frac{1}{2}, 0, \frac{7}{9}$ (d) π

2. > **3.** 13 **4.** -6.8 **5.** -4 **6.** 10

7. 10 **8.** 47 **9.** -160 **10.** 8 **11.** -30

12. 1 **13.** $\frac{17}{24}$ **14.** $\frac{2}{15}$ **15.** $\frac{7}{12}$ **16.** -27

17. -0.64 **18.** 33 **19.** 235 **20.** -2

21. Distributive Property

22. Multiplicative Inverse Property

23. Associative Property of Addition

24. Commutative Property of Multiplication

25. $\frac{2}{9}$ **26.** $2\cdot 2\cdot 2\cdot 3\cdot 3\cdot 3$

27. 58 feet per second **28.** $6.43

CHAPTER 2

Section 2.1 *(page 74)*

1. $7.55w$ **3.** $3.79m$ **5.** x **7.** m, n **9.** k

11. $4x, 3$ **13.** $6x, -1$ **15.** $\frac{5}{3}, -3y^3$ **17.** $a^2, 4ab, b^2$

19. $3(x + 5), 10$ **21.** $15, \frac{5}{x}$ **23.** $\frac{3}{x + 2}, -3x, 4$

25. 14 **27.** $-\frac{1}{3}$ **29.** $\frac{2}{5}$ **31.** 2π

33. 3.06 **35.** $y \cdot y \cdot y \cdot y \cdot y$

37. $2 \cdot 2 \cdot x \cdot x \cdot x \cdot x$ **39.** $4 \cdot y \cdot y \cdot z \cdot z \cdot z$

41. $a^2 \cdot a^2 \cdot a^2 = a \cdot a \cdot a \cdot a \cdot a \cdot a$

43. $-4 \cdot x \cdot x \cdot x \cdot x \cdot x \cdot x$

45. $-9 \cdot a \cdot a \cdot a \cdot b \cdot b \cdot b$

47. $(x + y)(x + y)$ **49.** $\left(\frac{a}{3s}\right)\left(\frac{a}{3s}\right)\left(\frac{a}{3s}\right)\left(\frac{a}{3s}\right)$

51. $2 \cdot 2 \cdot (a - b)(a - b)(a - b)(a - b)(a - b)$

53. $-2u^4$ **55.** $(2u)^4$ **57.** $(-a)^3b^2$

59. $(-3)^3(x - y)^2$ **61.** $\left(\frac{x + y}{4}\right)^3$ **63.** (a) 0 (b) -9

65. (a) 3 (b) 13 **67.** (a) 6 (b) 4

69. (a) 5 (b) 14 **71.** (a) 3 (b) -20

73. (a) 33 (b) 112

75. (a) 0 (b) Division by zero is undefined.

77. (a) $-\frac{1}{5}$ (b) $\frac{3}{10}$ **79.** (a) 0 (b) 11

81. (a) $\frac{15}{2}$ (b) 10 **83.** (a) 72 (b) 320

85. (a)

x	-1	0	1	2	3	4
$3x - 2$	-5	-2	1	4	7	10

(b) 3 (c) $\frac{2}{3}$

87. (a) $x + 6$ (b) 29 inches

89. $(n - 5)^2$, 9 square units

91. $a(a + b)$, 45 square units

93. (a) $\frac{3(4)}{2} = 6 = 1 + 2 + 3$

(b) $\frac{6(7)}{2} = 21 = 1 + 2 + 3 + 4 + 5 + 6$

(c) $\frac{10(11)}{2} = 55 = 1 + 2 + 3 + 4 + 5 + 6 + 7 + 8 + 9 + 10$

95. (a) 4, 5, 5.5, 5.75, 5.875, 5.938, 5.969; Approaches 6

(b) 9, 7.5, 6.75, 6.375, 6.188, 6.094, 6.047; Approaches 6

97. No. The term includes the minus sign and is $-3x$.

99. The product of an even number and an odd number $[n(n + 1)$, where $n \geq 1$, and $n(n - 3)$, where $n \geq 4]$ is even, so it divides evenly by 2. This will always yield a natural number.

101. 17 **103.** 10 **105.** 24 **107.** 12

109. Commutative Property of Multiplication

111. Distributive Property

Section 2.2 *(page 85)*

1. Commutative Property of Addition

3. Associative Property of Multiplication

5. Additive Identity Property

7. Multiplicative Identity Property

9. Associative Property of Addition

11. Commutative Property of Multiplication

13. Distributive Property

15. Additive Inverse Property

17. Multiplicative Inverse Property

19. Distributive Property

21. Additive Inverse Property, Additive Identity Property

23. $(-5r)s = -5(rs)$ Associative Property of Multiplication

25. $v(2) = 2v$ Commutative Property of Multiplication

27. $5(t - 2) = 5(t) + 5(-2)$ Distributive Property

29. $(2z - 3) + [-(2z - 3)] = 0$ Additive Inverse Property

31. $-5x\left(-\frac{1}{5x}\right) = 1, x \neq 0$ Multiplicative Inverse Property

33. $12 + (8 - x) = (12 + 8) - x$

Associative Property of Addition

35. $32 + 16z$ **37.** $-24 + 40m$ **39.** $90 - 60x$

41. $-16 - 40t$ **43.** $-10x + 5y$ **45.** $8x + 8$

47. $-24 + 6t$ **49.** $4x + 4xy + 4y^2$ **51.** $3x^2 + 3x$

53. $8y^2 - 4y$ **55.** $-5z + 2z^2$ **57.** $-12y^2 + 16y$

59. $-u + v$ **61.** $3x^2 - 4xy$

63. ab; ac; $a(b + c) = ab + ac$

65. $2a$; $2(b - a)$; $2a + 2(b - a) = 2b$

67. $16t^3, 3t^3$; $4t, -5t$ **69.** $4rs^2, 12rs^2$; $-5, 1$

71. $4x^2y, 10x^2y$; $x^3, 3x^3$ **73.** $-2y$ **75.** $-2x + 5$

77. $11x + 4$ **79.** $3r + 7$ **81.** $x^2 - xy + 4$

83. $17z + 11$ **85.** $z^3 + 3z^2 + 3z + 1$

87. $-x^2y + 4xy + 12xy^2$

89. $\frac{2}{x} + 8$ **91.** $\frac{11}{t} - 5t$

93. False. $3(x - 4) = 3x - 12$ **95.** True. $6x - 4x = 2x$

97. False. $2 - (x + 4) = -x - 2$

99. 416 **101.** 432 **103.** -224 **105.** 35.1

107. $12x$ **109.** $-4x$ **111.** $6x^2$ **113.** $-10z^3$

115. $9a$ **117.** $-\frac{x^3}{3}$ **119.** $-24x^4y^4$ **121.** $2x$

123. $13s - 2$ **125.** $-2m + 21$ **127.** $8x + 38$

129. $8x + 26$ **131.** $2x - 17$ **133.** $10x - 7x^2$

135. $3x^2 + 5x - 3$ **137.** $4t^2 - 11t$ **139.** $26t - 2t^2$

141. $\frac{x}{3}$ **143.** $\frac{5z}{4}$ **145.** $-\frac{11x}{12}$ **147.** $\frac{31x}{30}$

149. $5 + (3x - 1) + (2x + 5) = 5x + 9$

151. (a) $8x + 14$ (b) $3x^2 + 21x$

153. (a) Answers will vary. (b) $\frac{57}{2}$ square units

155. $x^2 + 50x$

157. $(6x)^4 = (6x)(6x)(6x)(6x) = 6 \cdot 6 \cdot 6 \cdot 6 \cdot x \cdot x \cdot x \cdot x$; $6x^4 = 6 \cdot x \cdot x \cdot x \cdot x$

159. The exponents of y are not the same.

161. Distributing -3 to each term of $(x - 1)$ would give $-3x + 3$, not $-3x - 3$.

$$\begin{aligned} 4x - 3(x - 1) &= 4x - 3(x) - 3(-1) \\ &= 4x - 3x + 3 \\ &= x + 3 \end{aligned}$$

163. 12 **165.** -11 **167.** $\frac{1}{80}$

169. (a) 4 (b) -5 **171.** (a) 9 (b) 12

Mid-Chapter Quiz *(page 90)*

1. (a) 0 (b) 10
2. (a) Division by zero is undefined. (b) 0
3. Terms: $4x^2, -2x$; Coefficients: $4, -2$
4. Terms: $5x, 3y, -z$; Coefficients: $5, 3, -1$
5. $(-3y)^4$
6. $2^3(x - 3)^2$
7. Associative Property of Multiplication
8. Distributive Property
9. Multiplicative Inverse Property
10. Commutative Property of Addition
11. $6x^2 - 2x$
12. $-12y - 18y^2 + 36$
13. $20y^2$
14. $-\frac{x^2}{5}$
15. $9y^5$
16. $\frac{10z^3}{21y}$
17. $y^2 + 4xy + y$
18. $\frac{3}{u} + 3u$
19. $8a - 7b$
20. $-8x - 66$
21. $8 + (x + 6) + (3x + 1) = 4x + 15$
22. (a) $\frac{x}{6}$ (b) 5 students

Section 2.3 *(page 101)*

1. d **2.** c **3.** e **4.** f **5.** b **6.** a

7. $x + 5$ **9.** $x - 25$ **11.** $x - 6$ **13.** $2x$

15. $\frac{x}{3}$ **17.** $\frac{x}{50}$ **19.** $\frac{3}{10}x$ **21.** $3x - 5$

23. $3(x - 5)$ **25.** $\frac{x}{5} + 15$ **27.** $|x + 4|$ **29.** $x^2 + 1$

31. A number decreased by 10

33. The product of 3 and a number, increased by 2

35. One-half of a number, decreased by 6

37. Three times the difference of 2 and a number

39. The sum of a number and 1, all divided by 2

41. One-half decreased by the quotient of a number and 5

43. The square of a number, increased by 5

45. $x(x + 3) = x^2 + 3x$ **47.** $x - (25 + x) = -25$

49. $x^2 - x(2x) = -x^2$ **51.** $\frac{8(x + 24)}{2} = 4x + 96$

53. $0.10d$ **55.** $0.06L$ **57.** $\frac{100}{r}$ **59.** $15m + 2n$

61. $t = 10.2$ years **63.** $t = 11.9$ years

65. One less than 2 times a number; 39 **67.** $a = 5, b = 4$

69.

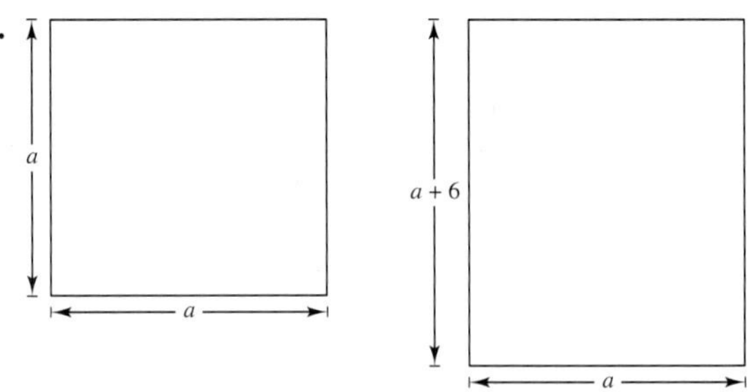

Perimeter of the square: $4a$ centimeters; Area of the square: a^2 square centimeters; Perimeter of the rectangle: $4a + 12$ centimeters; Area of the rectangle: $a(a + 6)$ square centimeters

71. 5.75 feet

73. $2 \cdot 2x + \frac{1}{2}[(9x - 4) - 2x] \cdot 2 = 11x - 4$

75. The annual cost (in dollars) is $12d$.

77. The perimeter is $5w$. **79.** The start time is missing.

81. The amount of the weekly earnings is missing, as well as either the number of hours worked or the hourly wage.

83. (a), (b), and (e)

85. $\frac{5}{3n}, \frac{5}{n} \cdot 3 = \frac{15}{n}$; The expression $\frac{3n}{5}$ is not a possible interpretation because the phrase "the quotient of 5 and a number" indicates that the variable is in the denominator.

87. 78 **89.** $\frac{3}{4}$ **91.** $\frac{23}{9}$

93. Commutative Property of Addition

95. Distributive Property

Section 2.4 *(page 111)*

1. (a) Not a solution (b) Solution
3. (a) Not a solution (b) Solution
5. (a) Not a solution (b) Solution
7. (a) Not a solution (b) Not a solution
9. (a) Solution (b) Not a solution
11. (a) Solution (b) Solution
13. (a) Not a solution (b) Not a solution
15. (a) Not a solution (b) Not a solution
17. (a) Not a solution (b) Solution
19. (a) Solution (b) Not a solution
21. (a) Solution (b) Not a solution
23. (a) Not a solution (b) Solution
25. (a) Solution (b) Not a solution

27.

$x - 8 = 3$	Original equation
$x - 8 + 8 = 3 + 8$	Add 8 to each side.
$x = 11$	Solution

29.

$\frac{2}{3}x = 12$	Original equation
$\frac{3}{2}\left(\frac{2}{3}x\right) = \frac{3}{2}(12)$	Multiply each side by $\frac{3}{2}$.
$x = 18$	Solution

31.

$5x + 12 = 22$	Original equation
$5x + 12 - 12 = 22 - 12$	Subtract 12 from each side.
$5x = 10$	Combine like terms.
$\frac{5x}{5} = \frac{10}{5}$	Divide each side by 5.
$x = 2$	Solution

33.

$2(x - 1) = x + 3$	Original equation
$2x - 2 = x + 3$	Distributive Property
$2x - 2 - x = x + 3 - x$	Subtract x from each side.
$2x - x - 2 = x - x + 3$	Group like terms.
$x - 2 = 3$	Combine like terms.
$x - 2 + 2 = 3 + 2$	Add 2 to each side.
$x = 5$	Solution

35.

$\frac{x}{3} = x + 1$	Original equation
$3\left(\frac{x}{3}\right) = 3(x + 1)$	Multiply each side by 3.
$x = 3x + 3$	Distributive Property
$x - 3x = 3x + 3 - 3x$	Subtract $3x$ from each side.
$x - 3x = 3x - 3x + 3$	Group like terms.
$-2x = 3$	Combine like terms.
$\frac{-2x}{-2} = \frac{3}{-2}$	Divide each side by -2.
$x = -\frac{3}{2}$	Solution

37. A number decreased by 6 is 32.
39. Twice a number increased by 5 is 21.
41. Ten times the difference of a number and 3 is 8 times the number.
43. The sum of a number and 1, all divided by 3, is 8.
45. $x + 12 = 45$ **47.** $\frac{x}{8} = 6$ **49.** $3x + 4 = 16$
51. $120 - 6x = 96$ **53.** $\frac{4}{x + 5} = 2$ **55.** 15 dollars
57. 150 pounds **59.** 6000 feet **61.** 240 centimeters
63. 13 **65.** 10 **67.** $x + 6 = 94$
69. $1044 + x = 1926$ **71.** $x + 45 = 375$
73. $\frac{x}{18} = 4.5$ **75.** $\frac{x}{3} = 1100$ **77.** $6 \cdot 4 \cdot h = 72$
79. $3x + 25 = 160$ **81.** $7x = 150 - 72$
83. $750{,}000 - 3D = 75{,}000$
85. $10a + 6\left(\frac{3}{4}a\right) = 986$
$\frac{29}{2}a = 986$
87. No, there is only one value of x, $\frac{b}{a}$, for which the equation is true.
89. Sample answer: The total cost of a shipment of bulbs is \$840. Find the number of cases of bulbs if each case costs \$35.
91. t^7 **93.** $15x$ **95.** $8b$ **97.** $x + 23$ **99.** $4x + 7$

Review Exercises *(page 118)*

1. $60t$ **3.** x **5.** a, b
7. Terms: $12y, y^2$
Coefficients: 12, 1
9. Terms: $5x^2, -3xy, 10y^2$
Coefficients: 5, -3, 10

11. Terms: $\frac{2y}{3}, -\frac{4x}{y}$

Coefficients: $\frac{2}{3}, -4$

13. $(5z)^3$ **15.** $(-3x)^5$ **17.** $6^2(b-c)^2$

19. (a) 5 (b) 5 **21.** (a) 4 (b) -2

23. (a) 0 (b) -7 **25.** (a) -3 (b) 6

27. Multiplicative Inverse Property

29. Commutative Property of Multiplication

31. Associative Property of Addition

33. $3(m^2n) = 3m^2(n)$ Associative Property of Multiplication

35. $(3x+8) + [-(3x+8)] = 0$ Additive Inverse Property

37. $4x + 12y$ **39.** $-10u + 15v$ **41.** $8x^2 + 5xy$

43. $a - 3b$ **45.** $2x + 6 - 4y$ **47.** $3x, 2x$

49. $10, -2$ **51.** $-2a$ **53.** $11p - 3q$

55. $\frac{15}{4}s - 5t$ **57.** $x^2 + 2xy + 4$ **59.** $3x - 3y + 3xy$

61. $3\left(1+\frac{r}{n}\right)^2$ **63.** $48t$ **65.** $45x^2$ **67.** $-12x^3$

69. $8x$ **71.** 183 **73.** 686 **75.** $5u - 10$

77. $5s - r$ **79.** $10z - 1$ **81.** $2z - 2$

83. $8x - 32$ **85.** $-2x + 4y$ **87.** $-\frac{x}{2}$ **89.** $\frac{17z}{10}$

91. (a) $6x + 12$ (b) $2x^2 + 12x$ **93.** $6n + 3$

95. $(4x)(16x) - x(6x) = 58x^2$

97. *Verbal model:*

Base pay per hour + Additional pay per unit · Number of units produced per hour

Algebraic expression: $8.25 + 0.60x$

99. $\frac{2}{3}x + 5$ **101.** $2x - 10$ **103.** $50 + 7x$

105. $\frac{x+10}{8}$ **107.** $x^2 + 64$ **109.** A number plus 3

111. A number decreased by 2, all divided by 3

113. $0.05x$ **115.** $625n$

117. Four more than 3 times a number; 64

119. (a) Not a solution (b) Solution

121. (a) Not a solution (b) Solution

123. (a) Solution (b) Not a solution

125. (a) Not a solution (b) Solution

127. (a) Solution (b) Solution

129.

$-7x + 20 = -1$	Original equation
$-7x + 20 - 20 = -1 - 20$	Subtract 20 from each side.
$-7x = -21$	Combine like terms.
$\frac{-7x}{-7} = \frac{-21}{-7}$	Divide each side by -7.
$x = 3$	Solution

131.

$x = -(x - 14)$	Original equation
$x = -x + 14$	Distributive Property
$x + x = -x + 14 + x$	Add x to each side.
$x + x = -x + x + 14$	Group like terms.
$2x = 14$	Combine like terms.
$\frac{2x}{2} = \frac{14}{2}$	Divide each side by 2.
$x = 7$	Solution

133. $x + \frac{1}{x} = \frac{37}{6}$ **135.** $6x - \frac{1}{2}(6x) = 24$

Chapter Test *(page 123)*

1. Terms: $2x^2, -7xy, 3y^3$

Coefficients: $2, -7, 3$

2. $x^3(x+y)^2$

3. Associative Property of Multiplication

4. Commutative Property of Addition

5. Additive Identity Property

6. Multiplicative Inverse Property

7. $3x + 24$ **8.** $20r - 5s$ **9.** $-3y + 2y^2$

10. $-36 + 18x - 9x^2$ **11.** $-a - 7b$ **12.** $8u - 8v$

13. $4z - 4$ **14.** $18 - 2t$ **15.** 6 **16.** -28

17. Division by zero is undefined. **18.** $\frac{1}{3}n - 4$

19. (a) Perimeter: $2w + 2(2w - 4) = 6w - 8$;

Area: $w(2w - 4) = 2w^2 - 4w$

(b) Perimeter: 34 units; Area: 70 square units

20. (a) $25m + 20n$ (b) \$110

21. (a) Not a solution (b) Solution

CHAPTER 3

Section 3.1 *(page 134)*

1. -6 **3.** 13 **5.** 4 **7.** -9

9. $5x + 15 = 0$ Original equation
$5x + 15 - 15 = 0 - 15$ Subtract 15 from each side.
$5x = -15$ Combine like terms.
$\frac{5x}{5} = \frac{-15}{5}$ Divide each side by 5.
$x = -3$ Simplify.

11. $-2x - 8 = 0$ Original equation
$-2x - 8 + 8 = 0 + 8$ Add 8 to each side.
$-2x = 8$ Combine like terms.
$\frac{-2x}{-2} = \frac{8}{-2}$ Divide each side by -2.
$x = -4$ Simplify.

13. Addition **15.** Multiplication **17.** 2 **19.** -7
21. 6 **23.** $-\frac{7}{3}$ **25.** 3 **27.** 2 **29.** 4 **31.** $\frac{2}{3}$
33. 30 **35.** -2 **37.** $\frac{5}{3}$ **39.** $\frac{5}{6}$ **41.** 2 **43.** $\frac{1}{3}$
45. -2 **47.** No solution **49.** 1 **51.** 4
53. -2 **55.** Infinitely many solutions **57.** $\frac{2}{5}$
59. $\frac{9}{2}$ **61.** 0 **63.** $\frac{2}{3}$ **65.** Infinitely many solutions
67. 2 **69.** No solution **71.** 75 centimeters
73. 20 inches $\times$ 40 inches
75. Yes. Subtract the cost of parts from the total to find the cost of labor. Then divide by 44 to find the number of hours spent on labor. $2\frac{1}{4}$ hours
77. 150 seats **79.** $5\frac{1}{2}$ hours **81.** 4
83. 35, 37 **85.** 51, 53, 55
87. The red box weighs 6 ounces. If you removed three blue boxes from each side, the scale would still balance. The Subtraction Property of Equality
89. False. Multiplying each side of the equation $3x = 9$ by 0 yields $0 = 0$. The equation $3x = 9$ has one solution, $x = 3$, and the result $0 = 0$ suggests that the equation has infinitely many solutions.
91. False. $(2m + 1) + 2n = 2m + 2n + 1 = 2(m + n) + 1$, which is odd.

93.

95.

97. (a) Solution (b) Not a solution
99. (a) Not a solution (b) Solution
101. (a) Not a solution (b) Not a solution

Section 3.2 *(page 144)*

1. 4 **3.** -5 **5.** $\frac{22}{5}$ **7.** 2 **9.** -10
11. 9 **13.** 3 **15.** 30 **17.** No solution **19.** -4
21. No solution **23.** $\frac{8}{5}$ **25.** 1 **27.** No solution
29. $\frac{5}{6}$ **31.** 1 **33.** 3 **35.** $-\frac{3}{2}$ **37.** $\frac{5}{2}$ **39.** $-\frac{2}{5}$
41. $\frac{35}{2}$ **43.** $-\frac{10}{3}$ **45.** No solution **47.** $\frac{1}{6}$ **49.** 50
51. $\frac{32}{5}$ **53.** 10 **55.** 10 **57.** 0 **59.** $\frac{16}{3}$ **61.** $\frac{4}{3}$
63. $\frac{4}{11}$ **65.** 6 **67.** 10^2 **69.** 10^3 **71.** 5
73. 7.71 **75.** 123 **77.** 3.51 **79.** 8.99
81. 4.8 hours **83.** 97 **85.** 25 quarts **87.** 2028
89. Because each brick is 8 inches long and there are n bricks, the width that is made up of bricks is represented by $8n$. Because there is $\frac{1}{2}$ inch of mortar between adjoining bricks and there are $n - 1$ widths of mortar, the width that is made up of mortar is represented by $\frac{1}{2}(n - 1)$. Because the width of the fireplace is 93 inches, the equation is $8n + \frac{1}{2}(n - 1) = 93$.
91. You could divide each side of the equation by 3.
93. Dividing by a variable assumes that it does not equal zero, which may yield a false solution.
95. $4x^6$ **97.** $5z^5$ **99.** $x - 4$
101. $-y^4 + 2y^2$ **103.** $\frac{17}{3}$ **105.** -9

Section 3.3 *(page 155)*

1. 62% **3.** 20% **5.** 7.5% **7.** 238%
9. 80% **11.** 125% **13.** $83\frac{1}{3}\%$ **15.** 105%
17. 0.125 **19.** 1.25 **21.** 0.085 **23.** 0.0075
25. $\frac{3}{8}$ **27.** $\frac{13}{10}$ **29.** $\frac{7}{500}$ **31.** $\frac{1}{200}$

	Percent	*Parts out of 100*	*Decimal*	*Fraction*
33.	40%	40	0.40	$\frac{2}{5}$
35.	7.5%	7.5	0.075	$\frac{3}{40}$
37.	63%	63	0.63	$\frac{63}{100}$
39.	15.5%	15.5	0.155	$\frac{31}{200}$
41.	60%	60	0.60	$\frac{3}{5}$

43. $37\frac{1}{2}\%$ **45.** $41\frac{2}{3}\%$ **47.** 45 **49.** 0.42
51. 2100 **53.** 132 **55.** 72% **57.** 2.75%

	Cost	*Selling Price*	*Markup*	*Markup Rate*
59.	\$26.97	\$49.95	\$22.98	85.2%
61.	\$40.98	\$74.38	\$33.40	81.5%
63.	\$69.29	\$125.98	\$56.69	81.8%
65.	\$13,250.00	\$15,900.00	\$2650.00	20%
67.	\$107.97	\$199.96	\$91.99	85.2%

	List Price	*Sale Price*	*Discount*	*Discount Rate*
69.	\$39.95	\$29.95	\$10.00	25%
71.	\$23.69	\$18.95	\$4.74	20%
73.	\$189.99	\$159.99	\$30.00	15.8%

	List Price	*Sale Price*	*Discount*	*Discount Rate*
75.	\$119.96	\$59.98	\$59.98	50%
77.	\$995.00	\$695.00	\$300.00	30.2%

79. \$544 **81.** \$3435 **83.** 9.5%

85. 4% **87.** \$37,380 **89.** 500 points

91. \$312.50 **93.** 0.107%

95. Media networks: \$14.91 billion
Parks and resorts: \$10.65 billion
Studio entertainment: \$7.455 billion
Consumer products: \$2.485 billion

97. (a) 3,246,000 (b) 98,000 (c) 119,000

99. If $a > b$, the percent is greater than 100%.
If $a < b$, the percent is less than 100%.
If $a = b$, the percent is equal to 100%.

101. False. $1\% = 0.01 \neq 1$

103. False. Because $68\% = 0.68$, $a = 0.68(50)$.

105. 0 **107.** (a) 7 (b) 16 **109.** $8x - 20$

111. -3 **113.** -12

Section 3.4 *(page 166)*

1. $\frac{4}{1}$ **3.** $\frac{1}{2}$ **5.** $\frac{17}{4}$ **7.** $\frac{2}{3}$ **9.** $\frac{9}{1}$ **11.** $\frac{32}{53}$

13. $\frac{2}{1}$ **15.** $\frac{2}{3}$ **17.** $\frac{1}{4}$ **19.** $\frac{7}{15}$ **21.** $\frac{2}{1}$ **23.** $\frac{3}{8}$

25. $\frac{3}{50}$ **27.** $\frac{4}{5}$ **29.** $\frac{3}{10}$ **31.** \$0.049 per ounce

33. \$0.073 per ounce **35.** a **37.** b **39.** a

41. 12 **43.** $\frac{10}{3}$ **45.** $\frac{175}{8}$ **47.** 16 **49.** $\frac{3}{16}$

51. $\frac{1}{2}$ **53.** 27 **55.** $\frac{14}{5}$ **57.** $\frac{2}{3}$ **59.** $\frac{26}{7}$ **61.** $\frac{20}{1}$

63. $\frac{3}{2}$ **65.** $\frac{100}{49}$ **67.** 16 gallons **69.** 250 blocks

71. \$2.10 **73.** 22,691 votes **75.** $46\frac{2}{3}$ minutes

77. 384 miles **79.** $12\frac{1}{2}$ pounds **81.** $\frac{5}{2}$

83. $6\frac{2}{3}$ feet **85.** 20% **87.** \$7983 **89.** \$0.94

91. No. It is also necessary to know either the number of men in the class or the number of women in the class.

93. Answers will vary. **95.** -122 **97.** -4

99. 8 **101.** 40.5 **103.** 93%

Mid-Chapter Quiz *(page 170)*

1. 6 **2.** 8 **3.** $\frac{19}{2}$ **4.** 0 **5.** $-\frac{1}{3}$ **6.** $\frac{35}{12}$

7. 36 **8.** $\frac{11}{5}$ **9.** 5 **10.** -2 **11.** 2.06

12. 51.23 **13.** 15.5 **14.** 42 **15.** 200%

16. 455 **17.** 10 hours

18. 6 square meters, 12 square meters, 24 square meters

19. 93 **20.** 16% **21.** 3 hours **22.** $\frac{225}{64}$

23. 26.25 gallons

Section 3.5 *(page 179)*

1. $h = \dfrac{2A}{b}$ **3.** $r = \dfrac{A - P}{Pt}$ **5.** $l = \dfrac{V}{wh}$

7. $C = \dfrac{S}{1 + R}$ **9.** $m_2 = \dfrac{Fr^2}{\alpha m_1}$ **11.** $b = \dfrac{2A - ah}{h}$

13. $a = \dfrac{2(h - v_0t)}{t^2}$ **15.** 100π cubic meters

17. 24 pounds per square inch **19.** \$49.59 **21.** 6%

23. 48 meters **25.** 16 hours **27.** $114.\overline{1}$ m/sec

29. 4 inches **31.** 2.5 meters **33.** (a) 15 feet (b) \$3240

35. 8.6% **37.** \$15,975 **39.** 1154 miles per hour

41. 0.17 hour **43.** 46 stamps at \$0.27, 54 stamps at \$0.42

45. Solution 2: 75 gallons; Final solution: 100 gallons

47. Solution 2: 5 quarts; Final solution: 10 quarts

49. $\frac{8}{7}$ gallons **51.** \$2000 at 7%, \$4000 at 9%

53. $1\frac{1}{5}$ hours **55.** Answers will vary.

57. 125 milliliters per hour

59. 16 dozen roses, 8 dozen carnations

61. (a)

Corn, x	Soybeans, $100 - x$	Price per ton of the mixture
0	100	\$200
20	80	\$185
40	60	\$170
60	40	\$155
80	20	\$140
100	0	\$125

(b) Decreases (c) Decreases
(d) Average of the two prices

63. $l = 13$ inches, $w = 5$ inches

65. 15 meters, 15 meters, 53 meters **67.** 28 miles

69. 3 hours on the first part, 1 hour and 15 minutes on the last part

71. Divide by 2 to obtain 90 miles per 2 hours. Divide by 2 again to obtain 45 miles per hour.

$$d = rt \Rightarrow \frac{d}{t} = r \Rightarrow \frac{180 \text{ miles}}{4 \text{ hours}} = 45 \text{ miles per hour}$$

73. Use $h = \dfrac{2A}{x + y}$ to find the height h of a trapezoid.

Use $x = \dfrac{2A}{h} - y$ to find the base x of a trapezoid.

Use $y = \dfrac{2A}{h} - x$ to find the base y of a trapezoid.

75. The circumference would double; the area would quadruple.
Circumference: $C = 2\pi r$, Area: $A = \pi r^2$
If r is doubled, $C = 2\pi(2r) = 2(2\pi r)$ and $A = \pi(2r)^2 = 4\pi r^2$.

77. (a) 7, 1 (b) 7, 1, -3 (c) 1.8, $\frac{1}{10}$, 7, -2.75, 1, -3 (d) None

79. (a) 9 (b) 9, -6 (c) -2.2, 9, $\frac{1}{3}$, $\frac{3}{5}$, -6 (d) $\sqrt{13}$

81. 9 **83.** 6 **85.** 16

Section 3.6 *(page 193)*

1. (a) Yes (b) No (c) Yes (d) No

3. (a) No (b) No (c) Yes (d) No

5. a **6.** e **7.** d **8.** b **9.** f **10.** c

11.

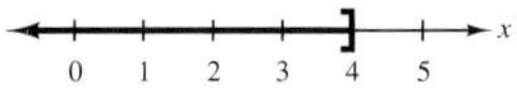

13.

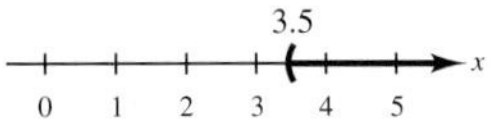

15.

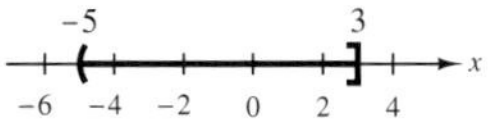

17.

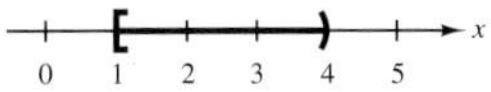

19.

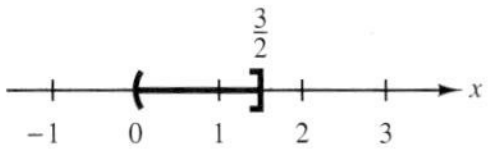

21.

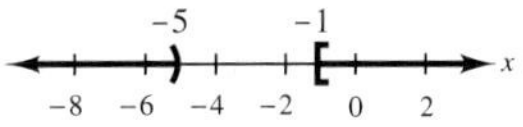

23.

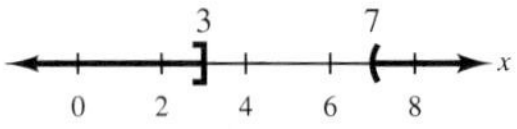

25. $-15 + x < -24$

27. Not equivalent **29.** Not equivalent

31. Equivalent **33.** Not equivalent

35. $x \geq 4$

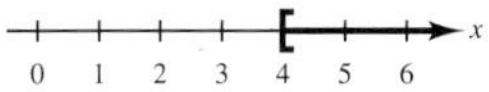

37. $x \leq 2$

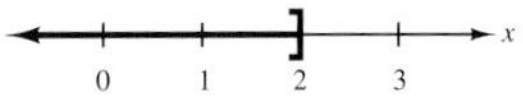

39. $x < 4$

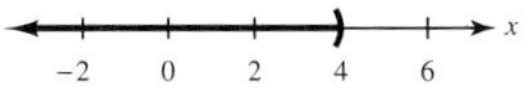

41. $x \leq -4$

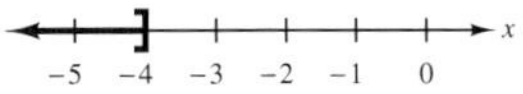

43. $x > 8$

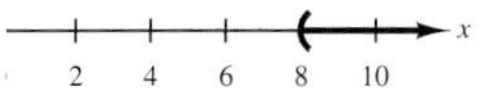

45. $x \geq 7$

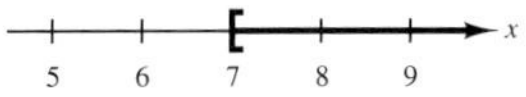

47. $x > 7.55$

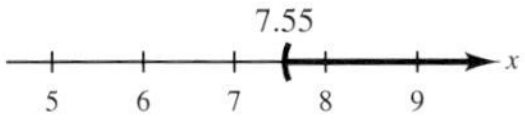

49. $x > -\frac{2}{3}$

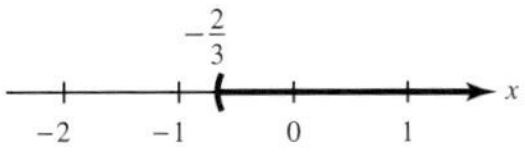

51. $x > \frac{9}{2}$

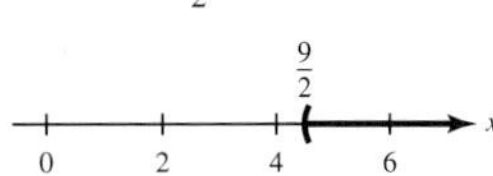

53. $x > \frac{20}{11}$

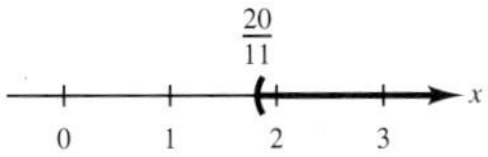

55. $x > \frac{8}{3}$

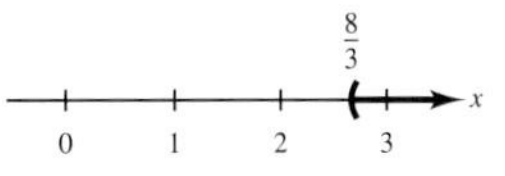

57. $x \leq -8$

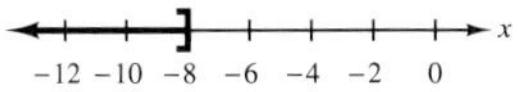

59. $x > -15$

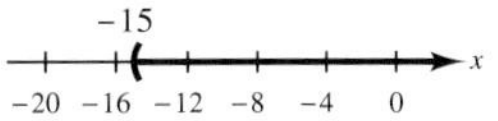

61. $\frac{5}{2} < x < 7$

63. $-3 \leq x < -1$

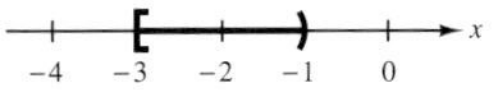

65. $-5 < x < 5$

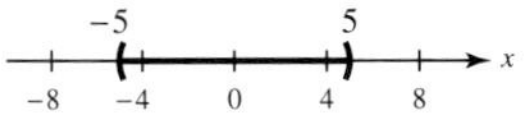

67. $-\frac{3}{2} < x < \frac{9}{2}$

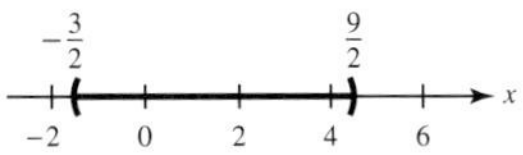

69. $1 < x < 10$

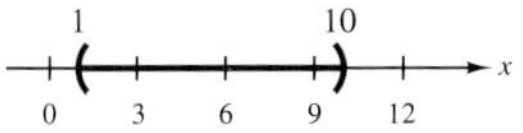

71. $-1 < x \leq 4$

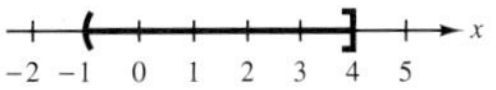

73. No solution

75. $-\infty < x < \infty$

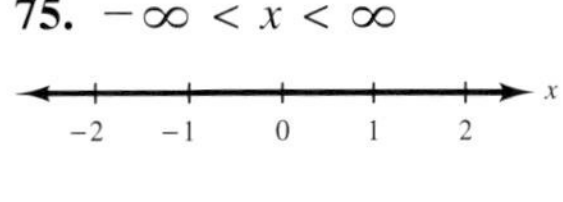

77. $x < -\frac{8}{3}$ or $x \geq \frac{5}{2}$

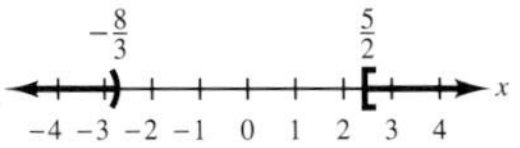

79. $y \leq -10$

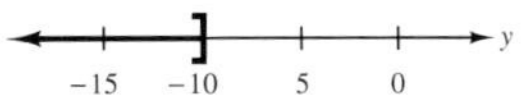

81. $-5 < x \leq 0$

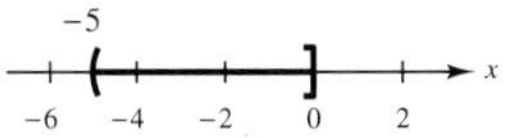

83. $x < -3$ or $x \geq 2$, $\{x|x < -3\} \cup \{x|x \geq 2\}$

85. $-5 \leq x < 4$, $\{x|x \geq -5\} \cap \{x|x < 4\}$

87. $x \leq -2.5$ or $x \geq -0.5$, $\{x|x \leq -2.5\} \cup \{x|x \geq -0.5\}$

89. $\{x|x \geq -7\} \cap \{x|x < 0\}$

91. $\left\{x|x > -\frac{9}{2}\right\} \cap \left\{x|x \leq -\frac{3}{2}\right\}$

93. $\{x|x < 0\} \cup \left\{x|x \geq \frac{2}{3}\right\}$

95. $x \geq 0$ **97.** $z \geq 8$ **99.** $10 \leq n \leq 16$

101. x is at least $\frac{5}{2}$. **103.** y is at least 3 and less than 5.

105. \$2600

107. The average temperature in Miami is greater than the average temperature in New York City.

109. 26,000 miles **111.** $x \geq 31$

113. $23 \leq x \leq 38$ **115.** $3 \leq n \leq \frac{15}{2}$

117. $12.50 < 8 + 0.75n$; $n > 6$ **119.** 1999, 2000, 2001

121. The multiplication and division properties differ. The inequality symbol is reversed if both sides of an inequality are multiplied or divided by a negative real number.

123. The solution set of a linear inequality is bounded if the solution is written as a double inequality. Otherwise, the solution set is unbounded.

125. $a < x < b$; A double inequality is always bounded.

127. $x > a$ or $x < b$; $x > a$ includes all real numbers between a and b and greater than or equal to b, all the way to ∞. $x < b$ adds all real numbers less than or equal to a, all the way to $-\infty$.

129. $<$ **131.** $=$ **133.** No; Yes

135. Yes; No **137.** $\frac{17}{2}$ **139.** $-\frac{1}{4}$

Section 3.7 *(page 203)*

1. Not a solution **3.** Solution

5. $x - 10 = 17$; $x - 10 = -17$

7. $4x + 1 = \frac{1}{2}$; $4x + 1 = -\frac{1}{2}$ **9.** $|3x| = 1$

11. $|2x| = 2$ **13.** $4, -4$ **15.** No solution

17. 0 **19.** $3, -3$ **21.** $4, -6$ **23.** $11, -14$

25. $\frac{4}{3}$ **27.** $\frac{17}{5}, -\frac{11}{5}$ **29.** 2 **31.** $\frac{15}{4}, -\frac{1}{4}$

33. 2, 3 **35.** $7, -3$ **37.** $\frac{1}{3}$ **39.** $\frac{1}{2}$

41. $|x - 4| = 9$ **43.** Solution

45. Not a solution **47.** $-3 < y + 5 < 3$

49. $7 - 2h \ge 9$ or $7 - 2h \le -9$ **51.** $-4 < y < 4$

53. $x \le -6$ or $x \ge 6$ **55.** $-7 < x < 7$

57. $-1 \le y \le 1$ **59.** $x < -16$ or $x > 4$

61. $-3 \le x \le 4$ **63.** No solution

65. $-104 < y < 136$ **67.** $-5 < x < 35$

69. $-\infty < x < \infty$

71. $-2 < x < \frac{2}{3}$

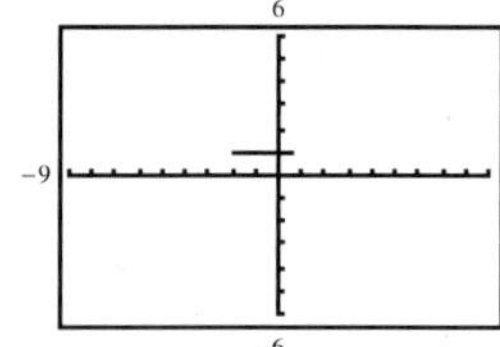

73. $x < -6$ or $x > 3$

75. $3 \le x \le 7$

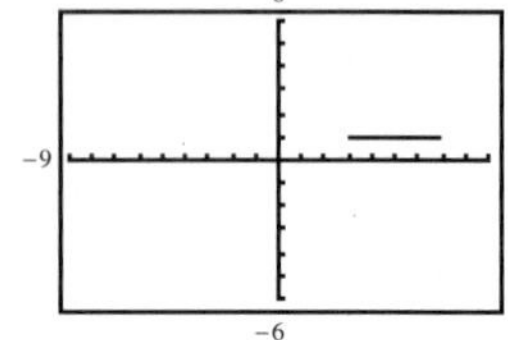

77. d **78.** c **79.** b **80.** a **81.** $|x| \le 2$

83. $|x - 19| > 2$ **85.** $|x| < 3$ **87.** $|2x - 3| > 5$

89.

Fastest: 41.826 seconds; Slowest: 42.65 seconds

91. (a) $|s - x| \le \frac{3}{16}$ (b) $4\frac{15}{16} \le x \le 5\frac{5}{16}$

93. All real numbers that are within one unit of 4

95. $|2x - 6| \le 6$ **97.** $4(n + 3)$

99. Selling price: \$112, Markup: \$32

101. $x > 20$ **103.** $x \ge 4$

Review Exercises *(page 208)*

1. 5 **3.** -4 **5.** $\frac{3}{5}$ **7.** 3 **9.** 3 **11.** 5

13. 4 **15.** 3 **17.** $\frac{4}{3}$ **19.** 20 **21.** 12 units

23. 80 meters $\times$ 50 meters **25.** 20 **27.** 6 **29.** 1

31. 7 **33.** $\frac{19}{3}$ **35.** 20 **37.** 3 **39.** $-\frac{1}{7}$

41. 23.26 **43.** 224.31 **45.** 3 hours

	Percent	*Parts out of 100*	*Decimal*	*Fraction*
47.	60%	60	0.60	$\frac{3}{5}$
49.	80%	80	0.80	$\frac{4}{5}$
51.	20%	20	0.20	$\frac{1}{5}$
53.	55%	55	0.55	$\frac{11}{20}$

55. 20 **57.** 400 **59.** 60% **61.** \$85.44

63. (a) 8.5% (b) 58.3% **65.** $\frac{1}{8}$ **67.** $\frac{4}{5}$

69. 24-ounce container **71.** $\frac{7}{2}$ **73.** $-\frac{10}{3}$ **75.** 9

77. \$133.33 **79.** \$151

81. $x = zs + m$

	Distance, d	*Rate, r*	*Time, t*
83.	520 mi	65 mi/hr	8 hr
85.	855 m	5 m/min	171 min
87.	3000 mi	60 mi/hr	50 hr

89. About 1108.3 miles **91.** 30 feet $\times$ 26 feet

93. \$475 **95.** 13 dimes, 17 quarters

97. $\frac{30}{11} \approx 2.7$ hours

99.

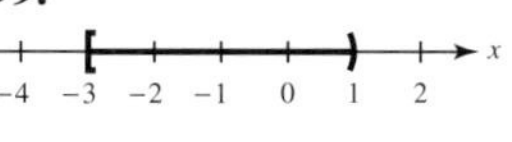

101.

103. $x > 4$

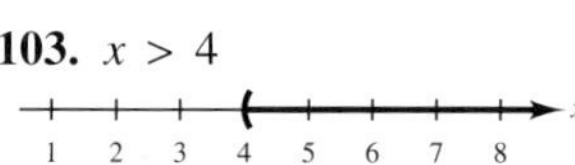

105. $x \le 6$

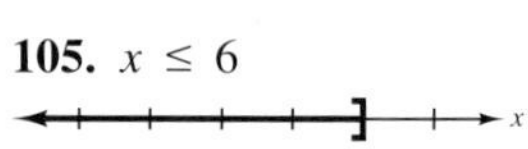

107. $y > -\frac{70}{3}$

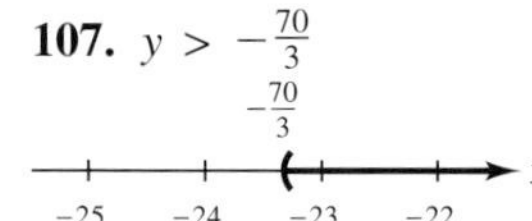

109. $-7 \le x < -2$

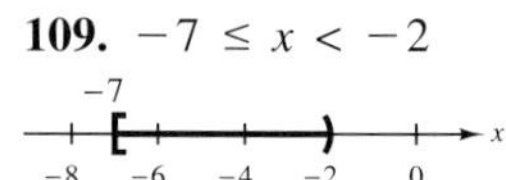

111. $-3 < x < 2$

113. At least \$800 **115.** $4, -\frac{4}{3}$ **117.** $0, -\frac{8}{5}$

119. $\frac{1}{2}, 3$ **121.** $x < 1$ or $x > 7$ **123.** $-4 < x < 4$

125. $b < -9$ or $b > 5$

127. $x \le 1$ or $x \ge 5$ **129.** $|x - 3| < 2$

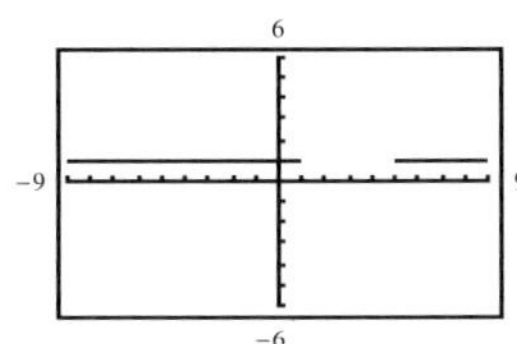

131.

Maximum: 116.6 degrees Fahrenheit

Minimum: 40 degrees Fahrenheit

Chapter Test *(page 212)*

1. -13 **2.** $\frac{21}{4}$ **3.** 7 **4.** 1 **5.** $-\frac{1}{3}$

6. -6 **7.** $5, -11$ **8.** $\frac{2}{3}, -\frac{4}{3}$ **9.** 11.03

10. $2\frac{1}{2}$ hours **11.** $31\frac{1}{4}\%$, 0.3125 **12.** 1200

13. 36% **14.** $\frac{5}{9}$; 2 yards = 6 feet = 72 inches

15. $\frac{12}{7}$ **16.** 5 **17.** 66 miles per hour

18. $\frac{36}{7} \approx 5.1$ hours **19.** $b = \dfrac{a}{p + 1}$ **20.** \$6250

21. $t \ge 8$ **22.** 25,000 miles

23. $x \ge 5$

24. $x > 1$

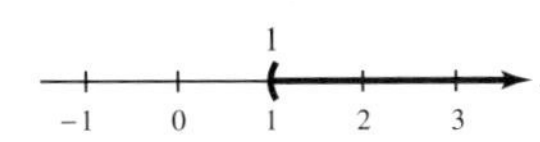

25. $-7 < x \le 1$

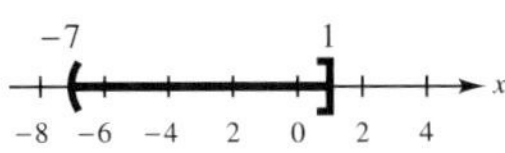

26. $-1 \le x < \frac{5}{4}$

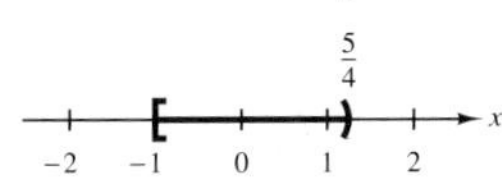

27. $1 \le x \le 5$

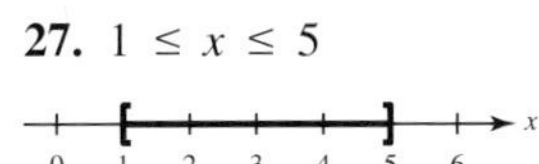

28. $x < -\frac{9}{5}$ or $x > 3$

Cumulative Test: Chapters 1–3 *(page 213)*

1. < **2.** 1200 **3.** $-\frac{11}{24}$ **4.** $-\frac{25}{12}$ **5.** 8

6. 14 **7.** 28 **8.** -30 **9.** $-\frac{11}{2}$

10. $3^3(x + y)^2$ **11.** $-2x^2 + 6x$

12. Associative Property of Addition

13. $15x^7$ **14.** $7x^2 - 6x - 2$ **15.** $-3x^2 + 18x$

16. 6 **17.** $\frac{52}{3}$ **18.** -7

19. $x \ge -7$

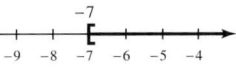

20. $\dfrac{15{,}000 \text{ miles}}{1 \text{ year}} \cdot \dfrac{1 \text{ gallon}}{30 \text{ miles}} \cdot \dfrac{\$3.00}{1 \text{ gallon}} \approx \$1500$ per year

21. $\frac{3}{4}$ **22.** \$920 **23.** \$57,000

CHAPTER 4

Section 4.1 *(page 223)*

1.

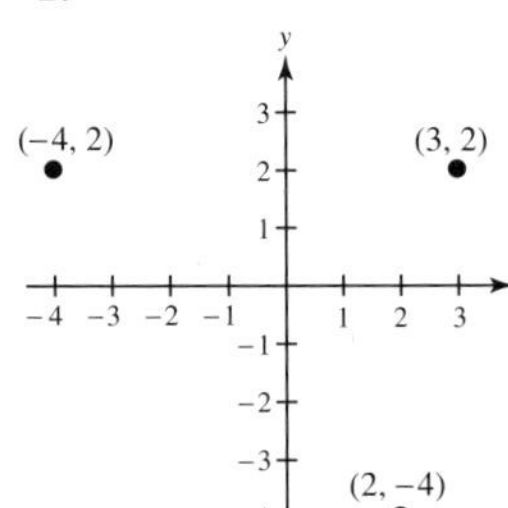

3.

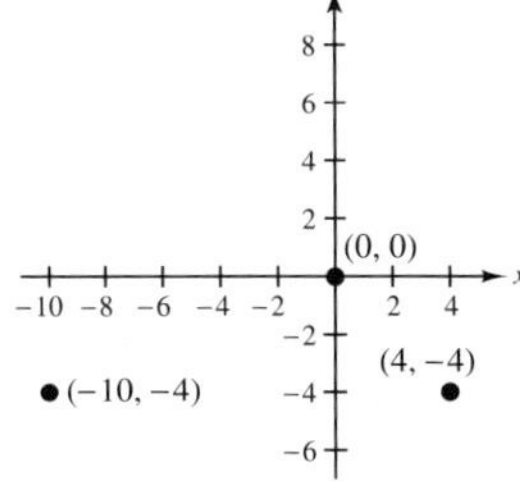

5.

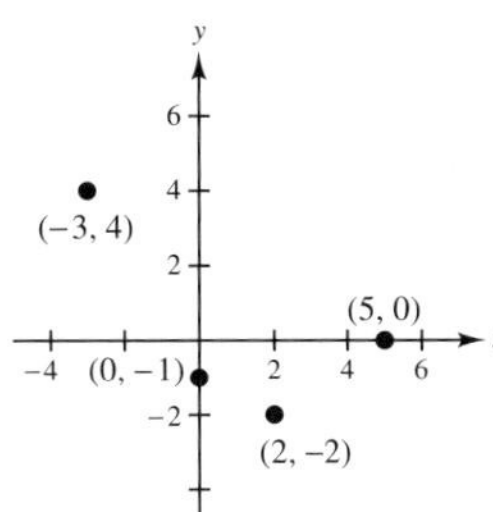

7.

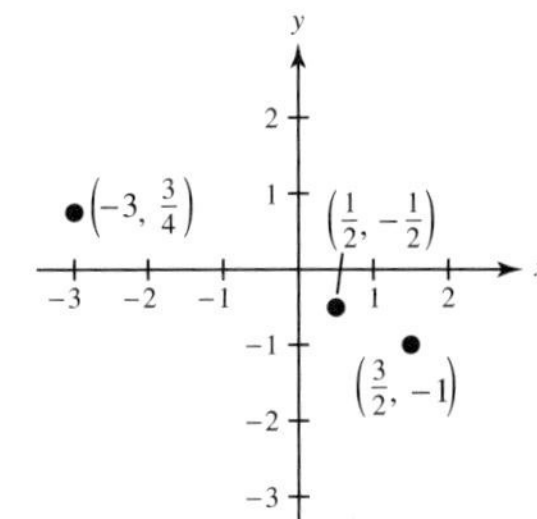

9.

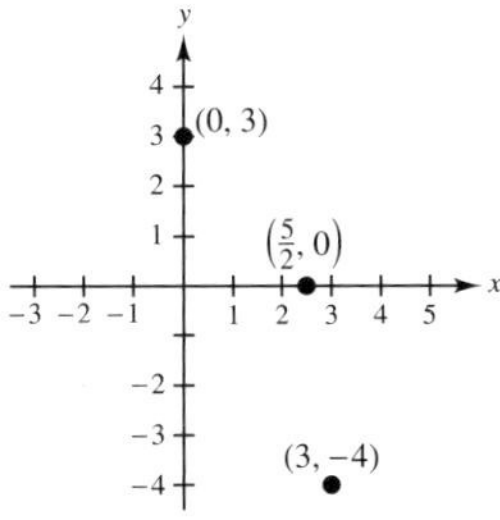

11. A: (5, 2), B: (−3, 4), C: (2, −5), D: (−2, −2)

13. A: (−1, 3), B: (5, 0), C: (2, 1), D: (−1, −2)

15. Quadrant II **17.** Quadrant III **19.** Quadrant III

21. Quadrant II or III **23.** Quadrant III or IV

25. Quadrant II or IV

27.

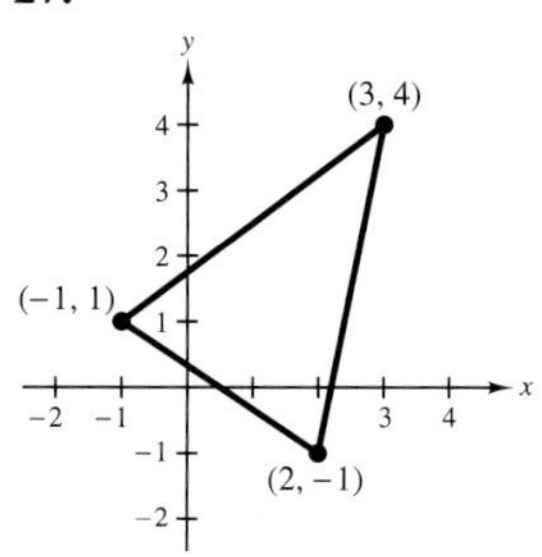

29.

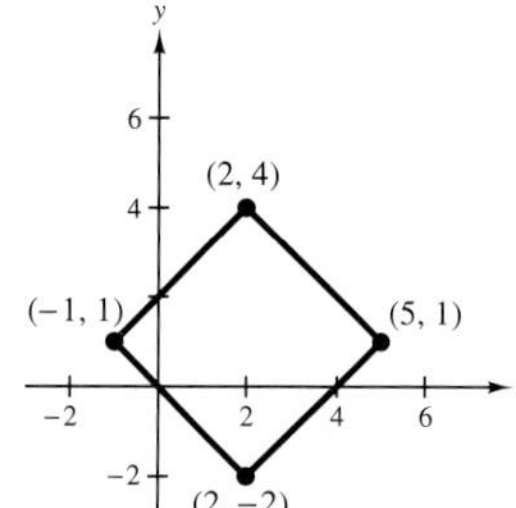

31.

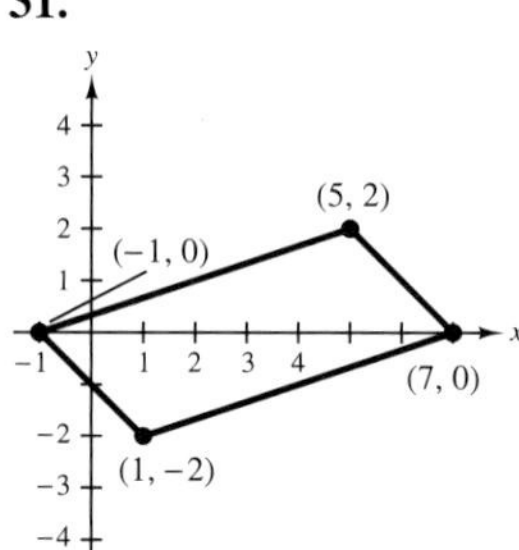

33.

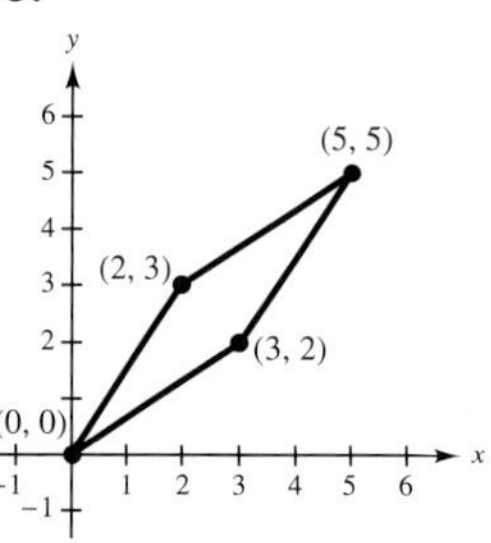

35.

x	-2	0	2	4	6
$y = 3x - 4$	-10	-4	2	8	14

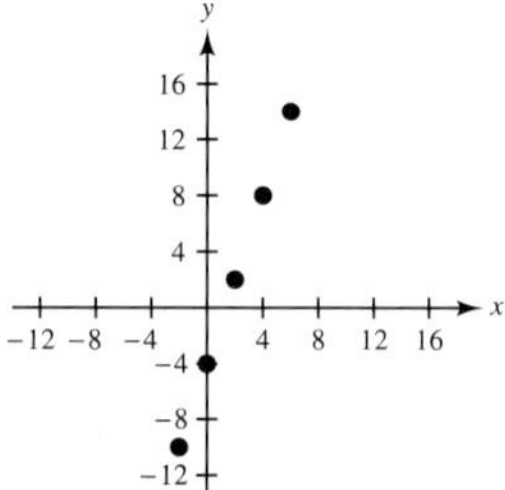

37.

x	-2	0	4	6	8
$y = -\frac{3}{2}x + 5$	8	5	-1	-4	-7

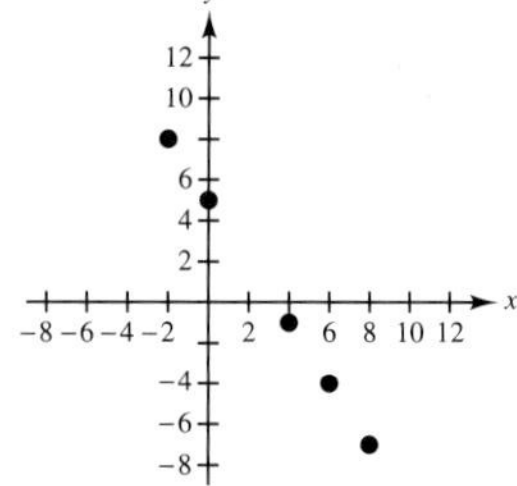

39.

x	-2	-1	0	1	2
$y = -4x - 5$	3	-1	-5	-9	-13

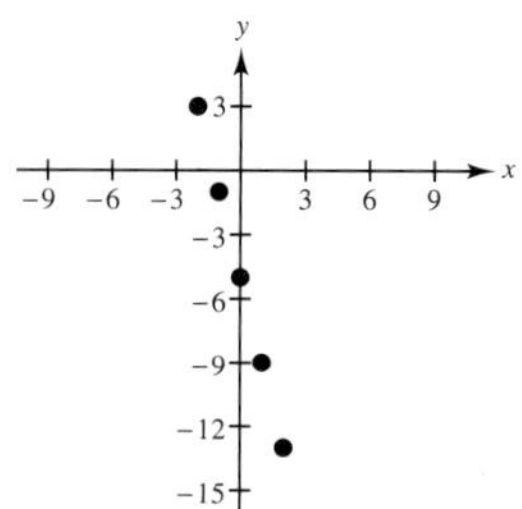

41. $y = -7x + 8$ **43.** $y = 10x - 2$

45. $y = 2x - 1$ **47.** $y = -\frac{1}{4}x + 2$ **49.** $y = \frac{4}{5}x - \frac{3}{5}$

51. (a) Solution (b) Not a solution (c) Not a solution (d) Solution

53. (a) Solution (b) Solution (c) Not a solution (d) Solution

55. (a) Not a solution (b) Solution (c) Solution (d) Not a solution

57. (a) Solution (b) Not a solution (c) Not a solution (d) Not a solution

59. (a) $\left(-\frac{4}{3}, 0\right)$ (b) $(4, 16)$ (c) $(-2, -2)$

61.

x	20	40	60	80	100
$y = 0.066x$	1.32	2.64	3.96	5.28	6.60

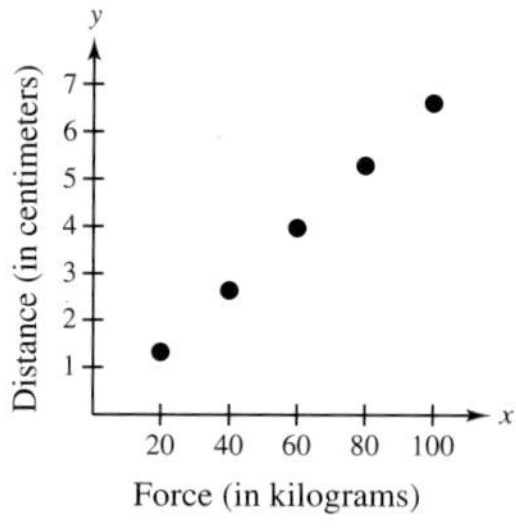

63. $y = 25x + 5000$

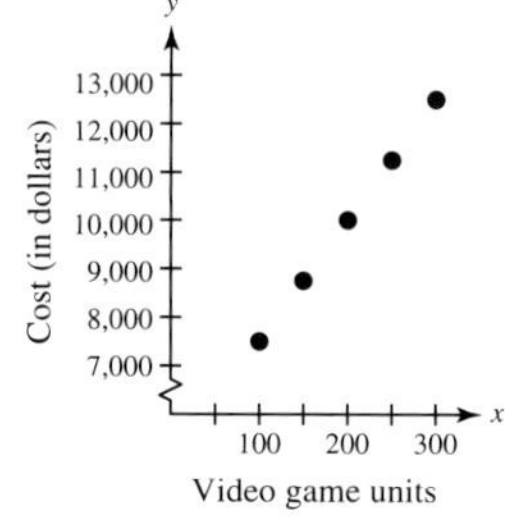

65. (a)

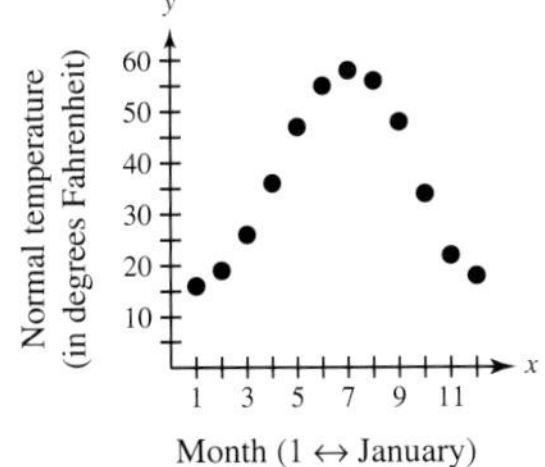

No, because there are only 12 months, but the temperature ranges from 16°F to 58°F.

(b) August

67. (a)

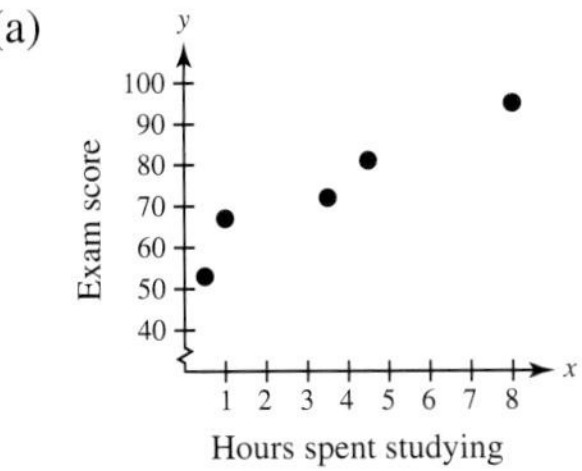

(b) Scores increase with increased study time.

69. \$1.20 **71.** \$0.30 **73.** \$30,750 **75.** 6%

77. 6.5% **79.** 65°F

81. (a) and (b)

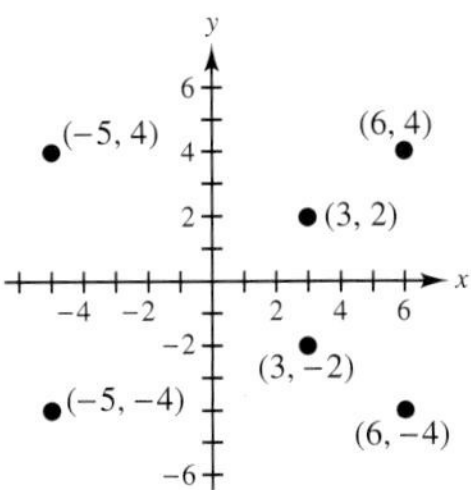

(c) Reflection in the x-axis

83. (6, 4)

85. No. The scales are determined by the magnitudes of the quantities being measured by x and y. If y is measuring revenue for a product and x is measuring time in years, the scale on the y-axis may be in units of \$100,000 and the scale on the x-axis may be in units of 1 year.

87. -10 **89.** 14 **91.** 6 **93.** 144 **95.** $x > -1$

97. $x < 4$

Section 4.2 *(page 234)*

1. g **2.** b **3.** a **4.** e **5.** h **6.** c

7. d **8.** f

9.

x	-2	-1	0	1	2
y	11	10	9	8	7

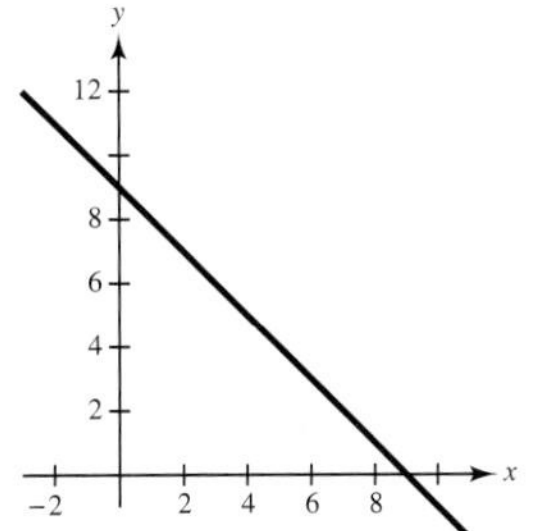

11.

x	-2	0	2	4	6
y	3	2	1	0	-1

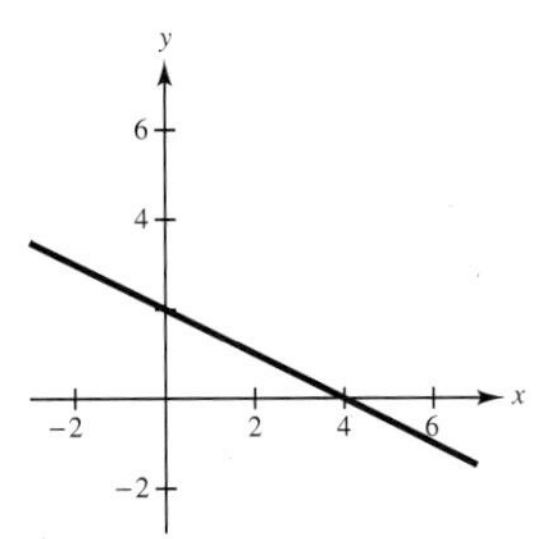

13.

x	-2	-1	0	1	2
y	7	4	3	4	7

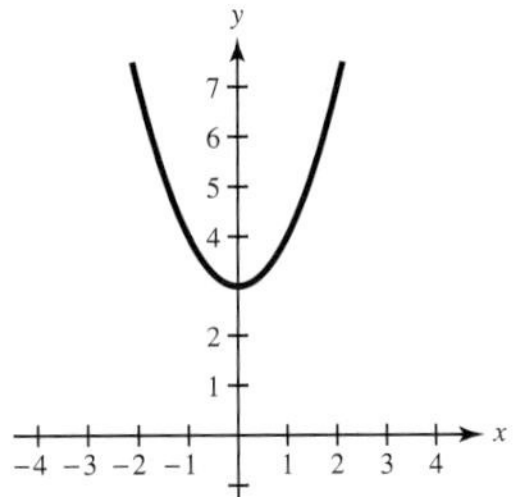

15.

x	-3	-2	-1	0	1
y	2	1	0	1	2

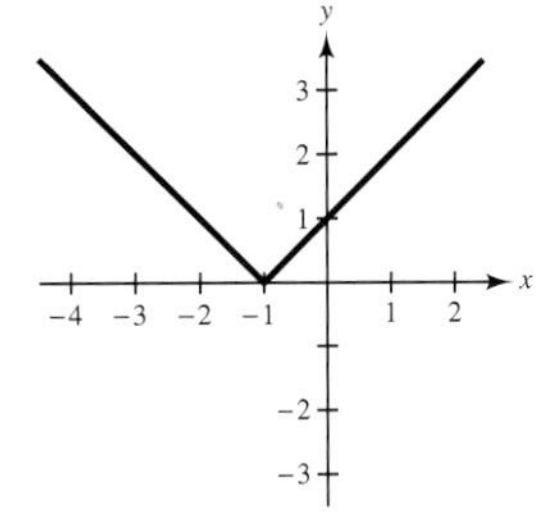

17. $(-2, 0), (0, 4)$ **19.** $(6, 0), (0, 2)$

21. $(-3, 0), (3, 0), (0, -3)$ **23.** $(-4, 0), (4, 0), (0, 16)$

25. $\left(\frac{7}{2}, 0\right), (0, 7)$ **27.** $(2, 0), (0, -1)$

29. $(1, 0), (0, -1)$ **31.** $(-1, 0), (0, -2)$

33. $\left(\frac{9}{2}, 0\right), \left(0, \frac{3}{2}\right)$ **35.** $(4, 0), (0, -6)$

37.

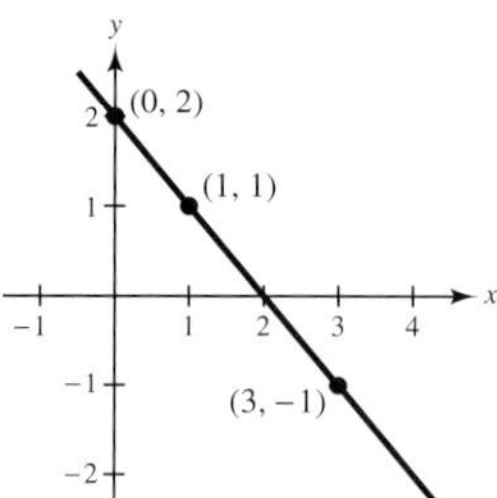

39.

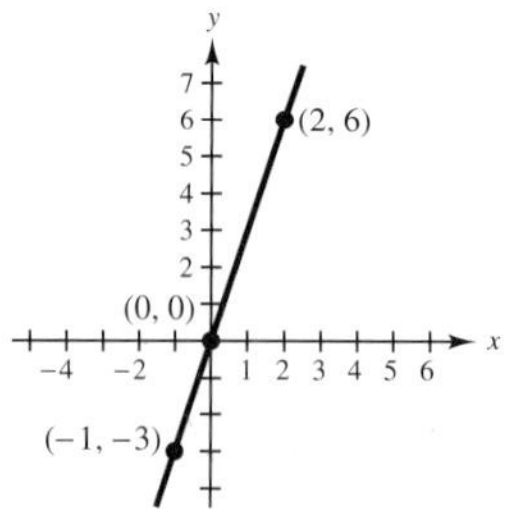

41.

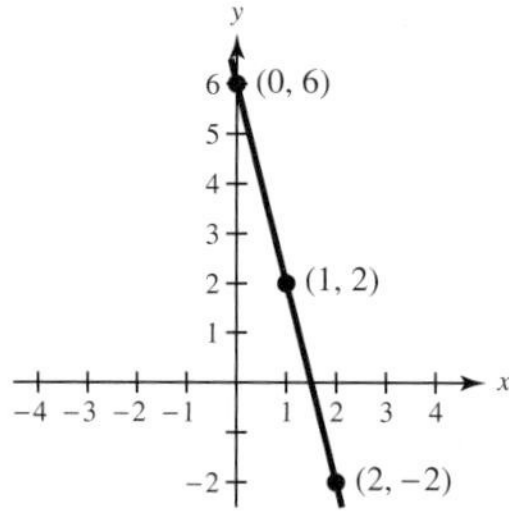

43.

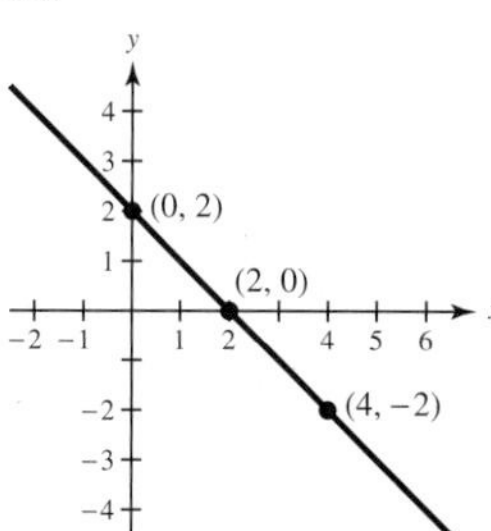

45.

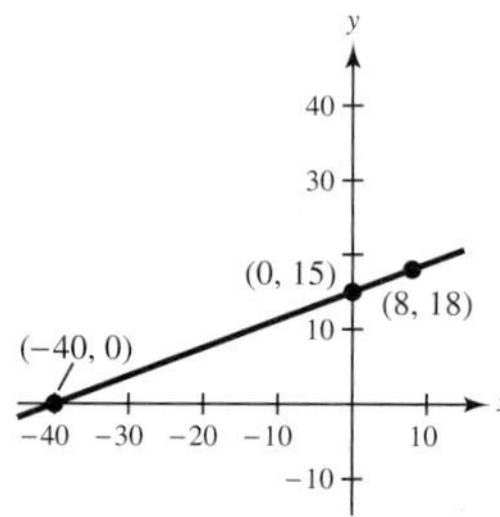

47.

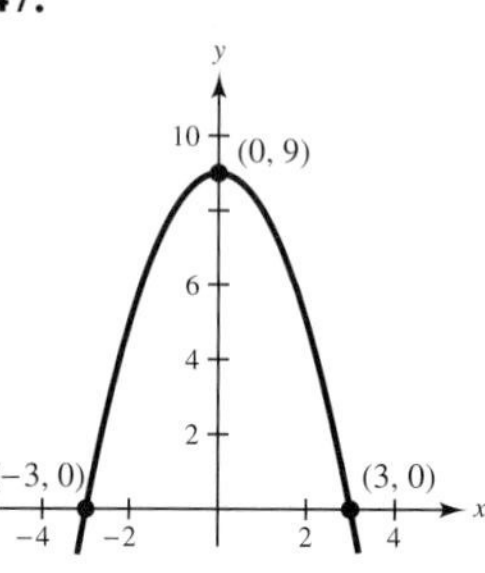

49.

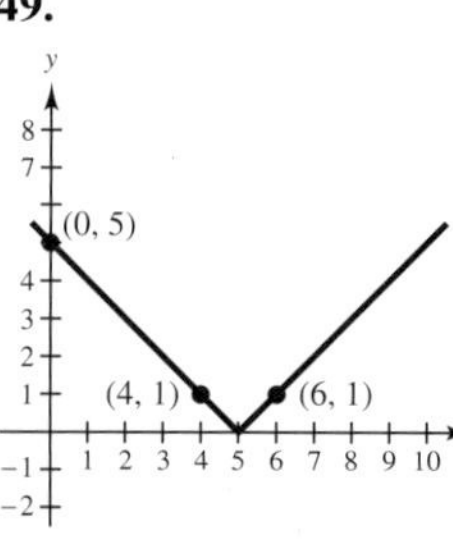

51.

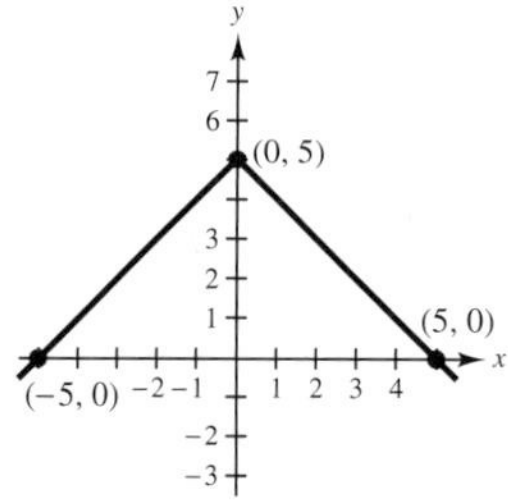

53.

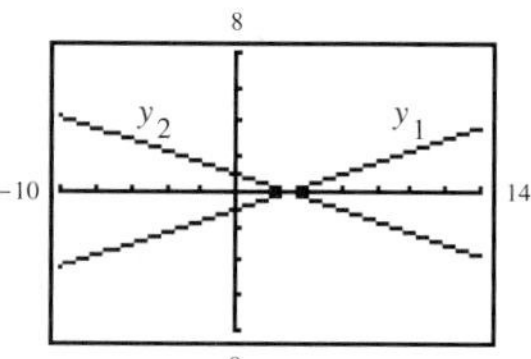

No

55.

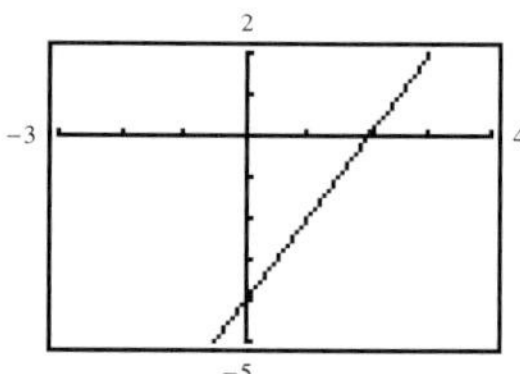

Yes; Distributive Property

57. No

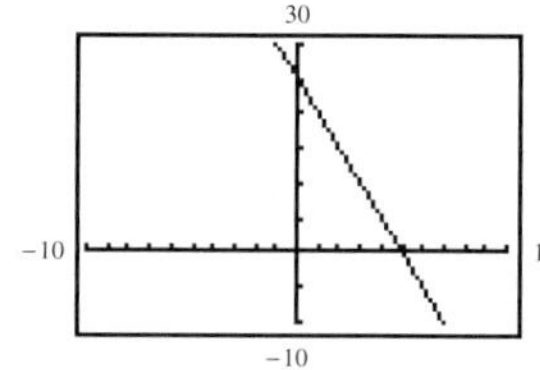

59. Yes

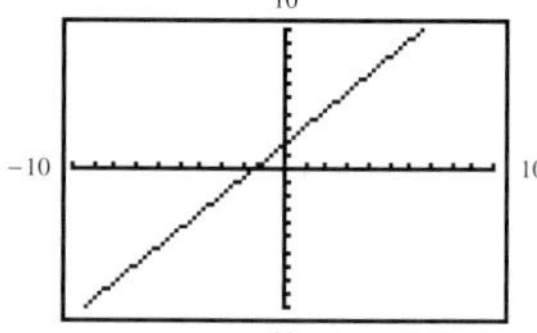

Xmin = -10
Xmax = 10
Xscl = 1
Ymin = -10
Ymax = 30
Yscl = 5

61. No

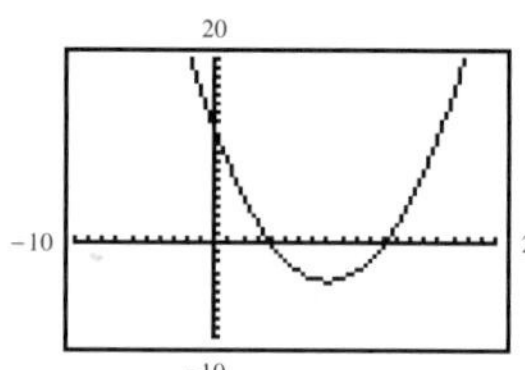

63. Yes

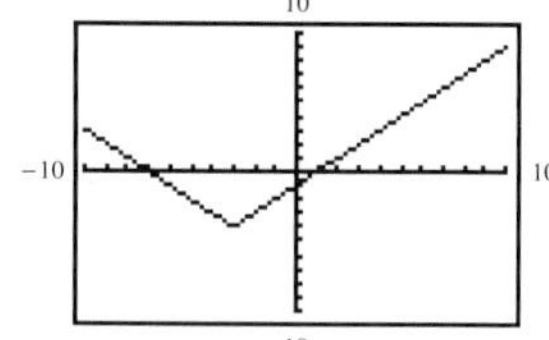

Xmin = -10
Xmax = 20
Xscl = 1
Ymin = -10
Ymax = 20
Yscl = 1

65. $y = 35t$

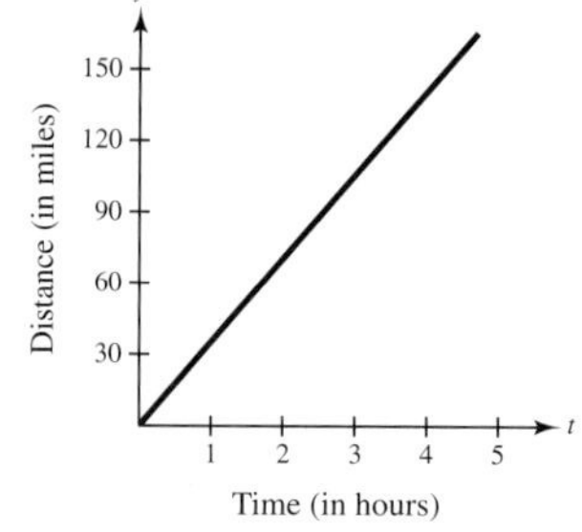

67. (a) $y = 1120 - 80x$

(b)

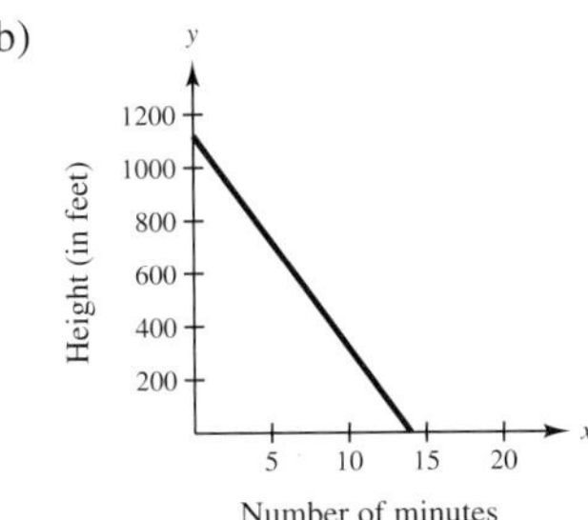

(c) (0, 1120); the initial height of the hot-air balloon

69. (a) (b) 77.4 years

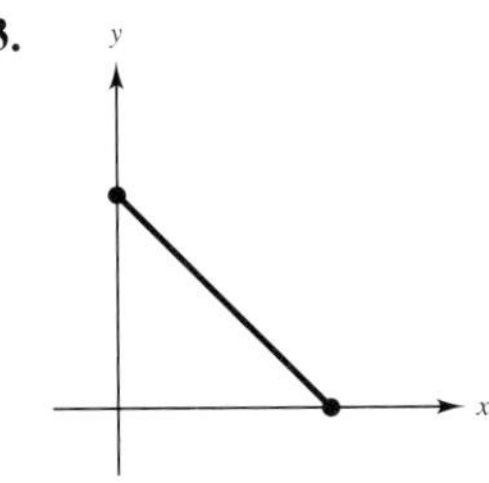

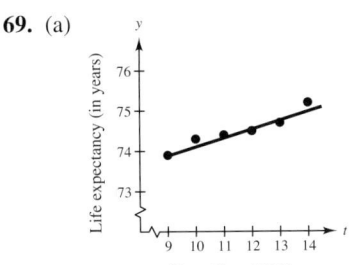

71. Yes. For any linear equation in two variables, x and y, there is a resulting value for y when $x = 0$. The corresponding point $(0, y)$ is the y-intercept of the graph of the equation.

73.

The distance between you and the tree decreases as you move from left to right on the graph. The x-intercept represents the number of seconds it takes you to reach the tree.

75. -13 **77.** 13 **79.** $-\frac{1}{6}$ **81.** $\frac{3}{10}$ **83.** $-\frac{36}{5}$

85. (a) Not a solution (b) Not a solution (c) Solution (d) Solution

87. (a) Not a solution (b) Not a solution (c) Not a solution (d) Not a solution

Section 4.3 *(page 243)*

1. Domain: $\{-4, 1, 2, 4\}$; Range: $\{-3, 2, 3, 5\}$

3. Domain: $\left\{-9, \frac{1}{2}, 2\right\}$; Range: $\{-10, 0, 16\}$

5. Domain: $\{-1, 1, 5, 8\}$; Range: $\{-7, -2, 3, 4\}$

7. Function **9.** Not a function **11.** Function

13. Not a function **15.** Not a function

17. Function **19.** Function **21.** Not a function

23. Function **25.** Function **27.** Function

29. Function **31.** Function **33.** Not a function

35. Not a function **37.** (a) 1 (b) $\frac{5}{2}$ (c) -2 (d) $-\frac{1}{3}$

39. (a) -1 (b) 5 (c) -7 (d) -2

41. (a) 49 (b) -4 (c) 1 (d) $-\frac{13}{8}$

43. (a) 8 (b) 8 (c) 0 (d) 2

45. (a) 6 (b) 6 (c) 66 (d) 11

47. (a) 1 (b) 15 (c) 0 (d) 0

49. (a) 4 (b) 0 (c) 12 (d) $\frac{1}{2}$

51. (a) -1 (b) 0 (c) 26 (d) $-\frac{7}{8}$

53. $D = \{0, 1, 2, 3, 4\}$ **55.** $D = \{-8, -6, 2, 5, 12\}$

57. $D = \{-5, -4, -3, -2, -1\}$

59. The set of all real numbers r such that $r > 0$

61. (a) $f(10) = 15$, $f(15) = 12.5$ (b) Demand decreases.

63. (a) 100 miles (b) 200 miles (c) 500 miles

65. High school enrollment is a function of the year.

67. $f(2001) = 15{,}000{,}000$

69. $P = 4s$; P is a function of s. If you make a table of values where $s > 0$, no first component has two different second components.

71. (a) Yes, L is a function of t. (b) $9 \le L \le 15$

73. Yes

Domain → *Range*
1 → 4
1 → 5
2 → 5
3 → 6

75. Yes

Domain → *Range*
1 → 6
2 → 7
3 → 8
4 → 9
5 → 9

77. $x < 0$ **79.** $85 \le z \le 100$ **81.** $-8, 8$

83. $-6, 6$ **85.** $-9, 1$ **87.** No solution

Mid-Chapter Quiz *(page 248)*

1.

2. Quadrant I or IV, or on the x-axis

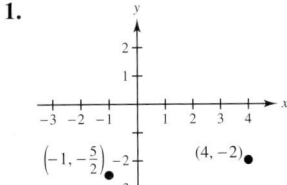

3. (a) Solution (b) Not a solution (c) Solution (d) Not a solution

4. 2000: $145 billion
2001: $165 billion
2002: $185 billion
2003: $205 billion
2004: $220 billion
2005: $230 billion

5. $(12, 0), (0, -4)$ **6.** $\left(\frac{2}{7}, 0\right), (0, 2)$

7.

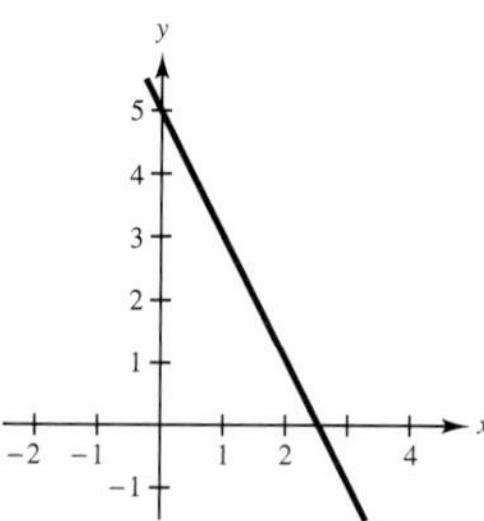

8.

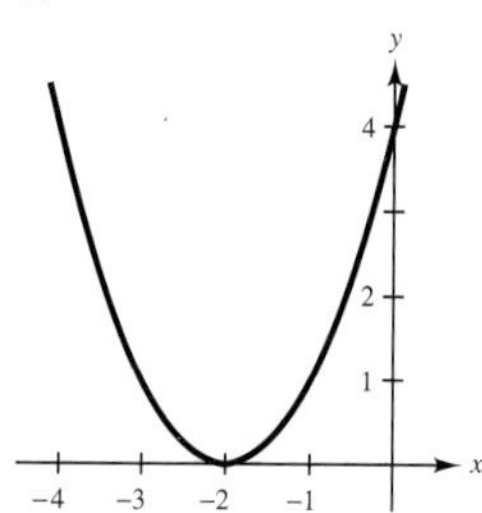

9.

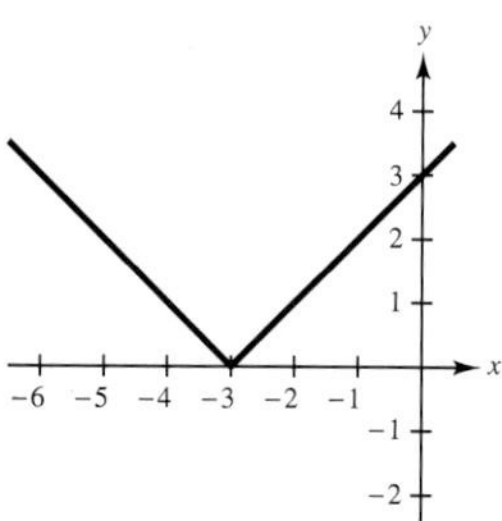

10. Domain: $\{1, 2, 3\}$; Range: $\{0, 4, 6, 10, 14\}$

11. Domain: $\{-3, -2, -1, 0\}$; Range: $\{6\}$

12. Not a function **13.** (a) 2 (b) -7

14. (a) 3 (b) -60 **15.** Domain: $\{10, 15, 20, 25\}$

16.

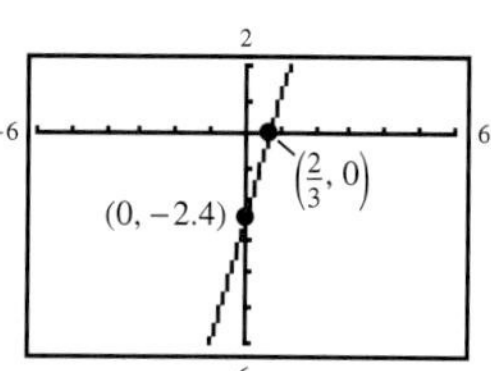

Substitute the coordinates for the respective variables in the equation and determine if the equation is true.

17. (a) $y = 2000 - 500t$

(b)

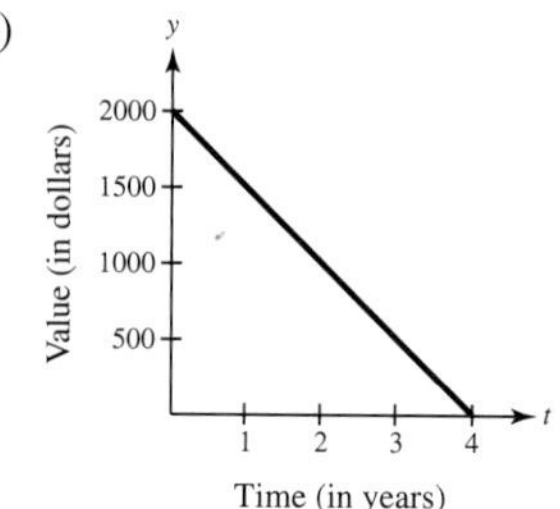

(c) $(0, 2000)$; the original value of the computer system

Section 4.4 *(page 258)*

1. 1 **3.** 0 **5.** $-\frac{1}{3}$

7. (a) L_2 (b) L_3 (c) L_4 (d) L_1

9.

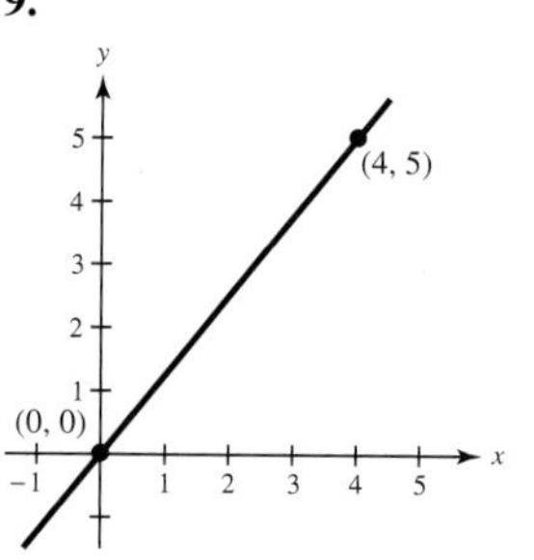

$m = \frac{5}{4}$; The line rises.

11.

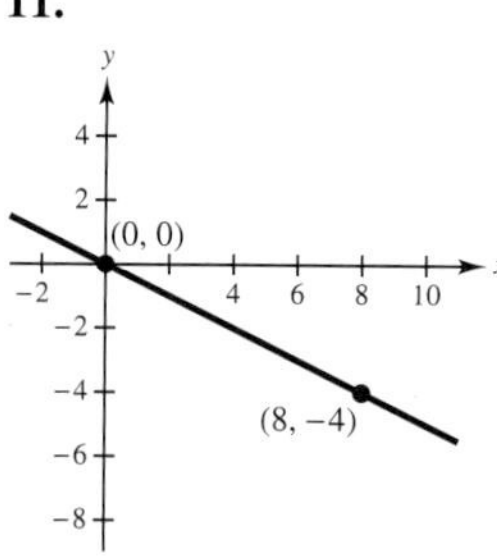

$m = -\frac{1}{2}$; The line falls.

13.

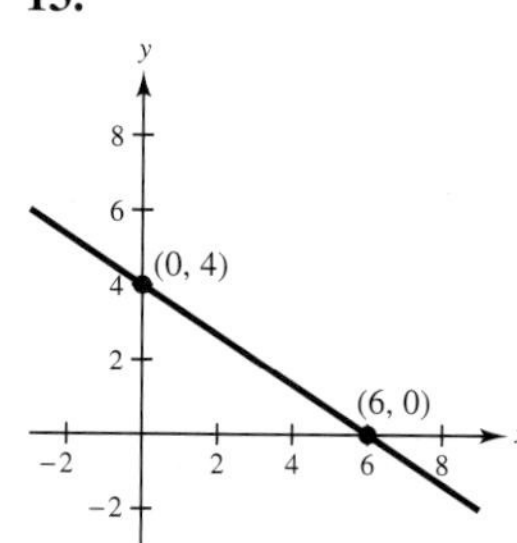

$m = -\frac{2}{3}$; The line falls.

15.

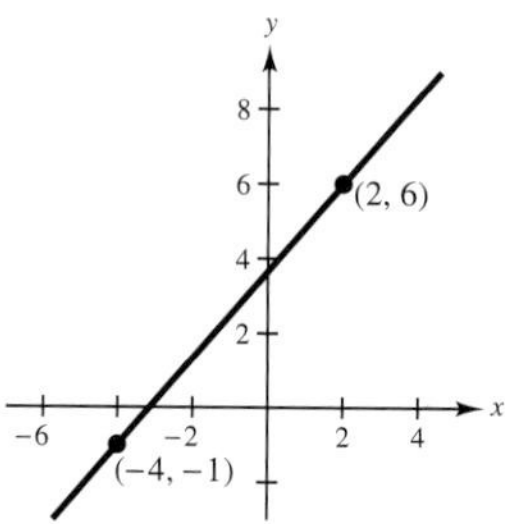

$m = \frac{7}{6}$; The line rises.

17.

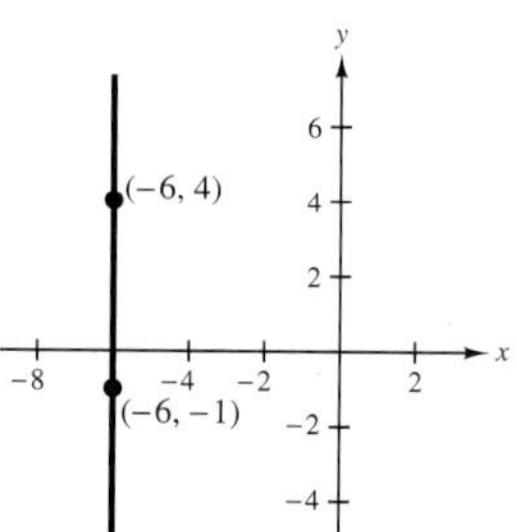

m is undefined.
The line is vertical.

19.

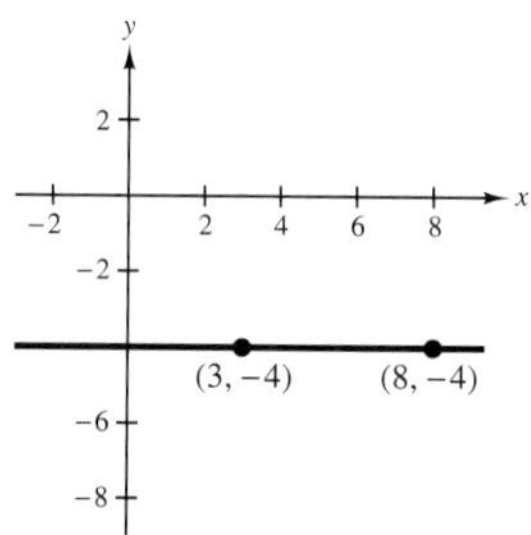

$m = 0$
The line is horizontal.

21.

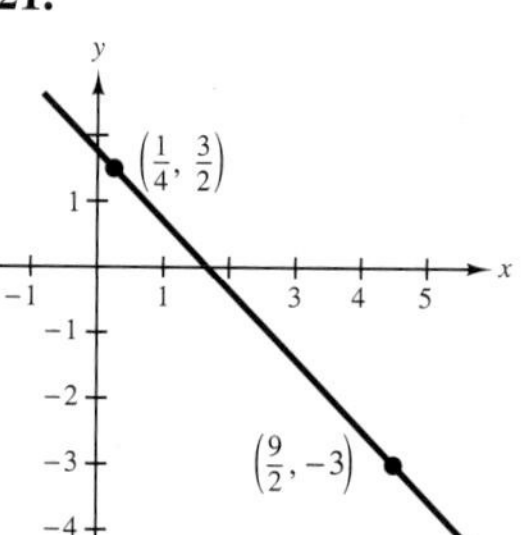

$m = -\frac{18}{17}$; The line falls.

23.

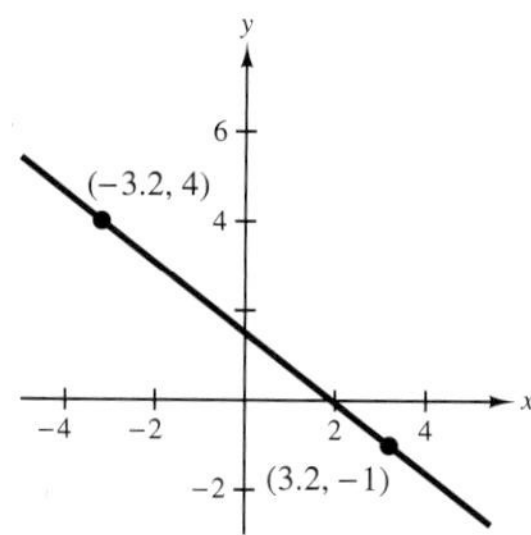

$m = -\frac{25}{32}$; The line falls.

25.

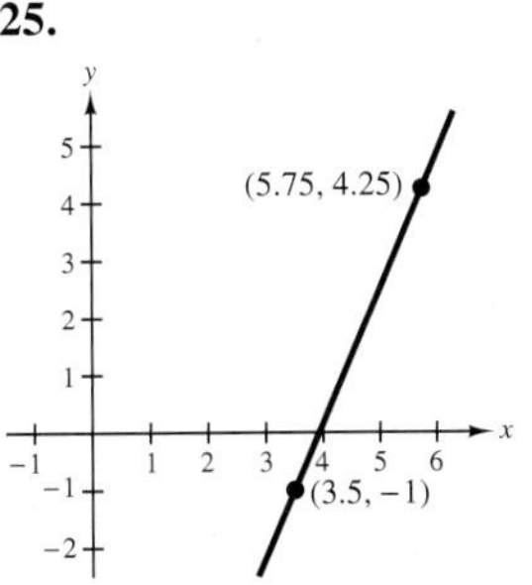

$m = \frac{7}{3}$; The line rises.

27.

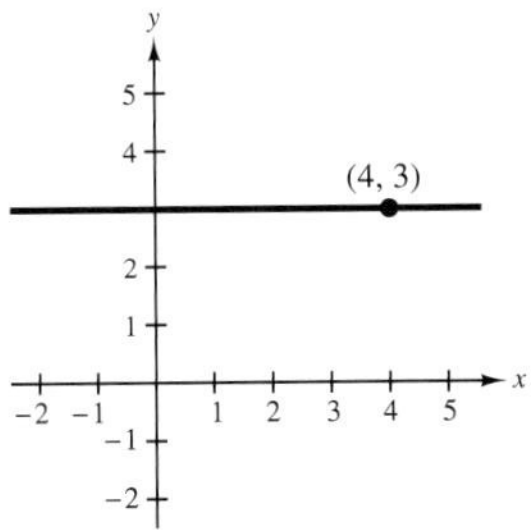

$m = 0$; The line is horizontal.

29. $m = -2$

x	-2	0	2	4
y	2	-2	-6	-10
Solution point	$(-2, 2)$	$(0, -2)$	$(2, -6)$	$(4, -10)$

31. $y = 15$ **33.** $y = -\frac{43}{2}$

35.

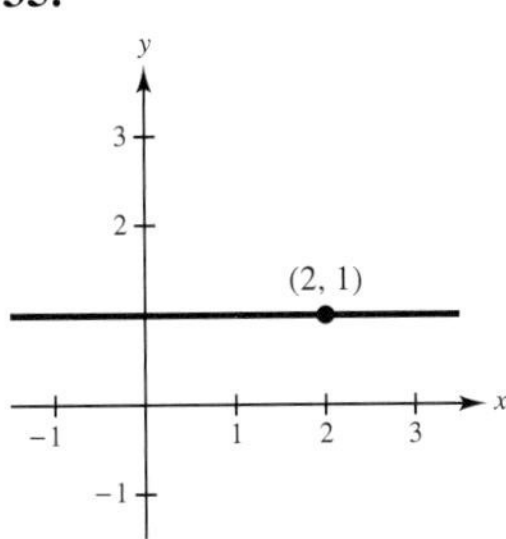

$(0, 1), (1, 1)$

37.

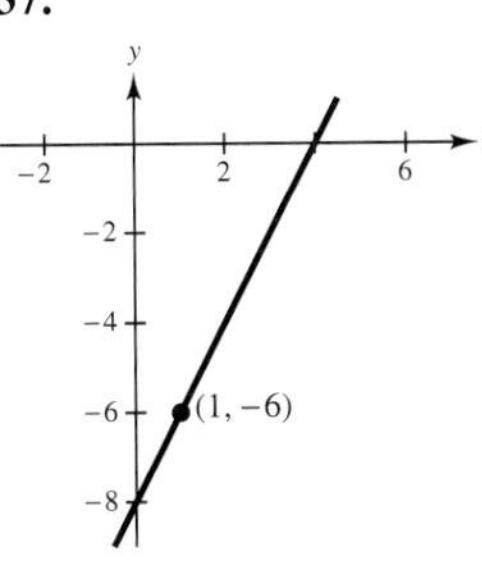

$(2, -4), (3, -2)$

39.

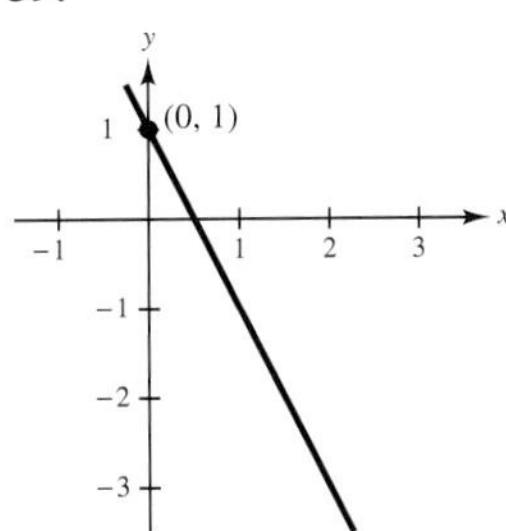

$(1, -1), (2, -3)$

41.

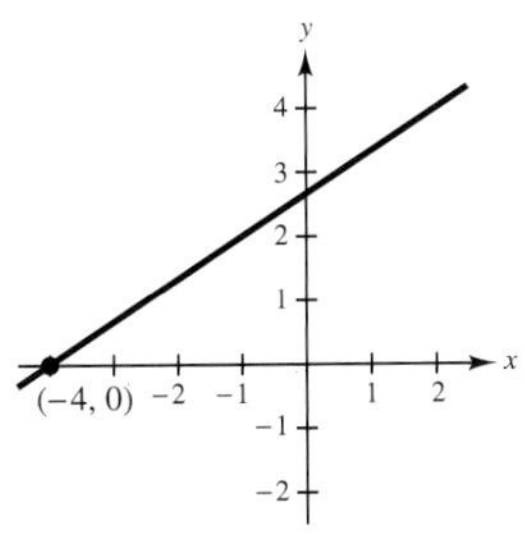

$(-1, 2), (2, 4)$

43.

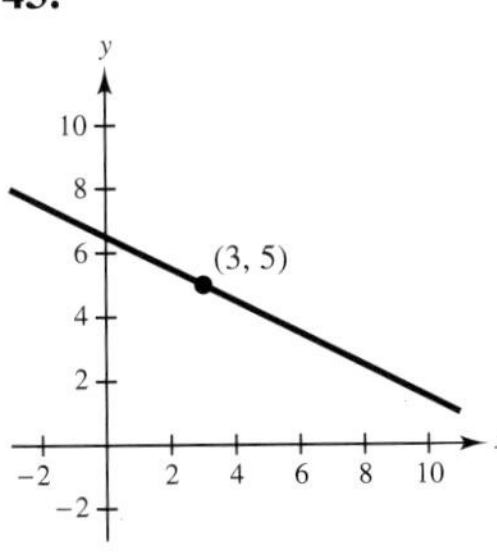

$(5, 4), (7, 3)$

45.

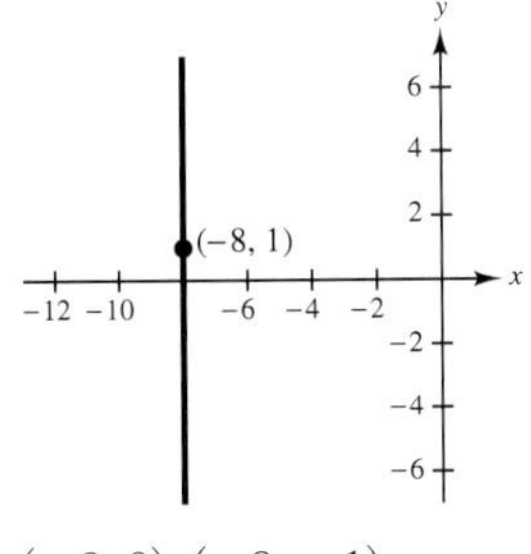

$(-8, 0), (-8, -1)$

47.

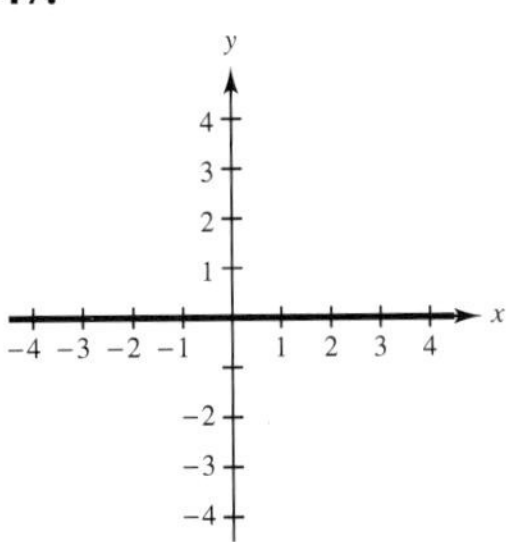

49.

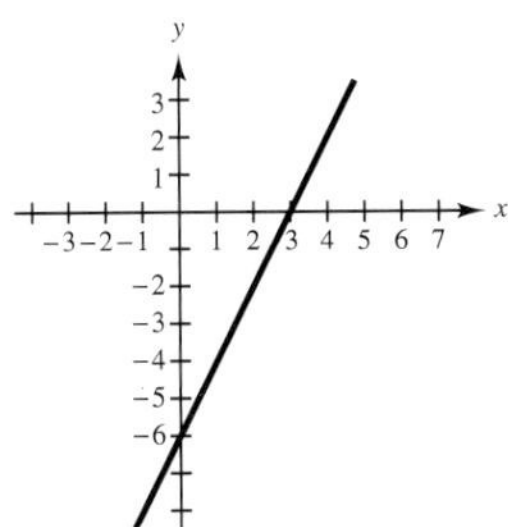

51.

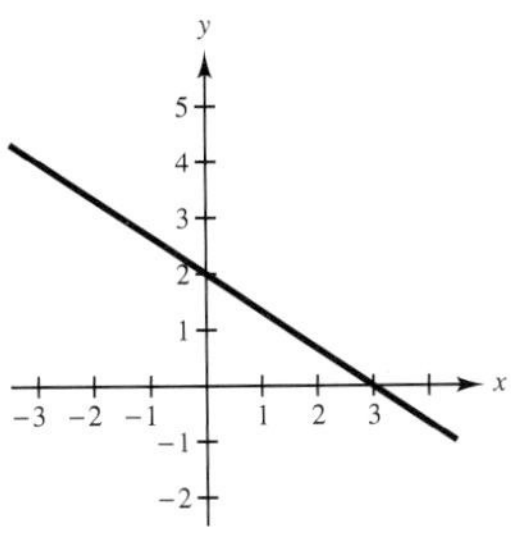

53.

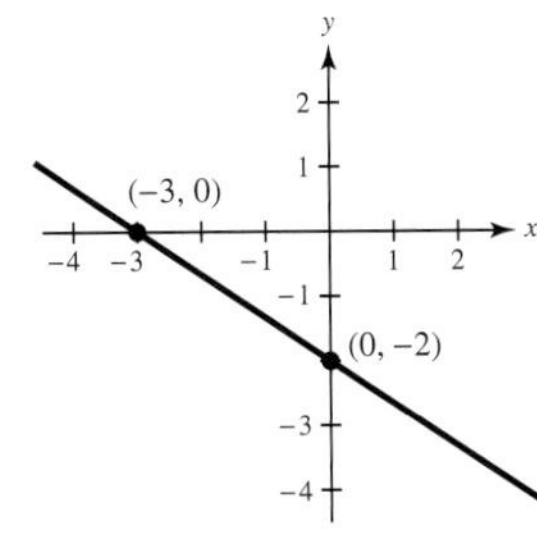

55.

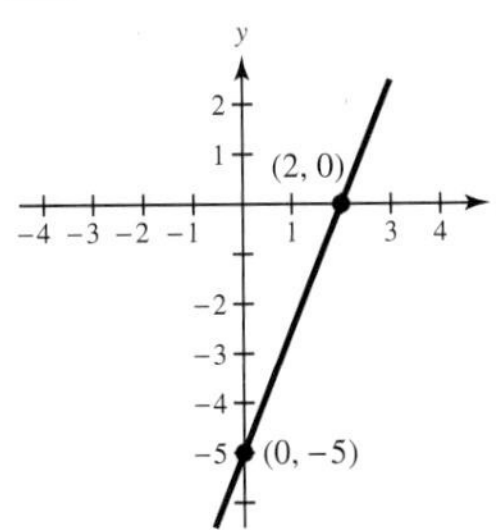

57.

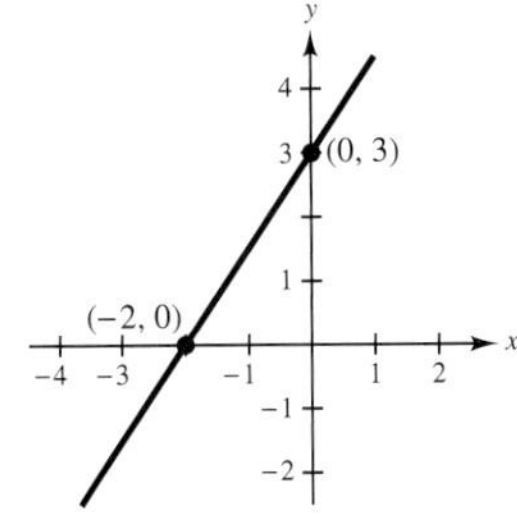

59. $y = x$

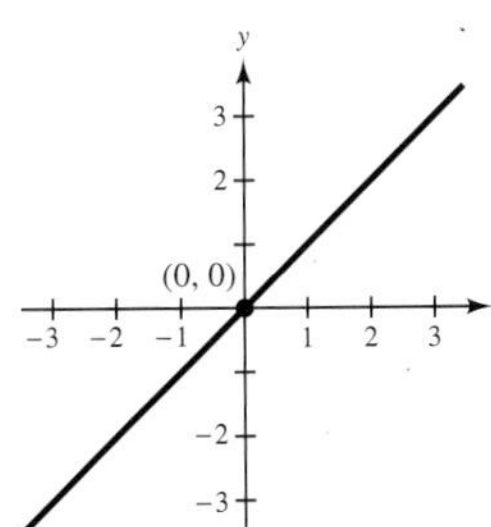

61. $y = -\frac{1}{2}x$

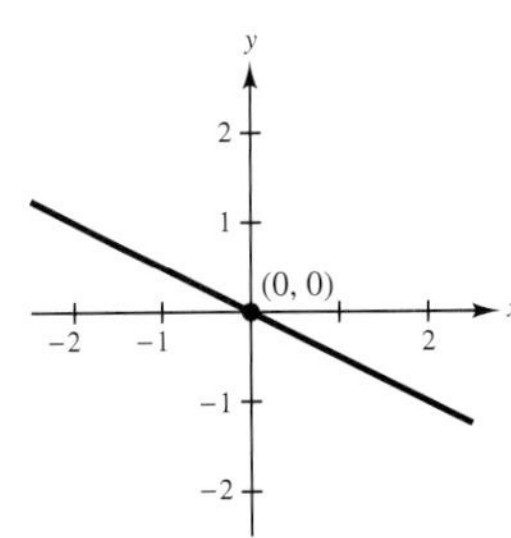

63. $y = 2x - 3$

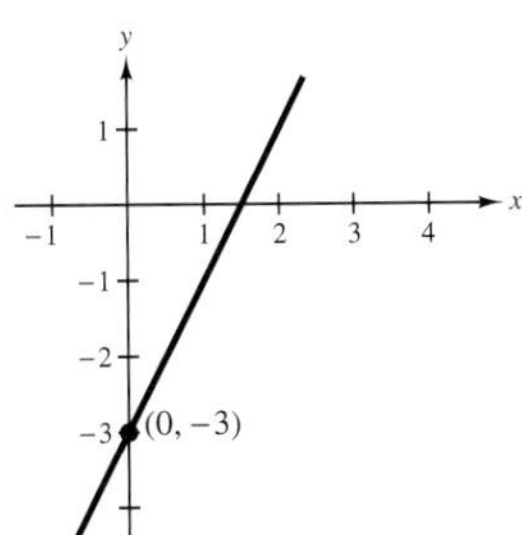

65. $y = \frac{1}{2}x + 1$

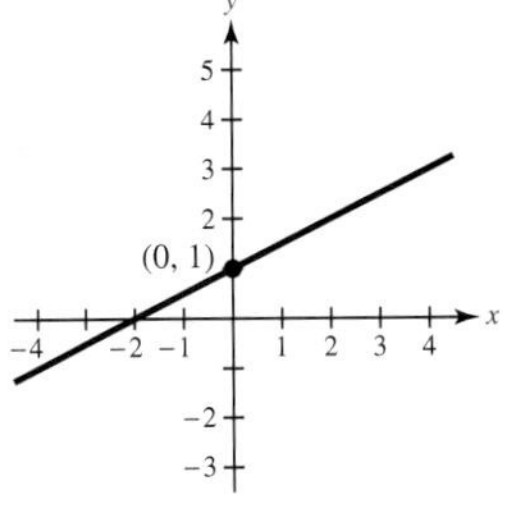

67. $y = \frac{1}{3}x - \frac{5}{2}$

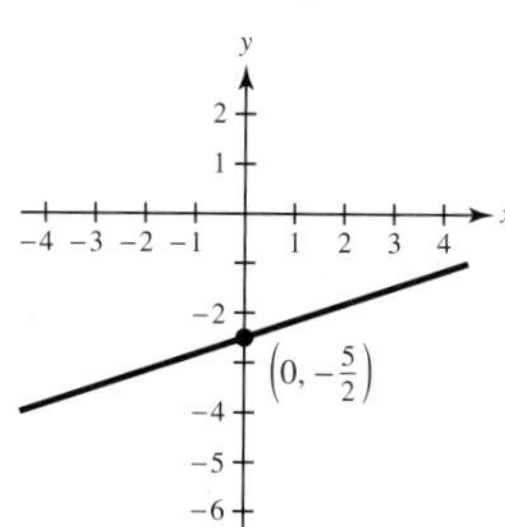

69. $y = \frac{3}{4}x + \frac{1}{2}$

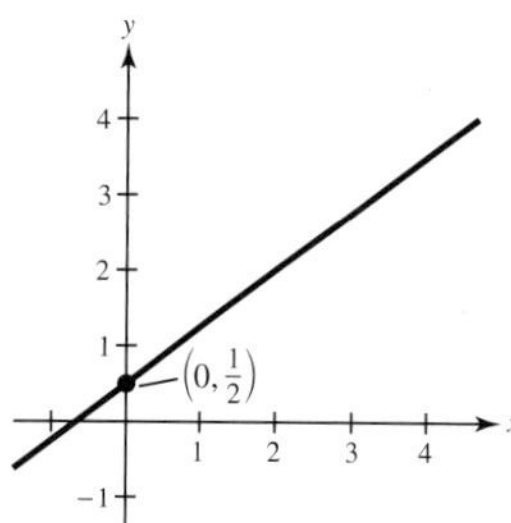

71. $y = -5$

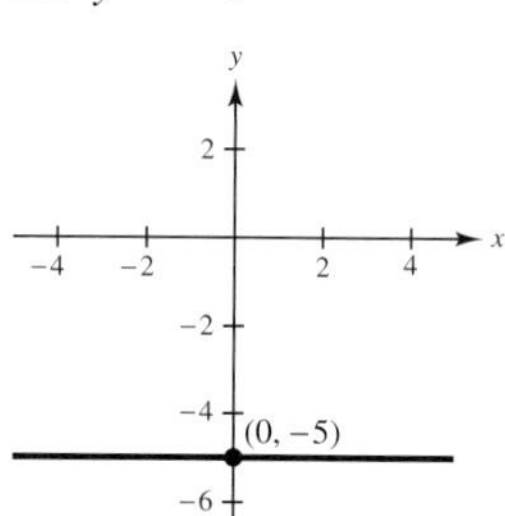

73.

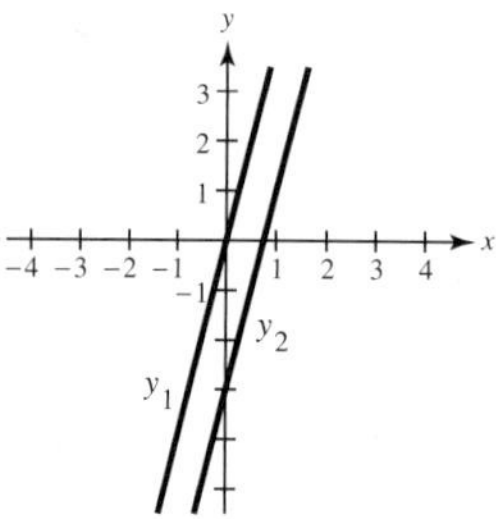

$m_1 = 4, (0, 0)$

$m_2 = 4, (0, -3)$

75.

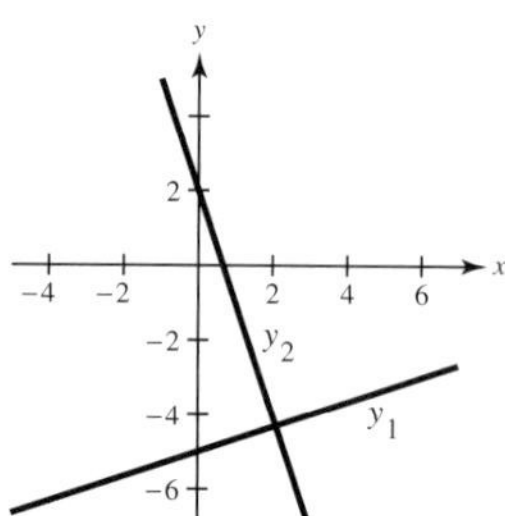

$m_1 = \frac{1}{3}, (0, -5)$

$m_2 = -3, (0, 2)$

77.

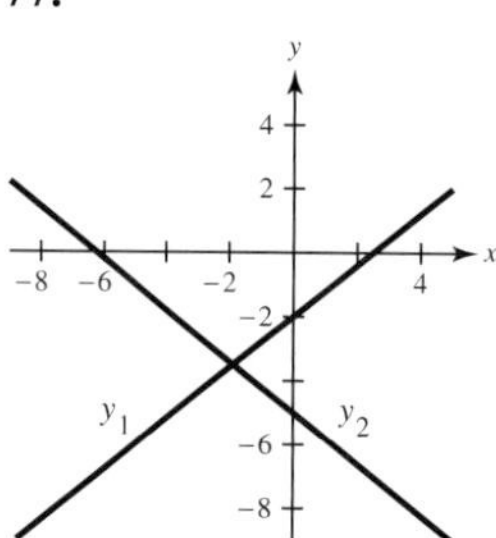

$m_1 = \frac{4}{5}, (0, -2)$

$m_2 = -\frac{4}{5}, (0, -5)$

79.

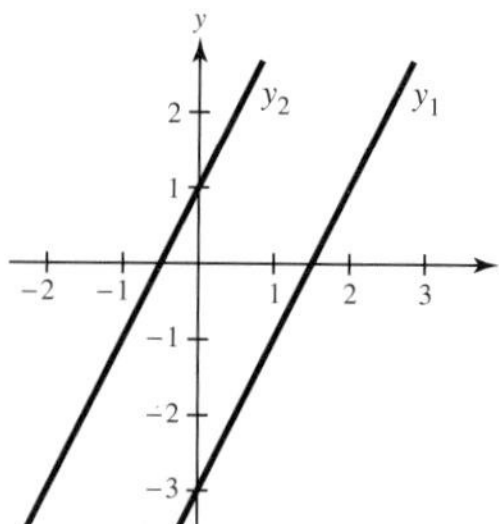

Parallel

81.

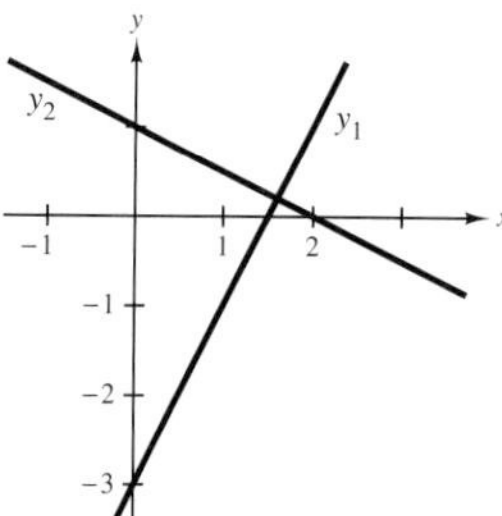

Perpendicular

83. Perpendicular **85.** Parallel **87.** $-\frac{2}{3}$ **89.** $\frac{1}{5}$

91. (a)

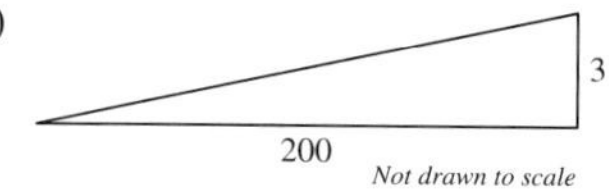

(b) $\frac{3}{200}$ (c) Yes; $\left|\frac{3}{100}\right| > \left|\frac{3}{200}\right|$

93. (a) 7, 55, 21, 58

(b) The slope is 35.25 and represents the average annual increase in the number of theatrical films released from 2002 to 2006.

95. (a) The slopes represent an increase of \$0.2 million in profit per year for P_1 and an increase of \$0.3 million in profit per year for P_2.

(b) $P_2 = 0.3t + 2.4$ (c) \$4.4 million, \$5.4 million

(d)

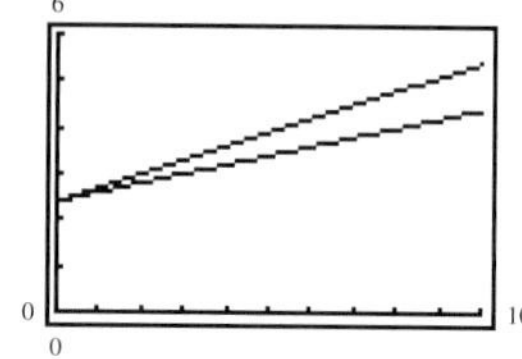

97. Sales increase by 76 units.

99. Sales remain the same.

101. Sales decrease by 14 units.

103. No. The slopes of nonvertical perpendicular lines have opposite signs. The slopes are the negative reciprocals of each other.

105. The slope

107. Yes. You are free to label either one of the points as (x_1, y_1) and the other as (x_2, y_2). However, once this is done, you must form the numerator and denominator using the same order of subtraction.

109. x^5 **111.** $-y^3$ **113.** $50x^5$

115. $x + 2$ **117.** $\left(\frac{1}{2}, 0\right), (0, -3)$

119. $\left(-\frac{3}{2}, 0\right), (0, -3)$ **121.** $(9, 0), \left(0, \frac{3}{2}\right)$

Section 4.5 *(page 270)*

1. $2x + y = 0$

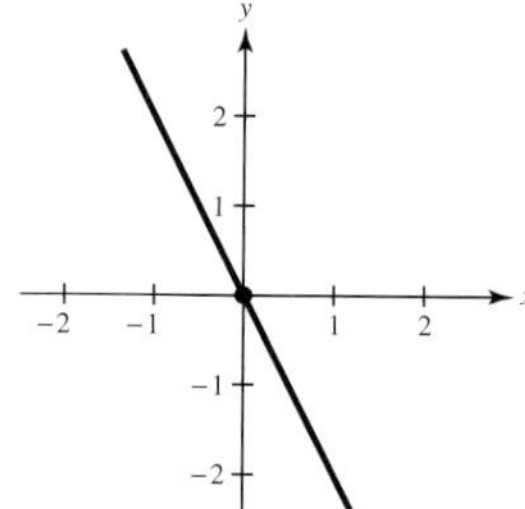

3. $x - 2y = 6$

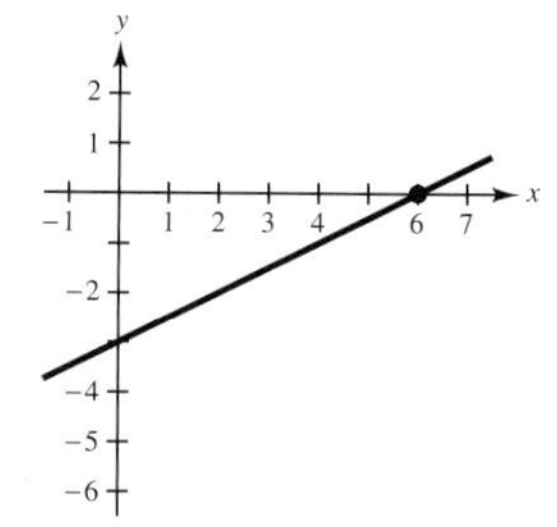

5. $2x - y = -5$ **7.** $x + 5y = -13$

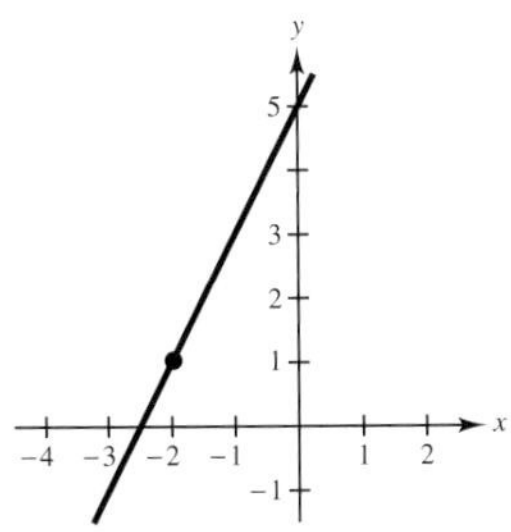

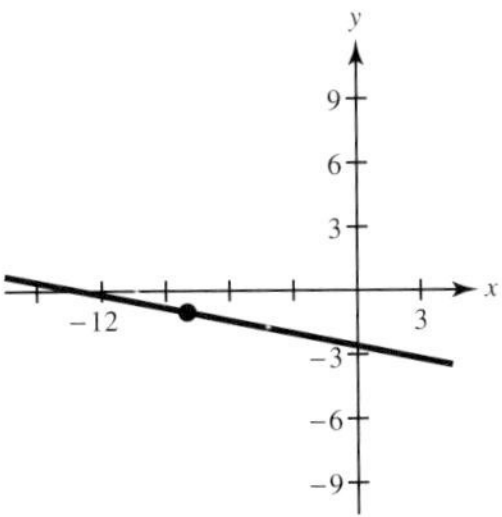

9. $y = -3$ **11.** $4x - 6y = -9$

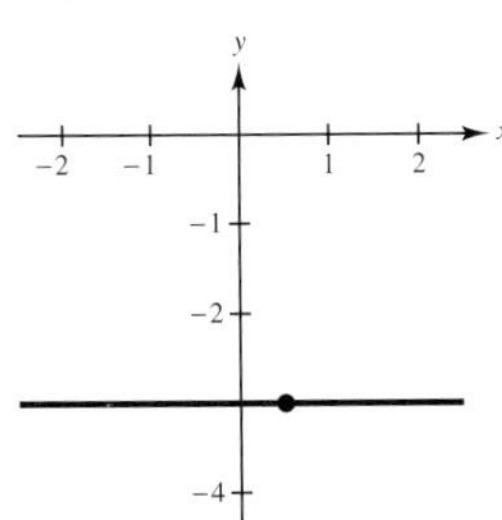

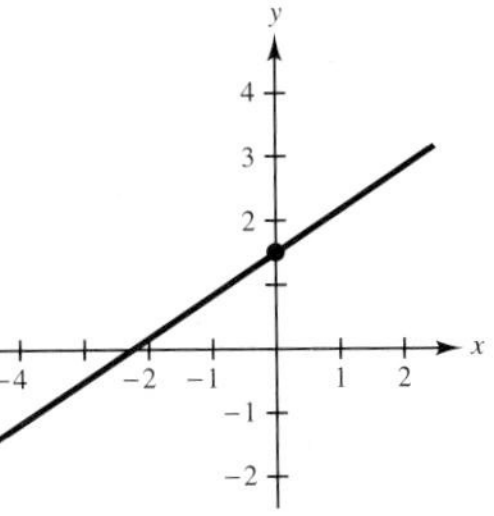

13. $4x + 5y = 28$

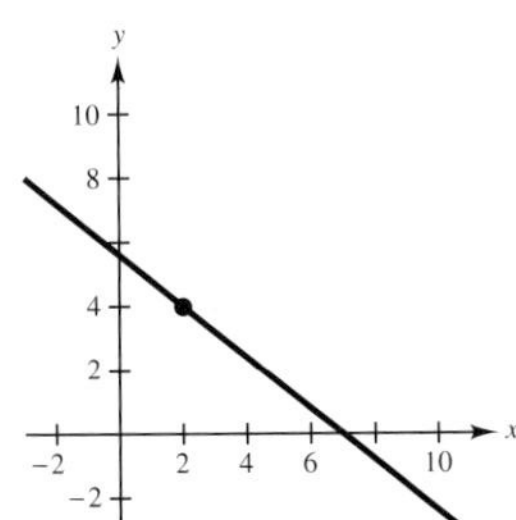

15. $y = 4x - 4$ **17.** $y = -3x - 3$ **19.** $y = -\frac{1}{3}x + 3$

21. $y = 4$ **23.** $y = -\frac{3}{4}x + 5$ **25.** $y = \frac{3}{8}x + \frac{7}{4}$

27. $\frac{3}{8}$ **29.** 5 **31.** $\frac{2}{3}$ **33.** 0 **35.** Undefined

37. $\frac{3}{2}$ **39.** $y = 3x - 1$ **41.** $y = -\frac{1}{2}x + 2$

43. $y - 2 = -\frac{1}{3}(x + 1)$

45. $y + 1 = \frac{1}{3}(x + 2)$ or $y - 1 = \frac{1}{3}(x - 4)$

47. $y = -x$ **49.** $y = 2x - 8$

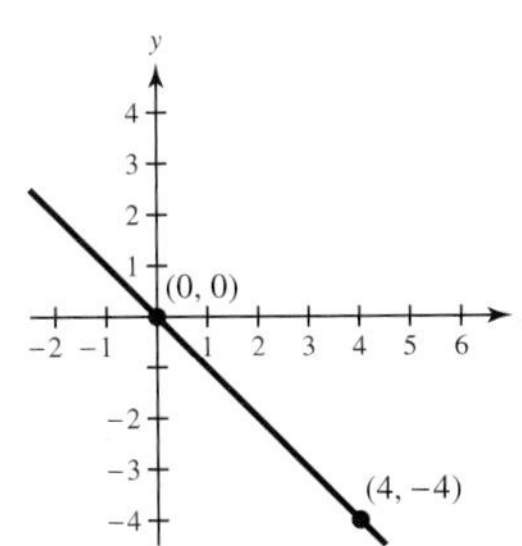

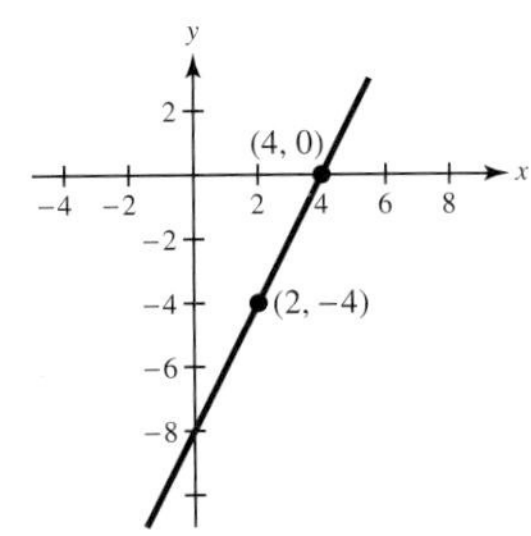

51. $y = -x + 1$ **53.** $y = \frac{1}{3}x + 4$

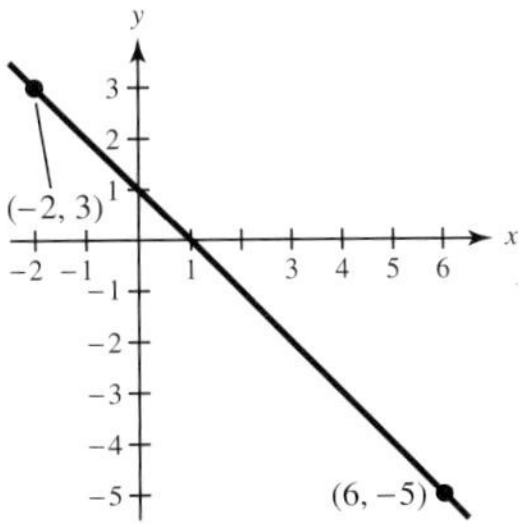

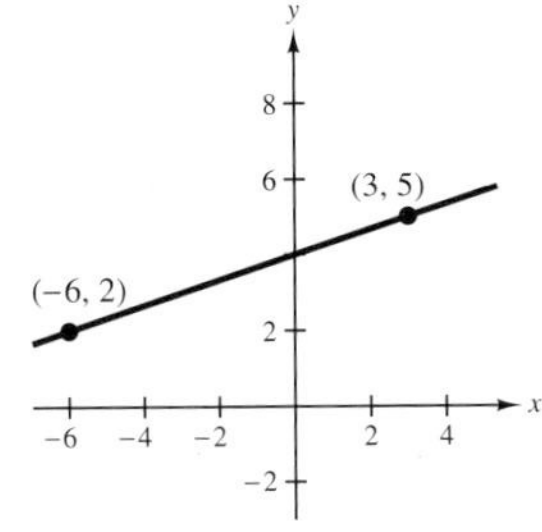

55. $x = 5$ **57.** $y = 4x - 11$

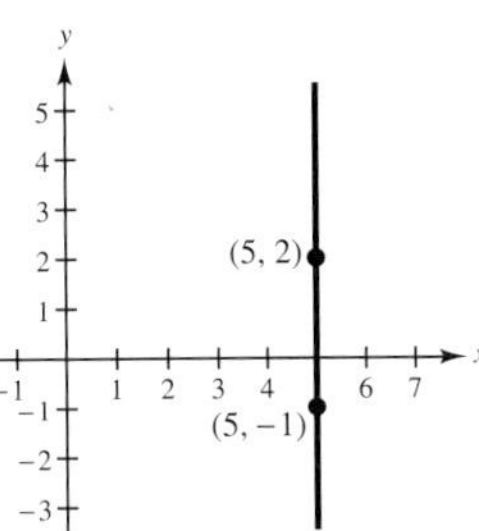

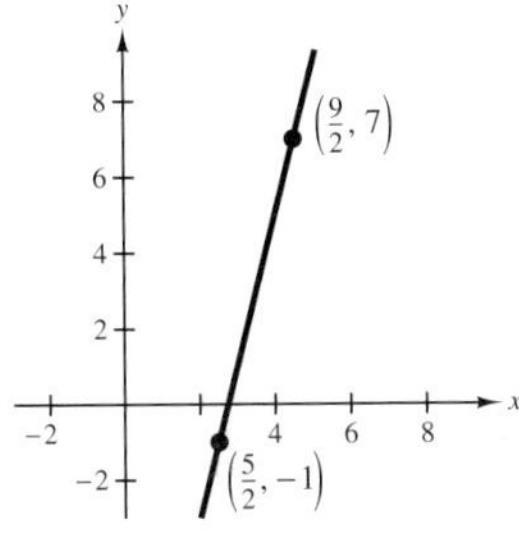

59. $x + y - 3 = 0$ **61.** $3x + 5y - 10 = 0$

63. $-3x + 2y - 20 = 0$ **65.** $-x + 4y + 9 = 0$

67. $3x + 5y - 31 = 0$ **69.** $8x + 6y - 19 = 0$

71. $6x + 5y - 9 = 0$

73. (a) $y = x - 1$ (b) $y = -x + 3$

75. (a) $y = -\frac{3}{4}x - 5$ (b) $y = \frac{4}{3}x + 20$

77. (a) $y = -2x + 5$ (b) $y = \frac{1}{2}x + \frac{5}{2}$

79. (a) $y = 0$ (b) $x = -1$

81. (a) $y = \frac{2}{3}x - \frac{11}{3}$ (b) $y = -\frac{3}{2}x + 5$

83. $x = -2$ **85.** $y = \frac{2}{3}$ **87.** $x = 4$

89. $y = -8$

91.

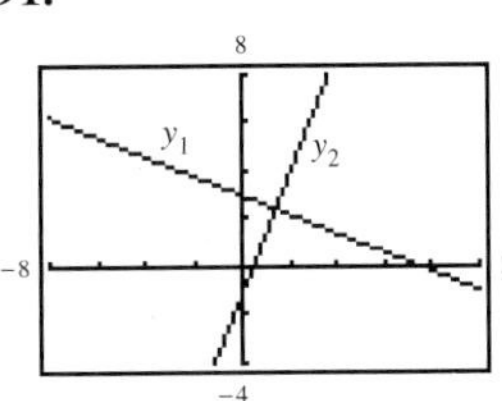

Perpendicular

93.

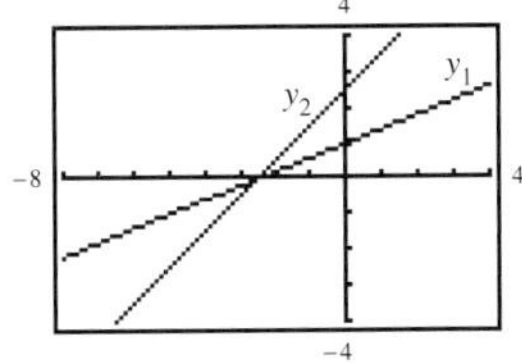

Neither

95.

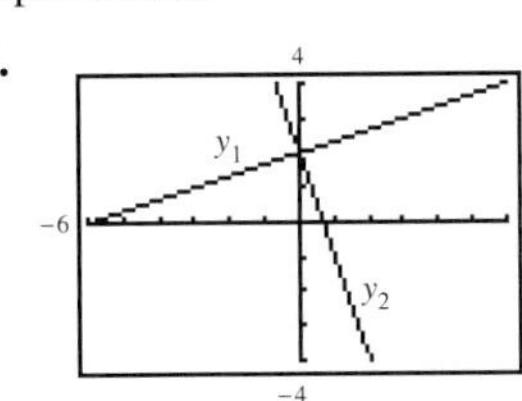

y_1 and y_2 are perpendicular.

97. $W = 2000 + 0.02S$

99. (a) $S = L - 0.2L$

(b)

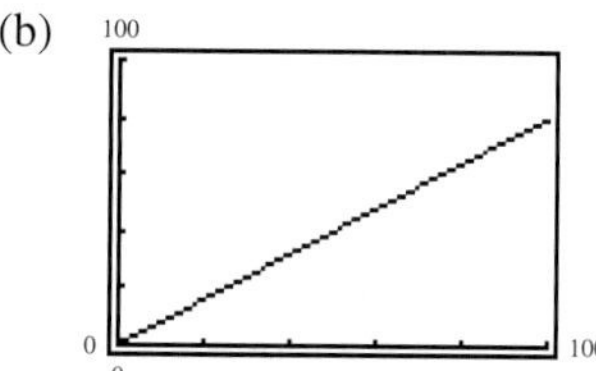

(c) $40, $39.98

101. (a) $V = 25{,}000 - 2300t$ (b) $18,100

103. 3 gallons per minute; $y = 3x + 30$

105. (a) (50, 580), (47, 625)

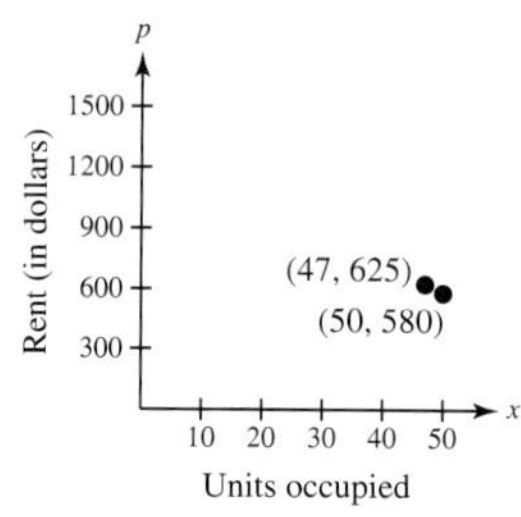

(b) $p = -15x + 1330$

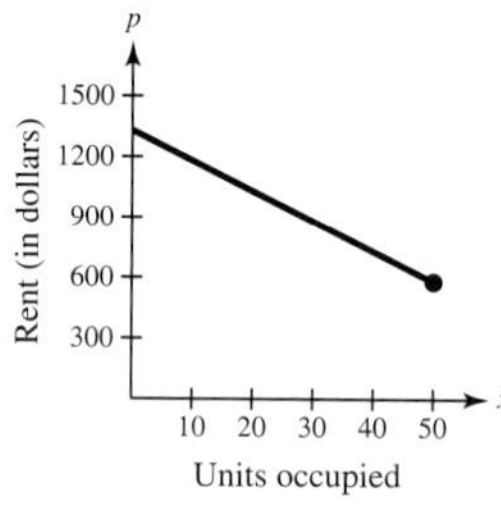

As the rent increases, the demand decreases.

(c) 45 units (d) 49 units

107. $s = 3.5t - 10.9$; $13.6 billion

109. (a) $V = 125t + 1665$

(b) $V = 4.5t + 124.5$

(c) $V = -2000t + 34{,}400$

(d) $V = -2300t + 61{,}100$

111. Yes. Because every line has one specific slope, any pair of points on the line will give you this slope.

113. Let $y = 0$ and solve for x.

$$y = mx + b$$
$$0 = mx + b$$
$$-b = mx$$
$$-\frac{b}{m} = x$$

115. The lines are parallel or they coincide.

117. $12 - 8x$ **119.** $x + 10$

121. $y = -3x + 4$ **123.** $y = \frac{4}{5}x + \frac{2}{5}$

125. 2 **127.** $-\frac{1}{7}$

Section 4.6 *(page 280)*

1. (a) Not a solution (b) Solution (c) Not a solution (d) Solution

3. (a) Solution (b) Solution (c) Solution (d) Not a solution

5. (a) Solution (b) Not a solution (c) Solution (d) Not a solution

7. (a) Solution (b) Solution (c) Solution (d) Solution

9. Dashed **11.** Solid **13.** b **14.** c **15.** d

16. a **17.** c **18.** a **19.** b **20.** d

21.

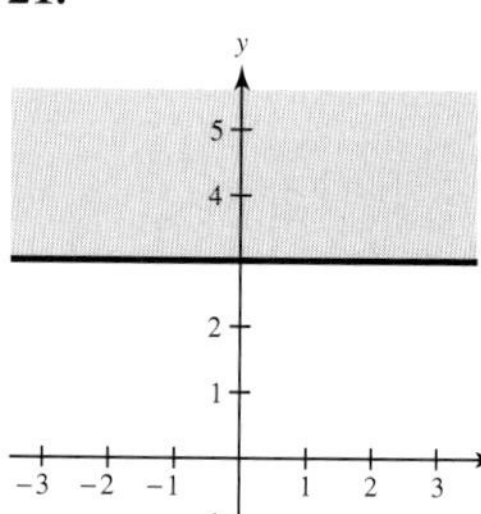

23.

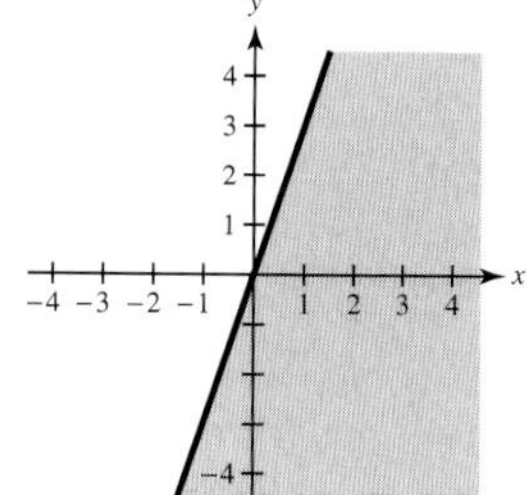

25.

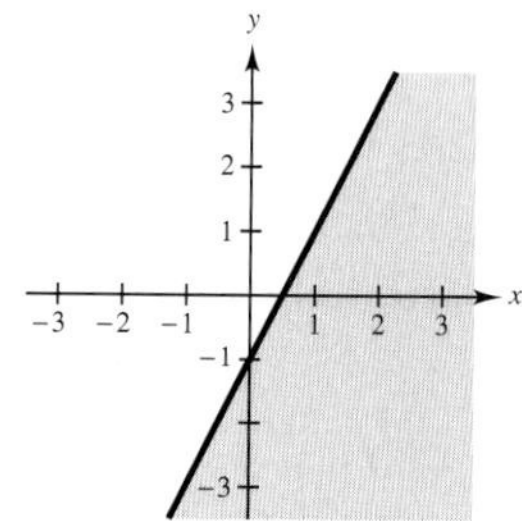

27.

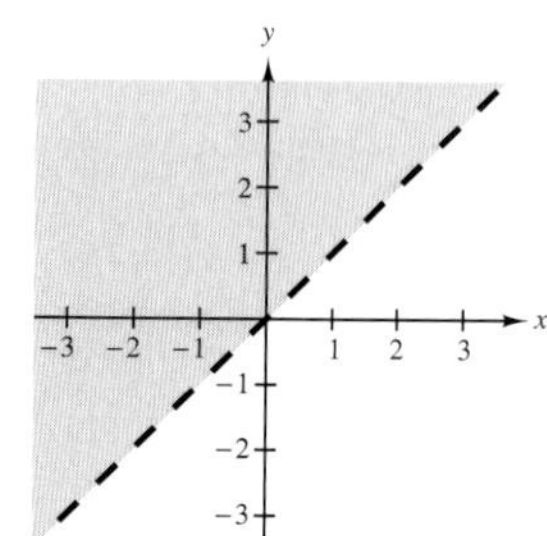

29.

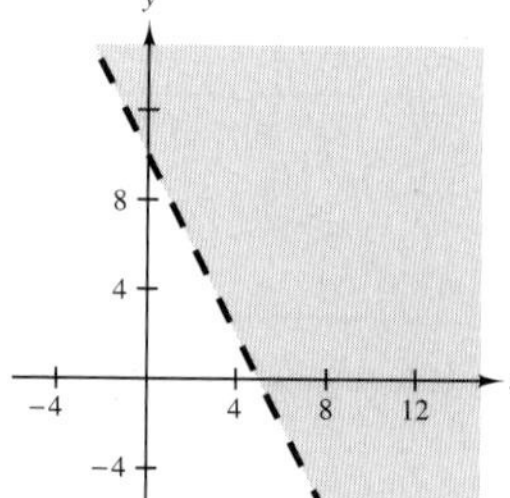

31.

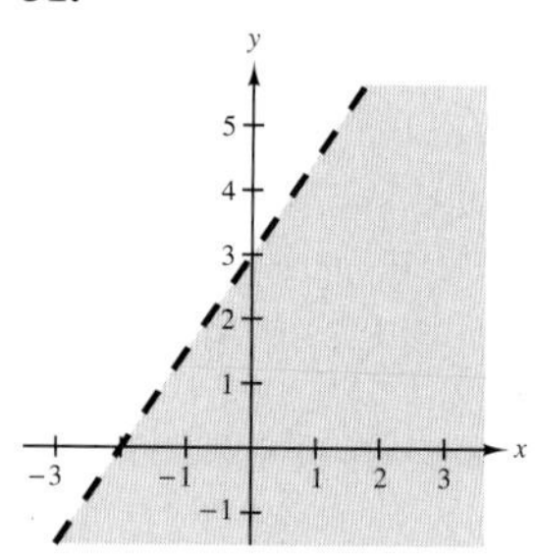

33.

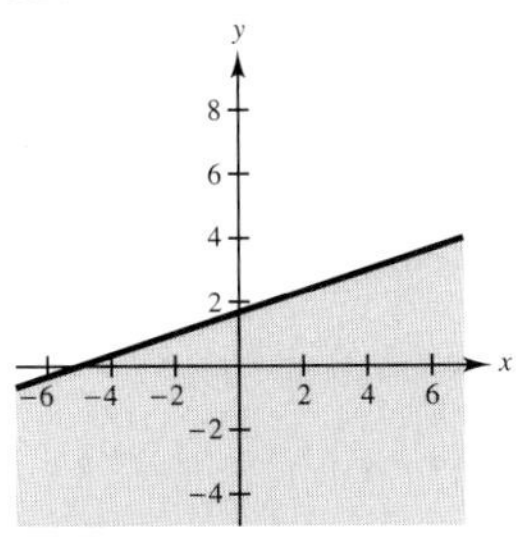

35.

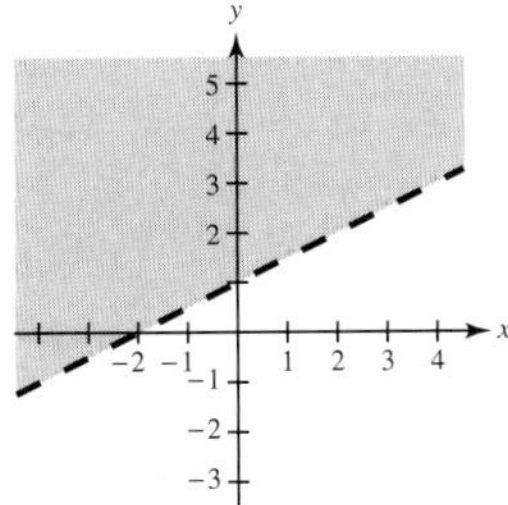

37.

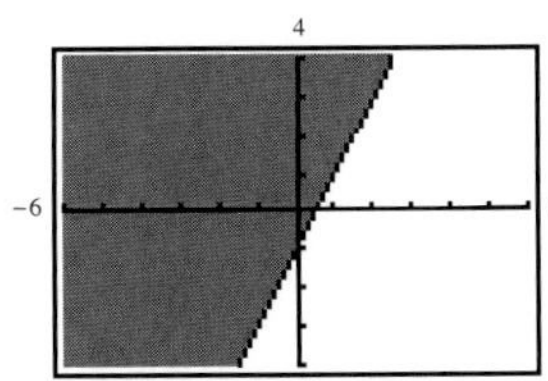

39.

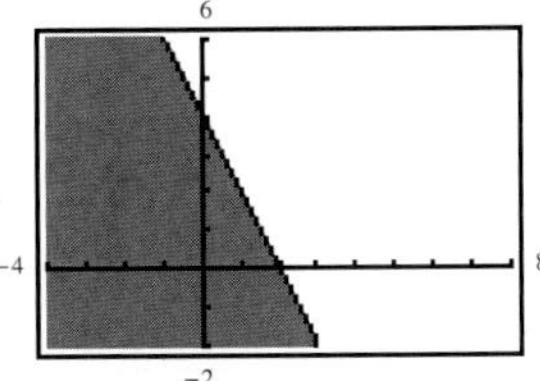

41.

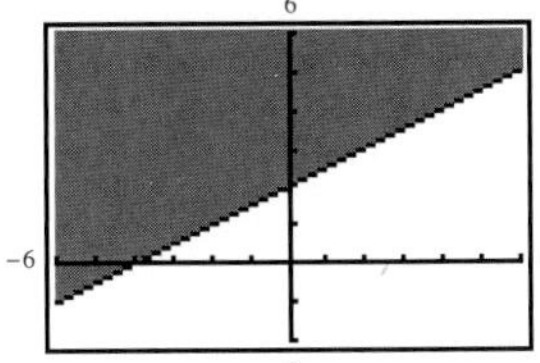

43.

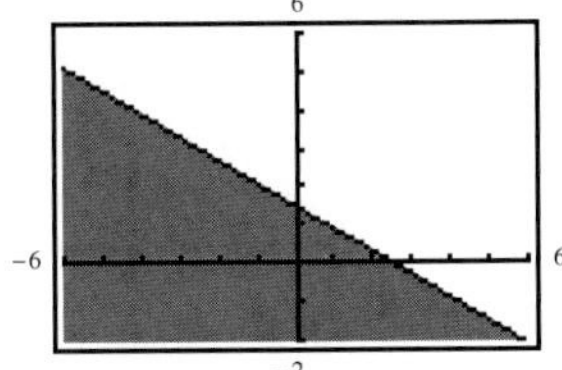

45. $y > 6x$

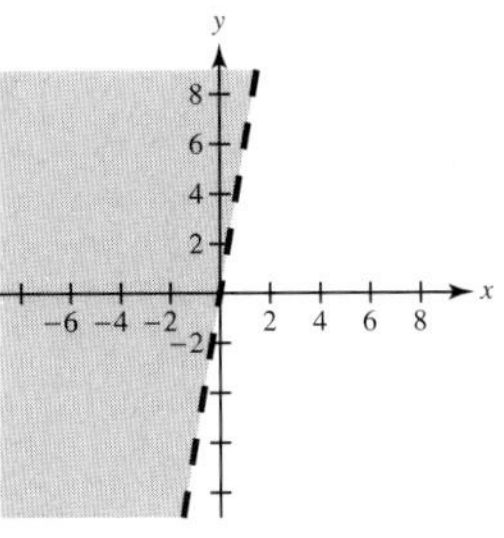

47. $x + y \geq 9$

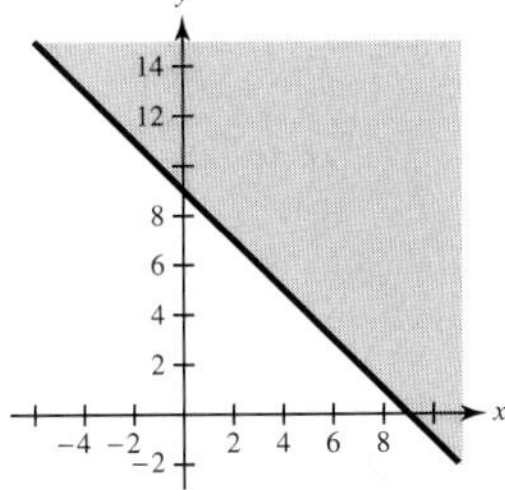

49. $y \leq x + 3$

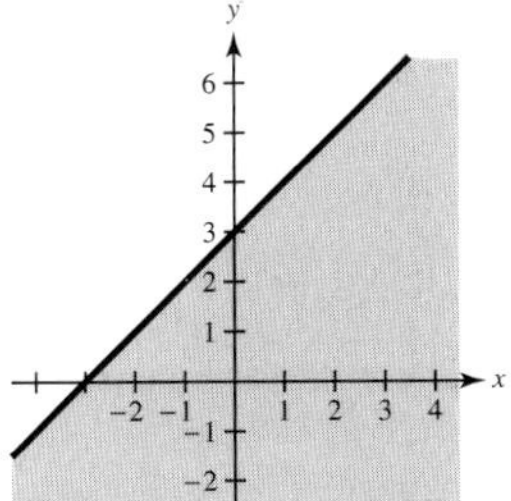

51. $y \geq 2$ **53.** $y \leq 2 - 2x$ **55.** $y < 2x$

57. (a) $y \leq 0.35x$, where x represents the total calories consumed per day and y represents the fat calories consumed per day.

(b)

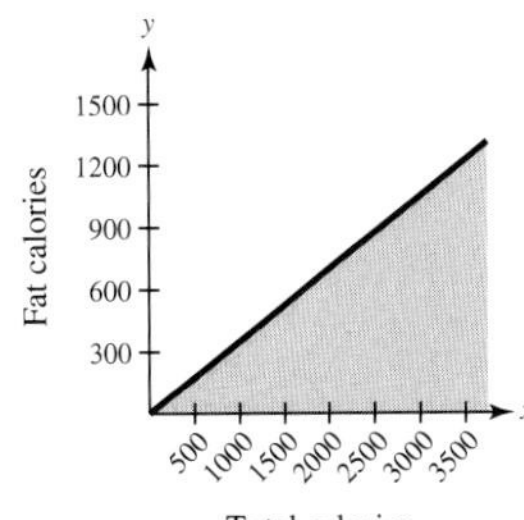

Sample answer: (1500, 500), (2000, 700), (2500, 800)

59. (a) $9x + 12y \geq 210$, where x represents the number of hours worked at the grocery store and y represents the number of hours worked mowing lawns.

(b)

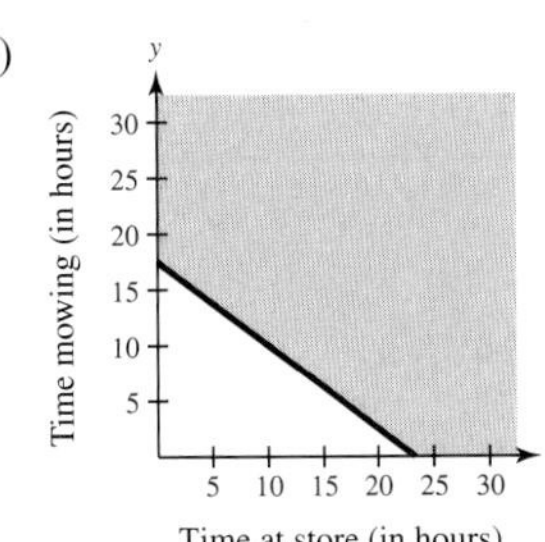

Sample answer: (2, 16), (15, 7), (20, 20)

61. (a) $t + \frac{3}{2}c \leq 12$, where t represents the number of tables and c represents the number of chairs.

(b)

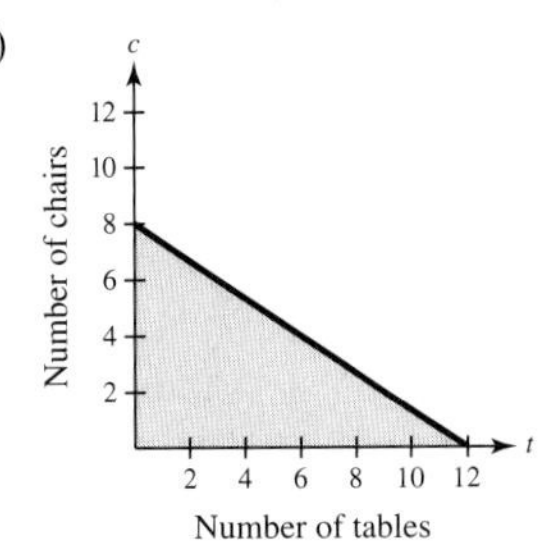

Sample answer: (5, 4), (2, 6), (0, 8)

63. $y > 0$

65. Yes. When you divide each side of $2x < 2y$ by 2, you get $y > x$.

67. To graph $x \geq 1$ (a) on the real number line, use a solid dot at 1 and shade the number line to the right of 1 to represent all points on a number line that are solutions of $x \geq 1$. To graph $x \geq 1$ (b) on a rectangular coordinate system, graph the corresponding equation $x = 1$, which is a vertical line. The points (x, y) that satisfy the inequality $x \geq 1$ are those lying on and to the right of this line.

69. $\frac{4}{9}$ **71.** $\frac{1}{6}$ **73.** $\frac{23}{24}$ **75.** $y = 3x - 7$

77. $y = -\frac{1}{4}x + 4$ **79.** $y = -9x + \frac{11}{2}$

Review Exercises *(page 286)*

1.

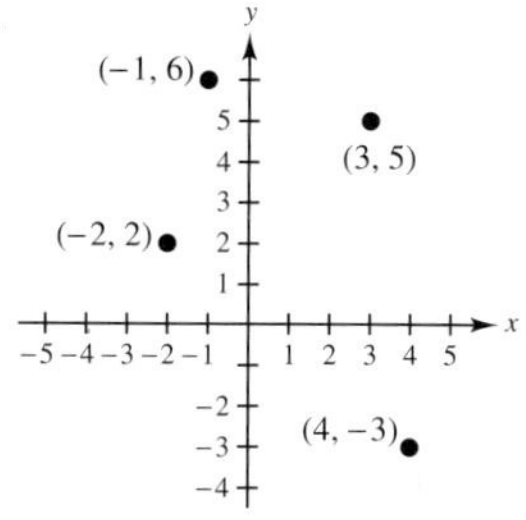

3. A: $(3, -2)$; B: $(0, 5)$; C: $(-1, 3)$; D: $(-5, -2)$

5. Quadrant II **7.** Quadrant II or III, or on the x-axis

9.

x	-1	0	1	2
$y = -\frac{1}{2}x - 1$	$-\frac{1}{2}$	-1	$-\frac{3}{2}$	-2

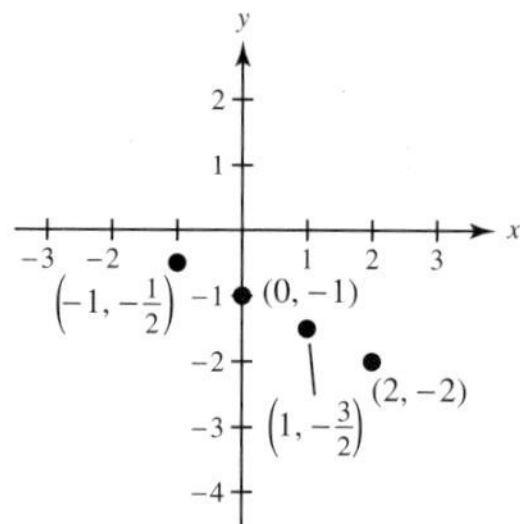

11. $y = -\frac{3}{4}x + 3$ **13.** $y = 3x - 4$ **15.** $y = \frac{1}{2}x - 4$

17. (a) Solution (b) Not a solution
(c) Not a solution (d) Not a solution

19. (a) Solution (b) Solution
(c) Not a solution (d) Solution

21. (a)

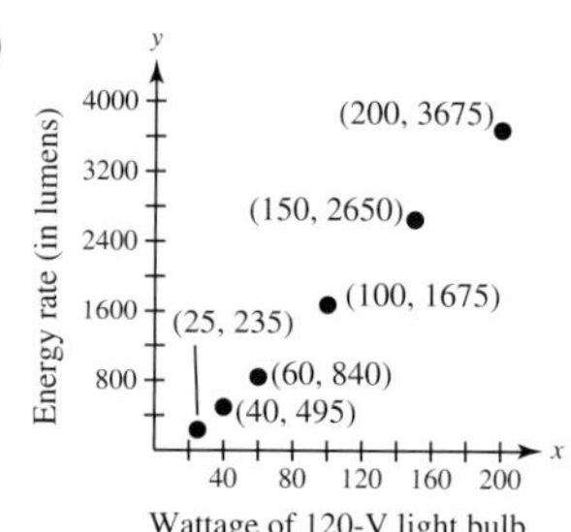

(b) Approximately linear

23.

x	-2	-1	0	1	2
y	3	0	-1	0	3

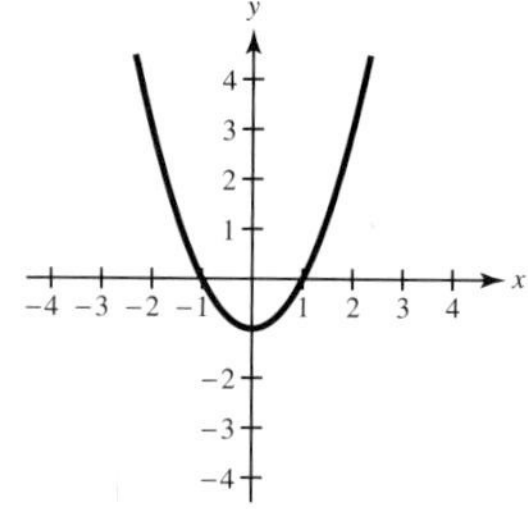

25.

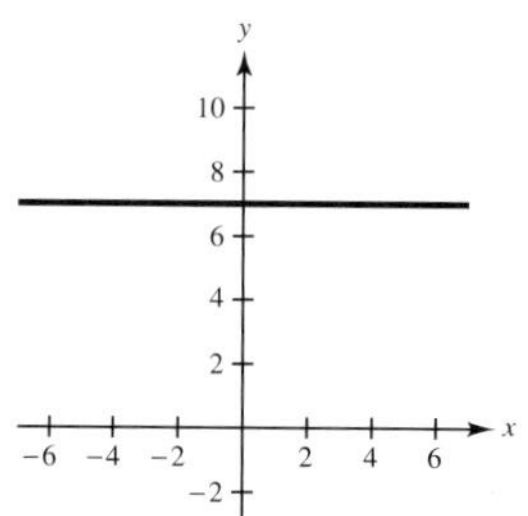

27.

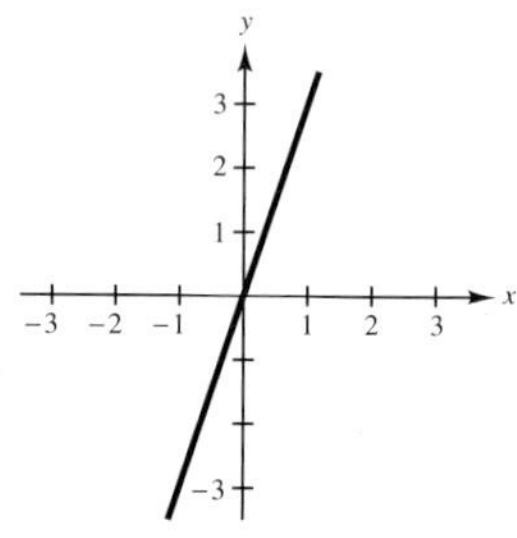

29.

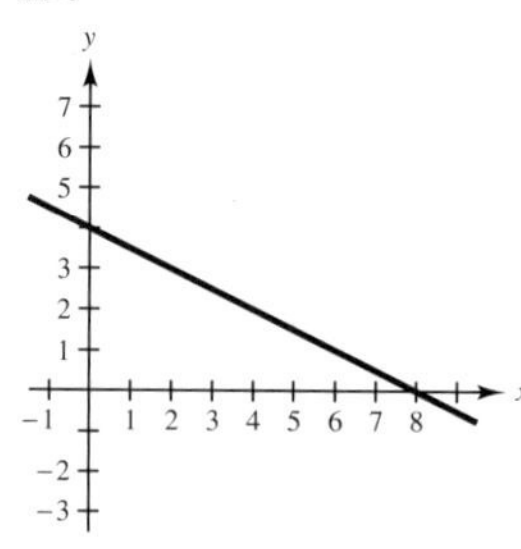

31.

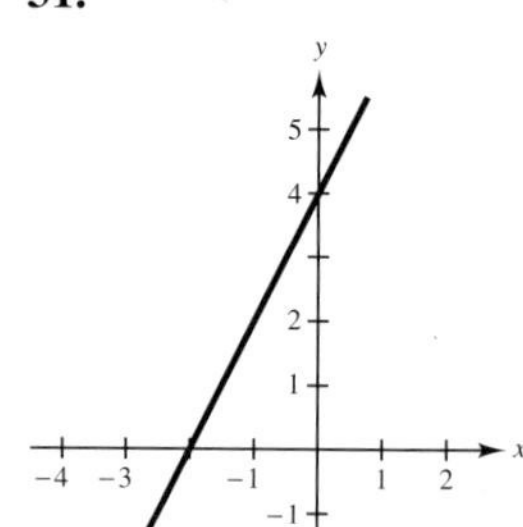

33.

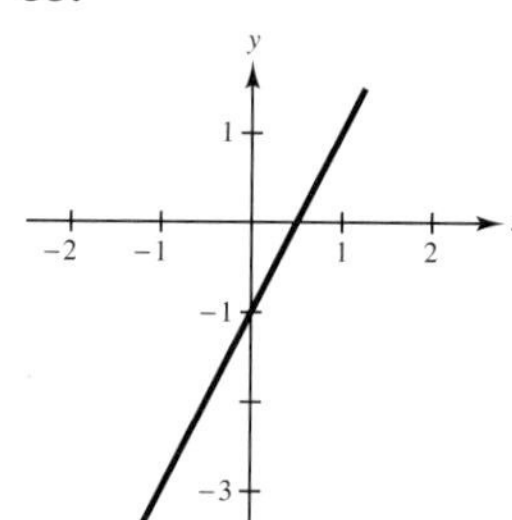

35.

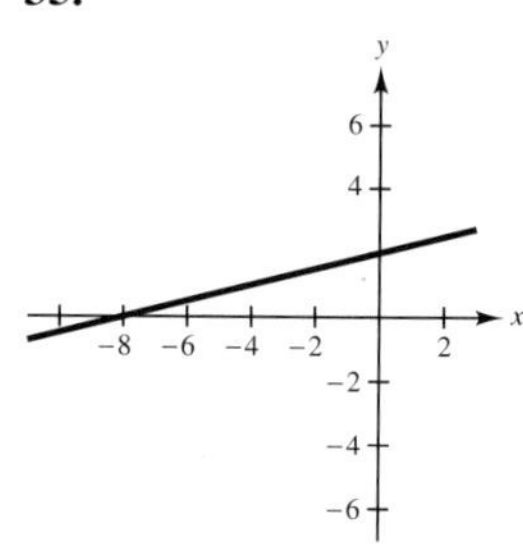

37.

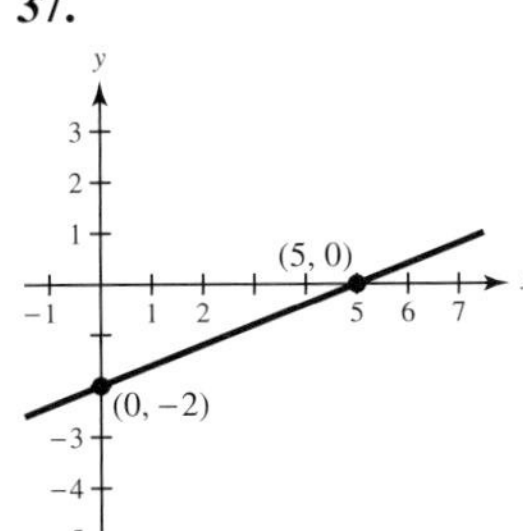

39.

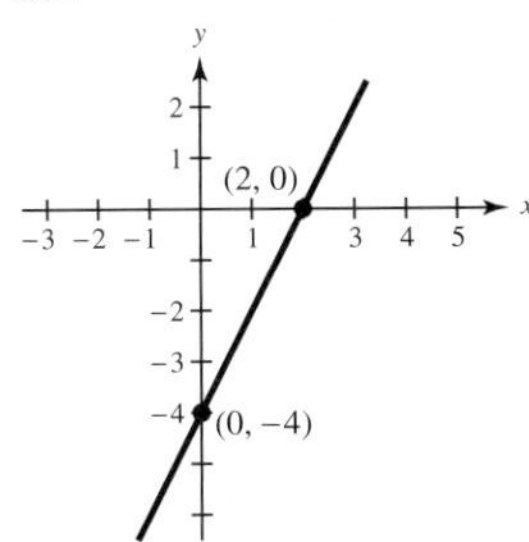

41.

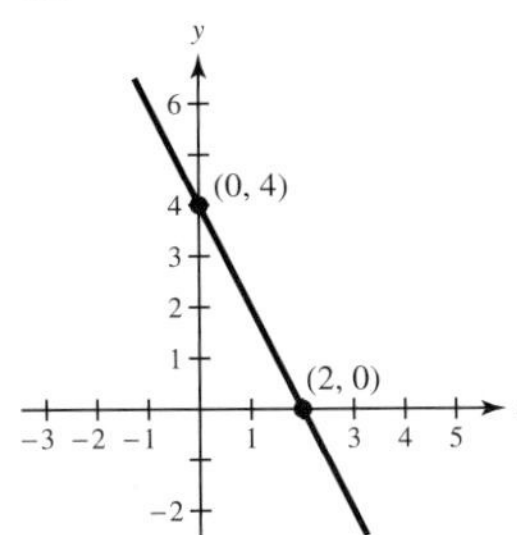

43. $C = 3x + 125$

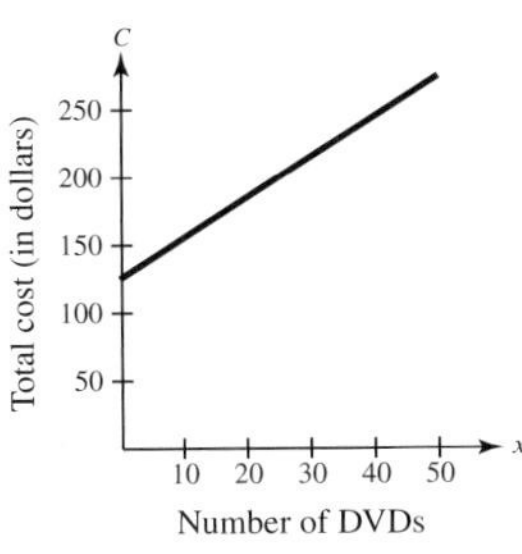

45. Domain: $\{-2, 3, 5, 8\}$; Range: $\{1, 3, 7, 8\}$

47. Domain: $\{-2, 2, 4\}$; Range: $\{-3, 0, 3, 4\}$

49. Function **51.** Function **53.** Not a function

55. (a) $-\frac{3}{4}$ (b) 3 (c) $\frac{15}{2}$ (d) -1

57. (a) 64 (b) 63 (c) 48 (d) 0

59. (a) 3 (b) 13 (c) 5 (d) 0

61. (a) 38 (b) 30 (c) 20

63. Domain: $\{1, 2, 3, 4, 5\}$ **65.** Domain: $\{-2, 0, 3, 4, 7\}$

67. $\frac{1}{2}$ **69.** 3

71.

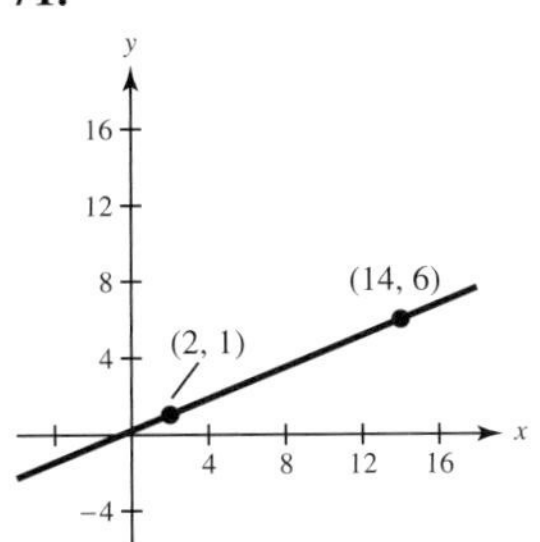

$m = \frac{5}{12}$; The line rises.

73.

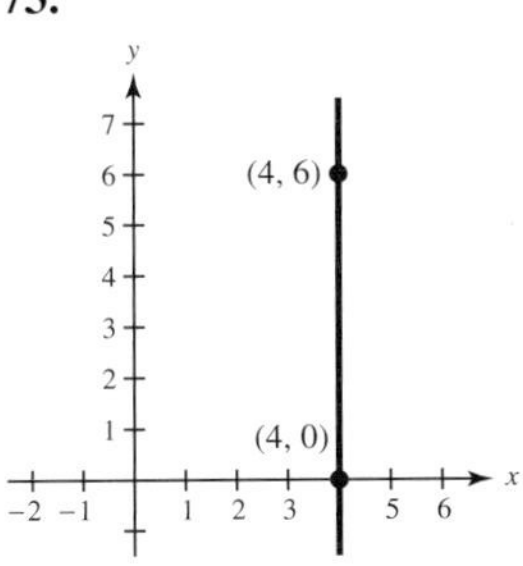

m is undefined; The line is vertical.

75.

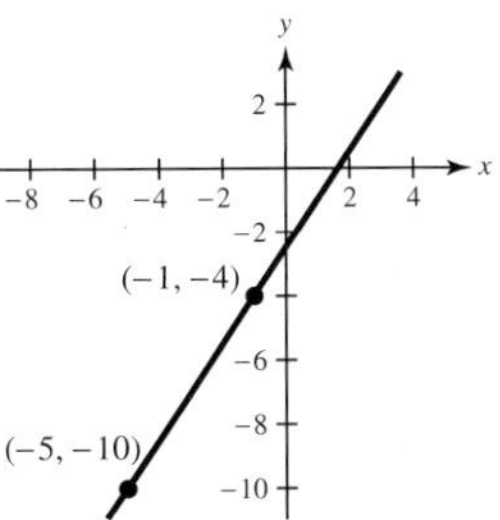

$m = \frac{3}{2}$; The line rises.

77. $\frac{2}{3}$

79. $y = -x + 6$

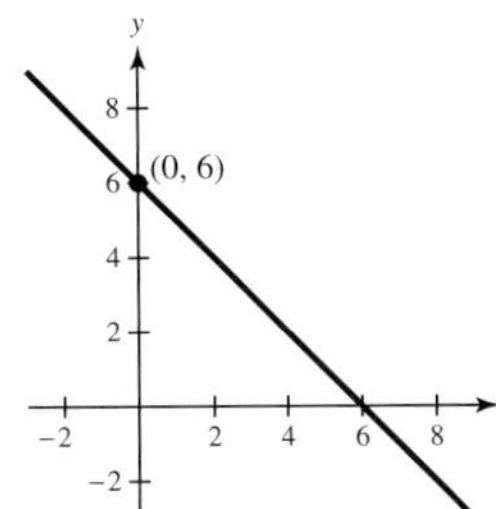

81. $y = 2x + 1$

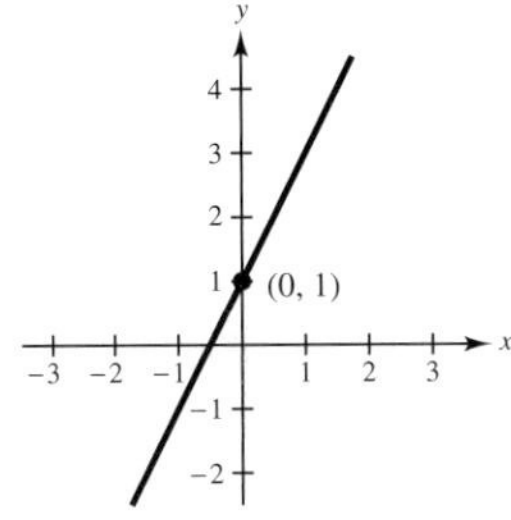

83. $y = -\frac{1}{2}x + 2$

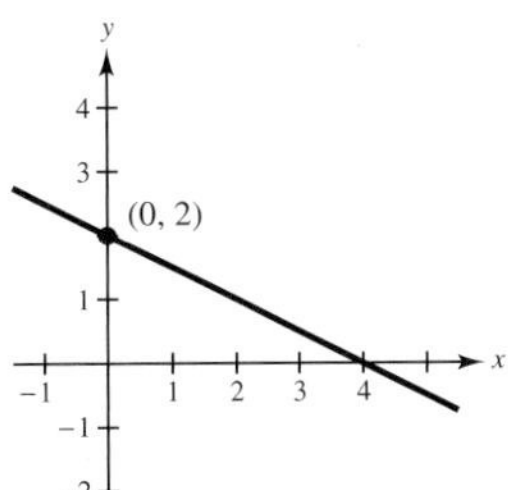

85. $y = \frac{2}{5}x + 1$

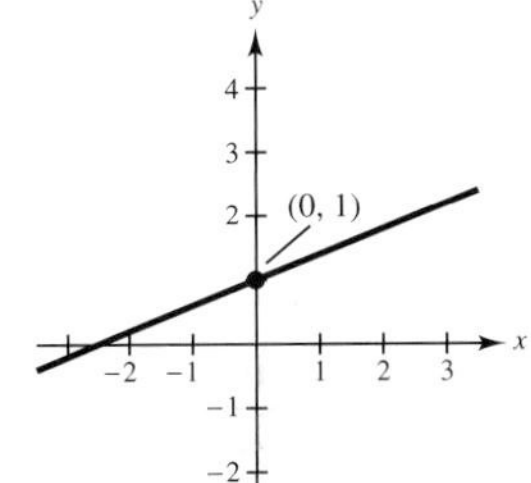

87. Parallel **89.** Neither **91.** $y = 2x - 9$

93. $y = -4x + 6$ **95.** $y = \frac{4}{5}x + 2$

97. $x = 3$ **99.** $x + 2y + 4 = 0$ **101.** $y - 8 = 0$

103. $-5x + 3y + 1 = 0$ **105.** $25x - 20y + 6 = 0$

107. (a) $y = x + 9$ (b) $y = -x - 3$

109. (a) $y = 4$ (b) $x = \frac{3}{8}$ **111.** $y = 5$ **113.** $x = 5$

115. $W = 5500 + 0.07S$

117. (a) Not a solution (b) Not a solution (c) Solution (d) Solution

119. (a) Not a solution (b) Solution (c) Solution (d) Not a solution

121.

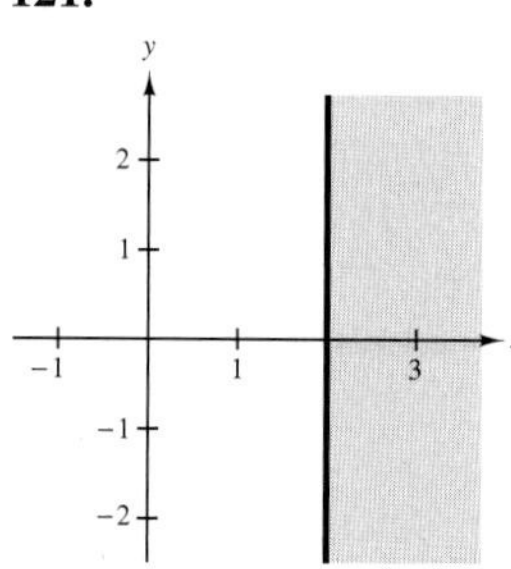

123.

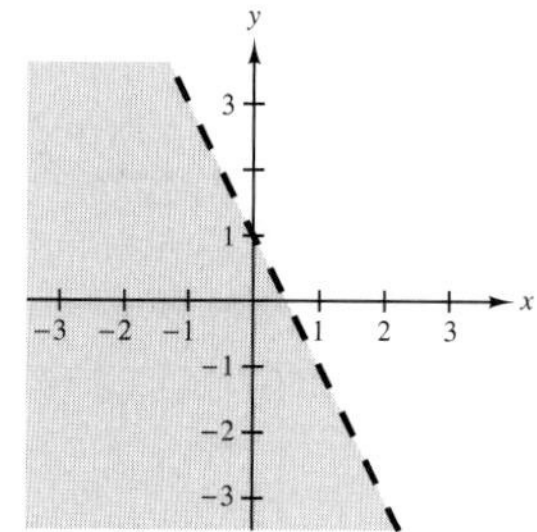

125.

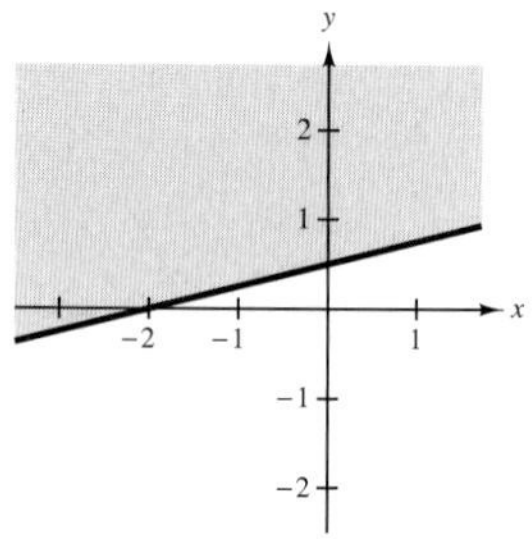

127. $y \le x + 1$

129. $2x + 3y \le 120$, where x represents the number of DVD players and y represents the number of camcorders.

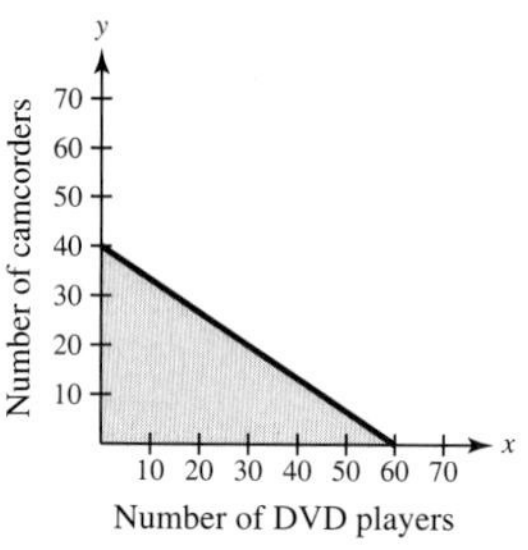

Sample answer: (10, 15), (20, 20), (30, 20)

Chapter Test (page 291)

1.

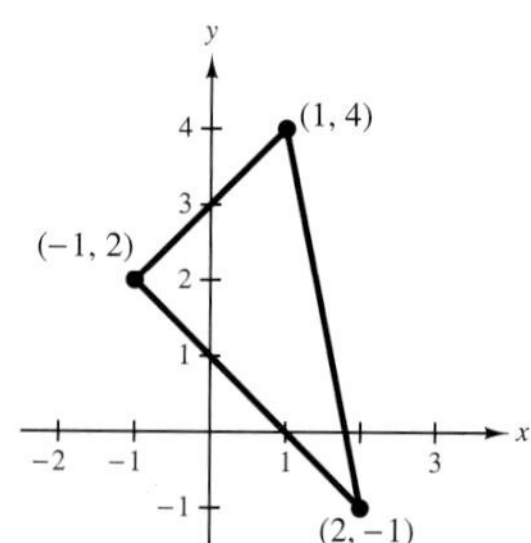

2. (a) Not a solution (b) Solution (c) Solution (d) Not a solution

3. 0 **4.** $(-2, 0), (0, 8)$

5.

x	-2	-1	0	1	2
y	2	-1	-4	-7	-10

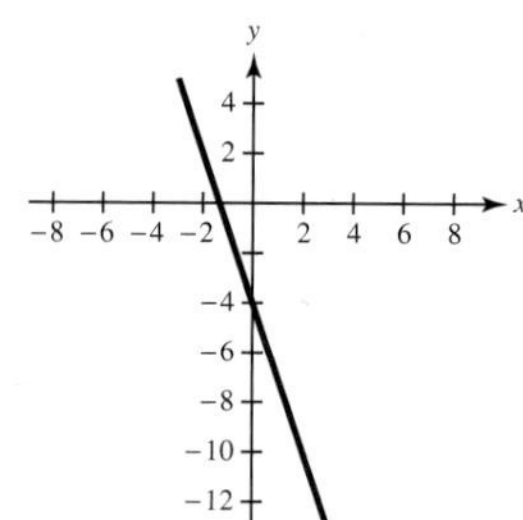

6.

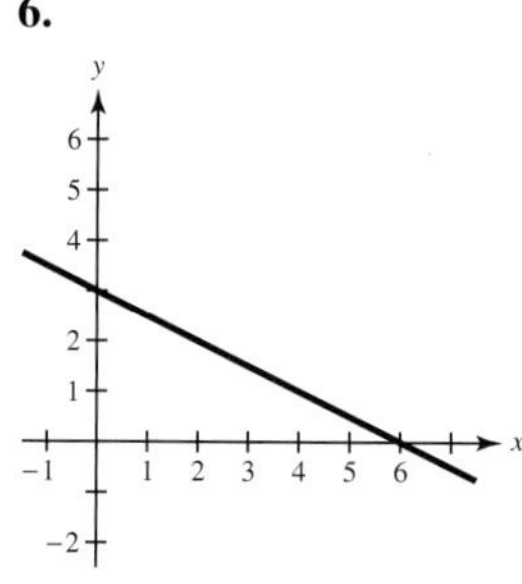

7.

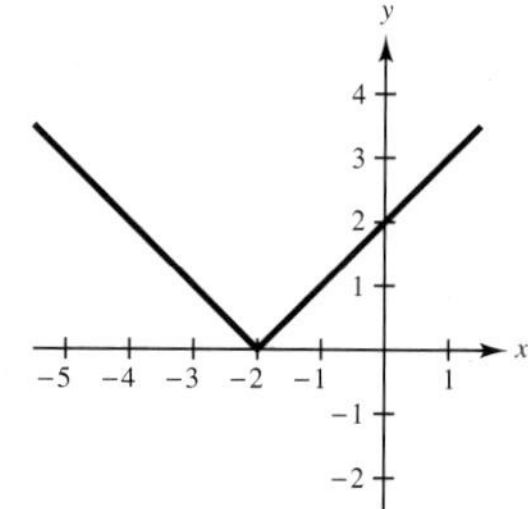

8.

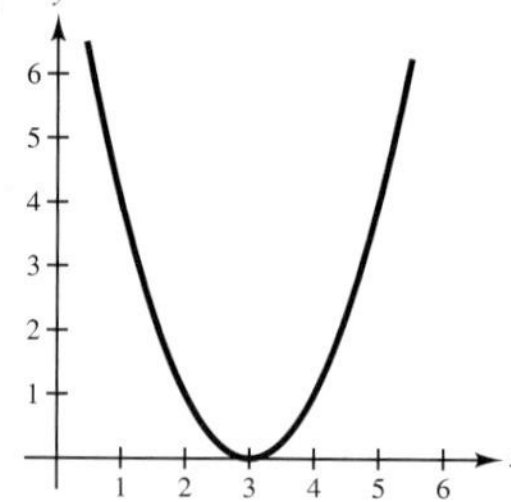

9. No, some input values, 0 and 1, have two different output values.

10. Yes, because it passes the Vertical Line Test.

11. (a) 0 (b) 0 (c) -16 (d) $-\frac{3}{8}$

12. $\frac{3}{14}$; $y = \frac{3}{14}x + \frac{15}{14}$

13. $(-2, 2), (-1, 0)$

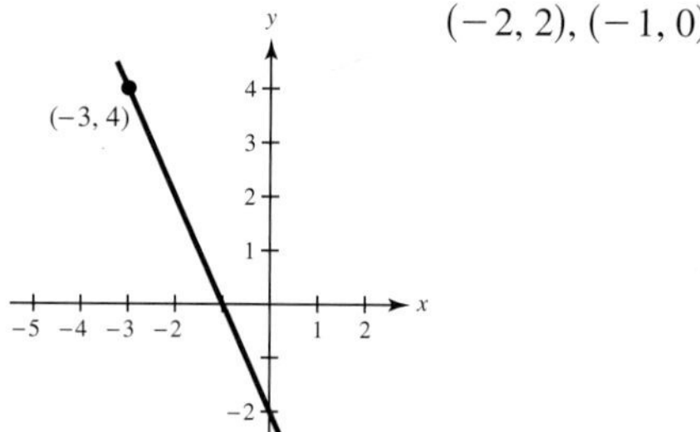

14. $-\frac{8}{7}$ **15.** $3x + 8y - 48 = 0$ **16.** $x = 3$

17. (a) Solution (b) Solution (c) Solution (d) Solution

18.

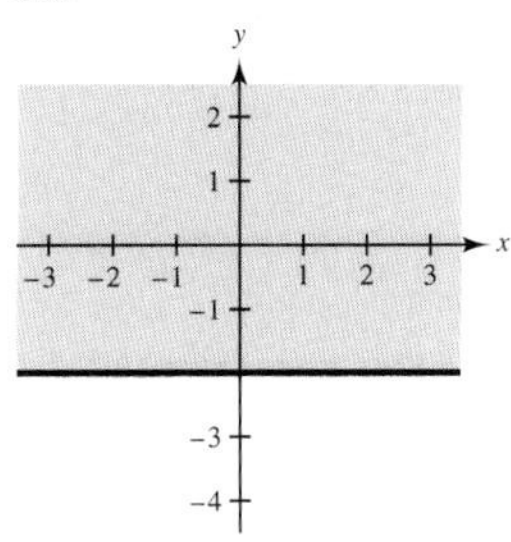

19.

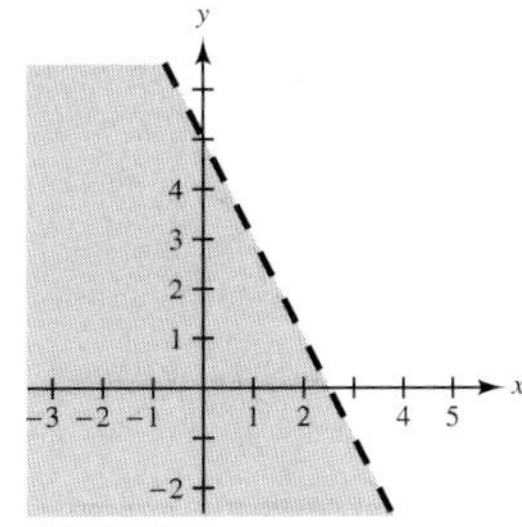

20.

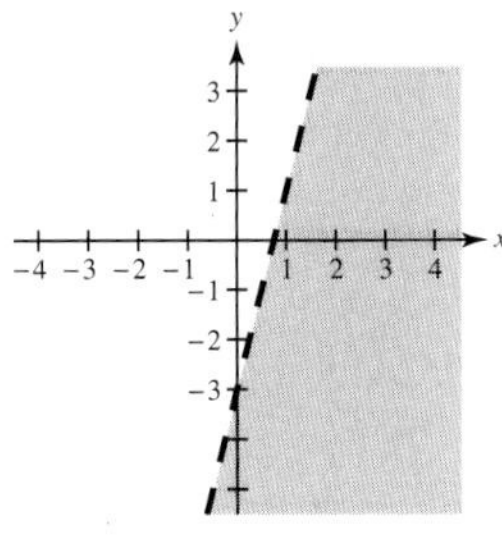

21. Sales are increasing at a rate of 230 units per year.

CHAPTER 5

Section 5.1 *(page 300)*

1. (a) $-3x^8$ (b) $9x^7$ **3.** (a) $-125z^6$ (b) $25z^8$

5. (a) $2u^3v^3$ (b) $-4u^9v$ **7.** (a) $-15u^8$ (b) $64u^5$

9. (a) $-m^{19}n^7$ (b) $-m^7n^3$

11. (a) $3m^4n^3$ (b) $3m^2n^3$ **13.** (a) $\frac{9x^2}{16y^2}$ (b) $\frac{125u^3}{27v^3}$

15. (a) $\frac{8x^4y}{9}$ (b) $-\frac{2x^2y^4}{3}$ **17.** (a) $\frac{25u^8v^2}{4}$ (b) $\frac{u^8v^2}{4}$

19. (a) $x^{2n-1}y^{2n-1}$ (b) $x^{2n-2}y^{n-12}$ **21.** $\frac{1}{25}$

23. $-\frac{1}{1000}$ **25.** 1 **27.** 64 **29.** -32 **31.** $\frac{3}{2}$

33. 1 **35.** 1 **37.** 729 **39.** 100,000 **41.** $\frac{1}{16}$

43. $\frac{1}{64}$ **45.** $\frac{3}{16}$ **47.** $\frac{64}{121}$ **49.** $\frac{16}{15}$ **51.** y^2 **53.** z^2

55. $\frac{7}{x^4}$ **57.** $\frac{1}{64x^3}$ **59.** x^6 **61.** $\frac{4a}{3}$ **63.** t^2

65. $\frac{1}{4x^4}$ **67.** $-\frac{12}{xy^3}$ **69.** $\frac{y^4}{9x^4}$ **71.** $\frac{10}{x}$ **73.** $\frac{x^5}{2y^4}$

75. $\frac{81v^8}{u^6}$ **77.** $\frac{b^5}{a^5}$ **79.** $\frac{1}{2x^8y^3}$ **81.** $6u$ **83.** x^8y^{12}

85. $\frac{2b^{11}}{25a^{12}}$ **87.** $\frac{v^2}{uv^2+1}$ **89.** b **91.** 4 **93.** 144

95. 1 **97.** $-\frac{4}{5}$ **99.** $\frac{1}{144}$ **101.** 3.6×10^6

103. 4.762×10^7 **105.** 3.1×10^{-4} **107.** 3.81×10^{-8}

109. 5.73×10^7 **111.** 9.4608×10^{12} **113.** 8.99×10^{-2}

115. 720,000,000 **117.** 0.0000001359

119. 34,659,000,000 **121.** 15,000,000

123. 0.00000000048 **125.** 6.8×10^5 **127.** 2.5×10^9

129. 6×10^6 **131.** 9×10^{15} **133.** 1.6×10^{12}

135. 3.46×10^{10} **137.** 4.70×10^{11} **139.** 4.43×10^{25}

141. 2.74×10^{20} **143.** 9.3×10^7 miles

145. 84,830,000,000,000,000,000,000 free electrons

147. 1.59×10^{-5} year $\approx$ 8 minutes

149. 8.99×10^{17} meters

151. $\$7.87 \times 10^{12} = \$7,870,000,000,000$

153. Scientific notation makes it easier to multiply or divide very large or very small numbers because the properties of exponents make it more efficient.

155. False. $\frac{1}{3^{-3}} = 27$, which is greater than 1.

157. The Product Rule can be applied only to exponential expressions with the same base.

159. The Power-to-Power Rule applied to the expression raises the base to the *product* of the exponents.

161. $6x$ **163.** $a^2 + 3ab + 3b^2$

165.

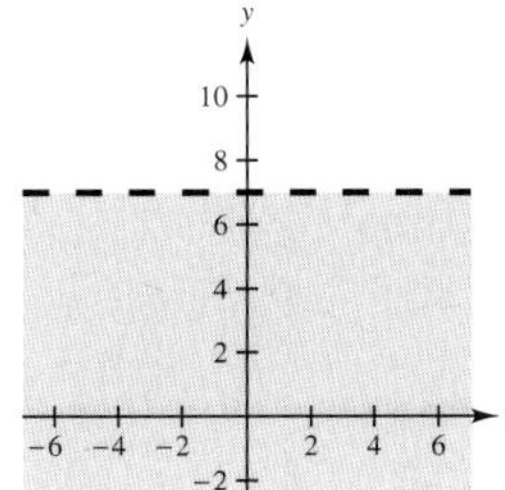

167.

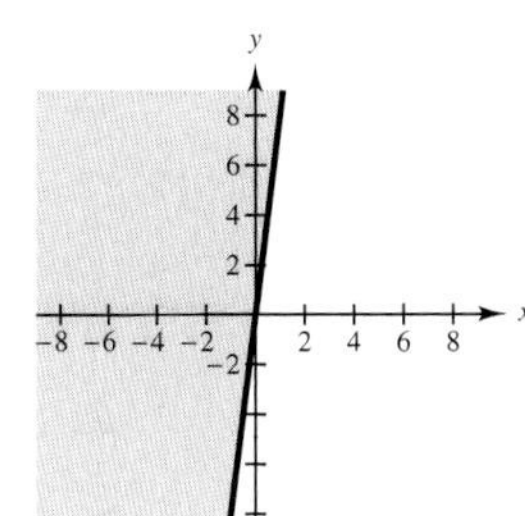

169.

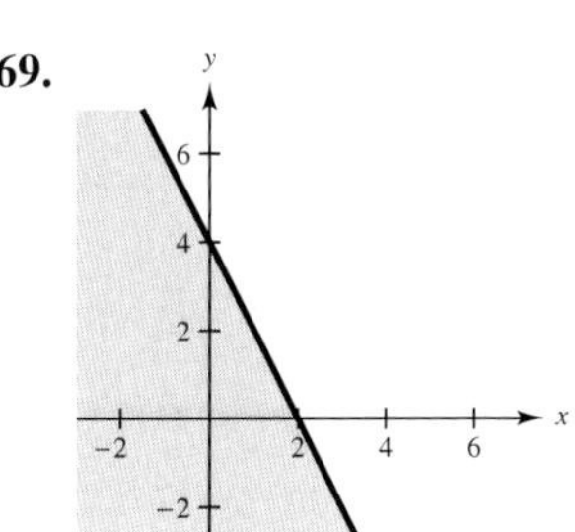

Section 5.2 *(page 310)*

1. Polynomial

3. Not a polynomial because the exponent in the first term is not an integer

5. Not a polynomial because the exponent is negative

7. Polynomial

9. Standard form: $12x + 9$
Degree: 1
Leading coefficient: 12

11. Standard form: $-5x^2 + 7x + 10$
Degree: 2
Leading coefficient: -5

13. Standard form: $-3m^5 - m^2 + 6m + 12$
Degree: 5
Leading coefficient: -3

15. Standard form: 10
Degree: 0
Leading coefficient: 10

17. Standard form: $-16t^2 + v_0t$
Degree: 2
Leading coefficient: -16

19. Binomial **21.** Monomial **23.** Trinomial

25. $5x^3 - 10$ **27.** $3y^2$ **29.** $x^6 - 4x^3 - 2$

31. $20w - 4$ **33.** $4z^2 - z - 2$ **35.** $\frac{3}{2}y^2 + \frac{5}{4}$

37. $1.6t^3 - 3.4t^2 - 7.3$ **39.** $2b^3 - b^2$ **41.** $4b^2 - 3$

43. $5x + 13$ **45.** $-x - 28$ **47.** $3x^2 + 2$

49. $2x^3 + 2x^2 + 8$ **51.** $3x^4 - 2x^3 - 3x^2 - 5x$

53. $4x^2 + 2x + 2$ **55.** $5y^3 + 12$

57. $9x - 11$ **59.** $x^2 - 2x + 2$ **61.** $-3x^3 + 1$

63. $-w^3 + w + 8$ **65.** $-3x^5 - 3x^4 + 2x^3 - 6x + 6$

67. $x - 1$ **69.** $-x^2 - 2x + 3$

71. $-2x^4 - 5x^3 - 4x^2 + 6x - 10$ **73.** $-2x^3$

75. $4t^3 - 3t^2 + 15$ **77.** $-x^3 + 9x^2 - x - 10$

79. $3x^3 + 4x + 10$ **81.** $-2x - 20$

83. $3x^3 - 2x + 2$ **85.** $2x^4 + 9x + 2$

87. $8x^3 + 29x^2 + 11$ **89.** $12z + 8$

91. $y^3 - y^2 - 3y + 7$ **93.** $4t^2 + 20$

95. $6v^2 + 90v + 30$ **97.** $10z + 4$

99. $2x^2 - 2x$ **101.** $21x^2 - 8x$

103. (a) $T = 0.007t^2 + 0.47t + 69.1,\ 10 \le t \le 15$

(b)
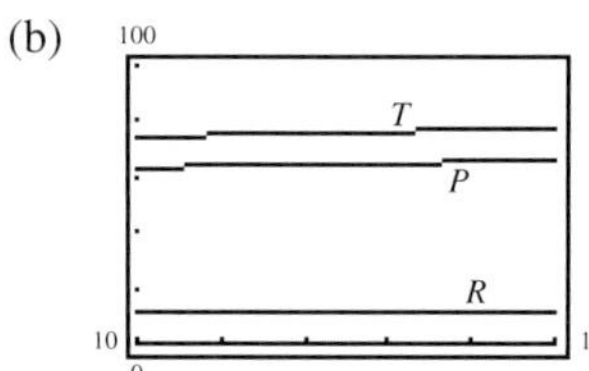

(c) Increasing; increasing; increasing

105. Two terms are like terms if they are both constant or if they have the same variable factor(s); Numerical coefficients.

107. Yes. A polynomial is an algebraic expression whose terms are all of the form ax^k, where a is any real number and k is a nonnegative integer.

109. 6 **111.** $-\frac{3}{5}$

113.
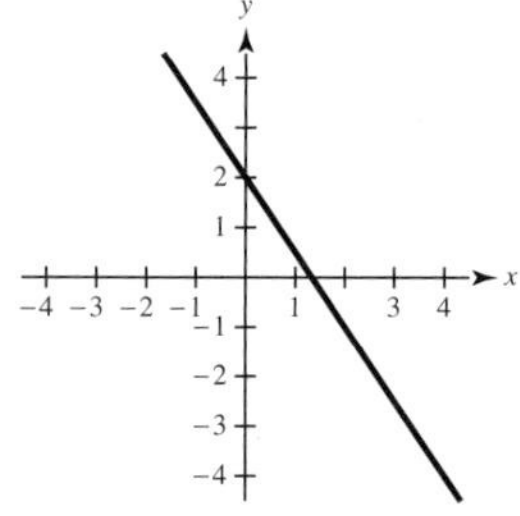

115. -5 **117.** 3.3714×10^{-15}

Mid-Chapter Quiz *(page 314)*

1. $256m^{12}n^8$ **2.** $72x^8y^5$ **3.** $-\dfrac{4}{3x^2y}$ **4.** $\dfrac{t}{12}$

5. $\dfrac{5}{x^2y^3}$ **6.** $\dfrac{3yz}{5x^2}$ **7.** $\dfrac{a^6}{9b^4}$ **8.** 1

9. 8.168×10^{12} **10.** 0.00000005021

11. Because the exponent of the third term is negative

12. Degree: 3
Leading coefficient: -4

13. $3x^5 - 3x + 1$ **14.** $y^2 + 6y + 3$

15. $-v^3 + v^2 + 6v - 5$ **16.** $3s - 11$

17. $3x^2 + 5x - 4$ **18.** $5x^4 + 3x^3 - 2x + 2$

19. $x^3 - x^2 + 5x - 19$ **20.** $x^2 - 2x + 6$

21. $5x^4 + 3x^3 - 4x^2 + 6$ **22.** $10x + 36$

Section 5.3 *(page 323)*

1. $-2x^2$ **3.** $4t^3$ **5.** $\frac{5}{2}x^2$ **7.** $6b^3$

9. $-y^2 + 3y$ **11.** $-x^3 + 4x$ **13.** $-6x^3 - 15x$

15. $24x^4 - 12x^3 - 12x$ **17.** $3x^3 - 6x^2 + 3x$

19. $2x^3 - 4x^2 + 16x$ **21.** $4t^4 - 12t^3$

23. $4x^4 - 3x^3 + x^2$ **25.** $-12x^5 + 18x^4 - 6x^3$

27. $30x^3 + 12x^2$ **29.** $12x^5 - 6x^4$ **31.** $x^2 + 7x + 12$

33. $3x^2 - 2x - 5$ **35.** $2x^2 - 5xy + 2y^2$

37. $15x^2 + 23x + 6$ **39.** $-8x^2 + 18x + 18$

41. $3x^2 - 5xy + 2y^2$ **43.** $3x^3 + 6x^2 - 4x - 8$

45. $4x^2 + 16x + 16$ **47.** $64x^2 + 32x + 4$

49. $15s + 4$ **51.** $-x^6 + 8x^3 + 32x^2 - 2x - 8$

53. $x^2 + 12x + 20$ **55.** $14x^2 - 31x - 10$

57. $x^3 + 3x^2 + x - 1$ **59.** $x^4 - 5x^3 - 2x^2 + 11x - 5$

61. $5x^3 - 8x^2 - 8$ **63.** $x^4 - 6x^3 + 5x^2 - 18x + 6$

65. $9x^4 - 12x^3 - 3x^2 - 4x - 2$ **67.** $x^2 + x - 6$

69. $8x^5 + 12x^4 - 12x^3 - 18x^2 + 18x + 27$

71. $3x^5 + 4x^3 + 7x^2 + x + 7$ **73.** $x^4 - x^2 + 4x - 4$

75. $-2x^4 - 9x^3 + 16x^2 + 11x - 12$

77. $x^3 - 6x^2 + 12x - 8$ **79.** $x^4 - 4x^3 + 6x^2 - 4x + 1$

81. $x^3 - 12x - 16$ **83.** $4u^3 + 4u^2 - 5u - 3$

85. $x^2 - 9$ **87.** $x^2 - 400$ **89.** $4u^2 - 9$

91. $16t^2 - 36$ **93.** $4x^4 - 25$ **95.** $16x^2 - y^2$

97. $81u^2 - 49v^2$ **99.** $x^2 + 12x + 36$

101. $t^2 - 6t + 9$ **103.** $9x^2 + 12x + 4$

105. $64 - 48z + 9z^2$ **107.** $16 + 56s^2 + 49s^4$

109. $4x^2 - 20xy + 25y^2$

111. $x^2 + y^2 + 2xy + 2x + 2y + 1$

113. $u^2 + v^2 - 2uv - 6u + 6v + 9$ **115.** $8x$

117. Yes **119.** $x^3 + 6x^2 + 12x + 8$

121. $(x^2 + 10x)$ square feet **123.** 4 feet × 4 feet

125. $x^2 + 5x + 4 = (x + 1)(x + 4)$

127. $(x + 2)^2 = x^2 + 4x + 4$; Square of a binomial

129. $(x + 4)(x + 5) = x^2 + 9x + 20$

131. (a) $T = 0.01054t^3 - 2.0236t^2 + 18.166t + 6364.16,$
$6 \leq t \leq 15$

(b) 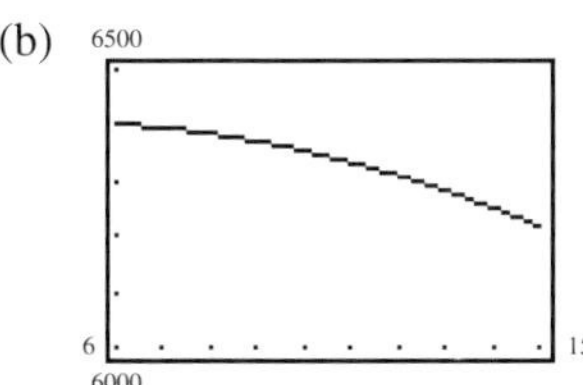

(c) Approximately 6251 million gallons

133. $500r^2 + 1000r + 500$

135. Multiplying a polynomial by a monomial is an application of the Distributive Property. Polynomial multiplication requires repeated use of this property.

$4x(x - 2) = 4x^2 - 8x$

137. The product of the terms of highest degree in each polynomial will be of the form $(ax^m)(bx^n) = abx^{m+n}$. So, the degree of the product is $m + n$.

139. False. $(x - 4)(x + 3) = x^2 - x - 12$

141. $15x - 7$ **143.** $-4x + 17$ **145.** 11.25

147. $33\frac{1}{3}\%$ **149.** 4 **151.** $\frac{15}{4}$

Section 5.4 *(page 334)*

1. $7x^2 - 2x,\ x \neq 0$ **3.** $-4x + 2,\ x \neq 0$

5. $m^3 + 2m - \dfrac{7}{m}$ **7.** $-10z^2 - 6,\ z \neq 0$

9. $v^2 + \frac{5}{2}v - 2,\ v \neq 0$ **11.** $x^3 - \frac{3}{2}x^2 + 3x - 2,\ x \neq 0$

13. $\frac{5}{2}x - 4 + \frac{7}{2}y,\ x \neq 0,\ y \neq 0$ **15.** $112 + \frac{5}{9}$

17. $215 + \frac{2}{3}$ **19.** $x - 5,\ x \neq 3$

21. $x + 10,\ x \neq -5$ **23.** $x - 3 + \dfrac{2}{x - 2}$

25. $x + 7,\ x \neq 3$ **27.** $5x - 8 + \dfrac{19}{x + 2}$

29. $4x + 3 - \dfrac{11}{3x + 2}$ **31.** $3t - 4,\ t \neq \dfrac{3}{2}$

33. $y + 3,\ y \neq -\frac{1}{2}$ **35.** $x^2 + 4,\ x \neq 2$

37. $3x^2 - 3x + 1 + \dfrac{2}{3x + 2}$ **39.** $2 + \dfrac{5}{x + 2}$

41. $x - 4 + \dfrac{32}{x + 4}$ **43.** $\dfrac{6}{5}z + \dfrac{41}{25} + \dfrac{41}{25(5z - 1)}$

45. $4x - 1,\ x \neq -\frac{1}{4}$ **47.** $x^2 - 5x + 25,\ x \neq -5$

49. $x + 2 + \dfrac{1}{x^2 + 2x + 3}$

51. $4x^2 + 12x + 25 + \dfrac{52x - 55}{x^2 - 3x + 2}$

53. $x^3 + x^2 + x + 1,\ x \neq 1$ **55.** $x^3 - x + \dfrac{x}{x^2 + 1}$

57. $7uv,\ u \neq 0,\ v \neq 0$ **59.** $3x + 4,\ x \neq -2$

61. $x + 3,\ x \neq 2$ **63.** $x^2 - x + 4 - \dfrac{17}{x + 4}$

65. $x^3 - 2x^2 - 4x - 7 - \dfrac{4}{x - 2}$

67. $5x^2 + 14x + 56 + \dfrac{232}{x - 4}$

69. $10x^3 + 10x^2 + 60x + 360 + \dfrac{1360}{x - 6}$

71. $0.1x + 0.82 + \dfrac{1.164}{x - 0.2}$ **73.** $(x - 3)(x + 4)(x - 2)$

75. $(x - 1)(2x + 1)^2$ **77.** $(x + 3)^2(x - 3)(x + 4)$

79. $5\left(x - \frac{4}{5}\right)(3x + 2)$ **81.** -8

83. 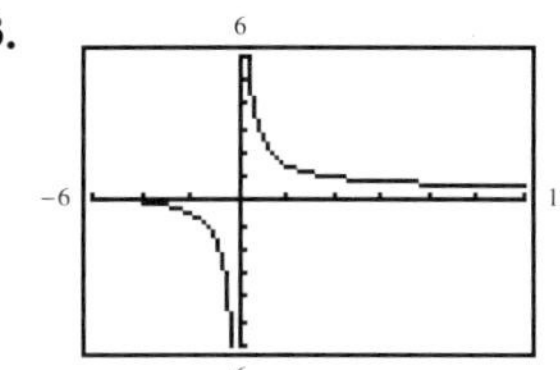

85. $x^{2n} + x^n + 4,\ x^n \neq -2$ **87.** $x^3 - 5x^2 - 5x - 10$

89. $f(k)$ equals the remainder when dividing by $(x - k)$.

k	$f(k)$	Divisor $(x - k)$	Remainder
-2	-8	$x + 2$	-8
-1	0	$x + 1$	0
0	0	x	0
$\frac{1}{2}$	$-\frac{9}{8}$	$x - \frac{1}{2}$	$-\frac{9}{8}$
1	-2	$x - 1$	-2
2	0	$x - 2$	0

91. $x^2 + 2x + 1$ **93.** $2x + 8$

95. x is not a factor of the numerator.

$$\frac{6x + 5y}{x} = \frac{6x}{x} + \frac{5y}{x} = 6 + \frac{5y}{x}$$

97. $\dfrac{x^2 + 4}{x + 1} = x - 1 + \dfrac{5}{x + 1}$

Dividend: $x^2 + 4$; Divisor: $x + 1$;
Quotient: $x - 1$; Remainder: 5

99.

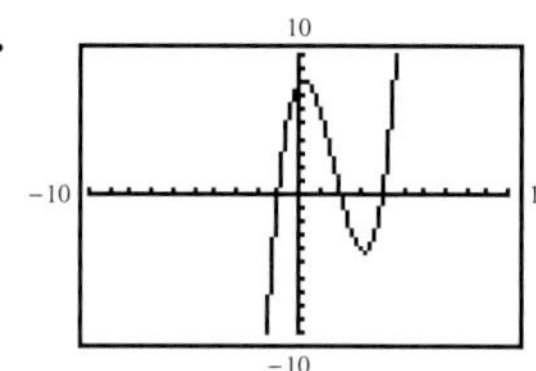

The x-intercepts are $(-1, 0)$, $(2, 0)$, and $(4, 0)$. The polynomials in parts (a), (b), and (c) are all equivalent.

101. $x < \frac{3}{2}$ **103.** $1 < x < 5$ **105.** $x \le -8$ or $x \ge 16$

107. Quadrant II or III **109.** Located on the y-axis

Review Exercises *(page 340)*

1. x^9 **3.** u^6 **5.** $-8z^3$ **7.** $4u^7v^3$ **9.** $2z^3$

11. $\dfrac{5g^2}{16}$ **13.** $144x^4$ **15.** $\frac{1}{72}$ **17.** $\frac{64}{27}$ **19.** $12y$

21. $\dfrac{2}{x^3}$ **23.** 1 **25.** $\dfrac{a^7}{2b^6}$ **27.** $\dfrac{4x^6}{y^5}$ **29.** $\dfrac{405u^5}{v}$

31. 5.38×10^{-5} **33.** 483,300,000 **35.** 3.6×10^7

37. 500

39. Standard form: $-5x^3 + 10x - 4$; Degree: 3; Leading coefficient: -5

41. Standard form: $5x^4 + 4x^3 - 7x^2 - 2x$; Degree: 4; Leading coefficient: 5

43. Standard form: $7x^4 + 11x^2 - 1$; Degree: 4; Leading coefficient: 7

45. $9x$ **47.** $5y^3 + 5y^2 - 12y + 10$

49. $7u^2 + 8u + 5$ **51.** $2x^4 - 7x^2 + 3$

53. $(4x - 6)$ units **55.** 4 **57.** $-6x^2 + 9x - 8$

59. $7y^2 - y + 6$ **61.** $3x^2 + 4x - 14$

63. (a) $-\frac{1}{2}x^2 + 14x - 15$

(b)

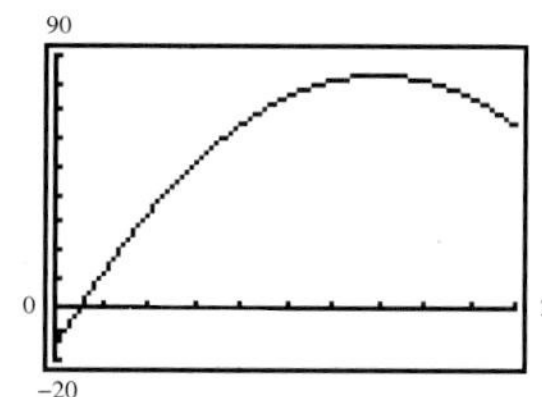

(c) \$83; When x is less than or greater than 14, the profit is less than \$83.

65. $3y^2 - 3y$ **67.** $-30y^2 + 42y^3$ **69.** $u^2 + 3u - 10$

71. $-14x^2 + 62x - 60$ **73.** $s^3 + s^2 - 15s + 9$

75. $4x^3 + 26x^2 - 8x - 10$

77. $15y^3 + 14y^2 + 19y + 36$

79. $x^4 - x^3 - 82x^2 + 124x - 24$

81. $27y^3 - 54y^2 + 36y - 8$

83. $(x^2 + 55x + 750)$ square meters **85.** $x^2 - 10x + 25$

87. $16 + 24b + 9b^2$ **89.** $r^2 - 49$ **91.** $16u^2 - 25v^2$

93. $2x^2 - \frac{1}{2}$, $x \ne 0$ **95.** $3xy - y + 1$, $x \ne 0$, $y \ne 0$

97. $2x^2 + \dfrac{4}{3}x - \dfrac{8}{9} + \dfrac{10}{9(3x - 1)}$ **99.** $x^2 - 2$, $x \ne \pm 1$

101. $x^2 - x - 3 - \dfrac{3x^2 - 2x - 3}{x^3 - 2x^2 + x - 1}$

103. $x^2 + 5x - 7$, $x \ne -2$

105. $x^3 + 3x^2 + 6x + 18 + \dfrac{29}{x - 3}$

107. $(x - 2)(x + 1)(x + 3)$

Chapter Test *(page 343)*

1. $\dfrac{x^4}{9y^6}$ **2.** $\dfrac{u^6}{16v^4}$ **3.** $\dfrac{3y^6}{x}$ **4.** $\dfrac{1}{4x^4y^2z^6}$

5. $-24u^9v^5$ **6.** $\dfrac{27x^6}{2y^4}$

7. (a) 1.5×10^{-4} (b) 80,000,000

8. The variable appears in the denominator.

9. Degree: 4; Leading coefficient: -3

10. $z^5 - 3z^3 + 5$ **11.** $2z^2 - 3z + 15$ **12.** $7u^3 - 1$

13. $-y^2 + 8y + 3$ **14.** $-6x^2 + 12x$

15. $x^2 - 14x + 49$ **16.** $4x^2 - 9$

17. $2z^3 + z^2 - z + 10$

18. $y^5 + 3y^4 + y^3 + 3y^2 - 4y - 12$ **19.** $2z^2 + \frac{1}{2}$

20. $-2x^2 + \frac{3}{2}$ **21.** $x^2 + 2x + 3$

22. $2x^2 + 9x + 17 + \dfrac{63}{2x - 4}$ **23.** $4x^2 - x$

24. $2x^2 + 11x - 6$ **25.** $P = x^2 - 47x - 150$

26. $x - 3$

CHAPTER 6

Section 6.1 *(page 351)*

1. z^2 **3.** $2x$ **5.** u^2v **7.** $3y^5z^4$ **9.** 1

11. $14a^2$ **13.** $x + 3$ **15.** $7x + 5$ **17.** $3(x + 1)$

19. $6(z + 6)$ **21.** $8(t - 2)$ **23.** $-5(5x + 2)$

25. $6(4y^2 - 3)$ **27.** $x(x + 1)$ **29.** $u(25u - 14)$

31. $2x^3(x + 3)$ **33.** No common factor

35. $2x(6x - 1)$ **37.** $-5r(2r^2 + 7)$

39. $4(3x^2 + 4x - 2)$ **41.** $25(4 + 3z - 2z^2)$

43. $3x^2(3x^2 + 2x + 6)$ **45.** $5u(2u + 1)$

47. $8a^3b^3(2 + 3a)$ **49.** $10ab(1 + a)$

51. $4xy(1 + 2x - 6x^3y^4)$ **53.** $-5(2x - 1)$

55. $-10(x + 300)$ **57.** $-(x^2 - 5x - 10)$

59. $-2(x^2 - 6x - 2)$ **61.** $(x - 3)(x + 5)$

63. $(q - 5)(y - 10)$ **65.** $(y + 4)(x^3 + y)$

67. $(a + b)(2a - b)$ **69.** $(x + 10)(x + 1)$

71. $(x + 3)(x + 4)$ **73.** $(x + 3)(x - 5)$

75. $(2x - 7)(2x + 7)$ **77.** $(2x + 1)(3x - 1)$

79. $(x + 4)(8x + 1)$ **81.** $(3x - 7)(3x + 2)$

83. $(x - 2)(2x - 3)$ **85.** $(t - 3)(t^2 + 2)$

87. $(2z + 1)(8z^2 + 1)$ **89.** $(x - 1)(x^2 - 3)$

91. $(x + 3)(x^2 - 6)$ **93.** $(y - 4)(ky + 2)$

95. $(x + 3)$ **97.** $(10y - 1)$ **99.** $(14x + 5y)$

101. $x + 1$ **103.** $6x^2$ **105.** $2\pi r(r + h)$

107. $kx(Q - x)$

109. There are no more common monomials that can be factored out.

111. Sample answers:

$2x^4 + 6x^3 - 4x^2 = 2x^2(x^2 + 3x - 2)$

$9x^2 + 33 = 3(3x^2 + 11)$

113. (a) Not a solution (b) Solution

115. (a) Solution (b) Not a solution

117. $\dfrac{9y^2}{4x^6}$ **119.** $\dfrac{1}{z^4}$ **121.** $3mn$ **123.** $x + 1$

Section 6.2 *(page 359)*

1. $(x + 1)$ **3.** $(y + 12)$ **5.** $(z - 3)$

7. $(x + 1)(x + 11)$; $(x - 1)(x - 11)$

9. $(x + 14)(x + 1)$; $(x - 14)(x - 1)$; $(x + 7)(x + 2)$; $(x - 7)(x - 2)$

11. $(x - 12)(x + 1)$; $(x - 1)(x + 12)$; $(x - 2)(x + 6)$; $(x - 6)(x + 2)$; $(x - 3)(x + 4)$; $(x - 4)(x + 3)$

13. $(x + 4)(x + 2)$ **15.** $(x + 5)(x - 3)$

17. $(x - 11)(x + 2)$ **19.** $(x - 7)(x - 2)$

21. $-(x - 5)(x + 3)$ **23.** $(u - 24)(u + 2)$

25. $(x + 10)(x - 7)$ **27.** $(x + 4)(x + 15)$

29. $(x - 9)(x - 8)$ **31.** Prime **33.** $(x - 9z)(x + 2z)$

35. $(x - 2y)(x - 3y)$ **37.** $(x + 5y)(x + 3y)$

39. $(a + 5b)(a - 3b)$ **41.** $4(x - 5)(x - 3)$

43. Prime **45.** $9(x^2 + 2x - 2)$ **47.** $x(x - 10)(x - 3)$

49. $3x(x + 4)(x + 2)$ **51.** $x^2(x - 2)(x - 3)$

53. $2x^2(x - 3)(x - 7)$ **55.** $x(x + 2y)(x + 3y)$

57. $-3x(y - 3)(y + 6)$ **59.** $2xy(x + 3y)(x - y)$

61. $\pm 9, \pm 11, \pm 19$ **63.** $\pm 2, \pm 34$

65. $\pm 12, \pm 13, \pm 15, \pm 20, \pm 37$

67. Sample answer: $2, -10$ **69.** Sample answer: 3, 4

71. Sample answer: $8, -10$

73. (a) $4x(x - 2)(x - 3)$;
This is equivalent to $x(4 - 2x)(6 - 2x)$, where x, $4 - 2x$, and $6 - 2x$ are the dimensions of the box. The model was found by expanding this expression.

(b)

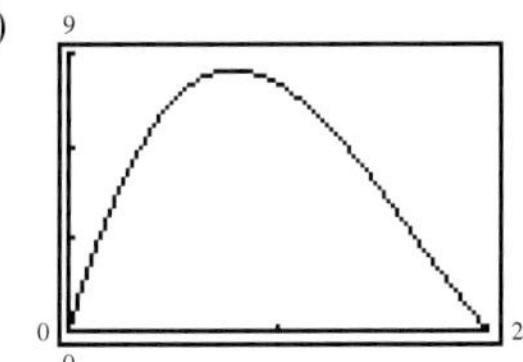

0.785 foot

75. 200 square units

77. (a) and (d); (a) Not completely factored; (d) Completely factored

79. The process is easier if c is a prime number because there will be only the prime number and 1 to test, whereas a composite number will require testing of more pairs of numbers.

81. 7 **83.** -1 **85.** -3 **87.** -2 **89.** $7t^2$

91. xy^2 **93.** $4xy^3$ **95.** $3xy$

Section 6.3 *(page 367)*

1. $(x + 4)$ **3.** $(t + 3)$ **5.** $(x + 2)$ **7.** $(x + 3)$

9. $(5a - 3)$ **11.** $(4z - 1)$ **13.** $(2x - 7)$

15. $(3a - 4)$ **17.** $(3t - 2)$

19. $(5x + 3)(x + 1)$; $(5x - 3)(x - 1)$; $(5x + 1)(x + 3)$; $(5x - 1)(x - 3)$

21. $(5x - 1)(x - 28)$; $(5x - 28)(x - 1)$; $(5x + 1)(x + 28)$; $(5x + 28)(x + 1)$; $(5x - 2)(x - 14)$; $(5x - 14)(x - 2)$; $(5x + 2)(x + 14)$; $(5x + 14)(x + 2)$; $(5x - 4)(x - 7)$; $(5x - 7)(x - 4)$; $(5x + 4)(x + 7)$; $(5x + 7)(x + 4)$

23. $(2x + 3)(x + 1)$ **25.** $(4y + 1)(y + 1)$

27. $(6y - 1)(y - 1)$ **29.** $(12x - 5)(x + 1)$

31. Prime **33.** Prime **35.** $(4s - 1)(2s - 3)$

37. $(x + 4)(4x - 3)$ **39.** $(3x - 2)(3x - 4)$

41. $(3u - 4)(6u + 7)$ **43.** $(5a - 2)(3a + 4)$

45. $(5t + 6)(2t - 3)$ **47.** $(5m - 4)(3m + 5)$

49. $(8z - 5)(2z - 3)$ **51.** $3x(2x - 1)$

53. $5y(3y - 8)$ **55.** $(u - 3)(u + 9)$

57. $2(v + 7)(v - 3)$ **59.** $-3(x^2 + x + 20)$

61. $3(z - 1)(3z - 5)$ **63.** $4(x - 2)(x + 1)$

65. $-x^2(5x + 4)(3x - 2)$ **67.** $x(3x^2 + 4x + 2)$

69. $6x(x - 4)(x + 8)$ **71.** $9u^2(2u^2 + 2u - 3)$

73. $-(2x - 9)(x + 1)$ **75.** $-(3x - 2)(x + 2)$

77. $-(6x + 5)(x - 2)$ **79.** $-(10x - 1)(6x + 1)$

81. $-(5x - 4)(3x + 4)$ **83.** $\pm 11, \pm 13, \pm 17, \pm 31$

85. $\pm 1, \pm 4, \pm 11$

87. $\pm 22, \pm 23, \pm 26, \pm 29, \pm 34, \pm 43, \pm 62, \pm 121$

89. Sample answer: $-1, -7$ **91.** Sample answer: $-8, 3$

93. Sample answer: $-6, -1$ **95.** $(3x + 1)(x + 1)$

97. $(7x - 1)(x + 3)$ **99.** $(3x + 4)(2x - 1)$

101. $(5x - 2)(3x - 1)$ **103.** $(3a + 5)(a + 2)$

105. $(8x - 3)(2x + 1)$ **107.** $(3x - 2)(4x - 3)$

109. $(u - 2)(6u + 7)$ **111.** $l = 2x + 3$

113. $2(x + 5) = 2x + 10$

115. (a) $y_1 = y_2$

(b)

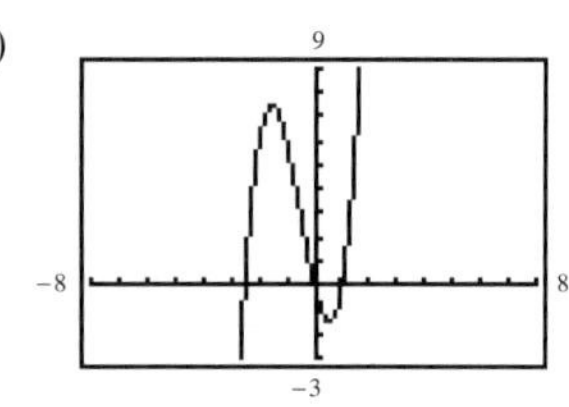

(c) $\left(-\frac{5}{2}, 0\right)$, $(0, 0)$, $(1, 0)$

117. The product of the last terms of the binomials is 15, not -15.

119. Four. $(ax + 1)(x + c)$, $(ax + c)(x + 1)$, $(ax - 1)(x - c)$, $(ax - c)(x - 1)$

121. Sample answer: $2x^3 + 2x^2 + 2x$

123. Factoring by grouping:

$$\begin{aligned} 6x^2 - 13x + 6 &= 6x^2 - (4x + 9x) + 6 \\ &= (6x^2 - 4x) - (9x - 6) \\ &= 2x(3x - 2) - 3(3x - 2) \\ &= (3x - 2)(2x - 3) \end{aligned}$$

$$\begin{aligned} 2x^2 + 5x - 12 &= 2x^2 + (8x - 3x) - 12 \\ &= (2x^2 + 8x) - (3x + 12) \\ &= 2x(x + 4) - 3(x + 4) \\ &= (x + 4)(2x - 3) \end{aligned}$$

$$\begin{aligned} 3x^2 + 11x - 4 &= 3x^2 + (12x - x) - 4 \\ &= (3x^2 + 12x) - (x + 4) \\ &= 3x(x + 4) - (x + 4) \\ &= (x + 4)(3x - 1) \end{aligned}$$

Preferences, advantages, and disadvantages will vary.

125. $3^2 \cdot 5 \cdot 7$ **127.** $5^2 \cdot 7 \cdot 13$ **129.** $9x^2 - 12x + 4$

131. $27y^2 + 81y + 24$ **133.** $(y + 5)(y - 3)$

135. $(x - 10)(x - 2)$

Mid-Chapter Quiz *(page 371)*

1. $(2x - 3)$ **2.** $(x - y)$ **3.** $(y - 6)$

4. $(y + 3)$ **5.** $3(3x^2 + 7)$ **6.** $5a^2b(a - 5b)$

7. $(x + 7)(x - 6)$ **8.** $(t - 3)(t^2 + 1)$

9. $(y + 6)(y + 5)$ **10.** $(u + 8)(u - 7)$

11. $x(x - 6)(x + 5)$ **12.** $2y(x + 8)(x - 4)$

13. $(2y - 9)(y + 3)$ **14.** $(3 + z)(2 - 5z)$

15. $(3x - 2)(4x + 1)$ **16.** $2s^2(5s^2 - 7s + 1)$

17. $\pm 7, \pm 8, \pm 13$; These integers are the sums of the factors of 12.

18. 16, 21; One pair of factors of c has a sum of -10.

19. m and n are factors of 6.

$(3x + 1)(x + 6)$ $(3x - 1)(x - 6)$
$(3x + 6)(x + 1)$ $(3x - 6)(x - 1)$
$(3x + 2)(x + 3)$ $(3x - 2)(x - 3)$
$(3x + 3)(x + 2)$ $(3x - 3)(x - 2)$

20. $10(2x + 8) = 20x + 80$

21.

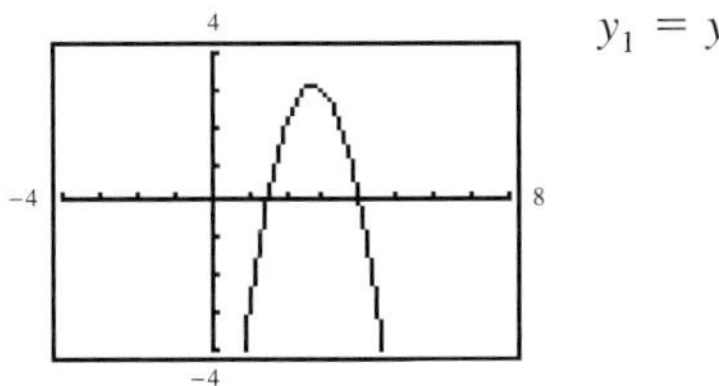

$y_1 = y_2$

Section 6.4 *(page 378)*

1. $(x + 3)(x - 3)$ **3.** $(u + 8)(u - 8)$

5. $(12 + x)(12 - x)$ **7.** $\left(u + \frac{1}{2}\right)\left(u - \frac{1}{2}\right)$

9. $\left(v + \frac{2}{3}\right)\left(v - \frac{2}{3}\right)$ **11.** $(4y + 3)(4y - 3)$

13. $(10 + 7x)(10 - 7x)$ **15.** $(x + 1)(x - 3)$

17. $-z(10 + z)$ **19.** $(6 + a)(2 - a)$

21. $(3y + 5z)(3y - 5z)$ **23.** $2(x + 6)(x - 6)$

25. $x(2 + 5x)(2 - 5x)$ **27.** $2y(2y + 5)(2y - 5)$

29. $(y^2 + 9)(y + 3)(y - 3)$

31. $(1 + x^2)(1 + x)(1 - x)$

33. $2(x^2 + 9)(x + 3)(x - 3)$

35. $(9x^2 + 4y^2)(3x + 2y)(3x - 2y)$ **37.** Yes **39.** No

41. Yes **43.** $(x - 4)^2$ **45.** $(x + 7)^2$ **47.** $\left(b + \frac{1}{2}\right)^2$

49. $(2t + 1)^2$ **51.** $(5y - 1)^2$ **53.** $\left(2x - \frac{1}{4}\right)^2$

55. $(x - 3y)^2$ **57.** $(2y + 5z)^2$ **59.** $(3a - 2b)^2$

61. ± 2 **63.** $\pm \frac{8}{5}$ **65.** ± 36 **67.** 9 **69.** 4

71. $(x - 2)(x^2 + 2x + 4)$ **73.** $(y + 4)(y^2 - 4y + 16)$

75. $(1 + 2t)(1 - 2t + 4t^2)$ **77.** $(3u - 2)(9u^2 + 6u + 4)$

79. $4(x - 2s)(x^2 + 2sx + 4s^2)$

81. $(3x + 4y)(9x^2 - 12xy + 16y^2)$ **83.** $4(x - 7)$

85. $u(u + 3)$ **87.** $5y(y - 5)$ **89.** $5(y + 5)(y - 5)$

91. $y^2(y + 5)(y - 5)$ **93.** $(x - 2y)^2$ **95.** $(x - 1)^2$

97. $(9x + 1)(x + 1)$ **99.** $2x(x - 2y)(x + y)$

101. $(3t + 4)(3t - 4)$ **103.** $-z(z + 12)$

105. $(t + 10)(t - 12)$ **107.** $u(u^2 + 2u + 3)$

109. Prime **111.** $2(t - 2)(t^2 + 2t + 4)$

113. $3(a + 2b)(a^2 - 2ab + 4b^2)$

115. $(x^2 + 9)(x + 3)(x - 3)$

117. $(x^2 + y^2)(x + y)(x - y)$

119. $(x + 1)(x - 1)(x - 4)$ **121.** $x(x + 3)(x + 4)(x - 4)$

123. $(2 + y)(2 - y)(y^2 + 2y + 4)(y^2 - 2y + 4)$

125. 441 **127.** 3599 **129.** $-(4t + 29)(4t - 29)$

131. $(x + 3)^2 - 1^2 = (x + 4)(x + 2)$

133. Box 1: $(a - b)a^2$; Box 2: $(a - b)ab$; Box 3: $(a - b)b^2$ The sum of the volumes of boxes 1, 2, and 3 equals the volume of the large cube minus the volume of the small cube, which is the difference of two cubes.

135. No. $(x + 2)(x - 2)$

137. No. $x^3 - 27 = (x - 3)(x^2 + 3x + 9)$ **139.** 4

141. -1 **143.** $\frac{5}{2}$ **145.** $(2x + 1)(x + 3)$

147. $(3m - 4)(2m + 5)$

Section 6.5 *(page 388)*

1. 0, 4 **3.** $-10, 3$ **5.** $-4, 2$ **7.** $-\frac{5}{2}, -\frac{1}{3}$

9. $-\frac{25}{2}, 0, \frac{3}{2}$ **11.** $-4, -\frac{1}{2}, 3$ **13.** 0, 5 **15.** $-\frac{5}{3}, 0$

17. 0, 16 **19.** 0, 3 **21.** ± 5 **23.** ± 4

25. $-2, 5$ **27.** 4, 6 **29.** $-5, \frac{5}{4}$ **31.** $-\frac{1}{2}, 7$

33. $-1, \frac{2}{3}$ **35.** 4, 9 **37.** 4 **39.** -8 **41.** $\frac{3}{2}$

43. $-2, 10$ **45.** ± 3 **47.** $-4, 9$ **49.** $-12, 6$

51. $-1, \frac{5}{2}$ **53.** $-7, 0$ **55.** $-6, 5$ **57.** $-2, 6$

59. $-5, 1$ **61.** $-2, 8$ **63.** $-13, 5$ **65.** 0, 7, 12

67. $-\frac{1}{3}, 0, \frac{1}{2}$ **69.** ± 2 **71.** $\pm 3, -2$ **73.** ± 3

75. $\pm 1, 0, 3$ **77.** $\pm 2, -\frac{3}{2}, 0$

79. $(-3, 0), (3, 0)$; The x-intercepts are solutions of the polynomial equation.

81. $(0, 0), (3, 0)$; The x-intercepts are solutions of the polynomial equation.

83.

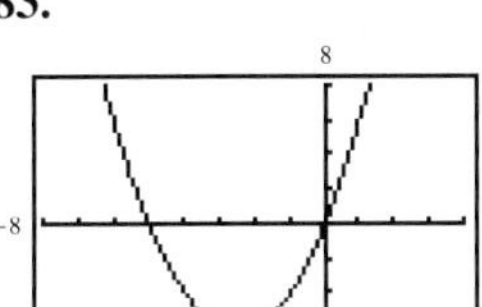

$(-5, 0), (0, 0)$

85.

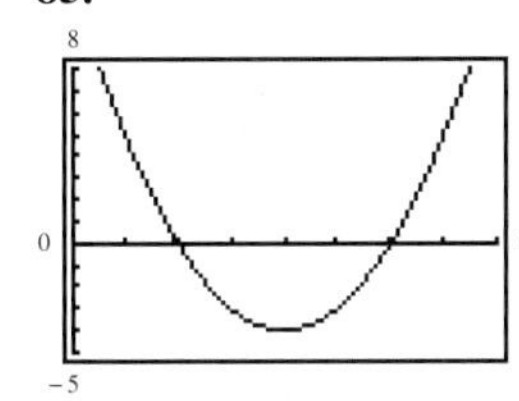

$(2, 0), (6, 0)$

87.

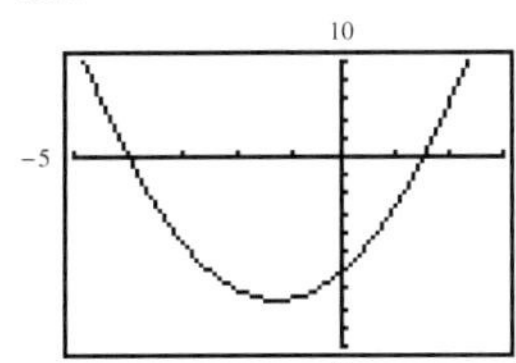

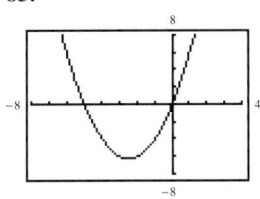

$(-4, 0), \left(\frac{3}{2}, 0\right)$

89.

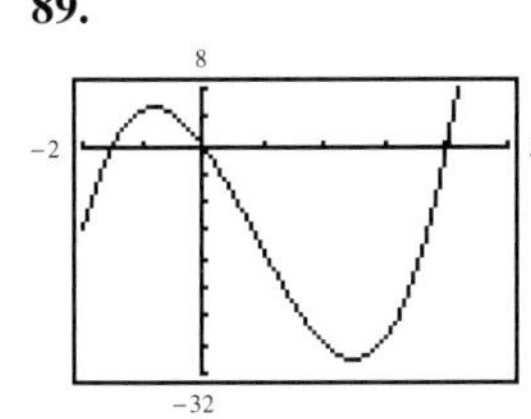

$\left(-\frac{3}{2}, 0\right), (0, 0), (4, 0)$

91. $-\dfrac{b}{a}, 0$ **93.** $x^2 - 4x - 12 = 0$ **95.** 15

97. 20 feet × 27 feet

99. Base: 6 inches; Height: 9 inches

101. 5 seconds **103.** 2 seconds **105.** 9.75 seconds

107. 40 units, 50 units

109. (a) $-6, -\frac{1}{2}$ (b) $-6, -\frac{1}{2}$ (c) Answers will vary.

111. (a) Length $= 5 - 2x$; Width $= 4 - 2x$; Height $= x$

Volume = (Length)(Width)(Height)

$V = (5 - 2x)(4 - 2x)(x)$

(b) $0, 2, \frac{5}{2}$; $0 < x < 2$

(c)

x	0.25	0.50	0.75	1.00	1.25	1.50	1.75
V	3.94	6	6.56	6	4.69	3	1.31

(d) 1.50

(e) 0.74

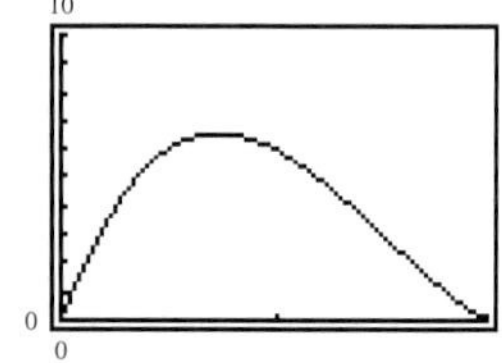

113. Maximum number: n. The third-degree equation $(x + 1)^3 = 0$ has only one real solution: $x = -1$.

115. When a quadratic equation has a repeated solution, the graph of the equation has one x-intercept, which is the vertex of the graph.

117. A solution to a polynomial equation is the value of x when y is zero. If a polynomial is not factorable, the equation can still have real number solutions for x when y is zero.

119. \$0.0625 per ounce **121.** \$0.071 per ounce

123. All real numbers x such that $x \neq -1$

125. All real numbers x such that $x \leq 3$

Review Exercises *(page 394)*

1. t^2 **3.** $3x^2$ **5.** $7x^2y^3$ **7.** $4xy$ **9.** $3(x - 2)$

11. $t(3 - t)$ **13.** $5x^2(1 + 2x)$ **15.** $4a^2(2 - 3a)$

17. $5x(x^2 + x - 1)$ **19.** $4(2y^2 + y + 3)$

21. $(p - 4)(p - 2)$ **23.** $x(3x + 4)$

25. $(x + 1)(x - 3)$ **27.** $(u - 2)(2u + 5)$

29. $(y + 3)(y^2 + 2)$ **31.** $(x^2 + 1)(x + 2)$

33. $(x + 3)(x - 4)$ **35.** $(x - 7)(x + 4)$

37. $(u - 4)(u + 9)$ **39.** $(x - 6)(x + 4)$

41. $(y + 7)(y + 3)$ **43.** $(b + 15)(b - 2)$

45. $(w + 8)(w - 5)$ **47.** $\pm 6, \pm 10$ **49.** ± 12

51. $(x - y)(x + 10y)$ **53.** $(y + 3x)(y - 9x)$

55. $(x + 2y)(x - 4y)$ **57.** $4(x - 2)(x - 4)$

59. $x(x + 3)(x + 6)$ **61.** $4x(x + 2)(x + 7)$

63. $3(x + 9)(x - 3)$ **65.** $-(3x + 5)(x - 1)$

67. $-(x + 10)(x - 5)$ **69.** $(3x + 2)(2x + 1)$

71. $(4y + 1)(y - 1)$ **73.** $(3x - 2)(x + 3)$

75. $(3x - 1)(x + 2)$ **77.** $(2x - 1)(x - 1)$

79. $\pm 2, \pm 5, \pm 10, \pm 23$ **81.** Sample answer: $2, -6$

83. $3(x + 6)(x + 5)$ **85.** $3(2y - 1)(y + 7)$

87. $3u(2u + 5)(u - 2)$ **89.** $4y(2y - 3)(y - 1)$

91. $2x(3x - 2)(x + 3)$ **93.** $3x + 1$

95. $(2x - 7)(x - 3)$ **97.** $(4y - 3)(y + 1)$

99. $(3x - 2)(2x + 5)$ **101.** $(7x + 5)(2x + 1)$

103. $(a + 10)(a - 10)$ **105.** $(5 + 2y)(5 - 2y)$

107. $3(2x + 3)(2x - 3)$ **109.** $(u + 3)(u - 1)$

111. $-(z - 1)(z - 9)$ **113.** $(x + 1)(x - 1)$

115. $3y(y + 5)(y - 5)$ **117.** $st(s + t)(s - t)$

119. $(x^2 + 9)(x + 3)(x - 3)$ **121.** $(x^2 + 4)(x - 2)$

123. $(x - 4)^2$ **125.** $(3s + 2)^2$ **127.** $(y + 2z)^2$

129. $\left(x + \frac{1}{3}\right)^2$ **131.** $(a + 1)(a^2 - a + 1)$

133. $(3 - 2t)(9 + 6t + 4t^2)$

135. $(2x + y)(4x^2 - 2xy + y^2)$ **137.** $0, 2$ **139.** $-\frac{1}{2}, 3$

141. $-10, -\frac{9}{5}, \frac{1}{4}$ **143.** $-\frac{4}{3}, 2$ **145.** $-3, \frac{1}{2}$

147. $-4, 9$ **149.** ± 10 **151.** $-4, 0, 3$ **153.** $0, 2, 9$

155. $\pm 1, 6$ **157.** $\pm 3, 0, 5$ **159.** $9, 11$

161. 45 inches $\times$ 20 inches **163.** 8 inches $\times$ 8 inches

165. 15 seconds

Chapter Test *(page 398)*

1. $9x^2(1 - 7x^3)$ **2.** $(z + 17)(z - 10)$

3. $(t - 10)(t + 8)$ **4.** $(3x - 4)(2x - 1)$

5. $3y(y - 1)(y + 25)$ **6.** $(2 + 5v)(2 - 5v)$

7. $(x + 2)(x^2 - 2x + 4)$ **8.** $-(z + 1)(z + 21)$

9. $(x + 2)(x + 3)(x - 3)$ **10.** $(4 + z^2)(2 + z)(2 - z)$

11. $(2x - 3)$ **12.** ± 6 **13.** 36

14. $3x^2 - 3x - 6 = 3(x + 1)(x - 2)$ **15.** $-4, \frac{3}{2}$

16. $-3, \frac{2}{3}$ **17.** $-\frac{3}{2}, 2$ **18.** $-1, 4$ **19.** $-2, 0, 6$

20. $\pm\sqrt{3}, -7, 0$ **21.** $x + 4$

22. 7 inches $\times$ 12 inches **23.** 8.875 seconds; 5 seconds

24. 24, 26 **25.** 300 feet $\times$ 100 feet

Cumulative Test: Chapters 4–6 *(page 399)*

1. Because $x = -2$, the point must lie in Quadrant II or Quadrant III.

2. (a) Not a solution (b) Solution
(c) Solution (d) Not a solution

3.

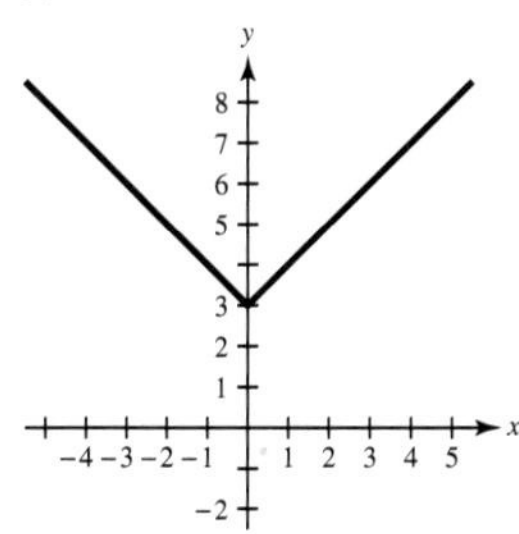

(0, 3)

4.

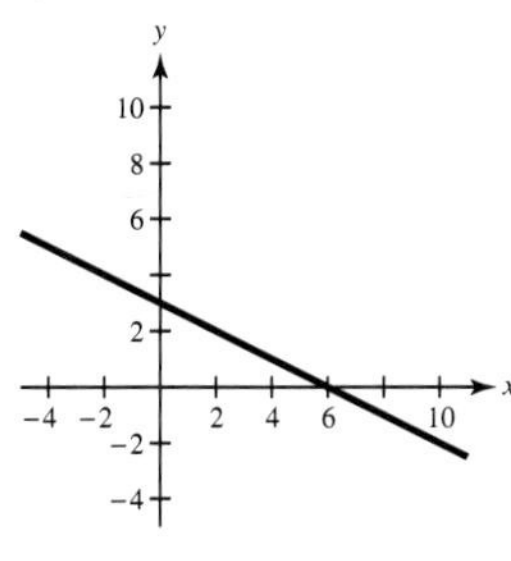

(6, 0), (0, 3)

5. Not a function

6. $(-2, 2)$; There are infinitely many points on a line.

7. $y = \frac{2}{5}x - \frac{3}{2}$

8.

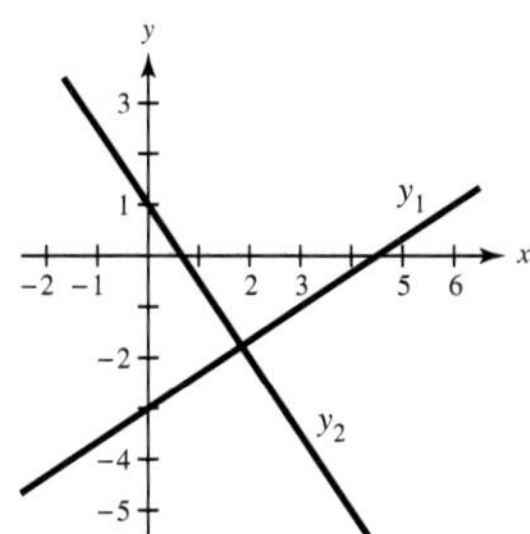

Perpendicular

9.

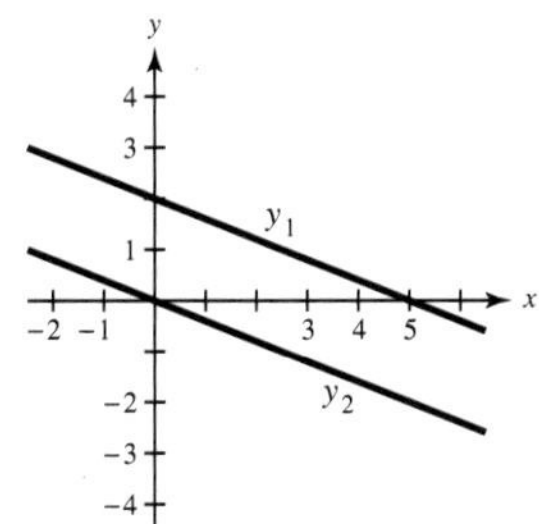

Parallel

10. $-5x^2 + 5$ 11. $-42z^4$ 12. $3x^2 - 7x - 20$

13. $25x^2 - 9$ 14. $25x^2 + 60x + 36$ 15. $x + 12$

16. $x + 1 + \frac{2}{x - 4}$ 17. $\frac{x}{54y^4}$ 18. $2u(u - 3)$

19. $(x + 2)(x - 10)$ 20. $x(x + 4)^2$

21. $(x + 2)^2(x - 2)$ 22. 0, 12 23. $-\frac{3}{5}, 3$ 24. $\frac{4}{x^2}$

25. $C = 150 + 0.45x$; \$181.50 26. 39,142 miles

CHAPTER 7

Section 7.1 *(page 409)*

1. $(-\infty, \infty)$ 3. $(-\infty, 3) \cup (3, \infty)$

5. $(-\infty, 9) \cup (9, \infty)$ 7. $(-\infty, -10) \cup (-10, \infty)$

9. $(-\infty, \infty)$ 11. $(-\infty, -3) \cup (-3, 0) \cup (0, \infty)$

13. $(-\infty, 0) \cup (0, 1) \cup (1, \infty)$

15. $(-\infty, -4) \cup (-4, 4) \cup (4, \infty)$

17. $(-\infty, 0) \cup (0, 3) \cup (3, \infty)$

19. $(-\infty, 2) \cup (2, 3) \cup (3, \infty)$

21. $(-\infty, -1) \cup \left(-1, \frac{5}{3}\right) \cup \left(\frac{5}{3}, \infty\right)$

23. (a) 1 (b) -8 (c) Undefined (division by 0) (d) 0

25. (a) 0 (b) 0 (c) Undefined (division by 0)
(d) Undefined (division by 0)

27. (a) $\frac{25}{22}$ (b) 0 (c) Undefined (division by 0)
(d) Undefined (division by 0)

29. $(0, \infty)$ 31. $\{1, 2, 3, 4, \ldots\}$ 33. $[0, 100)$

35. $x + 3$ 37. $3(x + 16)^2$ 39. $x(x - 2)$

41. $x + 3$ 43. $\frac{x}{5}$ 45. $x, x \neq 0$ 47. $\frac{6x}{5y^3}, x \neq 0$

49. $\frac{x - 3}{4x}$ 51. $x, x \neq 8, x \neq 0$ 53. $\frac{1}{2}, x \neq \frac{3}{2}$

55. $\frac{y + 7}{2}, y \neq 7$ 57. $\frac{1}{a + 3}$ 59. $\frac{x}{x - 7}$

61. $\frac{y(y + 2)}{y + 6}, y \neq 2$ 63. $\frac{y^2(y - 4)}{y - 3}, y \neq -4$

65. $-\frac{3x + 5}{x + 3}, x \neq 4$ 67. $\frac{x + 8}{x - 3}, x \neq -\frac{3}{2}$

69. $\frac{3x - 1}{5x - 4}, x \neq -\frac{4}{5}$ 71. $\frac{3y^2}{y^2 + 1}, x \neq 0$

73. $\frac{y - 8x}{15}, y \neq -8x$ 75. $\frac{5 + 3xy}{y^2}, x \neq 0$

77. $\frac{u - 2v}{u - v}, u \neq -2v$ 79. $\frac{3(m - 2n)}{m + 2n}$

81.

x	-2	-1	0	1	2	3	4
$\frac{x^2 - x - 2}{x - 2}$	-1	0	1	2	Undef.	4	5
$x + 1$	-1	0	1	2	3	4	5

$$\frac{x^2 - x - 2}{x - 2} = \frac{(x - 2)(x + 1)}{x - 2} = x + 1, \; x \neq 2$$

83. $\frac{x}{x + 3}, x > 0$ 85. $\frac{1}{4}, x > 0$

87. (a) $C = 2500 + 9.25x$ (b) $\overline{C} = \frac{2500 + 9.25x}{x}$
(c) $\{1, 2, 3, 4, \ldots\}$ (d) \$34.25

89. (a) Van: $45(t + 3)$; Car: $60t$ (b) $d = |15(9 - t)|$
(c) $\frac{4t}{3(t + 3)}$

91. π, $d > 0$ **93.** $\dfrac{1189.2t + 25{,}266}{-0.35t + 67.1}$, $1 \le t \le 5$

95. The rational expression is in simplified form if the numerator and denominator have no factors in common (other than ± 1).

97. $\dfrac{1}{x^2 + 1}$

99. The student incorrectly divided (the denominator may not be split up) and the domain is not restricted.

Correct solution: $\dfrac{x^2 + 7x}{x + 7} = \dfrac{x(x + 7)}{x + 7} = x,\ x \neq -7$

101. To write the polynomial $g(x)$, multiply $f(x)$ by $(x - 2)$ and divide by $(x - 2)$.

$g(x) = \dfrac{f(x)(x - 2)}{(x - 2)} = f(x),\ x \neq 2$

103. $\frac{3}{16}$ **105.** 1 **107.** $4a^4$

109. $-3b^3 + 9b^2 - 15b$

Section 7.2 *(page 419)*

1. x^2 **3.** $(x + 2)^2$ **5.** $u + 1$ **7.** $-(2 + x)$

9. $\dfrac{7}{3}$, $x \neq 0$ **11.** $\dfrac{s^3}{6}$, $s \neq 0$ **13.** $24u^2$, $u \neq 0$

15. 24, $x \neq -\dfrac{3}{4}$ **17.** $\dfrac{2uv(u + v)}{3(3u + v)}$, $u \neq 0$

19. -1, $r \neq 12$ **21.** $-\dfrac{x + 8}{x^2}$, $x \neq \dfrac{3}{2}$

23. $4(r + 2)$, $r \neq 3$, $r \neq 2$ **25.** $2t + 5$, $t \neq 3$, $t \neq -2$

27. $-\dfrac{xy(x + 2y)}{x - 2y}$ **29.** $\dfrac{(x - y)^2}{x + y}$, $x \neq -3y$

31. $\dfrac{(x - 1)(2x + 1)}{(3x - 2)(x + 2)}$, $x \neq \pm 5$, $x \neq -1$

33. $-\dfrac{x^2(x + 3)(x - 3)(2x + 5)(3x - 1)}{2(2x + 1)(2x + 3)(2x - 3)}$, $x \neq 0$, $x \neq \dfrac{1}{2}$

35. $\dfrac{(x + 3)^2}{x}$, $x \neq 3$, $x \neq \pm 2$ **37.** $\dfrac{x(x + 1)}{3(x + 2)}$, $x \neq -1$

39. $\dfrac{4x}{3}$, $x \neq 0$ **41.** $\dfrac{6}{x}$ **43.** $\dfrac{3y^2}{2ux^2}$, $v \neq 0$

45. $\dfrac{3}{2(a + b)}$ **47.** $\frac{4}{3}$, $x \neq -2$, $x \neq 0$, $x \neq 1$

49. $x^4y(x + 2y)$, $x \neq 0$, $y \neq 0$, $x \neq -2y$

51. $\dfrac{x - 4}{x - 5}$, $x \neq -6$, $x \neq -5$, $x \neq 3$

53. $\dfrac{x + 4}{3}$, $x \neq -2$, $x \neq 0$

55. $\dfrac{1}{4}$, $x \neq -1$, $x \neq 0$, $y \neq 0$

57. $\dfrac{(x + 1)(2x - 5)}{x}$, $x \neq -5$, $x \neq -1$, $x \neq -\dfrac{2}{3}$

59. $\dfrac{x^4}{(x^n + 1)^2}$, $x^n \neq -3$, $x^n \neq 3$, $x \neq 0$

61.

63. $\dfrac{w(2w + 3)}{6}$ **65.** $\dfrac{x}{4(2x + 1)}$ **67.** $\dfrac{\pi x}{4(2x + 1)}$

69. $Y = \dfrac{-0.696t + 8.94}{(-0.092t + 1)(0.352t + 15.97)}$, $1 \le t \le 6$

71. In simplifying a product of rational expressions, you divide the common factors out of the numerator and denominator.

73. The domain needs to be restricted, $x \neq a$, $x \neq b$.

75. The first expression needs to be multiplied by the reciprocal of the second expression, and the domain needs to be restricted.

$$\frac{x^2 - 4}{5x} \div \frac{x + 2}{x - 2} = \frac{x^2 - 4}{5x} \cdot \frac{x - 2}{x + 2}$$
$$= \frac{(x - 2)^2(x + 2)}{5x(x + 2)}$$
$$= \frac{(x - 2)^2}{5x},\ x \neq \pm 2$$

77. $\frac{9}{8}$ **79.** $\frac{13}{15}$ **81.** $-3, 0$ **83.** $\pm\frac{5}{2}$

Section 7.3 *(page 428)*

1. $\dfrac{3x}{2}$ **3.** $-\dfrac{3}{a}$ **5.** $-\dfrac{2}{9}$ **7.** $\dfrac{2z^2 - 2}{3}$ **9.** $\dfrac{x + 6}{3x}$

11. 1, $y \neq 6$ **13.** $\dfrac{1}{x - 3}$, $x \neq 0$ **15.** $\dfrac{1}{w - 2}$, $w \neq -2$

17. $\dfrac{1}{c + 4}$, $c \neq 1$ **19.** $-\dfrac{4}{3}$ **21.** $x - 6$, $x \neq 3$

23. $20x^3$ **25.** $36y^3$ **27.** $15x^2(x + 5)$

29. $126z^2(z + 1)^4$ **31.** $56t(t + 2)(t - 2)$

33. $2y(y + 1)(2y - 1)$ **35.** x^2 **37.** $u + 1$

39. $-(x + 2)$ **41.** $\dfrac{2n^2(n + 8)}{6n^2(n - 4)}, \dfrac{10(n - 4)}{6n^2(n - 4)}$

43. $\dfrac{2(x + 3)}{x^2(x + 3)(x - 3)}, \dfrac{5x(x - 3)}{x^2(x + 3)(x - 3)}$

45. $\dfrac{3v^2}{6v^2(v + 1)}, \dfrac{8(v + 1)}{6v^2(v + 1)}$

47. $\dfrac{(x - 8)(x - 5)}{(x + 5)(x - 5)^2}, \dfrac{9x(x + 5)}{(x + 5)(x - 5)^2}$ **49.** $\dfrac{-12x + 25}{20x}$

51. $\frac{7(a+2)}{a^2}$ **53.** $\frac{5(5x+22)}{x+4}$ **55.** $0,\ x \neq 4$

57. $\frac{3(x+2)}{x-8}$ **59.** $1,\ x \neq \frac{2}{3}$ **61.** $\frac{3(8v-3)}{5v(v-1)}$

63. $\frac{x^2-7x-15}{(x+3)(x-2)}$ **65.** $-\frac{2}{x+3},\ x \neq 3$

67. $\frac{5(x+1)}{(x+5)(x-5)}$ **69.** $\frac{4}{x^2(x^2+1)}$

71. $\frac{6}{(x-6)(x+5)}$ **73.** $\frac{4x}{(x-4)^2}$

75. $\frac{y-x}{xy},\ x \neq -y$ **77.** $\frac{2(4x^2+5x-3)}{x^2(x+3)}$

79. $-\frac{u^2-uv-5u+2v}{(u-v)^2}$ **81.** $\frac{x}{x-1},\ x \neq -6$

83.

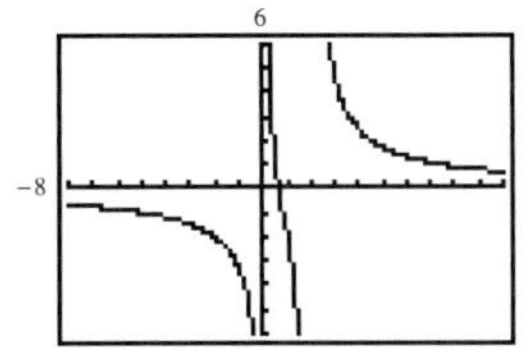

85. $\frac{5t}{12}$

87. $A = -4,\ B = 2,\ C = 2$

89. $T = \frac{633.50t^2 + 5814.42t + 13{,}159.79}{(0.183t+1)(0.205t+1)},\ 0 \le t \le 5$

91. When the numerators are subtracted, the result should be $(x-1)-(4x-11) = x-1-4x+11$.

93. Yes. The LCD of $4x+1$ and $\frac{x}{x+2}$ is $x+2$.

95. $2v+4$ **97.** x^2-2x-3 **99.** $(x-3)(x-4)$

101. $(a-6)(2a+3)$

Mid-Chapter Quiz *(page 432)*

1. $(-\infty, -1) \cup (-1, 0) \cup (0, \infty)$

2. (a) $\frac{1}{2}$ (b) $\frac{1}{2}$ (c) $-\frac{3}{2}$ (d) 0

3. (a) 0 (b) $\frac{9}{2}$ (c) Undefined (d) $\frac{8}{9}$

4. $\frac{3}{2}y,\ y \neq 0$ **5.** $\frac{2u^3}{5},\ u \neq 0, v \neq 0$

6. $-\frac{2x+1}{x},\ x \neq \frac{1}{2}$ **7.** $\frac{z+3}{2z-1},\ z \neq -3$

8. $\frac{5a+3b^2}{ab}$ **9.** $\frac{n^2}{m+n},\ 2m \neq n$ **10.** $\frac{t}{2},\ t \neq 0$

11. $\frac{5x}{x-2},\ x \neq -2$ **12.** $\frac{8x}{3(x+3)(x-1)^2}$

13. $\frac{2x^3y}{z},\ x \neq 0, y \neq 0$ **14.** $\frac{(a+1)^2}{9(a+b)^2},\ a \neq b, a \neq -1$

15. $\frac{2(x+1)}{3x},\ x \neq -2,\ x \neq -1$

16. $\frac{30}{x+5},\ x \neq 0,\ x \neq 1$

17. $\frac{4(u-v)^2}{5uv},\ u \neq \pm v$ **18.** $\frac{7x-11}{x-2}$

19. $-\frac{4x^2-25x+36}{(x-3)(x+3)}$ **20.** $0,\ x \neq 2,\ x \neq -1$

21. (a) $C = 25{,}000 + 144x$ (b) $\overline{C} = \frac{25{,}000 + 144x}{x}$

(c) \$194

Section 7.4 *(page 437)*

1. $\frac{1}{4}$ **3.** $6xz^3,\ x \neq 0,\ y \neq 0,\ z \neq 0$

5. $\frac{2xy^2}{5},\ x \neq 0,\ y \neq 0$ **7.** $-\frac{1}{y},\ y \neq 3$

9. $-\frac{5x(x+1)}{2},\ x \neq -1,\ x \neq 0,\ x \neq 5$

11. $\frac{x+5}{3(x+4)},\ x \neq 2$ **13.** $\frac{2(x+3)}{x-2},\ x \neq -3,\ x \neq 7$

15. $\frac{(2x-5)(3x+1)}{3x(x+1)},\ x \neq \pm\frac{1}{3}$

17. $\frac{(x+3)(4x+1)}{(3x-1)(x-1)},\ x \neq -3,\ x \neq -\frac{1}{4}$

19. $x+2,\ x \neq \pm 2,\ x \neq -3$

21. $-\frac{(x-5)(x+2)^2}{(x-2)^2(x+3)},\ x \neq -2,\ x \neq 7$ **23.** $\frac{y+4}{y^2}$

25. $-\frac{3x+4}{3x-4},\ x \neq 0$ **27.** $\frac{x^2}{2(2x+3)},\ x \neq 0$

29. $\frac{3}{4},\ x \neq 0,\ x \neq 3$ **31.** $\frac{5(x+3)}{2x(5x-2)}$

33. $y-x,\ x \neq 0,\ y \neq 0,\ x \neq -y$ **35.** $\frac{x+y}{x-y},\ x \neq 2y$

37. $-\frac{(y-1)(y-3)}{y(4y-1)},\ y \neq 3$ **39.** $\frac{20}{7},\ x \neq -1$

41. $\frac{1}{x},\ x \neq -1$ **43.** $\frac{x(x+6)}{3x^3+10x-30},\ x \neq 0,\ x \neq 3$

45. $\frac{y(2y^2-1)}{10y^2-1},\ y \neq 0$ **47.** $\frac{x^2(7x^3+2)}{x^4+5},\ x \neq 0$

49. $\frac{y+x}{y-x},\ x \neq 0,\ y \neq 0$ **51.** $\frac{y-x}{x^2y^2(y+x)}$

53. $-\frac{1}{2(h+2)},\ h \neq 0$ **55.** $\frac{11x}{60}$ **57.** $\frac{11x}{24}$

59. $\frac{b^2+5b+8}{8b}$ **61.** $\frac{x}{8}, \frac{5x}{36}, \frac{11x}{72}$

63. $\frac{R_1R_2}{R_1+R_2},\ R_1 \neq 0,\ R_2 \neq 0$

65. (a)

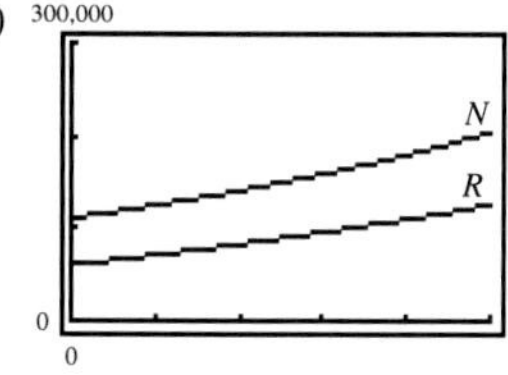

(b) $B = \dfrac{250(-487.42t^2 + 4510.08t + 60{,}227.5)}{3(-257.34t^2 + 1992.05t + 111{,}039.2)}$,

$0 \le t \le 5$

67. No. A complex fraction can be written as the division of two rational expressions, so the simplified form will be a rational expression.

69. In the second step, the parentheses cannot be moved because division is not associative.

$\dfrac{(a/b)}{b} = \dfrac{a}{b} \cdot \dfrac{1}{b} = \dfrac{a}{b^2}$

71. $72y^5$ **73.** $(3x - 1)(x + 2)$ **75.** $\dfrac{x}{8}$, $x \ne 0$

77. $(x + 1)(x + 2)^2$, $x \ne -2, x \ne -1$

Section 7.5 *(page 446)*

1. (a) Not a solution (b) Not a solution

(c) Not a solution (d) Solution

3. (a) Not a solution (b) Solution

(c) Solution (d) Not a solution

5. 10 **7.** 1 **9.** 0 **11.** 8 **13.** $-\frac{40}{9}$

15. $-3, \frac{8}{3}$ **17.** $-\frac{2}{9}$ **19.** $\frac{7}{4}$ **21.** $\frac{43}{8}$ **23.** 61

25. $\frac{18}{5}$ **27.** $-\frac{26}{5}$ **29.** 3 **31.** 3 **33.** $-\frac{11}{5}$

35. $\frac{4}{3}$ **37.** ± 6 **39.** ± 4 **41.** $-9, 8$ **43.** 3, 6

45. No solution **47.** -5 **49.** 8 **51.** 3 **53.** 5

55. $-\frac{11}{10}, 2$ **57.** 20 **59.** $\frac{3}{2}$ **61.** $3, -1$ **63.** $\frac{17}{4}$

65. $-3, 4$ **67.** (a) and (b) $(-2, 0)$

69. (a) and (b) $(-1, 0), (1, 0)$

71. (a)

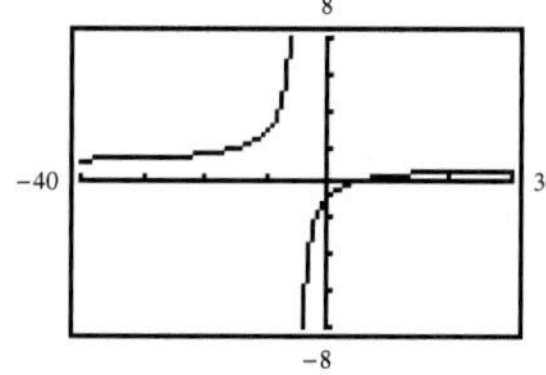

(b) $(4, 0)$

73. (a)

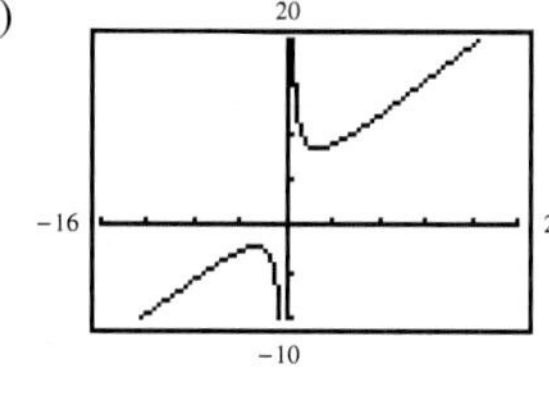

(b) No x-intercepts

75. (a)

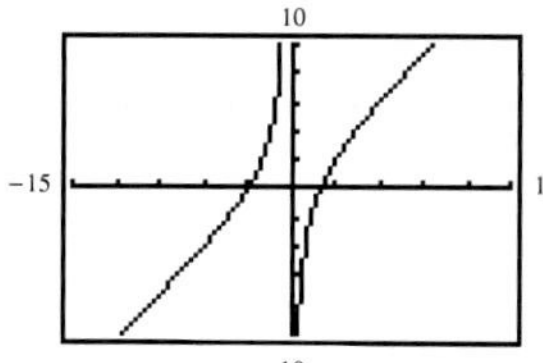

(b) $(-3, 0), (2, 0)$

77. -12 **79.** $\dfrac{x^2 + 2x + 8}{(x + 4)(x - 4)}$ **81.** $6, \dfrac{1}{6}$

83. 2.4 hours = 2 hours 24 minutes

85. 40 miles per hour **87.** 121 saves

89. An extraneous solution is a "trial solution" that does not satisfy the original equation.

91. When the equation is solved, the solution is $x = 0$. However, if $x = 0$, then there is division by zero, so the equation has no solution.

93. $(x + 9)(x - 9)$ **95.** $\left(2x - \frac{1}{2}\right)\left(2x + \frac{1}{2}\right)$

97. $(-\infty, \infty)$

Section 7.6 *(page 457)*

1. $I = kV$ **3.** $V = kt$ **5.** $u = kv^2$ **7.** $p = k/d$

9. $A = k/t^4$ **11.** $A = klw$ **13.** $P = k/V$

15. Area varies jointly as the base and the height.

17. Volume varies jointly as the square of the radius and the height.

19. Average speed varies directly as the distance and inversely as the time.

21. $s = 5t$ **23.** $F = \frac{5}{16}x^2$ **25.** $n = 48/m$

27. $g = 4/\sqrt{z}$ **29.** $F = \frac{25}{6}xy$ **31.** $d = 120x^2/r$

33.

x	2	4	6	8	10
$y = kx^2$	4	16	36	64	100

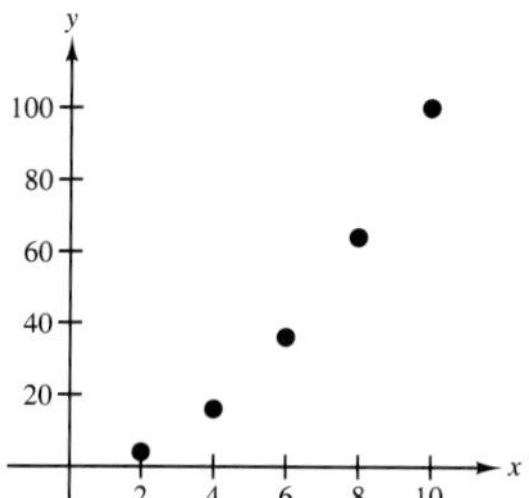

35.

x	2	4	6	8	10
$y = kx^2$	2	8	18	32	50

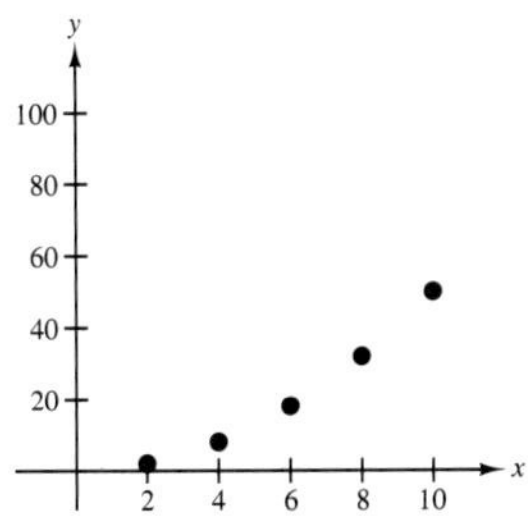

37.

x	2	4	6	8	10
$y = \frac{k}{x^2}$	$\frac{1}{2}$	$\frac{1}{8}$	$\frac{1}{18}$	$\frac{1}{32}$	$\frac{1}{50}$

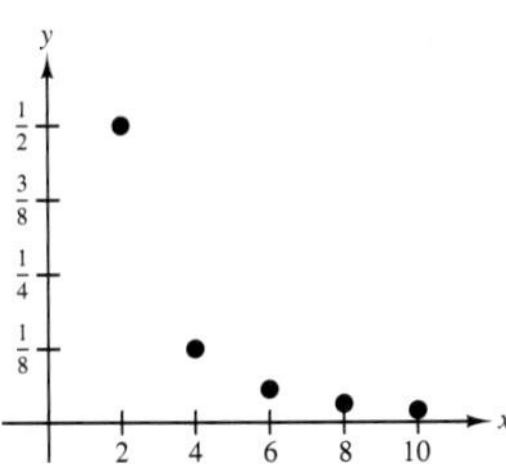

39.

x	2	4	6	8	10
$y = \frac{k}{x^2}$	$\frac{5}{2}$	$\frac{5}{8}$	$\frac{5}{18}$	$\frac{5}{32}$	$\frac{1}{10}$

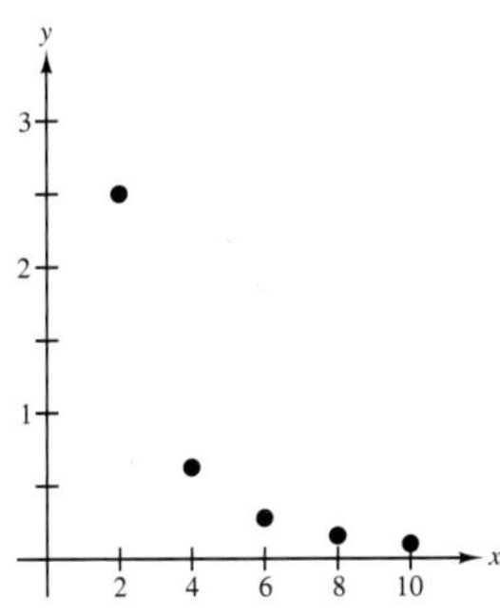

41. $y = k/x$ with $k = 4$

43. 7.5 miles per hour; 9 miles per hour **45.** 10 people

47. 180 minutes **49.** 85%

51. (a) $\{1, 2, 3, 4, . . .\}$

(b) 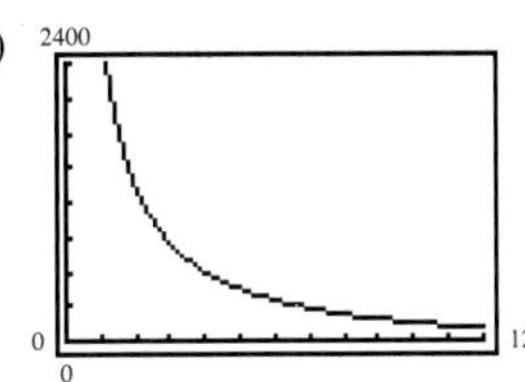

(c) 10d (d) 10d

53. \$5983; Price per unit

55. (a) 2 inches (b) 15 pounds **57.** 18 pounds

59. -128 feet per second **61.** 192 feet

63. 1350 watts **65.** 667 boxes **67.** $p = \frac{114}{t}$; 18%

69. \$6 per pizza; Answers will vary. **71.** \$22.50

73. (a) $P = \frac{kWD^2}{L}$

(b) Unchanged

(c) Increases by a factor of 8

(d) Increases by a factor of 4

(e) Decreases by a factor of $\frac{1}{4}$

(f) 3125 pounds

75. False. If x increases, then z and y do not both necessarily increase.

77. The variable y will be one-fourth as great. If $y = k/x^2$ and x is replaced with $2x$, the result is

$$y = \frac{k}{(2x)^2} = \frac{k}{4x^2}.$$

79. 6^4 **81.** $\left(\frac{1}{5}\right)^5$ **83.** $x - 7, \ x \neq -2$

85. $4x^4 - 2x^3, \ x \neq 3$

Review Exercises *(page 464)*

1. $(-\infty, 8) \cup (8, \infty)$ **3.** $(-\infty, \infty)$

5. $(-\infty, 1) \cup (1, 6) \cup (6, \infty)$ **7.** $(0, \infty)$

9. $\frac{2x^3}{5}, \ x \neq 0, \ y \neq 0$ **11.** $\frac{b - 3}{6(b - 4)}$ **13.** $-9, \ x \neq y$

15. $\frac{x}{2(x + 5)}, \ x \neq 5$ **17.** $\frac{1}{2}, \ x > 0$ **19.** $\frac{x}{3}, \ x \neq 0$

21. $\frac{y}{8x}, \ y \neq 0$ **23.** $12z(z - 6), \ z \neq -6$

25. $-\frac{1}{4}, \ u \neq 0, \ u \neq 3$ **27.** $20x^3, \ x \neq 0$

29. $\frac{125y}{x}, \ y \neq 0$ **31.** $\frac{1}{3x - 2}, \ x \neq -2, \ x \neq -1$

33. $\frac{x(x - 1)}{x - 7}, \ x \neq -1, \ x \neq 1$ **35.** $3x$ **37.** $\frac{4}{x}$

39. $-\dfrac{x-13}{4x}$ **41.** $\dfrac{5y+11}{2y+1}$ **43.** $\dfrac{7x-16}{x+2}$

45. $\dfrac{2x+3}{5x^2}$ **47.** $\dfrac{4x+3}{(x+5)(x-12)}$

49. $\dfrac{5x^3-5x^2-31x+13}{(x-3)(x+2)}$ **51.** $\dfrac{2x+17}{(x-5)(x+4)}$

53. $\dfrac{6(x-9)}{(x+3)^2(x-3)}$

55.

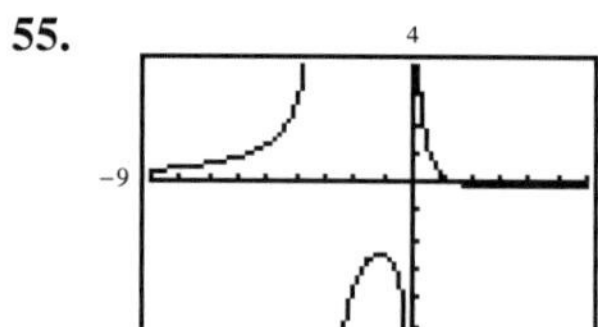

$y_1 = y_2$

57. $3x^2,\ x \neq 0$

59. $-\dfrac{1}{2},\ x \neq 0, x \neq 2$ **61.** $\dfrac{6(x+5)}{x(x+7)},\ x \neq \pm 5$

63. $\dfrac{3t^2}{5t-2},\ t \neq 0$ **65.** $x-1,\ x \neq 0, x \neq 2$

67. $\dfrac{-a^2+a+16}{(4a^2+16a+1)(a-4)},\ a \neq -4, a \neq 0$

69. 6 **71.** -120 **73.** $2, -\frac{3}{2}$ **75.** $\frac{36}{23}$ **77.** 5

79. $-4, 6$ **81.** $-\frac{16}{3}, 3$ **83.** $-4, -2$

85. No solution **87.** $-2, 2$ **89.** $-\frac{9}{5}, 3$

91. 16 miles per hour, 10 miles per hour **93.** 4 people

95. 8 years **97.** 150 pounds **99.** 2.44 hours

101. \$61.49; Answers will vary. **103.** \$922.50

Chapter Test *(page 469)*

1. $(-\infty, 1) \cup (1, 5) \cup (5, \infty)$ **2.** $-2,\ x \neq 2$

3. $\dfrac{2a+3}{5},\ a \neq 4$ **4.** $3x^3(x+4)^2$ **5.** $\dfrac{5z}{3},\ z \neq 0$

6. $\dfrac{4}{y+4},\ y \neq 2$ **7.** $\dfrac{14y^6}{15},\ x \neq 0$

8. $(2x-3)^2(x+1),\ x \neq -\frac{3}{2}, x \neq -1, x \neq \frac{3}{2}$

9. $\dfrac{x+1}{x-3}$ **10.** $\dfrac{-2x^2+2x+1}{x+1}$ **11.** $\dfrac{5x^2-15x-2}{(x-3)(x+2)}$

12. $\dfrac{5x^3+x^2-7x-5}{x^2(x+1)^2}$ **13.** $\dfrac{x^3}{4},\ x \neq -2, x \neq 0$

14. $-(3x+1),\ x \neq 0, x \neq \frac{1}{3}$

15. $\dfrac{(3y+x^2)(x+y)}{x^2y},\ x \neq -y$ **16.** $3x - 2 + \dfrac{4}{x}$

17. 16 **18.** $-1, -\frac{15}{2}$ **19.** No solution

20. $v = \frac{1}{4}\sqrt{u}$ **21.** 240 cubic meters

CHAPTER 8

Section 8.1 *(page 483)*

1. (a) Solution (b) Not a solution

3. (a) Not a solution (b) Solution

5. (a) Solution (b) Solution **7.** $(2, 0)$

9. $(-1, -1)$ **11.** Infinitely many solutions

13. No solution

15. $(1, 2)$

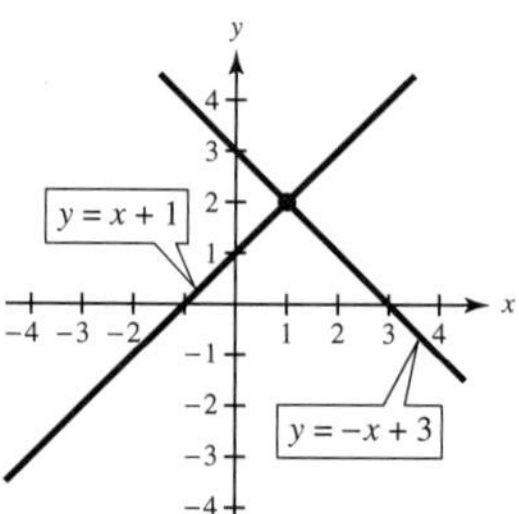

17. $(2, 0)$

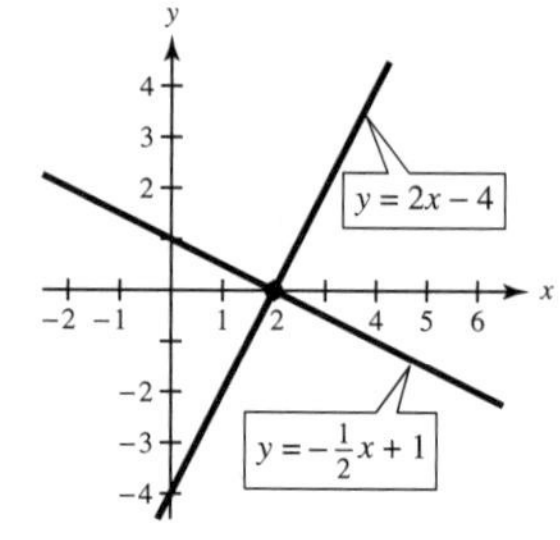

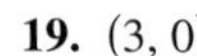

19. $(3, 0)$

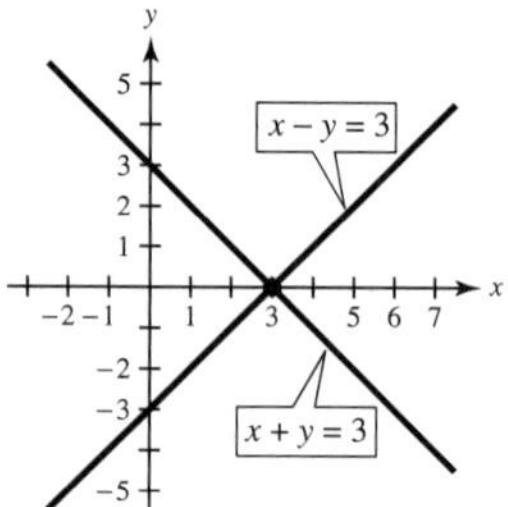

21. No solution

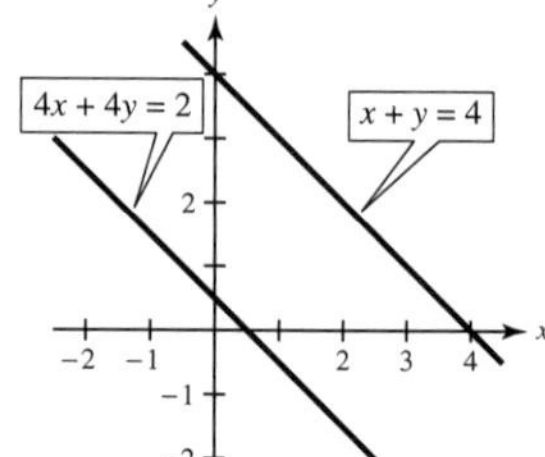

23. $(7, -2)$

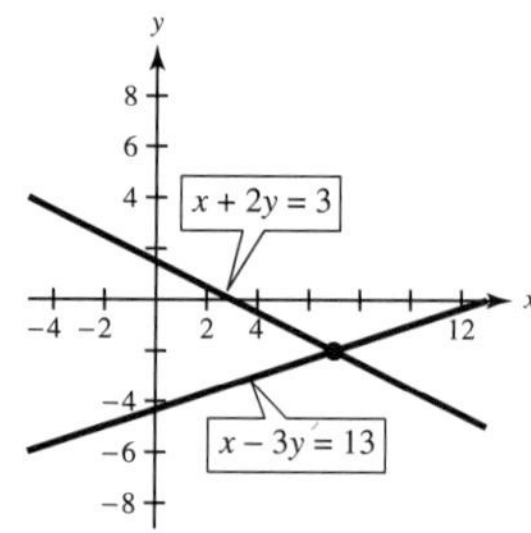

25. $(2, -1)$

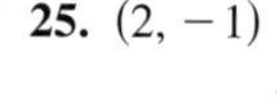

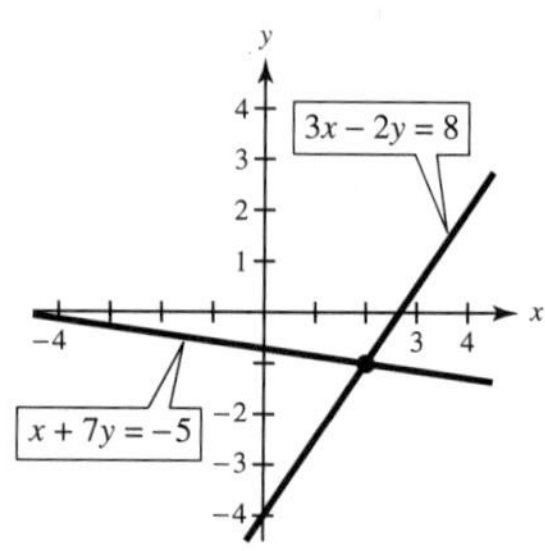

27. (1, 1)

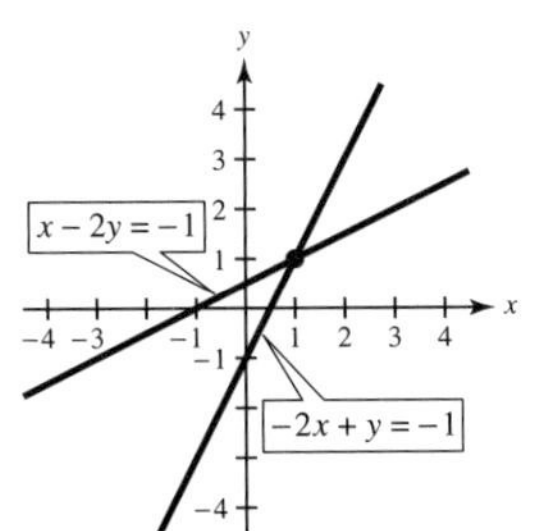

29. Infinitely many solutions

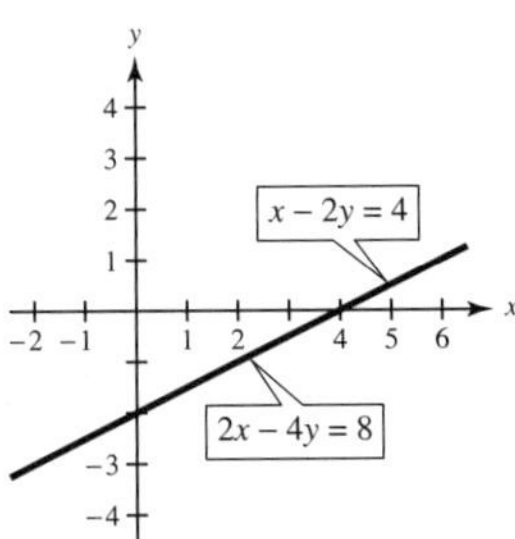

31. (0, −3)

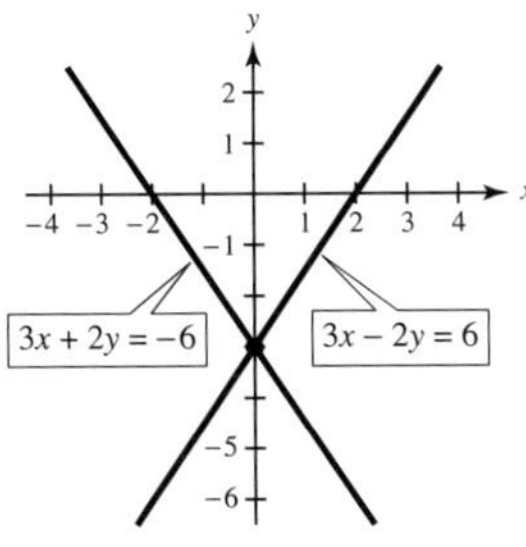

33. (5, 4)

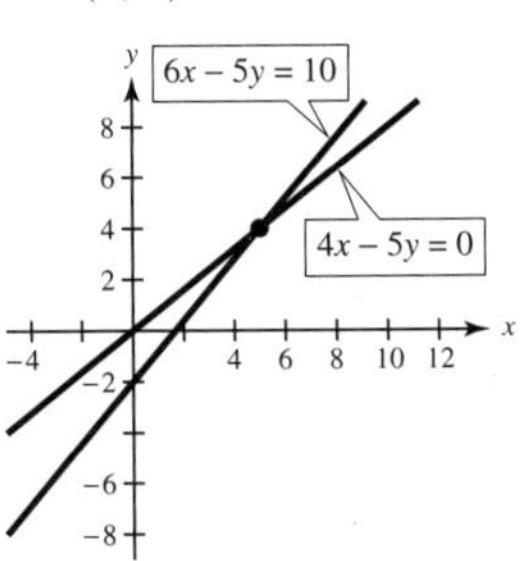

35. Infinitely many solutions

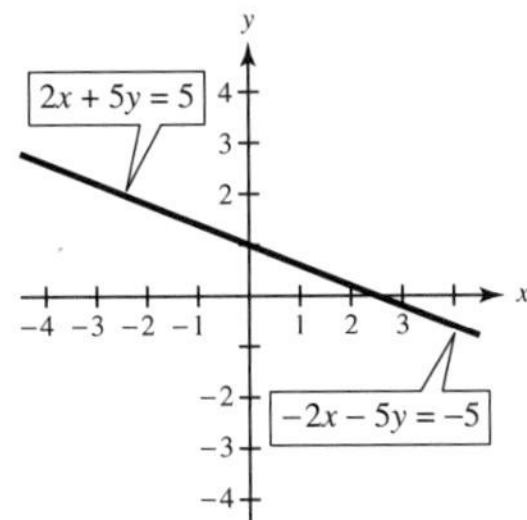

37. Infinitely many solutions

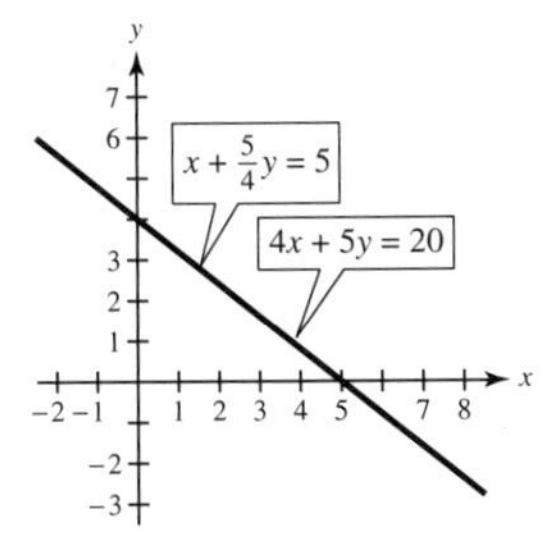

39. No solution

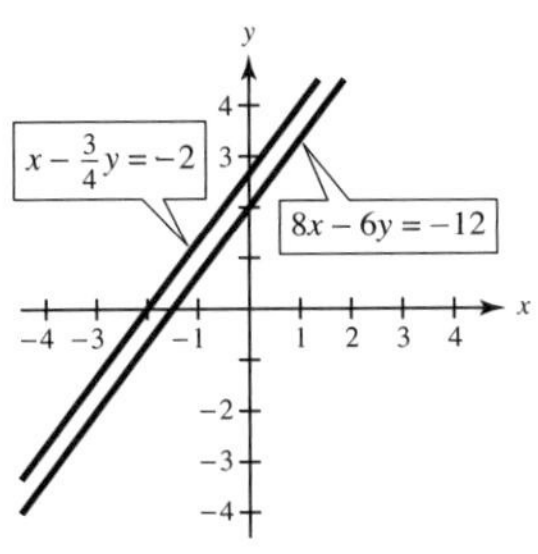

41. (2, 3)

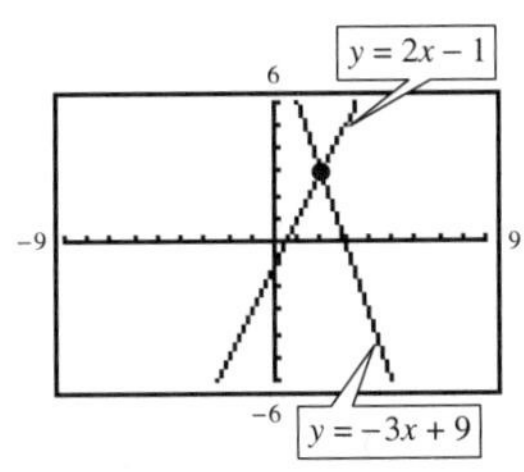

43. (3, 2)

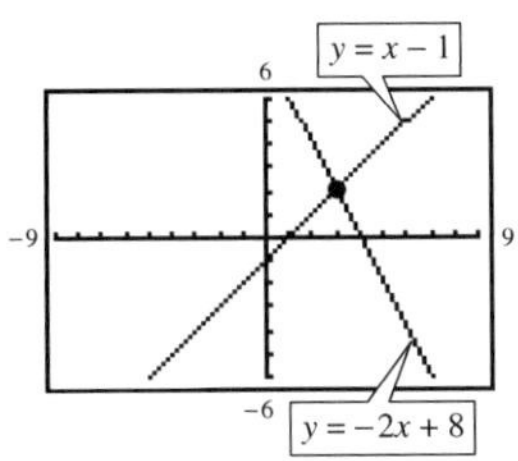

45. $y = \frac{2}{3}x + 4,\ y = \frac{2}{3}x - 1$; No solution

47. $y = \frac{1}{4}x + \frac{7}{4},\ y = \frac{1}{4}x + \frac{7}{4}$; Infinitely many solutions

49. $y = \frac{2}{3}x + \frac{4}{3},\ y = -\frac{2}{3}x + \frac{8}{3}$; One solution

51. $y = \frac{3}{4}x + \frac{9}{8},\ y = \frac{3}{4}x + \frac{3}{2}$; No solution

53. (1, 0) **55.** No solution

57. Infinitely many solutions **59.** (2, 3) **61.** (15, 5)

63. (4, −4) **65.** (0, 0) **67.** (2, 6) **69.** $\left(\frac{1}{2}, 3\right)$

71. (−3, 2) **73.** $\left(\frac{65}{72}, -\frac{1}{18}\right)$ **75.** $\left(\frac{5}{2}, -\frac{1}{2}\right)$ **77.** (0, 0)

79. (3, −2) **81.** Infinitely many solutions

83. $\left(\frac{5}{2}, 15\right)$ **85.** No solution **87.** (2, 0)

89. Infinitely many solutions **91.** (6, 0) **93.** $\left(\frac{5}{2}, \frac{3}{4}\right)$

95. (8, 4) **97.** $\left(\frac{115}{27}, \frac{157}{27}\right)$ **99.** $\left(\frac{18}{5}, \frac{3}{5}\right)$ **101.** (3, 0)

103. 11, 9

105. (a)

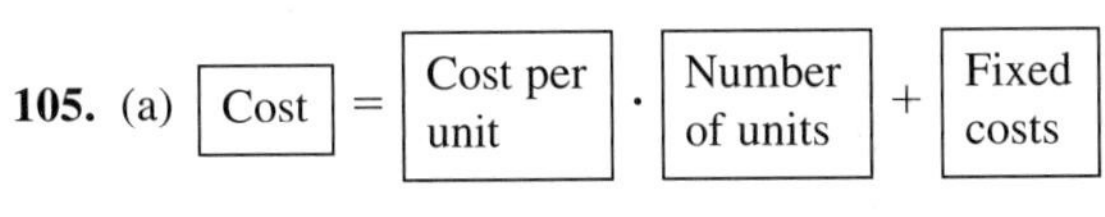

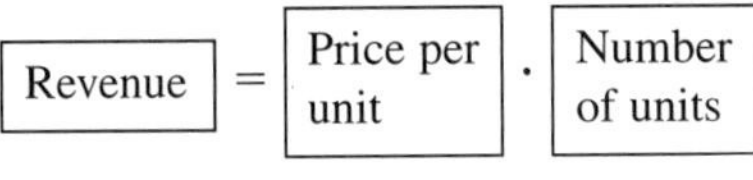

(b)

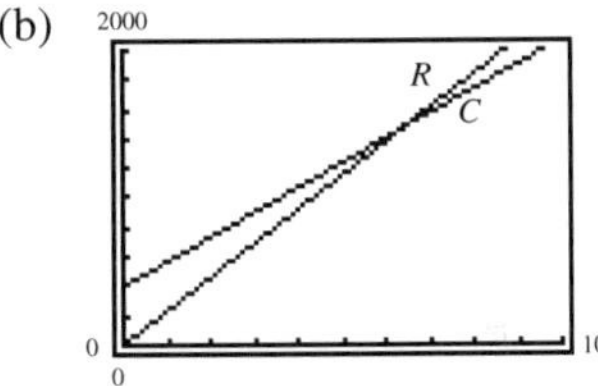

$x \approx 64$ units, $C = R \approx \$1472$; This means that the company must sell 64 feeders to cover their cost. Sales over 64 feeders will generate profit.

107. Because the slopes of the two lines are not equal, the lines intersect and the system has one solution: (79,400, 398).

109. 5%: \$10,000; 8%: \$5000 **111.** 2 adults

113. 100,000 miles **115.** $2x - y - 9 = 0$

117. A dependent system of linear equations is a consistent system in which two lines coincide (are identical), there are infinitely many points of intersection, the slopes are equal, and there are infinitely many solutions.

119. When you obtain a false result such as $-4 = 2$, then the system of linear equations has no solution.

121. The substitution method yields exact solutions.

123. Sample answer: $\begin{cases} x + y = 0 \\ x + y = 1 \end{cases}$

125. The solution should be checked in both equations in case the graph is not an accurate representation of the system of equations.

127. $b = 2$ **129.** $b = -\frac{1}{3}$ **131.** 1 **133.** $-\frac{1}{21}$

135. $-\frac{3}{2}$ **137.** $\frac{5}{11}$ **139.** 5 **141.** -13

Section 8.2 *(page 495)*

1. $(2, 0)$ **3.** $(-1, -1)$ **5.** $(8, 4)$ **7.** $(-4, 4)$

9. $(2, 1)$ **11.** $\left(\frac{21}{4}, -\frac{3}{2}\right)$ **13.** $(-2, 2)$ **15.** $(1, -2)$

17. $(-1, 1)$ **19.** $\left(\frac{7}{25}, -\frac{1}{25}\right)$ **21.** $(4, -1)$

23. $(4, -1)$ **25.** No solution

27. Infinitely many solutions **29.** $\left(\frac{1}{2}, 0\right)$ **31.** $\left(6, \frac{3}{2}\right)$

33. Infinitely many solutions **35.** $(8, 7)$ **37.** $(6, -4)$

39. $(2, -1)$ **41.** $(1, 1)$ **43.** $(5, 3)$ **45.** $(8, 4)$

47. $(-5, -3)$ **49.** $\left(\frac{15}{11}, \frac{15}{11}\right)$ **51.** $\left(\frac{4}{3}, \frac{4}{3}\right)$ **53.** $\left(1, -\frac{5}{4}\right)$

55. 15, 25 **57.** 48, 34

59. Two-point baskets: 7; Three-point baskets: 2

61. Student ticket: \$3; General admission ticket: \$5

63. \$4000

65. Private-lesson students: 7; Group-lesson students: 5

67. Yes, it is.

69. $y = \frac{1}{3}x + \frac{2}{3}$

71. Sample answer:

$$\begin{cases} 0.02x - 0.03y = 0.12 \\ 0.5x + 0.3y = 0.9 \end{cases}$$

Multiply each side of the first equation by 100 and multiply each side of the second equation by 10.

$$\begin{cases} 2x - 3y = 12 \\ 5x + 3y = 9 \end{cases}$$

73. Sample answer:

$$\begin{cases} y = 3x + 4 \\ y = -x - 8 \end{cases}$$

75.

$m = -\frac{8}{3}$

77.

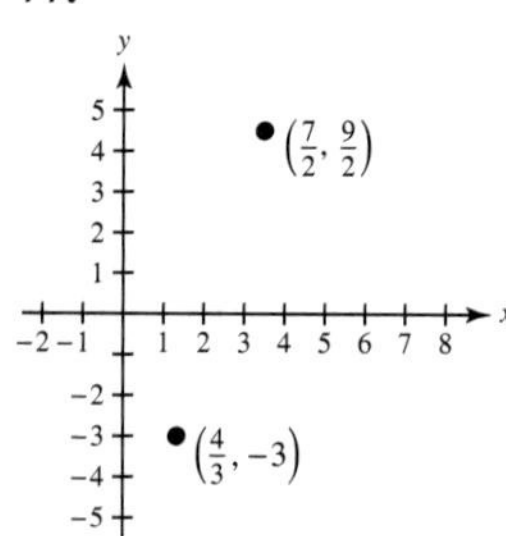

$m = \frac{45}{13}$

79.

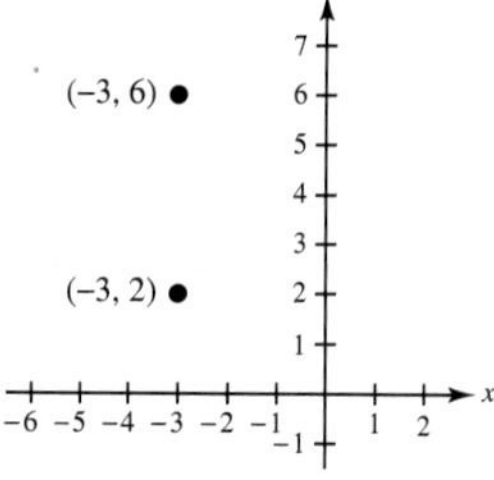

m is undefined.

81. $x \le 3$

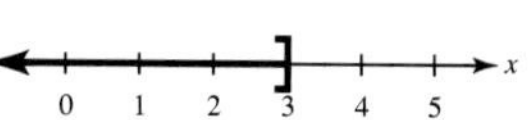

83. $x < 1$

85. $(5, 5)$ **87.** $(3, -1)$

Section 8.3 *(page 507)*

1. (a) Not a solution (b) Solution

(c) Solution (d) Not a solution

3. $(22, -1, -5)$ **5.** $(14, 3, -1)$

7. $\begin{cases} x - 2y = 8 \\ y = 14 \end{cases}$ Eliminated the x-term in Equation 2

9. Yes. The first equation was multiplied by -2 and added to the second equation. Then the first equation was multiplied by -3 and added to the third equation.

11. $(1, 2, 3)$ **13.** $(1, 2, 3)$ **15.** $(2, -3, -2)$

17. No solution **19.** $(-4, 8, 5)$ **21.** $(5, -2, 0)$

23. $\left(\frac{3}{10}, \frac{2}{5}, 0\right)$ **25.** $(-4, 2, 3)$ **27.** $\left(-\frac{1}{2}a + \frac{1}{2}, \frac{3}{5}a + \frac{2}{5}, a\right)$

29. No solution **31.** $(1, -1, 2)$

33. $\left(\frac{6}{13}a + \frac{10}{13}, \frac{5}{13}a + \frac{4}{13}, a\right)$ **35.** $\begin{cases} x + 2y - z = -4 \\ y + 2z = 1 \\ 3x + y + 3z = 15 \end{cases}$

37. $s = -16t^2 + 144$ **39.** $s = -16t^2 + 48t$

41. 88°, 32°, 60°

43. \$17,404 at 6%, \$31,673 at 10%, \$30,923 at 15%

45. \$4000 at 6%, \$8000 at 8%, \$3000 at 9%

47. 20 gallons of spray X, 18 gallons of spray Y, 16 gallons of spray Z

49. 84 hot dogs at \$1.50, 38 hot dogs at \$2.50, 21 hot dogs at \$3.25

51. (a) Not possible

(b) 10% solution: 0 gallons; 15% solution: 6 gallons; 25% solution: 6 gallons

(c) 10% solution: 4 gallons; 15% solution: 0 gallons; 25% solution: 8 gallons

53. 50 students on strings, 20 students in winds, 8 students in percussion

55. The solution is apparent because the row-echelon form is

$$\begin{cases} x = 1 \\ y = -3. \\ z = 4 \end{cases}$$

57. Three planes have no point in common when two of the planes are parallel and the third plane intersects the other two planes.

59. The graphs are three planes with three possible situations. If all three planes intersect in one point, there is one solution. If all three planes intersect in one line, there are an infinite number of solutions. If each pair of planes intersects in a line, but the three lines of intersection are all parallel, there is no solution.

61. $3x, 2; 3, 2$ **63.** $14t^5, -t, 25; 14, -1, 25$

65. $(4, 3)$ **67.** $(-2, 6)$

Mid-Chapter Quiz *(page 511)*

1. Not a solution, because substituting $x = 4$ and $y = 2$ into $3x + 4y = 4$ yields $20 = 4$, which is a contradiction.

2. Solution, because substituting $x = 2$ and $y = -1$ into both equations yields true equalities.

3. $(3, 2)$ **4.** $(4, 1)$ **5.** $(4, -1)$

6. $(-3, 11)$ **7.** $(2, 2)$

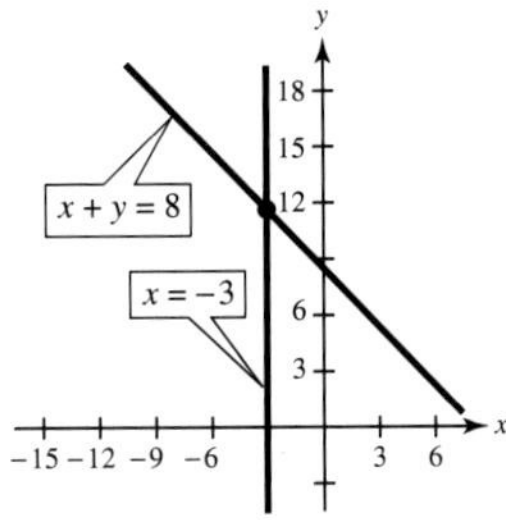

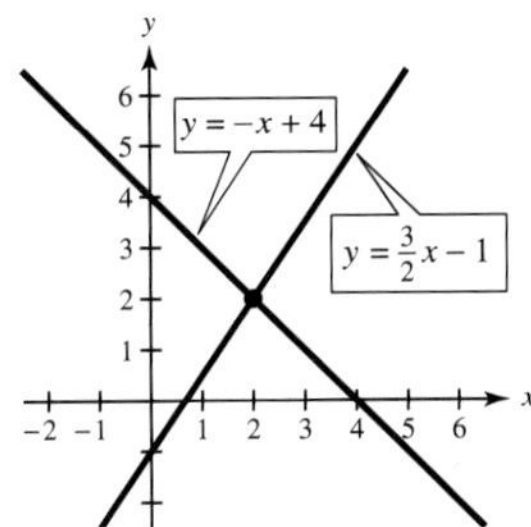

8. $(-1, 4)$ **9.** $(6, 2)$ **10.** $(3, 3)$

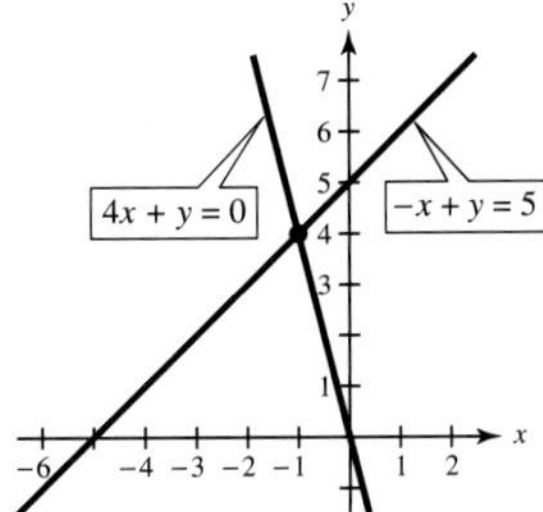

11. $\left(-\frac{1}{2}, 6\right)$ **12.** $(-2, 5)$ **13.** $\left(\frac{55}{23}, \frac{95}{23}\right)$

14. $\left(\frac{1}{2}, -\frac{1}{2}, 1\right)$ **15.** $(5, -1, 3)$

16. Sample answer:
$$\begin{cases} 2x + y = 7 \\ -5x - 2y = -13 \end{cases}$$

17. Sample answer:
$$\begin{cases} -x + y = -3 \\ 4x - 3y = 12 \end{cases}$$

18.

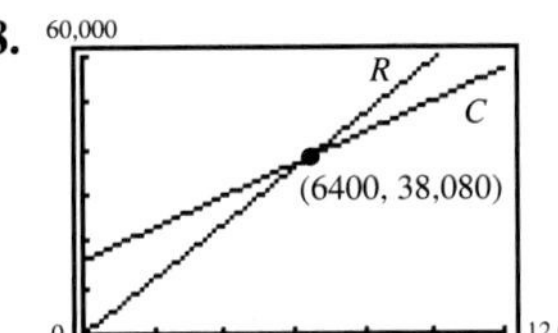

The point of intersection of the two functions, (6400, 38,080), represents the break-even point.

19. 45°, 80°, 55°

Section 8.4 *(page 520)*

1. 4×2 **3.** 2×2 **5.** 4×1 **7.** 1×1

9. 1×4

11. (a) $\begin{bmatrix} 4 & -5 \\ -1 & 8 \end{bmatrix}$ (b) $\left[\begin{array}{cc:c} 4 & -5 & -2 \\ -1 & 8 & 10 \end{array}\right]$

13. (a) $\begin{bmatrix} 1 & 1 & 0 \\ 5 & -2 & -2 \\ 2 & 4 & 1 \end{bmatrix}$ (b) $\left[\begin{array}{ccc:c} 1 & 1 & 0 & 0 \\ 5 & -2 & -2 & 12 \\ 2 & 4 & 1 & 5 \end{array}\right]$

15. (a) $\begin{bmatrix} 5 & 1 & -3 \\ 0 & 2 & 4 \end{bmatrix}$ (b) $\left[\begin{array}{ccc:c} 5 & 1 & -3 & 7 \\ 0 & 2 & 4 & 12 \end{array}\right]$

17. $\begin{cases} 4x + 3y = 8 \\ x - 2y = 3 \end{cases}$ **19.** $\begin{cases} x + 2z = -10 \\ 3y - z = 5 \\ 4x + 2y = 3 \end{cases}$

21. $\begin{cases} 5x + 8y + 2z = -1 \\ -2x + 15y + 5z + w = 9 \\ x + 6y - 7z = -3 \end{cases}$

23. $\begin{cases} 13x + y + 4z - 2w = -4 \\ 5x + 4y - w = 0 \\ x + 2y + 6z + 8w = 5 \\ -10x + 12y + 3z + w = -2 \end{cases}$

25. Interchange the first and second rows.

27. Multiply the first row by $-\frac{1}{3}$.

29. Add 5 times the first row to the third row.

31. $\begin{bmatrix} 1 & 1 & -4 & 2 \\ 0 & 4 & 5 & 5 \\ 0 & 0 & 8 & 3 \end{bmatrix}$ **33.** $\begin{bmatrix} 1 & -2 & 3 \\ 3 & 4 & 5 \end{bmatrix}$

35. $\begin{bmatrix} 1 & 4 & 3 \\ 0 & 0 & 0 \end{bmatrix}$ **37.** $\begin{bmatrix} 1 & 2 & 3 \\ 0 & 1 & 2 \end{bmatrix}$

39. $\begin{bmatrix} 1 & 0 & -\frac{7}{5} \\ 0 & 1 & \frac{11}{10} \end{bmatrix}$ **41.** $\begin{bmatrix} 1 & 1 & 0 & 5 \\ 0 & 1 & 2 & 0 \\ 0 & 0 & 1 & -1 \end{bmatrix}$

43. $\begin{bmatrix} 1 & -1 & -1 & 1 \\ 0 & 1 & 6 & 3 \\ 0 & 0 & 1 & 0.8 \end{bmatrix}$ **45.** $\begin{bmatrix} 1 & 1 & -1 & 3 \\ 0 & 1 & -4 & 1 \\ 0 & 0 & 0 & 0 \end{bmatrix}$

47. $\begin{cases} x - 2y = 4 \\ y = -3 \end{cases}$ $(-2, -3)$ **49.** $\begin{cases} x + 5y = 3 \\ y = -2 \end{cases}$ $(13, -2)$

51. $\begin{cases} x - y + 2z = 4 \\ y - z = 2 \\ z = -2 \end{cases}$ $(8, 0, -2)$ **53.** $(5, 1)$ **55.** $(1, 1)$

57. $(-2, 1)$ **59.** No solution **61.** $(2, -3, 2)$

63. $(2a + 1, 3a + 2, a)$ **65.** $(1, 2, -1)$ **67.** $(1, -1, 2)$

69. $(34, -4, -4)$ **71.** No solution

73. $(-12a - 1, 4a + 1, a)$ **75.** $\left(\frac{1}{2}, 2, 4\right)$ **77.** $\left(2, 5, \frac{5}{2}\right)$

79. \$800,000 at 8%, \$500,000 at 9%, \$200,000 at 12%

81. Theater A: 300 tickets; Theater B: 600 tickets; Theater C: 600 tickets

83. 5, 8, 20

85. 15 computer chips, 10 resistors, 10 transistors

87. Certificates of deposit: $250{,}000 - \frac{1}{2}s$

Municipal bonds: $125{,}000 + \frac{1}{2}s$

Blue-chip stocks: $125{,}000 - s$

Growth stocks: s

If $s = \$100{,}000$, then:

Certificates of deposit: \$200,000
Municipal bonds: \$175,000
Blue-chip stocks: \$25,000
Growth stocks: \$100,000

89. 3×5. There are 15 entries in the matrix, so the order is 3×5, 5×3, or 15×1. Because there are more columns than rows, the second number in the order must be larger than the first.

91. There will be a row in the matrix with all zero entries except in the last column.

93. The first entry in the first column is 1, and the other two are zero. In the second column, the first entry is a nonzero real number, the second number is 1, and the third number is zero. In the third column, the first two entries are nonzero real numbers and the third entry is 1.

95. -42 **97.** 26 **99.** $(4, -2, 1)$

Section 8.5 *(page 533)*

1. 5 **3.** 27 **5.** 0 **7.** 0 **9.** -24

11. -0.33 **13.** -24 **15.** -2 **17.** -30

19. 3 **21.** 0 **23.** 0 **25.** 102 **27.** -0.22

29. $x - 5y + 2$ **31.** 248 **33.** 105.625 **35.** 4.32

37. $(1, 2)$ **39.** $(2, -2)$ **41.** $\left(\frac{3}{4}, -\frac{1}{2}\right)$

43. Not possible, $D = 0$ **45.** $\left(\frac{2}{3}, \frac{1}{2}\right)$ **47.** $(-1, 3, 2)$

49. $\left(\frac{1}{2}, -\frac{2}{3}, \frac{1}{4}\right)$ **51.** Not possible, $D = 0$ **53.** $\left(\frac{22}{27}, \frac{22}{9}\right)$

55. $\left(\frac{51}{16}, -\frac{7}{16}, -\frac{13}{16}\right)$ **57.** -2 **59.** 16 **61.** 21

63. $\frac{31}{2}$ **65.** $\frac{33}{8}$ **67.** 16 **69.** $\frac{53}{2}$

71. 250 square miles **73.** Collinear

75. Not collinear **77.** Not collinear

79. $x - 2y = 0$ **81.** $7x - 6y - 28 = 0$

83. $9x + 10y + 3 = 0$ **85.** $16x - 15y + 22 = 0$

87. $I_1 = 1, I_2 = 2, I_3 = 1$ **89.** $I_1 = 2, I_2 = 3, I_3 = 5$

91. (a) $\left(\dfrac{13}{2k + 6}, \dfrac{3k - 4}{-2k^2 - 6k}\right)$ (b) $0, -3$

93. A square matrix is a matrix with the same number of rows and columns. Its determinant is simply a real number.

95. The determinant is zero. Because two rows are identical, each term is zero when expanding by minors along the other row. Therefore, the sum is zero.

97.

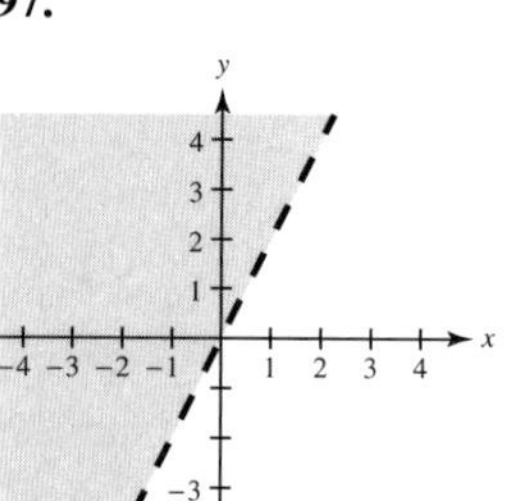

99.

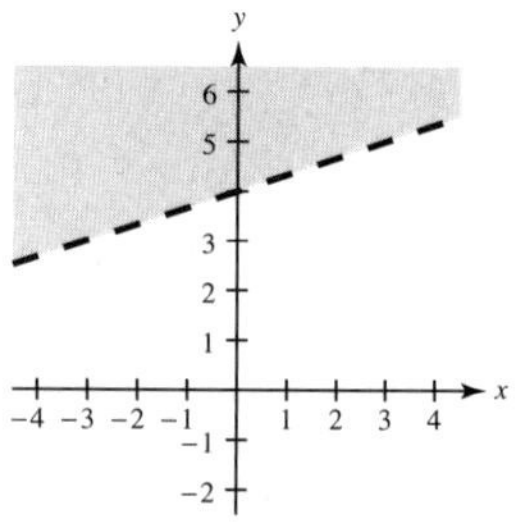

101. $x - 2y = 0$ **103.** $y = 2$ **105.** Function

107. Not a function

Section 8.6 *(page 542)*

1. d **2.** b **3.** f **4.** c **5.** a **6.** e

7. (a) Not a solution (b) Solution

9. (a) Not a solution (b) Not a solution

11.

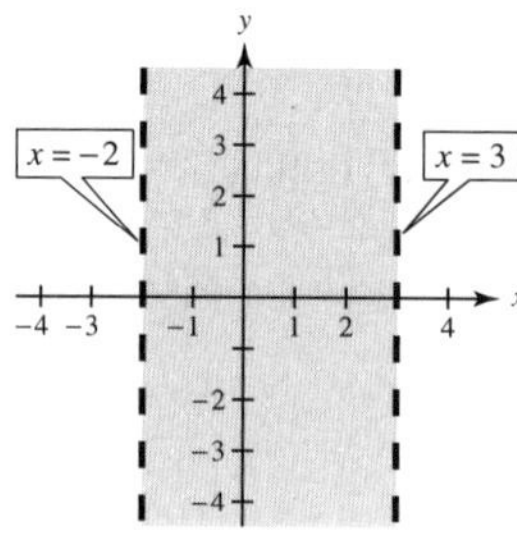

13.

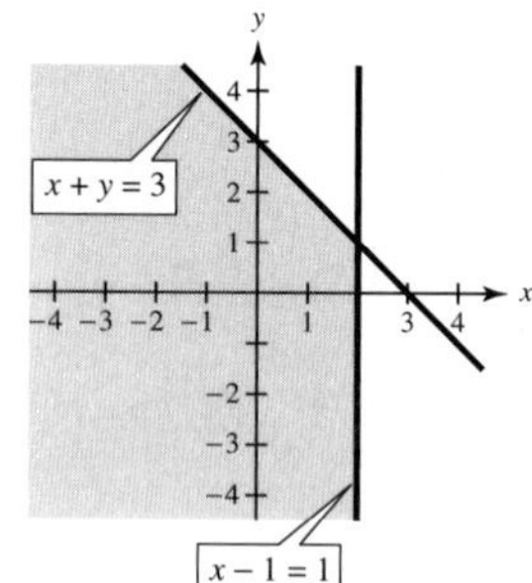

15.

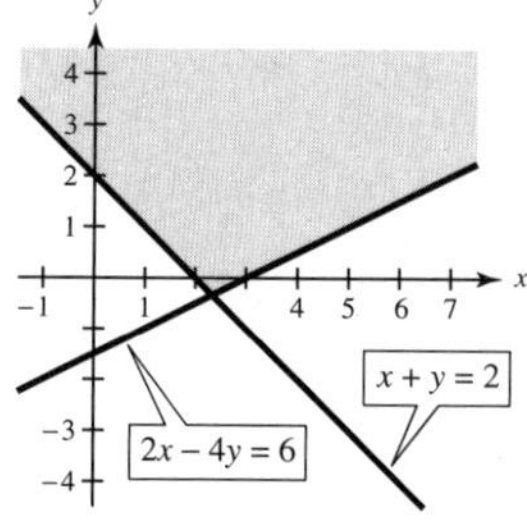

17.

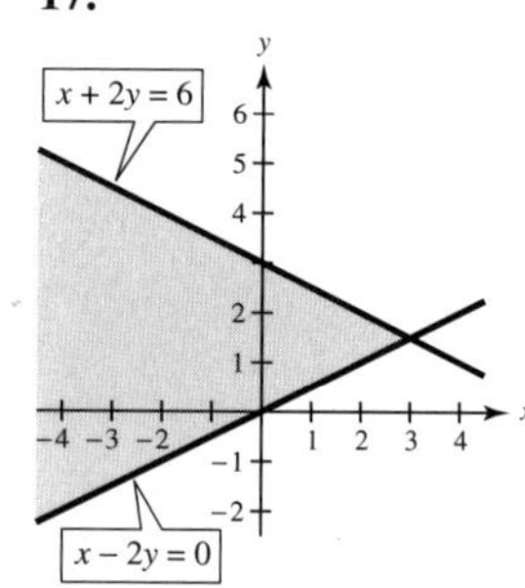

19.

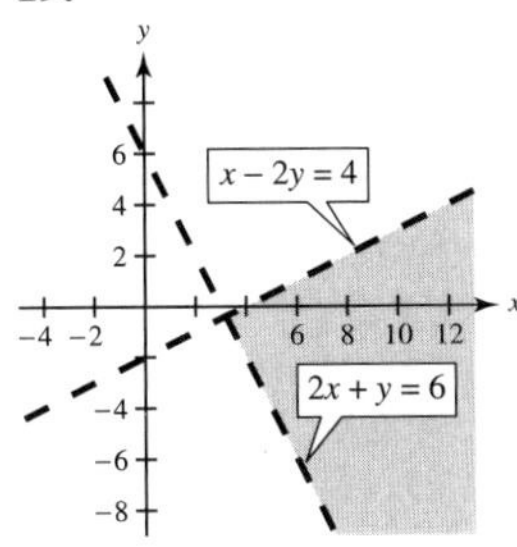

21.

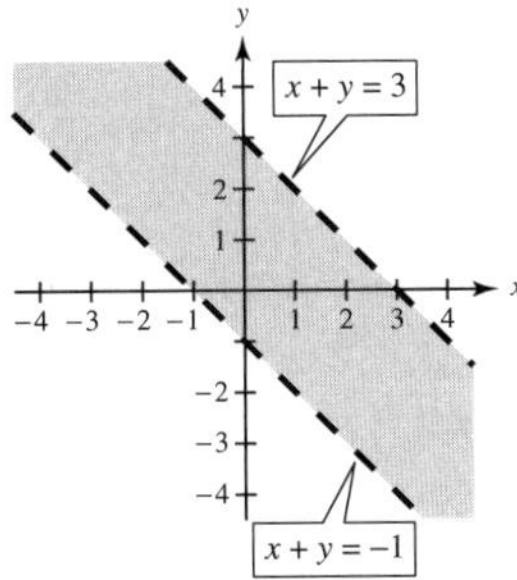
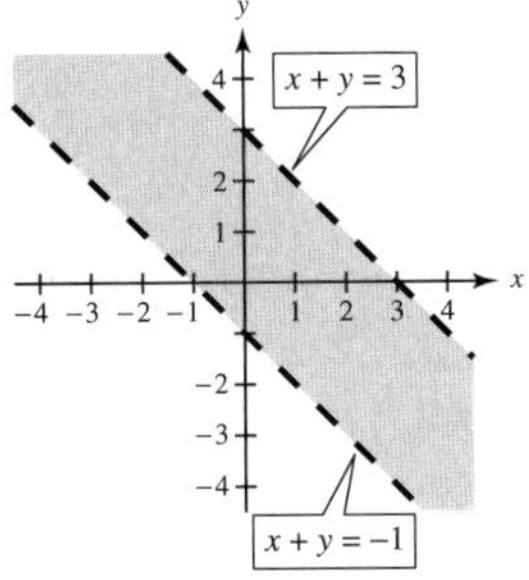
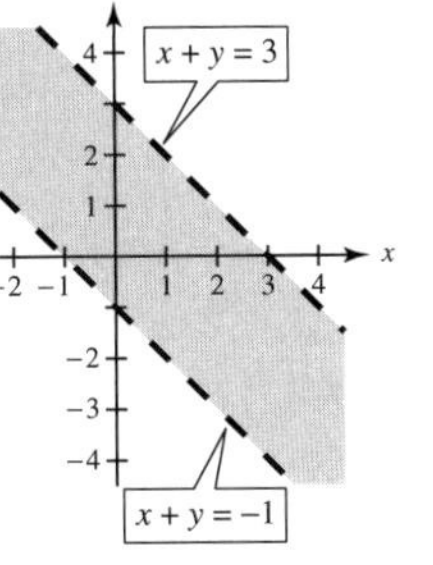

23.

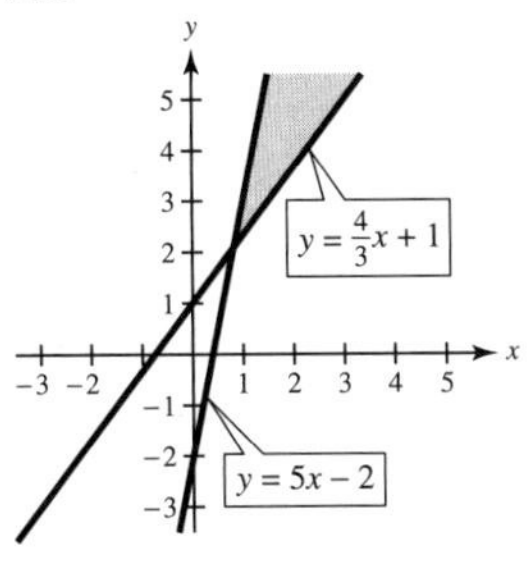

25.

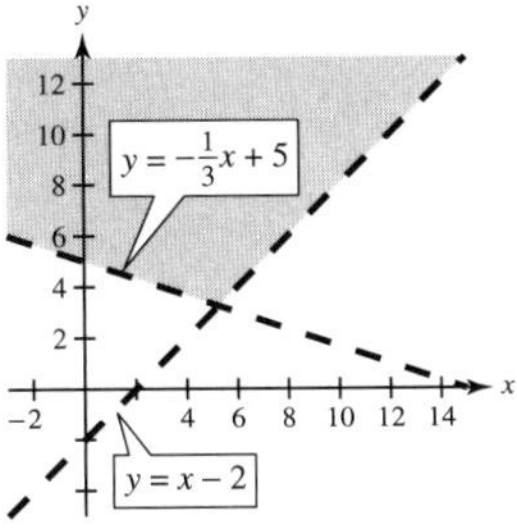

27.

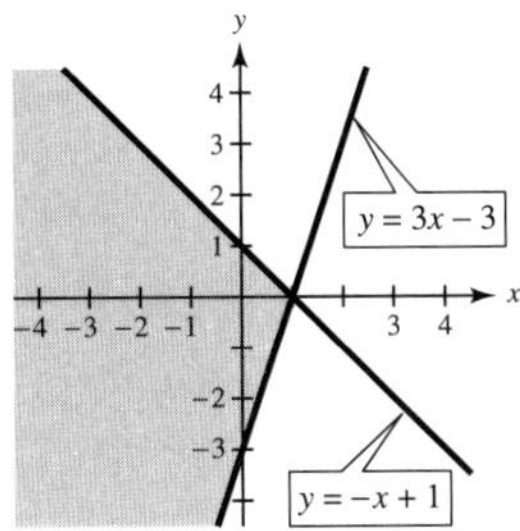

29.

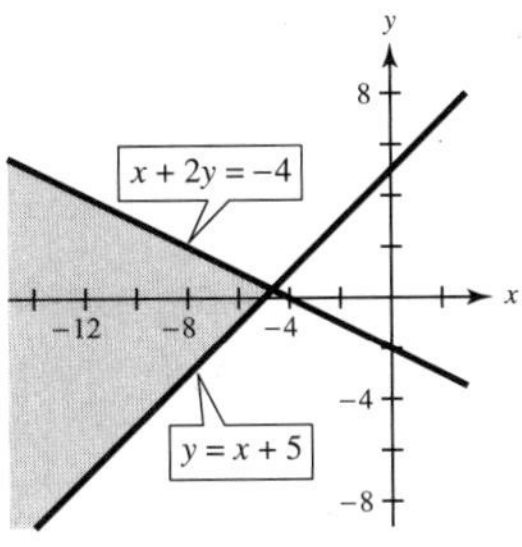

31.

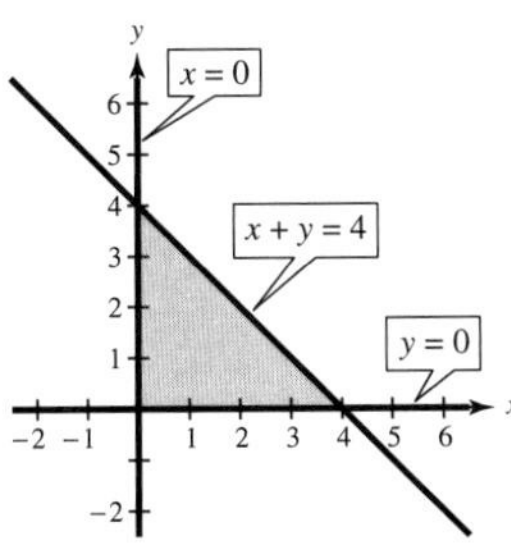

33.

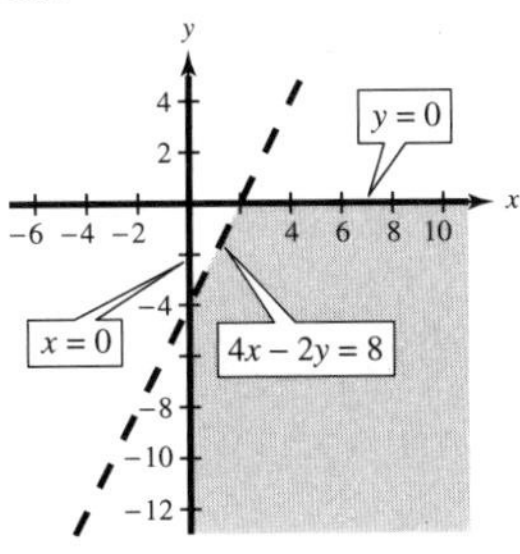

35.

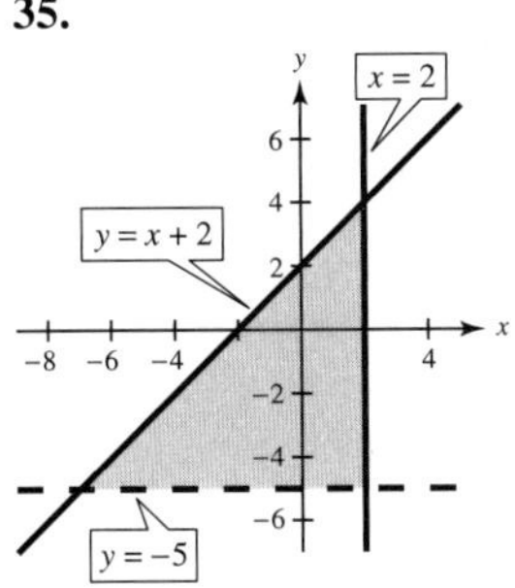

37.

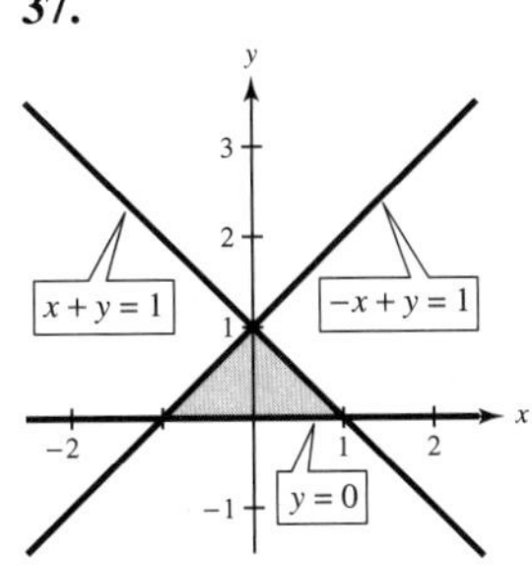

39. No solution

41.

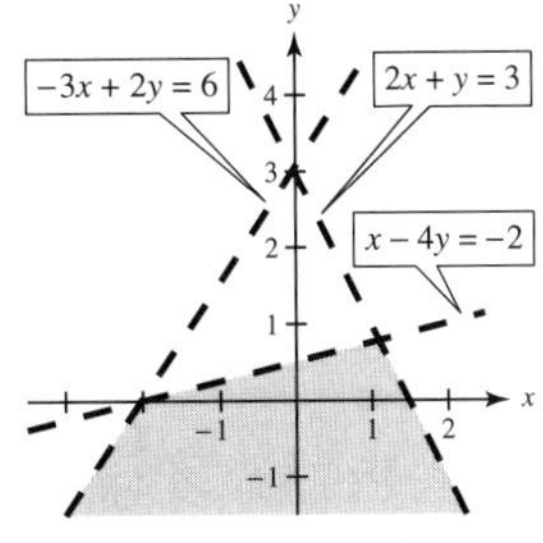

43.

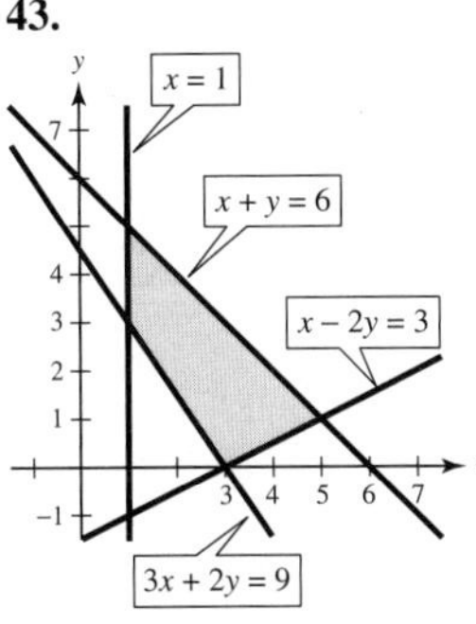

45.

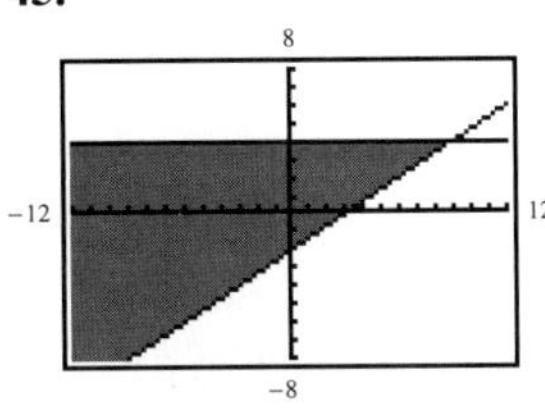

47.

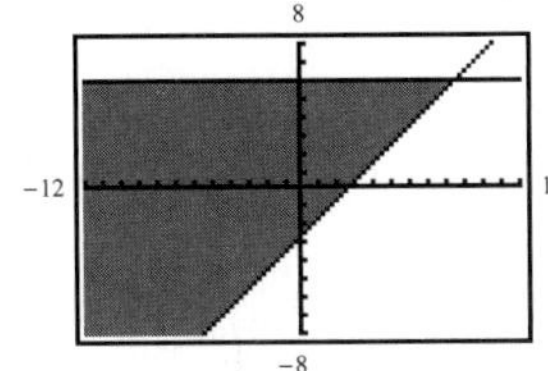

49.

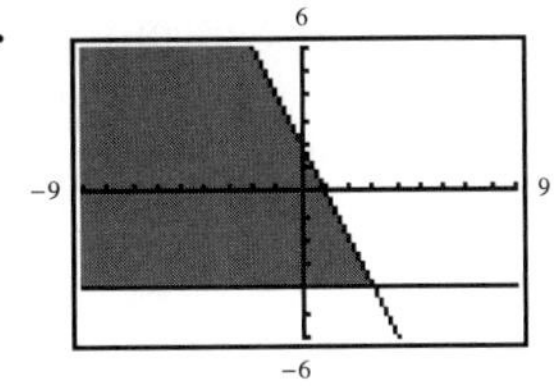

51. $\begin{cases} x \geq 1 \\ y \geq x - 3 \\ y \leq -2x + 6 \end{cases}$

53. $\begin{cases} x \geq -2 \\ x \leq 2 \\ y \geq -\frac{1}{2}x - 4 \\ y \leq -\frac{1}{2}x + 2 \end{cases}$

55. $\begin{cases} y \leq \frac{9}{10}x + \frac{42}{5} \\ y \geq 3x \\ y \geq \frac{2}{3}x + 7 \end{cases}$

57. $\begin{cases} x + \frac{3}{2}y \leq 12 \\ \frac{4}{3}x + \frac{3}{4}y \leq 16 \\ x \geq 0 \\ y \geq 0 \end{cases}$

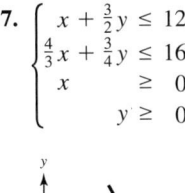

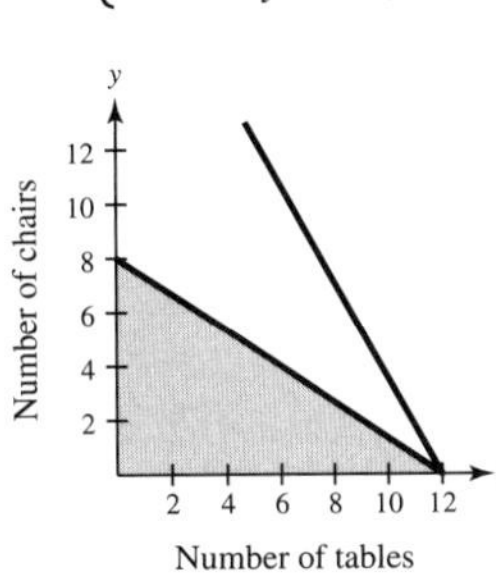

59. $\begin{cases} x + y \leq 40{,}000 \\ x \geq 10{,}000 \\ y \geq 2x \end{cases}$

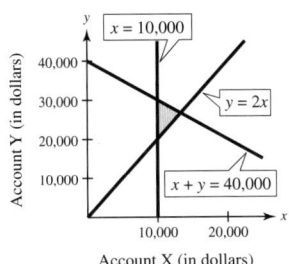

x = 10,000
y = 2x
x + y = 40,000
Account Y (in dollars)
Account X (in dollars)

61. $\begin{cases} 20x + 10y \ge 280 \\ 15x + 10y \ge 160 \\ 10x + 20y \ge 180 \\ x \ge 0 \\ y \ge 0 \end{cases}$

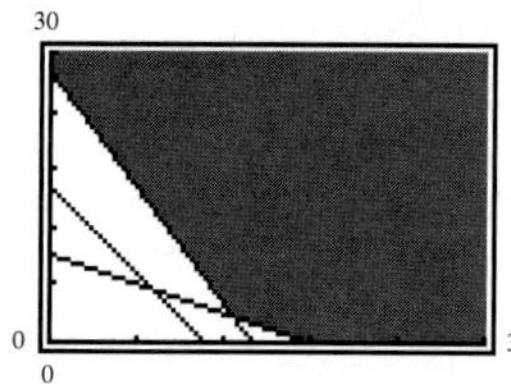

63. $\begin{cases} x + y \ge 15{,}000 \\ 30x + 45y \ge 525{,}000 \\ x \ge 8{,}000 \\ y \ge 4{,}000 \end{cases}$

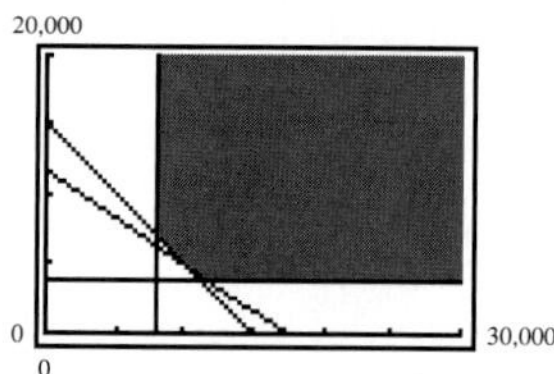

65. $\begin{cases} x \le 90 \\ y \le 0 \\ y \ge -10 \\ y \ge -\frac{1}{7}x \end{cases}$

67. The graph of a linear equation splits the xy-plane into two parts, each of which is a half-plane. The graph of $y < 5$ is a half-plane.

69. To determine the vertices of the region, find all intersections between the lines corresponding to the inequalities. The vertices are the intersection points that satisfy each inequality.

71. $\left(-\frac{1}{2}, 0\right), (0, 2)$ **73.** $(3, 0), (0, -1)$

75. $(-2, 0), (0, 2)$

77. (a) -10 (b) -5 **79.** (a) 0 (b) $6m - 4m^2$

Review Exercises *(page 548)*

1. (a) Solution (b) Not a solution

3. (a) Not a solution (b) Solution

5. $(1, 2)$ **7.** $(0, -3)$

9. $(5, 1)$

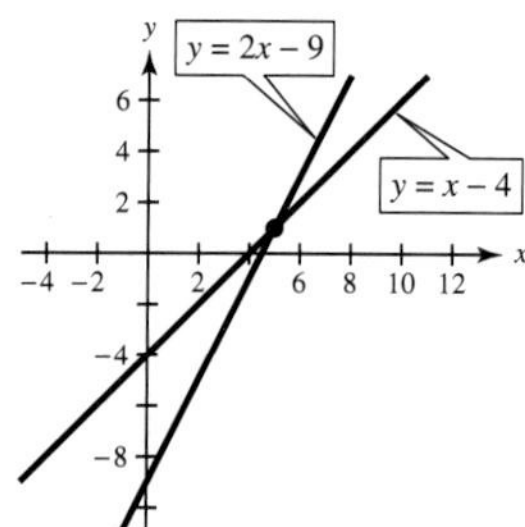

11. 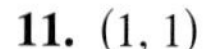 $(1, 1)$

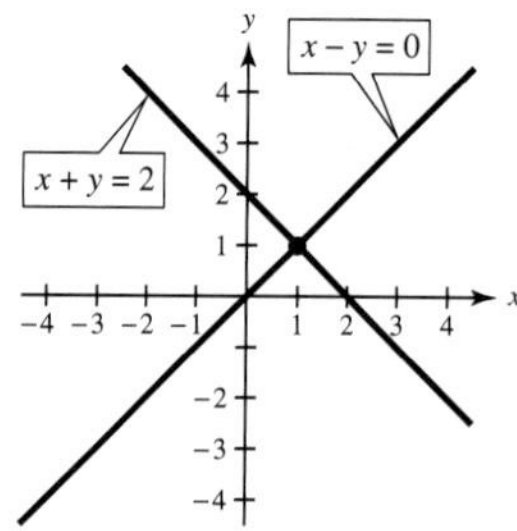

13. No solution

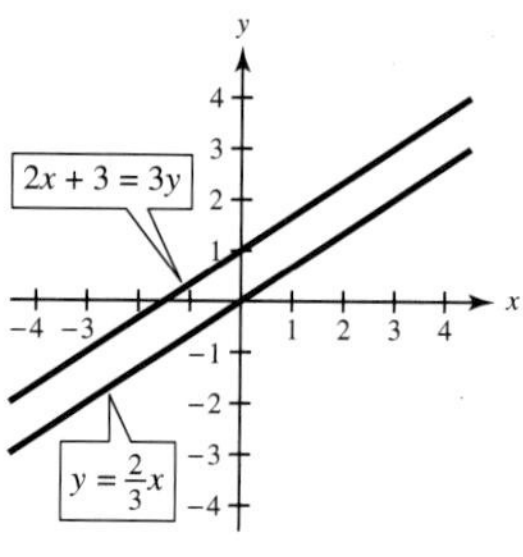

15. Infinitely many solutions

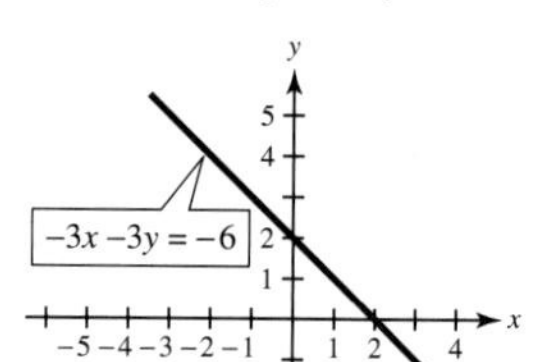

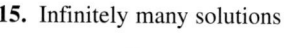

17. 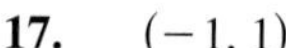 $(-1, 1)$

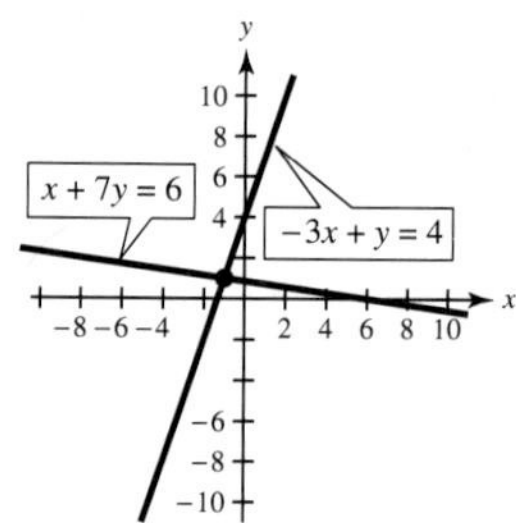

19. $(4, 8)$ **21.** $(4, 2)$ **23.** No solution **25.** $(4, -1)$

27. $\left(\frac{5}{2}, 3\right)$ **29.** Infinitely many solutions **31.** $(5, -5)$

33. 36, 16 **35.** 5%: \$8000; 10%: \$4000

37. 804 adult tickets and 706 children's tickets **39.** $(2, 1)$

41. $(10, -12)$ **43.** Infinitely many solutions

45. $\left(\frac{38}{35}, \frac{11}{35}\right)$ **47.** $\left(-\frac{1}{5}, \frac{7}{10}\right)$ **49.** $(5, 6)$ **51.** $(4, 2)$

53. $(-1, -1)$ **55.** $\left(4, -\frac{3}{2}\right)$

57. Infinitely many solutions **59.** $(-1, 3)$ **61.** $(10, 0)$

63. $(-3, 7)$ **65.** $\left(\frac{1}{3}, -\frac{1}{2}\right)$ **67.** $(3, 2, 5)$

69. $(0, 3, 6)$ **71.** $(2, -3, 3)$ **73.** $(0, 1, -2)$

75. \$8000 at 7%, \$5000 at 9%, \$7000 at 11%

77. 1×4 **79.** 2×3

81. (a) $\begin{bmatrix} 7 & -5 \\ 1 & -1 \end{bmatrix}$ (b) $\left[\begin{array}{cc|c} 7 & -5 & 11 \\ 1 & -1 & -5 \end{array}\right]$

83. $\begin{cases} 4x - y = 2 \\ 6x + 3y + 2z = 1 \\ y + 4z = 0 \end{cases}$ **85.** $(10, -12)$

87. $(0.6, 0.5)$ **89.** $(3, -6, 7)$ **91.** $\left(\frac{1}{2}, -\frac{1}{3}, 1\right)$

93. $(1, 0, -4)$ **95.** 10 **97.** -51 **99.** 1

101. $(-3, 7)$ **103.** $(2, -3, 3)$ **105.** 16 **107.** 7

109. Not collinear **111.** $x - 2y + 4 = 0$

113.

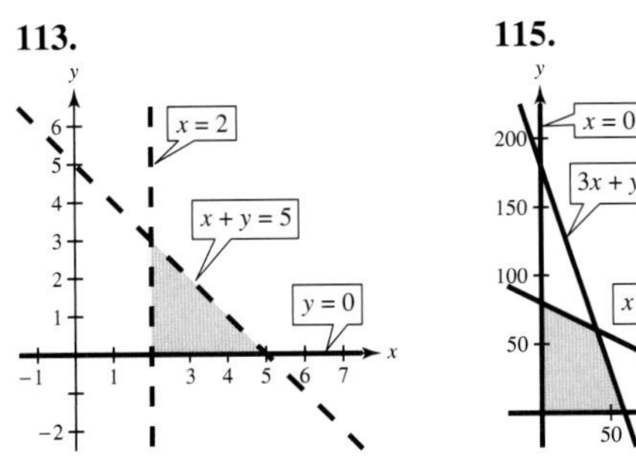

115.

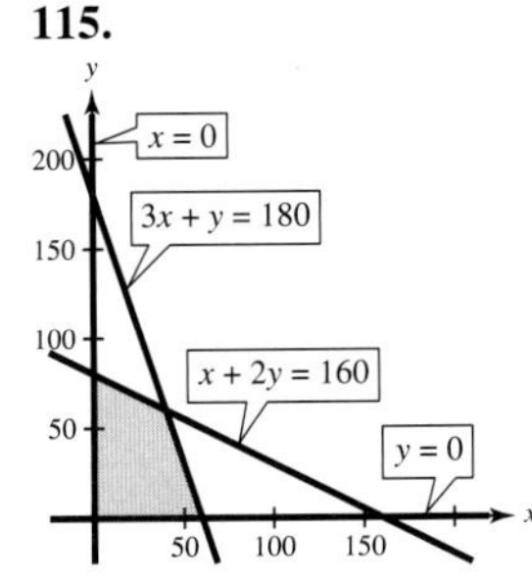

117.

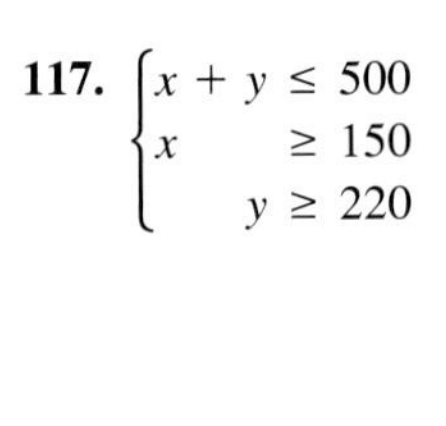

$$\begin{cases} x + y \le 500 \\ x \ge 150 \\ y \ge 220 \end{cases}$$

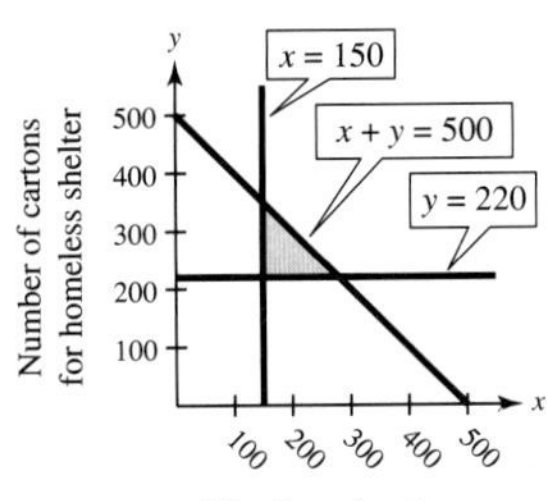

Chapter Test *(page 553)*

1. (a) Solution (b) Not a solution

2. $(3, 2)$ **3.** $(1, 3)$ **4.** $(2, 3)$ **5.** $(-2, 2)$

6. $(2, 2a - 1, a)$ **7.** $(-1, 3, 3)$ **8.** $(2, 1, -2)$

9. $\left(4, \frac{1}{7}\right)$ **10.** $(5, 1, -1)$ **11.** -16 **12.** 12

13.

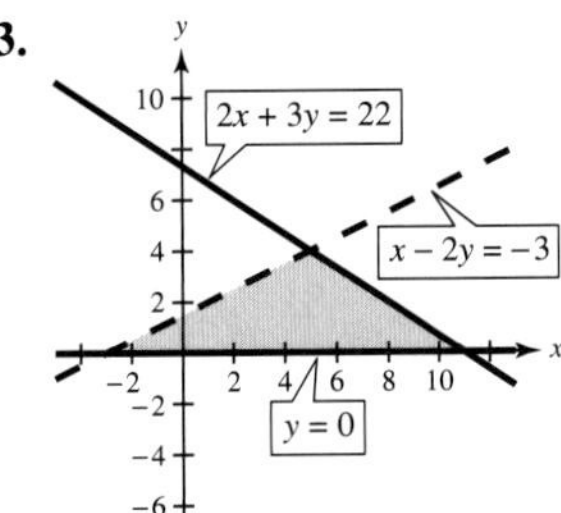

14. 50,000 miles

15. \$13,000 at 4.5%, \$9000 at 5%, \$3000 at 8%

16. $\begin{cases} 30x + 40y \ge 300{,}000 \\ x \le 9{,}000 \\ y \le 4{,}000 \end{cases}$

where x is the number of reserved seat tickets and y is the number of floor seat tickets.

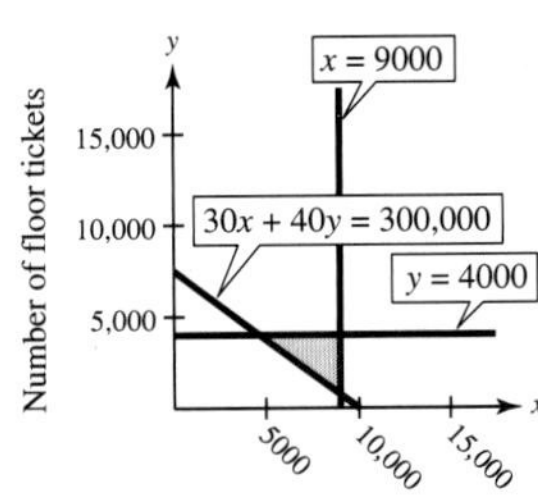

CHAPTER 9

Section 9.1 *(page 563)*

1. 8 **3.** -7 **5.** -3 **7.** Not a real number

9. Perfect square **11.** Perfect cube **13.** Neither

15. ± 5 **17.** $\pm\frac{3}{4}$ **19.** $\pm\frac{1}{7}$ **21.** ± 0.4 **23.** 2

25. $-\frac{1}{3}$ **27.** $\frac{1}{10}$ **29.** 0.1 **31.** Irrational

33. Rational **35.** Rational **37.** Irrational

39. 8 **41.** 10 **43.** Not a real number **45.** $-\frac{2}{3}$

47. Not a real number **49.** 5 **51.** -23 **53.** 5

55. 10 **57.** 6 **59.** $-\frac{1}{4}$ **61.** 11 **63.** -24

65. 3 **67.** -2

	Radical Form	*Rational Exponent Form*
69.	$\sqrt{36} = 6$	$36^{1/2} = 6$
71.	$\sqrt[4]{256^3} = 64$	$256^{3/4} = 64$

73. 5 **75.** -6 **77.** $\frac{1}{4}$ **79.** $\frac{1}{9}$ **81.** $\frac{4}{9}$ **83.** $\frac{3}{11}$

85. 9 **87.** -64 **89.** $t^{1/2}$ **91.** x^3 **93.** $u^{7/3}$

95. $x^{-1} = \dfrac{1}{x}$ **97.** $t^{-9/4} = \dfrac{1}{t^{9/4}}$ **99.** x^3

101. $y^{13/12}$ **103.** $x^{3/4}y^{1/4}$ **105.** $y^{5/2}z^4$ **107.** 3

109. $2^{1/3}$ **111.** $\dfrac{1}{2}$ **113.** $c^{1/2}$ **115.** $\dfrac{3y^2}{4z^{4/3}}$

117. $\dfrac{9y^{3/2}}{x^{2/3}}$ **119.** $x^{1/4}$ **121.** $y^{1/8}$ **123.** $x^{3/8}$

125. $(x + y)^{1/2}$ **127.** $\dfrac{1}{(3u - 2v)^{5/6}}$ **129.** 5.9161

131. 9.9845 **133.** 0.0367 **135.** 3.8158

137. 66.7213 **139.** 1.0420 **141.** 0.7915

143. (a) 3 (b) 5 (c) Not a real number (d) 9

145. (a) 2 (b) 3 (c) -2 (d) -4

147. (a) 2 (b) Not a real number (c) 3 (d) 1

149. $0 \le x < \infty$ **151.** $-\infty < x \le \frac{4}{9}$

153. $-\infty < x < \infty$ **155.** $-\frac{9}{2} \le x < \infty$

157. $0 < x < \infty$

159. Domain: $(0, \infty)$

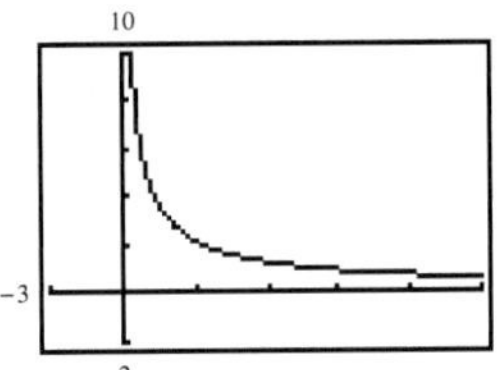

161. Domain: $(-\infty, \infty)$

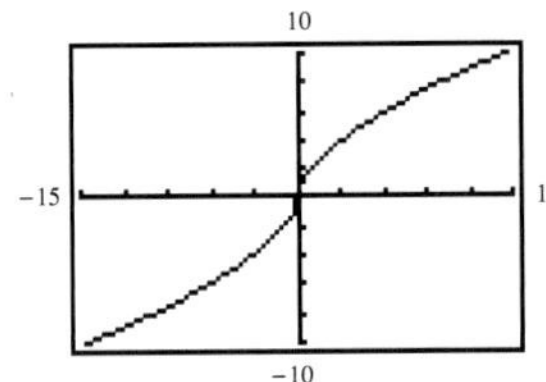

163. $2x^{3/2} - 3x^{1/2}$ **165.** $1 + 5y$ **167.** 0.128

169. 23 feet × 23 feet **171.** 104.64 inches

173. If a and b are real numbers, n is an integer greater than or equal to 2, and $a = b^n$, then b is the nth root of a.

175. No. $\sqrt{2}$ is an irrational number. Its decimal representation is a nonterminating, nonrepeating decimal.

177. 0, 1, 4, 5, 6, 9; Yes **179.** 5 **181.** $-\frac{4}{5}$

183. $s = kt^2$ **185.** $a = kbc$

Section 9.2 (page 572)

1. $3\sqrt{2}$ **3.** $3\sqrt{5}$ **5.** $4\sqrt{6}$ **7.** $3\sqrt{17}$

9. $13\sqrt{7}$ **11.** 0.2 **13.** $0.06\sqrt{2}$ **15.** $2\sqrt{5}$

17. $\frac{\sqrt{13}}{5}$ **19.** $3x^2\sqrt{x}$ **21.** $4y^2\sqrt{3}$ **23.** $3\sqrt{13}|y^3|$

25. $2|x|y\sqrt{30y}$ **27.** $8a^2b^3\sqrt{3ab}$ **29.** $2\sqrt[3]{6}$

31. $2\sqrt[3]{14}$ **33.** $2x\sqrt[3]{5x^2}$ **35.** $3|y|\sqrt[4]{4y^2}$ **37.** $xy\sqrt[3]{x}$

39. $|xy|\sqrt[4]{4y^2}$ **41.** $2xy\sqrt[5]{y}$ **43.** $\frac{\sqrt[3]{35}}{4}$

45. $|y|\sqrt{13}$ **47.** $\frac{4a^2\sqrt{2}}{|b|}$ **49.** $\frac{2\sqrt[5]{x^2}}{y}$ **51.** $\frac{3a\sqrt[3]{2a}}{b^3}$

53. $-\frac{w\sqrt[3]{3w}}{2z}$ **55.** $\frac{\sqrt{3}}{3}$ **57.** $\frac{\sqrt{7}}{7}$ **59.** $\frac{\sqrt[4]{20}}{2}$

61. $\frac{3\sqrt[3]{2}}{2}$ **63.** $\frac{\sqrt{y}}{y}$ **65.** $\frac{2\sqrt{x}}{x}$ **67.** $\frac{\sqrt{2}}{2x}$

69. $\frac{2\sqrt{3b}}{b^2}$ **71.** $\frac{\sqrt[3]{18xy^2}}{3y}$ **73.** $3\sqrt{5}$

75. 89.44 cycles per second **77.** $2\sqrt{194} \approx 27.86$ feet

79. $\sqrt{6} \cdot \sqrt{15} = \sqrt{90} = \sqrt{9} \cdot \sqrt{10} = 3\sqrt{10}$

81. $\left(\frac{5}{\sqrt{3}}\right)^2 = \frac{25}{3}$

No. Rationalizing the denominator produces an expression equivalent to the original expression, whereas squaring a number does not.

83. To find a perfect nth root factor, first factor the radicand completely. If the same factor appears at least n times, the perfect nth root factor is the common factor to the nth power.

85.

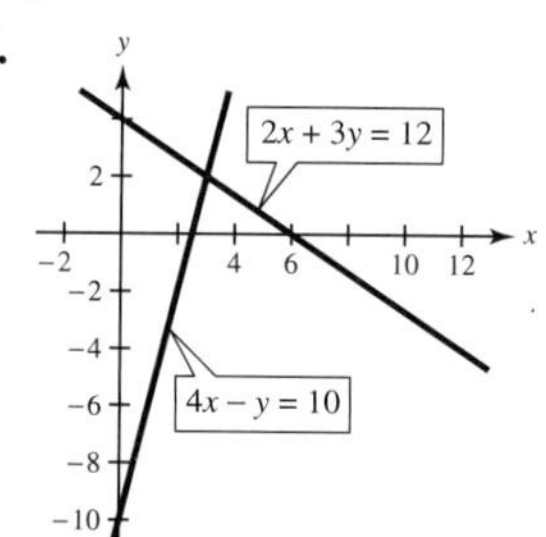

(3, 2)

87. No solution **89.** (4, −8, 10)

Section 9.3 (page 577)

1. $2\sqrt{2}$ **3.** $7\sqrt{6}$ **5.** Cannot combine **7.** $3\sqrt[3]{5}$

9. $13\sqrt[3]{y}$ **11.** $14\sqrt[4]{s}$ **13.** $9\sqrt{2}$

15. $4\sqrt[4]{5} - 7\sqrt[4]{13}$ **17.** $13\sqrt[3]{7} + \sqrt{3}$ **19.** $21\sqrt{3}$

21. $23\sqrt{5}$ **23.** $30\sqrt[3]{2}$ **25.** $12\sqrt{x}$ **27.** $13\sqrt{x+1}$

29. $13\sqrt{y}$ **31.** $(10 - z)\sqrt[3]{z}$ **33.** $(6a + 1)\sqrt{5a}$

35. $(x + 2)\sqrt[3]{6x}$ **37.** $4\sqrt{x - 1}$ **39.** $(x + 2)\sqrt{x - 1}$

41. $5a\sqrt[3]{ab^2}$ **43.** $3r^3s^2\sqrt{rs}$ **45.** 0 **47.** $\frac{2\sqrt{5}}{5}$

49. $\frac{9\sqrt{2}}{2}$ **51.** $\frac{5\sqrt{3y}}{3}$ **53.** $\frac{(3x + 2)\sqrt{3x}}{3x}$

55. $\frac{\sqrt{7y}(7y^3 - 3)}{7y^2}$ **57.** > **59.** < **61.** $12\sqrt{6x}$

63. $9x\sqrt{3} + 5\sqrt{3x}$

65. (a) $5\sqrt{10} \approx 15.8$ feet

(b) $400\sqrt{10} \approx 1264.9$ square feet

67. $T = -64 + 278.8\sqrt{t} - 214.1t + 51.8\sqrt{t^3}$; 45,756 people

69. No; $\sqrt{5} + (-\sqrt{5}) = 0$

71. Yes. $\sqrt{2x} + \sqrt{2x} = 2\sqrt{2x} = \sqrt{4} \cdot \sqrt{2x} = \sqrt{8x}$

73. (a) The student combined terms with unlike radicands; the radical expressions can be simplified no further.

(b) The student combined terms with unlike indices; the radical expressions can be simplified no further.

75. $\frac{3(z - 1)}{2z}$ **77.** $-3,\ x \neq 3$ **79.** $\frac{2v + 3}{v - 5}$

81. $\frac{15a^2}{2c^3},\ a \neq 0, b \neq 0$ **83.** $\frac{3(w + 1)}{w - 7},\ w \neq -1, w \neq 3$

Mid-Chapter Quiz (page 580)

1. 15 **2.** $\frac{3}{2}$ **3.** 7 **4.** 9

5. (a) Not a real number (b) 1 (c) 5

6. (a) 4 (b) 2 (c) 0

7. $-\infty < x < 0$ and $0 < x < \infty$ **8.** $-\frac{10}{3} \leq x < \infty$

9. $3|x|\sqrt{3}$ **10.** $2x^2\sqrt[4]{2}$ **11.** $\frac{2u\sqrt{u}}{3}$ **12.** $\frac{2\sqrt[3]{2}}{u^2}$

13. $5x|y|z^2\sqrt{5x}$ **14.** $4a^2b\sqrt[3]{2b^2}$ **15.** $4\sqrt{3}$

16. $3x\sqrt{7x},\ x \neq 0$ **17.** $3\sqrt{3} - 4\sqrt{7}$ **18.** $4\sqrt{2y}$

19. $7\sqrt{3}$ **20.** $4\sqrt{x + 2}$ **21.** $6x\sqrt[3]{5x^2} + 4x\sqrt[3]{5x}$

22. $4xy^2z^2\sqrt{xz}$ **23.** $23 + 8\sqrt{2} \approx 34.3$ inches

Section 9.4 (page 585)

1. 4 **3.** $3\sqrt{5}$ **5.** $2\sqrt[3]{9}$ **7.** 2 **9.** $3\sqrt{7} - 7$

11. $2\sqrt{10} + 8\sqrt{2}$ **13.** $3\sqrt{2}$ **15.** $12 - 4\sqrt{15}$

17. $y + 4\sqrt{y}$ **19.** $4\sqrt{a} - a$ **21.** $2 - 7\sqrt[3]{4}$

23. $\sqrt{15} - 5\sqrt{5} + 3\sqrt{3} - 15$ **25.** $8\sqrt{5} + 24$

27. $2\sqrt[3]{3} + 3\sqrt[3]{6} - 3\sqrt[3]{4} - 9$ **29.** -1 **31.** 29

33. $8 - 2\sqrt{15}$ **35.** $100 + 20\sqrt{2x} + 2x$

37. $45x - 17\sqrt{x} - 6$ **39.** $8x - 5$

41. $\sqrt[3]{4x^2} + 10\sqrt[3]{2x} + 25$ **43.** $y - 5\sqrt[3]{y} + 2\sqrt[3]{y^2} - 10$

45. $t + 5\sqrt[3]{t^2} + \sqrt[3]{t} - 3$ **47.** $x^2y^2(2y - |x|\sqrt{y})$

49. $4xy^3(x^4\sqrt[3]{x} + y\sqrt[3]{2x^2y^2})$ **51.** $x + 3$ **53.** $4 - 3x$

55. $2u + \sqrt{2u}$ **57.** $2 - \sqrt{5},\ -1$

59. $\sqrt{11} + \sqrt{3},\ 8$ **61.** $\sqrt{15} - 3,\ 6$

63. $\sqrt{x} + 3,\ x - 9$ **65.** $\sqrt{2u} + \sqrt{3},\ 2u - 3$

67. $2\sqrt{2} - \sqrt{4},\ 4$ **69.** $\sqrt{x} - \sqrt{y},\ x - y$

71. (a) $2\sqrt{3} - 4$ (b) 0 **73.** (a) 0 (b) $4 - 4\sqrt{3}$

75. $\dfrac{6(\sqrt{11} + 2)}{7}$ **77.** $\dfrac{7(5 - \sqrt{3})}{22}$ **79.** $\dfrac{5 + 2\sqrt{10}}{5}$

81. $\dfrac{\sqrt{6} - \sqrt{2}}{2}$ **83.** $2(2\sqrt{3} + \sqrt{7})$ **85.** $\dfrac{4\sqrt{7} + 11}{3}$

87. $\dfrac{2x - 9\sqrt{x} - 5}{4x - 1}$ **89.** $\dfrac{(\sqrt{15} + \sqrt{3})x}{4}$

91. $\dfrac{5\sqrt{t} + t\sqrt{5}}{5 - t}$ **93.** $4(\sqrt{3a} - \sqrt{a}), a \neq 0$

95. $\dfrac{3(x - 4)(x^2 + \sqrt{x})}{x(x - 1)(x^2 + x + 1)}$

97. $-\dfrac{\sqrt{u + v}(\sqrt{u - v} + \sqrt{u})}{v}$

99.

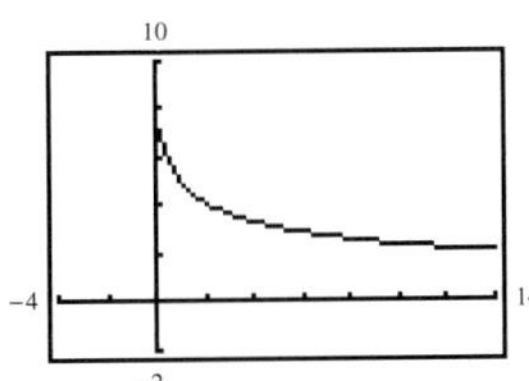

101.

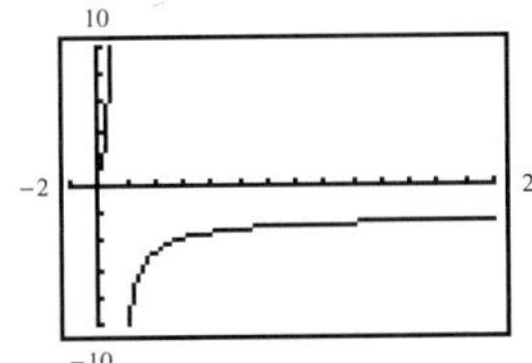

103. $\dfrac{2}{7\sqrt{2}}$ **105.** $\dfrac{10}{\sqrt{30x}}$ **107.** $\dfrac{4}{5(\sqrt{7} - \sqrt{3})}$

109. $\dfrac{y - 25}{\sqrt{3}(\sqrt{y} + 5)}$ **111.** $192\sqrt{2}$ square inches

113. $\dfrac{500k\sqrt{k^2 + 1}}{k^2 + 1}$

115. (a) If either a or b (or both) equal zero, the expression is zero and therefore rational.

(b) If the product of a and b is a perfect square, then the expression is rational.

117. $\sqrt{a} - \sqrt{b}$; $(\sqrt{a} + \sqrt{b})(\sqrt{a} - \sqrt{b}) = a - b$;
$\sqrt{b} - \sqrt{a}$; $(\sqrt{b} + \sqrt{a})(\sqrt{b} - \sqrt{a}) = b - a$;
When the order of the terms is changed, the conjugate and the product both change by a factor of -1.

119. 6 **121.** No solution **123.** $-12, 12$

125. $-5, 3$ **127.** $4|x|y^2\sqrt{2y}$ **129.** $2y\sqrt[4]{2x^2y}$

Section 9.5 *(page 595)*

1. (a) Not a solution (b) Not a solution
(c) Not a solution (d) Solution

3. (a) Not a solution (b) Solution
(c) Not a solution (d) Not a solution

5. 144 **7.** 49 **9.** 27 **11.** 49

13. No solution **15.** 64 **17.** 90 **19.** -27

21. $\frac{4}{5}$ **23.** 5 **25.** No solution **27.** 215

29. $\frac{577}{16}$ **31.** 4 **33.** No solution **35.** 7

37. -15 **39.** $-\frac{9}{4}$ **41.** 8 **43.** 1, 3 **45.** 1

47. $\frac{1}{4}$ **49.** $\frac{1}{2}$ **51.** 4 **53.** 7 **55.** 4

57. 216 **59.** $4, -12$ **61.** -16

63.

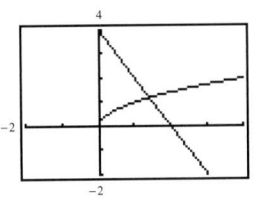

1.407

65.

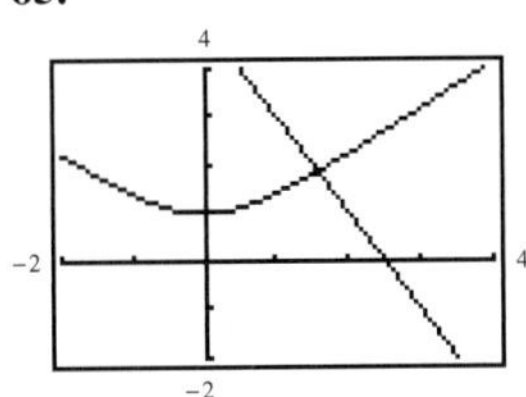

1.569

67.

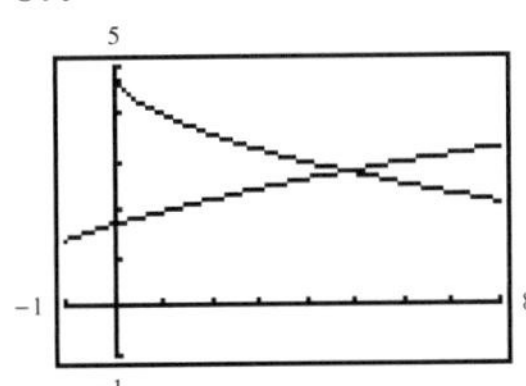

4.840

69.

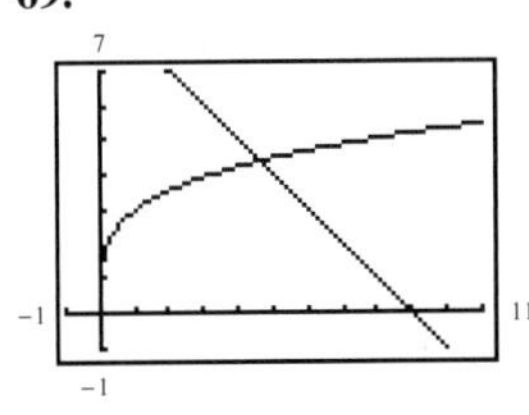

4.605

71.

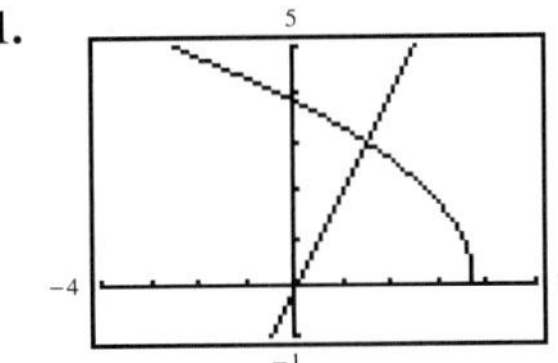

1.500

73. 25 **75.** 2, 6 **77.** $\frac{4}{9}$ **79.** 2, 6

81. 9.00 **83.** 12.00

85.

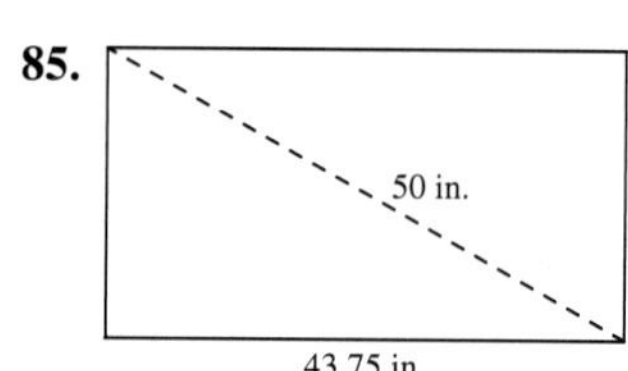

$\dfrac{25\sqrt{15}}{4} \approx 24.21$ inches

87. $2\sqrt{10} \approx 6.32$ meters **89.** 15 feet

91. 30 inches $\times$ 16 inches

93. $h = \dfrac{\sqrt{S^2 - \pi^2 r^4}}{\pi r}$; 34 centimeters

95. $r = \dfrac{\sqrt{\pi A}}{\pi}$

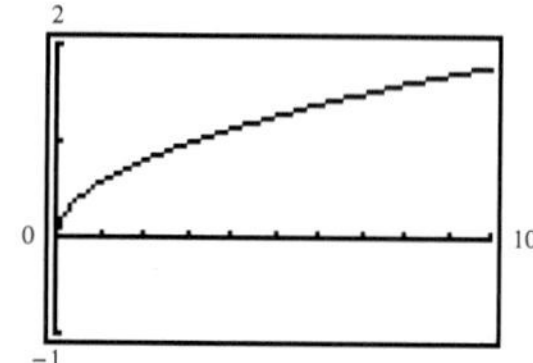

97. 64 feet **99.** $32\sqrt{5} \approx 71.55$ feet per second

101. 39.06 feet **103.** 1.82 feet **105.** 500 units

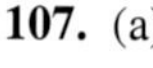

107. (a)

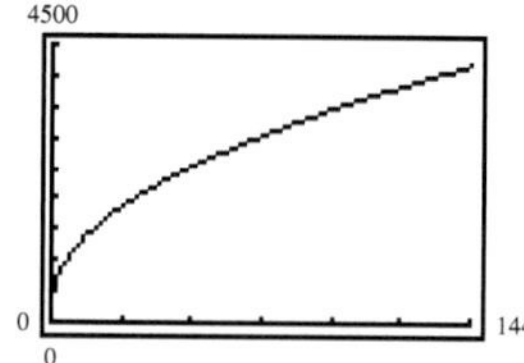

(b) 92 months

109. $(\sqrt{x} + \sqrt{6})^2 \neq (\sqrt{x})^2 + (\sqrt{6})^2$

111. Substitute $x = 20$ into the equation, and then choose any value of a such that $a \leq 20$ and solve the resulting equation for b.

113. Parallel **115.** Perpendicular **117.** (3, 2)

119. $a^{4/5}$ **121.** $x^{3/2}, x \neq 0$

Section 9.6 *(page 605)*

1. $2i$ **3.** $-12i$ **5.** $\frac{2}{5}i$ **7.** $-\frac{6}{11}i$ **9.** $2\sqrt{2}i$

11. $\sqrt{7}i$ **13.** 2 **15.** $\dfrac{3\sqrt{2}}{5}i$ **17.** $0.3i$ **19.** $10i$

21. $2i$ **23.** $3\sqrt{2}i$ **25.** $3\sqrt{3}i$ **27.** $-2\sqrt{6}$

29. $-3\sqrt{6}$ **31.** -0.44 **33.** $-3 - 2\sqrt{3}$

35. $-4\sqrt{5} + 5\sqrt{2}$ **37.** $4 + 3\sqrt{2}i$ **39.** -16

41. $-8i$ **43.** Equal **45.** Not equal

47. $a = 3, b = -4$ **49.** $a = 2, b = -3$

51. $a = -4, b = -2\sqrt{2}$ **53.** $a = 64, b = 7$

55. $10 + 4i$ **57.** $-14 - 40i$ **59.** $-14 + 20i$

61. $9 - 7i$ **63.** $3 + 6i$ **65.** $\frac{13}{6} + \frac{3}{2}i$

67. $-6.15 - 9.3i$ **69.** $-3 + 49i$ **71.** -36

73. 24 **75.** $-35i$ **77.** $27i$ **79.** -9

81. $-65 - 10i$ **83.** $20 - 12i$ **85.** $4 + 18i$

87. $-40 - 5i$ **89.** $-14 + 42i$ **91.** 9

93. $-7 - 24i$ **95.** $-21 + 20i$ **97.** $18 + 26i$

99. $-i$ **101.** 1 **103.** -1 **105.** i **107.** -1

109. 5 **111.** 68 **113.** 31 **115.** 100 **117.** 144

119. 4 **121.** 2.5 **123.** $-10i$ **125.** $-\frac{1}{5} + \frac{2}{5}i$

127. $2 + 2i$ **129.** $1 - 2i$ **131.** $-\frac{24}{53} + \frac{84}{53}i$

133. $\frac{6}{29} + \frac{15}{29}i$ **135.** $\frac{12}{13} - \frac{5}{13}i$ **137.** $-\frac{23}{58} + \frac{43}{58}i$

139. $\frac{9}{5} - \frac{2}{5}i$ **141.** $\frac{47}{26} + \frac{27}{26}i$ **143.** $\frac{14}{29} - \frac{35}{29}i$

145–147. (a) Solution and (b) Solution

149. (a) $\left(\dfrac{-5 + 5\sqrt{3}i}{2}\right)^3 = 125$

(b) $\left(\dfrac{-5 - 5\sqrt{3}i}{2}\right)^3 = 125$

151. (a) $1, \dfrac{-1 + \sqrt{3}i}{2}, \dfrac{-1 - \sqrt{3}i}{2}$

(b) $2, \dfrac{-2 + 2\sqrt{3}i}{2} = -1 + \sqrt{3}i,$

$\dfrac{-2 - 2\sqrt{3}i}{2} = -1 - \sqrt{3}i$

(c) $4, \dfrac{-4 + 4\sqrt{3}i}{2} = -2 + 2\sqrt{3}i,$

$\dfrac{-4 - 4\sqrt{3}i}{2} = -2 - 2\sqrt{3}i$

153. $2a$ **155.** $2bi$

157. Exercise 153: The sum of complex conjugates of the form $a + bi$ and $a - bi$ is twice the real number a, or $2a$.

Exercise 154: The product of complex conjugates of the form $a + bi$ and $a - bi$ is the sum of the squares of a and b, or $a^2 + b^2$.

Exercise 155: The difference of complex conjugates of the form $a + bi$ and $a - bi$ is twice the imaginary number bi, or $2bi$.

Exercise 156: The sum of the squares of complex conjugates of the form $a + bi$ and $a - bi$ is the difference of twice the squares of a and b, or $2a^2 - 2b^2$.

159. The numbers must be written in i-form first.

$\sqrt{-3}\sqrt{-3} = (\sqrt{3}i)(\sqrt{3}i) = 3i^2 = -3$

161. To simplify the quotient, multiply the numerator and the denominator by $-bi$. This will yield a positive real number in the denominator. The number i can also be used to simplify the quotient. The denominator will be the opposite of b, but the resulting number will be the same.

163. $-7, 5$ **165.** $-4, 0, 3$ **167.** 81 **169.** 25

Review Exercises *(page 610)*

1. -9 **3.** -4 **5.** $-\frac{3}{4}$ **7.** $-\frac{1}{5}$

9. Not a real number

	Radical Form	*Rational Exponent Form*
11.	$\sqrt[3]{27} = 3$	$27^{1/3} = 3$
13.	$\sqrt[3]{216} = 6$	$216^{1/3} = 6$

15. 81 **17.** Not a real number **19.** $\frac{1}{16}$

21. $x^{7/12}, x \neq 0$ **23.** $z^{5/3}$ **25.** $\dfrac{1}{x^{5/4}}$

27. $ab^{2/3}$ **29.** $x^{1/8}$ **31.** $(3x + 2)^{1/3},\ x \neq -\frac{2}{3}$

33. 0.0392 **35.** 10.6301

37. (a) Not a real number (b) 7

39. (a) -1 (b) 3 **41.** $\left(-\infty, \frac{9}{2}\right]$ **43.** $6u^2|v|\sqrt{u}$

45. $0.5x^2\sqrt{y}$ **47.** $2ab\sqrt[3]{6b}$ **49.** $\dfrac{\sqrt{30}}{6}$ **51.** $\dfrac{\sqrt[3]{4x^2}}{x}$

53. $\sqrt{145}$ **55.** $8\sqrt{6}$ **57.** $14\sqrt{x} - 9\sqrt[3]{x}$

59. $7\sqrt[4]{y+3}$ **61.** $3x\sqrt[3]{3x^2y}$ **63.** $6\sqrt{x} + 2\sqrt{3x}$ feet

65. $10\sqrt{3}$ **67.** $2\sqrt{5} + 5\sqrt{2}$ **69.** $3 - x$

71. $3 + \sqrt{7}; 2$ **73.** $\sqrt{x} - 20; x - 400$

75. $-\dfrac{(\sqrt{2} - 1)(\sqrt{3} + 4)}{13}$ **77.** $\dfrac{(\sqrt{x} + 10)^2}{x - 100}$

79. 32 **81.** No real solution **83.** 5

85. $-5, -3$ **87.** $\frac{3}{32}$ **89.** 8 inches × 15 inches

91. 2.93 feet **93.** 64 feet **95.** $4\sqrt{3}i$

97. $10 - 9\sqrt{3}i$ **99.** $\frac{3}{4} - \sqrt{3}i$ **101.** $15i$

103. $\sqrt{70} - 2\sqrt{10}$ **105.** $a = 10,\ b = -4$

107. $a = 4,\ b = 7$ **109.** $8 - 3i$

111. 25 **113.** $11 - 60i$ **115.** $-\frac{7}{3}i$

117. $\frac{9}{26} - \frac{3}{13}i$ **119.** $\frac{13}{37} - \frac{33}{37}i$

Chapter Test *(page 613)*

1. (a) 64 (b) 10 **2.** (a) $\frac{1}{25}$ (b) 6

3. $f(-8) = 7, f(0) = 3$ **4.** $\left[\frac{3}{7}, \infty\right)$

5. (a) $x^{1/3},\ x \neq 0$ (b) 25 **6.** (a) $\dfrac{4\sqrt{2}}{3}$ (b) $2\sqrt[3]{3}$

7. (a) $2x\sqrt{6x}$ (b) $2xy^2\sqrt[4]{x}$ **8.** $\dfrac{2\sqrt[3]{3y^2}}{3y}$

9. $\dfrac{5(\sqrt{6} + \sqrt{2})}{2}$ **10.** $6\sqrt{2x}$ **11.** $5\sqrt{3x} + 3\sqrt{5}$

12. $16 - 8\sqrt{2x} + 2x$ **13.** $3 + 4y$ **14.** 24

15. No solution **16.** 9 **17.** $2 - 2i$

18. $-16 - 30i$ **19.** $-8 + 4i$ **20.** $13 + 13i$

21. $\frac{13}{10} - \frac{11}{10}i$ **22.** 144 feet

Cumulative Test: Chapters 7–9 *(page 614)*

1. Domain $= \left(-\infty, \frac{3}{8}\right) \cup \left(\frac{3}{8}, \infty\right)$

2. $\dfrac{x(x + 2)(x + 4)}{9(x - 4)}, x \neq -4, 0$

3. $\dfrac{(x + 1)(x - 2)}{(x + 6)}, x \neq -4, -2, -1, 0$ **4.** $\dfrac{3x + 5}{x(x + 3)}$

5. $x + y, x \neq 0, y \neq 0, x \neq y$

6. (a) Solution (b) Not a solution **7.** b **8.** c

9. d **10.** a **11.** $(2, 1)$ **12.** $(3, -2)$ **13.** $(5, 4)$

14. $(0, 1, -2)$ **15.** $(1, -5, 5)$ **16.** $\left(-\frac{1}{5}, -\frac{22}{5}\right)$

17.

$-2x + y = 2$

$y = 0$

$x - 2y = 0$

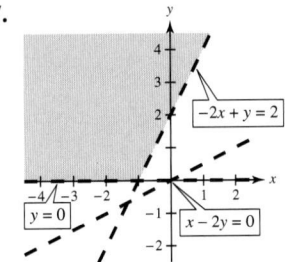

18. $-4 + 3\sqrt{2}i$ **19.** $21 + 20i$ **20.** $t^{1/2}$

21. $35\sqrt{5x}$ **22.** $6x^2 + 14\sqrt{6x} + 49$ **23.** $6 + 3\sqrt{5}$

24. $\frac{3}{20} + \frac{11}{20}i$ **25.** 2, 5 **26.** 2, 9 **27.** 16 **28.** 4

29. 128 feet; $d = 0.08s^2$ **30.** 50

31. Student tickets: 1035; Adult tickets: 400

32. \$20,000 at 8% and \$30,000 at 8.5%

33. $16(1 + \sqrt{2}) \approx 38.6$ inches **34.** About 66.02 feet

CHAPTER 10

Section 10.1 *(page 623)*

1. 6, 9 **3.** $-5, 6$ **5.** $-9, 5$ **7.** 8 **9.** $-\frac{8}{9}, 2$

11. 0, 3 **13.** 9, 12 **15.** $-\frac{9}{2}, 5$ **17.** 1, 6

19. $-\frac{5}{6}, \frac{1}{2}$ **21.** ± 7 **23.** ± 3 **25.** $\pm\frac{4}{5}$ **27.** ± 14

29. $\pm\frac{5}{2}$ **31.** $\pm\frac{15}{2}$ **33.** $-12, 4$ **35.** 2.5, 3.5

37. $2 \pm \sqrt{7}$ **39.** $-\dfrac{1}{2} \pm \dfrac{5\sqrt{2}}{2}$ **41.** $\dfrac{2}{9} \pm \dfrac{2\sqrt{3}}{3}$

43. $\pm 6i$ **45.** $\pm 2i$ **47.** $\pm \frac{\sqrt{17}}{3}i$ **49.** $3 \pm 5i$

51. $-\frac{4}{3} \pm 4i$ **53.** $-\frac{1}{4} \pm \sqrt{5}i$ **55.** $-3 \pm \frac{5}{3}i$

57. $1 \pm 3\sqrt{3}i$ **59.** $-1 \pm 0.2i$ **61.** $\frac{2}{3} \pm \frac{1}{3}i$

63. $-\frac{7}{3} \pm \frac{\sqrt{38}}{3}i$ **65.** $0, \frac{5}{2}$ **67.** $-4, \frac{3}{2}$ **69.** ± 30

71. $\pm 30i$ **73.** ± 3 **75.** $2 \pm 6\sqrt{3}$ **77.** $2 \pm 6\sqrt{3}i$

79. $-2 \pm 3\sqrt{2}i$

81.

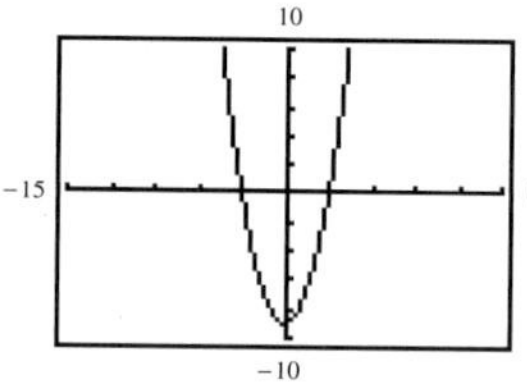

$(-3, 0), (3, 0)$;

The result is the same.

83.

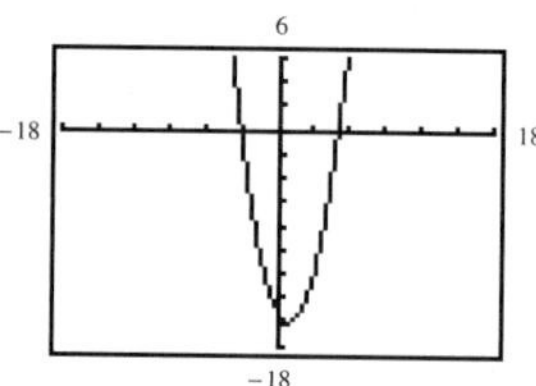

$(-3, 0), (5, 0)$;

The result is the same.

85.

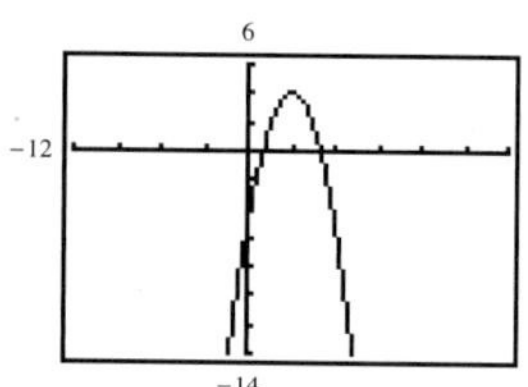

$(1, 0), (5, 0)$;

The result is the same.

87.

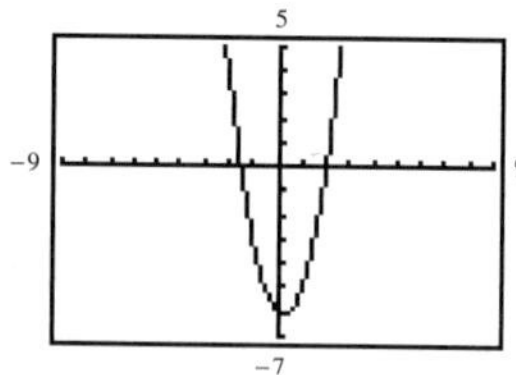

$(2, 0), \left(-\frac{3}{2}, 0\right)$;

The result is the same.

89.

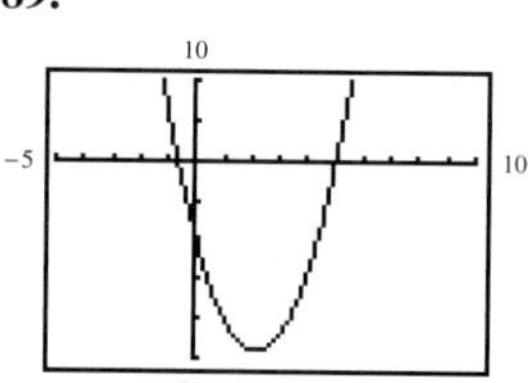

$\left(-\frac{2}{3}, 0\right), (5, 0)$;

The result is the same.

91.

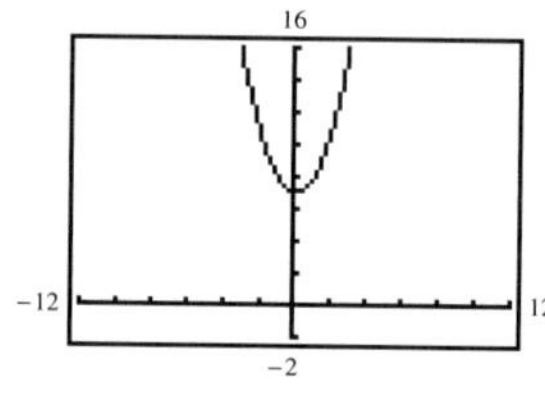

$\pm\sqrt{7}i$; complex solutions

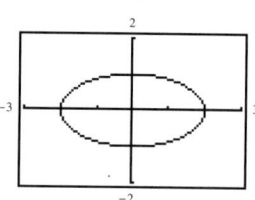

93.

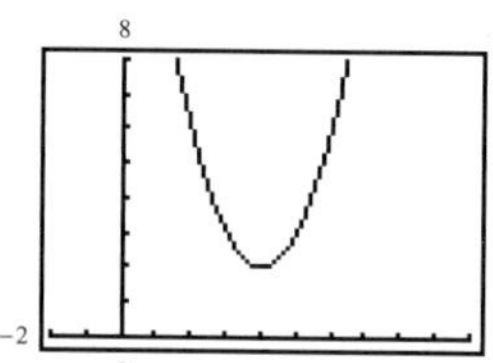

$4 \pm \sqrt{2}i$; complex solutions

95.

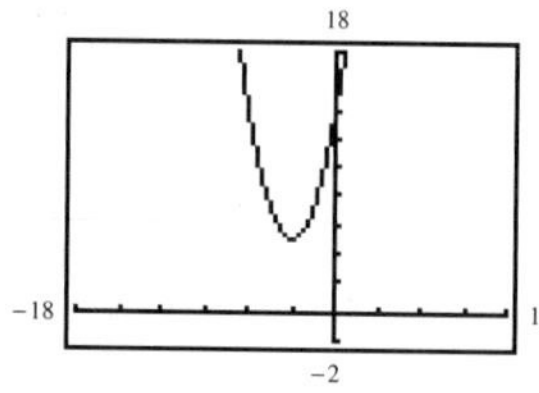

$-3 \pm \sqrt{5}i$; complex solutions

97. $f(x) = \sqrt{4 - x^2}$
$g(x) = -\sqrt{4 - x^2}$

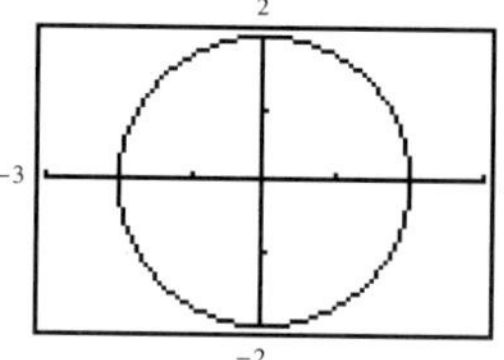

99. $f(x) = \frac{1}{2}\sqrt{4 - x^2}$
$g(x) = -\frac{1}{2}\sqrt{4 - x^2}$

101. $\pm 1, \pm 2$ **103.** $\pm\sqrt{2}, \pm\sqrt{3}$ **105.** $\pm 1, \pm\sqrt{5}$

107. 16 **109.** 4, 25 **111.** $-8, 27$ **113.** $1, \frac{125}{8}$

115. 1, 32 **117.** $\frac{1}{32}, 243$ **119.** 729 **121.** 1, 16

123. $\frac{1}{2}, 1$ **125.** $-1, \frac{4}{5}$ **127.** $\pm 1, 2, 4$ **129.** $\frac{12}{5}$

131. 120 feet **133.** 4 seconds

135. $2\sqrt{2} \approx 2.83$ seconds **137.** 9 seconds

139. 6% **141.** 2001

143. If $a = 0$, the equation would not be quadratic because it would be of degree 1, not 2.

145. Write the equation in the form $u^2 = d$, where u is an algebraic expression and d is a positive constant. Take the square root of each side of the equation to obtain the solutions $u = \pm\sqrt{d}$.

147. $x > 4$

149. $x \leq 5$

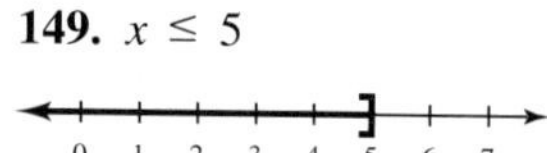

151. (3, 2, 1) **153.** $3\sqrt{3}$ **155.** $16\sqrt[3]{y} - 9\sqrt[3]{x}$

157. $(4n + 1)m^2\sqrt{n}$

Section 10.2 *(page 631)*

1. 16 **3.** 100 **5.** 49 **7.** $\frac{25}{4}$ **9.** $\frac{81}{4}$ **11.** $\frac{1}{36}$

13. $\frac{16}{25}$ **15.** 0.04 **17.** 0, 20 **19.** $-6, 0$

21. 0, 5 **23.** 1, 7 **25.** $-4, -3$ **27.** $-3, 6$

29. 3 **31.** $-\frac{5}{2}, \frac{3}{2}$

33. $2 + \sqrt{7} \approx 4.65$
$2 - \sqrt{7} \approx -0.65$

35. $-2 + \sqrt{7} \approx 0.65$
$-2 - \sqrt{7} \approx -4.65$

37. $-7, 1$

39. $6 + \sqrt{26} \approx 11.10$
$6 - \sqrt{26} \approx 0.90$

41. $-7, -1$ **43.** 3, 7

45. $-\frac{5}{2} + \frac{\sqrt{13}}{2} \approx -0.70$
$-\frac{5}{2} - \frac{\sqrt{13}}{2} \approx -4.30$

47. $3 \pm i$ **49.** $-2 \pm 3i$

51. $\frac{1}{2} + \frac{\sqrt{3}}{2}i \approx 0.5 + 0.87i$
$\frac{1}{2} - \frac{\sqrt{3}}{2}i \approx 0.5 - 0.87i$

53. $-\frac{7}{2} + \frac{\sqrt{5}}{2} \approx -2.38$
$-\frac{7}{2} - \frac{\sqrt{5}}{2} \approx -4.62$

55. $\frac{1}{3} + \frac{2\sqrt{7}}{3} \approx 2.10$
$\frac{1}{3} - \frac{2\sqrt{7}}{3} \approx -1.43$

57. $-\frac{3}{8} + \frac{\sqrt{137}}{8} \approx 1.09$
$-\frac{3}{8} - \frac{\sqrt{137}}{8} \approx -1.84$

59. $-2 + \frac{\sqrt{10}}{2} \approx -0.42$
$-2 - \frac{\sqrt{10}}{2} \approx -3.58$

61. $-\frac{3}{2} + \frac{\sqrt{21}}{6} \approx -0.74$
$-\frac{3}{2} - \frac{\sqrt{21}}{6} \approx -2.26$

63. $-\frac{1}{2} + \frac{\sqrt{10}}{2} \approx 1.08$
$-\frac{1}{2} - \frac{\sqrt{10}}{2} \approx -2.08$

65. $\frac{3}{10} + \frac{\sqrt{191}}{10}i \approx 0.30 + 1.38i$
$\frac{3}{10} - \frac{\sqrt{191}}{10}i \approx 0.30 - 1.38i$

67. $\frac{1}{3} + \frac{\sqrt{127}}{3} \approx 4.09$
$\frac{1}{3} - \frac{\sqrt{127}}{3} \approx -3.42$

69. $-\frac{5}{2} + \frac{\sqrt{17}}{2} \approx -0.44$
$-\frac{5}{2} - \frac{\sqrt{17}}{2} \approx -4.56$

71. $-\frac{5}{6} + \frac{\sqrt{47}}{6}i \approx -0.83 + 1.14i$
$-\frac{5}{6} - \frac{\sqrt{47}}{6}i \approx -0.83 - 1.14i$

73. $1 \pm \sqrt{3}$ **75.** $-2, 6$ **77.** $4 + 2\sqrt{2}$

79.

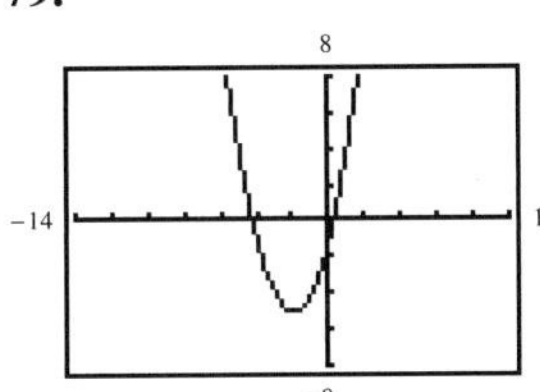

$(-2 \pm \sqrt{5}, 0)$

The result is the same.

81.

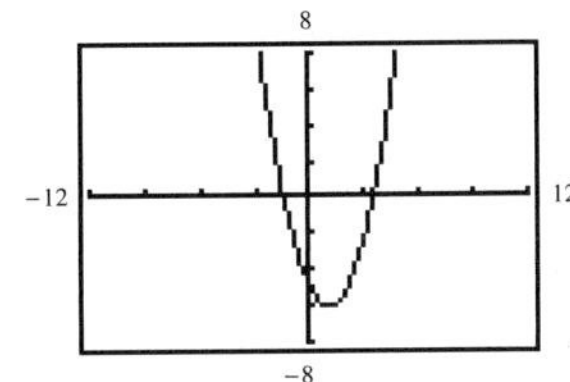

$(1 \pm \sqrt{6}, 0)$

The result is the same.

83.

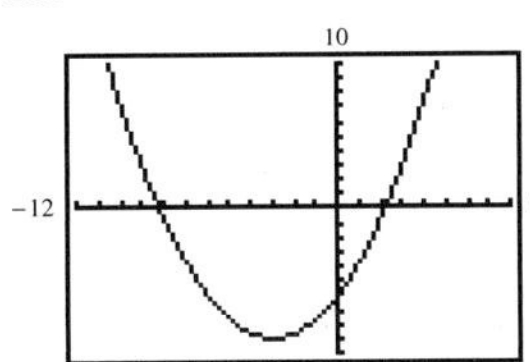

$(-3 \pm 3\sqrt{3}, 0)$

The result is the same.

85.

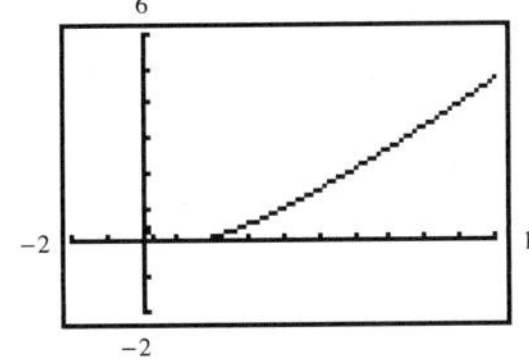

$(1, 0)$

The result is the same.

87. (a) $x^2 + 8x$ (b) $x^2 + 8x + 16$ (c) $(x + 4)^2$

89. 15 meters × $46\frac{2}{3}$ meters or 20 meters × 35 meters

91. 6 inches × 10 inches × 14 inches

93. 50 pairs, 110 pairs

95. Use the method of completing the square to write the quadratic equation in the form $u^2 = d$. Then use the Square Root Property to simplify.

97. (a) $d = 0$ (b) d is positive and is a perfect square. (c) d is positive and is not a perfect square. (d) $d < 0$

99. 150 **101.** 7 **103.** $11 + 6\sqrt{2}$

105. $\frac{4\sqrt{10}}{5}$ **107.** $\sqrt{a} \cdot \sqrt{b}$

Section 10.3 *(page 640)*

1. $2x^2 + 2x - 7 = 0$ **3.** $-x^2 + 10x - 5 = 0$

5. 4, 7 **7.** $-2, -4$ **9.** $-\frac{1}{4}$ **11.** $-\frac{3}{2}$

13. $-15, 20$ **15.** $1 \pm \sqrt{5}$ **17.** $-2 \pm \sqrt{3}$

19. $5 \pm \sqrt{2}$ **21.** $-\frac{3}{4} \pm \frac{\sqrt{15}}{4}i$ **23.** $-\frac{1}{3}, 1$

25. $-1 \pm \frac{\sqrt{10}}{2}$ **27.** $-\frac{3}{4} \pm \frac{\sqrt{21}}{4}$ **29.** $\frac{3}{8} \pm \frac{\sqrt{7}}{8}i$

31. $\frac{5}{4} \pm \frac{\sqrt{73}}{4}$ **33.** $\frac{1}{2} \pm \frac{\sqrt{13}}{6}$ **35.** $-\frac{1}{4} \pm \frac{\sqrt{13}}{4}$

37. $\frac{1}{5} \pm \frac{\sqrt{5}}{5}$ **39.** $-\frac{1}{5} \pm \frac{\sqrt{10}}{5}$

41. Two distinct complex solutions

43. Two distinct rational solutions

45. One repeated rational solution

47. Two distinct complex solutions **49.** ± 13

51. $-3, 0$ **53.** $\frac{9}{5}, \frac{21}{5}$ **55.** $-\frac{3}{2}, 18$ **57.** $-4 \pm 3i$

59. $\frac{13}{6} \pm \frac{13\sqrt{11}}{6}i$ **61.** $-\frac{8}{5} \pm \frac{\sqrt{3}}{5}$

63. $-\frac{11}{8} \pm \frac{\sqrt{41}}{8}$ **65.** $x^2 - 3x - 10 = 0$

67. $x^2 - 8x + 7 = 0$ **69.** $x^2 - 2x - 1 = 0$

71. $x^2 + 25 = 0$ **73.** $x^2 - 24x + 144 = 0$

75.

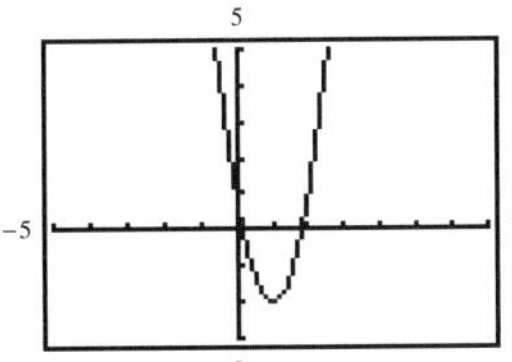

(0.18, 0), (1.82, 0)

The result is the same.

77.

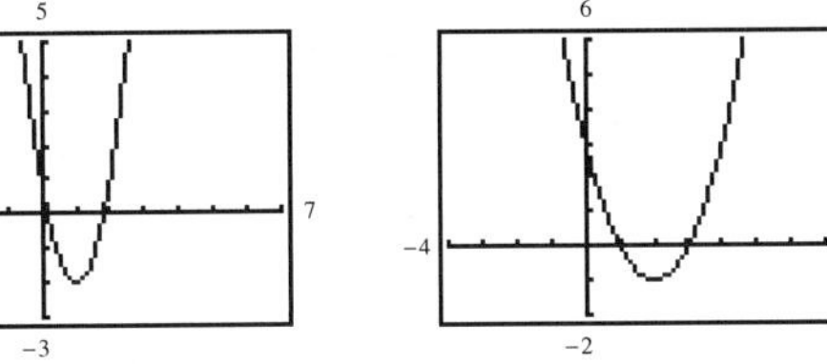

(1, 0), (3, 0)

The result is the same.

79.

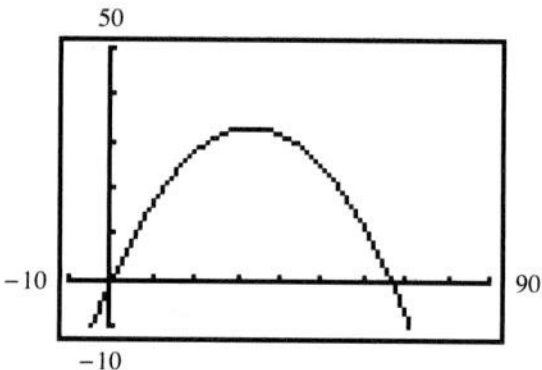

(0.20, 0), (66.47, 0)

The result is the same.

81.

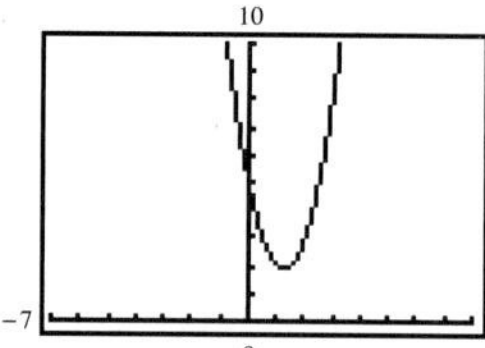

No real solutions

83.

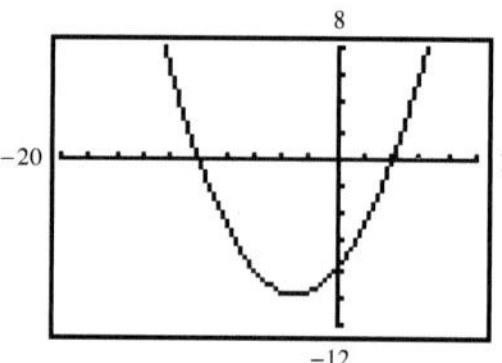

Two real solutions

85. $\frac{7}{4} \pm \frac{\sqrt{17}}{4}$ **87.** No real values **89.** $\frac{4}{3} \pm \frac{2\sqrt{13}}{3}$

91. $\frac{3}{2} + \frac{\sqrt{17}}{2}$ **93.** (a) $c < 9$ (b) $c = 9$ (c) $c > 9$

95. 5.1 inches × 11.4 inches

97. (a) 2.5 seconds (b) $\frac{5}{4} + \frac{5\sqrt{3}}{4} \approx 3.42$ seconds

(c) No. In order for the discriminant to be greater than or equal to zero, the value of c must be greater than or equal to -25. Therefore, the height cannot exceed 75 feet, or the value of c would be less than -25 when the equation is set equal to zero.

99. 2.15 or 4.65 hours

101. (a)

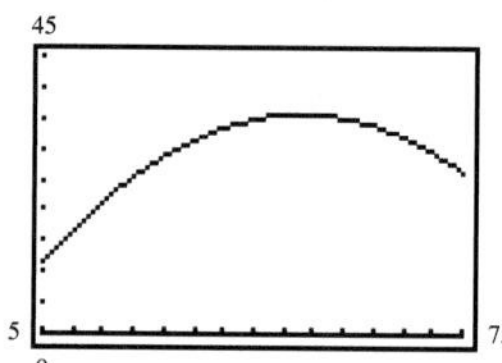

(b) 31.3 or 64.8 miles per hour

103.

	x_1, x_2	$x_1 + x_2$	x_1x_2
(a)	$-2, 3$	1	-6
(b)	$-3, \frac{1}{2}$	$-\frac{5}{2}$	$-\frac{3}{2}$
(c)	$-\frac{3}{2}, \frac{3}{2}$	0	$-\frac{9}{4}$
(d)	$5 + 3i, 5 - 3i$	10	34

105. The Square Root Property would be convenient because the equation is of the form $u^2 = d$.

107. The Quadratic Formula would be convenient because the equation is already in general form, the expression cannot be factored, and the leading coefficient is not 1.

109. When the Quadratic Formula is applied to $ax^2 + bx + c = 0$, the square root of the discriminant is evaluated. When the discriminant is positive, the square root of the discriminant is positive and will yield two real solutions (or x-intercepts). When the discriminant is zero, the equation has one real solution (or x-intercept). When the discriminant is negative, the square root of the discriminant is negative and will yield two complex solutions (no x-intercepts).

111. Collinear **113.** Not collinear

115.

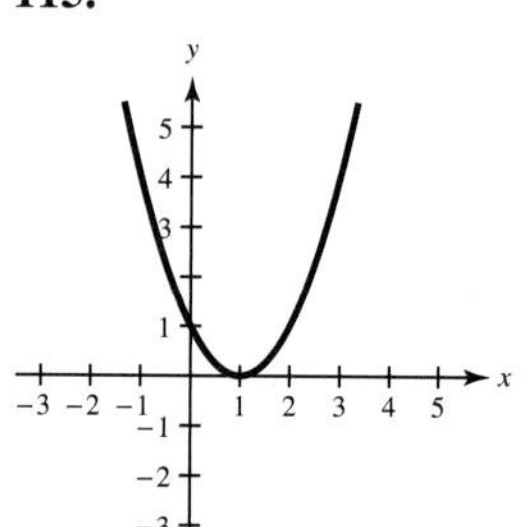

117.

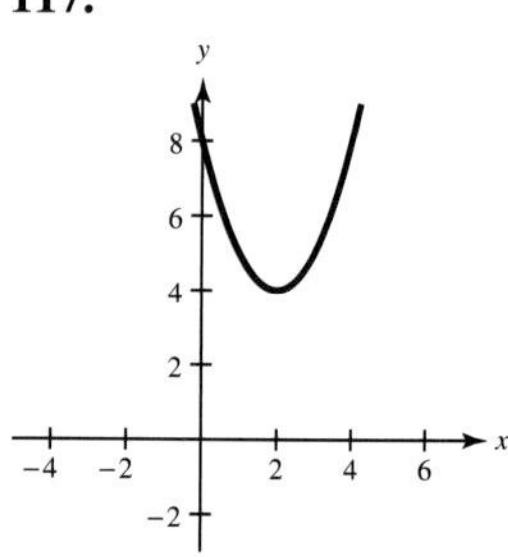

Mid-Chapter Quiz *(page 644)*

1. ± 6 **2.** $-4, \frac{5}{2}$ **3.** $\pm 2\sqrt{3}$ **4.** $-1, 7$

5. $-\frac{7}{2} \pm \frac{\sqrt{41}}{2}$ **6.** $-\frac{3}{2} \pm \frac{\sqrt{19}}{2}$ **7.** $-2 \pm \sqrt{10}$

8. $\frac{1}{4} \pm \frac{\sqrt{105}}{12}$ **9.** $-\frac{5}{2} \pm \frac{\sqrt{3}}{2}i$ **10.** $-2, 10$

11. $-3, 10$ **12.** $-2, 5$ **13.** $\frac{3}{2}$ **14.** $-\frac{5}{3} \pm \frac{\sqrt{10}}{3}$

15. 49 **16.** $\pm 2i, \pm\sqrt{3}i$ **17.** 1, 9

18. $\pm\sqrt{2}, \pm 2\sqrt{3}$

19.

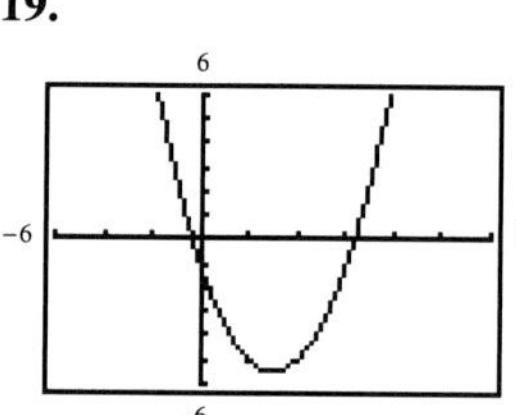

$(-0.32, 0), (6.32, 0)$

The result is the same.

20.

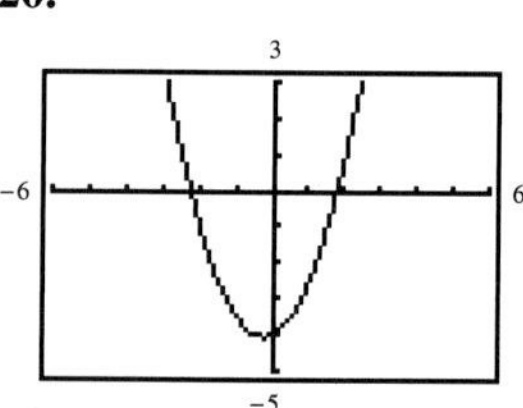

$(-2.24, 0), (1.79, 0)$

The result is the same.

21. 60 video games **22.** 35 meters × 65 meters

Section 10.4 *(page 650)*

1. e **2.** f **3.** b **4.** c **5.** d **6.** a

7. $y = (x - 1)^2 - 1, \ (1, -1)$

9. $y = (x - 2)^2 + 3$, $(2, 3)$

11. $y = (x + 3)^2 - 4$, $(-3, -4)$

13. $y = -(x - 3)^2 - 1$, $(3, -1)$

15. $y = -(x + 4)^2 + 21$, $(-4, 21)$

17. $y = 2\left(x + \frac{3}{2}\right)^2 - \frac{5}{2}$, $\left(-\frac{3}{2}, -\frac{5}{2}\right)$ **19.** $(4, -1)$

21. $(-1, 2)$ **23.** $\left(-\frac{1}{2}, 3\right)$ **25.** Upward, $(0, 2)$

27. Downward, $(10, 4)$ **29.** Upward, $(0, -6)$

31. Downward, $(3, 0)$ **33.** Downward, $(3, 9)$

35. $(\pm 5, 0)$, $(0, 25)$ **37.** $(0, 0)$, $(9, 0)$

39. $(-7, 0)$, $(1, 0)$, $(0, 7)$ **41.** $\left(\frac{3}{2}, 0\right)$, $(0, 9)$

43. $(0, 3)$ **45.** $\left(-\frac{3}{2} \pm \frac{\sqrt{19}}{2}, 0\right)$, $(0, 5)$

47.

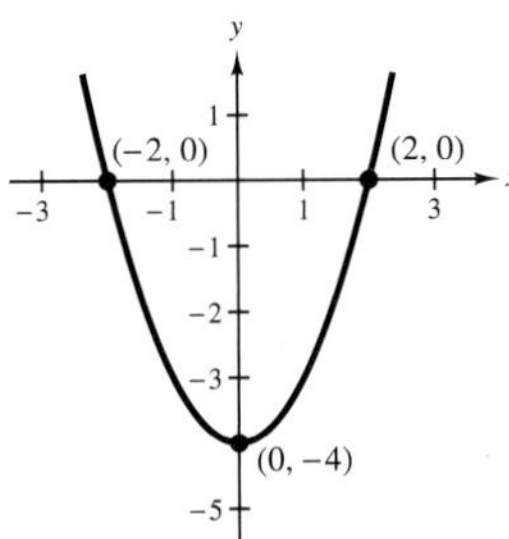

49.

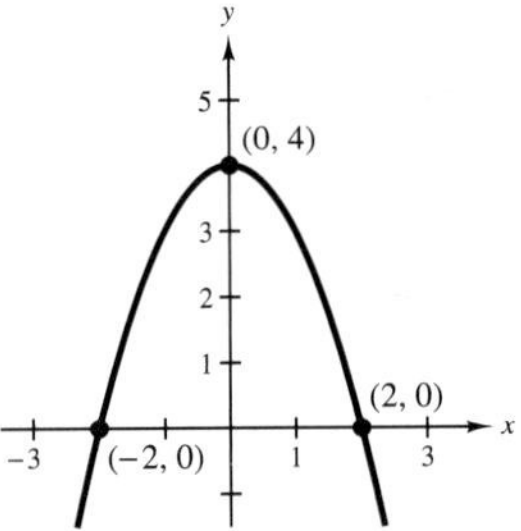

51.

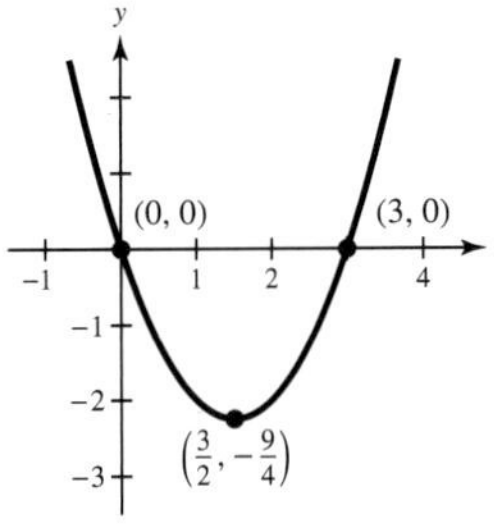

53.

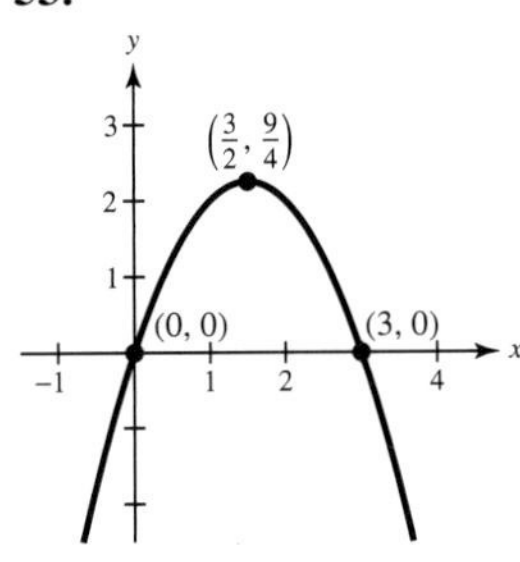

55.

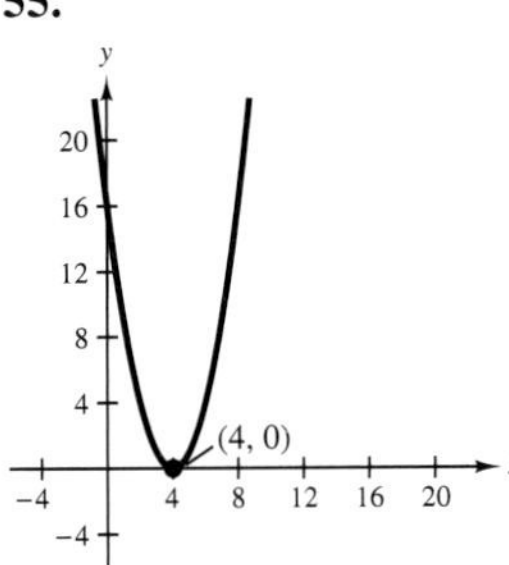

57.

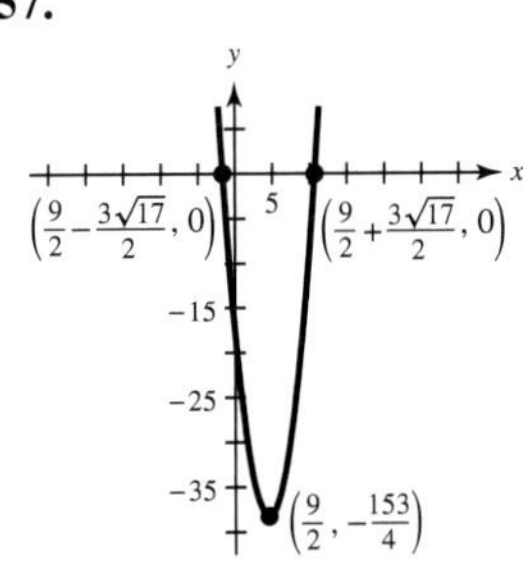

59.

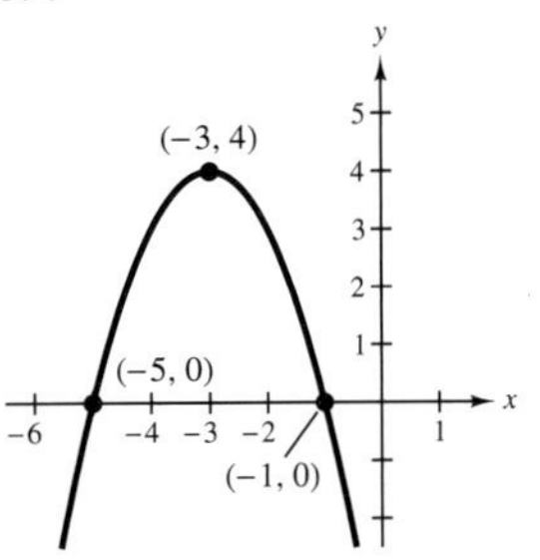

61.

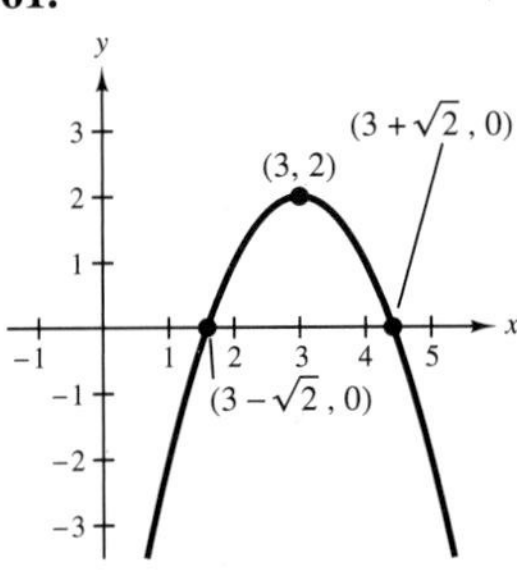

63.

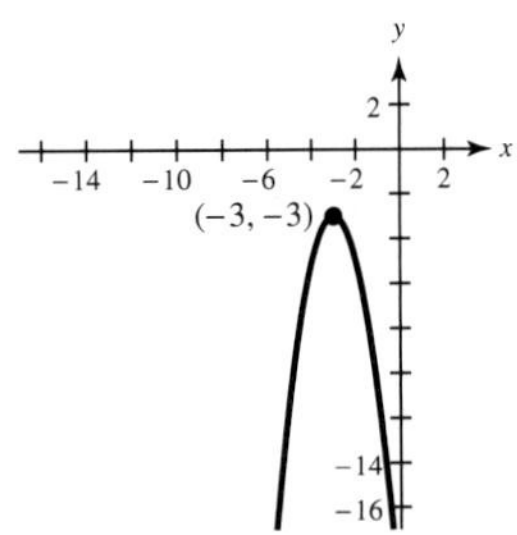

65.

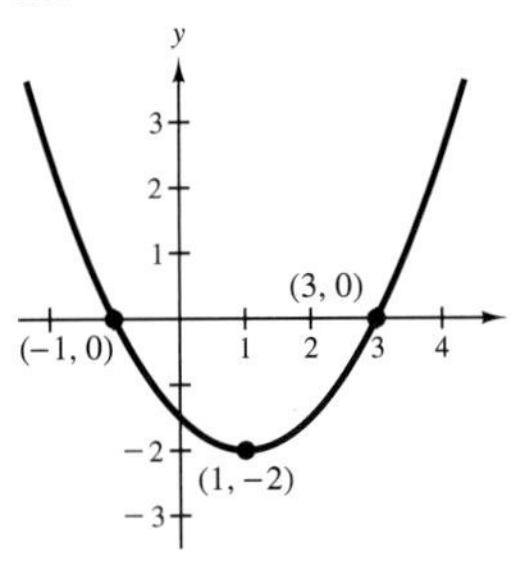

67.

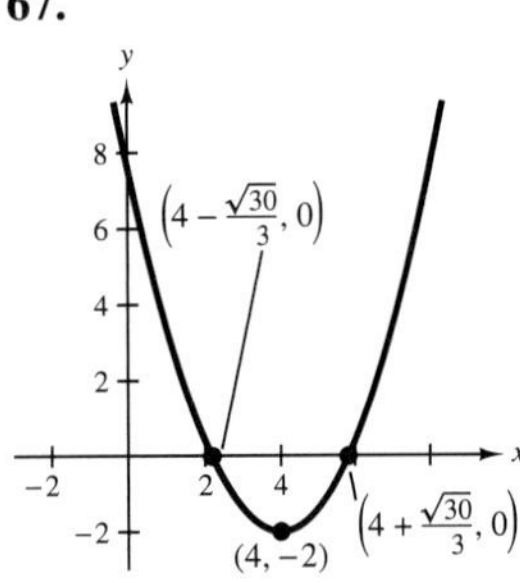

69.

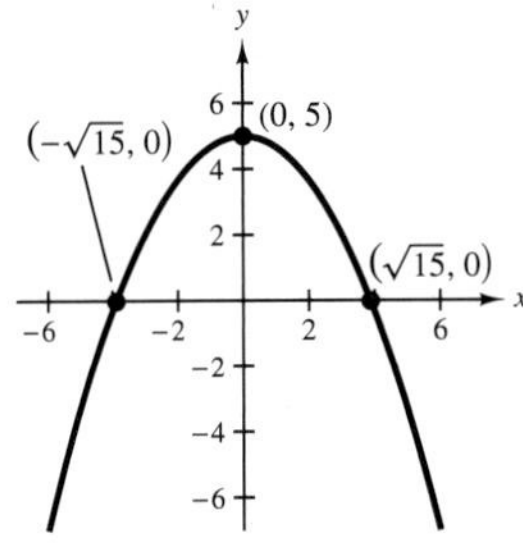

71. Vertical shift

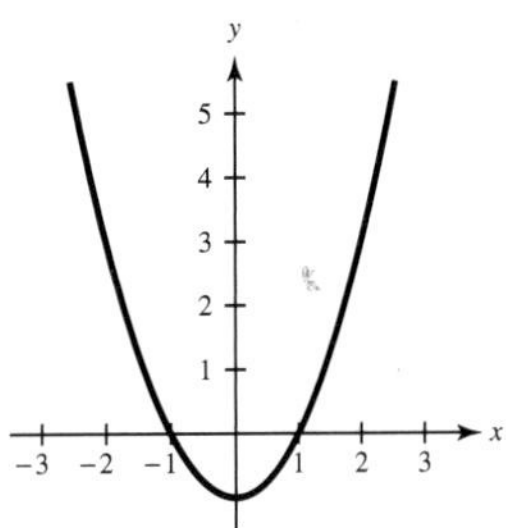

73. Horizontal shift

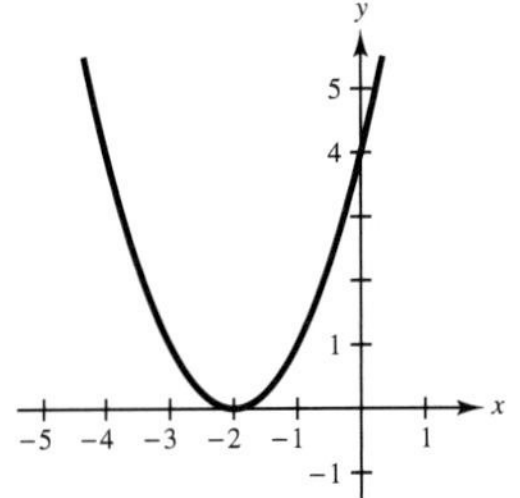

75. Horizontal shift and reflection in the x-axis

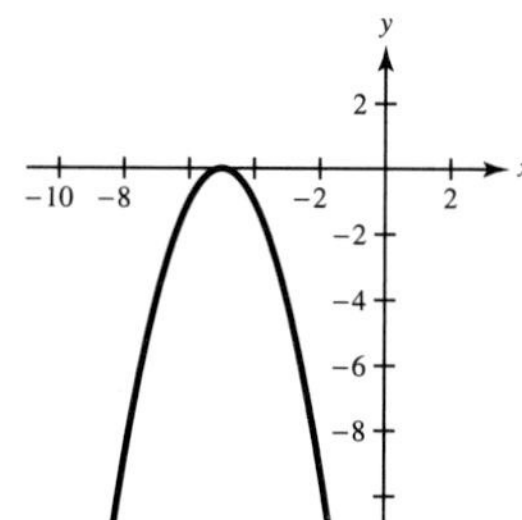

77. Horizontal and vertical shifts, reflection in the x-axis

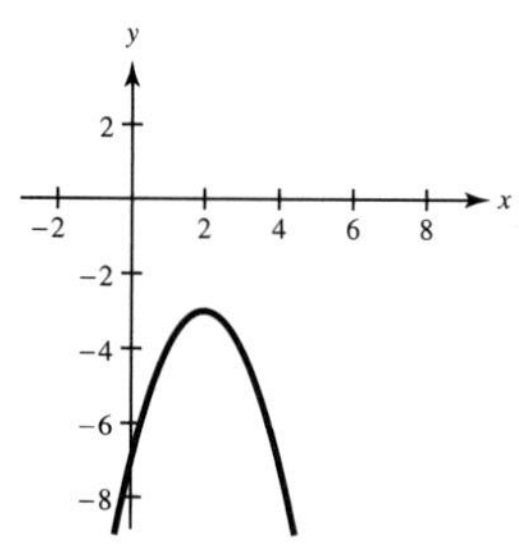

79.

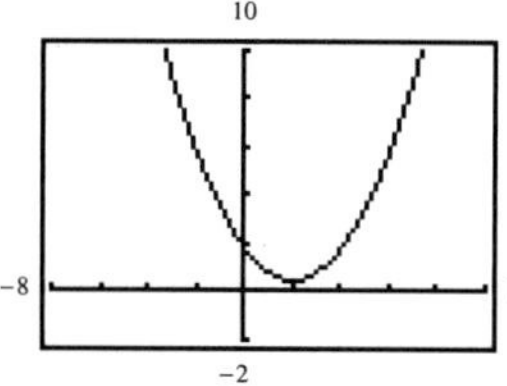

Vertex: (2, 0.5)

81.

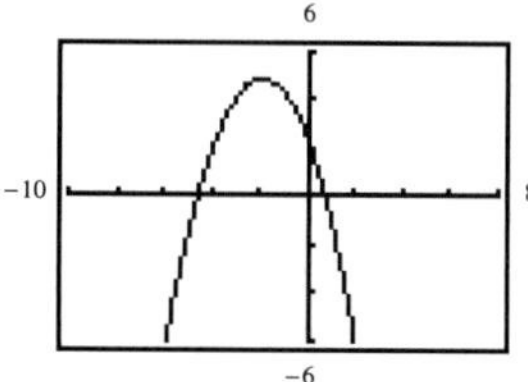

Vertex: (−1.9, 4.9)

83. $y = -x^2 + 4$ **85.** $y = 2(x + 1)^2 - 2$

87. $y = (x - 2)^2 + 1$ **89.** $y = (x - 2)^2 - 4$

91. $y = (x + 2)^2 - 1$ **93.** $y = \frac{2}{3}(x + 1)^2 + 1$

95. (a) 4 feet (b) 16 feet (c) $12 + 8\sqrt{3} \approx 25.9$ feet

97. (a) 3 feet (b) 48 feet (c) $15 + 4\sqrt{15} \approx 30.5$ feet

99. (a) 0 yards (b) 30 yards (c) 240 yards

101. 14 feet

103.

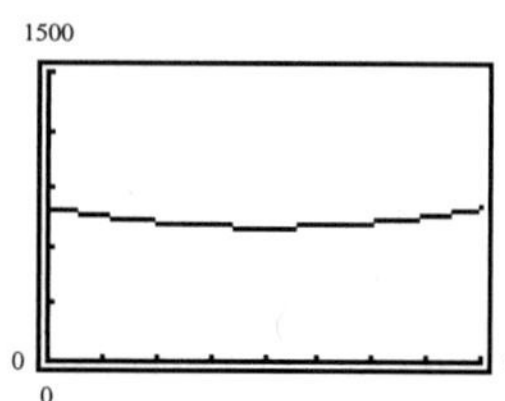

$x = 20$ when C is minimum.

105.

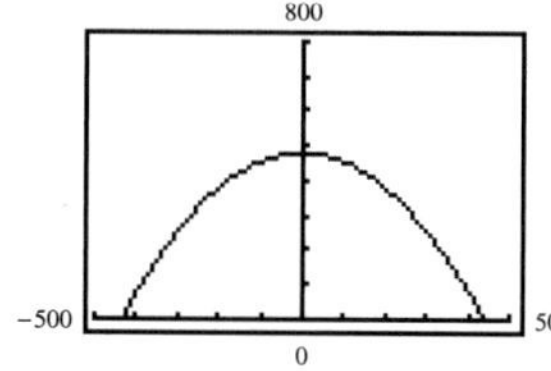

480 feet

107. $y = \frac{1}{60}x^2$

109. If the discriminant is positive, the parabola has two x-intercepts; if it is zero, the parabola has one x-intercept; and if it is negative, the parabola has no x-intercepts.

111. Find the y-coordinate of the vertex of the graph of the function.

113. $y = -\frac{1}{2}x$ **115.** $y = 2x$ **117.** $y = -\frac{11}{8}x + \frac{161}{16}$

119. $y = 8$ **121.** $8i$ **123.** $0.09i$

Section 10.5 *(page 661)*

1. 18 dozen, \$1.20 per dozen **3.** 10 DVDs at \$5 per DVD

	Width	*Length*	*Perimeter*	*Area*
5.	$1.4l$	l	54 in.	177.19 in.^{2}
7.	w	$2.5w$	70 ft	250 ft^2
9.	$\frac{1}{3}l$	l	64 in.	192 in.^{2}
11.	w	$w + 3$	54 km	180 km^2
13.	$l - 20$	l	440 m	12,000 m^2

15. 12 inches × 16 inches

17. 50 feet × 250 feet or 100 feet × 125 feet

19. No.

Area $= \frac{1}{2}(b_1 + b_2)h = \frac{1}{2}x[x + (550 - 2x)] = 43{,}560$

This equation has no real solution.

21. Height: 12 inches; Width: 24 inches **23.** 9.5%

25. 7% **27.** About 6.5% **29.** 15 people

31. 3.9 miles or 8.1 miles

33. (a) $d = \sqrt{h^2 + 100^2}$

(b)

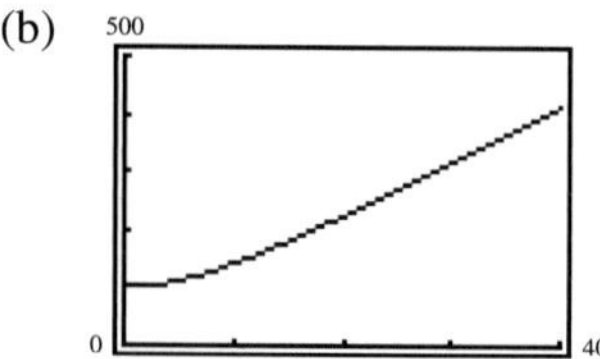

(c) $h \approx 173.2$ when $d = 200$.

(d)

h	0	100	200	300
d	100	141.4	223.6	316.2

35. 10.7 minutes, 13.7 minutes **37.** $3\frac{1}{4}$ seconds

39. About 9.5 seconds **41.** 4.7 seconds

43. (a) 3 seconds, 7 seconds (b) 10 seconds

(c) 400 feet

45. 13, 14 **47.** 12, 14 **49.** 17, 19

51. 400 miles per hour

53. 60 miles per hour or 75 miles per hour

55. (−6, 9), (10, 9)

57. (a) $b = 20 - a$; $A = \pi ab$; $A = \pi a(20 - a)$

(b)

a	4	7	10	13	16
A	201.1	285.9	314.2	285.9	201.1

(c) 7.9, 12.1

(d)

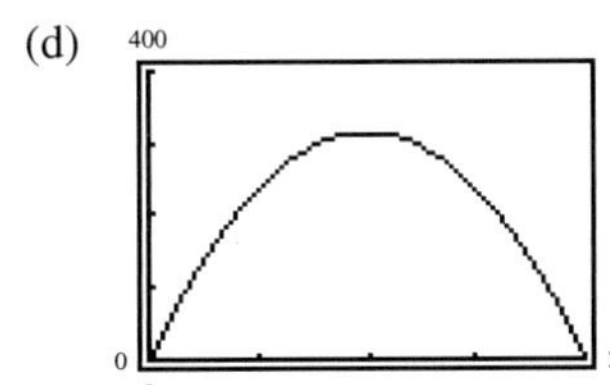

59. To solve a rational equation, each side of the equation is multiplied by the LCD. The resulting equations in this section are quadratic equations.

61. No. For each additional person, the cost-per-person decrease gets smaller because the discount is distributed to more people.

63. $x < -4$

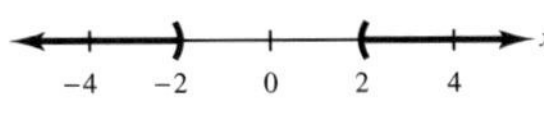

65. 0, 8

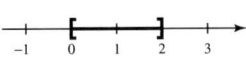

Section 10.6 *(page 672)*

1. $0, \frac{5}{2}$ **3.** $\pm\frac{9}{2}$ **5.** $-3, 5$ **7.** $1, 3$ **9.** $-3, \frac{5}{6}$

11. Negative: $(-\infty, 4)$; Positive: $(4, \infty)$

13. Negative: $(6, \infty)$; Positive: $(-\infty, 6)$

15. Negative: $(0, 5)$; Positive: $(-\infty, 0) \cup (5, \infty)$

17. Negative: $(-\infty, -2) \cup (2, \infty)$; Positive: $(-2, 2)$

19. Negative: $(-1, 5)$; Positive: $(-\infty, -1) \cup (5, \infty)$

21. $(0, 2)$

23. $[0, 2]$

25. $(-\infty, -2) \cup (2, \infty)$

27. $(-\infty, -2] \cup [5, \infty)$

29. $(-\infty, -4) \cup (0, \infty)$

31. $[-9, 4]$

33. $(-\infty, -3) \cup (1, \infty)$

35. No solution

37. $(-\infty, \infty)$

39. $(-\infty, 2 - \sqrt{2}) \cup (2 + \sqrt{2}, \infty)$

41. $(-\infty, \infty)$

43. No solution

45. $\left[-2, \frac{4}{3}\right]$

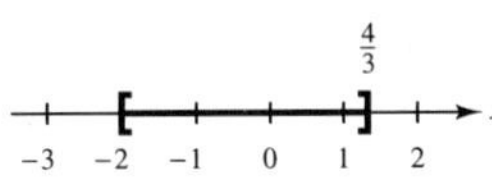

47. $\left(\frac{2}{3}, \frac{5}{2}\right)$

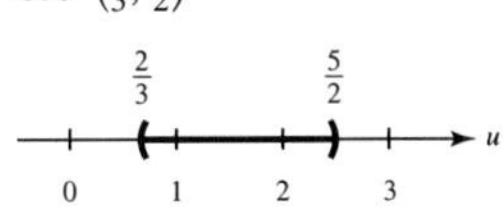

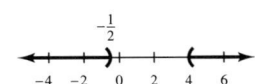

49. $\left(-\infty, -\frac{1}{2}\right) \cup (4, \infty)$

51. $-\frac{7}{2}$

53. No solution

55. $(-\infty, 2 - \sqrt{6}) \cup (2 + \sqrt{6}, \infty)$

57. $(-\infty, -9] \cup [-1, \infty)$

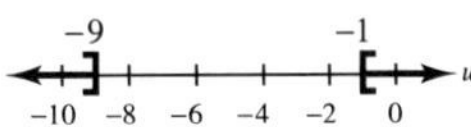

59. $(-2, 0) \cup (2, \infty)$

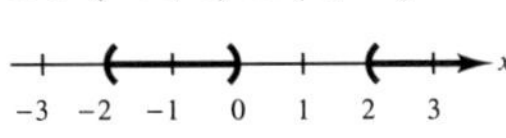

61.

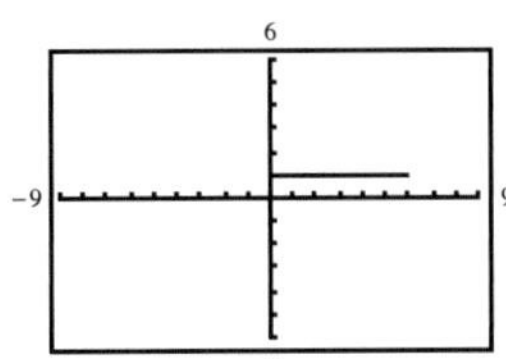

$(0, 6)$

63.

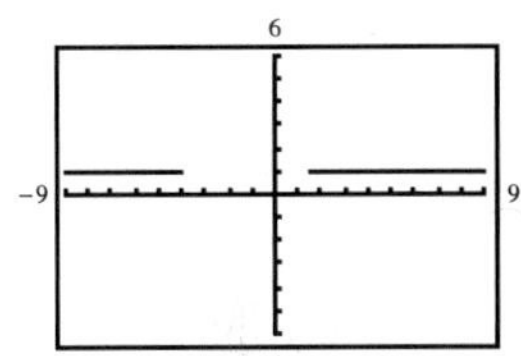

$(-\infty, -4) \cup \left(\frac{3}{2}, \infty\right)$

65.

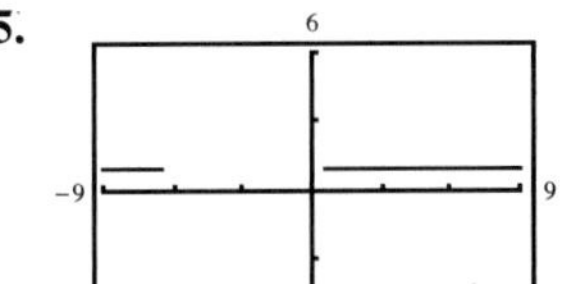

$(-\infty, -3 - 2\sqrt{3}] \cup [-3 + 2\sqrt{3}, \infty)$

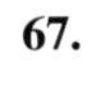

67.

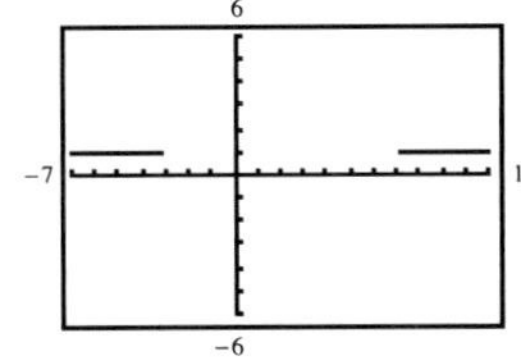

$(-\infty, -3) \cup (7, \infty)$

69.

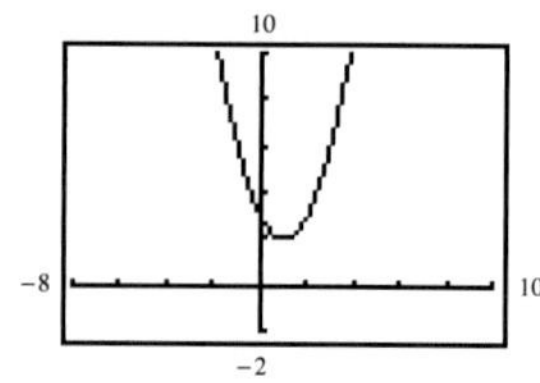

(a) $(-\infty, \infty)$ (b) $[-1, 3]$

71.

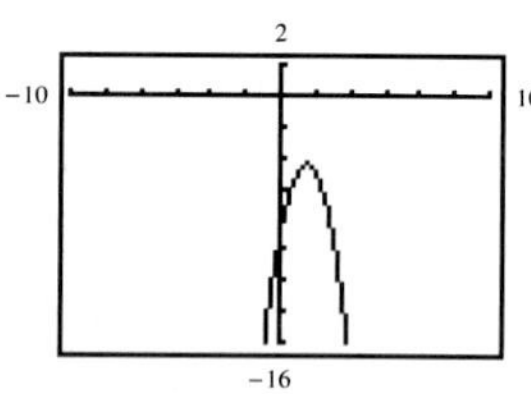

(a) $(-0.303, 3.303)$ (b) $(-\infty, \infty)$

73. 3 **75.** $0, -5$

77. $(3, \infty)$

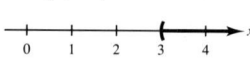

79. $(-\infty, 3)$

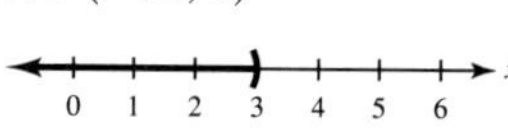

81. $[-2, 1)$

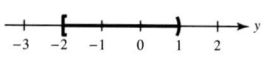

83. $(-\infty, -4) \cup (2, \infty)$

85. $(1, 4]$

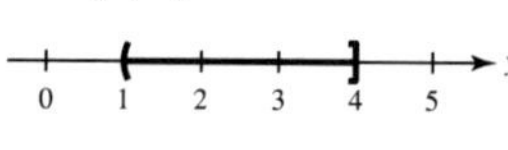

87. $\left(-\infty, \frac{1}{2}\right) \cup (2, \infty)$

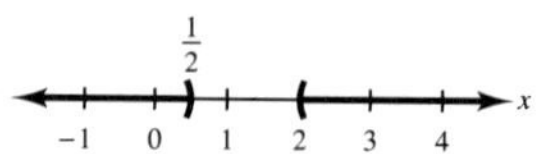

89. $\left[-2, -\frac{3}{2}\right)$

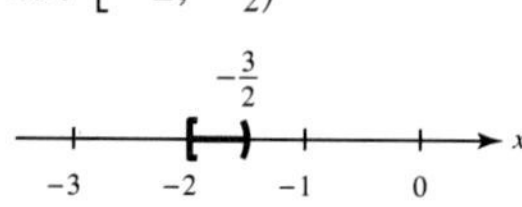

91. $(-1, 3)$

93. $\left(5, \frac{17}{3}\right)$

95. $\left(-2, -\frac{2}{5}\right)$

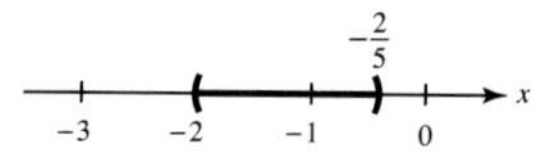

97. $(-\infty, 6) \cup [7, \infty)$

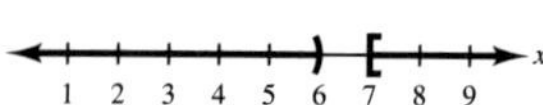

99.

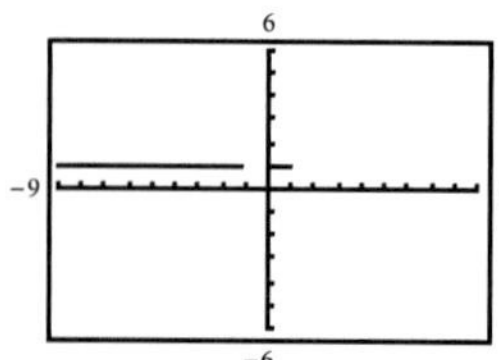

$(-\infty, -1) \cup (0, 1)$

101.

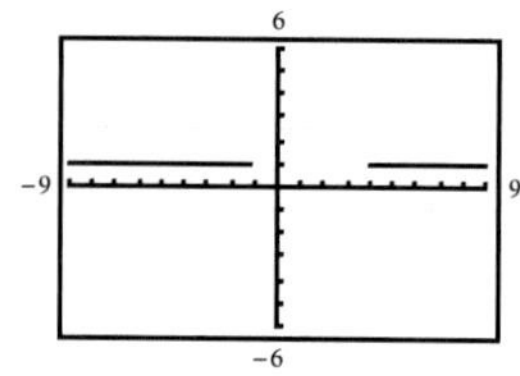

$(-\infty, -1) \cup (4, \infty)$

103.

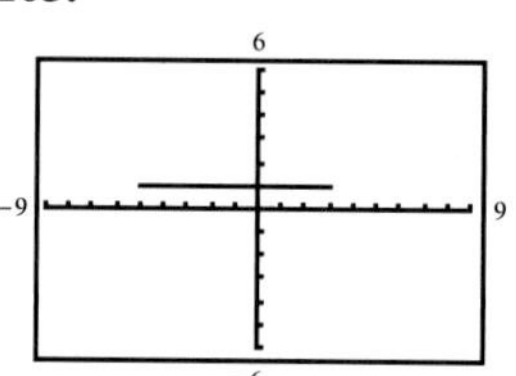

$\left(-5, \frac{13}{4}\right)$

105.

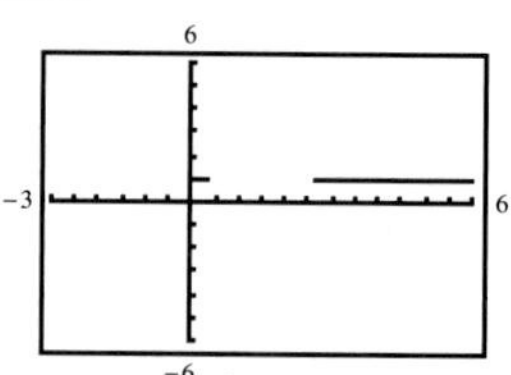

$(0, 0.382) \cup (2.618, \infty)$

107.

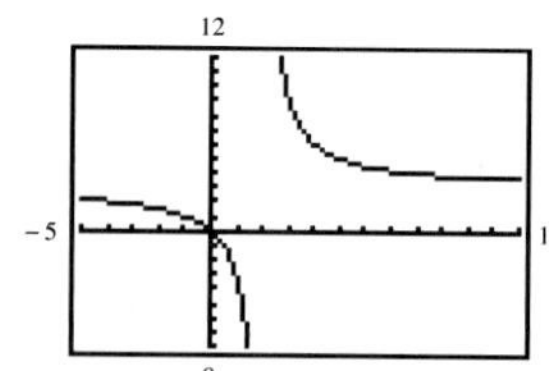

(a) $[0, 2)$ (b) $(2, 4]$

109.

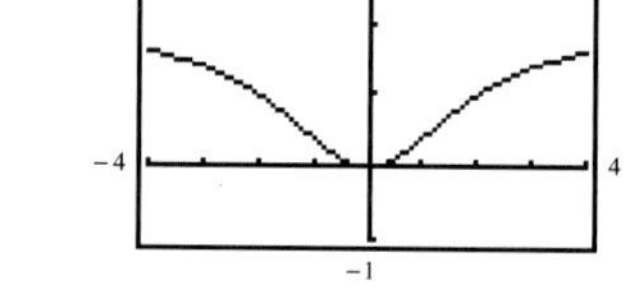

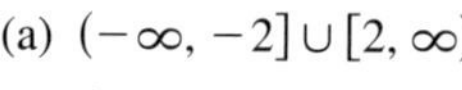

(a) $(-\infty, -2] \cup [2, \infty)$

(b) $(-\infty, \infty)$

111. $(3, 5)$ **113.** $r > 7.24\%$ **115.** $(12, 20)$

117. $90{,}000 \le x \le 100{,}000$

119. (a)

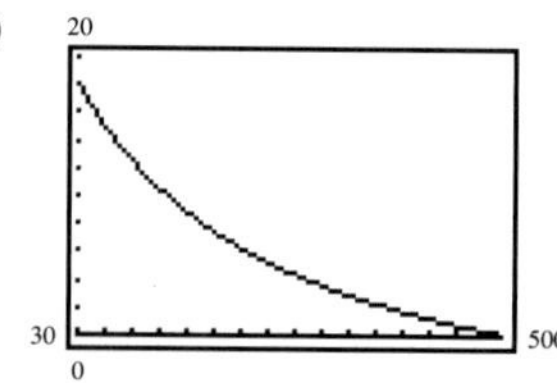

(b) $\frac{4225}{17} \approx 248.5$ minutes

121. The critical numbers of a polynomial are its zeros, so the value of the polynomial is zero at its critical numbers.

123. No solution. The value of the polynomial is positive for every real value of x, so there are no values that would make the polynomial negative.

125. $\dfrac{y^3}{2x^2},\ y \ne 0$ **127.** $\dfrac{(x-3)(x+1)}{4x^3(x+3)},\ x \ne -2$

129. $\dfrac{(x+4)^2}{3x(x-6)(x-8)}$ **131.** $\dfrac{1}{9}$ **133.** 79.21

Review Exercises *(page 678)*

1. $-12, 0$ **3.** ± 3 **5.** $-\frac{5}{2}$ **7.** $-9, 10$ **9.** $-\frac{3}{2}, 6$

11. ± 12 **13.** $\pm 2\sqrt{3}$ **15.** $-4, 36$ **17.** $\pm 11i$

19. $\pm 5\sqrt{2}i$ **21.** $-4 \pm 3\sqrt{2}i$ **23.** $\pm\sqrt{5}, \pm i$

25. $1, 9$ **27.** $1,\ 1 \pm \sqrt{6}$ **29.** $-343, 64$ **31.** 81

33. $\frac{225}{4}$ **35.** $\frac{1}{25}$

37. $3 + 2\sqrt{3} \approx 6.46;\ 3 - 2\sqrt{3} \approx -0.46$ **39.** $-4, -1$

41. $\frac{1}{3} + \frac{\sqrt{17}}{3}i \approx 0.33 + 1.37i;\ \frac{1}{3} - \frac{\sqrt{17}}{3}i \approx 0.33 - 1.37i$

43. $-7, 6$ **45.** $-\frac{7}{2}, 3$ **47.** $\frac{8}{5} \pm \frac{3\sqrt{6}}{5}$

49. One repeated rational solution

51. Two distinct rational solutions

53. Two distinct irrational solutions

55. Two distinct complex solutions

57. $x^2 + 4x - 21 = 0$

59. $x^2 - 10x + 18 = 0$ **61.** $x^2 - 12x + 40 = 0$

63. $y = (x - 4)^2 - 13$; Vertex: $(4, -13)$

65. $y = 2\left(x - \frac{1}{4}\right)^2 + \frac{23}{8}$; Vertex: $\left(\frac{1}{4}, \frac{23}{8}\right)$

67.

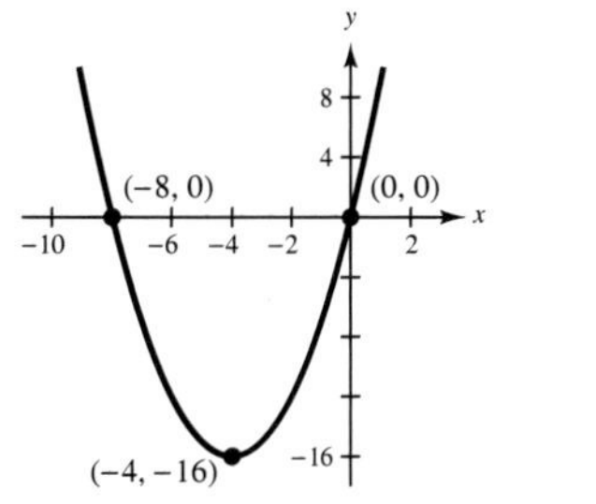

69.

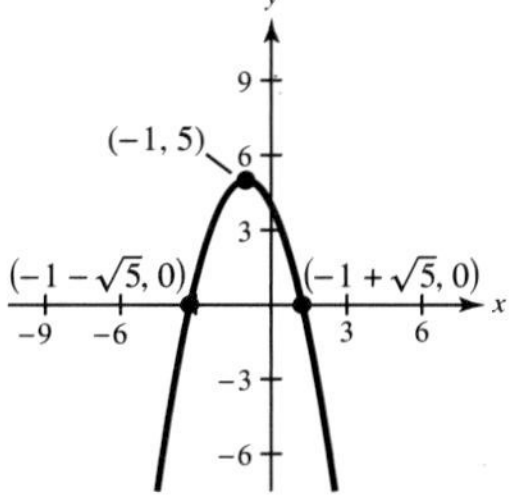

71. $y = 2(x - 2)^2 - 5$ **73.** $y = \frac{1}{16}(x - 5)^2$

75. (a)

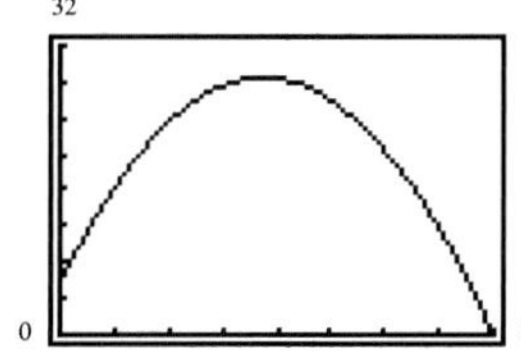

(b) 6 feet (c) 28.5 feet (d) 31.9 feet

77. 16 cars; \$5000 **79.** 5 inches × 17 inches

81. 15 people

83. $9 + \sqrt{101} \approx 19$ hours, $11 + \sqrt{101} \approx 21$ hours

85. $-7, 0$ **87.** $-3, 9$

89. $(0, 7)$

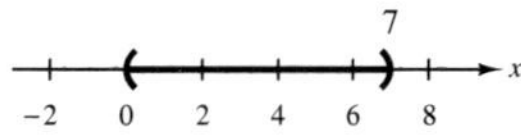

91. $(-\infty, -2] \cup [6, \infty)$

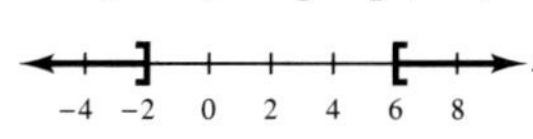

93. $\left(-4, \frac{5}{2}\right)$

95. $(-\infty, -3] \cup \left(\frac{7}{2}, \infty\right)$

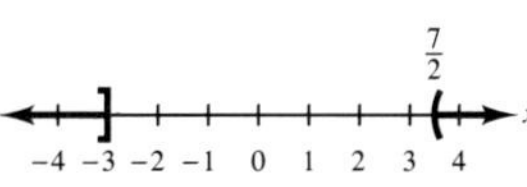

97. $(-4, 1)$ **99.** $(5.3, 14.2)$

Chapter Test *(page 681)*

1. 3, 10 **2.** $-\frac{1}{3}, 6$ **3.** 1.7, 2.3 **4.** $-4 \pm 10i$

5. $\frac{3}{2} \pm \frac{\sqrt{3}}{2}$ **6.** $1 \pm \frac{3\sqrt{2}}{2}$ **7.** $\frac{3}{4} \pm \frac{\sqrt{5}}{4}$ **8.** 1, 512

9. -56; A negative discriminant tells us the equation has two imaginary solutions.

10. $x^2 + 10x + 21 = 0$

11.

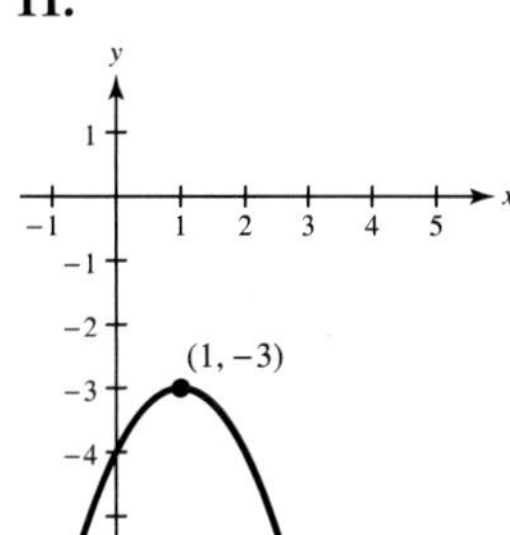

12.

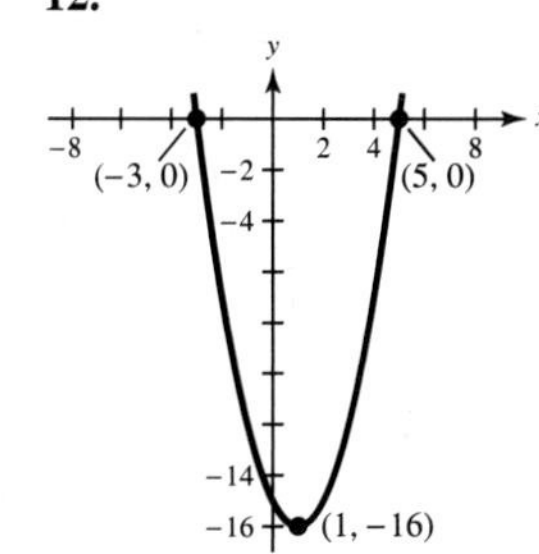

13. $(-\infty, -2] \cup [6, \infty)$

14. $(0, 3)$

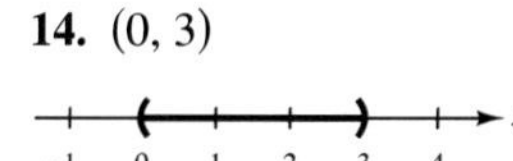

15. $[-1, 5)$

16. 8 feet × 30 feet

17. 40 members **18.** $\frac{\sqrt{10}}{2} \approx 1.58$ seconds

19. 35 feet, 120 feet

CHAPTER 11

Section 11.1 *(page 692)*

1. 3^{2x+2} **3.** e^2 **5.** $\frac{3}{e^{2x}}$ **7.** $-2e^x$ **9.** 9.739

11. 1.396 **13.** 107.561 **15.** 0.906

17. (a) $\frac{1}{9}$ (b) 1 (c) 3

19. (a) 0.455 (b) 0.094 (c) 0.145

21. (a) 500 (b) 250 (c) 56.657

23. (a) 1000 (b) 1628.895 (c) 2653.298

25. (a) 486.111 (b) 47.261 (c) 0.447

27. (a) 73.891 (b) 1.353 (c) 0.183

29. (a) 333.333 (b) 434.557 (c) 499.381

31.

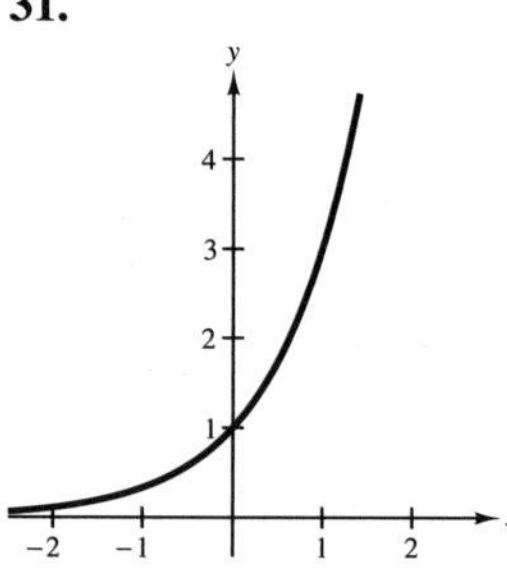

Horizontal asymptote: $y = 0$

33.

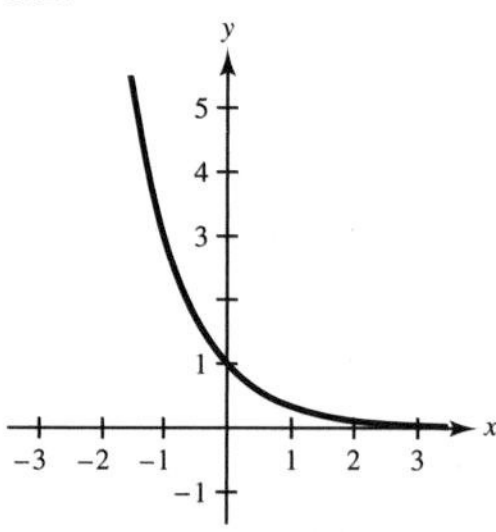

Horizontal asymptote: $y = 0$

35.

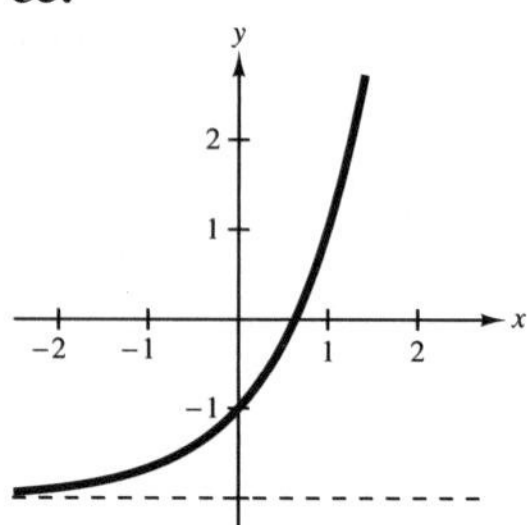

Horizontal asymptote: $y = -2$

37.

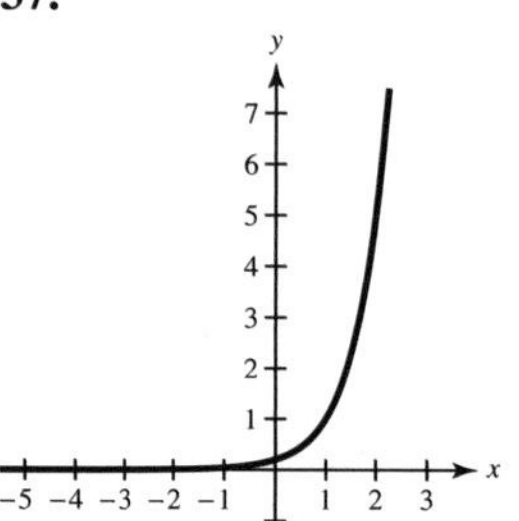

Horizontal asymptote: $y = 0$

39.

Horizontal asymptote: $y = 0$

41.

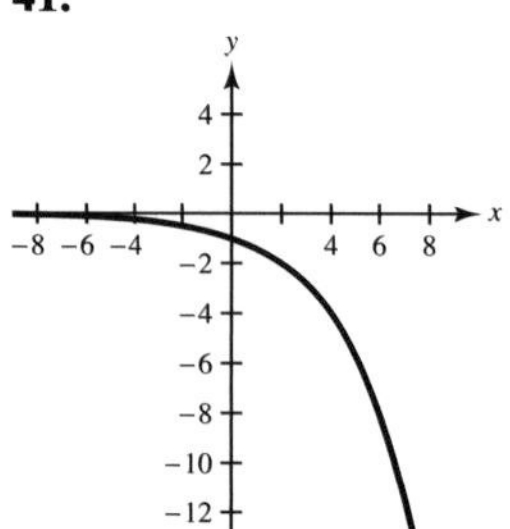

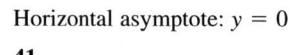

Horizontal asymptote: $y = 0$

43.

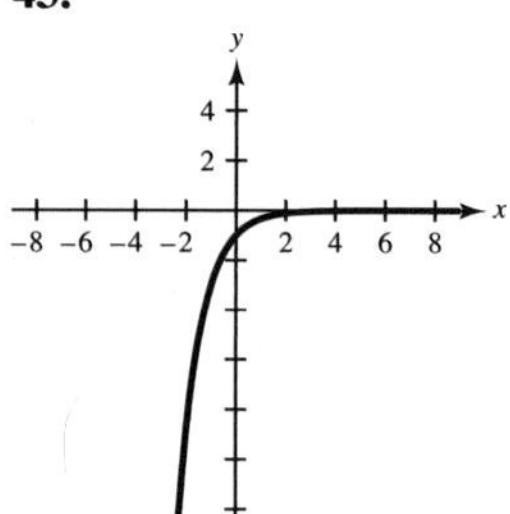

Horizontal asymptote: $y = 0$

45.

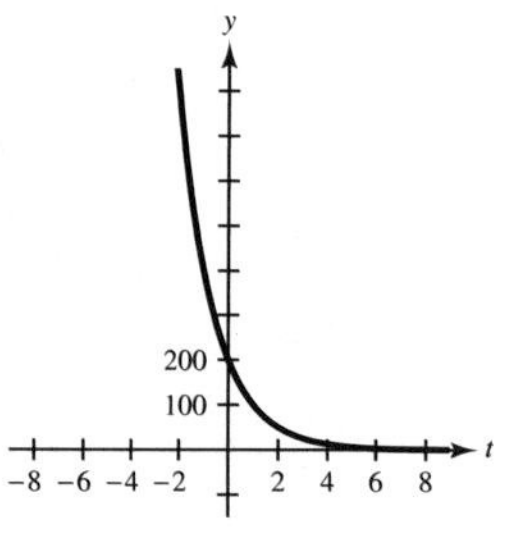

Horizontal asymptote: $y = 0$

47.

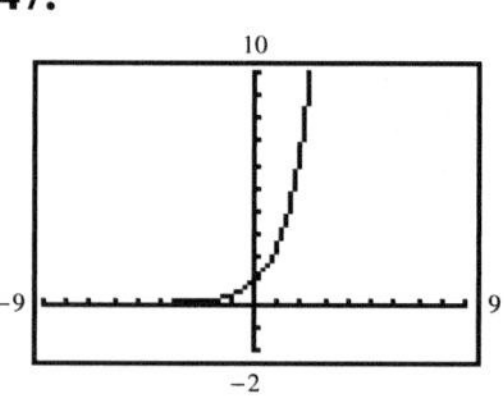

49.

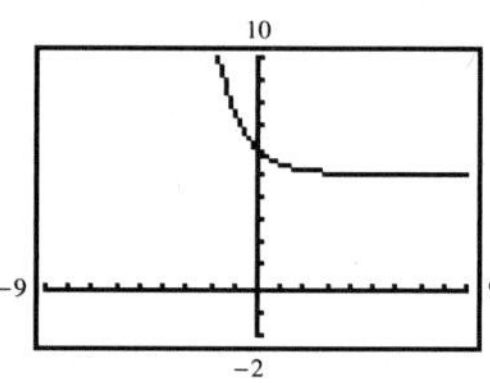

51.

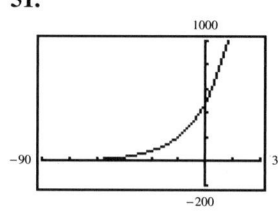

53.

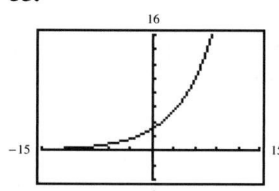

55.

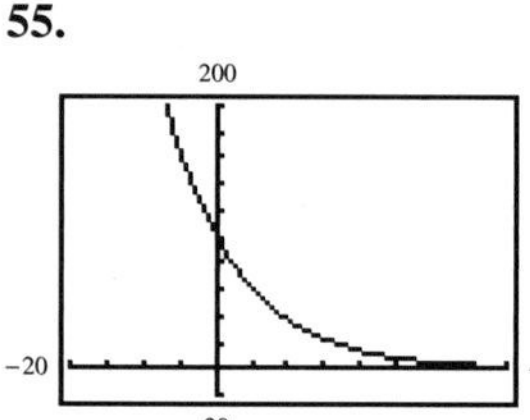

57.

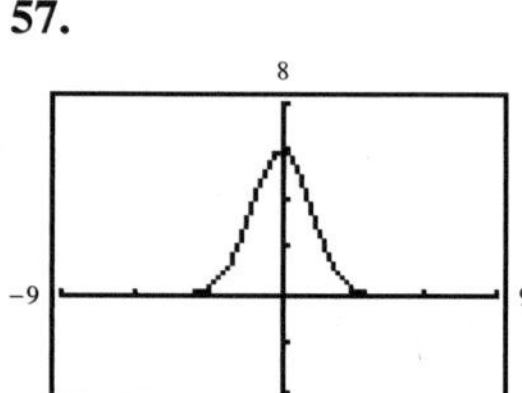

59. b **60.** a **61.** c **62.** d

63. Vertical shift

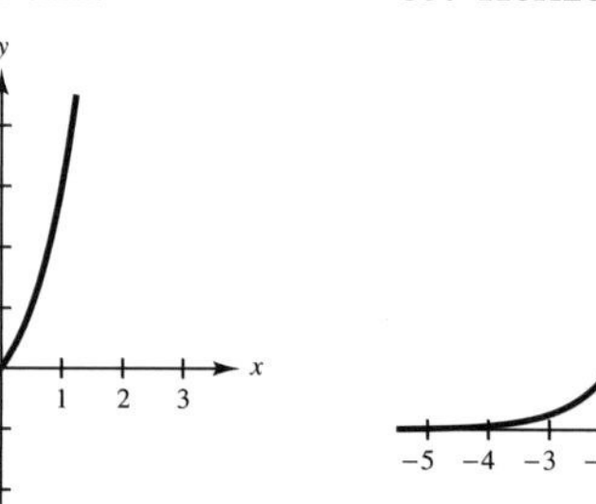

65. Horizontal shift

67. Reflection in the x-axis

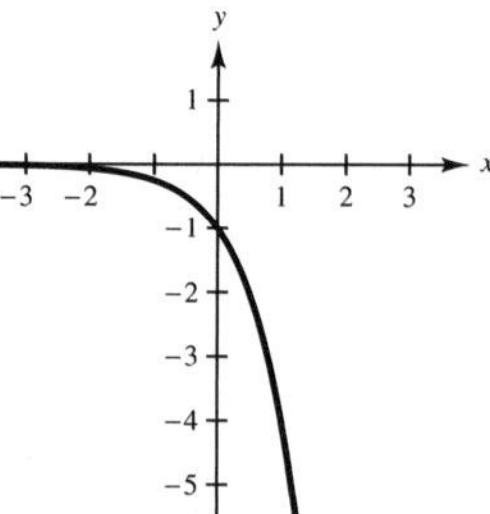

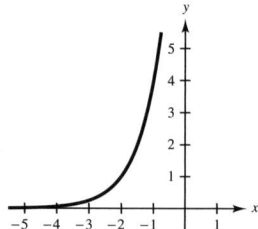

69. 2.520 grams **71.** \$6744.25

73.

n	1	4	12	365	Continuous
A	\$275.90	\$283.18	\$284.89	\$285.74	\$285.77

75.

n	1	4	12
A	\$4956.46	\$5114.30	\$5152.11

n	365	Continuous
A	\$5170.78	\$5171.42

77.

n	1	4	12
P	\$2541.75	\$2498.00	\$2487.98

n	365	Continuous
P	\$2483.09	\$2482.93

79.

n	1	4	12
P	\$18,429.30	\$15,830.43	\$15,272.04

n	365	Continuous
P	\$15,004.64	\$14,995.58

81. (a) \$22.04 (b) \$20.13

83. (a) \$80,634.95 (b) \$161,269.89

85. $V(t) = 16{,}000\left(\frac{3}{4}\right)^t$

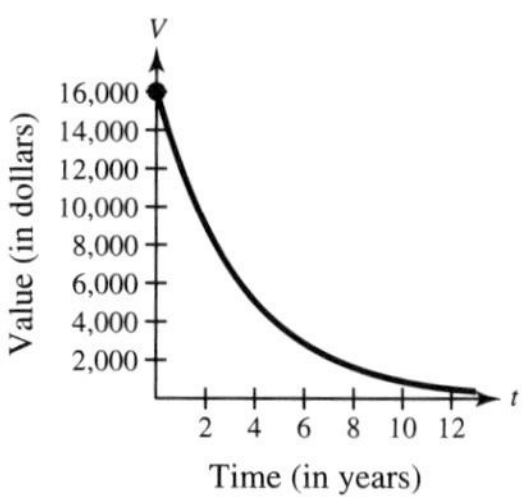

\$9000 after 2 years; \$5062.50 after 4 years

87. $f(x) = 1024\left(\frac{1}{2}\right)^x$; $f(8) = 4$ golfers

89. (a)

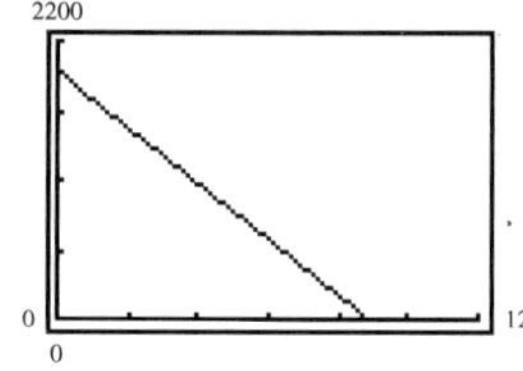

(b)

Time (in seconds)	0	10	20	30	40
Height (in feet)	2000	1731	1510	1290	1070

Time (in seconds)	50	60	70	80	90
Height (in feet)	850	630	410	190	0

(c) The height changes the most within the first 10 seconds because after the parachute is released, a few seconds pass before the descent becomes constant.

91. (a)

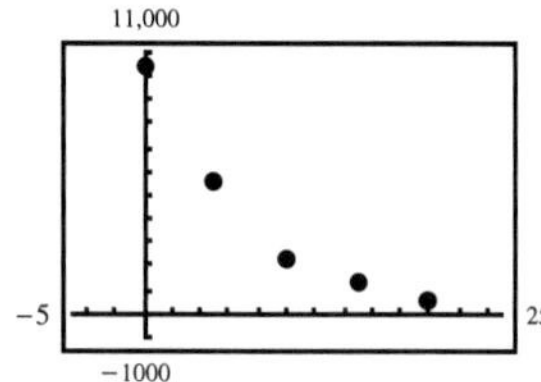

(b)

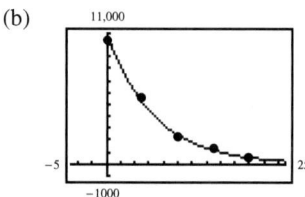

The model is a good fit for the data.

(c)

h	0	5	10	15	20
P	10,332	5583	2376	1240	517
Approx.	10,958	5176	2445	1155	546

(d) 3300 kilograms per square meter

(e) 11.3 kilometers

93. (a)

x	1	10	100	1000	10,000
$\left(1 + \frac{1}{x}\right)^x$	2	2.5937	2.7048	2.7169	2.7181

(b)

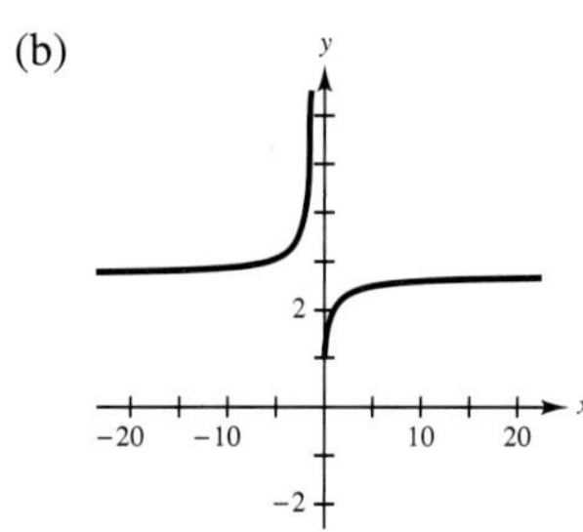

Yes, the graph appears to be approaching a horizontal asymptote.

(c) The value approaches e.

95. By definition, the base of an exponential function must be positive and not equal to 1. If the base is 1, the function simplifies to the constant function $y = 1$.

97. No; e is an irrational number.

99. When $k > 1$, the values of f will increase. When $0 < k < 1$, the values of f will decrease. When $k = 1$, the values of f will remain constant.

101. $[4, \infty)$

103.

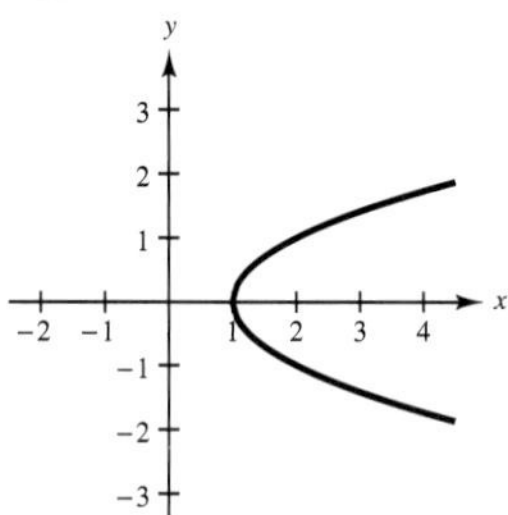

y is not a function of x.

Section 11.2 *(page 705)*

1. (a) $2x - 9$ (b) $2x - 3$ (c) -1 (d) 11

3. (a) $x^2 + 4x + 7$ (b) $x^2 + 5$ (c) 19 (d) 14

5. (a) $|3x - 3|$ (b) $3|x - 3|$ (c) 0 (d) 3

7. (a) $\sqrt{x + 1}$ (b) $\sqrt{x - 4} + 5$ (c) 2 (d) 7

9. (a) $\dfrac{x^2}{2 - 3x^2}, x \neq 0$

(b) $2(x - 3)^2, x \neq 3$

(c) -1

(d) 2

11. (a) -1 (b) -2 (c) -2 **13.** (a) -1 (b) 1

15. (a) 5 (b) 1 (c) 3 **17.** (a) 0 (b) 10

19. (a) $(f \circ g)(x) = 3x - 17$

Domain: $(-\infty, \infty)$

(b) $(g \circ f)(x) = 3x - 3$

Domain: $(-\infty, \infty)$

21. (a) $(f \circ g)(x) = \sqrt{x - 2}$

Domain: $[2, \infty)$

(b) $(g \circ f)(x) = \sqrt{x + 2} - 4$

Domain: $[-2, \infty)$

23. (a) $(f \circ g)(x) = x + 2$

Domain: $[1, \infty)$

(b) $(g \circ f)(x) = \sqrt{x^2 + 2}$

Domain: $(-\infty, \infty)$

25. (a) $(f \circ g)(x) = \dfrac{\sqrt{x - 1}}{\sqrt{x - 1} + 5}$

Domain: $[1, \infty)$

(b) $(g \circ f)(x) = \sqrt{-\dfrac{5}{x + 5}}$

Domain: $(-\infty, -5)$

27.

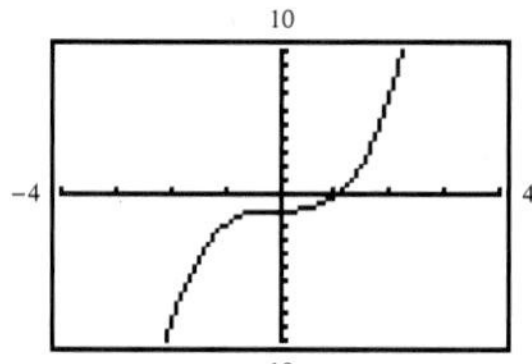

Yes

29.

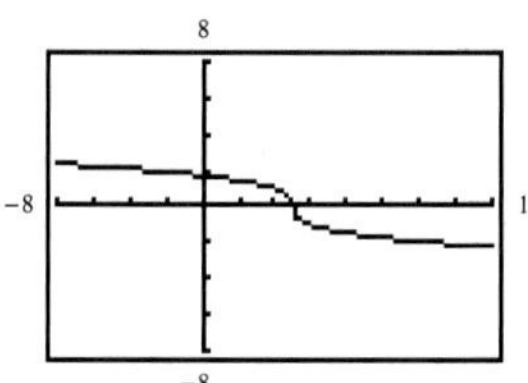

Yes

31.

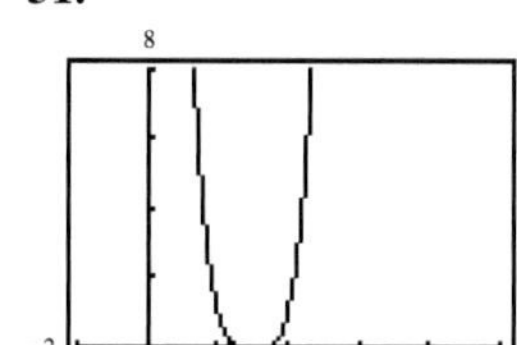

No

33.

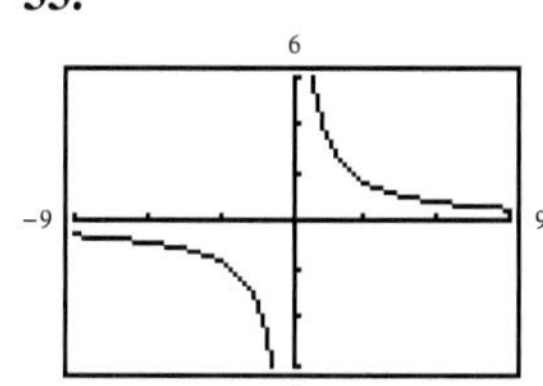

Yes

35. No **37.** Yes **39.** No

41. $f(g(x)) = f\left(-\frac{1}{6}x\right) = -6\left(-\frac{1}{6}x\right) = x$

$g(f(x)) = g(-6x) = -\frac{1}{6}(-6x) = x$

43. $f(g(x)) = f\left[\frac{1}{2}(1 - x)\right] = 1 - 2\left[\frac{1}{2}(1 - x)\right]$

$= 1 - (1 - x) = x$

$g(f(x)) = g(1 - 2x) = \frac{1}{2}[1 - (1 - 2x)] = \frac{1}{2}(2x) = x$

45. $f(g(x)) = f(x^3 - 1) = \sqrt[3]{(x^3 - 1) + 1} = \sqrt[3]{x^3} = x$

$g(f(x)) = g\left(\sqrt[3]{x + 1}\right) = \left(\sqrt[3]{x + 1}\right)^3 - 1$

$= x + 1 - 1 = x$

47. $f(g(x)) = f\left(\dfrac{1}{x}\right) = \dfrac{1}{(1/x)} = x$

$g(f(x)) = g\left(\dfrac{1}{x}\right) = \dfrac{1}{(1/x)} = x$

49. $f^{-1}(x) = \frac{1}{5}x$ **51.** $f^{-1}(x) = -\frac{5}{2}x$

53. $f^{-1}(x) = x - 10$ **55.** $f^{-1}(x) = 5 - x$

57. $f^{-1}(x) = \sqrt[9]{x}$ **59.** $f^{-1}(x) = x^3$

61. $g^{-1}(x) = x - 25$ **63.** $g^{-1}(x) = \dfrac{3 - x}{4}$

65. $g^{-1}(t) = 4t - 8$ **67.** Inverse does not exist.

69. $h^{-1}(x) = x^2,\ x \geq 0$ **71.** $f^{-1}(t) = \sqrt[3]{t + 1}$

73. $f^{-1}(x) = x^2 - 3,\ x \geq 0$

75. b **76.** c **77.** d **78.** a

79.

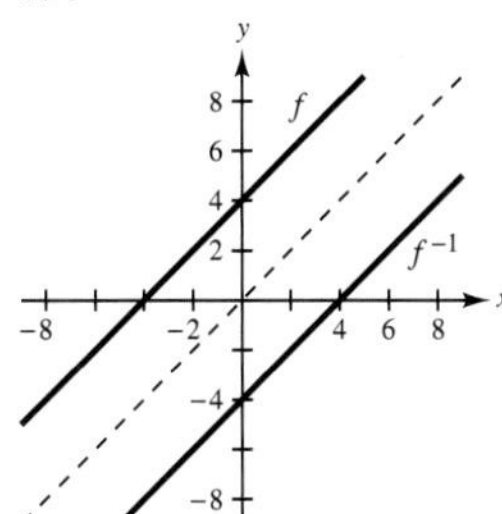

81.

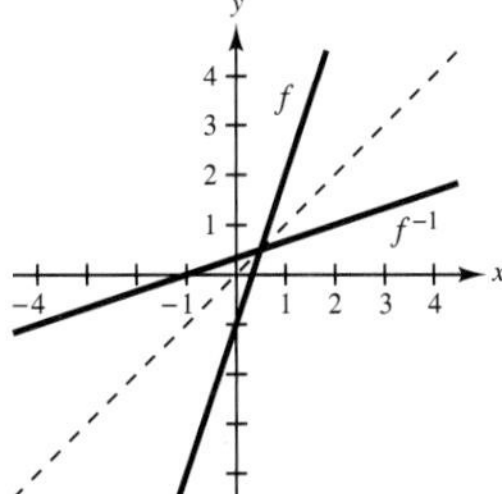

83.

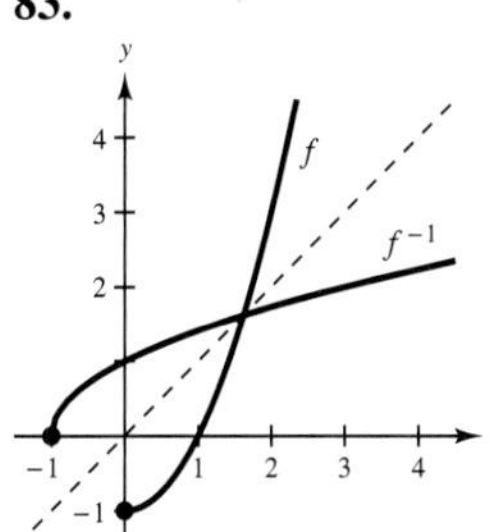

85.

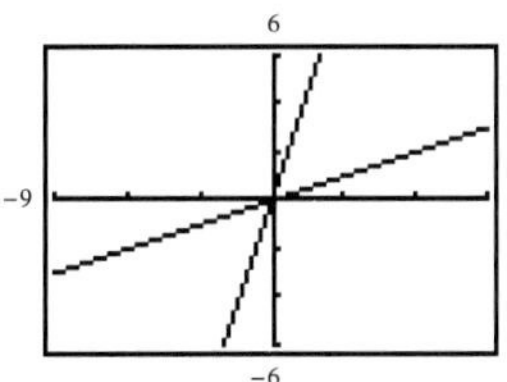

87.

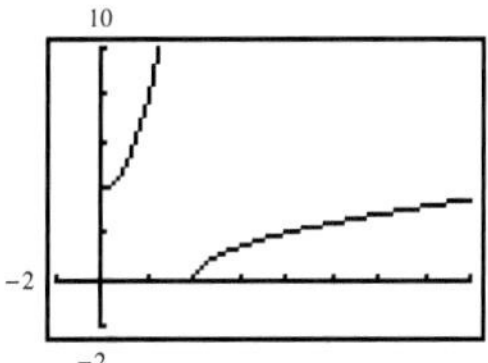

89.

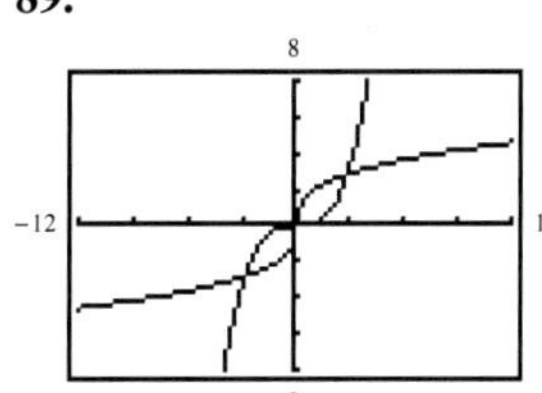

91.

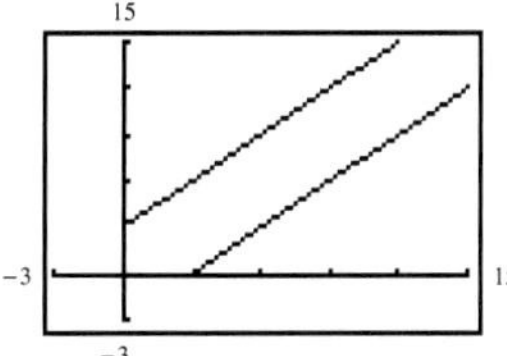

93. Domain of f: $x \geq 2$; $f^{-1}(x) = \sqrt{x} + 2$;
Domain of f^{-1}: $x \geq 0$

95. Domain of f: $x \geq 0$; $f^{-1}(x) = x - 1$;
Domain of f^{-1}: $x \geq 1$

97. $f^{-1}(x) = \frac{1}{2}(3 - x)$ **99.** c **101.** a

103. $p(s(x)) = 0.03x - 9000,\ x > 300{,}000$
This function represents the bonus earned for sales over \$300,000.

105. $A(r(t)) = 0.36\pi t^2$
Input: time; Output: area; $A(r(3)) = 10.2$ square feet

107. (a) $R = p - 2000$ (b) $S = 0.95p$
(c) $(R \circ S)(p) = 0.95p - 2000$;
5% discount followed by the \$2000 rebate
$(S \circ R)(p) = 0.95(p - 2000)$;
5% discount after the price is reduced by the rebate
(d) $(R \circ S)(26{,}000) = 22{,}700$; $(S \circ R)(26{,}000) = 22{,}800$
$R \circ S$ yields the smaller cost because the dealer discount is calculated on a larger base.

109. (a) $y = \frac{20}{13}(x - 9)$
(b) x: hourly wage; y: number of units produced
(c) $x \geq 9$ (d) 8 units

111. (a) $(f \circ g)(x) = 4x + 24$
(b) $(f \circ g)^{-1}(x) = \dfrac{x - 24}{4} = \dfrac{1}{4}x - 6$
(c) $f^{-1}(x) = \dfrac{1}{4}x$; $g^{-1}(x) = x - 6$
(d) $(g^{-1} \circ f^{-1})(x) = \dfrac{1}{4}x - 6$; The results are the same.
(e) $(f \circ g)^{-1}(x) = (g^{-1} \circ f^{-1})(x)$

113. True. If the point (a, b) lies on the graph of f, the point (b, a) must lie on the graph of f^{-1}, and vice versa.

115. False. $f(x) = \sqrt{x - 1}$; Domain: $[1, \infty)$;
$f^{-1}(x) = x^2 + 1,\ x \geq 0$; Domain: $[0, \infty)$

117. Interchange the coordinates of each ordered pair. The inverse of the function defined by $\{(3, 6), (5, -2)\}$ is $\{(6, 3), (-2, 5)\}$.

119. A function can have only one input for every output, so the inverse will have one output for every input and is therefore a function.

121. Reflection in the x-axis

123. Horizontal and vertical shifts **125.** $(6 + y)(2 - y)$

127. $-(u^2 + 1)(u - 5)$

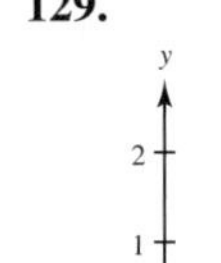

129.

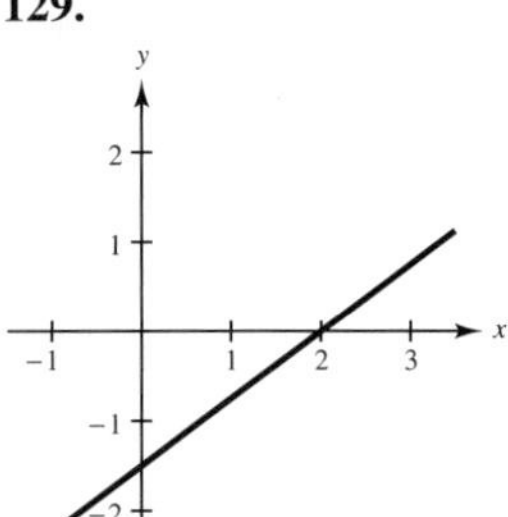

131.

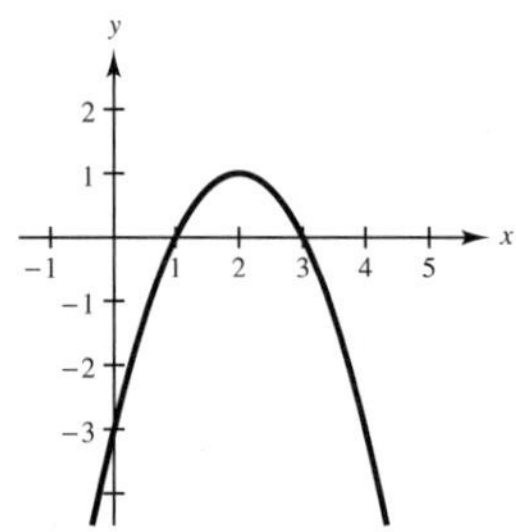

Section 11.3 *(page 718)*

1. $7^2 = 49$ **3.** $2^{-5} = \frac{1}{32}$ **5.** $3^{-5} = \frac{1}{243}$

7. $36^{1/2} = 6$ **9.** $8^{2/3} = 4$ **11.** $2^{2.4} \approx 5.278$

13. $\log_6 36 = 2$ **15.** $\log_5 \frac{1}{125} = -3$ **17.** $\log_8 4 = \frac{2}{3}$

19. $\log_{25} \frac{1}{5} = -\frac{1}{2}$ **21.** $\log_4 1 = 0$

23. $\log_5 9.518 \approx 1.4$ **25.** 3 **27.** 3 **29.** -4

31. -3 **33.** -4

35. There is no power to which 2 can be raised to obtain -3.

37. 0

39. There is no power to which 5 can be raised to obtain -6.

41. $\frac{1}{2}$ **43.** $\frac{3}{4}$ **45.** 4 **47.** 1.6232 **49.** -1.6383

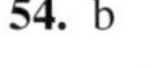

51. 0.7190 **53.** c **54.** b **55.** a **56.** d

57.

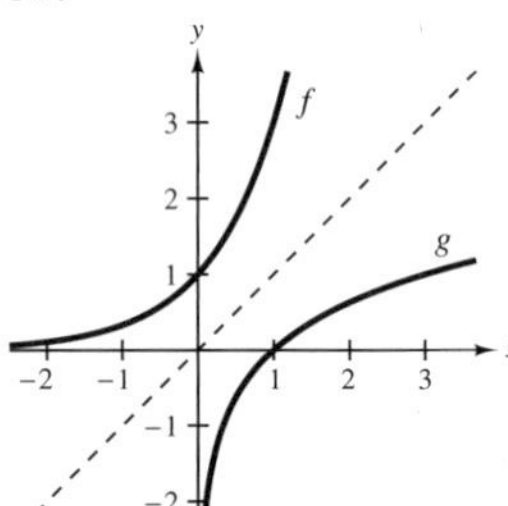

59.

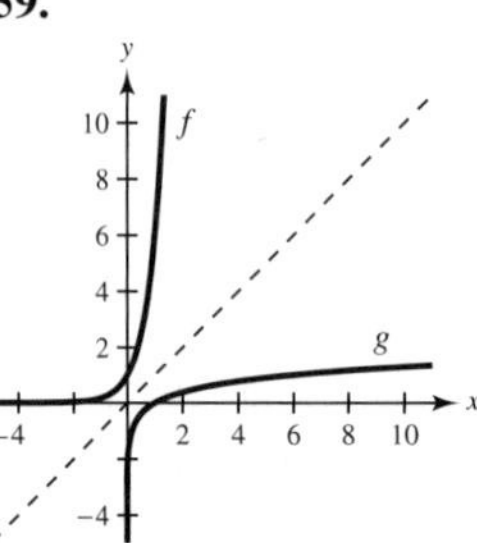

61. Vertical shift

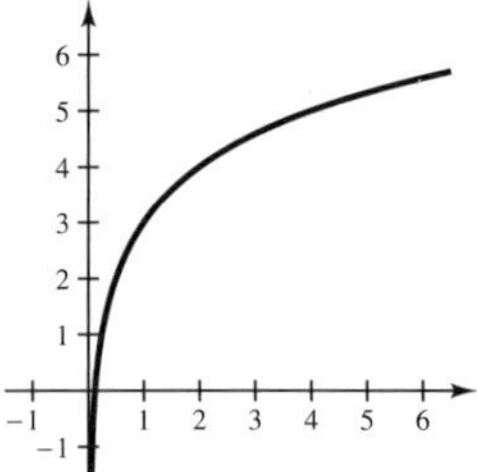

63. Horizontal shift

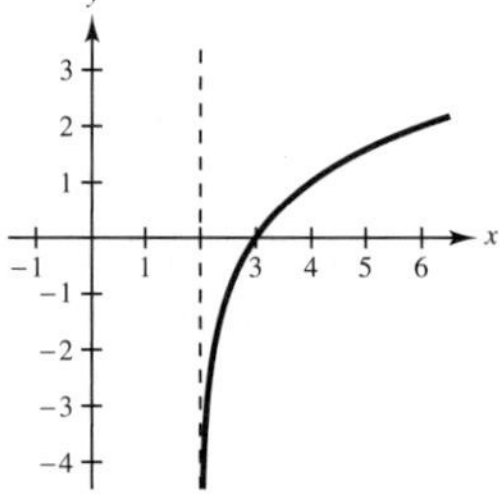

65. Reflection in the y-axis

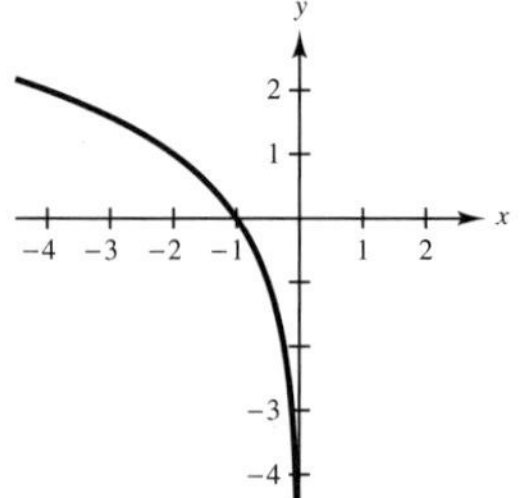

67.

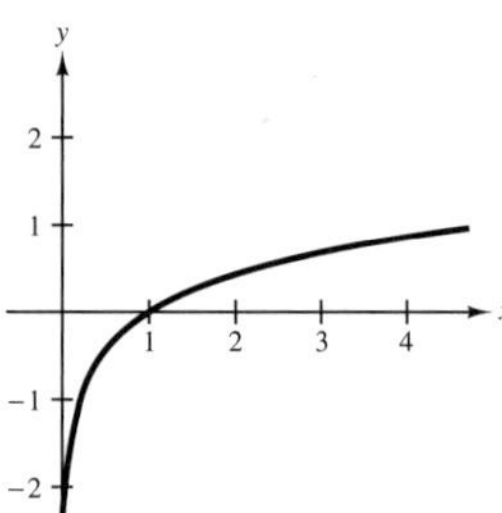

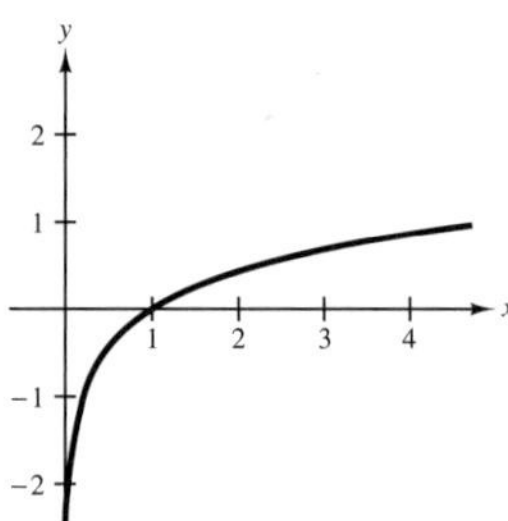

Vertical asymptote: $x = 0$

69.

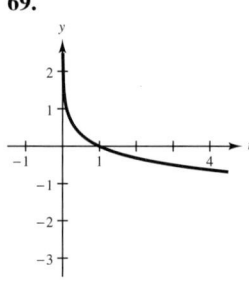

Vertical asymptote: $t = 0$

71.

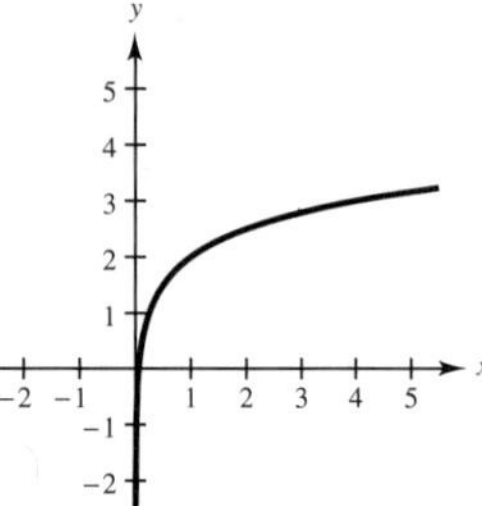

Vertical asymptote: $x = 0$

73.

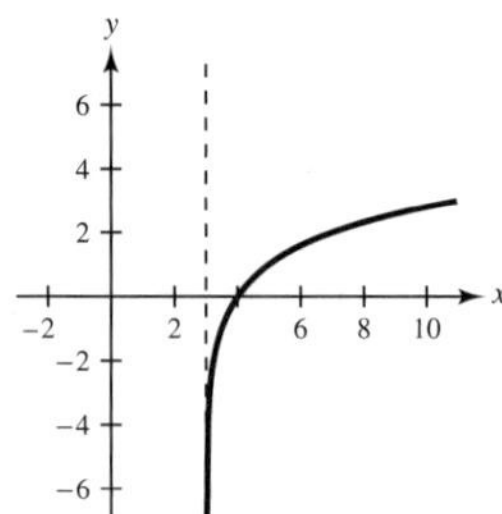

Vertical asymptote: $x = 3$

75.

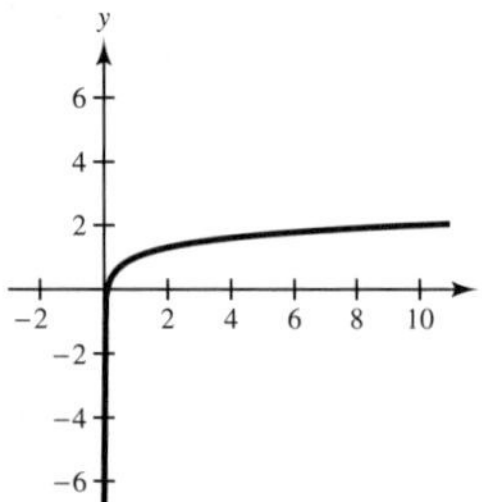

Vertical asymptote: $x = 0$

77. Domain: $(0, \infty)$

Vertical asymptote: $x = 0$

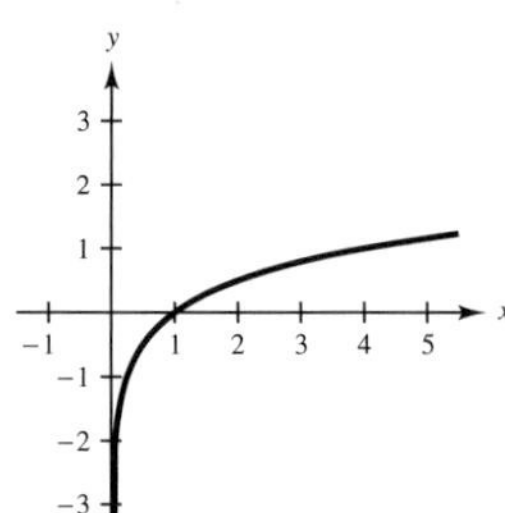

79. Domain: $(4, \infty)$

Vertical asymptote: $x = 4$

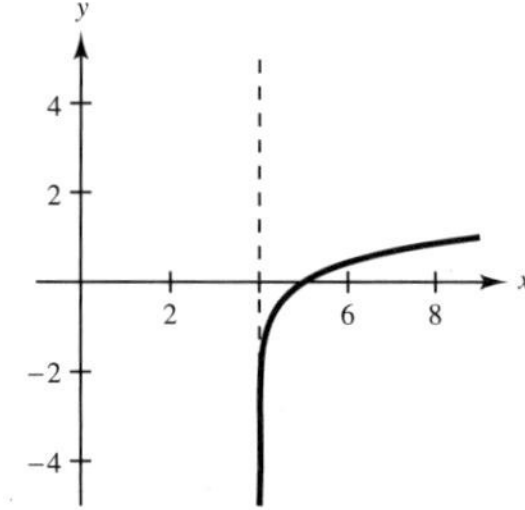

81. Domain: $(0, \infty)$

Vertical asymptote: $x = 0$

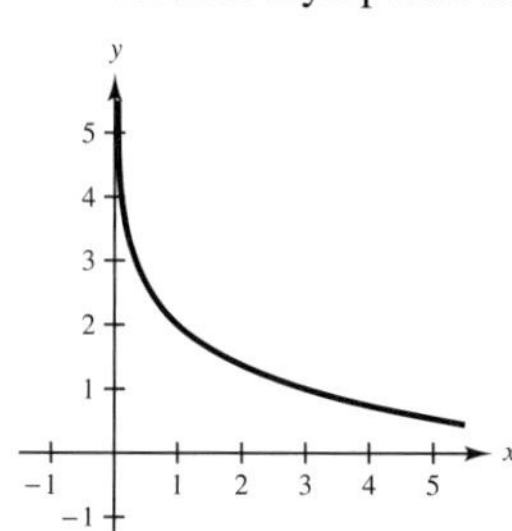

83. Domain: $(0, \infty)$

Vertical asymptote: $x = 0$

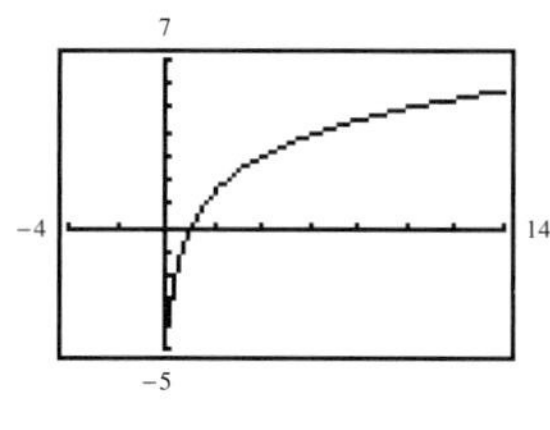

85. Domain: $(0, \infty)$

Vertical asymptote: $x = 0$

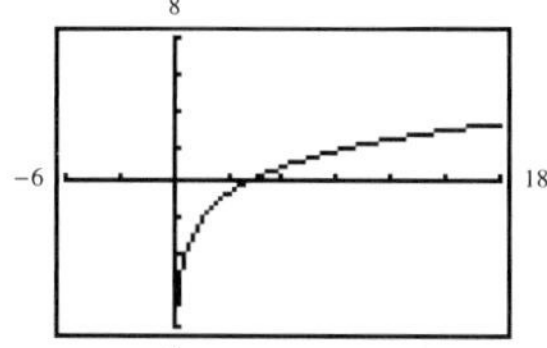

87. Domain: $(0, \infty)$

Vertical asymptote: $x = 0$

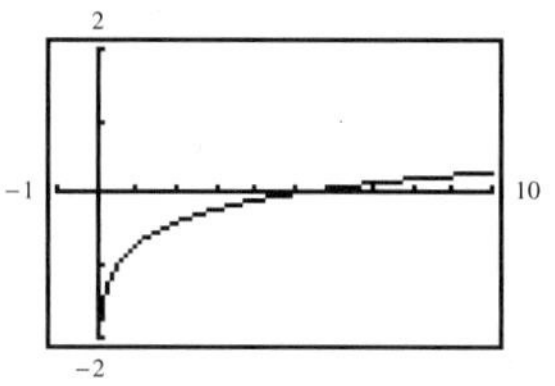

89. 3.6376 **91.** -1.8971 **93.** -1.1484 **95.** b

96. a **97.** d **98.** c

99. Vertical asymptote: $x = 0$

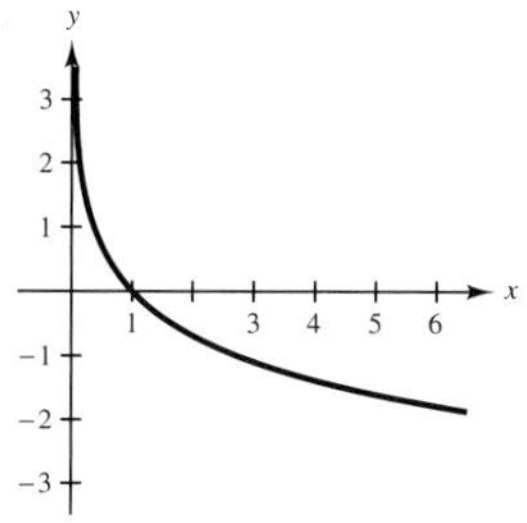

101. Vertical asymptote: $x = 0$

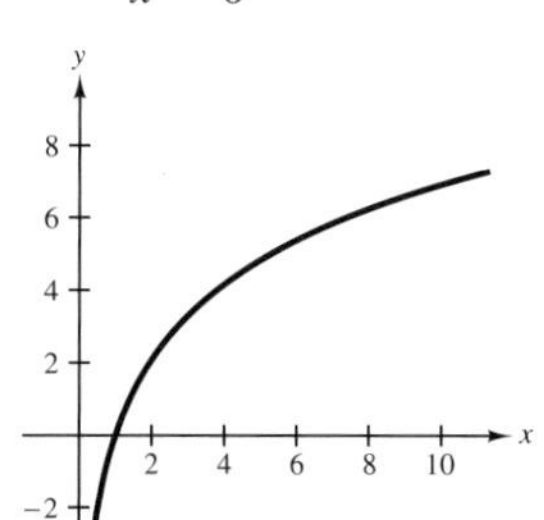

103. Vertical asymptote: $x = 0$

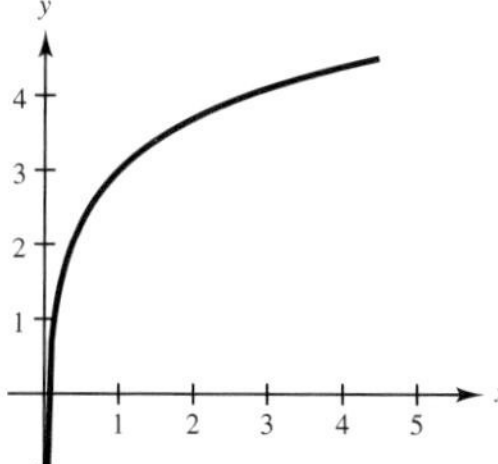

105. Vertical asymptote: $t = 4$

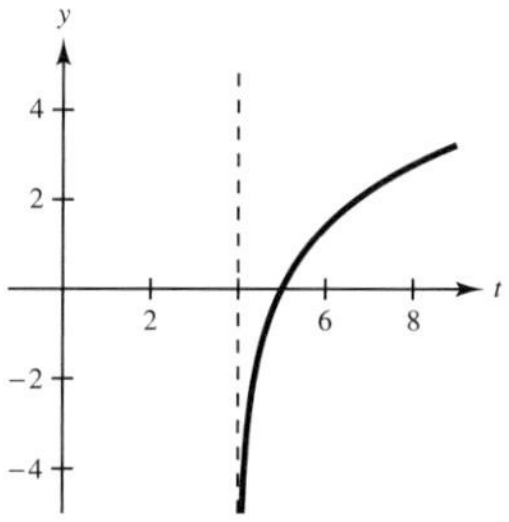

107.

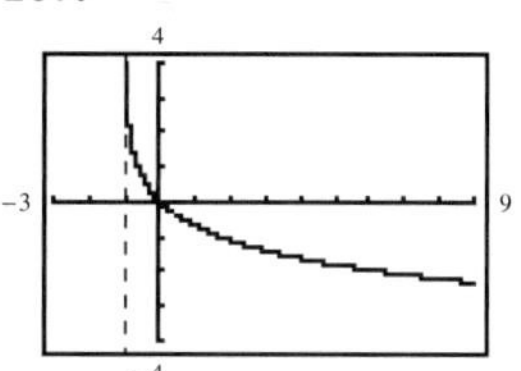

Domain: $(-1, \infty)$

Vertical asymptote: $x = -1$

109.

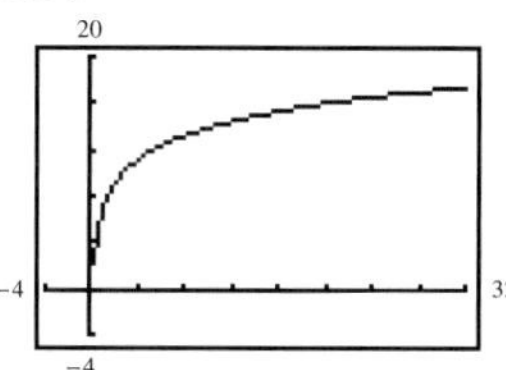

Domain: $(0, \infty)$

Vertical asymptote: $t = 0$

111. 1.6309 **113.** 1.6397 **115.** -0.4739

117. 2.6332 **119.** -2 **121.** 1.3481 **123.** 1.8946

125. 53.4 inches

127.

r	0.07	0.08	0.09	0.10	0.11	0.12
t	9.9	8.7	7.7	6.9	6.3	5.8

129. (a)

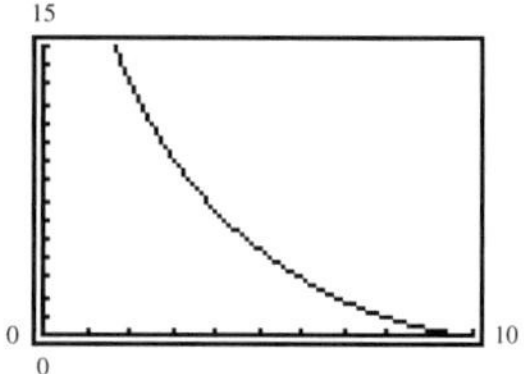

Domain: $(0, 10]$

(b) $x = 0$ (c) $(0, 22.9)$

131. $(0, \infty)$ **133.** $3 \le f(x) \le 4$ **135.** A factor of 10

137. Logarithmic functions with base 10 are common logarithms. Logarithmic functions with base e are natural logarithms.

139. A vertical shift or reflection in the x-axis of a logarithmic graph does not affect the domain or range. A horizontal shift or reflection in the y-axis of a logarithmic graph affects the domain, but the range stays the same.

141. $-m^{10}n^4$ **143.** $\frac{9x^3}{2y^2}$, $x \neq 0$

145. $19\sqrt{3x}$ **147.** $\sqrt{5u}$

Mid-Chapter Quiz *(page 722)*

1. (a) $\frac{16}{9}$ (b) 1 (c) $\frac{3}{4}$ (d) 1.540

2. Horizontal asymptote: $y = 0$

3.

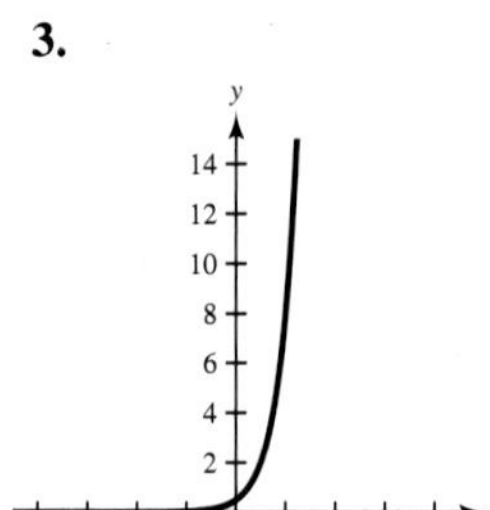

Horizontal asymptote: $y = 0$

4.

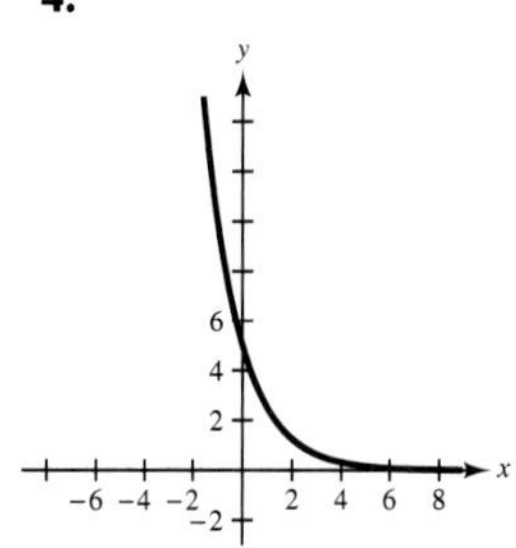

Horizontal asymptote: $y = 0$

5.

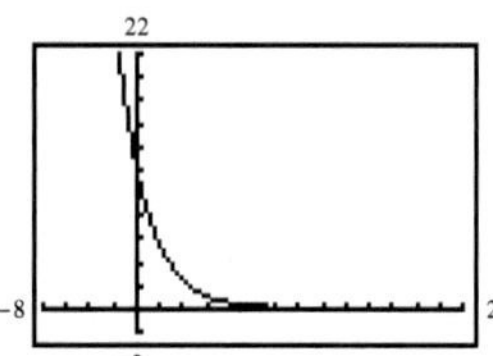

Horizontal asymptote: $y = 0$

6.

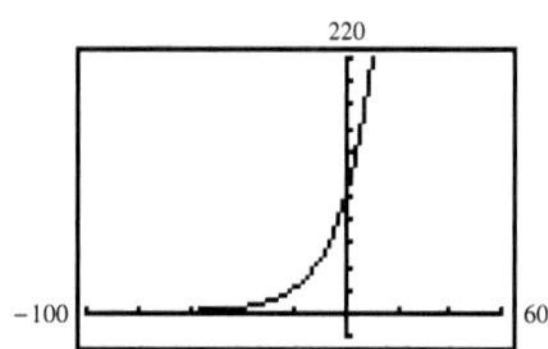

Horizontal asymptote: $y = 0$

7. (a) $2x^3 - 3$ (b) $(2x - 3)^3 = 8x^3 - 36x^2 + 54x - 27$
(c) -19 (d) 125

8.
$$f(g(x)) = 5 - 2\left[\tfrac{1}{2}(5 - x)\right] = 5 - 5 + x = x$$
$$g(f(x)) = \tfrac{1}{2}[5 - (5 - 2x)] = \tfrac{1}{2}(2x) = x$$

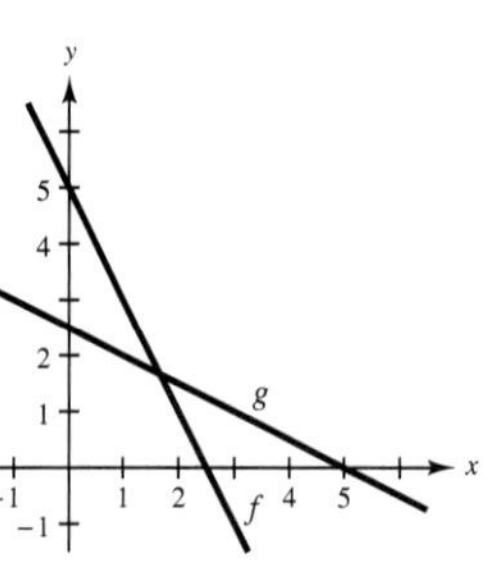

9. $h^{-1}(x) = \dfrac{x - 3}{10}$ **10.** $g^{-1}(t) = \sqrt[3]{2t - 4}$

11. $9^{-2} = \frac{1}{81}$ **12.** $\log_2 64 = 6$ **13.** 3

14.

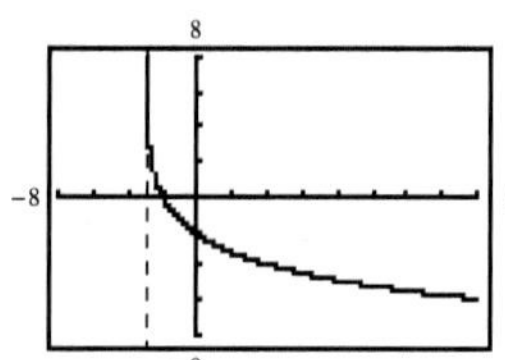

Vertical asymptote: $t = -3$

15.

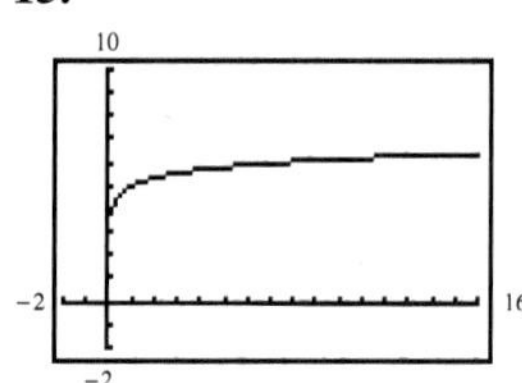

Vertical asymptote: $x = 0$

16. $h = 2, k = 1$ **17.** 6.0639

18.

n	1	4	12
A	\$2979.31	\$3042.18	\$3056.86

n	365	Continuous
A	\$3064.06	\$3064.31

19. 1.60 grams

Section 11.4 *(page 727)*

1. 3 **3.** -4 **5.** $\frac{1}{3}$ **7.** 0 **9.** -9 **11.** 2
13. 2 **15.** 3 **17.** 2 **19.** -3 **21.** 12 **23.** 1
25. 1.5000 **27.** 0.2925 **29.** 0.5283 **31.** 0
33. 2.1972 **35.** 0.5108 **37.** 1.9033 **39.** 8.7118

41. $-3 \log_4 2 = \log_4 2^{-3} = \log_4 \frac{1}{8}$

43.
$$\begin{aligned} -3\log_{10} 3 + \log_{10} \tfrac{3}{2} &= \log_{10} 3^{-3} + \log_{10}\tfrac{3}{2} \\ &= \log_{10}\tfrac{1}{27} + \log_{10}\tfrac{3}{2} \\ &= \log_{10}\left(\tfrac{1}{27}\cdot\tfrac{3}{2}\right) \\ &= \log_{10}\tfrac{1}{18} \end{aligned}$$

45. $-\ln\frac{1}{7} = \ln\left(\frac{1}{7}\right)^{-1} = \ln 7 = \ln\frac{56}{8} = \ln 56 - \ln 8$

47. $\log_3 11 + \log_3 x$ **49.** $\ln 3 + \ln y$ **51.** $2\log_7 x$

53. $-3\log_4 x$ **55.** $\frac{1}{2}(\log_4 3 + \log_4 x)$

57. $\log_2 z - \log_2 17$ **59.** $\frac{1}{2}\log_9 x - \log_9 12$

61. $2\ln x + \ln(y + 2)$ **63.** $6\log_4 x + 2\log_4(x + 7)$

65. $\frac{1}{3}\log_3(x + 1)$ **67.** $\frac{1}{2}[\ln x + \ln(x + 2)]$

69. $2[\ln(x + 1) - \ln(x + 4)]$ **71.** $\frac{1}{3}[2\ln x - \ln(x + 1)]$

73. $\ln x + 2\ln y - 3\ln z$ **75.** $\log_{12}\dfrac{x}{3}$ **77.** $\log_3 5x$

79. $\log_{10}\dfrac{4}{x}$ **81.** $\ln b^4$ **83.** $\log_5\dfrac{1}{4x^2}$ **85.** $\log_2 x^7z^3$

87. $\log_3 2\sqrt{y}$ **89.** $\ln\dfrac{x^3y}{z^2}$ **91.** $\ln x^4y^4$

93. $\ln\left(\dfrac{x}{x + 1}\right)^2$ **95.** $\log_4\dfrac{x + 8}{x^3}$ **97.** $\log_5\dfrac{\sqrt[3]{x + 3}}{x - 6}$

99. $\log_6\dfrac{(c + d)^5}{\sqrt{m - n}}$ **101.** $\log_2\sqrt[5]{\dfrac{x^3}{y^4}}$ **103.** $2 + \ln 3$

105. $1 + \frac{1}{2}\log_5 2$ **107.** $1 - 3\log_8 x$

109.

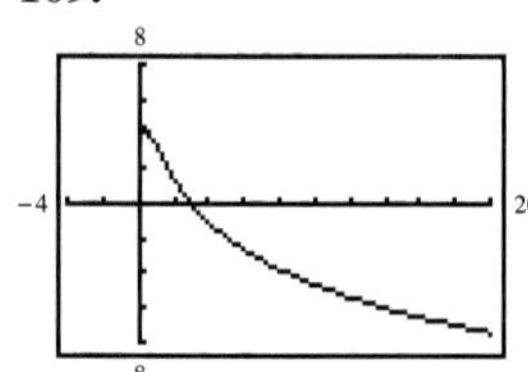

111.

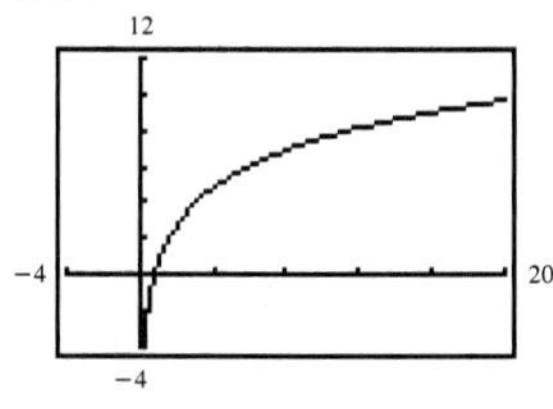

113. $B = 10(\log_{10} I + 16)$; 130 decibels

115. $E = \log_{10}\left(\dfrac{C_2}{C_1}\right)^{1.4}$

117. True; $\log_2 8x = \log_2 8 + \log_2 x = 3 + \log_2 x$

119. False; $\log_3(u + v)$ does not simplify.

121. True;

$$\begin{aligned} f(ax) &= \log_a ax \\ &= \log_a a + \log_a x \\ &= 1 + \log_a x \\ &= 1 + f(x) \end{aligned}$$

123. False; 0 is not in the domain of f.

125. False; $f(x - 3) = \ln(x - 3)$

127. True; $f(1) = 0$, so when $f(x) > 0$, $x > 1$.

129. Evaluate when $x = e$ and $y = e$.

131. $5 \pm 2\sqrt{2}$ **133.** $\frac{5}{3}$ **135.** 47

137. **139.**

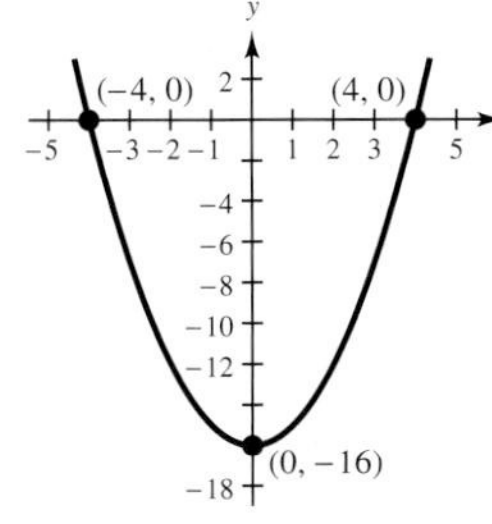

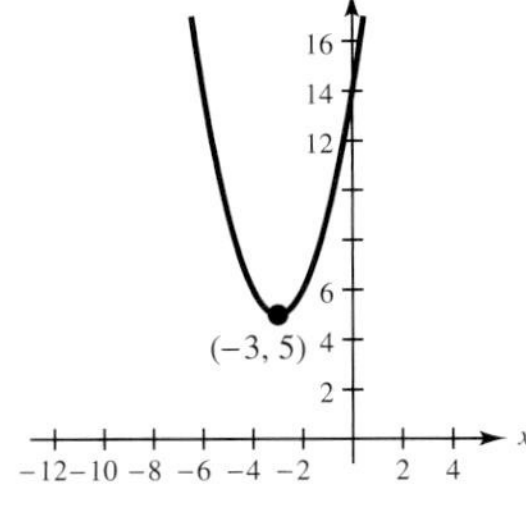

141. (a) $(f \circ g)(x) = \sqrt{x - 3}$

Domain: $[3, \infty)$

(b) $(g \circ f)(x) = \sqrt{x} - 3$

Domain: $[0, \infty)$

143. (a) $(f \circ g)(x) = \dfrac{5}{x^2 + 2x - 3}$

Domain: $(-\infty, -3) \cup (-3, 1) \cup (1, \infty)$

(b) $(g \circ f)(x) = \dfrac{5}{x^2 - 4} + 1 = \dfrac{x^2 + 1}{x^2 - 4}$

Domain: $(-\infty, -2) \cup (-2, 2) \cup (2, \infty)$

Section 11.5 *(page 737)*

1. (a) Not a solution (b) Solution

3. (a) Solution (b) Not a solution

5. (a) Not a solution (b) Solution

7. 3 **9.** -3 **11.** 4 **13.** 1 **15.** -2

17. -3 **19.** -6 **21.** 2 **23.** $\frac{22}{5}$ **25.** 6

27. -7 **29.** 4 **31.** No solution **33.** -7

35. $2x - 1$ **37.** $2x,\ x > 0$ **39.** 4.11 **41.** 1.31

43. 1.49 **45.** 0.83 **47.** 0.11 **49.** -3.60

51. -1.79 **53.** 3.00 **55.** -1.45 **57.** 35.35

59. 6.80 **61.** 12.22 **63.** 3.28 **65.** No solution

67. -1.04 **69.** 2.48 **71.** 0.90 **73.** 4.39

75. 0.39 **77.** 8.99 **79.** 9.73 **81.** 4.62

83. 0.10 **85.** 174.77 **87.** 2187.00 **89.** 6.52

91. 25.00 **93.** 0.61 **95.** ± 20.09 **97.** 3.00

99. -3.93 **101.** 2.83 **103.** 2000.00 **105.** 3.20

107. 4.00 **109.** 0.75 **111.** 5.00 **113.** 2.46

115. 3.33 **117.** 6.00

119. (1.40, 0) **121.** (21.82, 0)

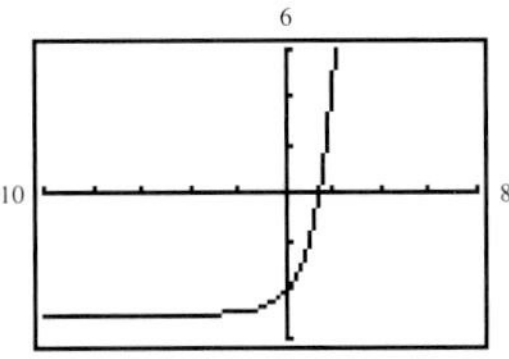

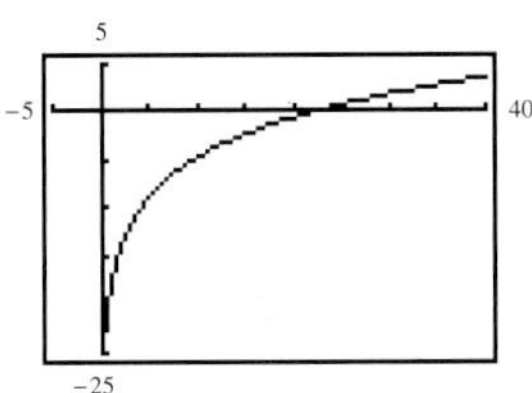

123. 5% **125.** 7.7 years

127. 10^{-8} watt per square centimeter

129. (a) 105°

(b)

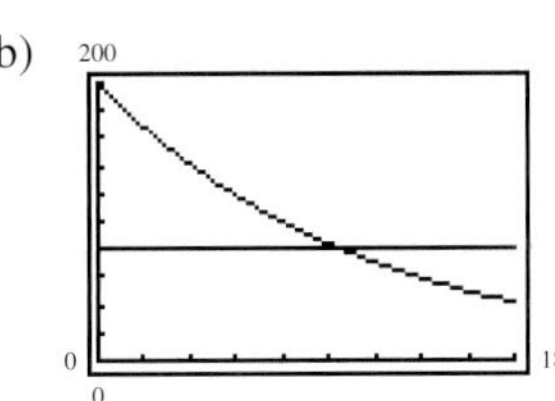

105°

131. (a) -0.144 (b) 6:24 P.M. (c) 90.2°F

133. 1.0234 and 1.0241 grams per cubic centimeter

135. $\log_a(uv) = \log_a u + \log_a v$; $\log_a \dfrac{u}{v} = \log_a u - \log_a v$;

$\log_a u^n = n \log_a u$

137. To solve an exponential equation, first isolate the exponential expression, then take the logarithm of each side of the equation and solve for the variable.

To solve a logarithmic equation, first isolate the logarithmic expression, then exponentiate each side of the equation and solve for the variable.

139. $\pm 5i$ **141.** $\pm\frac{4}{3}$ **143.** $\pm 2, \pm 3$

	Width	*Length*	*Perimeter*	*Area*
145.	$2.5x$	x	42 in.	90 in.^{2}
147.	w	$w + 4$	56 km	192 km^2

Section 11.6 *(page 747)*

1. 7% **3.** 9% **5.** 8% **7.** 9.27 years

9. 8.66 years **11.** 9.59 years **13.** Yearly

15. Continuous **17.** Quarterly **19.** 8.33%

21. 7.23% **23.** 6.14% **25.** 8.30%

27. No. Each time the amount is divided by the principal, the result is always 2.

29. \$1652.99 **31.** \$626.46 **33.** \$3080.15

35. \$951.23 **37.** \$5496.57 **39.** \$320,250.81

41. Total deposits: \$7200.00; Total interest: \$10,529.42

43. $k = \frac{1}{2} \ln \frac{8}{3} \approx 0.4904$ **45.** $k = \frac{1}{3} \ln \frac{1}{2} \approx -0.2310$

47. $y = 90e^{0.0082t}$; 110,000 people

49. $y = 286e^{0.0029t}$; 308,000 people

51. $y = 2589e^{0.0052t}$; 2,948,000 people

53. $y = 3834e^{0.0052t}$; 4,366,000 people

55. k is larger in Exercise 47, because the population of Aruba is increasing faster than the population of Jamaica.

57. 7.949 billion people

59. (a) $y = 73e^{0.4763t}$ (b) 9 hours

61. $y = 109{,}478{,}000e^{0.1259t}$, where $t = 0 \leftrightarrow 2000$; 562,518,000 users

	Isotope	*Half-Life (Years)*	*Initial Quantity*	*Amount After 1000 Years*
63.	^{226}Ra	1620	6 g	3.91 g
65.	^{14}C	5730	4.51 g	4.0 g
67.	^{239}Pu	24,100	4.2 g	4.08 g

69. 3.3 grams **71.** 7.5 grams **73.** \$15,204

75. The Chilean earthquake in 1960 was about 19,953 times greater.

77. The earthquake in Fiji was about 63 times greater.

79. 7.04 **81.** 10^7 times

83. (a)

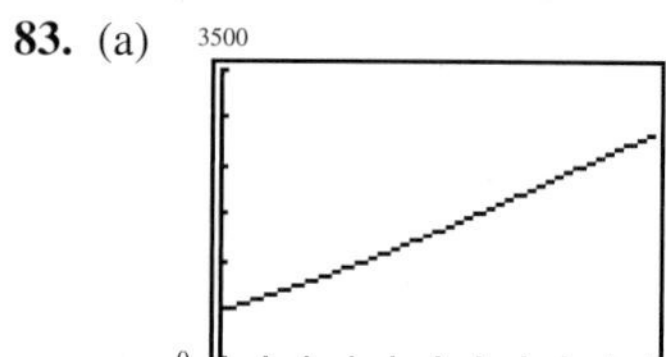

(b) 1000 rabbits

(c) 2642 rabbits

(d) 5.88 years

85. (a) $S = 10(1 - e^{-0.0575x})$ (b) 3314 pairs

87. If $k > 0$, the model represents exponential growth, and if $k < 0$, the model represents exponential decay.

89. When the investment is compounded more than once in a year (quarterly, monthly, daily, continuously), the effective yield is greater than the interest rate.

91. $\frac{7}{2} \pm \frac{\sqrt{69}}{2}$ **93.** $-\frac{3}{2} \pm \frac{\sqrt{33}}{6}$

95. $(4, \infty)$

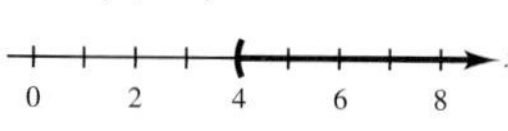

97. $(-\infty, -3) \cup (3, \infty)$

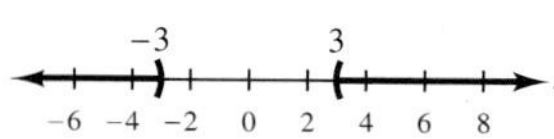

Review Exercises *(page 754)*

1. (a) $\frac{1}{64}$ (b) 4 (c) 16 **3.** (a) 5 (b) 0.185 (c) $\frac{1}{25}$

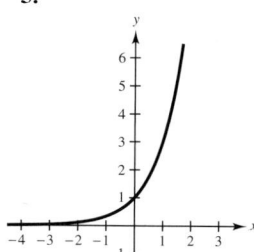

5.

Horizontal asymptote: $y = 0$

7.

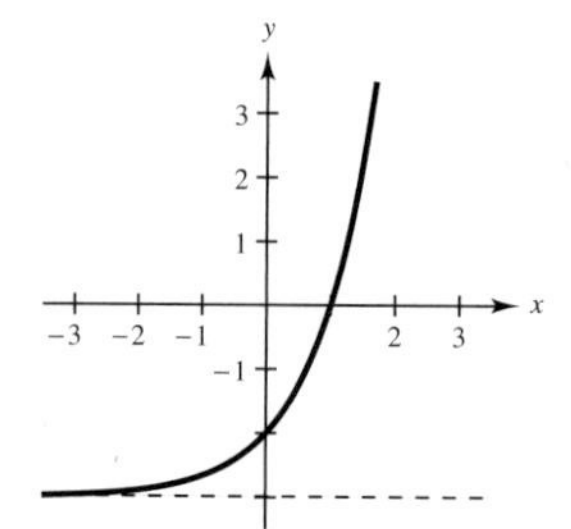

Horizontal asymptote: $y = -3$

9.

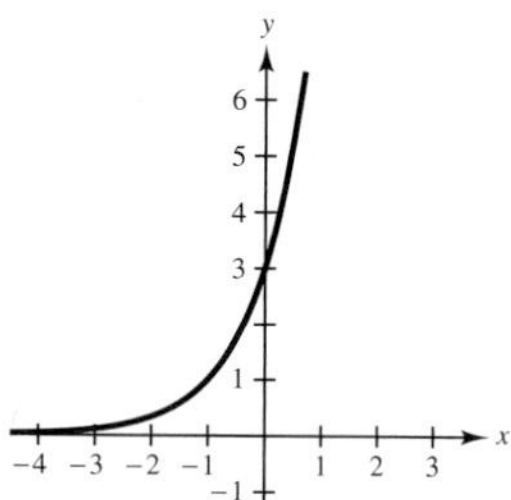

Horizontal asymptote: $y = 0$

11.

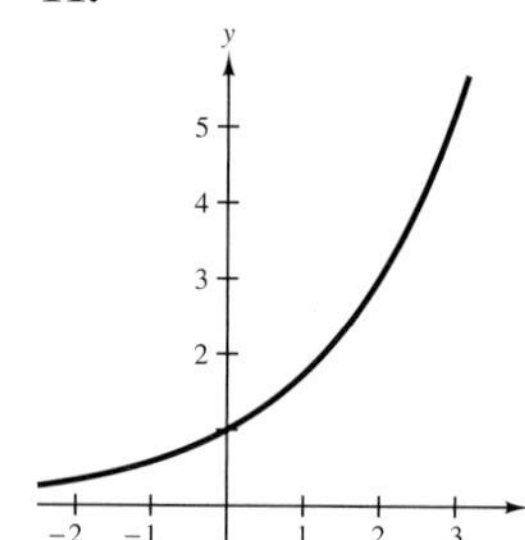

Horizontal asymptote: $y = 0$

13.

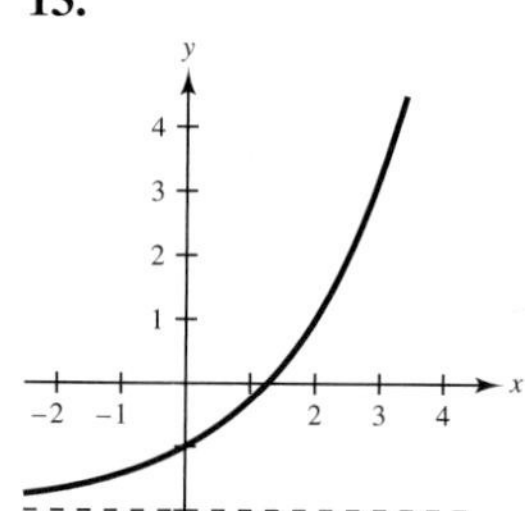

Horizontal asymptote: $y = -2$

15.

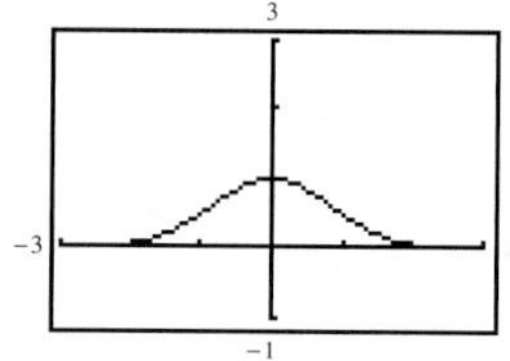

17.

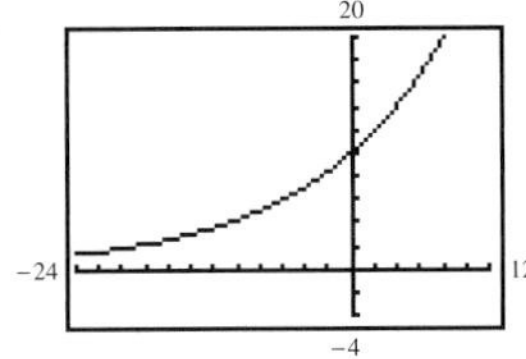

19. (a) 0.007 (b) 3 (c) 9.557×10^{16}

21.

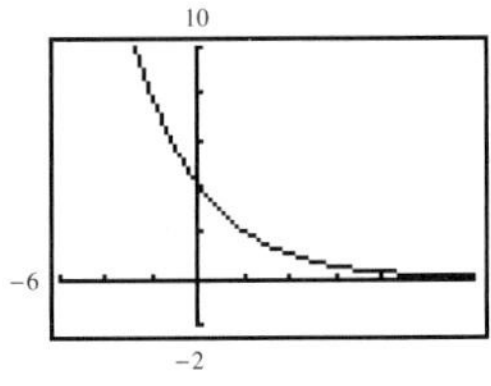

23.

25.

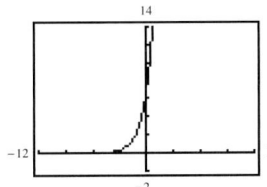

n	1	4	12
A	\$226,296.28	\$259,889.34	\$268,503.32

n	365	Continuous
A	\$272,841.23	\$272,990.75

27. 4.21 grams **29.** (a) 6 (b) 1 **31.** (a) 5 (b) −1

33. (a) $(f \circ g)(x) = \sqrt{2x + 6}$

Domain: $[-3, \infty)$

(b) $(g \circ f)(x) = 2\sqrt{x + 6}$

Domain: $[-6, \infty)$

35. No **37.** Yes **39.** $f^{-1}(x) = \frac{1}{3}(x - 4)$

41. $h^{-1}(x) = \frac{1}{5}x^2,\ x \geq 0$ **43.** $f^{-1}(t) = \sqrt[3]{t - 4}$

45.

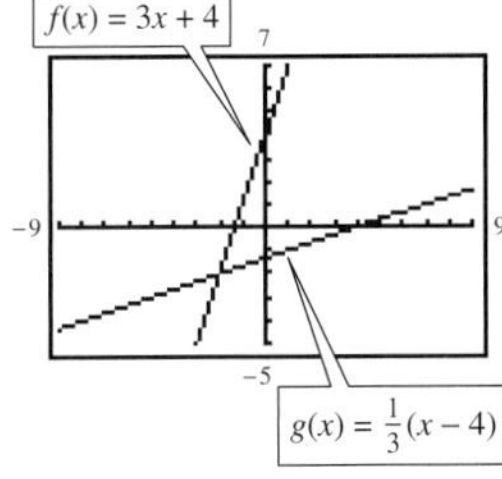

47.

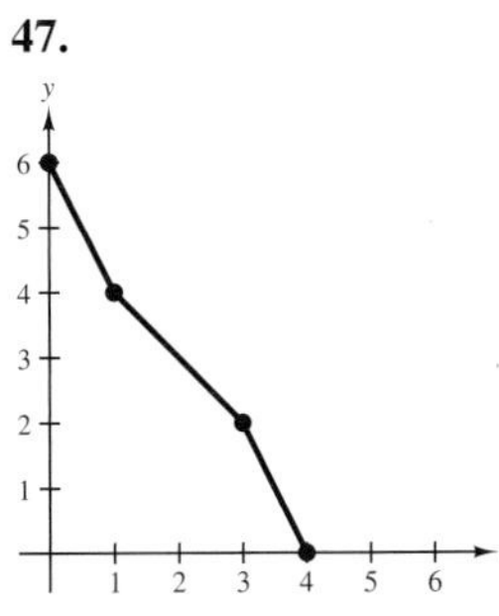

49.

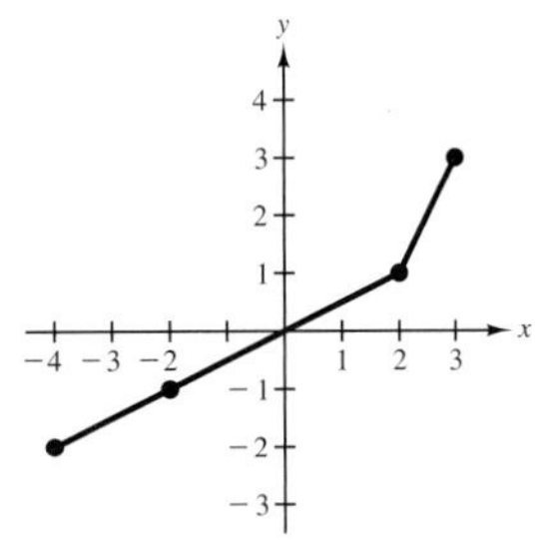

51. 3 **53.** −2 **55.** 6 **57.** 0

59.

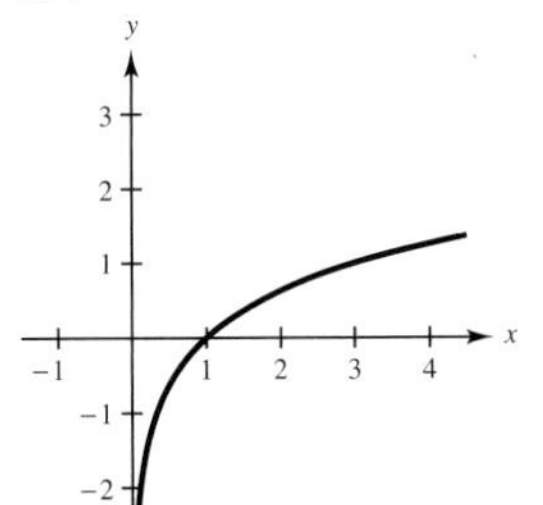

Vertical asymptote: $x = 0$

61.

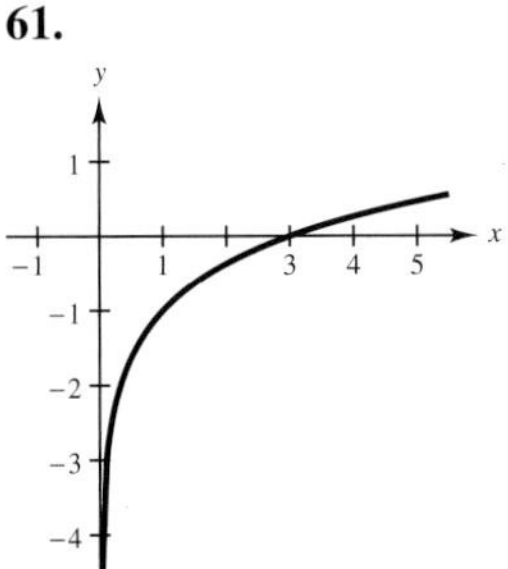

Vertical asymptote: $x = 0$

63.

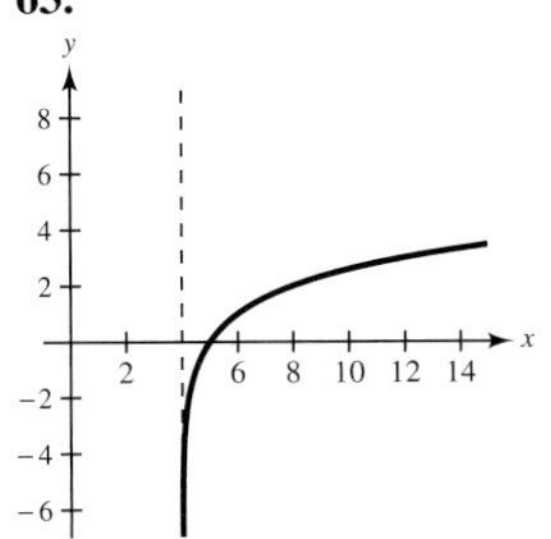

Vertical asymptote: $x = 4$

65. 3.9120

67.

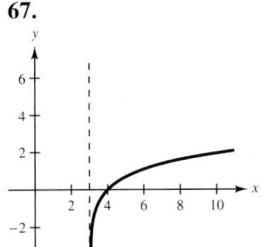

Vertical asymptote: $x = 3$

69.

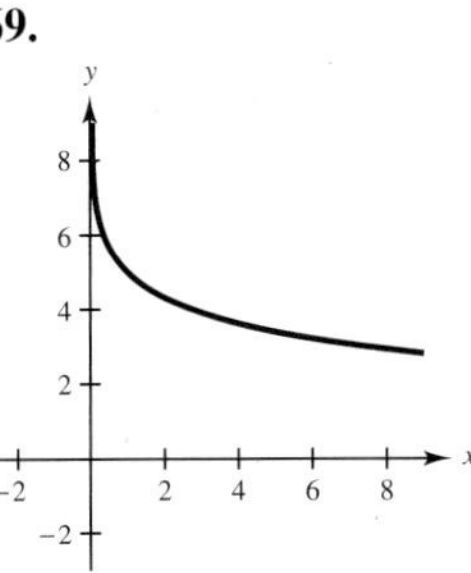

Vertical asymptote: $x = 0$

71. 1.5850 **73.** 2.4406 **75.** 1.7959 **77.** −0.4307

79. 1.0293 **81.** $\log_4 6 + 4 \log_4 x$ **83.** $\frac{1}{2} \log_5(x + 2)$

85. $\ln(x + 2) - \ln(x + 3)$

87. $\dfrac{1}{2}(\ln 2 + \ln x) + 5 \ln(x + 3)$ **89.** $\ln\left(\dfrac{1}{3y}\right)^{2/3},\ y > 0$

91. $\log_8 32x^3$ **93.** $\ln \dfrac{9}{4x^2},\ x > 0$

95. $\log_2\left(\dfrac{k}{k - t}\right)^4,\ k > t, k > 0$

97. $\ln(x^3y^4z),\ x > 0,\ y > 0,\ z > 0$

99. False; $\log_2 4x = \log_2 4 + \log_2 x = 2 + \log_2 x$

101. True **103.** True

105. $y = 0.83(\ln I_0 - \ln I)$; 0.20 **107.** 6

109. 1 **111.** 6 **113.** 5.66 **115.** 6.23

117. No solution **119.** 1408.10 **121.** 31.47

123. 4.00 **125.** 64.00 **127.** 11.57 **129.** 7%

131. 5% **133.** 7.5% **135.** 6.5% **137.** 5.65%

139. 7.71% **141.** 7.79%

	Isotope	*Half-Life (Years)*	*Initial Quantity*	*Amount After 1000 Years*
143.	^{226}Ra	1620	3.5 g	2.282 g
145.	^{14}C	5730	2.934g	2.6 g
147.	^{239}Pu	24,100	5 g	4.858 g

149. The earthquake in San Francisco was about 2512 times greater.

Chapter Test *(page 759)*

1. $f(-1) = 81$;
$f(0) = 54$;
$f\left(\frac{1}{2}\right) = 18\sqrt{6} \approx 44.09$;
$f(2) = 24$

2.

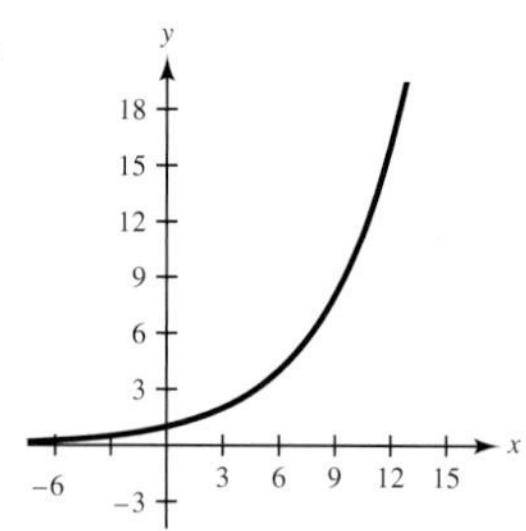

Horizontal asymptote: $y = 0$

3. (a) $(f \circ g)(x) = 18x^2 - 63x + 55$;
Domain: $(-\infty, \infty)$
(b) $(g \circ f)(x) = -6x^2 - 3x + 5$;
Domain: $(-\infty, \infty)$

4. $f^{-1}(x) = \frac{1}{9}(x + 4)$

5. $(f \circ g)(x) = -\frac{1}{2}(-2x + 6) + 3$
$= (x - 3) + 3$
$= x$
$(g \circ f)(x) = -2\left(-\frac{1}{2}x + 3\right) + 6$
$= (x - 6) + 6$
$= x$

6. -4

7. $g = f^{-1}$

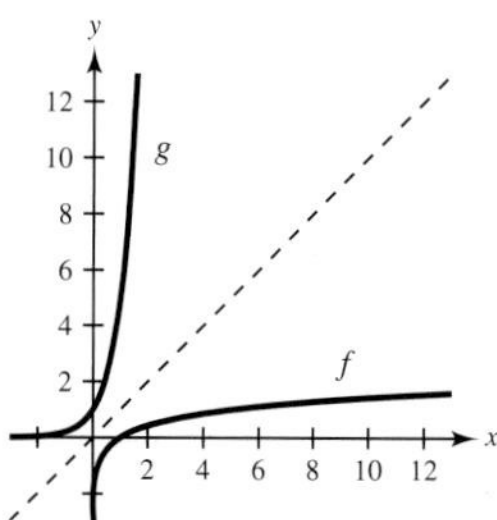

8. $\dfrac{2}{3} + \dfrac{1}{2}\log_8 x - 4\log_8 y$ **9.** $\ln \dfrac{x}{y^4},\ y > 0$

10. 32 **11.** 1.18 **12.** 13.73 **13.** 15.52

14. 5 **15.** 8 **16.** 109.20 **17.** 0

18. (a) \$8012.78 (b) \$8110.40 **19.** \$10,806.08

20. 7% **21.** \$4746.09 **22.** 600 foxes

23. 1141 foxes **24.** 4.4 years

CHAPTER 12

Section 12.1 *(page 769)*

1. e **2.** c **3.** d **4.** a **5.** f **6.** b

7. $x^2 + y^2 = 25$ **9.** $x^2 + y^2 = \frac{4}{9}$ **11.** $x^2 + y^2 = 36$

13. $x^2 + y^2 = 29$ **15.** $(x - 4)^2 + (y - 3)^2 = 100$

17. $(x - 6)^2 + (y + 5)^2 = 9$

19. $(x + 2)^2 + (y - 1)^2 = 4$

21. $(x - 3)^2 + (y - 2)^2 = 17$

23. Center: $(0, 0)$; $r = 4$ **25.** Center: $(0, 0)$; $r = 6$

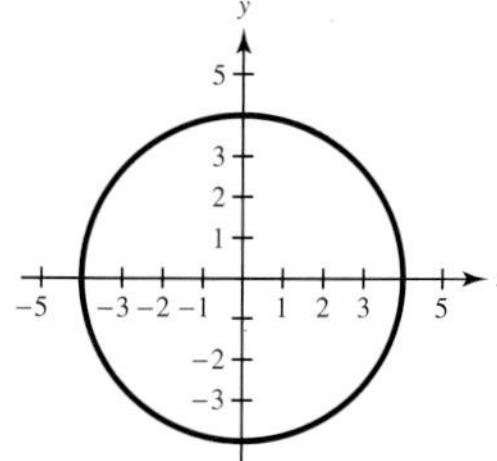

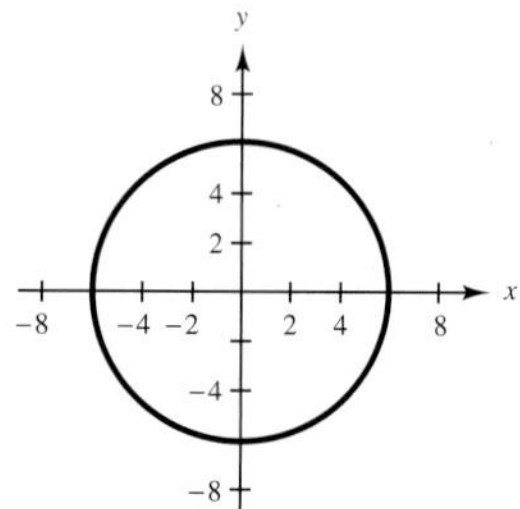

27. Center: $(0, 0)$; $r = \frac{1}{2}$ **29.** Center: $(0, 0)$; $r = \frac{12}{5}$

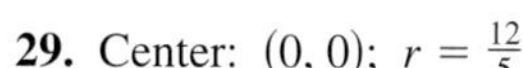

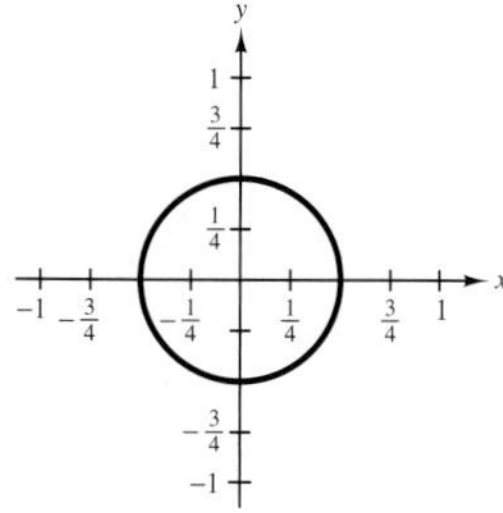

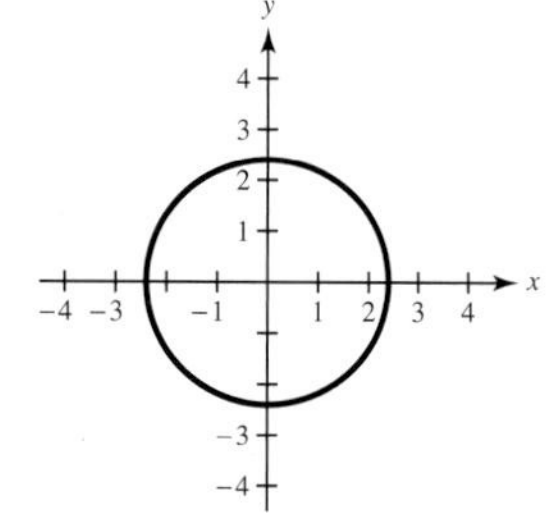

31. Center: $(-1, 5)$; $r = 8$

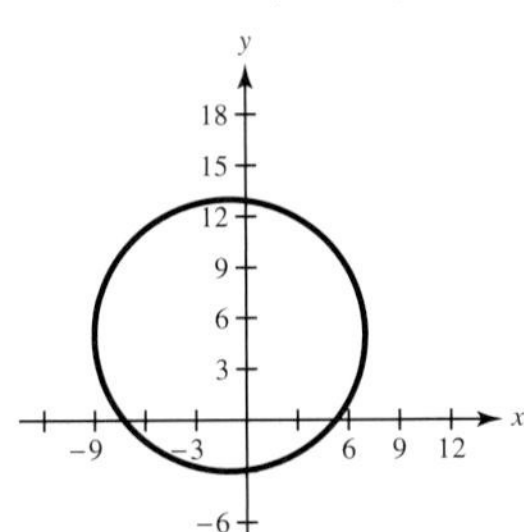

33. Center: $(2, 3)$; $r = 2$

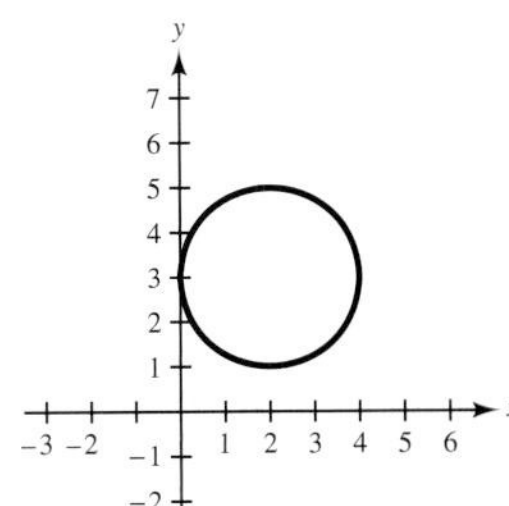

35. Center: $\left(-\frac{9}{4}, 4\right)$; $r = 4$

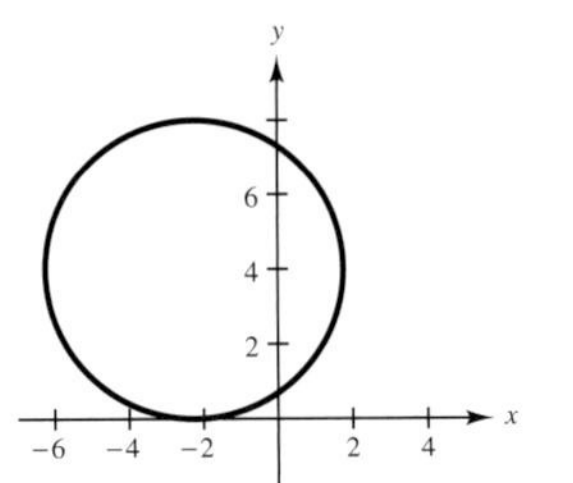

37. Center: $(2, 1)$; $r = 2$

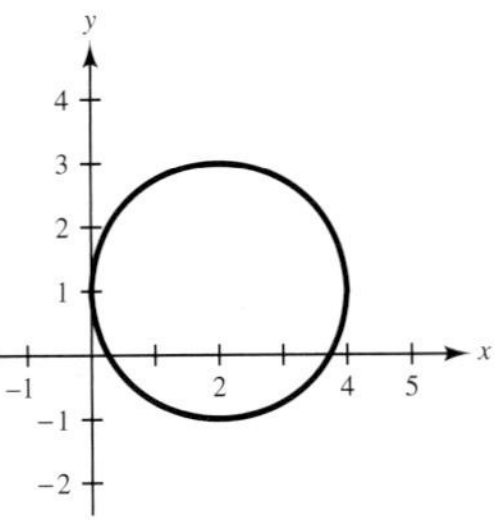

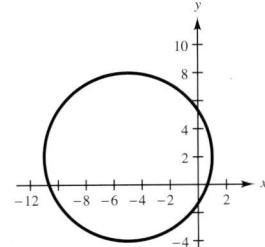

39. Center: $(-1, -3)$; $r = 2$

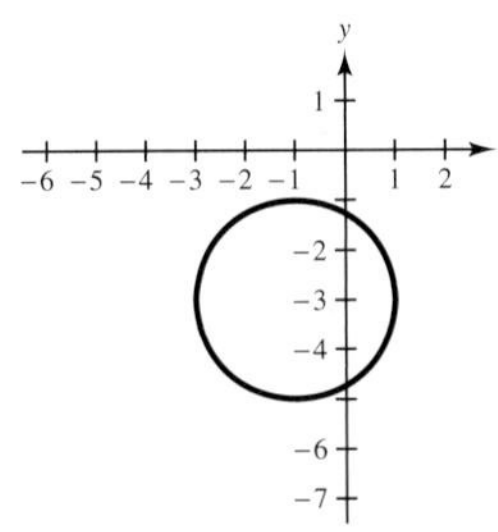

41. Center: $(-5, 2)$; $r = 6$

43.

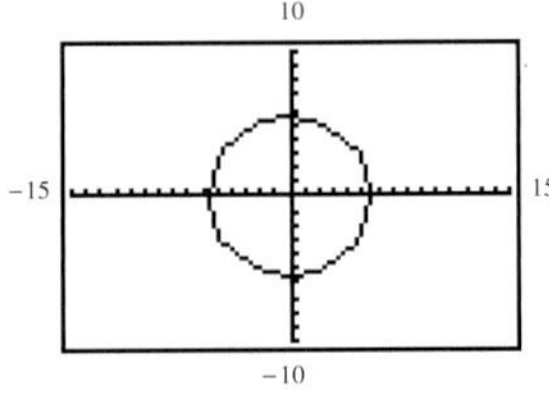

45.

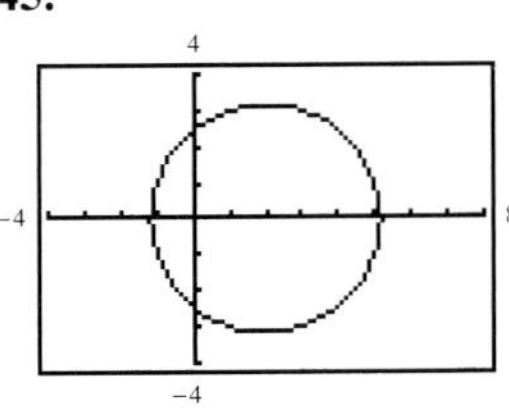

47. f **48.** a **49.** e **50.** c **51.** b **52.** d

53. $x^2 = \frac{3}{2}y$ **55.** $x^2 = -6y$ **57.** $y^2 = -8x$

59. $x^2 = 4y$ **61.** $y^2 = 24x$ **63.** $y^2 = 9x$

65. $(x - 3)^2 = -(y - 1)$ **67.** $y^2 = 2(x + 2)$

69. $(y - 2)^2 = -8(x - 3)$ **71.** $x^2 = 12(y + 4)$

73. $(y - 2)^2 = x$

75. Vertex: $(0, 0)$
Focus: $\left(0, \frac{1}{2}\right)$

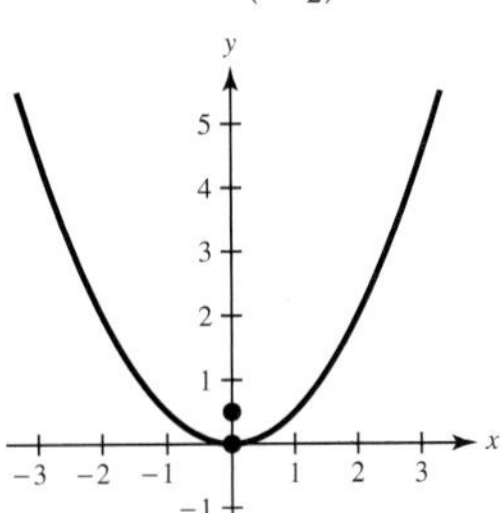

77. Vertex: $(0, 0)$
Focus: $\left(-\frac{5}{2}, 0\right)$

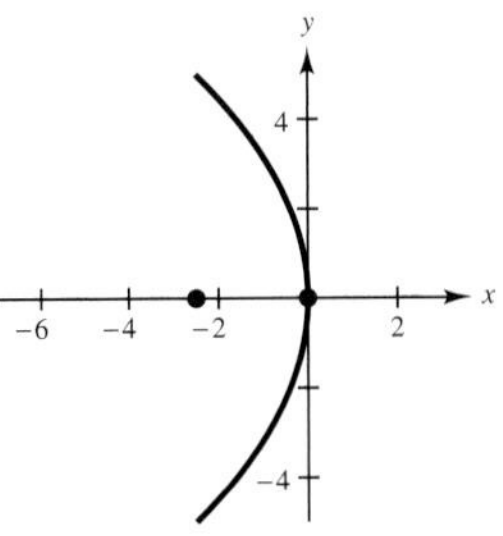

79. Vertex: $(0, 0)$
Focus: $(0, -2)$

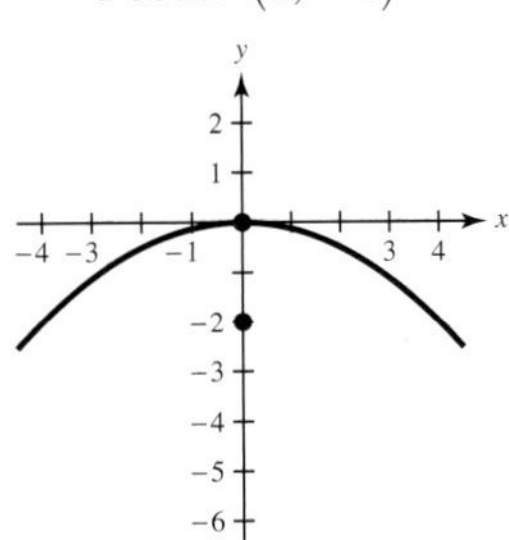

81. Vertex: $(1, -2)$
Focus: $(1, -4)$

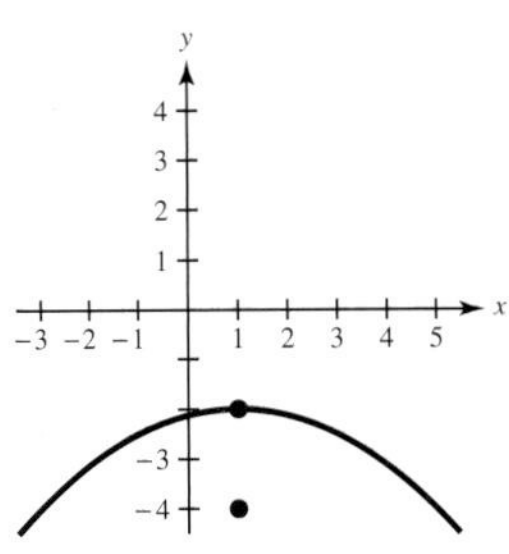

83. Vertex: $\left(5, -\frac{1}{2}\right)$
Focus: $\left(\frac{11}{2}, -\frac{1}{2}\right)$

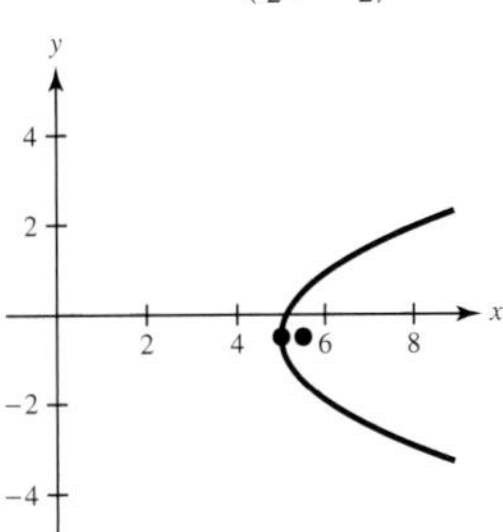

85. Vertex: $(1, 3)$
Focus: $\left(1, \frac{15}{4}\right)$

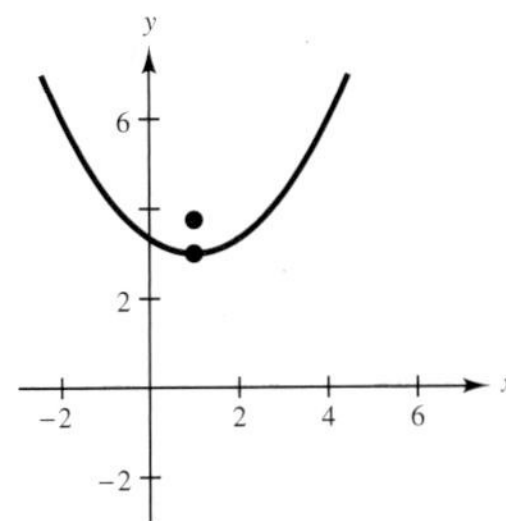

87. Vertex: $(-2, -3)$
Focus: $(-4, -3)$

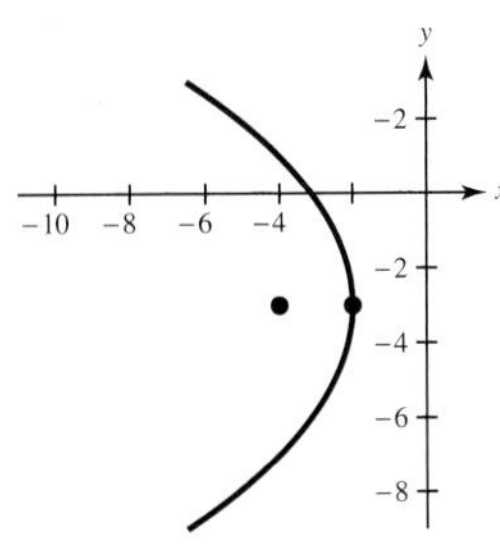

89. Vertex: $(-2, 1)$
Focus: $\left(-2, -\frac{1}{2}\right)$

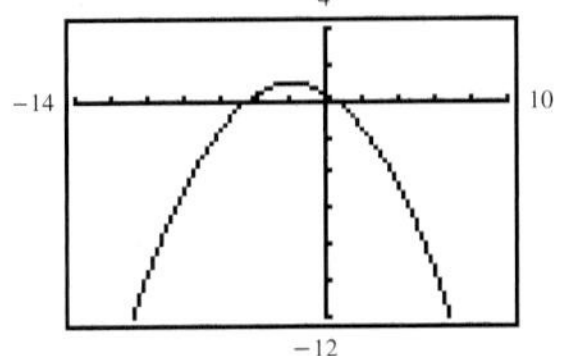

91. Vertex: $\left(\frac{1}{4}, -\frac{1}{2}\right)$
Focus: $\left(0, -\frac{1}{2}\right)$

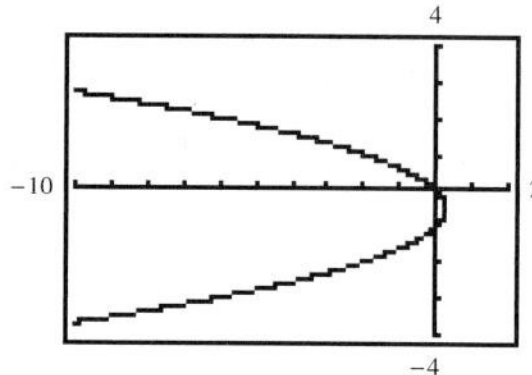

93. $x^2 + y^2 = 4500^2$

95. $(x - 10)^2 + (y - 5.5)^2 = 2.25$; 6.8 feet

97. (a) $x^2 = 180y$

(b)

x	0	20	40	60
y	0	$2\frac{2}{9}$	$8\frac{8}{9}$	20

99. (a)

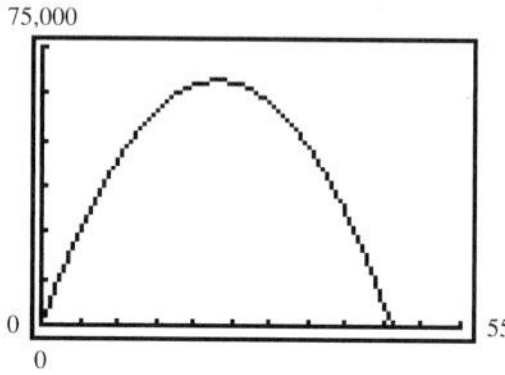

(b) 230 video game systems

101. (a) Answers will vary.

(b)

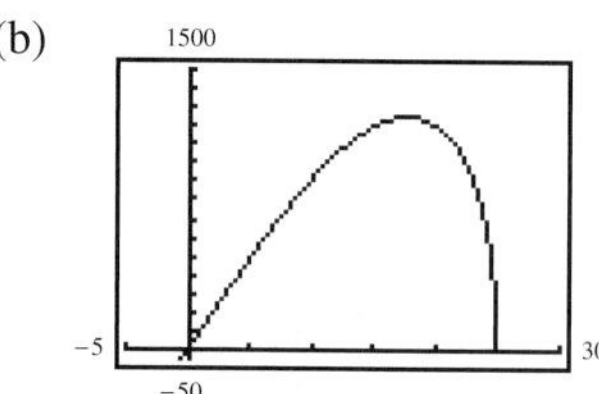

Maximum area when $x \approx 17.68$

103. No. The equation in standard form is $x^2 + (y - 3)^2 = 4$, which does represent a circle with center $(0, 3)$ and radius of 2.

105. No. If the graph intersected the directrix, there would exist points closer to the directrix than the focus.

107. $-2 \pm \sqrt{10}$ **109.** $-1, 3$ **111.** $-\frac{5}{4} \pm \frac{\sqrt{89}}{4}$

113. $10 \log_8 x$ **115.** $\ln 5 + 2 \ln x + \ln y$

117. $\log_{10} 6x$ **119.** $\ln \frac{x^3y}{9}$

Section 12.2 *(page 780)*

1. a **2.** f **3.** d **4.** c **5.** e **6.** b

7. $\frac{x^2}{16} + \frac{y^2}{9} = 1$ **9.** $\frac{x^2}{4} + \frac{y^2}{1} = 1$ **11.** $\frac{x^2}{9} + \frac{y^2}{36} = 1$

13. $\frac{x^2}{1} + \frac{y^2}{4} = 1$ **15.** $\frac{x^2}{9} + \frac{y^2}{25} = 1$

17. $\frac{x^2}{100} + \frac{y^2}{36} = 1$

19.

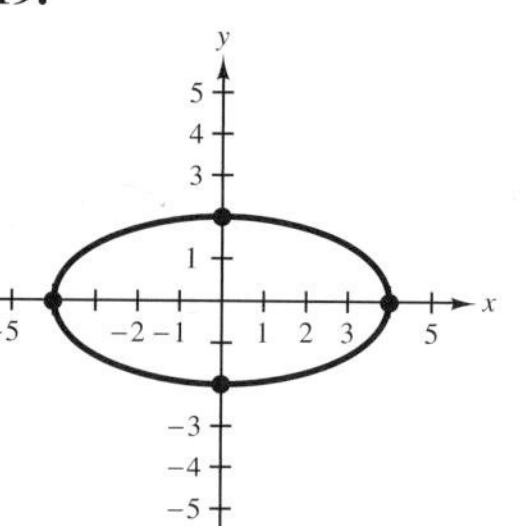

Vertices: $(\pm 4, 0)$
Co-vertices: $(0, \pm 2)$

21.

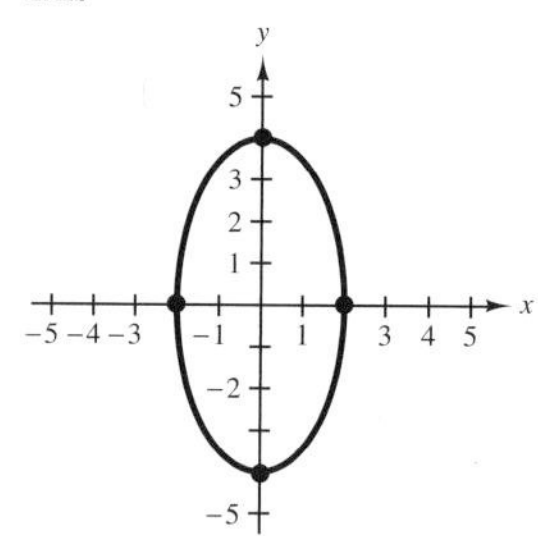

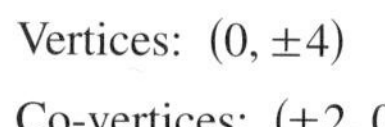
Vertices: $(0, \pm 4)$
Co-vertices: $(\pm 2, 0)$

23.

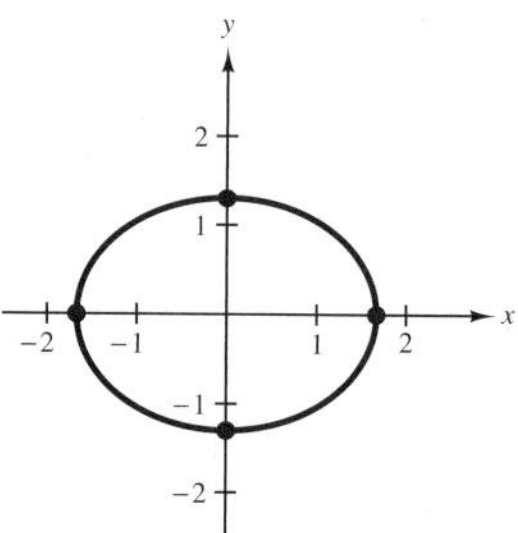

Vertices: $\left(\pm\frac{5}{3}, 0\right)$
Co-vertices: $\left(0, \pm\frac{4}{3}\right)$

25.

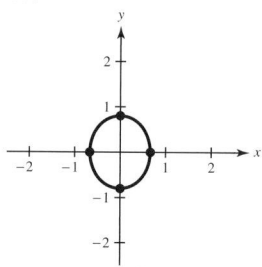

Vertices: $\left(0, \pm\frac{4}{5}\right)$
Co-vertices: $\left(\pm\frac{2}{3}, 0\right)$

27.

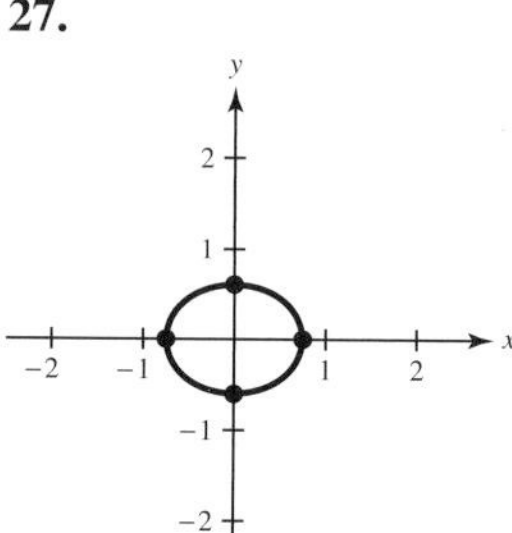

Vertices: $\left(\pm\frac{3}{4}, 0\right)$
Co-vertices: $\left(0, \pm\frac{3}{5}\right)$

29.

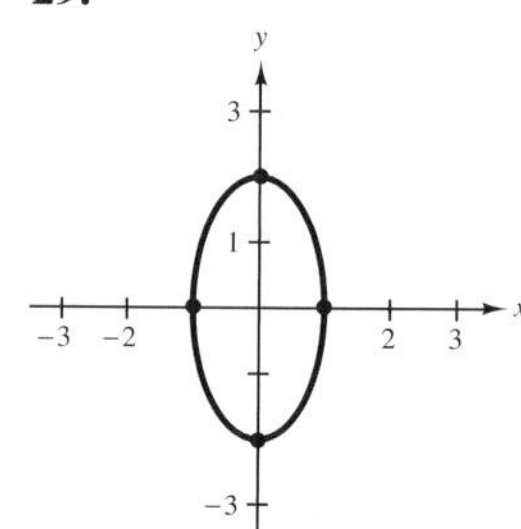

Vertices: $(0, \pm 2)$
Co-vertices: $(\pm 1, 0)$

31.

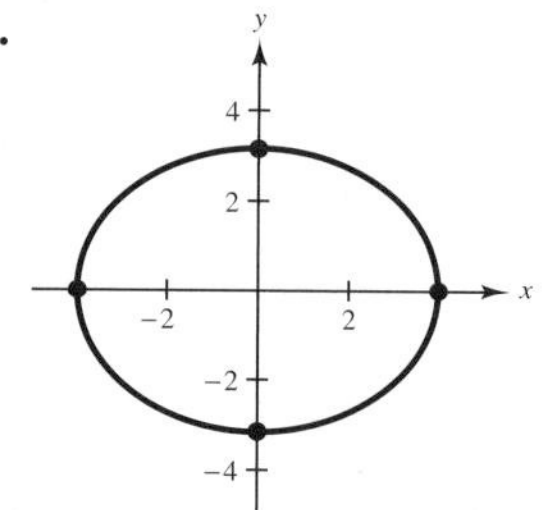

Vertices: $(\pm 4, 0)$
Co-vertices: $\left(0, \pm\sqrt{10}\right)$

33.

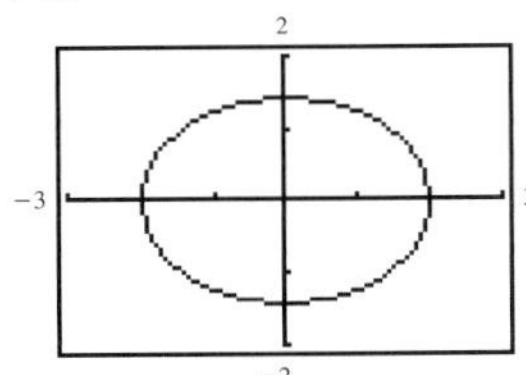

Vertices: $(\pm 2, 0)$

35.

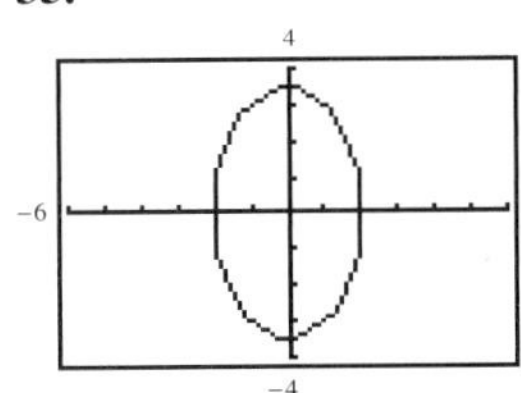

Vertices: $\left(0, \pm 2\sqrt{3}\right)$

37. $\dfrac{x^2}{1} + \dfrac{y^2}{4} = 1$ **39.** $\dfrac{(x-4)^2}{9} + \dfrac{y^2}{16} = 1$

41. Center: $(-5, 0)$

Vertices: $(-9, 0), (-1, 0)$

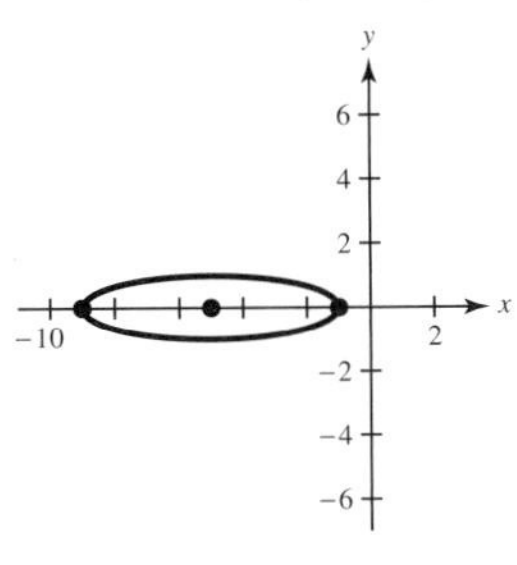

43. Center: $(1, 5)$

Vertices: $(1, 0), (1, 10)$

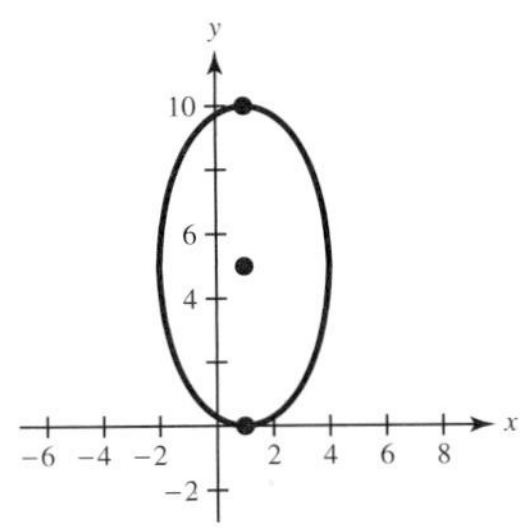

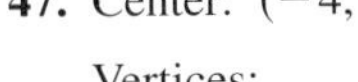

45. Center: $(2, -2)$

Vertices:

$(-1, -2), (5, -2)$

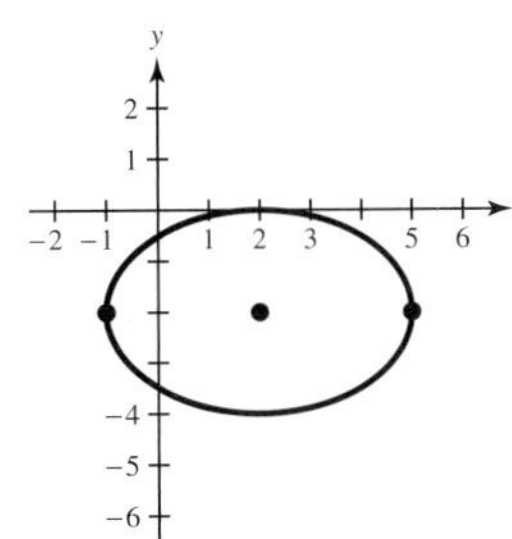

47. Center: $(-4, 1)$

Vertices:

$(-4, -3), (-4, 5)$

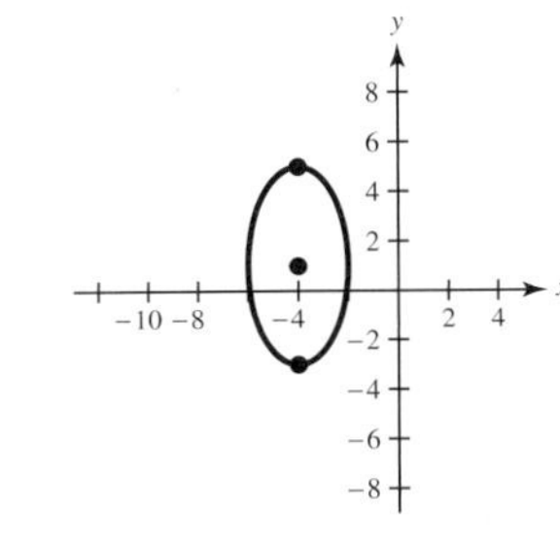

49. Center: $(-2, 3)$

Vertices:

$(-2, 6), (-2, 0)$

51. Center: $(4, -3)$

Vertices:

$(4, -8), (4, 2)$

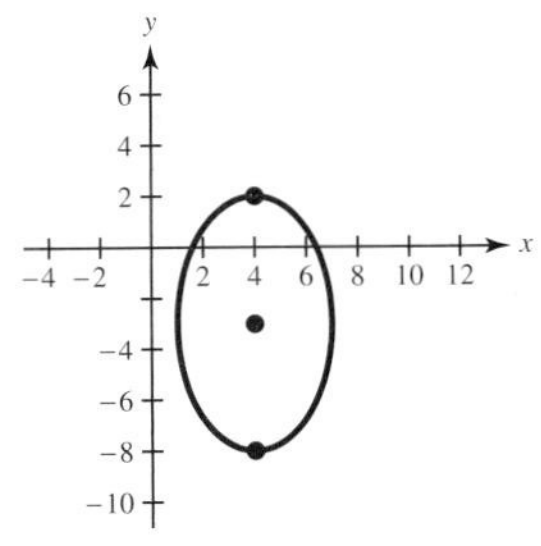

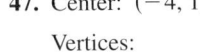

53. Center: $(2, 1)$

Vertices: $(-8, 1), (12, 1)$

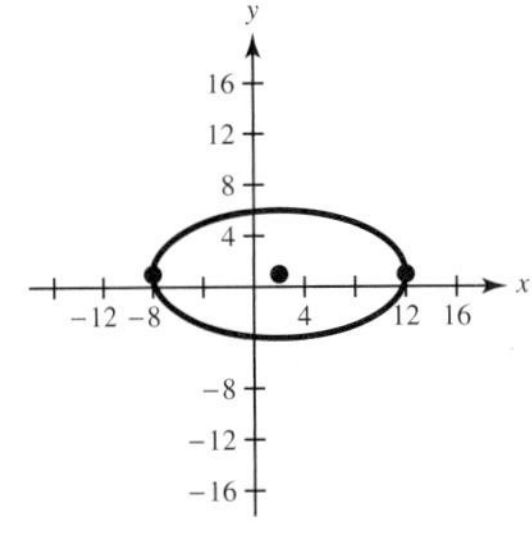

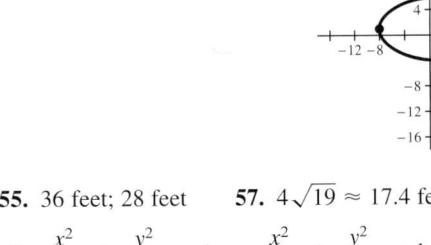

55. 36 feet; 28 feet **57.** $4\sqrt{19} \approx 17.4$ feet

59. $\dfrac{x^2}{615^2} + \dfrac{y^2}{290^2} = 1$ or $\dfrac{x^2}{290^2} + \dfrac{y^2}{615^2} = 1$

61. (a) Every point on the ellipse represents the maximum distance (800 miles) that the plane can safely fly with enough fuel to get from airport A to airport B.

(b) Airport A: $(-250, 0)$; Airport B: $(250, 0)$

(c) 800 miles; Vertices: $(\pm 400, 0)$

(d) $\dfrac{x^2}{400^2} + \dfrac{y^2}{\left(50\sqrt{39}\right)^2} = 1$

(e) $20{,}000\sqrt{39}\,\pi \approx 392{,}385$ square miles

63. A circle is an ellipse in which the major axis and the minor axis have the same length. Both circles and ellipses have foci; however, a circle has a single focus located at the center, and an ellipse has two foci that lie on the major axis.

65. The sum of the distances between each point on the ellipse and the two foci is a constant.

67. The graph of an ellipse written in the standard form

$$\frac{(x-h)^2}{a^2} + \frac{(y-k)^2}{b^2} = 1$$

intersects the y-axis if $|h| > a$ and intersects the x-axis if $|k| > b$. Similarly, the graph of

$$\frac{(x-h)^2}{b^2} + \frac{(y-k)^2}{a^2} = 1$$

intersects the y-axis if $|h| > b$ and intersects the x-axis if $|k| > a$.

69. (a) $f(-2) = 9$

(b) $f(2) = \frac{1}{9}$

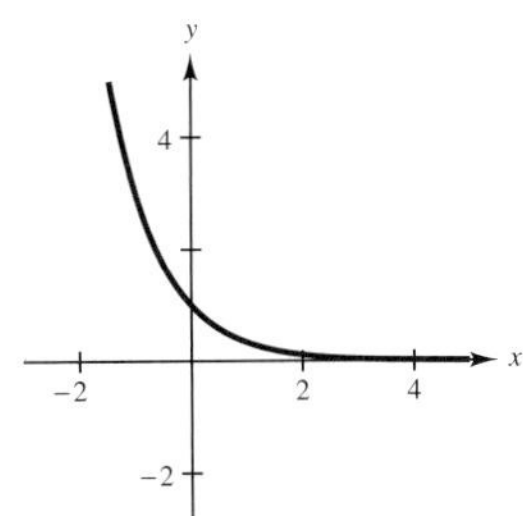

71. (a) $g(-1) \approx 3.639$

(b) $g(2) \approx 16.310$

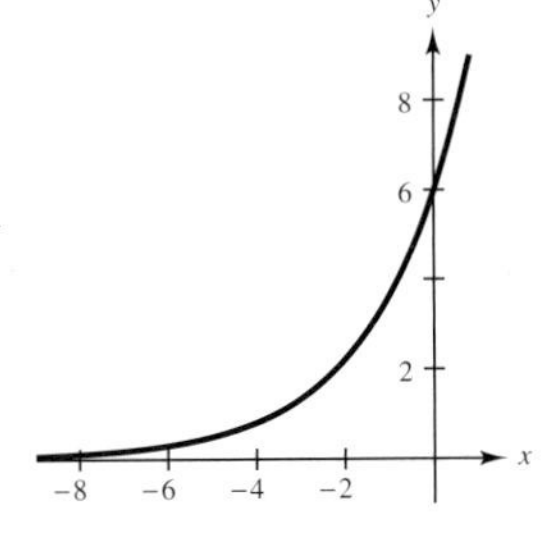

73. (a) $h(4) = 1$

(b) $h(64) = 2$

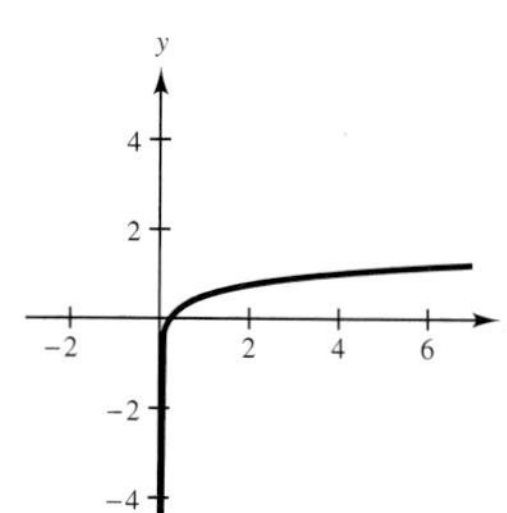

75. (a) $f(3)$ does not exist.

(b) $f(35) = \frac{5}{2}$

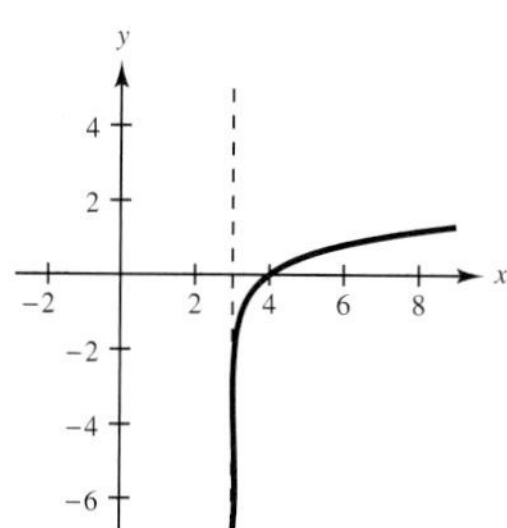

Mid-Chapter Quiz *(page 784)*

1. $x^2 + y^2 = 25$ **2.** $(y - 1)^2 = 8(x + 2)$

3. $\dfrac{(x + 2)^2}{16} + \dfrac{(y + 1)^2}{4} = 1$

4. $(x - 3)^2 + (y + 5)^2 = 25$

5. $(x - 2)^2 = -8(y - 3)$ **6.** $\dfrac{x^2}{36} + \dfrac{y^2}{100} = 1$

7. $x^2 + (y + 3)^2 = 16$; Center: $(0, -3)$; $r = 4$

8. $(x + 1)^2 + (y - 2)^2 = 1$; Center: $(-1, 2)$; $r = 1$

9. $(y - 3)^2 = x + 16$; Vertex: $(-16, 3)$; Focus: $\left(-\frac{63}{4}, 3\right)$

10. $(x - 4)^2 = -(y - 4)$; Vertex: $(4, 4)$; Focus: $\left(4, \frac{15}{4}\right)$

11. $\dfrac{(x - 2)^2}{9} + \dfrac{y^2}{36} = 1$

Center: $(2, 0)$

Vertices: $(2, -6)$, $(2, 6)$

12. $\dfrac{(x - 6)^2}{9} + \dfrac{(y + 2)^2}{4} = 1$

Center: $(6, -2)$

Vertices: $(3, -2)$, $(9, -2)$

13.

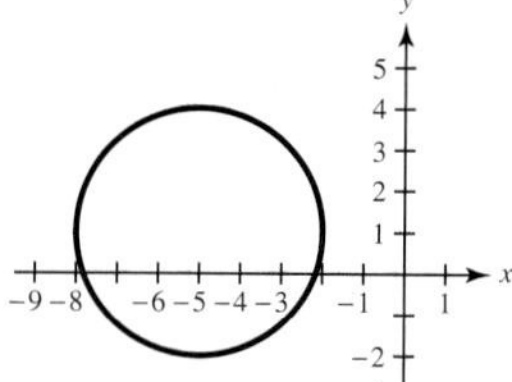

14.

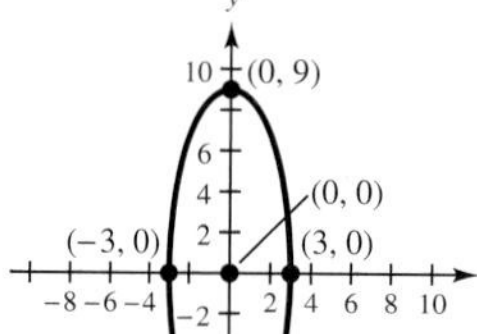

15.

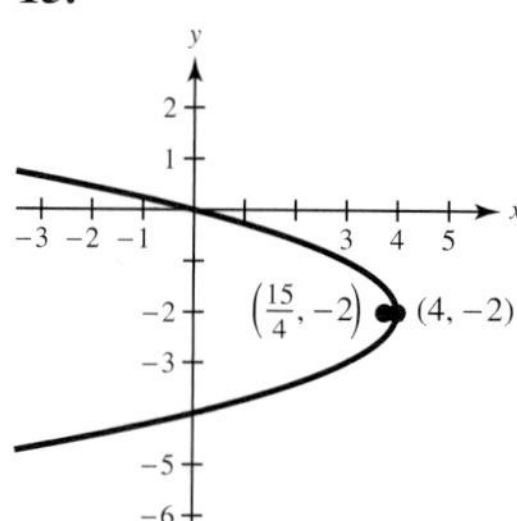

16.

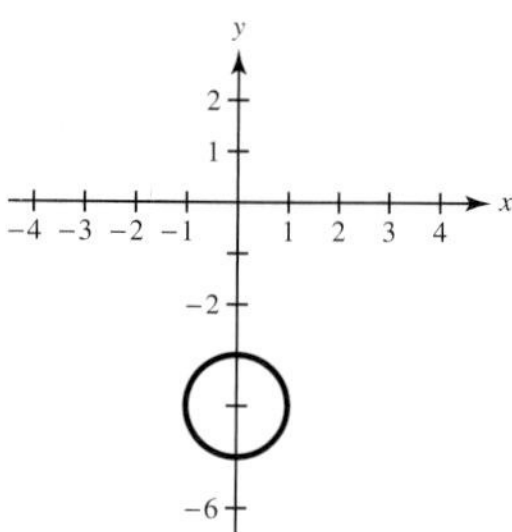

17.

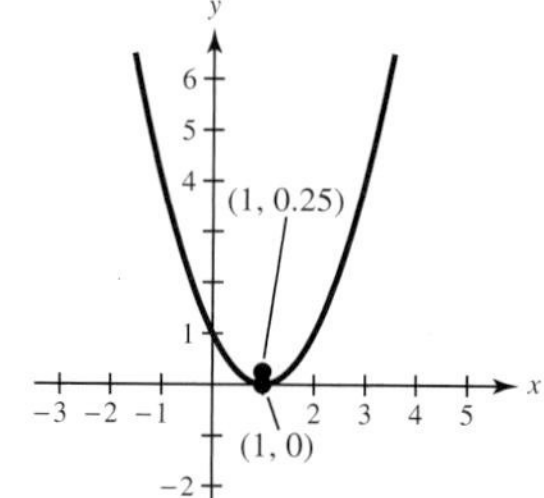

18.

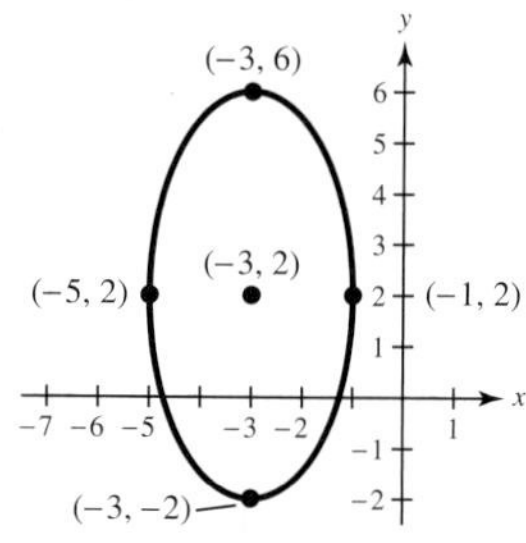

Section 12.3 *(page 790)*

1. c **2.** e **3.** a **4.** f **5.** b **6.** d

7.

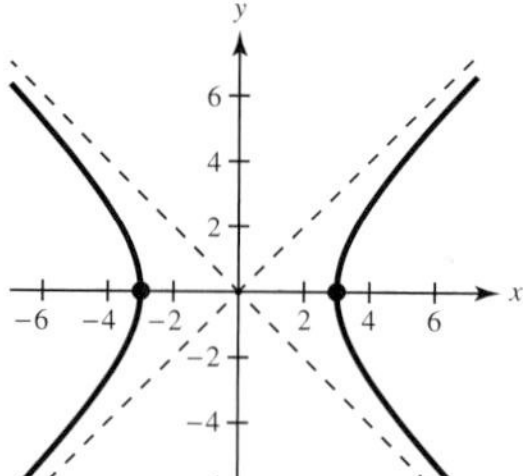

Vertices: $(\pm 3, 0)$

Asymptotes: $y = \pm x$

9.

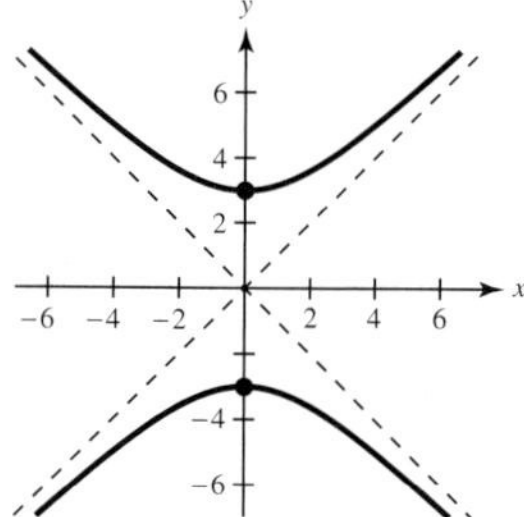

Vertices: $(0, \pm 3)$

Asymptotes: $y = \pm x$

11.

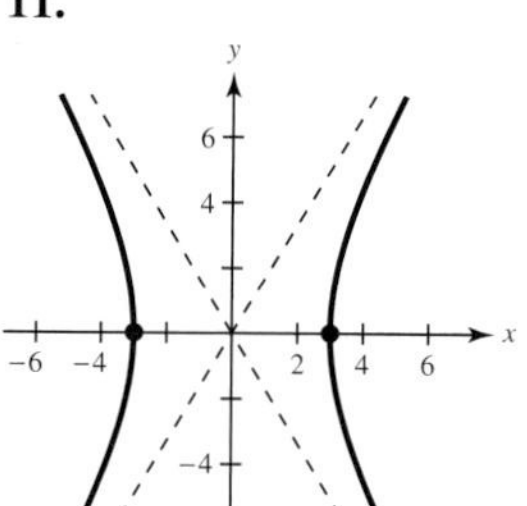

Vertices: $(\pm 3, 0)$

Asymptotes: $y = \pm \frac{5}{3}x$

13.

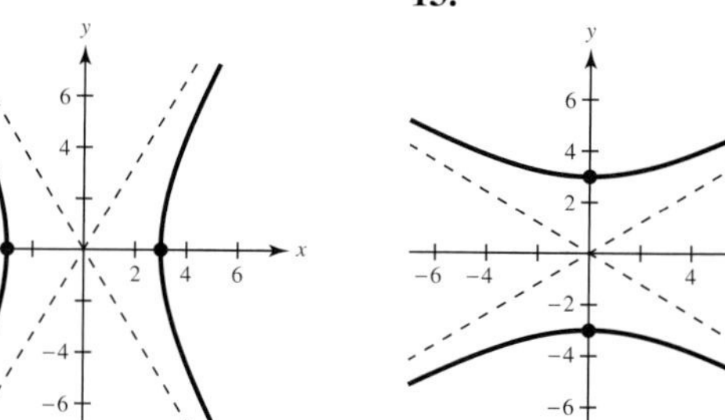

Vertices: $(0, \pm 3)$

Asymptotes: $y = \pm \frac{3}{5}x$

15.

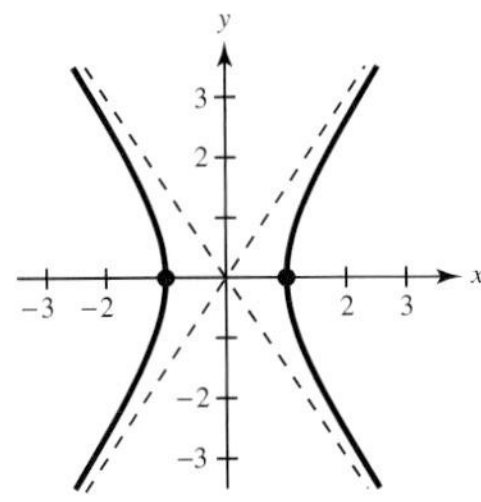

Vertices: $(\pm 1, 0)$

Asymptotes: $y = \pm\frac{3}{2}x$

17.

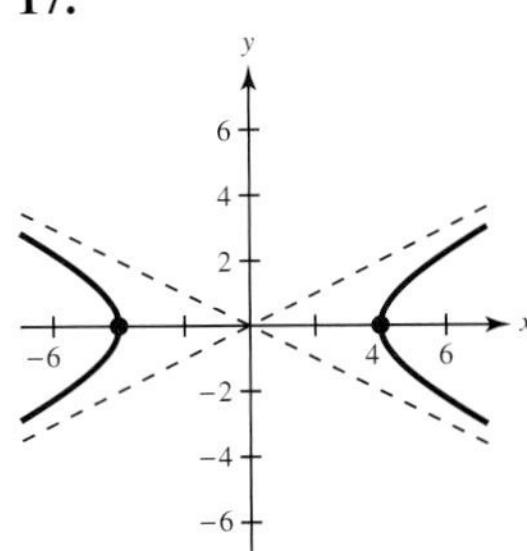

Vertices: $(\pm 4, 0)$

Asymptotes: $y = \pm\frac{1}{2}x$

19. $\dfrac{x^2}{16} - \dfrac{y^2}{64} = 1$ **21.** $\dfrac{y^2}{16} - \dfrac{x^2}{64} = 1$

23. $\dfrac{x^2}{81} - \dfrac{y^2}{36} = 1$ **25.** $\dfrac{y^2}{1} - \dfrac{x^2}{1/4} = 1$

27.

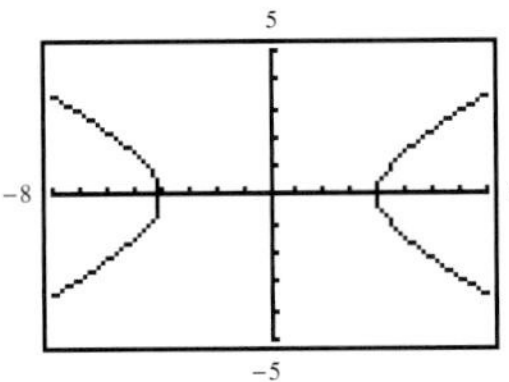

29.

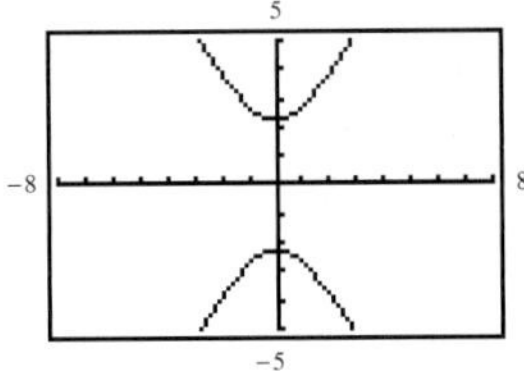

31. Center: $(3, -4)$

Vertices:

$(3, 1), (3, -9)$

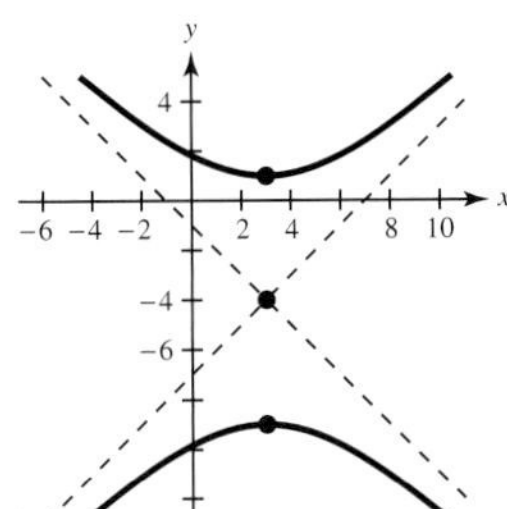

33. Center: $(1, -2)$

Vertices:

$(-1, -2), (3, -2)$

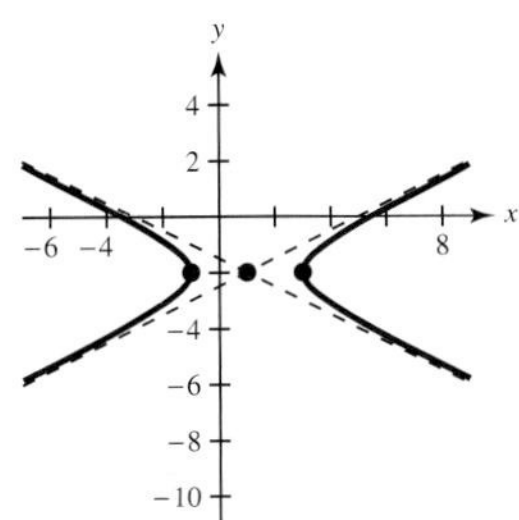

35. Center: $(2, -3)$

Vertices:

$(1, -3), (3, -3)$

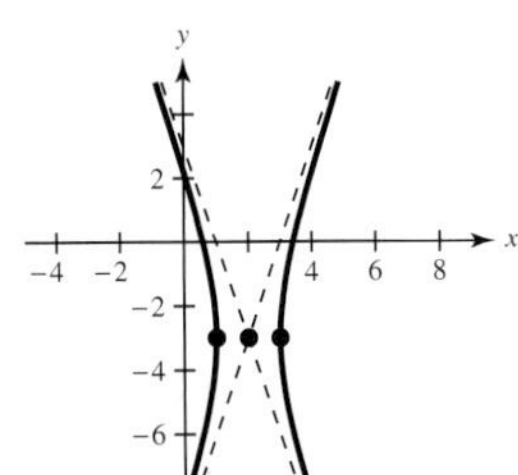

37. Center: $(-3, 2)$

Vertices:

$(-4, 2), (-2, 2)$

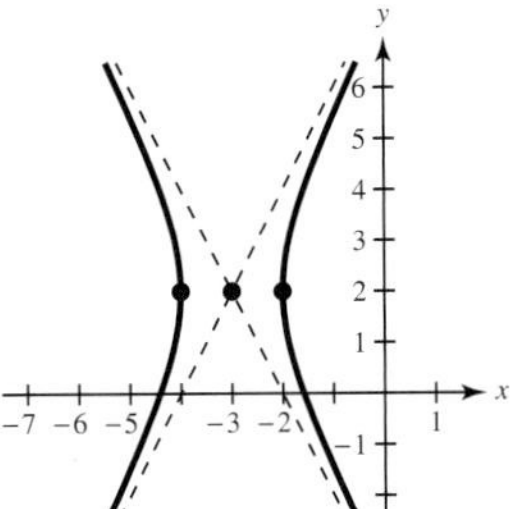

39. $\dfrac{y^2}{9} - \dfrac{x^2}{9/4} = 1$ **41.** $\dfrac{(x-3)^2}{4} - \dfrac{(y-2)^2}{16/5} = 1$

43. Ellipse **45.** Hyperbola **47.** Hyperbola

49. 110.3 **51.** (a) Left half (b) Top half

53. One. The difference in the distances between the given point and the given foci is constant, so only one branch of one hyperbola can pass through the point.

55. Consistent **57.** $\left(\frac{5}{2}, \frac{1}{2}\right)$

Section 12.4 *(page 800)*

1.

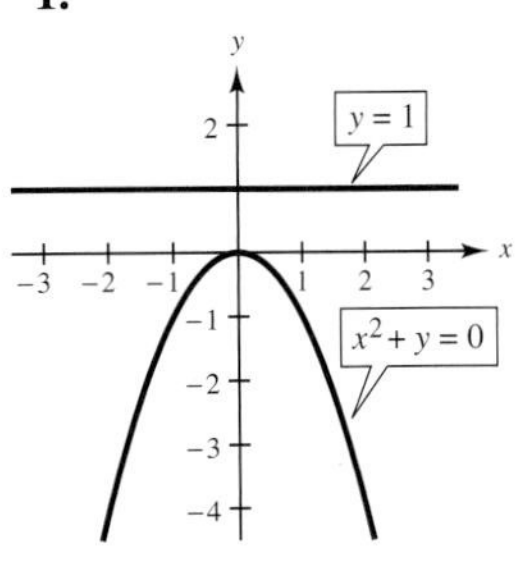

No real solution

3.

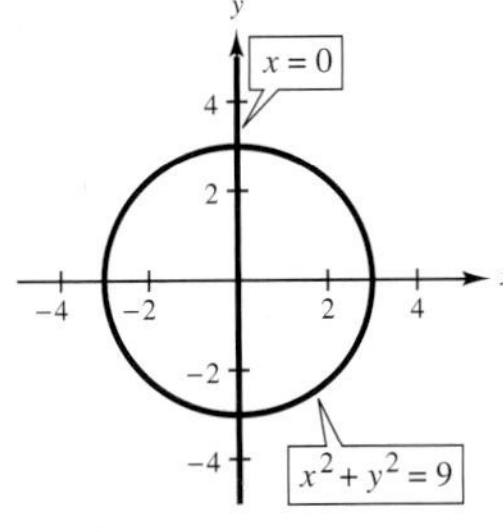

$(0, 3), (0, -3)$

5.

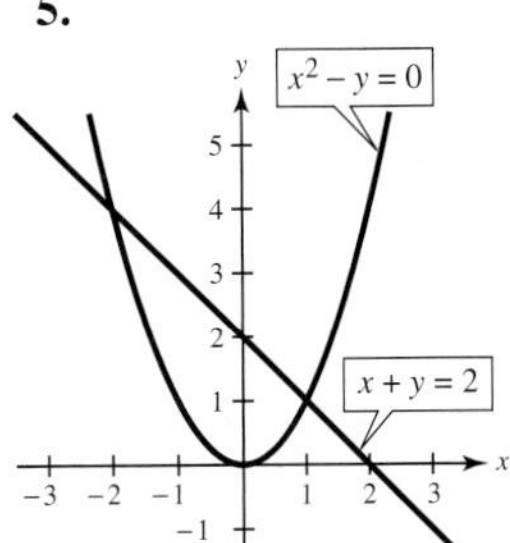

$(-2, 4), (1, 1)$

7.

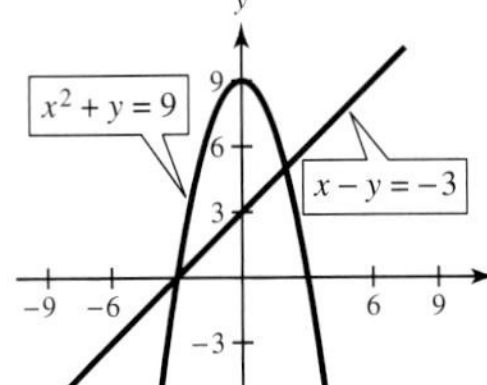

$(2, 5), (-3, 0)$

9.

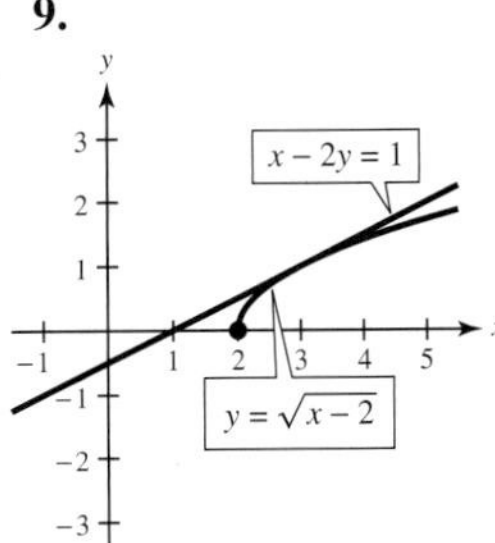

$(3, 1)$

11.

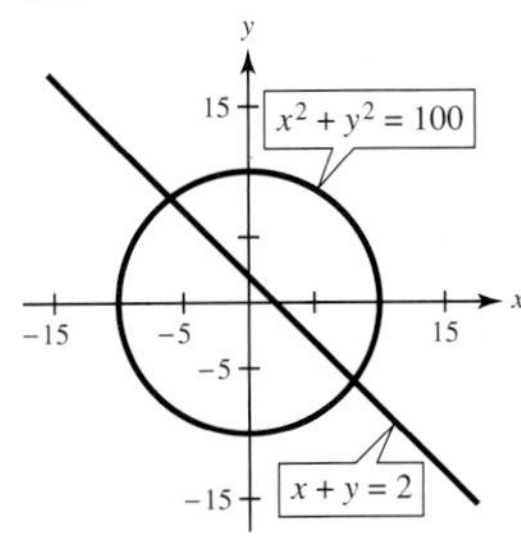

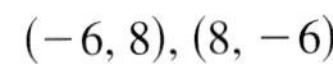

$(-6, 8), (8, -6)$

13.

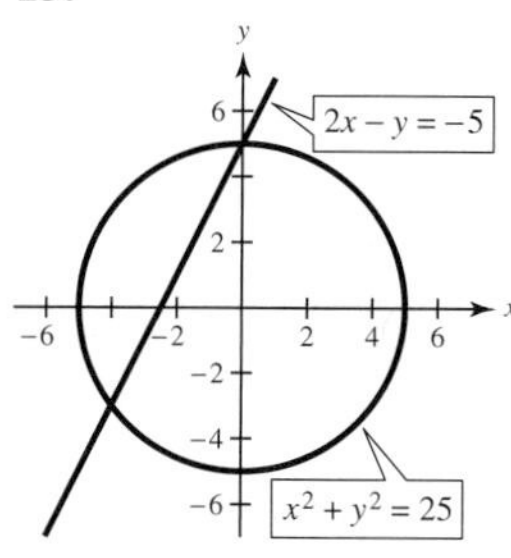

$(0, 5), (-4, -3)$

15.

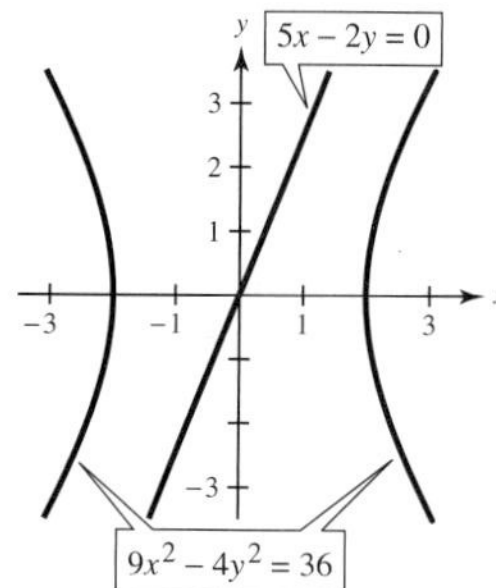

No real solution

17.

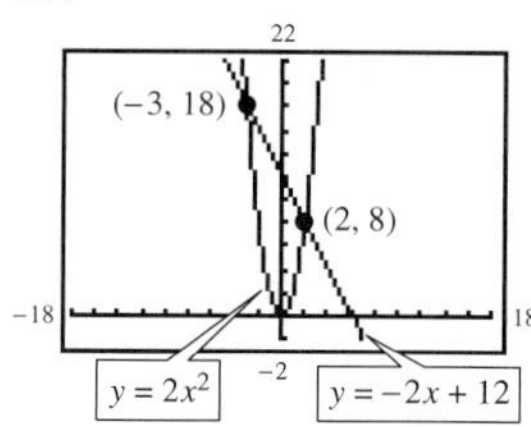

$(-3, 18), (2, 8)$

19.

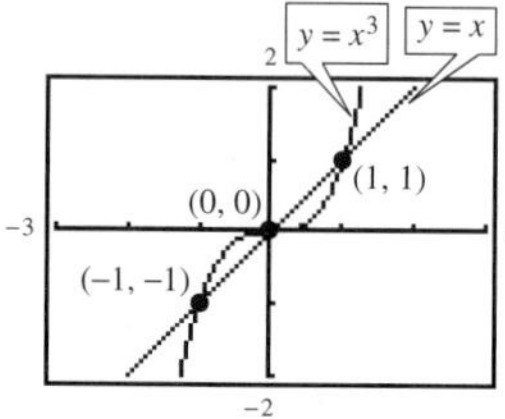

$(0, 0), (1, 1), (-1, -1)$

21.

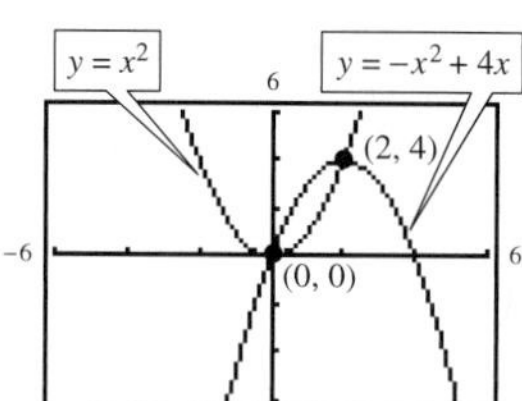

$(0, 0), (2, 4)$

23.

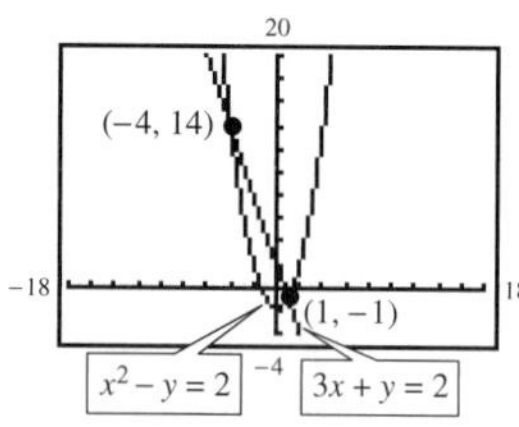

$(-4, 14), (1, -1)$

25.

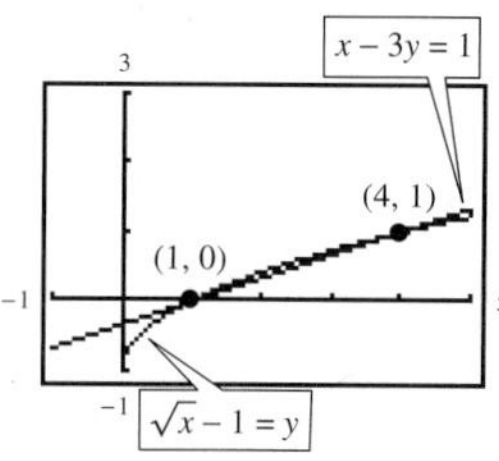

$(1, 0), (4, 1)$

27.

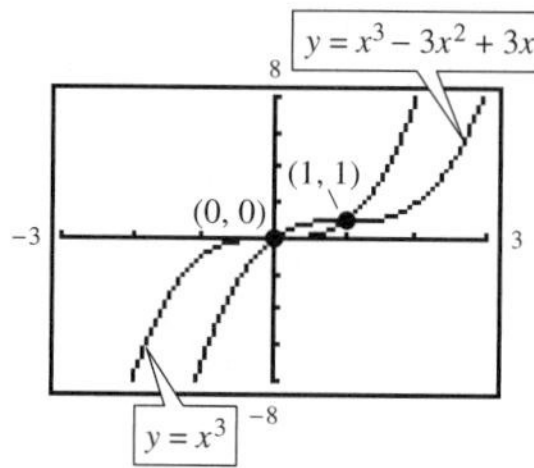

$(0, 0), (1, 1)$

29.

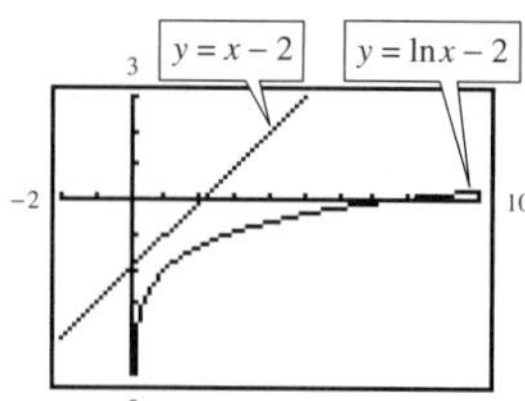

No real solution

31. $(1, 2), (2, 8)$ **33.** $(0, 5), (2, 1)$ **35.** $(0, 2), (2, 0)$

37. $(0, 5)$ **39.** $(0, 8), \left(-\frac{24}{5}, -\frac{32}{5}\right)$

41. $\left(-\frac{17}{2}, -6\right), \left(-\frac{7}{2}, 4\right)$ **43.** $\left(-\frac{9}{5}, \frac{12}{5}\right), (3, 0)$

45. $(-14, -20), (2, -4)$ **47.** $(-4, 11), \left(\frac{5}{2}, \frac{5}{4}\right)$

49. $(0, 2), (3, 1)$ **51.** No real solution

53. $(\pm\sqrt{5}, 2), (0, -3)$ **55.** $(0, 4), (3, 0)$

57. No real solution **59.** $(\pm\sqrt{3}, -1)$

61. $(2, \pm 2\sqrt{3}), (-1, \pm 3)$ **63.** $(\pm 2, \pm\sqrt{3})$

65. $(\pm 2, 0)$ **67.** $(\pm 3, \pm 2)$ **69.** $(\pm 3, 0)$

71. $(\pm\sqrt{6}, \pm\sqrt{3})$ **73.** No real solution

75. $(\pm\sqrt{5}, \pm 2)$ **77.** $\left(\pm\frac{2\sqrt{5}}{5}, \pm\frac{2\sqrt{5}}{5}\right)$

79. $(\pm\sqrt{3}, \pm\sqrt{13})$ **81.** 40 feet $\times$ 75 feet

83. 200 feet $\times$ 210 feet **85.** $(3.633, 2.733)$

87. Between points $\left(-\frac{3}{5}, -\frac{4}{5}\right)$ and $\left(\frac{4}{5}, -\frac{3}{5}\right)$

89. Solve one of the equations for one variable in terms of the other. Substitute that expression into the other equation and solve. Back-substitute the solution into the first equation to find the value of the other variable. Check the solution to see that it satisfies both of the original equations.

91. Two. The line can intersect a branch of the hyperbola at most twice, and it can intersect only one point on each branch at the same time.

93. -5 **95.** 9 **97.** 5 **99.** 4.564

101. 1.023 **103.** -0.632

Review Exercises *(page 806)*

1. Ellipse **3.** Circle **5.** Hyperbola

7. $x^2 + y^2 = 36$

9. Center: $(0, 0)$; $r = 8$

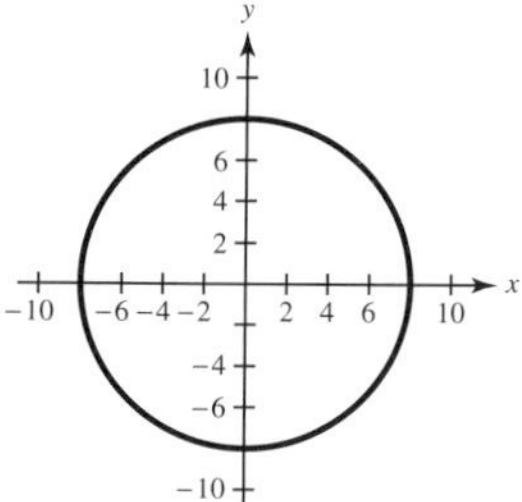

11. $(x - 2)^2 + (y - 6)^2 = 9$

13. Center: $(-3, -4)$; $r = 2$

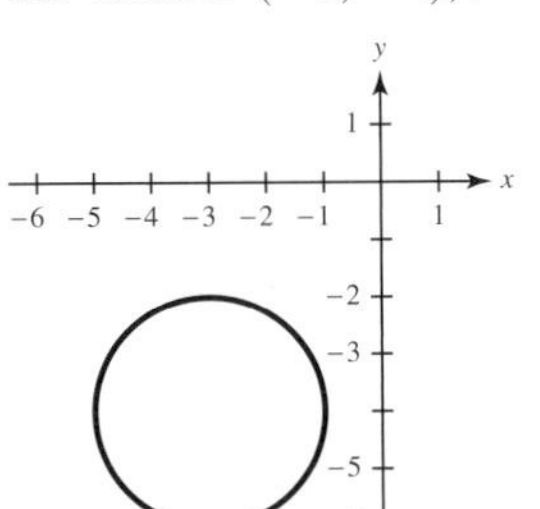

15. $y^2 = -8x$

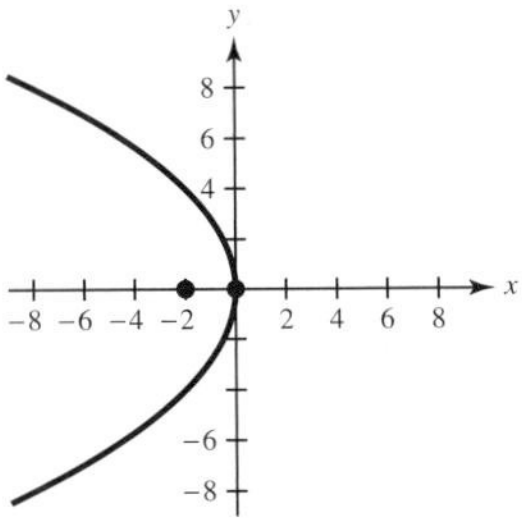

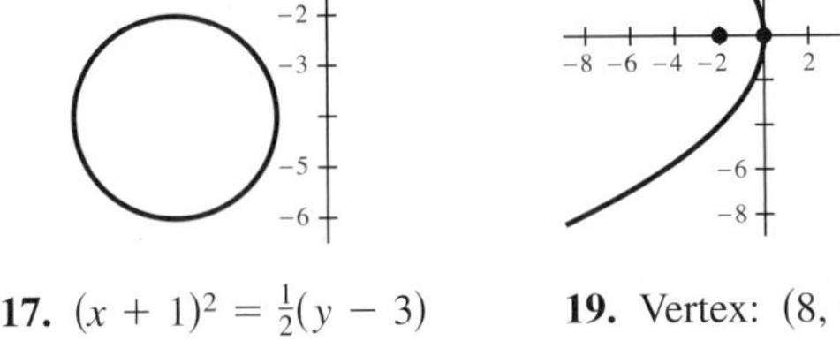

17. $(x + 1)^2 = \frac{1}{2}(y - 3)$

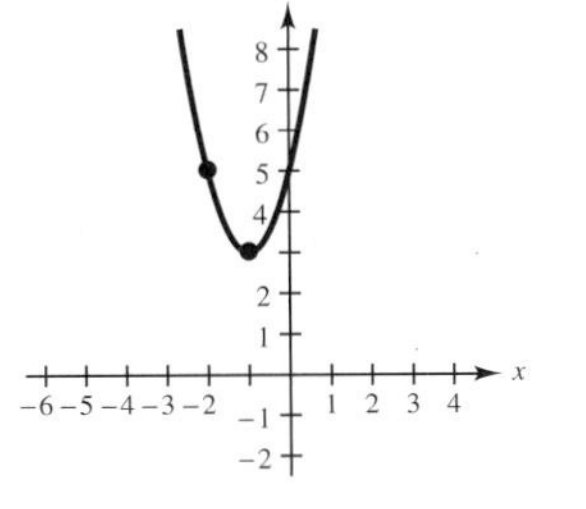

19. Vertex: $(8, -25)$
Focus: $\left(8, -\frac{49}{2}\right)$

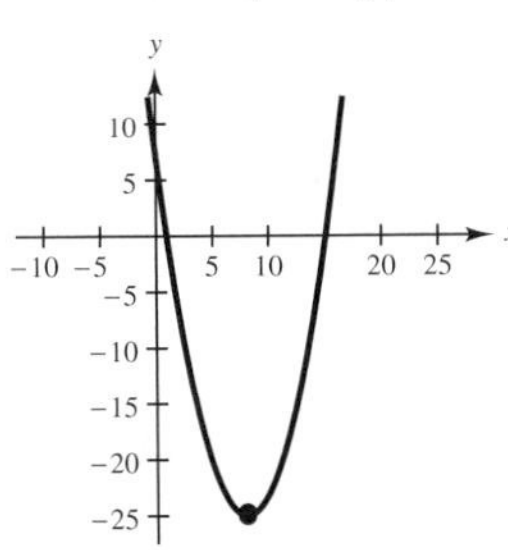

21. $\dfrac{x^2}{4} + \dfrac{y^2}{25} = 1$ **23.** $\dfrac{x^2}{4} + \dfrac{y^2}{9} = 1$

25.

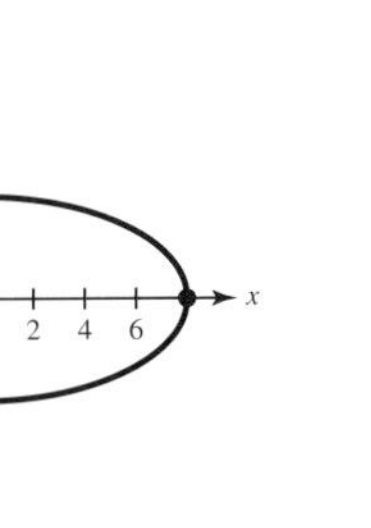

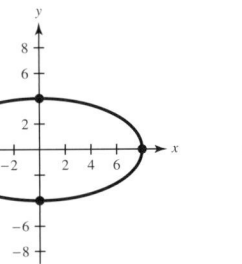

Vertices: $(\pm 8, 0)$
Co-vertices: $(0, \pm 4)$

27.

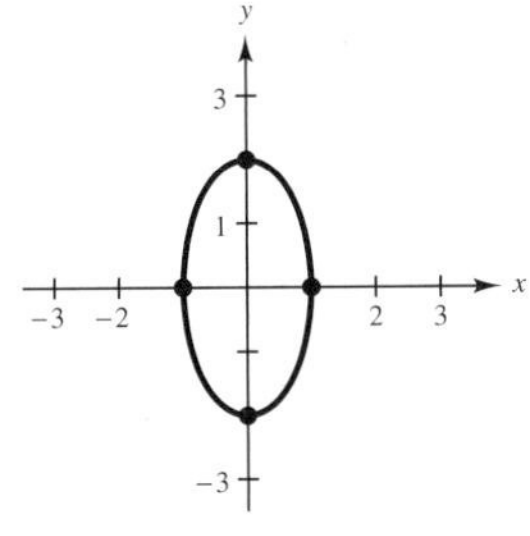

Vertices: $(0, \pm 2)$
Co-vertices: $(\pm 1, 0)$

29. $\dfrac{(x - 3)^2}{25} + \dfrac{(y - 4)^2}{16} = 1$ **31.** $\dfrac{x^2}{9} + \dfrac{(y - 4)^2}{16} = 1$

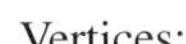

33. Center: $(-1, 2)$
Vertices:
$(-1, -4)$, $(-1, 8)$

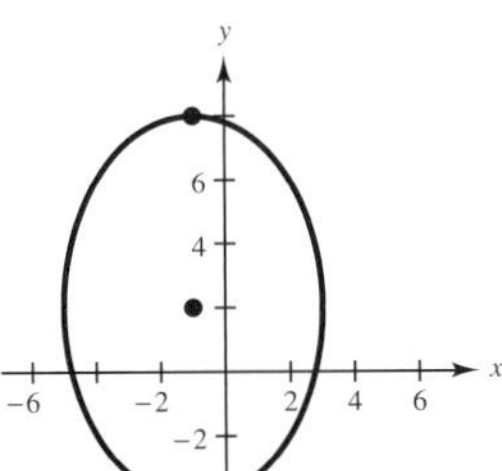

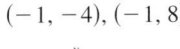

35. Center: $(0, -3)$
Vertices:
$(0, -7)$, $(0, 1)$

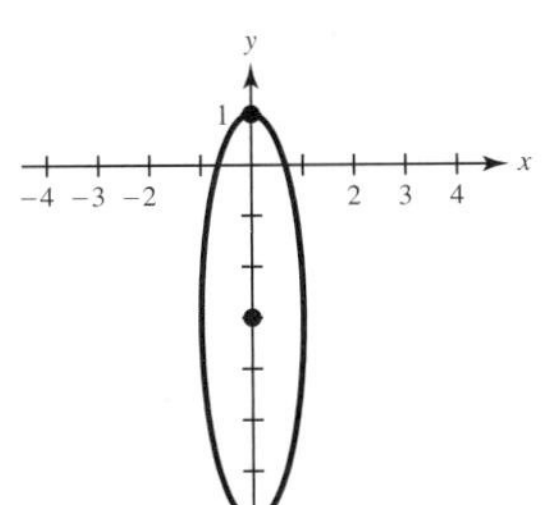

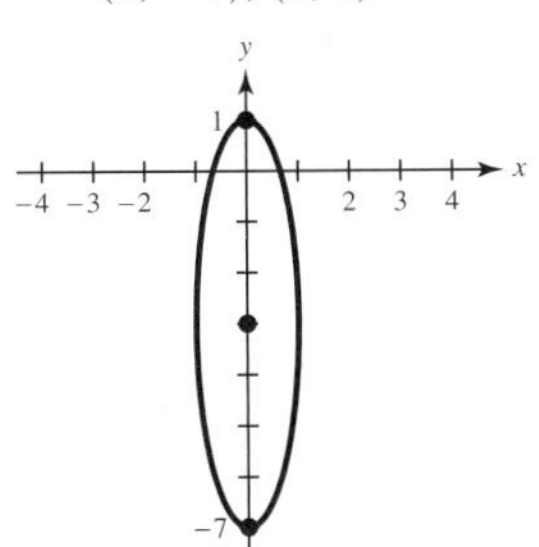

37.

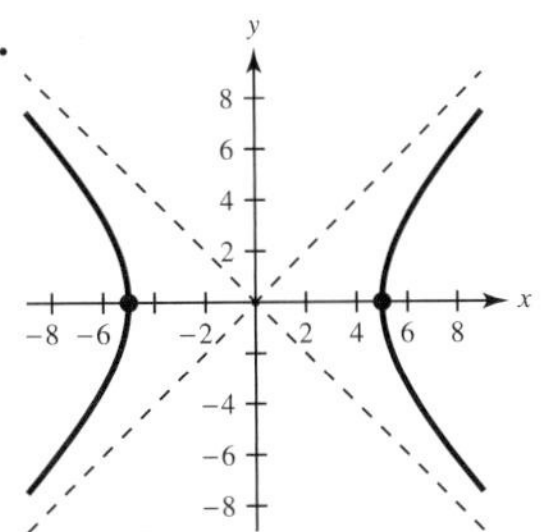

Vertices: $(\pm 5, 0)$
Asymptotes: $y = \pm x$

39.

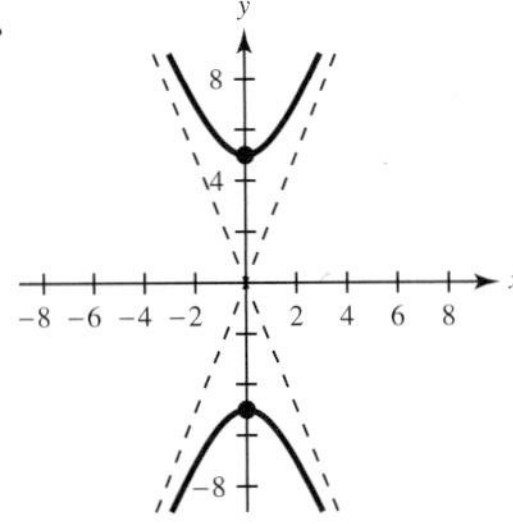

Vertices: $(0, \pm 5)$
Asymptotes: $y = \pm \frac{5}{2}x$

41. $\dfrac{x^2}{4} - \dfrac{y^2}{9} = 1$ **43.** $\dfrac{y^2}{64} - \dfrac{x^2}{100} = 1$

45. Center: $(3, -1)$
Vertices: $(0, -1)$, $(6, -1)$

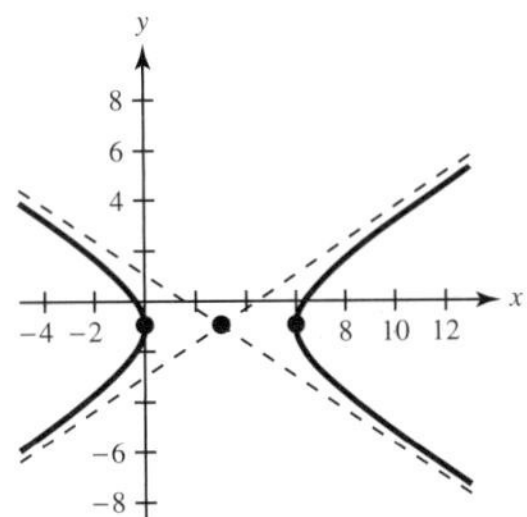

47. Center: $(4, -3)$
Vertices: $(4, -1)$, $(4, -5)$

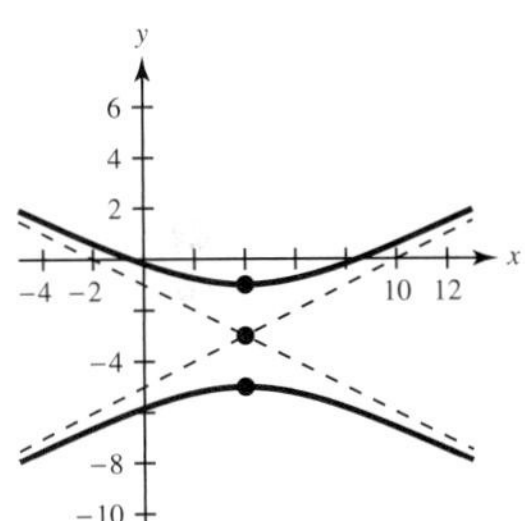

49. $\dfrac{(x + 4)^2}{4} - \dfrac{(y - 6)^2}{12} = 1$

51.

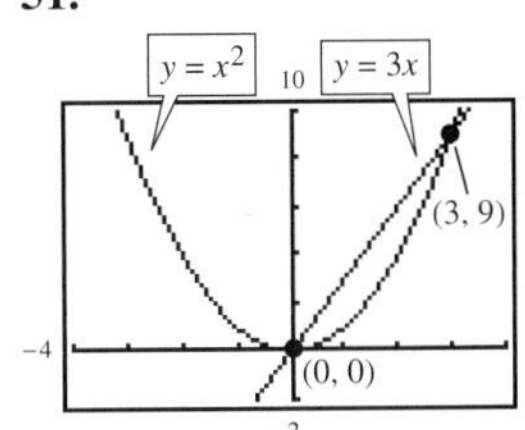

$(0, 0)$, $(3, 9)$

53.

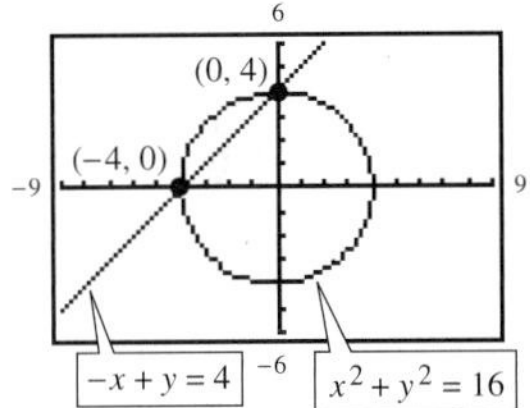

$(-4, 0)$, $(0, 4)$

55. $(-1, 5)$, $(-2, 20)$ **57.** $(-1, 0)$, $(0, -1)$

59. $(\pm 2, \pm 3)$ **61.** 6 centimeters $\times$ 8 centimeters

63. Piece 1: 38.48 inches; Piece 2: 61.52 inches

Chapter Test *(page 809)*

1. $(x + 2)^2 + (y + 3)^2 = 16$

2. $(x - 1)^2 + (y - 3)^2 = 9$

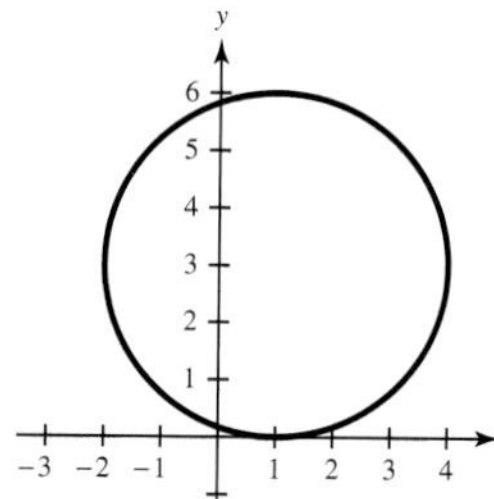

3. $(x + 2)^2 + (y - 3)^2 = 9$

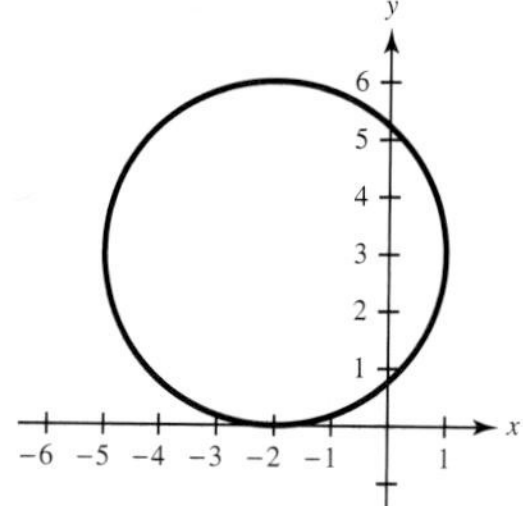

4. Vertex: $(4, 2)$; Focus: $\left(\frac{47}{12}, 2\right)$

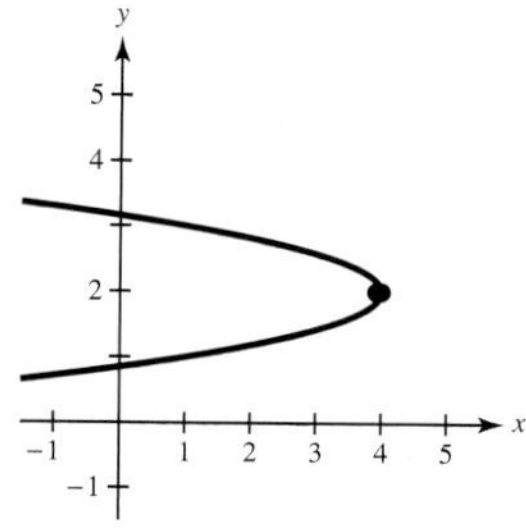

5. $(x - 7)^2 = 8(y + 2)$ **6.** $\dfrac{(x - 2)^2}{25} + \dfrac{y^2}{9} = 1$

7. Center: $(0, 0)$

Vertices: $(0, \pm 4)$

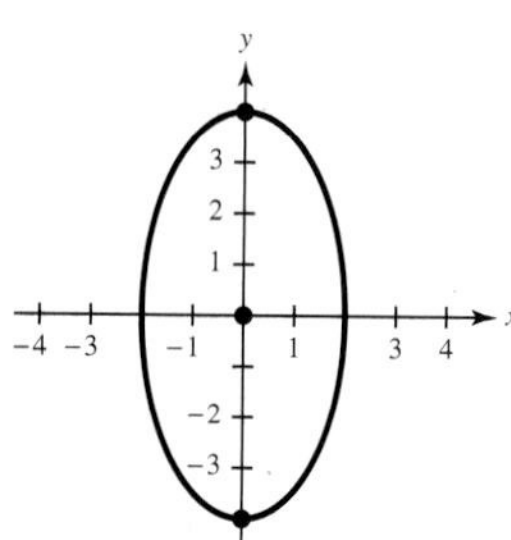

8. Center: $(1, 3)$

Vertices: $(1, -2), (1, 8)$

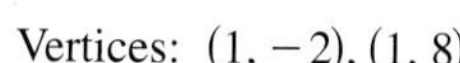

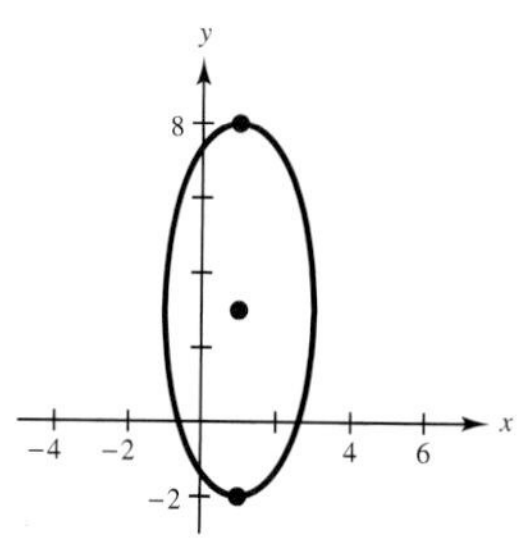

9. $\dfrac{x^2}{9} - \dfrac{y^2}{4} = 1$ **10.** $\dfrac{y^2}{25} - \dfrac{x^2}{4} = 1$

11. Center: $(3, 0)$

Vertices: $(1, 0), (5, 0)$

12. Center: $(4, -2)$

Vertices: $(4, -7), (4, 3)$

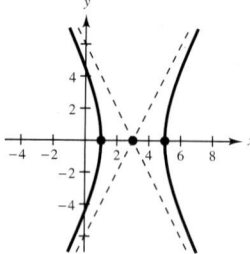

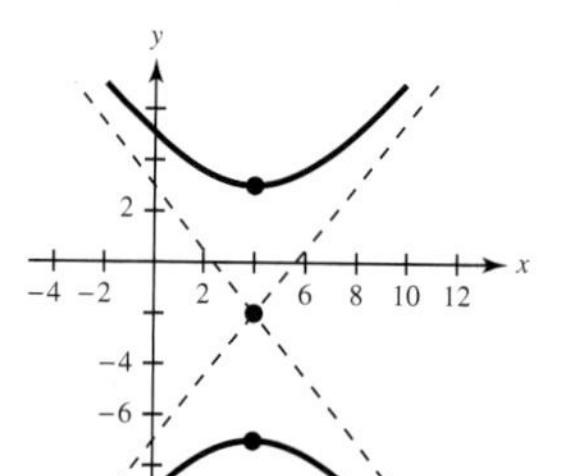

13. $(0, 3), (4, 0)$ **14.** $(\pm 4, 0)$

15. $(\sqrt{6}, 2), (\sqrt{6}, -2), (-\sqrt{6}, 2), (-\sqrt{6}, -2)$

16. $x^2 + y^2 = 5000^2$ **17.** 16 inches × 12 inches

Cumulative Test: Chapters 10–12 *(page 810)*

1. $-\frac{3}{4}, 3$ **2.** $-3, 13$ **3.** $5 \pm 5\sqrt{2}$

4. $-1 \pm \dfrac{\sqrt{3}}{3}$ **5.** $\pm\sqrt{3}, \pm\sqrt{5}$

6. $\left[-3, \dfrac{1}{3}\right]$

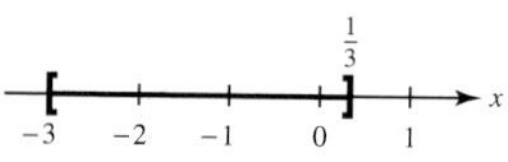

7. $\left(-\frac{4}{3}, \frac{1}{2}\right)$

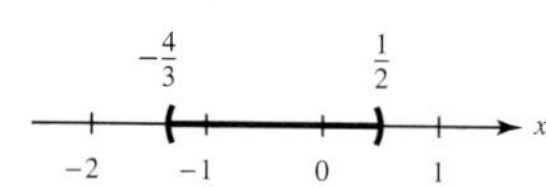

8. $x^2 - 4x - 12 = 0$

9. (a) $(f \circ g)(x) = 50x^2 - 20x - 1$; Domain: $(-\infty, \infty)$

(b) $(g \circ f)(x) = 10x^2 - 16$; Domain: $(-\infty, \infty)$

10. $f^{-1}(x) = -\frac{4}{3}x + \frac{5}{3}$

11. $f(1) = \frac{15}{2}; f(0.5) \approx 7.707; f(3) = \frac{57}{8}$

12. Horizontal asymptote: $y = 0$

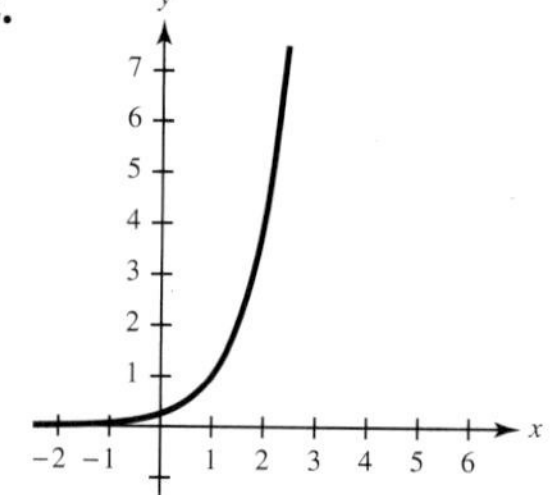

13. The graphs are reflections of each other in the line $y = x$.

14. Vertical asymptote: $x = 1$

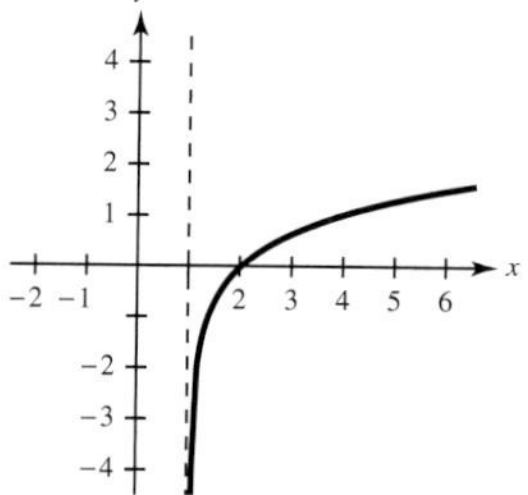

15. -2 **16.** $\log_2 \dfrac{x^3y^3}{z}$ **17.** $\dfrac{1}{2}\log_{10}(x + 1) - 4\log_{10} x$

18. 3 **19.** 12.182 **20.** 18.013 **21.** 0.867

22. \$34.38 **23.** 8.33% **24.** 19.8 years

25. $(x - 3)^2 + (y + 7)^2 = 64$

26. Vertex: $(5, -45)$

Focus: $\left(5, -\frac{359}{8}\right)$

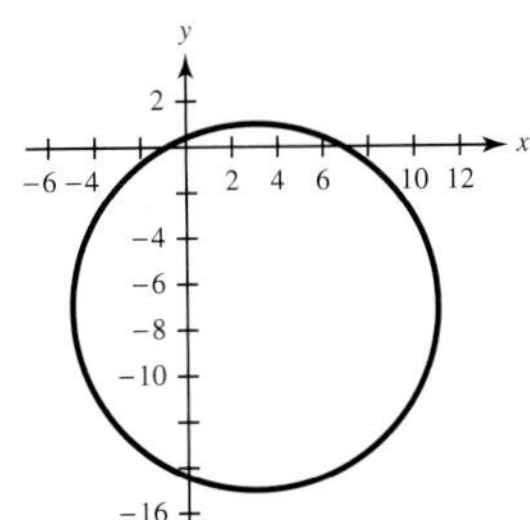

27. $\dfrac{(x + 3)^2}{4} + \dfrac{(y - 2)^2}{25} = 1$

28. Center: $(0, 0)$

Vertices: $(0, \pm 2)$

29. $\dfrac{y^2}{9} - x^2 = 1$

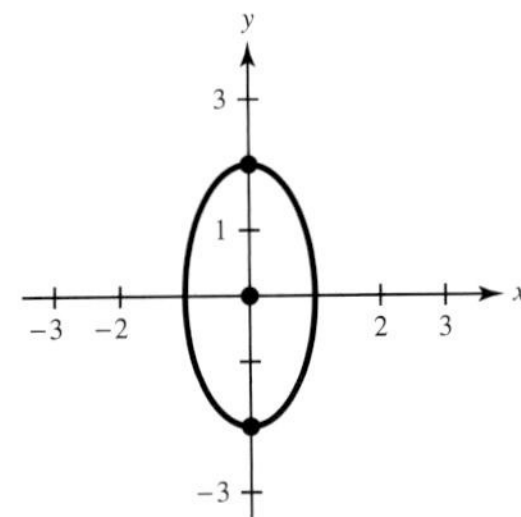

30. Center: $(0, 1)$

Vertices: $(\pm 12, 1)$

31. $(1, -1), (3, 5)$

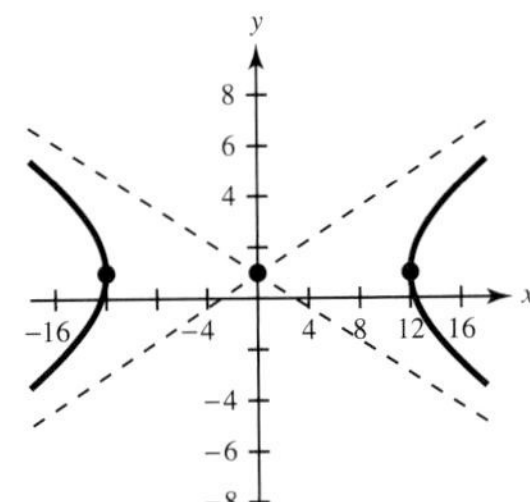

32. $(-4, \pm 1), (-1, \pm 2)$

33. 8 feet $\times$ 4 feet

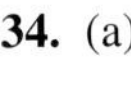

34. (a)

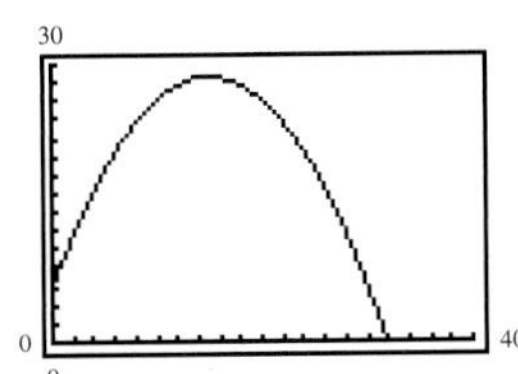

(b) Highest point: 28.5 feet; Range: $[0, 28.5]$

CHAPTER 13

Section 13.1 *(page 820)*

1. 2, 4, 6, 8, 10 **3.** $\frac{1}{4}, \frac{1}{16}, \frac{1}{64}, \frac{1}{256}, \frac{1}{1024}$

5. $-2, 4, -6, 8, -10$ **7.** $\frac{1}{4}, -\frac{1}{8}, \frac{1}{16}, -\frac{1}{32}, \frac{1}{64}$

9. 3, 8, 13, 18, 23 **11.** $1, \frac{4}{5}, \frac{2}{3}, \frac{4}{7}, \frac{1}{2}$ **13.** $\frac{3}{4}, \frac{2}{3}, \frac{9}{14}, \frac{12}{19}, \frac{5}{8}$

15. $-1, \frac{1}{4}, -\frac{1}{9}, \frac{1}{16}, -\frac{1}{25}$ **17.** $\frac{9}{4}, \frac{33}{16}, \frac{129}{64}, \frac{513}{256}, \frac{2049}{1024}$

19. 2, 3, 4, 5, 6 **21.** $0, 3, -1, \frac{3}{4}, -\frac{1}{4}$ **23.** -72

25. $\frac{31}{2520}$ **27.** 5 **29.** $\frac{1}{132}$ **31.** 53,130

33. $\dfrac{1}{n + 1}$ **35.** $n(n + 1)$ **37.** $2n$ **39.** b

40. d **41.** c **42.** a

43.

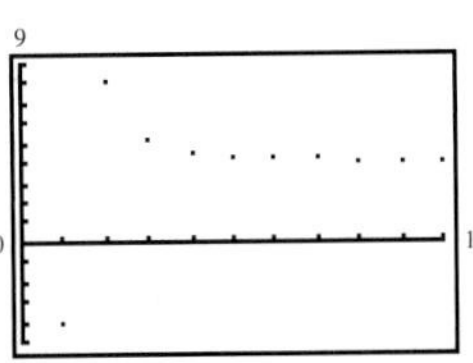

45.

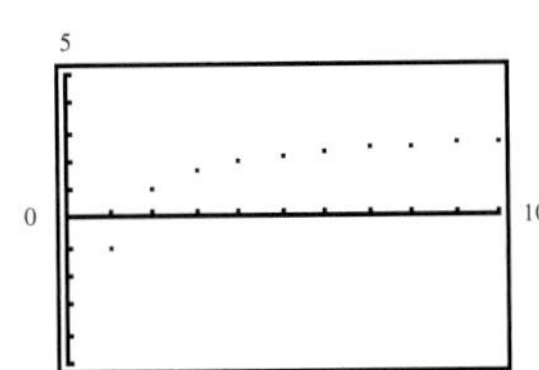

47.

49. $a_n = 2n - 1$

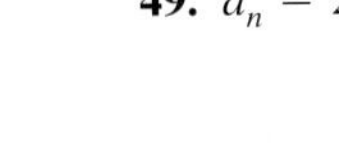

51. $a_n = 4n - 2$ **53.** $a_n = n^2 - 1$

55. $a_n = (-1)^{n+1}2n$ **57.** $a_n = \dfrac{n + 1}{n + 2}$

59. $a_n = \left(-\dfrac{1}{5}\right)^n$ **61.** $a_n = \dfrac{1}{2^{n-1}}$ **63.** $a_n = 1 + \dfrac{1}{n}$

65. $a_n = \dfrac{(-1)^n}{(n + 1)!}$ **67.** $S_1 = 7; S_2 = 16; S_6 = 72$

69. $S_2 = \frac{3}{2}; S_3 = \frac{11}{6}; S_9 = \frac{7129}{2520}$ **71.** 90 **73.** 77

75. 100 **77.** $\frac{3019}{3600}$ **79.** $\frac{15{,}551}{2520}$ **81.** -48 **83.** $\frac{8}{9}$

85. $\frac{182}{243}$ **87.** 2040 **89.** 852 **91.** 16.25

93. 6.5793 **95.** $\sum_{k=1}^{5} k$ **97.** $\sum_{k=1}^{6} 5k$ **99.** $\sum_{k=1}^{50} \dfrac{3}{1 + k}$

101. $\sum_{k=1}^{10} \dfrac{1}{2k}$ **103.** $\sum_{k=0}^{12} \dfrac{1}{2^k}$ **105.** $\sum_{k=1}^{20} \dfrac{1}{k(k + 1)}$

107. $\sum_{k=1}^{11} \dfrac{k}{k + 1}$ **109.** $\sum_{k=1}^{20} \dfrac{2k}{k + 3}$ **111.** $\sum_{k=0}^{6} k!$

113. (a) \$535, \$572.45, \$612.52, \$655.40, \$701.28, \$750.37, \$802.89, \$859.09

(b) \$7487.23

(c)

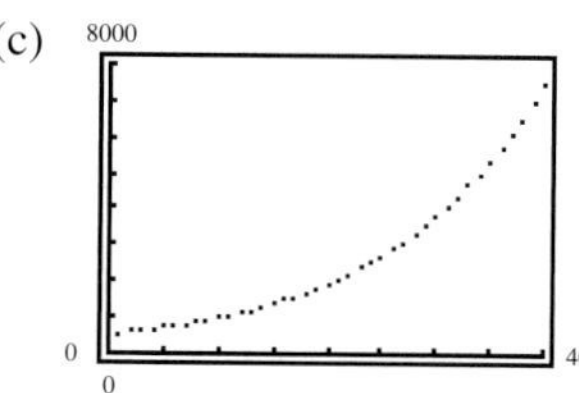

(d) Yes. Any investment earning compound interest increases at an increasing rate.

115. 36°, 60°, 77.1°, 90°, 100°, 108°

117. (a) \$13,500

(b) \$5695.31; No. The SUV loses $1 - \left(\frac{3}{4}\right)^3 = 1 - \frac{27}{64} \approx 58\%$ (a little more than half) of its value every three years.

119. A sequence is a function because there is only one value for each term of the sequence.

121. $a_n = 4n! = 4(1 \cdot 2 \cdot 3 \cdot 4 \cdot \cdots \cdot (n-1) \cdot n)$

$a_n = (4n)!$

$= 1 \cdot 2 \cdot 3 \cdot 4 \cdot \cdots \cdot 4(n-1) \cdot (4n)$

123. True

$$\sum_{k=1}^{4} 3k = 3 + 6 + 9 + 12$$

$$= 3(1 + 2 + 3 + 4) = 3\sum_{k=1}^{4} k$$

125. 9 **127.** -11

129. Center: $(0, 0)$; $r = 6$

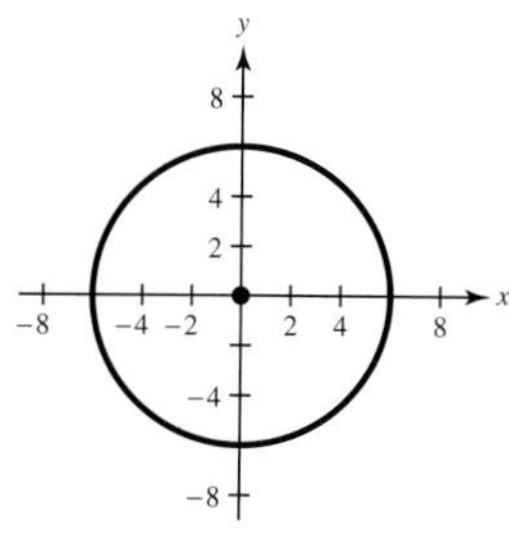

131. Center: $(-2, 0)$; $r = 4$

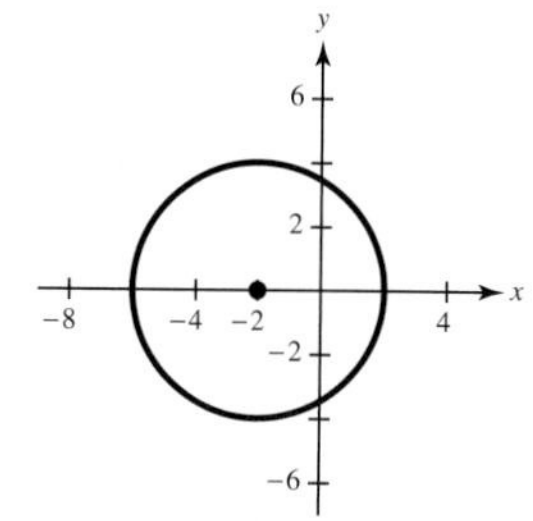

133. Vertex: $(0, 0)$

Focus: $\left(0, \frac{3}{2}\right)$

135. Vertex: $(0, -4)$

Focus: $(0, -6)$

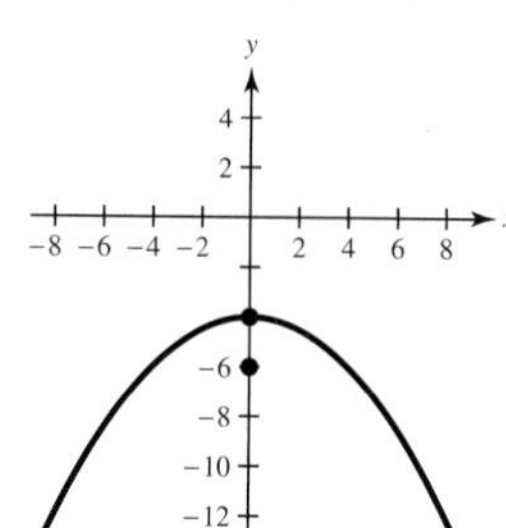

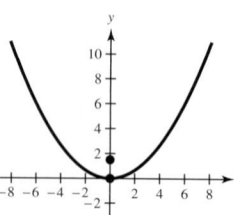

Section 13.2 *(page 829)*

1. 3 **3.** -6 **5.** -12 **7.** $\frac{2}{3}$ **9.** $-\frac{5}{4}$

11. 4 **13.** -3 **15.** $\frac{1}{2}$ **17.** Arithmetic, 2

19. Arithmetic, -2 **21.** Arithmetic, -16

23. Arithmetic, 0.8 **25.** Arithmetic, $\frac{3}{2}$

27. Not arithmetic **29.** Not arithmetic

31. Not arithmetic **33.** 7, 12, 17, 22, 27

35. 11, 15, 19, 23, 27 **37.** 20, 16, 12, 8, 4

39. 6, 11, 16, 21, 26 **41.** 22, 18, 14, 10, 6

43. 7, 10, 13, 16, 19 **45.** 6, 4, 2, 0, -2

47. $\frac{3}{2}, 4, \frac{13}{2}, 9, \frac{23}{2}$ **49.** $\frac{8}{5}, \frac{11}{5}, \frac{14}{5}, \frac{17}{5}, 4$ **51.** $4, \frac{15}{4}, \frac{7}{2}, \frac{13}{4}, 3$

53. $a_n = 3n + 1$ **55.** $a_n = \frac{3}{2}n - 1$

57. $a_n = -5n + 105$ **59.** $a_n = \frac{3}{2}n + \frac{3}{2}$

61. $a_n = \frac{5}{2}n + \frac{5}{2}$ **63.** $a_n = 4n + 4$

65. $a_n = -10n + 60$ **67.** $a_n = -\frac{1}{2}n + 11$

69. $a_n = -0.05n + 0.40$ **71.** 14, 20, 26, 32, 38

73. 23, 18, 13, 8, 3 **75.** $-16, -11, -6, -1, 4$

77. 3.4, 2.3, 1.2, 0.1, -1 **79.** 210 **81.** 1425

83. 255 **85.** 62,625 **87.** 35 **89.** 522

91. 1850 **93.** 900 **95.** 12,200 **97.** 243 **99.** 23

101. b **102.** f **103.** e **104.** a

105. c **106.** d

107.

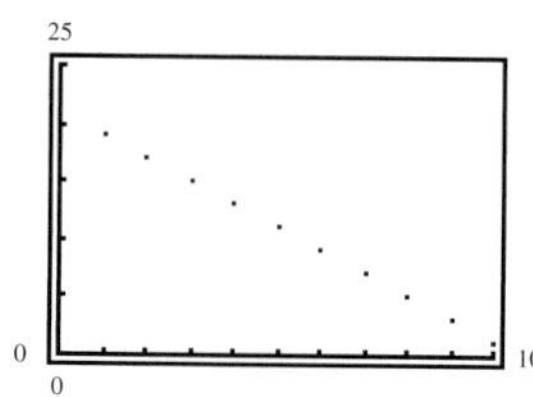

109.

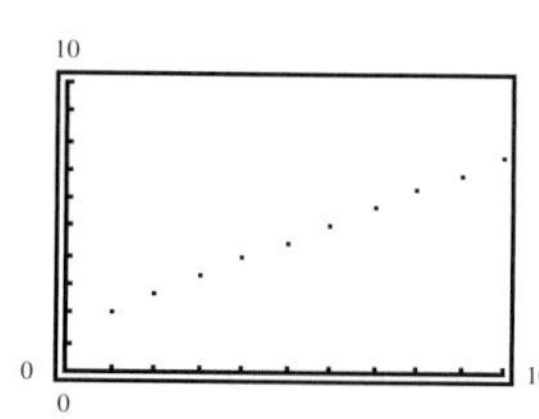

111.

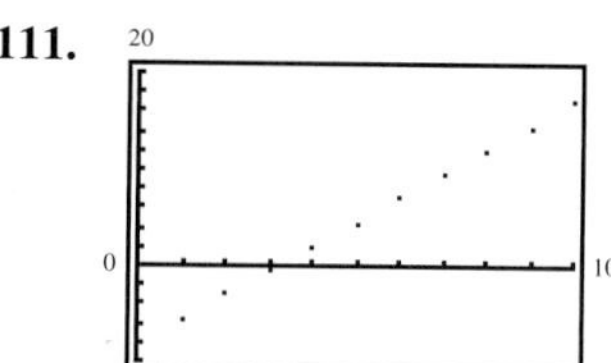

113. 9000

115. 13,120 **117.** 3011.25 **119.** 2850 **121.** 2500

123. \$369,000 **125.** \$25.43 **127.** 632 bales

129. 114 times **131.** 1024 feet **133.** 30.9 barrels

135. A recursion formula gives the relationship between the terms a_{n+1} and a_n.

137. Yes. Because a_{2n} is n terms away from a_n, add n times the difference d to a_n.

$a_{2n} = a_n + nd$

139. (a) 4; 9; 16; 25; 36

(b) No. There is no common difference between consecutive terms of the sequence.

(c) $\sum_{k=1}^{n}(2k - 1) = n^2$; 49

(d) Answers will vary.

141. Center: $(4, -5)$

Vertices: $(-1, -5)$, $(9, -5)$

143. Center: $(1, -3)$

Vertices: $(1, -6)$, $(1, 0)$

145. $\sum_{k=1}^{7}(k + 2)$ **147.** $\sum_{k=1}^{5}(3k + 9)$

Mid-Chapter Quiz *(page 834)*

1. 4, 8, 12, 16, 20 **2.** 7, 9, 11, 13, 15

3. 32, 8, 2, $\frac{1}{2}$, $\frac{1}{8}$ **4.** $-\frac{3}{5}$, 3, $-\frac{81}{7}$, $\frac{81}{2}$, -135 **5.** 100

6. 40 **7.** 87 **8.** 25 **9.** 40 **10.** 26

11. $\sum_{k=1}^{20}\frac{2}{3k}$ **12.** $\sum_{k=1}^{25}\frac{(-1)^{k+1}}{k^3}$ **13.** $\sum_{k=1}^{20}\frac{k-1}{k}$

14. $\sum_{k=1}^{10}\frac{k^2}{2}$ **15.** $\frac{1}{2}$ **16.** -6 **17.** $a_n = -3n + 23$

18. $a_n = -4n + 36$ **19.** 40,200 **20.** \$33,397.50

Section 13.3 *(page 840)*

1. 2 **3.** -1 **5.** $-\frac{1}{2}$ **7.** $\frac{1}{5}$ **9.** π **11.** 1.04

13. Geometric, $\frac{1}{2}$ **15.** Not geometric **17.** Geometric, 2

19. Not geometric **21.** Geometric, $-\frac{2}{3}$

23. Geometric, 1.1 **25.** 4, 8, 16, 32, 64

27. 6, 3, $\frac{3}{2}$, $\frac{3}{4}$, $\frac{3}{8}$ **29.** 5, -10, 20, -40, 80

31. -4, 2, -1, $\frac{1}{2}$, $-\frac{1}{4}$ **33.** 10, 10.2, 10.40, 10.61, 10.82

35. 10, 6, $\frac{18}{5}$, $\frac{54}{25}$, $\frac{162}{125}$ **37.** $\frac{3}{2}$, 1, $\frac{2}{3}$, $\frac{4}{9}$, $\frac{8}{27}$ **39.** $a_n = 2^{n-1}$

41. $a_n = 2(2)^{n-1}$ **43.** $a_n = 10\left(-\frac{1}{5}\right)^{n-1}$

45. $a_n = 4\left(-\frac{1}{2}\right)^{n-1}$ **47.** $a_n = 8\left(\frac{1}{4}\right)^{n-1}$

49. $a_n = 14\left(\frac{3}{4}\right)^{n-1}$ **51.** $a_n = 4\left(-\frac{3}{2}\right)^{n-1}$

53. $\frac{3}{256}$ **55.** $48\sqrt{2}$ **57.** 1486.02

59. $-\frac{40}{6561}$ **61.** $\frac{81}{64}$ **63.** 12 **65.** $\frac{256}{9}$

67. b **68.** d **69.** c **70.** a **71.** 1023

73. 772.48 **75.** 2.25 **77.** 2.47 **79.** 6.67

81. $-14,762$ **83.** 16.00 **85.** 13,120 **87.** 48.00

89. 1103.57 **91.** 26,886.11 **93.** 2 **95.** $\frac{6}{5}$

97. $\frac{10}{9}$ **99.** 32

101.

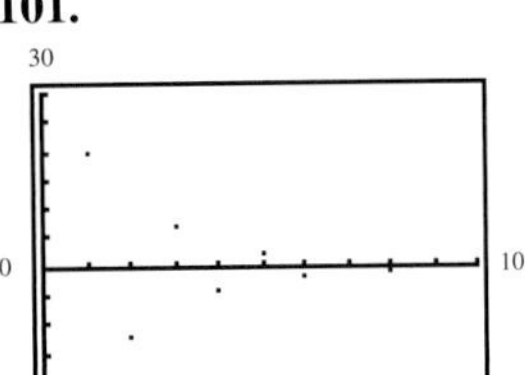

103.

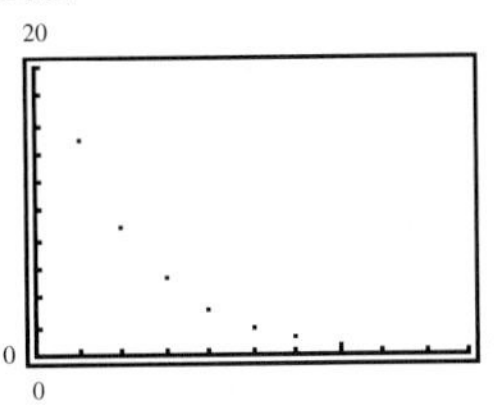

105. (a) There are many correct answers.

$a_n = 187{,}500(0.75)^{n-1}$ or $a_n = 250{,}000(0.75)^n$

(b) \$59,326.17 (c) The first year

107. \$3,623,993.23 **109.** \$9748.28 **111.** \$105,428.44

113. \$100,953.76

115. (a) \$5,368,709.11 (b) \$10,737,418.23

117. (a) $P = (0.999)^n$ (b) 69.4%

(c)

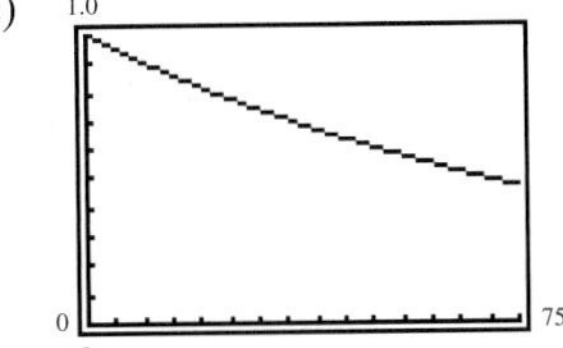

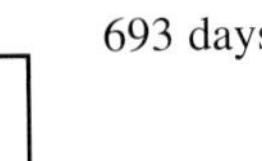

693 days

119. $\frac{37}{64} \approx 0.578$ square unit

121. 666.21 feet **123.** $a_1 = 12$. Answers may vary.

125. When a positive number is multiplied by a number between 0 and 1, the result is a smaller positive number, so the terms of the sequence decrease.

127. The nth partial sum of a sequence is the sum of the first n terms of the sequence.

129. $(-1, 2)$, $(2, 8)$ **131.** 8% **133.** 7.5%

135.

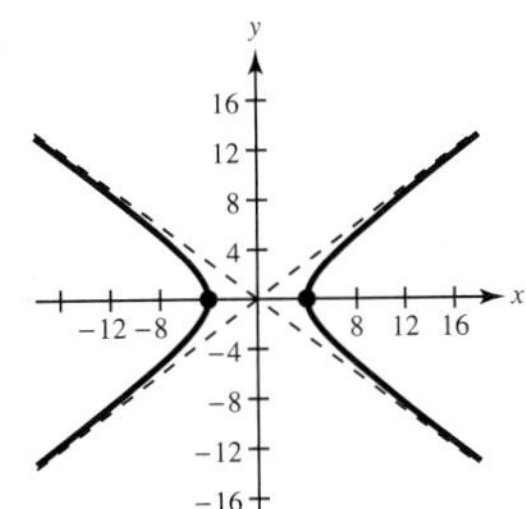

Vertices: $(\pm 4, 0)$

Asymptotes: $\pm\frac{3}{4}x$

Section 13.4 *(page 850)*

1. 15 **3.** 252 **5.** 1 **7.** 38,760 **9.** 38,760

11. 593,775 **13.** 2,598,960 **15.** 85,013,600

17. 15 **19.** 35 **21.** 70

23. $t^3 + 15t^2 + 75t + 125$

25. $m^5 - 5m^4n + 10m^3n^2 - 10m^2n^3 + 5mn^4 - n^5$

27. $243a^5 - 405a^4 + 270a^3 - 90a^2 + 15a - 1$

29. $64y^6 + 192y^5z + 240y^4z^2 + 160y^3z^3 + 60y^2z^4 + 12yz^5 + z^6$

31. $x^8 + 8x^6 + 24x^4 + 32x^2 + 16$

33. $x^6 + 18x^5 + 135x^4 + 540x^3 + 1215x^2 + 1458x + 729$

35. $u^3 - 6u^2v + 12uv^2 - 8v^3$

37. $81a^4 + 216a^3b + 216a^2b^2 + 96ab^3 + 16b^4$

39. $x^4 + \dfrac{8x^3}{y} + \dfrac{24x^2}{y^2} + \dfrac{32x}{y^3} + \dfrac{16}{y^4}$

41. $32x^{10} - 80x^8y + 80x^6y^2 - 40x^4y^3 + 10x^2y^4 - y^5$

43. $120x^7y^3$ **45.** $163{,}296a^5b^4$ **47.** 120

49. 54 **51.** 1.172 **53.** 510,568.785

55. $\frac{1}{32} + \frac{5}{32} + \frac{10}{32} + \frac{10}{32} + \frac{5}{32} + \frac{1}{32}$

57. $\frac{1}{256} + \frac{12}{256} + \frac{54}{256} + \frac{108}{256} + \frac{81}{256}$

59. The sum of the numbers in each row is a power of 2. Because the sum of the numbers in Row 2 is $1 + 2 + 1 = 4 = 2^2$, the sum of the numbers in Row n is 2^n.

61. The signs of the terms alternate in the expansion of $(x - y)^n$.

63. The first and last numbers in each row are 1. Every other number in the row is formed by adding the two numbers immediately above the number.

65. 390 **67.** 97,656

Review Exercises *(page 854)*

1. 8, 11, 14, 17, 20 **3.** $\frac{1}{2}, \frac{2}{5}, \frac{3}{8}, \frac{4}{11}, \frac{5}{14}$

5. 2, 6, 24, 120, 720 **7.** $\frac{1}{2}, \frac{1}{2}, 1, 3, 12$ **9.** $a_n = 3n + 1$

11. $a_n = \dfrac{1}{n^2 + 1}$ **13.** $a_n = -2n + 5$

15. $a_n = \dfrac{3n^2}{n^2 + 1}$ **17.** 28 **19.** $\dfrac{13}{20}$ **21.** $\displaystyle\sum_{k=1}^{4} (5k - 3)$

23. $\displaystyle\sum_{k=1}^{6} \frac{1}{3k}$ **25.** -5.5 **27.** 127, 122, 117, 112, 107

29. $2, \frac{7}{3}, \frac{8}{3}, 3, \frac{10}{3}$ **31.** $80, \frac{155}{2}, 75, \frac{145}{2}, 70$

33. $a_n = 4n + 6$ **35.** $a_n = -50n + 1050$

37. 486 **39.** 1935 **41.** 2527.5 **43.** 5100

45. 462 seats **47.** $\frac{5}{2}$ **49.** 10, 30, 90, 270, 810

51. $100, -50, 25, -\frac{25}{2}, \frac{25}{4}$ **53.** $4, 6, 9, \frac{27}{2}, \frac{81}{4}$

55. $a_n = \left(-\frac{2}{3}\right)^{n-1}$ **57.** $a_n = 24(3)^{n-1}$

59. $a_n = 12\left(-\frac{1}{2}\right)^{n-1}$ **61.** 8190 **63.** -1.928

65. 116,169.538 **67.** 4.556×10^{13} **69.** 8 **71.** 12

73. (a) There are many correct answers. $a_n = 120{,}000(0.70)^n$
(b) \$20,168.40

75. 321,222,672 visits **77.** 56 **79.** 1365

81. 91,390 **83.** 177,100 **85.** 10 **87.** 70

89. $x^4 - 20x^3 + 150x^2 - 500x + 625$

91. $125x^3 + 150x^2 + 60x + 8$

93. $x^{10} + 10x^9 + 45x^8 + 120x^7 + 210x^6 + 252x^5 + 210x^4 + 120x^3 + 45x^2 + 10x + 1$

95. $81x^4 - 216x^3y + 216x^2y^2 - 96xy^3 + 16y^4$

97. $u^{10} + 5u^8v^3 + 10u^6v^6 + 10u^4v^9 + 5u^2v^{12} + v^{15}$

99. $13{,}440x^4$ **101.** $-61{,}236$

Chapter Test *(page 857)*

1. $1, -\frac{3}{5}, \frac{9}{25}, -\frac{27}{125}, \frac{81}{625}$ **2.** 2, 10, 24, 44, 70 **3.** 60

4. 45 **5.** -45 **6.** $\displaystyle\sum_{k=1}^{12} \frac{2}{3k + 1}$ **7.** $\displaystyle\sum_{k=1}^{6} \left(\frac{1}{2}\right)^{2k-2}$

8. 12, 16, 20, 24, 28 **9.** $a_n = -100n + 5100$

10. 3825 **11.** $-\frac{3}{4}$ **12.** $a_n = 4\left(\frac{1}{2}\right)^{n-1}$ **13.** 1020

14. $\frac{3069}{1024}$ **15.** 1 **16.** $\frac{50}{3}$ **17.** 1140

18. $x^5 - 10x^4 + 40x^3 - 80x^2 + 80x - 32$

19. 56 **20.** 490 meters **21.** \$153,287.87

APPENDIX

Appendix B *(page A45)*

1. **3.**

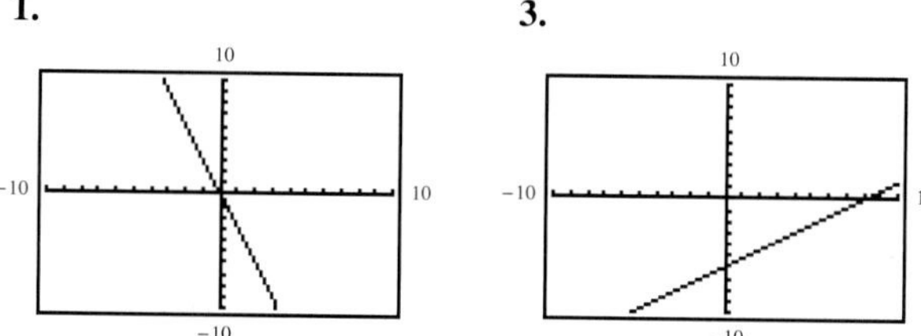

5. **7.**

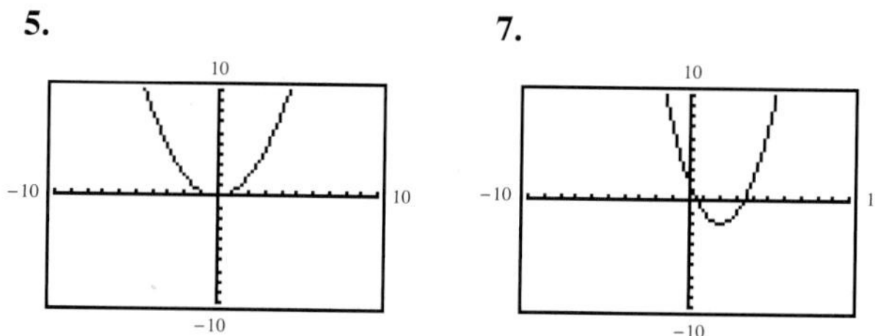

9. **11.**

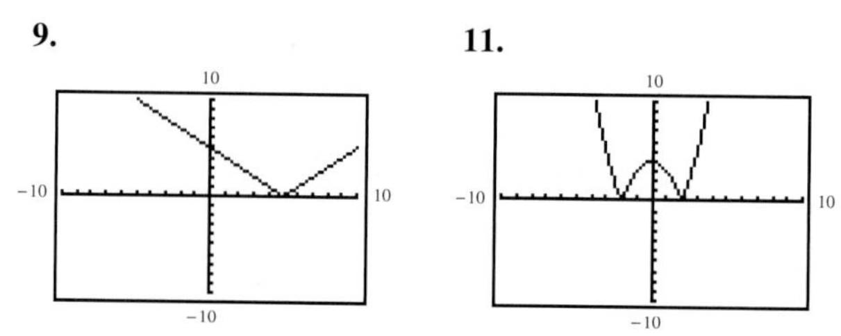

13.

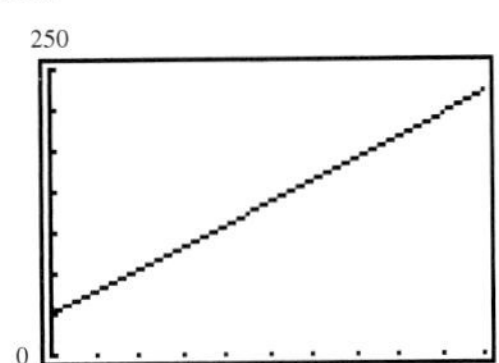

15.

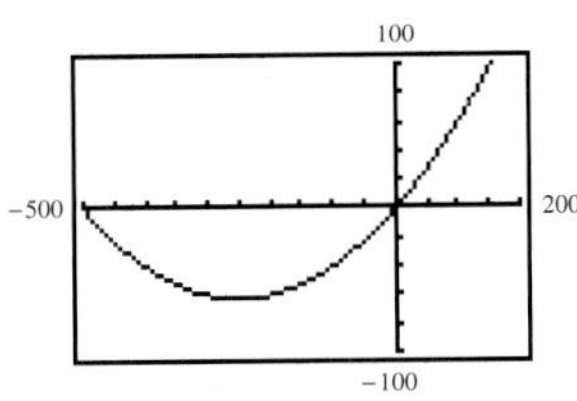

17. Sample answer:

Xmin = 4
Xmax = 20
Xscl = 1
Ymin = 14
Ymax = 22
Yscl = 1

19. Sample answer:

Xmin = -20
Xmax = -4
Xscl = 1
Ymin = -16
Ymax = -8
Yscl = 1

21.

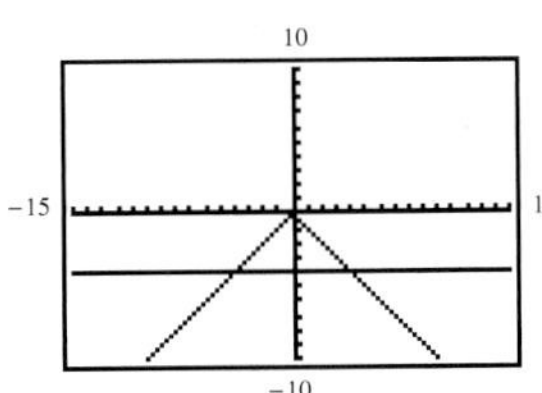

Triangle

23.

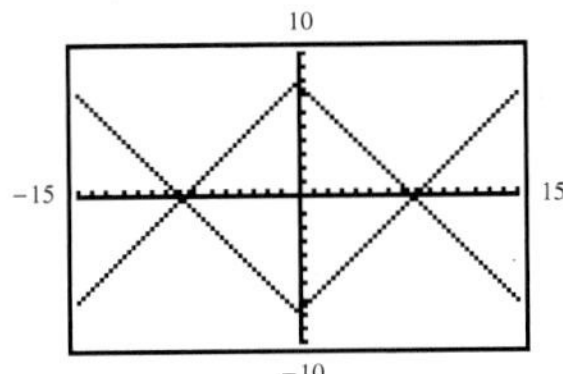

Square

25.

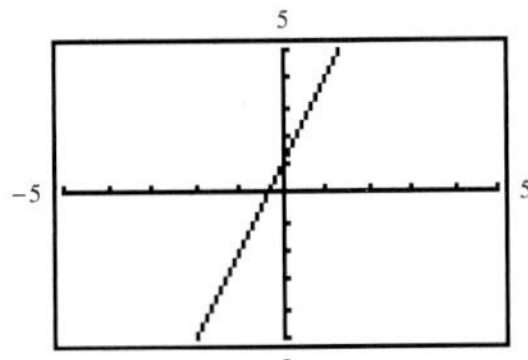

Yes, Associative Property of Addition

27.

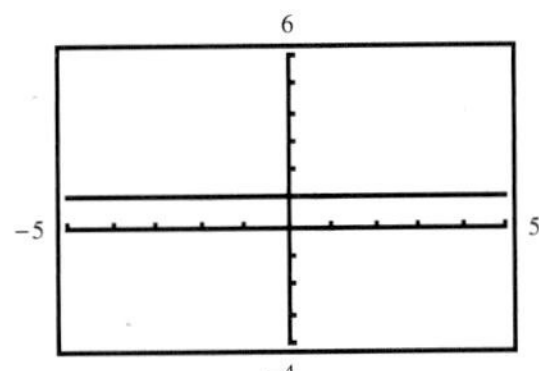

Yes, Multiplicative Inverse Property

29. $(-3, 0), (3, 0), (0, 9)$ **31.** $(-8, 0), (4, 0), (0, 4)$

33. $\left(\frac{5}{2}, 0\right), (0, -5)$ **35.** $(-2, 0), \left(\frac{1}{2}, 0\right), (0, -1)$

37.

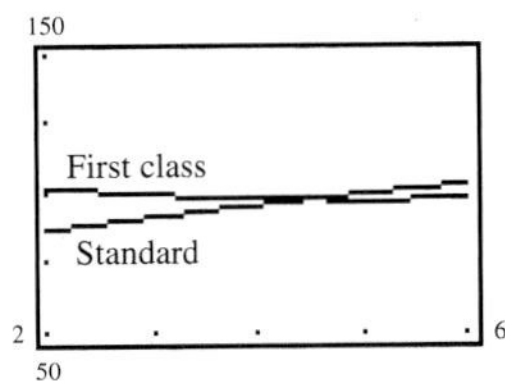

Index of Applications

Biology and Life Sciences

Business

Chemistry and Physics

Construction

Consumer

Geometry

Interest Rates

Miscellaneous

Time and Distance

U.S. Demographics

Index